StarBriefs Plus

StarBriefs Plus

A Dictionary of Abbreviations, Acronyms and Symbols in Astronomy and Related Space Sciences

Compiled by

André Heck
Strasbourg Astronomical Observatory, France

KLUWER ACADEMIC PUBLISHERS
DORDRECHT / BOSTON / LONDON

A C.I.P. Catalogue record for this book is available from the Library of Congress.

ISBN 1-4020-1925-4

Published by Kluwer Academic Publishers,
P.O. Box 17, 3300 AA Dordrecht, The Netherlands.

Sold and distributed in North, Central and South America
by Kluwer Academic Publishers,
101 Philip Drive, Norwell, MA 02061, U.S.A.

In all other countries, sold and distributed
by Kluwer Academic Publishers,
P.O. Box 322, 3300 AH Dordrecht, The Netherlands.

Printed on acid-free paper

Foreword

This compilation probably looks like one of the craziest things a human being could spend his or her time on. Yet nobody would wonder at someone taking a short walk every day – after some thirty years that person would have covered a surprisingly long distance. This is exactly the story behind this list, which appeared first as a few pages within the directory *StarGuides* (or whatever name it had at that time) and as a distinct sister publication since 1990. This edition includes about 60,000 new entries – which represents an increase of 43% compared with the previous edition (Heck 2000). See the review entitled *From Early Directories to Current Yellow-Page Services* (Heck 2003a) for a history of those resources.

The idea behind this dictionary is to offer astronomers and related space scientists practical assistance in decoding the numerous abbreviations, acronyms, contractions and symbols which they might encounter in all aspects of the vast range of their professional activities, including traveling. Perhaps it is a bit paradoxical, but if scientists quickly grasp the meaning of an acronym within their own specific disciplines, they will probably encounter more difficulties when dealing with adjacent fields. It is for this purpose that this dictionary might be most often used. Scientists might also refer to this compilation in order to avoid identifying a project by an acronym which has already too many meanings or confused definitions.

Hence *StarBriefs Plus*, the enlarged and fully updated version of a series of dictionaries, gathers together about 200,000 entries encountered in the literature of astronomy and space sciences, as well as of related fields such as aeronautics, aeronomy, astronautics, atmospheric sciences, chemistry, communications, computer sciences, data processing, education, electronics, energetics, engineering, environment, geodesy, geophysics, information handling, management, mathematics, meteorology, optics, physics, remote sensing, and so on.

Abbreviations, acronyms, contractions and symbols in common use and/or of general interest have also been included where appropriate. The traveling scientist has also not been forgotten, with the inclusion of the codes of locations, currencies and so on. Compared with the previous versions, new categories such as computer file extensions and major conferences have also been included. Separate sections are devoted to Greek letters, mathematical symbols, special signs and characters, as well as to entries with a numerical beginning.

We allowed an extensive number of technical acronyms to enter the list, because it is often difficult to find their explanations in other sources. Deviations from standard or 'official' acronyms are also listed if their frequency of occurrence in the literature justifies it – in such cases, the recommended explanation is flagged with an asterisk. Many such precisions have been added to this edition (about 1300 in total). Similarly, in numerous instances, complementary information has been offered between brackets (about 25,000 occurrences).

This compilation is essentially carried out in parallel with the permanent updating of the master files for *StarGuides*, a directory of astronomy, space sciences and related organizations of the world made available as a sister publication. In practice, all relevant abbreviations and acronyms encountered when scanning the literature received relating to the *StarGuides* directory are gathered here, the underlying principle being that they might also appear one day under the eyes of astronomers and related space scientists. We insist that these are 'real' abbreviations and acronyms, in other words that they are not constructed automatically from dictionaries as in some resources.

The associated directory *StarGuides Plus* (Heck 2003b) and its derived databases gather together all practical data available on associations, societies, scientific committees, agencies, companies, institutions, universities, etc., and more generally organizations, involved in astronomy and space sciences. Many other types of entries have also been included such as academies, advisory and expert committees, bibliographic services, consultants, data and documentation centres, dealers, distributors, funding agencies and organizations, journals, manufacturers, meteorological services, museums, norms and standards offices, planetariums, public observatories, publishers, research institutions in related fields, software producers and distributors, etc.

The master files for the present compilation of *StarBriefs Plus* have been made accessible as an on-line database at the *European Southern Observatory (ESO)* under the label *StarWords* (Albrecht & Heck 1994) and at the *Centre de Données astronomiques de Strasbourg (CDS)* as the database *StarBits* (Heck *et al.* 1994). They are now reachable on the web site of *Kluwer Academic Publishers* (http://www.wkap.nl/).

It should be noted that the manner of abbreviation is not always carried out in the same way in all countries. The use of full points seems to be one of the features which varies the most (for instance, in some countries they might be used after contractions whilst this would not be the case in other countries).

The meaning of abbreviations, acronyms, contractions and symbols should always be explained at their first occurrence in any document. It would be even more practical and helpful if authors and editors systematically tabulated them, with their explanations, at the end of the papers, chapters, reports, proceedings, books, etc., they produce. Such a policy would at least remove any ambiguity, since, as shown many times in the following pages, a given group of characters may have numerous meanings.

Socio-linguists might be interested in the following table giving the sixty most frequent terms (stop words and country names excluded) appearing in the acronym explanations out of a total of about 107,800 different words (at least two characters) for the whole dictionary (about 782,000 words in total).

Rank	Word	Freq.	Rank	Word	Freq.	Rank	Word	Freq.
1	System	11936	21	National	1808	41	Plan	1247
2	Data	5218	22	Time	1766	42	Language	1140
3	Control	4591	23	Engineering	1707	43	Automatic	1130
4	Information	4329	24	Association	1707	44	Access	1121
5	Management	3232	25	Interface	1601	45	Institute	1118
6	Test	3116	26	High	1588	46	Centre	1098
7	Program	2887	27	Digital	1475	47	Satellite	1083
8	Center	2650	28	Development	1421	48	Society	1081
9	Unit	2611	29	Integrated	1410	49	Review	1081
10	Research	2322	30	Office	1400	50	Maintenance	1038
11	International	2280	31	Equipment	1379	51	Mission	997
12	Network	2250	32	Systems	1358	52	Design	987
13	Analysis	2227	33	Facility	1345	53	Joint	972
14	Computer	2132	34	Operations	1326	54	European	971
15	Service	2024	35	Processing	1324	55	Standard	969
16	Support	1972	36	Communications	1320	56	Flight	969
17	Science	1870	37	Software	1289	57	Evaluation	968
18	Group	1870	38	Committee	1279	58	Area	966
19	Space	1851	39	Advanced	1278	59	Experiment	960
20	Technology	1839	40	Technical	1259	60	University	949

This frequency table, obtained from a significantly enlarged dictionary, confirms the results obtained in the previous edition (Heck 2000). It gives a pretty good picture of our everyday jargon, with all terms being English/American ones in spite of the fact that the dictionary is multilingual. 90% of the sixty most frequent terms of the previous edition remain among the current "Top 60". Six new terms moved in: 'Communications' (ranking now 36th, previously 64th) makes the third largest jump forward, following 'Plan' (now 41st, previously 84th) and mainly 'Evaluation' (now 57th, previously 107th). The other newcomers are 'Maintenance' (now 50th, previously 70th), 'Joint' (now 53rd, previously 78th) and 'Area' (now 58th, previously 69th). Conversely the terms 'New', 'Assembly', 'Remote, 'Energy', 'Optical', 'Power', and 'Laboratory' dropped behind the "Top 60" pack of the present version.

'Space' (ranking now 19th, previously 12th) is still among the twenty most frequent terms, whilst the first astronomy-related term, 'Observatory', comes well after the leading group (ranking 82nd) with 797 occurrences. 'Astronomical' and 'Telescope' are located around the 100th position with both 720 occurrences, while 'Solar' follows (ranking 108th) with 679 appearances. 'Physics', 'Astronomy' and 'Physical' are way behind with respectively 574, 511 and 228 occurrences only.

Significantly higher occurrences of words such as 'virtual' and derivatives were expected in this edition since the spreading of *virtual-reality* techniques and the multiplication of *virtual-observatory* projects, but all together they reach only about 500 appearances.

We also investigated how a frequency distribution would look when based on *concepts* or *notions* rather than on specific terms. For instance under 'Science', we included not only its plural 'Sciences', but also terms such as 'Scientific' and their equivalent in different languages. In a similar way, 'Test' covered words such as 'Testability', 'Testing', 'Tester', Testbed', 'Tests', and so on. The following table offers such a distribution for the 30 most frequent *concepts* in this dictionary.

Rank	Concept	Freq.	Rank	Concept	Freq.	Rank	Concept	Freq.
1	System	13701	11	Unit	3216	21	Time	2356
2	Control	5427	12	Research	2903	22	Automatic	2344
3	Data	5313	13	Service	2818	23	Technology	2221
4	Information	4450	14	Processing	2799	24	Engineering	2183
5	Center	4321	15	Analysis	2747	25	Group	2085
6	Program	4259	16	Network	2714	26	Technical	2068
7	Management	3904	17	International	2591	27	Association	2065
8	Test	3579	18	Space	2569	28	Support	2054
9	Science	3451	19	Integrated	2460	29	National	2029
10	Operations	3390	20	Computer	2440	30	Institute	1923

The overall results are roughly similar to those of the first table, with small fluctuations of ranks. Some differences are significant though. Thus 'Operations' (including 'Operational', etc.) moves from Rank 34 in the first table to Rank 10 in the second one, 'Processing' (including 'Processor', etc.) from Rank 35 to Rank 14, 'Automatic' (including 'Automated', etc.) from Rank 43 to Rank 22, 'Institute' (including 'Institution', etc.) from Rank 45 to Rank 30, and 'Technical' (including 'Technique', etc.) from Rank 40 to Rank 26. Conversely, the largest movement backwards is by 'Support', going from the 16th position to the 28th one.

For the purists, perhaps a couple of additional comments on the details are in order. Caesurae have been introduced as appropriately as possible, but the compact two-column presentation did not allow to always cut the words in full respect of the strict rules languages (such as French) have sometimes for this kind of operations. We had to reach a compromise between these and other constraints such as avoiding overlapping lines and making the right-hand edges of columns as regular as possible.

Acronyms and abbreviations appear sometimes fully capitalized, sometimes fully in low-case characters, or as a mixture of both. The per-default rule of this compilation has been to use the fully capitalized (upper-case) format. Versions of abbreviations and acronyms involving low-case characters have however been listed when appearing regularly in the literature.

Perhaps as a matter of interest to socio-linguists: as mentioned earlier, acronyms have been routinely extracted from the literature read by astronomers such as periodicals, proceedings, books, more generally publications of all kinds used in astronomy-related activities at large. As it can be expected, new acronyms have been frequently found in new projects, but also more frequently in astronomy-related colloquia than in symposia. This is probably due to the organization format of the respective conferences, symposia being more "formal" and/or dealing with scientific themes established since a longer time.

When compiling catalogues such as *StarGuides* and *StarBriefs*, one cannot help but be impressed by the very broad spectrum of disciplines to which astronomy and related space sciences are linked, and by the very rich variety of techniques applied in these fields.

Acknowledgements

The implementations, as databases, of the *Star*s Family* products by the European Space Agency, the European Southern Observatory, Strasbourg Astronomical Data Centre, and now Kluwer Academic Publishers have been strong incentives to continue and always improve these time-consuming compilations for the benefit of the best possible communication within the astronomical community, as well as between it and the outside world.

November 2003.

REFERENCES

Albrecht, M.A. & Heck, A.: 1994, StarWords – An On-line Database of Acronyms, Abbreviations, and Contractions in Astronomy, Space Sciences, and Related Fields, *Astron. Astrophys. Suppl.* **103** 471.

Heck, A.: 2000, StarBriefs 2001 – A Dictionary of abbreviations, acronyms, and symbols in astronomy, space sciences and other related fields, Kluwer Acad. Publ., Dordrecht, viii + 814 pp. (ISBN 0-7923-6510-0).

Heck, A.: 2003a, From Early Directories to Current Yellow-Page Services, in *Information Handling in Astronomy – Historical Vistas*, Ed. A. Heck, Kluwer Acad. Publ., Dordrecht, 183-205.

Heck, A.: 2003b, StarGuides Plus – A World-Wide Directory of Organizations in Astronomy and Related Space Sciences, Kluwer Acad. Publ., Dordrecht, x + 1140 pp. (ISBN 1-4020-1926-2).

Heck, A., Egret, D. & Ochsenbein, F.: 1994, StarWorlds – StarBits (announcement of two databases), *Astron. Astrophys. Suppl.* **108**, 447-448.

Author's address: **André HECK**
Observatoire Astronomique
11, rue de l'Université
F-67000 Strasbourg
France
Telephone: (+33)(0) 390 242 420
Telefax: (+33)(0) 388 508 772
WWW: http://vizier.u-strasbg.fr/~heck

Dictionary

The asterisks (*) indicate recommended explanations among several similar ones.

A

a About • Absorptivity • Acceleration • Accepté • Accepted • Acre • Acreage • Acting • Active • Actual • Address • Adjective • Advance • Advanced • Afternoon • Age • Akzeptiert • Alto • Am • Amateur • Ampere • An • Année • Anno • Anonymous • Answer • Ante • Antes • Are • Arrive • Arriving • At • Atto • Auf • Deceleration • Electron Annihilation Operator • Lattice Space of Si • Linear Absorption Coefficient • Mean Sound Absorption • Radiation Density Constant • Semi-Major Axis • Sonic Velocity • Specific Acidity • Specific Helmholtz Free Energy • Speed of Sound • Stefan-Boltzmann Radiation Constant • Surface Albedo • Thermal Diffusivity • Universal Affirmative • Year

Å Ångström

A Abell • Absolute • Absorbance • Absorption • Academician • Academy • Ace • Acre • Acreage • Acting • Active • Activity • Additional • Address • Adjutant • Adult • Advanced • Aerodynamic Reference Area • Affinity • Air • Albedo • Alp • Alpha (international alphabet) • Alt • Altesse • Alto • America • American • Ammeter • Ampere • Ampere-Turn • Amplitude • Analog • Analysis • Anchor • Anemic • Anno • Annual • Anode • Anonymous • Answer • Ante • Apogee • Appendix • April • Area • Ariel • Argon • Arrêté • Arroyo • Art • Arterial • Assistant • Associate • Atom • Atomic • Atomic Weight • August • Austral (currency) • Australia • Australian • Austria (international vehicle registration code) • Austria (postal country code) • Azimuth • Cross-Section • Distance Unit • Equivalent Absorption Area • Five Hundred (Roman numeral) • Helmholtz Free Energy • Helmholtz Function • Interstellar Extinction • Linear Current Density • Magnetic Vector Potential • Mass Number • Nuclear Number • Nucleon Number • Oort's Constant of Galactic Rotation • Optical Density • Permeance • Richardson Constant • Rotational Constant • Ten • Spectral Type • Standard Paper Size • Universal Affirmative • Work • 500 (Roman numeral)

aa Acting Appointment • Afar (ISO language code) • After Arrival • Always Afloat • Años • Approximate Absolute • Area Administrator • Arithmetic Average • Attendance Allowance • Author's Alteration

aA anderer Ansicht

A-A Air-to-Air

A/A Airplane Avionics • Air-to-Air • Analog-to-Analog • Angular Accelerometer

A&A Additions & Amendments • Advertise & Award • Astronomy and Astrophysics (periodical code)

AA Abbreviated Analysis • Absence Autorisée • Academic Affairs • Accelerated Assembly • Accelerometer Assembly • Access Authorization • Accreditation Authority • Achieved Ability • Achieved Availability • Acquisition Activity • Acte Authentique • Acting Appointment • Adjoint Administratif • Administrative Area • Administrative Assistant • Administrative Authority • Advance Agreement • Advertise and Award • Affirmative Action • Affordability Analysis • After Arrival • Age Allowance • Agent Administratif • Aircraft Availability • Airplane Avionics • Air-to-Air • Aitreya Aranyaka • Algorithmic Architecture • Allied Affairs • Altezas • Analysis of Alternatives • Angle of Arrival • Angular Accelerometer • Angular Aperture • Antiaircraft • Antiproton Accumulator • Application Association • Appointing Authority • Approval Authority • Approximate Absolute • Aptitude Area • Ardennes-Astronomie (France) • Area Administrator • Arithmetic Average • Arrival Angle • Artificial Agent • Arts Appliqués • Assembly Area • Assistant Administrator • Assistant Astronomer • Associate Administrator • Associate Astronomer • Associate in Accounting • Associate of Arts • Astrograph Astrometry • Astronautica Acta • Astronomical Application • Astronomy and Astrophysics • Atlas Agena • Atomic Absorption • Atomic Adsorption • Attack Assessment • Attendance Allowance • Audit Agent • Auswärtiges Amt • Author's Alteration • Auto-Analysis • Auto-Answer • Automatic Answer • Automobile Association • Autonomous Acquisition • Avenue of Approach • Size of Battery • Size of Cell

A&AA Astronomy and Astrophysics Abstracts (periodical code)

AAA Absolute Astronomical Accelerometer • Absorption Additivity Assumption • Academic Assistance Access • Accéléromètrie Astronomique Absolue • Acceptance Approving Authority • Achievement in Asia Award (OCPA) • Acta Astronomica Archive • Advanced Acoustic Array • Advanced Aircraft Analysis • Agricultural Adjustment Administration • Agrupación Astronómica Aragonesa (Spain) • Agrupación Astrónomos Aficionados (Venezuela) • Agrupación de Aficionados a la Astronomía (Uruguay) • Airframe Airworthiness Assurance • Allocation - Assessment - Analysis • Alternate Assembly Area • Amateur Astronomers Association (USA-NY) • Amateur Astronomers of Acadiana (USA-LA) • American Accounting Association (USA) • American Arbitration Association (USA) • American Automobile Association (USA) • Amor - Aten - Apollo • Anaa (Tuamotu Islands) (location identifier) • Anno Ante Christum • Apollo - Amor - Aten • Artifacts - Agents - Activities • Asociación Argentina de Astronomía (Argentina) • Assistant Associate Administrator • Association Astronomique d'Anjou (France) • Association Astronomique de l'Ain (France) • Association d'Astronomie de l'Artois (France) • Association of Amateur Astronomers (USA-TX) • Associazione Acquavivese Astrofili (Italy) • Astronaut-Actuated Abort • Astronomy and Astrophysics Abstracts • Aten - Apollo - Amor • Authentication, Authorization and Accounting • Authorization Accounting Activity • Author's Argumentation Assistant • Automated Airline Assignment • Automatic Amplitude Control • Avionics Air Assembly

AAAA Albany Area Amateur Astronomers (USA-NY) • American Association of Amateur Astronomers (USA-TX) • Ames Area Amateur Astronomers (USA-IA) • Ars Ad Astra Association (Switzerland) • Asociación Argentina Amigos de la Astronomía • Association des Astronomes Amateurs d'Auvergne (France) • Astronomische Arbeitsgemeinschaft Aalen eV (Germany) • Atlas Automatisé des Absorptions Atmosphériques • Automatized Atmospheric Absorption Atlas

AAAAA Anne Arundel Association of Amateur Astronomers (USA-MD)

AAAAT Association des Astronomes Amateurs Abitibi-Témiscamingue (Canada-QC)

AAACE American Association for Adult and Continuing Education (USA)

AAAD Amateur Astronomers Association of Delhi (India)

AAAE Alliance of Associations for the Advancement of Education (USA) • American Association for Adult Education (USA) • American Association of Academic Editors (USA) • American Association of Airport Executives (USA)

AAAF Association Aéronautique et Astronautique de France

AAAI American Association for Artificial Intelligence (USA) • Association of Australian Aerospace Industries

AAAL American Academy of Arts and Letters (USA)

AAAOA Amateur Astronomers As Outreach Ambassadors (project)

AAAP Amateur Astronomers Association of Pittsburgh (USA-PA) • Amateur Astronomers Association of Princeton (USA-NJ)

AAAPT Asian-African Association for Plasma Training

AAAS American Academy of Arts and Sciences (USA) • American Association for the Advancement of Science (USA) • Arizona Amateur Astronomy Society (USA-AZ) • Automated Airport Advisory Service

AAATE Association for Advancement of Assistive Technology in Europe

AAAV Amateur Astronomers' Association of Vadodara (India)

AAB Above And Beyond • Academic Advisory Board • Adaptive Abstract Builder • Administrative Appeal Board • All-to-All Broadcast • Arrabury (Australia-QLD) (location identifier) • Association d'Astronomie El-Battani (Algeria) • Associazione Astrofili Bolognesi (Italy) • Astronomischer Arbeitskreis Bamberg (Germany)

AABNCP Advanced Airborne National Command Post

AABU Al al-Bayt University (Jordan)

AABV Associazione Astrofili del Basso Vicentino (Italy)

AABW Antarctic Bottom Water

AAC Abbreviated Address Calling • Academic Advisory Council • Acquisition Advice Code • Acquisition/Alignment Control • Activity Address Code • Advanced Audio Coding • Aeronautical Administrative Communication • Agrupación Astronómica Cántabra (Spain) • Agrupación Astronómica Complutense (Spain) • Agrupación Astronómica de Córdoba (Spain) • Alachua Astronomy Club (USA-FL) • Alamogordo Astronomy Club (USA-NM) • Al Arish (Egypt) (location identifier) • All-Aspect Capability • Alumni Advisory Council • Amateur Astronomy Centre (UK) • Anglo-American Cataloguing (rules) • Angola Air Charter Ltd. • Apprentissage Anticipé de la Conduite • Asesores Astronómicos Cacereños (Spain) • Asset Allocation Committee • Associate Analysis Center • Associated Analysis Center • Association des Archivistes Canadiens (Canada) • Association of American Colleges (USA) • Associazione Astronomica Cassino (Italy) • Associazione Astronomica Cortina (Italy) • Astronomisches Abenteuer Camp (Germany) • Atlanta Astronomy Club (USA-GA) • ATM Access Concentrator • Atomic Absorption Coefficient • Attaché d'Administration Centrale • Auto Cruise Control • Automatic Amplitude Control • Automatic Answering Capability • Automatic Aperture Control • Automatic Approach Control • Autonomous Activation Condition

AACB Aeronautics and Astronautics Coordinating Board

AACC Airport Associations Coordinating Council • American Association of Clinical Chemistry (USA) • American Association of Community Colleges (USA) • American Automatic Control Council (USA) • Anglo-American Cataloguing Committee

AACCM Anglo-American Cataloguing Committee for Cartographic Materials

AACCS Analytical Approach to Candidate Constellation Selection

AACCUP Accrediting Agency of Chartered Colleges and Universities of the Philippines (formerly PAASCU)

AACE Airworthiness Assurance Center of Excellence • American Association of Cost Engineers (USA) • Asociación Argentina de Ciencias Espaciales (Argentina) • Association for Adult and Continuing Education (UK) • Association for the Advancement of Computing in Education

AACG American Association for Crystal Growth (USA)

AACGM Altitude-Adjusted Corrected Geomagnetic

AACI Amateur Astronomers of Central Iowa (USA-IA)

AACJC American Association of Community and Junior Colleges (USA)

AACM Active Access Control Module

AACO Arab Air Carriers Association • Association Astronomique de la Côte d'Or (France)

AACOBS Australian Advisory Council of Bibliographical Services

AACP Ambient Air Control Panel

AACPSM Actual Average Cost per Successful Mission

AACR Anglo-American Cataloguing Rules

AACRAO American Association of Collegiate Registrars and Admissions Officers (USA)

AACS Advanced Automatic Compilation System • Asynchronous Address Communications Systems • Attitude and Antenna Control System • Attitude and Articulation Control System • Automated-Access Control System • Auxiliary Attitude Control System

AACTE American Association of Colleges for Teacher Education (USA) • Asociación para el Avance de la Ciencia y la Tecnología en España (Spain)

AAC&U Association of American Colleges & Universities (USA)

AACU Association of American Colleges and Universities (USA)

AACUP Accrediting Agency of Colleges and Universities of the Philippines (formerly PAASCU)

AAD Ad-Dabbah (Sudan) (location identifier) • Admission And Disposition • Agrupación Astronómica de los Dolores (Spain) • Analog Alignment Diskette • Analysis and Application Division • Attitude Anomaly Detector • Australian-Antarctic Discordance

AADC Alaska Aerospace Development Corporation (USA-AK) • All-Application Digital Computer • Approach And Departure Control • Automated Area Distribution Center

AADEW Attack Assessment for Directed-Energy Weapon

AADI Asociación Argentina de Derecho Internacional (Argentina)

AADPS Algorithmic Architecture Data Processing System

AADR Applications Approval Design Review

AADS Ascent Air Data System

AADSF Advanced Automated Directional Solidification Furnace

AADZ Astronomsko-Astronauticko Drustvo Zagreb (Croatia)

AAE Abort Advisory Equipment • Above Aerodrome Elevation • Academy for Academic Excellence • Aerospace Auxiliary Equipment • African American English • Agrupación Astronáutica Española (Spain) • Allgemeine Anschalterlaubnis • American Association of Engineers (USA) • Ames Aerospace Encounter (USA-CA) • Annaba (Algeria) (location identifier) • Association for Adult Education (UK) • Association for Astronomy Education (UK) • Automatic Adaptive Equalization • Automatic Answering Equipment

AAEC Attitude Axis Emergency Control • Australian Atomic Energy Commission

AAED Advanced Airborne Expendable Decoy

AAEE American Academy of Environmental Engineers (USA) • American Association of Electrical Engineers (USA) • Asuntos Exteriores

AAEN Alternative Academic Educational Network

AAEO Association of Adult Education Organizations

AAES American Association of Engineering Societies (USA) • Automated Aeromedical Evacuation System

AAEU Aquatic Animal Experiment Unit

AAF Agrupación Astronómica de Fuerteventura (Canary Islands) • Annual Adjustment Factor • Antecedent Action Form • Apalachicola (USA-FL) (location identifier) • Association des Archivistes Français (France) • Associazione Astrofili Fiorentini (Italy) • Associazione Astronomica Frusinate (Italy) • Auxiliary Airfield

AAFB Alamogordo Air Force Base (USA-NM) (now HAFB) • Andrews Air Force Base (USA-MD)

AAFC Association Astronomique de Franche-Comté (France)

AAFE Advanced Applications Flight Equipment • Advanced Applications Flight Experiment (*) • Aeroassist Flight Experiment

AAFI Association des Anciens Fonctionnaires Internationaux

AAFIF Automated Air Facility Information File

AAfz Astrometriia i Astrofizika (periodical code)

AAG American Association of Geographers (USA) • Area Auditor General • Asociación Astronómica de Guatemala • Association of American Geographers (USA) • Associazione Astrofili Garfagnana (Italy)

AAGC Agrupación Astronómica de Gran Canaria (Canary Islands)

AAGG Associazione Astrofili Galileo Galilei (Italy)

AAGS American Association for Geodetic Surveying (USA) • Association of African Geological Surveys

AAH Agrupación Astronómica de Huesca (Spain) • Allocation pour Adulte Handicapé • Association d'Aide Humanitaire

AAHA Awaiting Action of Higher Authority

AAHam Astronomische Abhandlungen Hamburger Sternwarte (periodical code)

AAHE American Association for Higher Education (USA)

AAI Additional Authorized Item • Amateur Astronomers Inc. (USA-NJ) • Angle of Approach Indicator • Application to Application Interface • Arraias (Brazil-TO) (location identifier) • Arrival Aircraft Interval • Association Astronomique de l'Indre (France) • Associazione Astrofili Imolesi (Italy) • Astronomischer Arbeitskreis Ingolstadt eV (Germany) • Astronomy Acknowledgment Index • Authorized Active Inventory • Automated Acquisition Information • Avionics-ATC Integration

AAIB Air Accident Investigation Branch

AAIC Active Addressing Integrated Circuit • Aircraft Accident Investigation Committee

AAID Arithmetic Array Identification

AAIE American Association of Industrial Engineers (USA)

AAIL Applied Artificial Intelligence Laboratory

AAIM American Association of Industrial Management (USA)

AAIP American Association of Independent Publishers (USA)

AAIR Advanced Atmospheric Imaging Radiometer

AAIS American Association for Information Science (USA) • Associate Administrator for Information Systems • Automated Acquisition Information System

AAIT Air Accident Investigation Team

AAIzN Alma Ata Izdatel Nauka (periodical code)

AAJ Amateur Astronomers of Jackson (USA-MI) • Association of American Journalists (USA)

AAJA Afro-Asian Journalists' Association

AAJC American Association of Junior Colleges (USA)

AAK Aranuka (Kiribati) (location identifier) • Astronomischer Arbeitskreis Kassel eV (Germany)

AAL Above Aerodrome Level • Above Airport Level • Absolute Assembly Language • Account Access Layer • Acronym and Abbreviation List • Additional Authorization List • Additional Authorized List • Aero-Acoustic Levitation • Ålborg (Denmark) (location identifier) • Amateurs Astronomes du Luxembourg (Luxembourg) • Asian Accelerator Laboratory • Assign Affiliation List • Associated Absorption Line • Association of Assistant Librarians • Astronomische Arbeitsgruppe Laufen eV (Germany) • Astronomy Association of Lawrence (USA-KS) • Atelier Astronomie de Limoges (France) • ATM Adaptation Layer • Authorized Allowance List

AALC Association Astronomique du Loir-et-Cher (France)

AALCD Active Addressing Liquid Crystal Display

AALCP ATM Adaptation Layer Common Part

AALL American Association of Law Libraries (USA)

AALM ATM Adaptation Layer Management

AAM Absolute Angular Momentum • Advanced Access Mode • Advanced Analysis Module • Advanced Astro Master • Agrupación Astronómica de Madrid (Spain) • Agrupación Astronómica de Manresa (Spain) • American Association of Museums (USA) • Application Activity Model • Application Architectural Model • Architecture Assessment Methodology • Associazione Astrofili Mantovani (Italy) • Associazione Astronomica Milanese (Italy) • Atmospheric Angular Momentum • ATOS Application Module • Mala Mala (South Africa) (location identifier)

AAMA Area Airspace Management Authority

AAMAS Autonomous Agents and Multiagent Systems (conference)

AAMC Auto-Adaptive Multimedia Composer

AAMI Advanced Active Microwave Instrument • Association for the Advancement of Medical Instrumentation

AAMMH Annual Available Maintenance Man-Hours

AAMP Advanced Architecture Microprocessor

AAMPS Automated Aircraft Mission Planning System

AAMS Analytical Accounting Management System

AAMSI American Association for Medical Systems Informatics (USA)

AAMT Associazione Astronomica Monti della Tolfa (Italy)

AAMU Alabama A&M University (USA-AL)

AAMV Amateurastronomen Max Valier (Italy)

AAMVG Astronomisches Arbeitskreis Merkur/Venus Göttingen (Germany)

AAN Al Ain (UAE) (location identifier) • All Area Network

AANC Astronomical Society of Northern California (USA-CA)

AANEAS Anglo-Australian Near-Earth Asteroid Survey

AANT Autonomous and Advanced Navigation Technique

AANZ Astronautics Association of New Zealand

aaO am angeführten Ort • am angegebenen Ort

AAO Abastumani Astrophysical Observatory (Georgia) • Adjacent Arctic Ocean • Agrupación Astronómica d'Osona (Spain) • All At Once • Amtsanschlußorgan • Anaco (Venezuela) (location identifier) • Anglo-Australian Observatory • Approved Acquisition Objective • Astronaut Activities Office (NASA) • Authorized Acquisition Objective

AAOE Airborne Antarctic Ozone Experiment

AAOEPP Anglo-Australian Observatory - Epping (Australia-NSW)

AAofA Amateur Astronomers of Acadiana (USA-LA)

AAOL Association Against Obtrusive Lighting (Australia)

AAONw Anglo-Australian Observatory Epping Newsletter (periodical code)

AAOPr Anglo-Australian Observatory Preprint (periodical code)

AAOS Array Acousto-Optical Spectrometer

AAOSPS Anglo-Australian Old Stellar Populations Survey

AAP Abbreviated Acquisition Plan • Abbreviated Acquisition Program • Adjoint Administratif Principal • Advanced Acquisition Plan • Advanced Acquisition Planning • Advanced Application Program • Advanced Automation Program • Affirmative Action Plan • Affirmative Action Program • Agrupación Astronómica de la Palma (Canary Islands) • Airlock Adapter Plate • American Association of Publishers (USA) • Analog Autopilot • Apollo Applications Program • Applications Access Point • Area Air Picture • Assemble At Pad • Assign Alternate Parent

• Association of American Publishers (USA) • Associative Array Processor • Associazione Amici dei Planetari (Italy) • Astronomische Arbeitsgemeinschaft Pfaffenwinkel (Germany) • Astronomischer Arbeitskreis Pforzheim (Germany) • Atmospheric Analysis and Prediction • Australian Associated Press • Automatic Asset Placement • Houston Andrau (USA-TX) (location identifier)

AAPA American Association of Port Authorities (USA) • Association of Asia Pacific Airlines

AAPART Aviator Annual Proficiency And Readiness Test

AAPB Astronomy and Astrophysics Planning Board

AAPF Astronomy and Astrophysics Postdoctoral Fellowship (NSF)

AAPG American Association of Petroleum Geologists (USA)

AAPIER American Association for Public Information, Education and Research (USA)

AAPM American Association of Physicists in Medicine (USA)

AAPMH Annual Available Productive Man-Hours

AAPP Administraciones Públicas • Affirmative Action Program Plan • ATOVS and AVHRR Processing Package

AAPPS Association of Asian and Pacific Physical Societies

AAPR Angle Averaged Partial Redistribution

AAPS Advanced Antenna Prototype System • American Association for the Promotion of Science (USA) • American Association of Pharmaceutical Scientists (USA)

AAPSS American Academy of Political and Social Science (USA)

AAPT American Association of Physics Teachers (USA)

AApTr Astronomy and Astrophysics Transactions (periodical code)

AAQ Anapa (Russia) (location identifier) • Association des Archivistes du Québec (Canada-QC) • Astronomical Association of Queensland (Australia-QLD)

A&AR Astronomy and Astrophysics Review (periodical code)

AAR Administrative Element Report • Advanced Address Recognizer • After-Action Report • After-Action Review • Against All Risks • Aircraft Accident Report • Airport Acceptance Rate • Air-to-Air Receive • Air-to-Air Refueling • Århus (Denmark) (location identifier) • Association for Automated Reasoning • Association of American Railroads (USA) • Attaché d'Administration de la Recherche • Auto-Analysis Results • Automatic Adaptive Routing • Automatic Alternate Routing • Average Annual Rainfall

AARA Access and Amendment Refusal Authority • After-Action Reconstruction and Analysis

AARAO Ålborg Amateur Radio Astronomy Observatory (Denmark)

AARC Anglo-American Cataloguing Rules • Arctic and Antarctic Research Group • Auroral Atmospheric Radiance Code

AARCS Automatic Agent Radio Communications System

AARD Advanced Accelerator Research and Development

Aarg Aargang

AARG Astronomy and Astrophysics Research Group

AARI Advanced Airborne Remote Instrumentation • Arctic and Antarctic Research Institute (Russia)

AARIT Austrian Association for Research in Information Technology (Austria)

AARL Association for Academic and Research Libraries (USA)

AARMS Authorized/Assigned Resource Management System

AARN Australian Academic Research Network

AARNet Australian Academic and Research Network

AARNET Australian Academic Research Network

AARP American Association of Retired Persons (USA) • AppleTalk Address Resolution Protocol • Attaché d'Administration de la Recherche Principal

AARPS Air-Augmented Rocket Propulsion System

AARS Asian Association of Remote Sensing

AART Assessment And Recovery Team

AARTS Automated Audio Remote Test Set • Automated Audio Remote Test System

AARU Astronomy and Atmospheric Research Unit (USM)

A&ARv Astronomy and Astrophysics Review (periodical code)

A&AS Astronomy and Astrophysics Supplement Series (periodical code)

AAS Abingdon Astronomical Society (UK-England) • Abort Advisory System • Academy of Applied Science (USA) • Acta Astronomica Sinica • Advanced Antenna System • Advanced Automation System • Advisory and Assistance Service • African Academy of Sciences • Agrupació Astronòmica de Sabadell (Spain) • Airport Advisory Service • All-to-All Scatter • Alternate Architecture Study • American Academy of Sciences (USA) • American Aerospace Society (USA) • American Astronautical Society (USA) • American Astronomical Society (USA) • American Astronomical Society Meetings (periodical code) • Amphibious Assault Ship • Andersen Air Sampler • Andover Astronomical Society (UK-England) • Annual Assurance Statement • Apalapsili (Indonesia) (location identifier) • Archdale Astronomical Society (USA-NC) • Arithmetic Assignment Statement • Asian Accelerator School • Associate in Applied Science • Association Astronomique du Soissonnais (France) • Association des Archivistes Suisses (Switzerland) • Associazione Astrofili di Salò (Italy) • Associazione Astrofili Sardi (Italy) • Associazione Astrofili Segusini (Italy) • Associazione Astrofili Spezzini (Italy) • Astronomical Association - Sofia (Bulgaria) • Astronomischer Arbeitskreis Salzkammergut (Austria) • Athens Astronomical Society (USA-GA) • Atomic Absorption Spectrometer • Atomic Absorption Spectrometry • Atomic Absorption Spectrophotometry • Atomic Absorption Spectroscope • Atomic Absorption Spectroscopy • Auburn Astronomical Society (USA-AL) • Aurora Astronomical Society (USA-OH) • Austin Astronomical Society (USA-TX) • Australian Academy of Sciences • Automated Accounting System • Automatic Addressing System • Ayrshire Astronomical Society (UK-Scotland) • Azimuth Alignment System

AASA Association Astronomique Saint-Avertinoise (France)

AASANA Administración de Aeropuertos y Servicios Auxiliares a la Navegación Aérea

AASAP Applied Analysis Spectral Analysis Process

AASC Astronomy and Astrophysics Survey Committee (NAS)

AASCU American Association of State Colleges and Universities (USA)

AASE Airborne Arctic Stratospheric Expedition • Airborne Arctic Stratospheric Experiment

AASERT Augmentation Award for Science and Engineering Research Training

AASES ATA Aviation Safety Exchange System • Australian Amateur Space Engineering Society

AASG Association of American State Geologists (USA)

AASI Amateur Astronomical Society of Iran • Astronomical Association of Southern Illinois (USA-IL)

A'asia Australasia

AASIR Advanced Atmospheric Sounder and Imaging Radiometer • Advanced Atmospheric Sounding and Imaging Radiometer

AASM Associated African States and Madagascar • Association of American Steel Manufacturers (USA)

AASP All Arizona Star Party (USA-AZ) • American Association of Stratigraphic Palynologists (USA) • ASCII

Asynchronous Support Package

AASPB American Astronomical Society Photo Bulletin (periodical code)

AASPP Astronomy and Astrophysics Series (periodical code)

AASR Airport and Airways Surveillance Radar

A&ASS Astronomy and Astrophysics Special Supplement Series (periodical code)

AASS Astronomy and Astrophysics Supplement Series • Astronomy And Space Sciences • Asuntos Sociales

AASSREC Association of Asian Social Science Research Councils

AASTINO Automated Astrophysical Site Testing Infant Observatory

AASTO Australian Astrophysical Site Testing Observatory • Automated Astrophysical Site Testing Observatory (*)

AASTR Advanced Along Track Scanning Radiometer

AASTRA American Astronomical Society Teacher Resource Agent

AASU Armstrong Atlantic State University (USA-GA) • Attaché d'Administration Scolaire et Universitaire (France)

AASW American Association of Scientific Workers (USA)

AASZ Astronomical and Astronautical Society of Zadar (Croatia)

AaT Archives at the Telescope (system)

A&AT Astronomical and Astrophysical Transactions (periodical code)

AAT Achievement Anxiety Test • Administration de l'Assistance Technique • Administrative Appeals Tribunal • Advanced Altimeter Tracker • Advanced Antenna Technology • Advanced Application of Technology • Agrupación Astronómica de Tenerife (Canary Islands) • Altay (China-PRC) (location identifier) • Amateur-Astronomie-Tagung • Analogic Alias Table • Anglo-Australian Telescope • Apple Advanced Typography • Architecture Analysis Tool • Area Adaptive Thresholding • Art and Architecture Thesaurus • Associazione Astrofili Teatini (Italy) • Associazione Astrofili Thethys (Italy) • Associazione Astrofili Treutini (Italy) • Associazione Astrofili Triestini (Italy) • Attitude Acquisition Technique • Australian Antarctic Territory • Automatic Analog Test • Automatic Analog Tester • Average Access Time • Aviation-Applied Technology • Avis d'Arrêt de Travail

AATB Anglo-Australian Telescope Board

AATC Advanced Automatic Train Control • Automatic Air Traffic Control

AATD Automated Articulation Testing Device • Aviation Applied Technology Directorate

AATGS AAT Guide Star

AATM At All Times

AATMS Airborne Air Traffic Management System

AATO Army Air Transport Organisation

AATR Air-to-Air Transmit/Receive

AATS AMOS Acquisition Telescope System • Automated Architecture Tool Suite

AATSE Australian Academy of Technological Science and Engineering

AATSR Advanced Along-Track Scanning Radiometer

AATT Advanced Air Transportation Technologies (project)

AATU Automatic Antenna Tuning Unit

AAU Århus University (Denmark) • Asau (Samoa) (location identifier) • Association of African Universities • Association of American Universities (USA) • Attaché d'Administration Universitaire

AAUA Agrupación Astronómica Universitaria de Alicante (Spain)

aaub Antwoord Alstublieft

AAUF Associations Aéronautiques de l'Union Française

AAUI Apple Attachment Unit Interface

AAUP American Association of University Presses (USA) • American Association of University Professors (USA) • Association Astronomique Universitaire de Perpignan (France) • Association of American University Presses (USA) • Association of Australian University Presses (Australia)

AAUW American Association of University Women (USA)

AAV Action-Angle Variable • Agrupación Astronómica Vizcaina (Spain) • Alah (Philippines) (location identifier) • Associazione Astrofili Valdinievole (Italy) • Associazione Astrofili Valtellineri (Italy) • Associazione Astrofili Veneziani (Italy)

AAVD Automatic Alternate Voice/Data

AAVSO American Association of Variable Stars Observers (USA-MA)

AAVV Autores Varios

AAW Abbottabad (Pakistan) (location identifier) • Active Aeroeleastic Wing • Advanced Analyst Workstation • Aerospace and Astrophysics Working Group • Algemene Arbeidsongeschiktheidswet • All Around the World

AAWF Auxiliary Aviation Weather Facility

AAWORD Association of African Women for Research and Development

AAX Araxá (Brazil-MG) (location identifier) • Area of Absolute Exclusion

AAY Al Ghaydah (Yemen) (location identifier) • Always All Yours

AAZ Airport Approach Zone • Areas And Zones

a/b Airborne

ab Abkhazian (ISO language code) • About • Abridgment

aB Auf Bestellung

Ab Astatine

A/B Afterburner • Airborne • Aktiebolaget

AB Able Seaman • Abstract • Abstracting Board • Acceleration Boundary • Ada-Based • Adapter Booster • Address Bus • Advisory Board • Aft Bulkhead • Afterburner • Aharonov-Bohm (interferometer) • Air Base • Air Bearing • Airborne • Aitreya Brāhmana • Alberta (Canada) • Algemeen Bestuur • Alliance Balkanique • Allocated Baseline • Altbau • Anchor Bolt • Answer Back • Antibody • Archive Branch • As Before • Assembly Bill • Assez Bien • Assyro-Babylonian • Asymptotic Bias • Atomic Bremsstrahlung • Autobeacon • Autobiographic • Autobiography • Automated Bibliography • Avionics Bay • Bachelor of Arts

ABA Abakan (Russia) (location identifier) • America - Britain - Australia • American Bankers' Association (USA) • American Bar Association (USA) • American Book Award • American Booksellers Association (USA) • American - British - Australian • Amplifier - Buffer - Attenuator • Annual Budget Authorization • Applied Behavior Analysis • Appropriation and Budget Activity • Asociación Boliviana de Astronomía (Bolivia) • Associate in Business Administration • Australian Bankers' Association • Australian Booksellers Association

ABAA Appropriation and Budget Activity Account

ABAG Association of Bay Area Governments (USA-CA)

ABAIT Automatic Board Assembly, Inspection and Test

AbAO Abastumani Astrophysical Observatory (Georgia)

AbAOB Abastumanskaia Astrofizicheskaia Observatoriia Bulleten (periodical code)

ABARS Aggregate Backup And Recovery Support (*) • Application Backup And Recovery Support • Automatic Briefing And Research System

ABATS Automatic Bit Access Test System

Abb Abbaye • Abbey • Abbildung • Abbreviation • Abbuono

ABB Abbreviation • Abingdon (UK-England) (location identifier) • Asea Brown Boveri • Asociación Boliviana de Bibliotecarios (Bolivia) • Associação Brasileira de Bibliotecários (Brazil)

ABBA Automated Biomass Burning Algorithm

Abbr Abbreviate • Abbreviated • Abbreviation

Abbrev Abbreviate • Abbreviated • Abbreviation

ABBS Astronomical Bulletin Board System

ABC Academia Brasileira de Ciências (Brazil) • Acquisition Basic Course • Active Broadcast Control • Activity-Based Costing • Ada Browse Capability • Administrative and Billing Center • Advanced Bar Code • Advanced Biomedical Capsule • Advanced Blade Concept • After Bottom Center • Airborne Control • Air Business Contact • Airway - Breathing - Circulation • Alcoholic Beverage Control • Allocations for Budgetary Control • America - Britain - Canada • American Bibliographical Center (USA) • American Broadcasting Co. (USA) • American - British - Canadian • Analog Backup Computer • Animal Birth Control • Antenna Base Computer • Antwerp - Brussels - Charleroi (Belgium) • Apparel Business Control • Application Builder Class • Approach By Concept • Argentina - Brazil - Chile • Artifact-Based Collaboration • Aruba - Bonaire - Curaçâo • Asociación de Bibliotecarios de Chile • Astrometric Binaries Catalog • Atanasoff-Berry Computer • Atomic - Biological - Chemical • Audit Bureau of Circulation • Augmented Bibliographic Citation • Australian Bibliographical Centre • Australian Broadcasting Commission • Australian Broadcasting Corporation • Automatic Bandwidth Control • Automatic Bar Code • Automatic Binary Computer • Automatic Break Control • Automatic Brightness Control • Automation of Bibliography through Computerization • Axial Bay Camera

ABCA America - Britain - Canada - Australia • American - British - Canadian - Australian • Association of British Columbia Archivists (Canada-BC)

ABCANZ America - Britain - Canada - Australia - New Zealand

ABCAS Algorithm Beam Center and Shape

ABCB Air-Blast Circuit Breaker

ABCC Association of British Chambers of Commerce

ABCCC Airborne Command and Control Center

ABCD Asymmetry - Border - Color - Diversity

ABCE Automated Bar Code Evaluator

ABCF As-Built Configuration File

ABCG Abell Galaxy Clusters • Automation and Bar Code Group

ABCL As-Built Configuration List

ABCM Associação Brasileira de Ciências Mecânicas (Brazil)

ABCP Airborne Command Post • Associate Business Continuity Professional

ABCR As-Built Configuration Record

ABCS Advanced Business Communication via Satellite • Automated Budget Compilation System

Abd Abdicate • Abdicated • Abdomen • Abdominal

ABD Abadan (Iran) (location identifier) • Abdicate • Abdicated • Aboard • Access Block Diagram • Ada-Based Document • Administration de Banque de Données • Administration de Base de Données • Airborne Ballistics Division (NASA) • All But Dissertation (*) • All But Done • Allgemeine deutsche Biographie • Assign Bypass Destination • Association Belge de Documentation (Belgium) • Average Body Dose • Axial-Boom Drive

ABDA American - British - Dutch - Australian • Analog-Based Domain Analysis

ABDC After Bottom Dead Center

ABDDD Ada-Based Development Design Document

ABDL Automatic Binary Data Link

ABDP Association of British Directory Publishers (UK)

ABDPEA Association Belge pour le Développement Pacifique de l'Énergie Atomique (Belgium)

ABDS Accounting Budget and Distribution System

ABDT Answer Back Day and Time

ABE Acetone - Butanol - Ethanol • Advisory Board on Energy • Agent Building Environment • Air-Based Electronics • Air-Breathing Engine • Allentown (USA-PA) (location identifier) • Arm-Based Electronics • Associação Brasileira de Educação (Brazil) • Astrobiology Explorer • Automated Bar code Evaluator • Bethlehem (USA-PA) (location identifier) • Easton (USA-PA) (location identifier)

ABEC Aitken Bicentennial Exhibition Centre (Canada-New Brunswick) • Analog Backup Engine Control

ABEd Bachelor of Arts in Education

ABEDA Arab Bank for Economic Development in Africa

ABELF Association des Éditeurs Belges de Langue Française (Belgium)

ABEND Abnormal End

Aber Aberdeen (UK-Scotland) • Aberdonian

Aberr Aberration

ABERT Auto Bit Error Rate Test

ABES Air-Breathing Engine System

ABET Accreditation Board for Engineering and Technology

ABETS Airborne Beacon Electronic Test Set

ab ex Ab Extra

abf Adobe Binary Format (file extension)

Abf Abfahrt

ABF Abaiang (Kiribati) (location identifier) • Adobe Binary Font • Advanced Beam Former • Application Binary Format • Asset Balance File • Association des Bibliothécaires Français (France) • Association des Bibliothèques de France • Availability Balance File • Average Branching Factor

AbfG Abfallbeseitigungsgesetz

ABFM Airborne Field Mill Program

ABFN Adaptive Beam-Forming Network • Adjustable Beam-Forming Network

ABFT Algorithm-Based Fault Tolerance

Abg Abgeordneter

ABG Abingdon (Australia-QLD) (location identifier) • Anti-Blooming Gate • Arterial Blood Gas • Association Belge des Géomorphologues (Belgium)

ABGB Allgemeines Bürgerliches Gesetzbuch

Abgk Abgekürzt

Abh Abhandlung • Abhandlungen

ABH Actual Bodily Harm • Alpha (Australia-QLD) (location identifier)

AbhKP Abhandlungen der Königlichen Preußichen Akademie der Wissenschaften (periodical code)

ABHRS As Built Hardware Reporting System

ABI Abilene (USA-TX) (location identifier) • Ada Binding Interface • Adaptive Brain Interface • Advance Book Information • Advanced Baseline Imager • Advanced Book Information • American Biographical Institute (USA-NC) • Application Binary Interface • Area-Based Initiative • Artificial Biosensing Interface • Association des Bibliothèques Internationales • Associazione di Bibliotecari Italiani (Italy) • Avionics Bus Interface

ABICS Ada-Based Integrated Control System

ABIDES Automated Budget Interactive Data Environment System

ABIDS Automated Biometric Identification System

ABIEG Amtsblatt der Europäischen Gemeinschaften

ab init Ab Initio

ABIOS Advanced Basic Input/Output System

ABIR Agent-Based Information Retrieval • All-Band Intercept Receiver

ABIRA American Biographical Institute Research Association (USA)

ABIRD Aircraft-Based Infrared Detector

ABIST Automatic Built-In Self-Test

ABIT Advanced Built-In-Test • Airborne Information Transmission • Automatic Built-In-Test

ABJ Abidjan (Ivory Coast) (location identifier)

ABJPAA Association Belge des Journalistes Professionnels de l'Aéronautique et de l'Astronautique (Belgium)

abk Autobackup (file extension)

Abk Abkürzung

ABK Kabri Dar (Ethiopia) (location identifier)

ABKie Astronomische Beobachtungen auf der Sternwarte der Königlichen Christian-Albrechts-Universität zu Kiel (periodical code)

abl ablehnend

Abl Ablative • Amtsblatt

ABl Amtsblatt

ABL Adaptation Binary Load • Airborne Laser • Aircraft-Based Laser • Albedo • Allocated Baseline • Allocation Baseline • Ambler (USA-AK) (location identifier) • Astrometry-Based Luminosity • Atmospheric Boundary Layer • Atmospheric Burst Locator

ABLE Activity Balance Line Evaluation • Airborne Laser Evaluation • Amazon Boundary Layer Experiment • Arctic Boundary Layer Expedition • Arctic Boundary Layer Experiment (*) • Atmospheric Balloon-borne Lidar Experiment • Atmospheric Boundary Layer Expedition

ABLE ACE Airborne Laser Extended Atmospheric Characterization Experiment

ABLISS Association of British and Information Science Schools

ABLOS Algorithm Beam Line Of Sight

ABLS Association of British Library Schools • Bachelor of Arts in Library Science

abm Album (file extension)

Abm Abeam

ABM Acquisition and Business Management • Acquisition Bus Monitor • Adams - Bashforth - Moulton (method) • Advanced Bill of Materials • Apogee Boost Motor • Area Base Maintenance • Associate in Business Management • Asynchronous Balanced Mode • Atmosphère Basse et Moyenne • Atomic Beam Method • Automatic Batch Mixing • Bamaga (Australia-QLD) (location identifier)

ABMP Abastumani Blazar Monitoring System

ABMPS Automated Business Mail Processing System

aBN Amorphous Boron Nitride

Abn Aberdeen (UK-Scotland)

ABN Airborne • Albina (Suriname) (location identifier) • African Biosciences Network • Agrupación de Bibliotecas Nacionales (Spain) • Algemeen Beschaafd Nederlands • Australian Bibliographic Network

ABNCP Airborne National Command Post

ABNEI Association Belge des Négociants Exportateurs et Importateurs (Belgium)

ABNF Augmented Backus-Naur Form

Abnl Abnormal

Abnml Abnormal

ABNT Associação Brasileira de Normas Técnicas (Brazil)

ABO Aboisso (Ivory Coast) (location identifier) • Adiabatic Born-Oppenheimer • Annular Beam Oscillator

Abog Abogado

ABOL Advisor Business-Oriented Language

ABOM Ada Bit-Oriented Message • Australian Bureau of Meteorology

ABOMO Airborne Optic and Magnetic Observatory

ABOP Algemene Bond van Onderwijzend Personeel

ABOR Abort

ABOS Algemeen Bestuur voor Ontwikkelingssamenwerking

ABp Archbishop

ABP Absolutely Black Paint • Acoustic Backscatter Probe • Algemeen Burgerlijk Pensioenfonds • Alternate Bipolar • Answer Back Procedure • Arterial Blood Pressure • Associated Book Publishers • Association des Bibliothèques Pédagogiques • Atkamba (Papua New Guinea) (location identifier)

ABPA American Book Producers Association (USA) • Australian Book Publishers' Association (Australia)

ABPDD Ada-Based Product Design Document

ABPP Arecibo-Berkeley Pulsar Processor

ABPS Air-Breathing Propulsion System • Automatic Blood Pressure System

ABQ Albuquerque (USA-NM) (location identifier)

abr Adobe Brush (file extension)

Abr Abrégé • Abridge • Abridged • Abridgment • Abril • Abrogé

ABR Aberdeen (USA-SD) (location identifier) • Abridged • Abril • Air Breathing Rocket • Answer Bit Ratio • Area Border Router • Assigned Bypass Reception • Asymmetric Bit Rate • Automatic Baud Rate • Automatic Beacon Removal • Available Bit Rate

ABRACOS Anglo-Brazilian Amazonian Climate Observational Study

ABRAMS Automated Bidirectional Reflectance Measurement System

ABRC Advisory Board for the Research Councils (UK) • Advisory Board to Research Council • Archives and Business Records Centre • Association des Bibliothèques de Recherche du Canada

ABRD Automatic Baud Rate Detection

ABRES Antiballistic Reentry System

Abrev Abreviación • Abreviado

Abrév Abréviation

ABRIXAS A Broadband Imaging X-ray All-sky Survey

ABRS Automated Book Request System

ABRV Advanced Ballistic Reentry Vehicle

Abs Absatz • Absence • Absender • Absent • Absolu • Absolument • Absolute • Absolutely • Arsorbent • Abstract

ABS Absent Subscriber • Abus de Biens Sociaux • Abu Simbel (Egypt) (location identifier) • Acoustic Bubble Spectrometer • Acrylonitrile Butadiene Styrene • Activity-Based Statistics • Address Book Synchronization • Advanced Baseline Sounder • Agile Beam System • Airborne Bathymetric Survey • Air-Break Switch • Alerting Base Station • American Bureau of Shipping (USA) • American Business Systems Inc. • Ammonia Boiler System • Anti-Blockier-System • Anti-Blocking System • Antilock Braking System • Association des Bibliothécaires Suisses (Switzerland) • Astro-Biology Society (USA-CA) • Asymmetrically Bright Shell • Atomic Beam Source • Australian Bureau of Statistics • Automated Bid Set • Automated Briefing System • Automatic Backup System • Automatic Bibliographic Service • Aux Bons Soins de

ABSC Advanced Biological Safety Cabinet

Abschn Abschnitt

abse. re. Absente Reo

abse. feb. Absente Febre

abs. feb. Absente Febre

ABSM Answer Back Service Message

ABSMAG Absolute Magnitude

Absol Absolu • Absolument • Absolute • Absolutely

Absorp Absorption

abs. re. Absente Reo

ABSS Astronomical Bulletin Board System

ABST Abstention

ABSTI Advisory Board on Scientific and Technical Information

Abstr Abstract • Abstracted

Abs. Visc. Absolute Viscosity

ABSW Association of British Science Writers (UK)

Abt About • Abteilung

ABT Abort • About • Air-Bearing Table • Al-Baha (Saudi Arabia) (location identifier) • Assign Bulk Transfer • ATM Block Transfer • Automatic Battery Test • Average Burning Time

ABTA Association of Travel British Agents

ABTAC Australian Book Trade Advisory Committee

ABTad Akademiia Nauk Tadzhikskoi SSR Institut Astrofiziki Biulleten (periodical code)

ABTCU Active Baseplate Thermal Control Unit

ABTEMA Asociación Boliviana de Teledetección para el Medio Ambiente (Bolivia)

ABU Acceptance for Beneficial Use • Analog Backup • Asian Broadcasting Union • Asian-Pacific Broadcasting Union • Asociación de Bibliotecarios del Uruguay • Atambua (Indonesia) (location identifier)

ABUEN Asociación de Bibliotecas Universitarias y Especializadas de Nicaragua

ABUP Asociación de Bibliotecarios Universitarios de Paraguay

ABV Above • Absolute Value • Abuja (Nigeria) (location identifier) • Annual Buy Value

ABVA Algemene Bond Van Ambtenaren

ABVM Automated Best Value Model

ABVV Algemeen Belgisch Vakverbond

abw abweichend

ABW Abau (Papua New Guinea) (location identifier) • Algemene Bijstandswet • Antarctic Bottom Water • Apparent Bandwidth • Arctic Bottom Water • Aruba (ISO country code)

ABWR Advanced Boiling Water Reactor

ABX Albury (Australia-NSW) (location identifier)

ABY Albany - Dougherty County (USA-GA) (location identifier)

Abys Abyssinian

abz abzüglich

ABZ Aberdeen Dyce (UK-Scotland) (location identifier)

abzgl abzüglich

Abzw Abzweigstelle

a/c Account • Account Current • A Cuenta • Air Conditioning • Ao Cuidado de

àc À Compte

ac A Capo • Acre • Activity • Air Conditioning • Alternating Current • Anno Christi • Año Citada • Ante Christum • Ante Cibum • Anno Currente

aC Antes de Cristo • Avanti Cristo

Ac Acetate • Acetyl • Actinium • Altocumulus

A&C Addenda & Corrigenda

A/C Absolute Ceiling • Account • Account Current • Air Conditioner • Air Conditioning • Aircraft • Approach Control • Area Code • Associate Contractor • Current Account

AC Abstract Computer • Access Charge • Access Concentrator • Access Control • Accord Commercial • Accretion Column • Aceiro • Acquisition Center • Acre (Brazil) • Action Code • Activated Carbon • Active Centre • Active Component • Actual Cost • Actual Count • Address Control • Address Cycle • Adjudication Committee • Administrateur Civil • Administration Centrale • Advanced Camera • Advanced Composite • Advisory Circular • Advisory Committee • Advisory Council • Aerodynamic Center • Aerospace Center • Agent Comptable • Agent Consulaire • Agent Contractuel • Air Carrier • Air Conditioning • Air Control • Aircraft • Aircraft Carrier • Aire Corrigée • Allocation de Chômage • Alternating Current • Altocumulus • American Cartographer (journal) • Amorphous Carbon • Analog Computer • Analyse Chimique • Analysis Center • Analysis Console • Analysis Coordinator • Analyst Console • Angular Correlation • Année Civile • Année Courante • Anno Corrente • Annual Conference • Anomalous Cepheid • Anomalous Component • Ante Christum • Ante Cibum • Antenna Computer • Anticenter • Anticoincidence • Antigalactic Center • Appeal Court • Appellation Contrôlée • Application Context • Application Controller • Area Code • Area Coverage • Arm Class (galaxies) • Army Corps • Arrêté de Compte • Asociación Civil • Assegno Circolare • Assemblée Constituante • Assembly Complete • Astrographic Catalog • Astronomical Constant • Astronomical Council • Atelier Central • Atlas-Centaur • Atmospheric Chemistry • Atmospheric Corrector • Attenuated Continuum • Attitude Control • Audio Center • Audit Committee • Auracanian • Australia-California • Australian Capital Territory • Authentification Center • Authorization Center • Authorization Code • Author's Correction • Autocheck • Autocollimation • Autocorrelation • Automatic Checkout • Automatic Computer • Automatic Control • Automation and Communications • Automobile Club • Auxiliary Computer • Avocat-Conseil • Axiom of Choice • Azimuth Center

AcA Acta Astronomica (periodical code)

ACA Academic Cooperation Association • Acapulco (Mexico) (location identifier) • Acquisition Capability Assessment • Action Commerciale Appliquée • Adjacent-Channel Attenuation • Advanced Computer Application • Advisory Committee for Astronomy • After Contract Award • Agency for cultural Affairs • Air Clearance Authority • Air Control Authority • Airspace Control Authority • Airspace Coordination Area • Algorithme de Compression Avancé • Altitude Controller Assembly • ALMA Compact Array • American Cartographic Association (USA) • American Conservation Association (USA) • American Crystallographic Association (USA) • Annunciator Control Assembly • Application Control Architecture • Applications of Computer Algebra (conference) • Aspect Camera Assembly • Associate Contractor Administration • Associate Contractor Agreement • Association Ciel d'Anjou (France) • Association of Consulting Actuaries • Associazione Cernuchese Astrofili (Italy) • Astro-Club Aubagnais (France) • Astronomie Centre Ardenne (Belgium) • Astronomy Club of Akron (USA-OH) • Astronomy Club of Augusta (USA-NJ) • Asynchronous Communications Adapter • Attitude Controller Assembly • Audio Compression Algorithm • Australian Communications Authority • Author Co-citation Analysis • Automatic Channel Assignment • Automatic Circuit Assurance • Automatic Conference Arranger • Waco (USA-TX) (location identifier)

ACAA Administrative Committee on Academic Affairs • Asociación Cordobesa Amigos de la Astronomía (Argentina) • Automatic Chemical Agent Alarm

AcAAS Acta Aeronautica et Astronautica Sinica (periodical code)

AcAau Acta Astronautica (periodical code)

ACABQ Advisory Committee on Administrative and Budgetary Questions

ACAC Administration Centrale de l'Aviation Civile • African Civil Aviation Conference • Ancient City Astronomy Club (USA-FL)

ACACC Association des Cartothèques et Archives Cartographiques du Canada

ACACE Advisory Council for Adult and Continuing Education (UK)

ACACIA A Consortium for the Application of Climate Impact Assessments

AcAcu Acta Acustica (periodical code)

Acad Academia • Academic • Academical • Academician • Académie • Académique • Academy

ACAD Advanced Computer-Aided Design • Automatic Chemical Agent Detector

Acad. Fran. Académie Française

ACADS Association for Computed-Aided Design

Acad. Sci. Académie des Sciences

ACAE Australian Commission for Advanced Education

AcAer Acta Aerodynamica Sinica (periodical code)

ACAI Advanced Course on Artificial Intelligence

ACAIS Air Carrier Activity Information System

ACAL Academia de Ciencias de América Latina • Advanced Computer Architecture Laboratory

ACAMPS Automated Communications and Message Processing System

Acap Acapulco (Mexico)

ACAP Århus Centre for Advanced Physics (Denmark) • Advanced Composite Aircraft Program • Advanced Computer for Array Processing • Application Configuration Access Protocol • Application Control Access Protocol • Assessment Correlation and Analysis Process • Automatic Circuit Analysis Program

AcApS Acta Astrophysica Sinica (periodical code)

ACaPS Agile Cable Production Service

ACAPS Automated Chart Production System

ACAQ Advisory Committee on Air Quality

ACAR Angular Correlation of Annihilation Radiation

ACARD Advisory Council for Applied Research and Development (UK)

AcArm Acta Armamentarii (periodical code)

ACARS Aircraft Communications Addressing and Reporting System • ARINC Communications, Addressing And Reporting System • ARINC Communications And Reporting System • Automatic Communications And Reporting System

ACARU Association Canadienne des Administrateurs de Recherche Universitaire (Canada)

AcAS Acta Astronomica Supplementa (periodical code)

ACAS Academic Services Planning Unit • Advisory Conciliation and Arbitration Service • Airborne Collision Avoidance System • Aircraft Collision Avoidance System • Application Control Architecture Services • Atlantic Canada Airborne Sensing

AcASn Acta Astronomica Sinica (periodical code)

ACAST Advisory Committee on the Application of Science and Technology

ACATA Association for Computer-Assisted Text Analysis (USA)

ACATS Advisory Committee for Advanced Television Service • Airborne Chromatograph for Atmospheric Trace Species

ACAU Automatic Calling/Answering Unit

AcAuS Acta Automatica Sinica (periodical code)

ACAVC Advisory Committee on Agricultural and Veterinary Chemicals

ACAWS Advisory, Caution, And Warning System

ACB Access Control Block • Advanced Concepts Base • Air Circuit Breaker • Air Cushion Barge • Allocated Configuration Baseline • Application Control Block • Architectural Control Board • Asociación Cubana de Bibliotecarios (Cuba) • Association Canadienne des Bibliothèques (Canada) • Astro Club de Beauce (France) • Automatic Call Back • Bellaire - Antrim County (USA-MI) (location identifier)

ACBAR Arcminute Cosmology Bolometer Array Receiver

ACBCT Automatic Circuit Board Card Tester

ACBCU Association Canadienne des Bibliothèques de Collèges et d'Universités (Canada)

ACBD Association pour la Coopération entre Bibliothèques et Centre de Documentation (France)

ACBE Advisory Committee on Business and the Environment (UK)

ACBL American Contract Bridge League (USA)

ACBLF Association Canadienne des Bibliothèques de Langue Française (Canada)

ACBP Advanced Concepts Base Program

ACBSP Association of Collegiate Business Schools and Programs

Acc Accelerate • Accelerated • Acceleration • Accélération • Accélérer • Accent • Accept • Acceptance • Accepted • Accompanied • Accompaniment • Accord • Accordant • According • Account • Accountant • Accounted • Accumulate • Accumulation • Accumulator • Accumuler • Accusative

AcC Acta Cosmologica (periodical code)

ACC Academia de Ciencias de Cuba • Accelerate • Accelerated Case Closure • Acceleration • Accept • Acceptability • Acceptance • Acceptance Communications Controller • Access Control Center • Accounting Classification Code • Accra Kotoka (Ghana) (location identifier) • Accréditif • Accumulator • Acid Copper Chromate • Action Concertée Coordonnée • Active Clearance Control • Acute Cardiovascular Collapse • Adaptive Cruise Control • Administration Commerciale et Comptable • Administrative Committee of Coordination • Administrative Committee on Coordination • Administrative Coordinating Committee • Advanced Charge Composition • Advanced Communications Controller • Advanced Communications Course • Advanced Computer Communications • Aft Cargo Carrier • Aft Cargo Compartment • Air Control Center • Air Coordinating Committee • Airspace Control Center • Airspace Coordinating Center • Allocation Control Center • ALMA Coordination Committee • Alternate Control Center • Altocumulus Castellanus • Annual Capital Charge • Antarctic Circumpolar Current • Antarctic Conservation Committee • Antenna Control Computer • Approval Cause Code • Arab Cooperation Council • Architecture Control Committee • Architecture Coordination Council • Area Communication Controller • Area Control Center • Area Control Computer • Area Coordination Center • Argonne Code Center (USA-IL) • Array Control Computer • Assignment Classification Code • Ateliers de Constructions du Centre (France) • Atlantic Climate Change • ATM Cross-Connector • Attitude Control Computer • Australian Chamber of Commerce • Australian Copyright Council • Automatic Callback Calling • Automatic Carrier Control • Automatic Control Console • Automation Classification Code • Auxiliary Crew Compartment • Aviation Civile et Commerciale

ACCA Air Charter Carriers Association • Automatic Cloud Cover Assessment

Accad Accademia

ACCAD Advisory Committee on Climate Applications and Data

ACCAT Advanced Command and Control Architectural Testbed

ACCBD Advisory Committee on Conservation of Biological Diversity

ACCC Area Control Computer Complex • Arthur C. Clarke Centre (Sri Lanka) • Australian Competition and Consumer Commission

ACCD American Coalition of Citizens with Disabilities (USA) • Astrometric Charge-Coupled Device

ACCDS Area Control Center Display System

Acce Acceptance

ACCE Advisory Committee on Chemicals in the Environment • Augmented Continuous Control Evaluator

Accel Accelerate • Accelerated • Acceleration

Accél Accélération • Accélérer

ACCEL Automated Circuit Card Etching Layout

ACCEM Atmospheric Chemistry and Climate Explorer Mission

ACCENT Atmospheric Chemistry of Combustion Emissions Near the Troposphere

ACCES Advanced Cryptologic Carry-on Exploitation System • Atelier de Capitalisation des Connaissances et d'Exploitation du Savoir-Faire • Automated Command and Control Evaluation System

Access Accessory

ACCESS Advanced Calorimeter for Composition of Elements of the Space Station • Advanced Cosmic-ray Composition Experiment for the Space Station • All Children Can Explore the Solar System • Alliance for Collaborative Change in School Systems • APM-Cambridge Catalog of the Equatorial Southern Strip • Assembly Concept for Construction of Erectable Space Structures • Automatic Computer-Controlled Electronic Signaling System

ACCF Area Communications Control Facility • Automated Command and Control Facility

ACCH Associated Control Channel

Accid Accident • Accidental

ACCIMT Arthur C. Clarke Institute for Modern Technologies (Sri Lanka)

ACCIS Advisory Committee for the Coordination of Information Systems (UNO) • Automated Command and Control Information System

ACCM Advanced Cooperative Collection Management

ACCME Accreditation Council for Continuing Medical Education

ACCMPS Automated Command and Control Message Processing System

ACCMT Arthur C. Clarke Centre for Modern Technologies (Sri Lanka)

ACCN Audit Central Control Network

Accom Accommodate • Accommodation

ACCOM Accommodation

Accomp Accompanied • Accompaniment • Accompany • Accomplishment

ACCOR Accounting Office of Record

ACCORD AURA Coordinating Committee of Observatory Research Directors

ACCP Advanced Command and Control Program • Atlantic Climate Change Programme

Accred Accredited

Accrd Accrued

ACCS Adaptive Communication Control System • Advanced Communications Control System • Area Command and Control System • Automated Command and Control System • Automatic Calling Card Service

ACCSCT Accrediting Commission of Career Schools and Colleges of Technology (USA)

ACCSIC Atomic Collision Cross-Sections Information Center

ACCSQ ASEAN Consultative Committee for Standard and Quality

Acct Account • Accountant • Accounting

ACCT Account • Advisory Committee on Certification and Testing • Agence de Coopération Culturelle et Technique (France) • Ancillary Computer Compatible Tape • Application of Common Characteristics and Testability

ACCTG Accounting

ACCTPO Accountable Property Officer

ACCTSTR Accountable Strength

ACCU Academisch Computer Centrum Utrecht (Netherlands) • Audio Central Control Unit

Accum Accumulate • Accumulated • Accumulation • Accumulative

Accus Accusative

ACCV Asian Conference on Computer Vision

Accy Accountancy

ACCY Accountancy

Acd Accord

ACD Acandi (Colombia) (location identifier) • Accuracy Control Document • Acrobat Capture Document • Administrative Commitment Document • Administrative Control Desk • Administrative Control Document • Adobe Capture Document • Advanced Concept Division • Aeronautical Charting Division • Allgemeine Computer-Datenbank • Allocated Configuration Documentation • Analysis and Computation Division • Anode-Cathode Distance • Anticoincidence Detector • Architectural Control Document • Assistant Command Director • Atmospheric Chemistry Division • Audio Control Panel • Auto Cell Device • Autocommutateur Digital • Automated Call Distribution (*) • Automatic Call Directory • Automatic Call Distribution • Automatic Call Distributor

ACDA Advanced Continuous Descent Approach • Advisory Committee on Distinction Awards • Arms Control and Disarmament Agency (USA) • Asociación Columbiana De estudios Astronómicos (Columbia)

ACDAR Acoustic Radar

ACDAS Automatic Control and Data Acquisition System

ACDB Automated Cost Database

AC/DC Alternating Current/Direct Current

AC-DC Alternating Current/Direct Current

ACDCS Air Contract Data Collection System

ACDDS Advanced Cartographic Data Digitizing System

ACDF Access Control Decision Facility • Access Control Decision Function

ACDI Agence Canadienne de Développement International (Canada) • Asynchronous Communications Device Interface

ACDK Accretion Disk

ACDM Asymmetric Cable Distribution Modem

ACDMS Automated Control of a Document Management System

ACDNS Attitude Control, Determination, and Navigation System

ACDO Air Carrier District Office • Assistant Command Duty Officer

ACDP Area Controller Display Position

ACDPI Association for a Competitive Data Processing Industry (USA)

ACDR Actual Cost Detail Report

ACDS Active Control Data Set • Archival Catalog and Database System • Astronomical Catalogs Database System • Attitude Control and Determination System

ACDSB Archive, Catalogs, and Data Services Division (STScI)

Acdt Accident

ACDTI Advanced Concept Development and Technology Integration

ACE Accelerated Capital Enrichment • Acceptance Checkout Equipment • Access Connection Element • Access Control Encryption • Acer Computer Explorer • Acidification in Clouds Experiment • Acquisition Center of Excellence • Action Communautaire pour l'Environnement • Actuator Control Electronics • Ada Common Environment • Ada Compilation Environment • Adaptive Command Element • Adaptive Communications Element • Adaptive Communications Environment • Adaptive Contrast Enhancement • Added Capability Effort • Admission and Credential Evaluators • Adobe Certified Expert • Adult and Continuing Education • Advanced Camera for Exploration • Advanced Cartographic Environment • Advanced Cartographic Equipment • Advanced Communication Engine • Advanced Communications Enhancement • Advanced Communications Environment • Advanced Components Evaluation • Advanced Composition Explorer (*) • Advanced Computing Environment • Advanced Computing Equipment • Advanced Concept Exploration • Advanced Control Experiment • Advanced Cooled Engine • Advanced Cosmic-ray Explorer • Adverse Channel Enhancement • Advisory Centre for Education (UK) • Advisory Committee on Environment • Advisory Committee on the Environment • Aerosol Characterization Experiment • Affinity Capillary Electrophoresis • Aide au Conseil aux Entreprises • Aids to Communication in Education • Airborne Command Element • Airborne Communications and Electronics • Airborne Communications Electronics • Air Cargo Express Inc. • Air Control Element • Aircraft Configuration Equipment • Airspace Control Element • Airspace Coordination Element • Alpha Centaurids (meteor stream) • Alternate Command Element • Alternating Conditional Expectation • Altitude Control Electronics • American Council on Education (USA) • Americans Communicating Electronically • Analysis and Control Element • Analysis Command Element • Analytical Computing Environment • Animated Computer Education • Animateur de Club Expérimental • Application Construction Environment • Arctic Cloud Experiment • Army Corps of Engineers • Array Control Electronics • Arts, Computation and Engineering • Assemblée Consultative Européenne • Assistant Chief of Engineers • Assistant Corps Engineer • Association for the Conservation of Energy • Association of Conference Executives • Association of Consulting Engineers • Association of Cultural Exchange • Astronomical Cataloguing Environment • Astronomical Consultants and Equipment Inc. (USA-AZ) • Astronomy Catalog Extractor • Atmosphere and Climate Explorer • Atmospheric Chemistry Education • Atmospheric Climate Experiment • Atmospheric Compensation Experiment • Attitude Control Electronics • Attitude Control Engine • Australian College of Education • Automated Commercial Environment • Automatic Calibration and Equalization • Automatic Calling Equipment • Automatic Centroid Extractor • Automatic Checkout Equipment • Automatic Computing Engine • Automatic Computing Engineering • Auxiliary Communications Equipment • Award in Continuing Education • Lanzarote (Canary Islands) (location identifier)

ACEA Action Committee for European Aerospace • Association Canadienne pour l'Éducation des Adultes (Canada) • Association des Constructeurs Européens d'Automobiles

ACEB Association Canadienne des Écoles de Bibliothécaires (Canada)

ACEC Ada Compiler Evaluation Capability • Advisory Committee on Electromagnetic Compatibility • American Consulting Engineers Council (USA) • Area of Critical Environmental Concern • Asociación Canaria para la Enseñanza de las Ciencias (Spain) • Ateliers de Construction Électriques de Charleroi (Belgium)

ACECHEM Atmospheric Composition Explorer for Chemistry

ACEEE American Council for an Energy-Efficient Economy (USA)

ACEF Ada Compiler Evaluation Facility • Anisotropic Curvature Eikonal Flow

ACEID Asian Center for Educational Innovation and Development

ACEIT Automated Cost Estimating Integrated Tools

ACEL Air Crew Equipment Laboratory

AcEle Acta Electronica (periodical code)

ACELE Accord Centre-Européen de Libre-Échange • Association Centre-Européenne de Libre-Échange

ACELP Algebraic Code-Excited Linear Prediction

AcElS Acta Electronica Sinica (periodical code)

Acep Acepción

ACEP Advisory Committee on Export Policy • American College of Emergency Physicians (USA)

Acept Aceptación

ACER Advisory Committee on Environmental Resources • Australian Council for Educational Research

ACERSS Advanced Civilian Earth Remote Sensing System

ACeS Asia Cellular System

ACES Acceptance Checkout and Evaluation System • Acceptance Control Equipment Section • Access Control and Environmental System • Acoustic Containerless Experiment System • Acquisition Center Executive System • Active Control Equivalence Studies • Altus Cumulus Electrification Study • Atomic Clocks Ensemble in Space • Automated Computer Examination System • Automatic Checkout Equipment Sequencer

ACESJ Applied Computational Electromagnetics Society Journal (periodical code)

Acet Acetone

ACET Advisory Committee on Electronics and Telecommunication

ACEU Aerocontrol Electronics Unit

acf Agent Character (file extension)

ACF Académie Canadienne Française • Acceptance Check Flight • Access Control Facility • Access Control Field • Access Coordination Fee • Acquisition Career Field • Advanced Common Flightdeck • Advanced Communications Facility • Advanced Communications Function • Agent Character File • Alternate Command Facility • Alternate Communications Facility • Angular Correlation Function • Applications Control Facility • Area Computing Facilities • Area Confinement Facility • Area Control Facility • Arizona Carbon Foil Co. Inc. (USA-AZ) • Astro-Club de France • Asynchronous Communications Facility • Australian Conservation Foundation • Automated Control Function • Autocorrelation Function • Awareness Campaign Forum

ACFA Analyse des Coûts de la Formation Administrative • Asian Committee for Future Accelerators

ACFAS Association Canadienne-Française pour l'Avancement des Sciences

ACFCI Association des Chambres Françaises de Commerce et d'Industrie (France)

ACFD Applied Computational Fluid Dynamics

ACFG Automatic Continuous Function Generation

ACFHE Association of Colleges of Further and Higher Education

ACFM Alternating-Current Field Measurement

ACFNET Advanced Communications Function Network

ACFO Aircraft Certification Field Office

ACFP Advanced Computer Flight Plan

ACFPS Advanced Computer Flight Planning System

AcFre Acta Frequenza (periodical code)

ACFS Abbreviated Cost Form System • Advanced Computerized Forwarding System • Advanced-Concept Flight Simulator

Acft Aircraft

ACFT Aircraft

ACG Address Coding Guide • Administration Computing Group • Archival Collection Guide • Area Coordination Group • Arp's Compact Galaxy • Automatic Common Generator • Automatic Control Gear

AcGeP Acta Geophysica Polonica (periodical code)

ACGI Advanced Computer Generated Image • Asynchronous Common Gateway Interface

ACGIH American Conference of Governmental Industrial Hygienists (USA)

ACGR Annual Compound Growth Rate

ACGW Analytical Construction by Generalized Work

ACH Achat • Acheté • Air Changes per Hour • Altenrhein (Switzerland) (location identifier) • Association for Computer and the Humanities • Attempts per Circuit per Hour • Automated Clearinghouse

ACHA American Clearinghouse Association (USA) • Applied and Computational Harmonic Analysis (journal)

ACHAYA Asociación Chilena de Astronomía y Astronautica (Chile)

ACHE ADR Control and Housekeeping Electronics • Alabama Center for Higher Education (USA-AL)

ACHEFT Automated Clearinghouse Electronic Funds Transfer

ACHEX Aerosol Characterization Experiment

Achiev Achievement

Achievd Achieved

AcHPh Acta Helvetica Physica (periodical code)

ACHR American Society of Human Rights (USA)

ACI Accelerated Capability Initiative • Access Control Information • Action Concertée Incitative (*) • Action Coordonnée Incitative • Adjacent Channel Interference • Admission and Coding Information • Advanced Chip Interconnection • Age-Controlled Item • Airports Council International • Alarm Control Interface • Alderney - The Blaye (Channel Islands) (location identifier) • Alliance Coopérative Internationale • Allocated Configuration Identification • Allocated Configuration Item • Assign Call Inhibit • Association Cartographique Internationale • Aurigny - The Blaye (Channel Islands) (location identifier) • Authority to Change Information • Automated Carrier Interface • Automatic Card Identification • Autonomous Citation Index • Aviation Civile Internationale

ACIA Asynchronous Communications Interface Adapter

ACIAAT Advisory Committee on Instrumentation for the Anglo-Australian Telescope

ACIAS Automated Calibration Internal Analysis System • Automated Calibration Interval Analysis System

ACIBS Architecture for Computer Integrated Business Systems

ACIC Aeronautical Chart and Information Center • Australian Chemical Industry Council

ACiCh Astronomical Circular (periodical code)

ACID Acidification Chemistry Information Database • Acoustic Imaging Development • Aircraft Identification • Analysis Console for Intrusion Databases • Atomic - Consistent - Isolated - Durable • Atomicity - Consistency - Isolation - Durability • Automatic Classification and Interpretation of Data

ACIFS Automated Central Issue Facility System

ACIGA Australia Consortium for Interferometric Gravitational Astronomy

ACIL Automatic Controlled Instrument Landing

ACIM Advanced Communications Interface Module

ACIP Aerodynamic Coefficient Instrumentation Package • Air Characteristic Improvement Panel • Assembly Configuration and Integration Panel • Aviation Career Incentive Pay

ACIR Advanced Computational Infrastructure and Research • Advisory Commission on Intergovernmental Relations

ACIRI AT&T Center for Internet Research at ICSI (USA-CA)

ACIS Advanced Camera for Imaging Spectroscopy • Advanced CCD Imaging Spectrometer • American Committee for Interoperable Systems (USA) • American Council for International Studies (USA) • Automated Claim Information System • AXAF CCD Imaging Spectrometer

ACISD Advanced Computational and Information Sciences Directorate

ACISP Australian Conference on Information Security and Privacy

ACIT Atmospheric Čerenkov Imaging Technique

ACIU Advanced Central Interface Unit

ACJ Attitude Control Jet

Ack Acknowledge • Acknowledged • Acknowledgment

ACK Acknowledge • Acknowledged • Acknowledgment • Amsterdam Compiler Kit • Nantucket (USA-MA) (location identifier)

ACKED Acknowledged

Ackgt Acknowledgment

ACL Academia de Ciências de Lisboa (Portugal) • Access Control List • Acquisition Career Level • Ada Command Language • Address Correction Letter • Advanced CMOS Logic • Advanced Communications Link • Advanced Computer Laboratory • Advanced Computing Laboratory • Advisory Committee on Libraries • Agent Communication Language • Aguaclara (Colombia) (location identifier) • Aircraft Cabin Load • All Critical Learning • Allowable Cabin Load • Allowable Cargo Load • Allowable Container Load • Alternate Concentration Level • An Chomhairle Leabharlanna (Ireland) • Applications Control Language • Ascent Closed Loop • Association for Computational Linguistics • Asynchronous Communications Line • Authorized Consumption List • Automatic Cartridge Library

ACLAE Association Canadienne des Laboratoires d'Analyse Environnementale

ACLC Adaptive Communication Live Controller

ACLD Above Clouds • Advisory Council on Library Development

ACLI Ada Command Language Interpretation

ACLL Association for Literary and Linguistic Computing

ACLR Access Control, Logging and Reporting

ACLS Advanced Cardiac Life Support • Air Cushion Landing System • American Council of Learned Societies (USA) • Automatic Carrier Landing System • Automatic Control and Landing System

ACLT Actual Calculated Landing Time

ACLU American Civil Liberties Union (USA)

ACLV Accrued Leave

ACM Accelerometer • Acquisition Control Module • Additional Crew Member • Address Complete Message • Advanced Climate Model • Advanced Component Mounter • Advanced Concept Manager • Agenda des Conférences

Mathématiques (EMS) • Allowable Cost Model • Analyse des Correspondances Multiples • Anticollision Maneuver • Arica (Colombia) (location identifier) • Asbestos-Containing Material • Association for Computing Machinery • Astéroïdes - Comètes - Météores • Asteroids - Comets - Meteorites • Asteroids - Comets - Meteors • ATC Communications Management • Atmospheric Control Module • Audio-Compression Manager • Australian Chamber of Manufacturers • Authorized Controlled Material • Auto-Correlation Method • Automated Communications Monitor • Automated Contract Manager • Auxiliary Core Memory • Auxiliary Crew Member • Average Cost per Mission

ACMA Acquisition Career Management Advocate • Advanced Concept Manager • Air Cargo Management Association

ACMAD African Centre of Meteorological Applications for Development

ACMan Astronomical Contributions from the University of Manchester (periodical code)

ACMB Acquisition Career Management Board • Applications Configuration Management Board • Attic Cycladic Metamorphic Belt

AcMCS Acta Materiae Compositae Sinica (periodical code)

ACME Adaptive Collection Management Environment • Advanced Computer for Medical Research • Advanced Cosmic Microwave Explorer • Airborne Cloud Measurement Equipment • Association of Consulting Management Engineers • Attitude Control and Maneuvering Electronics

AcMec Acta Mechanica (periodical code)

AcMeS Acta Meteorologica Sinica (periodical code)

AcMet Acta Metallurgica (periodical code)

AcMik Acta Mikrochimica (periodical code)

ACMIO Arhangelsk Complex Magnetic Ionospheric Observatory (Russia)

ACMIS Automated Configuration Management/Integration System

ACML Association of Canadian Map Libraries

ACMLA Association of Canadian Map Libraries and Archives

ACM&M Acta Metallurgica et Materialia (periodical code)

ACMMS Acta Metallurgica et Materialia Supplement (periodical code)

ACMNS Association for Computational Machinery Network Services

ACMO Acquisition Career Management Office • Acquisition Career Management Officer • Agent Chargé de la Mise en Oeuvre • Aicraft Maintenance Officer

Acmp Accompany

ACMP Acquisition Career Management Program • Advisory Committee on Marine Pollution • Annual Cost Monitoring Plan • Applied and Computational Mathematics Program (ARPA)

AcMph Acta Universitatis Caroliae. Mathematica et Physica (periodical code)

ACMPP Annual Cost Monitoring Program Plan

ACMR Annual Cost Monitoring Report • Annual Course Monitoring Report

ACMRR Advisory Committee on Marine Resources Research

ACMS Administration Configuration Management System • Advanced Configuration Management System • Advanced Cost Management System • Aerosol Composition Mass Spectrometer • Aircraft Condition Monitoring System • Analytical Cost Management System • Applications Control and Management System • Applications Control Management System • Automated Census Mapping System • Automated Career Management System • Automated Communications Management System • Automated Configuration Management System • Avionics Control Management System

ACMSA Antarctic Center for Millimetric and Submillimetric Astrophysics

AcMSn Acta Mechanica Sinica (periodical code)

AcMSS Acta Mechanica Solida Sinica (periodical code)

ACMT Automated Configuration Management Tool

ACN Acceptance Change Notice • Accession Change Number • Account Control Number • Accounting Control Number • Action Control Number • Administrative Change Notice • Advance Change Notice • Advisory Committee on Networking • Airborne Classification Number • Airborne Communications Node • Aircraft Classification Number • Alkane Carbon Number • All Concerned Notified • Ante Christum Natum • Assign Commercial Network • Assignment Control Number • Association Canadienne de Normalisation (Canada) • Authorized Code Number • Ciudad Acuna (Mexico) (location identifier)

ACNA Access Customer Name Abbreviation • Arctic Institute of North America

ACNFP Advisory Committee on Novel Foods and Processes

ACNN Australian Conference on Neural Networks

ACNO Advisory Committed of National Organizations (USA) • Assistant Chief of Naval Operations

AcNum Acta Numerica (periodical code)

A&CO Activation & Checkout • Assembly & Checkout

ACO Abell - Corwin - Olowin (catalog) • Acceptance Checkout • Action Object • Activation and Checkout • Administrative Contracting Officer • Administrative Contract Office • Advanced Concepts Office • Aircraft Certification Office • Ant Colony optimization (method) • Ascona (Switzerland) (location identifier) • Assembly and Checkout • Assembly Checkout • Associate Contracting Officer

ACOA Adaptive Course Of Action • Alternate Course Of Action

ACOB Altair Complex-Object Benchmark

ACOC Alternate Command Operations Center • Area Communications Operations Center • Area Control Operations Center • Associated Contractor Originated Change • Autocollimating Optical Coupler

ACOG Atlanta Committee for the Olympic Games (USA-GA)

ACOH Advisory Committee for Operational Hydrology

ACOM Australian Community Ocean Model • Automated Centralized Operations and Maintenance • Automatic Centralized Operations and Maintenance

ACOMEN Advanced Computational Methods in Engineering (conference)

ACOMS Automated Collection Management System

ACONET Akademisches Computer Netz

ACONIT Association pour un Conservatoire de l'Informatique et de la Télématique (France)

ACONS AMOS Control Software

AcOpS Acta Optica Sinica (periodical code)

AcOpt Acta Optica (periodical code)

ACOR Administrative Contracting Officer Representative

ACORD Advisory Council On Research and Development

ACORN Acronym-Oriented Nut • Advanced Collaborative Open Resource Network • Associative Content Retrieval Network • Automated Conversion Routine • Automatic Checkout and Recording Network

ACoS Advanced Communication Services (workshop)

ACOS Advanced Computer-Oriented System • Advanced Coronal Observing System • Advanced Crystal Observation System • Advisory Committee on Safety • Arms Control Observation Satellite • Assistant Chief Of Staff • Attitude and Orbit Control System • Automated Commissary Operation System

ACOSM Australian Conference On Software Management

ACOSS Agence Centrale des Organismes de Sécurité Sociale (France)

ACOST Advisory Council on Science and Technology

ACOT Apple Classroom Of To-morrow

A&COTA Assembly & Checkout Technical Analysis

ACOTA Assembly and Checkout Technical Analysis

AcouL Acoutics Letters (periodical code)

Acoust Acoustique

ACP Abstract Cauchy Problem • Acceptance Checkout Procedure • Access Control Protocol • Action Point • Active Control Point • Activité Complémentaire des Personnels • Adaptive Control Process • Adjacent-Channel Power • Advanced Computational Processor • Advanced Computer Program • Advanced Computer Project • Advanced Conventional Pressure • Advanced Cooperative Project • Aerosol Collector and Pyrolyser • Aerosol Collector/Pyrolyser • Aerospace Computer Program • Affordable Composite for Propulsion • African - Caribbean - Pacific • Afrique - Caraïbes - Pacifique • Agence Comptable Principale • Airborne Command Point • Air Control Point • Airspace Control Plan • Algebra of Communicating Processes • Alternate Command Post • Alternating Current Plasma • American Center for Physics (USA-MD) • Analyse en Composantes Principales • Analytical Computer Program • Ancillary Control Program • Aniline Cloud Point • Antenna Control Program • Architecture and Capabilities Platform • Area Coordination Paper • Argus Communications Protocol • Array Control Processor • Asia - Caribbean - Pacific • Aspen Center for Physics (USA-CO) • Assemblée Consultative Provisoire • Assign Common Pool • Associate Computing Professional • Association Canadienne des Physiciens (Canada) • Association of Canadian Publishers (Canada) • Association of Contingency Planners • Astronaut Control Panel • Astronomical Computing Panel (UK) • Atmospheric Chemistry Program • Atypical Clustering Procedure • Audio Control Panel • Automated Command Post • Automated Communications Processor • Automatic Communications Processor • Auxiliary Communications Package • Auxiliary Control Panel • Auxiliary Control Process • Azimuth Change Pulse

ACPA Adaptive Controlled Phase Array • Association of Computer Programmers and Analysts (USA) • Audio Capture and Playback Adapter

ACPAQ Advisory Committee on Postadjustment Questions

ACPAR Angular Correlation of Positron-Annihilation Radiation

ACPB Acquisition Career Program Board

ACPC Austrian Center for Parallel Computation

ACPD Alternating-Current Plasma Detector • Alternating-Current Plasma Difference

ACPDP AC Plasma Display Panel

ACPE Association for Continuing Professional Education (USA)

ACPG Astronomical Community Participation Group

ACpH Air Changes per Hour

ACPH Air Changes Per Hour

AcPhA Acta Physica Austriaca (periodical code)

AcPhH Acta Physica Hungarica (periodical code)

AcPhy Acta Physica (periodical code)

ACPI Advanced Configuration and Power Interface • Advanced Configuration Power Interface

ACPL Assembly Component Parts List

ACPM Activity Career Program Manager • Associate Contractor Program Manager • Attitude Control Propulsion Motor

ACPO Associate Contractor Projects Office • Association of Chief Police Officers

ACPQMB Acquisition Processes Quality Management Board

ACPRB Acquisition Career Program Review Board

ACPS Attitude Control Propulsion System • Automated Contract Preparation System

AcPSc Acta Polytechnica Scandinavica Applied Physics Series (periodical code)

ACPSEM Australasian College of Physical Scientists and Engineers in Medicine

AcPSl Acta Physica Slovaca (periodical code)

AcPSn Acta Physica Sinica (periodical code)

Acpt Acceptance

ACPT Accept

ACPU Association Canadienne des Professeurs d'Université (Canada) • Auxiliary Computer Power Unit

ACPVI Analyse en Composantes Principales avec Variables Instrumentales

ACPY Accompany

acq Acquire • Acquisition • Acquittal

ACQ Acquisition • Action Courte Qualifiante • Advisory Committee on Quality • All Call Query

ACQE Acquisition and Control Query Executive

ACQN Acquisition

ACQW Asymmetric Coupled Quantum Well

ACR Absolute Cryogenic Radiometer • Access Control Register • Accounting Clearance Record • ACCS Capability Requirements • ACCS Change Requirements • Acronyme • Active Cavity Radiometer • Actual Cost Report • Advanced Capabilities Radar • Advanced Cell Routing • Advanced Concepts and Requirements • Advanced Concepts Requirements • Aerodrome Control Radar • Afloat Correlation System • Air Carrier • Air Control Radar • Aircraft Control Room • Air Crew • Air Crew Rescue • Airfield Control Radar • Allowance Change Request • Allowed Cell Rate • Alternate Carrier Routing • Alternate CPU Recovery • Anomalous Cosmic Ray • Antenna Coupling Regulator • Approach Control Radar • Araracuara (Colombia) (location identifier) • Assign Channel Reassignment • Astroclub Radebeul (Germany) • Astrometric Calibration Region • Attenuation-to-Crosstalk Ratio • Audio Cassette Recorder • Authorization Change Request • Automatic Card Reader • Automatic Character Reader • Automatic Character Recognition • Automatic Compression Regulator • Automatic Current Regulation • Auxiliary Computer Room • Avionics Computer Resource

ACRB Acquisition Career Record Brief • Acquisition Civilian Record Brief

ACRBC Acceptance, Checkout, Retest, and Backout Criteria

ACRC Advanced Computing Research Centre • Alternate Control Reporting Center • Area Calibration and Repair Center • Assured Crew Return Capability (now ACRV)

Acrd Accrued

ACRE Acquisition of Requirements • Advanced Clock/Ranging Experiment • Assessing Calculus Reform Efforts • Automatic Climatological Recording Equipment

ACRES Airborne Communications Relay Station • Australian Centre for Remote Sensing • Automated Change Request and Evaluation System

ACRF Advanced Computer Research Facility • Advanced Computing Research Facility • Atmospheric Cloud Radiative Forcing

ACRFNet Advanced Computing Research Facility Network

Acrg Acreage

AcRhe Acta Rheologica (periodical code)

ACRI Advanced Computer Research Institute • Advanced Computer Research International • Advanced Computing Research Institute (USA-NY)

ACRIM Active Cavity Radiometer for Irradiance Monitor • Active Cavity Radiometer for Irradiance Monitoring • Active Cavity Radiometer Irradiance Monitor (*) • Active Cavity Radiometer Irradiance Monitoring

ACRIMSAT Active Cavity Radiometer for Irradiance Monitor Satellite

ACRL Association of College and Research Libraries

ACRMP Automation Communications Resource Management Plan

ACRN Account Classification Reference Number • Accounting Classification Reference Number

Acron Acronym

ACROSS Automated Cargo Release and Operations Service System

ACRP Advanced Concepts Research Project

ACRRP Automated Contractor Responsibility Review Program

ACRS Accelerated Cost Recovery System • Across • Advanced Carbon Reactor Subassembly • Advisory Committee on Reactor Safeguards • African Council on Remote Sensing • Asian Conference on Remote Sensing • Astrographic Catalog of Reference Stars

ACRT Accelerated Crucible Rotation Technique

ACRV Advanced Crew Return Vehicle • Assured Crew Rescue Vehicle • Assured Crew Return Vehicle (formerly ACRC) (*) • Astronaut Crew Rescue Vehicle • Astronaut Crew Return Vehicle

a/cs Accounts

ACS Academic Computer Specialist • Acesso • Access • Access Control Set • Access Control System • Accounting Classification Structure • Accounting Control System • Accumulator Switch • Achinsk (Russia) (location identifier) • Acquisition and Command Support • Active Cancellation System • Active Control System • Actuator Control System • Ada Compilation System • Adaptive Code System • Adaptive Control System • Address Change Service • Advanced Camera for Surveys • Advanced Civil Schooling • Advanced Communications Service • Advanced Communications System • Advanced Composite Structure • Advanced Computer System • Advanced Connectivity System • Afloat Correlation System • Aft Crew Station • Airborne Control System • Air Charter Services • Air Conditioning Shelter • Air Conditioning System • Air Control System • Airspace Control System • ALMA Common Software • Alternating Current Synchronous • Altitude Control System • Ambient Cooling System • American Chemical Society (USA) • Amphibious Command Ship • Analog Computer System • Analysis Computer System • Antarctic Conservation Strategy • Anti-Coincidence System • Application Communications System • Arab Computer Society • Archive and Catalog System • Archived Catalog System • Area Common User • Asian Chemical Societies • Assembly Control System • Asset Control System • Asynchronous Communications Server • ATM Circuit Steering • Atmosphere Control and Supply • Atmosphere Control System • Attitude Control and Stabilization • Attitude Control System • Australian Computer Society (Australia) • Autocorrelation Spectrometer • Autograph Card Signed • Automated Cartography System • Automated Cartridge System • Automated Commercial System • Automated Communications Set • Automated Confirmation Service • Automated Control System • Automated Conversion System • Automatic Checkout System • Automatic Class Selection • Automatic Climate Station • Automatic Container System • Automatic Control System • Autopilot Control System • Auxiliary Cooling System • Auxiliary Core Storage

ACSA Acquisition and Cross-Servicing Agreement

ACSAC Annual Computer Security Application Conference

ACSAD Arab Centre for the Study of Arid Zones and Dry Lands

ACSAG Advanced Composite Supportability Advisory Group

ACSB American Ceramic Society Bulletin (periodical code)

ACSC American Ceramic Society Communications (periodical code) • American Council on Schools and Colleges (USA) • Applied Communications System Center • Association of Casualty and Surety Companies

ACSE Access Control and Signaling Element • Analyse et Conduite des Systèmes d'Exploitation • Application Control Service Element • Arctic Coastal and Shelf Ecosystem • Association Control Service Element • Association Control Service Entity • Association of Consulting Structural Engineers

ACSEP Aircraft Certification System Evaluation Program

ACSES Alberta Centre for Surface Engineering and Science (Canada-AB)

ACSET Advisory Committee on the Supply and Training of Teachers

ACSF Association des Centres Serveurs Français (France)

ACSGP Advisory Committee on Standards, Guidelines and Protocols

ACSI American Consumer Satisfaction Index • Association Canadienne des Sciences de l'Information (Canada)

ACSIA Aide à la Communication pour les Systèmes d'Information Automatisés

ACSIOM Applied Computing System Institute of Massachusetts Inc. (USA-MA)

ACSIR Advisory Council for Scientific and Industrial Research (UK)

ACSIS Advisory Committee on Standards Information Services • Auto-Correlation Spectrometer Imaging System

ACSJ American Ceramic Society Journal (periodical code)

ACSL Advanced Continuous Simulation Language

ACSM Advanced Composite Supportability Manager • American Congress of Surveying and Mapping (USA)

ACSN Advance Change Schedule Notice • Advance Change Study Notice • Advanced Change Study Notice

ACSNet Australian Computer Science Network

ACSOE Atmospheric Chemistry Studies in the Oceanic Environment

ACSP Advanced Control Signal Processor • Advisory Council on Scientific Policy (UK) • Asynchronous Connection Server Program

ACSR Aluminium Conductor Steel-Reinforced

ACSS Analog Computer Subsystem • Automated Colour Separation System

Acst Acoustic

ACST Access Time • Advisory Council on Science and Technology (Australia)

ACSTI Advisory Committee for Scientific and Technical Information (UK)

ACSU Alternating-Current Switching Unit

Acsy Accessory

ACSyS Arctic Climate System Study

Act Actif • Acting • Action • Activates • Active • Aktivität • Activité • Activity • Actor • Actual • Actuary

ACT Acquisition Coordinating Team • Accredited Compaq Technician • Acquisition Coarse Tracker • Acquisition - Control - Test • Acting • Activated • Active • Active Control Technology • Active Cooling Test • Active Cooling Teststand • Activity • Actor • Actriz • Actual • Actuator • Adaptive Current Tomography • Advanced Cargo Transport • Advanced Communications Technology • Advanced Communications Terminal • Advanced Composites Technology • Advanced Compton Telescope • Advanced Computational Technology • Advanced Computer Technique • Advanced Computer Technology • Advanced Concept Technology • Advanced Concepts and Technology • Advanced Corporation Tax • Advanced Crew Terminal • Advisory Council on Technology (UK) • Air Cargo Terminal • Air-Čerenkov Telescope • Air Control Team • Air-Cushion Technology • Algebraic Compiler and Translator • Alliance Commerciale Technologique • American College Test • Analog Convection Transducer • Analogical-Circuit Technique • Analysis Complexity Tool • Analysis Control Team • Analysis of Complexity Tool • Annual Change Traffic • Applied Coherent Technology • Applied Computerized Telephony • Area Capability Training • Association of Classroom Teachers • Association of Commercial Televisions • Astrographic Catalogue (Tycho) • Atmospheric-Čerenkov Technique • Atmospheric-Čerenkov Telescope (*) • Attitude Control Thruster • Audio Center Transmission • Australian Capital Territory • Automated Circuit Test • Automated Control of Trainees • Automatic Code Translation • Automatic Command Tracking • Automatic Continuous Tuning • Azimuth Control Torquer • Waco (USA-TX) (location identifier)

a cta A Cuenta

ACTA Advanced Civilian Technology Agency (USA) • American Carriers Telecommunications Association (USA) (*) • America's Carriers Telecommunications Association (USA)

ACTAp Automatic Control Theory Applications (periodical code)

ACTAS Advanced Communication Technologies And Services • Alliance of Computer-based Telephony Application Suppliers

AcTC Acta Theoretica Chimica (periodical code)

ACTC Associated Colleges of the Twin Cities (USA-MN)

ACTD Advanced Communications Technology Demonstration • Advanced Concept Technical Demonstration • Advanced Concept Technology Demonstration • Advanced Concept Technology Demonstrator • Attitude Control Torquing Device

ACTE Approval Committee for Telecommunications Equipment • Automatic Checkout Test Equipment

AcTeC Acta Technica (periodical code)

ACTF Advanced Components Test Facility

ACTFEL AC Thin-Film Electroluminescent

ACTFL American Council on Teaching Foreign Languages (USA)

Actg Acting

ACTG Acting

ACT GS Actual Ground Speed

ACTI Advisory Committee on Technology Innovation

ACTID Advanced Concepts Test Integration Directorate

ACTIM Agence pour la Coopération Technique et Industrielle en Météorologie

ACTION Assisting Carers using Telematics Interventions to meet Older person's Needs

Activ Activement • Activité

ACTIVE Advanced Control Technology for Integrated Vehicles • Auroral Composition and Toroidal Ion Velocity Experiment

Actl Actual

ACTLU Activate Logical Unit

ACTM Atmospheric Chemical Transport Model

Actnt Accountant

ACTO Action Officer • Association en Touraine et Centre-Ouest (France) • Automatic Computing Transfer Oscillator

ACTP Adobe Certified Training Provider • Advanced Concepts Technology Program • American College Testing Program

ACTPO Accountable Property Officer

ACTPU Activate Physical Unit

Actr Actuator

ACTR Actuator

ACTS Acoustic Control and Telemetry System • Acquisition Coarse Tracking System • Acquisition Control and Tracking System • Advanced Civil Transport Simulator • Advanced Communications Technologies and Services • Advanced Communications Technology Satellite (*) • African Center for Technology Studies • Area Communications Terminal System • Attitude Control and Translation System • Automated CDRL and Tracking System • Automated Centralized Ticketing System • Automated Circuit Test System • Automated Computer Time Service • Automatic Coin Telephone Service

ACTT Advanced Communications and Timekeeping Technology

ACTU Australian Council of Trade Unions

Actuat Actuating

Actv Activation • Activity

ACTV Active

Actvt Activate

Actvtd Activated

Actvty Activity

Acu Acustica (periodical code)

AC&U Association of Colleges & Universities

ACU Acceleration Compensation Unit • Access Control Unit • Achutupo (Panama) (location identifier) • Address Calculation Unit • Address Control Unit • Air Control Unit • Alarm Control Unit • Alternating Current Control Unit • Altocumulus • Analog Control Unit • Analog Conversion Unit • Annunciator Control Unit • Antenna Control Unit • Area Common User • Arithmetic and Control Unit • Array Control Unit • Association of Colleges and Universities • Association of Commonwealth Universities • Association of Computer Users • Attitude Control Unit • Autocalling Unit • Autocall Unit • Automatic Calling Unit • Automatic Call Unit • Automatic Client Update • Automatic Client Upgrade • Avionics Control Unit • Avionics Cooling Unit

ACUC Animal Care and Use Committee • Area Common Use Computer

ACUI Automatic Calling Unit Interface

ACUIM Atmospheric Correction Using Imbedded Models

ACUO Avionics Cooling Unit Operator

ACUP Association of Canadian University Presses

ACURA Association of Canadian Universities for Research in Astronomy (Canada)

Acus Acusativo

ACUS Administrative Conference of the United States (*) • Area Common User System • Automated Conference of the United States

Acúst Acústica

ACUT Association de Culture Universitaire et Technique

ACV Actual Cash Value • Air-Cushion Vehicle • Alternating Current Volt • Antenna Cable Vehicle • Arcata (USA-CA) (location identifier) • Astro Club Voyager (Italy) • Eureka (USA-CA) (location identifier)

ACVA Accidente Cerebrovascular Agudo

ACVC Ada Compiler Validation Capability • Ada Compiler Validation Capacity

ACVE Accelerometer Calibration Vibration Exciter

ACVEN Advisory Committee on Vehicle Emissions and Noise

ACVS Astrometric Catalogue of Variable Stars

ACVT Atmospheric Chemistry Validation Team

ACW Aircraft Control and Warning • Alternating Continuous Wave • Anticlockwise • Automatic Car Wash

ACWAR Agile Continuous Wave Acquisition Radar

ACWC Advisory Committee on Weather Control

ACWP Actual Cost of Work Performed

ACWS Actual Cost of Work Scheduled • Aircraft Control and Warning System

ACWT Average Customer Waiting Time

ACX Alpha Charge Exchange

ACXN Analog Call Extension

ACY Any Country • Atlantic City - Pomona (USA-NJ) (location identifier)

ACYC Anticyclonic

ACZ Aerodrome Control Zone

AC2F Automated Command and Control Facility

AC2P Advanced Command and Control Program

a/d A Dater de • A la Date de

ad Adapt • Adaptation • Adapter • Adverb • After Date • an dem • an den • an der • Ante Diem • auf der • Autograph Document

aD außer Dienst

Ad Adiabatic • Admis • Admission • Advantage • Adverb • Advert • Advertisement

A&D Admission & Disposition • Aerospace & Defense

A/D Acquisition/Discrimination • Aerodrome • After Date • Analog-Digital • Analog to Digital • Antenna Drive • Arm/Disarm • Arrival/Departure • Assembly/Disassembly

A-D Analog to Digital

AD Abbreviated Dialing • Absolute Deviation • Access Door • Accession Document • Accidental Damage • Accretion Disk • Accumulated Dose • Acquisition Director • Acte de Décès • Active Directory • Active Duty • Activity Description • Addendum • Adjoint Administratif • Administrateur Délégué • Administration de Données • Administration Déléguée • Administration Départementale • Administrative Decision • Administrative Department • Administrative Directive • Administrative Domain • Admission and Disposition • Advanced • Advanced Deployability • Advanced Deployment • Advanced Design • Advanced Development • Advisory Direction • Aerodynamic Decelerator • Aeronautical Data • Aerospace and Defense • After Date • Air Density • Air Directive • Air Dried • Airworthiness Directive • Albrecht Dürer (1471-1528) • Algerian Dinar (currency) • Algorithmic Differentiation • Alternate Days • Altitude Deviation • Alzheimer's Disease • Ambipolar Diffusion • Aménageur-Développeur • Ampere Demand • Amplifier Detector • Analog-to-Digital • Ancillary Device • Andorra (ISO country code) • Andorre (code ISO de pays) • Anno Domini • Anodal Deviation • Ante Diem • Antimatter Decelerator • Antiproton Decelerator • Aperture Door • Apparent Drift • Applicable Document • Application Development • Architectural Design • Archive Device • Area Director • Arithmetic Device • Art Director • As Drawn • Assembly District • Assembly Drawing • Assistant Director • Associate Director • Astrometry Department • Astronautics Division • Asymmetric Drift • Atomic Drive • Attachment Device • Attribute Directory • Auditory Display • Auto Dial • Auto Doubler • Autograph Document • Autogroup Definition • Automated Data • Automatic Dictionary • Automatic Differentiation • Automatic Display • Autonomous Domain • Average Deviation • Avis de Débit • Ayant Droit

A/D/A Analog to Digital to Analog

ADA Acceleration - Deceleration - Acceleration • Acquisition Decision Authority • Action Data Automation • Adana (Turkey) (location identifier) • Adaptive Diagnostics Authoring • Additional Drill Assembly • Administration Agent • Adobe Developers Association • Advanced Data Analysis • Advanced Delivery Alert • Advanced Disk Array • Aerial Delivery Adapter • Aeronautical Development Agency • After Date of Award • Alternative Discipline Agreement • American with Disabilities Act (USA) • Amplifier Detector Assembly • Array Drive Assembly • Asociazione Divulgazione Astrofisica (Italy) • Assign Directive Acknowledgment • Association Drômoise d'Astronomie (France) • Astronomy in Daily Art (network) • Atomic Development Authority • Automatic Data Acquisition • Automatic Document Analysis • Azimuth Drive Assembly

AdA&A Advances in Astronomy and Astrophysics (periodical code)

ADAAC Affiliated Distributed Active Archive Center

ADAAPS Aircraft Data Acquisition Analysis and Presentation System

ADAB Australian Development Aid Bureau

ADABAS Adaptable Database System

ADAC Acoustic Data Analysis Center • Allgemeiner Deutscher Automobil-Club (Germany) • Astronomical Data Analysis Center • Attitude Determination And Control • Automatic Direct Analog Computer

ADACE ALMA Data Analysis Center

ADACS Attitude Determination And Control System • Automatic Data Acquisition and Control System

ADAD After Development Award Date • Automatic-telephone Dialing-Announcing Device

ADADL Ada Design And Documentation Language

ADADS ASTER Data Archive and Distribution System

ADAES Advanced Data Assessment Expert System

ADAF Advection-Dominated Accretion Flow

ADAGE Avionics Domain Application Generation Environment

ADALT Advanced Radar Altimeter

ADAM Administrative Documentation And Management • Advanced Data Management • A Dynamic Attribute Manager • Analog Data Acquisition Module • Architecture Development Analysis Model • Art Design Architecture and Media • Astronomical Data Acquisition and Management • Astronomical Data Acquisition Monitor • Astronomical Data Analysis and Management • Automated Distributed Architecture Model • Automatic Data Acquisition Module • Automatic Distance and Angle Measurement

ADAMAT Ada Metric Analysis Tool

ADAMHA Alcohol, Drug Abuse, and Mental Health Administration (now SAMHSA)

ADAMS Administrative Documentation And Management System • Automated Design And Maintenance System • Automated Dynamic Analysis of Mechanical Systems

Adap Adapted

ADAP Advanced Data Analysis Program • Aerodynamic Data Analysis Program • Argos Data Acquisition Platform • Automated Data Analysis Program

AdApM Advances in Applied Mathematics (periodical code)

ADAPS Advanced Data Acquisition and Processing System • Atmospheric Data Acquisition and Processing System • Automatic Data Acquisition and Processing System • Automatic Display And Plotting System • AVHRR Data Acquisition and Processing System

ADAPSO Association of Data Processing Organizations

ADAPT Adaptive Diagnostic Approach to Performance Testing • Adoption of Automatically Programmed Tools • Advanced Design And Production Technologies (program) • Analog-Digital Automatic Program Tester • Architecture Design, Analysis and Planning Tool • Automatic Density Analysis Profile Technique

ADAPTS Advanced Data And Picture Transformation System

ADAR Advanced Design Array Radar • Airborne Data Acquisition and Registration

ADARF Adjoint Administratif de Recherche et Formation (France)

ADARS Adaptive Antenna Receive System

ADARTS Ada-based Design Approach for Real-Time Systems

AdAs Ad Astra (periodical code)

ADAS Advanced Data Acquisition System (NASA) • Aerospace Data Acquisition System • Airborne Data Acquisition System • Aircraft Departure and Arrival Sequencing • Altrincham and District Astronomical Society (UK) • Ancillary Data Analysis Software • Asiago-DLR Asteroid Survey • Astronomical Data Analysis Software • Astronomical Data Analysis System • Atmospheric Data Acquisition System • Atomic Data and Analysis Structure (package) • Automatic Data Acquisition System • Automatic Digital Acquisition System • Auxiliary Data Annotation Set

ADASS Astronomical Data Analysis Software and Systems

ADAT Automatic Data Accumulator and Transfer

ADATE Automatic Digital Assembly Test Equipment

AdAtS Advances in Atmospheric Sciences (periodical code)

ADATS Adjustable Diversity Acoustic Telemetry System • AMOS Data Transmission System

ADAVAIL Advise Availability

ADB Accidental Death Benefit • African Development Bank • Apple Desktop Bus • Applications of Databases (conference) • Asian Development Bank • Assign Database • Astronomical Database • Izmir - Adnan Menderes (Turkey) (location identifier)

ADBIS Advances in Databases and Information Systems (conference)

ADBK Astronomical Database Kernel

ADBS Arecibo Dual-Beam Survey • Association des Documentalistes et Bibliothécaires Spécialisés (France)

ADBU Association des Directeurs des Bibliothèques Universitaires (France)

AD&C Attitude Determination & Control

ADC Academic Development Committee • Accretion Disk Corona • Active Duty Commitment • Actual Direct Cost • Adaptive Data Compression • Add with Carry • Advanced Data Controller • Advanced Developing Country • Advanced Diamond Composite • Advances in Distributed Computing (talks) • Advice of Duration and Charge • Affiliated Data Center • Aide De Camp • Air Data Computer • American Digital Cartography • American Digital Cellular • Analog Detection Circuit • Analog Digital Combiner • Analog-to-Digital Conversion • Analog-to-Digital Converter (*) • Antenna Drive Cabinet • Apparent Diffusion Coefficient • Application Development Conference • Application Device Channel • Applied Direct Cost • Area Data Center • Area Damage Control • Area Distribution Center • Astrometric Differential Coordinates • Astronomical Data Center • Astronomy Data Center (*) • Asynchronous Data Communications • Atlantic Data Coverage • Atmospheric Dispersion Compensator • Atmospheric Dispersion Corrector • Attitude Determination and Control • Automated Distribution Control • Automatic Data Collection • Automatic Data Controller • Automatic Detect Classification • Automatic Digital Calculator • Automatic Drift Control • Average Depth of Compensation

ADCA Advanced Continuous Descent Approach

ADCAP Advanced Capability

ADCBu Astronomical Data Center Bulletin (periodical code)

ADCC Advanced Data Communication Control Procedure • Automatic Data Cataloguing and Characterization

ADCCC Astrophysics Data Center Coordinating Council

ADCCP Advanced Data Communications Control Procedure • Advanced Data Communications Control Protocol

AD&CD Advanced Displays & Controls Development

ADCD Advanced Displays and Controls Development

ADCI Automatic Display Call Indicator

ADCLS Advanced Data Collection and Location System • Automated Data Collection and Location System

AdCM Advanced Ceramic Materials (periodical code)

ADCN Attitude Determination, Control, and Navigation

ADCNS Attitude Determination, Control, and Navigation System

ADCO Alcohol and Drug Control Office

AdCoL Advanced Composites Letters (periodical code)

ADCOM Administrative Command • Aerospace Defense Command (now AFSPC)

ADCON Administrative Control

ADCOP Area Damage Control Party

ADCOS Advisory Committee for Space Sciences

A&DCP Analysis & Data Collection Plan

ADCP Acoustic Doppler Current Profiler • Acquisition and Distribution of Commercial Products • Analysis and Data Collection Plan

ADCR Applicable Document Contractual Record

ADCS Association pour le Développement de la Culture Scientifique (France) • Attitude Determination and Control System • Australasian Document Computing Symposium • Automated Data Conversion System • Automated Document Conversion Service • Automated Document Conversion System • Automatic Data Collection System • Automation Data Collection System

ADCT Adaptive Discrete Cosine Transform

add Adidit

Add Addenda • Addendum • Addition • Additional • Address • Addressed

ADD Accessibility and Dissemination of Data • Adaptive Defect Detection • Adaptive Device Driver • Addendum • Addis Ababa (Ethiopia) (location identifier) • Airstream Direction Detector • Analog Data Decommutator • Analog Data Digitizer • Architectural Design Document • Architecture Definition Document • Arctic Data Directory • Assistive Devices Division • Attention Deficit Disorder • Attitude Determination Device • Automatic Document Detection • Automatic Document Distribution

ADDA American Design Drafting Association (USA) • Australian Database Development Association

ADDAR Automatic Digital Acquisition And Recording

ADDAS Airborne Digital Data Acquisition System

ADDB Antarctic Digital Database

ADDDS Automatic Direct Distance Dialing System

ADDE Abstract Data Distribution Environment

ADDEF Association pour le Développement du Droit de l'Espace en France (France)

ADDER Algorithm Description, Development, Evaluation and Refinement

ADDF Advanced Data Distribution Facility

ADDISS Advanced Deployable Digital Imagery Support Segment • Advanced Deployable Digital Imagery Support System

Addit Additional

Addl Additional

ADDL Additional

ADDM Asynchronous Digital Data Modem

ADDMACS Automated Data Distribution Management And Control System

ADDMD Administrative Directory Management Domain

Addn Addition

ADDNET Acid Deposition Data Network

Addnl Additional

ADDO Assistant Deputy Director for Operations

ADDP Antarctic Digital Database Project

Addr Address

ADDR Address • Address Register

ADDS Air Deployable Delivery System • Antarctic Data Directory System • Applications Developmental Data System • Applied Digital Data System • Automatic Data Distribution System • Automatic Document Distribution Service • Aviation Digital Data Service

Addsd Addressed

Addv Addvection

ADE Actuator Drive Electronics • Ada Development Environment • Aden (Yemen) (location identifier) • Aerial Delivery Equipment • Aeronautical Development Establishment • Agency-wide Distributed Environment • Application Development Environment • Approved Data Element • Array Drive Electronics • Assign Digit Editing • Assistant Division Engineer • Authorized Data Element • Autodesk Data Extension • Automated Design Engineering • Automatic Directory Exchange • Avance et Développement Électronique

ADEA Age Discrimination in Employment Act (USA)

ADEB Association Des Éditeurs Belges (Belgium)

ADEC Association pour le Développement des Entreprises et des Compétences

ADECAM Association pour le Développement de la Communication Animée et Multimédia

ADEF Action de Développement d'Évaluation et de Formation • Association de Défense des Économiquement Faibles

ad effect Ad Effectum

ADEIE Associated Digital Error Improvement Equipment

ADEL Antenne de Diffusion des Échanges du Livre (France)

ADELE Articulation de Déploiement à Lames d'Enroulement

ADEME Agence de l'Environnement et de la Maîtrise de l'Énergie (France)

ADEN Association des Documentalistes de l'Éducation Nationale (France)

ADENA Asociación para la Defensa de la Naturaleza (Spain)

AdEOA Advances in Earth-Oriented Applications (periodical code)

ADEOS Advanced Earth Observation Satellite • Advanced Earth Observing Satellite (*) • Advanced Earth Observing Spacecraft

ADEP Association for the Development of Earthquake Prediction (Japan) • Association pour le Développement de l'Éducation Permanente (France)

ADEPA Association pour le Développement de la Production Automatisée

AdEPS Advances in Earth and Planetary Sciences (periodical code)

ADEPT Advanced Distributed Environment for Production Technology • Analysis-based Data Exchange for Publications and Training • Assisting in Deployment of Energy Practices and Technologies

ADEQUA Aide à la Décision de la Qualité

ADER Association pour le Développement de l'Enseignement et de la Recherche (France)

ADERA Association pour le Développement de l'Enseignement et de la Recherche en Aquitaine (France)

ADES Association de Documentation Économique et Sociale • Automatic Digital Encoding System

ADESS Analog Data Equipment Switching System • Automated Data Editing and Switching System

ADESSO Analytical Data Exchange and Storage System Online

ADEST Association pour le Développement et l'Étude de la Scientométrie et de la Technométrie (France)

ADETEM Association pour le Développement des Techniques d'Études du Marché (France)

ADETIM Association pour le Développement des Techniques des Industries Mécaniques (France)

ad eund Ad Eundem

ADF Abundance Distribution Function • Active Dark Filament • Adapter Description File • Age Distribution Function • Airborne Direction Finder • ALFA Deep Field • Algorithm Development Facility • Alliance for a Digital Future • Application-Dependent Format • Application Development Facility • Approved Deposit Fund • Asian Development Fund • Astrophysics Data Facility • Automatically-Defined Function • Automatic Direction Finder • Automatic Direction Finding • Automatic Display Finder • Automatic Document Feed • Automatic Document Feeder • Avian Development Facility • Avionics Development Facility

ADFA Australian Defence Force Academy

ADFCS Affordable Digital FBW Control System

ad fin Ad Finem

ADFRF Ames-Dryden Flight Research Facility (NASA) (formerly DFRF)

ADFTO Advanced Data Flow Technology Office

ADG Adrian - Lenawee County (USA-MI) (location identifier) • Advanced Deployment Group • Assistant Director General • Authority Derivation Graph • Automated Data Generation • Automated Deduction in Geometry (workshop)

AdGeo Advances in Geophysics (periodical code)

ADH Advanced Development Hardware • Aldan (Russia) (location identifier) • Alternative Decohering History • Antidiuretic Hormone • Armagh - Dunsink - Harvard (telescope) • Automatic Data Handling • Average Delay till Handling

ADHD Attention Deficit/Hyperactivity Disorder

ad h.l. Ad Hunc Locum

ADHS Analog Data Handling System

AdI Agência de Inovação (Portugal)

ADI Abstract Data Interface • Acceptable Daily Intake • Air Defense Initiative • Alternating Direction Implicit • Alternative Direct Implicit (scheme) • Analog-to-Digital Interface • Analytic and Design Investigation • Apollo Document Index (NASA) • Application Data Interchange • Application Data Interface • Approved Driving Inspector • Arandis (Namibia) (location identifier) • Archivage et Documentation Intelligents • Area of Dominant Influence • Association de Droit International • Associazione Dottorandi e Dottori di Ricerca Italiani (Italy) • Attitude Direction Indicator • Attitude Display Indicator • AutoCAD Device Interface • Autodesk Device Interface • Automatic Direction Indicator

ADIAF Association pour le Développement de l'Informatique dans l'Administration Française (France)

ADIC Automatic Data Interface Console

ADID Authority and Description Identifier

ADIE Acquisition Data Input Equipment

ADIF Aide à la Définition d'Intercepteurs Futurs • Analog Data Interchange Format • Antarctic Directory Interchange Format

ADIFOR Automatic Differentiation of Fortran

ADII Agency for the Development of Innovations and Inventions

ADIL Astronomy Data Image Library (periodical code) • Astronomy Digital Image Library

ad inf Ad Infinitum

ad init Ad Initium

ad int Ad Interim

ADION Association pour le Développement International de l'Observatoire de Nice (France)

ADIOS Adiabatic Inflow-Outflow Solution • Advection-Dominated Inflow-Outflow Solution • Automatic Digital Input-Output System • Automatic Digitizing of Interpretation Overlay System

ADIRC Association pour le Développement Informatique en Région Centre (France)

ADIRS Air Data and Inertial Reference System • Air Data Inertial Reference System • Air Data/Inertial Reference System

ADIRU Air Data/Inertial Reference Unit

ADIS Acquisition and Due-In System • Arctic Data Interactive System • Association de Diffusion de l'Iconographie Scientifique • Association for the Development of Instructional Systems (USA) • Automated Data Interchange System

ADIT Agence pour la Diffusion de l'Information Technologique (France) • Analog-Digital Integrating Translator • Analog-Digital Integration Translator • Assistant Dean for Information Technology

ADIVA Association des Distributeurs Informatiques à Valeur Ajoutée (France)

ADIZ Air Defense Identification Zone

Adj Adjacent • Adjective • Adjetivo • Adjoining • Adjoint • Adjourned • Adjudged • Adjunct • Adjust • Adjustable • Adjusted • Adjustment • Adjutant

ADJ Accrétion - Disques - Jets • Adjacent • Adjustment • Amman (Jordan) (location identifier)

Adjt Adjoint • Adjutant

ADK Adak Island (USA-AK) (location identifier) • Additional Decryption Key • Application Development Kit

ADKC Automatic Key Distribution Center

ADKMS Advanced Data and Knowledge Management System

ADL Acoustic Delay Line • Activities of Daily Living • Activity Definition Language • Ada Design Language • Address Data Latch • Adelaide (Australia-SA) (location identifier) • Advanced Development Laboratory • Advanced Distributed Learning • Advances in Digital Libraries (conference) • Aeronautical Data Link • Agent Description Language • Airborne Data Link • Apollo Documentation List (NASA) • Après la Durée Légale • Architecture Description Language • Arthur D. Little (USA-MA) • Assertion Definition Language • Atmospheric Diffuse Light • Au-delà de la Durée Légale • Authorized Data List • Automated Data Library • Automatic Data Link • Avionics Development Laboratory

ADLAT Adaptive Lattice

ad lib Ad Libitum

ADLIPS Automatic Data Link Plotting System

ad loc Ad Locum

ADLP Advanced Data Link Processor • Advanced Data Link Program • Automated Data Link Protocol

ADLS Airborne Doppler Lidar System

ADLY Arrival Delay

Adm Administración • Administratif • Administration • Administrative • Administrator • Admiraal • Admiral • Admiralty • Admission • Admitted

ADM Acquisition Decision Memorandum • Active Device Mount • Activity Data Method • Adaptive Delta Modulation • Add-Drop Multiplexer • Add-Drop Multiplexing • Administração • Administrador • Administratif • Administration • Administrative • Admiral • Advanced Demonstration Model • Advanced Development Model • Aeronautical Decision Making • AFXS Deployment Mechanism • Aggregate Data Manager • Air Data Management • Analysis/Design Model • Annual Delegate Meeting • Ardmore (USA-OK) (location identifier) • Arme de Destruction Massive • Arnowitt - Deser - Misner (formalism) • Arrow Diagram Method • Asynchronous Disconnect Mode • Atmospheric Dynamics Mission (ESA) • Automated Deployment Mechanism • Automatically Defined Macro • Automatic Display Mode • Automatic Distribution of Microfiche • Average Daily Membership

ADMA Aviation Distributors and Manufacturers Association

ADMACS Apple Document Management and Control System

ADMAPS Automated Document Management and Publishing System

ADMATCH Address Matching

ADMC Agriresidue Dry Matter Content

ADMCEN Administration Center

AdMD Administrative Management Domain

ADMD Administrative Directory Management Domain • Administrative Management Domain

ADMF Asynchronous Data Mover Facility

Admin Administered • Administrate • Administratif • Administrating • Administration • Administrative • Administrator

ADMINCEN Administration Center

ADMIRE Automatic Diagnostic Maintenance Information Retrieval

ADMIS Administrative Management Information System • Automated Data Management Information System

ADMIT African Drought Monitoring Integrative Technique • Automatic Drought Monitoring Integrative Technique

Admón Administración

ADMN Administration

Admor Administrador

ADMOS Automatic Device for Mechanical Order Selection

admov Admove • Admoveatur

ADMP Adjusted Monthly Program

Admr Administrador

ADMS Advanced Digital Modem System • Analog and Digital Monitoring System • Antarctic Data Management System • Automated Data Management System • Automatic Digital Message Switch

ADMSC Automatic Digital Message Switch Center

ADMSE Automatic Digital Message Switch Equipment

ADMSR Airborne Dual-frequency Microwave Radiometer-Scatterometer

ADMT Active Directory Migration Tool

AdMUM Advances in Mechanics Uspekhi Mekhaniki (periodical code)

ADMYTE Archivo Digital de Manuscritos Y Textos Españoles (Spain)

ADN Abbreviated Dialing Number • Acide Désoxyribonucléique • Advanced Digital Network • Agency Disclosure Notice • Allgemeiner Deutscher Nachrichtendienst • Andes (Colombia) (location identifier) • Anti-Dwarf Nova •

Area Distribution Node • ASTER Data Network • Automatic Digital Network • Automatic Dobson Network • Yemen (international vehicle registration code)

ADNDT Atomic Data and Nuclear Data Tables • Atomic Data and Nuclear Data Tables (periodical code)

ad neut Ad Neutralizandum

ADNFI Association des Nouvelles Formations d'Ingénieurs

ADNS Automated Data Network System • Automated Digital Network System

ADNX Access Digital Network Exchange • Adaptive Digital Network Exchange • Automated Digital Network Exchange

ADO Abstract Data Object • Active Data Object • ActiveX Data Object • Advanced Development Objective • Airport District Office • Andamooka (Australia-SA) (location identifier) • Anschlußdose • Arbitrary Digital Object • Assistant District Officer • Associate Director of Operations • Associate Disbursing Officer • Astronomy Department Observatory • Automated Delivery Order • Automotive Diesel Oil

ADOC Architecture Description and Operational Concept • Architecture Design and Operational Concept

ADOM Acid Deposition and Oxidant Model

ADONIS Adaptive Optics Near-Infrared System • AMOS Daytime Optical Near-infrared Imaging System • Article Delivery Over Network Information Systems • Automatic Digital Online Instrumentation System

ADOP Advanced Distributed Onboard Processor • Automated Design Optimization Program

ADOR Astrometry and Design Operations Review • Astronomy Design and Operations Review

ADOS African Doppler Observation Survey

ADOT Advanced Display Optimization Tool • Automatic Digital Optical Tracker

AdP Arbeitsgemeinschaft deutschsprachiger Planetarien

ADP Acceptance Data Package • Acceptance Documentation Package • Accountable Paper Depository • Administrative Data Processing • Advanced Data Processing • Advanced Design Program • Advanced Detection Processor • Advanced Development Plan • Advanced Development Program • Advanced Ducted Propeller • Advanced Ducted Pulsion • Advance Development Plan • Advance Development Program • Aéroports de Paris (France) • Airborne Data Processor • Air Data Package • Air Data Probe • Anuradhapura (Sri Lanka) (location identifier) • Application Development Platform • Application Distribution Protocol • Arbeitsgemeinschaft deutschsprachiger Planetarien • Architecture Development Process • Area Density of Particles • Associated Deep Penetration • Astrophysics Data Program • Atmospheric Dynamics Program • Attitude Determination Processor • Automated Data Process • Automated Data Processing • Automatic Data Processing • Automatic Data Processor • Peseta (Andorra) (ISO currency code)

ADPA Air Data Probe Assembly

ADPC Automated Data Processing Center • Automatic Data Processing Center

ADPCM Adaptive Delta Pulse Code Modulation • Adaptive Difference Pulse Code Modulation • Adaptive Differential Pulse Code Modulation (*)

ADPE Automated Data Processing Equipment • Automatic Data Processing Equipment • Auxiliary Data Processing Equipment

ADPF Automated Data Processing Facility • Automatic Data Processing Facility

ADPG Advances in Planetary Geology (periodical code)

AdPhy Advances in Physics (periodical code)

ADPL As Designed Parts List • Average Daily Patient Road

ADPLAN Advanced Planning

ADPLO Automatic Data Processing Liaison Officer

ADPLP Automatic Data Processing Laser Printer

ADPM Adaptive Delta Pulse Modulation • Automatic Data Processing Machine

ADPMIS Automatic Data Processing Management Information System

ADPMP Automatic Data Processing Master Plan

ADPP Automatic Data Processing Printer • Automatic Data Processing Program

ADPR Automated Data Processing Resources • Automatic Data Processing Resources

ADPS Acceptance Data Package System • Automated Data Processing System • Automatic Data Processing System • Automatized Differential Photometry System

ADPSC Automatic Data Processing Service Center

ADPSCP Automatic Data Processing Status and Control Panel

ADPSEC Automatic Data Processing Security

ADPSO Association of Data Processing Service Organizations • Automatic Data Processing Security Officer • Automatic Data Processing Selection Office

ADPSSEP Automatic Data Processing System Security Enhancement Program

ADPSSO Automatic Data Processing Systems Security Officer

Adptr Adapter

ADQ Audit on Data Quality • Kodiak (USA-AK) (location identifier)

ADQC Astronomical Data Quality Control

ADQCS Astronomical Data Quality Control Scientist

Adr Adresse

ADR Accelerated Delivery Request • Accident Data Recorder • Achievable Data Rate • Active Data Repository • Active Distributed Routing • Address • Adiabatic Demagnetization Refrigerator • Advanced Digital Recording • Adverse Drug Reaction • Advisory Route • Aeroporti Di Roma (Italy) • Aggregate Data Rate • Airborne Data Relay • Aircraft Damage Repair • Air Data Relay • Alternative Dispute Resolution • Analog Digital Recorder • Andrews (USA-SC) (location identifier) • Annonce Différée de Résultats • Anomaly Detection and Reporting • Applied Data Research • Architectural Design Review • ASTRA Digital Radio • Automated Demand Resolution • Automatic Data Relay

ADRA Administrative Dispute Resolution Act (USA)

ADRAC Automatic Digital Recording And Control

ADRAM Advanced Digital Range Modification

ADRD Automatic Data Rate Detection

ADRG Arc Digitized Raster Graphics • Automatic Digital Raster Graphics

ADRIES Advanced Digital Radar Imagery Exploitation System

ADRL Automatic Drawing Release List

ADRP Automated Data Reduction Program

Adrs Address • Addressee

ADRS A Departmental Reporting System • Analog-to-digital Data Recording System • Astrometry Data Reduction System • Automated Data Retrieval System • Automatic Data Reporting System • Automatic Discrepancy Reporting System • Automatic Document Request Service

ADRST Association pour le Développement de la Recherche Scientifique et Technique (France)

ADRT Analog Data Recorder Transcriber

ad s Ad Sectum

AdS Académie des Sciences

ADS Académie Des Sciences • Accelerator-Driven System • Accurately-Defined System • Acid Deposition System • Acquisition Data System • Acquisition Display System • Action Démocratique et Sociale • Active Directory Service • Activity Data Sheet • Ada Development System • Adaptive Diagnostics System • ADF Data System • Administrative Data System • Advanced Data Streaming • Advanced Data System • Advanced Degaussing System • Advanced Deployable System • Advanced Digital Switch • Advanced Digital System • Advanced Display System • Advanced Distributed Simulation • Advanced Distributive Simulation • Advanced Distribution System • Advanced Dressing Station • Advection-Dominated Source • Aerial Delivery System • After-Dinner Speech • Aide Décisive à la Sécurité • Airborne Data System • Air Data Sensor • Air Data System • Aitken Double Stars • American Dialect Society (USA) • Analog Distribution System • Angle Displacement Sensor • Angular Displacement Sensor • Application Development Software • Application Development Solution • Application Development System • Application-Driven Scalability • Archive Data Set • Ascent Design System • Aspect Determination System • Asset Data System • Associate Director for Science • Astrophysical Data System • Astrophysics Data System • Asynchronous Data Set • Attitude Determination System • Attitude Displacement System • Attitude Display System • Audio Distribution System • Authoritative Data Source • AutoCAD Development System • Auto-Debiting System • Autograph Document Signed • Automated Data Service • Automated Data System • Automated Dependent Surveillance • Automatic Data System • Automatic Decompression System • Automatic Dependent Surveillance • Automatic Depressurization System • Automatic Distribution System • Automatic Door Seal • Auxiliary Data System • Azimuth Determining System • Dallas Addison (USA-TX) (location identifier)

ADSA Alternate Demonstration Structural Article • Angular Displacement Sensor Assembly

ad saec Ad Saeculum

ADSAP Advise As Soon As Possible

ad sat Ad Saturandum

ADSB Automatic Dependent Surveillance Broadcast

ADSC Acquisition Data Systems Control • Acquisition Data Systems Controller • Address Status Changed • Adobe Document Structuring Convention • Automated Data Source Collection • Automatic Data Service Center • Automatic Digital Switching Center

ADSD Asiago Database of Spectroscopic Databases

ADSEP Advanced Separation Process • Automatic Data Set Editing Program

ADSERV Advertizing Service

ADSF Advanced Directional Solidification Furnace (*) • Automated Directional Solidification Furnace • Automatic Directional Solidification Furnace

ADSG Agents Du Service Général

ADSI Active Directory Services Interface • Analog Directory Services Interface • Analog Display Services Interface

A&DSL Administrative & Direct Support Logistics

ADSL ABR Digital Subscriber Line • Administrative and Direct Support Logistics • ASCA Deep Survey in the Lockman hole • Asymmetrical Digital Subscriber Line • Asymmetric Digital Subscriber Line (*) • Asymmetric Digital Subscriber Loop

ADSM ADSTAR Distributed Storage Manager • Automated Data Systems Manual

ADSN Accounting and Disbursing Station Number

ADSP Advanced Digital SAR Processor • Aerially Deployed Soil Penetrator • AppleTalk Data Stream Protocol

AdSpR Advances in Space Research (periodical code)

ADSR Administration of Designed Services Review • Attach - Decay - Sustain - Release

ADSRM Auto Dispatch System for Registered Mail

ADSRN ASLV Down Range Station

ADSS Acquisition Decision Support System • Advanced Digital Stereophotogrammetric System • Analysis of Digitized Seismic Signals • Atmospheric Density Specification Satellite • Automated Data Subsystem • Automated Decision Support System • Automatic Data Switching System • Automated Dictionary Support System

ADST Advanced Distributed Simulation Technology

ADSTAR Automatic Document Storage and Retrieval

adst. feb. Adstante Febre

ADSU Air Data Sensor Unit • ATM Data Service Unit

ADSUP Automated Data Systems Uniform Practices

ADT Abstract Data Type • Actual Departure Time • Ada (USA-OK) (location identifier) • Advanced Design Team • Advanced Development Technology • Advanced Display Technology • Advanced Display Terminal • Airborne Data Terminal • Air Data Transfer • Alaska Daylight Time • Alphanumeric Display Terminal • Anomalous Diffraction Theory • Application Data Type • Application Development Tool • Approved Departure Time • Architecture Development Team • Astronomy Diagnostic Test • Atlantic Daylight Time • Automatic Data Transfer • Automatic Data Translator • Automatic Data Transmission • Automatic Data Transmitter • Automatic Detection and Tracking • Automatic Digital Tester • Average Daily Traffic

ADTA Advance Deposit Trust Account • Aircraft Development Test Activity • Air Data Transducer Assembly • Aviation Development Test Activity

ADTAC Automatic Digital Tracking Analyzer Computer

ADTD Association of Data Terminal Distributors

ADTE Advanced Technology Demonstrator Engine

ADTF ACR Decrease Time Factor

ADTG Assign Digital Transmission Group

ADTN Administrative Data Transmission Network

ADTP Accelerated Development Test Program

ADTRS Advanced Tracking and Data Relay Satellite

ADTS Advanced Detection Technology Sensor • Automated Data and Telecommunications Service • Automated Digital Terminal System • Automatic Data and Telecommunications Service

ADTT Advanced Digital Television Technologies

ADTU ASCII Data Translation Utility • Automatic Digital Test Unit • Auxiliary Data Transfer Unit • Auxiliary Data Translator Unit

ADU Acceleration-Deceleration Unit • Accumulation and Distribution Unit • Air Data Unit • Alpha Demonstration Unit • Analog-Digital Umsetzer • Analog/Digital Unit • Analog Display Unit • Analog to Digital Unit (*) • Annunciator Display Unit • Application Development University • Arbitrary Data Unit • Ardabil (Iran) • Ascent Driver Unit • Assign Dial Up • Asynchronous Data Unit • Audio Distribution Unit • Automatic Dialing Unit • Auxiliary Display Unit • Azimuth Drive Unit

ADUA Administrative Directory User Agent

ADUrb Astronomy Department of the University of Illinois at Urbana (periodical code)

ad us Ad Usum

ADUS Avionics Data Utilization System

ad us. ext. Ad Usum Externum

ad v Ad Valorem

Adv Advance • Advanced • Advancement • Advent • Adverb • Adverbial • Adverbially • Adverbio • Adversaire • Adverse • Adversus • Advertisement • Advertising • Advice • Advise • Adviser • Advisory • Advocate

ADV Abstract Data View • Advance • Advanced • Advise • Advogado • Air Distribution and Ventilation •

Allgemeine Datenverarbeitung • Andover (UK-England) (location identifier) • Arbeitsgemeinschaft für Datenverarbeitung • Automatische Datenverarbeitung • Automatisierte Datenverarbeitung

ADVA Automatische Datenverarbeitungsanlage

ad val Ad Valorem

Advb Adverb

ADVCAP Advanced Capability

ADVCTN Advection

Adven Adventure

Advers Adversativo

ADVIS Advances in Information Systems (conference)

ADVISAR All Digital Video Imaging System for Atmospheric Research

ADVISOR Advanced Integrated Safety and Optimizing Computer

ADVN Advance

ADVOX Advanced Oxidation

ADVOXP Advanced Oxidation Process

ADVP Aide au Développement Vocationnel et Personnel

Adv. Pmt. Advance Payment

ADVPMT Advance Payment

AdvSpaRes Advances in Space Research

Advt Advertisement

ADVT Advanced Development Verification Test

Advtg Advertising

ADVY Advisory

AdW Akademie der Wissenschaften

ADW Advanced Design Workshop • Asymptotic Distorted Wave • Camp Springs (USA-MD) (location identifier)

ADWG Astronomical Database Working Group

ADWO Associate Director for WET Operations

ADWS Active Data Warehouse System • Automated Digital Weather Switch

ADX Advanced Development Experimental • Assign Digit Translation • Automatic Data Exchange • Automatic Digital Exchange • Saint Andrews - Leuchars (UK-Scotland) (location identifier)

ADXPS Angular Dependent X-ray Photoelectron Spectroscopy

ADY Alldays (South Africa) (location identifier)

ADZ Als Daar Zijn • Approach and Departure Zone • San Andrés Island (Colombia) (location identifier)

ae Aetatis

aE am Ende

A&E Accident & Emergency • Airframe & Engine • Allocation & Engineering • Analysis & Evaluation • Architects & Engineers • Architecture & Engineering

A/E Application Entity • Architect/Engineer

A-E Architect-Engineer • Architect Engineering • Architectural Engineering

AE Abbreviate-Evaluate • Above or Equal • Account Executive • Acid Ester • ACIS Extract • Acoustic Emission • Acoustoelectric • Acquisition Executive • ACS Engine • Actual Evapotranspiration • Adaptation Engine • Additional Entry • Address Extension • Adjoint d'Enseignement • Adult Education • Aeronautical Engineer • Aeronautical Engineering • Affaires Étrangères • Afrique Équatoriale • After Execution • Age Exemption • Aide-Éducateur • Airframe and Engine • Alfvén Eigenmode • All England • Almost Everywhere • American English • American Express • Amplitude Equation • Analog Electronics • Antenna Equipment • Application Engineer • Application Entity • Application Environment • Application Execution • Applications Engineer • Applications Explorer • Apportioned Effort • Architect-Engineer • Architects and Engineers • Architectural and Engineering • Architectural Engineering • Arithmetic Element • Arithmetic Expression • Artificial Evolution • Associate Editor • Associate in Education • Associate in Engineering • Association Européenne • Astronomische Eenheden • Astronomische Einheit • Atmosphere Explorer • Atmospheric Explorer • Atomic Energy • Attenuation Equalizer • Auroral Electrojet • Auto Estrada • Autoexposure • Automatic Exposure • Autumn Equinox • Auxiliary Equipment • Avis d'Expédition • Émirats Arabes Unis (code ISO de pays) • United Arab Emirates (ISO country code)

AEA Abemama Atoll (Kiribati) (location identifier) • Acoustic Emission Analysis • Advanced Extension Award • Aerial Experiment Association • Aft End Assembly • Agence Européenne pour l'Environnement • Alternative Engineering Assumption • American Electronics Association (USA) • Antenna Elevation Angle • Assignment Eligibility and Availability • Association of European Airlines • Atomic Energy Act (USA) • Atomic Energy Authority (UK) • Automatic Error Analysis • Automation Economic Analysis • Auxiliary Engine Assembly

AEAAC Association Éducative des Amateurs d'Astronomie du Centre (France)

AEAC Alternate Emergency Action Center

AeAm Aerospace America (periodical code)

AEAP Atmospheric Effects of Aviation Project

AEAT Agencia Estatal de la Administración Tributaria (Spain)

AEB Acquisition Executive Brief • Aft Equipment Bay • Agência Espacial Brasileira (Brazil) • Analog Expansion Bus • Analysis and Evaluation Board • Area Electricity Board • Associated Examining Board • Astronomy Education Board (AAS) • Auxiliary Equipment Building • Average Extent Burning

AEBP Atmospheric Equivalent Boiling point

AEBR Association of European Border Regions

AEC Acte d'État Civil • Additional Extended Coverage • Address Element Correction • Aft End Cone • Aft Events Controller • Airborne Electronic Equipment • ALMA Executive Committee • Altitude Engine Control • Alternate Exchange Carrier • American Express Co. • Application Essentiality Code • Architectural and Engineering Construction • Architecture - Engineering - Construction • Archived Exposure Catalog • Archive Exposure Catalog • Area Equipment Compounds • Association des Éditeurs Canadiens (Canada) • Association of Education Committees (UK) • Atomic Energy Commission • Atomic Energy Corporation • Atomic Energy Council • Atomic Energy of Canada • Australian Environmental Council • Automated Element Correction • Automatic Engine Control • Automatic Experiment Container • Automatic Exposure Control • Automatic Exposures Catalog • Auxiliary Equipment Cabinet

AECB Atomic Energy Control Board

AECE Asociación Española de Comercio Electrónico (Spain)

AeCh Aerospace China (periodical code)

AECI Agencia Española de Cooperación Internacional (Spain)

AECIC Affaires Étrangères et Commerce International du Canada

AECL Atomic Energy of Canada Ltd.

AeCM Aerospace Composites Materials (periodical code)

AECM Alternating Expectation Conditional Maximization

AECMA Association Européenne des Constructeurs de Matériel Aérospatial

AECO Advance Engineering Change Order

AECS Advanced Entry Control System • Advanced Environmental Control System • Area Entry Control System •

Atmospheric Effect Correction System • Automated Entry Control System

AECSE Association des Enseignants et Chercheurs en Sciences de l'Éducation (France)

AECSI ACM Electronic Community Services Initiative • ACM Electronic Community Strategic Initiative

AECT Association of Educational Communications and Technology

AECU Advanced Electronic Control Unit

AEd Associate in Education

AED Actual End Date • Advanced Electronics Design • ALGOL Extended for Design • Allocation Engineering Division • Analog Event Distributor • Analysis and Evaluation Division • Architectural Engineering Directorate • Association of Engineering Distributors • Atomic Emission Detector • Automated Engineering Design • Automatic External Defibrillator • Dirham (UAE) (ISO currency code)

AEDAF Asociación Española de Asesores Fiscales (Spain)

AEDB Apollo Engineering Documentation Board (NASA)

AEDC Antarctic Environmental Data Centre • Arnold Engineering Development Center (USA-TN)

AEDD Antarctic Environmental Data Directory

AEDDWG Antarctic Environmental Data Directory Working Group

AEDE Association Européenne des Enseignants

AEDI Audiovisual Event Desçription Interface

AEDIMA Asociación Española de Informática y Matemática Aplicada (Spain)

AEDMS Automated Engineering Data Management Section

AEDO Aeronautical Engineering Duty Officer

AEDP Astrometry and Engineering Data Processing

AEDS Analog Event Distributor System • Arctic Environmental Data System • Atomic Energy Detection System

AEDT Australian Eastern Daylight Time

AEDV Adaptive Electronic Document Viewer

AeDy Aerospace Dynamics (periodical code)

AeE Aeronautical Engineer

AEE Achieving Engineering Excellence • Administración Espacial Europea • Agence Européenne de l'Environnement • Airborne Evaluation Equipment • Alkaline Earth Element • Application Execution Environment • APPN Enterprise Extender • Architecture Électronique Embarquée • Association for Environmental Education • Atomic Energy Establishment

AEEA Association Européenne pour l'Enseignement de l'Astronomie

AEEC Aeronautical Electronics Engineering Committee • Airline Electronic Engineering Committee

AEEI Atmospheric Emitted Radiance Instrument • Autonomous Energy-Efficiency Index

AEELS Automatic Electronic Emitter Locating System

AeEn Aerospace Engineering (periodical code)

AEEN Agence Européenne pour l'Énergie Nucléaire

AEENM Association des Élèves de l'École Nationale de la Météorologie (France)

AEEP Affirmative Employment Program Plan • Association of Environmental Engineering Professors

AEERL Air and Energy Engineering Research Laboratory

AEES Association des Élèves des Écoles Spéciales

AEESCA Association for Engineering Education in South and Central Asia

AEESEAP Association for Engineering Education in Southeast Asia and the Pacific

AEF Aerodrome Environmental Federation (UK) • Aerospace Engineering Facility • Affaires Économiques et Financières • Air Expeditionary Force • Air Experience Flight • American Expeditionary Force • Arbeitsgemeinschaft Extraterrestrische Forschung (Germany) • Asociación Española de Fundaciones (Spain) • Average Emission Factor

AEFE Agence pour l'Enseignement du Français à l'Étranger • Association Éducation-Formation-Entreprise (France)

AeFS Aeronautical Fixed Service

AEFS American Express Financial Services (USA-MN)

aeg Ad Eundem Gradum • All Page Edges Gilt

AEG Active Element Group • Aek Godan (Indonesia) (location identifier) • Aeromedical Evacuation Group • Aircraft Evaluation Group • Air Experience Gliding • Allgemeine Elektrizitäts-Gesellschaft (Germany) • Antenna Evolution Group • Association of Engineering Geologists • Association of Exploration Geochemists • Asynchronous Exponential Growth

AEGE ASEAN Experts Group on the Environment

AEGEE Association des États Généraux des Étudiants d'Europe • Association des États Généraux des Étudiants Européens

AEGIS Advanced Electronic Guidance and Instrumentation System • An Existing General Information System • Australian Electronic Government Information Service

AEGS Association of European Geological Societies

AEH Abecher (Chad) (location identifier) • Antenna Effective Height

AEHF Advanced Extremely High Frequency

AEHP Atmospheric Electrical Hazards Protection

AEI Accretion-Ejection Instability • Action Éducative Innovante • Additional Equipment Interface • Aerial Exposure Index • Air Express International • Albert-Einstein-Institut (Germany) • Alternative Energy Institute • American Express International • Assign Equipment In • Association des Écoles Internationales • Average Efficiency Index • Azimuth Error Indicator

AeIBu Aerokosmicheski Issledovaniia Bulgariia (periodical code)

AEIC ASEAN Earthquake Information Center

AEIDC Arctic Environmental Information Data Center

AEIE Agrupación Europea de Interés Economico

AEIMEE Antarctic Environmental Implications of Mineral Experiments

AEIMP Apple Event Interprocess Messaging Protocol

AeJ Aeronautical Journal (periodical code)

AEJ Association of European Journalists

AEK All-Electric Kitchen • Aseki (Papua New Guinea) (location identifier)

Aekv Aerodinamika Kanalov i Ventiliatorov (periodical code)

AEL Adverse Effect Level • Aeronautical Engine Laboratory • Aeronautics Education Laboratory (USA-MD) • Aerospace Educational Laboratory • Aircraft Engine Laboratory • Albert Lea (USA-MN) (location identifier) • Allowance Equipment Listing • Associated European Laboratories • Authorized Equipment Listing • Automation Engineering Laboratory

AELE Association Européenne de Libre Échange

AELS Airborne Electronic Library System

A&ME Aeronautical & Mechanical Engineering

AEM Acoustical Emission Monitoring • Aeronautical and Mechanical Engineering • Airborne Electromagnetic • Analytical Electron Microscope • Analytical Electron Microscopy • Animal Enclosure Module (*) • Animal Exposure Module • Application Explorer Mission • Arsenal Exchange Model • ASEAN Economic Ministers • Assembly and External Maintenance • Astronomical Equipment

Maker • Astronomical Equipment Making • Astronomy Education Material • Atmospheric Explorer Mission • Automation Equipment Maintenance

AeMat Aerospace Materials (periodical code)

AEME Agence de l'Environnement et de la Maîtrise de l'Énergie (France)

AeMed Aerospace Medicine (periodical code)

AEMO Action Éducative en Milieu Ouvert

AeMP Aeronautical Meteorology Programme

AE&MP Ambassador Extraordinary & Minister Plenipotentiary

AEMP Association of European Management Publishers

AEMR Acquisition Executive Management Report

AeMS Aeronautical Mobile Service

AeMSS Aeronautical Mobile Satellite Service

AEMU Advanced Extravehicular Mobility Unit

aen Aeneus

AEN Agence pour l'Énergie Nucléaire

AENA Aeropuertos Españoles y Navegación Aérea (Spain)

AEng Associate in Engineering

AENOR Asociación Española de Normalización y Certificación (Spain)

AENSB Association de l'École Nationale Supérieure des Bibliothèques

AEO Acousto-Electrical Oscillator • Aioun - El Atrouss (Mauritania) (location identifier) • Alkaline Earth Oxide • All Engines Operating • All-European Observatory • An Element Of • Assistant Education Officer • Assistant Experimental Officer • Association of Exhibition Organizers

AEOP Adaptive Electro-Optical Processor

AEORMA Asociación Española para la Ordenación del Medio Ambiente (Spain)

AEOS Advanced Electro-Optical System

AE&P Ambassador Extraordinary & Plenipotentiary

AEP Accrued Expenditures Paid • Accrued Expenses Paid • Affirmative Employment Plan • Affirmative Employment Program • Annual Execution Plan • AppleTalk Echo Protocol • Application Environment Profile • Application Execution Platform • Aquatic Environmental Protection • Arbeitsgemeinschaft Extraterrestrische Physik (Germany) • Association Européenne de Productivité • Asymptotic Equipartition Property • Atmospheric Environment Programme • Buenos Aires - Jorge Newbery (Argentina-BA) (location identifier)

AEPB Astronomy Education Policy Board (AAS)

AEPC Asociación Española para el Progreso de las Ciencias (Spain)

AEPG Analysis and Experimentation Planning Group

AEPI Atmospheric Emissions Photo Imaging • Atmospheric Emissions Photometric Imaging

AEPIA Asociación Española Para la Inteligencia Artificial (Spain)

AEPM APACM Energy and Particle Monitor

AEPOC Association Européenne pour la Protection des Oeuvres Cryptées

AEPS Airline Employment Placement Service • Animation Éducative Périscolaire • Arctic Environmental Protection Strategy • Arctic Environment Protection Strategy • Average Expenditure Per Sortie

AEPSI Asociación Española de Provedores de Servicios de Internet (Spain)

AEPT Advanced Engine Propulsion Technology • Asociación Española de Prensa Técnica (Spain)

AEPUA Association d'Éducation Permanente des Universités de l'Atlantique

AEPUC Association pour l'Éducation Permanente dans les Universités du Canada

aeq Aequalis

AeQ Aeronautical Quarterly (periodical code)

AEQ Active Equalizer • ADC External Query • Award Evaluation Questionnaire

Aér Aéronautique

Aer Aerodrome • Aeronáutica • Aeronautics • Aeroplane

AER Academic Evaluation Report • Adler Sochi (Russia) (location identifier) • Antenna Equipment Room • Association of European Radios • Astronomy Education Research • Astronomy Education Review • Atmospheric and Environmental Research Inc. (USA-MA) • Atomic Energy Research • Azimuth - Elevation - Range • Sochi (Russia) (location identifier)

Aera Aeration

AERA American Educational Research Association (USA) • Annus Erat Regni Augusti • Automated En-Route ATC

AerAs L'Aéronautique et l'Astronautique (periodical code)

AERC Asociación Española de Radiodifusión Comercial (Spain)

AERE Association of Environmental and Resource Economists • Atomic Energy Research Establishment (UK)

AEREA Association of European Research Establishments in Aeronautics (now EREA)

AERI Atmospheric Emitted Radiance Instrument • Atmospheric Emitted Radiance Interferometer (*)

AERIE Airborne Equipment for Remote Imaging of the Environment

AERL Association of European Research Libraries

AERM Alignment Error Rate Monitor

AerMS L'Aerotecnica Missili e Spazio (periodical code)

AERN Association of European Research Networks • Astronomy Education Resource Network

AeRNS Aeronautical Radionavigation Service

AeRNSS Aeronautical Radionavigation Satellite Service

Aero Aerodrome • Aerodynamic • Aeronautic • Aeronautical • Aeronautics • Aeropuerto • Aerospace • Aerospace UK (periodical code)

AERODAS Aeromagnetic Data System

Aerodyn Aerodynamic

Aéron Aeronomie

Aeron Aeronautical • Aeronautics • Aeronomy

Aeronaut Aeronautical • Aeronautics

Aéron Aéronautique

AERONICA Aerolineas Nicaragüenses SA

AEROS Advanced Earth Resources Observational Satellite • Artificial Earth Research and Orbiting Satellite

Aerosp Aerospace • Aerospatial

Aérosp Aérospatial

AERP Aquatic Effects Research Program

AERQS Asiago-ESO/RASS QSO Survey

AerRv Revue Aérospatiale (periodical code)

AERS Access/Egress Roadway System • Advanced Earth Resource Satellite • Advanced ESA Resource Satellite • Association pour l'Expansion de la Recherche Scientifique

AerST Aerosol Science Technology (periodical code)

AERT Association for Education by Radiotelevision (USA)

AerTe Aerojet Technology (periodical code)

AE&S Air Equipment & Support

AES Accord Économique et Social • Acquisition Executive Service • Acuerdo Económico y Social • Ada Evaluation System • Advanced Encryption Standard • Advanced Energy System • Advanced Expert System • Advanced Extravehicular Suit • Aerospace Electrical Society • Aerospace Environmental Support • Ålesund (Norway) (location identifier) • Ahmedabad Earth Station (India) • Airborne Emission Spectrometer • Air Equipment and Support • Air Evaporation System • Analog Event System • Analysis and Evaluation Staff • Application Environment Service • Application Environment Specification • Application Environment Support • Artificial Earth Satellite • Association des Écrivains Scientifiques (France) • Astronomical Explorer Satellite • ATM Edge Switch • ATM Enhancement Service • Atmosphere Exchange System • Atmospheric Emission Spectrometer • Atmospheric Environmental Service • Atmospheric Environment Service • Atmospheric Exchange System • Atomic Emission Spectrography • Atomic Emission Spectrometer • Atomic Emission Spectrometry • Atomic Emission Spectroscopy • Atomic Emission Spectrum • Audio Engineering Society • Auger Electron Spectroscopy • Auger Electron Spectrum • Automated Enrollment System

AESA Active Electronically Scanned-array Antenna • Approximation Elimination Search Algorithm • ATM End-System Address • Atmospheric Effect of Stratospheric Aircraft

AESB Avionics Engineering Subboard

AeSc Collection of Papers on Aerospace Science (periodical code)

AESC American Engineering Standards Committee (USA) • Automatic Electronic Switching Center

AESE Airborne Electrical Support Equipment • Association of Earth Science Editors

AESF Association des Écrivains Scientifiques de France

AeSIJ Aeronautical Society of India Journal (periodical code)

AESIS Australian Earth Sciences Information System

AESL Aerospace Energy Systems Laboratory

AESO Advanced Engineering Supplemental Order

AESOP Advanced End-to-end Simulation of Onboard Processing • Aircraft - Environment - Situation - Operation - Personnel • AMOS Ephemeris Satellite Orbit Predictor • An Evolutionary System for Online Processing • An Extensible Symbolic Optics Package • Anti-Electron Suborbital Payload • Automated Engineering and Scientific Optimization Program

AESP Aerospace Education Services Project

AESS Automatic Electronic Switching System

AEST Australian Eastern Standard Time

Aesth Aesthetics

AESTM Association Européenne des Sciences et Techniques de la Mer

aet Aetatis

AET Active Edge Table • Actual Elapsed Time • Actual Evapotranspiration • Actual Exposure Time • Advanced Entry Training • Algorithm Evaluation and Test • Allakaket (USA-AK) (location identifier) • Application Entity Table • Application Entity Title • Approximate Exposure Time

AETA Association Européenne des Transporteurs Aériens

aetat Aetatis

AETB Advanced Expander Cycle Testbed • Alumina-Enhanced Thermal Barrier

AETC Air Education and Training Command

AET&CD Acquisition Education, Training & Career Development

AETCD Acquisition Education, Training, and Career Development

AETL Approved Engineering Test Laboratory

AetR Aller et Retour

AETR Advanced Engineering Test Reactor

AETT Association for Educational and Training Technology (UK)

AEU Accrued Expense Unpaid • Airborne Electronics Unit • Altitude Encoder Unit • American Ethical Union (USA) • Archiv für Elektronik und Übertragungstechnik • Asian Electronics Union • Auxiliary Electronics Unit • Avionics Electronics Unit • Avionics Equipment Unit

AEV Air Evacuation • Anti-Earth Viewing

AEVC Astronomy Education and Visitors Center (NRAO)

AEW Airborne Early Warning • Asset Evaluation Worksheet

AEW&C Airborne Early Warning & Control

AEWC Airborne Early Warning and Control

AEWG Astronomy Education Working Group (AAS)

AEWS Aircraft Early Warning System

AEX Alexandria (USA-LA) (location identifier) • Automated Execution

AExpr Astronomy Express (periodical code)

AEY Akureyri (Iceland) (location identifier) • Auger Electron Yield

AEZ Agro-Ecological Zone • Alkaline Earth Zeolite

a/f A Favor • Airfield

af Ad Finem • Advanced Freight • A Favor • Afrikaans (ISO language code) • Anno Futuro • Audio Frequency

aF alte Fassung

Af Afghani (currency) • Africa • African • Afrikaans • Aruban Florin (currency) • Assez Faible

A/F Airfield • Airframe • Air-to-Fuel • As Found

A&F Accounting & Finance

AF Abbreviated From • Académie Française • Active Filament • Address Field • Adult Fiction • Affaires Foncières • Affiliation • Afghanistan (ISO country code) • Africa • Aft Fuselage • Agent de Fabrique • Air Force • Airframe • Alliance Française • Allocations Familiales • Alta Frequenza • Alternating Field • Ancien Franc (France) • Anglo-French • Année Fiscale • Antiferromagnet • Antiferromagnetic • Appropriated Funds • Arrêté Fédéral • Assemblée Fédérale • Associate Fellow • Astrometric Field • Astronomical Foundation (USA-CA) • Asymptotically Flat • ATM Forum • Attenuation Factor • Attitude File • Audio Fidelity • Audio Frequency • Augmentation Force • Authentication Framework • Autofocus • Automatically Following • Automatic Filter • Auxiliary Flag • Availability Factor • Award Fee • Axial Flow

AFA Accelerated File Access • Action de Formation Alternée • Advanced Fuel Assembly • Afghani (Afghanistan) (ISO currency code) • Air Force Academy (USA) • Airframe Assembly • Analog Filter Assembly • Anonymous FTP Archive • Associate in Fine Arts • Association des Fournisseurs d'Accès (France) • Association Française d'Astronomie (France) • Association of Flight Attendants • Audio Frequency Amplifier • Automatic Fire Alarm • San Rafael (Argentina-MD) (location identifier)

AFAC Airborne Forward Air Controller • Association Française pour l'Assurance de la Qualité (France) • Association pour la Formation et l'Animation Culturelle (France) • Asynchronous Fast Adaptive Composite

AFACTS Automatic Facilities Test System

AFADS Automatic Force Adjustment Data System

AFAE Association Française des Administrateurs de l'Éducation (France)

AFAIAA Associate Fellow of the American Institute of Aeronautics and Astronautics

AFAIK As Far As I Know

AFAL Air Force Astronautical Laboratory

AFAM Associazione Friulana di Astronomia e Meteorologia (Italy)

AFAMS Automated Flying And Maintenance Scheduling

AFAO Approved Force Acquisition Objective

AFAQ Association Française pour l'Assurance de la Qualité (France)

AFAS Association Française pour l'Avancement des Sciences (France) • Automated Fuels Accounting System

Afb Afbeelding

AFB Air Force Base • Antifriction Bearing • Association Française des Banques (France)

AFBC Atmospheric Fluidized-Bed Combustor

A&FC Airworthiness & Flight Characteristics

AFC Active Flight Control • Administration and Finance Committee • Administrative and Finance Committee • Advanced Facer-Canceler • Aerodynamic Flight Control • African Laser Center • Aircraft Flight Correlator • Airworthiness and Flight Characteristics • Alkaline Fuel Cell • Analog to Frequency Converter • Analyse Factorielle des Correspondances • Application Foundation Class • Area Forecast Center • Area Frequency Coordinator • Automatic Flight Control (*) • Automatic Flight Controller • Automatic Font Change • Automatic Frequency Control • Automatic Frequency Controller

AFCAC African Civil Aviation Commission

AFCAS Automatic Flight Control Augmentation System

AFCC Air Force Communications Command

AFCE American Foundation for Continuing Education (USA) • Automatic Flight Control Equipment

ADCEA Armed Forces Communications and Electronics Association

AFCEE Association Française pour le Commerce et les Échanges Électroniques (France)

AFCET Association Française pour la Cybernétique Économique (France)

AFCF Allocation Forfaitaire pour Charges de Famille

AFCM Automatic Fine Cull Machine

AFCO All Fuel Cutoff • Automatic Fuel Cutoff

AFCOHT Association Française du Commerce Hors Taxes (France)

AFCOS Association Française des Conseils en Organisation Scientifique (France)

AFCR Automated Financial Control Register

AFCRL Air Force Cambridge Research Laboratory (USAF)

AFCS Adaptive Flight Control System • Advanced Facer Canceler System • Advanced Flight Control System • Automatic Flight Control System

AFCT Affect

AFCVI Analyse Factorielle des Correspondances avec Variables Instrumentales

Afd Afdeling

AfD Amt für Datenverarbeitung

A/FD Airport/Facility Directory

AFD Abbreviated Functional Description • Accelerated Freeze Dried • Accelerated Freeze Drying • Advance Function Printing • Aft Flight Deck • Agence Française de Développement (France) • Application Filing Date • Architecture Flow Diagram • Assign Fixed Directory • Assistant Flight Director • Automated File Designator • Automated Flaw Detector • Automatic File Distribution • Axial Flux Density • Axial Flux Difference • Port Alfred (South Africa) (location identifier)

AFDA Axial Flux Difference Alarm

AfDB African Development Bank

AFDC Aid for Families with Dependent Children • Aid to Families with Dependent Children

AFDE Aft Flight Deck Equipment

AFDET Association Française pour le Développement de l'Enseignement Technique (France)

AFDI Automatic Fault Detection and Isolation

AFDIE Association Française pour le Développement de l'Intelligence Économique (France)

AFDIN Association Française de Documentation et d'Information Nucléaires (France)

AFDK After Dark

AFDM Advanced Fluid-Dynamics Model • Ash Free Dry Mass

AFDO Aft Flight Deck Operator • Assistant Flight Dynamics Officer • Award Fee Determining Official

AFDR Ascent Flight Design Review

AFDS Automatic Flight Director System • Autonomous Free-flight Dispenser System

AFDW Active Framework for Data Warehousing • Ash-Free Dry Weight

AF&E Airframe & Engine

AFE Airframe and Engine • American Flight Echocardiograph • Analog Front End • Annual Frequency Estimate • Apple File Exchange • Aromatic Feature in Emission • Association Française de l'Éclairage (France)

AFEA Abbreviated Functional Economic Analysis • Association des Femmes et Étudiantes Africaines • Association Française pour l'Éducation des Adultes (France)

AFEAS Alternative Fluorocarbons Environmental Acceptability Study

AFEB Award Fee Evaluation Board

AFEC Association de Formation et d'Enseignement Continu • Association Francophone d'Éducation Comparée (France) • Award Fee Evaluation Committee

AFECEI Association Française des Établissements de Crédit et des Entreprises d'Investissements (France)

AFECTI Association Française des Experts de la Coopération Technique Internationale (France)

AFEJ Asian Forum of Environmental Journalists

AFeN Analoges Fernsprechnetz

AFEQ Adiabatic Fast Equilibration

AFERO Asia and the Far East Regional Office

AFES Automated Feature Extraction System • Avionik-Flug-Erprobungs-System

AFESD Arab Fund for Economic and Social Development

AFEUR Association Française d'Étude pour l'Union Européenne (France)

AFEWES Air Force Electronic Warfare Effectiveness Simulator

Aff Affaire • Affairs • Affectionate • Affiliate • Affiliated • Affirmatif • Affirmative • Affix

AFF Aberration-Free Focus • Acceptance and Ferry Flight • Affiliate • Affiliated • Autonomous Formation Flying • Colorado Springs (USA-CO) (location identifier)

AFFDL Air Force Flight Dynamics Laboratory (USAF)

Aff.Etr. Affaires Étrangères

AFFF Aqueous Film Forming Foam

Affil Affiliate • Affiliated • Affiliation

AFFIRM Association For Federal Information Resources Management (USA)

Affmo Afectísimo

AFFORD Affordability

AFFS Allocation de Formation de Fin de Stage

Afft Affidavit

Afg Afghanistan

AFG Administrative and Finance Group • Advanced Frame Grabber • Afghanistan (international vehicle registration code) • Afghanistan (ISO country code) • Analog Function Generator • Annual Funding Guidance • Antenna Field Gain • Arbeitsförderungsgesetz • Arbitrary Function Generator • Auto Force Generator • Automatic Force Generator

AFGE American Federation of Government Employees (USA) • Awards to Facilitate Geoscience Education

Afgh Afghanistan

Afghan Afghanistan

AFGL Air Force Geophysical Laboratory • Air Force Geophysics Laboratory (*) • Air Force Geophysics Laboratory Research Papers (periodical code)

AFGS Automatic Flight Guidance System

AFGWC Air Force Global Weather Center (USAF)

AFGWIS Air Force Global Weather Information System

AFH Antenna Feed Horn

AFHB Arbeitsgemeinschaft der Fachhochschulbibliotheken

AFI Affinity Index • Amalfi (Colombia) (location identifier) • American Film Institute (USA) • AppleTalk Filing Interface • Assistant Flying Instructor • Association Fiscale Internationale • Authority and Format Identifier • Authorized Flight Instructor • Automatic Fault Isolation

AFIA Association Française pour l'Intelligence Artificielle (France)

AFIAP Artiste de la Fédération Internationale de l'Art Photographique

AFID Association Française d'Information et de Documentation (France)

AFIGEO Association Française pour l'Information Géographique (France)

AFII American Federation of International Institutes (USA) • Association for Font Information Interchange (USA) • Association Française des Investissements Internationaux (France)

AFIIM Associate Fellow of the Institution of Industrial Managers

AFIJ Association pour Faciliter l'Insertion des Jeunes (France)

AFIM Antarctic Flight Information Manual

AFIMA Associate Fellow of the Institute of Mathematics and its Applications

AFIP Affichage des Fonctions d'Intérêt Prioritaire

AFIPS American Federation of Information Processing Societies (USA)

AFIRM Automated Fingerprint Image Reporting and Match

AFIRO Association Française d'Informatique et de Recherche Opérationnelle (France)

AFIS Aerodrome Flight Information Service • Association Française pour l'Information Scientifique (France) • Automated Fingerprint Identification System • Automated Flight Inspection System • Automatic Fingerprint Identification System

AFISS Airborne Flight Information Subsystem

AFIT Air Force Institute of Technology (USAF)

AFITE Association Française des Ingénieurs et Techniciens de l'Environnement (France)

AFITV Air Force Instrumented Test Vehicle

AFJQS Air Force Job Qualification Standard

AfK Arbeitsgemeinschaft für Kommunikationsforschung

AFK Away From Keyboard

Afl Aflevering

AFL Abone Field Level • Abstract Family of Languages • Afloat • Airborne Flying Laboratory • Alta Floresta (Brazil-MT) (location identifier) • American Federation of Labour (USA) • Association Française pour la Lecture • Astronomical First Light • Automatic Fault Location

AFLA Asian Federation of Library Associations

AFLC Air Force Logistics Command

AFLCD Antiferroelectric Light Emitting Diode

AFLCIO American Federation of Labour and Congress of Industrial Organizations (USA)

afld Airfield

AFLEX Automatic Flight Landing Experiment

AfRL Akademie für Raumforschung and Landesplannung (Germany)

Aflt Afloat

afm Adobe Font Metrics (file extension)

A/FM Arm Firing Mechanism

AFM Abrasive Flow Machining • Adobe Font Metrics • Advances in Fluid Mechanics (conference) • Aircraft Flight Manual • Airplane Flight Manual • Analog Frame Marker • Antiferromagnetic • Anti-Friction Metal • Asociación Filosófica Mexicana (Mexico) • Association Française de Mécanique (France) • Atomic Force Microscope • Atomic Force Microscopy • Audio Frequency Modulation • Automated Forward Modeling • Automatic Flight Management • Award Fee Monitor

AFMa Air Force Magazine (periodical code)

AFMC Air Force Material Command (USAF) • Asian Fluid Mechanics Congress

AFME Agence Française pour la Maîtrise de l'Énergie (France)

Afmo Afectísimo

AFMR American Foundation for Management Research (USA) • Anti-Ferromagnetic Resonance • Association pour la Formation en Milieu Rural (France)

AFMS Advanced Flight Management System • Automatic Flight Management System • Automatische Fernmeldende Meteorologische Station • Auxiliary Fuel Monitoring System

AFMSS Air Force Mission Supprt System

AFMWG Acquisition and Financial Management Working Group

AFN Alaska Federation of Natives • All For Nothing • Australian Fiducial Network • Average Failure Number • Jaffrey (USA-NH) (location identifier)

AFNIC Association Française pour le Nommage Internet en Coopération (France)

AFNIL Agence Francophone pour la Numérotation Internationale du Livre

AFNOR Association Française de Normalisation (France)

AFO Abort From Orbit • Accounting and Finance Office • Afton (USA-WY) (location identifier) • Announced Flight Opportunity • Announcement of Flight Opportunity • Astronomischer Freundeskreis Ostsachsen (Germany)

AFOAT Air Force Office of Atomic Test

AFOC Advanced Fiber Optic Coupler

AFOE Advanced Fiber Optic Echelle

AFOEV Association Française des Observateurs d'Étoiles Variables (France) • Bulletin de l'Association Française des Observateurs d'Étoiles Variables (periodical code)

AFOGEC Association pour la Formation Générale Et Continue (France)

AFOP Association Française des Industries de l'Optique et de la Photonique (France)

AFOPT Association Française des Opérateurs Privés en Télécommunications (France)

AFOS Active Fiber Optic Segment • Advanced Field Operations System • Antarctic Fiber Optic Spectrometer • Automated Forecasting Operational System • Automation of Field Operations and Services

AFOSC Asiago Faint Object Spectrograph and Camera

AFOSH Air Force Occupational Safety and Health

AFOSR Air Force Office of Scientific Research (USA-DC)

AFOTEC Air Force Test and Evaluation Center (USA)

AFP Actualité de la Formation Permanente • Adaptive Fuzzy Partition • Adiabatic Fast passage • Administradora de Fondos de Pensiones • Advanced Flexible Processor • Advanced Functional Programming • Advanced Function Presentation • Advanced Function Printing • Agence France-Presse • Air Force Program • Analyst Private File • Annual Financial Plan • Annual Funding Program • AppleTalk File Protocol • AppleTalk Filing Protocol • Approval for Full Production • Area Flight Plan • Ariane-Furnished Property • Attached Fortran Processor • Avis de Fortes Précipitations

AFPA Association pour la Formation Professionnelle des Adultes (France) • Automatic Flow Process Analysis

AFPAC Algebraic Frames for the Perception-Action Cycle (workshop)

AFPC Association pour la Formation Professionnelle Continue (France)

AFPD Authorization For Program Development

AFPDS Advanced Function Presentation Data Stream

AFPES Automated Final Polyelastic Sampling

AFPI Association Française des Professionnels de l'Internet (France) • Association pour la Formation et la Promotion dans l'Industrie (France)

AFPL Alladin Free Public License

AFPM Advanced Fluid Physics Module

AFPRO Air Force Plant Representative Office

AFPS Advanced File and Print Server

AFPU Association Française des Presses d'Université (France)

AFQA Air Force Quality Assurance

Afr Africa • African • Afrikaans • Afrique

A-Fr Anglo-French

AFr Anglo-French

AFR Accident Frequency Rate • Actual Fuel Remaining • Afore (Papua New Guinea) (location identifier) • Africa • Air Force Regulation • Air-Fuel Ratio • Allocation Formation Reclassement • Annualized Failure Rate • Automatic Field Recognition • Automatic Fingerprint Recognition • Automatic Format Recognition • Available For Release

AFRA Association Française de Régulation et d'Automatisme (France)

AFRAA African Airlines Association

AFRALTI African Advanced Level Telecommunications Institute (Kenya)

AFRAND African Foundation for Research And Development

AFRASEC Afro-Asian organization for Economic Cooperation

AFRB Award Fee Review Board

AFRC Agricultural and Food Research Council (UK) • Agriculture and Food Research Council (UK) • Automatic File Redundancy Checking

AFRE Alta Frequenza (periodical code)

AFREPEN African Energy Policy Research Network

AF-RF Audio Frequency - Radio Frequency

AFRG Australian Fusion Research Group

AFRHC Association Française de Recherche sur l'Histoire du Cinéma (France)

AFRI Association Française de Robotique Industrielle (France)

Afrik Afrikaans

AFRIMS Armed Forces Research Institute of Medical Sciences (USA)

AFRL Air Force Research Laboratory (USA)

AFRO Africa Regional Office

AFRP Approval for Full-Rate Production

AFRPL Air Force Rocket Propulsion Laboratory (USAF)

AFRSI Advanced Flexible Reusable Surface Insulation

AfS Amt für Standardisierung

AFS Active Federal Service • Advanced Flats Sorter • Air Force Station • AIRS Facility System • American Field Service • Andrew File System • Antenna Feeder System • Antenna Feed System • Arch Filament System • Ascent Flight System • Astronomisk Forening for Sydsjaelland (Denmark) • Atomic Fluorescence Spectrometry • Atomic Fluorescence Spectroscopy • Atomic Frequency Standard • Attestation de Formation Spécialisée • Audio Frequency Shift • Autogenic Feedback System • Automatic Flight System • Automatic Frequency Stabilization • Automation Feasibility Study • Axial Field Spectrometer

AFSA American Foreign Service Association (USA)

AFSAC Air Force Security Assistance Center (USA)

AFSATCOM Air Force Satellite Communications (USAF)

AFSB Air Force Studies Board (USAF)

AFSBO American Federation of Small Business Organizations (USA)

AFSC Aggregate Friction Seal Coat

AFSCF Air Force Satellite Control Facility

AFSCN Air Force Satellite Control Network

AFSIG Ascent Flight System Integration Group

AFSCME American Federation of State, County, and Municipal Employees (USA)

AFSCN Air Force Satellite Control Network (USAF)

Afsd Aforesaid

AFSK Automatic Frequency Shift Keying

AFSM Advanced Flats Sorting Machine • Automated Flats Sorting Machine

AFSPACECOM Air Force Space Command (Now AFSPC)

AFSPC Air Force Space Command

AFSRR Avionics Flight Systems Readiness Review

AFSRT Avionics Flight Systems Readiness Test

AFSS Automated Flight Service Station • Automated Fuel Service Station

AFSSE Agence Française de Sécurité Sanitaire Environnementale (France)

AFSTB Air Force Science and Technology Board (USAF)

AFSV Automatische Fernmelde-Speicher-Vermittlung

AFSWC Air Force Space Weather Center (USA)

Aft After • Afternoon

AFT Active Fine Tracker • Adapter Fault Tolerance • Aerodynamic Flight Test • After • Afutara (Solomon Islands) (location identifier) • Ahlfors Finiteness Theorem • Alert Fire Team • Ambiguity Function Technique • American Federation of Teachers (USA) • Annual Financial Target • Aschberg-Frühjahrs-Teleskoptreffen (Germany) • Association Française de Topographie (France) • Association Française du Transport (France) • Atmospheric Flight Test • Automated File Transfer • Automated Fund Transfer • Automatic Flight Termination • Automatic Fund Transfer • Axial-Flow Turbine

AFTA Acoustic Fatigue Test Article • Aft Frame Tilt Actuator

AFTAC Air Force Technical Application Center

AFTDS Automated Flight Test Data System

AFTE American Federation of Technical Engineers (USA) • Autogenic Feedback Training Experiment

AFTEL Association Française de Télématique (France)

AFTF Avionics Flight Test Facility

AFTI Advanced Fighter Technology Integration • Automated Interchange of Technical Information

AFTM Additional Full Time Manning

Aftn Afternoon

AFTN Aeronautical Fixed Telecommunication Network • Afternoon • Automated Fixed Telecommunication Network

AFTP Acceptance Functional Test Procedure • AGB Final Thermal Pulse • Anomymous File Transfer Protocol • APPC File Transfer Protocol

AFTRA American Federation of Television and Radio Artists (USA)

AFTS Aft Fuselage Thrust Structure • Audio Frequency Tone Shift

AFTT Association Française de Télétravail et de Téléactivités (France)

AFU Advanced Flying Unit • All Fouled-Up • Analog Feedback Unit • Assistance au Freinage d'Urgence

AFUE Annual Fuel Utilization Efficiency

AFUL Association Française des Utilisateurs de Linux (France)

AFUTT Association Française des Utilisateurs de Téléphone et de Télécommunications (France)

AFUU Association Française des Utilisateurs d'Unix (France)

AFUSIE Association Française des Utilisateurs de Systèmes d'Information Électroniques (France)

AFV Alternate Fuel Vehicle • Awaiting Fix Verification

AFVG Anglo-French Variable Geometry

AFVP Association Française des Volontaires du Progrès (France)

AFW Fort Worth - Alliance (USA-TX) (location identifier)

AFWA Air Force Weather Agency (USA)

AFWIN Air Force Weather Information Network (USA)

AFY Afyon (Turkey) (location identifier)

AFYVE Asociación Fonográfica Y Videográfica de España (Spain)

Afz Astrofizica (periodical code)

AFZ Acronym-Free Zone • Astro Fax Zirkular • Automobile-Free Zone

af2i Association Française des Intermédiaires en Information

aG auf Gegenseitigkeit

Ag Acting • Agent • Agreement • Antragsgegner • Argent • August • Silber • Silver

A-G Air-to-Ground

A/G Air-to-Ground

A&G Acquisition & Guidance • Acquisition & Guiding • Adquisición & Guiado • Astronomy and Geophysics (periodical code)

AG Accountant-General • Acoustic Gravity • Address Generation • Ad Gentes • Adjutant General • Administration Générale • Advisory Group • Agent du Gouvernement • Agent-General • Agglomération • Airglow • Aktiengesellschaft • Amstgericht • Andean Group • And Gate • Annales Geophysicae • Anti-Gas • Antigravity • Antigua & Barbuda (ISO country code) • Application Gateway • Applications Generator • Arbeitsgemeinschaft • Arbeitsgruppe • Arbetsgrupp • Arresting Gear • Artificial Gravity • Assemblée Générale • Astronomische Gesellschaft (Germany) • Attitude Gyroscope • Attorney-General • Auftraggeber • Ausführungsgesetz • Avocat Général • Auxiliary Gas • Barbuda (Antigua &_) (ISO country code)

AGA Adaptive Genetic Algorithm • Agadir (Morocco) (location identifier) • Agent Administratif • American Gas Association (USA) • Approche Globale d'Apprentissage • Arab Geologists Association • As Good As • Association of Government Accountants • Attitude Gyro Assembly

AGAA Association Guadeloupéenne d'Astronomes Amateurs (France) • Attitude Gyroscope Accelerometer Assembly • Automatic Gain Adjusting Amplifier

AGAb Astronomische Gesellschaft Abstract Series (periodical code)

AGAC American Guild of Authors and Composers (USA) • Archivo General de la Administración Central

AGACS Automatic Ground-to-Air Communications System

AGADES Accès Général à la Documentation Électronique Spécialisée

AGADS Advanced Graphics Avionics Display System

AGAGE Advanced Global Atmospheric Gases Experiment

AGAL Australian Government Analytical Laboratories

AGAP Attitude Gyroscope Accelerometer Package

AGAPE Andromeda Galaxy and Amplified Pixels Experiment (formerly) • Andromeda Gravitational Amplification Pixel Experiment (*)

AGARAS Antiglare - Antireflective - Antistatic

A-GARD Amphiphile Graded Autocatalysis Replication Domain (model)

AGARD Advisory Group for Aerospace Research and Development (NATO) • Amphiphile Graded Autocatalysis Replication Domain (model)

AGARF Agent Administratif de Recherche et Formation (France)

AGASA Akeno Giant Air Shower Array (Japan)

AGASC AXAF Guide and Aspect Star Catalog

AGASP Arctic Gas and Aerosol Sampling Programme

AGATA Advanced Gas Turbine for Automobiles

AGATE Advanced Gamma-ray Astronomy Telescope • Advanced Gamma-ray Astronomy Telescope Experiment • Advanced General Aviation Transport Experiment

AGB Accessory Gearbox • A Goog Brand • Any Good Brand • Allgemeine Geschäftsbedingungen • Association of Governing Board • Astronomische Gesellschaft Baden (Switzerland) • Astronomische Gesellschaft Bern (Switzerland) • Asymptotic Giant Branch • Augsburg (Germany) (location identifier)

AGBD Above Ground Biomass Density

AGC Abell Galaxy Cluster • Adaptive Gain Control • Administration Générale Civile • Advanced Graduate Certificate • Arecibo General Catalog • Association Géologique du Canada • Australian Geoscience Council • Automatic Gain Control • Pittsburgh - Allegheny County (USA-PA) (location identifier)

AGCA Automatic Ground-Controlled Approach

AGCD Administration Générale de la Collaboration au Développement

AGCFM Advisory Group on Computational Fluid Mechanics

AGCH Access Grant Channel

AGCI Aspen Global Change Institute

AGCL Answer Garden Command Language • Automatic Ground-Controlled Landing

AGCM Atmospheric General Circulation Model

AGCP Advanced Guidance and Control Processor

AGCPO Arctic Global Change Programme Office

AGCS Accord Général sur les Commerces de Services • Automatic Ground Control Station

AGCSF Anisotropic Generalized Curve Shortening Flow

AGCT Army General Classification Test

AGCU Air/Ground Cooling Unit • Attitude Gyro Coupling Unit

Agcy Agency

AGCY Agency

Agd Agreed

AGD Agreed • Anggi (Indonesia) (location identifier) • Atelier de Génie Didacticiel • Australian Geodetic Datum • Axial Gear Differential

AGDA Andøya Geophysical Data

AGDB African Geodetic Data Bank • Australian Geographic Database

AGDS American Gauge Design Standard • Automatic Graphic Digitizing System

AGE Aerospace Ground Equipment • Aircraft Ground Equipment • Air/Ground Equipment • Angle of Greatest Elongation • Applied General Equilibrium • Associated Ground Equipment • Associate in General Education • Association Générale des Étudiants • Automatic Ground Equipment • Automatic Guidance Electronics • Auxiliary Ground Equipment • Avionics Ground Equipment • Wangerooge (Germany) (location identifier)

AGED Advisory Group on Electron Device • Allocation de Garde d'Enfants à Domicile

AGEP Alliance for Graduate Education and the Professoriate

AGEX Address Generation/Execute Cycle

AGF Absorbing Galaxy Fraction • Agen (France) (location identifier) • Águas-Furtadas • All Glass Fiber • Arbeitsgemeinschaft der Großforschungseinrichtungen (Germany) • Association des Géographes Français (France) • Assurances Générales de France • Automated Gradient Furnace

AGFA Aktiengesellschaft für Anilinfabrikation (Germany)

AGFD Arizona Game and Fish Department (USA-AZ)

ag. feb. Aggrediente Febre

AGFS Aviation Gridded Forecast System

Agg Aggregate

AGG Aliased Gravity Grid • Allgemeine Geschäftsgrundsätze • Angoram (Papua New Guinea) (location identifier) • Astronomische Gesellschaft Graubünden (Switzerland) • Attributed Graph Grammar

AGGG Advisory Group on Greenhouse Gases

Aggl Agglomération

Aggr Aggregate

AGGS Accelerated Guided Gradient Search

Agh Afghani (currency)

AGH Hälsingborg Ängelholm (Sweden) (location identifier) • Helsingborg Angelholm (Sweden) (location identifier)

AGHF Advanced Gradient Heating Facility (*) • Advanced Gradient Heating Furnace

AGI Address Generation Interlock • Adjusted Gross Income • Advanced Ground Instructor • American Geographical Institute (USA) • American Geological Institute (USA) • Analytical Graphics Inc. (USA-PA) • Année Géophysique Internationale • Annual General Inspection • Application Graphics Interface • Applied Geophysics Institute • Association for Geographic Information (UK) • Authorized Ground Instructor • Auxiliary General Intelligence • Wageningen (Suriname) (location identifier)

AGICE Automatic Graphic Identification of Constitutive Equations

AGID Association of Geoscientists for International Development

AGIF Arbeitsgemeinschaft für Innerbetriebliche Fortbildung (Germany)

AGIL Airborne General Illumination Light

AGILE Astro-rivelatore Gamma e Immagini Leggero

AGIP Agenzia Generale Italiana Petroli (Italy) • Azienda Generale Italiana Petroli (Italy)

AGIRC Association Générale des Institutions de Retraite des Cadres (France)

AGIRE Agence d'Imagerie pour la Recherche et l'Enseignement

AGIRS Advanced Geostationary Infrared Sounder

AGIS Arctic Geographic Information System • Auto-Guiding and Imaging System • Automated Geographic Information System

agit Agitat • Agitatum

agit. ante sum. Agita Ante Sumendum

AGJ Aguni (Japan) (location identifier)

AGK Aktiver Galaktischer Kern • Allgemeine Geschäftskosten • Astronomische Gesellschaft Katalog • Kagua (Papua New Guinea) (location identifier)

AGKR AGK Reference Star

AGK2 Zweiter Katalog der Astronomischen Gesellschaft

AGL Above Ground Level • Above Ground Link • Absolute Ground Level • Allied Geophysical Laboratories • ASTRONET Graphics Library • Astronomische Gesellschaft Luzern (Switzerland) • Atelier de Génie Logiciel • Average Grade Level • Azores-Gibraltar Line • Wanigela (Papua New Guinea) (location identifier)

Aglm Agglomerate

AGM Active Grating Monochromator • Advisory Group Meeting • Alternate Graphics Mode • Angmagssalik (Greenland) (location identifier) • Annual General Meeting • Arithmetic-Geometric Mean • Association General Meeting • Automatic Global Matcher

AGMA American Gear Manufacturing Association (USA)

AGMS Arbeitsgemeinschaft Massenspektrometrie (Germany)

Agn Again

AGN Active Galactic Nucleus • Additive Gaussian Noise • African Gravity Network • Again • Angoon (USA-AK) (location identifier) • Archives Générales de la Nation

AGN&C Advanced Guidance, Navigation & Control

AGNC Advanced Guidance, Navigation and Control

AGND Analog Ground

AGNES A Graphical Notation Editing System

Ago Agosto

AGO Administrative Grant Officer • Angola (ISO country code) • Assemblée Générale Ordinaire • Astronomical and Geophysical Observatory • Astronomische Gesellschaft Oberwallis (Switzerland) • Australian Greenhouse Office • Automated Geophysical Observatory • Magnolia (USA-AR) (location identifier)

AGOALF Agostoalferes

AGOCG Advisory Group On Computer Graphics (UK)

AGOES Advanced Geosynchronous Observation Environment Satellite

AGOF Arbeitsgemeinschaft Online-Forschung

AGONET Automatic Geophysical Observatory Network

AGOR Accélérateur Groningen-Orsay • Auxiliary-General Oceanographic Research

AGOS Air-Ground Operations System

AGOSS Automated Ground Operations Scheduling System

AgOx Silver Oxide

AGP Accelerated Graphics Port • Acide Gras Polyinsaturé • Advanced Geosynchronous Platform • Advanced Graphics Port • A Gateway Protocol • Astronomy, Geophysics, and Planetary sciences • Automatic Guidance Programming • Aviation General Policy • Malaga (Spain) (location identifier)

AGPSP ASCA Galactic Plane Survey Project

AGQ Agrinion (Greece) (location identifier) • Arbeitsgemeinschaft Quantenoptik (Germany)

Agr Agreement • Agrégé • Agricultura • Agricultural • Agriculture

AGR Advanced Gas-cooled Reactor • Advanced Gas Reactor • Agra (India) (location identifier) • Agricultural • Agriculture • Antenna Group Refurbishment • Apollo Gamma Ray • Archive Generation Request • Archives Générales du Royaume (Belgium)

AGRE Active Geophysical Rocket Experiment

Agree Agreement

AGREE Advisory Group on Reliability of Electronic Equipment

AGRF Advanced Guidance Research Facility

AGRHYMET Agriculture-Hydrology-Meteorology Program

Agric Agricultural • Agriculture

Agricult Agricultural • Agriculture

AGRINET Agricultural Network (UK)

AGRIS Agricultural Information System

AgRISTARS Agriculture and Resources Inventory Surveys Through Aerospace Remote Sensing

Agron Agronomy

AGRS Acid-Gas Removal System • Active GPS Reference System • Airborne Gamma-Ray Spectrometer

AGS Abort Guidance Section • Abort Guidance System • Actual Ground Speed • Advance Guidance System • Aguascalientes (Mexico) • Airborne Ground Surveillance • Aircraft General Standard • Alternating Gradient Synchrotron • American Geographical Society (USA) • Analog Guidance Sensor • Antigravity Suit • Artificial Gravity Structure • Assistant General Secretary • Associate in General Studies • Assurance de Garantie des Salaires • ASTER Ground System • Asynchronous Gateway server • Augusta Bush (USA-GA) (location identifier) • Automatic Gain Stabilization

AGSAC Australian Gemini Science Advisory Committee

AGSC Australian Gemini Steering Committee

AGSE Astrophysical Grant in Support of Education • Astrophysics Grant Supplements for Education • Astrophysics Grant Supplement to Education • Aviation Ground Support Equipment

AGSO Australian Geological Survey Organization

AGSS American Geographical and Statistical Society (USA)

AGSSB Atmospheric, Geophysical and Space Sciences Branch (PAGASA)

Agst Against

AGST Applied General Systems Theory

Agt Agent • Agreement

AGt Anschaltgerät

AGT Above Ground Test • Above Ground Testing • Adaptive Guidance Throttling • Advanced Gas Turbine • Agent • Agent Technique • Association fo Geology Teachers • Astronomische Gesellschaft Tagung • Attractive-Game Theory • Average Generation Time • Ciudad del Este - Alejo García (Paraguay) (location identifier)

AGTDC Accord Général sur les Tarifs Douaniers et le Commerce

AGTM Archive Generation Termination Message

AGTS Automatic Gyro Test Set

AGU Address-Generation Unit • Aerospace Ground Unit • Aguascalientes (Mexico) (location identifier) • American Geophysical Union (USA) • Automatic Ground Unit

AGV Acarigua (Venezuela) (location identifier) • Assurance Générale de Vie • Automated Guided Vehicle • Automatic Guided Vehicle • Avion à Grande Vitesse

AGVA Anglo-German Variable Geometry

AGVS Automatically Guided Vehicle System

AGVT Advanced Ground Vehicle Technology

AGW Acoustic-Gravity Wave • Acquisition - Guiding - Wavefront sensing • Actual Gross Weight • Agnew (Australia-QLD) (location identifier) • Allowable Gross Weight • Atmospheric Gravity Wave

AGWP Absolute Global Warming Potential

AGX Agatti Island (India) (location identifier)

Agy Agency

AGY Agency • Argyle Downs (Australia-WA) (location identifier)

AGZ Actual Ground Zero • Aggeneys (South Africa) (location identifier)

AGZO Astronomische Gesellschaft Zürcher Oberland (Switzerland)

AGZU Astronomische Gesellschaft Zürcher Unterland (Switzerland)

Ah Ampere-Hour

A-H Ampere-Hour

A&H Accident & Health

A/H Altitude/Height

AH Acoustic Holography • Adaptive Hypermedia • Adresse Habituelle • After Hatch • Aft Hatch • Air Heater • Alternate Headquarters • Ampere-Hour • Angle Horaire • Angulo Horario • Anno Hebraico • Anno Hegirae • Application Header • Aromatic Hydrocarbon • Artificial Horizon • AstroHubs • Attitude Hold • Authentication Header • Available Hours • Avant-Hier

AHA Adaptive Hypermedia Architecture • American Historical Association (USA) • American Hospital Association (USA) • Arbeitsgemeinschaft Hildesheimer Amateursastronomen • Automated Historical Archive • Okinawa Naha (Japan) (location identifier)

AHAC AEOS-Haleakala Atmospheric Characterization

AHAF Advanced Hypervelocity Aerophysics Facility

AHAM Adaptive Hypermedia Application Model • Association of Home Appliance Manufacturers

AHARS Attitude And Heading Reference System

AHAS Allegheny Highlands Astronomical Society (USA-PA) • Automated Historical Archives Set • Automated Historical Archives System

AHAT Advanced Hardened Avionics Technology

AHB Abha (Saudi Arabia) (location identifier)

AHBC Apparent Horizon Boundary Conditions

AHC Aircraft Handling Characteristics • Animal Health Committee • Automated Head Count • Automatic Hold-off Command • Herlong Amedee (USA-CA) (location identifier)

AHCC Asynchronous Health Care Communications

A&HCI Arts & Humanities Citation Index (ISI)

AHCI Arts and Humanities Citation Index (ISI)

AHCIET Asociación Hispanoamericana de Centros de Investigación y Estudios de Telecomunicaciones

AHCPR Agency for Health Care Policy and Research (USA)

AHCS Advanced Helicopter Control System • Advanced Hybrid Computer System

Ahd Ahead

AHD Advanced Hardware Development • Ahead • Ardmore Downtown (USA-OK) (location identifier) • Arrowhead • Arunah Hill Days (USA-MA)

AHDAS Adaptable Hydrologic Data Acquisition System

AHDL Analog Hardware Descriptive Language

AHDP Advanced H/W Development Baseline

AHDS Arts and Humanities Data Service

AHE Adaptive Histogramme Equalization • Associate in Home Economics • Association for Higher Education • Automated History Entry

AHEAD Animal Health/Emerging Animal Diseases

AHERA Asbestos Hazard Emergency Response Act (USA)

AHF Adaptive High Frequency • American Heritage Foundation (USA) • Anhydrous Hydrogen Fluoride • Antihemophilic Factor • Arapahoe (USA-NE) (location identifier) • Attitude History File

AHFA Adaptive High Frequency Appliqué

AHFC Adaptive High Frequency Controller

AHFG ATM-attached Host Functional Group

AHG Ad Hoc Group • Altschuldenhilfegesetz

AHH Amery (USA-NE) (location identifier)

AHI Amahai (Indonesia) (location identifier) • American Health Institute (USA) • Artificial Horizon Indicator

AHIMA American Health Information Management Association (USA)

AHISS Arecibo HI Strip Survey

AHJ Authority Having Jurisdiction

ahl Ad Hunc Locum

AHL Aishalton (Guyana) (location identifier) • Angulo Horario Local

AHM Acutely Hazardous Metal • Adaptive Hypermedia • Ampere-Hour-Meter • Amsterdam Hypermedia Model • Automatic Hold Mode

AH&MA American Hotel & Motel Association (USA)

AHMA American Hotel and Motel Association (USA)

AHMM Aspect Hidden Markov Model

AHMR Apartment House Mail Receptacle

AHN Archivo Histórico Nacional • Athens (USA-GA) (location identifier)

A/H/O Abort/Hold/Orbit

AHO Alghero (Italy) (location identifier)

AHOF Astronaut Hall Of Fame (USA-FL)

AHOI Allgemeines Hochschul-Orientierungs- und Informationssystem

AHP Absorption Heat Pump • Air Horsepower • Analytical Hierarchy Process

AHPCA Advanced High-Performance Computing Activity • Advanced High-Performance Computing Application

AHPCC ARPA High-Performance Computer Center

AHPCP Affordable High-Performance Computing Prospect

AHPCRC Army High-Performance Computation Research Center • Army High-Performance Computing Research Center

AHPL A Hardware Programming Language

AHQ Air Headquarters • Army Headquarters

AHR Ablative Heat Rate • Acceptable Hazard Rate • After Heat Removal • Alternate Headquarters

AHRA Audio Home Recording Act (USA)

AHRB Arts and Humanities Research Board (UK)

AHRI Arts and Humanities Research Institute

AHRIS Advanced High-Resolution Imaging Spectrometer

AHRS Attitude and Heading Reference System • Attitude, Heading and Reference Systems • Attitude Heading Reference System

AHRV Advanced High-Resolution Visible (imager)

AHS Adaptive Hypermedia System • Adult Health Study • Aft Heat Shield • Ahuas (Honduras) (location identifier) • Allgemeinbildende Höhere Schule (Austria) • Anno Humanae Salutis • Annual Housing Survey • Automated Highway System

AHSB Authority, Health, and Safety Branch

AHSE Assembly and Handling Support Equipment

AHSJ American Helicopter Society Journal (periodical code)

AHST Advanced Human Support Technology

AHStatG Außenhandelsstatistikgesetz

AHT Absorption Heat Transformer • Amchitka (USA-AK) (location identifier) • ATM Header Translator • Attitude History Table • Average Holding Time

AHTTI Association pour l'Histoire de Télécommunications et des Techniques de l'Information (France)

AHU Air Handling Unit • Al Hoceima (Morocco) (location identifier)

ahv Ad Hanc Vocem

AHW Als Het Ware • Atomic Hydrogen Welding

AHWG Ad Hoc Working Group

AHWGA Ad Hoc Working Group on Archiving

AHWGB Ad Hoc Working Group on Broadcasting

AHWGC Ad Hoc Working Group on Consumers

AHWGD Ad Hoc Working Group on Distribution

AHWGE Ad Hoc Working Group on Education

AHWGG Ad Hoc Working Group on Growth

AHWGH Ad Hoc Working Group on Health

AHWGI Ad Hoc Working Group on Imagery

AHWGP Ad Hoc Working Group on Production

AHWGR Ad Hoc Working Group on Research

AHWGT Ad Hoc Working Group on Teaching

AHWGX Ad Hoc Working Group on Exchange

AHWGY Ad Hoc Working Group on Youth

AHX Avionic Heat Exchanger

AHY Ambatolahy (Madagascar) (location identifier)

AHZ Alpe d'Huez (France) (location identifier)

ai Ad Interim

A&I Abstracting & Indexing • Accident & Indemnity • Analysis & Integration

A/I Actuator/Indicator

AI Access Interface • Acquisition Instruction • Action Intégrale • Action Item • Activity Index • Adaptation Interdite • Adapter Interface • Admiralty Islands • Adresse Incomplète • Agenda Item • Airspeed Indicator • Allocation d'Insertion • Allocation Intermédiaire • Altesse Impériale • Altimeter Indicator • Altitude Indicator • American Institute • Amnesty International • Analog Input • Anguilla (ISO country code) • Anno Inventionis • Ainti-Icing • Air Interface • Application Information • Application Interface • Arbitrage International • Area of Influence • Area of Interest • Artificial Insemination • Artificial Intelligence • Asociación de Internautas (Spain) • Assignment Instruction • Assistant Ingénieur • Assistant Instructor • Associate Investigator • Associative Ionization • Astronomical Institute • Asymmetry Index • Attitude Indicator • Audio Interface • Authentication Information • Automatic Identification • Automatic Indexing • Automatic Input • Avalanche Injection • Azimuth Indicator

AIA Aeronautical Industries Association (USA) • Aerospace Industries Association (USA) • Air Intelligence Agency • Alliance (USA-NE) (location identifier) • American Industries Association (USA) • American Institute of Aeronautics (USA) • American Institute of Architects (USA) • American Insurance Association (USA) • Anguilla (ISO country code) • Application Installation Aid • Applications Integration Architecture • Associazione Industrie Aerospaziali (Italy) • Atmospheric Imaging Assembly • Automated Image Analysis • Automatic Information Acquisition • Automation Interoperability Appraisal • Avionics Integration Area

AIAA Aircraft Industries Association of America • American Institute of Aeronautics and Astronautics (USA) • Area of Intense Aerial Activity

AIAAJ American Institute of Aeronautics and Astronautics Journal (periodical code)

AIAB Aerospace Industries Association of Brazil

AIAC Aerospace Industries Association of Canada • Artificial Intelligence Advisory Committee

AIACP Atelier Industriel de l'Aéronautique de Cuers-Pierrefeu (France)

AIAG Automotive Industry Action Group

AIAI Artificial Intelligence Applications Institute (UK-Scotland)

AIAO Aerospace Industry Association of Oregon (USA-OR)

AIAS Airborne Integrated Antenna System

AIASA Annual Integrated Assessment of Security Assistance

AIASJ American Institute of Aeronautics and Astronautics Student Journal (periodical code)

AIB Accident Investigation Board • Accidents Investigation Branch • Acquisition Interface Buffer • Action Item Browser • Aminoisobutyric • Anita Bay (USA-AK) (location identifier) • Aromatic Infrared Band • Associazione Italiana Biblioteche (Italy)

AIBD Asian Institute for Broadcasting Development

AIBS American Institute of Biological Sciences (USA)

AIC Academia de la Investigación Científica (Mexico) (now AMC) • Account Identifier Code • Accretion-Induced Collapse • Achromatic Interfero-Coronagraph • Acrobat Interapplication Communication • Advanced Integrated Circuit • Aeronautical Information Circular • Airok (Marshall Islands) (location identifier) • Akaike Information Criterion • Alfén Ion Cyclotron • Allowance Item Code • American Institute of Chemists (USA) • Analog Interface Circuit • Antenna Interface Controller • Applied Informatics and Communications (conference) • Architecture Integration Committee • Artificial Intelligence Center • Assigned Interaction Class • Assignment Information Card • Assignment Instruction Code • Astro Imaging Center • Astronomer In Charge • Audio Intercommunications Center • Automated Intelligence Correlation • Automatic Information Center • Automatic Intercept Center

AICA Arizona-Idaho Conservation Act (USA) • Associazione per l'Informatica ed il Calcolo Automatico (Italy)

AICAS Astronomical Institute of the Czech Academy of Sciences (Czech Republic)

AICB Association Internationale Contre le Bruit

AICC Action Item Control Card • Automatic Intercept Communications Controller

AICCSA ACS/IEEE International Conference on Computer Systems and Applications

AICDT Astronomer In Charge Discretionary Time

AICE American Institute of Chemical Engineers (USA)

AIChE American Institute of Chemical Engineers (USA)

AICMA Association Internationale des Constructeurs de Matériel Aérospatial

AICP Aerodynamic Instrumental Coefficient Package

AICPA American Institute of Certified Public Accountants (USA)

AICPS Advanced Integrated Collective Protection System

AICRIP All India Coordinated Programme for Rice Improvement

AICRO Association of Independent Contract Research Organizations

AICS Action Item Closeout Sheet • American Institute for Computer Science (USA) • Artificial Intelligence - Cognitive Science (conference) • Association Internationale du Cinéma Scientifique

AICU Astronomical Institute of the Comenius University (Slovakia)

AICV Applications of Invariance in Computer Vision (workshop)

AID Abbreviated Item Description • Accident - Incident - Deficiency • Action Item Directive • Action Item Disposition • Acute Infection Disease • Aeronautical Inspection Directorate • Agence Internationale pour le Développement • Agency for International Development • Algebraic Interpretive Dialogue • Allgemeiner Informationsdienst • Analog Input Differential • Anderson (USA-IN) (location identifier) • Architecture Interconnection Diagram • Association Internationale de Développement • Association Internationale pour le Développement • Audit Item Disposition • Automated Interaction Detector • Automatic Indexing of Documents • Automatic Interaction Detection

AIDA Advanced Integrated-circuit Design Aid • Aerosol Interactions and Dynamics in the Atmosphere (chamber) • ALIS Internal Data Administration • Alternatives for International Document Availability • Arecibo Initiative on Dynamics of the Atmosphere • Artificial Intelligence Discrimination Architecture • Associazione Italiana per la Documentazione Avanzata (Italy) • Astronomical Image Data Archive • Attention - Interest - Desire - Action • Automatic In-flight Data Acquisition

AIDAA Associazione Italiana di Aeronautica e Astronautica (Italy)

AIDAB Australian International Development Assistance Bureau

AIDAS Advanced Instrumentation and Data Analysis System

AIDB Analyst Information Database

AIDBP Antarctic Ice Drifting Buoy Programme

AIDC Aeronautic Industry Development Centre • Australian Industry Development Corporation

AIDDE Ames' Interactive Dynamic Display Editor

AIDE Aerospace Installation Diagnostic Equipment • Automated Integrated Design Engineering • Automated Intrusion Detection Environment • Automated Inventory Distribution and Exchange

AIDES Automatic Image Data Extraction System

AIDMO Arab Industrial Development and Mining Organization

AIDOS Automatisiertes Informations- und Dokumentationssystem

AIDJEX Arctic Ice Deformation Joint Experiment • Arctic Ice Dynamics Joint Experiment (*)

AIDP Associate of the Institute of Data Processing

AIDS Abort Inertial Digital System • Accident Information Display System • Administrative Information Data System • Advanced Image Digitizing System • Advanced Integrated Data System • Advanced Integrated Display System • Airborne Integrated Data System • Aircraft Integrated Data System • Astronomical Information Distributed Service • Automated Information and Documentation System • Automated Information Data System • Automated Information Display System • Automated Information Dissemination System • Automated Information Distribution System • Automatic Integrated Debugging System

AIE Acceptance Inspection Equipment • Accident - Incident - Événement • Ada Integrated Environment • Agence Internationale de l'Énergie • Agencia Internacional de la Energía • Aide Individuelle Exceptionnelle • Aiome (Papua New Guinea) (location identifier) • American Institute of Engineers (USA) • Asociación de Intérpretes y Ejecutantes (Spain) • Association for International Education • Associazione Italiana degli Editori (Italy) • Author Index Entry

AIEA Agence Internationale de l'Énergie Atomique

AIEE Associate of the Institution of Electrical Engineers • Association des Instituts d'Études Européens

AIEP Additive Inverse Eigenvalue Problem

AIER Allied Interface Engineering Request

AIERI Association Internationale des Études et Recherches sur l'Information

AIESEC Association Internationale des Étudiants en Sciences Économiques et Commerciales

AIESI Association Internationale des Écoles des Sciences de l'Information

AIEWS Advanced Integrated Electronic Warfare System

aif Audio Interchange Format (file extension)

AIF Adjusted Impact Factor • Algorithm Integration Facility • Allocation d'Insertion et de Formation • Annual Impact Factor • Asociación Internacional de Fomento • Assis (Brazil-SP) (location identifier) • Associazione per l'Insegnamento della Fisica (Italy) • Astro-Info-Flow • Astronomical Instrumentation Facility • Audio Interchange Format • Automated Installation File

AIFC Audio Interchange File Compressed

AIFF Audio Interchange File Format

AIFS Australian Integrated Forecast System

AIFT Association Interprofessionnelle pour la Formation Technique (France)

Aig Aiguille

AIG Address Identifier Group • Address Indicating Group • Aerosol-Ionization Gas • Analog Indicator Group • Association Internationale de Géodésie • Astronomical Information Gateway • Astronomisches Institut Graz (Austria) • Astronomy Interest Group • Yalinga (Central African Rep.) (location identifier)

AIGA American Institute of Graphic Arts (USA) • Association Internationale de Géomagnétisme et d'Aéronomie

AIGG Association Internationale de Géodésie et de Géophysique

Aigle Aiguille

AIGO Australian Interferometric Gravitational-wave Observatory • Australian International Gravitational Observatory

AIH Acoustical Imaging and Holography (periodical code) • All In Hands • American Institute of Hydrology (USA)

AIHA American Industrial Hygiene Association (USA) • American International Health Alliance (USA)

AIHEP Artificial Intelligence in High-Energy Physics

AIHS Académie Internationale d'Histoire des Sciences • Association Internationale d'Hydrologie Scientifique

AII Accord Inter-Institutionnel • Advanced Interactive Interface • Alisabieh (Djibouti) (location identifier) • Analogical and Inductive Inference (workshop) • Antarctic Infrared Interferometer • Astronomical Information Infrastructure • Automated Item Identification

AIIA Associazione Italiana per l'Intelligenza Artificiale (Italy) • Australian Information Industry Association • Australian Institute of International Affairs

AIIC Association Internationale des Interprètes de Conférences

AIIDS Analyst Interpreter Interactive Display System

AIIE American Institute of Industrial Engineers (USA)

AIIM Association for Information and Image Management (USA)

AIIMS All India Institute of Medical Sciences

AiInt Air International (periodical code)

AIIOC Advanced Imagery Interpretation Orientation Course

AIIP Advanced Industry Information Package • Association of Independent Information Professionals (USA)

AIIS Advanced Infrared Imaging Seeker • Automated Information Infrastructure System

AIISA Automated Ice Information System for Arctic

AIJ Activities Implemented Jointly

AIK Aid In Kind • Aiken (USA-SC) (location identifier) • Aircraft Installation Kit • Artificial Intelligence Knowledge

AIKR Artificial Intelligence Knowledge Representation

Ail Aileron

AIL Action Item List • Ailigandi (Panama) (location identifier) • Année Internationale du Livre (1972) • Asset Interchange Language • Assign Interswitch Link • Association of International Libraries • Authoring In the Large

AILAS Automatic Instrument-Landing Approach System

AILIS Association of International Librarians and Information Specialists

AILS Advanced Integrated Landing System • Airborne Information for Lateral Spacing • ARC Imaging Library Server • Automatic Instrument Landing System

AIM Acquisition Interface Management • Acquisition Interface Module • Adaptive Image Manager • Advanced Industrial Management • Advanced Informatics in Medicine • Advanced Information Memory • Advanced Information Module • Advanced Instrumentation and Measurements • Advanced Instrument Manufacturing • Advanced Integrated Management • Aeronautical Information Manual • Aeronomy of Ice in the Mesosphere • Aerospace Industries Malaysia • African Index Medicus • Ailuk Island (Marshall Islands) (location identifier) • Airman's Information Manual • Alternative Investment Market • American Institute of Management (USA) • American Institute of Mathematics (USA-CA) • Amplitude Intensity Modulation • Analog Intensity Modulation • Analytical Ion Microscope • Analytical Ion Microscopy • Antenna Interface Module • AOL Instant Messenger • Apogee Injection Module • Application Interpreted Model • Approximate Intertial Manifold • Approximate Inverse Mapping • Artificial Ionospheric Mirror • Association Internationale de Météorologie • Associative Index Method • Asteroid Investigation with Microspacecraft • Astrometric Interferometry Mission • Astronomisches Institut Münster (Germany) • Astronomy Internet Master's (degree) • Asynchronous Interface Module • ATM Inverse Multiplexer • ATM Inverse Multiplexing • Atmosphere - Ionosphere - Magnetosphere • Attitude and Information Management • Audio Indicate Muted • Auroral Ionospheric Mapper • Australian Institute of Management • Authoring Instructional Material • Automated Information Management • Automated Input Methodology • Automated Internal Management • Automated Inventory Management • Automatic Immission Monitoring • Automatic Information Management

AIMA As Interest May Appear

AIMAS Associazione Italiana di Medicina Aeronautica e Spaziale (Italy)

AIMC Asociación para la Investigación de Medios de Comunicación (Spain)

AIMCC Association of Ice Monitoring Contractors of Canada

AIMCR Alternative Internal Management Control Review

AIMCS Airborne Investigations of Mesoscale Convective Systems

AIMD Aviation Intermediate Maintenance Department • Avionics Integrated Maintenance Department

AIME American Institute of Mechanical Engineers (USA) • American Institute of Mining Engineers (USA) • Association for Information Media and Equipment • Association of the Institute of Mechanical Engineers • Autonomous Integrity Monitored Extrapolation

AIMES Automated Inventory Management Evaluation System

AIMI Aviation Intensive Management Item

AIMIA Australasian Interactive Multimedia Industry Association

AIML Astronomical Instrument Markup Language

AIMNET Acquisition Information Management Network

AIMP Anchored Interplanetary Monitoring Platform

AIMPA Association Internationale de Météorologie et de Physique de l'Atmosphère

AIMLIS Automatic Intelligent Material Location and Identification System

AIMO Association of Indian Meteor Observers (India)

AIMR Airborne Imaging Microwave Radiometer

AImS Acousto-optic Imaging Spectrometer

AIMS Acousto-optic Imaging Spectrometer • Acquisition Information Management System • Advanced Information Management System • Advanced Integrated Maintenance System • Advanced Integrated Millimeter-wave Subassembly • AFGL Interactive Meteorological System • Agile Infrastructure for Manufacturing Systems • Airborne Integrated Mapping System • Aircraft Identification Monitoring System • Aircraft Integrated Monitoring System • Aircraft Information Management System • Altitude Identification and Measuring System • Analytical Imagery Matching System • Applied Information Management System • Australian Institute for Marine Sciences • Auto Indexing Mass Storage • Automated Image Measuring Software • Automated Image Measuring System • Automatic Image Matching System • Automated Industrial Management System • Automated Information and Management System • Automated Information Management System • Automated Infrastructure Management System • Automated Instructional Management System • Automated Inventory Management System • Automatic Identification and Monitoring System

AIMSA Artificial Intelligence: Methodology, Systems, and Applications • (conference)

AIMT Artificial-Intelligence-based Maintenance Tutor

AIMTB Artificial Intelligence Modules Testbed

AIMUX ATM Inverse Multiplexing

AIMW Advanced International Management Workshop

AIN Advanced Intelligent Network • Approved Item Name • Assembly Identification Number • Wainwright (USA-AK) (location identifier)

AINA Arctic Institute of North America

AInd Anglo-Indian

AINSE Australian Institute for Nuclear Science and Engineering

AiO Assistent in Opleiding

AIO Activities - Interests - Opinions • Airborne Infrared Observatory • Application Interoperability • Arecibo Ionospheric Observatory (Puerto Rico) • Assistent In Opleiding • Asynchronous Input/Output • Atlantic (USA-IA) (location identifier) • Auroral Imaging Observatory

AIOCS Automated Industry and Occupation Coding System

AIOD Automatic Identificated of Outward Dialing • Automatic Identificated Outward Dialing • Automatic Identification Outward Dialing • Automatic Identified Outward Dialing

AION Alphabetical Index Of Names

AIOP All-In-One Procedure • Analog Input/Output Package • Association Internationale d'Océanographie Physique • Association Internationale d'Orientation Professionnelle

AIP Action Incitative Programmée (CNRS) • Adaptive Interferometer Processor • Advanced Information Processing • Advanced Instrumentation Payload • Advanced Instrumentation Platform • Aeronautical Information Publication • Agence Internationale de Presse • Ailinglapalap Island (Marshall Islands) (location identifier) • Airborne Ion Probe • Aircraft Instrumented Payload • Airport Improvement Program • Algorithm Intercomparison Programme • Algorithm Intercomparison Project • American Institute of Physics (USA) • Application Input Processor • Approval In Principle • Approved In Principle • Archival Information Package • Archive Information Package • Associate Investigator Program • Association of Internet Professionals • Astronomical Image Processing • Astrophysikalisches Institut Potsdam (Germany) • Atmosphere-Independent Propulsion • Australian Institute of Physics • Automated Implementation Plan • Automatic Information Processing • Avionics Improvement Package • Avionics Improvement Plan • Avionics Integration Plan • Azimuthal Intensity Profile

AIPC Alternate Indentured Product Code • American Institute of Physics Conference Series (periodical code) • Asociación Iberoamericana de Periodistas Científicos • Association Internationale des Palais des Congrès • Automatic Information Processing Center

AIPCN Association Internationale Permanente des Congrès de Navigation

AIPCR Association Internationale Permanente des Congrès de la Route

AIPE Apple Image Processing Educator

AIPEA Association Internationale Pour l'Étude des Argiles

AIPES Automated Intermediate Polyelastic Sampling

AIPG American Institute of Professional Geologists (USA)

AIPLA American Intellectual Property Law Association (USA)

AIPLF Association Internationale des Parlementaires de Langue Française

AIPO American Institute of Public Opinion (USA) • Australian Industrial Property Organization

AIPPI Association Internationale pour la Protection de la Propriété Industrielle

AIPS Advanced Integrated Propulsion System • Advanced Intrapulse Processor System • Association Internationale de Prévoyance Sociale • Astronomical Image Processing System • Automated Information Processing System • Automated Intelligence Processing System • Automatic Indexing and Proofreading System

AIPU Association Internationale des Professeurs d'Universités

AIPULF Association Internationale des Presses Universitaires de Langue Française

AIQ All Is Quiet • Analyst Interest Queue

AIR Action Item Report • Adaptive Intercommunication Requirement • Additive Increase Rate • Advanced Information Retrieval • Aero International Regional • After Initial Release • Airborne Imaging Radar • All India Radio • American Indian Reservation (USA) • American Institute for Research (USA) • Ancillary Items Rack • Annals of Improbable Research • Aripuanã (Brazil-MT) (location identifier) • Artist In Residence • Association Internationale des Registres • Automated Image Retrieval • Automatic Indexing and Retrieval

AIRAPT Association Internationale pour la Recherche Avancée en Pneumatique Technique

AIRC Association of Independent Radio Companies

AIRCAM Aerospace Infrared Camera

AirCo Air Conditioning • Air et Cosmos (periodical code)

AirE Aircraft Engineering (periodical code)

AIRE AOR/IER Resource Estimator • Astrophysical Integrated Research Environment

AIREA American Institute of Real Estate Appraisers (USA)

AIREP Aircraft Report

AIRES Advanced Imagery Requirements Exploitation System • Airborne Infrared Echelle Spectrometer • Automatic Inverted Recording Echo Sounder

AIRF Additive Increase Rate Factor

AIRG Applied Imaging Research Group

AIRH Association Internationale de Recherches Hydrauliques

AIRIS Alaska Integrated Resource Inventory System

AIRK Area Interswitch Rekeying Key • Assigned Interarea Rekeying Key

AIRMS Advanced Infrared Measuring System

AIRO Antarctic Infrared Observatory • Associazione Italiana di Ricerca Operativa (Italy)

AIROBICC Airshower Observation By Integrating Čerenkov Counters

AIRP Airborne Instrumentation Research Project • Association Internationale des Relations Publiques

Airpl Airplane

Airpt Airport

AIRS Acoustic-Imaging Recognition System • Advanced Inertial Reference Sphere • Advanced Infrared Radiation Sounder • Advanced Infrared Radiometer Sounder • Advanced Infrared Sounder • Aerometric Information Retrieval System • Aircrew Incident Reporting System • Antarctic International Radiometric Survey • Atmospheric Infrared Sounder • Auroral Imaging Remote Sensor • Auroral Ionospheric Remote Sensor • Automatic Image Retrieval System • Automatic Information Retrieval System

AIRSAR Airborne Synthetic Aperture Radar

AirSp Air and Space (periodical code)

AIRSS Airborne Infrared Linescan System

AIRST Advanced Infrared Search and Track

AIRT Autonomous Infrared Technology

AIRTO Association of Independent Research and Technology Organizations

AIRV Assigned Interarea Rekeying Variable

AIS Academic Instruction School • Accident Investigation Site • Accountable Instruction System • Accounting Information System • Action Information System • Action Item Sheet • Active Interference Suppression • Active Isolation System • Adaptation et Intégration Scolaire • Address Information System • Administrative and Information Services • Administrator for Information Systems • Advanced Image Slicer • Advanced Imaging System • Advanced Indication System • Advanced Information System • Advanced Instructional System • Advanced Ionospheric Spectrometer • Advance In Schedule • Adversary Information System • Aeronautical Information Service • Agreed Industry Standard • Airborne Imaging Spectrometer • Alarm Indication Signal • Alarm Indicator Signal • All-sky Imaging Survey • American Information Systems Inc. (USA-IL) • American Interplanetary Society (USA) • Analog Instrumentation System • Analysis Instrument Set • Answer-Indicative Sentence • Application Information System • Application Interface and Services • Arbeitsgruppe für Integrierte Schaltungen • Architecture Integration Study • Archival Information System • Arizona Imager/Spectrometer • Arorae Island (Kiribati) (location identifier) • Artificial Immune System • Association Internationale de Sociologie • Attitude Indicating System • Audio Intercommunications System • Australian Information Service • Authoring In the Small • Automated Identification System • Automated Information Security • Automated Information System • Automated Instrumentation System • Automated Intelligence System • Automatic Intercept System • Autonomous Intelligent System • Avionics Integrated Support

AISA Association Internationale pour la Sécurité Aérienne • Astronomical Institute of the Slovak Academy

AISAO Astrofizicheskie Issledovaniia (periodical code)

AISAS Astronomical Institute of the Slovak Academy of Sciences (Slovakia)

AISB Association of Imaging Service Bureaus

AISC Academic Information Services Committee • Agence Internationale de Satellites de Contrôle • Airborne Imaging Spectrometer Campaign • American Institute of Steel Construction • Artificial Intelligence and Symbolic Computation (conference) • Assessment and Information Service Centre • Association of Independent Software Companies (USA)

AISE Association Internationale des Sciences Économiques • Automated Information System Equipment

AISES American Indian Science and Engineering Society (USA)

AISGC Automatic Information System of Geodesy and Cartography

AISH Association Internationale des Sciences Hydrologiques

AISI American Iron and Steel Institute (USA)

AISIST Australian Institute of Spatial Information Sciences and Technology

AISJ Association Internationale des Sciences Juridiques

AIS&LUS All India Soil & Land Use Survey

AISLUS All India Soil and Land Use Survey

AISM Association Internationale de Signalisation Maritime • Association Internationale des Sociétés de Microbiologie

AISMC Artificial Intelligence and Symbolic Mathematical Computing (conference)

AISO Automated Information Security Officer

AISof Astrofizicheskie Issledovaniia Sofia (periodical code)

AISP Association Internationale des Sciences Politiques • Association of Information Systems Professionals (USA)

AISPIT Association Internationale de Séismologie et de Physique de l'Intérieur de la Terre

AISPRT Automated Information Security Police Review Team

AISR Academic Information Services and Research • Applied Information Systems Research • Astrophysics Information Systems Research

AISRC Automated Information Systems Review Council

AISRP Applied Information Systems Research Program (NASA)

AISS Adversary Information Systems Security • African Institute of Space Science • Association Internationale de la Science du Sol • Association Internationale de la Sécurité Sociale • Australian International Space School • Automated Information Systems Security

AISSAP Automated Information Systems Security Assessment Program

AISSP Adversary Information Systems Security Program • Automated Information Systems Security Policy • Automated Information Systems Security Program

AIST Abbreviated Integration System Test • Advanced Interferometric System Technology • Agency of Industrial Science and Technology (Japan) • Astrometric Artificial Satellite Telescope • Automatic Informational Station

AISTR Advanced Interferometric System Technology Requirement

AISTS Associazione Italiana della Stampa Tecnica e Scientifica (Italy)

AI&T Algorithm Integration & Test • Assembly, Integration & Testing

AIT Acceptor Idle Table • Active Imaging Testbed • Advanced Identification Technique • Advanced Imaging Tracker • Advanced Individual Training • Advanced Information Technology • Advanced Institutional Training • Advanced Intelligent Tape • Agency for Instructional Technology • Agency for Instructional Television • Aide Technique • Airborne Intelligence Technician • Aitutaki (Cook Islands) (location identifier) • Algorithmic Information Theory • Algorithm Integration and Test • Algorithm Integration Team • American Institute in Taiwan • Analysis and Integration Team • Application/Infrastructure Tool • Applications Interface Test • Artificial Ionospheric

Turbulence • Asian Institute of Technology (UNU) • Assembly, Integration and Test • Assembly Integration Test • Assembly - Integration - Testing • Assistance Technique d'Ingénieur • Associazione Italiana Telerilevamento (Italy) • Astrometric and Imaging Telescope • Astrometric Imaging Telescope (*) • Astronomisches Institut Tübingen (Germany) • Atmospheric Interceptor Technology • Autoignition Temperature • Autoignition Test • Automated Identification Technology • Automated Information Transfer • Automatic Imaging Telescope • Automation Integration Technology • Automotive Information Test • Average Instruction Time

AITA Asociación Internacional de Transporte Aéreo • Association Internationale des Transports Aériens

AITAL Asociación Internacional de Transporte Aéreo Latinoamericano

AITC Airport Intermodal Transit Connector • Association Internationale des Traducteurs de Conférence • Association of Investment Trust Companies

AITI Automated Interchange of Technical Information

AITMP Arctic Ice Thickness Monitoring Project

AITP Aerospace Industry Technology Program • Arctic Ice Thickness Project • Association of Information Technology Professionals (USA) • Audit Information Transfer Protocol

AITS Action Item Tracking System • Adopted Information Technology Standard • Advanced Information Technology Service • Advanced Information Technology System • Advanced Information Transmission System

AITT Acadia Institute for Teaching and Technology • Algorithm Integration and Test Team

AIU Abort Interface Unit • Access Interface Unit • Accident Investigation Unit • Advanced Interface Unit • Antenna Interface Unit • AOCS Interface Unit • Association Internationale des Universités • Atiu Island (Cook Islands) (location identifier) • Attaché d'Intendance Universitaire • Auxiliary Interface Unit • Avionics Interface Unit

AIUB Astronomisches Institut der Universität Bern (Switzerland)

AIUP Acceptable Internet Usage Policy

AIUTA Association Internationale des Universités du Troisième Âge

AIUW Astronomisches Institut der Universität Würzburg (Germany)

AIV Aliceville - George Downer (USA-AL) (location identifier) • Assemblage - Intégration - Vérification • Assembly - Integration - Verification • Association Internationale de Volcanologie • Automatische Informationsverarbeitung

AIVS Automatisiertes Informationsverarbeitungssystem

AIW APPN Implementer's Workshop • Arbeitskreis für Internationale Wissenschaftskommunikation • Astronomisches Institut Würzburg (Germany) • Atlantic Intercoastal Waterway

AIWEX Arctic Internal Wave Experiment

AIWM American Institute of Weights and Measures (USA)

AIWS Artificial Intelligence Workshop

AIX Advanced Interactive Executive • Alternate Image Exploitation • Artificial Intelligence Executive

AIY Atlantic City - Bader (USA-NJ) (location identifier)

AIZ Aerodrome Information Zone • Kaiser - Lake of the Ozarks (USA-MO) (location identifier)

AI2 Advanced Image Intensification • Asociación de Ingenieros en Informática (Spain)

Aj Ajouté

A/J Antijam

AJ Acte Judiciaire • Administrative Judge • Antijamming • Apojove • Assembly Jig • Astronomical Journal (periodical code) • Auberge de Jeunesse

AJa Astronomischer Jahresbericht (periodical code)

AJA Adjoint Administratif • Ajaccio - Campo dell Oro (France) (location identifier)

AJAASS Arab Journal for Astronomy And Space Sciences

AJAP Adjoint Administratif Principal

AJAS Association des Jeunes Amis de la Science (France)

AJB Astronomischer Jahresbericht

AJCC Alternate Joint Communications Center

AJCL American Journal of Computational Linguistics (periodical code)

AJCON Automated Job Control

AJCS Automated Job Card System

AJE Asociación de Jovenes Empresarios (Spain)

AJF Al-Jouf (Saudi Arabia) (location identifier) • Jouf (Saudi Arabia) (location identifier)

AJFP Adaptive Joint Force Package

AJI Association Juridique Internationale

AJIT Aides aux Jeunes pour l'Innovation Technologique

AJJ Akjoujt (Mauritania) (location identifier)

AJL Aizawl (India) (location identifier)

AJM Abrasive Jet Machining

AJN Anjouan (Comoros) (location identifier)

AJO Al-Jouf (Yemen) (location identifier)

AJP Advanced Joint Planning • American Journal of Physics (periodical code)

AJPAA Association des Journalistes Professionnels de l'Aéronautique et de l'Astronautique (France)

AJPO Ada Joint Program Office

AJQ Assignment and Justification Questionnaire

AJR Apport Journalier Recommandé • Arvidsjaur (Sweden) (location identifier)

AJS Abreojos (Mexico) (location identifier)

AJSE Arabian Journal of Science Engineering (periodical code)

AJSPI Association des Journalistes Scientifiques de la Presse d'Information (France)

AJT Adjoint Technique

AJU Aracaju (Brazil-SE) (location identifier)

AJY Agades (Niger) (location identifier)

AK Accadian • Adaptive Kernel • Alaska (USA) • Alphabetischer Katalog • Apogee Kick • Arbeitskreis • Automatic Keying

aka Also Known As

AKA Active Keel Actuator • Also Known As • Anisotropic Kinetic Alpha • Ankang (China-PRC) (location identifier) • Arbeitskreis Asteroiden (Germany) • Ausschuß zur Koordinierung der Auslandsarbeit

Akad Akademie

AKAF Arbeitskreis Astronomie Freiburg eV (Germany)

AKAM Advanced Knowledge Area Module • Arbeitskreis Astrometrie

AkAtP Akademia Athenon Praktika (periodical code)

AKB Application Knowledge Base • Atka (USA-AK) (location identifier)

AKC Asymmetric Key Cryptography

AKCLIS Australian Key Centre for Land Information Studies

AKD Akola (India) (location identifier) • Automatic Key Distribution

AKDC Automatic Key Distribution Center

AKE Akieni (Gabon) (location identifier) • Arbeitskreis Energie (Germany)

AKF Arbeitskreis Festkörperphysik (Germany) • Kufrah (Libya) (location identifier)

AKG Akademie für Kunst und Geschichte (Germany) • Anguganak (Papua New Guinea) (location identifier)

AKI Akiak (USA-AK) (location identifier)

AKJ Asahikawa (Japan) (location identifier)

AKK Akhiok (USA-AK) (location identifier)

AKL Auckland (New Zealand) (location identifier)

AKM Apogee Kick Motor • Applied Knowledge Management • Arbeitskreis Meteore (Germany) • Automated Key Management • Automatic Key Management • Axial k-Means (algorithm) • Zakouma (Chad) (location identifier)

AKMC Automatic Key Management Center

AKN King Salmon (USA-AK) (location identifier)

AKNLDG Acknowledge

AKO A Kind Of • Akron - Washington County (USA-CO) (location identifier)

AKP ALMA Key Personnel • Anaktuvuk Pass (USA-AK) (location identifier) • Arbeitskreis Planetenbeobachter (VdS)

AKPA Association of Korean Physicists in America

AKQ Astraksetra - Gunung Batin (Indonesia) (location identifier)

AKR Akure (Nigeria) (location identifier) • Auroral Kilometric Radiation

AKS Algorithmic Key Selection • Aluminium - Potassium - Silicon • Anormale Kosmische Strahlung • Auki (Solomon Islands) (location identifier)

AKSAS Alaska State Accounting System

AKT Akrotiri (Cyprus) (location identifier)

AkTeK Akustika i Ul trazvukovaia Tekhnika (periodical code)

AktG Aktiengesetz

AKU Aksu (China-PRC) (location identifier)

AKUP Association of Korean University Presses (Korea-ROK)

AKV Akulivik (Canada-QC) (location identifier)

AKW Atomkraftwerk • Authors and Key Words

AKWIC Altered Keyword In Context • Authors and Keywords In Context

AKX Akt'ubinsk (Kazakhstan) (location identifier)

AKY Akyab (Myanmar) (location identifier) • Sittwe (Myanmar) (location identifier)

AkZh Akusticheskii Zhurnal (periodical code)

al Alia • Alii • Allée

Ál Álava (Spain)

Al Alemán • Alfvén Number • Alinéa • Allée • Aluminium

A/L Administration/Logistics • Airlock • Approach and Landing • Autoland

A&L Approach & Landing

AL Abraham Lincoln (1809-1865) • Absence Légale • Access Line • Access Link • Access Lock • Acquisition Letter • Acquisition Logistician • Acquisition Logistics • Acronym List • Address List • Administration Légale • Administrative Letter • Administrative Logistics • Airlock • Alabama (USA) • Alagoas (Brazil) • Alameda • Albania (international vehicle registration code) • Albania (ISO country code) • Albanie (code ISO de pays) • Alberta (Canada) • Algeria • Algonquin • Alignment • Altai • América Latina • Analytical Limit • Anglo-Latin • Année-Lumière • Anno Lucis • Antistokes Luminescence • Appleton Laboratory (UK) (see RAL) • Applications Layer • Approach and Landing • Après Livraison • Arab League • Arbitrated Loop • Artificial Life • Aschgraues Licht • Ashen Light • Assemblée Législative • Assembly Language • Assistance Légale • Astronautics Laboratory • Astronomical League (USA) • Attachment Level • Attenuation Length • Attestation Légalisée • Autograph Letter • Autoland

Ala Alabama (USA) • Alanine

ALA Actual Launch Azimuth • All Letters Answered • Alma Ata (Kazakhstan) (location identifier) • Almaty (Kazakhstan) (location identifier) • Alpha-Linolenic Acid • American Librarians Association (USA) • American Library Association (USA) (*) • American Lung Association (USA) • Asia and Latin America • Associate in Liberal Arts • Association Lotoise d'Astronomie (France) • ATM Link Accelerator • Authorized Landing Area • Authors League of America • Automobile Legal Association

ALABOL Algorithmic And Business-Oriented Language

ALACE Autonomous Lagrangian Circulation Explorer

ALACT Asociación Latinoamericana de Ciencias Fisiológicas

ALADA Asociación Latinoamericana de Derecho Aeronáutico-espacial

ALADDIN Atmospheric Layer And Density Distribution of Ions and Neutrals

ALADI Asociación Latinoamericana De Integración

ALADIN Atmospheric Laser Doppler Instrument

ALAE Atmospheric Lyman Alpha Emission • Atmospheric Lyman Alpha Experiment

ALAGE Asociación Latinoamericana de Geofísica Espacial

ALALC Asociación Latinoamericana de Libre Comercio

ALAN Application Layer Active Networking

ALAP AppleTalk Link Access Protocol • As Late As Possible • As Low As Possible • As Low As Practicable • As Low As Practical

ALARA As Low As Reasonably Achievable

ALARM Airborne Lidar Agent Remote Monitor

ALARME Alerte aux Risques Météorologiques Exceptionnels

ALARP As Low As Reasonably Practicable

Alas Alaska (USA)

ALAS Advanced Liquid Axial Stage • Approach/Landing Autopilot System • Atmospheric Laboratory for Application Science

ALASAT Air-Launched Antisatellite

ALASKA Arctic Lands And Shelves - Key Assessments

ALASS Air Launch Atmospheric Sounding System • Alternate Low-Altitude Surveillance System

ALAT Aviation Légère de l'Armée de Terre (France)

alb Albus

Alb Albania • Albanian • Alberta (Canada) • Albion

ALB Academic Libraries of Brooklyn (USA-NY) • Acceptance Log Book • Active Longitude Belt • Adaptive Load Balancing • Albania (ISO country code) • Albanie (code ISO de pays) • Albany (USA-NY) (location identifier) • Analog Loopback • Asthenosphere-Lithosphere Boundary • Automated Logbook

Alba Alberta (Canada)

Albac Albacete (Spain)

ALBO Automatic Line Buildout

Albq Albuquerque (USA-NM)

ALBS Air-Launched Balloon System

alc À la Carte

Alc Alcalde • Alcohol • Alcool

ALC Accounting Legend Code • Administration and Logistics Center • Administrative and Logistics Center • Advanced Logistics Center • Aft Load Controller • Agency Location Code • Air Logistics Center • Alicante - El Altet (Spain) (location identifier) • Alphabet Code • American Language Course • Analog Leased Circuit • Area Logistics Command • Arithmetic and Logic Circuits • Artificial

Luminous Cloud • Assembler Language Code • Assembly Language Coding • Assured Logistics Communications • Attached Laboratory Centre • Audio Load Compensator • Augmented Landing Site • Automatic Level Control • Automatic Level Controller • Automatic Light Control • Automatic Light Controller • Automatic Load Control • Automatic Load Controller

ALCA Aft Load Controller Assembly • Acuerdo de Libre Comercio de América • Área de Libre Comercio de las Américas • Automatic Level-Control Assembly

ALCAE Association of Librarians of Colleges of Advanced Education (USA)

ALCANUS Alaska - Canada - United States

ALCATRAZ A Large Catalog of Atmospheric Transparencies for Relative Astronomical Zeroing

ALCC Airborne Launch Control Center • Air Launch Control Center

ALCG Analog Line Conditioner Group • Analog Line Conditioning Group

Alch Alchemy

AlCl Aluminium Chloride

ALCOA Aluminium Co. of America

ALCOM Advanced Liquid Crystalline Optical Materials • Algebraic Compiler • ALGOL Compiler

ALCOMA Algebraic Combinatorics and Applications (conference)

ALCON All Concerned • Astronomical League Convention

ALCOP Alternate Command Post

ALCOR Astronomical League Correspondent • Astronomical Lighting Control Region

ALCORSS Australian Liaison Committee On Remote Sensing by Satellite

ALCP Area Local Control Panel

ALCPT American Language Course Placement Test

ALCS Airborne Launch Control System • Augmented Longitudinal Control System • Authors' Licensing and Collecting Society (UK)

ALCTS Association for Library Collections and Technical Services

Ald Alderney (Channel Islands)

ALD Accounting Line Designator • Acquisition Logistics Division • Affection de Longue Durée • Alerta (Peru) (location identifier) • Analog Line Driver • Arc-Length Difference • Ariane Lift Derivative • Assign Local Distribution • Assign Local Duplication • Astronomy from Large Databases • At a Later Date • Automatic Line Drawing • Available-to-Load Date

ALDC Adaptive Lossless Data Compression • Allowance List Depreciation Guide • Automatic Load and Drive Control

ALDF Advanced Lightning Detection Finder

ALDHU Asociación Latino-americana para los Derechos Humanos

ALDO Activity-Level-Dependent Operation

ALDOC Arab League Documentation Centre

ALDP Advanced Launch Deployment Program • Advanced Launcher Development Programme • Automatic Language Data Processing

ALDS Automatic Lightning Detection System

ALDT Administrative and Logistics Delay Time • Administrative Logistics Downtime • Average Logistics Delay Time • Aldeamento

ALE Accord de Libre Échange • Additional Living Expense • Address Latch Enable • Agence Locale pour l'Emploi • Airborne Laser Experiment • Air Liaison Element • Airport Lighting Equipment • Alpine (USA-TX) (location identifier) • Annual Loss Estimate • Annual Loss Expectancy • Application Link Enabling • Arbitrary Lagrange Euler • Arbitrary Lagrangian Eulerian (code) • Association de la Libre Entreprise • Association de Libre-Échange • ATM Link Enhancement • Atmospheric Lifetime Experiment • Atomic Layer Epitaxy • Automated Link Establishment • Automatic Link Establishment

ALEC Alternate Local Exchange Carrier • Alternative Local Exchange Carrier

ALECSO Arab League Educational, Cultural and Scientific Organization

ALEGE Asociación Latino-Americana de Geofísica Espacial

ALENA Accord de Libre-Échange Nord-Américain

ALENEX Algorithm Engineering and Experimentation (workshop)

ALEPH Apparatus for LEP Physics

ALERT Acute Launch Emergency Reliability Tip • Acute Launch Emergency Restraint Tip • Arctic Long-term Environmental Research Transect • Association pour L'Étude des Risques du Travail • Augmented Learning Environment and Renewable Teaching (project) • Automated Local Evaluation in Real Time

ALES Automated Land Evaluation System

ALEXIS Array of Low-Energy X-ray Imaging Sensors • Atmospheric Lidar Experiment In Space

AlF Aluminium Fluoride

ALF Acoustic Levitating Furnace • Acoustic Levitation Furnace • Airborne Laser Fluoroscope • Alert Log File • Aloft • Alta (Norway) (location identifier) • Annual Licence Fee • Application Level Framing • Automatic Letter Facer • Auxiliary Landing Field

ALFA Adaptive Optics with Lasers for Astronomy • Aéroports de Langue Française Associés • Air-Lubricated Free Attitude • Arecibo L-band Feed Array (Puerto Rico) • Astronomical Low-Frequency Array

ALFE Advanced Liquid Feed Experiment • Advanced Liquid Fuel Experiment • Analog Line Front End

a/Lfg An Lieferung

ALFG Additive Lagged-Fibonacci Generator

ALFI Association des Laboratoires Français Indépendants (France)

ALFOSC Andalucía Faint Object Spectrograph and Camera

AlFr Alta Frequenza (periodical code)

ALFS Airborne Low-Frequency Sonar • Automated Leave Form System

ALFT Action-Level Fault Tolerance • Approach and Landing Flight Test

Álg Álgebra

Alg Algebra • Algebraic • Algèbre • Algébrique • Alger (Algeria) • Algeria • Algerian • Algiers (Algeria)

ALG Algebra • Alger (Algeria) (location identifier) • Algorithm • Along • Application Layer Gateway • Autonomous Landing Guidance

AlGaAs Aluminium Gallium Arsenide

ALGEC Algorithmic Language for Economic Calculations

Alger Algeria • Algerian

ALGHNY Allegheny

ALGI Address-Linked Geological Inventory

ALGO Algorithm

ALGOL Algorithmic-Oriented Language

ALH Advanced Light Helicopter • Advanced Liquid Hydrogen • Alarm Handler • Albany (Australia-WA) (location identifier) • Allan Hills (Antarctica)

ALI Accelerated Lambda Iteration (technique) • Alice (USA-TX) (location identifier) • American Law Institute (USA)

• American Library Institute (USA) • Annual Limit on Intake • Approximate Lambda Iteration • Approximate Lambda-operator Iteration • Array Level Interface • Automatic Location Identification • Automatic Logic Implementation

ALIA Australian Library and Information Association (Australia)

ALIAS Aircraft Laser Infrared Absorption Spectrometer

ALIBI Adaptive Location of Internetworked Bases of Information

Alic Alicante (Spain)

ALICE Adventists Libraries Information Cooperative • A Large Ion Collider Experiment • A Large Isotopic Composition Experiment • Algorithme d'Injection de Courants Électriques • Algorithm for Listing, Counting, and Enumeration (conference) • Analyse des Liquides Critiques dans l'Espace • Array-Limited Control Electronics • Array-Limited Instrument Control Environment • Artificial Linguistic Internet Computer Entity

ALID Automated Library Issue Document

Align Alignment

ALIGN Alignment

ALIM Asynchronous Line Interface Module

ALIO Activity-Level-Independent Operation

ALIRA Atacama Large Infrared Array

ALIS Advanced Life Information System • Auroral Large Imaging System • Automated Land Information System • Automated Library and Information System

ALISE Association of Library and Information Science Education

ALISSA Atmosphère par Lidar Sur Saliout

ALIT Automatic Line Insulation Testing

ALIVARS Algol-Like Irregular Variable Star

ALIVE Artificial Life Interactive Video Entertainment

ALJ Alexander Bay (South Africa) (location identifier) • Australian Law Journal

Alk Alkali • Alkaline

ALK Alkalinity • Asela (Ethiopia) (location identifier)

ALKS Adaptive Least k-th Squares

All Allegato

ALL Additional Authorization List • Airborne Laser Laboratory • Albenga (Italy) (location identifier) • Lek (Albania) (ISO currency code)

ALLC Association for Literary and Linguistic Computing

Alld Allemand

ALLEA All European Academies

Alleg Allegory

allg allgemein

allgA allgemeine Ansicht

Alloc Allocate

Allow Allowance

ALLS Advanced Laser Light Source

ALLSTAR Aeronautics Learning Laboratory for Science, Technology And Research

Alfz Alférez

Alm Almacén • Almanac • Almería (Spain)

ALM Advanced Lightning Mapper • Alamogordo (USA-NM) (location identifier) • Almirante • AppWare Loadable Module • Asset and Liability Management • Assurance de Longue Maladie • Asynchronous Line Multiplexer • Audiolingual Method

ALMA Atacama Large Millimeter Array (Chile) • Automated Logistics Management Activity

ALMDA Airlock Multiple-Docking Adapter

ALMIS Acquisition Logistics Management Information System

ALMRS Automated Land and Mineral Record System

ALMS Airborne Laser Mapping System • Analytic Language Manipulation System • Automated Library Management System • Automated Logistics Management System

Almte Almirante

AlN Aluminium Nitride

ALN Alton (USA-IL) (location identifier) • Asynchronous Learning Network

ALNOT Alert Notice

ALO Advanced Lunar Operation • Allocate • Alternate Launch Officer • Annonce Légale Obligatoire • Astronomical Latitude Observatory • At Least One • Authorized Level of Organization • Automatic Lock-On • Waterloo (USA-IA) (location identifier)

ALOC Administrative/Logistics Operations Center

ALOE AppleTalk Low Overhead Encapsulation

ALOFT A Language Oriented for Flight Testing • A Low Frequency orbiting Telescope • AMOS Large Optical Test Facility

ALOHA Aerial Location of Hazardous Atmospheres (model) • A Long-term Oligotrophic Habitat Assessment • Astronomy Librarians Of Hawaii Association (USA-HI)

Aloj Alojamiento

ALOMAR Arctic Lidar Observatory for Middle Atmosphere Research

ALOPE Airborne Lidar Oceanographic Probing Experiment

ALOS Advanced Land Observing Satellite • Average Length Of Stay

ALOT Advanced Large Optics Technology • Apple Library Of To-morrow (USA-CA)

Alp Alpes • Alpin • Alpine

ALP Accélération par Laser de Particules • Advanced Leadership Programme • Advanced Logistics Planning • Advanced Logistics Program • Advanced Lunar Project • Advancement of Librarianship Programme (IFLA) • Aleppo Neirab (Syria) (location identifier) • Algebraic and Logic Programming (workshop) • Analyst Language Processor • Anomalous Luminous Phenomenon • Approval for Limited Production • Assembly Language Processor • Assembly Language Program • Automated Language Processing • Automated Learning Process • Automated Library Program • Automatic Linearity Plot • Auxiliary Line Printer

ALPA Acquisition Logistics and Product Assurance • Air Line Pilots Association (USA)

ALPAC Automated Language Processing Advisory Committee • Automatic Language Processing Advisory Committee (*)

ALPAL A Livermore Physics Applications Language

ALPC Adaptive Link Power Control

ALPEX Alpine Experiment

Alph Alphabetical • Alphabétique

Alpha Alphabetical • Alphabétique

ALPHA Airborne Lidar Plume and Haze Analyzer • Automatic Literature Processing, Handling and Analysis

ALPO Association of Lunar and Planetary Observers (USA-CA)

ALPOJ Association of Lunar and Planetary Observers Journal (periodical code)

ALPS Accidental Launch Protection System • Addresses, Lists and Proposals System • Advanced Linear Programming System • Advanced Liquid Propulsion System • Aggregate Level Simulation Protocol • Airborne Laser Polarization Sensor • Alternative Launch-Point System •

Approach and Landing Procedures Simulator • Assembly Line Processing System • Associated Logic Parallel System • Automated Language Processing System • Automated Launch Processing System • Automated Library Processing System • Automated Library Publishing System • Automated Logistic Publishing System • Automated Logistics Planning System • Automatic Linear Positioning System

ALPSP Association of Learned and Professional Society Publishers

Alq Alquimia

ALQ Alegrete (Brazil-RS) (location identifier) • Association des Librairies du Québec (Canada-QC)

ALQDS All Quadrants

alr Aliter

ALR Action Leakage Rate • Advanced Logic Research Inc. • Alexandra (New Zealand) (location identifier) • American Law Report • Antenne Linéaire Remorquée • Art Lost Register • Automated Leave Report

ALRI Acute Lower Respiratory Infection

ALRR Active Longitude Rotation Rate

ALRS Airborne Laser Ranging System

ALRSS Advanced Land Remote Sensing System

ALRT Alert

ALRU Automatic Line Record Update

ALRUTS Automated Line Replaceable Unit Tracking System

ALS Absorption-Line System • Accident Localization System • Action Lead Sheet • Active Line State • Ada Language System • Adjacent Link Station • Administrative and Legal Service • Advanced Landsat System • Advanced Launching System • Advanced Launch System (*) • Advanced Letter Sorter • Advanced Life Support • Advanced Light Source • Advanced Limb Scanner • Advanced Logistics System • Airborne Lidar System • Airborne Link Segment • Air-Land-Sea • Airlock System • Alamosa (USA-CO) (location identifier) • Alternate Launch System • Alternate Line Service • Alternative Landing Site • American Lunar Society (USA-PA) • Annual Linux Showcase (conference) • Application Layer Structure • Approach and Landing System • Approach Lighting System • Attitude Limiting System • Australian Landsat Station • Autograph Letter Signed • Automated Library System • Automated List Service • Automatic Landing System • Automatic Level Setting • Azimuth Laying Set

ALSA Advanced Linux Sound Architecture • Astronaut Life Support Assembly

Alsat Alsatian

ALSATER Réseau Alsacien de Télécommunications de la Recherche (France)

ALSCC Apollo Lunar Surface Closeup Camera • Ariane Launch Sales Control Committee

ALSE Ada Language System Extension • Astronaut Life Support Equipment • Availability of Logistics Support Elements • Available Logistics Support Element

ALSEC All Sectors

al seg Al Segno

ALSEP Apollo Lunar Surface Experiment Package

ALSF Area Logistics Support Facility

ALSMA Automated Logistics System Management Activity

ALSOR Air-Launched Sounding Rocket

ALSP Aggregate Level Simulation Protocol

ALSPP Atmospheric and Land Surface Processes Project

ALSS Acquisition and Logistic Support System • Advanced Life Support System • Advanced Logistic Support Site • Airlock Support System • ASCA Large Sky Survey

ALST Alaska Standard Time

ALSTG Altimeter Setting

Alt Alter • Alteration • Alternate • Alternatif • Alternative • Alternator • Altesse • Altimeter • Altitude • Alto

ALT Acquisition Lead Time • Active Leap Technology • Administrative Lead Time • Alenquer (Brazil-PA) (location identifier) • Algorithmic Learning Theory (workshop) • Alter • Alternate • Alternate Leave Time • Altimeter • Altitude • Approach and Landing Test • Automated Language Translation

Alta Alberta (Canada)

ALTA Active Long-Term Archive • Alberta (Canada) • Alberta Large-area Time-coincidence Array

ALTAC Algebraic Transistorized Automatic Computer

ALTAIR ARPA Long-range Tracking And Instrumentation Radar • Automatic Logical Translation And Information Retrieval

alt-az Altazimuth

altaz Altazimuth

Atld Altitude

alt. dieb. Alternis Diebus

Alte Almirante

Altér Altération

Alter Alteration

Altern Alternativement

ALTG Analog Line Trunk Group • Auxiliary Line Termination Group

alt. hor Alternis Horis

ALTHQ Alternate Headquarters

Altm Altimeter • Altmark

ALTM Airborne Laser Terrain Mapper

ALTMS Automated Library Technical Manual System

Altn Alternate • Alternative

ALTN Alternate

ALTNLY Alternately

alt.noct. Alternis Noctibus

Altntr Alternator

Altnv Alternative

ALTR Approach and Landing Test Requirement

Altrd Altered

Altrn Alternation

ALTS Aided Laser Tracking System • Airborne Laser Terrain Scanner • Automatic Laser Tracking System

ALTV All-Light Television

ALU Alula (Somalia) (location identifier) • Analog Line Terminal Unit • Analog Line Termination Unit • Arithmetic and Logic Unit • Arithmetic/Logic Unit

ALUE Admissible Linear Unbiased Estimator • Arizona Land Use Experiment

Alum Aluminium • Alumna • Alumnae • Alumni • Alumnus

ALV Advanced Launch Vehicle • Advanced Logistics Vehicle • Algemene Ledenvergadering • Andorra la Vella (Andorra) (location identifier) • Apprentissage au Long de la Vie • Australian Launch Vehicle • Auswärtiger Leihverkehr • Autonomous Land Vehicle

ALVINN Autonomous Land Vehicle In a Neural Net

ALW Aard- en Levenwetenschappen • Allowance • Walla Walla (USA-WA) (location identifier)

ALWF Actual Wind Factor

ALWR Advanced Light-Water Reactor

ALX Alexander City - Thomas C. Russell (USA-AL) (location identifier)

Aly Alley

ALY Alexandria - El Nouzha (Egypt) (location identifier) • Alley

ALZ Alitak (USA-AK) (location identifier)

a/m Above Mentioned

am Amber • Amharic (ISO language code) • Ante Meridiem • Attometre

aM am Main

A/m Amperes per Meter

Am America • American • Americium • Amiral • Metallic-Line A-type Star

A&M Administration & Management • Agricultural & Mechanical

A/M Area to Mass • Auto/Manual

AM Access Manager • Accounting Management • Acknowledgment Message • Acknowledge Message • Acquisition Management • Acquisition Manager • Active Matrix • Active Message • Active Monitor • Actuator Mechanism • Adaptation Model • Address Management • Adhesion Model • Adjacent Matrix • Administrative Memorandum • Administrative Module • Aeromedical Monitor • Agent de Maîtrise • Agile Mirror • Aide-Mémoire • Airline Mileage • Airlock Module • Air Mail • Alert Message • Allocation Mensuelle • Alpes-Maritimes (France) • Altazimuth Mount • Alternate Medicine • Amazonas (Brazil) • American • Amplitude Modulated • Amplitude Modulation • Angular Momentum • Anno Mundi • Annus Mirabilis (1666) • Ante Meridiem • Antenna Management • Aperture Masking • Application Manager • Application Mediation • Application Mediator • Archive Manager • Area Manager • Armenia (ISO country code) • Arménie (code ISO de pays) • Arp-Madore (catalog) • Arrêté Ministériel • Assertional Module • Asset Management • Assistant Manager • Associate Member • Associative Memory • Assurance Management • Assurance Mutuelle • Asynchronous Mode • Audit Manager • Automated Mapping • Automatic Monitoring • Auxiliary Memory • Auxiliary Module • Avis de Mutation • Award of Merit • Master of Arts

AMA Absolute Measurement Accuracy • Active Matrix Array • Against Medical Advice • Agent Management Assistant • Amarillo (USA-TX) (location identifier) • American Management Association (USA) • American Marketing Association (USA) • American Medical Association (USA) • American Microfilm Association (USA) • Ancient and Medieval Astronomy • Area Management Activity • Area Minimum Altitude • Articulated Mirror Assembly (*) • Articulating Mirror Assembly • Association Malmédienne d'Astronomie (Belgium) • Associazione Marchigiana Astrofili (Italy) • Attitude Measurement Assembly • Australian Medical Association • Automatic Message Accounting • Automatic Memory Allocation • Automatic Message Accounting

AMAC ALMA Management Advisory Committee

AMAD Activity Median Aerodynamic Diameter • Airborne Missions and Applications Division • Aircraft-Mounted Accessory Drive • Auxiliary-Mounted Accessory Drive

AMAE American Museum of Atomic Energy (USA)

AMAG Austria Metal AG

Amal Amalgamated

AMAL Aeronautical Medical Acceleration Laboratory • Authorized Medical Allowance List

AMANDA Antarctic Muon And Neutrino Detector Array

AMANDDA Automatic Messaging And Directory Assistance

AMAP Academic Map Analysis Package • Advanced Metric Analysis Program • Arctic Monitoring and Assessment Programme

AMAPS Apogee Motor Assembly with Paired Satellites

AMARC Aerospace Maintenance And Regeneration Center • Automatic Message Accounting Recording Center

AMARS Automatic Message Address Routing System

AMARV Advanced Maneuverable Reentry Vehicle

AMAS Advanced Millimeter-wave Atmospheric Sounder • Association Marseillaise d'Astronomie (France) • Automated Materiel Acquisition System • Aviation Management Accounting System

AMASE Astrophysics Multispectral Archive Search Engine

AMASIC Autonomous Mapping by Analytic Sensor and Intelligence Correlation

AMASS Airport Movement Area Safety System • Archive Management And Storage System

AMAST Algebraic Methodology And Software Technology (conference)

Amat Amateur

AMATF Arctic Monitoring and Assessment Task Force

AMATIS Automated Meteorological Terminal Information Service

AMATPS Automatic Message Accounting Teleprocessing System

AMATYC American Mathematical Association of Two-Year Colleges (USA)

amb Ambulance

Amb Ambassade • Ambassadeur • Ambassador • Ambiguo • Ambulance • Ambulatory

AMB Aircraft Maintenance Branch • Air Mission Brief • Airways Modernization Board • Aluminous Mare Basalt • Ambassador • Ambilobe (Madagascar) (location identifier) • Ambulance • Ambulatory

AMBAC Asociación Mexicana de Bibliotecarios (Mexico)

Ambas Ambassador

AMBD Automated Maintenance Database

AMBER Advanced Multiple Beam Equalization Radiography • Astronomical Multiple Beam Recombiner

AMBF Asset Master Balance File

Ambig Ambiguous

AMBIT Algebraic Manipulation By Identity Translation • Acronym May Be Ignored Totally

AMBL Ambulance

AMBLADS Advise Method, Bill of Lading, and Date Shipped

AMC Academia Mexicana de Ciencias (Mexico) (formerly AIC) • Acceptable Means of Compliance • Acquisition Method Code • Active Motion Control • Adaptive Modulation and Coding • Adjoint Model Compiler • Aerospace Maintenance Cell • Aerospace Manufacturers' Council • Aerospace Manufacturing Center • Aft Motor Controller • Airport Mail Center • All Major Cards • Alternate Master Clock • Amorphous Carbon • Am Timan (Chad) (location identifier) • Année Mondiale des Communications (1983) • Appalachian Mountain Club • Archives and Manuscripts Control • Area Management Committee • Associate Members' Committee • Association Minéralogique du Canada • Associative Memory Computer • Associative Memory Controller • Automatic Manual Control • Automatic Measuring Complex • Automatic Message Counting • Automatic Mission Control • Automatic Mixture Control • Axial Meridian Circle

AMCA Aft Motor Controller Assembly

AMCD Addressograph - Multigraph - Copier - Duplicator • Annular Momentum Control Device

AMCEN African Ministerial Conference on the Environment

AMCIT American Citizen

AMCM Advanced Missions Cost Model

AMCO Assistant Monitor and Control Operator

AMCP Advanced Multimedia Content Processing (conference)

AMCR Automated Master Charge Record • Automatic Magnetic Card Reader

AMCS Attitude Measurement and Control System • Attitude Monitoring and Control System

AMCT Air Movement Control Team

AM-CW Amplitude-Modulated Continuous Wave

AMCW Amplitude-Modulated Continuous Wave

Amd Amend • Amendment

AMD Acid Mine Drainage • Active Matrix Display • Administration Management Domain • Administrative Management Division • Advanced Micro Devices Inc. • Aerospace Medical Division • Age-Metallicity Degeneracy • Age-related Macular Degeneration • Ahmedabad (India) (location identifier) • Air Movement Data • Air Movement Designator • Amend • Amended • Amendment • Angular Momentum Deficit • Antarctic Master Directory • Application Missions Division • Applied Mathematics Division • Appropriate Minimum Degree • Asset Management and Disposition • Assign Memory Display • Asteroid Meteoroid Detector • Automatic Message Display • Auxiliary Memory Drum • Average Monthly Demand • Dram (Armenia) (ISO currency code)

AMDAPS Automatic Meteorological Data Acquisition and Processing System

AMDAR Aircraft Meteorological Data Relay

AMDAS Avionics Maintenance Data Analysis System

AMDC Advanced Manufacturing Development Center

AMDCS Assurance Management Document Control System

AMDDRCS Automated Message Distribution, Determination, Reproduction, and Collation System

AMDE Analyse des Modes de Défaillance et de leurs Effets

AMDEA Association of Manufacturers of Domestic Electrical Appliances

AMDF Average Magnitude Difference Function

AMDG Ad Majorem Dei Gloria

AMDIC Associazione (Astronomica) Madonna di Campiglio (Italy)

AMDIS Atomic and Molecular Data Interactive System

AMDL Abstract Machine Description Language

AMDO Articulated Motion and Deformable Objects (workshop)

AMDP Aircraft Maintenance Delayed for Parts

AMDRFM Advanced Monolithic Digital Radio Frequency Memory

AMDS Advanced Missions Docking System • Advanced Monitor Display System • Amplitude Modulation Data System • Automatic Message Distribution System

Amdt Amendment

AME Abort Maneuver Evaluator • Accord Monétaire Européen • Acuerdo Monetario Europeo • Adaptive Mixture of Experts • Advanced Master of Education • Aerospace Management Element • Air Mobility Element • Airspace Management Element • Alternate Mission Equipment • Alto Molocue (Mozambique) (location identifier) • Amplitude Modulation Equivalent • Antarctic Marine Ecosystem • Antenna-Mounted Electronics • Application Model Editor • Association of Municipal Engineers (USA) • Astronaut Maneuvering Equipment • Asynchronous Modem Eliminator • Attitude Measurement Electronics • Attitude Measurement Error • Authorized Medical Examiner • AutoCAD Modeling Extension • Automatic Measuring Engine • Automatic Microfiche Editor • Auxiliary Mission Equipment

AMEC Aft Master Events Controller • Agent-Mediated Electronic Commerce (workshop) • Association of Mining and Exploration Companies

AMeDAS Automated Meteorological Data Acquisition System

AMEDAS Automated Meteorological Data Acquisition System

AMEDE Amélioration du Mécanisme de Déploiement

AMEGE Antarctic Marinen Engineering Geological Expedition

AMEL Active-Matrix Electroluminescent (display)

Am. Emb. American Embassy

Amend Amendment

Amendt Amendment

Amér América

Amer America • American

Amér Amérique

Amer. Ind. American Indian

AmerInd American Indian

Amer. Sp. American Spanish

AmerSp American Spanish

Amer. Std. American Standard

AMES Accent on Mathematics and Engineering Science • Aeronautical and Mechanical Engineering Sciences • Automatic Message Entry System

AMET Academy of Maritime Education and Training • Agent-Mediated Electronic Trading (workshop) • Automatic Meteorological (system) • Automatic Meteorology

AmEx American Express

AMEX American Express • Atlantic Meteorology Experiment • Australian Monsoon Experiment

AMF Abort Motor Facility • Adaptive Matched Filter • Aide-Mémoire Fiscal • Aircraft Maintenance Facility • Airmail Field • Airmass Factor • Airport Mail Facility • Ama (Papua New Guinea) (location identifier) • Apogee Motor Fire • Apogee Motor Firing (*) • Arab Monetary Fund • Area Maintenance Facility • Arrangement Multifibre • Aspect Moderator Framework • Astronaut Memorial Foundation • Automatic Maneuvering Flap • Automatic Micro Furnace • Automatic Mirror Furnace • Autorité des Marchés Financiers (France)

AMFABS Advanced Maintenance Free Aircraft Battery System

AMFB Acquisition Management Functional Board

AMFM Automated Manufacturing Facilities Management • Automated Mapping Facilities Management

Amg Among

AMG Activation Management Group • Ada Management Group • Algebraic Multigrid • Amboin (Papua New Guinea) (location identifier) • Amplitude Modulation Generator • Angle of Middle Gimbal • Antenna Mast Group • Applied Mathematics Group • Architecture Management Group • Arzneimittelgesetz (Germany) • Australian Map Grid • Automatic Magnetic Guidance

AMGBS Archiv Meteorologie Geophysik und Bioklimatologie Serie Meteorologie und Geophysik (periodical code)

AMGS Acceleration Monitoring Guidance System • Advanced Meteorological Ground Station

Amh Amharic

AMH Arba Mintch (Ethiopia) (location identifier) • Automated Message Handler • Automated Message Handling • Automatic Message Handling

AMHS Automatic Message Handling System

AmI Ambient Intelligence

AMI Accidental Motor Ignition • Accord Multilatéral sur les Investissements • Active Microwave Instrument • Active Microwave Instrumentation • Acuerdo Marco Interconfederal • Acute Myocardial Infarction • Advanced Microwave Instrument • Aerospace Materials Information •

Agence Maritime Internationale • Air Mercury International • Air Mileage Indicator • Alternate Mark Inversion • American Megatrends Inc. (USA) • Analyses et Mesures Industrielles • Apogee Memory Image • Apogee Motor Igniter • Application Manager Interface • Application of Metrics in Industry • Arcminute Microkelvin Imager • Assurance Maladie-Invalidité • Astrophysique et Méthodes Interférométriques • Audit Manager Infrastructure • Axial Meridian Circle • Mataram (Indonesia) (location identifier)

AMIA American Medical Informatics Association (USA) • Atelier Mobile d'Initiation à l'Astronomie (Canada) • Australian Map Industry Association

AMIAP Associate Member of the Institution of Analysts and Programmers

AMiBA Array for Microwave Background Anisotropy

AMIBE Applications Multimodales pour Interfaces et Bornes Évoluées

AMIC Acquisition Management Information Center • Aerospace Materials Information Center • Aircraft Mishap Investigation Course • Australian Mining Industry Council • Automated Management Information Center • Automatic Management Information Center

AMICA Asteroid Multiband Imaging Camera • Automatic Multispectral Image Compaction Algorithm

AMICIST Auroral Microphysics and Ion Conics: Investigation of Space and Time

AMICO Art Museum Image Consortium • Asteroids, Meteorites, Impacts and their Consequences (conference)

AMIDE Advanced Multimedia Information Dissemination Environment

AMIDS Advanced Multispectral Image Descriptor System • Audit Monitoring and Intrusion Detection System • Aviation Maintenance Integrated Diagnostics System

AMIE Airglow Measurements of Infrared Emissions • Assimilative Mapping of Ionospheric Electrodynamics • Asteroid Microimager Experiment • Automated Methane Instrument Evaluation

AMIERE Associate Member of the Institute of Electronic and Radio Engineers

AMIF Applied Mathematics for Industrial Flow • Automated Map Information File

AMIGAS Advanced Meteorological Image and Graphics Analysis System

AMIGO American Inter-hemispheric Geobiosphere Organization (USA)

AMIGOS Access Method for Indexed Data Generalized for Operating System

Am. Ind. American Indian

AMIP Aircraft Modification Integration Program • Atmospheric Model Intercomparison Project

AMIR Advanced Microwave Imaging Radiometer

AMIRA Australian Mining Industry Research Association

AMIS Accounting Management Information System • Acquisition Management Information System • Advanced Management Information System • Advanced Microwave Imaging Sensor • Audio Message Interchange Specification • Australian Municipal Information System • Automated Maintenance Information System • Automated Management Information System • Aviation Management Information System

AMISAR Active Microwave Imager Synthetic Aperture Radar

AMISR Advanced Modular Incoherent Scatter Radar

AMIX American Information Exchange

AMJ Academy of Management Journal (periodical code) • Almenara (Brazil-MG) (location identifier) • April - May - June • Assemblée Mondiale de la Jeunesse

AmJC American Journal of Cardiology (periodical code)

AmJPh American Journal of Physics (periodical code)

AmJS American Journal of Science (periodical code)

AMK Aircraft Modification Kit • Anti-Misting Kerosene • Durango Animas (USA-CO) (location identifier)

AML Abandoned Mine Land • Acquisition Material List • Acute Myeloid Leukemia • Amplitude Modulated Link • Angular Momentum Loss • Applied Mathematics Laboratory • Approved Materials List • ARC Macro Language • Array Machine Language • Astronomical Markup Language • Asymmetric Maximum Likelihood • Puerto Armuellas (Panama) (location identifier)

AMLC Asynchronous Multiline Controller

AMLCD Active-Matrix Liquid Crystal Display

AMLI Algebraic Multilevel Iteration

AMLLV Advanced Multipurpose Large Launch Vehicle

AMLP Amplitude Modulation Link Program

AMLR Antarctic Marine Living Resources

AMLS Advanced Manned Launch System • Airspace Management Liaison Section • Master of Arts in Library Science

AMM Adaptive Mathematical Model • Aircraft Maintenance Manual • Amman - Queen Alia (Jordan) (location identifier) • Analog Modem Module • And Much More • Année Mondiale des Mathématiques • Anomalous Magnetic Moment • Antarctic Micrometeorite • Application Model Manager • ASEAN Ministerial Meeting • Association Médicale Mondiale • Auroral Multiscale Mission • Automatic Maintenance Monitor • Autorisation de Mise sur le Marché

AMMA Advanced Memory Management Architecture

AMME Automated Multimedia Exchange

AMMES Automated Multimedia Exchange System

AMMH Annual Maintenance Man-Hours

AmMin American Mineralogist (periodical code)

AMMIS Aerospace Manpower Management Information System • Automated Manpower Management Information System

AMML Automated Microbial Metabolism Laboratory

AMMM Autonomous Magnetic Momentum Management

AMMOS Advanced Multimission Operations System

AMMP Advanced Manned Missions Program • African Magnetic Mapping Project • Approved Modernization and Maintenance Program • Approved Modification and Maintenance Program

AMMPEI Annual Maintenance Man-hours Per Equipment Item

AMMR Advanced Multichannel Microwave Radiometer • Aircraft Multichannel Microwave Radiometer • Algebraic Methods for Model Reduction (meeting)

AMMS Advanced Microwave Moisture Sounder • Advanced Multichannel Microwave Sensor • Airborne Microwave Moisture Sounder • Allowable Materiel Management System • Automated Message Management System • Automatic Multimode Mass Spectrometry

AMMSA Alliance for Microgravity Materials Science and Applications (USA)

AMMT Automated Multimedia Terminal

AMN Alma - Gratiot Community (USA-MI) (location identifier) • Atomic Mass Number

AMND Ammended

AMNH American Museum of Natural History (USA-DC)

AMNIP Adaptive Man-machine Non-arithmetical Information Processing

AMO Acquisition Management Office • Acquisition Management Organization • Air Mass Zero • Air Movement Designator • Alpha Monocerotids (meteor stream) • Alternate Molecular Orbital • Area Maintenance Office •

Atomic - Molecular - Optical • Automation Management Office • Automation Management Officer • Automation Management Organization • Mao (Chad) (location identifier)

AMOB Automatic Meteorological Oceanographic Buoy

AMOC Aircraft Maintenance Officer Course • Asian Mobile Computing Conference

AMOHAS Aerospace Medicine and Occupational Health Advisory Subcommittee

AMOHD Aerospace Medicine and Occupational Health Division

AMOLF Atoom- en Molecuul Fysica (FOM Institute)

AMON ATM Monitoring

AMOO Aerospace Medical Operations Office

AMOOS Advanced Maneuvering Orbit-to-Orbit Shuttle

AMOP Amplitude Modulation On Pulse • Arctic Marine Oil Program • Atomic, Molecular and Optical Physics

AMOR Advanced Meteor Orbit Radar

AMORE Arctic Mid-Ocean Ridge Expedition

AMORS Australian Modular Optical Remote Sensor

AMORSA Atmospheric and Meteorological Ocean Remote Sensing Assembly

Amort Amortization

AMOS Acoustic Meteorological Observing System • Advanced Mechanical and Optical Systems (Belgium) • Aerodrome Meteorological Observing System • Aerospace Maintenance and Operational Status • Air Force Maui Optical Site (USA-HI) • Air Force Maui Optical Station (USA-HI) (*) • Antarctic Meteorological Observing System • Assembly Management Operating System • Associative Memory Organizing System • Australian Meteorological and Oceanographic Society • Automated Materials Ordering System • Automated Meteorological Observation Station • Automated Multiobject Spectrograph • Automated Multiobject Spectrometer • Automated Multiple Object Spectrograph • Automatic Magnetic Observatory System • Automatic Message Operating Scheduling • Automatic Meteorological Observation Station • Automatic Meteorological Observation System • Automatic Meteorological Observing Station (*) • Automatic Meteorological Observing System

AMOSS Airline Management and Operations Support System

AMOTS Airborne Meteorological and Optical Test Support

AMOW Amplitude Modulated Orderwire

AMOWG Astronomy Management Operations Working Group • Astrophysics Management Operations Working Group

amp Ampere • Amplitude

Amp Amperage • Ampere • Amplified • Amplifier • Amplify • Amplitude

AMP Accelerometer Monitoring Program • Access Module Processor • Acquisition Management Plan • Acquisition Master Plan • Active Measurement Program • Active Monitor Present • Adaptation Mathematical Processor • Adjusted Month Program • Adjusted Monthly Program • ADPS Master Plan • Advanced Management Program • Advanced Maneuvering Program • Advanced Maneuvers Package • Advanced Materials and Processes (periodical code) • Advanced Microprocessing • Advanced Microprocessor • Aeroelastic Model Program • Aerospace Media Publishing • Alarm Manager Process • Alliance for Minority Participation • Ampanihy (Madagascar) (location identifier) • Angular Measurement Precision • Apparent Motion Parameter • Amplifier • Area Mail Processing • ARGOS Meteorological Platform • Aseptic Maintenance by Pressurization • Association of Media Producers (USA) • Associative Memory Processor • Asymmetric Multiprocessing • Automated Message Processing • Automated Mission Planning • Average Mean Pressure • Average Month Program • Axially Magnetized Plasma

AMPAS Academy of Motion Picture Arts and Sciences (USA-CA) • Analytical Maintenance Program Analysis System

AMPC Area Mail Processing Center • Automatic Message Processing Center

AMPCSS Automatic Master Paralleling Controller Switching System

AMPD Applied Materials Physics Division

AMPDP Acquisition Management Professional Development Program

AMPE Automated Message Processing Equipment • Automated Message Processing Exchange

AMPES Automated Message Processing Exchange System

Amph Amphibious

AMPI Annual Mean Percentage Increase

AMPIC Axiomatic Machine Process Inference Confirmer

Ampl Amplified • Amplifier • Amplify • Amplitude

AMPL Advanced Microprocessor Programming Language • A Macro Programming Language • A Manuscript Preparation Language

AMPLE Advanced Multipurpose Language

AMPLG Amplidyne Generator

AMPLTD Amplitude

AM/PM Amplitude Modulation/Phase Modulation

AMPN Amplification

AMPO Astronaut Memorial Planetarium and Observatory (USA-FL)

AMPOS Aircraft Maintenance Procedure Optimization System

AMPP Advanced Microprogrammable Processor • Arctic Meteorology Photographic Probe

AMPQC Atomic and Molecular Physics and Quantum Chemistry

AMPR Advanced Microwave Precipitation Radiometer • Aeronautical Manufacturer Planning Report • Aeronautical Manufacturer Progress Report

AMPRS Automated Management and Progress Reporting System

amps Amperes

Amps Amperes

AMPS Active Microwave Pressure Sounder • Adaptive Mode Planning System • Advanced Maneuvering Propulsion System • Advanced Meteorological Processing System • Advanced Mission Planning System • Advanced Mobile Phone Service • Advanced Mobile Phone Standard • Advanced Mobile Phone System • Assembly of Mathematical and Physical Sciences • Association for Modeling, Planning, and Simulation • Atmospheric and Magnetospheric Plasma-in-Space (Shuttle) • Atmospheric and Magnetospheric Plasma System • Automated Message Processing System • Automated Meter Posting System • Automated Mission Planning System • Automated Mobility Processing System • Automatic Mapping and Planning System • Automatic Material Processing System • Automatic Message Processing System

AMPSK Amplitude Modulation Phase Shift Keying

AMPSSO Automated Message Processing System Security Office • Automated Message Processing System Security Officer • Automatic Message Processing System Security Office • Automatic Message Processing System Security Officer

AMPT Advanced Maneuvering Propulsion Technology

AMPTE Active Magnetospheric Particle Tracer Explorer

AMQ Action Modulable Qualifiante • Aircraft Maintenance Qualification • Ambon (Indonesia) (location identifier)

AMQP Aircraft Maintenance Qualification Program

AMQUA American Quaternary Association (USA)

AMR Abnormal Mission Routine • Absolute Maximum Rating • Academy of Management Review (periodical code) • Adaptive Mesh Refinement • Advanced Material Request • Advanced Microwave Radiometer • Advanced Micro Research • Aerospace Medical Research • Age-Metallicity Relation • Altimetry Microwave Radiometer • Anisotropic Magnetoresistance • Area-to-Mass Ratio • Arno (Marshall Islands) (location identifier) • Automated Management Report • Automated Meter Reading • Automatic Membership Registration • Automatic Message Recording • Automatic Message Registering • Automatic Message Routing

AMRA Adaptive Mesh Refinement for Astrophysics • Advanced Multimode Radar Altimeter

AMRAD Automatic Message Routing And Distribution

AMRC Accès Multiple à Répétition de Codes • Acquisition Method Reason Code • American Manufacturing Research Consortium (USA) • Antarctic Meteorological Research Center

AMRF Accès Multiple par Répartition en Fréquence • Automated Manufacturing Research Facility

AMRIR Advanced Medium Resolution Imagery Radiometer • Advanced Medium Resolution Infrared Radiometer

Amrit Amritsar (India)

AMRL Adjusted Main Ring Length • Aeronautical and Maritime Research Laboratory (Australia-VIC)

AMROC American Rocket Co. (USA)

AMRPD Applied Manufacturing Research and Process Development

AMRS Airborne Microwave Radiometer System • Automated Mail Retrieval Service

AMRSA Atmospheric and Meteorological Remote Sensing Assembly

AMRV Astronaut Maneuvering Research Vehicle

AMRVI Advanced Medium Resolution Visible Imager

AmS Amateur Service

AMS Aberdeen Mechanical Society (UK-Scotland) • Accelerator Mass Spectrometer • Accelerator Mass Spectrometry • Accelerator Mass Spectroscopy • Acceptable Means of Compliance • Access Method Service • Accident-Monitoring System • Account Management Specialist • Account Management System • Acoustic Measurement System • Acquisition Management System • Actuation Mechanism System • Acute Mountain Sickness • Adaptive Mode Switch • Address Management System • Advanced Management System • Advanced Manned Spacecraft • Advanced Mapping System • Advanced Memory System • Advanced Meteorological Satellite • Advanced Meteorological System • Advanced Mission Studies • Advanced Mobility System • Advertising Management System • Aeronautical Material Specification • Aerospace Material Specification • Airborne Maritime Surveillance • Airborne Measurement System • Aircraft Multiband Scanner • Airlock Module Station • Airmass • Alert Management System • Alpha Magnetic Spectrometer • Alternate Mailing System • Alternate Mission Scientist • Altitude Measurement System • American Materiel Standard • American Mathematical Society (USA) • American Meteorological Society (USA) • American Meteor Society (USA) • Amsterdam Schiphol (Netherlands) (location identifier) • Anchor Macroscopic Semantic • Anisotropy of Magnetic Susceptibility • Anomalous-Mass Star • Anomaly Management System • Antenna Mast Set • Antimatter Spectrometer • Apollo Mission Simulator • Applied Mathematical Sciences • Archive Management System • Arctic Marine Science • Army Map Service • Articulated Mirror System • Artificial Moon Satellite • Asset Management System • Assistance Médico-Sociale • Associative Memory System • Atmospheric Monitor System • Attached Manipulator System • Attitude Measurement Sensor • Attitude Measurement System • Audiovisual Multimedia Services • Auto-Manual System • Automated Meteorological Station • Automated Maintenance System • Automated Management System • Automated Manifest System • Automated Mapping System • Automated Material System • Automated Measurement System • Automated Message Switch • Automated Meteorological System • Automated Microbial System • Automatic Management Switch • Automatic Meteorological System • Autonomous Message Switch • Auxiliary Memory Set

AMSA Area Maintenance Support Activity • Associazione Maremmana Studi Astronomici (Italy)

AMSAS AOA Meteorological Satellite Analysis System

AMSAT Amateur Satellite

AMSB Anomaly Mediated Supersymmetry Breaking

AMSBu American Meteorological Society Bulletin (periodical code)

AMSC Acquisition Management Systems Control • Acquisition Method Suffix Code • American Mobile Satellite Corporation (USA) • Attitude Measurement and Control System

AMSCAT Airborne Microwave Scatterometer

AmSci American Scientist (periodical code)

AMSD Adjusted Mapping Support Data • Advanced Mirror System Demonstrator

AMSDL Acquisition Management Source Data List • Acquisition Management Systems Data List

AMSDS Automated Management Supporting Data System

AMSE Asymptotic Mean Squared Error

AMSEE Association Méditerranéenne des Sciences de l'Environnement et de l'Espace (France)

AMSF Area Maintenance Supply Facility

AMSGA Association of Manufacturers and Suppliers for the Graphic Arts (UK)

AMSIE Annual Meeting and Science Innovation Exposition (AAAS)

AMSL Above Mean Sea Level

AMSO Arab Organization for Standardization and Metrology

AMSOE Automated Mobility Schedule Of Events

AMSR Advanced Mechanically Scanned Radiometer • Advanced Microwave Scanning Radiometer (*) • Advanced Microwave Sounding Radiometer

AMSRC Acquisition Management Systems Review Committee

AmSS Amateur Satellite Service

AMSS Advanced Manned Space Simulator • Advanced Meteorological Sounding System • Advanced Multimission Sensor System • Aeronautical Mobile Satellite Service • Airborne Multispectral Scanner • Aircraft Multichannel Scanner System • ASCA Medium Sensitivity Survey • Automatic Meteorological Sensor Station • Automatic Meteorological Sensor System

AMSSA Assured Mission Support Space Architecture

AM/SSB Amplitude-Modulated Single Sideband

AMSSB Amplitude-Modulated Single Sideband

Amst Amsterdam (Netherlands)

AMST Advanced Medium STOL Transport

AMSTS Automatic Multiparameter Semiconductor Test Set

AMSU Advanced Microwave Sounder Unit • Advanced Microwave Sounding Unit (*) • Advanced Microwave Sound Unit

Amt Amount

AMT Accelerated Mission Testing • Acquisition Management Team • Active Matrix Technology • ACTS Mobile Terminal • Adaptive Matched Filter • Address Mapping

Table • Aerial Mail Terminal • Air Mail Transfer • Alternative Minimum Tax • Amata (Australia-NT) (location identifier) • Amount • Amplitude-Modulated Transmitter • Analog Magnetic Tape • Angular Mapping Transformation • Apogee Motor Timer • Apple Media Tool • Area Management Team • Associate in Mechanical Technology • Astrograph Moon Time • Automatic Moon Tracking • Automated Monitoring Telescope • Automatic Motor Tester • Auxiliary Master Tape • Available Machine Time • Aviation Maintenance Technician

AMTA Advanced Multicolor Tracker for AMOS • Association for Machine Translation in the Americas (USA)

AMTAS Automatic Modal Tuning and Analysis System

AMTD Affordability and Manufacturing Technology Demonstration • Automatic Magnetic-Tape Dissemination

AMTEC Alkali Metal Thermal to Electric Converter • Association des Média et de la Technologie en Éducation au Canada

AMTEX Airmass Transformation Experiment • Airmass Transportation Experiment

AMTF Acoustic Model Test Facility

AMTFT Active Matrix Thin Film Transistor

AMTI Area Moving Target Indicator

AMTRAC Amphibian Tractor • Amphibious Tractor

AMTRAK American Travel by Track

AMTRAN Automatic Mathematical Translator

AMTS Active Maintenance Training Simulator • Advanced Meteorological and Temperature Sounder • Advanced Meteorological Temperature Study • Advanced Mobile Telephone System • Advanced Moisture and Temperature Sounder (*) • Advanced Moisture Temperature Sounder • Association Mondiale des Travailleurs Scientifiques • Automatic Maritime Telecommunications System

AMU Aircraft Maintenance Unit • Aligarh Muslim University (India) • Amanab (Papua New Guinea) (location identifier) • Antenna Matching Unit • Arab Maghreb Union • Associated Midwestern Universities (USA) • Astronaut Maneuvering Unit • Atomic Mass Unit • Attitude Match Update • Audio Management Unit • Auxiliary Memory Unit • Average Monthly Usage

AMUS Automated Manpower Utilization System

AMUSS Association des Musées des Sciences de Strasbourg (France)

AMV Amderma (Russia) (location identifier) • Armored Maintenance Vehicle

AMVER Automated Mutual Vessel Rescue System

AMVT Advanced Medium Vehicle Technology

AMW Active Microwave • Ames (USA-IA) (location identifier) • Angular Momentum Wheel • Average Molecular Weight

AMWA Area Microwave Assembly

AMWG Architecture Methodology Working Group

AMX Ammaroo (Australia-NT) (location identifier) • Automatic Message Exchange

AMY Ambatomainty (Madagascar) (location identifier) • Ampere-Modulated Yielding

AMZ Ardmore (New Zealand) (location identifier) • Astro Mail Zirkular

AM0 Air Mass Zero

AM2 Amperes per Meter Squared

an Anno

An Actinon • Annam • Annum • Anonymous • Answer

A/N Account Name • Advice Note • Alphanumeric • As Needed

AN Above Named • Accession Number • Access Network • Access Node • Acid Number • Airman • Air Navigation • Alphanumeric • Ammonium Nitrate • Analog Non-secure • Anglo-Norman • Angola • Annunciator • Antarctica • Antilles Néérlandaises (code ISO de pays) • Antilles (Netherlands_) (ISO country code) • Applications Notice • Assemblée Nationale • Associated Number • Astronautics Notice • Astronomische Nachrichten (periodical code) • Atmospheric Neutrino • Atomic Number • Audible Noise • Auftragnehmer • Netherlands Antilles (ISO country code)

ANA Abridged Nautical Almanac • Absence Non-Autorisée • Adaptive Null Antenna • Advanced Network Applications • American Newspaper Association (USA) • Anaheim (USA-CA) (location identifier) • Article Numbering Association • Association of National Advertisers • Automatic Number Analysis • Automatic Number Announcement

ANABAD Asociación Nacional de Archiveros, Bibliotecarios, Antiquarios y Documentalistas (Spain)

ANAC Automatic Number Announcement Circuit

AnaCh Analytical Chemistry (periodical code)

ANACOM Analog Computer

ANADAT Analog Data Tabulation

ANADP Association of North American Directory Publishers (USA)

ANAE Académie Nationale de l'Air et de l'Espace (France)

Anag Anagram

AnaIn Analytical Instrumentation (periodical code)

Anal Analogie • Analogique • Analogous • Analogy • Analyse • Analyser • Analysis • Analyst • Analytic • Analytical • Analytique • Analyze • Analyzer

ANAL Advise - Negate - Arrange - Legislate • Analysis

Analyt Analytical

ANAM Agence Nationale des Aérodromes et de la Météorologie

ANAO Australian National Audit Office

AnAp Annales d'Astrophysique (periodical code)

ANAP Associaçã Nacional dos Artistas Plásticos • Association Narbonnaise d'Astronomie Populaire (France)

ANARE Australian National Antarctic Research Establishment • Australian National Antarctic Research Expeditions

ANARS Autonomous Navigation and Attitude Reference System

ANAS Adaptation in Natural and Artificial Systems • Azerbaijan National Academy of Sciences

ANASA Azerbaijan National Aerospace Agency

ANASAC ALMA North American Science Advisory Committee

ANASE Association des Nations de l'Asie du Sud-Est

Anat Anatomia • Anatomical • Anatomie • Anatomique • Anatomist • Anatomy

ANATRAN Analog Translator

ANB Akademia Nauk Belarusi (Belarus) • American National Biography • Anniston (USA-AL) (location identifier) • Atmospheric Neutrino Background

ANBEF Association Nationale des Bibliothèques d'Expression Française (Belgium)

AnBel Annales de l'Observatoire Royal de Belgique (periodical code)

AnBes Annales de l'Observatoire de Besançon (periodical code)

ANBF Australian National Beamline Facility

AnBos Annals of the Bosscha Observatory (periodical code)

ANBRE Analog Biomechanical Recorder

AnBru Annales de l'Observatoire Royal de Bruxelles (periodical code)

ANBS Asian Network of Biological Sciences

Anc Ancien • Ancient

ANC Academia Nacional de Ciencias (Mexico) • Access Node Control • Acid-Neutralizing Capacity • Active Noise Cancellation • Active Noise Control • Active Nutation Control • Advanced Nuclear Computer • All Number Calling • Alternate Network Connectivity • Anchorage (USA-AK) (location identifier) • Assistant Network Controller • Automatic Noise Control

ANCA Australian Nature Conservation Agency

AnCap Annals of Cape Observatory (periodical code)

ANCB Academia Nacional de Ciencias de Bolivia

ANCCAC Australian National Committee on Computation and Automatic Control

ANCD Automated Network Control Device

ANCE Agence Nationale pour la Création d'Entreprises (France) • Association Nationale du Commerce Extérieur

AnCGH Annals of Royal Observatory, Cape of Good Hope (periodical code)

Anch Anchorage • Anchored

ANCIC Australian National Copyright Information Centre

ANCOSPAR Australian National Committee on Space Research

ANCOVA Analysis of Covariance

ANCS Automatic Net Control System

ANCSA Alaska Native Claims Settlement Act (USA)

Anct Ancient

ANCT Alternate Network Control Terminal

and Andante

And Andalucía (Spain) • Andromeda-Andromedae (constellation)

AND Administración Nacional de Puertos • Alphanumeric Display • Anderson (USA-SC) (location identifier) • Andorra (international vehicle registration code) • Andorra (ISO country code) • Andorre (code ISO de pays)

ANDE Alphanumeric Display Equipment

AnDea Annals of Dearborn Observatory (periodical code)

ANDEP Agence Nationale pour le Développement de l'Éducation Permanente (France)

ANDES Advanced NOAO Database Expert System

ANDF Architecturally Neutral Distribution Format • Architecture Neural Distribution Format • Architecture Neutral Development Format • Architecture Neutral Distribution Format • Application Neural Distribution Format

ANDI Applied Neutron Diffraction for Industry • Automated Non-Destructive Inspector

ANDMS Advanced Network Design and Management System

ANDS Advanced Navigation Display System

AnDud Annals of the Dudley Observatory (periodical code)

ANDVT Advanced Narrowband Digital Voice Terminal

Andz Anodize

ANE Acuerdo Nacional de Empleo • Aeronautical and Navigational Electronics • Angers (France) (location identifier)

ANEC American Nuclear Energy Council (USA) • Averaged Null Energy Condition

AnEdi Annals of the Royal Observatory, Edinburgh (periodical code)

ANEEL Agência Nacional de Energia Eléctrica (Brazil)

ANELE Asociación Nacional de Editoriales de Libros de Enseñanza (Spain)

ANEORP Arcetri Near-Earth Object Recovery Program

Aner Aneroid

AnErg Ergänzungshefte zu den Astronomischen Nachrichten (periodical code)

ANEXGOVT At No Expense to Government

ANF Adult Non-Fiction • Antofagasta - Cerro Moreno (Chile) (location identifier) • Automatic Number Forwarding

ANFC Annual Natural Forest Cleared

ANFE Aircraft Not Fully Equipped

ANFIAC Association Nationale pour la Formation Artistique et Culturelle (France)

ANFIS Adaptive Neuro-Fuzzy Inference System

ANFR Agence Nationale des Fréquences Radioélectriques (France) • Algorithms with Neglecting Functional Residuals

Ang Angle • Angular

AnG Annales de Géophysique (periodical code)

ANG Air National Guard (USA) • Angle • Angoulême (France) (location identifier) • Guilder (Netherlands Antilles) (ISO currency code)

ANGAS Advanced Nuclear Gamma-ray Analysis System

ANGB Air National Guard Base (USA)

AnGeo Annales Geophysicae (periodical code)

Angew Angewandt

Angl Angle • Angleterre • Anglican • Anglicized

Anglr Angular

ANGO Adviescommissie Niet-Gebonden Onderwijsonderzoek

AnGVP Annales du Bureau des Longitudes, Gauthier-Villars, Paris (periodical code)

Anh Anhalt • Anhang

ANH Anuha Island (Solomon Islands) (location identifier)

AnHar Annals of the Astronomical Observatory of Harvard College (periodical code)

ANH&MRC Australian National Health & Medical Research Council

Anhyd Anhydrous

ani Animation (file extension)

Ani Animation

ANI Access Network Interface • Accord National Interprofessionnel • Advanced Networking Infrastructure • Agência de Noticias de Informações • Agência de Noticias e Informações • Agencia Nacional de Informaciones • Agência Nacional de Informações • Aniak (USA-AK) (location identifier) • Authorized - Not Implemented • Automatic Number Identification

ANIBAL Abaissement du Niveau de Bruit des Avions Légers

ANIF Automatic Number Identification Failure

AnIHP Annales de l'Institut Henri Poincaré (periodical code)

ANIL Association Nationale d'Information sur le Logement (France)

ANIPLA Associazione Nazionale Italiana per l'Automazione (Italy)

ANIR Advanced Networking Infrastructure and Research • Automated Network Initialization and Reconfiguration

ANISN Associazione Nazionale Insegnanti di Scienze Naturali (Italy)

ANITA A New Inspiration To Arithmetic • Antarctic Impulsive Transient Antenna

AnITPM Antarctic Ice Thickness Measurement Programme

ANJ Asociação Nacional dos Jornais (Brazil) • Zanaga (Congo) (location identifier)

Ank Ankunft

ANK Alphanumeric Keyboard • Ankara (Turkey) (location identifier)

Anl Anlage

ANL Accademia Nazionale dei Lincei (Italy) • Active-Node List • Agence Nationale du Logiciel (France) • Andulo

(Angola) (location identifier) • Argonne National Laboratory (USA-IL) • Australian National Library • Automatic Noise Limiter • Automatic Noise Limiting

ANLAR Accademia Nazionale dei Lincei - Atti Rendiconti (Scienze Fisiche) (periodical code)

ANLCI Agence Nationale de Lutte Contre l'Illettrisme (France)

AnLei Annals of Leiden Observatory (periodical code)

Anlg Analog

AnLow Annals of the Lowell Observatory (periodical code)

ANLP Applied Natural Language Processing

Anlr Annular

ANLST Analyst

AnLun Annals of Lund Observatory (periodical code)

ANLYS Analysis

Anm Anmerkung

A/NM Administrative Network Management

ANM Académie Nationale de Médecine • Adaptation and Negotiation Module • Administrative Network Management • Answer Message • Antalaha (Madagascar) (location identifier) • Associated News Media • Automated Network Management • Automated Network Manager

ANMC American National Metric Council (USA)

AnMet Annalen Meteorologie (periodical code)

AnMN Antarctic Meteorite Newsletter (periodical code)

ANMP Account Network Management Program

ANMS Automated Network Monitoring System • Automated Notice to Mariners System

Ann Annales • Annals • Année • Annex • Annexe • Anni • Anno • Annonce • Annual • Annuel • Annuelle • Annuity • Annulation

ANN Adaptive Non-linear Network • All-figures Number Now • Annette Island (USA-AK) (location identifier) • Annals • Annotations • Architecture for Neural Networks • Artificial Neural Network

ANNA Air Force - NASA - Navy - Army

ANNE Artificial Neural Network Emulator

AnNic Annales de l'Observatoire de Nice (periodical code)

Anniv Anniversary

ANNLY Annually

Annot Annotate • Annotated • Annotation • Annotator

AnnRev A & A Annual Review of Astronomy and Astrophysics

Annu Annual • Annuel • Annuelle

Annul Annulment

ANNX Annex

ANO Above Named Officer • Air Navigation Officer • Air Navigation Order • Alphanumeric Output • Angoche (Mozambique) (location identifier) • Any Other • Automatic Network Operations

ANOC Alternate Network Operations Control • Association des Nouveaux Opérateurs de Câble (France) • Authorized Notice Of Change

ANODE Auroral Nitric Oxide Dayglow Experiment

ANOK Adult Next Of Kin

ANOM Authorized Network Operations Manager

Anon Anonyme • Anonymous • Anonymously

AnORB Annales de l'Observatoire Royal de Belgique (periodical code)

ANote Astronomical Notes University of Gothenburg Sweden (periodical code)

ANOV Analysis Of Variance

ANOVA Analysis Of Variance

AnP Annalen der Physik (periodical code)

ANP Actual Navigation Performance • Algemeen Nederlands Persbureau (Netherlands) • Annapolis Lee (USA-MD) (location identifier)

ANPA Agenzia Nazionale per la Protezione dell'Ambiente (Italy) • American Newspaper Publishers Association (USA)

AnPar Annales de l'Observatoire de Paris (periodical code)

ANPC Advanced Navigation and Positioning Corporation

AnPCS Annales de Physique - Colloquium Supplement (periodical code)

ANPE Agence Nationale Pour l'Emploi (France) • Association Nationale des Professionnels de l'Enseignement (France)

AnPh Annales de Physique (periodical code)

AnPhy Annales of Physics (periodical code)

ANPIT Agency for the National Program on Information Technology

ANPRM Advanced Notice of Proposed Rulemaking

ANQ Angola - Tri-State Steuben County (USA-IN) (location identifier) • Archives Nationales du Québec (Canada-QC)

Anr Another

ANR Active Noise Reduction • Advanced Networking Research • Advanced Network Routing • Another • Antwerpen Deurne (Belgium) (location identifier) • Assign Net Rekey • Automatic Network Routing • Average Normalized Rank

ANRA Air Navigation Radio Aid

ANRAC Aids Navigation Radio Control

ANRC American National Red Cross (USA) • Australian National Research Council

AnRE Annual Review of Energy (periodical code)

ANRE Advanced Nuclear Rocket Engine • Agency of Natural Resources and Energy

AnRev Annual Review (periodical code)

AnRFM Annual Review of Fluid Mechanics (periodical code)

ANRI Automatic Net Radio Interfacing

AnRMS Annual Review of Materials Science (periodical code)

ANRT Association Nationale de la Recherche Technique (France)

ans Ansi (file extension)

Ans Answer

ANS Advanced Network and Services Inc. • Advanced Network Services • Advanced Neutron Source • Air Navigation System • American National Standard • American Nuclear Society (USA) • Andahuaylas (Peru) (location identifier) • ANSI • Answer • Array Normal Spin • Artificial Nervous System • Artificial Neural System • Astronomical Netherlands Satellite • Astronomische Nederlandse Satelliet • Australian Neuroscience Society • Autograph Note Signed • Automatic Navigation System • Automatic Nervous System • Autonomous Navigation System

ANSA Advanced Network System Architecture • Advanced Null Steering Array • Agenzia Nazionale Stampa Associata (Italy)

ANSARI Astrophysical Network Space Archiving Research Institute

ANSBK Answer Back

ANSBR American National Standard for Bibliographic References

ANSC American National Standards Committee (USA)

ANSCII American National Standard Code for Information Interchange

ANSCR Alphanumeric System for Classification of Recordings

ANSCSOCR American National Standard Character Set for Optical Character Recognition

ANSE Association Nationale pour les Stages à l'Étranger (France) • Autonomous Nervous System Experiment

ANSEI Asociación Nacional de la Seguridad de Entornos Informáticos (Spain)

ANSER Actuated Nose Strakes for Enhanced Rolling

AnShO Annals of Shanghai Observatory (periodical code)

ANSI American National Standards Institute (USA)

ANSIC Aerospace Nuclear Safety Information Center (USA)

ANSL Australian National Standards Laboratory

ANSMET Antarctic Search for Meteorites

ANSORC Analog System for Optical Reading of Characters

ANSP Academy of Natural Sciences of Philadelphia (USA-PA)

ANSR Advanced Neutron Source Reactor

ANSS Aeronautical Navigation Satellite System

ANSSR Aerodynamically Neutral Spin-Stabilized Rocket

Anst Anstalt

ANSTC Australian Nuclear Science and Technology Council

ANSTI African Network of Scientific and Technological Institutions

ANSTJ Association Nationale Science Techniques Jeunesse (France)

ANSTO Australian Nuclear Science and Technology Organization

ANSTP Austrian National Space Technology Programme

AnStr Annales de l'Observatoire de Strasbourg (periodical code)

Ant Antarctica • Antenna • Antérieur • Anterior • Anticuado • Antigua • Antiguo • Antillas • Antlia-Antliae (constellation) • Antonym

ANT Actor-Network Theory • Antenna • Antenna Noise Temperature • Antilles Néérlandaises (code ISO de pays) • Antilles (Netherlands_) (ISO country code) • Articulated Navigation Testbed • Autonomous Navigation Technology • Netherlands Antilles (ISO country code) • Sankt Anton (Austria) (location identifier)

Antarc Antarctica

ANTARES Antarctic Research • Astronomy with a Neutrino Telescope and Abyss Environmental Research • Astrophysics Network for Training and Research

ANTAS Automatische Naß-Trocken Ablagerungssammler

ANTC Advanced Networking Test Center • Advanced Network Test Center

AnTec Annals Tectonicae (periodical code)

AnTel Annales des Télécommunications (periodical code)

Anten Antenna • Antenny (periodical code)

Anthol Anthology

Anthrop Anthropological • Anthropology

Anthropol Anthropological • Anthropology

ANTI Automated Near-Term Improvement

Antífr Antífrasis

Antig Antigua

antilog Antilogarithm

ANTIOPE Acquisition Numérique et Télévisualisation d'Images Organisées en Pages d'Écriture

Antiq Antiquarian • Antiquary • Antiquity

Antk Antarktika (periodical code)

AnTok Annals of Tokyo Astronomical Observatory (periodical code)

ANTOSTRAT Antarctic Offshore Acoustic Stratigraphy

AnTou Annales de l'Observatoire de Toulouse (periodical code)

Antrop Antropología

ANTS Algorithmic Number Theory Symposia • Automatic Nitrogen Transfer System • Autonomous Nano Technology Swarm

Antw Antwerp (Belgium) • Antwort

Antwpn Antwerpen (Belgium)

ANU Antigua (Leeward Islands) (location identifier) • Australian National University (Australia-ACT)

ANUATC Australian National University Astrophysical Theory Centre

ANULAS Australian National University Landsat Analysis System

ANV Anvik (USA-AK) (location identifier)

ANVAR Agence Nationale pour la Valorisation de la Recherche (France)

ANVIS Aviator's Night Vision Imaging System

AnVNG Angestellstenversicherungs-Neuregelungsgesetz

ANVR Algemene Nederlandse Vereniging van Reisbureaus

ANW Advanced NetWare • Ainsworth (USA-NE) (location identifier)

ANWC Alternate National Warning Center

ANWD Alphanumeric Wall Display

AnWie Annalen der Universitäts-Sternwarte Wien (periodical code)

ANWR Arctic National Wildlife Refuge

Anx Annex

ANX Anexo • Annex • Automatic Network Exchange • Automotive Network Exchange

ANY Anthony (USA-KS) (location identifier)

Anz Anzeiger

ANZ Angus Downs (Australia-NT) (location identifier)

ANZAAS Australian and New Zealand Association for the Advancement of Science

ANZEC Australian and New Zealand Environment Council

ANZECC Australian and New Zealand Environment and Conservation Council

ANZi Beobachtungs-Zirkular der Astronomischen Nachrichten (periodical code)

ANZIAM Australia and New Zealand Industrial and Applied Mathematics

ANZES Australian and New Zealand Scientific Exploration Society

ANZLIC Australia-New Zealand Land Information Council

ANZONE Antarctic Zone

ANZUK Australia - New Zealand - United Kingdom

ANZUS Australia - New Zealand - United States

a/o Account of

a/O An der Oder

aO An der Oder

Ao Arroyo

A-O Acousto-Optical

A/O Account Of • Analog Output • And Others

AO Abgabenordnung • Abnormal Occurrence • Abort Output • Abstract Object • Access Opening • Accompanying Object • Accountant Officer • Accounting Office • Account Of • Acousto-Optical • Action Office • Action Officer • Adaptive Optics • Additional Observation • Administrative Office • Administrative Officer • Administrative Official • Administrative and Overhead • Aerial Observer • Airline Operator • All Other • Allotment Order • Always On • Among Others • Amplifier Output • Analog Output • And Others • Angola (ISO country code)

• Annonce d'Opportunité • Anno Ordinis • Announcement of Opportunity • Appel d'Offre • Arctic Oscillation • Area Office • Area of Operations • Arecibo Observatory (Puerto Rico) • Arecibo Occultation • Assembly Order • Assembly Outline • Assisté par Ordinateur • Associated Object • Associate Office • Atmosphere-Ocean • Atomic Orbital • Atomic Oxygen • Attribute Object • Authorizing Official • Automated Operation • Autres Objets

AoA Analysis of Alternatives • Angle of Attack

AOA Abbreviated Operational Assessment • Abort Once Around • Adaptive Optics Associates Inc. (USA-MA) • Administration On Aging (USA) • Airborne Optical Adjunct • Airport Operating Area • Airport Operators' Association • Allen Osborne Associates • American Ordnance Association (USA) • Angle Of Arrival • Angle Of Attack • Aroa (Papua New Guinea) (location identifier) • Assessment Of Alternatives • Associazione Ogliastrina di Astronomia (Italy) • At Or Above

AOAC Association of Official Analytical Chemists (USA-VA)

AOARD Asian Office of Aerospace Research and Development

AOAS Advanced Oceanic Automation System • Arkansas-Oklahoma Astronomical Society (USA)

AOB Adaptive Optics Bonnette • Advanced Operational Base • Alternate Operational Base • Angle Of Bearing • Annanberg (Papua New Guinea) (location identifier) • Annual Operating Budget • Any Other Business • Approved Operating Budget • At Or Below

AOBI Any Other Bit of Information

AOBP Arctic Ocean Buoy Program

A&OC Attitude & Orbit Control

AOC Adaptive Optics Coronagraph • Advanced Operational Capability • Advice Of Charge • Aeronautical Operational Control • Aircraft Operations Center • Airline Operations Center • Air Officer Commanding • Air Operations Center • Air Operator's Certificate • Altenburg (Germany) (location identifier) • Annual Operating Cost • Anodic Oxide Coating • Appellation d'Origine Contrôlée • Apportionment Of Credit • Approved Operational Concept • Area Of Concern • Area Operations Center • Array Operations Center (NRAO) • Asociación de Operadores de Cable (Spain) • Assumption Of Control • Astronomical Observatory Copenhagen (Denmark) • Attitude and Orbit Control • AURA Oversight Committee • Automated Operations and Control • Automatic Operation Control • Automatic Output Control • Automatic Overload Control • Auxiliary Output Chip • Average Operating Cost • Award Of Contract

AOCA Amis de l'Observatoire des Creusets à Arbaz (Switzerland)

AOCB Any Other Competent Business

AOCC Astronomical Observatory of Campo catino (Italy) • Atomic-Orbital Close-Coupling (method)

AOCE Apple Open Collaboration Environment • Apple Open Collaborative Environment (*) • Attitude and Orbit Control Electronics

AOCI Accredited Off-Campus Instruction • Airborne Ocean Color Imager Spectrometer

AOCM Advanced Ocean Color Monitor

AOCMS Attitude and Orbit Control and Measurement System

AOCP AVCS Onboard Computer Program

AOCR Advanced Optical Character Reader • Advanced Optical Character Recognition

AOCRD Acceptance and Operational Checkout Requirements Document

AOCRS African Organization for Cartography and Remote Sensing

AOCS Active Optics Control System • Altitude and Orbit Control System • Arctic Ocean Climate Station • Attitude and Orbital Control System • Attitude and Orbit Control System (*) • Attitude Orbit Control System • Automated Orbit Control System • Autonomous Onboard Commanding System • Autonomous Onboard Command System

AOCSI Adaptive-Optics Compensated Speckle Interferometry

AOCSN Arctic Ocean Climate Station Network

AOCU Analog Orderwire Control Unit • Attitude and Orbit Control Unit

AOD Abort Of Dialing • Abou Deia (Chad) (location identifier) • Acousto-Optic Deflector • Acousto-Optic Device • Acquisition Of Data • Advanced Object Data • Aerosol Optical Depth • Analog Output Differential • Applied Optical Detection • Area Of Display • Area-Oriented Distribution • As Of Date • Assistant Operations Director • Atmosphere and Ocean Dynamics • Audio-On-Demand

AO&DA Advanced Optics & Display Application

AODA Advanced Optics and Display Application

AODC Acridine Orange Direct Count • Australian Oceanographic Data Centre

AODMD Acousto-Optic Delay-and-Multiple Device

AODP Adaptive Optics Development Program

AODPA Acquisition Orbit Determination Program Assembly

AODV Ad-hoc On-demand Distance Vector (protocol)

AoE Airport of Entry

AOE Application Operating Environment • At Our Expenses

AOEC Aero-Optical Evaluation Center

AOELT Adaptive Optics for Extremely Large Telescopes

AOES Advanced Orbit/Ephemeris System

AOET Atomic Oxygen Exposure Tray

AOF Absolute Open Flow • Afrique Occidentale Française (obsolete) • Anode Oxide Film • Archive Operations Facility • Authoring On the Fly

AOFEE Advanced Orbiter Flight Evaluation Experiment

AOFP Absolute Open-Flow Potential

AofS Academy of Sciences

AOFSCE Absolute Oil Fluorescence Spectral Conversion Efficiency

AOFTM Arecibo Observatory Fourier Transform Machine

AOG Abstract Object Generic • Aircraft On Ground • Anshan (China-PRC) (location identifier) • Architecture Oversight Group • Auxiliary Output Generator

AOGC Arctic Ocean Grand Challenge

AOGCM Atmosphere-Ocean General Circulation Model

AOGO Advanced Orbiting Geophysical Observatory

AOH Ancient Ocean Hypothesis • Available Operating Hour • Average Operating Hour • Lima - Allen County (USA-OH) (location identifier)

AOI Additional Operational Issue • Ancona (Italy) (location identifier) • Angle Of Incidence • Announcement of Opportunity Instruments • Any Other Information • Arab Organization for Industrialization • Area Of Interest • Associate Office Infrastructure • Associazione Ottica Italiana (Italy) • Astrometric Optical Interferometer • Automated Optical Inspection • Automatic Optical Inspection

AOIB All-Optical Imaging Brassboard

AOIC Additional Operational Issues and Criteria

AOIP Association des Ouvriers en Instruments de Pression (France)

AOIPS Atmospheric and Oceanographic Image Processing System • Atmospheric and Oceanographic Information Processing System

AOIRT Automated Operational Information Requirements Tool

AOIS Alps Observation and Information System • Automatic Optical Inspection System

AOIV Automatically Operated Inlet Valve

AOJ Aomori (Japan) (location identifier)

AOK All Okay • Astronomical Observatory Kitab (Uzbekistan) • Astrooptik Kohler (Switzerland) • Karpathos (Greece) (location identifier) • Kwanza (Angola) (ISO currency code)

AOL Absent Over Leave • Airborne Oceanographic Lidar • Airborne Optical Laser • America On Line • Application-Oriented Language • Area Of Launch • Astronomy On Line • Paso de los Libres (Argentina-CR) (location identifier)

AOLO Advanced Orbital Launch Operation

AOLR Amplifier Open Loop Response

AO&M Administration, Operation & Maintenance

AOM Acousto-Optical Modulator • Acousto-Optic Modulator • Adaptive Optics Module • Administration - Operation - Maintenance • Alignment Object Map • Amorphous Organic Matter

AOMD Amended Operator and Maintainer Decision

AOML Atlantic Oceanographic and Meteorological Laboratory (NOAA)

AOMPS Automatic Outgoing Message Processor System

AOMS Advanced Optical Monitoring System • Area Organizational Maintenance Shop

AON Active Optical Network • Act Of Nature • All Optical Network • Arona (Papua New Guinea) (location identifier) • New Kwanza (Angola) (ISO currency code)

AONBS Asian Oceania Network of Biological Sciences

AOO Altoona (USA-PA) (location identifier) • American Oceanographic Organization (USA) • Analyse Orientée Objets • Any Other Object • Approche Orientée Objet • Area Of Operations

AOODB Active Object-Oriented Database

AOP Additive Operational Product • Additive Operational Program • Additive Operational Project • Advanced Observing Program • Advanced Operating Platform • Airborne Optical Platform • Annual Operating Plan • Annual Operating Program • Application Output Processor • Approved Operating Plan • Approved Operating Program • Aspect-Oriented Paradigm • Aspect-Oriented Programming (*) • Assembly Operations Plan • Association of Online Professionals • Astrophysikalisches Observatorium Potsdam (Germany) • Atmospheric Optical Parameter

AOPA Aircraft Owners and Pilots Association (USA)

AOPC Atmospheric Observations Panel Committee (GCOS)

AOPI Any Other Piece of Information

AOPL Advance Order Parts List

AOPM Airline Operations Planning Model

AOPP Atmospheric, Oceanic, and Planetary Physics

AOPR Area Of Prime Responsibility

a. o. Prof. außerordentlicher Professor

AOPS Airborne Operations • Automatic Office Protocol Standard

AOQ Average Outgoing Quality

AOQL Average Outgoing Quality Limit

a/or And/Or

AOR Abstract Object Repository • Accumulated Operating Results • Advice Of Rights • Alor Setar (Malaysia) (location identifier) • Angle Of Reflection • Annual Operating Requirement • Antenna Ohmic Resistance • Application Owning Region • Arctic and Offshore Research • Area Of Responsibility • Astronomical Observation Request • Atlantic Ocean Region • Authorized Organizational Representative

AORA Atlantic Ocean Recovery Area

AORD Astronaut Operations Requirement Document

AORIS Arctic and Offshore Research Information System

AORS Abnormal Occurrences Reporting System • Acousto-Optical Radio Spectrograph

A&OS Atmospheric & Oceanic Sciences

AOS Academic Operating System • Acoustic Overwatching Sensor • Acousto-Optical Spectrograph • Acousto-Optical Spectrometer • Acousto-Optical Spectroscopy • Acousto-Optic Spectrograph • Acousto-Optic Spectrometer • Acousto-Optic Spectroscopy • Acquisition Of Signal • Active Optical Sensor • Activity Operating Schedule • ACTS Optical System • Adaptive-Optics System • Add-On Stabilization • Add Or Subtract • Advanced Operating System • Advanced Orbiting System • Advice Of Shipment • Airborne Optical Sensor • Airborne Optical Surveillance • Airborne Optical System • Algebraic Operating System • All Over the Shop • Alternate Operator Service • Aluminium Oxide Sphere • Amateur Observers' Society (USA-NY) • Amook (USA-AK) (location identifier) • Amount Of Something • Amplifier Output Stage • Area Of Specialization • ASTER Operations Segment • Astronomical Observatory Satellite • Astronomiczne Obserwatorium Szerokościowe (Poland) • Atmospheric and Oceanic Sciences • Australian Optical Society • Automated Office System • Automatic Observing System • Azimuth Orientation System

AOSB Arctic Ocean Sciences Board

AOSC Alternate Operational Support Center

AOSD Aspect-Oriented Software Development • Atomic Object State Diagram

AOSE Agent-Oriented Software Engineering

AOSGE Arctic Ocean System in the Global Environment

AOSIS Association of Small Island States

AOSL Authorized Organizational Stockage List

AOSN Autonomous Ocean Sampling Network

AOSO Advanced Orbiting Solar Observatory

AOSP Active Optics Simulation Program • Advanced Onboard Signal Processor • Automatic Operating and Scheduling Program

AO&SS Atmospheric, Oceanic & Space Sciences

AOSS Active Optics Simulation System • Airborne Oil Surveillance System • Atmospheric, Oceanic and Space Sciences • AURA's Observatory Support Services • Automated Office Support System

AOST Association des Opérateurs de Services de Télécommunications (France)

AOT Actual Operating Time • Adaptive Optics Team • Aeroassisted Orbital Transfer • Aerosol Open Test • Aerosol Optical Thickness • Agent-Oriented Technique • Alignment Optical Telescope • Allowed Outage Time • Assignment-Oriented Training • Astronomical Observation Template • Astronomical Observing Template • Atmospheric Optical Thickness • Auxiliary Output Tester • Average Operation Time • Avionics Operating Time • Avionics Overall Test

AOTD Active Optical Target Detector

AOTF Acousto-Optical Tunable Filter (*) • Acousto-Optic Tunable Filter

AOTP Abbreviated Outline Test Plan

AOTS Advanced On-the-job Training System • Advanced Orbital Test Satellite

AOTV Aero-assisted Orbital Transfer Vehicle • Aero-assisted Orbit Transfer Vehicle

AOU Area Of Uncertainty • Astronomiska Observatoriet Uppsala (Sweden) • Attopeu (Laos) (location identifier) • Azimuth Orientation Unit

AoV Analysis of Variance

AOV Annonce d'Opportunité de Vol

AOVEF Arecibo Observatory Visitor and Educational Facility (Puerto Rico)

AOVP Arctic Ocean Variability Project

AOW Algemene Ouderdomswet • All Over the World • Asian-Oceanic Workshop • Asia-Oceania Workshop

AOX Adsorbable Organic Halogen

AOZ Aircraft Operating Zone

Ap Apartado • Aparte • Apelación • Apparent • Approach • April • Astrofizika (periodical code) • Astrophysics (periodical code) • Autopilot • Peculiar A-type Star

A&P Administrative & Personnel • Airframe & Powerplant • Asia & Pacific • Attitude & Pointing

A/P Accounting Period • Account Paid • Accounts Payable • Active/Passive • Additional Premium • Airport • Autopilot

AP Above Proof • Absolute Parallelism • Academic Press (now HBJ) • Accelerometer Package • Access Panel • Access Permit • Access Point • Accident Professionnel • Accompte Provisionnel • Accounting Period • Account Payable • Accreditation Panel • Acoustic Processor • Acquisition Plan • Acquisition Policy • Action Potential • Active Prominence • Additional Premium • Additional Program • Adjective Phrase • Adjunct Processor • Administration Publique • Administrative Point • Administrative Procedure • Advanced Placement • Advanced Post • Advance Party • Advance Pay • Aerosol Particle • Affranchissement Postal • After Peak • Agence de Placement • Agence de Presse • Agency Procedure • Aided Pilotage • Airborne Platform • Aire Projetée • Air Patrol • Air Pollution • Air Pressure • Air Publication • Alkaline Permanganate • All Points • All Purposes • Amapá (Brazil) • Ammonium Perchlorate • Analog Processor • Analysis Package • Analysis Processor • Analytical Plotter • Annales de Physique • Année Préparatoire • Annual Production • Anomalous Propagation • Antenna Pattern • Ante Prandium • Antipersonnel • Antiproton • Apartado Postal • Application • Application Part • Application Process • Application Processing • Application Processor • Application Profile • Application Program • Application Project • Application Protocol • Applied Physics • Appliqué • À Protester • Arecibo Pulsar • Argument Programming • Arithmetic Progression • Array Processor • Arrêté Préfectoral • Asia-Pacific • As Prescribed • As Purchased • Ascent Phase • Assembly Platform • Assistance Publique • Associated Press • Association Professionnelle • Associative Processor • Assurance Produit • Astrometric Photometer • Astrophysical Parameter • Atmospheric Physics • Atmospheric Pressure • Atomic Power • Atomic Powered • Attached Processor • Attendant Position • Attitude and Pointing • Attitude Processor • Austrittspupille • Authority to Pay • Authority to Purchase • Author's Proof • Automated Procedure • Automated Processing • Automatic Processing • Automatic Programming • Autopilot • Avant-Projet • Avant-Propos • Average Payable • Average Power • Average Price • Average Projection • Avionics Package • Avis Préalable • Pakistan (international aircraft registration code)

APA Abort Programmer Assembly • Adaptive Packet Assembly • Additional Personal Allowance • Administrative Procedure Act (USA) • Advanced Pilot's Aid • Advanced Processor Assembly • Advance of Pay and Allowances • Aerodromes Protection Agency (UK) • Alkaline Phosphatase Activity • Allocation Personnalisée à l'Autonomie (France) • Allowance for Program Adjustment • All Points Addressable • Amateur Press Association (USA) • American Philosophical Association (USA) • American Physicists Association (USA) • American Planning Association (USA) • American Press Association (USA) • American Psychological Association (USA) • Amplified Piezo-Actuator • Amplitude Phase Analysis • Analysis Plan for Assessment • Annular Phased Array • Antarctic Protected Area • Antenna Pattern Analyzer • Antenna Phase Angle • Appropriation Procurement Account • Appropriation Purchase Account • Arithmetic Processing Accelerator • Arithmetic Processor Accelerator • Asociación Paraguaya de Autores (Paraguay) • Asociación Peruana de Astronomía (Peru) • Associate in Public Administration • Associate Principal Astronomer • Association of Patent Attorneys • Associative Principle for Addition • Augmented Predictive Analyzer • Australian Postgraduate Award • Austria Presse Agentur • Automatic Photographic Analysis • Axial Pressure Angle • Denver - Arapahoe County (USA-CO) (location identifier)

APAA Asian Patent Attorneys Association • Associação Portuguesa de Astrónomos Amadores (Portugal)

ApAc Applied Acoustics (periodical code)

APAC Aerial Photogrammetry, Analysis and Cartography • Antenna Pointing Angle Change • Asian Particle-Accelerator Conference • Asociación Peruanan para el Avance de la Ciencia (Peru) • Attaché Principal d'Administration Centrale • Australian Partnership for Advanced Computing • Australia Pacific Airports Corporation

APACHE Antarctic Plateau Anisotropy Chasing Experiment

APACM Atmospheric Physical and Chemical Monitor • Atmospheric Physics and Chemistry Monitor

APACS Association for Payment Clearing Services

APAD Application-Adaptable Database System • Asynchronous Packet Assembler/Disassembler

APADE Automated Purchase and Accounting Data Entry • Automation of Procurement and Accounting Data Entry

APADS Advanced Precision Airborne Delivery System • Automatic Programming and Data System

APAE Attached Payload Accommodation Equipment

APAFO Advanced Particles And Fields Observer

APAG Atlantic Policy Advisory Group (NATO)

APAI Advanced Planning Acquisition Information

APAL Array Processor Assembly Language

APALS Autonomous Precision Approach and Landing System

APAM Array Processor Access Method

ApAN Apatity Akademiia Nauk SSSR (periodical code)

APAN Asia-Pacific Advanced Network

APANA Australian Public Access Network Association

APAP Amis du Planétarium d'Aix-en-Provence (France)

APAPI Abbreviated Precision Approach Path Indicator

APAQS Asia-Pacific Conference on Quality Software

APAR Absorbed Photosynthetically Active Radiation • Authorized Program Analysis Report • Automatic Program Analysis Report • Automatic Programming and Recording

APAREN Address Parity Enable

APARS Automated Printing And Retrieving System

Apart Apartment

APAS Action de Promotion d'Accompagnement et de Soutien • ADABAS Performance Analysis System • Aerodynamic Performance and Sizing Program • Androgynous Peripheral Assembly System • Astrophysical, Planetary and Atmospheric Sciences • Attività de Preparazione, Accompagnamento e Sostegno • Automated Program for Aerospace-Vehicle Synthesis • Automated Programmable Assembly System • Automatic Performance Analysis System

APASEG All-party Parliamentary, Astronomy, and Space Environment Group (UK)

APAST Association Africaine pour l'Avancement des Sciences et des Techniques

APASU Attaché Principal d'Administration Scolaire et Universitaire

APAT Acquisition Process Action Team • Atmospheric Pressure and Ambient Temperature

APATS Acquisition Planning And Tracking System • Antenna Pattern Test System

APAU Asociación de Profesores de Astronomía de Uruguay • Attaché Principal d'Administration Universitaire

APB Acquisition Program baseline • All-Points Bulletin • Apolo (Bolivia) (location identifier) • Approval Program Baseline • Avionics Planning Baseline

APBAD Associação Portuguesa de Bibliotecários, Arquivistas e Documentalistas (Portugal)

APBI Advanced Planning Briefing for Industry

Apble Appreciable

APBO Accumulated Postretirement Benefit Obligation

APBS Automated Property Book System

APC Abbreviated Performance Characteristics • Accelerated Provisioning Concept • Accelerometer Pulse Converter • Accounting Processing Code • Account Processing Code • Acquisition Position Category • Acquisition Professional Community • Activity Processing Code • Adaptive Payload Carrier • Adaptive Power Control • Adaptive Predictive Coding • Advance Procurement Change • Advance Procurement Check • Advanced Process Control • Advanced Programming Course • Advanced Propulsion Comparison • Advanced Propulsion Cooling • Advertising Policy Council • Aerial Port Commander • Aeronautical Planning Chart • Aeronautical Public Correspondence • Aft Power Controller • Agency Program Coordinator • Airborne Particle Counter • Aircraft-Pilot Coupling • Airplane-Pilot Coupling • Air Pollution Control • American Personal Communications • American Power Conversion • Angle Polished Convex • Antarctic Place-names Committee • Antenna Pattern Correction • Application Program Control • Approach • Approach Power Compensator • Arbeitsplatzcomputer • Area Positive Control • Armored Personnel Carrier • Aspirin - Phenacetin - Caffeine • Association des Planétariums du Canada • Association for Progressive Communications (USA) • Association for Progressive Computing • Asteroid Photometric Catalog • Attitude Processing Center • Automated Postal Center • Automated Processing Center • Automatic Performance Control • Automatic Phase Control • Automatic Pitch Control • Automatic Power Control • Automatic Predicate Calculus • Automatic Public Convenience • Autonomous Payload Control • Autonomous Payload Controller • Autopilot Controller • Napa County (USA-CA) (location identifier)

APCA Aft Power Controller Assembly • Air Pollution Control Association (USA)

APCAJ Air Pollution Control Association Journal (periodical code)

APCAPS Automated Payroll, Cost, and Personnel System

APCATS Asian-Pacific Conference on Aerospace Technology and Science

APCB Air Pollution Control Board

APCC Alternate Processing and Correlation Center

APCCE Asian Pacific Confederation of Chemical Engineering

APCCI Assemblée Permanente des Chambres de Commerce et d'Industrie (France)

APCCIRN Asian-Pacific Coordinating Committee for International Research

APCD Air Pollution Control District

APCE Agence Pour la Création d'Entreprises (France)

APCF Advanced Protein Crystallization Facility (*) • Automated Protein Crystallization Facility

APCG Advanced Protein Crystal Growth • ATM Pilot Coordination Group

APCH Approach

APCHG Approaching

APCI Agence pour la Promotion de la Création Industrielle (France)

APCM Aerial Photography Contract Management • Assemblée Permanente des Chambres de Métiers (France)

APCN Anno Post Christum Natum • Asia-Pacific Cable Network

APCO Association for Public (safety) Communication Officials

APCOM Application Program to Communication

APCOPPLSRF Analysis and Program for Calculation of Optimum Propellant Performance for Liquid and Solid Rocket Fuels

ApCpH Attempts per Circuit per Hour

APCPH Attempts Per Circuit Per Hour

APC-REX Autonomous Payload Control Rocket Experiment

APCREX Autonomous Payload Control Rocket Experiment

APCS Active-Passive Correlation Spectroscopy • Aeronautical Production Control System • Alberta Photogrammetric Control System • Atmosphere Pressure Control System • Attitude and Pointing Control System • Automated Project Control System

APCST Asia and Pacific Council for Science and Technology

APC-STEX Autonomous Payload Control Satellite Experiment

APCSTEX Autonomous Payload Control Satellite Experiment

APCTP Asia Pacific Center for Theoretical Physics (ROK)

APCUG Association of Personal Computers Users Groups (USA)

APCVD Atmospheric Pressure Chemical Vapor Deposition

apd Aldus Printer Description (file extension)

Apd Approved

APD Absolute Pointing Drift • Acoustic-Phonetic Decoding • Activity Precedence Diagram • Adjustable Pitch Device • Advanced Planning and Design • Advanced Planning Document • Advanced Program Development • Advance Planning Document • Aeronautical Propulsion Division • Agencia de Protección de Datos (Spain) • Aide Publique au Développement • Air Particulate Detector • Aldus Printer Description • American Places Dictionary • Amplitude Probability Distribution • Analog Processing and Decommutation • Analog-to-Pulse Duration • Angular Position Digitizer • Application Program Designation • Area Passive Dosimeter • Area Postal Director • Area Postal Directory • Asian Pacific Division • Astronomical Pocket Diary • Automated Products Division • Auxiliary Power Distribution • Avalanche Photodiode • Avalanche Photodiode Detector • Avant-Projet Définitif • Avant-Projet Détaillé • Average Particle Diameter • Average Percentage Damage • Average Pore Diameter • Average Power Dissipation

APDA Apple Programmers and Developers Association (USA) • Arctic Precipitation Data Archive

APDC Acquisition Professional Development Council • Associação Portuguesa para o Desenvolvimento das Comunicações (Portugal) • Association Pour le Développement des Communications (France)

APDE Application Protocol Development Environment

APDEC Application for Partial Differential Equations Center

APDF Astronomical Data Processing Facility

APDL Ada Programming Design Language • Algorithmic Processor Description Language

APDM Amended Program Decision Memorandum

Apdo Apartado

APDOOR Aperture Door

APDP Acquisition Professional Development Program

APDS Advanced Personnel Data System • Advanced Planning Data Sheet • Analog Processor Discriminator Station • Automated Personnel Data System • Automated Postal Directory Service

APDU Application Protocol Data Unit

APE Absolute Pointing Error • Accompagnement Personnalisé pour l'Emploi • Active Phasing Equipment • Active Phasing Experiment • Activité Principale Exercée • Advanced Production Engineering • Airborne Polar Experiment • Alfvén Propulsion Engine • Allocation Parentale d'Éducation • All Paths Explorer • Alpha - Proton - Electron • Arithmetic Processing Extension • Asemblea Parlamentaria Europea • Assemblée Parlementaire Européenne • Atmospheric Physics Experiment • Atomic Photoelectric Effect • Aurora Photography Experiment • Automatic Photomapping Equipment • Automatic Positioning Equipment • Available Potential Energy • Available Power Efficiency • San Juan Aposento (Peru) (location identifier)

APEA Antenna Pattern Error Analysis

APEC Acquisition Planning Executive Council • Advisory Panel on Environmental Change • Agence Pour l'Emploi des Cadres (France) • All-Purpose Electronic Computer • Asia-Pacific Economic Cooperation • Association Pour l'Emploi des Cadres (France) • Astrophysical Plasma Emission Code • Autonomous Robot Control and Simulation

APED Astrophysical Plasma Emission Database

APEJ Aide au Premier Emploi des Jeunes (France) • Association for Physics Education in Japan

APEM Association du Personnel de la Météorologie

ApEn Applied Energy (periodical code) • Approximate Entropy

ApEnM Applied Environmental Microbiology (periodical code)

APEP Administrative and Professional Exchange Program

Aper Aperture

APERAL Air Pollution Effects Research At Lancaster

APERT Aerospace Programme for Education, Research and Technology

APES Advanced Processor Emulator System • Advanced Protective Eyewear System • Automated Polyelastic Sampling

APEX Active Plasma Experiment • Advanced Photovoltaic and Electronics Experiment • Advanced Purchase Excursion • Antiproton Experiment • Arctic Polynya Experiment • Ariane Payload Experiment • Astrophysical Plasmadynamic Explorer • Atacama Pathfinder Experiment • Athena Precursor Experiment

APF All Plastic Fiber • Analyseur de Plasma Froid • Analyst Private File • Appropriated Funds • Approximate Partition Function • Asia-Pacific Forum • Atomic Partition Function • Attitude Parameters File • Authorized Program Facility • Naples (USA-FL) (location identifier)

APFA Association of Professional Flight Attendants (USA)

APFCS Automatic Power-Factor-Control System

APFIM Atom-Probe Field-Ion Microscope • Atom-Probe Field-Ion Microscopy

APFO Aerial Photography Field Office

APFR Automatic Programmable Film Reader

APFS Apparent Places of Fundamental Stars

APFTI Airline Passenger Fair Treatment Initiative

APFTQ Association des Producteurs de Films et de Télévision du Québec (Canada-QC)

APG Aberdeen Phillips (USA-MD) (location identifier) • Accelerating Population Growth • Acquisition Planning Guide • Algorithmic Pattern Generator • All Paths Generator • Antenna Power Gain • Application Performance Group • Application Program Generator • Asia-Pacific Group • Astrometric-Photometric Group • Atlas of Peculiar Galaxies • Automated Password Generator • Automated Program Generator • Azimuth Pulse Generator

APGC Air Proving Ground Command

APGERN Asian-Pacific Global Environmental Research Network

Aph Aphetic • Aphorism

APh Astroparticle Physics (periodical code)

APH Attitude Pointing History • Bowling Green - Camp A.P. Hill (USA-VA) (location identifier)

APHA American Public Health Association (USA)

APHAZ Aircraft Proximity Hazard

APHELIE Astronomie, Physique, Élaboration, Instrumentation et Observation

APHID Atmospheric Phase Inference Device

APHIS Animal and Plant Health Inspection Service

APhO Asia Physics Olympiad

APHT Advanced Physical Test

api Application Program Interface (file extension)

API Access Programming Interface • Acoustic Properties Instrument • Acquisition Program Indicator • Acquisition Program Integration • Advanced Procurement Information • Advance Publication Information • Aerial Photointerpretation • Aide aux Projets d'Innovation • Air Photo Interpretation • Air Pollution Index • Air Position Indicator • Alphabet Phonétique International • American Petroleum Institute (USA) • Angle Position Indicator • Antenna Position Indicator • Apiay (Colombia) (location identifier) • Application et Promotion de l'Innovation • Application Program Interface (*) • Application Programmer Interface • Application Programmer's Interface • Application Programming Interface • Applications Portability Interface • A Priori Importance • Association Phonétique Internationale • Association pour la Promotion de l'Image (France) • Atmospheric Pressure Ionization • Automatic Payment of Invoices • Automatic Priority Interrupt

API&A Acquisition Policy, Integrity & Accountability

APIA Acquisition Policy, Integrity and Accountability

APIAN Advanced Propulsion Integration of Aerodynamics and Noise

APIB Application Program Integration Board

APIC Acquisition Process Improvement Campaign • Acquisition Procurement Improvement Campaign • Advanced Programmable Interface Controller • Advanced Programmable Interrupt Controller • Antecedent Precipitation Index Catchment • Automatic Power Input Controller • Automatic Programming Information Centre (UK)

APICS American Production and Inventory Control Society (USA)

APID Application Identifier • Applications Process Identification • Applications Process Identifier

APIF Automated Process Information File

APIG Applied Physics Interest Group

APII Asia-Pacific Information Infrastructure

ApInv Astrophysical Investigations (periodical code)

APIO Advanced Programming Interoperability

APIRD Authorized Procurement Information Requirements Description

APIRL Authorized Procurement Information Requirements List

APIRM Application Programming Interface Reference Manual

APIS Advanced Passenger Information System • Aerial Photography Information System • Antarctic Pack Ice Seals • Application Program and Information System

APISP Association des Professeurs d'Initiation aux Sciences Physiques (France)

APIT Association des Professeurs d'Initiation Technologique (France)

APIU Adaptable Programmable Interface Unit • Adaptive Programmable Interface Unit

ApJ Astrophysical Journal (periodical code)

APJ Agent de Police Judiciaire

APJA Agent de Police Judiciaire Adjoint

ApJL Astrophysical Journal Letters (periodical code)

ApJS Astrophysical Journal Supplement Series (periodical code)

APK Amplitude Phase Keying • Apataki (Tuamotu Islands) (location identifier) • Astronaut Preference Kit

APKC Achromatic Phase Knife Coronagraph

ApKF Apatity Kol skii Filial AN SSSR (periodical code)

APKIM Actual Plate Kinematic Model

Apl Aplicado • April

ApL Astrophysical Letters (periodical code)

APL Adult Performance Level • Advanced Parts Language • Advanced Programming Language • Advance Procurement List • Aide Personnalisée au Logement (France) • Allowance Parts List • Alternating Polarization Laser • Applicable Price List • Applications Program Library • Applied Physics Laboratory • Applied Physics Letters • Approved Parts List • Approved Products List • A Programming Language • Array Processing Language • Assembly Part List • Assembly Programming Language • Association for Programmed Learning • Associative Programming Language • Automated Programming Language • Average Picture Level • Nampula (Mozambique) (location identifier)

APLA American Patent Law Association (USA)

APLAC Analysis Program Linear Active Circuit • Asia Pacific Laboratory Accredition Cooperation

ApLC Astrophysical Letters Communications (periodical code)

APLCN Appalachian

APLDI APL Data Interface

APLE Average Power Laser Experiment

APLF Association des Planétariums de Langue Française • Association of Patent Law Firms (USA)

APLM Applications Program Library Maintenance

APLMF Asia-Pacific Legal Metrology Forum

APLO Automation Planning and Liaison Office

APLS Automated Propellant Loading System

APLSS Attached-Payload Space Station • Attached-Payload Support System

APLU Automatic Program Loading Unit

APM Acquisition Program Manager • Acquisition Project Manager • Additional Packet Mode • Advanced Power Management • Aerosol Physics Module • Airborne Particulate Matter • Air Particulate Monitor • Air Pollution Meteorologist • Akustische Plattenmode • Alarm Panel Monitor • Analog Panel Meter • Antenna Pointing Mechanism • Antenna Positioning Mechanism • Appended Pressurized Module • Area Prediction Model • Ascent Particle Monitor • Assistant Paymaster • Assistant Program Manager • Assistant Project Manager • Associate Program Manager • Associate Project Manager • Association des Professeurs de Mathématiques (France) • Associative Principle for Multiplication • Atmospheric Particle Monitor • Attached Pressurized Module (now CAL) • Automated Plate Machine • Automatic Photographic Measuring • Automatic Plate Measurement • Automatic Plate Measuring • Automatic Printing Machine

ApMaC Applied Mathematics Computation (periodical code)

APMAC Asia-Pacific Market Awareness Committee

ApMaL Applied Mathematics Letters (periodical code)

ApMaM Applied Mathematics and Mechanics (periodical code)

ApMaO Applied Mathematics and Optimization (periodical code)

ApMat Applied Mathematics (periodical code)

APMC Advanced Program Management Course

APME Associative Processor Microelectronic Element

APMG Assistant Postmaster General • Associate Postmaster General

APMI Aerial Photography Micrographic Index • Area Precipitation Measurement Indicator

ApMic Applied Microbiology (periodical code)

APMIS Automated Project Management Information System • Automatic Project Management Information System

APML Assistant Program Manager for Logistics

ApMM Applied Mathematics and Mechanics (periodical code)

APMM Association of Policy Market Makers

APMMRI Automatic Point Marking, Measuring and Recording Instrument

APMP Acid Precipitation Mitigation Programme • Asia-Pacific Metrology Programme • Association of Proposal Management Professionals

APMQ Association des Producteurs en Multimédia du Québec (Canada-QC)

APMR Adaptive Pointable Multibeam Radar

ApMRv Applied Mechanics Reviews (periodical code)

APMS Advanced Power Management System • Advances in Production Management Systems • Automated Performance Measurement System • Automated Performance Measuring System • Automated Project Management System

APMSM Assembly Phase Maintenance Simulation Model

Apmt Appointment

ApMT Applied Microgravity Technology (periodical code)

APMT Advanced Planetary Mission Technology • Antenna Pattern Measurement Test • Asia Pacific Mobile Telecommunications

APN Abatement Priority Number • Access Point Name • Advances in Petri Nets (conference) • All Pass Network • Alpena (USA-MI) (location identifier) • Asia-Pacific Network • Augmented Proportional Navigation • Authorized Part Number

APNIC Asian Pacific Network Information Center

ApNM Applied Numerical Mathematics (periodical code)

ApNr Astrophysica Norvegica (periodical code)

Apo Apogee

APO Access Protection Option • Advanced Planner and Optimizer • Amatérska Prohlídka Oblohy (Czech Republic) • Analysis Procedure Object • Apache Point Observatory (USA-NM) • Apartado (Colombia) (location identifier) • A Part Of • Appointed Post Office • Army Post Office • Australian Patent Office • Austrian Patent Office • Automated Purchase Order • Automatic Purchase Order

APOA Arctic Petroleum Operators Association

Apóc Apócope

APOD Astronomy Picture Of the Day

APOE Airport Of Embarkation

Apog Apogee

APOLLO Apache Point Observatory Lunar Laser-ranging Operation • Article Procurement with Online Local Ordering

APOM Adjusted Program Objective Memorandum • Association des Producteurs d'Oeuvres Multimedia (France)

APOMA American Photo-Optical Manufacturers Association (USA) • American Precision Optics Manufacturers Association (USA)

APON ATM over Passive Optical Network

APOP Authenticated Post Office Protocol • Authentication Post Office Protocol

Apops Apostrophe

ApOpt Applied Optics (periodical code)

APOR Advisory Panel for Operations Research

APOS Advanced Polar Orbiting Satellite

APOTA Automatic Positioning Telemetering Antenna

App Apparat • Apparatus • Appareil • Apparent • Apparently • Appeal • Appearance • Appendice • Appendix • Applicable • Application • Applied • Apply • Appointed • Appointment • Apprehend • Apprentice • Approach • Approval • Approve • Approved • Approximate • Approximately • Approximation

AP&P Advanced Plans & Programs

APP Advanced Planetary Probe • Advanced Plans and Programs • Advanced Procurement Package • Advanced Procurement Plan • Advanced Project Planning • Aerospace Propulsion Products • Agence pour la Protection des Programmes (France) • Agency Peculiar Property • Air Parcel Post • Amyloid Precursor Protein • Analog Preprocessor • Analytic Pointing Performance • Antenna Position Programmer • Apollo Applications Programme • Apparatus • Application • Applications Portability Profile • Apprehend • Approach • Appropriation • Asapa (Papua New Guinea) (location identifier) • Association of Professional Photographers (UK) • Associative Parallel Processor • Astrophysics Payload • Atelier Pédagogique Personnalisé • Atmospheric Physics Program • Atmospheric Pressure Perturbation • Automated Photographic Patrol • Automatic Page Positioning • Automatic Plate Processor • Automatic Position Planning • Auxiliary Power Package • Auxiliary Power Plant

APPA American Public Power Association (USA)

APPAG Association Pour un Planétarium dans l'Agglomération Grenobloise (France)

APPAM Association for Public Policy Analysis and Management (USA)

AppAn Applicable Analysis (periodical code)

Appar Apparatus • Apparent • Apparently

APPC Advanced Peer-to-Peer Communication • Advanced Program-to-Program Communication • Asia Pacific Physics Conference

Appd Approved

APPE Association for Practical and Professional Ethics

APPEAL Academic Press Print and Electronic Access License • Asia-Pacific Programme of Education for All

ApPEC Astroparticle Physics - European Coordination

APPECS Adaptive Pattern Perceiving Electronic Computer System

Append Appendice • Appendix

APPEX Advanced Power Projection and Execution

APPF Automated Payload Processing Facility

ApPhA Applied Physics A - Solids and Surfaces (periodical code)

ApPhC Applied Physics – Communications (periodical code)

ApPhL Applied Physics – Letters (periodical code)

ApPhy Applied Physics (periodical code)

APPI Acquisition Pollution Prevention Initiative • Advanced Peer-to-Peer Internetworking • Advanced Planning Procurement Information

Appl Appeal • Applicable • Applicant • Application • Applied • Apply

APPLE Advanced Propulsion Payload Effects • Applied Parallel Programming Language Experiment • Ariane Passenger Payload Experiment • Associative Processor Programming Language Evaluation

APPLES ACS Pure Parallel Lyman-alpha Emission Survey

Applic Application

Appl.Op. Applied Optics

Appl.Opt. Applied Optics

APPM Accident Prevention Program Manager • Avant-Projet de Plan de Masse

Appmt Appointment

Appn Appropriation

APPN Advanced Peer-to-Peer Networking • Appropriation

APPNTAM Advanced Peer-to-Peer Networking Topology and Accounting Manager

ApPP Astrophysics and Plasma Physics

ApPPL Applied Physics B – Photophysics Laser Chemistry (periodical code)

Appr Apprentice • Approved • Approximately

APPR Approach

Appro Approbation • Approval

Approp Appropriate • Appropriation

APPROV Approval

Approx Approximate • Approximately • Approximation

APPROX Approximately

Apprp Appropriate

Apprx Approximate

Apps Appendices • Applications

APPS Advanced Planning and Packaging Support • Advanced Planning Program Scheduling • Aerosol Physical Properties of the Stratosphere • Analytical Photogrammetric Positioning System • Analytical Photogrammetric Processing System • Application Portability Profile Standards • Applications • Asynchronous Parallel Pattern Search • Atmospheric Pressure Plasma System • Australian Physiological and Pharmacological Society • Automated Package Processing System • Automated Past Performance System • Automated Publications and Printing System • Automated Publications Preparation System • Automated Publications Production System • Auxiliary Payload Power System

APPSR Aerial Photography Progress Summary Report

Appt Appartement • Appoint • Appointed • Appointing • Appointment

Appd Appointed

Appx Appendix

AP2I Acquisition Pollution Prevention Initiative

AP3 Acquisition Pollution Prevention Program

APQ Absolutely - Positively - Quality • Actual Pointing Quality • Arapiraca (Brazil-AL) (location identifier) • Associação Portuguesa para a Qualidade (Portugal)

APQCO Asian-Pacific Quality Control Organization

APQF Aerial Photography Quad File

Apr April • Aprile

APR Acoustic Paramagnetic Resonance • Acquisition Procurement Request • Active Prominence • Active Prominence Region • Adaptive Pointing Radar • Advanced Production Release • Advance Parts Release • Agency Procurement Request • Agency Progress Report • Airborne Profile Recorder • All-Purpose Room • Alternate Path

Retry • Annual Percentage Rate • Annual Planning Report • Annual Program Review • Annual Progress Report • Annual Purchase Rate • Antenna Position Recorder • April • April River (Papua New Guinea) (location identifier) • Arbeitsplatzrechner • Area Planning Report • Atmospheric Processes Research • Audiovisual Production Requirements • Automatic Pattern Recognition • Automatic Power Reserve • Automatic Pressure Relief • Automatic Programming and Recording

APRA Australian Postgraduate Research Award

APRB Acquisition Plan Review Board • Adaptive Pattern Recognition Processing

APRC Anno Post Roman Conditam

Aprch Approach

APRD Atmosphere Particulate Radioactivity Detector

APRF Aerophysics/Propulsion Research Facility • Automated Password Retrieval Facility

APRIA Asian-Pacific Rim Institute for Astrophysics

APRM Asian-Pacific Regional Meeting • Automatic Position Reference Monitor • Average Power Range Monitor

APRO Aerial Phenomena Research Organization

APRODI Association pour la Promotion et le Développement Industriel (France)

Aprox Aproximadamente

APRP Adaptive Pattern Recognition Processing

APRS Acoustic Position Reference System • Acquisition Position Review System • Applied Physics Research Section • Australian Pattern Recognition Society (Australia) • Automatic Position Reference System

APRSS Australian Photogrammetric and Remote Sensing Society

Aprt Airport

APRT Advanced Productivity Research and Technology

Aps Apus-Apodis (constellation)

A&PS Administration & Program Support

APS Accelerometer Parameter Shift • Accessory Power Supply • Acquisition Processor Subsystem • Active Pixel Sensor • Active Protective System • Address Program Support • Adjustment Procedure Software • Adjustment Processing System • Administration and Program Support • Advanced Photon Source • Advanced Photosynthetic System • Advanced Photo System • Advanced Pixel Sensor • Advanced Planning and Scheduling • Advanced Planning System • Advanced Power System • Advanced Printing Service • Advanced Propellant System • Advanced Propulsion System • Aerodynamic Particle Sizer • Aft Propulsion System • Aiming and Pointing System • Airborne Pointing System • Airborne Power Source • Air Pollution Syndrome • Airport Planning Standard • Air Pressure Switch • Albuquerque Public Schools (USA-NM) • Alphanumeric Photocomposer System • Alpha Particle Spectrometer • Alternate Propulsion Source • Altitude Proximity Sensor • American Philosophical Society (USA) • American Physical Society (USA) • American Physiological Society (USA) • American Polar Society (USA) • Amplifier Power Supply • Anápolis (Brazil-GO) (location identifier) • Ångström Pyrheliometric Scale • Angular Position Sensor • Antenna Pointing System • Appearance Potential Spectroscopy • Arbeitsplatzsystem • Arc-Plasma Spraying • Area Parameter Specification • Array Processor Software • Array Processor System • Ascent Propulsion System • Assembly Programming System • Astronomy and Planetary Science • Astrophysical and Planetary Sciences • Asynchronous Production System • Asynchronous Protocol Specification • Atmospheric Pollution Sensor • Atmospheric Pressure Sensor • Attitude Propulsion System • Australasian Planetarium Society • Australian Public Service • Automated Planning System • Automated Plate Scanner • Automated Process Software • Automated Procurement System • Automated Production System • Automated Publishing System • Automatic Page Search • Automatic Patching System • Automatic Phasing System • Automatic Pilot System • Automatic Pointing System • Automatic Processing System • Automatic Program System • Automatic Protection Switch • Automatic Protection Switching • Auxiliary Power Supply • Auxiliary Power System • Auxiliary Program Storage • Auxiliary Propulsion System • Auxiliary Publication Service • Avant-Projet Sommaire • Average Particle Size

APSA American Political Science Association (USA)

APSAT Asia-Pacific Satellite

APSB Advanced Planning Systems Branch • Advanced Program Systems Branch • Astronomy and Planetary Science Board

APSC Asia Pacific Space Centre Pty Ltd (Christmas Island)

APSCC Asia-Pacific Satellite Communication Council

ApScR Applied Scientific Research (periodical code)

APSE Ada Programming Support Environment

ApSF Amperes per Square Foot

APSF Amperes Per Square Foot • Australian Pacific Science Foundation

APSG Asia-Pacific Space Geodynamics

APSGP Asia-Pacific Space Geodynamics Project

ApSI Amperes Per Square Inch

APSI Advanced Propulsion System Integration • Aircraft Propulsion System Integration • Amperes Per Square Inch

APSIN Alaska Public Safety Information Network

APSK Amplitude Phase-Shift Keying

APSO Asian Physical Society • Audio Precision System One

APSP Array Processor Subroutine Package

ApSpE Applied Spectroscopy (periodical code)

APSR Aerial Photography Summary Record

APSRS Aerial Photography Summary Record System

Ap&SS Astrophysics & Space Sciences (periodical code)

ApSS Applied Surface Science (periodical code)

APSS Advanced Planetary Spacecraft System • Applied Physics A - Solids and Surfaces (periodical code) • Atmospheric Pressure Supply System • Automated Program Support System • Automatic Planning and Scheduling System

APST Association of Professional Scientists and Technologists

APSTM Analytical Photon Scanning Tunneling Microscope • Analytical Photon Scanning Tunneling Microscopy

ApSup Applied Superconductivity (periodical code)

Apt Airport • Apartment • Aptitude

AP&T Advanced Physics & Technology

APT Acquisition - Pointing - Tracking • Acquisition Position Type • Address Pass Through • Address Path Through • Advanced Parallel Technology • Advanced Passenger Train • Advanced Patent Technique • Advanced Personnel Training • Advanced Physics and Technology • Advanced Platform Technology • Advanced Postal Teller • Advanced Processing Testbed • Advanced Process Technology • Airborne Pointer Tracker • Air Passenger Terminal • Airport • Alliance for Public Technology (USA-DC) • All-Picture Transfer • Analog Program Tape • Analyseur de Plasma Thermique • Apartamento • Apartment • Applatissement • Application Productivity Tool • Application Table • Applied Potential Tomography • Architectural Power Tool • Area of Priority Treatment • Asia-Pacific Telecommunity • Asia Pacific Telescope • Association of Professional Translators (USA) • Astronaut Preference Test • Astronomers Planning Tool • Astronomer's

Proposal Tool • Augmented Programming Training • Automated Patrol Telescope • Automated Photometric Telescope • Automated Postal Teller • Automated Programming Tool • Automatically Programmed Tool • Automatic Photoelectric Telescope • Automatic Photometric Telescope • Automatic Picture Taking • Automatic Picture Transmission (*) • Automatic Position Telemetering • Automatic Programmed Tool • Automatic Programming of Tools • Automation Planning and Technology • Autonomous Picture Transmission • Autorisation Provisoire de Travail • Jasper - Marion County (USA-TN) (location identifier)

Aptdo Apartado

APTE Asociación de Parques Científicos y Tecnológicos de España (Spain) • Automated Production Test Equipment • Automatic Production Test Equipment

ApTI Actions per Time Interval

APTI Actions Per Time Interval

APTIC Air Pollution Technical Information Center

Apto Apartamento

APTP Arithmetic Proficiency Training Program • Asia-Pacific Technology Program

Apts Apartments

APTS Aerial Profiling of Terrain System • Automatic Picture Transmission System

APTU Aerodynamic and Production Test Unit • Aerodynamic and Propulsion Test Unit • African Postal and Telecommunications Union

APU Academic Planning Unit • Accessory Power Unit • Acute Psychiatric Unit • Apucarana (Brazil-PR) (location identifier) • Arab Postal Union • Arithmetic Processing Unit • Assessment of Performance unit • Automated Program Understanding • Auxiliary Power Unit • Auxiliary Propulsion Unit • Azusa Pacific University (USA-CA)

APUC Association des Presses Universitaires Canadiennes (Canada) • Average Procurement Unit Cost

APUHS Automatic Program Unit High Speed

APULS Automatic Program Unit Low Speed

APUPA Alien - Penumbral - Umbral - Penumbral - Alien

APUS Auxiliary Power Unit System

APUT Auxiliary Power Unit Test

Apv Approve

APV Accounts Payable Voucher • Ahorro Previsional Voluntario • Algemene Politieverordening • Apple Valley (USA-CA) (location identifier)

Apvd Approved

Apvl Approval

APW Abel-Poisson Wavelet • Apia Faleolo (Samoa) (location identifier) • Apparent Polar Wander • Application and Practice Workshop • Approximation of Plane Waves • Augmented Plane Wave

APWA American Public Welfare Association (USA) • American Public Works Association (USA)

APWL Automatically Processed Wire List

APWP Actual Price of Woek Performed • Apparent Polar Wander Path

APWR Advanced Pressurized Water Reactor

Apx Appendix

APX Alpha Proton X-ray • Approximately • Arapongas (Brazil-PR) (location identifier)

APXS Alpha Proton X-ray Spectrometer

APY Alto Parnaíba (Brazil-MA) (location identifier)

APZ Zapala (Argentina) (location identifier)

Aq Aqua • Aqueous • Aquatic

AQ Accomplishment Quotient • Accumulator-Quotient (register) • Achievement Quotient • Acquisition • Action Qualifiante • Adaptive Quantisation • Air Quality • Antarctica (ISO country code) • Antarctique (code ISO de pays) • Any Quality • Assurance de Qualité • Astronomy Quarterly • Astrophysical Quantities • Astroqueyras (France) • Attainment Quotient • Authorized Qualifyer

AQA Agence de la Qualité de l'Air • Air Quality Act (USA) • Air Quality Assessment • American Quaternary Association (USA) • Approved Quality Assurance • Araraquara (Brazil-SP) (location identifier) • Assessment and Qualifications Alliance (UK)

AQAB Air Quality Advisory Board

AQAFO Aeronautical Quality Assurance Field Office

AQAG Air Quality Advisory Group

AQAHWCI Algorithm Quality Assurance Hardware Configuration Item

AQAP Allied Quality Assurance Provision • Allied Quality Assurance Publication

aq. bull. Aqua Bulliens

AQC Air Quality Control • Automatic Quality Control • Automatic Quench Calibration • Automatic Quench Correction • Automatic Quench Compensation

aq. cal Aqua Calida

AQCB Automated Quote Contract Billing

AQCD Air Quality Criteria Document

AQCEN Accès à la Qualité, à la Certification et à la Normalisation

AQCL Analytical Quality Control Laboratory

aq. com. Aqua Communis

aq. comm. Aqua Communis

AQD Additional Qualification Designator • Air Quality Data

aq. dest. Aqua Destillata

aq. ferv. Aqua Fervens

aq. frig. Aqua Frigida

AQDHS Air Quality Data Handling System

AQE Absolute Quantum Efficiency • Actual Quadrant Elevation • Airman Qualifying Examination

AQF Advanced Quick Fix

AQG Anqing (China-PRC) (location identifier)

AQH Average Quarter Hour

AQI Air Quality Index • Qaisumah (Saudi Arabia) (location identifier)

AQIS Air Quality Information System

AQJ Aqaba (Jordan) (location identifier)

Aql Aquila-Aquilae (constellation)

AQL Acceptable Quality Level • Acceptance Quality Level • Advanced Query Language • Advanced Quick Look • A Query Language • Astronomical Query Language • Astronomy Query Language • Average Quality Limit

AQLT Acquisition Lead Time

AQM Additive Quark Model • Air Quality Management • Air Quality Model • Air Quality Monitoring • Ariquemes (Brazil-RO) (location identifier) • Atmospheric Quality and Modification

AQN Actual Quantity Needed • Azimuthal Quantum Number

AQO Air Quality Organization • Automatic Quality Organization

AQP Acquisition Quality Plan • Advanced Qualification Programme • Air Quality Plan • Airworthiness Qualification Plan • Arequipa - Rodriguez Ballón (Peru) (location identifier)

AQTD Airworthiness Qualification Test Directorate

AQPU Association Québécoise des Presses Universitaires (Canada-QC)

aq. pur. Aqua Pura

AQQPRI Amended Qualitative and Quantitative Personnel Requirements Information

Aqr Aquarius-Aquarii (constellation)

AQR Acceptable Quality Rate • Approximate and Qualitative Reasoning

AQS Advanced Quality System • Air Quality Standard • Air Quality System • Approximate Quadratic Search • Saqani (Fiji) (location identifier)

Aqstn Acquisition

AQT Acceptable Quality Test • Analytical Query Time • Association Québécoise de Télédétection (Canada-QC) • Aviation Qualification Test

aq. tep. Aqua Tepida

Aque Aqueduct

Aqued Aqueduct

AQY Alyeska (USA-AK) (location identifier)

a/r Accounts Receivable

ar Arabic (ISO language code) • Argentina

aR Am Rhein

Ár Árabe

Ar Arabia • Arabian • Arabic • Aragón (Spain) • Argon • Arrival • Arrive • Arrived • Arrives • Arroyo • Aryl Group

A/R Accept/Refuse • Accounts Receivable • Accusé de Réception • All Risks • Anno Regni • As Required • At the Rate of • Autoland/Rollout • Azimuth/Range

A&R Adapter & Rotator • Assault & Robbery • Assembly & Recycle • Astronomie und Raumfahrt (periodical code) • Automation & Robotics

AR Abteilungsrechner • Acceleration Request • Acceptance Readiness • Acceptance Review • Access Rate • Access Register • Accomplishment Ratio • Account Receivable • Accusé de Réception • Achievement Ratio • Acid-Resisting • Acknowledgment of Receipt • Acquisition Radar • Acquisition Reform • Acrylic Rubber • Action Register • Action Request • Active Radar • Active Radical • Active Region • Active Route • Activity Report • Acute Rejection • Adapter Rotator • Address Register • Administrative Ruling • Advanced Reactor • Advanced Robotics • Advance Release • Advice of Receipt • Advisory Report • Aerodynamic Report • Agent Report • Age Replacement • Airborne Relay • Aircraft Rocket • Air Reconnaissance • Air Refueling • Alarm Report • Aller et Retour • Aller-Retour • Allocated Reserve • All Risks • Alternate Route • Altesse Royale • Anna Regina • Annual Register • Annual Report • Annual Return • Annual Review • Annual Reviews Inc. • Anode Reaction • Anomalous Refraction • Anomaly Report • Anti-Reflection • Anti-Reflective • Application Region • Applied Research • Apportionment Request • Approved for Release • April • Arabia • Archival Research • Archival Researcher • Argentina (ISO country code) • Argentine (code ISO de pays) • Arkansas (USA) • Arrêté Royal • Arrival and Return • Arrives • Arriving • Ascención Recta • Aspect Ratio • As Required • Assembly and Recycle • Assembly and Repair • Association Régionale • Astronomy-Related • Atmosphere Revitalization • Augmented Reality • Automatic Resources • Automation and Robotics • Autonomous Region • Autoregression • Autoregressive • Average Rating • Average Revenue • Avionics Requirements • Avis de Réception • Avril

Ara Ara-Arae (constellation)

ArA Arkiv for Astronomi (periodical code)

ARA Abbreviated Registered Address • Academic and Research Affairs • Accelerated Readiness Analysis • Activité à Responsabilité Académique • Address Register Area • Advanced Radar Altimeter • Advanced Rigid Array • Aerospace Research Association • Airborne Radar Approach • Analyse du Rayonnement Atmosphérique • Angular Rate Assembly • Apple Remote Access • AppleTalk Remote Access • Assigned Responsible Agency • Associate of the Royal Academy • Association for Research Astronomy • Association of Research Astronomers (USA) • Associazione Reggiana Astrofili (Italy) • Associazione Romana Astrofili (Italy) • Atmospheric Radiation Analysis • Attitude Reference Assembly • Attribute Registration Authority • Austral (Argentina) (ISO currency code) • Automatic Reimbursement Authority • Autoresonant Accelerator • Auxiliary Recovery Antenna • Average Response Amplitude • New Iberia - Acadiana (USA-LA) (location identifier)

ARA&A Annual Review of Astronomy & Astrophysics (periodical code)

ARAA Annual Review of Astronomy and Astrophysics

Arab Arabia • Arabian • Arabic

AraBN Arab Biosciences Network

ARABSAT Arab Satellite Communication Organization

ARAC Aerospace Research Applications Center • Army Radar Approach Control • Atmospheric Release Advisory Capability • Aviation Rulemaking Advisory Committee

ArAco Archives Acoustics (periodical code)

ARAD Average Response Amplitude Data

Arag Aragonés

ARAG Antireflective-Antiglare • Atmospheric Radiation Analysis Group

ARAGO Advanced Robotic Agile Observatory

ARAK Arakelian (catalog)

ArAku Archiwum Akustiki (periodical code)

ARAL Automatic Record Analysis Language

ARALL Aramide Aluminium Laminate

ARALR Association Régionale d'Astronomie pour le Languedoc-Roussillon (France)

ARAM Automated Reliability/Availability/Maintainability

ARAMCO Arabian American Co.

ARAMIS Adaptation du Radar À la Météorologie Infra-Synoptique • Antarctic Remote-sensing Aerial-photography and Mapping Information System • Antarctic Research in Applied Meteorology, Imaging and Sounding • Atmospheric Radiation And Minor Species

ARAMP Arctic Remote Autonomous Measurement Platform

ARAP Age-Rotation-Activity Paradigm • AppleTalk Remote Access Protocol • Association de Recherche et d'Action Pédagogique • Astronaut Rescue Air Pack • Automated Reliability Assessment Program

ArApM Archive of Applied Mechanics (periodical code)

ARAR Accident Risk Assessment Report • Applicable or Relevant and Appropriate Requirement • Associazione Ravennate Astrofili Rheyta (Italy)

ARARAT Advanced Radiometric Ray Tracer

ARAS Antireflective - Antistatic • Atomic Resonance Absorption Spectroscopy

ARAT Avion de Recherche Atmosphérique et de Télédétection

Arauc Araucano

ArAuT Archiwum Automatyki i Telemekhaniki (periodical code)

Arb Arbeit • Arbeiter • Arbitrer • Arbitrageur • Arbitraire • Arbitrary • Arbitration • Arbitrator

ARB Accreditation Review Board • Acquisition Review Board • Adaptive Rate-Based • Address Record Buffer • Address Recorder Buffer • Aerospace Research Bulletin (AIAA) • Air Registration Board • Airworthiness Review Board • Analysis Review Board • Ann Arbor (USA-MI) (location identifier) • Arbitrager • Assign Receive Bypass • Award Review Board

ArBei Astrophysics Reports of the Beijing Astronomical Observatory (periodical code)

ARBG Acquisition Reform Benchmarking Group

ArbGG Arbeitsgerichtsgesetz

ARBIRS Advanced Region-Based Image Retrieval System

ArBuM Archiwum Budowy Maszyn (periodical code)

ArbZG Arbeitszeitgesetz

ArbZRG Arbeitszeitrechtsgesetz

arc Arcus

Arc Arcade • Arcaico

A&RC Application & Resource Control

ARC Academic Regulations Committee • Accelerating Rate Calorimeter • Accounting Requirements Code • Acquisition Review Council • Action de Recherche Concertée • Active Radar Calibration • Active Radar Calibrator • Active Radical Combustion • Adjustment Reply Code • Administrative Radio Conference • Advanced Radial Camera • Advanced Reentry Concept • Advanced Remote Control • Advanced Research Center • Advanced RISC Computing • Advanced Rotor Concept • Aeronautical Research Council • Aerospace Remote Calculator • Aerospace Research Chamber • Aggregation of Red Cells • Aide au Recrutement de Cadres • Airborne Radio Communications • Alberta Research Council (Canada) • Alternate Routes to Computing (program) • Altitude Rate Command • American Red Cross (USA) • Ames Research Center (NASA) • Amplitude and Risetime Compensation • Analog Response Conditioner • Analyzer - Response - Controller • Annual Relative Citedness • Antarctic Research Center • Antireflection Coating • Application and Resource Control • Applied Research Corporation • Arcade • Archive • Archived • Arcsecond Raster Chart • Arctic Radiation and Chemistry (experiment) • Arctic Research Commission • Arctic Village (USA-AK) (location identifier) • Area Reprogramming Capability • Area Resources Center • Assign Remote Connection • Association pour la Recherche Cognitive (France) • Astronomy Research Consortium (USA) • Astrophysical Research Consortium (USA) • Astrophysics Research Center • Atelier de Recherche et de Création • Atlantic Research Corporation • ATM Research Consortium • Atmospheric Radiance Code • Attached Resources Computer • Attached Resources Computing • Australian Research Council • Automated Rendezvous and Capture • Automatic Rate Control • Automatic Ratio Control • Automatic Reconfiguration Criterion • Automatic Relay Calculator • Automatic Remote Control • Automatic Responsivity Correction • Auxiliary Roll Control • Average Resonance Capture • Average Response Computer

ARCA Advanced RISC Computing Architecture

ARCAD Arctic Auroral Density

ARCADE Absolute Radiometer for Cosmology, Astrophysics, and Diffuse Emission • Ampere Remote Control Access Data Entry • Architectural Computer-Aided Design and Evaluation • Automatic Radar Control And Data Equipment

ARCAL Aircraft Radio Control of Aerodrome Lighting

ARCANE Action de Recherche sur la Circulation en Atlantique Nord-Est

ARCB Arcebispo

ARCC Analysis of Recent Climate Change

ARCCOH Asian Regional Coordinating Committee on Hydrology

ARCDR Altimetry and Radiometry Composite Data Record

ARCFPCS Algorithm with Residual Cancellation in Functionals and at Points of Control Switching

Arch Archaic • Archaism • Archbishop • Archery • Archipel • Archipelago • Archipiélago • Architect • Architectural • Architecture • Archive • Archives • Archivo

Archbp Archbishop

Archel Archipel

ARCH Architecture • Argonne/Chicago Development Corporation • Articulated Computer Hierarchy • Auto-Regressive Conditional Heteroscedasticity

Archaeol Archaeological • Archaeology

Archbp Archbishop

ARCHIMEDES Advanced Research in Communications using Highly Inclined orbits for Mobile Experiments and Demonstration with European Satellites

ARCHISS Archival Climatic History Survey

ARCHV Archive

ARCISS Association of Research Centres in the Social Sciences (UK)

ARCL Allowable Residual Contamination Level

arcmin Minute of Arc

ARCN Attached Resource Computer Network

ARCnet Attached Resource Computing Network • Attached Resource Computer Network

ArCom Archivum Combustionis (periodical code)

arcos Arc Cosine

ARCOS Area Correlation for Determination of Object Surfaces

arcot Arc Cotangent

ARCOVA American Research Corporation of Virginia (USA-VA)

ARCQUERY Archive Query

ARCS Achievement Rewards for College Scientists (foundation) • Acquisition Radar and Control System • Admiralty Raster Chart Service • Aft Reaction Control System • Altitude Rate Command System • Architecture of Computing Systems (conference) • Arcminute Radio Clusterlens Search • Associate of the Royal College of Science • Atmospheric Radiation and Cloud Station • Attached Resources Computer System • Austrian Research Centers Seibersdorf (Austria) • Automated Radar Calibration System • Automated Reusable Components System • Average Rapidness of Citation Successions

arcscs Arc Cosecant

arcsec Arc Second • Second of Arc

arcsin Arc Sine

ARCSS Arctic Systems Science

ARCST Associate of the Royal College of Science and Technology

arctan Arc Tangent

ARCU American-to-Russian Converter Unit

ARCUS Arctic Research Consortium of the United States

Ard Ardennes (France)

ARD Abort Region Determinator • Absolute Radiation Detector • Acute Radiation Disease • Acute Respiratory Disease • Aeronautical Research Division • Agence Régionale de Développement (France) • Alor Island (Indonesia) (location identifier) • Analyse Relationnelle de Données • Annonce de Résultats Différés • Arbeitsgemeinschaft der öffentlich-rechtlichen Rundfunkanstalten der Bundesrepublik Deutschland (Germany) • Arrière Droit • ASCII Region Definition • Assistant Resident Director • Atmospheric Recovery Demonstrator • Atmospheric Reentry Demonstrator (*) • Automatic Reentry Demonstrator • Automatic Release Date • Average Response Data

ARDA Advanced Research and Development Activity • Advanced Research and Development Agency • Analog Recording Dynamic Analyzer

ARDB Arctic Runoff Database

ARDC Air Research and Development Command (USA)

ARDEMI Association pour la Recherche et le Développement de l'Enseignement Multimédia et Interactif (France)

ARDES Automatic Recorder and Data Evaluation System

ARDF Aerial Radio Direction Finding • Automatic Radio Direction Finding

ARDIS Advanced Radio Data Information Service

ARDL Average Readiness Days Lost

ARDME Automatic Radar Data Measuring Equipment

ARDMS Automated Route Design and Management System

ARDS Advanced Range Data System • Advanced Remote Display Station • Automated Resources Directory Service

Ardt Arrondissement

ARDU Aircraft Research and Development Unit

ARE Acoustic Radiation Element • Acquisition Reform Executive • Activated Reactive Evaporation • Admiralty Research Establishment (UK) • Aerothermal Reentry Experiment • Aide à la Recherche d'un Emploi • Algebraic Riccati Equation • All Routes Explorer • Antenna Range Equipment • Arab Republic of Egypt • Arecibo (Puerto Rico) (location identifier) • Array Regulator Electronics • Assemblée des Régions d'Europe • Asymptotic Relative Efficiency • ATM Routing Engine • Automated Responsive Environment • Automatic Record Evaluation • Auxiliary Rocket Engine • Émirats Arabes Unis (code ISO de pays) • United Arab Emirates (ISO country code)

AREA Angle Resolving Energy Analyzer

AREAL Atmospheric Research and Exposure Assessment Laboratory

AREAS Area Resource Analysis System

AREC Air Resource Element Coordinator

ARED Aperture Relay Experiment Definition

AREDOR Association de Recherche et d'Exploitation de Diamant et d'Or

AREG Antarctic Research Evaluation Group

ARELAS Arequipa Laser Station (Peru)

ArEle Archiwum Elektrotechniki (periodical code)

ARELS Association of Recognized English Language Services

ArElU Archiv Elektronik und Übertragungstechnik (periodical code)

AREN Australian Regional Geodetic Network

ARENTS ARPA Environmental Test Satellite

ARep Astronomy Reports (periodical code)

AREP Atmospheric Research and Environment Programme

AREPS Annual Review of Earth and Planetary Sciences (periodical code)

AREPT Agent Report

ARES Advanced Railroad Electronic System • Advanced Research and Experiments in Sensing • Advanced Research EMP Simulator • Amateur Radio Emergency Services • Architectural Reasoning for Embedded Software (workshop) • Association pour la Recherche sur l'Environnement Spatial • Automated Risk Evaluation System • Automated Runway Evaluation System • Automatic Record Evaluation System

ARESCE Association Royale des Étudiants en Sciences Commerciales et Économiques

ArEVO Arbeitsentgeltsverordnung

ARF Abort - Retry - Fail • Acaricuara (Colombia) (location identifier) • Active Region Filament • Acute Respiratory Failure • Aeronautical Research Foundation • Aerospace Recovery Facility • Airport Reservation Function • Alarm Reporting Function • Ancillary Response File • Ancillary Response Function • Anticipated Receive File • Aquatic Research Facility • Automatic Reporting Feature

ARFA Allied Radio Frequency Agency

ARFDS Automatic Reentry Flight Dynamics Simulator

ARFE Association des Régions Frontalières d'Europe • Association des Régions Frontalières Européennes

ARFF Airport Rescue and Firefighting

ARFM Annual Review of Fluid Mechanics (periodical code)

ARfS Astronomer Royal for Scotland

Arg Argent • Argentina • Argentine • Argentinian

Årg Årgang

ARG Accident Response Group • Aeronautical Research Group • Air Rescue Group • Argentina (ISO country code) • Argentine (code ISO de pays) • Arrière Gauche • Attributed Relational Graph • Walnut Ridge (USA-AR) (location identifier)

Argen Argentina • Argentine

Argent Argentina • Argentine

ARGO Astrophysical Radiation and Gamma Observatory • Automatic Ranging Grid Overlay

ARGOS Advanced Remote Gaseous Oxides Sensor • Advanced Research and Global Observation Satellite

ARGOSI Application-Related Graphics and OSI Standards Integration

ARGUS Advanced Real-time Gaming Universal Simulation • Automatic Routine Generating and Updating System

a/Rh Am Rhein

a. Rh. Am Rhein

ARH Arkhangelsk (Russia) (location identifier)

ARHP Association Rule Hypergraph Portioning

ARHS Attitude and Heading Reference System

Ari Aries-Arietis (constellation)

ARI Academic Research Infrastructure • Accelerated Research Initiative • Access Rule Index • Access to Research Infrastructure • Acute Respiratory Infection • Addressee Routing Indicator • Address Recognized Indicator • Advanced Research Initiative • Airborne Remote Instrumentation • Annual Recurrence Interval • Annual Reviews Inc. (USA-CA) • Appareil Respiratoire Isolant • Arica Chacalutta (Chile) (location identifier) • Astronomisches Rechen-Institut (Germany) • Astronomy Research and Instrumentation • Automated Readibility Index • Automatic Return Item • Automotive Radio Information • Autonomous Research Institute

ARIA Advanced Range Information Aircraft • Advanced Range Instrumentation Aircraft (*) • Aeroflot Russia International Airlines • Aeroflot Russian International Airlines • American Risk and Insurance Association (USA) • Atmospheric Response In Aurora • Automated Requirements Integration and Analysis

ARIAT Acquisition Reform Implementation Assessment Team

ARIB Association of Radio Industries and Businesses (USA)

ARIC Acid Rain Information Centre

ARICS Accounts Receivable Invoice Control System

ARID Adaptive Recursive Interpolated Differential

ARIDPCM Adaptive Recursive Interpolated Differential Pulse Code Modulation

ARIES Airborne Research Integrated Experiments System • Antarctic Reception Imagery for Environmental Studies • Applied Resource Image Exploitation System • Arizona (infrared) Imager and Echelle Spectrograph • Association de Recherche, d'Information et d'Échanges Européens • Astronomical Radio Interferometric Earth Survey • Astronomical Radio Interferometric Earth Surveying • Astronomy Resources for Intercurricular Elementary Science • Atmospheric Research Information Exchange Study • ATM Research and Industrial Enterprise Study • Australian Resources Information and Environment Satellite • Authentic Reproduction of an Independent Earth Satellite

ARIGC Astronomisches Rechen-Institut General Catalogue

ARIL Automated Return Items List • Automatic Return Items List

ARIMA Autoregressive Integrated Moving Average

ARIN Aerospace Research Information Network • American Registry of Internet Numbers (USA)

ARINC Aeronautical Radio Inc.

ARIPO African Regional Industrial Property Organization

ARIS Advanced Range Instrumentation Ship • Aggregate Route-based IP Switching • Alabama Resource Information System • Arizona Resource Information System • Atomic Reactor In Space • Attitude and Rate Indicating System • Australian Resources Information System • Authority - Responsibility - Information - Support • Automatic Recording Infrared Spectrometer • Automatisches Retrieval Informationssystem

ARISAL Association Régionale des Ingénieurs et des Scientifiques d'Alsace (France)

ARISE Advanced Radio Interferometry between Space and Earth • Agricultural Resources Inventory and Survey Experiment • A Reusability Infrastructure for Software Engineering

ARISS Amateur Radio on the International Space Station

Arist Aristotle (384-322 BC)

ARIST Agence Régionale d'Information Scientifique et Technique (France) • Annual Review of Information Science and Technology

Aristoph Aristophanes (c. 445 - c. 380 BC)

ARISTOTE Association de Réseaux Informatiques en Système Totalement Ouvert et Très Élaboré

ARISTOTELES Application and Research Involving Space Techniques Observing The Earth fields from Low-Earth-orbit Spacecraft • Application and Research Involving Space Technologies Observing The Earth's field from Low Earth orbit Satellite

Arit Aritmética

Arith Arithmetic • Arithmetical • Arithmetician

Ariz Arizona (USA)

ARJ Admission Reject • Admission Rejected • Arso (Indonesia) (location identifier)

ARJD Admission Rejected

Ark Arkansas (USA)

ARK Arusha (Tanzania) (location identifier)

ARKod Annual Report of the Kodaikanal Observatory (periodical code)

ARL Acceptable Reliability Level • Access Rights List • Adaptive Representation through Learning • Adjusted Ring Length • Admiralty Research Laboratory • Aeronautical Research Laboratory • Aerospace Research Laboratory • Air Resources Laboratory • AKZO Research Laboratories • Applied Research Laboratory • Arctic Research Laboratory • Arly (Burkina Faso) (location identifier) • Army Research Laboratory (USA) • Association of Research Libraries • Astrophysics Research Laboratory • Atelier de Raisonnement Logique • Australian Radiation Laboratory • Authorized Retention Level • Average Remaining Lifetime • Average Run Length

ARLA Annual Review of Language Acquisition • Association of Residential Letting Agents

ARLC Area Resource Learning Center

ARLL Advanced Run Length Limited

ARLM Airborne Reconnaissance Low-Multifunction

ARLO Acoustics Research Letters Online

ARLTAB Army Research Laboratory Technical Assessment Board

Arm Armagh (UK-Northern Ireland) • Armenia • Armenian • Armorican

ArM Master of Architecture

ARM Aachen Robot Manipulator • Accelerated Reply Mail • Account Resource Management • Acorn RISC Machine • Active Rule Mechanism • Ada Requirements Methodology • Adjustable-Rate Mortgage • Advanced RISC Machine • Airborne Remote Mapping • Alarm • Algorithmic Remote Manipulation • All-Risk Management • Amateur Radio Monitor • Angular Resolution Measure • Anhysteretic Remanent Magnetization • Annotated Reference Manual • Application Reference Model • Application Response Measurement • Applied Research Management • Applied Research Methodology • Arcsecond Raster Map • Area Radiation Monitor • Armagh (Starlink internal code) • Armenia (ISO country code) • Arménie (code ISO de pays) • Armidale (Australia-NSW) (location identifier) • Asynchronous Response Mode • Atmospheric Radiation Measurement • Atmospheric Radiation Monitoring • Atomic-Resolution Microscope • Atomic-Resolution Microscopy • Automated Route Management • Automatic Resource Management • Automatic Resource Manager • Automatic Restart Management • Automatic Restart Manager • Availability - Reliability - Maintainability

ARMA Accumulator-Reservoir Manifold Assembly • Alarmstelle für Radioaktivität an der Meteorologischen Anstalt • Association of Records Managers and Administrators (USA) • Autoregressive and Moving Average • Autoregressive Moving Average (*)

ARMADE Automation and Robotics for Microgravity Applications Demonstrator

ArMAF Ark. Mat. Astr. Fys. (periodical code)

ArMAM Archives Mechanics Archiwum Mechaniki Stosowanej (periodical code)

ARMAR Airborne Rain Mapping Radar

ARMC Advanced Resource Management Course

ARMCAS Arctic Radiation, Microphysics, Clouds and Aerosols Study

ARMD ATAPI Removable Media Device

ARME Automated Réseau Measuring Equipment

Armen Armenia • Armenian

ArMeS Archiwum Mechaniki Stosowanej (periodical code)

ARMIP Accounting and Reporting Management Improvement Program

ARMM Analysis and Research of Methods for Management • Automated Retroactive Minimal Moderation

ARMMS Automated Reliability and Maintainability Measurement System • Automatically Reconfigurable Modular Multiprocessing System • Automatically Reconfigurable Modular Multiprocessor System

ArmOL Armagh Observatory Leaflet (periodical code)

A&RMOWG Astronomy & Relativity Mission Operations Working Group

ARMS Absolute Radiometric Measurements in Space • ADPE Resources Management System • Advanced Real-time Metric System • Advanced Receiver Model System • Advanced Retrieval Mail System • Advanced Robotics Manipulation System • Aerial Radiological Measurements and Survey • Aerial Radiological Measuring System • Application of Remote Manipulators in Space • Automated Records Management Study • Automated Records Management System • Automated Requirements Management System • Automated Resources Management System • Automated Retrieval Mail Service • Automated Retrieval Mail System • Automated Review Management System • Automatic Radiation Monitoring System • Automatic Receiving and Measuring System • Automatic Remote Manned System • Automation Resource Management System • Aviation Reconfigurable Manned Simulator

ARMSAT Atmospheric Radiation Measurement Satellite

ARMU Addressable Remote Multiplexer Unit

ARMZ Armazém

ARN Access Remote Node • Acide Ribonucléique • Action and Reply Notice • Additional Reference Number • Aeronautical Radio Navigation • Alteration Request Number • Astronomical Research Network • Atmospheric Radio Noise • ATS Route Network • Autorité de Régulation Nationale • Stockholm Arlanda (Sweden) (location identifier)

ARND Around

ARNEE ASEAN Region Network on Environmental Education

ARNEWS Acid Rain National Early Warning System

ARNFM Annual Review of Numerical Fluid Mechanics (periodical code)

Arngmt Arrangement

ARNICA Arcetri Near-Infrared Camera

ARNO Algorithm for Radio Network Optimization (project)

ARO Acquisition Reform Office • Acquisition Requirements Office • Advanced Research Objective • After Receipt of Order • Airport Reservation Office • Air Radio Officer • Air Request Net Operator • Algonquin Radio Observatory (Canada) • Apogee Robotic Observatory • Applied Research Objective • Arboletas (Colombia) (location identifier) • Arizona Radio Observatory (USA-AZ) • Army Research Office (USA) • Asian Regional Organization • Automatic Range Only • Autonomous Robotic Observatory • Auxiliary Readout

AROC Ariane Recoverable Orbital Carrier

AROD Airborne Radar Orbital Determination • Airborne Range and Orbit Determination • Airborne Ranging and Orbit Determination • Airborne Remotely-Operated Device • Airborne Remote Operated Device

ARODS Airborne Radar Orbital Determination System • Airborne Ranging and Orbit Determination System

AROF Ambiguity Resolution On the Fly

AROM Alterable Read-Only Memory

AROS African Regional Organization for Standardization • Automated Regional Operations System

AROU Aviation Repair and Overhaul Unit

ARP Acquisition Requirements Package • Adaptive Routing Protocol • Address Resolution Program • Address Resolution Protocol • Adjustable-Rate Preferred • Advanced Reentry Program • Advanced Research Project • Aerodrome Reference Point • Aerospace Reference Project • Airport Reference Point • Alternativa Racional a la Pseudociencias • Alternative Release Procedure • Antarctic Research Program • Antenna Radiation Pattern • Aragip (Papua New Guinea) (location identifier) • Arctic Research Plan • As-Run Procedure • Atmospheric Research Program • ATV Rendezvous Predevelopment • Automatic Reimbursement Program • Peso (Argentina) (ISO currency code)

ARPA Advanced Research Projects Agency (USA) (formerly DARPA) • Arctic Research and Policy Act (USA) • Automatic Radar Plotting Aid

ARPANet Advanced Research Projects Agency Network

ARPANSA Australian Radiation Protection and Nuclear Safety Agency (Australia)

ARPAS Automated Resource Planning and Analysis System

ARPC Automated Radar Prediction Capability

ARPCS Atmospheric Revitalization Pressure Control System

ARPD Advanced Research Program Directive

ARPDP Association of Rehabilitation Programs in Data Processing

ARPDS Auto-Regressive Power Density Spectrum

ARPEGE Action de Recherche à Petite Échelle et Grande Échelle

ARPEGES Astronomie Radio Planétaire, Extragalactique, Galactique et Stellaire (Observatoire de Paris-Meudon)

ARPES Angle-Resolved Photoelectric Spectroscopy • Angle-Resolved Photoelectron Spectroscopy • Angle-Resolved Photoemission Spectroscopy (*) • Angular-Resolved Photoemission Spectroscopy

ARPESH Accurate and Reliable Prototype Earth Sensor Head

ARPIC Aerospace Radioisotope Power Information Center

Arpk Airpark

ARPL Adjust Requested Privilege Level

ARpM Average Revenue per Message

ARPM Adgeziia Rasplavov i Paika Materialov (periodical code) • Average Revenue Per Message

ARPMIS Automated Resource and Plan Management Information System

ARPR Annual Reviews Preprints and Reprints • Automatic Radar Pattern Recognition

ARPS Advanced Radar Processing System • Advanced Radioisotope Power System • Advanced Regional Prediction System • Australasian Radiation Protection Society • Automated Resource Projection System

ArPSp Archiwum Procesow Spalania (periodical code)

Arpt Airport

ARPT Advanced Rocket Propulsion Technology

ARPU Average Revenue Per User

ARPV Advanced Remotely Piloted Vehicle

Arq Arquetipo • Arquitecto • Arquitectura

ARQ Annual Review Questionnaire • Arauquipa (Colombia) (location identifier) • Arquitecto • Authorized Requisition Quantity • Automated Response to Query • Automatic Repeat-Request • Automatic Request

Arqto Arquitecto

Arqueol Arqueología

Arr Arrange • Arranged • Arrangement • Arranger • Arrecife • Arrêté • Arrival • Arrive • Arrived • Arrivée • Arriver • Arrives • Arrondissement

ARR Accounting Rate of Return • Advanced Requirements Review • Advanced Restricted Report • Advanced Robotics Research • Advance Release Record • Aerospace Rapid Response • Aerospace Rescue and Recovery • Airborne Radio Relay • Allowance Requirement Register • Alto Rio - Senguerr (Argentina) (location identifier) • Andøya Rocket Range (Norway) • Anno Regni Reginae • Anno Regni Regis • Annual Requisition Review • Antenna Radiation Resistance • Antenna Rotation Rate • Anti-Repeat Relay • Archive Readiness Review • Arrangement • Asynchronous Reply Required • ATLO Readiness Review • Attributes for Representing Relationships • Automatic Repeat Request • Automatic Rerouting

ARRA Announced Retransmission Random Access

ARRC Aerospace Rescue and Recovery Center

ARRCC Analysis of Rapid and Recent Climate Change

ARRDATE Arrival Date

Arrdt Arrondissement

ARRE Autoread Relocation

ARRECAD Arrecadação

Arrest Arresting

ARRF Automatic Recording and Reduction Facility

ARRG Aerospace Rescue and Recovery Group

Arrgt Arrangement

ARRI Automation and Robotics Research Institute (USA-TX)

ARRL American Radio Relay League (USA)

ARRM Arruamento

ArRMA Archive of Rational Mechanics and Analysis (periodical code)

Arron Arrondissement

Arrond Arrondissement

ARRQ Automatic Repeat Request

ARRS Advanced Rescue and Recovery System • Aerospace Rescue and Recovery Service • Air Recovery and Rescue Service • American Roentgen-Ray Society (USA) • Attitude-Referenced Radiometer Study

ARRSTC Asian Regional Remote Sensing Training Centre

ARRT Absolute Reaction Rate Theory

ARRUNK Arrival Unknown

ArS Archives des Sciences (periodical code)

ARS Action Request System • Active Repeater Satellite • Acute Radiation Syndrome • Address Recognition System • Address Resolution Server • Advanced Reconnaissance Satellite • Advanced Reconnaissance System • Advanced Record System • Advanced Recovery System • Advanced Reentry System • Advanced Rescue System • Advanced Righting System • Advanced Rocket System • Aerial Reconnaissance and Security • Aerial Reconnaissance and Surveillance • Aeronautical Research Scientist • Aerospace Research Satellite • Agricultural Research Service • Airborne Radiometer Scanner • Air Rescue Service • Air Revitalization System • Alpha Ray Spectrometer • Alternate Route Selection • Aménagement des Rythmes Scolaires • American Radium Society (USA) • American Rescue Service (USA) • American Rocket Society (USA) (now AIAA) • Analog Recording System • Analog Ring Sampler • Angular Rate Sensor • Antenna Receiving Station • Aragarças (Brazil-GO) (location identifier) • Archive and Retrieval System • Area Scan • Artificial Reality System • Assembleo Regionale Siciliana (Italy) • Atmosphere Revitalization Section • Atmosphere Revitalization System • Atmospheric Revitalization System • Atmospheric Radiance Study • Attitude Reference System • Audio Response System • Automated Reconciliation System • Automated Relay System • Automated Response System • Automated Retrieval System • Automatic Recording Spectrometer • Automatic Route Selection • Autonomous Robotic System

ARSA Airport Radar Service Area • Automated Radioxenon Sample Analyzer

ARSAS Agricultural Remote Sensing Analyzing System

ARSC African Remote Sensing Council • Alberta Remote Sensing Centre • Analog Rotation Speed Control • Arctic Region Supercomputing Center (USA-AK) • Association des Revues Scientifiques et Culturelles (Belgium)

ARSDA Advanced Radiation Space Defense Application

ARSE Alpha Ray Spectrometric Equipment

arsec Inverse Secant

ARSEC Association of Roman Catholics for Space Exploration and Colonization (USA-MA)

ARSFSS Advanced Reduced Scale Fuel System Simulator

arsin Inverse Sine

ARSIP Accuracy, Reliability and Supportability Improvement Program

ARSL Atmospheric Rendezvous Space Logistics

ARSLOE Atlantic Remote Sensing Land/Ocean Experiment

ARSM Algebraic Reynolds Stress Model

ARSO African Regional Standards Organization

ARSOM Académie Royale des Sciences d'Outre-Mer (Belgium)

ARSP Aerospace Research Support Program • African Remote Sensing Programme • Analog Radar Signal Processor • Applied Remote Sensing Program

ARSR Air Route Surveillance Radar

Art Article • Artículo • Articulo • Artificial • Artikel • Artist

ART Abteilung Reprotechnik • Accelerated Ray Tracing • Adaptive Refinement Tree • Adaptive Resonance Theory • Address Resolution Table • Admissible Rank Test • Advanced Research and Technology • Advanced Research Technologies • Advanced Resolution Technology • Advanced Rotorcraft Technology • Airborne Radiation Thermometer • Alarm Response Team • Algebraic Reconstruction Technique • Allgemeine Relativitätstheorie • Alpha-Recoil Track • Analysis and Replay Tool • Animated Reconstruction of Telemetry • Another Ray Tracer • Applied RISC Technology • Architecture Review Team • Arithmetic Reasoning Test • Article • Assessment Recovery Team • Audible Ringing Tone • Automated Reasoning Tool • Automated Request Transmission • Automatic Radiotheodolite • Automatic Range Tracker • Automatic Range Tracking • Automatic Reporting Telephone • Autorité de Régulation des Télécommunications (France) • Avionics Research Testbed • Watertown (USA-NY) (location identifier)

ARTA Apple Real-Time Architecture • Ariane Research and Technology Accompaniment • Association pour la Recherche et la Technologie en Alsace (France)

ARTAS ATC Radar Tracker And Server

ARTB Advanced Radar Testbed

ARTC Air Route Traffic Control • Annual Regional Telecommunications Conference

ARTCC Air Route Traffic Control Centre

ARTDB Active, Real-Time, and Temporal Databases (workshop)

ARTE Ada Run Time Environment

ARTEMIS Advanced Relay and Technology Mission Satellite (*) • Advanced Relay Technology Mission • Advanced Research and Technology Mission • African Real-Time Environmental Monitoring Information System • African Real-Time Environmental Monitoring using Imaging Satellites • Antimatter Research Through the Earth Moon Ion Spectrometer

ARTEP Ariane Technology Experiment Platform

ARTES Advanced Research in Telecommunications Systems (ESA)

ARTEWG Ada Runtime Environment Working Group

ARTFL American Research on the Treasury of French Language

ARTI Advanced Range Technology Initiative (NASA) • Advanced Rotorcraft Technology Integration

ARTIC A Real-Time Interface Coprocessor

Artif Artificial • Artificiel

ARTIM Automated Reasoning Tool for Information Management

ARTIS Airborne Real-Time Imaging System • Airborne Real-Time Instrumentation System

ARTIST Astronomy-Related Teacher In-Service Training

ARTK Ada Real-Time Kernel

ARTL Ada Runtime Library

ARTO Advanced Radiation Technology Office

ARTOOS Advanced Radiation Technology Office Optical Support

ARTRAC Advanced Range Testing, Reporting and Control • Advanced Real-Time Range Control

ARTS Adapted Real-Time Strategy • Advanced Radar Terminal System • Advanced Radiofrequency Tracking System • Advanced Remote Tracking Station • Annual Research Task Summary • Aspects of Real-Time Systems (workshop) • Asynchronous Remote Takeover Server • Automated Radar Terminal System • Automated Radar Tracking System • Automated Remote Tracking System

• Automated Requirements Traceability System • Automated Resource Tracking System • Automated Restricted Tracking System • Automated Retrieval and Tracking System • Automatic Remote Tracking System • Autonomous Range and Target Simulator

ArtSa Artificial Satellites (periodical code)

ARTSE Ada Run-Time Support Environment

ARTSIP Adaptive Real-Time Strategies for Image Processing

ArTSp Archiwum Termodynamiki i Spalania (periodical code)

ARTSS Advanced Real-Time Simulation System

ARTT Aménagement de la Réduction du Temps de Travail • Asynchronous Remote Takeover Terminal

ARTTel Automatic Request Transmission by Telephone

ARTU Automatic Range Tracking Unit

ARU Academic Resource Unit • Access/Relay Unit • Address Recognition Unit • Altitude Reference Unit • Analog Remote Unit • Analog Response Unit • Araçatuba (Brazil-SP) (location identifier) • Astrophysical Research Unit • Attitude Reference Unit • Audio Response Unit

ARUNK Arrival Unknown

ARUPS Angle-Resolved Ultraviolet Photoemission Study

ARV Aeroballistic Reentry Vehicle • Aerospace Research Vehicle • American Revised Version • Arctic Research Vessel • Ascent and Reentry Vehicle • Minocqua (USA-WI) (location identifier)

ARVAD Armagh (Starlink internal code)

ARVISA Advanced Real-time Vision System and Architecture

ARVO Association for Research in Vision and Ophthalmology

ARW Advanced Research Workshop • Arad (Romania) (location identifier) • Associated Random Walk • Atmospheric Radio Wave • Attitude Reaction Wheel

ARWAT Automated Resources Workload Assessment Toolkit

ARX Asbury Park (USA-NJ) (location identifier) • Automatic Retransmission Exchange • Autoregressive with External input

ARXPS Angle-Resolved X-ray Photoelectron Spectroscopy

ARY Ararat (Australia) (location identifier)

ARZ Autorestricted Zone • N'Zeto (Angola) (location identifier)

AR5I/F Ariane 5 Interface

a/s Au Sujet de • Aux Soins de

as Aanstaande • Assamese (ISO language code) • As Stated

As Altostratus • Ampere Second • Arsen • Arsenic • Asia • Asian • Asiatic • Asociación • Astronomische Schemering • Asymmetrical

A/S Account Sales • Acquisition Sensor • Advanced Supplementary • After Sight • Airspeed • Aktieselskab • Aktjeselskab • Alongside • Analysis/Synthesis • Arm/Safe • Ascent Stage • Auxiliary Stage

A&S Accident & Sickness • Alignment & Stabilization

AS Absorbing Substrate • Academy of Science • Access Switch • Access System • Accident de Service • Accumulator Sampler • Acquisition Section • Acquisition Strategy • Action Spécifique • Active Service • Actual Start • Adapter Section • Adaptive Secondary (mirror) • Additional Source • Additional Stars • Address State • Add-Subtract • Administrative Support • Admittance Spectroscopy • Advanced Schottky • Advanced Subsidiary • Advanced System • Aerial Sensor • Aerial Surveillance • Aerospace • Aerospace Standard • Aerospace Study • Aft Shroud • Agent de Service • Air Sortie • Airspeed • Air Station • Air-to-Surface • Alignment and Stabilization • Aller Simple • All Sections • All Source • Almost Surely • Alternate Site • Altostratus • Aluminium Silicate • American Samoa (ISO country code) • American Standard • Ammeter Switch • Ammunition Ship • Analog Secure • Analog Signal • Analog State • Angle Sidéral • Anglo-Saxon • Angulo Sidéreo • Anno Salutis • Anno Salvatoris • Anodic Stripping • Answer Supervision • Antenna System • Anti-Scintillation • Antisolar • Antispoof • Antispoofing • Anti-Sun • Aperture Synthesis • Apogee Stage • Applicability Statement • Application System • Applied Science • Applied Statistics • Archive Server • Area Security • Area Served • Area Source • Artificial Satellite • Ascent Stage • Asia • Assistance Service • Assistance Sociale • Assistante Sociale • Assistant Secretary • Associate in Science • As Stated • Astroclimatic Station • Asynchronous System • Atmosphere and Space • At Sight • Attitude Set • Augmentation System • Australian Standard • Authentification Server • Authorization System • Automated Storage • Automatic Skin • Automatic Sprinkler • Autonomous System • Auxiliary Stage • Samoa (American_) (ISO country code) • Samoa Américaine (code ISO de pays)

ASA Abort Sensor Assembly • Academic Staff Assembly • Acetylsalicylic Acid • Acoustical Society of America • Adapter Service Area • Ada Software Alliance • Adjustable Shock Absorber • Advanced System Architecture • Advanced System Avionics • Aerospace States Association • Aerosurface Amplifier • Aerosurface Servoamplifier • Agency System Administrator • Agent Systems and Applications (symposium) • Air Starline AG • All Source Analysis • Alternate Structure Approach • American Sociological Association (USA) • American Standards Association (USA) (now ANSI) • American Statistical Association (USA) • American Student Association (USA) • Amplifier and Switch Assembly • Antarctic Support Associates • Antenna Subassembly • Apodized Square Aperture • Apprentissage Symbolique Automatique • Area Software Analysis • Assab (Ethiopia) (location identifier) • Association Sammielloise d'Astronomie (France) • Association Scientifique d'Astronomie (Algeria) • Association Sologne Astronomie (France) • Associazione Astrofili Salentina (Italy) • Astroclub Solaris Aarau (Switzerland) • Astronomical Society of Australia • Astronomical Society of the Atlantic (USA-FL) • Atomic Scientist Association (UK) • Atmospheric Spectroscopy Application • Atomic Sphere Approximation • Austrian Space Agency • Automatic Steering Antenna • Automatic Systems Analysis • Azimuth Servo Assembly

ASAA Asian Studies Association of Australia • Association Saranaise des Astronomes Amateurs (France)

ASAAS Asymmetric Stress Analysis of Asymmetric Solids

ASAC Active Satellite Attitude Control • Aerodynamic Surfaces Assembly and Checkout • All Source Analysis Center • ALMA Science Advisory Committee • Altostratus and Altocumulus • Antarctic Science Advisory Council • Application-Specific Architecture Compilation • Australian Scientific Advisory Committee

AsAcS Astronautica Acta Supplement (periodical code)

ASACS Advanced Secure Access Control System

ASAD Ames Spatial Auditory Display (NASA)

ASADA Automated Smoke/Aerosol Detection Algorithm

ASAE American Society for Aerospace Education (USA) • American Society of Agricultural Engineers (USA) • American Society of Association Executives (USA)

AsAer Astronautics and Aeronautics (periodical code)

ASAFI Asociación de Astronomos Aficionados de Colombia

ASAIHL Association of Southeast Asian Institutions of Higher Learning

ASAJ Acoustical Society of America Journal (periodical code)

ASAL Arid and Semi-Arid Land

ASAMC Advanced Software Acquisition Management Course

ASAMS Automatic Structure Analysis of Mass Spectra

ASANCA Advanced Study of Active Noise Control in Aircraft

ASAP Academic Strategic Alliances Program (USA) • Active Scattering Aerosol Spectrometer • Advanced Scientific Array Processor • Advanced Sensor Analysis Program • Advanced System Analysis Package • Aerospace Safety Advisory Panel • AFGL Statistical Analysis Programs • Airborne Science and Applications Program • Airport Scenario Analysis Package • Amsterdam Special Action Programme • Analog System Assembly Pack • Any Service - Any Port • Any Service to Any Port • Ariane Structure for Auxiliary Payload • As Soon As Possible • Australian Science Archive Project • Automated Shipboard Aerological Program • Automated Statistical Analysis Program • Automatic Self-Allocating Processor • Automatic Statistical Analysis Program • Automatic Switching And Processing

ASAPTS Automated Scheduling And Product Tracking System

ASAR Advanced Synthetic Aperture Radar

ASARS Advanced Synthetic Aperture Radar System • Airborne Search-And-Rescue System

ASAS Active Scattering Aerosol Spectrometer • Advanced Solid Axial Stage • Advanced Solid Axial System • Advanced Solid-state Array Sensor • Advanced Solid-state Array Spectrometer • Advanced Solid-state Array Spectroradiometer • Aerodynamic Stability Augmentation System • Airborne Separation Assurance System • All-Sky Astronomical Survey • All-Sky Automated Survey • All-Source Analysis System • Australian Science and Application Spacecraft • Aviation Safety Analysis System

ASASAC Asociación de Astrónomos Autodidactos de Colombia

ASASP Active Scattering Aerosol Spectrometer Probe

ASAT Anti-Satellite • Automatic Spares Analysis Technique • Avionics Situation Awareness Trainer

ASATS Anti-Satellite System

Asb Asbestos

AsB Anschlußbereich

ASB Adad Source Builder • Administration and Storage Building • Airlock Stowage Bag • Air Safety Board • Altitude Sensor Bypass • Analysis and Simulation Branch • Archenhold-Sternwarte Berlin (Germany) • Ashkhabad (Turkmenia) (location identifier) • Associated Standardisation Body • Associated Standardising Bodies • Asymmetrical Sideband • Audit Service Bureau • Australian Space Board • Automated Status Board

ASBC Advanced Solid-Block Correlator

ASBCA Armed Services Board of Contract Appeals

ASBH American Society for Bioethics and Humanities (USA)

Asbl Assemble

ASBL Association Sans But Lucratif

ASBMR American Society for Bone and Mineral Research (USA)

ASBR Academically-Separated Budgeted Research

ASBU Arab States Broadcasting Union

AsBVSt Anschlußbereichsvermittlungsstelle

asc ascii (file extension)

Asc Ascend • Ascenseur • Ascenso • Ascent

ASc Associate in Science

ASC Accelerometer Signal Conditioner • Accounting Service Center • Accredited Standards Committee • ACE Science Center (USA-CA) • Adams State College (USA-CO) • Additional Specialty Code • Advanced Scientific Computer • Advanced Simulation Center • Advanced Structural Component • Advanced System Concept • Aerodynamic Surface Control • Aeronautical Systems Command • Aerospace Planning Chart • Aerospace Static Converter • Aerosurface Control • Agence Spatiale Canadienne (Canada) • Agnes Scott College (USA-GA) • Airlock Support Component • Albany State College • Allocation Spécifique de Conversion • Allowance Source Code • All-Sky Camera • American Satellite Co. • American Security Council (USA) • American Society of Cartographers (USA) • American Society of Cinematographers (USA) • American Society of Cybernetics (USA) • American Standards Committee (USA) • Analog Signal Converter • Analog Signal Correlator • Analog Strip Chart • Apache Survival Coalition (USA) • Applied Science Corporation • Applied Superconductivity Conference • Area Safety Coordinator • Area Surveillance Center • Arizona State College (USA-AZ) • Arlington State College • Array Stowage Container • Ascención (Bolivia) (location identifier) • ASCII • ASEAN Standing Committee • Assembly Shortage Control • Association of Systematics Collections • Association Sportive et Culturelle • Associative Structure Computer • Astronautical Society of Canada • Astronautics Support Center • Astronomy Survey Committee • Astro Space Center (Russia) • Asynchronous Communication Procedure • Aughey Spark Chamber • Australian Seismological Centre • Australian Space Council • Authorized Support Center • AUTODIN Switching Center • Automated Software Engineering • Automatic Scan Counter • Automatic Selectivity Control • Automatic Sensitivity Control • Automatic Slope Control • Automatic Spacecraft Controller • Automatic Switching Center • Automatic Synchronized Control • Automatic System Control • Automatic System Controller • Automation Security Committee • Autonomous Spacecraft Controller • Auxiliary Storage Cassette • Auxiliary Switch Closed • AXAF Science Center (USA-MA)

AsCA Asian Crystallographic Association

ASCA Advanced Satellite for Cosmology and Astrophysics (Japan) (formerly Astro-D) • Asian Crystallographic Association • Association for Science Cooperation in Asia

ASCADS Aft Station Command And Data System

ASCAN Astronaut Candidate

ASCAP American Society of Composers, Authors and Publishers (USA)

ASCAS Automated Security Clearance Approval System • Automated Security Clearance Authentication System

ASCAT Advanced Scatterometer • Analog Self-Checking Automatic Tester

ASCATT Advanced Scatterometer

ASCB Address Space Control Block • American Society of Cell Biology (USA) • Avionics Standard Communication Bus

ASCC Academia Sinica Computing Center (China-ROC) • Aeronautical Satellite Communications Center (USA) • Air Standardization Coordination Committee • All-Sky Compiled Catalogue • Asia-Pacific Satellite Communication Council • Astronomical Society of the Central Coast (USA-CA) • Automatic Sequence-Controlled Calculator

ASCD ACTS Systems Concept Development • All-Source Correlated Database • American Society of Computer Dealers (USA) • Association for Supervision and Curriculum Development

ASCDB All-Source Correlated Database

ASCDS AXAF Science Center Data System

ASCE Abrupt Space Charge Edge • Advanced Solar Coronal Explorer • Advanced Spectroscopic and Coronagraphic Explorer • Airlock Signal Conditioning Electronics • American Society of Civil Engineers (USA) • Association Service Control Element

ASCEND After-School Center for Exploration and New Discovery • Agenda of Science for Environment and Development

ASCHOT Astrophysical Schmidt Orbiting Telescope

ASCENT Assembly System for Central Processor

ASCF Ames Satellite Communications Facility

ASCH American Society for Circumpolar Health (USA)

AsChE American Society of Chemical Engineers (USA)

ASCI Advanced School for Computing and Imaging (Netherlands) • Advanced Simulation and Computing Initiative

ASCID Altered State of Consciousness Induction Device

ASCII American Standard Code for Information Interchange

ASCI Accelerated Strategic Computing Initiative (USA) • Advanced Speech Cell Item • Association Canadienne des Sciences de l'Information (Canada)

ASCL Analog Scan Control Logic • Astrophysics Source Code Library

ASCLA Association of Specialized and Cooperative Library Agencies

ASCM Advanced Spaceborne Computer Module

ASCMN Area Signal Center Main Code

ASCN Ascension

ASCNET Australian Computer Science Network

ASCO Aboard Sensing Control Unit • Arab satellite Communications Organization • Asian Science Communicators' Organization • Automatic Sustainer Cutoff

ASCOM Astronomy Common Object Model (initiative)

ASCON Associated Contractor • Automated Switched Communications Network

ASCORE Automatic Shipboard Checkout and Readiness Equipment

ASCOT Accelerated Simulation of Charge particle Orbits in a Tokamak • Airborne Stabilized Cooled Optical Telescope • Atmospheric Studies in Complex Terrain • Automatic Scanning-system for Corpus-Oriented Tasks

ASCP American Society of Consulting Planners (USA) • Attitude Set Control Panel • Aviation System Capacity Plan

ASCPL Astro Scientific Centre Pte Ltd (Singapore)

ASCR Advanced Scientific Computing Research • Analog Strip Chart Recorder • Asymmetrical Semiconductor Controlled Rectifier

ASCS Advanced Stirling Conversion System • Aerospace Sector Certification Scheme • Aft Shroud Cooling System • Air Support Control Section • Area Surveillance Control System • Association des Sociétés de Commercialisation de Services (France) • Atmosphere Storage and Control System • Atmospheric Storage and Control Section • Attitude and Spin Control System • Attitude Stabilization Control System • Automated Storage Control System • Automated Scan Control System • Automatic Stabilization and Control System

ASCSP Australia Space Centre for Signal Processing

ASCT Association Sportive et Culturelle Toussaintaise (France) • Auxiliary Satellite Control Terminal

ASCU Association of State Colleges and Universities (USA) (now AASCU) • Automatic Scanning Control Unit

AScW Association of Scientific Workers (UK)

ASD Academic Support Division • Access Security Device • Accreting Single Dwarf (star) • Accurate Space Derivative • Additional Site Data • Adjustable Speed Drive • Advanced Systems Development • Advanced Systems Directorate • Aeronautical Systems Division • Aerospace Services Division • After-Sales Department • Aircraft Situation Display • Air Situation Display • Air Support Division • Allentown School District (USA-PA) • Alternate Source Development • Analysis and Simulation Division • Analysis of Spectral Data • Andros Town (Bahamas) (location identifier) • Application Support Division • Application Systems Developer • Arbre Senonique de Décision • Archive Software Developer • Arc Supérieur Droit • Assign Software Developer • Assistant Secretary of Defense • Association for Software Design • Association Suisse de Documentation (Switzerland) • Astronomy Support Department • Astro Socio Dynamics • Astrostation Demmin (Germany) • Atmospheric Sciences Division • Atomic Spectra Database • Automated Seamless Dispatch • Automated Software Distribution • Automated Structural Design • Automated System Demonstration • Automatic Synchronized Discriminator

ASDA Accelerate-Stop Distance Available • Atomic and Space Development Authority • Australian Standard Data Archive

ASDAC Advanced Software Development And Commercialization

ASDAR Aircraft-to-Satellite Data Relay

AsDB Asian Development Bank

ASDC Aeronomy and Space Data Center • Alternate System Design Concept • Atmospheric Science Data Center • Australian Space Development Conference • Automated Source Data Collection

ASDCS Automated Source Data Collection System

a/s de Aux Soins de

ASDE Aerosurface Driver Electronics • Airport Surface Detection Equipment

ASDEC Automatic Selection of Digital Electronic Computers

ASDEM Automatic Sample Detection and Efficiency Monitoring

ASDEX Axially Symmetric Divertor Experiment

ASDG Automated Systems Development Group

ASDI Analog Simple Data Interface • Arctic Sustainable Development Initiative • Australian Space Data Infrastructure • Automatic Selective Dissemination of Information

ASDIS Automated Sample Data Instrumentation System • Automated Sampled Data Information System

ASDM Aerosurface Driver/Monitor • Apollo-Soyuz Docking Module • Automated System Design Methodology

ASDMS Autonomous Spacecraft Data Management System

ASDO Accredited Standards Developing Organization

ASDP Abbreviated System Decision Paper • Accelerated Software Development Process

ASDPSIM Advanced System Data Processing Simulation

ASDRS Automatic Spectrum Display and Recognition System

ASDS Air Situation Display System • Astronomical Software and Documentation Service (*) • Astronomical Software Directory Service (*) • Astronomical Software Documentation Service

ASDSP AppleTalk Secure Data Stream Protocol • Application-Specific Digital Signal Processor

ASDSRS Automatic Spectrum Display and Signal Recognition System

ASDT Accumulated Standardized Data Tape

ASDTIC Analog Signal to Discrete Time Interval Converter

ASDU Aurora Software Development Unit

ASDWG Astro Socio Dynamics Working Group

AS&E American Science & Engineering

ASE Accredited Systems Engineer • Active Seismic Experiment • Administrative Support Equipment • Advanced Sensor Exploitation • Advanced Software Engineering • Advanced Space Engine • Advanced Systems Engineering • Aerospace Support Equipment • Agence Spatiale Européenne • Air and Space Europe (periodical code) • Airborne Search Equipment • Airborne Support Environment • Airborne Support Equipment • Air-Sea Experiment • Airspace Support Equipment • Allowable Steering Error • American Science and Engineering • American Society of Educators (USA) • American Society of Engineers

(USA) • American Stock Exchange • Amplified Spontaneous Emission • Annular Solar Eclipse • Application Service Element • Application Software Entity • Application Specific Environment • Applied Service Element • Aspen (USA-CO) (location identifier) • Associate of the Society of Engineers • Association for Science Education • Association of Space Explorers • Association Socio-Éducative • Astronomical Society of Edinburgh (UK-Scotland) • Atmospheric Structure Experiment • Attitude Sensor Electronics • Audio Support Equipment • Automated Software Engineering • Automated Stabilization Equipment • Automatic Stabilization Equipment • Automatic Support Equipment • Automotive Stirling Engine • Autonomous Sciencecraft Experiment • Avionics Simulated Environment • Snowmass (USA-CO) (location identifier)

ASEA American Solar Energy Association (USA) • Association of South East Asia

ASEAMS Association of Southeast Asian Marine Scientists

ASEAN Association of Southeast Asian Nations

ASEAP Association of Southeast Asian Publishers

ASEAUS Association of South East Asian University Students

ASEB Aeronautics and Space Engineering Board

ASEC Applied Solar Energy Corporation

ASECNA Agence pour la Sécurité de la Navigation Aérienne en Afrique et à Madagascar

ASED Application Specific Environment Description • Avionics System Engineering Division

ASEDEM Asociación Española de Empresas de Multimedia (Spain)

ASEE American Society for Engineering Education (USA)

ASEER Aerospace and Environmental Education Research

ASEF Advanced Sensor Evaluation Facility • Aircrew System Evaluation Facility

ASEL Aircraft Single Engine Land • Automatic Space Ecological Laboratory

ASEM Advanced Space Environment Monitor • Application Specific Electronic Module • Assembly of Station by Extravehicular Methods

ASEP Advanced Science Education Program • Advanced Scientific-computer Enhancement Project • Advanced Skills Education Program • Automated Signal Excess Prediction • Automatic Signal Execution and Processor

ASEPS Astronomical Studies of Extrasolar Planetary Systems • Automated Signal Excess Prediction System

ASER Amplification by Stimulated Emission of Radiation

ASERL Association of Southeastern Research Libraries (USA)

ASES American Solar Energy Society (USA) • Aviation Safety Exchange System

ASESS Aerospace Environment Simulation System

ASET Advanced Sensor Evaluation and Test • Astronomical Society of East Texas (USA-TX) • Australian Society for Educational Technology • Author System for Education and Training • Automated Security Enhancement Tool

ASEU Advanced Sensor Emulator Unit

ASF Accelerometer Scale Factor • Actively Star-Forming • Active Streaming Format • Additional Safety Factor • Additional Secondary Factor • Additional Selection Factor • Administrative Support Facility • Advanced Simulation Facility • Advanced Simulation Framework • Area Sampling Frame • Aerospace Security Force • Air Safety Foundation (USA) • Alaska SAR Facility • Amperes per Square Foot • Amplitude Spread Function • Angle Selection Function • Angular Selection Function • Applicant Supply File • Area Sampling Frame • Arithmetic Statement Function • Astounding Science Fiction • Astrakhan (Russia) (location identifier) • Astronomes Sans Frontières • Astronomical Society of Frankston (Australia-VIC) • Atmospheric Science Facility • Atmospheric Stabilization Framework • Atomic Scattering Factor • Atomic System Function • Automatic Store and Forward • Auxiliary Service Facility • Auxiliary Supporting Feature • Aviation Support Facility

ASFA American Science Film Association (USA)

ASFAS Academia Scientiarum Fennica Annales Series Physica (periodical code)

ASFE Accelerometer Scale Factor Error

ASFG Actively Star-Forming Galaxy • Atmospheric Sound-Focusing Gain

ASFIP Accelerometer Scale Factor Input Panel

ASFIR Active Swept Frequency Interferometer Radar

ASFIS Aquatic Sciences and Fisheries Information System

ASFM Algebraic Stress/Flux Model • Astronomical Society of Fort McMurray (Canada-AB)

ASFS Adelaide Space Frontier Society (Australia-SA)

ASFSE American-Swiss Foundation for Scientific Exchange

ASFT Abbreviated System Functional Test

ASFTS Auxiliary System Function Test Stand

Asg Assign

ASG Access Service Group • Acquisition and Show Guide • Acting Secretary-General • ADAM Support Group • Ada Structure Graph • Advanced Systems Group • Aeronautical Standards Group • Aggregate Specialization Graph • Antenna Steering Group • Approaching Side Gap • Area Support Group • Ashburton (New Zealand) (location identifier) • Assessment Subgroup • Assistant Secretary-General • Astronomical Society of Georgia • Astronomical Society of Glasgow (UK-Scotland) • Außensteuergesetz • Australian Spaceport Group • Auxiliary Status Generator • Avionics Systems Group

ASGA Advanced Solar GaAs Array • Association des Services Géologiques Africains • Australian Standard Geographical Classification

ASGB Aeronautical Society of Great Britain

Asgd Assigned

ASGH Astronomical Society of Greater Hartford (USA-CT)

ASGI Astronomical Science Group of Ireland (Northern Ireland)

ASGLS Advanced Space-Ground Link System

Asgmt Assignment

Asgn Assign

ASGPD Attitude Set and Gimbal Position Display

ASGRG Australasian Society for General Relativity and Gravitation (Australia)

ASGSB American Society for Gravitational and Space Biology (USA)

ASGTE Action Spécifique "Grands Télescopes Étrangers" (France)

A&SH Astronomie & Sciences Humaines

ASH Action on Smoking and Health (USA) • Alternate Support Headquarters • Anelastic Spherical Harmonic (code) • Astronomical Society of Haringey (UK) • Astronomical Society of Harrisburg (USA-PA) • Astronomical Society of the Hunter (Australia) • Avion Spatial HERMES • Nashua Boire (USA-NH) (location identifier)

ASHE A Simple Hypertext Editor

ASHOE Airborne Southern Hemisphere Ozone Experiment

Ashp Airship

ASHPC Advanced Simulation and High-Performance Computing

ASHRA Action Spécifique Haute Résolution Angulaire (France) • ASHRAE Journal (periodical code)

ASHRAE American Society of Heating, Refrigerating and Air-conditioning Engineers (USA)

AS&I Assembly, Surveillance & Inspection

ASI Adapter Support Interface • Additional Skill Identifier • Advanced Science Instrument • Advanced Scientific Instrument • Advanced Studies Institute • Advanced Study Institute • Aerospace Static Inverter • Agenzia Spaziale Italiana (Italy) • Airspeed Indicator • Ajustement Structurel Intensif • Alarm Status Indicator • All-Sky Imager • All-Source Intelligence • Alphabetical Subject Indexing • Alternate Space Inversion • Amended Shipping Instructions • American Safety Institute (USA) • American Society of Indexers (USA) • Analytic Subject Index • Appui Social Individualisé • Artemis Society International (USA-CA) • Articulated Subject Index • Asian Statistical Institute • Assembly, Surveillance and Inspection • Assign Switch Initialization • Association de Solidarité Internationale • Astronomical Society of India • Asynchronous/Synchronous Interface • Atmospheric Structure Instrument • Atmospheric Structure Investigation • Augmented Spark Igniter (*) • Augmented Spark Indicator • Augmented System Igniter • Augmented System Ignition • Australian Science Index • Aviation Safety Inspector • Axial Scientific Instrument • Axial Shape Index • Azimuth Speed Indicator • Georgetown Wideawake (Saint Helena) (location identifier)

ASIA Action Spécifique d'Initiatives Académiques • Advanced Simulation Interoperable Architecture

ASIAA Academia Sinica's Institute of Astronomy and Astrophysics (China-ROC)

ASIAC Australian Surveying Industry Advisory Committee

ASIAS Astronomy Science Information and Abstract Service

ASIB NASA Advanced Studies Institute - Series B (periodical code)

ASIC Application-Specific Integrated Circuit (*) • Application-Specific Integrated Circuitry • Area Security Information Center • Australian Standard Industry Classification • NASA Advanced Studies Institute - Series C (periodical code)

ASICC Astro Space Informational Computing Centre (MAO) • Australian Space Industry Chamber of Commerce

ASI&CS Acoustic Senor Integration & Control System

ASICS Acoustic Senor Integration and Control System

ASID Access, Searching and Indexing Directories • Archival Storage Identifier

ASIDIC Association for Scientific Information and Dissemination Centres • Association of Scientific Information Dissemination Centres

ASIDS Abbreviated Status Information Display System • Australian Space Industry Development Strategy

ASIES Academia Sinica's Institute of Earth Sciences (China-ROC)

ASII Air-Sea-Ice Interaction • American Science Information Institute (USA)

ASIL American Society of International Law (USA) • Application-Specific Integrated Lens

ASIM Automated Security Incident Measurement

ASIMS Automated Security Incident Measurement System

ASIOE Associated Support Item Of Equipment

ASIP Advanced System Improvement Plan • Aircraft Structural Integrity Program • All-Sky Imaging Photometer • All-Source Imagery Processor

ASIPI Asociación Interamericana de la Propiedad Industrial

ASIR Avionics System Integration Rig

ASIRT Association for Safe International Road Travel

ASIS Abort Sensing and Implementation System • Ada Semantic Interface Specification • Airborne Spectroradiometric Imagery System • American Society for Industrial Security (USA) • American Society for Information Science (USA) • Application Software Installation Server • Automated Sea Ice Segmentation

ASIST Advanced Scientific Instruments Symbolic Translator • American Society for Information Science and Technology (USA)

ASIT Adaptable Surface Interface Terminal • Advanced Security and Identification Technology

ASIV Automatisiertes System für Informationsverarbeitung

ASIZ Advanced Sea Ice Zone

ASJ Amami O-shima (Japan) (location identifier) • Astronomical Society of Japan

ASJC American Society of Journalists and Composers (USA)

ASK Acquisition of Strategic Knowledge • Actively Shared Knowledge • American School of Kuwait • Amplitude Shift Key • Amplitude Shift Keying • Analog Select Keyboard • Anomalous State of Knowledge • Anschlußleitung • Assisted Search for Knowledge • Astronaut Survival Kit • Available Seat Kilometer • Yamoussoukro (Ivory Coast) (location identifier)

ASKA Automatic System for Kinematic Analysis

ASKC Astronomical Society of Kansas City (USA-MO)

ASKE Association for Skeptical Enquiry (UK)

ASKI All-Sky Imager

ASKS Automatic Station Keeping System

ASL Above Sea Level • Adaptive Speed Levelling • Adjusted Stock Level • Advanced Schottky Logic • Albuquerque Seismological Laboratory (USA-NM) • Allowable Supply List • Alternate Storage Location • American Sign Language • Applied Science Laboratory • Approved Source List • Arbeitskreis Sternwarte Lübeck (Germany) • Arithmetic Shift Left • Assault • Assessment of Student Learning • Assigned Stockage Level • Assigned Stockage List • Assign Switch Locator • Association for Symbolic Logic • Astigmatic Spectral Line • Astronomical Society of London (UK) (now RAS) • Astronomisches Sommerlager • Astro Space Locator (project) • Atmospheric Sciences Laboratory • Atmospheric Surface Layer • Authorized Stockage Level • Authorized Supply Level • Automated Soft Lander • Automatic Systems Laboratories Ltd. (UK) • Average Sortie Length • Average Staffing Level • Azimuth Steering Line • Marshall - Harrison County (USA-TX) (location identifier)

ASLA Application Service Level Agreement

ASLC Adaptive Sidelobe Canceler • Astronomical Society of Las Cruces (USA-NM)

ASLE Adjusted Single Loss Expectancy • Association for the Study of Literature and Environment

ASLEEP Automatic Scanning Low-Energy Electron Probe

ASLEP Apollo Surface Lunar Experiments Package

ASLI Astronomical Society of Long Island (USA-NY)

ASLK Algemene Spaar- en Lijfrentekas

ASLLR Accademia di Scienze e Lettere Istituto Lombardo (periodical code)

ASLM Apple Shared Library Manager

ASLO American Society of Limnology and Oceanography (USA) • Australian Scientific Liaison Office

Aslt Assault

ASLT Advanced Solid Logic Technology

ASLU Antenna Select Logic Unit

ASLV Advanced Small Launch Vehicle • Augmented Satellite Launch Vehicle

Asm Assemble

ASM Abstract State Machine • Accelerated Surface Mail • Acquisition Security Manager • Accretion Stream Mapping • Actuator Signal Management • Administrative Support Manual • Administrator of Session Monitors • Advanced Servomanipulator • Aerospace Structural Material

• Airspace Management • Agrégats Stables Minimums • Algebraic Stress Model • Algemene Salarismaatregelen • Algorithmic State Machine • All-Sky Monitor • American Samoa (ISO country code) • American Society for Metals (USA) • Angular Second Moment • Angular Second Momentum • Antenna Switching Matrix • Aperture Synthesis Map • Application of Software Management • Application Specific Module • Applied Simulation and Modeling • Applied Software Measurement • Appointment and Scheduling Module • Approximate String Matching • Asmara (Ethiopia) (location identifier) • Assembler • Assistant Sales Manager • Assistant Stage Manager • Assistant Station Master • Association for Systems Management (USA) • Association Set Manager • Astrometric Sky Mapper • Astronomical Seeing Monitor • Astronomical Site Monitor • Asynchronous State Machine • Attached Sync Maker • Audiovisual Multimedia Services • Austrian Society for Aerospace Medicine (Austria) • Automated Scheduling Message • Automated Scheduling Module • Automated Storage Management • Automatic Sorting Machine • Automatic Switch Monitor • Auxiliary Storage Manager • Auxiliary Storage Module • Available Seat Mile • Azimuth Scan Mirror • Samoa (American_) (ISO country code) • Samoa Américaine (code ISO de pays)

ASMA Aerospace Medical Association • Antarctic Specially Managed Area • Automatic Segment Matching Algorithm

ASMC Alternate System Monitoring Center • AUTODIN Station Maintenance Console

ASMD Atmospheric Sciences Modeling Division

ASME American Society of Magazine Editors (USA) • American Society of Mechanical Engineers (USA)

ASMES Advanced Synchronous Meteorological Satellite

ASMET African Satellite Meteorology Training

ASMI Aircraft Surface Movement Indicator

ASMIEG Advanced Structural Materials Information Exchange Group

ASMO Arab Standards and Metrology Organization

ASMP American Society of Magazine Photographers (USA) • American Society of Media Photographers (USA) • Asymmetric Multiprocessing

ASMPC Automated Staff Message Processing Center

ASMS Advanced Synchronous Meteorological Satellite • American Society for Mass Spectrometry (USA) • Atmosphere Sensing and Maintenance System • Automated Systems Management System

ASMSS Airborne Stereo Multispectral Sensor

ASMU Automatically Stabilized Maneuvering Unit

ASMW Amt für Standardisierung Meßwesen und Warenprüfung (Germany)

Asn Asparagine • Assign

ASN Abstract Syntactical Notation • Abstract Syntax Notation • Access Stack Node • Accountable Station Number • Acknowledged Sequence Number • Advanced Shipping Notice • Advance Ship Notice • AeroSpace Network • Allotment Serial Number • AMOS Star Number • Army Service Number • Assigned Serial Number • Association • Association Suisse de Normalisation (Switzerland) • Astronomical Society of Nevada (USA-NV) • Autonomous System Number • Autorité de Sûreté Nucléaire (France) • Average Sample Number • Talladega (USA-AL) (location identifier)

ASNAP Automatic Steerable Null Antenna Processor

Asnau Astronautik (periodical code)

ASNAV Association Nationale pour l'Amélioration de la Vue (France)

ASNE American Society of Naval Engineers (USA) • American Society of Newspaper Editors (USA)

ASNH Astronomical Society of New Haven (USA-CT)

ASNNE Astronomical Society of Northern New England (USA-ME)

ASNORA Association Normande d'Astronomie (France)

ASNSW Astronomical Society of New South Wales (Australia-NSW)

ASNT American Society of Non-destructive Testing (USA) • Astronomical Society of North Texas (USA-TX)

ASNY Astronomical Society of New York (USA-NY)

ASO Administrative Service Office • Administrative Service Only • Advanced Solar Observatory • Ajustement Structurel Ordinaire • Algemeen Secundair Onderwijs • Amateur Satellite Observers (USA-AR) • Application Service Object • Area Safety Officer • Area Supply Officer • Asosa (Ethiopia) (location identifier) • Astronomía Sigma Octante (Bolivia) • Astronomical Search for Origins • August - September - October • Australian Space Office (now ASPU) • Australian Survey Office • Auxiliary Switch Open • Aviation Safety Office • Aviation Safety Officer

Asoc Asociación

ASOC Air Sovereignty Operation Center • Alternate Systems Operation Control • Antarctic and Southern Ocean Coalition

ASOCON Asian Soil Conservation

ASOD Astronomical Society of the Desert (USA-CA)

ASOP Automatic Scheduling and Operating Program

ASOPS Astronomical Search for Origins and Planetary Systems

ASOS Automated Seismological Observation System • Automated Structure Optimization System • Automated Surface Observing System • Automatic Storm Observation Service • Automatic Surface Observing System • Auto Surface Observing System

ASOSV Amateur Satellite Observers of Southeast Virginia (USA-VA)

ASOVAC Asociación Venezolana para el Avance de la Ciencia (Venezuela)

asp Active Server Page (file extension)

Asp Aspartic Acid

ASP Abstract Service Primitive • Abstract Syntax Processor • Accepté Sous Protêt • Acquisition Strategy Panel • Acquisition Strategy Plan • Acquisition Strategy Process • Acquisition Streamlining Plan • Acquisition Support Program • Acquisition System Protection • Active Server Page • Active Splitting Point • Activity Scheduling Processor • Activity Scheduling Program • Adjunct Service Point • Advanced Signal Processor • Advanced Simulation Program • Advanced Stokes Polarimeter • Advanced Study Program • Aerospace Plane • Aerosurface Position • Aggregated Switch Procurement • Airborne Science Program • Aircraft Servicing Platform • Alarm Signal Processor • Alberta Star Party (Canada-AB) • Alice Springs (Australia-NT) (location identifier) • All-Service Processor • All-Source Production • American Selling Price • American Society of Photogrammetry (USA) • Analog Signal Processor • Annual Service Practice • Anti-Solar Point • AppleTalk Session Protocol • Applet Security Profile • Application Service Provider • Application Support Professional • Appui Spécifique Personnalisé • Area Support Programmer • Arizona Search for Planets • Arrival Sequencing Program • Aspirante • Assistance Scolaire Personnalisée • Association of Shareware Professionals (USA) • Association-Storing Processor • Associative Structures Package • Astronautics Standard Practice • Astronomical Society of the Pacific (USA-CA) • Astronomy Spacelab Payload • Atmosphere Sounding Projectile • Atmospheric Sensor Package • ATM Service Provider • Attached Support Processor • Attitude Sensor Package • Attitude Sensor Payload (*) • Augmented Support Period • Augmented Support Plan • Auroral Scanning Photometer • Authorized Service Provider • Automated

Schedule Procedure • Automated Sneak Program • Automated Spooling Priority • Automatic Sample Processor • Automatic Schedule Procedure • Automatic Servo Plotter • Automatic Switching Panel • Automatic Synthesis Program • Auto Sequence Program • Auto Sequence Programming • Auxiliary Spacecraft Power

ASPA Antarctic Specially Protected Area • Association of South Pacific Airlines • Association pour la Surveillance de la Pollution Atmosphérique

ASPAC Asian and Pacific Council

ASPB Astronomical Society of the Palm Beaches (USA-FL)

ASPC Accepté Sous Protêt pour Compte • Asia Pacific Space Centre (Christmas Island) • Astronomical Society of the Pacific Conference Series (periodical code)

ASPCA American Society for Prevention of Cruelty to Animals (USA)

ASPCS Astronomical Society of the Pacific Conference Series

ASPE Advances in Solar Physics Euroconference • American Society for Precision Engineering (USA)

ASPEI Association of South Pacific Environmental Institutions

ASPEN Asian Physics Education Network • Automatic System for Performance Evaluation of the Network

ASPERA Analyzer of Space Plasma and Energetic Atoms • Automatic Solar-system Particle Experiment with Rotating Analyzer • Automatic Space Plasma Experiment with a Rotating Analyser

ASPERT Analyse Stratégique, Prospective et Évaluation de la Recherche et de la Technologie (France)

Asph Asphalt

ASPH Adaptive Smooth Particle Hydrodynamics

ASPI Advanced SCSI Programming Interface • Advanced SCSI Protocol Interface • Aerosurface Position Indicator • Australian Space Policy Institute

ASPIC Apparatus for Surface Physics and Interfaces at CERN

ASPIN Arizona State Public Information Network

ASPIP Arab Society for the Protection of Industrial Property

ASPJ Airborne Self-Protection Jammer

ASPK Accompanying Spare Parts Kit

ASPL Advanced Space Propulsion Laboratory (JSC)

ASPLD Application-Specific Programmable-Logic Device

ASPLOS Architectural Support for Programming Languages and Operating Systems (conference)

ASPN Algorithm for Synthesis of Polynomial Networks

ASPO Acquisition System Protection Office • Acquisition System Protection Officer • American Society of Planning Officials (USA) • ATMOS Science Program Office • Automated Systems Program Office

ASPOC Active Satellite Potential Control • Active Spacecraft Potential Control (*)

ASPOL A Simulation Process Oriented Language

ASPP Acquisition Strategy Planning Process • Acquisition Systems Protection Plan • Acquisition Systems Protection Program • Advanced Satellite Products Project • Alloy-Steel Protective Plating • American Society of Picture Professionals (USA) • American Society of Plant Physiologists (USA) • Antenna Solar Panel Positioner • Atmospheric and Space Plasma Physics

ASPRO Academic Staff Public Representative Organization

ASPRS American Society of Photogrammetry and Remote Sensing (USA)

ASPRT Adjusted Sequential Probability Ratio Test

ASPRv Astrophysics and Space Physics Reviews (periodical code)

ASPS Action Spécifique de Physique Stellaire • Advanced Signal Processing System • Annular Suspension and Pointing System

ASpScROW Astronomy, Space Sciences and Related Organizations of the World (directory) (now StarGuides)

Aspt Aspirant

ASPT Advanced Satellite Products Team (NOAA)

ASPU Academic Services Planning Unit • Australia Space Policy Unit (formerly ASO)

ASPWG Acquisition Systems Protection Working Group

ASQ Administrative Science Quarterly (periodical code) • American Society for Quality (USA) • Austin (USA-NV) (location identifier) • Automated Software Quality

ASQC American Society for Quality Control (USA)

ASQP Airline Service Quality Performance

ASR Acceptance Summary Report • Access Service Request • Accumulators Shift Right • Acquisition Strategy Report • Activation Status Register • Active Surge Region • ADA Software Repository • Address Space Register • Advanced Special Receiver • Advanced Systems Requirement • Advances in Space Research • Airport Surveillance Radar • Air-Sea Rescue • Alphabetisches Sachregister • Alternate Supply Route • Alternative System Review • Altimeter Setting Region • American Sociological Review (periodical code) • AMOS Spectral Radiometer • Analog Shift Register • Annual Status Report • Annual Summary Report • Approved System Requirement • Arithmetic Shift Right • Array Switching Regulator • Astronomy Space and Radio • Astrophysics and Space Research • Atmospheric Sounding Radiometer • Atmospheric Sound Refraction • Automatic Send and Receive • Automatic Send-Receive • Automatic Speech Recognition • Automatic Speech Recognizer • Automatic Step Regulator • Automatic Storage and Retrieval • Auxiliary Sender/Receiver • Auxiliary Status Reply • Available Supply Rate • Avionics System Review • Kayseri (Turkey) (location identifier)

AS&RA Astrophysics Software & Research Aids

ASRA ADP Systems Resources Analysis • Advanced Systems Research Aircraft • Applied Science and Research Application • Automatic Stereo Recording Amplifier

ASRAS Astronomy Section of the Rochester Academy of Sciences (USA-NY)

ASRB Alpine Surface Radiation Budget • Astronomy Space and Radio Board

ASRC Astronomical Society of Rowan County (USA-NC) • Atmospheric Sciences Research Center

ASRD Age - Sex - Rank - Diagnosis • Aircraft Shipment Readiness Date

AsRDS Astronomy Research and Development Section (PAGASA)

ASRE American Society of Refrigerating Engineers (USA)

ASREQ As Required

ASRF Asian Synchrotron Radiation Facility

ASRI Asher Space Research Institute (Israel) • Australian Space Research Initiative • Australian Space Research Institute

ASRL Atmospheric Sciences Research Laboratory

ASRLO Australian Scientific Research Liaison Office

ASRM Abort Solid Rocket Motor • Advanced Solid Rocket Motor • Antenna System Readiness Monitor

ASRO Asociatia de Standardizare din Romania (Romania) • Astronaut Rover

ASRP Australian Synchrotron Research Programme

ASRRC Avionics System Requirement Review Conference

ASRS Airborne Satellite Receiving Station • Aircraft Suitability Reporting System • Automated Shuttle Requirements System • Automated Support Requirements System • Automatic Storage and Retrieval System • Auxiliary Support Reaction System • Aviation Safety Reporting System

ASRT Academy of Scientific Research and Technology

Ass Assegno • Assemblée • Assembly • Assistant • Association • Associazione • Assorted • Assurance

A&SS Alignment & Stabilization System

ASS Accessory Supply System • Acquisition Sun Sensor • Advanced Sensor System • Advanced Space Station • Aerosol Sampling System • Aerospace Surveillance System • Airborne Support System • Airlock Support System • Alignment and Stabilization System • Allocation Spécifique de Solidarité • Altitude Sensing System • Analog Simulation System • Analog Sun Sensor • Antenna Support Structure • Antenna Support System • Assistante Sociale • Astrophysics and Space Science (periodical code) • Automatic Space Station • Automatic Stabilization System • Automatic Switching System • Axisymmetric Spiral (field) • Axisymmetric Spiral Structure

ASSA Astronomical Society of South Australia • Astronomical Society of Southern Africa • Austrian Solar and Space Agency (now ASA) • Automated Support System Architecture

ASSAB Académie Serbe des Sciences et des Arts - Bulletin (periodical code)

ASSAS Advanced Solid-State Array Sensor

ASSC Accounting Standards Steering Committee • Age Structure of Scientific Collaboration • Association for the Scientific Study of Consciousness

ASSE American Society of Sanitary Engineers (USA) • Automation System for Scientific Experiments

ASSEDIC Association pour l'Emploi Dans l'Industrie et le Commerce (France)

ASSESS Airborne Science Shuttle Experiments System Simulation • Airborne Science Spacelab Experiments System Simulation • Assessment

ASSET Abstracts of Selected Solar Energy Technology • Advanced Systems Synthesis and Evaluation Technique • All-Source Satellite Evaluation Tool • American Society of Scientific and Engineering Translators (USA) • Asset Source for Software Engineering Technology • Association for Schools Science, Engineering and Technology • Astronomical Society of SE Texas (USA-TX) • Automated Software Support Entry System

ASSF Acta Societatis Scientiarum Fennicae (periodical code)

ASSFA Acta Societatis Scientiarum Fennicae - Series A (periodical code)

ASSH Académie Suisse des Sciences Humaines (Switzerland) • Advanced Space Systems Hardening

ASSI Active Stabilization in Stellar Interferometry • Additional Special Skill Indicator • Applications Statistiques Scientifiques Informatiques • Atmospheric amd Solar (EUV) Spectrometer Instrument

ASSIA Applied Social Science Index and Abstracts

Assigt Assignment

Assim Assimilate

ASSIST Advanced System Support and Information System Technology • Automated Special Security Information Systems Terminal • Automatic System of Secondary Index Selection Technique

ASSL Astrophysics and Space Science Library (periodical code)

ASSM Académie Suisse des Sciences Médicales (Switzerland) • Aerodynamics of Solid Segmented Motors • Automated Space Service Module

Assmt Assessment

Assn Association

ASSN Académie Suisse des Sciences Naturelles (Switzerland) • Accademia Svizzera di Science Naturali (Switzerland)

ASSNA Action Spécifique pour la Simulation Numérique en Astrophysique (France)

ASSO Assistant System Security Officer

Assoc Associate • Associated • Association • Associé • Associer

Assocn Association

ASSOP Analytical Space Surveillance Object Processor

Assoz Associazione

ASSP Aerosol Scattering Spectrometer Probe • Aerospace Systems Security Program • Analog Scanner Signal Processor • Application-Specific Standard Product • Architecture for Survivable Systems Processing • Auxiliary Scattering Spectrometer Probe

ASSPU Advanced Sensor Signal Processor Unit

ASSR Agreed Set of Security Rules • Autonomous Soviet Socialist Republic (now obsolete)

ASSS Aerospace Systems Safety Society • Automated Service Station System

Ass/t Assistant

Asst Assist • Assistant

ASST Académie Suisse des Sciences Techniques (Switzerland) • American Society for Steel Treating (USA) • Analytic Software Sizing Tool • Assistant

Asstd Assorted

ASSTD Advanced Satellite System Technology Demonstration

ASSU Ambulance de Secours et de Soins d'Urgence • Annual Scientific Supplement to Urania (periodical code)

ASSW Astronomical Society of the South West (Australia)

Assy Assembly

ASSY Assembly

ast Assistant (file extension)

Ast Antragsteller • Astronomy (periodical code) • Asturiano • Asturias (Spain)

ASt Außenstelle

AST Abort-Scan Table • Absolute Space-Time • Abstract Syntax Tree • Accademia delle Scienze di Torino (Italy) • Active Solar-Type • Adaptive Structuration Theory • Add-Subtract Time • Administrative-Supply Technician • Advanced Sensor Technology • Advanced Simulation Technology • Advanced Software Technology • Advanced Solar Telescope • Advanced Survivability Technology • Advanced Switch Test • Advanced System Technology • Aerospace, Science and Technology • Aerospace Technologist • Agent de Service Technique • Agent des Services Techniques • Airborne Surveillance Testbed • Airbus Super Transporter • Aircraft System Trainer • Airlock Systems Test • Alaskan Standard Time • Albert (Wong) - Safi (Qreshey) - Thomas (Yuen) (company) • Algorithm Support Team • All-Systems Test • Alternate Stand Talker • Analytical Simulation Team • Antarctic Submillimeter Telescope • Apparent Sidereal Time • Architectural Selection Tool • Assembly and Structure Test • Assign Secondary Traffic • Assistant • Assured Shorthold Tenancy • ASTER Science Team • Astoria (USA-OR) (location identifier) • Astrometry • Astrometry Science Team • Astronomical Society of Tasmania (Australia-TAS) • Astronomy Science Team • Astro Star Tracker • Asynchronous System Trap • Asynchronous Trap • Atlantic Standard Time • Atmospheric Surveillance Technology • Atomized Suspension Technique • At Same Time • Automatic Spectroscopic Telescope • Average Sampling Time

AStA Allgemeine Studentenausschuß (Germany)

ASTA Advanced Software Technology and Algorithms • Aerospace Technologies of Australia • Airport Surface Traffic Automation • Allgemeiner Studenten-Ausschuß (Germany) • American Society of Travel Agents (USA) • Arab Scientists and Technologists Abroad • Arrivals See Time Average • Association of Short-circuit Testing Authorities • Astronomical Time Series Analysis • Auckland Science Teacher's Association (New Zealand)

ASTAP Advanced Statistical Analysis Program

Astab Astable

ASTB Astronomiska Sällskapet Tycho Brahe (Sweden)

ASTC Advanced Satellite Tracking Center • Advanced Science and Technology Center • Advanced Science-Technology Center • Advanced Semiconductor Technology Center • AlliedSignal Technical Services Corporation (USA) • Association of Science and Technology Centers (USA-DC)

ASTD Advanced Space Technology Division • Area Scale Temperature Display

AstDyS Asteroids Dynamic Site

ASTE Action Spécifique "Télescopes Étrangers" (INSU) • Advanced Strategic/Tactical Expendable • Aerospace Systems Test Environment • American Society of Test Engineers (USA) • American Society of Tool Engineers (USA) • Association pour le développement des Sciences et des Techniques de l'Environnement (France) • Atacama Submillimeter Telescope (Chile)

ASTEC Aerospace System Test and Evaluation Complex • Australian Science and Technology Council

ASTED Avancement des Sciences et des Techniques de la Documentation

ASTEG All Systems Test Equipment Group

ASTEP Algorithm Simulation Test and Evaluation Program • Astrobiology Science and Technology for Exploring Planets (program)

ASTeR Audio System for Technical Reading

ASTER Advanced Spaceborne Thermal Emission and Reflection (radiometer) • Assessment Tools for the Evaluation of Risk • Atmosphere-Surface Turbulence Exchange Research • Audio System for Technical Readings

ASTERIX Analyse Stéréoscopique par Radar à Impulsions X

ASTEX Atlantic Stratocumulus Transition Experiment

ASTF Aeropropulsion System Test Facility

ASTG Aerospace Test Group

AstHe Astronomical Herald (periodical code)

ASTI Advanced Science and Technology Institute • Area of Special Tourist Interest

ASTID Astrobiology Science and Technology Instrument Development

ASTIP Army Scientific and Technical Information Program

ASTIS Arctic Science and Technology Information Service • Australian Science and Technology Information Service

AstL Astronomy Letters (periodical code)

ASTL Advanced Space Technology Laboratory • Automated SPARCS Command Link

ASTM American Society for Testing and Materials (USA) • American Society for Testing Materials (USA) (*) • American Standards and Test Method

ASTMS Association of Scientific, Technical and Managerial Staff • Association of Specialist Technical Organizations (UK)

ASTOR Airborne Stand-Off Radar • Area Stand-Off Radar

ASTOS Association of Specialized Technical Organizations for Space

ASTOVL Advanced Short Takeoff and Vertical Landing

ASTP Advanced Sensor Technology Program • Advanced Space Technology Program • Advanced Space Transportation Program • Advanced Systems and Technology Programme • Apollo-Soyuz Test Project

Astr Astronomer • Astronomical • Astronomy

ASTR Addition, Subtraction, Timing and Ratio • Aerospace Systems Test Reactor • Astronome • Automated Software Trouble Report

ASTRA Advanced Space Technologies for Robot Applications • Application of Space Techniques Relating to Aviation • Applied Space Technology Regional Advancement • Association in Scotland to Research into Astronautics • Associazione per lo Studio e la Ricerca Astronomica (Italy) • Astronome Adjoint • Astronomical and Space Techniques for Research on the Atmosphere • Astronomical Society of the Toms River Area (USA-NJ) • Astronomical Space Telescope Research Assembly • Automated Scale Transition • Automated Spectrophotometric Telescope Research Associates (project) • Automatic Satellite Tracking Antenna • Automatic Sorting, Testing, Recording Analysis

ASTRAL Alliance for Strategic Token Ring Advancement and Leadership • Analog Schematic Translator to Algebraic Language

Astr. Ap. Astronomy and Astrophysics

ASTRE Accéléromètre Sensible Triaxial Électrostatique • All-Sky Supernova and Transient Explorer • Analyse Statistique des Transferts Radiatifs dans les Écoulements • Astrophysiciens à la Recherche de l'Entreprise (France) (now defunct)

ASTREX Advanced Space Technology Research Experiment

Astrf L'Astrofilo (periodical code)

ASTRF Agent des Services Techniques de Recherche et de Formation

ASTRI Analytical Stereorestitution Instrument

ASTRID Advanced Stars: a Tool for Relating Information and Data • Agence de Soutien des Technologies de la Recherche Industrielle et du Développement (France) • Association Scientifique et Technique pour la Recherche en Information Documentaire (Belgium)

ASTRIS Astrofili della Società di Telecomunicazioni del Raggruppamento IRI-STET (Italy)

Astro Astronaute • Astronautique • Astronomer • Astronomical • Astronomie • Astronomique • Astronomy

AST/RO Antarctic Submillimeter Telescope/Remote Observatory

ASTRO Air Space Travel Research Organization • Air Support To Regular Operations • Antarctic Submillimeter Telescope and Remote Observatory • Artificial Satellite Time and Ratio Orbit • Ascent Trajectory Optimization • Astronomers in the Schools Teaching Resource Organization

Astrol Astrología • Astrologie • Astrologique • Astrology

Astron Astronaute • Astronautique • Astronomer • Astronomía • Astronomical • Astronomie • Astronomique • Astronomy

ASTRON Stichting Astronomisch Onderzoek in Nederland (Netherlands)

Astronaut Astronautical • Astronautics • Astronautique

Astrophys Astrophysical • Astrophysics • Astrophysique

ASTS American Space Transport System • Associazione della Stampa Tecnica e Scientifica (Italy) • Avionics System Test Specification

Ast. T. Astronomical Time

ASTU Automatic Systems Test Unit

ASTUTE Automated System for Thesaurus Updating, Testing and Editing

ASU Administration Scolaire et Universitaire • Administrative Service Unit • Administrative Support Unit • Airborne Sensing Unit • Air Separation Unit • Albany State University (USA-GA) • Allocation de Salaire Unique • American Students Union (USA) • Angelo State University (USA-TX) • Approval for Service Use • Approved for Service Use • Area Service Unit • Arizona State University (USA-AZ) • Arkansas State University (USA-AR) • Association pour la Statistique et ses Utilisations (France) • Astronomical Server URL • Astronomy Study Unit • Asunción - Silvio Pettirossi (Paraguay) (location identifier) • Asynchron/Synchronumsetzer • Automatic Switching Unit

ASUP AUTODIN Switch Upgrade Project

ASUR Airborne Submillimeter Radiometer

ASUT Association Suisse de Utilisateurs du Téléphone

ASV Advanced Safety Vehicle • Aerospace Vehicle • Airline Schedule Vendor • Amboseli (Kenya) (location identifier) • American Standard Version • Analog Secure Voice • Angle Stop Valve • Anode Supply Voltage • Anodic Stripping Voltammetry • Arabic Vowelized Simplified • Arithmetic Simple Variable • Atlas Stellarum Variabilium • Automatic Self-Verification • Automatic Shuttle Valve • Autonomous Space Vehicle

ASVC Automatic Secure Voice Communication

ASVD Analog Simultaneous Voice/Data • Asynchronous Simultaneous Voice and Data

ASVIP American Standard Vocabulary for Information Processing

ASVP Ada Simulated Validation Program

ASVS Advanced Space Vision System

ASVT Address Space Vector Table • Applications Systems Verification and Transfer • Applications Systems Verification Test

ASW Acoustic Surface Wave • Address and Synchronization Word • Amorphous Solid Water • Application Software • Association of Scientific Workers (UK) • Astronomical Society of Wakabadai (Japan) • Aswan (Egypt) (location identifier) • Augmented Spherical Wave • Auxiliary Switch • Average Surface Wind

ASWA Astronomical Society of Western Australia Inc. (Australia-WA)

ASWG Ad hoc Science Working Group • American Steel Wire Gauge • Architecture Systems Working Group

ASX Answer-Stock Exchange • Ashland (USA-WI) (location identifier)

ASY Ashley (USA-ND) (location identifier) • Assembly

Asym Asymmetric • Asymmetry

Asymp Asymptote

Async Asynchronous

ASYNC Asynchronous

AS3AP ANSI SQL Standard Scalable And Portable

at Austria

At Ampere-turn • Astat • Astatine • À Temps • Atención • À Terme • Atmosphere • Atom • Atomic • Attainable • Attitude • Attorney

A/T American Terms

A&T Acceptance & Transfer • Acquisition & Technology • Administration & Training • Assembly & Test • Assembly & Turnaround

AT Absolute Threshold • Abstract Type • Accelerator Technology • Acceptance Tag • Acceptance Test • Access Tandem • Access Time • Access Tower • Access Type • Accident de Trajet • Accident du Travail • Accidente de Trabajo • Achievement Test • Acquisition Team • Action Taken • Action Time • Admission Temporaire • Advanced Technology • Advanced Trainer • Agent Technique • Aide Technicien • Air Temperature • Air Traffic • Air Transport • Alert Time • Aliveness Test • Allocation Temporaire • Alta Tensione • Alternative Technology • Ambient Temperature • Ampere-Turn • Analog Tape • Analysis Time • Anisotropy Telescope • Annual Tour • Annual Training • Apparent Time • Appearance Time • Application Technology • Appropriate Technology • Aptitude Technique • Arrêt de Travail • Arrival Time • Assembly and Test • Assembly Truss • Astrometry Team • Astronomical Time • Asynchronous Transmission • Atlantic Time • Atomic Time • Attainment Target • Attention • Attenuation • Auroral Time • Australia Telescope • Austria (ISO country code) • Automated Telecommunications • Automatic Telescope • Automatic Test • Automatic Testing • Automatic Ticketing • Automatic Translation • Autorisation Temporaire • Autothrottle • Autriche (code ISO de pays) • Auxiliary Tank • Auxiliary Telescope • Auxiliary Timer

A&TA Assembly & Test Area

ATA Abort Time Assembly • Academic Travel Abroad • Actual Time of Arrival • Additional Training Assembly • Advanced Technologies and Applications • Advanced Technology Attachment • Advanced Test Accelerator • Aerodynamic Testing Alliance • Aft Test Article • Airport Traffic Area • Allen Telescope Array (USA-CA) • American Translators' Association (USA) • American Transportation Association (USA) • Ammonia Tank Assembly • Annual Training Assistance • Anta (Peru) (location identifier) • Antarctica (ISO country code) • Antarctique (code ISO de pays) • Assembly and Test Area • Assembly Test Article • Associate Technical Aide • Associazione Technica dell'Automobile • Associazione Tuscolana di Astronomia (Italy) • AT Attachment • Atlantic Treaty Association • Automatic Track Acquisition • Automatic Tracking Antenna • Automatic Trouble Analysis • Auto Terrain Avoidance • Auxiliary Timer Assembly • Avionics Test Article • Azimuth Torquer Amplifier

ATAC Abbreviated Transportation Accounting Classification • Advanced Technology Advisory Committee • Aérodynamique des Tuyères et Arrière-Corps • Air Transport Advisory Council • Air Transport Association of Canada • ALMA Technical Advisory Committee • Applied Technology Advanced Computer • Association des Transporteurs Aériens par Charter • Australian Time Assignment Committee (AAO)

ATACS Advanced Telecommunications Abuse Control System

ATAFS Automated Training Analysis and Feedback System

ATAG Authoring Tool Accessibility Guidelines

ATAI Association de Transport Aérien International

ATAL Agent Theories, Architectures, and Languages (workshop) • Alternate Transoceanic Abort Landing

ATALA Association pour le Traitement Automatique des Langues

ATAM Architecture Trade-off Analysis Method • Association for Teaching Aids in Mathematics

ATAO Annals of Tokyo Astronomical Observatory

ATAP Any Time - Any Place

ATAPI Advanced Technology Attachment Packet Interface

ATAR Advanced Technology Assessment Report

ATARR Advanced Turbine Aerothermal Research Rig

ATARS Advanced Tactical Air Reconnaissance System

ATAS Advanced Technology Alert System • Airspace and Traffic Advisory Service • Automatic Test Analysis System • Automatic Three-Axis Stabilization • Automatic Tracking Antenna System

ATB Airborne Testbed • All Terrain Bike • All Trunks Busy • Association of Tropical Biology • Atbara (Sudan) (location identifier) • Automated Ticket and Boarding • Average Time Burning

ATBD Algorithm Theoretical Basis Document

ATBM Average Time Between Maintenance

ATC Ablative Thrust Chamber • Ablative Thrust Control • Absolute Time Command • Acoustical Test Chamber • Action Taken Code • Active Thermal Control • Active Transfer Command • Active Turbulent Combustion • Adaptive Transform Coding • Address Translation Cache • Adiabatic Toroidal Compressor • Advanced Technology Center • Affordability Through Commonality • After Top Center • Aircraft Target Computer • Air Temperature Control • Air Traffic Control • Air Traffic Controller • Air Training Command • All Things Considered • Applied Technology Council • Arthur's Town (Bahamas) (location identifier) • Assistant Test Conductor • Astrophysical Theory Centre • Atmosphere Technology Division • ATM Transfer Capability • Australian Telecommunications Commission • Authority To Construct • Authorization To Copy • Authorized Training Centre • Automated Text Categorization • Automatic Target Cueing • Automatic Temperature Compensation • Automatic Temperature Control • Automatic Threshold Circuit • Automatic Tracking Control • Automatic Traffic Control • Automatic Train Control • Automatic Transmission Control • Automatic Tuning Control • Automation Training Center • Average Total Cost

ATCA Attitude Translation Control Assembly • Australia Telescope Compact Array

ATCAA Air Traffic Control Assigned Airspace

ATCAS Air Traffic Control Automation System

ATCBI Air Traffic Control Beacon Interrogator

ATCC Airborne Transmitter Control Center • Air Traffic Control Center • Automated Telecommunications Center

ATCCC Air Traffic Control Command Center

ATCCS Air Traffic Command and Control System

ATCCSS Air Traffic Control Communications Switching System

ATCD Advanced Technology Concept Demonstration

ATCE Ablative Thrust Chamber Engine

ATCF Air Traffic Control Flight

Atch Attach

ATCH Attach • Attachment

Atchd Attached

ATCIS Automated Tool Control and Inventory System

ATCL Air Traffic Control Line

ATCM Antarctic Treaty Consultative Meeting

ATCO Air Taxi Commercial Operator • Air Traffic Coordinating Office • Air Traffic Control Officer • Ambient Temperature Catalytic Oxidizer • Astronomie en Touraine et Centre-Ouest (France) • Astronomie, Techniques et Communications (France) (*) • Automated Test and Checkout • Automatic Telephone Control Office

ATCOS Atmospheric Composition Satellite

ATCP Absolute Time Command Processor • Antarctic Treaty Consultative Party • AppleTalk Control Protocol

ATCRB Air Traffic Control Radar Beacon

ATCRBS Air Traffic Control Radar Beacon System

ATCS Active Thermal Control System • Air Traffic Communications Station • Air Traffic Control Section • Air Traffic Control Specialist • Air Traffic Control System • Automatic Test Control System • Auxiliary Telescope Control Software

ATCSCC Air Traffic Control System Command Center

ATCSR Alternate Transmitted Chirp Slope Reversal

ATCSS Air Traffic Control Signaling System

ATCT Airport Traffic Control Tower • Air Traffic Control Tower (*)

ATD Above Threshold Detachment • Actual Time of Departure • Address Transition Detector • Advanced Technology Demonstrator • Advanced Technology Demonstration • Advanced Technology Development • Advanced Training Division • Aerospace Technology Division • Air Tasking Directive • Allocation Temporaire Dégressive • Alpha-Track Detector • Amphibious Transport Dock • Associated Technical Document • Associative Template Data-flow • Asynchronous Time Division • À Titre Définitif • Atmospheric Technology Division • Atoifi (Solomon Islands) (location identifier) • Automated Training Division • Automatic Tape Degausser • Automatic Teaching Device

ATDA Alternate Target Docking Adapter • Augmented Target Docking Adapter

ATDB Aerothermal Data Block • Aerothermodynamic Data Book

ATDBA Association Tunisienne des Documentalistes, Bibliothécaires et Archivistes (Tunisia)

ATDC After Top Dead Center

ATDD Asynchronous Time Diversity Device • Atmospheric Turbulence and Diffusion Division

ATDE Advanced Technology Demonstrator Engine

ATDF Archival Tracking Data File • Automated Test Development Folder

ATDI Appropriate Technology Development Institute

ATDL Adaptive Targeting Data Link • Atmospheric Turbulence and Diffusion Laboratory

ATDLG Advanced Technology Demonstration Laser Gyro

ATDM Adaptive Time Division Multiplexer • Asynchronous Time Division Multiplexing (*)

ATDN Advanced Technology Demonstration Network

ATDnet Advanced Technology Demonstration Network

ATDOS Advanced Technique for Developing Operational Software

ATDP Academic Talent Development Program • Acceptance Test Data Package • Advanced Technology Demonstration Plan • Attention Dial Pulse

ATDRS Advanced Tracking and Data Relay Satellite

ATDRSS Advanced Tracking and Data Relay Satellite System

ATDT Attention Dial Tone • Automated Testability Decision Tool

ATE Accéléromètre Triaxial Électrostatique • Acceptance Test Equipment • Actual Time En-route • Advanced Technological Education • Advanced Technology Engine • Aerospace Test Equipment • Aerospace Training Equipment • Airborne Test Equipment • Antlers (USA-OK) (location identifier) • Area Test Equipment • Artificially-Triggered Erection • ASCII Table Extension • Association of Teachers Educators • Associativity Test Equation • A Temporal Element • A Temporal Entity • Auslandstätigkeitserlaß • Automated Terminal Exchange • Automated Test Equipment (*) • Automatic Test Engineer • Automatic Test Equipment

ATEC Automated Technical Control • Automated Test Control

ATEGG Advanced Turbine Engine Gas Generator

ATEL Atelier

ATEM Analytical Transmission Electron Microscope • Analytical Transmission Electron Microscopy

AtEn Atomnaia Energiia (periodical code)

AtEnv Atmospheric Environment (periodical code)

ATEP Advanced Technical Education Program

ATER Attaché Temporaire d'Enseignement et de Recherche • Attaché Temporaire d'Études et de Recherche • Automatic Testing, Evaluation and Reporting

ATERP Arctic Terrestrial Environmental Research Program

ATERS Automatic Test Equipment Reporting System

AtERv Atomic Energy Review (periodical code)

ATES Alcatel Thomson Espace (France)

ATEX Atlantic Tropospheric Experiment

ATF Accelerator Test Facility (BNL) • Acceptance Test Facility • Acoustic Test Facility • Advanced Tactical Fighter • Advanced Toroidal Facility • Aerodrome Traffic Frequency • Algebraic Technological Function • ALMA Test Facility • Ambato Chachoan (Ecuador) (location identifier) • Anatomical Transfer Function • Anstalt für Tieftemperaturforschung (Germany) • Antenna Test Facility • Application Transport Failover • Astrometric Telescope Facility • Astronomical Test Facility • Atmospheric Transmission Factor • Attention To File • Automated Telescope Facility • Automatic Telescope Facility • Automatic Terrain Following • Automatic Tracking Feature • Avionics Test Facility • French Southern Territories (ISO country code) • Southern Territories (French_) (ISO country code) • Territoire Français Austral (code ISO de pays)

ATFJ Atelier pour le Travail et la Formation des Jeunes

ATFM Air Traffic Flow Management

ATFOS Alignment and Test Facility for Optical Systems

ATFP AppleTalk Filing Protocol

ATFS Association of Track and Field Statisticians

ATFT Affine Toda Field Theory

ATG Address Translating Gateway • Advanced Technology Group • Aerospace Test Group • Antenna Test Group • Antenna-Transceiver Group • Antigua & Barbuda (ISO country code) • Assign Trunk Group • Automatic Test Grading • Automatic Test Guide • Automatic Timeline Generator • Barbuda (Antigua &_) (ISO country code)

ATGIA Accurate Tidal Gravimetry In Antarctica

ATGS Advanced Terminal Guidance System

Ath Athlete • Athletic

ATH Abbreviated Trouble History • Above The Horizon • Assign Thresholds • Astronomie-Treff Hückelhoven (Germany) • Athens (Greece) (location identifier) • Athinai (Greece) (location identifier) • Attention Hang-up • Auxiliary Telescope Hall

ATHDFS Autralia Telescope - Hubble Deep Field South (survey)

ATHE Association of Teachers in Higher Education

Athen Athenian

ATHENA Acquisition des données de Topographie de surface et d'Hydrologie dans l'Est Nord Atlantique

Athl Athlete • Athletic

ATHL Airborne Takeoff and Horizontal Landing

ATi Astronomisk Tidsskrift (periodical code)

ATI Above Threshold Ionization • Action Thématique Innovante (INSU) • Advanced Technical Information • Advanced Technologies and Instrumentation • Advanced Technology Insertion • Advanced Technology Instrumentation • Advanced Telecommunications Institute • Air Technical Index • Allocation Temporaire d'Invalidité • Apparent Thermal Inertia • Application Transport Interface • Area Time Integral • Artigas (Uruguay) (location identifier) • Asociación de Técnicos de Informática (Spain) • Assistant Technique à l'Industrie • Association of Technical Institutions • Automated Technical Information • Automatic Test Inhibit • Automation of Technical Information • Average Total Inspection • Azimuth - Time - Intensity

ATIA Advanced Techniques for Information Analysis

ATIC Access To Interpersonal Communication • Advanced Thin Ionization Calorimeter • Aerospace Technical Intelligence Center • Analog Telephony Interface Card • Avionics Test and Integration Complex

ATICA ACIS Telemetry Interpreter for Calibration Analysis

ATICCA Australian Tertiary Institution Consulting Companies Association

ATID Advanced Technical Interoperability Demonstration

ATII Advanced Technique for Imagery Interpretation

ATILO Administration Tribunal of the International Labour Office

ATIM Advanced Technology Insertion Module

ATIP Advanced Technology Insertion Program • Airborne Telemetry Instrument Package • Asian Technology Information Program • Autonomous Telemetry and Instrumentation Package

ATIRCM Advanced Tactical Infrared Countermeasure (*) • Advanced Threat Infrared Countermeasure

ATIRS Automated Test Incident Recording System

ATIS Advanced Technical Information Support • Advanced Technical Information System • Aerodrome Terminal Information Service • Agence de Traitement des Images Satellitaires • Alliance for Telecommunications Industry Solution (USA-DC) • Appropriate Technology Information Service • Atherton Tools Integration Service • Automated Terminal Information Service • Automatic Telecommunications Information System • Automatic Telescope Instruction Set • Automatic Telescope International Standard • Automatic Terminal Information Service • Automatic Terminal Information System • Automatic Transmitter Identification System

ATJ Antsirabe (Madagascar) (location identifier)

ATJAM ASME Transactions - Journal of Applied Mechanics (periodical code)

ATJDS ASME Transactions - Journal of Dynamic Systems (periodical code)

ATJEG ASME Transactions - Journal of Engineering, Gas Turbines and Power (periodical code)

ATJEl ASME Transactions - Journal of Electronic Packaging (periodical code)

ATJEM ASME Transactions - Journal of Engineering Materials and Technology (periodical code)

ATJEn ASME Transactions - Journal of Engineering Power (periodical code)

ATJEP ASME Transactions - Journal of Engineering Power (periodical code)

ATJFE ASME Transactions - Journal of Fluids Engineering (periodical code)

ATJHT ASME Transactions - Journal of Heat Transfer (periodical code)

ATJLT ASME Transactions - Journal of Lubrication Technology (periodical code)

ATJSE ASME Transactions - Journal of Solar Energy and Engineering (periodical code)

ATJTr ASME Transactions - Journal of Tribology (periodical code)

ATJTu ASME Transactions - Journal of Turbomachinery (periodical code)

ATJVA ASME Transactions - Journal of Vibration Acoustics (periodical code)

Atk Attack

ATK Andrew Toolkit • Atqasuk (USA-AK) (location identifier) • Attack • Available Ton-Kilometer • Average Ton-Kilometer

Atl Atlantic • Atlas

AT&L Acceptance Test & Logistics

ATL Above The Line • Acceptance Test and Logistics • Active Template Library • Active Time List • Actual Total Loss • Adjusted Termination Level • Advanced Technology Laboratory • Alternate Target List • Analog Threshold Logic • Annotated Task List • Applications Terminal Language • Artificial Transmission Line • Atlanta Hartsfield (USA-GA) (location identifier) • Atlantic • Atlas • Automated Tape Library • Automatic Tape Library • Automatic Test Line • Avionics Test Laboratory

ATLA American Trial Lawyers Association (USA)

ATLAS AAO Tunable Littrow Articulated Spectrograph • Abbreviated Test Language for All Systems • Abbreviated Test Language for Avionics Systems • Advanced Technology LADAR System • Airborne Tunable Laser Absorption Spectrometer • Analysis of Thrust Structure Loads And Stresses • Anglo-Australian Tunable Littrow Astronomical Spectrograph • Argonne Tandem-Linear Accelerator System • Atmospheric Laboratory for Applications and Science (*) • Atmospheric Laboratory for Applications in Science • A Toroidal LHC Apparatus • Automatically Tuned Linear Algebra Subroutines • Automatic Tape Label Assignment System • Automatic Thin-Layer Analytical System

ATLAST Augment the Teaching of Linear Algebra through Software Tools

ATLC Atlantic

ATLH Atalho

ATLID Atmospheric Lidar

ATLIS Automated Technical Library Information System

ATLO Acceptance Test and Launch Operations • Assembly, Test and Launch Operations

ATLOS Advanced Launcher Trajectory Optimization Software

ATLS Advanced Trauma Life Support • Automated Tape Library Support

AtlSV Specola Astronomica Vaticana Atlas Stellarum Variabilium (periodical code)

AtlVS Atlas Poiskovykh Kart Peremennykh Zvezd (periodical code)

at. m. Atomic Mass

atm Adobe Type Manager (file extension) • Standard Atmosphere

At/m Ampere-Turns per Meter

Atm Atmosphere • Atmosphère • Atmospheric • Atmosphérique

ATM Accroissement Total Moyen • Adaptive Triangular Mesh • Adobe Typeface Manager • Adobe Type Manager (*) • Advanced Targeted Marketing • Advanced Telescope Mission • Advanced Trauma Management • Aerodynamic Twisting Moment • Airborne Thematic Mapper • Airborne Topographic Mapper • Air Terminal Management • Air Traffic Management • Air Turbine Motor • Altamira (Brazil-PA) (location identifier) • Amateur Telescope Maker • Amateur Telescope Making • Analog-to-Time Module • Antenna Test Model • Apollo Telescope Mount • Approved Testing Material • Ascent Trajectory Model • Assign Traffic Metering • Association of Teachers of Management • Association of Teachers of Mathematics • Asynchronous Time Mode • Asynchronous Transfer Mode (*) • Asynchronous Transmission Manager • Asynchronous Transmission Mode • Atmosphere • Atmospheric Transmission at Microwaves • Atmospheric Transmission Model • Atmospheric Transport Model • At The Moment • Automated Teller Machine • Automatic Teller Machine (*) • Automatic Test Method • Auxiliary Tape Memory

ATMA Accroissement Total Moyen Annuel • Amateur Telescope Maker's Association (USA-WA)

ATMAB Association Technique Maritime et Aéronautique - Bulletin (periodical code)

ATMAC Air Traffic Management Automated Center

ATMAP Atmospheric Tides Middle Atmosphere Program

ATMARP Asynchronous Transfer Mode Address Resolution Protocol

ATMAS Association Technique Maritime et Aéronautique Session (periodical code)

ATME Atmospheric Transmission Measurement Equipment • Automatic Transmission Measuring Equipment

AtmEn Atmospheric Environment (periodical code)

ATMES Atmospheric Model Evaluation Study

ATMH Automatic Text Message Handling

ATMHS Automatic Text Message Handling System

ATMIS Atmospheric and Meteorology Instrument System • Atmospheric Moisture Intercomparison Study

ATMJ Amateur Telescope Making Journal

ATMOB Amateur Telescope Makers Of Boston (USA-MA)

Atmos Atmosphere • Atmosphere (periodical code)

ATMOS Atmospheric Trace Molecular Spectroscopy • Atmospheric Trace Molecule Spectroscopy (*) • Atmospheric Trace Molecules Observed by Spectroscopy (*) • Atmospheric Trace Molecules Observing Spectrometer • Atmospheric Trace Molecules Spectroscopy • Atmospheric Turbulence Measurement and Observation System • Automated Tactical Meteorological Observing System • Automatic Testing Multiple Operating System

Atmosph Atmosphere • Atmosphère • Atmospheric • Atmosphérique

ATMOSPHERE Advanced Techniques and Models of System Production in a Heterogeneous, Extensible and Rigorous Environment

ATMP Acceptance Test Management Plan • Ascend Tunnel Management Protocol

AtmRe Atmospheric Research (periodical code)

AT&MS Advanced Technology & Mission Studies (NASA)

ATMS Advanced Technology Microwave Sounder • Advanced Traffic Management System • Air Traffic Management System • Assumption-based Thruth Maintenance System • Atmospheric Turbulence Measuring Set • Automated Technical Manual System • Automatic Test Management System • Automatic Transmission Measuring System

AtmTe Atmospheric Technology (periodical code)

Atn Atención

ATN Actual Test Number • Adaptive Tactical Navigation • Aeronautical Telecommunications Network • Amateurastronomische Tagung an Niederrhein (Germany) • Amateur Television Network (USA) • AMOS Test Number • Arc Tangent • Astrogeophysical Transmission Network • Attention • Augmented Transition Network • Automatic Telescope Network • Namatanai (Papua New Guinea) (location identifier)

ATNC Agent Transmissible Non Conventionnel

ATNF Australia Telescope National Facility

at. no. Atomic Number

Ato Atento

AtO Atmosphere and Ocean (periodical code)

ATO Abort To Orbit • Action Technical Order • Actual Time Over • Advanced Technologies Operation • Airborne Test Operator • Air Tasking Order • Air Task Order • Antipodal Transient Observatory • Assisted Takeoff • Athens - Ohio University (USA-OH) (location identifier) • Atmospheric Test Orbit • Australian Tax Office • Automatic Train Operation • Automatic Turn-On

ATOC Acoustic Thermometry of Ocean Climates • Average Total Operating Cost

ATOG Automated Telescope Offset Guider

ATOLL Acceptance Test Of Launch Language • Analysis of the Tropical Oceanic Lower Layer

ATOM Analog Tree-Organized Multiplexer • Astronomical Telescope Orientation Mount • Automatic Telescope for Optical Monitoring • Automatic Topographic Mapper

ATOMIC Action of Transport On Minor Constituents • Astrophysical Transient Observatory Multiple Imaging Camera

ATOMS Air Traffic Operations Management System • Automated Technical Order Management System

ATOP Ada Technology Operational Planning

ATOS Administratif, Technique, Ouvrier et de Service • Advanced Technology Operations System • Automated Tasking Order System • Automated Technical Order System • Automatisation Totale de l'Observation de la Surface • Automatisation Totale de l'Observation de Surface

ATOVS Advanced TIROS Operational Vertical Sounder

ATOX Atomic Oxygen

ATP Absolute Time Processor • Accelerated Training Program • Acceptance Test Plan • Acceptance Test Procedure • Acceptance Test Program • Accroissement Total Périodique • Acquisition - Tracking - Pointing • Action Thématique Programmée (CNRS) • Activation Test Program • Active Training Period • Adenosine Triphosphate • Adult Treatment Panel • Advanced Technical Payroll • Advanced Technology Park • Advanced Technology Program • Advanced Telescope Project • Advanced Turboprop • Affectation à Titre Provisoire • Aid Trade Provision • Airborne Tracking and Pointing • Aitape (Papua New Guinea) (location identifier) • All Tests Pass • Amateur Turned Professional • AppleTalk Transaction Protocol • Array Transform Processor • Astronautics Test Procedure • Astrophysical Theory Program • Astrophysics Theory Program (NASA) • Augmented Thrust Propulsion • Authority To Participate • Authority To Proceed • Authorization To Proceed • Author - Translation - Publisher • Automated Test Pattern • Automated Test Plan • Automated Test Procedure • Automated Text Processing • Automated Tow Placement • Automatic Test Program • Automatic Text Processing • Automatic Train Protection • Autorisation de Transfert Préalable • Available To Promise

ATPA Accroissement Total Périodique Annuel • Alpha Temperature Probe Assembly

ATPDC Atomic Transition Probabilities Data Center

ATPDU Acknowledge Transport Protocol Data Unit

ATPG Automatic Test Path Generator • Automatic Test Pattern Generation • Automatic Test Program Generation • Automatic Test Program Generator

ATPL Air Transport Pilot's Licence

ATPO Assistant Technical Project Officer • Associate Technical Project Officer

ATPS Advanced Translator Processing System • Analysis of Training Performance System • AppleTalk Printing Services • Atmosphere Temperature Profile Sounder • Automated Test Planning System

ATPSE Acquisition, Tracking, and Pointing System Emulator

ATPU Advanced Tactical Power Unit

ATQ Amritsar (India) (location identifier) • AppleTalk Queue • AppleTalk Transition Queue

ATR Above Threshold Reporting • Acceptance Test Report • Accès de Tiers au Réseau • Access Time of Reading • Accidental Trigger Rate • Achievement of Test Readiness • Advanced Technology Radar • Advanced Test Reactor • Advanced Thermal Reactor • Aided Target Recognition • Air Tasking Request • Air Traffic Regulation • Air Transport Rack • Air Transport Radio • Air Transport Rating • Air Transport Regulation • Air Turbo Rocket • Ambient Temperature Recorder • Analog Tape Recorder • Angle - Time - Range • Annual Training Requirement • Answer To Reset • Antenna Test Range • Anti-Transmit Receive • Assistant Task Representative • Assistant Technique Régional • Atar (Mauritania) (location identifier) • Atlantic Tracking Range • Attenuated Total Reflectance • Attenuated Total Reflection • Attribute • Australian Telecommunication Research (periodical code) • Automated Target Recognition • Automatic Target Recognition • Automatic Terminal Recognition • Automatic Thrust Restoration • Autonomous Target Recognition

ATRAN Automatic Terrain Recognition And Navigation

ATRAS Automated Travel Record Accounting System

ATRI Advanced Thermionics Research Initiative

ATRIB Average Transfer Rate of Information Bits

ATRIMA As They Respective Interests May Appear

ATRIP Advancement of Teaching and Research in Intellectual Property

ATRIS Automatic Test Resource Information Systems

ATRR Acceptance Test Readiness Review

ATRRL Advanced Technology Reusable Rocket Launcher

ATRS Automated Training Records System • Automated Travel Records System • Automated Trouble Reporting System

ATRT Atelier Thématique de Réadaptation par le Travail

ATRWG Automatic Target Recognition Working Group

ATS Absolute Temperature Scale • Abstract Test Suite • Academic Trainer System • Accelerator Test Stan • Acceptance Test Specification • Acquisition and Tracking System • Acquisition Track Sensor • Administrative Telephone Service • Administrative Telephone System • Administrative Terminal System • Advanced Technical Skill • Advanced Technology Satellite • Advanced Telecommunications Services • Advanced Telescope System • Advanced Terminal System • Advanced Training System • Aerothermal Structure • Airborne Telemetry System • Air Traffic Service • Air Traffic System • Ambulatory Telemetry System • Aménagement du Temps Scolaire • American Trading Syndicate (USA-CA) • Angle Tracking System • Analoge Teilnehmerschaltung • Analog Tone Signal • Annual Training Site • Antarctic Treaty System • Antenna Test Station • Antique Telescope Society (USA-PA) • Apple Terminal Services • Application des Technologies Spatiales • Applications and Technology Satellite • Applications Technology Satellite • Artesia (USA-NM) (location identifier) • Assembly Trusses and Structure • Assign Terminal Service • Astronomical Time Switch • Asynchronous Task Storage • Atlantic Tracking Ship • Atlantic Tracking Station • Atmospheric Sounder • Atomic Time Scale • At The Suit of • Attitude Thrustor System • Attitude Transfer System • A Tutorial System • Automated Tasking System • Automated Terminal System • Automated Test Set • Automated Test System • Automated Text System • Automated Tilting System • Automated Time Standard • Automated Trade System • Automated Travel System • Automated Tuning System • Automatic Telemetry System • Automatic Terminal System • Automatic Test Set • Automatic Test System • Automatic Transcribing System • Automatic Transfer Switch • Auxiliary Telescope System • Auxiliary Territorial Service • Aviation Transport Services • Axial Thrust System • Ayudante Técnico Sanitario (Spain) • Schilling (Austria) (ISO currency code)

ATSA Aerotransporte SA

ATSB Australian Transport Safety Bureau

AtSc Atmospheric Science (periodical code)

ATSC Advanced Television Systems Committee • AlliedSignal Technical Services Corporation (USA-MD) • Automated Telecommunications Systems Center

ATSCM Antarctic Treaty Special Consultative Meeting

ATSDR Agency for Toxic Substances and Disease Registry (USA)

ATSE Atacama Submillimeter Telescope Experiment

ATSF Avion de Transport Supersonique du Futur

ATSIM Acquisition and Track Simulation

ATsir Astronomicheskii Tsirkulyar (periodical code)

ATSIS Automated Technical-Specification Information System

ATSM Automated Technique for Spacecraft Monitoring

ATSOCC Applications Technology Satellite Operations Control Center

ATSORA Air Traffic Service Outside Regulated Airspace

ATSP Airborne Tracking Solar Photometer • Asymmetric Traveling Salesman Program

ATSR Activity Time Status Report • Along-Track Scanning Radiometer

ATSS Acquisition and Tracking Subsystem • Automatic Test Support System • Aviation Training Support System

ATSSI Advanced Technology Solar Spectroscopic Imager

ATSM Automatic Telecommunications System Security Manager

AtST Atomisation Spray Technology (periodical code)

ATST Advanced Technology Solar Telescope

ATSU Air Traffic Services Unit

Att Attach • Attache • Attaché • Attached • Attempt • Attempted • Attend • Attendance • Attended • Attending • Attente • Attention • Attitude • Attorney

AT&T American Telephone & Telegraph Co. (USA)

ATT Accelerated Test Technology • Acceptance Test Team • Acceptance Thermal Testing • Advanced Technology Telescope • Advanced Transport Technology • Advanced Transport Telematics • All Thrust Termination • Aménagement du Temps de Travail • Assign Terminal Type • Astrometric Technology Telescope • Astronomical Tele-Telescope • Astronomischer Tausch- und Trödeltreff (Germany) • Asynchronous Time Trap • Atmautluak (USA-AK) (location identifier) • Attenuation • Atterrissage Tout Temps • At This Time • Attitude • Authorization To Test • Automatic Target Tracker • Automatic Target Tracking • Average Task Time

Atta Atenta

ATTA Advanced Terrain Tracking Altimeter • Atom Trap Trace Analysis

ATTAC Advanced Target/Threat Assessment Code

ATTAP Advanced Turbine Technology Application Program

ATTAS Advanced Technology Testing Aircraft System • Advanced Transport Technology Aircraft System • Advanced Transport Testing Aircraft System

ATTC Automatic Transmission Test and Control

ATTCH Attach • Attachment

ATTCS Advanced Takeoff Thrust Control System

ATTD Advanced Technology Transfer Demonstration

Atte Atentamente

ATTE Automatic Transistor Test Equipment

ATTEN Attenuation

ATTIC Alternative Technology Treatment Information Center

ATTIS AT&T Information System

ATTM AEG Telefunken Technische Mitteilungen (periodical code) • Authorization To Transfer Material

Attmte Atentamente

Attn Attention

ATTN Attention

Attnd Attend • Attended

Attndt Attendant

Atto Atento

ATTOM Acoustic Travel Time Ocean Current Monitor

ATTR Average Time To Repair

ATTRAS Automatic Telemetry Tracking Antenna System

Attrib Attribute • Attributed to • Attribution • Attributive • Attributively

ATTRIB Attribute

ATTS Automated Test Technology Standard • Automatic Telemetry Tracking System • Automatic Test Technology Standard

ATTTJ ATT Technical Journal (periodical code)

ATTW Association of Teachers of Technical Writing

Atty Attorney

ATTY Attorney

Atty. Gen. Attorney General

ATU ADSL Transceiver Unit • Advanced Training Unit • American Technological University (USA) • Antenna Tuning unit • Arab Telecommunication Union • Arkansas Tech University (USA-AR) • Attu Island - Casco Cove (USA-AK) (location identifier) • Audio Terminal Unit • Automated Translation Unit • Automatic Tracking Unit • Autonome Transfer Unit • Auxiliary Test Unit

ATUC African Trade Union Confederation • Australia Telescope Users Committee • Average Total Unit Cost

ATUG Australian Telecommunications Users Group

ATV Advanced Technology Validation • Advanced Television • Advanced Test Vehicle • All-Terrain Vehicle • Amateur Television • Arbeidstijdverkorting • Ariane Transfer Vehicle • Associated Television • Ati (Chad) (location identifier) • Atmospheric Test Vehicle • Automate à Transitions Valuées • Automated Transfer Vehicle • Automatic Threshold Variation

ATVA Advanced Technology Visible Array

ATVC Ascent Thrust Vector Control • Ascent Thrust Vector Controller • Automatic Thrust Vector Control

ATVCD Ascent Thrust Vector Control Driver

ATVCS Automatic Thrust Vector Control System

ATVEF Advanced Television Enhancement Forum

ATVT Alignment/Thermal Verification Tool

ATW Appleton (USA-WI) (location identifier) • Atomic Weight • Menasha (USA-WI) (location identifier) • Neenah (USA-WI) (location identifier)

ATWDDS Automated Terminal Weather Dissemination Display System

ATWG Avionics Test Working Group

At. Wt. Atomic Weight

ATX Advanced Technology Extended • Atbasar (Kazakhstan) (location identifier)

ATY Watertown (USA-SD) (location identifier)

ATZ Airport Traffic Zone • Assiut (Egypt) (location identifier)

au Audio (file extension) • Australia

Au Author • Gold • Or

AÜ Automatische Übersetzung

AU Access Unit • Accounting Unit • Administrative Unit • Administrative User • Administrator Unit • African Union (formerly OAU) • Agrégé d'Université • Alignment Unit • All Up • Alternate Use • American University (USA-DC) • Analyzable Unit • Ångström Unit • Août • Apollo Unit • Arbitrary Unit • Archival User • Arithmetic Unit • Astronomical Unit • Atomic Unit • Auburn University (USA-AL) • Audio • August • Australia (ISO country code) • Australie (code ISO de pays) • Author • Authorized User • Auxiliary Unit

AUA Aruba - Reina Beatrix (Netherlands Antilles) (location identifier) • Associations des Universités Africaines • Associations des Universités Arabes • Associations des Universités de l'Atlantique

AUAGA All-Union Astronomical-Geodetic Association (U-SSR) (obsolete)

AUAI Association for Uncertainty in Artificial Intelligence

AUASS Arab Union for Astronomy and Space Sciences

AUB Aft Utility Bridge • Alstublieft • American University of Beirut (Lebanon) • Itauba (Brazil-MT) (location identifier)

AUBas Astronomisches Institut der Universität Basel (periodical code)

AUBC Association of Universities of the British Commonwealth

AUC Ab Urbe Condita • Academic User Committee • American University in Cairo (Egypt) • American University of the Caribbean • Anno Urbis Conditae • Arauca (Colombia) (location identifier) • Area Under Curve (*) • Area Under the Curve • Australian Universities Commission • Authentification Center • Average Unit Cost

AUCBE Advisory Unit for Computer-Based Education (UK)

AUCBM Arab Union for Concrete and Building Materials

AUCC Association des Universités et Collèges du Canada • Association of Universities and Colleges of Canada

Auct Auctorum

AUCV Advanced Unmanned Cargo Vehicle

Aud Audience • Audiencia • Audit • Auditor • Auditory

AUD Allocation Unique Dégressive • Audible • Audio • Audit • Auditing • Augustus Downs (Australia-QLD) (location identifier) • Automatic Data Processing • Dollar (Australia) (ISO currency code)

AUDACIOUS Automatic Direct Access to Information with the Online UDC System

AUDE Association des Utilisateurs de Détecteurs Électroniques (France)

AUDF Automated Unit Development Folder

AUDIST Agence Universitaire de Documentation et d'Information Scientifique et Technique

AUDIT Automated Data Input Terminal

AUDIX Audio Information Exchange

AUE Abu Rudeis (Egypt) (location identifier) • Acta Unica Europea • Acte Unique Européen • Association des Universitaires d'Europe

AUEF Association Universités-Entreprises pour la Formation

AUEL Automated Unit Equipment List

AUF Agence Universitaire de la Francophonie • Authentication Function • Authorized User File

Aufl Auflage

AufzV Aufzugverordnung

Aug Augmentation • Augmentative • Augmented • Augmenter • August

AÜG Arbeitnehmer-Überlassungsgesetz

AUG Administrative Unit Group • Application User Group • Arbeit - Umwelt - Gesundheit • August • Augusta (USA-ME) (location identifier)

Augm Augmentative

AUH Abu Dhabi (UAE) (location identifier)

AUI Access Unit Interface • Akhawayn University in Ifrane (Morocco) • Asociación de Usuarios de Internet (Spain) • Associated Universities Inc. (USA-DC) • Association des Utilisateurs d'Internet (France) • Attached Unit Interface • Attachment Unit Interface • Attachment Universal Interface • Attentive User Interface • Aua Island (Papua New Guinea) (location identifier) • Auxiliary Unit Interface

AUIF Access Unit Interface Female

AUIM Access Unit Interface Male

AUIS Analog Input-Output Unit

Auj Aujourd'hui

AUJ Active Universal Joint • Ambunti (Papua New Guinea) (location identifier)

AuJPA Australian Journal of Physics - Astrophysical Supplement (periodical code)

AuJPh Australian Journal of Physics (periodical code)

AuJPS Australian Journal of Physics - Supplement (periodical code)

AUK Alakanuk (USA-AK) (location identifier)

AUL Astronomy User's Library • Aur Island (Marshall Islands) (location identifier) • Average Useful Life

AULC Arizona Universities Library Consortium (USA-AZ)

Aum Aumentativo • Aumento

AUM Adaptive User Model • Animal Unit Months • Asset Utilization Model • Association Universitaire de Mécanique • Austin (USA-MN) (location identifier) • Aviation Unit Maintenance

AuMM Australian Meteorological Magazine (periodical code)

AuMSJ Australian Mathematical Society Journal (periodical code)

aun Absque Ulla Nota

AUN Auburn (USA-CA) (location identifier)

AUNT Automatic Universal Translator

AUO Active User Object • Administratively Uncontrollable Overtime • African Unity Organization • Area Utilization Officer • Auburn Opelika (USA-AL) (location identifier)

AUP Aberdeen University Press • Acceptable Use Policy • Actual Unit Price • Agaun (Papua New Guinea) (location identifier) • American University of Paris (France) • Appropriate Use Policy • Astroparticle and Underground Physics

AUPC Average Unit Procurement Cost

AUPELF Association des Universités Partiellement ou Entièrement de Langue Française

AUPG American University Publishers Group (USA)

AUPU Average Usage Per User

AUQ Atuona (Marquesas Islands) (location identifier)

Aur Auriga-Aurigae (constellation)

AuR Astronomie und Raumfahrt

AUR All Up Round • Alpha Aurigids (meteor stream) • Aurillac (France) (location identifier) • Automated Underreporter

AURA Association of Universities for Research in Astronomy (USA-DC) • Atmospheric Ultraviolet Radiance Analyzer

AURBO Aurora Borealis

AURELHY Analyse Utilisant le Relief pour l'Hydrométéorologie

AURIO Auroral Imaging Observatory

AURL Aggregate Uniform Resource Locator

AUROC Association des Utilisateurs du Réseau OCLC

AURP AppleTalk Update-based Routing Protocol

AurPh Auroral Physics (periodical code)

AURS Automated Unit Reference Sheet

Aus Australia • Australian • Austria • Austrian

AUS Army of the United States (USA) • Austin - Robert Mueller (USA-TX) (location identifier) • Australia (international vehicle registration code) • Australia (ISO country code) • Australie (code ISO de pays)

AUSAC Arecibo Users and Science Advisory Council

AUSDEC Australasian Spatial Data Exchange Centre

AUSEANET Austroasian Network

AUSF American Urban Star Fest (USA-NV)

Ausg Ausgabe

AUSINET Australian Information Network

AUSLAS Austrian Laser Station

AuslG Ausländergesetz

AUSLIG Australian Survey and Land Information Group • Australian Surveying and Land Information Group (*)

AUSM Advection Upstream Splitting Method

AusSI Australian Society of Indexers (Australia)

Aust Australia • Australian • Austria • Austrian

AUST American University of Sciences and Technology (Lebanon)

Austr Australia • Australian • Austria • Austrian

Austral Australasia • Australasian • Australia • Australian

Aut Auteur • Author • Autograph • Automatique • Automne • Automobile • Autonomy • Autorisation • Autriche • Autumn

AUT Advanced User Terminal • Aristotelian University of Thessaloniki (Greece) • Association of University Teachers (UK) • Atauro (Indonesia) (location identifier) • Austria (ISO country code) • Automated Ultrasonic Testing • Autriche (code ISO de pays)

Auth Authentic • Author • Authoress • Authority • Authorization • Authorize • Authorized

AUTH Authentification • Authorized

AUTHGR Authority Granted

Auto Automatic • Automobile • Automotive

AUTO Automated • Automatic • Automation

Autobiog Autobiographical • Autobiography

AUTOBUS Automated Budget System

AUTODIN Automatic Digital Network

AUTODN Automatic Digital Network

AUTOEXEC Automatic Execution

AUTOFIB Automatic Fibre

Autog Autograph

Autogr Autographe

Autom Automatica (periodical code) • Automatique

Automov Automovilismo

Auton Autonomous

AUTONAV Automatic Navigation

AUTONET Automatic Network

AUTOPATH Automated (flight) Path

AUTOPIC Automatic Personal Identification Code

AUTOSCRIPT Automatic System for Composing, Revising, Illustrating, and Phototypesetting

AUTOSPEC Automated Specification

Autostr Autostrade

AUTOVON Automatic Voice Network

Autp Autopista

Autref Autrefois

AUTOSTRT Automatic Starter

AUTOSTRTG Automatic Starting

AUTOVON Automatic Voice Network

AUTRAN Automatic Translator

AUTS Automatic Update Transaction System

Autv Autovía

AUU ATM User-to-User • Aurukun Mission (Australia-QLD) (location identifier)

AUUA Americas UNIVAC Users Association

AUV Auslandumzugskosten-Verordnung • Autonomous Underwater Vehicle

AUVS Automated Unit Vacancy System

AUW Airframe Unit Weight • All-Up Weight • Authorized Unpriced Work • Wausau (USA-WI) (location identifier)

AUWS Automatic Unmanned Weather Station

aux Auxiliary (file extension)

Aux Auxiliary

AUX Araguaína (Brazil-TO) (location identifier) • Auxiliary

AUXF Auxiliary Frame

Auxil Auxiliary

AUY Aneityum (Vanuatu) (location identifier)

AUZ Aurora (USA-IL) (location identifier)

av Ad Valorem • Annos Vixit

Áv Ávila (Spain)

Av Avance • Avant • Avenida • Avenue • Average • Avocat • Avoirdupois • Avril

A&V Assembly & Verification

A/V Ad Valorem • Air Vehicle • Audiovisual

AV Abnormal Voltage • Acid Value • Acoustic Velocity • Actual Value • Acuité Visuelle • Ad Valorem • Aerial Vehicle • Air Vehicle • Allgemeine Verfügung • Antipodal Void • Antivirus • Aperture Value • Arteriovenous • Artificial Viscosity • Asset Value • Asset Visibility • Astronomicheskii Vestnik • Atharvaveda • Atrioventicular • Audiovisual • Augmented Virtuality • Authenticity Verification • Authorized Version • Auxiliary Vector • Avenida • Avenue • Average Value • Average Variability • Aviation • Avionics • Avis de Versement • Avis de Virement • Avoirdupois

AVA Adaptive Vehicular Antenna • Adirondack Video Astronomy (USA-NY) • Aerodynamische Versuchsanstalt • Agenda de la Vie Astronomique • Asociación Valenciana de Astronomía (Spain) • Astronomische Vereinigung Aarau (Switzerland) • Astronomische Vereinigung Albstadt (Germany) • Astronomische Vereinigung Augsburg (Germany) • Astrophysics Visualization Archive • Attribute Value Assertion • Audiovisual Aid • Audiovisual Authoring

AVAC Antelope Valley Astronomy Club (USA-CA)

AVADS Automated Vacancy Announcement Distribution System

Avail Availability • Available

AVAIL Available

Aval Available

AVARS Audiovisual Analysis and Retrieval System

AVASI Abbreviated Visual Approach Slope Indicator

AVB Allgemeine Versicherungsbedingungen • Analog Video Bandwidth • Aviano (Italy) (location identifier)

AVBATTS Aviation Battery Standardization

Avbl Available

AVBPA Audio- and Video-based Biometric Person Authentication (conference)

AVBS Absolute Value Bit Synchronizer

av. C. Avanti Cristo

AVC Accident Vasculaire Cérébral • Additional Voluntary Contribution • Ahorro Voluntario de Cesantia • Assurance of Voluntary Compliance • Audio Visual Connection • Automatic Valve Control • Automatic Vent Control • Automatic Vibration Control • Automatic Voice Control • Automatic Voltage Control • Automatic Volume Control • Average Variable Cost

AVCG Automatic Vapor Crystal Growth

AVCS Advanced Vidicon Camera System • Attitude and Velocity Control System

Avd Avenida • Avoid

AVD Acoustic Variable Density • Advanced Vocational Degree • Air Vehicle Detection • Alternate Voice/Data • Alternating Voice and Data • Anode Voltage Drop • Astronomischer Verein Dortmund (Germany) • Automatic Voice Data • Automatic Voltage Digitizer • Avant Droit • Aviation Training Device

Avda Avenida

AVDA Associazione Vigevanese Divulgazione Astronomica (Italy)

AVDO Aerospace Vehicle Distribution Office

avdp Avoirdupois

AVDS Audiovisual Display System

AVDTG Analog Via Digital Trunk Group

AVDU Audiovisual Display Unit

Ave Avenue • Average

AVE Aeronautical Vehicle Equipment • Aerospace Vehicle Electronics • Aerospace Vehicle Equipment • Airborne Vehicle Electronics • Airborne Vehicle Equipment • Alta Velocidad • Analisis de Variabilidad Estelar • Asociación de Variabilistas de España (Spain) • Atmospheric Variability Experiment • AutoCAD Visual Extension • Automatic Virtual Environment • Automatic Volume Expansion • Avenue • Average

AVEN Avenue

AVENU Avenue

AVERT Automatic Verification, Evaluation and Readiness Tester

AVES Adaptive Virtual Echelle Spectrograph

AVESCA Aerovías Especiales de Carga Ltda.

AVest Astronomicheskii Vestnik (periodical code)

AVET Atomno Vodorodnaia Energetika i Tekhnologiia (periodical code)

AVF Astronomische Vereinigung Frauenfeld • Avoriaz (France) (location identifier)

Avg Average

AVG Allgemeine Verlagsgesellschaft • Amateurastronomische Vereinigung Göttingen (Germany) • Ambulatory Visit Group • American Volunteers Group • Angestelltenversicherungsgesetz • Approximation de la Vallée Généralisée • Audiovisual Group • Auvergne (Australia-NT) (location identifier) • Avant Gauche • Average • Avion à Grande Vitesse

AVGB Astronomischer Verein der Grafschaft Bentheim (Germany)

Avge Average

AVGR Ames Vertical Gun Range

AVGSE Average String Error

AVGWE Average Word Error

AvH Alexander von Humboldt (1769-1859)

AVH Air Vehicle Handler • Almenvitenskapelige Høyskole (Norway) • Astronomischer Verein Hoyerswerda eV (Germany)

AVHRR Advanced Very High-Resolution Radiometer

avi Audio Video Interleaved (file extension)

AVI Advanced Visual Interface • Air Velocity Index • Arbitrary Variant of an Image • Audio Video Interlaced • Audio Video Interleave • Audio Video Interleaved (*) • Audio Visual Interleaved • Automatic Vehicle Identification • Ciego de Avila (Cuba) (location identifier)

Aviac Aviación

AVIC Aviation Industries of China (China-PRC)

AVICA Association des Villes Câblées (France)

AVID Advanced Visual Information Display • Automated Verification Information Database

AVIM Aviation Intermediate Maintenance

AVIRG Audiovisual Information Research Group

AVIRIS Airborne Visible and Infrared Imaging Spectrometer • Airborne Visible/Infrared Imaging Spectrometer

AVIS Advanced Visual Information System • Audiovisual Information System • Automatic Visual Inspection System

AVISH Air Vehicle Inventory Status and History

AVISO Archivage, Validation et Interprétation des Satellites Océanographiques • Archiving, Validation and Interpretation of Satellite Oceanography

AVISS Astronomicheskii Vestnik Issledovaniia Solnechnoi Sistemy (periodical code)

AVISURS Aerospace Vehicle Inventory, Status and Utilization Reporting System

AVIT Audiovisual Instructional Technology

av. J-C Avant Jésus-Christ

av. JC Avant Jésus-Christ

AVK Arvaikheer (Mongolia) (location identifier) • Astronomische Vereinigung Karlsruhe (Germany) (now AVKa) • Astronomische Vereinigung Kärntens (Austria) • Astronomische Vereinigung Kreuzlingen (Switzerland) • Audio Video Kernel

AVKa Astronomische Vereinigung Karlsruhe (Germany) (formerly AVK)

AvKos Aviatsiia i Kosmonavtika (periodical code)

AVL Address Validity • Adel'son - Vel'skii - Landis (tree) • Advanced Visualization Laboratory • Amtsverbindungsleitung • Angle Versus Length • Asheville (USA-NC) (location identifier) • Assign Variable Location • Audiovisual Language • Automatic Vehicle Location • Avionics Verification Laboratory • Hendersonville (USA-NC) (location identifier)

Avlbl Available

AVLDS Automatic Vehicle Location and Dispatch System

AVLIS Atomic Vapor Laser Isotope Separation

AVLN Automatic Vehicle Location and Navigation

AVLSI Advanced Very-Large-Scale Integration

AVM Accelerated Value Model • Audiovisual Modulator • Automatic Vehicle Monitoring • Automatic Vending Machine

AVMI Automated Video Maintenance Information

AVMIS Automated Vehicle Management Information System

AVML Asiatic Vegetarian Meal

AVMS Automatic Vehicle Monitoring System

Avn Aviation

AVN Adobe Virtual Network • Astronomische Vereinigung Nürtingen eV (Germany) • Avenue • Aviation • Avignon Caumont (France) (location identifier)

AVNA Automatic Vector Network Analyzer

AVNIR Advanced Visible and Near Infrared Radiometer

AVNL Automatic Video Noise Limiter

AVNUE Avenue

AVO Ada Validation Office • Algemeen Vormend Onderwijs • Analog Voice Orderwire • Astronomical Virtual Observatory • Astronomischer Verein Olten (Switzerland) • Astronomischer Verein Ortenau (Germany) • Astrophysical Virtual Observatory • Available Vehicle Occupancy • Average Vehicle Occupancy • A View Of • Avoid Verbal Orders • Avon Park (USA-FL) (location identifier)

AVOSS Aircraft Vortex Spacing System

AVOU Analog Voice Orderwire Unit

AVOW Analog Voice Orderwire

AvP Average Precision

AVP Adaptive Video Processor • Address Verification Pulse • Agression sur la Voie Publique • Application Validation Project • Audiovisual Processor • Authorized Vendor Program • Wilkes Barre (USA-PA) (location identifier) • Scranton (USA-PA) (location identifier)

AVPD Antivirus Product Developers

AvPE Avtomatizatsiia Proektirovaniia Elektronike (periodical code)

AVPN ATM Virtual Private Network

Avr Aviator • Avril

AVR Acquisition Validation Report • Arbeitskreis Volkssternwarte Recklinghausen (Germany) • Automatic Voice Relay • Automatic Voice Recognition • Automatic Voltage

Regulation • Automatic Voltage Regulator • Automatic Volume Recognition • Avril • Axial Velocity Ratio

AVRA Automatic Visual Range Assessor

AVRCS Attitude-Velocity-Reaction Control System

AVRG Automated Virtual Reality Generator

AVRIST Association pour la Valorisation des Relations Internationales Scientifiques et Techniques (France)

AVRP Audiovisual Recording and Presentation

AVRS Astronomischer Verein Remscheid eV (Germany)

AVS Advanced Visualization System • Advanced Visual System • Aerospace Vehicle System • Airborne Video Surveillance • Altitude Vertical Scale • American Vacuum Society (USA) • Application Visualization Software • Application Visualization System • Artificial Visual System • Audiovisual System • Automated Verification System • Autonomous Vision System • Average Speed

AVSA Associazione Valdostana Scienze Astronomiche (Italy)

AVSC Audiovisual Support Center

AvSeM Aviation, Space, and Environmental Medicine (periodical code)

AVSL Association of Vision Science Libraries

AVSM Auxiliary Video Switching Matrix • Auxiliary Video Switch Matrix

AVSO Allgäuer Volkssternwarte Ottobeuren (Germany)

AVSOA American Association of Variable Stars Observers Abstracts (periodical code)

AVSOM American Association of Variable Stars Observers Monographs (periodical code)

ASVSOQ American Association of Variable Stars Observers Quarterly Reports (periodical code)

AVSOR American Association of Variable Stars Observers Reports (periodical code)

AvSP Aviation Safety Programme (NASA)

AVSS Audio Video Support System • Automated Voice Switching System

AVSSE Atmospheric Variability Severe Storms Experiment

AVST Advanced Video Simulation Technique

AVSTC Apple Valley Science and Technology Center (USA-CA)

AVSTOVL Advanced Short Takeoff and Vertical Landing

AVSVB Associazione Veneta Osservatori di Stelle Variabili Bulletin (periodical code)

Avt Avant

AVT Acceptance Vibration Testing • Active Voice Technology • Advanced Vehicle Technology • Algorithm Verification Test • All Volatile Treatment • Analog Voice Telephony • Applied Vehicle Technology • Applied Voice Technology • Astronomische Vereinigung Toggenburg (Switzerland) • Astronomische Vereinigung Tübingen (Germany) • Audiovisual Tutorial • Automatic Video Tracker

AvTek Aviatsionnaia Tekhnika (periodical code)

AvTel Avtomatika i Telemekhanika (periodical code)

Avtme Avtometriia (periodical code)

Avtn Aviation

AVTRS Automated Vehicle Time Recording System

AVTS Advanced Visual Technology System

AVU African Virtual University • Artificial Vision Unit • Avu Avu (Solomon Islands) (location identifier)

AVUS Automobilverkehrs- und Übungstraße (Germany)

AVUSD Apple Valley Unified School District (USA-CA)

AVV Allgemeine Verwaltungsvorschrift • Avalon (Australia-VIC) (location identifier)

AVVEU Altitude and Vertical-Velocity Electronic Unit

AVVI Altitude and Vertical-Velocity Indicator

AVVID Architecture for Voice, Video and Integrated Data (Cisco)

AVW Tucson - Avra Valley (USA-AZ) (location identifier)

A&VWG Assembly & Verification Working Group

AVWG Assembly and Verification Working Group

AVWM Astronomische Vereinigung West-München (Germany)

AvWST Aviation Week and Space Technology (periodical code)

AVX Catalina Island - Avalon Bay (USA-CA) (location identifier)

AVZ Analysen- und Vorhersagenzentrale • Astronomische Vereinigung Zürich (Switzerland)

A/W Actual Weight • Antenna Window

AW Above Water • Accompanying Wave • Actual Weight • Administrative Weight • After Web • Airlock Wall • Airworthiness • Alfén Wave • All Water • All Weather • Antwort • Armstrong-Whitworth • Aruba (ISO country code) • Assembly Workstand • Astronomical Watch • Atomic Weight • Attack Warning • Axial Width • Schreiner Airways BV

AWA Advise When Able • African West Air • All Wave Antenna • Argonne Wakefield Accelerator • Audio Wave Analyzer • Awassa (Ethiopia) (location identifier) • Away Without Authorization

AWAA Attack Warning and Attack Assessment

AWAC Airborne Warning And Control

AWACS Airborne Warning And Control System

AWAD Assembly Work Authorization Document

AWAGS Australia-Wide Array of Geomagnetic Stations

AWAPS Advanced Weather Analysis and Prediction System

AWAR Area Weighted Average Resolution

AWARD All Weather Arrival and Departure

AWAS Automated Work Authorization System

AWB Airway Bill • Akademie der Wissenschaften zu Berlin (Germany) • Awaba (Papua New Guinea) (location identifier)

AWBE Automatic Weather Broadcast Equipment

AWC Arbitrary Waveform Composer • Aviation Weather Center

AWCS Agency-Wide Coding Structure • Automated Wafer Cartridge System • Automatic Warning and Control System

AWCU Association of World Colleges and Universities (now AWE)

Awd Award

AWD African Wave Disturbance • Aniwa (Vanuatu) (location identifier) • Award

AWDAAS Automated Weather Data Acquisition and Archiving System

AWDL Aviation Weather Development Laboratory

AWDS Automated Weather Distribution System

AWE Adaptable Windowing Environment • All-Weather Electronics • Alowe (Gabon) (location identifier) • Association for World Education (formerly AWCU) • Atomic Weapons Establishment (UK)

AWEA American Wind Energy Association (USA)

AWES Area Weapons Effects System

AWF Aging Waste Facility

AWFE Advanced Work Flow Engine

AWFQ Adaptive Weighted Fair Queuing

AWFS Automatisierten Waldbrandfrühwarnsystem

AWG Acceptance Working Group • Accreditation Working Group • Activation Working Group • Advanced Waveform Generator • Advanced Wavefront Generator • American Wire Gauge • Analysis Working Group • Arbitrary Waveform Generator • Architectural Working Group • Arrayed Waveguide Grating • Astronomy Working Group • Astrophysics Working Group • Außenwirtschaftsgesetz • Aviation Week Group • Guilder (Aruba) (ISO currency code)

AWGAA Australian Working Group for Antarctic Astronomy

AWGN Additive With Gaussian Noise

AWH Alfvén Wave Heating • Awareh (Ethiopia) (location identifier)

AWHFY Are We Having Fun Yet

AWI Accommodation Weight Investigation • Air Water Interface • Alfred Wegener Institute • Assembly Work Instruction • Automated Work Instruction

AWIN Aviation Weather Information

AWIPS Advanced Weather Interactive Processing System • Automated Weather Interactive Processing System

AWIS Airport Weather Information System • Archive Writer Interface Software • Association for Women In Science (USA)

AWISE Association for Women in Science and Engineering

AWJ Abrasive Water Jet

Awk Awkward

AWK Aho - Weinberger - Kernighan (language) • Awkward • Wake Island (US Minor Outlying Islands) (location identifier)

AWL Absent With Leave • Absent Without Leave • Administrative Weight Limitation • All-Weather Landing • Application Wrapper Layer • Atmospheric Waves Lidar • Augmented Wire List • Automated Wire List • Average Work Load

AWLS All-Weather Landing System

AWM Advanced Warehouse Management • Advanced Wave Machine • Association for Women in Mathematics (USA) • Automatic Writing Machine • Awaiting Maintenance • West Memphis (USA-AR) (location identifier)

AWMA Air and Waste Management Association

AWMS Advanced Welding Manufacturing System

AWN Alton Downs (Australia-SA) (location identifier) • Automated Weather Network • Automatic Weather Network

AWO Adult Warrant Officer • Attack Warning Order

AWOC Aegean Workshop on Computing

AWOG All-Weather Operations Group

AWOL Absent Without Leave

AWON Stichting Aardwetenschappelijk Onderzoek (Netherlands)

AWOS Absent Without Leave • Automated Weather Observation System

AWP Annual Work Plan • At Work Printing • Austral Downs (Australia-NT) (location identifier) • Aviation Weather Processor • Awaiting Parts

AWPA American Word Processing Association (USA)

AWPDS Attack Warning Processing and Display System

AWPG Aviation Weather Products Generator

AWPSP Association Wallonne du Personnel des Services Publics (Belgium)

AWPV Automated Weld Process Verification

AWQ Ambient Water Quality

AWR Advanced Weather Radar • Airborne Weather Radar • Automated Work Request • Awar (Papua New Guinea) (location identifier)

AWRA American Water Resources Association (USA)

AWRE Atomic Weapons Research establishment (UK) • Autowrite Relocation

AWRS Airborne Weather Reconnaissance System

AWS Acoustic Well Sounding • Advanced Warning System • Advanced Workstations and Systems • Air Weather Service • Air Weather Station • All-Weather System • Alternate Work Schedule • Alternating Work Schedule • Alternative Work Schedule • Altitude Warning System • American Weather Service (USA) • American Welding Society (USA) • Analyst Workstation • Arbitrary Waveform Synthesizer • Astroclimatic Weather Station • Astronaut Workstation • Automated Warehouse System • Automated Wiring System • Automatic Warning System • Automatic Weather Station

AWSK Amsterdaamse Weer- en Sterrenkundige Kring (Netherlands)

AWSO All-Weather Surface Observations

AWSR Air Weather Service Regulation

AWSt Automatische Wetterstation

AW&ST Aviation Week & Space Technology

AWT Abstract Windowing Toolkit (*) • Abstract Windows Toolkit • Actual Work Time • Advanced Windowing Toolkit • Average Working Time

AWTB All-Weather Testbed

AWU Ambulatory Work Unit • Atomic Weight Unit

AWV Atmospheric Wind Velocity

AWVVO Außenwirtschaftsverordnung

AWWA American Water Works Association (USA)

AWY Airway

AWZ Ahwaz (Iran) (location identifier)

Ax Axial • Axiom

AX American Express • Analyst Executive • Architecture Extended • Attack Experimental • Automatic Transmission

AXA Anguilla (Leeward Islands) (location identifier)

AXAF Advanced X-ray Astrophysical Facility • Advanced X-ray Astrophysics Facility (now CXO) (*)

AXAF-I AXAF Imaging

AXAFI AXAF Imaging

AXAF-S AXAF Spectroscopy

AXAFS AXAF Spectroscopy

AXB Alexandria Bay (USA-NY) (location identifier)

AXBT Airborne Expendable Bathythermograph

AXC Additional Extended Coverage • Aramac (Australia-QLD) (location identifier)

AXCP Airborne Expendible Current Profiler

AXCTD Autonomous - Expendable Conductivity - Temperature Depth

AXD Alexandroupolis (Greece) (location identifier) • Auxiliary Drum

Axe Annexe

AXEL Astrophysical Experimental Library

AXF Advanced X-ray Facility

AXG Algona (USA-IA) (location identifier)

AXGAM All-sky X-ray Gamma-ray Astronomy Monitor

AXI Automated X-ray Inspection • Axiom

AXIS Atmospheric X-ray Imaging Spectrometer

AXK Ataq (Yemen) (location identifier)

AXL Alexandria (Australia-NT) (location identifier)

AXM Advanced X-ray Monitor • Armenia (Colombia) (location identifier)

AXML Article Extensible Markup Language

AXN Alexandria (USA-MN) (location identifier)

AXP Anomalous X-ray Pulsar • Spring Point (Bahamas) (location identifier)

AXR Arutua (Tuamotu Islands) (location identifier)

AXRC Auxiliary Receiver

AXS Access • Alpha-backscatter/X-ray-fluorescence Spectrometer • Alpha-proton X-ray-fluorescence Spectrometer • Alpha-proton X-ray Spectrometer • Alpha X-ray Spectrometer • Altus (USA-OK) (location identifier) • ASCA X-ray Source

AXSND Auxiliary Sender

AXT Akita (Japan) (location identifier) • Anomalous X-ray Pulsar

AXU Axum (Ethiopia) (location identifier)

AXUVD Absolute Extreme Ultraviolet Diode

AXV Wapakoneta - Neil Armstrong (USA-OH) (location identifier)

AXX Angel Fire (USA-NM) (location identifier)

ay Aymara (ISO language code)

AY Alderney (Channel Islands) • Apportionment Year • Assyrian

AYA And Yet Again • Association of Yale Alumni (USA-CT) • Ayapel (Colombia) (location identifier)

AYB Accessory Bulletin

AYC Accessory Change • Ayacucho (Colombia) (location identifier)

AYD Alroy Downs (Australia-NT) (location identifier)

AYE At Your Expenses • Fort Devens (USA-MA) (location identifier)

AYG Yaguara (Colombia) (location identifier)

AYH Alconbury (UK) (location identifier) • American Youth Hostels

AYHA American Youth Hostels Association (USA)

AYI Yari (Colombia) (location identifier)

AYK Arkalyk (Kazakhstan) (location identifier)

AYL Anthony Lagoon (Australia-NT) (location identifier)

AYLI As You Like It

AYN Anyang (China-PRC) (location identifier)

Ayo Arroyo

AYO Ayolas (Paraguay) (location identifier)

AYP Ayacucho (Peru) (location identifier)

AYQ Ayers Rock (Australia-NT) (location identifier)

Ayr Ayrshire (UK)

AYR Ayr (Australia-QLD) (location identifier)

AYS Waycross - Ware County (USA-GA) (location identifier)

AYT Antalya (Turkey) (location identifier) • Are You There?

Ayto Ayuntamiento

AYU Aiyura (Papua New Guinea) (location identifier)

AYW Ayawasi (Indonesia) (location identifier)

AYZ Amityville Zahns (USA-NY) (location identifier)

az Azerbaijani (ISO language code) • Azimuth • Azure

aZ auf Zeit

Az Aktenzeichen • Azimuth

AZ Absolute Zero • Age Zero • Aktenzeichen • Altitude Zero • Arizona (USA) • Azerbaijan (ISO country code) • Azimuth • Azinhaga • Aztec

AZB Amazon Bay (Papua New Guinea) (location identifier)

AZD Yazd (Iran) (location identifier)

AZE Azerbaijan (ISO country code)

AZ-EL Azimuth-Elevation

AZEL Azimuth-Elevation

Azer Azerbaijan

AZG Apatzingan (Mexico) (location identifier)

AZh Astronomicheskii Zhurnal (periodical code)

AZH Astronomisches Zentrum Halberstadt (Germany)

AZI Abu Dhabi - Bateen (UAE) (location identifier)

AZK Alphabetischer Zentralkatalog

Azm Azimuth

AZM Astronomisches Zentrum Magdeburg (Germany) • Manat (Azerbaijan) (ISO currency code)

AZN Andizhan (Uzbekistan) (location identifier)

Azo Azores

AZO Allgemeine Zollordnung • Arbeitszeitordnung • Kalamazoo (USA-MI) (location identifier)

AZOP Antarctic Zone Of Peace

AZP Mexico City - Atizapan (Mexico) (location identifier)

AZR Adrar (Algeria) (location identifier)

AZS Absolute Zero State • Astronomisches Zentrum Schkeuditz (Germany) • Automatic Zero Set

AZT Zapatoca (Colombia) (location identifier)

AZZ Ambriz (Angola) (location identifier)

A0 Standard Paper Size

A1 Standard Paper Size

A10 Standard Paper Size

A2 Algorithmic Architecture • Botswana (international aircraft registration code) • Standard Paper Size

A2A Administration to Administration

A2B Administration to Business

A2C Administration to Customer

A2ET Avionics Architecture Evaluation Tool

A2LA American Association for Laboratory Accreditation (USA)

A2S2 Astronomy And Space Sciences

A3 Affordable Acquisition Approach • Astronomes Amateurs Aixois (France) • Aurora Astronomical Association (USA-CO) • Standard Paper Size • Tonga (international aircraft registration code)

A3C Action Concertée Chambre de Combustion

A3I Accelerated Advance Acquisition Initiative

A3IE Association des Ingénieurs de l'Institut d'Informatique d'Entreprise

A3RI Antarctic Astronomy and Astrophysics Research Institute

A4 Astronomische Arbeitsgemeinschaft Aalen eV (Germany) • Standard Paper Size

A4O Oman (international aircraft registration code)

A5 Bhutan (international aircraft registration code) • Standard Paper Size

A6 Standard Paper Size • UAE (international aircraft registration code) • United Arab Emirates (international aircraft registration code)

A7 Qatar (international aircraft registration code) • Standard Paper Size

A8 Standard Paper Size

A9 Standard Paper Size

A9C Bahrain (international aircraft registration code)

A° Arroyo

B

b Bachelor • Bajo • Bale • Bar • Barn • Bass • Battle • Bay • Beam • Bedroom • Before • bei • beim • Billion • Bin • Binding Factor • Bis • Bit • Black • Blend • Blended • Book • Born • Bottom • Bound • Breadth • Brother • Brotherhood • Burgers Vector • By • Collision Parameter • Ecliptic Latitude • Galactic Latitude • Heliocentric Latitude • Impact Parameter • Magnetic-Field Strength • Mobility Ratio • Phonon Annihilation Operator • Semi-Minor Axis • Wien Displacement Law Constant

B Baai • Bachelor • Bacillus • Bad • Baht (currency) • Bahia • Baía • Baie • Baja • Balboa (currency) • Ban • Band • Banque • Barrage • Baryon Number • Base • Bass • Baud • Baumé • Bay • Beatus • Beau • Bel • Belgium (international vehicle registration code) • Belgium (postal country code) • Bien • Billion • Biological • Bishop • Black • Block • Blue • Board • Bois • Bolívar (currency) • Boliviano • Book • Bor • Born • Boron • Boundary Conditions • Bowen Ratio • Bravo (international alphabet) • Breadth • Breakdown Potential • Brewster • Brief • British • Broadband • Brother • Brotherhood • Bucht • Building • Bulb • Bulletin • Byte • China (PRC) (international aircraft registration code) • Ecliptic Latitude of the Sun • Eleven • Magnetic Flux Density • Magnetic Induction • Mass Defect • Oort's Constant of Galactic Rotation • PRC (international aircraft registration code) • Rotational Constant • Secondary Road • Spectral Type • Susceptance • Three Hundred (Roman numeral) • 300 (Roman numeral)

ba Bashkir (ISO language code)

Ba Bahía • Bath • Bathroom • Barium • Baryum

B/A Bank Angle • Barometric Altimeter • Boost/Abort • Booster Adapter • Buffer Amplifier

B&A Boat & Aircraft

BA Bachelor of Arts • Bagages Accompagnés • Bahía (Brazil) • Balancing Allowance • Bangladesh • Bank Acceptance • Bank Angle • Bantu • Barometric Altimeter • Bas Âge • Base Assembly • Basic Agreement • Basic Authorization • Basisanschluß • Basses-Alpes (France) • Behavior Aggregate • Belief Agent • Bibliography of Astronomers • Bilan Annuel • Binary Add • Blind Approach • Block Adjustment (technique) • Blockausgleichung • Boolean Algebra • Booster Adapter • Bordereau d'Achats • Bosnia-Hercegovina (ISO country code) • Bosnie-Herzégovine (code ISO de pays) • Bourse d'Apprentissage • Breathing Air • Breathing Apparatus • Bridge Amplifier • British Academy • British America • British Association (UK) • Bronze Age • Budget Activity • Budget Analysis • Budget Authority • Budget Authorization • Buenos Aires (Argentina) • Buffer Amplifier • Bulletin d'Admission • Bureau Auxiliaire • Bureau d'Affrètement • Bureau d'Archives • Bus Automaton • Bus Available

BAA Bachelor of Applied Arts • Bialla (Papua New Guinea) (location identifier) • Blanket Authorization Agreement • British Airport Authority (UK) • British Astronomical Association (UK) • Broad Agency Announcement • Broad Area Announcement • Broadband Active Analyzer • Budget Activity Account • Buffalo Astronomical Association (USA-NY) • Burgenländische Amateurastronomen (Austria) • Business Area Analysis

BAAA Boletín de la Asociación Argentina de Astronomía (periodical code)

BAAE Bachelor of Aeronautical and Astronautical Engineering

BAAL British Association for Applied Linguistics (UK)

BAAM Basic Administration And Management

BAAN Budget Authorization Account Number

BAAPG Bulletin of the American Association of Petroleum Geologists (periodical code)

BAAS Billing/Accounting Application Service • British Association for the Advancement of Science (UK) • Broad Area Surveillance System • Bulletin of the American Astronomical Society (periodical code)

Bab Babylon

BAB Booster Assembly Building • Bring A Bottle • Budget Advisory Board • Bundesautobahn • Bureau pour l'Automatisation des Bibliothèques (France) • Marysville Beale (USA-CA) (location identifier)

BABAS Bay Area Book Awareness Service (USA-CA)

BABEL Broadcasting Across the Barriers of European Languages

BABI Basaltic Achrondite Best Initial

BABOGAS Balloon-Borne Gaschromatograph

BABS Beam Approach Beacon System • Blind Approach Beacon System

BABT British Approval Board for Telecommunications (UK)

Bac Baccalauréat (France)

BAC Baccalauréat (France) • Balanced Asynchronous Class • Barranca de Upia (Colombia) (location identifier) • Barometric Altitude Control • Basic Attitude Control • Beijing Astrophysics center (China-PRC) • Beleid Advies Commissie • Billing - Accounting - Charging • Binary-Analog Conversion • Binary Asymmetric Channel • Blood-Alcohol Concentration • Blueshifted Absorption Component • Boeing Aerospace Co. • Boeing Aircraft Co. • Booster Assembly Contractor • Bordeaux Astrographic Catalogue • Bristol Astronomy Club (UK) • British Aerospace Corporation (UK) • Broad Acoustic Category • Budget At Completion • Buffer Access Card • Bulletin d'Activité Cyclonique • Bureau d'Archives Collectives

BACA Beleid Advies Commissie Aardwetenschappen (Netherlands)

BACAS Belgian Academy Council of Applied Sciences (Belgium)

BACE Basic Automatic Checkout Equipment • Booster Adapter Control Electronics

Bach Bachelor

BACH Banc (de tests) d'Attitude à Configuration Hybride • Beautiful Aperiodic Crystal of Harmony

BACIS Budget Accounting Information System

BACK Balloon Air Čerenkov (technique)

Backgrd Background

BACL Beleid Advies Commissie Levenwetenschappen (Netherlands)

BACODINE BATSE Coordinates Distribution Network

BACP Bandwidth Allocation Control Protocol

BACS Bank Automated Clearing System (UK) • Bibliographic Access and Control System • Body Axis Coordinate System

Bact Bacterial • Bacteriology • Bacterium

BACT Best Available Control Technology

BACV Budget At Completion Variance

Bad Badajoz (Spain)

BAD Banque Africaine de Développement • Biological Activity Database • Broken As Designed • Shreveport Barksdale (USA-LA) (location identifier)

BADAS Binary Automatic Data Annotation System • Blackpool And District Astronomical Society (UK)

BADC Beijing Astronomical Data Centre (China-PRC) • Binary Asymmetric Dependent Channel • British Atmospheric Data Centre (UK)

BADEA Banque Arabe de Développement Économique en Afrique

BADG British Aerospace Dynamics Group

BADPG Berlin Akademie Verlag Ergebnisse Plasmaphysik und Gaselektronik (periodical code)

BAdW Bayerische Akademie der Wissenschaften (Germany)

BAe British Aerospace

BAE Bachelor of Aeronautical Engineering • Barcelonnette (France) (location identifier) • Beacon Antenna Equipment • Biblioteca de Autores Españoles • Bureau of American Ethnology

BAeA British Aerospace Australia

BAEF Belgian-American Educational Foundation

BAES British Aviation Enthusiasts Society (UK)

BAeSC British Aerospace Space and Communications

BAF Bacterial Air Filter • Barnes Absorption Filter • Baseline Asset File • Bâtiment d'Assemblage Final • Benefield Anechoic Facility (USA-CA) • Biological Air Filter • Biological Air Filtering • Westfield Barnes (USA-MA) (location identifier)

BAFA Brevet d'Aptitude aux Fonctions d'Animateur (France)

BAFD Brevet d'Aptitude aux Fonctions de Directeur (France)

BAFI Base des Agents Financiers

BAFO Best And Final Offer • British Air Forces of Occupation

BAFöG Bundesausbildungsförderungsgesetz

BaF2 Barium Fluoride

Bag Bagages • Baggage

BAG Baggage • Baguio (Philippines) (location identifier) • Branche Asymptotique des (étoiles) Géantes • Brigade Anti-Gang (France) • Budget Activity Group • Budget Aggregate Group • Bundesamt für Gesundheit (Germany) • Bundesamt für Gesundheitswesen (Switzerland) • Bundesarbeitsgericht (Germany)

BAGBI Booksellers' Association of Great Britain and Ireland

BAGI Backscatter Absorption Gas Imaging

Bah Bahamas

BAH Bahrain Muharraq (Bahrain) (location identifier) • Basic Adaptive Hardware • Basic Allowance for Housing

BAHC Biospheric Aspects of the Hydrological Cycle

BAI Bande Aromatique Infrarouge • Barometric Altitude Indicator • Bearing Altitude Indicator • Beneficios Antes de Impuestos • Beratende Ausschuß der Industriephysiker (Germany) • Budget Advisory Board • Buenos Aires (Costa Rica) (location identifier)

BAIC Barring All Incoming Calls • Binary Asymmetric Independent Channel

BAICz Bulletin of the Astronomical Institutes of Czechoslovakia (periodical code)

BAID Boolean Array Identifier

BAIE British Association of Industrial Editors (UK) • Bureau d'Accueil et d'Information sur l'Emploi

BAIL Boundary And Interior Layers (conference)

BAILER British Association for Information and Library Education and Research (UK)

BAIM Baseline Advanced Industrial Management

BAIR Breathing Air

Ba. Is. Bahamas Islands

BAIT Berkeley Automatic Imaging Telescope (USA-CA) (now KAIT)

BAJ Bachelor of Arts in Journalism • Bali (Papua New Guinea) (location identifier)

bak Backup (file extension)

BAK Backup • Baku (Azerbaijan) (location identifier) • Bildaufklärung • Binary Adaptation Kit

BAKSA Bahagian Kajian Sains Angkasa (Malaysia)

BAKT Basisanschluß-Konzentrator

Bal Balance • Balanced • Baluchistan

BAL Balance • Basic Assembly Language • Batman (Turkey) (location identifier) • Blood Alcohol Level • Boîte Aux Lettres • Broad Absorption Line

BALAAG Broad Absorption Line Radio Galaxy

BALAST Balloon Astronomy

BALI Backweb Authoring Language Interface

BALIS Bavarian Agriculture Information System (Germany)

Balk Balkan

BALM Block and List Manipulator

Baln Balnéaire

BALO Bulletin d'Annonces Légales Officielles

BAlp Basses-Alpes (France)

BALPA British Airline Pilots Association (UK)

BALQ Broad Absorption Line Quasar

BALQSO Broad Absorption Line Quasi-Stellar Object

BALR Broad Absorption Line Region

BALS Broad Absorption Line System

BALSA Brown University Algorithm Simulator and Animator

BALSAMINE Ballons Surpressurisés pour l'Analyse de la Mousson Indienne d'Été

Balt Baltic • Baltimore (USA-MD)

BALT Barometric Altitude

BaltA Baltic Astronomy (periodical code)

BALUN Balanced/Unbalanced

BAM Bachelor of Applied Mathematics • Bacteriology and Applied Microbiology • Baikal - Amur - Magistral (railway) • Balloon Anisotropy Measurement • Basic Access Method • Basic Access Mode • Basic Acoustic Modeling • Battle Mountain - Lander County (USA-NV) (location identifier) • Bifunctional Adaptive Mesh • Bipartite Agreed Materiel • Bit Assignment Matrix • Boyan Action Module • Brackett Alpha Measurement • Bradley Aberration Method • Bundesanstalt für Materialforschung (Germany) • Burst Amplitude Modulation

BAMBI Better Access to Manuscripts and Browsing of Images

BAMIN Biology/Applied Mathematics Instruction Model

BamKV Kleine Veröffentlichungen der Remeis-Sternwarte Bamberg (periodical code)

BAMM Balloon Altitude Mosaic Measurement • Bidirectional Asymmetric Multipoint-to-Multipoint

BAMN By Any Means Necessary

BAMO Polyazidomethyloxetane

BAMR Bipartite Agreed Materiel Requirement

BamVe Veröffentlichungen der Remeis-Sternwarte Bamberg (periodical code)

BAMS Bulletin of the American Meteorological Society (periodical code)

BAMx Basisanschluß-Multiplexer

BAN Base Académique des Nomenclatures • Base Aéronavale • Basic Access Node • Basongo (Zaire) (location identifier) • Boundary-Access Node • Budget Allocation Notice • Bulgarska Akademiia na Naukite (Bulgaria) • Bulletin of the Astronomical Institutes of the Netherlands (periodical code)

BANANA Build Absolutely Nothing Anywhere Near Anyone

BANCS Bell Application Network Control System

BAND Boston Area NEXRAD Demonstration • Bulgarska Akademiia Naukite Doklady (periodical code)

BANG Broadband Active Network Generation

Banl Banlieue

Banq Banque

BANR Board on Agriculture and Natural Resources

BANS Bulletin of the Astronomical Institutes of the Netherlands, Supplement Series (periodical code)

BANZARE British, Australian, New Zealand Antarctic Research Expedition

BAO Bachelor of Arts in Obstetrics • Ban Mak Khaen - Udorn (Thailand) (location identifier) • Beijing Astronomical Observatory (China-PRC) • Bisei Astronomical Observatory (Japan) • Boulder Atmospheric Observatory (USA-CO) • Byurakan Astronomical Observatory (Armenia)

BAOC Barring All Outgoing Calls

BAORB Bulletin Astronomique de l'Observatoire Royal de Belgique (periodical code)

BAOSS Beijing Astronomical Observatory Supernova Survey

bàp Billets à Payer

Bap Baptist • Baptized

BAP Baibara (Papua New Guinea) (location identifier) • Bandwidth Allocation Protocol • Basaltic Achondrite Parent • Baseline Activity Profile • Basic Assembler Program • Basic Assembly Program • Best Adaptive Path • Best Air Penetration • Binary Age Postulate • Biotechnology Action Programme • Bipartite Advisory Publication • Bottom Agitation Process • Branch Arm Piping • Branche d'Activité Professionnelle • Brevet d'Aptitude Physique • British Association of Planetaria (UK) • Bulletin d'Alerte aux Précipitations

BAPE Balloon Atmospheric Propagation Experiment • Bipartite Agreed Plan of Engineering

BAPI Business Applications Programming Interface

BAPM Bidirectional Asymmetric Point-to-Multipoint

BAPMoN Background Air Pollution Monitoring Network (*) • Baseline Air Pollution Monitoring Network

BAPP Bidirectional Asymmetric Point-to-Point

BAPS Baseline Air Pollution Station (Australia-TAS) • Berg Analysis and Prediction System • Bulletin de l'Académie Polonaise des Sciences (periodical code) • Bulletin of the American Physical Society (periodical code)

Bapt Baptist • Baptized

BAPT Bundesamt für Post und Telekommunikation (Germany)

BAPTA Bearing And Power Transfer Assembly

BAPU Bureau d'Aide Psychologique Universitaire

BAQ Barranquilla - Ernesto Cortissos (Colombia) (location identifier)

bàr Billets à Recevoir

Bar Barometer • Barometric • Baron • Barones • Baronne • Barrel • Barrister

BAR Baker Island (USA-AK) (location identifier) • Ballistocraft for Acceleration Research • Band Area Ratio • Bandwidth Adaptive Reencoding • Barão • Barometer • Base Address Register • Bimonthly Activity Report • Broad Area Review • Bucket Activation Ratio • Buffer Address Register

Barb Barbados

BARB Ballast Aerating Retrieval Boom • Board Arbiter • Bulletin de l'Academie Royale de Belgique (periodical code)

Barc Barcelona (Spain)

BARC Bay Area Research Center (USA-CA) • Bhabha Atomic Research Centre (India) • Block Adaptive Rate Controlled

BARD Bay Array Regional Deformation (USA-CA) • Binational Agricultural Research and Development

BARE Basic Architecture Receive Element

Barg Bargain

BARN Biological and Agricultural Reactor Netherlands

Baro Barometer

BARO Barometer

BARON Baronesa

BARP Bundesamt für Raumplanung (Switzerland)

Barr Barrister

BARRNet Bay Area Regional Research Network (USA-CA)

BARS Backup Attitude Reference System • Barred And Ringed Spirals • Baseline Accounting and Reporting System • Bistatic Auroral Radar System • Budget Analysis Reporting System

BARSC British Association of Remote Sensing Companies (UK)

BaRT Bayesian Reasoning Tool

BART Bay Area Rapid Transit (USA-CA) • Billing Analysis Reporting Tool

BARTOC Bus Advanced Real-Time Operational Control

BARTS Bell Atlantic Regional Timesharing

BARWAN Bay Area Research Wide Area Network

bas BASIC (file extension)

BAS Bachelor of Applied Sciences • Balalae (Solomon Islands) (location identifier) • Ballarat Astronomical Society (Australia-VIC) • Baltimore Astronomical Society (USA-MD) • Barbados Astronomical Society • Barnard Astronomical Society (USA-TN) • Baseline Assessment Strategy • BASIC • Basic Absorption Sample • Basic Activity Set • Basic Allowance for Subsistence • Bedford Astronomical Society (UK) • Birmingham Astronomical Society (UK) • Birmingham Astronomical Society (USA-AL) • Bit-rate Allocation Signal • Boise Astronomical Society (USA-ID) • Bolton Astronomical Society (UK) • Boolean Assignment Statement • Bradford Astronomical Society (UK) • Braintree Astronomical Society (UK) • Branch And Save • Breathing Air System • Breckland Astronomical Society (UK-England) • Brevard Astronomical Society (USA-FL) • Bridgewater Astronomical Society (UK) • Bridport Astronomical Society (UK) • Brisbane Astronomical Society (Australia-QLD) • Bristol Astronomical Society (UK) • British Antarctic Survey • Broadcast Access Server • Broadcast Auxiliary Service • Budget Allocation Summary • Bulgarian Academy of Sciences • Bundesamt für Schiffsvermessung (Germany) (now BSH) • Bureau d'Aide Sociale • Burst Alert Station

BASc Bachelor of Applied Sciences

BASC Board on Atmospheric Science and Climate (NRC) • Brouwer Award Selection Committee (AAS)

BASD Ball Aerospace Systems Division • Basic Active Service Date

BASE Beam Aspect Seeker Evaluation • Beaufort and Arctic Storms Experiment • Beta Alumina Solid Electrolyte • Bipolar Advanced Silicon for Europe • British Aerospace Systems and Equipment

BASF Badische Anilin- und Sodafabrik

BASG Ball Aerospace Systems Group

BASH Bird Aircraft Strike Hazard • Born-Again Shell

BASI Bulletin of the Astronomical Society of India (periodical code) • Bureau of Air Safety Investigation

BASIC Basic Algebraic Symbolic Interpretive Compiler • Basic Automatic Stored Instruction Computer • Bay Area

Shared Information Consortium (USA-CA) • Beginner's All-purpose Symbolic Instruction Code • British-American Security Information Council

BASIL Boeing Automated Systems Information and Logging

BASIN Basic Studies on Air glow, Smog and Inversion

BASIS Bank Automated Service Information System • Bibliothekarisch Analytisches System zur Informationsspeicherung • Budgetary And Scheduling Information System • Burst and All-Sky Imaging Survey • Burst Arcsecond Imaging and Spectroscopy

BASJE Bolivian Air Shower Joint Experiment

BASM Built-in Assembler

BASOPS Base Operations

BASR Branch And Save Register

BASS Backup Avionics System Software • Base Augmentation Support Set • Base Automated Service Store • Base Solaire Sol • Broad Area Surveillance System • Broadband Array Spectrograph System • Broadband Array Spectrometer System (*) • Bulk Avalanche Semiconductor Switch

BAST Best Available and Safest Technology • Board on Army Science and Technology

BASV Baade-Swope Variable (star)

BASW Baade-Swope • British Association of Social Workers (UK)

BASYS Basic System

bat Batch Job (file extension)

Bât Bâtiment

Bat Batavia • Battery

BAT Back Arc Tectonics • Barretos (Brazil-SP) (location identifier) • Basic Air Temperature • Batch • Battery • Beam Assembly Teleoperator • Best Available Technology • Biochemical Automated Technology • Blind Approach Training • Block Address Translation • Brevet d'Aptitude Technique • British Antarctic Territory • Buffer Alias Table • Bundesangestelltentarifvertrag (Germany) • Bureau d'Assistance Technique • Burst Alert Telescope

BATC Ball Aerospace and Technologies Corporation (USA-CO) • Beijing - Arizona - Taipei - Connecticut

BATE Base Activation Test Equipment

BATEA Best Available Technology Economically Available

BATES Battlefield Artillery Tactical Engagement System

BATF Blind Approach Training Flight

BATGE Biosphere-Atmosphere Trace Gas Exchange • Biosphere-Atmosphere Tropical Global Exchange

Bath Bathroom

BAth Bulletin of the Astronomical Institute of Athens (periodical code)

BATH Best Available True Heading

BATO Balloon-Assisted Takeoff

Bâts Bâtiments

BATS Basic Additional Teleprocessing Support • Bayesian Analysis of Time Series • BDT Acquisition Telescope System • Billing Action Tracking System • Biosphere-Atmosphere Transfer Scheme • Broadband Antenna Test System

BATSE Burst And Transient Source Experiment

Batt Battalion • Battery

BATT Battery

Bau Bureau

BAU Bande d'Arrêts d'Urgence • Baryon Asymmetry of the Universe • Bauru (Brazil-SP) (location identifier) • Business As Usual

BAUA Business Aircraft Users Association

BAUEM Bundesamt für Übermittlungstruppen (Switzerland)

bàv Bon à Vue

Bav Bavaria (Germany) • Bavarian

BAV Ballistic Accuracy Verification • Baotou (China-PRC) (location identifier) • Berliner Arbeitsgemeinschaft für Veränderliche Sterne eV (Germany) • Breitband-Anschlußvermittlungsstelle • Bundesaufsichtsamt für das Versicherungswesen (Germany) • Bundesdeutsche Arbeitsgemeinschaft für Veränderliche Sterne eV (Germany) • Buque de Alta Velocidad

BAVMM Berlin Akademie Verlag Shriftenreihe des Zentralinstituts Mathematik und Mechanik (periodical code)

BAVSM Berliner Arbeitsgemeinschaft für Veränderliche Sterne - Mitteilungen (periodical code)

BAVSO Bulletin of the American Association of Variable Stars Observers (periodical code)

BAVSR Berliner Arbeitsgemeinschaft für Veränderliche Sterne - Rundbrief (periodical code)

baw Bis auf Weiteres

BAW Bayerische Akademie der Wissenschaften (Germany) • Biawonque (Gabon) (location identifier)

BAWMN Bayerische Akademie Wissenschaften mathematisch naturwissenschaftliche Klasse Sitzungsberichte (periodical code)

BaWü Baden-Württemberg (Germany)

BAX Barnaul (Russia) (location identifier)

Bay Bayern (Germany) • Bayonet

BAY Baia Mare (Romania) (location identifier) • Bayer

Bayer Bayerisch

BAYS British Association Youth Section

BAZ Barbelos (Brazil-AM) (location identifier) • Buchloer Astronomisches Zirkular

bb Books

B/B Baseband • Breadboard

B&B Bed & Breakfast

BB Backbone • Ball Bearing • Banc-Balance • Band-by-Band • Barbade (code ISO de pays) • Barbados (ISO country code) • Bare Base • Baseband • Bayesian Bound • Bed and Breakfast • Behavior-Based • Bénéfice Brut • Bescherming Burgerbevolking • Big Bang • Billet de Banque • Black Body • Blattbereich • Blowback • Blue Belly • Blue Book • Brassboard • Breadboard • Bright Band • Broad Band • Building Block • Bulletin Board • Bulletin de Bagages • Bureau de Bienfaisance • Bureau of Budget • Business Board • Double Black

BBA Bachelor in Business Administration • Balanced Budget Act (USA) • Balmaceda (Chile) (location identifier) • Blue Book Abstract • Broadband Antenna • Bucket Brigade Algorithm

BBAA Back Bay Amateur Astronomers (USA-VA)

BBAO Black Birch Astronomical Observatory (New Zealand)

BBAR Broadband Anti-Reflective (coating)

BBAS Balloon-Borne Astronomical Studies • Base Budget Automation System • Big Bend Astronomical Society (USA-TX)

BBASS Budget Backup And Support System

BBauG Bundesbaugesetz

BBB Beans - Bacon - Biscuits • Bed, Breakfast and Bath • Benson (USA-MN) (location identifier) • Better Business Bureau • Big Blue Bump • Treble Black • Triple Black (pencil)

BBC Backup Bus Controller • Base Band Cabinet • Base Band Convertor • Basic Bibliographic Citation • Bay City (USA-TX) (location identifier) • Beam-to-Beam Correlation • Before Bottom Center • Before Business Clearance • Big Bang Cosmology • Blockhouse Battery Charger • Breadboard Camera • British Broadcasting Corporation (UK) • Broadband Bearer Capability • Broadband Conducted

BBCEn BBC Engineering (periodical code)

BBCI Biophysics and Biochemistry Citation Index

BBCNC British Broadcasting Corporation Networking Club

BBD Bivariate Brightness Distribution • Black-Body Detection • Block Branch diagram • Brady Curtis (USA-TX) (location identifier) • Bucket Brigade Device • Bundesbaudirektion (Germany) • Dollar (Barbados) (ISO currency code)

BBDC Before Bottom Dead Center

BBE Background Block Error • Big Bell (Australia-WA) (location identifier) • Big Building East

BBED Broad-Band Energy Distribution

BBEM Bed, Breakfast, and Evening Meal

BBER Background Block Error Ratio

BBergG Bundesberggesetz

BBesG Bundesbesoldungsgesetz

BBF Balloon-Borne Filter • Beijing Book Fair (China-PRC) • Berlin Brandenburg Flughafen (Germany) • Bivariate Brightness Function • Bulletin des Bibliothèques de France • Burlington (USA-MA) (location identifier) • Bursty Bulk Flow

BBFR Balloon-Borne Filter Radiometer

BBFW Broadband Fixed Wireless

BBG Butaritari (Kiribati) (location identifier)

BBGKY Born - Bogoliubov - Green - Kirkwood - Yvon (hierarchy)

BBH Barth (Germany) (location identifier)

BBHIR Broadband Hemispheric Infrared Radiometer

BBHSR Broadband Hemispheric Solar Radiometer

BBI Berlin Brandenburg International (Germany) • Bhubaneswar (India) (location identifier) • Broadband Interactive • Broadband Interferometer • Broadband Interferometry • Bureau des Brevets et Inventions

BBIAT Be Back In A Tick

BBiG Berufsbildungsgesetz

BBIR Broadband Infrared

BBJ Ball-Bearing Joint • Bitburg (Germany) (location identifier) • Boeing Business Jet

BBK Breadboard Kit • Kasane (Botswana) (location identifier)

bbl Barrel

BBL Babolsar (Iran) (location identifier) • Barrel • Basic Business Language • Branch Back and Load

BBLP Broadband Linear Polarization

BBLT Bus Block Transfer

BBM Battambang (Cambodia) (location identifier) • Beam-Bending Magnet • Blocking and Bracing Material • Books By Mail • Breadboard Model • Break Before Make • Building Block Monochromator • Bulk Business Mail

BBMF Balloon-Borne Magnet Facility

BBMIP Branch-Bound Mixed Integer Programming

BBML Baby Meal

BB&N Bolt, Beranek & Newman Inc.

BBN Backbone Network • Bario (Malaysia) (location identifier) • Big Bang Nucleosynthesis (*) • Big Building North • Black Body Nucleosynthesis • Bloomberg News Radio • Bolt, Beranek and Newman Inc.

BBNS Big Bang Nucleosynthesis • Broadband Network Services

BBO Berbera (Somalia) (location identifier) • Beta Barium Borate • Booster Burnout

BBOL Building-Block-Oriented Language

BBORD Bulletin Board

BBP Bembridge (Isle of Wight) (location identifier) • Binary Backoff Procedure • Broadband Photometer • Building Block Principle

BBPC Broadband Polarizing Cube

BBPS Behavior-Based Personal System

BBQ Barbecue • Barbuda (Leeward Islands) (location identifier)

BBQC British Board of Quality Control (UK)

BBR Backbone Ring • Balloon-Borne Radio • Band-to-Band Registration • Basse Terre (Guadeloupe) (location identifier) • Black-Body Radiation • Broadcast-Radiated

bbrg Ball Bearing

BBRS Balloon-Borne Radio System

bbs Bulletin Board System (file extension)

BBS Baseline Build Standard • Big Building South • Blackbush (UK) (location identifier) • Bright Star Spot • Bulletin Board Server • Bulletin Board Service • Bulletin Board System

BBSAG Bedeckungsveränderlichen-Beobachter der Schweizerischen Astronomischen Gesellschaft (Switzerland) • Bulletin der Bedeckungsveränderlichen-Beobachter der Schweizerischen Astronomischen Gesellschaft (periodical code)

BBSes Bulletin Board Systems

BBSO Big Bear Solar Observatory (USA-CA)

BBSP Balloon-Borne Solar Pointer

BBSRC Biotechnology and Biological Sciences Research Council (UK)

BBSS Balloon-Borne Sounding System

BBT Basal Body Temperature • Berberati (Central African Rep.) (location identifier) • Big-Bang Theory • Broadband Technology

BBTT Berlin-Brandenburger Teleskoptage (Germany)

BBU Bit Buffer Unit • Bucureşti Baneasa (Romania) (location identifier)

BBUSM Babes Bolyai Universitas Studia Series Physica (periodical code)

BBUSS Babes Bolyai Universitas Studia Series Mathematica (periodical code)

BBV Bereby (Ivory Coast) (location identifier)

BBW Big Building West • Broken Bow (USA-NE) (location identifier) • Bundesamt für Bildung und Wissenschaft (Switzerland)

BBX Blue Bell (USA-PA) (location identifier)

BBXRT Broad Band X-Ray Telescope

BBY Bambari (Central African Rep.) (location identifier)

BBZ Zambezi (Zambia) (location identifier)

bc Before Christ • Between Centers • Blind Copy

B/C Bench Check • Birth Certificate

B&C Boller & Chivens

BC Bachelor of Chemistry • Bachelor of Commerce • Back Connected • Back Course • Bad Character • Bad Check • Bail Commercial • Baisse Comprise • Baisse des Cours • Baja California (Mexico) • Balance Commerciale • Baldone Catalogue • Ballistic Camera • Ballistic Centre • Ballistic Coefficient • Bar Code • Base Construction • Baseline Configuration • Basic Control • Battery Charge • Battery Charger • Battery Conditioner • Beam Collimator • Bearer Capability • Bearing Change • Beat Cepheid • Beco • Before Christ • Before Computer • Begin Construction • Bell Cord • Bench Check • Bénéfice Commercial • Between Centers • Bibliothekcentralen • Big Crunch • Billet Collectif • Binary Calculator • Binary Code • Binary Counter • Birth Certificate • Black Carbon • Black Cluster • Blind Copy • Bloch Curve • Block Check • Blue Core • Board of Control • Body-Centered • Bolometric Correction • Bolt Center • Bon de Caisse •

Bon de Commande • Bordereau de Coupons • Borocarbon • Bose Condensation • Boston College (USA-MA) • Bottom Center • Bottom Current • Boundary Condition • Bound-Complex (mechanism) • Bourse de Commerce • Branch Code • Brevet de Capacité • Brigade Criminelle • British Columbia (Canada) • British Commonwealth • British Council • Broadcast • Broadcast Control • Broad Classification • Bubble Chamber • Bubble Combustion • Budget Center • Budgeted Cost • Bulkhead Connector • Bulletin de Commande • Buoyancy Compensator • Bureau Central • Bureau de Chômage • Bureau de Contrôle • Bureau de Correspondance • Bureau of the Census (USA) • Bus Controller • Bus Coupler • Business Class • Business Component • Busy Condition • Butyl Chloride

Bca Boca

B/CA Business and Commercial Aviation

BCA Bachelor of Commerce and Administration • Baracoa (Cuba) (location identifier) • Base Contracting Automated (system) • Benefit/Cost Analysis • Bilderberg Continuum Atmosphere • Binary Collision Approximation • Board of Contract Appeal • Booster Change Assembly • British Crystallographic Association (UK) • Bureau Canadien des Archivistes (Canada) • Bureau Commun d'Architecture • Bureau of Canadian Archivists • Business and Commercial Aviation • Business Council of Australia • Buyoant Cable Antenna

BCAA Business Clearance Approval Authority

BCAAS Berks County Amateur Astronomical Society (USA-PA)

BCAC Barker Coded Alternating Code • British Conference on Automation and Computation

BCAG Boeing Commercial Airplane Group

BCAR British Civil Airworthiness Requirement

BCAS Base Contracting Automated System

BCB Baseline Configuration Board • Battery Control Building • Binary Code Box • Blacksburg - Virginia Tech (USA-VA) (location identifier) • Broadcast Band

BCBG Bon Chic, Bon Genre

BCC Base Communications Center • Basic Computer Concept • Beacon Control Console • Bear Creek (USA-AK) (location identifier) • Bellevue Community College (USA-WA) • Bibliographic Compilation Catalogue • Binary Control Character • Blind Carbon Copy • Blind Courtesy Copy • Block Check Character • Block Check Count • Block Control Character • Blocked Call Cleared • Body-Centered Cubic • Booster Countdown Controller • Brevard Community College (USA-FL) • British Copyright Council (UK) • Budget Classification Code • Bureau Central de Compensation • Bureau Central de Comptabilité • Business Cooperation Centre

BCCH Broadcast Control Channel

BCCI Bulletin de la Chambre de Commerce et d'Industrie

BCCT Break Control Command Transducer

BCCU Battery Charge Control Unit

BCD Bacolod (Philippines) (location identifier) • Bad Conduct Discharge • Bar Coded Decimal • Baseline Change Directive • Baseline Concept Definition • Baseline Concept Description • Baseline Concept Document • Baseline Configuration Description • Baseline Configuration Document • Basic Calibrated Data • Beam Commuter Device • Behind Completion Date • Bibliographic Content Designator • Binary-Coded Data • Binary-Coded Decimal • Blocked Call Delay • Blue Compact Dwarf (galaxy) • Bottom Collider Detector • Budget Change Document

BCDC Binary-Coded Decimal Counter • Business Cycle Dating Committee

BCDG Blue Compact Dwarf Galaxy

BCDI Bureau Central de Documentation et d'Information

BCDIC Binary-Coded Decimal Interchange Code

B-CDMA Broadband Code-Division Multiple Access

BCDMA Broadband Code-Division Multiple Access

BCDMS Bologna - CERN - Dubna - Munich - Saclay (collaboration)

BCDS Backup Control Data Set

Bce Balance

BCE Bachelor of Chemical Engineering • Bachelor of Civil Engineering • Backup Control Electronics • Banca Centrale Europea • Banco Central Europeo • Banque Centrale Européenne • Barium Cloud Experiment • Base Civil Engineer • Baseline Cost Estimate • Beam Collimation Error • Before Christian Era • Before Common Era • Biblioteca Central de Educação (Brazil) • British Commonwealth and Empire • Bryce (USA-UT) (location identifier) • Bubble Chamber Experiment • Burn Control Element • Bus Control Electronics • Bus Control Element

BCEAO Banque Centrale des États de l'Afrique de l'Ouest

BCEE Bureau Commun d'Études et d'Enquête

BCEFM Business, Cost Estimating, and Financial Management

BCEN Banque Commerciale pour l'Europe du Nord

BCEOM Bureau Central pour les Équipements d'Outre-Mer

BCET Balance Contrast Enhancement Technique

BCF Bandpass Crystal Filter • Basic Calibration File • Basic Control Frequency • Beam Correction Factor • Biconcentration Factor • Billet de Chemin de Fer • Billion Cubic Feet • Bouca (Central African Rep.) (location identifier) • Bridge Control Facility

BCFHT Bulletin of the Canada-France-Hawaii Telescope (periodical code)

BCFS Backup Flight Control System

BCFSK Binary Code Frequency Shift Keying

BCG Bacille Calmette-Guérin • Bemichi (Guyana) (location identifier) • Biconjugate Gradient • Block-Connected Graph • Blue Compact Galaxy • Bright Cluster Galaxy • Brightest Cluster Galaxy (*)

BCGC Best Case Government Cost

BCGM Biconjugate Gradient Method

bch Body-Centered Hypercubic

Bch Branch • Bunch

BCh Bachelor of Chemistry

BCH Baker - Campbell - Hausdorff (formula) • Baucau - English Madeira (Indonesia) (location identifier) • Beach • Binary-Coded Hexadecimal • Binary-Coded Hollerith • Bose - Chaudhuri - Hocquenheim (coding) • Bureau Central des Hôpitaux

Bche Bouche

BChE Bachelor of Chemical Engineering

Bchon Bouchon

BCI Barcaldine (Australia-QLD) (location identifier) • Baseline Configuration Identification • Battery Condition Indicator • Binary-Coded Information • Bit Count Integrity • Brain-Computer Interface • Broadcast Interference • Budgetary Cost Information

BCID Best Candidate Item Database

BCIE Banco Centroamericano de Integración Economico

BCIR Bureau Climatologique Interrégional

BCI Bureau Central de Séismologie

BCIU Battery Control and Interface Unit • Bolometer Computer Interface Unit • Bus Control Interface Unit

BCJ Baca Grande (USA-CO) (location identifier)

BCJS Buffer Control Junction Switch

BCK Back • Backwards • Bolwarra (Australia-QLD) (location identifier)

BCKG Backing

Bcl Bicyclette

BCL Barra Colorado (Costa Rica) (location identifier) • Basic Contour Line • Batch Command Language • Binary Closed Loop • Binary-Coded Language • Binary Cutter Location • Boulesteix Cluster (catalog) • Burroughs Common Language

BCLA Beam Combination Laboratory Area (VLT)

BCLF Base Cardiac Life Support

BCM Bacau (Romania) (location identifier) • Baseline Correlation Matrix • Back Course Marker • Baseline Correlation Matrix • Basic Capacity Module • Basic Compatibility Module • Basic Control Monitor • Beam Change Monitor • Beam Charge Monitor • Become • Beyond Capability of Maintenance • Beyond Capacity of Maintenance • Block Coded Modulation • Brevet de cadre de Maîtrise • Bright Cluster Member • Brightest Cluster Member • Bureau Central Météorologique • Bureau de Contrôle Médical • Business Clearance Memorandum • Business Control Monitor

BCMC Block Change Management Council

BCMG Becoming

BCMN Bureau Central de Mesures Nucléaires (Belgium)

BCMR Boîtier de Commande des Mécanismes et Réchauffeurs

BCMT Bureau Central de Magnétisme Terrestre

Bcn Beacon

BCN Backbone Concentrator Node • Backward Congestion Notification • Barcelona - El Prat (Spain) (location identifier) • Base Centrale des Nomenclatures • Beacon • Boarding Control Number • Breakdown Control Number • British Commonwealth of Nations • Broadband Communication Network • Bureau Control Number

BCNet Business Cooperation Network

BCNF Boyce-Codd Normal Form • Bus Controller Navigation Filter

BCNRC Bulletin of the Canadian National Research Council of Canada (periodical code)

BCNRS Bulletin Signalétique du Centre National de la Recherche Scientifique (periodical code)

BCNU Be Seeing You

BCNZ Broadcasting Corporation of New Zealand

Bco Banco

BCO Base Central Office • Base Contracting Office • Belgisch Comite voor Optica (Belgium) • Binary-Coded Octal • Booster Cutoff • Jinka (Ethiopia) (location identifier)

BCOB Broadband Connection-Oriented Bearer

BCoE Backgrounds Center of Expertise

BCOE Bench Checkout Equipment

BCOI British Central Office of Information (UK)

Bcp Beaucoup

BCP Base Centrale de Pilotage • Base Communications Processor • Base Comprehensive Plan • Base Condemnation Percent • Base Control Program • Baseline Change Proposal • Basic Calcium Phosphate • Basic Call Process • Basic Celestial Phenomenon • Benchmark Control Point • Best Commercial Practice • Best Current Practice • Bibliothèque Centrale de Prêt (France) • Binary Control Protocol • Break Cloud Procedure • Brevet de Compagnon Professionnel • Bridge Control Protocol • Bridging Control Protocol • Brilliant Computer Products • Budget Change Proposal • Bulk Copy Program • Byte-Controlled Protocol • Byte Control Protocol

BCPL Basic Combined Programming Language (*) • Basic Commercial Pilot's Licence • Basic Computer Programming Language

BCPM Berkeley-CalTech Pulsar Machine

BCPN Business Customer Premises Network

B/C/P/S Base/Post/Camp/Station

BCPST Biologie - Chimie - Physique - Sciences de la Terre

BCQ Boat Crew Qualification

BCR Bar Chart Report • Bar Code Reader • Base Condemnation Rate • Baseline Change Request • Baseline Configuration Review • Battery Charge Regulator • Benefit/Cost Ratio • Benefit Cost Research • Block Coordinate Relaxation • Bôca do Acre (Brazil-AM) (location identifier) • Branch Constitutive Relation • Branched-Chain Reaction • Budget Change Request • Bureau Communautaire de Référence (EU) • Business Communications Review • Byte Count Register

BCRA Business Clearance Reviewing Authority

BCrAO Bulletin Crimean Astrophysical Observatory (periodical code)

BCRC British Columbia Research Council (Canada) • Bureau Central de Recherches Criminelles

BCRCI Brigade Centrale de Répression de la Criminalité Informatique (France)

BCRD Basic Consolidated Requirements Document • Budget Civil de Recherche et de Développement (France)

BCRS Barycentric Celestial Reference Frame • Beleidscommissie Remote Sensing (Netherlands)

BCRST Bibliography on Cold Regions Science and Technology

BCRT Binary-Coded Range Time • Binary Code Range Time (*) • Bright Cathode-Ray Tube

BCRTS Binary-Coded Range Time Signal

BCS Bachelor of Chemical Science • Bachelor of Commercial Science • Backscatter Cross-Section • Backup Control System • Bar Code Sender • Bar Code Sorter (*) • Bardeen-Cooper-Schrieffer (theory) • Baseline Comparative System • Baseline Comparison System • Basic Catalog Structure • Basic Combined Set • Basic Communications Services • Basic Contract Specification • Basic Control System • Batch Change Supplement • Beam Control System • Behavioral and Cognitive Sciences • Belle Chasse (USA-LA) (location identifier) • Bent Crystal Spectrometer • Binary Compatibility Standard • Block Check Sequence • Block Control Signal • Boeing Computer Services • Boîtier de Contrôle de Sauvegarde • Boston Computer Society (USA-MA) • Bragg Crystal Spectrometer • Bragg Curve Spectroscopy • Brightest Cluster Sample • British Computer Society (UK) • Broadband Cable System • Broadcast Control Station • Bulk Change Supplement • Bus Controller Simulator • Business Communications Services • Business Components System

BCSAIL British Columbia Spatial Archive and Interchange Format

BCSB Berkeley Center for Structural Biology (USA-CA)

BCSD Business Council for Sustainable Development

BCSCO Bismuth-Calcium-Strontium-Copper Oxide

BCSD Blank Corrected Sample Data

BCSF Bureau Central Sismologique Français (France)

BCSI Business Customer Satisfaction Index

BCSM Basic Cal State Model

BCSR Basic Commercial Security Requirement

BCSS Banque Carrefour de la Sécurité Sociale (Belgium) • Bociny Computer Support Services

BCSSS British Columbia Space Sciences Society (Canada-BC)

Bcst Broadcast

BCST Board on Chemical Sciences and Technology

BCSVS Bibliographic Catalogue of Suspected Variable Stars

BCT Bandwidth Compression Technique • Barycentric Coordinated Time • Boca Raton (USA-FL) (location identifier) • Body Centered Tetragonal • Bureau Central de

Transmission • Bus Configuration Table • Business Cordless Telephone

BCU Basic Computer Unit • Battery Control Unit • Battery Coolant Unit • Bauchi (Nigeria) (location identifier) • Bench Checkout Unit • Bibliothèque Cantonale et Universitaire (Switzerland) • Binary Computing Unit • Block Cycle Update • Branch Control Unit • Buffer and Conversion Unit • Buffer Control Unit • Bus Control Unit • Bus Coupler Unit

BCV Belmopan (Belize) (location identifier) • Brightness Contrast Value • Business Continuation Volume

BCVS Basic Conversational Verb Set • Bibliographic Catalogue of Variable Stars

BCVT Basic Class Virtual Terminal

BCW Buffer Control Word

BCWP Budget Costs for Work Performed • Budgeted Cost of Work Performed

BCWR Budgeted Cost of Work Remaining

BCWS Budget Costs for Work Scheduled • Budgeted Cost of Work Scheduled

BCX Beloreck (Russia) (location identifier)

BCY Bulchi (Ethiopia) (location identifier)

BCZ Bickerton Island (Australia-NT) (location identifier)

b/d Barrels per Day • Brought Down

bd Bis Die

bD Bidécadaire

Bd Band • Baud • Board • Bold • Bond • Boulevard • Bound • Broad • Bundle

B/D Bank Draft • Bills Discounted • Binary to Decimal • Brought Down • Bulge-to-Disk

BD Bachelor of Divinity • Backward Difference • Bahrain Dinar (currency) • Balmer Discontinuity • Banc de Dynalpie • Band • Bande Dessinée • Bangladesh (international vehicle registration code) • Bangladesh (ISO country code) • Bank Draft • Barrière Douanière • Base de Datos • Base de Données • Baseline Description • Baud • Beam Deflection • Beam Diameter • Bidirectional • Big Deal • Bilan Définitif • Billet à Découvert • Bills Discounted • Binary Decoder • Binary Digit • Binary Discrete • Binary Divide • Bit Density • Blowing Dust • Board • Bon de Douane • Bonner Durchmusterung • Boot Descriptor • Booster Development • Bordereau Descriptif • Brief Description • Brown Dwarf (star) • Building Distributor • Buiten Dienst • Burning Disk • Bus Director

BDA Backup Drive Amplifier • Base de Données Académiques • Base de Données d'Amas • Battle Damage Assessment • Beacon Drive Amplifier • Bermuda • Bermuda Kindley (Bermuda) (location identifier) • Bermuda Tracking Station • BIOS Data Area • Blast Danger Area • Block Decoder Assembly • Bomb Damage Assessment • Brazilian Decimetric Array • Business Data Analysis • Business Domain Architecture • Hamilton (Bermuda) (location identifier)

BDACS BDM Data Acquisition and Conversion System

BDAM Basic Direct-Access Method

BDAP Banque de Données Analysées et Prévues

BDAS Bendigo District Astronomical Society Inc. (Australia-VIC)

BDAT Best Demonstrated Available Technology

BDB Banque de Données Bibliographiques • Base Development Board • Big Dumb Booster • Bundaberg (Australia-QLD) (location identifier)

BDBS Base de Données Bathymétriques du SHOM

BDC Backgrounds Data Center (USA-DC) • Backup Digital Computer • Backup Domain Controller • Barra do Corda (Brazil-MA) (location identifier) • Beijing Data Center (China-PRC) • Bidirectional Converter • Binary Decimal Counter • Blood Donor Center • Bottom Dead Center • Brown Dwarf Catalog • Brown Dwarf (star) Candidate • Business Development Committee

BDCF Baseline Data Collection Facility

BDCR Baseline Document Change Request

Bdcst Broadcast

BDCU Base de Données Charge Utile

BdD Banque de Données • Base de Données

BDD Badu Island (Australia-QLD) (location identifier) • Binary Decision Diagram • Banque De Données • Baseline Definition Document • Baseline Description Document • Binary Decision Diagram • Binary Digital Data • Binary-to-Decimal Decoder • Blanket Delivery date • Brown Dwarf Desert • Burst Detection Dosimeter • Burst Detector Dosimeter

BDDB Drown Dwarf Database

Bde Brigade

BDDE Bulletin de la Direction Départementale de l'Équipement (France)

BDDN Broadcast Data Dissemination Network • Broadcast Data Distribution Node

BDDPI Banque de Données du Droit de la Propriété Industrielle (INPI)

BDDR Banque de Données Documentaires Réparties

BDDS Broadcast Data Distribution System

Bde Bände • Brigade

BDE Base de Données Extragalactiques • Baudette (USA-MN) (location identifier) • Bidirectional Electron-event • Block Diagram Editor • Borland Database Engine • Boundary of Diffuse Encroachment • Bureau Des Élèves • Business Development Expense • Buyer-Designated Equipment

BDEAC Banque de Développement des États de l'Afrique Centrale

BDEC Bendix Field Engineering Corporation

BDES Bibliographic Data Exchange System

BdF Bibliothèque de France

BDF Backward Difference Formula • Backward Differencing Formula • Backward Differential Formula • Backward Differential Function • Bibliothèque De France • Bi-Directional Factor • Binary Distribution Format • Bitmap Distribution Format • Block Data Format • Bradford Rinkenberger (USA-IL) (location identifier) • Bulk Data Frame

BDFA Basic Daily Food Allowance

Bdg Binding

BDG Base de Données Géographiques • Binding (USA-UT) (location identifier) • Blanding (USA-UT) (location identifier) • Blue Dwarf Galaxy • Bridge • Bureau de la Documentation et de Gestion

BDH Bandar Lengeh (Iran) (location identifier)

BDI Bande Diffuse Interstellaire • Base de Données Intelligente • Bearing Deviation Indicator • Binary Digital Image • Bird Island (Seychelles) (location identifier) • Black Digital Interface • Both Dates Included • Both Days Included • Bourse de Doctorat pour Ingénieur • Budget Data and Integration • Buffered Direct Injection • Bundesverband der Deutschen Industrie (Germany) • Bureau de Documentation et d'Information • Burundi (ISO country code)

BDIN Book Dealer Identification Number

BDIS Boîtier de Distribution

BDJ Banjarmasin (Indonesia) (location identifier) • Brans - Dicke - Jordan (theory)

BDK Bondoukou (Ivory Coast) (location identifier)

Bdl Bundle

BdL Bureau des Longitudes (France)

BDL Banque de Données Locales • Baseline Demonstrator Laser • Basic Design Language • Battery Data Link • Boundary Double Layer • Bundle • Bureau Des Longitudes (France) • Hartford - Bradley (USA-CT) (location identifier) • Springfield - Bradley (USA-MA) (location identifier) • Windsor Locks - Bradley (USA-CT) (location identifier)

BDLC Burroughs Data Link Control

Bdle Bundle

BDLI Bundesverband der Deutschen Luftfahrt-, Raumfahrt- und Ausrüstungindustrie eV (Germany)

BDM Background Debug Mode • Bandirma (Turkey) (location identifier) • Baryon-Dominated Model • Baryonic Dark Matter • Base de Données Météorologiques • Binary Delta Modulation • Binary Digital Multiplier • Bone Densitometer • Bound Density Maxima (algorithm) • Branch Delegates' Meeting • Bureau de Documentation Minière • Bureau Des Méthodes

BDMI Bibliographic Dictionaries Master Index

BDMM Basic Differential Multiplier Method

BDMS Baseline Data Management System

BDN Badin (Pakistan) (location identifier) • Bell Data Network • Bright Dark Nebula • Bulk Data Network

BDO Bandung (Indonesia) (location identifier) • Basic Delivery Order • Bipartite Development Objective • Blanket Delivery Order

BDOS Basic Disk Operating System

BDP Base Development Plan • Base Development Planning • Baseline Data Package • Bhadrapur (Nepal) (location identifier) • Biblioteca di Documentazione Pedagogica (Italy) • Bibliothèque de Documentation Pédagogique (France) • Bibliothèque Départementale de Prêt (France) • Bourse de Demi-Pension • Breakdown Pressure • Burst Detection Processor • Burst Detector Processor • Business Data Processing

BDPA Black Data Processing Association • Bureau pour le Développement de la Production Agricole

BDPI Base Data Processing Installation • Baseline Data Package Integration

BDPME Banque de Développement des Petites et Moyennes Entreprises (France)

BDPT Brown Dwarf Prototype

BDPU Bubble, Drop and Particle Unit (*) • Bubble Droplet Particle Unit

BDQ Vadodara (India) (location identifier)

BDR Baseline Design Review • Basic Data Record • Battery Discharge Regulator • Bidirectional Reflectance • Border • Bridgeport (USA-CT) (location identifier) • Build Design Review • Bis Device Request

BDRF Bidirectional Reflectance Function • Bidirectional Reflective Factor

Bdrm Bedroom

BDRM Bedroom

BDRN Broadcast Data Receiver Node

Bdry Boundary

bds Bis in Die Sumendus

Bds Boards • Bundles

BDS Barbados (international vehicle registration code) • Base de Données Scientifiques • Basic Data Service • Beijing Document Service (China-PRC) • Binary Digital Search • Biographical Dictionary of Scientists • Bioreactor Demonstration System • Bioreactor Development System • Briefing and Display System • Brindisi (Italy) (location identifier) • Budget Development System • Bulgarian Committee for Standardization and Metrology • Bulk Data System • Bundle Delivery System • Burnham's Double Stars • Business Development Specialist

BDSF Bidirectional Scattering Function

BDSG Bundesdatenschutzgesetz (Germany)

BDSP Basic Data Set Project

BDSS Base de Données Sédimentologiques du SHOM • Brown Dwarf Spectroscopy Survey

BDST British Double Summer Time

BD/T Beam Director/Tracker

BDT Barycentric Dynamical Time • Barycentric Dynamic Time • Basa de Dados Tropical • Base Deck to Tape • Base de Données Technologiques • Beam Deflection Tube • Beam Director/Tracker • Bidirectional Proton-event • Bilinear Dual Transform • Billing Data Transmitter • Binary Deck-to-Tape • Block Data Transfer • Bon Du Trésor • Brans-Dicke Theory • Brittle-Ductile Transition • Bulk Data Transfer • Bureau de Développement des Télécommunications • Gbadolite (Zaire) (location identifier) • Taka (Bangladesh) (ISO currency code)

BDTC British Dependent Territory Citizen

BDU Bardufoss (Norway) (location identifier) • Baryon-Dominated Universe • Basic Display Unit • Battery Display Unit • Bundesverband Deutscher Unternehmensberater (Germany) • Bus Data Unit • Business Development Unit

B-du-R Bouches-du-Rhône (France)

BDus Byulleten Instituta Astrofiziki Dushanbe (periodical code)

BdV Bund der Vertriebenen (Germany)

BDV Basis-Datenverarbeitung • Breakdown Voltage • Breitband-Durchgangsvermittlungsstelle • Moba (Zaire) (location identifier)

BDW Bandwidth • Bedford Downs (Australia-WA) (location identifier) • Block Description Word

BDWS Biological Detection and Warning System

Bdx Bordeaux (France)

BDX Broadus (USA-MT) (location identifier)

Bdy Boundary

BDY Bandon State (USA-OR) (location identifier) • Body

BDZ Baindoung (Papua New Guinea) (location identifier)

be Belgium • Bezüglich • Belarussian (ISO language code) • Bielorussian (ISO language code) (obsolete)

Be Beryllium

Bé Baumé • Breveté

B&E Beginning & Ending • Breaking & Entering

B/E Bill of Exchange

BE Bachelor of Education • Bachelor of Engineering • Back End • Baldwin Effect • Bande Étroite • Bank of England • Banlieue Est • Barkhausen Emission • Basic Encyclopaedia • Beacon Explorer • Beam Expander • Before Execution • Belarus • Belgique (code ISO de pays) • Belgium (ISO country code) • Bell End • Below or Equal • Benin • Best Estimate • Beyond Einstein • Bharat Electronics • Bibliografía Española (Spain) • Bill of Exchange • Binary Encounter • Binding Edge • Binding Energy • Biocomplexity in the Environment (NSF) • Blue Edge • Board of Education • Bon d'Électricité • Bon d'Épargne • Booster Engine • Bordereau d'Envoi • Bordereau d'Escompte • Boron Epoxy • Bose-Einstein • Bourse d'Études • Bradford Engineering • Branch if Equal • Breaker End • Brevet Élémentaire • Brevet d'Études • Brilliant Eyes • British Element • British Embassy • Budget Estimate • Bulk Encryption • Bureau d'Enregistrement • Bureau d'Études • Bursty Erection

BEA Bereina (Papua New Guinea) (location identifier) • Binary Encounter Approximation • Bond Energy Algorithm • Bond Energy Analysis • Bright Earth Avoidance (region) • Budget Enforcement Act (USA) • Budget Estimate Agreement • Bureau à l'Énergie Atomique • Bureau des Enquêtes et Accidents • Bureau of Economic Analysis

BEAA Bureau d'Enquête sur les Accidents d'Aviation (Switzerland)

BEAC Banque des États de l'Afrique Centrale

BEAD Beginning Extended-Area Descriptor

BEAF Beam Expander Alignment Function

BEAM Brain Electrical Activity Mapping

BEAMS Base Engineer Automated Management System • Becoming Enthusiastic About Math and Science

BeamtVG Beamtenversorgungsgesetz

BeAP Beryllium Ammonium Perchlorate

BEAP Banc d'Essai des Accélérateurs à Poudre

BEAR Barents Euro-Arctic Region • Beacon and Auroral Research • Bonus Extension and Retraining Program

bearb Bearbeitet

BEARS Beam Array Receiver System • Berkeley EUV Airglow Rocket Spectrometer

BEAST Background Emission Anisotropy Scanning Telescope

BEAT Best Execution Analysis Tabulation • Brevet Élémentaire d'Aptitude Technique

BEATEP Brevet d'État d'Animateur Technicien de l'Éducation Populaire (France)

Beau Bordereau

BEB Benbecula (UK) (location identifier) • Binary-Encounter-Bethe (model) • Binary Exponential Backoff • Bureau d'Études du Bâtiment

BEBA Bolometer Electronics Box Assembly

BEBC Big European Bubble Chamber

Bec Because

BEC Backup Event Controller • Base Environment Coordinator • Base Extension Course • Basic Emergency Care • Beginning of Equilibrium Cycle • Bose-Einstein Condensate (*) • Bose-Einstein Condensation • Bose-Einstein Correlation • Brevet d'Enseignement Commercial • Brevet d'Études Commerciales • Business Executive Council • Business-oriented Electronic Commerce • Convertible Franc (Belgium) (ISO currency code) • Wichita Beech (USA-KS) (location identifier)

BECAT Background Estimation Convolution And Threshold

BECAUSE Benchmark of Concurrent Architectures for Use in Scientific Engineering

BECC Burst Error Correcting Coder

BECI Bibliografía Española de Ciencias de la Información (Spain)

BECMG Becoming

BECN Backward Error-Congestion Notice • Backward Error-Congestion Notification • Backward Error-Correction Notification • Backward Explicit Congestion Notice • Backward Explicit Congestion Notification

BECO Booster Engine Cutoff

BECTA Bureau d'Exploitation et de Coordination des Transports Aériens

BECU Billion European Currency Units

Bed Bedeutend

BEd Bachelor of Education

BED Bachelor of Education • Bedford Hanscom (USA-MA) (location identifier) • Bibliothèque d'Étude et de Documentation • Biologically Effective Dose • Blue-Emitting Diode • Box External Data • Budget Execution Direction • Budget Execution Document

BEDAC Burst Error Detection And Correlation

BEDIS Booktrade EDI Standards

BEDO Burst Extended Data Out

Beds Bedfordshire (UK-England) • Bedrooms

BEDT Bisethylenedithia

BEE Bachelor of Electrical Engineering • Beagle Bay (Australia-WA) (location identifier) • Bio-Environmental Engineer

BEEC Binary Error Erasure Channel

BEEd Board on Engineering Education

BEEF Big Explosive Experimental Facility (USA) • Business and Engineering Enriched Fortran

BEEL Biological Equivalent Exposure Limit

BeElM Beskontaktnye Elektricheskie Mashiny (periodical code)

BEEM Ballistic Electron Emission Spectroscopy

BEEP Bureau Européen de l'Éducation Populaire

BEES Board on Energy and Environmental Systems

BEET Best Estimated Evaluation Trajectory

Bef Before

BEF Background Equivalent Flux • Ballast Efficiency Factor • Band Elimination Filter • Bluefields (Nicaragua) (location identifier) • Building Entrance Facility • Franc (Belgium) (ISO currency code)

BEff Baldwin Effect

BEFS Berkeley Extreme and Far-ultraviolet Spectrometer

Beg Beginning

BEG Bach - Escher - Gödel (trilogy) • Beograd (Yugoslavia-Serbia) (location identifier) • Bright Elliptical Galaxy • Budget Estimate Guidance • Bureau d'Études Géologiques

Begr Begründet

BEH Benton Harbor (USA-MI) (location identifier)

BeHAP Beryllium Hydride Ammonium Perchlorate

BeH2 Beryllium Hydride

BEI Back-scattered Electron Imaging • Banque Européenne d'Investissement • Beica (Ethiopia) (location identifier) • Best Estimate of Impact • Brevet d'Enseignement Industriel (France)

Beibl Beiblatt

beif Beifolgend

BeiGe Beiträge zur Geophysik (periodical code)

beigeb Beigebunden

beil Beiliegend

BEIR Biological Effects of Ionizing Radiation

BeiRe Beijing Astronomical Observatory Reprints (periodical code)

Beitr Beitrag • Beiträge

BEJ Berau (Indonesia) (location identifier)

Bek Bekanntmachung

BEK Rae Bareli (India) (location identifier)

Bel Belarussian • Belgian • Belgium

BEL Basic Equipment List • Belém (Brazil-PA) (location identifier) • Belfast (Starlink internal code) • Belgique (code ISO de pays) • Belgium (ISO country code) • Bell • Broad Emission Line • Financial Franc (Belgium) (ISO currency code)

BELC Broad Emission Line Cloud

Belg Belgian • Belgium

BELGOSPACE Belgian Space Industry Association

BELINDIS Belgian Information and Dissemination Service

BELLCORE Bell Communications Research

BELR Broad Emission Line Region

BELST Belarussian Committee for Standardization, Metrology and Certification (Belarus)

Bem Bemerkung

BEM Bachelor of Engineering of Mines • Big Emerging Market • Biological Environment Monitor • Bossembele (Central African Rep.) (location identifier) • Boundary Element Method • Bug-Eyed Monster

BEMA Biogenic Emissions in the Mediterranean Area

BEMAR Backlog of Essential Maintenance And Repair

BEMO Base Equipment Management Office

BEMS Bioelectromagnetics Society • Building Energy Management System

Ben Bengali

BEN Basic Encyclopaedia Number • Benghazi Benina (Libya) (location identifier) • Benin (ISO country code) • Broadband Electronic Noise • Broadband Electrostatic Noise • Bureau d'Études Nucléaires

BENCHMARC Biodegradation of Environmental Chemicals Modelled with Aquatic, Relative-rate Coefficients

Bend Bendigo (Australia-VIC)

Benef Benefice • Beneficiary

BENELUX Belgium - Netherlands - Luxembourg

Benf Beneficiario

Beng Bengal • Bengali

BENHS British Entomology and Natural History Society (UK)

BENS Bounded Error Navigation System

BENT Beginning Evening Nautical Twilight

BEO Base Engineer Officer • Bilan - Évaluation - Orientation • Biologie - Energie - Ökologie • Boulder Extension Office (SWRI) • Newcastle Belmont (Australia-NSW) (location identifier)

Beob Beobachter • Beobachtung

BEOP Best Estimate of Orbital Parameters

BEOS Bremen Development of Orbital Systems

BEP Bachelor of Engineering Physics • Backward Error Propagation • Basic Emergency Plan • Bellary (India) (location identifier) • Best Efficiency Point • Best Erection Profile • Biomolecular Engineering Programme • Bit Error Probability • Bon d'Épargne Progressif • Break-Even Point • Brevet d'Études Professionnelles (France) • Budget Execution Plan • Bureau Exécutif Permanent • Bureau of Engraving and Printing

BEPC Beijing Electron-Positron Collider (China-PRC) • Brevet d'Études du Premier Cycle

BEPD Basic Entry Pay Date

BEPG Broad Economic Policy Guidelines (EU)

BePhA Beiträge zur Physik Atmosphäre (periodical code)

BEPI Bureau d'Études et de Participation Industrielle

BePl Beiträge Plasmaphysik (periodical code)

BEPP Brevet Élémentaire de Pilote Privé

BEPS Biological Experiments Per Second

Beq Bequeath

BEQ Best Estimated Quantity • Bury Saint Edmunds - Honington (UK-England) (location identifier)

Beqt Bequest

ber berichtig

Ber Berber • Bericht • Berlin (Germany)

BER Basic Encoding Rule • Berlin (Germany) (location identifier) • Binary Encoded Representation • Bit Error Rate • Bit Error Ratio • Budget Execution Review

Berb Berber • Berbère • Berberisco

BERC Bit Error Rate Counter

BERD Banque Européenne de Reconstruction et de Développement

BERDI Balkan Environmental Research and Development Institute (Greece)

Bergstr Bergstraße

BERI Bureau d'Études et de Réalisations Industrielles

Ber. Is. Bermuda Islands

Berks Berkshire (UK-England)

Berl Berlin (Germany)

Berm Bermuda

BERM Basic Encyclopaedic Redundancy Media

BERMS Boreal Ecosystem Research and Monitoring Sites (project)

BERNET Berlin Netz (Germany)

BERP Bit Error Rate Performance

BERR Bit Error Rate Record

BERT Basic Energy Reduction Technology • Best Error Ratio Tester • Bit Error Rate Test • Bit Error Rate Tester

BERTS Basic Exchange Radio Telecommunications Service • Bit Error Rate Test Set

BErzGG Bundeserziehungsgeldgesetz

Bes Besonders

BES Bachelor of Engineering Science • Bachelor of Environmental Studies • Banc d'Essais Site • Basic Energy Sciences • Beam Emission Spectroscopy • Best Erection Schedule • Bioengineering and Environmental System • Booster Exhaust Stream • Bose-Einstein Spectrum • Bose-Einstein Statistics • Bourse de l'Enseignement Secondaire • Brest Guipavas (France) (location identifier) • Brevet d'Enseignement Social • British Ecological Society (UK) • British Electrophoresis Society (UK) • Budgeted End Strength • Budget Estimate Submission • Budget Execution System • Business Expansion Scheme

BeschFG Beschäftigungsförderungsgesetz

BESD Baseline Extraction and Sky Deglitching

BESEP Base Electronics Systems Engineering Plan

BESEX Bering Sea Experiment

BESI Budget Execution System Installation • Bus Electronic Scanning Indicator

BeSN Be Star Newsletter (periodical code)

BESR Board on Earth Sciences and Resources

BESS Balloon-Experiment with Superconducting Solenoid (*) • Balloon-Experiment with Superconducting Spectrometer • Biomedical Experiment Scientific Satellite • Biomedical Experiment Support Satellite • Bit Error Sensing System • Bottom Environmental Sensing System

BESSY Berlin Electron Storage Synchrotron • Berliner Elektronenspeicherringgesellschaft für Synchrotronstrahlung

Best Bestellung • Bestimmt

BEST Baryonic Extragalactic Structure Tracer • Bathymetric Estimator Search Technique • Battery Energy Storage Test • Berlin Exoplanet Search Telescope • Bids Evaluation Support Tool • Bilan Énergétique du Système Tropical • Board on Environmental Sciences and Technology • Booster Exhaust Study Test • British Expertise in Science and Technology • Business Electronic System Technique • Business Equipment Software Technique

Best.Nr. Bestellnummer

Bet Between

BET Baseline Evaluation Test • Best Estimate of Trajectory • Best-Estimate Trajectory • Bethel (USA-AK) (location identifier) • Bit-Enhanced Technology • Black Entertainment Television • Budget Énergie Transport • Brunauer - Emmett - Teller (method) • Bureau d'Études Techniques

BETA Ballistic Exploitation and Target Acquisition • Billion-channel Extraterrestrial Array • Bureau d'Économie Théorique et Appliquée

BETECH Boundary Element Technology

BETEL Broad Exchange over Transeuropean Links

Betr Betreff • Betreffend • Betreffs • Betrifft

Betrbf Betriebsbahnhof

BETRS Basic Exchange Telecommunications Radio Service

BetrVG Betriebsverfassungsgesetz

BETS Building Energy Technical Survey • Business Ethics Teaching Society

BETSCE Brilliant Eyes Ten-Kelvin Sorption Cryocooler Experiment

BETT Best Estimate of Trajectory Tape

Betw Between

BEU Bedourie (Australia-QLD) (location identifier) • BENELUX Economic Union • Bulk Encryption Unit • Bureau of Energy Utilization

BEUC Bureau Européen des Unions de Consommateurs

BeurkG Beurkundungsgesetz

Bev Bevel • Bevölkerung

BeV Billion Electronvolts

BEV Beer Sheba (Israel) (location identifier) • Billion Electron Volts • Blacksburg Electronic Village • Bundesamt für Eich- und Vermessungswesen (Austria)

BEW Beira (Mozambique) (location identifier)

BewG Bewertungsgesetz

BEX Benson (UK-England) (location identifier) • Broadband Exchange

BEXEC Budget Execution

BEXR Basic Exchange Radio

BEY Beirut (Lebanon) (location identifier)

bez Bezahlt • Bezüglich

Bez Bezeichnung • Bezirk

BEZ Beru (Kiribati) (location identifier)

BEZS Bandwidth Efficient Zero Support

bezw Beziehungsweise

b/f Bring Forward • Brought Forward

bf Base Frequency • Boldface • Bona Fide

Bf Bahnhof • Brief

B/F Background/Foreground • Blade Fold • Bring Forward • Brought Forward

BF Backface • Back Feed • Back Focal • Background Field • Backup Force • Bad Flag • Bailleur de Fonds • Ballast Factor • Banque de France • Bassa Frequenza • Basse Fréquence • Beam Forming • Beat Frequency • Bebaute Fläche • Belgian Franc (currency) • Bellman-Ford • Bénéfice Forfaitaire • Berthing Feet • Best Fit • Binary Fraction • Black Forest (Germany) • Body Flap • Boldface • Bottom Face • Boundary Function • Boyfriend • Branching Function • Breathing Frequency • Brevet Français • Broadband Filter • Brought Forward • Burkina Faso (ISO country code) • Bypass Filter • Franc (Belgium) (currency)

BFA Bachelor of Fine Arts • Burkina Faso (ISO country code)

BFAP Blocking Factor Analysis Program

BFAS Basic File Access System

B'fast Breakfast

B/FAST Breakfast

BFAST Breakfast

BFB Bang For the Buck • Blue Fox Bay (USA-AK) (location identifier) • Boundary First Born • Business Finance Branch

BFBI Brute Force and Bloody Ignorance

bfc Briefcase Document (file extension)

BFC Backup Flight Computer • Backup Flight Control • Block Framing Character • Bloomfield (Australia-QLD) (location identifier) • Blue Featureless Continuum • Body Flap Control • Bohr Frequency Condition • Bold Face Capital

BFCE Banque Française du Commerce Extérieur

BFCO Band Filter Cutoff

BFCS Backup Flight Control System

BFCU Blade Fold Control Unit

BFD Back Focal Distance • Best Fit Decreasing • Boolean Function Designator • Bradford (USA-PA) (location identifier) • Budget Formation Directive

BFDK Before Dark

BFE Base Flood Elevation • Beam Forming Electrode • Bielefeld (Germany) (location identifier) • Blacker Front End • Blue Fundamental Edge • Buyer-Furnished Equipment

BFEC Bendix Field Engineering Corporation (USA-MD)

BFEM Balloon Flight Engineering Model

BFET Ballistic Field-Effect Transistor

bff Binary File Format (file extension)

BFF Binary File Format • Buffered Flip-Flop • Bund Freischaffender Forodesigner (Germany) • Scottsbluff (USA-NE) (location identifier)

BfFS Bundesanstalt für Flugsicherung (Germany)

Bfg Briefing

BFG Binary Frequency Generator • Brute Force Gyro • Bullfrog Basin (USA-VT) (location identifier)

BfGR Bundesanstalt für Geowissenschaften und Rohstoffe (Germany)

BFGS Broyden - Fletcher - Goldfarb - Shanno (method)

BFH Bitfehlerhäufigkeit • Bundesfinanzhof (Germany)

BFHZ Bayerisch-Französisches Hochschulzentrum

BFI British Film Institute (UK) • Broadband Filter Imager • Brute Force and Ignorance • Seattle Boeing (USA-WA) (location identifier)

BFID Boolean Function Identifier

BFII Banking, Finance and Insurance Information

BFJ Ba (Fiji) (location identifier)

BFK Denver Buckley (USA-CO) (location identifier)

BFL Back Focal Length • Bakersfield (USA-CA) (location identifier) • Back Focal Length • Bonded Film Lubrication • Buffered FET Logic • Bundesamt für Landestopographie (Switzerland)

BfLR Bundesanstalt für Landeskunde und Raumentwicklung (Germany)

bfm Unix Font Metrics (file extension)

BFM Basic Flight Maneuver • Best Fit Mass (model) • Best Fit Model • Blocking Flow Method • Body Fluid Measurement • Boson-Fermion Model • Bowen Fluorescence Mechanism • Budget Financial Manager • Bus Functional Model • Business and Financial Manager • Mobile (USA-AL) (location identifier)

BFMMP Bureau Fédéral des Mesures et Métaux

Bfn Bloemfontein (South Africa)

BFN Beam-Forming Network • Bloemfontein (South Africa) (location identifier) • Bye For Now

BFO Beat Frequency Oscillation • Beat Frequency Oscillator • Black Forest Observatory (USA-CO) • Buffalo Range (Zimbabwe) (location identifier)

BFOC Bayonet Fiber Optic Connector

BFOSC Bologna Faint Object Spectrometer and Camera

BFP Beaver Falls (USA-PA) (location identifier) • Best-Fitting Paraboloid • Binary Floating Point • Buyer-Furnished Part

BFPDDA Binary Floating Point Digital Differential Analyzer

BFPOP British Forces Post Office

BFQ Bid For Questions

Bfr Before • Buffer

BFr Belgian Franc (currency)

BFR Bedford - Virgil I. Grissom (USA-IN) (location identifier) • Before • Belgian Franc (currency) • Biennial Flight Review • Big Fast Router • Block Format Recording • Franc Belge (currency)

BFRL Basic Facilities Requirements List

BFRP Boron Fiber-Reinforced Plastic

BfS Bundesanstalt für Strahlenschutz (Germany)

BFS Backup Flight Software • Backup Flight System • Band Filter Set • Belfast (UK-Northern Ireland) (location identifier) • Best Fit Sphere • Breadth First Search (method) • Brute Force Supply • Bureau Fédéral de Statistiques

BFSK Binary Frequency Shift Keying

BFSL Best Fit Straight Line

BFSP Black Forest Star Party (USA-PA)

BFSS Backup Flight Software System

BFT Beaufort County (USA-SC) (location identifier) • Binary File Transfer • Budget de Fonctionnement Total

BFTA Bulk Fuel Tank Assembly

BFU Bengbu (China-PRC) (location identifier) • Beratungsbüro für Umwelt • Betriebsführungsumsetzer • Bundesstelle für Flugunfalluntersuchung (Germany) • Büro für Flugunfalluntersuchungen (Switzerland) • Burst-Forming Unit

BFW Broadband Fixed Wireless • Bundesfachverband der Wohnungsverwalter (Germany) • Bundesverband Freier Wohnungsunternehmer (Germany)

BFWSG Braunschweig Friedrich Vieweg und Sohn GmbH (periodical code)

BFX Bafoussam (Cameroon) (location identifier)

BFY Budget Fiscal Year

b/g Bonded Goods

bg Bulgaria • Bulgarian (ISO language code)

Bg Background • Bag • Berg • Bourg • Brug • Budget

B&G Blood & Guts • Buildings & Grounds

BG Back Gear • Background • Band Gap • Big Grain • Billet de Groupe • Black Galaxy • Block Group • Blood Group • Bologna Galactic • Border Gateway • Branch Group • Bridgestone • Brigadier General • Bulbe Galactique • Bulgaria (international vehicle registration code) • Bulgaria (ISO country code) • Bulgaria (postal country code) • Bulgarie (code ISO de pays) • Bureau de Garantie • Bureau Géographique • Burg • Bürgerliches Gesetzbuch

BGA Ball Grid Array • Behavior Genetics Association (USA) • British Geophysical Association (UK) • Bucaramanga (Colombia) (location identifier) • Bundesverband des (Deutsches) Groß- und Außenhandels (Germany)

BGAAC Blue Grass Amateur Astronomical Club (USA-KY)

BGB Booue (Gabon) (location identifier) • Bürgerliches Gesetzbuch

BGBGM Berlin Gebrüder Bornträger Geoexploration Monographs Series (periodical code)

BGBl Bundesgesetzblatt

BGC Bank Giro Credit • Biogeochemical Cycle • Blue Globular Cluster • Bragança (Portugal) (location identifier) • Bright Galaxy Catalog • Burnham General Catalog

BGCS Baangebruiks Controle Systeem

BGD Bangladesh (ISO country code) • Belgische Geologische Dienst (Belgium) • Borger (USA-TX) (location identifier)

Bge Barrage • Beige

BGE Bach - Gödel - Escher (trilogy) • Backus-Gilbert Estimate • Block Gaussian Elimination • Bologna Galactic Extended • Brainbridge - Decatur County (USA-GA) (location identifier) • Branch if Greater or Equal

BGeod Bulletin Géodésique (periodical code)

Bger Bouger

BGF Bangui (Central African Rep.) (location identifier) • Boursier du Gouvernement Français (France) • Brutto-Geschoßfläche • Bruttogrundfläche

BGG Background Galaxy • Bongouanou (Ivory Coast) (location identifier) • Booster Gas Generator • Brightest Group Galaxy

BGGC Beijing Group of Galaxy and Cosmology (China-PRC)

BGGM BGS Global Geomagnetic Model

BGH Boghe Abbaye (Mauritania) (location identifier) • Bundesgerichtshof (Germany)

BGHT Bight

BGI Barbados (location identifier) • Bibliographie Géographique Internationale • Borland Graphics Interface • Bureau Gravimétrique International

BGJ Borgarfjördur Eystri (Iceland) (location identifier)

BGK Berliner Gesamtkatalog • Bhatnagar - Gross - Krook (lattice) • Big Creek (Belize) (location identifier)

BGL Baglung (Nepal) (location identifier) • Below Ground Level • Lev (Bulgaria) (ISO currency code)

BGLS Bragg Grating Laser Sensor

Bgm Bürgermeister

BGM Binghamton (USA-KY) (location identifier) • Buffered Gate Modulation • Endicott (USA-KY) (location identifier) • Johnson City (USA-KY) (location identifier)

BGMP Border Gateway Multicast Protocol

BGN Began • Begin • Board on Geographic Names (USA) • Brüggen (Germany) (location identifier)

BGNW Benutzergruppe Netzwerke

BGO Bergen (Norway) (location identifier) • Bismuth Germanate • Bismuth Germanate Oxide • Bismuth Germanium Oxide • Bismuth Germinate • Burke-Gaffney Observatory (Canada-NS)

BGP Bongo (Gabon) (location identifier) • Border Gateway Protocol • Borland Gateway Protocol • Boundary Gateway Protocol • Bulletin Grand Public • Byzantine Generals Problem

BGQ Big Lake (USA-AK) (location identifier)

BGR Bangor (USA-ME) (location identifier) • Bulgaria (ISO country code) • Bulgarie (code ISO de pays) • Bundesanstalt für Geowissenschaften und Rohstoffe (Germany)

BGS Backup Guidance System • Big Spring - Webb (USA-TX) (location identifier) • Blue-Green Survey • Bright Galaxy Sample • British Geological Survey (UK) • Bundesgrenzschutz (Germany) • Burgs

BGSU Bowling Green State University (USA-OH)

Bgt Bight • Bought

BGT Bagdad (USA-AZ) (location identifier) • Biorack Ground Team • Branch if Greater Than

BGU Bangassou (Central African Rep.) (location identifier) • Basic Generation Unit • Ben Gurion University (Israel)

BGUID Basic Generation Unit Interface Device

BGV Background Vortex • Bento Gonçalves (Brazil) (location identifier) • Boost Glide Vehicle

Bgw Bergwerk

BGW Baghdad (Iraq) (location identifier) • Bienaymé - Galton - Watson (process)

BGX Bagé (Brazil) (location identifier)

BGY Bergamo (Italy) (location identifier) • Milano - Orio al Serio (Italy) (location identifier)

BGZ Braga (Portugal) (location identifier)

b/h Behind

bh Bihari (ISO language code)

B-H Binary to Hexadecimal

BH Bahrain (ISO country code) • Bahrein (code ISO de pays) • Bâtiment Hydrographique • Belize (international vehicle registration code) • Bellman-Harris (process) • Black Hole • Block Handler • Block Header • Block House • Bosnia-Hercegovina • Breather Hose • Burlington House • Bushongo • Busy Hour

BHA Bahia de Caraquez (Ecuador) (location identifier) • Base Helix Angle • Bibliographie de l'Histoire de l'Art • Butt-Head Astronomer

BHAD Black Hole Accretion Disk

B'ham Birmingham (UK-England)

BHarO Bulletin of Harvard Observatory (periodical code)

BHAS Black Hills Astronomical Society (USA-SD) • Brighton and Hove Astronomical Society (UK-England)

BHB Bar Harbor (USA-ME) (location identifier) • Black Hole Binary • Blue Horizontal Branch

BHC Beam-Heated Cathode • Benzene Hexochloride • Black Hole Candidate • Born-Haber Cycle • Bulk Handling Center • Bullhead City (USA-AZ) (location identifier) • Busy Hour Call • Laughlin (USA-NV) (location identifier)

BHCA Busy Hour Call Attempt

BHCIG British Human-Computer Interaction Group (UK)

Bhd Berhad • Bulkhead

BHD Belfast City (Northern Ireland) (location identifier) • Dinar (Bahrain) (ISO currency code)

BHE Biharmonic Equation • Blenheim (New Zealand) (location identifier)

B'head Birkenhead (UK-England)

Bhf Bahnhof

BHF Background Heat Flux • Bahia Cupica (Colombia) (location identifier)

BHG Bauherrengemeinschaft • Berlinhilfsgesetz • Blue Hertzsprung Gap • Brus Laguna (Honduras) (location identifier) • Bureau de l'Hygiène Générale

BHH Bisha (Saudi Arabia) (location identifier)

BHI Bahia Blanca - Commandante Espora (Argentina-BA) (location identifier) • Balbus-Hawley Instability • Branch if Higher • British Humanities Index • Bureau Hydrographique International

BHIS Branch if Higher or Same

BHJ Bhuj (India) (location identifier)

BHK Bukhara (Uzbekistan) (location identifier)

BHL Bahia Angeles (Mexico) (location identifier) • Bursty High Latitude

B-HLI Broadband Higher-Layer Information

BHLI Broadband Higher-Layer Information

Bhm Birmingham (UK-England)

BHM Birmingham (USA-AL) (location identifier)

BHMO Blue Hill Meteorological Observatory (USA)

BHMP Black Hole Mass Function

BHN Beihan (Yemen) (location identifier) • Brinell Hardness Number

BHNC Bridge-function Hypernetted Chain

BHND Behind

BHO Bâtiment Hydro-Océanographique • Bhopal (India) (location identifier) • Black Hole Observatory (Germany)

bhp Brake Horsepower

BHP Bacteriohopanepolyol • Basic Hydrogen Peroxide • Bhojpur (Nepal) (location identifier) • Brake Horsepower

BHQ Broken Hill (Australia-NSW) (location identifier)

BHR Bahrain (ISO country code) • Bahrein (code ISO de pays) • Bharatpur (Nepal) (location identifier) • Block Handler Routine

BHRA British Hydraulics Research Association (UK)

BHRC Budget and Human Resources Committee

bHROS Benchmounted High-Resolution Optical Spectrograph

BHS Bahamas (ISO country code) • Base Heat Shield • Bathurst Raglan (Australia-NSW) (location identifier) • Bessel-Hankel Space • Black Hole Safari • Black Hole System • British Hydrology Society (UK) • Bulletin of Hispanic Studies (periodical code) • Burke High School (USA-NE)

BHSXT Black Hole Soft X-ray Transient

BHT Bipolar Heterojunction Transistor • Bottom-Hole Temperature • Branch History Table • Brighton Downs (Australia-QLD) (location identifier) • Butylated Hydroxytoluene

Bhu Bhutan

BHU Banaras Hindu University (India) • Bhavnagar (India) (location identifier)

BHV Bahawalpur (Pakistan) (location identifier)

BHVAD Birmingham (Starlink internal code)

BHvaO Hvar Observatory Bulletin (periodical code)

BHW Base Hazardous Waste • Sargodha Bhagatanwala (Pakistan) (location identifier)

BHWG Bird Hazard Working Group

BHX Bank Holiday Excepted • Birmingham (UK-England) (location identifier)

BHXB Black Hole X-ray Binary

BHXN Black Hole X-ray Nova

BHXRT Black Hole X-Ray Transient

BHXT Black Hole X-ray Transient

BHY Beihai (China-PRC) (location identifier)

BHZ Belo Horizonte - Pamphula (Brazil-MG) (location identifier)

BH1 Bâtiment Hydrographique de Première Classe

BH2 Bâtiment Hydrographique de Deuxième Classe

bi Bislama (ISO language code)

Bi Bipolar • Bismuth • Wismut

BI Baccalauréat International • Backbone Interface • Background Information • Background Investigation • Back-Illuminated • Back-side Illuminated • Balnicity Index • Basic Index • Binary Input • Base d'Imposition • Basic Instinct • Bénéfice d'Inventaire • Biens Immobiliers • Big Island • Billet Individuel • Binary • Binary Input • Binary Interaction • Blanking Input • Bordereau d'Inscription • Branch Immaterial • Brevet d'Invention • Briefing to Industry • Brief Introduction • Buffer Index • Built In • Bulk Issue • Bulletin d'Information • Bureau d'Instruction • Burn In • Burundi (ISO country code) • Business Intelligence • Bus Interface

BIA Basic Information Analysis • Bastia Poretta (France) (location identifier) • Blue Ice Area • Bomb Impact Assessment • Boost, Insertion and Abort • Booster Interstage Assembly • British Island Airways Ltd. • Bureau of Indian Affairs (USA) • Burned In Address • Business Impact Analysis • Bus Interface Adapter

BIAC Business Industry Advisory Committee

BIAPS Battery Inverter Accessory Power Supply

BIAS Bakosurtanal Image Analysis System • Better International Air Service

BIAst Bjull. Inst. Astrofizikii (periodical code)

BIATEX Biosphere-Atmosphere Exchange

bib Bibe • Bibliography (file extension) • BibTEX (file extension)

Bib Bible • Biblical • Bibliothèque

BIB Backward Indicator Bit • Baidoa (Somalia) (location identifier) • Balanced Incomplete Block • Bibliography • Biographical Information Bank • Blocked Impurity Band • Board for International Broadcasting • Bus Interface Board • Bulletin d'Information des Bibliothèques

BIBA Bachelor in International Business Administration • Bilingualer Basisanschluß

BIBB Bundesinstitut für Berufsbildung

BIBD Balanced Incomplete Block Design • Blocked Impurity Band Detector

BIBEX Biomass Burning Experiment

BIBIB Back-Illuminated Blocked Impurity Band

BIBID Bibliographic Identifier

Bibl Biblical • Bibliographer • Bibliographical • Bibliographie • Bibliographique • Bibliography • Biblioteca • Bibliothèque

Biblio Biblioteca

Bibliogr Bibliografía • Bibliographer • Bibliographical • Bibliographie • Bibliographique • Bibliography

BIBLIS Bibliotheeks Informatie Systeem

BIBNET Bibliographic Network

BIBO Bounded-Input Bounded-Output

BIBSYS Bibliotek System

BIC Barium Ion Cloud • Barring Incoming Calls • Battery Interconnecting Cable • Bayesian Information Criterion • Bénéfices Industriels et Commerciaux • Biased Ion Collector • Big Creek (USA-AK) (location identifier) • Binary Correlator • Book Industry Communications • Bulletin d'Information Cyclonique • Bureau de l'Information et de la Communication • Bureau International du Cinéma • Business Information Conference • Bus Interface Circuit • Bus Interface Controller • Byte Input Control

BICA Bureau International de Chimie Analytique

BICAA Bénéfices Industriels, Commerciaux, Agricoles et Artisanaux

Bicarb Bicarbonate

BICC Bearer Independent Call Control • Berne International Copyright Convention • Biomedical Information Communication Center • Bureau International des Chambres de Commerce

BICCG Block Incomplete Choleski Conjugate Gradient (method)

BICDS Bulletin d'Information du Centre de Données de Strasbourg (periodical code)

BICE Balloon-borne Infrared Carbon Explorer • Board on Infrastructure and Constructed Environment

BICEP Basic Image Classification and Enhancement Package

BICEPT Book Indexing with Context and Entry Points from Text

BICER Baikal International Centre for Ecological Research

BICFET Bipolar Inversion-Channel Field-Effect Transistor

BiCG Biconjugate Gradients (algorithm)

BICI Book Item and Contribution Identifier • Broadband Intercarrier Interface • Bureau International du Commerce et de l'Industrie

BICM Bilan Individuel de Compétences Managériales

BiCMOS Bipolar Complementary Metal Oxide Silicon

BICS Burroughs Inventory Control System

BICSE Board on International Comparative Studies in Education

bid Bis in Die

BID Bachelor of Industrial Design • Balance Identification Code • Banco Interamericano de Desarollo • Banque Interaméricaine de Développement • Basic Iterative Deconvolution • Bis In Die • Blind Iterative Deconvolution • Block Island (USA-RI) (location identifier) • Böblingen Instruments Division (HP) • Built-In Diagnostics • Business Improvement District • Business Incubator Development • Business Integrated Division

BIDA Bureau International des Droits d'Auteurs

BIDC Barbados Investment and Development Corporation

BIDDS Basic Information Digital Distribution System

BIDE Basis Identity Data Element • Bulletin International de Découvertes et d'Explorations

BIDF Bureau International de Documentation Fiscale

BiDi Bidirectional

BIDKA Bibliotheek-, Informatie- en Documentatiecentrum van de Kristelijke Arbeidersbeweging (Belgium)

BIDM Basic Interoperability Data Model

BIDP Bunyip Internet Directory Project

BIDR Basic Image Data Record

BIDS Backup Image Display System • Backup Information Display System • Bangladesh Institute of Development Studies • Bath Information and Data Service (UK) • Bath ISI Data Service (UK) • Bendix Integrated Data System • Biological Integrated Detection System • Broadband Integrated Distributed Star

Bie Bergerie

BIE Bachelor of Industrial Engineering • Base Interface Equipment • Beatrice (USA-NE) (location identifier) • Boundary Integral Equation • Bureau International d'Électrothermie • Bureau International de l'Éducation • Bureau International des Expositions • Bureau of Industry Economics

BIEF Banque d'Information sur les États Francophones

BIEL Bundesverband Internationaler Express- und Logistiekdienste (Germany)

Bien Biennial

BIET British Institute of Engineering Technology (UK) • Bureau International de l'Enseignement Technique • Business Initiatives, Education and Training

BIF Banded Iron Formation • Basic In Flow • Benchmark Interchange Format • Beta-Induced Fluorescence • Building Information Folder • Bureau d'Information sur la Formation • Bus Identification Field • Bus Interface Function • El Paso - Biggs (USA-TX) (location identifier) • Franc (Burundi) (ISO currency code)

BIFADEC Board for International Food and Agricultural Development and Economic Cooperation

BIFET Bipolar Field-Effect Transistor

BIFR Board for Industrial and Financial Reconstruction

BIG Beliefs in Government • Big Delta (USA-AK) (location identifier) • Biophysics Interdepartmental Group • Brazilian Institute of Geography and Statistics • Byurakan-IRAS Galaxy (sample)

BIGFET Bipolar Insulated Gate Field-Effect Transistor

BIGIS Bendix Interactive Geographic Information System

BIGRAT Bicentennial Gamma-Ray Telescope

BIGS Bonn International Graduate School (Germany) • Booster Inertial Guidance System

BIH Bishop (USA-CA) (location identifier) • Board on International Health • Bosnia-Hercegovina (ISO country code) • Bosnie-Hercégovine (code ISO de pays) • Built-In Hold • Bulletin d'Information Hebdomadaire • Bureau International de l'Heure (France)

BII Background Illumination Intensity • Basic Issue Item • Bikini Atoll (Marshall Islands) (location identifier) • Binary Independence Indexing

BIICC Bureau International d'Information des Chambres de Commerce

BIIEA Bureau International d'Information sur les Éphémérides Astronomiques

BIIL Basic Impulse Insulation Level • Basic Impulse Isolation Level • Basic Issue Item List

BIIS Binary Interchange of Information and Signaling

Bij Bijouterie

BIJ Biliau (Papua New Guinea) (location identifier) • Bureau d'Information pour la Jeunesse

Bijl Bijlage

Bijv Bijvoorbeeld

Bijz Bijzonder

BIK Biak (Indonesia) (location identifier)

BIL Band Interleaved by Line • Basic Insulation Level • Bâtiment d'Intégration Lanceur (CSG) • Billings (USA-MT) (location identifier) • Block Input Length • Braille Institute Library

Bilat Bilatéral

BILBO Built-In Logic Block Observer

BILD Bureau International de Liaison et de Documentation

BILE Balanced Inductor Logical Element

BiLuf Bildmessung und Luftbildwesen (periodical code)

Bim Bimestrale • Bimestre

BIM Backplane Interface Module • Balanced Implicit Multiway • Before Image • Beginning of Information Marker • Best In Match • Bimestriel • Bimini (Bahamas) (location identifier) • Binary Ionic Mixture • Biomedicine • Bit Image Memory • Boundary Integral Method • British Institute of Management (UK) • Bulletin Individuel de Mouvement • Bus Interface Model

BIMA Berkeley-Illinois-Maryland Array • Berkeley-Illinois-Maryland Association (*) • British Interactive Multimedia Association (UK)

BIMAC Bistable Magnetic Core

BIMAF Boletín del Instituto de Matemática Astronómica y Fósica de Córdoba (periodical code)

Bimest Bimestriel

BIML Bureau International de Métrologie Légale

BIMOS Bipolar Metal-Oxide Semiconductor

BIMS Browse Image Management System • Bus Ion Mass Spectrometer (PV)

BImSchG Bundesimmissionsschutzgesetz

bin Binary (file extension)

Bin Binary

BIN Bamiyan (Afghanistan) (location identifier) • Barrier Intrinsic-N+ • Basic Information Network • Bell Information Network • Billing Information Number • Binary • Bureau International de Normalisation • Business Information Network

BINAC Binary Automatic Computer

BINAS Biosafety Information Network and Advisory Servers

Bind Binding

BIND Berkeley Internet Name Daemon • Berkeley Internet Name Domain (*)

BINM Bureau of International Narcotics Matters

BINOMEXP Binomial Expansion

BINR Basic Intrinsic Noise Ratio

BINS Boreal Institute of Northern Studies

BInvG Bundesinvestitionsgesetz

BIO Basic Interoperability • Bedford Institute of Oceanography (Canada-NS) • Bilbao Sondica (Spain) (location identifier) • Biography • Biological • Biology • Bureau of International Organizations

BIOAPI Biometrics Application Programming Interface

BIOC Break Into Other Computers

Biochem Biochemical • Biochemist • Biochemistry

Biogr Biographer • Biographical • Biographie • Biographique • Biography

Biol Biología • Biological • Biologist • Biology

BIOM Buffer Input-Output Memory

BIOMASS Biological Investigation of Marine Antarctic Systems and Stocks

BIOME Biological Imaging and Observational Mission to Earth

Bion Bionika (periodical code)

BION Believe It Or Not

BIOP Blatant Interruption of Operational Potentialities • Bulletin Intérieur de l'Observatoire de Paris • Bureau pour l'Information, l'Orientation et le Perfectionnement

BIOPAR Bifurcating Optical Pattern Recognizer

Bioquím Bioquímica

BIOR Business Input-Output Rerun

BiOS Biomedical Optics Society

BIOS Basic Input/Output System • Biological Investigations Of Space • Biological Satellite

BioSc BioScience (periodical code)

BIOT British Indian Ocean Territory • Built-In Operational Test

BIOTAM Built-In Operational Test Analytic Method

BIP Background Information Package • Balanced In Plane • Balloon Interrogation Package • Band Interleaved by Pixel • Base Interoperability Protocol • Baseline Implementation Plan • Basic Interface for Parallelism • Bâtiment d'Intégration des Propulseurs (CSG) • Biens et Intérêts Privés • Bit-Interleaved Parity • Bruttoinlandsprodukt • Budget Increment Package • Bulimba (Australia-QLD) (location identifier) • Bureau d'Information et de Propagande • Bureau d'Insertion et du Partenariat • Bureau International de Presse

BIPAR Bistatic Parasitic Radar

BIPE Bureau d'Information et de Prévision Économique

BIPM Bureau International des Poids et Mesures (France)

BIPO Basic Information Protection Office

BIPP Bonn International Physics Program • British Institute for Professional Photography (UK)

BIpS Billion Instructions per Second

BIPS Backup Impact Prediction Software • Backup Impact Prediction System • Bayesian Inference for the Physical Sciences • Billion Instructions Per Second • Bulletin of the Israel Physical Society (periodical code)

BIPT Business-Integrated Process Team

BIQ Biarritz Anglet Bayonne (France) (location identifier)

BIQUUE Best Invariant Quadratic Uniformly Unbiased Estimate

BIR Band Irradiance Response • Based on Information Revealed • Benefit/Investment Ratio • Binary Independence Retrieval • Biratnagar (Nepal) (location identifier) • Birmingham (Starlink internal code) • Bit Receiver • Board of Inspection and Survey • British Institute of Radiology (UK) • Browser Information Request • Bulletin d'Informations Rapides • Bureau of Immigration Research • Bureau of Industry Research • Bus Interface Receiver

BIRA Belgische Instituut voor Ruimte Aëronomie (Belgium)

BIRAC Beijing Infrared Array Camera

BIRAMIS Bilan Radiatif de l'Atmosphère par Micro-accélérométrie Spatiale

BIRAP Balloon Infrared Astronomy Platform

BIRCH Balanced Iterative Reducing and Clustering using Hierarchies

BIRD Banco Internacional para la Reconstrucción y el Desarollo • Banque Internationale pour la Reconstruction et le Développement • Binational Industrial Research and Development • Browsing Interface for the Retrieval of Documents

BIRDE Banque Internationale pour la Reconstruction et le Développement Économique

BIRF Banco Internacional para la Reconstrucción y Fomento

BIRG Bright Infrared Galaxy

Birm Birmingham (UK-England)

BIRM Beijing Institute of Radiometrology and Measurement (China-PRC) • Bulletin de l'Institut Royal Météorologique de Belgique (periodical code)

BIRMD Balanced Incomplete Repeated Measures Design

BIRN Biomedical Informatics Research Network

BIRS Banff International Research Station (Canada-AB) • Basic Indexing and Retrieval System • Basic Information Retrieval System • Binary Infrared Stars • Bright Infrared Stars

Bis Bissextile

BIS Bachelor of Science in Information Systems • Bande Intermédiaire Satellite • Bank for International Settlements • Basic Information Sheet • Beam-Induced Stripping • Bibliographic Information System • Bibliotheks-Informations-System • Bien de Interés Cultural • Binary Synchronous • Biocide Injection System • Bio-geographic Information System • Bisei Imaging Spectrograph • Bismarck (USA-ND) (location identifier) • Blind Information System • Board of Inspection and Survey • Border Intermediate System • Bremsstrahlung Isochromat Spectroscopy • British Interplanetary Society (UK) • Building Information Schedule • Building Information Society • Built-In Simulation • Bulk Insured Services • Bureau of Indian Standards (India) • Bureau of Information Science • Business Information System • Byurakan-IRAS Stars (sample)

BISA Basic Information System Acquisition • Belgian Institute for Space Aeronomy (Belgium)

BISAC Book Industry System Advisory Committee (USA)

BISAM Basic Indexed Sequential Access Method

BISAR Bistatic Array

BISCA Bolzano International Schools on Cognitive Analysis (Italy)

B-ISDN Broadband Integrated Services Digital Network

BISDN Broadband Integrated Services Digital Network

BISE Best Index Slope Extraction

BISHOP Better Image Software with Help On-line for Programs

BISMC Basic Information System Management Center

BISO Board on International Scientific Organizations (USA)

BiSON Birmingham Solar Oscillation Network (*)

BISON Birmingham Solar Oscillation Network

BISP Basic Information System Plan • Business Information System Program

BISS Basic Installation Sensor System • Basic Intrusion Security System • Berkeley Interactive Statistical System

BISSCAN Binary Identification System from S-Curve Analysis

BIST Baseline Integrated Satellite Test • Built-In Self-Test

BISUP Broadband Integrated Services User Part

BISY British Industrial Space Year • British Industry in Space Year

BiSync Binary Synchronous Communication

BISYNC Binary Synchronous Communication

Bit Binary Digit

BIT Bachelor of Industrial Technology • Baitadi (Nepal) (location identifier) • Behavior and Information Technology • Bilateral Investment Treaty • Binary Digit • Binary Information Transfer • Born-Infeld Theory • Broadband Infrastructure Technology • Browser Information Transfer • Built-In Test • Built-In Testing • Built-In Training • Bureau International du Travail (Switzerland) • Business Information Technology

BITA Byulleten Instituta Teoreticheskoi Astronomii (Leningrad) (periodical code)

BitBlT Bit Block Transfer

BITBLT Bit Block Transfer

BITC Basic Information Transfer Center • Basic Instructor Training Course

BITE Built-In Test Equipment • Built-In Training Equipment

BITFTP BITNET FTP

BITNET "Because It's Time" Network

BITNIC BITNET Network Information Center

BITon Boletín del Instituto de Tonantzintla y Tacubaya (periodical code)

BiTraC Bidirectional Transfer Channel

BITS Basic Information Transfer System • Biological Information Technology System • Biweekly Indicator Tracking System • Book Integrity Token System • Boot Integrity Token System • Budget Impact Tracking System • Building Integrated Timing Supply

BIU Basic Information Unit • Basic Interface Unit • Bibliothèque Interuniversitaire • Bildudalur (Iceland) (location identifier) • Buffer Interface Unit • Bureau International des Universités • Bus Interface Unit

BIV Biometric Identity Verification • Bivouac • Bria (Central African Rep.) (location identifier) • Built-In Variance

BIW Billiluna (Australia-WA) (location identifier)

BIX Biloxi Keesler (USA-MS) (location identifier) • Binary Information Exchange • Byte Information Exchange

BIXS Beta-ray-Induced X-ray Spectrometry

BIY Bisho (South Africa) (location identifier)

BIZ Bimin (Papua New Guinea) (location identifier) • Business

Bi2S3 Bismuth Sulfide

Bj Baujahr

BJ Bachelor of Journalism • Benin (ISO country code) • Bonne Journée • Bubble-Jet

BJA Bejaia (Algeria) (location identifier)

BJAT Bonne Journée À Tous

BJC Broomfield Jeffco (USA-CO) (location identifier) • Bubble Jet Color

BJCEB British Joint Communications Electronics Board (UK)

BJD Bakkafjördur (Iceland) (location identifier)

BJF Batsfjord (Norway) (location identifier) • Bundesverband Jugend und Film (Germany)

BJG Bolaang (Indonesia) (location identifier)

BJH Bajhang (Nepal) (location identifier)

BJHS British Journal for the History of Science

BJI Bemidji (USA-MN) (location identifier)

BJJ Wooster - Wayne County (USA-OH) (location identifier)

BJK Benjina (Indonesia) (location identifier)

BJL Banjul (Gambia) (location identifier)

BJM Bujumbura (Burundi) (location identifier)

BJN Bajone (Mozambique) (location identifier)

BJO Bermejo (Bolivia) (location identifier) • British Journal of Ophthalmology

Bjr Bonjour

BJR Bahar Dar (Ethiopia) (location identifier) • Blockhouse Junction Rack

BJS Beijing (China-PRC) (location identifier) • Peking (China-PRC) (location identifier)

BJT Bipolar Junction Transistor

BJU Bajura (Nepal) (location identifier) • Bob Jones University (USA-SC)

BJW Bajawa (Indonesia) (location identifier)

BJX León-Guanajuato - Del Bajio (Mexico) (location identifier)

BJY Beograd Batajnica (Yugoslavia-Serbia) (location identifier)

BJZ Badajoz - Talavera la Real (Spain) (location identifier)

bk Backup (file extension)

Bk Back • Backwardation • Bank • Barrack • Berkelium • Black • Block • Book • Break

BK Backup • Benutzerklasse • Betriebskosten • Breitbandkabel • Breitbandkommunikation • Build Kit • Bundeskanzler • Bürokommunikation

BKA Bundeskriminalamt (Germany) • Moscow Bykovo (Russia) (location identifier)

BKB Bikaner (India) (location identifier)

Bkble Bookable

Bkbndr Bookbinder

BKC Benzalkonium Chloride • Buckland (USA-AK) (location identifier)

Bkcy Bankruptcy

Bkd Blackboard • Booked

BKD Blocked • Breckenridge - Stephens County (USA-TX) (location identifier)

Bkdn Breakdown

BKE Baker (USA-OR) (location identifier)

BKEP Boosted Kinetic Energy Penetrator

bkf Backup (file extension)

BKF Brooks Lake (USA-AK) (location identifier)

Bkfst Breakfast

Bkg Banking • Booking • Bookkeeping • Breakage

BKG Background • Banking • Bundesamt für Kartographie und Geodäsie (Germany)

Bkgd Background

BKGG Bundeskindergeldgesetz

BKH Kekaha - Barking Sands (USA-HI) (location identifier)

BKI Kota Kinabalu (Malaysia) (location identifier)

BKJ Boke (Guinea) (location identifier)

BKK Bangkok Donmuang (Thailand) (location identifier)

Bkkg Book-Keeping

BKL Cleveland - Burke Lakefront (USA-OH) (location identifier)

bklr Black Letter

Bklt Booklet

Bklyn Brooklyn (USA-NY)

BKM Bakalalan (Malaysia) (location identifier)

BKN Birni Nkoni (Niger) (location identifier) • Broken

BKNO3 Boron-Potassium Nitrate

BKO Bamako (Mali) (location identifier) • Barkhausen-Kurz Oscillator

BKoAS Bulletin of Korean Astronomical Society (periodical code)

BKobO Bulletin of the Kobe Marine Observatory (periodical code)

bKOM Broadband Kilometric (radiation)

BKP Backup • Barkly Downs (Australia-QLD) (location identifier)

Bkpg Bookkeeping

Bkpr Bookkeeper

Bkpt Bankrupt

BKQ Blackall (Australia-QLD) (location identifier)

BKR Bokoro (Chad) (location identifier)

Bkrpt Bankrupt

Bks Banks • Barracks • Books

BKS Basic Kernel System • Bengkulu (Indonesia) (location identifier) • Broadcast Key Station • Broad Kolmogorov Spectrum

BKSP Backspace

Bkt Basket • Bracket • Bucket

BKT Blackstone (USA-VA) (location identifier) • Camp Pickett - Blackstone (USA-VA) (location identifier)

BKU Betioky (Madagascar) (location identifier)

BKW Beckley (USA-WV) (location identifier)

BKX Brookings (USA-SD) (location identifier)

Bky Berkeley (USA-CA)

BKY Berkeley (USA-CA) • Bukavu (Zaire) (location identifier)

BKZ Bukoba (Tanzania) (location identifier)

b/l Bill of Lading

bl Barrel • Bel

Bl Bale • Black • Blatt • Block • Blue • Boulevard

B-L Bill of Lading

B&L Bausch & Lomb

B/L Baseline • Bill of Lading

BL Backlit • Bande Large • Barrel • Baseline • Basic Load • Beam Line • Bell • Bend Line • Between Layers • Billet de Logement • Bill of Lading • Bioluminescence • Bit Line • Black Letter • Blank Line • Bloch Line • Bloco • Blowing • Blue Lightning • Blue Loop (star) • Bon de Livraison • Bottom Layer • Boundary Layer • Bowling Lane • Boy's Life • British Library • Budget Line • Bulge Luminosity • Bulleted List • Bureau de Location • Busy Line • By-Laws

BLA Barcelona (Venezuela) (location identifier) • Base de Lancement Ariane (CSG) • Base Level Assessment • BL Lacertae (object) • Blocking Acknowledgment • Puerto de la Cruz (Venezuela) (location identifier)

BLabR Bell Laboratories Record (periodical code)

BLACS Basic Linear Algebra Communication Subprogram • Basic Linear Algebra Communication Subroutine

BLADE Basic Level Automation of Data through Electronics • Bell Laboratories Automatic Device

BLADES Bell Laboratories Automatic Device System

Bladz Bladzijde

BLAGN Broad-Line Active Galaxy Nucleus

BLAI Brown Leaf Area Index

BLAISE British Library Automated Information Service

BLAM Binary Logarithmic Arbitration Method

Blas Blasón

BLAS Basic Linear Algebra Subprogram • Basic Linear Algebra Subroutine • Blackburn Leisure Astronomy Section (UK)

BLAST Balloon-borne Large-Aperture Stratospheric Telescope • Basic Local Alignment Search Tool • Bates Large Acceptance Spectrometer Toroid • Battlefield Laser Acquisition Sensor Test • Blocked Asynchronous/Synchronous Transmission • Burst Locations with an Arcsecond Telescope

BLASTSP Blocked Asynchronous Transmission Software Protocol

BlazD BLAZAR Data (periodical code)

BLB Balboa (Panama) (location identifier)

BLBD Binary Light Beam Deflector

Blc Blanc

BLC Bali (Cameroon) (location identifier) • Baseline Configuration • BL Lac (object) Candidate • Boundary Layer

Control • British Library Catalogue • Broad Line Cloud • Broad Line Component

BLCE Base Level Commercial Equipment • Baseline Calibration Equipment

Blche Blanche

BLCIP British Library Cataloguing In Publication

BLCMP Birmingham Libraries Cooperative Mechanisation Project

BLCS Base Level Contingency Simulation

BLCTD British Library Computing and Telecommunications Department

bld Bload (file extension)

Bld Boldface • Boulevard

BLD Baseline Design • Beam-Lead Device • Below Limit of Detection • Blade • Boulder City (USA-NV) (location identifier) • Build

BLDB Bilingual Lexical Database

BLDC Brushless Digital Converter

Bldg Building

BLDG Building

Bldgs Buildings

BlDok Bulgarska Akademiia Nauk Doklady (periodical code)

Bldr Builder

BLDRS Baseline Data Reduction Software

BLDSC British Library Document Supply Centre (UK)

BLDUP Buildup

BLE Bahadur-Lazersfeld Expansion • Baseline Environment • Binary Logic Element • Borlänge (Sweden) (location identifier) • Box-fitting Least Squares • Branch if Less or Equal

BLEDE Backscattered Laser Energy Digitizing Equipment

BLEG Broad Line Emitting Gas

BLEIS Blazar + ESO Imaging Survey (project)

BLER Block Error Rate

BLERT Block Error Rate Test • Block Error Rate Tester

BLEU Belgo-Luxembourg Economic Union

BLF Bivariate Luminosity Function • Bluefield (USA-WV) (location identifier) • Bluff • Boundary Layer Flow • Busy Lamp Field • Busy Line Field • Princeton (USA-WV) (location identifier)

BLFS Bluffs

Blg Belge

BLG Belaga (Malaysia) (location identifier) • Binary Line Generalization • Blast Load Generator • Bundesleistungsgesetz

BlGeo Bulgarska Akademiia Naukite Geofizichni Institut – Izvestiia (periodical code)

BlGSp Bulgarsko Geofizichno Spisanie (periodical code)

BLH Band-Limited Hiss • Blythe (USA-CA) (location identifier)

BLH&SS British Library Humanities & Social Sciences

BLI Bellingham (USA-WA) (location identifier) • Budget Line Item • Business Logic Interface

BLID Budget Line Item Description

BLII Base Level Information Infrastructure

BLIMP Balloon-Launched Integral Model Payload • Boundary Layer Integral Matrix Procedure

BLIN Budget Line Identification Number • Budget Line Item Number

BLING Bladed Ring

BLIP Background-Limited Infrared Photoconductor (*) • Background-Limited Infrared Photodetection • Background-Limited Infrared Photodetector • Background-Limited Infrared Photography • Background-Limited Instrument Performance • Background Limiting Performance • Basic Landsat Image Processing • Boundary Layer Instrumentation Package

BLIR Bilingual Information Retrieval • Boundary Layer Illumination and Radiance

BLIS Baffle/Liner Interface Seal • Basic Load Inspection System • Bell Laboratories Interpretive System • Bibliographic Information System • Boundary Layer Instrumentation System

BLISK Bladed Disk

BLISS Balloon-borne Laser In Situ Sensor • Basic Language for the Implementation of System Software • Beowulf for LISA Source Simulation • Bibliographic and Library Information Search Service • British Library Information Sciences Service

BLIT Block Image Transfer

BLJ Batna (Algeria) (location identifier)

BlJMH Bulgarian Journal of Meteorology and Hydrology (periodical code)

BlJPh Bulgarian Journal of Physics (periodical code)

Blk Balkon • Black • Blank • Block • Bulk

BLK Black • Blackpool (UK-England) (location identifier) • Block

Blkd Bulkhead

Blkg Blanking • Blocking

Blks Blocks

BLKT Blanket

BLK/WHT Black/White

BLL Below Lower Limit • Billund (Denmark) (location identifier) • BL Lac (object)

BLLE Balanced Line Logical Element

BLLI Broadband Low Layer Information

BLLS Boundary Layer Lidar System

BLM Basic Language Machine • Belmar - Monmouth County (USA-NJ) (location identifier) • Bi-Linear Modeling • Boundary Layer Model • Bureau of Land Management (USA)

BLMC Buried Logic Macrocell

BLMI Broadband Local Management Interface

BLML Bland Meal

Bln Balloon

BLN Backbone Link Node • Benalla (Australia-VIC) (location identifier)

BLNAF British Library Name Authority File

BLNBS British Library National Bibliographic Service

BLNSA British Library National Sound Archive

BLNT Broadband Local Network Technology

BLO Below • Below Clouds • BL Lacertae Object • Blönduós (Iceland) (location identifier) • Broad-Lined Object

BLOb Binary Large Object

BLOB Binary Large Object

Bloc Blockade

BLOS Beyond Line Of Sight • Branch if Lower Or Same

BLOW Booster Lift-Off Weight

BLP Balmer Line Profile • Baryon Load Problem • Basic Launch Plan • Bellavista (Peru) (location identifier) • Best Lasting Penetration • Best Linear Predictor • Boundary Layer Physics • Boundary Layer Profile • Bypass Label Processing

BLPG Buried Long-Period Grating

BLPP Base Level Planning Process

BLPU Basic Land and Property Unit

BLQ Backlog Queue • Bologna (Italy) (location identifier)

BLQAM Blackman Quadrature Amplitude Modulation

BLQSO Broad Line Quasar

BLR Bangalore (India) (location identifier) • Baseline Restorer • Belarus (ISO country code) • Bielorussia (ISO country code) (obsolete) • Bélarus (code ISO de pays) • Biélorussie (code ISO de pays) (obsolete) • Boiler • Boucle Locale Radio • Boundary Layer Radar • Broad Line Region

BLR&DD British Library Research & Development Department

BLRDD British Library Research and Development Department

BLRG Broad-Line Radio Galaxy

BLRIC British Library Research and Innovation Centre (UK)

B-LRIP Beyond Low-Rate Initial Production

BLRIP Beyond Low-Rate Initial Production

BLRO Broad Line Recombination Object

BLS Bachelor of Library Science • Balanced Line System • Balloon Launching Station • Band-Limited Signal • Baseline Spectroscopy • Basic Linked System • Bell Laboratories Survey • Bell Log System • Bern - Lötschberg - Simplon (Switzerland) • Bidimensional L System • Block Level Sharing • Bollon (Australia-QLD) (location identifier) • Borne Libre-Service • Botswana - Lesotho - Swaziland • Bottom Left Side • Boundary Layer Separation • Brillouin Light Scattering • Bundle of Lines of Sight • Bureau of Labor Statistics

BLSI Boolean Latent Semantic Indexing

BLSN Blowing Snow

BlSpi Bulgarska Akademiia Naukite Spisanie (periodical code)

BLSR Bidirectional Line-Switched Ring

BLSRIS British Library Science Reference and Information Service

Blt Built

BLT Bacon, Lettuce and Tomato • Basic Language Translator • Basic Leadership Training • Blackwater (Australia-QLD) (location identifier) • Block Transfer • Boundary Layer Transition • Branch Less Than • Branch Liaison Team • Briefing and Liaison Team • Bundesamt für Landestopographie (Switzerland) • Bureau de Liaison des Traitements

BlTeK Bulgarska Akademiia Naukite Institut Tekhnicheska Kibernetika Izvestiia (periodical code)

Bltn Bulletin

BlTse Bulgarska Akademiia Naukite Tsentralna Laboratoriia Geodeziia Izvestiia (periodical code)

BLU Basic Link Unit • Basic Logic Unit • Blue Canyon (USA-CA) (location identifier)

BLUE Best Linear Unbiased Estimator

BLUP Best Linear Unbiased Prediction

BLUUE Best Linear Uniformly Unbiased Estimate

BLV Belleville Scott (USA-IL) (location identifier) • Bleed Valve • Busy Line Verification

BLVA Bayerisches Landesvermessungsamt (Germany)

Blvd Boulevard

BLVD Boulevard

Blvrd Boulevard

Blw Below

BLW Below • Waimanalo Bellows (USA-HI) (location identifier)

BLWN Band-Limited White Noise

BLWR Blower

BLX Basic Launch Complex • Belluno (Italy) (location identifier)

BLY Belmullet (Ireland) (location identifier)

Blz Bladzijde

BLZ Belize (ISO country code) • Blantyre Chileka (Malawi) (location identifier) • Boundary Layer Zone

Blzd Blizzard

BLZD Blizzard

bm Bitmap Graphics (file extension)

Bm Bearer Mobile

B/M Bench Maintenance • Bench Mark • Buffer/Multiplexer • Bulletin Maritime

BM Banco Mundial • Banque Mondiale • Baryonic Matter • Basal Metabolism • Battle Management • Baumasse • Beam Monitor • Bench Maintenance • Bench Mark • Bending Magnet • Bermuda (ISO country code) • Bermudes (code ISO de pays) • Bibliothèque Municipale • Bimensuel • Binary Multiply • Black Marker • Black Moon • Body-Mounted • Boltzmann Machine • Bon Marché • Boundary Marker • Boyer-Moore • Branch Manager • Branch Material • Branch Memorandum • Brevet de Maîtrise (France) • British Museum (UK) • Broker Manager • Brownian Motion • Bubble Memory • Buffer Manager • Buffer Module • Bulletin Mensuel • Bulletin Météorologique • Business Manager

BMA Balanced Magnetic Amplifier • Body-Mounted Accelerometer • Brevet des Métiers d'Arts • British Medical Association (UK) • Bulk Mail Acceptance • Bulletin Météorologique d'Alerte • Bulletin Météorologique d'Avertissement • Stockholm Bromma (Sweden) (location identifier)

BMaA Bundesministerium für auswärtige Angelegenheiten

BMAA Bays Mountain Amateur Astronomers (USA-TN) • Bucks-Mont Astronomical Association (USA-PA)

BMAED Board on Manufacturing and Engineering Design

BMAG Body-Mounted Attitude Gyro

BMai Bulletin de la Station Astrophotographique de Mainterne (periodical code)

BMAN Broadband Metropolitan-Area Network

BMAP Background Measurement and Analysis Program • Biological Monitoring and Abatement Program • Buffer Map

BMAR Backlog, Maintenance And Repair

BMAS Body-Mounted Accelerometer System

BMath Bachelor of Mathematics

BMATT Briefcase Multimission Advanced Tactical Terminal

BMAU Bulk Mail Acceptance Unit

BMB Base Maintenance Building • Bench Maintenance Building • Biomedical Belt • Bumba (Zaire) (location identifier) • Business Management Board

BMBF Bundesministerium für Bildung und Forschung (Germany)

BMBW Bundesministerium für Bildung und Wissenschaft (Germany)

BMC Battelle Monte Carlo • Binary Magnetic Core • Block Multiplexer Channel • Bone Mineral Content • Brigham City (USA-UT) (location identifier) • British Museum Catalogue • Broadcast/Multicast Control • Bulk Mail Center • Bulk Motion Comptonization

BMCD Blind Mate Connector Bracket

BMCS Branch Movement Control System

BMCT Begin Morning Civil Twilight

BMC3 Battle Management Command, Control, and Communications

BMD Bangladesh Meteorological Department • Belo (Madagascar) (location identifier) • Beryllium Mirror Demonstrator • Biomedical • Bone Mineral Density • Boonshoft Museum of Discovery (USA-OH) • Business Management Department (STScI) • Business Management Division • Dollar (Bermuda) (ISO currency code)

BMDADS Biomedical Data Analysis and Display System

BMDP Biomathematical Data Processing • Biomedical Package

BMDS Bite Maintainability Data Summary

BME Bachelor of Mechanical Engineering • Bachelor of Mining Engineering • Bench Maintenance Equipment • Bench-Mounted Echelle • Biomedical Engineering • Bipolar Mode Ejection • Born-Mayer Equation • Broome (Australia-WA) (location identifier) • Business Mail Entry

Bment Bâtiment

BMEP Brake Mean Effective Pressure

BMES Biomedical Engineering Society

BMEU Business Mail Entry Unit

BMF Background Magnetic Field • Background Mean Field • Bakouma (Central African Rep.) (location identifier) • Basic Main Frame • Basic Mobile Facility • Beryllium Machining Facility • Boron Metal Fiber • Bundesministerium für Finanzen • Business Master File

BMFT Bundesministerium für Forschung und Technologie (Germany)

BMG Badan Meteorologi dan Geofisika (Indonesia) • Bemessungsgrundlage • Bloomington (USA-IN) (location identifier) • Budget and Manpower Guidance • Bundesmietengesetz • Bundesministerium für Gesundheit

BMGR Branch Manager

BMH Base Message Host • Bomai (Papua New Guinea) (location identifier)

BMI Backplane Message Interface • Backplane Message Interfacer • Baltimore Museum of Industry (USA-MD) • Battelle Memorial Institute • Bismaleimide • Bloomington Normal (USA-IL) (location identifier) • Body-Mass Index • Branch if Minus • Bravais-Miller Index • Bundesministerium des Innern (Germany) • Business Management Intern

BMIC Bus Master Interface Controller

BMIL Budget Milestone

BMIS Basic Management Information System • Budget Management Information System

BMJ Baramita (Guyana) (location identifier) • British Medical Journal (periodical code) • Bundesministerium für Justiz (Germany)

BMK Borkum (Germany) (location identifier)

BML Basic Machine Language • Berlin (USA-NH) (location identifier) • Bidder Mailing List • British Museum Library • Bulk Material Length • Business Management Layer

BMLS Balloon Microwave Limb Sounder

BMM Berührungslose Mittendickenmeßgerät • Biological Macromolecule • Bitam (Gabon) (location identifier) • Bohr and Mottleson Model

BMMD Body Mass Measuring Device

BMMP Branch Modernization Management Plan

BMN Bamerny (Iraq) (location identifier) • Bulletin Météorologique National

BMNH British Museum of Natural History (UK)

BMNT Begin Morning Nautical Twilight • Beginning Morning Nautical Twilight

BMO Bhamo (Myanmar) (location identifier) • Black Moshannon Observatory (USA-PA) • Bond Molecular Orbital • British Meteorological Office (UK) • Business Machine Operator

BMOA Business Memorandum Of Agreement

BMOC Big Machine On Campus • Big Man On Campus

B'mouth Bornemouth (UK-England)

bmp Bitmap Graphics (file extension)

BMP Bad Manufacturing Practice • Basic Multilingual Plane • Batch Message Processing • Best Management Practice • Best Manufacturing Practice • Binary Millisecond Pulsar • Bitmap Picture • Bloc des Moyens Provisoires • Blue Metal-Poor (star) • Branch Modernization Plan • Break Mean Power • Brampton Island (Australia-QLD) (location identifier) • Burst Mode Protocol

BMPA Broadband Microwave Power Amplifier

BMPCOE Best Manufacturing Practice Center Of Excellence

BMPR Bimonthly Progress Report

BMPT Bundesministerium für Post und Telekommunikation (Germany) • Buren - Mosset - Parmentier - Toroni (painters)

BMQ Bamburi (Kenya) (location identifier)

BMR Baltrum (Germany) (location identifier) • Basal Metabolism Rate • Bearingless Main Rotor • Bulletin Météorologique Régional • Bureau of Mineral Resources • Business Management Review

BMRC Bureau of Meteorology Research Centre (Australia)

BMS Bachelor of Marine Science • Background Mapping Sensor • Background Measurement Satellite • Balanced Magnetic Switch • Bandwidth Management Services • Bandwidth Manager Service • Basic Management System • Basic Mapping Service • Basic Mapping Support • Basic Meteorological Service • Battle Management Simulation • Below Minimum Standards • Bias Management System • Biomedical Monitoring System • Board on Mathematical Sciences • Body Monitoring System • Bondi - Metzner - Sachs (group) • Boston Museum of Science (USA-MA) • Broadcast Message Server • Broadcast Microwave System • Broad Meteor Stream • Brumado (Brazil-BA) (location identifier) • Building Management Science • Bulletin Mensuel de Statistiques • Bulletin Météorologique Spécial • Burst Measuring System • Business Mailer Support • Business Management System • Business Modeling System • Bypass Monochrome Signal

BMSP Binary Millisecond Pulsar

BMT Bargmann - Michel - Telegdi (equation) • Basic Motion Test • Beaumont (USA-TX) (location identifier) • Beginning of Magnetic Tape • British Mean Time • Brooklyn-Manhattan Transit (New York)

BMTA Backbone Message Transfer Agent

BMTC Bureau of Meteorology Training Centre

BMTE Bright Magnitude Theoretical Excess

BMTS Bench Maintenance Test Set

BMU Battery Management Unit • Bermuda (ISO country code) • Bermudes (code ISO de pays) • Best Matching Unit • Best Match Unit • Bima (Indonesia) (location identifier) • Bundesumweltministerium (Germany) • Bus Monitor Unit

BMUX Block Multiplexer

BMV Ban Me Thuot - Phung-Duc (Viet Nam) (location identifier) • Bulletin Météorologique de Vigilance • Bundesministerium für Verkehr

BMVA British Machine Vision Association (UK)

BMVC Biologically-Motivated Computer Vision (workshop)

BMVg Bundesministerium für Verteidigung (Germany)

BMW Bayerische Motoren-Werke • Beam Width • Bordj Badji Mokhtar (Algeria) (location identifier) • Brera Multiscale Wavelet (project)

BMwA Bundesministerium für wirtschaftliche Angelegenheiten

BMWF Bundesministerium für Wissenschaft und Forschung

BMWi Bundesministerium für Wirtschaft

BMWV Bundesministerium für Wissenschaft und Verkehr (Austria)

BMX Big Mountain (USA-AK) (location identifier)

BMY Belep Island (New Caledonia) (location identifier)

BMZ Bamu (Papua New Guinea) (location identifier) • Baumassenzahl • Brandmeldezentrale • Bundesministerium für wissenschaftliche Zusammenarbeit (Germany)

bn Bangla (ISO language code) • Bengali (ISO language code)

Bn Baan • Basin • Billion • Born • Brown

BN Ballistic Number • Bank Note • Becklin-Neugebauer (object) • Belief Network • Biblioteca Nacional • Bibliothèque Nationale • Binary Number • Boron Nitride • Boundary Node • Bridge Number • Brunei (ISO country code) • Bull Nose • Bureau de Normalisation • Bureau National • But Not

BNA Backbone Network Architecture • Base-Neutralized Acid • Basic Network Availability • Block Numbering Area • British North America • Burroughs Network Architecture • Nashville (USA-TN) (location identifier)

BNAE Bureau de Normalisation de l'Aéronautique et de l'Espace (France)

BNAP Basic Hydrological Network Assessment Project

BnB Branch and Bound

BNB Boende (Zaire) (location identifier) • British National Bibliography

BNC Baby N Connector • Base-Neutralizing Capacity • Bayonet - Neill - Concelman (connector) • Bayonet Nut Collector • Bénéfice Non-Commercial • Beni (Zaire) (location identifier) • Boisson Non Comprise • British National Connector • British National Corpus

BNCAR British National Committee for Antarctic Research (UK)

BNCC Base Network Control Center • British National Committee for Chemistry (UK)

BNCE Banque Nationale pour le Commerce Extérieur

BNCF Biblioteca Nazionale Centrale di Firenze (Italy)

BNCI Banque Nationale pour le Commerce et l'Industrie

BNCIHP Belgisch Nationaal Comité voor het Internationaal Hydrologisch Programma

BNCOD British National Conference On Databases

BNCSR British National Committee on Space Research (UK)

BNCT Boron Neutron Capture Therapy

BND Bandar Abbas (Iran) (location identifier) • Beam Normalization Detector • Bend • Bénéfices Non Déclarés • Bound • Bundesnachrichtendienst (Germany) • Buy Nothing Day • Dollar (Brunei) (ISO currency code)

BNDRY Boundary

BNDT Base Nationale des Données Topographiques (Canada)

Bne Borne

BNE Branch if Not Equal • Brisbane (Australia-QLD) (location identifier) • But Not Exceeding

BNEC Belgian National ESO Committee (Belgium) • British Nuclear Energy Conference

BNES British Nuclear Energy Society (UK)

BNF Backus-Naur Form (*) • Backus Normal Form • Backus Normal Formalism • Baranof (USA-AK) (location identifier) • Bibliographie Nationale Française • Bibliothèque Nationale de France (*) • Big-Name Fan • Boolean Normal Form • Boron Nitride Fiber • British Nuclear Fuels (UK)

BNFL British Nuclear Fuels Ltd. (UK)

BNG Banning (USA-CA) (location identifier) • Blue Nucleated Galaxy • Branch No Group

BNH Hartford Barnes (USA-CT) (location identifier)

BNI Before Next Interaction • Benin City (Nigeria) (location identifier) • Bibliografia Nazionale Italiana • Broadband-to-Narrowband Interface

BNIAB British National Institute of Agricultural Botany (UK)

BNIST Bureau National de l'Information Scientifique et Technique

BNJ Bonn (Germany) (location identifier)

BNK Ballina (Australia-NSW) (location identifier) • Bank • Bereichsnetzknoten

Bnkg Banking

BNKL Becklin - Neugebauer - Kleinmann - Low (region)

BNL Barnwell County (USA-SC) (location identifier) • Bibliothèque Nationale du Luxembourg • Brookhaven National Laboratory (USA-NY)

Bnls Boneless

Blnx Benelux

BNM Bibliotheca Neerlandica Manuscripta (Netherlands) • Bodinumu (Papua New Guinea) (location identifier) • Bureau National de Métrologie (France)

BNMS Bus Neutral Mass Spectrometer (PV)

BNN Boundary Network Node • Broadcast News Navigator • Brønnøysund (Norway) (location identifier)

BNO Baksan Neutrino Observatory (Russia) • Becklin-Neugebauer Object • Burns (USA-OR) (location identifier)

BNP Bannu (Pakistan) (location identifier) • British National Bibliography • Bruto Nationaal Produkt

BNPC Bathymetric Navigation Planning Chart

BNPD Bivariate Normal Probability Density

BNQ Baganga (Philippines) (location identifier) • Bas Niveau de Qualification • Bibliothèque Nationale du Québec (Canada-QC) • Bureau de Normalisation du Québec (Canada-QC)

Bnr Burner

BNR Banfora (Burkina Faso) (location identifier) • Bloomberg News Radio

BNRID Basic Net Radio Interface Device

BNRSC British National Remote Sensing Centre (UK)

BNS Bachelor of Natural Science • Backbone Network Services • Barinas (Venezuela) (location identifier) • Bathymetric Navigation System • Beacon Numerical Service • Bibliothèque Nationale Suisse (Switzerland) • Billed Number Screening • Binary Number System • Bomb Navigation System • Braille 'N Speak • Buyer No Seller

BNSC British National Space Centre (UK)

Bnsh Burnish

BNSI Barbados National Standards Institution

BNT Baker-Nunn Telescope • Broadband Network Termination • Bundi (Papua New Guinea) (location identifier)

BNTH Beneath

BNU Basic Networking Utilities • Bibliothèque Nationale et Universitaire (France) • Blumenau (Brazil-SC) (location identifier)

BNUS Bibliothèque Nationale et Universitaire de Strasbourg (France)

BNV Belgische Natuurkundige Vereniging (Belgium) • Boana (Papua New Guinea) (location identifier)

BNW Boone (USA-IA) (location identifier)

BNX Banja Luka (Bosnia-Hercegovina) (location identifier)

BNY Bellona (Solomon Islands) (location identifier)

Bnz Bronze

BNZ Banz (Papua New Guinea) (location identifier)

b/o Brought Over

bo Bolivia • Tibetan (ISO language code)

Bo Barrio

B/O Booster/Orbiter • Breakout • Brought Over • Burnout • Buyer's Option

B-O Binary to Octal

BO Back Order • Bâtiment Océanographique • Beating Oscillator • Beat Oscillator • Bench Order • Billet à Ordre • Binary Output • Binary to Octal • Blackout • Blanket Order • Blocking Oscillation • Blocking Oscillator •

Body Odor • Boletín Oficial • Bolivia (ISO country code) • Bolivie (code ISO de pays) • Booster Orbiter • Born-Oppenheimer • Box Office • Braeside Observatory (USA-AZ) • Branch Office • Breakout • Broker's Order • Bulle Opérationnelle • Bulletin Officiel • Burnout • Business Object • Butcher-Oemler (effect) • Buyer's Option

BOA Baja Obszervatórium Alapítvány (Hungary) • Basic Object Adapter • Basic Operating Agreement • Basic Order Agreement • Basic Ordering Agreement • Big Optical Array • Big Orbital Array • Bipolar Operational Amplifier • Board Of Advisors • Boma (Zaire) (location identifier) • Bright Object Aperture • British Optical Association (UK) • Broad Ocean Area • Building Omission Area

BOAC Basic Officer Advanced Course • Billed Office Address Code • Billing Office Address Code

BOAD Banque Ouest-Africaine de Développement

BOAMP Bulletin Officiel des Appels du Marché Public

BOAO Bohyunsan Optical Astronomy Observatory (Korea-ROK)

BOAS Bright Object Alert System

BOATS Budget Organization And Tracking System

BoB Break-out Box

BOB Beginning Of Business • Best On Best • Best Orthogonal Basis • Bijzondere Opsporingsbrigade • Boliviano (Bolivia) (ISO currency code) • Bolometer Box • Bora Bora (Society Islands) (location identifier) • Break-Out Box • Broker of Observation Blocks • Bureau Of the Budget • Business Opportunities Bulletin (periodical code)

BOBCATSSS Barcelona - Oslo - Budapest - Copenhagen - Amsterdam - Tampere - Sheffield - Szombathely - Stuttgart (symposium)

BOBDN Best Orthogonal Basis Denoising

BOBeo Bulletin de l'Observatoire Astronomique de Belgrade (periodical code)

BOC Barring Outgoing Calls • Base Operations Center • Base Operations Contract • Basic Operating Company • Basic Operational Capability • Basic Operator Console • Beginning Of Cycle • Bell Operating Co. • Beoordelingscommissie (Netherlands) • Best Operational Capability • Bimodal Optical Computer • Block-Oriented Computer • Blowout Coil • Bocas del Toro (Panama) (location identifier) • Boîte d'Outils de Communication • Bureau Of Census • Byte Output Control

BOCA Borland Object Component Architecture

BOCCA Board of Coordination of Civil Aviation

BOCES Board Of Cooperative Educational Services

BOCL Business-Object Communication Language

BOCO Buying Office Contracting Official

BOCOL Basic Operating Consumer-Oriented Language

BOCS Business Object Component Set

Bod Bodega

BoD Board of Directors

BOD Bandwidth On Demand • Base Operations Division • Basic Operational Data • Beneficial Occupancy Date • Bid Opening Date • Biochemical Oxygen Demand • Biological Oxygen Demand • Board Of Directors • Bordeaux Mérignac (France) (location identifier) • Bright-Object Detector • Buy-Off Date

BODACC Bulletin Officiel des Annonces Civiles et Commerciales

BODC British Oceanographic Data Centre (UK)

BOE Basis Of Estimate • Blanket Open End • Board Of Education • Boletín Oficial del Estado • Bottom Of Edge • Boundji (Congo) (location identifier) • Break Of Entry

BOEN Bulletin Officiel de l'Éducation Nationale (France)

BOF Basic Oxygen Furnace • Beginning Of File • Beurre - Oeufs - Fromage • Bias Oscillator Frequency • Binary Oxide Film • Bird Of a Feather • Washington Bolling (USA-DC) (location identifier)

BofB Back of Board

BofE Bank of England

BOFROST Bonner Frozen Spin Target

BOFS Biogeochemical Ocean Flux Study • Bird-Of-a-Feather Session

BoG Board of Governors

BOG Board Of Governors • Bogotá Eldorado (Colombia) (location identifier) • Breadboard Operational Guide

Boh Bohemia • Bohemian

BOH Back Of House • Bottom Of Hill • Bournemouth (UK-England) (location identifier)

BoI Board of Inspection

BOI Basic Operator Interface • Basis Of Issue • Board Of Inspection • Boise (USA-ID) (location identifier) • Branch Output Interrupt • Break Of Integrity • Business Object Interface

BOIC Barring Outgoing International Calls

BOIFD Basis Of Issue Feeder Data

BOIG Basic Order of Issue Guidance

BOIP Basis of Issue Plan

BOIPFD Basis of Issue Plan Feeder Data

BOIR Based On Information Revealed

BOJ Bourgas (Bulgaria) (location identifier)

BOK Brookings State (USA-OR) (location identifier)

bol Bolus

Bol Boletín • Bolivia • Bolivian

BOL Bally Kelly (UK) (location identifier) • Bearing Only Launch • Beginning Of Life • Beginning Of Line • Bolivia (ISO country code) • Bolivie (code ISO de pays) • Business-On-Line

BOLD Bibliographic Online Display • Blood Oxygenation Level Dependent • Boeing On-Line Data

Boll Bollettino

BoLMe Boundary-Layer Meteorology (periodical code)

BOLT Broadband Observatory for the Localization of Transients

BOLTS Broad Area Optical Lightning Telescope Sensor

Bom Bombay (India)

BoM Begin of Message • Bureau of Meteorology

BOM Base Operations Manager • Basic Operating Monitor • Beginning Of Measurement • Beginning Of Message • Beginning Of Mission • Beginning Of Month • Bill Of Materials • Bill-Oriented Message • Binary Order of Magnitude • Bombay (India) (location identifier) • Born-Oppenheimer Method • Bureau Of Meteorology • Bureau Of Mines • By Other Means • Byte Order Mark

BOMA Building Owners and Managers Association (USA) • Business Object Management Architecture

Bomb Bombay (India)

BOMB Bombeiros

BOMEX Barbados Ocean Mixing Experiment • Barbados Oceanographic and Meteorological Experiment

BOMI Bill Of Materiel Indicator

BOMP Base Organization and Maintenance Processor

BOMS Bollettino delle Opere Moderne Straniere

BON Better Object Notation • Bonaire Flamingo (Netherlands Antilles) (location identifier)

BonD Bonner Durchmusterung (periodical code)

BOND Bandwidth On Demand

BONeS Block-Oriented Network Simulator

BONT Broadband Optical Network Termination

BONTT Bergen-Oslo Nuclear Theory Team (Norway)

Boo Bootes-Bootis (constellation)

BOO Bid Opening Officer • Bodø (Norway) (location identifier) • Business Operations Office

BOOde Odesskij Gosudarstvennyi Universitet im. I. I. Mechnikova Byulletin Astronomicheskoi Observatorii (periodical code)

BOOK Bibliographic Online Organized Knowledge • Built-in Orderly Organized Knowledge

BOOM Binocular Omni-Orientation Monitor (*) • Binocular Omni-Oriented Monitor

BOOMERANG Balloon Observations of Millimetric Extragalactic Radiation and Geophysics

BOOST Broadband Object-Oriented Service Technology

BOOTES Burst Observer and Optical Transient Exploring System

BOOTP Bootstrap Protocol

BOP Balance Of Payments • Balance Of Power • Baseline Operations Plan • Basic Operation Plan • Basic Overall Polarity • Basic Oxygen Process • Beginning Of Period • Bill Of Process • Binary Output Program • Bit-Oriented Protocol • Bouar (Central African Rep.) (location identifier) • Building Optimization Program • Bureau Of Printing • Business Opportunity Page • Byte-Oriented Protocol

BOPA Balance Of Payments Act (USA)

BOPACE Boeing Plastic Analysis Capability for Engines

BOPI Bulletin Officiel de la Propriété Industrielle (France)

BOpS Billion Operations per Second

BOPS Billion Operations Per Second

BOPul Bulletin of Pulkovo Observatory (periodical code)

BOQ Bachelor Officers' Quarters • Boku (Papua New Guinea) (location identifier)

BOQT Bit Optimized Query Technology

Bor Borough

BOR Basic Output Record • Belfort Fontaine (France) (location identifier) • Board Of Review • Body Of Revolution • Borough • Bureau Of Reclamation

BORAM Block-Oriented Random Access Memory

BOREAS Boreal Ecosystem-Atmosphere Study (*)

BORG Best Overall Results Generator

BORIS BOREAS Information System

Boro Borough

BORPQU Borland Pro Quattro

BORQU Borland Quattro

Borr Borrador

BORSCHT Battery - Overvoltage - Ringing - Supervision - Coding - Hybrid - Testing

BörsG Börsengesetz

BOS Background Operating System • Backup Operating System • Base Operating Support • Base Operations Support • Basic Operating System • Behörden und Organisationen mit Sicherheitsaufgaben • Binary Object Sequence • Board Of Studies • Book Order and Selection • Boston Logan (USA-MA) (location identifier) • Bright-Object Sensor • Business Office Supervisor

BOSC Base Operations Support Contract

BOSEX Baltic Open Sea Experiment

BOSO Ballistic Orbital Support Operations

BOSP Bulletin Officiel du Service des Prix

BOSS Background Optical Suppression Sensor • Behavior of Offshore Structures • Big Occulting Steerable Satellite • Billing and Order Support System • Block-Oriented System Simulation • Business-Oriented Software System

BOST Baselined Operational System Test

Bot Botanic • Botánica • Botanical • Botanist • Botany • Bottle • Bottom • Bought

BOT Back On Topic • Beam On Target • Beginning Of Table • Beginning Of Tape • Beginning Of Track • Beginning Of Transaction • Board Of Trade • Board Of Trustees • Boset (Papua New Guinea) (location identifier) • Bottom • Build - Operate - Transfer • Bureau des Objets Trouvés • Bureau d'Organisation Technique • Bureau d'Orientation Technique • Bureau Of Trade

BOTA Board On Testing and Assessment

BOTor Biuletyn Obserwatorium Astronomicznego Uniwersytetu M. Kopernika w Toruniu (periodical code)

BOTT Boletín de los Observatorios Tonantzintla y Tacubaya (periodical code)

BOTTM Bottom

BOU Bourges (France) (location identifier)

Boul Boulevard

BOUL Boulevard

BOULV Boulevard

Bourg Bourgmestre

BOV Bias Output Voltage • Boang (Papua New Guinea) (location identifier) • Board Of Visitors

BOW Bag Of Words • Bartow (USA-FL) (location identifier) • Basic Operating Weight • Bill Of Work • Bright Object Warning

BOX Borroloola (Australia-NT) (location identifier)

BOY Bobo Dioulasso (Burkina Faso) (location identifier)

BOZ Bozoum (Central African Rep.) (location identifier) • Bureau Onderwijszaken

b/p Blueprint • Bills Payable

Bp Beaucoup • Bon pour • Peculiar B-type Star

B&P Bid & Proposal • Budgetary & Planning

B/P Bills Payable • Blood Pressure • Blueprint

BP Back Penetration • Back Plane • Back Pressure • Back Projection • Back Propagation • Bald Patch • Bande Passante • Bandpass • Barber Pole • Barometric Pressure • Base Pair • Base Plate • Base Point • Base Pointer • Basic Pay • Basic Protocol • Basis Point • Basis Pursuit • Basse Pression • Battery Package • Before the Present • Below Proof • Between Perpendiculars • Bibliothèque Publique • Bid and Proposal • Biens Propres • Big Point • Bills Payable • Binding Post • Birth Place • Bit Pattern • Black Planet • Block Parity • Blood Pressure • Blueprint • Board President • Boiler Plate • Boiling Point • Boîte Postale • Bon à Payer • Bonum Publicum • Boost Phase • Boulevard Périphérique • Brennpunkt • Brevet Professionnel • Bright Point • British Patent • British Petroleum • Brutal Penetration • Budgetary and Planning • Budget Program • Budget Project • Bulk Posting • Bulk Processing • Bulletin de Paye • Bureau de Placement • Bureau de Poste • Bureau du Personnel • Business Process • Buttock Plane • Bypass

BPA Bachelor of Professional Arts • Bahnpostamt • Balanced Parametric Amplifier • Bandpass Amplifier • Basic Pressure Altitude • Basic Pricing Agreement • Basic Probability Assignment • Beam Plasma Amplification • Bethpage Grumman (USA-NY) (location identifier) • Bioshield Power Assembly • Blanket Purchase Agreement • Blanket Purchase Authorization • BNL Plant Analyzer • Board on Physics and Astronomy • Bonnes Pratiques Astronomiques • Branch Planning Analysis • Brookhaven Plant Analyzer • Budget Project Account • Buenas Prácticas Astronómicas • Business Process Analyst • Business Publications Audit

BPAA Battle Point Astronomical Association (USA-WA)

BPAC Budget Program Account Code • Budget Program Activity Code • Budget Project Activity Code

BPAD Bisynchronous Packet Assembler/Disassembler

BPAM Basic Partitioned Access Method

BPANZ Book Publishers Association of New Zealand

BPAS Bulletin of Pure and Applied Science (Physics) (periodical code)

bpb Bank Post Bills

B/PB Boost/Postboost

BPB BIOS Parameter Block • Boridi (Papua New Guinea) (location identifier) • Business Process Base

Bpc Bishopric

BPC Back Pressure Control • Bamenda (Cameroon) (location identifier) • Beam Pipe Calorimeter • Biomass Production Chamber • Blood Pressure Control • Book Prices Current • Boost Protective Cover • Brevet Professionnel Commercial • Bureau de Pilotage et de Coordination • Business Practices Course • Butyl Pyridium Chloride

BPCA Bioshield Pyrotechnic Control Assembly

BPCD Barrels Per Calendar Day

BPCF Bandpass Crystal Filter

BPCO Bronchopneumopathie Chronique Obstructive

BPCS British Post Consultancy Service (UK)

BPCST Biologie - Physique - Chimie - Sciences de la Terre

BPD Bapi (Papua New Guinea) (location identifier) • Barrels Per Day • Baseline Program Document • Baseline Program Documentation • Basic Planning Document • Beam-Plasma Discharge • Beam-Pulse Detector

BPDF Bidirectional Polarization Distribution Function

BPDM Budget Phase Data Menu

BPDN Basis Pursuit Denoising

BPDU Bridge Protocol Data Unit

BPe Bardeen-Petterson Effect

BPE Bachelor of Physical Education • Bardeen-Petterson Effect • Best Preliminary Estimate • Bit-Plane Encoding • Boiling Point Elevation

BPED Basic Pay Entry Date

BPEI Bascom Palmer Eye Institute (USA-FL)

BPES Brutal Intermediate Polyelastic Sampling

BPF Bandpass Filter • Batuna (Solomon Islands) • Berkeley Packet Filter • Best Path First • Bilan Pédagogique et Financier • Bon Pour Francs • Bottom Pressure Fluctuation • Bypass Filter

BPG Barra do Garças (Brazil-MT) (location identifier) • Baseline Process Guide • Bordeaux Planetary Group (France) • Break Pulse Generator • Broadcasting Press Guild • Business Process Group

BPGC Beam-Plasma-Gas Coupling

BPh Bachelor of Philosophy

BPH Bachelor of Public Health • Bislig (Philippines) (location identifier)

bpi Bits per Inch

BPI Best Penetration Inclination • Bibliothèque Publique d'Information (France) • Big Piney Marbleton (USA-WY) (location identifier) • Boost Phase Intercept • Breakthrough Productivity Initiative • Brevet Professionnel de l'Informatique • Bright Point Index • Bureau de Presse et d'Information • Bureau of Public Inquiries • Business Process Improvement • Bytes Per Inch

BPID Blueprint Phase Implementation Directive • Blueprint Phase Implementation Directory • Book Physical Inventory Difference

BPIE Base Procured Investment Equipment

BPI/R Business Process Improvement/Reengineering

BPIT Business Processes and Information Technology

BPJ Balanced Pressure Joint • Best Professional Judgment

BPKM Billion Passenger Kilometer

BPKT Basic Programming Knowledge Test

Bpl Birthplace

B/PL Book Plate

BPL Boston Public Library (USA-MA) • Branch if Plus • Brevet Professionnel de Librairie

BPLA Beijing Proton Linear Accelerator (China-PRC)

BPLR Bipolar

bpm Beats per Minute • Bits per Minute

BPM Background Processing Management • Barrels Per Minute • Basic Parameters Module • Basic Program Monitor • Batch Processing Monitor • Beam Positioning Magnet • Beam Position Monitor • Beam Profile Monitor • Beam Propagation Method • Beats Per Minute • Bipolytropic Model • Bits Per Minute • Blood Pressure Monitor • Bound Printed Matter • Bruce Proper Motion (survey) • Business Performance Management • Business Process Management • Business Process Manager • Business Program Manager • By-Product Material

BPMS Blood Pressure Measuring System

BPN Balikpapan (Indonesia) (location identifier) • Bibliography and Program Notes • Blanket Purchase Number • Breakdown Pulse Noise

BPNN Back Propagation Neural Network

BPO Base Post Office • Blanket Purchase Order • Blood Program Office • Borland Pascal with Objects • Business Process Object • Business Process Orientation • Business Process Outsourcing • Business Program Office

BPOA Basic Project Order Agreement • Blanket Purchase Order Agreement

BPOM Breadboard on Passive Optic Network

BPON Broadband Passive Optical Network

bpp Bits per Pixel

BPP Best Penetration Profile • Binary Packing Problem • Bits per Pixel • Black Patch Panel • Boarding Pass Printer • Bon Pour Pouvoir • Bounded-error Probabilistic Polynomial • Brevet de Pilote Professionnel • Bridge Port Pair • Brokered Private Peering • Bubble Pulse Period • Business Planning Process

BPPAC Burning Plasma Program Advisory Committee

BPPBS Biennial Planning, Programming, and Budgeting System

BPPBES Biennial Planning, Programming, Budgeting, and Execution System

BPQ Buying Power Quota

BPR Back-Pressure Regulator • Bardeen-Petterson Radius • Base Repair Percentage • Battery Powered Recorder • Béton de Poudres Réactives • Big Pulkovo Radiotelescope (Russia) • Bimonthly Progress Report • Biological and Physical Research (office) (NASA) • Bottom Pressure Recorder • Bureau des Personnels et du Rectorat • Bureau of Public Roads • Business Process Redesign • Business Process Reengineering (*) • Bypass Ratio

BPRC Beard Polar Research Center

BPRF Bullet Proof

BPRM Breit-Pauli R Matrix

BPRR Budget and Program Resource Review

BPRS Bulk Parcel Return Service

bps Bits per Second

BpS Bytes per Second

B/PS Boxy/Peanut-Shaped (galaxy bulge)

BPS Banque de Programmes et de Services • Basic Programming System • Battery Power Supply • Battery Power System • Beats Per Second • Belarussian Physical Society • Belgian Physical Society • Best Penetration Schedule • Binary (stars) Population Synthesis • Biomass Production System • Boost Propulsion System • Boundary Plasma Sheet • Boxy-Peanut-Shaped (galaxy bulge) • Budget Preparation System • Bulk Processing System • Bureau of Product Standards (Philippines) • Bureau of Professional

Standards • Bytes Per Second • Porto Seguro (Brazil-BA) (location identifier)

bpsi Bits per Square Inch

BPSI Budget Preparation System Installation

BPSK Binary Phase Shift Key • Binary Phase Shift Keying (*) • Bi-Phase Shift Key • Bi-Phase Shift Keying

BPSN Budget Project Symbol Number

BPSS Basic Packet-Switching Service • Biopack Subsystem • Budget Program Service System

bpt Bits per Track • Boiling Point

BPT Bandpass Transformer • Beaumont - Jefferson County (USA-TX) (location identifier) • Best Practice Technology • Body Point • Boiling Point • Port Arthur (USA-TX) (location identifier)

BPTF Black Patch and Test Facility

BPU Base Production Unit • Basic Processing Unit • Basic Production Unit • Beppu (Japan) (location identifier) • Branch Prediction Unit • Branch Processing Unit

BPV Bipolar Violation • Bipropellant Valve

BPX Boundary Point of Exit • Broadband Packet Exchange • Burning Plasma Experiment

BPY Besalampy (Madagascar) (location identifier)

BPZ Below Primary Zone

Bq Becquerel

BQ Best Qualified • Billet de Quai • Blue Quasar • Branch Qualification

BQA Baler (Philippines) (location identifier) • Best Questions and Answers

BQD Browse Quality Data

BQE Bubaque (Guinea-Bissau) (location identifier)

BQF British Quality Foundation (UK)

BQH Biggin Hill (UK) (location identifier)

BQK Brunswick Glynco (USA-GA) (location identifier)

BQL Basic Query Language • Boulia (Australia-QLD) (location identifier)

BQN Aguadilla (Puerto Rico) (location identifier)

BQO Bouna (Ivory Coast) (location identifier)

BQP Bounded-error Quantum Polynomial

BQQ Barra (Brazil-BA) (location identifier)

BQR Bonus Qualité Recherche • Boqueirão

BQS Berkeley Quality Software • Blagoveshchensk (Russia) (location identifier) • Bright QSO Survey • Bright Quasar Survey

BQT Brest (Belarus) (location identifier)

BQU Port Elizabeth (Saint Vincent & Grenadines) (location identifier)

Bque Banque

Bquin Bouquin

BQV Gustavus Bartlett (USA-AK) (location identifier)

BQW Balgo Hills (Australia-WA) (location identifier)

b/r Bills Receivable

br Branch • Brazil • breit • Breton (ISO language code)

Br Bearing • Bedroom • Birr (currency) • Branch • Brass • Brazil • Breite • Breton • Brevet • Breveté • Brick • Bridge • Brief • Brig • Britain • British • Broché • Brom • Brome • Bromine • Bronze • Brother • Brown • Brücke • Brume • Brut • Bruto

B-R Beobachtung-Rechnung

B/R Backup/Restore • Bills Receivable • Blue/Red • Bridges/Router • Brücken/Router

BR Background Reading • Bacteriorhodopsin • Bad Register • Bairro • Bank Rate • Baseline Review • Basic Research • Basic RPC • Bedroom • Begin Row • Belief Revision • Bibliothèque Royale (Belgium) • Bills Receivable • Binary Relation • Biorack • Biorichness • Biosphere Reserve • Bit Rate • Bit Reverse • Blind Reflex • Block Release • Block Replacement • Book of Reference • Book Review • Booster Regulator • Boundary Representation • Branch • Branching Ratio • Branch Report • Brazil (international vehicle registration code) • Brazil (ISO country code) • Break Request • Bremsstrahlung Radiation • Brésil (code ISO de pays) • Bridge • Bridging • Briefing Room • Bright Region • British Rail (UK) • Browse • Buffer Register • Bundesrat (Germany) • Bureau of Reclamation • Buriat

BRA Barreiras (Brazil-BA) (location identifier) • Basic Rate Access • Basic Research Area • Bee Research Association • Bench Replaceable Assembly • Boston Redevelopment Authority (USA-MA) • Bradford (Starlink internal code) • Brazil (ISO country code) • Brésil (code ISO de pays)

BRAA Bearing - Range - Altitude - Aspect

BRAC Base Realignment And Closure • Basic Rest-Activity Cycle

Brad Bradford (UK-England)

BRAD British Rates And Data

BRAG Brandeis Radio Astronomy Group (USA-MA)

BRAHMS Broad Range Hadron Magnetic Spectrometer

BRAIN Basic Research in Adaptive Intelligence and Neurocomputing • Basic Research in Aircraft Interior Noise

BRAINIAC Brain-Imitating Almost-automatic Computer

Br.Am. British America

BRAM Broadcast Recognition Access Method • Bulletin Régional d'Alerte Météorologique • Bulletin Régional d'Avertissement Météorologique

BRAN Broadband Radio Access Network

BRAO Bundesrechtsanwaltsordnung

BRAP Broadcast Recognition with Alternating Priorities

Bras Brasil

BRAS Baton Rouge Astronomical Society (USA-LA) • Black River Astronomical Society (USA-OH)

BRASS Berkeley Reconfigurable Architecture, Systems and Software

BRAT Basic RISC Architecture Timer • Budget/Readiness Analysis Technique

BRAVO Back-end Redesign for Added Value for Observers

BRAVOS Business Risk And Value of Operations in Space

Braz Brazil • Brazilian

Brazz Brazzaville (Congo)

BRB Barbade (code ISO de pays) • Barbados (ISO country code) • Barreirinhas (Brazil-MA) (location identifier) • Be Right Back • Broad Band

Br.C. British Columbia (Canada)

BR/C Budget Request/Change

BRC Below Regulatory Concern • Below Regulatory Control • Blue Ribbon Contractor • Bright Rimmed Cloud • Budget Review Committee • Business Reply Card • San Carlos de Bariloche (Argentina-RN) (location identifier)

BRCC Blue Ribbon Contractor Certificate

Brch Branch

Brck Brücke

BRCL Blue Ribbon Contract List • Blue Ribbon Contractor List

BRCS Basic Reference Coordinate System

BRD Baseline Requirements Document • Biological Resources Division (USGS) • Blank Recording Disc • Bonded Rolled Diaphragm • Booster Requirements Document • Brainerd (USA-MN) (location identifier) • Breadboard • Bundesrepublik Deutschland (Germany)

Brdcst Broadcast

BRDF Bidirectional Reflectance Distribution Function (*) • Bidirectional Reflective Distribution Function

Brdg Bridge

BRDGE Bridge

BRDL Biomedical Research and Development Laboratory

BRDT Bayesian Reliability Demonstration Test

BRE Bistatic Radar Equation • Bremen Neuerland (Germany) (location identifier) • Bridge Relay Encapsulation • British Research Establishment • Bureau de Rapprochement des Entreprises (France) • Business Reply Envelope • Cruzeiro (Brazil) (ISO currency code)

BREBS Bowen Ratio Energy Balance System

BREC Battery Regulator and External Converter

BReg Bundesregierung

BREJ Bipolar Rotating Episodic Jet

BREM Boeing Radiation Exposure Model

BRER Board on Radiation Effects Research

BRESEX Brazilian Remote Sensing Experiment

BRESM Bureau de Recherches et d'Études SHOM/Météo-France

Brev Brevet • Breveté • Brevetto

BREW Biospherics Research on Emissions from Wetlands

Brf Brief

BRF Band Rejection Filter • Basic Research Fund • Below Rig Floor • Benchmark Report Format • Best Replacement Factor • Bidirectional Reflectance Factor • Blind Relevance Feedback • Bradford (UK-England) (location identifier) • Bridge Relay Function • Brief

BRFI Bidirectional Reflectance Field Instrument

Brg Bearing

BRG Bearing • Bridge • Bright Red Galaxy • Bright Rimmed Globule • Budget Review Group • Business Research Group • Whitesburg (USA-KY) (location identifier)

BRGC Binary Reflected Gray Code

BRGG Bureau de Recherches Géologiques et Géophysiques

BRGGM Bureau de Recherches Géologiques, Géophysiques et Minières

BRGM Bureau de Recherches Géologiques et Minières (France)

BRH Brahman (Papua New Guinea) (location identifier)

BRHR Basic Research and Human Resources

bRi Bulk Richardson Number

Br. I. British India • British Isles

BRI Banque des Règlements Internationaux • Bari (Italy) (location identifier) • Bartol Research Institute (USA-DE) • Basic Rate Interface (*) • Basic Rate Interference • Basic Rate ISDN • Brain Response Interface • Bristol (Starlink internal code) • Bruttorauminhalt • Building Research Institute (Japan) • Bureau de Renseignements Internationaux

BRIAN Barrier Reef Image Analysis

BRIC Biological Research In Canisters • Bureau de Recherche pour l'Innovation et la Convergence (France) • Bureau de Recherches Industrielles et Commerciales

BRICS Basic Research In Computer Sciences

BRIDGE Biotechnology, Research for Innovation, Development and Growth in Europe

BRIEF Basic Reconfigurable Interactive Editing Facility

Brig Brigade • Brigadier

BRIG Brigadeiro

BRIL Baseline Requirements Item List • Baseline Resource Item List • Budget Receipt/Issuance Log

BRIM Biosphere Reserve Integrated Monitoring • Bridge-Router Interface Module

BRIMS Budgeted Resource Information Management Source

Brisb Brisbane (Australia-QLD)

BRISCA Bristol Ice and Snow Cover Algorithm

Brist Bristol (UK-England)

Brit Britain • Britannia • British • Briton

BRITE Basic Rate Interface Transmission Extension • Basic Research in Industrial Technologies for Europe • Bright Radar Indicator Terminal Equipment

Brit.Mus. British Museum

Brit.Pat. British Patent

BRJ Bright (Australia-VIC) (location identifier)

Brk Brake • Brick

BRK Bourke (Australia-NSW) (location identifier) • Brake • Break • Brook

Brkg Breaking

BRKN Broken

Brkr Breaker

BRKS Breaks • Brooks

Brkt Bracket

BRKT Bracket

Brl Barrel

BRL Ballistics Research Laboratory • Bendix Research Laboratories • Beyond Richardson-Lucy • Building Restriction Line • Burlington (USA-IA) (location identifier) • Real (Brazil) (ISO currency code)

BRLO Broad Recombination Line Object (*) • Broad Recombination of Line Objects

BRM Barometer • Barquisimeto (Venezuela) (location identifier) • Baseline Reference Mission • Basic Reference Model • Basic Remote Module • Binary Rate Multiplier • Biological Research Module • Bio Research Module • Booster Recovery Module • Bureau of Radio Management (China-PRC) • Business Redesign Methodology • Business Reply Mail

BRMAS Business Reply Mail Accounting System

BRMIC Behavior Research, Methods, Instruments and Computers (periodical code)

BRMP Background Radiation Monitoring Programme

Br.n. Broché Neuf

Brn Brown

BRN Bahrain (international vehicle registration code) • Bern Belp (Switzerland) (location identifier) • Berne Belp (Switzerland) (location identifier) • Brazilian New Cruzado (ISO currency code) • Brunei (ISO country code)

BRNCH Branch

Brng Bearing • Burning

Bro Brother • Brotherhood

BrO Bromine Monoxide

BRO Breault Research Organization Inc. (USA-AZ) • Brother • Brownsville (USA-TX) (location identifier)

Broc Brochure

BROGS Belgian Royal Observatory Communications Series Geophysics Series (periodical code)

Brom Bromide

BROOM Ballistic Recovery Of Orbiting Man

Bros Brothers

Brosch Broschiert

BROTS Beneficial Rays Of The Sun

BRP Basic Research Plan • Biaru (Papua New Guinea) (location identifier) • Biological Reclamation Process • Blue Ribbon Panel • Bulletin Régulier de Précipitations

BRPA Basic Research and Proposal Announcement

BRPM Blue Ribbon Program Monitor • Breath Rate Per Minute

BRPP Blue Ribbon Program Panel

BRPRS Blue Ridge Parkway Regional Stargaze (USA-NC)

BRPW Bulk Revenue Pieces and Weights

BRQ Brno (Czech Rep.) (location identifier)

BRR Barra (UK) (location identifier)

BRRL Broad Radio Recombination Line

BRRS Banana River Repeater Station

BRRU Business Regulation Review Unit

BRS Balloon Radio System • Bandwidth Reservation System • Barycentric Reference System • Baseline Resource Scheduler • Basic Research Symposium • Baurechtsammlung • Bibliographic Reference Service • Bibliographic Research Service • Bibliographic Retrieval Service (*) • Bibliographic Retrieval System • Biennial Reporting System • Big Red Switch • Binary Relational Store • Binary Ring Sequence • Block Recorder System • Bottom Right Side • Break Request Signal • Bristol Lulsgate (UK-England) (location identifier) • Broadcast Radiation Station • Budget Review System • Bureau of Resource Sciences • Buried Ridge Structure

BRSC Beijing Remote Sensing Corporation (China-PRC)

BRSL Balasore Rocket Launching Station (India)

Brst Burst

Brt Bright • Brightness • Brut

BrT Brevet Technique

BRT Base Repair Time • Base Resistance Transistor • Bathurst Island (Australia-NT) (location identifier) • Belgische Radio en Televisie (Belgium) • Binary Run Tape • Bradford Robotic Telescope • Bright • Brightness • Bruttoregistertonne

BRU Battery Regulator Unit • Branch Registration Unit • Brunei (international vehicle registration code) • Brussel Zaventem (Belgium) (location identifier) • Brussels Zaventem (Belgium) (location identifier) • Bruxelles Zaventem (Belgium) (location identifier) • Bureau de Renseignements Universitaires • Bus Regulator Unit

Brunsw Brunswick

Brux Bruxelles (Belgium)

BRV Backup Recovery Vector • Ballistic Reentry Vehicle • Bremerhaven (Germany) (location identifier)

BRW Barrow Wbas (USA-AK) (location identifier) • Brake Release Weight

BrWK Brennstoff Wärme Kraft (periodical code)

BRWM Board on Radioactive Waste Management

brx Browse Index (file extension)

Brx Brussels (Belgium)

BRX Barahona (Dominican Rep.) (location identifier)

BRY Bardstown Samuels (USA-KY) (location identifier)

Brz Brazil • Bronze

BRZ Borotou (Ivory Coast) (location identifier)

BRZN Bibliotheksrechenzentrum Niedersachsen (Germany)

b/s Bill of Sale

Bs Bags • Bales • Bas • Bolivars (currency) • Boliviano (currency)

B/S Balance Sheet • Behind Schedule • Bill of Sale • Bits per Second

BS Bachelor of Science • Backplane Segment • Backspace • Backup System • Bahamas (international vehicle registration code) • Bahamas (ISO country code) • Balance Sheet • Base Section • Base Station • Bashkir • Battery Simulator • Baudhāyana Sulbasūtra • Below Specification • Be Specific • Beam Splitter • Beam-Splitting • Beam Steering • Beam Stop • Bearer Service • Bibliography Software • Bibliothèque Spécialisée • Bill of Sale • Binary Scale • Binary Star • Binary State • Binary Subtract • Bit Selection • Bit-Sliced • Bit Synchronization • Black Star • Block Specification • Blowing Snow • Blue Storm • Blue Straggler • Bon de Sortie • Booking System • Bootstrap • Bordereau de Situation • Bow Shock • Brevet de Spécialité • Brevet Supérieur • Bright Spot • Bright Star • Bright Stars (catalog) • British Standard • Broadcasting Satellite (Japan) • Broadcasting Service • Broadcast Satellite (Japan) • Broadcast Source • Broadcast Station • Broad Structure • Browse Service • Building Section • Bulletin de Salaire • Bureau of Standards • Bureau of Statistics • Burgerlijke Stand • Bus Switch

BSA Backwards Stepping Aggregation • Base Service Area • Base Support Agreement • Basic Service Agreement • Basic Serving Agreement • Basic Standardization Agreement • Battery Solar Array • Bibliographical Society of America (USA) • Biological Safety Adviser • Bipartite Standardization Agreement • Bit Sync Acquisition • Board of Scientific Affairs • Bohr-Sommerfeld Atom • Bon de Souscription d'Actions • Bon de Souscription en Actions • Bossaso (Somalia) (location identifier) • Boy Scouts of America • Brookhaven Science Associates • Budget Subactivity • Burst Spectrum Analyzer • Business Software Alliance • Bus Solar Array

BSAA Bachelor of Science in Applied Arts

BSAdv Bachelor of Science in Advertising

BSAE Bachelor of Science in Aeronautical Engineering

BSAF Bids Solicited As Follows

BSAM Basic Sequential Access Method • Basic Software Acquisition Management

BSAMC Basic Software Acquisition Management Course

BSAP Base Station Application Port

Bs. As. Buenos Aires (Argentina)

BSAS Barnard Seyfert Astronomical Society (USA-TN)

Bsb Brisbane (Australia-QLD)

BSB Brain-State-in-the-Box • Brasília (Brazil-DF) (location identifier) • Brassboard • British Satellite Broadcasting Ltd.

BSBA Bachelor of Science in Business Administration

BSBB Baryon-Symmetric Big bang

BSBD Biblio Service Data Bank

BSc Bachelor of Science

B/SC Brake Skid Control

BSC Bachelor of Science • Bahia Solano (Colombia) (location identifier) • Base Station Controller • Base Support Contractor • Basic Sequential Access Method • Beam Splitter Chamber • Beam Steering Computer • Bergen Scientific Centre (IBM) • Bibliographical Society of Canada • Billing Status Code • Binary Symmetric Channel • Binary Synchronous Communication • Biological Safety Committee • Bi-Synchronous Communication • Bit Scan Command • Blue Spheroid Candidate • Boyan Script • Brake Skid Control • Brake Skid Controller • Brevet Supérieur Commercial • Bright Source Catalog • Bright Stars Catalog • British Steel Corporation (UK) • Business Service Center

BSCA Binary Synchronous Communications Adapter • Boresight Camera Assembly

BSCCO Bismuth Strontium Calcium Copper Oxide

BSCE Bachelor of Science in Civil Engineering

BSCG Bisei Spaceguard Center (Japan)

BSChE Bachelor of Science in Chemical Engineering

BSCM Bisynchronous Communications Module

BSCN Bit Scan

BSCom Bachelor of Science in Communications

BSCP British Standard Code of Practice

BSCS Bachelor of Science in Computer Science

BSCW Basic Support for Cooperative Work

BSD Baccalauréat du Second Degré • Baoshan (China-PRC) (location identifier) • Ballistic Systems Division • Baseline System Description • Basic Standard Definition • Beam Steering Device • Bergedorfer Spektral Durchmusterung • Berkeley Software Design Inc. (USA-CO) • Berkeley Software Distribution (*) • Berkeley Standard Distribution • Berkeley System Distribution • Bit Storage Density • Board of Sustainable Development • Bright Surge on Disk • British Space Development • Bulk Storage Device • Business System Development • Dollar (Bahamas) (ISO currency code)

BSDAM Basic Sequential Data Access Method

BSDC Binary Symmetric Dependent Channel

BSDF Bidirectional Scatter Distribution Function

BSDI Berkeley Software Design Inc. (USA-CO) • Berkeley Software Design International (USA-CO) (*)

BSDIA Brazilian Symposium on Document Image Analysis

BSDL Boresight Dotum Line

BSDP Booster Stage Discharge Pressure

BSDS Biological Standoff Detection System • Boot Strap Data Set

BSE Bachelor of Science in Education • Bachelor of Science in Engineering • Back-Scattered Electron • Base Station Equipment • Base Support Element • Basic Service Element • Bâtiment de Stockage des EAP (CSG) • Bâtiment de Stockage des Étages (CSG) • Bethe-Salpeter Equation • Binary Star Evolution • Booster Systems Engineer • Broadcast Satellite Experiment • Sematan (Malaysia) (location identifier)

B/sec Bits per Second

BSEc Bachelor of Science in Economics

BSEC Base Security Executive Council • Black Sea Economic Cooperation • Brevet Supérieur d'Études Commerciales

BSEd Bachelor of Science in Education

BSED Ball Systems Engineering Division

BSEE Bachelor of Science in Electrical Engineering

BSEEA Iasi Institutul Politechnic Buletinul Sectia Electrotechnica Electronica Automatizari (periodical code)

BSEEng Bachelor of Science in Electrical Engineering

BSElE Bachelor of Science in Electronic Engineering

BSEng Bachelor of Science in Engineering

BSEP Bachelor of Science in Engineering Physics

BSES Bachelor of Science in Engineering Sciences • Brevet Supérieur de l'Enseignement Supérieur

BSET Base Security Evaluation Team

BSF Bachelor of Science in Forestry • Backspace File • Back Surface Field • Baseline System Functional • Berliner Sternfreunde eV (Germany) • Best Slit Throughput • Binational Science Foundation • Bit Scan Forward • Pohakuloa Bradshaw (USA-HI) (location identifier)

BSFC Brake Specific Fuel Consumption

BSFR Back Surface Field Reflector (*) • Back Surface Field with Reflection • Back Surface Filled Reflector

BSFS Baseline System Functional Specification

BSG Bata (Equatorial Guinea) (location identifier) • Beam Steering Group • Bénéficiaire de Soins Gratuits • Bit Sync Generator • Blue Supergiant (star) • British Standard Gauge • Bundessozialgericht (Germany)

BSGC Bisei Spaceguard Center (Japan)

BSGDG Brevetè sans Garantie du Gouvernement

BSGE Bachelor of Science in General Engineering

BSH Brighton (UK-England) (location identifier) • Bundesamt für Seeschifffahrt und Hydrographie (Germany)

BSHG Bundessozialhilfegesetz

BSHK Biophysical Society of Hong Kong (China-PRC)

BSHM Bâtiment de Soutien de Haute Mer • British Society for the History of Mathematics (UK)

BSHS British Society for the History of Science (UK)

BSI Base Support Installation • Basic Shipping Information • Basic Shipping Instruction • Battery Status Indicator • Béton Spécial Industriel • Bibliographical Star Index • Blairsville (USA-PA) (location identifier) • Booster Situation Indicator • Branch and Store Instruction • British Standards Institution (UK) • Broken Scale Invariance • Bundesamt für Sicherheit in der Informationstechnik (Germany) • Bureau de Statistique Industrielle

BSIC Base Station Identity Code • Binary Symmetric Independent Channel

BSIR Bachelor of Science in Industrial Relations

BSIRA British Scientific Instrument Research Association (UK)

BSIT Bachelor of Science in Industrial Technology • Base Systems Integration Testbed

BSJ Bachelor of Science in Journalism • Bairnsdale (Australia-VIC) (location identifier)

Bsk Basket

BSK Biostack • Biskra (Algeria) (location identifier)

Bskt Basket

Bs/L Bills of Lading

BSL Bachelor of Science in Law • Bachelor of Science in Linguistics • Bâle Mulhouse (Switzerland) (location identifier) • Basel Mulhouse (Switzerland) (location identifier) • Basic System Listing • Basle Mulhouse (Switzerland) (location identifier) • Biot-Savart Law • Blackwell Science Ltd. • Bright Source List • Bright Surge on Limb • British Sign Language • Mulhouse Bâle (France) (location identifier)

BSLS Bachelor of Science in Library Science

BSM Austin Bergstrom (USA-TX) (location identifier) • Backup Setup Message • Basic Storage Module • Basic Sustainment Material • Basic System Memory • Basic System Module • Beam-Steering Mechanism • Boîtier de Servitude Moteur • Booster Separation Motor • Branch Sales Manager

BSMD Bulk Store Memory Device

BSME Bachelor of Science in Mechanical Engineering

BSMF Bit Sync Matched Filter

BSML Bioinformatic Sequence Markup Language

BSMM Bidirectional Symmetric Multipoint-to-Multipoint

BSMMT IASI Institutul Politechnic Buletinul Sectia Matematica Mecanica Teoretica Fizica (periodical code)

Bsmnt Basement

B/SMPL Bits per Sample

BSMPL Bits per Sample

BSMSP Bernouilli Society for Mathematical Statistics and Probability (ISI)

Bsmt Basement

BSMT Basement • IASI Institutul Politechnic Buletinul Sectia Mecanica Tehnica (periodical code)

BSMTP Batched Simple Mail Transfer Protocol

BSN Bachelor of Science in Nursing • Backward Sequence Number • Basic Symbol Number • Basis Systeme Netzwerk • Bossangoa (Central African Rep.) (location identifier) • Business Service Network

BSNE Bachelor of Science in Nuclear Engineering

BSNI Barbados National Standards Institution

BSNR Backward Sequence Number Received

Bsns Business

BSO Bacteria-Shaped Object • Ballon Stratosphérique Ouvert • Basco (Philippines) (location identifier) • Basic Security Option • Beroeps Secundair Onderwijs • Big

Sky Observatory (Canada-AB) • Blue Stellar Object • Bright Stellar Object • Broad System of Ordering • Budget Submitting office • Buro voor Systeemontwikkeling BV (Netherlands)

Bson Buisson

BSOP Bipartite Standing Operating Procedure

BSP Bachelor of Science in Pharmacy • Backup Service Provider • Base Support Plan • Beam Splitter • Beam Steering Processor • Beam-Switched Position • Bensbach (Papua New Guinea) (location identifier) • Bibliographical Service Provider • Binary Space Partition • Biodiversity Support Program • Biospecimen Sharing Program • Blackwell Scientific Publications • Boolean Satisfiability Problem • Bulk Synchronous Parallel • Bulk Synchronous Parallelism • Burroughs Scientific Processor • Business Strategy Panel • Business Systems Planning

BSPA Bachelor of Science in Public Administration

BSPM Bidirectional Symmetric Point-to-Multipoint

BSPP Bidirectional Symmetric Point-to-Point

BSPS Billion Samples Per Second

bspw beispielweise

BSQ Band Sequential • Binary Sequential (file) • Bisbee (USA-AZ) (location identifier)

BSR Back Surface Reflector • Basic System Release • Basra (Iraq) (location identifier) • Bâtiment de Soutien de Région • Beyond Sensor Range • Bistatic Spaceborne Radar • Bit Scan Reverse • Bit Slippage Rate • Blood Sedimentation Rate • Board of Standard Review • Bragg Scattering Region • Brevet de Sécurité Routière • Bulk Shielding Reactor • Byte Status Register

BSRAM Burst Static Random Access Memory

BSRC Behavioral Science Research Centre

BSRF Basic System Reference Frequency • Beijing Synchrotron Radiation Facility (China-PRC) • Biological Space Research Facility

BSRL Boeing Scientific Research Laboratory

BSRM Booster Solid-Rocket Motor

BSRN Baseline Surface Radiation Network

BSROC Belgium Space Remote Operations Center

BSRSL Bulletin de la Société Royale des Sciences de Liège (periodical code)

BSRT Bachelor of Science in Radiological Technology

BSS Bachelor of Secretarial Science • Bachelor of Social Science • Backup System Service • Balanced Salt Solution • Balloon-borne Superconducting Selenoid • Balsas (Brazil-MA) (location identifier) • Bare Strange Star • Base Station System • Basic Safety Standard • Basic Service Set • Basic Space Sciences • Basic Synchronized Set • Beam Steering System • Bigelow Sky Survey • Billing Support System • Bisymmetric Spiral (field) • Bisymmetric Spiral Structure • Block Started by Symbol • Blue Straggler Star • Borehole Seismic System • Bottom-Side Sinusoidal • Breakthrough Science Society (India) • Bright Star Spot • British Standards Specification • British Sundial Society (UK) • Broadband Switching System • Broadcasting Satellite Service • Broadcast Satellite Service • Broler-Supplied System • Business Support System • Byurakan Second Survey

BSSA Bachelor of Science in Secretarial Administration • Bulletin of Seismological Society of America (periodical code)

BSSAP Base Station System Application Part

BSSC Bit Synchronizing and Signal Conditioner • Board for Software Standardization and Control • Board of Software Standard and Control

Bsse Basse

BSSE Bachelor of Science in Secondary Education • Basic Set Superposition Error • Brazilian Symposium on Software Engineering

BSSGP Base Station System Global Positioning

BSSI Bulk Solar System Initial

BSSMAP Base Station System Management Application Part

BSSR BSSR Institut Teplo i Massoobmena Vsesoiuznaia Konferentsiia Teplomassoobmenu Minsk Belarussian SSR Preprint (periodical code)

BSSS Bachelor of Science in Secretarial Studies • Bachelor of Science in Social Science

BST Backward-pointing Star Tracker • Barium Strontium Titanate • Base Shop Test • Basic Services Terminal • Beam Steering Transducer • Beam-Switching Tube • Belgian Space Training (Belgium) • Bid-Sample Testing • Binary Synchronous Transmission • BioServe Space Technologies (USA-CO) • Bost (Afghanistan) (location identifier) • Brevet Supérieur Technique • British Standard Time • British Summer Time • Bureau de Standardisation des Télécommunications • Byggstandardiseringen i Sverige (Sweden)

BStaO Byulleten Stalinabadskoi Astronomicheskoi Observatorii (periodical code)

BStatG Bundesstatistikgesetz (Germany)

Bstb Bistable

BStBl Bundesteuerblatt (Germany)

BSTC Biotechnology Specimen Temperature Controller

BSTF Best Shop Test Facility • Business Systems Tast Force

Bstg Bastogne (Belgium)

BSTI Bangladesh Standards and Testing Institution

BST&IE Bachelor of Science in Trade & Industrial Education

BSTJ Bell System Technical Journal

Bstr Booster

BSTR Booster

BSTS Best Ship Test Station • Boost Surveillance and Tracking System • Boost Surveillance Tracking System • Broadband Series Test System

BSU Ball State University (USA-IN) • Basankusu (Zaire) (location identifier) • Base Spatial Unit • Basic Sounding Unit • Basic Structural Unit • Beam Steering Unit • Bench Scale Unit • Blood Supply Unit • Boise State University (USA-ID) • Bulk Storage Unit

BSV Boot Sector Virus • Bosset (Papua New Guinea) (location identifier)

BSVC Broadcast Switched Virtual Connection

BSVit Bollettino della Società dei Variabilisti Italiani (periodical code)

BSW Basic Software • Boswell Bay (USA-AK) (location identifier) • British Standard Whitworth • Business Software

BSX Bassein (Myanmar) (location identifier)

BSY Bardera (Somalia) (location identifier) • Busy

BSYNC Binary Synchronous Communication

BSZ Bartletts (USA-AK) (location identifier)

BS2 Basic Space Sciences

b/t Between

Bt Bâtiment • Beat • Benefit • Bent • Billet • Boat • Bought • Brevet • Brut • Bukit

B-T Business-Technology

BT Bachelor of Teaching • Background Test • Ballistic Transport • Basic Training • Bassa Tensione • Basse Tension • Bathythermograph • Bathythermography • Begin Text • Beijing Time • Bibliothèque de Travail • Bit Test • Board of Trade • Body Text • Bon à Tirer • Booster Trailer • Botswana • Boutan (code ISO de pays) • Brevet de Technicien • Brevet Technique • Brigade Territoriale •

Brightness Temperature • British Telecom (UK) • Broadband Terminal • Broader Term • Bulk Transfer • Bulletin Technique • Bundestag (Germany) • Bureau de Transport • Bureau Technique • Burn Time • Burst Tolerance • Business Telecommunications • Bus Terminator • Buthan (ISO country code)

BTA Ballistic Trajectory Aggregate • Basic Trading Area • Basse Tension Aérienne • Bertoua (Cameroon) (location identifier) • Best Technical Approach • Better Than Average • Binary Tables Agreement (FITS) • Bolshoi Teleskop Azimutal'nij (Russia) • Booster Track Accuracy

BTAC Binary Tree Algebraic Computation • Branch Target Address Cache • Branch Target Address Code

BTAIM Be That As It May

BTAM Basic Tape Access Method • Basic Telecommunications Access Message • Basic Telecommunications Access Method (*) • Basic Terminal Access Method

BTB Bathythermograph Buoy • Believed-To-Be • Betou (Congo) (location identifier) • Block Transfer Bus • Branch Target Buffer • Burst Time Plan

BTC Batticaloa (Sri Lanka) (location identifier) • Before Top Center • Bench Test Console • Best Technical Course • Big Throughput Camera • Binary Time Code • Bit Test and Complement • Bit Time Counter • Block Transfer Controller • Block Truncating Coding • Breakthrough Curve • British Transport Commission (UK) • Bus Tie Contractor

Btca Biblioteca

BTCAM Big Throughput Camera

BTCF Base Technical Control Facility

BTCI Biotechnology Citation Index

BTCM Bent Triangular Crystal Monochromator

BTD Balanced Tape Drive • Binary To Decimal • Boston Transportation Department (USA-MA) • Breath Testing Device • Brief Task Description • Brightness Temperature Difference • Brunette Downs (Australia-NT) (location identifier) • Burn To Depletion

BTDAM Basic Telecommunications Data Access Method

BTDC Back of Top Dead Center • Before Top Dead Center

BTDL Basic Transient Diode Logic

bte Binte

Bte Boîte

Bté Breveté

BTE Bachelor of Textile Engineering • Bench Test Equipment • Better Than Expected • Boltzmann Transport Equation • Bonthe (Sierra Leone) (location identifier) • British Telecommunications Engineering (periodical code) • Broadband Terminal Equipment • Bulk Tape Eraser

BTECC Book Trade Electronic Communications Committee

BTEM Beam Test Engineering Model

BTESS Brevet Technique d'Études Supérieures Spécialisées

BTEX Benzene-Toluene-Ethylbenzene-Xylene

BTF Baryonic Tully-Fisher (law) • Baseline Test Facility • Battery Test Facility • Beam Transfer Function • Bench Test Fixture • Biotechnology Facility • Bold Typeface • Bountiful - Salt Lake (USA-UT) (location identifier)

BtG Betreuungsgesetz

BTG Batangafo (Central African Rep.) (location identifier) • Bon de Transport Gratuit • Broadband Transmit Group

Bth Bath • Bathroom • Berth

BTH Bath • Batu Besar (Indonesia) (location identifier) • Below The Horizon • Beyond The Horizon

BTHP Béton à Très Hautes Performances

BThU British Thermal Unit

BTI Balanced Technology Initiative • Barter Island (USA-AK) (location identifier) • Baseline Technology Improvement • Beginning Tape Label • Bench Test Instruction • Bridged Tap isolator • Business Travel International

BTIC Business and Technological Information Centre

BTJ Banda Aceh (Indonesia) (location identifier)

Btk Buttock

BTK Bijzonder Tijdelijk Kader • Bratsk (Russia) (location identifier)

Btl Bottle

BTL Backplane Transceiver Logic • Basic Termination Liability • Battle • Battle Creek (USA-MI) (location identifier) • Beacon Tracking Level • Beam Transport Line • Bell Telephone Laboratories • Below The Line • Better Than Life • Between Layers • Bread - Tomato - Lettuce

BTLB Block Translator Look-aside Buffer

BTLC Bureau Technique de Liaison et de Coordination

BTLS Basic Tape Library Support

Btm Bottom

BTM Bayerisches Teleskopmeeting (Germany) • Benchmark Timing Methodology • Bench Test Model • Bottom • Brevet Technique des Métiers • Bromotrifluoromethane • Butte (USA-MT) (location identifier)

BTMC Bell Telephone Manufacturing Co. (USA)

BTMS Body Temperature Measuring System

Btn Baton • Button

BTN Bennettsville (USA-SC) (location identifier) • Between • Bhutan (ISO country code) • Billing Telephone Number • Boutan (code ISO de pays) • Ngultrum (Bhutan) (ISO currency code)

BTO Basic Technology Option • Beam Transfer Optics • Botopasie (Suriname) (location identifier) • Branch Transportation Office • Brief Task Outline

BTOA Binary To ASCII

BtoB Back to Back

BTOBD Be There Or Be Dead

BTOBS Be There Or Be Square

BToIT Bulletin of the Tokyo Institute of Technology (periodical code)

BTok Bulletin of the Tokyo Astronomical Observatory (periodical code)

Bton Bouton

BTP Bachelor of Town Planning • Back-Traffic Protection • Batch Transfer Program • Bâtiment et Travaux Publics • Behind The Plane • Branch Technical Position • Branch Transfer Program • Broader Term Partitive • Built To Print • Butler Graham (USA-PA) (location identifier)

BTQ Butare (Rwanda) (location identifier)

BTR Baton Rouge - Ryan (USA-LA) (location identifier) • Better • Bimetric Theory of Relativity • Bit Test and Reset • Block Tape Recorder • Boolean Topic Representation • Brevet de Technicien de la Représentation • Burn Time Remaining • Bus Transfer

BTRL British Telecom Research Laboratory (UK)

Btry Battery

BTRY Battery

BTS Balloon Transport System • Base Telecommunications System • Base Transceiver Station • Batch Terminal Simulator • Battery Test Set • Beacon Tracking System • Beam Transport System • Bench Test Specification • BIH Terrestrial System • Bit Test and Set • Boolean Time Sequence • Bratislava - M.R. Stefanik (Slovakia) (location identifier) • Brevet de Technicien Supérieur (France) • Bureau of Technical Security • Bureau of Transportation Statistics • Business-Technology Strategy • Business Transaction Services

BTSA Brevet de Technicien Supérieur Agricole (France)

BTSCC Block Transfer Serial Crate Controller

BTSF Biotechnology Strategic Forum

BTSHD Block Transfer Serial Highway Driver

BTSS Basic Time-Sharing System

BTT Back To Traffic • Beginning-to-Tape Test • Bettles (USA-AK) (location identifier) • Branch Training Team

BTTF Blue Taurus Tunable Filter

Btu British Thermal Unit

BTU Basic Terminal Unit • Basic Transmission Unit • Bintulu (Malaysia) (location identifier) • British Thermal Unit • Bus Terminal Unit

BTUM Braunschweig Technische Universität - Mitteilungen (periodical code)

BTV Beacon Test Van • Bulletin Télévisé • Burlington (USA-VT) (location identifier)

BTW Batu Licin (Indonesia) (location identifier) • Belasting over de Toegevoegde Waarde • Between • By The Way

BTWC Biological and Toxic Weapons Convention

Btwn Between

BTWN Between

BTX Benzene-Toluene-Xylene • Betoota (Australia-QLD) (location identifier) • Bildschirmtext • British Telecom Exchange

Bty Battery

BTY Battery • Beatty (USA-NV) (location identifier)

BTZ Banadas - Teitelboin - Zanelli (black hole) • Bildungs- und Tagungszentrum (DWD) • Bursa (Turkey) (location identifier)

Bu Bureau • Bushel • Butyl Group

B/U Backup • Backup Unit • Build-Up

BU Backup • Base Unit • Bibliothèque Universitaire • Binding Unit • Birmingham University (UK-England) • Black Universe • Bollettino Ufficiale • Boston University (USA-MA) • Bottom Up • Branch Unit • Break-Up • Branch Unit • Brihadāranyaka Upanishad • Brown University (USA-RI) • Budgetary Unit • Buildup • Bulgaria • Burma (ISO country code) • Bushman • Business Unit • Buzzer

BUA Booster Umbilical Assembly • Buka (Papua New Guinea) (location identifier)

BUAAJ Beijing University Aeronautics and Astronautics Journal (periodical code)

BUAF Big Ugly ASCII Font

BUAG Big Ugly ASCII Graphic

BuAIC Bulletin of the Astronomical Institutes of Czechoslovakia (periodical code)

BuAkK Budapest Akademiai Kiado (periodical code)

BUAN Baseline Upper Air Network

BuARB Bulletin de l'Académie Royale de Belgique (periodical code)

BUB Boston University of Brussels (Belgium) • Burwell (USA-NE) (location identifier)

BUBL Bulletin Board for Libraries

BUC Backup Computer • Backup Control • Bangor University College (UK-Wales) • Burketown (Australia-QLD) (location identifier)

Bucks Buckinghamshire (UK-England)

BUCLD Boston University Conference on Language Development

BUCOE Boulder Upslope Cloud Observation Experiment

BUCS Backup Computer System • Backup Control System

Bud Budget

BUD Beneficial Use Date • Budapest Ferihegy (Hungary) (location identifier) • Budget

BUDC Backup Digital Computer

Budpst Budapest (Hungary)

BUDS Backup Digital System

BUE Buenos Aires (Argentina-BA) (location identifier)

BUEC Backup Emergency Communications

BUECE Backup Emergency Communications Equipment

BuENS Osaka University Bulletin Series Engineering Natural Sciences (periodical code)

BUES Budapest University of Economic Sciences (Hungary)

BUF Backup Facility • Buffalo (USA-NY) (location identifier) • Buffer

BUFC Backup Flight Controller

BUFF Buffer

BUFORA British Unidentified Flying Objects Research Association (UK)

BUFR Binary Universal Form for Representation (of data)

BUG Benguela (Angola) (location identifier) • Bloomsburg University Goniometer

BUGS Backup Guidance System • Bayesian (interface) Using Gibbs Sampling

BUH Bucharest (Romania) (location identifier) • Bucureşti (Romania) (location identifier)

BUI Bokondini (Indonesia) (location identifier)

BUIC Backup Intercept Control • Backup Interceptor Control • Bureau Universitaire d'Information sur les Carrières

BuIPS Bulletin of the Israel Physical Society (periodical code)

BUJ Boussaada - Ain Eddis (Algeria) (location identifier)

BUK Albuq (Yemen) (location identifier) • Burman Kyat (ISO currency code)

BUKG Bundesumzugskostengesetz (Germany)

BUKS Bibliotheeks Uitleen- en Katalogus Systeem (Netherlands)

Bul Bulgarian • Bulletin

BUL Bulletin • Bulolo (Papua New Guinea) (location identifier)

Bulg Bulgaria • Bulgarian

Bull Bulletin

BUM Broad Upshifted Maximum • Butler (USA-MO) (location identifier)

BUMA Bureau voor Muziek-Auteursrecht

BuMat Bulletin Mathématique (periodical code)

BUN Buenaventura (Colombia) (location identifier) • Bgultrum (Bhutan) (ISO currency code)

BUNCH Burroughs - Univac - NCR - Control Data - Honeywell

BUO Burao (Somalia) (location identifier)

BUOU Backup Optical Unit

bup Backup (file extension)

BUP Bachillerato Unificado Polivalente (Spain) • Backup Plate • Bathinda (India) (location identifier) • British United Press

BUPT Beijing University of Posts and Telecommunications (China-PRC)

BUQ Bulawayo (Zimbabwe) (location identifier)

Bur Bureau • Buried • Burma • Burmese

BUR Backup Rate • Burbank (USA-CA) (location identifier) • Buried • Burma (international vehicle registration code)

Burg Burgalés • Burgess • Burgemeester • Burgomaster • Burgos

Burl Burlesque

BUrlG Bundesurlaubsgesetz

Burm Burma • Burmese

BURP Backup Rate of Pitch

BURR Backup Rate of Roll

BURS Bradford University Remote Sensing Ltd. (UK)

BURU Bottom Up Review Update

BURY Backup Rate of Yaw

Bus Bushel • Business

BUS Backup Structure • Batumi (Georgia) (location identifier) • Beilstein Unique Sequence • Broadcast and Unknown Server • Bureau Universitaire de Statistiques • Business

BUSCA Bonn University Simultaneous Camera

BUSD Blow Up/Settle Down

BUSDSP Bureau Universitaire de Statistiques et de Documentation Scolaires et Professionnelles

Bush Bushel • Bushing

BUSS Balloon-borne Ultraviolet Stellar Spectrometer • Beneficial User Shipping System

BUST Baksan Underground Scintillation Telescope (Russia)

But Butter • Button

BUT Basic Unit Training • Brevet Universitaire de Technologie • Burtonwood (UK-England) (location identifier)

BUTE Butte

BUTEC Bureau Technique

BUU Basic Unit User • Basic User Unit • Boltzmann - Ühling - Uhlenbeck (theory) • Buyo (Ivory Coast) (location identifier)

BUV Backscatter Ultraviolet • Bella Union (Uruguay) (location identifier)

BUVS Backscatter Ultraviolet Spectrometer

BUW Baubau (Indonesia) (location identifier)

BUX Bunia (Zaire) (location identifier)

Buy Buyer • Buying

BUY Bunbury (Australia-WA) (location identifier)

BUZ Bushehr (Iran) (location identifier)

BuZa Buiten Zaken

bv Bijvoorbeeld

B/V Bleed Valve

BV Back View • Balanced Voltage • Bamberg Variable (star) • Basic Vehicle • Basic Version • Bauvorschrift • Bene Vale • Berechnungsverordnung • Besloten Vennootschap • Best Value • Bijvoorbeeld • Bleed Valve • Boîte avec Valeur (déclarée) • Bonnet Valve • Book Value • Booster Vehicle • Boost Vehicle • Bouvet Island (ISO country code) • Bred Vector • Brightness Value • Bulletin de Visite • Bulletin de Vote • Bulletin of Verification • Bureau Voucher • Île Bouvet (code ISO de pays)

BVA Beauvais (France) (location identifier) • Billing Validation Application • Bureau des Vins et Alcools • Bureau de Voyages Aériens

BVAC Beaver Valley Astronomy Club (USA-PA)

BVAS Boston University Astronomical Society (USA-MA)

BVAT Before Value-Added Tax

BVB Boa Vista (Brazil-RR) (location identifier)

BVBA Besloten Vennootschap met Beperkte Aansprakelijkheid

BVC Boa Vista (Cape Verde) (location identifier) • Billing Validation Center • Boost Voltage Converter

BVCP Banyan Vines Control Protocol

BVCT Best Value Contracting Tool

BVD Beacon Video Digitizer • Beaver Inlet (USA-AK) (location identifier) • Binnenlandse Veiligheidsdienst

BVE Barotropic Vorticity Equation • Binocular Visual Efficiency • Bivariate Exponential • Breadboard Verification Equipment • Brive-la-Gaillarde (France) (location identifier) • Bureau de la Vie Étudiante

BVF Brunt-Väisälä Frequency • Bua Dama (Fiji) (location identifier)

Bvg Beverage

BVG Berlevåg (Norway) (location identifier) • Berliner Verkehrsgesellschaft (Germany) • Bundesversorgungsgesetz

BVH Basic Video Handler • Vilhena (Brazil-RO) (location identifier)

BVI Birdsville (Australia-QLD) (location identifier) • Blade-Vortex Interaction • British Virgin Islands • Bundesverband für Investmentgesellschaften (Germany)

BVIM Base Visual Information Manager

BVISC Base Visual Information Support Center

BVLS Battery-Voltage Limit System

BVK Bundesverdienstkreuz

BVM Bayerische Volkssternwarte München eV (Germany) • Belgische Vereniging voor Microscopie (Belgium) • Belmonte (Brazil-BA) (location identifier) • Best Value Manufacturing • Boundary Value Method

BVN Breitbandkabel-Verteilnetz • Bureau Voucher Number

BVO Bartlesville (USA-OK) (location identifier) • Bundesverdienstorden

BVol Bulletin of Volcanology (periodical code)

BVP Beacon Video Processor • Blood Volume Pulse • Bolovip (Papua New Guinea) (location identifier) • Booster Vacuum Pump • Boundary Value Problem • British Visitor's Passport • Bureau de Vérification de la Publicité

BVPA Boundary Value Problem Approach

BVPN Broadband Virtual Private Network

BVPS Beacon Video Processing System • Booster Vacuum Pump System

BVQI Bureau Veritas Quality International

BVR Beyond Visual Range • Beyond Visual Recognition • Brava (Cape Verde) (location identifier)

BVRO Belgische Vereniging voor Raketonderzoek (Belgium)

BVRS Breadboard Visual Reference System

BVS Basic Variable Set • Bergische Volkssternwarte Wuppertal eV (Germany) • Best Value Selection • Best Value Supplier • Best Variable Selection • Bibliothekverbundsystem • Branch VME System • Breves (Brazil-PA) (location identifier) • Broadcast Video Server • Buoyant Venus Station • Bureau of Vital Statistics

BVSP Basaltic Volcanism Study Project

BVSS Banc de Validation du Système Satellite

Bvt Brevet • Brevetted

BVT Basic Vehicle Timing • Bouvet Island (ISO country code) • Île Bouvet (code ISO de pays)

BVU Brightness Value Unit • Buena Vista University (USA-IA)

BVV Belgische Vereniging voor Verlichtingskunde (Belgium)

BVVSA Bratislava Veda Vydavatelstvo Slovenskej Akademie Vied (periodical code)

BVW Batavia Downs (Australia-QLD) (location identifier)

BVX Batesville (USA -AR) (location identifier)

BVY Beverly (USA-MA) (location identifier)

BVZ Beverly Springs (Australia-WA) (location identifier)

b/w Black and White

B-W Baade-Wesselink (method)

B&W Beam & Whiteside • Black & White

B/W Black and White • Black to White

BW Baade's Window • Baade-Wesselink (method) • Bandwidth • Beam Width • Before Web • Bitte Wenden! • Black and White • Body Weight • Botswana (ISO country code) • Bridge Wire • Bulk Water • Burgerlijk Wetboek • Business Week • Bus Watcher

BWA Backward Wave Amplifier • Bent Wire Antenna • Bhairawa (Nepal) (location identifier) • Botswana (ISO country code) • Broadband Wireless Access

BWB Barrow Island (Australia-WA) (location identifier) • Blended Wing Body

BWC Backward Wave Converter • Bandwidth Compression • Basic Weight Controller • Biological Weapons Convention (1972) • Brawley (USA-CA) (location identifier) • Broadband Waveguide Circulator • Bureau Weather Control

Bwd Backward

BWD Backward • Basic Wind Data • Brownwood (USA-TX) (location identifier)

BWE Bandwidth Expansion • Billet de Weekend • Braunschweig (Germany) (location identifier)

BWER Bounded Weak Echo Region

BWF Barrow-in-Furness (UK-England) (location identifier)

BWFl Bruttowohnfläche

BWFR Black and White Film Recorder

BWG Beam Waveguide • Bipartite Working Group • Birmingham Wire Gauge • Bowling Green - Warren County (USA-KY) (location identifier)

BWH Butterworth (Malaysia) (location identifier)

BWI Baltimore Washington International (USA-MD) (location identifier) • British West Indies

BWiN Breitbauwissenschaftsnetz

BWJ Bawan (Papua New Guinea) (location identifier)

BWK Belt Weather Kit • Bol (Croatia) (location identifier)

BWL Blackwell (USA-OK) (location identifier)

BWLF Better White Light Filter (Coronado)

BWM Backward Wave Magnetron • Bandwidth Manager • Block-Write Mode • Bowman (USA-ND) (location identifier) • Broadcast Warning Message

BWN Bandar - Seri Begawan (Brunei) (location identifier) • Bergmann - Wagoner - Nordtvedt (theory)

BWO Backup While Open • Backward Wave Oscillator • Balakovo (Russia) (location identifier)

BWOS Backward Wave Oscillator Synchronizer

BWOT Backward Wave Oscillator Tube

BWP Bewani (Papua New Guinea) (location identifier) • Pula (Botswana) (ISO currency code)

BWPA Backward Wave Power Amplifier

BWQ Brewarrina (Australia-NSW) (location identifier) • Buzz Word Quotient

BWR Bandwidth Ratio • Bandwidth Reservation • Boiling Water Reactor

BWRD Backward

BWS Base Weather Station • Beaufort Wind Scale • Black-White Symmetry • Blaine (USA-WA) (location identifier) • Building Wire Standardization

BWT Backward Wave Tube • Bohr-Wheeler Theory • Broadcast Warning TWX • Burnie Wynyard (Australia-TAS) (location identifier)

BWTP Basalt Waste Isolation Project

BWTS Bandwidth Test Set

BWU Bankstown (Australia-NSW) (location identifier)

BWV Bach Werke-Verzeichnis • Back Water Valve

BWY Woodbridge (UK-England) (location identifier)

Bx Box • Bordeaux (France)

BX Base Exchange • Basic Exchange • Box • Bus Extender

BXA Bogalusa - George R. Carr (USA-LA) (location identifier)

BXB Babo (Indonesia) (location identifier)

BXC Boxborough (USA-MA) (location identifier)

BXD Bade (Indonesia) (location identifier)

BXE Bakel (Senegal) (location identifier)

BXFM Binaire X de Faible Masse

BXI Boundiali (Ivory Coast) (location identifier) • Bus Extender Interface

BXK Buckeye (USA-AZ) (location identifier)

Bxl Bruxelles

BXL Blue Lagoon (Fiji) (location identifier)

BXM Batom (Indonesia) (location identifier)

BXN Bodrum Imsik (Turkey) (location identifier)

BXO Bissau (Guinea-Bissau) (location identifier)

BXP Border Check Point

BXR Basic Expander • Bureau for External Relations

BXS Borrego Springs (USA-CA) (location identifier)

BXT Bontang (Indonesia) (location identifier)

BXU Butuan (Philippines) (location identifier)

BXV Breiddalsvík (Iceland) (location identifier)

BXX Borama (Somalia) (location identifier)

By Barony • Byte

BY Babylon • Base Year • Basic Year • Bayern (Germany) • Belarus (ISO country code) • Bélarus (ISO country code) • Best Year • Biélorussie (code ISO de pays) (obsolete) • Billion Years • Byelorussia (ISO country code) (obsolete) • Budget Year

BYA Billion Years Age • Boundary (USA-AK) (location identifier)

BYam Bulletin of the Yamagata University (periodical code)

BYB Baya (Oman) (location identifier) • Dibaa (Oman) (location identifier)

BYC Yacuiba (Bolivia) (location identifier)

BYD Beidah (Yemen) (location identifier) • Beyond

Bye Byelorussia (obsolete) • Byelorussian (obsolete)

BYG Buffalo (USA-WY) (location identifier)

BYH Blytheville (USA-AR) (location identifier)

BYI Burley (USA-ID) (location identifier) • Rupert (USA-ID) (location identifier)

BYK Bouake (Ivory Coast) (location identifier)

BYL Bella Yella (Liberia) (location identifier)

BYM Bayamo (Cuba) (location identifier)

BYN Bayankhongor (Mongolia) (location identifier)

Bynd Beyond

Byo Bulawayo (Zimbabwe)

BYO Backyard Observatory • Bring Your Own • Build Your Own

BYOB Bring Your Own Beer • Bring Your Own Booze • Bring Your Own Bottle

BYOP Bring Your Own Popcorn

BYOS Bring Your Own Snacks

Byp Bypass

BYP Bypass

BYPA Bypass

BYPS Bypass

BYQ Bunyu (Indonesia) (location identifier)

BYr Billion Years

BYR Rouble (Belarus) (ISO currency code)

BYS Fort Irwin - Bicycle Lake (USA-CA) (location identifier) • Biélorussie (code ISO de pays) (obsolete) • Byelorussia (ISO country code) (obsolete)

BYT Bantry (Ireland) (location identifier)

BYU Bayoo • Bayou • Bayreuth (Germany) (location identifier) • Brigham Young University (USA-UT)

BYUP Brigham Young University Press

ByuPr Byurakanskaia Astrofizicheskaia Observatoriia Preprints (periodical code)

ByuRe Byurakanskaia Astrofizicheskaia Observatoriia Reprints (periodical code)

BYW Blakely Island (USA-WA) (location identifier)

BYX Baniyala (Australia-NT) (location identifier)

Byz Byzantine

Bz Benzene • Benzoyl • Bezirk

B-Z Belouzhov-Zhabotinskii (oscillator)

BZ Battle Zone • Belize (ISO country code) • Belouzhov-Zhabotinskii (oscillator) • Below the Zone • Blandford-Znajek (effect) • British Zone • Buffer Zone • Byzantine

BZA Bonanza - San Pedro (Nicaragua) (location identifier) • Breathing Zone Air

BZC Búzios (Brazil-RJ) (location identifier)

BZD Balranald (Australia-NSW) (location identifier) • Dollar (Belize) (ISO currency code)

BZE Belize City - Goldston (Belize) (location identifier)

BZG Bydgoszcz (Poland) (location identifier)

bzgl bezüglich

BZI Balikesir (Turkey) (location identifier)

BZK Briansk (Russia) (location identifier)

BZL Barisal (Bangladesh) (location identifier)

BZM Bergen op Zoom - Woensdrecht (Netherlands) (location identifier)

BZN Bozeman Gallatin (USA-MT) (location identifier)

BZO Bolzano (Italy) (location identifier)

BZP Bizant (Australia-QLD) (location identifier)

Bzr Buzzer

BZR Béziers (France) (location identifier)

BZRG Bundeszentralregistergesetz

BZS Basic Zirconium Sulphate • Washington - Buzzards Point (USA-DC) (location identifier)

BZT Bundesamt für Zulassungen in der Telekommunikation (Germany) • Nrazoria - Hinkles Ferry (USA-TX) (location identifier)

BZU Buta (Zaire) (location identifier)

BZV Brazzaville (Congo) (location identifier)

bzw beziehungsweise

BZW Blizzard Warning

BZX Basic Zone Expansion

BZY Bel'cy (Moldova) (location identifier)

BZZ Brize Norton (UK) (location identifier)

B0 Paper Size

B1 First Bologna Survey • Paper Size

B10 Paper Size

B2 Paper Size • Second Bologna Survey

B2B Business-to-Business

B2C Business-to-Consumer

B2FH Burbidge - Burbidge - Fowler - Hoyle (theory)

B2G Business-to-Government

B2X Binary to Hexadecimal

B3 Paper Size

B4 Paper Size

B5 Paper Size

B6 Paper Size

B7 Paper Size

B8ZS Binary Eight Zero Substitution

B8 Paper Size

B9 Paper Size

C

c/ Calle • Cargo • Cuenta

c About • Calorie • Candle • Capacity • Caput • Carbohydrate • Carbon • Carton • Case • Cathode • Cent • Centavo (currency) • Centi • Centigrade • Centime (currency) • Centimeter • Centre • Century • Chairman • Chairperson • Chapter • Characteristic Velocity • Charm • Chief • Child • Cibus • Circa • Circiter • Circum • Cirrus • City • Cloudy • Cognate • Cold • Colour • Como • Compact • Compound • Concentration • Congius • Constant • Consul • Contra • Contrast • Convection • Copper • Copy • Copyright • Corps • Court • Cube • Cubic • Cum • Curie • Currency • Cycle • Exhaust Velocity • Molecular Velocity Vector • Specific Heat Capacity • Speed of Light in Vacuum • Velocity of Light • Velocity of Sound

C/ Calle • Cargo • Cuenta

C Acceleration Correction • Cabo • Caldo • Caliente • Calle • Calorie • Cambrian • Campo • Canada (international aircraft registration code) • Candle • Cap • Capacitance • Cape • Capo • Carbon • Carbone • Carboniferous • Cargo • Carrier • Cathode • Celestron • Celsius • Cent • Centavo (currency) • Center • Centesimo • Centigrade • Centime • Centre • Centrifugal Force • Centro • Centum • Century • Chapter • Charge Conjugation • Charlie (international alphabet) • Charm Quantum Number • Chaud • Chemical • Chief • Circling • Circulaire • Circular • Cité • Ciudad • Clock • Closed • Cloudy • Coated • Code • Coefficient • Cold • Collector • College • Colon • Color Index • Common • Commonwealth • Complement • Complete • Compliance • Computer • Conceptual • Conductivity • Confidential • Congress • Conservative • Consonant • Constant • Contact • Control • Convective Transport • Core • Cosine • Costa • Coulomb • Council • Count • Country • County • Coverage Factor • Cuba (international vehicle registration code) • Cuesta • Cumulus • Cycle • Cytidine • Euler Number • Heat Capacity • Hundred (Roman numeral) • Kohlenstoff • Molecular Concentration • Parallactic Angle • Programming Language • Quantity of Electricity • Rotational Constant • Set of Complex Numbers • Spectral Type • Viscous Damping • 100 (Roman numeral)

ca Cache Data (file extension) • Canada • Catalan (ISO language code) • Cathode • Centare • Centiare • Circa

Ca Cable • Calcium • Canada • Canadian • Carré • Carretera • Compagnia • Companhia • Compañía • Cuesta

C/A Capital Account • Cash Account • Closest Approach • Coarse Acquisition (*) • Contract Administrator • Corrective Action • Course Acquisition • Credit Account • Cuenta de Ahorros • Current Account

C&A Calibration & Analysis • Certification & Accreditation

CA Cabin Attendant • Cable • Calar Alto (Spain) • Calculateur Analogique • Calculateur Arithmétique • Calibrated Altitude • Calibration Adjust • California (USA) • Canada (ISO country code) • Capital Account • Capital Asset • Carrier Aircraft • Catamarca (Argentina) • Cell Arrival • Cellular Automata • Cellular Automaton • Central America • Centre d'Apprentissage • Centrifugal Atomization • Certificat Administratif • Certificat d'Aptitude • Certificate Authority • Certification Authority • Certified Authorities • Change Agreement • Channel Access • Chargé d'Affaires • Chartered Accountant • Chemical Abstracts • Chief Accountant • Chiffre d'Affaires • Chromospherically Active • Chronological Age • City Attorney • Civil Affairs • Civil Aviation • Clear and Add • Clear Aperture • Clerical Assistant • Closest Approach • Closing Angle • Cloud Analysis • Coarse Acquisition • Coarse Alignment • Code Available • Collision Avoidance • Command Authority • Commercial Activities • Commercial Agent • Common Applications • Communications Activities • Comparative Advantage • Competitor Analysis • Completion of Assembly • Compressed Air • Comptroller of Accounts • Computer-Aided • Computer Animation • Computer Assembly • Computers and Automation • Condition Action • Conditional Access • Cone Angle • Configuration Audit • Connection Approval • Connection Assistant • Connector Architecture • Conseil d'Administration • Constant Amplitude • Construction Analysis • Construction Authorization • Consular Agent • Continental Arctic • Contingency Abort • Continuous Averaging • Contour Anomaly • Contract Administration • Contract Administrator • Contract Advice • Contract Analysis • Contract Authorization • Contract Authority • Contract Award • Contractor Assessment • Contractor Assisted • Contractual Agreement • Contrat d'Adaptation • Contrat d'Apprentissage • Control Account • Control Area • Control Assembly • Control Authority • Controlled Atmosphere • Controller of Accounts • Controlling Authority • Controlled Atmosphere • Convening Authority • Cooperative Agreement • Coordination Action • Cornerstone Agent • Coronagraphic Apodizer • Correct or Amplify • Corrective Action • Correspondence Analysis • Corriente Alterna • Cost Account • Cost Analysis • Coulomb Acceleration (model) • Coulomb Approximation • Courant Alternatif • Cours d'Appel • Course Alignment • Court of Appeal • Court of Auditors • Cover Aft • Credit Account • Crisis Alerting • Critical Analysis • Criticality Analysis • Cryogenic Assembly • Cryptanalysis • Current Account • Current Assets • Current Awareness • Cylindrical Anode

CAA Cambridge Astronomical Association (UK-England) • Canadian Authors' Association (Canada) • Canadian Aviation Authority (Canada) • Career Advancement Award • Catacamas (Honduras) (location identifier) • Cedar Amateur Astronomers (USA-IA) • Central Arid Asia • Centre à Autonomie d'Acheminement • Chinese Arctic Administration • Chinese Astronomy Abstracts • Circular Aperture Antenna • Civil Aeronautics Administration • Civil Aeronautics Authority • Civil Air Augmentation • Civil Aviation Agency • Civil Aviation Authority (UK) • Civilian Agency Acquisition • Clean Air Act (USA) • Command Arrangement Agreements • Committee for Astronomy and Astrophysics • Committee on Astronomy and Astrophysics • Competent Authority Approval • Composite Audio Amplifier • Computational Aeroacoustics • Computer-Aided Animation • Computer Amplifier Alarm • Computer-Assisted Accounting • Context-Aware Appliance • Controlled Access Area • Cost Account Authorization • Cour Administrative d'Appel • Coût d'Acquisition des Abonnés • Crypto Access Authorization • Cuyahoga Astronomical Association (USA-OH)

CAAA Clean Air Act Amendment (*) • Clean Air Amendment Act (USA) • Composers, Authors, and Artists of America

CAAB Club d'Astronomes Amateurs de Bienne (Switzerland)

CAAC Capital Area Astronomy Club (USA-MI) • Charlotte Amateur Astronomers Club (USA-NC) • Civil Aviation Administration of China (China-PRC) • Civilian Agency Acquisition Council • Club Astrónomos Aficionados del Cerro (Uruguay) • Club d'Astronomie Alpha Centaure (France) • Commission Académique d'Action Culturelle

CAACM Club des Astronomes Amateurs Centre Mauricie (Canada-QC)

CAACTD Comité Asesor sobre la Aplicación de la Ciencia y la Tecnología al Desarollo (ONU)

CAADRP Civil Aircraft Airworthiness Data Recording Program

CAAE Canadian Association for Adult Education (Canada) • Certificat d'Aptitude à l'Administration des Entreprises

CaAer Canadian Aeronautics and Space Journal (periodical code)

CAAFU Civil Aviation Authority Flight Unit (UK)

CAAIS Computer-Assisted Action Information System

CAAL Club des Astronomes Amateurs de Laval (Canada-QC) • Club des Astronomes Amateurs de Longueuil (Canada-QC)

CAALIS Computer-Aided Assembly-Line Inspection System

CAAMI Caisse Auxiliaire d'Assurance Maladie-Invalidité (Belgium)

CAAO Center for Astronomical Adaptive Optics

CAAP Composition and Acidity of Asian Precipitation

CAAR Compressed Air Accumulator Rocket

CAAS Central Arkansas Astronomical Society (USA-KS) • Chinese Association for the Advancement of Science • Civil Aviation Authority of Singapore • Commission Académique d'Action Sociale • Computer-Assisted Acquisition System • Club des Astronomes Amateurs de Sherbrooke (Canada-QC) • Contracted Advisory and Assistance Services • Crimean Amateur Astronomical Society (Ukraine)

CAAT Centro Astronómico del Alto Turia (Spain) • Computer-Aided Assembly Tool • Computer-Assisted Audit Techniques • Coordinamento Associazioni delle Astrofili Toscane (Italy)

CAATS Canadian Automated Air Traffic System • Compact Airborne Automatic Tracking System

CAAW Cercle des Astronomes Amateurs de Waterloo (Belgium)

Cab Cabin • Cabinet • Cable

CAB Cabin • Cabinda (Angola) (location identifier) • Cabinet • Calculateur Arithmétique Binaire • Canadian Association of Broadcasters (Canada) • Centralized Accounting and Billing • Centro Atomico Bariloche (Argentina) • Centro de Astrobiologia • Cercle Astronomique de Bruxelles (Belgium) • Change Analysis Board • Change Appeals Board • Change Authorization Board • Circolo Astrofili Bergamaschi (Italy) • Circolo Astrofili Bolzano (Italy) • Circuit Assurance Block • Citizens Advice Bureau • Civil Aeronautics Board • Civil Aviation Board • Climate Applications Branch • Command Answer Back • Commonwealth Agricultural Bureau • Communication Accelerator Board • Computer Address Bus • Corrective Action Board • Cost Analysis Brief • Cost Audit Board • Current Awareness Bulletin

CABE Computer-Aided Business Engineering

CABEI Central American Bank for Economic Integration

CABERS Command Automated Budget Execution and Retrieval System

CABS Carrier Access Billing System • Chromospherically Active Binary Star • Cockpit Airbag System • Command Automated Budget System • Committee on Automated Bibliographic Services (UK) • Computerized Annotated Bibliography System • Crashworthy Airbag System

CABSCAT Catalog of Chromospherically Active Binary Stars

Các Cáceres (Spain)

CaC Cost at Completion

CAC Cable Access Cover • Call Accepted • Call Admission Control • Cancellation Administrative Charges • Capabilities Assessment Center • Carrier Access Code • Cascavel (Brazil-PR) (location identifier) • Casualty Area Command • CECOM Acquisition Center • Center for Advanced Computation • Central Air Conditioning • Central Arizona College (USA-AZ) • Centre d'Action Culturelle • Centro Atomico Constituyentes (Argentina) • Channel Activity Command • Chemical Abstracts Condensates • Circuit Administration Center • Ciudad de los Artes y de las Ciencias • Clear All Channels • Climate Analysis Center • COBE Analysis Center • Codex Alimentarius Commission • Colonial Astronomy Club (USA-GA) • Command and Activity Controller • Command And Control • Committee on Atmospheric Chemistry • Community Access Center • Community Advisory Committee • Complete Address Constant • Compressed Aeronautical Chart • Computer Advisory Committee • Computer-Aided Classification • Computer-Aided Control • Computer-Assisted Classification • Concentration - Asymmetry - Clumpiness • Concordia Astronomy Club (USA-NJ) • Connection Admission Control • Consumer Advisory Council • Consumidor A Consumidor • Contract Audit Coordinator • Contract Award Committee • Control Analysis Center • Conversational Associative Category • Cooperation And Coordination • Correction Action Committee • Cost At Completion • Coupling Among Classes • Crisis Action Cell • Crossroads Astronomy Club (USA-TX) • Cumberland Astronomy Club (USA-MD) • Current Actions Center • Curriculum Advisory Committee • Customer Access Connection • Customer Advisory Council

CACA Computer-Aided Circuit Analysis

CACAE Commonwealth Advisory Committee on Advanced Education (Australia)

CACB Center Aisle Connector Bracket • Compressed-Air Circuit Breaker

CACC Communications And Configuration Console • Computer Application Control Code

CACD Collins Avionics and Communications Division • Computer-Aided Circuit Design

CACDS Centre for Advanced Computer and Decision Support

CACE Central Advisory Council for Education (UK) • Computer-Aided Control Engineering

CACEAS Computer-Assisted Circuit Engineering and Allocation System

CACES Computer-Aided Cost Estimating Support

CACGP Commission on Atmospheric Chemistry and Global Pollution (IAMAP)

CACHE Computer Aids for Chemical Engineering Education

CACID Computer-Aided Concurrent Integral Design

CACM Central Advisory Council for the Ministry • Central American Common Market • Communications of the Association for Computing Machinery

CACMI Comité Africain pour la Coordination des Moyens d'Information

CACO Community Activities Coordination Office • Corporate Administrative Contracting Officer

CACOFD Caribbean Congress of Fluid Dynamics

CACON Cargo Container

CACP Central Arbitration Control Point

CACS Canadian Association of Computer Sciences (Canada) • Center for Advanced Computer Studies • Centralized Alarm Control System • Chinese Association of Computer Science (China) • Computer-Aided Communications System • Computer-Assisted Command System • Contract Audit Closing Statement

CACSD Computer-Aided Control System Design

CACST Central Advisory Council for Science and Technology (UK)

CACSUSS Canadian Advisory Committee on Scientific Utilization of Space Station

CACT Carrier-Aided Code Tracking

CACTUS Capteur Accélérométrique Capacitif Triaxial Ultra-Sensible

CACUL Canadian Association of College and University Libraries (Canada)

CAC&W Continental Aircraft Control & Warning

c.-à-d. C'est-à-Dire

càd C'est-à-Dire

Cád Cádiz (Spain)

Cad Cadauno • Cadmium

CAD Cadillac (USA-MI) (location identifier) • Cartridge-Activated Device • Cash Against Documents • Catalog of Astrometric Data • Center for Astronomical Data • Civil Aviation Department • Club d'Astronomie de Drummondville (Canada-QC) • Club d'Astronomie de Duttlenheim (France) (now defunct) • Collecte Automatique des Données • Collective Address Designator • Comité d'Aide au Développement • Comité d'Assistance au Développement • Command Action Directive • Commercial Aircraft Development • Computation and Analysis Division • Computer Access Device • Computer Adaptor Display • Computer-Aided Design • Computer-Aided Design (periodical code) • Computer-Aided Detection • Computer-Aided Dispatching • Computer-Aided Drafting • Computer-Aided Drawing • Computer-Assisted Design • Computer-Assisted Device • Computer-Assisted Diagnosis • Computer-Assisted Dialog • Computer-Associated Device • Condemnation At Depot • Condemned At Depot • Consolidated Acquisition Directive • Consultatiebureau voor Alcohol en Drugs • Containment Atmosphere Dilution • Contract Administration Data • Contract Appeals Division • Contract Award Date • Coronary Artery Disease • Cost Analysis Division • Course Administrative Data • Cylindrical Algebraic Decomposition • Dollar (Canada) (ISO currency code)

CADA Commission d'Accès aux Documents Administratifs • Computer-Aided Design and Analysis • Computer-Assisted Data Acceptance • Computer-Assisted Distribution and Assignment

CADAF Colorado Astronomical Data Analysis Facility (USA-CO)

CADAM Computer-Augmented Design And Manufacturing

CADAP Computer-Aided Dentrite Analysis Program

CADAR Computer-Aided Design, Analysis and Reliability

CADARS Catalog of Apparent Diameters and Absolute Radii of Stars

CaDAS Cleveland and Darlington Astronomical Society (UK)

CADAS Comité des Applications De l'Académie des Sciences • Conseil pour les Applications De l'Académie des Sciences (France) (*)

CADAT Computer-Aided Design Analysis Tool • Computer-Aided Design And Test

CADBIT Computer-Aided Design for Built-In Test

CADC Canadian Astronomical Data Centre (Canada) • Canadian Astronomy Data Centre (Canada) (*) • Canadian Astrophysical Data Centre (Canada) • Central Air-Data Computer • Colour Analysis Display Computer

CADD Computer-Aided Design and Development • Computer-Aided Design and Drafting • Computer-Aided Drafting and Design

CAD&DA Computer-Aided Design & Design Automation • Computer-Assisted Design & Design Automation

CADDA Computer-Aided Design and Design Automation • Computer-Assisted Design & Design Automation

CADDAC Central Analog Data Distributing and Computing System

CADDE Central Automatic Digital Data Encoder

CADDET Centre for Analysis and Dissemination of Demonstrated Energy Technologies

CADDMAS Computer-Assisted Dynamics Data Monitoring and Analysis System

CADDR Change Address

CADDS Computer-Aided Design and Drafting System • Computer-Aided Design Drafting System

CAD/E Computer-Aided Design and Engineering

CADE Computer-Aided Design Engineering • Computer-Aided Design and Evaluation • Computer-Aided Document Engineering • Computer-Assisted Data Evaluation • Controller/Attitude-Direct Electronics

CADEC Counter-Air Directed Energy Capability

CADEP Computer-Aided Design of Electronic Products

CADES Computer-Assisted Data Entry System • Computer-Assisted Data Evaluation System

CADESS Computer-Aided Data Exchange Standards/Specifications

CADET Computer-Aided Design and Electrical Test • Computer-Aided Design of Experiment Translator • Course of Action Display and Evaluation Tool

CADETS Classroom-Aided Dynamic Educational Time-Sharing

CADF Commutated Antenna Direction Finder • CRT Automatic Direction Finding

CADFISS Computation And Data Flow Integration Subsystem

CADH Communications And Data Handling

CADI Canadian Advanced Digital Ionosonde • Central Apollo Data Index (NASA) • Centre Algérien d'Information et de Documentation (Algeria) • Certificat d'Aptitude à la Documentation Informatisée • Computer Access Device Input

CADIAP Comment And Data Integration And Printing

CADIC Centro Austral de Investigaciones Científicas (Argentina) • Computer-Aided Design of Integrated Circuits

CADIS Calar Alto Deep Imaging Survey • Central And Distributed (data acquisition and processing) Integrated System • Computer-Aided Design Information System

CADIST Centre d'Acquisition De l'Information Scientifique et Technique • Centre d'Acquisition et de Diffusion de l'Information Scientifique et Technique (France) • Centre d'Acquisition et D'Information Scientifique et Technique

CADITS Coupling And Dynamics of the Ionosphere-Thermosphere System

CADIZ Canadian Air Defense Identification Zone

CADL Communications And Data Link • Critical Application Distribution List

CADM Configuration Administration and Data Management

CADMAP Computer-Aided Drafting, Mapping, and Photogrammetry

CADMAT Computer-Aided Design, Manufacture And Test • Computer-Aided Design, Manufacturing, And Testing

CADMINI Computer Administrative Instruction

CADMOS Centre d'Aide au Développement de la Microgravité et aux Opérations Spatiales

CADMP Computer-Aided Data Management Procedure

CADN Cellular Access Digital Network

CADNET Chemical Agent Detector Network

CADO Computer Access Device Output

CADOSS Computer-Aided Development Of Standards/Specifications

CADPO Communications And Data Processing Operation

CADPOS Communications And Data Processing Operation System

CADQAD Computer-Aided Development of Quality Assurance Data

CADR Computer-Aided Design Reliability • Contract Administration Data Record

CADRE Center for Aerospace Doctrine, Research and Education • Centre d'Animation, de Développement et de Recherche en Éducation • Computer-Aided Data Retrieval and Evaluation • Coupling And Dynamics of Regions Equatorial • Current Awareness and Document Retrieval for Engineers

CADRG Compressed Arc Digital Raster Graphics

CADRIS Computer-Aided Design and Reuse with Intelligent Support

CADRS Concern And Deficiency Reporting System

CADRT Computer-Aided Dead Reckoning Tracer

CADS Cartridge-Actuated Devices • Central Air Data System • Chemical Agent Detecting Solution • Code Abuse Detection System • Command And Data Simulator • Computer-Aided Debriefing System • Computer-Aided Design System • Computer Anomaly Detection System • Computer-Assisted Dispatching System • Computer-Assisted Display System • Concept And Design Study • Control - Accessibility - Disposal - Storage • Control Authority Data Structure

CADSI Communications and Data Systems Integration

CADSS Combined Analog-Digital Systems Simulator • Complex Applications Design and Support Systems

CADU Channel Access Data Unit • Control And Display Unit • Controller And Data Unit (*)

CADV Combined Alternate Data/Voice

Cae Caelum-Caeli (constellation)

CAE Calculateur Arithmétique Électronique • Canadian Academy of Engineering (Canada) • Chinese Academy of Engineering (China-PRC) • Client Application Enabler • Cóbrese Al Entregar • College of Advanced Education • Columbia (USA-SC) (location identifier) • Committee on Advanced Education • Common Applications Environment • Compositeur - Auteur - Éditeur • Component Acquisition Executive • Computer Aeroelastics • Computer-Aided Education • Computer-Aided Engineering • Computer-Assisted Entry • Computer-Assisted Engineering • Computer-Assisted Enrollment • Computer-Assisted Estimating • Conseil des Affaires d'État • Control Actuation Electronics • Corrective Action Engineer

CAEA Climate and Atmospheric Environment Activity

CAEAL Canadian Association for Environmental Analytical Laboratories (Canada)

CAECAL Computer-Aided Engineering for Cargo Accommodation and Location

CAEDE Computer-Aided Engineering and Design for Electronics

CAEDS Computer-Aided Emulation Design System

CAEFF Center for Advanced Engineering of Fibers and Films (USA-SC)

CAEG Computer-Aided Engineering Graphics

CAEI Certificat d'Aptitude à l'Enseignement des Inadaptés (France)

CAEIU Conversion Actuator Electrical Interface Unit

CAELEP Certificat d'Aptitude à l'Enseignement dans les Lycées Professionnels (France)

CAeM Commission for Aeronautical Meteorology (WMO)

CAEM Comité d'Assistance Économique Mutuelle • Conseil d'Assistance Économique Mutuelle

CAEN Computer-Aided Engineering Network

CAEO Coalition of Adult Education Organizations

CAEPE Centre d'Achèvement et d'Essais des Propulseurs et Engins

Caer Caernarvonshire (UK-England)

CaER Customer and Employee Relations

CAER Community Awareness and Emergency Response • Concours d'Accès aux Échelles de Rémunération

Caern Caernarvonshire (UK-England)

Caerns Caernarvonshire (UK-England)

CAERPA Concours d'Accès aux Échelles de Rémunération des Professeurs Agrégés

Caes Caesar

CAES Center for Advanced Educational Services • Comité d'Action et d'Entraide Sociales • Component Acquisition Executives • Compressed-Air Energy Storage

CAESAR CCD Airborne Experimental Scanner for Applications in Remote Sensing

CAESR Cold Atmospheric Emission Spectral Radiometer

CAET Certificat d'Aptitude à l'Enseignement Technique (France) • Computer-Assisted Electrical Test

CAETI Computer-Assisted Education and Training Initiative

CAETS Council of Academies of Engineering and Technological Sciences (USA-DC)

CAEU Council of Arab Economic Unity

CAEX Computer-Aided Exploration

CAF Caisse d'Allocations Familiales (France) • Call Attach Facility • Canadian Air Force • Carauari (Brazil-AM) (location identifier) • Central African Federation • Central African Rep. (ISO country code) • Chemical Analysis Facility • Clean Air Facility • Clean Assembly Facility • Cleared As Filed • Command Advisory Function • Core Assembly Facility • Cost And Freight • Cost - Assurance - Freight • Coût - Assurance - Fret • République Centrafricaine (code ISO de pays)

CAFA Centre Académique de Formation et d'Administration (France)

CAFAP Consumer Affairs Financial Assistance Program

CAFB Certificat d'Aptitude aux Fonctions de Bibliothécaire (France)

CAFBC Circulating Atmospheric Fluidized Bed Combustion

CAFC Court of Appeals for the Federal Circuit

CAFE Cassegrain Fibre Environment • Centre d'Accueil des Français sur FUSE (France) • Computer-Aided Film Editor • Conditional Access For Europe • Corporate Average Fuel Economy

CAFED Computer-Aided File of Engineering Drawings

CAFF Conservation of Arctic Flora and Fauna

CAFFE Computer-Aided Fast Fabrication Exploration

CAFIE Centre d'Autoformation et d'Ingénierie Éducative (France)

CAFIT Computer-Assisted Fault Isolation Test • Computer-Assisted Fault Isolation Testability

CAFM Commercial Air Freight Movement

CAFO Central Analysis and Forecast Office

CAFOC Centre Académique à la Formation Continue (France)

CAFOE Circulaire de l'Association Française d'Observateurs d'Étoiles Variables (periodical code)

CAFOS Calar Alto Faint Object Spectrograph

CAFS Content-Addressable File Store

CAFTA Concours Annuel des Fabricants de Télescopes Amateurs

CAG Cagliari (Italy) (location identifier) • Canadian Association of Geographers (Canada) • Canopus Acquisition Gate • Change Analysis Group • Civil Affairs Group • Civilian Advisory Group • Clothing Advisory Group • Column Address Generator • Commercial Arbitration

Group • Common Avionics Group • Competitiveness Advisory Group • Comptroller and Auditor General • Computer-Aided Gauging • Computing Applications Group • Concept Action Group • Contrôle Automatique de Gain • Cost Advisory Group

CAGC Coded Automatic Gain Control

CAGD Computer-Aided Geometric Design • Computer-Assisted Geometric Design

CAGE Central Arctic Geological Expedition • Commercial And Government Entity • Composition of Aleatorically Generated Elements • Computerized Aerospace Ground Equipment • Contractor And Government Entity

CAgM Commission for Agricultural Meteorology (WMO)

CAGR Civil Advanced Gas-Cooled Reactor • Commercial Advanced Gas-Cooled Reactor • Compound Average Growth Rate • Compounded Average Growth Rate

CAGS Canadian Association of Graduate Schools (Canada) • Certificate of Advanced Graduate Study

Cah Cahier • Cahiers

CAH Va Mau (Viet Nam) (location identifier)

CAHA Centro Astronómico Hispano-Alemán (Spain)

CAHD Computer-Aided Hypermedia Development

CAHDE Computer-Aided Hypermedia Development Environment

CAHE Core Auxiliary Heat Exchanger

C/AHRS Compass/Attitude and Heading Reference System

CAHRS Compass/Attitude and Heading Reference System

CAHS Comprehensive Automation of the Hydrometeorological Service

CAHT Chiffre d'Affaires Hors Taxes

Cai Caithness (UK-Scotland)

CAI Cairo (Egypt) (location identifier) • Calcium-Aluminium-rich Inclusion • Call Assembly Index • Canadian Aeronautical Institute (Canada) • Centre d'Analyse de l'Information • Centre d'Analyse des Images (France) • Close Approach Indicator • Co-Authorship Index • Coded-Aperture Imaging • Color Alteration Index • Comisión Asesora para la Investigación • Comité Arctique International • Common Air Interface • Communications Action Identifier • Compression After Impact • Compression of Astronomical Images • Computer-Administered Instruction • Computer-Aided Instruction • Computer-Analog Input • Computer-Assisted Instruction • Computer-Assisted Interviewing • Computer Associates International • Confederation of Australian Industry • Configuration Acceptance Instruction • Configuration Audit Inspection • Current-Awareness Information

CAIA Conference on Artificial Intelligence for Applications • Context-Aware Information Access

CAIC Computer-Assisted Indexing and Classification • Computer-Assisted Instruction Center

CAICYT Centro Argentino de Información Científica y Tecnológica (Argentina) • Comisión Asesora de Investigación Científica y Técnica (Spain)

CAID Clear Accountability In Design • Computer-Aided Industrial Design • Contractor Accountability In Design • Control Authority Identification • Control Authority Identifier

CAIDA Cooperative Association for Internet Data Analysis (USA-CA)

CAIF Command Automated Installation File

CAIG Cost Analysis Improvement Group

CAIL Conditional Attributed Interaction Lyndenmaier System

CAILE Comité sur l'Accès à l'Information et la Liberté d'Expression (Canada)

CAIM Computer-Assisted Instruction Management

CAIMS Canadian Applied and Industrial Mathematics Society (Canada) • Computer-Aided Instruction Management System

CAIN Camara Infraroja • Comet and Asteroid Information Network • Computer Algebra Information Network • Conflict Archive on the Internet

CAI/O Computer Analog Input/Output

CAIP Computer Aids for Industrial Productivity • Computer Analyses of Images and Patterns (conference) • Computer-Assisted Indexing Program • Consortium of Affiliates for International Programs

CAIR Content Analysis and Information Retrieval

CAIRN Collaborative Advanced Internet Research Network

CAIRP Centro Argentino para la Información y Refutación de la Pseudociencia (Argentina)

CAIR Committed Aggregate Information Rate

CAIRS Central Automated Inventory and Referral System • Commercial Activities Inventory and Reporting System • Computer-Assisted Information Retrieval System • Computer-Assisted Interactive Resources Scheduling • Computer-based Automatic Information Retrieval System

CAIS Canadian Association for Information Science (Canada) • Casualty Assistance Information System • Central Abstracting and Indexing Services • Central Abstracting and Information Services • Centre for Advanced Instrumentation Systems • Chemical Agent Identification Set • Collocated Automatic Intercept System • Common Airborne Integrated System • Common APSE Interface Set • Common APSE Interface Specification • Computer-assisted Action Information Service

CAISc Canadian Association for Information Science (Canada)

CAISE Conference on Advanced Information Systems Engineering

CAISMI Centro per l'Astronomia Infrarossa e lo Studio del Mezzo Interstellare (Italy)

CAISMS Computer-Assisted Instruction Study Management System

CAISS Computer-Assisted Information Support System

CAISSE Computer-Aided Information System on Solar Energy

CAIT Center for Advanced Information Technology • Center for Applied Information Technology • Computer-Aided Instruction Terminal • Computer-Aided Instruction Trainer

Caith Caithness (UK-Scotland)

CAITS Computerized Automatic Inertial Test Set

CAIV Cost as An Independent Variable

CAIVRS Credit Alert Interactive Voice Response System

CAJ Canaima (Venezuela) (location identifier)

CaJCh Canadian Journal of Chemistry (periodical code)

CaJES Canadian Journal of Earth Sciences (periodical code)

CaJPh Canadian Journal of Physics (periodical code)

CaJRS Canadian Journal of Remote Sensing (periodical code)

CAJUN CD-ROM Acrobat Journals Using Networks

CAK Akron Canton (USA-OH) (location identifier) • Castor - Abbott - Klein (flow)

CAKT Computer-Assisted Keyboard Training

cal Calando • Calendar (file extension) • Calorie • Calvus

Cal Calcutta (India) • Calendar • Calibrate • Calibrated • Calibration • Calibre • California (USA) • Called • Kilocalorie

CAL Calculated Average Life • Calendar • Caliber • Calibrate • Calibration • Calibration Abstract Layer • Calibre • Campbeltown (UK-England) (location identifier) • Canadian Astronautics Ltd. (Canada) • Canister Airlock

• Central Association of Libraries (USA-CA) • Client Access License • Columbia Astrophysics Laboratory • Columbus Attached Laboratory (formerly APM) • Command Asset List • Command Authorization List • Common Assembler Language • Communications Application Language • Component Action List • Computer-Aided Learning • Computer Aided Logistics • Computer-Assisted Learning • Computer Animation Language • Confined Area Landing • Contract Advice Letter • Contractor Alert List • Conversational Algebraic Language • Cornell Aeronautical Laboratory • Coupling of Atmospheric Layers (conference) • Creative Applications Laboratory

CALA Club d'Astronomie de Lyon-Ampère (France) • Chinese American Librarians Association • Computer-Aided Loads Analysis • Concours d'Accès à la Liste d'Aptitude

CALAS Computer-Assisted Language Analysis System

Calç Calçada

Calc Calculate • Calculated • Calculation • Calculus • Calcutta (India)

CALC Calculate • Calculated • Calculus As a Laboratory Course • Customer Access Line Charges

CALCE Computer-Aided Life Cycle Engineering

CALCRE Comité des Auteurs en Lutte Contre le Racket de l'Édition (France)

Cald Calculated

CALD Caldeira

CalDB Calibration Database

CALDB Calibration Database

CALEEP California Earthquake Education Program

Calg Calculating • Calgary (Canada-AB)

CALGO Collected Algorithms (ACM)

Calib Calibrate • Calibrated • Calibration

Calibr Calibrate • Calibrated • Calibration

CALIBRE Common Access to Libraries in Europe

CALICO Computer-Assisted Language Instruction Consortium

Calif California (USA)

Caligr Caligrafía

CALIM Consortium of Academic Libraries in Manchester (UK)

CALIMS Computer-Assisted Library and Information System

CALISCE Computer-Aided Learning and Instruction in Science and Engineering (conference)

CALL Center for Applied Linguistics and Languages (Australia-QLD) • Computer-Aided Language Learning • Computer-Aided Learning Laboratory • Computer-Assisted Language Learning • Computer-Assisted Learning Laboratory • Computer-Augmented Loft Lines

CALM COBOL Automatic Language Modifier • Collected Algorithms for Learning Machines • Committee for Administrative and Legal Matters • Computer-Aided Load Manifest (system) • Computer-Assisted Library Mechanization • Conservation And Land Management • Contra-Almirante

CALMIS Calibration Management Information System

CALMS Computer-Aided Load Manifest System

Caln Calculation

CALN Computer-Assisted Learning Network

CALP Computer-Assisted Learning Package

CALPERS California Public Employees Retirement System (USA-CA)

CALRS Centralized Automatic Loop Reporting System

CALS Cloud and Aerosol Lidar System • Comprehensive Analytical Laboratory Services • Comprehensive Automated Learning System • Computer-aided Acquisition and Logistics Support • Computer-aided Acquisition, Logistics and Support • Computer-aided Acquisition Logistics Support • Computer-Aided Large Stage • Computer-Aided Lifecycle Support • Computer-Aided Logistics and Support • Computer-Aided Logistics Support • Computer-Aided Logistic System • Continuous Acquisition and Life-cycle Support • Current Awareness Literature Search • Current Awareness Literature Service

CALSA Computer-Aided Logical Support Analysis

CALSIP Computer-aided Acquisition and Logistic Support Implementation Plan

CALST Cornell Aeronautical Laboratory Shock Tunnel

CALT Clearance Altitude • Contract Administrative Lead Time • Control Altitude

CalTech California Institute of Technology (USA-CA)

CALTF Combined Air Lift Task Force

CALTRANS California Department of Transportation (USA-CA)

CALUTRON California University Cyclotron

Calv Calvinist

CAL/VAL Calibration/Validation

Cam Cambodia • Cambodian • Cambridge • Camelopardalis (constellation) • Camera • Cameroon • Camino

CAM Cambridge (Starlink internal code) • Camera • Caminho • Camiri (Bolivia) (location identifier) • Cancellation of Amplitude Modulation • Carrier Aircraft Modification • Carte À Mémoire • Cascade Access Method • Cell Adherence Molecule • Cellular Adhesion Molecule • Cellular Automata Machine • Cellular Automata Model • Central Address Memory • Centre d'Aviation Météorologique • Centrifuge Accommodation Module • Cercle Astronomique Mosan (Belgium) • Channel Access Method • Checkout and Automatic Monitoring • Chemical Agent Monitor • Circolo Astrofili di Milano (Italy) • Circular Area Method • Civil Aeronautics Manual • Cockpit Area Microphone • Coded Aperture Mask • Collision Avoidance Maneuver • Commercial Air Movement • Commercial Assets Mobilization • Common Access Method • Common Access Model • Communication Access Method • Communication - Advertising - Marketing • Compass Angle Measurement • Complementary Alternative Medicine • Complementary and Alternative Medicine (*) • Computational and Applied Mathematics • Computer Achievement Monitoring • Computer Address Matrix • Computer-Aided Machining • Computer-Aided Makeup • Computer-Aided Management • Computer-Aided Manufacture • Computer-Aided Manufacturing • Computer-Aided Mapping • Computer Annunciation Matrix • Computer-Assisted Makeup • Computer-Assisted Maintenance • Computer-Assisted Manufacture • Computer-Assisted Manufacturing • Computerized Anatomical Man • Concentrator Access Module • Consolidated Assessment Model • Containment Atmospheric Monitoring • Content-Addressable Memory (*) • Content-Addressed Memory • Contingency Analysis Model • Continuous Air Monitor • Contract Audit Manual • Control Access Manager • Control Account Manager • Coronagraphic Apodizer Mechanism • Coronagraphic Apodizer Mechanizer • Cost Account Management • Cost Account Manager • Coupled-Angular Mode • Crisis Action Module • Customer Account Monitor • Customer Asset Management

CAMA Central Arctic Management Area • Centralized Automatic Message Accounting • Civil Aerospace Medical Association

CAMAC Computer-Aided Measurement And Control • Computer-Automated Measurement And Control

CAMAR Common Aperture Multifunction Aeeay Radar

CAMARC Computer-Aided Movement Analysis in a Rehabilitation Context

CAMAS Central Automatic Message Accounting System

Camb Cambridge

CAMB Civilian Acquisition Management Branch

Cambs Cambridgeshire (UK-England)

CAMC Carlsberg Automatic Meridian Circle

CAMCOS Computer-Assisted Maintenance and Control System

CAMD Center for Advanced Microstructures and Devices • Computer-Aided Mechanical Drawing • Computer-Aided Molecular Design

CAME Computer-Aided Manufacturing Environment • Consejo de Ayuda Mutua Económica

CAMEC China Aerospace Machinery and Electronics Corporation (China-PRC)

CAMEO Characterizations Automatically Made and Edited Online • Computer-Aided Management of Emergency Operations • Computer-Assisted Management of Emergency Operations

CAMERA Control And Monitoring Equipment Resource Allocation

CAMEX Computer-Assisted Map Exercise • Convection and Moisture Experiment

CAMG Central Administration Management Group

CAMI Civil Aeromedical Institute

CaMILE Collaborative and Multimedia Interactive Learning Environment

CAMILE Concerted Action on Management Information for Libraries in Europe

CAMIS Center for Advanced Medical Informatics at Stanford • Customs Accounting and Management Information System

CAML Consequential Angular Momentum Loss

CAMM Computer-Aided Maintenance Management • Computer-Aided Manufacturing Management • Computer-Aided Molecular Modeling • Computer-Assisted Maintenance Management • Computer-Assisted Manufacturing Management • Computer-Assisted Molecular Modeling

CAMMICE Charge And Mass Magnetospheric Ion Composition Experiment

CAMMIS Computer-Assisted Map Maneuver System

CAMMS Computer-Aided Maintenance Management System • Computer-Assisted Map Maneuver Simulation

CAMMU Cache/Memory Management Unit

CAMO Contracting and Acquisition Management Office

CAMOS Committee on Atomic, Molecular, and Optical Sciences

CAMP Campeche (Mexico) • Campo • Code Applications and Maintenance Program • Cold Arctic Mesopause Project • Command Automation Master Plan • Common Administrative Management Program • Compiler for Automatic Machine Programming • Computer-Aided Message Processing • Computer-Assisted Mathematics Program • Computer-Assisted Message Preparation • Computer-Assisted Movie Production • Controls And Monitoring Processor

CAMPS Computer-Aided Mission Planning System • Computer-Assisted Mission Planning System

CAMPUS Center Automated Manpower and Project Update System

Camr Camera

CAMR Camera • Conférence Administrative Mondiale des Radiocommunications • Conferencia Administrativa Mundial de Telecomunicaciones

CAMRAS Computer-Assisted Mapping and Records Activities System

CamRe Cambridge Observatories Reprints (periodical code)

CAMRSS Center for Autonomous and Man-controlled Robotic and Sensing Systems

CAMS Cabin Atmosphere Monitoring System • Calibrated Airborne Multispectral Scanner • Caminhos • Canadian Applied Mathematics Society (Canada) • Cast Application Mining Suite • Center for Advanced Mathematical Sciences • Center for Applied Mathematical Sciences (USA-CA) • Central Atmosphere Monitoring System • Collection Automated Management System • Collision Activation Mass Spectrum • Computer-Aided Maintenance System • Computer-Aided Manipulation System • Control And Monitoring System • Core Automated Maintenance System • Crisis Action Management System

CAMSP Center for Atomic, Molecular and Surface Physics (UWA)

CAMST Community Action on Magnetic-Storage Technology

CAMU Corrective Action Management Unit

Can Canada • Canadian • Canal • Canarias • Canberra (Australia-ACT) • Cancel • Cancelled • Cañon • Canton • Canyon

CA/N Convertisseur Analogique/Numérique

CAN Campus Area Network • Canada (ISO country code) • Cancel • Central Asian Network • Certification Analysis Network • Committee of an Advisory Nature • Computer-Aided Navigation • Computer Algebra Nederland • Computer Algebra Netherlands (*) • Controller Area Network • Convertisseur Analogique/Numérique • Cooperative Agreement Notice • Cost Account Number • Customer Access Network • Guangzhou (China-PRC) (location identifier)

CANA Centro Astronomico Neil Armstrong (Italy)

Canad Canada • Canadian

CANAMEX Canadian - American - Mexican

CANARI Code pour l'Analyse par Navier-Stokes de l'Aérodynamique d'une Roue Isolée

CANARIE Canadian Network for the Advancement of Research, Industry and Education (Canada)

Canc Cancel • Cancellation • Cancelled

CANC Cancelled

Cand Candelabra • Candidate

CAND Computer-Aided Network Design

CANDI Combined Analogue Digital Integration • Commercial And Non-Development Item

CANDLE Center for the Advancement of Natural Discoveries using Light Emission

CANDU Canadian Deuterium-Uranium

CANE Chemical and Nuclear Environment Report

CANEL Connecticut Advanced Nuclear Engineering Laboratory (USA-CT)

CANEX Canadian Experiment

CanF Canadian French

Can. Fr. Canadian French

CANGAROO Collaboration between Australia and Nippon for a Gamma-Ray Observatory in the Outback

Can. I. Canary Islands (Spain)

CANM Communications in Applied Numerical Methods (periodical code)

CANOPUS Canadian Auroral Network for the Origin of Plasma in Earth Neighborhood Program Unified Study

CANP Civil Aviation Notification Procedure

Can. Pac. Canadian Pacific

CANS Computer-Assisted Network Scheduling

Cant Canterbury (UK-England) • Cantería • Cantilever • Cantonese

CANTCO Cannot Comply

Cantil Cantilever

CANTO Caribbean Association of National Telecommunication Organizations

CANTPR Cannot Process

CANTPRO Cannot Process

CANTRAN Cancel in Transmission

CANUS Canada - United States

Canv Canvas

CANV Club Astronomie Nature du Valromey (France)

CANZ Canada - Australia - New Zealand

CAO Cagliari Astronomical Observatory (Italy) • Canadian Association of Optometrists (Canada) • Career Access Opportunity • Case Administering Office • Casualty Assistance Office • Central Accounting Office • Central Aerological Observatory • Central Applications Office • Central Astronomy Observatory (Russia) • Centro Astronomico Orione (Italy) • Change of Appointing Office • Chief Accountant Officer • Chief Administrative Officer • Circuit Activation Order • Classification Advisory Officer • Clayton (USA-NM) (location identifier) • Club d'Astronomie d'Ottignies (Belgium) • Cold Air Outbreak • Collateral Action Officer • Commissie Antarctisch Onderzoek (Netherlands) • Commission for Asia and Oceania • Computer-Aided Office • Computer-Aided Optimization • Conception Assistée par Ordinateur • Contract Administration Office • Contract Administration Officer • Control Authority Office • Cost Administration Office • County Advisory Officer • Coupled Atmosphere-Ocean • Crimean Astrophysical Observatory

CAOC Combined Air Operations Center • Constant Axial Offset Control

CAOE Contractor-Acquired Operational Equipment

CAOI Catalog of Astronomical Object Identifications

CAOIS Control Authority Office Information System

CAOQ Center for Astronomical Observing Quality (USA-CO)

CAORF Computer-Aided Operation Research Facility

CAOS Code for Adaptive Optics Systems • Comité Académique des Oeuvres Sociales • Completely Automatic Operational System • Computer-Aided Observation System • Computer-Aided Organic Synthesis

CAOW Cable Orderwire

cap Capillatus • Caption (file extension)

Cap Capacitance • Capacitor • Capacity • Capiat • Capital • Capitale • Capitalize • Capitán • Capitol • Capitolo • Capitúlo • Capitulum • Capricornus-Capricorni (constellation) • Captain • Caput

CAP Alpha Capricornids (meteor stream) • Cable Access Point • Cable Access Protocol • Calorimeter Analogue Processor • Canadian Association of Physicists (Canada) • Capability • Capability Assurance Program • Capacitor • Capacity • Cap Haitien (Haiti) (location identifier) • Capitão • Capture • Card Assembly Program • Carrier Amplitude Phase • Carrierless Amplitude and Phase • Carrierless Amplitude-Phase (modulation) • Cartridge Access Port • CCMS Application Program • Center for Astronomy and Physics • Central Arbitration Point • Centralized Assignment Procedure • Centre d'Analyse et de Programmation • Centre d'Archivage et de Prétraitement • Certificat d'Aptitude Pédagogique • Certificat d'Aptitude Professionnelle (France) • Channel Access Protocol • Civil Air Patrol • Cloud Atmospheric Parameter • Club Astronómico del Paraguay • Codice di Avviamento Postale • Coefficient of Author Position • Collective Action Plan • Combinatorial Auction Problem • Command Acceptance Pattern • Commission Administrative Paritaire • Committee on Applied Physics • Common Agricultural Policy • Communications Access Processor • Common Air Picture • Communications Access Processor • Communications Alternative Provider • Communications Application Platform • Competitive-Access Provider • Complex Autonomous Payload • Component Approval Process • Composer - Author - Publisher • Computer Address Panel • Computer-Aided Planning • Computer-Aided Printing • Computer-Aided Production • Computer-Aided Publishing • Computer Algebra and Parallelism (workshop) • Computer Application Program • Computer-Assisted Printing • Computer-Assisted Production • Computer-Assisted Publishing • Concurrent Algorithmic Programming • Configuration and Alarm Panel • Configuration Audit Plan • Connection Approval Program • Conservation Assessment Program • Consolidated Aerial Port • Consolidated Appeals Process (UNO) • Contamination Analysis Program • Continuous Audit Program • Contract Administration Plan • Contractor-Acquired Property • Contractor Assessment Program • Contrat d'Achat Privilégié • Control and Alarm Panel • Control and Audio Panel • Control and Authorization Process • Control Assembly Program • Controlled-Access Production • Convertible Adjustable Preferred • Cooperative Agreement Program • Corporate Affiliates Program • Corporate Automation Plan • Corrective Action Plan • Cost Account Package • Cost Account Plan • Cost Analysis Plan • Council on Advanced Programming • Crew Activity Plan • Crisis Action Planning • Crisis Action Procedure • Critical Acquisition Position • Critical Angle Profile • Cryogenic Associative Processor • Cryotron Associative Processor • Current Approved Program • Current Assessment Plan • Customer Administration Panel

CAPA Commission Administrative Paritaire Académique (France) • Computer-Aided Plastic Application

CAPAC Caisse Auxiliaire de Paiements des Allocations de Chômage (Belgium) • Composers', Authors', and Publishers' Association of Canada

CAPAL Computer and Photographic Assisted Learning

CAPAN Capacity Analyzer

CAPAS Comité de l'Académie Pour les Applications de la Science (Belgium)

CAPAV Committee on Atmospheric Problems of Aerospace Vehicles

CAPCA Canadian Advanced Protein Crystallization Apparatus

CAPCHE Component Auto-Programmed Checkout Equipment

CAPCOM Capsule Communicator

CAPD Commission Administrative Paritaire Départementale (France) • Computer Assistance to Persons with Disabilities • Computing to Assist Persons with Disabilities

CAPDS Chemical Agent Point Detector System

CaPE Convection and Precipitation Experiment

CAPE Capability And Performance Evaluation • Centre d'Analyse de Politique Économique • Certificat d'Aptitude au Professorat des Écoles (France) • Communications Automatic Processing Equipment • Community of Agile Partners in Education (USA) • Computer-Aided Parametric Estimating • Computer-Aided Process Engineering • Computer-Aided Production Engine • Computer-Aided Project Engineering • Computer Applications in Production and Engineering • Connecting Asia-Pacific to Europe • Consortium for the Advancement of Physics Education • Contractor Assessment Product Evaluation • Convective Available Potential Energy

CAPEB Confédération de l'Artisanat et des Petites Entreprises du Bâtiment (France)

CAPER Cleft Accelerated Plasma Experimental Rocket • Collaborative Applications for Project-based Education • Computer-Aided Pattern Evaluation and Recognition

CAPES Capability Evaluation System • Certificat d'Aptitude au Professorat de l'Enseignement Secondaire (France) • Certificat d'Aptitude Pédagogique à l'Enseignement Secondaire • Certificat d'Aptitude Préparatoire à l'Enseignement Secondaire • Certificat d'Aptitude Professionnelle l'Enseignement Secondaire • Coordenação de Aperfeiçoamento de Pessoal de Nível Superior (Brazil)

CAPET Certificat d'Aptitude au Professorat de l'Enseignement Technique (France) • Certificat d'Aptitude Professionnelle l'Enseignement Technique

CAPFI Certificat d'Aptitude Professionnelle aux Fonctions de l'Informatique

CAPG Control of Application Process Group

CAPH Client Available Position Hours

CAPI Common Application Program Interface • Common Application Programmers Interface • Common Application Programming Interface • Comprehensive Authority, Power and Influence • Computer-Assisted Personal Interviewing • Computer-Assisted Photo-Interpretation • Cryptographic Application Programming Interface

Capit Capitalist

CAPJ Certificat d'Aptitude à la Profession de Journaliste

CAPL Capillary Pump Loop • Coastal Anti-Pollution League • Critical Acquisition Position List

CAPLP Certificat d'Aptitude au Professorat de Lycée Professionnel (France)

CAPM Capital Asset Pricing Model • Certified Associate of Project Management • Computer-Aided Patient Management • Computer-Aided Project Management

CapMi Royal Observatory Cape Mimeogram (periodical code)

Capn Captain

CAPN Commission Administrative Paritaire Nationale

CAPO Canadian Army Post Office (Canada)

CAPP Canadian Association of Professional Physicists (Canada) • Center for Accelerator and Particle Physics • Center for Applied Parallel Processing • Committee on Astronomy and Public Policy (AAS) • Computer-Aided Process Planning • Computer-Aided Production Planning • Computer-Aided Project Planning • Computer-Assisted Process Planning • Computer-Assisted Production Planning • Content-Addressable Parallel Processor

CAPPE Centre for Applied Philosophy and Public Ethics (Australia-ACT)

CAPPI Constant Altitude Plan Position Indicator

CAPPS Checkout, Assembly and Payload Processing Services • Contract Appraisal System

CAPR Capability Request • Capability Requirement

CAPRI Capacitance Rate Integrator • Captive Reset Ignitor • Card And Printer Remote Interface • Computational Analysis Programming Interface • Computer-Aided Personal Reference Index • Computerized Advance Personnel Requirements and Inventory

Capric Capricorn

CAPRICE Cosmic Antiparticle Ring-Imaging Čerenkov Experiment

Caps Capitals • Capsule

CAPS Cabin Air Pressure System • Call Attempts Per Second • Capitals • Captive Animals Protection Society • Cassette Programming System • Cassini Plasma Spectrometer • Catalogue Automatisé de Publications en Série • Cell Atmosphere Processing System • Center for Analysis and Prediction of Storms (USA) • Center for Analysis of Particle Scattering • Centralized Assignment Procedures System • Centralized Automated Payment System • Centralized Automatic Payment System • CESG Assisted Products Scheme • Common Attitude Pointing System • Communications/Aural Protective System • Computer-Assisted Problem Solving • Configuration Analysis and Projection System • Corporate Accounts Processing System • Cost Accounting Practices Statement • Crew Activity Planning System • Crew Altitude Protection System

CAPSAIS Certificat d'Aptitude au Professorat Spécialisé d'Adaptation et d'Intégration Scolaires (France)

CAPSI CalTech Precollege Science Initiative

CAPSIM Captive Simulation

CAPSS Computer-Aided Planning and Scheduling System

Capt Captain • Caption

CAPT Certificat d'Aptitude au Professorat Technique • Computer-Assisted Procedures Trainer • Conversational Parts Programming Language

CAPTAC Conférence des Administrations des Postes et Télécommunications de l'Afrique Centrale

CAPTAIN Computer-Aided Processing and Terminal Access Information • Covariance Analysis Program for the Study of Augmented Inertial Navigators

CAPTERM Curation and Analysis Planning Team for Extraterrestrial Materials

CAPU Central Administration Planning Unit • Crew Altitude Protection Unit

CAPUST Capacité Universitaire Scientifique et Technique (France)

Capy Capacity

CAQ Caucasia (Colombia) (location identifier)

CAQA Computer-Aided Quality Assurance

Car Carat • Cargo • Carina-Carinae (constellation) • Carrefour

CAR Capital Authorization Request • Cardiff (Starlink internal code) • Caribou (USA-ME) (location identifier) • Center for Aerospace Research • Central African Republic • Centre d'Analyse et de Recherche • Certification Analysis Requirement • Certification Approval Request • Certification Audit Report • Channel Address Register • Chargé d'Administration de la Recherche • Civil Airworthiness Regulation • Cloud Absorption Radiometer • Cloudtop Altitude Radiometer • Club Astronomique Rochefortois (Belgium) • Collectieve Arbeidsvoorwaardenregeling • Collisionally Activated Reaction • Command Action Response • Command Assessment Review • Commission Administrative Régionale • Committed Access Rate • Communications Automation Review • Compounded Annual Rate • Compounded Annual Return • Computer-Assisted Reproduction • Computer-Assisted Research • Computer-Assisted Retrieval • Conditional Autoregression • Condition And Recommendation • Configuration Analysis Report • Configuration And Acceptance Review • Configuration Audit Review • Content of Address Register • Contract Administration Report • Contract Authorization Request • Contractual Action Request • Control Authority Ratio • Corporación Autónoma Regional • Corporation for Astronomical Research • Corrective Action Record • Corrective Action Reply • Corrective Action Report • Corrective Action Request • CSIRO Atmospheric Research • Current Action Report • Customer Acceptance Review • Customer Account Representative • Customer Affairs Representative

CARA California Association for Research in Astronomy (USA-CA) • Center for Astrophysical Research in Antarctica (*) • Centre for Astrophysical Research in the Antarctic • Chemical Assessments and Related Activities • Combined Altitude Radar Altimeter • Current Aerospace Research Activities

CARAFE Cloud, Aerosol, and Radiation Arctic Field Experiment

CARAIB Carbon Assimilation In the Biosphere

CARAM Content-Addressable Random Access Memory

CARAN Centre d'Accueil et de Recherche des Archives Nationales (France)

Carb Carbon • Carbonate • Carburetor

CARB California Air Resources Board (USA-CA) • Central Acquisition Review Board • Central Arbiter

CARC Chemical Agent Resistant Coating

CARCN Central American Regional Communications Network

CARCSLR Career Councillor

Card Cardinal

CARD Cartographic Representation of Data • Center for Astrophysics Research Discussion • Channel Allocation and Routing Data • Comet and Asteroid Rendezvous Docking • Compact Automatic Retrieval Device • Compact Automatic Retrieval Display • Computer-Aided Research and Development • Constraints And Restriction Document • Cost Analysis Requirements Description • Cost Analysis Requirements Document

CARDA Computer-Aided Reliability Data Analysis

CARDIF Cardiff (Starlink internal code)

CARDME Concerted Action for Research on Demand Management in Europe

Cards Cardiganshire (UK-Wales)

CARDS Climatological Aerological Reference Data Set • Computer-Aided Reliability Data Systems • Concept And Requirements Definition Study

CARE Career Awareness Resource Education • Center of Applied Research in Education • Computed-Aided Requirements Engineering • Computed-Aided Risk Evaluation • Cooperative for American Relief Everywhere • Customers Are Really Everything

CAREM Centre Armoricain de Recherche en Environnement (France)

CARES Cartographic Retrieval Service • Computer-Assisted Retrieval Service

CARF Central Altitude Reservation Facility • Central Altitude Reservation Function

CARIAN Centre for Advanced Research in International Affairs in the Netherlands

Carib Caribbean

CARIB Caribbean

CARICOM Caribbean Community

CARID Customer Acceptance Report Item Disposition • Customer Acceptance Review Item Disposition

CARIFTA Caribbean Free Trade Area

CARIN Center for Applied Research on Information Networking

CARINA Caribbean Institute of Astronomy (Trinidad)

CARIRI Caribbean Research Institute (Trinidad & Tobago)

CARIS Computer-Aided Resource Information System • Computerized Agricultural Research Information Service

CARISMA Corrections to Applied Research Laboratories Ion-Sputtering Mass Analyzers

CARJ Clube de Astronomia do Rio de Janeiro (Brazil-RJ)

CARL California Academic and Research Libraries (USA-CA) • Canadian Association of Research Libraries (Canada) • Capacité d'Absorption des Radicaux Libres • Chemical Algorithm for Reticulation Linearization • Club Astronomique de la Région Lilloise (France) • Colorado Alliance of Research Libraries (*) • Colorado Association of Research Libraries • Compressor Aero Research Laboratory

CARLA Code Actuated Random Load Apparatus

CARM Customer Accommodation Requirement Model

CARMA Combined Array for Research in Millimeter-wave Astronomy • Community Aerosol and Radiation Model Atmosphere • Continuous-time Autoregressive Moving Average • Controlled Auto-Regressive Moving Average

Carms Carmarthenshire (UK-Wales)

CARN Competitive Applied Research Network

Carns Caernarvonshire (UK-Wales)

CarOB Carter Observatory Bulletins (periodical code)

Carol Carolingian

Carp Carpathian • Carpenter • Carpentry • Carpintería

CARP Cache Array Routing Protocol • Calculated Air Release Point • Case Routing Program • Comisión Administrativa del Río de la Plata • Computed Air Release Point • Computer-Assisted Regional Planning • Contingency Alternate Routing Plan • Contingency Alternate Routing Program • Contract Award Release Package • Contract Award Review Panel • Corrective Action Review Process • Crisis Action Response Plan

CARPS Collaborative Astrophysical Research Project System

Carr Carrera • Carretera • Carriage • Carrier

CARR Customer Acceptance Readiness Review • Customer Accommodation Requirement Representative

CarRe Carter Observatory Reprints (periodical code)

Carr. Fwd. Carriage Forward

CARRS Coherent Anti-Stokes Resonance Raman Scattering • Cost Analysis Resource Reference System

CARS Cable Television Relay Service • Cable Television Relay Station • Center for Applications of Remote Sensing • Climate Applications Referral Service • Climate Applications Referral System • Coherent Anti-Stokes Raman Scattering (*) • Coherent Anti-Stokes Raman Spectrum • Coherent Anti-Stokes Raman Spectroscopy • Combined Altimeter Radiometer Sensor • Common Automated Recovery System • Community Antenna Relay Service • Community Antenna Relay System • Computer-Aided Reference Service • Computer-Aided Routing System • Computer-Assisted Reference Selection • Computer Audit Retrieval System • Computerized Automotive Reporting System • Configuration Accounting and Reporting System • Consolidated Acquisition Reporting System • Control And Release System • Corrective Action Reporting System

CARSO Carnegie Southern Observatory (now Las Campanas Observatory) (Chile)

CARSS Center for Atmospheric and Remote Sensing Studies

Cart Cartonné

CART Cartridge • Center for Advanced Research and Telecommunications • Central Automatic Reliability Tester • Central Automatic Replenishment Technique • Centralized Automatic Recorder and Tester • Classification And Regression Tree • Cloud and Atmospheric Radiation Testbed • Cloud And Radiation Testbed • Compact Array Radiolocation Technique • Complete And Ready for Test • Computer Analysis and Response Team • Computerized Automatic Rating Technique • Conditional Assembly Release and Transfer • Conformal Array Radar Technology • Custom Asynchronous Receiver/Transmitter

CARTE Center for Advanced Research and Technology for Education • Centre d'Applications et de Recherches en Télédétection • Collaboration on Aeronautical Research and Technology in Europe

CARTEL Centre d'Applications et de Recherches en Télédétection (Canada-QC)

Carth Carthage

Cartog Cartographer • Cartography

CARTS Common Ada Run-Time System • Controlled Atmosphere Reliable Transportation System

CARTT Centre Académique de Ressources et de Transfert de Technologie

CARU Central Administration Resource Unit • Comisión Administrativa del Río Uruguay

CARVE Criticality - Accessibility - Recuperability - Vulnerability - Effect

CARVER Criticality - Accessibility - Recuperability - Vulnerability - Effect - Recognizability

cas Castellanus • Comma-delimited ASCII Data (file extension)

Cas Cassegrain • Cassiopeia-Cassiopeiae (constellation) • Castle • Casual • Casualty

CAS Calculated Airspeed • Calibrated Airspeed • California Academy of Sciences (USA-CA) • Calumet Astronomical Society (USA-IN) • Canadian Astronautical Society (Canada) • Canberra Astronomical Society (Australia-ACT) • Canterbury Astronomical Society (New Zealand) • Cardiff Astronomical Society (UK) • Casablanca (Morocco) (location identifier) • Case Analysis System • Cassette • Casualty Assessment System • Catalog Archive Server • Center for Aerospace Structures • Center for Astrophysical Sciences • Central Administrative Services • Central Alarm Station • Centralized Attendant Services • CERN Accelerator Schools • Certificate of Advanced Studies • Change Analysis Section • Channel-Associated Signaling • Charlottesville Astronomy Society (USA-VA) • Chemical Abstracts Service • Chesmont Astronomical Society (USA-PA) • Chester Astronomical Society (UK) • Cheyenne Astronomical Society (USA-WY) • Chicago Academy of Sciences (USA-IL) • Chicago Air Shower Array (USA-IL) • Chicago Astronomical Society (USA-IL) • Chinese Academy of Sciences (China-PRC) • Chinese Astronomical Society (China-PRC) • Cincinnati Astronomical Society (USA-OH) • Civil Aviation Security • Clean Air System • Coarse Alignment Servo • Cockermouth Astronomical Society (UK) • Cognizant Area Supervisor • Collision Avoidance System • Columbus Astronomical Society (USA-OH) • Column Address Select • Column Address Strobe • Combined Antenna System • Command Augmentation System • Command Automated System • Command Automation System • Commission for Atmospheric Sciences (WMO) • Communicating Applications Specification • Communications Applications Specification • Communications Analysis Section • Communications Antenna Sleeve • Communications Applications Specification • Communications Applications Standard • Community Authorization Service • Competitor Analysis System • Complex Adaptive System • Computational Aeroscience • Computed Airspeed • Computer-Aided Search • Computer-Aided Software • Computer-Aided Styling • Computer-Aided Supportability • Computer-Aided System • Computer Algebra System • Computer-Assisted Search • Computer-Assisted Survey • Computerized Acquisition System • Computerized Autodial System • Conception de l'Architecture du Système • Conditional Access System • Configuration Accounting System • Contract Administration Services • Contract Administration Surcharge • Contract Administrative Services • Contract Analysis System • Control Augmentation System • Cooperative Application Satellite • Corps Administratif Supérieur • Cosmo Astronomical Society (USA-FL) • Cost Accounting Standard • Cost Accounting System • Cotswold Astronomical Society (UK) • Council on Atmospheric Studies • Crew Accommodation System • Crisis Action Support • Crisis Action System • Criticality Alarm System • Croydon Astronomical Society (UK) • Cumberland Astronomical Society (USA-TN) • Current Awareness Services • Current Awareness Situation

CASA Center for Advanced Spatial Analysis • Center for Advanced Studies of Accelerators • Center for Advanced Study in Astronomy • Center for Astrophysics and Space Astronomy • Centi Astro-Space Activities (USA-PA) • Central And South America • Centre of Advanced Study in Astronomy (India) • Chef Adjoint de Service Administratif • Chicago Air Shower Array (USA-IL) • Construcciones Aeronáuticas SA (Spain) • Controller Automation Spacing Aid • Cost Analysis Strategy assessment

CASAC Clean Air Scientific Advisory Committee

CASAFA Commission on the Application of Science to Agriculture, Forestry and Aquaculture

CASAT Computer-Assisted System Approach to Training

CAsB Clube de Astronomia de Brasília (Brazil)

CASB Catalogs And Surveys Branch • China Association for Science and Business (China-ROC) • Cost Accounting Standards Board

CASBAA Cable And Satellite Broadcast Association of Asia

CASC Cataloguing And Standardization Center • Center for Applied Scientific Computing • China Aerospace Corporation (China-PRC) • Club Astronomique de Saint-Claude (France) • Commercial Activities Steering Committee • Contract Administration Service Component

CASCA Campanas All-Sky Survey • Canadian Astronomical Society - Société Canadienne d'Astronomie (Canada)

CASCADE Computer-Aided Sonar Classification And Data Extraction

CASCADES Calculs et Analyses de Seuils Capables d'Annonces des Débordements Soudains • Calculs et Analyses de Signaux Capables d'Annonces des Débordements Soudains

CASCDS Cascades

CASCO Conformity Assessment Committee (ISO)

CASD Chemical Analysis Solutions Division (HP) • Computer-Aided Structural Design • Computer-Aided System Design

CASDE Computer-Assisted Software Development Environment

CASDN Comité d'Action Scientifique de la Défense Nationale (France)

CASE Care Application Service Element • Center for Advanced Studies in the Environment • Christian Association of Stellar Explorers (USA-AR) • Circum Arctic Structural Events • Civilian Access to Space Effort • Coherent Active Source Experiment • Committee on Academic Science and Engineering (USA) • Common Access Switching Equipment • Common Application Service Element • Computer-Aided Software Engineering • Computer-Aided Support Equipment • Computer-Aided System Engineering • Computer-Aided System Evaluation • Computer-Assisted Software Engineering • Computer-Assisted Systems Engineering • Computer-Assisted Systems Evaluation • Computer-Automated Software Engineering • Computer-Automated Support Equipment • Computer-Automated System Engineering • Confederation for the Advancement of State Education • Connecticut Academy of Science and Engineering (USA-CT) • Cooperative Award in Science and Engineering • Coordinating Agency for Supplier Evaluation • Correlation of Alpine Structural Events • Council for Advancement and Support of Education • Council for Advancement of Secondary Education (US)

CASEA Computer-Assisted System Engineering and Analysis

CASES Compilers, Architecture and Synthesis for Embedded Systems (conference)

CASF China Association of Science Foundations (China-PRC)

Cash Cashier

CASH Catalog of Available and Standard Hardware

CASI Canadian Aeronautics and Space Institute (Canada) • Canadian Aeronautics and Space Institute Transactions (periodical code) • Center for Aerospace Information (USA-VA) • Compact Airborne Spectrographic Imager (*) • Compact Airborne Spectrographic Imagery • Compact Airborne Spectrographic Instrument • Computer-Assisted Self-Interviewing

CASIA Center for Analysis of Satellite Interference with Astronomy

CASIC Computer-Assisted Survey Information Collection

CASID Committee for Aviation and Space Industry Development (China-ROC)

CASIMIR CalTech Submillimeter (and far-infrared) Mixing Receiver • Catalyst Studies for Industry through Micrograsvity Research

CASIN Centre for Applied Studies in International Negotiations

CASIS Center for the Advanced Study of Information Systems

CASJ Canadian Aeronautics and Space Journal (periodical code)

CASJQ Canadian Aeronautics and Space Journal Quarterly (periodical code)

CASL Calibration Service Laboratory • California Association of School Libraries (USA-CA) • Crosstalk Application Scripting Language

CASLEO Complejo Astronómico El Leoncito (Argentina-SJ)

CASLIN Czech And Slovak Library Information Network

CASLIS Canadian Association of Special Libraries and Information Services (Canada)

CASO Cancellation Addendum Sales Order

CASOAR Chambre Adaptée à la Simulation à l'Oxygène Atomique pour les Revêtements

CASOP Crisis Action Standing Operating Procedure

CASP Cambridge Arctic Shelf Programme • Capability Support Plan • Center for Advanced Space Propulsion • Center for Applied Solar Physics • Center for Astronomy and Space Physics • Change Assessment Summary Package • Communications Applications Specific Processor • Computer-Aided Search Planning • Computer-Assisted Search Planning • Computerized Administrative Software Program • Contracting and Acquisition Support Plan • Crew Activities Scheduling Program

CASPAR Cambridge Analog Simulator for Predicting Atomic Reactions

CASPEC Cassegrain Echelle Spectrograph • Cassegrain Spectrograph

CASPIA Computer-Assisted System for the Processing of International Applications

CASPIR Cryogenic Array Spectrometer for the Infrared • Cryogenic Array Spectrometer/Imager

CASPR Configurable Architecture for Smart Pixel Research

CASR Cost and Schedule Reporting System

CasRe Case Western Reserve University Warner and Swasey Observatory Reprints (periodical code)

CASRE Computer-Aided Software Reliability Estimation

CASREP Casualty Report

CASRN Chemical Abstracts Service Registry Number

CASRS Countdown and Status Receiving Station • Countdown and Status Receiving System

Cass Cassegrain (focus)

CASS Center for Advanced Space Studies • Center for Astronomy and Space Sciences • Center for Astrophysics and Space Sciences • Center for Atmospheric and Space Sciences • Center for Automated Space Science • Chinese Academy of Social Sciences (China-PRC) • Coarse Analog Sun Sensor • Coding Accuracy Support System • Common Applications Support Software • Computer-Assisted Search Service • Computerized Algorithmic Satellite Scheduler • Conditional Access Subsystem • Conférence des Académies Scientifiques Suisses (Switzerland) • Consolidated Automated Support System • Consolidated Automatic Support System • Copyright Accounting and Settlement System

CASSAD Centre for African Settlement Studies and Development

CASSI Chemical Abstracts Service Source Index • Custom Astronomical Support Services Inc. (USA)

CASSIOPE Computer-Aided System for Scheduling, Information and Operation of Public transport in Europe

CASSIS Classification and Search Support Information System • Classified and Search Support Information System

CASSM Content Addressable Segment Sequential Memory

CASSP Computer-Aided Support for Strategic Planning

CASST Commercial Aviation Safety Strategy Team

CASSY Cockpit Assistant System

Cast Castellano • Castle • Castilian • Castilla (Spain)

CAST Castelo • Casting And Solidification Technology • Center for Application of Sciences and Technology • Centre d'Actualisation Scientifique et Technique (France) • CERN Axion Solar Telescope • China Association for Science and Technology • Chinese Academy of Space Technology • Commercial Aviation Safety Team • Common Access Security Terminal • Computer-Aided Search Technique • Computer-Aided Software Testing • Computer-Aided Software Translator • Computer-Aided Structural Technology • Computer-Aided Systems Theory • Computer-Assisted Scanning Technique • Computer-Assisted Scheme Training • Computer-Assisted Self-Training • Computer-Assisted Software Test • Computer-Assisted Software Testing • Computerized Automatic Systems Tester • Cooperation in Applied Science and Technology • Cost Analysis Support Team

CASTD Committee on Application of Science and Technology to Development

CASTI Canter for the Analysis of Science and Technical Information

CASTLe CfA-Arizona Space Telescope Lens (survey)

CASTLES CfA-Arizona Space Telescope Lens Survey

CASTNET Clear Air Status and Trends Network

CASTOR Chaîne d'Acquisition, de Surveillance et de Traitement des Observations Radar

CASTS Countdown And Status Transmission System

CASU Cambridge Astronomical Survey Unit • Conseiller Administratif des Services Universitaires • Conseiller d'Administration Scolaire et Universitaire (France) • Consolidated Administrative Support Unit

CASW Council for the Advancement of Scientific Writing

Cat Catalán • Catalog • Cataluña • Cataract • Catégorie • Category • Catena • Catenate

CAT Cable Tester • Calling Attention To • Carburetor Air Temperature • Catalog • Category • Category Alias Table • Cat Island (Bahamas) (location identifier) • Celestial Atomic Trajectory • Cellular Automata Transform • Centralized Automatic Testing • Centre d'Aide par le Travail • Centre for Advanced Technology • Centre for Alternative Technology • Centro de Astrofísica Teórica • Cercle Astronomique de Tournai (Belgium) • Čerenkov Array in Themis (France) • Čerenkov Atmospheric Telescope (*) • Certificat d'Aptitude Technique • Civil Air Transport • Classical Analytic Technique • Classroom Assignment Technique • Cleanup And Treatment • Clear-Air Turbulence • Client Availability Time • Close Approach Tool • Club Astronomique Tarbais (France) • Cockpit Automation Technology • College of Advanced Technology • Columbus Automation Testbed • Comisión para Asignación de Tiempo • Comité de Asignación de Tiempos • Comité para la Asignación de Tiempos • Command Action Team • Committee on Applications and Technology • Common Assessment Task • Common Authentication Technology • Communications And Tracking • Compile And Test • Complementary Analysis Team • Component Advanced Technology • Composition Automatique de Textes • Comprehensive Analysis Tool • Computed Axial Tomography • Computer Adaptive Test • Computer-Aided Teaching • Computer-Aided Technology • Computer-Aided Telescope • Computer-Aided Test • Computer-Aided Testing • Computer-Aided Tomography • Computer-Aided Topography • Computer-Aided Trading • Computer-Aided Training • Computer-Aided Transcription • Computer-Aided Translation • Computer-Aided Typesetting • Computer Annex Trailer • Computer-Assisted Teaching • Computer-Assisted Technology • Computer-Assisted Telescope • Computer-Assisted Testing • Computer-Assisted Thermography • Computer-Assisted Tomography • Computer-Assisted Trading • Computer-Assisted Training • Computer-Assisted Transcription • Computer-Assisted Translation • Computer-Assisted Typesetting • Computer-Automated Terminal • Computer Automatic Testing • Computerized Axial Tomography (*) • Computer of Average Transients • Concatenate • Conditionally Accepted Tag • Consolidated Atomic Time • Contractor Acceptance Test • Control and Analysis Tool • Control and Assessment Team • Control Attenuator Timer • Controlled Attenuator Timer • Coordinated Analysis of the Thermosphere • Cornell Atacama Telescope • Correlation Analysis Tool • Cosmic Anisotropy Telescope • Coudé Auxiliary Telescope • Craft Access Terminal • Crisis Action Team • Curbline Access Test

Cata Catalog

CATA Computers And Their Applications (conference)

Catal Catalan

CATAM Computer-Aided Teaching of Applied Mathematics

CATC Carrier Air Traffic Control • Computer-Assisted Technical Control

CATCC Carrier Air Traffic Control Center

CATCH Couplage avec l'Atmosphère en Conditions Hivernales

CATD Crisis Action Team Director

CATE Computer-Aided Test Equipment • Computer-Aided Test Engineering • Computer-Assisted Threat Evaluation • Computers and Advanced Technology in Education (conference) • Contrat d'Aménagement du Temps de l'Enfant

CATEG Category

CATEX Category Exclusion

Cath Cathedral • Cathode

CATI Centre Académique de Traitement de l'Information • Colorado Advanced Technical Institute (USA-CO) • Colorado Advanced Technology Institute (USA-CO) • Computer-Aided Telephone Interviewing • Computer-Assisted Telephone Inquiry • Computer-Assisted Telephone Interviewing

CATIA Cabine Téléphonique Intégrée à l'Architecture • Computer-Aided Three-Dimensional Interactive Application

CATIE Centro Agronomico Tropical de Investigación y Enseñanza

CATIES Common Aperture Technique for Imaging Electro-Optical Sensors

CATIS Common Applications and Tools Integration Services • Computer-Aided Technical Information System

CATLAS Centralized Automatic Trouble Locating and Analysis System

CATM Cashless Automatic Teller Machine • Cellular Asynchronous Transfer Mode • Computer-Aided Technical Management

CATNIP Computer-Assisted Technique for Numerical Indexing Purposes

CATO Comet-Asteroid Transition Object • Compiler for Automatic Teaching Operation

Catóp Catóptrica

Catópr Catóptrica

CATOX Catalytic Oxidation

CATP Computer-Aided Typesetting Process

CATR Compact Antenna Test Range

CATREF Combination and Analysis of Terrestrial Reference Frames

CATREP Crisis Action Team Report

CATS Canadian Association for Translation Studies (Canada) • Carload And Truckload Shipment • Change Action Tracking System • Cheap Access To Space • Chicago/Argonne Terabyte System • Committee on Applied and Theoretical Statistics • Common Automatic Test System • Communications And Tracking System • Comprehensive Analytical Test System • Computer-Accessed Telemetry System • Computer-Aided Teaching System • Computer-Aided Telemetry System • Computer-Aided Test System • Computer-Aided Theatrical Store • Computer-Aided Troubleshooting • Computer-Assisted Test Shop • Computer-Assisted Test System • Computer-Assisted Trading System • Computer-Assisted Training System • Computer-Assisted Troubleshooting System • Computer-Automated Test System • Computing And Telecommunications Services • Conference And Travel Services • Conformance Abstract Testing Suite • Core Analysis Tools • Correspondence Action Tracking System • Credit Accumulation Transfer Scheme

CATSAT Cooperative Astrophysics and Technology Satellite

CATSS Computer-Aided Training Support System

CATT Center for Advanced Technology in Telecommunications • Centre d'Accueil et de Traitement Téléphonique • Combined Arms Tactical Trainer

CATTB Component Advanced Technology Testbed

CATU Command And Telemetry Unit

CATV Cable And Television • Cable Television • Common Antenna Television • Community Access Television • Community Antenna Television (*)

CATVS Community Antenna Television System

CATWG Calibration Target Working Group

Cau Caucasian

CAU Calculatrice Arithmétique Universelle • Caruaru (Brazil-PE) (location identifier) • Cell Adapter Unit • Centre d'Apprentissage Universitaire • Channel Adapter Unit • Coarse Alignment Unit • Command Acquisition Unit • Command Activation Unit • Command Acquisition Unit • Command Activation Unit • Complex Arithmetic Unit • Concentrator Access Unit • Control Access Unit • Control Arithmetic Unit • Controlled Access Unit • Converter Amplifier Unit • CORDIC Angle Unit • CPU Access Unit • Crypto Adapter Unit • Crypto Ancillary Unit • Crypto Authorization Unit • Crypto Auxiliary Unit • Cryptographic Ancillary Unit • Cryptographic Auxiliary Unit • Customer Acquisition Unit

CAUB Common Area User Block

CAUCA Código Aduanero Uniforme Centroamericano

CAUCE Canadian Association for University Continuing Education (Canada) • Coalition Against Unsolicited Commercial E-mail

CAUL Council of Australian University Libraries (Australia)

CAUP Canadian Association of University Presses (Canada) • Centro de Astrofísica da Universidade do Porto (Portugal)

CAURA Canadian Association of University Research Administrators (Canada)

Caus Causation • Causative

CAUSE College And University System Exchange

CAUT Canadian Association of University Teachers (Canada) • Caution

CAUV Centro de Astronomia Ulisses Villela (Brazil)

Cav Cavalier • Cavaliere • Caveat • Caveau • Cavity

CAV Cazombo (Angola) (location identifier) • Ceiling And Visibility • Česka Akademie Ved (Czech Republic) • Chiefland Astronomy Village (USA-FL) • Circolo Astrofili Veronesi (Italy) • Club d'Astronomie de Villemur (France) • Commercial Asset Visibility • Composite Analog Video • Computer-Aided Verification • Constant Air Volume • Constant Angular Velocity • Coordinate - Anticipate - Verify

CAVAD Cambridge (Starlink internal code)

CAVAL Cavaleiro

CAVE CAVE Automatic Virtual Environment • Cellular Authentication of Voice Encryption • Computer Access for Vision and Employment • Computer-Augmented Video

Education • Computer-Automated Virtual Environment • Contractual Aspect of Value Engineering

CAVERN Computer Access for Vision and Employment Research Network

CAVI Centre Audiovisuel International

CAVIAR Computer Aided Visitor Information and Retrieval

CAVM Circumpolar Arctic Vegetation Mapping

CAVOK Cloud And Visibility OK

CAVORT Coherent Acceleration and Velocity Observations in Real Time

CAVT Constant Absolute Vorticity Trajectory

CAVU Ceiling And Visibility Unlimited • Clear And Visibility Unlimited

CAW Campos - Bartolomeu Lisandro (Brazil-RJ) (location identifier) • Certification Authority Workstation • Channel Address Word • Club Astro de Wittelsheim (France) • Crisis Action Weather

CAWAS Coventry and Warwickshire Astronomical Society (UK-England)

CAWG Coaxial Adapter Waveguide • Communications Architecture Working Group • Customer Accommodation Working Group

CAWIA Commission on Atomic Weights and Isotopic Abundances (IUPAC)

CAWS Commercial Analyst Workstation • Common Aviation Weather System

CAX Carlisle (UK-England) (location identifier) • Community Automatic Exchange • Computer-Assisted Exercise

Cay Cayenne (French Guyana) • Cayman Islands

CAY Cayenne (French Guyana) (location identifier) • Centre Astronomique des Yvelines (France) • Centro Astronómico de Yebes (Spain)

CAZ Cobar (Australia-NSW) (location identifier)

Cb Columbium • Cumulonimbus

C-B Center of Balance • Chemical-Biological

C/B Center of Balance • Chemical/Biological • Circuit Breaker • Cost/Benefit

CB Cable Budget • Call Back • Campus Backbone • Cantilever Beam • Cape Breton (Canada) • Carrier Band • Carte Bancaire • Carte Blanche • Carte Bleue • Cash Book • Cataclysmic Binary • Cell Broadcast • Census Bureau • Center of Buoyancy • Central Bureau • Change Board • Change Bulletin • Character-Based • Chemical and Biological • Chemical-Biological • Chlorobromomethane • Chubut (Argentina) • Circuit Breaker • Circumbinary • Citizens' Band • Close Binary (star) • Column Builder • Common Base • Common Battery • Common Bench • Communications Base • Comparator Buffer • Compass Bearing • Component-Based • Component Broker • Condition Bit • Conduction Band • Conference Bridge • Configuration Baseline • Constant Bandwidth • Contact Binary (star) • Continental Breakfast • Continuous Breakdown • Control Board • Control Break • Control Building • Coordination Board • Cost Benefit • County Borough • Covalent Bonding • Cumulonimbus • Crystal Ball • Currency Bond

CBA Center for Backyard Astrophysics • Center for Basement Astrophysics (USA) • Central Baffle Assembly • CMM-Based Appraisal • Collective Bargaining Agreement • Colliding Beam Accelerator • Commonwealth Broadcasting Association • Communication Burden Assessment • Corner Bay (USA-AK) (location identifier) • Cosmic Background Anisotropy • Cosmological Baryon Asymmetry • Cost-Benefit Analysis

CBAA Comité Belge des Astronomes Amateurs (Belgium)

CBAIPI CMM-Based Appraisal for Internet Process Improvement

Cbal Counterbalance

CBAL Counterbalance

CBAM Content-Based Access to Multimedia

CBARS Classified Baseline Accounting and Reporting System

CBAS Cape Breton Astronomical Society (Canada-NS) • Command Budget Automated System

CBASSE Commission on Behavioral and Social Sciences and Education

CBAT Central Bureau for Astronomical Telegrams (USA-MA) • College Board Achievement Test

CBB Cluster-Based Browsing • Cochabamba (Bolivia) (location identifier) • Commission Belge de Bibliographie (Belgium) • Consumer Buying Behavior • Contract Budget Base • Contract Budget Baseline • Contractor's Budget Baseline

CBBD Centre Belge de la Bande Dessinée (Belgium)

CBBS Computer-Based Bibliographic Search • Computerized Bulletin Board System

CBC Call Broker Client • Campus Backbone Cable • Canadian Broadcasting Corp. (Canada) • Caribbean Broadcasting Co. • Cell Broadcast Center • Change Board Commitment • Cherrabun (Australia-WA) (location identifier) • Cipher Block Chain • Cipher Block Chaining • Closed Brayton Cycle • Complete Blood Count • Congé de Bilan de Compétences • County Borough Council • Coupled Biogeochemical Cycle • Cross-Border Coverage • Current Book Contents

CBCB Center for Bioinformatics and Computational Biology

CBCC Conviction By Civil Court

CBCH Cell Broadcast Channel

CBCP Certified Business Continuity Professional

CBCR Change Board Commitment Record • Channel Byte Count Register

CBCS Common Baseline Circuit Switch • Commonwealth Bureau of Census and Statistics • Computer-Based Conference System

C-BD C-Band

C/BD Chemical/Biological Defense

CBD Candidate Brown Dwarf (star) • Car Nicobar (India) (location identifier) • Cash Before Delivery • C-Band • Center of Business District • Central Business District • Centrale Bibliotheekdienst (Netherlands) • Centre de Bibliographie et Documentation • Chemical Bath Deposition • Chemical-Biological Defense • Citrate - Bicarbonate - Dithionite • Climatologie - Base de Données • Commerce Business Daily (periodical code) • Complementary But Disjoint • Component-Based Development • Compton Backscatter Densitometer • Condemned Below Depot • Constant Bit Density • Convention on Biological Diversity (1992) • Crossed-Beam Display • Cypher Block Chaining

CBDE Carbonated Beverage Dispenser Evaluation • Center for Business Development and Entrepreneurship

CBDi Component-Based Development and Integration

CBDS Calibration Database System • Centre Belge de Documentation Scientifique (Belgium) • Collection Box Decontamination System • Connectionless Broadband Data Service

CBDTPA Consumer Broadband and Digital Television Promotion Act (USA)

CBe Coubomb-Bethe (approximation)

CBE Cell Broadcast Equipment • Chemical Beam Epitaxy • Chemical - Biological - Environmental • Coherent Backscatter Enhancement • Collisionless Boltzmann Equation • Command Budget Estimate • Companion of the British Empire • Computer-Based Education • Condensat Bose-Einstein • Connector Bracket Experiment • Convention sur le Brevet Européen • Costs, Budgeting and Economics • Council for Basic Education • Cumberland (USA-MD) (location identifier)

CBeA Coubomb-Bethe Approximation

CBEC Canadian Book Exchange Centre

CBECI Centre Belge des Échanges Culturels Internationaux (Belgium)

CBED Convergent-Beam Electron Diffraction

CBEL Cambridge Bibliography of English Literature

CBEM Computer-Based Educational Material

CBEMA Computer Business Equipment Manufacturers Association (USA)

CBEP Community-Based Environmental Protection

CBER Center for Biologics Evaluation and Research

CBERS China-Brazil Earth Resources Satellite • Chinese-Brazilian Earth Resource Satellite

CBET Computer-Based Education and Training

CBEVE Central Bureau for Educational Visits and Exchanges

CBF Central Board of Finance • Cerebral Blood Flow • Cold Brittle Point • Council Bluffs (USA-IA) (location identifier) • Current Basis Function

CBFS Cesium Beam Frequency Standard • Command Budget Formulation System

CBG Cambridge (UK-England) (location identifier) • Computer-Based Gaming

CBGA Carpathian Balkan Geological Association • Ceramic Ball Grid Array

CBH Bechar (Algeria) (location identifier)

CBHS Chemical Biological Hardened Shelter

CBI Cape Breton Island • Caribbean Basic Initiative • Center for Biological Informatics • Center for Business Information • Central Bureau of Identification • Commission Baleinière Internationale • Competency-Based Instruction • Complementary Binary • Complete Background Investigation • Compound Batch Identification • Computer-Based Information • Computer-Based Instruction • Confederation of British Industries (UK) • Confidential Business Information • Cosmic Background Imager • Cumulative Book Index

Cbien Combien

CBIL Common and Bulk Items List • Common Bulk Items List

CBIOS Compatibility Basic Input/Output System

CBIPO Custom Build Installation Productivity Option

CBIQS Content-Based Indexing and Querying System

CBIR Content-Based Image Retrieval • Content-Based Information Retrieval

CBIRF Chemical and Biological Incident Response Force

CBIS Central Bureau Information System • Centre de Rectification des Images Spatiales • Computer-Based Information System • Computer-Based Instructional System

CBIT Computer-Based Interactive Training • Continuous Built-In Test

C-BITE Continuous Built-In Test Equipment

CBITE Continuous Built-In Test Equipment

CBITS Computer-Based Instruction Training System

CBIV Computer-Based Interactive Videodisk

CBJ Cabo Rojo (Dominican Rep.) (location identifier)

CBK Checkbook • Clearback (signal) • Colby (USA-KS) (location identifier)

cbl COBOL (file extension)

Cbl Cable

CBL Calculator-Based Laboratory • Ciudad Bolivar (Venezuela) (location identifier) • Cloud Boundary Layer • COBOL • Commercial Bill of Lading • Common Business Language • Community-Based Learning • Computer-Based Learning • Convective Boundary Layer

CBLT Character Block Transfer

CBM Centre for Bibliographic Management • Coalbed Methane • Columbus (USA-MS) (location identifier) • Common Bathing Mechanism • Communications Buffer Memory • Condition-Based Maintenance • Conduction Band Mission • Confidence Building Measure • Core Business Manual • Cubic Meter

CBMIS Computer-Based Management Information System

CBMS Computer-Based Mail System • Computer-Based Medical System • Computer-Based Message System • Conference Board of Mathematical Sciences

CBMU Current Bit Monitor Unit

cBN Cubic Boron Nitride

CB&N Chemical Biological & Nuclear

CBN Centre Bibliographique National • Chemical Biological and Nuclear • Cirebon (Indonesia) (location identifier) • Computerized Bibliographic Network • Content-Based Navigation • Cooperative Base Network • Cubic Boron Nitride

CBNM Central Bureau for Nuclear Measurements

CBNS Corporate Backbone Network Switch

Cbo Colombo (Sri Lanka)

CBO Career-Broadening Objectives • Comité Belge d'Optique (Belgium) • Community-Based Organization • Computer Burst Order • Congressional Budget Office (USA) • Continuous Bitstream Oriented • Cooperative Business Object • Cotabato (Philippines) (location identifier)

CBOE Chicago Board of Options Exchange (USA-IL) • Coherent Backscattering Opposition Effect

CB-OHS Component-Based Open Hypermedia System

CBOHS Component-Based Open Hypermedia System

CBOI Complete Basis Of Issue

CBOIP Complete Basis Of Issue Plan

CBOT Chicago Board Of Trade (USA-IL)

CBP Coimbra (Portugal) (location identifier) • Collateral Benefits Plan • Commission for Bibliography and Publications • Connector Bracket Power • Convergent Billing Platform • Cool Bottom Processing • Corporate Business Plan • Cosmic Background Polarization

CBPC Center Beam Candlepower

CBPF Centro Brasileiro de Pesquisas Fisicas (Brazil)

CBPL Capability-Based Program List

CBPO Consolidated Base Personnel Office

CBPQ Corporation des Bibliothécaires Professionnels du Québec (Canada-QC)

CBPS Chemical and Biological Protective Shelter

CBPSS Chemical and Biological Protective Shelter System

CBQ Calabar (Nigeria) (location identifier) • Class-Based Queuing

CBR CalTech Baseband Recorder • Canberra (Australia-ACT) (location identifier) • Case-Based Reasoning • Centraal Bureau voor de Rijvaardigheidsbewijzen (Netherlands) • Change Board Register • Chemical, Bacteriological, and Radiation • Chemical, Bacteriological, and Radiological (warfare) • Chemical - Biological - Radioactive • Cloud Base Recorder • Clustering Based on Rules • Code Coverage Analysis • Combined Compression and Analysis • Community-Based Research • Concurrent Budget Resolution • Constant Bit Rate • Content-Based Retrieval • Continuous Bit Rate • Cosmic Background Radiation • Crude Birth Rate • Current Budget Resolution

CBRG Cost Benefit Review Group

CBRL Cell Biology Research Laboratory

CBRM Centre Belge de Recherches Mathématiques (Belgium)

CBRS Coherent Bremsstrahlung Radiation Spectrum • Computer-Based Reference Service • Concept-Based Requirements System

CBRT Centroid Bias Radio Test

CBS Cabimas (Venezuela) (location identifier) • Call Broker Server • Capital Budget System • Case Blue Stars • Cauchy - Buniakowskii - Schwarz (inequality) • Centraal Bureau voor de Statistiek • Central Bibliographic Service • Central Bibliographic System • Central Broadcasting System • Central Bureau of Statistics • Channel Block Sequence • Close Binary System • Cluster-Based Search • Coherent Bremsstrahlung-radiation Spectrum • Columbia Broadcasting System • Commission for Basic Systems (WMO) • Committed Burst Size • Common Base Station • Common Basic Services • Community Base Station • Connector Bracket Signal • Continuing Balance System • Contract and Business Services • Cost Breakdown Structure

CBSA Cargo Bay Stowage Assembly • Component-Based Software Architecture

CBSD Celestial Background Scene Description • Component-Based Software Development

CBSE Component-Based Software Engineering

CBSR Census Bureau Shipment System

CBSS Close-Binary Supersoft Source • Council of the Baltic Sea States

CBST Current Bibliography on Science and Technology

cbt Computer-Based Training (file extension)

CBT Case Builder Tool • Catumbela (Angola) (location identifier) • Cesium Beam Tube • Code-Based Testing • Computer-Based Test • Computer-Based Trainer • Computer-Based Training • Conduction-Band Tail • Core-Based Tree • Cross-Bar Transceiver

CBTR Carrier Bit Timing Recovery

CBTS Cesium Beam Time Standard

CBU Cluster Bomb Unit • Cluster Box Unit • Commodity Business Unit • Completely Built-Up • Conference Bridge Unit

CBUD Computer-Based User Documentation

C-BUS Control Bus

CBUS Control Bus

CBV Cabin Bleed Valve • Cerebral Blood Volume • Coban (Guatemala) (location identifier)

C/BW Chemical/Biological Warfare

CBW Call Bandwidth • Chemical and Biological Warfare • Constant Bandwidth • Convert Byte to Word • Critical Bandwidth

CBWFQ Class-Based Weighted Fair Queuing

CBX Common Branch Extender • Company Branch Exchange • Computer Branch Exchange • Computer-controlled Branch Exchange • Computerized Branch Exchange • Condobolin (Australia-NSW) (location identifier)

CBY Canobie (Australia-QLD) (location identifier)

CBZ Cabin Creek (USA-AK) (location identifier) • Coastal Boundary Zone

c/c Compte Courant • Cuenta Corriente

cc Carbon Copy • Centuries • Chapters • Cicero • Copies • Cross-Couple • Cubic Centimeter

Cc Cirrocumulus

C&C Candidate & Calendar • Collection & Classification • Command & Control • Communication & Control • Computer & Communications • Computing & Communications • Control & Coordination • Creativity & Cognition (conference)

C-C Carbon-Carbon • Center-to-Center • Computer-to-Computer • Cross-Country

C/C Carbon/Carbon • Combustion Chamber • Compte Courant • Configuration Control • Conto Corrente

CC Calçada • Calibration Curve • Call Confirm • Call Control • Candidate Country • Carbon-Carbon • Carbon Composite • Carbon Copy • Card Code • Carriage Control • Cartographic Catalog • Casa de Campo • Cascade Correlation • Cash Credit • Cashier's Check • Casilla de Correo • Caucasus • Center Commander • Central Commission • Central Committee • Central Component • Central Computer • Central Console • Central Control • Centre de Calcul • Cepstrum Coefficient • Chamber of Commerce • Chance for Citedness • Change Course • Channel Command • Channel Controller • Charged Current • Chart Convergency • Cheap Chip • Chemical Control • Chemically Consistent • Chief Clark • Chile (international aircraft registration code) • Chronometer Correction • Chronometric Cosmology • Circuit Court • Circulating Copy • Cirrocumulus • City Centre • City Council • City Councillor • Civil Court • Classification Code • Clearance Center • Clearing Center • Close Control • Close-Coupled • Close Coupling • Closed Circuit • Closing Coil • Cloud to Cloud • Cluster Controller • Coarse Control • Cocos (Keeling) Islands (ISO country code) • Code Civil • Code Control • Code Converter • Código Civil • Coefficient of Contingency • Coefficient of Correlation • Collaborative Coefficient • Collision Course • Color Center • Color Code • Color Compensation • Color Contrast • Color Conversion • Color Correction • Column Chromatography • Combination Center • Combined Cycle • Combustion Chamber • Comité Central • Comité de Coordination • Command and Control • Command Center • Command Computer • Command Conference • Commercial Consumable • Commercial Control • Common Carrier • Common Collector • Common Control • Common Criteria • Communication and Control • Communication Center • Communications Central • Communications Control • Communications Coordinator • Community Council • Compact Companion • Compass Course • Compatibility Change • Compensating Current • Competitive Category • Compilation Catalog • Compilation Control • Compiled Catalog • Compiler Construction • Complete Controllability • Component Commander • Composite Capability • Compte Courant • Computational Chemistry • Computation Center • Computer Calculator • Computer Code • Computer Community • Computer Complex • Concave • Conceptual Clustering • Concurrency Control • Condition Code • Configuration Control • Connection Confirmed • Conseil Constitutionnel • Constant Current • Consular Corps • Container Code • Contamination Control • Continuation Clause • Continuity Cell • Continuity Check • Continuous Clocking • Continuous Current • Conto Corrente • Contra Credit • Contract Completion • Control Center • Control Character • Control Computer • Control Console • Contrôle Continu • Convective Collapse • Conventional Conflict • Conversion Complete • Cool Component • Coordinate Converter • Coordinating Center • Coordination Center • Copie Conforme • Corner Cube • Corps Consulaire • Correction Coil • Corriente Continua • Cost Center • Countdown Counter • County Clerk • County Commissioner • County Council • County Court • Country Code • Country to Country • Coupled Channel • Courant Continu • Cours Complémentaire • Courtesy Car • Courtesy Copy • Credit Card • Crew Certified • Crew Compartment • Criteria Check • Cross-Connect • Cross-Correlation • Cubic Centimeter • Current Contents (periodical) • Current Cost • Cursor Control • Cushion Craft • Cusp Cap • Îles Cocos Keeling (code ISO de pays) • Keeling Islands (Cocos_) (ISO country code)

CCA Cadre Communautaire d'Appui • Call Connexion Agent • Call Control Agent • Canadian Cartographic Association (Canada) • Canadian Coalition for Astronomy (Canada) • Canadian Council of Archives (Canada) • Canonical Circulation Analysis • Carrier-Controlled Approach • CENELEC Certification Agreement • Center for Computing and Automation • Center for Cost Analysis • Central Control Analysis • Centre Culturel de l'Astronomie (France) • Centre de Calcul Analogique • Chromated Copper Arsenate • Circuit Card Assembly • Circuit Court of Appeals

• Close Coupling Approximation • Cloud Chamber Analysis • Cluster-Cluster Agglomeration • Collaborative Compression Architecture • Comité de Coordination des Actions • Command and Control Agency • Command Center Automation • Commission de Contrôle des Assurances (France) • Commission on Crystallographic Apparatus • Command Channel Assignment • Common Cryptographic Architecture • Communications Carrier Assembly • Competitive Contracting Act (USA) • Component Cost Analysis • Composite Cell Address • Computability and Complexity in Analysis (workshop) • Computer Corporation of America • Conceptual Communication Area • Configuration Control Article • Conseil Canadien des Archives (Canada) • Consumers' Cooperative Association • Container Control Activity • Contamination Control Area • Continental Control Area • Contingency Capabilities Assessment • Continuum Computer Architecture • Contract Change Authority • Contract Change Authorization • Contract Construction Agent • Coolant Control Assembly • Copy Control Association • County Court of Appeals • Critical Characteristics Audit • Cross-Correlation Analysis • Cryogenic Cooler Assembly • Current Cost Accounting • Customer Control Access • Fort Chaffee (USA-AR) (location identifier)

CCAA Center for Computer-Aided Analysis • Common Cabin Air Assembly • Comunidades Autónomas (Spain) • Contract Clearance Approval Authority • Contract Clearance Approving Authority

CCAB Canadian Circulations Audit Board (Canada) • Consultative Committee of Accounting Bodies

CCAD Center for Computer-Aided Design • Central American Commission on the Environment and Development • Comisión Centro-Americana de Ambiente y Desarollo • Consolidated Contract Administration Data

CCAF Call Control Agent Function • Circolo Culturale di Farra d'Isonzo (Italy)

CCAFS Cape Canaveral Air Force Station (USA-FL)

CCAG Cahier des Clauses Administratives Générales

CCAL Certified Contractor Access List

CCAM Canadian Congress of Applied Mechanics • Commission for Climatology and Applications of Meteorology • Conversational Communication Access Method

CCAMLR Commission for Conservation of Antarctic Marine Living Resources • Convention for Conservation of Antarctic Marine Living Resources

CCAMS Computer-Controlled Antenna Measurement System

CCAP Cable Coupling Analysis Program • Cahier des Charges Administratives Particulières • Coastal Change Analysis Program • Commercial Complex Autonomous Payload • Communications Control Application Program • Competitive Cooperative Agreement Program

CCAPS Circuit Card Assembly and Processing System

CCAQ Consultative Committee on Administrative Questions

CCAR Cost Change Action Request

CCARS Comprehensive Cost And Requirement Report

CCAS Cape Canaveral Air Station (USA-FL) • Cape Cod Astronomical Society (USA-MA) • Center of Curvature Alignment Sensor • Center of Curvature Alignment System (HET) • Central Coast Astronomical Society (USA-CA) • Centre Communal d'Action Sociale • Chester County Astronomical Society (USA-PA) • Cleveland County Astronomical Society (USA-NC) • Contingency Contract Administration Services • Convention for the Conservation of Arctic Seals

CCASS Construction Contract Appraisal Support System

CCAT Centre de Coordination et d'Appui Technique

CCATS Command, Communications And Telemetry Systems • Communications, Command And Telemetry Systems

CCASTD Comité Consultatif sur l'Application de la Science et de la Technique au Développement

CCAT Comité de Coordination de l'Assistance Technique (ONU)

CCAUV Comité Consultatif de l'Acoustique, des Ultrasons et des Vibrations (BIPM)

CCB Call Control Block • Catalogue de Compilation Bibliographique • Change Control Board • Channel Command Block • Circuit Concentration Bay • Civil Cooperation Bureau • Clock Control Board • Clock Correction Board • Collective Catalogue Belgium • Command Control Block • Configuration Change Board • Configuration Control Board • Configuration Coordination Board • Contract Change Board • Upland Cable (USA-CA) (location identifier)

CCBD Change Control Board Directive • Configuration Control Board Directive

CCBL Collect Commercial Bill of Lading

CCBR Configuration Control Board Request

CCBS Change Control Board Summary • Committee on Capacity Building in Science • Completion of Calls to Busy Subscriber

CCBT Centre de Coordination Bibliographique et Technique

CCBX Computer-Controlled Branch Exchange

C-CC Carbon-Carbon Composite

CCC Canadian Chamber of Commerce (Canada) • Canadian Climate Centre (Canada) • Canadian Committee on Cataloguing (Canada) • Cape Communications Control (NASA) • Carbon-Carbon Composite • Cargo Category Code • Carriage Control Character • Catalogue de Compilation Critique • Cayo Coco (Cuba) (location identifier) • CCATS Command Controller • CDMA Cluster Controller • CEN Certification Committee • Central Classification Committee • Central Compact Core • Central Computer Complex • Central Control Commission • Central Control Complex • Central Cool Component • Change and Configuration Control • Change Current Controller • Chemical Contamination Control • Chronological Coverage Code • Civilian Conservation Corps (USA) • Classified Control Clerk • Código Cuenta Cliente • Cold and Climate Change • Comité de Certification du CEN • Command and Communications Center • Command and Control Center • Command Center Complex • Command - Control - Communications • Command Control Console • Communication Center Console • Communications Command Center • Communications Control Center • Community of Common Concern • Compact Central Core • Compact Computer Core • Complex Control Center • Computer Checkout Console • Computer Command Control • Computer Communications Console • Computer Communications Converter • Computer Control Center • Computer Control Communication • Computer Control Complex • Concourse Computer Center • Conference Control Channel • Configuration Control Document • Conseil Consultatif Conjoint • Conseil de Coopération Culturelle • Consolidated Command Center • Contaminant Control Cartridge • Contiguity-Constrained Clustering • Contract Category Code • Controller Checkout Console • Converged Close Coupling • Convergent Close Coupling (theory) • Cooling Core Cluster • Coordinate Conversion Computer • Copyright Clearance Center (USA) • Core Containment Container • Correlator Control Computer • Cosmic Censorship Conjecture • Countercurrent Chromatography • Crew Coordination Concept • Critical Compilation Catalog • Critical Control Circuit • Cross-Cultural Communications • Cryogenic Current Comparator • Cusp Cap Collar • Customs Cooperation Council • Cyclic Check Character

CCCB Configuration Change Control Board • Contractor Configuration Control Board

CCCC Computer Command and Control Company • Computerized Conferencing and Communications Center (USA-NJ)

CCCD Crew-Centered Cockpit Design • Crew Compartment Configuration Drawing

CCCE Closed-Cycle Cryogenic Equipment

CCCF Central Committee on Communications Facilities • Chaos Computer Club de France

CCCG Canadian Conference on Computational Geometry

CCCGH Centro Científico y Cultural de Geografía e Historia (Ecuador) (now CCIGM)

CCCH Central Computer Complex Host • Common Control Channel

CCCI Case Closure Certificate Inventory • Command - Control - Communications - Intelligence

CCCII Chinese Character Code For Information Exchange

CCCL Complementary Constant Current Logic

CCCM Canadian Climate Centre Model • Centro de Control de Calidad y Metrología • Council Committee on Consumer Matters

CCCMMMR Closed-Chest Cardiac Massage and Mouth-to-Mouth Resuscitation

CCCN CDIAC Computing System Network

CCCNT Conseil Consultatif Canadien sur les Normes de Télécommunications • Country Code to Country Name

CCCO Committee on Climate Changes and the Ocean • Committee on Climatic Changes and the Ocean (*) • Corning Community College Observatory

CCCP Centro de Control de Contaminación del Pacífico (Colombia)

CCCR Communication and Control Command Requirement

CCCS Central Control Computer System • City Carrier Cost System • Common Core Character Set • Conseil Canadien des Centres des Sciences

CCCSN Command, Control, and Communications System Network

CCCT Computer, Communications and Control Techniques (conference)

CCCU Crew Compartment Cooling Unit

CCCUNY City College of City University of New York (USA-NY)

CC-D Current Contents on Diskette

CCD Calcite Compensation Depth • Calibration Curve Data • Cambridge Crystallographic Database (UK) • Camouflage, Concealment and Deception • Cash Concentration and Disbursement • Census County Division • Central Command Decoder • Central Computing Department • Charge-Coupled Detector • Charge-Coupled Device (*) • Checkout Command Decoder • Cold-Cloud Duration • Color-Color Diagram • Command and Control Display • Command Case Directive • Common Category Database • Computer Center Division • Computer-Controlled Display • Configuration Change Directive • Configuration Control Document • Conseil de Coopération Douanière • Constant Change Display • Construction Completion Date • Consumer Computer Device • Control Center DORIS • Convective Cloud Differential (method) • Convection to Combat Desertification (1994) • Countercurrent Distribution • Course Control Document • Critical Component Demonstration • Current Contents on Diskette • Los Angeles - Century City (USA-CA) (location identifier)

CCDA CCD Astronomy (periodical code) • Centre Canadien de Données Astronomiques (Canada-BC) • Commercial Chemical Development Association • Commission pour la Coordination de la Documentation Administrative (France)

CCDB Carbon Content of Dead Biomass • Committee on Chemical Databases • Common Core Database in Biotechnology

CCDC Cambridge Crystallographic Data Centre • Central Control and Display Console

CCDF Complementary Cumulative Distribution Function

CCDIS Crustal Dynamics Data Information Station

CCDL Cross-Channel Data Link

CCDM Catalogue des Composantes de Doubles et Multiples • Cold Collisionless Dark Matter • Comité Consultatif pour la Définition du Mètre (BIPM) (now CCL) • Consultative Committee on the Definition of the Meter • Control Command and Data Management

CCDN Corporate Consolidated Data Network

CCDP Cisco Certified Design Professional • Climate Change Detection Project

CCDR Contract Cost Data Report • Contractor Cost Data Report • Contractor Cost Data Reporting • Contractor Critical Design Review

CCDS Center for the Commercial Development of Space • Comité Consultatif pour la Définition de la Seconde (BIPM) (now CCTF) • Console Control and Distribution System • Consultative Committee for the Definition of the Second

CCDU Control, Communications, and Display Unit

C&CE Command & Control Electronics

CCE Centre de Calcul Électronique • Centrifuge Control Electronics • Centro de Calculo Electrónico • Centro de Computação Electrônica • Certificat de Compétence en Entreprise • Change Compositive Explorer • Charge Collection Efficiency • Charge Composition Explorer • Chartered Civil Engineer • Chesler-Cram Equation • Comité Central d'Entreprise • Comité Consultatif d'Électricité (BIPM) (now CCEM) • Command and Control Element • Command and Control Evaluation • Command Center Element • Command Control Equipment • Commercial Construction Equipment • Commission des Communautés Européennes • Communications Control Equipment • Component Cost Estimate • Compound-Cycle Engine • Computer Command Engineer • Computer - Communications - Electronics • Configurable Crypto Engine • Configuration Control Element • Conseil des Communes d'Europe • Conseiller du Commerce Extérieur • Container Control Element • Contingency Communications Element • Continuing Criminal Enterprise • Continuous Comprehensive Evaluation • Continuous Control Evaluator • Corporate Component Engineering • Cryocooler Electronics • Current-Carrying Element • Saint-Martin - Grand Case (French Antilles) (location identifier)

CCEB Combined Communications Electronics Board

CCEC Command and Control Engineering Center

CCEE Countries of Central and Eastern Europe

CCEF Command Center Evaluation Facility

CCEI Composite Cost Effectiveness Index • Coordinating Committee on Engineering Information (USA)

CCEL Centre Canadien de l'Échange du Livre (Canada)

CCEM Comité Consultatif d'Électricité et de Magnétisme (BIPM)

CCEMRI Comité Consultatif pour les Étalons de Mesure des Rayonnements Ionisants (BIPM) (now CCRI)

CCEP Comprehensive Clinical Evaluation Program

CCES Chemically-Consistent Evolutionary Synthesis • Cold-Cathode Emission Spectroscopy • Command Center Evaluation System • Contingency Communications Extension Switch

CCETT Centre Commun d'Études de Télédiffusion et de Télécommunications (France)

CCEU Council on Continuing Education Units (USA)

CCEV Command Center Experimental Version

ccf Hundred Cubic Feet

CCF Call Control Function • Carcassonne Salvaza (France) (location identifier) • Catalogue Collectif de France • Central Clearance Facility • Central Computer Facility • Central Computing Facility • Central Control Facility • Centralized Control Facility • Change Control Form • Chondrite Carbonnée Fission • Collection Coordination Facility • Common Cause Failure • Common Communication Format • Communications Control Facility • Compartmented Communications Facility • Component Cleaning Facility • Configuration Control Form • Consolidated Computer Facility • Contingency Control Facility • Contrôle en Cours de Formation • Converter Compressor Facility • Correctional Custody Facility • Cross-Correlation Function • Cumulative Count Function • Current Calibration File • Current Crest Factor • Custom Control Factory

CCFA Capillary and Connective Flow Apparatus • Comité des Constructeurs Français d'Automobiles (France)

CCFD Complementary Cumulative Frequency Distribution

CCFE Coordinating Committee on Fusion Energy

CCFF Cape Canaveral Forecast Facility (USA-FL) • Crew Compartment Fit and Function

CCFL Counter-Current Flow Limit

CCFM Centre Canadien de Fusion Magnétique (Canada) • Cryogenic Continuous Film Memory

CCFP Centre de Coordination de la Formation Professionnelle • Consultative Committee for the Fusion Programme

CCFSS Collection Coordination Facility Support System

CCFT Captive Carry Flight Test • Cold-Cathode Fluorescent Tube • Controlled Current Feedback Transformer

CCFY Current Contractor Fiscal Year

CCG Carl-Cranz-Gesellschaft (Germany) • Cartesian Coordinate Grid • Catalogue de Compilation Général • Central Cluster Galaxy • Centre Canadien de Géomatique (Canada) • Certification Consultative Group • Change Control Group • Classical Conjugate Gradient (method) • Command and Control Group • Computer Coordination Group • Constant Current Generator • Country Commercial Guide • Crane (USA-TX) (location identifier) • Crisis Coordinating Group • Crisis Coordination Group

CCGA Communications Control Group Assembly

CCGCR Centre for Climate and Global Change Research (Canada)

CCGE Canadian Council on Geographic Education (Canada)

CCGP Command and Control Group

CCGS Canadian Coast Guard College (Canada-NS)

CCGT Combined Cycle Gas Turbine

CCGTM Commonwealth Consultative Group on Technology Management

cch Corel Chart (file extension)

CCH Channel Check • Chile Chico (Chile) (location identifier) • Colegio de Ciencias y Humanidades • Commercial Clearinghouse • Connections per Circuit per Hour • Control Channel • Cubic Capacity of Holds

CCHAS Center for Computer High Assurance System

CCHP Constant-Conductance Heat Pipe

CCHS Crew Compartment Heat Shield

CCHSC Conference on Compilers and High-Speed Compilation

CCHX Component Cooling Heat Exchanger

CCI Canadian Copyright Institute (Canada) • Card Computer Interface • Centre de Calcul Interfacultaire • Centro di Calcolo Interfacolta • Chambre de Commerce et d'Industrie • Chambre de Commerce Internationale • Charge Conjugation Invariance • Charge-Coupled Imager • Charging Contrast Image • Client Communication Interface • Comissão Central de Informática • Comité Científico Internacional • Command Control Interface • Commercial and Customer Intelligence • Common Client Interface • Compétence Complémentaire en Informatique • Compliant - Compatible - Interoperable • Component Communication Infrastructure • Component Cost Index • Composite Configuration Item • Computer Carrier Interrupt • Computer Crime Initiative • Computer Crime Investigation • Computer Crime Investigator • Concôrdia (Brazil-SC) (location identifier) • Contactless Chipcard Interface • Contract Change Identification • Contrôle et Commande des Instruments • Controlled Cryptographic Item • Cost Control Item

CCIA Comité de Coordination pour l'Information et l'Automatisme • Computer and Communications Industry Association (USA) • Console Computer Interface Adapter

CCIB Call Completion - Internet Busy • Command and Control Interoperability Board • Command Center Integration Branch

CCIC Cold Climatic Information Center (USA) • Committee on Computing, Information and Communications

CCID Comité Consultatif pour l'Information et la Documentation • Communities Committee for Information and Documentation (EU)

CCIDES Command and Control Interactive Display Experimentation System

CCIE Cisco Certified Internetwork Expert

CCIF Centre de Calcul Interfacultaire

CCIFQ Centre de Coopération Interuniversitaire Franco-Québécoise

CCIG Centre Canadien d'Information sur les Géosciences • Cold Cathode Ion Gauge

CCIGM Centro Cultural del Instituto Geográfico Militar (Ecuador) (formerly CCCGH)

CCIL Continuously-Computed Impact Line

CCIM Command Computer Input Multiplexer • Common Communications Interface Model

CCIMP Central Course Information Management Project

CCIMS Central Course Information Management System

CCINU Comité Commun de l'Information des Nations Unies (ONU)

CCIP Central Community Investment Plan • Certificat de Capacité d'Insertion Professionnelle (France) • Chambre de Commerce et d'Industrie de Paris (France) • Cisco Certified Internetwork Professional • Committee on Communications and Information Policy • Continuously-Computed Impact Point

CCIPL Colloquium on Current Issues in Programming Languages

CCIR Center for Communication and Information Research • Comité Consultatif International des Radiocommunications • Consultative Committee on International Radiocommunications

CCIRN Coordinating Committee for Intercontinental Research Networking • Coordinating Committee for Intercontinental Research Networks (*) • Coordinating Committee for International Research Networking

CCIS Center for Computer and Information Services • Command and Control Information System • Command and Control Interswitch Signaling • Command, Control, and Information System • Command Control Information System • Common Channel Interoffice Signaling • Computer-Controlled Interconnect System • Coordinating Committee for Information Services

CCISD Centre de Coopération Internationale en Santé et en Développement

CCIST Commission Consultative sur l'Information Scientifique et Technologique (Canada)

CCIT Cable Communications for Information Transfer

CCITT Comité Consultatif International Télégraphique et Téléphonique (now ITU) • Command, Control, and Intelligence Technical Test • Consultative Committee for International Telegraph and Telephone • Consultative Committee on International Telegraph and Telephone (*) • Consultative Committee on International Telegraphy and Telephony

CCIW Canadian Centre for Inland Waters (Canada)

CCJ Calicut (India) (location identifier) • Circuit Court Judge

CCK Cell Culture Kit • Cocos Islands (location identifier) • Cocos (Keeling) Islands (ISO country code) • Communications Commission of Kenya • Complementary Code Keying • Îles Cocos Keeling (code ISO de pays) • Keeling Islands (Cocos_) (ISO country code)

CCl Commission for Climatology (WMO)

CCL Calibration Control Language • Cambridge Consultants Ltd. • Canadian Circumpolar Library (Canada) • Carrier Common Line • Chinchilla (Australia-QLD) (location identifier) • Closed-Circuit Loop • Closed Conference Loop • Comité Consultatif des Longueurs (BIPM) • Command and Control Laboratory • Command Connection Language • Command Control Language • Commerce Control List • Commodity Control List • Commonality Candidate List • Common Command Language • Communications Command Language • Communications/Computer Link • Communications Control Language • Composite Cell Logic • Computer Control Loading • Configuration Control Logic • Connection Control Language • Constraints in Computational Logics (conference) • Continuous Communications Link • Control Center Line • Council for the Central Laboratory • Coupled Cavity LINAC • Critical Characteristics List • Critical Components List • Cursor Control Language

CCLB Carbon Content of Living Biomass

CCLC Carrier Common Line Charge

CCLI Course, Curriculum and Laboratory Improvement

CCLID Charge Coupled Linear Imaging Device

CCLKWS Counterclockwise

CCLMS Concurrent Computing Laboratory for Materials Simulations

CCLN Council for Computerized Library Networks (USA)

CCLO Coordinating Council of Library Organizations (USA)

CCLRC Council for the Central Laboratory of the Research Councils (UK)

CCls Court of Claims

CCLS Computer-Controlled Launch Set • Consolidated Contractor Logistics Support

ccm cc:Mail (file extension)

CCM Cell Culture Module • Center for Composite Materials • Central Case Management • Central Case Manager • Central Customer Management • Central Customer Manager • Centre de Contrôle et de Mission • Centrum voor Constructie en Mechatronica (Netherlands) • Chromatographie en Couches Minces • Climate Community Model • Close Control Module • Comité Consultatif pour la Masse (BIPM) • Commission Centrale des Marchés • Committee on Challenges of a Modern Society • Common Configuration Method • Communications Controller Multichannel • Communications Control Module • Community Climate Model • Company-Conformed Name • Computer Control Module • Consejo Cultural Mundial • Constant Current Modulation • Continuous Core Measurements • Controlled Carrier Modulation • CORBA Component Model • Coronagraph Camera Module (NGST) • Cost Centre Manager • Counter-Countermeasure • County College of Morris (USA-NJ) • Coupled Cluster Method • Crew Cargo Module • Criciúma (Brazil-SC) (location identifier) • Cross-Connect Multiplexer • Cross-Correlation Method • Cross-Country Movement • Current Confinement Model

CCMA Comité de Compradores de Material Aeronáutico • Commission Consultative Mixte Académique • Converging Computing Methodologies in Astronomy (ESF) • Crew Correctable Maintenance Action

CCMC Community Coordinated Modeling Center

CCMD Canadian Centre for Management Development (Canada) • Commission Consultative Mixte Départementale • Continuous-Current Modulating Device • Continuous-Current Monitoring Device • Crew Correctable Maintenance Demand

CCME Contract Change Mass Estimate

CCMI Comité Consultatif sur les Météorites et les Impacts (Canada)

CCMIP Commission de Contrôle des Mutuelles et des Institutions de Prévoyance (France)

CCMM Centre Contre les Manipulations Mentales (France) • Course Curriculum Model Manager

CCMMP Committee on Condensed-Matter and Materials Physics (BPA)

CCMP Coordinating Committee on the Moon and Planets

CCMS Capacity-Constrained Machine Scheduling • Central Critical Monitoring System • Checkout Control and Monitor System • Command Control and Monitor System • Commission on the Challenges of Modern Society • Committee on the Challenges of Modern Society (*) • Committee on the Challenges to Modern Society • Computer Center Management System • Contract Change Management System

CCMSS Computer-Controlled Microfilm Search System

CCMTC Command and Control Maintenance Team Complex

CCMU Computer Controller Multiplexer Unit

CCN Campus Computing Network • Case Control Number • Catalogue Collectif National (France) • Center for Civic Networking • Central Command Network • Centre Climatologique National • Chakcharan (Afghanistan) (location identifier) • Cisco Communications Networks (now AVVID) • Classification Change Notice • Cloud Condensation Nucleus • Cluster Control Node • Command Control Number • Common Communications Network • Community Care Network • Computing, Communications and Networking • Configuration Change Negotiation • Configuration Change Notice (*) • Configuration Control Number • Conseil Canadien des Normes • Contract Change Negotiation • Contract Change Notice • Contract Change Notification • Contract Change Number • Contract Completion Notice • Convention Collective Nationale • Cordless Communications Network • Cyclomatic Complexity Number

CCNA Cisco Certified Network Associate

CCNAA Coordination Council for North American Affairs

CCNC Common Channel Network Controller • Computer-Communications Network Center

CCNDT Canadian Council for Non-Destructive Technology (Canada)

CCNE Comité Consultatif National d'Éthique (France)

CCNH Calçadinha

CCNNU Comité de Coordination et de Normalisation des Nations Unies (ONU)

CCNP Cisco Certified Network Professional

CCNPS Catalogue Collectif National des Publications en Série (France)

CCNR Completion of Call on No Reply • Consultative Committee for Nuclear Research

CCNS Computer-Controlled Navigation System

CCNT Centre de Contrôle Nacelle-Télescope

CCNU Chloroethylcyclohexylnitrosourea

CCNUMA Cache Coherent Non-Uniform Memory Access

CCNY City College of New York (USA-NY)

CC&O Customer Cost & Obligation

CCO Call Connected • Campus Computing Organization • Carimagua (Colombia) (location identifier) • Case Control Office • Catalogue Collectif des Ouvrages • Central Coding Office • Central Compact Object • Central Control Officer • Circuit Control Office • Classified Control Officer • Communications Collection Outstation • Communications Control Office • Compact Central Object • Computational Cosmology Observatory • Configuration Change Order • Contingency Contracting Officer • Contract Change Order • Contrôle-Commande Opérationnelle • Convective Core Overshoot • Countdown Control Officer • Current-Controlled Oscillator • Customer Contracting Officer • Customer Cost and Obligation • Crystal-Controlled Oscillator

CCOC Consolidated Contingency Operations Center

CCOD Current Contents On Disquette

CCOE Catalogue Collectif des Ouvrages Étrangers

CCOF Complementary Columbus Orbital Facility

CCOH Corrosive Contaminants, Oxygen, and Humidity

CCOHS Canadian Centre for Occupational Health and Safety (Canada)

CCOL Coordinating Committee on the Ozone Layer

CCOM Centro Comercial

CCOP Carnot Coefficient Of Performance

CCORE Centre for Cold Ocean Resources Engineering

CCOSM Chemistry and Circulation Occultation Spectroscopy Mission

CCOW Control Channel Order Wire

CCP Canadian Climate Programme • Catalogue Collectif des Périodiques • Center Console Panel • Central Cluster Panel • Central Communication Processor • Central Contract Point • Central Control Point • Certificate in Computer Programming • Certified Computing Professional • Change Control Package • Change Control Panel • Change Control Point • Charge Capacitance Probe • Circulation Control Point • Civil Code of Procedure • Code of Civil Procedure • Collaborative Computational Project • Combined Command Post • Comisión del Canal de Panamá • Comité Consultatif de Photométrie (BIPM) (now CCPR) • Comité de Concertation Pluridisciplinaire • Command Control Panel • Command Control Post • Commercial Change Proposal • Commission on Comparative Planetology • Common Conversion Point • Communications Career Program • Communications Checkpoint • Communications Configured Package • Communications Control Program • Communications Control Package • Communications Control Panel • Communications Control Processor • Community College of Philadelphia (USA-PA) • Compression Control Protocol • Compte Chèque Postal • Compte Courant Postal • Computer Control Panel • Concepción - Carriel Sur (Chile) (location identifier) • Conference on Computational Physics • Configuration Change Plan • Configuration Change Point • Configuration Control Panel • Configuration Control Phase • Configuration Control Plan • Configuration Control Point • Console Command Processor • Consolidated Cryptologic Program • Contact Capacity Potential • Container Consolidation Point • Contamination Control Plan • Contingency Communications Package • Continuity Check Request • Continuous Correlation Processing • Conto Corrente Postale • Contract Change Proposal • Control Computerizado de Procesos • Controlled Components Program • Conventional Contingency Plan • Coordinated Commentary Programming • Corrosion Control Plan • Cosmological Constant Problem • Cost Control Program • Court of Common Pleas • Cross-Check Procedure • Credit Card Purchase • Critical Compression Pressure • Cubic-Close Packed • Current Cell Rate

CCPA Cloud Chamber Photographic Analysis • Commission Consultative Paritaire Académique

CCPB Catálogo Colectivo del Patrimonio Bibliográfico

CCPC Central Control Processing System • Civil Communications Planning Committee • Communications Computer Programming Center • Control Center Programming Center

CCPCB Control Command Power Control Board

CCPD Centre de Coopération Policière et Douanière • Coupling Capacitor Potential Device • Cross-Correlation Peak Distribution

CCPDS Command Center Processing and Data System • Command Center Processing and Display System

CCPDSR Common Center Processing and Display System Replacement • Communications Center Processing and Display System Replacement

CCPE Canadian Council of Professional Engineers (Canada)

CCPEP Center for Chemistry and Physics of the Earth and Planets (formerly CEPI)

CCpFS Ceskoslovensky Casopis pro Fyziku Sekce (periodical code)

CCPI Consultative Committee on Public Information

CCPIT China Council for the Promotion of International Trade (China-PRC)

CCPL Command and Control Product Line • Computer Center Program Library

CCPM Command Career Program Manager • Commission Consultative Paritaire Mixte

CCPN Commission Consultative Paritaire Nationale

CCPO Central Civilian Personnel Office • Consolidated Civilian Personnel Office

CCPOQ Consultative Committee on Programme and Operational Questions

CCPP Catálogo Colectivo de Publicações Periódicas • Committee on Computers and Public Policy (ACM) • Communications-Computer Program Plan

CCPPSP Communications-Computer Program Plan Support Program

CCPR Comité Consultatif de Photométrie et de Radiométrie (BIPM)

CCPS Central Control Processing System • Command/Control Pressure Suit • Contingency Communications Parent Switch

CCPSD Colloquium on Combining Paradigms for Software Development

CCPSDR Command Center Processing and Display System Replacement

CCPT Cockpit Control Position Transducer

Ccpto Concepto

CCQ Cachoeira (Brazil-BA) (location identifier) • Cross-Crew Qualification

CCQJC Christopher Columbus Quincentenary Jubilee Commission

CCQM Comité Consultatif pour la Quantité de Matière (BIPM)

C&CR Command & Control Reports

CCR Call Completion Rate • Cargo Compatibility Review • Cassette Camera Recorder • Center for Cell Research (USA-PA) • Center for Claim Resolution • Center for Climatic Research • Central Contractor Registration • Central Control Room • Centre Commun de Recherche • Centre de Contrôle Régional • Centro Comune de Ricerche • Closed-Circuit Radio • Cognitive Coherence Relation • Combined Contractor Registration • Command Conference Room • Command Control Receiver • Commission of Civil Rights • Commission on Civil Rights • Commitment, Concurrency and Recovery (protocol) • Common Center of Research • Communicating Cells Ratio • Complex Chemical Reaction • Complex Control Room • Computer and Computation Research • Computer Character Recognition • Computer Communications Research

• Computer-Controlled Retrieval • Concord (USA-CA) (location identifier) • Configuration Change Request • Configuration Control Record • Console Card Reader • Consumer-Controlled Reconfiguration • Continuous Catalyst Regeneration • Contract Change Record • Contract Change Release • Contract Change Request • Contractor Change Request • Contractor Change Review • Contractor Cost Reduction • Control Center Rack • Convention Collective Régionale • Conversation Control Record • Corrected Compass Rose • Cube-Corner Reflector • Current Cell Rate • Customer-Controlled Reconfiguration

CCRA Cape Canaveral Reference Atmosphere (USA-FL)

CCRB Configuration Control Review Board

CCRC Communications Control and Restoral Center

CCRE Conseil des Communes et Régions d'Europe

CCRF Consolidated Communication Recording Facility

CCRI Comité Consultatif de Recherches en Informatique • Comité Consultatif des Rayonnements Ionisants (BIPM) • Community College of Rhode Island (USA-RI)

CCRIA Comité Consultatif de Recherches en Informatique et Automatique

CCRL Computer Crime Research Laboratories

CCRP Continuously Computed Release Point

CCRRDT Comité Consultatif Régional de la Recherche et du Développement Technologique

CCRS Canada Centre for Remote Sensing • Centre Canadien de Ressources Spatiales (Canada) • Cluster Center Radio Source

CCRSE Commitment, Concurrency and Recovery Service Element

CCRST Comité Consultatif de la Recherche Scientifique et Technique (France)

C-CS Communication Computer Systems

C&CS Command & Control System

CCS Calculus for Communicating Systems • Calculus of Communicating Systems • Camera Control System • Caracas - Simon Bolivar (Venezuela) (location identifier) • Carrier Cost System • Carrington Coordinate System • Case Control System • Center for Computational Sciences • Central Checkout System • Central Command Schedule • Central Computer and Sequencer • Central Computer Services • Central Control Section • Central Control Software • Central Control Station • Central Control System • Centre de Calcul de Strasbourg (France) • Centre de Contrôle des Satellites • Centre de Contrôle Spécialisé • Centre de Culture Scientifique • Centre du Cadran Solaire (France) • Centro de Control de Satélites • Change Commitment Summary • Change Control System • Charge Control System • Classified Catalog System • Classified Control System • Classroom Communications System • Clock Comparator System • Collective Call Sign • Combinatorics and Computer Science (conference) • Command and Communications System • Command and Control Segment • Command and Control Set • Command and Control System • Command and Coordination Set • Command Control Segment • Command Control System • Comments Collection System • Commitment Control System • Common Carrier System • Common Channel Signaling • Common Command Set • Common Communications Service • Common Communications Support • Common Control Software • Communications and Control Segment • Communications Computer Systems • Communications Control Set • Communications Control System • Complex Control Set • Computer and Communications Security (conference) • Computer Command System • Computer Consulting Service • Computer Core Segment • Computing Classification Scheme • Computing Classification System • Computing Consulting Service • Concurrency Control Service • Confidential Cover Sheet • Configuration Control Secretariat • Console Communication System • Container Control Site • Contamination Control System • Continuous Commercial Service • Continuous Composite Servo • Contract Control System • Control and Communications System • Control Center System • Contrôle, Commande et Servitude • Controlled Combustion System • Controlled Component Synthesis • Cool Carbon Stars • Cooperative Credit Society • Coördinatiecommissie Synchrotronstraling • Cryocooler System • Custom Computer System

CCSA Centre de Calcul Scientifique de l'Armement • Command and Control Support Activity • Common Control Switching Arrangement

CCSAIT Colorado College Student Astronomy Instrument Team

CCSB Configuration Control Subboard

CCSC Canadian Council of Science Centres (Canada) • Central Connecticut State College (USA-CT) • Centre de Calcul de Strasbourg-Cronenbourg (France) (defunct) • Command and Control Systems Center • Commercial Computer Security Centre (UK) • Consortium for Computing in Small Colleges (USA) • Coordinating Committee for Satellite Communications • Customer Support Call Center

CCSD Cellular Circuit-Switched Data • Centre pour la Communication Scientifique Directe • Command Communications Service Designator • Computer-Compatible Signal Data • Control Communications Service Designator • Couple-Cluster Single and Double

CCSDS Consultative Committee for Space Data Systems • Consultative Committee for Standard Data Services

CCSE Comité pour la Compétitivité et la Sécurité Économique

CCSEM Centro di Cultura Scientifica Ettore Majorana (Italy) • Computer-Controlled Scanning Electron Microscope • Computer-Controlled Scanning Electron Microscopy

CCSF CalTech Concurrent Supercomputing Facility (USA-CA) • City College of San Francisco (USA-CA)

CCSG Computer Components and System Group • Consolidated Contingency Steering Group

CCSI California Commercial Spaceport Inc. (USA-CA) • Cisco Certified Systems Instructor • Communications-Computer Systems Integration

CCSID Coded Character Set Identification • Coded Character Set Identifier

CCSIL Command and Control Simulation Interface Language • Command Control Software Interface Language

CCSIO Communications-Computer Systems Integration Office

CCSK Cycle Code Shift Keying • Cyclic Code Shift Keying

CCSM Cross-Correlogram Spectral Matching

CCSMO Command and Control Systems Management Office

CCSN Community College of Southern Nevada (USA-NV) • Core-Collapse Supernova

CCSNe Core-Collapse Supernovae

CCSO Command and Control Systems Office • Command and Control Systems Organization • Communications-Computer Systems Officer • Computing and Communicating Services Office

CCSP Communications Concentrator Software Package • Consolidated Computer Security Program • Contingency Contracting Support Plan

CCSPE Commission de Coordination des Sciences de la Planète et de l'Environnement (CNES)

CCSQ Commission des Cadrans Solaires du Québec (Canada-QC) • Consultative Committee on Substantive Questions

CCSR Carbon-Carbon Survivable Radiator • Communications-Computer Systems Requirement • Crew Compartment Stowage Review

CCSRB Communications Computer Systems Requirements Board

CCSRD Communications Computer Systems Requirements Document

C-CSS Communication Computer Systems Security

CCSS Command and Control Switching System • Command Commodity Supply System • Common Channel Signaling System • Communication Computer Systems Security • Computer Crime and Security Survey • Control Center Standard Software • Coordination and Contract Summary Sheet

CCSSC Coca-Cola Space Science Center (USA-GA)

CCST Center for Computer Sciences and Technology • Centre Culturel Scientifique et Technique • Centre de Culture Scientifique et Technique • Coordination Committee for Science and Technology

CCSTG Computer-Controlled Small Tool Polisher

CCSTI Centre de Culture Scientifique, Technique et Industrielle

CCSU Central Connecticut State University (USA-CT) • Configuration Control and Sensing Unit

CCSV Comité de Coordination du Service Volontaire (France)

CCT Captive Carry Test • Cauchy Convergence Test • Central Control Terminal • Centre Canadien de Télédétection (Canada) • Centre de Compétences Techniques • Cerebral Circulation Time • Clean Coal Technology • Code Carrier Tracker • Colonia Catriel (Argentina-RN) (location identifier) • Comité Consultatif de Thermométrie (BIPM) • Command and Control Technology • Command and Control Transmitter • Command/Control Transmitter • Common Customs Tariff • Communications Control Team • Complaint Control Terminal • Complex Coordination Test • Computational Convexity Technique • Computer-Compatible Tape • Computer-Controlled Typesetting • Consolidated Change Table • Constant Current Tokamak • Constant Current Transformer • Consultative Committee on Thermometry • Contact Charge Transfer • Contingency Communications Team • Continuous Cooling Transformation • Continuous-Current Tokamak • Controlled Clinical Trial • Correlated Color Temperature • Crew Compartment Trainer • Crystal-Controlled Transmitter • Current Completion Time

CCTA Central Computer and Telecommunications Agency (UK) • Commission de Coopération Technique pour l'Afrique • Computer Communications and Telecomms Agency • Coordinating Committee of Technical Assistance

CCTAC Computer Communications Trouble Analysis Center

CCTC Central Command Transmitter Console • Command and Control Technical Center

CCTCF Communication Circuit Technical Control Facility

CCTDA Cockpit Control Thrust Drive Actuator

CCTDE Compound Cycle Turbine-Diesel Engine

CCTF Comité Consultatif du Temps et des Fréquences (BIPM) • Command and Control Test Facility • Consultative Committee for Time and Frequencies

CCTG Cahier des Clauses Techniques Générales • Constant Concentration Tracer Gas

CCTI Centre de Calcul et de Traitement de l'Information • Control Center Technology Interchange

ccTLD Country Code Top Level Domain

CCTP Cahier des Clauses Techniques Particulières

CCTR Cassette Compact Toroid Reactor • Chromosphere-Corona Transition Region

CCTS Comité de Coordination des Télécommunications par Satellite

CCTT Command and Control Team Trainer

CCTV Closed-Circuit Television

CCU Cable Control Unit • Calcutta - Dum Dum (India) (location identifier) • Camera Control Unit • Cardiac Care Unit • Cell Culture Unit • Central Computer Unit • Central Control Unit • Certificate of Clearance for Use • Change Control Unit • Channel Control Unit • Charge Control Unit • Chart Comparison Unit • Chung Cheng University (China-ROC) • Cluster Control Unit • Coastal Carolina University (USA-SC) • Comité Consultatif des Unités (BIPM) • Comité Consultatif des Universités • Command Center Upgrade • Command Control Unit • Common Control Unit • Communication and Consultation Unit • Communication Control Unit • Communications Carrier Umbilical • Communications Carrier Unit • Communications Concentrator Unit • Communications Control Unit • Communications Coupling Unit • Compact Computer Unit • Computer Control Unit • Conversion Computer Unit • Convolutional Coding Unit • Coronary-Care Unit • Coupling Control Unit • Crewman Communications Umbilical • Crew Communications Unit • Crew Compartment Umbilical

CCUFB Centre de Coopération Universitaire Franco-Bavarois

CCUMC Consortium of College and University Media Centers

CCURV Cable-Controlled Underwater Recovery Vehicle

CCUS Chamber of Commerce of the United States

CCV Cash Collection Voucher • Chamber Coolant Valve • Control-Configured Vehicle • Coolant Control Valve • Craig Cove (Vanuatu) (location identifier)

CCVA Chamber Coolant Valve Actuator

CCVD Combustion Chemical Vapor Deposition

CCVPS Computer-Controlled Vent and Pressurization System

CCVR Centre de Calcul Vectoriel pour la Recherche

CCVS Central Clearance Verification System • COBOL Compiler Validation System • Current-Controlled Voltage Source

CCVT Coupling Capacitor Voltage Transformer

CCW Cable Cutoff Wavelength • Carbon Copy for Windows • Channel Command Word • Channel Control Word • Component Cooling Water • Continuous Carrier Wave • Continuous Composite WORM • Counterclockwise • Cowell (Australia-SA) (location identifier)

CCWCP Coordinating Committee for the World Climate Programme

CCWO Commercial Communications Work Order

CCWP Classified Computer Working paper

CCWS Component Cooling Water System

CCX Cáceres (Brazil-MT) (location identifier)

CCY Charles City (USA-IA) (location identifier)

CCZ Chub Bay (Bahamas) (location identifier) • Command and Control Zone

c/d Carried Down • Certificate of Deposit

cd Candela • Cum Dividendo

Cd Cadmium • Card • Ciudad • Cord • Could

C-D Convergent-Divergent

C&D Cease & desist • Command & Decision • Communication & Data • Control & Display

C/D Certificate of Deposit • Code/Debug • Command/Data select • Construction/Development • Consular Declaration • Control Data • Countdown • Cup/Disc • Customs Declaration

CD Cable Driver • Cable Duct • Calendar Day • Calibration Device • Call Deflection • Call Director • Campus Distributor • Canadian Dollar • Candidate Detection • Capacitive Discharge • Capacitor Diode • Card • Carried Down • Carrier Density • Carrier Detect • Carrier Detection • Case Designator • Cash Discount • Center Director • Center Distance • Central Detector • Central District • Certificate of Deposit • Change Directive • Change Directory • Channel Designator • Chapter Director • Chart

Distance • Chemiluminescence Detection • Chemistry Division • Chilldown • Chromatic Dispersion • Circuit Description • Circuit Diagram • Circular Dichroism • Circular Dispersion • Civil Defense • Classification Décimale • Classification of Defects • Clearance Delivery • Climatological Data • Clock Driver • Closing Date • Cloud Detection • Cluster-Directed • Coded Diphase • Coefficient of Drag • Cold Drawn • Collaborative Document • College Diploma • Collision Detection • Color Display • Column • Comité de Dirección • Command Decoder • Committee Draft • Common Digitizer • Common Direction • Communication and Data • Community Development • Compact Disk • Compound Document • Computational Differentiation • Computational Domain • Concept Definition • Concept Development • Conference on Disarmament • Confidential Document • Congressional District • Connection Director • Consiglio Direttivo • Constant Delay • Constant Density • Contact Discontinuity • Content Dependent • Contour Dynamics • Contract Definition • Contract Design • Contrast Detail • Control and Display • Control Data • Convection-Diffusion • Convective-Dispersive • Converging-Diverging • Converted Distance • Cool Disk • Coordinate Data • Coordinated Development • Coordination Document • Coordinating Draft • Córdoba (Argentina) • Córdoba Durchmusterung • Core Damage • Core-Dominated • Corporate Design • Corporate Development • Corps Diplomatique • Così Detto • Countdown • Critical Dimension • Criticality Designator • Criticality Dosimeter • Cross-Disperser • Crystal Diode • Crystal Driver • Current Density • Current Dollars • Current-Driven • Current Driver

cda CD-Audio (file extension)

CDA Call Data Accumulator • Case Directive Amendment • Catalog Data Activity • Catalogue des Données Astrométriques • Central Design Activity • Central Design Agency • Central Design Authority • Centro de Datos Astronómicos • Centro de Divulgação da Astronomia (Brazil) • Cesium Dihydrogen Arsenate • Chandra Data Archive • Circuit Distribution Assembly • Cognitive Decision Aiding • Columbus Design Authority • Combinatorial Data Analysis • Command and Data Acquisition • Command Data Administration • Commander's Decision Aid • Common Data Access • Common Data-casting Architecture • Communication Decency Act (USA) • Compound Documents Architecture • Computer Design Activity • Computer Design and Applications • Conceptual Design Activity • Consolidated Disposition Authority • Continuous Descent Approach • Contractual Document Action • Control and Data Acquisition • Convolutional Decoder Assembly • Cooinda (Australia-NT) (location identifier) • Coordinated Data Analysis • Coordinating Design Authority • Cosmic Dust Analyzer • Craft Digital Assistant • Critical Design Activity • Critical Design Audit

CDAA Central Data Analysis Area • Circularly Disposed Antenna Array

CDAC Composite Design and Analysis Code • Cosmology Data Analysis Center

Cdad Ciudad

CDAd Component Data Administrator

CDAF Convection-Dominated Accretion Flow

CDAG Computational Directed Acyclic Graph

CDAI Centre de Documentation Aéronautique Internationale (France)

CDAP Centrally-Directed Audit Program • Civil Damage Assessment Program • Climatic Data Analysis Program

CDAPS Central data Acquisition Processing System

CDAS Central Data Acquisition System • Central Decision/Analysis Support • Cleethorpes and District Astronomical Society (UK) • Climate Data Assimilation System • Cognitive Decision Aiding System • Command and Data Acquisition Station • Commission Départementale d'Action Sociale (France) • Configuration Design Analysis Simulation • Configuration Development Analysis Simulation • Core Design Analysis Simulation

CDATS Chemical Detection and Alarm Training System

CDAW Coordinated Data Analysis Workshop

CDB Calibration Database • Candidate Detection Buffer • Caribbean Development Bank • Carrier Development Board • Central Data Bank • Central Data Buffer • Central Database • Cold Bay (USA-AK) (location identifier) • Command Database • Command Data Buffer • Command Descriptor Block • Common Database • Common Data Bus • Commonwealth Development Bank • Communications Database • Compiler Database • Comprehensive Database • Computer Data Bus • Congested District Board (Ireland) • Constraint Database • Corporate Database • Cosmic Diffuse Background • Current Data Bit

Cdbd Cardboard

CDBFR Common Data Buffer

CDBM Current Diffusive Ballooning Mode

CDBMS Climate Database Management System

CDBN Column-Digit Binary Network

CDBS Calibration Database Software • Calibration Database System • Cluster Database System

CdC Carte du Ciel

CDC Call Directing Character • Call Directing Code • Capsule Drive Core • Career Development Course • Cedar City (USA-UT) (location identifier) • Center for Disease Control • Central Digital Computer • Central Directory Component • Central Disease Center • Central Document Control • Central Drift Chamber • Centrale De Commutation • Centre De Contrôle • Centre de Documentation Chimique • Centro de Documentação Cientifica • Centro de Documentación Científica • Ceramic Disk Capacitor • Chemical Data Center • Chemical Decontamination Center • Chromatic Dispersion Coefficient • Classified Document Control • Climate Diagnostics Center (NOAA) • Cloud Droplet Concentration • Coalition to Diversify Computing (ACM) • Coaxial Directional Coupler • Code Décimal Cyclique • Code Directing Character • Cold-Drawn Copper • Comisión de Documentación Científica • Command and Data-handling Console • Command Data Channel • Command Decoder Coaxial • Common Development Cycle • Common Distributed Change • Computer Development Center • Computer Display Channel • Concensus Development Conference • Conditioned Diphase Converter • Configuration Data Control • Confined Detonating Cord • Consolidated Duplicating Center • Context-Dependent Clustering • Contract Data Coordination • Control Data Corporation • Control Distribution Center • Convective Day Category • Core Data Center • Cost Determination Committee • Count - Double Count • Countdown Clock • Critical Density of Current • Cryogenics Data Center • Crystal Data Center • Crystallographic Data Center • Curriculum Development Center

CDCA Centre de Dépouillement des Clichés Astronomiques

CD&CC Central Data & Cataloguing Center

CDCC Central Data and Cataloguing Center • Centro de Divulgação Cientifica e Cultural

CDCD Cargo Design Control Drawing

CDCE Central Data Conversion Equipment

CdCF Cahier des Charges Fonctionnelles

CDCF Cahier Des Charges Fonctionnelles • Cosmic Dust Collection Facility • Cosmic Dust Collector Facility

CDCFL Capacitator Diode-Coupled FET Logic

CDCI Continuous Duration and Constant Impact

CDCL Capillary-Driven Cooling Loop • Collective Distributive Constraint Language

CDCM Carbon Dioxide Concentrating Module

CDCN Command Document Control Number • Contract Data Change Notice • Contract Document Change Notice

CDCP Center for Disease Control and Prevention • Central Data Collection Point • Command Display and Control Processor

CDCR Center for Communication and Documentation Research (USA-OH) • Conceptual Design and Cost Review

CDCS Central Data Collection System • Centre for Development Cooperation Services • Customer Depot Complaint System

CDCT Centro de Documentação Científica e Técnica • Centro de Documentación Científica y Tecnológica

CDCTE Centro de Documentación Científica, Técnica y Económica

CDCU Communications Digital Control Unit

CDD Central Data Dictionary • Central Data Display • Central Dispersion Drop • Centre De Documentation • Certificate of Disability for Discharge • Change Drive and Directory • Charge Density Distribution • Charge/Discharge Device • Charging/Discharging Device • Chart Distribution Data • Circumplanetary Dust Dynamics • Classification Décimale de Dewey • Closing the Digital Divide (conference) • Coded Decimal Digit • Color Data Display • Column Density Distribution • Come Dovevasi Dimostrare • Command Decoding and Distribution • Commission du Développement Durable (ONU) • Common Data Dictionary • Common Data Directory • Compact-Disk Drive • Competitive Design/Development • Computer-Directed Drawing • Concept and Development Definition • Conceptual Design Document • Contrat à Durée Déterminée • Cosmic Dust Detector

CD-DA Compact Disk - Digital Audio

CDDA Compact Disk - Digital Audio

CDDB Central Demand Database • Common Diagnostic Database

CDDC Center Data Descriptions Catalog

CDDF Central Data Distribution Facility • Column Density Distribution Function

CDDI Computer-Directed Drawing Instrument • Copper-Distributed Data Interface

CDDIS Crustal Dynamics Data Information Station • Crustal Dynamics Data Information System (*)

CDDN China Digital Data Network (China-PRC)

CDDP Cataloguing Distribution Data Processing • Centre Départemental de Documentation Pédagogique (France)

CDDS Cluster Data Distribution System

CDDT Countdown Demonstration Test

Cde Conde

CDE Caledonia (Panama) (location identifier) • CR-ROM Edition • Centre de Documentation Européenne • Certification in Data Education • Collège Doctoral Européen • Commercial Derivative Engine • Commission for Development and Exchange • Commission on Development and Exchange • Commission pour le Développement et les Échanges • Common Data Environment • Common Desktop Environment • Compact Disk Erasable • Complex Data Entry • Computational Design Engine • Concept Demonstrator Engine • Conde • Content Directory Entry • Continuous Distribution of Ellipsoids • Convection-Dispersion Equation • Cooperative Development Environment • Cost Driver Effect

CDEC Commission Départementale d'Équipement Commercial (France)

CDEDS Collection and Distribution of Ecological Data by Satellite

C. de E. Consejo de Estado

C. de F. Chemin de Fer

CDEFI Conférence des Directeurs d'Écoles et Formations d'Ingénieurs (France)

CDEK Computer Data Entry Keyboard

CDEN Conseil Départemental de l'Éducation Nationale (France)

CDEP Concept Development and Educational Programme

CDER Centre de Développement des Énergies Renouvelables

CDES Computer Data Entry System • Conversational Data Entry System • Crossed-Dispersed Echelle Spectrograph

cdf Cumulative Distribution Function

CdF Collège de France

CD&F Class Determination & Finding

CDF Cable Distribution Frame • California Department of Forestry (USA-CA) • Capodimonte Deep Field • Catalog Data File • Central Database Facility • Central Data Facility • Central Discontinuance Fund • Chandra Deep Field • Channel Definition Format • Charte des Droits Fondamentaux • Circuit Design Fabrication • Class Determination and Finding • Collider Detector at Fermilab (*) • Collider Detector Facility • Combined Data File • Combined Distributing Frame • Combined Distribution Function • Comma-Delimited Format • Command Display Facility • Common Data Form • Common Data Format • Communications Data Field • Community Development Fund • Complete Disjunctive Form • Computer-Derived Format • Concurrent Design Facility • Confined Detonating Fuse • Conventional Data Format • Cool-Down Facility • Core-Damage Frequency • Cortina d'Ampezzo - Fiames (Italy) (location identifier) • Cost Documentation Format • Criticality Difficulty Frequency • Cumulative Damage Function • Cumulative Density Function • Cumulative Distribution Function • Custom-Defined Function • Cutoff Decrease Factor • Cyberspace Description Format

CDFA Component Functional Data Administrator

CDFCHB Command Data Format Control Handbook

CDFIF Communications Direction-Finding Interface

CDFL Capacitator Diode-coupled FET Logic

CDFN Chandra Deep Field North

CDFNT Cold Front

CDFS CD-ROM File System • Chandra Deep Field South • Compact Disk File System • Core Driven Fan Stage

CDF&TDS Circuit Design, Fabrication & Test Data Systems

CDFTDS Circuit Design, Fabrication, and Test Data Systems

cd. fwd. Carried Forward

Cdg Cardiganshire (UK-Wales) • Coding

CDG Capacitance Diaphragm Gauge • Capacitor Diode Gate • Central Design Group • Central Dominant Galaxy • Centrally Dominant Galaxy • Circular Diffraction Grating • Coder-Decoder Group • Compact Dwarf Galaxy • Competitive Development Group • Concept Development Group • Cosmic Diffuse Gamma-ray • Paris - Charles de Gaulle (France) (location identifier)

CDGF Conseil de Discipline de la Gestion Financière (France)

C&DH Command & Data Handling • Communications & Data Handling • Control & Data Handling

CDH Camden (USA-AR) (location identifier) • Command and Data Handling • Communications and Data Handling • Constant Delta Height • Constant Differential Height • Control and Data Handling

CDHC Command and Data Handling Console

CDHF Central Data Handling Facility • Cluster Data Handling Facility • Coordinated Data Handling Facility

C&DHS Command & Data Handling System

CDHS Command and Data Handling System • Comprehensive Data Handling System

CD-I Common Disk Interactive • Compact Disk Interactive

CDI Capacitive Discharge Ignition • Cargo Disposition Instructions • Centre de Documentation et d'Information

• Centre des Impôts (France) • Centre for the Development of Industry • Centre pour le Développement Industriel • Change Direction Indicator • Circle Digit Identification • Classification - Discrimination - Identification • Coder-Decoder Indicator • Collector Diffusion Isolation • Coloured Digital Image • Command Data Interface • Command Detector Interface • Command Display Indicator • Commercial Delta Inc. • Common Disk Interactive • Compact Disk Interactive • Compressed Data Interface • Computer-Directed Instrument • Concept Development Investigation • Constrained Data Item • Contract Definition Instruction • Contrat à Durée Indéterminée • Control Direction Indicator • Core Data Item • Course Deviation Indicator • Critical Decision Issue • Cumulative Damage Index

CDIA Centre de Documentation et d'Information de l'Assurance • Certified Document Imaging Architect

CDIAC Carbon Dioxide Information Analysis Center

C-DIC Circular Differential Interference Contrast

CDIC Carbon Dioxide Information Center • Circular Differential Interference Contrast

CDID Contract Data Item Description

CDIF CASE Data Interchange Format • Compound Document Interchange Format

CDIG Contract Data Input Group

CDII Contrat à Durée Indéterminée Intermittent

CDIM Current Diffusive Interchange Mode

CDIP Centre de Documentation et d'Information Pédagogique

CDIS Comité Directeur de l'Information Statistique • Command and Data Interface Simulator

CDIT Centre de Documentation et d'Information Technique • Centro de Documentación e Información Tecnológica

c. div. Cum Dividend

CDIVS Core Data Item Verification System

CDJ Conceição do Araguaia (Brazil-PA) (location identifier)

CDK Cedar Key - Lewis (USA-FL) (location identifier) • Control Development Kit • Custom Development Kit • Cyberspace Developer Kit

CDL Cable Delay Line • California Digital Library (USA-CA) • Candle (USA-AK) (location identifier) • Carbon Dioxide Laser • Case Definition Language • Central Development Laboratory • Client Display Library • Cloud Detection Lidar • Coaxial Diode Limiter • Comité De Liaison • Command Downlink • Commercial Driver's Licence • Common Data Link • Common Diode Limiter • Common Display Logic • Component Definition Language • Component Description Language • Computer Description Language • Computer Design Language • Contract Data List • Control Development Language • Coronal Dividing Line • Course Description Language • Current-carrying Double Layer • Cyberspace Development Kit

CDLB Carbon Dioxide Laser Beam

CDLC Cellular Data Link Control

CDLI Common Data Link Interface

C&DM Configuration & Data Management

CDM Cable Driver Modem • Canonical Data Model • Case Development Module • Category Discrimination Method • C Data Manager • Central Data Management • Centre Départemental de la Météorologie • Centre Départemental Météorologique • Chandra Data Model • Chemistry Data Management • Clean Development Mechanism • Code Division Modulation • Code Division Multiplexing • Cold Dark Matter • Collaborative Decision Making • Collider Dipole Magnet • Commission on Dynamic Meteorology • Common Data Model • Communications Data Manager • Conceptual Data Model • Configuration and Data Management • Configuration Data Manager • Configuration Design Module • Content Data Model • Contractor Data Manual • Contrat de Durée Maximale • Control Development Methodology • Córdoba Durchmusterung • Core Discharge Monitor • Critical Damping Mode • Crypto Development Module

CDMA Code Division Multiple Access

CDMD Configuration Data Manager Database

CDMF Collaborative Document Management Facility • Commercial Data Masking Facility

CDMGM Cable Driver Modem Group Buffer

CD-MO Compact Disk - Magneto-Optical

CDMO Compact Disk Magneto-Optical

CDMP Configuration and Data Management Procedure

CDMR Cyclic Data Management Routine

CDMS Central Data Management System • Climate Data Management System • Coherent Doppler Measurement System • Command and Data Management System • Command Data Management System • Communications and Data Management System • Communications Data Management System • Computer-aided Data Management System • Configuration Definition Management System • Contract Data Management System • Contractor Data Management System • Control and Data Management System • Cryogenic Dark Matter Search • Cryospheric Data Management System • Cyberspace Data Monitoring System

CDMTS Configuration/Data Management Tracking System

CDMU Central Data Management Unit • Command and Data Management Unit

Cdn Canadian

CDN Camden Woodward (USA-SC) (location identifier) • Canada (international vehicle registration code) • Canadian • CDR Discrepancy Notice • Central Data Recording • Centre de Documentation Nationale • Certificat De Navigabilité • Chicago Daily News • City-Datennetz • Coded Decimal Notation • Comparison Dark Nebula • Compressed Dial Number • Conceptual Design Review • Consolidated Data Network • Content Delivery Network • Content Distribution Networking • Coordinations - Definitions - Notations • Customer-Dedicated Network

CDNC Cloud Drop Number Concentration • Computer-aided Design Numerical Control

Cdnes Condiciones

CDNNET Canadian Network

CDNR CDR Discrepancy Notice Record

CDNS Contingency Data Network System

CDNU Control Display Navigation Unit

Cdo Commando • Consulado

CDO Central Dense Overcast • Central Disbursing Officer • Centre de Documentation Ophtalmologique (France) • Collaborative Data Object • Comité des Directeurs d'Organismes • Command Duty Officer • Commando • Community Dial Office • Complex Duty Officer • Continuous Distribution of Oscillators • Continuous Duty Overnight • Counter District Office • Cradock (South Africa) (location identifier)

CDOM Colored Dissolved Organic Matter

CDOS Chief Data and Operations Scientist • Comité Départemental des Oeuvres Sociales • Concurrent Disk Operating System • Customer Data and Operations System (*) • Customer Data Operations System

CDOT Colloidal Disorder-Order Transition

CD-P Compact Disk - Photographic

C&DP Control & Data Processor

CDP Calorimeter Digital Processor • Candidate Development Program • Career Development Plan • Census Designated Place • Central Data Processing • Central Data Processor • Central Distribution Panel • Centralized Data

Processing • Centre de Données Planétaires (France) • Certification Data Processing • Certified Data Processor • Charge des Droits du Passé • Checkout Data Processor • Cisco Discovery Protocol • Closest Discrete Pairwise • Color Display Processor • Command and Data Processor • Command Data Processor • Command, Data and Processing • Committee of Directors of Polytechnics • Common Depth Point • Common Document Pool • Communications Data Processor • Community Development Project • Compact Disk - Photographic • Compressor Discharge Pressure • Computational Data Processing • Computer Data Processing • Concept Definition Phase • Concept Definition Proposal • Concept Demonstration Program • Concept Development Plan • Condensed Dispersed Phase • Conditioned Diphase • Confidence Development Program • Constant Deviation Prism • Contract Definition Phase • Control Data Panel • Control and Data Processor • Control and Display Panel • Cooperative Development Program • Correlated Data Processor • Cost Data Plan • Critical Decision Point • Critical Design Parameter • Crustal Dynamics Project • Cuddapah (India) (location identifier)

CdPA Called Party Address

CDPA California Desert Protection Act (USA) • Classified Data preparation Area

CDPC Central Data Processing Computer • Commercial Data Processing Center • Computation and Data Processing Center

CDPD Cellular Digital Packet Data

CDPF Central Data Processing Facility

CDPI Command, Data Processing and Instrumentation

CDPIC Caisson à Dépression de la Partie Inférieure du Corps

CDPIE Command Data Processing Interface Equipment

CDPIS Command, Data Processing and Instrumentation System

C/DPM Cost/Design Performance Management

CDPM Cost-Design Performance Management

CdPN Called Party Number

CDPP Centre de Données de la Physique des Plasmas (France)

CDPS Communications Data Processing System • Computing and Data Processing Society (Canada)

CDPSK Coherent Differential Phase-Shift Keying

CDPU Central Data Processing Unit

CDQ Core-Dominated QSO • Core-Dominated Quasar • Croydon (Australia-QLD) (location identifier)

CDQR Critical Design and Qualification Review

cdr Corel Draw (file extension)

Cdr Commander

CdR Cahier de Recettes

CD-R Compact Disk Recordable

CDR Call Detailed Record • Call Detail Record • Call Detail Recording • Cargo Delivery Receipt • Centrale Des Revues • Centrality-Density Ratio • Chadron (USA-NE) (location identifier) • Cleaning Decontamination Request • Climate Data Record • Column Density Ratio • Command Destruct Receiver • Commander • Common Data Record • Common Data Representation • Compact Disk Recordable • Concept and Design Review • Concept Design Review • Conceptual Design Requirement • Conceptual Design Review • Configuration Design Review • Consolidated Data Report • Consortium De Réalisation • Constant Dose Range • Content of Decrement Register • Content of the Decrement of a Register • Continuous Data Recording • Continuous Data Replication • Contract Data Requirement • Cooperative Development Research • Coordinator of Data Requirements • Cost Detail Report • Countdown Deviation Request • Crash Damage Repair • Critical Design Review • Crude Death Rate • Cumulative Data Report • Current Design Reusable • Current Directional Relay

CDRA Carbon Dioxide Removal Assembly

CDRAM Cached Dynamic Random Access Memory

CDRB Critical Design Review Baseline

CDRC CIESIN Data and Research Center

CDRD Carbon Dioxide Research Division • Computational and Data Reduction Division • Contract Data Requirements Document

CDRG Catastrophic Disaster Response Group

CDRH Center of Devices and Radiological Health

CDRL Contract Data Requirements List • Contractor Data Requirements List

CDRLQ Core-Dominated Radio-Loud Quasar

CD-ROM Compact-Disk - Read-Only Memory

CDROM Compact-Disk Read-Only Memory

CDRP Carbon Dioxide Research Program • Certification for Disaster Recovery Planners • Coherent Dispersion Removal Processor

CDRR Contract Documentation Requirements Record

CDRS Computer Data Recording System • Container Design Retrieval System • Contour Drawing Rating Scale • Contract Data Requirements Schedule • Control and Data Retrieval System

CDRSC Cross-Domain Resource

CD-RTOS Compact-Disk - Real-Time Operating System

CDRTOS Compact-Disk Real-Time Operating System

CD-RW Compact Disk Rewritable

CdS Cadmium Sulfide

C&DS Command & Data Simulator

CDS Capability Design Specification • Cataloguing Distribution Service • Cathode Dark Space • Cavity Digitizing System • Cell Directory Service • Center for Development Studies • Central Data Station • Central Data System • Centralized Data System • Centre de Données de Strasbourg (France) (*) • Centre de Données Stellaires (France) • Centre for Development Studies • Centre for Disaster Studies (China-PRC) • CERN Document Server • Characteristic Damage State • Cheap Data Center • Chemical Discriminator System • Childress (USA-TX) (location identifier) • Chromatography Data System • Circuit Data Service • Circuit Data Sheet • Circular Data Stamp • Circumstellar Dust Shell • Cleaning and Deicing System • Climate Data Set • Climate Data Service • Closeout Door System • Cockpit Development Station • Cockpit Dynamic Simulator • Coded Doppler Sonography • Cold-Drawn Steel • Collision Detector System • Color Data System • Color Difference Signal • Comité Directeur Scientifique • Command and Data Simulator • Command and Data System • Command and Decision System • Command Direct Support • Communications and Data System • Compact Digital Switch • Compact Disk Single • Compatible Duplex System • Complex Data System • Compliance Data System • Comprehensive Display System • Compressed Data Storage • Computer Data Switchboard • Computer Data System • Computer Design and Simulation • Computer-Drive System • Computer Duplex System • Computerized Delivery Sequence • Computerized Dispersive Spectroscopy • Computerized Documentation System • Conceptual Design Study • Configuration Design System • Configuration Development System • Congressional Data Sheet • Congressional Descriptive Summary • Consolidated Direct Support • Container Delivery System • Continuing Design Service • Continuous Dynamical System • Contractor Data Sheet • Control and Data System • Control and Display System • Control and Dynamical System • Control Data Set • Control Data System • Control Display System • Control Distribution System • Contrôle Dynamique de Stabilité • Controller Select Direction • Coronal Diagnostic Spectrometer

• Correlated Double Sample • Correlated Double Sampler • Correlated Double Sampling • Cortex Dealer System • Coupling Display System • Cryogenic Dilution System • Current Directory Structure

CDSA Common Data Security Architecture • Condessa

CDSB Concise Dictionary of Scientific Biography

CD&SC Communications Distribution & Switching Center

CDSC Communications Distribution and Switching Center

CDSCC Canberra Deep Space Communications Complex (Australia-ACT)

CDSCP Cell Directory Service Command Program

CDSE Computer-Driven Simulation Environment

CDSF Central Data Services Facility • Commercially-Developed Space Facility • Customer Data Service Facility

CDSI Committee on the Dissemination of Scientific Information • Continuous Duration and Sporadic Impact

CDSL Computer Data System Laboratory • Consumer Digital Subscriber Line

CDSLR Crustal Dynamics Satellite Laser Ranging

CDSP Core Digital Signal Processing

CDSPH Chemo-Dynamical Smoothed Particle Hydrodynamics

CDSR COBE Science Data Room • Contract Data Status Report • Controlled Deployment Specular Reflector • Cost Data Summary Report

C&DSS Communication & Data Subsystem

CDSS Communication and Data Subsystem • Compressed Data Storage System • Contract Data Surveillance System • Cost data Summary Sheet • Customer Decision Support System

CDSSI Common Data Sharing System Infrastructure

CDST Central Daylight Saving Time • Centre de Documentation Scientifique et Technique (CNRS)

Cdt Cadet • Commandant • Cordialement

CdT Cadmium Telluride • Chef de Travaux

CDT Call Data Transmitter • Capabilities Description Table • Catalogs and Databases Team • Cell Delay Tolerance • Center for Democracy and Technology • Central Data Terminal • Central Daylight Time • Centre de Documentation Technique • Centro de Diagnostico y de Tratamiento • Centro de Documentación Técnica • Chargeable Downtime • Class Descriptor Table • Command Descriptor Table • Common Data Transistor • Compressed Data Tape • Compressor Discharge Temperature • Computer Data Transmission • Computer Data Transmitter • Computer-Directed Training • Computer Display Terminal • Concept Design Team • Concrete Data Type • Conduct • Configuration Data Table • Connaissance des Temps (periodical code) • Connectionless Data Transmission • Conseiller en Développement Technologique • Constrained Delaunay Triangulation • Contract Development Test • Contractor Demonstration Test • Control Data Terminal • Control Departure Time • Control Differential Transformer • Control Display Terminal • Controlled Departure Time • Coordinate Data Transmission • Corporación de Desarollo Tecnológico • Countdown Demonstration Test • Countdown Time • Craft - Design - Technology • Crew Duty Time

CDTA Capital District Transportation Authority

CDTAC Composite Damage Tolerance Analysis Code

CDTCP Communication and Data Transmission Control Program

CDTD Centre de Distribution de Travail à Domicile

CdTe Cadmium Telluride

CDT&E Contractor Development Test & Evaluation

CDTE Centre de Développement des Techniques d'Enseignement • Contractor Development Test and Evaluation

CDTED Compressed Digital Terrain Elevation Data

CDTI Centre Départemental du Traitement de l'Information • Centro para el Desarollo Tecnológico e Industrial • Centro para el Desarollo Tecnológico Industrial (Spain) • Cockpit Display of Traffic Information

CDTL Common Data Translation Language

CDTP Centre de Documentation Technique et Pédagogique (Belgium)

CDTR Commercial Training Device Requirement • Common Data Transfer Record

CDTS Computer-Directed Training System

CD-TV Commodore Dynamic - Total Vision • Compact Disk - Television

CDTV Commodore Dynamic - Total Vision • Compact Disk - Television

CDU Calculatrice Digitale Universelle • Camden (Australia-NSW) (location identifier) • Central Display Unit • Centre de Documentation Universitaire • Clasificación Decimal Universal • Classification Décimale Universelle • Cockpit Display Unit • Command Data Unit • Command Decoder Unit • Command Detector Unit • Command Distribution Unit • Computer Display Unit • Control and Diagnostic Unit • Control and Display Unit • Control Data Unit • Control Display Unit • Countdown Display Unit • Counter Display Unit • Coupling Data Unit • Coupling Display Unit • CRT Display Unit • Current Data Unit • Customer Data Unit

CDUA Conservation District Usage Application

CD-V Compact Disk Video

CD/V Concept Definition/Validation

CDV CARMA Data Viewer • Carte De Visite • Cartographic Data Visualizer • Cell Delay Variation • Committee Draft for Voting • Compact Disk - Video • Compressed Digital Video • Concept Definition Vehicle • Cordova - Mile 13 (USA-AK) (location identifier) • Current Domestic Value

CDVT Cell Delay Variation Tolerance

CDW Caldwell Wright (USA-NJ) (location identifier) • Car Damage Waiver • Centrifugally Driven Wind • Charge Density Wave • Chilled Drinking Water • Circumpolar Deep Water • Cold Drinking Water • Collision Damage Waiver • Command and Display Workstation • Command Data Word • Command Word • Computer Data Word • Continuum Distorted Wave • Control Data Word • Critical Design Walkthrough

CDWG Controls and Display Working Group • Countdown Working Group

CD-WO Compact-Disk - Write Once

CDWO Compact-Disk - Write Once

CDWR Chilled Drinking Water Return

CDWT Critical Design Walkthrough

CDX Central Data Exchange • Compound Index • Control Differential Transmitter

CD-XA Compact-Disk - Extended Architecture

CDXA Compact-Disk Extended Architecture

CDY Cagayan de Sulu (Philippines) (location identifier)

Cdz Cádiz (Spain)

CDZ Compact Disk Zone

CdZnTe Cadmium Zinc Telluride

ce Caveat Emptor • Correo Electrónico

Ce Cer • Cerium • Cérium

C-E Communications-Electronics • Communications-Equipment

C&E Ciel et Espace (periodical code)

C/E Calculation/Experiment • Cost Effectiveness

CE Cache Enable • Calculateur Électronique • Campaign Evaluation • Capillary Electrophoresis • Carbon Equivalent • Cargo Element • Ceará (Brazil) • Celestial Equator

• Cellular Encoding • Celtic • Central Engine • Centre of Effort • Change Engineer • Change Evaluation • Channel Element • Channel Extension • Chapman-Enskog • Chargé d'Enseignement • Charge Exchange • Chartered Engineer • Chemical Energy • Chemical Engineer • Chief Engineer • Chip Enable • Christian Era • Chromospheric Emission • Circular Error • Circumstellar Envelope • Civil Engineer • Civil Engineering • Clean Enclosure • Clear End • Clear Entry • Clinical Evaluation • Close Encounter • Coarse Erection • Coil Experiment • Collision Elimination • Combined Electronics • Combustion Efficiency • Comitato Ejecutivo • Comité d'Entreprise • Comité d'Établissement • Command Element • Commercial Engineer • Commission Européenne • Common Emitter • Common Entrance • Common Envelope • Common Environment • Common Era • Communauté Européenne (now UE) • Communications Electronics • Commutator End • Compact Elliptical • Compass Error • Complement Edge • Comprehensive Evaluation • Compression Engine • Computer Engineer • Computer Entry • Computing Efficiency • Concept Evaluation • Concept Exploration • Concurrent Engineering • Conducted Emission • Conference End • Conformité Européenne • Connection Element • Connection Endpoint • Conseil de l'Europe • Conseil d'Établissement • Conseil Économique • Conseiller d'Éducation • Consiglio d'Europa • Consumer Electronics • Contamination Experiment • Continuing Education • Continuous Evaluation • Contrast-Enhanced • Control Element • Control Engineer • Convective Envelope • Conversion Efficiency • Convert Enable • Cooperative Engagement • Corps of Engineers • Correlation Energy • Correlation Environment • Cost Effectiveness • Cost Element • Cost Estimate • Coulomb Explosion • Council of Europe • Cours Élémentaire • Crew Equipment • Critical Examination • Cryocooler Electronics • Current Efficiency • Current Estimate • Current Expendable • Curvature Equation • Cusp Extension • Customer Engineer

CEA Canadian Exporters' Association (Canada) • Cause-and-Effect Analysis • Center for EUV Astrophysics (USA-CA) • Central Electricity Authority • Central Electronics Assembly • Chemical Engineering Abstract • Circular Error Average • Clube Estudiantil de Astronomia (Brazil) • Combined Electronics Assembly • Comité Electrotécnico Argentino (Argentina) • Comité Européen des Assurances • Commissariat à l'Énergie Atomique (France) • Commission Économique pour l'Afrique • Committee on Education in Astronomy • Communications Electronics Activity • Computational Electromagnetics and Acoustics • Conference of Educational Associations • Control Electronics Assembly • Control Element Assembly • Consorzio Engineering Aeroportuale • Cost Effective Analysis • Cost Effectiveness Analysis (*) • Council of Economic Advisers • Council of Educational Advance • County Education Authority • Wichita Cessna (USA-KS) (location identifier)

CEAA Center for Editions of American Authors • Council of European-American Associations

CEAAL Centro de Estudios Astronômicos de Alagoas (Brazil)

CEAB Canadian Engineering Accreditation Board (Canada)

CEABA Chemical Engineering and Biotechnology Abstracts

CEAC Commission Européenne de l'Aviation Civile • Cost Economic Analysis Center • Cost Evaluation and Analysis Center

CEADEN Centro de Estudios Aplicadas para el Desarollo Nuclear

CEAEO Commission Économique pour l'Asie et l'Extrême-Orient

CEAL Civil Engineering Automation Library

CEAM Center for Exposure Assessment Modeling • Coarse Earth Acquisition Mode • Concerted European Action on Magnets • Cost Effectiveness Analysis Methodology

CEAMIG Centro de Estudios Astronômicos de Minas Gerais (Brazil-MG)

CEAO Communauté Économique de l'Afrique de l'Ouest • Communauté Économique pour l'Afrique de l'Ouest

CEAREX Coordinated Eastern Arctic Experiment • Coordinated Eastern Arctic Regional Experiment

CEAS Computer Engineering and Application Society • Confederation of European Aerospace Societies

CEASE Campaign for the Elimination of Acronyms in Scientific Exchange

CEATM Cost Effectiveness At The Margin

CEATS Central European Air Traffic Services

CEB Cebu (Philippines) (location identifier) • Central Electricity Board • Central Electronics Building • Centre Électronique de Brest (France) • Change Evaluation Board • Comité Électrotechnique Belge (Belgium) • Communications Electronics Board • Configured Engine Bay

CEBAF Continuous Electron Beam Accelerator Facility (now TJNAF)

CEBAS Closed Equilibrated Biological Aquatic System

CEBC China-EC Biotechnology Center

CEBEDOC Centre Belge de Documentation (Belgium)

CeBIT Centrum für Büro- und Informationstechnik (Germany)

CEBM Centre Européen de Biologie Moléculaire • Conférence Européenne de Biologie Moléculaire (*)

CEBO Coordinated EISCAT and Balloon Observations

CEC California Energy Commission • Canadian Electrical Code • Capillary Electrochromatography • Capsule End Cover • Cation Exchange Capacity • Central Electronic Complex • Central European Country • Central Executive Committee • Centre d'Expertise Clientèle • Centre Éducatif et Culturel • Centre for Educational Communication (India) • Civil Engineering Corps • Commission of the European Communities • Commonwealth Education Conference • Complex Equipment Contract • Compound Elliptical Concentrator • Computer Electronic Complex • Computer Engineering Center • Confédération des Étudiants Chercheurs (France) • Consumer-oriented Electronic Commerce • Contractor Employment Compliance • Contrat Emploi Consolidé • Control Electronics Container • Control Encoder Coupler • Controlled Element Computer • Cooperative Engagement Capability • Coordinating European Council • Crescent City (USA-CA) (location identifier) • Cryogenic Engineering Conference

CECA Communauté Européenne du Charbon et de l'Acier • Confederación Española de Organizaciones Empresariales (Spain)

CECAM Centre Européen de Calcul Atomique et Moléculaire

CECC CENELEC Electronic Components Committee • Central European Candidate Country (EU)

CECDC Cost Estimate Control Data Center

CECE Confederación Española de Centros de Enseñanza (Spain)

CECED Collaborative Environment for Concurrent Engineering Design • Conseil Européen de la Construction Électrodomestique

CECIMO Comité Européen de Coopération des Industries de la Machine-Outil

CECL Cascade Rmitter-Coupled Logic

CECLES Centre Européen de Construction de Lanceurs d'Engins Spatiaux • Conseil Européen pour la Construction de Lanceurs d'Engins Spatiaux (now merged with CERS into ASE)

CECM Centre for Experimental and Constructive Mathematics (Canada-BC) • Composite Engineering Change Memorandum

CeCO Chief e-Commerce Officer

CECO Center Engine Cutoff • Cost Estimate Change Order

CECOM Communications and Electronics Command

CECORS Civil Engineering Contract Reporting System

CECP Compatibility Engineering Change Proposal • Country Extended Code Pages

CECS Center for Energy and Climate Solutions (USA-VA) • Centro de Estudios Científicos del Sur (Chile) • Centro de Estudios Científicos de Santiago (Chile) • Civil Engineering Computing System

CECSD Citizens for Energy Conservation and Solar Development

CE/D Concept Exploration/Definition

CED Capacitance Electronic Disc • Capacitive Electronic Disk • Cederblad (catalog) • Ceduna (Australia-SA) (location identifier) • Centre d'Édition et de Documentation • Centre Européen de Documentation • Cohesive Energy Density • Collins English Dictionary • Committee for Economic Development • Communauté Européenne de Défense • Computer Entry Device • Concept Exploration and Definition • Concurrent Engineering Directorate • Cost Estimate Dispersion • Cost Estimating Data • Council for Economic Development

CEDA Centro de Estudios de Astronomía (Venezuela)

CEDAM Combined Electronic Display And Map

CeDAR Center for Database Access Research

CEDAR Center of Excellence for Document Analysis and Recognition • Central European Environmental Data Request • Cost Evaluation and Data Archive • Coupling, Energetics and Dynamics of Atmospheric Regions

CEDAS Cluster EFW Data Analysis

CEDC Centre d'Études, de Documentation et de Coordination • Cyclic Error Detection Code

CEDD Committee on Exchange of Digital Data

CEDDA Center for Experiment Design and Data Analysis

CEDE Centre d'Études et de Documentation Européennes • Centro de Estudios sobre el Desarollo Económico

CEDEA Communauté Économique Des États d'Afrique

CEDEAO Communauté Économique Des États de l'Afrique de l'Ouest

CEDEFOP Centre Européen pour le Développement de la Formation Professionnelle

CEDEP Centre Européen D'Éducation Permanente

CEDES Centre d'Études et de Documentation pour l'Enseignement Supérieur

CEDEX Courrier d'Entreprise à Distribution Exceptionnelle

CEDF Conception - Étude - Développement - Fabrication

CEDH Convention Européenne des Droits de l'Homme • Cour Européenne des Droits de l'Homme

CEDI Centre Européen de Documentation et d'Information • Committee on Educational Development and Innovation

CEDIME Centre de Diffusion des Méthodes d'Enseignement

CEDIS Centre d'Études de la Documentation et de l'Information Scolaire

CEDISA Centro de Diseño de Sistemas Automatizados

CEDM Control Element Drive Motor

CEDO Centre for Educational Development Overseas

CEDP Centre d'Enseignement et de Diffusion Pédagogique

CEDR Center for Earth Dynamics Research

CEDRE Calculs d'Écoulements Diphasiques Réactifs pour l'Énergétique • Centre de Documentation, de Recherche et d'Expérimentation • Centre d'Études et de Recherches • Contamination Et Dégradation des Revêtements dans l'Espace • Coopération pour l'Évaluation et le Développement de la Recherche

CEDRO Centro Español de Derechos Reprográficos (Spain)

CEDT Centre for Electronics Design and Technology

C&EE Central & Eastern Europe

CEE Center for Excellence in Education • Central and Eastern Europe • Central and Eastern European • Central Excess Emission • Centre d'Études de l'Emploi • Centro de Estudios Espaciales • Certificate of Extended Education • Cherepovets (Russia) (location identifier) • Civil and Environmental Engineering • Comité Estatal de Estadísticas • Commercial Equivalent Equipment • Commission Économique pour l'Europe • Common Entrance Examination • Common Execution Environment • Communauté Économique Européenne (now UE) • Communications Electronics Element • Comprehensive Environmental Evaluation • Computer Ergonomics Evaluation • Computers and Electrical Engineering (periodical code) • Comunità Económica Europea • Contrat d'Échange Express • Council of Environment Education

CEEA Communauté Européenne de l'Énergie Atomique

CEEAC Communauté Économique des États de l'Afrique Centrale

CEEB College Entrance Examination Board

CEEC Central and Eastern European Countries • Council for European Economic Education

CEED Centre for Economic and Environmental Development • Centre for Economic and Environment Development

CEEF Controlled Ecological Experiment Facility

CEEFI Convention Établissement-Entreprise de Formation Intégrée

CEEI Center for Energy and Environmental Information • Centre Européen d'Études et d'Information • Comité Européen d'Entreprise et d'Innovation

CEEJ Canadian Electrical Engineering Journal (periodical code)

CEEL Core Electron Energy Loss

CEELS Center for Environmental Education and Lake Science (Japan)

CEEM Centre for Energy and Environmental Management

CEEP Centre Européen des Entreprises Publiques • Centre Européen d'Études de Population • Comité Européen de l'Entreprise Publique • Composite Entry Evaluation Program

CEER Council of European Energy Regulators

CEERS Command Excess Equipment Redistribution System

CEES Committee on Earth and Environmental Sciences

CEESAR Centre Européen d'Études Socio-économiques et Accidentologiques des Risques

CEESP Conselho Estadual de Educação de São Paulo (Brazil-SP)

CEF Cable Entrance Facility • Centralized Environment Facility • Chicopee (USA-MA) (location identifier) • Civil Engineering File • Common Equipment Facility • Common Exchange Format • Computer Execute Function • Confederación Española de Fundaciones (Spain) (now AEF) • Contrat Emploi-Formation • Crystalline Electric Field • Springfield Westover (USA-MA) (location identifier)

CEFD Current Estimate Date Fix

CEFI Centre d'Économie et de Finance Internationale • Centre d'Études sur les Formations d'Ingénieurs • Comité d'Études sur les Formations d'Ingénieurs

CEFIC Conseil Européen des Fédérations de l'Industrie Chimique

CEFOAM Checkout Equipment For Onboard Automatic Maintenance

CEFP Centre d'Études et de Formation Pédagogiques

CEFSR Committee for Evaluating the Feasibility of Space Rocketry

CEFTA Central European Free Trade Agreement

CEG Chester (UK-England) (location identifier) • Chief Engineering Group • Civil Engineering Group • Collège d'Enseignement Général • Collins English-German • Combined Environment Group • Common Equipment Group • Computer Education Group • Continuous Edge Graphics • Council for Excellence in Government

CEGB CEGB Research (periodical code) • Central Electricity Generating Board (UK)

CEGEP Collège d'Enseignement Général et Professionnel (Canada)

CEGOS Commission d'Études Générales de l'Organisation Scientifique

CEGP Close-in Extrasolar Giant Planet

CEGRN Central Europe GPS Reference Network

CEGT Conseil de l'Enseignement Général et Technique

CEH Centre for Ecology and Hydrology

CEHT Camera Electrical High Tension

CEI Center for Environmental Information • Central European Initiative • Centre d'Études Industrielles • Cercle Europe Ingénieurs • Channel-Extending Intermediary • Chiang Rai (Thailand) (location identifier) • Cloudtop Entrainment Instability • Collisional Equilibrium Ionization • Collisional Equilibrium Ionized (plasma) • Comitato Elettrotecnico Italiano (Italy) • Commercial-Economic Information • Commission Électrotechnique Internationale • Committee for Engineering Information • Communauté des États Indépendants • Communication - Education - Information • Communications Electronics Installation • Communications-Electronics Instruction • Comparably Efficient Interconnection • Computer-Extended Instruction • Conducted Electromagnetic Interference • Configuration End Item • Connected Element Interferometer • Connection Endpoint Identifier • Contract End Item • Cost Effectiveness Index • Coulomb Explosion Imaging • Council of Engineering Institutions (UK) • Critical Employment Indicator

CEIA Canadian Environmental Industry Association (Canada)

CEIAC Coastal Engineering Information Analysis Center

CEIB Computer Equipment Information Bureau (USA)

CEIC Caribbean Environmental Information Center • Chemical Industry Education Centre • Commission on Electronic Information and Communications

CEICN Centre Européen d'Information pour la Conservation de la Nature

CEIF Council of European Industrial Federations

CEIE Centre d'Études et d'Information sur l'Enseignement

CEII Configured End Item Identification

Ceil Ceiling

CEIN Contract End Item Number

CEIPI Centre d'Études Internationales sur la Propriété Industrielle

CEIR Corporation for Economic and Industrial Research

CEIS Centre d'Initiation à l'Enseignement Supérieur • Combined Elements and Integrated Systems • Contract End Item Specification • Corporate Executive Information System

CEIST Centre Européen d'Informations Scientifiques et Techniques

CEIT Cargo Equipment Integrated Test • Cost Engineering Integrated Tool • Cost Engineering Integration Tool • Crew Equipment Interface Test • Crew Equipment Integration Test

CEJ Chernigov (Ukraine) (location identifier) • Counter Electrojet • Cour Européenne de Justice

CEJP Central European Journal of Physics (periodical code)

CEK Chelyabinsk (Russia) (location identifier) • Computer Entry Keyboard

Cel Celebrate • Celebrated • Celebration • Celsius • Celestial

CEL Capacitatively-Enhanced Logics • Cape Eleuthera (Bahamas) (location identifier) • Carbon Equilibrium Loop • Celescope • Celestial • Célula • Certificat Européen de Langue • Chemical Enrichment Law • Contractor Experience List • Contrat d'Emploi Local • Conversational Extensible Language • Coronal Emission Line • Cosine Emission Law • Critical Experiment Laboratory • Cryogenic Engineering Laboratory

CELC Commonwealth Education Liaison Committee

CELE Center for Evolutionary Language Engineering (Netherlands)

CELENV Cellule d'Environnement

CELESCOPE Celestial Telescope

CELESTE Čerenkov Low-Energy Sampling and Timing Experiment (Themis)

CELG Compact Emission-Line Galaxy

CELI Computer-Enhanced Learning Initiative

CELIA Centre Lasers Intenses et Applications (France) (formerly PMOH)

CELIAS Charges, Elements and Isotope Analysis System (SOHO)

CELJ Conference of Editors of Learned Journals

CELLS California Educational Linkage in the Life Sciences

Cel.Mech. Celestial Mechanics

CELP Codebook Excited Linear Predictive (coding) • Code-Excited Linear Prediction

CELPA Centro de Experimentación y Lanzamiento de Proyectiles Autopropulsados

Cels Celsius

CELS Continuous Extended Laser Spark • Controlled Ecological Life Support

CELSS Closed Ecological Life Support Systems • Close-Ecology Life Support System • Controlled Ecological Life Support Systems (*)

Célt Céltico

Celt Celtic • Celtique

CELT California Extremely Large Telescope (USA-CA) • Centre for Effective Learning and Teaching • Classified Entries in Lateral Transposition • Coherent Emitter Location Technique • Coherent Emitter Location Testbed

Celtolat Celtolatino

CELV Commercial Expandable Launch Vehicle • Complementary Expandable Launch Vehicle

Cem Cement • Cemetery

CEM Camera Electronics Module • Cascade Exiton Model • Center for Electromechanics • Central (USA-AK) (location identifier) • Centro Español de Metrología (Spain) • Channel Electron Multiplier • Chemical Evolutionary Model • Chromospheric Evaporation Model • Circular Electric Mode • Color Evaporation Model • Column Emission Measure • Communications - Electronics - Meteorological • Compatibilité Électromagnétique • Complex Electromechanical • Computational Electromagnetics • Concept Evaluation Model • Concurrent Engineering Methodology • Configuration Element Management • Constrained Energy Minimization • Contact Electronic Mailbox • Continuous Emission Measure • Continuous Emission Monitoring • Control Electronics Module • Conventional Enhancement Modification • Cost and Effectiveness Method • Cost Engineering Model • Cost Estimating Method • Coulomb Explosion Method • Crew Escape Module • Cumulus Ensemble Model

CEMA Consumer Electronics Manufacturers Association • Council for Economic Mutual Assistance • Council of Economic and Mutual Assistance

CEMAC Committee of European Associations of Manufacturers of Active Electronic Components • Communauté Économique et Monétaire d'Afrique Centrale

CEMAGREF Centre National du Machinisme Agricole, du Génie Rural et des Eaux et Forêts (France)

CEMAP Centre Européen de Mathématique Appliquée

CEMAS Civil Engineering Materials Acquisition System

CEMC Communications-Electronics Management Center • Continental Electronics Manufacturing Co.

CEMD Center for Electrophotonic Materials and Devices

CeMDA Celestial Mechanics and Dynamical Astronomy (periodical code)

CeMec Celestial Mechanics (periodical code)

CEMES Communications, Electronics, and Meteorology Equipment Status

CEMF Counter-Electromotive Force

CEMISE Common Management Information Service Element

CEMM Compaq Expanded Memory Manager

CEMN Centre d'Enseignement de la Météorologie Nationale (France)

CEMOS Central European Motivation Seminar

CEMP Carbon-Enhanced Metal-Poor (star) • CCAMLR Ecosystem Monitoring Program

CEMR Conférence Européenne des Ministres de la Recherche • Contractor Estimating Methods Review • Council of European Municipalities and Regions

CEMS Central Electronic Management System • Centre d'Équipement en Matériel Scientifique • Community of European Management Schools • Constituent Electronic Mail System • Conversion Electron Mössbauer Spectroscopy • Conversions-Elektronen-Mössbauer Spektroskopie

CEMT Center for Electronic Messaging Technologies • Command Equipment Management Team • Conférence Européenne des Ministres des Transports

Cen Centaurus-Centauri (constellation) • Center • Central • Centre • Century

CEN Center • Center for Electronic Materials • Centre d'Étude de la Neige • Centre d'Études Nucléaires • Centre Européen de Normalisation • Ciências Exactas e Naturais • Ciudad Obregón (Mexico) (location identifier) • Comité Estatal de Normalización • Comité Européen de Normalisation (*) • Comité Européen des Normes • Communications - Electronics - Navigation • Computational Electronics and Nanoelectronics • Cygnus Egg Nebula • RAL (Starlink internal code)

CeNA Cellular Network Analyzer

CENA Centre d'Études de la Navigation Aérienne (France) • Climatologie Étendue des Nuages et des Aérosols • Coopération d'Étalonnage Nord-Américaine

CENAM Centro Nacional de Metrología (Mexico)

CENB Centre d'Études Nucléaires de Bruyères-le-Châtel (France)

CENBG Centre d'Études Nucléaires de Bordeaux Gradignan (France)

CENCER Comité Européen de Normalisation - Certification

CEND Compton-Emission Neutron Detector

CENDIC Centro Nacional de Documentación e Información Científica

CENDL Chinese Evaluated Nuclear Data Library (China-PRC)

CENDOC Centralized Documentation

CENELEC Comité Européen de Normalisation Électrotechnique (*) • Comité Européen des Normes Électrotechniques

CENFAR Centre d'Études Nucléaires de Fontenay-aux-Roses (CEA)

CENG Centre d'Études Nucléaires de Grenoble (CEA)

CENIC Centro Nacional de Investigaciones Científicas

CENIDE Centro Nacional de Investigaciones para el Desarollo de la Educación (Spain)

CEnIT Center for Entrepreneurship in Information Technology (USA-CA)

CENL Council of European National Libraries

CENNA Convention on Early Notification of a Nuclear Accident

CENR Committee on Environmental and Natural Resources • Committee on the Environment and Natural Resources

Cens Censor • Censorship

CENS Centre d'Études Nucléaires de Saclay (CEA)

CenSIS Centre for Spatial Information Studies

Cent Cantavo (currency) • Center • Centesimo (currency) • Centigrade • Centigrado • Centime • Central • Centre • Centrifugal • Centro • Centum • Century

CENT Center

CENTAR Code for Extended Non-linear Transient Analysis of Reactors

Centig Centigrade

CENTO Central Treaty Organization

CENTORA Centre des Technologies Optiques en Rhône-Alpes (France)

CEnTp Engineering Thermophysics China (periodical code)

CENTR Center • Central

CENTREX Central Exchange

CENTX Central Telephone Exchange

CEO Catalyzed Electrochemical Oxidation • Celestial Ephemeris Origin • Center for Earth Observation • Centre Européen de l'Orgue • Chief Education Officer • Chief Engineer's Office • Chief Executive Officer • Combination of Expert Opinions • Comité Européen de l'Outillage • Communications-Electronics Officer • Confederation of Employee Organizations • Corporate Executive Officer • Waco Kungo (Angola) (location identifier)

CEOI Central European Olympiad in Informatics

CEONet Canadian Earth Observation Network

CEOP Conditional End Of Page • Coordinated Enhanced Observing Period

CEOPS Civil Engineering Operational Planning System

CEOS Committee on Earth Observation Satellites • Corporate Equal Opportunity Specialist

CEOSR Center for Earth Observing and Space Research (GMU)

CEOT Communications Equipment Operator Trainer

Cep Cepheus-Cephei (constellation)

CEP Cable Entrance Panel • Camera Evaluation Program • Capability Evaluation Plan • Career Enhancement Plan • CEA Event Pipeline • Celestial Ephemeris Pole • Center for Environmental Prediction • Central East Pacific • Centre d'Éducation Professionnelle • Centre d'Exposition Permanente • Certificat d'Études Primaires • Certificat d'Études Professionnelles • Circle End Point • Circle of Equal Probability • Circular Error Probability • Circular Error Probable (*) • Civil Emergency Planning • Civil Engineering Package • Civil Engineering Plan • Codigo de Endereço Postal • Color Evaluation Program • Comité Ecumenico de Proyectos • Committee for Environmental

Protection • Committee on Environmental Protection • Common Electronics Parts • Community Education Programme • Component Error Propagation • Comprehensive Event Pipeline • Computed Ephemeris Position • Computer Entry Punch • Concepción (Bolivia) (location identifier) • Concept Evaluation Phase • Concept Evaluation Plan • Concept Evaluation Program • Concept Experimentation Program • Concept Exploration Program • Conduction Electron Polarization • Configuration Evaluation Plan • Connection End Point • Consejo Electoral Provisional • Constraint Exploiter Process • Contextual Enhancement Processor • Continuous Estimation Program • Continuous Evaluation Program • Contract Estimating and Pricing • Contract Execution Plan • Contract Evaluation Plot • Contrat d'Études Prospectives • Cooperative Engagement Processor • Council on Economic Priorities • Cylindrical Electrostatic Probe

CEPA Canadian Environmental Protection Act (Canada) • Centre d'Études de Physiologie Appliquée (France) • Commonwealth Environment Protection Agency • Coupled Electron Pair Approximation

CEPAC COSTEP/ERNE Particle Analyser Collaboration

CEPAL Comisión Económica (de las Naciones Unidas) Para América Latina (y el Caribe)

CEPD Committee on Education and Professional Development

CEPE Centre d'Écologie et de Physiologie Énergétique (France) • Compagnie d'Électricité et de Piézo-Électricité • Cylindrical Electrostatic Probe Experiment

CEPES Centre Européen pour l'Enseignement Supérieur • Comité Européen pour le Progrès Économique et Social

CEPEX Central Equatorial Pacific Experiment

CEPG Command Execution Planning Group

CEPGL Communauté Économique des Pays des Grands Lacs

CEPH Centre d'Étude du Polymorphisme Humain (France)

CEPHAG Centre d'Études des Phénomènes Aléatoires Géophysiques

CEPI Center for Earth and Planetary Interiors

CEPII Centre d'Études Prospectives et d'Informations Internationales (France)

CEPIS Council of European Professional Informatics Societies

CEPMART Centre for Promotion of Mathematical Research in Thailand

CEPME Crédit d'Équipement des Petites et Moyennes Entreprises (France)

CEPMMT Centre Européen de Prévision Météorologique à Moyen Terme

CEPO Center Education Program Officer • Certified Exchange Point Operator

CEPP Committee on Elementary Particle Physics

CEPPAD Comprehensive Energetic Particle and Pitch Angle Distribution • Comprehensive Energetic Particle Pitch Angle Distribution

CEPr Centre d'Essai des Propulseurs

CEPR Centre for Economic Policy Research • Compromising Emanation Performance Requirement • Concept Experimentation Program Review

CEPS Case Evaluation Profile System • Center for Earth and Planetary Studies (USA-DC) • Centre d'Étude et de Prospective Stratégique (France) • Centre for European Policy Studies

CEPT Comité d'Études Pédagogiques et Techniques • Comprehensive Electrical Performance Test • Conférence Européenne des Postes et Télécommunications

CEPUP Chemical Education for Public Understanding Program

CEQ Cannes (France) (location identifier) • Cost Estimate Quality • Council on Environmental Quality

CEQA California Environmental Quality Act (USA)

Cer Ceramic

CER Cell Error Rate • Cell Error Ratio • Centre des Énergies Renouvelables • Ceramic Electrochemical Reactor • Certification Evaluation Review • Character Error Rate • Charge Exchange Recombination • Cherbourg (France) (location identifier) • Civil Engineering Report • Closer Economic Relation • Command and Expenditure Report • Communications and External Relations • Complete Engineering Release • Component Engineering Request • Comprehensive Evaluation Report • Consultative Environmental Review • Contactless Electroreflectance • Corotation Eccentricity Resonance • Cost Estimating Relationship

CERA Centre d'Étude et de Recherche pour l'Automobile (France)

CERAG Centre d'Études et de Recherches Appliquées à la Gestion (France)

CERAP Central Radar Approach

CERB Concurrent Engineering Review Board

CERBM Centre Européen de Recherche en Biologie et en Médecine

CERC Center for Energy Research Computation • Central Engine Room Control • Concurrent Engineering Review Center

CERCA Centre de Recherche en Calcul Appliqué

CERCG Centre d'Études et de Réalisations Cartographiques et Géographiques (France)

CERCLA Comprehensive Environmental Response, Compensation, and Liability Act (USA)

CERCLIS Comprehensive Environmental Response, Compensation and Liability Information System (USA)

CERCO Comité Européen de Responsables de la Cartographie Officielle

CERCOA Centre d'Étude et de Recherche en Chimie Organique Appliquée

CERD Catalog of Environmental Resource Data

CERDU Central European Research and Development Unit

CERE Computer Entry and Readout Equipment

CEREG Centre d'Études et de Recherches Éco-Géographiques (France)

CEREGE Centre Européen de Recherche et d'Enseignement des Géosciences de l'Environnement (France)

CEREM Centre d'Études et de Recherches sur les Matériaux (France)

CEREN Centre d'Études et de Recherches sur l'Énergie (France)

CEReS Center for Environmental Remote Sensing

CERES California Environmental Resources Evaluation System • Center for Educational Resources • Center for Research Support • Centre d'Études et de Recherches des Enseignements Supérieurs • Clouds and Earth Radiant Energy System • Clouds Earth Radiant and Energy System (experiment) • Consortium for European Research on Extragalactic Surveys • Crop Environment Resource Synthesis

CERESI Cellule de Recherche en Sciences de l'Information

CERESIS Centro Regional de Seismología

CERET Centre Européen de Réflexion et d'Étude en Thermodynamique

CERF Central Emergency Revolving Fund (UNO) • Compresseur d'Études et de Recherches Fondamentales

CERFA Community Environmental Response Facilitation Act (USA)

CERFACS Centre Européen de Recherche et de Formation Avancée en Calcul Scientifique (France)

CERFnet California Education and Research Federation Network

CERGA Centre d'Études et de Recherches en Géodynamique et Astrométrie (France) (*) • Centre d'Études et de Recherches Géodynamiques et Astronomiques (France)

CERGOP Central Europe Regional Geodynamics Project

CERI Center for Educational Research and Innovation • Center for Environmental Research Information • Centre d'Enseignement et de Recherche en Informatique (France) • Centre d'Études des Relations Internationales

CERIF Common European Research Information Format

CERII Centre d'Études et de Recherche sur l'Information et l'Innovation

CERIS Centre d'Étude des Relations Internationales et Stratégiques (Belgium)

CERIST Centre d'Études et de Recherche d'Information Scientifique et Technique

CERL Civil Engineering Research Laboratory • Computer-based Education Research Laboratory • Consortium of European Research Libraries • Construction Engineering Research Laboratory • Corvallis Environmental Research Laboratory

CERM Coordinated Emergency Response Measures

CERMA Centre d'Études et de Recherches en Médecine Aérospatiale (France)

CERMAV Centre de Recherche sur les Macromolécules Végétales (France)

CERMET Ceramic-to-Metal

CERMICS Centre d'Enseignement et de Recherche en Mathématiques, Informatique et Calcul Scientifique

CERN Centre Européen pour la Recherche Nucléaire (*) • Conseil Européen pour la Recherche Nucléaire

CERP Center for Educational Research and Practice (Japan) • Centre Européen des Relations Publiques • Confédération Européenne des Relations Publiques • Cooperative Ecological Research Project

CERPEMA Centre de Recherche sur les Propriétés Électroniques de Matériaux Avancés (Canada-QC)

CERPET Centre de Recherches Pédagogiques pour les Enseignements Technologiques (France)

CERPS Consolidated and Expenditure Reporting and Processing System

CERR Centre d'Écologie des Ressources Renouvelables (France)

Cerraj Cerrajería

CERRO Control European Regional Research Organization

CERS Carreer Evaluation and Reporting System • Carrier Evaluation and Reporting System • Centre Européen de Recherches Spatiales (now merged with CECLES into ASE) • Commission pour l'Éducation, la Recherche et la Science • Compact Extragalactic Radio Source • Cost-Estimating Relationship

CERSI Centre d'Études et de Recherche en Sciences de l'Information (France)

Cert Certificate • Certificated • Certification • Certified • Certify

CERT Centre d'Études et de Recherche de Toulouse (ONERA) • Centre d'Études et de Recherche Technique • Certification • Combined Environmental Reliability Test • Committee on Energy Research and Technology • Computer Emergency Response Team

CERTA Centre d'Expertise, de Réponse et des Traitements des Attaques (informatiques) (France)

CERTI Centro Regional de Tecnologia em Informatica

Certif Certificate • Certificated

CERTIF Certification

Cert. Inv. Certified Invoice

CES Center for Energy Studies • Central Electronics System • Centre d'Études Sociologiques • Certificat d'Études Spéciales • Certificat d'Études Supérieures • Cessnock (Australia-NSW) (location identifier) • Chemically Enhanced Self-sputtering • Circuit Emulation Service • Civil Engineering Support • Closed Ecological System • Coast Earth Station • Collège d'Enseignement Secondaire • Combined Earth Sensor • Comité d'Environnement Social • Comité Économique et Social • Comité Électrotechnique Suisse (Switzerland) • Committee on Earth Sciences • Committee on Earth Studies • Commonwealth Employment Service • Communications Equipment Support • Community Energy System • Computational Engineering and Science • Computer Engineering Service • Concept Evaluation Study • Confédération Européenne des Syndicats • Conference for European Statisticians • Conference for European Statistics • Conical Earth Sensor • Connection Event Sequence • Conseil de l'Enseignement Supérieur • Conseil Économique et Social • Consejo Económico y Social • Constant Elasticity of Substitution • Consumer Electronics Show (USA-NV) • Consumers Electronic Show • Contrat Emploi-Solidarité • Control Electronics Section • Control Electronics System • Controlled Environmental System • Controlled Environment System • Cooperative Extension Service • Coordinated Evaluation System • Corpus Encoding Standard • Cost Effectiveness Study • Cost Element Structure • Cost-Estimating System • Coudé Echelle Spectrograph • Coudé Echelle Spectrometer • Crew Escape System • Cryospheric Environment Simulator

CESA Central Electronics Subassembly • Commission des Études Spatiales Amont • Conseil Économique et Social d'Alsace (France) • Construcciones Españolas de Sistemas Aéronauticas (Spain)

CESAC Communications Electronics Scheme Accounting Control

CESAME Centre Étatique de Simulation Avancée et de Modélisation Électromagnétique (France)

CESAP Commission Économique et Sociale pour l'Asie et le Pacifique (ONU)

CESAR Canadian Expedition to Study Alpha Ridge • Capsule Escape and Survival Applied Research • Center for Engineering Science Advanced Research (USA-TN) • Center for Engineering Systems Advanced Research • Central European Satellite for Advanced Research • Committee on Educational Strategy And Resource

CESB Centre d'Études Supérieures de la Banque (France)

CESBIO Centre d'Études Spatiales de la Biosphère (France)

CESC Conferencia Europea de Seguridad y Cooperación

CESDIS Center of Excellence for Space Data and Information Sciences (USA-MD)

CESDS Central Environmental Satellite Data System

CESE Center for Earth Science and Engineering • Civil Engineering Support Equipment • Coalition for Earth Science Education • Coalition of Earth Science Education • Communications Equipment Support Element

CESEPUP Committee on Science, Engineering and Public Policy

CESG Communications and Electronics Security Group • Communications Electronics Security Group • Communications Equipment Support Group • Cryogenic Electrically Suspended Gyroscope

CESGA Centro de Supercomputación de Galicia (Spain)

CESHR Center for Electrochemical Systems and Hydrogen Research

CESI Centre d'Études Supérieures Industrielles

CESID Centro Superior de Información para la Defensa (Spain)

CESLA Cost Estimate Logistic Support Analysis

CESLS Constant Energy Synchronous Luminescence Spectrometry

CESN Canadian Electronic Scholarly Network (Canada)

CESO Communications-Electronics Staff Officer • Council of Engineering and Scientific Organizations

CESP Centre de Supercomputació de Catalunya (Spain) • Centre d'Étude des Supports Publicitaires (France) • Civil Engineering Support Plan • Civil Engineering Support Program • Command Element Specific Program

CESPG Civil Engineering Support Plan Group

CESQG Conditionally Exempt Small Quantity Generator

CESR Center for Earth System Research • Centre d'Étude Spatiale des Rayonnements (France) • Communications-Electronics Systems Requirement • Conduction Electron Spin Resonance • Conseil Économique et Social Régional (France) • Contractor Estimating System Review • Cornell Electron-positron Storage Ring (USA-NY)

CESRA Committee of European Solar Radio Astronomers

CESRB Communications-Electronics Systems Requirements Board

CESRD Communications-Electronics Systems Requirements Document

CESSDA Council of European Social Science Data Archives

CEST Center for Exploitation of Science and Technology • Central European Saving Time • Central European Summer Time • Communications Gillies Inc. Electronic Systems in Transportation (periodical code)

CESTA Centre d'Étude des Systèmes et des Technologies Avancées

CESU Cooperative Ecosystems Studies Unit (USA)

Cet Cetus-Ceti (constellation)

CET Capsule Elapsed Time • Central Earth Terminal • Central European Time • Centre d'Études Techniques • Certificat Élémentaire Technique • Cholet - Le Pontreau (France) (location identifier) • Civilian Engineering Team • Clutter Equivalent Target • Cockpit Evaluation Team • Collège d'Enseignement Technique • Combined Effects Test • Combined Element Test • Combined Environment Test • Commission on Education and Training • Common Efficiency Team • Common External Tariff • Compte Épargne-Temps • Concern for Europe's To-morrow • Concurrent Engineering Team • Conseiller d'Enseignement Technologique • Constant Elasticity of Transformation • Contract Engineering Task • Contractor Engineering Task • Corrected Effective Temperature • Cost Evaluation Team • Council for Educational Technology • Creative and Emerging Technology • Cumulative Elapsed Time

CETA Carbon Emissions Trajectory Assessment • Centre d'Étalonnage Agréé • Chinese-English Translation Assistance • Comprehensive Education and Training Administration • Comprehensive Employment and Training Act (USA) • Contractor Employee Technical Assistance • Corrosion Evaluation and Test Area • Cost-Effective Training Analysis • Crew and Equipment Translation Aid (*) • Crew Equipment Translation Aid

CETAF Consortium of European Taxonomic Facilities

CETD Concept Exploration and Technology Demonstration • Cross-Enterprise Technology Development

CETDP Cross-Enterprise Technology Development Program (NASA)

CETEL Cycle d'Enseignement de la Télédétection

CETEX Ceramic Tile Experiment • Committee on Extraterrestrial Exploration • Contamination by Extraterrestrial Exploration • Contamination of Extraterrestrial Exploration

CETF CalTech Environment Task Force • Critical Evaluation Task Force

CETH Center for the Electronic Texts in the Humanities

CETI Communication with Extraterrestrial Intelligence

CETIM Centre Technique des Industries Mécaniques (France)

CETIS Centre Européen de Traitement de l'Information Scientifique

CETM Collège d'Enseignement Technique Maritime • Customized Electronic Technology Magazine

CETO Centro de Ciências e Tecnologias Ópticas

CETP Collaborative for Excellence in Teacher Preparation (NSF)

cet. par. Ceteris Paribus

Cetr Cetrería

CETR Center for Explosives Technology Research

CETS Chemical Energy Transmission System • Commission on Engineering and Technical Systems • Conference Européenne des Télécommunications par Satellite • Contract Engineering and Technical Services • Contractor Engineering and Technical Services

CETSP Contractor Engineering Technical Services Program

CEU Camera Electronics Unit • Central Electronic Unit • Clemson - Oconee County (USA-SC) (location identifier) • Command Execute Unit • Commercial Entertainment Unit • Commission of the European Union • Computer Electronics Unit • Continuing Education Unit • Control Electronics Unit • Cooling Equipment Unit • Coupler Electronics Unit

CEUS Central and Eastern United States

CEV Centre d'Essais en Vol (France) • Connersville Mettle (USA-IN) (location identifier) • Controlled Environment Vault • Corona Extinction Voltage • Cryogenic Explosive Valve

CEVAT Combined Environmental, Vibration, Acceleration, Temperature

CEVI Common Equipment Voltage Indicator

CEVP Countermeasures Evaluation and Validation Project (NASA)

CEVT Contingency Extravehicular Transfer

CEVU Conseil des Études et de la Vie Universitaire

CEW Crestview - Bob Sikes (USA-FL) (location identifier)

CEWES Corps of Engineers Waterways Experiment Station

CEWG Communications Electronics Working Group

CEWS Communications Early Warning System

CEX Chena Hot Springs (USA-AK) (location identifier) • Cyclic Executive

Cey Ceylon

CEY Murray - Calloway County (USA-KY) (location identifier)

Ceyl Ceylon

CEZ Cortez (USA-CO) (location identifier)

CE1 Close Encounters of the First Kind

CE2 Close Encounters of the Second Kind

CE3 Close Encounters of the Third Kind

CE4 Close Encounters of the Fourth Kind

c/f Carried Forward • Cost and Freight

cf Confer • Cubic Foot • cum figuriis

Cf Californium • Confer

C&F Circulation & Fulfillment • Cost & Freight

C/F Center Frequency

CF Calibration Factor • Call Forwarding • Canada (international aircraft registration code) • Cannot Find • Capacity Factor • Cape Faint • Carbon Film • Carbon Filtered • Carbon Fraction • Carboxylfluorescine • Career Field •

Carriage Forward • Carried Forward • Carrier-Free • Carrier Frequency • Carry Flag • Carry Forward • Catalog of Faint (objects) • Cathode Follower • Causeway Ferry • Center Focus • Center Forward • Center Frequency • Center of Flotation • Central African Republic (ISO country code) • Central File • Central Formation • Centrally Funded • Centrifugal Force • Centripetal Force • Centro de Fundaciones (Spain) (now AEF) • Certainty Factor • Change in Formula • Change to Frequency • Characteristic Function • Chemin de Fer • Chèque Formation • Circuit Finder • Circulation and Fulfillment • Cloud Fraction • Cluster Feature • Coherence Function • Coin First • Cold Front • Collaborative Filtering • Color Filter • Color Function • Communication Factor • Communications Facility • Communications Flight • Commutation Factor • Comorian Franc (currency) • Compartmented Facility • Compartment Facility • Compensation Fee • Completion Force • Composition Filter • Computers and Fluids (periodical code) • Concentration Factor • Concept Facility • Concept Feasibility • Concept Formulation • Concrete Floor • Confidence Factor • Congé de Formation • Consequence of Failure • Constant Frequency • Consumption Factor • Content Filtering • Context-Focused • Context Free • Continuous Flow • Contract Formulation • Contribution Function • Control Facility • Control Function • Controlled Facility • Convergence Function • Conversion Factor • Cooling Flow • Coordination Function • Copy Furnished • Corner Fitting • Correction Factor • Correlation Factor • Corresponding Fellow • Cost and Freight • Count Forward • Coupling Facility • Crew Factor • Crowding Factor • Crystal Field • Cubic Feet • Cumulative Frequency • Customer Facility • République Centrafricaine (code ISO de pays)

CfA Center for Astrophysics

CF&A Chief of Finance & Accounting

CFA Call Forward Area • Centre de Formation d'Apprentis (France) • Centrifugal Fast Analyzer • Chartered Financial Analyst • Chief of Finance and Accounting • Closed Form Analysis • Cloud Fractional Absorptance • Coffee Point (USA-AK) (location identifier) • Cognizant Field Activity • Communauté Financière Africaine • Communauté Française d'Afrique • Communications Functional Area • Compact Flash Association • Concept Feasibility Analysis • Confirmatory Factor Analysis • Congé de Fin d'Activité • Continuous Flow Analysis • Controlled Firing Area • Correspondence Factor Analysis • Correspondence Factorial Analysis • Cross-Field Amplifier • Cross-Fielded Amplifier • Current Files Area

CFAA Computer Fraud and Abuse Act (USA)

CFAC Constant-Frequency Alternating Current

CFAD Contoured Frequency by Altitude Diagrams

CFADC Controlled-Fusion Atomic Data Center

CFAE Contractor-Furnished Aeronautical Equipment • Contractor-Furnished Airborne Equipment • Contractor-Furnished Aircraft Equipment

CFAI Centre de Formation à l'Apprentissage Industriel

CFAM Cogeneration Feasibility Analysis Model

CfAO Center for Adaptive Optics

CFAO Conception de la Fabrication Assistée par Ordinateur • Conception et Fabrication Assistées par Ordinateur (*) • Contrôle de la Fabrication Assistée par Ordinateur

CfAPr Center for Astrophysics Preprints (periodical code)

CfAR Center for Automation Research

CFAR Constant False Alarm Rate

CFAS Cape Fear Astronomical Society (USA-NC) • Center For Architecture and Standards • Central Florida Astronomical Society (USA-FL)

CFAT Carnegie Foundation for the Advancement of Teaching

CFB Calibration Fiber • Call Forwarding Busy • Carey Foster Bridge • Catalina Flying Boat • Change Franc Pacifique • Cipher Feedback • Circulating Fluid Bed • Configurable Function Block • Contract Financial Brief

CFBC Circulating Fluid Bed Combustion

CFBS Canadian Federation of Biological Societies (Canada)

CFC Carbon Fiber Composite • Carbon Fiber Compound • Central Field Camera • Central Fire Control • Central Flow Control • Centre de Formation Continue • Centre Français du Copyright (France) • Chlorinated Fluorocarbon • Chlorofluorocarbon • Circulating Fluidized Combustion • Combined Federal Campaign (USA) • Comisión Federal de la Competencia (Mexico) • Comité Français de Cartographie • Complex Facility Console • Component Finance Center • Congés de Fin de Carrière • Conseiller en Formation Continue (France) • Continuous Flow Cryostat • Cube à Faces Centrées • Culler - Facer - Canceler

CFCA Communications Fraud Control Association

CFCB Computer Format Control Buffer

CFCC Census Feature Class Code • Contract File Content Checklist • Coupling Facility Control Code

CFCE Centre Français du Commerce Extérieur (France)

CFCF Central Flow Control Facility

CfCP Center for Cosmological Physics (USA)

CFCS Central Flow Control Service

cfd Cubic Feet per Day

CFD Bryan Coulter (USA-TX) (location identifier) • Career Field Designate • Career Field Designation • Centre de Fabrication et de Distribution • Common File Device • Composite Fluffy Dust (model) • Computational Fluid Dynamics • Continuous Flow Diffusion • Control Flow Diagram • Cross-Functional Development • Cubic Feet per Day • Cumulative Frequency Distribution • Current Force Design

CFDA Catalog of Federal Domestic Assistance (USA) • Component Functional Data Administrator • Computational Fluid Dynamics Analysis • Cross-Functional Development Assignment

CFDAA CFD Advances Applications (periodical code)

CFDCC Computational Fluid Dynamics Community Club

CFDF Canada-France Deep Field

CFDP Calibration File Derivation Procedure • Coherent File Distribution Protocol

CfDS Campaign for Dark Skies

CFDS Centralized Fault Display System

CFDT Cartesian Full Disk Telescope

CFE Cathode Flicker Effect • Center For Engineering • Central Framework Executive • Centre de Formalités des Entreprises (France) • Clermont-Ferrand Aulnat (France) (location identifier) • College of Further Education • Consolidated Front End • Contractor-Furnished Equipment • Control Fuel Element • Control Function Equipment • Controlled Flash Evaporation • Critical Function Equipment • Customer-Furnished Equipment

CFEA Collective Front-End Analysis

CFEM Combined Fates and Effects Model

CFEN Certificat de Fin d'Études Normales • Contractor-Furnished Equipment Notice

CFEPP Comité Français pour l'Étude des Phénomènes Paranormaux (France)

CFES Certificat de Fin d'Études Secondaires • Continuous Flow Electrophoresis System

CFESP Certificat de Fin d'Études Secondaires Professionnelles

CFET Centrifuge-based Flight Environment Trainer

CFETP Career Field Education and Training Plan

CFF Canfunfo (Angola) (location identifier) • Carbon Fraction of Fuel • Cartographic Feature File • Chemins de Fer

Fédéraux (Switzerland) • Columbus Free Flyer • Commission on Fossil Fuels • Committee on Finance and Fundraising • Compact Font Format • Compressible Flow Facility • Conical Flow Field • Counter Flip-Flop • Critical Flicker Frequency • Critical Fusion Frequency

CFFT Columbus Free Flying Laboratory

CFFTP Canadian Fusion Fuels Technology Project

cfg Configuration Data (file extension)

CFG Caine Farber and Gordon • Certificat de Formation Générale • Cienfuegos (Cuba) (location identifier) • Computing Facilities Group • Configuration • Constant-Frequency Generator • Context-Free Grammar

CFGRS CalTech Faint Galaxy Redshift Survey

CFGSRV Configuration Server

CFGT Chinese Future Giant Telescope • Comité Français des Grands Télescopes (France) (*) • Conseil Français des Grands Télescopes (France)

CFGU Common Frequency Generation Unit

cfh Cubic Feet per Hour

CFH Carefree Handling • Clifton Hills (Australia-SA) (location identifier) • Cubic Feet per Hour

CFHT Canada-France-Hawaii Telescope

CFHTLS Canada-France-Hawaii Telescope Legacy Survey

CfI Center for Inquiry

CFI CAD Framework Initiative • Camfield (Australia-NT) (location identifier) • Canada Foundation for Innovation • Canadian Foundation for Innovation • Card Format Identifier • Certified Flight Instructor • Chief Flying Instructor • Cloud-Free Interval • Comprehensive Flare Index • Contractor Final Inspection • Contractor-Furnished Information • Corporación Financiera Internacional • Cost - Freight - Insurance • Court of First Instance • Critical Fault Identifier

CFIA Component Failure Impact Analysis

CFIB Corporate Functional Integration Board

CFI&I Center For Integration & Interoperability

CFII Center For Integration and Interoperability

CFIRB Cosmic Far-Infrared Background

CFIS Conflict-Free Instrument Schedule

CFISU Canadian Foundation for the International Space University (Canada)

CFIT Controlled Flight Into Terrain

CFIUS Committee on Foreign Investment in the United States

CFK Carbonfaserverstärkte Kunststoffe • Chloorfluorkoolwaterstof • Color Find Key

cfl Corel Flow (file extension)

CFL Calibrated Focal Length • Close Focus Lens • Color-Flavor Locked • Compact Fluorescence Lamp • Compact Fluorescent Lamp • Compact Fluorescent Lighting • Competent Familiarity Loading • Context-Free Language • Control Flag • Courant - Friedrichs - Levy (condition)

CFLOS Cloud-Free Line Of Sight

CFLP Colloquium on Functional and Logic Programming

CFLS Canada-France Legacy Survey

CFLSE Critical Fluid Light Scattering Experiment

CFLSI Critical Fluid Light Scattering Instrument

cfm Confirm • Confirmed • Cubic Feet per Minute

Cfm Confirm • Confirmation

C/F/M Corrosion/Film/Moisture

CFM Candle Flames in Microgravity • Cathode Follower Mixer • Center Frequency Modulation • Chlorofluoromethane • Closed Flux Memory • Code Fragment Manager • Collège Français de Métrologie (France) • Common Functional Model • Computer Facilities Management • Configuration Management • Conformally Flat Model • Contingency For Movement • Contractor Financial Management • Contractor-Furnished Material • Contractor-Furnished Materiel • Copy For Me • Cryogen-Free Magnet • Crystal Frequency Multiplier • Cubic Feet per Minute • Customer-Furnished Material

CFMA Central Financial Management Activities • Centre de Formation National d'Avignon (France)

CFMF Crip Flow Management Facility

CFMI Circular Fluorescent Microscope Illuminator

CFMO Construction and Facilities Management Office

CFMS Central Financial Management System • Centralized Fuels Management System • Chained File Management System

CFMU Central Flow Management Unit

CFMW Case Financial Management Worksheet

CFN Carrickfinn (Ireland) (location identifier) • Center for Functional Nanostructures (Germany) • Confine • Donegal (Ireland) (location identifier)

CFNR Call Forwarding - No Reply

CFO Calling For Orders • Central Facilities Operations • Central Forecasting Office • Ceramics Fiber Optics • Channel For Orders • Chief Financial Officer • Chief Flight Operation • Cognizant Field Office • Comité Français d'Organisation • Commissioning and Fitting Out • Complex Facility Operator • Confreza (Brazil-MT) (location identifier) • Critical Flashover Voltage

CFOA Chief Financial Officer's Act (USA)

CFONB Comité Français d'Organisation et de Normalisation Bancaire (France)

CFOP Club des Fibres Optiques Plastiques

CFOR Call Forwarding On Reply • Conversational Fortran

CFOV Center of the Field Of View

cFP Core Fundamental Plane

CFP Call For Papers • Call For Participation • Carpentaria Downs (Australia-QLD) (location identifier) • Carrier Frequency Pulse • Center of Filtering and Plotting • Centre de Formation des Professeurs • Centre de Formation Pédagogique • Centre de Formation Permanente • Centre de Formation Professionnelle • Certificat de Formation Professionnelle • Certified Financial Planner • Certified Financial Planning • Code of Federal Regulations (USA) • Coefficient of Fractional Parentage • Cold Frontal Passage • Cold Front Passage • Combined Filter and Plot • Communauté Française du Pacifique • Communauté Financière du Pacifique • Computerized Flight Plan • Computers - Freedom - Privacy • Concept Formulation Package • Concept Formulation Process • Conceptual Flight Profile • Contractor-Furnished Product • Contractor-Furnished Property • Franc (French Pacific Islands) (ISO currency code)

CfPA Center for Particle Astrophysics (USA-CA)

cfpd Cubic Feet per Day

cfph Cubic Feet per Hour

CFPL Computational Fluid Physics Laboratory

CFpM Cubic Feet per Minute

CFPO Canadian Forces Post Office (Canada)

CFPP Centre de Formation Professionnelle et de Perfectionnement • Centre de Formation Professionnelle Permanente

CFPPA Comité Français de Pilotage du Plan d'Action

cfps Cubic Feet per Second

CFPs Calls For Papers

CFPS Collaborative Flight Plan Services

CFPT Center for Particle Tracking • Centre de Formation des Personnels Territoriaux (France) • Centre de Formation des Professeurs Techniques (France) • Centre de Formation Professionnelle et Technique • Cold Fusion Performance Tuning

Cfr Chauffeur • Confronta

CFR Caen Carpiquet (France) (location identifier) • Call For Release • Cambridge Fast Ring • Carbon-Film Resistor • Carbon Formation Reactor • Catastrophic Failure Rate • Centre des Faibles Radioactivités (France) • Centre for Field Research • Cluster Formation Rate • Code of Federal Regulations (USA) • Commercial Fast Reactor • Computerized Facial Recognition • Computerized Fault Reporting • Connectionist Knowledge Representation • Constant Frame Rate • Contractor Funds Report • Coupling Facility Receiver • Cumulative Failure Rate

CFRD Confidential, Formerly Restricted Data

CRFE Contract Financial Requirements Estimate

CFRM Coupling Facility Resource Management • Coupling Facility Resource Management • Coupling Facility Resource Manager

CFRO Comité Français de Recherche Opérationnelle

CFRP Carbon Fibre Reinforced Plastic • Conceptual Framework for Reuse Planners • Conceptual Framework for Reuse Processes

CFRS Canada-France Redshift Survey • Computerized Fault Reporting System

cfs Coste - Flete - Seguro • Cubic Feet per Second

CFS Caching File System • Carrier Frequency Shift • Cassegrain Feed System • Center For Standards • Center Frequency Stabilization • Central Frequency Synthesizer • Cesium Feed System • Chronic Fatigue Syndrome • Coffs Harbour (Australia-NSW) (location identifier) • Coherent Frequency Synthesizer • Cold-Finished Steel • Collection and Forwarding System • Combination Frequency Signal • Combined File Search • Committee on Fundamental Science • Common File System • Compact Flat Spectrum • Completely Finished Set • Component Failure Summary • Composite Feed System • Computerized Forwarding System • Conflict-Free Schedule • Consolidated Front Section • Contract Field Services • Contract Financial Status • Coste - Flete - Seguro • Coudé Feed Spectrograph • Coupling Facility Sender • Cryogenic Fluid Storage • Cubic Feet per Second

CFSC Community Family Support Center • Cryogenic Fluid Storage Container

CFSD Citizens For Space Demilitarization • Critical Function Sequence Diagram

CFSE Crystal-Field Stabilization Energy

CSFK Coherent Frequency Shift Keying

CSFM Communication Finite State Machine

CFSP Common Foreign and Security Policy (EU)

CfSR Center for Solar Research (NJIT)

CFSR Center For Solar Research (NJIT) • Contract Funds Status Report • Contract Funds Summary Report • Contract Funds Status Review

CFSS Combined File Search Strategy

CFSSI Centre de Formation à la Sécurité des Systèmes d'Information

CFST Context-Free Syntactical Translator

CFSTI Clearinghouse for Federal Scientific and Technical Information

Cft Craft

CfT Call for Tender

CFT Captive Flight Test • Captive Flight Trainer • Centre de Formation Technologique • Chirp Fourier Transform • Clifton Morenci (USA-AZ) (location identifier) • Cluster Fault Tolerance • Cognitive Flexibility Theory • Common Facilities Test • Companion Fault Tolerance • Complement Fixation Test • Complex Fourier Transform • Conformal Field Theory • Continuous Fourier Transport • Contract Field Team • Contract Field Technician • Contractor Field Team • Contractor Flight Test • Coudé Feed Telescope • Cross-File Transfer • Cross-Flow Turbine • Cross-Functional Team • Crystal Field Theory

CFTA Centre de Formation Technique Agréé

CFTC Commodity Futures Trading Commission • Commonwealth Fund for Technical Cooperation

CFTE Critical Fluid Thermal Equilibration

Cftmn Craftsman

CFU Call Forwarding Unconditional • Control Functional Unit • Corfu (Greece) (location identifier) • Kerkyra (Greece) (location identifier)

CFUR Continuous Flow Unstirred Reactor

CFUVM Call Forwarding Unconditional to Voice Mailbox • Club Français des Utilisateurs de VM

CFV Call For Votes • Carrier-Festverbindung • Coffeyville (USA-KS) (location identifier) • Continuous Flow Ventilation

CFVME Critical Fluid Viscosity Measurement Experiment

CFW Call Forward

CFWP Central Flow Weather Processor

CFWSU Central Flow Weather Service Unit

CFWU Central Flow Weather Unit

CFX Central Funds Exchange

CFY Clarify • Company Fiscal Year • Current Fiscal Year (*)

CFZ Critical Focus Zone • Critical Friendly Zone

CFZF Commercial Float Zone Facility • Commercial Float Zone Furnace

cg Centigram

Cg Center of Gravity • Congress

C&G Control & Guidance

CG Case Galaxy • Center of Gravity • Chairman's Guidance • Character Generator • Churyumov-Gerasimenko (comet) • Cloud to Ground • Cluster of Galaxies • Coast Guard • Cold Grain • Cometary Globule • Command Guidance • Commanding General • Committee on Governance • Communications Gateway • Compact Galaxy • Compact Group • Comptroller General • Computational Geometry • Computer-Generated • Computer Graphics • Computers and Games (conference) • Conceptual Graph • Condensed Graph • Congo (ISO country code) • Conjugate Gradient • Consolidated Guidance • Consul General • Contadora Group • Contrainte Grammaticale • Control Gate • Control Group • Coolant Gas • Cool Gas • Coordonnées Généralisées • Coulomb-Glauber • Cumulated Gain

CG&A Computer Graphics & Applications

CGA Call Global Area • Canadian Gas Association (Canada) • Canonical Genetic Algorithm • Centre de Gestion Agréé • Coast Guard Academy • Color Graphics Adapter • Compressed Gas Association • Comptabilité et Gestion Administrative • Computer Graphics Adapter • Computer Graphics and Applications • Craig (USA-AK) (location identifier) • Cultural Grant Aid

CGAC China General Aviation Corporation

CGAM Centre for Global Atmospheric Modeling • Condensed Graph Abstract Machine

CGAS Coast Guard Air Station • Cooled Grating Array Spectrometer

CGAU Cabin Gas Analysis Unit

CGB Circuit Group Blocking • Convert Gray to Binary • Cosmic Gamma-ray Background • Cuiabá - M. Rondon (Brazil-MT) (location identifier)

CGBA Circuit Group Blocking Acknowledgment • Commercial Generic Bioprocessing Apparatus

CGC Canadian Geoscience Council (Canada) • Cape Gloucester (Papua New Guinea) (location identifier) • CEMS

Graduate Conference • Clebsch-Gordan Coefficient • Climate and Global Change • Coast Guard College • Color Glass Condensate • Color Graphics Converter • Comité de Gestion et de Coordination • Command Guidance Computer • Commission Géologique du Canada • Committee on Global Change • Confédération Générale des Cadres • Continuous Global Coverage • Coût Global Client

CGCD Crossed-Grid Charge Detector

CGCG Catalog of Galaxies and Clusters of Galaxies

CGCM Coupled General Circulation Model

CGCP Canadian Global Change Programme • Climate and Global Change Program

CGCT Compagnie Générale de Constructions Téléphoniques

CGCU Centre de Gestion de la Charge Utile

CGD Changde (China-PRC) (location identifier) • Climate and Global Dynamics • Committee for Geophysical Data • Coplanar Grid Detector

CGDA Cross-link Gimbal Drive Assembly

CGDB Central Geographic Database

CGDI Canadian Geospatial Data Infrastructure

CGDS Computer Graphics Display System

Cge Carriage • Charge

CGE Cambridge (USA-MD) (location identifier) • Capillary Gel Electrophoresis • Chadwick-Goldhaber Effect • Common Graphics Environment • Compagnie Générale d'Électricité • Compagnie Générale des Eaux • Computable General Equilibrium • Conférence des Grandes Écoles (France) • Conseil Général des Études

CGED Committee on Geophysical Environmental Data

CGEN Character Generator

CGER Caisse Générale d'Épargne et de Retraite • Commission on Geosciences, Environment and Resources

CGES Center for Global Environmental Studies

CGF Cleveland - Cuyahoga County (USA-OH) (location identifier) • Color Guard Fiber • Command Generation Function • Computer-Generated Force • Computer-Generated Form • Crystal Growth Facility • Crystal Growth Furnace • Czech Government Flying Service

CGG Casiguran (Philippines) (location identifier) • Centro Galileo Galilei (CNAA) • Commission on Global Governance • Compact Group of Galaxies • Compagnie Générale de Géophysique • Compensated Geothermal Gradient • Contingency Gravity Gradient

CGGTTS CCDC Group on GPS Time Transfer Standards

CGH Cape of Good Hope (South Africa) • CHON Grain Halo • Computer-Generated Hologram • Computer-Generated Holographic • Concept Generalization Hierarchy • São Paulo - Congonhas (Brazil-SP) (location identifier)

CGHA Castel Gondolfo H-Alpha

CGHES Closed Geothermal Heat Exchange System

cgi Common Gateway Interface (file extension)

CGI Calibration Group Identifier • Canadian Glacier Inventory • Cape Girardeau (USA-MO) (location identifier) • Common Gateway Interface • Computer-Generated Image • Computer-Generated Imagery • Computer-Generated Imaging • Computer-Generated Interface • Computer Graphics Interface • Corrugated Galvanized Iron

CGIA Centre for Geographical Information and Analysis

CGIAR Consultative Group on International Agricultural Research

CGIC Canadian Geoscience Information Centre (Canada)

CGIM Computer Graphics and Imaging (conference)

CGIP Computer Graphics and Image Processing • Computer Graphics Image Processing (periodical code)

CGIS Canadian Geographical Information System • Continental Geographical Information System

CGIVS Computer-Generated Image Visual System

CGJ Chingola (Zambia) (location identifier)

CGK Jakarta Soekarno-Hatta (Indonesia) (location identifier)

CGL Charge Generation Layer • Circling Guidance Light • Commercial General Liability • Complex Ginzburg-Landau • Comprehensive General Liability • Continuous Gas Laser • Corrected Geomagnetic Latitude

CGLE Complex Ginzburg-Landau Equation

CGLM Classical General Linear Model

cgm centigram • Computer Graphics Metafile (file extension)

CGM Camiguin Mambajao (Philippines) (location identifier) • Circumgalactic Medium • Computer Graphics Materiel • Computer Graphics Metafile • Conjugate Gradient Method • Corrected Geomagnetic (coordinates) • Cylindrical Grating Monochromator

CGMO Coast and Geomagnetic Observatory

CGMP Cisco Group Management Protocol • Current Good Manufacturing Practice

CGMRF Compound Gaussian Markov Random Field

CGMS Coordination Group on Meteorological Satellites • Coordination on Geostationary Meteorological Satellites (*)

CGMW Catalog of Galaxies behind the Milky Way • Commission for the Geological Map of the World

CGN CCSDS Ground Network • Centre de Gravité National • Cologne Bonn Konrad Adenauer (Germany) (location identifier) • Concentrator Group Number • Köln Bonn Konrad Adenauer (Germany) (location identifier)

Cgo Cargo

CGO Canadian Gemini Office • Cargo • Code Generation and Optimization (symposium) • Cruz Gonzales O-stars) • Zhengzhou (China-PRC) (location identifier)

CGOM Consultative Group on Ocean Mapping

CGP Canonical Genetic Programming • Central Grounding Point • Character Graphics Printer • Chittagong (Bangladesh) (location identifier) • Color Graphics Printer • Command Generation Program • Committee on Gravitational Physics • Control Ground Point

CGPA2 Second General Catalog of stars observed with Photoelectric Astrolabes

CGPJ Consejo General del Poder Judicial (Spain)

CGPL Conversational Graphical Programming Language

CGPM Conférence Générale des Poids et Mesures

CGPME Confédération Générale des Petites et Moyennes Entreprises (France)

CGPO Canadian Gemini Project Office (Canada)

CGPS Canadian Galactic Plane Survey • Compiling Genetic Programming System

CGQ Changchun (China-PRC) (location identifier)

CGR Campo Grande (Brazil-MS) (location identifier) • Captured Gamma Ray • Carbon Glass Resistor • Compound Growth Rate

CGRF Canadian Geomagnetic Reference Field

CGRI Commissariat Général aux Relations Internationales

CGRM Computer Graphics Reference Model

CGRO (Arthur Holley) Compton Gamma-Ray Observatory

CGROSSC (Arthur Holley) Compton Gamma-Ray Observatory Science Support Center

CGRS Central Gyro Reference System • Compact Gamma-Ray Spectrometer

CGRT Carbon-Glass Resistance Thermometer

cgs Centimeter - Gram - Second

C&GS Coast & Geodetic Survey (USA) • Commercial & Government Systems

CGS Canadian Geotechnical Society (Canada) • Center for Government Studies (USA-CA) • Centimeter - Gram - Second • Central Geological Survey • Centre de Gestion des Servitudes • Centre de Gestion Scientifique • Centro di Geodesia Spaziale (Italy) • Civil GPS Service • Coast and Geodetic Survey (USA) • Coast Guard Station • College Park (USA-MD) (location identifier) • Common Ground Segment • Common Ground Sensor • Common Ground Station • Common Ground System • Communications Gateway System • Computer Ground Station • Conjugate Gradients Squared (algorithm) • Continuous Grain Silicon • Cooled Grating Spectrometer • Core Ground Segment • Cryogenic Grating Spectrometer

CGSA Computer Graphics Structural Analysis

CGSB Canadian General Standards Board (Canada) • Canadian Government Specifications Board (Canada)

CGSC Chinese Geodetic Star Catalog

CGSD Computer Graphics Systems Development

CGSE Common Ground Support Equipment • Customer Ground Support Equipment

CGSEL Common Ground Support Equipment List

CGSIC Civil GPS Service Interface Committee

CGSPO Combined Gas Steam Plant Optimization

CGSS Cryogenic Gas Storage System

CGT Capital Gains Tax • Ceramic Gas Turbine • Chinguitti (Mauritania) (location identifier) • Closed Gas Turbine • Compagnie Générale Transatlantique • Compressed-Gas Turbomotor • Consolidated Ground Terminal • Corrected Geomagnetic Time

CGTE Global Changes in Terrestrial Ecosystems

CGTI Conseil Général des Technologies de l'Information (France)

CGTO Contracted Gaussian-Type Orbital

CGU Canadian Geophysical Union (Canada) • Circuit Group Unblocking • Ciudad Guayana (Venezuela) (location identifier) • Claremont Graduate University (USA-CA) • Complex Grammar Unit • Computer Graphics Unit

CGUA Circuit Group Unblocking Acknowledgment

CGV Caiguna (Australia-WA) (location identifier) • Computer-Generated Voice • Conditions Générales de Vente • Critical Grid Voltage

CGW Computer Graphics World • Continuous Gravitational Wave • Corning Glass Works • Customer Gateway

CGWIC China Great Wall Industry Corporation (China-PRC)

CGWT Cylindrically Guided Wave Technique

CGX Chicago - Meiggs Field (USA-IL) (location identifier)

CGY Cagayan de Oro (Philippines) (location identifier)

CGZ Casa Grande (USA-AZ) (location identifier)

ch/ Cheque

ch Candle-Hour • Change • Choke • Clipper Header (file extension) • Cosinus Hyperbolique • Switzerland

Ch Chain • Chair • Chairman • Chaldron • Chamber • Chambre • Chancellor • Channel • Chapel • Chapelle • Chapitre • Chaplain • Chapter • Charge • Charges • Chart • Château • Chaussée • Check • Chemical • Chemin • Chemistry • Cheval-vapeur • Chief • Child • Children • Chile • Chilean • China • Chinese • Choir • Choke • Methyladyne

C&H Cosmic & Heliospheric

C/H Camera Head • Central Heating

CH Call Hold • Candle-Hour • Carnegie Hall • Case Hardening • Central Heating • Centre d'Hydrographie • Certified Hardware • Chaco (Argentina) • Chairman • Channel • Chapel • Character • Character Height • Chickasaw • Chief Hydrologist • Chile • China • Condritic-Porous • Chondrule • Chuvash • Clearinghouse • Collision Hypothesis • Common Hardware • Compass Heading • Conductor Head • Confederatio Helvetica • Confédération Helvétique • Control Heading • Control House • Coronal Hole • Court House • Critical Height • Cursor Horizontal • Customs House • Methylidyne • Suisse (code ISO de pays) • Switzerland (international vehicle registration code) • Switzerland (ISO country code) • Switzerland (postal country code)

Cha Chameleon-Chameleonis (constellation)

CHA Chattanooga (USA-TN) (location identifier)

ChA&A Chinese Astronomy & Astrophysics (periodical code)

CHAC Computer History Association of California (USA-CA)

CHAD Code to Handle Angular Data

CHAID CHI-squared Automatic Interaction Detector

ChaIRN Chamaeleon Infrared Nebula

Chal Chaldron • Chaleur • Challenge

CHAM Chemical Abstract Machine

CHAMA Commission for History of Ancient and Medieval Astronomy

CHAMADE Change Management in Distributed Environment

Chamb Chambers

CHAMEX Cloud Height And Motion Experiment

CHAMMP Computer Hardware, Advanced Mathematics and Model Physics

ChaMP Chandra Multiwavelength Project

CHAMP Carbon Heterodyne Array of the MPIfR • Challenging Mini-satellite Payload • Comet Halley Active Monitoring Program

CHAMPs Charge Massive Particles

CHAMPS Configurable Hardware for Multidisciplinary Projects in Space

Chan Chancellor • Chancery • Channel

CHAN Channel

Chanc Chancellor • Chancery

ChAOS Chicago Adaptive Optics System

CHAOS Canadian High Acceptance Orbit Spectrometer • Chapel Hill Astronomical and Observational Society (USA-NC) • Cornell High-order Adaptive Optics Survey

Chap Chapitre • Chaplain • Chapter

CHAP Challenge Authentication Protocol • Challenge Handshake Authentication Protocol (*) • Channel Processor

CHAPS Clearinghouse Automatic Payments System

Chapt Chapter

Char Character • Characteristic • Characterize • Charter

CHAR Character

CHARA Center for High-Angular Resolution Astronomy (USA-GA)

Charact Characterize

CHARC Characteristics

CHARGE Chemical Analysis of Released Gas Experiment

CHARM Camera High-Angular-Resolution Module • Composite High-Altitude Radiation Model

CHARME Comité pour l'Histoire de l'Armement (France)

CHARON Calern High-Angular-Resolution Optical Network (France)

chart Charta

chart. cerat. Charta Cerata

Chas Chassis

CHAS Chagrin Valley Astronomical Society (USA-OH) • Containment Heat Removal System

CHASE Chamber Association of Small Enterprises • Coronal Helium Abundance Skylab Experiment

Chât Château

CHAT Conversational Hypertext Access Technology

Chb Cherbourg (France)

CHB Chilas (Pakistan) (location identifier) • Core Helium Burning

CHBDL Common High-Bandwidth Data Link

Chbre Chambre

Chbres Chambres

Chbrs Chambers

ChC Choke Coil

CHC Chance • Christchurch (New Zealand) (location identifier)

CHCF Component Handling and Cleaning Facility

CHCI Communications Hardware Client Item • Communications Hardware Communications Interface • Communications Hardware Configuration Item

CHCK Channel Check

CHCP Change Code Page

CHCS Cabin Humidity Control System • Composite Health Care System

Chd Chaud • Chord

ChD Doctor of Chemistry

CHD Chandler (USA-AZ) (location identifier) • Coronary Heart Disease

Chde Chaude

Ch. de F. Chemin de Fer

chdir Change Directory

CHDIR Change Directory

CHDL Computer Hardware Description Language

CHDM Cold and Hot Dark Matter • Cold plus Hot Dark Matter

CHDU Current Header Data Unit

ChE Chemical Engineer • Chief Engineer

CHE Caherciveen Reenroe (Ireland) (location identifier) • Cargo Handling Equipment • Channel End • Chemical Engineer • Chemistry • Container Handling Equipment • Suisse (code ISO de pays) • Switzerland (ISO country code)

CHEAF Centrum voor Hoge-Energie Astrofysica (Netherlands)

CHEASE Cubic Hermite Element Axisymmetric Static Equilibrium

CHeB Core Helium Burning

CHEC Channel Evaluation and Call

Chem Chemical • Chemically • Chemin • Chemist • Chemistry

CHEM Charge - Energy - Mass • Chemicals - Health - Environment - Me • Chemicals - Health - Environment - Medicine

ChemE Chemical Engineer

CHEMFET Chemical-sensing Field Effect Transistor

ChemG Chemikaliengesetz

CHEMRAWN Chemical Research Applied to World Needs

CHEMS Charge Energy Mass Spectrometer

CHEMTREC Chemical Transportation Emergency Center (USA)

ChEnC Chemical Engineering Communications (periodical code)

ChEnN Chemical Engineering News (periodical code)

ChEnS Chemical Engineering Science (periodical code)

CHEOPS Chemical Operations System • Chemistry of Ozone in the Polar Stratosphere • Chemistry Of the Polar Stratosphere

CHEP Computing in High-Energy Physics

CHEPS Center for Higher Education Policy Studies

CHERPS Computerized Hypergol Emergency Response Protocol System

CHERS Charge Exchange Recombination Spectroscopy

Ches Cheshire (UK-England)

CHES Cryptographic Hardware and Embedded Systems (workshop)

CHESS Cornell High-Energy Synchrotron Source (USA-NY)

CHEST Combined Higher Education Software Team

CHET Commercial Heavy Equipment Transporter

CHeX Confined Helium Experiment

CHEX Committee on Human Exploration

Chf Chief

CHF Calibration History File • Chemical Heritage Foundation • Chinhae (Korea-ROK) (location identifier) • Congestive Heart Failure • Continuous Heating Furnace • Coupled Hartree-Fock • Critical Heat Flux • Franc (Liechtenstein & Switzerland) (ISO currency code)

CHFET Complementary Heterojunction Field-Effect Transistor

CHFN Change Finger

Chg Change • Charge

CHG Centre Hospitalier Général • Change • Chaoyang (China-PRC) (location identifier) • Charge

Chgd Charged

Chge Change • Charge

ChGeo Chemical Geology (periodical code)

Chges Changes

Chgment Changement

Chgov Changeover

CHGR Charger

CHGRP Change Group

ChGSE Chiral Gaussian Symplectic Ensemble

CHH Chachapoyas (Peru) (location identifier)

Chteau Château

Chtx Châteaux

Chi Chicago (USA-IL) • China • Chinese

CHI Catalog Handling Interface • Challenge Identification • Chicago (USA-IL) (location identifier) • Coaxial Helicity Injection • Computer Horizon Inc. • Computer-Human Interaction • Computer-Human Interface • Computer-Human Interfacing

CHIA Cumulative Hydrologic Impact Assessment

Chic Chicago (USA-IL)

CHIC Community Health Information Classification

Chich Chichester

CHICOS California High School Cosmic-ray Observatory (USA-CA)

CHIDP Condritic-Porous Interplanetary Dust Particle

CHIEF Combined Helmholtz Integral Equation Formulation

ChIEJ Chinese Institute of Engineers Journal (periodical code)

CHIH Chihuahua (Mexico)

CHII Compact HII (region)

Chil Chilean

CHIL Current Hogging Injection Logic

CHILD Cognitive Hybrid Intelligent Learning and Development • Computer Having Intelligent Learning and Development

CHILL CCITT High-Level Language

Chim Chimie • Chimique

Chin China • Chinese

CHIN Canadian Heritage Information Network (Canada) • Community Health Information Network

CHIO Cultural Heritage Information Online

CHIP Character-Handling Interface Processor • Chromospheric Helium Imaging Photometer • Chromospheric HI Imaging Photometer • Computerized Homework In Physics • Constraint-Handler In PROLOG • Constraint-Handling In PROLOG

ChIPS Chandra Imaging and Plotting Software (*) • Chandra Imaging and Plotting System

CHIPS Clearinghouse Interbank Payments System • Common Hardware Intelligence Processing System • Cosmic Hot Interstellar Plasma Spectrometer

CHIPSat Cosmic Hot Interstellar Plasma Spectrometer Satellite

CHIRLAM Chemical High-Resolution Limited Area Model

CHIRP Confidential Hazardous Incident Reporting Programme • Confidential Human Incidence Reporting Programme • Confidential Human Incidence Reporting Project

CHIS Chiapas (Mexico) • Computerized Hospital Information System

CHISSA Center for High Integrity System Software Assurance

ChJ Chief Justice

CHJ Chipinge (Zimbabwe) (location identifier)

ChJA Chinese Journal of Aeronautics (periodical code)

ChJAA Chinese Journal of Astronomy and Astrophysics (periodical code)

ChJIR Chinese Journal of Infrared Research (periodical code)

ChJL Chinese Journal of Lasers (periodical code)

ChJNP Chinese Journal of Nuclear Physics (periodical code)

ChJS Chinese Journal of Semiconductors (periodical code)

ChJSS Chinese Journal of Space Science (periodical code)

chk Checkdisk Recovery (file extension) • Temporary Document (file extension)

Chk Check

CHK Check • Chickasha (USA-OK) (location identifier)

ChkB Check Bit

chkdsk Check Disk

CHKDSK Check Disk

Chkr Checker

chl Configuration History Log (file extension)

Chl Chloride • Chloroform

Ch-L Chef-Lieu

CHL Central Helium Liquefier • Certified Hardware List • Challis (USA-ID) (location identifier) • Channel • Chef-Lieu • Chile (ISO country code) • Chili (code ISO de pays) • Cited Half Life

ChLC Cholesteric Liquid Crystal

Chld Chilled

Chlo Chloride • Chloroform

CHLW Commercial High-Level Waste

Chm Chairman

CHM Capital Humano y Movilidad • Chemistry • Chimbote (Peru) (location identifier) • Commission on History of Meteorology (AMS) • Constructable Habitat Module • Cooperative Hypermedia

Chmbr Chamber

CHMBR Chamber

CHMC Commission on the History of Modern Chemistry (IUHPS)

CHMI Czech Hydrometeorological Institute

CHML Child Meal

CHMM Chairman • Continuous Hidden Markov Model

Chmn Chairman

CHMMN Commission Hydrographique de la Méditerranée et de la Mer Noire

Chmn Chairman

chmod Change Mode

CHMOF Change Mode

Chn Channel

CHN China (PRC) (ISO country code) • Chine (RPC) (code ISO de pays) • Chonju (Korea-ROK) (location identifier) • Hydrogen Cyanide • PRC (ISO country code) • République Populaire de Chine (code ISO de pays)

Chng Change

CHNO Hydrogen Isocyanate

CHNS Hydrogen Isothiocyanate

CHO Aldehyde • Charlottesville (USA-VA) (location identifier)

CHOA Corporate Home Office Audit

Choc Chocolate

CHOC Change of Operational Control

CHOICE Chemical Online Information Center

CHOL Common High Order Language

CHON Carbon - Hydrogen - Oxygen - Nitrogen (particle)

CHOP Change of Operational Procedures • Countermeasures Hands-On Program • Cryogenic HEMT Optimization Program

CHOPS Chief of Operations

CHORUS CERN Hybrid Oscillation Research Apparatus

CHOWN Change Owner

chp Ventura Chapter (file extension)

Chp Chairperson • Champ • Chopper

C&HP Cosmic & Heliospheric Physics

CH&P Crew Habitability & Protection

CHP California Highway Patrol • Catalog Handling Plus • Center for the History of Physics (USA-MD) • Certified Health Professional • Channel Path • Chapter • Chemical Heat Pump • Circle Hot Springs (USA-AK) (location identifier) • Combined Heat and Power

ChPAS Republic of China National Science Council Proceedings Applied Sciences (periodical code)

CHPC Center for High-Performance Computing

Ch. Pd. Charges Paid

ChPhL Chinese Physics Lasers (periodical code)

ChPhy Chinese Physics (periodical code)

CHPI Chung Hua Polytechnic Institute (China-ROC)

Ch. Ppd. Charges Prepaid

CHPPM Center for Health Promotion and Preventive Medicine

Chps Champs

CHPS Characters Per Second

Chq Cheque • Chèque

CHQ Chania (Greece) (location identifier) • Cheque

chr ASCII Character Data (file extension)

CHR Cafés - Hôtels - Restaurants • Centre Hospitalier Régional • Chapter • Character • Character Handling Routine • Chateauroux (France) (location identifier) •

Chronicle • Committee on Human Rights • Cooper-Harper Rating

CHRD Channel Ready

Chrg Charge

Chrge Charge

Chrger Charger

Chrgr Charger

CHRIS Chemical Hazard Response Information System • Compact High-Resolution Imaging Spectrometer

CHRISP Coronagraphic High-Resolution Infrared Spectrograph

CHRISTA Cryogenic Infrared Spectrometers and Telescopes for the Atmosphere

Chrm Chairman

ChRM Characteristic Remanent Magnetization

CHRMA Civilian Human Resource Management Agency

CHRNS Center for High-Resolution Neutron Scattering

Chrnx Charneux (Belgium)

CHROMEX Chromosomes Experiment

Chron Chronicle • Chronique • Chronograph • Chronological • Chronologically • Chronologique • Chronology • Chronometer • Chronomètre • Chronometry

Chronol Chronological • Chronologically • Chronologique • Chronology

CHRP Calcul et Réseaux à Haute Performance • Common Hardware Reference Platform

CHRU Centre Hospitalier Régional Universitaire

CHRUS Centre for Housing Research and Urban Studies

Chrs Chambers

Chrst Characteristic

ChRST China Rept Sci Technol (periodical code)

Chrt Chart

Chs Chapters

CHS Calibration Heat Source • Canadian Hydrographic Service (Canada) • Carlsbad High School (USA-NM) • Change Sign • Chapters • Charleston (USA-SC) (location identifier) • Comité d'Hygiène et de Sécurité • Common Hardware and Software • Common Hardware Suite • Common Hardware Systems • Community Health Service • Compact Helical System • Cooperative Hypermedia Server • Cooperative Hypertext Server • Cross-Hole Seismic • Cylinder Head Sector

ChSAJ Chinese Society of Astronautics Journal (periodical code)

ChSBu Chinese Science Bulletin (periodical code)

CHSCT Comité d'Hygiène, de Sécurité et des Conditions de Travail (France)

CHSF Cargo Hydrazine Service Facility

ChSMJ Chinese Society of Mechanical Engineers Journal (periodical code)

CHSP Certified Health Security Professional • Common Hardware Software Program

CHSSI Common HPC Software Support Initiative

ChSST Chinese Space Science Technology (periodical code)

CHSTR Character String

CHT Cathode Heating Time • Chatham Island (New Zealand) (location identifier) • Cylinder Head Temperature

Chtg Charting

CHU Celsius Heat Unit • Centigrade Heat Unit • Centre Hospitalier Universitaire • Chuathbaluk (USA-AK) (location identifier) • Current Header Unit

ChUDA Chile Universidad Departamento de Astronomia Publicaciones (periodical code)

ChUI Character User Interface

CHUM Chart Updating Manual

CHUNET Swiss University Network

CHUNNEL Channel Tunnel

CHUR Chondritic Uniform Reservoir

CHV Chaves (Portugal) (location identifier)

CHVC Compact High-Velocity Cloud

CHW Command Header Word • Constant Hot Water • Jiuquan (China-PRC) (location identifier)

CHWS Chilled Water System

Chx Choix • Choux

CHX Cabin Heat Exchanger • Changuinola (Panama) (location identifier) • Condensing Heat Exchanger

CHW Chilled Water • Cold Heading Wire • Cold and Hot Water

CHX Condensing Heat Exchanger

CHy Commission for Hydrology (WMO)

CHY Choiseul Bay (Solomon Islands) (location identifier) • Columbia - Harvard - Yale

CHZ Chiloquin (USA-OR) (location identifier) • Circumstellar Habitable Zone • Continuously Habitable Zone

CH2O Formaldehyde

CH3Cl Methyl Chloride

CH3CN Methyl Cyanide

CH4 Methane

c/i Certificate of Insurance

Ci Cirrus • Curie

C+I Compatibility + Interoperability

C&I Checkout & Integration • Commerce & Industry • Commercial & Industrial • Communication & Instrumentation • Compatibility & Interoperability • Cost & Insurance • Criteria & Indicators

C/I Carrier-to-Interference • Carrier-to-Intermodulation • Certificate of Insurance

CI Call Indicator • Call Intrusion • Canary Islands • Carbonaceous (chondrite of type) I • Card Input • Caroline Islands • Cascade Ionization • Case Identifier • Cast Iron • Catalog Interoperability • Cédula de Identidad • Cell Identification • Center of Impact • Central Interval • Centre d'Information • Certification Inspection • Chain Index • Channel Islands (*) • Chemical Inspectorate • Chemical Ionization • Chemi-Ion • Chief Inspector • Chief Instructor • Chimalateco • Christmas Island • Circuit Intégré • Circuit Interrupter • Circulaire Interministérielle • Cirrus • Civilian Instructor • Classical Imaging • Clock In • Close-In • Code Instrument • Coefficient of Intelligence • Coeficiente Intelectual • Collisional Ionization • Color Index • Color Interior • Combustion Index • Command Information • Commercial Item • Common Interface • Commonwealth Institute • Communication and Instrumentation • Communications Interface • Compass and Inclinometer • Competitive Intelligence • Component Interface • Compound Interferometer • Compression-Ignition • Computational Intelligence • Computer Indicator • Concentration Index • Confidence Interval • Configuration Information • Configuration Inspection • Configuration Interaction • Configuration Item • Congestion Indicator • Congiunzione Inferiore • Consommation Intermédiaire • Consular Invoice • Containment Integrity • Content Indicator • Context Independent • Continuous Improvement • Continuous Injection • Continuous Interlock • Contract Initiation • Contract Item • Contractor Inquiry • Contrast Index • Control Indicator • Control Interface • Control Interval • Converging Inflow • Coordinate Index • Coordinating Installation • Coronal Index • Coronagraphic Imager • Corporate Identity • Corporate Information • Corrosion Inhibitor • Cost Index • Côte d'Ivoire (international vehicle registration code) • Côte d'Ivoire (code ISO de pays) • Courrier International • Course Indicator • Course Interpreter • Course Invalid •

Crédit d'Impôts • Crew Interface • Critical Item • Cubic Inch • Cumulative Index • Curriculum and Instruction • Customer Installations • Ivory Coast (international vehicle registration code) • Ivory Coast (ISO country code)

Cia Compagnia • Companhia • Compañía

CIA Call Initiate Answer • CAM Interactive Analysis (software) • Cash In Advance • Centrale Informatie Afdeling (Netherlands) • Central Intelligence Agency (USA) • Central Inventory Accounting • Centralized Information Alert • Centre d'Intelligence Artificielle • Certificate Issuing and Authentication • Change Impact Assessment • Chemical Industries Association • Chemical Institute of Australia • Citation Index Activity • Collectif Inter-Associatif • Collection Interface Agent • Collisionally-Induced Absorption • Collision-Induced Absorption • Communications Interface Assembly • Communications Interrupt Analysis • Competitive Intelligent Agent • Computer Industry Association (USA) • Computer Interface Adaptor • CONE Interface Adapter • Conference Interface Agent • Confidentiality - Integrity - Availability • Configuration Identification Accounting • Conseil International des Archives • Control Interface Assembly • Controller Interface Assembly • Cooperative Information Agent • Cooperative Information Agents (workshop) • Core Instability Accretion • Coronal Imaging Assembly • Coronographe Interférentiel Achromatique • Current Instruction Address • Roma Ciampino (Italy) (location identifier)

CIAA Centro de Investigación Astronómico de Alicante (Spain) • Coordinator of Inter-American Affairs

CIAB Conseil International des Agences Bénévoles

CIAC Centro de Información Ambiental del Caribe • Computer Incident Advisory Capability • Computer Incident Advisory Council

CIAD Climate Impact Assessment Division

CIADT Comité Interministériel de l'Aménagement du Territoire (France)

CIAE Canadian Institute for Adult Education (Canada) • Chinese Institute of Atomic Energy (China-PRC)

CIAHG Copyright Issues Ad Hoc Group

CIAI Commerce and Industry Association Institute

Cial Commercial

CIALS Confédération Internationale des Arts, des Lettres et des Sciences

CIAM Congrès International d'Architecture Modern • Cooperative Institute for Applied Meteorology

CIAME Comité Interprofessionnel pour l'Automatisation et la Mesure Électronique (France)

CiAO Cimini Astronomical Observatory (Italy)

CIAO Chandra Interactive Analysis of Observations (package) • Conception Industrielle Assistée par Ordinateur • Coronagraphic Imager with Adaptive Optics • Critical Infrastructure Assurance Office

CIAP Central Intelligence Architecture Plan • Climatic Impact Assessment Program

CIAPS Customer Integrated Automated Purchasing System

CIAR Canadian Institute for Advanced Research (Canada-ON) • Center for Inter-American Relations

CIArb Chartered Institute of Arbitrators

CIAS Circuit Inventory and Analysis System • Conference of Independent African States

CIASE China Institute of Aeronautic System Engineering • Computer Institute for Applications in Science and Engineering

CIAT Comité Interministériel d'Aménagement du Territoire (France) • Contrat Interministériel d'Aménagement du Territoire

CIATF Comité International des Associations Techniques de Fonderie

CIATI Centre Interacadémique de Traitement de l'Information • Centro de Investigación y Asistencia Técnica a la Industria

CIAX Cassegrain Instrument Auto Exchanger

cib Cibus

CIB Catalina Island (USA-CA) (location identifier) • Central Intelligence Board • Centre d'Information Bibliographique • Centre International Bibliographique • Centrum voor Informatiebeleid (Netherlands) • Change Impact Board • Change Implementation Board • Classification Internationale des Brevets • COBOL Information Bulletin • Cold Interface Board • Combined Information Bureau • Command Input Block • Control Interface Block • Controlled Image Base • Controlled Imagery Base • Controller Image Base • Cosmic Infrared Background

CIBA Control Instrumentation for Balloon

CiBAA Circular of the British Astronomical Association (periodical code)

CIBR Cosmic Infrared Background Radiation

CIBS Common Integrated Broadcast Service

CIBTC China International Book Trading Corporation

cic Culpa in Contrahendo

Cic Cicero (106-43 BC)

C-I-C Cloud-In-Cell

CIC Cable In Conduit • Canadian Institute of Chemistry (Canada) • Cargo Integration Contractor • Carrier Identification Code • Central Integrated Checkout • Centre d'Investigation Clinique • Centro de Información Científica • Centro Inter-dipartimentale della Communicazione • Chemical Information Center • Chemical Institute of Canada • Chico (USA-CA) (location identifier) • China Internet Corporation (China-PRC) • Chip Implementation Center • CIGNET Information Center (USA-MD) • Circuit Identification Code • Cloud-In-Cell • Combined Intelligence Center • Comet Impact Campaign • Comisión de Investigaciones Científicas • Comité International de Coordination • Commander In Chief • Command Information Center • Command Input Coupler • Command Interface Control • Committee on Information and Communications • Committee on Institutional Cooperation • Commonwealth Information Centre • Communications Instructor Console • Communications Intelligence Corporation • Communications Interface Control • Communications Interface Controller • Community Inversion Code • Compensated Ion Chamber • Complex Integrated Circuit • Computer Instruments Corporation • Computer Interface Control • Computerized Information Center • Computers, Information and Communications • Computing, Information and Communications • Configuration and Interface Controller • Connectivity Industry Consortium • Connector Integrated Cell • Consumer Information Catalog • Consumers Interest Council • Contamination Instrument Complement • Content Indicator Code • Control and Information Center • Control Inteligente Computerizado • Controller In Charge • Coordination and Information Center • Core Image Converter • Coronado Instrument Group Inc. (USA-AZ) • Coronagraphic Imaging Camera • Creative Incentive Coalition • Crew Interface Coordinator • Critical Item Code • Critical Item Consolidation • Customer Identification Code

CICA Canadian Institute of Chartered Accountants • Center for Innovative Computer Applications • Center for Innovative Computing Applications • Competition In Contracting Act (USA) • Council of International Civil Aviation • Customer Identification Control and Accounting

CICADA Central Instrumentation Control And Data Acquisition • Configurable Instrumentation Control And Data Acquisition (*)

CICAP Comitato Italiano per il Controllo delle Affermazioni sul Paranormale (Italy)

CICAR Cooperative Investigations of the Caribbean and Adjacent Regions

CICAS Contractor Information Computer Assistance System

CICATS Computer Industry Coalition for Advanced Television Service

CICC Cable-In-Conduit Conductor • Cargo Integration Control Center • Center of the International Cooperation for Computerization

CICCR Configuration Information Control Center Requirement

CICE Centre International de Calcul Électronique (ONU)

CICESE Centro de Investigación Científica y de Educación Superior de Ensenada (Mexico)

CICF Customer Information and Control File

CICG Centre International des Conférences de Genève (Switzerland) • Centre Interuniversitaire de Calcul de Grenoble (France)

CICH Centro de Información Científica y Humanística (UNAM)

CICHE Committee for International Cooperation in Higher Education

CICI Confederation of Information Communication Industries

CICIAM Committee for the International Conferences on Industrial and Applied Mathematics

CICID Comité Interministériel de la Coopération Internationale et du Développement (France)

CICIS Chemicals In Commerce Information System

CICM Communications Interface Control Module

CICOM Centre d'Information et de Communication

CICP Committee to Investigate Copyright Problems • Communications Interface Control Processor

CICPP Central Integrated Checkout Program Plan

CICR Comité International de la Croix-Rouge

CICRIS Cooperative Industrial and Commercial Reference and Information Service

CICRP Centre Interuniversitaire de Calcul de la Région Parisienne (France)

CICS Center for Intelligent Control Systems • Central Integrated Checkout System • Commercial Industrial Contract System • Cooperative Institute for Climate Studies • Core Instrument Control System • Customer Information Control System • Customer Interface Control System • Custom Information Control System

CICT Centro de Información Científica-Ténica • Computing, Information and Communications Technology • Conseil International du Cinéma et de la Télévision

CICU Central Interface Converter Unit • Commission Internationale de la Classification Universelle • Communications Interface Control Unit • Computer Interface Conditioning Unit • Computer Interface Control Unit

CICYT Comisión Interministerial de Ciencia y Tecnología

CID Cable Interconnection Diagram • Cabling Interface Drawing • Campus Infrastructure Division • Capability Integration Division • Channel Identifier • Character Identifier • Charge-Injection Device • Ciller Identification • Cedar Rapids (USA-IA) (location identifier) • Center for Information and Documentation • Central Institute of the Deaf • Centre d'Information et de Documentation • Centre d'Informatique Documentaire • Centre International de Documentation • Centro de Investigación Documentaria • Centro de Investigación y Desarollo • Change In Design • Channel Identification • Characteristic Item Description • Charge Injection Device • Chemical Information Division • Cidade • Circular Intensity Difference • Civil Investigative Demand • Collection of Indexable Data • Collisionally-Induced Dissociation • Collision-Induced Dissociation • Commercial Item Description • Commercial Item Drawing • Communications Implementation Directive • Component Identification • Comprehensive Interior Design • Computer-assisted Imaging Device • Computer-Integrated Design • Computer-Integrated Documentation • Computer Interface Device • Configuration Identification • Configuration - Installation - Distribution • Connection Identification • Context Identification • Converted In Distance • Criminal Investigation Department • Criminal Investigation Division • Cubic Inch Displacement • Current-Image Diffraction • Customer Interaction Data • Iowa City (USA-IA) (location identifier)

CIDA Canadian International Development Agency (Canada) • Centre d'Informatique et de Documentation Automatique • Centro de Investigaciones de Astronomía (Venezuela) • Centro de Investigaciones y Desarollo de la Armada (Spain) • Cometary and Interstellar Dust Analyzer • Comité Intergouvernemental du Droit d'Auteur

CIDA-E Centro de Investigación y Difusión Aeronáutico-Espacial (Uruguay)

CIDAE Centro de Investigación y Difusión Aeronáutico-Espacial (Uruguay)

CIDAS Centro Italiano Divulgazione Astronomica e Spaziale (Italy)

CIDAV Centro Internacional de Documentación Audiovisual

CIDCW Calling Identification on Call Waiting

CIDE Centre d'Information et de Documentation sur l'Enseignement • Centro de Investigación y Desarollo Espacial • Centro de Información y Documentación de Empresas • Centro de Investigación y Desarollo de la Educación • Comisión de Inversión y Desarollo Económico • Communications Information Data Exchange

CIDEP Chemically Induced Dynamic Electron Polarization

CIDESA Centre International de Documentation Économique et Sociale Africaine

CIDEX Courrier Individuel à Distribution Exceptionnelle

CIDF Control Interval Definition Field

CIDH Comisión Interamericana de Derechos Humanos • Commission Interaméricaine des Droits de l'Homme

CIDIE Committee of International Development Institutions for the Environment

CIDIM Cooperating Intelligent System for Distributed Management

CIDJ Centre d'Information et de Documentation pour la Jeunesse

CIDL Canadian Initiative on Digital Libraries (Canada) • Configuration Item Data List • Control Interface Document List

CIDM Client, Interoperability, and Data Management

CIDN Change In Drawing Notice

CIDNP Chemically Induced Dynamic Nuclear Polarization

CIDOC Center for Intercultural Documentation

CIDP Centre International de Documentation Parlementaire (Switzerland) • Control Instrument Data Processor

CIDR Classless Interdomain Routing • Classless Internet Domain Routing • Contract Inventory Delinquency Report • Critical Intermediate Design Review

CIDRE Circulation Intelligente de Dossiers Répartis • Convention d'Indemnisation Directe et de Recours

CIDS CEPEX Integrated Data System • Circular Intensity Differential Scattering • Circularity and Inner Diameter Station • Collateral Information Display System • Command Information Display System • Commercial Intrusion Detection System • Configuration Item Development Specification • Contracting Information Database System • Contractor Software Design System • Critical Item Development Specification

CIDSC Configuration Item Development Specification Circuit • Contractor In-house Data Support Capability

CIDSE Consolidated Integrated Development Support Environment

CIDSP Centre d'Information des Données Socio-Politiques

CIDST Comité d'Information et de Documentation Scientifique et Technique

CIDT Centre d'Information et de Documentation Technique

CIDU Centre d'Information et de Documentation Universitaire

CIDX Chemical Industry Data Exchange

Cie Compagnie

CIE Cargo Integration Equipment • Centre d'Information Éducative • Centre for International Economics • Cesium Ion Emission • Chinese Institute of Electronics • Chinese Institute of Engineers • Ciclo de Información Espacial • Clothing and Individual Equipment • Collie (Australia-WA) (location identifier) • Collisional Ionization Equilibrium • Comité Inter-Entreprises • Commissariat International de l'Éclairage • Commission Indépendante d'Experts • Commission Internationale de l'Éclairage • Common Ion Effect • Communication - Information - Education • Communications Interface Equipment • Compagnie Internationale de l'Éclairage • Computer-Integrated Engineering • Confédération Internationale des Étudiants • Contrat Initiative-Emploi (France) • Controlled Internal Extension

CIEA Conseil International d'Éducation des Adultes • Committee on International Environmental Activities

CIEC Citizens Internet Empowerment Coalition • Commission Internationale de l'État Civil • Conseil International des Études Canadiennes (Canada)

CIECC Consejo Interamericano para Educación, Ciencia y Cultura

CIEE Council of International Educational Exchange

CIEEM Center for Integrated Electronics and Electronics Manufacturing (USA-NY)

CIEF Capillary Isoelectric Focusing • Conseil International d'Études Francophones

CIEG Common Information Exchange Glossary

CIEM Comité International pour l'Exploration de la Mer • Computer-Integrated Engineering and Manufacturing • Conseil International pour l'Exploration de la Mer

CIEMAT Centro de Investigaciones Energéticas, Medioambientales y Tecnológicas (Spain)

CIEMS Catalog of Information Exchange and Message Standards

Cien Ciencia • Científico

CIEPS Centre International d'Enregistrement des Publications en Série

CIES Centre International des Étudiants et Stagiaires (*) • Centre International des Étudiants Stagiaires

CIESIN Consortium for International Earth Science Information Network

CIESM Commission Internationale pour l'Exploration Scientifique de la Mer Méditerranée

CIESP Centro de Estudos de Informática e Educação de São Paulo (Brazil-SP)

CIETT Confédération Internationale des Entreprises de Travail Temporaire

cif Ventura Chapter Information (file extension)

CIF Candidate Item File • Capacitor Input Filter • Cargo Integration Facility • Cell In Flight • Cell In Frames • Central Index File • Central Information Facility • Central Information File • Central Instrumentation Facility • Central Integration Facility • Centro Internacional de Física • Chifeng (China-PRC) (location identifier) • Cloning Inhibiting Factor • Código de Identificación Fiscal (Spain) • Cohesive Intermolecular Force • Color Interior Film • Common Image Format • Common Interchange Format • Common Intermediate Format • Computer Integrated Factory • Computer Interface Function • Configuration and Interface Controller • Congé Individuel de Formation • Conseil International des Femmes • Consolidated Instrumentation Facility • Constructed Impact Factor • Contrast Improvement Factor • Contrat Industriel de Formation • Corrected Impact Factor • Cost - Insurance - Freight • Crédit Impôt-Formation • Crystallographic Information File • Crystallographic Input Format • Customer Information File

CIFAX Ciphered Facsimile

CIFC Cost - Insurance - Freight - Commission

CIFCI Cost - Insurance - Freight - Commission - Interest

CIFE Colleges and Institutes for Further Education • Conseil International du Film d'Enseignement • Cost - Insurance - Freight - Exchange

CIFEG Centre International pour la Formation et les Études Géologiques (France)

CIFI Cost - Insurance - Freight - Interest

CIFJ Centre International du Film pour la Jeunesse

CIFKU Conditional Instability of the First Kind Upside-down

CIFMA Centro Internacional de Física y Matemáticas Aplicadas (Mexico)

CIFOR Center for International Forestry Research

CIFRE Convention Industrielle de Formation par la Recherche

CIFS Committee for the International Freedom of Scientists (APS) • Common Internet File System • Consorzio Interuniversitario per la Fisica Spaziale (Italy) • Copenhagen Institute for Future Studies (Denmark)

CIG Cable Integrity Group • Canadian Institute of Geomatics (Canada) • Centro de Investigaciones Geoténicas • Chemical Ion Generator • Climate Interest Group • Clumpy Irregular Galaxy • Comité International de Géophysique • Communications and Interface Group • Communications Interface Group • Compound Interlace Grating • Computer Image Generation • Computer Image Generator • Computerized Interactive Graphics • Conférence Intergouvernementale • Connector Integration Group • Connector Integrity Group • Coordinate Indexing Group • Craig Moffat (USA-CO) (location identifier) • Customer Interface Group

CIGAR Channel Inspection and Gauging Apparatus of Reactors

CIGB Commission Internationale des Grands Barrages

CIGBP Complementary Insulated Gate Bipolar Transistor

CIGIS Comprehensive Institute of Geotechnical Investigations and Surveying

CIGMA Cooperative Institute for Geoscience Management and Applications

CIGNET Cooperative International GPS Network

CIGR Commission Internationale du Génie Rural

CIGRE Conférence Internationale des Grands Réseaux Électriques

CIGREF Club Informatique des Grandes Entreprises Françaises (France)

CIGS Computer-Integrated General Scene • Cooled Infrared Grating Spectrometer • Copper Indium Gallium Diselenide

CIGSS Common Imagery Ground/Surface System

CIGTF Central Inertial Guidance Test Facility

CIH Certificate in Industrial Health • Changzhi (China-PRC) (location identifier) • Coronographe Interférentiel Hybride

CIHS Conseil de l'Informatique Hospitalière et de Santé

CI&I Compatibility, Interoperability & Integration

CII Center for Industrial Innovation (USA-NY) • Change Installation Instruction • Chartered Insurance Institute •

Compatibility - Interoperability - Integration • Computer-Integrated Instruction • Configuration Identification Index • Configuration Item Identification • Configuration Item Identifier • Configuration Item Index • Current Impact Index

CIIA Canadian Institute of International Affairs (Canada)

CIIBA Comité Interministériel pour l'Informatique et la Bureautique dans l'Administration (France)

CIIC Controlled Inventory Item Code

CIID Centro Internacional de Investigaciones para el Desarollo • Commission Internationale des Irrigations et du Drainage

CIIDS Commercial Interior Intrusion Detection System

CIIF Common Imagery Interoperable Facility

CIIL Configured Items Identification List • Control Interface Intermediate Language

CIIR Center for Intelligent Information Retrieval (USA-MA) • Central Institute for Industrial Research

CIIRC Centro Internacional para Investigación sobre Recursos Costeros (Spain)

CIIS Compatibility - Interoperability - Integration - Security • Comprehensive Interactive Information Strategy

CIISTU Commissione Italiana per l'Insegnamento della Scienze della Terra e dell'Universo (Italy)

CIIWG Communications, Implementation, and Installation Working Group

CIJ Cobija (Bolivia) (location identifier)

CIJB Cradle Interface Junction Box

CIJE Current Index to Journals in Education

CIJEDAE Comisión Interamericana de Juristas Expertos en Derecho Aéreo y Espacial

CIK Chalkyitsik (USA-AK) (location identifier) • Cryptographic Ignition Key • Crypto Ignition Key

CIKM Conference on Information and Knowledge Management

CIL Common Interface Language • Common Intermediate Language • Communications Interface Link • Complex Intermediate Language • Component Integration Laboratories • Computer Interpreter Language • Configuration Inspection Log • Contracting Information Letter • Contract Inspection Log • Contrat d'Initiative Locale • Council (USA-AK) (location identifier) • Critical Items List

CILB Commission Internationale de Lutte Biologique

CILET Circular Letter

CILF Conseil International de la Langue Française

CI/LI Corrosion Inhibitor/Lubricity Improver

CILI Corrosion Inhibitor/Lubricity Improver

CILN Condensation de l'Information en Langue Naturelle

CILS Computer Information Library Service • Continuous Integrated Logistics System

CILT Center for Innovative Learning Technologies (USA)

CIM Cable Interface Module • Center for Information Management • Cimitarra (Colombia) (location identifier) • Collection Insertion Map • Command Interface Module • Commercial Information Management • Common Information Model • Common Interface Model • Communications Improvement Memorandum • Communications Interface Module • Compaq Insight Manager • CompuServe Information Manager • Computer Input from Microfilm • Computer Input Matrix • Computer Input Media • Computer Input Microfilming • Computer Input Multiplexer • Computer Input on Microfilm • Computer-Integrated Manufacturing • Computer Integration Manufacturing • Continuous Image Microfilm • Corporate Information Management • Council on Information Management • Critical Item Monitor • Cryogenic Transfer Module • Crystal Impedance Meter

CIMASI Conférence Internationale sur les Mathématiques Appliquées et les Sciences de l'Ingénieur

CIME Centro Internazionale Matematico Estivo • Comité Intergouvernemental pour les Migrations Européennes • Computer Integrated Manufacturing and Engineering • Conseil International des Moyens d'Enseignement • Consejo Intergubernamental para las Migraciones Europeas • Council of Industry for Management Education • Cursus Intégré pour la Mobilité des Étudiants

CIMEX Convective Interfacial Mass Experiment

CIMH Caribbean Institute for Meteorology and Hydrology (formerly CMI) • Commission Internationale pour la Météorologie Historique

CIMI Centre d'Information sur les Méthodes d'Instruction • Computer Interchange of Museum Information • Consortium for the Interchange of Museum Information

CIMIC Center for Information Management and Connectivity (USA-NJ)

CIML Comité International de Métrologie Légale • Contract Item Material List

CIMM Canadian Institute of Mining and Metallurgy (Canada)

CIMMS Cooperative Institute for Mesoscale Meteorological Studies

CIMN Channel Identifier and Message Number

CIMNE Centro Internacional de Métodos Numéricos en Ingeniería (Spain)

CIMNET Center for Information Management Network

CIMO Commission des Instruments et des Méthodes d'Observation • Commission des Instruments et Méthodes d'Observation • Commission for Instruments and Methods of Observation (WMO)

CIMOS Center for Intelligent Manufacturing of Semiconductors (USA-CA)

CIMP Climatic Index Monitoring Program • Commission on Igneous and Metamorphic Petrogenesis • Common Information Management Protocol • Corporate Information Management Plan

CIMS Chemical Information Management System • Chemical Ionization Mass Spectrometry • Consolidated Information Management System • Corporate Information Management System

CIMSS Cooperative Institute for Meteorological Satellite Studies

CIMU Compatibility Integration Mock-Up

CIN Cargo Increment Number • Carrier Input • Carroll (USA-IA) (location identifier) • Center Information Network • Change Identification Number • Change Identifier Number • Change Incorporation Notice • Change Instrumentation Notice • Chart Information Network • Chemical Industry Notes • Circuit Identification Number • Code Identification Number • Comité International de Normalisation • Command Information Network • Content Identifier Number • Contract Item Number • Cooperative Information Network • Corporate Internet • Course Identification Number

CINA Comisión Internacional de Navegación Aérea

CINAA Cyclic Instrumental Neutron Activation Analysis

CinC Commander in Chief

CINC Combined Intelligence Center • Commander in Chief

CINCAT Capacity Increase through Computer Assistance Tools

CIND Computer Index of Neutron Data

CINDA Chrysler Integrating Numerical Difference Analyzer

CINDAS Center for Information and Numerical Data Analysis and Synthesis (USA-IN)

CINDOC Centro de Información y Documentación Científica (CSIC)

CINE Comet Impact Network Experiment

Cinem Cinematografía

CINES Centre Informatique National de l'Enseignement Supérieur (France)

Cinn Cincinnati (USA-OH)

CINOF Campaign for Integrated Observations of Solar Flares

CINR Central Institute for Nuclear Research

CINS Command Information Network Supplement • Command Information Network System • Cryogenic Inertial Navigation System

CINT Call Interception • Center for Integrated Nanotechnologies (USA-NM)

CINTEL Computer Interface for Television

CINTEX Catalog Interoperability Experiment

CINTS Control Information Navigation and Training System

CINU Centre d'Information des Nations Unies (ONU)

CINTEX Catalog Interoperability Experiment

CINVESTAV Centro de Investigación y de Estudios Avanzados (Mexico)

CIO Calibrated Individual Observation • Catalog of Infrared Observations • Central Imagery Office (USA) • Central Information Office • Central Input/Output • Centre d'Information et d'Orientation • Centro de Investigaciones en Optica • Chief Information Office • Chief Information Officer • Chief Investment Officer • Comité International d'Organisation • Comité International Olympique • Comité Interorganismes • Command-Issuing Office • Commission International d'Optique • Common Item Order • Compagnie des Indes Orientales • Computer Interoperability • Congress of Industrial Organizations • Continuous Improvement Opportunity • Conventional International Origin

CIOB Central Indian Ocean Basin

CIOCS Communications Input and Output Control System • Communications Input/Output Control System

CIOD Cast Iron Outside Diameter

CIOH Centro de Investigaciones Oceanográficas e Hidrográficas (Colombia)

CIOMS Council for International Organizations of Medical Sciences

CION Centre Informatique de l'Observatoire de Nice

CIOP CAMAC Input-Output Processor • Centro de Investigaciones Opticas • Committee on International Organizations and Programs • Common Interface Operating Procedure

CIOS Conseil International pour l'Organisation Scientifique

CiP Computers in Physics

CIP Candidate Interoperability Plan • Capability Improvement Program • Capital Investment Plan • Capital Investment Program • Capsule Internal Programmer • Carrier Identification Parameter • Carte d'Identité Professionnelle • Cast Iron Pipe • Catalog Interoperability Protocol • Cataloguing In Publication • Celestial Intermediate Pole • Cellule d'Information Préventive • Cellule Information Prévention • Central Information Processor • Centralized Information Processing • Centre d'Information de la Presse • Change Implementation Plan • Change Improvement Plan • Channel Interface Processor • Chipata (Zambia) (location identifier) • Classroom Instruction Program • Clustering Initial Pattern • Cold Isostatic Pressing • Color Inkjet Printer • Combined Instrument Panel • Combined Interoperability Program • Command Information Program • Command Instrument Processor • Command Interface Port • Commands In Progress • Commercial Instruction Processor • Commercially Important Person • Committee on International Programs • Common Indexing Protocol • Common Industrial Policy • Common Input Processor • Common Integrated Processor • Communications Implementation Plan • Communications Improvement Plan • Communications Improvement Program • Communications Interface Panel • Communications Interface Processor • Compatible Independent Peripheral • Component Improvement Program • Component Industry Promotion • Compressor Inlet Pressure • Computer Image Processing • Computer Information Processing • Computer-Investitions-Programm • Computers In Physics (periodical) (now CiSE) • Consolidated Imagery Program • Consolidated Intelligence Program • Constant Injection Pressure • Contour Interpolation Program • Contract Information Processor • Contrat d'Insertion Professionnelle • Control and Interface Panel • Convergence and Implementation Plan • Coordinated Instrument Package • Coronal/Interplanetary Physics • Cost Improvement Plan • Critical Infrastructure Protection • Critical Intelligence Parameter • Critical Item Program • Cryogenic Insulation Program • Crypto-Ignition Plug • Current Injection Probe • Current In the Plasma • Customer Information Package

CIPA California Institute for Physics and Astrophysics (USA-CA) • Chartered Institute of Patent Agents • Classified Information Procedure Act (USA) • Contributions for International Peacekeeping Activities

CIPE Center for Image Processing in Education (USA-AZ) • Central Institute for Physics of the Earth • Committee on Computers In Physics Education • Concurrent Image Processing Executive • Conseil Interministériel de Planification Économique

CIPFS Critical Item Product Function Specification

CIPI Canadian Intellectual Property Institute (Canada) • Change In Procurement Instructions

CIPIA Critical Infrastructure Protection and Information Assurance

CIPM Comité International des Poids et Mesures

CIPO Canadian Intellectual Property Office (Canada) • Comité International pour la Protection des Oiseaux (now BirdLife)

CIPP Communications Individual Positioning Program • Context - Input - Process - Product

CIpR Cubic Inches Per Revolution

CIPR Commission Internationale de Protection contre les Radiations • Concept In Process Review • Contractor Insurance/Pension Review • Cubic Inches Per Revolution

CIPRA Commission Internationale pour la Protection des Alpes

CIPS Canadian Information Processing Society (Canada) • Centralized Information Processing System • Cesium Ion Propulsion System • C Image Processing System • Continuous Ice Particle Sampler • Customer-Initiated Payment System

CIPSG Canadian Information Processing Society Graphics Interface (periodical code)

CIPSH Conseil International de Philosophie et de Sciences Humaines • Conseil International pour la Philosophie et les Sciences Humaines (*)

CIPSO Commercial Internet Protocol Security Option • Common Internet Protocol Security Option

CIPT Cost-Integrated Product Team

CIPW Cross - Iddings - Pirsson - Washington

CIPWG Critical Infrastructure Protection Working Group

CIQ Chiquimula (Guatemala) (location identifier)

CIQM Centre for Information Quality Management

Cir Circa • Circinus-Circini (constellation) • Circle • Circonscription • Circuit • Circular • Circulation • Circumference • Circus • Cirugía

CIR Cairo (USA-IL) (location identifier) • Call Initiate Request • Cargo Integration Review • Central Intelligence

Report • Central Intensity Ratio • Centralized Inventory and Requirement • Circa • Circle • Circular • Circulate • Color Infrared • Comité Interrégional • Commission on Industrial Relations • Committed Information Rate • Compagnie Industrielle Radioélectrique • Computational Information Retrieval (workshop) • Computer-Integrated Repair • Computerized Information Retrieval • Computerized Inspection Report • Configuration Inspection Report • Configuration Investigation Review • Contractor Information Report • Conversational Information Retrieval • Corotating-Interacting Region • Corotating Interaction Region (*) • Corotation-Inclination Resonance • Corotation Inclined Resonance • Cost Information Report • Council on Industrial Relations • Crédit d'Impôt Recherche • Critical Implementation Review • Critical Information Request • Critical Information Requirement • Critical Item Report • Critical Item Review • Current Instruction Register

CIRA Center for Information Revolution Analyses • Centre Interministériel de Renseignements Administratifs (France) • Centro Italiano Ricerche Aerospaziale (Italy) • Conference on Industrial Research Associations • Cooperative Institute for Research in the Atmosphere • COSPAR International Reference Atmosphere

CIRAC Canadian Institute for Research in Atmospheric Chemistry (Canada)

CIRANO Centre Interuniversitaire de Recherche en Analyse des Organisations

CIRB Cosmic Infrared Background • Create Interrupt Request Block

Circ Circa • Circle • Circuit • Circular • Circulation • Circumference • Circumstance • Circus

CIRC Central Information Reference and Control • Circle • Circular • Circular Reference • Circulation Input Recording Center • Circumvalación • Circumvention • Cross-Interleaved Reed-Solomon Code

CIRCE Centre Interrégional de Calcul Électronique • Complex Instruments Ranking with a new Computational Environment

CIRCL Circle

CIRCLE Champollion Infrared Comet Lander Experiment

Circum Circumference

CIRCUM Circumference

CIRD Cargo Integration Review Dry (run)

CIRE Collaborative to Integrate Research and Education • Comité Interministériel pour la Réforme de l'État (France)

CIRED Centre International de Recherche sur l'Environnement et le Développement

CIRES Cooperative Institute for Research in Environmental Sciences (USA-CO)

CIRF Centralized Intermediate Repair Facility • Computer-Integrated Repair Facility

CIRG Critical Incident Response Group

CIRH Centro de Investigación de Recursos Hidrobiológicos (Nicaragua)

CIRHS Critical Items and Residual Hazards List

CIRI Central India Research Institute (India)

CIRIS Central Inertial Reference Instrumentation System • Completely Integrated Range Instrumentation System • Completely Integrated Reference Instrumentation System • Cryogenic Infrared Radiance Instrumentation for Shuttle

CIRK Common Interswitch Rekeying Key

CIRLA Cheasapeake Information Research Library Alliance

CIRM Celestial Infrared Mapping • Celestial Infrared Measurement • Comité International Radio-Maritime

CIRMES Centre for Integrated Resource Management and Environmental Science

CIRMM Centre Interuniversitaire de Rencontres Mathématiques de Marseille (France)

CIRMS Celestial Infrared Measurement System

CIRP Conference on Infrared Physics • Critical Issue Resolution Plan

CIRPAS Center for Interdisciplinary Remotely-Piloted Aircraft Studies

CIRPASS Cambridge Infrared Panoramic Survey Spectrograph

CIRRIS Cryogenic Infrared Radiance Instrumentation for Shuttle • Cryogenic Infrared Radiance Instrument for Shuttle

CIRRUS Cloud Infrared Radiometer for UnESS

CIRS Center for Information and Research Services • Circles • Composite Infrared Spectrometer (Cassini) • Computerized Information Retrieval Service • Contractor Inventory Redistribution System • Corotating Interacting Regions

CIRSI Cambridge Infrared Survey Instrument

CIRST Centre Interministériel de la Recherche Scientifique et Technologique (France) • Centre Interuniversitaire de Recherche sur la Science et la Technologie

CIRT Computer Incident Response Team

CIRTV Centre International de Radio et Télévision

CIRTEVS Compact Infrared Television System

CIRUISEF Conférence Internationale des Responsables d'Universités et Instituts Scientifiques d'Expression Française

CIRUS Consortium for Industrial Research in the Use of Space

CIRV Common Interswitch Rekeying Variable

CIS Cadre Interview Score • Canadian Ice Service (Canada) • Canadian Institute of Surveying (Canada) • Canton Island (Kiribati) (location identifier) • Capillary Ion Source • Card Information Structure • Cargo Integration Schedule • CASE Integration Service • Cataloguing In Source • CDIAC Information System • Cells In Space • Center for Imaging Science • Center for Interdisciplinary Studies (UK) • Center for International Studies • Centre for Information Science • Central Integration Site • Cesium Ion Source • Change Impact Summary • Character Instruction Set • Chemical Information Services • Chemical Information System • Client Information Server • Client Information System • Cluster Ion Spectrometer • Cluster Ion Spectrometry • Commentator Information System • Common Information System • Commonwealth of Independent States • Communications and Information System • Communications and Instrumentation System • Communications Information System • Communications Interface System • Community Information System • Community Innovation Survey • Compact Image Sensor • Compact Interactive Standard • Compensated Image System • Compensated Imaging System • Component Identification Sheet • Complementary Information System • Compressed Image Sequence • CompuServe Information Service • Computer Information System • Computerized Information System • Configuration Information System • Congressional Information Service • Contact Image Sensor • Contractor Information Submittal • Contractor Information System • Contractor Initial Support • Contrat d'Intervention sur Site • Conventional Inertial System • Cooperative Information System • Coordinate Indexing System • Copper Indium Diselenide • Corporate Information System • Cost Information System • Countryside Information System • Critical Information System • Cryogenic Imaging Spectrometer • Cryogenic Instrumentation System • Cryogenic Interferometer/Spectrometer • Cue Indexing System • Current Information Selection • Customer Information System • Customer Interaction Software • Customer Interaction System • Customs Information System • Cycles In Service

CISA Circular Index Sequential Access

CISAC Center for International Security and Cooperation (USA-CA) • Committee on International Security and Arms Control (NAS) • Confédération Internationale des Sociétés d'Auteurs et de Compositeurs

CISAM Compressed Index Sequential Access Method

CISAS Centro Interdipartimentale di Studi e Attività Spaziali (Italy)

CISC Complex Instruction Set Computer • Complex Instruction Set Computing

CISCO Cooled Infrared Spectrograph and Camera for OHS

CISD Critical Incident Stress Debriefing

CISDA Centre for Information Society Development in Africa

CiSE Computer in Science and Engineering (periodical)

CISE Centro de Investigaciones y Servicios Educativos • Computer and Information Science and Engineering

CISET Committee on International Science, Engineering and Technology

CISF Centralized Integration Support Facility • Communication Services Industrial Fund

CISG Communauté Industrielle Spatiale de Guyane (French Guyana)

CISI Comité Interministériel pour la Société de l'Information (France)

CISIG Geographic Information Systems Interest Group

CISIR Ceylon Institute of Scientific and Industrial Research

CISK Conditional Instability of the Second Kind (*) • Convective Instability of the Second Kind

CISM Canadian Institute of Surveying and Mapping (Canada) • Centre International pour les Sciences Mécaniques • Certified Information Security Manager • Coupled Ionosphere Scintillation Model

CISNAR Commission for Integrated Survey of Natural Resources

CISO Chief Information Security Officer

CISP Committee on International Space Programs • Cooperative Interactive Storyboarding Prototyping

CISPD Consolidated Information Services Planning Document

CISPR Comité International Spécial des Perturbations Radioélectriques

CISR Centre for Information Science Research • Comité International Spécial sur les Perturbations Radioélectriques • Commercial International Security Requirements • Configuration Item and Status Report

CISS Centaur Integrated Support Structure • Center for Information Systems Security • Comprehensive Integrated Security System • Conseil International des Sciences Sociales • Cooperation of ISO Serendipity Survey

CISSP Certified Information Security Systems Professional

CISSV Circolare Interna della Sezione Stelle Variabili dell'Unione Astrofili Italiani (periodical code)

CIST Centre d'Information sur les Sciences de la Terre • Centre International pour la Science et la Technologie • Computer-Aided Surgery and Therapy • Conference on Information Systems and Technology

CISTI Canadian Institute for Scientific and Technical Information (Canada)

CISTL Configuration Item Software Team Leader

CISTOD Confederation of International Scientific and Technological Organizations for Development

Cit Citation • Cited • Citizen • Citrate

CIT Calibration Implementation Team • California Institute of Technology (USA-CA) • Carnegie Institute of Technology • Case Institute of Technology (USA-OH) • Center for Innovative Technology (USA-VA) • Central Institute of Technology • Centro de Información Técnica • Chemical Instrumentation Test and Evaluation • Chimkent (Kazakhstan) (location identifier) • Circumstellar Imaging Telescope • Citation • Combined Interrogator/Transponder • Combustor Inlet Temperature • Comité International de Télévision • Command Interface Test • Communication and Information Technology • Compact Ignition Tokamak • Compression In Transit • Compressor Inlet Temperature • Computer-Illustrated Text • Computer-Integrated Telephoning • Computer-Integrated Telephony • Computer Interface Technology • Computer Interface Terminal • Configuration Identification Table • Consumable Item Transfer • Contact Ion Thruster • Convection-Induced Turbulence • Counselor In Training • Country In Transition • Cranfield Institute of Technology (UK) • Critical Item Tag

CITA Canadian Institute for Theoretical Astrophysics (Canada) • Center on Information Technology Accommodation • Commercial-Industrial-Type Activity

CITAB Computer Instruction and Training Assistance for the Blind

CITAC Cooperation on International Traceability in Analytical Chemistry

CITADEL Citations and Delivery

CITC Community Imagery Training Committee • Computing Information Technology Center

CITCE Comité International de Thermodynamique et de Cinétique Électrochimique

CITE Capsule Integrated Test Equipment • Cargo Integration Test Equipment • Centre d'Information sur les Techniques d'Enseignement • Centre for International Technology and Education (UK-England) • Chemical Instrument Test and Evaluation • Classification Internationale des Types d'Éducation • College of Information Technology and Engineering (USA-WV) • Computer Integration Test Equipment • Conservatoire de l'Informatique et de la Télématique (France) • Controller Input Test Equipment • Council on Technology Education

CITED Copyright In Transmitted Electronic Documents

CITEL Committee for Inter-American Telecommunications • Conferencia Interamericana de Telecomunicaciones

CITEP Colorado Io Torus Emissions Package

CITERE Convention Industrielle et Technique d'Études et de Recherche à l'Étranger

CITES Convention on International Trade of Endangered Species

CITF Commercial-Industrial Type Function

CITHAR Calculateurs pour une Informatique en Temps Réel et Hétérogène dans des Architectures en Réseau

CITI Centre Interuniversitaire de Traitement de l'Information • Charlotte Institute for Technology Innovation (USA-NC) • Classification Internationale Type pour l'Industrie • Columbia Institute for Teleinformation • Common Internetworking Topology Initiative • Computer Internet Telephony Integration • Confédération Internationale des Travailleurs Intellectuels • Corporate Information Technology Infrastructure

CITIC China International Trust and Investment Corporation

CITIS Centralized Integrated Technical Information Service • Centralized Integrated Technical Information System • Contractor Integrated Technical Information Service • Contractor Integrated Technical Information System

CITM Chubu Institute Technology Memoirs (periodical code)

CITO Centraal Instituut voor Toetsontwikkeling • Communications and Information Technology Ontario (Canada-ON)

CITP Capillary Isotachophoresis

CITRAL Commission on International Trade Law

CITS Center for Information Technology Services • Central Integrated Test System • Command Information Transfer System

CITT Computerintegrierte Telefonie und Telematik

CITTI Colorado Institute for Technology Transfer and Implementation (USA-CO)

CIU Cell Interface Unit • Central Information Unit • Central Interpretation Unit • Central Interface Unit • Channel Interface Unit • Cockpit Interface Unit • Command Interface Unit • Communications Interface Unit • Computer Interface Unit • Control Indicator Unit • Control Interface Unit • Controller Interface Unit • Sault Sainte Marie - Chippewa County (USA-MI) (location identifier)

CIUC Consejo Internacional de Uniones Científicas

CIUI Crustal Intelligent User Interface

CiUO Circular of the Union Observatory Johannesburg (periodical code)

CIUP Centro de Informática da Universidade do Porto (Portugal) • Cité Internationale Universitaire de Paris (France)

CIUS Conseil International des Unions Scientifiques

CIUT Columbus IVA Underwater Testing

Civ Civil • Civilian • Civilization

CIV Cell Interarrival Variation • Central International Visibility • Centrum voor Informatieverwerking • Chomley (USA-AK) (location identifier) • Civil • Civilian • Colegio de Ingenieros de Venezuela • Commission Internationale du Verre • Conversation Interface Verb • Corona Inception Voltage • Côte d'Ivoire (code ISO de pays) • Critical Ionization Velocity • Ivory Coast (ISO country code)

CIVR Computer and Interactive Voice Response • Configuration Item Verification Review

CIVT Cargo Interface Verification Test

CIVU Cargo Integration Verification Unit

CIW Canouan Island (Grenadine Islands) (location identifier) • Carnegie Institution of Washington (USA-DC) • Ceramic Insulated Wire • Combined Iterative Weight

CIWG Cargo Integration Working Group • Catalog Interoperability Working Group • Communications Interoperability Working Group

CIX Chiclayo (Peru) (location identifier) • Commercial Internet Exchange • Compulink Information Exchange • Computer Information Exchange • Computing Information Exchange

CIXS Compact Imaging X-ray Spectrometer

CIY Comiso (Italy) (location identifier)

CIZ Coari (Brazil-AM) (location identifier)

CIZA Clusters In the Zone of Avoidance

cj Conjectural

C&J Collection & Jamming • Communications & Jamming

CJ CalTech - Jodrell Bank (survey) • Chief Justice • Conjugation and Parity • Core Jet • Cosmic Jet • Cour de Justice • Court of Justice

CJ&A Class Justification & Approval

CJA Cajamarca (Peru) (location identifier) • Class Justification and Approval

CJAA Chinese Journal of Astronomy and Astrophysics (periodical code)

CJB Coimbatore (India) (location identifier) • Congressional Justification Book

CJC Calama (Chile) (location identifier) • Clean Japan Centre

CJChE Canadian Journal of Chemical Engineering (periodical code)

CJCR Chinese Journal Citation Reports (periodical code)

CJD Candilejas (Colombia) (location identifier) • Centre des Jeunes Dirigeants • Correlation Job Description

CJDA Critical Joint Duty Assignment

Cje Corretaje

CJE Cour de Justice Européenne

CJECE Canadian Journal of Electrical Computer Engineering (periodical code)

CJF CalTech - Jodrell Bank Flat-spectrum (survey)

CJFET Complementary Junction Field Effect Transistor

CJH Chiko Lake (Canada-BC) (location identifier)

CJI Crafton Island (USA-AK) (location identifier)

CJIS Criminal Justice Information Service • Criminal Justice Information System

CJK Chinese - Japanese - Korean

CJL Chitral (Pakistan) (location identifier)

CJLI Command Job Language Interpreter

CJLTP Chinese Journal of Low-Temperature Physics (periodical code)

CJMA Communications Junction Module Assembly

CJN El Cajon (USA-CA) (location identifier)

CJO Cambridge Journals Online • Chemical Journals Online

CJP Club Jean Perrin (France) • Communications Jamming Processor • Complete Joint Penetration • Cornu-Jellet Prism • Cultureel Jongerenpaspoort

CJPE Certified Java Programmer Examination

CJR Career Job Reservation

CJS Ciudad Juarez (Mexico) (location identifier)

CJTCS Chicago Journal of Theoretical Computer Science

CJTF Combined Joint Task Force

CJU Cheju (Korea-ROK) (location identifier)

CJW Coplanar Joint Wing

CJ1 First CalTech - Jodrell Bank (survey)

CJ2 Second CalTech - Jodrell Bank (survey)

Ck Cask • Check • Cook • Cork • Creek

CK Check • Cometary Knot • Compartment Key • Containerized Kitchen • Cook Islands (ISO country code) • Îles Cook (code ISO de pays)

CKA Cherokee Kegelman (USA-OK) (location identifier)

ckb C Keyboard Mapping (file extension)

CKB Clarksburg (USA-WV) (location identifier)

Ckbd Corkboard

CKBO Classical Kuiper Belt Object

CKC Cherkassy (Ukraine) (location identifier)

CKD Complete Knocked Down • Completely Knocked Down • Count Key Data • Crooked Creek (USA-AK) (location identifier)

CKDC Chemical Kinetics Data Center

CKE Clear Lake (USA-CA) (location identifier) • Cumulus Kinetic Energy

CKF Chronic Kidney Failure

CKG Chongqing (China-PRC) (location identifier) • Cooperative Key Generation

CKH Chokurdah (Russia) (location identifier)

CKI Croker Island (Australia-NT) (location identifier)

CKIP Chinese Knowledge Information Processing

CKIPG Chinese Knowledge Information Processing Group (China-PRC)

CKK Ckerokee (USA-AR) (location identifier)

CKL Compromised Key List

CKM Cabibbo - Kobayashi - Masakawa (matrix) • Certified Knowledge Manager • Charged Kaons at the Main (injector) (experiment) • Clarksdale Fletcher (USA-MS) (location identifier)

CKN Crookston (USA-MN) (location identifier)

CKO Centrum voor Klimaatonderzoek (Netherlands) • Chief Knowledge Officer • Contingency Contracting Officer • Cornélio Procópio (Brazil-PR) (location identifier)

CKP Cayley-Klein Parameter • Control Knowledge Pool

Ckpt Cockpit

CKPT Checkpoint • Cockpit

CKR Commercial Key Recovery • Crane Island (USA-WA) (location identifier)

CKRA Cape Kennedy Reference Atmosphere

CKS Carajás (Brazil-PA) (location identifier) • Checks

CKSM Check Sum

Ckt Circuit

CKT Circuit

CKU Cordova (USA-AK) (location identifier)

CKV Clarksville (USA-TN) (location identifier)

ckw Clockwise

CKW Clockwise • Chlorkohlwasserstoff

CKX Chicken (USA-AK) (location identifier)

CKY Conakry Gbessia (Guinea) (location identifier)

CKZ Canakkale (Turkey) (location identifier)

c/l Cash Letter

cl Centiliter • Chile • Common LISP (file extension) • Cum Laude

Cl Calle • Canal • Chlor • Chlore • Chlorine • Claim • Clarinet • Class • Classe • Classification • Clause • Clearance • Clerk • Climb • Clos • Close • Closet • Closing • Closure • Cloth • Cluster • Clutch • Council

C/L Cash Letter • Centre Line • Checklist • Cross-Link

C&L Commerce & Labor • Control & Line

CL Cable Length • Calculateur Logique • Calendar Line • Car Load • Carnegie Library (USA) • Cathode Luminescence • Cathodoluminescence • Centerline • Centerline Light • Central Laboratory • Central Line • Chain Length • Channel • Checklist • Chemical Laser • Chemiluminescence • Chief of Logistics • Child-Langmuir • Chile (ISO country code) • Chili (code ISO de pays) • Circular Letter • Civil Law • Clark Lake • Class • Clear • Clearance Limit • Close • Closed • Closed Loop • Cluster • Coding Language • Coefficient of Lift • Collimating Lens • Command Language • Common Law • Common LISP • Competitive Learning • Compiler Language • Complete Linearization • Computational Linguistics • Computational Logic • Condition aux Limites • Condition Lever • Condition Limite • Confidence Level • Confidence Limit • Connectionless • Control Language • Control Leader • Control Level • Control Logic • Conversion Loss • Convertible Lens • Cristal Líquido • Critical List • Current Layer • Current Loop • Cygnus Loop • Sri Lanka (international vehicle registration code)

CLA Camera Lens Adaptor • Canadian Library Association (Canada) • Carry Lookahead Adder • Cash Limited Allocation • Catch Limit Algorithm • Center for Laser Applications (UTSI) • Center Line Average • Clear and Add • Cloud Analysis • Comilla (Bangladesh) (location identifier) • Commission for Latin America • Communications Line Adapter • Communications Link Analyzer • Conjugated Linoleic Acid • Control Logic Array • Copyright Licensing Agency (UK) • Couche Limite Atmosphérique • Cover Layer Assembly • Cross-Levelling Agreement

CLAC Centre de Lecture et d'Animation Culturelle

CLAES Cryogenic Limb Array Etalon Spectrometer

CLAF Centro Latino-Americano de Física

CLAIMS Class Assignee Index Method Search

CLAIR Center for Library And Information Resources • Clevers Leaf Area Index by Reflectance

CLAIRE Clear Air Infrared Radiation Experiment

CLAM Camera for a Lander on Mercury • Completed Leasing Action Message • Computational Linear Algebra Machine

CLAM-D Descent Camera for a Lander on Mercury

CLAM-S Surface Camera for a Lander on Mercury

CLAMS Classification Automatique de Données Multispectrales • Cleared Lane Marking System

CLAN Cordless Local Area Network

CLANS Computerized Link Analysis System

CLAR Canadian Large Adaptor Reflector (Canada) • Centre for Latin American Research • Channel Local Address Register • Communications Loading Analysis Report

CLARA Computer Load And Resources Analysis

CLARCS Copyright Licensing Agency's Rapid Clearance Service (*) • Copyright Licensing Agency's Rapid Clearance System

CLARET Cloud Lidar And Radar Exploratory Test

CLAROM Club pour les Actions de Recherche sur les Ouvrages en Mer

CLAS Calibration Laboratory Assessment Service • California Learning Assessment System • CEBAF Large Acceptance Spectrometer • China Lake Astronomical Society (USA-CA) • Classification • Communications Link Analyzer System • Computerized Loads Analysis System • Computer Library Applications Service

CLASP Comprehensive Logistics Automated Support Program • Computer Language for Aeronautics and Space Programming • Computer Launch And Separation Problem

Class Classic • Classical • Classification • Classified • Classify

CLASS Canadian Land Surface Scheme (Canada) • Center for Lidar and Atmospheric Sciences Students • Classification Literature Automated Search Service • Client Access to Systems and Services • Coherent Lidar Airborne Shear Sensor • College Library Access and Storage System • Comprehensive LSA Automated Support System • Computer-based Laboratory for Automated School Systems • Computerized Labelling and Address Sequence System • Computerized Literature Access Search Service • Cooperative Library Agency for Systems and Services • Cosmic Lens All-Sky Survey • Cross-chain Loran Atmospheric Sounding System • Cryogenic Liquid Acquisition, Storage and Supply • Cumulative and Linguistic Advanced Search System • Customized Launch Support Services • Custom Local-Area Signaling Services

CLASSI Classification System for Information

CLASSMATE Computer Language to Aid and Stimulate Scientific, Mathematical, and Technical Education

C&LAT Cargo & Loading Analysis Table

CLAT Cargo and Loading Analysis Table

CLAWS Classify, Locate and Avoid Wind Shear • Coherent Launch Atmospheric Wind Sounder

CLB Cape Lookout Bight (USA-NC) • Cash Limited Budget • Castlebar (Ireland) (location identifier) • Club • Common Logic Board • Community Land Bank • Configurable Logic Block • Constant Level Balloon

CLBA Centre de Lancement de Ballons Atmosphériques (France)

cLBV Candidate Luminous Blue Variable (star)

CLC Change Letter Control • Clean Lakes Clearinghouse • Clear Carry • Clear Confirmation • Clearlake (USA-TX) (location identifier) • Color Laser Copier • Combinatorial Logic Circuit • Command Line Controller • Commonwealth Liaison Committee • Computer Liaison Committee • Control Launch Center • Couche Limite Convective • Course Line Computer • Critical Load Cycle

CLCC Ceramic Leadless Chip Carrier • Closed-Loop Congestion Control • Closed-Loop Continuity Check

CLCD Contemporary Life Cycle Development

CLCE Communications Link Characterization Equipment

Cl-Cl Cluster-Cluster

ClCl Cluster-Cluster

CLCPW Capacitively-Loaded Coplanar Waveguide

CLCR Controlled Letter Contract Reduction

CLCS Checkout and Launch Control System • Chinese Language Computer Society • Consequence Limiting Control System • Current - Logic - Current - Switching

CLCSE Center for Life Cycle Software Engineering

CLCW Command Link Control Word

Cld Called • Cancelled • Cleared • Cloud • Colored • Cooled • Could

CLD Called • Carlsbad (USA-CA) (location identifier) • Central Line Data • Chômeur de Longue Durée • Clear Direction • Cloud • Colour-Luminosity Diagram • Command Language Directive • Communications Logistics Depot • Computer Logic Demonstrator • Congé de Longue Durée • Contrat de Longue Durée • Cost Laid Down • Cumulative Luminosity Distribution • Current-Limiting Device

CLDA Control Logic and Drive Assembly

CLDAP Connectionless Lightweight Directory Access Protocol

CLDB Common Land Database

CLDC Centre de Loisirs et de Diffusion Culturelle • Connected Limited Device Configuration

CLDCC Ceramic Leaded Chip Carrier

CLDG Closed Loop Degaussing

CLDP Centre Local de Documentation Pédagogique • Convertible Laser Designation Pod

Clds Clouds

CLDS Canada Land Data System • Clouds

CLDST Closed Loop Dynamic Stability Test

C/LDT Create/Load Distribution Tape

CLDT Create/Load Distribution Tape

Cldy Cloudy

ClDy Climate Dynamics (periodical code)

CLE Centre de Loisirs Éducatifs • Cleveland Hopkins (USA-OH) (location identifier) • Comité Local d'Éducation • Comité Local de l'Emploi • Communication - Lecture - Écriture • Continuous Legal Education • Control Law Electronics • Council of Legal Education • Customer-Located Equipment

CLEA Civilian Law Enforcement Agency • Comité de Liaison Enseignants Astronomes (France) • Contemporary Laboratory Experiences in Astronomy • Council of Local Education Authorities

CLEAPSE Consortium of Local Education Authorities for the Provision of Science Equipment

CLEAR Closed-Loop Evaluation And Reporting • Compiler, Executive Program, Assembler Routines • Components Life Evaluation And Reliability • Coronagraph and Low-Emissivity Astronomical Reflector • Coronagraphic and Low-Emissivity Astronomical Reflector (*)

CLEARS Cornell Laboratory for Environmental Applications of Remote Sensing

CLEC Competitive Local-Exchange Carrier

CLEF Comité du Livre Électronique Français • Commercially Licensed Evaluation Facility • Cross-Language Evaluation Forum

CLEI Common-Language Equipment Identification

CLEM Cargo Lunar Excursion Module • Closed Loop Eigenvector Matrix

CLEO Clear Language for Expressing Orders • Conference on Lasers and Electro-Optics • Control Library for Engineers and Operators (GBT)

CLEOPATRA Cloud Experiment Oberpfaffenhofen And Transports

CLEP College Level Examination Program

CLF Capacitive Loss Factor • Center-to-Limb Function • Central Laser Facility (UK) • Chilean Investment Unit (ISO currency code) • Civilian Labor Force • Clear Lake Facility • Cliff • Cluster Luminosity Function • Control and Limit File • Control Laser Facility • Controller - Load Forward • Control Lyapunov Function • Critical Link Factor • Cumulative Luminosity Function

CLFIC Center Launch and Flight Instrumentation Center

CLFS Cliffs

CLFST Cambridge Low-Frequency Synthesis Telescope

CLFT Classical Local Field Theories

Clg Ceiling • Cooling

ClG Cluster of Galaxies

CLG Ceiling • Central Laboratory for Geodesy • Coalinga (USA-CA) (location identifier) • Collaborative Linkage Grant • Company Limited by Guarantee • Compile, Load and Go

CLGEE Comité de Liaison des Géomètres-Experts Européens

CLH Coolah (Australia-NSW) (location identifier)

CLI Calling Line Identification • Calling Line Identity • Call-Level Interface • Canada Land Inventory • Cellular Line Interface • Certified Lotus Instructor • C Language Interface • Clear Indication • Clear Interrupt • Client Library Interface • Clintonville (USA-WI) (location identifier) • Coefficient of Luminous Intensity • Coherent Laser Illumination • Command-Language Interpreter • Command-Line Interface • Command Line Interpreter • Commission Locale d'Information • Common Language Infrastructure • Communications Line Interface • Communications Link Interface • Compare Logical Immediate • Contractor Line Item • Contrat Local d'Insertion • Coronal Loop Interaction • Cost-of-Living Index • Critical Line Item • Cumulative Leakage Index

CLIB Clementine Lunar Image Browser

CLIBS Colinear Laser Ion Beam Spectroscopy

CLIC Centre Local d'Information et de Coordination • Collisionneur Linéaire du CERN • Command Language for Interrogating Computers • Communications Library for Industrial Codes • Compact Linear Collider (CERN)

CLICOM Climate Computing

CLICS Categorical Logic In Computer Science

CLID Calling Line Identification • Course Length In Days

CLIF Computational Linguistics in Flanders (Belgium) • Concept Lattice Interchange Format

CLIH Course Length In Hours

CLIIP Common Law Institute of Intellectual Property

Clim Climate • Climatic

CLIM Common LISP Interface Manager • Consolidated Link Layer Management • Course Length In Months

CLIMAP Climate Applications Project • Climate: Long-range Interpretation, Mapping, And Prediction

CLIMAT Climatology

CLIMEX Climatic Extremes

Clin Clinic • Clinical

CLIN Community Learning and Information Network • Contract Line Item Number

CLINS Climatic Laboratory Instrumentation System

CLIODN Committee for Liaison with International Organizations and Developing Nations

CLIP Calling-Line Identification Parameter • Calling-Line Identification Presentation (*) • Cancel Launch In Progress • Cirrus Lidar Probe • Classical Internet Protocol • Combined Laser Instrumentation Package • Compiler Language for Information Processing • Computer Launch Interference Problem • Configuration Logistic Information Program • Crew Loads Instrument Pallet • Croydon Library Internet Project (UK)

CLIPRT Comité Local d'Information et de Prévention des Risques Technologiques (France)

CLIPS C Language Integrated Production System • Climate Information and Prediction Services (WMO) • Communications Link Interface Planning System

CLIR Calling Line Identification Restriction • Cross-Language Information Retrieval

CLIRCAM Claremont-Riverside-Infrared Camera

CLIS College of Library and Information Services • Contract Line Item Status

CLISP Conversational List Processing

CLIST Command List

CLIVAR Climate Variability

CLIW Course Length In Weeks

CLJ Cluj (Romania) (location identifier)

CLJA Closed Loop Jumper Assembly

Cljo Complejo

Clk Clerk • Clock

CLK Clerk • Clinton (USA-OK) (location identifier) • Clock • Hong Kong - Chek Lap Kok (China-PRC)

Clkw Clockwise

Clkws Clockwise

CLL Bryan (USA-TX) (location identifier) • Chief of Legislative Liaison • Chronic Lymphocytic Leukemia • Collaborative Learning Laboratory • College Station (USA-TX) (location identifier) • Corrected Line Length • Critical Load Level

CLLI Common-Language Location Identification

Cllj Callejón

CLLM Consolidated Link Layer Management

Cllr Councillor

Clm Column

CLM Career Limiting Move • Care Logic Module • Cell-Loss Margin • Central Meridian Longitude • Circumlunar Mission • Congé de Longue Maladie • Conversational Line Multiplexer • Council of Logistics Management • Port Angeles (USA-WA) (location identifier)

CLMC Central Logistics Management Center • Chemical Laser Mode Control

CLMS Component and Library Management System

Cln Clean

CLN Carolina (Brazil-MA) (location identifier) • Central Library Network

CLNA Cooled Low-Noise Amplifier

CLNAP Connectionless Network Access Protocol

Clnc Clearance

CLNIP Connectionless Network Interface Protocol

ClNO3 Chlorine Nitrate

CLNP Connectionless-mode Network Protocol • Connectionless Network Protocol

CLNR Cleaner

CLNS Connectionless-mode Network Service • Connectionless Network Service • Cornell Laboratory of Nuclear Studies (USA-NY)

Clnt Coolant

Clo Closet • Clothing

ClO Chlorine Monoxide

CLO Cali (Colombia) (location identifier) • Chapter Liaison Officer • Chief Liaison Officer • Civil Liaison Officer • Clothing • Club Liaison Office • Communications Liaison Officer • Computer Lock-On • Concentric Line Oscillator • Congressional Liaison Office • Consular Liaison Officer • Contrat Local d'Orientation • Contrat Logiciel • Country Liaison Officer • Crystal Lake Observatory (USA)

Cloir Couloir

Cloirs Couloirs

CLONES Central Location On-line Entry System

ClONO2 Chlorine Nitrate

CLORA Club des Organismes de Recherche Associés (France)

Clos Closet

CLOS Common LISP Object System • Coordination of Land Observing Satellites

CLOT Combined Loads Orbiter Test

CLOUD Cosmics Leaving Outdoor Droplets (experiment)

CLOUDS Cloud Logic to Optimize Use of Defense Systems • Continuous Longitude Observations of Ultracool Dwarfs (collaboration)

clp Clipboard (file extension)

CLP Cell Loss Priority • Cell Loss Probability (*) • Clarks Point (USA-AK) (location identifier) • Clipboard • Club Laser de Puissance (France) • Command Language Processor • Concurrent Logic Programming • Constraint Logic Programming • Continuous Line Plotter • Contract Laboratory Program • Cornell List Processor • Corporate Learning Program • Couche Limite Planétaire • Peso (Chile) (ISO currency code)

CLPA Common Link Pack Area

Clpbd Clapboard

CLPC Computerized Label Printing Center

Clpr Clapper

CLPS Calibration Lamp Power Supply

CLPX Cold Land Processes Experiment

CLQ Clear Request • Colima (Mexico) (location identifier)

Clr Clear • Color • Cooler • Couleur • Councillor

CLR Calipatria (USA-CA) (location identifier) • Canistered Traffic Balloon • Cell Loss Ratio • Circuit Layout Record • Clear • Combined Line and Recording • Command - Leadership - Resource • Common Language Runtime • Computer Language Recorder • Computer Language Research • Coronal-Line Region • Council on Library Resources • Councillor • Current-Limiter Resistor • Current-Limiting Resistor

CLRB Cost Limit Review Board

CLRC Central Laboratory of the Research Councils (UK) • Circuit Layout Record Card • Country Logistics Readiness Center

Clrg Clearing

CLRG Clearing

Clrm Classroom

CLRO Clark Lake Radio Observatory (USA-CA)

CLRP Central Laboratory for Radiological Protection • Command Logistics Review Program

Clrs Colors • Couleurs

CLRS Central Launch and Recovery Station • Clear and Smooth

CLRTAP Convention on Long-Range Transportation of Anthropogenic Pollutants

cls Class (file extension) • Clear Screen

CLS Case Late Stars • Case Law Summary • Center for Language Studies • Central Level Summary • Certificate in Library Science • Chehalis Centralia (USA-WA) (location identifier) • Cislunar Space • Classical Least Squares • Clear and Subtract • Clear Screen • Client System •

Closed Loop Support • Closed Loop System • Close Lunar Satellite • Cloud Lidar System • Collecte et Localisation par Satellite • Command and Launch System • Commercial Launch Services • Committee on Life Sciences • Common Language System • Concept Learning System • Connectionless Server • Constant Level Signal • Constrained Least Squares • Contingency Landing Site • Contract Logistics Support • Contractor Logistics Support • Cycle Line State

CLSA Canada Land Use Monitoring Program

ClSCC Cleveland State Community College (USA-TN)

Clsd Closed

CLSES Closed Loop Solvent Extraction System

CLSF Communications Logistics Support Facility • Connectionless Service Function

Clsg Closing

CLSI Computer Library Service International

CLSID Class Identifier

CLSM Compact Letter Sorting Machine • Confocal Laser Scanning Microscope • Confocal Laser Scanning Microscopy • Conical Log Spiral Mobile • Crew Life Support Monitor

CLSO Contingency Landing Support Officer

CLSP Center for Language and Speech Processing

CLSR Cooled Limb Sounding Radiometer

CLSS Combined Logistics Support System • Computerized Literature Searching Service

CLSSA Cooperative Logistics Supply Support Arrangement

CLSX Closed Loop Support Extended

Clsy Classify

CLT Calculated Landing Time • Cellular Logistics Team • Center for Learning and Teaching • Central Limit Theorem • Charlotte (USA-NC) (location identifier) • Closed-Loop Telemetry • Closed-Loop Test • Common Leader Training • Communications Line Terminal • Computational Learning Theory • Computer Language Translator • Container Load Trailer • Culture

CLTB Communications Link Testbed

C&LTE Cost and Lead Time Estimate

CLTE Coefficient of Linear Thermal Expansion • Cost and Lead Time Estimate

CLTG Center Level Transaction Group

Clthg Clothing

CLTP Connectionless Transport Protocol

CLTR Cross-Language Text Retrieval

CLTS Clear Task Switch • Closed Loop Tracking System • Connectionless Transport Service • Cryogenic Linear Temperature Sensor

CLTU Command Link Transmission Unit

CLTV Closed-Loop Television

CLU California Lutheran University (USA-CA) • Central Logic Unit • Collimator Lens Unit • Columbus (USA-IN) (location identifier) • Combination Logic Unit • Command and Launch Unit • Command Logic Unit • Constant Linear Unit • Cryptographic Logic Unit

CLUD Conceptual Clustering with Usage Dependency

CLUE Čerenkov Light Ultraviolet Experiment • Clustering Environment • Common LISP User-interface Environment • Compensated Laser Ultrasonic Evaluation

CLUI Command Language User Interface • Command Line User Interface

CLUMP Canada Land Use Monitoring Program

CLUP Contrat de Licence Utilisateur Final

CLUSIF Club de la Sécurité Informatique Français (France)

CLUT Color Look-Up Table • Computer Logic Unit Tester

CLV Caldas Novas (Brazil-GO) (location identifier) • Center-Limb Variation • Center-to-Limb Variation • Clear Voice • Constant Linear Vector • Constant Linear Velocity • Corrected Loss Variance

CLW Clearwater (USA-FL) (location identifier) • Cloud Liquid Water • Contaminated Liquid Waste • Correction Location Word

Clws Clockwise

CLX Clorinda (Argentina-FO) (location identifier) • Common LISP X-interface

CLY Calvi Sainte-Catherine (France) (location identifier)

CLZ Calabozo (Venezuela) (location identifier) • Craft Landing Zone • Cushion Landing Zone

cm Carat Métrique • Causa Mortis • Centimeter • Centimetric

c/m Call of More

Cm Curium

C/M Communications Multiplexer

C-M Color-Magnitude (relation) • Control-Monitor

C&M Care & Maintenance • Controlling & Monitoring • Control & Monitor

CM Calibrated Magnification • Calibration Magnification • Calibration Module • Cambodia • Camera Module • Cameroon (ISO country code) • Cameroun (code ISO de pays) • Candidate Material • Career Management • Cargo Manager • Carnegie Monitor • Cartographic Material • Case Manager • Celestial Mechanics • Center of Mass • Center of Momentum • Centimeter • Central Maine (USA-ME) • Central Memory • Central Meridian • Certificated Master • Certified Master • Chairman's Memorandum • Chambre des Métiers • Change Management • Chargé de Mission • Chef de Mission • Chrome-Molybdenum • Circulaire Ministérielle • Circular Measure • Circulation Manager • Classification Management • Classification Model • Classified Message • Cloud Medium • Collection Management • Collection Manager • Color-Magnitude • Combined Method • Combustion Modification • Combustion Module • Commander's Manual • Command Module • Commercial Middeck • Common Market • Common Meter • Common Mode • Communications Management • Communications Manager • Communications Modem • Communications Multiplexer • Complementary Medicine • Complex Mixture • Composite Material • Computer Malfunction • Computer Modern (font) • Computer Module • Computing Media • Conceptual Model • Condition Monitoring • Confidential Memorandum • Configuration Management • Configuration Manager • Configuration Monitor • Congé de Maladie • Connection Machine • Connectionnist Model • Construction and Machinery • Construction Management • Consumable Management • Contamination Mode • Contraction Mapping • Contract Management • Contract Modification • Contract Monitor • Contrast Medium • Control and Monitoring • Contrôle Médical • Control Mark • Control Mode • Control Modem • Control Monitor • Coordination Meeting • Core Meltdown • Core Memory • Corporate Membership • Correction Memorandum • Corrective Maintenance • Corresponding Member • Cost Monitoring • Counterfeit Materiel • Countermeasure • Cours Magistral • Cours Moyen • Credit Memorandum • Crew Module • Crisis Management • Critical Mass • Cruise Mode • Cursor Movement • Cuthill-McKee (algorithm)

Cma Cima

CMa Canis Major - Canis Majoris (constellation)

CMA Calcium Magnesium Acetate • Canadian Medical Association (Canada) • Canadian Micrographic Association (Canada) • Canadian Museums Association (Canada) • Cash Management Account • Cell Management Agent • Census Metropolitan Area • Centrally-Managed Allocation • Central Management Agency • Central Mapping Agency • Central Mapping Authority • Centre de

Mathématiques Appliquées • Centre de Modane Avrieux (ONERA) • Centre for Mathematical Analysis • Centro Meteorológico Antártico • Cercle Méridien Automatique • Certificate of Management Accounting • Channel Multiplier Array • Chef de Mission Adjoint • Chemical Manufacturers' Association • Chemical Manufacturing Association • China Meteorological Administration (China-PRC) • Círculo Meridiano Automático • Classified Mail Address • Collection Management Authority • Command Modulator Assembly • Commercial Mailing Agent • Committee on Minorities in Astronomy • Communications Managers Association (USA) • Complex Modulus Apparatus • Computer Management Association (USA) • Computer-Marked Assignment • Concert Multithread Architecture • Configuration Management Accounting • Conical Monopole Antenna • Conseil Mondial de l'Alimentation • Consistent Multifluid Advection • Contact Making Ammeter • Contact Motion Analysis • Contract Management Assistant • Contrat Matériel • Control - Monitor - Alarm • Cunnamulla (Australia-QLD) (location identifier) • Cylindrical Mirror Analyser

CMAA Central Missouri Astronomical Association (USA-MO) • Communications Mission Area Analysis

CMAC Centralized Maintenance and Administration Center • Cerebellar Model Articulation Controller

CMACS Central Monitor And Control System

CMAG Configuration Management Advisory Group

CMAI Chemical Material Acquisition Initiative

CMAIS Configuration Management Automated Information System

CMAK Conical Monopole Antenna Kit

CMAL Controlled Multiple-Address Letter

CMAM Canadian Middle Atmosphere Model • Commercial Middeck Augmentation Module • Computational Methods in Applied Mathematics • Condition Monitoring and Asset Management

CMAME Computer Methods in Applied Mechanics and Engineering (periodical code)

CMAN Configuration Management

CMAO Calcul Mathématique Assisté par Ordinateur • Contract Management Assistance Officer

CMAP Climate Modeling, Analysis and Prediction • Climatic Modeling and Applications Program

CMAR Configuration Management Action Request

CMAS Central Montana Astronomy Society (USA-MT) • Computational Methods in Applied Sciences • Confédération Mondiale des Activités Subaquatiques • Configuration Management Applications Service • Configuration Management Applications Software • Control Monitoring Automated System • Council for Meteorology and Atmospheric Sciences (India)

CMAT Compatible Materials • Coupled Middle Atmosphere and Thermosphere

CMA2 Computations for Metaphors, Analogy, and Agents (workshop)

CMB Caractères Magnétiques en Bâtonnets • Career Management Board • Chemical Mass Balance • Collection Management Board • Colombo Katunayake (Sri Lanka) (location identifier) • Configuration Management Board • Configuration Management Branch • Configuration Management Bulletin • Console Message Buffer • Core-Mantle Boundary • Corrective Maintenance Burden • Cosmic Microwave Background • Crossed Molecular Beam

CMBP Cosmic Microwave Background Polarization

CMBR Cosmic Microwave Background Radiation

CMBT Cosmic Microwave Background Telescope

CMC Call Management Center • Camocim (Brazil-CE) (location identifier) • Canadian Meteorological Centre (Canada) • Carbon Matrix Composite • Carlsberg Meridian Catalogue • Carlsberg Meridian Circle • Center for Macromolecular Crystallography • Central Maintenance Computer • Central Massive Core • Central Methodist College (USA-MO) • Ceramic Matrix Composite • Certificate Management Consultant • Chromosphere Magnetic Canopy • Code for Magnetic Characters • Coherent Multichannel • Collection Management Center • Combined Meteorological Committee • Command Management Center • Command Module Computer • Common Mail Call • Common Messaging Call • Communications Management Configuration • Communications Mode Control • Compact Molecular Cloud • Complement Carry • Complex Monitor and Control • Composite à Matrice Céramique • Computer-Mediated Communication • Conservation Monitoring Centre • Constant Mean Curvature • Construction Maintenance Center • Convolve - Multiply - Convolve • Cooperative Multimodal Communications • Cost Monitoring Coordinator • Crew Module Computer • Crisis Management Center • Crisis Management Council • Current Mode Coupler • Customized Multimedia Connection • Cylindrical Mesh Cathode

CMCC Central Mission Control Centre • Classified Material Control Center

CMCD Centro de Missão de Coleta de Dados • Cylindrical Mesh Cathode Device

CMCI CompuMath Citation Index • Computer and Mathematics Citation Index

CMCIM Categorical Methods for Concurrency, Interactions, and Mobility (conference)

CMCIRS Computerized Map Cataloguing and Information Retrieval System

CMCK Centre Médico-Chirurgical de Kourou (French Guyana)

CMCL Command Management Control List

CMCO Classified Material Control Officer

CMCRL Consolidated Master Cross-Reference List

CMCS Case Management Control System • Centralized Maintenance Computer System • Cluster Mission Control System • Communications Monitoring and Control System • Complex Monitor Control System • COMSAT Maritime Communications Satellite • Construction Management Control System

CMCSM Center for Mathematical and Computational Science in Medicine

CMCTL Current Mode Complimentary Transistor Logic

cmd Commands (file extension)

Cmd Command • Commendation

CMD Catalog Management Data • Catalogued Metadata • Centralized Message Distribution • Central Meridian Distance • Centre Mondial de Données • Chargé de Mission Diplomatique • Charge Modulation device • Circuit Mode Data • Collection Management and Dissemination • Color-Magnitude Diagram • Color Monitor Device • Color Monitor Display • Command • Commanded • Commendation • Compatibility Mode Driver • Contract Management District • Contract Management Division • Cootamundra (Australia-NSW) (location identifier) • Core Memory Drive • Core Memory Driver • Cryogenic Magnetic Detector • Current Meter Digitizer

CMDAC Current-Mode Digital-to-Analog Converter

CMDB Composite Modified Double Base • Configuration Management Database • Consolidated Manpower Database

CMDE Chargé de Mission de Défense Économique (France)

CMDF Combined Main Distribution Frame • Command Masking Data Facility

Cmdg Commanding

CMDL Climate Monitoring and Diagnostics Laboratory (*) • Climate Monitoring Diagnostics Laboratory

CMDM Control Multiplexer/Demultiplexer

CMDN Catalog Management Data Notification

Cmdo Comodoro

Cmdr Commander

CMDR Card Maintenance and Data Recorder • Commander • Command Reject

Cmdre Commodore

CMDRG Chislennye Metody Dinamike Razrezhennykh Gazov (periodical code)

CMDS Centralized Message Data System • Cockpit Management Display System • Commands • Configuration Management Data System • Consortium for Materials Development in Space • Computer Misuse Detection System

Cmdt Comandante • Commandant

CMDT Commandant • Conférence Mondiale de Développement des Télécommunications (UIT)

Cmdte Comandante

Cme Centime

CME Camera Module Electronics • Center for Media Education • Centre Médico-Éducatif • Chemically Modified Electrode • Circuit Multiplexing Equipment • Ciudad del Carmen (Mexico) (location identifier) • Coastal and Marine Ecology • Commercial Measuring Equipment • Common Mode Error • Community Modeling Effort • Component Management Entity • Computer Memory Element • Conférence Mondiale de l'Énergie • Conformance Management Entity • Conseil Mondial de l'Énergie • Continuing Medical Education • Contract Manpower Equivalent • Coronal Mass Ejection

CMEA Cellular Message Encryption Algorithm • Council for Mutual Economic Assistance

CMEM Canonical Multievent Model • Computational Methods and Experimental Measurements (conference)

CMEO Coronal Mass Ejection Onset

CMEP Comité Mixte d'Évaluation et de Prospective

CMER Center for Management and Economic Research

CMERI Central mechanical Research Institute

CMES Common Modular Elint System

CMET Command Management Engineering Team • Consolidated Master Edit Table

CMEX Center for Mars Exploration (NASA)

cmf Corel Metafile (file extension)

CMF Cancelling Magnetic Feature • Carbon Monofluoride • Career Management Field • Cartesian Mapping Function • Case Master File • Cask Maintenance Facility • Central Maintenance Facility • Central Microstructure Facility • Chambéry (France) (location identifier) • Chip-Matched Filter • CICS Monitor Facility • Cloud Mass Function • Cluster Mass Function • Cockpit Motion Facility • Coherent Memory Filter • Command Management Framework • Common Mode Failure • Comoving Frame • Complex Maintenance Facility • Comprehensive Management Facility • Compressed Magnetic Field • Configuration Management Facility • Conjugate Matched Filter • Conseil des Marchés Financiers (France) • Consolidated Maintenance Facility • Containerized Maintenance Facility • Core-Mantle Friction • Core Mass (probability) Function • Crisis Management facility • Cross-Modulation Factor • Current Master File • Cylindrical Magnetic Film

CMFA Common-Mode Failure Analysis

CMFD Computational Magneto-Fluid Dynamics

CMFE Civilian Manpower and Funding extract

CMFG Communications Management Functional Group

CMG Collaboration in Mathematical Geosciences • Commercial Middeck Galley • Commission for Marine Geology • Computer Management Group • Computer Measurement Group • Control Moment Gyro • Control Monitor Group • Corumbá (Brazil-MS) (location identifier)

CMGA Control Moment Gyro Assembly

CMGD Centre of Marine Geological Data

CMGEA Control Moment Gyro Electrical Assembly

CMGIA Control Moment Gyro Inverter Assembly

CMGR Committee on Microgravity Research

CMGS Control Moment Gyro System

CMH Columbus - Port Columbus (USA-OH) (location identifier)

CMHAS Crayford Manor House Astronomical Society (UK)

CMHMM Continuous Mixture Hidden Markov Model

CMi Canis Minor - Canis Minoris (constellation)

CMI Cable Microcell Integrator • Caribbean Meteorological Institute (now CIMH) • Cell-Mediated Immunity • Certain Meteoritical Inclusion • Champaign (USA-IL) (location identifier) • Chr. Michelsen Institute (Norway) • Classified Military Information • Clay Mathematics Institute (USA-MA) • Coded Mark Inversion • Comité Maritime International • Comité Météorologique International • Common Management Information • Computer-Managed Instruction • Continuous Measurable Improvement • Copyright Management Information • Correlative Measurement Investigator • Crop Moisture Index • Cryomagnetics Inc. • Cyclotron Maser Instability • Czech Metrology Institute (Czech Republic) • Urbana (USA-IL) (location identifier)

CMIA Cash Management Improvement Act (USA)

CMIB Controlled Multispectral Image Base

CMIC Computer Microfilm International Corporation

CMID Certificat Médical Initial Descriptif

CMIF Carrier Management Individual File • Common Module Integration Facility • Core Module Integration Facility • CWI Multimedia Interchange Format

CMIL Circular Mail

CMILT Coût Moyen Incrémental de Long Terme

C/Min Counts per Minute • Cycles per Minute

CMIO Communications Material Issuing Office

CMIP Carrier Management Information Protocol • Common Management Information Protocol (*) • Common Management Interface Protocol • Cost Management Improvement Program • Cost Methods Improvement Program

CMIQ Centro Méxicano de Información Química (Mexico)

CMIR Centre Météorologique Interrégional • Corotating Merged Interaction Region

CMIS Command Management Information System • Commitment Management Information Service • Committee Members Information System • Common Management Information Service • Common Management Information System • Common Module Integration Simulator • Computer-Managed Instruction System • Computer Management Information System • Computer Microimage System • Computer-oriented Management Information System • Configuration Management Information System • Connection Machine Instruction Set • Control Monitor Isolation System • Corporate Management Information System

CMISE Common Management Information Service Element

CMIT Common Management Information Transmission • Configuration Management Interoperability Test

CMITU Centre Mondial d'Informations Techniques et d'Urbanisme

CMIX Commercial Materials ITA Experiment

CMJ Chi Mei (China-ROC) (location identifier)

CMJN Cyan - Magenta - Jaune - Noir

CMK Club Makokola (Malawi) (location identifier)

Cml Commercial

CML Camooweal (Australia-QLD) (location identifier) • Central Meridian Longitude • Chemical • Chemical Markup Language • Chemical - Medical - Logistic • Cognitive

Modeling Language • Collimated Monochromatic Light • Combined Membership List • Commands and Measurement List • Common Machine Language • Computer-Managed Laboratory • Computer-Managed Learning • Computer-Management Learning • Conceptual Modeling Language • Concurrent Metalanguage • Conference Mailing List • Conversational Modeling Language • Core Mass-Luminosity (relation) • Coupled Map Lattice • Current-Mode Logic • Customized Markup Language

CMLOPS Chemical Operations

CMLR Core-Mass-Luminosity Relation

CMLS Computer Multiple Listing Service

CMM Capability Maturity Model (*) • Cargo Movement Module • Carmelita (Guatemala) (location identifier) • Centre de Météorologie Marine (France) • Centre de Météorologie Maritime (*) • Certified Medical Manager • Coherent Microwave Memory • Color Management Module • Command Memory Management • Commission for Marine Meteorology (WMO) • Commission for Maritime Meteorology • Compact Master Manipulator • Condition-Monitored Maintenance • Coordinate Measurement Machine • Coordinate Measuring Machine (*) • Core Mechanical Mockup • Core Memory Module

CMMA Clothing Monetary Maintenance Allowance

CMMACS CSIR Centre for Mathematical Modeling and Computer Simulation

CMMDA Command Module Multiple-Docking Assembly

CMME Compton's Multimedia Encyclopaedia

CMMF Component Maintenance and Mockup Facility

CMMI Capability Maturity Model Integration

CM/MM Collection Management/Mission Management

CMMP Commodity Management Master Plan • Condensed Matter and Materials Physics

CMMR Common Mode Rejection Ratio

CMMS Computerized Maintenance Management Software • Computerized Maintenance Management System

CMMTL Composites Materials and Manufacturing Technology Laboratory

CMMU Cache/Memory Management Unit

Cmn Commission

CMN Casablanca - Mohammed V (Morocco) (location identifier) • Centre Météorologique National • Cerium Magnesium Nitrate • Cerous Magnesium Nitrate • Common • Configuration Management Network • Constructions Mécaniques de Normandie (France) • Contract Management Network • Contract Master Network • Crewman • Customer Management Network

CMNC Commence

CMNH Cleveland Museum of Natural History (USA-OH)

CMNPS Canadian Minimum Navigation Performance Specification (Canada)

Cmnr Commissionner

CMNS Connection Mode Network Service

CMNSW Common Software

Cmnt Comment

CMNT Comment

CMO Canada Map Office • Caribbean Meteorological Organization • Central Management Office • Central Meteorological Observatory • Chief Marketing Officer • Chief Meteorological Officer • Chief of Mission Operations • Circuit Management Office • Civil-Military Operations • Class Mix Option • Collateralized Mortgage Obligation • Collection Management Office • Collection Management Organization • Common Main Objective • Community Missions Office • Compact Massive Object • Configuration Management Office • Configuration Management Officer • Configuration Management Organization • Contract Management Office • Contract Management Officer • Contracts Management Office • Convoy Movement Order • Coordinating and Monitoring Office • Coordination and Monitoring Office • Obbia (Somalia) (location identifier)

CMOD Centre Mondial d'Orientation Documentaire • Course Module

CMOI Contrôle et Mesures Optiques pour l'Industrie

CMOP Configuration Management Operation Plan

CMOPE Confédération Mondiale des Organisations de la Profession Enseignante

CMORE Civilian Manpower Obligations Resources

CMOS Canadian Meteorological and Oceanographic Society (Canada) • Carbon Molybdenum Steel • Cargo Movement Operating System • Complementary Metal-Oxide Semiconductor • Complementary Metal-Oxide Silicon

CMOSM Configuration Management Operating Systems Manual

CMOT CMIP Over TCP • Communication Management Over TCP/IP

CMOV Conditional Move

Cmp Compromise

CMP Camp • Celestial Mapping Program • Cellular Multiprocessing • Central Meridian Passage • Central Modeling Programme • Centre de Maintien à Poste • Centre de Mise à Poste • Centre Médico-Psychologique • Chemical Mechanical Polishing • Chemical Mechanical Processing • Christa McAuliffe Planetarium (USA-NH) • Climate Modeling Programme • Coastal Mapping Program • Code des Marchés Publics • Command and Monitor Panel • Command Module Pilot • Common Message Parser • Common Mid-Point • Communications and Media Policy • Communications Message Processor • Communiqué Météorologique de Presse • Compare • Comprehensive Monitoring Program • Computer • Computer Modeling Package • Computer Model Package • Condensed Matter Physics • Configuration Management Panel • Configuration Management Plan • Configuration Management Procedure • Configuration Management Program • Congestion Monitoring Period • Console Message Processor • Contrast Mode Photometer • Controlled Materials Production • Corrective Maintenance Performed • Correlation of Molecular Pairs • Cost of Maintaining Project • Cost Monitoring Program • Course Management Plan • Coût Moyen Pondéré • Cross-Media Publishing • Custom Mass Production • Santana do Araguaia - Campo Alegre (Brazil-PA) (location identifier)

CMPA Canadian Magazine Publishers Association (Canada)

Cmpd Compound

CmpEn Composites Engineering (periodical code)

CMPF Core Mass Probability Function

CMPh Condensed Matter Physics

CMPL Complete • Completion

Cmplmt Complement

Cmplte Complete

Cmpltn Completion

CmpMa Composites Manufacturing (periodical code)

Cmpn Companion

Cmpns Compensate

Cmpnt Component

CMPP Centre Médico-Psycho-Pédagogique • Configuration Management Program Plan

Cmpr Compare

CMPR Compare

CMPRT Compartment

CMPS Cluster Mission Planning System • Command Module Procedures Simulator • Compare String • Computer, Mathematical, and Physical Sciences • Concept Master Plan System

Cmpsg Compensating

Cmpsn Compensation

Cmpst Composite

CmpSt Composite Structures (periodical code)

CmpST Composites Science Technology (periodical code)

Cmpt Compute

Cmptd Computed

CMPTD Computed

Cmptg Computing

Cmptr Computer

CMPTR Computer

CMQ Center for Measurement Quality • Clermont (Australia-QLD) (location identifier)

CMQA Configuration Management/Quality Assurance

CMR Call Modification Request • Cameroon (ISO country code) • Cameroun (code ISO de pays) • Cargo Movement Requirement • Catalytic Membrane Reactor • Cell Misinsertion Rate • Cell Misinterpretation Rate • Cellule Multifonctionnelle Régionale • Center for Marine Resources • Center Management Review • Center Materials Representative • Central Meter Reading • Chargé de Mission de la Recherche • Cockpit Management Resources • Collosal Magnetoresistance • Colmar (France) (location identifier) • Color-Magnitude Relation • Color-Magnitude Relationship • Colossal Magnetoresistance • Commander • Commercial Mail Relay • Common Mode Rejection • Communications Monitoring Report • Communications Moon Relay • Computer Modern Roman • Configuration Management Representative • Configuration Management Review • Contract Management Review • Contractor Management Reserve • Contractor Management Review • Countdown Modification Request • Country Management Review • Critical Manufacturing Review • CSIRO Marine Research

CMRA Chemical Marketing and Resources Association • Commercial Mail Receiving Agency • Commercial Mail Receiving Agent

CMRB Contract Modification Review Board • Contractor Material Review Board

CMRDI Central Metallurgical Research and Development Institute

CMRN Cooperative Meteorological Rocket Network

CMRP Configuration Management Requirements Plan

CMRR Common Mode Rejection Ratio

CMRS Calibration/Measurement Requirements Summary • Center for Magnetic Reconnection Studies (USA-IA) • Centre Météorologique Régional Spécialisé • Commercial Mobile Radio Services • Configuration Management Requirements Standard

CMRST Committee on Manpower Resources for Science and Technology

cms Cras Mane Sumendus

CM&S Communications Maintenance & Storage

CMS Cable Management System • Call Management System • Canadian Mathematical Society (Canada) • Card Management Service • Cargo Movement System • Catalog Management System • Center for Mathematical Sciences • Center for Measurement Standards • Central Maintenance System • Central Materials Service • Central Materials Supply • Centre de Mécanique Spatiale • Centre de Météorologie Spatiale (France) • Centre Médico-Social • Centre Mobile de Sécurité • Cepstral Mean Subtraction (method) • Change Management System • Channel Module Shelf • Circuit Maintenance System • Civil and Mechanical Systems • Cockpit Management System • Coded Mask System • Code Management System • Color-Magnitude Stereogram • Color Management System • Command and Monitor System • Command Management System • Command Module Simulator • Common Management Service • Common Mapping System • Common Mode Signal • Communications Maintenance and Storage • Communications Management System • Communiqué Météorologique Spécial • Community Management Staff • Compact Muon Solenoid • Comparative Media Study • Compiler Monitor System • Complement Management and Selection • Complement Management System • Component Model Synthesis • Component Mode Synthesis • Computer Management System • Configuration Management Subgroup • Configuration Management System • Container Management System • Contact Management Software • Content Management System • Contingency Mutual Support • Contract Master Schedule • Contractor Maintenance Services • Contractor Maintenance Support • Controlled Mode Security • Control Monitor Set • Conventional Monitoring System • Conversational Monitoring System • Conversational Monitor System • Conversation Monitoring System • Conversation Monitor System • Core Module Simulator • Corrective Maintenance System • Cost Measurement System • Cost Monitoring Specialist • Cours de Mathématiques Spéciales • Courseware Management System • Crisis Management System • Cross-Memory Service • Current-Mode Switching • Customer Management System • Scusciuban (Somalia) (location identifier)

CMSA Consolidated Metropolitan Statistical Area

CMSC Consolidated Mission Support Center

CMSCD Communication Mode Selector Control Device

Cmsd Commissioned

CMSDG Configuration Management System Development Group

CMSE Candidate Materials Space Exposure

CMSEP Contractor Management System Evaluation Program

Cmshft Camshaft

CMSI Command Monitor System Interface • Crisis Management System Interface

CMSITRI Centre for Measurement Standards of the Industrial Technology Research Institute (China-ROC)

CMSM Committee on Marine Surveying and Mapping

Cmsn Commission

Cmsnd Commissioned

CMSP Concept Model Spectropolarimeter • Configuration Management Support Plan

CMSQ Communications Maintenance Squadron

Cmsr Commissioner

CM&SS Contractor Maintenance & Supply Support

CMSS Canadian Mobile Servicing System • Commission on Molecular Structure and Spectroscopy • Configuration Management Support System • Crisis Management Support System

CMST Collection Management Support Tool

CMSU Central Missouri State University (USA-MO)

Cmt Comment

CMT Cadmium Mercury Telluride • Cametá (Brazil-PA) (location identifier) • Career Management Training • Carlsberg Meridian Telescope (Canary Islands) • Cellular Mobile Telephone • Centroid - Moment - Tensor • Code Matching Technique • Collective Monitored Training • Colossal Magneto-Resistant (material) • Comisión del Mercado de Telecomunicaciones (Spain) • Common Maintenance Truck • Compressible MHD Turbulence • Computer Memory Tester • Connection Management • Continuous Media Toolkit • Continuum Mechanics and Thermodynamics (periodical code) • Core Makeup Tank • Corrected Mean Temperature

CMTA Centre Multifonction de Télécommunications Avancées

CMTBCF Control Mean Time Between Critical Failures

CMTBDE Control Mean Time Between Downing Events

CMTD Common Maintenance Training Device

Cmte Comandante • Committee

CMTI Celestial Moving Target Indicator

CMTK Common Mapping Tool Kit

CMTRK Configuration Management Tracking (system)

Cmts Comments

CMTS Comments • Compliance Monitoring and Tracking System • Computerized Maintenance Test System • Computer Maintenance Test Set

CMTTRF Control Mean Time To Restore Function

CMTTRS Control Mean Time To Restore System

CMTU Cartridge Magnetic Tape Unit

CMU Carnegie Mellon University (USA-PA) • Centralized Markup Unit • Central Management Unit • Central Michigan University (USA-MI) • Chiang Mai University (Thailand) • Clear Memory Utility • Clock Monitor Unit • Command and Monitoring Unit • Common-Mode Voltage • Communications Management Unit • Construction and Maintenance Unit • Control and Monitor Unit • Core Memory Unit • Cotisation Maladie Universelle • Couverture Maladie Universelle • Kundiawa (Papua New Guinea) (location identifier)

CMUA Certificate Management User Agent

cmv Corel Move (file extension)

CMV Cloud Motion Vector • Common Mode Voltage • Coromandel (New Zealand) (location identifier)

CMVA Coordinated Millimeter VLBI Array

CMVAS Configuration Management Verification Accounting System

CMVC Configuration Management Version Control • Cooperative Model-View Controller

CMVM Contact-Making Voltmeter

CMW Camaguey (Cuba) (location identifier) • Circular Magnetic Wave • Cloud-Motion Wind • Compartmented Mode Workstation • Continuous Microwave

CMwA Computers and Mathematics with Applications (periodical code)

CMWG Configuration Management Working Group

CMX Crisis Management Exercise • Hancock (USA-MI) (location identifier)

CMY Civilian Man-Year • Cyan - Magenta - Yellow • Sparta McCoy (USA-WI) (location identifier)

CMYB Cyan - Magenta - Yellow - Black

CMYK Cyan - Magenta - Yellow - Black

CMZ Caia (Mozambique) (location identifier) • Calcium - Magnesium - Zirconium • Central Molecular Zone

CM2 Contingency Management Computer Model

cn China (PRC) • Cone • Cras Nocte • PRC

C/N Carbon/Nitrogen • Carrier/Noise • Circular Note • Consignment Note • Contract Note • Cover Note • Credit Note

CN Canadian National • Carbon-Nitrogen • Change Notice • Channel • China (PRC) (ISO country code) • Chine (RPC) (code ISO de pays) • Circular Note • Classical Nova • Code Napoléon • Column • Combined Nomenclature • Cometary Nebula • Comité de Normalización • Comité National • Commande Numérique • Command Network • Common Name • Common Network • Communications Network • Compound Nucleus • Condensation Nucleus • Cone • Contract Number • Contractor Notification • Control Number • Coordination Number • Core Network • Corporate Network • Correction Notice • Correspondent Node • Current Node • Customer Network • Cyanogen Radical • Morocco (international aircraft registration code) • PRC (ISO country code) • République Populaire de Chine (code ISO de pays)

CN/A Convertisseur Numérique/Analogique • Customer Name/Address

CNA Canadian National Archives (Canada) • Cananea (Mexico) (location identifier) • Cascaded Network Architecture • Centre National de l'Automatisation (France) • Certified NetWare Administrator • Certified Network Administrator • Certified NetWare Analyst • Certified Novell Administrator • Certified Nursing Assistant • Chemical Notation Association • Code Not Allocated • Comisión Nacional de Astronomía • Communications Network Application • Communications Network Architecture • Computer and Network Architecture • Computer Network Attack • Convertisseur Numérique/Analogique • Copper-Nickel Alloy • Corporate Network Accreditor • Cosmic Noise Absorption • Customer Name and Address

CNAA Community Needs Assessment Approach • Concours National de l'Apprentissage Artisanal (France) • Consorzio Nazionale per l'Astronomia e l'Astrofisica (Italy) • Council for National Academic Awards (UK)

CNAD Conference of National Armaments Directors

CNAE Comisión Nacional de Actividades Espaciales (Argentina)

CNAF Caisse Nationale des Allocations Familiales (France)

CNAM Caisse Nationale d'Assurance Maladie (France) • Conservatoire National des Arts et Métiers (France)

CNAME Canonical Name

CNAMTS Caisse Nationale d'Assurance Maladie des Travailleurs Salariés (France)

CNAP Conseil National des Astronomes et Physiciens (France)

CNAPS Connected Network of Adaptive Processors System • Coprocessing Node Architecture for Parallel Systems

CNAR Compound Net Annual Rate

CNAV Caisse Nationale d'Assurance Vieillesse (France)

CNB Comité National Belge (Belgium) • Coonamble (Australia-NSW) (location identifier)

CNBOS Comité National Belge de l'Organisation Scientifique (Belgium)

CNBPHI Comité National Belge pour le Programme Hydrologique International

Cnc Cancer-Cancri (constellation)

CNC Cámara Nacional de Comercio • Central Navigation Computer • Centre National de la Cinématographie (France) • Centre National de la Consommation (France) • Coconut Island (Australia-QLD) (location identifier) • Comité National de Chimie • Compact Nuclear Cluster • Computerized Numerical Control • Computerized Numerical Controlled • Computerized Numerically Controlled • Computer Numerical Control • Computer Numerical Controller • Computer Numerically Controlled (*) • Computer Numeric Control • Condensation Nucleus Counter • Configurable Network Computing • Contrôle Numérique par Ordinateur

CNCA Conseil National des Clients Aériens (France)

CNCB Computer Networks and Communications Branch

CNCC Country Name to Country Code • Customer Network Control Center

CNCE Command Nodal Control Element • Communications Network Control Element • Communications Nodal Control Element

CNCI Control, Network, and Computational Intelligence

CNCIM Centre National de Coordination et de l'Indexation Matières

CNCIS Commission Nationale de Contrôle des Interceptions de Sécurité

Cncl Conceal • Council

CNCL Cancel • Common Network Command Language

Cncld Cancelled • Concealed

CNCL Commission Nationale de la Communication et des Libertés

Cnclr Councillor

CNCPI Chambre Nationale des Conseils en Propriété Industrielle (France)

CNCPRST Centre National de Coordination et de Planification de la Recherche Scientifique et Technique (Morocco)

CNCPT Concept

CNCPTS Concepts

Cncr Concurrent

CNCR Characterization of Neurospora Circadian Rythms

CNCS Cold Neutron Chopper Spectrometer

CNCT Consejo nacional de Ciencia y Tecnología (Mexico)

Cnctrc Concentric

Cncv Concave

Cñd Cañadón

Cnd Conduit

CND Cannot Duplicate • Centre National de Documentation • Circumnuclear Disk • Computer Network Defense • Condition • Constanta (Romania) (location identifier) • Contrôle Non-Destructif

Cñda Cañada

Cndct Conduct

CNDCT Centro Nacional de Documentación Científica y Tecnológica (Bolivia)

CNDCTE Centro Nacional de Documentación Científica, Ténica y Económica (Uruguay)

Cndctr Conductor

Cndcty Conductivity

CNDEP Centre National de Documentation sur l'Enseignement Privé (France)

CNDI Commercial Non-Development Item

CNDL Centre National de Dépôt et de Labellisation (France)

CNDN Canadian

CNDO Complete Neglect of Differential Overlap

CNDP Centre National de Documentation Pédagogique (France) • Communications Network Design Program

Cnds Condensate

CNDS Centro Nazionale di Documentazione Scientifica (Italy)

CNDST Centre National de Documentation Scientifique et Technique (Belgium) • Centro Nazionale di Documentazione Scientifico-Tecnica (Italy) • Conseil National de la Documentation Scientifique et Technique

CNDT Centro Nazionale di Documentazione Tecnica (Italy)

Cndtn Condition

CNDWI Critical Nuclear Defense Weapon Information

CNE Campus Network Environment • Canon City (USA-CO) (location identifier) • Certified NetWare Engineer • Certified Network Engineer • Certified Novell Engineer • Comisión Nacional de Energía • Comité National d'Évaluation • Communications Network Emulator • Concentrateur Numérique Éloigné • Contrôle National des Emplois

CNEA Comisión Nacional de Energía Atomica (Argentina) (now CONEA)

CNEAF Coal, Nuclear, Electric, and Alternate Fuels

CNEAI Comisión Nacional Evaluadora de la Actividad Investigadora (Spain)

CNECO Change Notice Engineering Change Order

CNED Center for Non-conventional Energy Development • Centre National d'Éducation à Distance (France) • Centre National d'Enseignement à Distance (France) (*)

CNEI Conseil National des Entreprises d'Insertion (France)

CNELG Compact Narrow Emission-Line Galaxy

CNEN Comissão Nacional de Energia Nuclear (Portugal) • Comitato Nazionale per l'Energia Nucleare (Italy) • Comité National d'Études Nucléaires

CNEP Centre National d'Études à Distance (France)

CNER Comité National d'Évaluation de la Recherche (France) • Conseil National des Économies Régionales (France)

CNES Centre for Neural and Evolutionary Systems (UK) • Centre National d'Études Spatiales (France)

CNESER Conseil National de l'Enseignement Supérieur Et de la Recherche (France)

CNESR Conseil National de l'Enseignement Supérieur et de la Recherche (France)

CNET Centre National d'Études des Télécommunications (France) • Conseil National de l'Enseignement Technique (France)

CNETC Conseil National de l'Enseignement Technique par Correspondance (France)

CNETP Consolidated New Equipment Training Plan

CNEXO Centre National pour l'Exploitation des Océans (France)

cnf Configuration (file extension)

CNF Belo Horizonte - Confins (Brazil-MG) (location identifier) • Belo Horizonte - Tancredo Neves (Brazil-MG) (location identifier) • Canadian Neutron Facility (Canada) • Configuration • Conjunctive Normal Form • Coordinator Nationale Faciliteit (Netherlands)

CNFA Comité National Français d'Astronomie (France)

CNFE Composite Network Front End

CNFL Confluence

CNFGG Comité National Français de Géodésie et de Géophysique (France) (*) • Conseil National Français de Géodésie et de Géophysique (France)

Cnfig Configuration

CNFM Comité National de Formation en Microélectronique (France)

CNFP Centrul National de Formare a Personalulu (Romania) • Commission Nationale de Formation Permanente (France)

CNFPT Centre National de Formation Publique Territoriale (France)

CNFRS Comité National Français de Radioélectricité Scientifique (France) • Compact Nuclear Flat-Spectrum Radio Source

CNFS Chinese National Foundation of Sciences (China-PRC)

CNFWMP Canadian Nuclear Fuel Waste Management Program (Canada)

CNG Calling • Charge Number Grouping • Cognac Parvaud (France) (location identifier) • Compressed Natural Gas

CNGPS Chinese Navigation Global Positioning System

CNGS CERN Neutrinos to Gran Sasso (experiment)

CNH Claremont (USA-NH) (location identifier)

CNHS Coupled Natural and Human System

CNI Calcul Numérique Intensif • Carte Nationale d'Identité • Center for Networked Information • Centre National de l'Informatique • Certified NetWare Instructor • Certified Novell Instructor • Changhai (China-PRC) (location identifier) • Coalition for Networked Information • Commissariat National à l'Informatique • Committee for Nuclear Information (USA) • Common Network Interface • Communications - Navigation - Identification • Consejo Nacional de Investigación • Coulomb-Nuclear Interference

CNIAE Conseil National de l'Insertion par Activité Économique (France)

CNIB Centro Nazionale di Informazioni Bibliografique (Italy)

CNIC Centre National de l'Informatique Chimique (France) • Centro Nacional de Investigaciones Científicas • Computer Network and Information Center • Computer Network Information Center

CNICT Centro Nacional de Información Científica y Ténica (Venezuela) • Consejo Nacional de Investigaciones Científicas y Ténicas (Argentina)

CNIDC Centro Nacional de Información y Documentación Científica (Spain)

CNIDCYT Centro Nacional de Información y Documentación Científica Y Tecnológica (Peru)

CNIDE Centre National d'Information et de Documentation Économiques (Algeria)

CNIDR Clearinghouse for Networked Information Discovery and Retrieval

CNIDS CommerceNet/Nielsen Internet Demographic Survey

CNIDST Comité National de l'Information et de la Documentation Scientifique et Technique (France)

CNIE Comisión Nacional de Investigaciones Espaciales (Argentina)

CNIES Centro Nacional de Información Económica y Social (Mexico)

CNIG Centre National de l'Information Géographique (France) • Centro Nacional de Informação Geográfica (Portugal) • Conseil National de l'Information Géographique (France)

CNIL Commission Nationale de l'Informatique et des Libertés (France) (*) • Communications, Navigation, and Identification Laboratory

CNIN Centraal Normalisatie Instituut in Nederland (Netherlands)

CNIP Calling Name Identification Presentation • Center for Networked Information and Publishing

CNIPA Committee of National Institutes of Patent Agents

CNIPTG Communications, Networks and Information Processing Theory Group

CNIR Calling Name Identification Restriction • Centre National d'Information Routière (France) • Connected Name Identification Restriction

CNIS Centre National d'Information Statistique (France)

CNISF Conseil National des Ingénieurs et Scientifiques de France

CNIT Centre National des Industries et des Techniques (France)

CNIU Communications/Navigation Interface Unit • Customer Network Interface Unit

CNJ Cloncurry (Australia-QLD) (location identifier)

CNK Circumstances Not Known • Common Noun Keyword • Concordia Blosser (USA-KS) (location identifier)

CNKD Circumnuclear Keplerian Disk

Cnl Cancel

CNL Canadian National Library (Canada) • Carrier Noise Level • Central National Library • Circuit Net Loss • Committee on National Libraries (UK) • Commonwealth National Library (Australia) • Computer Newsletter • Sindal (Denmark) (location identifier)

CNLA Council of National Library Associations (USA)

CNLD Cancelled

CNLH Canelha

CNLIA Council of National Library and Information Associations (USA)

CNLLS Complex Non-Linear Least Squares

CNLP Connectionless Network Layer Protocol

CNLS Center for Non-Linear Science (LANL) • Complex Non-linear Least Squares

CNM Carlsbad (USA-NM) (location identifier) • Centre National de Métrologie • Centro Nacional de Metrología (Mexico) • Cold Neutral Material • Cold Neutral Medium (*) • Cold Nuclear Medium • Comité National de Mathématique • Communications Network Management • Communications Network Manager • Cool Neutral Medium • Customer Network Management

CNMA Communication Network for Manufacturing Applications

CNME Centre National des Mouvements d'Énergie (France) • Communications in Numerical Methods in Engineering (periodical code)

CNMI Communications Network Management Interface

CNMS Canadian National Measurement System (Canada) • Communications Network Management System

CNMV Comisión Nacional del Mercado de Valores (Spain)

CNN Cable News Network • Cellular Neural Network • Commercial News Network • Composite Network Node • Conference on Neural Networks

Cno Corno

CNO Carbon - Nytrogen - Oxygen • Chief of Naval Operations • Chief Nursing Officer • Chino (USA-CA) (location identifier) • Collaborateur Non Officiel

CNOC Canadian Network for Observational Cosmology (Canada) • Commission on Nomenclature in Organic Chemistry (IUPAC)

CNOFS Communication Navigation Outgae Forecast System

CNOT Controlled Not (gate)

CNOUS Centre National des Oeuvres Universitaires et Scolaires (Francc)

CNP Candidate Nomination Proposal • Centre National de Prêt • Committee on Nuclear Physics • Connection Not Possible • Connexion Non-Pertinente • Conseil National des Programmes • Customized Networking Platform • Eastgreenland (Greenland) (location identifier)

CNPCN Comité National pour la Protection du Ciel Nocturne (France)

CNPF Conseil National du Patronat Français (France)

CNPI Communication, Navigation, and Position Integration

CNPq Conselho Nacional de Pesquisas (Brazil)

CNPR Centre National de Promotion Rurale (France) • Channel Precedence

CNPS Centre National des Professions de Santé (France) • Common Network Planning Software • Compact Nuclear Power Source • Conseil National de la Politique Scientifique (Belgium)

CNQ Conselho Nacional da Qualidade (Portugal) • Corrientes - Camba Punta (Argentina-CR) (location identifier)

Cnr Corner

CNR Carrier-to-Noise Ratio • Chanaral (Chile) (location identifier) • Change Notice Request • Circumnuclear Ring • Committee on Natural Resources • Complex Node Representation • Composite Noise Rating • Conseil National de la Recherche • Conservatoire National de Région (France) • Consiglio Nazionale delle Ricerche (Italy) • Coordination Nationale des Réseaux (France) • Corner • Cosmic-Ray Nucleus • Council of National Representatives

CNRC Conseil National des Recherches du Canada

CNRD China's Natural Resources Database

CNRET Centre for Natural Resources Energy and Transport

CNRF Cold Neutron Research Facility

CNRG Comité National de Recherches Géomorphologiques

CNRI Corporation for National Research Initiatives

CNRL Communications and Navigation Research Laboratory

CNRM Centre for Natural Resources Management • Centre National de Recherches Météorologiques (France) • Committee on Nuclear Risk Management

CNRS Centralized Network Reference Service • Centre National de la Recherche Scientifique (France)

CNRT Centre National de Recherche Technologique

CNRVA Comité National de Reconnaissance et de Validation des Acquis (France)

CNS Cairns (Australia-QLD) (location identifier) • Canadian Nuclear Society (Canada) • Catalogue of Nearby Stars • Center for Non-linear Science (UNT) • Central Navigation Support • Central Nervous System • Certified Novell Salesperson • Channel Networking Service • Cognitive and Neural System • Cognitive Neuroscience Society • Cold Neutron Source • Comet News Service • Commercial Network Survivability • Common Number System • Communications, Navigation and Surveillance • Communications Network Simulator • Communications Network System • Complementary Network Service • Complimentary Network Service • Composite Nuclear Spectrum • CompuServe Network Services • Computer Network System • Computing and Network Services • Context Name Service • Continuous • Custom News Service

CNSA Centro Nacional de Sanidad Ambiente (Spain) • China National Space Administration (Chine-ROC)

CNSC Center for Non-Linear Studies

CNSF Coalition for National Science Funding (USA) • Cornell National Supercomputer Facility (USA-NY)

CNSFR Circumnuclear Star-Forming Region

CNSG Catalogue of Nearby Stars from the Ground • Comité National des Sciences Géologiques

CNSH Catalogue of Nearby Stars from Hipparcos

CNSL Console

Cnsld Consolidate

Cnsltnt Consultant

CNSNS Comisión Nacional de Seguridad Nuclear y Salvaguardias (Mexico)

Cnsp Conspicuously

CNSR Canadian Network for Space Research (Canada) • Comet Nucleus Sample Return

CNSS Core Nodal Switching System • Council of the National Seismic Systems (USA)

CNSTAT Committee on National Statistics

Cnstr Canister

Cnt Count

CNT Canadian National Telegraph • Carbon Nanotube • Celestial Navigation Trainer • Centre National de Télédétection • Charata (Argentina-CH) (location identifier) • Computer Network Technology • Continue • Contract • Coordinated Network Testing

CNTBT Communications Network Testbed • Comprehensive Nuclear Test Ban Treaty

Cntd Contained • Continued

CNTE Centre National de Télé-Enseignement (France)

CNTER Center

CNTG Computer Network Transaction Group

Cntgcy Contingency

CNTI Centre National de Transfert de l'Information

Cntl Control

CNTL Control

Cntn Contain

Cntor Contactor • Contractor

Cntr Container • Contribute • Contribution

CNTR Center • Counter

CNTRCT Contract

Cntrf Centrifugal

CNTL Central

CNTRL Control • Controller

Cntrln Centerline

CNTRLS Controls

CNTRS Centers

CNTS Centre National des Techniques Spatiales

CNTV Consejo Nacional de Televisión

CNTY Country

CNU Chanute - Martin Johnson (USA-KS) (location identifier) • Chungnam National University (Korea-ROK) • Conseil National des Universités (France) • CSM Navigation Update

CNUAH Centro de las Naciones Unidas para los Asentamientos Humanos

CNUCE Centro Nazionale Universitario di Calcolo Elettronico (Italy)

CNUCED Conférence des Nations Unies sur le Commerce Et le Développement

CNUED Conférence des Nations Unies sur l'Environnement et le Développement

CNUL Council on National University Libraries

CNUSC Centre National Universitaire Sud de Calcul (France)

CNUST Conférence des Nations Unies pour la Science et la Technique

cnv Data Conversion (file extension)

CNV Canavieiras (Brazil-BA) (location identifier) • Centre National du Volontariat (France) • Conventional • Conversion

CNVA Conseil National de la Vie Associative (France)

Cnvc Convenience

Cnvr Conveyor

Cnvt Convert

CNVT Convert

Cnvtb Convertible

Cnvtd Converted

Cnvtr Converter

Cnvx Convex

CNWRA Centre for Nuclear Waste Regulatory Analyses

CNX Certified Network Expert • Chiang Mai (Thailand) (location identifier)

CNY Moab (USA-UT) (location identifier) • Yuang Ren-Min-Bi (China-PRC) (ISO currency code)

CNYITP C.N. Yang Institute for Theoretical Physics (USA-NY)

CNYN Canyon

CNYTDO Central New York Technology Development Organization (USA-NY)

CNZ Cangamba (Angola) (location identifier)

c/o Care Of • Carried Over • Cash Order • Change Over • Consist of

co Company • Corsican (ISO language code)

C/o Care Of • Carried Over

Co Camino • Cerro • Cobalt • Colombia • Colorado (USA) • Company • Compagnie • Continuity • Continuous • Corsican (ISO language code) • County • Cowling Number • Kobalt

C&O Concept & Objectives

C/O Carbon-to-Oxygen Ratio • Care Of • Cash Order • Certificate of Origin • Changeout • Checkout • Contamination/Overpressure • Cutoff

CO Cabinet Office • Call Offer • Carbon Monoxide • Cardiac Output • Carried Over • Cash Order • Catalogue d'Observation • Central Office • Change Order • Channel Overflow • Checkout • Chief Officer • Chief Operator • Clerical Officer • Clock Oscillator • Clock Out • Collaborateur Officiel • Colombia (international vehicle registration code) • Colombia (ISO country code) • Colombie (code ISO de pays) • Color • Colorado (USA) • Colour • Column • Combined Operations • Comité d'Organisation • Commanding Officer • Command Operations • Command Output • Commissioned Officer • Communications Officer • Compact Object • Company • Complete Observability • Compteur ordinal • Compte Ouvert • Computer Operator • Conceptual Object • Connection-Oriented • Conscientious Objector • Conseiller d'Orientation • Console Operator • Contingency Operations • Contracting Officer • Contrat d'Orientation • Convert Out • Co-Occurrence • Coordinated Organizations • Correction Officer • Correction Optics • Coupled Oscillator • Crystal Oscillator • Current Oscillation • Cutoff • Cut Out

CoA Comments on Astrophysics

COA Center Operations Area • Central Operating Agency • Central Operating Authority • Certificate Of Authority • Certificate of Operating Authority • Change Of Address • Change Of Assignment • Change Over Acknowledgment • Coherent Optical Array • Columbia (USA-CA) (location identifier) • Commission On Accreditation • Common Optics Assembly • Condition On Admission • Confirmation Of Arrival • Constant Output Amplifier • Course Of Action • Current Operating Allowance

COAA Centro de Observação Astronômica do Algarve (Portugal)

COAAT Course Of Action Assessment Tool

COACH Cognitive Adaptive Computer Help

COACK Component-Oriented Advanced Control Kernel

COADC Committee for Oceanographic Advice to Developing Countries

COADS Comprehensive Ocean-Atmosphere Data Set

COAE Centre d'Orientation et d'Action Éducative • Consultation d'Orientation et d'Action Éducative

COAH Coahuila (Mexico)

COAL Customer Order Acceptance List

COALA Calibration for Ozone by Atmospheric Link Acquisitions

COAM Customer-Owned and Maintained

COAME Customer-Owned and Maintained Equipment

COAN Comptroller Office Automation Network

CoAnk Communications of the Department of Astronomy of Ankara University (periodical code)

CoAp Comments on Astrophysics (periodical code)

CoAP Co-occurrence Agreement Probability

COAP Center for Ocean Analysis and Prediction

COAPEC Coupled Ocean-Atmosphere Processes and European Climate

COAPI Coordinamento degli Osservatori Astronomici Popolari Italiani (Italy)

COARDS Cooperative Ocean-Atmosphere Research Data Service

COARE Coupled Ocean-Atmosphere Response Experiment

CoArm Contributions from the Armagh Observatory (periodical code)

COARS Command On-line Accounting And Reporting System

COAS Coarse Optical Alignment Sight • Council of the Organization of American States • Crewman Optical Alignment Sight • Crew Optical Alignment Sight (*) • Customer Order Acceptance Summary

CoAsi Contributions dell'Osservatorio Astrofisica dell'Università di Padova in Asiago (periodical code)

COASST Committee On the Application of Satellite and Space Technology

COAST Cache On A Stick • Cambridge Optical Aperture Synthesis Telescope • Computer Operations, Audit and Security Technology • Controller's Assistant • Course Of Action Support Tool • Course Of Action System Tool

COAT Coherent Optical Adaptive Technique • Corrected Outside Air Temperature

CoAth Contributions from the Research and Computing Center Academy of Sciences of Athens - Series I Astronomy (periodical code)

Coax Coaxial

COAX Coaxial

cob COBOL Code (file extension)

COB Change Order Board • Chip On Board • Classical Over-the-Barrier (model) • Clip Onboard • Close Of Business • Collocated Operating Base • Columbus Operations Board • Command Operating Budget • Commission des Opérations de Bourse (France) • Communications Office Building • Contenu Optimal des Bases • Coolibah (Australia-NT) (location identifier) • Cosmic Optical Background • Cryogenic Optical Bench • Currently On Board

COBA Communications Operating in Built-up Areas

COBAE Comissão Brasileira de Atividades Espaciais (Brazil)

COBALT Collocation-Based Language Translator

COBE Command Operating Budget Estimate • Cosmic Background Explorer

COBIM Collocated Operating Base Information Model

COBIS Command Operating Budget Information System • Computer-Based Instruction System

COBM Central-Office-Based Multiplexing

COBOL Common Business-Oriented Language

CoBos Contributions of Bosscha Observatory (periodical code)

cobq Cum Omnibus Bonis Quiescat

CoBRA Computerized Bibliographic Record Actions

COBRA CalTech Owens (Valley) Broadband Reprogrammable Array • Coherent Bragg Rod Analysis • Compatible Onboard Ranging • Complete Oil Breakdown Rate Analyzer • Computerized Bibliographic Record Actions • Computerized Boolean Reliability Analysis • Cooperation in Bureau, Research and Administration • Cosmic Background Radiation Anisotropy

COBRAS Cosmic Background Radiation Anisotropy Satellite

CoBrn Contributions of the Public Observatory and Planetarium in Brno (periodical code)

COBS Command Operating Budget System

COBY Current Operating Budget Year

CoC Center of Curvature • Certificate of Compliance • Committee of Council (ESO)

COC Certificate Of Competency • Certificate Of Completion • Certificate Of Compliance • Certificate Of Conformance • Certification Of Compliance • Chain Of Custody • Chamber Of Commerce • Chart Overlay Copier • Circuit Order Control • Close - Open - Close • Coded Optical Character • Code Of Conduct • Combined Operations Center • Command Operations Center • Computer Operations Center • Computer Output Committee • Concordia (Argentina-ER) (location identifier) • Cone Of Confusion

COCA Clearinghouse On Computer Accommodation

CoCam Contributions of Cambridge Observatory (periodical code)

COCAST Council for Overseas Colleges of Arts, Science, and Technology

COCC Central Operations Command Center • Contractor's Operational Control Center

COCESNA Central American Corporation for Air Navigation Services

COCF Centre for Our Common Future • Connection-Oriented Convergence Function

coch Cochlear

COCHISE Cold Chemi-excited Infrared Simulation Experiment • Cosmological Observations at Concordia with High-sensitivity Instrument for Science Extraction

COCI Committee On Chemistry and Industry • Consortium On Chemical Information

COCKTAIL Collisions Of Clouds Keep Tree Algorithm In Labor (code)

COClA Catalog of Open-Cluster Ages

CoCo Configuration Control

CO-CO Contractor Owned - Contractor Operated

COCO Coaxial-Collinear • Collaborative Computing • Contractor Owned - Contractor Operated • Coordinator Of Chain Operations

CoCoi Comunicações do Observatorio Astronômico da Universidade de Coimbra (periodical code)

CoCom Control and Computers (periodical code)

COCOM Coordination Committee

CoCoMo Constructive Cost Model

COCOMO Constructive Cost Model • Cost Construction Model

COCONA Conseil de Coopération Nord-Atlantique

COCONUDS Coordinated Constellation of User-Defined Satellites

COCOS Constraint Optimization and Constraint Satisfaction (conference)

COCOT Customer-Owned Coin-Operated Telephone

COCP Customer Order Control Point

COCPFF Conventional Ordinance Consumption Plan Factor File

COCS Camera Override Control System

COCSP Contingency Operational Contracting Support Plan • Contingency Operational Contracting Support Program

cod Fortran Compiled Code (file extension)

Cód Código

Cod Codex • Codicil • Codification

CoD Córdoba Durchmusterung

COD Carrier Onboard Delivery • Cash On Delivery • Center Operations Directorate • Centre for Orbit Determination • Chamber Of Deputies • Code • Cody (USA-WY) (location identifier) • Coherent Optical Device • Collect On Delivery • Computer-Originated Document • Concise Oxford Dictionary • Confirmation Of Delivery • Connection-Oriented Data • Contract Operations Data • Converted Out Distance • Cooperative Opportunities Document • Correction Of Deficiencies • Cost of Operations Division • Crack Opening Displacement • Current Operation Division

CODA Chemical Online Data Analyzer • Conservation Option for Decision Analysis

CODAB CODATA Bulletin (periodical code)

CODAG Combined Diesel And Gas • Cosmic Dust Aggregation (experiment)

CODAN Carrier-Operated Device Anti-Noise

CoDAO Contributions of the Dominion Astrophysical Observatory (periodical code)

CODAP Client-Oriented Data Acquisition Program • Comprehensive Occupational Data Analysis Program • Computer Documentation of Anomalous Phenomena • Control Data Assembly Program

CODAR Coastal Ocean Dynamics Application Radar • Comprehensive Occupational Data Analysis Program • Correlation Detection And Ranging • Correlation Display Analyzing And Recording

CoDAS Condensed Matter Directed Alerting Service

CODAS Customer-Oriented Data System

CODASYL Conference on Data System Languages

CODAT Composition and Dynamic Analysis Tools (package)

CODATA Committee on Data for Science and Technology • Confederation of Design and Technology Associations

CODC Canadian Oceanographic Data Centre (Canada)

CODCF Central Office Data Connecting Facility

Codd Codices

CODD Central Online Data Directory

CoDDO Contributions of the David Dunlap Observatory (periodical code)

CODE Canadian Organization for Development through Education (Canada) • Centre for Orbit Determination in Europe • Client/Server Open Development Environment • Coastal Ocean Dynamics Experiment • Common Data Environment • Common Operating Digital Environment • Cooperative Olympus Data Experiment

CoDec Coder/Decoder • Compression/Decompression

CODEC Coder/Decoder • Compression/Decompression • Compressor/Decompressor

CODED Computer-Oriented Design of Electronic Devices

CODEL Communications Definition Language

CODER Composite Document Effective Retrieval • Composite Document Expert Retrieval • Composite Document Extended Retrieval

CODES Common Digital Exploitation System • Computer Design and Evaluation System • Computerized Onsite Data Entry System

CODEST Comité de Développement Européen pour la Science et la Technologie • Comité pour le Développement Européen de la Science et de la Technologie • Committee for the Development of European Science and Technology

CODEVER Code Verification

CODEVI Compte pour le Développement Industriel (France)

CoDEX Chemo-Dynamical Evolution of Galaxies

CODF Crystallite Orientational Distribution Function

CODI Cloud Ozone Dust Imager

CODIAC Centralized Or Distributed Integrated Access Control

CoDIAK Concurrent Development, Integration, and Application of Knowledge

CODIC Color Difference Computer • Computer-Directed Communications

CODIF Composition and Distribution Function

CODIL Control Diagram Language

CODIS Controlled Digital Simulator

CODISC Centre Opérationnel de la Direction de la Sécurité Civile (France)

CODMAC Committee on Data Management And Computation (*) • Committee on Data Management And Computing • Committee On Data Management, Archiving and Computing

CODN Component Operational Data Notice

CODOC Code Documentaire

CODOG Combined Diesel Or Gas

CODOT Classification of Occupations and Directory of Occupational Titles

CODP Coherent Optical Dynamic Processor

CoDR Concep Design Review • Conceptual Design Review

CODS Classified Output Distribution System

CoDun Contributions from the Dunsink Observatory (periodical code)

CoE Center of Excellence • Center of Expertise • Council of Europe

COE Center Of Excellence • Center Of Expertise • Central Office Equipment • Certificate Of Entitlement • Certification Of Equivalency • Chief Of Engineers • Chief Operating Engineer • Code Of Ethics • Coeur d'Alene (USA-ID) (location identifier) • Cognizant Operations Engineer • Command Operating Environment • Commission On Ecology • Commitments - Obligations - Expenditures • Committee On ECDIS • Committee On Education • Common Operating Environment • Complete Operating Equipment • Concept Of Engineering • Cooperative Effort Experience • Core Operating Environment • Corps Of Engineers • Cost Of Electricity • Cost Of Energy

COEA Cost and Operational Effectiveness Analysis • Cost Operational Effectiveness Analysis

Co-Ed Co-Editor

CoEd Co-Editor

COED Computer-Operated Electronic Display • Computer-Oriented Electronic Display • Concise Oxford English Dictionary

COEES Central Office Equipment Engineering System

Coef Coefficient

Coeff Coefficient

Coéff Coéfficient

COEFF Coefficient

COEI Component Of End Item

Coel Coelum (periodical code)

COEL Chef des Opérations de l'Ensemble de Lancement • Committee on the Origins and Evolution of Life

COEM Commercial Original Equipment Manufacturer • Component Environmental Model

CoEnCo Committee for Environmental Conservation

COER Central Office Equipment Report

COES Committee for Operations and EGSE Standardisation • Committee for Operations and EGSE Standards • Committee on Open Electronic Standard

COESA Committee On Extension for the Standard Atmosphere

COF Central Operating Facility • Classroom Of the Future • Cocoa Patrick (USA-FL) (location identifier) • Columbus Orbital Facility • Computer Operations Facility • Computer-Optimized Fabrication • Construction Of Facilities • Contract Operations File • Correlator Output Format • Cost Of Facilities

CofA Certificate of Airworthiness

COFAC Construction Facility

CofC Chamber of Commerce

CoFC Centennial of Flight Commission (USA)

COFC Container On a Flat Car

COFCO Configuration Control

COFD Contracting Officer Final Decision

COFDM Coded Orthogonal Frequency Division Multiplex • Coded Orthogonal Frequency Division Multiplexing

CofE Certificate of Experience

COFECUB Comité Français d'Évaluation de la Coopération Universitaire avec le Brésil (France)

COFF Common Object File Format • Cut Off

CofG Center of Gravity

COFI Checkout and Fault Isolation

COFIDA Comité des Fonds Internationaux de Droits d'Auteurs

COFIL Core File

CoFl Combustion and Flame (periodical code)

COFM Correlation Of Forces and Means

CofP Center of Pressure

COFR Certificate Of Flight Readiness • Certification Of Flight Readiness

CoFra Contributions of the CNR Astrophysics Laboratory in Frascati (periodical code)

COFRAC Comité Français d'Accréditation (France)

COFRAD Central-Office Frame Relay Access Device

COFRS Chinese Operational Earth Resources Satellite

CofS Chief of Staff

COFS Coastal and Ocean Flux Study • Coastal Ocean Forecasting System

CofT Certificate of Test

CoFuR Cost of Fulfillment of Request

COFUSI Comité Français des Unions Scientifiques Internationales (France)

COFW Certificate Of Flight Worthiness

cog Center of Gravity

Cog Cognate • Cognizant

CoG Center of Gravity • Curve of Growth

COG Center Of Gravity • Centralized Ordering Group • Cluster Of Galaxies • Collaborative Open Group • Computer Operations Group • Condoto (Colombia) (location identifier) • Congo (ISO country code) • Continuity Of Government • Council Of Governments • Course Over Ground • Curve Of Growth

COGB Certified Official Government Business

COGENE Committee on Genetic Experimentation

COGENT Compiler and Generalized Translator

COGEODATA Committee on Geological Data

COGEODOC Commission on Geological Documentation

COGEP Confined One Gluon Exchange Potential

COGI Centre d'Organisation Générale pour l'Informatique (France)

CoGIS Cooperative Geographic Information System

COGO Coordinate Geometry

COGS Combined Operations Ground System • Computer-Oriented Geological Society • Continuous Orbital Guidance System • Cost Of Goods Sold

COGSIS Committee on Government Statistics and Information Services (USA)

COGUANOR Comisión Guatemalteca de Normas (Guatemala)

COH Cash On Hand • Connection Overhead • Cooch Behar (India) (location identifier)

COHC China Ocean Helicopter Corporation

COHMAP Cooperative Holocene Mapping Project

COHO Coherent Oscillator • Coordinated Heliospheric Observations

COHP Computerized Occupational Health Program

COHRA Cambridge Observing Handbooks for Research Astronomers (CUP)

COHS Collaborative Open Hypermedia Systems

COHSI Cambridge OH Suppression Instrument

Co-I Co-Investigator

CoI Co-Investigator

COI Center for Optoelectronics and Imaging • Central Office of Information • Certificate of Origin and Interest • Charge-Ordered Insulator • Chief Of Installation • Class Of Interface • Coast Orbital Insertion • Cocoa - Merritt Island (USA-FL) (location identifier) • Co-Investigator • Comisión Oceanográfica Intergubernamental • Comité

Olympique International • Commission de l'Océan Indien • Commission Océanographique Intergouvernementale • Commission Océanographique Internationale • Communication Operations Instruction • Community Of Interest • Composite Optics Inc. (USA-CA) • Conflict Of Interest • Connection-Oriented Interconnection • Contemporary Operational Issues • Contingency Orbit Insertion • Cost Of Illness • Course Of Instruction • Critical Operational Issue

CoIAB Contributions de l'Institut d'Astrophysique de Paris - Série B (periodical code)

CoIAP Contributions de l'Institut d'Astrophysique de Paris - Série A (periodical code)

COIC Canadian Oceanographic Identification Centre (Canada) • Careers and Occupational Information Center • Critical Operational Issues and Criteria

COIE Centro de Orientación, Información y Empleo

COIF Charities Official Investment Fund

COIL Chemical Oxygen-Iodine Laser

COIM Checkout Interpreter Module

COIN Committee On Information Needs • Common Operators Interface Nucleus

COINS Commercial Operation Integrated System • Committee On Improvement of National Statistics • Community Online Intelligent Network System • Computer and Intelligence Sciences • Computer-Operated Instrument System • Contractor Invoice Service

CoIoa Contributions from the Laboratory of Astronomy - Ioannina (periodical code)

COIP Centre d'Orientation Interprofessionnel • Contemporary Operational Issues Process

COIR Content-Oriented Information Retrieval

COIS Canals Otoliths Integration Studies • Computation-Oriented Information System

COISTD Confédération des Organisations Internationales Scientifiques et Techniques pour le Développement

COJ Coonabarabran (Australia-NSW) (location identifier)

COK Cochin (India) (location identifier) • Cook Islands (ISO country code) • Îles Cook (code ISO de pays)

COKE CORBA Keyword

CoKit Contributions from the Kitt Peak National Observatory (periodical code)

CoKon Communications of Konkoly Observatory (periodical code)

CoKwa Contributions from the Kwasan and Hida Observatories (periodical code)

CoKyo Contributions from the Institute of Astrophysics - Kyoto (periodical code)

col Multiplan Spreadsheet (file extension)

Col Cola • Colección • Colegio • Collect • Collected • Collection • Collector • College • Collegiate • Colombia • Colombian • Colon • Colonia • Colonial • Colony • Color • Colorado (USA) • Colored • Columba-Columbae (constellation) • Column • Columna

COL Carry-On Laboratory • Checkout Language • Coherent Optical Laser • Colima (Mexico) • Collision • Coll Island (UK) (location identifier) • Colombia (ISO country code) • Colombie (code ISO de pays) • Column • Communications-Oriented Language • Computer-Oriented Language • Cosmos-On-Line • Cost Of Living

COLA Center for Ocean-Land-Atmosphere (studies) (USA-MD) • Collision Avoidance • Committee On Library Automation • Compact Objects in Low-luminosity AGNs • Compact Objects in Low-powered AGNs • Constant-Output Level Adapter • Cost-Of-Living Adjustment • Cost-Of-Living Allowance

COLAGE Conferencia Latinoamericana de Geofísica Espacial

colat Colatus

COLD Computer Output to Laser Disk

COLDS Cargo Off-Load and Discharge System • Common Opto-electronic Laser Detection System

COLDSAT Cryogenic On-orbit Liquid Depot Storage, Acquisition and Transfer

Colec Colección

Colect Colectivo

colent Colentur

colet Coletur

COLI Company Owned Life Insurance

CoLic Contributions of Lick Observatory (periodical code)

COLICO Committee On Library Cooperation (Ireland)

COLIDAR Coherent Light Detection And Ranging

COLINGO Compile Online and Go

CoLIS Conceptions of Library and Information Science

COLIS Council On Library and Information Services

Coll Collateral • Colleague • Collect • Collected • Collection • Collective • Collector • College • Collège • Collegiate • Collègue • Collaboration • Colloque • Colloquial • Colloquium

COLL Collect • Collection • Collective • Collision

Collab Collaborate • Collaboration • Collaborator

COLLAGE Collaborative Analysis and Graphics Environment • Collision and Aggregation Experiment

Collat Collateral • Collaterally

Collect Collective • Collectively

COLLIDE Collision Into Dust Experiment

COLLNET Collaborative Network

COLLOG Columbus Logbook

Colloq Colloque • Colloquial • Colloquialism • Colloquially • Colloquium

Collr Collector

collun Collunarium

collut Collutorium

collyr Collyrium

Colm Column

COLM Column

Colo Colorado (USA)

colog Cologarithm

COLOG Computer Logic (conference)

Colon Colonial

CoLon Communications of the University of London Observatory (periodical code)

Coloq Coloquial

color Coloretur

COLOS Columbus Operating System

CoLou Contributions of the Louisiana State University Observatory (periodical code)

COLP Connection Line Presentation

CoLPL Communications of the Lunar and Planetary Laboratory (periodical code)

COLPT Collection Point

COLR Connection Line Restriction

Cols Columns

COLS Communications for Online Systems • Container Off Load System

COLT Central Office Line Tester • Communications Line Terminator • Computerized Online Testing • Computer-Oriented Language Translator • Control Language Translator

COLTRIMS Cold Target Recoil - Ion Momentum Spectroscopy

com Command Executable (file extension)

Com Coma Berenices - Comae Berenices (constellation) • Comedy • Comercio • Comisión • Comité • Comma • Command • Commander • Commentary • Commerce • Commercial • Commission • Commissioner • Committee • Commodore • Common • Commonality • Commonly • Commonwealth • Commune • Communicate • Communicated • Communication • Community • Común • Comunidad

CoM Center of Mass • Chief of Mission • Continuation of Message

COM Center Of Mass • Central Office of Measures • Change Order Modification • Character-Oriented Message • Checkout Mode • Checkout Operations Manual • Chief Of Mission • Coleman (USA-TX) (location identifier) • Collection Operations Management • Color Octet Model • Coma Berenicids (meteor stream) • Command • Commander • Commercial • Commissary Operating Manual • Common Object Model • Common Object Modeling • Common Organization of the Market • Communications • Comores (code ISO de pays) • Comoros (ISO country code) • Compact Optronic Mast • Complex Organic Mixture • Component Object Model • Computerized Operations Management • Computer Operations Manager • Computer Operations Manual • Computer Output Management • Computer Output Microfilm • Computer Output Microfiche • Computer Output Microform • Computer Output on Microfilm • Computer Output on Microfiche • Computer Output on Microform • Concept Of Maintenance • Constant Of Motion • Continuation Of Message • Cyclic Object Methodology

CoMA Comet Mass Analyzer

COMA Conserving Momentum Approximation • Copenhagen Opacities for Model Atmospheres

COMAC Continuous Multiple-Access Collator

COMADAB Columbus Marketing Database

COMANDOS Construction and Management of Distributed Open Systems

ComAp Comments on Astrophysics (periodical code)

COMAP Consortium for Mathematics and its Applications

COMAS Central Office Maintenance and Administration System • Combined Orbital Maneuvering and Abort System

COMAT Computer-Assisted Training

Comb Combination • Combine • Combined • Combining • Combustible • Combustion

COMB Combine • Combined • Combiner • Combustion • Console-Oriented Model Building

Combl Combustible

COMBO Classifying Objects by Medium-Band Observations • Correlation Of Miss Between Orbits

ComBr Communication Broadcasting (periodical code)

Combu Combustion

COMCAT Character-Oriented Message Catalog

CoMcD Contributions from the McDonald Observatory (periodical code)

COMCEN Communications Center

ComCon Command Console

COMCON Communications Control

Comd Command • Commander

COMD Command • Component Modeling

COMDAC Command - Display - Control

COMDEG Communications Degradation

COMDEX Communications Development Exposition • Computer Dealers Exposition

Comdg Commanding

Comdr Commander

Comdt Commandant

COMED Combined Map and Electronic Display

COMEDI Commerce Electronic Data Interchange

COMEND Comendador

COMES Comité Mexicano para el Eclipse Solar (Mexico) • Commissariat à l'Énergie Solaire

COMESA Common Market for Eastern and Southern Africa

COMET Coherent Electromagnetic-Energy Transmission • Combined Optical Measurements Experimentation Test • Combined Optical Measurements Experiment Team • Command Executive Translator • Commercial Experiment Transporter • Committee for Middle East Trade • Computer-Operated Management Evaluation Technique • Cooperative Operational Meteorology Education and Training (program) • Coordinated Multimedia Explanation Testbed • Cornell Macintosh Terminal Emulator

COMETE Commission de Mémoire Technique

COMETS Career-Oriented Module to Explore Topics in Science • Commission d'Éthique pour les Sciences • Computer-Operated Multifunction Electronic Test System

COMETT Community in Education and Training for Technology

COMEX Commodity Exchange • Communications Exercise • Communications Experiment

COMFAC Communications Facility

COMI Comité d'Orientation des Moyens Informatiques • Communications Interface

COMICS Cooled Mid-Infrared Camera and Spectrometer

CoMil Contributions of Milano Observatory (periodical code)

COMIND Cognitive Match Interface Design

COMINT Communications Intelligence

COMIREX Committee on Imagery Requirements and Exploitation

COMIS Cartographic Operations Management Information System • Coded Mask Imaging Spectrometer • Committee Meeting Information System

COMIX Cooperative Multiuser Interface to X-window

Coml Commercial

COML Commercial

COMLO Compass Locator

Comm Command • Commander • Comment • Commentary • Commerce • Commercial • Commercially • Commission • Commitment • Committee • Common • Commonwealth • Communicate • Communication • Community • Commutator

COMM Commercial • Commodity • Communications

COMMANDOS Construction and Management of Distributed Open Systems

Commd Commissioned

COMMEL Communications/Electronics

COMMEN Compiler Oriented for Multiprogramming and Multiprocessing Environments

COMMEND Computer-aided Mechanical Engineering Design

COMMEX Communications Exercise

COMMINT Communications Intelligence

Commn Commission

Commnd Commissioned

COMMO Communications Officer

ComMP Communications in Mathematical Physics (periodical code)

COMMRI Communication Routing Indicator

COMMS Communications • Communications System

COMMSEC Communications Security

Commun Communication • Community

Commy Commissary

CommZ Communications Zone

COMMZ Communications Zone

COMn n-th Serial Port

COMNAP Council of Managers of National Antarctic Programmes

ComNet Communications Network

COMNET Communications Network

COMOD Comodoro

COMOPS Combined Operations • Communications Operating Summary

Comp Compañía • Companion • Comparative • Comparativo • Comparator • Compare • Comparing • Comparison • Compass • Compatible • Compensate • Compensating • Compensation • Compete • Competitive • Competitor • Compilación • Compilador • Compilation • Compile • Compiled • Compiler • Compiling • Complement • Complete • Component • Compose • Composed • Composer • Composite • Composition • Compositor • Compound • Compounded • Comprehensive • Compression • Comprising • Compromise • Compromiso • Computation • Compute • Computer • Computer (periodical code) • Computing

CoMP Contributions to Mineralogy and Petrology (periodical code)

COMP Compare • Compass • Compatibility • Compensate • Competitive • Complete • Component • Composite • Composition • Computer • Conceptually Oriented Mathematics Program • Controlled Optics Modeling Package

COMPAC Computer Program for Automatic Control

COMPACT Commercial Product Acquisition Team • Compatible Algebraic Compiler and Translator • Computer Planning and Control Technique

Compar Comparative • Comparison

COMPAR Comparison

COMPARE Computerized Performance and Analysis Response Evaluator • Console for Optical Measurement and Precise Analysis of Radiation from Electronics

COMPAS Coastal Ocean Management, Planning and Assessment System • Computer-Oriented Metering Planning and Advisory System

ComPASS Common Planning And Scheduling System

COMPASS Catalogs of Objects and Measured Parameters from All-Sky Surveys • Common Muon and Proton Apparatus for Structure and Spectroscopy (CERN) • Common Operational Modeling, Planning, And Simulation Strategy • Communications Planning Assistant • Compiler-Assembler • COMPTEL Processing and Analysis Software System • Computer-Aided Subject Search • Computerized Movement Planning And Status System • Computerized Optimization Model for Predicting and Analyzing Support Structure • Computerized Procurement And Supply System • Cosmic Microwave Polarization At Small Scales

Compd Compound

COMPEL Compute Parallel

COMPEN Compensator

COMPES Contingency Operations, Mobility Planning and Execution System

COMPFEC Compressed with Forward Error Correction

COMPG Composite Group

ComPh Computers and Physics (periodical code)

Compl Complement • Complément • Complete • Completed • Compliment • Complimentary

COMPL Complete • Completed

COMPLAN Communications Plan

COMPLD Completed

COMPLEX Committee on Planetary and Lunar Exploration (NAS)

COMPLIP Computation of Manpower Programs by Linear Programming

COMPLIS Collaboration for Measurements using a Pulsed Laser Ion Source

Complt Complaint

CompM Computational Mechanics (periodical code)

Compn Compensate

COMPNT Component

Compo Composites (periodical code) • Composition

COMPO Component

COMPOOL Communications Pool

Compr Compression • Compressive

COMPR Compress • Compressed • Compressor

COMPRD Compressed

COMPRESS Computer Research, Systems and Software

COMPROG Computer Program

Comps Compliments

Compt Compartment • Comptroller

COMPT Compartment • Comptroller • Controlled Optics Modeling Package

COMPTEL Compton Telescope

Comptr Comparator • Comptroller

Compu Computing (periodical code)

COMPUSEC Computer Security

Comr Commissioner

COMRAA Committee on the Organization and Management of Research in Astronomy and Astrophysics (NRC)

COMREP Communications Report

COMS Collection Objectives Management System • Computer-based Operations Management System • Contamination Monitoring System

COMSAT Commercial Satellite • Communications Satellite (*)

COMSCEE Committee on Science in Central and Eastern Europe

ComSE Computing Systems in Engineering (periodical code)

COMSEC Communications Security

COMSER Commission on Marine Science and Engineering Research

COMSIM Communications Simulation (system) • Communications Simulator

COMSIT Communications Situation

COMSL Communication Simulation Language

Comsnd Commissioned

ComSo Communications Society (periodical code)

COMSOFT Consortium for the Management of Software Technologies

ComSp Commercial Space (periodical code)

COMSR Communications Support Requirements

COMSS Coastal Oceans Monitoring Satellite System (now ERS)

ComST Composites Science Technology (periodical code)

COMSTAC Commercial Space Transportation Advisory Committee

COMSTAT Communications Status

CoMStOC Coronal Magnetic-Structures Observing Campaign

COMSY Commissary

COMTI Common Object Model Transaction Integrator

COMTR COMSAT Technical Review (periodical code)

COMTRAN Commercial Translator

CoMtW Contributions from the Mount Wilson Observatory (periodical code)

Comunic Comunicación

COMUTA Comité des Usagers du Transport Aérien (France)

COMZ Communications Zone

COM1 First Serial Port

COM2 Second Serial Port

COM3 Third Serial Port

COM4 Fourth Serial Port

con Congestus

Con Concentration • Concerning • Concerto • Conclusion • Confidence • Conjunx • Connection • Conservative • Consolidate • Consolidated • Consol • Consul • Contingency • Continue • Continued • Contra • Control • Controlled • Convenience • Conversation

CON Concentrator • Concord (USA-NH) (location identifier) • Condensation • Cónego • Connect • Console • Contact • Contactor • Contention • Contingency • Continuous Orthonormalization • Contrast • Control • Controlled • Convention • Conventional

CONA Comité Oceanográfico Nacional (Chile)

CONACYT Consejo Nacional de Ciencia y Tecnología (Mexico)

CONAE Comisión Nacional de Actividades Espaciales

CONAFE Consejo Nacional de Fomento Educativo

CONAM Consejo Nacional del Ambiente (Peru)

CONAMA Comisión Nacional del Medio Ambiente (Chile)

Conbd Contributed

Conc Concentrate • Concentrated • Concentrating • Concentration • Concentric • Concerning • Concesivo • Concrete

CONCAM Continuous Camera

CONCAP Construction Capability

Concd Concentrated

CONCERT Combined Neutron Center for European Research and Technology • Conformity Certification

Concg Concentrating

CONCISE COSINE Network's Central Information Service for Europe

Concl Conclude • Concluded • Concluding • Concludingly • Conclusion

Concn Concentration

CONCN Concentration

CONCORDE Connected Car-Oriented Entity

Concr Concrete

CONCYTEC Consejo Nacional de Ciencia y Tecnología (Peru)

Cond Condense • Condensed • Condenser • Condensing • Condicional • Condition • Conditional • Conditioning • Conditioner • Conduct • Conducted • Conducting • Conductivity • Conductor

COND Condenser • Condition • Conditioner • Conductor

CONDORS Convective Diffusion Observed by Remote Sensors

Condr Conductor

CONDR Conditioner

CONE Combined Neutral and Electron • Common OSI Networking Environment • Cryogenic Orbital Nitrogen Experiment

CONEA Comisión Nacional de Energía Atomica (Argentina) (formerly CNEA)

CONEE Comisión Nacional del Espacio Exterior

CONES Consejo Nacional de Educación Superior (Bolivia)

conf Configuration Data (file extension)

Conf Confédération • Confer • Conference • Confidence • Confidential • Configuration • Confirmación • Confirmed • Conformance

CONF Conference • Confidence • Confidential • Confirmation

CONFDC Confidence

Confed Confederacy • Confederate • Confederation

Confid Confidential

Config Configuration

CONFIG Configuration

CONFU Conference on Fair Use

cong Congius

Cong Congress • Congressional

CONG Congress

CONGEN Consul General

Congr Congruent

Coni Conical

CONICA Coudé Near-Infrared Camera

CONICET Consejo Nacional de Investigaciones Científicas y Técnicas (Argentina)

CONICIT Consejo Nacional de Investigaciones Científicas y Tecnológicas (Venezuela) • Consejo Nacional para Investigaciones Científicas y Tecnológicas (Costa Rica)

CONICOR Consejo de Investigaciones de Córdoba (Argentina-CO)

CONICYT Comisión Nacional de Investigación Científica y Tecnológica (Chile) • Consejo Nacional de Investigaciones Científicas y Técnicas (Uruguay) • Consejo Nacional de Investigaciones Científicas y Tecnológicas (Venezuela)

CONIDA Comisión Nacional de Investigaciones y Desarollo Aerospacial (Peru)

CONIE Comisión Nacional de Investigaciones del Espacio (Spain)

CONIRIS Calibrated Optical and Near Infrared Imaging System

Conj Conjugation • Conjunción • Conjunction • Conjunctive • Conjunto

CoNLL Conference on Natural Language Learning

CONMAINT Consolidated Maintenance

CoNMx Contributions of the Observatory of New Mexico State University - Las Cruces (periodical code)

Conn Connect • Connecté • Connected • Connecter • Connecticut (USA) • Connecting • Connection • Connector • Connexion • Connotation

CONN Connect • Connector

CONNEX Connection Matrix

CONOP Contingency Operations

CONOPS Concept of Operations • Contingency Operations • Continuing Operations • Continuity of Operations • Continuous Operations • Control Operations

CONP Connected Name Presentation • Connection-Oriented Network Protocol

CONPLAN Concept Plan • Contingency Plan

ConPh Contemporary Physics (periodical code)

CONPS Carbon - Oxygen - Nitrogen - Phosphorus - Sulfur

CONQUEST Concurrent Querying Space and Time

CONRAIL Consolidated Rail Corporation (USA)

CONREP Contingency Report

CONRIM Council on Northern Resources Information Management

Cons Consecrated • Consecutive • Consequence • Conséquence • Conservation • Conservative • Conservatoire • Conservatorium • Conservatory • Consider • Consideration • Considération • Considérer • Considering • Consist • Consistance • Consistency • Consisting • Consolidate • Consolidated • Consonant • Constable • Constant • Constante • Constitution • Constitutional • Construction • Consul • Consult • Consultant • Consulter • Consulting

CONS Connection-Oriented Networking Session • Connection-Oriented Network Service • Conselheiro • Console • Construct

CONSAL Congress of Southeast Asian Libraries

CONSCAN Conical Scan • Conical Scanning

Consec Consecutive

Conseq Consequence

Conséq Conséquence

CONSERT Comet Nucleus Sounding Experiment by Radiowave Transmission (Rosetta)

Conserv Conservatoire • Conservatorium • Conservatory

Consid Consideration

CONSOIL Conservation Of Natural Resources through Sharing Of Information Layers

Consol Consolidated

Conspec Construction Specification

Const Constable • Constant • Constante • Constitución • Constitution • Constitutional • Construct • Constructed • Construction • Constructor

Constl Constitutional

Constr Constraint • Construct • Construction

CONSTR Construct • Construction

CONSUL Control Subroutine Language

CONSULT Consultório

Cont Contabilidad • Contact • Contador • Contain • Container • Containing • Content • Contents • Continent • Continental • Continuación • Continuation • Continue • Continued • Continuous • Continuum • Contour • Contra • Contract • Contraction • Control • Controller

CONT Constant • Construct • Constructed • Container • Contingency • Continue • Continued • Continuous • Contractor • Control • Controller

CONTAC Central Office Network Access

CONTAM Contaminate • Contamination

Contbg Contributing

Cont'd Continued

Contd Contained • Continued

CONTDVD Continental Divide

Contemp Contemporary

Contg Containing

CONTG Continuing

CONTGY Contingency

CONTIG Contiguous

Contin Continued • Continuous

CONTLY Continuously

Contn Continuation

CONTONE Continuous Tone

CONTOUR Comet Nucleus Tour

Contr Contracción • Contract • Contracted • Contraction • Contrary • Contrast • Contrasted • Contribution • Control • Controller

CONTR Contract • Contractor • Control • Controller

Contracc Contracción

CONTRAIL Condensation Trail

CONTRAN Control Translator

Contrib Contributed • Contributing • Contribution • Contributor

CONTU Commission on New Technological Uses

CONUS Continental United States

Conv Convection • Convective • Convenient • Convention • Conventional • Conversation • Conversion • Convert • Converter • Converting • Convertible • Convocation

CONV Convection • Converge • Convergence • Convergent • Conversion • Converter

CONVEL Constant Velocity

Convl Conventional

Convn Convection

COO Cargo Operations Officer • Centre d'Océanographie Opérationnelle • Centre d'Orbitographie Opérationnelle • Centre Opérationnel d'Orbitographie • Chief Operating Officer • Computerondersteund Onderwijs • Concept Of Operations • Cotonou (Benin) (location identifier)

COOH Carboxyl Group

COOL Chorus Object-Oriented Layer • Common Object-Oriented LISP • Concurrent Object-Oriented Language • Control Object-Oriented Language

COOLLAR Cryogenic On-Orbit Long-Life Active Refrigerator

Coop Cooperativa • Coopérative • Cooperative

COOP Concurrent Object-Oriented Programming • Continuity Of Operations Plan • Continuity Of Operations Procedures • Cooperation • Cooperative • Craft-Of-Opportunity Program

COOPLOG Cooperative Logistics

COOR Centre for Organizational and Operational Research • Coordination

CoORB Communications de l'Observatoire Royal de Belgique (periodical code)

Coord Coordinador • Coordinate • Coordinated • Coordinating • Coordination • Coordinator

COORD Coordinate • Coordinator

COORDWG Coordination Working Group

COOS Chemical Orbit-to-Orbit Shuttle • Coastal Ocean Observing System

COOSRA Canadian Offshore Oil Spill Research Association (Canada)

COOTS Conference on Object-Oriented Technologies and Systems

CoOxf Communications from the University Observatory Oxford (periodical code)

Cop Copernican • Copia • Copper • Copyright • Copyrighted

COP Calibrated Output Processing • Celescope Optical Package • Center of Pressure • Central Operations Program • Certificat d'Orientation Professionnelle • Chane-Over Point • Character-Oriented Protocol • Coefficient Of Performance • Coherent Optical Processor • Combinatorial Optimization Problem • Combined Operations Plan • Command Operating Program • Command Operation Procedure • Commitment - Obligation - Procurement • Committee for Operations and Plans • Common Online Package • Common Operating Picture • Common Operational Picture • Common Operational Plan • Communication Output Printer • Community Of Practice • Community-Oriented Policing • Composite Operations Plan • Computer Operations Parameters • Computer Optimization Package • Computer Optimization Program •

Concept Of Operations • Conference Of the Parties • Conseiller d'Orientation - Psychologue • Consistent Operating Picture • Constabulary On Patrol • Constituent Object Parser • Contingency Operations Plan • Continuation Of Pay • Continuous Operation Program • Co-Orbiting Platform • Copperstown (USA-NY) (location identifier) • Count-Oriented Protocol • Craft of Opportunity Program • Crash-Out Package • Cross-Order Profile • Current Operating Plan • Current Operating Procedure • Customer-Orienting Program • Customer-Owned Property • Peso (Colombia) (ISO currency code)

COPA Child Online Protection Act (USA) • Columbus Payload • Coordinated Observations of Polar Electrodynamics • Council On Postsecondary Accreditation • Crisis Operations Procedures Analysis

COPAC Colegio Oficial de Pilotos de Aviación • Continuous Operation Production Allocation and Control

COPAD Contractor-Operated Parts Depot

COPAI Colegio Oficial de Pilotos de la Aviación Comercial

COPANIT Comisión Panameña de Normas Industriales y Técnicas

COPANT Comisión Panamericana de Normas Técnicas

COPARS Contractor-Operated Parts Store

COPARSS Committee On Practical Applications of Remote Sensing from Space

COPAT Comisión Panamericana Técnica

COPC Combined Operations Planning Committee

COPD Chronic Obstructive Pulmonary Disease

COPE Carbon Monoxide Pollution Experiment • Command Overview and Program Execution • Communications-Oriented Processing Equipment • Computer-Oriented Project Engineering • Co-Orbiting Platform Elements • Custodian Of Postal Effects

COPEARMS Coordinating Project for Electronic Author's Rights Management Systems

COPEM Columbus Precursor Environment Monitor

COPEPP Committee on Public Education and Public Policy

CoPer Contributions of Perkins Observatory (periodical code)

COPERS Commission Préparatoire Européenne de Recherche Spatiale (now defunct)

COPES Committee on Program Evaluation and Support • Conceptually Oriented Program in Elementary Science • Control Program for Engineering Synthesis • Cooperating Expert System

CoPhC Computer Physics Communications (periodical code)

COPI Computer-Oriented Programmed Instruction

COPICAT Copyright Ownership Protection for Computer-Assisted Training

COPICS Copyright Office Publication and Interactive Cataloguing System

COPIDAB Columbus Pictures Data Bank

COPL Centre d'Optique, Photonique et Laser • Core Persistent Object Library

COPLAC Council of Public Liberal Arts Colleges (USA)

COPO Chief Of Personnel Operations

Copol Copolarized

COPOLCO Consumer Policy Committee (ISO)

CoPP Contributions to Plasma Physics (periodical code)

COPPA Children's Online Privacy Protection Act (UK)

COPPER Consolidation of Pay and Personnel

COPPUL Council of Prairie and Pacific University Libraries (USA)

Copr Copyright

COPR Concept of Operations Review

CopRe Copenhagen University Observatory Reprints (periodical code)

CoPri Contributions from the Princeton University Observatory (periodical code)

CoPS Complex Product System

COPS Canadian Optics and Photonics Society (Canada) • CFHT Optical PDCS Survey • Chrysler Optical Processing Scanner • Circuit Order Preparation System • Circuit Order Processing System • Coastal Ocean Prediction Systems • Coherent Optical Processing System • Committee On Physical Security • Common Open Policy Service • Communications Operational Planning System • Complex Operations Per Second • Computerized Oracle and Password System • Computer Oracle and Password System • Computer-Oriented Partial Sum • Computer Output Processing Service • Concept-Oriented Programming System • Current Operations Support

COPSK Coded Octal Phase Shift Keying

COPT Convection Profonde Tropicale

Coptr Copartner

COPTRAN Communication Optimization Program Translator

Copul Copulativo

COPUOS Committee on the Peaceful Uses of Outer Space (UNO)

COPUS Committee On the Public Understanding of Science (UK)

COPV Composite Overwrapped Pressure Vessel

COPYME Confederación de la Pequeña y Mediana Empresa (Spain)

COQ Certificate Of Qualification • Certification Of Qualification • Choibalsan (Mongolia) (location identifier) • Cost Of Quality • Coût d'Obtention de la Qualité

COQC Certificate Of Quality Compliance

Cor Corner • Corollaire • Corpus • Correct • Corrected • Correction • Correlation • Correlative • Correspondence • Correspondent • Corresponding • Corrigendum

CoR Committee of Region

COR Center for Operations Research • Centrale Ondernemingsraad • Central Office of Record • Centre d'Opérations du Réseau • Circular Of Requirements • Coherent Optical Receiver • Cold Optics Radiometer • Commande Optimale du Rotor • Committee Of Region • Committee Of the Regions (*) • Component Overhaul and Repair • Concentric Orbit Rendezvous • Continental Operating Range • Contracting Officer's Representative • Córdoba - Pajas Blancas (Argentina-CD) (location identifier) • Corner • Correction • Corrigendum • Cut-Off Rigidity

CORA Coherent Optical Radar Amplifier • Conditioned-Reflex Analog • Conditioned-Response Analog

CoRad Communications from the Radcliffe Observatory - Pretoria (periodical code)

CORAD Correlation Radiometer

CORAL Class-Oriented Ring-Associative Language • Computer Online Real-time Applications Language

CORALS Complete Optical radio Absorption Line System

CORALz Compact Radio source At Low Redshift

CORAM Centre d'Observation et de Recherche Astronomique Mandeville (Belgium)

CORAVEL Correlation Radial Velocities

CORB Communications de l'Observatoire Royal de Belgique (periodical code)

CORBA Common Object Request Broker Architecture

Córd Córdoba

Cord Cordillera

CorD Córdoba Durchmusterung (periodical code)

CORD Computer with Online Remote Devices • Coordinated Operational Requirements Document

CORDIC Coordinate Rotation Digital Computer

CORDIS Community Research and Development Information Service (EU)

CORDS Center for Orbital and Reentry Debris Studies • Coherent On-Receive Doppler System • Copyright Office Registration and Deposit System

CORE Center for Operations Research and Econometrics (Belgium) • Central Operation of Resources for Educators • Chemical Online Retrieval Experiment • Coherent Optical Receiver Engineering • Cold Ocean Resources Engineering • Common Operational Research Equipment • Complex Operating Rule Engine • Computer-Oriented Reporting Efficiency • Contingency Response • Continuous Observation of the Rotation of the Earth • Contractor Renormalization • Cost-Oriented Resource Estimating • Council of Registrars • Cross-disciplinary Optics Research and Education

CORECT Collaborative Requirements Capture • Committee on Renewable Energy, Commerce and Trade

COREDOC Core Document

CoREN Corporation for Research and Enterprise Network

COREP Combined Overload Repair Control

COREPER Comité des Représentants Permanents

CORES Computer-assisted Order Routing and Execution System

CORETA Centre d'Organisation de Recherches et d'Expertises en Technologies Avancées

CORF Committee on Radio Frequencies

CoRHO Contributions of Rosemary Hill Observatory (periodical code)

CORI Celestial Observers of Rhode Island (USA-RI)

CORIA Complexe de Recherche Interprofessionnelle en Aérothermisme

CORINE Coordination of Information on the Environment

CORKS Computer-Oriented Record Keeping System

CORLAT Collection of Reference Laterite

CORM Council for Optical Radiation Measurements (USA)

CorMaSS Cornell-Massachusetts Slit Spectrograph

Corn Cornish • Cornwall (UK)

CORN Computer Resources Nucleus

CoROE Communications from the Royal Observatory Edinburgh (periodical code)

Corol Corollary

Coroll Corollary

CORONAS Complex Orbital Near-Earth Observations of the Solar Activity

COROT Convection-Rotation • Convection-Rotation à Transits (planétaires)1

Corp Corporal • Corporate • Corporation

Corpn Corporation

CORPS Computing, Organizations, Policy and Society

Corr Correct • Corrected • Correctement • Correction • Corrective • Correctly • Correlate • Correlating • Correlation • Correlative • Correlator • Correlation • Correlative • Corrente • Correspond • Correspondence • Correspondent • Corresponding • Corridor • Corrigenda • Corriger • Corrugated • Corrupt • Corrupted • Corruption

CoRR Computing Research Repository

CORR Correction

CORRD Corrected

Correg Corregidor

Correl Correlative

Corresp Correspondence • Corresponding

Corrup Corrupción

Cors Corsica • Corsican

CORS Canadian Operational Research Society (Canada) • Catalytic Oxygen Removal System • Continuously Operating Reference Station • Corners

CORSA Cosmic Ray Satellite

CORSAIR Coastal Optical Remote Sensing Airborne Radiometer • Computer-Oriented Reference System for Automatic Information Retrieval

CoRSE Communications of the Royal Society of Edinburgh (periodical code)

CORSE Chabot Observatory Remote Sensing Experiment • Conference on Remote Sensing Education

CORSPERS Committee on Remote Sensing Programs for Earth Resource Survey

Corspnd Correspond

CORSS Coordination of Ocean Remote Sensing Satellites

CORSSAC Civil Operational Remote Sensing Satellite Advisory Committee

CORTA Cellule Opérationnelle de Régulation du Trafic Aérien

CorTi Tirada Aparte del Observatorio Astronomico de Córdoba (periodical code)

CORTREAT Corrosion Treatment

CORTS Convert Range Telemetry System

CORU Candidate Orbital Replaceable Unit

CoRut Contributions from the Rutherford Observatory - New York (periodical code)

cos cosine

Cos Companies • Counties

CoS Class of Service

COS Carbonyl Sulfide • Carry-on Oxygen System • Cash On Shipment • Cassette Operating System • Celestial Observation Satellite • Central Office of Statistics • Certificate Of Service • Changeover Switch • Checkout Station • Chief Of Section • Chief Of Staff • Chief Of State • Chief Of Station • Chief Of Supply • Civilian Occupational Specialty • Class Of Service • Cloud Observing System • Coefficient d'Occupation des Sols • Colorado Springs - Peterson (USA-CO) (location identifier) • Combinatorial Organic Synthesis • Common Operating System • Communication Operation Station • Communications Operating System • Communications Segment • Compatibility Operating System • Compatible Operating System • Component Oversight • Conceptual Operational System • Concurrent Operating System • Console Operating System • Convergence Sublayer • Corporation for Open Systems • Cosine • Cosmic Origins Spectrograph • Cray Operating System • Critical Occupational Specialty • Crystal Observation System

co. sa. Come Sopra

COSA Colorado Office of Space Advocacy (USA-CO)

COSAC Cometary Sampling And Composition (experiment) • Communications Sans Connexions

COSAG Combined Steam And Gas

COSAL Consolidated Shipboard Allowance List

COSAMC Commission for Special Applications of Meteorology and Climatology

COSAR Compression Scanning Array Radar

COSATI Committee On Scientific And Technical Information

COS-B Cosmic gamma-ray Satellite B

COSB Central Operations and Systems Branch • Cosmic gamma-ray Satellite B

COSC Chabot Observatory and Science Center (USA-CA) (now CCSC) • Computer Operations Service Center

COSCAT Coherent Scatter

COSCL Common Operating System Control Language

COSCQ Central Organization for Standardization and Quality Control (Iraq)

COSD Composite Object State Diagram

COSDIF Cost Differential

COSE Combined Office Standard Environment • Committee on Optical Science and Engineering • Committee On Science Education • Common Open Software Environment • Common Open Systems Environment • Common Operating Software Environment • Cooperative Operating System Environment

cosec Cosecant

CoSEC Collaborative Sun-Earth Connector

cosech Hyperbolic Cosecant

COSEMS Comprehensive Operational Support and Evaluation Model for Space

COSEPP Committee on Science, Engineering and Public Policy

COSEPUP Committee on Science, Engineering and Public Policy

COSETI Columbus Optical Search for Extraterrestrial Intelligence

COSG Consolidated Operations Steering Group

CoSGI Contributions of the Slovak Geophysical Institute (periodical code)

cosh Hyperbolic Cosine

COSH Charte des Organisations Sociales et Humanitaires

COSI Center of Science and Industry • Closeout System Installation • Combiner Simulator • Comet Simulation

COSIC Computer Security and Industrial Cryptography

COSIMA Cometary Secondary Ion Mass Analyzer (Rosetta) • Crystallization of Organic Substances In Microgravity for Applications

COSINE Cooperation for OSI Networking in Europe

COSIP Computer Open System Implementation Program

COSIS Care Of Supplies In Storage

COSIT Conference On Spatial Information Theory

CoSka Contributions of Skalnate Pleso Observatory (periodical code)

CoSkL Contributions of Skalnate Pleso Observatory - Letters (periodical code)

CoSkS Contributions of Skalnate Pleso Observatory - Supplement (periodical code)

COSM Center for the Origin and Structure of Matter (USA) • Checkout Servicing and Maintenance

COSMAS Core Strategic Measurements for Atmospheric Science

COSMIC Carnegie Observatories Spectroscopic Multislit and Imaging Camera • Common System Main Interconnecting • Computer Software Management and Information Center • Constellation Observing System for Meteorology, Ionosphere, and Climate

CoSMO Consolidated Supercomputing Management Office

COSMOS Centralization Of Supply Management Operations • Computer System for Mainframe Operations • Cosmologically Significant Mass Oscillation Search (Fermilab)

COSMOSOMAS Cosmological Structures On Medium Angular Scales

Cosmog Cosmogony • Cosmographical • Cosmography

Cosmogr Cosmografía

COSMOS Computer-Oriented System for Management Order Synthesis • Computer System for Mainframe Operations • Console-Oriented Statistical Matrix Operator System • Coordinates, Size, Magnitude, Orientation and Shapes

COSMT Czech Office for Standards, Metrology and Testing (Czech Republic)

COSNA Composite Observing System for the North Atlantic

co. so. Come Sopra

COSOD Conference on Scientific Ocean Drilling

COSOMA Copyright Society of Malawi

COSP Centre d'Orientation Scolaire et Professionnelle

CosPA Cosmology and Particle Astrophysics

COSPAC Cost/Schedule Planning And Control

COSPAR Committee on Space Research

COSPEC Correlation Spectrometer

COSPIN Cosmic-ray and Solar-charged Particles Investigation

COSPO Community Open Source Program Office

COSPUP Committee on Science and Public Policy

COSQ Cost On Software Quality

COSR Contract Officer Security Representative

COSRAD Coastal Ocean Surface Radar

CosRe Cosmic Research (periodical code)

COSRIMS Committee on Support of Research in the Mathematical Sciences

COSRO Conical Scan on Receive Only

Coss Consules

COSS Common Object Services Specification • Common Object System Specification • Contractor-Operated Support Site • Conventional Ordnance Status System • Customer-Oriented Strategic System

COSSA CSIRO Office of Space Science and Applications

COSSAC Chief Of Staff to Supreme Allied Commander

COSSAM Codice per la Sintesi Spettrale nelle Atmosfere Magnetiche

COSSC Compton Observatory Science Support Center

CosSe Cosmic Search (periodical code)

Cost Costante

COST Committee for Overseas Science and Technology • Committee on Science and Technology • Contracting Officer Support Tool • Coopération en Science et Technologie • Cooperation On Science and Technology • Coopération Scientifique et Technique • Cost-Oriented Systems Technique

Costa Costanilla

CoStA Communications from the University Observatory - St Andrews (periodical code)

COSTA Cost Accounting

COSTAR Corrective Optics Space Telescope Axial Replacement (*)

CoSte Contributions University of Steward Observatory - Tucson (periodical code)

COSTED Committee on Science and Technology in Developing Countries

COSTEP Comprehensive Suprathermal and Energetic Particle (analyzer) (SOHO)

CoStr Computers and Structures (periodical code)

COSTR Collaborative Solar-Terrestrial Research

COSY Checkout Operating System • Correlation Spectroscopy

cot Cotangent

CoT College of Technology

COT Central Office Terminal • Central Outer Tracker • Change Of Temperature • Checkout Time • Class Of Traffic • Columbus Operations Team • Committee On Toxicology • Computer-Operated Terminal • Computer - Output - Typesetting • Consistent Orbit Tracing • Cotulla (USA-TX) (location identifier) • Critical Optical Technology • Customer Office Terminal • Customer-Operated Terminal • Cyclooctatetraene • Cyclooctatetraenyl

cotan Cotangent

COTAS Commission pour la Technologie et les Applications des Sciences

COTASP Council of Technology and Allied Sciences of the Philippines

COTD Common Operator Training Device

COTE Checkout Terminal Equipment

COTF Contractor-Operated Test Facility

coth Hyperbolic Cotangent

CoThe Contributions from the Astronomical Department of the University of Thessaloniki (periodical code)

COTIF CEDAR Optical Tomographic Imaging Facility • CEDAR Optical Tomography of the Ionosphere Facility

COTM Circuit-Oriented Transfor Mode • Communications On The Move

CoTok Contributions from the Department of Astronomy University of Tokyo (periodical code)

CoTol Contributions of Cerro Tololo Inter-American Observatory (periodical code)

COTP Connection-Oriented Transport Protocol

COTR Contracting Officer's Technical Representative

COTS Cargo Offload and Transfer System • Checkout Test Set • Committee On Technical Security • Commercial Off-The-Shelf (*) • Commodities Off-The-Shelf • Connection-Oriented Transport Service

COTSS Classified Operational Telecommunications Switching System

Cott Cottage

COTT Central Organization for Technical Training • Control Of Transmission Test

Cou Coudé

COU Cable Orderwire Unit • Columbia (USA-MO) (location identifier) • Concept of Operation and Utilization • Curso de Orientación Universitaria (Spain)

CoUCL Contributions de l'Université Catholique de Louvain (periodical code)

Coun Council • Councillor • Counselor

COUNT Computer-Operated Universal Test

COUP Consolidated Operations and Utilization Plan

Cour Courant

Cov Covariance • Coventry (UK-England) • Cover • Covered • Covering

COV Calculus Of Variation • Coefficient Of Variation • Committee Of Visitors • Composé Organique Volatile • Concept-Oriented View • Corona Onset Voltage • Counter Obstacle Vehicle • Covariance • Covilha (Portugal) (location identifier) • Cutoff Value • Cutoff Valve • Cutoff Voltage

COVAM Capture Orbit Vehicle Assembly Mode

COVE Collaborative Object-oriented Virtual Environment • Coronal Velocity Explorer

COVENIN Comisión Venezolana de Normas Industriales (Venezuela)

COVIRA Computer Vision in Radiology

CoVis Collaborative Visualization

CoVVO Contributions of Van Vleck Observatory (periodical code)

COW Cell-site On Wheel • Cluster Of Workstations • Committee on Water • Copy On Write • Council of the Wise

COWAR Committee on Water Research

CoWas Contributions from the Washburn Observatory of the University of Wisconsin (periodical code)

CoWat Contributions of the University of Waterloo Observatory (periodical code)

COWAT Coaxial Waveguide Transformer

COWD Carbon/Oxygen White Dwarf (star)

COWI Commissie Wetenschappelijke Informatieverzorging (Netherlands)

CoWro Contributions from the Wroclaw Astronomical Observatory (periodical code)

COx Carbon Oxide

COX Congo Town (Bahamas) (location identifier)

COY Company • Coolawanyah (Australia-WA) (location identifier)

COYOTES Coordinated Observations of Young Objects from Earthbound Sites

COZ Costanza (Dominican Rep.) (location identifier)

COZAC Conservation Zone Advisory Committee

CO2 Carbon Dioxide • Consideration Of Others

CO2EL Carbon Dioxide Electrolysis

CO2T Consideration Of Others Training

C° Cerro

cp Copy

cP Centipoise

Cp Camp • Candlepower • Capability Index • Carriage Paid • Compare • Copy

C&P Capabilities & Procedures • Characteristics & Performance • Collection & Processing • Contracts & Pricing

C/P Cartesian to Polar • Charter Party • Cold Plate • Command Post • Coronagraph/Polarimeter

CP Bolivia (international aircraft registration code) • Caixa Postal • Calibration Procedure • Call for Proposals • Cambridge Pulsar • Camp • Campus Police • Canadian Pacific • Canadian Press • Candlepower • Capability Package • Cape • Cape Photographic • Cape Province (South Africa) • Capillary Plate • Cardinal Point • Card Punch • Career Program • Cargo Projects • Carriage Paid • Carrier Platform • Carte Perforée • Casella Postale • Case Postale • Case Processor • Cathodic Protection • Center of Pressure • Center Point • Centrally Procured • Central Peak • Central Point • Central Processing • Central Processor • Certificate Policy • Certification Program • Chamber Pressure • Change Package • Change Point • Change Proposal • Character Printer • Charge Parity • Checkpoint • Chemically Peculiar • Chemically Pure • Chemical Physics (periodical code) • Chemische Physik • Circuit Package • Circularly Polarized • Circular Pitch • Circular Polarization • Circular Pure • Civil Procedure • Clinical Practice • Clock Phase • Clock Pulse • Close Protection • Code of Practice • Code of Procedure • Code Postal • Codice Penale • Código Postal • Coefficient of Performance • Coherent Processor • Code Postal • Código Postal • Cold Penetration • Cold Plate • Colegio Público • Colis Postal • Collection Point • Collective Protection • College Park (USA-MD) • Color Printing • Command Processor • Command Post • Command Pulse • Commission Permanente • Common Part • Common Picture • Common Point • Common Program • Communications Processor • Communications Program • Community Program • Completed Path • Compteur de Programme • Computer Program • Concentration Polarization • Concurrent PROLOG • Condensation Product • Conditional Processor • Conditional Proof • Conference Paper • Conference Proceedings • Conference Publication • Configurable Profile • Confirmation Page • Congés Payés • Congressional Presentation • Conjugation Parity • Connector Panel • Connexion Pertinente • Conseiller

Pédagogique • Conservation Park • Console Processor • Consolidation Point • Constant Potential • Constant Pressure • Constrained Procedure • Constraint Programming • Construction Permit • Contact Point • Content Particle • Content Provider • Continuous Path • Continuous Penetration • Contracting Party • Contracts and Pricing • Contributed Presentation • Contribution Profile • Control Panel • Control Platform • Control Point • Control Post • Control Procedure • Control Processor • Control Program • Coordinated Procurement • Copilot • Copy • Copy Protected • Core Polarization • Core Program • Correction Processor • Cosmological Principle • Cost and Performance • Cours Préparatoire • Cox Process • Crab Pulsar • Cracking Pressure • Crack Propagation • Critical Path • Crossover Point • Cross-Polarization • Cross-Product • Current Paper • Customized Processor • Cyberspace Protocol • Cycle Préparatoire • Cyclic Protocol • Phosphorus Carbide

CPA Cape Palmas (Liberia) (location identifier) • Cardiopulmonary Arrest • Center for Particle Astrophysics • Central Planning Area • Central Pulse Amplifier • Centre de Pédagogie Active • Centro de Pesquisas Antárticas • Certified Public Accountant • Cessation Progressive d'Activités • Change Proposal Analyst • Chairman's Program Assessment • Chartered Public Accountant • Chief of Public Affairs • Chirped Pulse Amplification • Circularly Polarized Antenna • Circularly Polarized Array • Circular Probable Area • Classe Préparatoire à l'Apprentissage • Classification of Products by Activity • Closest Point of Approach • Coherent Potential Approximation • Cold-Plasma Analyzer • Command Project Acceptance • Commission on Preservation and Access • Commutative Principle for Addition • Conjugation-Parity Asymmetry • Constant Potential Accelerator • Contingency Planning Aid • Contract Price Adjustment • Conventional Plate Adjustment • Cost Planning and Appraisal • Critical Parametric Analysis • Critical Path Analysis • Cross-Program Activity • Curved-Plate Analyzer

CPAA Charge Particle Activation Analysis • China Patent Agents Association (China-PRC)

CPAB Corrosion Prevention Advisory Board

CPAC Cambridge Planck Analysis Center (USA-MA) • Castle Point Astronomy Club (UK-England) • Civilian Personnel Activity Center • Civilian Personnel Administration Center • Civilian Personnel Advisory Center

CPAF Cost Plus Award Fee

CPAG Commission on Preservation and Access Glossary

CPAL Computer Processing on Asian Languages

CPAM Caisse Primaire d'Assurance Maladie (France) • Columbus Payload Accommodation Handbook • Communications in Pure and Applied Mathematics (periodical code) • Continental Polar Air Mass

CPAN Circumpolar Protected Area Network • Comprehensive Perl Archive Network

CPAO Conception de Programme Assistée par Ordinateur

CPAP Chef de Projet Ariane Production • Control Parameter Assembly Program

CPAR Construction Productivity Advanced Research • Contractor Performance Assessment Report

CPARS Contractor Performance Assessment Reporting System

CPAS Cellular Priority Access Service • Central Procurement Accounting System • Centre Public d'Aide Sociale

CPAT Critical Process Assessment Tool

CPAX Computerized Private Automatic Exchange

CPB Capabilities - Programming - Budgeting • Capurgana (Colombia) (location identifier) • Centraal Planbureau • Central Planning Bureau • Channel Program Block • Charged Particle Beam • Civilian Personnel Bulletin • Control Processor Board • Corp. for Public Broadcasting • Cross-Platform Bridge

CPBX Computerized Private Branch Exchange

CPC Canadian Polar Commission (Canada) • Calculatrice à Programmes par Cartes • Cape Photographic Catalogue • Card Programmed Calculator • Career Planning Center • Cargo Processing Contract • Center for Personal Communication • Center for Plant Conservation • Central Pennsylvania Consortium (USA-PA) • Central Planning Center • Central Processing Center • Central Processing Complex • Central Processing Console • Central Processor Complex • Central Product Classification • Centrifugal Partition Chromatography • Characteristics Properties Code • Chiron Perihelion Campaign • Chopped Photometric Channel • Civilian Personnel Circular • Climate Prediction Center • Clock Pulse Control • Clock-Pulsed Control • Coastal Patrol Craft • Coated Paper Copier • Code of Professional Conduct • Collaborative Product Commerce • Color Pack Camera • Column Parity Check • Column Position Counter • Comité du Programme et de la Coordination • Commission Professionnelle Consultative • Common Product Classification • Communications Processing Center • Complementary Protection Certificate • Compound Parabolic Concentrator • Computer Physics Communications • Computer Power Center • Computer Print Control • Computer Process Control • Computer Program Command • Computer Program Component • Computer Program Configuration • Computer Programming Concept • Condensation Particle Counter • Confederación de la Producción y del Comercio • Conference Planning Committee • Conseiller Pédagogique de Circonscription • Constant Point Calculation • Constant Pressure Chart • Contact Postcode • Contract Progress Control • Control Pressure Conditioner • Control Processing Center • Control Processing Complex • Control Processing Console • Corps Propulseur Chargé • Corrosion Prevention and Control • Creative Programming Consultants (USA-TX) • Critical Propulsion Component • Current Production Code • Customer Premises Cabling • Cyclic Permutation Code • Cyclic Program Control • Cyclic Program Counter • San Martin de los Andes - Chapelco (Argentina-NE) (location identifier)

CPCA Conférence Permanente des Coordinations Associatives (France) • Consolidated Procurement Cost Avoidance

CPCB Crew Procedures Control Board

CPCC Communications Processor Conversion Center • Computer Program Configuration Component

CPCEI Computer Program Contract End Item

CPCG Commercial Protein Crystal Growth

CPCI Commission Permanente de Concertation pour l'Industrie (France) • Computer Program Change Instruction • Computer Program Configuration Item • Computer Program Configured Item • Cross-Pointer Course Indicator

CPCL Computer Program Change Library • Computer Program Control Library

CPCM Certified Professional Contracts Manager

CPCN Command Post Communications Network • Conférence des Présidents du Comité National

CPCO Central Port Call Office

CPCOM Central Processing Communications

CPCP Center for Planetary Chemistry and Physics • Center for the Physics and Chemistry of the Planets

CPCPC Commission Paritaire Consultative des Personnels Contractuels

CpCpH Connections per Circuit per Hour

CPCPH Connections Per Circuit Per Hour

CPCR Computer Program Change Request • Contractor Packaging Capability Review • Crew Procedures Change Request

CPCS Check Processing Control System • Common Part Convergence Sublayer • Common Part of Convergence

Sublayer • Communication Payload Checkout System • Conversion Process Controller System • Cost Planning and Control System

CPCSB Computer Program Configuration Subboard

CPCSE Check Point Certified Security Expert

CPCT Center for Photoinduced Charge Transfer

CPCU Computer Power Conditioning Unit

CPC2 Second Cape Photographic Catalogue

Cpd Compound

CpD Cards per Day • Cycles per Day

C/PD Cost/Pricing Data

CPD Cape Photographic Durchmusterung • Cards Per Day • Center Program Director • Central Postal Directory • Central Processing and Distribution • Change Plot Device • Chemical Property Data • Chemical Propulsion Division • Circuit Protection Device • Circumprimary Disk • Civilian Personnel Director • Coaxial Power Divider • Cold Plasma Detector • Collector Potential Depression • Color-Position Diagram • Commercial Product Development • Commercial Program Development • Common Packet Data • Common Picture Display • Common Position Date • Communications Planning and Development • Competitive Professional Development • Compound • Computer Program Development • Congressional Presentation Document • Conseiller Pédagogique Départemental • Consolidated Procurement Directive • Consolidated Programming Document • Contact Potential Difference • Continuing and Professional Development • Continuing Professional Development • Coober Pedy (Australia-SA) (location identifier) • Cosmic-Particle Detector • Course Pointing Device • Crew Passive Dosimeter • Crew Procedures Division • Cumulative Probability Distribution • Cycles Per Day

CPDB Characteristics and Performance Database • Common Picture Database

CPDC Centre Pédagogique de Documentation et de Communication • Command Processor Distributor Control • Command Publishing Distribution Center • Committee on the Participation of Developing Countries • Continuing Professional Development Committee

CPDD Conceptual Project Design Description

CPDDS Computer Program Detail Design Specification

CPDF Central Personnel Data File

CPDL Controller-Pilot Data Link

CPDLC Controller-to-Pilot Data Link Communication

CPDP Centre Privé de Documentation Pédagogique • Computer Program Development Plan • Computer Program Development Program • Continuing Professional Development Plan

CPDR Contractor's Preliminary Design Review

Cpds Compounds

CPDS Computer Program Design Specification • Computer Program Detailed Specification • Computer Program Development Specification • Crew Procedures Documentation System

Cpdt Cependant

CPDT Center for Professional Development and Training

CPDU Central Power Distribution Unit • Command Pulse Distribution Unit

CPE Campeche (Mexico) (location identifier) • Cape • Cathodic Protection Equipment • Centrally Planned Economy • Central Processing Element • Central Processing Equipment • Central Programmer and Evaluator • Centre de Prospective et d'Évaluation • Certified Professional Evaluator • Certified Processing Element • Chief Program Engineer • Chief Project Engineer • Circadian Periodicity Experiment • Circular Probable Error • Cloud Processing Equipment • Collective Protection Equipment • Comètes - Planètes - Exoplanètes • Commission Paritaire de l'Emploi • Commission Paritaire d'Établissement (France) • Communauté Politique Européenne • Communications Privacy Equipment • Communications Processor Element • Computer Performance Evaluation • Computer Peripheral Equipment • Conseiller Principal d'Éducation • Constant Phase Element • Consumer Premise Equipment • Consumer-Provided Equipment • Continuing Professional Education • Contractor Performance Evaluation • Conventional Planning and Execution • Convergence Protocol Entity • Coopération Politique Européenne • Cosmic-Particle Event • Customer Premises Equipment • Customer-Provided Equipment

CPEF Centre Permanent d'Entraînement et de Formation

CPEG Certificat Préparatoire aux Études de Gestion

CPEI Computer Program End Item

CPEM Conference on Precision Electromagnetic Measurements

CPEN Centre Permanent d'Étude la Nature (Belgium)

CPEP Contemporary Physics Education Project • Contractor Performance Evaluation Plan

CPER Contrat de Plan État-Région (France)

CPERT Continuous Professional Education Round Table

CPES Center for Power Electronics Systems (USA) • Checked Polyelastic Sampling • Chemistry, Physics, and Environmental Sciences • Contractor Performance Evaluation System • Crew Procedures Evaluation Simulator • Crew Procedures Evaluator Simulator

CPET Charged Particle Electrostatic Thruster • Contractor Performance Evaluation Team

CPEUG Computer Performance Evaluation Users Group

CPEV Customer Premises Equipment Voice

CpF Cost per Flight

CPF Calibration Product File • Cargo Processing Facility • Central Processing Facility • Central Provident Fund • Cepu (Indonesia) (location identifier) • Change Parameter File • Civilian Personnel Flight • Command Propagation Facility • Common Publishing Format • Computerized Physical Fit • Context Passing Format • Contractor Performance Factor • Control Program Facility • Cost Per Flight • Critical Point Facility • Cumulative Percentage Frequency

CPFF Cost Plus Fixed Fee

CPFL Contingency Planning Facilities List

CPFM Composition and Photodissociative Flux Measurement

CPFSK Continuous Phase Frequency Shift Keying

CPG Calibration Proposal Generation • Call Progress • Candidate Pass Generator • Carmen de Patagones (Argentina-BA) (location identifier) • Catalog of Paired Galaxies • Catalog of Pairs of Galaxies • Central Processing Group • Central Processor Group • Certificat de Physique Générale • Change Planning Group • Clock Pulse Generator • Coarse-Pointing Gimbal • Communications Processor Group • Contingency Planning Guidance

CPGA Catalog of Peculiar Galaxies and Associations • Ceramic Pin Grid Array

CPGE Classes Préparatoires aux Grandes Écoles (France)

CPGS Compiling Genetic Programming System

cph Cards per Hour • Characters per Hour • Counts per Hour • Cycles per Hour

Cph Cost Per Hour

CPH Calls Per Hour • Central Powerhouse • Copenhagen (Denmark) (location identifier) • Cycles Per Hour • København (Denmark) (location identifier)

CPhD Cape Photographic Durchmusterung (periodical code)

CPHE Comets in the Post-Halley Era

cpi Characters per Inch • Clock per Instruction • Code Page Information • Cycles per Instruction

CpI Cost per Instruction

CPI Carbon Preference Index • Center of Programmed Instruction • Central Patent Index • Centre de Première Intervention • Centro Público de Investigación • Channel Port Index • Channel Protocol Interpreter • Characters Per Inch • Charge Particle Instrument • Chemical Patent Index • Clock Per Instruction • Clock Pulse Interval • Code de la Propriété Intellectuelle (France) • Code-Page Information • Code-Page Interface • Coherent Processing Interval • Committee for Public Information • Committee on Public Information • Common Part Indicator • Common-Path Interferometry • Common Programming Interface • Communications Processing Interface • Communications Processor and Interface • Comprehensive Plasma Instrument • Comprehensive Plasma Investigation • Computer PBAX Interface • Computer-Prescribed Instruction • Conception de Produits Industriels • Conference Papers Index • Consumer Price Index • Continuous Process Improvement (*) • Continuous Progress Improvement • Contractor Preliminary Inspection • Converging Partial Indicators (method) • Corte Penal Internacional • Cost Performance Index • Cost Per Instruction • Cost Plus Incentive • Cour Pénale Internationale • Crash Position Indicator • Crédits de Politique Industrielle • Cross-Pointer Indicator • Current Physics Index

CPIA Centro Paraguayo de Informaciones Astronómicas (Paraguay) • Close-Pair Interstitial Atom

CPIC Charged Particles Information Center (USA) • China Patent Information Center (China-PRC) • Common Programming Interface for Communications • Computer Program Integration Contract • Computer Program Integration Contractor

CPIF Character Position In Frame • Cost Plus Incentive Fee

CPIL Critical Played Items List

CPILS Correlation Protection Integrated Landing System

CPIM Communications Processor and Interface Module

CPIN Computer Program Identification Number • Computer Program Identification Numbering

CPINS Computer Program Identification Numbering System

CPIO Copy In and Out

CPIP Chemical Physics Interdisciplinary Program • Club des Professionnels de l'Industrie du Photon (France) • Computer Program Implementation Process

CPIPI Commission pour la Protection Internationale de la Propriété Industrielle

CPIPT Cost and Performance Integrated Product Team

CPIR Corporate Preparedness Industrial Representative

CPIRSS Catalog of Positions of Infrared Stellar Sources

CPIS Canadian Polar Information System (Canada) • Center for Planning and Information Systems • Comprehensive Planning Information System • Contingency Planning Information System

CPIT Contract Price of Item Terminated

CPIU Central Payload Interface Unit

CPIWI Customer Premise Inside Wire Installation

CPJI Cour Permanente de Justice Internationale

CPK Corey - Pauling - Koltun (model)

cpl Control Panel (file extension)

Cpl Caporal • Corporal • Couple

CpL Characters per Line

CPL Callable Personal Librarian • Capillary Pumped Loop • Central Public Library • Chaparral (Colombia) (location identifier) • Characters Per Line • Chemical Physics Letters (periodical code) • Circularly Polarized Light • Combined Programming Language • Commercial Pilot's Licence • Commercial Price List • Common Program Language • Common Programming Language • Common Pulse Line • ComPASS Planning Language • Computer Program Library • Consolidated Parts List • Contractor Procurement List • Contract Policy Letter • Control Panel • Conversational Planning Language • Conversational Programming Language • Corporal • Couple • Crash Program Locator • Credential Privilege List • Critical Path Length • Current Privilege Level • Customer Priority List

CPLC Commission du Personnel des Laboratoires Conventionnés

Cpld Coupled

CPLD Complex Programmable Logic Device • Coupled

CPLEE Charged Particle Lunar Environment Experiment

Cplet Complet

Cplète Complète

CPLF Certificat Pratique de Langue Française

Cplg Coupling

CPLG Coupling

CPLI Calling Party Line Identification

CPLOM Center for Production, Logistics and Operations Management

Cplr Coupler

Cplry Capillary

CPLT Complete • Copilot

cpm Cards per Minute • Characters per Minute • Counts per Minute • Cycles per Minute

CpM Cost per Minute

CP/M Control Program for Microcomputers • Control Program Monitor

CPM Capsule Positioning Mechanism • Career Program Manager • Cargo Propulsion Module • Cathode Pulse Modulation • Cell Performance Monitor • Center Planning Manager • Center Program Manager • Central Path Method • Central Processing Module • Central Processor Module • Change Proposal Manager • Characters Per Minute • Chemical Propulsion Module • Civilian Personnel Manual • Combinatorial Pattern Matching • Command Processor Module • Common Particular Meter • Common Proper Motion • Compton (USA-CA) (location identifier) • Computer Performance Management • Computer Programming Manual • Computer Program Module • Computer Project Manager • Conference Preparatory Meeting • Continuous Phase Modulation • Contractor Performance Measurement • Control Processor Module • Control Program for Microcomputers • Control Program Monitor • Conversational Program Module • Cost and Performance Measurement • Cost Performance Measure • Cost Per Minute • Cost Per Thousand • Counts Per Minute • Coût Par Millier • Critical-Path Method • Critical-Path Movement

CPMA Centre de Projets en Microélectronique Avancée (France)

CPMAS Cross-Polarization Magic Angle Spinning

CPMB Common Proper Motion Binary

CPMC Contractor Performance Measurement Course

CPME Charged Particle Measurement Experiment

CPMF Computer Program Maintenance Facility • Cost Plus Management Fee

CPMIS Consolidated Personnel Management Information System

Cpmm Characters per Millimeter

CPMM Characters Per Millimeter

CPMOH Centre de Physique Moléculaire Optique et Hertzienne (France) (now CELIA)

CPMP Carrier Performance Measurement Plan • Circum-Pacific Man Project • Computer Program Management

Plan • Crew Procedures Management Plan • Crew Products Management Plan

CPMR Computer Program Management Review

CPMS Career Planning Management System • Climatological Prediction by Model Statistics • Computer-assisted Photogrammetric Mapping System • Computer Performance Management System • Contractor Property Management System • Control Point Management Service

CPMST Certificat Préparatoire à la Maîtrise de Sciences et Techniques (France)

CPMT Components, Packaging and Manufacturing Technology

Cpn Copenhagen (Denmark) • Coupon

CPN Calling Party Number • Cape Rodney (Papua New Guinea) (location identifier) • Causal Probabilistic Network • CCSDS Principal Network • Contract Payment Notice • Corps de Propulseur Nu • Customer Premises Network • Customer Premises Node

CPND Charged Particle Nuclear Data

CPNE Commission Paritaire Nationale de l'Emploi (France)

Cpnhgn Copenhagen (Denmark)

CPNI Consumer Proprietary Network Information

CPNS Customer Preferred Name System

Cpntr Carpenter

Cpo Campo

CPO Canberra Planetarium and Observatory (Australia-ACT) (now CSDO) • Cancel Previous Order • Capilla Peak Observatory (USA-NM) • Catalytic Partial Oxidation • Central Pennsylvania Observers (USA-PA) • Central Planning Office • Central Post Office • Central Procurement Office • Central Project Office • Charged Particle Optics • Chief Petty Officer • Chief Post Office • Chinese Patent Office (China-PRC) • Circular Parking Orbit • Civilian Personnel Office • Civilian Personnel Officer • Cloud Physics Observatory • Command Pay Office • Community Post Office • Complete Partial Order • Complete Provisions Only • Compulsory Purchase Order • Controlled Precision Oscillator • Copiapo (Chile) (location identifier) • Core Project Office

CPODA Contention Priority-Oriented Demand Assignment

CPOL Communications Procedure-Oriented Language

CPOS Civilian Personnel Occupational Standards

CPOZ Compressed Profile Ozone

CPP Cable Patch Panel • Calling Party Pays • Camouflage Pattern Printing • Cancellation Protection Plan • Capacitive Phase Probe • Capital Purchase Program • Caractères Par Pouce • Card Punching Printer • Centralized Postage Payment • Central Processing Point • Centre Permanent Professionnel • Certificat de Pratique Professionnelle • Chemical Processing Plant • Citations Per Paper • Citations Per Publication • Civilian Pay Plan • Cloud Photopolarimeter • Clustered Pearson Predictor • Columbus Polar Platform • Comet Physical Properties Package • Commercial Practices Program • Communications Protocol Processor • Competitive Phased Pricing • Controllable Pitch Propeller • Convective Protoplanet • Corps Propulseur Protégé • Critical Path Planning • Current Perpendicular to the Plane • Current Purchasing Power

CPPA Canadian Periodical Publishers Association (Canada) • Centralized Parts Procurement Agency • Cosmological Pressureless Parabolic Advection

CPPC Career Program Policy Committee • Core Project Planning Committee • Cost Plus Percentage of Cost

CPPL Command Processor Parameter List

CPPM Center for Process Plant Management • Centre de Physique des Particules de Marseille (France)

CPPP Compressed Point-to-Point Protocol

CPPS Comisión Permanente del Pacífico Sur • Commission Permanente pour le Pacifique Sud • Computer Programming Performance Specification • Computer Program Product Specification • Critical Path Planning and Scheduling

Cpq Compaq

CPQ Campinas (Brazil-SP) (location identifier) • Comissão Permanente para a Qualificação

cpr Cycles per Revolution

CPR Canadian Pacific Railway (Canada) • Cardiopulmonary Rescucitation • Cardiovascular Pulmonary Rescucitation • Card Puncher and Reader • Carrier Performance Rating • Casper (USA-WY) (location identifier) • Cell Phone Radiation • Cellular Packet Radio • Center for Public Resources Inc. • Centre de Programmation Récréative • Centre Pédagogique Régional • Chairman's Program Recommendation • Circularly Polarized Ratio • Circular Polarization Ratio • Civilian Personnel Regulation • Clock Pulse Repeater • Cloud Physics Radiometer • Cloud Profiling Radar • Color-Period Relation • Command Performance Report • Command Performance Review • Committee of Permanent Representatives • Common Position Reference • Compact Positioning Reporting (algorithm) • Computer-based Patient Record • Computerized Patient Record • Computer Personal Research • Consolidated Progress Report • Continuous Plankton Recorder • Contractor Performance Report • Contract Procurement Request • Contract Progress Review • Cost Performance Report • Courseware Problem Report • Critical Problem Report • Customer Payload Requirement • Cut Paraboloidal Reflector • Cycle Pressure Ratio

CPRA Compressed Pulse Radar Altimeter

CPRC Central Price Regulation Committee

CPRDB Comprehensive Pacific Rainfall Database

CPRI Computerized Patient Record Institute

Cpris Compris

CPRM Copy Protection for Recordable Media

Cprs Compress

CPRS Central Policy Review Staff • Cloud Profiling Radar System

Cprsd Compressed

Cprsn Compression

Cprsr Compressor

CPRU Construction Planning and Research Unit

cps Cards per Second • Centipoise • Characters per Second • Counts per Second • Cycles per Second

CpS Connections per Second

CPS Camera Power Supply • Caractères Par Seconde • Carte de Professionnel de Santé • Cartographic Production System • Cathode Potential Stabilized • Center for Policy Studies • Central Pain Syndrome • Central Plasma Sheet • Central Police Station • Central Power Supply • Central Power System • Central Processing System • Centrum voor Plasmafysica en Stralingstechnologie • Certificate Practice Statement • Certification Practice Statement • Certified Professional Secretary • Characters Per Second • Chicago Public Schools (USA-IL) • Circuit Package Schematic • Close-Packed Sphere • Coarse Pointing System • Collection and Processing System • Collective Protecting Shelter • Color Picture Signal • Combustion and Plasma System • Commercial Processing System • Commission Permanente de Standardisation (France) • Commonwealth Public Service • Communications Processing System • Competitive Prototyping Strategy • Complementary Polar Satellite • Computer and Peripheral System • Computer Power Supply • Computer Process System • Computer Programming Service • Computer Program Specification • Computer Program System • Connections Per Second • Contour Plotting System • Contractor Performance Support • Contractor Performance System • Contractor Profile System • Contractor Purchasing System • Control Power Supply • Control Power

Switch • Control Pressure System • Conversational Programming System • Conversation Program System • Conversion Program System • Copy Processing System • Core Processing System • Counts Per Second • Crew Procedures Simulator • Critical Path Schedule • Critical Path Scheduling • Critical Path System • Current Population Survey • Current Program Status • Cycle Processor System • Cycles Per Second • Saint Louis - Bi-State Parks (USA-MO) (location identifier)

CPSC Consumer Product Safety Commission • Contingency Planning Support Capability

CPSD Corrected Pattern Standard Deviation • Cross-Power Spectral Density

Cpse Counterpoise

CPSE Common Payload Support Equipment • Complex Problem Solving Environment • Crew and Passenger Support Equipment

CPSG Common Power Supply Group

CPSK Cathodic Protection Survey Kit • Coherent Phase Shift Keyed

CPSM Continuous Phase Shift Modulation • Cyclic Pulse System Modulation

CPSMA Commission on Physical Sciences, Mathematics and Applications

CPSO Cataloguing Policy and Support Office • Cumberland Plateau Seismological Observatory

CPSR CalTech-Parkes-Swinborne Recorder • Computer Professionals for Social Responsibility • Contractor Procurement System Review • Contractor Purchasing System Review • Contract Procurement System Review

CPSS Call Processing Subsystem • Central Processing Subsystem • Cold Plate Support Structure • Common Program Support System • Control Packet Switching System • Cooperative Program in Space Science • Critical Phase System Software

CPSSP Comité Professionnel des Stocks Stratégiques Pétroliers (France)

CPSST Cross-Product Sea Surface Temperature

CPST Comité pour la Politique Scientifique et Technique • Commission on Professionals in Science and Technology • Committee to Promote Science and Technology

CPSU Central Processor Subunit

Cpt Capitain • Cockpit • Comptant • Counterpoint

CpT Cost per Thousand

CPT Cape Town - D.F. Malan (South Africa) (location identifier) • Captain • Cargo Processing Technician • CCI - PS - TRI • Central Planning Team • Centre de Perfectionnement Technique • Centre de Physique Théorique • Charged-Particle Telescope • Charge - Parity - Time • Clock, Programming and Timing • Cockpit Procedures Trainer • Coherent Population Trapping • Combination Photoelectric Telescope • Command Pass Through • Command Post Terminal • Comparison Test • Comprehensive Performance Test • Computer Program Tape • Computer Program Test • Contractor-Provided Training • Control Position Transducer • Control Power Transformer • Crisis Planning Team • Critical Path Technique • Critical Process Team • Cryogenic Pressure Transducer • Crystal Pressure Transducer • Curriculum Practical Training

CPTA Comité Technique Paritaire Académique (France) • Computer Programming and Testing Activity • Core Propulsion Test Article

Cptán Capitán

CPTC Central Processor Test Console • Computer Performance Technical Center

CPTD Comité Technique Paritaire Départemental (France) • Comprehensive Plan for Training devices

Cpte Capitaine • Compte

CPT&E Computer Program Test & Evaluation

CPTE Computer Program Test and Evaluation • Computer Program Test Equipment

CPTEC Centro de Previsão de Tempo e Estudos Climáticos

CPTED Crime Prevention Through Environmental Design

CPTF Common Patch and Test Facility

CPTP Computer Program Test Plan • Computer Program Test Procedure • Coordinated Project Technology Programme

CPTPR Computer Program Test Procedure

CPTR Compact Payload Test Range • Computer Program Test Report

CPTS Computer Program Test Specification

CPTYC Committee on Physics in Two-Year Colleges

CpU Cost per Unit

CPU Central Packaging Unit • Central Processing Unit (*) • Central Processor Unit • Centre de Première Urgence • Collective Protection Unit • Command Processor Unit • Committee on the Physics of the Universe (NRC) • Communications Processing Unit • Communications Processor Unit • Computer Peripheral Unit • Computer Printer Unit • Computer Processing Unit • Computer Program Update • Computer Publishing Unit • Conférence des Présidents d'Universités • Contract Postal Unit • Control Processing Unit • Cururupu (Brazil-MA) (location identifier)

CPUAX Central Processing Unit - Arithmetic Extended

CPUC Current Procurement Unit Cost

CPUID Central Processing Unit Identifier

Cpv Capoverso

CPV Campina Grande - João Suassuna (Brazil-PB) (location identifier) • Cape Verde (ISO country code) • Cap Vert (code ISO de pays) • Centre de Programmation Végétation • Chlorinated Polyvinyl • Chlorure de Polyvinyle • Command Post Vehicle • Common Pressure Vessel • Common Procurement Vocabulary

CPVC Chlorinated Polyvinyl Chloride

CPW Center for Public Works • Certified Public Weigher • Coplanar Waveguide

CPWG Center Projects Working Group • Communications Procedures Working Group

cpx Corel Presentation Exchange (file extension)

CPX Charged Pigment Xerography • Common Picture Exchange • Culebra (Puerto Rico) (location identifier)

CPXe Common Picture Exchange Enhancement

CPY Copy

CPZ Contingency Planning Zone

CP2 Contractor Performance Certification Program

cq Casu Quo

CQ Cercle de Qualité • Change of Quarters • Color-book Query • Concept Query • Conditionally Qualified • Contrat de Qualification • Current and Qualified

CQA Construction Quality Assurance • Contract Quality Assurance

CQAO Contrôle de Qualité Assisté par Ordinateur

CQAP Construction Quality Assurance Plan

CQAT Combined Qualification and Acceptance Testing

CQC Contractor Quality Control • Critères - Qualité - Contrôles

CQCM Controlled Quartz Crystal Microbalance • Cryogenically-cooled Quartz Crystal Microbalance

CQCT Centre Québécois de Coordination de la Télédétection

CQDR Critical Qualification Design Review

CQE Central Quasi-Emission • Certified Quality Engineer • Cognizant Quality Engineer

CQF Calais (France) (location identifier)

CQFD Ce Qu'il Fallait Démontrer

CQFP Ceramic Quad-Flatpack

CQG Classical and Quantum Gravity

CQGra Classical Quantum Gravity (periodical code)

CQGrS Classical Quantum Gravity Supplement (periodical code)

CQI Continuous Quality Improvement

CQIM Configuration and Quality Information Management

CQIP Continuous Quality Improvement Process

CQL Computers and the Quality of Life • Contractual Quantization Limit

CQM Colloquium • Control Quality Monitor • Cost Quality Management

CQMP Clinical Quality Management Program

CQP Cape Flattery (Australia-DLD) (location identifier) • Contrat de Qualification Professionnelle

CQR Contract Quality Requirement

CQS Call Queue Service • Costa Marques (Brazil-RO) (location identifier)

CQSSP Completed Quasar, Satellite and Star Positioning • Coupled Quasar, Satellite and Star Positioning (*)

CQT Capacitor Qualification Test • Caquetania (Colombia) (location identifier)

CQTP Capacitor Qualification Test Program

c/r Company's Risk

cr Con Riserva • Costa Rica • Crux • Cum Rights

Cr Chrom • Chrome • Chromium • Commander • Councillor • Cour • Create • Created • Creation • Credit • Creditor • Creek • Crown • Cruise • Cruiser

C/R Command/Response • Commutation Rate • Converter/Regulator • Counter Rotating

C&R Circumvention & Recovery

CR Calculated Risk • Calendrier Républicain • Call Reference • Camera Ready • Cameroon • Card Reader • Carriage Return • Carrier Route • Carrier's Risk • Carrington Rotation • Cash Receipt • Cathode Ray • Cathode Reaction • Central Region • Central Registry • Central Repository • Čerenkov Radiation • Certification Recommendation • Certification Request • Certification Requirement • Challenge Ring • Change Recommendation • Change Record • Change Report • Change Request • Channel Rate • Channel Reliability • Character Reader • Chargé de Recherche • Circuit Reliability • Citations per Researcher • Cited Research • Civis Romanus • Clarification Report • Clarification Request • Classification Research • Classroom • Close Range • Cold Region • Collisional-Radiative • Comb Response • Command Register • Command-Response • Commutation Rate • Compact Ridge • Company's Risk • Complete Redistribution • Complete Regression • Compression Ratio • Compte-Rendu • Comptes-Rendus • Comptes-Rendus de l'Académie des Sciences (periodical code) • Computer Resources • Computing Reviews • Concentration Ratio • Conception Review • Conditioned Reflex • Conditioned Response • Condition Recognition • Conference Report • Configuration Review • Confirm Response • Conflict Resolution • Congo Red • Congressional Record • Connection Request • Consciousness-Raising • Conseil Régional • Conservation Reserve • Constrained Realization • Consultant Report • Continuing Resolution • Contractor-Recommended • Contractor Report • Contractor's Response • Contract Report • Controlled Rectifier • Control Relay • Control Room • Coordinate Reader • Coordinate Registration • Corner Reflector • Corotation • Corotation Radius • Corotation Resonance • Correlation Radiometer • Correlation Recognition • Corrientes (Argentina) • Cosmic Ray • Costa Rica (international vehicle registration code) • Costa Rica (ISO country code) • Cost Reimbursement • Counter-Rotating • Counter-Rotation • Count Reverse • Country Representative • Credential Record • Credit • Credit Rating • Credit Report • Creek • Crew • Critical Ratio • Critical Resource • Croatia • Cross-Reference • Crystal Rectifier • Current Rate • Current Regulator • Current Relay • Curvature Radiation • Customer's Request • Cyclic Reduction • Cyclotron Resonance • Portugal (international aircraft registration code)

CrA Corona Australis - Coronae Australis (constellation) • Corona Austrina - Coronae Austrinae (constellation)

CRA Carbon Rod Atomizer • Cassegrain Reflector Antenna • Catalog Recovery Area • Centre de Recherches Atmosphériques (France) • Centro di Ricerche Aerospaziale • Command Relationship Agreement • Commission Régionale de l'Apprentissage • Composite Research Aircraft • Computer Research Association • Computing Research Association (USA) • Conflict Resolution Advisory • Consiglio per le Ricerche Astronomiche • Continuing Resolution Act (USA) • Continuing Resolution Authority • Contrat de Réinsertion par Alternance • Coordinated Regional Assessment • Cooperative Research Agreement • Coordination Review Authority • Corrosion-Resistant Alloy • Cosmic-Ray Altimeter • Council for Research in Astronomy • Craiova (Romania) (location identifier) • Crew Reception Area

CRAA Corporation for Research Amateur Astronomy (USA-CA)

CRAAE Centro de Radio-Astronomia e Aplicações Espaciais (Brazil)

CRAAG Centre de Recherches en Astronomie, Astrophysique et Géophysique (Algeria)

CRAAM Centro de Rádio-Astronomia e Astrofísica Mackenzie (Brazil-SP) (formerly GRAM)

CRAB Compte-Rendu d'Activité des Bâtiments

CRAC Career Research Advisory Centre (UK) • Cellule Rectorale d'Action Culturelle • Computer Resource Acquisition Course

CRAD Cooperative Research and Development

CRADA Cooperative Research And Development Agreement • Cooperative Research And Development Award

CRADO Collaborative Research And Development Operation

CRADP Cooperative Research and Development Plan

CRADT Conférence Régionale d'Aménagement et de Développement du Territoire (France)

CRAF Comet Rendezvous and Asteroid Flyby • Committee on Radio Astronomy Frequencies

CRAFT Computerized Relative Allocation of Facilities Technique • Cooperative Research Action For Technology • Cray Research Adaptive Fortran

CRAG Contractor Risk Assessment Guide

CRAL Centre de Recherche Astronomique de Lyon (France)

CRAM Caisse Régionale d'Assurance Maladie (France) • Card Random Access Memory • Challenge-Response Authentication Mechanism • Communication Routing And Metering • Compte-Rendu Automatisé Mensuel • Computerized Reliability Analysis Method • Control Random Access Memory

CRAMP Computerized Routing for Automated Mail Processing

CRAND Cosmic-Ray Albedo Neutron Decay

CRANE Cosmic-Ray Nuclear Experiment

CrAO Crimean Astrophysical Observatory (Ukraine)

CRAP Cercle de Recherche et d'Action Pédagogique • Cleveland Regional Association of Planetariums (USA-OH)

CRAPE Central Register of Aerial Photography for England

CraR Polskie Towarzystwo Milosnikow Astronomii (periodical code)

CRAR Critical Reliability Action Report

CraRe Cracow Observatory Reprints (periodical code)

CRAS CIMSS Regional Assimilation System • Comet Rapid Announcement Service • Comptes-Rendus de l'Académie des Sciences (France) • Cost Reduction Alternative Study

CRASA Comptes-Rendus de l'Académie des Sciences - Vie Académique (periodical code)

CRASB Comptes-Rendus de l'Académie des Sciences - Série B Physique (periodical code)

CRASG Comptes-Rendus de l'Académie des Sciences - La Vie des Sciences (periodical code)

CRASM Comptes-Rendus de l'Académie des Sciences - Mathématiques (periodical code)

CRASP Comptes-Rendus de l'Académie des Sciences - Mécanique Chimie (periodical code)

CRAST Coordinate Refinement Analytical Strip Triangulation

CRATE Centimetric Radiometer Absolute Temperature

CRAWS Cold Region Automatic Weather Station

CRAZI Count Routine Applied to Zero Input

CrB Corona Borealis - Coronae Borealis (constellation)

CRB Central Radio Bureau • Certification Record Brief • Change Review Board • Collarenebri (Australia-NSW) (location identifier) • Command Review Board • Concept Review Board • Configuration Review Board • Contract Review Board • Correspondence Review Board • Cost Review Board • Cramer-Rao Bound

CRBE Conversational Remote Batch Entry

CRBP Cold Regions Bibliography Project

CRC Camera-Ready Copy • Canada Research Chairs (programme) • Cancer Research Campaign • Candidate Review Committee • Career Resource Center • Carriage Return Contact • Cartago (Colombia) (location identifier) • Cataloguing Rules Committee • Central Reporting Center • Centre Régional de Coordination • Centre Régional de la Consommation (France) • Centre Réseau Communication • Centro de Rastreiro e Controle • Chambre Régionale des Comptes • Chargé de Recherche Contractuel • Circuit Routing Chart • Civil Rights Commission • Class - Responsibility - Collaborator • Climate Research Committee • Code de Redondance Cyclique • Colón (Costa Rica) (ISO currency code) • Color Ratio Composite • Column Reference Code • Combination Research Centre • Common Reference Code • Communications Relay Center • Communications Research Centre • Compact Radio Core • Computer Resources Center • Computer Resources Council • Contractor-Recommended Code • Contractor-Recommended Coding • Control and Reporting Center • Cooperative Research Centre • Coordinating Research Council • Cost Reduction Curve • Counter-Rotating Core • Cour Régionale des Comptes • Crisis Response Cell • Cumulative Results Criterion • Cyclical Redundancy Character • Cyclic Redundancy Character • Cyclic Redundancy Check • Cyclic Redundancy Checking • Cyclic Redundancy Code

CRCB Computer Resource Control Board

CRCC Center for Research on Concepts and Cognition • Combined Rescue Coordination Center • Consolidated Record Communications Center

CRCD Combined Research-Curriculum Development • Coordinated Research and Curriculum Development

CRCG Center for Research in Computer Graphics • Common Routing Connection Group

CRCH Central Register and Clearinghouse • Central Reprint Clearinghouse

CRCI Chambre Régionale de Commerce et d'Industrie (France) • Computer Resource Configured Item

CRCIM Czech Research Consortium for Informatics and Mathematics (Czech Rep.)

CRCL Circle

CRCLE Circle

CRCLMP Computer Resource Life Cycle Management Plan

Crclt Circulate

Crcltd Circulated

CRCM Conference on Recent Crustal Movements • Commission on Recent Crustal Movements

Crcmf Circumference

CRCS Cassegrain Rotator Control System • Clinical Record Cover Sheet

CRCSS Cooperative Research Centre for Satellite Systems (QUT)

CRCT Centre de Recherche sur la Combustion Turbulente • Centre de Recherche sur la Culture Technique • Congé pour Recherches ou Conversion Thématique

Crctn Correction

CRCU Crypto Remote Control Unit

CRCV Circunvalação

CRCW Concurrent-Read - Concurrent-Write

crd Windows Card (file extension)

CRD Capacitor-Resistor Diode • Cardfile • Cavity-Ring-Down (technique) • Centre de Recherche Documentaire • Change Request Disposition • CODATA Referral Database • Color Rendering Dictionary • Comodoro Rivadavia (Argentina-CB) (location identifier) • Complete Redistribution • Computer Readout Device • Confidential Restricted Data • Congressional Reference Division • Cosmic-Ray Detector • Cost Reporting Document

CRDA Contractor Research and Development Agreement • Converging Runway Display Aid • Cooperative Research and Development Agreement

CRDAS Cavity Ring Down Absorption Spectroscopy

CRDB Communications Requirements Database • Contingencies Resources Database • Cost Risk Database

CRDD Calibrated Raw Detector Data • Calibrated Reconstructed Detector Data (*)

CRDDP Centre Régional de Documentation et de Diffusion Pédagogique

CRDEC Chemical Research, Development, and Engineering Center

CRDF Cathode-Ray Direction Finding • Civilian Research and Development Foundation (USA)

CRDI Centre de Recherches pour le Développement International

CRDL Critical Requirement Deficiency List

CRDM Control Rod Drive Mechanism

CRDP Centre Régional de Documentation Pédagogique (France) • Certification Review Data Package • Computer Resources Development Plan

CRDS Calligraphic Raster Display System • Cavity Ring Down Spectroscopy • Contribution au Remboursement de la Dette Sociale (France)

CRe Cosmic-Ray Electron

C&RE Cost & Resource Estimate

CRE Cation-Responsive Electrode • Cauchy-Riemann Equation • Center for Renewable Energy • Chemical Release Experiment • Cold Readout Electronics • Commission de Régulation de l'Électricité (France) • Computer Reprogramming Equipment • Conférence des Recteurs Européens • Contrat de Retour à l'Emploi • Controlled Requirement Expression • Control Reporting Element • Correlation - Regression - Effectiveness • Corrosion-Resistant • Cosmic-Ray Exposure • Cross-Range Error • Cryogenic Readout Electronics • Cumulative Radiation Effect • Myrtle Beach - Grand Strand (USA-SC) (location identifier)

CREA Centre de Recherches et d'Études Appliquées • Cosmic-Ray Exposure Age

CREAM Cognitive Reality and Error Analysis Method • Cosmic-Radiation Effect and Activation Monitor (*) • Cosmic-Ray Effect and Activation Monitor • Cosmic-Ray Effect and Activation Monitoring • Cosmic-Ray Energetics And Mass

Crec Creciente

CREDAS Centro Regional de Datos Satelitales

CREDIF Centre de Recherche et d'Étude pour la Diffusion du Français

CREDO Comprehensive Report Examination and Display Option • Cosmic-Ray Effects and Dosimetry

CREEM Centre de Recherche et d'Expérimentation pour l'Enseignement Mathématique

CREF Centre Rural d'Éducation et de Formation (France) • Computer-Ready Electronic File • Connection Refused

CREFAL Centro Regional de Educación Fundamental para la América Latina

CREG Concentrated Range Extension with Gain

CREI Centro Regional de Educación en Informática • Centro Regional para la Enseñanza de la Informática

CREID Conversational Response for ETL Image Database

CREIM Canadian Rockies Ecosystem Interagency Management (Canada)

CREIT Centre de Recherches et d'Études en Informatique et Télématique

CREM Centre de Recherche sur l'Enseignement des Mathématiques • Corporate Real Estate Management

CREMADE Centre de Recherches de Mathématique de la Décision (France)

CREME Cosmi-Ray Effects on Microelectronics

CREN Chemical Resources and Environmental Needs • Computer Research Education Network • Consortium for Research and Education Network • Corporation for Research and Educational Networking

CREOL Center for Research and Education in Optics and Lasers • Center for Research in Electro-Optics and Lasers

CREP Centre Régional d'Éducation Populaire • Contract Repair Enhancement Process • Contract Repair Enhancement Program

CREPA Centre Régional d'Éducation Populaire pour Adultes (France)

CREPE Cosmic-Ray Emulsion Plastic Equipment

CREPUQ Conférence des Recteurs Et Principaux des Universités du Québec (Canada-QC)

CRERI Centre de Recherche en Économie Régionale et Industrielle (France) • Centre of Empirical Research in International Relations

CRERS Crew Rescue and Equipment Retrieval System

Cres Crescent

CRES Centre for Research and Environmental Studies • Centre for Research on Environmental Systems • Centre Régional d'Énergie Solaire • Command Readiness Exercise System • Corrosion-Resistant Steel • Crescent

CRESALC Centro Regional para la Educación Superior de América Latina y el Caribe

CRESER Centre Régional de l'Enseignement Supérieur Et de la Recherche

CRESNA Centre de Recherches pour l'Étude des Substances Naturelles (Belgium)

CRESP Consortium for Risk Evaluation with Stakeholder Participation • Coordinated Research and Environmental Surveillance Project

CRESS Center for Research in Earth and Space Science • Center for Research in Experimental Space Science • Center for Research in Social Systems

CRESST Cryogenic Rare Event Search with Superconducting Thermometers

CREST Career-Related Experience in Science and Technology • Cenozoic Reef Evolution in Space and Time • Center for Research and Education in Science and Technology (India) • Center for Research Excellence in Science and Technology • Center for Research in Environmental Science and Technology • Centre de Recherche en Évaluation Sociale des Technologies • Comité de la Recherche Scientifique et Technique • Core Research for Evolutional Science and Technology • Creativity in Science and Technology • Curriculum Resources in Energy Science and Technology

CRESST Center for Research and Studies on Strategy and Technology

CRESTech Centre for Research in Earth and Space Technology (Canada-ON) (merger of ISTS and WCGR)

CRESTEL Club de Réflexion et d'Études sur les Systèmes de Télécommunications (France)

CRESU Cinétique de Réaction en Écoulement Supersonique Uniforme

CRET Centre de Recherches et d'Échanges Technologiques • Cathode-Ray Electron Tube

CRF Cable Retransmission Facility • Calibration Resource File • Capacitor Resonance Frequency • Capital Recovery Factor • Carnot (Central African Rep.) (location identifier) • Cathode-Ray Furnace • Celestial Receiver Frame • Celestial Reference Frame • Cell Relay Function • Cell Relaying Function • Central Repair Facility • Central Retransmission Facility • Channel Reassignment Function • Cloud Radiation Forcing • Cloud Radiative Forcing • Combustion Research Facility • Compressor Research Facility • Computer Readable Form • Computer Resources Focus • Concentrator Relay Function • Connection-Related Function • Consolidated Revenue Fund • Content Retrieval Filter • Corotating Frame • Correspondence Routing Form • Corticotropin Releasing Factor • Cosmic-Ray Flux • Crisis Response Force • Cross-Reference File

CRFC Centre Régional de Formation Continue

CRFMAES Centre Régional de Formation des Maîtres de l'Enseignement Spécialisé

CRFP Centre Régional de Formation Professionnelle

Crg Carriage

CRG Centre de Recherche en Gestion • Centre de Recherches Géophysiques • Change Review Group • Collaborative Research Grant • Collaborative Research Group • Communications Relay Group • Compressed Rastor Graphics • Computer Resources Group • Control Rate Gyro • Correspondence Review Group • Cosmic-Ray Gas • Cost Report Generator • Jacksonville Craig (USA-FL) (location identifier)

CRGP Converse Reduced Game Property

CRH Cherribah (Australia-QLD) (location identifier)

CRHD Convective Radiation Hydrodynamics

CRHI Centre de Recherche en Histoire de l'Innovation

CRHST Centre de Recherche en Histoire des Sciences et des Techniques

CRI Cambridge Research and Instrumentation (USA-MA) • Centraal Rekeninstituut (Netherlands) • Centrale de Règlements Interbancaires (France) • Centre de Rattrapage Intégré • Centre de Réception des Images • Centre de Recherche en Informatique • Centre de Ressources Informatiques • Centre Relai Innovation • Certification Requirements Index • Climate Research Institute • Clinical Research Initiative • Collective Routing Indicator • Color Rendering Index • Color Reproduction Index • Color Reversal Intermediate • Comité de Recherches en Informatique • Contrôle des Règlements Interbancaires • Corporate Research Institute • Cosmic-Ray Institute • Cosmic-Ray Intensity • Cosmic-Ray Interstellar (model) • Costa

Rica (ISO country code) • Cray Research Inc. • Criterion-Referenced Instruction • Croix Rouge Internationale • Crooked Island (Bahamas) (location identifier) • Cross-Reference Index • Crown Research Institute (New Zealand) • Cyclotron Resonance Feature

CRIA Center for Research on Information Access

CRIB Computer Resources Information Base

CRIBABB Centro Regional de Investigaciones Basicas y Applicadas Bahia Blanca (Argentina)

CRIB Computerized Resources Information Bank

CRIC Centre de Recherche Industrielle du Québec (Canada-QC) • Centre for Research on Innovation and Competition (UK) • Copyright Research and Information Center

CRICCS Center for Real-time, Intelligent, Complex Computer System

CRICR Centre Régional d'Information et de Coordination Routière (France)

CRICyT Centro Regional de Investigaciones Científicas y Tecnológicas (Argentina-MD)

CRID Centre de Recherche et d'Information pour le Développement

CRIE Cosmic-Ray Isotope Experiment

CRIEPI Central Research Institute of Electric Power Industry (Japan)

CRIES Comprehensive Resource Inventory and Evaluation System

CRIF Cargo Routing Information File

CRIG Capacitor Rate-Integrating Gyro

CRIHAN Centre de Ressources Informatique de Haute-Normandie (France)

Crim Criminal

CRIM Centre de Recherche Informatique de Montréal (Canada-QC) • Commercial Refrigerator Incubator Module

Criminol Criminologist • Criminology

CRIN Centre de Recherches en Informatique de Nancy (France) • Club Recherche et Industrie

CRIP Collective Routing Indicator Processing

CRIPTE Calcul sur Réseaux des Interactions Perturbatrices en Topologie Électromagnétique

CRIRES Cryogenic Infrared Echelle Spectrograph • Cryogenic Infrared Echelle Spectrometer (*)

CrIS Cross-track Infrared Sounder

CRIS Calibration Recall and Information System • Carrier Route Information System • Catania Relativistic Ion Studies • Center for Research on the Information Society • Centre de Rectification des Images Spatiales (CNES) • Clinical Radiology Imaging System • Command Retrieval Information System • Communications Resource Information Sharing • Computerized River Information System • Computer-Related Information Symposium • Computer Resources Integrated Support • Connector Removal and Installation Sheet • Cosmic-Ray Isotope Spectrometer • Current Research Information System

CRISAP Centro de Referencia y Información de Suelos de Amazonia Peruana (Peru)

CRISD Computer Resource Integrated-Support Data • Computer Resource Integrated-Support Document • Computer Resource Integration Support Document

CRISL Contract Repair Initial Support List

CRISM Compact Reconnaissance Imaging Spectrometer for Mars

CRISP Centre for Remote Imaging, Sensing and Processing • C-machine Reduced Instruction Set Processor • Computer Reconstructed Images from Scene Photographs • Computer Resources Integrated Support Plan • CONTOUR Remote Imager/Spectrograph • CONTOUR Remote Imaging Spectrograph • Counter Rotating Integrated Shrouded Propfan • Critical Surface Spectrometer • Cross-Industry Standard Process • Cryptographic Reduced Instruction Set Processor • Crypto-Reduced Instruction Set Processor

CRISTA Cryogenic Infrared Spectrometers and Telescopes for the Atmosphere

Crit Criterion • Critic • Critical • Critically • Criticism • Critique • Criticized

CRIT Critical

CRITO Center for Research on Information Technology and organization

CRITT Centre Régional d'Innovation et de Transfert Technologique (France)

CRIWG Computer Resources Interoperability Working Group

CRJ Call Record Journal • Coorabie (Australia-SA) (location identifier)

CRJE Conversational Remote Job Entry

Crk Cork (Ireland) • Crank

CRK Cable Repair Kit • Cosmic-Ray Killer • Creek • Luzon Island - Clark (Philippines) (location identifier)

CRL Cambridge Research Laboratories • Cartographic Research Laboratory • Cathode-Ray Lamp • Central Reference Library • Central Regional Laboratory • Central Research Library • Certificate Revocation List • Certification Revocation List • Certified Record Librarian • Certified Reference Librarian • Certified Reporting Limit • Charleroi Gosselies (Belgium) (location identifier) • Common Rail Launcher • Communications Research Laboratory • Complex Record Language • Computing Research Laboratory • Contrat de Recherche Libre • Custody Receipt Listing • Customer Requirements List

CRLC Circulation

CRLCF Circulation Facility

CRLCMP Computer Resources Life Cycle Management Plan

CRLF Carriage Return + Line Feed

CRLIS Current Research in Library and Information Science

CRLJ Communications Research Laboratory Journal (periodical code)

CRLRv Communications Research Laboratory Review (periodical code)

CRLSE Conference Release

CRM Camera-Ready Materiel • Camera-Ready Mechanical • Camera-Ready Mechanics • Catarman (Philippines) (location identifier) • Cell Rate Margin • Central Rocky Mountains (USA) • Centre de Recherche des Macromolécules • Certified Reference Material • Chandra Radiation Model • Change Request Manager • Change Request Material • Character Recognition Module • Chemical Release Module • Chemical Remanent Magnetization • Citizen Relationship Management • Cloud Resolving Model • Cockpit Resource Management • Collection Requirements Management • Collisional-Radiative Model • Command Record Manager • Comment Resolution Meeting • Communications Resource Manager • Compte-Rendu de Matériel • Computer Resources Management • Conceptual Reference Mission • Connection Request Mode • Continuous Risk Management • Control and Reproducibility Monitor • Core Restraint Mechanism • Count Rate Meter • Courtesy Reply Mail • Crew Resource Management • Criterion-Referenced Measurement • Customer Relationship Management • Customer Resource Management • Cyclotron Resonance Maser

CRMA Cyclic-Reservation Multiple Access

CRMB Collection Requirements Management Board

CRMC Communication Resources Management Centre

CRMD Centre de Recherche en Météorologie Dynamique • Computer Resources Management Data

CRMHT Centre de Recherche sur les Matériaux à Haute Température (France)

CRMI Committed-Rate Measurement Interval

CRMP Coastal Resources Management Project • Collection Requirements Management Plan • Collection Requirements Management Program • Computer Resources Management Plan • Computer Resources Management Program

CRMS Computer Resources Management System

CRMT Computer Resources Management Team • Computer Resources Management Technology

Crn Crane • Crown

CRN Cable Routing Notation • Canadian Road Network • Cellular Radio Network • Centre de Recherche Nucléaire • Change Review Notice • Conflict Resolution Network • Contract Revision Number • Coronal Rain • Cosmic-Ray Nuclei (experiment) • Cosmic-Ray Nucleus • Cromarty (UK-Scotland) (location identifier) • Crown

CRNL Chalk River Nuclear Laboratories

Crnmtr Chronometer

Cro Croatian

CRO Career Referral Office • Cathode-Ray Oscilloscope • Cathode-Ray Oscillograph • Chemical Release Observation • Chief Recruiting Officer • Community Relations Officer • Component Repair and Overhaul • Compulsory Rights Order • Contract Research Organization • Control Room Operator • Corcoran (USA-CA) (location identifier)

CROAP Conception et Réalisation d'Outils d'Aide à la Programmation

Croat Croatia • Croatian

CROC Computer Review and Orientation Course

Crom Cromarty (UK-Scotland)

CROM Capacity Read-Only Memory • Control Read-Only Memory

CROMOS Cryogenic Multiobject Spectrometer

Cronol Cronología

CROP Cosmic-Ray Observatory Project (USA-NE) • Customized Reduction Of Products

CROS Computerized Reliability Organization System • Crystalline Operating System

CROSS Combined Radio/Optical Space Surveillance • Cross-Support Service

CROSSPATE Coordinative Retrieval of Selectively Sorted Permuted Analogue-Title Entries

CROUS Centre Régional des Oeuvres Universitaires et Scolaires (France)

CrO2 Chromium Oxide

CRp Cosmic-Ray Proton

CRP Card Reader Program • Center for Responsive Politics • Central Receiving Point • Central Repair Point • Centre de Recherche Pédagogique • Centre de Recherche Public • Centre de Rééducation Professionnelle • Centre de Ressources Pédagogiques • Change Review Panel • Climate Research Program • Command and Reporting Post • Command Readiness Program • Common Relevant Picture • Completion Report • Compulsory Reporting Point • Computer Reset Pulse • Configuration Requirements Processing • Contract Requirements Package • Control and Reporting Point • Control and Reporting Post • Coordinated Reconnaissance Plan • Coordinated Research Programme • Corpus Christi (USA-TX) (location identifier) • Cosmic-Ray Particle • Cost Reduction Program • Cross-Reference Project • Current Research Profile

CRPA Centre de Recherches en Physique de l'Atmosphère • Controlled-Radiation Pattern Antenna • Controlled-Radiation Patterned Antenna • Controlled-Reception Pattern Antenna • Controlled-Reception Patterned Antenna • Croatian Radio Protection Association

CRPD City Real Property Database

CRPE Centre de Recherches en Physique de l'Environnement (France) • Concours de Recrutement des Professeurs des Écoles

CRPG Centre de Recherches Pétrographiques et Géochimiques (France)

CRPHT Centre de Recherches sur la Physique des Hautes Températures

CRPL Central Radio Propagation Laboratory (USA) • Cosmic-Ray Physics Laboratory

CRPP Center for Research in Particle Physics • Centre de Recherches en Physique des Plasmas

CRPPH Committee on Radiation Protection and Public Health

CRPS Centre Régional de Pédagogie Spéciale

crPSO Crystallographic Point Symmetry Operation

Crpt Carpet

CRPZ Centre Régional de Prévisions de Zone (France)

CRQ Call Request • Caravelas (Brazil-BA) (location identifier) • Command Response Queue • Compte-Rendu Quotidien • Corequisite

CRR Capabilities and Requirements Review • Capabilities Requirements Review • Cargo Readiness Review • Case Reconciliation Review • Case Requisition Review • Ceres (Argentina-SF) (location identifier) • Commissioning Results Review • Computer Run Report • Constant-Retard Ratio • Critical Requirements Review • Cyclic Redundancy Rate

CRRC Construction Requirements Review Committee

CRRE Combined Radiation and Release Experiment

CRREL Cold Regions Research and Engineering Laboratory

CRRERIS Commonwealth Regional Renewable Energy Resources Information System

CRRES Combined Release and Radiation Effects Satellite

CRRL Carbon Radio Recombination Line

CRRP Contractor Responsibility Review Program

crs Conversion Resource (file extension)

Crs Cours • Course

CRS Calibration Recall System • Calibration Requirements Summary • Call Relay Service • Candidate Review Subcommittee • Capability Reporting System • Cavity-Ringdown Spectrometer • Cavity-Ringdown Spectroscopy • Celestial Reference Frame • Cell Relay Service • Center for Remote Sensing • Central Referral System • Centro de Referencia de Salud • Certified Repair Station • Character Recognition System • Child Restraint System • Cluster Red Sequence • Coastal Radio Station • Cold-Rolled Steel • Command Relay Section • Command Retrieval System • Committee of Remote Sensing • Communications Relay Set • Comprehensive Retrieval System • Compressed Raw Storage • Computerized Reservation System • Computerized Retrieval Service • Computer Resources Support • Configuration Report Server • Congressional Research Service (USA) • Container Recovery System • Content Replication System • Contract Research Staff • Control and Reporting Service • Controlled Random Search • Cooperative Reporting System • Corporate Review Structure • Correctional Reporting System • Corsicana (USA-TX) (location identifier) • Cosmic Radio Source • Cosmic-Ray Shower • Cosmic-Ray Spectrum • Cosmic-Ray System • Course • Crew Rescue System • Cross • Current Research Summary

CRSC Carrier-Relay Spacecraft • Center for Remote Sensing and Cartography • Center for Research in Scientific

Communication • Center for Research in Scientific Computation

Crscnt Crescent

CRSCNT Crescent

CRSDA Community Recreation and Skill Development Activity

CRSDP Codar Remote Sensing Demonstration Programme

CRSE Centre de Recherche des Sciences Économiques • Course

CRSENT Crescent

CRSEST Crescent

CRSF Cyclotron Resonance Scattering Feature

CRSFD Cross-Feed

CRSI Ceramic Reusable Surface Insulation

CRSM Configuration Register and Status Multiplexer

Crsn Corrosion

CRSN Centre de Recherche en Sciences Naturelles

CRSNG Conseil de Recherche en Sciences Naturelles et Génie

CRSNT Crescent

CRSP Central Receiving and Shipping Point • Commercial Remote Sensing Program • Contract Repair Supply Procedure • Cooperative Research Support Project • Core Research Support Program • Cryogenic Spectrograph • Cryogenic Spectrometer

CRSR Center for Radiophysics and Space Research (USA-NY) • Cursor

CRSS Cambridge-ROSAT Serendipity Survey • Canadian Remote Sensing Society (Canada) • Chemically Rigidized Space Structure • Commercial Remote Sensing Satellite • Commercial Remote Sensing System • Critical Resolved Shear Stress

Crssng Crossing

CRSSNG Crossing

Crst Crescent

CRST Crescent • Crest • Customer Requirement Support Team

CRSTI Canadian Remote Sensing Training Institute (Canada)

Crsv Corrosive

Crsvr Crossover

CRSVR Crossover

crt Certificate (file extension)

Crt Courant • Court • Crater-Crateris (constellation)

CRT Cathode-Ray Tube • Centre de Ressources Technologiques • Centre Régional de Télécommunications • Chinese Remainder Theorem • Circuit Requirement Table • Complex Ridge Terrain • Composite Rate Tax • Composite Readiness Test • Computer Remote Terminal • Computer Research Technology • Computer Response Time • Conformal Radiotherapy • Contract Repair Team • Cosmic-Ray Telescope • Cosmic-Ray Tracking • Counter Recovery Time • Court • Criterion-Referenced Test • Criterion-Referenced Testing • Crossett (USA-AR) (location identifier)

CRTAM Computer Real-Time Access Method

CRTBT Centre de Recherches sur les Très Basses Températures (France)

CRTC Canadian Radio-television and Telecommunications Commission (Canada) • Cathode-Ray Tube Controller

CRTD Command Real-Time Display

CRTEAN Centre Régional de Télédétection des États de l'Afrique du Nord

CRTF Central Receiver Test Facility

Crtg Cartridge • Crating

CRTG Cartridge

CRTI Computer Real-Time Interrupt

Crtkr Caretaker

Crtl Courtil

CRTL C Run-Time Library

CRTO Cathode-Ray Tube Oscillograph

CRTP Compressed Real-Time Protocol

CRTR Crater

Crts Courtes

CRTS Centre Royal de Télédétection Spatiale (Morocco) • Cathode-Ray Tube Shield

CRTST Centre de Recherches Transdisciplinaires sur les Sciences et les Techniques

CRTT Cathode-Ray Tube Tester • Computer Resources Technology Transition

CRTU Combined Receiving and Transmitting Unit • Cryogenic Radiator Test Unit

CRTV Command Receiver Test Van • Composite Reentry Test Vehicle

Cru Crux-Crucis (constellation)

CRU Card Reader Unit • Carriacou Island (Windward Islands) (location identifier) • Central Registration Unit • Climatic Research Unit • Combined Rotating Unit • Comité Réseau des Universités • Command and Response Unit • Command Relay Unit • Communications Register Unit • Computer Resource Unit • Control Radio Unit • Control Register User

CRUC Catalèg de Revistes de les Universitats Catalanes

CRUD Computer Resource Unit Data • Corrosive, Radioactive, Undetermined Deposit • Create - Retrieve - Update - Delete

CRUDDES Crustal Dynamics Database Expert System

CRUE Center for Research on Unexpected Events • Conferencia de los Rectores de la Universidades Españolas (Spain)

CRUI Cenferenza dei Rettori della Università Italiane (Italy)

CRUME Committee for Research in Undergraduate Mathematics Education

CRUS Cruise

CRUSK Center for Research on Utilization of Scientific Knowledge

CRUSO Coordinated Request and User Support Office

CRUX Cosmic-Ray Upset Experiment

Crv Corvus-Corvi (constellation) • Curve

CRV Cargo Return Vehicle • Centre de Recherche en Volcanologie (France) • Coding Rule Violation • Compte-Rendu du Vent • Computer-Readable Version • Constant Radial Velocity • Crew Rescue Vehicle • Crew Return Vehicle • Crotone (Italy) (location identifier) • Cryptography Verification

CRW Channel Record Word • Charleston (USA-WV) (location identifier) • Collective Robotics Workshop • Control Read/Write • Crew

CRWG Computer Resources Worling Group

CRWI Coalition for Responsible Waste Incineration

CRWMS Civilian Radioactive Waste Management System

CRX Corinth - Roscoe Turner (USA-MS) (location identifier)

CRY Carlton Hill (Australia-WA) (location identifier)

Cryo Cryogenic • Cryogenics (periodical code)

CRYO Cryogenic

CRYOFD Cryogenic Flexible Diode

Crypto Cryptographic • Cryptography

CRYSP Crystal Sample Package

CRYSYS Cryospheric System • Cryospheric System Study

Cryst Crystal • Crystalline • Crystallized • Crystallography

CRYSTAL Cirrus Regional Study of Tropical Anvils and (cirrus) Layers

Crystallog Crystallography

Crystd Crystallized

Crystn Crystallization

CRZ Chardzhou (Turkmenia) (location identifier) • Close Reconnaissance Zone

CRZT Cruzamento

c/s Call Sign • Cases • Cycles per Second

cs Case • Census • Come Supra • Consul • Czech (ISO language code)

cS Centistoke

Cs Cäsium • Cesium • Césium • Cirrostratus

C&S Calendaring & Scheduling • Casualty & Survivability • Computers & Systems • Computing & System

C/S Call Sign • Chief of Staff • Client/Server • Concentrator/Spectrometer • Cost/Schedule • Counts per Second • Course/Speed • Cycles per Second

C-S Client-Server • Cox-Stewart

CS Calcium Silicate • Calendaring and Scheduling • Calibration System • Call Sign • Call Signal • Camera Site • Canadian Standard • Canal Sémaphore • Capability Set • Capital Stock • Captain Starlink • Carbon Monosulfide • Carbon Steel • Card Station • Carrier Selection • Cast Steel • Catalog Subgroup • Center Section • Centraal Station • Central Secretariat • Central Source • Central Station • Central Storage • Central Supply • Certificate in Statistics • Change of Status • Change Status • Channel Spacing • Charge Shuffling • Chartered Surveyor • Checkout Station • Check Sum • Chemical Society • Chief of Staff • Chief Scientist • Chief Secretary • Chip Select • Chirp Scaling • Circuit Switch • Circuit Switching • Circumstellar • Circumstellar Shell • Cirrostratus • City Surveyor • Civil Servant • Civil Service • Classified Sensitive • Classifier Syntax • Classifier System • Clear and Subtract • Clearing Server • Clear to Send • Client Server • Clock Synchronization • Close Support • Code Scheme • Code Segment • Cognitive Science • Cognitive System • College of Science • Collision Strength • Color Scanner • Color Specification • Colour Start • Comité Scientifique • Command Set • Command System • Commercial Service • Commercial Standard • Common Set • Common Software • Common Subsequence • Communications and Security • Communication Segment • Communication Services • Communications Satellite • Communications Security • Communications Server • Communications Software • Communications Station • Communications Support • Communications Switcher • Communications System • Community Service • Comparison Side • Component Specification • Composite Service • Composite Superconductor • Compton Scattering • Computer Science • Computer Section • Computer Simulation • Computer Software • Computer System • Computing Science • Conditioned Stimulus • Conducted Susceptibility • Confederate States • Configuration server • Confined Space • Confinement Spare • Confinement System • Congiunzione Superiore • Conjugate Structure • Conseil Scientifique • Constant Source • Constant Speed • Consumables Status • Contact Scientist • Continental Shelf • Contingency Sample • Contractor Share • Contractor Support • Contract Specialist • Controlled Space • Control Sample • Control Scanner • Control Section • Control Segment • Control Set • Control Signal • Control Slip • Control Star • Control Station • Control Strobe • Control Switch • Convergence Sublayer • Convergent Stereoscopic • Coolant System • Cooperative Society • Core Segment • Core Shift • Core System • Cornerstone • Corporate Source • Correct Selection • Correlador Solar • Cosine-Sine • Cost Sharing • County Seat • Coupled States • Cox-Stewart • Crew Simulator • Crew Station • Crew System • Critically Sensitive • Critical Sensitive • Critical Source • Critical Surface • Crossover Switch • Cross-Scan • Cross-Section • Cross-Strapping • Cubic Spline • Current Science • Current Sensor • Current Series • Current Source • Current Strength • Curve Shortening • Cusp Shortening • Cycle Sequence • Czechoslovakia (ISO country code) (now obsolete) • Czech Republic (international vehicle registration code) • Portugal (international aircraft registration code) • Slovakia (international vehicle registration code)

C/S/A Command, Services, and Agencies

CSA Calendaring and Scheduling API • Cambridge Scientific Abstracts • Canadian Space Agency (Canada) • Canadian Standards Association (Canada) (*) • Canadian Standards Authority (Canada) • Carbazole Sulphonic Acid • Carrier Serving Area • Center for Space Astrophysics • Central Study Area • Character Scan or Alternate • Charge-Sensitive Amplifier • Chinese Society of Astronautics • Classical Simulated Annealing • Client Server Agent • Client Server Architecture • Client Services Agent • Client Services Approach • Cloud Size Analyzer • Colonsay Island (UK) (location identifier) • Coarse Sensor Array • Commercial Space Act (USA) • Commission Spécialisée d'Astronomie • Common Service Area • Common Storage Area • Common Support Application • Common System Area • Communications Service Authorization • Compensation System Analyst • Compression Service Adapter • Computer Science Association • Computer Sciences of Australia • Computer Security Act (USA) • Computer System Analyst • Computing Society of America • Confederate States of America • Configuration Status Accounting • Conical Scan Antenna • Conseil Supérieur de l'Audiovisuel (France) • Contextual Sequences Analysis • Contractor Support Area • Contract Specification Amendment • Contrat Support Application • Control Switching Assembly • Covariance Structure Analysis • Cryogenic Society of America (USA) • Current Source Amplifier • Cyclic Strain Attenuator

CSAA Committee on Space Astronomy and Astrophysics

CSAB Computing Sciences Accreditation Board

CSAC Census Statistical Area Committee (USA) • Command Studies Advisory Committee • Command Studies Advisory Council • Computer Science Accreditation Commission

CSACS Centralized Status, Alatm and Control System

CSAD Capsule Systems Advanced Development • Computer System Address Directory • Computer System Authorization Directory

CSAER Configuration Status Accounting and Engineering Records

CSAGI Comité Scientifique de l'Année Géophysique Internationale • Comité Spécial de l'Année Géophysique Internationale (*)

CSAIN Canadian Substance Abuse Information Network (Canada)

CSAIO Chef du Service Académique d'Information et d'Orientation

CSALT Canadian Society for the Advancement of Legal Technology (Canada)

CSAM Circular Sequential Access Memory • Coarse Sun Acquisition Mode • College of Science And Mathematics • Computer Security for Acquisition Managers • Computer Systems and Applied Mathematics (workshop) • Crinkled Single Aluminized Mylar

CSAO CAS Shaanxi Astronomical Observatory (China-PRC)

CSAP Call Service Attendant Processing • Computer Security Accreditation Plan • Control System Analysis Program

CSAPI Common Speller Application Programming Interface

CSAR Canadian Space Agency Rocket • Center for Simulation for Advanced Rockets • Channel System Address

Register • Communications Satellite Advanced Research • Configuration Status Accounting Report • Custom Scanned Absolute Response

CSAS Colorado Springs Astronomical Society (USA-CO) • Command and Stability Augmentation System • Computerized Status Accounting System • Configuration Status Accounting System

CSASG Contributions of the Slovak Academy Sciences Geophysical Institute (periodical code)

CSASP Classical Scattering Aerosol Spectrometer Probe

CSASS Client/Server Application Support Service

CSAT Center for Space and Advanced Technology • Combined Systems Acceptance Test • Crew Station Automation Technology

CSATI Centro Studi ed Applicazione sulle Tecnologie dell'Informazione

CSAU Circuit Switched Access Unit

CSB Calcium Silicate Vrick • Caransebes (Romania) (location identifier) • Change Schedule Board • Chinese Science Bulletin (periodical code) • Closely Spaced Basing • Close Support Bridge • College of Saint Benedict (USA-MN) • Common-Sense-Based • Complex Support Building • Computer Support Base • Conseil Supérieur des Bibliothèques • Customer Service Branch

CSBC China Ship Building Corp. (China-ROC)

CSBCS Carrier Sequence Bar Code Sorter

CSBM Committee on Space Biology and Medicine

CSBTS China State Bureau of Technical Supervision (China-PRC)

c&sc Capitals & Small Capitals

csc Corel Script (file extension) • Cosecant

CSc Candidate of Sciences

C/SC Cost/Schedule Control

CSC Canadian Society for Chemistry (Canada) • Canas (Costa Rica) (location identifier) • Carnegie Science Center • Carrier-Smoothed Code • Cathode Strip Chamber • Center for Scientific Computing • Center for the Study of Communications • Central Switching Concept • Cercle Scientifique de Charleroi (Belgium) • Circuit Switch Control • Circuit Switching Center • Circuit Switching Controller • Civil Service Commission • Coastal Service Center (NOAA) • Commercial Space Center • Commission de la Sécurité des Consommateurs (France) • Common Signaling Channel • Commonwealth Science Council • Communications Switchboard Console • Communications Systems Center • Communications Systems Control • Compact Spherical Coil • Complex Support Controller • Comprehensive Self-Check • Computer Science Conference • Computer Science Corporation • Computer Security Center • Computer Service Center • Computer Software Component • Computer Software Configuration • Computer System Center • Computer System Controller • Computer Systems Command • Conical-Shaped Charge • Conseil Scientifique Consultatif • Construct Service Compiler • Consumer Service Card • Contingency Support Center • Control Signal Code • Conventional Standoff Capability • Conventional Systems Committee • Convention for Safe Containers • Core Support Capability • Cost - Schedule - Control • Critical Success Chain • Critical System Characteristic • Customer Service Control

CSCA Catalog of Star Clusters and Associations • Computer-Supported Collaborative Argumentation

CSCB Contractor's Summary Cost Breakdown

CSCC Cleveland State Community College (USA-TN) • Communications System Category Code • Concurrent Supercomputing Consortium • Consulting Services Contract Coordinator

CSCCC Consolidated System Command Control Center

CSCCE Contingency Satellite Configuration Control Element

CSCD Character Set Computer Development • Chinese Science Citation Database • Control System Command Decoder

CSCE Canadian Society for Civil Engineering (Canada) • Common Student Computing Environment • Communications System Control Element • Conference on Security and Cooperation in Europe • Conférence sur la Sécurité et la Coopération en Europe

CSCG Communications Security Control Group

csch Hyperbolic Cosecant

CSCHE Canadian Society for Chemical Engineering (Canada)

CSCI Center for Satellite Control Initiative • Commercial Satellite Communications Initiative • Commercial Satellite Communications Interconnection • Computer Software Configuration Interface • Computer Software Configuration Item • Current Source Current Intensive

CSCL Computer-Supported Collaborative Learning • Computer-Supported Cooperative Learning • Computer Support of Collaborative Learning (conference)

CSCR Classification System for Computing Reviews

C/SCS Cost/Schedule Control System

CSCS Central Security Control System • Centro Svizzero di Calcolo Scientifico (Switzerland) • Component Support Cost System • Contractor Satellite Control Site

CSCSAT Commercial Synchronous Communication Satellite

C/SCSC Cost/Schedule Control System Criteria

CSCSI Canadian Society for Computational Studies of Intelligence (Canada)

CSCSO Circuit Switch Call Service Operator

CSCT Canadian Society for Chemical Technology (Canada)

CSCU Conseil Supérieur des Corps Universitaires (France)

CSCW Computer-Supported Collaborative Work • Computer-Supported Cooperative Work (*) • Contract Start Date • Cresswell Downs (Australia-NT) (location identifier) • Crystal Size Distribution • Cumulative Sum Diagram

CSD Cellular Security Device • Chartered Society of Designers • Chemical Systems Division • CICS System Definition • Circuit Switched Data • Circumsecondary Disk • Civil Service Department • Cold Shutdown • Commission on Sustainable Development (UNO) • Common Subdivision • Comparative Soot Diagnostics (*) • Complément de Service Donné • Computational Structural Dynamics • Computer Security Day • Computer Service Directive • Computer Services Department • Computer System Directive • Computer System Director • Computer Systems Division • Concepts and Studies Division • Conference on Sustainable Development • Configuration Standardization Document • Constant-Speed Drive • Contract Start Date • Controlled-Slip Differential • Control System Development • Cooperative Soot Diagnostics • Core-Shift Drive • Corrective Service Disk • Corrective Service Diskette • Crew Systems Division • Current Sheet Disruption • Current Source Density

CSDA Computational Statistics and Data Analysis • Continuous Slowing Down Approximation

CSDB Central Specialized Design Bureau (Russia)

CSDC Circuit Switched Digital Capability • Cluster Science Data Centre • Commonwealth Spatial Data Committee • Communications Systems Design Center

CSDD Conceptual System Design description • Control Systems Development Division

CSDF Central Source Data File • Crew Station Design Facility

CSDG Common Software Development Group

CSDI Computer Software Distribution Item • Cross-Spectral Dispersion Imaging

CSDL Charles Stark Draper Laboratory (MIT) • Conceptual Scheme Definition Language

CSDM Computer Systems Diagnostic Manual

CSDN Circuit-Switched Data Network

CSDO Canberra Space Dome and Observatory (Australia-ACT)

CSDP Command Supply Discipline Program

CSDR COBE Science Data Room

CSDRA Centre for Spatial Data Research and Applications

CSDRBL Considerable

CS&DS Customer Service & Sales District

CSDS Cluster Science Data System • Contractor Software Design System • Controlled Speed Drive Starter • Critical Space Debris Sphere • Customer Service and Sales District • C3 Software Development and Support

CSDT Computer Software Data Tape • Computer Software Data Type

CSDU Centre de Stockage des Déchets Ultimes

Cse Course

CS&E Computational Science & Engineering (*) • Computer Science & Engineering

CSE Camel Service Environment • Center for Science and Engineering • Center for Software Engineering • Center for Software Evaluation • Center for Statistical Education • Centraal Schriftelijk Eindexamen • Central Signal Establishments • Centre de Suivi Écologique • Certificate of Secondary Education • Certified System Engineer • Chief System Engineer • Circumstellar Envelope • Client Server Environment • Cluster Soft Excess • Cognitive Systems Engineering • Cold Start Entry • Colloquium on Software Engineering • Commission des Spécialistes d'Établissement • Common Support Element • Common Support Equipment • Commonwealth Service of Employment (Australia) • Communications Security Element • Communications Security Establishment (Canada) • Communications Support Element • Communications Systems Engineer • Computational Science and Engineering • Computer Science and Engineering • Computer Support Equipment • Conférence Spatiale Européenne • Configuration Switching Equipment • Conseil Supérieur de l'Éducation • Contrat de Suivi d'Exploitation • Control Section Electronics • Control Systems Engineering • Core Storage Element • Coupled-channel Schrödinger Equation • Crested Butte (USA-CO) (location identifier) • Critical Skills Engineer • Cryostat Servicing Equipment

CSEA Civil Service Employees Association

CSEAR Centre for Social and Environmental Accounting Research

CSEC Centre for the Study of Environmental Change • Computer Security Evaluation Center

CSECT Control Section

CSEDI Cooperative Studies of the Earth's Deep Interior

CSEE Compagnie de Signaux et d'Équipements Électroniques (France) • Computer Science and Electrical Engineering • Conference on Software Engineering Education

CSEET Conference on Software Engineering Education and Training

CSEG Canadian Society for Exploration Geophysicists (Canada) • Command Support Equipment Group

CSEL Circuit Switch Select Line • Consolidated Support Equipment List

CSELI Centro Studi E Laboratori Telecomunicazioni (Italy)

CSEM Centre Sismologique Euro-Méditerranéen • Centre Suisse d'Électronique et de Microélectronique (Switzerland) • Climatologie Sol-Espace et Modélisation • Continuous Sensitivity Equation Method

CSEMS Computer Science, Engineering and Mathematics Scholarship

CSEN Conseil Supérieur de l'Éducation Nationale (France)

CSEOL Center for the Study of Evolution and Origin of Life (IGPP)

CSEP Communications System Engineering Plan

CSER Centre for Satellite Engineering Research

CSERD Computational Science Education Reference Desk

CSERGE Centre for Social and Economic Research on the Global Environment

CSES Center for the Study of Earth from Space

CSET College of Science, Engineering and Technology • Computer Security Education and Training

CSETI Center for the Study of Extraterrestrial Intelligence

CSETWG Computer Security Education and Training Working Group

CsF Cesium Fluoride

CSF Call Service Function • Central Serials File • Central Simulation Facility • Central Simulation Form • Central Supply Facility • Central Support Facility • Central Switching Facility • Cerebrospinal Fluid • Chaos - Solitons - Fractals (periodical code) • Character Scan or Fail • Chi-Squared Function • Circumnuclear Star Formation • Coding Sequence Finder • Columbus Simulation Facility • Community Support Framework • Companion Star Fraction • Computer System Facility • Configuration State Function • Constant Star Formation • Continuum Surface Force • Contractor Support Facility • Contract Supply Facility • Contrast Sensitivity Function • Cost Sensitivity Factor • Creil (France) (location identifier) • Crew Storage Facility • Critical Success Factor • Curve Shortening Flow • Customer Support Facility

CSFA Canadian Scientific Film Association (Canada)

CSFC Cruise Specific Fuel Consumption

CSFD Canadian Symposium on Fluid Dynamics

CSFH Cosmic Star Formation History

CSFP Contemporary Star Formation Process

CSFR Color Strip Film Recorder • Committee on Scientific Freedom and Responsibility • Continuous Scan Film Recorder

CSFS Committee on Sciences for Food Security • Control Section Flight Software

CSFT Combined System Functional Test

CSFW Computer Security Foundation Workshop

Csg Casing

CSG Calibration Signal Generator • Centre Spatial Guyanais (French Guyana) • Clock Sequence Generator • Columbus (USA-GA) (location identifier) • Communications Strategy Group • Complex System Guideline • Composite Support Group • Computer Systems Group • Concept Synthesis Group • Constructive Solid Geometry • Consulting Services Group • Consumer Support Group • Contribution Sociale Généralisée • Coordinating Subgroup • Cryptologic Support Group

CSGD Chinese Solar-Geophysical Data

CSGE Common Software Engineering Environment

CSGF Computational Science Graduate Fellowship

CSGS Computer Systems Group Software

CSH Calcium Silicate Hydrate • Cape Sarichef (USA-AK) (location identifier) • Communications Soft Hat • Console Signal Handler • Convective Stratiform Heating • Critical Systems Heuristics

CSHARP Cooperative Student High-Altitude Rocket Payload

CSHELL Cryogenic Echelle Spectrometer

CSHKP Carmichael - Sturrock - Hirayama - Kopp - Pneuman (model)

CSHP Conseil Supérieur d'Hygiène Publique

CSHPM Canadian Society for the History and Philosophy of Mathematics (Canada)

CSHPS Canadian Society for the History and Philosophy of Science (Canada)

CsI Caesium Iodide

CSI Carrier-Scale Internetworking • Cartographie par Satellite et Informatique • Casino (Australia-NSW) (location identifier) • Catalogue of Stellar Identifications • Centre de Sociologie de l'Innovation • Cercle Scientifique Interfacultaire • Channel Status Information • Channel Status Indicator • Circuit Switch Interface • Cité des Sciences et de l'Industrie • Coastal Studies Institute (USA) • Code Structure Identifier • Coelliptic Sequence Initiation • Coil Side Injection • College of Staten Island (USA-NY) • Command Sequence Introducer • Commercial Sites Index • Commercial Systems Integration • Common Software Interface • Common System Interface • Compact Source Iodide • Computational Sciences and Informatics • Computer Science Institute • Computer Science Instructor • Computer Security Institute (USA) • Computer Services Index • Computer Society of India • Computer-Supported Instruction • Computer Synthesized Image • Computer System Integration • Concentric Sequence Initiation • Configuration Subitem • Conseil de Sécurité Intérieure (France) • Consolidated Serviceable Inventory • Consolidated Software Inventory • Consorzio Servizi Informatici • Consumer Satisfaction Index • Contractor Simulator Instruction • Control and Structural Integrity • Control Servo Input • Control Signal Indicator • Control Structure Interaction • Cost Since Inception • Cost Since Inspection • Cost System Indicator • Crew Software Interface • Critical Safety Item • Critical Success Index • Critical Sustainability Item • Customer Satisfaction Index • Customer Service Index • Customer Source Inspection

CSIA Compound Specific Isotope Analysis

CSIAM Chinese Society for Industrial and Applied Mathematics

C/SiC Carbon/Silicium Carbide

CSIC Computer System Interface Circuit • Consejo Superior de Investigaciones Científicas (Spain)

CSICOP Committee for the Scientific Investigations of Claims of the Paranormal (USA-NY)

CSID Calling Station Identification • Call Subscriber Identification • Convergence Source-Image Distortion • Current System Interface Drawing

CSIDA CSIRO System for Interactive Data Analysis

CSIDC Computer Society International Design Competition (IEEE)

CSIESR Comité des Services Informatiques de l'Enseignement Supérieur et de la Recherche

CSIF Communications Service Industrial Funds

CSII Configuration Similarity Iterative Improvement

CSIIS Canadian Sea Ice Information System

CSIL Customer Service Information Line

CSILE Computer-Supported Intentional Learning Environment (*) • Computer-Supported Intentional Learning Experiment

CSIM Cosmic Simulator • Custom Spectral Irradiance Measurement

CSIN Communications Security Interface Net • Computer Systems Information Network

CSIO Central Science Instruments Organization (India) • Computer Systems Integration Office

CSIOE Common Support Items Of Equipment

CSIP Committee for the Scientific Investigations of the Paranormal • Component Sponsored Investment Program • Continental Scale International Project

CSIPG Circuit Switch Interface Planning Guide

CSIR Civil Society Internet Rights Project • Commonwealth Scientific and Industrial Research (South Africa) • Computer Software Integration Review • Computer Systems Integration Review • Council for Scientific and Industrial Research (South Africa) • Council of Scientific and Industrial Research (India) • Crew Software Interface Review

CSIRO Commonwealth Scientific and Industrial Research Organization (Australia)

CSIRONET CSIRO Network

CSIS Center for Secure Information Systems (CMU) • Center for Strategic and International Studies • Centralized Storm Information System • Centre for Spatial Information Systems

CSISO Conditioned Single Input - Single Output

CSISRS Cross-Section Information Storage and Retrieval System

CSIST Chung Shan Institute of Science and Technology (Taiwan)

CSIT Computational Science and Information Technology • Customer Service Improvement Team

CSIU Core Segment Interface Unit

CSJ Cape Saint Jacques (Viet Nam) (location identifier) • Chemical Society of Japan

CSJAG Crew Station Joint Advisory Group

CSJWG Crew Station Joint Working Group

Csk Cask

CSK Cap Skirring (Senegal) (location identifier) • Cathodic Survey Kit • Committee of Standardization of the DPRK

CSL California State Library (USA-CA) • Casal • Centre Spatial de Liège (Belgium) • Chemical Sewer Line • Clock Sync Logic • Code Selection Language • Command Sequence Language • Command Signal Limiter • Common Security Label • Common Specification Language • Compact Shuttle Lidar • Computer Science Laboratory • Computer Science Library • Computer Science Logic • Computer Sensitive Language • Computer Simulation Language • Computer Status Light • Computer Structure Language • Computer System Language • Computer Systems Laboratory • Console • Context Sensitive Language • Context Switch Latency • Continuous Spontaneous Localisation • Control and Simulation Language • Control and Status Logic • Crew Systems Laboratory • Critical Sequence List • Current Switch Logic • San Luis Obispo (USA-CA) (location identifier)

CSLA China Satellite Launch Agent • Communications Security Logistics Activity • Communications Security Logistics Agency

CSLC Coherent Sidelobe Canceler

CSLCP Computer Security Life Cycle Process

CSLF Carbon Stars Luminosity Function • Conseil Supérieur de la Langue Française

CSLG Campus-wide Software License Grants

CSLI Civilian Space Leadership Initiative

CSLIN Contract Subline Item Number

CSLIP Compressed Serial Line Interface Protocol

CSLM Client Service Level Monitor • Coarsening in Solid-Liquid Mixtures (facility) • Confocal Scanning Laser Microscope

CSLO Canadian Scientific Liaison Office (Canada)

CSLP Cooperative Satellite Learning Program

CSLR Continuous Spoken Language Recognition

CSM Camera - Speaker - Microphone • Cargo System Manual • Circumstellar Matter • Circumstellar Medium • Climate System Model • Climate System Modeling • Climate System Monitoring • Clinton (USA-OK) (location identifier) • Coaxial Switching Matrix • College of Science and Mathematics • Color-Singlet Model • Combined Symbol

Matching • Cometary Stirring Model • Command and Service Module • Command Service Module • Commission for the Synoptic Meteorology • Committee for Standardization and Metrology • Commodity Support Manager • Common Support Module • Communications Services Manager • Communications Storage Manager • Company Stage Manager • Compartmented Security Mode • Component and Supplier Management • Comprehensive System Model • Computational Structural Mechanics • Computer Software Management • Computer Status Matrix • Computer Systems Manager • Computer Systems Manual • Conseil Supérieur de la Météorologie • Continuous Sheet Memory • Convexity - Symmetry - Maximum • Cumberland Science Museum (USA-TN) • Curvature Sensing Method • Cyclotron Scattering Model

CSMA Carrier-Sense Multiple Access • Committee on the Status of Minorities in Astronomy (AAS) • Computational Structural Mechanics and Applications (periodical code)

CSMA/CA Carrier-Sense Multiple Access with Collision Avoidance

CSMA/CD Carrier-Sense Multiple Access with Collision Detection

CSMC Communications Systems Management and Control

CSMD Computer Science and Mathematics Division

CSMDS Customer Services Management Data System

CSME Canadian Society of Mechanical Engineering (Canada) • Computational Science, Mathematics and Engineering

CSMEE Committee for Support of Mathematicians from Eastern Europe • Committee on Science, Mathematics and Engineering Education

CSMF Communications Satellite Monitoring Facility

CSMM Critical Stream Meteoroid Mass • Cryogenic Scan Mirror Mechanism

CSMNR Complex Studies and Mapping of Natural Resources

CSMOL Control Station Manual Operating Level

CSMP Climate System Modeling Program • Climate System Modeling Project • Common Software Management Plan • Continuous System Modeling Program • Control System Modeling Program

CSMPA Ankara University, Faculty of Sciences Communications - Series Maths, Physics and Astronomy (periodical code)

CSMPS Computerized Scientific Management Planning System

CSMR Course Spatial-resolution Multispectral Radiometer

CSMS Combined Support Maintenance Shop • Communications and Systems Management Segment • Computer Services Management Symposium • Computing Science - Mathematics - Statistics • Consolidated Support Maintenance Shop • Customer Support Management System

CSMT Carlos Sanchez Magro Telescope • Center for Space Microelectronics Technology • Circuit Switching Magnetic Tape

CS-MUX Circuit-Switched Multiplexer

CSMUX Circuit-Switched Multiplexer

CSN Card Select Number • Caribbean Satellite Network • Carson City (USA-NV) (location identifier) • Central Satellite Numérique • Centro di Scienze Naturali • Channel Sequence Number • Circuit Switched Network • Circuit Switching Network • Comité Supérieur de Normalisation • Concentrateur Satellite Numérique • Consejo de Seguridad Nuclear • Content Storage Network • Control Symbol Number • Coopérant du Service National (France) • Current Stock Number • Customer Source Number • Cycles Since New

CSNA Classification Society of North America (USA-IL)

CSNAS Computer-Supported Network Analysis System

CSNE Coopérant du Service National en Entreprise (France)

CSNet Computer Science Network

CSNET Computer Science Network

CSNF Common Source Noise Figure

CSNH Casalinho

CSNI Committee on the Safety of Nuclear Installations • Communications Systems Network Interoperability • Czech Standards National Institute (Czech Republic)

CSNLP Cognitive Science of Natural Language Processing

CSNP Complete Sequence Numbers PDU

CSNR Comet Nucleus Sample Return

CSNSM Centre de Spectroscopie Nucléaire et de Spectrométrie de Masse (France)

CSNW Client Service for NetWare

Cso Corso

CSO CalTech Submillimeter Observatory (USA-HI) • Case Stellar Object • Central Selling Organization • Central Services Organization • Central Statistical Office • Chained Sequential Operation • Chief Scientific Officer • Chief Security Officer • Chief Signal Officer • Chief Staff Officer • Closely-Spaced Object • Cognizant Security Office • Color Separation Overlay • Commonwealth Scientific Office • Communications Service Order • Communications Support Organization • Communications System Office • Community Service Order • Compact Symmetric Object • Complex Safety Office • Complex Safety Officer • Composite SWING Object • Computer Security Office • Computer Security Officer • Computer Service Office • Computer Service Officer • Computer Systems Office • Computer Systems Officer • Consolidated Space Operations • Contracting Security Office • Contracting Security Officer • Customer Service Office

CSOAR Colorado Student Ozone Atmospheric Rocket

CSOB Calibration Science and Operations Branch • Command Systems Operations Branch • Crew System and Operations Branch

CSOC Consolidated Space Operations Center (USA-CO) • Consolidated Space Operations Contract • Consolidated Space Operator's Center • Consolidated Support Operations Center • Construction Special Operations Center • Customer Support Operations Center

CSOCR Code Sort Optical Character Reader

CSOD Command Systems Operations Division

CSOG Communications Security Operational Group

CSOM Computer Software Operator's Manual • Computer System Operator's Manual • Confocal Scanning Optical Microscope

CSOP Carl Sagan Observatory Project (Mexico) • Crew Systems Operating Procedures

CSOR Computer Security Objects Register

CSOT Constant Source Operator Terminal

C/S/P Cost/Schedule/Performance

C/SP Cost/Schedule Performance

CS&P Competitive Sourcing & Privatization

CSP Call Service Position • Canada Scholarship Program • Cape Spencer (USA-AK) (location identifier) • Carbon Sheet Pump • Carbon Sheet Pumping • Catégorie Socio-Professionnelle • Center for Space Physics • Center for Space Power (USA-TX) • Central Signal Processor • Central Supply Point • Centro di Supercalcolo Piemonte (Italy) • Certified Service Provider • Certified Special Processor • Certified Systems Professional • Character Shape Player • Chiefland Star Party (USA-FL) • Classical Sensor Package • Coder Sequential Pulse • Coherent Signal Processing • Coherent Signal Processor • Collaborative Special Project • Commence Search Point • Commercial Space Package • Commercial Subroutine Package • Commercial Support Program • Committee on Science Policy • Common SAR Processor • Common Signal Processor

• Communicating Sequential Process • Communications Satellite Program • Communications Security Publication • Communications Sequential Process • Communications Service Provider • Communications Support Processor • Communications System Processor • Competitive Sourcing and Privatization • Component Service Provider • Composants et Systèmes de Précision (France) • Composite Stellar Population • CompuCom Speed Protocol • Computer Simulation Program • Computer Support Program • Concurrent Spare Parts • Conformal Space Projection • Conseil Scientifique et Pédagogique • Constraint Satisfaction Problem • Contact Support Processor • Contingency Support Package • Continuous Sampling Program • Contracting Support Plan • Contract Strategy Paper • Control and Status Panel • Controlled Storage Point • Control Signal Processor • Control Switching Point • Convention de Site et de la Profession • Cooperative Statistical Program • Corporate Strategies and Programs • Cost - Schedule - Performance • Council for Science Policy • Council for Scientific Policy • Crisis Staffing Procedure • Cryptographic Service Provider • Cryptographic Support Package • Cryptographic Support Processor • Cutting Stock Problem

CSPAA Conférence de Solidarité des Pays Afro-Asiatiques

CSPAAR Center for Space Plasma, Aeronomy and Astrophysics Research (UAH)

CSPAR Center for Space Plasma and Aeronomical Research (USA-AL)

C/SPC Cost/Schedule Performance Criteria

CSPC Call service Position Console • Conditional Stored Program Command • Cost and Schedule Planning and Control • Cost Schedule Performance Criterion

C/SPCS Cost/Schedule Planning and Control Specification

CSPDN Circuit Switched Public Data Network

CSPE Communications Systems Planning Element

CSPEI Computer Software Product End Item

CSPF Critical Single-Point Failure

CSPG Canadian Society of Petroleum Geologists (Canada) • Common Source Power Gain

CSPI Coherent State Path Integral

CSPISE Center for Stochastic Processes in Science and Engineering

CSPL Certified Special Processor List

CSPM Cercle des Sciences Physiques et Mathématiques • Communications Security Publication Memorandum

CSPN Central Star of Planetary Nebula

CSPNe Central Stars of Planetary Nebulae

CSPO Communications Satellite Project Office • Computer Security Program Office • Control Systems Procurement Office

CSPP Collaborative Special Project and Programme • Computer Systems Policy Project • Crew Scheduling and Phasing Plan

CSPR Committee on Scientific Planning and Review (ICSU)

C/SPS Cost/Schedule Performance System

CSPS Coherent Signal Processing System • Cost Schedule Performance System

CSPSW Center for Solar Physics and Space Weather (IACS)

CSPT Channel Status Pointer Table

CSPU California State Polytechnic University (USA-CA) • Core Segment Processing Unit

CSPW Computer-Supported Personal Work

CSQ Command Sequencer • Creston (USA-IA) (location identifier)

C/SR Cost/Schedule Report

CSR Casuarito (Colombia) (location identifier) • Cell Sequenced Ratio • Cell Switch Router • Center for Space Research • Center for Subatomic Research (Canada) • Central Serials Record • Centre de Supervision du Réseau • Certification Status Report • Certified Shorthand Reporter • Change Status Report • Check Signal Return • Check Status Register • Check Status Reply • Check Status Report • Circuit Switch Routing • Civil Service Regulation • Civil Service Retirement • Clementine Spectral Reflectance • Command Status Register • Commercial Security Requirement • Communications Satellite Relay • Communications Service Request • Compact Stellar Remnant • Compensation System Review • Complément de Service Reçu • Complete Spatial Randomness • Comprehensive Spending Review • Concept to Ship Review • Conical Scan Radiometer • Continuous Speech Recognition • Contract Status Report • Control and Status Register • Controlled Supply Rate • Control Status Register • Cost - Schedule - Risk • Crew Station Review • Customer Service Representative • Customer Signature Required • Customer Support Room

CSRA Centro de Sensoriamento Remoto da Amazonia (Brazil) • Civil Service Reform Act (USA) • Comité Scientifique pour les Recherches Antarctiques

CSRAM Custom Static Random Access Memory

CSRB Computer Systems Requirements Board

CSRC Canadian Space Resource Centre (Canada) • Content Source • Crash-Safety Research Centre

CSRD Comprehensive School Reform Demonstration (program) • Computer System Requirement Document

CSRDF Civil service Retirement and Disability Fund

CSRF Common Source Routing File • Control Systems Research Facility

CSRFS Common Source Routing File System

CSRG Catalogue of Southern Ringed Galaxies • Computer Science Research Group • Computer Systems Research Group

CSRI Chemistry and Soil Research Institute

CSRL Common Strategic Rotary Launcher • Conceptual Structures Representation Language

CSRO Consolidated Standing Route Order • Contingency Standing Route Order

CSRP Computers and Software Review Panel

CSRR Component Software Requirements Review

CSRS Canadian Spatial Reference System • Centre Suisse de Recherches Scientifiques (Switzerland) • Civil Service Retirement System • Committee Services and Research Section • Continuous Speech Recognition System • Cooperative State Research Service

CSRSR Center for Space and Remote Sensing Research

CSRT Conseil Supérieur de la Recherche et de la Technologie

CSRTEP Circuit Switch Routing Task Execution Plan

css Cascading Style Sheet (file extension)

CS&S Computer Sciences & Statistics

CSS Capability Support System • Card and Socket Services • Cargo Storage and Securing • Carrier Smart Switch • Cascading Style Sheet • Cassilândia (Brazil-MS) (location identifier) • Catalina Sky Survey • Catalog of S Stars • Center for Space Science • Center for Spatial Studies • Central Security Service • Central Site System • Central Structure Store • Certificat de Sécurité-Sauvetage • Channel Signaling System • Channel Subsystem • Character String Scanner • Circuit Switching Station • Circuit Switching System • Clock Subsystem • Coarse Sun Sensor • Cockpit System Simulator • Coded Switch System • Cognitive Science Society • Coherent Signal Simulator • Combined Service Support • Command Specific Server • Commercial Satellite Survivability • Common Semantics and Syntax • Common Support System • Commonwealth Superannuation Scheme • Communications Security System • Communications Subset • Communications Subsystem • Communications Support Segment • Communications Support System • Communications Switching System • Communications Systems Segment • Compact Steep Spectrum • Complete Statistical System • Component Scheduling System • Computational Support Section • Computer Sciences and Statistics • Computer Science School • Computer Search Services • Computer Service Support • Computer Service System • Computer Subsystem • Computer Support System • Computer System Simulator • Connecticut Space Center (USA-CT) • Consolidated Support System • Constant Surveillance Service • Contact Start-Stop • Content Scramble System (*) • Context Scramble System • Contingency Support Staff • Continuous System Simulator • Contractor Support Services • Contract Subsystem • Contract Support System • Control Stick Steering • Control Subsystem • Conventional Serial Stagnation • Conversational Software System • Cooperative Spacecraft System • Coordination de la Sécurité au Sol • Core Segment Simulator • Council for Science and Society • Crew Safety System • Cryogenic Support System • Customer Switching System

CSSA Center for Space Science and Astrophysics • Circuit Switch Summary Alarm • Coarse Sun Sensor Assembly • Computer Society of South Africa • Computer Software Support Agency • Configuration Similarity Simulated Annealing

CSSAR Centre for Space Science and Applied Research

CSSB Compatible Single Sideband

CSSC Canadian Strategic Software Consortium (Canada) • Chabot Space and Science Center (USA-CA) (formerly COSC) • Coded Switch Set Control • Coded Switch Set Controller • Computer System Security Office • Computer System Security Officer • Computer System Support Center

CSSD Command Support System Demonstrator • Communications Software Support Division • Communications System Status Display

CSSE Center for Space Science and Exploration (USA-NM) • Computer Science and Software Engineering • Control System Simulation Equipment

CSSEC Computer Systems Support and Evaluation Command

CSSH Congress of the Social Sciences and Humanities

CSSI Computer Security System Interpretation

CSSIN Conseil Supérieur de la Sûreté des Installations Nucléaires (France)

CSSIP Centre for Sensor Signal and Information Processing

CSSL Communications and Space Science Laboratory • Continuous Systems Simulation Language

CSSM Center System Safety Manager • Centre for the Subatomic Structure of Matter (Australia-SA) • Client-Server Systems Management

CSSMRS Computerized Site Security Monitor and Response System

CSSO Computer Special Security Office • Computer Special Security Officer • Computer Systems Security Office • Computer Systems Security Officer

CSSP Circuits, Systems and Signal Processing (periodical code) • Committee on Solar and Space Physics • Computer Security Support Program • Council of Scientific Society Presidents • Cutting and Skiving Stock Problem

CSSPAB Computer System Security and Privacy Advisory Board

CSSPM Center System Safety Program Manager

CSSPPT Commission Supérieure du Service Public des Postes et Télécommunications (France)

CSSPT Common Supply Support

C/SSR Cost/Schedule Status Report

CSSR Center for Solar and Space Research • Communication Systems Segment Replacement • Contractor System Status Review • Contract Summary Status Report • Cost Schedule Status Report

CSSS Carrier Smart Switch Services • Circuit Switch Switching Shelter • Colorado Springs Space Society (USA-CO)

CSST China Space Solar Telescope • Commission Spécialisée des Sciences de la Terre • Computer System Science Training

CSSTE Centre for Space Science and Technology Education

CSSTEAP Centre for Space Science and Technology Education for Asia and Pacific

CSSU Converter Simulator Signal Unit

Cst Coast • Constant

CST Cadre Spécial Temporaire • Capsule Systems Test • Carlos Sánchez Telescope • CARMA Software Toolkit • Castaway (Fiji) (location identifier) • Caustic Surface Trace • Central Standard Time • Centre Spatial de Toulouse (France) • Certified Survey Technician • Channel Status Table • Character Set Technology • Classification on Science and Technology • Coast • Code Sort Translator • College of Science and Technology • Combined Station and Tower • Combined Systems Test • Combustion Science and Technology (periodical code) • Comité Supérieur de la Télématique • Communications Subtest • Compact Standard Transponder • Complex Safety Technician • Computational Structure Technology • Computer Systems and Telematics • Constant • Contract Supplemental Tooling • Control System Test • Convective-Stratiform Technique • Conventional Stability Test • Convulsive Shock Therapy • Council for Science and Technology • Countdown Simulation Test • Crew Software Trainer • Crew Station Trainer • Crew Systems Trainer • Culture Scientifique et Technique • Customer Support Team

CSTA Canterbury Science Teachers' Association • Computer Sciences Technicolor Associates • Computer-Supported Telecommunications Applications • Computer-Supported Telephony Applications • Computer Support Telecommunications Application • Crew Software Training Aid

CSTAG Centre for the Study of Telematics and Governance (UK-Scotland)

CSTAR Center for Space Transportation and Applied Research • Classified Scientific and Technical Aerospace Report • Continuous Stellar-Tracking Attitude Reference

CSTB Centre Scientifique et Technique du Bâtiment (France) • Computer Science and Telecommunications Board

CSTC Computer Science Teaching Center • Computer Security Technology Center • Consolidated Satellite Test Center • Consolidated Space Test Center

CSTCR Consolidated Space Test Center Regulation

C/STCS Cost/Schedule Technical Control System

CSTCTO Country Standard Time Compliance Technical Order

CSTD Centre of Science and Technology for Development (*) • Committee on Science and Technology for Development • Computer Sciences and Telecommunications Division

Cste Constante

CsTe Cesium Tellurium

CSTE Certified Software Test Engineer

CSTEA Center for the Studies of Terrestrial and Extraterrestrial Atmospheres

CSTEE Comité Scientifique sur la Toxicité, l'Écotoxicité et l'Environnement

CSTF Central Support Technical Facility

Cstg Casting

CSTG Computer Systems Technical Group • Coordination of Space Techniques for Geodesy and Geodynamics

CSTGAG Coordination of Space Techniques for Geodesy And Geophysics

CSTI Center for Scientific and Technological Information • Civil Space Technology Initiative (USA) • Committee for Scientific and Technical Information • Council of Science and Technology Institutes • Council of Scientific and Technical Institutes • Culture Scientifique, Technique et Industrielle

CSTIN China Science and Technology Information Network (China-PRC)

CSTL Chemical Science and Technology Laboratory • Coastal

CSTM Custom Spectral Transmission Measurement

CSTO Computer Systems Technology Office • Country Standard Technical Order

CSTOL Customer System Test and Operations Language

CSTP Columbus Solar-Terrestrial Physics • Committee for Science and Technology Policy • Committee on Science and Technology Policy

CSTPC Chinese Science and Technology Publications and Citations (database)

CSTR Canister • Committee on Solar-Terrestrial Research • Computer Science Technical Report • Continuous-flow Stirred Tank Reactor

CSTS Centro di Studio per le Telecomunicazioni Spaziali (Italy) • Combined System Test Stand • Commercial Space Transportation Study • Cryogenic Storage and Transfer System

CSTSS Combined Systems Test Simulator Set

CSTU Combined Systems Test Unit

CSTVRP Computer Security Technical Vulnerability Reporting Program

CSU Calibration Source Unit • California State University (USA-CA) • Central Statistical Unit • Central Switching Unit • Channel Service Unit • Channel Sharing Unit • Channel Synchronizer Unit • Circuit Switching Unit • Collège Scientifique Universitaire • Colorado State University (USA-CO) • Command Sensor Unit • Communications Switching Unit • Composante Sol-Utilisateur • Computer Software Unit • Conseil Supérieur des Universités • Constant Speed Unit • Contol Selection Unit • Conventional Signals Upgrade • Conventional Speed Unit • Customer Service Unit • Customer Set-Up • Customer Support Unit • Santa Cruz do Sul (Brazil-RS) (location identifier)

CSUCA Consejo Superior de Universidades de Centroamérica

CSUDH California State University, Dominguez Hills (USA-CA)

CSUF California State University, Fullerton (USA-CA)

CSUH California State University, Hayward (USA-CA)

CSUL Colorado State University Libraries (USA-CO)

CSULB California State University, Long Beach (USA-CA)

CSUM Call Summary

CSUMB California State University, Monterey Bay (USA-CA)

CSUN California State University, Northridge (USA-CA) • Commission for Symbols, Units and Nomenclature (IUPAC)

CSUS California State University, Sacramento (USA-CA)

CSUSB California State University, San Bernardino (USA-CA)

CSUSM California State University, San Marcos (USA-CA)

csv Comma-Separated Values (file extension)

CSV Catalogue of Suspected Variables (stars) • Cathodic Stripping Voltammetry • Characteristic Statistical Value • Circuit-Switched Voice • Comma-Separated Values • Comma-Separated Variables • Common Services Verb • Control Step Verification • Corona Start Voltage • Crossville (USA-TN) (location identifier)

CSVC Core Sample Vacuum Container

CSVL Control Secure Voice Link

CSVS Canadian Space Vision System • Catalogue of Suspected Variable Stars (periodical code)

CSW Certified Social Worker • Channel Status Word • Colliding Stellar Wind • Common Software • Communications Switching • Compartment Stowage Worksheet • Continuous Seismic Wave • Contract Specialist Workbook

CSWA Canadian Science Writer's Association (Canada) • Committee on the Status of Women in Astronomy (AAS)

CSWG Circumpolar Seabird Working Group • Communications Systems Working Group • Cost and Scheduling Working Group • Crew Station Working Group

CSWP Committee on the Status of Women in Physics (ASP) • Common Software Processing

Cswy Causeway

CSWY Causeway

CSX Cayman Stock Exchange (Cayman Islands) • Changsha (China-PRC) (location identifier)

CSXT Civilian Space Exploration Team

CSY Cheboksary (Russia) (location identifier)

CSZ Chilean Special Zone • Clearly-Separated Zones

CS2 Carbon Disulfide

ct Carat • Centum

Ct Canton (China-PRC) • Caught • Cent • Certificate • Circuit • Connecticut (USA) • Content • Count • County • Courant • Court • Credit • Current

C&T Chips & Technologies • Ciel et Terre (periodical code) • Ciência & Tecnologia • Clothing & Textile • Collection & Textile • Command & Telemetry • Communications & Tracking • Contingency & Training

C/T Conference Terms • Control and Timing

CT Cable Test • Cable Testing • Cable Transfer • Calendar Time • Call Transfer • Canal de Transit • Cape Town (South Africa) • Captive Test • Case-Telescoped • Cassette Tape • Center Tap • Center Thickness • Central Time • Centre Technique • Centro Técnico • Čerenkov Telescope • Certificated Teacher • Certification Test • Certification Testing • Certified Teacher • Chef Technicien • Chemical Transfer • Church-Turing (thesis) • Cipher Text • Circuit Theory • Clock Time • Coarse Track • Cognitive Technology • Cognizant Test • Collective Task • Collective Training • College of Technology • Colour Type • Combined Test • Combined Testing • Comissão Técnica • Command Table • Command Transmitter • Commercial Translator • Commercial Traveller • Commission Technique • Common Tier • Communications Technician • Communications Technology • Communications Terminal • Compact Tension • Compact Toroid • Comparison Test • Compatibility Test • Compatibility Testing • Compliance Test • Compliance Testing • Component Test • Component Testing • Compressed Term • Computational Technique • Computed Tomography • Computed Tomography • Computerized Tomography • Computer Technology • Computer Telephony • Computer Terminal • Conformance Test • Conformance Testing • Connecticut (USA) • Connectivity Table • Conseiller Technique • Constrained Transport • Contract Team • Contractor Test • Contractor Testing • Contrôle Technique • Controlled Term • Control Telemetry • Control Transformer • Copyright Treaty • Cordless Telephone • Cordless Telephony • Corona-Thread • Corporate Technology • Correlation Tracker • Counter Timer • Court • Crawler-Transporter • Critical Temperature • Cryogenic Trapping • Cryptological Technician • Cryptology Technician • Current Transformer • Cursor Table • Customer Test • Customer Testing • Cycle Time • Cyclic Time • Cypher Text

cta Cum Testamento Annexo

Cta Caleta • Cuenta • Cuesta

CTA Canadian Target Assembly • Caribbean Technical Assistance • Catania (Italy) (location identifier) • Center

Test Authority • Central Technical Activity • Centro Técnico Aeroespacial • Chef de Travaux Adjoint • Chemical-Technical Assistant • Chief Technical Adviser • Clothing Table of Allowance • Cognitive Task Analysis • Comité de la Télématique Anonyme • Command and Telemetry Assembly • Commission for the Teaching of Astronomy (IAU) • Commodities Trading Adviser • Common Table of Allowance • Compatibility Test Area • Component Test Area • Computational Technology Area • Computer Technology Associates • Computer Traders Alliance • Consolidated Table of Allowance • Consolidated Training Activities • Constructive Technology Assessment • Contractor Technical Assistance • Contractual Technical Assistance • Contrat de Télé-Assistance • Control Area • Controlled Thrust Assembly • Controlled Time of Arrival • Cryogenic Telescope Assembly

CTAA Comité Technique d'Attribution des Aides

CTAC Central Technology Advisory Committee • Cetyltrimethylammonium Chloride • Computational Techniques and Applications Conference • Computer Tester Analyzer Controller

CTAD Conference on Trade And Development

CTAF Common Traffic Advisory Frequency

CTAG Canadian Time Allocation Group (Canada)

CTAK Cipher Text Autokey

CTAM Continental Tropical Air Mass

CTAMN Centre de Télédétection et d'Analyse des Milieux Naturels

CTAN Comprehensive TEX Archive Network

CTAP Communications Technology Assessment Process

CTAR Critical Threat Assessment Report

CTAS Center/Tracon Automation System • Central Texas Airborne Systems (USA-TX) • Central Texas Astronomical Society (USA-TX)

CTB Calibration Test Box • Centraal Technische Bibliotheek (Netherlands) • Central Terminal Building • Cipher-Type Byte • Commonwealth Telecommunications Board • Communications Testbed • Communications Toolbox • Complex-to-Basic • Comprehensive Test Ban • Control Testbed • Council tax Benefit • Cut Bank (USA-MT) (location identifier)

CTBA Centre Technique du Bois et de l'Ameublement • Composite Triple Beat Abstand

CTBL Cloud-Topped Boundary Layer

CTBP Center for Theoretical Biological Physics (UCSD)

CTBT Comprehensive Test Ban Treaty

CTBTO Comprehensive Test Ban Treaty Organization

CTC Camera, Timing and Control • Canadian Transport Commission • Capsule Test Conductor • Carbon Tetrachloride • Catamarca (Argentina-CA) (location identifier) • CCATS Telemetry Controller • Center for Technology Commercialization • Centralized Traffic Control • Central Technical Catalog • Central Test Center • Channel-To-Channel • Chief Test Conductor • City Technology College • Climate Test Chamber • Closed Timelike Curve • Command Transmitter Controller • Committee for Technical Cooperation • Communications Technology Center • Community Technology Center • Compañía de Teléfonos de Chile • Compatibility Test Capsule • Computer Technology Center • Consolidated Test Center • Constant Torque Compensation • Contract Target Cost • Cooke Transit Circle • Cornell Theory Center (USA-NY) • Coste de Transición a la Competencia • Cost To Complete • Cost To Completion • Counter/Timer Circuit • Counter-Timer Control • Crew Training Centre • Crew Training Complex

CTCA Channel To Channel Adaptor

Ctcas Características

CTCC Consolidated Telecommunications Center • Critical Technology Coordinating Committee

CTCF Channel Technical Control Facility

CTCI Classification Type pour le Commerce International • Contract Technical Compliance Inspection

CTCM Critical Technical Characteristics Matrix

CTCMS Center for Theoretical and Computational Materials Science

CTCP Client-To-Client Protocol • Communications and Transmission Control Program • Configuration Test Change Period • Contract Task Change Proposal • Customer-To-Customer Protocol

CTCPEC Canadian Trusted Computer Product Evaluation Criteria

CRCS Central Thermal Control System

C&TCS Communications & Tracking Checkout System

CTCS Category Theory and Computer Science (conference) • Communications and Tracking Checkout System

CTCSS Continuous Tone Coded Squelch System

Ct. Ct. Compte Courant

CTCU Correlation Tracker and Control Unit

Ctd Coated • Continued • Created

CTD Cell Transfer Delay • Centrale Technische Dienst • Central Tracking Detector • Centre for Telecommunications Development • Certified Test Data • Charge-Transfer Device • Chemical Technology Division • Classified Telephone Directory • Computing and Telecommunications Division • Conductivity - Temperature - Depth • Consecutive Triple Difference • Cost To Date • Critical Technology Development • Cumulative Trauma Disorder • Current - Temperature - Density

CTDC Chemical Thermodynamics Data Center • Control Track Direction Computer

CTDCS Common Test Data Collection System

CTDF Contracting Technical Data File

CRDH Command and Telemetry Data Handling

CTDIF Cambridge Tabular Data Interchange Format

CTDP Centre de Traitement Doris/Poséidon • Communications and Tracking Data Processor

CTDR Commercial Training Device Requirement

C/TDS Count/Time Data System

CTDS Code Transition Data System

CTDU Cross-link Transponder Data Unit

Cte Commandante • Corriente

CT&E Contractor Test & Evaluation

CTE Carti (Panama) (location identifier) • Center for Teaching Excellence • Central Timing Equipment • Centre de Télé-Enseignement • Certification Test and Evaluation • Charge Transfer Effect • Charge Transfer Efficiency (*) • Ciencias de la Tierra y del Espacio (periodical code) • Coefficient of Thermal Expansion • Commercial Test Equipment • Common Test Environment • Component Test Equipment • Concurrent Test Environment • Content - Transfer - Encoding • Contractor Test and Evaluation • Contractual Technical Evaluation • Cross-Track Error • Cruise Thermal Efficiency • Customer Terminal • Customer Trading Exchange

CTEA Copyright Term Extension Act (USA) • Cost and Training Effectiveness Analysis

CTEC Certified Technical Education Center

CTED Civilian Training, Education, and Development

CTEI Centre de Traitement et d'Exploitation des Images • Cloud-Top Entrainment Instability

CTEIP Central Test and Evaluation Investment Program

CTEM Conventional-Transmission Electron Microscope • Conventional-Transmission Electron Microscopy

CTEP Camera Tube Evaluation Program

Cter Center

CTER Command Terminal

CTERM Communications Terminal

CTERMP Communications Terminal Protocol

CTeS Corporate Telecommunications Strategies

Ctf Certificate • Certify

CTF Capital de Temps Formation • Central Test Facility • Central Training Facility • CLIC Test Facility (CERN) • Closure Task Force • Coal-Tar Fuel • Collective Training Facility • Combined Task Force • Combined Test Force • Commit To Flight • Common Technical Framework • Common Test Facility • Communauté des Télévisions Francophones • Community Task Force • Consolidated Task Force • Consolidated Test Facility • Contrast Transfer Function • Control Test Flight • Core Test Facility • Correction To Follow • Counting Test Facility

ctFAM Combined Through-Fleet Assignment Model

CTFC Central Time and Frequency Control

CTFE Chlorotrifluoroethylene

CTFM Continuous-Transmission Frequency Modulated

CTFO Correct Tension Friction Optimization (system)

Ctg Cartridge • Coating • Cutting

CTG Cartagena (Colombia) (location identifier) • Center for Technology in Government (USA) • Central Training Group • Centrum voor Technische Geowetenschappen (Netherlands) • Chemical Topological Graph • CICS Transaction Gateway • Commonality Task Group • Communications Transmission Group

Ctge Cartage • Cartridge • Cottage

CTGY Category

CTH Chalmers Tekniska Göteborg (Sweden) • Cloud Top Height • Coatesville - Chester County Carlson (USA-PA) (location identifier) • Comité Technique d'Homologation • Compact Toroidal Hybrid

CTI Camera Timing Indicator • CCD Transit Instrument • Centre des Technologies de l'Information • Centre des Télécommunications et de l'Informatique • Centre de Traitement de l'Information • Centre de Traitement Informatique • Centre Technique Industriel • Charge Transfer Inefficiency • Climate Technology Initiative • Coherent Toroidal Interconnect • Comité des Titres d'Ingénieur • Command Technical Inspection • Commission des Titres d'Ingénieur (France) • Compile Time Instrumentation • Complaint Type Investigation • Component Test Instruction • Computational Testbed for Industry • Computer Telephone Integration • Computer Telephony Integration • Computer-Telescope Interface • Consumer Technology Index • Contract Technical Instructor • Couplage Téléphonie-Informatique • Critical Technical Issue • Critical Technology Institute • Critical Transportation Item • Cryogenic Thermal Imager • Cuito Canavale (Angola) (location identifier) • Current Technology Index

CTIA Capacitance Transimpedance Amplifier • Capacitative Transimpedance Amplifier • Cellular Technology Industry Association (USA-DC) (*) • Cellular Telecommunications and Internet Association (USA-DC) • Cellular Telecommunications Industry Association (USA-DC)

CTIC Contractor Technical Information Code

CTID Communications Transmission Identifier

CTIDP Common Technical Interface Design Plan

CTIF Cellule de Traitement des Informations Financières

CTIM Coupled Thermosphere-Ionosphere Model

CTIO Cerro Tololo Interamerican Observatory (Chile) • Coatings Technology Integration Office (USAF)

CTIOPI Cerro Tololo Interamerican Observatory Parallax Investigation

CTIR Computation Theory and Information Retrieval

CTIS Center for Trade and Investment Services • Comprehensive Telemetry Indexing System • Computerized Test Information Stand • Crawler-Transporter Intercom System

CTIT Center for Telematics and Information Technology

CTIU Command and Telemetry Interface Unit

CTIV Centre de Traitement des Images de Végétation

CTJ Central-Tendency Journal

CTK Canton (USA-SD) (location identifier) • Composite Toolkit

CTKP Composite Toolkit Program

ctl Control Data (file extension)

Ctl Central • Control

CTL Calibration Task List • Canoga Test Laboratory • Central • Central Technical Library • Charleville (Australia-QLD) (location identifier) • Checkout Test Language • Command Talking Language • Complementary Transistor Logic • Complex Text Layout • Component Team Leader • Component Test Laboratory • Construction Technical Letter • Constructive Total Loss • Control • Controller Tape Location • Control Transfer Logic • Core Transistor Logic • Critical Task List • Critical To Launch

Ctlry Cutlery

Ctlst Catalyst

CTLV Center-To-Limb Variation

CTLZ Control Zone

ct/m Counts per Minute

CTM Cargo Traffic Message • Cell Transfer Matrix • Centre de Transfert des Microtechniques • Centrifugal Twisting Moment • Chemical Tracer Model • Chemical Transport Model • Chemistry/Transport Model • Chetumal (Mexico) (location identifier) • Ciências e Tecnologias dos Materiais • Ciencia y Tecnologia del Mar • Circuit Transfer Mode • Collapsible Tube Mast • Combined Transforms Method • Combustion Theory Modeling (periodical code) • Command Training Manager • Communications Terminal Module • Community Trademark (EU) • Component Test Motor • Computerized Text Management • Computerized Tomographic Myelography • Continuum Storm • Contract Technical Manager • Cordless Terminal Mobility • Critical Thermal Maximum • Crystalline Transitional Material • Current Transformation Matrix

CTMA Center Traffic Management Advisor

CTMC Classical Trajectory Monte Carlo (method) • Communications Terminal Module Controller

CTMFM Comisión Técnica Mixta del Frente Marítimo

CTMP Channel Traffic Movement Plan

CTMS Carrier Transmission Measuring System

CTMT Containment

ctn Cotangent

Ctn Carton

CTN Cable Termination Network • CALS Test Network • Canadian Technology Network • Carton • Centre des Technologies Nouvelles (Belgium) • Certification Test Network • Comité Technique de Normalisation • Commercial Telecommunications Network • Cooktown (Australia-QLD) (location identifier) • Corporate Telecommunications Network

CTNE Compañía Telefónica Nacional de España (Spain)

Ctnr Container

CTNS Center for Theology and the Natural Sciences (USA-CA)

Cto Conto

CTO Call Transfer Outside • Calverton - Peconic River (USA-NY) (location identifier) • Cancelled To Order • Central Technical Order • Centre de Topographie des Océans (France) • Charge-Transforming Operator • Chief Technical Officer • Chief Technology Officer • Circular Terminal Orbit • Cognizant Technical Official • Cognizant Transportation Office • Control Tower Operator

CTOCU Central Technical Order Control Unit

CTOF Charge Time Of Flight

CTOL Cataloguing Tools Online • Conventional Takeoff and Landing

CTOM Centralized Technical Order Management

CTOS Cassette Tape Operating System • Computerized Tomography Operating System • Converging Technologies Operating System

CTP Capacitor Test Program • Carutapera (Brazil-MA) (location identifier) • Center for Theoretical Physics • Characteristic Technical Parameter • Charge-Transfer Photography • Charge Transforming Parameter • Chief Test Pilot • Chinese Translation Pattern • Christmas Tree Pattern • Closed-Time Path • Clustering Terminal Pattern • Coded Telemetry Processor • Comité Technique Paritaire • Command Translator and Programmer • Commission Technique Paritaire • Commit Target Point • Common Transport Policy • Communications Tasking Plan • Communications Terminal Processor • Communications Timing Procedure • Compiled Translation Process • Composite Theoretical Performance • Computer To Plate • Computer To Press • Computer To Print • Confidence Test Program • Configuration Testing Protocol • Conseil Technique Paritaire • Consent To Publish • Consolidated Telecommunications Program • Construction Test Procedure • Conventional Terrestrial Pole • Coordinated Test Plan • Coordinated Test Program • Core Technology Programme • Critical Technical Parameter • Critical Test Parameter • Cyclic Time Processor

CTPA Coax-to-Twisted-Pair Adapter • Commission Technique Paritaire Académique

CTPB Carboxyl-Terminated Polybutadiene

CTPC Comité Technique Paritaire Central

CTPD Commission Technique Paritaire Départementale • Crew Training and Procedures Division

CTPL Central Technical Publications Library

CTPM Commission Technique Paritaire Ministérielle

CTPN Comité Technique et Pédagogique National (France) • Commission Technique Paritaire Nationale (France)

CTPP Central Tactical Processing Program

CTPS Commit Target Point Selection

CTQ Calán-Tololo Quasar (catalog) • Santa Vitória do Palmar (Brazil-RS) (location identifier)

Ctr Center • Contour • Contribution • Contributor • Culture • Cutter

CTR Capital Type Rehabilitation • Cattle Creek (Australia-NT) (location identifier) • Center • Centre de Traitement Régional • Centre de Traitements Régionaux • Certification Test Requirement • Certification Test Review • Certified Test Record • Certified Test Result • Civil Tiltrotor • Classical Transition Region • Common Technical Regulation • Computer Technology Research • Consolidate Time Rate • Continuous Technology Refreshment • Contractor Task Report • Contract Technical Representative • Contractual Technical Report • Control • Controlled Thermonuclear Reaction • Controlled Thermonuclear Research • Controller • Cooperative Threat Reduction • Coordinate Transformation Restoration • Core Transistor Register

CTRC Colorado Technical Reference Center (USA-CO)

Ctre Centre

CTRF Conventional Terrestrial Reference Frame

Ctrfgl Centrifugal

Ctrg Centering

Ctrl Central

CTRL Control

Ctrlr Controller

Ctrlrs Controllers

Ctro Centro

CTRR Component Test Results Review

CTRS Centers • Component Test Requirements Specification • Conditional Term Rewriting System (workshop) • Conventional Terrestrial Reference System

Ctrst Contrast

CTRW Continuous-Time Random Walk

ct/s Counts per Second

Cts Centavos • Centimes • Cents • Certificates • Counts • Crates

C&TS Communications & Tracking System

CTS Cable Telemetry System • Cable To Shore • Calán-Tololo Survey • Calorimeter Thermal Sink • Canadian Technology Satellite • Captive Trajectory System • Carrier Test Switch • Cassegrain Twin Spectrograph • Center for Theoretical Study • Centralized Translation System • Central Timing System • Centro Turistico Studentesco • Certification Test Specification • Certified Test Results • Cesium Time Standard • Change Tracking System • Characteristic Time Scale • Charge-Transfer Spectrum • Chemical and Transport Systems • Chirp Transform Spectrometer • Circuit Test Set • Clear To Send • Cloud Top Scanner • Coherent Thought Sequencing • Collective Thomson Scattering • Command and Telemetry System • Committee on the Teaching of Science • Common Transport Semantics • Communications and Tracking System • Communications Technology Satellite • Communications Terminal Synchronous • Communications Test Station • Communications Transmission System • Compatibility Test Suite • Compatible Time Sharing • Compatible Timesharing System • Component Test Set • Computerized Tomographic Scanner • Computerized Training System • Computer Test Set • Computer Transition Systems (Australia-VIC) • Computer Typesetting • Computer Typing System • Conformance Testing Service • Conformance Test Service • Console Test Set • Contact Test Set • Context Thesaurus System • Contingency Transfer System • Continuous Tracking Survey • Contractor Technical Service • Contractor Technical Support • Contrat de Télé-Services • Conventional Terrestrial System • Conversational Terminal System • Conversational Time Sharing • Conversation Time Sharing • Coordinate Transformation System • Corporate Technology Strategies • Cosmic Top Secret • Counts • Courier Transfer Station • Course Training Standards • Courts • Crawler Transportation System • Cryogenic Temperature Sensor • Cryogenic Temperature Source • Culler - Tiler - Singler • Sapporo Chitose (Japan) (location identifier)

CTSA Computer-Supported Telephony Application

CTSB Critical Task Selection Board

CTSC Center Technical Services Contractor • Computer Technical Services Contractor

CTSD Computer Test Sequences Document

CTSF Centralized Training Systems Facility • Central Technical Support Facility • Central Test Support Facility • Consolidated Testing and Support Facility

CTSG Central Timing Signal Generator

C&TSS Communications & Tracking Subsystem

CTSS Communications & Tracking Subsystem • Compatible Time-Sharing System • Computer and Telecommunications Systems Staff • Conversational Time-Sharing System • Cray Time-Sharing System

CTSSB Critical Task Site Selection Board

CTST Commercial Space Transportation Study

CTSWG Consolidated Training Support Working Group

cTT Classical T Tauri (star)

CTT Capital Transfer Tax • Card-to-Tape Tape • Cartridge Tape Transport • Centre for Technology Transfer • Channel Transfer Table • Chemical Thermodynamic Table • Church-Turing Thesis • Classical T Tauri (star) •

Cloud Top Temperature • Combined Test Team • Command Transfer Tape • Common Task Test • Common Task Training • Common Test Training • Communications Tactical Terminal • Comprehensive Tracing and Tracking • Contractor Technical Test • Crew Transfer Tunnel • Le Castellet (France) (location identifier)

CTTB Checkout Techniques Testbed

CTTC Cartridge Tape Transport Controller • Copper To The Curb

Cttee Committee

CTTF Central Technical Testing Facility • Central Test and Training Facility

CTTH Copper To The Home

CTTI Center for Technology Transfer and Integration

CTTL Center for Technology in Teaching and Learning

CTTMP Center for Transport Theory and Mathematical Physics

CTTP Critical Technical Test Parameter

cTTS Classical T Tauri Star

CTTY Console Teletype

CTU Cabin Telecommunications Unit • Capsule Test Unit • Captive Test Unit • Cargo Transport Unit • Caribbean Telecommunication Union • Cartridge Tape Unit • Centigrade Thermal Unit • Central Telemetry Unit • Central Terminal Unit • Central Thermal Unit • Central Timing Unit • Chengdu (China-PRC) (location identifier) • Command and Telemetry Unit • Command Telemetry Unit • Communications Terminal Unit • Compatibility Test Unit • Component Test Unit • Custom, Tradition and Usage • Czech Technical University (Czech Republic)

CTUG Chinese TEX Users Group

CTUS Customs Territory of the United States

CTV Cable Television • Cargo Transfer Vehicle • Channel Television • Color Television • Commercial Television • Compatibility Test Van • Computerisierte Textverarbeitung • Constant Tangential Velocity • Controlled Test Vehicle • Control Test Vehicle • Crew Transfer Vehicle • Crew Transport Vehicle (*)

CTVA Canadian Transportable VLBI Antenna

CTW Coronal Transient Wave • Cottonwood (USA-AZ) (location identifier)

CTWG Calibration Targets Working Group

Ctwlk Catwalk

Ctwt Counterweight

CTWT Counterweight

ctx Ciphertext (file extension)

CTX Center for Technical Excellence • Clear To Transmit • Combined Training Exercise • Cortland (USA-NY) (location identifier)

Cty County

CTY Center for Talented Youth • Console Teletype • County • Cross City (USA-FL) (location identifier)

CTZ Clinton - Sampson County (USA-NC) (location identifier) • Control Traffic Zone

c/u Cada Uno

cu Basque (ISO language code) • Cuba • Cubic

Cu Copper • Cuivre • Cumulus • Kupfer

C-U Command-Unique

CU Calibration Unit • Calibration Uplinks • Cambridge University (UK-England) • Camera Unit • Central Unit • Centre d'Usinage • Centre Universitaire • Centroiding Unit • Certificat d'Université • Certificat d'Urbanisme • Charge Utile • Chinese University • Cité Universitaire • Ciudad Universitaria • Close-Up • Code Unit • Coefficient of Utilization • Colorado University (USA-CO) • Columbia University (USA-NY) • Combining Unit • Command Unit • Common Unit • Common User • Communications Unit • Computer Unit • Computing Unit • Concerned Unit • Construction Unit • Control Unit • Cornell University (USA-NY) • Correlation Unit • Cryptographic Unit • Crystal Unit • Cuba (international aircraft registration code) • Cuba (ISO country code) • Cubic Capacity • Cumulus

CUA Catholic University of America (USA-DC) • Ciudad Constitución (Mexico) (location identifier) • Commercial Utilization Area • Common User Access • Computer Users Association • Corporate Unit Agreement

CUAO Copenhagen University Astronomical Observatory (Denmark)

CUAS Cambridge University Astronomical Society (UK-England) • Champaign-Urbana University Astronomical Society (USA-IL) • Common Use Alternative Statement • Computer Utilization Accounting System

Cub Cubic • Cubicle

CUB Central Utility Building • Columbia Owens (USA-SC) (location identifier) • Commonality Usage Board • Cuba (ISO country code) • Cubic • Cursor Backward

CUBE Concertation Unit for Biotechnology in Europe • Cosmic Ultraviolet Background Experiment

CUBIC Caribbean Uniform Building Code • Cosmic Unresolved Background Imaging Camera • Cosmic Unresolved Background Instrument using CCDs

CUBOL Computer Usage Business-Oriented Language

CUBRC Calspan-University of Buffalo Research Center (USA-NY)

CUBS Coronal Ultraviolet Berkeley Spectrometer

CUC Common User Compatible • Common User Contract • Computer Usage Control • Crystal Unit Cell • Cucuta (Colombia) (location identifier)

CUCAP Cambridge University Committee on Aerial Photography

CUCCA Columbia University Center for Computing Activities (USA-NY)

CUCI Clasificación Uniforme para el Comercio Internacional

CUCRS Commonwealth Users Committee on Remote Sensing

CUCV Cargo Utility Carrier Vehicle • Commercial Utility Carrier Vehicle

CUD Caloundra (Australia-QLD) (location identifier) • Computer Underground Digest • Cursor Down

CUDAT Common User Data Terminal

CUDF Call User Data Field

CUDI Centro Universitario de Documentación e Información

CUDIX Common User Data Information Exchange • Common User Digital Exchange • Common User Digital Information Exchange

CUDIXS Common User Digital Information Exchange System

CUDN Common User Data Network

CUDOS Continuously Updated Dynamic Optimizing System

CUDSS Canada-UK Deep SCUBA Survey • Canada-UK Deep Submillimeter Survey (*)

CUE Cité Universitaire Européenne • Cognizant User Engineer • Collaborative Ukrainian Experiment • Command Uplink Electronics • Command Use Equipment • Committee on Undergraduate Entrance • Common Usage Equipment • Communication Unit Executor • Computer Update Equipment • Computer Updating Equipment • Configuration Utilization Evaluation • Cuenca (Ecuador) (location identifier)

CUED Cambridge University Engineering Department

CUEJ Centre Universitaire d'Enseignement et de Journalisme (France) • Centre Universitaire Européen du Journalisme (France)

Cuen Cuenca (Spain)

CUES Centre Universitaire d'Enseignement Supérieur • Colleges and Universities Entrance Scheme • Computer Utility Educational System

cuf C Utilities Form (file extension)

CUF Capacity Utilization Factor • Command Unit Function • Common University Fund • Common User Follow-on • Cross-Utilization File • Cuneo (Italy) (location identifier) • Cursor Forward

CUFOS Center for Unidentified Flying Object Studies

CUFS Common User Fax System

cu. ft. Cubic Foot

CUG Closed User Group • Control Unit Group • Cudal (Australia-NSW) (location identifier) • C Users' Group • Orange Cudal (Australia-NSW) (location identifier)

CUGR Cargo Utility GPS Receiver

CUH Cushing (USA-OK) (location identifier)

CUHK Chinese University of Hong Kong (China-PRC)

CUI Centre Universitaire d'Informatique • Centre Universitaire International • Change User Interface • Character-oriented User Interface • Character User Interface • Close-Up Imager • Common Usage Item • Common User Interface • Controlled Unclassified Information • Currillo (Colombia) (location identifier)

CUIC Consejo Internacional de Uniones Científicas

CUIL Common Usage Item List

cu. in. Cubic Inch

CUIO Centre Universitaire d'Information et d'Orientation • Cellule Universitaire d'Information et d'Orientation

CUIP CPS UC Internet Project

Cuis Cuisine

CUIS Columbus Users Information System • Columbus Utilization Information System

CUISAE Canadian Universities Institute for Space and Atmospheric Environments (Canada)

CUISP Columbus Users Information System Prototype

CUITN Common User Installation Transport Network

cuj Cujus

CUJ Culion (Philippines) (location identifier)

CUJT Complementary Unijunction Transistor

CUK Caye Caulker (Belize) (location identifier)

cul Cursor Library (file extension)

CUL Cambridge University Library • Command Uplink • Common Use List • Common User Language • Cornell University Library • Culiacan (Mexico) (location identifier) • See You Later

CULER Cryogenic Upper (atmosphere) Limb Emission Radiometer

CULP Computer Usage List Processor

CULR Common Upper Layer Requirements

Cult Cultivé • Cultural • Culture • Culturel

CULT Common User Land Transportation

Cum Cumulative

CUM Central-Unit Memory • Cumana (Venezuela) (location identifier) • Cumulative

Cumb Cumberland (UK-England)

CUMD Continuous Update Memory Display

CUMP Central-Unit Memory Programmer

CUN Cancún (Mexico) (location identifier)

CUNUMBR Control Unit Number

CUNY City University of New York (USA-NY)

CUO Caruru (Colombia) (location identifier) • Copenhagen University Observatory (Denmark)

CUP California University of Pennsylvania (USA-PA) • Cambridge University Press • Canadian University Press • Carupano (Venezuela) (location identifier) • Columbia University Press (USA-NY) • Commonality Usage Proposal • Common User Processor • Common User Product • Communications User Program • Comprehensive Use Plan • Cursor Position • Custom and Practice • Peso (Cuba) (ISO currency code)

CUPA College and University Professional Association • Comparative Understanding of Planetary Atmospheres

CUPC Canadian Undergraduate Physics Conference

CUPE Cranfield Unit for Precision Engineering

CUPID Completely Universal Processor I/O Design • Conversation-mode Utilities for Processing Image Data • Cosmic Ultraviolet Polarimetric Imaging Device

CUPLE Comprehensive Unified Physics Learning Environment

CUPR Catholic University of Puerto Rico

CUPRI Cornell University Portable Radar Interferometer

CUpS Connection Updates per Second

CUPS Connection Updates Per Second • Consolidated Uniform Payroll System • Consortium for Upper-level Physics Software

CUQ Coen (Australia-QLD) (location identifier)

CUQS Chile-UK Quasar Survey

cur Cursor (file extension)

Cur Currency • Current

CUR Council on Undergraduate Research • Curaçao Hato (Netherlands Antilles) (location identifier) • Currency • Current • Cursor

CURDIF Centre Universitaire de Recherche et Développement Informatique

CURDS Centre for Urban and Regional Development Studies (UK)

CUREA Consortium for Undergraduate Research and Education in Astronomy

CURES Computerized Unsatisfactory Report Engineering System • Computer Utilization Reporting System

CURI Comité d'Utilisateurs des Ressources Informatiques

CURL Consortium of University Research Libraries

Curr Currency • Current • Curriculum

CURR Centre for Urban and Regional Research • Current

CURRI Centre Universitaire Régional de Ressources Informatiques

Currt Current

CURS Central United States Registry (USA)

CURSA Catalog Utilities for Reporting, Selecting and Arithmetic

Curt Current

CURTS Common User Radio Transmission System

CURV Cable-Controlled Underwater Recovery Vehicle • Commandable Underwater Recovery Vehicle • Curve

CUS Catalog Update Server • Central User Service • Clean-Up System • Coherent Unit Set • Columbus (USA-NM) (location identifier) • Common User Support • Common User System • Conférence Universitaire Suisse (Switzerland) • Continental US

CUSA Centrifugal Urine Separator Assembly

CUSBL Cumulus Under Stratocumulus Boundary Layer

CUSE Committee on Undergraduate Science Education

CUSH Comité des Utilisateurs du SHOM (France)

CUSI Club des Utilisateurs de Systèmes Informatiques (France) • Configurable Unified Search Index • Configurable Unified Search Interface (*)

CUSIP Committee for Uniform Security Identification Procedures

CUSP Compton Upscattering Process • Cyclotron Upscattering Process

CUSPEA China-United States Physics Examination and Application (program)

CUSR Central United States Registry (USA)

Cust Custodian • Custody • Customer • Customs

Custod Custodian

CUSUM Cumulative Sum (chart)

CUSV Commercial Utility Cargo Vehicle

CUT Canonical Unit of Time • Circuit Under Test • Class Under Test • Control Unit Terminal • Control Unit Tester • Coordinated Universal Time • Cross-Utilization Training • Cutral (Argentina-NE) (location identifier)

CUTE Clarkson University Terminal Emulator • Common Use Terminal Equipment • Constrained and Unconstrained Test Environment • Constrained and Unconstrained Testing Environment

CUTLASS Collaborative UK Twin Located Auroral Sounding System

CUTS Computer Users' Tape System • Customer Usage Tracking System

CUU Chihuahua (Mexico) (location identifier) • Computerunterstützter Unterricht • Conference on Universal Usability • Cursor Up

CUV Casigua (Venezuela) (location identifier)

CUVIT Canadian Ultraviolet Imaging Telescope

CUVRA Characteristics of Ultraviolet in the Alps

CUW Cube Cove (USA-AK) (location identifier)

CUX Corpus Christi - Cuddihy (USA-TX) (location identifier)

CUY Cue (Australia-WA) (location identifier)

CUZ Cuzco (Peru) (location identifier)

cv CodeView (file extension) • Coefficient of Variance • Cras Vespere

C-V Capacitance-Voltage

C/V Crew/Vehicle • Customer/Vendor

CV Calibrated Volume • Call Variable • Calorific Value • Cape Verde (ISO country code) • Cap Vert (code ISO de pays) • Cardiovascular • Cargo Variant • Carrier Vehicle • Cataclysmic Variable (star) • Cavallo Vapore • Cavalos Vapor • Cave • Central Version • Chaotic Variability • Check Valve • Chèque de Voyage • Cheval-Vapeur • Chief Value • Circuit Virtuel • Clandestine Vulnerability • Classification Value • Coefficient of Variation • Collective Vector • Command Vehicle • Command Verification • Common View • Complex Visibility • Computer Vision • Configuration Verification • Constant Voltage • Constant Volume • Continuously Variable • Contribution Value • Conventional Vehicle • Convertible • Córdoba Variable (star) • Core vehicle • Cost Variance • Counter Voltage • Cove • Crecimiento Vegetativo • Crew/Vehicle • Cross-Validation • Cruise Vehicle • Cryptographic Variable • Cryptovariable • Crystal Violet • Curriculum Vitae • Customer/Vendor • Cyclic Voltammetry

CVA Central Valley Astronomers (USA-CA) • Central Verification Activity • Cerebrovascular Accident • Clandestine Vulnerability Analysis • Communications Vulnerability Assessment • Pittsburgh Civic (USA-PA) (location identifier)

CVAC Catawba Valley Astronomy Club (USA-NC) • Comox Valley Astronomy Club

CVAF Cold Vapor Atomic Fluorescence • Common VTOC Access Facility • Core Vehicle Assembly Facility

C/VAM Compass/Vertical Angle Measurement

CVAM Compass/Vertical Angle Measurement

CVAP Cataclysmic Variable (stars) Alert Programme

CVAS Chagrin Valley Astronomical Society (USA-OH) • Chattaboochee Valley Astronomical Society (USA) • Chippewa Valley Astronomical Society (USA-OH) • Configuration Verification and Accounting System

CVB Canonical Valence Bond • Chungribu (Papua New Guinea) (location identifier)

CVC Capacitance Voltage Characteristics • Card Value Center • Card Verification Code • Cesium Vapor Cathode • Circular Velocity Curve • Cleve (Australia-SA) (location identifier) • Coherent Video Cancellation • Current-to-Voltage Converter • Current Voltage Characteristics • Customer Value Contractor

CVCDU Coded Virtual Channel Data Unit

CVCF Constant Voltage and Constant Frequency

CVCM Collected Vacuum Condensible Materials • Collected Volatile Condensible Materials

CVCP Committee of Vice-Chancellors and Principals (UK)

CVCS Chemical and Volume Control System

CVD Carte Virtuelle Dynamique • Cash Versus Documents • Cave Direito • Central Velocity Dispersion • Čerenkov Viewing Device • Chemical Vapor Deposition • Chromatic Variation in Dispersion • Come Volevasi Dimostrare • Compact Video Disk • Congressional Visit Day (USA) • Coupled Vibration Dissociation • Current Voltage Diagram

CVDA Commercial Vapor Diffusion Apparatus

CVDE Collaborative Virtual Design Environment

CVDF Cumulative Velocity Dispersion Function

CVDR Cockpit Video Data Recorder

CVDV Coupled Vibration Dissociation Vibration

CVE Cave Esquerdo • Centrale Verwerkingseenheid • Collaborative Virtual Environment • Common Vulnerabilities and Exposures • Complete Vehicle Erector • Cooperative Virtual Environment • Covenas (Colombia) (location identifier) • Escudo (Cape Verde) (ISO currency code)

CVEI Convertisseur de l'Électronique Image

CVEL Cross-Velocity

CVEsc Escudo of Cape Verde (currency)

CVF Central Visual Field • Circular Variable Filter • Command Verified File • Compressed Volume File • Continuously Variable Filter • Courchevel (France) (location identifier)

CVFR Controlled Visual Flight Rules

CVFS Cesium Vapor Feed System • Circular Variable Filter Spectrometer

CVG Certificat de Valeur Garantie • Cincinnati - Northern Kentucky (USA-OH) (location identifier) • Compressor Variable Geometry

CVGA Color Video Graphics Adapter • Color Video Graphics Array

CVGIP Computer Vision, Graphics and Image Processing (periodical code)

CVH Caviahue (Argentina-NE) (location identifier) • Common VTOC Handler

CVI Caleta Olivia (Argentina-SC) (location identifier) • Carbon Vapor Impregnation • Centrum voor Informatieverwerking (Netherlands) • Certified Vendor Information • Chemical Vapor Infiltration • Conditional Voluntary Indefinite • Configuration Verification Index • Configuration Verification Inspection • Counterflow Virtual Impactor • Crew/Vehicle Interface • Current Variable Inductor • Customer/Vendor Interface

CVIA Computer Virus Industry Association

CVIB Compressor Vibration

CVIC Conditional Variable Incremental Computer

CVIM Collaboration, Visualization and Information Management • Configurable Vision Input Module

CVJ Constant Velocity Joint • Cuernavaca (Mexico) (location identifier)

CVL Cape Vogel (Papua New Guinea) (location identifier) • Centre de Vacances et de Loisirs • Configuration Verification List • Copper Vapor Laser

CVLF Compact Very Low Frequency • Consolidated Very Low Frequency

CVLP Chinese Visiting Lecturer Programme

CVLS Central Vacuum Loading System

CvM Cramer - von Mises (test)

CVM Cash Value Machine • Centre de Veille Météorologique • Chromospheric Variability Monitor • Ciudad Victoria (Mexico) (location identifier) • Cluster Variation Method • Computational Vision Model • Control Valve Module

CVn Canes Venatici - Canum Venaticorum (constellation)

CVN Clovis (USA-NM) (location identifier) • Conversion

Cvntl Conventional

CVO Albany (USA-OR) (location identifier) • Canadian Virtual Observatory • Chief Veterinary Officer • China Virtual Observatory (China-PRC) • Corvallis (USA-OR) (location identifier)

CVOPS C-based Virtual Operating System

CVOS C Virtual Operating System

cvp Cover Page (file extension)

CVP Circuit Virtuel Permanent • Collaborative Virtual Prototyping • Commissioning and Verification Phase • Content Vectoring Protocol • Cost Volume Profit

CVPE Collaborative Virtual Prototyping Environment

CVPP Command and Validation Peripheral Processor

CVPR Computer Vision and Pattern Recognition

CVQ Carnarvon (Australia-WA) (location identifier)

Cvr Cover

CVR Centerline Vertical Restraint • Change Verification Record • Change Verification Report • Cockpit Voice Recorder • Command Voltage Regulator • Computer Voice Response • Configuration Verification Review • Constant Velocity Recording • Constant Voltage Reference • Cover • Culver City - Hughes (USA-CA) (location identifier) • Crystal Video Receiver

CVROS Compact Video Rate Optical Scanner

CVRP Computer System Vulnerability Reporting Program

CVRR Computer Vision and Robotics Research

CVRS Central Vehicle Registration System • Cockpit Voice Recording System

Cvrsn Conversion

cvs Canvas Graphics (file extension)

CVS Cargo Visibility System • Cellular Visual Microprocessor • Clovis Cannon (USA-NM) (location identifier) • Comisión Vela Solar • Commercial vendor Services • Common Video System • Concurrent Version System • Corrigé des Variations Saisonnières • Coves

CVSD Continuously Variable Slope Data • Continuous Variable Slope Delta

CVSDM Continuous Variable Slope Delta Modification • Continuous Variable Slope Delta Modulation

CVSE Conferentie voor Veiligheid en Samenwerking in Europa

CVSF Cowichan Valley Star Finders (Canada-BC)

CVSRF Crew-Vehicle Systems Research Facility

CVSS Cuyahoga Valley Space Society (USA-OH)

Cvt Convertible

CVT Canonical Variational Theory • Centroidal Voronoi Tessellation • Chemical Vapor Transport • Color Video Tape • Command Vetting Translator • Communications Vector Table • Concept Verification Test • Concept Verification Testing • Configuration Verification Test • Constant Variable Transmission • Constant Voltage Transformer • Continuing Vocational Training • Continuously Variable Transmission • Controlled Variable Time • Convert • Cosmic Virial Theorem • Coventry (UK-England) (location identifier) • Cray Visualization Toolkit • Current Value Table

CVTE Chemical Vapor Transport Experiment • Crystals by Vapor Transport Experiment • Crystals of Vapor Transport Experiment

CVTS Compressed Video Transmission Service

CVU Calibration Verification Unit • Constant Voltage Unit • Corvo Island (Portugal) (location identifier)

CVV Cryostat Vacuum Vessel • Cryo Vacuum Vessel

CVW CodeView for Windows • Collaborative Virtual Workplace

CVX Critical Velocity Experiment • Critical Viscosity of Xenon

CVZ Constant Viewing Zone • Continuous Viewing Zone • Continuous Visibility Zone

c/w Complete with

C&W Cable & Wireless • Caution & Warning • Control & Monitoring

C/W Carrier Wave • Caution & Warning

CW Call Waiting • Carrier Wave • Caution and Warning • Center Wavelength • Chandler Wobble • Character Width • Chemical Warfare • Chemical Weapon • Chi-Writer • Clean Water • Clear Write • Clockwise • Close Workstation • Cockcroft-Walton • Cold-Worked • Colliding Wind • Color Width • Combined Weight • Command Word • Commercial Weight • Community Workshop • Complete With • Composite Wave • Computer Warfare • Computer Week • Computing Week • Constant Watch • Continuous Wave • Continuous Waveform • Control Word • Conventional Wisdom • Curie-Weiss

CWA Cash With Acceptance • Caution and Warning Advisory • Caution and Warning Annunciator • Center Weather Advisory • CEN Workshop Agreement • Chemical Warfare Agent • Civil Works Administration • Clean Water Act (USA) • Clean Work Area • Closed World Assumption • Collective Worling Agreement • Constant-Wattage Autotransformer • Contractor Warranty Administration • Controlled Work Area • Core-Wing Anomaly • Mosinee (USA-WI) (location identifier) • Stevens Point (USA-WI) (location identifier) • Wausau - Central Wisconsin (USA-WI) (location identifier)

CWAIP Centre for Wavelets, Approximation and Information Processing (Singapore)

CWAN Coalition Wide Area Network

CWAR Continuous Wave Acquisition Radar

CWARC Canadian Workplace Automation Research Centre (Canada)

CWAS Central Wyoming Astronomical Society (USA-WY) • Contractor's Weighted Average Share

CWB Central Weather Bureau • Colliding Wind Binary (model) • Curitiba - Afonso Pena (Brazil-PR) (location identifier)

CWBS Consolidated Work Breakdown Structure • Contractor Work Breakdown Structure • Contract Work Breakdown Structure • Cost Work Breakdown Structure

CWC Central Water Commission • Centre for Wireless Communications (Singapore) • Change Working Directory • Chemical Weapons Convention (1993) • Chernovtsy (Ukraine) (location identifier) • Closed Waste Container • Condensed-Water Content • Cyclone Warning Centre

CWCS Celestial World Coordinate System

CWD Cable Wiring Diagram • Change Working Directory • Convert Word to Double (word) • Cool White Dwarf (star) • Cyclotron Wave Device

CWDB Close White Dwarf Binary

CWDE Center for World Development Education

CWDM Coarse Wave Division Mutiplex • Coarse Wave Division Mutiplexing • Coarse Wavelength Division Mutiplexer

CWDS Cyclone Warning Dissemination System

CWDW Convert Word to Double Word

CWE Caution and Warning Electronics • Caution and Warning Equipment • Current Work Estimate • Current Working Estimate

CWEA Caution and Warning Electronics Assembly

C'wealth Commonwealth

Cwealth Commonwealth

CWEAR Critical-Work Element Audit Review

CWES Cloud Wind Estimation System

CWEU Caution and Warning Electronics Unit

CWF Coal-Water Fuel • Composite Wave Filter • Cornwell-Weisskopf Formula

CWFM Continuous Wave Frequency-Modulated

CWFS Calibration Wave-Front Sensor

CWG Callaway Gardens (USA-GA) (location identifier) • Closed Waveguide • Commonality Working Group • Communications Working Group • Computer Working Group • Constant Wear Garment • Coordination Working Group • Costing Working Group

CWI Call Waiting - Internet • Centralized Warning Indicator • Centrum voor Wiskunde en Informatica (Netherlands) • CICS Web Interface • Clinton (USA-IA) (location identifier) • Collection-Wide Information • Continuous Wave Illuminator

CWIB Call Waiting - Internet Busy

CWIC Classification by Weighted Iterative Clustering

CWIF Continuous Wave Intermediate Frequency

CWIFS Compact Wide Field Sensor

CWIK Chemical World Index Key

CWIN Cockpit Weather Information Needs

CWIP Camera with Workstation for Image Processing • Construction Work In Progress

CWIS Campus-Wide Information Service • Campus-Wide Information System • Community-Wide Information Service • Current Weather Instrument System

CWISP Campus-Wide Information System Protocol

CWIT CDF Windowing Image Tool

cwk Claris Works (file extension)

CWKB Complex Wentzel-Kramers-Brillouin (method)

CWL Cardiff Wales (UK-Wales) (location identifier) • Center Wavelength • Continuous Wave Laser

CWLM Caution and Warning Limit Module

CWLS Conventional Weighted Least Squares

Cwlth Commonwealth

CWM Center for Waste Management • Common Warehouse Model

CWMTU Cold-Weather Materiel Test Unit

CWN Contract Work Notification

CWO Capital Work Order • Cash With Order • Chief Warrant Officer • Communications Watch Officer • Continuous-Wave Oscillator • Mineral Wells - Fort Wolter (USA-TX) (location identifier)

CWON Collection Work Order Number

CWOT Complete Waste Of Time

CWP Campbellpore (Pakistan) (location identifier) • Causeway Pier • Central Weather Processor • Circulating Water Pump • Contractor's Working Party • Contractor Work Plan • Controller Working Position • Cumulative Weight Percent • Current Workspace Pointer

CWPI Configuration Work Package Item

CWpM Correct Words per Minute

CWPM Correct Words Per Minute

CWPS Council on Wage and Price Stability

CWQ Center for Water Quality

CWR Calm Water Ramp • Civil Work Requirements • Colliding Wind Region • Continuous-Wave Radar • Cowarie (Australia-SA) (location identifier)

CWRS Core Water-Recycling System

CWRU Case Western Reserve University (USA-OH)

CWS Cartographic Workstation • Caution and Warning Status • Caution and Warning System • Center Island (USA-WA) (location identifier) • Center Work System • Chilled Water Supply • CICS Web Support • Clinician Workstation • Collaborative Work Service • Collision Warning System • Contingency Workstation • Continuous Wave Spectroscopy • Cooling Water System • Crew Workstation • Customer Warning System • Cyclone Warning System

CWSF Canada-Wide Science Fair • Chemical Waste Storage Facility

CWSS Cold-Weather Sleep System

CWSU Caution and Warning Status Unit • Center Weather Service Unit • Central Weather Service Unit • Condensate Water Servicing Unit

cwt Hundredweight

CWT Call Waiting • Center Wing Tank • Code Walk-Through • Continuous Wavelet Transform • Cowra (Australia-NSW) (location identifier)

CWTS Centrum voor Wetenschaps- en Technologiestudies (Netherlands) • China Wireless Telecommunications Standard

CWU Composite Weighted Unit

CWVC Campus-Wide Virtual Computer

CWVW Cloud and Water Vapor Wind

CWW Corowa (Australia-NSW) (location identifier)

CWY Clearway

Cx Complex • Convex

CX Cancelled • Central Exchange • Charge Exchange • Christmas Island (ISO country code) • Coaxial • Color Exterior (film) • Control Transmitter • Cycle Crossover • Île de Noël (code ISO de pays) • Uruguay (international aircraft registration code)

CXA Caicara (Venezuela) (location identifier)

CXB Cosmic X-ray Background • Cox's Bazar (Bangladesh) (location identifier)

CXBR Cosmic X-ray Background Radiation

CXC Chandra X-ray Center • Chitina (USA-AK) (location identifier)

CXCDS Chandra X-ray Center Data System

CXF Coldfoot (USA-AK) (location identifier)

CXH Vancouver - Coal Harbour (Canada-BC) (location identifier)

CXI Christmas Island (Kiribati) (location identifier) • Common X-windows Interface

CXJ Caxias do Sul - Campo dos Bugres (Brazil-RS) (location identifier)

CXL Calexico (USA-CA) (location identifier) • Cancelled

CXMS Conversion X-ray Mössbauer Spectroscopy

CXN Candala (Somalia) (location identifier)

CXO Chandra X-ray Observatory (formerly AXAF) • Conroe - Montgomery County (USA-TX) (location identifier)

CXOU Chandra X-ray Observatory Unidentified (source)

CXP Cilacap (Indonesia) (location identifier)

CXQ Christmas Creek (Australia-WA) (location identifier)

CXR Carrier • Christmas Island (ISO country code) • Cygnus-X Ridge • Île de Noël (code ISO de pays)

CXRB Cosmic X-Ray Background

CXS Charge Exchange Spectroscopy

CXT Charters Towers (Australia-QLD) (location identifier)

CXY Cat Cay (Bahamas) (location identifier)

cy Welsh (ISO language code)

Cy Capacity • Century • City • County • Currency • Cyanide • Cycle

CY Calendar Year • Chypre (code ISO de pays) • Copy • Cubic Yard • Current Year • Cycle • Cyprus (international vehicle registration code) • Cyprus (ISO country code)

CYA Les Cayes (Haiti) (location identifier)

CYB Cayman Brac (Cayman Islands) (location identifier)

Cyber Cybernetics

Cyc Cycle • Cycling

CYC Caye Chapel (Belize) (location identifier) • Compact Young Cluster • Cycle • Cyclonic

CYCLONE Cyclotron de Louvain-la-Neuve (Belgium)

CYE Crystal Lake (USA-PA) (location identifier)

CYF Chefornak (USA-AK) (location identifier)

Cyg Cygnus-Cygni (constellation)

CYG Corryong (China-ROC) (location identifier)

CYI Chiayi (China-ROC) (location identifier)

CYKOS Chromospherically Young, Kinematically Old Star

Cyl Cylinder • Cylindrical

CYL Coyoles (Honduras) (location identifier) • Cylinder

CYLINA Cyberspace Leveraged Intelligent Agent

CYM Cayman Islands (ISO country code) • Chatham (USA-AK) (location identifier) • Îles Cayman (code ISO de pays)

CYMK Cyan - Yellow - Magenta - Black

Cyn Canyon

CYN Canyon

CYO Cayo Largo del Sur (Cuba) (location identifier) • Choose Your Own

Cyp Cyprian • Cyprus

CYP Calbayog (Philippines) (location identifier) • Cape York Peninsula (Australia-QLD) • Chypre (code ISO de pays) • Cyprus (ISO country code) • Pound (Cyprus) (ISO currency code)

CYR Colonia (Uruguay) (location identifier)

CYRIC Cyclotron and Radioisotope Center

CYS Cheyenne (USA-WY) (location identifier)

CYSA Cape York Space Agency Pty. Ltd. (Australia)

CYSB Cape York Space Base (Australia-QLD)

CYSF Cape York Space Facility (Australia-QLD)

CYSP Cape York Space Port (Australia-QLD)

CYT Yakataga (USA-AK) (location identifier)

CYTED Ciencia Y Tecnología para el Desarollo

CYU Cuyo (Philippines) (location identifier)

CYX Cherskiy (Russia) (location identifier)

CYZ Cauayan (Philippines) (location identifier)

cz Czech Republic

Cz Cenozoic • Czech

CZ Canal Zone (Panama) • Clear Zone • Conservation Zone • Control Zone • Convection Zone • Czechoslovakia (now obsolete) • Czech Republic (ISO country code) • République Tchèque (code ISO de pays)

CZA Chichén Itzá (Mexico) (location identifier)

CZB Cruz Alta - Carlos Ruhl (Brazil-RS) (location identifier)

CZC Copper Center (USA-AK) (location identifier)

CZCS Coastal Zone Color Scanner • Coastal Zone Color Scanning

Cze Czech

CZE Capillary Zone Electrophoresis • Coro (Venezuela) (location identifier) • Czech Republic (ISO country code) • République Tchèque (code ISO de pays)

CZF Cape Romanzof (USA-AK) (location identifier)

CZH Corozal (Belize) (location identifier)

CZJ Corazón de Jesús (Panama) (location identifier)

CzJPh Czechoslovak Journal of Physics (periodical code)

CZK Cascade Locks (USA-OR) (location identifier) • Koruna (Czech Republic) (ISO currency code)

CZL Constantine (Algeria) (location identifier)

CZM Coastal Zone Management • Cozumel (Mexico) (location identifier)

CZMA Coastal Zone Management Act (USA)

CZMG Coastal Zone Management Group

CzMJ Czechoslovak Mathematical Journal (periodical code)

CZMS Coastal Zone Management Subgroup

CZN Chisana (USA-AK) (location identifier)

Czo Cozzo

CZO Chistochina (USA-AK) (location identifier)

CZP Cape Pole (USA-AK) (location identifier)

CZR Columnar Zone Radius

CZS Cruzeiro do Sul (Brazil-AC) (location identifier)

CZT Cadmium Zinc Telluride • Carrizo Springs (USA-TX) (location identifier)

CZU Corozal (Colombia) (location identifier)

CZW Czestochowa (Poland) (location identifier)

CZX Changzhou (China-PRC) (location identifier)

CZY Cluny (Australia-QLD) (location identifier)

CZZ Campo (USA-CA) (location identifier)

C2 Command and Control • Nauru (international aircraft registration code)

C2C Channel-to-Channel • Consumer-to-Consumer

C2CDM Command and Control Core Data Model

C2CN Command and Control Computer Network

C2CS Command and Command Communications System

C2D Character-to-Decimal • Command and Control Disruption • Core to Disk

C2DA Command and Control Decision Aid

C2DATE Command and Control Decision Aids Test Environment

C2DBS Command and Control Databases

C2E Command and Control Element • Command Center Element • Continuous and Comprehensive Evaluation

C2FAA Command and Control Functional Area Analysis

C2FAC Command and Control Facility

C2FACRP Command and Control Functional Analysis and Consolidation Review Panel

C2FMO Command and Control for Mobile Operations

C2GCR Centre for Climate and Global Change Research

C2I Command, Control, and Information • Command, Control, and Intelligence

C2IC Command and Control Integration Council

C2IE Command and Control Information Element • Command and Control Information Exchange

C2IL Command and Control Integration Language

C2IP Command and Control Initiatives Program • Command, Control, and Intelligence Processor

C2IPS Command and Control Information Processing System

C2IS Command and Control Information System

C2ISM Command and Control Information Systems Manager

C2ISMO Command and Control Information Systems Management Office

C2MA Command and Control Mission Area

C2MAA Command and Control Mission Area Analysis

C2MADP Command and Control Mission Area Development Plan

C2MO Command and Control for Mobile Operations

C2MS Command and Control Modernization Methodology

C2MUG Command and Control Microcomputer Users Group

C2P Centre de Compétence en Projection Plasma • Command and Control Planning • Command and Control Processor • Command and Control Protect • Command and Control Protection

C2PWG Command and Control Protect Working Group

C2S Co-alignment/Co-phasing System • Command and Control Support • Command and Control System • Command - Control - Status

C2SA Command and Control Support Activity

C2SID Command and Control System Integration Directorate • Command and Control System Integration Division

C2SIP Command and Control Systems Interoperability Program

C2SPMO Command and Control Systems Program Management Office

C2SPR Command and Control Systems Program Review

C2SRS Command and Control Strength Reporting System

C2T Cable Chaining Technology • Command and Control Technology

C2T2 Commercial Communications Technology Testbed

C2T2L Commercial Communications Technology Testbed Laboratory

C2V Command and Control Vehicle

C2VR Centre de Calcul Vectoriel pour la Recherche

C2VTC Command and Control Video Teleconference

C2W Command and Control Warfare

C2X Character-to-Hexadecimal

C3 Andorra (international aircraft registration code) • Clsuter Command Control • Command, Control and Communications • Communications, Command and Control • Consultation, Command and Control • Correlator Control Computer

C3A Club Ajaccien des Amateurs d'Astronomie (France) • Command - Control - Communications - Automation

C3AD Computer-Aided Aerospace Analysis and Design

C3AG Command, Control, and Communications Advisory Group

C3I Command, Control, Communication and Information • Command, Control, Communications and Intelligence • Command, Control, Computing and Intelligence

C3IA Command, Control and Communication Acquisition

C3IC Coalition, Coordination, Communications and Integration Center

C3IRM Command, Communications and Control Information Resource Management

C3MP Command, Control and Communications Master Plan

C3P2 Command, Control and Communications Program Plan

C3RIS Command, Control and Communications Systems Readiness Information System

C3S Command, Control and Communications System

C3SMP Command, Control and Communications Systems Master Plan

C3T Concept - Capability - Consensus - Technology

C4 Center for Clouds, Chemistry and Climate • Comet Coma Chemical Composition • Command - Control - Communications - Computers

C4ADA Command, Control, Communications, and Computers Automated Decision Aid

C4H2 Diacetylene

C4I Command - Control - Communications - Computers - Intelligence

C4ID Command, Control, Communications, Computers, and Intelligence Dissemination

C4ISA Command, Control, Communications, Computers, and Intelligence Support Activity

C4ISP Command, Control, Communications, Computers, and Intelligence Support Plan

C4ISR Command - Control - Communications - Computer - Intelligence - Surveillance - Reconnaissance

C4I2 Command, Control, Communications, Computers, Intelligence and Information • Command, Control, Communications, Computers, Intelligence and Interoperability

C4I2T Command, Control, Communications, Computers, and Intelligence/Information Technology

C4RDP Command, Control, Communications, and Computers Requirements Definition Program

C4S Center for Command and Control Communication Systems • Command, Control, Communications and Computer Systems

C4SMP Command, Control, Communications, and Computers Systems Master Plan

C4SSTAG Command, Control, Communications, Computer Systems Security Training • Advisory Group

C4TA Command, Control, Communications, and Computers Technical Architecture

C5 Gambia (international aircraft registration code)

C6 Bahamas (international aircraft registration code)

C6H2 Triacetylene

C9 Mozambique (international aircraft registration code)

c/- Care of • Currency

D

d Apparent Diameter • British Penny • Da • Date • Daughter • Day • Dead • Deceased • Deci • Decision • Decree • Deep • Degeneracy of Vibrational Mode • Degree • Delete • Deliver • Delivery • Delta • Dem • Denarius • Density • Depart • Depth • Deputy • Der • Desert • Dextro • Deuteron • Día • Dialect • Dialectal • Diameter • Died • Dime • Diminutif • Diminutive • Diminutivo • Dinar (currency) • Discharge • Distance • Distance Modulus • Dividend • Dollar • Dose • Down • Drachma (currency) • Droit • Droite • Lattice Plane Spacing • Pence • Penny • Relative Density • Surface Drift • Thickness

D Absorbed Dose • Da • Dag • Daily • Dalasi (currency) • Damen • Damping Constant • Darcy • Data • Day • Debye-Waller Factor • December • Deck • Declassify • Declination • Décret • Deep • Deliver • Delta (international alphabet) • Democrat • Democratic • Demonstration • Demy • Densité • Density • Department • Depot • Depression • Depth • Deputy • Desirable • Desired • Destroyer • Detached • Detectivity • Deus • Deuterium • Deutschland • Devonian • Diameter • Diffusion Coefficient • Diffusive Transport • Digital • Dimension • Dimensional • Dinar (currency) • Diode • Dioptre • Directing • Director • Diretto • Discrete • Dispersion • Dissociation Energy • Distinguished • Doctor • Dogana • Dom • Don • Dong (currency) • DOS • Dose • Douane • Double • Down • Drag • Droit • Droite • Drop Diameter • Dutch • Dynamic • Electric Displacement • Electric Flux Density • First Derivative • Five Hundred (Roman numeral) • Fractal Dimension • Germany (international aircraft registration code) • Germany (international vehicle registration code) • Germany (postal country code) • Linear Diameter • Mass Diffusion Coefficient • Mean Elongation • Optical Density • Route Départementale • Telescope Aperture • Total Drift • 500 (Roman numeral)

d/a Days after Acceptance • Documents against Acceptance

d-a Digital to Analog

da Danish (ISO language code) • Déca • Deka

d/A Documentos contra Aceptación

Da Aerospace Drift • Danish • Daughter • Day

D-A Digital to Analog

D/A Days after Acceptance • Decoder Activate • Delivery on Acceptance • Deposit Account • Digital to Analog • Documents against Acceptance • Documents for Acceptance

DA Algerian Dinar (currency) • Dakota • Damage Assessment • Danger Area • Data Accountability • Data Acquisition • Data Adapter • Data Administration • Data Administrator • Data Analysis • Data Augmentation • Data Automation • Data Available • Days after Acceptance • Decimal Add • Decimal-to-Analog • Decision Aid • Decision Altitude • Decision Analysis • Decoder Activate • Deflection Amplifier • Delayed Action • Delta Amplitude • Density Altitude • Département d'Astrophysique • Department of Administration • Department of Agriculture • Department of Astronomy • Department of Astrophysics • Department of the Army • Deployed Assembly • Deployment Assembly • Deposit Account • Deputy Administrator • Deputy Assistant • Derecho Administrativo • Derived Address • Descent Advisor • Descriptor Area • Design Authority • Design Authorization • Design Authorship • Design Automation • Desk Accessory • Destination Address • Destructive Analysis • Detector Assembly • Developing Activity • Developing Agency • Development Assistance • Development Authorization • Dictionary of Americanisms • Differential Algebra • Differential Analyzer • Digital-to-Analog • Dinar (currency) • Dip Angle • Diplexer Assembly • Direct Access • Direct Action • Direct Ascent • Direction de l'Administration • Direction of Administration • Directorate of Administration • Director of Administration • Directory Agent • Directory Assistance • Discrete Address • Display Active • Dissociative Attachment • Distributing Authority • Distribution Assembly • District Administrator • District Attorney • Divergence Angle • Docking Adapter • Documents against Acceptance • Documents Attached • Documents for Acceptance • Doesn't Answer • Dona • Don't Answer • Doppelader • Double Action • Double Aisle • Double Amplitude • Draft Addendum • Drift Angle • Drive Actuator • Drive Assembly • Dummy Antenna • Dust Analyzer • Dutch Antilles

DAA Data Access Arrangement • Data Acquisition Application • Data Analysis Assistant • Data Analysis in Astronomy • Data Automation Activity • Data Availability Acknowledgment • Decimal Adjust for Addition • Decision Aid Application • Delegation of Approval Authority • Departamento de Astronomía y Astrofísica • Designated Accreditation Authority • Designated Accrediting Authority • Designated Approval Authority • Designated Approving Authority • Designed Assembly Area • Detailed Analysis of Alternatives • Devon Astronomical Association (UK) • Digital/Analog Adapter • Digital Automatic Acquisition • Direct Access Arrangement • Display Alternate Area • Division Administrative Assistant • Fort Belvoir - Davison (USA-VA) (location identifier)

DAAC Data Active Archive Center • Distributed Active Archive Center (*) • Division des Archives Audiovisuelles et Cartographiques

DAACS Dump Analysis And Consulting System

DAAD Deutscher Akademischer Austauschdienst (Germany)

DAAES Diplôme d'Aptitude à Accéder à l'Enseignement Supérieur (France)

DAAIS Danger Area Activity Information Service

DAAK Deutsch-Amerikanisches Akademisches Konzil

DAAM Dynamic Associative Access Memory

DAAR Day Air-to-Air Refueling

DAAS Data Acquisition and Analysis System • Department of Astronomy and Atmospheric Sciences

DAAT Distributed Architecture Analysis Tool

DAATS Digital Analog Automatic Test System

DAB Data Acquisition Bus • Data Announcement Bulletin • Datenanschaltbaugruppe • Daytona Beach (USA-FL) (location identifier) • Defense Acquisition Board • Dictionary of American Biography • Diffuse And Bright • Diffusion Absorption Band • Digital Audio Broadcast • Digital Audio Broadcasting • Display Attention Bit • Distributeur Automatique de Billets • Docking And Berthing

DABBLE Definitive Atmospheric Buoyant Boundary Layer Experiment

DABRK Day Break

DABS Demand-Assigned Bandwidth Service • Discrete Address Beacon System

DABU Data Bus Unit

DAC Data Acquisition and Control • Data Acquisition Camera • Data Acquisition Chassis • Data Acquisition Computer • Data Acquisition and Control • Data Acquisition Controller • Data Analysis and Control • Data Analysis Center • Data Analysis Console • Data Analysis Control • Data Archiving Centre • Data Assembly Centre • Date of Award of Contract • Days After Contract • Deductible Average Clause • Degenerated Angel Cross • Delayed Action Command • Demodulator and Access Controller •

Design Analysis Cycle • Design Augmented by Computer • Design Automation Conference • Development Assistance Committee • Dhaka (Bangladesh) (location identifier) • Diamond Anvil Cell • Digital Audio Copy • Digital-to-Analog Converter • Digital-to-Analytical Conversion • Digital-to-Audio Converter • Digital Arithmetic Center • Direct Access Capability • Direct-Access Computing • Direct Air Cycle • Directors Advisory Committee • Disaster Assistance Center • Discrete Absorption Component • Discretionary Access Control • Display Analysis Console • Distance Amplitude Correction • District Access Coordinator • Document Availability Code • DSMC Analysis Code • Dual Attached Concentrator • Dual Attachment Concentrator

DACA Days After Contract Award • Division des Archives Cartographiques et Audiovisuelles

DACB Data Acquisition and Control Buffer • Data Adapter Control Block

DACBU Data Acquisition and Control Buffer Unit • Digital Acquisition and Control Buffer Unit

DaCC Display and Computer Console

DACC Direct-Access Communications Channels

DACCS Digital Access Crossconnect System

DACD Digital Audio Compact Disk

DACE Department of Adult and Continuing Education • Design and Analysis of Computer Experiments

DACEMS Data Communications Equipment Monitoring and Switching

DACG Discriminant Analysis with Constellation Graph

DACI Direct Adjacent Channel Interference

DACK DMA Acknowledgment

DACM Data Adapter Control Mode

DACO Divisional Administrative Contracting Officer

DACON Digital-to-Analog Converter

DACOR Data Correction

DACP Data Acquisition and Control Processor • Data Acquisition, Control, and Processing

DACR Data Assimilation Research Center

DACS Data Acquisition and Control System • Data Acquisition Control System • Data and Analysis Center for Software • Dedicated Address Computation Stage • Design and Artists Copyright Society (UK) • Digital Access and Crossconnect Service • Digital Access and Crossconnect System • Digital Access Crossconnect Switch • Digital Access Crossconnect System • Digital Acquisition and Control System • Digital Auto-Correlation Spectroscopy • Digital-to-Analog Conversion System • Directory Assistance Charging System • Discrete-Address Communication System • Divert and Attitude Control System • Divert Attitude Control System

DACST Department of Art, Culture, Science and Technology (South Africa)

DACT Data Automated Communications Terminal • Digital Automated Communications Terminal • Disposable Absorption Collection Trunk

DACU Data Acquisition and Control Unit • Digital Azimuth Control Unit • Digital-to-Analog Converter Unit

D&AD Design & Art Direction

DA&D Data Acquisition & Distribution • Discrimination Analysis & Design

DAD Da Nang (Viet Nam) (location identifier) • Data Access Diagram • Data Acquisition and Distribution • Data Acquisition Division • Deployment Assistance Device • Deputy Assistant Director • Design and Art Direction • Design Approval Data • Desktop Application Director • Detector Array Development • Diode-Array Detector • Double Amplitude Displacement • Draft Addendum • Driveability Analysis Data • Dual Air Density • Due At Department

DADA Distributed Astronomical Data Archive

DADAS Directory Assistance Direct Access Service

DADB Data Analysis Database

DADC Destination Area Distribution Center • Digital Air Data Computer • Digital Audio Disk Corporation

DADD Data Archived and Distribution Service

DADE Data Acquisition and Decommutation Equipment • Data Acquisition and Documentation Equipment • Dual Air Density Explorer

DADEE Dynamic Analog Differential Equation Equalizer

DADEMS Data Administration Data Element Management System

DADLO Démarche d'Amélioration Des Logiciels

DADMS DMA Automated Distributed Management System

DADP Devil's Advocate Decision Program

DADS Data Acquisition and Distribution System • Data Archive and Distribution Service • Data Archive and Distribution System (*) • Déclaration Annuelle de Données Sociales (France) • Digital Audio Distribution System • Digital Assisted Database System • Digital Automated Data System • Distributed Analysis for Decision Support • Document Availability and Distribution Service • Dual Air Density Satellite • Dynamic Analysis and Design of Systems • Dynamic Analysis and Design System

DADSM Direct-Access Device Space Management • Direct-Access Device Space Manager

DAdW Deutsche Akademie der Wissenschaften (Germany)

DA&E Data Analysis & Evaluation

DAE Daparizo (India) (location identifier) • Data Acquisition Equipment • Data Analysis and Evaluation • Data Automation Equipment • Department of Atomic Energy • Dictionary of American English • Differential Algebraic Equation (*) • Differential and Algebraic Equation • Digital Audio Extraction • Distributed Application Environment • Distributed Automation Edition • Dump Analysis and Elimination • Dynamic Asian Economies

DAEC Data Analysis and Evaluation Center • Département d'Astrophysique Extragalactique et de Cosmologie (Observatoire de Paris-Meudon)

DAEE Departamento de Águas e Energia Eléctrica

DAEF Data Analysis and Evaluation Facility

DAEI Direction des Affaires Économiques et Internationales

DAEMON Disk And Execution Monitor

DAEO Designated Agency Ethic Official

DAES Division des Actions Éducatives et de la Scolarité

DAET Délégué Académique à l'Enseignement Technique (France)

DAEU Diplôme d'Accès aux Études Universitaires (France)

DAF Data Acquisition Facility • Data Analysis Facility • Daup (Papua New Guinea) (location identifier) • Described As Follows • Destination Address Field • Direction des Affaires Financières • Directory Access Function • Discard At Failure • Dissolved Air Flotation • Document Absolute Frequency

DAFC Digital Automatic Flight Control • Digital Automatic Frequency Control

DAFCO Délégation Académique à la Formation Continue (France)

DAFCS Digital Automatic Flight Control System

DAFE Division des Affaires Financières et des Études

DAFI Documents And Forms Inventory

DAFM Discard-At-Failure Maintenance

DAFNE Double Annular Factory for Nice Experiments

DAFT Data Acquisition Frequency Table • Digital-to-Analog Function Table

dag Decagram

DAG Barstow Daggett (USA-CA) (location identifier) • Daggett Barstow (USA-CA) (location identifier) • Data Analysis Group • Data Authentication Group • Decision Action Group • Development Assistance Group • Direct Acyclic Graph • Directed Acyclic Graph • Direction de l'Administration Générale • Direktrufnetzabschlußgerät • Documentation Advisory Group

DAGC Delayed Automatic Gain Control • Digital Automatic Gain Control • Drawing And Graphic Committee

DAGEFI Division des Affaires Générales et Financières

DAGEN Division des Affaires Générales de l'Éducation Nationale

DAGH Distributed Adaptive Grid Hierarchy

DAGIC Division des Affaires Générales Internationales et de la Coopération

DAGS Dartmouth Institute for Advanced Graduate Studies • Data Analysis and Graphing Software • Dynamic Airspace Management System

DAGt Datenanschlußgerät

DAH Dathina (Yemen) (location identifier)

DAHAP Data Acquisition, Handling and Analysis Plan • Digital Archive of Historical Astronomy Pictures

DAHNI Demonstrator on the Access Hypermedia Non-visual Interface

DAHW Deployable Adaptive Hardware

DAI Damage Assessment Indicator • Darjeeling (India) (location identifier) • Data Adapter Interface • Data-stream Access Interface • Days After Installation • Deferred Acceptance Item • Dictionary of Astronomical Instrumentation • Deployment Action Item • Digital Audio Interface • Discrete-Activity Indicator • Dissertation Abstracts International • Distributed Artificial Intelligence

DAIC Data Acquisition and Instrument Control

DAID Data Applications and Interactive Display

DAII Distributed Adaptive Information Infrastructure

DAIM Data Analysis Information Memorandum • Dynamic Active Index Matrix

DAIR Direct Altitude and Identity Readout • Direct Attitude and Identity Readout • Driver Aid, Information and Routing • Dynamic Allocation Interface Routine

DAIRS Dial Access Information Retrieval System • Distributed Aperture Infrared System

DAIS Data Adapter Interface Software • Data Avionics Information System • Digital Airborne Imaging Spectrometer • Digital Avionics Information System • Digital Avionics Instruction Set • Distributed Applications and Interoperable Systems (conference) • Distributed Automatic Intercept System • Document Analysis and Information System

DAISY Data Acquisition and Interpretation System • Data Acquisition Information System • Data Acquisition Integrated System • Disposal Automated Information System • Double-precision Automatic Interpretive System

DAIU Digital-to-Analog Interface Unit

DAJ Dauan Island (Australia-QLD) (location identifier) • Direction des Affaires Juridiques

DAJL Direction de l'Animation de la Jeunesse et des Loisirs

Dak Dakota (USA)

DAK Double-Adding KNMI

dal Decaliter

DAL Dallas - Fort Worth - Love (USA-TX) (location identifier) • Data Abstraction Layer • Data Accessions List • Data Access Language • Data Access Layer • Data Access Line • Data Acquisition Language • Data Acquisition List • Data-Aided Loop • Data Analysis Laboratory • Data Analysis Language • Database Access Language • Dedicated Access Line • Detached Address Label • Director of Acquisition Logistics • Disk Access Lockout • Dissemination Authority List • Distribution Authority List • Drahtlose Anschlußleitung • Drawing Assembly List

DALATS Data Logging And Transmission System

DALC Divided Access Line Circuit

DALF Diplôme Approfondi de Langue Française

DALGT Daylight

DALI Dispositif d'Accès et de Lecture de l'Inventaire • Document And Library Integration

DALO Disconnect At Lift-Off

DALR Dry Adiabatic Lapse Rate

DALS Data Acquisition Logging System

DALY Disability-Adjusted Life Year

dam Decameter

DAM Damascus (Syria) (location identifier) • Data Access Manager • Data Access Method • Data Acquisition and Monitoring • Data Addressed Memory • Data Analysis Management • Data Association Message • Decameter • Decametric • Decametric Radiation • DECT Authentification Module • Definition, Analysis and Mechanization • Descriptor Attribute Matrix • Detection And Mapping • Deutsche Akademie für Metrologie (Germany) • Device Adapter Module • Diagnostic Acceptability Measure • Digital-to-Analog Multiplier • Digitized Analog Magnetogram • Direct-Access Memory • Direct-Access Method • Distributed Associative Memory • Double Alumunized Mylar • Driver Amplifier Module • Dynamic Address Mapping • Dynamisch Adaptive Mehrfachträger

DAMA Dark Matter (search) • Data Administration Management Association • Demand Access Multiple Assignment • Demand-Assigned Multiple Access • Demand Assignment Multiple Access

DAMAp Département Atomes et Molécules en Astrophysique (Observatoire de Paris-Meudon)

DAME Data Acquisition and Monitoring Equipment • Distance Azimuth Measuring Equipment

DAMGM Direction des Affaires Maritimes et des Gens de la Mer (France)

DAMI Designated Airworthiness Maintenance Inspection

DAML DARPA Agent Markup Language • Digital Added Main Line

DAMMPS Damping And Metal Matrix for Precision Structure

DAMOP Division of Atomic, Molecular and Optical Physics (APS)

DAMOYR Day - Month - Year

DAMPS Digital Advanced Mobile Phone Service

DAMPT Department of Applied Mathematics and Theoretical Physics

DAMR Dynamic Adaptive Mesh Refinement • Dynamically Adaptive Mesh Refinement

DAMRI Département des Applications de la Métrologie des Rayonnements Ionisants (France)

DAMS Deployable Automated Maintenance System • Direct-Access Management System

DAMT Distributed Applications Monitor Tool

DAMTP Department of Applied Mathematics and Theoretical Physics

DAMUS Data Management and User Services

Dan Danish

DAN Danville (USA-VA) (location identifier) • Data Analyzer • Data Analysis Network • Data Availability Notice • Deutsches Astronetz (Germany) • Deutsches Astronomisches Netz (Germany)

DANDE Despin Active Nutation Damping Electronics

DANEOPS DLR - Archenhold Near-Earth Objects Prediscovery Survey

DANET Data Administration Network

DANIDA Danish International Development Agency • Danish International Development Assistance

DANOF Département d'Astronomie Fondamentale (Observatoire de Paris-Meudon)

DANTE Delivery of Advanced Network Technology to Europe

DANTIC Délégation Académique aux Nouvelles Techniques d'Information et de Communication (France)

DaNzKo Datennetzkoordinator

DAO Data Access Object • Data Acquisition Object • Data Assimilation Office (GSFC) • Defense Acquisition Organization • Department Administrative Order • Department - Agency - Organization • Dessin Assisté par Ordinateur • Dial Assist Operator • Disbursing Accounting Officer • Disc At Once • District Advisory Officer • Documentation Assistée par Ordinateur • Dominion Astrophysical Observatory (Canada-BC) • Duly Authorized Officer

DAOG Data Archive Operations Group

DAOPHOT DAO Photometric Package

DAP Daily Activity Plan • Darchula (Nepal) (location identifier) • Data Access Point • Data Access Processor • Data Access Protocol • Data Access Provider • Data Acquisition and Processing • Data Acquisition Plan • Data Acquisition Platform • Data Acquisition Processor • Data Acquisition Programme • Data Administration Program • Data Analysis Plan • Data Analysis Program • Data Associated Parameter • Data Automation Panel • Data Automation Proposal • Day of Arrival at POE • Deformation of Aligned Phases • Delayed Alpha Particle • Demande d'Admission Préalable • Descent Analysis Program • Design Analysis Phase • Designated Acquisition Program • Detail Assembly Panel • Deterministic Adaptable Priority • Developer Assistance Program • Diastolic Arterial Pressure • Differential Accelerometer Package • Digital Area Photometer • Digital Assembly Program • Digital Audio Player • Digital Autopilot • Direct Access Program • Directed Audit Program • Direction de l'Administration et du Personnel • Director of Administrative Planning • Directory Access Protocol • Display Adjust panel • Distant Aiming Point • Distributed Array Processor • Disturbed Astronomer Path • Division Advisory Panel • Do All Possible • Do Anything Possible • Document Application Profile • Documents Against Payment • Donor-Acceptor Pair • Double-Amplitude Peak • Draw A Person • Dynamic Adaptive Modeling

DAPA Direction de l'Architecture et du Patrimoine (France) • Distribution And Pricing Agreement

DAPAF Data Processing And Analysis Facility

DAPF Data Analysis and Processing Facility

DAPFS Direct-Ascent Powered-Flight Simulation

DAPHNE Document Application Processing in a Heterogeneous Network Environment

DAPI Developers Application Programming Interface

DAPIE Developers Application Programming Interface Extension

DAPM Domain Analysis Process Model

DAPMO Data Administration Program Management Office

DAPN Directional Antenna Phasing Network

DAPNAD Deterministic Adaptive Prioritized Network Access Delay

DAPNIA Département d'Astrophysique, de Physique Nucléaire et d'Instrumentation Associée

DAPP Data Acquisition and Processing Program

DAppSc Doctor of Applied Science

DAPR Digital Automatic Pattern Recognition

DAPS DCS Automated Processing System • Decommutation And Processing System • Direct-Access Programming System • Display and Acquisition Processing System

DAPSK Differential Amplitude Phase Shift Keying

DAPU Data Acquisition and Processing Unit

DAQ Data Acquisition

DAR Daily Activity Report • Dar-Es-Salaam (Tanzania) (location identifier) • Data Access Register • Data Acquisition Rack • Data Acquisition Recorder • Data Acquisition Request • Data-Aided Receiver • Data Automation Requirement • Daughters of the American Revolution • Decay At Rest • Decision Analysis Report • Defense Acquisition Regulation • Designated Agency Representative • Designated Airworthiness Representative • Deutscher Akkreditierungsrat (Germany) • Deviation Approval Request • Deviation Authorization Request • Differentiation with Asymmetrical Reinforcement • Digital Audio Recording • Digital Autopilot Requirement • Direct Absorption Receiver • Direct-Access Recorder • Direct Access Restore • Distortion Adaptive Receiver • Division of Atmospheric Research • Doppler Acoustic Radar • Drawing Analysis Record • Driver-Augmented Readout

DARA Deutsche Agentur für Raumfahrtsangelegenheiten GmbH (Germany)

DARAI Deutsche Arbeitsgemeinschaft für Raumfahrt und Astronomie Information (Germany)

DARC Data Acquisition and Reports Control • Delayed-Action Radar Calibrator • Département d'Astrophysique Relativiste et de Cosmologie (Observatoire de Paris-Meudon) • Description - Acquisition - Restitution - Correlation • Dirac Atomic Relativistic Code • Direct Access Radar Channel • Direct Access Radar Control • Document Archive Controller

DARD Data Acquisition Requirements Document

DARDC Device for Automatic Remote Data Collection

DARE Data Archival and Retrieval Enhancement • Data Automatic Reduction Equipment • Data Automation Research and Experimentation • Data Rescue • Demand And Resource Evaluation • Dictionary of American Regional English • Document Abstract Retrieval Equipment • Documentation Automated Retrieval Equipment • Doppler Automatic Reduction Equipment • Drug Abuse Resistance Education

DARES Data Analysis and Reduction System • Data Retrieval System

DARHF Dual-Axis Radiographic Hydrodynamic Facility

DARHT Dual-Axis Radiographic Hydrodynamic Test

DARI Database Application Remote Interface

DARIS Diffusion et Accès Rapide à l'Information Scientifique

DARLI Digital Angular Readout by Laser Interferometry

DARM Data Acquisition Room

DARMIS Data Requirements Management Information System

DARMS Data Automation Resource Management System • Digital Alternate Realization of Musical Symbols • Drifting Automatic Radiometeorological Station

DARN Dual Auroral Radar Network

DARO Days After Receipt of Order • Defense Airborne Reconnaissance Office (USA)

DARP Direct Access RTCS Pointer

DARPA Defense Advanced Research Projects Agency (USA) (now ARPA)

DARPMI Direction de l'Action Régionale et de la Petite et Moyenne Industrie (France)

DARR Delft Atmospheric Research Radar (Netherlands)

DARS Data Accumulating and Reporting Sheet • Data Acquisition Recording System • Data Acquisition and Reduction Specification • Data Acquisition and Reduction

System • Decommutation And Readout System • Digital Acoustic Receiver System • Digital Adaptive Recording System • Digital Attitude and Rate System • Digital Attitude Reference System • Digital Audio Radio Service

DART Daily Automatic Rescheduling Technique • Data Analysis and Review Team • Data Analysis Recording Tape • Data and Remote Control Terminal • Data Retrieval Tool • Decision-Aided Rescheduling Tool • Decomposed Ammonia Radioisotope Thruster • Deficiency Analysis Ranking Technique • Deployable Automatic Relay Terminal • Deutsche Amateur-Radioastronomische Tagung (Germany) • Dexterous Anthropomorphic Robotic Testbed • Dial-A-Ride-Transportation • Digital Audio Reconstruction Technology • Direct Access Repository Toolkit • Direct Advisory of Recorded Transactions • Directional Automatic Realignment of Trajectory • Director And Response Tester • Disaster Assistance Response Team • Display for Advanced Research Technology • Dual Anamorphic Reflector Telescope • Dual Axis Rate Transducer • Dublin Area Rapid Transit (Ireland) • Dust Accumulation and Removal Tool • Dynamic Acoustic Response Trigger • Dynamically Adaptive Receiver Transmitter • Dynamic Analysis and Replanning Tool • Dynamic Analytical Replanning Tool • Dynamic Analytic Replanning Tool

DARTE Discrete Angle Radiative Transfer Equation

DARTFire Diffusive and Radiative Transport in Fire

DARTS Data Archive and Transmission System • Destination Acceptance Reporting and Tracking System • Digital Automated Radar Tracking System • Digital Azimuth Range Tracking System • Dynamics Algorithms for Real-Time Simulation

DARWIN Design Assessment of Rotors With Inspection

DAS Dairo Air Services • Dakota Astronomical Society (USA-ND) • Data Access Security • Data Acquisition Station • Data Acquisition System • Data Analysis Software • Data Analysis Station • Data Analysis System • Data Archive Server • Data Assimilation System • Data Automation System • Data Availability Schedule • Dataset Attribute Structure • Decimal Adjust for Subtraction • Deflection Amplifier System • Delaware Astronomical Society (USA-DE) • Denver Astronomical Society (USA-CO) • Department of Administrative Services • Delivered Alongside Ship • Density Analysis System • Deputy Assistant Secretary • Design, Analysis, and Simulation • Detailed Activity Schedule • Development Advisory Service • Dial Access System • Differential Absorption and Scattering • Digital Address System • Digital Analog Simulator • Digital Angiography Subtraction • Digital Announcement System • Digital Autocorrelation Spectrometer • Digital Avionics System • Dimer Adatom Stacking (model) • Dipole Antenna System • Direct Access Service • Direct Access Subscriber • Direct Access System • Direct Admissions System • Direct Air Support • Direct Attached Storage • Directory Access Service • Directory Assistance System • Distributed Authentication Service • Distributor And Scanner • Disturbance Analysis System • DME-based Azimuth System • Document Analysis System • Documentation Accountability Sheet • Doppler Acoustic Sounder • Double Algebraic Sum • Dual-Attached Station • Dual-Attach Station • Dual Attachment Station • Dundee Astronomical Society (UK-Scotland) • Dust-Acoustic Structure • Dwingeloo Autocorrelation Spectrometer • Dynamically Assigned Socket

DASA Deutsche Aerospace AG (now part of EADS) • Dual Aerospace Servo Amplifier • Dual Aerosurface Servo Amplifier

DASAT Datenübertragung über Satellit

DASC Deep Atmosphere Spectral Camera

DASD Data Acquisition Support Document • Direct-Access Storage Device • Directly Accessible Storage Device

DASDL Data And Structure Definition Language

DASE Differential Absorption and Scattered Energy • Differential Absorption and Scattering Equipment

DASF Direct-Access Storage Facility

DASGal Département d'Astronomie Stellaire et Galactique (Observatoire de Paris-Meudon)

DASH Data Analysis Software Hierarchy • Data Analysis System Hierarchy • Deployable Automation Support Host • Digital Audio-recorder with Stationary Head • Display And Sight Helmet • Distributed Analysis Software Hierarchy • Distributed Analysis System Hierarchy

DASI Degree Angular Scale Interferometer • Digital All-Sky Imager • Digital Array Scanned Interferometer

DASL Data Access System Language • Data Approximation Subroutine Library

DA&SM Data Acquisition & Systems Maintenance

DASM Data Acquisition System Manager • Design and Analysis Support Model

DASO Demonstration And Shakedown Operations

DASOC Disturbance Accommodation Standard-Deviation Optimal Controller (HST)

DASOP Département d'Astronomie Solaire et Planétaire (Observatoire de Paris-Meudon)

DASP Discrete Analog Signal Processing

DASR Data Acquisition Statistical Recorder • Digital Airport Surveillance Radar • Direct Air-to-Satellite Relay

DASS Data Analysis Support Staff • Data Archive and Science Support • Defensive Aids Subsystem • Demand Assignment Signaling and Switching • Digital Acoustic Sensor Simulator • Direct Access Secondary Storage • Direction des Affaires Sanitaires et Sociales • Distributed Authentication Security Service • Doppler Acoustic Sounding System • Double Aperture Scintillation Sensor

DASSS Decentralized Automated Service Support System

DASTCOM Database of Asteroids and Comets

DAS3 Decentralized Automated Service Support System

dat Data (file extension)

Dat Date • Datif • Dative • Dativo • Datum

DAT Daily Activity Team • Data • Data Acceptance Test • Data Acquisition Terminal • Data Acquisition Test • Data Analysis Tool • Data-bus Analysis Tool • Datong (China-PRC) (location identifier) • Déclaration d'Achèvement des Travaux • Deployment Action Team • Design Acceptance Test • Design Approval Test • Design, Assembly and Test • Desktop Analysis Tool • Detail Assembly Template • Detailed Assessment Tool • Development Acceptance Test • Digital Archival Tape • Digital Audio Tape • Direct-Ascent Threat • Disconnect Actuating Tool • Disk Array Technology • Docking Alignment Target • Dummy Alias Table • Duplicate Address Test • Dynamic-Address Translation

DATA Decision Aids for Target Aggregation • Defense Air Transport Administration (USA) • Direct-Access Terminal Application • Distribution and Automation Technology Advancement

DATAC Data Analog Computer

DATACOM Data Communications

DATAMS Data Analysis, Test and Mission Software

DATAN Data Analysis

DATAR Délégation à l'Aménagement du Territoire et à l'Action Régionale (France) • Digital Autotransducer And Recorder

DATDE Direction de l'Aménagement du Territoire et du Développement Économique (France)

DATE Decision Aids Test Environment • Dedicated Access to (X.25) Transport Extension • Design, Automation, and Test in Europe (conference) • Dynamics, Acoustics, and Thermal Environment

DATEL Data Telecommunications

DATEX Data Exchange

DAtF Deutsches Atomforum (Germany)

DATICO Digital Automatic Tape Intelligence Checkout

DATINT Data Intelligence

DATIS Digital Airborne Topographic Imaging System

DATOS Detection And Tracking Of Satellites

DATP Detailed Acceptance Test Procedure

DATPE Direction de l'Aménagement du Territoire et de la Protection de l'Environnement (France)

DATPG Digital Automatic Test Pattern Generator

DATR Design Acceptance Test Report • Design Approval Test Report

DATS Data Acquisition and Transmission System • Detail Acceptance Test Specification • Dynamic Accuracy Test Set • Dynamic Accuracy Test System

DATSS Development And Test Support System

DATTS Data Acquisition, Telecommand and Tracking Station • Data Acquisition, Tracking and Telecommand Station

DATU Data Adapter Termination Unit • Direct Access Testing Unit

DATV Digitally-Assisted Television

Dau Daughter

DAU Daru (Papua New Guinea) (location identifier) • Data Acquisition Unit • Data Adapter Unit • Data Arithmetic Unit • Delta Aurigids (meteor stream) • Digital Adapter Unit • Digital Amplifier Unit • Digital-Analog-Umsetzer • Display Adapter Unit • Document Administratif Unique • Dümmster Anzunehmender User

DAUF Degenhardt Amateur User Facility (USA-TN)

DAV Data Above Voice • Data Available • Data Valid • David (Panama) (location identifier) • Deutscher Autorenverband (Germany) • Digital Audio/Video

DAVC Delayed Automatic Volume Control

DAVD Directional Acceleration Vector-Driven • Directional Acceleration Vector-driven Displacement

DAVE Do Anything Very Easily

DAVES Data/Analog Voice Expansion Shelf

DAVI Drowned Ancient Volcanic Island • Dynamic Antiresonant Vibration Oscillator

DAVIC Digital Audio and Video Council • Digital Audio Video Council • Digital Audiovisual Council

DAVID Digital Audio/Video Interactive Decoder • Distributed Access View Integrated Database

DAVP Doctrinal Audiovisual Program

DAW Data Analysis Workshop • Deutsche Akademie der Wissenschaften (Germany) • Digital-Analog-Wandler • Digital Audio Workstation • Document Archive Writer • Dry Active Waste • Dust-Acoustic Wave

DAWCLM Data Automation Workload Control and Library Maintenance

DAWG Data Acquisition Working Group • Data Analysis Working Group • Delivery Accuracy Working Group

DAWID Device for Automatic Word Identification and Discrimination

DAWMS Deep Attack Weapons Mix Study

DAWN Digital Automatic Weather Network

DAWNS Digital Automatic Weather and NOTAM System

DAWS Data Analysis Workstation • Digital Advanced Wireless Service

DAX Daxian (China-PRC) (location identifier) • Deutsche Aktienindex

DAY Dayton - James M. Cox (USA-OH) (location identifier)

DAZ Darwaz (Afghanistan) (location identifier)

DAZD Double Anode Zener Diode

db Database (file extension)

dB Decibel

Db Dobra (currency)

D-B Decimal to Binary

D/B Daybook • Decimal to Binary • Disk to Bulge • Docking/Berthing

DB Dagelijks Bestuur • Daily Bulletin • Daimler-Benz • Dark Band • Data Bank • Database • Database Browser • Data Block • Data Buffer • Data Bus • Datenbank • Daybook • Deadband • Decibel • Delayed Broadcast • Design Baseline • Design Bureau • Designierte Brücke • Detached Binary (star) • Detailed Balance • Detection Buffer • Deutsche Bibliothek (Germany) • Deutsche Bundesbahn (Germany) • Deutsche Bundesbank (Germany) • Diffuse Background • Diffusion-Bonded • Diffusion Bonding • Digital Block • Diode Bus • Direct Beam • Direct Budget • Directing Board • Direct Broadcast • Dirty Beam • Disparition Brusque • Display Buffer • Disposition Board • Distribution Board • Distribution Box • Document Base • Double Base • Double Bed • Double Biased • Double Bottom • Double Break • Drifting Buoy • Dunkle Bänder • Duschbad • Dry Bulb • Dynamic Braking

d/b/a Doing Business As

dBa Decibel Adjusted

DBA Adjusted Decibel • Dalbandin (Pakistan) (location identifier) • Dassault Aviation Belgique (Belgium) • Database Accelerator • Database Access • Database Accessory • Database Administration • Database Administrator • Datenbankadministrator • Derwent Biotechnology Abstracts • Design Basis Accident • Diode Box Assembly • Direct Budget Authority • Doctor of Business Administration • Doing Business As • Doppelbesteuerungsabkommen • Double-Bend Achromat • Drum Buffer Assembly

DBAAM Disk Buffer Area Access Method

DBAC Database Administration Center

DBAd Database Administrator

DBAF Database Access Facility

DBAM Database Access Method • Database Access Module

DBAO Digital Block AND-OR (gate)

DBAS Database Administration System

DBASI Digital Barometer Altimeter Setting Indicator

DBAT Database Access Tool

DBAU Doing Business As Usual

DBAWG Database Administration Working Group

DBB Database Browser • Detector Back Bias • Detector Balanced Bias • Disk Black Body • Dynamic Bass Boost

DBC Database Computer • Database Controller • Data Buffer Coupler • Data Bus Control • Data Bus Coupler • Diameter Bolt Circle • Differential Box Counting • Digital-to-Binary Converter • Distributed Batch Controller

DBCC Database Consistency Check

DBCL Database Command Language

DBCO Digital Block Clock Oscillator

DBCP Drifting Buoy Cooperation Panel

DBCR Database Change Request

DBCS Delivery Bar Code Sorter • Double-Byte Character Set

DBCU Data Bus Control Unit

DBd Disparition Brusque Dynamique

DBD Database Description • Database Descriptor • Database Design • Database Document • Database Driver • Dead Band • Detailed Block Diagram • Detailed Budget Decision • Dhanbad (India) (location identifier) • Double-Base Diode • Double Beta Decay

DBDA Database Design Aid • Design Basis Depressurization Accident

DB-DB Database to Database

DBDB Digital Bathymetric Database

DBDD Database Design Document

DBDF Data Block Definition File

DBDL Database Definition Language

DBDQ Distributed Queue Dual Bus

DBDR Database Design Review

DBDS Database Design Specification

DBDU Digital Bargraph Display Unit

DBE Database Element • Database Engine • Database Entry • Data Bus Element • Data Bus Enable • Deep Basin Experiment • Design-Basis Event • Deutsche biographische Enzyklopädie • Digital Backend • Digital Based Exciter • Disadvantaged Business Enterprise

DBEATS Dispatch Payable Both Ends All Time Saved

DBELTS Dispatch Payable Both Ends on Lay Time Saved

DBES Database and Expert System Program

dbf dBase (file extension)

DBF Database File • dBase Format • Demodulator Band Filter • Design Basis Fault • Digital Beam Forming • Direction des Bibliothèques de France

DBFF Digital Block Flip-Flop

DBFN Data Base File Number • Data Bus File Number

DBFW Digital Fly-By-Wire

dbg Debugging Data (file extension)

DBG Database Generation • Database Generator • Data Bus Group • Bebug • Deutsche Bunsengesellschaft (Germany) • Diversified Business Group

DBGF Database Generation Facility

DBGMP Data Bus Generation and Maintenance Package

DBGS Database Generation System

DBH Database Handler • Diameter at Breast Height

DBHS Database Handling System

DBI Database Interface • Data Bus Interface • Decomposition-Based Induction • Design Basic Incident • Deutsches Bibliotheksinstitut (Germany) • Digital Bus Interface • Division of Biological Infrastructure (NSF) • Drag Balance Instrument • Dull But Important

DBIA Data Bus Interface Adapter • Data Bus Isolation Amplifier • Digital Block Inverter Amplifier

DBIF Database Interface

DBIL Database Input Language

DBIMA Database of Images

DBIN Description Bibliographique Internationale Normalisée

DBIO Database Input/Output

DBIS Discoveries and Breakthroughs Inside Science

DBIU Data Bus Interface Unit

dbk Database Backup (file extension)

Dbk Drawback

DBK Data Bank • Database Kernel • Dominant Battlespace Knowledge

Dbl Double

DBL Damage Before Launch • Database Language • Database List • Database Load • Deuterium Burning Limit

Dblr Doubler

dbm Database (file extension)

dBM Decibels above one Milliwatt

D/BM Docking/Berthing Mechanism

DBM Database Machine • Database Management • Database Manager • Database Mining • Data Buffer Memory • Data Buffer Module • Data Bus Monitor • Debra Marcos (Ethiopia) (location identifier) • Decibel Meter • Décision Budgétaire Modificative • Detailed Balancing Model • Diploma in Business Management • Docking Berthing Mechanism

DBMC Data Bus Monitor/Controller • Destination Bulk Mail Center

DBMD Database Maintenance Daemon • Digital Burst Message Device

DBME Database Management Environment

DBMIST Direction des Bibliothèques, des Musées et de l'Information Scientifique et Technique (France)

DBML Database Management Language • Diabetic Meal

DBMP Database Management Plan

DBMR Database Management Report

DBMS Database Management Software • Database Management System (*)

DBMV Digital Block Multivibrator

Dbn Durban (South Africa)

DBN Data Bus Network • Design By Numbers • Dublin (USA-GA) (location identifier)

DBNN Difference Boosting Neural Network

DBO Dual Beam Oscilloscope • Dubbo (Australia-NSW) (location identifier)

DBOI Development Basis Of Issue

DBOMP Database Organization and Maintenance Processor

DBOR Database Of Record

DBOS Disk-Based Operating System

dBP Decibels above one Picowatt

DBP Debepare (Papua New Guinea) (location identifier) • Design Baseline Program • Deutsche Bundespost (Germany) • Division of Biological Physics

DBPA Decentralized Blanket Purchase Agreement

DBPC Database Processing Center

DBPL Database Programming Language

DBPSK Differential Binary Phase Shift Keying

DBpSM Decibels per Square Meter

DBPSM Decibels Per Square Meter

DBQ Database Query • Document-Based Query • Dubuque (USA-IA) (location identifier)

DBQL Database Query Language

DBR Database Retrieval • Datenbankrechner • Deterministic Bit Rate • Development Baseline Review • Distributed Bragg Reflector • DOS Boot Record • Dual-Band Radar

DBRA Dynamic Bandwidth Resource Allocation

DBRAD Database Relational Application Directory

DBRC Database Recover Control

DBRD Dépense Brute de Recherche et de Développement

Dbre Diciembre

DBRITE Digital Bright Radar Indicator Tower Equipment

DBRM Database Resource Management

dBRN Decibels above Reference Noise

DBRN Data Bank Release Notice

DBRTD Double Barrier Resonance Tunneling Diode

dbs Database (file extension)

DBS Database Server • Database Specification • Database Storage • Database System • Data Bus Simulator • Deep Brain Stimulation • Deep Brain Stimulator • Définition des Besoins du Système • Demand Broadcast System • Design Build Standard • Dibromostyrene • Diffraction

Back-Scattering • Digital Book System • Digital Broadcasting by Satellite • Direct-Broadcasting Satellite • Direct-Broadcasting Service • Direct-Broadcast Satellite • Direct-Broadcast Service • Direct-Broadcast System • Diversity-Based Summarization • Diversity-Based Summarizer • Docking Berthing System • Doppler Beam Sharpening • Doppler Beam Swinging • Double Beam Spectrophotometer • Double-Beam Switch • Dual-Beam Spectrograph • Dubois (USA-ID) (location identifier) • Duplex Bus Selector

DBSA Database and System Administration

DBSC Digital Block Slave Clock

DBSCAN Density-Based Spatial Clustering of Applications with Noise

DBSM Database System Manager • Decibels per Square Meter

DBSP Double-Based Solid Propellant

DBST Digital Block Schmitt Trigger • Double British Summer Time

dbt Database Text (file extension)

DBt Disparition Brusque Thermale

DBT Database Transfer • Database Transmission • Debra Tabor (Ethiopia) (location identifier) • Design Built Team • Deutsche Bundespost Telekom (Germany) • Disparition Brusque Thermale • Double-Base Transistor • Dual-Buffer Tritation • Dynamique et Bilan de la Terre

DBTF Database Task Force

Dbtfl Doubtful

DBTG Database Task Group

DBTT Ductile-to-Brittle Transition Temperature

dBU Decibel Unit

DBU Data Bus Unit • Deutsche Bundesstiftung Umwelt (Germany) • Digital Bus Unit • Direct Business Unit • Disk Buffer Utilization

DBUR Data Bank Update Request

DBUT Database Update Time

dBV Decibels above one Volt • Deutscher Bibliothekverbund (Germany) • Dubrovnik Cilipi (Croatia) (location identifier)

DBV Database Variable

DBVS Datenbankverwaltungssystem

dBW Decibel reference to one Watt

DBW Data Bus Wire • Design Bandwidth • Differential Ballistic Wind

DBWA Design By Walking Around

DBWR Database Writer

DBX Digital Branch Exchange • Distributed Branch Exchange

DBY Dalby (Australia-QLD) (location identifier)

DB2I Database-to-Database Interface

dc Duty Cycle

dC Después de Cristo • dopo Cristo • Dwarf Carbon (star)

D&C Display & Console • Display & Control • Distribution & Control

D/C Displays/Controls • Down-Converter

DC Da Capo • Dairy Cattle • Damage Control • Data Camera • Data Cell • Data Center • Data Channel • Data Check • Data Classifier • Data Code • Data Collection • Data Communication • Data Compressor • Data Concentrator • Data Control • Data Conversion • Data Coordinator • Dead Center • Dearborn Catalog • Death Certificate • Decimal Classification • Decimal Currency • Decision Conferencing • Decreto • Dedicated Circuit • Default Colour • Define Constant • Definitive Contract • Degré de Condensation • Degree Centigrade • Degree of Condensation • Delayed Coincidence • Delete Character • Delta Clipper • Dense Component • Deployment Configuration • Deputy Chief • Deputy Commissioner • Deputy Consul • Deputy Counsel • Derecho Civil • Design Change • Design Contractor • Desk Checking • Detector Controller • Developed Country • Developing Country • Development Center • Development Characteristic • Development Cost • Device Control • Device Coordinates • Diagnostic Centre • Dialogue Controller • Dictionary of Canadianisms • Dienstcommissie • Differential Calculus • Differential Correction • Diffuse Cloud • Digital Clock • Digital Code • Digital Comparator • Digital Computer • Diners Club • Diplomatic Corps • Direct Cite • Direct Command • Direct Current • Direct Cycle • Directional Control • Disconnect Confirm • Disk Controller • Display and Console • Display and Control • Display Computer • Display Console • Display Control • Display Controller • Display Coupler • Displays/Controls • Disque Compact • Dissimilarity Coefficient • Dissipative Collapse • Distributed Center • Distributed Component • District Chief • District Commissioner • District Council • District of Columbia (USA) • Division Chief • Double Column • Double Compact • Double Contact • Double Coronagraph • Drift Chamber • Drift Correction • Dry Cell • Dry Chemical • Duty Cycle • Golden Air Flyg AB (obsolete?)

DC&A Data Collection & Analysis

DCA Data Capture Agent • Data Collection Assignment • Data Correction Amplifier • Debit-Credit Account • Decade Counting Assembly • Defense Communications Agency (USA) • Deflection Coil Amplifier • Delta Cancrids (meteor stream) • Department of Civil Aviation • Design Change Authorization • Detailed Configuration Accounting • Detector Control Assembly • Detrended Correspondence Analysis • Development Cost Avoidance • Digital Command Assembly • Digital Communications Associates Inc. (USA-GA) • Direct-Current Amplifier • Director of Community Activities • Distributed Communications Architecture • Distributed Computing Architecture • Distribution Control Assembly • Document Content Architecture • Document Control Architecture • Double Conversion Adapter • Double Coulomb Approximation • Double Cyclotron Absorption • Drift Correction Angle • Droplet Combustion Apparatus • Dynamic Channel Allocation • Ronald Reagan Washington National (US-A-DC) (location identifier)

DCAA Defense Contracting Audit Agency (USA) • Dual Automatic Answer Device

DCAC Defense Communication Agency Circular • Define and Control Airplane Configuration

DCAF Distributed Console Access Facility

DCAG Data Collection and Analysis Group

DCAI Direct-Current Analog Input

DCAJ Direction des Contrats et des Affaires Juridiques

DCAM Data Communications Access Method • Department of Civil Aviation and Meteorology (Qatar) • Direct Chip Attach Module • Dynamique et Climat de l'Atmosphère Moyenne

DCAMIP Data Center for Atomic and Molecular Ionization Processes

DCAMPS Deployable Computer-Aided Mission Planning System

DCAO Digital Card And-Or Gate

DCAP Data Capture And Presentation • Dynamic Controls and Analysis Package

DCAR Danish Centre for Atmospheric Research • Data Collection Analysis Request • Design Corrective Action Report • Discrepancy and Corrective Action Report

DCARS Data Collection And Reduction System

DCAS Database Commitment Accounting System • Data Collection and Analysis System • Defense Contract Administration Service

DCASMA Defense Contract Administration Service Management Area

DCASR Defense Contract Administration Service Region

DCASS Digital Cartographic Software System

DCAT Developmental Comparative Active Telescope

DCATS Deployable Contract Action Tracking System

DCATT Data Certification And Technology Transfer

DCAWG Data Collection and Analysis Working Group

DCB Damage Control Book • Data Channel Bank • Data Control Block • Data Control Board • Data-set Control Block • Device Control Block • Design Change Board • Design Control Board • Development Component Baseline • Device Control Block • Differential Code Bias • Digital Continuum Backend • Disk Coprocessor Board • Double Cantilever Beam • Drawing Configuration Baseline • Dynamic Component Bulletin

DCBS Doubly-Compressed Bit Slices

DCBU Digital Conference Bridge Unit

DCBX Distributed Computerized Branch Exchange

DCC Damage Control Center • DAQ and Control Computer • Data Collection Center • Data Communications Channel • Data Communications Computer • Data Compression Conference • Data Computation Complex • Data Condition Code • Data Control Console • Data Country Code • Dedicated Control Centre • Denial Cause Code • Descriptive Cataloguing Committee • Design Change Control • Design Concept Change • Development Control Center • Device Control Character • Differential Chain Code • Digital Command Control • Digital Communications and Control • Digital Communications Control • Digital Communications Controller • Digital Compact Cassette • Digital Control Computer • Direct Cable Connection • Direct Client Connection • Direct Commercial Contract • Discrete Cell Control • Discrimination and Control Computer • Display Channel Complex • Display Combination Code • Display Computer Control • Display Control Console • Distributed Call Center • Distribution Control Center • District Communications Center • Division Computer Center • Documentation Coordinating Committee • Document Control Center • Dynamic Component Change

DCCC Data Communication Control Character

DCCF Discrete Cross-Correlation Function • Distance Classifier Correlation Filter

DCCH Dedicated Control Channel • Digital Control Channel

DCCI Distributed Computing Communications Interface • Distributed Computing Configuration Item

DCCLA Direct Current Closure Adapter

DCCO Defense Commercial Communications Office (USA) • Digital Card Clock Oscillator

DCCP Design Change Control Program • Digital Computer Control Panel • Distributed Call Control Protocol

DCCR Display Channel Complex Rehost

DCCS Data Communications Control System • Digital Camera Control System • Digital Command Communications System • Digital Comparator Correlator System • Digital Crossconnect System • Discontiguous Shared Segments • Distributed Command and Control System • Dynamic Cell Culture System

DCCU Data Communications Control Unit • Data Correlation Control Unit • Decommutator Conditioning Unit • Digital Cluster Control Unit • Digital Command and Control Unit • Digital Communications and Control Unit

Dcd Decode

DCD Data Carrier Detect • Data Collection Device • Data Compression Device • Decimal Classification Division • Deflection Coil Drive • Design Change Document • Design Control Document • Desired Completion Date • Differential Current Density • Dimensional Control Drawing • Diode - Capacitor - Diode • Document Content Definition • Document Control Division • Double Channel Duplex • Double-Crystal Diffraction • Drain-Current Drift • Dual Command Decoder • Dynamic Computer Display

DCDB Digital Cadastral Database • Digital Cartographic Database • Domain Controller Database

DCDD Design Characteristics Definition Document

DCDG Diode-Capacitor-Diode Gate

DCDL Digital Control Design Language

DCDM Digitally Controlled Delta Modulator

DCDOS Deputy Chief Data and Operations Scientist

Dcdr Decoder

DCDS Digital Cartographic Data Standard • Digital Control Design System • Digital Countdown Display System • Distributed Computing Design System

DCDSTF Digital Cartographic Data Standards Task Force

DCDU Data Collection and Distribution Unit

DCE Data and Control Electronics • Data Carrier Equipment • Data Circuit-terminating Equipment • Data Circuit-termination Equipment • Data Collection Effort • Data Collection Event • Data Communications Equipment • Data Compression Equipment • Data Conversion Equipment • Department of Conservation and Environment • Deployable Cargo Element • Deployment Control Electronics • Despin Control Electronics • Detector Control Electronics • Device Characteristics Evaluation • Device Control Entry • Diffusion-Convection Equation • Digital Communications Equipment • Digital Communications Experiment • Diploma in Chemical Engineering • Direction des Centres d'Expertise • Distributed Communication Environment • Distributed Computer Environment • Distributed Computing Environment (*) • Distributed Computing Equipment • Doctor of Civil Engineering • Document-Centered Approach • Domestic Credit Expansion • Dossier de Consultation des Entreprises • Drive Control Equipment • Droplet Combustion Experiment • Dust Control Equipment

DCEA Digital Control Electronics Assembly

DCED Département de la Coordination, de l'Exploitation et du Développement • Distributed Computing Environment Daemon

DCEP Data Collection and Evaluation Plan

DCES Data Collection and Evaluation System • DSS Communications Equipment System

DCESS Danish Center for Earth System Science (Denmark)

DCF Data Capture Facility • Data Channel Filter • Data Collection Form • Data Communication Facility • Data Compression Facility • Data Control Facility • Data Correlation Facility • Data Count Field • Differential Counts Function • Digital Cartographic File • Discounted Cash Flow • Discrete Correlation Function • Dispersion-Compensated Fiber • Distributed Computing Facility • Distributed Coordination Function • Document Composition Facility • Document Control File • Dominica Cane Field (Dominican Rep.) (location identifier) • Driver Configuration File • Dynamic Coercive Force

DCFEM Dynamic Crossed-Field Electron Multiplication

DCFF Digital Card Flip-Flop

DCFG Direct-Current Free Gyro

DCFL Direct Coupled FET Logic

DCFO Data Collection and Field Observations

DCFP Draft Concept Formulation Package • Draft Concept Formulation Paper • Dynamic Crossed-Field Photomultiplier

DCFS Data Communications Functional Support

DCG Data Control Group • Data Coordination Group • Definite Clause Grammar • Design Control Group • Dichromated Gelatin • Diode-Capacitor Gate • Direct-Current Generator • Discounted Cumulated Gain • Discounted Cumulative Gain (*) • Double-Current Generator • Dwarsverband Commissie Gewasbescherming

DGCE Digital Command Guidance Electronics

DCGT Direction et Contrôle des Grands Travaux

DCH D Channel Handler • Dedicated Channel • Direct Containment Heating

Dcha Derecha

DCHCI Distributing Computing Hardware Configuration Item

Dche Douche

DChE Doctor of Chemical Engineering

Dcho Derecho

DCI Data Capture Interface • Data Collection Initiative • Data Collection Integrator • Decimomannu Rafsu (Italy) (location identifier) • Defense Capability Initiative • Definite Clause Grammar • Definite Gaussian Grammar • Deliverable Contract Item • Dénomination Commune Internationale • Desorption Chemical Ionization • Dielectric Constant Indicator • Digital Clock Indicator • Digital Constant Image • Digital Constant Imaging • Direct Computer Input • Director of Central Intelligence • Display Control Interface • Documentation Change Instruction • Documentation, Control, and Integration • Document de Contrôle d'Interface • Domestic Collaborative Index • Double Column Inch • Dual Channel Interchange • Ductile Cast Iron • Dwarsverbandcommissie Informatica (Netherlands) • Rafsu Decimomannu (Italy) (location identifier)

DCIA Digital Card Inverting Amplifier

DCIB Data Communication Input Buffer

DCID Development Change In Design

DCILM Direct Computer Input Load Module

DCIM Display Control Interface Module

DCIP Data Correction Indicator Panel

DCIS Defense Criminal Investigative Service (USA) • Distribution Construction Information System • Downrange Computer Input System

DCIST Directory of Computerized Information in Science and Technology

DCIU Digital Control and Interface Unit

D-CIXS Demonstration of a Compact Imaging X-ray Spectrometer

DCK Dahl Creek (USA-AK) (location identifier)

Dckg Docking

DCKNG Docking

DCKP Direct Current Key Pulsing

Dcl Declaration

DCL Data Characterization Language • Data Checklist • Data Control Language • DEC Command Language • DEC Control Language • Declaration • Delphi Common LISP • Depôt Central du Livre (France) • Designate Command Line • Design Capability Line • Designer Choice Logic • Detailed Checklist • Detector Characterization Laboratory • Device Clear • Dickens - Currie - Lucey (catalog) • Digital Command Language • Digital Computer Laboratory • Digital Control Language • Digital Control Logic • Direct Communications Link • Direct-Coupled Logic • Doctor of Civil Law • Document Change List • Dynamic Characteristic Load

DCLC Drift Cyclotron Loss Cone

DCld Dark Cloud

DCLE Data Conversion Language Engine

DCLI Direct Connection Line Interface

DCLIR Dead Cat Lying In the Road

DCLM Department of Command, Leadership amd Management • Department of Conservation and Land Management

Dclr Decelerate

DCLS Data Collection and Location System • Dual Channel Line Scanner

DCLU Digital Carrier Line Unit

DCLZ Data Compression Lempel-Ziv (algorithm)

D/CM Diagnostic and Condition Monitoring

DCM Castres (France) (location identifier) • Data and Configuration Management • Data and Control Management • Data Channel Multiplexer • Data Circulation Model • Data Communication Mission • Data Communications Manager • Data Communications Module • Data Communications Multiplexer • Data Control Module • Decommutator Control Memory • Deputy Chief of Mission • Device Communication Manager • Diagnostic-Controlled Modem • Diagramme Couleur-Magnitude • Differential Code Modulation • Digital Carrier Module • Digital Cartographic Model • Digital Circuit Module • Digital City Map • Dimension Constraint Manager • Direction Cosine Matrix • Display and Control Module • Distributed Computer Model • Distributed Computing Manager • Double-Chirped Mirror • Double Crystal Monochromator • Duty Committee Member • Dwarsverbandcommissie Materialonderzoek (Netherlands)

DCMAO Defense Contract Management Area Operations

DCMB Development Configuration Management Board

DCMC Defense Control Management Center • Denis Catalogue towards Magellanic Clouds

DCMD Dose Calculation for Multiple Data

DCME Digital Circuit Multiplexing Equipment • Digital Circuit Multiplication Equipment

DC-MEF Disque Compact à Mémoire Fixée

DCMF Data Capture and Management Facility

DCMP Division of Condensed Matter Physics

DCMS Data Communications Management System • Data Control Multiplex System • Digital Capacitance Measuring System • Digital Communications Management System • Distributed Call Measurement System • District Cost Monitoring Specialist

Dcmsn Decommission

Dcmsnd Decommissioned

DCMT Danish-Chinese Meridian Telescope

DCMU Digital Concentrator Measurement Unit

DCMV Digital Card Multivibrator

DCN Data Center Network • Data Change Notice • Data Communications Network • Design Change Notice • Development Change Notice • Directeur de Commande Numérique • Direction des Constructions Navales (France) • Discontinued Component Notice • Distributed Computer Network • Document Change Notice • Document Control Number • Draft Change Notice • Drawing Change Notice

DCNA Data Communication Network Architecture

DCNP Document Change Notice Proposal

DCO Damage Control Officer • Data Collection Order • Data Control Office • Demande Clinique en Oxygène • Deputy Commanding Officer • Detailed Checkout • Dial Central Office • Digital Central Office • Digital Controlled Oscillator • Document Change Order • Drawing Change Order

d. col Double Column

DCOM Data Communications • Distributed Component Object Model

DCOMP Division of Computational Physics

DCOP Detailed Checkout Procedure • Displays, Controls and Operation Procedure

DCOS Data Communication Output Selector • Deputy Chief Of Staff • Downrange Computer Output System

dcp Data Code Page (file extension)

DC&P Data Collection & Processing

DCP Data Collecting Platform • Data Collection and Processing • Data Collection Platform • Data Collection Program • Data and Command Processor • Data Communication Processor • Data Communication Protocol • Decision Coordinating Paper • Decision Coordinating Plan • Decision Coordination Paper • Decreasing Correlation Period • Default Configuration Procedure • Degree Completion Program • Delayed Command Processor • Depot Condemnation Percentage • Design Change Package • Design Change Proposal • Design Criteria Plan • Development Concept Paper • Development Cost Plan • Device Control Protocol • Diffuse Cloud Phase • Digital Cartographic Production • Digital Clock Pulse • Digital Communications Processor • Digital Communications Protocol • Digital Computer Processor • Digital Computer Programming • Direct Coupled Plasma • Direct Coupled Pump • Direct-Coupling Photometer • Direct-Current Plasma • Display Control Panel • Display Control Program • Distributed Collaborative Planning • Distributed Communications Processor • Division of Chemical Physics • Drug Control Programme • DTS Calibration Parameter • Duplex Central Processor

DCPEI DEU Control Program End Item • Display Control Program End Item

DCPG Digital Clock Pulse Generator

DCPI Digital Continuous Profiling Infrastructure

DCPL Distributed Control Programming Language

DCPLS Data Collection Platform Location System

DCPRS Data Collecting Platform Retransmission System

DCPS Data Collection and Processing System • Data Communications Protocol Standard • Data Compression Processing System • Digital Cartographic Production Segment • Digitally-Controlled Power Source • District of Columbia Public Schools (USA-DC) • Dynamic Crew Procedures Simulator

DCPSK Differential Coherent Phase Shift Keyed • Differentially Coherent Phase Shift Keyed

DCPSP Direct-Current Power Supply Panel

DCPU Display Control Power Unit

DCQ Daily Contract Quality

DCQM Digital Circuit Quality Monitor

DCR Daily Communication Report • Data Change Request • Data Collection Recorder • Data Collection Request • Data Conversion Receiver • Data Coordinator and Retriever • Decatur Hi-Way (USA-IN) (location identifier) • Decision Circuit Reception • Decrease • Dedicated Control Room • Design Certification Review • Design Change Record • Design Change Request • Design Characteristic Review • Design Clarification Review • Design Conception Review • Design Concept Report • Design Concept Review • Design Concern Report • Design Criteria Report • Detector Count Rate • Differential Chromatic Refraction • Differential Colour Refraction • Differential Correlation Radiometer • Digital Coded Radar • Digital Continuum Receiver • Digital Conversion Receiver • Direct Connecting Receptacle • Direct Conversion Receiver • Direct Cost Reimbursement • Disposition of Contract Request • Distributed Computing Revolution • Document Change Record • Document Change Request • Document Change Review • Dual Channel Radiometer • Dynamic Color Rendition

DCRA Dynamic Communications Resource Allocation

DCRB Design Change Review Board

DCRE Degenerate Configuration of Reflective Equilibrium

DCRG Decreasing

DCRI Direction de la Coopération et des Relations Internationales

DCRR Design Change Requirement Record

DCRS Data Collection and Reduction System • Danish Centre for Remote Sensing

DCRT Division of Computer Research and Technology

DC&S Display & Control System

DCS Damage Control System • DASD Calculation Services • Data Characterization Scheme • Data Classification System • Data Collecting System • Data Collection Site • Data Collection System • Data Communications System • Data Conditioning System • Data Control Sequencer • Data Control Services • Data Control System • Data Conversion System • Data Cycle System • Defense Communications System • Defined Context Set • Deflection Coil Set • Delayed Coincidence Spectroscopy • Department of Computer Science • Department of Computing Science • Depot Computer System • Deputy Chief of Staff • Design Change Schedule • Design Change Summary • Design Communication System • Design Control Specification • Design Criteria Specification • Desktop Colour Separation • Despin Control System • Detailed Checkout Specification • Detector Controller Simulator • Detector Control Software • Detector Control System • Development of Communications Systems • Diagnostic Control System • Differential Cross-Section • Digital Camera System • Digital Cartography Section • Digital Cell Site • Digital Cellular Switch • Digital Cellular System • Digital Color Separate • Digital Command System • Digital Communications Services • Digital Communications System • Digital Computer System • Digital Control Signal • Digital Control Station • Digital Control System • Digital Crossconnect System • Direct Commercial Sales • Direct Connect System • Direct-Current Sensor • Direction Culturelle et Sociale • Dispatch Critical System • Display and Control System • Display Control System • Distributed Communications System • Distributed Computer System • Distributed Computing System • Distributed Control System • Doctor of Commercial Science • Document Control Sheet • Document Control Software • Document Control System • Doncaster Finningley (UK-England) (location identifier) • Double Compton Scattering • Double Correlated Sampling • Dual Checkout Station

DCSA Direct-Current Servo Amplifier

DCSC Digital Card Slave Clock

DCSCE Downsized Communications Systems Control Element

DCSCU Dual Capability Servo Control Unit

DCSi Diversified Computer Systems Inc. (USA-CO)

DCSI Data and Control Signal Interface • Data Channel Status Indicator

DCSO Deputy Commander for Space Operations

DCSP Design Controlled Spare Part • Digital Call Service Position • Digital Control Signal Processor

DC/SR Display and Control/Storage and Retrieval

DCSR Data Control Status Report

DCSS Digital Communications Satellite System • Digital Communications Subsystem • Digital Conference Switching System • Digital Conferencing and Switching System

DCSSI Direction Centrale de la Sécurité des Systèmes d'Information

DCST Data Collection System Tape • Data Collection System Transponder • Digital Card Schmitt Trigger • Direction de la Coopération Scientifique et Technique

DCSTS Data Collection Storage and Transmission System

DCSU Differential Corrected Spectral Unit • Digital Computer Switching Unit

dct Dictionary (file extension)

Dct Direct • Document

DCT Database Configuration Tool • Data Collection Technician • Data Communications Terminal • Data Congestion Threshold • Data Conversion Transmitter • Decimal Code Translator • Dedicated Carbon Tax • Design - Code - Test • Destination Control Table • Developmental Certification Test • Development Certification Testing • Device Control Table • Dictionary • Digital Carrier Trunk • Digital Communications Terminal • Digital Cordless Telephony • Direct • Direct Cosine Transform • Discipline Coordination Team • Discrete Cosine Transform • Dissector Camera Tube • Distributed Computer Telephony • Divisão de Ciências de Terra • Document Control Ticket • DSS Communications Terminal • Duncan Town (Bahamas) (location identifier) • Dynamic Circuit Timing

DCTI Direction Centrale du Traitement de l'Information

DCTL Direct-Coupled Transistor Logic

DCTM Direct-Current Torque Motor

DCTN Defense Commercial Communications Network

DCTP Developmental Certification Test Plan

DCTS Digital Coordinate Transformation System • Dimension Custom Telephone Service

DCTT DSN Compatibility Test Trailer

DCTV Digital Color Television

DCU Data-Cache Unit • Data Collection Unit • Data Communications Unit • Data Command Unit • Data Computer Unit • Data Control Unit • Date Conversion Utility • Decade Counting Unit • Decatur Pyor (USA-AL) (location identifier) • Decimal Counting Unit • Deployment Control Unit • Detection and Control Unit • Detector Control Unit • Digital Coefficient Unit • Digital Computer Unit • Digital Conference Unit • Digital Control Unit • Digital Counting Unit • Direct Control Unit • Dispenser Control Unit • Display and Control Unit • Display Controller Unit • Display Control Unit • Distribution Control Unit • Drum Control Unit • Dublin City University (Ireland) • Dynamic Component Upgrade

DCUPS Densely-Centered Uniform-P Search

DCUTL Direct-Coupled Unipolar Transistor Logic

DCV Demain Chez Vous • Digital Coded Voice • Direct-Current Volt • Directional Control Valve • Document Cut-off Value

DCVR Direct-Current Voltage Reference • Direct-Current Voltage Regulator

DCVZ Denver Convergence Vorticity Zone

DCW Data Control Word • Digital Chart of the World • Distributed Communities on the Web (workshop)

DCWS Debris Collision Warning System • Detection, Control and Warning System • Directional Control and Warning System

DCWV Direct-Current Working Volt

DC-X Delta Clipper Experimental

DCX Delta Clipper Experimental • Direct Current Experiment • Double Charge Exchange

DC-XA Delta Clipper Experimental Advanced

DCXA Delta Clipper Experimental Advanced

DCXRD Double Crystal X-Ray Diffraction

DC2 Distributed Command and Control

dd De Dato • Días • Disk Doubler (file extension) • Dono Dedit

d/d Delivered

d/D Depth-to-Diameter (ratio)

D/d Days after Date • Delivered

Dd Dated • Dedicated • Delivered

D/D Days after Date

D&D Decontamination & Decommissioning • Design & Development • Drug & Disease free

DD Damage Done • Data Definition • Data Depository • Data Description • Data Descriptor • Data Device • Data Dictionary • Data Directory • Data Discrepancy • Data Display • Data Distillery • Days • Day's Date • Days after Date • Days after Delivery • Deadline Date • Decimal Divide • Decoder Driver • De Dato • Dedicated Display • Deep Down (*) • Deep-Drawn • Degree-Day • Delayed Delivery • Delete Description • Demand Draft • Depacketization Delay • Departure Date • Deputy Director • Deputy Directorate • Description Description • Design and Development • Design Deviation • Destructive Dilemna • Detailed Design • Dewey Decimal (classification) • Differential Doppler • Digital Data • Digital Display • Direct Draw • Directional Discontinuity • Directives Documentation • Director's Discretionary • Disc Diameter • Disconnecting Device • Dishonorable Discharge • Disk Drive • Dispersion Direction • Doctor of Divinity • Document Delivery • Document Director • Domain Decomposition • Domain Defined • Donnée de Définition • Dossier de Définition • Double Deck • Double Degenerate • Double Demy • Double Density • Dry Dock • Due Date

DDA AAS Division on Dynamical Astronomy Meetings (periodical code) • Data Delivery Acknowledgment • Data Descriptive Area • Data Distribution Assembly • Decommissioning and Disposal Analysis • Dedicated Digital Access • Demand Deposit Account • Demand Deposit Application • Deputy Director for Administration • Designated Development Activity • Designated Development Agency • Development Display Assembly • Digital Differential Analyzer • Digital Directory Assistance • Digital Driver Amplifier • Direction des Développements Ariane • Discrete Dipole Approximation • Distributed Data Access • Division Design Analysis • Division on Dynamical Astronomy (AAS) • Domain-Defined Attribute • Drawing Departure Authorization • Dual Digital Audio

DDAE Dossier de Demande d'Autorisation d'Exploiter (France)

DDAM Dynamic Design Analysis Method

DDAP Digital Diagnostics And Prognostics

DDAS Digital Data Acquisition System • Digital Data Archives System

DDASS Direction Départementale swa Affaires Sanitaires et Sociales (France)

DDB Data Display Board • Data Display Buffer • Decision Database • Delayed Direct Broadcast • Depot Data Bank • Design Data Block • Device-Dependent Bitmap • Device Descriptor Block • Digital Data Buffer • Distributed Database • Double Declining Balance • Dynamic Database

DDBMS Distributed Database Management System

DDBP Dedicated Database Processor

DDBS Derived Database System • Descriptor Database System • Display Database System • Distributed Database System

DD&C Data Dictionary & Catalog

DDC Data Display Center • Data Display Central • Data Dissemination Center • Data Distribution Center • Data Documentation Cost • Day Display Component • Deck Decompression Chamber • Defense Documentation Center • Delivery Distribution Center • Design and Development Contract • Dewey Decimal Classification • Diffraction Dilution Coefficient • Digital Data Cell • Digital Data Channel • Digital Data Communications • Digital Data Control • Digital Data Conversion • Digital Data Converter • Digital Display Console • Digital Display Converter • Digital Document Connection • Direct Data Channel • Direct Department Calling • Direct Digital Control • Direction du Développement et de la Coopération (France) • Directed Digital Control • Discipline Data Center • Display Data Channel • Disruptive Discharge Cleaning • District Data Center • District Development Community • Division Data Center • Document Development and Control • Dodge City (USA-KS) (location identifier) • Double-Doped Crystal • Dual Diversity Comparator • Dynamic Data Channel

DDCE Digital Data Conversion Equipment

DDCFG Data Display and Control Functional Group

DDCM Differential Decay Curve Method

DDCMP Digital Data Communications Message Protocol

DDCP Draft Development Concept Paper

DD&CS Dedicated Display & Control System

DDCS Dedicated Display and Control System • Digital Database Connection Service • Digital Data Calibration System • Digital Data Communications Set • Direct Digital Control System • Distributed Database Connection Services • Distributed Data Connection Services • Double Differential Cross-Section

DDCU DC-to-DC Converter Unit

DD/D Data Dictionary/Directory

DDD Data Definition Document • Data Description Document • Dat, Dicat, Dedicat • Date Deficiency Discovered • Date Discrepancy Discovered • Deadline Delivery Date • Désétatisation - Défiscalisation - Dérégulation • Design Definition Document • Desired Delivery Date • Detailed Data Display • Detailed Design Document • Deutscher Depeschen-Dienst (Germany) • Digital Daemon Disciple • Digital Data Distributor • Digital Design Document • Digital Diagnostic Diskette • Digital - Digital - Digital • Digital Display Driver • Dimension - Durée - Dispersion • Direct Distance Dial • Direct Distance Dialing • Display Decoder Drive • Distributed Document Database • Dono Dedit Dedicavit • Drug Delivery Device • Duplexed Display Distributor

DDDA Decimal Digital Differential Analyzer

DDDDD Demand Drive Direct Digital Dissemination

DDDL Digital Data Downlink

DD/DS Data Dictionary/Directory System

DDDS Digital Data Display System • Digital Data Distribution System • Dynamic Delegation Discovery System

DDDU Digital Decoder Driver Unit

DDE Data Dictionary Entry • Decentralized Data Entry • Delaunay Density Estimator • Direct Data Entry • Direct Digital Encoder • Direction Départementale de l'Équipement (*) • Director Design Engineering • Distributed Data Entry • Distributive Data Environment • Division Départementale de l'Équipement • Dwight David Eisenhower (1890-1969) • Dynamic Data Exchange

DDEM Delaunay Density Estimator Method

DDEML Dynamic Data Exchange Management Library • Dynamic Data Exchange Manager Library

DDEN Délégué Départemental de l'Éducation Nationale (France)

DDES Data Display Expert System • Digital Data Editing System • Direct Data Entry Station

DDF Data Description Field • Data Description File • Data Descriptive File • Data Disposition Facility • Data Distribution Facility • Descriptive Data File • Dielectric Dissipation Factor • Digital Distribution Frame • Director's Discretionary Fund • Display and Debriefing Facility • Document Description File • Dynamic Data File • Dynamic Data Format • Dynamic Data Formatting

DDFT Design - Development - Fabrication - Testing

DDG Dandong (China-PRC) (location identifier) • Data Display Generator • Deputy Director-General • Digital Data Generator • Digital Data Group • Digital Display Generator • Dual Display Generator

DDGE Digital Display Generator Element

DDGI Deutscher Dachverband für Geoinformation (Germany)

DDGL Device-Dependent Graphic Layer

DDH Digital Display Handling • Distributed Document Handling

DDHA Digital Data Handling Assembly

DD.HH. Derechos Humanos

DDHS Digital Data Handling System

DDHU Digital Data Handling Unit

ddi Disk Doubler Image (file extension)

DDI Data Display Indicator • Daydream Island (Australia-QLD) (location identifier) • Dedicated Display Indicator • Delayed Dynamic Instability • Delivery Distribution Indicator • Depth Deviation Indicator • Device Driver Interface • Digital Display Indicator • Diplôme de Docteur Ingénieur • Direct Dial In • Direct Dialing In • Direct Dialing Inwards • Direct Dial Inwards • Directed Drawing Instrument • Discrete Data Input • Discrete Digital Input • Driver Display Interface • Dynamic Data Interchange

DDICT Data Dictionary

DDID Digital Data Interface Device

DDIF Digital Document Interchange Format

d.d. in d. De Die in Diem

DDIP Division de Documentation et d'Information Pédagogique

DDIS Data Depository Index System • Depository Distribution Information System • Document Data Indexing Set

DDIST Data Distribution

DDJ Digital Differencing Junction

DDK Device Development Kit • Device Driver Kit • Disk Driver Kit • Driver Development Kit

ddl Degré de Liberté

DDL Daily Delivery List • Database Description Language • Data Definition Language • Data Description Language • Data Descriptive Language • Data Design Language • Data Distribution Laboratory • Data Distribution List • Data Documentation List • Data Downlink • Data Drawing List • Delay-Locked Loop • Description Definition Language • Detector Development Laboratory • Dialogue Description Language • Differential Delay Line • Differential Distribution Law • Digital Data Link • Digital Data Logger • Digital Design Language • Diode-Diode Logic • Direct Downlink • Direct Drive Linear • Directed Distribution List • Dispersive Delay Line • Display Description Language • Distributed Design Language • Documentation Distribution List • Document Data Listing • Document Description Language • Document Distribution List • Double Delay Line • Dual Data Link

DDLC Data Description Language Committee

DDLCN Distributed Double Loop Computer Network

DDLI Direct Data Link Interconnectivity

DDLT Diagnostic Decision Logic Table

ddm Différence de Marche (optique)

DDM Data Demand Module • Data Display Module • Data Display Monitoring • Decaying Dark Matter • Délégué Départemental de Météo France • Deputy Demonstration Manager • Design Decision Memorandum • Desktop Document Manager • Difference in Depth of Modulation • Digital Data Modem • Digital Display Machine • Digital Display Makeup • Discrete Data Management • Distributed Data Management • Distributed Data Manager • Distributed Design Methodology • Dodoima (Papua New Guinea) (location identifier) • DOME Data Manager

DDMA Disk Direct Memory Access

DDMan Data Display Module Alphanumeric

DDMg Data Display Module Graphical

DDML Document Definition Markup Language

DDMP Domain Decomposition Message Passing

DDMS Deputy Daily Mission Specialist • Digital Data Measuring System • Digital Desertification Monitoring System

DDMWG Data Distribution And Marketing Working Group

DDM3d Data Display Module 3D Graphical

DDN Data Delivery Notice • Data Dissemination Network • Dedicated Digital Network • Defense Data Network • Delta Downs (Australia-QLD) (location identifier) • Design Decision Notice • Digital Data Network • Direct Dial Network • Distributed Data Network • Distributed Data Node • Documentation Development Notification

DDNA Divided Difference Noise Analysis

DDNC Direct Digital Numerical Controller

DDNMC Data Dissemination Network Management Centre

DDNS DEC Distributed Name Service • Dynamic Domain Naming Service • Dynamic Domain Naming System

DDO Data-Driven Object • David Dunlap Observatory (Canada-ON) • Director's Discretionary Observation • Discrete Data Output • Discrete Digital Output • Dynamically Distributed Overlay • Dynamic Data Object

DDOCE Digital Data Output Conversion Equipment

DDOP Differential Dilution Of Precision

DDORe David Dunlap Observatory Reprints (periodical code)

DDoS Distributed Denial of Service

DDOS Distributed Denial Of Service

DDOV Digital Data Over Voice

ddp Device Driver Profile (file extension)

DDP Data Distribution Panel • Data Distribution Plan • Data Distribution Program • Datagram Delivery Protocol • Declaration of Design and Performance • Deep Discount Price • Demand Development Period • Deployment Detailed Planning • Design Data Package • Design Development Plan • Detailed Demonstration Plan • Digital Data Processing • Digital Data Processor • Digital Doppler Processor • Director's Discretionary Program • Distributed Data Processing • Distributed Data Processor • Dorado (Puerto Rico) (location identifier)

DDPAT Distributed Database Performance Analysis Tool

DDPC Digital Data Processing Center • Digital Data Processing Complex

DDPE Digital Data Processing Equipment

DDPF Dedicated Display Processing Function

DDPO Defense Dissemination Program Office

DDPS Data-Directed Programming System • Digital Data Processing System • Discrimination Data Processing System • Distribution Data Processing System

DDPT Disk Drive Parameter Table

DDPU Digital Data Processing Unit

DD&R Data Display & Review

DDR Daily Demand Rate • Data Definition Record • Data Description Record • Data Descriptive Record • Data Direction Register • Data Discrepancy Report • Data Display and Review • Data Display Room • Data Distribution Requirement • Design Development Record • Design Development Review • Detail Design Review • Deutsche Demokratische Republik (now merged into BRD) • Developer Documentation Requirement • Development Discrepancy Report • Dial on Demond Routing • Digital Data Receiver • Direct Drive Rotary • Double Data Rate • Double Drift Region • Dynamic Device Reconfiguration

DDR&E Deputy Director for Research & Engineering • Director of Defense Research & Engineering (DoD)

DDRE Design Development Research and Engineering

DDRG Double-Double Radio Galaxy

DDRH Digital Data Recording Head

DDRI Design Drafting Reference Information

DDRP Direct/Delayed Response Project

DD&RR Data Display & Review Room

DDRS Digital Data Recording System

DDRT Data Distribution Review Team

Dds Dedans

DDS Data Design Specification • Data Dialog System • Data Display Set • Data Display System • Data Disposition System • Data Dissemination System • Data Distribution System • Dataphone Digital Service • Deputy Directorate of Science • Descent Design System • Design Data Sheet • Detailed Design Specification • Development Data Sheet • Dewey Decimal System • Digital Dataphone Service • Digital Data Service • Digital Data Servo • Digital Data Storage • Digital Data System • Digital Display Scope • Digital Display System • Digital Drafting System • Digital Dynamics Simulator • Direct Digital Service • Direct Digital Synthesis • Direct Digital Synthesizer • Discipline Data System • Display and Debriefing System • Display Data Set • Distant DRAGN Survey • Distributed Database Service • Distributed Database System • Distributed Data Service • Distributed Data System • Distributed Directory Service • Division of Digital Systems • Documentation Distribution System • Document Delivery Service • Document Development System • Domestic Data Service • Drug Delivery System • Dry Drop of Sweat • Dual Driver Service • Dust Detection System • Dust Detector System

DDSC Direction de la Défense et de la Sécurité Civile (France)

DDSM Digital Data Switching Matrix

DDSOT Digital Daily System Operability Test

DDSR Deleted Data Space Release • Dynamic Data Space Release

DDSRV Document Data Server

DDSS Digital Data Switching System

DDS&T Deputy Director for Science & Technology

DDST Deputy Director for Science and Technology • Direction Départementale des Services Techniques (France)

DDSU Digital Data Storage Unit

DD&T Design, Development & Test

DDT Data Description Table • Debye Dipole Theory • Deflagration - Detonation - Transition • Department of Development and Tourism • Design Data Transmittal • Design, Development and Test • Design Development Test • Diagnostic Decision Table • Dichloro-Diphenyl-Trichlorethane • Digital Data Terminal • Digital Data Transmission • Digital Data Transmitter • Digital Debugging Tape • Digital Demodulation Technique • Director's Discretionary Time • Distributed Decision (aid) Tool • Domain Definition Table • Don't Do That • Doppler Data Translator • Double Deflection Tube • Dynamic Debugging Technique • Dynamic Display Tester

DDTC Director's Discretionary Time Committee

DDTD Digital Data Transmission Device

DDT&E Design, Development, Testing & Engineering • Design, Development, Testing & Evaluation

DDTE Design - Development - Test - Engineering • Design - Development - Test - Evaluation • Digital Data Terminal Equipment • Direction Départementale du Travail et de l'Emploi (France)

DDTEFP Direction Départementale du Travail, de l'Emploi et de la Formation Professionnelle (France)

DDTESM Digital Data Terminal Equipment Service Module

DDTF Docking Dynamics Test Facility • Dynamic Docking Test Facility

DDTP Director's Discretionary Time Proposal

DDTS Digital Data Transmission System • Direct Dial Telephone System • Distributed Data and Telecommunications System • Distributed Defect Tracking System • Dynamic Docking Test System

DDU Dadu (Pakistan) (location identifier) • Data Display Unit • Data Distribution Unit • Decommutator Distribution Unit • Dedicated Display Unit • Destination Delivery Unit • Diagnostic Display Unit • Digital Data Unit • Digital Display Unit • Digital Distributing Unit • Digital Distribution Unit • Direct Deploying Unit • Disc Data Unit • Disk Drive Unit • Display and Debug Unit • Display Driver Unit

DDV Dark Dense Vegetation • Datendirektverbindung • Dialog Data Validation

DDVA Délégation Départementale à la Vie Associative (France) • Délégué Départemental à la Vie Associative (France)

DDVN Direct Dedicated Voice Network

DDVR Displayed Data Video Recorder

DDW D-Density Wave • Digital Data Warfare

DDWG Digital Data Working Group

DDX Digital Data Exchange • Distributed Data Exchange • Double Differential Cross-Section

de German (ISO language code) • Germany • Individual Drift Error

dE Diffuse Elliptical • Dwarf Elliptical (galaxy)

De Total Drift Error

D&E Demonstration & Exercise

DE Allemagne (code ISO de pays) • Condor Flugdienst GmbH (obsolete?) • Damage Estimation • Damage Expectancy • Dassault Électronique • Data Element • Data Encoder • Data Entry • Data Extraction • Decay • December • Décembre • Decision Element • Decision Error • Deflection Error • Delaware (USA) • Delay Equalizer • Demandeur d'Emploi • Denmark • Descent Engine • Design Engineering • Design Evaluation • Detection Efficiency • Detector Efficiency • Detector Electronics • Development Estimate • Développement Exploratoire • Device End • Diesel-Electric (engine) • Diesel Engine • Digital Electronics • Digital Element • Digital Encoder • Digital Engineering • Diplôme d'État • Directed Energy • Direct Electrolysis • Direct Encounter • Direct Excitation • Directory Enquiry • Directory Entry • Discard Eligibility • Discard Eligible • Disconnection Event • Display Electronics • Display Entity • Dissociative Excitation • Distance Education • Distant Element • District Engineer • Doctor of Engineering • Document Expansion • Dokumentationseinheit • Doppler Extractor • Dose Equivalent • DOS Extender • Double Elephant • Double End • Double Entry • Double-Excited • Draw End • Driver Education • Dynamics Explorer • Germany (ISO country code)

dea Aerospace Drift Error

Dea Dean

DEA Data Encryption Algorithm • Data Envelopment Analysis • Data Exchange Agreement • Data Exchange Annex • Defense Exchange Agreement • Délégation aux Études et aux Audits • Department of Environmental Affairs • Deployed Electronics Assembly • Detector Electronics Assembly • Dielectric Analysis • Diffusione Edizioni Angloamericane (Italy) • Digital Electronics Assembly • Diploma de Estudios Avanzados (Spain) • Diplôme d'Études Approfondies (France) • Director of Educational Activities • Directory-Enabled Application • Display Electronics Assembly • Divisão de Electrônica Aeroespacial • Division des Études Avancées • Doctrine Evaluation Activity • Drug Enforcement Administration • Drug Enforcement Agency (USA) • Dynamo Electric Amplifier

DEAA Downeast Amateur Astronomers (USA-ME)

DeaCo Dearborn Observatory Contributions (periodical code)

DEACT Deactivate

DEAG Data Extraction and Analysis Group

DEAL Data Entry Application Language • Decision Evaluation and Logic

DEAR Department of Energy Acquisition Regulation

deb Debug (file extension)

DEB Data Extent Block • Debrecen (Hungary) (location identifier) • Debris • Déclaration d'Échange de Biens • Detached Eclipsing Binary (star) • Detector Electronics Box • Digital Electronics Box • Digital European Backbone • Division of Environmental Biology (NSF) • Dynamical Ephemeral Body

DEBIE Debris In-orbit Evaluator

DEBIS Distributed Expert-Based Information System

DEBITS Deposition of Biogeochemically Important Trace Species

DEBS Digital Electron Beam Scanner

DEBRA Diffuse Extragalactic Background Radiation

dec Decena • Decimeter • Declination • Decrement

Dec Deceased • Deceleration • December • Decimal • Decimeter • Decision • Declaration • Declare • Declared • Declination • Decode • Decoded • Decoder • Decoding • Decompose • Decomposition • Decrease • Decreased • Decreasing • Decrement • Decrescendo

Déc Décembre

DEC Decatur (USA-IL) (location identifier) • December • Décembre • Declination • Decode • Decoder • Decrease • Decrement • Department of Environmental Conservation • Device Clear • Digital Echo Cancellation • Digital Equipment Corporation • Digital Evaluation Computer • Direct Energy Converter • Division des Examens et Concours • Division Environmental Coordinator • Double Electron Capture

DECA Department of Education and Cultural Affairs • Descent Engine Control Assembly • Digital Electronic Control Assembly

Decaf Decaffeinated

DECAL DEC Author Language

DECAS Direction des Entreprises Commerciales, Artisanales et de Services (France)

DECB Data Event Control Block

DECC Digital European Cordless Communications

Decd Deceased

DECDEE Designing Centers for Distance Education in Estonia

DECF Diplôme d'Études Comptables et Financières

DECIMAL Decision Making in Libraries

DECIMALS Dynamical, Electrodynamical and Chemical Interactions near the Mesopause at Arctic Latitudes during Summer

dECL Differential Emitter-Coupled Logic

DECL Declaration • Declassified • Declassify • Deliverable Equipment Checklist • Direct Energy Conversion Laboratory

Decn Decision

DECN Decision

DECNET Digital Equipment Corporation Network

DECNET E-net DEC Engineering Network

DECODE Dual-channel EUV Coronal Dynamics Explorer

DECOM Decommutate • Decommutator

DECOMATE Delivery of Copyright Materials to End-users

Decomp Decompose • Decomposed • Decomposition

DECOMP Decompiler

Decompn Decompression

Decon Decontaminate • Decontamination

Decontn Decontamination

DECOR Digital Electronic Continuous Ranging

DECPU Digital Enhanced Central Processing unit

Decr Decrease

DECR Decrease

DECS Department of Education, Culture and Sports • Diplôme d'Études Comptables Supérieures

DECSAI Departamento de Ciencias de la Computación e Inteligencia Artificial (Spain)

DECT Digital Enhanced Cordless Telecommunications • Digital European Cordless Telecommunications (*)

DECU Data Exchange Control Unit • Development and Educational Communication Unit • Digital Engine Control Unit

DECUS Digital Equipment Computer Users' Society

Ded Dedicate • Dedicated • Deduce • Deduct • Deduction

DEd Doctor of Education

DED Dark-Emitting Diode • Data Element Definition • Data Element Dictionary • Data Element Directory • Data Entry Device • Data Entry Display • Dedicate • Dedicated • Dehra Dun (India) (location identifier) • Deputy Executive Director • Design Engineering Discrepancy • Dynamic Entity Database • Dynamic Ergodic Director

DEDA Data Entry and Display Assembly

DEDB Data Entry Database

de d. in d. De Die in Diem

DEDL Data Element Description List

DEDP Data Entry and Display Channel • Document d'Examen des Droits à la Pension • Dossier d'Examen des Droits à la Pension

DEDS Data Entry and Display System • Digital Enhancement Database System • Digital Error Detection System

DEDSL Data Entity Dictionary Specification Language

DEE Datenendeinrichtung • Dexterous End Effector • Digital Evaluation Equipment • Digital Events Evaluator • Diploma in Electrical Engineering

DEEC Digital Electronic Engine Control

DEEM Dynamic Environmental Effects Model

DEEP Deep Extragalactic Evolutionary Probe • Directly Elected European Parliament • Division des Études, de l'Évaluation et de la Prospective

DEES Dynamic Electromagnetic Environment Simulator

DEET Department of Employment, Education and Training

DEETYA Department of Employment, Education, Training and Youth of Australia

def Definition (file extension)

Def Default • Defect • Defectif • Defective • Defectivo • Defend • Defended • Defense • Defensive • Deferred • Deficiency • Deficient • Deficit • Define • Defined • Definite • Definition • Definitive • Deflate • Deflation • Deflect • Deflected • Deflection • Defunct • Defunctus

Déf Définition

DEF Data Exchange Format • Default • Defend • Definition • Delivery Evaluation Factor • Development and Evaluation Facility • Display Evaluation Form • Duplicate Emergency File

DEFA Digital Event Filter and Address • Diplôme d'État aux Fonctions d'Animation (France)

DEFCAL Deflection Calibration

DEFCS Digital Electronic Flight Control System

DEFD Demandeur d'Emploi en Fin de Droit

DefEl Defense Electronics (periodical code)

Defek Defektoskopiia (periodical code)

DEFI Data Entry File Inquiry • Documentation Économique et Financière Informatisée

DEFIL Data Edit File

Defl Deflate • Deflation • Deflect • Deflected • Deflection • Deflector

DEFL Deflect

DEFM Demandeur d'Emploi en Fin de Mois • Demographic and Economic Forecasting Model

DEFR Développement Exploratoire de Fusion de Renseignements

DEFRAG Defragment

DEFT Direct Electronic Fourier Transform • Discrete Electronic Fourier Transform • Dynamic Error-Free Transmission

Deg Degré • Degree

DEG Degenerate Electron Gas • Degree • Derecho Especial de Giro • Directorate of Environmental Geology • Document Evaluation Guidelines • Dwarf Elliptical Galaxy • Dynamique des Espaces Géographiques

DEGA Deutsche Gesellschaft für Akustik (Germany)

de g. à dr. De Gauche à Droite

DEGt Datenendgerät

DEH Decorah (USA-IA) (location identifier) • Directorate of Engineering and Housing • Director of Engineering and Housing

DEHE Directorate General of Higher Education

DEHI Double Exposure Holographic Interferometry

DeHyZ Deutsche Hydrographische Zeitschrift (periodical code)

DEI Data Exchange Interface • Deliverable End Item • Denis Island (Seychelles) (location identifier) • Descent Imager • Design, Engineering, and Integration • Design Engineering Identification • Design Engineering Inspection • Development Engineering Inspection • Diffraction-Enhanced Imaging • Dispositif d'Étalonnage Interne • Division des Études et de l'Information • Dutch East Indies

D/EIB Designate/Encoder Interface Buffer

DEIC Differential Equations in Industry and Commerce

DEIMOS Deep Extragalactic Imaging Multiobject Spectrograph • Deep Imaging Multiobject Spectrograph

DEINT Digital Enhanced Interface

DEIS Design Engineering Inspection Simulation • Design Engineering Inspection Simulator • Design Evaluation and Inspection Simulator • Draft Environmental Impact Statement

Dek Deken

DEK Dansk Elektroteknisk Komite (Denmark) • Data Encryption Key • Donald E. Knuth

del Delineator • Delineavit

Del Delaware (USA) • Delegación • Delegate • Delegated • Delegation • Delete • Deleted • Deletion • Delhi (India) • Delineation • Deliver • Delivered • Delivery • Delphinus-Delphini (constellation)

DEL Data Entry Language • Decode-Encode Language • Delay • Delete • Delhi - Indira Gandhi (India) (location identifier) • Deliver • Delivery • Deorbit - Entry - Landing • Deployable Equipment List • Deploying Equipment List • Diode Électroluminescente • Direct Exchange Line

Deld Delivered

DELD Demandeur d'Emploi de Longue Durée

Dele Delete

DELE Delete

Deleg Delegación • Delegate • Delegation

DELEX Data Elements Lexicon

DELF Diplôme Élémentaire de Langue Française

DELHP Division on Education, Labor, and Human Performance

DELICAT Data Enhancement of Library Catalogs

delin Delineavit

DELM Department of the Environment and Land Management

DELO Diagonal Element Lambda Operator

DELOG Design Engineering Logistics

DELP Department of the Environment, Land and Planning

DELPHI Detector with Lepton Photon and Hadron Identification

DELSTR Delete String

Delt Delineator

DELT Dynamic Environmental Laboratory Test

DELTA Description Language for Taxonomy • Detailed Labor and Time Analysis • Determination Effective Levels of Task Automation • Developing European Learning through Technological Advance • Distance Education and Learning Telecommunications Applications • Dual Entry Large Altitude (chamber)

Delv Deliver • Delivered

Dely Delivery

dem Demonstration (file extension)

Dem Demand • Demanded • Democrat • Democratic • Demodulate • Demodulated • Demodulator • Demonstrate • Demonstrated • Demonstration • Demonstrative

DEM Data Energy Modernization • Dembidollo (Ethiopia) (location identifier) • Demonstration • Density Emission Measure • Department of Environmental Management • Differential Emission Measure • Digital Electronics Module • Digital Elevation Map • Digital Elevation Mapping • Digital Elevation Matrix • Digital Elevation Model • Digital Event Marker • Discrete Event Marker • Dynamic Enterprise Modeller • Mark (Germany) (ISO currency code)

DEMA Direct E-mail Marketing Association

DEMARC Demarcation

DEMAT Department of Educational Media and Technology

DEMD Digital Message Entry Device

DEMIL Demilitarization • Demilitarize • Demilitarized

DEMIRM Département de Matière Interstellaire et Radioastronomie Millimétrique (Observatoire de Paris-Meudon)

DEMKO Danmarks Elektriske Materielkontrol (Denmark)

DEMM Database Evolution and Meta-Modeling (workshop) • Division of Educational Media Management

Demo Demonstration

DEMO Demonstration

Demod Demodulator

DEMOD Demodulation • Deployment Model

Demon Demonstrate • Demonstrative

DEMON Diagnostic Evaluation Monitor • Diminishing Error Method of Optimization for Networks

DEMO Diffuse Metal-Oxide Semiconductor

DEMP Dispersed Electromagnetic Pulse

DEMR Department of Energy, Mines, and Resources (Canada)

DeMS Deployment Management System

DEMS Digital Electronic Message Service • Digital Error Monitoring System

DEMUX Demultiplex • Demultiplexer

DEMVAL Demonstration and Evaluation

Den Denmark • Denomination • Denotation • Denote • Denoted • Density • Denver (USA-CO)

DEN Data Element Number • Denver Stapleton (USA-CO) (location identifier) • Directory-Enabled Network • Directory-Enabled Networking • Document Enabled Networking

DEng Doctor in Engineering

DEngS Doctor in Engineering Science

DENI Department of Education of Northern Ireland

DENIS Deep Near-Infrared Survey

DENR Department of Environment and Natural Resources (Philippines)

Dens Density

DENS Demande d'Emploi Non-Satisfaite

DEO Data Exploitation Operator • Dearborn - Hyatt Regency (USA-MI) (location identifier)

DEOF Double End Of File

D/EOM Directorate of Earth Observation and Microgravity • Director of Earth Observation and Microgravity

DEORB Deorbit

DEOS Deeply Embedded Outburst Star • Delft Institute for Earth-Oriented Space Research (Netherlands) • Digital Engine Operating System • Dynamic Earth Observation System

DEOSS Demonstration of Electro-Optical Surveillance Sensor

Dep Depart • Departs • Department • Departure • Dependency • Dependent • Deponent • Deportes • Depose • Deposed • Deposit • Depósito • Depot • Deputy

Dép Départ • Département • Député

DEP Data Element Profile • Data Entry Panel • Data Exchange Program • Data Extraction Point • Dedicated Experiment Processor • Deep • Deflection Error Probable • Delayed Entry Program • Delft Electronic Products • Deparizo (India) (location identifier) • Department • Département de l'Économie Publique • Department of Environmental Protection • Department of Environment and Planning • Departs • Departure • Deployed • Depot Element Profile • Depth • Deputy • Descanse En Paz • Design Engineering Program • Dielectrophoresis • Direction de l'Évaluation et de la Prospective • Disfonctionnement Érectile Permanent • Display Electronics Package • Division of Educational Programs • Draft Equipment Publication

DEPA Danish Environmental Protection Agency • Defense Evaluation and Research Agency

DEPART Departamento

DEPI Differential Equations Pseudocode Interpreter

Depl Deplete • Depleted • Depletion

DEPL Deploy • Deployment

DEPLAN Deployment Plan

DepLIS Department of Library and Information Services

DEPMAS Deployment Management System

Depn Dependent

DEPOS Depósito

Depr Depreciation • Depression

DEPRESS Depressurize

DEPS Dernier Entré - Premier Sorti • Differential Empirical Population Synthesis • Division of Engineering and Physical Sciences

DEPSK Differential-Encoded Phase Shift Keying (*) • Differential Encoding Phase Shift Keying

DEPSO Departemental Standardization Office

Dept Department • Deponent • Deputy

Dépt Département

DEPT Department

DEPTH Design Evaluation for Personnel, Training and Human (factors)

DEPTM Draft Equipment Publication Technical Manual

Depto Departamento

Depts Departments

DEPULP Département d'Éducation Permanente de l'Université Louis Pasteur (France)

DEPUS Departemental User

DEQ Department of Environmental Quality • Dose Equivalent • Double-Ended Queue

DEQPSK Differentially-Encoded Quadrature Phase Shift Keying

DEQUE Double-Ended Queue

Dér Dérivatif • Dérivé

Der Derecha • Derecho • Derivado • Derivative • Derive • Derived

DER Data Entity Relationship • Declining Error Rate • Defective Equipment Review • Derim (Papua New Guinea) (location identifier) • Designated Engineering Representation • Designated Engineering Representative • Design Error Report • Direction des Études et des Recherches • Documentation Error Report • Document Error Report • Drawing Error Report

DERA Defence Evaluation and Research Agency (UK) • Droits des Citoyens dans leurs Relations avec l'Administration (France)

DERAD Département de Radioastronomie Décimétrique (Observatoire de Paris-Meudon)

DERAT Département d'Études et de Recherches en Aérothermodynamique

Derby Derbyshire (UK-England)

DERF Dynamical Extended Range Forecasting

DERG Data Extraction and Reduction Guide

DERI Département d'Études et de Recherche en Informatique

Deriv Derivation • Derivative • Derive • Derived

DERL Derived Emergency Reference Level

DERM Digital Elevation and Radiometry Model

DERO Département d'Études et de Recherches en Optique (ONERA)

DEROD Data Entry and Readout Device

DERP Defective Equipment Replacement Program

DERRA Department of Extragalactic Research and Relativistic Astrophysics

DERSI Développement des Relations Spatiales Industrielles

DERV Diesel-Engine Road Vehicle

Des Desert • Desierto • Design • Designate • Designated • Designation • Designer

DES Data Element Standardization • Data Encryption Standard (*) • Data Encryption System • Data Entry Sheet • Data Exchange System • Data Exploitation System • Data Extension Segment • Data Extraction Segment • Deep Ecliptic Survey • Delhi Earth Station (India) • Department for Education and Science • Department of Education and Science • Descend • Descent • Description • Design Environmental Simulator • Design Expansion System • Desroches (Seychelles) (location identifier) • Dessinateur • Destination • Destination End Station • Destination End System • Detached Eddy Simulation • Deutsches Elektronen Synchrotron • Development Enquiry Service • Dielectronic Satellite • Differential Energy Spectrum • Differential Equation Solver • Digital Editing Station • Digital Encryption Standard • Digital Expansion System • Diplôme d'Études Spécialisées • Diplôme d'Études Supérieures • Directed Energy Starfire • Direct Entry System • Direction de l'Enseignement Supérieur • Directorate of Evaluation and Standardization • Discrete Event System • Dismounted Extension Switch • Distributed Energy Storage • Documentation Économique et Sociale

DESA Data Encryption Standard Algorithm • Defense Evaluation Support Agency • Direction de l'Enseignement Supérieur Administratif

Desap Desaparecida

DESAT Desaturated • Desaturation

Desc Descarga • Descend • Descendant • Descenso • Descent • Describe • Described • Description

DESC Defense Electronics Supply Center • Descent • Digital Evaluation Solving Computer • Direction d'Études Spécialisées Complémentaires • Display ESR SAR Coverage • Dutch Experiment Support Centre (Netherlands)

DeScE Defense Science Electronics (periodical code)

Descr Describe • Description

Descto Descuento

DESCW Display ERS Swath Coverage for Windows

DESD Department for Economics and Social Development

DESE Diplôme d'Études Supérieures Économiques

DESG Diplôme d'Études Scientifiques Générales

Desgl Desgleichen

DESHE Distributed Expert System Hosting Environment

Desid Desideratum

Desig Designate

DESIG Designate

DESIRE Design Information Recovery Environment • Development of a European Service for Information on Research and Education

Desp Dispatch • Dispatched

DESPA Département de Recherche Spatiale (Observatoire de Paris-Meudon)

Desp Despectivo

DESPOT Design Performance Optimization

DESR Division de l'Enseignement Supérieur et de la Recherche

DèsS Docteur ès Sciences

DESS Diplôme d'Études Supérieures Spécialisées (France)

DESSA Distressed and Expatriate Scientists and Scholars from Africa

DèsSc Docteur ès Sciences

Dest Destination • Destroy • Destroyed

DEST Destination • Diplôme d'Études Supérieures Techniques • Direction de l'Enseignement Supérieur des Télécommunications (France)

DESTEC Département des Sciences et Techniques de l'Expression et de la Communication

Destn Destination

DESTU Diplôme d'Études Scientifiques et Techniques Universitaires • Diplôme d'Études Supérieures Technologiques Universitaires

DESU Diplôme d'Études Supérieures Universitaires

Desus Desusado

DESY Deutsches Elektronen-Synchrotron (Germany)

Det Detach • Detached • Detachment • Detail • Detailed • Detainee • Detect • Detected • Detection • Detector • Determination • Determine • Determined • Determiner

Dét Détaché • Détachement • Détacher • Détail • Détaillé • Détailler • Détecté • Détecter • Détection • Détermination • Déterminé • Déterminer

DET Data Entry Terminal • Data Evaluation Terminal • Decision Error Tradeoff • Déclaration d'Entrée sur le Territoire • Density Equilibration Timescale • Design Evaluation Test • Detach • Detail • Detailed • Detection Error Tradeoff (*) • Detection Evaluation Tradeoff • Detector • Detent • Detroit City (USA-MI) (location identifier) • Device Execute Trigger • Digital Event Timer • Direct Energy Transfer • Directory Entry Table • Displaced Equipment Training • Division of Educational Technology • Dust Erosion Tunnel • Dynamic Environment and Terrain

DETA Dielectric Thermal Analysis • Diethylene Triamine

DETEC Defensive Technology Evaluation Code • Département de l'Environnement, des Transports, de l'Énergie et de la Communication (Switzerland)

DETEL Département de l'Environnement, des Transports, de l'Énergie et des Communications (Switzerland)

DETEMER Detection of Electromagnetic Emission Regions

Deter Determinado

DETF Double Ended Tuning Fork

DETIR Defensive Technology Information Repository

DETL Digital Emulation Technology Laboratory

Detn Determination

DETN Detection

DETP Displaced Equipment Transition Plan

DETR Department of the Environment, Transport and the Regions (UK) • Detector

DETS Digital Element Test Set

DETYA Department of Education, Training and Youth Affairs (Australia)

DEU Allemagne (code ISO de pays) • Data Encoder Unit • Data Encryption Unit • Data Entry Unit • Data Exchange Unit • Development Engineering Unit • Digital Electronics Unit • Digital Evaluation Unit • Display Electronics Unit • Germany (ISO country code)

DEUCE Digital Electronic Universal Computing Engine

DEUG Diplôme d'Enseignement Universitaire Général • Diplôme d'Études Universitaires Générales (France)

DEUP Diplôme d'Études Universitaires Professionnalisées (France)

DEUST Diplôme d'Études Universitaires en Sciences et Techniques • Diplôme d'Études Universitaires Scientifiques et Techniques (France)

DEUT Data Encoder Unit Transmitter

dev Device Driver (file extension)

Dev Develop • Developed • Developer • Developing • Development • Developmental • Deviate • Deviated • Deviation • Device • Devon (UK-England) • Devonshire (UK-England)

Dév Développé • Développer • Développement

DEV Data Element Value • Develop • Developed • Developing • Development • Deviation • Device

DEVA Design for Validation • Development Acceptance • Development-Validation

DEVCO Development Committee

Devel Development

DEVN Deviation

Devon Devonshire (UK-England)

DEVOS Deep Extragalactic VLBI-Optical Survey

Devp Develop

Devpt Development

DEVS Driver's Enhanced Vision System

DEVSIS Development Science Information System

DEW Directed Energy Weapon • Distance Early Warning • Distant Early Warning

DEWIZ Distant Early Warning Identification Zone

dex Power of Ten

DEX Data Exchange • Dexterity

DEXA Database and Expert-system Applications (conference)

Dez Dezember • Dezembro

DEZ Deirezzor (Syria) (location identifier) • Deutsches Energiezentrum (Germany) • Dezembro

d/f Días Fecha

D/F Design/Fabrication • Detached Flight • Direction Finder • Direction Finding

D&F Determination & Finding

DF Damage Free • Data Facility • Data Field • Data Flag • Data Flow • Data Fusion • Dead Freight • Dean of Faculty • Decimal Fraction • Decontamination Factor • Deep Field • Default Forwarding • Deflection • Degree of Freedom • Density Functional • Derechos Fundamentales • Describing Function • Destination Field • Detection Fraction • Development Flight • Development Forward • Device Flag • Dielektrische Festkörper • Diesel Fuel • Dirac-Fock • Direct Flight • Direct Flow • Direction Finder • Direction Finding • Disappearing Filament • Disassembly Facility • Discrimination Filter • Disk File • Dispersion Factor • Display Function • Disposition Form • Dissipation Factor • Distribution Factor • Distribution Frame • Distribution Function • Distrito Federal • Djibouti Franc (currency) • Document Frequency • Don't Fragment • Dose Factor • Double Feeder • Double Flag • Double Focusing • Double Foolscap • Draft • Drawing File • Droits Fondamentaux • Dust Formation • Duty Factor • Duty Free • Dynamical Friction

DFA Data Flow Analyst • Department of Foreign Affairs • Describing Function Analyzer • Design For Automation • Deterministic Finite Automaton • Detrended Fluctuation Analysis • Development Fund for Africa • Digital Fault Analysis • Digital Frequency Analysis • Direction Finding Antenna • Discriminant Function Analysis • Dominant Feature Analysis • Dual-Frequency Altimeter • DXI Frame Address • Dynamic Force Analysis

DFAC Drag-Free and Attitude Control

DFACS Drag-Free and Attitude Control System

DFAD Digital Feature Analysis Data • Digital Feature Attribute Data

DFAE Département Fédéral des Affaires Étrangères (Switzerland)

DFAIT Department of Foreign Affairs and International Trade

DFAT Department of Foreign Affairs and Trade • Destination Final Acceptance Test

DFAX Digital Facsimile

DFB Datenfernschaltbaugruppe • Digital Filter Board • Distributed Feedback (*) • Distribution Feedback • Distribution Fuse Board • Driftwood Bay (USA-AK) (location identifier)

DFBLD Distribution Feedback Laser Diode

DFBS Digitized First Byurakan Survey

DFBSC-LD Dispersion Feedback Semiconductor Laser Device

DFBSCLD Dispersion Feedback Semiconductor Laser Device

DFBW Digital Fly-By-Wire

DFC Data Feed Client • Data Field Channel • Data Filter Controller • Data Flow Control • Data Format Converter • Data Format Converter • Département des Finances et de la Comptabilité • Design Field Change • Diagnostic Flow Chart • Directions For Conducting • Disk File Control • Divergence-Free Constraint • Division Financière et Comptable • Dossier de Fabrication et de Contrôle • Dust-Free Chamber

DFCD Data Format Control Document

DFCE Divergence-Free Constraint Equation

DFCF Dispersion-Flattened Compensated Fiber

DFCG Directeur Financier et de Contrôle de Gestion

DFCR Deputy Functional Chief Representative

DFCS Digital Flight Control Software • Digital Flight Control System (*)

DFD Data Flow Diagram • Data Function Diagram • Design For Discard • Deutsches Fernerkundungsdatenzentrum (Germany) • Digital Flight Display • Digital Frequency Discriminator • Digital Frequency Display • Direct Fringe Detection • Displaced Frame Difference • Division of Fluid Dynamics

DFDC Disk File Descriptor Control

DFDL Dual-Frequency Doppler Lidar • Dual-Frequency Dye Laser

DFDR Digital Flight Data Recorder

DFDS Data File Distribution System

DFDSS Data Facility Data-set Services

DFDT Dynamic Fault Diagnosis Technique

DfE Department for Education

DFE Dark Frame Error • Data Flow Editor • Data Flow Engineer • Debye-Falkenhagen Effect • Decision Feedback Equalizer • Department For Education • Designated Federal Entity • Designed For the Environment • Desktop Functional Equivalent • Direction-Finding Equipment • Directorate of Facilities Engineering • Distributed Framework Executive • Dynamic Finite Element

DfEE Department for Education and Employment (UK)

DFELL Duke Free Electron Laser Laboratory (USA-NC)

DFEO Diplôme de Fin d'Études Obligatoires

DFF Dispersion-Flattened Fiber

DFFR Dynamic Forcing Function Report

DFFS Distance From Face Space

DFG Data Flow Graph • Deutsche Forschungsgemeinschaft (Germany) • Digital Function Generator • Diode Function Generator • Discrete Frequency Generator • Display Format Generator • Distinguished Foreign Guest

DFGI Deutsches Geodätisches Forschungsinstitut (Germany)

DFGII Deutsches Geodätisches Forschungsinstitut - Abteilung I (Germany)

DFGS Digital Flight Guidance System

DFGt Datenfernschaltgerät

DFH Dong Fang Hong

DFHSM Data Facility Hierarchical Storage Manager

DFI Dark Field Illumination • Data Field Identifier • Defiance (USA-OH) (location identifier) • Design - Fabricate - Install • Developmental Flight Instrument • Developmental Flight Instrumentation • Development Flight Instrumentation • Development Flight Integration • Digital Facility Interface • Direct Foreign Investment • Direction des Finances • Direction Finding Indicator • Domain Format Identifier • Double Fourier Interferometry

DFIB Data Function Information Block

DFID Department For International Development (UK)

DFI PLT Development Flight Instrumentation Pallet

DFIR Data Flow Instrument Responsible

DFIS Digital Facsimile Interface System • Dual Filament Ion Source

DFK Digitale Flurkarte

Dfl Deflating

DFL Data Field Line • Data File Line • David Florida Laboratory (Canada) • Daily Flight Log • Decommutator Format Load • Decreto con Fuerza de Ley • Deterministic Fragment Loss • Display Formatting Language

DFLD Distribution-Free Logic Design

DFLS Direction Finding and Location System

DFM Data Filter Manager • Data Flow Management • Data Flow Manager • Defence Mapping Agency (USA) • Deputy for Financial Management • Design For Maintainability • Design For Manufacturability • Design For Manufacture • Design For Manufacturing • Digital Frequency Meter • Dispositif de Formation Modulaire • Dual-Frequency Method • Dust Flux Monitor

DFMA Design For Manufacturing and Assembly

DFMI Dust Flux Monitor Instrument

DFMO Division Financial Management Office • Doppler Filter Mixer-Oscillator

DFMR Dual Frequency Microwave Radar

DFMS Digital Facility Management System

DFMW Dust Flux Monitor Whipple (shield)

DFN Deutsches Forschungsnetz (Germany)

DFNT Dipartimento di Fisica Nucleare e Teorica

DFO Data Flow Operations • Department of Fisheries and Oceans • Director of Flight Operations • Disk File Optimizer • Dynamic Feedback Optimization • Dynamics Functional Objective

DFOS Danish Faint Object Spectrograph

DFOSC Danish Faint Object Spectrograph and Camera

DFOT Doi - Fugugita - Okamura - Tarusawa (catalog)

DFP Data Facility Product • Davidon - Fletcher - Powell (method) • Délégation à la Formation Professionnelle • Digital Flat Panel • Diode Flat Pack • Dipole Flat Plate • Display Function Processor • Distributed Functional Plane • Drumduff (Australia-QLD) (location identifier) • Dry Film Processor • Dry Filter Processing • Dynamic Flow Parameter

DFPDN Digital Fixed Public Data Network

DFPL Data Flow Programming Language

Dfr Defrost

DFR Decreasing Failure Rate • Defrost • Digital Flight Recorder • Direction Finding Receiver • Document Filing and Retrieval • Drag-Free Reference • Dual-Frequency Receiver • Dust-Free Room

DFRC Dryden Flight Research Center (USA-CA) (formerly DFRF)

DFRD Data Format Requirements Document

DFRF Dryden Flight Research Facility (NASA) (now DFRC)

DFRL Differential Relay

DFRNL Differential

DFS Daily Flow System • Dansk Fysisk Selskab (Denmark) • Data Flow Software • Data Flow System • Decision Fusion Simulator • Demodulation and Frame Synchronizer • Depth First Search (algorithm) • Deutsche Flugsicherung (Germany) • Deutscher Fernmelde Satellit (Germany) • Development Flight Satellite • Digital Field System • Digital Frequency Synthetizer • Direct Flow Sampler • Directional Finding System • Direction Finding System • Disk Filing System • Display Formatting System • Distributed File Service • Distributed File System • Document Finding System • Donnerstag - Freitag - Samstag • Doppler Frequency Shift • Dynamic Flight Simulator

DFSA Designated Finance Support Activity

DFSG Datenfernschaltgerät

DFSK Direct Frequency-Shift Keying • Double Frequency-Shift Keying

DFSM Dispersion Flatted Single Mode

DFSMS Data Facility Storage Management System

DFSO Division Finance Support Officer

DFSR Detailed Functional System Requirement

DFSS Debt-for-Science Swap • Depth From Shading and Stereo

Dft Draft • Drift

DFT Density Functional Theory • Design Feasibility Test • Design For Test • Design For Testability • Diagnostic Function Test • Digital Filtering Technique • Direct Fourier Transform • Discrete Fourier Transform • Discrete Fourier Transformation • Distributed Function Terminal • Distribution Functional Terminal • Document File Transfer • Down Flow Tent • Draft • Dual-Flow Turbine

Dftg Drafting

DFTNIC Data Flow Technology Network Information Center

Dftr Deflector

DFTS Dispersive Fourier Transform Spectrometer • Dispersive Fourier Transform Spectroscopy

DFTUZ Departamento de Física Teórica de la Universidad de Zaragoza (Spain)

DFÜ Datenfernübertragung

DFU Data File Utility

DFUS Diffuse

DFV Datenfernverarbeitung

DFVLR Deutsche Forschungs- und Versuchsanstalt für Luft- und Raumfahrt (Germany) (now DLR)

DFW Dallas - Fort Worth (USA-TX) (location identifier) • DASD Fast Write

DFWG Data Flow Working Group

DFWM Degenerate Four Wave Mixing

DFWMAC Distributed Foundation Wireless Media Access Control

DFyT Derecho Financiero y Tributario

DFZ Dislocation Free Zone

dg Decigram • Dei Gratia

D/G Dust/Gas Ratio • Dwarf/Giant (star)

DG Data General • Data Good • Data Group • Dei Gratia • Density Gradient • Deo Gratias • Dependence Graph • Design Guide • Deutsche Grammophon • Differential Generator • Differential Geometry • Diode Gate • Directeur Général • Directional Gyroscope • Direction Générale • Directorate General • Director General • Direttore Generale • Direzione Generale • Displacement Gyro • Display Generator • Display Group • Documentation Group • Dorschner-Görtler (catalog) • Double Groove • Drafting Group • Dwarf Galaxy

DGA Dangriga (Belize) (location identifier) • Délégation Générale pour l'Armement (France) • Demi-Grand Axe • Direcção Geral do Ambiente (Portugal) • Direct Graphics Access • Distributed Genetic Algorithm

DGAC Direction Générale de l'Aviation Civile (France)

DGACM Direction Générale de l'Aviation Civile et de la Météorologie (Central African Rep.)

DGAFP Direction de l'Administration Générale et de la Fonction Publique (France)

DGaO Deutsche Gesellschaft für angewandte Optik (Germany)

DGB Danger Bay (USA-AK) (location identifier) • Disk Gap Band

DGC Data Gathering Document • Data General Corporation • Degahbur (Ethiopia) (location identifier) • Deutsche Gesellschaft für Chronometrie (Germany) • Direct Geodetic Constraint

DGCA Directorate General of Civil Aviation

DGCCRF Direction Générale de la Concurrence, de la Consommation et de Répression des Fraudes (France)

DGCI Discrete Geometry for Computer Imagery (workshop)

DGCL Direction Générale des Collectivités Locales (France)

DGCM Direct Geodetic Constraint Method • Dynamic Global Vegetation Model

DGCP Direction Générale de la Comptabilité Publique (France)

DGCR Direcção Geral de Desenvolvimento Regional (Portugal)

DGCS Downsized Ground Control Station

DGD Decision Guidance Document • Dotation Globale de Décentralisation • Dynamic Gravity Detector

DGDA Downlink Gimbal Drive Assembly

DGDAS Davenport Group Draft Advisory Standard

DGDDI Direction Générale des Douanes et des Droits Indirects (France)

DGDF Digital Geospatial Data File • Director General Discretionary Fund

DGDR Double Giant Dipole Resonance

DGDT Downsized Ground Data Terminal

DGE Davisson-Germer Experiment • Délégation Générale à l'Espace • Deutsche Gesellschaft für Ernährung (Germany) • Digital Ground Equipment • Direcção Geral de Energia (Portugal) • Division of Graduate Education (NSF) • Doppler Gravity Experiment • Dotation Globale d'Équipement • Dust and Gas Envelope • Mudgee (Australia-NSW) (location identifier)

DGED Direction Générale des Études et de la Documentation

DGEF Diffraction Grating Evaluation Facility

DGEFP Direction Générale de l'Emploi et de la Formation Professionnelle

DGEMP Direction Générale de l'Énergie et des Matières Premières (France)

DGENORS Direction Générale de l'Enseignement Non-Obligatoire et de la Recherche Scientifique (Belgium)

DGES Direction Générale de l'Enseignement Scolaire

DGESR Direction Générale des Enseignements Supérieurs et de la Recherche

DGF Digital Gamma Finder • Direction Générale des Finances • Dotation Globale de Fonctionnement • Douglas Lake (Canada-BC) (location identifier)

DGfB Deutsche Gesellschaft für Biophysik (Germany)

DGFI Deutsches Geodätisches Forschungsinstitut (Germany)

DGfK Deutsche Gesellschaft für Kartographie (Germany)

DGfS Deutsche Gesellschaft für Sprachwissenschaft (Germany)

DGG Daugo (Papua New Guinea) (location identifier) • Department of Geology and Geophysics • Deutsche Geologische Gesellschaft (Germany) • Deutsche Geophysikalische Gesellschaft (Germany) • Deutsche Glastechnische Gesellschaft (Germany) • Dynamic Gravity Generator • Dysprosium Gallium Garnet

DGGE Denaturing Gradient Gel Electrophoresis

DGGMNT Deutsche Gesellschaft für Geschichte der Medizin, Naturwissenschaft und Technik (Germany)

DGGMR Directorate General of Geology and Mineral Resources

DGH Dotation Globale Horaire

DG-HZ Delayed Gratification Habitable Zone

DGHZ Delayed Gratification Habitable Zone

DGI Department of Geographic Information • Direct Gasoline Injection • Directional Gyroscopic Indicator • Direction Générale des Impôts (France)

DGICYT Dirección General de Investigación Científica y Técnica (Spain)

DGIS Direct Graphics Interface Standard • Directoraat Generaal Internationale Samenwerking (Netherlands)

DGIWG Digital Geographic Information Working Group

D-GIX Distributed Global Internet Exchange

DGIX Distributed Global Internet Exchange

DGK Deutsche Geodätische Kommission (Germany) • Deutsche Gesellschaft für Kristallographie (Germany)

DGKGN Deutsche Geodätische Kommission Gravity Network (periodical code)

Dgl Dergelijke • Dergleichen • Desgleichen

DGL Diffuse Galactic Light • Distinguished Guest Lecturer • Doped Glass Laser • Douglas (USA-AZ) (location identifier)

DGLF Délégation Générale à la Langue Française

DGLR Deutsche Gesellschaft für Luft- und Raumfahrt (Germany)

DGM Deputy General Manager • Deutsche Gesellschaft für Materialkunde (Germany) • Digital Ground Modeling • Digital Group Multiplexer • Directional Gyro Mode • Discontinuous Galerkin Method

DGMA Digital Group Multiplexer Assembly

DGMP Deutsche Gesellschaft für Medizinische Physik (Germany)

DGN Dahlgren (USA-VA) (location identifier) • Degrees Grid North • Dirección General de Normas (Mexico) • Domestic Geographic Name

DGO Directional Gyro Operation • Director General's Office • Durango (Mexico) • Durango - Guadalupe Victoria (Mexico) (location identifier)

DGON Deutsche Gesellschaft für Ortung und Navigation (Germany)

DGOR Deutsche Gesellschaft für Operations Research eV (Germany)

DGP Data Generating Program • Datagram Packet • Daugavpils (Latvia) (location identifier) • Digital Graphic Product • Dirección General de la Policía • Direction Générale de la Production • Distributed Genetic Programming

DGPA Department of General Physics and Astronomy

DGPC Dave's Genetic Programming Code • Direction Générale de la Programmation et de la Coordination

DGPF Deutsche Gesellschaft für Photogrammetrie und Fernerkundung

D-GPS Differential Global Positioning System

DGPS Data Gathering and Processing System • Differential Global Positioning System

DGPST Differential Global Positioning System Technology

DGPT Direction Générale des Postes et Télécommunications

DGQ Deutsche Gesellschaft für Qualität

Dgr Degrease

DGR Dargaville (New Zealand) (location identifier) • Deductible Gift Recipient • Designated Government Representative • Dwarf-to-Giant Ratio

DGRB Diffuse Gamma-Ray Background

DGRC Digital Government Research Center (USA)

DGRCST Direction Générale des Relations Culturelles, Scientifiques et Techniques

DGRF Definitive Geomagnetic Reference Field

DGRST Direction Générale à la Recherche Scientifique et Technique (France) • Direction Générale de la Recherche Scientifique et Technologique

DGRT Direction Générale de la Recherche et de la Technologie

Dgrx Dangereux

dgs Diagnostics (file extension)

Dgs Degaussing

DGS Data Gathering System • Data Ground Station • Degaussing System • Degrees Grid South • Density Gradient Sendimentation • Deployable Ground Station • Deployable Ground System • Deutsche Gesellschaft für Standardisierung (Germany) • Digital Ground Station • Digital Graduate Student • Dirección General de Seguridad • Display Generation System • Distributed Graphics System • Distributed Graph Storage • Document-Generating System

DGSC Defense General Supply Center (USA)

DGSE Direction Générale de la Sécurité Extérieure (France)

DGSI Direction Générale des Services Informatiques • Direction Générales des Stratégies Industrielles (France) • Dow Geochemical Services Inc.

DGSM Directorate General for Specifications and Measurements

DGSNR Direction Générale de la Sûreté Nucléaire et de la Radioprotection (France)

DGSRD Directorate-General of Scientific Research and Development

DGSS Deep Groth Strip Survey

Dgt Digit

DGT Deutsche Große Teleskop (Germany) • Digit • Digital Geographic Technology • Dirección General de Tributos (Spain) • Directional Gamma-ray Telescope • Direction Générale des Télécommunications • Direction Générale Technique • Directorate General of Telecommunications • Division de la Gestion et des Traitements • Dumaguete (Philippines) (location identifier)

Dgtl Digital

DGTL Digital

DGTMM Dirección General de Territorio Marítimo y Marina Mercante (Chile)

DGU Dedougou (Burkina Faso) (location identifier) • Digital Gateway Unit • Digital Graphics Unit • Directional Gyro Unit • Display Generator Unit

DGV Dark Gray Varnish • Doppler Global Velocimetry

DGVM Dynamic Global Vegetation Model

DGW Douglas - Converse County (USA-WY) (location identifier)

DGWK Deutsche Gesellschaft für Warenkennzeichnung (Germany)

DGZ Designated Ground Zero • Desired Ground Zero

DGZfP Deutsche Gesellschaft für Zerstörungsfreie Prüfung (Germany)

dh das heißt • Dirham (currency)

D/H Deuterium/Hydrogen

D-H Decimal to Hexadecimal

DH Dahomean • Damage History • Data Handling • Data History • Dead Heat • Decametric-Hectometric • Decision Height • Decoherent History • Default Hierarchy • De Havilland • Dekameter-Hectometric • Derechos Humanos • Design Handbook • Design History • Direction Indicator • Dirham (currency) • Double Heterostructure • Double Hydrant • Drill Hole • Droits de l'Homme • Dynamically Hot • Moroccan Dirham (currency)

DHA Department of Humanitarian Affairs • Dependent Housing Area • Design Hazards Analysis • Destination Hardware Address • Dhahran (Saudi Arabia) (location identifier) • Diffie-Hellamn Algorithm

DHAE Department of Home Affairs and Environment

Dhc Doctor Honoris Causa

DHC Dark-Haloed Crater • Data Handling Center • Docteur Honoris Causa

DHCN Daily Historical Climate Network

DHCP Dynamic Host Configuration Program • Dynamic Host Configuration Protocol (*) • Dynamic Host Control Protocol

DHCU Data Handling and Control Unit

DHD Durham Downs (Australia-QLD) (location identifier)

DHDC Don Harrington Discovery Center (USA-TX)

DHDD Digital High-Definition Display

DHDS Data Handling and Display System

DHE Data Handling Equipment • Debye-Hueckel Equipment • Double Hop Experiment

DHEA Dihydroepiandrosterone

DHEI Diplôme des Hautes Études Internationales

DHEW Department of Health, Education, and Welfare

DHF Abu Dhabi - Al Dhafra (UAE) (location identifier) • Data Handling Facility

DHG Dotation Horaire Globale • Double Henyey-Greenstein (function)

DHH Doppelhaushälfte

DHHCS Department of Health, Housing and Community Services

DHHS Department of Health and Human Services (USA)

DHI Deutsches Hydrographisches Institut (Germany) (now BSH) • Dhangarhi (Nepal) (location identifier) • Disk-Halo Interaction

DHL Désastre Humanitaire pour les Localités • Dhala (Yemen) (location identifier) • Dutch Holographic Laboratories • Dynamic Head Loading

DHM Departamento de Hidrometeorologia • DEVISE Hypermedia • Dharamshala Gaggal (India) (location identifier)

DHME Departamento de Hidrometeorologia do Estado

DHMM Discrete Hidden Markov Model

DHMO Dihydrogen Monoxide

Dhmy Dehumidify

DHN Dothan (USA-AL) (location identifier)

DHO Damped Harmonic Oscillator

DHP Developed Horsepower

DHPE Data Hardware Project Engineer

DHPI Data Handling Plug-In

DHR Decay Heat Removal • Den Helder - De Kooy (Netherlands) (location identifier) • Diagramme Hertzsprung-Russell

DHRD Department of Housing and Regional Development

DHRM Dexter Hypermedia Reference Model • Dexter Hypertext Reference Model

DHS Data Handling System • Department of Homeland Security (USA) • Department of Human Services • Director of Health Services • Discrete Horizon Sensor • Division of History of Science • Double Heterostructure

DHSD Duplex High-Speed Data

DHSH Department of Human Services and Health

DHSS Data Handling Subsystem • Department of Health and Social Security (UK)

DHT Dalhart (USA-TX) (location identifier) • Directory Hash Table • Discrete Hartley Transform • Distributed Hash Table

DHTE Dynamic-Mapping Hypertext Engine

dHTML Dynamic Hypertext Markup Language

DHTML Dynamic Hypertext Markup Language

DHU Data Handling Unit • Detector Head Unit

DHV Dangerously High Violence • Deutscher Hochschulverband (Germany) • Doors - Hatches - Vents

DHW Dengle - Hart - Weinberg (planetary nebula) • Domestic Hot Water • Double-Hung Window

DHX Dump Heat Exchange • Dump Heat Exchanger

Dhyr Dehydrator

di das ist • Dat is • De-ice • Dienstags • Dinsdag

dI Dwarf Irregular (galaxy)

Di Didymium • Diameter • Dienstag

D&I Development & Implementation

D/I Distinctive Insignia • Drop and Insert

DI Daily Inspection • Data In • Data Input • Data Integration • Data Integrator • Data Interchange • Data Interface • Data Item • Dat Is • Daily Inspection • Decrement Interrupt • Deep Impact • Délégation à l'Informatique (France) • Demand Indicator • Department of the Interior • Dépense Intérieure • Descent Image • Descent Imager • Design Interface • Desk Instruction • Destination Index • Detachment Index • Developed Item • Developmental Item • Development Integration • Diameter • Dienstag • Dienstags • Differential Interferometry • Diffusion Internationale • Digital Image • Digital Input • Diode for Isolation • Diplomatic Immunity • Diplôme d'Instituteur • Diplomingenieur • Directeur de l'Informatique • Direct Impact • Direct-Injected • Direct Injection • Direction Indicator • Direction Informatique • Directivity Index • Dirty Image • Discrete Input • Disk Instability • Display Interface • Distance Indicator • Distributed Intelligence • District Inspector • Disturbance Index • Divisional Inspector • Documentation Integrator • Document Identifier • Document Imaging • Doppler Imaging • Double Imperial • Drill Instructor • Driving Issue • Dynamic Instability • Dynamic Interface

dia Diagraph (file extension)

Dia Diagram • Dialect • Diameter

DIA Darmstadt Indexing Approach • Date Impact Analysis • Date of Initial Appointment • Defense Intelligence Agency • Denver International Airport (USA-CO) • Diameter • Difference Image Analysis • Differential Image Analysis • Digital Interface Adapter • Direct Interaction Approximation • Document Indexing Approach • Document Interchange Architecture • Due In Asset

DIAD Drum Information Assembler/Dispatcher

DIADS Digital Image Analysis and Display System • Digital Image Analysis Display System • Digital Integrated Air Defense System

Diag Diagnose • Diagnosis • Diagonal • Diagram

DIAG Diagnostics • Diagonal • Diagram

DIAGD Diagnostic Daemon

DIAGN Diagnose

Dial Dialect • Dialecte • Dialectal • Dialectic • Dialéctica • Dialectical • Dialectique

DIAL Department of Information on the Astronomical Life • Diagnostic Instrumentation and Analysis Laboratory (USA-MS) • Differential-Absorption Lidar • Digital-Image Analysis Laboratory • Direct Information Access Line • Disability Information Awareness Line • Display Interactive Assembly Language

DIALCOM Dialed Communications

Dialect Dialectal

Diam Diameter • Diamètre

DIAM Data Independent Accessing Model • Data Independent Access Model • Defense Intelligence Agency Manual

DIAN Digital Analog

DIANA Data and Information Available Now in Africa • Digital Integrated Access Network Arrangement • Direct Information Access Network for Africa

DIANE Direct Information Access Network for Europe

DIANNE Direct Inversion based on Artificial Neural Networks

Diaph Diaphragm

DIARAD Dual Irradiance Absolute Radiometer

DIAS Defense Automatic Integrated System • Digital Image Analysis System • Distributed Intelligent Actuator and Sensor • Dublin Institute for Advanced Studies (Ireland) • Dynamic Inventory Analysis System

DIAW Dust-Ion-Acoustic Wave

dib Device-Independent Bitmap (file extension)

DIB Data Input Box • Data Input Bus • Data Interface Buffer • Déchet Industriel Banal • Device-Independent Bitmap • Dibrugarh (India) (location identifier) • Dictionary of International Biography • Diffuse Interstellar Band • Direct Inward Dialing • Directory Information Base • Dual Independent Bus

DIBI Device-Independent Backup Interface

DIBOC Distributed Input Buffering with Output Control

DIBOL DEC's Business-Oriented Language

DIBS Deskbook Information Breakdown Structure • Diffuse Interstellar Bands • Dynamic Infrared Background System

dic Dictionary (file extension)

Dic Dicembre • Diciembre

DIC Data Information Center • Data Input Clerk • Data Input Console • Data Insertion Converter • Data Interface

Control • Data Item Category • Defence Industries Council (UK) • Departamento de Investigaciones Científicas • Dictionary • Differential Interference Contrast • Digital Input Computer • Digital Integrated Circuit • Digital Integrating Computer • Digital Interface Controller • Dili (Zaire) (location identifier) • Diploma of Imperial College (UK) • Direct Image Copy • Direction de l'Information et de la Communication • Dissolved Inorganic Carbon • Distributed Information Centre • Documentation and Information Centre • Document Identifier Code • Double Ionization Cell • Double Ionization Chamber

DICAP Direct-Current Circuit Analysis Program

DICAPS Document Imaging and Computer-Assisted Publication System

DICB Data Interface Control Board

Dicc Diccionario

DICC Digital Interface Code Converter

DICCG Deflated Incomplete Choleski Conjugate Gradient (method)

DICD Draft Interface Central Document

DICE DAO Instrument Control Environment • DARPA Initiative in Concurrent Engineering • Defense Initiative for Concurrent Engineering • Digital Integrated Circuit Element • Digital Intercontinental Conversion Equipment • Direct Intercommunication Experiment • Direct Inter-establishment Communication Experiment • Direct Inter-establishment Communication in Europe • Distributed Interactive Collaboration Environment • Dual Imaging Čerenkov Experiment • Dynamic Integrated Climate Economy

DICEF Digital Communications Experimental Facility

DICIFER Digital Interactive Complex for Image Feature Extraction and Recognition

DICO Data Information and Coordination Office

DICS Data Interface Control Scientist • Digital Image Correction System

Dict Dictation • Dictator • Dictionary

DICT Dirección de la Información Científica y Técnica

DICU Data Interface Control Unit • Digital Interface and Control Unit • Display Interface Control Unit

DICW Data Integrity Control Workstation

DID Damage - Injuries - Death • Data Information Delivery • Data Information Description • Data Ingest Distribution • Data Input Display • Data Item Description • Data Item Descriptor • Destination Identification • Digital Image Data • Digital Image Document • Digital Information Display • Digital Interactive Decoder • Direct Inbound Dialing • Direct Injection Diesel • Direct Inward Dial • Direct Inward Dialing • Display Interface Device • Double Isotope Derivative • Drum Information Display • Dust Impact Detection

DIDA Dynamic Instrumentation Digital Analyzer

DIDAC Digital Data Acquisition • Dutch ISO Data Centre (Netherlands)

DIDAK Digital Image Data Analysis Karlsruhe

DIDC Data Input and Distribution Card • Data Input Display Console

DIDD Dynamic Integrated Data Display

DIDDS Dynamic Integrated Data Display System

DIDF Dual Input Describing Function

DIDL Digital Integrated Design Language

DI/DO Data Input / Data Output

DIDOM Distribution de la Décision dans les Organisations Matricielles

DIDOT Digitizing and Designing Of Type

DIDS Decision Information Display System • Decision Information Distribution System • Defense Integrated Data System • Digital Information and Display System • Digital Information Display System • Distributed Information Delivery System • Distributed Intrusion Detection System • Domestic Information Display System • Donor-Impurity Density of States

DIDSY Dust Impact Detection System

DIE Antsiranana (Madagascar) (location identifier) • Diploma in Industrial Engineering • Documentation Information Element

dieb. alt. Diebus Alternus

DIEBRA Digital Image Enlarging Balanced Reconstruction Algorithm

dieb. secund. Diebus Secundis

dieb. tert. Diebus Tertius

Diel Dielectric

DIES Data Interpretation and Evaluation System • Digital Image Enhancement System

DIESA Department of International and Social Affairs

DIET Desorption-Induced by Electronic Transitions

DIEU Digital Interface Electronics Unit

dif Data Interchange Format (file extension)

Dif Differ • Difference • Different • Differential

DIF Data Integration Function • Data Interchange Format • Data Interface Facility • Daily Impact Factor • Decay In Flight • Descriptive Item File • Development Integration Facility • Difference • Difficulty - Importance - Frequency • Diffuse • Digital Interface Frame • Directory Interchange Format • Discrete Increment Filter • Display Information Facility • Document Interchange Format

DIFAR Directional Finding And Ranging

DIFAS Digital Ice Forecast and Analysis System

DIFAX Digital Facsimile

Diff Differ • Difference • Different • Differential • Difficult • Difficulty • Diffraction

DIFF Differential

DIFM Due In From Maintenance

Dig Digest • Digit • Digital • Digitalize • Digitalized

DIG Deputy Inspector-General • Designated Industry Group • Diffuse Ionized Gas • Digital • Digital Image Generation • Digital Imposition Geometry • Direction Informatique Généralisée • Disabled Income Group • Dwarf Irregular Galaxy

DIGACC Digital Guidance And Control Computer

DIGBE Diffuse Galactic Background Experiment

DIGCOM Digital Computer

DIGENOR Dirección General de Normas y Sistemas de Calidad

DIGEST Digital Geographic Exchange Standard

DIGI Deutsche Interessengemeinschaft Internet eV (Germany) • Digital

DIGICOM Digital Communications System

DIGITEF Diffusion par l'Informatique de Graphiques, d'Images et de Textes pour l'Enseignement du Futur

DIGITIP Direction Générale de l'Industrie, des Technologies, de l'Information et des Postes (France)

DIGL Device-Independent Graphics Layer

DIGMAP Digital Map

DIGON Digitales Ortsnetz

DIGPM Diphase Group Modem

DIGRAPH Directed Graph

DIGS Delta Inertial Guidance System • Device-Independent Graphics Service • Digital Inertial Guidance System

DIH Discrete Input High

DIHIDRONAV Dirección de Hidrografía y Navegación (Peru)

DIHK Deutsche Industrie- und Handelskammer (Germany) (formerly DIHT)

DIHT Deutsche Industrie- und Handelstag (Germany) (now DIHK)

DII Diode Ion Injector • Doppler Imaging Inversion • Dynamic Invocation Interface

DIIC Dialectically Isolated Integrated Circuit

DIID Data Independent Interface for Databases

DIIG Digital Information Infrastructure Guide

DIIP Direct Interrupt Identification Port

DIIPS Digital Interactive Image Processing System

DIIR Digital Image-Intensifier Radiography

DIIS Directorate for Intelligence and Information Security

DIIT Décès, Invalidité ou Incapacité Temporaire

Dij Dijon (France)

DIJ Délégation à l'Insertion des Jeunes • Dijon (France) (location identifier)

DIK Data - Information - Knowledge • Dickinson (USA-ND) (location identifier)

Dil Dilute • Diluted • Dilution

DIL Deliverable Items List • Digital Integration Laboratory • Digitization Integration Laboratory • Dili (Indonesia) (location identifier) • Discrete Input Low • Distributed Interaction Language • Dual In-Line • Dynamic Imaging Library

DILGEA Department of Immigration, Local Government and Ethnic Affairs

DILHR Department of Industry, Labor and Human Relations

DILIB Document and Information Library

Diln Dilution

DILO Deep Insertion - Large Opening

DILSA Distributed Integrated Logistics Support Analysis

dim Dimidius

Dim Dimanche • Dimension • Diminish • Diminuendo • Diminutive

DIM Data In Memory • Data Input Message • Data Input Module • Data Interface Module • Data In the Middle • Decimetric • Decimetric Radiation • Department of Immigration and Multicultural Affairs (Australia) • Description, Installation and Maintenance • Design Impact Meeting • Design Information Manual • Design Interface Manual • Design Interface Meeting • Differential Image Motion • Differential Inverse Method • Diffuse Ionized Medium • Digital Image Matching • Digital Input Module • Dimbokro (Ivory Coast) (location identifier) • Diminish • Diminished • Diminishing • Diminution • Diplôme d'Ingénieur-Maître • Discrete Interface Module • Disk Instability Model • Distributed Information Management • Distributed Information Manager • Document Image Management • Dust Impact Monitor

DIMA Direct Imaging Mass Analyzer

DIMACS Discrete Mathematics and Theoretical Computer Science

DIMAP Distributed Image Management And Projection

DIMAPS Digital Image Manipulation Analysis and Processing System • Digital Image Manipulation And Processing System

DIMD Dichroic Medium Dispersion Mode

DIMDI Deutsches Institut für Medizinische Dokumentation und Information (Germany)

DIME Development of Integrated Monetary Electronics • Direct Memory Execute • Dual Independent Map Encoding

Dimen Dimensions (periodical code)

DIMEN Dimension

DIMES Delfts Interfacultair Instituut voor Micro-Elektronica en Submicrontechnologie (Netherlands) • Diffuse Microwave Emission Survey • Digital Image Manipulation and Enhancement System

DiMF Dreier in Metallfolie

DIMIAS Digital Interactive Multiimage Analysis System

Dimin Diminish • Diminuendo • Diminutive

dIML Digitale Internationale Mietleitung

DIMM Differential Image Motion Monitor • Diplôme en Ingénierie Multilingue et Multilangue • Dual In-line Memory Module

DIMS Data and Information Management System • Data Information and Manufacturing System • Data Input Management System

DIMT Diagonal Intrinsically Multiply Transitive

DImTe Display Imaging Technology (periodical code)

DIMWG Data and Information Management Working Group

Din Dinar (currency) • Dining • Dinner • Dinsdag

DIN Data Identification Number • Deutsche Industrie-Normen (Germany) • Deutsches Institut für Normung (Germany) • Dien Bien Phu (Viet Nam) (location identifier) • Device Identification Number • Digital Imaging Network • Digital Network • Direct Injection Nebulizer • Disaster Information Network • Document Identification Number

DINA Detector of Ions and Neutral Atoms • Direct Noise Identification • Distributed Information Network Architecture

DINAMEL Dirección Nacional de Metrologia Legal (Uruguay)

DIng Doctor of Engineering

DINS Digital Inertial Navigation System

DINU Dual Inertial Navigation Unit

Dio Diode

DIO Dark Irregular Object • Data Input/Output • Device Input/Output • Digital Input/Output • Digital Interoperability • Diomede Island (USA-AK) (location identifier) • Distinguished Information Office • Division Information Officer

DIOB Digital Input/Output Buffer

DIOCU Digital Input/Output Control Unit

DIODE Designing Inorganic-Organic Devices (network) • Détermination Immédiate d'Orbites par DORIS Embarquée • Digital Input/Output Display Equipment

DIOI Digital Input/Output Interface

DION Documentair Informatiesysteem voor Onderwijsliteratuur in Nederland

DIOnAS Data Ingest and Online Access System

Dióp Dióptrica

DIOP Digital Input/Output Package

Diopt Diopter

Dióptr Dióptrica

DIOS Direct Input/Output System

Dip Diploma • Diputación

DIP Data In-line Package • Data Input Processor • Data Integrated Plan • Data Integration Program • Debtor In Possession • Deep Inelastic Process • Designated Inspection Point • Design Improvement Program • Detailed Integration Plan • Dialup Internet Protocol • Diapaga (Burkina Faso) (location identifier) • Digital Image Processing • Digital Image Processor • Digital Imaging Processing • Digital Incremental Plotter • Digital Information Pheronome • Digital Instrumentation Programmer • Dimensional Inspection Planner • Diploma • Direct Imaging Plotter • Directories In Print (Gale) • Display Information Processor • Display Input Processor • Display Interface Processor • Dissemination Information Package

• Distributed Information Processing • Distribution and Ingest Peripheral • Document Image Processing • Document Improvement Programme • Document Interchange Protocol • Drop and Insert Point • Dual In-line Package • Dual In-line Pin

DIPADA Définition - Identification - Preuve - Acceptation - Désir - Action

DipAM Diploma in Applied Mechanics

DIPAOS Direction des Personnels Administratifs, Ouvriers et de Service

DipAppSc Diploma of Applied Science

DipCE Diploma in Civil Engineering

DipChemEng Diploma in Chemical Engineering

DipEd Diploma in Education

DipEl Diploma in Electronics

DipEng Diploma in Engineering

DIPES Distributed and Parallel Embedded Systems (conference)

Dipl Diplexer • Diploma • Diplomat • Diplomatic

DIPL Display Initial Program Load • Drain-Induced Barrier Lowering

DipLib Diploma of Librarianship

Dipl.-Ing. Diplomingenieur

DipLSc Diploma of Library Science

Diplxr Diplexer

DipMechE Diploma of Mechanical Engineering

DIPORS Digital Image Processing Of Remotely Sensed Data

DipPA Diploma in Public Administration

DIPPA Digital Parallel Processing Array

DIPPR Design Institute for Physical Property Data (USA)

DIPR Departmental Industrial Plant Reserve • Detailed In-Process Review

DIPS Development Information Processing System • Digital Image Processing System • Digital Imagery Processing System • Digital Information Processing System • Display Image Processing System • Display Information Processor System • Distributed Internet Patent Service • Dynamic Isotope Power System

DipT Diploma in Teaching

DipTech Diploma in Technology

DIQ Delivered Image Quality • Divinópolis (Brazil-MG) (location identifier)

dir Directory (file extension) • Dirham (currency)

Dir Dirección • Direct • Directed • Directeur • Direction • Directional • Directive • Director • Directorate • Directory • Direttore • Direzione

DIR Data Item Requirement • Daytime Infrared • Department of Industrial Relations • Design Implementation Review • Design Information Release • Design Information Request • Developer Inhibitor Release • Device Identification Register • Diagnostics Responder • Direct • Direct Information Reading • Direct Information Recording • Direction • Direction Interrégionale • Director • Directorate • Directory • Dire Dawa (Ethiopia) (location identifier) • Disassembly and Inspection Report • Discipline-oriented Information Retrieval • Division of Intramural Research • Document Information Retrieval • Documentation Information Record • Document Information Record

DIRA Data Information Retrieval in Astronomy • Distribution Information Retrieval in Astronomy

DiRaG Dinamika Razrezhennykh Gazov (periodical code)

DIRB Diffuse Infrared Background

DIRBE Diffuse Infrared Background Experiment

DIRC Detection of Internally Reflected Čerenkov (light) • Digital Interrelay Communication • Din International Research Center (USA-NM)

DIRCM Directed Infrared Countermeasure

DIRCN Décennie Internationale pour la Réduction des (effets des) Catastrophes Naturelles

DIRD Dépense Intérieure de Recherche et de Développement

DIRECT Digital Rectifier

DIREE Direct Integration of Relaxed Einstein Equations

DIREN Direction Régionale de l'Environnement (France)

DIREP Difficulty Report

Dir.Gen. Director General

Dir.-Genl. Director-General

DIRM Directorate for Information and Resource Management

DIRPA Directorate of Personnel and Administration

dIrr Dwarf Irregular (galaxy)

DIRS Damage Information Recording System • Dialog-Information Retrieval System • Digital Image Rectification System • Digital Image Registration System • Distributed Information Retrieval System • Dynamic Information Retrieval System

DIRSP Dynamic Infrared Scene Projector

DIRT Dust Infrared Toolbox • Dusty Infrared Test

DIRTS Data and Information Related To Seismology

dis Distribution (file extension)

Dis Discharge • Disciple • Discipline • Disconnect • Discontinue • Discount • Dispense • Distance • Distant • Distribute

DIS Data and Information System • Data Information System • Data Input System • Data Interface System • Data Interpretation System • Daylight Imaging System • Deep Imaging Survey • Deep Inelastic Scattering • Defense Industrial Security • Defense Investigative Service • Design Improvement Study • Designing Interactive Systems (conference) • Development Information System • Device Interface System • Digital Image Stabilizing • Digital Imaging Spectrophotometry • Digital Imaging Spectroscopy • Digital Imaging Subset • Digital Imaging System • Digital Instrumentation System • Digital Integration System • Diplôme Interdisciplinaire de Spécialisation • Diplôme Interuniversitaire de Spécialisation • Director of Installation Support • Disabled In Service • Disagree • Discussion • Dispersive Interferometric Spectroscopy • Display • Distributed Information System • Distributed Interactive Simulation • Distributed Interactive System • Divisional Information System • Documentation Index System • Document Imaging System • Document Inquiry and Storage • Document Inquiry System • Doppler Inertial System • Double Imaging Spectrograph • Draft Information Standard • Draft International Standard • Dynamic Impedance Stabilization • Loubomo (Congo) (location identifier)

DISA Data Interchange Standards Association (USA) • Defense Information Systems Agency • Defense Information Systems Agency Information (periodical code) • Deployable Information Systems Architecture • Direct Inward Station Access • Direct Inward System Access • Directory Information Service Application

Disabld Disabled

DISAC Digital Simulator And Computer

DISAM Direct and Index Sequential Access Method

DISAO Designated Independent Senior Acquisition Official

DISAP Disapproved

Disassm Disassemble

Disassy Disassembly

DISASSY Disassembly

DISABL Disable

Disc Disciple • Discipline • Disconnect • Disconnected • Disconnecter • Discontinue • Discontinued • Discount •

Discover • Discover (periodical code) • Discovered • Discoverer • Discovery • Discrete

DISC Delay In Separation Code • Differential Scattering • Digital Information System CD-ROM • Digital Stereo Correlation System • Discone • Disconnect • Discrete • Distributed Computing (symposium)

Disch Discharge • Discharged

DISCH Discharge

DISCO Design and Implementation of Symbolic Computation Systems (symposium) • Direct-Image Stabilized Camera Option • Direct SCF Optimization • Division Information Systems Control Officer • Dual Irradiance and Solar Constant Orbiter

DISCOM Digital Selective Communications • Division Support Command

DISCON Discrepancy In Shipment Conformation

DISCOS Database and Information System Characterizing Objects in Space • Distributed Computing Solution • Disturbance Compensation System

Discr Discriminate • Discriminated • Discrimination

DISCR Director for Industrial Security Clearance Review

Disct Discount

DISCUS Delaware Innovative Science/Math Collaborative for Undergraduate Success • Digital Self-healing Control for Upgraded Safety

DISD Data and Information Systems Division

DISEL Diseño e Ingenieria de Sistemas Electrónicos

DISENG Disengage

DISGM Diphase Supergroup Modem

DISH Design and Implementation of Software for Historians

DISIDS Display and Information Distribution System

DISIMP Device Independent Software for Image Processing

DiSis Dinamicheskie Sistemy (periodical code)

DISMAC Digital Scene Matching Area Correlation

DISME Distrito de Meteorologia

DISN Defense Information Systems Network • Digital Integrated Services Network

DISOSS Distributed Office Support System • Distributed Office Systems Support

Disp Dispatcher • Dispense • Dispenser • Disperse • Dispersion • Display • Displacement • Disposición

DISP Directory Information Shadowing Protocol • Dispenser • Display • Displacement • Display • Disposition • Drug Investigation Support Program

DISPAC Domestic and International Scientific Planning and Cooperation

DISPERS Distributed Personnel System

DISPI Dispersed Position-coded Intensity

Displ Displaced • Displacement • Displays (periodical code)

Dispo Disponible • Disposition

Disr Disruption

DISR Descent Imager and Spectral Radiometer • Descent Imager/Spectral Radiometer • Disruption

Diss Dissipation

DISS Diffractive Interstellar Scintillation

DISSPLA Display Integrated Software System and Plotting Language

Dist Distance • Distant • Distinguish • Distinguished • Distort • Distorted • Distortion • Distribute • Distributed • Distribution • District

DIST Daewoo Institute of Science and Technology (Korea-ROK) • Délégation à l'Information Scientifique et Technique (France) • Department of Industry, Science and Technology (Australia) • Direction de l'Information Scientifique et Technique • Distance • Distribution • District • Documentation et Information Scientifique et Technique

DISTB Direction de l'Information Scientifique et des Bibliothèques

Distn Distortion

DISTN Direction à l'Information Scientifique et aux Technologies Nouvelles (France)

DISTNB Direction de l'Information Scientifique, des Technologies Nouvelles et des Bibliothèques (France)

Distr Distribución • Distribute • Distributed • Distribution • Distributivo • Distributor • District • Distrikt

DISTA Distribution Authority

DISTR Distributor

Dists Districts

DISUM Daily Intelligence Summary

Disyunt Disyuntivo

DI&T Design, Integration & Test

DIT Data Ingest Technician • Data Inquiry Terminal • Décès ou Incapacité de Travail • Deployable Internetworking Test • Deployed Internetworking Test • Depth in Inheritance Tree • Design, Integration and Test • Detector Integration Time • Detroit Institute of Technology (USA-IL) • Development Interoperability Testing • Directeur de l'Informatique et des Télécoms • Directory Information Tree • Display Index Table • Divisão de Informação Técnica • Division Informatique et Telecommunications • Division of Information Technology • Documentation Implementation Team • Documentation Industrielle et Technique • Double Income Tax • Dublin Institute of Technology (Ireland) • Draft Initiation Time • Dynamic Integrated Test

DITAC Department of Industry, Technology And Commerce

DITDOS Distributed Inventory Tracking and Data Ordering Specifications

DITEN Direction de l'Information et des Technologies de l'Éducation Nationale (France)

Dithm Dithmarschen (Germany)

DITMO Development, Integration, Test, and Mission Operations

DITN Direction de l'Information et des Technologies Nouvelles

DITR Department of Industry, Technology and Resources • Deutsches Informationszentrum für Technische Regeln (Germany)

DITS Digital Imagery Terminal System • Digital Imagery Transmission System • Digital Information Transfer System • Distributed Instrumentation Tasking System

DITU Digital Interface Test Unit

DITY Do It Yourself

DIU Data Interface Unit • Data Information Unit • Dedicated Interface Unit • Detector Interface Unit • Digital Indoor Unit • Digital Input Unit • Digital Insertion Unit • Digital Interface Unit • Diu (India) (location identifier) • Document Interchange Unit

Div Divergence • Diverse • Diversion • Diverter • Divide • Dividend • Divided • Divine • Divinity • Division • Divisional • Divisione • Divisor

DIV Data In Virtual • Délégation Interministérielle à la Ville (France) • Distance d'Itakura Vectorielle • Divide • Division • Divo (Ivory Coast) (location identifier) • Dynamic Imagery Viewer

DiVA Distance Visibility Algorithm

DIVA Département d'Information sur la Vie Astronomique • Deutsches Interferometer für Vielkanalphotometrie und Astrometrie (Germany) • Dialogprogramm zur Versuchsdaten-Analyse • Dicyanovinyl Anisole • Digital Input Voice Answerback • Digital Video Application • Distributeur de

Vignettes d'Affranchissement • Double Interferometer for Visual Astrometry

DIVE Direct Interface Video Extension • Distributed Interactive Virtual Environment

DIVIMP Divertor Impurity

Divn Division

Divnl Divisional

DIVOLI Diviseur Optique de Ligne

DIVOT Digital-to-Voice Translator

DIVOTS Data Input Voice Output Telephone System

DIW Defense Information Warfare • Department of Industrial Works • Detection - Identification - Warning • Deutsche Institut für Wirtschaftsforschung (Germany) • D-Inside Wire

DIWS Digital Imagery Workstation

DIX DEC/Intel/Xerox • Deutsche Immobilienindex (Germany)

DIXRD Delft Instruments X-Ray Diffraction (Netherlands)

DIY Diyarbakir (Turkey) (location identifier) • Do It Yourself

dJ der Jüngere • dieses Jahres

Dj Djebel

DJ Date Julienne • Deceptive Jamming • Deutsche Justiz (Germany) • Dinner Jacket • Disk Jockey • District Judge • Djibouti (ISO country code) • Doctor Juris • Doctor of Law • Dust Jacket

DJA Djougou (Benin) (location identifier) • Dose Journalière Autorisée

DJB Jambi (Indonesia) (location identifier)

DJC Dependent Job Control

DJCE Diplôme de Juriste Conseil d'Entreprise

DJD Degenerative Joint Disease

DJDE Dynamic Job Definition Entry

DJE Deflected Jet Exhaust • Djerba (Tunisia) (location identifier)

DJF December - January - February • Franc (Djibouti) (ISO currency code)

DJG Deutsch-Japanische Gesellschaft (Germany) • Djanet (Algeria) (location identifier)

DJH Date Julienne Héliocentrique

DJI Djibouti (ISO country code) • Dual Jet Injection

DJIA Dow Jones Industrial Average

DJJ Jayapura (Indonesia) (location identifier)

DJM Date Julienne Modifiée • Djambala (Congo) (location identifier) • Dual Jet Model

DJN Delta Junction (USA-AK) (location identifier) • Deutsch-Japanisches Netzwerk

DJNR Dow Jones News Retrieval

DJO Daloa (Ivory Coast) (location identifier)

DJU Degré-Jour Unifié • Djúpivogur (Iceland) (location identifier)

DJV Deutscher Journalisten Verband (Germany)

DJVA Direction de la Jeunesse et de la Vie Associative

dk Denmark

Dk Dark • Deck • Dock

DK Danemark (code ISO de pays) • Dark • Data and Knowledge • Deck • Denmark (international vehicle registration code) • Denmark (ISO country code) • Denmark (postal country code) • Dezimal-Klassifikation • Dielektrizitätskonstante • Dienstekennung • Display Keyboard • Docking

DKB Decimal Keyboard • Distributed Knowledge Base

DKBMS Data and Knowledge Base Management System

DKC Design Knowledge Capture

DKD Deutsche Kalibrierdienst (Germany) • Deutsches Komitee für Dokumentation (Germany) • Distributed Knowledge Discovery

DKE Digital Knowledge Environment

DKES Drift Kinetic Equation Solver

DKET Deployable Ku-band Earth Terminal

dkg Dekagram

DKG Deutsche Keramische Gesellschaft (Germany)

DKI Deutsche Kommission für Ingenieur-Ausbildung (Germany) • Drift Kink Instability • Driver-Kernel Interface • Dunk Island (Australia-QLD) (location identifier)

DKK Dunkirk (USA-NY) (location identifier) • Krone (Denmark) (ISO currency code)

dkl Dekaliter

dkm Dekameter

DKM Dreikoordinatenmeßgerät

DKMM Deutsches Komitee für Meeresforschung und Meerestechnik (Germany)

DKNF Domain Key Normal Form

Dkr Dansk Krone (currency)

DKR Dakar (Senegal) (location identifier) • Dansk Krone (currency)

DKRZ Deutsches Klimarechenzentrum (Germany)

dks Decastere

DKS Dikson (Russia) (location identifier) • Direct Keying System • Distribution Key Share • Dote - Kivelson - Schwartz (model)

DKT Dahl-Kirkam Telescope

DKUUG Danish Unix-systems User Group (Denmark)

DKV Docker River (Australia-NT) (location identifier)

DKW Don't Know Why

DKZ Datenkonzentrator • Dienstkennzahl • Digitales Kennzeichen

dl Deciliter

d/L Decreto Ley

D/L Data Link • Demand Loan • Deorbit/Landing • Down Link • Down List • Down Load

DL Dale • Daresbury Laboratory (UK-England) • Data Language • Data Level • Data Link • Data List • Data Logging • Daylight • Dead Letter • Deadline • Dead Load • Décret-Loi • Decreto-Lei • Definition List • Delay Line • Deorbit/Landing • Depósito Legal • Dépôt Légal • Description List • Description Logics • Descriptive List • Design Language • Detection Limit • Developed Length • Development Left • Diesel • Diffraction Limited • Diffusion Locale • Digital Library • Diode Laser • Diode Logic • Directive de Lancement • Direct Line • Disk Light • Display Level • Display Line • Distance Learning • Distinguished Lecture • Distinguished Lecturer • Distributed List • Distribution Line • Distribution List • Dividing Line • Doctor of Laws • Documentation Library • Document Librarian • Document Log • Dose Level • Double Layer • Double Ledger • Double Lobe • Downlink • Download • Drawing List • Driver's Licence • Driving Licence • Dynamic Logic

DLA Damped Ly-alpha Absorber • Damped Ly-alpha Absorption • Damped Lyman Alpha • Data Link Address • Declination of Launch Asymptote • Defense Logistics Agency (USA) • Delay • Delay Line Assembly • Dérivé Léger Ariane • Developmental Leadership Assessment • Diffuse Limited Aggregation • Diffusion-Limited Aggregation (*) • Division de la Logistique Administrative • Division of Library Automation • Documentation - Libraries - Archives • Douala (Cameroon) (location identifier) • Double Logarithmic Approximation

DLAI Documentation, Libraries and Archives Infrastructure

DLAS Damped Lyman-alpha Absorption System • Damped Lyman-Alpha System

DLAT Destructive Lot Acceptance Testing

DLB Dalbertis (Papua New Guinea) (location identifier) • Digital Line Backend • Dynamic Load Balance

DLBI Differential Long-Baseline Interferometer (PV) • Differential Long-Baseline Interferometry

DLC Dalian (China-PRC) (location identifier) • Data Link Control • Data Link Controller • Date Limite de Consommation • Delay Line Control • Diamond-Like Carbon • Diamond-Like Coating • Digital Lease Circuit • Digital Library Coordinator • Digital Line Carrier • Digital Line Concentrator • Digital Logic Circuit • Digital Loop Carrier • Direction des Lycées et des Collèges • Direct Lift Control • Distributed Loop Carrier • District Library Center • Downlink Communications • Dual-Layer Capacitor • Duplex Line Control • Duplex Line Controller

DLCC Diamond-Like Carbon Coating

DLCF Data Link Control Field

DLCI Data Link Channel Identifier • Data Link Connection Identifier

DLCN Distributed Loop Computer Network

DLCP Data Link Control Panel

DLCQ Discretized Light-Cone Quantization

DLCS Data Line Concentration System • Data Link Communications System • Delay Line Control Software • Digital Line Conditioning System

DLCU Digital Line Carrier Unit

Dld Delivered

DLD Data Link Decoder • Deadline Date • Display List Driver • Download • Geilo Dagali (Norway) (location identifier)

DLDED Division Level Data Entry Device

DLDSS Digital Legacy Data Storage System

DLE Data Line Equalizer • Data Link Equipment • Data Link Escape • Delta Leonids (meteor stream) • Density and Luminosity Evolution • Digital Live Editor • Dôle Travaux (France) (location identifier)

DLEB Double Lined Eclipsing Binary (star)

DLEC Digital Library Education Collaboratory

DLED Dedicated Line Encryption Device • Dedicated Loop Encryption Device • Digital Line Encryption Device • Digital Loop Encryption Device

DLEP Distributed List Expansion Point

DLESE Digital Library for Earth System Education

DLF Data List File • Degraded Laminate Failure • Del Rio - Laughlin (USA-TX) (location identifier) • Diffraction Limited Focusing • Dopolavoro Ferroviario

DLG Digital Line Graph • Dillingham (USA-AK) (location identifier)

DLGE Digital Line Graph Enhanced

DLGRM DLGR Magnetofluiddynamics (periodical code)

DLH Direct Labour Hour • Duluth Superior (USA-MN) (location identifier)

DLI Dalat Lienkhang (Viet Nam) (location identifier) • Data Link Interface • Digital Library Initiative • Dilute Lens Imager • Direct Liquid Introduction • Division de la Logistique Immobilière

Dlib D-Lib Magazine (periodical code)

DLIRIM Diffraction-Limited Infrared Imager

DLIS Department of Library and Information Science • Department of Library and Information Studies

DLIU Data Link Interface Unit

DLJS Document-Level Java Script

DLK Data Link • Dulkaninna (Australia-SA) (location identifier)

DLKG Dual Loop Key Generator

dll Dynamic Link Library (file extension)

DLL Data Linked Library • Data Link Layer • Data Link Library • Delay Lock Loop • Digital Local Loop • Dillon (USA-SC) (location identifier) • Direction des Livres et de la Lecture • Disclosure Limitation List • Dynamically Linked Libraries • Dynamically Loadable Library • Dynamic Linked Library • Dynamic Link Library (*) • Dynamic Load Library

DLM Dalaman (Turkey) (location identifier) • Deep-Layer Mean • Delay Line Memory • Depot Level Maintenance • Digital Landscape Model • Digital Logic Module • Dimanche - Lundi - Mardi • Diphase Loop Modem • Direct Logistics Maintenance • Distributed List Member • Distributed Lock Manager • Down Link Manager • Dynamic Link Module

DLMF Digital Library of Mathematical Functions (NIST)

DLMPS Division of Logic, Methodology and Philosophy of Science

DLMS Digital Landmass System

DLN Dillon (USA-MT) (location identifier) • Distance Learning Network • Double Length Number

DLNR Department of Land and Natural Resources

DLO Dead Letter Office • Dienst Landbouwkundig Onderzoek (Netherlands) • Director of Launch Operations • Disk-Like Object • Distribution List Owner • Division Liaison Officer • Document-Like Object • Dolomi (USA-AK) (location identifier) • Dual Loading Operation

DLOC Division Logistical Operations Center

DLOGLR Deutsche Lilienthal Oberth Gesellschaft für Luft- und Raumfahrt (Germany)

DLOR Downward Light Output Ratio

DLOS Disturbed Line Of Sight • Double Line Of Sight

DLP Data Link Processor • Data Listing Program • Deep Lasting Penetration • Digital Light Processing • Digital Light Processor • Directional Langmuir Probe • Distance Learning Program • Distinguished Lectureship Program • Dual Laser Plasma

DLPDU Data Link Protocol Data Unit

DLPI Data Link Provider Interface

DLPMA Diphase Loop Modem Assembly

DLPO DoD Landsat Program Office

DLPP Data Link Preprocessor

DLPR Dual Linear Polarization Radar

DLPS Data Link Processing System • Data Link Processor System

DLPU Data Link Processor Unit

DLQ Deck Landing Qualification

Dlr Dealer • Dollar

DLR Data Link Receiver • Dependent Logical Requester • Depot Level Repairable • Deutsche Forschungsanstalt für Luft- und Raumfahrt eV (Germany) • Deutsches Zentrum für Luft- und Raumfahrt eV (Germany) (*) • Docklands Light Railway (UK-England) • DOS LAN Requester

DLRL Diffraction-Limited Raman Laser

DLRNa DLR Nachrichten (periodical code)

DLRP Data Link Reference Point

Dlrs Dollars

Dlrx Douloureux

DLS Damped Lyman-alpha System • Data Link Set • Data Link Simulator • Data Link Support • Data Link Switching • Data Link System • Data Loader Simulator • Data Loader System • Data Logging System • Deep Lens Survey • Department of Library Services • Department of Library Studies • Dependent LU Server • Depot Logistics System • Differential Light Scattering • Digital Link Service • Digital Logic System • Diphasique Lagrangien

Structuré • Dirty Little Secret • Distributed Link Service • Distributed Load Sharing • DME-based Landing System • Dock Landing Ship • Doctor of Library Science • Domestic Substances List • Doppler Lidar Sounder • Dynamic Light Scattering • Dynamic Limb Sounder • Dynamic Load Simulator • The Dalles (USA-OR) (location identifier)

DLSc Doctor of Library Science

DLSC Daily Living Science Center (USA-LA) • Data Link to a Satellite Channel • Defense Logistics Service Center

DLSM Data Link Summary Message

DLSP Diffraction-Limited Spectropolarimeter • Diffraction-Limited Stokes-Polarimeter • Distribution List Submit Permission

DLSw Data Link Switching

Dlt Delete

DLT Data Line Translator • Data Link Terminal • Data Link Translator • Data Loop Transceiver • Decision Logic Table • Depletion Layer Transistor • Digital Library Technology • Digital Linear Tape • Direct Linear Transform • Distributed Language Translator • Downlist • Dual Language Translation

Dltd Deleted

DLTM Data Link Terminal Module • Data Link Test Message • Digital Line Terminating Module • Digital Line Termination Module

DLTR Data Link Terminal Repeater • Data Link Transmission Repeater

DLTS Data Link Test Set • Deep-Level Transient Spectroscopy

DLTU Digital Line Termination Unit • Digital Line/Trunk Unit

DLU Data Link Upgrade • Dependent Logical Unit • Development Laboratory Unit • Digital Line Unit • Digitizer Logic Unit • Display Logic Unit • Downlink Unit

DLUO Date Limite d'Utilisation Optimale

DLUR Dependent Logical Unit Requester

DLUS Dependent Logical Unit Server

DLV Delissaville (Australia-NT) (location identifier)

DLVA Detector Logarithmic Video Amplifier

Dlvd Delivered

DLVO Darjaguin - Landau - Verwey - Overbeek (model)

DLVQ Dynamic Learning Vector Quantization

Dlvy Delivery

DLVY Delivery

DLWG Departmental Level Working Group • Department Level Working Group

DLWI Deformational Longwave Instability

Dlx Deluxe

Dly Daily • Delay

DLY Delay • Dillons Bay (Vanuatu) (location identifier)

DLZ Dalanzadgad (Mongolia) (location identifier) • Dynamic Launch Zone

dm Decimeter • Decimetric

dM dieses Monats

Dm Data Mobile • Dios Mediante

D/M Demodulate/Modulate • Demodulation/Modulation • Demodulator/Modulator

D&M Detection & Monitoring • Driving & Maintenance

DM Dam • Dark Marking • Dark Matter • Data Management • Data Manager • Data Manual • Data Mark • Data Master • Data Memory • Data Mining • Data Model • Data Modulation • Data Monitor • Debugging Mode • Decimal Multiply • Decision Maker • Decision-Making • Decision Memorandum • Deep Monitoring • Definitional Module • Deformable Mirror • Delay Modulation • Delta Modulation • Demand Meter • Demonstration Manager • Demonstration Model • Demonstration Motor • Density Meter • Deployment Module • Depot Maintenance • Descriptive Method • Design Manual • Design Margin • Design Memorandum • Detached Module • Detection and Monitoring • Detector Module • Deutsche Mark (currency) • Development and Maintenance • Development Milestone • Development Model • Development Motor • Diagonal Mirror • Dialectic Method • Dialog Manager • Dialogue Mode • Dienstmerkmal • Differential Mode • Dipole Moment • Direct Mail • Director of Management • Disassembly Manual • Discard Message • Disconnect Mode • Dispersion Measure • Dispersion Measurement • Display Manager • Distributed Management • Distributed Memory • District Manager • Docking Mechanism • Docking Module • Docteur en Médecine • Doctor of Mathematics • Domain Manager • Domain Model • Dominica (ISO country code) • Dormant Mode • Double Mode • Double Module • Downshifted Maximum • Drafting Manual • Draft Manual • Dual Mode • Durchmusterung • Dynamical Meteorology

DMA Data Management Administrator • Data Management Analysis • Data Management Assembly • Data Management Association • Decision-Making Aid • Deep Map Assistant • Defense Mapping Agency (USA) • Depot Maintenance Activity • Designated Maintenance Agency • Designated Market Area • Differential Mobility Analyzer • Direct Marketing Association • Direct Memory Access • Direct Memory Address • Direct Memory Addressing • Division Maintenance Area • Document Management Alliance • Document Management Attribute • Dominica (ISO country code) • Drive Motor Assembly • Drum Memory Assembly • Dynamic Mechanical Analysis • Dynamic Memory Access • Tucson - Davis Monthan (USA-AZ) (location identifier)

DMAAC Defense Mapping Agency Aerospace Center

DMAC Data Management Analysis Center • Data Management and Analysis Center • Direct Memory Access Controller

DMACS Distributed Manufacturing Automation and Control Software

DMAE Degeneración Macular Asociada a la Edad

DMAG Depot Maintenance Activity Group

DMAHTC Defense Mapping Agency Hydrographic and Topographic Center

DMAP Data Management and Analysis Plan • DECT Multimedia Access Profile • Depot Maintenance Activation Plan • Digital Mapping • Distributor Management Application Package

DMAPI Data Management Application Programming Interface

DMAPS Design, Manufacturing And Producibility Simulation

DMAR Deferred Maintenance And Repair

DMAS Data Management and Archive System • Des Moines Astronomical Society (USA-IA) • Digital Multiplexed Audio System • Distributed Management Accounting System

DMAT Data Management and Archive Team • Digital Module Automatic Tester

DMATC Defense Mapping Agency Topographic Center

DMath Doctor of Mathematics

DMB Database Management Branch • Data Management Branch • Datum Marker Buoy • Digital Multimedia Broadcast • Digital Multipoint Bridge • Dzhambul (Kazakhstan) (location identifier)

DMC Dark Molecular Cloud • Database Maintenance Component • Data Management Center • Data Management Channel • Data Management Component • Data Management Computer • Data Mode Control • Data Multiplexer Control • Depot Maintenance Concept • Diffuse

Molecular Cloud • Digital Microcircuit • Digital Monitor Computer • Dirección Meteorológica de Chile • Direct Maintenance Costs • Direct Manufacturing Costs • Direct Monte Carlo • Direct Multiplexer Channel • Discrete Memoryless Channel • Disk Memory Controller • Distributed Master Control • Divisão de Mecânica e Controle • Double-Mode Cepheid • Drought Monitor Centre • Drought Monitoring Centre • DSIF Monitor and Control • Dwarf Molecular Cloud

DMCA Digital Millennium Copyright Act (USA)

DMCF Deservicing, Maintenance, and Checkout Facility

DMCh Dirección Meteorológica de Chile

DMCI Data Management Component Interface

DMCL Device Media Control Language

DMCM Double-density Modular Core Memory

DMCS Data Management and Control System • Data Management Control System • Direct Monte Carlo Simulation • Domestic Mail Classification Scheme

DMD Dark Matter Disk • Data Management Division • Data Measurement Device • Data Message Device • Data Monitor and Display • Demodulate • Département Maintenance et Documentation • Deployment Management System • Devices Management Directorate • Dienstag - Mittwoch - Donnerstag • Differential Metabolic Display • Differential Mode Delay • Digital Message Device • Digital Methods Division • Digital Micromirror Device • Direct Manipulation Device • Directory Management Device • Directory Management Domain • Division Manager Designator • Doomadgee Mission (Australia-QLD) (location identifier) • Dual Mode Display

DMDCP Dual-Mass Dynamic Cone Penetrometer

DMDD Distributed Multiplexing - Distributed Demultiplexing

DMDF Digital Map Data Format

DMDP Data Maintenance Diagnostic Program

DMDS Data Management Display System • Dimethyl Disulfide

DME Department of Mines and Energy • Diagnostic Monitor Executive • Digital Multiplex Equipment • Dimethyl Ether • Diploma in Mechanical Engineering • Direction de la Maîtrise d'Étude • Direct Measurements Explorer • Directorate of Maintenance Engineering • Display Major Equipment • Distance Measurement Equipment • Distance-Measuring Equipment • Distributed Management Environment • Dropping Mercury Electrode • Dynamic Mission Equipment • Moscow Domodedovo (Russia) (location identifier)

DMEA Damage Modes and Effects Analysis

DMED Digital Message Entry Device

DMEMP Distributed Memory Explicit Message Passing

DMES Digital Message Entry System • Dimethylethoxis

DMF Dansk Matematisk Forening (Denmark) • Data Management Facility • Data Mining Facility • Depot Maintenance Facility • Digital Multiplexing and Formatting • Dimethylformamide • Distribution Media Floppy • Distribution Media Format • Division des Moteurs de Fusées • Drug Master File • DSIF Maintenance Facility

DMFA Drustvo Matematikov, Fizikov in Astronomov Slovenije (Slovenia)

DMfAD Distributed Monitoring for Active Directory

dmg Disk Image (file extension)

Dmg Damage

DMG Data Management Group • Deutsche Meteorologische Gesellschaft (Germany) • Digital Map Generator • Digital Media Group

DMGD Dark Moon Geophysical Day

DGMS Digital Map Generation System

DMH Dual Mode Hydrazine

DMI Danish Maritime Institute • Danish Meteorological Institute • Danmarks Meteorologiske Institut (Denmark) • Deferred Maintenance Item • Definition of Management Information • Deployment Mapping Instrument • Desktop Management Information • Desktop Management Interface • Diagnostic Mathematics Inventory • Digital Multiplexed Interface • Direct Manipulation Interface • Direct Memory Interface • Distance Measuring Instrument • Dual Mode Interceptor

DMIA Dual Multiplexer Interface Adapter

DMIC Deputy Meteorologist-In-Charge

DMICB Daily Mission Integration Control Board

DMICS Distributed Management Information and Control System

DMIDP Dark Matter Inner Density Profile

DMIF Distributed Multiuser Information Fusion • Dynamic Multiuser Information Fusion

DMII Design, Manufacture and Industrial Innovation

DMILL Durci Mixte sur Isolant Logico-Linéaire

DMIR Designated Manufacturing Inspection Representative

DMIS Data Management Information System • Dimensional Measuring Interface Specification • Directorate for Management Information System • Dynamic-Mapping Information System

DMIST Direction des Musées et de l'Information Scientifique et Technique

DMJ Defense Management Journal (periodical code)

DMK Dynamic Management Kit

DML Database Management Language • Data Macro Language • Data Management Language • Data Manipulation Language • Data Manipulation Logic • Decision and Modeling Language • Depot Maintenance Level • Digitized Message Link • Dronning Maud Land (Antarctica) • Dual Mode Laser • Dynamic Multilevel

DMLA Dégénérescence Maculaire Liée à l'Âge

DMLCC Design and Manufacturing of Low-Cost Composites

DMLE Distributed Multimedia Learning Environment

DMLS Dissecting Microscope Lighting System • Doppler Microwave Landing System • Dynamical Method of Least Squares

DMM Design Methodology Manager • Digital Multimeter • Directorate for Materiel Management • Disaster Monitoring and Mitigation • Division of Materiel Management • Domestic Mail Manual • Don't Mock Me • Dynamic Memory Management

DMMG Displacement Method Matrix Generator

DMMH Direct Maintenance Man-Hours

DMMM Division of Materials, Mechanics, and Manufacturing

DMMO Déclaration des Mouvements de Main d'Oeuvre

DMMP Distributed-Memory Microprocessor

DMMS Dynamic Memory Management System

DMMV Deutscher Multimedia Verband (Germany)

Dmn Dimension • Dimensional

DMN Data Multiplexing Network • Deming (USA-NM) (location identifier) • Dimethylnitrosamine • Direction de la Météorologie Nationale • Don't Move Now

DMNA Dimethylnitrosamine

DMNH Denver Museum of Natural History (USA-CO)

DMNS Denver Museum of Nature and Science (USA-CO)

dmo Demonstration (file extension)

DM&O Data Management & Operations

DMO Data Management Office • Data Management Officer • Data Management Organization • Diffuse Massive Object • Direction de la Maîtrise d'Ouvrage • Direct Marketing Organization • Directory Maintenance Official • Distributed Management Object • Documentation Management Officer • Domain Management Organisation • Sedalia (USA-MO) (location identifier)

DMOC DSPSE Mission Operations Center (USA-VA)

DMOD Demodulation

DMON Discrete Monitoring

DMOS Diffusion Metal Oxide Semiconductor • Diffusive Mixing of Organic Solutions • Double-diffused Metal Oxide Semiconductor

dmp Dump (file extension)

DMP Daily Maintenance Pack • Daily Management Plan • Data Management Plan • Data Management Pointer • Data Management Processor • Data Merger Processor • Data Migration Plan • Deployable Maintenance Platform • Deployment and Maintenance Platform • Diagnostic and Monitoring Protocol • Dimethylphthalate • Direct Manipulation • Disk-storage Management Program • Division of Materials Physics • Documentation Material Processed • Dot Matrix Printer • Drive Management Protocol • Dump

DMPC Distributed Memory Parallel Computer

DMPCS Deep Mission Planning and Control Station

DMPI Daily Mean Percentage Increase • Designated Mean Point of Impact • Desired Mean Point of Impact

DMPN Dictionary of Minor Planet Names

Dmpr Damper

DMPS Deep Mission Planning Station

DMQS Display Mode Query and Set

Dmr Dimmer

DMR Data Management Routine • December • Defective Materiel Report • Detailed Mission Requirements • Differential Microwave Radiometer (*) • Diffuse Microwave Radiometer • Digital Microwave Radio • Digital Module Radio • Division of Materials Research (NSF) • Dynamic Modular Replacement • Ohamar (Yemen) (location identifier)

DMRC Directorate of Meteorology Research Committee

DMRG Density-Matrix Renormalization Group

DMRS Data Management and Retrieval System

DMRT Dense Medium Radiative Transfer

DMS Daily Mission Schedule • Daily Mission Scientist • Dark Matter Sphere • Database Management System • Data Management Service • Data Management Software • Data Management System • Data Measuring System • Data Memory Set • Data Mining Survey • Data Monitoring System • Data Multiplexer Subunit • Data Multiplex System • Decision-Making System • Dedicated Message System • Deep Multicolor Survey • Dense Meteor Stream • Dense Multiprocessor System • Deployed Maintenance System • Deputy Mission Scientist • Design Maintenance System • Design Management System • Desktop Mapping System • Development Management System • Dewar Main Shell • Diagnostic Monitor Software • Differential Maneuvering Simulator • Digital Management System • Digital Mapping System • Digital Map System • Digital Mass Storage • Digital Message Switch • Digital Message System • Digital Microwave System • Digital Modem Server • Digital Multiplexed System • Digital Multiplexing Synchronizer • Digital Multiplex Switch • Digital Multiplex System • Dimethyl Sulfide • Diminishing Manufacturing Source • Directeur de la Mission Satellite • Disk Monitor System • Display Management System • Distance Measuring System • Distribution Management System • Division of Mathematical Sciences • Docking Mechanism System • Docking Module System • Documentation Management System • Documentation of Molecular Spectroscopy • Document Management Service • Document Management System • Doppler Measuring System • Drawing Management System • Drift Magnetic Separator • Drum Memory System • Dutch Meteor Society (Netherlands) • Dynamic Micromagnetic Structure • Dynamic Motion Simulator • Dynamic Motion System

DMSA Double and Multiple Systems Annex

DMSB Disposable Molecular Sieve Bed

DMSD Data Model Specification Document • Digital Multistandard Decoding

DMSF Data Management Systems Facility

DMSH Diminish

DMSI Data Management Segment Interface

DMSK Differential Minimum Shift Keying

DMSO Defense Modeling and Simulation Office (USA) • Dimethylsulphoxide

DMSP Data Management Summary Processor • Defense Meteorological Satellite Program • Depot Maintenance Support Plan • Dichroic Microspectrophotometer • Distributed Mail Service Protocol • Dual Mode Speech Processor

DMSS Data Management System Simulator • Data Multiplex Subsystem • Dedicated Mission Support System • Defense Meteorological Satellite System • Distributed Mission Support System

Dmst Demonstration

DMST Division of Military Science and Technology

Dmstn Demonstration

Dmstr Demonstrator

DM&T Data Management & Training

DMT Dark Matter Telescope • Data Management and Training • Data Management Team • Design Manufacture and Test • Desired Merge Total • Deutsche Meteorologentagung (Germany) • Developmental Maintenance Testing • Diagonal Multiply Transitive • Diamantino (Brazil-MT) (location identifier) • Digital Message Terminal • Digital Multitone • Dimensional Management Training • Direct Modulation Technique • Disaster Management Team • Discrete Multitone • Dispersive Mechanism Test • Distributed Mission Training

DMTC Digital Message Terminal Computer

DMTD Digital Message Transfer Device

DMTDSG Digital Message Transfer Device Subgroup

DMTF Desktop Management Task Force • Distributed Management Task Force

DMTG Data Manipulation Task Group

DMTH Data Manager Threshold

DMTI Digitized Moving Target Indicator

DMTPS Digital Magnetic Tape Plotting System

DMTS Digital Module Test Set

DMTT Dual Magnetic Tape Transport

DMTU Default Maximum Transmission Unit • Digital Magnetic Tape Unit

DMU Data Management Unit • Data Manipulation Unit • Data Measurement Unit • Data Migrator Utility • Data Multiplexer Unit • Data Multiplex Unit • Decision-Making Unit • De Montfort University (UK-England) • Detached Mail Unit • Digital Management Unit • Digital Master Unit • Digital Message Unit • Digital Monitor Unit • Dimapur (India) (location identifier) • Directly Managed Unit • Disk Memory Unit • Dual Maneuvering Unit

dmv Door Middel van

DMV Department of Motor Vehicles (USA) • Deutsche Matematiker Vereinigung eV (Germany) • Digital Motion Video

DMW Digital Media Warehouse • Dissimilar-Metal Weld

DMWG Data Management Working Group • Data Migration Working Group • Depot Maintenance Work Requirement

DMX Data Multiplex • Decision-Making Exercise • Dual Motorway Extension

DMXDA Demultiplexer Data

DMY Day - Month - Year

DMZ Data Monitoring Zone • Demilitarized Zone

Dn Don • Down • Dozen • Dyne

D/N Debit Note

DN Dark Nebula • Dark Nucleus • Data Name • Data Net • Data Number • Date Number • Day and Night • Decimal Number • Dedicated Network • Delayed Neutron • Delivery Notification • De Novo • Design Note • Despacho Normativo • Diffusion Nationale • Digital Non-secure • Digital Number • Directorate Notice • Directory Name • Directory Number • Discrepancy Notice • Dispatch Note • Distinguished Name • Distrito Nacional • Document Number • Domain Name • Down • Dwarf Nova

Dna Doña

DNA Datennetzabschlußgerät • DEC Network Architecture • Defense Nuclear Agency • Designated National Agency • Designated National Authority • Desoxyribonucleic Acid • Deutscher Normenausschuß (Germany) • Did Not Answer • Did Not Attend • Digital Network Architecture • Directional Neighborhoods Approach • Direction de la Navigation Aérienne • Directorio Nacional Argentino • Division de la Navigation Aérienne • Document-Enabled Network • Does Not Answer • Does Not Apply • Downstream Network Address • Okinawa Kadena (Japan) (location identifier)

DNAD Director of the NRAO ALMA Division

DNAE Datennetzabschlußeinrichtung

DNAEE Departamento Nacional de Águas e Energia Eléctrica (Brazil)

DNAG Decade of North American Geology

DNANS Digital Network Architecture Naming Service

D/NAPS Day/Night Adverse weather Pilotage System

DNAT Diplôme National des Arts et Techniques

DNB Deutsches Nachrichtenbüro (Germany) • Diagnostic Neutral Beam • Dictionary of National Biography (UK) • Dunbar (Australia-QLD) (location identifier)

DNBA Diplôme National des Beaux-Arts (France)

DNC Danane (Ivory Coast) (location identifier) • Day/Night Capability • Delayed-Neutron Counter • Delayed-Neutron Counting • Depreciation of National Capital • Did Not Come • Digital Nautical Chart • Direct Numerical Control • Direct Numerically Controlled • Distributed Numerical Control • Dynamic Network Controller

DNCC Deployed Network Control Center • Distributed Network Control Center • Domain Network Control Center

DNCP DECNET Control Protocol

DNCRI Division of Networking and Communications Research and Infrastructure

DNCS Data Network Control Station • Day/Night Camera System

DNCT Direction Nationale de la Cartographie et de la Topographie

DND Department of National Defence • Department of National Defense • Direct Numerical Diagonalization • Do Not Disturb • Do Not Duplicate • Dundee (UK-Scotland) (location identifier)

DNDO Do Not Disturb Override

Dne Douane

DNe Dwarf Novae

DNE Dallas North (USA-TX) (location identifier) • Daily Negative Effect • Day/Night Effect • Digital News Editor • Does Not Exist • Do Not Erase • Dynamic Node Exchange

DNEL Disconnect Non-Essential Loads

DNES Department of Non-conventional Energy Sources

DNESYS Drainage Network Extraction System

DNF Derna (Libya) (location identifier) • Did Not Finish • Disjunctive Normal Form

DNG Datennetzabschlußgerät • Donnée Numérique Géographique

DNH Department of National Heritage (UK) • Dunhuang (China-PRC) (location identifier)

DNHR Dynamic Non-Hierarchical Routing

DNI Desktop Network Interface • Detailed Network Interface • Documento Nacional de Identidad • Do Not Invite • Wad Medani (Sudan) (location identifier)

DNIC Data Network Identification Code

DNID Data Network Identity

DNIF Duty Not Involving Flying

DNIS Databases in Networked Information Systems (workshop) • Dialed Number Identification Service

DNK Danemark (code ISO de pays) • Denmark (ISO country code) • Deutscher National Komitee (Germany) • Dnepropetrovsk (Ukraine) (location identifier)

DNKZ Datennetzkontrollzentrum

DNL Augusta Daniel (USA-GA) (location identifier) • Detector-Noise Limited • Differential Non-Linearity • Director of Naval Laboratories • Discipline Non-Linguistique • Do Not List • Dynamic Noise Limiter

DNLB Danmarks Natur- og Laegevidenskabelige Bibliotek

DNLK Downlink

DNLS Derivative Non-Linear Schrödinger

DNLT Downlist

DNM Damped Newton Method • DEC Network Manager • Delayed Neutron Monitor • Denham (Australia-WA) (location identifier) • Dirección Nacional de la Meteorología • Direct Nucleus Measurement

DNMI Det Norske Meteorologiske Institutt (Norway)

DNMR Dynamic Nuclear Magnetic Resonance

DNMS Distributed Network Management System

DNN Dalton (USA-GA) (location identifier) • Dialectic Neural Network

DNNA Digital Neural Network Architecture

DNO Descending-Node Orbit • Dianópolis (Brazil-TO) (location identifier) • Distributed Networks Office • Do Not Open • Dwarf Nova Oscillation

DNOM Direction of Naval Oceanography and Meteorology

DNOS Distributed Network Queuing System

DNP Dang (Nepal) (location identifier) • Document Non Pertinent • Do Not Print • Do Not Publish • Dynamic Nuclear Polarization

DNPA Data Numbering Plan Area

DNPG Digital Network Product Group

DNQ Deniliquin (Australia-NSW) (location identifier) • Did Not Qualify • Do Not Quote

DNR Dauernutzungsrecht • Department of Natural Resources • Détection par Neutrons Rapides • Deutscher Naturschutzring eV (Germany) • Dialed Number Recorder • Digital Network Radio • Digital Noise Reduction • Dinard Pleurtuit (France) (location identifier) • Domain Name Resolver • Domain Name Routing • Do Not Rescucitate • Do Not Return

DNRC Department of Natural Resources and Conservation

DNRD Dépense Nationale de Recherche et de Développement

DNRS Day/Night Range Sight

Dns Downs

DNS Decimal Number System • Denison (USA-IA) (location identifier) • Dense • Departure Nuclear Stage • Did Not Show up • Digital Nervous System • Digital Network Subscriber • Direct Network Subscriber • Direct Numerical Simulation • Directory Name Service • Discrete Network Simulation • Distributed Network Supervisor • Domain Name Server • Domain Name Service • Domain Name System • Domain Naming Service • Domain Naming System • Doppler Navigation Sensor • Doppler Navigation System • Double Neutron Star

DNSC Data Network Service Center

DNSLP Downslope

DNSO Domain Name Supporting Organization

DNSSEC Domain Name System Security

DNSTRM Downstream

DNSVT Digital Narrowband Secure Voice Terminal

DNSW Dark Skies Northwest (USA)

DNT Desired Non-merge Total • Détection par Neutrons Thermiques • Durée Normale du Travail • Santa Ana Downtown (USA-CA) (location identifier)

DNTS Diplôme National de Technologie Spécialisé

DnU Differentsial'nye Uravneniia (periodical code)

DNU Did Not Use • Dinangat (Papua New Guinea) (location identifier) • Do Not Use

DNV Danville (USA-IL) (location identifier)

DNVT Digital Narrowband Voice Terminal • Digital Non-secure Voice Telephone • Digital Non-secure Voice Terminal

DNW Dedicated Network • Deutsch-Niederländischer Windkanal • Duits-Nederlandse Windtunnel

DNX Dinder Galegu (Sudan) (location identifier) • Dynamic Network X-connect

DNZ Denizli Cardak (Turkey) (location identifier)

d/o Daughter of

do Ditto • Dominican Rep. • donnerstags

dO Der Obige

Do Ditto • Donnerstag

D&O Description & Operations • Doctrinal & Organizational

D-O Decimal to Octal

D/O Delivery Order

DO Daily Operations • Data Organizer • Data Out • Data Output • Dearborn Observatory (USA-IL) • Decision Opportunity • Defined Order • Delivery Order • Denominación de Origen • Dépense Ordinaire • Deputy Organizer • Derived Operand • Description and Operations • Design Objective • Digital Ortho • Digital Output • Directeur des Opérations • Direct Object • Direct Optimized • Director of Operations • Director's Office • Disbursing Officer • Discrete Output • Disponibilité d'Office • Dissolved Oxygen • Distributed Object • District Office • District Officer • Ditto • Divisional Office • Divisional Officer • Document-Ordered • Dominican Rep. (ISO country code) • Dominion Observatory (Canada) • Donnerstag • Donnerstags • Drawing Office • Duty Officer • République Dominicaine (code ISO de pays)

DOA Date Of Availability • Dead On Arrival • Department Of Administration • Department Of Aviation • Deputy Orbital Analyst • Direction Of Arrival • Direct Obligation Authority • Display-Oriented Application • Dissolved-Oxygen Analyzer • Doany (Madagascar) (location identifier) • Dutch Occultation Association (Netherlands) • Dynamic Obstacle Avoidance

DOAFT Data Organization and Access Focus Team

DOAMS Distant-Object Attitude-Measuring System

DoArm Akademiia Nauk Armianskoi SSR Doklady (periodical code)

DOAS Differential Optical Absorption Spectrometer • Differential Optical Absorption Spectroscopy (*) • Differentielle Optische Absorptionspektroskopie • Dual Optical Absorption Spectroscopy

DoAze Akademiia Nauk Azerbaidzhanskoi SSR Doklady (periodical code)

DOB Data Output Box • Date Of Birth • Deployable Optical Bench • Depth Of Burst • Dobo (Indonesia) (location identifier)

DoBel Akademiia Nauk BSSR Doklady (periodical code)

DOBIS Dortmunder Bibliotheks Informations System

doc Word Document (file extension)

Doc Doctor • Document • Documentation • Documento

DoC Department of Commerce • Department of Communications

DOC Daily Operating Cost • Data, Operations and Control • Data Operations Center • Data Operations Console • Data Operations Control • Data Operations Controller • Data Optimizing Computer • Data Output Channel • Date Of Change • Decimal to Octal Conversion • Deepsky Observer's Companion • Department Of Commerce • Department Of Communications • Department Of Conservation • Descend On Course • Designated Operational Coverage • Designed Operational Capability • Design Operation Capability • Desired Operational Capability • Direct Operating Cost • Director Of Contracting • Discipline Operations Center • Disque Optique Compact • Dissolved Organic Carbon • Distributed Object Computing • Document • Documentation • Document-Oriented Computing • Dornoch (UK-England) (location identifier) • Drift-Orbit Correction • Dynamic Overload Control

DOCA Date Of Change of Accountability • Date Of Current Appointment • Distributed Object Computing Architecture

DOCASE Distribution and Objects in CASE

DOCDEL Document Delivery

DOCED Document Edition

DocEng Doctor of Engineering

DOCEX Document Exploitation

DOCi Documentation des Observateurs - Circulaire (periodical code)

DOCKET Document and Code Knowledge Elicitation Toolset

DOCP Degree Of Circular Polarization

DOCS Designated Operational Capability Statement • Disk-Oriented Computer System • Display Operator Console System • Document Organization and Control System • Documents

DOCSIS Data Over Cable System Interface Specification

DOCSV Data Over Circuit Switched Voice

Docu Document • Documentary • Documentation

DOCUP Document Unique de Programmation

DOCUS Display-Oriented Computer Usage System

DoD Department of Defense • Depth of Discharge

DOD Date Of Death • Department Of Defense • Department of Ocean Development • Depth Of Discharge • Deutsches Ozeanographisches Datenzentrum (Germany) • Development Operations Division • Dial On Demand • Digital Object Delivery • Digital Optical Data • Digital Optical Disk • Direct Outward Dialing • Division Organisation et Développement • Dodoma (Tanzania) (location identifier) • Drops On Demand

DODAS Digital Oceanographic Data Acquisition System

DoDIIS Department of Defense Intelligence Information System

DoDISS Department of Defense Index of Specifications and Standards

DODISS Department Of Defense Index Specifications and Standards

DODL DOCASE Design Language

DODM Data Orderwire Diphase Modem

DO-DO Drive-On - Drive-Off

DODR Digital Original Data Record

DODS Database Object Design Studio • Definitive Orbit Determination System • Different Orbitals for Different Spins • Distributed Oceanographic Data System

DODT Design Option Decision Tree

DoE Department of Education • Department of Energy • Design of Experiments

DOE Department Of Education • Department Of Electronics • Department Of Energy • Department Of the Environment • Depending On Experience • Depends On Experience • Design Of Experiments • Diffractive Optical Element • Distributed Objects Everywhere • Djoemoe (Suriname) (location identifier)

DOEM Designated Official for Environmental Matters

DOEn Department Of Environment

DOES Directorate Of Evaluation and Standardization • Disk-Oriented Engineering System

DOF Degree Of Freedom • Department Of Finance • Diffractive Optical Frame • Direction Of Flight • Displacement Of Fluids • Distributed Object Framework • Dora Bay (USA-AK) (location identifier)

DOFIC Domain-Originated Functional Integrated Circuit

DoG Difference of Gaussians

DOG Depth Of Gating • Difference Of Gaussian • Dongola (Sudan) (location identifier) • Dot On the Ground • Drei-Objekt-Grenze

DOGE Doris Orbitography and Geopotential Evaluation

DOGS DGFI Orbit and Geodetic (parameter estimation) Software • Dwingeloo Obscured Galaxy Survey

DOH Department Of Health (USA) • Dirección de Obras Hidráulicas • Discrete Output High • Doha (Qatar) (location identifier)

DOHC Double Overhead Camshaft • Dual Overhead Camshaft

DoI Department of Interior

DOI Degree Of Interest • Department of Industry • Department of the Interior • Descent Orbit Insertion • Differential Orbit Improvement • Digital Object Identifier • Directeur de l'Organisation de l'Information • Direction de l'Organisation de l'Information • Direction des Opérations Informatiques • Direct Orbit Insertion • Director Of Information • Division Of Infrastructure • Document Object Identifier • Document-Oriented Interface • Doini (Papua New Guinea) (location identifier) • Domain Of Interpretation

DOIAP Documentation des Observateurs - Institut d'Astrophysique de Paris (periodical code)

DOIM Directorate Of Information Management • Director Of Information Management

DOIP Dial Other Internet Providers

DOIS Delivery Operations Information System

DoIT Department of Information Technology

DO-IT Disabilities, Opportunities, Internetworking and Technology

DOIT Disabilities, Opportunities, Internetworking and Technology

DOITT Department Of Information, Technology and Telecommunications

DoJ Department of Justice

DOJ Department Of Justice

DOK Deutsches Optisches Komitee (Germany) • Donetsk (Ukraine) (location identifier)

DokAN Doklady Akademia Nauk Minerologia USSR (periodical code)

DOKM Diagonal Optokinetic Nystagmus

Dol Dollar

DoL Department of Labor

DOL Daily Operating Log • Day Of Launch • Department Of Labor • Deauville Saint-Gatien (France) (location identifier) • Department Of Labor • Directorate Of Logistics • Director Of Logistics • Discrete Output Low • Dispersed Operating Locations • Display-Oriented Language • Division Of Labor • Dollar • Dynamic Object Library • Dynamic Octal Load

DOLA Department Of Land Information

DOLARS Departmental Online Accounting and Reporting System • Digital Off-Line Automatic Recording System

DOLCE Digital Online Cryptographic Equipment

DOLE Detection Of Laser Emissions

DOLP Date Of Last Posting • Date Of Last Procurement • Degree of Linear Polarization

DOLR Decentralized Object Location and Routing

Dols Dollars

Dom Domain • Domenica • Domestic • Domicile • Dominant • Domingo • Dominica • Dominican • Dominion

DOM Data Organization and Management • Data Output Message • Data Output Module • Data Output Multiplexer • Date Of Manufacture • Day Of Month • Deo Optimo Maximo • Département d'Outre-Mer (France) • Description, Operation and Maintenance • Development Operations Manager • Digital Ohmmeter • Dirty Old Man • Discrete Ordinate Method • Dispersed Organic Material • Dissolved Organic Material • Dissolved Organic Matter • Distributed Object Management • Document Object Model • Dominica - Melville Hall (Dominica) (location identifier) • Dominican Rep. (international vehicle registration code) • Dominican Rep. (ISO country code) • République Dominicaine (code ISO de pays)

DOMAIN Distributed Operating Multiaccess Interactive Network

DOMAINS Distributed Open Management Architecture In Networked Systems

DOMC Dynamically Optimized Monte Carlo (method)

DOME Distributed Object Management Environment

DOMES Directory Of MERIT Sites

DOMF Distributed Object Management Facility

Domin Dominican

DOMINO Demonstrationszentrum für Optische Meßtechnik und Integrierte Optik (Germany)

DOMIS Directory of Materials Information Sources

DOMM Distributed Object Management Model

DOMS Distributed Object Management System

DOMSAT Domestic Satellite

don Donec

DON Delivery Order Notice • Department Of the Navy • Disque Optique Numérique • Dissolved Organic Nitrogen • Document Order Number • Dos Lagunas (Guatemala) (location identifier)

DONA Decentralized Open Network Architecture

DONUT Direct Observation of the Nu Tau • Documentation Offered with No Usability Testing

DOO Daily Operations Order • Dorobisoro (Papua New Guinea) (location identifier)

DOOD Deductive and Object-Oriented Databases (conference)

DOOL Distributed Object-Oriented Language

doorg Doorgang

DOORS Dynamic Object-Oriented Requirements System

DOP Daily Observing Plan • Daily Operating Plan • Data Object Processed • Data Operational Program • Degree Of Polarization • Degree Of Protection • Dept Of Penetration • Dermo-Optical Perception • Designated Overhaul Point • Detailed Operating Procedure • Developing-Out Paper • Development Options Paper • Dilution Of Precision • Directory Operational Protocol • Dolpa (Nepal) (location identifier) • Doppler • Driver-Operated Plug • Drop-Off Point • Peso (Dominican Rep.) (ISO currency code)

DOPAA Description Of Proposed Action and Alternatives

DOPAOS Direction de l'Organisation des Personnels Administratifs, Ouvriers et de Service

DOPIE Department Of Primary Industries and Energy

DOPLID Doppler Lidar

DOPLIGHT Doppler Lightning

DOPP Delivery Order Performance Plan

D/OPS Directorate of Operations

DOPS Digital Optical Protection System • Division de l'Organisation de la Programmation Scolaire

DOQ Digital Orthophoto Quadrangle • Digital Orthoquad

DOQQ Digital Orthophoto Quarter Quadrangle

Dor Dorado-Doradus (constellation) • Dorian • Doric

DOR Daily Operating Report • Daily Operational Report • Date Of Rank • Date Of Request • Design Objective Reliability • Differenced One-way Range • Digital Optical Recording • Director of Operational Requirements • Direct Oxide Reduction • Domain Of Responsibility • Dori (Burkina Faso) (location identifier) • Due Out Release

DORAN Doppler Range And Navigation

DORAS Development of Operational Retrieval Algorithm from Spectra

DORCA Dynamic Operational Requirements and Cost Analysis

DORIS Détermination d'Orbite et Radiopositionnement Intégré par Satellite • Development Of Reasoning In Science • Doppler Orbitography and Radiopositioning Integrated by Satellite

DorPo Dornier Post (periodical code)

DORRAN Distributed Ordering, Reporting, Researching, and Accounting Network

Dors Dorset (UK-England)

DORS Data Output Recording System • Departementaal Overleg Remote Sensing (Netherlands)

dor. s. tr. Doré sur Tranche

DORT Daily Operational Readiness Test

dos DOS Text (file extension)

dOS Direct Opacity Sampling

DoS Denial of Service • Department of State

D.O/S Distributed Operating System

DOS Date Of Separation • Day Of Supply • Denial Of Service • Density Of States • Department Of Space • Department Of State • Digital Operation System • Digital Orthophoto System • Dios (Papua New Guinea) (location identifier) • Discrete Ordinates System • Discrete Orthonormal Sequence • Disk Operating System (*) • Distributed Operating System • Division de l'Organisation Scolaire • Division of Optics and Spectroscopy • Domain Ontology Server

DOSE Dictionary Of Substances and their Effects • Dynamics Of the Solid Earth

DOSEM Disk Operating System Emulation

DOSES Development of Statistical Expert Systems

DOSF Distributed Office Support Facility

DOSIM Dosimeter

DOSLI Department Of Survey and Land Information

DOSS Decision-Oriented Scheduling System • Department Of State Services • Dexterous Orbiter Servicing System

DoSSR Akademiia Nauk SSSR Doklady (periodical code)

DOST Department Of Science and Technology

DOSV Deep Ocean Survey Vehicle

DOSX DOS Extender

Dot Dotation

DoT Department of Tourism • Department of Transport

DOT Data and Operations Team • Database for Objects in Time-domain • Data Observe Time • Data Output Tape • Day Of Training • Deep Ocean Transponder • Deep Offshore Technology • Department Of Overseas Trade • Department Of Telecommunications • Department Of Transportation • Department Of Treasury • Deployment Operations Team • Designating Optical Tracker • Design Optimization Tool • Digital Optical Tape • Digital Output Timer • Directly Observed Therapy • Director Of Training • Doctrine - Organization - Training • Dutch Open Telescope (formerly OTT)

DOTA Deployable Optical Telescope Assembly

DOTAC Department Of Transport And Communications

DoTad Akademiia Nauk Tadzhikskoi SSR Doklady (periodical code)

DOTC Dansk Organisation for Test og Certificering (Denmark) • Data Observing Testing Console

DOTD Director Of Training Development

DOT&E Director, Operational Test & Evaluation

DOTE Director of Operational Test and Evaluation

DOTIS Département d'Orientation Technique et d'Information Scientifique

DOTM Due Out To Maintenance

DOTS Data Ordering and Tracking System • Data Oversight and Transfer System • Digital Office Timing Supply • Dynamic Ocean Tracking System • Dynamic Ocean Track System

Dott Dottore

DOTT Detection Of Tornadic Thunderstorms

DOT&V Director, Operational Test & Verification

DOTV Director of Operational Test and Verification

DOU Daily Orbit Update • Data Object Unprocessed • Dourados (Brazil-MS) (location identifier)

Doubl Doublet

DoUkr Akademiia Nauk Ukrains koi RSR Dopovidi (periodical code)

DoUzb Akademiia Nauk Uzbekskoi SSR Doklady (periodical code)

DOV Data Over Voice • Director of Orbital Verification • Discreet Operational Vehicle • Dispatch Of Values • Dover Cheswold (USA-DE) (location identifier)

DOVAC DERA-ONERA Vibro-Acoustic Collaboration

DOVAP Doppler Velocity And Position

DOVE Data Over Voice Equipment

DOVF Dependent Overflow

DOW Data Orderwire • Date Of Withdrawal • Day Of Week • Digital Orderwire • Direct Overwrite

DOWB Deep Operating Workboard

DOWS Directional Ocean Wave Spectrum

DOX Direct Operational Exchange • Dongara (Australia-WA) (location identifier)

DOY Day Of Year

doz Description Out Zip (file extension)

Doz Dozen

DOZ Dolmetscherschule Zürich (Switzerland) • Dozen

dp Daily Planner (file extension) • Dot Pitch

d/P Documentos contra Pago

Dp Deep • Department • Departamento

D-P Double-Peaked

D&P Development & Production • Drain & Purge

D/P Data Pack • Définitif/Provisoire • Delivery against Payment • Delivery on Payment • Documents against Payment

DP Damp-Proof • Damp-Proofing • Data Pack • Data Package • Data Pattern • Data Policy • Data Printer • Data Processing • Data Processor • Data Product • Data Production • Davis-Putnam • Decision Package • Decision Point • Deep • Deep Penetration • Definitive Parameter • Deflection Plate • Degree of Polimerization • Delayed Procurement • Deliberate Planning • Delivery Point • Delta Pressure • Demand Priority • Demarcation Point • Demonstration Plan • Departure Point • Deployable Payload • Deployment Platform • Depreciation Percentage • Depth • Description Pattern • Design Phase • Design Proof • Destination Port • Developmental Priority • Development and Production • Development Phase • Development Plan • Development Proposal • Development Prototype • Dew Point • Dial Pulse • Dial Pulsing • Diametral Pitch • Diametrical Pitch • Differential Pressure • Diffuse Penetration • Digital Photogrammetry • Direction de la Prévision • Directione Propria • Direction Pédagogique • Director of Personnel • Direct Playback • Disabled Person • Disk Pack • Displaced Person • Display Panel • Display Postscript • Display Processing • Display Processor • Disruptive Pattern • Distributed Publishing • Distribution Point • Distrito Postal • Division President • Doctor of Philosophy • Document Pertinent • Domus Procerum • Dot Pitch • Double Paper • Double Penetration • Double Play • Double Pole • Double Precision • Drafting Panel • Draft Proposal • Drain and Purge • Drill Purpose • Drum Processor • Dual Purpose • Durable Penetration • Durable Press • Dynamic Programming

D&PA Design & Product Assurance

DP&A Data Processing & Analysis

DPA Chicago - Du Page County (USA-IL) (location identifier) • Data Processing Activity • Data Processing Agency • Data Processing Algorithm • Data Processing Area • Data Processing Assembly • Deferred Payment Account • Delegation of Procurement Authority • Demand Protocol Architecture • Demi-Petit Axe • Departamento de Pesquisa e Aplicações • Department of Physics and Astronomy • Department of Political Affairs • Design and Product Assurance • Destructive Physical Analysis • Detector Position Angle • Deutsches Patentamt (Germany) • Deutsche Press-Agentur (Germany) • Diphenylamine • Diploma in Public Administration • Direction des Personnels Administratifs • Direct Piezo-Actuator • Direct Pricing Agreement • Division Planning Authorization • Doctor of Public Administration • DoD Protocol Architecture • Double-Precision Arithmetic

DPAC Dedicated Plant Assignment Center • Direction des Programmes pour l'Aviation Civile (France)

DPAD Directeur de Projet d'Animation et de Développement (France)

DPA&E Director, Program Analysis & Evaluation

DPAE Director of Program Analysis and Evaluation

DPAH Direct Product Actual Hour

DPAI Division du Patrimoine et des Actions Immobilières

DPAIS Data Processing Advanced Information System

DPAM Demand Priority Access Method

DPANS Draft Proposed American National Standard

DPAOS Direction des Personnels Administratifs, Ouvriers et de Service

DPAP Demand Priority Access Protocol

DPAREN Data Parity Enable

DPARS Data Processing Automatic Record Standardization

DPAS Digital Patch and Access System • Door Peninsula Astronomical Society (USA-WI)

DPAT Dynamic Program Analysis Tool

DPATE Direction des Personnels Administratifs, Techniques et d'Encadrement

DPASV Differential Pulse Anodic Stripping Voltammetry

DPB Data Plotting Board • Data Processing Board • Data Processing Branch • Deposit Pass Book • Dewpoint Temperature • Document Processing Branch • Drive Parameter Block

DPBC Delivery Point Bar Code

DPBIT Data Processor Built-In Test

DPBO Division Property Book Officer

DPBX Digital Private Branch Exchange

DPC Database Promotion Centre (Japan) • Data Processing Center • Data Processing Central • Data Processing Control • Deadtime-Primbsching Correlation • Decision Point Cluster • Dedicated Personal Communicator • Destination Point Code • Differential Pressure Control • Digital Phase Comparator • Digital Pressure Converter • Digital Process Controller • Diplôme du Premier Cycle • Direct Program Control • Display Power Control • Domestic Policy Council • Dual Payload Carrier • Dual Power Converter

DPCA Data Processing Control Area • Director of Personnel and Community Activities • Displaced Phase Centre Antenna • Dual Port Communications Adapter

DPCC Data Processing Control Center • Data Processing Control Console • Differential Pulse Code Modulation

DPCI Domain PC Interconnect

DPCL Dedicated Payload Communication Link

DPCM Delta Pulse Code Modulation • Differential Pulse Code Modulation • Digital Pulse Code Modulation

DP&CS Data Processing & Communications Security

DPCS Data Processing & Communications Security • Deep Photographic Cluster Survey • Distributed Process Construction System

DPCSD Department for Policy Coordination and Sustainable Development

DPCT Device Performance Characteristics Table

DPCTE Data Processor and Computer Test Equipment

DPCU Digital Processing and Control Unit

DPD Data Processing Department • Data Processing Division • Data Procurement Document • Data Project Directive • Data Protection Directive • Debrecen Photoheliographic Data • Detailed Procedures Description • Digital Plane Driver • Digital Phase Difference • Digital Product Definition • Direction de la Programmation et du Développement • Discrete Probability Distribution

DPDLD Detailed Program Design Language Document

DPDM Differential Pulse Duration Modulation • Double Pulse Duration Modulation

DPDS Data Processing and Display System • Data Processing Design System • Distributed Processing and Display System

DPDT Double-Pole Double-Throw • Dual-Pole Dual-Throw

DPDU Direction de la Programmation et du Développement Universitaire

DPE Data Parity Error • Data Path Extender • Data Processing Electronics • Data Processing Element • Data Processing Environment • Data Processing Equipment • Data Processing Extract • Delayed Pulse Extraction • Demand Processing Unit • Development Project Engineer • Dieppe (France) (location identifier) • Digital Processing Electronics • Digital Processing Environment • Direction des Personnels d'Encadrement • Direction des Personnels Enseignants • Distributed Processing Environment

DPEK Differential Phase-Exchange Keying

DPEO Deputy Program Executive Officer

DPES Direction des Personnels de l'Enseignement Supérieur • Distributed Polyelastic Sampling

DPESR Direction des Personnels de l'Enseignement Supérieur et de la Recherche

DPET Data Processing Equipment Trainer

DPF Data Private Facility • Data Processing Facility • Data Production Facility • Dense Plasma Focus • Differential Pressure Feedback • Directional Potential Function • Double Point Failure • Dust Particle Facility • Dynamic Pressure Feedback

DPFC Double-Pole Front Connected

DPFO Data Processing Field Office

DPFT Data Processing Focus Team

Dpg Damping

DPG Data Processing Group • Data Processor Group • Dedicated Packet Group • Deutsche Physikalische Gesellschaft (Germany) • Deutsche Plannungsgesellschaft (Germany) • Digital Pattern Generator • Dual-Purpose Gyroscope • Dugway Michael (USA-UT) (location identifier)

DPGS Data Product Generation System

DPh Doctor of Philosophy

DPH Department of Public Health • Director of Public Health

DPhG Département de Physique Générale

DPhil Doctor of Philosophy

DPHS Diphase

dpi Dots per Inch (file extension)

DPI Data Processing Installation • Department of Primary Industries • Department of Public Information • Derivative Port Identifier • Detail Program Interrelationship • Deuterium Pellet Injector • Digital Process Instrument • Digital Products Inc. • Digital Pseudorandom Inspection • Distributed Protocol Interface • Divisão de Processamento de Imagens • Dot Pitch Integer • Dots per Inch • Double Pendulum Interferometry

DPID Direction des Personnels d'Inspection et de Direction

DPIE Department of Primary Industries and Energy

DPIR Data Processing and Information Retrieval

DPIV Digital Particle Image Velocimeter

DPK Deer Park (USA-NY) (location identifier) • Deutsche Patentklassifikation (DPA)

DPKO Department of Peacekeeping Operations

Dpl Diplomat • Diplomatic • Duplex

DPL Data Processing Laboratory • Data Processing Language • Data Programming Language • Davis - Putnam - Loveland • Declared Parts List • Degaussed Product List • Descriptor Privilege Level • Development Problem Log • Digital Power Line • Diplomat • Dipolog (Philippines) (location identifier) • Distributed Processing Library • Distributed Program Link • Document Processing Language • Dual Polarization Lidar • Dublin Public Libraries (Ireland)

DPLCS Digital Propellant Level Control System

DPLE Declarative Programming Languages in Education

DPLL Digital Phase-Locked Loop

DPLM Data Product Line Manager • Deputy Program Manager for Logistics

DPLO Deep Penetration - Large Opening

DPLOA Draft Proposed Letter Of Agreement

DPLR Doppler

DPLY Deployment

DPM Data Processing Machine • Data Processing Management • Data Processing Manager • Data Processing Modeling • Data Processing Module • Decentralized Production Management • Deflectable Photomultiplier • Deputy Prime Minister • Deputy Program Manager • Deputy Project Manager • Development Planning Memorandum • Development Preliminary Manual • Dichroic Parametric Mirror • Digital Panel Meter • Digital Phase Modulation • Digital Plotter Map • Digital Prominence Monitor • Dinamika i Prochnost Mashin (periodical code) • Dissemination Program Manager • Domaine Public Maritime • Double Projection Model • Draft Presidential Memorandum • Drop Physics Module (*) • Drop Physics Molecule • Dual Parton Model • Dual-Port Memory

DPMA Data Processing Management Association (USA) • Deutsches Patent und Markenamt (Germany) • Direction du Personnel, de la Modernisation et de l'Administration • Distributive Principle of Multiplication over Addition

DPM&C Department of Prime Minister & Cabinet

DPMC Département de Physique de la Matière Condensée • Disk Phase-Mask Coronagraph

DPMG Deputy Postmaster General

DPMI DOS Protected Mode Interface

DPMIS Data Processing Management Information System

DPML Deputy Program Manager for Logistics

DPMMS Department of Pure Mathematics and Mathematical Statistics

DPMO Defects Per Million Opportunities

DPMR Deputy Project Manager Resources • Director of Personnel Manpower Resources

DPMS Dahlgren Polar Monitoring Service (USA) • Data Project Management System • Display Power Management Signaling • Display Power Management System • DOS Protected Mode Services

DPMT Deputy Project Manager Technical

DPN Data Packet Network • Datenpaketnetzwerk • Deepen • Deutsches Provider Netzwerk (Germany) • Dual Processing Node

DPNG Deepening

DPNH Dinitrophenylhydrazine

DPNSS Digital Private Network Signaling System

DPO Data Phase Optimization • Data Processing Operation • Designated Post Office • Devonport (Australia-TAS) (location identifier) • Digital Phosphorous Oscilloscope • Digital Processing Oscilloscope • Direction Par Objectifs • Disaster Preparedness Office • Distinct Point of Origin • Distributing Post Office • District Pay Office

DPOB Date and Place Of Birth

DPODP Double-Precision Orbit Determination Program

DPOP Data Processing Operating Procedure

DPOSS Digital Palomar Observatory Sky Survey • Digitized Palomar Observatory Sky Survey (*)

DPP Data Patch Panel • Data Processing Plan • Data Project Plan • Day of Pay Period • Decision Process Pattern • Dedicated Procurement Program • Deferred Payment Plan • Delivery Planning Package • Demand Priority Protocol • Deployment/Pointing Panel • Detailed Project Plan • Development Program Plan • Differential Pulse Polarography • Digital Preprocessor • Director of Public Prosecutions • Direct Product Profitability • Distributed and Parallel Processing • Distributed Production Program • Dynamic Passage Partition

DPPA Double Pumped Parametric Amplifier

DPPE Data Processing Project Engineer

DPPI Data Processing Plug-In • Designing Pleasurable Products and Interfaces (conference)

DPPO Development - Production - Prove-out

DPPR Direction de la Prévention de la Pollution et des Risques (France)

DPQ Double-Precision Quantity

DpR Degrees per Revolution

DPR Data Pedigree Review • Data Processing Request • Data Protection Register • Debrecen Photoheliographic Results • December Progress Review • Definition Phase Review • Degrees Per Revolution • Democratic People's Republic • Department of Professional Regulation • Design Problem Report • Development Planning Review • Development Progress Review • Direction of Public Relations • Document Problem Report • Dual Precipitation Radar

DPRI Disaster Prevention Research Institute (Japan)

DPRK Democratic People's Republic of Korea (North Korea)

DPRP Disaster Prevention and Recovery Plan

DPRS DECT Packet Radio Service • Docking Probe Retraction System

DpS Degrees per Second • Dividend per Share

DP&S Data Processing & Software

DPS AAS Division of Planetary Science Meetings (periodical code) • Data Preparation System • Data Processing and Software • Data Processing Set • Data Processing Standard • Data Processing System • Data Production Specialist • Decision Package Set • Deep Public Survey • Degrees Per Second • Delegated Production System • Delivery Point Sequence • Denpasar-Bali - Ngurah Rai (Indonesia) (location identifier) • Descent Power System • Descent Propulsion System • Design and Performance Specification • Design and Procedure Standard • Detector Power Supply • Development Processing System • Diagnostic Problem Solver • Differential Power Switch • Digisonde Portable Sounder • Digital Phase Shifter • Digital Photogrammetric Station • Digital Photogrammetric System • Digital Photogrammetry System • Digital Plotter System • Digital Power Supply • Digital Processing System • Digital Production System • Diode Phase Shifter • Diplôme Professionnel Spécialisé • Director of Personal Services • Director of Postal Services • Direct Personnel Support • Disk Programming System • Display Postscript • Display Power Supply • Display Processing System • Distributed Processing System • Disturbance Propagation System • Dividend Per Share • Division of Planetary Science • Doctor of Public Service • Document Preparation System • Document Processing System • Drifting Pulsation Structure • Droit Préférentiel de Souscription • Dust Protection System • Dynamic Personalization Server • Dynamic Power Spectrum

DPSC Data Processing Service Center

DPSCA Department of Political and Security Council Affairs

DPSD Data Processing Systems Division • Digital Publications Standard Development

DPSH Direct Product Standard Hour

DPSK Differential Phase Shift Keying • Digital Phase Shift Keying

DPSM Diode Phase Shifter Module • Double Passage Soustractif Multicanal

DPSN Department of Physics of Stars and Nebulae • Deployable Packet Switched Node

DPSO Data Products Standards Office • Deep Penetration - Small Opening • Discovery Program Support Office

DPSPAR Digital Processing Systems Personal Animation Recorder

DPSR Data Processing Service Request • Data Processing System Requirements

DP&SS Data Processing & Software System

DPSS Data Processing and Services System • Data Processing and Software System • Data Processing Services System • Data Processing Software System • Data Processing Subsystem • Data Processing Switching System • Data Processing System Simulator • Data Product Screening System

DPSST Data Product Screening Software Team

DPST Double-Pole Single-Throw • Dual-Pole Single-Throw

Dpstk Dipstick

DPSU Downstream Processing Unit

Dpt Département • Department • Deponent • Deposit • Depot

DPT Data Punched Tape • Delayed Picture Transmission • Descent Performance Test • Design Proof Test • Diphteria - Pertussis - Tetanus • Director of Plans and Training • Distributed Processing Technology • Division of Production Technology • Drive Parameter Table • Drive Parameter Tracking • Duty Performance Test • Dynamic Packet Transport • Dynamic Pressurization Test

DPTH Depth

Dpto Departamento

DPtoTP Display Coordinates to Tablet Coordinates

DPTS Dartmouth/Princeton Timing System

Dpty Deputy

DPU Data Parallel Unit • Data Path Unit • Data Processing Unit • Digital Processing Unit • Display and Processing Unit • Distributed Processing Unit • Distributed Processing Utility • Droit de Préemption Urbain • Dual Processor Unit • Dumpu (Papua New Guinea) (location identifier)

DPV Delivery Point Verification • Depressurization Valve • Differential Pulse Voltammetry • Distributed Partition View

DPVSt Datenpaketvermittlungsstelle

DPW Dedicated Professional Workforce • Department of Public Works • Digital Photogrammetric Workstation • Directorate of Public Works • Director of Public Works

DPWB Diensten voor de Programmatie van het Wetenschapsbeleid

DPWF Dedicated Professional Workforce

DPWG Dapa Processing Working Group

DPWR Data Process Work Request

DPWS Digital Photogrammetric Workstation

Dpx Duplex

DPX Duplex • Duplexer

DPY Deploy

DQ Data Qualification • Data Qualifier • Data Quality • Definite Quantity • Dernier Quartier • Design Qualification • Destination Queue • Direct Question • Discussion Question • Dispersion Quotient • Disqualified • Disqualify • Duque • Dynamic Query • Fiji (international aircraft registration code)

DQA Data Quality Assessment • Diagonal Quadratic Approximation

DQC Data Quality Control • Definite Quantity Control • Dynamic Quality Control

DQCM Data Quality Control Monitor

DQDB Distributed-Queue Dual Bus • Dual Queue Dual Bus

DQE Detective Quantum Efficiency • Detector Quantum Efficiency (*) • Devis Quantitatif Estimatif • Differential Quantum Efficiency

DQF Data Quality File • Data Query Function

DQH Distributed Query Handler

DQI Data Quality Index • Digital Quartz IMU • Digital Quartz Inertial

DQL Database Query Language • Data Query Language

DQM Data Quality Manager • Digital Quality Monitor • Distributed Queue Management

DQMCS Data Query and Manipulation Common Service

DQO Data Quality Objective

DQP Designated Qualified Person • Dès Que Possible • Diode Qualification Program

DQPSK Differential Quadratic Phase Shift Keying • Differential Quadrature Phase Shift Keying • Differential Quaternary Phase Shift Keying

DQQ Digital Quarter Quadrangle

DQR Design Qualification Requirement

DQS Distributed Queuing System • Domain Queuing System • Duques

DQSA Duquesa

DQSR Data Quality Survey Report

DQT Demand Quarter Tally • Design Qualification Test • Diode Qualification Test

DQTP Design Qualification Test Plan • Diode Qualification Test Program

Dque Duque

DQV Data Quality Verification

dqy Database Query (file extension)

dr Drachma (currency) • Dropper

Dr Debtor • Directeur • Director • Docteur • Doctor • Dokter • Doktor • Dollar • Door • Drachma (currency) • Draht • Drain • Dram • Drama • Draw • Drawer • Drawn • Dreef • Drève • Drill • Drive • Driver • Droit • Druk • Drum

D/R Deposit Receipt • Digital to Resolver • Direct or Reverse • Disassemble/Reassemble

DR Daily Report • Daily Review • Data Ready • Data Receive • Data Received • Data Receiver • Data Record • Data Recorder • Data Reduction • Data Register • Data Report • Data Repository • Data Request • Data Requester • Data Requirement • Dead Reckoning • Debug Register • Decay Ratio • Decision Review • Decommissioning Review • Decreto Reglamentar • Deficiency Report • Definitively Relevant • Délégation Régionale • Delivery Rate • Demi-Reliure • Departmental Report • Designated Router • Design Requirement • Design Review • Desktop Resource • Detailed Report • Detection Radar • Development Report • Development Right • Deviation Range • Deviation Ratio • Dielectronic Recombination • Digital Receiver • Digital Research • Differential Rate • Differential Rotation • Diffusion Régionale • Dilution Refrigerator • Dining Room • Directeur de Recherche • Direction de la Recherche • Direction Régionale • Disaster Recovery • Disconnect Request • Discount Rate • Discrepancy Record • Discrepancy Report • Dispatch Reliability • Dispersion Relation • Disposition Record • Disposition Report • Dissociative Recombination • Distinguished Reviewer • Distributive Recursion • District Railway • District Registry • Doctor • Documentation Requirement • Document Report • Document Retrieval • Dominican Republic • Door • Double Royal • Doutor • Down Range • Drachma (currency) • Drafting Request • Draft Recommendation • Draft Release • Drifting • Drift Rate • Drive • Drizzle and Rain • Dutch Reformed • Dynamic Range • Dynamic Relaxation

Dra Draco-Draconis (constellation)

DR&A Data Reduction & Analysis • Data Requirements & Analysis

DRA Data Rate Adapter • Data Reduction Agent • Data Reduction and Analysis • Data Reformatter Assembly • Data Release Authorization • Data Request Acknowledgment • Data Requirements and Analysis • Data Resource Adapter • Dead Reckoning Analyzer • Decision Risk Analysis • Defence Research Agency • Deployment and Retraction Assembly • Design Review Activity • Design Review Agreement • Digital Read-in Assembly • Digital Recorder Analyzer • Discontinuous Reinforced Aluminium • Document Release Authorization • Doutora • Drum Read Amplifier • Mercury - Desert Rock (USA-NV) (location identifier)

DRAAG Design Review And Acceptance Group

DRAC Data Reduction and Analysis Cell • Direction Régionale de l'Aviation Civile • Direction Régionale des Affaires Culturelles (France) • Distributed Read Address Counter

DRACO Dynamics, Reconnection And Configuration Observatories (mission)

D/RADEX Digitized Radar Experiment

DRAE Diccionario de la Real Academia de la Lengua Española

DRAG Doctrine Review and Approval Group

DRAGN Double Radiosource Associated with Galactic Nuclei

DRAGNs Double Radiosource in Active Galactic Nuclei

DRAI Dead Reckoning Analog Indicator

DRAL Daresbury and Rutherford Appleton Laboratories (UK) (now CCLRC)

DRAM Detection-Radar Automatic Monitoring • Digital Random Access Memory • Dynamic Random Access Memory (*)

DRAMA Data Review Analysis and Monitoring Aid • Digital Radio And Multiplexer Acquisition

DRAMS Data Reduction And Management System • Digital Recording And Measuring System

DRANS Data Reduction and Analysis System

DRAO Daeduk Radio Astronomy Observatory (Korea-ROK) (now TRAO) • Dominion Radio Astrophysical Observatory (Canada-BC)

dr. ap. Dran (Apothecaries)

DRAPE Data Recording And Processing Equipment • Digital Recording And Playback Equipment

DRASC Diffusion Raman Anti-Stokes Cohérente

DRAT Data Reduction Analysis Tape • Data Requirements Analysis Tool • Demonstration Reliability Acceptance Test

dr. av. Dran (Avoirdupois)

DRAW Direct Read After Write

DRB Data Review Board • Decimal Register Binary • Derby (Australia-WA) (location identifier) • Design Requirements Baseline • Design Review Board • Discrepancy Review Board

DRBTM Dynamic Radiation Belt Test Model

DRC Daily Relative Citedness • Damage-Risk Criterion • DAMA Remote Controller • Data Recording Camera • Data Recording Control • Data Recording Controller • Data Recovery Component • Data Reduction Center • Data Reduction Compiler • Data Reduction Complex • Data Return Capsule • Data Return Container • Deployment Readiness Condition • Depot Repair Cycle • Design Rule Check • Digital Reality Creation • Directeur de Recherche Contractuel • Direct-Reading Compass • Direct Reporting Command • Dirico (Angola) (location identifier) • Disability Review Council • Discrete Rate Command

DRCC Department of Cosmic Rays and Cosmology

DRCCE Directeur de Recherche Contractuel de Classe Exceptionnelle

DRCE Directeur de Recherche de Classe Exceptionnelle • Direction Régionale du Commerce Extérieur (France)

DR&CG Data Reduction & Computer Group

DRCG Data Reduction and Computer Group • Dressler Cluster of Galaxies

Drch Drachma (currency)

DRCI Direction Régionale du Commerce et de l'Industrie

DRCS Data Receiver and Command Sender • Digital Radio Concentrator System • Dynamically Redefinable Character Set

Drct Direct

DRCT Depot Repair Cycle Time • Direct

DRCVR Digital Receiver

DRD Data Reading Device • Data Recording Device • Data Requirements Description • Data Requirements Document • Demand Return and Disposal • Design Requirements Document • Detailed Requirements Document • Document Requirement Description • Dorunda Station (Australia-QLD) (location identifier) • Drum Read Driver

DRDA Distributed Relational Data Architecture • Distributed Relational Database Access • Distributed Relational Database Architecture (*)

DRDBMS Distributed Relational Database Management System

DRDC Detector Research and Development Committee

DRDM Distributed Relational Data Manager

DRDS Dynamic Reconfiguration Demonstration System

DRDW Direct Read During Write

DRe Dead Reckoning Error

D/RE Disassemble/Reassemble Equipment

DRE Data Reduction Equipment • Dead Reckoning Equipment • Department of Resources and Energy • Differential Riccati Equation • Directional Reservation Equipment • Direction Régionale de l'Équipement • Direct Recording Electronic (machine) • Display Remoting Equipment • Distributed Real-time and Embedded (system) • Diversity Reception Equipment • Downrange Error • Drummond Island (USA-MI) (location identifier) • Ducted Rocket Engine

DREAM Database on Rare Earths At Mons (University) • Definitive Realization of Energy and Matter

DRED Data Routing and Error Detecting • Deferred Requisitioning of Engineering Data • Deferred Requisitioning of Engineering Drawings • Detection Radar Environment Display • Direction de la Recherche et des Enseignements Doctoraux • Direction de la Recherche et des Études Doctorales (*)

DREE Direction des Relations Économiques Extérieures (France)

DREN Defense Research and Education Network (USA)

DrEng Doctor of Engineering

DREO Defence Research Establishment Ottawa (Canada-ON)

DRES Defence Research Establishment Suffield (Canada-ON) • Direction des Relations avec l'Enseignement Supérieur • Direct Reading Emission Spectrograph • Diversity Resources and Educational Services

DRET Direction de la Recherche et de la Technologie • Direction des Recherches et Études Techniques

DREWS Direct Readout Equatorial Weather Satellite

DRF Data Recovery Field • Data Reduction Facility • Data Reporting Form • Data Request Form • Data Requirement Form • Desorientation Recovery Facility • Deutsche Rettungsflugmacht (Germany) • Digital Raster File • Digital Radio Frequency • Digital Research Facility • Disaster Response Force • Discrepancy Report Form • Distinguished Research Fellowship • Doctoral Records File • Documentation Requisition Form • Document Relative Frequency • Drift River (USA-AK) (location identifier)

DRFC Direction Régionale de la Formation Continue

DRFM Digital Radio Frequency Memory

DRFP Direction Régionale à la Formation Professionnelle • Draft Request For Proposals

DRFT Drift

Drftd Drifted

DRG Data Record Generator • Data Reduction Group • Deering (USA-AK) (location identifier) • Defense Research Group • Deutsche Röntgengesellschaft (Germany) • Diagnostic-Related Group • Digital Random Generator • Digital Ranging Generator • Digital Raster Graphic • Directional Receiver Group

DRGP Digital Representation of Graphic Products

DRGS Direct Readout Ground Station

DRH Département des Ressources Humaines • Digital Readout Head • Directeur des Ressources Humaines • Direction des Ressources Humaines

DRHAF Direction des Ressources Humaines et des Affaires Financières

Dr. hc Doctor Honoris Causa

DRI Data Rate Indicator • Data Recording Interface • Data Reduction Interpreter • Data Representation Interface • Dead-Reckoning Indicator • Defense Reinvestment Initiative • Defense Research Internet • De Ridder - Beauregard Parish (USA-LA) (location identifier) • Desert Research Institute (USA-NV) • Design Research Institute (USA-NY) • Differential Refraction Instrument • Diffuse-Reflectance Infrared Fourier Transform • Digital Research Inc. • Direction des Relations Internationales • Directly Riemann Integrable • Distributed Recovery Intelligence

DRIBU Drifting Buoy

DRIC Direction des Relations Internationales et de la Coopération

DRID Direct Readout Image Dissector

DRIE Deep Reactive Ion Etched • Deep Reactive Ion Etching (*)

DRIFT Diversity Receiving Instrumentation For Telemetry

DRIG Digital Rate-Integrating Gyro

DRIL Directed Item Return List

Dr.-Ing. Doktor der Ingenieurwissenschaft

DRIP Data Reduction Input Program

DRIPS Dynamic Real-Time Information Processing System

DRIR Direction Régionale de l'Industrie et de la Recherche (France) • Direct Readout Infrared • Direct Readout Infrared Radiometer

DRIRE Direction Régionale de l'Industrie, de la Recherche et de l'Environnement (France)

DRIRU Dry Rotor Inertial Reference Unit

DRISS Digital Read-In Subsystem

DRIU Damage Repair Instruction Unit

DRIV Drive

DRIVE Dedicated Road Infrastructure for Vehicles in Europe

DRK Display Request Keyboard

DRL Data Reduction Laboratory • Data Requirements Language • Data Requirements List • Data Required to Load • Data Retrieval Language • Daytime Running Lamp • Decision Representation Language • Design Review List • Diagnostic Reference Level • Digital Resource Librarian • Document Requirements List • Double-Rack Laboratory • Dynamically Reconfigurable Logic

DRLH Directed Recursive Labelnode Hypergraph

DRLMS Digital Radar Landmass Simulation

DRLT Database Report Language Team

DRM Data Records Management • Data Relay Mode • Data Replication Manager • Data Resource Management • Département des Recherches Météorologiques • Design Reference Mission • Design Reference Model • Design Requirements Manual • Design Review Meeting • Detector Response Matrix • Detrital Remanent Magnetization • Digital Radiometer • Digital Range Machine • Digital Range Meter • Digital Rights Management • Digital Road Map • Diploma in Resource Management • Directorate for Resource Management • Director for Resource Management • Discipline Requirements Manager • Distinct Region Model • Drafting Room Manual • Drama (Greece) (location identifier) • Drawing Requirements Manual

DRMD Deployment Requirement and Manning Document

DRMI DOS Routed Mode Interface

DRMO Defense Reutilization Marketing Organization

DRMP Design Reference Mission Profile

DRMS Data Resource Management System • Distance Root Mean Square

DRMU Digital Remote Measurement Unit

Drn Drawn

DRN Data Record Number • Data Reference Number • Dewan Riset Nasional (Indonesia) • Dirranbandi (Australia-QLD) (location identifier) • Document Release Notice • Document Revision Notice • Drain

DRNN Diagonal Recurrent Neural Network • Dynamic and Recurrent Neural Network • Dynamical Recurrent Neural Network (*)

DRO Daily Routine Order • Data Request Output • Datenverarbeitung-Rationalisierung-Organisation • Design Requirements Overview • Destructive Readout • Dielectric Resonator Oscillator • Digital Readout • Digital Revision Overlay • Direct Readout • Disposal Release Order • Document Release Order • Domestic Route Order • Double Resonant Oscillator • Durango - La Plata (USA-CO) (location identifier) • Purgatory (USA-CO) (location identifier)

DROG Drogue

DROID Distributed Readout Imaging Device

DROM Decoder Read-Only Memory

DROO Digital Readout Oscilloscope

DROP Data Printout Program

DROPPS Distribution and Role Of Particles in the Polar Summer

DROS Direct Readout Satellite • Disk-Resident Operating System

DROWS Direct Readout Weather Satellite

DRP Data Reduction Procedure • Data Reduction Program • Data Restoration Program • Dead Reckoning Plotter • DECNET Routing Protocol • Deficiency Report Program • Densest Random Packing • Digital Recording Process • Digital Routing Protocol • Disaster Recovery Plan • Distribution and Replication Protocol • Dividend Reinvestment Plan

DRPP Data Relay Preparatory Programme

DRPM Direct Reporting Program Manager

DRPR Date Repair Part Required • Drawing Practices

DRR Data Recorder/Reproducer • Data Redundancy Removal • Design Readiness Review • Design Release Review • Design Request Review • Design Requirements Review • Digital Radio-Relay • Digital Receiver Replacement • Document Release Record • Durrie (Australia-QLD) (location identifier)

DRRB Data Requirements Review Board

DRRC Devices Requirements Review Committee

DRRD Design Requirements Review Disposition

Dr. rer. nat. Doctor of Natural Science

DRRF Documentation Retrieval Request Form

DRRS Drawing Research and Request System

DRRT Délégation Régionale à la Recherche et à la Technologie (France)

DRS Data Rate Select • Data Rate Selector • Data Reaction System • Data Receiving Station • Data Recording Set • Data Recording System • Data Reduction Software • Data Reduction System • Data Relay Satellite • Data Relay Station • Data Relay System • Data Requirement Specification • Data Retrieval System • DCP Retransmission System • Deficiency Reporting System • Design Requirements Sheet • Design Requirements Specification • Dielectric Relaxation Spectroscopy • Digitales Richtfunksystem • Digital Radar Simulator • Digital Radio System • Digital Range Safety • Digital Readout System • Digital Receivers • Digital Receiver Station • Digital Recording System • Direct Receiving Station • Direct Reception System • Disassembly/Reassembly Station • Discrepancy Reporting System • Display Resource (file extension) • Dispute Resolution Specialist • Disturbance Reduction System • Documentation Retrieval System • Document Registration System • Document Retrieval System • Donning and Recharge System • Doppler Radar System • Downlink Receiving Station • Dresden Klotzsche (Germany) (location identifier) • Driver Resource • Drives • Dynamic Reflectance Spectroscopy

DRSA Data Recording System Analyst

DRSAM Dynamic Random-Sequential Access Method

DRSCS Digital Range Safety Command System

Drsg Dressing

DRSG Digital Recorder Signal Generator

DRSI Destination Station Routing Indicator

DRSN Drifting Snow

DRSS Data Recording Subsystem • Data Relay Satellite System

Drt Droit

DRT Data Receive Terminal • Data Relay Terminal • Data Relay Transponder • Data Review Technician • Dead Reckoning Tracer • Decision Response Time • Del Rio (USA-TX) (location identifier) • Deployment Readiness Test • Design Reference Test • Design Reference Timeline • Design Reference Trajectory • Design Review Team • Device Rise Time • Diagnostic Rhyme Test • Digital Readout Timer • Diplôme de Recherches Technologiques • Direction de la Recherche et de la Technologie (France) • Direction des Relations de Travail • Discourse Representation Theory

DRTC Designated Reporting Technical Control

Drte Droite

DRTE Direction Régionale du Travail et de l'Emploi

DRTEFP Direction Régionale du Travail, de l'Emploi et de la Formation Professionnelle

DRTL Diode Resistor Transistor Logic

DRTM Data Relay and Technology Mission (*) • Data Relay Technology Mission

DRTS Data Relay and Tracking Satellite • Data Relay Test Satellite • Detecting, Ranging and Tracking System

DRTSS Data Relay and Tracking Satellite System

DRU Data Reference Unit • Data Retrieval Unit • Digital Radio Unit • Digital Range Unit • Digital Recovery Unit • Digital Register Unit • Digital Remote Unit • Digital Retiming Unit • Direct Reporting Unit • Drive Unit • Drummond (USA-MT) (location identifier) • Dynamic Reference Unit

DRUB Digital Remote Unit Buffer

DRUC Disposition Record Unsatisfactory Condition

DRUI Délégation aux Relations Internationales Universitaires

DRUID Dynamic Rules for User Interface Design

DRUM Deep Reflection of the Upper Mantle

DRUMS Defeasible Reasoning and Uncertainty Management System • Detailed Revision/Update of Message Standards

Dr. und Vrl. Druck und Verlag

Dr. u. Vrl. Druck und Verlag

drv Driver (file extension)

DRV Daily Reference Value • Dark Red Varnish • Democratic Republic of Viet Nam • Drive • Driver

DRVID Differenced Range Versus Integrated Doppler

Drvr Driver

DRVS Differential Radial Velocity Spectroscopy

DRVT Downsized Remote Video Terminal

DRW Darwin (Australia-NT) (location identifier) • Draw • Drawing

DRWG Data Reduction Working Group

drx Drachma (currency)

DRZ Deep Reconnaissance Zone

Drzl Drizzle

DRZL Drizzle

d/s Days' Sight

Ds Dans • Decistere • Destro • Dysprosium • Element of Path

dS Dwarf Spheroidal (galaxy)

D&S Davis & Sanford Co. (USA-NY) • Demand & Supply • Display & Storage

D/S Days after Sight • Deploy/Stow • Descent Stage • Digital to Synchro • Downsize • Dynamic/Static

DS Danish Standards • Dansk Standard (Denmark) • Dansk Standardiseringsråd (Denmark) • Dark Spot • Database Specification • Data Scanning • Data Segment • Data Send • Data Server • Data Set • Data Sheet • Data Specialist • Data Standard • Data Storage • Data Strobe • Data Synchronization • Data System • Date of Service • Daylight Saving • Decimal Subtract • Decision Sheet • Decontamination Solution • Deep Space • Deep Survey • Defect Size • Deficiency System • Define Storage • Délégué Syndical • Delivery Schedule • Demonstration Science • Dempster-Schafer (rule) • Department of State • Department Scientifique • Depleted Sample • Deputy Secretary • Descanso Semanal • Descent Stage • Descent System • Design Sheet • Design Specification • Design Standard • Detached Subgiant • Detail Specification • Devoir Surveillé • Diagnostic Software • Dictionary of Scientists • Diesel Ship • Diffractive System • Diffuse Surface • Digital Scanner • Digital Selection • Digital Signal • Digital Switch • Diode for Suppression • Diode Switch • Diplomatic Security • Directed Search • Directing Staff • Directionally Solidified • Directory Service • Direct Sequence • Direct Sequencing • Direct Signal • Direct Support • Discipline Scientist • Disclosure Statement • Disconnect Switch • Discover • Discovery Science • Disk Star • Disk Storage • Display System • Distributed Services • Distributed Software • Distributed System • Distribution System • Docking System • Doctor of Science • Doctrine Sponsor • Document Signed • Domestic Service • Dornier System • Double-Sided • Double Star • Draft Standard • Draw Start • Dressing Station • Drifting Structure • Driver Seated • Drive System • Drum Storage • Dünne Schicht • Dust Storm • Duty Scientist • Dwarf Spheroidal (galaxy) • Dynamical System

DSA Data Service Adapter • Data Systems Administration • Defense Supply Agency • Deployable Solar Array • Demonstration Structural Article • Deputy for Systems Acquisition • Designated Support Activity • Design Schedule Analysis • Destination Software Address • Destination Subarea • Deutsche Schüler Akademie (Germany) • Dial Service Assistance • Dial System Assistance • Diffusive Shock Acceleration • Digital Signal Analyzer • Digital Signature Algorithm • Digital Spectrum Analyzer • Dimensionally Stabilized Anode • Directory Service Agent • Directory System Agent • Directory System Alert • Discrete Sample Analyzer • Disk Storage Assembly • Display System Activity • Distant Support Area • Distributed Systems Architecture • Doppler Spectrum Analyzer • Dynamic Scalable Architecture • Dynamic Signal Analyzer

DSAA Division of Science and Academic Affairs

DSAC Data Synthetic Analysis Center

DSAD Data System Authorization Directory • Data Systems and Analysis Directorate

DSALS Decelerated Step Approach and Landing System

DSAM Direct Sequential Access Method

DSAO Data Sources in Asian-Oceanic Countries • Design Structuring and Allocation Optimization

DSAP Data Systems Automation Program • Destination Service Access Point

DSAR Daily Significant Activity Report • Data Sampling Automatic Receiver

DSAS DAMUS Satellite Archive System • Density Scalable Atomic Sphere • Digital Sun Angle Sensor • Distributed Systems Application Services

DSASS Digital Sun Angle Sensor System

DSAT Diagnostic Service Analysis Testing

DSAZ Deutsche-Spanische Astronomisches Zentrum (Spain)

DSB Defense Science Board • Device Status Byte • Dictionary of Scientific Biography • Digital Signaling Buffer • Diplôme Supérieur des Bibliothèques • Direct Satellite Broadcasting • Direct-Sound Broadcast • Direct-Sounder Broadcast • Direct-Sounding Broadcast • Disable • Dispute Settlement Body • Document Status Bulletin • Double Sideband • Double Strand Break

DSBAM Double-Sideband Amplitude Module

DSBD Diameter versus Surface Brightness Diagram

DSBM Designated Subnet Bandwidth Manager

DSBNF Double Sideband Noise Figure

DSBSC Double Sideband Suppressed Carrier

Dsbl Disable

Dsbld Disabled

DSBS Digital Sound Broadcasting Satellite

dsc Discard (file extension)

DSc Doctor of Science

DSC Albanian Direction of Standards • Data Separator Card • Data Services Center • Data Synchronizer Channel • Debye-Sears Cell • Decision Support Center • Dedicated Signal Conditioner • Delaware State College (USA-DE) • Deliberate Security Compromise • Delivery Source Code • Depot Support Concept • Detector Spacecraft (XEUS) • Differential Scanning Calorimeter • Differential Scanning Calorimetry • Differential Signal Control • Digital Scan Converter • Digital Selective Calling • Digital Setting Circle • Digital Source Collector • Digital-to-Synchro Converter • Direction de la Sécurité Civile (France) • Direct Satellite Communications • Disconnect • Distant Station Connected • Distribution of Stockage Code • Documentation Standards Committee • Document Service Center • Document Structure Convention • Document Structuring Convention • Document Supply Center • Drift Scan Camera • Dschang (Cameroon) (location identifier) • Dynamic Sequential Control • Dynamic SQL Caching • Dynamic Standby Computer

DScA Docteur en Sciences Appliquées

DSCB Data Set Control Block

DSCC Deep Space Communication Complex

DScEng Doctor of Science in Engineering

DSCF Destination Sectional Center Facility • Dispersion-Slope-Compensated Fiber

DSchG Denkmalschutzgesetz (Germany)

DSCHG Discharge

D/SCI Directorate of Science • Director of Science

DSCI Data Channel Status Indicator

DSCIM Display System Computer Input Multiplexer

DSCN Design and Simulator Computation Network

Dscont Discontinue

Dscontd Discontinued

DSCP Differential Service Code Point

DSCR Direction de la Sécurité et de la Circulation Routières (France)

Dscrm Discriminate

Dscrmd Discriminated

Dscrmr Discriminator

DSCS Defense Satellite Communications System • Desk Side Computer System • Digital Satellite Compression Service • Digital Simulator Computer System

DSCT Delta Scuti (star) • Directional Solidification of Cadmium Telluride • Double Secondary Current Transformer

DScTech Doctor of Technical Science

DSCU Digital Computer Switching Unit

D&SD Deployment & Support Division

DSD Dansk Selskab for Datalogi (Denmark) • Dark Surge on Disk • Data Status Display • Data Storage Device • Data Structure Diagram • Data System Designator • Data Systems Division • Digital System Design • Direct Stream Digital • Disk Storage Device • Display Subscriber Directory • Double-Sided Disk • Dual-Scan Display • La Désirade (Guadeloupe) (location identifier)

DSDB Digital Sound and Data Broadcasting

DSDC Direct Service Dial Capability

DSDD Double-Side - Double Density (*) • Double-Sided - Double Density

DSDI Digital Secure Data Interface • Digital Simple Data Interface

DSDIC Design Support for Distributed Industrial Control

DSDIE Digital Simple Data Interface - Enhanced

DSDL Data Storage Definition Language • Data Storage Description Language • Double-Side Double-Layer

DSDM Delamination Simulation by Damage Mechanism • Drop Site Database Manager • Dynamic System Development Method

DSDN Distributed Software Development Network

DSDO Data System Design Office

DSDP Deep-Sea Drilling Project

DSDS Data Storage and Distribution System • Data Systems Dynamic Simulator • Decision Support Display System • Digital Synchro Data Source • Document Survey Data Sheet

DSDT Discrete Space and Discrete Time

DSDU Data Storage Distribution Unit

DSE Data Set Experiment • Data Set Extension • Data Storage Encoder • Data Storage Equipment • Data Structure Editor • Data Switching Equipment • Data Switching Exchange • Data System Equipment • Data System Experiment • Data Systems Engineering • Debye-Sears Effect • Decision Support Environment • Depot Support Equipment • Deputy for Software Evaluation • Derecho del Trabajo • Dessie (Ethiopia) (location identifier) • Developmental Supportability Engineering • Development Support Equipment • Development Support Environment • Dimensionally Stable Electrode • Diplôme en Sciences Économiques • Dirección de Seguridad del Estado • Direction pour les Systèmes et l'Espace (ONERA) • Directory Services Element • Distributed Single-layer Embedded • Distributed Systems Environment • Division of Solar Energy • Division Support et Environnement • Dossier Social Européen

DSEA Data Storage Electronics Assembly • Display Station Emulation Adapter

DSEC Director of Security

DSECT Dummy Section

DSEE Domain Software Engineering Environment

DSEG Data Systems Engineering Group

DSEO Data Systems Evaluation Office

DSES Division of Social and Economic Studies

DSEST Direct Support Electrical System Test

DSESTS Direct Support Electrical System Test Set

DSETS Direct Support Electrical Test System

DSF Data Scanning and Formatting • Data Stream Formation • Deep Space Exploration Society (USA-CO) • Delivery Sequence File • Design Safety Factor • Deputy Secretary General • Digital Signal Generator • Directional Solidification Furnace • Directory System Function • Disappearing Solar Filament • Disk Storage Facility • Dispersion-Shifted Fiber • Dispersion-Strengthened Ferritic • Distributed Simulation Framework • Document Status File

DSFC Direct Side-Force Control

DSFG Deep-Sky Field Guide

DSFT Distributed System Fault Tolerance

DSG Data Systems Group • Datenschutzgesetz (Austria) • Default Slot Generator • Defense Systems Group • Digital Signal Generator • Dilasag (Philippines) (location identifier) • Dynamique des Systèmes Gravitationnels

Dsgn Design • Designer

DSGN Design

DSH Dye-Sensitized Heterojunction

DSHD Double-Sided - High Density • Double-Side - High Density (*)

DSI Data System Integration • Data Systems Inquiry • Délégation aux Systèmes d'Information • Delivered Source Instruction • Destin (USA-FL) (location identifier) • Detailed Spectrum Investigation • Development Sciences Inc. • Differential Speckle Interferometry • Diffusion Sélective de l'Information • Digital Speech Interpolation • Digital Speech Interpolator • Directeur des Systèmes d'Information • Direction des Services Informatiques • Direction des Systèmes d'Information • Directorate of Scientific Information • Discrete Smooth Interpolation • Discrete Surface Integral • Diseminación Selectiva de Información • Distribution System Integration • Division des Services Informatiques • Division of Science Information • Dual and Sequential Ignition • Dynamic Skeleton Interface

DSIC Departamento de Sistemas Informáticos y Computación • Documentation and Scientific Information Centre

DSID Data Set Identification • Divergence Source-Image Distortion

DSIF Deep Space Instrumentation Facility

DSIM Distributed Shared Image

D-SIMM Dual Single In-line Memory Module

DSIMM Dual Single In-line Memory Module

DSIN Direction de la Sûreté et des Installations Nucléaires (France)

DSIO Directeur des Systèmes d'Information et d'Organisation

DSIP Development Support and Integration Program

DSIPS Digital Satellite Image Processing System

DSIPT Dissipate

DSIPTG Dissipating

DSIR Department of Scientific and Industrial Research

DSIRnet New Zealand Government Network

DSIS Decision Support Information System • Deep-Structure Indexing System • Distributed Statistical Information System • Distributed Support Information Standard

DSIT Directeur des Systèmes d'Information et de Télécommunications • Direction des Systèmes d'Information et de Télécommunications

DSJ Defense Science Journal (periodical code)

dsk Desktop (file extension)

DSK Datenschutzkommision (Germany) • Dera - Ismail Khan (Pakistan) (location identifier)

Dsl Diesel

DSL Dark Side Limb • Daru (Sergio Leone) (location identifier) • Data Set Label • Data Simulation Language • Data Structures Language • Design Specification List • Digital Simulation Language • Digital Subscriber Line • Display Switch Locator • Distributed Service Level • Distributed Service Logic • Document Summary List • Domain-Specific Language • Drawing and Specification Listing • Dynamic Simulation Language

DSLAM Digital Subscriber Line Access Multiplexer

DSLOC Delivered Source Line Of Code

DSLS Division of Space Life Sciences

Dslv Dissolve

Dslvd Dissolved

dsm Digital Sound Module (file extension)

DSM Data Source Matrix • Data Status Message • Data Storage Memory • Data Systems Manager • Data Systems Modernization • Decision Support Matrix • Decision Synchronization Matrix • Dedicated Security Mode • Deep Space Maneuver • Deep Space Measurement • Demand-Side Management • Department of Standards Malaysia • Des Moines (USA-IA) (location identifier) • Detector Structural Model • Development Support Manager • Development System Manager • Diagnostic and Statistical Manual • Digital Select Matrix • Digital Select Module • Digital Simulation Model • Digital Support Matrix • Digital Surface Model • Digital Switching Module • Digital System Model • Direction of Systems Management • Directorate of Servicing and Maintenance • Direct-Solution Method • Direct Support Maintenance • Discrete and Serial Module • Distinguished Service Medal • Distributed Shared Memory • Distributed System Management • Distribution Storage Management • Distribution System Management • Downhill Simplex Method • Dynamic Scattering Mode • Dynamic Situation Model • Dynamic Stiffness Matrix

DSMA Digital Sense Mutiple Access

DSMAC Digital Scene Matching Area Correlation

DSMC Data Systems Management Course • Defense System Management College • Direct Simulation Monte Carlo

DSMCC Digital Storage Media Command and Control

DSMCS Document Structure Management Common Service

Dsmd Dismissed

DSMD Discount Schedule and Marketing Data

DSML Directory Services Markup Language

DSmP Digital Signal Microprocessor

DSMP Defense Meteorological Satellite System (USA)

DSMRD Distribution of Sporadic Meteor Radiants Density

DSMS Deep Space Mission Support

DSMSC Dynamic System Modeling, Sensing and Control

DSMSL Document Style Semantics and Specification Language

dsn Design (file extension)

DSN Data Set Name • Data Source Name • Data Steering Network • Deep-Space Network (NASA) • Delta Scuti Network • Dependable Systems and Networks (conference) • Derived Services Network • Dewan Standardisasi Nasional (Indonesia) • Digital Secure Network • Digital Services Network • Digital Subscriber Network • Digital Switched Network • Digital Switching Network • Digital System Network • Diminishing Source Notice • Distributed System Network • Dongsheng (China-PRC) (location identifier) • Standardization Council of Indonesia

DSNA Dictionary Society of North America

DSNP Digital Signal Noise Processing

DSNT Distant

DSO Dark-Sky Observatory (USA-NC) • Dark-Sky Observer • Data Set Optimizer • Data Systems Office • Deep-Sky Object • Deep-Sky Observer • Deputy Safety Officer • Designated Senior Official • Design Secondary Objective • Detailed Secondary Objective • Detailed Supplementary Objective • Development System Office • Digital Storage Oscilloscope • Direct System Output • Display Switching Oscilloscope • Distinguished Service Order • Distributed Shared Object • District Staff Officer • Dynamic Shared Object

DSOB Data Systems Operations Branch

DSOC Discipline Science Operations Center

DSOF Dispersion-Shifted Optical Fiber

D-SOFT D-Shaped Optical Fiber Transmission

DSOFT D-Shaped Optical Fiber Transmission

DSOM Distributed System Object Model • Distributed Systems: Operations and Management (workshop)

DSOP Draft Statement of Principles

dsp Decessit Sine Prole

DSp Datenspeicher

DSP Data Source Panel • Data Source Provider • Deep Space Probe • Defense Support Program • Densité Spectrale de Puissance • Department of State Planning • Departure Sequencing Program • Deployable Solar Panel • Desktop Storage Pack • Despacho • Differential Scanning Polarimetry • Digital Sense Processor • Digital Signal Process • Digital Signal Processing • Digital Signal Processor • Digital Spectropolarimeter • Digital Speech Processor • Digital Strip Printer • Direction de la Stratégie et des Programmes • Directory Service Protocol • Directory Synchronization Protocol • Directory System Protocol • Display Simulation Program • Display System Protocol • Disposal • Distributed System Program • Domain Specific Part • Dominating Set Problem • Doppler Signal Processing • Doppler Spectrum Processor • Durable/Stable Penetration • Dynamic Site Publisher • Dynamic Support Program • Dynamics of the Solar Photosphere

DSPE Digital Signal Processing Experiment • Division of Scientific Personnel and Education

DSpec Design Specification

dSph Dwarf Spherical (galaxy)

DSph Dwarf Spheroidal (galaxy)

Dspl Display • Disposal

Dspld Displayed

dSph Dwarf Spheroidal

DSPL Display System Programming Language

Dsplcd Displaced

DSPM Designated System Program Manager • Designated System Project Manager • Detailed System Performance Model

Dspn Disposition

DSPN Digital Secure Packet Network • Direct Sequence Pseudo-Noise

Dspo Disposal

DSPO Defense Support Project Office

DSP&P Director for Security Plans & Programs

DSPP Director for Security Plans and Programs

DSPRM Digital Signal Processing Resource Manager • Digital Signal Processor Resource Manager

DSPRMI Digital Signal Processing Resource Manager Interface • Digital Signal Processor Resource Manager Interface

DSPS Digital Signal Processing System

DSPSE Deep Space Program Science Experiment

DSPT Département Scientifique, Pédagogique et Technique

DSPU Downstream Physical Unit

DSPW Digital Stereo Photogrammetric Workstation

DSQ Développement Social des Quartiers

DSQD Double-Sided - Quad Density

DS&R Data Storage & Retrieval

DSR Daily Status Report • Dansk Selskab for Rumfartsforskning (Denmark) • Data Scanning and Routing • Data Set Ready • Data Signaling Rate • Data Specification Request • Data Storage and Retrieval • Data Survey Report • Datenstationsrechner • Design Status Review • Device Status Register • Device Status Report • Digitales Satellitenradio • Digital Satellite Radio • Digital Shift Register • Digital Signaling Rate • Digit Storage Relay • Direct Server Return • Direction of Scientific Research • Director of Scientific Research • Direct Storage Recorder • Disaster Survey Report • Discriminating Selector Repeater • Display System Replacement • Division of Solar Research • Document Search and Retrieval • Document Status Report • Document Summary Record • Dolby Sound Reduction • Dynamic Service Register • Dynamic Spatial Reconstruction • Dynamic Spatial Reconstructor • Dynamic Storage Region

DSRA Document Summary Record Address

DSRC Dedicated Short-Range Communications • Digital Short-Range Communications

Dsrd Desired

DSRD Depot Support Requirements Document • Draft Systems Requirements Document

DSREDS Digital Storage and Retrieval Engineering Data System

DSRI Danish Space Research Institute • Destination Station Routing Indicator • Digital Standard Relational Interface

DSRM Director of Safety Risk Management

DSRMC Defense Systems Review Military Communications (periodical code)

DSRR Digital Short-Range Radio

DS&RS Document Storage & Retrieval System

DSRS Document Storage and Retrieval System

DSRSG Deep South Regional Star Gaze (USA)

DSRT Deep Sea Reversing Thermometer • Dynamic Soft Real-Time (scheduler)

DSRV Deep Submergence Rescue Vehicle

DSS Daily Science Schedule • Dakota Space Society (USA-ND) • Dark-Sky Switzerland • Data Server System • Data Set Service • Data Simulation System • Data Source Selector • Data Station Selector • Data Storage Set • Data Storage System • Data Support Section • Data Support System • Datensichtstation • Decision Support Shell • Decision Support System • Deductive Subsystem • Deep Seismic Sounding • Deep Sky Survey • Deep Space Station • Deep Spectroscopic Survey • Deep Survey Spectrometer • Delft Sensor System • Deliverable Survival Shelter • Demonstration Sensor System • Dense Stellar System • Department of Science Service • Department of Social Security • Department of Social Services • Department of Supplies and Services • Department of Supply and Services • Department Store System • Department Summary Schedule • Deployable Sensor System • Deployed Supply System • Deutsche Satellitensysteme GmbH (Germany) • Device Simulation and Simulator • Diagnostic Simulation System • Digital Satellite Services • Digital Satellite System • Digital Signal Standard • Digital Signal Synchronizer • Digital Signature Standard • Digital Simulator System • Digital Sky Survey • Digital Small Switch • Digital Storage System • Digital Subset • Digital Subscriber Service • Digital Subscriber System • Digital Subsystem • Digitized Sky Survey • Directory and Security Services • Direct Station Selection • Direct Supply Support • Direct Support System • Discrete-Source Scan • Distributed Sample Scrambler • Distributed Small Satellite • Distributed Snuffer System • Distributed Standard System • Distribution Standard System • Documentation Support Services • Document Search System • Document Status Sheet • Document Storage System • Dornier Satellitensysteme • Double Spot System • Drum Storage System • Dynamic Simulation System • Dynamic Steady State • Dynamic Support System

DSSA Domain-Specific Software Architecture • Dynamic Security Surveillance Agent

DSSB Double Single-Sideband

DSSC Deep Space Station Complex • Distributed Services Steering Committee

DSSD Double Sided - Single Density • Double Side - Single Density (*)

DSSE Development Software Support Environment • Development System Software Environment

DSSEP Dermatomal Somatosensory Evoked Potential • Developmental Software Support Environmental Plan

DSSI Digital Standard Systems Interconnect • Digital Storage Systems Interconnect

DSSL Document Semantics and Specification Language

DSSM Data Standards and Structures Manual • Dedicated Solar Sortie Mission • Dynamic Subsystem Management

DSSMOF Dispersion-Shifted Single Mode Optical Filter

DSSN Delta Scuti Star Newsletter (periodical code) • Disbursing Station Symbol Number

DSSO Data System Support Organization

D/SSP Directorate of Space Station and Platforms • Director of Space Station and Platforms

DSSP Direct Supply Support Point • Double-Sided Strip Detector

DSSR Deep-Space Surveillance Radar • Digital Short-Range Radio

DSSS Direct-Sequence Spread Spectrum

DSSSD Double-Sided Si Strip Detector

DSSSL Document Style, Semantics, and Specification Language

Dst Disturbance Short Time

DST Dark Streak • Data Source Terminal • Data System Test • Datenstation • Daylight Saving Time • Decision Support Template • Deep-Sky-Tagung (Germany) • Deep-Space Transponder • Department of Science and Technology • Destination • Detailed System Test • Developer Solutions Team • Digital Storage Tape • Dimensional Special Tooling • Direction de la Surveillance du Territoire (France) • Director of Supplies and Transport • Direct-Sounding Transmission • Direct Support Team • Discrete Sine Transform • Discrete Symmetry Transform • Distance • Distribution of Sign Times • Disturbance Storm Time (*) • Documentation Scientifique et Technique • Domeless Solar Telescope • Domeless Solar Tower • Double Solar Telescope • Double Summer Time • Downsized Tester • Dual Standard Transponder • Dunkle Striche • Dunn Solar Telescope (USA-NM) (formerly VTT) • Dynamic Surface Tension

DSTA Diplôme Scientifique de Technologie Avancée

DSTE Data Subscriber Terminal Equipment • Digital Subscriber Terminal Equipment • Dossiers Statistiques du Travail et de l'Emploi

DSTF Delta Spin Test Facility

DSTI Direction des Services et des Techniques Informatiques • Direction Science-Technologie-Industrie (OCDE) • Division of Scientific and Technical Information

DSTL Defence Science and Technology Laboratory (UK)

Dstlt Distillate

Dstltd Distillated

Dstn Destination

DSTN Double Supertwisted Nematic (*) • Dual-Scan Twisted Nematic

Dstng Distinguish

Dstngd Distinguished

DSTO Defence Science and Technology Organization (Australia)

DSTP Data Self-Test Program • Data Space Transport Protocol

DStR Deutsches Steuerrecht

DSTR Destroy

D/STS Directorate of Space Transportation Systems • Director of Space Transportation Systems

DSTS Deep-Space Tracking System

DSTU Digital Signal Transfer Unit • Draft Standard for Trial Use • State Committee for Standardization, Metrology and Certification (Ukraine)

DSTV Direct Support Team Vehicle

DSU Data Service Unit • Data Storage Unit • Data Support Unit • Data Synchronization Unit • Data Synchronizer Unit • Decision Storage Unit • Decoder Switching Unit • DEU Switching Unit • Développement Social Urbain • Digital Service Unit • Digital Signaling Unit • Digital Standard Update • Digital Storage Unit • Digital Switching Unit • Directory Search Unit • Direct Support Unit • Disk Storage Unit • Drum Storage Unit

DSUP Defense System Upgrade Program (*) • Defensive System Upgrade Program

DSUR Data Storage Unit Receptacle

DSUWG Data Systems Users Working Group

DSV Dansville (USA-NY) (location identifier) • Deep Submergence Vehicle • Design - Specification - Verification • Diameter Spherical Volume • Digital Sum Variation • Digital System Validation

DSVD Digital Simultaneous Voice/Data

DSVF Data System Verification Facility

DSVIS Design, Specification, and Verification of Interactive Systems (workshop)

DSVM Distributed Shared Virtual Memory

DSVT Digital Secure Voice Terminal • Digital Subscriber Voice Telephone • Digital Subscriber Voice Terminal

DSW Data Status Word • Device Status Word • Dirac Scattered Wave

DSWC Disposable Solid Waste Cask

DSX Digital Signal Crossconnect • Digital System X-connect • Distributed System Executive

DS2 Decision Support Display System • Second Dark Spot (Neptune)

DS3 Deep Space Surveillance System

DS4 Decentralized Supply Service Support System

dt Deutsch

dT Thymidine

Dt Doit

D/T Double Theodolite

D&T Detection & Tracking • Development & Technology

DT Daily Telegraph • Damage-Tolerant • Dark Time • Data Technician • Data Terminal • Data Transfer • Data Translator • Data Transmission • Data Transmit • Data Type • Dateitransfer • Daylight Time • Decay Time • Decision Table • Decision Tree • Deduction Theorem • Definition Term • Delayed Time • Delay Time • Delete Translation • Delirium Tremens • Delta Time • Demi-Traitement • Demonstration Test • Department of Transportation • Department of Treasury • Deployment Transaction • Description Title • Destructive Testing • Detective • Deuterium-Tritium • Deutsche Telekom (Germany) • Developer Testing • Developmental Test • Developmental Tester • Developmental Testing • Development and Technology • Development Test • Development Testing • Device Type • Differential Time • Digital Technique • Digital Tracker • Digital Trunk • Digit Tube • Diplôme Technique • Direction de la Technologie • Direction Technique • Discrimination Technique • Disk Tape • Disk Torque • Dispersion Time • Displaced Threshold • Disposición Transitoria • Distance Test • Division Technique • Document Type • Domaine Technologique • Double Throw • Double Time • Downlink Transmitter • Downtime • Downtown • Drift Tube • Drop Tank • Dwell Time • Dynamic Tester • Dynamic Time

DTA Damage Tolerance Analysis • Data • Data Terminal Adapter • Daily Testability Analyzer • Decision Tree Analysis • Delta (USA-UT) (location identifier) • Design To Affordability • Deutsche Technische Akademie (Germany) • Development Test Article • Differential Thermal Analysis • Diplôme Technique Avancé • Directed Training Assistance • Disk Transfer Area • Distributing Terminal Assembly • Dual Telephone Adapter

DTAB Dispatching Table

DTAC Dynamically-Typed Access Control

DTAD Digital Terrain Analysis Data

DTAG Deutsche Telekom AG (Germany)

DTAM Document Transfer And Manipulation

DTAMS Data Transmission And Message System

DTAP Direct Transfer Application Port

DTARS Digital Transmission And Routing System

DTAS Data Transmission And Switching • Digital Test Access System • Digital Transmission And Switching

DTB Danmarks Tekniske Bibliotek • Data Terminal Buffer • Data Transfer Bus • Decimal To Binary • Diplôme Technique de Bibliothécaire

DTBA Date To Be Advised

DT&C Department of Transport & Communications

DTC Data Terminal Controller • Data Transfer Cartridge • Data Transmission Center • Data Transport Computer • Dead Time Correction • Decision Threshold Computer • Defense Trade Control • Delivery Term Code • Department of Transport and Communications • Desert Training Center • Design-Time Control • Design To Cost • Desktop Computer • Desktop Conferencing • Development Test Center • Differential Thermal Calorimetry • Digital Tape Conversion • Digital Technical Control • Digital Television Controller • Digital Trunk Controller • Display Test Chamber • Display Timing Control • Dispositif de Transfert de Charges • Distributed Terminal Controller • Distributed Transaction Coordinator • Domaine des Technologies-Clés • Domain Technology Committee • DSIF Telemetry and Command • Dynamic Traffic Control

DTCD Department of Technical Cooperation for Development

DTCP Digital Transmission Content Protocol

DTCR Data Transfer and Certification Record

DTCS Data Transmission and Control System • Digital Test Command System

DTCU Data Transmission Control Unit

DTCW Data Transfer Command Word

dtd Document Type Definition (file extension)

Dtd Dated

DTD Datadawai (Indonesia) (location identifier) • Data Tag Definition • Data Transfer Device • Data Transport Device • Data Type Definition • Deputy Test Director • Detailed Test Description • Digital Terrain Data • Digital Topographic Data • Digitized Terrain Database • Directorate of Technical Development • Director of Technical Development • Document Tag Description • Document Type Declaration • Document Type Definition (*) • Document Type Description

DTDMA Distributed Time Division Multiple Access

DTDP Different Time - Different Place

DT&E Demonstration, Testing & Evaluation • Developmental Test & Evaluation • Development Test & Evaluation

DTE Daet - Camarines Norte (Philippines) (location identifier) • Data Terminal Equipment • Data Terminating Equipment (*) • Data Transfer Equipment • Data Transmission Equipment • Data Transmission Exchange • Data Transmit Exchange • Data Transmitting Equipment • Deliverable Test Equipment • Detective Quantum Efficiency • Developmental Test and Evaluation • Developmental Test Equipment • Development Test and Evaluation • Development - Test - Evaluation • Digital Television Equipment • Digital Terminal Equipment • Digital Tone Extractor • Digital Transmission Equipment • Direct To Earth • Distributed Transaction Equipment • Domain and Type Enforcement • Down To Earth • Dumb Terminal Emulator

DTEA Developmental Training Effectiveness Analysis

DTED Digital Terrain Elevation Data (*) • Digital Terrain Evaluation Data

DTel Deutsche Telekom (Germany)

D/TEL Directorate of Telecommunications • Director of Telecommunications

DTEM Digital Terrain Elevation Mapping

DTEMS Digital Terrain Elevation Mapping System

DTEO Defence Test and Evaluation Organisation

DTF Daily Transaction File • Data Transfer Facility • Data Transmission Factor • Data Transmission Feature • Data Transmission Function • Data Transmittal Form • Define The File • Demonstration Test Flight • Development and Test Facility • Development Test Facility • Dial Tone First • Dielectric Thin Film • Digital Tape Format • Direct Tully-Fisher (relation) • Distributed Test Facility • Distributed Transaction Facility • Domain Task Force

Dtg Dienstag

DTG Date-Time Group • Desktop Guide • Digital Tape Generation • Digital Transmission Group • Digital Trunk Group • Direct Trunk Group • Display Transmission Generator • Distance To Go • Dry Tuned Gyro • Dynamically-Tuned Group

DTH Death Valley (USA-CA) (location identifier) • Direct To Home • Down The Hill

DThPT Diploma in Theory and Practice of Teaching

DTI Danish Technological Institute (Denmark) • Delivery Ticket Invoice • Department of Trade and Industry (UK) • Development - Technology - Innovation • Development Test Instrumentation • Direction des Transmissions et de l'Informatique • Display Technologies Inc.

DTIC Defense Technical Information Center (USA)

DTID Disposal Turn-In Document

DTIF Digital Transmission Interface Frame

DTJ Digital Technical Journal (periodical code)

DTK Desired Track

Dtl Detail

DTL Data-Time Logged • Dead Time Log • Decomposition Type List • Designated Transit List • Desktop Library • Detroit Lakes (USA-MN) (location identifier) • Detroit Testing Laboratory (USA-IL) • Dialog Tag Language • Dictionary Term Lookup • Diode Transfer Logic • Diode-Transistor Logic

DTLCC Design To Life Cycle Cost

Dtld Detailed

DTLN Dateline

DTLS Descriptive Top-Level Specification

DTLU Digital Trunk and Line Unit

DTM Data Transfer Mechanism • Data Transfer Module • Data Transmission Medium • DEC Test Manager • Deflection Transformation Matrix • Delay Timer Multiplier • Department of Terrestrial Magnetism • Design Test Model • Desktop Manufacturing • Deterministic Turing Machine • Developmental Test Model • Development Test Model • Diamond Turning Machine • Digital Terrain Map • Digital Terrain Model • Digital Terrain Modeling • Dortmund (Germany) (location identifier) • Dual Tome Multiple (frequency) • Dynamic Test Model • Dynamic Transfer Mode

DTMAG Dual Technique Magnetometer

DTME Diplôme Technologique de Management Européen

DTMF Data Tone Multiple Frequency • Dial Tone Modulated Frequency • Dial Tone Multifrequency • Dual Tone Multifrequency (*) • Dual Tome Multiple Frequency

DTML Document Template Markup Language

DTMO Design Test and Mission Operation • Development, Test and Mission Operation

DTMOP Development Test Measures Of Performance

DTMP Data Technical Management Panel

DTMS Development, Test, and Mission Support • Digital Test Measurement System • Digital Test Monitoring System • Digital Test Monitor System

DTN Data-flow Technology Nederland BV • Data Transmission Network • Designated Transport Node • Directeur Technique National • Direction Technique Nationale • Shreveport Downtown (USA-LA) (location identifier)

DTNI Direction Technique Nationale Informatique

DTNM Dirección de Tecnología, Normalización y Metrología

Dto Descuento • Distrito

DT&O Data, Test & Operation

DTO Daily Tasking Order • Data Takeoff • Data, Test and Operation • Descuento • Design Test Objective • Detailed Test Objective • Developmental Test Objective • Development Test Objective • Digitally-Tuned Oscillator • Digital Terminal Operator • Digital Testing Oscilloscope • Direct To Office • Direito • Division Transportation Office • Division Transportation Officer • Documentation Technique Officielle

DTOA Differential Time Of Arrival

DTOE Draft Table of Organization and Equipment

DTOG Depth To Generator

DTOL Digital Test-Oriented Language

DTOP Digital Topographic (data)

Dtor Director

DTOSC Design To Operations and Support Costs

DTP Darwin Technology Package • Datagram Transport Protocol • Data Transfer Process • Data Transmission Process • Design To Price • Desktop Publishing • Detailed Test Plan • Detailed Test Procedure • Detail Test Plan • Développement Technique Probatoire • Development Test Plan • Digital Transfer Phase • Diode Test Program • Diphteria - Tetanus - Pertussis • Directory Tape Processor • Distributed Transactional Processing • Distributed Transaction Process • Distributed Transaction Processing • Dolph-Tchebyscheff Pattern • Dynamic Test Panel • Dynamic Testing Program

DTPL Domain Tip Propagation Logic

DTPM Distributed Transaction Processing Middleware • Distributed Transaction Processing Monitor

DTQE Detection Quantum Efficiency

DTR Daily Total Rate • Data Terminal Ready • Data Transfer Rate • Data Transfer Receptacle • Data Transrecorder • Decatur Island (USA-WA) (location identifier) • Dedicated Token Ring • Definite-Time Relay • Delay - Throughput - Reliability • Desktop Reproduction • Detailed Test Report • Development and Test Resource • Diffusion-Transfer Reversal • Digital Tape Recorder • Distribution Tape Reel • Diurnal Temperature Rate • Document Transmittal Record • Draft Technical Report • Dynamic Transaction Routing

DTRA Defense Threat Reduction Agency (USA)

Dtrbd Daughterboard

DTRD Development Test Requirement Document

DTRE Data Type Refinement Environment

DTREM Dust, Thermal and Radiation Engineering Measurement

DTRF Data Transmittal and Routing Form

DTRIC Dielectric Totally Internally Reflecting Concentrator

DTRM Development and Technology Review Mission • Development Technology Review Mission • Dual Thrust Rocket Motor

DTRR Developmental Test Readiness Review

Dtrs Distress

DTRS Developmental Test Readiness Statement • Development Test Requirement Specification • Dryden Technical Report Server

DTRT Deteriorate

DTS Data Terminal Set • Data Test Station • Data Transfer System • Data Transmission System • Data Transport System • Date Time Satisfied • Dedicated Transmission Service • Defense Technology Study • Detailed Test Specification • Development Test Satellite • Development Test Sensor • Development Test Set • Digitale Teilnehmerschaltung • Digital Telemetry System • Digital Termination Service • Digital Terrain System • Digital Theater System • Digital Tracking System • Digital Transmission System • Diplomatic Telecommunications Service • Diplomatic Telecommunications System • Display Terminal Service • Distinguished Teaching Scholar • Distributed Time Server • Distributed Time Service • Droits de Tirages Spéciaux • Dynamic Test Station • Dynamic Test System

DTSC Department of Toxic Substances Control

Dtsch Deutsch

DTSL Data Transfer Systems Laboratory

DTSP Digitized Training Support Package

DTSR Digital Temporary Storage Recorder • Digital Temporary Storage Recording

DTSS Dartmouth Time-Sharing System • Digital Topographic Support System

DTST Defense Technology Study Team

DTT Data Throttling Time • Data Transfer Timing • Design Test Tool • Design Thermal Transient • Desktop Trainer • Detroit (USA-MI) (location identifier) • Direction des Transports Terrestres • Display Terminal Type • Doctrine - Tactics - Training • Double Taxation Treaty

dTTL Differential Transistor-Transistor Logic

DTTU Data Transmission Terminal Unit

DTTV Digital Terrestrial Television

DTU Data Takeoff Unit • Data Tape Unit • Data Terminal Unit • Data Transfer Unit • Data Transmission Unit • Development Test Unit • Diffusive Time Unit • Digital Tape Unit • Digital Telemetry Unit • Digital Test Unit • Digital Transfer Unit • Digital Transmission Unit • Direct To User • Documents Techniques Unifiés

DTUPC Design To Unit Price Cost • Design To Unit Production Cost

D'ture Departure

DTUTF Digital Tape Unit Test Facility

DTV Data Translation and Verification • Desktop Video • Development Test Vehicle • Digital Television • Digital to Television • Digitvizuális Távcsö • Driver Thermal Viewer • Dynamic Test Vehicle

DTVC Data Transmission and Verification Converter • Desktop Video Conferencing • Digital Transmission and Verification Converter

DTVE Digital Television Element

DTV&I Development, Test, Verification, & Integration

DTVI Development, Test, Verification, and Integration

DTVM Differential Thermocouple Voltmeter

DTW Desktop Workstation • Detroit - Wayne County (USA-MI) (location identifier) • Dynamic Time Warping

DTWA Dual Trailing Wire Antenna

DTWG Data Tools Working Group

Dtx Douteux

DTX Discontinuous Transmission

Dtzd Dutzend

du Duplicatus

dU der Unterzeichnete

Du Duke • Dutch

DÜ Datenübertragung • Dienste-Übergang

DU Data Unit • Decision Unit • Delhi University (India) • Dependable Undertaking • Depleted Uranium • Detector Unit • Digital Unit • Diplôme d'Université • Diplôme Universitaire • Disk Usage • Display Unit • Distribution Unit • Dobson Unit • Doctor of the University • Documentation Unit • Do Until • Dredge-Up • Dust • Dutch

DUA Demande d'Utilisation d'Ariane • Deterministic Uncertainty Analysis • Digital Uplink Assembly • Directory User Agent • Durant Eaker (USA-OK) (location identifier)

DUAL Dynamic Universal Assembly Language

DUAP Dual Use Application Program

DUAT Direct User Access Terminal

DUATS Direct User Access Terminal Service

DUAU Data Update and Access Unit

Dub Dubious • Dubitans • Dublin (Ireland)

DUB Dublin (Ireland) (location identifier)

DUBBLE Dutch-Belgian Beamline Experiment

DUBCS Delivery Unit Bar Code Sorter

DUBS Dual-bus Unidirectional Broadcast System

DUC Data User's Committee • Demand Used in Computation • Dense Upper Cloud • Digital Uplink Command • Display Use Chain • Document Understanding Conference • Duncan Halliburton (USA-OK) (location identifier) • Dutch Utilization Centre (Netherlands)

DUCMA Dust Counter and Mass Analyzer

DUCS Déclaration Unifiée de Cotisations Sociales (France)

DUD Doesn't Use Derivatives • Dunedin (New Zealand) (location identifier)

DUDL Déclaration Universelle des Droits Linguistiques

DudOR Dudley Observatory Reports (periodical code)

DÜE Datenübertragungseinrichtungen

DUE Déclaration Unique d'Embauche (France) • Division of Undergraduate Education (NSF) • Dundo (Angola) (location identifier)

DUEL Data Update Edit Language

DUER Design Unit Engineering Report

DUES Diplôme Universitaires d'Études Scientifiques

DUF Diffusion Under Film • Duck (USA-NC) (location identifier) • Dynamic Update Facility

DUG Deep Under Ground • Douglas (USA-AZ) (location identifier)

DUI Data Use Identifier • DELPHI User Interface • Diplôme Universitaire d'Initiation • Directory User Interface • Driving Under Influence • Driving Under the Influence

DUITS Digital - Ubiquitous - Interoperable - Transparent - Secure

DUJ Digital User Journal • Dubois (USA-PA) (location identifier)

DUK Dukuduk (South Africa) (location identifier)

Dulc Dulcis

DULCE Demonstration Unit for Low-order Cryogenic Etalon

DULCO Diplôme Unilingue des Langues et Civilisations Orientales

DUM Dumai (Indonesia) (location identifier) • Dummy

DUMAND Deep Underwater Muon And Neutrino Detection • Deep Underwater Muon And Neutrino Detector

DUMC Dutch Union Map Catalogue

Dun Dundee (UK-Scotland)

DUN Dial-Up Networking • Dundas (Greenland) (location identifier)

DUNCE Dial-Up Network Connection Enhancement

DUNDEE Down Under Doppler and Electricity Experiment

DUNE Dust (measurements) Near Earth (mission)

DUniv Doctor of the University

DunOP Dunsink Observatory Publications (periodical code)

DunRe Dunsink Observatory Reprints (periodical code)

DUNS Data Universal Numbering System

Duo Duodecimo

DUO Decision Unit Overview • Disk Unseen Object

Dup Duplicate

DUP Data User Part • Data User Programme • Data Utilization Programme • Déclaration d'Utilité Publique • Décret d'Utilité Publique • Disk Utility Program • Dredge-Up • Duplicate

Dupdo Duplicado

Dupe Duplicate

Dupl Duplicado • Duplicate

Dupld Duplicated

DUPLEX Dust-Prominent Longitudinally Extended (PPN)

Duplx Duplex

Duplxr Duplexer

DUPS Datenübertragungsprogrammsystem

DUQ Duncan Quam (Canada-BC) (location identifier)

Dur Duration • Durban (South Africa) • Durham (UK-England)

DUR Duration • Durban - Louis Botha (South Africa) (location identifier) • Durham (Starlink internal code)

DURC During Climb

DURD During Descent

DURG During

Durn Duration

DURS Durham/UKST Redshift Survey

DÜS Datenübertragungssystem

DUS Data Utilization Station • Design Unit Specification • Diplôme Universitaire Scientifique • Düsseldorf Lohausen (Germany) (location identifier) • Düsseldorf Rhein-Rhur (Germany) (location identifier)

DUSO Dutch User Support Organization

ÜST Datenübertragungssteuerung

DUST Datenumsetzerstelle

Dut Dutch

DUT Delft University of Technology (Netherlands) • Democretian University of Thrace (Greece) • Device Under Test • Diplôme Universitaire de Technologie (France) • Dutch Harbor (USA-AK) (location identifier)

DUTA Display Unit Test Assembly

DUTE Digital Universal Test Equipment

DUUD Direct Unit Uplight Density

DUV Damaging Ultraviolet • Data Under Voice • Deep Ultraviolet

DUVAD Durham (Starlink internal code)

DUVE Diffuse Ultraviolet Experiment

DUVS Deep Ultraviolet Survey

DUWIR Dual Wavelength Infrared Imager

DUWL Dual Use Wireless LAN

DUX Designing the User Experience (conference)

DUZ Deutsche Universitätszeitung

d/v Días Vista

dv Deo Volente

D/V Demonstration/Validation

D&V Demonstration & Validation • Design & Validation

DV Daily Value • Data Valid • Data Visualization • Datenverarbeitung • Defective Vision • Déficient Visuel • Delta Velocity • Demonstration/Validation • Deo Volente • Designated Verification • Designee for Verification • Design Verification • De Vaucouleurs • Development • Differential Velocity • Differential Voltage • Digital Video • Direct View • Direct Vision • Discrete Variational (method) • Discrete Velocity • Discrimination Value • Distinguished Visitor • Divide • Double Vision • Driver • Driver's Viewer • Dump Valve

DVA Datenverarbeitungsanlage • Deutsche Verlagsanstalt (Germany) • Deva (Romania) (location identifier) • Differential Velocity Aberration • Differential Voltage Amplifier • Directed Vector Autocorrelation • Direct Vector Autocorrelation • Direct Video Access • Distance Vector Algorithm • Distant Vector Algorithm • Diverse Vector Area • Document Validation Audit • Dynamic Visual Acuity

DVAA Delaware Valley Amateur Astronomers (USA-PA)

DVAL Demonstration and Validation

DVB Digital Video Broadcast • Digital Video Broadcasting

DVC Desktop Video Conferencing • Device • Digital Video Cassette • Digital Video Compression • Digital Video Conferencing • Digital Voice Communication

DVCCS Differential Voltage-Controlled Current Source

DVCS Deutscher Verband Chartered Surveyors (Germany)

D/VD Data/Voice Data

DVD Data/Voice Data • Datenverarbeitungsdienst • Delta Velocity Display • Detailed Velocity Display • Digital Versatile Disk • Digital Video Disk • Digital Video Drive • Direct Vendor Delivery • Direct Versatile Drive • Divide

DVDP Dry Valley Drilling Program

DVDR Density versus Velocity Divergence Relation

DVDS Digital Video Display System

DVE Datenverarbeitungseinrichtung • Design Verification Experiment • Differential Vector Equation • Digital Video Effect • Distributed Virtual Environment • Driver's Viewer Enhanced • Driver's Vision Enhancer • Durée de Vie d'Eddington

DVED Direct View Encoding Desk

DVF Digital Visual Fortran • Displacement Vector Field

DVFO Digital Variable-Frequency Oscillator

DVFR Day Visual Flight Rule • Defense Visual Flight Rule

D/VG Directional/Vertical Gyros

DVG Datenverarbeitungsgesellschaft • Deutsche Vakuum Gesellschaft (Germany) • Digital Video Generator • Digital Video Graphics

DVGW Deutscher Verein des Gas- und Wasserfaches (Germany)

DVH Durée de Vie de Hubble

dvi Device Independent (file extension)

DVI Data/Voice Integration • Datenverarbeitungsinstitut • Device-Independent • Difference Vegetation Index • Differential Vegetation Index • Digital Video Imaging • Digital Video Interaction • Digital Video Interactive • Digital Video Interface • Digital Visual Interface • Direct Voice Input • Disque Vidéo Interactif

DVIP Designated Very Important Person • Digital Video Integrator and Processor • Distinguished Very Important Person

DVITS Digital Video Imagery Transfer System • Digital Video Imagery Transmission System

DvJ Dag van het Jaar

Dvl Develop

DVL Data/Voice Logger • Deutsche Versuchanstalt für Luftfahrt (Germany) (now DLR) • Devils Lake (USA-ND) (location identifier) • Direct Voice Line • Direct Voice Link

DVLP Develop

Dvlpd Developed

DvM Dag van de Maand

DVM Design validation Model • Digital Velocity Meter • Digital Video Microscope • Digital Voltmeter • Discontinuous Variational Method • Displaced Virtual Machine • Displayed Virtual Machine • Doctor of Veterinary Medicine

DVMA Direct Virtual Memory Access

DVMG Digital Vector Magnetograph

DVMRP Distance Vector Multicast Routing Protocol

DVN Davenport (USA-IA) (location identifier)

DVO Davao Mati (Philippines) (location identifier) • Decimal Voltage Output • Delta Velocity Off • Delta Velocity On • Digital-Video Object • Direct View Optics • Durchführungsverordnung

DVOM Digital Voltohmmeter

DVOR Doppler VHF Omnidirectional Range

DVOW Digital Voice Orderwire

DVP Data Validation Program • Davenport Downs (Australia-QLD) (location identifier) • Design Verification Program • Digital Video Plotter • Digital Video Processing • Digital Video Processor • Digital Video Producer • Dynamic Virus Protection

DVPDF Dry Vacuum Pump Discharge Filter

DVPF Dry Vacuum Pump Filter

DVQ Distinguished Visitors Quarters

Dvr Driver

DVR Daly River (Australia-NT) (location identifier) • Datenübertragungsvorrechner • Digital Video Recorder • Digital Video Recording • Direct Voltage Readout • Discrete-Variable Representation • Document Validation Report • Driver

DVRS Digital Voice Recorder System

DVRT Differential Variable Reluctance Transducer

DVS Datenverarbeitungssystem • Datenvermittlungssystem • Daten - Video - Sprache • Daily Visits Statistics • Delta Velocity System • Design Verification Specification • Digital Voice System • Digital Voltage Source • Direct Via Satellite • Distributed Virtual System • Document Vector Space • Dome Ventilation System • Doppler Velocity Sensor

DVSt Datenvermittlungsstelle

DVST Direct View Storage Tube

DVSZ Datenverarbeitungs-Servicezentrum (Germany)

Dvt Devant • Deviate

DVT Deep Vein Trombosis • Delta Voice Terminal • Design Verification Test • Design Verification Testing • Deutsche Verband Technisch-Wissenschaftlicher Vereine (Germany) • Digital Video Tape • Document Verification Team • Phoenix - Deer Valley (USA-AZ) (location identifier)

DVTC Desktop Video Conferencing

Dvtd Deviated

DVTE Design Verification Test Experimental

DVTR Digital Video Tape Recorder

DVU Delta Velocity Ullage • Design Verification Unit

DVV Downward Vertical Velocity

DVVD Dual Vision Viewing Device

DVVP Development, Verification and Validation Plan

DvW Dag van de Week

DVWG Deutsche Verkehrswissenschaftliche Gesellschaft (Germany)

DVX Digital Voice Exchange

DVZ Datenverarbeitungszentrum

d/w Dust Wrapper

dw Deelswoord • Dienstwillig

Dw Dwarf

D-W Germany (formerly West Germany) (postal country code)

D/W Dock Warrant

D&W Detection & Warning

DW Data Warehouse • Data Warehousing • Data Word • Dead Weight • Delivery Weight • Developed With • Diabatic Warming • Direct Weighted • Dishwasher • Distilled Water • Distorted Wave • Double Weight • Drum Write • Dual Wheel • Dwingeloo

DWA Data Warehouse Administrator • Distorted Wave Approximation • Dwangwa (Malawi) (location identifier)

DWAC Distributed Write Address Counter

DWAPS Defense Warehousing Automated Processing System

DWARF Dwingeloo/Westerbork Astronomical Reduction Facility (Netherlands)

DWB Data Word Buffer • Soalala (Madagascar) (location identifier)

DWBA Distorted Wave Born Approximation

DWBer Dauerwohnberechtigter

DWC Digital Wireless Communications

DWCD Drinking Water Criteria Document

DWD Deutscher Wetterdienst (Germany) • Double White Dwarf (star) • Dynamic Weather Display

DWDD Domain Well Dynamic Displacement

DWDF Direct Wave Direction Finding

DWDM Dense Wavelength Division Multiplexer (*) • Dense Wavelength Division Multiplexing (*) • Digital Wave Division Multiplexer • Digital Wave Division Multiplexing

DWE Delivery With Equipment • Der Wohnungseigentümer • Doppler Wind Experiment • Dynamic Wavelength Equalizer

DWEE Discrete Wind Emission Element

dwf Drawing Web Format (file extension)

DWF Dayton Wright (USA-OH) (location identifier) • Distributed Workstation Feature

dwg Drawing (file extension)

Dwg Drawing

DWG Data Working Group • Digital Waveform Generator • Discipline Working Group • Documentation Working Group • Drawing

DWH Data Warehousing • Don't Walk Here • Houston - David Wayne Hooks (USA-TX) (location identifier)

DWI Data Word In • Descriptive-Word Index • Descriptor Word Index • Digital Words Institute (USA-FL) • Driving While Intoxicated • Dutch West Indies (now Netherlands Antilles)

DWIA Distorted Wave Impulse Approximation

DWIM Diffuse Warm Ionized Medium • Do What I Mean

DWiN Deutsches Wissenschaftsnetz (Germany)

DWIPS Digital Weather Imaging Processing System

DWIS Do What I Say

DWIT Do What I Think

DWL Dominant Wavelength • Doppler Wind Lidar

DWMT Discrete Wavelet Multitone

Dwn Drawn

DWN Down • Oklahoma City Downtown (USA-OK) (location identifier)

DWNDFTS Downdrafts

DWNLD Download

DWO Development Work Order • Direct Writing Oscillograph

DWP Decommissioning Work Plan • Department of Water and Power • Digital Wave Processing • Digital Wave Processor (Cluster)

DWPNT Dew Point

DWQT Deferred Write Queue Threshold

Dwr Drawer

DWR Dauerwohnrecht • Design With Reuse • Development Work Request • Dewar • Digital Wired Recorder • Disaster Warning Receiver • Doppler Weather Radar

DWS Debris Washing System • Demineralized Water System • Design Work Study • Detailed Work Statement • Detector Workstation • Development Work Statement • Disaster Warning Satellite • Disaster Warning System (*) • Doppler Wind Sensor • Doppler Wind Sounder • Dynamic Wavelength Slicing • Orlando - Walt Disney World (USA-FL) (location identifier)

DWSG Direct Write Scene Generator

dwt Pennyweight

DWT Dead Weight Tonnage • Dead Weight Tons • Discrete Wavelet Transform • Dog Wags Tail

DWTS Digital Wideband Tracking System

DWV Dielectric Withstanding Voltage • Duty White Uniform

DWX Distorted Wave (approximation) with Exchange

DWZ Dat Wil Zeggen • Directie Wetenschap en Technologie

Dx Diagnosis • Duplex

DX Daily Exposure • Data Exchange • Data Explorer • Data Extraction • Direct Exchange • Directory Exchange • Distance • Duplex • Dynamic Experimentation • Long Distance

DXA Deux-Alpes (France) (location identifier) • Direct Exchange Activity

DXB Diffuse X-ray Background • Drawing Exchange Binary • Drawing Interchange Binary • Dubai (UAE) (location identifier)

DXC Data Exchange Control • Digital Crossconnect • Drawing Exchange Control

DXD Digital By Digital • Dixie (Australia-QLD) (location identifier)

dxf Drawing Interchange Format (file extension)

DXF Data Exchange File • Data Exchange Format • Digital Exchange Format • Drawing Exchange Format • Drawing Interchange File • Drawing Interchange Format

DXI Data Exchange Interface • Direct Exchange Item

DXP Data Exchange Protocol • Digital Experiment Processor • Digital X-ray Processor

DXR Danbury (USA-CT) (location identifier) • Deep X-Ray

DXRB Diffuse X-Ray Background

DXRBS Deep X-ray Radio Blazar Survey

DXS Data Exchange Standard • Data Exchange System • Deep Extragalactic Survey • Diagnostic X-ray Spectrometry • Diffuse X-ray Scattering • Diffuse X-ray Spectrometer • Diffuse X-ray Spectrometry

DXT Data Extract

dy Demy

d/y Delivery

Dy Delivery • Deputy • Dysprosium

DY Benin (international vehicle registration code) • Benin (ISO country code) • Dahomey (ISO country code) • Day • Dockyard

DYA Dysart (Australia-QLD) (location identifier)

DYANA Dynamics-Adapted Network for Atmosphere • Dynamics-Adapted Network for the Atmosphere • Dynamik Adaptiertes Netz von Atmosphärer Netzstationen

DyAtO Dynamics of Atmospheres and Oceans (periodical code)

DYB Dynamic Breaking

Dyd Dockyard

DyeRe Dyer Observatory Reprints (periodical code)

DYF Do Yourself In

DYG Dayong (China-PRC) (location identifier) • Dynamic Young Galaxy

DYL Direct Yaw Line • Doylestown (USA-PA) (location identifier)

DYM Diamantina Lakes (Australia-QLD) (location identifier)

dyn Dyne

Dyn Dynamic • Dynamics • Dynamo • Dynamometer • Dynastic • Dyne

DyN Día y Noche

DYN Dynamic

DYNAPAC Dynamic Plasma Analysis Code

DynCo Dynamics and Control (periodical code)

DYR Anadyr (Russia) (location identifier)

DYS Abilene Dyess (USA-TX) (location identifier)

DYSC Discovery Young Scientist Challenge

DYSIM Dynamic Simulator

DYTEC Dynamique de la Terre er Évolution du Climat

DYU Dushanbe (Tajikistan) (location identifier)

DYW Daly Waters (Australia-NT) (location identifier)

dz Buthani (ISO language code)

Dz Dozen • Drizzle

DZ Algeria (international vehicle registration code) • Algeria (ISO country code) • Algérie (code ISO de pays) • Defense Zone • Doppelzentner • Dozen • Drizzle • Dropping Zone

DZA Algeria (ISO country code) • Algérie (code ISO de pays) • Dzaoudzi (Comoros) (location identifier)

DZB Deutsche Zentralbibliothek (Germany)

DZC Dual-Zone Chamber

DZCh Deutscher Zentralausschuß für Chemie (Germany)

DZD Dinar (Algeria) (ISO currency code)

DZI Codazzi (Colombia) (location identifier)

DZN Dzezkazgan (Kazakhstan) (location identifier)

DZNM Croatian State Office for Standardization and Metrology

DZO Durazno (Uruguay) (location identifier)

DZU Dazu (China-PRC) (location identifier)

D-Zug Durchgangszug

DZug Durchgangszug

D0 Day Zero

D2 Angola (international aircraft registration code) • Data Distribution • Degrade and Destroy • Denial and Destruction • Disrupt and Destroy

D2C Decimal to Character

D2CO Doubly Deuterated (form of) Formaldehyde

D2D Disk-to-Disk

D2T2 Dye Diffusion Thermal Transfer

D2X Decimal to Hexadecimal

D3 Decide - Detect - Deliver • Defending the Digital Domain (project) • Defense - Description - Designation • Demand-Driven Dispatch • Detection - Designation - Discrimination

D3A Decide - Detect - Deliver - Access

D4 Cape Verde (international aircraft registration code) • Design Data Dictionary and Directory

D6 Comoros Islands (international aircraft registration code)

D8 Group of Eight Developing Countries

StarBriefs Plus

E

e/ Entre • Envío

e Base of Natural Logarithms • Eccentricity • Edition • Educated • Elder • Eldest • Electric • Electricity • Electromotive • Electron • Elementary Charge • End • Engineer • Engineering • Entrance • Época • Equatorial Conformation • Erlang • Error • Euler Number • Excellence • Excellent • Internal Energy • Polarization Vector • Positron • Relative Elongation • Specific Internal Energy • Transcendental Number • Unit Coordinate Vector • Wet Air

E Color Excess • Earl • Earth • East • Easter • Eastern • Eccentric Anomaly • Eccentricity • Echo (international alphabet) • Edinburgh (UK-Scotland) • Efficiency • Egypt • Egyptian • Eigentum • Eingeladen • Einsteinium • Einwohner • Electric Field • Electric Field Intensity • Electric Field Strength • Electrode Potential • Electrometric Effect • Electromotive Force • Electronic • Electronics • Elevation • Elevation Angle • Elevator • Elimination Reaction • Elliptical Galaxy • Elongation • Emalangeni (currency) • Eminence • Éminence • Ended • Energy • Engineer • Engineering • England (UK) • English • Enhanced • E-number • Environment • Eötvös • Equation of Time • Equator • Equatorial • Equivalent Hours • Erdgeschoß • Erg • Erlang • Error • España • Essential • Est • Estacionamiento • Este • Estimated • European • Evening • Exa • Examination • Excellent • Expected • Expressway • Extensive • Illuminance • Internal Energy • Irradiance • Kinetic Energy of an Electron • Modulus of Elasticity • Radiative Flux • Spain (international vehicle registration code) • Spain (postal country code) • Total Probable Error • Two Hundred Fifty (Roman numeral) • Unit Matrix • Universal Negative Categorical Proposition • Voltage • Young Modulus • 250 (Roman numeral)

ea Each • En Andere • En Anderen

eA Emissivity Area

Ea Estancia

E/A Ephemeris/Attitude

E&A Engineering & Acquisition • Evaluation & Assistance

EA Each • East Africa • Eastern Arctic • Economic Adviser • Economic Analysis • Educational Affairs • Effective Action • Effective Address • Effective Agent • Effective Address • Effective Area • Electronic Attack • Electronics Assembly • Elemental Analyzer • Elevation Angle • Emergency Access • Emergency Action • Emergency Area • En Andere • En Anderen • Endangerment Assessment • Engagement Area • Engineering Analysis • Engineering Authority • Ente Autonomo • Enumeration Area • Environment Agency • Environmental Agent • Environmental Analysis • Environmental Assessment • Erreur d'Alimentation • Ethylacrylate • Evaluation Authority • Event Analyzer • Evolution Artificielle • Evolutionary Acquisition • Evolutionary Algorithm • Executing Agency • Executing Authority • Executive Agency • Executive Agent • Executive Assistant • Expedited Knowledge • Expenditure Authority • Experimental Area • Experiment Assembly • Expert Agent • Expert Assistant • Exposure Adjust • Exposure Age • Extended Address • Extended Area • Extended Attribute

EAA Eagle (USA-AK) (location identifier) • École des Arts Appliqués • Electo-Acoustic Assembly • Encyclopaedia of Astronomy and Astrophysics • Engineer in Aeronautics and Astronautics • Entretien Annuel d'Activités • Ethylene Acrylic Acid • Excitatory Amino-Acid • Experimental Aircraft Association (USA) • Export Administration Act (USA)

EAAA Escambia Amateur Astronomers Association (USA-FL) • European Association of Advertising Agencies

EAAE European Association for Astronomy Education (Germany)

EAAI European Association of Aerospace Industries

EAAM Effective Atmospheric Angular Momentum • European Association of Aerospace Manufacturers

EAAP European Association for Animal Production

EAAS Euro-Asian Astronomical Society (Russia)

EAB Abbse (Yemen) (location identifier) • Editorial Advisory Board • Electronic Address Book • Emergency Action Book • Emergency Air Breathing • Enterprise Access Builder • European ALMA Board • Exclusion Area Boundary • Executive Advisory Board • Extended Attribute Byte • External Access Box • Extra-Anatomic Bypass

EABI Embedded-Applications Binary Interface

EABPP Explosive Air Blast Propagation and Prediction

EABRD Electrically-Activated Band Release Device

EAC Early Analysis Capability • East African Community • Educational Advisory Committee • Emergency Action Cell • Emergency Action Committee • Emergency Action Console • Emergency Action Coordinator • Employee Action Circle • Empresa A Consumidor • Energy-Absorbing Capacity • Engineering Accreditation Commission • Engineering Advisory Committee • Equipment Activity Controller • Estimate At Completion • European Accreditation Committee • European Accreditation of Certification • European Aerospace Conference • European Astronauts Centre (ESA) • European Atomic Commission • Evaluation and Analysis Center • Expect Approach Clearance • Expect Approach Control • Experimental Apparatus Canister • Experiment Apparatus Container • Extended Attitude Control

EACC Electronic Asset Control Center • Error Adaptive Control Computer

EACCS EUMETSAT Advisory Committee of Cooperating States

EACE Early Analysis Capability Enhanced • Equinox Atmospheric Continuity Experiment • European Association of Cognitive Ergonomics

EACH European Academies Clearinghouse

EACL Emergency Action Checklist • Énergie Atomique du Canada Limitée (Canada) • European Association for Computational Linguistics

EACN Equivalent Alkane Carbon Number

EACP European Area Communications Plan

EACRO European Association of Contract Research Organisations

EACT Emergency Action and Coordination Team

ead Eadem

EAD Earliest Arrival Date • École des Arts Décoratifs • Electrically Alterable Device • Encoded Archival Description • Enseignement À Distance • Entered Active Duty • Entry on Active Duty • European Aerospace Database • Evaluation and Analysis Division • Extended Active Duty

EADA Escuela de Alta Dirección y Administración (Spain)

EADAS Engineering and Administrative Data Acquisition System

EADB East African Development Bank

EADI Electronic Altitude Director Indicator • Electronic Attitude Direction Indicator (*) • Electronic Attitude Director Indicator • Electronic Attitude Display Indicator • European Association of Development Institutes

EADN European Astrophysical Doctoral Network

EADS Engineering Analysis and Data System • Environmental Assessment Data System • European Aeronautic, Defense and Space Co. (merging of DASA and Aerospatiale-Matra)

EADT Experiment Agency Data Tape

EADTU European Association of Distance Teaching Universities

EAE Each And Every • Emae (Vanuatu) (location identifier) • Emergency Action Element • Empresa A Empresa • Enquête Annuelle d'Entreprise

EAEC European Atomic Energy Community

EAEEIE European Association for Education in Electrical and Information Engineering

EAEG European Association of Exploration Geophysicists • Extended Asynchronous Exponential Growth

EAEMP European Agency for the Evaluation of Medical Products

EAEO Equal Access End Office

EAERE European Association of Environmental and Resource Economists

EAES European Atomic Energy Society

EaEvS Earth and Evolution Sciences (periodical code)

EAF Effort Adjustment Factor • Electric-Arc Furnace • Emergency Action File • Engineering Analysis Facility • Engineering Analysis Function • Equivalent Availability Factor

EAFB Edwards Air Force Base (USA-CA)

EAFS Effective Aerial Film Speed

EAFT European Association For Terminology

EAG Engine Advisory Group • Experimental Astrophysics Group • External Advisory Group

EAGA East Asian Growth Accord

EAGB Early Asymptotic Giant Branch (star) • Expéditions Astrophysiques et Géophysiques Belges (Belgium)

EAGE Electrical Aerospace Ground Equipment • European Association of Geoscientists and Engineers

EAGIL Expert Associates Guide to Industrial Learning

EAGLE Enhanced Application of Ground Laser Equipment • EOS Atmospheric Global Lidar Experiment • European Association for Grey Literature Exploitation

EAGLES Expert Advisory Group on Language Engineering Standards

EAGS European Association of Exploration Geophysicists

EAHIL European Association for Health Information and Libraries

EAHP École d'Application des Hauts Polymères

EAI Engine Air Inlet • Enseignement Animé Interactif • Enterprise Application Integration • External Authoring Interface

EAIA École Avancée d'Intelligence Artificielle

EAIC European Aerospace Industry Council

EAID Engine Auxiliary Inlet Door • Equipment Authorization Inventory Data • Equipment Authorization Inventory Document

EAIE European Association for International Education

EAIM End Article Item Manager

EAIR Enterprise Analysis Integration Review

EAIS European Astronomical Information Service

EAITC External Affairs and International Trade Canada

EAITech Euro-American Institute of Technology

EAJ Engineering Academy of Japan

EAJA Equal Access to Justice Act (USA)

EAJNM Executive Agent for Joint Network Management

EAK Kenya (international vehicle registration code)

EAL Échelle Atomique Libre • Electromagnetic Amplifying Lens • Emergency Action Level • Entry Access List • Entry Authority List • Equipment Airlock • ESA Attached Laboratory • European Accreditation of Laboratories • European Astronomical Landscape

EALG Entschädigungs- und Ausgleichsleistungsgesetz

EALT Earliest Anticipated Launch Time

EAM Earth Acquisition Mode • Effective Angular Momentum • Electrical Accounting Machine • Electronic Accounting Machine • Embedded Amplitude Mask • Embedded-Atom Method • Emergency Action Message • Eurocontrol Airspace Model • Event Analysis Matrix • Nejran (Saudi Arabia) (location identifier)

EAMA East-Asian Meeting on Astronomy

EAMAC École Africaine pour la Météorologie et l'Aviation Civile

EAMAS Emergency Action Message Authentication System

EAMEDPM Electronic Accounting Machine and Electronic Data Processing Machine

EAMPAG European Astronaut Medical Psychological Advisory Group

EAMPS Extended Advanced Mobile Phone Service

EAMREA Environmental Impact of Mineral Exploration-Exploitation in Antarctica

EAMT European Association for Machine Translation

EAN Effective Atomic Number • Equivalent Atomic Number • European Academic Network • European Article Number • Wheatland Phifer (USA-WY) (location identifier)

EANA European Astrobiology Network Association

EANCO Emergency Action Non-Commissioned Officer

EANDC European-American Nuclear Data Committee

EANS Emergency Action Notification System • European Article Numbering System

EAO Emergency Action Office • Emergency Action Officer • Engelhardt Astronomical Observatory (Kazakhstan) • Enseignement Assisté par Ordinateur • Étude Assistée par Ordinateur • Extrême Adaptive Optics

EAOI Earth's Atmospheric Optical Instability

EAON European Asteroidal Occultation Network • Except As Otherwise Noted

EAOS Easy Access Ordering System • Environment Australia Online Service

EAP Basel Mulhouse (Switzerland) (location identifier) • East Asia and Pacific • Educational Awareness Program • Effective Air Path • Emergency Action Plan • Emergency Action Procedure • Employee Assistance Program • Enterprise Architecture Planning • Environmental Affairs Program • Environmental Assessment Panel • Environment Assistance Programme • EOSDIS Advisory Panel • Equivalent Air Pressure • Étage d'Accélération à Poudre • Evolutionary Acquisition Phase • Experimental Activity Proposal • Experimental Aircraft Project • Extensible Authentication Protocol • Mulhouse Bâle (France) (location identifier)

EAPA European Alliance of Press Agencies

EAPBX Electronic Automatic Private Branch Exchange

EAPC Euro-Atlantic Partnership Council

EAPE Eddy Available Potential Energy

EAPEE European Association for Political and Evolutionary Economy

EAPG European Association of Petroleum Geoscientists

EAPGE European Association of Petroleum Geoscientists and Engineers

EApJL Electronic Astrophysical Journal Letters

EAPLS European Association for Programming Languages and Systems

EAPOE Extensible Authentication Protocol Over Ethernet

EAPROM Electrically Alterable Read-Only Memory

EAPS Earth, Atmospheric and Planetary Sciences • Engine Air Particle Separator • Euro-Asian Physical Society

EAR Early Acquisition Radar • Electronically Agile Radar • Employee Attitude Research • Energy-Absorbing Resin • Engineering Acceptance Request • Engineering Action Request • Engineering Analysis Report • Engineering Analysis Review • Enterprise Application Repository • Entity - Attribute - Relationship • Environmental Analysis Report • Excess Absolute Risk • Export Administration Regulation • External Access Register • Kearney (USA-NE) (location identifier)

EARA Equipment Authorizations Review Activity • European Association for Research in Astronomy

EARB Engineering Action Request Review Board

EARCOM East Africa Regional Remote Sensing Management Committee

EARG Environmental Adaptation Research Group • Estación Astronómica de Río Grande (Argentina-TF)

EARL Easy Access Report Language • Electronic Access to Resources in Libraries • Electronically Accessible Russian Lexicon

EARMA European Association of Research Managers and Administrators

EARN European Academic and Research Network • European Academic Research Network • European Asteroid Research Node

EAROM Electrically Alterable Read-Only Memory

EAROS Electrically Alterable Read-Only Store

EARP Environmental Assessment Review Process

EARS Electronic Access to Reference Services • Environmental Analysis and Remote Sensing • External Audio Receptor System

EARSC European Association of Remote Sensing Companies

EARSEC European Airborne Remote Sensing Capabilities

EARSeL European Association of Remote Sensing Laboratories

EARSEL European Association of Remote Sensing Laboratories

EARSF European Airborne Remote Sensing Facility

EARSO European Airborne Remote Sensing Consortium

EarthCARE Earth Clouds, Aerosol and Radiation Explorer

EARTO European Association of Research and Technology Associations (Belgium)

EARTS En-route Automated Radar Tracking System

EAS Earth and Atmospheric Sciences • Earth's Artificial Satellite • Eastbay Astronomical Society (USA-CA) • ECS Advertising System • Electrical Accommodation System • Electron Acoustic Structure • Electron-beam Acoustic Structure • Electronic Animation System • Emergency Action System • Engine Actuation System • Engineering Administrative System • Engineering Analysis System • Enterprise Application Server • Environmental Analyses Section • Equipment Accountability System • Equivalent Airspeed • Error Analysis Study • European Astronomical Society • Everett Astronomical Society (USA-WA) • Evolutionary and Adaptive System • Executive and Administrative Schedule • Exeter Astronomical Society (UK) • Expense Assignment System • Experiment Assurance System • Expert Address Selection • Extended Announcement System • Extended Area Service • Extensive Air Shower • Extensive Atmospheric Shower • San Sebastian (Spain) (location identifier)

EASA European Academic Software Award • European Air Safety Authority (*) • European Aviation Safety Authority

EASB European Astronauts Selection Board

EASD Equal Access Service Date

EASE East Auroral Snapshot Experiment • Easy Access Screen Entry • Electrical Automatic Support Equipment • Electronically-Assisted Solicitation Exchange • Empirical Assessment in Software Engineering • Engineering Applications Support Programme • European Association of Science Editors • Evaporation - Accretion - Self-Enrichment • Experimental Assembly of Structures in EVA • Expert Activity Scheduling Environment • Eye-Assisted Selection and Entry

EASEP Early Apollo Scientific Experiments Package

EASF Enhanced All-Source Fusion

EASI European Academic Supercomputing Initiative • Expanded Additional Skill Identifier • Expert Analysis System Interface

EASIT European Association for Software Access and Information Transfer

EASIZ Ecology of the Antarctic Sea Ice Zone

EASL Engineering Analysis and Simulation Language

EASLS East African Scientific Literature Service

EASOE European Arctic Stratospheric Ozone Experiment

EASS European Astronomy and Space Service

EASSIS European Astronomy and Space Sciences Service

EASST European Association for the Study of Science and Technology

East Eastern

EAST East Australian Standard Time • Euro-African Satellite Telecommunications • European Advanced Software Technology • European Assistance for Science and Technology Ltd. Co. (UK)

EASTPAC Eastern Pacific

EASTT Environmental Advisory Service for Transfer of Technology

EASy Evolutionary and Adaptive System

EASY Early Acquisition System • Electronic Application System • Emergency Action System • Enterprise Assessment System • Evolutionary and Adaptive System

EAT Earliest Arrival Time • Encoder Address Translator • Environmental Acceptance Test • Environnement, Aménagement et Télédétection • Estimated Approach Time • Expected Approach Time • Extended Alias Table • External Air Transport • Tanzania (international vehicle registration code) • Wenatchee (USA-WA) (location identifier)

EATA Enhanced AT-bus Attachment

EATCHIP European Air Traffic Control Harmonization and Integration Programme • European ATC Convergence Harmonisation and Implementation Programme

EATCS European Association for Theoretical Computer Science • External Active Thermal Control System

EATF Externally Applied Thermal Field

EATMP European Air Traffic Management Programme

EATMS European Air Traffic Management System

EATR Extravehicular Assist Teleoperator Robot

EATS Extended Area Test System

EATWG Executive Agency Transfer Working Group

EAU Eau Claire (USA-WI) (location identifier) • Economic Analysis Unit • Émirats Arabes Unis • Erase All Unprotected • Extended Arithmetic Unit • Uganda (international vehicle registration code)

EAV Effective Angular Velocity • End of Active Video • Engine Assembly Vehicle

EAVO Enseignement Assisté par Audiovisuel et Ordinateur • External Air Vehicle Operator

EAVS Emergency Action Voice System • Équipe Académique de la Vie Scolaire

EAW Electron-Acoustic Wave • Equivalent Average Word

EAWAG Eidgenössische Anstalt für Wasserversorgung, Abwasserreinigung und Gewässerschutz (Switzerland)

EAWEP East Asia and Western Pacific

EAWS Engineering Analysis Workstation

EAX Electronic Automatic Exchange • Environmental Automatic Extension

EAZ Endgeräte-Auswahlziffer • Tanzania (international vehicle registration code)

EAZO Energy Active Zone of the Ocean

E/B Equipment Bay

EB Eastbound • Eclipsing Binary (star) • Economy and Business • Eigenbewegung • Electric Boat • Electricity Board • Electron Beam • Electron Bombarded • Electron Bombardment • Electron Bremsstrahlung • Electronic Beam • Electronic Book • Electronics Bay • Ellerman Bomb • Emergency Box • Empirical Bayes • Employee Benefits • Encapsulation Boundary • Encyclopaedia Britannica • Energy per Bit • Equatorial Band • Equatorial Belt (Jupiter) • Equipment Bay • Erstbezug • Exabyte • Executive Board • Executive Body • Executive Bulletin • Expendable Booster • Extragalactic Background

EBA École des Beaux-Arts • Elba Island (Italy) (location identifier) • Electronics Bay Assembly • Exponentiated Born Approximation

EBAA European Business Aviation Association

EBAD École des Bibliothécaires, Archivistes et Documentalistes

EBAE European Bureau for Adult Education

EBAM Electron-Beam- Accessed Memory

EBAR Edited Beyond All Recognition

EBB Electronic Bulletin Board • Elegant Breadboard • Entebbe Kampala (Uganda) (location identifier)

EBBR Energy-Balance Bowen Ratio

EBBS Electronic Bulletin Board System

EBBT Equivalent Black-Body Temperature

EBC EISA Bus Controller • Electron Beam Coating • Electron Beam Control • Electronics Begin Countdown • Emulated Buffer Computer • European Bibliographical Center • Executive Business Center

EBCCD Electron-Beam Charge-Coupled Device • Electron-Bombarded Charge-Coupled Device

EBCD Extended Binary-Coded Decimal

EBCDIC Extended Binary Coded Decimal Interchange Character • Extended Binary Coded Decimal Interchange Code (*)

EBCG European Biotechnology Coordination Group

EBCi Eclipsing Binaries Circulars (periodical code)

EBCI Explicit Backward Congestion Indication

EBCOT Embedded Block Coding with Optimized Truncation

EBCS Eastern Boundary Containment System

EBCT Electron Beam Computed Tomography

ebd Ebenda

EBD Eclipsing-Binary (star) Distance (method) • Electronic Breaking Distribution • El Obeid (Sudan) (location identifier) • Equivalent Binary Digit

EBDI Electronic Business Data Interchange

EBE Excédent Brut d'Exploitation • Except British Commonwealth • Extraterrestrial Biological Entity

e-beam Electron Beam

EBEB Europäisches Büro für Erwachsenenbildung

EBER Electron-Beam Electroreflectance

EBES European Board for Edifact Standards

EBF Electronic Book Format • Emergency Bug Fix • Enduring Business Framework

EBG El Bagre (Colombia) (location identifier) • Electron Beam Generator • Electron Beam Gun • Escher - Bach - Gödel (trilogy)

EBGR Extragalactic Background Radiation

EBI Enterprise Business Integration • Equivalent Background Input • European Bioinformatics Institute • Extended Background Investigation • Extended BIOS Interface • Extensive Background Information • Extensive Background Investigation

EBIC Electron Beam Induced Current • Electron Beam Induction Current • European Business Information Centre

EBICON Electron Bombardment Induced Conductivity

EBIM Ethernet Bridge Interface Module

EBIN Eindhovens Bibliotheek Informatie Netwerk

EBIP European Biotechnology Information Project

EBIR Electron Beam Image Recorder

EBIS Electron Beam Ion Source • ESO Budget Information System • European Biotechnology Information Service

EBIT Earnings Before Interest and Taxes • Electron Beam Ion Trap • European Broadband Interconnection Trial

EBITDA Earnings Before Interest, Taxes, Depreciation and Amortization

EBJ Esbjerg (Denmark) (location identifier)

EBK Einstein - Brillouin - Keller (method) • Emergency Backup Key

EBL Electron Beam Lithography • Explanation-Based Learning • Extragalactic Background Light

EBLIDA European Bureau of Library, Information and Documentation Associations

EBLP Estimated Best Linear Predictor

EBM El Borma (Tunisia) (location identifier) • Electrical Bridgman Method • Electro-Beam Machine • Electro-Beam Machining • Energy Balance Model • Entity-Based Model • Estimation Before Modeling • Ethernet Bridge Module • Evidence-Based Medicine • Extended Boolean Model

EBMD Electron Beam Mode Discharge

eBMF e-Business Model Framework

eBMH e-Business Model Handbook

eBML e-Business Modeling Language

EbN East by North

EBN East By North • Ebon (Marshall Islands) (location identifier) • Ericsson Business Network • European Backbone Network • European Business Network • Extended Border Node

EBND Eastbound

EBNF Extended Backus-Naur Form (*) • Extended Backus-Naur Formalism • Extended Backus Normal Form

EBNN Explanation-Based Neural Network

EBO Ebon (Marshall Islands) (location identifier) • Embedded Base Organization • Engineering Buy-Off • Equivalent Bond Orbital

EBOF Enterprise Business Objects Facility

EBONE European Backbone

EBOP Eclipsing Binary Orbit Program • Enhanced BATSE Occultation Package

EBP Electron Beam Process • Electron Beam Pumped • Electronic Book Player • Electronics Base Plate • Enduring Business Process • European Books in Print • EUV Bright Point • Exhaust Back Pressure

EBPA Electron Beam Parametric Amplifier

EBPP Effelsberg-Berkeley Pulsar Processor • Electronic Bill Payment and Presentment

EBr Encyclopaedia Britannica

EBR Baton Rouge Downtown (USA-LA) (location identifier) • Electron Beam Recorder • Electron Beam Recording • Electron Beam Regulator • Employee Badge Reader • Epoxy Bridge Rectifier • Experimental Breeder Reactor • Extended Boot Record • Extended-partition Boot Record • Extragalactic Background Radiation

EBRD European Bank for Reconstruction and Development

EBRF Extended Bidirectional Reflectance Factor

EBRS Enterprise Backup Recovery System

EbS East by South

EBS East By South • Eclipsing Binary Star • Economics and Business Studies • Electron Beam Semiconductor • Electron Beam System • Electron-Bombarded Semiconductor • Electron-Bombarded Silicon • Electronically-Built Society • Electronically-controlled Braking System • Electronic Band Spectra • Electronic Brain Storming • Emergency Breathing System • Emergency Broadcast System • Entrée Banalisée Synchrone • Equipment Breakdown Structure • European Business School • Excess Burst Size • Executive Briefing System • Webster City (USA-IA) (location identifier)

EBSA European Biophysical Societies Association

EBSI École de Bibliothéconomie et des Sciences de l'Information (Canada-QC) (*) • École des Bibliothèques et des Sciences de l'Information (Canada-QC)

EBT Electron Beam Technique • Electron Beam Test • Electronic Benefits Transfer • Electronic Book Technologies • Electronic Book Technology • Elementary Burst Theory • Elmo Bumpy Torus • Enduring Business Theme • Extended Blue Tail

EBTE Electronics Bay Test Equipment

EBTP École du Bâtiment et des Travaux Publics

EBTR Electronic Bearing-Time Recorder

EBU Eddy Break-Up • Electronics Bay Unit • European Broadcasting Union • Saint-Étienne Bouthéon (France) (location identifier)

EBV Electron Beam Vaporization

EBW Ebolowa (Cameroon) (location identifier) • Effective Bandwidth • Electron Beam Welding • Electron Bernstein Wave • Exploding Bridge Wire • Explosive Bridge Wire

EBWR Experimental Boiling-Water Reactor

EBX Electronic Branch Exchange

e/c En Cuenta

ec Ecuador • Exempli Causa

Ec Ecuador

E/C Encoder/Coupler

E&C Engineering & Commissioning

EC Early-type Contact (binary star) • Earth Convergency • Earth Coverage • Earth Curvature • East Caribbean • East Coast • Echo Cancellation • Echo Canceler • Echo Cancelling • École Centrale • Economy Class • Ecuador (international vehicle registration code) • Ecuador (ISO country code) • Eddy Correlation • Eddy Current • Edinburgh-Cape (survey) • Educational Committee • Education Committee • Effective Concentration • Elastic Center • Electoral Committee • Electrochemical • Electrolytic Corrosion • Electron Capture • Electron Cyclotron • Electronic Calculator • Electronic Commerce • Electronic Computer • Electronics Controller • Elemental Carbon • Element Contractor • Embedded Channel • Embedded Computer • Emergency Commission • Emulsion Chamber • Enabling Command • Encoder-Coupler • Engine Change • Engine Control • Engine Cutoff • Engineer Corps • Engineering Change • Engineering Corps • Engineering Council • Enrichment Cell • Entry Course • Environmental Capacity • Environmental Compliance • Environmental Control • Environment Canada • Équateur (code ISO de pays) • Equipment Change • Equipment Custodian • Erase Character • Error Checking • Error Control • Error Correcting • Error Correction • Essentiality Code • Ethylene Carbonate • Eurocheque • European Commission • European Communities • European Community • Europese Commissie • Events Controller • Events Coupler • Evolutionary Computation • Exchange Carrier • Executive Committee • Executive Council • Expansive Classification • Experiment Canister • Experiment Computer • Extended Control • Extended Coverage • Extension Course • External Compton (scattering) • Spain (international aircraft registration code)

ECA Early Comparability Analysis • Earth Central Angle • Earth-Crossing Asteroid • East Tawas - Emmet County (USA-MI) (location identifier) • École de Commerce et d'Administration • Economic Commission for Africa • Economic Cooperation Administration • Economic Cooperation Agreement • Eigenvalue Change Analysis • Electron-Cyclotron Absorption • Electronic Commerce Acquisition • Electronic Component Assembly • Electronic Computer Assembly • Electronic Control Amplifier • Electronics and Computer Assembly • Electronics and Control Assembly • Electronics Control Assembly • Embodied Conversational Agent • Engine Control Assembly • Engineering Change Analysis • Engineering Change Authorization • Epoxy Curing Agent • European Cockpit Association • European Common Allocation • European Court of Auditors • European Crystallographic Association • Event - Condition - Action • Export Credit Agency

ECAB Electronic Current Awareness Bibliography

ECAC Earth-Crossing Asteroids and Comets • Electromagnetic Compatibility Analysis Center (USA) • Electronic Compatibility Analysis Center • European Civil Aviation Conference

ECACC European Centre for Animals Cell Cultures

ECAD Electronic Computer-Aided Design • Electronics and Computer Applications Division • Electronics Computer-Aided Design

ECAFE Economic Commission for Asia and the Far East (now ESCAP)

ECAI European Conference on Artificial Intelligence

Ecal Electronic Calibration

ECAL East Coast Alert Landing • Electronic Channel Alignment • Enjoy Computing And Learn • European Conference on Artificial Life

ECALPO European Computer-Aided Learning Program

ECAM Electronic Centralized Aircraft Monitor • Electronic Centralized Aircraft Monitoring • Electronic Centralized Aircraft Monitoring • Energy Conservation And Management • European Conference on Application of Meteorology • Extended Communications Access Method

ECAMP Environmental Compliance Assessment and Management Programme • European Conference on Atomic and Molecular Physics

ECAN Electronic Calibration And Normalization

ECAO Environmental Criteria and Assessment Office

ECAP Electronic Circuit Analysis Program • Electronic Customer Access Program • Évaluation de Compétences et Acquis Professionnels

ECARP European Computational Aerodynamic Research

ECAS Eight-Color Asteroid Survey • European Courses in Advanced Statistics • Experiment Computer Application Software • External Compliance and Assessment System

ECASS Electronically-Controlled Automatic-Switching System • Export Control Automated Support System

ECAT Electronic Card Assembly and Test • Electronic Commerce Acquisition Team • Electronic Commerce Action Team • European Centre for Automatic Translation

ECB Educational Communications Board • Electronic Code Book • Electronic Components Board • Engineering Change Board • Entry Control Block • Environment Coordination Board • European Central Bank • European Chemical Bureau • Event Control Block • Event Control Buffer • Event Controlled Block

ECBC Extended Cross-Border Coverage

ECBRS Enhanced Concept-Based Requirements System

Ecc Eccentric • Eccetera

ECC Earth Continuity Conductor • Electrochemical Concentration Cell • Electron Capture to the Continuum • Electronically-Controlled Coupling • Electronic Calibration Center • Electronic Command Center • Electronic Computer Center • Electronic Cuff Checklist • Element Command Center • Element Control Center • Elliptical-Curve Cryptography • Elliptical-Curve Cryptosystem • Embedded Control Channel • Emergency Communications Center • Emergency Control Center • Emergency Core Coolant • Emergency Core Cooling • Engineering Change Control • Engineering Critical Component • Enter Cable Change • Equipment Category Code • Error Check and Correcting • Error Check Code • Error Checking and Correcting • Error Checking and Correction • Error Code Correction • Error-Correcting Code • Error Correction and Control • Error Correction Circuitry • Error Correction Code • Estimated Correction Cost • Estimated Cost at Completion • EUROCONTROL Experimental Centre • European Communication Council • European Coordinating Committee • European Coordination Committee • European Crystallographic Committee • Evacuation Control Center • Exercise Control Center • Experimental Computer Complex • Experiment Concept Committee • Extended Core Configuration • Extinct Comet Candidate

ECCA Electronic Component Checkout Area

ECCAI European Coordinating Committee on Artificial Intelligence

ECCAP Electromagnetic Compatibility Circuit Analysis Program • Electromagnetic Compatibility Computer Analysis Proposal • Electromagnetic Compatibility Computerized Analysis Program

ECCB Engineering Change Control Board • Executive Configuration Control Board

ECCC Electronic Community Coordinating Committee • Electronic Command and Control Center

ECCCS Emergency Command Control Communications System

ECC&D Embedded Computer Control & Display

ECCD Electron-Cyclotron Current Drive • Electronic Cockpit Control Device • Embedded Computer Control and Display

ECCE Early Command Center Experiment

ECCI Electronic Commerce of Component Information

ECCIS Engineer Command and Control Information System

Eccl Ecclesiastic

Eccles Ecclesiastic

ECCM Electronic Counter-Counter Measure • Equilibrium Cloud Condensation Model

ECCN European Climate Computing Network • Export Commodity Control Number • Export Control Classification Number

ECCO Estimating the Circulation and Climate of the Ocean • European Communities Communications Office • Extremely-heavy Cosmic-ray Composition Observatory

ECCOMAS European Community on Computational Methods in Applied Sciences

ECCP European Climate Change Programme • Experiment Configuration Control Panel

ECCRN Emergency Command Control Radio Network

ECCS Emergency Core-Cooling System • Engineer Command and Control System

ECCV European Conference on Computer Vision

ECD Early Closing Day • Effective Cutoff Diameter • Electric Control Drive • Electrochromic Device • Electron Capture Detection • Electron Capture Detector • Electron Current Detector • Electronic Cash Disbursement • Electronic Commerce Directive (EU) • Electronic Control Display • Electrostatic Charge Decay • Engineering Change Document • Engineering Control Drawing • Enhanced Color Display • Entry Corridor Display • ESA Computer Department • Estimated Completion Date • Extragalactic Cepheid Database

ECDA Engine Control Development Area • Experiments on Cosmic Dust Analogues • Extensible Compound Document Architecture

ECDB Engineering Chemistry Database

ECDC Economic Cooperation among Developing Countries • Electrochemical Diffused Collector • Engineering Configuration Data Control • External Countdown Clock

ECDH Elliptical Curve Diffie Hellman (system)

ECDIN Environmental Chemical Data and Information Network

ECDIS Electronic Chart Display and Information System

ECDL Elliptical Curve Discrete Logarithm • European Computer Driving Licence • European Conference on Digital Libraries

ECDO Electronic Community Dial Office

ECDR Electrostatic Corona Discharge Reactor • Experiment Conceptional Design Review

ECDS Enhanced Container Delivery System

ECDU Electrical Coupling Display Unit • Electronic Coupling Display Unit • Enhanced Control and Display Unit

E&CE Electrical & Computer Engineering

ECE Eastern and Central Europe • Economic Commission for Europe • Electrical and Computer Engineering • Electrical Checkout Equipment • Electron-Cyclotron Emission • Element Characteristics Equation • Elements of Communications Equipment • Energy Conservation Engine • Engineering Capacity Exchange • Experiment Checkout Equipment • External Consultant Examiner

ECECS Electrical and Computer Engineering and Computer Science

ECEF Earth-Centered Earth-Fixed

ECEH European Centre for Environment and Health

ECEMP Electron-Caused Electromagnetic Pulse

ECF Echo Frame • Effective Cloud Fraction • Effective Cutoff Frequency • Electron Cyclotron Frequency • Electronic Contracts Folder • Element Circuit Function • Energy Conversion Factor • Engine Calibration Facility • Engineering Computer Facility • Enhanced Connectivity Facility • Entry Control Facility • Equipment Calibration Facility • Equivalency Capability File • European Coordinating Facility • Experimental Calibration Facility • Extended-Care Facility • Externally Caused Failure

ECFA European Committee for Future Accelerators

ECFM Eddy Current Flow Meter

ECFY Engineering Careers Foundation Year

ECG Ecosystem Conservation Group • Electrocardiogram • Electrocardiograph • Electrocardiography • Electrocatalytic Gas • Electrochemical Grinding • Electroepitaxial Crystal Growth • Elizabeth City (USA-NC) (location identifier) • Emergency Coordination Group • Exercise Control Group

ECGC Electron Capture Gas Chromatograph

ECGI Error-Correcting Grammatical Inference

Ech Échelon

ECH Echuca (Australia-VIC) (location identifier) • Electron Cyclotron Harmonic (wave) • Electron Cyclotron Heating

ECHAM European Centre Hamburg Model

ECHB End of Core Hydrogen Burning

ECHIVAL European (project) on Climatic and Hydrological Interactions between Vegetation, Atmosphere and Land

ECHO European Commissions Host Organisation • European Community Humanitarian Office • Experiment for Coordinated Heliospheric Observations • Extraction and Classification of Homogeneous Objects

ECHOS Enhanced Coach and Bus Occupant Safety

ECHP European Community Household Panel

ECHR European Convention on Human Rights

ECHT European Conference on Hypertext

ECI Earth-Centered Inertia • Earth-Centered Inertial (*) • Eau Chaude Industrielle • Échanges Culturels Internationaux • École Centrale d'Informatique • Education - Communication - Information • Electrical Circuit Interrupter • Electronic Control Instrumentation • Electronic Conversion Interface • Electronic Course Identification • Employment Cost Index • Energy-Cost Indicator • Engineer Contract Instruction • Engineering Change Instruction • Equipment Characteristic Information • Equipment Configuration Item • European Cooperation in Informatics • European Corpus Initiative • Extension Course Institute • External Call Interface

ECIA European Council of Information Associations

ECIC Electronic Commerce In Contracting • Electronic Commerce Information Center • Electronic Communications Implementation Committee

ECICS European Customs Inventory of Chemical Substances

ECID En-Route Computer Identification

ECIDC Educational Cooperation between Industrialized and Developing Countries

ECIE Executive Council on Integrity and Efficiency

ECIL Electronic Corporation of India Ltd.

ECIP Electronic Cataloguing in Publications • Energy Conservation Investment Program • European Community Investment Partner

ECIR Étoile Carbonnée Infrarouge

ECIS Earth-Centered Inertial System • Eindhoven Center for Innovation Studies (Netherlands) • Error Correction Information System • European Committee for Interoperable Systems • European Community Information Service • European Conference on Information Systems • Extragalactic Card Index System • Extremely Cold Infrared Source

ECISS European Committee for Iron and Steel Standardization

ECIT Enhanced Communication Interface Terminal

ECITC European Committee for Information Technology Certification

ECJ European Court of Justice

ECK Emergency Communication Key

ECKD Extended Count - Key - Data

ECL École Centrale de Lyon (France) • Effluent Concentration Limit • Electronic Circuit Logic • Embedded-Constraint Language • Emergency Cooling Limit • Emitter Control Logic • Emitter-Coupled Logic • End Communication Layer • Engine Control Lever • English Comprehension Level • Equipment Component List • Execution Control List • Executive Control Language • Existing Chemicals List

ECLA Economic Commission for Latin America • European Classification • European Community Library Association

ECLAC Economic Commission for Latin America and the Caribbean

ECLAIR European Collaborative Linkage of Agriculture and Industry through Research • Évaluation de Critères Liés pour l'Automatisation des Interrogations et des Recherches • Extensible Class Library for Information Retrieval

Ecli Eclipse • Ecliptic

ECLIPS Experimental Cloud Lidar Pilot Study (*) • Experimental Cloud Lidar Project Study

ECLIS Economic and Cultural Landscape Information System

ECLO Emergency Locust Operations Center • Emitter-Coupled Logic Operator

ECLRv Electrical Communication Laboratories Review (periodical code)

ECLS Environmental Control and Life Support • Environmental Control and Life-support System

ECLSS Environmental Control and Life Support System

ECLT English Comprehensive Level Test

ECM Electric Coding Machine • Electrochemical Machining • Electromagnetic Countermeasure • Electromechanical Counting Mechanism • Electron-Cyclotron Maser • Electronic Control Module • Electronic Countermeasure • Ellipsoid Collector Mirror • Emulation Control Module • Endpoint Congestion Management • Energy Conservation Measure • Energy Conversion Management (periodical code) • Engine Condition Monitoring • Engineering Change Management • Enterprise Commerce Management • Enterprise Commercial Management • Entitlement Control Message • Entity Coordination Management • Environmental Corrosion Monitor • Error Correcting Memory • Error Correction Mode (*) • European Common Market • European Crystallographic Meeting • Expectation Conditional Maximization • Extended Core Memory • Extracloud Medium • Extra Crew Member

ECMA East Coast Magnetic Anomaly • European Computer Manufacturers' Association

ECMAST European Conference on Multimedia Applications, Services and Techniques

ECMB Energy Center Management Board

ECMC Emergency Crisis Management Center • Enhanced Crisis Management Capability

ECME Electron Cyclotron Maser Emission

ECMEA European Conference of Meteorological Experts for Aeronautics

ECMI European Consortium for Mathematics in Industry

ECML Electronic Commerce Modeling Language • European Conference on Machine Learning

ECMOG Erie County Mobile Observing Group (USA-PA)

ECMP ECS Configuration Management Plan • Equal Cost Multipath • European Configuration Management Plan

ECMR Electronic Calibration Maintenance Record

ECMS Electronic Copyright Management System • Engine Condition Monitoring System • Enhanced Crisis Management System

ECMU Extended Core Memory Unit

ECMWF European Centre for Medium-range Weather Forecasting • European Centre for Medium-range Weather Forecasts (UK) (*)

ECN Electronic Communications Network • Energieonderzoek Centrum Nederland • Engineering Change Notice • Environmental Change Network • Environmental Communications Network • Epoxy - Cresol - Novolak • Equipment Control Number • Ercan (Cyprus) (location identifier) • European Centres and Networks • European Consulting Network • European Cosmological Network • European Cosmology Network (*) • Explicit Congestion Notification

ECNE Enterprise-Certified NetWare Engineering

ECNOD ESA Computer and Network Operations Department

Eco Ecological • Ecology

ECO Earth-Crossing Object • Ecological • Ecology • Economic Change Order • Economic Cooperation Organization • Effective Competitive Opportunity • Electron-Coupled Oscillator • Electronic Central Office • Electronic Checkout • El Encanto (Colombia) (location identifier) • Emergency Control Officer • Energy Conservation Opportunity • Engine Combustion • Engine Cutoff • Engineering Change Order • Entry Control Office • Equipment Control Officer • Etscorn Campus Observatory (USA-NM) • Europe Centrale et Orientale • Expected Cost of Observing

ECOA Equal Credit Opportunity Act (USA)

ECOD European Consortium for Ocean Drilling

ECODU European Control Data Users

ECOGAS European Council of General Aviation Support

ECOHSE European Centre for Occupational Health, Safety, and the Environment

Ecol Ecología • Ecological • Ecology

ECOLE European Collaborative Learning Environment

E-COM Electronic Computer-Originated Mail

ECOM Electronic Computer-Originated Mail • En-route Communications • ESA Costing Model

ECOMARGE Ecology of the Continental Margins

ECOMET European Cooperation in Meteorology

Econ Economia • Economic • Economics • Economist • Economy

ECOOP European Conference on Object-Oriented Programming

ECOPS European Committee for Ocean and Polar Sciences • European Committee on Ocean and Polar Sciences

ECOR Engineering Commission on Ocean Resources • Engineering Committee on Oceanic Resources (*)

ECORS Étude des Continents et Océans par Réflexion Sismique

ECOS ESA Costing Software • Experiment Computer Operating System

ECOSAT Étude du Climat par Observation Spatiale de l'Atmosphère de la Terre

ECOSOC Economic and Social Council

ECOSSE Educational Channel Out of Scotland Servicing Europe

ECOT Electronic Community Of Teachers

ECOWARM European Committee for Water Resources Management

ECOWAS Economic Community of West African States

ECP Écart Circulaire Probable • École Centrale de Paris (France) • Effective Core Potential • Electrical Codes of Practice • Emergency Change Package • Emergency Change Proposal • Emergency Control Procedure • Encryption Control Protocol • Engineering Change Proposal • Engineering Control Program • Enhanced Capabilities Port • Entry Control Port • Environmental Choice Program • Environmental Control Panel • Equipment Change Proposal • Equipment Collecting Point • European Common Position • European Common Proposal • European Conference on Planning • Executive Cellular Processor • Executive Control Program • Exercise Control Plan • Explicitly-Coded Program • Extended Capabilities Port • External Communications Processor

ECPA Electronic Communications Privacy Act (USA) • European Commission on Preservation and Access

ECPD Engineers Council for Professional Development (now ABET)

ECPE European Centre for Public Enterprise

ECPF Electromagnetic Containerless Processing Facility • Encuesta Continua de Presupuestos Familiares

ECPI Electronic College of Process Innovation

ECPIU Electronic Circuit Plug-In Unit

ECPM École de Chimie, Polymères et Matériaux (France)

ECPN Electronic Commerce Processing Node

ECPR Electrically Calibrated Pyroelectric Radiometer • European Consortium for Political Research

ECPSA Elliptical Curve Digital Signature Algorithm

ECQ Executive Core Qualification

ECQAC Electronic Components Quality Assurance Committee

Écr Écrire • Écrit • Écriture

ECR Earth Centered Rotating • Efficient Customer Response • El Charco (Colombia) (location identifier) • Electron-Cyclation Radiation • Electron-Cyclotron Resonance (*) • Electron-Cyclotron Response • Electronic Cash Register • Electronic Control Relay • Embedded Computer Resources • Emergency Conference Room • Engineering Change Request • Enhanced Carrier Route • Equipment Change Record • Estimate Change Request • Executive Control Routine • Expected Cost of Repair • Experimental Control Processor • Experiment Change Record • Experiment Concept Review • Extragalactic Cosmic Ray

ECRA Engineering Change Request Authorization • Environmental Cleanup Responsibility Act (USA)

ECRC Electronic Commerce Resource Center • Engineering Contract Requirements Committee • European Computer Research Centre

ECRH Electron-Cyclotron Resonance Heating

ECRIN Échange et Coordination Recherche-Industrie

ECRM Electronic Customer Relationship Management

ECRQP Equality Constraints Recursive Quadratic Programming

ECRS Economic and Emergency Reserve Stock

ECRU Enhanced Command Response Unit • Environmental Consulting and Research Unit (CSIRO)

ECRV Exponentially Corrected Random Variable

ECS Eau Chause Sanitaire • École Chez Soi • École des Carrières Supérieures • Edgeworth-Cramer Series • Effective Cross-Section • Electrical and Communications Systems • Electrical Control Station • Electrochemical Society • Electron-Capture Spectroscopy • Electronically Collimated System • Electronic Charting System • Electronic Chart System • Electronic Commerce Service • Electronic Composing System • Electronic Control Switch • Electronic Control System • Electronic Courier Service • Electronic Crossconnect System • Electronic Customer Service • Electronics Control System • Elevation Coverage System • Embedded Computer Support • Embedded Computer System • Emergency Coolant System • Emergency Cooling System • Enable Control System • Enclosure Control System • Encryption Control Signal • Energy Conversion System • Energy Corrected Sudden • Engine Control System • Engineering Change Support • Engineering Change System • Engineering for Complex Systems • Enterprise Computer Solutions • Envelope Circumstellaire • Environmental Control Shroud • Environmental Control System • EOSDIS Core System • Equipment Concentration Site • European Chemical Society • European Committee for Standardization • European Communications Satellite • Exact Cubic Search • Executive Communication Systems • Executive Control System • Exercise Control System • Experimental Communications Satellite • Extended Computer Service • Extended Core Storage • Exterior Communications System • External Compton Scattering • Newcastle Mondell

(USA-WY) (location identifier) • Sucre (Ecuador) (ISO currency code)

ECSA European Computing Services Association • Exchange Carrier Standards Association • Extended Common Service Area

ECSC European Coal and Steel Community • European Conference on Satellite Communications

ECSCW European Conference on Computer-Supported Cooperative Work

ECSD Enhanced Circuit-Switched Data

ECSE Experimental Computer Science and Engineering

EcSJ Electrochemical Society Journal (periodical code)

ECSL European Centre for Space Law

ECSMR European Conference on Software Maintenance and Engineering

ECSN European Climate Support Network

ECSO European Council of Skeptical Organizations

ECSQARU European Conference on Symbolic and Quantitative Approaches to Reasoning and Uncertainty

ECSQAU European Conference on Symbolic and Quantitative Approaches to Uncertainty

ECSR Enhanced Commercial Security Requirements

ECSS European Cooperation for Space Standardisation • Executive Control Subordinate System • Extendable Computer System Simulator

ECST European Conference on Science and Religion

ECSTASY Economical Storage and Access System

ECSU Eastern Connecticut State University (USA-CT)

ECS2 Executive Control Subordinate System

ECS2C Executive Control Subordinate System Concept

ECT Économie des Changements Techniques • Eddy Current Testing • Electro-Convulsive Therapy • Embedded Computer Technology • Emergency Communication Terminal • Emission-Computed Tomography • Emission-Computerized Tomography • Emulsion Chamber Technology • Engineer Center Team • Engineering Center Team • Environment Control Table • Equatorial Crossing Time • Estimated Completion Time • Experiment Command Tape • Exposure Control Technique

ECTA Electronics Component Test Area • European Communities Trademark Association

ECTEI École Centrale des Techniciens de l'Électronique et de l'Informatique

ECTF Edinburgh Centre for Tropical Forests • Enterprise Computer Telephony Forum • European Community Telework Forum

ECTL Electronic Communal Temporal Lobe

ECTP European Conference on Thermophysical Properties

ECTRA European Committee for Telecommunications Regulatory Affairs

ECTS European Computer Trade Show • European Conference on Telecommunications by Satellite • European Credit Transfer System

ECU Earth-Centered Universe • Ecuador (ISO country code) • EISA Configuration Utility • Electrical Control Unit • Electrical Conversion Unit • Electronic Control Unit • Electronic Conversion Unit • Electronic Coupling Unit • Electronics Control Unit • Engine Change Unit • Engine Control Unit • Environmental Change Unit • Environmental Conditioning Unit • Environmental Control Unit • Équateur (code ISO de pays) • Equipment Control Unit • Error Correction Unit • European Currency Unit • Exercise Control Unit • Experiment Control Unit • Extreme Close-Up

Ecua Ecuador

Ecuad Ecuador

ECUMN European Conference on Universal Multiservice Networks

ECUP European Copyright User Platform

ECV Earthquake Controller - Vertical • École de Communication Visuelle • Electric Cargo Vehicle • Enhanced Customer View

ECVAM European Centre for Validation of Alternate Methods

ECVFP Expanded Charted Visual Flight Procedures

ECVV European Conference on Computer Vision

ECW Electron Cyclotron Wave • Enhanced Compressed Wavelet

ECWA East Central Wisconsin Astronomers (USA-WI) • Economic Commission for Western Asia

ECWAT European Consortium for Web Assurance Trust

ECWG Emergency Communications Working Group • Evaluation Coordination Working Group

ECWMT Europese Centrum voor Weersverwachtingen op Middellange Termijn

ECWS European Centre for Work and Society • Extreme Cold Weather Shelter

ECWSS Extreme Cold Weather Sleep System

EC2 En-route Command and Control

EC3SE Enhanced Command, Control, Communications, Survivability and Endurance

ed En Dergelijk • Ex Divident

Éd Éditer • Éditeur • Édition

Ed Edición • Edile • Edilizia • Edinburgh (UK-Scotland) • Edit • Edited • Editing • Editio • Edition • Editor • Educated • Education

E/D Edge Distance • Enable/Disable

E&D Engineering & Development • Evaluation & Development

ED Earth Distance • Eastern District • École Doctorale • Edge Distance • Education Department • Effective Date • Effective Dose • Election District • Electromagnetic Dissociation • Electron Device • Electron Diffraction • Elevation Data • End Delimiter • Ending Delimiter • Energy Division • Engineering and Development • Engineering Data • Engineering Demonstration • Engineering Design • Engineering Development • Engineering Directive • Engineering Directorate • Engineering Division • Entry Date • Enumeration District • Envelope Delay • Environmental Damage • Equivalent Dose • Erase Display • Erection Duration • Error Density • Error Detecting • Estimated Date • European Date • European Datum • Evaluation and Development • Event Designator • Events Dispatcher • Evolutionary Design • Evolutionary Development • Ex Dividend • Executive Director • Expedited Data • Experimental Design • Experimental Detector • Exploratory Development • Explosive Device • Extended Definition • Extended Disk • External Device • Extra Dividend • Extrahigh Density • Extralow Dispersion

EDA Early Departure Authorized • Economic Development Administration • Edge Damage Analysis • Edna Bay (USA-AK) (location identifier) • Electronic Design Automation • Electronic Differential Analyzer • Electronic Display Assembly • Electronic Document Access • Elevation Difference Accuracy • Elevation Drive Assembly • Embedded Document Architecture • Engineering Design Activity • Environmental Damage and its Assessment • Error Detector Assembly • Error and Dispersion Analysis • Estimated Date of Arrival • Estimated Date of Availability • Expert Diagnostic Analyzer • Exploratory Data Analysis • Explorative Datenanalyse

EDAC Electromechanical Digital Adapter Circuit • Error Detecting And Correcting • Error Detection And Correction

EDACS Enhanced Digital Access Communications System

EDAIS Electronic Document Automated Information System

EDAM Electronic Design And Manufacturing

EDAP Environmental Decision Alignment Process

EDAPS Electronic Data Processing System

EDAR Electronic Design Analysis Review

EDARC Enhanced Direct Access Radar Channel

EDARF Engineering Data Activity Record File

EDAS Enhanced Data Acquisition System

EDASRE Engineering Drawings Automated Storage and Retrieval Equipment

EDASRES Engineering Drawings Automated Storage and Retrieval System

EDATE Effective Date

EDAU Extended Data Acquisition unit

EDAX Energy-Dispersive Analysis of X-rays

EdB Bachelor of Education

EDB Economic Development Board • Educational Data Bank • Eldebba (Sudan) (location identifier) • Electronic Development Board • Elektronische Datenverarbeitung • Embrittlement Database • End of Data Block • Energy Database • Engineering Database • Entry Database • Environmental Data Book • Environmental Data Buoy • Ethylene Dibromide • Experience Database • Extended Dynamic Backtracking • Extensional Database • External Database • External Data Bus

EDBD Environmental Database Directory

EDBMS EXOSAT Database Management System

EDBS Expert Database System

EDBT Extending Database Technology (conference)

EDC Early Developmental Care • Economic Development Committee • Economic Development Conveyance • Edinburgh District Council (UK-Scotland) • Educational Development Center • Education in Developing Countries • Effective Date of Change • Effective Dielectric Constant • Einstein Data Center • Electrochemical Depolarized Cell • Electron Distribution Curve • Electronic Data Communications • Electronic Desk Calculator • Electronic Digital Computer • Electronic Draft Capture • Electrostatic Dust-Cyclotron (wave) • Emergency Decontamination Center • Energy Data Collection • Energy Distribution Curve • Engineering and Development Center • Engineering Data Center • Engineering Design Centre • Engineering Design Change • Enhanced Data Correction • Enterprise Development Center • Environmental Data Center • EOS Data Center • EROS Data Center • Error Detecting Code • Error Detection and Correction • Error Detection Code • Error Detection/Correction • Estimated Date of Completion • Euroavia Design Contest • EUROLAS Data Centre • European Data Center • European Defense Community • European Dialing Code • European Documentation Centre • External Database Coordination • External Data Coordination • External Data Criteria • External Device Code • Extradark Colour

EDCARS Engineering Data Computer-Assisted Retrieval System • Engineering Data Computer Automated Retrieval System • Engineering Data Control Automated Retrieval System

EDCAS Equipment Designer's Cost Analysis System

EDCATS Education Division Computer-Aided Tracking System (NASA)

EDCC Edinburgh-Durham Cluster Catalogue • Electronic Digital Control Computer • Engineering Drawing Control Center • European Dependable Computing Conference

EDCD Electronic Digital Control Device

EDCEN Education Center

EDCL Electric-Discharge Convection Laser

EDCOM Editor and Compiler

E&DCP Evaluation & Data Collection Plan

EDCP Engineering Design Change Proposal • Evaluation and Data Collection Plan

EDCPF Environmental Data Collection and Processing Facility

EDCS Electrical Distribution and Control System • Error Detection and Correction System • EURECA Dedicated Control System • Evolutionary Design of Complex Systems • Exportable Data Collection System

EDCSA Effective Date of Change of Strength Accountability

EDCT Expected Departure Clearance Time

EDCTP European Developing Countries Trials Programme

EDCW External Device Control Word

Edd Ediderunt • Editiones

EdD Doctor of Education

EDD Earliest Delivery Date • Earliest Departure Date • Electronic Data Display • Electronic Direct Debit • Electronic Document Delivery • Electronic Document Distribution • Envelope Delay Distortion • Environmental Data Directory • Erldunda (Australia-NT) (location identifier) • Estimated Delivery Date • Estimated Departure Date • Exchange of Digital Data • Expected Date of Delivery

EDDA Electronic Digital Data Acquisition

EDDC Extended Distance Data Cable • EURECA Data Distribution System

EDDD Electronic Document Development and Delivery

EDDF Error Detection and Decision Feedback

EDDI Extended Disk Driver Interface

EDDIC Experimental Design Development Integration Center

EDDP Engineering Development and Design Plan

EDDR Electron Dipole-Dipole Reservoir

EDDS Electronic Document Delivery Service • Engineering Data Support System • Experiment Data Distribution System

EDE Edenton (USA-NC) (location identifier) • Effective Dose Equivalent • Electrical Design Engineer • Elliptical Differential Equation • Emitter Dip Effect • Estimated Duration of Exposure • Exact Differential Equation

EDED Employee Development and Education Division • Error Detection Encoder-Decoder

EDELWEISS Expérience pour Détecter Les WIMPs en Site Souterrain

EDEN Encouragement au Développement d'Entreprises Nouvelles • Environmental Data European Network

EDEW Enhanced Distant Early Warning

EDEX Electrodynamics Explorer

EDF Anchorage Elmendorf (USA-AK) (location identifier) • Earliest Deadline First • ECS Development Facility • Edifício • Électricité de France • Electron Distribution Function • Empirical Distribution Function • Engineering and Development Facility • Engineering Data File • Estimated Demand Factor • European Development Fund • Event Data Format • Event Definition File • Experiment Data Facility • Expertise Distribution Function • Extensible Description Formalism • External Delay Factor

EDFA Erbium-Doped Fibre Amplifier

EDFE EVA Development Flight Experiment

EDFG Edge Device Functional Group

EDFL École Polytechnique Fédérale de Lausanne (Switzerland)

EDFP Engineering Data For Provisioning

EDFR Effective Date of Federal Regulation

EDFT EVA Demonstration Flight Test

EDG Edgewood Weide (USA-MD) (location identifier) • Educational Development Group • Emergency Diesel Generator • European Data Grid • Exploratory Development Goal

EDGAR Electronic Data Gathering, Analysis and Retrieval

EDGC Engineering Data Guidance Conference

EDGE Electronic Data Gathering Equipment • Electronic Document Gathering Environment • Enhanced Data rates for Global Evolution • Enhanced Data rates for GSM Evolution (*) • Enhancing Diversity in Graduate Education • Environmental Design Guidance for Evaluation • Explorer of Diffuse Galaxy Emission

EDGES Electronic Data/Guidelines for Element Survivability

EDGT Elevation Data Editing Terminal

EDH EPIC Data Handling • Error Detection and Handling

EDHE Experimental Data Handling Equipment

EDHS ECS Data Handling System

EDI Échange de Données Informatisé • Échange de Données Informatisées (*) • Edinburgh (Starlink internal code) • Edinburgh Turnhouse (UK-Scotland) (location identifier) • Eidgenössisches Departement des Innern (Switzerland) • Electron Drift Instrument • Electronic Data Interchange (*) • Electronic Data Interfacing • Electronic Dissemination of Information • Electronic Document Interchange • Engineering Data Interface • Environnement de Développement Intégré • Equivalent-Damage Index • ESA-Developed Instrument • Explicitly Defined Intron • Externally Dispersed Interferometer

EDIA Electronic Data Interchange Association

EDIC Engineering Data Identification and Control

EDICT Electronic Data Interchange Catalog Transaction • Engineering Data Interchange Catalog Transaction

EDID Extended Display Identification

EDIES EDC Digital Image Enhancement System • EROS Digital Image Enhancement System

Edif Edifício

EDIF Electronic Data Interchange Format • Electronic Design Interchange Format

EDIFACT Electronic Data Interchange For Administration, Commerce and Transportation

eDIG Extraplanar Diffuse Ionized Gas

EDIINT Electronic Data Internet Integration

EDIL Electronic Data Interchange between Libraries • Electronic Document Interchange between Libraries

EDILIBE Electronic Data Interchange for Libraries and Booksellers in Europe

EDIMS Electronic Data Interchange Messaging System

Edin Edinburgh (UK-Scotland)

Ed. in Ch. Editor in Chief

EDIPAS Engineering Data Interactive Presentation and Analysis System

EDIPS EDC Digital Image Processing System • EROS Digital Image Processing System

EDIS Educational Documentation and Information System • Electrical Distribution Information System • Electronic Data Imaging System • Electronic Document Information System • Engineering Data Information System • Environmental Data and Information Service (now merged with NESS into NESDIS) • Environmental Data Information Service • Erderkundungs- Daten und Informations System (Germany)

EDisCS ESO Distant Cluster Survey

EDISMC Electronic Data Interchange Standards Management Committee

Edit Edited • Éditer • Éditeur • Editing • Édition • Editor • Editore • Editorial

EDIT Electronic Dissemination of Information Technology • Engineering Data Isolation Technique • Engineering Development Integration Test • Error Deletion by Iterative Transmission

EDITAR Electronic Digital Tracking and Ranging

EDITOR ERTS Data Interpreter and Tenex Operations Recorder

EDIV Engineering Design Intent Verification

EDK El Dorado (USA-KS) (location identifier)

EdL Édition de Luxe

EDL Edit Decision List • Eldoret (Kenya) (location identifier) • Electrical Double Layer • Electric Discharge Laser • Electrodeless Discharge Lamp • Electronics Development Laboratory • Engineering Data List • Engineering Development Laboratory • Entry - Descent - Landing • Equatorial Dust Lane • Extended Disk Life • Extensor Digitorum Longus

EDLC Ethernet Data Link Control

EDLN Engineering Development Logic Network

EDLS Engineering Drawing Library System • Enhanced Dynamic Load Sensor • Enhanced Dynamic Load Server

EDLU Encoder and Decoder Lander Unit

EdM Master of Education

EDM Electrical Discharge Machined • Electrical Discharge Machining • Electric Dipole Moment • Electric Discharge Machining • Electrodischarge Machining • Electron Discharge Machining • Electronic Distance Measurement • Electronic Distance Measuring • Electronic Distance Meter • Electronic Document Management • Electronic Drafting Machine • Embryo Development Module • Energy-Density Model • Engineering Data Management • Engineering Demonstration Model • Engineering Design Model • Engineering Development Mission • Engineering Development Model • Enhanced Data Mode • Enterprise Data Management • Enterprise Data Model • Enterprise Desktop Manager • Enterprise-wide Data Model • Environmental Descriptor Manager • Equipment Deadlined for Maintenance • La Roche sur Yon (France) (location identifier)

EDMA Engineering Data Management Agency • Extended Direct Memory Access

EDMAS Environmental Dose Modeling and Analysis System

EDMCC European Conference on Distributed Memory Computing

EDMD Electronic Document Message Directory

EDMF Electric Dipole Moment Function • Extended Data Management Facility

EDMI Electronic Distance Measuring Instrument

EDMICS Engineering Data Management Information and Control System

EDMIS Education Management Information System

EDMM Erasable Disk Mass Memory

Edmn Edmonton (Canada-AB)

EDMO Engineering Data Management Officer

EDMP ECS Data Management Plan • Engineering Data Management Plan • Engineering Drawing Management Program

EDMS Electronic Data Management System • Electronic Document Management System • Engine Distress Monitoring System • Engineering Data Management System • Engineering Document Management System • Expert Dialogue Management System • Extended Data Management System

Edn Edition

EDN Edition • Electronic Design Notebook • Electronic Document and Notebook • Emergency Data Network •

Environmental Data Network • European Distribution Network • Experimental Data Network

EDNA European Data Network Agency

Ednbhg Edinburgh (UK-Scotland)

EDNES Earth Data Network for Education and Scientific Exchange

EDNS Expected Demand Not Supplied

Edo Estado

EDO Effective Diameter of Objective • Engineering Distributed Objects (workshop) • Engineering Duties Only • Engineering Duty Officer • Enhanced Data Output • Executor Director of Operations • Extended Data Out • Extended-Duration Orbiter

EDOC Effective Date Of Change • Enterprise Distributed Object Computing (conference)

EDODRAM Extended Data Out Dynamic Random Access Memory

EDOF Electronic Document Order Form

EDOMP Extended Duration Orbiter Medical Project

EDORAM Extended Data Out Random Access Memory

EDOS Effective Date Of Supply • Enhanced Disk Operating Sequence • Enhanced Disk Operating System • EOS Data and Operations System • Expected Date Of Supply • Extended Disk Operating System

EDP Early Packet Discard • Earth Dynamics Program • Éditions - Diffusion - Presse (France) • Educational Data Processing • Electricidade De Portugal • Electron Density Profile • Electron Diffraction Pattern • Electronic Data Processing • Electronic Data Processor • Electronic Display Panel • Electronic Documentation Project • Embedded Data Processor • Engineering Data Package • Engineering Design Proposal • Engineering Development Phase • Entry Date Patent • Environmental Definition Program • Environmental Differential Program • Equivalent Density Profile • Estimated Date of Publication • Event Display Process • Executive Development Program • Expedite Departure Path • Experimental Data Processing • Experiment Data Processing

EDPC Electronic Data Processing Center

EDPE Electronic Data Processing Equipment

EDPF Electronic Data Processing Facility

EDPICS Experience - Diagram - Principles - Investigate - Check - Summarize

EDPIS Electronic Data Processing and Information System

EDPITAF Educational Development Projects Implementing Task Force

EDPL Engineering Development Processor Laboratory

EDPM Electronic Data Processing Machine • Every Damm Problem Multiplied

EDPO Einstein Data Products Office

EDPRT Expert Diagnostic, Predictive, and Resolution Tools

EDPS Electronic Data Processing System • Engineering Data Pipeline System • Equipment Distribution Planning Studies • Experimental Data Processing System • Experiment Data Processing System

EDPT Electronic Data Processing Test • Enhanced Drive Parameter Table

EDPU Electrical Power Drive Unit

EDQ Erandique (Honduras) (location identifier)

EDR Early Data Release • Edward River (Australia-QLD) (location identifier) • Effective Data Rate • Effective Downwind Report • Elastic Recoil Detection • Electrodermal Reaction • Electronic Data Reader • Electronic Data Reading • Electronic Dictionary Research • Electronic Document Report • Elektronische Drosselklappenregelung • Engineering Data Record • Engineering Design Release • Engineering Design Review • Engineering Drawing Review • Environmental Data Record • Equivalent Direct Radiation • European Drawer Rack • Evolutionary Decision Review • Experiment Data Record

EDRAM Enhanced Dynamic Random Access Memory • Extended Dynamic Random Access Memory

EDRC Engineering Design Research Center

EDRE Emergency Deployment Readiness Exercise

EDRIS Engineering Data Requisition and Index System

EDRS Enforcement Document Retrieval System • Engineering Data Reproduction System • Engineering Data Retrieval System • European Data Relay Satellite

EDRSAM Extended DRSAM

EDRSS European Data Relay Satellite System

EDRT Effective Date of Release from Training

Eds Editions • Editors

EdS Educational Specialist • Einstein - de Sitter (cosmology)

EDS Earth Data System • Education Specialist • Electrical Distribution System • Electron Devices Society • Electronic Data Switching • Electronic Data System • Electronic Data Systems (USA-MA) • Electronic Dip Switch • Electronic Display System • Electronic Document Storage • Electronic Drive System • Emergency Deorbit System • Emergency Detection System • Energy-Dispersive Spectrometer • Energy-Dispersive Spectrometry • Energy-Dispersive Spectroscopy • Energy-Dispersive Spectrum • Engine Diagnostic System • Engineering Data System • Engineering Design Simulation • Engineering Design System • Engineering Development Schedule • Entry Data System • Environmental Data Service • Environmental Data System • Error Detection System • Estimated Date of Separation • European Distribution System • Exchangeable Disk Store • Executive Database System • Executive Data System • Executive Display System • EXOSAT Database System • Experiment Display System • Explosive Detection System • Explosive Device System • Express Delivery Service

EDSAC Electronic Data Storage Automatic Computer • Electronic Delay Storage Automatic Computer • Electronic Discrete Sequential Automatic Computer

EDSAT Educational Satellite

EDSC Early Deployment System Concept • Electronic Data Service Center

EDSD Electronic Document Segment Directory

EDSG Energy Demand Steering Group

EDSGC Edinburgh/Durham Southern Galaxy Catalogue

EDSI Enhanced Small Device Interface

EdSIG Education Special Interest Group

EDSL Extended Digital Subscriber Line

EDSP European Defence and Security Policy

EDSR Electronic Document Storage and Retrieval • Engineering Data Storage and Retrieval

EDSS Environment Decision-making Support System • Equipment Deployment and Storage System • Expert Decision Support System

EDST Eastern Daylight Saving Time

EDSX Electronic Digital Signal X-connect

EDT Eastern Daylight-saving Time • Eastern Daylight Time • Effective Date of Training • Electric Discharge Tube • Electronic Data Transmission • Electronic Theodolite System • Eligible Device Table • Energy Design Technique • Energy Dissipation Test • Engineer Design Test • Engineer Design Testing • Engineering Development Test • Estimated Delivery Time • Estimated Departure Time • Externally Directed Team • Extra Dispersion Triplet

EDTA Ethylene Diamine Tetraacetic Acid

EDTCC Electronic Data Transmission Communications Central • Electronic Data Transmission Control Center

Edtn Edition

EDTP Engineering Design Test Plan

Edtr Editor

EDTR Experimental, Developmental, Test and Research

EDTS Electronic Data Transfer System • Expanded Data Transfer System

EDTV Enhanced Definition Television • Extended Definition Television

Edu Education • Educational

EDU EDOS Data Unit • Education • Electrical Drive Unit • Electronic Display Unit • Engine Diagnostic Unit • Engineering Development Unit • EPIC Data Unit • Event Detection Unit • Exchange Data Unit • Experimental Development Unit

Educ Educate • Educated • Education • Educational

EDUSAT Educational Satellite

EDV Electronic Depressurization Valve • Electronic Depressurizating Valve • Elektronische Datenverarbeitung

EDVAC Electronic Discrete Variable Automatic Computer

EDVM Elektronische Datenverarbeitungsmaschine

EDVS Elektronische Datenverarbeitungssystem

EDVZ Elektronische Datenverarbeitungszentrum

EDW Edwards (USA-CA) (location identifier) • Elementary Data Word • Engineering Data Writer

EDX Energy Dispersive X-ray • En-route Data Exchange • Enterprise Data Exchange • Event-Driven Executive • Exoatmospheric Discrimination Experiment

EDXA Energy Dispersive X-ray Analysis

EDXD Energy Dispersive X-ray Diffraction

EDXF Energy Dispersive X-ray Fluorescence

EDXRD Energy Dispersive X-Ray Diffraction

EDXRFA Energy Dispersive X-Ray Fluorescence Analysis

EDXS Energy Dispersive X-ray Spectrometer • Energy Dispersive X-ray Spectroscopy

EDZ Emission Density Zoning • Europäisches Dokumentationszentrum

e&e Each & Every

ee Estonia

E-E End to End

E/E Electrical to Electrical

E&E Electrical & Electronic • Elektromashinostroenie i Elektrooborudovanie (periodical code) • Evasion & Escape

EE Early English • Edge to Edge • Electrical Engineer • Electrical Engineering • Electrodynamic Explorer • Electron Emission • Electronic Editing • Electronics Enclosure • Electronics Engineer • Electronics Engineering • Elements of Expenses • Elevator Effectiveness • Emergency Engineering • Emergency Establishment • Employment Exchange • Encircled Energy • End Effector • Endeinrichtung • End to End • Engagement Effectiveness • Engineering Effectiveness • Engineering Evaluation • Environmental Education • Environmental Engineering • Errors Excepted • Estonia (ISO country code) • Estonia (postal country code) • Estonie (code ISO de pays) • Execution Environment • Experimental Evaluation • Exponential Elevation • Extended Edition • Extended Enterprise • Eye and Ear

EEA Economic Espionage Act (USA) • Electromagnetic-Environment Analysis • Electron Electrostatic Analyzer • Electronic Engineering Association • Électronique, Électrotechnique et Automatique • Électronique et Électrotechnique Automatique • Embedded Electronic Architecture • Equal Employment Act (USA) • Essential Elements of Analysis • Établissement d'Enseignement Adapté • European Economic Area • European Economic Association • European Environment Agency • Extended Euclidian Algorithm

eeb Equation Editor Bar (file extension)

EEB Eccentric Eclipsing Binary (star) • European Environmental Bureau (*) • European Environment Bureau

EEBD Emergency Escape Breathing Device

EEC East European Country • East European Craton • Electronic Engine Control • Elements of Expenses Code • End of Equilibrium Cycle • Energy-Energy Correlation • Engineering and Education Center • Engineering Education Center • Equipment Essentially Code • Eurocontrol Experimental Centre • European Economic Community (now EU) • European Extragalactic Collaboration • Explosive-Emission Cathode • Extended Error Correction

EECA Electrostatic Energy-Charge Analyzer

EECF Earthnet ERS-1 Central Facility • ESRIN ERS Central Facility ((*))

EECL Emitter-Emitter Coupled Logic • Equivalent Electrical Cable Length

EECOM Electrical, Environmental, and Communications

EECR Electronic Entities Cataloguing Rules • Extreme Energy Cosmic Ray

EECS Electrical-Energy Conservation System • Electrical Engineering and Computer Science • Electronic Entry Control System

EECT End of Evening Civil Twilight • End-to-End Call Trace

EED Education and Evaluation Division • Effective Equivalent Dose • Electroexplosive Device • Electron Energy Distribution • Encircled Energy Diameter • Essentials Elements of Data • Exo-Earth Discoverer • Extracted Engineering Data • Needles (USA-CA) (location identifier)

EEE Electrical - Electronic - Electromechanical • Électrique - Électromécanique - Électronique • Electromagnetic Environment Effect • Electromagnetic-Environment Experiment • Electronic - Electrical - Electromechanical • Ephemeris Enhancement Endeavor • Espace Économique Européen • Essential Elements of Evaluation

EEEA Électriciens - Électrotechniciens - Électroniciens - Automaticiens

EEEC Electromagnetic Energy Environment Criteria

EEED Electronic and Electrical Engineering Division

EEEE Electronic Encyclopaedia of Examples and Exercises

EEEU End Effector Electronics Unit

EEF Encircled Energy Function • Engineering Employers Federation • Enhanced Emission Feature • Equivalent-Energy Function

EEFAE Efficient and Environmentally Friendly Aero Engine • Efficient and Environmentally Friendly Aircraft Engine

EEFI Essential Elements of Friendly Information

EEFSU Eastern Europe and Former Soviet Union

EEFT End-to-End Force Tracking

EEG Electroencephalogram • Electroencephalograph • Electroencephalographic • Electroencephalography • Electronics Engineering Group • Étiquette Électronique de Gondole • Europese Economische Gemeenschap

EEGS Environmental and Engineering Geophysical Society

EEHLL Entry Emulator High-Level Language

EEHO Either End Hop Off

EEI Edison Electric Institute • Electrical Engineer Inspector • Electroexplosive Initiator • Energy and Environment Information • Equipment-to-Equipment Interface • Espacio Europeo de Investigación • Essential Elements of Information • Estación Espacial Internacional • Exo-Earth Imager • External Environment Interface

EEIC Elements of Expenses Investment Code • Elevated Electrode Integrated Circuit • Environmental Education and Information Committee

EEID EOSDIS External Interface Document

EEIG European Economic Interest Group (*) • European Economic Interest Grouping

EEIS End-to-End Information System

EEJ Equatorial Electrojet

EEK Eek (USA-AK) (location identifier) • Kroon (Estonia) (ISO currency code)

EEL Electrical Equipment List • Electron Energy Loss • Epsilon Extension Language

EELG Extended Emission Line Gas

EELL Equivalent Electrical Lobe Length

EELR Extended Emission Line Region (*) • Extensive Emission Line Region

EELS Electron Energy Loss Spectroscopy • Electronic Emitter Location System • Emergency Escape Lighting System • European Evaluation of Language Systems

EELT Exploded Extremely Large Telescope

EELV Evolutionary Expendable Launch Vehicle • Evolved Expendable Launch Vehicle (*) • Evolving Expendable Launch Vehicle

EEM Earth Entry Module • Electrical Engineering Model • Electronic Equipment Monitoring • External Expansion Module

EEMA European Electronic Messaging Association

EEMBC Embedded Microprocessor Forum Benchmark Consortium

EEMIN European Environment Monitoring and Information Network

EEMOC European EOSDIS Mission Operations Center

EE&MP Envoy Extraordinary & Minister Plenipotentiary

EEMRL Equivalent Electrical Main Ring Length

EEMS Electrostatic Engine Monitoring System • Enhanced Expanded Memory Specification • Enhanced Extended Memory Specification • European Environment Monitoring Satellite • Evaluation of Expanded Memory System

EEN Brattleboro (USA-VT) (location identifier) • Extended Enterprise Network • Keene (USA-NH) (location identifier)

EENR Economic Evaluation of Natural Resources

EENT End Evening Nautical Twilight • Eye - Ear - Nose - Throat

EEO Elliptical Earth Orbit • End-to-End Operation • Equal Employment Office • Equal Employment Officer • Equal Employment Opportunity • European Express Organization

EEOA Equal Employment Opportunity Agency

EEOC Equal Employment Opportunity Commission

EEOO Equal Employment Opportunity Office • Equal Employment Opportunity Officer

EEOS European Earth Observation System • European Earth Observing System

EEP Early Experimental Phase • Earth Equatorial Plane • East Equatorial Pacific • Eastern Equatorial Pacific • East European Programme • Einstein Equivalence Principle • Electron-Excited Particle • Electronic Equipment Package • Elliptical Error of Probability • Explicit Equivalence Principle

EEPC European Educational Programme Coordinator

EEPG European Engineering Planning Group

EEPROM Electrically Erasable Programmable Read-Only Memory

EER Electronic Equipment Representative • Energy Efficiency Ratio • Engineering and Economic Research • Enhanced Early Retirement • Enhanced Edge Recognition • Equal Error Rate • Equilibrium Equivalent Radon • Estonian Crown (currency) • Explosive Echo Ranging • Extended Entity Relationship • External Engineering Review

EERI Earthquake Engineering Research Institute

EERIE École pour les Études et la Recherche en Informatique et Électronique (France)

EERM Établissement d'Études et de Recherches Météorologiques (France)

EERO European Environmental Research Organization

EEROC Expedited Essential Required Operational Capability

EEROM Electrically Erasable Read-Only Memory

EERP Extended Enterprise Resource Planning

EERSA East Europe Reports on Scientific Affairs (periodical code)

EERST East Europe Reports on Science and Technology (periodical code)

Eerw Eerwaarde

EES Earth and Environmental Sciences • Earth Elevation Sensor • Earth Exploration Satellite • Ejection Escape Suit • Electrical Environmental System • Electroexplosive System • Electron Energy Spectroscopy • Electron Energy Spectrum • Emergency Ejection Suit • End-to-End System • Engineering Education Scheme • Engineering Environment Services • Environment Effects Statement • Escape Ejection Suit • Escrowed Encryption Standard • Escrow Encryption Standard • European Economic Space • European Economic Sphere • European Exchange System • Expendable Explorer Spacecraft

EESA Electron Electrostatic Analyzer

EESAT Environmental Education, Science and Technology

EESD Earth Environmental Science Database

EESI Earth Environment Satellite Initiative • Earth Environment Space Initiative

EESMB Electrical and Electronics Standards Management Board

EESS Earth Exploration Satellite Service • Emergency Egress Severance System • Emergency Egress Slide System • Engineering Education Science Scheme

EEST Enhanced Ethernet Serial Transceiver • ESAnet Engineering Support Team

EET Eastern European Time • Edge Enhancement Technology • Education Equivalency Test • Electronic Engine Thrust • Engineering Evaluation Test • Entry Elapsed Time • Equator Earth Terminal • Equivalent Exposure Time • ESA Earth Terminal • Estimated Elapsed Time • Estimated Exposure Time

EETB Electronic/Electrical Termination Building

EETE European Training Technology Event

EETF Electronic Equipment Test Facility

EEU Electronic Equipment Unit • Experiment Environmental Unit • Experiment Environment Unit • Extravehicular Excursion Unit

EE.UU. Estados Unidos

EEV English Electric Valve

EEVC European Enhanced Vehicle-safety Committee

EEVL Edinburgh Engineering Virtual Library

EEVP End-to-End Virtual Path

EEVPN End-to-End Virtual Private Network

EEVT Electrophoresis Equipment Verification Test

EEWS Emergency Egress Window System

EEX Electronic Egg Exchange

EExSc Earth and Extraterrestrial Sciences (periodical code)

EEZ Electronic Exclusion Zone • Exclusive Economic Zone

Ef Efectos

E&F Equipment & Facilities

E/F Educator/Facilitator

EF Early Finish • Edge to Face • Eisenbahnfähre • Electronic Field • Electronic Filing • Electronic Format • Electronic Frontier • Elementary Function • Élement Fini • Elevation Finder • Emergency Fund • Emission Factor • Emitter Follower • Empty Field • Energy-Filtered • Enrichment Factor • Enterprise Framework • Equilibrium Flow • Érection Finale • Error Factor • Event Foreground • Expedited Forwarding • Experimental Flight • Explicit Forwarding • Exposed Facility • Extended Filament • Extrafine

EFA Early Findings Analysis • Engineering Flight Activity • Engine Fault Analysis • Eurofighter Aircraft • European Fighter Aircraft • European Food Authority (EU) • Experiment Flight Application • Extended File Attribute

EFAA Examen de Fin d'Apprentissage Artisanal

EFAC Enhanced Forward Air Controller

EFAM Extreme Floods Alarm Mission

EFAR Early Distant (galaxies program)

EFAS Electronic Flash Approach System • Enhanced Fault Alarm System • En-route Flight Advisory Service • Experimental Fiction Analysis System

EFATO Engine Failure After Takeoff • Engine Failure At Takeoff

EFB Eight Fathom Bight (USA-TX) (location identifier) • European Federation of Biotechnology

EFC Electronic Fee Collection • Electronic Filing Cabinet • Equivalent Full Charge • Estimated Final Cost • Expect Further Clearance • Extreme Forward Calorimeter

EFCC Externally Fired Combined Cycle

EFCE European Federation of Chemical Engineering

EFCI Explicit Forward Congestion Indication • Explicit Forward Congestion Indicator

EFCS Earth-Fixed Coordinate System • Electronic Filmless Camera System • Electronic Flight Control System • Extended Faceted Classification Schema

EFD Earliest Finish Date • Early Fault Detection • Electric Fields Detector • Electronic Fill Device • Electronic Forms Designer • End Frame Delimiter • Energy Flux Density • Engineering Flow Diagram • Excess Facilities Disposition • External Floppy Disk • External Floppy Drive • Houston Ellington (USA-TX) (location identifier)

EFDA European Federation of Data (processing) Associations • European Fusion Development Agreement

EFDD External Floppy Disk Drive

EFDP European Federation of Data Processing

EFE Einstein's Field Equation • Electronic Field Experiment • Engineering Feasibility Effort

EFEI Escola Federal de Engenharia de Itajuba (Brazil)

EFET École Française d'Enseignement Technique (France)

Eff Effect • Effective • Effetto • Efficace • Efficiency

EFF Eggleton - Faulkner - Flannery (formalism) • Efficiency • Electronic Frontier Foundation • Essential Function Failure • Experimental Forecast Facility

EFFECT Exchange Format for Electronic Components and Texts

EFFF Electric Field Flow Fractionation

Effy Efficiency

EFG Earth Fixed Geocentric (coordinates) • Efogi (Papua New Guinea) (location identifier) • Electric-Field Gradient • Entscheidungen der Finanzgerichte • Evidence Flow Graph

EFGC École Française de Gestion Commerciale (France)

EFGF Epitaxial Ferrite Garnet Film

EFH Earth Far Horizon • Einfamilienhaus • Engine Flight Hour

EFI Electromechanical Frequency Interference • Electronic Fuel Injection • Electronics For Imaging • Enrico Fermi Institute (USA-IL) • Error Frame Indicator • Error-Free Interval • Exploding Foil Initiator

EFIA Electric Field Intensity Analysis

EFIC European Fuel Information Centre

EFIE Electric Field Integral Equation

EFILA European Forum for Implementors of Library Applications

EFIN Environmental Financing Information Network

EFIS Electronic Flight Information System (*) • Electronic Flight Installation System • Electronic Flight Instrumentation System • Electronic Flight Instrument System • ESA Financial and Invoicing System

EFIT Engineer - Furnish - Install - Test

EFK Newport (USA-VT) (location identifier)

Efl Effluent

EFL Effective Focal Length • Emitter Follower Logic • Engineering First Light • English as a Foreign Language • Equivalent Focal Length • Executive Flight Level • Extended Fortran Language • External Financial Limit • Kefallinia (Greece) (location identifier)

EFLA Education Film Library Association (USA) • European Feeder Link Antenna

EFLC European Foundation for Library Cooperation

EFM Eight-to-Fourteen Modulation • Energy Flux Model • Ethernet in the First Mile • Extended Functionality Module

EFMD European Foundation for Management Development

EFMS Executive Financial Management System • Experimental Flight Management System • Expert Fault Management System

EFO East Fork (USA-AK) (location identifier) • Error - Freak - Oddity

EFOG Enhanced Fiber-Optic Guided

EFOIA Electronic Freedom Of Information Act (USA)

EFOSC ESO Faint Object Spectrograph and Camera

EFOV Effective Field Of View

EFP Éducation et Formation Professionnelle • Effective Fragment Potential • Einstein - Fokker - Planck (equation) • Electronic Field Production • Emulating Floating Point • Error-Free Performance • ESA-Furnished Property • European Fusion Programme

EFPA Educational Film Producers' Association (UK)

EFPD Effective Full-Power Day

EFPH Effective Full-Power Hour • Equivalent Full-Power Hour

EFPISS Extrinsic Fabry-Pérot Interferometric Strain Sensor

EFPOP Electrical Field Poling by Optical Polarimetry

EFPY Effective Full-Power Year

EFQ Estimated Fuel Quantity

EFQM European Foundation for Quality Management

EFR Emerging Flux Region • Enhanced Full Rate • Equipment Facility Requirements • Estimated Fuel Remaining • European Fast Reactor • Experimental Fast Reactor

EFRAP Electronic Feeder Route Analysis Program

EFRAPS Extracted Feature Rectification and Processing System

EFRR Executive Flight Readiness Review

EFS Earth-Fixed System • Économie Familiale et Sociale • Eesti Füüsika Selts (Estonia) • Electronic Filing System • Electronic Frequency Selection • End Frame Sequence • End-of-Frame Sequence • Enhanced Filing System • Enhanced Flight Screener • Epoch Folding Search • Erdfunkstelle • Error-Free Seconds • European Forward Study • Extensible File System

EFSM Ethernet/FDDI Switching Module • Extended Finite State Machine

EFSN European Federation of Scientific Networks

EFSPI European Federation of Statisticians in Pharmaceutical Industry

EFSS Eastern Formosa Summer School

EFSSS Engine Failure Sensing and Shutdown System

EFSY ESA Financial System

Eft Efecto

EFT Earliest Finish Time • Early Finish Time • École de Formation Technique • Effective Field Theory • Electronic Field Trip • Electronic Funds Transfer • End-to-end Force Tracking • Engineering Feasibility Testing • Enhanced File Transfer • Euro File Transfer • Expected Flare of Temper

EFTA École de Formation Technique de l'Aéronautique • European Free Trade Association

EFTF European Frequency and Time Forum

EFTO Encrypted For Transmission Only

EFTPoS Electronic Funds Transfer at Point of Sale

EFTPOS Electronic Funds Transfer at Point Of Sale

EFTS Electronic Funds Transfer System • Elementary Flying Training School

EFTSU Equivalent Full Time Student Unit

EFU Energetic Feed Unit

EFUV Extreme and Far Ultraviolet

Efvo Efectivo

EFW Electric Fields and Waves (experiment) (Cluster) • Energy From Waste • Jefferson (USA-IA) (location identifier)

EFX Effects

EFY End of Fiscal Year

EFZG Entgeltfortzahlungsgesetz

eg Egypt • Exempli Gratia • Ex Grege • For Example

eG Eingetragene Genossenschaft

Eg Egypt • Egyptian

EG Eggen • Egypt (ISO country code) • Égypte (code ISO de pays) • Electronographic • Elliptical Galaxy • Encyclopaedia Galactica • Erdgeschoß • Europäische Gemeinschaft • European Group • Europeiska Gemenskaperna • Europese Gemeenschap • Execution Guidance • Expert Group • Exponentiated Gradient • External Gradient

EGA Earth Gravity Assist • Earth-Grazing Asteroid • Eingabegerät für Alphanumerik • Electron Generator Assembly • Engati (Papua New Guinea) (location identifier) • Enhanced Graphics Adapter • Enhanced Graphics Array • Evolved Gas Analyser • Evolved Gas Analysis • Extended Graphics Adapter • Extended Graphics Array

EGAD Earth Gravity Assist Date • Electrical (power) Generation And Distribution • Extended Generation And Distribution

EGAL European Group of Astronomy Librarians

EGAO Einführungsgesetz zur Abgabenverordnung

EGARD Environmental Gamma-ray And Radon Detector

EGAS European Group on Atomic Spectroscopy • European Guaranteed Access to Space

EGB Educación General Básica • Electronic Gimbal Box • Escher - Gödel - Bach (trilogy)

EGBGB Einführungsgesetz zum Bürgerlichen Gesetzbuch

EGBGR Extragalactic Background Radiation

EGC Bergerac (France) (location identifier) • École de Gestion et de Commerce • École du Génie Civil • Education in Global Change • Elliptical Galaxies to Clusters (sequence) • Engine Gimbal Control • Enhanced Group Call • Experiment Ground Computer

EGCAT Extragalactic Catalog

EGCM Eddy-resolving General Circulation Model

EGCS Environmental Generation and Control System

EGD Electrogravimmetric Device • Evolved Gas Detection

EGDW Engineering Graphics Display Workstation

EGE Enhanced Greenhouse Effect • Epsilon Geminids (meteor stream) • European Group on Ethics • Vail - Eagle County (USA-CO) (location identifier)

EGEAS Electric Generating Expansion Analysis System

EGF Electrical Grapple Fixture • Epidermal Growth Factor

EG&G Edgerton, Germershausen & Grier

EGG European GOCE Gravity (consortium) • Evaporating Gaseous Globule

EGGF European Guidance and Guarantee Fund

EGH Electrogastrogram • Europäischer Gerichtshof

EGHIIR Extragalactic Giant HII Region

EGI Embedded GPS/Inertial • Exhaust Gas Ignition • Extended Gaussian Image • Valparaiso Duke (USA-FL) (location identifier)

EGIC Éditeur Graphique Interactif de Calculs

EGIL Electrical General Instrumentation and Lighting

EGInsO Einführungsgesetz zur Insolvenzordnung

EGIS European Geographical Information System

EGJ Extragalactic Jet

EGKS Europese Gemeenschap voor Kolen en Staal

EGL Extragalactic Light • Neghelli (Ethiopia) (location identifier)

EGLE Extreme Gravitational Lensing Event

EGLS Estimated Generalized Least Squares

EGM Earth Gravitational Model • Eggen Proper Motion • Electronic Governor Module • Electrostatics of Granular Material • Estudio General de Medios • Extraordinary General Meeting • Sege (Solomon Islands) (location identifier)

eGmbH Eingetragene Genossenschaft mit Beschränkter Haftpflicht

EGMS Escuela de Geología, Minas y Geofísica (UCV)

EGMT Elapsed Greenwich Mean Time

EGN Eingabegerät für Numerik • Einzelgebührennachweis • Enhanced Generic Naming • Geneina (Sudan) (location identifier)

EGNOS European Geostationary Navigation Overlay Service • European Geostationary Navigation Overlay System

EGO Belgorod (Russia) (location identifier) • Eccentric Geophysical Observatory • Économie - Gouvernements - Organisations • Edit Graphics Object • European Gravitational Observatory (Italy) • EUVE Guest Observer • Experimental Geophysical Orbiting • Extragalactic Object

EGOC EUVE Guest Observer Center

EGOCS EUVE Guest Observer Center Software

EGOP European Group for Optics and Photonics

EGORD Experiment Ground Operations Requirements Document

EGOS European Group on Ocean Stations

EGOWG EUVE Guest Observer Working Group

EGP Eagle Pass - Maverick County (USA-TX) (location identifier) • Enhanced Graphics Processor • European Glaciological Programme • Experimental Geodetic Payload • Extensive Gateway Protocol • Exterior Gateway Protocol (*) • External Gateway Protocol • Extrasolar Giant Planet • Pound (Egypt) (ISO currency code)

EGPGV Eurographics workshop on Parallel Graphics and Visualization

EGPM European Global Precipitation Mission

EGPRS Enhanced General Packet Radio Service

EGPWS Enhanced Ground Proximity Warning System

Egr Egregio

EGR Earned Growth Rate • Electronic Governor Regulator • Embedded GPS Receiver • Enhanced Gate Reader • Exhaust Gas Recirculation

EGREP Extended Global Regular Expression Print

EGRET Energetic Gamma-Ray Experiment Telescope

EGRS Extragalactic Radio Source

EGRU Enhanced Grid Reference Unit

EGS Earth Gateway Station • Earth Ground Station • Egilsstadir (Iceland) (location identifier) • Electron and Gamma-ray Shower • Electron Gamma-ray Shower • EOS Ground System • European Geophysical Society • Expeditionary Ground Station • Experimental Geodetic Satellite

EGSE Electrical Ground Support Equipment

EGMS Enhanced Ground Station Module

EGSMA Egyptian Geological Survey and Mining Authority

EGSO European Grid of Solar Observations

EGSP Electronics Glossary and Symbol Panel

EGStGB Einführungsgesetz zum Strafgesetzbuch

EGT Einstein-invariant Gauge Theory • Elapsed Ground Time • Estimated Ground Time • European Geotraverse • Exhaust Gas Temperature

EGTT Enhanced GPS Telemetry Transmitter

EGTV Efficient Global Transaction for Video

EGV Eagle River (USA-WI) (location identifier) • External Gelation of Uranium

EGW Edge Gateway • Effective Gravitational Wave • Electro Gas Welding • Equipment Ground Wire • Erste Gemeinnützige Wohnungsgesellschaft (Austria)

EGX Egegik (USA-AK) (location identifier)

EGY Egypt (ISO country code) • Égypte (code ISO de pays)

Egypt Egyptian

EH Ehrenhalber • English Heritage • Era Hispánica • Event Header • Evolvable Hardware • Sahara Occidental (code ISO de pays) • Sahara (Western_) (ISO country code) • Western Sahara (ISO country code)

EHA Earth Horizon Angle • Electro-Hydraulic Actuator • Electro-Hydrostatic Actuator • ESO Historical Archive

EHAPE Environmental Health Action Plan for Europe

EHAZ Equatorward-moving High-latitude Activity Zone

EHB Electro-Hydraulic Brake • Electronic Handbook • Extended Horizontal Branch • Extreme Horizontal Branch (*)

EHBO Eerste Hulp Bij Ongelukken

EHC Effective Heat Capacity • Electrical Heating Control • Electro-Hydraulic Control • Environmental Health Committee • Explosive Hazard Classification

EHCR Electronic Health Care Record

EHD Elastohydrodynamique • Electrohydrodynamical • Electrohydrodynamically • Electrohydrodynamics

EHDC EMP-Hardened Dispersal Communications

EHDM Enhanced Hierarchical Development Methodology

EHe Extreme Helium (star)

EHE Extended HII Envelope • Extraordinary Hall Effect • Extremely High Energy

EHECR Extremely-High-Energy Cosmic Ray

EheG Ehegesetz

EHEI École des Hautes Études Internationales

Ehem Ehemalig • Ehemals

EHEP Experimental High-Energy Physics

E-HEPNET European High-Energy Physics Network

EHEPNET European High-Energy Physics Network

EHESS École des Hautes Études en Sciences Sociales (France)

EHF Elevated HF Antenna • Encoding Header Field • European Headache Federation • Extremely High Frequency

EHFA Extremely High Frequency Appliqué

EHFR Extremely High Frequency Radio

EHI Eurohandelsinstitut

EHIA Environmentally High Impact Area

EHIHS Expert System Information Handling System

EHII Extended HII (region)

EHKP Einheitliche Höhere Kommunikationsprotokolle

EHL Effective Half-Life • Elastohydrodynamic Lubrication • El Bolsón (Argentina-RN) (location identifier) • Electron-Hole Liquid

EHLLAPI Emulator High-Level Language Applications Programming Interface (*) • Enhanced High-Level Language Applications Programming Interface

EHLSC European High-Level Scientific Conference

EHM Cape Newenham (USA-AK) (location identifier) • Electron Holographic Microscope • Extremely Hazardous Material

EHMD Expanded Helmet Mounted Display

EHMO Extended Hückel Molecular Orbital

EHO Environmental Health Office • Environmental Health Officer

EHOT External Hydrogen/Oxygen Tank

Ehp Equivalent Horsepower

EHP Effective Horsepower • Electric and Hybrid Propulsion • Electric Horsepower • Electron-Hole Pair • Electron-Hole Plasma • Equivalent Horsepower • Error Handling Package • Estimated Horsepower

EHPO Enhanced High-Performance Option

EHR Educational and Human Resources • Education and Human Resources • Electronic Health Record • Electronic Human Resource

EHS Earth Horizon Scanner • Earth Horizon Sensor • Environmental Health Service • Environmental Health Standard • Euro-Hermespace • European Home System • European Hybrid Spectrometer • Extrahigh Strength • Extremely Hazardous Substance

EHSD Existing Hardware/Software Database

EHSI Electronic Horizontal Situation Indicator

EHT East Hartford - Rentschler (USA-CT) (location identifier) • Electrical High Tension • Electrically Heated Thruster • Electrothermal Hydrazine Thruster • End-to-end Hardware Testbed • End-to-end Hardware Testing • Erector Holding Time • Extrahigh Tension

EHTO European Health Telematics Observatory

EHTR Emergency Highway Traffic Regulation

EHTS Emacs-based Hypertext System

EHTU Enhanced Handheld Terminal Unit • Expanded Handheld Terminal Unit

EHU Environmentally Homogeneous Unit • Extension Header Unit

EHV Electric and Hybrid Vehicle • Electro-Hydraulic Valve • Extrahigh Voltage • Extremely High Velocity

EHW Evolvable Hardware • Extremely Hazardous Waste

EHX Experiment Heat Exchanger

EHz Exahertz

E&I Engineering & Installation

E-I Electronic Installation • Engineering-Installation

EI Earth Interface • East Indian • East Indies • Electrical Insulation • Electromagnetic Interaction • Electromagnetic Interference • Electronic Installation • Electronic Integration • Electronics Interface • Electron Impact • Electron Ionization • Emission Index • Employee Involvement • End Injection • End Item • Energy Intake • Engineering Index • Engineering Information • Engineering Installation • Engineering Instruction • Engineering Investigation • Enterprise Integration • Entity Identifier • Entreprise Individuelle • Entry Interface • Environmental Impact • Exercise Item • Exposure Index • Extended Interface • External Input • External Interconnect • External Interface • External Interrupt • External Irradiation • Ireland (international aircraft registration code)

EIA Economic Impact Assessment • Education Initiative in Astronomy • Eia (Papua New Guinea) (location identifier) • Electrical Industries Association • Electrical Integration Assembly • Electronic Industries Association (USA) • Energy Information Administration (DOE) • Engineering In Astronomy • Environmental Impact Analysis • Environmental Impact Appraisal • Environmental Impact Assessment • Environment Institute of Australia • Equatorial Ionization Anomaly • Equatorial Ionospheric Anomaly • European Information Association • Experimental and Integrative Activities

EIAC Environmental Information and Analysis Center (USA)

EIAF Expanded Inward Access Feature

EIAJ Electronics Industry Association of Japan

EIAM École Internationale des Affaires et du Management

EIAO Enseignement Intelligemment Assisté par Ordinateur

EIAP Environmental Impact Analysis Process • Environmental Impact Analysis Program • Extended Interoperable Application Profile

EIARD European Initiative for Agricultural Research for Development

EIASN End Item Assembly Sequence Number

EIAT End Item Acceptance Test

EIB Electrical Interface Box • Engineering Information Bureau • Europäische Investitionsbank • European Investment Bank • Evaporating Infalling Body

EIC Earned Income Credit • Editor-In-Chief • Educational Information Center • Education - Information - Communication • Electronic Imaging Center • Electrostatic Ion Cyclotron • Element Identification Code • Employer Identification Code • End Item Code • End Item Contract • Engineer-In-Charge • Engineer-In-Chief • Engineering and Integration Contractor • Engineering Information Center • Engineering Installation Center • Engineering Institute of Canada • Environment Information Centre • Equatorial Infrared Catalogue • Equipment Identification Center • Equipment Identification Code • Euro Info Centre • Experimental Intercom • External Inverse Compton

EICAS Engine Indicating and Crew Alerting System • Engine Indicating and Crew Alert System • Engine Indication and Crew Alerting System • Engine Indication and Crew Alert System • Engine Instrument and Crew Alerting System • Engine Instrument and Crew Alert System (*)

EICC Edinburgh International Congress Centre (UK-Scotland) • Emergency Information and Coordination Center • Euro Info Correspondance Centre

EICD Experiment Interface Control Document

EICIT Electrical Interface Continuity/Isolation Test

EICM Electronic Information and Communication Module • Employer's Inventory of Critical Manpower

EICMC Enterprise Integration Corporate Management Center

EICOS European Initiative for Communicators in Science

EICS Energetic Ion Composition Spectrometer

EICW Electrostatic Ion Cyclotron Wave

EID Earliest Induction Date • Electron-Induced Disordering • End Item Delivery • End Item Description • End Item Documentation • Engineering Installation Division • Entry Identification Number • Evolution of Interstellar Dust • Experiment Interface Document

EIDE Enhanced Integrated Drive Electronics (*) • Enhanced Integrated Drive Electronics • Extended Integrated Device Electronics

EIDERS Externally Ionized Disks in the Environs of Radio Sources

EIDES ETL Image Database for Experimental Studies

EIDF Electronic Intercept and Direction Finding

Eidg Eidgenössisch

EIDOSTAS European Infrastructure for Document Supply in Technology and Applied Sciences

EIDP End Item Data Package

EIDR External Intermediate Design Review

EIDS Electronic Information Delivery Service • Electronic Information Delivery System • Electronic Information Display System • Electronic Integrated Display System • ESA IUE Data Server

EIDSO Engineer Information and Data Systems Office

EIE Electron Impact Excitation • Elutable Ion Exchange • Eniseysk (Russia) (location identifier) • Equilibrium Isotope Effect • Equipment Inventory Evaluation • European Industrial Engineering

EIEB Enterprise Integration Executive Board

EIED Engine Instrument Electronics Display

EIEIO Environmentally Integrated Effects Initiatives for Operations

EIES Electronic Information Exchange System

EIF Electronic Interface Format • Electron Impact Fluorescence • European Internet Foundation • European Investment Fund • External Interface • External Interface File

EIFA Element Interface Functional Analysis

EIFOV Effective Instantaneous Field Of View

EIFS Economic Impact Forecast System • Equalized Integral Field Spectroscopy

Eig Eigenlijk

EIG Engineering and Installation Group • Exchange Information Group

EIGRP Enhanced Interior Gateway Routing Protocol • Enhanced Internet Gateway Routing Protocol

Eigtl Eigentlich

EigZ Eigenheimzulage

EigZulG Eigenheimzulagengesetz

EIH Einasleigh (Australia-QLD) (location identifier) • Electronic Information Handling

EII Early Industry Involvement • Electron Impact Ionization • Électronique et Informatique Industrielle • Environmental Investigations Instruction

EIIA European Information Industry Association

EIII Electrical Industry Information Institute

EIISO Edison International Infrared Space Observatory (ESA)

Eil Eiland • Eilanden

EIL Electronics Integration Laboratory • Electron Injection Laser • Emploi d'Initiative Locale • Fairbanks Eielson (USA-AK) (location identifier)

EILIS Electronic Integrated Library and Information Service

EIM Electronic Image Management • Electron Imaging Management • End Item Manager • Engineering Information Management • Environmental Information Management • Ethernet Interconnection Module • Excitability Inducing Material

EIMB Engelhardt Institute for Molecular Biology (Russia)

EIMI Environmental Interactions Measurement Investigation

EIMS Employment Information in the Mathematical Sciences • Enterprise Integrated Management System

EIMSURS Equipment Inventory, Multiple Status, and Utilization Reporting System

EIMU Environmental Information Management Unit

EIN Educational Information Network • Education Information Network • Eindhoven (Netherlands) (location identifier) • Electronic Information Network • Employer Identification Number • Engineering Information Network • Enterprise Information Network • Environmental Information Network • Equipment Identification Number • Equivalent Input Noise • European Informatics Network

E-in-C Engineer in Chief

EinC Engineer in Chief

E. Ind. East Indian • East Indies

EINECS European Inventory of Existing Commercial Chemical Substances

EINET Enterprise Integration Network

EINETS Enterprise Integration Network System

Einf Einführung

EinfG Einführungsgesetz

Einl Einleitung

EINS European Information Network Service

Einschl Einschließlich

Einw Einwohner

EIO Enhanced Input/Output • Evaluation and Integration Office • Extended Interaction Oscillator • Extended Interoperability

EIOL European Infrastructure for Open Learning

EIONet Environmental Information and Observation Network (EU)

EIOOTS European In-Orbit Operations Technology Symposium

EIOP Extended Input/Output Processor

EIP Early Intervention Project • Easy Introduction and Penetration • Economic Inventory Procedures • Electronic Information Publishers • End Item Parameter • Engineering Installation Package • Engineering Installation Plan • Engineering Internship Program • Enterprise Implementation Plan • Environmental Interface Processor • Equipment Improvement Project • Evolutionary Implementation Plan • Excess Inventory Program • Experiment Implementation Plan • Extended Internet Protocol

EIPA European Institute of Public Administration

EIPD Everything In Place Date

EIPF End Item Parameter File

EIPL Environnement Intégré de Production de Logiciel

EIPO European IGAC Project Office

EIPPCB European Integrated Pollution and Prevention and Control Bureau

EIPS Electronic Information Products and Services • Electron Impact Point Source

EIPT Échelle Internationale Pratique de Température • Executive Integrated Product Team

EIQ End Item Quantity

EiR Engineering in Reverse

EIR Earned Income Relief • End Item Requirement • Environmental Impact Report • Equipment Identity Register • Equipment Improvement Recommendation • Equipment Improvement Report • Excess Information Rate • Experiment Integration Review • Extended Information Rate • External Information Requirement • External Intelligence Requirement • Extreme Infrared • Ireland (international vehicle registration code)

EIRB Extragalactic Infrared Background

EIRDT External Interface Rights Description Table

EIRENE European Integrated Radio Enhanced Network

EIRI Energy Information Resource Inventory (DoE)

EIRMA European Industrial Research Managers Association

EIRO European Industrial Relations Observatory • European Intergovernmental Research Organisation

EIRP Earth Instantaneous Radiated Power • Effective Isotropically Radiated Power • Effective Isotropic Radiated Power (*) • Equivalent Isotropically Radiated Power • Equivalent Isotropic Radiated Power

EIRR External Independent Readiness Review

EIRS Equipment Inventory Reporting System

EIRSG European Information Retrieval Specialist Group

EIRT Executive Independent Review Team

Eis Eisern

EIS Beef Island (British Virgin Island) (location identifier) • Earth Imaging System • Economic Information Service • Economic Information System • Effluent Information System • Electrical Integration System • Electronic Information Security • Electronic Information Service • Electronic Instrument System • Ellipsoïde Isotherme Singulier • E-mail Interface Server • Employee Information System • End Item Specification • Energy Information System • Engineering Information Service • Engineering Information System • Engineering of Intelligent Systems • Enterprise Information System • Enterprise Integration System • Environmental Imaging Spectrometer • Environmental Impact Statement • Environmental Impact Study • Environmental Information Service • Environmental Information System • Environment-Induced Scattering • ESO Imaging Survey • EUV Imaging Spectrometer • Evaluation Information System • Executive Information Summary • Executive Information System • Expanded Inband Signaling • Explosives Inspection Section • Extended Instruction Set • Tortola (British Virgin Island) (location identifier) • Virgin Gorda (British Virgin Island) (location identifier)

EISA Extended Industry Standard Adapter • Extended Industry Standard Architecture (*)

EISAC European Imaging Spectrometer Aircraft Campaign

EISc Economics and Information Sciences

EISCAT European Incoherent Scatter

EISE Extendible Integration Support Environment

EISF Escape Integral Source Function

EISG Experimental Investigation of Spacecraft Glow

EISMINT European Ice Sheet Modeling Initiative

EISN Environmental Information and Support Network

EISO Edison Infrared Space Observatory • Environmental Information System Office (ORNL)

EISP Eastern Iowa Star Party (USA-IA) • Embedded Information Security Product • End Item Sale Price • Engineering and Installation Support Plan

EISS Economic Impact Study System • Electronic Integrated Sensor Suite • European Intelligence Support System • Executive Information Systems Seminar

EISTI École Internationale des Sciences du Traitement de l'Information

EIT Échelle Internationale de Température • Economy In Transition • Electrical Impedance Tomography • Electromagnetically-Induced Transparency • Électronique - Informatique - Télécommunications • Encoded Information Type • Engineer In Training • Enterprise Integration Technologies • Entry Interface Time • Equatorial Ionosphere-

Thermosphere • EUV Imager Telescope • EUV Imaging Telescope (*)

EITA European Information Technology Association

eITB Electron Internal Thermal Barrier

EITF Emerging Issue Task Force

EITI Emergent Information Technologies Inc.

EITO European Information Technology Observatory

EIU Economic Intelligence Unit • Economist Intelligence Unit • Electronic Interface Unit • Engine Interface Unit • Ethernet Interface Unit • Event Identification Unit • (The) Economist Intelligence Unit

EIUF European ISDN User Forum

EIV Error-In-Variable

EIVE Electrical Interface Verification Equipment

EIVM Elektronische Informationsverarbeitung Maschine

EIVT Electrical and Instrumentation Verification Test • Electrical Interface Verification Test • Electronic Installation Verification Test

EIW Engineering Installation Wing • Enhanced Imagery Workstation

EIWB End Item Workbook

EIWG Earth Imaging Working Group • Experimentation Integration Working Group

EIX Electrochemical Ion Exchange

EIY Ein Yahav (Israel) (location identifier)

EJ Electronic Journal • Exajoule • Ireland (international aircraft registration code)

EJA Barrancabermeja (Colombia) (location identifier)

EJAM European Journal of Applied Mathematics

EJASA Electronic Journal of the Astronomical Society of the Atlantic

EJB Enterprise Java Beans

EJC Electronic Journal of Communications • Engineering Joint Council

EJCSC European Joint Committee of Scientific Cooperation

Ejem Ejemplar • Ejemplo

EJFET Enhancement-mode Junction Field-Effect Transistor

EJH Wedjh (Saudi Arabia) (location identifier)

EJIC Education and Job Information Center

EJIS European Journal of Information Systems (periodical code)

EJMF European Journal of Mechanics - Fluids (periodical code)

EJMS European Journal of Mechanics - Solids (periodical code)

Ejn Ejection

EJN Electronic Journal Navigator (Blackwell)

EJO Electronic Journals Online

EJP Effacement des Jours de Pointe • Exo-Jovian Planet

EJPh European Journal of Physics (periodical code)

EJRS Emergency Joint Reporting System

EJRT Engineering Job Release Ticket

EJS Electric Jet System

EJT Extended Joint Test

Ejtr Ejector

Ejud Ejusdem

EJV External Jugular Vein

EK Armenia (international aircraft registration code) • Eastman Kodak • Edgeworth-Kuiper (belt) • Eerste Kwartier • Eerstkomend • E-Kernel • Eskimo • Event Clock • Event Kernel

EKA Eureka Murray (USA-CA) (location identifier)

EKAW European Knowledge Acquisition Workshop

EKB Edgeworth-Kuiper Belt • Ekibastuz (Kazakhstan) (location identifier)

EKBO Edgeworth-Kuiper Belt Object

EKC Eastman Kodak Co.

EKD Electronic Key Distribution • Elkedra (Australia-NT) (location identifier) • Evangelische Kirche in Deutschland

EKE Eddy Kinetic Energy • Ekereku (Guyana) (location identifier) • Enhanced Kinetic Energy • Equatorial Kerr Effect

EKEN Education and Knowledge Exchange Network

EKF Extended Kalman Filter

EKG Effective Kilogram • Electrocardiogram • Électrocardiogramme • Elektrokardiogramme

EKH Emerson - Klein - Haslam (algorithm)

EKI Elkhart (USA-WI) (location identifier)

EKIO Experience - Knowledge - Information - Object

EKM Empirical Kinetic Model

EKMS Electronic Key Management System

EKN Elkins (USA-WV) (location identifier)

EKNF Elementary Key Normal Form

EKO Edgeworth-Kuiper Object • Elko (USA-NV) (location identifier)

EKS Electrokinetic Surveying • Electronic Keyboard System • Energy Kernel System • Europäische Kulturstiftung

EKT Envelope Kanalteiler • Eskilstuna (Sweden) (location identifier)

EKTS Electronic Key Telephone Set

EKU Eastern Kentucky University (USA-KY)

EKV Exoatmospheric Kill Vehicle

eKW Equivalent Kilowatt

EKW Equivalent Kilowatt

EKX Elizabethtown (USA-KY) (location identifier)

EKZ Europäisches Kulturzentrum (Switzerland)

el Greek (ISO language code)

El Elected • Electo • Electric • Electrical • Électricité • Electricity • Électrique • Electroluminescent • Electron • Électron • Electronic • Électronique • Element • Elevated • Elevation • Elongation

E/L Entry/Landing

EL Earth Limb • Education Level • Elastic Limit • Electrical Laboratory • Electroluminescence • Electroluminescent • Electronics Laboratory • Electron Lattice • Element • Element Layer • Elevation • Emitter Locator • Engineering Laboratory • Epitaxial Layer • Equipment Laboratory • Erase Line • Europhysics Letters (periodical code) • Excess Luminosity • Executive Level • Extremely Low • Liberia (international aircraft registration code)

ELA Eagle Lake (USA-TX) (location identifier) • Effective Launch Azimuth • Effective Learning Adviser • Eight-Letter Acronym • Electronic Learning Aid • Electron Linear Accelerator • Engineering - Logistics - Acquisition • Ensemble de Lancement Ariane

ELAB Electronic Laboratory

ELAC Economically Less Advantaged Country • European Lidar Airborne Campaign

ELACS Extended Life Attitude Control System

ELAG European Library Automation Group

ELAIS European Large Area Infrared Survey • European Large Area ISO Survey (*)

ELAN Emulated Local Area Network • Environment in Latin-America Network • Error Logging Analysis

ELAP EtherTalk Link Access Protocol

Elas Elasticity

ELAS Earth Laboratory Applications Software

Elast Elastic (periodical code)

ELAWS Engineering Logistics Automated Workflow System

Elb Elbow

ELB El Blanco (Colombia) (location identifier) • Electrical Load Bank • Emergency Locator Beacon • Environment Liaison Board

ELBA Emergency Location Beacon

ElBah Elektrische Bahnen (periodical code)

ELBBO Extended Life Balloon-Born Observatory

ELBL Extended Line-By-Line

ELC Earth Limb Clutter • Echo Limiting Contour • Eclipsing Light Curve • Elcho Island (Australia-NT) (location identifier) • Electronic Learning Community • Electron Loss to Continuum • Embedded Linking and Control • Emission-Line Cloud • Emission-Line Cluster • End of Line Code • Environmental Law Centre • Environment Law Centre • Environment Liaison Centre • Execute Launch Command • Executive Leadership Course • Extended Life Cooler

ELCA Earth Control Landing Assembly

ELCAS Elevated Causeway System

ELCD Electrolytic Conductivity Detector

ELCI Environment Liaison Centre International

ELCID European Living Conditions Information Directory

ELCO Environnement Limité Concentrique Ordonné

ElCom Electronics Communications of Japan (periodical code)

ELCP Emission-Line Coronal Photometer • Extended Linear Complementary Problem

Elctd Electrode

Elctlt Electrolyte

Elctrn Electron

Elctrochem Electrochemical

Eld Elder • Eldest

ELD Earth Launch Date • Economic Load Distribution • El Dorado (USA-AR) (location identifier) • Electroluminescent Display • Electronic Literature Directory • Electron Landau Damping • Emitter Location Data • Excedente de Libre Disposición • Extended Low-Density • Extralong Distance

ELDA European Language resources Distribution Agency

ELDO European Launcher Development Organization (merged now with ESRO into ESA)

ELDORA Electra Doppler Radar

ELDP Executive Leadership Development Program • European Lake Drilling Programme

ELDS Editorial Layout Display System

ELDV Electrically-Operated Depressurization Valve

ELDWIM Extended Low-Density Warm Ionized Medium

Ele Electo

ELE Earth Limb Experiment • Element • El Real (Panama) (location identifier) • Extinction Level Event

Élec Électricité • Électrique • Électron • Électronique

Elec Election • Elector • Electoral • Electric • Electrical • Electrically • Electrician • Electron • Electronic

ELEC Electrical • Electronics

Elecm Electromagnetics (periodical code)

ELECOM Electronic Computer

Élect Élection • Électoral • Électricité • Électrique • Électron • Électronique

Elect Election • Electoral • Electric • Electrical • Electrician • Electricity

Électr Électricité • Électrique • Électron • Électronique

Electr Electric • Electrical • Electricidad • Electricity • Electronical • Electronics

ELECTRA Enhanced Light Efficiency Cophasing Telescope Resolution Actuator

Electromagn Electromagnetismo

Electrón Electrónica

Électron Électronique

Electron Electronical • Electronics

ELED Edge-emitting Light-Emitting Diode

ELEGANT Electron Gamma-ray Neutrino Telescope

Elek Elektronika (periodical code)

Elekt Elektrisch • Elektrizität

Elektr Elektrisch • Elektrizität • Elektronisch • Elektronik

Elektron Electronisch • Elektronik

Elem Elektromekhanika (periodical code) • Element • Elementary

ELES Expanded Liquid Engine Simulation

ELETE Electric Engine Transfer Experiment

Elev Elevate • Elevation • Elevator

ELEV Elevation

ELF Early Lunar Flare • El Fasher (Sudan) (location identifier) • European Laser Facility • Executable and Linking Format • Extensible Language Facility • Extension Language Facility • Extralow Frequency • Extremely Low Frequency

ELFEXT Equal Level far End Crosstalk

ELFIN European Laminar Flow Investigation

ELFS Extremely Long Focal-length Spectrograph

ELG Education Librarians Group • El Golea (Algeria) (location identifier) • Emission-Line Galaxy • Engineering Log Generator • European Launching Group • European Liaison Group • European Library Group • Executive Level Group • Executive Level Guidance • Extended Local Group

ELGRA European Low-Gravity Association

ELGS Emission Line Galaxy Sample

ELH North Eleuthera (Bahamas) (location identifier)

ELI Elim (USA-AK) (location identifier) • Emitter Location and Identification • English Language Interpreter • Environmental Law Institute • Eurocell Library and Interfaces

ELIA European League of Institutes of Arts

ELIAS Earth Limb Infrared Atmospheric Structure • Entry Level Interactive Applications System

Elim Eliminate

ELIN Equipment Line Number • Exhibit Line Item Number

ELINCS European List of Notified Chemical Substances

ELINOR Electronic Information Online Retrieval

ELINT Electromagnetic Information • Electromagnetic Intelligence • Electronic Intelligence

ELIP Electrostatic Latent Image Photography

ELIR Emission Location Information Report

ELISE Electronic Library Image Service for Europe

ELITE Electric Insertion Transfer Experiment • Electron Isotope Telescope • European Laboratory of Intelligent Techniques Engineering • Extended Long-range Integrated Technology Evaluation

ELJ El Recreo (Colombia) (location identifier)

ELK Elk City (USA-OK) (location identifier)

ElL Electronics Letters (periodical code)

ELL Ellisras (South Africa) (location identifier) • English Language Laboratory

ElLC Electrotechnical Laboratory Circulars (periodical code)

ELLC Enhanced Logical Link Control

Ellip Elliptical

ElLR Electrotechnical Laboratory Researches (periodical code)

elm Element

ELM Ecological Land Management • Edge-Localized Mode • Elaboration Likelihood Model • Electronic Mail • Elmira Corning (USA-NY) (location identifier) • Employee and Labor (relations) Manual • Experimental Logistics Module (*) • Experiment Logistics Module • Extended Lunar Mission

ELMC Electrical Load Management Center

ELMF European Large Magnetic-field Facility

ELMO Electronic Map Ordering

ElMod Elektronnoe Modelirovanie (periodical code)

ELMS Earth Limb Measurement Satellite • Elastic-Loop Mobility System • Electrical Load Management System • Engineering Lunar Model Surface • EOS Microwave Limb Sounder • Experimental Library Management System

ELN Ellensburg Bowers (USA-WA) (location identifier)

ELNET European Library Network

Elng Elongate

ELNGT Elongate

Elo Elocution • Eloquence

ELO Eldorado (Argentina-MI) (location identifier) • Emerging Local Operator • Emission-Line Object • Enabling Learning Objective • European Limitless Office

ELOC Executable Line Of Code

ELOCARS Electro-Optical Control Analysis and Reporting System

ELOCATS Electro-Optical Collection Analysis and Targeting System

ELOISE European Land-Ocean Interaction Studies • European Large Orbiting Instrumentation for Solar Experiments

Elong Elongate • Elongated • Elongation

E. Long. East Longitude

ELOPTINT Electro-Optical Intelligence

ELOR Extended Lunar Orbital Rendezvous

ELOS Electronic Line Of Sight • Experimental Land Observation System • Extended Line Of Sight

ELOT Hellenic Organization for Standardization (Greece)

Elp Elliptical

ElP Electronics Power (periodical code)

ELP Earth-Like Planet • El Paso (USA-TX) (location identifier) • Emission Line Project • English Language Program • Environmental Leadership Program • Equivalent Local Potential • European Librarians and Publishers • Extensions of Logic Programming (workshop)

ELPC Electroluminescent-Photoconductive (device)

ELPH Elliptical Head

ELPOA Étoile Laser Polychromatique pour l'Optique Adaptative

ELPP Equal Level Patch Panel

ElPro Electronic Progress (periodical code)

ELQ Gassim (Saudi Arabia) (location identifier)

ELQP Extended Linear-Quadratic Program

ELR Emission-Line Region • Engineering Laboratory Report • Environmental Lapse Rate • Environment Lapse Rate • Environment Lapse Rate • Exceptional Leave to Remain • Extended-Line Region • Extralong Range

ELRA European Language Resources Association

ELRAD Earth Limb Radiance

ELRM Employee and Labor Relations Manual

ELS Earth Landing System • Earth Limbb Sensor • Eastern Launch Site • East London (South Africa) (location identifier) • Easy Link to Spot (satellite) • Eggen - Lynde-Bell - Sandage (model) • Electronic Library System • Electron Lodgement Service • Electron Spectrometer • Elevon Load System • Emission Line Star • Emitter Location System • Energy-Loss Spectroscopy • Entry Level System • Environment Labelling Scheme • Error-Likely Situation • Exportable Logistics System • Extended Low State • External Lighting System

ELSA Electronic Library SGML Application • Electronic Lobe-Switching Antenna • Ensemble Logiciel de Simulation en Aérodynamique • Environmental Life Support Assembly • Ethical, Legal and Social Aspects • Ethics - Law - Social Aspects • European Laboratory for Structural Assessment • European Law Students' Association

ELSC Earth Landing Sequence Controller

ELSEC Electronic Security

ELSET Element Set

Elsewh Elsewhere

ELSI Enhanced Longwave Spectrometer/Imager

ELSIE Electronic Signaling and Indicating Equipment • Electronic Speech Information Equipment

ELSIR Electronic Letter Sorting and Indicating Equipment

ELSPA European Leisure Software Publishers Association

ELSS Emergency Life-Support System • EVA Life-Support System

ELSSE Electronic Sky Screen Equipment

ELSW Elsewhere

ELSy Experimental-Laserkreisel-System

ELT Early Launch Terminator • Effective Lagrangian Theory • Electronic Light Table • El Tor (Egypt) (location identifier) • Emergency Locating Transmitter • Emergency Locator Transmitter (*) • Emission Line Tool • English Language Teaching • English Language Training • Extrapolated Launch Time • Extreme Late-Type (object) • Extremely Large Telescope

ElTA Elektronnaia Tekhnika Avtomatike (periodical code)

ELTIS European Local Transport Information Service

ELU El Oued (Algeria) (location identifier)

ELV Earth Launch Vehicle • Electrically-Operated Valve • Elfin Cove (USA-AK) (location identifier) • Enclosed-frame Low Voltage • End-of-Life Vehicle • Expendable Launch Vehicle • Extralow Voltage

ELVA Emergency Low-Visibility Approach

ELVIL European Legislative Virtual Library

ELVIRA Electronic Library and Visual Information Research Association

ELVIS Early Light Vision Isomorphic Singular (system) • Enhanced Link Visual information System • Expandable Launch Vehicle Integrated Services • Experimenter's Laboratory for Visualized Interactive Science

Elvn Elevation

ElW Elektrizitätswerk

ELW Earth Launch Window • Einliegerwohnung • Ellamar (USA-AK) (location identifier)

ELWRT Electrowriter

ELWW European Laboratory Without Walls

Ely Easterly

ELY Ely (USA-NV) (location identifier)

Elz Elzevieren

ELZ Wellsville (USA-NY) (location identifier)

Em Emanation • Eminent

E&M Ear & Mouth • Earth & Magnet • Electrical & Mechanical • Elektrotechnik und Maschinenbau (periodical code)

E/M Engineering Model • Escape Motor

E-M Expectation-Maximization (algorithm)

EM Earth-Moon • Ecosystem Management • Education Manual • Effective Mass • Ejecta Mobility • Electrical and Mechanical • Electrical Model • Electric Module • Electromagnetic • Electromechanical • Electromotive • Electromodulation • Electronic Mail • Electronic Module • Electronic Monitoring • Electron Microscope • Electron Microscopy • Electron Model • Element Manager • Emergency Maintenance • Emission Measure • Emphasized • End Matched • End of Media • End of Medium • End (user) Manual • Energy Management • Engineering Management • Engineering Manual • Engineering Memorandum • Engineering Model • Engineer of Mines • Enhanced Monitoring • En Mano • Enterprise Manager • Enterprise Model • Environmental Modeling • Environmental Module • Environment Management • Equatorial Mount • Equipment Maintenance • Equipment Manager • Equipment Module • Escape Motor • Estimation Module • Estrada Municipal • Evaluation Model • Event Management • Exception Monitor • Execution Monitor • Executive Manager • Expanded Memory • Expectation-Maximization (algorithm) • Experienced Member • Experiment Module • Express Mail • External Memorandum • External Memory

EMA East-Asian Meeting on Astronomy • East Midlands (UK-England) (location identifier) • École des Mines d'Alès (France) • Effective Medium Approximation • Egyptian Meteorological Authority • Électromagnétisme et Méthodes d'Analyse • Electromechanical Actuation • Electromechanical Actuator • Electromechanical Analysis • Electronic Mail Association • Electronic Messaging Association (*) • Elektronische Mitteilungen zur Astronomiegeschichte • Engineering and Manufacturing Aid • Enhanced Monitoring Array • Enterprise Management Architecture • Enterprise Modular Array • Envelope Manufacturers Association (USA) • Environmental Management Agency • Equipment Maintenance Allowance • European Monetary Agreement • Exponential Moving Average • Extended Memory Addressing • Extended Memory Autocoder • External Mount Assembly

EMAC Electronic Media, Arts, and Communication (RPI) • Emergency Management Advisory Committee • Emergency Message Action Center • Ethernet Media Access Control • Europe Market Awareness Committee

EMACH Extended Maximum Average Correlation Height

EMACS Editing Macros • Editor Macros • Engine Monitoring And Control System

EMAD Engine Maintenance And Disassembly • Equatorial Middle Atmosphere Dynamics

EMAE EPIC MOS Analogue Electronics

EMAEE European Meeting on Applied Evolutionary Economics

e-mail Electronic Mail

EMail Electronic Mail

EMALS Electromagnetic Aircraft Launch System

EMAP Electronic Mapping • Environmental Monitoring and Analysis Program • Environmental Monitoring and Assessment Program

EMaPS European (boards) for Marine and Polar Science

EMARAC Electromagnetic Memory And Research Arithmetical Calculator

EMAS Eco-Management and Audit Scheme • Edinburgh Multiaccess System • Equipment Marking Accounting System

EMAT Electromagnetic Acoustic Transducer

EMATS Emergency Message Automatic Transmitting System • Emergency Mission Automatic Transmission System • Equipment Manipulation And Transportation System • Equipment Manipulation And Transport System • Experiment Manipulation And Transportation System • Experiment Manipulation And Transport System

EMAV Electromagnetic Relief Valve

Emb Embajada • Embajador • Embankment • Embark • Embassy

EMB Electromagnetic Bias • Electronic Material Bulletin • Elektromagnetische Beeinflussung • Embaixador • Embark • Embedded • Environmental Management Bureau • Extended Memory Block • San Francisco - Embarkadero (USA-CA) (location identifier)

EMBARC Electronic Mail Broadcast to A Roaming Computer

EMBD Embedded

EMBDD Embedded

EMBL European Molecular Biology Laboratory

EMBnet European Molecular Biology Network

EMBO European Molecular Biology Organization

Embriol Embriología

EMC Early Mid-Course • École des Métiers de la Communication • Educational Media Council (USA) • EISCAT Magnetospheric Cross • Electromagnetic Capability • Electromagnetic Cleanliness • Electromagnetic Compatibility (*) • Electromagnetic Conduction • Electromagnetic Control • Electromagnetic Coupling • Electronic-Mail Connection • Electronic Maintenance Complex • E-Mail Connection • Emergency Message Change • Energy Management Centre • Energy Management Company • Engineering Mockup Critical (experiment) • Enhanced Machine Controller • Enhanced Memory Chip • Enhanced Multicoating • Enterprise Monitoring and Coordination • Environmental Modeling Center • Environmental Module Computer • Environmental Monitoring and Compliance • Environment Management Committee • Equivalent Mission Cycle • European Muon Collaboration • Executive Management Course • Explicit Multiplexing Capability • Extended Math Coprocessor

EMCA Express Mail Corporate Account

EMCAB Electromagnetic Compatibility Advisory Board

EMCC Eight-Man Crew Capability • Electromagnetic Code Consortium • Emergency Mission Control Center • European Molecular Computing Consortium

EMCD Electromechanical Control Diagram

Emcee Master of Ceremonies

EMCF European Monetary Cooperation Fund

EMCFA Electromagnetic Compatibility Frequency Analysis

EMCH EPIC MOS Camera Head

EMCI Engineering Model Configuration Inspection

EMCO Early Mid-Course Option

EMCOF European Monetary Cooperation Fund

EMCOM Emission Control Measure

EMCON Emission Control

EMCONS Emission Control System

EMCP Electromagnetic Compatibility Program

EMCPP Electromagnetic Compatibility Program Plan

EMCR EPIC MOS Control and Recognition

EMCS Electromagnetic Compatibility Standardization • Energy Management and Control System • Energy Monitoring and Control System • Energy Monitoring Control System • Environmental Monitoring and Control System • EPIC MOS Camera System • Event Management Control System

EMCSC Ettore Majorana Centre for Scientific Culture

EMCT EMC Technology (periodical code) • Extended-duration Metabolic Control Test

EMCTP Electromagnetic Compatibility Test Plan • Electromagnetic Compatibility Test Program

EMCU EPIC MOS Camera Unit

E&MD Engineering & Manufacturing Development

EMD Earth Mover's Distance • Electric-Motor Driven • Electrolytic Manganese Dioxide • Electronic Map Data • Electronic Map Display • Electronic Music Distribution • Elliptical Mass Distribution • Emerald (Australia-QLD) (location identifier) • Emission Measure Distribution • Engineering and Management Development • Engineering and Manufacturing Development • Engineering Manufacturing-Design • Engineering - Manufacturing - Development • Entry Monitor Display • Environmental Management Division • Equilibrium Mode Distribution • Exploitation Maximale des Données

EMDB Equipment Maintenance Database

EMDI Energy Management Display Indicator

EMDM Enhanced Multiplexer/Demultiplexer

EMDR Eye Movement Desensitization and Reprocessing

EMDS EURECA Mimic Diagram System • European Meteorological Data System

EMDSO Eastern Missouri Dark-Sky Observers (USA-MO)

EMDU EPIC MOS Data Unit

EME Earth - Mars - Earth • Earth - Moon - Earth (*) • Electromagnetic Effect • Electromagnetic Energy • Electromagnetic Environment • Emden (Germany) (location identifier) • Equipment Management Element

EMEA Europe - Middle East - Africa

EMEC Electromagnetic Effects Capability • Electromagnetic Effects Compatibility • Enhanced Master Events Controller

EMEGS Electromagnetic Environment Generation System

EMEO Examen Médical d'Embauche Obligatoire

EMEP European Monitoring and Evaluation Programme

Emer Emergency • Emeritus

EMER Emergency

EMERAUDE Élaboration de Modèles d'Étude, de Recherche et d'Application, Utilisant des Domaines Étendus

Emerg Emergency

EMERG Emergency

EMES Earth Monitoring Educational System • Electrical, Mechanical, and Environmental Systems

EMETF Electromagnetic Environmental Test Facility

EMEX Equatorial Mesoscale Experiment

emf Enhanced Metafile Format (file extension)

EMF Earth Magnetic Field • Electromagnetic Field • Electromagnetic Force • Electromagnetic Frequency • Electromotive Force • Electronics Manufacturing Facility • Elektromagnetische Felder • Energy Modeling Forum • Enhanced Metafile Format • European Multimedia Forum • Evaporated Metal Films Corporation (USA-NY)

EMG Ecological Management Group • Educational Management Group • Electromagnetogram • Electromyogram • Electromyograph • Electromyographic • Electromyography • Empangeni (South Africa) (location identifier) • Executive Management Group

EMGCM European Mars General Circulation Model

EMH Efficient Market Hypothesis • Erector Maximum Height • External Message Handler

EMHAZ Electromagnetic Hazard

EMHD Electron Magnetohydrodynamics

EMI Earthquakes and Megacities Initiative • Electrical and Musical Industries • Électrodynamique des Milieux Ionisés • Electromagnetic Interface • Electromagnetic Interference (*) • Electromotive Interference • Electronic Money Issuer • Emergency Management Institute • Emirau (Papua New Guinea) (location identifier) • Empirical Mutual Information • Ernst-Mach-Institut (Germany) • European Monetary Institute • Expérience de Mort Imminente • External Machine Interface • External Memory Interface • External Muon Identifier

EM&IAS Electromechanical & Integrated Avionics System

EMIAS Electromechanical and Integrated Avionics System

EMIC Electromagnetic Interference Control • Electromagnetic Ion Cyclotron • Emission Microscope • Emission Microscopy

EMICE Electromagnetic Interference Control Engineer

EMICP Electromagnetic Interference Control Plan

EMIDS Experiment for the Management of Information Data Systems

EMIF ESCON Multiimage Facility

EMIH Electromagnetic Interference Hardening

EMIM Expected Mutual Information Measure

EMIP European Market for Infrastructural Projects

EMIR Environmental Management and International Restrictions (conference) • Espectrógrafo Multiobjeto Infrarojo • European Mathematical Information Retrieval • European Microgravity Research • European Multilingual Information Retrieval

EMIRS Electrochemically Modulated Infrared Spectroscopy

EMIS Earth Moon Ion Spectrometer • Educational Management Information System • European Mathematical Information Service (EMS) • Express Mail International Service

EMISM Electromagnetic Interference Safety Margin

EMITF Electromagnetic Interference Test Facility

EMITS Electronic Mail for Invitations to Tender Software (*) • Electronic Mailing of Invitations to Tender Software

EMK Emergency Medical Kit • Emmonak (USA-AK) (location identifier)

eml Electronic Mail (file extension)

EML Electromagnetic Launcher • Electronic Mail • Element Management Layer • Enhanced Member List • Equipment Maintenance Log • Equipment Modification List • Expanded Markup Language • Expected Measured Loss

eMLPP Enhanced Multilend Precedence and Preemption

EMLS EOS Microwave Limb Sounder

EMM Electromagnetic Measurement • Electromechanical Mockup • Electronic Mail Message • Enterprise Mail Manager • Entitlement Management Message • État-Major de la Marine • Ethernet Management Module • Expanded Memory Manager • Expert Management Module • Express Mail Manifesto • Extended Memory Management • Extended Memory Module • Kemerer (USA-WY) (location identifier)

EMMA Early Micrometeorites Accretion • Electron Microscopy and Microanalysis

EMMCVPR Energy Minimization Methods in Computer Vision and Pattern Recognition (workshop)

EMMI ESO Multimode Instrument

EMML Expectation Maximization Maximum Likelihood

EMMS Electronic Mail and Messaging System • Electronic Music Management System • Engine Maintenance and Management Service

EMMU Extended Memory Management Unit

EMN Engineering Management Network • Entreprise Multinationale • Extended Meta-Note • Nema (Mauritania) (location identifier)

EMNLP Empirical Methods in Natural Language Processing (conference)

EMO Eagle Mountain Observatory (USA-NY) • Edgeworth-Kuiper Object • Éducation en Milieu Ouvert • Electronic Microorganism • Emo (Papua New Guinea) (location identifier) • Environmental Management Office • Equipment Management Office • Evolutionary Multicriterion Optimization

EMOC EOSDIS Mission Operations Center • EOS Mission Operations Center

EMOD Erasable Memory Octal Dump

EMON Exception Monitoring

EMOP Équipe Mobile d'Ouvriers Professionnels

EMOS Earth Mean Orbital Speed • ECMWF Meteorological Operational System

EMOT European Management and Organisations in Transition

EMOWG EOS Mission Operations Working Group

emp Emplastrum • Ex Modo Praescripto

Emp Empennage • Emperor • Empire • Empress

EM&P Earth, Moon and Planets (periodical code)

EMP Earth, Moon and Planets (periodical code) • Earth - Moon - Planets • École des Mines de Paris (France) • Ecological Monitoring Programme • Electromagnetic Pulse • Electromechanical Power • Electron Microprobe • Emporia (USA-KS) (location identifier) • End of Month Payment • Engineering Master Plan • Enhanced Monitoring Period • Environmental Management Plan • Environmental Management Programme • Environmental Monitoring Plan • Environmental Monitoring Programme • Environment Management Plan • Environment Management Program • Ephemerides of Minor Planets • Equipment Mounting Plate • Erasable Memory Program • Event Manager Process • Experimentation Master Plan • Extremely Metal-Poor

EMPA Eidgenössischer Materialprüfungs- und Versuchsanstalt (Switzerland)

EMPACT Electrons Muons Partons Air Core Toroid

EMPAR European Multipurpose Phased Array Radar

EMPB European Multiprotocol Backbone

EMPC Electrostatic Molecular Potential Contour

EMPDB Electronic Materials Properties Data Bank

Empf Empfohlen

EMPGS Electromechanical Power Generation System

EMPIRE Early Manned Planetary-Interplanetary Round Trip Experiment

EMPIS Engineering Materials and Processes Information Service • Environmental Management and Planning Information System

Empl Emplacement • Employ • Employee • Employment

Empld Employed

EMPM Electromagnetic Particle Model

EMPR Emprendimento

EMPS Elevation Matrix Processing System • En-route Maintenance Processor System • Extremely-Metal-Poor Star

EMPST Energy - Matter - Personality - Space - Time

EMQ Erreur Moyenne Quadratique • Event Manager Queue

EMR Electromagnetic Radiation • Electromagnetic Response • Electromechanical Relay • Electromechanical Research • Electronic Medical Record • Energy, Mines and Resources (Canada) • Engineering Management Review • Engine Mixed Ratio • Engine Mixture Ratio • Enhanced Metafile Record • Environmental Management Report • Environmental Management Review • Equipment Maintenance Record • Executive Management Review • Expanding Molecular Ring

EMRC European Medical Research Council

EMRDF Electronic Mail Run Data File

EMRE Electromagnetic Radiation Effect

EMREL Emission Release

EMRH Electromagnetic Radiation Hazard

EMRL Equipment Maintenance Requirements List

EMRO Electromagnetic Radiation Operations

EMRP Effective Monopole Radiated Power

EMRS Electronic Marketing Reporting System • Emergency Management Requirements Study • European Materials Research Society

EMRTC Energetics Materials Research and Testing Center (USA-NM)

ems Enhanced Menu System (file extension)

EMS Edinburgh Mathematical Society (UK-Scotland) • Effluent Management System • Electromagnetic Monitoring System • Electromagnetic Spectrum • Electromagnetic Surveillance • Electromagnetic Susceptibility • Electro-Mechanical Systems • Electronic Mail Service • Electronic Mail System • Electronic Maintenance Shop • Electronic Maintenance System • Electronic Manuscript Standard • Electronic Manuscript Submission • Electronic Meeting System • Electronic Message Service • Electronic Message System • Electron Multiplex Switch • Electronic Mail System • Electronic Management System • Element Management System • Embessa (Papua New Guinea) (location identifier) • Emergency Management System • Emergency Medical Service • Energy Management System • Engineering Master Schedule • Engineering Measurement System • Engine Monitoring System • Enhanced Memory Services • Enhanced Messaging Service • Enregistreur Magnétique Spatial • Enterprise Management System • Enterprise Messaging Server • Entry Monitoring System • Entry Monitor System • Environmental Management System • Environmental Modeling Simulation • Environmental Monitoring Satellite • ENVISAT Mission System • Equipment Management System • Erector Maximum Strength • Error Mean Square • Error Message Service • European Macroseismic Scale • European Mathematical Society • European Meteorological Service • European Mobile System • European Monetary System • Europees Monetair Systeem • Evolving Magnetic Structure • Expanded Memory Services • Expanded Memory Specification • Expanded Memory Support • Expanded Memory System • Expanding Molecular Shell • Experimental Monitoring Satellite • Experiment Mounting Structure • Exposure Management System • Express Mail Service • Extended Memory Specification • Extended Memory System • Extended Meteor Stream

EMSA Electron Microscopy Society of America

EMSAW En-route Minimum Safe Altitude Warning

EMSC European Mediterranean Seismological Centre • Experiment Mission Support Center

EMSD Efficient Mail Submission and Delivery

EMSE École Nationale Supérieure des Mines de Saint-Étienne (France)

EMSG Energy Management Steering Group

EMSI Electromagnetic Sensing and Interpretation • European Manned Space Infrastructure

Emsl Emsland (Germany)

EMSL Environmental Molecular Sciences Laboratory • Environmental Monitoring Systems Laboratory • European Microwave Signature Laboratory

Emsn Emission

EMSP Enhanced Modular Signal Processor

EMSPS European Mobility Scheme for Physics Students

EMSS Einstein Medium-Sensitivity Survey (*) • Emergency Mission Support System • Emulated Multispectral Scanner • Enhanced Mobile Satellite Services • Extended Medium-Sensitivity Survey

EMST Extended Magic-Set Transformation

EMSU Environmental Meteorological Support Unit

EMSW21 Enhancing the Mathematical Workforce in the 21st Century (NSF)

EMT Effective-Mass Theory • Effective Medium Theory • Elapsed Maintenance Time • Electrical Metallic Tubing • Element Management Tool • El Monto (USA-CA) (location identifier) • Embedded Management Tool • Emergency Management Team • Emergency Medical Technician • Emergency Medical Treatment • End of Magnetic Tape • Energy Momentum Tensor • Engineering and Management Team • Enseignement Manuel et Technique • Equipment Management Tool • Équivalent Mégatonnique • Éducation Manuelle et Technique • European Mathematical Trust • Évaluation en Milieu de Travail • External Momentum Trap

EMTA Engineering Marine Training Authority

EMTE Electromagnetic Test Environment

EMTED Electromagnetic Test and Evaluation Data System

Emtr Emitter

emu Emulation Data (file extension)

EMU Eastern Michigan University (USA-MI) • Economic and Monetary Union • Electrical Multiple Unit • Electromagnetic Unit • Electronic Message Unit • Elevation Measuring Unit • Engineering Model Unit • Engine Management Unit • European Monetary Union • European Monetary Unit • EVA Maneuvering Unit • Extended Memory Unit • Extravehicular Maneuvering Unit • Extravehicular Mobility Unit (*)

EMUG European MAP User Group

emuls Emulsio

EMUX Electronic Multiplexer • Electronic Multiplexing

EMV Electromagnetic Vehicle • Electromagnetic Vulnerability • Elektromagnetische Verträglichkeit • Europay - Mastercard - Visa

EMVC EPIC MOS Voltage Converter

EMVG Elektromagnetisches Verträglichkeitsgesetz (Germany)

EMVICT Engineering Model Verification and Interface Compatibility Test

EMVU Elektromagnetische Verträglichkeit zur Umwelt

EMW Erector Maximum Weight

EMWA Engineering Minor Work Authorization

EMWAC European Microsoft Windows Academic Centre

EMX Electronic Mobile Exchange • El Maiten (Argentina-CB) (location identifier) • Enterprise Mail Exchange • Enterprise Messaging Exchange

EMY Elder - Middle-ager - Younger • El Minya (Egypt) (location identifier)

En Energy • Enero • Engineer • English (ISO language code)

EN École Navale • École Normale • Économie Nationale • Éducation Nationale • Electrical Notebook • Electronic Notebook • El Niño • Emission Nebula • Emissora Nacional • End Node • Enforcement Notification • England • Estrada Nacional • Europastandard • European Network • European Norm • Evaluation Notice • Exception Noted • Extension Node • External Network • Extrapolation Number

ENA Eastern North America • École Nationale d'Administration (France) (*) • École Nationale d'Affichage • Energetic Neutral Atom • Enterprise Network Architecture • European Neuroscience Association • Extended Network Architecture • Kenai (USA-AK) (location identifier)

ENAA Encontro Nacional de Astronomia e Astrofísica (Portugal) • Epithermal Neutron Activation Analysis

ENAC École Nationale de l'Aviation Civile (France) • Energetic Neutral Atom Camera • Expanded National Agency Check

ENACEOS Energetic Neutral Atom Camera for Earth-Orbiting Satellite

ENACS ESO Nearby Abell Cluster Survey

Enam Enamel

ENAM European and National Astronomy Meeting • European North Atlantic Margin • Exotic Nuclei and Atomic Masses

ENAN Education Native American Network (USA)

ENAP Energetic Neutral Atom Precipitation • Évaluation du Niveau d'Aptitudes Professionnelles

EnAt Energia es Atomtechnika (periodical code)

ENAV Electronic Navigation • En-route Navigational Aid

ENB Eneabba West (Australia-WA) (location identifier) • Engineering Notebook • Escuela Nacional de Bibliotecarios (Argentina) • Extragalactic Neutrino Background

ENBA École Nationale des Beaux-Arts • Escuela Nacional de Biblioeconomía y Archivonomía (Mexico)

Enbl Enable

ENBL Enable

ENBW Equivalent Noise Bandwidth

enc Encoded (file extension) • Encrypted

Enc Enclos • Enclosed • Enclosure • Encoder • Encuadernación • Encyclopaedia

EnC Energy Conversion (periodical code)

ENC École Nationale des Chartes (France) • Eisenhower National Clearinghouse (USA-OH) • Electronic Navigational Chart • Enclosed • Encode • Encoded • Equivalent Noise Charge • Estimated Number of Citations • Nancy (France) (location identifier)

ENCA Enterprise Network Computing Architecture

Encap Encapsulate • Encapsulated

Enc. Brit Encyclopaedia Britannica

ENCC Environmental Noise Control Committee

Encd Encased • Encode

ENCD Electronic Navigation Chart Data

Encdd Encoded

ENCDR Encoder

ENCG Évaluation du Niveau des Connaissances Générales

Encl Enclose • Enclosed • Enclosing • Enclosure

ENCORE Evolutionary Computations Repository

ENCP Évaluation du Niveau des Connaissances Professionnelles

ENCPT Encrypt

ENCS Existing and New Chemical Substance

ENCT École Nationale des Cadres Techniques (France)

Encuad Encuadernación

Ency Encyclopaedia • Encyclopaedic

Ency.Brit Encyclopaedia Britannica

Encyc Encyclopaedia • Encyclopaedic

Encyc.Brit Encyclopaedia Britannica

Encycl Encyclopaedia • Encyclopaedic

End Endorse • Endorsed • Endorsement • Endorsing

eND Enhanced Navigation Display

END Enid Vance (USA-OK) (location identifier) • Expert National Détaché

ENDAK End of Acknowledgment

ENDEC Encoder/Decoder

ENDEX End of Exercise • Environmental Data Index (NOAA)

ENDF Evaluated Neutron Data File • Evaluated Nuclear Data File

ENDG Ending

ENDIF Enterprise Network Data Interconnectivity Family

ENDO Endoatmospheric

ENDOC Environmental Documentation (EU)

ENDOR Electron Nuclear Double Resonance

Endow Endowment

ENDP Exception to National Disclosure Policy

ENDPLAN Endurability Plan

ENDR Endurance • Entretien Non-Directif de Recherche

ENDS European Nuclear Documentation System

Endvr Endeavor (periodical code)

Ene Enero

E/NE Effective/Non-Effective

ENE East-Northeast • East-Northeastern • Ende (Indonesia) (location identifier) • Est-Nord-Est

ENEA Energia Nucleare e Energie Alternative (Italy) • European Nuclear Energy Agency

ENEC European Nuclear Energy Community

EnEng Environmental Engineering (periodical code)

ENEP Escuela Nacional de Estudios Profesionales (Mexico)

Energ Energetika (periodical code)

ENERN East-Northeastern

ENEWD East-Northeastward

ENF End-Notched Flexure • Enfermeira • Enontekio (Finland) (location identifier) • Européen Non-Français • Event Notification Facility

ENFI Electronic Networks For Interaction

ENFIA Exchange Network Facility for Interstate Access

ENFJ Extraverted Intuitive Feeling Judging

EnFM Engineering Fracture Mechanics (periodical code)

ENFOL Evanescent Near-Field Optical Lithography

ENFP Extraverted Intuitive Feeling Perceiving

eNFS Enhanced Network File System

Eng Engine • Engineer • Engineering • England (UK) • English • Engrave • Engraver • Engraving

ENG Electronic News Gathering • Electronystagmogram • Engenheiro • Engine • Engineer • Engineering • England

Enga Engage

ENGA Engage

EngAn Engineering Analysis (periodical code)

EngCo Engineering Computations (periodical code)

Engg Engineering

Engin Engineer • Engineering

Engl England (UK) • English

ENGO Environmental Non-Governmental Organization

Engr Engineer • Engineering • Engraved • Engraver • Engraving

Engrg Engineering

ENGRG Engineering

Engrv Engrave

Engy Energy

ENH Earth Near Horizon • Enhancement • Enshi (China-PRC) (location identifier)

ENHA Electronic Newsletter for the History of Astronomy

ENI École Nationale d'Ingénieurs (France) • El Nido (Philippines) (location identifier)

ENIAC Electronic Numerical Integrator Analyzer and Computer • Electronic Numerical Integrator And Calculator (*) • Electronic Numerical Integrator And Computer

ENIC École Nouvelle d'Ingénieurs en Communications (France)

ENID EGC Network Identity • Endnote Identification

ENIM École Nationale d'Ingénieurs de Monastir (Tunisia)

ENIS Electronic Networks and Information Sharing • European Nuclear Information Service

ENIU Enhanced Network Interface Unit

Enj Enjuiciamiento

ENJ El Naranjo (Guatemala) (location identifier)

Enk Enkelvoud

ENK Enniskillen - Saint Angelo (Northern Ireland) (location identifier)

Enl Enlarge • Enlarged • Enlargment • Enlisted

ENL Centralia (USA-IL) (location identifier) • Enlarge • Enlarged

Enlg Enlarge

ENLR Extended Narrow-Line Region (*) • Extremely Narrow-Line Region

ENM École Nationale de la Météorologie (France) • Enterprise Network Management

ENMAN Engine Management

ENN Nenana (USA-AK) (location identifier)

ENNA Établissement National de Navigation Aérienne

ENNS European Neural Network Society

ENO Encarnación - Teniente Prim Alarcón (Paraguay) (location identifier) • Essentially Non Oscillatory • European Northern Observatory (Canary Islands)

ENOS El Niño Oscilação Sul

ENP École Nationale de la Photographie (France) • Escuela Nacional Preparatoria

ENPA Energetic Neutral Particle Analyzer

ENPBW Equivalent Noise Power Bandwidth

ENPI Energetic Neutral Particle Imager

ENPC École Nationale des Ponts et Chaussées (France)

ENPT École Nationale des Postes et Télécommunications (France)

ENQ Coronel E. Soto Cano (Honduras) (location identifier) • Enquire • Enquiry

ENQA European Network for Quality Assurance

EnQL English-like Query Language

ENR Énergie Naturellement Renouvelable • Engineering New Record • En Route • Enterprise Network Roundtable • Equivalent Noise Resistance • Excess Noise Ratio

Enrgz Energize

ENRIC Environment and Natural Resources Information Center

ENRICH European Network for Research In global Change • European Network for Research on global Change

ENRM Environment and Natural Resources Management

Ens Ensign

ENS École Normale Supérieure (France) • Emergency Notification System • Enschede Twente (Netherlands) (location identifier) • Ensemble • Ensign • Enterprise Networking Services • Enterprise Network Services • Enterprising Messaging System • European Nervous System • Extended Novel Services

ENSA European Neutron Scattering Association

ENSAAMA École Nationale Supérieure d'Arts Appliqués et des Métiers d'Art (France)

ENSAE École Nationale Supérieure de l'Aéronautique et de l'Espace (France)

ENSAI École Nationale de la Statistique et de l'Analyse de l'Information (France) • École Nationale Supérieure de l'Analyse de l'Information (France)

ENSAIS École Nationale Supérieure des Arts et Industries de Strasbourg (France)

ENSAM École Nationale Supérieure des Arts et Métiers (France)

ENSAR Établissement National des Sciences Agricoles et Rurales (France)

ENSB École Nationale Supérieure des Bibliothécaires (France) (now ENSSIB) (*)

ENSC École Normale Supérieure de Chimie (France)

ENSCI École Nationale Supérieure de Création Industrielle (France)

ENSCP École Nationale Supérieure de Chimie de Paris (France)

ENSDF Evaluated Nuclear Structured Data File

ENSE Enterprise Network Strategies Europe

ENSEEIHT École Nationale Supérieure d'Électrotechnique, d'Électronique, d'Informatique et d'Hydraulique de Toulouse (France)

ENSERB École Nationale Supérieure d'Électricité et de Radio de Bordeaux (France)

ENSERG École Nationale Supérieure d'Électronique de Grenoble (France)

ENSES École Normale Supérieure de l'Enseignement Secondaire (France)

ENSET École Normale Supérieure de l'Enseignement Technique (France)

ENSG École Nationale des Sciences Géographiques (France)

ENSGI École Nationale Supérieure de Génie Industriel (France)

ENSG École Nationale des Sciences Géographiques (France)

ENSI École Nationale Supérieure d'Ingénieurs (France) • Equivalent Noise Sideband Input

ENSICA École Nationale Supérieure d'Ingénieurs de Constructions Aéronautiques (France)

ENSIETA École Nationale Supérieure des Ingénieurs des Études et Techniques d'Armement

ENSIMEV École Nationale Supérieure d'Ingénieurs en Mécanique et Énergétique de Valenciennes (France)

ENSIP Engine Structural Integrity Program

ENSITM École Nationale Supérieure des Industries Textiles de Mulhouse (France)

ENSJF École Normale Supérieure de Jeunes Filles (France)

ENSL École Normale Supérieure de Lyon (France)

ENSM École Nationale Supérieure des Mines (France)

ENSMA École Normale Supérieure de Mécanique et d'Aéronautique (France)

ENSMP École Normale Supérieure des Mines de Paris (France)

ENSO El Niño Southern Oscillation

ENSPS École Nationale Supérieure de Physique de Strasbourg (France)

ENSPTT École Nationale Supérieure des Postes et Télécommunications (France)

ENSS European Navigation Satellite System • Experiment Network Surveillance System • Exterior Nodal Switching System

ENSSIB École Nationale Supérieure des Sciences de l'Information et des Bibliothèques (France)

EnST Environmental Science Technology (periodical code)

ENST École Nationale Supérieure des Télécommunications (France)

ENSTA École Nationale Supérieure des Techniques Avancées (France)

ENSURE Expedited Non-Standard Urgent Requirements for Equipment

Ent Enterprise • Entrance • Entry

ENT Ear - Nose - Throat • Engineering and Technology • Eniwetok Island (Marshall Islands) (location identifier) • Entrada • Entropy • Equivalent Noise Temperature • European Negotiating Team

ENTC Engine Negative Torque Control

ENTCS Electronic Notes in Theoretical Computer Science (periodical code)

ENTEL Empresa Nacional de Telecomunicaciones (Chile)

ENTICE Energetic Transiron Composition Experiment

ENTIS Engineering Terrain Information System

ENTJ Extraverted Intuitive Thinking Judging

Entlo Entresuelo

ENTP Extraverted Intuitive Thinking Perceiving

Entom Entomología

Entr Enter • Entrance

EnTR Energy Technology Review (periodical code)

ENTR Entire • Entrance

Entro Entropie (periodical code)

EntSchG Entschädigungsgesetz

entspr entsprechend

Entw Entweder

EntwHStG Entwicklungshilfe-Steuergesetz

EntwLStG Entwicklungsländer-Steuergesetz

ENU Enugu (Nigeria) (location identifier)

EnUK Energy UK (periodical code)

Enum Enumerate • Enumeration

ENUR Enterprise Network Users Roundtable

ENUWAR Environmental consequences of Nuclear War

env Environment (file extension)

Env Envelope • Environ • Envoi • Envoyer

ENV Envelope • Europäische Normvorschlag • Wendover (USA-UT) (location identifier)

ENVI Environment for Visualizing Images

ENVIR Environment

Environ Environment • Environmental • Environnement

ENVIRON Environment

ENVISAT Environmental Satellite

ENVL Envelope

Envr Environment • Environmental

ENW Kenosha (USA-WI) (location identifier)

ENX Europe Network Exchange

ENY Yan'an (China-PRC) (location identifier)

Enz Enzovoort • Enzovoorts

Enzkr Enzkreis (Germany)

eo Esperanto (ISO language code) • Ex Officio

Eö Eötvös

E&O Engineering & Operations • Errors & Omissions

E/O Electrical to Optical • Electro-Optical • Engineering/Operations

E-O Electro-Optical • Electro-Optics

EO Earth Observation • Earth Observing • Earth Orbit • Earth Orbiter • Earth-Orbiting • Earth Orientation •

Édition Originale • Education Officer • Einstein Observatory • Electronic Operation • Electronics Unit • Electro-Optic • Electro-Optical • Electro-Optics • Elliptical Orbit • Emergency Operation • Enabling Objective • End Office • Engineering and Operations • Engineering Order • Engine-Out • Equal Opportunity • Erasable Optical • Escape Orbit • Executive Officer • Executive Order • Ex Officio • External Output • Eyes Only

EOA Early Operational Assessment • Early Operations Assessment • End Of Address • Equal Opportunity Advisor • Examination - Opinion - Advice • Extension Of Application

EOAC Early Operational Assessment Capability • Earth Observation Advisory Committee

EOAE Earth-Orientated Applications Experiment

EOAP Earth Observations Aircraft Program

EOARD European Office of Aerospace Research and Development

EOAS Engineering Order Accountability System

EOAST Earth-Oriented Applications and Space Technology (periodical code)

EoB End of Bus

EOB Electronic Order of Battle • Electro-Optical Breadboard • End Of Block • Executive Office Building • Expense Operating Budget • Explanation Of Benefits • Extensible Optical Bench

EOBT Estimated Off-Block Time

Eoc Eocene

EOC Early Operational Capability • Earth Observation Center • Earth Orientation Catalog • Earth Orientation Centre • Edge Of Coverage • Electronics Operations Center • Electron Optical Camera • Electro-Optical Center • Electro-Optics Camera • Electro-Optics Center • Element Operations Center • Emergency Operating Center • Emergency Operations Center • Employee-Owned Company • End Of Carrier • End Of Charge • End Of Construction • End Of Contents • End Of Contract • End Of Conversation • End Of Conversion • End Of Course • End Of Cycle • Engine Order Capability • Engine-Out Capability • Enhanced Operational Capability • EOS Operations Center • Equal Opportunities Commission • Equivalent Operational Capability • ESPRIT Optical Computing • Executive Oversight Committee • Executive Oversight Council • Extended Operating Cycle

EOCA Estación de Observación de Calar Alto (Spain)

EOCAP Earth Observations Commercial Applications Program (*) • Earth Observations Commercialization Application Program • Earth Observing Commercialization and Applications Program

EOCB Electro-Optical Circuit Board

EOCCT End-Of-Course Comprehensive Test

E/OCM Electromagnetic/Optical Countermeasure

E-OCM Electro-Optical Countermeasure

EOCM Electro-Optical Countermeasure

EOCR Engineering Order Control Record

EOCY End Of Calendar Year

Eod Eodem

EoD End of Discharge

EOD Electro-Optic Device • Élément Optique Diffractif • End Of Data • End Of Day • End Of Dial • End Of Discussion • Engineering Operations Division • Entry On Duty • Erasable Optical Disk • Estimated On Dock • Estimated Operational Date • Every Other Day • Expected Occupancy Date • Explosive Ordnance Disposal

EODB End Of Data Block

EODC Earth Observation Data Centre

EODET Early Orbit Determination

Eod.Loc. Eodem Loco

EODMS Earth Observation Data Management System

Eod.Op. Eodem Opere

EODRS Earth Observation Data Retrieval System

EODS Electro-Optic Direction Sensor

E&OE Errors & Omissions Excepted

EOE Early Operational Experimentation • Earth Observation Experiment • Earth Orbit Ejection • Earth Orbit Equipment • Electronic Order Exchange • End Of Everything • Equal Opportunity Employer • Errors and Omissions Excepted

EOEC End Of Equilibrium Cycle

EOED Earth Orbit Escape Device

EOEP Earth Observation Envelope Programme (ESA) • Earth Observation Experimental Programme

EOF Earth Orbital Flight • Emergency Operating Facility • Empirical Orthogonal Function • End Of File • Enterprise Object Framework • Equal Opportunities Form • Expected Operations Forecast • Experiment Operations Facility

EOFS Erasable Optical File System

EOFY End Of Fiscal Year

EOG Electrooculogram • Electrooculograph • Electrooculographic

EOH Emergency Operations Headquarters • Engineering Open House • Engine Overhaul • Equipment On Hand • Medellín - Enrique Olaya Herrera (Colombia) (location identifier)

EOHP Except Otherwise Herein Provided

EOHT External Oxygen and Hydrogen Tank

EoI Expression of Interest

EOI Earth Orbit Insertion • Eday (UK) (location identifier) • Electro-Optical Industries Inc. (USA-AZ) • End Of Integration • End Of Interrupt • End Or Identify • Escuela de Organización Industrial (Spain) • Expression Of Interest

EOIM Evaluation of Oxygen Interaction with Materials

EOIS Earth Observing Information System • Electro-Optical Imaging System

EOJ End Of Job

EOK Keokuk (USA-IA) (location identifier)

EOL Earth Observation Laboratory • Earth Orbit Launch • End Of Life • End Of Line • End Of Link • End Of List • Europe Online • Expected Operational Life • Expression-Oriented Language

EOLB End Of Line Block

EOLC Effective Outdoor Lighting Council (USA-TX)

EOLM Electro-Optical Light Modulator

EOLR Electrical Objective Loudness Rating

EOLS EINSTEIN Online Service

EOLT End Of Logical Tape

EOM Earth Observation Mission (now ATLAS) • Earth Orbital Mission • Electro-Optical Modulator • End Of Medium • End Of Memory • End Of Message • End Of Mission • End Of Month • End of Observation Marker • Engineering Operations Manual • Enterprise Object Model • Equation Of Motion • European Optical Module • Every Other Month • External Optical Modulator • Extractable Organic Material • Extractable Organic Matter

EOMF End Of Minor Frame

EON End Of Number • EOSAT Online Network • ESO Outreach Network

EONR European Organization for Nuclear Research

EONS Environments Of Nearby Stars

EOO End Of Order

EOOE Erreur Ou Omission Exceptée

EOOW Engineering Officer Of the Watch

EOP Earth Observations Program • Earth Orbit Plane • Earth Orientation Parameters • Emergency Operating Procedure • Emergency Oxygen Pack • End of Output • End Of Part • End Of Period • End Of Process • End Of Program • End Output • Épreuve d'Organisation des Points • Equal Opportunity Program • Evaluator Operational Plan • Examen d'Orientation Professionnelle • Examination Of Product • Executive Office of the President • Experiment of Opportunity Program

EOPAG ERS-1 Operations Planning Advisory Group

EOPAP Earth and Ocean Physics Applications Program

EOPB Earth Observation Programme Board

EOPC Electro-Optical Phase Change

EOPD Equipment-Oriented Publication Database

EOPF End Of Powered Flight

EOPG École et Observatoire de Physique du Globe (France) (now EOST)

EOPGS École et Observatoire de Physique du Globe de Strasbourg (France)

EOPM Electro-Optic Phase Modulation • Electro-Optic Phase Modulator

EOPP Earth Observation Preparatory Programme

EOPR Engineering Order Purchase Request • Engineering Order Purchase Requisition

EOPS ERBE Operational Processing System

EOQ Earth Observation Quarterly • Economic Order Quantity • End Of Quarter • European Organization for Quality

EOQC European Organization for Quality Control

EOQL End Of Qualified Life

EoR Epoch of Reionization

EOR Earth Observing Radar • Earth Orbit Rendezvous • Either-Or Region • El Dorado (Venezuela) (location identifier) • Electro-Optical Research • Element Of Resource • End Of Record • End Of Reel • End Of Refresh • End Of Routine • End Of Routing • End Of Runway • Epoch Of Recombination • Epoch Of Reionization • Equal Opportunity Representative • Equipment Operationally Ready • Estimate Of Recuperability • Eureca Occultation Radiometer • EUV Opacity Rocket • Exclusive Or • Expedite Operations Requirement

EORC Earth Observation Research Center

EORCU European Ozone Research Coordinating Unit

EORM Enhanced Object Relationship Model

EORQ Engineering Order Request for Quotation

EORSAT Electronic Ocean Reconnaissance Satellite • ELINT Ocean Reconnaissance Satellite

EORTC European Organisation for Research and Treatment of Cancer

EoS Equation of State

EOS Earth Observation Satellite • Earth Observatory Satellite • Earth Observation System • Earth Observing System • Earth Orbital Shuttle • Earth Orbiting System • Egyptian Organization for Standardization (Egypt) • Electrical Overstress • Electro Optical System • Electro Optic System • Electrophoresis Operations in Space • Electrostatic Overstress • Ellery Open System • Emergency Oxygen System • Employed Operational State • Employee Opinion Survey • End Of Screen • End Of Selection • End Of Service • End Of String • End Of Sweep • Equal Opportunity Specialist • Equation Of State • Erasable Optical Storage • ERIN Online Service • Estación de Observación Solar (Mexico-SON) • European Optical Society • Extended Operating System • Neosho (USA-MO) (location identifier)

EOSAT Earth Observation Satellite Corporation

EOSC Earth Observations Science Committee • Engineering Operations Support Center • Equal Opportunities Subcommittee

EOSCAT Einstein Observatory Source Catalog

EOSD Earth Observation for Sustainable Development • Electro-Optical Systems Design (periodical code)

EOSDIS Earth Observing System Data and Information System • Earth Observing System Data Information System (*)

EOSH Electro-Optical Sensor Hardening

EOSL Earth Observations Sciences Ltd. (UK)

EOSOC Earth Orbiting Satellite Observers Club (USA)

EOSP Earth Observing Scanning Polarimeter

EOSRAM Earth Observing System Regional Amazon Model

EOSS Earth Observation Satellite System • Earth Orbital Space Station • Electro-Optic Sensor System • Engineering Operational Sequencing System

EOSSMS Earth Observing System Space Measurement System

EOSSRP Earth Observing System Scientific Research Program

EOST École d'Organisation Scientifique du Travail (France) • École et Observatoire et des Sciences de la Terre (France)

EOSTAG Earth Observation Scientific and Technical Advisory Group

EOSTr EOS Transactions (periodical code)

EOT Early On-orbit Test • Earth-Observed Time • Employment and Operational Training • End Of Table • End Of Tape • End Of Task • End Of Test • End Of Text • End Of Tour • End Of Track • End Of Train • End Of Transaction • End Of Transfer • End Of Transmission • Engineering Operations Trainer

EOTAD Electro-Optics Target Acquisition and Designation

EOTC European Organization for Testing and Certification

EOTD Enhanced Observed Time Difference

EOT&E Early Operational Test & Evaluation

EOTE Early Operational Test and Evaluation

EOTF Engineering Order To Follow

EOTS Electro-Optical Threat Sensor • Electro-Optical Tracking System

EOU Eastern Oregon University (USA-OR) • End Of User

EOUT Engine Out

EoV End of Visibility

EOV Economic Order Van • End Of Visibility • End Of Volume • Explanation Of Votes

EOW Empty Operating Weight • End Of Week • End Of Word • Energy Over Weight • Engineering Orderwire • Every Other Word

EOWG Earth Observation Working Group • Engineering and Operations Working Group

EOWL Exploded Overwhelmingly Large (telescope)

EOY End Of Year

EOZ Elorza (Venezuela) (location identifier)

ep Editio Princeps

E/P Electron/Proton

EP Early Picture • Earth Probe • Easy Projection • Echo Protocol • École Polytechnique • Edge Preserving • Editio Princeps • Effective Pitch • Elbow Pitch • Electrically-conducting Polymer • Electrically Polarized • Electrical Power • Electric Power • Electric Propulsion • Electronic Prevention • Electronic Printer • Electronic Protect • Electronic Protection • Electronic Publishing • Electronic Pulse • Electrophotographic • Electrostatic Precipitator • Element Probability • Elliptical Potential • Emergency Preparedness • Empty Pattern • Emulation Program • Endpaper • End Point • Energetic Particle • Enfermedad

Profesional • Engineering Personnel • Engineer Personnel • Enseignement Programmé • Entrepiso • Entreplanta • Entry Point • Environmental Park • Environmental Protection • Environment Protection • Equipment Publication • Equivalence Principle • Equivalent Person • Equivalent Personnel • Érection Permanente • Érection Progressive • Érection Pulsative • Error Protocol • Estimated Position • Établissement Public • European Parliament • European Plan • European Portuguese • Evaluation Package • Evoked Potential • Evolutionary Programming • Execution Planning • Exit Pupil • Exoplanet • Expanded Polystyrene • Experience Point • Experiment Planning • Extended Plan • Extended Play • External Pilot • Extraction Procedure • Extrasolar Planet • Extraterrestrische Physik • Extreme Pressure • Iran (international aircraft registration code)

EPA Earth Polar Axis • Economic Planning Agency • Economic Price Adjustment • Edge-Preserving Adaptive • Educational and Public Affairs • Educational Paperback Association (USA) • Educational Priority Area • Electron-Positron Accumulator • Electron-Positron Annihilation • Electron Probe Analyzer • Electro-optical Polarization Analyzer • Employment Protection Act (USA) • Encuesta de Población Activa (Spain) • Energetic Particles Analyzer • Energy-Performance Assessment • Engineering Personnel Authorization • Engineering Project Authorization • Enhanced Performance Architecture • Environmental Protection Agency (USA) • Environmental Protection Authority • Environment Planning Authority • Environment Protection Agency (*) • Equatorial Pitch Angle • Equilibrium Point Analysis • Estimated Plan of Action • Établissement Public à caractère Administratif • Établissement Public d'Aménagement • Europäischer Patentamt • European Photochemistry Association • Exoatmospheric Penetration Aid • Explicitly Parallel Algorithm • Extended Planning Annex

EPABX Electronic Private Automated Branch Exchange • Expanded Private Area Branch Exchange

EPAC East Pacific • Economic Planning Advisory Group • Energetic-Particle Anisotropy and Composition (experiment) (*) • Energetic-Particle Composition • Engineering Physics Activity Club • European Particle Accelerator Conference • Experimental Program Advisory Committee • Exploration Program Advisory Committee

EPACML Environmental Protection Agency Composite Model for Landfills

EPACT Energetic Particle Acceleration, Composition and Transport

EPAD Electrically Powered Actuation Design • Electrically Powered Actuator Design • Entry Pad

EPAIRS Environmental Protection Agency Information Retrieval System (USA)

EPALL Emergency Preparedness At Local Level

EPAM Electron, Proton, and Alpha Monitor • Elementary Perceiver And Memorizer

EPAO Educational and Public Affairs Office

EPAPS Electronic Physics Auxiliary Publication Service (AIP)

EPAQ Electronic Parts of Assessed Quality

EPAS Electron and Proton wide-Angle Spectrometer • Energetic-Particle Anisotropy Spectrometer • Exercise Production and Analysis System • Expanded Preaward Survey

EPAT Earliest Probable Arrival Time

EPB Electronic Picture Book • Emergency Power Buildup • Equivalent Passband • Equivalent Pension Benefit • ESS Parallel Benchmark • Eucrite Parent Body • Europäischer Beschäftigungspakt

EPBM Elliptical Polarized Bending Magnet

EPBX Electronic Private Branch Exchange

EPC Earth Prelaunch Calibration • Easy Processing Channel • École Par Correspondance • Edge Punched Card • Edinburgh Portable Compiler • Editorial Policy Committee • Editorial Processing Center • Educational Policy Committee • Educational Publishing Council • Electrical Power Conditioner • Electronic Parts Control • Electronic Power Conditioner • Electronic Program Control • Environmental Planning Commission • Environmental Pollution Control • Environmental Protection Committee • Equipment Procurement Committee • Error Protection Code • Étage à Propulsion Cryotechnique • Étage Principal Cryotechnique • Étude Personnelle Chiffrée • European Paleoclimate • European Patent Convention • Event-driven Process Chain • Executive Policy Committee • External Power Contractor • Extrasolar Planetary Candidate

EPCA Energy Policy and Conservation Act (USA) • Extended Principle Component Analysis

EPCC Edinburgh Parallel Computing Centre (UK-Scotland) • Environment Policy Coordinating Committee

EPCDC Electrical Power Conditioning, Distribution, and Control

EPCE EPIC p-n Control Electronics

EPCH EPIC p-n Camera Head

EPCI Établissement Public de Coopération Intercommunale

EPCO Emergency Power Cutoff • European Planetary and Cometary Observers

EPCP Étude du Plateau Continental Polaire

EPCRA Emergency Planning and Community Right-to-know Act (USA)

EPCS EPIC p-n Camera System • Experiment Physics Control System • Experiment Point Control System

EPCSCP Établissement Public à Caractère Scientifique, Culturel et Professionnel (France)

EPCU Édifice de Préparation de la Charge Utile • Ensemble de Préparation des Charges Utiles (CSG) • EPIC p-n Camera Unit

EPD Earliest Possible Date • Earliest Practical Date • Early Packet Discard • Electrical Power Distribution • Electric Power Distribution • Emergency Preparedness Division • Emergency Procedures Document • Energetic-Particle Detector (*) • Energetic Plasma Detector • Energy-charged Particle Detector • Engineering Product Definition • Engineering Project Description • Environmental Protection Division • ESA Publications Division • Estimated Propagation Delay • Etch-Pit Density • European Pollen Database • Excluded from Public Domain

EPDA Education Professions Development Act (USA)

EPDB Environmental Protection Database • Experiment Power Distribution Box

EPD&C Electrical Power Distribution & Control

EPDC Electrical Power Distribution and Control

EPD&CS Electrical Power Distribution & Control System

EPDCS Electrical Power Distribution and Control System

EPDG Execution Plan Data Generation

EPDH EPIC p-n Data Handling

EPDM Enterprise Product Data Management • Ethylene-Polypropylene-Diene-Monomer

EPDS Earth Probe Data Set • Earth Probe Data System • Electrical Power and Distribution System • Electrical Power Distribution System • Electronic Processing and Dissemination System • ERS-1 Products Distribution Service

EPDT Electronics, Photonics, and Device Technologies

EPDU Electrical Power Distribution Unit • EPIC p-n Data Unit

EPE Editorial Project for Education • Emergency Power Engineering • Energetic Particles Experiment • Energetic Particles Explorer • Enhanced Privacy and Encryption • Environmental Protection Expenditure • European Partnership for Education • European Power Electronics

EPEA EPIC p-n Event Analyser • Experiment Pointing Electronic Assembly

EPER Environmental Processes and Effects Research

EPERR Evaluation Package Evaluation Readiness Review

EPES Établissement Privé d'Enseignement Supérieur • Extended Polyelastic Sampling

EPEST Enhanced Project Evaluation and Scheduling Technique

EPEX Extreme Phenomena Explorer

EPF Electric Penetration Feedthrough • Emission-Phase Function • Environmental Planning Function • Europa Pathfinder • Excess Pair Fraction • Exposed Payload Facility

EPFL École Polytechnique Fédérale de Lausanne (Switzerland)

EPFS Equipment Pool for Field Spectroscopy

EPFZ École Polytechnique Fédérale de Zurich (Switzerland)

EPG Electrical Power Generator • Electronic Program Guide • Electronic Proving Ground • Emergency Procedure Guideline • Eminent Persons Group • Engineering Proposal Generation • Engineering Proving Ground • European Paleoclimatology Glaciology • European Participating Government • Experiment Planning Guide • Weeping Water - Browns (USA-NE) (location identifier)

EPGC Earth Processes and Global Changes

EPGCS Electric Power Generation and Control System

EPGS École de Physique du Globe de Strasbourg (France) • Electric Power Generating System • Electric Power Generation System

EPGWS Enhanced Proximity Ground Warning System

Eph Ephemeris

EPH Emergency Planning Handbook • Ephrata (USA-WA) (location identifier)

EPHC Eastern Pacific Hurricane Center

EPHIN Electron Proton Helium Instrument

EPHOS European Procurement Handbook for Open Systems

EPI Eagle Picher Industries • Earth Path Indicator • Echo Planar Imaging • Effective-Pair Interaction • Electrical Patents Index • Electronic Position Indicator • Energetic Particles Investigation • Engineering Process Improvement • Environmental Productivity Index • Environnement de Programmation Intégré • Epi (Vanuatu) (location identifier) • Équipement de Protection Individuelle • ESOC Precipitation Index • Evolutionary Program Induction • Expanded Position Indicator • External Presentation Interface • External Protocol Interface • Extreme Physical Information

EPIc Extrasolar Planet Interferometric (survey)

EPIC Earth-Pointing Instrument Carrier • Earth Polychromatic Imaging Camera • Educational Policy Information Center • Efficient Pyramid Image Coder • Electronic Privacy Information Center (USA) • Energetic Particle and Ion Composition • Engineering Performance Information Center • Engineering Physics Information Center • Equatorial Processes Including Coupling • Établissement Public à caractère Industriel et Commercial • European Photon Imaging Camera • Explicit Planetary Isentropic Coordinate

EPICA European Project for Ice Coring in Antarctica

EPIcS Extrasolar Planet Interferometric Survey

EPICS Electrolysis Performance Improvement Concept Study • Experimental Physics and Industrial Control System

EPIE Educational Products Information Exchange

Epil Epilogue

EPIOTIC ESPRIT Projects In Optical Techniques of Interface Characterization

EPIP Evolutionary Phase Implementation Plan

EPIRB Emergency Position-Indicating Radio Beacon (*) • Emergency Position-Indicator Radio Beacon

EPIRSS Extended Performance Infrared Search Set

EPISD El Paso Independent School District (USA-TX)

Epit Epitaph • Epitome

EPITA École Pour l'Informatique et les Techniques Avancées (France)

EPJ European Physical Journal

EPK Episkopi (Cyprus) (location identifier)

EPL Early Programming Language • Effective Privilege Level • Electrical Power Level • Electronic Program Language • Emergency Power Level • Emitter Parameters List • Encoder Programming Language • Engineering Parts List • Engineering Propulsion Laboratory • Environmental Protection Limit • Épinal (France) (location identifier) • Equipment Parts List • Eruptive Prominence on Limb • Evaluated Products List • Exact Power Law • Excess Path Length • Exciter Power Logic • Expected Power Level • Experiment Payload • Extensible Programming Language

EPLD Electrically Programmable Logic Device • Erasable Programmable Logic Device (*)

EPLE Établissement Public Local d'Enseignement

EPLRS Enhanced Position Location Reporting System

EPLV Enseignement Précoce des Langues Vivantes

epm En Propria Mano

EPM Earth Pointing Mode • Electric Printing Machine • Electronic Postmark • Electronic Protective Measure • Electron-Probe Microanalysis • Energetic Particle Mode • Energetic Particle Monitor • Engineering Procedure Memorandum • Engineering Project Manager • Enhanced Presentation Manager • Enterprise Process Management • Environmental Planning and Management • Environmental Project Manager • Escape Probability Method • European Physiology Module • European Project Manager • Expanding Photosphere Method (*) • Expanding Photosphere Model • Extent of Presence Metaphor • External Polarization Modulation • External Power Module • External Protection Material

EPMA Electron Probe Microanalyser • Electron-Probe Microanalysis

EPMC Estimación de Parámetros por Mínimos Cuadrados • Executive Program Management Course

EPMIS Emergency Preparedness Management Information System

EPMR Electric Power Margin Report

EPMS Electronic Protection and Mapping System • Engineering Performance Management System

EPN Elemental Point Number • Epena (Congo) (location identifier) • Établissement Public National • European Planetarium Network • European Pulsar Network

EPNdB Effective Perceived Noise Decibel

EPNI European Premier Network Integrator

E&PO Education & Public Outreach

E/PO Education/Public Outreach

EPO Earthnet Programme Office • Earth Parking Orbit • Education and Public Outreach • Electronic Publishing Only • Electronic Purchase Order • Element Project Office • Emergency Power Off • Engineering Placement

Office • Enseignement Programmé par Ordinateur • Enterprise Profit Organization • Entry Post Office • EOS Project Office • Error Performance Objective • Europäische Patentorganisation • European Patent Office (*) • European Patent Organisation

EPOC Educational Point Of Contact • Employee Participation in Organisational Change • External Payload Operations Center

EPOCH European Programme On Climatology and Hazards

EPOCS Eastern Pacific Ocean Climate Study • Equatorial Pacific Ocean Climate Studies

EPOF Ethernet Passive Optical Fiber

EPOG Eye Point Of Gaze

EPONA Energetic Particle Onset Admonitor

EPOP Electric Propulsion Orbital Platform • European Polar Orbiting Platform • European Polar Platform

EPOPE Étude de Processus dans l'Ocean Pacifique Équatorial

EPOQUE European Parliament On-line Query • European Patent Office Query (system) (*)

EPOS Earth Parameter and Orbit System • Earthquake Phenomena Observation System • Effelsberg Pulsar Observation System • Electronic Point Of Sale • Engineering and Project-management Oriented Support System • Enhanced Preferred Observation Schedule • European Polarstern Study • European Project On the Sun • European Proximity Operations Simulator

EPOSE Electronic Point Of Sale - Enhanced

EPP Electric Power Plant • Electronic Power Plant • Emergency Preparedness Plan • Emploi - Poste - Personnel • Energetic Particle Package • Engineering Professional Program • Enhanced Parallel Port • Enterprise Portal Plan • Environment Protection Programme • Essential Product Program • European Polar Platform • Experimental Physics Project • Extended Planning Period

EPPI Electronic Programmed Procurement Information

EPPL Environmental Planning and Programming Language

EPPP Electrical Power Plant Programme • Electronic Publishing Promotion Project

EP&PS Electrical Power & Pyrotechnics System

EPPS Electrical Power and Pyrotechnics System • Electronic Page Printing System • Energetic Particles and Plasma Stability

EPPSS Electrical Power/Pyro Sequential System

EPPT European Printer Performance Test

EPQ Economic Production Quantity

EPR East Pacific Rise • Education and Public Relations • Einstein - Podolsky - Rosen (paradox) • Electrical Pressure Regulator • Electronic Patient Record • Electronic Patient Report • Electronic Proposal Review • Electronic Publishing Resource • Electron Paramagnetic Resonance • Emergency Personnel Record • Engineering Purchase Request • Engine Pressure Ratio • Enseignement Primaire Rénové (Belgium) • Equipment Performance Report • Esperance (Australia-WA) (location identifier) • Essential Performance Requirement • Estimated Probability of Relevance • Établissement Public Régional • Ethylene Propylene Rubber • European Passive Reactor • European Pressurized Reactor • European Pressurized-water Reactor • EUVE Problem Reporting • Executive Progress Review • Exhaust-Pressure Ratio • Explorer Platform Retrieval

EPRB Einstein - Podalsky - Rosen - Bohm (experiment) • Emergency Position-indicating Radio Beacon

EPRCA Enhanced Proportional Rate Control Algorithm

EPRG Engineering Psychology Research Group

EPRI Electric Power Research Institute (USA-CA) • Electron Paramagnetic Resonance Imaging

EPRIJ EPRI Journal (periodical code)

EPRN Emergency Program Release Notice

EPROM Electrically Programmable Read-Only Memory • Erasable and Programmable Read-Only Memory • Erasable Programmable Read-Only Memory

EPRR Evaluation Package Readiness Review

eps Encapsulated Postscript (file extension)

EpS Earnings per Share

EPS Earnings Per Share • Earth - Planets - Space • Electrical Power System • Electronic Power Shift • Electronic Proposal Submission • Electronic Publishing System • Electrostatic Plotting Software • Electrostatic Positioning System • Embedded Processor System • Emergency Power System • Empirical Population Synthesis • Encapsulated Postscript • Engineered Performance Standard • Enhanced Press-Schechter (theory) • Ensemble Prediction Scheme • Ensemble Propulsif Stockable • Entreprise de Portage Salarial • Equilibrium Problem Solver • Établissement Psychiâtrique Spécialisé • Étage à Propergols Stockables (*) • Étage Propulsif Stockable • EUMETSAT Polar System • Euregio Publiekscentrum voor Sterrenkunde (Netherlands) • European Physical Society • European Pointing System • European Polar Satellite • European Polar System • European Project Scientist • Evolutionary Population Synthesis • Exchangeable Production System • Exercise Production System • Experimental Power Supply • Experimental Publication System • Experiment Pointing System • Extended Press-Schechter (formalism) • Extended Protocol Specification • Extensible Programming System

EPSAT Estimation des Précipitations par Satellite • Étude des Précipitations par Satellite

EPSCOR Experimental Program to Stimulate Competitive Research

EPSCS Enhanced Private Switched Communications Service • Enhanced Private Switched Communications System

EPSD Electronics and Power Sources Directorate

EPSEX Electric-Propulsion Space Experiment

epsf Encapsulated Postscript Format (file extension)

EPSF Encapsulated Postscript File • Encapsulated Postscript Format • Expenditure Per Sortie Factor

EPSHOM Établissement Principal du Service Hydrographique et Océanographique de la Marine (France)

EPSI Encapsulated Postscript Interchange

EPSIG Electronic Publishing Special Interest Group

E&PSL Earth and Planetary Science Letters (periodical code)

EPSL Earth and Planetary Science Letters • Emergency Power Switching Logic

EPSP Excitatory Postsynaptic Potential • Experiment Power Switching Panel

EPSQ Electronic Personnel Security Questionnaire

EPSRC Engineering and Physical Sciences Research Council (UK)

EPSS Electrical Power Subsystem • Electrical Power Supervision System • Electronic Performance Support System • Enhanced Packet Switched Service • Experimental Packet-Switching System

EPST Établissement Public à caractère Scientifique et Technologique

EPSTF Electrical Power System Test Facility

EPT Échelle Provisoire de Température • Effective Performance Time • Electrostatic Printing Tube • Eliptamin (Papua New Guinea) (location identifier) • Emergency

Procedure Trainer • Environmental Proof Test • Environment Policy and Technology • Ethylene-Polypropylene-Terpolymer • Excess-Profits Tax • Experiment Parameter Table • External Pipe Thread

EPTCE Electrical Power Thermal Control Electronics

EPTCS Electrical Power Thermal Control System

EPTE Existed Prior To Entry

EPTF Engineering Prototype Test Facility

EPTOMS Earth Probe Total Ozone Mapping Spectrometer

EpTU Events per Time Unit

EPTU Events Per Time Unit

EPU Electric Power Unit • Elliptical Polarized Undulator • Emergency Power Unit • Energy Process Unit • Epoxy-Urethane • Europäisches Patentübereinkommen • European Payments Union • European Picture Union

EPÜ Europäisches Patentübereinkommen

EPub Electronic Publishing

EPUSP Escola Politécnica da Universidade de São Paulo (Brazil)

EPUU Enhanced PLRS User Unit

EPV Endpoint Vendor • Ensemble de Prise de Vues • Exclusion Principle Violating

EPVAC Electronic Publishing Volunteer Advisory Committee

EPVC EPIC p-n Voltage Converter

EPVT Extended Private Storage Area

EPW Electrical Pulse Width

EPWAT Equatorial Plasma Wave Turbulence

EPWG Experiment Planning Working Group

EPY Expanded Planning Yard

EPZ Emergency Planning Zone • Export Processing Zone

Eq Equal • Equality • Equate • Equation • Équation • Equator • Equatorial • Équatorial • Équipement • Equipment • Equitable • Équitable • Equity • Equivalence • Équivalence • Equivalent • Équivalent

EQ Earthquake • Educational Quotient • Elementary Query • Emotional Quotient • Encephalization Quotient • Environmental Quality • Equal • Equality • Equalization • Equalizer • Equation • Equatorial • Équatorial • Equipment • Equipment Qualification • Equivalence • Équivalence • Example Query • External Inquiry

EQA Easy Questions and Answers • European Quality Alliance • European Quality Association • European Quality Award

EQATE Electronic Quality Assurance Test Equipment

EQC Executive Quality Council • External Quality Control

EQD Equipment Designator

EQE Effective Quadrant Elevation • Event Queue Element

EQEC European Quantum Electronics Conference

EQF Embedded Query Facility • Embedded Query Function

EQI Environmental Quality Index • Exhaust Quality Index

EQIMG Equipment Industrial Management Group

EQIP Environmental Quality Information Panel

Eql Equal • Equally

EQLIPSE Evaluation and Quality in Library Performance Systems

Eqlz Equalize • Equalizer

EQM Écart Quadratique Moyen • Electrical Qualification Model • Engineering and Qualification Model • Engineering Qualification Model • Environmental Quality Modeling • Evaluation and Qualification Model

eqn Equation (file extension)

Eqn Equation

EQN Equation • Équation

EQNET European Quality Network

Eqns Equations

EQO Environmental Quality Objective • Environment Quality Objective

EQP Equipment • Examen de Qualification Professionnelle

Eqpt Equipment

EQR Equipment Qualification Record

EQS Edinburgh QSO Survey • Edinburgh Quasar Survey • Environmental Quality Standard • Esquel (Argentina-CB) (location identifier) • European Quality System

EQT Engine Qualification Test • Environmental Qualification Test

Equ Equuleus-Equulei (constellation)

Equat Equatorial

Equil Equilibrium

Equip Equipment • Equipped

Équip Équipement

EQUIP Equipment • Equipment Usage Information Programme

EQUIS Equatorial Ionospheric Studies

Equit Equitación

Equiv Equivalent

Équiv Équivalent

EQUS Electronic Questionnaires System

EQV Equivalent

Eqvlt Equivalent

EqW Equivalent Width

EQWF Equatorward-Facing

er Elder

Er Erbium

E-R Entity-Relationship

E&R Engineering & Reliability • Evasion & Recovery

ER Early Return • Earth Resources • Earth Rotation • Eastern Range • East Reading (UK-England) • East River (USA-NY) • Echo Ranging • Edge Recombination • Edge Router • Effectiveness Report • Efficiency Report • Effort Remaining • Elastic Reality • Electrical Resistance • Electronic Reconnaissance • Electron Reflectometer • Electroreflectance • Electrorheological • Elongation Rate • Emergency Receiver • Emergency Request • Emergency Rescue • Emergency Room • Emission Region • End Row • Energy Research • Engineer Regulation • Engine Room • Enhanced Radiation • Enhanced Reactivation • Entity-Relation • Entity-Relationship • Entre Rios (Argentina) • Environmental Report • Environmental Requirement • Equatorial Red (survey) • Equatorial Ring • Équilibre Radiatif • Équipe de Recherche • Equipment Readiness • Equipment Repair • Equipment Room • Equivalence Ratio • Érection Rapide • Érection Répétitive • Eritrea (ISO country code) • Error • Error Register • Error Report • Érythrée (code ISO de pays) • Established Reliability • Evaporation Residue • Evolutionary Robotics • Expanded Response • Experiment Representative • Explanation Report • Explicit Rate • Explicit Route • Extended Range • Extended Reliability • Extended Result • External Relations • Moldova (international aircraft registration code)

ERA Early Retirement Adjustment • Earned Run Average • Earth Rotation Angle • Educational Recording Agency (UK) • Electrical Replaceable Assembly • Electronic Reading Automation • Electron-Ring Accelerator • Element - Relationship - Attribute • Emergency Relief Administration • Employee Reimbursement Account • Entités - Relations - Attributs • Entity - Relationship - Attribute • Environmental Resources Analysis • Equal Rights Amendment (USA) • Équipe de Recherche Associée • Erigavo (Somalia) (location identifier) • European Research Area • European Robot Arm • European Robotic Arm (*) • Exobiology and Radiation Assembly • Exobiology Radiation Assembly • Extended Registry Attribute • External Robotic Arm

ERAA European Regional Airline Association

ERAB Energy Research Advisory Board

ERAC European Radio Astronomy Club

ERAD Eliminate Record Administration Duplication • Enroute Radar

ERAF Earth Resources Aircraft Facility

ERAM Earth Resources Applications Mission • European Regional Astronomy Meeting

ERAMS Environmental Radiation Ambients Monitoring System

ERAP Earth Resources Aircraft Program • Earth Resources Aircraft Project • Emergency Readiness Assurance Plan

ERAR Error Return Address Register

Eras Erasmus (1466-1536)

ERAS Electronic Routing and Approval System

ERASME Étude Radar des Sols et de la Mer

ERASMUS European Action Scheme for the Mobility of University Students • European Robotics, Automation, Simulation and Mission Utilization Support

ERAST Environmental Research Aircraft and Sensor Technology

ERATO Étude d'un Rotor Aéroacoustiquement Optimisé • Exploratory Research for Advanced Technology

ERAU Embry-Riddle Aeronautical University

ERB Earth Radiation Budget • Editorial Review Board (*) • Emergency Release Board • Engineering Reference Branch • Engineering Review Board (*) • Engineering Review Body • Equipment Review Board • Ernabella (Australia-SA) (location identifier) • Executive Resource Board • Expendable Rocket Booster • Experiment Review Board

ERBD Équipe de Recherche sur les Bases de Données

ERBE Earth Radiation Backscatter Experiment • Earth Radiation Budget Experiment (*)

ERBES Earth Radiation Budget Explorer Satellite

ERBF Earth Resources Browse Facility

ERBI Earth Radiation Budget Instrument

ERBS Earth Radiation Budget Satellite

ERBSS Earth Radiation Budget Satellite System

ErbSt Erbschaftsteuer

ErbStDV Erbschaftsteuer-Durchführungsverordnung

ErbStG Erbschaftsteuergesetz

ERC Earth Rate Compensation • Earth Re-entry Capsule • Earth Return Capsule • Economic Research Council • Educational Reference Center • Educational Resource Center • Educator Resource Center • Electrical Rule Checker • Electronics Research Center • Electronics Research Council • Embedded Real-time Computer • Emergency Relocation Center • Employee Recognition Center • Employment Rehabilitation Centre • Energy Research Corporation • Engineering Readiness Center • Engineering Research Center • Engineering Research Council • Equatorial Ring Current • Equipment Readiness Code • Erzincan (Turkey) (location identifier) • European Radiocommunications Committee • European Research Conference • European Research Council • Evaluation Review Committee • Excessive Requirement Cost • Executive Review Committee • Executive Review Council • Expérience de Repérage de Champ • Explicit Route Control

ERCA European Research Courses on Atmospheres

ERCB Energy Resources Control Board

ERCC Expendability - Recoverability - Reportability Category

Ercg Erecting

ERCH Electron Resonance Cyclotron Heating

ERCIM European Research Consortium for Informatics and Mathematics

ERCM Egghe-Rousseau Core Membership • Engineering Requirements and Change Management

ERCN Educator Resource Center Network (NASA)

ERCO European Radio Communications Office

ERCOS Expérience Rayonnements Cosmiques

ERCR Electronic Retina Computing Reader

ERCS Emergency Response Computer System • Emergency Rocket Communications System

Erct Erection

ERD Berdyansk (Ukraine) (location identifier) • Ecological Research Division • Edited Raw Data • Elastic Recoil Detection • Electronic Reference Document • Electronic Report Distribution • Elementary Requirements Document • Emergency Response Division • Emergency Return Device • Entity-Relation Diagram • Entity-Relationship Diagram • Environmental Radiation Data • Equipment Readiness Date • Equivalent Residual Dose • Error Recording Device • Estimated Receipt Date • Estimated Release Date • Evolutionary Requirement Definition • Experiment Requirements Document • External Requirements Document • Extrapolated Response Dose

ERDA Elastic Recoil Detection Analysis • Energy Research and Development Administration (USA) • Exchange Regional Development Assistant

ERDAM Energy Research and Development Administration Manual

ERDAS Earth Research Data Analysis System • Earth Resources Data Analysis System (*) • Earth Resources Digital Analysis Software • Earth Resources Digital Analysis System

ERDC Earth Resources Data Center • Energy Research and Development Centre • Engineer Research and Development Center (US Army)

ERDE Engineering Research and Development Establishment

ERDF European Regional Development Fund

ERDIC Energy Research, Development, and Information Centre

ERDL Engineering Research and Development Laboratory

ERDOS Expérience Radiations Dosimétrie

ERDS Emergency Response Data System

ERE Effective Resolution Element • Electric Rocket Engine • Émission Rouge Étendue • Energy Release Efficiency • Entity Relationship Editor • Erave (Papua New Guinea) (location identifier) • Expediente de Regulación de Empleo • Extended Red Emission • Extent of Reaction Effect • Extremely Red Emission

EREA École Régionale d'Enseignement Adapté • European Research Establishments in Aerospace (formerly AEREA)

Erec Erection

EREC Error Resilient Entropy Coding

EREG Extracted Register

EREM Equipment Reduction Evaluation and Management

EREP Earth Resources Experimental Package • Environmental Recording Editing and Printing • Environment Recording Event Program

erf erforderlich

ERF Edge Response Function • Electronic Repair Facility • Electron Rest Frame • End of Radioactive Fallout • Energy Rejection Filter • Erfurt (Germany) (location identifier) • European Research Foundation • Extended Range Freighter

ERFA European Radio Frequency Agency

ERFAK Emergency Response Fly-Away Kit

ERFDP Earth Resources Flight Data Processor

ERFEN Estudio Regional del Fenomeno El Niño

ERFIS Emergency Response Facility Information System

ERFPI Extended Range Floating Point Interpretive (system)

ERFU Environmental Restoration and Facilities Upgrade

Erg Ergänze!

ERG Electrical Resistance Gauge • Electroretinogram • Electroretinograph • Electroretinography • Energy and Resources Group • Executive Review Group • Extremely Red Galaxy

ERGO European Research Gateway Online

ERGOM European Research Group On Management

Ergon Ergonomics

ERGS En-Route Guidance System

ERH Errachidia (Morocco) (location identifier) • Europäischer Rechnungshof

ErholNutzG Erholungsnutzungsrechtsgesetz

Eri Eridanus-Eridani (constellation)

ER&I Engine Receiving & Inspection

ERI Earthquake Research Institute (Japan) • Educational Research Information • Engine Receiving and Inspection • Erie (USA-PA) (location identifier) • Eritrea (ISO country code) • Érythrée (code ISO de pays) • EVN Reliability Index • Extravehicular Reference Information

ERIA Estudios y Realizaciones en Informatica Aplicada

ERIC Educational Research and Information Center • Educational Research Information Center • Educational Resources Information Center • Educational Resources Information Clearinghouse • Energy Rate Input Controller • Environmental Research and Information Centre

ERICA Experiment on Rapidly Intensifying Cyclones over the Atlantic

ERIM Earth Resource Information Management • Environmental Research Institute of Michigan (USA-MI)

ERIN Earth Resources Information Network • Environmental Resources Information Network

ERIPS Earth Resources Interactive Processing System

ERIS Earth Resources Information System • Earth Resources Inventory System • Emergency Response Information System • Energy Research Information System • Environmental Resources Information System • Exoatmospheric Reentry System

ERISA Employee Retirement Income Security Act (USA) • Employment Retirement Income Security Act (USA) (*)

ERISTAR Earth Resources Information Storage, Transformation, Analysis, and Retrieval

Erit Eritrea

ERK Electronic Record Keeping

ERKE Excess Rotational Kinetic Energy

ERKN Explicit Runge-Kutta-Nyström (method)

Erl Erlang • Erlaß • Erläuterung

ERL Earth Resources Laboratory • Echo Return Loss • Electronic Reference Library • EMSS Radio-Loud (quasar) • Energy Recovery Linear (accelerator) • Energy Research Laboratory • Environmental Research Laboratory • Error Reference Line • Event Record Log • Excess Review Listing • Extended Relational Language • Extragalactic Reference Link

ERLAP European Reference Laboratory for Air Pollution

ERLAS Earth Resources Laboratory Application Software

ERLL Enhanced Run Length Limited

ERM Earth Radiation Mission • Earth Resource Mapping • Earth Return Module • École Royale Militaire • Electronic Records Management • Electronic Rights Management • Empirical Risk Minimization • Enterprise Resource Management • Entity-Relationship Model • Environmental Resource Management • EPIC Radiation Monitor • Erechim - Comandante Kraemer (Brazil-RS) (location identifier) • Ethernet Router Module • Events Resourcing Model • Exact Repeat Mission • Exchange Rate Mechanism • Executive Review Meeting • Explicit Rate Marketing • Extended Relational Model • External Relations and Marketing • External Resource Manager

ERMC External Relations and Marketing Committee

ERMD EPIC Radiation Monitor Detector

ERME EPIC Radiation Monitor Electronics

ERMES European Radio Messaging System

ERMP Effective Records Management Project • Environment Management and Review Program

ERMS Environmental Resources Mapping System • EPIC Radiation Monitor System

ERN Eirunepé (Brazil-AM) (location identifier) • Engineering Release Notice • Enterprise Resource Network • Explicit Route Number

ERNA European Recoil (separator) for Nuclear Astrophysics

ERNE Energetic and Relativistic Nuclei and Electrons (SOHO)

ERNET Educational Research Network

ERNIE Electronic Random Number Indicator Equipment • Electronic Random Numbering and Indicating Equipment

ERNN External Recurrent Neural Network

ERNS Emergency Response Notification System

ERO Early Release Observation • Eldred Rock (USA-AK) (location identifier) • Electronic Return Originator • Emergency Repair Overseer • Emergency Response Organization • Energy Research Office • Engineering Release Operation • Engineering Release Order • Engine Running Offload • Enhanced Reliability Option • European Radiocommunications Office • European Regional Organization • European Research Office • Executive Records Office • Explicit Route Object • Extended Range Operations • Extremely Red Object

EROC Ecological Rate Of Change

EROM Erasable Read-Only Memory

EROMM European Register of Microform Masters

EROPS Extended-Range Operations

EROS Earth Resources Observation Satellite • Earth Resources Observation System • Earth Resources Observing System • EISCAT Real-time Operating System • European Rivers Outflow Studies • Expérience de Recherche d'Objets Sombres • Experimental Reflection Orbital Shot

ERP Earth Resources Program • Earth Rotation Parameters • Educational Reimbursement Program • Education and Public Relations • Effective Radiated Power • Elevated Release Point • Elevator Reference Plane • Emergency Requirements Plan • Emitted Radiated Power • Energy Radiation Pattern • Engineer Regulating Point • Engineer Requirement Plan • En-route Reporting Point • Enterprise Resource Planning • Environmental Research Park • Episodic Response Project • Eppawala Rock Phosphate • Equipment Reception Party • Error Recovery Package • Error Recovery Procedure • Établissement Recevant du Public • European Recovery Programme • Event-Related Potential • Evoked Response Potential • Extended Range Proximity • Extensional Recursive Processing • Exterior Routing Protocol • Eye Reference Point

ERPG Emergency Response Planning Guidelines

ERPLD Extended Range Phase-Locked Demodulator

ERPS Equipment Release Priority System • Experimental Rocket Propulsion Society (USA-CA)

ERPSL Essential Repair Parts Stockage List

err Error Message (file extension)

ERR Electrical Resistance Requirement • Energy Release Rate • Engineering Release Record • Engineering Review Report • Engineering Revision Request • Engine Requirements Review • Equivalent Range Rate • Errol (USA-NH) (location identifier) • Error • Excess Relative Risk

ERRAC European Rail Research Advisory Council

ErrC Error Correction

ERRC Expendability, Recoverability, and Repair Capability • Expendability, Recoverability, and Repairability Category • Expendability, Recoverability, and Repairability Code • Expendability, Recoverability, Recoverability Category

ERRI European Rail Research Institute (Netherlands)

Erron Erroneous • Erroneously

ERRP Equipment Replacement and Refurbishment Plan

ERRRIS E-Region Rocket/Radar Instability Study

ERRSAC Eastern Regional Remote Sensing Applications Center

ERRT Error Rate Relative to Truncation

Ers Erase • Erased

ERS Earnings-Related Supplement • Earth Recovery System • Earth Remote Sensing • Earth Resources Satellite • Earth Resources Survey • Economic Research Service • Electric Resource System • Electronic Receipt Settlement • Element Rotation Stand • Emergency Radio Service • Emergency Relocation Site • Emergency Response System • Engineering Release System • Enterprise Router Switch • Entrée Réservée Synchrone • Entry and Recovery Simulation • Environmental Research Satellite • Erase • Error Reporting System • Escadron de Reconnaissance Stratégique • Estimated Release Schedule • European Remote Sense • European Remote Sensing • Extended Review Session • External Reference Specification • Extragalactic Radio Source • Windhoek Eros (Namibia) (location identifier)

ERSAC Environmental Remote Sensing Applications Centre

ERSAF Environment Remote Sensing Analysis Facility • European Remote Sensing Aircraft Facility

ERSAL Environmental Remote Sensing Applications Laboratory

ERSAR Earth Resources Synthetic Aperture Radar

ERSC Environmental Remote Sensing Centr • Eurimage, Radarsat International, Spot Image Consortium • European Research Standing Committee

ERSCOS European Remote-sensing Satellite Central User Service

ERSD Estimated Return to Service Date

ERSDAC Earth Remote Sensing Data Analysis Center • Earth Resources Satellite Data Analysis Center (Japan)

ERSDIS Earth Resources Satellite Data and Information Center

ERSE Erase

ERSI Elastomeric Reusable Surface Insulation • Environmental and Remote Sensing Institute

ERSIS Earth Resources Spectral Information System

ERSISS European Remote-sensing Satellite Interface Subset

ERSOS Earth Resources Survey Operational System

ERSP Earth Resources Survey Program • Electronic Reservation Services Provider • Epipolar Rectified Stereo Pair

ERSR Equipment Reliability Status Report

ERSS Earth Resources Survey Satellite

ERSSP European Remote Sensing Satellite Programme

ERST European Report on Science and Technology (periodical code)

ERSTI European Report on Science and Technology Indicators

ERSUN Environmental Remote Sensing Unit

ERT Earth Received Time • Earth Resources Technology • Educational Requirements Test • Education Round Table • Emergency Response Team • Environmental Response Team • Environment Round Table • Équipe de Recherche Technologique • Equipment Repair Time • Erdenet (Mongolia) (location identifier) • European Round Table • Excess Retention Tax • Execution Reference Time • Extended Range Transmitter

ERTBP Elliptical Restricted Three-Body Problem

ERTCS Emergency Recirculation and Tritium Cleanup System

ERTE Exact Radiative Transfer Equation

ERTG Economic Radioisotope Thermoelectric Generator

ERTMS European Railway Traffic Management System

ERTS Earth Resources Technology Satellite (later called Landsat)

ERTU Egyptian Radio and Television Union

ERU Earth Rate Unit • Emergency Reaction Unit • Emergency Recovery Unit • Energy Research Unit • Équipe de Réponse aux Urgences • Error Return Update • Erume (Papua New Guinea) (location identifier)

ERUS ERS User Services

ERV Earth Return Vehicle • Emergency Rendezvous • English Revised Version • Experimental Reentry Vehicle • Expiratory Reserve Volume • Kerrville (USA-TX) (location identifier)

ERVS Early Retirement and Voluntary Severance (scheme)

Erw Erweitert

ERW Electronic Resistance Welding

Erweit Erweitert

ERX Electronic Remote Switching

ERY Early

ERZ Erzurum (Turkey) (location identifier)

Erzgeb Erzgebirge (Germany)

es Spain • Spanish (ISO language code)

Es Einsteinium • Esempio

E/S Engine Speed • Entrées/Sorties • Equipment Section

E&S Earth & Space (periodical code) • Engineering & Science • Engineering & Scientific • Engineers & Scientists • Evans & Sutherland Computer Corporation (USA-UT)

ES Earth Science • Earth Station • Earth Sensor • Echo Sounding • Echo Suppression • École Secondaire • Économique et Social • Eddington-Sweet (circulation) • Education Specialist • Éducation Surveillée • Elastic Scattering • Eldest Son • Electrical Sounding • Electric Starting • Electrochemical Society • Electromagnetic Switching • Electronic Stabilization • Electronic Storage • Electronic Structure • Electronic Support • Electronic Switch • Électronique Séquentielle • Electron Synchrotron • Electrostatic • Ellipsoidal Shell • El Salvador (international vehicle registration code) • Embedded System • Employment Strategy • Enstelle • End Strength • End System • Energy Spectrum • Engineering and Science • Engineering Specification • Engine Speed • Enriched Sample • Enseignement Supérieur • Enter Shadow • Entwicklungssynthese • Environmental Shelter • Equal Section • Equipment Section • Equipment Shelf • Equipment Specialist • Érection Souple • Error Second • Escape System • Eskimo • Espagne (code ISO de pays) • Espíritu Santo (Brazil) • Estonia (international aircraft registration code) • Étoile Symbiotique • Euler Sum • Evolutionary Synthesis • Evolution Strategy • Executive Secretary • Exercise Scenario

• Exercise Server • Exercise Specialist • Expanded Storage • Experimental Station • Experiment Segment • Expert System • Exposed Section • External Subscriber • Extra Segment • Spain (ISO country code)

ESA Ecological Society of America (USA) • EFTA Surveillance Authority • Electrical Surge Arrestor • Electronically Scanned Array • Electronically Scanning Antenna • Electronically Scanning Array • Electronically Steerable Antenna • Electronically Steered Array • Electronic Static Amplifier • Electrostatic Analyzer • Electrostatic Sector Analyzer • Electrostatic Spherical Analyzer • Emergency Safe Altitude • Endangered Species Act (USA) • End System Administrator • Energy Sensor Array • Energy System Analysis • Engineering Sciences and Applications • Engineering Supply Area • Engineering Support Activity • Engineering Support Area • Engineering System Assembly • Enhanced Situation Awareness • Enterprise Storage Automation • Enterprise Systems Architecture • Entomological Society of America (USA) • Environmentally Sensitive Area • Environmental Sensor Assessment • Equatorial South America • Esa'Ala (Papua New Guinea) (location identifier) • European Sociological Association • European Space Agency • European Symposium on Algorithms • European System of Accounts • Evening Sector Arc • Excited State Absorption • Expiration of Service Agreement • Explosive Safe Area • Extended Service Area

ESAA École Supérieure des Arts Appliqués

ESAAB Energy System Acquisition Advisory Board

ESABu European Space Agency Bulletin (periodical code)

ESAC Earth Sciences Advisory Committee (ESA) • Electronic Systems Assistance Center • Employé de Service Administratif et Commercial • Enterprise-wide Security and Audit Conference • Environmental Studies Association of Canada • European Science Advisory Committee (ALMA)

ESAD Earth Science and Applications Division • École Supérieure d'Art et de Design (France) • Electronic Safe and Arm Device

ESADE Escuela Superior de Administración y Dirección de Empresas (Spain)

ESADS Earth Science and Applications Data Systems

ESAE École Supérieure d'Administration des Entreprises (France)

ESAEU Examen Spécial d'Accès aux Études Universitaires

ESAGB Early Super Asymptotic Giant Branch

ESAJ European Space Agency Journal (periodical code)

ESAL Entidad Sin Ánimo de Lucro

ESAM Earth Sensor Assembly Module • Energy Saving Assessment Model • Environmental Science And Management

ESANET European Space Agency Network

ESANN European Symposium on Artificial Neural Networks

ESAP ECLS Systems Assessment Program

ESAR Electromagnetic Spectrum Allocation Request • Electronically-Scanned Array Radar • Electronically-Steerable Array Radar • Exercice - Symbole - Assemblage - Règle • Extended Subsequent Application Review

ESARS Expert System for Accessing Remotely Sensed (data)

ESAS Enhanced Situational Awareness System • Extended Source Analysis Software

ESAST European Space Agency Scientific Technical Review (periodical code)

ESAT École Supérieure des Arts et Techniques (France) • ERB Solar Analysis Tape

ESATAN ESA Thermal Analysis Network

ESATS Electrical Subassembly Test Set

ESAW Engineering Societies in the Agent Worlds (workshop)

ESB Ankara Esenboga (Turkey) (location identifier) • B-Type Spanish Peseta (ISO currency code) • Electrical Stimulation of the Brain • Electricity Supply Board • Electric Storage Battery • Emerging Small Business • Empire State Building (USA-NY) (*) • Empirical Surface Brightness (method) • Empty State Building (USA-NY) (earlier nickname for (*)) • Endside Bus • Engineering Standards Board • Engineering Support Branch • Erweiterter Sonderbereich • Essential Switching Box • Exaspération Subite et Brutale • Executive Safety Board

ESBS École Supérieure de Biotechnologie de Strasbourg (France)

ESBT Experimental S-Band Terminal

Esc Escalera • Escalier • Escape • Escompte • Escrow • Escudo (currency) • Escultura

ESC École Supérieure de Chimie • École Supérieure de la Communication • Economic and Social Committee • Economic and Social Council • Educational Scientific Computer • Electronic Security Command • Electronic Still Camera • Electronic Structural Correlator • Electronic Switching Center • Electronic Systems Center • Electronic Systems Command • Electrostatic Compatibility • Engineering Service Circuit • Engineering Support Center • Environmental Support Center • EPIC Science Console • Equipment Serviceability Criteria • Equipment Subcommittee • Escada • Escanaba (USA-MI) (location identifier) • Escape • Escudo (currency) • ESIS Steering Committee • Étage Supérieur Cryogénique • European Seismological Commission • European Space Conference • Euro Space Center (Belgium) • Evergreen State College (USA-WA) • Executive Steering Committee • Expedited Site Characterization • Extended Core Storage

ESCA Electron Spectroscopy for Chemical Analysis • Engenharia de Sistemas de Controle e Automação • European Speech Communication Association

Escand Escandinavo

ESCAP Economic and Social Commission for Asia and the Pacific (UN)

ESCAPE Experiment of the Sun for Complementing the ATLAS Payload for Education

Escarp Escarpment

ESCAT Emergency SEcurity Control Air Traffic

ESCB European System of Central Banks

ESCC European Space Components Coordination • European Supernova Cosmology Consortium • Experiment Scoring and Control Centre • External Stress Corrosion Cracking

ESCCD Electrical Short-Circuit Current Decay

ESCDR Escadaria

ESCES Experimental Satellite Communication Earth Station

ESCI Environmental Science Citation Index

ESCIN European Science Communications and Information Network

Escl Escalator • Escalinata

ESCM Extended Services Communications Manager

ESCNH Escadinha

ESCO Educational, Scientific, and Cultural Organization • European Satellite Consulting Organization

ESCON Enterprise Systems Connection

ESCP Earth Science Curriculum Project

ESCPI École Supérieure de Chimie et de Physique Industrielle

ESCR Escritório • External Standard Channel Ratio

ESCS Electronic Spacecraft Simulator • Emergency Satellite Communications System • Enterprise Strategy Control Structure

ESCSI Embedded Small Computer System Interface

ESCT École Supérieure des Cadres et Techniciens • European Supersonic Civil Transport

ESCULT Escultor

ESCWA Economic and Social Commission for Western Asia

ESD Earliest Start Date • Eastsound (USA-WA) (location identifier) • Echo-Sounding Device • Ecologically Sustainable Development • Educational School District • Educational Service District • Educational Systems Division • Electromagnetic Shower Detector • Electronically-Stabilized Data • Electronic Software Distribution • Electronic Spectroscopic Diffraction • Electronic Systems Division • Electronics Systems Division • Électronique Serge Dassault • Electron-Stimulated Desorption • Electrostatic Discharge • Electrostatic Sensitive Device • Electrostatic Spray Deposition • Emergency Shutdown • Emission Standards Division • Energy Spectral Density • Engineering Services Department • Engineering Systems Division • Engine Shutdown • Enseignement du second Degré • Environmental Sciences Division • Estimated Shipping Date • European Standard Drawer • Experiment Systems Division • Exploitation Support Data • External Symbol Dictionary

ESDA Earth Sciences Data Standards • Electronic System Design Automation • Electrostatic Deposition Analysis

ESDB Endangered Species Database • Extended Source Database

ESDD Earth Science Data Directory

ESDI Enhanced Small Device Interface (*) • Enhanced Standard Device Interface • Enhanced System Device Interface • Extended Small Device Interface • Extended Small Disk Interface

ESDIM Earth System Data and Information Management

ESDIS Earth Science and Distributed Information System • Earth Science Data and Information System • Earth Science Data Information System

ESDOC European Science Data and Operations Centre

ESDP European Spatial Development Perspective • External Sensor Data Processing

ESDS Electrostatic Discharge Sensitive • Entry Sequenced Data Set

ESDT Earth Science Data Type

ESDU Engineering Sciences Data Unit • Enhanced Standalone Display Unit • Event Storage and Distribution Unit

E&E Electrical & Systems Engineering

ESE Earth Science Enterprise (NASA) • East-Southeast • East-Southeastern • Electrical and Systems Engineering • Electrical Support Equipment • Electronic Security Environment • Electronic Sonar Equipment • Electronic Support Equipment • Electron Spin Echo • Engineering Support Equipment • Engineer Support Equipment • Enhanced Spectrum Efficiency • Ensenada (Mexico) (location identifier) • Est-Sud-Est • Experiment Support Equipment • Extreme Scattering Event

ESEA Elementary and Secondary Education Act (USA)

ESEC European Software Engineering Conference

ESECA Energy Supply and Environmental Coordination Act (USA)

ESEE European Society for Engineering Education • Exhaustive Set of Exclusive Events

ESEM Environmental Scanning Electron Microscope • European Society for Engineering and Medicine • Evaluation of Space Environment with Materials

ESEO École Supérieure d'Électronique de l'Ouest (France)

ESEP Earth System Evolution Program • Engineer and Scientist Exchange Program • Enhanced Service Engineering Plan • European Science Exchange Programme

ESERD Electronic Support Equipment Recommendation Data

ESERN Eastern

ESES Environmentally Sound Energy System • ESA Software Engineering Standards

ESEU Examen Spécial d'Entrée à l'Université

ESEWD East-Southeastward

ESEX Electric Space Experiment

ESF Alexandria Esler (USA-LA) (location identifier) • Earth Surface Find • Earth Surface Fixed • Economic Support Fund • Edge Spread Function • Éducateurs Sans Frontières • Emergency Support Function • Engineering Safety Feature • Equatorial Spread-F • Établissement Scientifique Fédéral (Belgium) • Eureka Software Factory • European Science Foundation • European Social Fund • Evans Solar Facility (NSO) • Explosive Safe Facility • Extended Service Frame • Extended Superframe • Extended Superframe Format

ESFC Equivalent Specific Fuel Consumption

ESFJ Extraverted Sensing Feeling Judging

ESFON European Star Formation Network

ESFP Extraverted Sensing Feeling Perceiving

ESFRI European Strategy Forum on Research Infrastructure

ESFU Enhanced Standard Format Unit

ESG Earth System Grid • École Supérieure de Gestion • Education Support Grant • Electronic Security Group • Empirical Study Group • Enseignement Secondaire Général • Executive Steering Group • Experimental Study Group • Mariscal Estigarribia (Paraguay) (location identifier)

ESGAC European Space and Ground Astronomy Coordination

ESGEO ESRO Geostationary Earth Orbiting Satellite

ESGN Electrically-Suspended Gyro Navigation

ESGP Earth Science Geostationary Platform

Esgr Esgrima

ESGT École Supérieure des Geomètres et des Topographes (France)

esh Extended Shell (file extension)

ES&H Environmental Safety & Health • Environment, Safety & Health

ESH Earth System History • Environmental Safety and Health • Environment, Safety and Health • Equivalent Solar Hour • Equivalent Standard Hour • Sahara Occidental (code ISO de pays) • Sahara (Western_) (ISO country code) • Shoreham by Sea (UK-England) (location identifier) • Western Sahara (ISO country code)

ESHL EROS SMC HAeBe Candidate

ESHOE Electrically Switchable Holographic Optical Element

ESHP Equivalent Shaft Horsepower

ESI Earth Sciences Information • Earth Space Institute (France) • Earth Systems Institute • Echelle Spectrograph and Imager • Echellette Spectrograph and Imager • École des Sciences de l'Information • École Supérieure d'Informatique • Economic Strategy Institute (USA) • Electrical System Integration • Electronic Share Information • Electrospray Ionization • Ellery Systems Inc. (USA-CO) • End System Identifier • Engineering and Scientific Interpreter • Enhanced Serial Interface • Enseignement Secondaire Inférieur • Environmental Sensitivity Index • Environmental Severity Index • Environment Sensitivity Index • Escadron Spécial d'Intervention • Especially Sensitive Information • Espinosa (Brazil-MG) (location identifier) • Essential Sustainment Item • Ethernet

Serial Interface • European Satellite Industry • European Software Institute (Spain) • Experimental Systemic Initiative • Exploration Science Institute (USA-TX) • Externally Specified Index • External System Interface • Extremely Sensitive Information

ESIAC Electronic Satellite Imaging Analysis Console

ESIC Earth Science Information Center • Environmental Science Information Center

ESID EOSDIS Systems Interface Document

ESIE Elementary, Secondary, and Informal Education

ESIEA École Supérieure d'Informatique Électronique Automatique (France)

ESIEE École Supérieure d'Ingénieurs en Électrotechnique et Électronique (France)

ESIG Environmental and Society Impact Group • European SMDS Interest Group

ESIHS Expert System Information Handling System

ESIIP Environmental Services Interface Implementation Plan

ESIM Earth Science Information Manager • École Supérieure d'Ingénieurs de Marseille (France)

ESIMS Engineering Services Information Management System

ESIMUD European Space Industry Multimedia Database

ESIN Earth Science Information Network

ESIO Earth Science Information Office

ESIP Earth Science Information Partner • European Software Innovation Prize

ESIPP Earth Science Imaging Processing Package

ESIRS Expert System Information Retrieval System

ES-IS End System to Intermediate System

ESIS End System to Intermediate System • ESCAP Statistical Information System • European Social-science Information System • European Space Information System • European Survey of Information Society

ESIST European Society and its Interactions with Science and Technology

ESIWG Element and Systems Integration Working Group

ESJ École Supérieure de Journalisme (France)

ESJWG Earth Sciences Joint Working Group

Esk Eskimo

ESK Eskisehir (Turkey) (location identifier)

ESKAC European Square Kilometer Array Consortium

ESL Electronic Software Licensing • Electronic Systems Laboratory • Elista (Russia) (location identifier) • Emergency Staffing Level • Energy Sciences Laboratories • Engineering Societies Library • English as a Second Language • ESA Simulation Language • European Small Launcher • European System Language • Expected Search Length • Expected Significance Level • Extended Shelf Life

ESLAB European Space Laboratory

ESLIT Entrance Slit

ESLOC Equivalent Source Line Of Code

ESLRF Eye-Safe Laser Range Finder

ESLSCA École Supérieure Libre des Sciences Commerciales Appliquées (France)

ESM Earth Search Mode • École Supérieure de Métrologie • Edible Structural Material • Electronic Signal Monitoring • Electronic Support Measure • Electronic Surveillance Measurement • Emission Sensing Monitor • Energy Storage Module • En Sus Manos • Enterprise Storage Management • Environmental Systems Monitor • Esmeraldas (Ecuador) (location identifier) • Executive Secretaries Meeting • Executive Staff Meeting • Experimental Support Module • Experiment Support Module • Experiment Systems Monitor • Extended Sum-rule Model • External Storage Module

ESMAP Energy Sector Management Assistance Programme

ESMDM Enhanced Space-Station Multiplexer/Demultiplexer

ESME École Supérieure de Mécanique et d'Électricité

ESMF Efficient Separable Median Filter

ESMN European Solar Magnetometry Network

ESMO Earth Science Mission Operations

ESMOS European Stratospheric Monitoring Station

ESMP ECS Schedule Management Plan

ESMR Electrically Scanned Microwave Radiometer • Electrically-Scanning Microwave Radiometer • Emergency Specialized Mobile Radio • Enhanced Specialized Mobile Radio • Expert Systems Message Router

ESMS Environmental and Special Measurements System

ESMT École Supérieure Multinationale des Télécommunications

Esn Essential

ESN Earth Science Network • Easton (USA-MD) (location identifier) • Educationally Subnormal • Electronic Security Number • Electronic Serial Number • Electronic Switched Network • Emergency Service Number • English-Speaking Nation • Enhanced Service Node • Enterprise Storage Network • EOSDIS Science Network • European Science Notes (periodical code)

ESNG Engineering Subworking Group

Esntl Essential

ESNU Enseignement Supérieur Non-Universitaire (Belgium)

ESO École Supérieure d'Optique • Educación Secundaria Obligatoria (Spain) • Embarkation Staff Officer • Emergency Security Option • Emergency Services Officer • Enseñanza Secundaria Obligatoria • Entry Server Offering • Espanola (USA-NM) (location identifier) • European Southern Observatory • Event Sequence Override • Expanded Storage Only • Extended Security Option

ESOAR European Southern Observatory Annual Report (periodical code)

ESOB Environmental State Operating Budget • European Southern Observatory Bulletin (periodical code)

ESOC Emergency Supply Operation Center • European Space Operations Centre (ESA) • EUVE Science Operations Center (USA-CA) • Extrasolar Oort Cloud

ESOF European Science Open Forum

ESOH Environment, Safety and Occupational Health • Equipment and Supplies On Hand

ESOL English for Speakers of Other Languages

ESOMAR European Society for Opinion and Marketing Research

ESOMe European Southern Observatory Messenger (periodical code)

ESOP Electron Scattering Off Partons • Employee Share Ownership Plan • Employee Stock Ownership Plan • Enhanced Switch Operating Position • Enhanced Switch Operating Procedure • Enhanced Switch Operations Program • European Symposium on Occultation Projects • European Symposium On Programming

ESORICS European Symposium On Research In Computer Security

ESOSP European Southern Observatory Scientific Preprints (periodical code)

ESOSR European Southern Observatory Scientific Reports (periodical code)

ESOT Employee Stock Ownership Trust

ESOW Engineering Statement Of Work

Esp Espace • Espagne • España • Español • Especial • Especially • Especialmente • Esperanto • Esplanade

ESP Early Support Planning • Early Support Program • East Stroudsburg - Birchwood-Pocono (USA-PA) (location identifier) • Economic Simulation Program • Electric Submersible Pump • Electronic Schematic Processor • Electronic Shock Protection • Electronic Sort Processor • Electronic Stability Program • Electronic Standard Procedure • Electronic Steering Programme • Electroshock Protection • Electron Spin Polarization • Electron Stream Potential • Elsevier Science Publishers • Emergency Special Program • Empirical Studies of Programers • Employee Suggestion Program • Emulation Sensing Processor • Encapsulating Security Payload (*) • Encapsulating Security Protocol • Encapsulation Security Payload • Encryptor Support Package • Energetic Solar Particle • Energetic Storm Particle • Energy Savings Performance • Energy Simulation Package • Energy Sum Processor • Engineering and Scientific Processor • Engineering Simulation Program • Engineer Supply Point • English for Special Purposes • English for Specific Purposes • Enhanced Scalar Processing • Enhanced Serial Port • Enhanced Service Provider • Enhanced Support Package • En-route Spacing Program • Enseignement Secondaire Professionnel • Enterprise Server Platform • Enterprise Storage Platform • Entry Schedule for Periodicals • Environmental Sensing Platform • Equipment Survey Program • ESO Slice Project • Esoteric Scholarly Publication • Espagne (code ISO de pays) • Essential Service Protection • Estimated Selling Price • European Society of Photobiology • Evolutionary Spiral Process • Exchangeable Sodium Percentage • Executive System Planning • Expandable Stored Program • Expanding Space Paradigm • Expanding Spread Profile • Experiment Sensing Platform • Extended Self-contained PROLOG • Extended Service Plan • Extended Service Program • Extended Storage Provision • Extended Surface Photometry • Extended Support Plan • Extrasensory Perception • Extrasolar Planet • Extrasolar Planetary • Extravehicular Support Pack • Peseta (Spain) (ISO currency code) • Spain (ISO country code)

ESPACE European System for Particle and Astrophysics Cybereducation • Évaluation et Suivi de la Production Agricole en fonction du Suivi et de l'Environnement

E-SPAN European SPAN

ESPAN European SPAN

ESPAR Electronically-Steerable Phased Array Radar

ESPAS ESO Search for Potential Astronomical Sites

ESPC Energy Saving Performance Contract • European Space Power Conference

ESPCI École Supérieure de Physique et de Chimie Industrielles (France)

ESPE Escuela Politécnica del Ejercito

Espec Especial • Especially

ESPEO École Supérieure des Procédés Électroniques et Optiques

ESPF Experimental SAR Processing Facility

ESPG ESR Software Project Group • Experimental Space Plasma Group (UNH)

ESPI Electronic Speckle Pattern Interferometry • Extrasolar Planetary Imager • Extrasolar Planet Imager

ESPIN Emergency Spare Parts Information Network • European Space Information Network

Espl Esplanada

ESPL Extensible Structure Processing Language

ESPM Escola Superior de Propaganda e Marketing

ESPN Entertainment Sports Network

ESPOD Electronic Systems Precision Orbit Determination

ESPOL Escuela Superior Politécnica del Litoral (Ecuador) • Executive System Problem-Oriented Language

ESPP Employee Share Purchase Plan

ESPRIT European Strategic Programme for Research in Information Technology • European Strategic Programme for R&D in Information Technology (*)

ESPS Entropic Signal Processing System • Experiment Segment Pallet Simulator

ESPT État de Stress Posttraumatique

Esq Esquina • Esquire

ESQ Esquerdo • Esquire

Esq° Esquerdo

Esqr Esquire

ESR Early Site Review • Early Strategy Review • Earth Science Review • Earth Space Review • Educational System Reform • Effective Sunrise • EISCAT Svalbard Radar • Electrical Substitution Radiometer • Electron Spin Resonance • El Salvador (Chile) (location identifier) • Emergency Sun Re-acquisition • Engineering Support Request • Enseignement Secondaire Rénové (Belgium) • Equipment Status Report • Equivalent Series Resistance • Equivalent Service Round • Event Service Routine • Evolutionary Scheduling and Routing • Experimental Storage Ring

ESRA European Safety and Reliability Association

ESRANGE European Sounding Rocket Launching Range • European Space Range

ESRC Economic and Social Research Council (UK) • Engineering and Services Readiness Center • Environmental Science Research Center • European Science Research Council

ESRD Equipment Shipment Ready Date

ESRF Environmental Studies Revolving Funds • European Synchrotron Radiation Facility

ESRI Earth Sciences and Resources Institute • Environmental Systems Research Institute

ESRIN European Space Research Institute

ESRM Enterprise Storage Resource Management

ESRO European Space Research Organization (merged now with ELDO into ESA)

ESRP European Supersonic Research Program

ESRR Early Site Review Report

ESRS European Synchrotron Radiation Society

ESRU Earth Satellite Research Unit

ESRv Earth Science Reviews (periodical code)

Ess Essence • Essential • Essentially • Essex (UK-England)

ES&S Engineering Services & Safety

ESS Early Supply Support • Earth and Space Sciences • Earth System Science • Effective Sunset • Efficient Sensing System • Electronic Scoring System • Electronic Security System • Electronic Signaling System • Electronic Standards System • Electronic Surveillance System • Electronic Switching System • Elementary Science Study • Emergency Survival System • Energy Storage System • Engineered Safety System • Engineering and Software Services • Engineering Services and Safety • Engineering Support System • Engineering System Solution • Enseignement Secondaire Supérieur • Enterprise Security Server • Enterprise Storage Server • Environmental Simulator System • Environmental Stress Screening • Environmental Support System • Environment, Science and Society • Equipment Support Section • Equipment Support System • Escrow Encryption Standard • ESO Sculptor Survey • Essen (Germany) (location identifier) • Essential • Ethernet Segment Switch • European Simulation Symposium • European Social Survey • European Spallation Source • European Statistical System • Evolutionarily Stable Strategy • Executive Support System • Executive Systems Software • Expendable Second Stage • Experiment Support System • Experiment System Simulator • Expert Support System • Expert System Shell • Exploitation Support Segment • Exploitation Support

System • Exploration of the Solar System • Extended Self-Similarity • Extended Service Set • External Suspension System

ESSA Electrically Steerable Spherical Array • Electronic Switching Spherical Array • Electrostatic Surge Arrestor • Environmental Science Services Administration (USA) • Environmental Science Services Agency • Environmental Survey Satellite

ESSAC Earth System Science and Applications Advisory Committee

ESSAT European Society for the Study of Science and Theology

ESSC Earth System Science Center • Earth System Sciences Committee • Edmonton Space and Science Centre (Canada-AB) • Electronic Sustainment Support Center • Engineering Support Services Contract • European Space Science Committee

ESSCO Electronics Space Systems Co.

ESSD Electronic Sensors and Systems Division • Environmentally Sound and Sustainable Development • European Space Science Directory • Executive Software Systems Directorate

ESSDE European Space Software Development Environment

ESSE Earth System Science Education

ESSEC École Supérieure des Sciences Économiques et Commerciales (France)

ESSENTIAL European Systems Strategy for the Evolution of New Technology In Advanced Learning

ESSEX EUV Solar Spectrograph Explorer • Experimental Solid-State Exchange

ESSG Engineering Strategic Studies Group

ESSI European Software and Systems Initiative

ESSII Expert System for Satellite Image Interpretation

ESSIR European Summer School on Information Retrieval

ESSIS Earth and Space Science Information System • Earth System Science Information System

ESSIT ECS Science Software Integration Team

ESSL Engineering and Scientific Subroutine Library

ESSLLI European Summer School in Logic, Language, and Information

ESSLR Eye-Safe Simulated Laser Rangefinder

ESSM Equip - Supply - Service - Maintain • Expert System Simulation Model

ESSN Earth System Science Network

ESSO Earth Science Support Office • European Software Support Office

ESSOPE Expert System for Spacecraft Operations Planning and Execution

ESSP Earth System Science Pathfinder • Earth System Science Program • Educational Space Simulations Project • Enchanted Skies Star Party (USA-NM)

ESSS Electronic Single Support Strategy • Evolutionary Space Surveillance System • External Stores Support System

ESST European Studies in Science and Technology

ESSX Electronic Switching System Exchange

Est Established • Establishment • Estate • Estatuto • Estimate • Estimated • Estimation • Estimator • Estonia • Estonian • Estrecho • Estuary • Estudio

ESt Einkommensteuer • Eisenbahn Station

ES&T Environmental Science & Technology • Environment, Science & Technology

EST Earliest Start Time • Earth - Satellite - Target • Eastern Standard Time (*) • Eastern Summer Time • École Secondaire et Technique • Effective Straight-line Trajectory • Electronic Storage Tube • Electronic Support and Training • Electroshock Therapy • Electroshock Treatment • Ellipsoidal Shell Tokamak • Emergency Service Team • Emergency Support Team • Engineering String Test • Engineering Support and Technologies • En-route Support Team • Enseignement Secondaire Technique • Environmental System Test • Environmental System Testing • Environment, Science and Technology • Establish • Estate • Estherville (USA-IA) (location identifier) • Estimate • Estimation • Estonia (ISO country code) • Estonie (code ISO de pays) • Expanded Service Test • Expanding Spacetime • Expanding Spacetime Theory • Expert System Tools • Expressed Sequence Tag

ESTA Engineering Support Task Authorization • Escape System Test Article • European Science and Technology Assembly

Estab Establish • Established • Establishment

ESTAB Estabelecimento

ESTAC Estação

Estad Estadística

ESTAE Extended Specify Task Abnormal End

ESTAR Electronically Scanned Thinned Array Radiometer (*) • Electronically Steered Thinned Array Radiometer • External Servicing Technology Testbed for Automation and Robotics

Estát Estática

Estb Establish • Established • Establishment

ESTB EGNOS System testbed

ESTC Ethiopian Science and Technology Commission (Ethiopia)

EStDV Einkommensteuer-Durchführungsverordnung

ESTE Electrical Special Test Equipment

ESTEC European Space Research and Technology Centre (ESA)

ESTEEM Earth Science Teachers Exploring Exemplary Materials

ESTF Electronic System Test Facility

Estg Estimating

EStG Einkommensteuergesetz

EStH Einkommensteuerhinweise

ESTI European Science Teachers Initiative

ESTJ Extraverted Sensing Thinking Judging

ESTL Electronic Systems Test Laboratory • European Space Tribology Laboratory

ESTMP European Space Technology Master Plan

Estn Estimation

ESTO Earth Science and Technology Organization (Japan) • Electronic Systems Technology Office • Engineering Service Test Office • European Science and Technology Observatory

ESTOL EOS Standard Text and Operation Language

ESTP Extraverted Sensing Thinking Perceiving

EStR Einkommensteuerrichtlinien

ESTR Engineering and Science Tape Recorder • Environment, Science and Technology Resources • Estrada

ESTRACK European Space Tracking (ESA)

EStRG Einkommensteuerreformgesetz

ESTRU Environmental Science and Technical Research Unit

ESTS Electronic System Test Set • Estates

ESTT École Supérieure Technique des Transmissions

ESTU Extended Stretched Transuranium (WNSL)

ESU Electronic Storage Unit • Electrostatic Unit • Element Supervisor Unit • Emporia State University (USA-KS) • Environmental Simulator Unit • Equal Size Use • Eyepiece Storage Unit

ESUS ECS Scientist User Survey

ESV Earth Satellite Vehicle • Emergency Shutdown Valve • Emergency Shutoff Valve • Erstschichtversagen • Essential Service Value • Établissement Scolaire Virtuel • Experimental Safety Vehicle

ESVA Earth Station Verification and Acceptance

ESVS Escape Suit Ventilation System

ESW Easton State (USA-WA) (location identifier) • East-Southwest • Engine Status Word • Error Status Word • Existing Software Workbench

ESWG Electronic Services Working Group (ACM)

ESWGTT Euroscience Working Group on Technology Transfer

ESYS Expert System

ESZB Europäisches System der Zentralbanken

Ét Étage

et Estonian (ISO language code) • Ethiopia

Et Ethil

E&T Education & Training

ET Early Time • Earth Terminal • Eastern Time • École Technique • Edge Thickness • Edge-Triggered • Educational Television • Education and Training • Effectiveness Team • Effective Temperature • Egypt (international vehicle registration code) • Eigenvalue Transform • Elapsed Time • Electrical Time • Electrical Transcription • Electric Telegraph • Electric Typewriter • Electron Transfer • Electron Transition • Electro-Thermal • Element Test • Elevated Temperature • Elongated Tokamak • Embedded Trainer • Embedded Training • Emerging Technology • Emptying Time • Enabling Technology • Energy Transfer • Engineering Test • Engineering Testing • Enhanced Technology • Enhanced Telephone • Enhancement Technology • Enterprise Team • Ephemeris Technology • Ephemeris Time • Equation of Time • Equivalent Term • Equivalent Training • Erection Time • Érection Totale • Escape Tower • Est • Estimated Time • Ethiopia (international aircraft registration code) • Ethiopia (ISO country code) • Éthiopie (code ISO de pays) • European Technology • Evapotranspiration • Event Timer • Exchange Termination • Excitation Transport • Execution Territory • Exoplanet Tracker • Expansion Turbine • Extended Test • Extended Threat • Extended Time • External Tank • Extraterrestrial

ETA Earth Trojan Asteroid • Educational Television Association (UK) • Eesti Teaduste Akadeemia (Estonia) • Electronic Test Automation • Electronic Travel Aid • Emanation Thermal Analysis • Energetic Transient Array • Engineering Test Article • Estação Terrena de Alcântara (Brazil) • Estimated Time of Acquisition • Estimated Time of Appearance • Estimated Time of Arrival • Eta Aquarids (meteor stream) • European Teachers' Association • Expected Time of Arrival • Experimental Test Accelerator • Explosive Transfer Assembly • External Tank Attachment

Étab Établissement

ETAC EMASS Technical assistance Center • Environmental Technical Applications Center • Environment Technical Advisory Committee • Environment Technical Applications Center

ETACS Extended Total Access Communications System

ETADS Enhanced Transportation Automated Data System

ETAGO End Tag Open

ETAI Electronic Transactions on Artificial Intelligence

et al Et Alia • Et Alibi • Et Alii

ETAM Employés - Techniciens - Agents de Maîtrise • Entry Telecommunications Access Method

ETAN European Technology Assessment Network

ETANN Electrically Trainable Analog Neural Network

ETAP Education, Training, and Awareness Program • Estación de Tratamiento de Agua Potable • Expanded Technical Assistance Program • Experimental Technical Assessment Panel

ETAS Effective True Airspeed

ETATF Eesti NSV Teaduste Akadeemia Toimetised Fuusika Matemaatika (periodical code)

ETB Birr (Ethiopia) (ISO currency code) • Eclipse Testbed • End of Transmission Block • Engineering Test Basis • Equipment Transfer Bag • West Bend (USA-WI) (location identifier)

ETBE Ethyl Tertio Butyl Ether

etc Et Caetera

ETC Earth Terminal Complex • Earth Terrain Camera • École Technique Coopérative • Electronic Test Centre • Electronic Time Clock • Electronic Toll Collection • Electron Transport Chain • Electrothermal Chemical • Energy Technology Center • Engineering and Training Center • Engineering Training Course • Enhanced Terminal Capability • Enhanced Throughput Cellular • Estação Terrena de Cuiabá (Brazil) • Estimated Time of Completion • Estimated Time to Complete • Estimate To Complete • Et Caetera • European Test Conference • European Translation Centre • European Travel Commission • Explosive Transient Camera • Exposure Time Calculator • Extended Time Counter • Extraterrestrial Civilization

ETCA Equivalente a Tiempo Completo Anual • Établissement Technique Central de l'Armement (France) • Études Techniques et Constructions Aérospatiales (Belgium)

ETCB End of Transmission Control Block

ETCC Extraterrestrial Century Club (award)

ETCD Équipement de Terminaison de Circuit de Données

ETCG Elapsed Time Code Generator

ETCO Emergency Traffic Coordinating Officer • Equipment Transfer/Change Order • External Tanks Corporation

ETCS European Train Control System

ETD Effective Transfer Date • Electrical Terminal Distributor • Electronic Theses and Dissertations • Electronic Transfer Device • End-Transaction Descriptor • Entwicklung Technische Dokumentation • Estimated Time of Departure • Estimated Turnover Date • Etadunna (Australia-SA) (location identifier) • Evolutionary Terminal Defense • Expected Time of Departure • Experimentation and Test Directorate • Extension Trunk Dialing • External Tank Door

ETDC East Tennessee Discovery Center (USA-TN)

ETDE Energy Technology Data Exchange

ETDMA Extended Time Division Multiple Access

ETDs Electronic Theses and Dissertations

ET&E Engineering Test & Evaluation

ETE Éditions Techniques Européennes (France) • Electronic Test Equipment • Emploi Temporaire pour Étudiant • End-To-End • Engenharia e Tecnologia Espacial • Engineering Test and Evaluation • Engineering Test Equipment • Equivalent Telephone Erlang • Estimated Time En-route • Evacuation Time Estimate • Evolution of Terrestrial Ecosystems • Expected Time En-route • Genda Wuha (Ethiopia) (location identifier)

ETEC Expendable Turbine Engine Concept

ETEDB External Thermal Environment Database

ETEEM End-To-End Engagement Model • End-To-End Engineering Model

ETES Éditions Techniques, Économiques et Sociales

ETET End-To-End Tester

ETEX European Tracer Experiment

ETF Educational Trust Fund • Electrical Test Facility • Electronic Test Facility • Electronic Transfer of Funds •

Electronic Trip Fare • Electron Translation Factor • Electrothermal Feedback • Elgin Test Facility (NASA) • Engineered Test Facility • Engineering Test Facility • Engine Test Facility • Environmental Task Force • Estimated Time of Flight • European Technology Facility • European Training Foundation • European Transfer Format • Exchange Traded Fund

ETFD Electronic Toll Fraud Device

ETFE Ethylene-Tetrafluoroethylene

ETFRN European Tropical Forest Research Network

ETFSP Enhanced Technology Flight Segment Prototype

Étg Étang

ETG Electronics Training Group • Electron Temperature Gradient • Enhanced Target Generator • European Tripartite Group

ETGTS Electronic Text and Graphics Transfer System

Eth Ether • Ethical • Ethics • Ethiopia • Ethiopian

ETH Eidgenössische Technische Hochschule (Switzerland) • Elat (Israel) (location identifier) • Ethiopia (international vehicle registration code) • Extraterrestrial Hypothesis • Falkland Islands (ISO country code) • Îles Malouines (code ISO de pays) • Malvinas (ISO country code)

ETHICS Effective Technical and Human Implementation of Computer-based Systems

Ethnog Ethnography

Ethnol Ethnological • Ethnology

ETHP Equivalent Thrust Horsepower

ETHZ Eidgenössische Technische Hochschule Zürich (Switzerland)

ETI Economic and Technical Intelligence • Economic and Technological Intelligence • Economic Thickness of Insulation • Economic Transition and Integration • Éditions Techniques et Industrielles • Elapsed-Time Indicator • Environmental Technology Initiative • Epner Technologies Inc. (USA-NY) • Estimated Time of Interception • Expertcenter for Taxonomic Identification • Extraterrestrial Intelligence

ETIC Electronic Technical Information Center • Environmental Technical Information Center • Estimated Time In Commission

Etim Etimología

ETIS Elapsed Time Indicator System • Environmental Technical Information System • ESA Tele-Invoicing System (now EFIS) • European Telecommunications Informatics Services

ETK Embedded Toolkit

ETKD Electronic Transfer Keying Device

ETKM Every Test Known to Man

ETKU Electronic Transfer Keying Unit

ETL Educational Technology Language • Electrotechnical Laboratory • Ending Tape Label • Ending Time Label • Endorsed Tools List • Engineering Technical Letter • Enhanced Transceiver Logic • Environmental Technology Laboratory • Environmental Test Laboratory • Équilibre Thermodynamique Local • Extraction - Transformation - Loading • Extract - Transform - Load

ETLA Equivalent Two-Level Atom • Extended Three-Letter Acronym • Extended Time Limited Area

ETLOW External Tank Lift-Off Weight

ETLS Executable TEX Language Specification

ET&M Electronic Test & Maintenance

ETM ECS Technical Manager • Elapsed-Time Meter • Electrical Thermal Model • Electronically Transmitted Message • Electronic Technical Manual • Electronic Test and Maintenance • Electronic Test and Measurement • Electronic Transmission • Electron-Trapping Material • Engineering Test Model • Enhanced Thematic Mapper • Enhanced Timing Module • Évapotranspiration Maximale • Exonération du Ticket Modérateur • Exportable Training Material • Extension Training Material

ETMA Efficient Turbulence Model for Aeronautics

ETMD Extended Thematic Mapper Derivative

ETMG Electronic Trajectory Measurements Group

ETMS Education Training Management System • Enhanced Traffic Management System

ETN Eastland (USA-TX) (location identifier) • École Technique Normale • Educational Teleconferencing Network • Electronic Tandem Network • Electronic Telecommunications Network • Equipment Table of Nomenclature

ETNO European Telecommunications Network Operators

Etnogr Etnografía

Etnol Etnología

EtO Ethylene Oxide

ETO Earth-To-Orbit • Electronic Trading Opportunity • Engineering Task Order • Espace de la Transformation en Ondelettes • Estimated Takeoff • Estimated Time Off • Estimated Time in Operations • Estimated Time of Operations • Estimated Time Operational • Ethylene Oxide • European Telecommunications Office • European Transport Organization • Express Transportation Order • Extended Time Observation

ETOC Estimated Time Of Correction

EtoE End to End

ETOL ESA Test and Operations Language • European Test and Operations Language • European Test Operations Language

ETOM Electron-Trapping Optical Memory

ETOP Education and Training in Optics and Photonics

ETOPIM Electrical Transport and Optical Properties of Inhomogeneous Media (conference)

ETOPS Extended Twin-Engine Operations

ETOS Earthquake and Tsunami Observation System • Extended Tape Operating System

ETP Electronic Technical Publishing • Engineering Test Phase • Engineering Test Procedure • Engineering Test Program • English Translation Pattern • Enquête Technique Préalable • Environmental Test Plan • Equal Time Point • Equal Treatment Property • Equipment Test Plan • Équivalent à Temps Plein • Equivalent Top Product • Estimated Turnaround Point • Estimated Turning Point • European Training Programme • Évapotranspiration Potentielle • Exploitants de Télécommunications Publiques • Exploration Technology Program • Exportable Training Package • Extrathin Pellicle

ETPD Emerging Technology Program Database

ETPL Endorsed Tempest Products List

E. Tpo. Ecuación de Tiempo

ETPS Electronic Technical Publication System

ETR Early Token Release • Eastern Test Range • Educational Technology Review (AACE) • Education and Training • Effective Transmission Rate • Element Test Review • Embedded Training Requirement • Engineering Tape Recorder • Engineering Tape Recording Data • Engineering Test Reactor • Engineering Test Request • Enhanced Temperature Region • Estimated Time of Repair • Estimated Time of Return • ETSI Technical Report • Évapotranspiration Réelle • Experimental Test Reactor • Experimental Thermonuclear Reactor • Export Traffic Release

ETRA Eye Tracking Research and Applications (conference)

ETRD Engineering Tape Recorded Data

ETRDL ERCIM Technical Reference Digital Library

ETRE Elektrosviaz Telecommunications Radio Engineering Telecommunications (periodical code)

ETRF European Travel Research Foundation

Étrgr Étranger

ETRI Electronics and Telecommunications Research Institute (Korea-ROK)

ETRO Estimated Time to Return to Operations

ETROD Eastern Test Range Operations Directive

ETRP Education and Training Programme

ETRR Environmental Test Readiness Review • Export Traffic Release Request

ETRS European Terrestrial Reference System

Éts Établissements

E&TS Engineering & Technology Support

ETS Earth Thermal Storage • Econometric Time Series • Educational Testing Service • Edwards Test Station (NASA) • Electrical Test Set • Electrical Test Setup • Electronic Tandem Switching • Electronic Telegraph System • Electronic Translation System • Electronic Translator System • Electron Triggering System • Emergency Temporary Standard • Energy Transfer System • Engineering and Technical Services • Engineering and Technical Support • Engineering and Technology Support • Engineering Technical Services • Engineering Technical System • Engineering Test Satellite • Enterprise (USA-AL) (location identifier) • Environmental Technical Specification • Environmental Test Specification • EOSDIS Test System • EOS Test System • Equinox Transition Study • Estimated Time of Separation • Ethernet Terminal server • ETSI Technical Standard • European Telecommunication Standard • European Teleprocessing System • European Test Services • Executable Test Suite • Experimental Technical Satellite • Experimental Test Site • Experiment Test System • Expertise Transfer System • Expiration of Term of Service • Extended Target Track • External Tank System

ETSA Electronic Tandem Switching Administration • Electron Transport System Activity

ETSACI Electronic Tandem Switching Administration Channel Interface

ETSC European Transport Safety Council

E&TSD Engineering & Technology Support Department

ETSD Engineering and Technology Support Department

ETSEP External Tank Separation

ETSF Éditions Techniques et Scientifiques Françaises (France)

et seq Et Sequens • Et Sequentia

et seqq Et Sequentes • Et Sequentia

ETSI European Telecommunications Standards Institute

ETSO European Transmission System Operators

et sqq Et Sequentes • Et Sequentia

ETSS Engineering and Technical Service Specialist • Entry Time-Sharing System • Experimental Time-Sharing System • Extended Training Service Specialist • External Tank Separation System

ETSSP Electronic Tandem Switching Status Panel

ETSU East Tennessee State University (USA-TN)

ETT Education and Technology Transfer • Empresa de Trabajo Temporal • Entreprise de Travail Temporaire • Environmental Technology Team • ESTEC Telescience Testbed • Estimated Time of Track • Estimated Travel Time

ETTC Educational Technology Training Center • European Test and Telemetry Conference

ETTD Équipement de Terminaison de Traitement de Données • Équipement Terminal de Transmission de Données

ETTF Estimated Time To Fix • Extreme Temperature Transition Furnace

ETTRT European Transactions on Telecommunications-Related Technologies (periodical code)

ETU Easy To Use • Elementary Time Unit • Engineering Technical Unit • Enhanced Terminology Upgrade • Experimental Test Unit

ETUA European Teleconferencing User Association

ETUC European Trade Union Confederation

ETUD External Tank Umbilical Disconnect

et ux Et Uxor

ETV Educational Television • Eigentümerversammlung • Electrical Test Vehicle • Electronic Tape Vaulting • Electrothermal Vaporization • Electrothermal Vaporizer • Engine Test Vehicle • Estimated Time Value • Extraterrestrial Visitation

ETVA External Tank Vent Arm

ETVM Electrostatic Transistorized Voltmeter

ETVX Entry - Task - Validation - Exit

ETW Eigentumswohnung • End of Tape Warning • Error Time Word • European Transsonic Windtunnel

ETX End of Text

Ety Etymology

Etym Etymological • Etymology

Etymol Etymological • Etymology

ETZ Metz-Nancy-Lorraine (France) (location identifier)

ETZL European Time Zone Logistics

eu Basque (ISO language code)

Eu Euler Number • Europium

E-U États-Unis

EU Economic Updating • Einstein Universe • Electrical Unit • Electronics Unit • Emotional Universe • Encyclopaedia Universalis • End User • Engineering Unit • Ephemeris Uncertainty • États-Unis • Europäische Union (formerly EWG) • Europe • European Union (formerly EEC) • European University • Europese Unie • Europe Unie • Execution Unit • Expander Unit • Experimental Unit

EUA Early User Appraisal • Estados Unidos de América • États-Unis d'Amérique • Eua (Tonga) (location identifier) • European University Association • External User Access

EUATMS European Unified Air Traffic Management System

EUB Energy Use Budget • Essential User Bypass

Euc Euclid (c. 330 - c. 275 BC)

EUC End-User Certificate • End-User Computing • Engineering Unit Conversion • Erasmus User Centre (ESTEC) • Eucla (Australia-WA) (location identifier) • Extended Unix Code

EUCAG End-User Computing Assistance Group

EUCD European Union Copyright Directive

EUCE End-User Computing Equipment

EUCLID European Conference of Library and Information Educational Research • European Cooperation for the Long term In Defence

EUCMG End-User Computing Management Group

EUCOS European Composite Observing System

EUCREM European Cloud Resolving Modeling

EUCREX European Cirrus Radiation Experiment • European Cloud Radiation Experiment

EUDC Extraurban Driving Cycle

EUDISED European Documentation and Information System for Education

EUDPD European Union Data Privacy Directive

EUE Early User Evaluation • Early User Experiment • Early User Experimentation • Eureka (USA-NV) (location identifier)

EUF Equivalent Unavailability Factor • Eufaula Weedon (USA-AL) (location identifier)

EUG Eugene (USA-OR) (location identifier) • European Union of Geosciences

EuGH Europäischer Gerichtshof

EUI Easy User Interface • EGRET Unidentified (source) • End-User Interface • Energy Utilization Index • Energy Utilization Intensity • European University Institute (Italy)

EUIT Early User Innovative Test • Educational Uses of Information Technology

EUL Economic Useful Life

EULA End User License Agreement • End User Licensing Agreement

EUM End User's Manual • Neumünster (Germany) (location identifier)

EUMAC European Modeling of Atmospheric Constituents

EUMD Extended Unit Manning Document

EUMETNET European Meteorological Network

EUMETSAT European Meteorological Satellite Organization

EUN Egyptian Universities Network • Laayoune (Morocco) (location identifier)

EUNEASO European Near-Asteroid Search Observatories

EUnet European Unix Network

EUNICA Extensible Universal Control of Appliance

EUNIS Extreme Ultraviolet Normal Incidence Spectrograph

EUO Paratebueno (Colombia) (location identifier)

EUP Edinboro University Planetarium (USA-PA) • End User Proxy • English Universities Press (UK) • Equipment Usage Profile

EUPEN European Physics Education Network

EUPRIO European Universities Public Relations and Information Officers (association)

EUPRO European Union of Physics Research Organisations

Eur Europe • European • Européen

EUR Erasmus Universiteit Rotterdam (Netherlands) • Euro (ISO currency code) • Europe

EURAB European Research Advisory Board

EURACA European Air Carriers Association

EURACS European Radar Cross-Section

EURAD European Acid Deposition

EURADOS European Radiation Dosimetry

EURAM European Research on Advanced Materials

EURASEP European Association of Satellite Experiments on Pollution

EURASIP European Association for Signal Processing

EURCORES European Science Foundation/Collaborative Research (programme)

EURD Espectrógrafo Ultravioleta para la Radiación Difusa

EURECA European Retrievable Carrier

EUREF European Reference Frame

EUREKA European Research Coordination Agency

EURESCO European Research Conference

EURET European Research for Transport

EURILIA European Initiative in Library and Information in Aerospace

EURIPA European Information Providers Association

EURIS European Information Service

EURISY European Association for the International Space Year • Europe in Space Year

EURL Entreprise Unipersonnelle à Responsabilité Limitée

EURO European Regional Office

EUROCAE European Organization for Civil Aviation Electronics

EUROCAL European Conference on Computer Algebra

EUROCASE European Council of Applied Sciences and Engineering

EUROCAST European Computer-Aided Systems Theory (workshop)

EUROCEAN European Oceanographic Association

EUROGI European Umbrella Organization for Geographic Information

EUROHEPNET European High-Energy Physics Network

EuroHORC European Heads Of Research Councils

EUROHORCS European Union Research Organizations Heads Of Research Councils

EUROLAS European Laser Consortium

EUROMECH European Mechanics Committee

EUROMET European Meteorological Education and Training • European Metrology

EUROPA Expanding Unidirectional Ring of Pages (webring)

EuroPlaNet European Planetarium Network

EuroPVM European Parallel Virtual Machine (conference)

EUROS European Retrievable Orbiting System

EUROSEP European Association of Scientists in Environmental Pollution

EUROSIM European Real-time Operations Simulator

EUROSPACE European industrial Space study group

EUROVISION European Television

EURYI European Young Investigator (award)

EUS Eastern United States • End User Support

EUSC European Utility Support Centre

EUSJA European Union of Science Journalists Associations

EUSO Extreme Ultraviolet Spatial Observatory • Extreme Universe Space Observatory (*)

EUSP European University at Saint Petersburg (Russia)

EUSS Extended Update and Support Service

EUT Early User Test • End-User Training • Equipment Under Test • European Union Treaty

EUT&E Early User Test & Evaluation • Early User Test & Experimentation

EUTE Early User Test and Evaluation • Early User Test and Experimentation

EUTELSAT European Telecommunications Satellite Organization

EUTP Early User Test Plan • Enhanced Unshielded Twisted Pair

EUUG European Unix User Group

EUV Extreme Ultraviolet

EUVE Extreme Ultraviolet Explorer

EUVI Extreme Ultraviolet Imager

EUVL Extreme Ultraviolet Light

EUVIP Extreme Ultraviolet Imaging Photometer

EUVITA Extreme Ultraviolet Imaging Telescope Array

EUVS Extreme Ultraviolet Spectrograph

EUVSAT Extreme Ultraviolet Satellite

EUVT Extreme Ultraviolet Telescope

EUX Saint Eustatius (Netherlands Antilles) (location identifier)

ev En Volgende

eV eingetragener Verein • Electronvolt • Elektronvolt

Ev Éventuel • Éventuellement

E&V Evaluation & Validation

EV Earned Value • Earth Viewing • Edited Volume • Eerstvolgend • Electrical Vehicle • Electronic Vision • Enforcement Vector • Engineering Validation • English Version • Enhanced Version • Entry Vehicle • En Ville • Era Volgare • Escort Vehicle • Etagenverteiler • Étoile Variable • Expandable Vehicle • Expanded Volume • Experimental Version • Experiment Version • Expert Visit • Exposure Value • Extravehicular

EVA Earned-Value Analysis • Échelle Visuelle Analogique • Economic Value Added • Electronic Valve Actuator • Electronic Velocity Analyzer • Empfangsverteilanlage • Entrée dans la Vie Active • Europese Vrijhandelsassociatie • Evadale (USA-TX) (location identifier) • Extended Visit Authorization • Extravehicular Activity • Extreme Value Analysis

Evac Evacuate • Evacuated • Evacuation

Évac Évacuation • Évacué • Évacuer

EVAC East Valley Astronomy Club (USA-AZ) • Evacuation • Extravehicular Activity Communications

Eval Evaluate • Evaluated • Evaluation

Éval Évaluation • Évalué • Évaluer

EVAL Earth-Viewing Application Laboratory • Evaluate • Evaluation

EVAN Expandable Van • Expansible Van

Evap Evaporate • Evaporated • Evaporation • Evaporator

Évap Évaporateur • Évaporation • Évaporé • Évaporer

Evapn Evaporation

EVAS Enhanced Vortex Advisory System • Extravehicular Activity System

EVATA Extravehicular Activity Translational Aid

EVBA Equivalent Vector Boson Approximation

EVC Engine Valve Control • Enhanced Video Connector • Extravehicular Communications

EVCA European Venture Capital Association • Evolving Cellular Automata

Evce Evidence

EVCF Eastern Vehicle Checkout Facility

EVCS Emergency Voice Communications System • External Voice Communications System • Extravehicular Communications System

EVCSD Electron Vision Corporation of San Diego (USA-CA)

EVCU Extravehicular Communications Umbilical

EVD Early Version Demonstration • Economische Voorlichtingsdienst • Eva Downs (Australia-NT) (location identifier) • External Visual Display

EVDE External Visual Display Equipment

Eve Evening

EVE Espace Vidéo Européen • European Videotelephony Experiment • EUV Variability Experiment • Evenes (Norway) (location identifier) • Evening • Evolvable View Environment • Extensible VAX Editor

EVES Emergency Voice Evacuation System

EVest Elektrotehniski Vestnik (periodical code)

EVF Equipment Visibility File

EVFU Electronic Vertical Format Unit

Evg Evening

EVG Sveg (Sweden) (location identifier)

EVGA Extended Video Graphics Adapter • Extended Video Graphics Array

EVH Evans Head (Australia-NSW) (location identifier)

EVI Economic Value of Information

Evid Evidence • Evident

Évid Évidence • Évident

E-ViEWS Emergency Vehicle Early Warning and Safety (system)

EViEWS Emergency Vehicle Early Warning and Safety (system)

EVIL Extended Video Interactive Language

EVINCO Evaluation and Interviewing Committee

EVIRI Enhanced Visible and Infrared Imager

EVL Electronic Visualization Laboratory • Event List

EVLA Expanded Very Large Array

EVLDF Evaluation Data File

EVLSS Extravehicular Life Support System

EVM Earned Value Management • Earth-Viewing Module • Electronic Voltmeter • Entry Vocabulary Module • Error Vector Magnitude • Eveleth (USA-MN) (location identifier) • External Verification Module

EVMS Earned Value Management Standard • Earned Value Management Support • Earned Value Management System • Eastern Virginia Medical School (USA-VA)

EVMSO Earned Value Management Support Office

EVMU Extravehicular Mobility Unit

EVN Einzelverbindungsnachweis • European VLBI Network • Yerevan (Armenia) (location identifier)

Evng Evening

EVNPC European VLBI Network Programme Committee

EVO Engineering Verification Order • Extravehicular Operation

EVOF Extreme-Velocity Outflow

Evol Evolution • Evolutionary

EVOM Electronic Voltohmmeter

EVOP European Volcanological Project • European Volcanology Project (*)

EVP Elastoviscoplastic • Entrer dans la Vie Professionnelle • Equilibrium Vapor Pressure • Executive Vice President • Experimental Version Prototype • Expert Visits Programme • External Viewing Platform

EVPA Early VLBI Polarization Axis • Electric Vector Position Angle • Experimental Version performance Assessment • Extragalactic Variable Position Angle

EVPI Expected Value of Perfect Information

EVPSN European Vehicle Passive Safety Network

EVR Electronic Video Recording • Electronic Voltage Regulator • Electrovideo Recorder

EVRIS Étude de la Variabilité, de la Rotation et des Intérieurs Stellaires

EVS Échelle Verbale Simple • Enhanced Vision System • Equipment Visibility System • Ethics and Value Studies • European Voluntary Service • External Visibility System • Extravehicular Suit • National Standards Board of Estonia

EVSC Extravehicular Suit Communications

EVSM Electrostatic Voltmeter System

EVSS Extravehicular Space Suit

EVSt Endvermittlungsstelle

EVSTC Extravehicular Suit Telecommunications

EVSU Extravehicular Space Unit

evt Event Log (file extension)

EVT Environmental Verification Testing • ESO Video Team • Eulerian Virial Theorem • Experiment Verification Test • Extravehicular Transfer

Evtl Éventuel • Eventuell

Evtlmt Éventuellement

EVÜ Einzelverbindungsübersicht

EVU Energieverwertungsunternehmen

EVUA European VPN User Association

EVV Evansville (USA-IN) (location identifier)

EVW Evanston (USA-WY) (location identifier)

EVX Electronic Voice Exchange • Évreux (France) (location identifier)

Evy Every

EVz Endstellenverzweiger • Endverzweiger

Ew Einwohner

E/W Energy/Weight

EW Belarus (international aircraft registration code) • Early Warning • East-West • Edwards-Wilkinson • Eigentumswohnung • Einheitswert • Electronic Warfare • Electroweak • Entrance Window • Equivalent Width • Estonia (international vehicle registration code) • European Workshop • Evanescent Wave

EWA Early Warning Assessment • End Warning Area • Estimated Warehouse Arrival

EWAM Extended Window Addressable Memory

EWAN Emulator Without A Name • Enterprise Wide-Area Network

EWAT Electronic Warfare Advanced Technology

EWB Fall River (USA-MA) (location identifier) • New Bedford (USA-MA) (location identifier)

EWC East West Center • Equatorial Wave Campaign • European Works Council • European Writers Congress

EWCBR European Workshop in Case-Based Reasoning

EWCS European Wideband Communications System

EWD Eastward • East/West Database (workshop) • Extended Warnier Diagram

EWDOS ECOOP Workshop on Distributed Object Security

EWDT Early Warning Data Transmission

EWE Emergency Window Escape • Ewer (Indonesia) (location identifier) • Extreme Weather Event

EWEB Eugene Water and Electricity Board (USA-OR)

EWF Extrazellulare Waschflüßigkeit

EWG Environmental Working Group • Europäische Wirtschaftsgemeinschaft (now EU) • Executive Working Group

EWGS European Weather Graphics System

EWI Education With Industry • Electronic Work Instruction • E-mail Waiting Indication • Enarotali (Indonesia) (location identifier) • Europäisches Währungsinstitut • Europäisches Wirtschaftinstitut

eWiC Electronic Workshop in Computing

EWiV Entscheidungen zum Wirtschaftsrecht

EWIV Europäische Wirtschaftliche Interessenvereinigung

EWL Equivalent Working Length • Evaporated Water Loss

EWLR European Workshop on Learning Robots

EWK Extent of World Knowledge • Newton (USA-KS) (location identifier)

EWM Engineering Wringout Meeting

EWMA Exponentially Weighted Moving Average

EWMS Electronic Warfare Management System

EWN Enterprise-Wide Network • New Bern (USA-NC) (location identifier)

EWNLG European Workshop on Natural Language Generation

EWO Educational Welfare Officer • Electronic Warfare Officer • Engineering Work Order • Essential Work Order • Ewo (Congo) (location identifier)

EWON European Work Organization Network

EWOS European Workshop for Open Systems

EWP Employee Wellness Program

EWQ Exceptionally Well Qualified

EWR Early-Warning Radar • Engineering Work Request • Europäischer Wirtschaftsraum • New York - Newark (USA-NJ) (location identifier)

EWS Early Warning System • Elektronisches Wählsystem • Emergency Water Supply • Employee-Written Software • Engineering Workstation • Europäisches Währungssystem • European Warranty System • European Workgroup on SGML • European Working-group on SGML • European Workstation • Expert Workbench System

EWSBS Electroweak Symmetry Breaking Scale

EWSK East-West Station Keeping

EWSL European Working Session on Learning

EWSP Elektronisches Wählsystem Paketvermittlung • European Workshop on Planning

EWSPT European Workshop on Software Process Technology

EWST European Week for Science and Technology

EWSTC European Week for Scientific and Technological Culture

EWTA Enterprise-Wide Technical Architecture

EWU Eastern Washington University (USA-WA)

EWW Extended Workweek

EWY Newbury Greenham (UK-England) (location identifier)

e. Wz. eingetragenes Warenzeichen

ex Exit

Ex Examination • Examine • Examined • Examiner • Example • Excellent • Except • Excepted • Exception • Excess • Exchange • Exclude • Excluding • Exclusive • Excursion • Execute • Executed • Executive • Exemple • Exempt • Exeter (UK-England) • Export • Express • Extra • Extended

EX Electronics Experimental • Erection Exercise • Example • Exchange • Execute • Executive • Exercise • Experimental • Express • Extract • Kyrgyzstan (international aircraft registration code)

ExA Experimental Astronomy (periodical code)

EXA Execution Agency

EXAFS Extended X-ray Absorption Fine Structure

Exag Exaggerate • Exaggeration

EXAGT Executive Agent

Exam Examination • Examine • Examined • Examinee • Examiner

Examd Examined

Examg Examining

Examn Examination

EXAMPLE Executive Automated Materiel Planning Equipment

EXAMS Exposure Analysis Modeling System

ExAO Extreme Adaptive Optics

EXAP Extended Arithmetic Processor

ex aq. Ex Aqua

EXAS Experiment on Autonomous SAR

ex b. Ex Bonus

exc Excudit

Exc Excellence • Excellency • Excellent • Excellentie • Except • Excepted • Exception • Excess • Exchange • Excursion

EXC Excellent • Experiment Computer • Extragalactic Astronomy and Cosmology (NSF)

ExCA Exchangeable Card Architecture

EXCA Exchangeable Card Architecture

EXCAP Exercise Capability

EXCC Exercise Control Center

EXCEDE Excitation by Electron Deposition

EXCG Exercise Control Group

Exch Exchange • Exchequer

EXCH Exchange

Excl Exclamation • Exclamativo • Exclude • Excluded • Excluding • Exclusif • Exclusive

Exclam Exclamación • Exclamation • Exclamatory

Exclvmt Exclusivement

ExCo Executive Committee

EXCO Executive Council

EXCOM Executive Committee • Extended Communications

EXCON Executive Control • Exercise Control

EXCP Except • Executive Channel Program

Excsv Excessive

Exctr Exciter

excud Excudit

Exd Examined

EXD Examined

EXDAMS Extended Debugging And Monitoring System

EXDB Experimental Database

EXDIR Executive Director

EXDIS Exclusive Distribution

Ex. Doc. Executive Document

exe Executable Module (file extension)

ExE Event by Event

EXE Executable • Execute • Execution

Exec Execute • Execution • Executive • Executor

EXEC Executable • Executive

EXECCTL Execution Control • Executive Control

EXELFS Electron Energy Loss Fine-Structure Spectroscopy

Exempl Exemplaire

EXEMSI Experimental European Manned Space Infrastructure

Exer Exercise

EXER Exercise

Exes Expenses

EXES Echelon-Cross-Echelle Spectrograph

EXEVAL Exercise Evaluation • External Evaluation

ExFl Experiments in Fluids (periodical code)

EXFOR Experimental Force

ex. g. Exempli Gratia

Exgl Extragalactic

ex. gr. Exempli Gratia

Exh Exhaust • Exhibit • Exhibition

EXH Exchange Hardware • Exhibit

Exhib Exibit • Exhibition

Exhbn Exhibition

EXI Excursion (USA-AK) (location identifier)

ExInEd Exploration In Education

ex int. Ex Interest

EXIST Energetic X-ray Imaging Survey Telescope

EXITE Energetic X-ray Imaging Telescope Experiment

Exkl Exklusive

ex lib. Ex Libris

ExM Experimental Mechanics (periodical code)

EXM Examination • Exmouth Gulf (Australia-WA) (location identifier)

EXMP Experimentation Master Plan

EXN Ex New

EXNOR Exclusive NOR

ExNPS Exploration of Neighboring Planetary Systems

EXNPS Exploration of Neighboring Planetary Systems

EXO Enriched Xenon Observatory • Executive Office • Exoatmospheric • Exosat (catalog) • Experiment Operator

ex off. Ex Officio

Exor Executor

EXOR Exclusive OR

EXORD Execute Order • Exercise Order

EOS Exosphere Satellite

EXOSA EXOSAT Express (periodical code)

EXOSAT European X-ray Observatory Satellite

ex p. Ex Parte

Exp Expand • Expanded • Expansion • Expect • Expected • Expecting • Expedition • Expenses • Experience • Experiment • Experimental • Experimented • Experimenting • Expert • Expiration • Expire • Expired • Expires • Explanation • Explain • Explained • Explaining • Expletivo • Exponential • Export • Exportation • Exported • Exporter • Expose • Exposed • Exposing • Exposure • Express • Expression • Expulsion

EXP Expandable • Expect • Expected • Expense • Experiment • Exponent • Exponential • Expressway

EXPC Expect

Expdn Expedition

Exped Expedite

Expen Expendable

EXPEN Expendable

Exper Experiment • Experimental

Expl Exemplar • Explain • Explanation • Explanatory • Exploration • Explore • Explored • Explorer • Explosion • Explosive

Expld Explode

Expln Explosion

EXPLODE Expanding List in the OSI Directory Environment

Exploit Exploitation

Expn Exposition

EXPN Expand

Expnt Exponent • Exponential

Expo Exposition

Exp. O. Experimental Order

EXPORT Exoplanetary Observational Research Team

Expr Expresión • Express

EXPR Experiment Related • Expressway

ExpRes Experiment Results

EXPRES Experimental Research in Electronic Submission

EXPRESS Expedited Processing of Experiments to Space Station • Experimental Reentry Space System • Experiment Reusable Space System

EXPRESSO Experiment for Regional Sources and Sinks of Oxidants

EXPRIM Expert pour la Recherche d'Images

EXPS Expanding Squares

Expsr Exposure

Expt Experiment

Expte Expediente

Exptl Experimental

EXPW Expressway

Expy Expressway

EXPY Expressway

Exr Executor

EXR Exclusive Rights

Exs Expenses

EXS Expansion Shelf • Expenses

EXSAS Extended Scientific Analysis System

ExSC Executive Standards Council

ExSEL Excellence in Science and Engineering Leadership

EXSTA Experimental Station

Exstg Existing

Ex Stre. Ex Store

EXSUM Executive Summary

ext Extension (file extension)

Ext Extend • Extended • Extending • Extension • Extent • Exterior • External • Externally • Extinct • Extinguish • Extra • Extract • Extraction • Extreme

ExT Experimental Techniques (periodical code)

EXT Exeter (UK-England) (location identifier) • Experiment Terminal • Extend • Extension • Exterior • External • Extinguished

Extd Extend • Extended

EXTD Extend

EXTEC Exploration science and Technology

Exten Extended • Extension

EXTER External

EXTEV External Evaluation

ExTFS Experimental Thermal Fluid Science (periodical code)

EXTGH Extinguish

Exting Extinguishing

Extm Extreme

Extn Extension • External

EXTN Extension

Extnr Extinguisher

Extr Exterior • Extract • Extraction (periodical code) • Extracto • Extrait • Extraordinary • Extremadura (Spain)

EXTRA Experiment for X-ray Timing and Relativistic Astrophysics • Export Transportation

EXTRAP Extrapolate

ExTRAS Experiment on Timing, Ranging and Atmospheric Soundings

Extraterr Extraterrestre • Extraterrestrial • Extraterrestrische • Extraterritorial

Extraterrit Extraterritorial

EXTRM Extreme

EXTRN External Reference

Extns Extensions

EXTNSN Extension

EXTS Extensions

EXTSV Extensive

EXTV European Experimental Test Vehicle

Exx Examples

EXU Execution Unit

EXUG European X User Group

Ex Whf. Ex Wharf

Ex Whse. Ex Warehouse

EXX Examples

Exz Exzellenz

EXZ Excessive Zero

EY Elder-Younger • Execution Year • Extended Year • Tajikistan (international aircraft registration code)

EYA End of Year Analysis

EYC Express Yourself Clearly

EYL Yelimane (Mali) (location identifier)

Eylt Eyelet

EYP El Yopal (Colombia) (location identifier)

Eypc Eyepiece

EYR Yerington (USA-NV) (location identifier)

EYS Eliye Springs (Kenya) (location identifier)

EYW Key West (USA-FL) (location identifier)

EZ Easy • Economische Zaken • Empowerment Zone • Entrainment Zone • Equatorial Zone • Ephemeridenzeit • Erection Zone • Exclusion Zone • Extraction Zone • Turkmenistan (international aircraft registration code)

EZB Effective Zone Boundary • Europäische Zentralbank • Exclusive Zone Boundary

EZCO Erection Zone Control Officer

EZE Buenos Aires - Ezeiza (Argentina-BA) (location identifier) • Buenos Aires - Ministro Pistarini (Argentina-BA) (location identifier)

EZHR Equivalent Zenithal Hourly Rate

eZIF Electrically-activated Zero Insertion Force

EZJ Easy Job

EZMW Europäisches Zentrum für Mittelfristige Wettervorhersagen

EZn Equatorial Zone - Northern Component (Jupiter)

EZR Expanded ZIP Retrofit

EZs Equatorial Zone - Southern Component (Jupiter)

EZS Elazig (Turkey) (location identifier)

EZU Europäische Zahlungsunion

EZW Embedded Zero-tree Wavelet (algorithm)

E2 Electromagnetic Effect • Exponential Elevation

E2A Ever Expanding Acronym

e2e End-to-End

E2E End-to-End

E2PROM Electrically Erasable Programmable Read-Only Memory

E2S Expert Software Systems (Belgium)

E3 Electromagnetic Environmental Effect • Electronic Entertainment Exhibition (USA-CA) • End-to-End Encryption

E3A Electronic Environment Effects Analysis

E3CP Electromagnetic Environmental Effects Control Program

E3I École d'Ingénieur en Informatique pour l'Industrie (France)

E3RB Electromagnetic Environmental Effect Requirement Board

E3RC Electromagnetic Environmental Effect Research Center

F

f Distribution Function • Earth Flattening • Fac • Facing • Faible • Fair • Farad • Farthing • Father • Fathom • Feet • Female • Féminin • Feminine • Feminino • Femto • Feuillet • Fiant • Fiat • Fiction • Field • Fine • Firm • Flat • Flattening • Fluid • Focal Length • Folgende • Foliación • Folio • Follow • Following • Follow Up • Foot • Fondé • For • Force • Form • Formula • Fortran (file extension) • Founded • Fraction • Franc • Frequency • From • Fugacity • Function of • Guilder (currency) • Mixture Fraction • Packing Fraction • Specific Helmholtz Function • Terrestrial Flattening • Thermal Utilization Factor • True Anomaly

F Electric Field Strength • Fach • Faculté • Faculty • Fahrenheit • Failure • Fair • False • Family • Farad • Faraday Constant • Farthing • Fast • Father • Fathom • February • Federation • Fellow • Female • Féminin • Feminine • Fen • Fermé • Fermi • Ferrovia • Field • Filament • Filial • Filter • Final • Fine • Firm • Firth • Flag • Flash • Flat • Fleet • Flight • Florijn • Fluor • Fluorine • Flux • Flux Density • Focal Length • Fog • Folio • Follow • Following • Follow Up • Force • Forint (currency) • Fort • Forte • Forty (Roman numeral) • Forward • Fourier (transform) • Foxtrot (international alphabet) • Franc (currency) • France (international aircraft registration code) • France (international vehicle registration code) • France (postal country code) • Frauen • Freddo • Free Energy • Freighter • French • Frequency • Frère • Friday • Frigate • Frio • Frío • Froid • Front • Fuel • Function of • Fundamental • Fuze • Helmholtz Function • Hyperfine Structure Quantum Number • Kinematic Flux • Magnetomotive Force • Nodal Forces • Quantum Number of Total Angular Momentum • Radiant Flux • Rotational Term • Spectral Type • Thrust • 40 (Roman numeral)

f&a Fore & Aft

fa Farsi (ISO language code) • Persian (ISO language code)

fA Femtoampere

Fa Factura • Firma • Florida (USA)

F/A Fabrication and Assembly • Failure Analysis • Fuel/Air • Fuel Assembly

F&A Fabrication & Assembly • Facilities & Administration

FA Fabrication and Assembly • Facial Analysis • Facility • Factor Analysis • Faculty Advisor • Failure Analysis • False Alarm • Family Allowance • Fatty Acid • Feasibility Assessment • Feedback Analysis • Field Activities • Field Allowance • Filter Array • Filtered Air • Final Approach • Final Assembly • Finance Act (USA) • Financial Adviser • Fine Art • Fire Alarm • First Aid • First Article • Fixed Asset • Flight Acceptance • Flight Aft • Flight Attendant • Flowing Afterglow • Fluffy Aggregate • Fluoranthene • Focus Amplifier • Force Analyzer • Forced Air • Foreign Agent • Free Agent • Free Aperture • Free Atmosphere • Fuel Assembly • Full Adder • Fully Automatic • Functional Analysis • Functional Area • Fuse Alarm

FA/A Functional Analysis/Allocation

FAA Fachmesse für Amateur-Astronomen (Germany) • False Alarm Avoidance • Faranah (Guinea) (location identifier) • Federación Argentina de Académicos (Argentina-BA) • Federal Aviation Administration (USA) • Federal Aviation Agency (USA) • Fellow of the Australian Academy • Foreign Assistance Act (USA) • Foreign Assistance Agency • Föreningen för Astronomi och Astronautik (Sweden) • Formalin Acetic Alcohol • Forum der Amateur-Astronomie • Forward Alighting Area • Forward Assembly Area • Free of All Average • Functional Area Analysis • Fund Administering Activity • Furnace Atomic Absorption

FAAA Final Acquisition Action Approach • Final Acquisition Action Approval

FAAAS Fellow of the American Academy of Arts and Sciences • Fellow of the American Association for the Advancement of Science

FAAB Frequency Allocation Advisory Board

FAAC Ford Amateur Astronomy Club (USA-MI)

FAAO Fellow of the American Academy of Optometry

FAAP Federation of Accrediting Agencies of the Philippines

FAAR Forward Area Alerting Radar

FAAQ Fédération des Astronomes Amateurs du Québec (Canada-QC)

FAAS Flame Atomic Absorption Spectroscopy • Foreign Affairs Administrative Support • Foreign Area Administrative Support • Forum for Amateur Astrospectroscopy • Forward Area Alerting System

FAAT Fully Analytical Aerotriangulation

fàb Franco à Bord

Fab Fabric • Fabricate • Fabricated • Fabrication

FÀB Franco À Bord

FAB Fabrica • Fabricate • Fabrication • Fast Atom Bombardment • Field Assistance Branch • Finance And Budget • First-Aid Box • Flight Assignment Baseline • Forward Avionics Bay • Fraction Actually Burned • French - American - British

Fabbr Fabbrica

FABEP Field Activity Basic Emergency Plan

FABF Fraction of Agriresidue Burned in Fields

FABITH Förderverband Astronomische Bildung in Thüringen (Germany)

FABIUS Fabrication Automatique de Bibliographies et d'Index Utilisant des Syntagmes

FABMS Fast Atom Bombardment Mass Spectroscopy

FABP Family of Alternative Bandit Processes

Fabr Fabricate • Fabrication

FABU Fuel Additive Blender Unit

FABx Fire Alarm Box

Fac Facade • Facial • Facility • Facsimile • Factor • Factory • Facultad • Faculté • Faculty

FAC Facilities • Facility • Failure Analysis Committee • Family Assistance Center • Fast Adaptive Composite • Fast As Can • Feature Analysis Code • Federal Acquisition Circular • Federal Affairs Committee • Field Accelerator • Field-Aligned Current • File Access Code • Final Approach Course • Fonds d'Aide et de Coopération • Force Accounting Code • Forecast At Completion • Forward Air Control • Forward Air Controller • Frequency Allocation Committee • Fresneau Astrographic Catalogue • Functional Account Code • Functional Area Chief • Functional Area Code

FACA Federal Advisory Committee Act (USA)

FACB Field Acquisition Bundle

FACC Feature Attribute Coding Catalog • Feature and Attribute Coding Catalog • Ford Aerospace Communications Corporation • Fundo de Apoio Comunidade Científica

FACCS Federation of Analytical Chemistry and Spectroscopy Societies (USA-TN)

FACD Foreign Area Customer Dialing

FACE Facilities And Communications Evaluation • Feature Analysis Comparison and Evaluation • Field-Alterable Control Element • Field Ancillary Computer Effort • Fine Attitude Control System • Florida Area Cirrus Experiment • Florida Area Cumulus Experiment • Forests Absorbing Carbon Dioxide Emission • Framework for Academic Cooperation in Europe • Free-Air CO Enrichment

FACELIF Flowing Atmospheric Chemistry Experiment with Laser-Induced Fluorescence

FACES Fortran Automatic Code Evaluation System

FACET Future Automated Commercial Environment Team

FACI First Article Configuration Inspection

Facil Facility

FACIT Fast Automatic Conversion with Integrated Tools • Formal Approaches to Computing and Information Technology

FACM Florida Association of Cadastral Mappers (USA-FL) • Functiones et Approximatio Comentarii Mathematici (periodical code)

FACNET Federal Acquisition Computer Network

FACO Fabrication and Acceptance Checkout • Factory Acceptance Checkout • Factory Assembly and Checkout • Final Assembly Checkout

FACP Fellow of the American College of Physicians • Forward Air Control Point • Forward Air Control Post • Fully Automated Computer Program

FACR First-Article Configuration Review • Fourier Analysis and Cyclic Reduction

FACRP Functional Analysis and Consolidation Review Panel

Facs Facsimile

FACS Facilities Assignment and Control System • Feature Attribute Coding Standard • Feature Attribute Coding System • Federation of Asian Chemical Societies • Fellow of the American College of Surgeons • Finance And Control System • Fine Attitude Control System • Fluorescence-Activated Cell Sorter • Forward Area Control System • Fully Automatic Compiling System • Functional Area Control System

FACSFAC Fleet Area Control and Surveillance Facility

FACSI Federal Advisory Council on Scientific Information (USA)

Facsim Facsimile

Fact Factura • Facture

FACT Factor Analysis Chart Technique • Factory • Fast Actuated Control Thruster • Federation Against Copyright Theft (UK) • First-Address Control Table • Flexible Automatic Circuit Tester • Flight Acceptance Composite Test • Foundation for Advanced Computer Technology • Fully Automated Cataloguing Technique • Fully Automatic Compiler-Translator • Fully Automatic Computer-Translator • Fully Automatic Compiling Technique

FACTS Facilities Assets Catalog and Tracking System • FDF Automated Control and Tracking System • Fingerprint Analysis and Criminal Tracing System • Flexible Astronomical Catalogue Transport System • Florida Atlantic Coast Transport Study • Fortran Analytical Cross-Reference Tabulation System

FACZ Forward Airspace Coordination Zone

FAD Fast Automatic Differentiation • Feasible Arrival Date • Fernsprechansagedienst • Fiber Actuator Device • Field Authorization Document • Filter Analog-to-Digital • Final Acceptance Date • Final Approach Display • Final Archive Definition • Flight Assessment Delta • Floating Add • Force/Activity Designator • Formation À Distance • Fraction Actually Degraded • Free Air Delivered • Funding Action Directive • Funding Allowance Document • Funding Authorization Document

FADABEF Fédération des Associations de Bibliothécaires, Archivistes et Documentalistes des États Francophones

FADBEN Fédération des Associations de Documentalistes Bibliothécaires de l'Éducation Nationale (France)

FADC Fast Analog Digital Converter • Flash Analog Digital Converter • Final Archive Definition Committee

FADEC Full-Authority Digital Electronic Control • Full-Authority Digital Engine Control (*) • Full-Authority Digital Engine Controller

FADS Fast Area Digitizing Scanner • Future Astronomical Data System

FADU File Access Data Unit

FAE Faroe Islands (location identifier) • Faroese • Field Application Engineer • Final Approach Equipment • Foyer d'Action Éducative • Functional Area Element

FAEP Fédération des Associations d'Éditeurs de Publications

FAES Flame Atomic Emission Spectroscopy

FAF Face À Face • Final Access Format • Final Approach Fix • Final Assembly Facility • Final Assembly Fixture • First Aerodynamic Flight • Follow-up for Billing Action • Fort Eustis (USA-VA) (location identifier) • Functional Area Evaluator

FAFB Falcon Air Force Base (USA-CO)

FAFR Focal Array-Fed Reflector

FAFS Falcon Air Force Station • Flame Atomic Fluorescence Spectroscopy

FAFTT Fonds d'Assurance Formation du Travail Temporaire

F&AG Finance & Administration Group

FAG Fagurholmsmyri (Iceland) (location identifier) • Fernmeldeanlagengesetz • Finance and Administration Group

FAGC Fast Automatic Gain Control

FAGS Federation of Astronomical and Geophysical Services • Fellow of the American Geographical Society

Fah Fahrenheit

FAH Farah (Afghanistan) (location identifier) • Final Anthropic Hypothesis • Fournisseur d'Applications Hébergées

FAHQMT Fully-Automatic High-Quality Machine Translation

FAHQT Fully-Automatic High-Quality Translation

Fahr Fahrenheit

Fahrb Fahrbahn

FAI Fail As Is • Fairbanks (USA-AK) (location identifier) • Federal Acquisition Institute • Fédération Aéronautique Internationale • Field-Aligned Irregularity • First Article Inspection • Fonds Autoroute de l'Information (France) • Fonds d'Aide à l'Innovation (France) • Fournisseur d'Accès à Internet • FREJA Auroral Imager • Fresh-Air Inlet

FAIA Functional Area Information Analysis

FAIAA Fellow of the American Institute of Aeronautics and Astronautics

FAIFE Free Access to Information and Freedom of Expression

FAIL Failure

FAIM Fast Aerial Image Model • Fully Analytic Ionospheric Model

FAIMS Financial Accounting and Information Management System

FAIP Fellow of the Australian Institute of Physics • First Assignment Instructor Pilot

FA/IPT First Article/Initial Production Test

FAIPT First Article/Initial Production Test

FAIR Fabrication, Assembly and Inspection Record • Failure Analysis Information Retrieval • Fairness and Accuracy In Reporting • Fast Access Information Retrieval • Federal Activities Inventory Reform • Federal Agricultural Improvement and Reform (USA) • Filled-Aperture

Infrared (telescope) • Financial Accounting and Inventory Record • Fly-Along Infrared • Fully-Automatic Information Retrieval • Functional Analysis Integration Review • Fundamentals of Artificial Intelligence Research (workshop)

FAIRS Federal Aviation Information Retrieval System (USA) • Fully Automatic Information Retrieval System

FAISAT Final Analysis Inc. Satellite

FAITC Foreign Affairs and International Trade Canada

FAJ Fajardo (Puerto Rico) (location identifier) • Filtered Accumulated Jitter • Fonds d'Aide aux Jeunes

Fak Faktura • Fakultät

FAK Fachnormenausschuß Klassifikation • False Island (USA-AK) (location identifier) • File Access Key • Freight All Kinds

FAKYR Fachbereich-Kybernetik-Retrieval-system

FAL Facility Area Layout • File Access Listener • Finite Automation Language • First Approach and Landing • Frequency Allocation List • Roma - Falcon State (USA-TX) (location identifier)

FALCON Fission-Activated Laser Concept • Frequency Agile Low Coverage Netted (method)

Falk.I. Falkland Islands

Falk.Is. Falkland Islands

FALOP Forward Area Limited Observation Plan

FALP Flowing Afterglow Langmuir Probe • Frequency Agile Laser Protection

FALT Final Altitude

FALTRAN Fortran to Algol Translator

Fam Familia • Familiar • Family

FAM Faculdade Anhembi Morumbi (Brazil) • Familiarization • Family • Farmington (USA-MO) (location identifier) • Fast Access Memory • Fast Action Method • Fast Auxiliary Memory • File Access Method • Five-Axis Mount • Fleet Assignment Model • Flight Assurance Manager • Full-Aperture Metrology • Fully Adaptive Method • Fully Adaptive Multigrid • Functional Area Manager

FAMAF Faculdad de Matemáticas, Astronomía y Física

FAMAS Flutter And Matrix Algebra System

FAME Fizeau Astrometric Mapping Explorer • Fluids And Materials Experiment • Forest Assessment and Monitoring Environment • Formatting And Multiplexing Equipment • Full-sky Astrometric Mapping Explorer

FAMIS Federal Aviation Management Information System (USA) • Financial Accounting Management Information System • Financial And Management Information System • Fulfillment And Member Information System • Foreign Affairs Management Information System

FAMJT Familiarization Job Training

FAMOS Flexible Automatisierte Montagesysteme • Flight Acceleration Monitor Only System • Floating-gate Avalanche-injection Metal-Oxide Semiconductor • Future of Atomic, Molecular, and Optical Sciences

FAMOUS Flexible Automation Monitoring and Operation User Station

FAMP Facilities Acquisition Management Plan

FAMPS Failure Analysis and Maintenance Planning System

FAMS FDF Automated Monitoring System • File Attribute Management System • Fuel Automated Management System

FAMSCO First Article Master Scheduling Conference

FAMT Fully Automatic Machine Translation

FAMU Fuel Additive Moisture Unit

FAN Family Area Network • Farsund (Norway) (location identifier) • Federal Aviation News • Functional Area network

FANE Fonds d'Action Nationale Européen

FANGIO Feedback Analysis for GCM Intercomparison and Observations

FANS Fully Automatic Navier-Stokes • Future Air Navigation System

FAO Fabrication Assistée par Ordinateur • Faro (Portugal) (location identifier) • Field Audit Office • Finance and Accounting Office • Finance and Accounting Officer • Finish All Over • Flight Activities Officer • Food and Agriculture Organization (UNO) • Foreign Area Officer • For the Attention Of • Foyer d'Accueil et d'Orientation

FAOB Federation of Asian and Oceanian Biochemists

FAOBMB Federation of Asian and Oceanian Biochemists and Molecular Biologists

FAP Fair Access Policy • False-Alarm Probability • Fast Atmospheric Pulsation • Field Application Panel • File Access Protocol • Filtre À Particules • Final Approach Plane • Final Approach Point • Final Approval Plan • Final Approved Program • Finance and Accounting Policy • First Aid Post • Flight Activity Performance • Floating-point Arithmetic Package • Fly-Away Package • Food and Agriculture Project • Fortran Assembly Program • Franc d'Avaries Particulières • Freedom-Attached Payload • Frequency Allocation Panel • Frequency Assignment Problem • Functional Area Plan • Future ATM Profile

FAPAE Federación de Asociaciones de Empresas Productoras de Audiovisuales de España (Spain)

FAPAM Fédération d'Astronomie Populaire du Midi (France)

FAPCAS Failure Analysis and Preventive/Corrective Action System

FAPE Fund for Assistance to Private Education

FAPES Force Augmentation Planning and Execution System

FAPESP Fundação de Amparo à Pesquisa do Estado de São Paulo (Brazil)

FAPI Family Application Program Interface

FAPIN Fabry-Pérot Interferometer

FAPM Functional Activity Program Manager

FAPR Formal and Applied Practical Reasoning (conference)

FAPS Fault Analysis Projection System • Filtered Accumulated Phase Slope

FAPSS Foreign Affairs Planning and Scheduling System

faq Frequently Asked Question (file extension)

FAQ Fair Average Quality • Foire Aux Questions • Frequently Asked Question

FAQL Frequently Asked Question List

FAQS FIRST-APM QSO Survey

FAR Failure Analysis Report • False Alarm Rate • Fargo (USA-ND) (location identifier) • Faster-than-Average Rotation • Federal Acquisition Regulation (USA) • Federal Aviation Regulation (USA) • Field Action Request • Field Activity Report • Final Acceptance Review • Final Assessment Report • Final Assessment Review • Fine Angular Resolution • First Assessment Report • Flight Acceptance Review • Flight Analysis Review • Force d'Action Rapide • Forward Acquisition Radar • Fuelflow to Airflow Ratio • Functional Analysis Review • Functional Area Requirement • Fund for Astrophysical Research • Funding Acquisition Request

FARA Federal Acquisition Reform Act (USA)

FARADA Failure Rate Data

FARC Federal Archives and Record Center (USA)

FARCOS Fast Adaptive Radio Communications System

FARE Fluid Acquisition and Resupply Experiment • Formation et Accompagnement à la Recherche d'Emploi

Farm Farmacia

FARM Farmaceútica • Fast Archive Recovery Method • Functional Area Record Manager

FARNET Federation of Academic Research Networks • Federation of Advanced Research Networks • Federation of American Research Networks (USA) (*)

FARR Frequency Agile Rain Radar

FARS Facility for Atmospheric Remote Sensing • Fast Acquisition Receiver System • Fatal Accident Reporting System • Federal Acquisition Regulation System (USA) • Financial Accounting and Reporting System

FAS Facilité d'Ajustement Structurel • Failure Analysis Section • Family of Alternative Superprocesses • Faskrudsfjördur (Iceland) (location identifier) • Fault Alarm System • Feature Analysis Section • Federation of American Scientists (USA) • Federation of Astronomical Societies (UK) • Field Alert Status • Final Approach Segment • Flame Absorption Spectroscopy • Flexible Assembly System • Flight Analysis System • Flight Application Software • Fonds d'Action Sociale • Fonds d'Aide Sociale • Foothills Astronomical Society (USA-SC) • Force Accounting System • Force Analysis Simulation • Forsyth Astronomical Society (USA-NC) • Forward Acquisition Sensor • Forward Acquisition System • FOS Analysis System • Frame Alignment Signal • Free Alongside Ship • Full Approximation Scheme • Full Authorization System • Functional Account Symbol • Functional Acquisition Specialist • Functional Address Symbol • Functional Address System • Functional Area Simulator • Fund Allowance System

FASA Federal Acquisition Streamlining Act (USA)

FASAB Federal Accounting Standards Advisory Board (USA)

FASAC Foreign Applied Sciences Assessment Center

FASAS Federation of Asian Scientific Academies and Societies

FASB Financial Accounting Standards Board (USA)

FASBMB Federation of African Societies of Biochemistry and Molecular Biology

Fasc Fascicolo • Fascicule • Fascículo • Fasciculus

FASc Fellow of the Indian Academy of Sciences

FASC Forward Area Signal Center

FASCOS Flight Acceleration Safety Cutoff System

FASE Federation of Acoustical Societies in Europe • Fundamental Approaches to Software Engineering (conference) • Fundamentally Analyzable Simplified English

FASEB Federation of American Societies for Experimental Biology (USA)

FASED Facilitation Awards for Scientists and Engineers with Disabilities (USA)

FASFID Fédération des Associations et Sociétés Françaises d'Ingénieurs Diplômés (France)

FASIC Function and Algorithm-Specific Integrated Circuit

FASINEX Frontal Air-Sea Interaction Experiment

FASIP Fresham/Alumni Summer Internship Program

FASIT Fully Automatic Syntactically-based Indexing of Text

FASMS Forecast/Allocation Submission Management System

FASN Fringe Amplitude Signal-to-Noise (ratio)

FASO Forward Airfield Supply Organisation

FASP Facility for Automated Software Production • Faculdades Associadas de São Paulo (Brazil) • Fly-Away Sensor Package • Functional Area Strategic Plan

FASR Frequency-Agile Solar Radiotelescope

FASRC Federation of Arab Scientific Research Councils

FASS Frequency Agile Signal Simulation • Functional Area Segment Specification

FASST Flexible Architecture Standard System Technology • Functional And System Support Team • Future of Archives Services at Space Telescope (study)

FAST Facility for Automatic Sorting and Testing • Factor Analysis System • Failure Analysis by Statistical Techniques • Fast Acquisition Search and Track • Fast Agent System Timer • Fast Algorithm for the Satisfaction Test • Fast All Sky Telescope • Fast Auroral Snapshot (explorer) • Fast Automatic Shuttle Transfer • Federal Acquisition Services for Technology • Federation Against Software Theft (UK) • Federazione delle Associazioni Scientifiche e Tecniche (Italy) • Field Assistance in Science and Technology • Field Assistance Support Team • Field Assist and Support Team • File Analysis and Selection Technique • File And Storage Technologies (conference) • Filled Aperture Smart Telescope • Final Approach Spacing Tool • First Application System Test • Five-hundred-meter Aperture Spherical Telescope • Flexible Algebraic Scientific Translator • Flow Analysis Software Toolkit • Forecasting and Assessment in Science and Technology • Formal Autoindexing of Scientific Texts • Formula And Statement Translator • Forward Area Support Team • Forward Area Support Terminal • Fourier Amplitude-Sensitivity Test • Frame ATM over SDH Transport • Frequency Agile Search and Track • Frontline Astrophysics for School Teachers • Functional Academic Skills Training • Functional Analysis System Technique • Function Analysis System Technique • Fundamental Astronomy by Space Techniques • Future Aircraft Systems Testbed • Future Avionics Systems Testbed

FASTAR Frequency Angle Scanning, Tracking, And Ranging

FASTEX Faint Standard Extension • Fronts and Atlantic Storm Track Experiment

FASTHAT Fast High-Accuracy Tunable Receiver

FASTI Fast Access to Systems Technical Information

Fastnr Fastener

FASTR Fast Term Recognizer

FASTS Federation of Australian Scientific and Technological Societies (Australia)

FASTT Flagstaff Astrometric Scanning Transit Telescope (USA-AZ)

FASU Filter And Shutter Unit

FAT Factory Acceptance Test • Factory Acceptance Trial • Fast Automatic Transfer • File Access Table • File Allocation Table • File Attribute Table • Final Acceptance Test • Final Acceptance Testing • Final Acceptance Trial • Final Assembly Test • First Article Test • First Article Testing • Fixed Analyzer Transmission • Flight Attitude Table • Fonds des Accidents du Travail (Belgium) • Foot Access Test • Foreign Alias Table • Foreign Area Translation • Formula Assembler Translator • Free Air Temperature • Fresno (USA-CA) (location identifier) • Functional Area Training

FATAL Fully Automatic Test Algebraic Language

FATCAT Frequency And Time Circuit Analysis Technique

FATE Formation of Aerosols and their Transformation over Europe

Fath Fathom

FATIMA Fast Analysis Technique for the Inversion of Magnetic Atmospheres

FATMA Frequency And Time Multiple Access

FATP Future Astronaut Training Program

FATR First Article Test Report

FATS Factory Acceptance Test Specification • Faint Asteroid Taxonomy Survey • Fault Tracking System • Fortran Automatic Timing System

Fatt Fattura

FAU Fixed Access Unit • Florida Atlantic University (USA-FL)

Faub Faubourg

FAUST Far Ultraviolet Space Telescope

Fav Favor • Favorable • Favorite

FAV Fakarava (Tuamotu Islands) (location identifier) • Favorite

FAVE Fonds d'Amélioration de la Vie Étudiante

FAW Financial Analysis Worksheet • Frame Alignment Word

FAWG Flight Assignment Working Group

FAWN Florida Automated Weather Network

FAWS Flight Advisory Weather Service

Fax Facsimile

FAX Facsimile

FAY Fayetteville (USA-NC) (location identifier)

FAZ Frankfurter Allgemeine Zeitung (Germany)

Fb Fabrik

FB Fachbereich • Faint Blue • Farley-Buneman (wave) • Flare Brightness • Franc Belge • Feedback • Feedback-Based • Fiber Burst • Field Bulletin • Film Badge • Filterbank • Final Braking • Fishbone-like Burst • Flaring Branch • Flat Bar • Fleet Broadcast • Force Builder • Forschungsbericht • Forward-Backward • Fourgon Blindé • Fraction Burned • Frame Buffer • Freight Bill • Functional Baseline • Fuze Block

FBA Fellow of the British Academy • Fixed Block Architecture

FBAC Fort Bend Astronomy Club (USA-TX)

FBAQ Flexible Block Adaptive Quantization

FBAS Fixed Base Aft Station

FBB Fast Break-Bulk • Fast-Burn Booster • Flyback Booster

FBC Fallen Building Clause • Faster - Better - Cheaper • Fluidized-Bed Combustion • Furnished By Contractor

FBCA Federal (PKI) Bridge Certification Authority (USA)

FBCF Formation Brute de Capital Fixe

FBCG Faint Blue Compact Galaxy

FBCR Faster, Better, Cheaper Report (NASA)

FBCS Fellow of the British Computer Society • Fixed Base Crew Station

FBCW Fallen Building Clause Waiver

Fbd Freeboard

FBD Faizabad (Afghanistan) (location identifier) • Functional Binary Decomposition

FBE Fédération Bancaire Européenne • Fernbetriebseinheit • Francisco Beltrão (Brazil-PR) (location identifier) • Free Buffer Enquiry • Fundamental Blue Edge

FBF Frankfurt Book Fair (Germany)

FBFT Flow Bias Functional Test

Fbg Faubourg

FBG Faint Background Galaxy • Faint Blue Galaxy • Fiber Bragg Grating • Fort Bragg (USA-NC) (location identifier)

FBI Federal Bureau of Investigations (USA)

F-BIDR Full-resolution Basic Image Data Record

FBIDR Full-resolution Basic Image Data Record

FBIRC Four Band Infrared Circuit

FBIS Foreign Broadcast Information Service

Fbk Firebrick

FBK Fairbanks - Fort Wainwright (USA-AK) (location identifier)

FBL Faribault (USA-MN) (location identifier) • Fly-By-Light • Functional Baseline

FBLA Future Business Leaders of America

FBLPY Fraction of Blocks Lost Per Year

FBLS Faculty of Biomedical and Life Sciences

FBM Fixed Block Management • Foot Board Measure • Fractional Brownian Motion • Frame Buffer Module • Fuzzy Bag Model • Fuzzy Bitmap • Lubumbashi (Zaire) (location identifier)

FBML Francis Bitter Magnet Laboratory

FBN Federal Base Network (USA) • Feedback Network • Fictitious Business Name • Flaring Branch Noise

FBNCC Fallback Network Control Center

F/B/O For the Benefit Of

FBO First Bound Object • Fixed-Base Operation • Fixed-Base Operator • Flaring Branch Oscillation • Foggy Bottom Observatory (USA-NY) • For the Benefit Of

FBOA Fellow of the British Optical Association

FBP Filtered Back-Projection • Final Boiling Point • Free Boundary Problem

FBPCS Federation of Behavioral, Psychological and Cognitive Sciences

FBPS Four-Band Photometric System

FBQ FIRST Bright Quasar

FBQS FIRST Bright Quasar Survey

Fbr Fiber • Fibre

FBR Fast Breeder Reactor • Fireball Radius • Fluidized-Bed Reactor • Fort Bridger (USA-WY) (location identifier)

FBRAM Frame Buffer Random Access Memory

Fbrbd Fireboard

Fbro Febrero

FBS Faint Blue Star • Fallback Switch • Feedback Signal • Feedback System • Fiber Bundle Switch • Firefighter Breathing System • First Byurakan Survey • Flare Build-up Study • Fluidized Bed Combustor • Forward-Based System • Frontal Boundary Study • Function-Based Sizing • Funkbasisstation

FBSR Feedback Shift Register

FBS2 Second Conference on Faint Blue Stars

FBS3 Third Conference on Faint Blue Stars

FBT Feedback Technology • Fringe Benefit Tax

FBTS Financial Bundle Trading System

FBu Burundi Franc (currency)

FBU Oslo Fornebu (Norway) (location identifier)

FBV Feuerbeschau • Field Base Visit • Frequency Bias Value • Fuel Bleed Valve

FBW Far Blue Wing • Fly-By-Wire • Fractional Bandwidth

FBY Fairbury (USA-NE) (location identifier) • Future Budget Year

FBZ Free Blocking Zone

fc Fin Courant • Foot-Candle

F&C Full & Change

F/C Fire Control • Fit Check • Flight Control • Flight Controller • Flow Control • Fuel Cell

FC Fachgruppe • Facilitated Communication • Facility Control • Facultad de Ciencias • Fail Closed • False Color • Feature Count • Featureless Continuum • Feedback Control • Ferrite Core • Fiber Channel • Fiber Connector • Fidei Commissum • Field Camera • Field Circular • Field Completion • Field Component • Field-Cooled • Fieri Curavit • Fifth Column • File Cabinet • File Code • File Compare • File Conversion • File Copy • Finance Code • Finance Committee • Financial Code • Financial Consultant • Financial Cryptography • Finding Chart • Fine Control • Fire Control • First Class • Fission Cell • Fit Check • Fixed Cooling • Fixed Cost • Flight Clearance • Flight Computer • Flight Control • Flight Controller

• Flight Critical • Floating Causeway • Floating Craft • Follow Copy • Foot-Candle • For Cash • Force Control • Formation Complémentaire • Formation Continue • Forward Checking • Fraction of Components • Fragment Code • Frame Character • Frame Control • Framing Control • Franck-Condon (factor) • Fuel Cell • Fuel Consumption • Full Course • Fully Coated • Fully Contained (event) • Functional Category • Functional Component • Functional Core • Function Code • Fund Code • Funnel Cloud

Fca Fábrica

FCA Fibre Channel Association • Field Change Analysis • Figure Control Actuator • Fire Control Area • Flexible Ceramic Armor • Flight Critical Aft • Flight Control Actuator • Flow Control Assembly • Fluids Control Assembly • Forward Controller Assembly • Frequency Control Analysis • Frequency Control and Analysis • Fuel Control Assembly • Full Cost Accounting • Functional Compatibility Analysis • Functional Configuration Analysis • Functional Configuration Assessment • Functional Configuration Audit • Functional Core Adapter • Fund Cite Authorization • Kalispell (USA-MT) (location identifier)

FCAA Field Contact Administration Activity

FCAC Forward Control and Analysis Center • Frequency Control and Analysis Center

FCAD Five College Astronomy Department (USA-MA)

FCAF Flight Crew Accommodation Facility

FCAG Facultad de Ciencias Astronómicas y Geofísicas

FCAGLP Facultad de Ciencias Astronómicas y Geofísicas de La Plata (Argentina-BA)

FC-AL Fibre Channel - Arbitrated Loop

FC/AL Fibre Channel/Arbitrated Loop

FCAL Fibre Channel - Arbitrated Loop

FCAM Ferroelectric Content-Addressable Memory

FCAP Facility Capacity • Fault - Configuration - Accounting - Performance • Flight Control Applications Program

FCAPS Fault - Configuration - Accounting - Performance - Security (management area)

FCAPT Four College Automated Photoelectric Telescope

FCAR Formation de Chercheurs et Aide à la Recherche • Free of Claim for Accident Reported

FCAS Fairfield County Astronomical Society (USA-CT) • Foreign Contract Administration Services

FCASI Fellow of the Canadian Aeronautics and Space Institute

FCAST Forecast

FCB Faster - Cheaper - Better • Ficksburg (South Africa) (location identifier) • File Control Block • Forms Control Buffer • Fracture Control Board • Fresh Cut Basalt • Functional Configuration Baseline

fcc Face-Centered Cubic

FCC Face-Centered Cubic • Facility Control Console • False Colour Composite • Federal Communications Commission (USA) • Federal Computer Conference • Federal Construction Council • Feynman Computer Center (Fermilab) • File Carbon Copy • Fits-Class Certificate • Flat Conductor Cable • Flight Communications Center • Flight Control Center • Flight Control Computer • Flight Coordination Center • Flight Crew Compartment • Fluid Catalytic Cracking • Fornax Cluster Catalog • Forward Command Channel • Four College Consortium (USA)

FCCA Floating Current-to-Current Amplifier

FCCAPT Four College Consortium Automatic Photoelectric Telescope

FCCB Field Configuration Control Board

FCCC Framework Convention on Climate Change

FCCE Fondos de Cohesión de las Comunidades Europeas

FCCH Frequency Correction Channel

FCCN Fundação para a Computação Cíentífica Nacional (Portugal)

FCCP Firm Contract Cost Proposal

FCCPO Federal Contract Compliance Programs Office

FCCS Forward Cabin Control Station

FCCSET Federal Coordinating Council for Science, Engineering and Technology (USA) (now NSTC)

FCCU Fluid Catalytic Cracker Unit

FCD Fecal Collection Device • Flight Control Division • FOS Command • Functional Capabilities Document • Functional Configuration Documentation • Function Class Decomposition

FCDB Flight Control Data Bus

FCDC Flexible Confined Detonating Cord • Flight Control Data Concentrator • Fundamental Constants Data Center

FCDR Facility Construction Design Requirements • Final Critical Design Review

FCDS Fleet Consolidated Data Set

FCE Factory Checkout Equipment • Faculty of Civil Engineering • Fast-Cycle Executive • Fire Control Element • Flexible Critical Experiment • Flight Control Equipment • Flight Crew Equipment • Fondo de Cultura Económica • Forward Control Element

FCEF Flight Crew Equipment Facility

FCEI Facility Contract End Item • Facility Contractor End Item

FC-EL Fibre Channel - Enhanced Loop

FC/EL Fibre Channel/Enhanced Loop

FCEL Fibre Channel - Enhanced Loop • Flight Crew Equipment Laboratory

FCELP Facultad de Ciencias Exactas de la Plata (Argentina-BA)

fcém Force Contre-Électromotrice

FCF Faculdad de Ciencias Físicas • Fax Control Field • First Captive Flight • Flight Critical Forward • Fluids Combustion Facility • Flux Conversion Factor • Free Cash Flow • Functional Check Flight • Fusion Critical Frequency

FCFM Flight Combustion Facility Monitor

FCFOV Full Coded Field Of View

FCFS First Come - First Served

FCFT Fixed Cost - Fixed Time

Fcg Facing

FCG Fire Control Group • Foreign Clearance Guide • Functional Coordinating Group • Fundação Calouste Gulbenkian (Portugal)

FCGA Fédération des Centres de Gestion Agréés (France)

FCGS Freight Classification Guide System

FCH Flight Controllers Handbook • Fresno Chandler (USA-CA) (location identifier)

FCHL Flight Control Hydraulics Laboratory

FCHP Fixed Conductance Heat Pipe

FCHR Functional Cost-Hour Report

FCI Fiber-Channel Interface • Fire Control Information • Flight Control Indicator • Flight Control Initialization • Flight Control Integration • Flight Crew Integration • Flight Critical Item • Flux Changes per Inch • Fondation Canadienne pour l'Innovation (Canada) • Force Concept Inventory • Forward Call Indicator • Fractally Compressed Image • Frame Copied Indicator • Functional Configuration Identification • Functional Configuration Inspection

FCIA Fiber Channel Industry Association (USA)

FCIC Federal Crop Insurance Co. • Fellow of the Chemical Institute of Canada

FCIF Flight Crew Information File • Formal Context Interchange Format • Full Common Intermediate Format

FCIL Formation Complémentaire d'Initiative Locale

FCIM Flexible Computer-Integrated Manufacturing • Flight Control Interface Module

FCIO Facilities Contract Issuing Office

FCIP Fiber Channel over Internet Protocol • Flight Cargo Implementation Plan

FCIR False Colour Infrared

FCIRT Fermilab Computer Incident Response Team

FCIT First Computer Interface Tester

FCJ Foreign Criminal Jurisdiction

FCKW Fluorchlorkohlenwasserstoffe

FCL Facility (security) Clearance • F-Center Laser • Feedback Control Loop • Feeder Control Logic • Final Coordination Line • Fire Coordination Line • Flight Crew License • Flight Crew Licensing • Fluid Coolant Loop • Format Control Logic • Foundation Class Layer • Freon Coolant Line • Freon Coolant Loop • Full Container Load • Functional Capability List • Fuzzy Controlled Learning

FCLA Florida Center for Library Automation (USA-FL)

FCLC Fibre Channel Loop Community

FCLS Factory Controlled Launch Support

FCLT Freeze Calculated Landing Time • Functional Central Limit Theorem

FCM Fast Component Mounter • Feature Change Map • Finite Cluster Model • First-Class Mail • Flight Contamination Monitor • Flow Control Module • Fond Cosmologique en Micro-ondes • Fondo Común Municipal • Fondo Cósmico de Microondas • Fuzzy Cognitive Map • Minneapolis - Flying Cloud (USA-MN) (location identifier)

FCMC Facility Control and Monitor Console

FCMCM Fuzzy C-Means Clustering Method

FCMD Fire Command

FCMS Flight Crew Mission Simulator • Fuel Control and Monitoring System • Functional Configuration Management System

FCMU Foot-Controlled Maneuvering Unit

FCMV Fuel-Consuming Motor Vehicle

FCN Field Change Notice • Free Core Nutation • Function

f. co. Fair Copy

Fco Franco

FCO Facility Control Office • Facility Control Officer • Federal Coordinating Officer • Field Change Order • Final Checkout • Fire Control • Flight Crew Operations • Foreign and Commonwealth Office (UK) • Fraction of Carbon Oxidized • Frequency-Controlled Oscillator • Full Cutoff • Functional Checkout • Rome - Leonardo da Vinci - Fiumicino (Italy) (location identifier)

FCOB Flight Control Operations Branch • Flight Crew Operations Branch

FCOD Flight Control Operations Directorate • Flight Crew Operations Directorate

FCOH Flight Controllers Operational Handbook • Flight Control Operations Handbook

FCOL Faculty of Communications and Open Learning

FCOS Flight Computer Operating System • Flight Control Operating System • Flight Control Operational Software

Fcp Foolscap

FCP Facilities Construction Plan • Facilities Criteria Plan • Failure Correction Panel • Fatigue-Crack Propagation • Federal Cataloguing Program • Fibre Channel Protocol • File Control Processor • Final Control Point • Firm Cost Proposal • Flat Concurrent PROLOG • Flight Control Procedure • Flight Correction Proposal • Fonds Commun de Placement • Forest Canopy Profile • Foundation Computing Technology • Franck-Condon Principle • Franck-Condon Pumping • Fuel Cell Plant • Fuel Cell Power • Fuel Cell Propellant • Function Control Package • Fusion and Correlation Processor

FCPA Federal Chamber of Patent Attorneys • Fractional Cutting-Plane Algorithm

FCPC Federal Computer Products Center • Field Coil Power Conversion • First Class Phone Card • Flight Control Primary Computer

FCPE Fonds Commun de Placement d'Entreprise (France)

FCPGA Flip Chip Pin Grid Array

FCPh Fundamentals of Cosmic Physics (periodical code)

FCpI Flux Changes per Inch

FCPI Flux Changes Per Inch • Fonds Commun de Placement pour l'Innovation (France)

FCPLR Formation Complémentaire Par La Recherche

F&CPP Foreign & Colonial Parcel Post (UK)

FCPR Fonds Commun de Placement à Risques

FCPS Fairfax County Public Schools (USA-VA) • Fuel Cell Power System • Fuel Consumption Projection System

FCPU Facility Central Processing Unit

FCQ Fourier Correlation Quotient (method)

FCR Facility Change Request • FIFO Control Register • Final Configuration Review • Final Costs Recovery • Fire Control System • First Customer Release • Flight Configuration Review • Flight Control Room • France Câbles et Radio • Functional Capability Requirement • Functional Career Representative • Fuse Current Rating

FCRA Fecal Collection Receptacle Assembly

FCRAO Five College Radio Astronomy Observatory (USA-MA)

FCRB Future Change Review Board

FCRC Federal Contract Research Center (USA) • Federated Computing Research Conference

F&CRe Flower and Cook Observatory Reprints (periodical code)

FCRN Fund Classification Reference Number

FCRSWG Forestry Canada Remote Sensing Working Group

FCRT Flight Cathode Ray Tube

FCRU Facilities Control Relay Unit

Fcs Francs

FCS Colorado Springs - Butts (USA-CO) (location identifier) • Faceted Classification Schema • Fast Control System • Fecal Containment System • Federal Catalog System • Federal Communications System (USA) • Federation of Communication Services • Feedback Control System • Fermion - Chern - Simons (theory) • Fiber Channel Standard • Field Citation Score • File Compression Software • Finally Came to my Senses • Financial Control System • Fine Calibration Source • Finite Coordinate Space • Fire Control Section • Fire Control System • First Customer Shipment • Flat Crystal Spectrometer • Flight Control Simulator • Flight Control System • Flight Crew System • Foreign Commercial Service • Forward Control Section • Foundations of Computer Security (conference) • Frame Check Sequence • Frequency Control Symposium • Frequency Conversion System • Front Crew System • Fuels Capabilities System • Full Communications Service

FCSC Federal Conversion Support Center • Fermilab Computer Security Coordinator • Flight Control Secondary Computer

FCSD Flight Control Systems Department • Flight Crew Support Division

Fcsg Focusing

FCSHS Fédération Canadienne des Sciences Humaines et Sociales (Canada)

FCSIR Flight Control System Integration Rig

Fcsle Forecastle

FCSM Flight Combustion Stability Monitor

FCSP Fiber Channel Switched Protocol

FCSS Flight Control System Simulator • Fornax Cluster Spectroscopic Survey • Fuel Cell Servicing System

FCST Federal Council for Science and Technology (USA) • Forecast

FCSTF Fuel Cell System Test Facility

FCT Facer Canceler Table • Fast Current Transformer • Federal Capital Territory • Filament Center Tap • File Control Table • Film-Cooled Tailpipe • Fire Control Team • First Coding Theorem • Fixed Conversion Time • Flight Controller Trainer • Flight Control Team • Flight Crew Trainer • Flux-Corrected Transport • Foreign Comparative Test • Foreign Comparative Testing • Fractional Component Thereof • Fuel Cell Test • Fuel Consumption Theorem • Function • Fundação para a Ciência a Tecnologia (Portugal) • Fundamentals of Computation Theory (conference) • Future Capability Team • Yakima (USA-WA) (location identifier)

FCTB Flight Crew Training Building

FCTF Fuel Cell Test Facility

FCTR Flight Controller Training Room

FCTS Flight Crew Trainer Simulator

Fctn Function

Fctnl Functional

Fcty Factory

FCU Fault-Containment Unit • Fluid Checkout Unit • Fluid Conditioning Unit • Focal-plane Control Unit • Frequency Control Unit • Fuel Control Unit

FCUL Faculdad de Ciências de la Universidad de Lisboa (Portugal)

FCV Flow Control Valve • Fuel Cell Vehicle

FCW Federal Computer Week (journal) • Format Control Word

Fcy Fancy

FCY Forrest City (USA-AR) (location identifier)

f&d Fill & Drain

f/d Focal Distance

fd Fortran Declaration (file extension)

Fd Feed • Filed • Fiord • Fjord • Flight Deck • Focal Distance • Ford • Forward • Found • Founded • Free Delivery • Free Discharge • Free Dispatch • Fund

F&D Facilities & Design • Fill & Drain • Findings & Determinations

F/D Face of Drawing • Field of Drawing • Fill/Drain • Flight Director • Focal Length to Diameter

FD Failure Definition • Failure Density • Fast Delivery • Fault Detection • Federal District • Feed • Feeder Data • Fermi-Dirac • Fictitious Domain • Field • Filament Darkness • File Description • File Descriptor • Fill and Drain • Fill Draw • Finite Difference • Fire Department • Fiscal Day • Flight Day • Flight Deck • Flight Demonstration • Flight Director • Flight Dynamics • Floor Distributor • Floppy Disk • Floppy Drive • Focal Distance • Focused Dispatch • Forbush Decrease • Force Development • Fraction Dead • Free Delivery • Freeze Dried • Frequency Domain • Full Development • Full Disk • Full Duplex • Fully Digital • Functional Dependency • Functional Description • Functional Designator • Function Designator

FDA Facility Disposal Area • Fault Detection and Annunciation • Final Design Approval • Flight Deck Assembly • Food and Drug Administration (USA) • Functional Data Administrator • Functional Data Analysis

FDAC Flight Demonstration of ASTA Concepts

FDAd Functional Data Administrator

FDAI Flight Director Attitude Indicator

FDAO Flight Dynamics Analysis Operation

FDAS Frequency Distribution Analysis Sheet

FDAT Final Development Acceptance Test

FDAU Flight Data Acquisition Unit

FDB Facility Database • Fahrenheit Dry Bulb • Federated Database • Field Dynamic Braking • Fit Database • Fuse Disconnection Buffer

Fdbk Feedback

FDBM Functional Database Manager

FDBS Federated Database System

FdC Fin de Citation

FDC Facility Design Criteria • Fast Device Controller • Fire Detection Center • First-Day Cover • Flight Data Center • Floppy-Disk Controller • Fluorosensor Data Correlator • Functional Data Coordinator • Functional Design Criteria

FDCD Fluorescence-Detected Circular Dichroism

FDCF Flight Design Control Facility

FDCI Fixed Duration and Constant Impact

FDCS Flight Deck Communications System • Fully Dimensional Cross-section

FDCT Forward Discrete Cosine Transform • Forward Discrete Cosine Transformation

FDD Facility Description Document • Final Delivery Date • Flight Dynamics Division • Floppy-Disk Drive • Force Design and Development • Formation De Développement • Frequency Division Duplex • Functional Decomposition Design • Functional Description Document • Functional Design Document

FDDA Four-Dimensional Data Assimilation

FDDB Function Designator Database

FDDF Flux Decrement Distribution Function

FDDI Fiber Digital Data Interface • Fiber Digital Device Interface • Fiber Distributed Data Interface (*)

FDDL Field Data Description Language • Frequency-Division Data Link

FDDO Feedback-Directed and Dynamic Optimization (workshop)

FDDS Flag Data Display System

FDE Fault Detection Electronics • Field Decelerator • Filter Driver Electronics • Finite Difference Equation • Flight Data Entry • Fluid Dynamics Experiment • Fond Diffus Extragalactique • Førde (Norway) (location identifier) • Full Duplex Ethernet • Functional Data Element • Functional Differential Equation

FDEP Flight Data Entry and Printout • Formatted Data Entry Program

FDES Fonds de Développement Économique et Social

fdf Forms Data Format (file extension)

FdF Fondation de France

FDF First Drop Flight • Flight Data File • Flight Dynamics Facility • Form Definition File • Form Description File • FORS Deep Field • Fort de France (Martinique) (location identifier)

FDFI Fault Detection - Fault Isolation

FDFS File Descriptor File System

FDFU Federation of Documentary Film Units

Fdg Funding

FDH Fixed Dynamical Heating • Fluid Dynamics and Hydraulics • Friedrichshafen (Germany) (location identifier)

FDHD Floppy Disk High-Density • Floppy Drive High-Density

FDHS Fast Data Handling Simulator

FD&I Failure Detection & Isolation

FDI Failure Detection and Isolation • Failure Detector Indicator • Fast Delivery Image • Fault Detection and Identification • Fault Detection and Isolation • Feeder Distribution Interface • Field Data Investor • Flight Director Indicator • Flight Director Instrument • Foreign Direct Investment • Forum des Droits sur l'Internet • Fourier Doppler Imaging • Frequency Domain Interferometry

FDIC Federal Deposit Insurance Corporation (USA)

FDIIR Fault Detection, Isolation, Identification, and Recompensation

FDIJ Fonds Départemental pour l'Initiative des Jeunes (France) • Foreign Direct Investment Jurisprudence

FDIO Flight Data Input/Output

FDIOC Flight Data Input/Output Center

FDIOR Flight Data Input/Output Remote

FDIP Full Development and Initial Production

FDIR Failure Detection, Isolation, and Recovery • Fault Detection, Identification, and Recovery • Fault Detection and Isolation Recovery • Fault Diagnosis, Isolation, and Recovery

FDISK Fixed Disk • Format Disk

FDIU Full Device Interface Unit

FDK Flight Display Keyboard • Frederick (USA-MD) (location identifier)

FDL Facility Data Link • Fast Delay Line • Fault Detection and Localization • File Definition Language • Fixed Document Language • Flight Director Loop • Flight Dynamics Laboratory • Formatted Definition Language • Form Definition Language • Forum on Design Languages

FDLC Final Design Load Cycle

FDLN Feed Line

FDLP Federal Depository Library Program (GPO)

FDLS Dault Detection Localization System

FDM Feasibility Demonstration Model • Fill and Drain Module • Finite Difference Method • Flight Data Manager • Flight Design Manager • Fluid Dark Matter • Formal Development Methodology • FOS Data Management • Frequency Data Multiplexer • Frequency Division Multiplex (*) • Frequency Division Multiplexer (*) • Frequency Division Multiplexing (*) • Frequency Domain Multiplex • Frequency Domain Multiplexer • Frequency Domain Multiplexing • Full Descriptive Method • Functional Data Manager • Fuzzy Dark Matter

FDMA Frequency Demand Multiple Access • Frequency-Division Multiple Access (*)

FDMDA Full Diameter Motorized Door Assembly

FDMJC Fédération Départementale des Maisons de Jeunes et de la Culture (France)

FDMP Full Duplex Message Protocol

FDMS Flight Data Management System

Fdn Foundation

FDN Filtre de Densité Neutre • Fixed Dialing Number

FDNR Frequency-Dependent Negative Resistance

FDNU Format-Dependent Non-Uniformity

Fdo Firmado

FDO Fee Determination Official • Fee Determining Official • Fiche Descriptive d'Organisation • Flexible Deterrent Option • Flight Deck Officer • Flight Director's Office • Flight Dynamics Officer • For Declaration Only • Foreign Disclosure Office • Foreign Disclosure Officer

FDOA Frequency Difference Of Arrival

FDOC Fraction of Degradable Organic Carbon

FDOR Flight Design Operations Review

FDP Fast Data Processor • Fast Delivery Processor • Fast Delivery Product • Fast Digital Processor • Feasibility Demonstration Program • Federal Demonstration Partnership (USA) • Federal Demonstration Project • Field Development Program • Field Distortion Pattern • Flight Data Processing • Flight Design Panel • Forecast Demonstration Project • Foreign Duty Pay • Formation De Perfectionnement • Four Dipole Problem • Fourniture de Documents Primaires • Frequency-Dependent Polarization • Full Disk Patrol • Full Dog Point • Functional Definition Process • Funded Delivery Period

FDPA Frequency-Dependent Position Angle

FDPC Federal Data Processing Center

FDPE Functional and Declarative Programming in Education (conference)

FDPO Foreign Disclosure Policy Office

FDPP Framework Demonstration Projects Program

FDPR Flight Data Processor Replacement

Fdr Feeder • Finder • Founder

FDr Fire Door

FDR Failure Detection and Recovery • False Discovery Rate • Fast Dump Restore • Final Data Report • Final Definition Review • Final Design Review • Fit Data Record • Flight Data Recorder • Flight Data Recording • Flight Definition Requirement • Flight Dynamics Room • Formal Design Review • Franklin Delano Roosevelt (1882-1945) • Frederick (USA-OK) (location identifier) • Frequency Domain Reflectometry • Functional Design Requirement • Functional Design Review • Future Data Radio • Future Design Review • Future Digital Radio

FDRD Faculty Director of Research Development • Flight Definition and Requirements Directive • Flight Definition and Requirements Document • Flight Determination and Requirements Directive

FRDF Financial Data Records Folder

FDRI Flight Director Rate Indicator

FDRP First Destination Reporting Point

FDRR Flexible Disc Recorder Reproducer

FDRS Flight Data Recording System • Functional Description Requirements Specification

FDRT Flexible Digital Receiving Terminal

Fdry Foundry

Fds Fonds

FDS Facility Data Sheet • Facility Documentation System • Failure Detection Electronics • Failure Dynamics Software • Faraday Dark Space • Fast Delivery Service • Fast Diode Switch • Fault Detection System • Fermi-Dirac Statistics • Final Draft Study • Finance Disbursing Section • Financial Data System • Fire Detection and Suppression • Fire Detection System • Fixed Distribution System • Flare Detection System • Flexible Display System • Flight Data System • Flight Design and Scheduling • Flight Design System • Flight Director System • Flight Display System • Flight Dynamics Simulator • Flight Dynamics Software • Flight Dynamics Support • Flight Dynamics System • Fluid Distribution System • Flux Difference Splitting • Formal Data Structure • Forward Direct Support • Frame Difference Signal • Fuels Dispensing System • Full Development System • Functional Design Specification

FDSC Failure Definition and Scoring Criteria • Flagstaff Dark Skies Coalition (USA-AZ) • Flight Dynamics Simulation Complex (*) • Flight Dynamics Situation Complex

FDSCU Fluid Distribution System Control Unit

FDSE Full Duplex Switched Ethernet

FDSL Fixed Directory Subscriber List

FDSMM Fluid Dynamics and Solidification of Metallic Melts

FDSN Federation of Digital Seismographic Networks

FDSR Final Design Status Review

FDSS Fault Detection Subsystem

FDSSS Flight Deck Status and Signaling System

FDSV Flight Demonstration Space Vehicle

FDSWG Future Data Systems Working Group

FDT Failure Diagnostic Team • Fault Detection System • First Destination Transportation • Flexible Digital Terminal • Flux-limited Diffusion Theory • Formal Description Technique • Full Duplex Teletype • Fully Developed Turbulent

FDTD Finite-Difference Time Domain

FDT&E Follow-on Development Test & Evaluation

FDTE Final Development Test and Evaluation • Follow-on Development Test and Evaluation

FDTR Full Duplex Token Ring

FDTS Field Data Tracking System

FDTSP Foreign Disclosure and Technical Security Plan

FDU Bandundu (Zaire) (location identifier) • Flexible Disk Unit • Flight Development Unit • Flight Divider Unit • Floppy Drive Unit • Fluid Distribution Unit • Frequency Determining Unit • Frequency Doubling Unit

FDUL Fixed Directory Unit List

FDV Fault Detection Verification • Fill and Drain Valve • Flowfield Dependent Variation

FDVDI Fiber-Distributed Video/Voice Data Interface

FDW Faculty Development Workshop • Forward Destination Word

FDx Full Duplex

FDX Full Duplex

FDY Findlay (USA-OH) (location identifier)

fe Fecit

Fe Eisen • Fer • Iron

F&E Facilities & Equipment • Facility & Equipment • Facility & Environment • Forschung & Entwicklung

F/E Forschung/Entwicklung • Full/Empty

FE Facilities Engineer • Facilities Engineering • Facility and Environment • Failure Equation • Farbexzeß • Far East • Far Encounter • Faulted Edge • Feature Extraction • February • Février • Fiberglass Epoxy • Field Elevation • Field Emission • Field Engineer • Field Expedient • Field Experiment • Fill End • Final Erection • Fine Erection • Finite Element • Fire Extinguisher • First Edition • Flexible Erection • Flight Engineer • Flight Engineering • Flight Examiner • Flight Execution • Flight Executive • Flight Experiment • Fly's Eye • Force Execution • Foreign Editor • For Example • Format Effector • Forschung und Entwicklung • Fossil Energy • Frequent Editor • Front End • Full Erection • Functional Element • Functional Entity • Fundable Expenses • Fundamentals of Engineering • Further Education

FEA Failure Effects Analysis • Federal Editors Association (USA) • Federal Energy Administration (USA) • Fetlar (UK) (location identifier) • Finite-Element Analysis • Flight Experiment Apparatus • Fluid Experiment Apparatus • Fluid Experiment Assembly • Front End Amplifier • Front End Analysis • Front End Assembly • Functional Economic Analysis • Functional Entity Action

FEAAS Federation of East Anglian Astronomical Societies (UK-England)

FEAC Far-End Alarm and Control

FEACO Fédération Européenne des Associations de Conseils en Organisation

FEANI Fédération Européenne d'Associations Nationales d'Ingénieurs

FEANTSA Fédération Européenne des Associations Nationales Travaillant avec les Sans-Abri

FeAp Fernsprech-Apparat

FEAP Finite-Element Analysis Program

FEARO Federal Environmental Assessment Review Office

FEAST Fast Ethernet using Advanced SMC Technology • Finite Element Analysis for Structural Test • Finite Element Analysis of Structures

FEAT Final Engineering Acceptance Test • Fine Error Averaging Test • Formal Evaluation Acceptance Test • Frequency of Every Allowable Term

Feb Febbraio • Febrero • February

FEB Fair Employment Board • Falling Evaporating Body • February • Fédération des Éditeurs Belges (Belgium) • Fédération des Entreprises Belges (Belgium) • Filter Electronics Box • Flight Evaluation Board • Fluidized Ejecta Blanket • Forward Equipment Bay • Fourth Entry Battery • Franklin Electronic Book • Functional Electronic Block • Functional Energy Block • Sanfebagar (Nepal) (location identifier)

Febb Febbraio

FEBE Far-End Bit Error • Far-End Block Error

FEBEL Faint Extended Broad Emission Line

FEBS Federation of European Biochemical Societies

Feby February

fec Fecit

FEC Fast Ethernet Channel • Federal Election Commission (USA) • Federal Express Corp. • Field Engineering Change • Fine Erection Complete • Fondation Européenne de la Culture • Foreign Exchange Carrier • Foreign Exchange Certificate • Foreign Exchange Cost • Forward Error Collection • Forward Error Control • Forward Error Correction (*) • Forward Events Controller • Forwarding Equivalence Class • Forwarding Equivalent Class • Front-End Computer • Front-End Controller

FECA Federal Employees Compensation Act (USA)

FECAI Federal Electronic Commerce Acquisition Instruction (USA)

FECAT Federal Electronic Commerce Acquisition Team (USA)

FECB File Extended Control Block • Foreign Exchange Control Board

FECC Federal Emergency Communications Coordinator (USA)

FECH Federación de Estudiantes de la Universidad de Chile (Chile)

FECM Fonds Européen de Coopération Monétaire

FECN Forward Explicit Congestion Notice • Forward Explicit Congestion Notification

FECOM Fonds Européen de Coopération Monétaire

FECN Forward Error Code Notification • Forward Explicit Congestion Notification

FECR Final Experiment Concept Review

FECS Federation of European Chemical Societies

FEC2 Forward Entry Command and Control

Fed Federal • Federalist • Federate • Federated • Federating • Federation

Féd Fédéral • Fédéraliste • Fédération • Fédéré • Fédérer

FED Far End Data • Federal • Fernsprechhauptanschluß für Direktruf • Field-Effect Diode • Field-Emission Device • Field-Emission Display (*) • Field-Emitter Display • Final Estimation of Data • Flight Events Demonstration • Fondo Europeo de Desarollo • Fonds Européen de Développement • Forward Entry Device • Frame Encapsulator/Decapsulator • Front End Driver

FEDAC Federal Education Data Acquisition Council (USA) • Forward Error Detection And Correction

FEDB Failure Experience Database • Flight Execution Database

FEDC Fusion Energy Design Center

FEDCAC Federal Computer Acquisition Center (USA)

FEDCIRC Federal Computer Incident Response Capability (USA)

FEDD Family of Electronic Deceptive Devices

FEDEP Federal Execution and Development Process

FEDER Fonds Européen de Développement Régional

FEDGE Finite Element Data Generation

FEDI Failure Experience Data Interchange

FEDIA Fédération d'Études des Infrastructures Alternatives (France)

FEDIT Federación Española de Entidades de Innovación y Tecnología (Spain)

FEDLINK Federal Library and Information Network

Fedn Federation

FEDP Facility and Equipment Design Plan

FEDR Facilities Engineering and Design Requirements

FEDS Federal Employment Data Service (USA)

FEDSTD Federal (telecommunications) Standard (USA)

FEDSTRIP Federal Standard Requisitioning and Issues Procedures (USA)

FEE Failure Effects Evaluation • Feature Extraction Environment • Federal Environmental Executive (USA) • Fédération des Étudiants Étrangers • Fédération Européenne des Experts-comptables • French Echocardiograph Equipment • Front End Electronics • Full Exit Entry • Functional Execution Element

FEED Field Exploitation of Elevation Data

FEEDS Facilities Engineering Expert and Diagnostic System

FEEE Fédération pour l'Éducation à l'Environnement en Europe

FEEMS Facilities Engineering Equipment Maintenance System

FEEP Field Emission Electric Propulsion

FEER Fast Eigensolution Extraction Routine • Fundamental Equilibrium Exchange Rate

FEES Front End Edit System

FEF Fédération des Étudiants Francophones • FITS Embedded Function • Flight Engineering Facility • Fondation Européenne pour la Formation • Fuel Examination Facility

FEFI Flight Environmental Fault Isolation

FEFO First Ended - First Out

FEFSC Finite Exposure Fractional Seeing Change

FEG Fergana (Uzbekistan) (location identifier) • Field-Emission Gun

FEGLI Federal Employees Group Life Insurance (USA)

FEHB Federal Employees Health Benefits (USA)

FEHBP Federal Employees Health Benefits Program (USA)

FEI Facility Engineering Item • Faculdade de Engenharia Industrial • Federation of the Electronics Industry • Financial Executive Institute • Flexible External Insulation • Fonds Européen d'Investissement • Functionality Emulative Increment

FEICRO Federation of European Industrial Cooperative Research Organizations

FEID Flight Equipment Interface Design • Flight Equipment Interface Device • Functional Engineering Interface Device

FEIN Federal Employer Identification Number (USA)

FEIP Facility and Equipment Improvement Program

FEIS Final Environment Impact Statement

FEISEAP Federation of Engineering Institutions of Southeast Asia and the Pacific

FEIT Fujitsi Enhanced Imaging Technology

FEJE Facilities Engineer Job Estimation

FEK Ferkessedougou (Ivory Coast) (location identifier) • Forskningsbibliotekernes Edb-Kontor (Denmark) • Frequency Exchange Keying

FEL First-Element Launch • Free-Electron Laser • Fürstenfeldbruck (Germany) (location identifier) • Fysich Elektronisch Laboratorium (Netherlands) • Fysich en Elektronisch Laboratorium

FELASI Federación Latin-Americana de Sociedades de Física

FELASOFI Federación Latin-Americana de Sociedades de Física

FELI Free Electron Laser Institute

FeLiX Fe-Line region X-ray

FELIX Free Electron Laser for Infrared Experiment • Front End of Laser Interceptor Transmitter

Fell Fellow • Fellowship

Felr Feeler

fém Force Électromotrice

Fem Female • Féminin • Feminine

FEM Field Effect Modified • Field Electron Microscope • Field Electron Microscopy • Field Emission Microscope • Field Emission Microscopy • Field Engineering Maintenance • Field Evaluation Model • Field Evaluation Modeling • Finite-Element Method • Finite-Element Model • Finite-Element Modeling • Flight Engineering Model • Flight Engineering Modeling • Flyable Engineering Model • Flyable Engineering Modeling • Fonds pour l'Environnement Mondial • Force Effectiveness Measure • Force Électromotrice • Foro Económico Mundial • Free Electron Maser • Front End Module • Functional Electrical Model • Functional Electrical Modeling

FEMA Federal Emergency Management Agency (USA) • Federation of European Meteor Observers

FEMC Flight Experiment Management Center

FEMCPL Facilities and Environmental Measurement Component Parts List

FEMF Foreign Electromotive Force

FeMik Finommechanika Mikrotechnika (periodical code)

FEMIS Fondation Européenne des Métiers de l'Image et du Son

FEML Funded Environmental and Morale Leave

FEMMS Frontiers of Electron Microscopy in Materials Science

FEMO Finite Element Modeling Optimization • Free-Electron Molecular Orbital

FEMP Federación Española de Municipios y Provincias (Spain) • Federal Energy Management Program • Free Energy Minimization Procedure

FEMR Foreign Excess Materiel Return

FEMS Federation of European Microbiological Societies

FEN Fédération de l'Éducation Nationale • Fenómeno El Niño • Fernando de Noronha (Brazil-FN) (location identifier) • Frequency Emphasizing Network

FENP Front-End Network Processor

FENS Federation of European Neuroscience Societies

FEO Federal Environment Office • Few Entry Object • Field Engineering Office • Field Engineering Order • Fin d'Études Orientées

FEOC Fixed Element Operations Center

FEORP Federal Equal Opportunity Recruitment Program (USA)

FEP Faculté d'Éducation Permanente • Fast Evolutionary Programming • Feature Extraction Processor • Federation of European Publishers • Finite Element Program • Flight Experiment Program • Floral Ethel Propane • Flow Enhancing Protocol • Fluorinated Ethene Propene • Fluorinated Ethylene Propylene (*) • Fonction d'Étalement de Point (*) • Fonction d'Étalon de Point • Forecast

Expenditure Plan • Formal Evaluation Phase • Format Evolution Process • Free Endpaper • Freeport Albertus (USA-IL) (location identifier) • Front-End Processor

FEPA Fair Employment Practices Act (USA)

FEPC Fair Employment Practices Commission (USA) • Fair Employment Practices Committee • Flight Equipment Processing Contractor • Format Evolution Process Committee

FEPCA Federal Employees Pay Comparability Act (USA)

FEPG Federal Engineering Planning Group • Federal European Planning Group

FEPI Front-End Programming Interface

FEPS Facility and Equipment Planning System • Far Encounter Planet Sensor • Formation and Evolution of Planetary Systems

FEPT Fédération Européenne de la Presse et de la Télévision

FER Federación Europea de Redes • Fergusons Gulf (Kenya) (location identifier) • Field Engineering Representative • Final Evaluation Report • Force Exchange Ratio • Fractional Exchange Ratio

FERA Federal Emergency Relief Administration (USA)

FeRAM Ferroelectric Random Access Memory

FERC Federal Energy Regulatory Commission (USA)

FERD Facility and Equipment Requirements Document • Forum Européen de la Recherche et du Développement • Fuel Element Rupture Detection

FERF Far-End Receive Failure • Far-End Remote Failure • Far-End Reporting Failure

FERIAS Forest Environment and Resource Information Analysis System

FERMI Front-End Readout Microsystem

FERN Forest Ecosystems Research Network • Further Education Research Network

FernVSt Fernvermittlungstelle

FEROS Fiber-fed Extended Range Optical Spectrograph

FERPA Family educational Rights and Privacy Act (USA)

Ferr Ferrocarriles

FERRO Far Eastern Regional Research Office

FERS Federal Employees Retirement System (USA) • Fédération Européenne des Réseaux Scientifiques

FERT Fortitudo Eius Rhodum Tenuit

Frev Fervens

FES Facilities Engineering Services • Factor Evaluation System • Fast Erect System • Feature Extraction Segment • Fédération des Éditeurs Suisses (Switzerland) • Field Emission Spectroscopy • Final Environmental Statement • Fine Error Sensor • Fine Error Signal • Fire and Emergency Services • Flame Emission Spectrometry • Flame Emission Spectroscopy • Flash Evaporator System • Flight Element Set • Fluids Experiments System • Forced Entry Switch • Force Entry System • Force Extension Switch • Fondation Européenne de la Science • Free Energy Simulation • Functional Electrical Simulation • Fuzzy Expert System

FESA Finite Element Structural Analysis

FESAC Fusion Energy Sciences Advisory Committee (USA)

FESC First EUVE Source Catalog

FESD Flight Experiment Specification Documentation

FESEM Field Emission Scanning Electron Microscope • Field Emission Scanning Electron Microscopy

FESG Forschungseinrichtung Satellitengeodäsie (TUM)

FESI Federación Española de Sociedades de la Informática (Spain)

FESIC Fédération des Écoles Supérieures d'Ingénieurs et des Cadres (France)

FESIP Fast External Satellite Image Processing

FESL Failure Effects Summary List

FESP Flight Execution Support Plan

FESR Finite Energy Sum Rules

FESS Facilities and Engineering Services Section (Fermilab) • Facility Engineering Supply System • Flight Experiment Shielding Satellite

Fest Festival • Festivo

FEST Foundation for Education, Science and Technology (South Africa)

FESTIP Future European Space Transportation Investigation Programme

FESWG Flight Experiment Specification Working Group

FESWMS Finite Element Surface Water Model System

FESWS Feature Extraction Workstation

FET Federal Estate Tax • Federal Excise Tax • Field-Effect Transistor (*) • Field Electron Transistor • Firewall Evaluation Team • Flight Elapsed Time • Follow-on Equipment Training • Fossil-Energy Technology • Fremont (USA-NE) (location identifier) • Future and Emerging Technologies

FETEM Field Emission Transmission Electron Microscope • Field Emission Transmission Electron Microscopy

FETV Full Envelope Thrust Vectoring

FeU Fern Universität (Germany)

FEU Filter Exchange Unit • Flight Equivalency Unit • Flight Evaluation Unit • Further Education Unit

FEUQ Fédération Étudiante Universitaire du Québec (Canada-QC)

Fév Février

Fev Fevereiro

FEV Far End Voice • Fevereiro • Février • Fisheye View • Flight Experiment Vehicle • Forced Expiratory Volume • Front Expansion Velocity

FEVARO Friedrich Ebert Visual and Radiotelescope Observatory (Germany)

FEVE Ferrocarriles Españoles de Vía Estrecha (Spain)

FEW Cheyenne Warren (USA-WY) (location identifier)

FEWG Flight Evaluation Working Group

FEWS Famine Early Warning System • Follow-on Early Warning System • Follow-on-Earth Warning System

FEX Field Exercise

FExt Fire Extinguisher

FEXT Far-End Crosstalk

FEZ Fez (Morocco) (location identifier) • Freizeit- und Erholungszentrum (Germany)

ff Faisant Fonction • Fecerunt • Feuillets • folgende • Folios • Following • Forms • fortfolgende • Fortissimo • Thick Fog

F&F Fire & Forget

F-F Face to Face • Flip-Flop

F/F Fill/Full • Frame Format • Full/Fill

FF Failure Factor • Far Field • Fast Forward • Fellows • Fellows • Filling Factor • Find File • Fitting Factor • Fixed Focus • Finish to Finish • First Fit • Fixed Focus • Flat Field • Flight Forward • Flip-Flop • Force Field • Ford Foundation • Forecasting Funding • Form Feed • Fortsetzung Folgt • Forward Fuselage • Free Fall • Free Flight • Free Flyer • Free Format • Free-Free • French Franc • Frozen Flow • Fuel Flow • Full Field • Full Functional • Functional Flow

FFA Fast Fourier Analyzer • Federal Facilities Agreement • Flygtekniska Försörksanstalten (Sweden) • For Further Assignment • Free Fall Acceleration • Free-Free Absorption • Free From Alongside • Functional Failure Analysis • Kill Devil Hills (USA-NC) (location identifier)

FF.AA. Fuerzas Armadas

FFAG Fixed-Field Alternating Gradient

FFAST Forest Fire Advanced System Technology

FFB Federal Financing Bank (USA) • Fourier Filtered-Back (algorithm)

FFBD Functional Flow Block Diagram

FFC Father Function Coefficient • Federal Facilities Council • Field File Custodian • Final Flight Certification • First-Flight Capability • Flagstaff Field Center (NIST) • Flip-Flop Complimentary • Floor-Fractured Crater • Foreign Funds Control • Free-Flyer Centre • Free From Chlorine

FFCA Federal Facility Compliance Act (USA)

FFCAPT Four College Consortium Automatic Photometric Telescope

FFCB Fédération Française de Coopération entre Bibliothèques (France)

FFCC Ferrocarriles • Free-Flyer Control Centre

FFCF First Flight Centennial Commission (USA)

FFD Fairford (UK-England) (location identifier) • First Fit Decreasing • Fixed-Format Display • Free Form Diagram • Functional Flow Diagram • Function Flow Diagram

FFDC First Failure Data Capture

FFDF Fluid Physics and Dynamics Facility

FFDP Far Field Diffraction Pattern

FFE Fluorescence par Faisceau d'Électrons • Forced Fault Entry • Furnitures - Fittings - Equipment

FFEL Förderung der Forschung in Entwicklungsländern

FFEM Freeze-Fracture Electron Microscope • Freeze-Fracture Electron Microscopy

FFES Fossil Free Energy System

FFEU Free-Flow Electrophoresis Unit

fff Fortississimo

FFF Field Flow Fractionation • Final Factory Functional • Fission - Fusion - Fission • Force-Free Field • Form - Fit - Function • Forschungsförderungsfonds (Austria)

FFFF Flow Field Flow Fractionation

FFFP Film Forming Fluoroprotein

FFFT Forced Flow Flamespreading Test

FFG Forschungsförderungsgesetz

FFGF Fill Frame Generation Function

FFH Fast Frequency Hopping

FFHR Force-Free Helical Reactor • Fusion-Fission Hybrid Reactor

FFI Find Further Information • Fine Field Imager • For Further Information • Forsvarets Forskningsinstitutt (Norway) • Frame Format Identification • Fraunhofer Filling In • Free From Infection • Fuel Flow Indicator • Full Field Investigation

FFIB Fuel Filter Impending Bypass

FFIRN Field Format Index Reference Number

FFIS Federal Facilities Information System

FFIT Form - Fit - Interoperability - Transportability

FFK Fixed Function Key

FFL Fairfield (USA-IA) (location identifier) • Fédération Française du Livre (France) • Flip-Flop Latch • Front Focal Length

FFLOP Field Fresnel Lens Optical Platform

Ffly Faithfully

FFM Fergus Falls (USA-MN) (location identifier) • Flats Forwarding Module • Fuel Flow Monitor • Free-Fall Machine • Free-Flying Module

FFMED Fixed Format Message Entry Device

FFMJC Fédération Française des Maisons de Jeunes et de la Culture (France)

FFMSR Federal Financial Management System Requirement

FFN Feedforward Neural

ffo Folios

FfO Forschungsinstitut für Optik

FFO Dayton Patterson (USA-OH) (location identifier) • Flux-Flow Oscillator • Fully Functional Operations

FFOL FDDI Follow-On LAN

FFon Faisant Fonction

FFOS Forward Flying Observation System

F&FP Force & Financial Plan

FFP Fabrication - Falsification - Plagiarism • Face-à-Face Pédagogique • Falsification - Fabrication of Results - Plagiarism • Faser - Fabry - Pérot (filter) • Fast Fluorescence Photometry • Federal Financial Participation • Fédération de la Formation Professionnelle (France) • Fifth Framework Programme (EU) • Firmed Fixed Price • Firm Fixed Price • Fixed Fee Procurement • Force and Financial Plan • Frequent Flyer Programme • Fresh Frozen Plasma • Fully Functioning Process

FFPE Front Focal Plane Enclosure

FFPI Flip-Flop Position Indicator

FFPIF Firm Fixed Price plus incentive Fee

FFQS FIRST Faint Quasar Survey

FFr French Franc (currency)

FFR Factory Flight Readiness • Fission-Fusion Ratio • Fixed Frequency Receiver • Flight Feasibility Review • Force Feasibility Review • Forward-Firing Rocket • Frame Forwarding Rate • Full Frequency Recording

FFRD Flip-Flop Relay Driver

FFR&DC Federally-Funded Research & Development Center (USA)

FFRDC Federally-Funded Research and Development Center (USA)

FFRR First Flight Readiness Review • Full-Frequency Range Recorder • Full-Frequency Range Recording

FFRRV Free-Flight Rotorcraft Research Vehicle

FFS Fast File System • Fast Fine Structure • Fat Fingered Syndrome • Federal Financial System (USA) • Fee For Service • Flame Fluorescence Spectroscopy • Flash File System • Force-Free Surface • Formatted File System • Free Fall Sensor

FFSA Fédération Française des Sociétés d'Assurance (France) • Field Functional System Assembly

FFSAC Field Functional Systems Assembly and Checkout

FFSD Failed Fuel Surveillance and Diagnosis

FFSK Fast Frequency Shift Keying

FFSM Fast Fourier Synoptic Mapping

FFSP Full Function Signal Processor

FFSS Full-Frequency Stereophonic Sound

FFST First Failure Support Technology

FFT Factory Functional Test • Fast Fourier Transform • Fast Fourier Transformation • File Format Table • Final Form Text • Flats Forwarding Terminal • For Further Transfer • Frame Format Table • Frankfort - Capital City (USA-KY) (location identifier) • Free Flight Test • Full Fuselage Trainer

FFTB Final Focus Test Beam

FFTD Finite Fourier Transform Definition

FFTF Fast Flux Test Facility • Final Focus Test Facility

FFTH Fiber To The Home

FFTO Free-Flying Teleoperator

FFTS Freight Forwarder Tracking System

FFTU Forward Freight Terminal Unit

FFTW Fastest Fourier Transform in the West (routines)

FFU Formatted Field Unit • Futaleufu (Chile) (location identifier)

FFV Fédération Française de Voile (France) • Field Failure Voltage • Flexible Fuel Vehicle • Focus Field Variation

FFW Failure-Free Warranty

FFwd Fast Forward

FFWF Fonds zur Förderung der Wissenschaftlichen Forschung

Ffy Faithfully

FFZ Free Fire Zone

fG freiwillige Gerichtsbarkeit

Fg Faubourg • Fine Grain • Fog • Fully Good

F/G Felsite/Granite • Fiberglass

FG Fachgremium • Faktengruppe • Feature Group • Federal Government • Fermi Gas • Fibonacci Generator • Fifth Generation • Filament Ground • File Gap • Filter Gate • Filtergewinde • Finance Group • Finanzgericht • Fine Grain • Fine-Grained • Fiscal Guidance • Fixed Gear • Flag • Floated Gyro • Floating Gate • Fluorescence Gated • Flux Gradient • Fog • Foreground • Forward Gate • Frais Généraux • Frame Grabber • Frame Ground • freiwillige Gerichtsbarkeit • French Guiana • Fully Good • Functional Group • Function Generator

FGA Foreign General Average • Free of General Average

FGAN Forschungsgesellschaft für Angewandte Naturwissenschaften (Germany)

FGB First Giant Branch • Flattened Gaussian Beam • Functional Energy Block (in Russian) • Functional Group Code

FGBA Fluids Generic Bioprocessing Apparatus

FGC Fast Guide Camera • Fifth Generation Computer • Fixed Gain Control • Flat Galaxy Catalogue

FGCC Federal Geodetic Control Committee (USA)

FGCS Federal Geodetic Control Subcommittee (USA) • Fifth Generation Computer Systems

Fgd Forged

FGD F Derik (Mauritania) (location identifier) • Flux Gas Desulphurisation

FGDC Federal Geographic Data Commission (USA) • Federal Geographic Data Committee (USA) (*) • Federal Geospatial Data Clearinghouse

FGE Fine Guidance Electronics • Flare Genesis Experiment

FGEF Federal Geographic Exchange Format (USA)

FGEP Fixed Ground Entry Point

FGFC Fixed Gas-Filled Capacitor

FGG Foreground Galaxy

FGGE First GARP Global Experiment

FGH Field - Goldsmith - Habing (theory) • Finanzgerichtshof • Flexible Gyro Header • Fourier Grid Hamiltonian

FGHA Flexible Gyro Header Assembly

FGHC Flat Guarded Horn Clause

FGI Apia Fagali'i (Samoa) (location identifier) • Fédération Graphique Internationale • Financial Guarantee Insurance • Finnish Geodetic Institute (Finland)

FGIPC Federation of Government Information Processing Councils (USA)

FGJ Forschungsgesellschaft Joanneum (Germany)

FGK Forschungsgesellschaft Kunststoffe (Germany)

FGKL Fifth Generation Kernel Language

FGKRL Fifth Generation Knowledge Representation Language

FGL Force Generation Level • Fourth Generation Language • Fox Glacier (New Zealand) (location identifier)

FGM Fast-Growth Material • Fiscal Guidance Memorandum • Fluxgate Magnetometer • Force Generation Model • Foreign Government Memorandum

Fgn Foreign

FGN Flying Ghost Nebula • Foreign • Fundamental Geodetic Network

FGNP Field Generated Normal Point

FGO Fderik (Mauritania) (location identifier) • Finanzgerichtsordnung

FGP Faktengruppenprofil • Fetch - Generalize - Project

FGPA Fluctuating Gunn-Peterson Approximation

FGREP Fixed Global Regular Expression Print

FGRF Functional Genomics Research Facility

FGS Fellow of the Geological Society • Fine Guidance Sensor • Fine Guidance System • Fixed Ground Segment • Fixed Ground Station • Florida Geological Survey (USA-FL) • Forschungsgruppe Satellitengeodäsie

FGSAA Fine Guidance Sensor Access Area

FGSB Fixed Ground Station Backup

FGSMRAO France-Germany-Spain Millimeter Radio Astronomy Observatory

FGSS Fine Guidance Sensor System • Flexible Guidance Software System

Fgt Freight

FGt Fernschaltgerät

FGT Federal Geographic Technology (USA) • Federal Gift Tax • First Generation Testbed • Formal Government Testing • Fundação Gomes Teixeira (Portugal) • Future Giant Telescope

FGU Fangatau (Tuamotu Islands) (location identifier)

FGUG Federal Government Users Group (USA)

FGV Ferry à Grande Vitesse

FGW Filter Grating Wheel

FGWA Filter-Grating Wheel Assembly

FGWE First Global Weather Experiment

Fgy Foggy

fh Fiat Haustus

F/H Freehold

FH Fachhochschule • Family Housing • Feed Horn • Field Headquarters • Fire Hose • Fire Hydrant • Firsthöhe • Flex Hose • Flexible Hose • Flight Hour • Fog Horn • Frequency Hopping • Fore Hatch • Forward Hatch • Frame Handler • Frequency Hop • Frequency Hopper • Frequency Hopping

FHA Fault Hazard Analysis • Federal Highway Administration (USA) • Federal Home Administration • Federal Housing Administration (USA) • Flexible Header Assembly • Foreign Humanitarian Assistance • Fuel-Holding Area • Fully-Heated Area

FHAI Federal Housing Authority Insurance

FHB Family Hold Back • Field Horizontal Branch (star)

FHBD Fachhochschule für Bibliothekswesen und Dokumentation (Germany)

FHC Flight Half Coupling • Fuel Handling Cell

FHCR Federated Health Care Record

FHD Farben-Helligskeits-Diagramm • First-Hand Distribution • Fixed-Head Disk • Flame Hydrolysis Deposition • Free Hand Drawing • Function Hierarchy Diagram

FHDTC Fixed-Head Disk Transfer Channel

FHF Flächenhelligskeit-Fluktuation • First Horizontal Flight

FhG Fraunhofer Gesellschaft (Germany)

FHHS Forum for History of Human Science

FHI Fritz-Haber-Institut (Germany) • Fuji Heavy Industries (Japan)

FHIL Forbidden High-Ionization Line

FHLC Faint High-Latitude Carbon (star)

FHLCS Faint High-Latitude Carbon Star

Fhld Freehold

FHLMC Federal Home Loan Mortgage Corporation

FHLT Force High-Level Terminal

FHMA Frequency Hop Multiple Access • Frequency Hopping Multiple Access

FHMUX Frequency Hopping Multiplexer

FHNC Fermi Hypernetted Chain Theory

FHO Fachhochschule Offenburg (Germany) • Failed Handover • Fenton Hill Observatory (USA-NM)

F'hold Freehold

fhp Friction Horsepower

FHP First Handling Pieces • Flying Hour Program • Forum on the History of Physics • Friction Horsepower • Fuel High Pressure

FHR Fire Hose Rack

F&HS Family & Housing Service

FHS Fachhochschule • Family and Housing Service • Field Halo Star • Flight Hardware Simulator • Foran High School (USA-CT) • Forward Heat Shield • Frequency Hopping Signal • Fuel Handling System

FHSF Fixed-Head Storage Facility

FHSG Family Housing

FHSS Forward Heat-Shield Separation • Frequency-Hopping Spread Spectrum

FHST Fixed-Head Star Tracker

FHT Fahrenheit • Fast Hartley Transform • Fast Hough Transform • Field Handling Trainer • File Handling Tool • Formal Hardware Testing • Frequency-Hopping Telemetry

FHTC Fixed High-Temperature Capacitor

FHTS Fluid Heat Transport System

FHU Fort Huachuca (USA-AZ) (location identifier) • Sierra Vista (USA-AZ) (location identifier)

FHVC Fixed High-Volt Capacitor

FHW Fachhochschule Wiesbaden (Germany) • First-Hand Witness • Flare of Holly Wrath

FHWA Federal Highway Administration (USA)

FHy Fire Hydrant

FHZ Fakahina (Tuamotu Islands) (location identifier) • Forest Habitable Zone

fi Finland • Finnish (ISO language code)

F&I Forschung im Ingenieurwesen (periodical code)

FI Fachinformation • Facility Inspection • Faculty Index • Faroe Islands • Falkland Islands • Fault Identification • Fault Isolation • Faulty Instrument • Field Intensity • Field Ionization • Fiji Islands • Final Issue • Finland (ISO country code) • Finlande (code ISO de pays) • Fire Insurance • Fixed Interval • Flare Index • Flexible Interconnect • Flight Instructor • Flight Instrumentation • Flow Indicator • Flow Injection • Force Integration • Forecast International (USA-CT) • Foreign Intelligence • For Instance • Formal Inspection • Formation Intégrée • Free In • Fréquence Intermédiaire • Front-Illuminated • Front-side Illuminated • Fuel Injected • Full Internet

FIA Fault Isolation Analysis • Federal Insurance Administration • Fédération Internationale d'Automatique • Fédération Internationale de l'Automobile • Financial Inventory Accounting • Flame Ionization Analyser • Flight Integration Assessment • Flow Injection Analysis • Flow Injection Analyzer • Force Integration Analysis • Freedom of Information Act (USA) • Free In Air • Free Interstitial Atom • Full Interest Admitted • Fully Illuminated Area • Functional Interoperability Architecture • Future Imagery Architecture

FIAB Fédération Internationale des Associations de Bibliothécaires • Flight Integration Assessment Board

FIABS Financial Inventory Accounting and Billing System

FIAC Foire Internationale d'Art Contemporain • Foire Internationale des Arts Contemporains

FIAF Federación Internacional de Archivos de Film

FIAI Fédération Internationale des Associations d'Instituteurs

FIAMS Flinders Institute for Atmospheric and Marine Studies

FIANN Fédération Internationale des Associations Nationales de Normalisation

FIAP Fellow of the Institution of Analysts and Programmers • Forum on Industrial and Applied Physics

FIAR Fabbrica Italiana Apparecchiature Radioelettiche • Failure Investigation Action Report

FIAS Formation Internationale Aéronautique et Spatiale

FIASc Fellow of the Indian Academy of Sciences

FIAT Fabbrica Italiana Automobili Torino • Federación Internacional de Archivos de Televisión • Field Inspection and Acceptance Test

fib Fibratus

Fib Fiber • Fibre • Fibro

FIB Fast Information Block • Flight Information Bulletin • Focused Ion Beam • Forward Indicator Bit • Forwarding Information Base

FIBA Full-Interoperability-Based Architecture

FIBC Filtered Image Brightness Corrected

FIBEX First International BIOMASS Experiment

FIBRE Friendly Inputting of Bibliographic Records

Fic Fiction • Fictional • Fictitious

FIC Fault Isolation Code • Federal Identifier Code • Federal Information Center • Federal Information Commission • Field Installed Connector • Fire Cove (USA-AK) (location identifier) • Fleet Installed Cost • Flight Information Center • Flight Information Controller • Florida Imaging Centre (Canada-BC) • Fogarty International Center • Fonds Interministériel de Coopération • Frequency Interference Control

FICA Federal Insurance Contributions Act (USA)

FICCDC Federal Inter-Agency Coordinating Committee on Digital Cartography (USA)

FICCI Federation of Indian Chambers of Commerce and Industry

FICO Fingerprint Image Conversion Operation

FICOME Fédération Interprofessionnelle de la Communication d'Entreprise (France)

Ficon Fiber Connecting

FICS Fault Isolation Checkout System • Financial Information Control System • Fingerprint Image Capture System • Forecasting and Inventory Control System

Fict Fiction • Fictitious

FICU Fonds International de la Coopération Universitaire

Fid Fidelity • Fiduciary

FID Failure Identification • Fault Isolation Demonstration • Fault Isolation Detection • Fédération Internationale d'Information et de Documentation • File Identification • Final Instrument Definition • Fishers Island (USA-NY) (location identifier) • Flame Ionization Detector • Flammen Ionisationsdetektor • Flight Implementation Directive • Format Identifier • Free Induction Decay

FIDAC Film Input to Digital Automatic Computer • Fireball Data Center (IMO)

FIDAE Feria Internacional del Aire y del Espacio (Chile)

FIDDO Focused Investigation of Document Delivery Options

FIDE Fully Integrated Data Environment • Fuzzy Inference Development Environment

FIDEA Federación Interamericana de Educación de Adultos

FIDES Forecaster's Intelligent Discussion Expert System

FIDEX Floating Ice Detection Experiment

FIDH Fédération Internationale des (ligues des) Droits de l'Homme

FIDIT Fonds d'Intervention pour le Développement Industriel et Technologique

FIDLE FIFE Doppler Lidar Experiment

FIDNet Federal Instrusion Detection Network (USA)

FIDO Facility for Integrated Data Organizations • Filing, Imaging and Distributed Objects • Flight Dynamics Officer • Flight Inspection District Office • Fog, Intense, Dispersion Of • Fog Investigation Dispersal Operations

FIDOM Fonds d'Investissement dans les Départements d'Outre-Mer (France)

FIDP Fellow of the Institute of Data Processing

FIDS Facility Intrusion Detection System

F&IE Facilities & Industrial Engineering

FIE Facilities and Industrial Engineering • Fair Isle (UK) (location identifier) • Ferric Ion Free • Flight Instrumentation Engineer • Fly-In Echelon • Fourier Integral Estimate • Formaldehyde-Induced Fluorescence • Frontiers In Education (conference)

FIEC Formation des Investigateurs aux Essais Cliniques

FIED Fédération Interuniversitaire de l'Enseignement à Distance

FIEE Fédération des Industries Électriques et Électroniques • Fellow of the Institution of Electrical Engineers

FIEEC Fédération des Industries Électriques, Électroniques et de Communication (France)

FIEEE Fellow of the Institute of Electrical and Electronics Engineers

FIEI Fellow of the Institution of Engineers in Ireland

FIEJ Fédération Internationale des Éditeurs de Journaux

FiELD Finite Element Luminaire Design

FIELD Foundation for International Environmental Law and Development

FIEP Federación Internacional de Editores de Periódicos

F&IER Forschung im Ingenieurwesen - Engineering Research (periodical code)

FIERA Fast Imager Electronic Readout Assembly

FIES Fiber-coupled Echelle Spectrograph

FIESTA Fast Instrument Echelle Spectrograph with Triple Arm

FIEU Fonds International d'Échanges Universitaires

FIEV Fédération des Industries des Équipements pour Véhicules (France)

fif Fractal Image Format (file extension)

FIF Failure Indicating Fuse • Fiber Instrument Feed • Fractal Image Format

FIFCLC Fédération Internationale des Femmes de Carrières Libérales et Commerciales

FIFDU Fédération Internationale des Femmes Diplômées des Universités

FIFE First ISLSCP Field Experiment

FIFI Far Infrared Fabry-Pérot Interferometer • Field Imaging Far-Infrared • Flexible Ideal Format for Information

FIFI LS Field Imaging Far-Infrared Line Spectrometer

FIFILS Field Imaging Far-Infrared Line Spectrometer

FIFM Fonds Interministériel de la Formation et de Modernisation (France)

FIFO First In - First Out • Flight Inspection Field Office • Floating Input - Floating Output

FIFOFS First In - First Out File System

FIFRA Federal Insecticide, Fungicide, and Rodenticide Act (USA)

fig Figura

Fig Figurado • Figurative • Figuratively • Figuré • Figure

FIG Fédération Internationale des Géomètres • Figure • Flight Inspection Group • Floating Integrating Gyro • FORTH Interest Group • Fria (Guinea) (location identifier) • Friends of Internet in Greece

FIGARO French Italian Gamma-Ray Observatory

FIGS Fabry-Pérot Infrared Grating Spectrometer • Figure Shift

FIH Kinshasa N'Djili (Zaire) (location identifier)

FII Federal Information Infrastructure • Federal Item Identification • Fédérations Internationale des Interprètes

FIICPI Fédération Internationale des Ingénieurs-Conseillers en Propriété Industrielle

FIIG Federal Item Identification Guide

FIIGS Federal Item Identification Guide System

FIIGMO Forget It, I've Got My Orders

FIIM Fédération Internationale des Ingénieurs Municipaux • Fellow of the Institution of Industrial Managers

FIIN Federal Item Identification Number

FIIT Flash Internal Information Transfer

FIIWG Facility Integration and Installation Working Group

FIJPA Fédération Internationale des Journalistes Professionnels de l'Aéronautique

FIJPAA Fédération Internationale des Journalistes Professionnels de l'Aéronautique et de l'Astronautique

FIK Field Ionization Kinetics • Finke (Australia-NT) (location identifier)

Fil Filament • Fillet • Fillister • Filosofia • Filter • Filtrate

FIL Filament • Fillmore (USA-UT) (location identifier) • Foire Internationale du Livre • Formation - Information - Liaison • Fuel Injection Line

FiLA Five-Letter Acronym

FILAO Fiber Link to Adaptive Optics

Filat Filatelia

FiLCHeRS Falsifiability - Logic - Comprehensiveness - Honesty -Replicability - Sufficiency

FILDR Federal Item Logistics Data Record

FILE Feature Identification and Location Equipment • Feature Identification and Location Experiment • Future Identification and Location Experiment

FILER File Information Language Executive Routine

FILG Filing

Filip Filipinas

FILM Far Infrared Line Mapper • Foire Internationale du Livre de Montréal (Canada-QC)

FILO First In - Last Out

Filol Filología

FILOS Far-Infrared Limb Observing Spectrometer

FILS Federal Information Locator System (USA) • Flarescan Instrument Landing System

Filt Filter

FILX File Index

FIM Facility Information Message • Facing Identification Mark • Failure Indication Module • Fault Isolation Meter • Fédération des Industries Mécaniques (France) • Festival International de Météorologie • Field Inspection Manual • Field Instruction Memorandum • Field Intensity Meter • Field Ion Microscope • Field Ion Microscopy • Filter Interest Manager • Finnish Mark (ISO currency code) • Flight Information Manual • Framework Information Model • Full Individual Membership • Functional Information Manager • Functional Integration Management • Markka (Finland) (ISO currency code)

FIMA Fault Isolation Maintainability Analysis • Fellow of the Institute of Mathematics and its Applications

FIMATE Factory-Installed Maintenance Automatic Test Equipment

FiMek Fizicheskaia Mekhanika (periodical code)

FIML Full Information Maximum Likelihood

FiMos Akademiia Nauk SSSR Otdelenie Obshchei Fiziki i Astronomii Nauchnaia Sessiia Moscow (periodical code)

FIMR Finnish Institute of Marine Research

FIMS Facility Information Management System • Far-ultraviolet Imaging Spectrograph • Fast Ion Mass Spectrometer • Fault Isolation and Monitoring System • Fellow of the Institute of Mathematical Statistics • Field Intensity Measuring System • Field Ionization Mass Spectrometry • Field-Ion Mass Spectrometer • Field-Ion Mass Spectrometry • Financial Information Management System • Fine-split Infrared Multispectral Scanner • Forms Interface Management System • FORS Instrumental Mask Simulator • Framework for Integrated Manufacturing System • FUV Imaging Spectrograph

Fin Final • Finance • Finances • Financial • Financier • Finis • Finish • Finland • Finnish

FIN Federal Item Name • Final • Finance • Financial • Finland (international vehicle registration code) • Finland (ISO country code) • Finland (postal country code) (formerly SF) • Finlande (code ISO de pays) • Finschhafen (Papua New Guinea) (location identifier) • Fish Information Network • Focused Information Network • Future International Network

FINA Following Items Not Available

FINAC Fast Interline Non-native Automatic Control

Finan Financial

FIND Fault Isolation by Nodal Dependence • Federal Item Name Directory

FINDAS FIRST Integrated Network and Data Archive System

FINDS Flexible Information Dissemination System • Foundation for the International Non-governmental Development of Space

FINE First In the Nation in Education

FINGIS Finnish Geographical Information System

FINI Fine Imager

FINITO Fringe-tracking Instrument Nice-Torino

FINNARP Finnish Antarctic Research Program

FINNIDA Finnish International Development Agency

FINO Flight Inspection National Office

FINPLAN Financial Plan

FINRACS File of Normalized Radar Cross-Sections

FINS Fixed Imaging Navigation Sensor

Fin.Sec. Financial Secretary

FIO File Input/Output • Foreign Intelligence Office • Foreign Intelligence Officer • For Information Only • Fortran Input/Output • Fourier Integral Operator • Free In and Out • Frequency In and Out • Furnished and Installed by Others

FIOC Final Initial Operational Capability • Fluoroscopic Intraoperative Cholangiogram • Frame Input/Output Control • Frame Input/Output Controller • Full Initial Operational Capability

FIOP Fortran Input-Output Package

FIOpt Fiber Integrated Optics (periodical code)

FIOS Free In and Out and Stowed

FIOT Free In and Out and Trimmed

FIP Facility Interface Processor • Fair Information Practice • Fairly Important Person • Far-Infrared Photometer • Fault Isolation Procedure • Federal Information Processing • Field Inspection Procedure • File Processor Buffering • First Ionization Potential • Foam In Place • Formation Interne des Personnels • Forum on International Physics • Free Instrument Package • Fuel Injection Pump • Future Impact Point

FIPA Foundation for Intelligent and Physical Agents (USA)

FIPC Financial Information Processing Center

FIPEX Flux Probe Experiment

FIPLF Fédération Internationale de la Presse de Langue Française

FIPP Fédération Internationale de la Presse Périodique

FIPR Federal Information Processing Resource

FIPRAP Federal Information Processing Resource Acquisition Plan

FIPS Facilities Inventory and Planning System • Federal Information Processing Standard (USA) • Federal Information Processing System • Federal Information Publication Standard (USA) • Fisheries Image Processing System • Flagstaff Image Processing System • Flight Inspection Positioning System • FOC Image Processing System • Freiburg Image Processing System

FIPSE Fund for the Improvement of Postsecondary Education (USA)

FIQ Fast Interrupt Request

FIR Facility Installation Review • Facility Interference Review • Far Infrared (*) • Far-Infrared Radiation • Fast Infrared • Fault Isolation Routine • Fern Infrarot • Field Intensity Receiver • Financial Inventory Report • Financial Item Reporting • Finite Impulse Response • Finite Input Response • First-Impression Report • Flight Incident Recorder • Flight Information Region • Flight Information Report • Flight Information Requirement • Flight Inspection Report • Fonds d'Intervention pour les Rapaces (France) • Forbidden - Intercombination - Resonance (lines) • Force d'Intervention Rapide • For Immediate Release • Fuel Indicator Reading • Functional Input Report • Functional Intelligence Ratio • Functional Item Replacement • Future Issue Requirement

FIRAN Force d'Intervention Rapide Alerte Nucléaire

FIRAS Far-Infrared Absolute Spectrophotometer

FIRB Far-Infrared Background

FIRBE Far-Infrared Balloon-borne Experiment

FIR&D Focused Independent Research & Development

FIRD Far-Infrared Detector • Fault Isolation Requirement Document

FIRE Far Infrared Experiment • Fast Imaging Ranicon Experiment • Film Image Recorder • Fire In Resource and Environment • First ISCCP Regional Experiment • Flare Infrared Experiment • Flight Investigation of the Reentry Environment • Force d'Intervention Rapide Électricité • Fusion Ignition Research Experiment

FIREC Fédération Internationale des Rédacteurs en Chef

FIRES Faint Infrared Extragalactic Survey

FIREX Free-flying Imaging Radar Experiment

FIRG Far-Infrared Galaxy • Force d'Intervention Rapide Gaz

FIRGA Far Infrared GaAs Array • Far-Infrared Gallium Arsenide (detector)

FIRI Force d'Intervention Rapide Inondations

FIRLF Far-Infrared Luminosity Function

FIRLS Fast Iteratively Reweighted Least Squares

FIRM Factory Information Resource Management

FIRMPC Federal Information Resources Management Policy Council

FIRMR Federal Information Resource Management Regulation

FIRMS Forecasting Information Retrieval of Management System • Forest Information Resource Management System • Fourier Ion Resonance Mass Spectrometer

FIRNV Force d'Intervention Rapide Neige/Verglas

FIRO Far-Infrared Observation

FIRP Far-Infrared Photometer • Far-Infrared Pointer • Federal Internetworking Requirements Panel (USA)

FIRPO Fast Infrared Photometer for Occultations

FIRPP Far-Infrared Pointer Package

FIRS Far-Infrared Spectrometer • Far-Infrared Survey • Federal Information Relay Service (USA) • File Interrogation and Reporting System • Fourier Transform Infrared Sounder • Free-text Information Retrieval System • Future Information Retrieval System

FIRSA Far-Infrared Stressed Array

FIRSDIMS Flexible Interactive Remote Sensing Data Information and Management System

FIRSSE Far Infrared Sky Survey Experiment

FIRST Faint Images of the Radio-Sky at Twenty-cm • Far-Infrared and Submillimeter Telescope (*) • Far-Infrared Search and Track • Far-Infrared Space Telescope (now HSO) • Fast Interactive Radio System Tool • Fast Interactive Retrieval System Technology • Financial Information Reporting System • First Image of the Radio Sky Telescope • First Imager of the Radio Sky at Twenty-cm (survey) • Flexible Information Retrieval System for Text (USA) • For Inspiration and Recognition of Science and Technology (competition) • Forum of Incident Response and Security Team

FIRTECH Formation des Ingénieurs par la Recherche Technique

Fís Física

FIS Fachinformationssystem • Facilities Inventory Study • Facility Interface Sheet • Fahrtgastinformationssystem • Far-Infrared Spectrometer • Far-Infrared Surveyor • Fault Isolation System • Federal Information System • Federated Information System • Field Information System • Field Installation Simulator • Filipino Inventors Society • Flight Information Service • Flight Instrumentation System • Floating-point Instruction Set • Fondation Internationale pour la Science • Forest Information System • Free In Store • Friendly Information System

FISA Facility Inspection Shakedown Acceptance • Foreign Intelligence and Surveillance Act (USA)

Fisc Fiscal • Fiscality

FISc Far-Infrared Scanner

FISC Foundation for International Scientific Coordination

FISCH Fibre-Schmidt

FISDW Field-Induced Spin-Density Wave

FISH First In - Still Here • Fluorescence In Situ Hybridization

Fisiol Fisiología

FISLIB Fortran Interactive Subroutine Library

FISp Far-Infrared Spectrometer

FISS Fédération Internationale des Sociétés Scientifiques

FISSP Federal Information Systems Support Program

FiSSR Akademiia Nauk Tadzhikskoi SSR Fiziko Tekhnicheskii Institut Dushanbe Tadzhik SSR Akademiia Nauk Tadzhikskoi SSR Doklady (periodical code)

FIST Factory Integrated System Test • Fault Injection Security Tool • Fault Isolation by Semi-automatic Techniques • Final Integrated System Test • Financial Information System Tool • First In - Still There • First Integrated System Test • Forward Integrated Support Team • Fully Integrated System Test • Functional Integrated System Test

FISTA Flying Infrared Signatures Technology Aircraft

FISTNI Fédération Internationale des Sociétés pour le Traitement Numérique de l'Information

FISU Fil In Signal Unit

fit File Index Table (file extension) • Flexible Image Transport (file extension)

FIT Fabrication, Integration and Test • Failure In Time • Fault Isolation Test • Fault Isolation Tree • Federal Income Tax • Federal Insurance Tax • Fédération Internationale des Traducteurs • Field Integration and Test • Field Investigation Team • File Information Table • Final Integration and Test • First Indication of Trouble • First Ionization Time • Fixed Interval Timer • Fizeau Interferometer Testbed • Fizeau Interferometry Testbed • Florida Institute of Technology (USA-FL) • Formation Intégrée au Travail • Forschungsgesellschaft für Informationstechnik (Germany) • Foundation for International Training • Free and Independent Traveller • Free of Income Tax • Frequent Independent Traveller • Front Intertropical • Functional Integration Test

FITA Fault Isolation Test Adapter

FITCE Fédération des Ingénieurs des Télécommunications de la Communauté Européenne

FITH Fire-In-The-Hole

FITI Fédération Internationale pour le Traitement de l'Information

FITL Fiber In The Loop

FITNR Fixed In The Next Release

FITR Final Integrated Technology Review

FITS Fault Identification and Test System • File Image Transfer System • Flexible Image Transport System (*) • Flexible Interchange Transport System • Functional Interpolating Transformational System • Functional Interpolating Transform System

FITSIO Flexible Image Transport System Input/Output

FITT Full Immersion Telepresence Testbed

FITW Federal Income Tax Withholding

FIU Fingerprint Identification Unit • Florida International University (USA-FL) • Forward Interpretation Unit

FiUkr Akademiia Nauk Ukrain'skoi RSR Dopovidi Seriia Fiziko Matematichki (periodical code) ta Tekhnichni Nauki (periodical code)

FIV Five Finger (USA-AK) (location identifier) • Fonds Interministériel pour la Ville (France) • Fuel Isolation Valve

FIUO For Internal Use Only

FIX Federal Information Exchange • Federal Interagency Exchange • Federal Interconnection Exchange • Federal Internet Exchange

Fiz Fizika (periodical code)

FIZ Fachinformationszentrum • Fitzroy Crossing (Australia-WA) (location identifier)

FizAO Akademiia Nauk SSSR Fizika Atmosfery i Okeana (periodical code)

FizAS Fizika Aerodispersnykh Sistem (periodical code)

FizAt Fiziki Atmosfery (periodical code)

FizCh Fiziki i Chemii Seria Fizika (periodical code)

FizEl Fizicheskaia Elektronika (periodical code)

FizGV Fizika Goreniia i Vzryva (periodical code)

FizKM Fiziko Khimicheskaia Mekhanika Materialov (periodical code)

FizKO Fizika i Khimiia Obrabotki Materialov (periodical code)

FizMM Fizika Metallov i Metallovedenie (periodical code)

FizMS Fizika Mnogochastichnykh Sistem (periodical code)

FizNT Fizika Nizkikh Temperatur (periodical code)

FizPl Fizika Plazmy (periodical code)

FizTT Fizika Tverdogo Tela (periodical code)

FizTV Fizika i Tekhnika Vysokikh Davlenii (periodical code)

FizZS Fizika Zhidkogo Sostoianiia (periodical code)

FI3G Forum International de l'Instrumentation et de l'Information • Géographiques

fj Fiji (ISO language code)

Fj Fjord

F-J Faber-Jackson (relation)

FJ Faber-Jackson (relationship) • Fast Jet • Fidji (code ISO de pays) • Fiji (ISO country code)

FJAG Framework Joint Activity Group

FJCC Fall Joint Computer Conference

Fjd Fjord

FJD Dollar (Fiji) (ISO currency code)

FJEP Foyer de Jeunes et d'Éducation Populaire

FJI Federación de Jovenes Investigadores (Spain) • Fidji (code ISO de pays) • Fiji (international vehicle registration code) • Fiji (ISO country code)

FJO Fundació Joan Oró (Spain)

FJOB Federal Job Opportunities Board (USA)

FJR Al-Fujairah (UAE) (location identifier) • Faber-Jackson Relationship

Fk Fork

FK Falkland Islands (ISO country code) • Function Key • Fundamental Katalog • Îles Malouines (code ISO de pays) • Malvinas (ISO country code)

FKA Formerly Known As

FKB Flight Display Keyboard

FKFO Fonds voor Kollektief Fundamenteel Onderzoek (Belgium)

FKH Fakenham Sculthorp (UK-England) (location identifier)

FKI Kisangani (Zaire) (location identifier)

FKJ Fukui (Japan) (location identifier)

FKL Franklin - Chess Lamberton (USA-PA) (location identifier)

FKM Funkkanalmodem

FKN Franklin (USA-VA) (location identifier)

FKO Franz-Keldysh Oscillation

FKP Förderverein Kieler Planetarium eV (Germany) • Pound (Falkland Islands) (ISO currency code)

FKQ Fak Fak (Indonesia) (location identifier)

FKr Faroese Krone (currency)

FKS Franz-Kroller-Sternwarte (Austria) • Fukushima (Japan) (location identifier)

FKSI Fourier-Kelvin Stellar Interferometer

FK3 Fundamental Katalog Nr. 3

FK4 Fundamental Katalog Nr. 4

FK5 Fundamental Katalog Nr. 5

FK6 Fundamental Katalog Nr. 6

fl Florin (currency) • Floruit • Guilder (currency)

fL Foot-Lambert

Fl Flanders • Flashing • Flat • Flemish • Fleuve • Floor • Florijn • Fluid • Flush • Guilder

F/L Forward Link

FL Fail • Far Left • Feed Line • Fermi Liquid • Fiber Link • Fine Lock • First Light • Flaring Loop • Flashing Light • Flemish • Flexible • Flight Leg • Flight Level • Floor • Florida (USA) • Flow Line • Flying Logo • Focal Length • Fonction de Luminosité • Foot Lumen • Force Level • Foreign Language • Format Line • Forward Link • Fraction Living • Franc Luxembourgeois (currency) • Full Lift • Full Load • Functional Limitation • Function Library • Funzione di Luminosità • Fuzzy Logic • Liechtenstein (international vehicle registration code) • Liechtenstein (postal country code)

Fla Florida (USA)

FLA Fiducial Light Assembly • Film Lens Antenna • Firm Line Altitude • Flexible Local Architecture • Flight Article • Florencia (Colombia) (location identifier) • Force Level Advisor • Foreign Launch Assessment • Four-Letter Abbreviation • Four-Letter Acronym • Future Large Aircraft

FLAAS Flexible Airborne ATN System

FLAC Fast Lagrangian Analysis of Continua • Field Level Access Control

FLACC Full Level ALGOL Checkout Compiler

FLAD Fundação Luse-Americana para o Desenvolvimento (Portugal)

FLAG Fiberoptic Link Around the Globe • Finite-Length-scale Anti-Gravity • Flemish Aerospace Group (Belgium) • Flexible Assignment Generator

FLAGE Flexible Lightweight Agile Guided Experiment

FLAI Fuzzy Logic in Artificial Intelligence (conference)

FLAIR Fibre-Linked Array-Image Reformatter • Fibre-Linked Astronomical-Image Reformatter • Food-Linked Agro-Industrial Research • Functional Language Articulated Interactive Resource

Flam Flamenco

FLAM Fault Location And Monitoring • Force Level Application Management

FLAME Fire Logistics Airborne Mapping Equipment

FLAMES Fibre Large-Array Multielement Spectrograph • Fibre Large-Array Multielement Spectrography

FLAMR Forward Looking Advanced Multilobe Radar

FLAP Fatal Light Awareness Program (Canada) • First Level Adaptive Program

FLAPF Federación Latinoamericana de Productores de Fonogramas (y Videogramas)

FLAPIR FLIR and Lidar Atmospheric Propagation in the Infrared

FLAPS Flight Application Software

FLAPW Full-potential Augmented Plane Wave

FLAR Fault Location And Repair • Forward-Looking Airborne Radar

FLAS Fuels Logistical Area Summary

FLASCO Faculdad Latinoamericana de Ciencias Sociales

FLASH Fast Low-Angle Shot • Fiber-Linked Astrophysical Spectrograph of Heidelberg • Fluorescence in Air Showers (experiment)

FLAT Families for Learning Accurate Theories (USA)

flav Flavus

FLB Faint Light Bridge • Floriano Cangapara (Brazil-PI) (location identifier) • Forward Logistics Base

FLC Facility Security Clearance • Falls Creek (Australia-VIC) (location identifier) • Federal Laboratory Consortium • Federal Library Committee • Ferro-electric Liquid Crystal • Force Level Control • Foreign Language Committee • Forward Load Control • Frequency and Load Controller • Friedmann-Lemaire Cosmology • Fuzzy Logic Controller

FLCA Forward Load Control Assembly • Forward Load Controller Assembly

FLCB Frequency and Load Controller Box

FLCC Force Level Command and Control • Force Level Control Capability

FLCD Ferroelectric Liquid Crystal Display • Force Level Command

FLCE Force Level Control Element

FLCF Force Level Control Function

FLCS Force Level Control Station • Force Level Control System

Fld Field • Fluid

FLD Fault Logic Diagram • Field • First-Level Destination • Flux Limited Diffusion • Fond du Lac (USA-WI) (location identifier) • Fraunhofer Line Discriminator • Functional Line Diagram

FLDC Fixed-Length Don't Care

Fldg Folding

FLDG Fielding

FLDMS Flied Maintenance Shop

fl. dr. Fluidram

FLDS Fields • First Level Decomposition and Specification

Fldt Floodlight

FlDy Fluid Dynamics (periodical code)

FlDyR Fluid Dynamics Research (periodical code)

FlDyT Fluid Dynamics Transactions (periodical code)

Fle Flemish

FLE Fiber Link Extender • Flare-Like Event • Forward Logistics Element • Petersburg - Fort Lee (USA-VA) (location identifier)

FLEA Flux Logic Element Array

FLECSE Flexible Environment for Collaborative Software Engineering

FLEEP Flying Lunar Excursion Experimental Platform

Flem Flemish

FLEP Fédération Laïque d'Éducation Populaire (France) • Foyer Laïc d'Éducation Populaire • Funded Legal Education Program

FLES Français Langue Étrangère et Seconde

FLETC Federal Law Enforcement Training Center

FLEVO Facility for Liquid Experimentation and Verification in Orbit

Flex Flexible

FLEX Flexible • Force Level Execution

FLF Fine Line Filter • Fixed-Length Field • Flensburg Schaferhaus (Germany) (location identifier) • Foul Language Filter • Future Large Facility

FLFS Fast and Large File System

Flg Flagging • Flashing • Flooring • Flying • Following

FLG Falling • Flag • Flagstaff (USA-AZ) (location identifier)

FLH Flotta (UK) (location identifier) • Fluorescence Line Height

FLHO Federal Lands Highway Office (USA)

FLI Fast Lyapunov Indicator • Fault Location Indicator • Flateyri (Iceland) (location identifier) • Fluorescence Line Imager • Font Library (TEX) • Force Level Information • Fractional Lunar Illumination • Free Language Indexing

FLIC Fault Location Indicating Console • Florida Imaging Center (USA-FL)

FLICC Federal Library and Information Center Committee (USA)

FLIDAR Fluorescence Lidar

FLIER Fast Low-Ionization Emission Region

FLIERS Fast Low-Ionization Emission Regions

FLIH First-Level Interrupt Handler

FLIHTT Flexible Intermodal Horizontal Transhipment Technique

FliIn Flight International (periodical code)

FLIL Functional Law of Iterated Logarithm

FLILO Flying Infrared for Low-level Operations

FLIM Forest-Light Interaction Model

FLIMAP Fast Laser Imaging Mapping And Profiling

FLIN Federal Library and Information Network

FLINN Fiducial Laboratories for an International Natural-sciences Network • Fiduciary Laboratory for an International Natural-sciences Network

FLIP Film Library Instantaneous Presentation • Flexible Image Processing • Flight Information Publication • Flight Instruction Procedures • Floating Indexed Point • Floating Index Point • Floating Instrument Platform • Flow Interferogram Processing

FLIPS FITS Large Image Processing Software • Flight Information Processing System • Future Language Information Processing System

FLIR Forward-Looking Infrared • Forward-Looking Infrared Radar

FLIRS Forward-Looking Infrared Radar System

FLIRT Fortran Logical Information Retrieval Technique

FLIS Fast Local Internet Service • Federal Logistic Information System

FLITE Federal Legal Information Through Electronics • Flying Laboratory for Integrated Testing and Evaluation • Free-flyer Laser Interferometer Technology Experiment

FLITECAM First-Light Infrared Test Experiment Camera

FLIZ F-Layer Irregularity Zone

FLJ Falls Bay (USA-AK) (location identifier)

FLK Falkland Islands (ISO country code) • Îles Malouines (code ISO de pays) • Malvinas (ISO country code)

FLKS Force Level Knowledge System

FLL Fixed Loss Loop • Flux Locked Loop • Fort Lauderdale (USA-FL) (location identifier) • Frequency-Locked Loop • Frequency Lock-Loop • Fresnel-Like Lens • Hollywood (USA-FL) (location identifier)

FLLG Frustum Lifting Lug Kit

Flli Fratelli

FLLK Frustum-Lifting Lug Kit

Flm Flame

FLM Federal Land Manager • Filadelfia (Paraguay) (location identifier) • Flight-weighted Laser Module • Fluorescent Light Module • Functional Level Management

Flmb Flammable

FlMSR Fluid Mechanics Soviet Research (periodical code)

FLN Florianópolis - Hercilio Luz (Brazil-SC) (location identifier) • Fuel Line

flo Floccus

FLO Fault Location Oscillator • First-Look Observation • First Lunar Outpost • Flight Line Operator • Florence (USA-SC) (location identifier) • Foreign Liaison Office • Foreign Liaison Officer • Functional Line Organization

FLODAC Fluid Operated Digital Automatic Computer

FLOGEN Flow Generator

FLOH Flugzeuglidar für Ozeanographie und Hydrologie

FLOLS Fresnel Lens Optical Landing System

FLOP Floating Octal Point • Floating-point Operation • FUSE Log of Observation Problems • Future Launcher Optimal Performances

FLOpS Floating-Point Operations per Second

FLOPS Floating-Point Operations Per Second (*) • Functional and Logic Programming Symposium

FloorCo Floor Cooling

flor Floruit

Flor Florence (Italy) • Florentine • Florida (USA)

FLOR Future Lunar Objectives and Requirements

FLORENCE Flux Océaniques Restitués par bilan Énergétique des Couches limites à moyenne Échelle

FLOS Fixed Line Of Sight • Flight Level Orientation Scheme

FLOSIAN Flow Simulation Analysis

Flot Flotation

FLOT Flotation

FLOTRAN Flowcharting Fortran

FLOTUS First Lady Of The United States

FLOWS FAA-Lincoln Operational Weather Study

fl. oz. Fluid Ounce

FLP Facility Layout Problem • Fast Link Pulse • Fault Location Panel • First-Level Processor • Flag Link Protocol • Flap • Flippin (USA-AR) (location identifier) • Fuzzy Logic Pack

FLPL Fortran List Processing Language

FLPP Future Launcher Preparatory Programme (ESA)

FLPS Flight Load Planning System • Flight Load Preparation System

FLPT Floating Point

Flr Failure • Filler • Floor • Florijn

FLR Failure • Field Line Resonance • Filed Line Resonance • Finite Larmor Radius • Firenze - Amerigo Vespucci (Italy) (location identifier) • Firenze Peretola (Italy) (location identifier) • Flare • Flight Log Recorder • Florence - Amerigo Vespucci (Italy) (location identifier) • Florence Peretola (Italy) (location identifier) • Forward-Looking Radar

FLRA Federal Labor Relations Authority (USA)

fl. rt. Flow Rate

FLRW Friedmann - Lemaître - Robertson - Walker (model)

FLRY Flurry

FLS Falls • Fault Locator System • Federated Component System • Feeder Link Station • Fields Logistics System • Finance Ledger System • First Line Supervisor • First-Look Survey • Flight Surgeon • Flinders Island (Australia-TAS) (location identifier) • Flow Switch • Force Level Simulation • Forward Logistics Site • Front Line State • Future Launching System • Fuzzy Logic System • Off-Line System

FLSA Fair Labor Standards Act (USA)

FLSC Final Logistics Support Concept • Flexible Linear Shaped Charge • Flexible Linear Shaped Cord

FLSF Future Large-Scale Facility (*) • Future Large Space Facility

FlSt Flurstück

FlStNr Flurstücknummer

Flt Flashlight • Flight • Float

FLT Fault • Fault Location Technology • Fermat's Last Theorem • First Level Trigger • Flat • Flat (USA-AK) (location identifier) • Flight • Floating • Full Line Terminal

FLTC Fault Controller • Flexible Long-Term Contracting

Fltg Floating

FLTNO Flight Number

FLTP Future Launcher Technology Programme

Fltr Filter

FLTR Filter

Flts Flats

FLTS Flats • Flight Line Test Set

FLTSATCOM Fleet Satellite Communications (USN)

FLTWT Flightweight

FLU Fault Location Unit • First Line Unit • New York - Flushing (USA-NY) (location identifier)

Fluc Fluctuant • Fluctuate • Fluctuated • Fluctuating • Fluctuation

FLUC Fluctuating

Fluor Fluorescent • Fluoridation • Fluoride

FLUOR Fiber-Linked Unit for Optical Recombination • Fluorescent

FluQ Fluids Quarterly (periodical code)

Fluid Fluidika (periodical code)

FLUREX Fluorescence Experiment

FLV Foreign Leave • Foreign Liaison Visit • Fort Leavenworth (USA-KS) (location identifier)

FLW Fault Location Word • Fellow • Flores Island (Azores) (location identifier) • Follow

FLWO Fred Lawrence Whipple Observatory (USA-AZ)

Flwp Followup

FLX Fallon (USA-NV) (location identifier)

FLY Finley (Australia-NSW) (location identifier)

FLZ Flight Level Zone

fm Fathom • Femtometre • Fermi • Festmeter • Fiat Mistura • Framemaker (file extension)

Fm Farm • Fermium • From • Metallic-Line F-type Star

F&M Feinwerktechnik und Meßtechnik (periodical code) • Force & Moment

FM Facility Maintenance • Facility Management • Facility Manager • Facility Module • Factory Management • Factory Model • Fault Management • Fax Modem • Ferritic and Martensitic • Ferro-Magnetism • Field Magnet • Field Maintenance • Field Manual • Field Marshall • Field Modification • Field Monitor • Field Monitoring • Figure of Merit • File Maintenance • Financial Management • Financial Manager • Fine Measurement • Fine Mese • First Mold • Fiscal Month • Fischer Matrix • Flight Manual • Flight Model • Floor Manager • Flow Meter • Fluid Management • Fluid Module • Focusing Mount • Folding Mirror • Force and Moment • Force Module • Foreign Mission • Formal Method • Format • Forms Management • Forward Motion • Foundation Manager • Foundation Member • Fracture Mechanism • Fréquence Modulée • Frequency Management • Frequency Meter • Frequency Modulated • Frequency Modulation • Frequency Monitor • Frequency Multiplex • From • Full Match • Full Moon • Functional Manager • Functional Module • Funzione di Massa • Micronesia (ISO country code) • Micronésie (code ISO de pays)

FMA Failure Mode Analysis • February - March - April • Field Modification Activity • Final Monitor Aid • Flight Margin Assessment • Foreign Media Analysis • Formosa - El Pucu (Argentina-FO) (location identifier) • Forward Maintenance Area • Fragment Mass Analyzer • Frequency Management Agency • Frequency Map Analysis • Fuzzy Medial Axis

FMAC Facility Maintenance And Control • Fargo-Moorhead Astronomical Club (USA-MN) • Financial Management Advisory Committee • Foreign Member-At-Large

FMAD Fluid Management And Management

FMAG Fluxgate Magnetometer

FMAHTS Flight Manifest And Hardware Tracking System

Fman Foreman

FMAPE Fast Maximum A Posteriori with Entropy

FMAR Final Mission Analysis Review

FMAS Fault Management Application Service • Foreign Media Analysis System

fmb File Manager Bar (file extension)

FMB Federal Maritime Board (USA) • Financial Management Board

FMBCR Flat Mail Bar Code Reader

FMBS Frame Mode Bearer Service

FM&C Financial Management & Comptroller

FMC Federal Manufacturers Code (USA) • Federal Maritime Commission (USA) • Ferrite Memory Core • Financial Management and Comptroller • First Magnitude Corporation (USA-WY) • Five Mile (USA-AK) (location identifier) • Fixed Mobile Convergence • Flexible Monte Carlo • Flight Management Computer • Food Machinery Corporation • Foreign Materiel Catalog • Formation et Mobilité des Chercheurs • Formation Médicale Continue • Forward Motion Compensation • Forward Motor Controller • Frequency Management Center • Frequency-Modulated Cyclotron • Fuel Management Computer • Full Mission Capable • Fully Mission Capable • Fully Multicoated

FMCA Forward Motor Controller Assembly

FMCAD Formal Methods in Computer-Aided Design (conference)

FMCF First Manned Captive Flight

FMCS Factory Management and Control System • Federal Mediation and Conciliation Service (USA) • Flight Management Computer System • Freight Movement Control System

FMCT Fissile Material Cutoff Treaty

FMCW Frequency-Modulated Continuous Wave • Frequency Modulation Continuous Wave

Fmd Formed

FMD Ferrous Metal Detector • Fields Metadata • Financial Management Division • Fixed-Module Drive • Forward Multiplicity Detector • Frequency Modulation Discriminator • Function Management Data

FMD&C Flight Mechanics, Dynamics & Control

FMDC Flight Mechanics, Dynamics, and Control

FMDM Flexible Multiplexer/Demultiplexer • Frequency Modulation Deviation Meter

FMDS Field Maintenance Data System • Flight Management Data System • Flight Model Discharge System

FMDT Frequency Modulated Differential Transducer

FME Failure Modes and Effects • Feature Manipulation Engine • Fort Meade - Tipton (USA-MD) (location identifier) • Frequency-Measuring Equipment

FMEA Failure Modes and Effects Analysis • Failure Mode Effects Analysis (*)

FMEC Forward Master Events Controller

FMECA Failure Mode and Effects Criticality Analysis • Failure Modes, Effects and Criticality Analysis

FMEIS Force Management Execution Information System

FMEP Foreign Materiel Exploitation Program

FMES Failure Modes Effects and Analysis • Full Mission Engineering Simulation

FMET Functional Management Engineering Team

FMEU Fonds Mondial d'Entraide Universitaire

FMF Facility Master File • Free Molecular Flow • Fuzzy Membership Function

FMFB Frequency Modulation with Feedback

FMFIA Federal Manager's Financial Integrity Act (USA)

FMFR Four-Minus-Five Restoral

FMFT Four-Minus-Five Test

FMG Ferrovia Monte Generoso (Switzerland) • Flamingo (Costa Rica) (location identifier) • Franc Malgache (currency) • Frequency Modulation Generator

FMGEC Flight Management Guidance and Envelope Computer • Flight Management Guidance Envelope Computer

FMGIS Flight Management Geographic Information System

FMGS Flight Management and Guidance System • Flight Management Guidance System

FMH Falmouth Otis (USA-MA) (location identifier)

FMHS Formal Message Handling System

FMI Faculty for Mathematics and Informatics • Failure Mode Indicator • Fermilab Main Injector • Finnish Meteorological Institute (Finland) • Fixed Mobile Integration • Flexible Modular Interface • Flow Measurement and Indication • Fluorescent Microthermographic Imaging • Focused Michelson Interferometer • Fonction de Masse Initiale • Fondo Monetario Internacional • Fonds Monétaire International • Force Module Identifier • Forest Management Institute • Free Motion Impedance • Functional Management Inspection • Funzione di Massa Iniziale • Kalemie (Zaire) (location identifier)

FMIC Frequency Modulation and Interference Control • Frequency Modulation Interference Control

FMICS Formal Methods for Industrial Critical System (ERCIM Working Group)

FMICW Frequency-Modulated Interrupted Continuous Wave

FMID Force Module Identifier

F-MIDR Full-resolution Mosaicked Image Data Record

FMIDR Full-resolution Mosaicked Image Data Record

FMIM Foreign Mass Injection Method

FMIP Financial Management Improvement Program

FMIS Facilities Management Information System • Financial Management Information System • Force Management Information System

fmk Fortran Makefile (file extension)

Fmk Finnish Mark (currency)

FMk Finnish Mark (currency)

FMK Fast Moving Knot • Field Modification Kit

FMKs Fast Moving Knots

Fml Formal

FML Fault Message Line • File Manipulation Language • Final Materials List • Flamsteed • Force Module Library

Fmla Formula

FMLIS Federal Mineral Lands Information System (USA) • Fuzzy Multispectral Landcover Information System

FMM Fast Multipole Method • Ferro-Magnetic Metal • Financial Management Manual • Flight Manual Manager

FMMC Fixed Mylar Metallized Capacitor

FMME Foreign Medical Materiel Exploitation

FMMUX Frequency Modulated Multiplexer

Fmn Formation

FMN Farmington (USA-NM) (location identifier) • Firme Multinationale • FOS Command Management

FMNBNA Frequency Modulation and Narrowband Noise Analyzer

FMNOC Fleet Numerical Meteorology and Oceanography Center

FMO Fast Moving Object • Field Maintenance Office • Financial Management Office • Financial Management Officer • Fiscal Management Office • Fiscal Management Officer • Flatland Meteorological Observatory • Flight Medical Officer • Flux de Matière dans l'Océan • Forms Management Officer • Frequency Management Office • Frequency Management Officer • Fuel Management Office • Fuel Management Officer • Future Method of Operations • Münster (Germany) (location identifier)

FMOCR Flat Mail Optical Character Reader

FMOD File Modify

FMOF First Manned Operational Flight • First Manned Orbital Flight

FMOFEVA First Manned Orbital Flight with Extravehicular Activity

FMOFPL First Manned Orbital Flight with Payload

FMOI Fédération Mondiale des Organisations d'Ingénieurs

FMOODS Formal Methods for Open Object-based Distributed Systems (workshop)

FMOP Frequency Modulation On Pulse

FMOS Fiber Multiobject Spectrograph

FMP Facility Management Plan • Finance and Management Policy • Financial Management Plan • Flight Manual Program • Fonds des Maladies Professionnelles (Belgium) • Force Module Package • Frequency Management Plan • Functional Maintenance Procedure

FMPAC Fixed Multipurpose Application Console

FMPCS Fixed Mechanization Process Control System

FMPEC Financial Management Plan for Emergency Conditions

FMPM Family-Manned Planetary Mission

FMPP Federal Merit Promotion Program (USA)

FMPS Fast Multidimensional Processing System • Fortran Mathematical Programming System • Functional Mathematical Programming System

FMPT First Materials Processing Test

FMP&TA Formal Methods in Programming & Their Applications (conference)

FMQ Frequency-Modulated Quartz • Fully Mission Qualified

Fmr Former

FMR Failure and Malfunction Report • Fair Market Rent • False Match Rate • Fast Magnetic Rotator • Ferromagnetic Resonance • Field Modification Request • Financial Management Regulation • Financial Management Report • Financial Management Review • Follow Me Roaming • Frequency-Modulated Radar • Frequency-Modulated Receiver • Functional Management Review • Function Maximum Rate

FMRB Forschungs- und Meßreaktor Braunschweig (Germany) • Functional Management Review Board

FMRC Factory Mutual Research Corporation

FMRD Flight Mission Rules Document

fMRI Functional Magnetic Resonance Imaging

FMRI Functional Magnetic Resonance Imaging

FMRL Functional Machine Representation Language

Fmrly Formerly

FMRS Forecast Movement Requirements System • Fuels Management Requirements System

FMS Facilities Management System • Facility Management System • False Memory Syndrome • Fast Matrix Solver • Fast Multiplexer System • Fault Management System • Fecal Management System • File Maintenance System • File Management Supervisor • File Management System • Filter Management System • Final Monthly Schedule • Final Multiple Score • Financial Management Service • Financial Management System • Fixed Mobile Substitution • Flexible Machining System • Flexible Manufacturing System • Flexible Measuring System • Flight Management System (*) • Flight Managing System • Flight Motion Simulator • Flow Measuring System • Fluid Management System • Flux Monitoring System • Focused Meteor Stream • Food Management System • Force Management System • Force Module System • Forces Modeling and Simulation • Formal Message Service • Forms Management System • Fort Madison (USA-IA) (location identifier) • Frequency-Mixer Stage • Frequency Management System • Frequency Measuring System • Frequency Monitoring Station • Frequency Monitoring System • Frequency Multiplier Storer • Functional Master Schedule • Fusion Management Service

FMSC Federal Manual for Supply Cataloguing

FMSF False Memory Syndrome Foundation

FMSP Formal Methods in Software Practice (workshop) • Frequency Modulation Signal Processor

FMSS Financial Management System Software

FMST Frequency Mass Spectrometer Tube

fmt Format (file extension)

FMT Facilities Maintenance Team • Field Maintenance Team • Field Maintenance Technician • Financial Management Team • Fine Multithreading • Flare Monitor Telescope • Flight Management Team • Format • Frequency-Modulated Transmitter • Friction Measurement Test • Full Machine Translation • Full Mission Trainer

FMTA Flash Mass Thermal Analysis

FMTP File Management Transaction Processor • First Materials Processing Test

FMTS Field Maintenance Test Set • Field Maintenance Test Station

FMTV Frequency Modulation Television

FMU Facility Management Unit • File Migration Utility • Financial Management Unit • Flight Monitor Unit • Force Measurement Unit • Francis Marion University (USA-SC) • Fuel Management Unit • Functional Mock-Up

FMUX Flexible Multiplexer

FMV Fast-Moving Vehicle • Full Motion Video

FMVSS Federal Motor Vehicle Safety Standards (USA)

Fmw Firmware

FMX Fire and Maneuver Exercise • Frequency Modulation Transmitter • Frequency Multiplexing

FMY Fort Myers (USA-FL) (location identifier)

FN False Negative • Fernando de Noronha (Brazil) • Fernmeldenetz • Finance • Financial Network • First Name • Flight Nurse • Flux Number • Footnote • Force Network • Foreign National • Forêt Noire

FNA Fast Neutral-particle Analyzer • For Necessary Action • Freetown Lungi (Sierra Leone) (location identifier) • Frequency Network Analyzer

FNAC Future Needs in Astronomical Computing

FNADT Fonds National de l'Aménagement du Territoire (France)

FNAEP Fédération Nationale des Associations d'Éducation Populaire (France)

FNAK Fondation Nationale Alfred Kastler (France)

FNAL Fermi National Accelerator Laboratory (USA-IL)

FNAP Fortran Network Analysis Program

FNARS FEMA National Radio System (USA)

FNATH Fédération Nationale des Accidentés du Travail et des Handicapés

FNARS FEMA National Teletype System (USA)

FNAUT Fédération Nationale des Associations d'Usagers de Transports (France)

FNAVS FEMA National Voice System (USA)

FNB Food and Nutrition Board

FNBDT Future Narrowband Digital Terminal

FNBR Fast Neutron Breeder Reactor

FNC Fast Neutron Cavity • Federal Networking Council (USA) • Fiber Nickel Cadmium • Forward Neutron Calorimeter • Funchal (Madeira Island) (location identifier)

FNCAC Federal Networking Council Advisory Committee (USA)

FNCF Fédération Nationale des Centres de Formation (France)

Fnct Function

Fnd Found

FND Facility Need Date • Fast Neutron Dose • FGS Neutral Density (filter)

FNDAE Fonds National pour le Développement des Adductions d'Eau (France)

Fndd Founded

FNDH Foreign National Direct Hire

Fndr Founder

FNDR Fondo Nacional de Desarollo Regional

FNDUT Fédération Nationale des Diplômés Universitaires de Technologie (France)

FNDVA Fonds National pour le Développement de la Vie Associative (France)

Fne Fine

FNE Fane (Papua New Guinea) (location identifier) • Fiscalía Económica Nacional • Fonds National de l'Emploi (France)

FNEP Fondation Nationale Entreprise et Performance (France)

FNET French UUCP Network

FNF First Normal Form

FNG Fada Ngourma (Burkina Faso) (location identifier)

FNI Nîmes Arles Camargue (France) (location identifier)

FNIC Fonds National Interconsulaire de Compensation (France) • Food and Nutrition Information Center

FNIH Foreign National Indirect Hire

FNIU Fiber Network Interface Unit

FNJ Pyongyang (Korea-DPRK) (location identifier)

FNK Fin Creek (USA-AK) (location identifier)

FNL Faint NEO List • Final • Fort Collins - Loveland (USA-CO) (location identifier)

FNLN Fine Line

FNM Fast Multipole Method • Financial Network Manager

FNMA Federal National Mortgage Association

FNMOC Fleet Numerical Meteorological and Oceanography Center (USN)

FNMT Fábrica Nacional de Moneda y Timbre (Spain)

FNNPE Federation of Nature and National Parks of Europe

FNOC First Numerical Oceanography Center • Fleet Numerical Oceanographic Center (USN) (*) • Fleet Numerical Operations Center • Fleet Numeric Oceanography Center

FNOES Fondation Nationale pour l'Organisation d'Expéditions Scientifiques (Belgium)

FNP Front-end Network Processor • Fusion Point

FNPA Foreign Numbering Plan Area

FNPO For NASA Personnel Only

FNPS Fédération Nationale de la Presse Spécialisée (France) • Fonds National de la Promotion Sociale (France)

FNR False Non-match Rate • Fonds National de la Recherche (Luxembourg) • Funter Bay (USA-AK) (location identifier)

FNRI Food and Nutrition Research Institute

FNRS Fonds National de la Recherche Scientifique (Belgium)

FNS Failure Notification Sheet • Failure Notification System • Fonds National Suisse (Switzerland) • Food and Nutrition Service • Fortified Navier-Stokes

Fnsh Finish

fnt Font (file extension)

Fnt Front

FNT Failure Notification Telex • Fermat Number Transform • Flint (USA-MI) (location identifier) • Font • Front

FNTGNS Frontogenesis

FNTLYS Frontolysis

FNU First Name Unknown

FNV Federatie Nederlandse Vakverenigingen (Netherlands)

FNWC Fleet Numerical Weather Center (USN)

f/o Full Out

fo Faeroese (ISO language code) • Folio

Fo Fourier Number

F/O Fiber Optics • Follow On • Fuel/Oxidizer

FO Fail Open • Fail Operating • Fail Operation • Fail Operational • Falsified Observation • Faroe Islands (ISO country code) • Fast Operate • Fast Operating • Federal Official • Fellowship Office • Fiber Optics • Field Office • Field Officer • Field Operations • Filter Output • Finance Officer • Firm Offer • Firm Order • First Officer • First Overtone • Flash Override • Flat Oval • Flight Operation • Foldout • Follow-On • Force Operations • Foreign Object • Foreign Office • Formatted Object • Formatting Object • Formosa (Argentina) • For Orders • Forward Observer • Fragment Offset • Fridays Only • Functional Objective • Furnace Oil • Îles Faroë (code ISO de pays)

FOA Failed On Arrival • Fiber-Optic Amplifier • Field Operating Agency • Figl-Observatorium für Astrophysik (Austria) • Foula (Shetland Islands) (location identifier) • Future Offensive Aircraft

FOAC Federal Office Automation Center (USA)

FOAD Formation Ouverte et À Distance (France)

FOAF Friend Of A Friend

FOAS Future Offensive Air System

FOB Federal Office Building • Fiber Optics Board • Flight Operations Building • Fort Bragg (USA-CA) (location identifier) • Forward Operating Base • Fractional Orbit Bombardment • Free On Board • Freight On Board • Friend Of Bill • Fresh Off the Boat • Fuel On Board • Full Operational Base

FOBC Fiber Optics Bulkhead Connector

FOBE First-Overtone Blue Edge

FOBS Fractional Orbit Bombardment System • Fractional Orbit Bombing System

Foc Focal • Focus

F&OC Full & Open Competition

FOC Factory Outlet Center • Faint Object Camera • Fiber Optic Cable • Fiber Optic Communication • Final Operating Capability • Final Operational Capability • Final Operational Configuration • Finance Office Code • Fleet Operational Capability • Flight Operating Costs • Flight Operations Center • Focus • Follow-On Contract • Foreign and Overseas Commission (UK) • Foreign Object Check • Form Of Occurrence • Fraction Of Circle • Free Of Charge • Full and Open Competition • Full Operating Capability • Full Operational Capability (*) • Full Operations Capability • Future Operating Capability • Fuzhou (China-PRC) (location identifier)

FOCA Federal Office of Civil Aviation • Fiber-Optic Cable Assembly • Focal Objective Corrective Astigmatism

FOCAL Facility for Optical Calibration At Liège (Belgium) • Formula Calculator • Forum Of Continuing Astronomical Liaison

FOCAM Faint Object Camera

FOCAP Fibre-Optic Coupled Aperture Plate

FOCAS Faint Object Camera And Spectrograph • Faint Object Classification and Analysis System • Fiber Optic Communications for Aerospace Systems

FOCAT-S Photographic Catalog of the Southern Hemisphere

FOCC Flight Operations Control Center

FOCES Fibre-Optic-coupled Cassegrain Echelle Spectrograph

FOCI Fisheries Oceanography Coordinated Investigations • Foreign Owned, Controlled, or Influenced • Foreign Ownership, Control or Influence

FOCIS Forest Classification and Inventory System

FOCL Fibre-Optic Communications Line

FOCP Fiber-Optic Cable Plant

FOCPAT Fiber-Optic Controlled Phased Array Technology

FOCpS Force Calculations per Second

FOCPS Force Calculations Per Second

FOCR Final Operational Concept Review

FOCS Fiber-Optic Cable System • Fiber Optic Chemical Sensor • Fiber-Optic Current Sensor • Force Calculations per Second • Foundations Of Computer Science (symposium)

FOCSI Fiber-Optic Control System Integration

FOCUS Fast Orbit Computation Utility Software • Federation on Computing in the United States • Forum Of Control Data Users • Fotoproduction of Charm: Upgraded Spectrometer • Frontiers in Optical Coherent and Ultrafast Science (project) • Future Operations Concepts for Utility Systems

FOD Fax On Demand • Fiber Optical Device • Field Operations Division • Finger Of Death • First Order Design • Flight Operations Directorate • Flight Operations Division • Foreign-Object Damage • Formation Ouverte à Distance • Fort Dodge (USA-IA) (location identifier) • Free Of Damage

FODA Feature-Oriented Domain Analysis

FödG Fördergebietsgesetz

FODL Fiber-Optic Data Link

FODMS Fiber-Optic Data Multiplexing System

FODO Foundations of Data Organization (conference)

FODS Faint Object Detection System • Fiber-Optic Demonstration System

FODT Fiber-Optic Data Transmission

FOE Fiber Optic Echelle • Field Operational Evaluation • Final Operational Evaluation • Flight Operations Engineer • Follow-On Evaluation • Follow-on Operational Evaluation • Friends Of the Earth • Topeka Forbes (USA-KS) (location identifier)

FOEB Flight Operations Evaluation Board

FOER Fiber-Optic Ethernet Repeat

FoF Friend of Friend

FOF FEP Output Format • Field Observing Facility • Field Of Fire • First Orbital Flight • First Operational Flight • Flight Operations Facility • Force On Force • Friends Of Friends

FOFCH First Order Flux Conservative Hyperbolic

FOFE Fiber Optic Front End

FOFL Follow-On Fiber LAN

FOFP Fiber-Optic Face Plate

FOFW Fiber-Optic Field Wire

FOG Faintly Observable Galaxy • Far Outer Galaxy • Fiber-Optic Guided • Fiber-Optic Gyroscope • First Osborne Group • Flight Operations Group • Foggia (Italy) (location identifier) • Forecast Generator • Forschungsorganisationsgesetz • Fourth Order Gravity • Frequency Offset Generator • Fiber Optics Guidance

FOGA Foundations of Genetic Algorithms

FoGeo Fotointerpretacia w Geografii (periodical code)

FOGN First-Order Gravity Net

FOGT First-Order Gradient Technique

FOH First-Order hold • Forced Outage Hour

FOHM Fundamental Open Hypermedia Model

FOHMD Fiber-Optic Helmet-Mounted Device • Fiber-Optic Helmet-Mounted Display

FOI Fault Detection Isolation • Field Operations Installation • Follow-On Interceptor • Freedom Of Information

FOIA Freedom Of Information Act (USA)

FOID Fibre Optic Image Dissector

FOIH Flight Operations Integration Handbook

FoIKS Foundations of Information and Knowledge Systems (symposium)

FOIL File-Oriented Interpretive Language • First - Outer - Inner - Last • Freedom Of Information Legislation

FOIM Field Office Information Management

FOIMS Field Office Information Management System

FOIPA Freedom Of Information/Privacy Act (USA)

FOIPP Freedom Of Information and Protection of Privacy (act)

FOIR Freedom Of Information Request

FOIRL Fiber-Optic Interrepeater Link

FOIS Formal Ontology Information Systems (conference)

FOIU Fiber-Optic Interface Unit

FOJT Formal On-the-Job Training

FOK Westhampton - Suffolk County (USA-NY) (location identifier)

fol Folio • Folium

Fol Folio • Folium • Follow • Followed • Following

FOL Fiber-Optic Laser • Fiber-Optic Light • Fiber-Optic Link • Folio • Forward Operating Location • Friends Of LISA

FoLA Four-Letter Acronym

FOLACL Federation of Local Authority Chief Librarians

FOLAN Fiber Optic Local Area Network

FOLBC Fiber-Optic Loop-Back Connector

FOLD Federally-Owned Landsat Data (USA)

Folg Following

FOLG Fiber Optics Laser Gyro

Foll Follow • Followed • Following

FOLPEN Foliage Penetration

fols Folios

FOLS Family Of Loudspeakers

FOM Facilitate Other Maintenance • Fault Of Management • Federation Objective Model • Federation Object Model • Fiber-Optic Modem • Field Offset Mirror • Field Operations Manual • Field Operations Memorandum • Figure Of Merit • Foumban (Cameroon) (location identifier) • Full Operation and Maintenance • Stichting voor Fundamenteel Onderzoek der Materie (Netherlands)

FOMA Facilities Operation and Maintenance Agreement • Facility Operation and Maintenance Activity

FOMAE Follow-On Management Application and Evaluation

FOMAU Fiber Optic Medium Access Unit • Fiber Optic Medium Attachment Unit

FOME Flight Observation Measurement Experiment

FoMLaDO Foundations of Models and Languages for Data and Objects (workshop)

FOMM Functionally-Oriented Maintenance Manual

FOMOS Fiber-Optics Multiobject Spectrograph

FOMR Flight Operations Management Room

FOMTE Forme, Orientation et Mouvement de la Terre dans l'Espace

fon Font • Phone (file extension)

Fon Fonética • Fonología

FON Fiber Optic Network • Font • Functional Operational Number • Phone

Fond Fondamenta • Fondamental • Fondation • Fondazione

FONDAP Fondo Nacional de Desarollo de Areas Prioritarias (Chile)

FONECON Phone Conversation • Telephone Conference • Telephone Conversation

FONEMA Fast Omnidirectional Non-scanning Energy Mass Analyser

FONPA Finding Of No Practicable Alternative

FONSI Finding Of No Significant Impact

FONTX Forbid One-Node Tree Crossover

FOO Flight Of Opportunity • Frequency of Optimum Operations • Numfoor (Indonesia) (location identifier)

FOOBAR Fooled up Beyond All Recognition

FOOL Foundations of Object-Oriented Languages (workshop)

FOOS Function-Oriented Organizational Structure

FOOT Follow-On Operational Test

FOP Facility Operating Procedure • Facility Operations Plan • Fiber Optics Probe • Financial Operating Plan • First Off Production • Flight Operations Panel • Flight Operations Plan • Follow-On Production • Forest Park - Morris (USA-GA) (location identifier) • Formation par Objectifs Pédagogiques • Frame-Oriented Protocol

FOPC First-Order Predicate Calculation

FOPG Flight Operations Planning Group

FoPh Foundations of Physics (periodical code)

FOPI First-Order Polynomial Interpolator

FOPMO Friends Of the Pine Mountain Observatory (USA-OR)

FOPP First-Order Polynomial Predictor • Follow-On Parts Production

FOPPR Flight Operations Plans, Procedures and Rules

FOPS Financial Operating System • Flight Operations Planning Schedule • Flight Operations Procedure System

FOPSS Flight Operation Product Scheduling System

FOPT Fiber-Optic Photo Transfer

FOQ Formation des Ouvriers Qualifiés

FOQA Flight Operations Quality Assurance

for Fortran (file extension)

For Foreign • Foreigner • Fornax-Fornacis (constellation)

FOR Field Of Reduction • Field Of Regard • File Owning Region • Flight Operations Review • Force • Forced Outage Rate • Foreign • Fortaleza - Pinto Martins (Brazil-CE) (location identifier) • Fortran • Free On Rails

Forc Forcing

FORCAST Faint Object Camera for the SOFIA Telescope

FORCS FAA Operational Reporting Communication System

FORD Fix Or Repair Daily • Flight Operations Requirement Document • Found On Road Dead

FördG Fördergebietsgesetz

FORE First Overtone Red Edge

Foregd Foreground

FOREGS Forum of European Geological Surveys

FORG Forging

FORGEN Force Generation

FORL Far Off Resonance Lattice

Form Formation • Formula • Formulate • Formulated • Formulation • Formule • Formuler

FORM Formation

FORMAC Formula Manipulation Computer

FORMATS FDF Orbital and Mission Aids Transformation System

Formn Foreman • Formation

FORNN Forenoon

FORP Fiber Optic Respiratory Plethysmograph

FORS Faint Object Red Spectrograph • Fiber-Optic Rate Sensor • Finance Office Research Section • Focal Reducer/Spectrograph

Forsch Forschung

FORSTAT Force Status

Fort Fortification • Fortified

FORT Foyer Optimisé pour le Radiotélescope

FORTE Fast On-orbit Recording and Transient Events • Fast On-orbit Recording of Transient Events (*)

FORTH Foundation for Research and Technology - Hellas (Greece)

FORTLP Fortran Linear Programming

FORTRAN Formula Translator

FORTRESS Framework for Operating, Recommending and Testing a Range of European Security Services

Forts Fortsetzung

FORTWIHR Forschungsverband für technisch-wissenchaftliches Hochleistungsrechnen (Germany)

FORUM Formula for Optimizing through Real-Time Utilization of Multiprogramming

FOS Faint Object Spectrograph • Family Of Services • Family Of Systems • Fiche d'Organisation des Services • Field Of Science • File Organization System • Flight Operations Security • Flight Operations Segment • Flight Operations Support • Follow-On Spares • Follow-On Support • Forrest (Australia-WA) (location identifier) • Fortran Operating System • Forward-looking Optronic System • Fragmentary Order System • Free On Station • Free On Steamer • Full Operating System • Full Operational Status

FOSA First-Order Smoothing Approximation • Flight Operations Support Annex

FOSD Functional Operational Sequence Diagram

FOSDIC Film Optical Sensing Device for Input to Computers

FOSE Federal Office Systems Exposition

FOSF Field Observing Support Facility

FOSI Format Option Specification Instance • Formatting Output Specification Instance • Formatted Output Specification Instance • Frontiers Of Space Imaging • Functional Output Specification Instance

FOSMS Facility Operations Scheduling and Monitoring System

FOSO Flight Operations Scheduling Office

FOSOL Florian's Own Statistically Oriented Language

FOSP Flight Operations Support Personnel • Friends Of Staerkel Planetarium (USA-IL)

FOSR Flexible Optical Solar Reflector

FOSS Family Of Systems Study • Fiber-Optic Sensor System • Fore Optic Support Structure • Full Option Science System • Future of Space Sciences

FoSSaCS Foundations of Software Science and Computation Structure (conference)

FOSSIL Fide/Opus/Seadog Standard Interface Layer

FOST Flight Operations Support Team

FOSTER Flight Opportunities for Science Teacher Enrichment • Flight Opportunities for Science Teacher Environment

FOT Faint-Object Telescope • Fiber-Optic Terminal • Fiber-Optic Topology • Fiber-Optic Transceiver • Fifth-Order Theory • Final Observation Tape • Fleet Ocean Tug • Flight Operations Team • Flight Operations Trainer • Flow Object Tree • Follow-on Operational Test • Follow-on Test • Follow-on Training • Forster (Australia-NSW) (location identifier) • Fourier Optical Transform • Fraction Of Time • Free Of Tax • Free On Truck • Frequency of Optimal Traffic • Frequency of Optimum Traffic • Frequency of Optimum Transmission • Fundamental Overtone

FOTC Fiber Optic Tow Cable

FOTD Fiber-Optic Towed Decoy

FOT&E Final Operational Test & Evaluation • Follow-on Operational Test & Evaluation • Follow-On Test & Evaluation • Functional Operational Test & Evaluation

FOTE Final Operational Test and Evaluation • Follow-on Operational Test and Evaluation • Follow-On Test and Evaluation • Functional Operational Test and Evaluation

FOTLAN Fiber-Optic Tactical Local Area Network

FOTM Functionally-Oriented Technical Manual

FotO Friends of the Observatory

FOTRA Free Of Tax to Residents Abroad

FOTS Fiber Optic Terminal System • Fiber Optic Transmission System • Follow-On Technical Support

FOTSLH Fiber Optic Transmission System Long-Haul

FOU Fougamou (Gabon) (location identifier)

FOUDRE Fourniture de Documents sur Réseau Électronique

Found Foundation • Founded

FOUO For Official Use Only

FOUPA Fiber-Optic Universal Port Adapter

FOURBI Fast Object/User Retrieval and Bibliographic Index • Follow-Up Relational Base for Images

FoV Field of View

FOV Face Of Venus • Family Of Vehicles • Field Of View • Field Of Vision • First Orbital Vehicle

FOVEA Field Of View Emissivity Area

FOVIA Fotómetro Visible Adaptado

FOW Finite Orbit Width

FOWG Fiber-Optic Wave Guide • Flight Operations Working Group

FOX Fiber Optic Transceiver • Field Operational X.500 • Fox (USA-AK) (location identifier)

FOY Foya (Liberia) (location identifier)

fp Forte-Piano

Fp Finite Part • Frontispiece

F-P Fabry-Pérot • Finson-Probstein • Fokker-Planck (model)

F/P Fabry-Pérot • Focal Plane

F&P Fluids & Propulsion • Functional & Performance (test)

FP Fabry-Pérot • Faceplate • False Positive • Fast Poll • Fast Processor • Feature Point • Federal Parliament • Federal Publication • Feedback Positive • Fiber Passive • Fictitious Point • Field Potential • File Processor • Film Pack • Filter Paper • Final Point • Financial Plan • Fine Paper • Fine Pointer • Fine Pointing • Fire Plug • Firing Point • First Performance • First Performed • Fission Product • Fixed Point • Fixed Price • Flash Point • Flat Panel • Flexible Processor • Flight Plan • Flight Program • Floating Point • Fluids and Propulsion • Fluorescent Particle • Focal Plane • Focal Point • Fokker-Planck (model) • Fonction Publique • Foolscap • Footpath • Foot-Pound • Force Package • Force Protection • Forest Park • Formación Profesional • Formation Permanente • Formation Professionnelle • Format Prefix • Forte-Piano • Forward Pass • Forward Perpendicular • Framework Programme • Freeze Point • Freezing Point • French Polynesia • Frequency Panel • Fresh Paragraph • Frontispiece • Front Page • Front Penetration • Fuel Pressure • Fuel Pump • Full Period • Full Point • Full Power • Full Problem • Full Propagation • Fully Paid • Functional Path • Functional Programming • Functional Proponent • Functional Protocol • Function Path • Function Point • Fundamental Plane • Fusing Point

FPA Failure Probability Analysis • Final Power Amplifier • First Point of Aries • Fixed Plant Adapter • Flat-Plane Array • Flat-Plate Antenna • Flat-Plate Array • Flexible Production Automation • Flight Path Analysis • Flight Path Angle • Flight Plan Approval • Floating-Point Accelerator • Floating-Point Arithmetic • Fluid Penetration Aid • Fluid Processing Apparatus • Focal-Plane Array • Focal-Plane Assembly • Force Planning Analysis • Foreign-Policy Advisor • Foreign Press Association • Formation Professionnelle des Adultes • Formula Pricing Agreement • Formula Pricing Arrangement • Fractal Pixon Axis • Free of Particular Average • Funding Program Advice

FPAA Field Programmable Analog Array • Flat-Plate Array Antenna • Flight Path Analysis Area

FPAC Flight Path Analysis and Command

FPAF Fixed Price Awarded Fee

FPAG Fundamental Physics Advisory Group (ESA)

FPAP Flight Plan Approval Package

FPAS Faculty of Pure and Applied Science • Focal Plane Assembly Shroud • Fully Polynomial Approximation Scheme

FPATC Focal Plane Array Test Chamber

FPAU Fringe Phase Alignment Unit

FPB Flight Process Board • Foreign Press Bureau • Fractal Pixon Basis • Fuel Preburner • Ful Problem

FPBD Functional Plan Block Diagram

FPBOV Fuel Preburner and Oxidizer Valve

F&PC Finance & Program Control

FPC Facility Power Control • Federal Port Controller • Federal Power Commission (USA) • Federal Publishers Committee (USA) • Final Processing Center • Finance and Program Control • Fixed-Point Calculation • Fixed-Precision Capacitor • Fixed-Price Call • Fixed-Price Contract • Fixed-Price Contractor • Fixed-Program Computer • Flexible Printed Circuit • Flight Path Control • Flight Processor Card • Floating-Point Calculation • Flow Process Chart • Flow Proportional Counter • Fluids Pressure Control • Focal Plane Chopper • Foreign Program Control • Formation Professionnelle Continue • For Private Circulation • Forward Power Controller • Front Panel Control • Functional Process Code

FPCA Forward Power Control Assembly • Forward Power Controller Assembly • Functional Programming and Computer Architecture (conference)

FPCB Field-Programmable Circuit Board

FPCC Fixed Polycarbonate Capacitor • Fixed Printed Circuit Capacitor

FPCE Floating-Point C Extension

FPCH Foreign Policy Clearinghouse

FPCS Fast Profiling Current Source • Focal Plane Crystal Spectrometer • Freezing-Point Calibration Standard • Full-Page Composition System

FPCT FITS Product Conformance Tester

FPD Fault Protection Device • Flame Photometric Detector • Flat-Panel Display • Flight Projects Directorate • Full-Page Display • Full-Power Day • Functional Purchase Description

FPDA Focal Plane Deck Assembly

FPDB Force Planning Database

FPDC Federal Procurement Data Center (USA)

FPDD Final Project Design Description

FPDE Focal Plane Drive Electronics

FPDIS Fabry-Pérot Doppler Imaging System

FPDS Federal Procurement Data System

FPDT Forecast Products Development Team

FPE Factory Process Engine • Final Prediction Error • Fixed Price with Escalation • Floating-Point Engine •

Fonds de Propriété d'État (France) • French Postural Experiment • Functional Program Element

FPEG Fast Pulse Electron Gun • Fundamental Plane of Elliptical Galaxies

FPEPA Fixed Price with Economic Price Adjustment

FPES Frozen Polyelastic Sampling

FPF Facility Parameter Field • Fine Pointing Facility • First Ply Failure • Fixed-Price Firm • Fluid Physics Facility • Force Package File • Full Power Frequency

FPG Fixed-Pattern Generator • Förderkreis Planetarium Göttingen eV (Germany)

FPGA Field Programmable Gate Array

fph Frames Per Hour

FPH Flight Procedures Handbook • Fondation pour le Progrès de l'Homme • Frames Per Hour • Free Phone

FPHA Federal Public Housing Authority

FPHB Flight Procedures Handbook

FPI Fabry-Pérot Interferometer • Fabry-Pérot Interferometry • First Periodic Inspection • Fixed-Price Incentive • Fixed-Price Initiative • Floating-Point Interface • Fluorescent Penetrant Inspection • Focal Plane Imager • Focused Process Improvement • Formación de Personal Investigador • Formation Professionnelle Initiale • Fuel Pressure Indicator • Functional Process Improvement

F&PIA Force & Program Impact Assessment

FPIA Force and Program Impact Assessment

FPIC Field Programmable Interconnect • Field Programmable Integrated Circuit • Field Programmable Interconnect Chip • Fixed price Incentive Contract

FPID Field Programmable Interconnect Device

FPIF Fixed Price Incentive Fee • Fixed Price Incentive Firm

FPIII Third Framework Programme

FPIP Floating Point Interpretive Program • Focal-Plane Instrument Package

FPIS Forward Propagation by Ionospheric Scatter

FPIT Field Preinstallation Test

FPITS Fabry-Pérot Inverse Transform Spectrometer

FPIV Fourth Framework Programme

FPK Fixed-Position Keyboard

FPKI Federal Public Key Infrastructure (USA)

Fpl Fireplace

FPL Federal Public Library (USA) • Ferry-Porter Law • Fetch Pointer List • Field-Programmable Logic (workshop) • Final Parts List • Final Protective Line • Flight Plan • Flight Propulsion Laboratory • Frequency Phase Lock • Full-Performance Level • Full Power Level • Full Power Load • Functional Programming Language • Fuzzy Programming Language

FPLA Field-Programmable Logic Array

FPLC Franklin Pierce Law Center

FPLD Field Programmable Logic Device

FPLE Functional Programming Languages in Education (symposium)

FPLMTS Future Public Land Mobile Telecommunication Service • Future Public Land Mobile Telecommunication System

FPLOE Fixed-Price Level Of Effort

FPLP Fortran List-Processing Language

fpm Feet per Minute

F&PM Fuel & Power Management

FPM Facilities Planning Module • Facility Power Monitor • Faculté Polytechnique de Mons (Belgium) • Fast-Page Mode • Federal Personnel Manual • Federal Procurement Manual • Feet Per Minute • Fiber Pulling in Microgravity • File Protect Memory • Finite Population Mean • First Polar Mission • Flashes Per Minute • Fluence Profile Monitor • Fluid Physics Module • Flux Particle Monitor • Folding Platform Mechanism • Formation Paritaire Mixte • Frames Per Minute • Frequency Position Modulation • Fuel and Power Management • Functional Planning Matrix • Functional Project Manager

FPMA Focal Plane Mount Array

FPMH Failures Per Million Hours

FPMIS Federal Personnel Management Information System (USA)

FPMR Federal Property Management Regulation (USA)

FPMs Faculté Polytechnique de Mons (Belgium)

FPN Federal Part Number • Fixed Pattern Noise

FPNNC Fédération des Parcs Naturels et Nationaux d'Europe

FPO Fast Photometric Observation • Field Post Office • Field Project Office • Fleet Post Office • For Positioning Only • For-Profit Organization • Freeport (Bahamas) (location identifier) • Future Projects Office

FPOE First Port Of Entry

FPOV Fuel Preburner Oxidizer Valve

FPP Facility Power Panel • Family Planning Programme • Fantasy-Prone Personality • Fixed Path Protocol • Fixed Point Program • Floating-Point Processor • Focal Plane Package • Focal Plane Polarimeter • Freon Pumped Package • Freon Pump Package

FPPA Field Programmable Processor Array

FPPC Fusion Physics Planning Committee

FPPL Function and Procurement Parts List

FPPO Fox Park Public Observatory (USA-MI)

FPPR Fixed Price with Price Revision

FPPS Full Page Phototypesetting System

FPPU Floating-Point Processor Unit

FpR Feet per Revolution

F&PR Functional & Performance Requirement

FPR Failure/Problem Report • Fan Pressure Ratio • Federal Procurement Regulation (USA) • Feet Per Revolution • Final Program Review • Final Progress Report • Final Project Review • Final Proposal Revision • Fixed-Point Representation • Flat-Plate Radiometer • Flight Performance Reserve • Floating-Point Register • Floating-Point Routine • Forest-Penetrating Radar • Fort Pierce (USA-FL) (location identifier) • Forward Pricing Rate • Full Power Response • Full Propellant Requirement • Functional and Performance Requirement • Functional Process Release • Funded Procurement Request

FPRA Forecasting, Planning and Resource Allocation • Forward Pricing Rate Agreement

FPRAS Fully Polynomial Randomized Approximation Scheme

Fprf Fireproof

FPRF Flexible Pseudo-Relevance Feedback

FPRP Forward Pricing Rate Proposal

FPRR Forward Pricing Rate Recommendation

F&PRS Functional & Performance Requirements Specification

FPRS Federal Property Resources Service • Functional and Performance Requirements Specification

FPRT Forward Pricing Review Team

FPRV Flight Product Requirements Verification

fps Feet per Second • Foot - Pound - Second • Frames per Second

FPS Fabry-Pérot Spectrometer • Fast Packet Switch • Fast Packet Switching • Favorite Picture Selection • Feet Per Second • Fiber Protection Switch • Financial Planning System • Financial Product System • Fine Pointing Simulation • Fine Pointing Simulator • Fixed Position Search •

Flights Per Second • Floating-Point System • Focal Plane Structure • Focus Projection and Scan • Focus Projection and Scanning • Foot - Pound - Second • Formation aux Premiers Secours • Forum on Physics and Society • Forward Power Supply • FOS Planning and Scheduling • Frames Per Second • Full-Pressure Suit

FPSA Focal Plane Structure Assembly

FPSI Formatting Presentation Specification Instance

FPSNW File and Print Server for NetWare

FPSO Flight Projects Support Office

FPSP Focal Plane Signal Processor

FPSR Flight Planning and Storage Review

fpsps Feet per Second per Second

FPSS Fine-Pointing Sun Sensor • Flat-Panel System Simulator • Focal-Plane Star Sensor • Focal-Plane Subsystem • Focal-Plane Support Structure

FPSTU Full-Pressure Suit Training Unit

FPS2 Feet Per Second Squared

FPT Female Pipe Thread • Field Patch tape • Finite Population Total • Fixed Price Tenders • Follow-on Production Test • Follow-on Production Testing • Forced Perfect Termination • Freezing Point • Full Power Trail • Full Production Training • Functional Performance Test • Fusion Plasma Theory

FPTE Facility Portable Test Equipment

FPTS Forward Propagation by Tropospheric Scatter

FPU First Production Unit • Floating-Point Unit • Focal Plane Unit • Fuel Purification Unit

FPV Fifth Framework Programme • Floating-Point Validating • Flow Proportioning Valve

FPWG Financial Processes Working Group • FSD Planning Working Group

FPY Financial Plan Year • Perry Foley (USA-FL) (location identifier)

FPYD Fission Product Yield Data

FQ Facteur de Qualité • Fair Queuing • Fall Quarter • Fiscal Quarter • Flight Qualification • Formal Qualification • Fused Quartz

FQ&A Following Question & Answer

FQA Fastener Quality Act (USA) • Final Quality Assurance • Fixed Quality Assurance • Following Question and Answer • Full Quality Assurance

FQAS Flexible Query Answering System • Flexible Query Answering Systems (conference)

FQAT Formal Qualification Acceptance Test

FQD Field Queuing Device

FQDN Fully Qualified Domain Name

FQF First Qualification Flight

FQHE Fractional Quantum Hall Effect

FQHN Fully Qualified Host name

FQI Federal Quality Institute (USA) • Flight Qualification Instrumentation • Fuel Quantity Indicator

FQIL Fuel Quartz Incandescent Lamp

FQIS Fuel Quantity Indicating System

FQL Functional Query Language

FQLFE Functional Query Language Front End

FQM Four-Quadrant Multiplier

FQO Forced Quantum Oscillator (method)

FQ&P Flight, Quality & Performance

FQP Flight - Quality - Performance • Fundamental Questions Program

FQPA Flight Quality Photomultiplier Assembly

FQPR Frequency Programmer

FQQPRI Final Qualitative and Quantitative Personnel Requirements Information

FQR Flight Qualification Recorder (*) • Flight Qualification Review (*) • Formal Qualification Review • Functional Qualification Review

FQRNT Fonds Québécois de la Recherche sur la Nature et les Technologies (Canada-QC)

FQT Formal Qualification Test • Formal Qualification Testing • Functional Qualification Test • Functional Quality Testing • Frequent

FQTR Flight Qualification Tape Recorder

FQTV Frequent Traveller

fr Folio Recto • Franc (currency) • Frei • French (ISO language code)

Fr Father • Fragment • Frame • Franc (currency) • Français • France • Francium • Franco • Frase • Fratelli • Frater • Frau • Free • Freitag • Freitags • French • Frequency • Frequent • Frère • Friar • Friday • From • Front • Froude Number

F/R Flared Rudder • Folio Reference

FR Face Recognition • Failure Rate • Failure Report • Falling Rapidly • Fanaroff-Riley (classification) • Faraday Rotation • Faroe Islands (international vehicle registration code) • Faroe Islands (postal country code) • Far Right • Fast Release • Fault Reporting • Federal Register • Federal Republic • Field Reversing • File Received • File Register • Finalité de la Recherche • Final Release • Final Report • Final Review • Firing Room • Fixed Ratio • Fixed Routing • Flame Retardant • Flat Rack • Flat Rate • Fletcher-Reeves • Flight Readiness • Flight Recorder • Flight Reliability • Flight Rule • Flux Randomization • Focal Reducer • Force-Reflective • Foreign Reimbursement • Frame • Frame Rate • Frame Relay • France (code ISO de pays) • Frei • Frequency Range • Frequency Rate • Frequency Response • Froude Number • Full Range • Full Rate • Full Resolution • Fully Reusable • Functional Requirement • Functional Review • Funding Request • Fusion Reactor

fra Factura • Fractus

FRA Federal Railroad Administration (USA) • Federal Records Act (USA) • Field Replaceable Assembly • Fixed Radio Access • Formation de Recherche Associée • Forward Rate Agreement • Forward Repair Activity • France (code ISO de pays) • Francfort (Germany) (location identifier) • Frankfurt (Germany) (location identifier) • Full Range Analyzer

FRAB Functional Requirement Authentication Board

Frac Fractional

FRAC Flint River Astronomy Club • Fonds Régional d'Actions Culturelles (France) • Fonds Régional d'Aide au Conseil (France) • Fonds Régional d'Art Contemporain (France) • Forward Reaction Attitude Control • Fracção • Fractionator Reflux Analog Computer

FRACA Failure Reporting Analysis and Corrective Action • Failure Reporting And Corrective Action

FRACAS Failure Reporting Analysis and Corrective Action Summary • Failure Reporting Analysis and Corrective Action System • Filter Response Analysis for Continuously Accelerating Spacecraft

FRACS Forward Reaction Attitude Control System

FRACTAL Fractional Dimensional

FRAD Frame Relay Access Device • Frame Relay Assembler/Disassembler

FRADE Formation et Recherche Appliquée pour le Développement Économique

FRAFP Fonds Régional de l'Apprentissage et de la Formation Professionnelle (France)

FRAFPC Fonds Régional de l'Apprentissage et de la Formation Professionnelle Continue (France)

Frag Fragile • Fragment • Fragmentation • Fragmento

FRAG Fragile • Fragment • Fragmentation

FRAGNET Fragmented Network

FRAM Fast Random Access Memory • Ferroelectric Random Access Memory (*) • Ferroelectronic Random Access Memory • Fine-Resolution Antarctic Model

FRANEC Frascati Raphson Newton Evolutionary Code

Fränk Fränkische

FRANK Frequency Regulation and Network Keying

FRAP Free Route Airspace Project

FRAS Fellow of the Royal Astronomical Society • Frame Relay Access Support

FRASC Front Range Astronomical Supercluster (USA-CO)

FRAT Free Radical Assay Technique

FRATT Fonds Régional d'Aide au Transfert de Technologie (France)

Fraud Fraudulent

FRB Failure Review Board • Fast Rise Balloon • Federal Reserve Bank (USA) • Federal Reserve Board (USA) • Forbes (Australia-NSW) (location identifier) • Functional Redundancy Checking • Functional Review Board

Frbd Freeboard

FRBS Frame Relay Bearer Service

FRC Facility Responsible Centre • Facility Review Committee • Federal Radio Commission (USA) • Federal Records Center (USA) • Fiber-Reinforced Composite • Field-Reversed Configuration • Final Reproducible Copy • Financial Review Committee • Fixed Radio Control • Flight Readiness Certification • Flight Research Center • Focal Reducing Camera • França (Brazil-SP) (location identifier) • Fresnel Reflection Coefficient • Front Range Consortium (USA-CO) • Full Route Clearance • Functional Redundancy Check • Functional Residual Capacity • Functional Residue Capacity • Fusion Research Center

FRCB Flight Rules Change Board • Flight Rules Control Board

FRCC Federal Research Contract Center (USA)

FRCD Floating-Rate Certificate of Deposit

FRCI Fibrous Refractory Composite Insulation

FRCN Fundamental Reference and Calibration Network

FRCP Facility Remote Control Panel • Fellow of the Royal College of Physicians

FRCS Federal Reserve Communications System (USA) • Fellow of the Royal College of Surgeons • Forward Reaction Control System

FCRT Fixed Record Communications Teletypewriter

Frd Friend • Froid

FRD Facilities Requirements Document • Failure Rate Data • Flight Requirements Document • Focal Ratio Degradation • Ford • Formerly Restricted Data • Foundation for Research Development (South Africa) (now NRF) • Friday Harbor (USA-WA) (location identifier) • Functional Reference Device • Functional Requirements Description • Functional Requirements Document

FRDB Failure Rate Database

FRDI Flight Research and Development Instrumentation

FRDM Finite Range Droplet Model

FRDR Flight Readiness Design Review

FRDS Federal Reporting Data System • Flight Records Data System • Fords

FRDT Function Rights Description Table

Fre Fracture • Freitag • French

FRE Fera Island (Solomon Islands) (location identifier) • Fichier Rectoral des Établissements (France) • Flight-Related Element • Fonds de Revitalisation Économique • Formation de Recherches en Évolution (France) • Format Request Element • Fundamental Red Edge

FRECOPA French Cooperative Package • French Cooperative Payload

FRed Focal Reducer

FRED Fast Rise and Exponential Decay • Fast Rise/Exponential Decay • Figure Reader Electronic Device • Figure Reading Electronic Device • Frame Editor • Front End Directory • Front-End to Disk

Free Freeway

FREESTAR Fast Reaction Experiment Enabling Science Technology Applications and Research

Freew Freeware • Freeway

FREGATE French Gamma-ray Telescope (HETE-2)

freiw Freiwillige

Fréq Fréquence

Freq Frequency • Frequent • Frequentative • Frequently • Frequenz (periodical code)

FREQ Frequency

FRERP Federal Radiological Emergency Response Plan (USA)

FRESA Frame Relay End-System Address

FRESCO Full Resolution Coadder • Full Resolution Survey Coaddition

FRESIP Frequency of Earth-Sized Inner Planets

FRESS File Retrieval and Editing System

FREX Fonds Régional à l'Exportation • Fridays Excepted

FRF First Reduction File • Flight Readiness Firing • Frame Relay Forum • Franc (France) (ISO currency code) • Frankfurt Rhein-Main (Germany) (location identifier) • Frequency Response Function • Frequency Reuse Factor

FRFA Final Regulatory Flexibility Analysis

FRFC Fonds de la Recherche Fondamentale Collective (Belgium)

FRFCIM Fonds de la Recherche Fondamentale Collective d'Initiative Ministérielle (Belgium)

FRFI Flug Revue Flugwelt International (periodical code)

FRFS Feeder Request For Service

FRFT File Ready For Transfer • Flight Readiness Firing Test • Fractional Fourier Transform

FRG Family Readiness Group • Farmingdale (USA-NY) (location identifier) • Federal Republic of Germany • Floated-Rate Gyro • Focused Research Group • Forge

FRGS Fellow of the Royal Geographical Society • Forges

Frgt Freight

FRGU Frequency Reference Generator Unit

FRH Fly Runway Heading • French Lick (USA-IN) (location identifier)

Fri Fribourg (Switzerland) • Friday

FRI Faranoff-Riley class I • Focused Research Initiative • Forest Research Institute • Forest Resource Inventory • Fort Riley (USA-KS) (location identifier) • Friday

FRIA Fellow of the Royal Irish Academy

Fric Friction • Frictional

FRICC Federal Research Internet Coordinating Committee (USA)

FRICS Fellow of the Royal Institution of Chartered Surveyors

Frict Friction • Frictional

Friesl Friesland

FRII Faranoff-Riley class II

FRINGE File and Report Information Processing Generator • Frontiers of Interferometry in Germany (team)

FrInJ Franklin Institute Journal (periodical code)

FRIS Forest Resource Information System

FRISC Fast Reduced Instruction Set Computer

Frisco San Francisco (USA-CA)

FRISCO Framework of Information System Concepts (working group)

FRISPI Fast Recording Imager and Spectrophotometric Instrument

FRIT Financial Restructuring Implementation Team

FRJ Fréjus Saint-Raphaël (France) (location identifier)

FRJD Forward Reaction Jet Driver

Frk Frøken • Fröken

FRK Fork • Fregate Island (Seychelles) (location identifier)

FRKS Forks

Frl Fräulein

FRL Forli (Italy) (location identifier) • Frame Reference Line • Frame Representation Language • Fräulein • Fuselage Reference Line

Frld Foreland • Freehold

Frm Frame • From

FRM Facility Requirements Manager • Fairmont (USA-MN) (location identifier) • Faraday Rotation Measure • Farm • Fast Resource Management • Fast Rotating Model • Fiber-Reinforced Material • Film-Reading Machine • FOS Resource Management • Foucault Rotating Mirror • Functional Reference Model

FRMA Functional Requirements Management Activity • Functional Requirements Management Agency

FRMAC Federal Radiological Monitoring and Assessment Center (USA)

FRMAP Federal Radiological Monitoring and Assessment Center Plan

FRMG Forming

FRMJC Fédération Régionale des Maisons de Jeunes et de la Culture (France)

FRMR Frame Reject

FRMS Frequency Record Management System • Frequency Resource Management System

FRMT Format

FRN Federal Register Notice • Floating-Rate Note • Force Requirement Number • Forskningsrådsnämmden • Fort Richardson (USA-AK) (location identifier) • Funding Resource Number

Frnc Furnace

FRNC Flame Retardant Non Corrosive

Frng Fringe

FRNI Frame Relay Network Interworking

FRNT Front

FRO Facilities Responsible Officer • Faculty Resource Officer • Faroe Islands (ISO country code) • Fire Risk Only • Fleurs Radioastronomical Observatory (Australia-NSW) • Flexible Response Option • Floro (Norway) (location identifier) • Focal Reduction Optics • Îles Faroë (code ISO de pays)

FROA Frame Relay Over ATM

FroCos Frontiers of Combining Systems (workshop)

FROD Functionally-Related Observable Difference

FROG Fédération Orcéenne en Géosciences (France) • Free-flight Rocket Over Ground • Frequency Resolved Optical Gating

Front Frontispiece

FRONT Fronteira • Frontière

FRONTIERS Forecasting Rain Optimized using New Techniques of Interactively Enhanced Radar and Satellite

Frontis Frontispiece

FROST Fast Robotic Observatory System for Transients • First Regional Observing Study of the Troposphere

FRP Facility Requirements Panel • Fast Reservation Protocol • Feature Recognition Processor • Federal Radionavigation Plan (USA) • Fiber-Reinforced Plastic • Filament-Reinforced Plastic • Flag Register Processing • Fresh Water Bay (USA-AK) (location identifier) • Fuel-Reprocessing Plant • Full-Rate Production • Functional Review Process • Fuselage Reference Plane

FRPA Fixed Radiation Pattern Antenna • Fixed Reception Pattern Antenna

FRPC Fast Reaction Procedure Card

FRPCC Federal Radiological Preparedness Coordinating Committee (USA)

FRP&D Full-Rate Production & Deployment

FRPD Full-Rate Production and Deployment

FRPDR Full-Rate Production Decision Review

Frpf Fireproof

FRPG Fantasy Role Playing Game

FRPH Frame Relay Protocol Handler

FRpI Flux Reversals per Inch

FRPI Flux Reversals Per Inch

Frpl Fireplace

FRQ Feramin (Papua New Guinea) (location identifier) • Frequent

FRR Facility Readiness Review • Failure and Rejection Report • Failure and Replacement Report • Failure Reporting Review • Fast Repetition Rate • FCA Review Report • Federal Reuse Repository • Federatie van Russische Republieken • Fédération des Républiques Russes • Federation of the Russian Republics • Field to Resolution Ratio • Flight Readiness Review • Force de Réaction Rapide • Formal Requirements Review • Front Royal - Warren County (USA-VA) (location identifier) • Full Reimbursement Rate • Functional Recovery Routine

FRRB Fast Rise Reflective Balloon

FRRID Flight Readiness Review Item Description • Flight Readiness Review Item Disposition

FRRPA Forest and Rangeland Renewable Planning Act (USA)

FRRS Frequency Resource Record System

Frs Francs (currency)

FRS Face Recognition Software • Faint Reference Star • Fast Retrieval Storage • Federal Reserve System (USA) • Fellow of the Royal Society (UK) • Figure Rating Scale • File Replication System • Filtered Rayleigh Scattering • First Readiness State • Fixed Resource Session • Flexible Route Selection • Flores (Guatemala) (location identifier) • Fragility Response Spectrum • Frame Relay Service • Freight Rate Specialist • Fully Responsive System • Functional Requirement Specification • Functional Respiratory System • Function Repository Server

FRSAS Fast-Response Solar-Array Simulator

FRSI Felt Reusable Surface Insulation • Flexible Reusable Surface Insulation • Frame Relay Service Interworking

FRSIS Fuzzy Relational Soil Information System

FRSL Fellow of the Royal Society of Literature

FRSM Fonds de la Recherche Scientifique Médicale (Belgium)

FRSQ Flat Radio Spectrum Quasar

FRSS Federal Register Search System • Fellow of the Royal Statistical Society • Flat Radio Spectrum Source • Flight Reference Stabilization System

FRSSCS Frame Relay Service Specific Convergence Sublayer

FRST Forest • Frost

FRSU Frame Relay Service Unit

Frt Freight • Front

FRT Fast Reaction Team • Fast Rotating Target • File Received Timeout • Film Register Treaty • Flight Readiness Test • Flight Readiness Training • Fonds de la Recherche et de la Technologie • Fonds de la Recherche Technologique • Frequency Response Test • Functional Requirements Test

FrTel France Telecom

FRTS Frame Relay Traffic Shaping

FRTT Fixed Round-Trip Time

FRU Bishkek (Kyrgyzstan) (location identifier) • Field-Replaceable Unit

FRUF Fédération des Résidences Universitaires de France (France)

FRUS Foreign Relation of the United States

FRUSA Flexible Rolled-Up Solar Array

FRV Final Rendezvous

FRW Far Red Wing • Francistown (Botswana) (location identifier) • Friedmann - Robertson - Walker (model)

FRWAY Freeway

FRWB Fully Reusable Winged Booster

FRWF Forecast Wind Factor

Frwk Framework

Frwy Freeway

FRWY Freeway

Fry Freeway

FRY Ferry • Former Republic of Yugoslavia • Fryeburg (USA-ME) (location identifier)

FRRY Ferry

Frz Freeze

FRZ Freeze • Fritzlar (Germany) (location identifier)

FRZN Frozen

Frzr Freezer

FRZR Freezer

fs Facsimile • Foot-Second • Functional Schematic

F/s Frames per Second

F/S Full Size

FS Fachschule • Facial Synthesis • Factor of Safety • Factor Space • Fail Safe • Fail to Sync • Faint Standard (star) • Faire Suivre • Far Side • Fast Screening • Fast Slew • Fast Switching • Fault Summary • Fax Software • Feasibility Study • Federal Specification • Federal Standard • Federal Survey • Fermi Surface • Fernschreiben • Ferrovie dello Stato (Italy) • Festschrift • Field Security • Field Separator • Field Service • Field Standard • Field Station • Field Stop • Field System • File Segment • File Separator • File Server • File Start • File System • Fill Start • Final Settlement • Final Speed • Financial Secretary • Financial Section • Financial Statement • Fine Structure • Finger Server • Finish to Start • Fin-Stabilized • Fire Station • Fire Suppression • First Stage • First Step • Fiscal Service • Fixed Service • Fixed Silo • Flexible Sustainment • Flight Safety • Flight Segment • Flight Simulator • Flight Spare • Flight Standard • Flight Structure • Flight System • Floating Sign • Float Switch • Fluid System • Flux Sheet • Flying Saucer • Flying Status • Foreign Service • Forest Service • Forward Shell • Forward Shock • Fraction of Savings • Frame Scan • Frame Status • Frame Switching • Franc Suisse • Freon Servicer • Frequency Shift • Frequency Stability • Frequency Standard • Frequency Switching • Frontal Strength • Full Scale • Full service • Full Size • Full Standard • Functional Schedule • Functional Schematic • Functional Selector • Functional Specification • Function Set • Function Specification • Fuselage Station • Future System

FSA Family Separation Allowance • Farm Services Agency • Fast Simulated Annealing • Federal Security Agency (USA) • Field Support Office • Final Site Acceptance • Final Stage Acceptance • Financial Services Authority (USA) • Financial Supervisory Agency (Japan) • Firewall Switching Agent • Finite State Automation • Flexible Solar Array • Fonction de Sélection Angulaire • Force Structure Assessment • Foreign Service Allowance • Foreign Service Availability • Foreign Systems Acquisition • Formal Safety Assessment • Fournisseur de Systèmes Applicatifs • Free Support Area • Frequency Stability Analyzer • Frequency Standard Assembly • Fuel Storage Area • Functional Support Agreement • Functional System Analyzer • Functional System Application • Functional Systems Audit

FSAA Flight Simulator for Advanced Aircraft

FSAD Force Structuring Analysis Division

FSAM Factor Space Access Method • Federal Storage Area Management • Fine Sun Acquisition Mode • Flexible Solar-Array Mechanism

FSAN Full Service Access Network

FSAO Family Services and Assistance Officer

FSAR Final Safety Analysis Report

FSAS Flight Service Automation System • Fuel Savings Advisory System

FSAT Field Site Acceptance Test • Full-Scale Aerial Target • Functional Strong Approximation Theorem

FSB Formation Scientifique de Base • Frequency Selective Bolometer • Frontside Bus

Fsbl Feasible • Fusible

FSc Full Scale

FSC Facility Security Clearance • Faint-Source Catalogue • Fairchild Space Co. • Fault Simulation Comparator • Federal Stock Classification • Federal Supply Catalog (USA) • Federal Supply Category (USA) • Federal Supply Classification (USA) • Federal Supply Code (USA) • Federal Supreme Court • Fernbank Science Center (USA-GA) • Figari (France) (location identifier) • File Sequence Code • Final Systems Check • Finance Subcommittee • Financial Supervisory Commission • FIRST Science Centre • First-Stage Conduit • Fixed Silicon Capacitor • Flat-Spectrum Component • Flight System Control • Florida Southern College (USA-FL) • Fluid Storage Container • Foreign Service Credits • Forest Stewardship Council • Frame Synchronization Clock • Frequency Shift Converter • Frequent Set Counting • Fuel Sulfur Content • Fuel Systems Capability • Full Spectrum Command • Full Systems Capable • Functional Standing Committee

FSCAP Flight Safety Critical Aircraft Part

FSCB Fielded Software Control Board

FSCC Facility Security Clearance Certificate • Fastbus Smart Crate Controller • First-Stage Conduit Container

FSCI Frequency Space Characteristic Impedance

FSCL Frequency Sensitive Competitive Learning

FSCM Federal Supply Classification Manufacturer • Federal Supply Code for Manufacturers (USA) (*)

FSCP Feasibility Study Change Proposal • Fire Sensor Control Panel • Flight safety Critical Part • Flight Support Computer Program

FSCR Faint Source Catalogue Reject (file) • Final System Concept Review

FSCRC Flow-on Small Computer Requirements Contract

FSCS Flight Simulation Computing System

FSCU Frequency Select Control Unit

FSD Feasibility Study Document • File System Descriptor • File System Driver • Financial Services Department • First Story Detection • Fixed Shroud Duplex • Fixed Shutter Dome • Flight Simulation Division • Fonds Social de Développement • Fuel Supply Depot • Full-Scale Deflection • Full-Scale Development • Fully Significant Difference • Functional Sequence Diagram • Functional

Statement Document • Functional System Description • Sioux Falls (USA-SD) (location identifier)

FSDA Fastbus Silicon Data Accumulator

FSDB Faint-Source Database • Flare Stars Database

FSDC Federal Statistical Data Center (USA) • Fiber-Supported Droplet Combustion

FSDIE Fonds de Solidarité et de Développement des Initiatives Étudiantes

FSDKO Formalisation of Satellite Design Knowledge for Operations

FSDO Flight Standards District Office • Flight Standards District Officer

FSD&P Flight Systems Design & Performance

FSDP First Sharp Diffraction Peak • Flight Systems Design and Performance

FSDPS Flight Service Data Processing System

FSDPSK Filtered Symmetric Differential Phase-Shift Keying

FSDR Final Site Description Review • Final System Design Review • Functional System Design Requirement

FSDS File System Descriptor Set • Foreign Satellite Data Service • Foreign Satellite Data Support

FSDT Functional System Design Team

FSDU Frame Synchronization and Decommutation Unit • Frequency Synthesizer and Distribution Unit

Fse Française

FSE Facilities System Engineer • Facility Support Element • Facility Support Equipment • Factory Support Equipment • Fast Software Encryption • Feuille de Soins Électronique • Field Support Equipment • Fill Start Entry • Final State Effect • Financial Services Authority (UK) • Flight Segment Engineer • Flight Simulation Engineer • Flight Support Equipment • Flight Systems Engineering • Fondo Social Europeo • Fonds Social Européen • Fonds Structurel Européen • Foundation of Software Engineering (USA-CA) • Foundations of Software Engineering (conference) • Foyer Socio-Éducatif • Fractionally Spaced Equalizer • Full-Scale Engineering • Full-Screen Edit • Full-Screen Editor • Functional System Element

FSEA Fellow of the Society of Engineering Associates

FSEC Fairchild Space and Electronics Co. • Federal Software Exchange Catalog • Federal Software Exchange Center (USA)

FSECO First-Stage Engine Cutoff

FSED Full-Scale Engineering Development

FSEN Future Small Extension Node

FSEP Federal Software Exchange Program (USA)

FSET FIRE Science Experiment Team

FSF Filtered Science File • First Static Firing • Fixed Sequence Format • Fragmented Signature File • Free Software Foundation • Front Surface Flat • Fundamental System of Functions

FSFDP Free Standing and Film Deposited Photodiode

FSFM Full Screen - Full Motion

FSFS Fuzzy Sets and Fuzzy Systems (conference)

FSG Family Support Group • Federal Stock Group • Federal Supply Group • Flexible Space Garment • Forschungseinrichtung Satellitengeodäsie • Full Self-Gravity • Fully Self-Gravitating • Funding Subgroup • Funding Subject Group

FSGC Florida Space Grant Consortium (USA-FL)

FSGH Fused Sphere Guided Homotopy

FSGS Flare/Shallow Glide Slope

FSH First-Stage Hydraulics • Fixed Superhard • Front-Side Halo

FSI Facility-class Science Instrument (SOFIA) • Federal Stock Item (USA) • Final State Interaction • Final Systems Installation • Flight Simulator Instructor • Flight Software Integration • Focus Software Inc. (USA-AZ) • Foreign Service Institute • Fort Sill - Henry Post (USA-OK) (location identifier) • Forum sur la Société de l'Information • Functionally Significant Item • Functional Simulation Installation • Functional System Interface

FSIC Feature Space Iterative Clustering • Forward Sensor Interface and Control

FSII Fuel System Icing Inhibitor

FSIM Functional Simulator

FSIOP File Server Input/Output Processor

FSIP Flight Standardisation Improvement Programme

FSIPM Formation Supérieure d'Ingénieurs en Physique des Matériaux

FSIRR Filter Scanning Infrared Radiometer

FSIS First-Stage Ignition System • Flight Safety Information System • Functional Support Information System

FSIT Formation Supérieure d'Ingénieurs en Télécommunications

FSIWG Flight System Interface Working Group

FSJ Feedback Summing Junction

FSK Feuerwehrschlüsselkasten • Fort Scott (USA-KS) (location identifier) • Frequency Shift Key • Frequency Shift Keying

FSKD Fuzzy Systems and Knowledge Discovery (conference)

FSKM Fuzzy Segmental K-Means

FSL Flexible Service Logic • Flight Simulation Laboratory • Flight Systems Laboratory • Fluid Science Laboratory • Fonds de Solidarité Locale (France) • Forecast Systems Laboratory • Foreign Science Library • Foreign Service Local • Formal Semantic Language • Fossil Downs (Australia-WA) (location identifier) • Full System Listing • Fundamental Straight Line

FSLLN Functional Strong Law of Large Numbers

FSLP First Spacelab Payload

FSLT First Sea Level Test

FSM Fast-Steering Mirror • Fédération Syndicale Mondiale • Field-Strength Meter • Fine Steering Mirror • Finite-State Machine • Firmware Support Manual • First Servicing Mission • First-Stage Motor • First Surface Mirror • Flight Schedule Monitor • Flight Simulator Mode • Flight Simulator Model • Flight Status Message • Flight System Mockup • Fluid Science Laboratory • Flying-Spot Microscope • Flying-Spot Microscopy • Food Service Management • Formal Specification Maturity • Fort Smith (USA-AR) (location identifier) • Forward Scatter Meter • Free Speech Movement • Frequency Shift Modulation • Frequency-Switched Mode • Fuel Supply Module • Functional Surveillance Manager • Micronesia (ISO country code) • Micronésie (code ISO de pays)

FSMA Functionally-Supported Maintenance Activity

FSMC First-Stage Motor Container

FSMMS Frequency Stability Module Management System

FSMO Flexible Single Master Operation

FSMOS File System Managed Object Specification

FSMR Fixed Simultaneous Multibeam Radar

FSMS File Storage Management System

FSMSC Federal Software Management Support Center (USA)

FSMWI Free Space Microwave Interferometer

FSN Federal Stock Number (USA) • Federation Stock Number • File System Navigator • Fiscal Station Number • Foreign Service National • Fort Sheridan - Haley (USA-IL) (location identifier) • French-Speaking Nation • Frequency Subset Number • Full Service Network

FSNCO Flight Safety Non-Commissioned Officer

FSO Facility Security Office • Facility Security Officer • Field Support Office • Flight Safety Officer • Foreign Service Officer • Frequency Sweep Oscillator • Fuel Supply Office • Full-Scale Output • Functional Supplementary Objective

FSOC Fairchild Satellite Operation Center • Flight Safety Officer Course

FSOH Flight Support Operations Handbook

FSOP Dield Standard Operating Procedure

FSOQ Frequency Shift Offset Quadrature

FSP Facility Security Profile • Facility Support Plan • Fault Summary Page • File Server Process • File Service Protocol • Flat Salary Payroll • Flight Safety Plan • Flight Scheduling Precedence • Flight Strip Printer • Food Stamp Program • Foreign Service Pay • Forschungschwerpunkt • Free Scientific Publishing • Frequency Shift Pulsing • Fuel Storage Point • Full-Scale Production • Full-Spectrum Processing • Functionality - Security - Performance • Saint-Pierre (Saint-Pierre & Miquelon) (location identifier)

FSPC Federal Science Policy Council (USA)

FSPD Freeze Speed Parameter

FSPL Free Space Path Loss

FSPM Formal Security Policy Model

FSQ Flat-Spectrum Quasar • Full Spectrum Quantization

FSQL Fuzzy Structured Query Language

FSR Facilities Support Requirement • Facultaire Studenraad • Faint Source Reject (file) • Fast Slew Rate • Feedback Shift Register • Fermi Selection Rule • Fielded System Review • Field Service Regulation • Field Service Representative • Final Software Review • Final System Release • Final System Review • Final System Run • Fin Stabilized Rocket • Fixed Sample Rate • Flight Safety Research • Flight Safety Routine • Flight Simulation Report • Flight Solar Reflectometer • Flight Specific Requirements • Flight Summary Report • Flight Support Request • Flight Support Room • Flux Sensitive Resistor • Folio Search and Replace • Force Sensitive Resistor • Frame Synchronized Ring • Free Spectral Range • Free System Resources • Frequency Scan Radar • Frequency Shift Receiver • Frequency Shift Reflector • Frozen Student Record • Full-Scale Record • Fund Summary Record

FSRA Fund Summary Record Address

FSRB Flight Safety Review Board

FSRD Federal Support of Research and Development (USA)

FSRG Feedback Shift Register Generator

FSRL Flat-Spectrum Radio Loud (quasar)

FSRM First-Stage Rocket Motor

FSRQ Flat-Spectrum Radio Quasar

FSRR Flight Software Readiness Review • Flight System Readiness Review • Flight System Recording System

FSRS Flat Spectrum Radio Source • Flight System Recording System • Frequency Selective Receiver System • Functional Security Requirements Specification

FSRT Firm Scheduled Return Time • Flight Systems Redundancy Test

FS&S Flight Systems & Servicing

FSS Facilities and Support Services • Faint-Source Survey • Feature Subset Selection • Federal Security Service (Russia) • Federal Supply Schedule (USA) • Federal Supply Service (USA) • Federal Supply System (USA) • Field Spectrometer System • Field Storage Site • Fine Sun Sensor • Fin Stabilizer System • Fire Suppression System • First-Stage Separation • Fixed Satellite Service • Fixed Service Structure • Flight Safety System • Flight Service Station • Flight Servicing System • Flight Support Station • Flight Support Structure • Flight Support System • Flight Systems Simulator • FOC Safing Sequence • Fokker Space and Systems • Font Selection Scheme • Fornax Spectroscopic Survey • Forres Kinloss (UK-Scotland) (location identifier) • Forward Supply System • Frame Storage System • Freitag - Samstag - Sonntag • Frequency Selective Surface • Frequency Surveillance Station • Frequency Surveillance System • Friday - Saturday - Sunday • Full-Scale Station • Full-text Storage System • Fully Separate Subsidiary • Functional Subsystem • Functional Support Services • Functional System Specification

FSSC Fellows and Students Selection Committee (ESO) • Field Software Support Center • Frequency Shifted Subcarrier

FSSD First-Stage Separation Device • Foreign Service Selection Date

FSSDE Formalisation of Standards for Satellite Data Exchange

FSSE Fixed Segment Support Equipment • Fixed Site Support Equipment • Flight Sensor Stimuli Equipment • Forward Service Support Element

FSSG Forward Service Support Group

FSSM Flexible Second Surface Mirror • Formal Specifications Strategies Maturity

Fssn Fission

FSSO Flight Science Support Office

FSSP Forward Scattering Spectrometer Probe • Fuel System Supply Point

FSSR Flight System Software Request • Flight System Software Requirement • Functional Subsystem Software Requirement

FSSS Future Strategic Systems Study

FSSTD Financing System for Science and Technology for Development (UNO)

FS&T Federal Science & Technology

FST Facility Science Team • Field Service Team • Field Star Table • Field Support Team • Field Support Terminal • File Status Table • Fine Star Tracker • Finite Sampling Time • Finite State Technology • FIRE Science Team • Fixed Service Tower • Flat Square Tube • Flight Support Team • Fort Stockton - Pecos County (USA-TX) (location identifier) • Forward-pointing Star Tracker • Forward Support Team • Free Space Transfer • French Summer Time • Frequency Shift Transmission • Frequency Shift Tuning • Full-Scale Test • Full-Scale Tunnel • Full-Scale Turbulence • Full Spectrum Theory • Full Spectrum of Turbulence (theory) • Functional Simulator and Translator • Functional System Test

FSTB Flight System Testbed

FSTC Financial Services Technology Consortium • Foreign Science and Technology Center (USA)

FSTD Flight Simulation Test Data

FSTE Factory Special Test Equipment

FSTI Federal Scientific and Technical Information (USA)

FSTJ Fujitsu Scientific Technical Journal (periodical code)

Fstnr Fastener

FSTS Federal Secure Telephone Service (USA) • Federal Secure Telephone System (USA) • Frequency Standard Test Set • Future Space Transportation System

FST&TCS Foundations of Software Technology & Theoretical Computer Science (conference)

FSTV Full-Scale Test Vehicle

fSU Former Soviet Union

FSU Fast Service Upgrade • Finance Support Unit • FLIR Sensor Unit • Florida State University (USA-FL) • Fonds de Solidarité Universitaire • Former Soviet Union • Fort Sumner (USA-NM) (location identifier) • Forward Support Unit • Freja System Unit • Freon Servicing Unit • Frequency Synthesizer Unit • Friedrich-Schiller-Universität (Germany) • Fringe Sensing Unit • Fringe Sensor Unit •

Frostburg State University (USA-MD) • Full-Scale Unit • Functional Storage Unit

FSV Final Stage Vehicle • Flight Space Vehicle • Frequency Selective Voltmeter

FSVS Faint Sky Variability Survey • Future Secure Voice System

FSW Flight Software • Forward Swept Wings • Friction Stir Welding

FSWG Flight Science Working Group • Flight Systems Working Group

FSY Factor of Safety Yield

FSz Full Size

FS3 Future Security Strategy Study

ft Feet • Fiat • Foot

Ft Feet • Forint (currency) • Fort

F/T Full Time

F&T Fire & Theft • Fuel & Transportation

FT Factory Test • Failure Tolerance • Far Term • Fast Track • Fast Train • Faulker Telescope (USA-HI) • Fault Tolerance • Feed Through • Feet • Field Terminator • Field Test • File Transfer • File Transferred • Filing Time • Financial Times • Fine Tracker • Fischer-Tropsch • Fixed Transmitter • Flight Team • Flight Test • Flight Time • Flow Tube • Flux Tube • Follow Through • Foot • Formal Training • Fort • Fourier Tachometer • Fourier Transform • Fraction of Time • France Telecom • Free Troposphere • Frente • Frequency Tracker • Frequent Traveller • Full Terms • Full Text • Full Time • Full-Tracked • Fully Tracked • Functional Test

FTA Failure To Appear • Fast Time Analysis • Fatigue Test Article • Fault Tree Analysis • Flight Test Article • Flip-Tick Architecture • Free-Trade Agreement • Futuna Island (Vanuatu) (location identifier)

FTAA Free Trade Area of the Americas

FTAG Field Theoretical Aspects of Gravity (workshop)

FTAM File Transfer, Access and Maintenance • File Transfer, Access and Management (*) • File Transfer, Access and Manipulation • File Transfer and Access Method • File Transfer And Management • File Telecommunications Access Method • File Transfer and Access Method • File Transfer And Management

FTAS Fast Time Analyzer System

FTB Film Transfer Boom • FITS Table Browser • Flying Testbed • Frequency Time Base • Front-To-Back (algorithm)

FT&C Formal Training & Certification

FTC Facilities Technical Criteria • Fast Time Constant • Federal Trade Commission (USA) • Fiber to the Curb • Flight Test Center • Flight Test Conductor • Florida Test Center • Formal Training and Certification • Fourier Transform Collimator • Fractal Transform Compression • Frequency Time Control

FTCCD Frame Transfer Charge Coupled Device

FTCP Flight Test Change Proposal

FTCR Functional Test Change Request

FTCS Fault-Tolerant Computing System • Forward Time Centered Space

Ftd Fitted

FTD Flight Test Direction • Flight Test Directive • Flight Test Director • Flight Training Device • Foreign Technical Department • Foreign Technology Division • Formal Technical Document • Foundational Technical Development • Freight Traffic Division • Functional Test Data • Functional Test Demonstration • Functional Track Directory

FTDMA Frequency and Time Division Multiple Access

FTDR Flight Test Data Recorder

FTDS Formal Training Data System

Fte Fonte

FTE Facility Training Equipment • Factory Test Equipment • Flight Test Engineer • Flight Test Equipment • Flight Test Evaluation • Flux Transfer Event • Fonte • Forced Test End • Forschung und Technologische Entwicklung • Full-Time Employee • Full-Time Employment • Full-Time Equivalent • Functional Test Equipment

FTEC Fluid and Thermal Energy Conversion (conference)

FTEE Full-Time Equivalent Employee

FTF Face To Face • Federal Telecommunications Fund (USA) • Field Test Facility • File Transfer Facility • Flared Tube Fitting • Floating Track Following • Functional Test Flight

FTFD Field Test Force Director

Ftg Fitting • Footing

FTG Final Trunk Group

FTGE Frontage

fth Fathom

FTH Fourier Transform Heterodyne

fthm Fathom

FTHR Farther • Further

FTI Fitiuta (American Samoa) (location identifier) • France Télécom Interactive • Frequency Time Indicator • Frequency Time Intensity

FTIP Flight Test Interface Panel

FTIR Fourier Transform Interferometer • Fourier Transform Interferometer Radiometer • Fourier Transform in the Infrared

FTIS Flight Test Instrumentation System • Fourier Transform Infrared Spectrometer • Fourier Transform Infrared Spectroscopy • Fourier Transform Interferometer Spectrometer • Function Test Information Sheet

FTIU Flight Test Interface Unit

FTK Fort Knox - Godman (USA-KY) (location identifier) • Freight Ton Kilometer

FTL Farran Technology Ltd. (Ireland) • Faster Than Light • Federal Telecommunications Laboratory (USA) • Food - Transportation - Lodging • Fortuna Ledge (USA-AK) (location identifier) • Full-Term License

FTLP Fourier Transformed Line Profile

FTLRS French Transportable Laser Ranging System

FTLS Final Top Level Statistics • Formal Top-Level Specification

FTM File Transfer Management • Flat Technology Monitor • Flat Tension Mask • Flight Test Manual • Flight Training Mission • Fonction de Transfert de Modulation • Force/Torque Module • FOS Telemetry • Fourier Transform Microwave (spectroscopy) • Freight Ton-Mile • Frequency Time Modulation • Full-Time Manning

FTMC Fault-Tolerant Memory Control • Frequency and Time Measurement Counter

FTMP Fault-Tolerant Multiprocessor

FTMS Flexible Test and Measurement Systems

FTN Fiber Transmission Network

FTNEA File Transfer Network Architecture

FTO Fellow Traveller Orbit • Final Technical Offer • Flight Test Objective • Flight Test Operation • Flight Training Organization • Flying Training Organization • Foreign Technology Office • Foreign Training Officer • Fourier Transform Operator • Functional Test Objective

FTOC Future Table Of Contents

FTOH Flight Team Operations Handbook • Flight Test Operations Handbook

FTOS Flight Termination Ordnance System

FTP Fault-Tolerant Processor • Federal Test Procedure • Fiber Termination Panel • Field Task Proposal • File

Transfer Program • File Transfer Protocol (*) • File Transport Protocol • First-order Theorem Proving (workshop) • Fixed-Term Plan • Fixed Throttle Point • Flare Time Profile • Flight Test Plan • Flight Test Procedure • Flight Test Program • Foiled Twisted Pair • Full Throttle Position • Full-Time Permanent • Full-Time Personnel • Functional Test Procedure • Functional Test Program • Functional Test Progress • Function Test Progress

FTPD File Transfer Protocol Daemon

FTPF Flight and Training Planning Facility

FTPP Fault-Tolerant Parallel Processor

FTR Federal Telecommunications Recommendation (USA) • Federal Travel Regulation (USA) • Fermi-surface Traversal Resonance • Final Technical Report • Final Test Rack • Financial Transaction Register • Flight Termination Report • Flight Test Report • Flight Test Requirement • Flow-Tube Reactor • Formal Technical Review • Functional Test Report • Functional Test Requirement • Functional Throughput Rate

FTRB Flight Test Review Board

FTRC Federal Telecommunications Records Center (USA)

FTRD Flight Test Requirements Document • Functional Test Requirements Document

FTRIA Flow and Temperature Removable Instrument Assembly

FTRM Flight Test Request Memorandum

FTRQ Fine Tuning Request

FTRTFT Formal Techniques in Real-Time and Fault-Tolerant (systems) (symposium)

ft/s Feet per Second

FTS Far-Term Study • Far-Term System • Fault Tolerance System • Federal Technology Service (USA) • Federal Telecommunications Service (USA) • Federal Telecommunications System (USA) • Federal Telephone System (USA) • Fiber-optic Test Set • Fiducial Transfer System • Field Test Set • File Transfer Service • File Transfer System • Fine Track Sensor • Flight Telemetry Subsystem • Flight Telerobotic Servicer • Flight Termination System • Flight Test Station • Flight Test Support • Flight Test System • Foot Switch • Fourier Transform Spectrograph • Fourier Transform Spectrometer • Fourier Transform Spectrometry • Fourier Transform Spectroscopy • Fourier Transform System • Free Text Search • French Technology Survey • Frequency and Timing System • Frequency Time Schedule • Frequency Time Standard • Full-Time Support • Full Turbulence Simulation • Full Turbulence Spectrum • Functional Test Specification

FTSA Fault-Tolerant System Architecture • Flight Termination System Adapter • Flight Termination System Approval

FTSC Fault-Tolerant Spacecraft Computer • Federal Telecommunication Standards Committee (USA)

FTSDBM Fourier Transform Spectrometer Database Management

FTSDTF Flight Telerobotic Servicer Demonstration Test Flight

FTSE Finite Temperature Schrödinger Equation

FTSP Federal Telecommunication Standards Program (USA)

FTSS Flight Test Simulation Station

FTT Field Test Telescope • File Transfer Timeout • Fischer-Tropsch Type • Fractal Transform Template • Full-Time Temporary • Fully Threaded Tree

FttA Fiber to the Amplifier

FTTA Federal Technology Transfer Act (USA) • Fiber To The Amplifier

FttB Fiber to the Building

FTTB Fiber To The Building

FttC Fiber to the Curb

FTTC Fiber To The Curb

FTTCab Fiber To The Cabinet

FttD Fiber to the Desk

FTTD Fiber To The Desk • Full-Time Training Duty

FttE Fiber to the Exchange

FTTE Fiber To The Exchange

FttH Fiber to the Home

FTTH Fiber To The Home

FttL Fiber to the Loop

FTTL Fiber To The Loop

FttO Fiber to the Office

FTTO Fiber To The Office

FTTP Full-Time Temporary Personnel

Fttr Fitter

FttR Fiber to the Radio

FTTR Fiber To The Radio

FttT Fiber to the Town

FTTT Fiber To The Terminal • Fiber To The Town

FTU FITS Translation Utility • Flight Test Unit • Florida Technological University (USA-FL) • Formal Training Unit • Fort Dauphin (Madagascar) (location identifier) • Freeman Time Unit • Freight Terminal Unit • Frequency Transfer Unit • Fuel Transfer Unit

FTUS Full-Time Unit Support

FTUSS Full-Time Unit Support System

FTV Flight Television • Flight Test Vehicle • Functional Technology Validation • Functional Test Vehicle • Masvingo (Zimbabwe) (location identifier)

FTVE Functional Technology Validation Experiment

FTW Fort Worth - Meacham (USA-TX) (location identifier) • Forward Traveling Wave • Future Technology Workshop

FTWG Flight Test Working Group

FTX Fault-Tolerant Unix • Field Training Exercise • Owando (Congo) (location identifier)

Fty Factory

FTY Atlanta - Fulton County (USA-GA) (location identifier)

FTZ Fernmeldetechnisches Zentralamt (Germany) • Forschungs- und Technologiezentrum • Free Trade Zone

Fu Fuel Mass • Fuse

FU Family Unit • Field Unit • Flight Unit • Follow-Up • Fouled Up • Freie Universität • Fuel • Functional User • Fuse • Fuze

FUA Follow-Up Amplifier • Fuel Use Act (USA)

FUAC Functional Area Code

FUACE Fédération Universelle des Associations Chrétiennes d'Étudiants

FUB Facility Utilization Board • Forward Utility Bridge • Fulleborn (Papua New Guinea) (location identifier)

FUBAR Fouled Up Beyond All Recognition • Fouled Up Beyond All Repair

FUCAM Faculté Universitaire Catholique de Mons (Belgium)

FUD Fear - Uncertainty - Doubt • Field Use Descriptor • Field Use Designator • First-Use Date • Follow-Up Detector

FUDD Follow-Up Detection Device • Functional Description Document

FUDR Failure and Usage Data Report

FUE First Unit Equipped • Fuerteventura (Canary Islands) (location identifier) • Full-Up Evaluation • Puerto del Rosario (Canary Islands) (location identifier)

FUEGOS Fibre Unit for European General Optical Spectroscopy

FUES Follow-up and Evaluation Section

FUFOR Fund for Unidentified Flying Object Research

FuFSt Funkfeststation

FUG Fuyang (China-PRC) (location identifier)

FUGI Future of Global Interdependence

Fuhld Fuseholder

FUI Federal Unemployment Insurance • File Update Information • FOS User Interface

FUJ Fukue (Japan) (location identifier)

FUK Fukuoka (Japan) (location identifier)

Ful Fulcrum

FUL Fondation Universitaire Luxembourgeoise (Belgium) • Fullerton (USA-CA) (location identifier)

FULBI Fédération des Utilisateurs de Logiciels pour Bibliothèques

FUM File Utilities Module • Fuma (Papua New Guinea) (location identifier) • Functional User's Manual

FUMO Functional Model

FUN Funafuti (Tuvalu) (location identifier)

Func Functional

FUNC Function • Functional

FUNCT Functional

Fund Fundación • Fundador • Fundamental

FUND Fundação

FUNDAP Fundação do Desenvolvimento Administrativo (Brazil)

FUNDP Facultés Universitaires Notre-Dame de la Paix (Belgium)

FUNET Finnish University Network

FUNI Frame-based User-to-Network Interface

Funl Funnel

FUNY Free University of New York (USA-NY)

FUO Fever of Unknown Origin • Follow-Up Output • Fuoshan (China-PRC) (location identifier)

FUOR FU Orionis (-type object)

FUORS FU Orionis Star

FUP Facility Utilization Plan

FUPAZ Frente Unido Pro Astronomía Zuliana (Venezuela)

Fur Furlong • Further

FUR Follow-Up Report • Furlong

FuRE Funkrufempfänger

FuRK Funkrufkonzentrator

Furl Furlough

Furn Furnace • Furnish • Furnished • Furniture

FURS Follow-Up Reporting System

FuRSe Funkrufsender

FuRVSt Funkrufvermittlungstelle

Fus Fuselage

FUS Focused Ultrasound • Focused Ultrasound Surgery • Full-Use Standard • Fuselage

FUSE Far-Ultraviolet Spectroscopic Explorer • Future Ubiquitous Service Environment

FUSL Facultés Universitaires Saint-Louis (Belgium)

Fuslg Fuselage

FUSLG Fuselage

FUSP Far-Ultraviolet Spectropolarimeter

FUSRAP Formerly Utilized Sites Remedial Action Program

Fut Future • Futuro

FUT Flight Unit Test • Futuna (Wallis & Futuna Islands) (location identifier)

FUTA Federal Unemployment Tax Act (USA)

FuTelG Funktelefongerät

Futur Future Spring (periodical code)

FuÜ Funkübertragung

FÜV Fernmelde - Überwachungs - Verordnung

FUV Far Ultraviolet

FUVITA Far Ultraviolet Imaging Telescope Array

FUVOP Far Ultraviolet Objective Prism

FuVSt Funkvermittlungstelle

fv File Viewer • Folio Verso

fV Femtovolt

F/V Fishing Vessel • Frequency/Voltage

FV Fachverband • Festverbindung • Fiducial Volume • Field Validation • Finite Volume • Fix Validator • Flight Vehicle • Flight Version • Flow Valve • Flux Valve • Front View • Full Voltage • Functional Validation

FVAP Federal Voting Assistance Program (USA)

FVAS Fox Valley Astronomical Society (USA-IL) • Fraser Valley Astronomical Society (Canada-BC)

FVC Forced Vital Capacity • Frequency-to-Voltage Converter

FVDR Flightdeck Video Data Recorder

FVDS Fuel Vent and Drain System

FVE Finite-Volume Element

FVF First Vertical Flight

FVIPS First Virtual Internet Payment System

FVK Faserverstärkte Kunststoffe

FVM Facility Verification Module • Finite-Volume Method

FVMS Fluid Volume Measurement System

FVP Feasibility Validation Program • Feedback Verification Procedure • Flash Vacuum Pyrolysis • Fluid Velocity Potential

FVQ Fuzzy Vector Quantization

FVS Flight Vehicle Structure • Flux Vector Splitting • Foreign Visits System

FVSat Festverbindung über Satellit

FVSEP Formal Verification System Evaluation Program

FVSL Full Viscous Shock Layer

FVSR File Validation Status Report

FVSS Fuel Vapor Scrubber System

FVSt Fernvermittlungstelle

FVSW Fox Valley Skywatchers (USA-IL)

FVT Field Verification Test • Final Version Test • Flash Vacuum Thermolysis • Full Video Translation • Full Volume Test • Functional Validation Test

FVU Functional Verification Unit

FVV Facility Verification Vehicle • Feste Virtuelle Verbindung • Fill and Vent Valve • First Vehicle Verification

FVW Fortran Visual Workbench

Fw Fußweg

F/W Filter Wheel • Firewall • Firmware • Freeware

FW Face Width • Falle Wave • Fernsprech-Wählnetz • Field Weakening • Filter Wheel • Firewall • Firmware • Fiscal Week • Fixed Wing • Flag Word • Forecaster Workstation • Formula Weight • Forschung im Weltraum • Fourier-Wiener • Freeware • Fresh Water • Fringeworthy • Fuel Wasting • Full Wave • Full Width

FW&A Fraud, Waste & Abuse

FWA Federal Works Agency • Feuerwehranfahrt • First Word Address • Fixed-Wing Aircraft • Fixed Wireless Access • Fluorescent Whitening Agent • Fort Wayne - Baer (USA-IN) (location identifier) • Forward Wave Amplifier • Fraud - Waste - Abuse • Full-Wave Amplifier

FWAS Fort Wayne Astronomical Society (USA-IN) • Fort Worth Astronomical Society (USA-TX)

F'way Farmway

FWB Farenheit Wet Bulb • Four Wheel Brake • Four Wheel Braking

FWBA Full-Wave Balanced Amplifier

FWBR Full-Wave Bridge Rectifier

FWC Filament-Wound Case • Full Well Capacity

FWCA Fish and Wildlife Coordination Act (USA)

FWCS Flight Watch Control Station

Fwd Foreword • Forward • Forwarded

FWD Forward • Four-Wheel Drive • Free World Dialup • Front-Wheel Drive

FWDD Forwarded

Fwdg Forwarding

FWER Family-Wise Error Rate

FWF Fixed Word Format • Förderung der Wissenschaftlichen Forschung • Free Widget Foundation

FWG Facilities Working Group • Facility Working Group • Financial Working Group • FITS Working Group • Flexible Waveguide • Functional Working Group

FWGA Folded Waveguide Antenna

FWH Flexible Wire Harness • Flexible Working Hours • Fort Worth - Carswell (USA-TX) (location identifier)

FWHE Full Width at Half Energy

FWHH Full Width Half Height

FWHIP Full Width between Half-Intensity Points

FWHM Full Width at Half Maximum (*) • Full Width at Half Meridian

FWHP Full Width at Half Peak

FWI Federation of West Indies • Flash With Information • French West Indies

FWIE Fish and Wildlife Information Exchange

FWIW For What It's Worth

FWL Farewell (USA-AK) (location identifier) • Fixed Word Length • Forward Working Limit

FWM Filter Wheel Mechanism • Food Web Model • Fort William (UK-Scotland) (location identifier) • Four-Wave Mixing

FWO Facility and Waste Operations • Fonds voor Wetenschappelijk Onderzoek - Vlaanderen (Belgium)

FWOP First Window Of Protection

FWP Field Work Proposal • Framework Programme

FWPC Federal Wireless Policy Committee (USA)

FWPCA Federal Water Pollution Control Act (USA) • Federal Water Pollution Control Administration (USA)

FWPD Firewall Product Developers

FWPO Four-Wave Parametric Oscillation

FWR Fish and Wildlife Resources • Full-Wave Rectification • Full-Wave Rectifier

FWRO Federal Wireless Review Office (USA)

FWRW Fixed-Wing/Rotary-Wing

FWS Federal Wage System (USA) • Filter Wedge Spectrometer • Final Work Statement • Fish and Wildlife Service • Forward Swept Wing

FWST Flight Weight Static Test

FWT Filtered Wavelet Transform • Forum für Wissenschaft und Technik • Forward Wave Tube • Free Water Tritium • Full Wavelet Transform

FWTS Federal Wireless Telecommunications Services (USA)

FWUF Federal Wireless Users' Forum (USA)

FWV Fixed-Wing Vehicle • Full Wave

Fww Flugwetterwarte

FWW Flugwetterwarte • Food, Water and Waste

FWWM Food, Water and Waste Management

FWWMS Food, Water and Waste Management System

Fwy Freeway

FWY Freeway

FWZ Feuerwehrzufahrt

FWZI Full Width at Zero Intensity

Fx Fracture • Fractured

FX Facsimile • Foreign Exchange • France (Metropolitan_) (ISO country code) • Fridays Excepted • Metropolitan France (ISO country code)

FXCH Floating-point Exchange

Fxd Fixed

FXD Fixed

FXE Fort Lauderdale - Executive (USA-FL) (location identifier)

Fxg Fixing

FXM Flaxman Island (USA-AK) (location identifier)

FXO Cuamba (Mozambique) (location identifier) • Foreign Exchange Office

FXP Flaring X-ray Pulsar

FXR Flash X-Ray

FXS Foreign Exchange Service • Foreign Exchange Station • Foreign Exchange Subscriber • Fully Expanded System

FXT Fast X-ray Transient • Foreign Exchange Service

Fxtr Fixture

FXU Fixed-point Unit

FXX France (Metropolitan_) (ISO country code) • Metropolitan France (ISO country code)

FXY Forest City (USA-IA) (location identifier)

fy Frisian (ISO language code)

Fy Ferry

FY Financial Year • Fiscal Year • Full Year

FYA For Your Attention

FYBP Fiscal-Year Base Plan

FYC For Your Consideration

FYCP Fiscal-Year Corporate Plan • Future-Year Corporate Plan

FYDO Fiscal-Year Design Objective

FYDP Fiscal-Year Development Plan

FYE For Your Eyes • Full-Year Equivalent

FYEO For Your Eyes Only

FYF For Your File

FYI For Your Information

FYIAP Five-Year Interoperability Assurance Plan

FYIG For Your Information and Guidance

FYIP Five-Year Instrumentation Plan • Future-Year Instrumentation Program

FYJ For Your Journal

FYL For Your Log

FYM Fayetteville (USA-TN) (location identifier)

FYMOP Five-Year Master Objectives Program

FYMP Five-Year Materiel Program

FYMTP Five-Year Master Test Plan

FYN Fuyun (China-PRC) (location identifier)

FYO Find Your Own • Fiscal-Year Option

FYP Fiscal-Year Plan • Five-Year Plan • Five-Year Program

FYPA Fiscal-Year Planning Assessment

FYPB Five-Year Planning Base

FYPP Five-Year Procurement Program

FYR Fiscal Year • Former Yugoslav Republic

FYROM Former Yugoslav Republic Of Macedonia

Fys Fysica

FYS Fixed Yaw Stabilizer • For Your Signature

FYT Faya (Chad) (location identifier) • For Your Trash

FYTD Fiscal Year To Date

FYTP Five-Year Test Plan • Five-Year Test Program • Future-Year Test Plan • Future-Year Test Program

FYTQ Fiscal-Year Transition Quarter

FYU Fort Yukon (USA-AK) (location identifier)

FYV Fayetteville (USA-AR) (location identifier)

FZ Float Zone • Fracture Zone • Franc Zone • Freezing • Free Zone • French Zone • Fuze

FZDZ Freezing Drizzle

FZE Fuse

FZFG Freezing Fog

FZJ Forschungszentrum Jülich (Germany)

FzK Forschungszentrum Karlsruhe (Germany)

FZK Forschungszentrum Karlsruhe (Germany)

FZO Filton (UK-England) (location identifier)

FZP Fresnel Zone Plate

FZRA Freezing Rain

F-Zug Fernschnellzug

FZug Fernschnellzug

FZY Fuzzy

F3 Form - Fit - Function

F3I Form-Fit-Function Interface

f77 Fortran 77 (file extension)

G

g Acceleration of Free Fall • Degeneracy • Density of Vibrational Modes • Gale • Gallon • Garade • Gas • Gaseous • Gauche • Gauge • Gender • General • Generally • Genitive • Geographical • Gerade • g-factor • Gluon • Gold • Good • Government • Grade • Grain • Gram • Grand • Grav • Gravitational Acceleration • Gravity • Gravity Constant • Great • Green • Grid • Gros • Groschen (currency) • Grosse • Guide • Guinea • Gun • Osmotic Coefficient • Specific Gibbs Free Energy • Statistical Weight • Surface Gravity • Variance of Mixture Fraction • Vibrational Term

G/ Giro

G Conductance • Flexibility • Four Hundred (Roman numeral) • Gain • Galerie • Garage • Gauche • Gauss • Geld • General • Generator • Geometry • German • Germany • Gesetz • Gibbs Function • Giclas (catalog) • Giga • Giornale • Giorno • Go • Golf (international alphabet) • Golfe • Golfo • Good • Gora • Gourde (currency) • Gram • Grand • Gravitational Constant • Gravity • Green • Grid • Ground • Growth Factor • Guarani (currency) • Guernsey • Gulden (currency) • Gulf • Gunong • Gunung • Gust • Modulus of Rigidity • Newtonian Constant of Gravitation • Phase Factor • Reciprocal Lattice Vector • Shear Modulus • Spectral Type • Storm • United Kingdom (international aircraft registration code) • United Kingdom (international vehicle registration code) • Weight • 400 (Roman numeral)

ga Gaelic (ISO language code) • Gas Amplification • Glide Angle • Graphic Ammeter • Irish (ISO language code)

Ga Billion Years • Gallium • Georgia (USA)

G&A General & Accounting • General & Administration • General & Administrative • Gyro & Accelerometer

G/A General Average • Ground to Air

G-A Ground to Air

GA Gabon (ISO country code) • Gain of Antenna • Gambia • Gas-phase Affinity • Gate Array • Gated Attenuation • Gear Assembly • Gemeinschafts-Antenne • General Accounting • General Agent • General Agreement • General Alert • General Arrangement • General Assembly • General Assignment • General Authority • General Availability • General Average • General Aviation • Generating Application • Génération d'Adresse • Generic Algorithm • Genetic Algorithm • Geographic Access • Geographical Association • Geographic Analysis • Georgia (USA) • Glide Angle • Global Analysis • Global Array • Global Assessment • Go Ahead • Go Around • Government Agency • Grammaire d'Attributs • Gran Atractor • Grand Attracteur • Grant Aid • Graphics and Administration • Grapple Adapter • Great Attractor • Ground Antenna • Growing Amplitude • Gyro Assembly

GAA Gain Adjuster Adapter • General Agency Agreement • General Aviation Activity • Générateur Atomique Alpha • Government Auditing Agency • Gruppo Astrofili Astigiani (Italy) • Guamal (Colombia) (location identifier)

GAAA General Aviation Activity and Avionics

GAAC German-American Academic Council

GAAE Grupo Astrofili Arthur Eddington (Italy)

GAANN Graduate Assistance in Areas of National Need

GAAP General Aviation Action Plan • Generally Acceptable Accounting Principles • Generally Accepted Accounting Principles (*) • Geological Analysis Aid Package

GAAREX Global Atmospheric Aerosol Radiation Experiment

GAARS Global Atmospheric and Aerosol Radiation Study

GAART Government Astronomy Administration Round Table

GaAs Gallium Arsenide

Gab Gabon

GAB Gabbs (USA-NV) (location identifier) • Gabinete • Gabon (ISO country code) • Générateur Atomique Beta • Gesellschaft für Astronomische Bildung • Gewestelijk Arbeidsbureau • Group Asynchronous Browsing • Group Audio Bridging • Guichet Automatique Bancaire

GABA Gamma-Aminobutyric Acid

GABE Groupement d'Association à But Éducatif

GABEK Ganzheitliche Bewältigung erfaßter Komplexität

GAC General Advisory Committee • General Area Coverage • Geographical Area Code • Geological Association of Canada • Global Accessibility Cone • Global Area Coverage • GMRT Array Combiner • Goddard Astronomy Club (USA-MD) • Gracias (Honduras) (location identifier) • Greenbelt Astronomy Club (USA-MD) • Gross Available Capacity • Ground Attitude Control • Group Audit Coordinator • Group-Average Clustering • Grumman Aerospace Corporation (USA-NY) • Gruppo Amici del Cielo (Italy) • Gustavus Adolphus College (USA-MN)

GACCOS Galactic Anticenter CO Survey

GACE Grupo de Astronomia y Ciencias del Espacio

GACEP Global Aerosol Climatology and Effects Program

GACIAC Guidance and Control Information Analysis Center

GA&CS Ground Acquisition & Command Station

GACS Ground Acquisition and Command Station

GACU Ground Air Conditioning Unit • Ground Avionics Cooling Unit

GAD Gadsden (USA-AL) (location identifier) • Générateur Atomique Delta • Geocentric Axial Dipole • German Astronomical Directory • Gruppo Astronomia Digitale

GADB General Attribute Database

GADG Gravity or Acceleration Due to Gravity

GADL Ground-to-Air Data Link

GADO General Aviation District Office

GADS Geodata Analysis and Display System • Geographic Analysis and Display System • Global Aerosol Data Set

Gae Gaelic

G&AE General & Administrative Expenses

GAE Gabes (Tunisia) (location identifier) • General and Administration Expenses • General American English • Gibbs Adsorption Equation • Global Alfvén Eigenmode

GAEC Grumman Aircraft Engineering Corporation (USA-NY)

Gaél Gaélico

GAF Gafsa (Tunisia) (location identifier) • Generic Application Frame • Gesellschaft für Angewandte Fernerkundung (Germany) • Grupo de Acción Financiera

GAFB General Airframe Bulletin

GAFC General Airframe Change

GAFD Geophysical and Astrophysical Fluid Dynamics

GAFIA Getting Away From It All

GAFOR General Aviation Forecast

GAFPIC Goddard Astronomical Fabry-Pérot Imaging Camera

GAFS General Accounting and Finance System

G/A/G Ground to Air to Ground

GAG Gage (USA-OK) (location identifier) • Gemini Acquisition and Guiding • Générateur Atomique Gamma • Genetic Algorithm Group • Genetic Algorithms Group • Georg-Agricola-Gesellschaft (Germany) • Gross Available Generation • Gruppo Astrofili Genovesi (Italy)

GAGAS Generally Accepted Government Auditing Standard

GAGE Global Atmosphere Gases Experiment

GAH Gayndah (Australia-QLD) (location identifier)

GAHC Group-average Agglomerative Hierarchical Clustering

GAHF Grapple Adapter Handling Fixture

GAHQ Groupe Astronomique Hague Querqueville (France) (formerly GAQ)

GAI Gaithersburg - Montgomery County (USA-MD) (location identifier) • Gemeenschappelijk Antenne-Inrichting • General Accounting Instruction • Gesellschaft für Angewandte Informatik • Gibbs Adsorption Isotherm • Global Ambient Intelligence • Guaranteed Annual Income

GAIA Geographic Access Information and Analysis • Global Astrometric Interferometer for Astrophysics • Graphical Astronomy and Image Analysis

GAIIA Global Alliance of Information Industry Associations

GAIL Glide Angle Indicator Light

GAIM Global Analysis, Interpretation and Modeling

GAIN Global Analysis and Information Network • Graphic Aids for Investigating Networks • Groupe d'Aide à l'Insertion

GAIS Global Area Information System • Global Area Intelligent Search • Gross Attributable Income

GAJ Yamagata (Japan) (location identifier)

GAK Gakona (USA-AK) (location identifier) • Gemeenschappelijk Administratiekantoor • Générateur Atomique Kappa • Göteborgs Astronomiska Klubb (Sweden)

gal Gallon

Gal Galerie • Galerij • Galicia (Spain) • Gallon

GAL Galena (USA-AK) (location identifier) • Galeria • Gas Analysis Laboratory • Gate Array Logic • Générateur Atomique Lambda • Generic Array Logic • Great-Attractor-Like (object) • Grow And Learn

GALAXY General Automatic Luminosity And XY (measuring machine) • Gigabit Astronomical Large Array Cross-connect

GALCIT Guggenheim Aeronautical Laboratory - California Institute of Technology (USA-CA)

GALD Greatest Axial Linear Dimension

GALE Generic Area Limitation Environment • Genesis of Atlantic Lows Experiment

GALEX Galaxy Evolution Explorer

Gall Gallego • Gallery • Gallon • Galloway (UK-Scotland)

GALLEX Gallium Experiment

GALP Good Automated Laboratory Practice

gals Gallons

Gals Gallons

Galv Galvanic • Galvanize • Galvanized • Galvanometer

Galvnm Galvanometer

Galy Galley

Gam Gambia

GAM Gambell (USA-AK) (location identifier) • Game • Geodesic Acoustic Mode • Geographical Analysis Machine • Geostationary Airglow Monitor • GPS-Aided Munition • Graphics Access Method • Groupe d'Activation Moteur • Groupe d'Assistance Mutuelle • Gruppo Astrofili Massesi (Italy) • Gruppo Astrofili Menkalinan (Italy) • Guaranteed Annual Minimum • Guide Acquire Module

GAMA General Aircraft Manufacturers Association • General Aviation Manufacturers Association (*) • General Aviation Manufacturing Association • Graphics-Assisted Management Application • Group of Amateur Astronomers (Czech Rep.)

GAMAC Groupement pour l'Avancement des Méthodes d'Analyse des Contraintes (France)

GAMBLE Game-Based Learning

GAME Generic Applications and Monitors Environment • GEWEX Asian Monsoon Experiment • Groupe d'Étude de l'Atmosphère Météorologique • Groupe pour l'Amélioration des Méthodes d'Enseignement

GAMES Gravity And Magnetic Earthprobe Studies • Gravity And Magnetic Earth Surveyor • Gravity And Magnetic Experiment Satellite

GAMETAG Global Atmospheric Measurements Experiment on Tropospheric Aerosols and Gases

GAMM Gesellschaft für Angewandte Mathematik und Mechanik (Germany)

GAMMA General Abstract Model for Multiset Manipulation • Génération Automatique des Maillages et Méthodes d'Adaptation • GSSS Automated Measuring Machine • Guide-star Automatic Measuring Machine

GAMP Global Atmospheric Measurement Program

GAMS General Algebraic Modeling System • Guide to Available Mathematical Software

GAMTA General Aviation Manufacturers and Traders Association (UK)

GaMuM Gesellschaft angewandte Mathematik und Mechanik (periodical code)

GaN Gallium Nitride

GAN Gan Island (Maldives) (location identifier) • Generalized Activity Network • Generating and Analyzing Network • Global Area Network

GANE GPS Attitude and Navigation Experiment

GANIL Grand Accélérateur National d'Ions Lourds (France)

GAO Gemini Adaptive Optics • General Accounting Office (USA) • Générateur Atomique Oméga • Gesellschaft für Angewandte Organisation • Gothard Astrophysical Observatory (Hungary) • Government Accounting Office • Guantanamo - Los Canos (Cuba) (location identifier)

GAOES Gunma Astronomical Observatory Echelle Spectrograph

GaP Gallium Phosphide

GAP Gas Analysis Package • Gas Analysis Program • General Assembly Programme • General Aviation Propulsion • Generic Address Parameter • Generic Address Profile • Geographic Application Program • Geostationary Atmospheric Profiler • Giant Aperiodic Crystal • Global Access Program • Global Activity Plan • Glycidil Acide Polymer • GOAL Automatic Procedure • Good Astronomical Practices • Government Available Property • Graduate Assistance Program • Grant Assistance Program • Group Activation Pack • Gruppo Astrofili di Palermo (Italy) • Gruppo Astrofili Pavese (Italy) • Gruppo Astrofili Peloritani (Italy) • Gruppo Astrofili Pesarese (Italy) • Gruppo Astrofili Piceni (Italy) • Guidance-Acquisition Package • Guildhall Automation Project (UK) • Gusap (Papua New Guinea) (location identifier)

GAPA Guichet d'Affranchissement Postal Automatique

GAPC Global Alternative Propulsion Center (GM)

GAPE Great Alaska and Pacific (VLBI) Experiment

GAPEWS Graphic Air Picture Early Warning System

GAPEX Ground-based Atmospheric Profiling Experiment

GAPF Groupe d'Astronomie Populaire Flammarion (Belgium)

GApFD Geophysical and Astrophysical Fluid Dynamics (periodical code)

GAPHYOR Gas Physics Orsay (database)

GAPI Gateway Application Programming Interface • Gestion Automatisée des Personnels Intérimaires

GAPL Group Assembly Parts List

GAPM Global Atmospheric Prediction Model

GAPP Geometric Arithmetic Parallel Processor • Geometric Array Parallel Processor • Global Alliance Partnership Program • Groupe d'Aide Psychopédagogique

GAPRA Groupement Astronomique Populaire de la Région d'Antibes (France)

GApS Giga-Actions per Second • Billions of Actions per Second

GAPS Billions of Actions Per Second • Gaseous Antiparticle Spectrometer • Giga-Actions Per Second • Global Assimilation and Prediction System

GAQ Gao (Mali) (location identifier) • Groupe Astronomique de Querqueville (France) (now GAHQ)

Gar Garage

GAR Garagem • Garaina (Papua New Guinea) (location identifier) • Government Analysis Report • Groupe d'Astrophysique Relativiste • Growth Analysis and Review • Gruppo Astrofili di Rozzano (Italy) • Gruppo Astrofili Romani (Italy) • Guided Aircraft Rocket

GARAG Genetic Algorithms Research and Application Group

GARAGe Genetic Algorithms Research and Application Group

GARCH Generalized Autoregressive Conditional Heteroscedasticity

Gard Garden

GARD General Address Reading Device • Group for Advanced Receiver Development

GARE Groupement des Animateurs Responsables de Formation

gARF Grating Ancillary Response File

GARL Group Action Request List

GARP Generic Attribute Registration Protocol • Genetic Algorithm for Rule Production • Global Atmosphere Research Program • Global Atmospheric Research Program

GARR Gruppo Armonizzazione delle Reti per la Ricerca (Italy)

GARS Geological Applications of Remote Sensing • Geostationary Archival and Retrieval System • GOES Archival and Retrieval System

GARSOM Garchinger Sonnenmodell

Gart Garten

GARTEUR Group for Aeronautical Research and Technology in Europe

gas Gasoline

Gas Gasoline

GAS Gallium Arsenide • Garissa (Kenya) (location identifier) • Gas Acquisition System • Gas Analyzer System • General Aviation Simulator • Generalized Applications System • Générateur Atomique Sigma • Genesee Astronomical Society (USA-MI) • Geomagnetic Aspect Sensor • Getaway Special • Giant Air Shower • Gibraltar Astronomical Society • Gisborne Astronomical Society (New Zealand) • Gislaved Astronomiska Sällskap (Sweden) • Graduate Assessment Survey • Groupe Astronomie de Spa (Belgium) • Guilford Astronomical Society (UK-England)

GASA Graduate Astronomy Students Association

GASAT Gender And Science And Technology

GASB Governmental Accounting Standards Board

Gasc Gascoigne

GASC Graphics and Software Communication System

GASCA Generic Avionics Scalable Computing Architecture

GASCAN Getaway Special Canister

GASER Gamma-ray Laser

GASIL General Aviation Safety Information Leaflet

GASL General Activity Simulation Language

GASP Galaxy Surface Photometry • Gammas At South Pole • Gemeinsame Außen- und Sicherheitspolitik • General Activity Simulation Program • General Automated Scheduling Program • Generalized Academic Simulation Program • Georgia Astronomy in State Parks (USA-GA) • Gettysburg Automated Stellar Photometer • Global Air Safety Program • Global Air Sampling Program • Global Assimilation and Prediction System • Global Atmospheric Sampling Program • Graded Assessment in Science Project • Grand Accelerated Space Platform • Graphic Applications Subroutine Package • Gravity-Assisted Space Probe • GSSS Archive Support Project • GSSS Astrometric Support Package • Guide-star Astrometric Support Program • Guide-star system Astrometric Support Package

GASPAR Groupe Académique de Soutien et de Prévention pour les Adolescents à Risques (France)

GASPI Gimbal Angle, Surface Position Indicator

GASPRA Garden State Planetarium Resource Association

GASS GAIA System Simulator • Globus Access to Secondary Storage • Ground Analysis Subsystem

GASSATA Genetic Algorithm for Simplified Security Audit Trail Analysis

GAST Globally Averaged Surface Temperature • Greenwich Apparent Sidereal Time

GAT Gap Tallard (France) (location identifier) • General Alias Table • General Analysis Technique • General Application Template • Generalized Algebraic Translator • Gifted And Talented • Göttinger Astronomische Tage (Germany) • Government Acceptance Test • Greenwich Apparent Time • Ground-to-Air Transmission • Ground-to-Air Transmitter • Groupe d'Activation Tuyère • Grupo de Astrofísica Teórica

GATAC General Assessment Tri-dimensional Analog Computer

GATAM General Automation Telecommunication Access Method

GATAN Gesellschaft Aerosolforschung Tagung über Aerosole Naturwissenschaften (Germany)

GATB General Aptitude Test Battery

GATC Graphic Arts Technical Committee

GATD Graphical Analysis of Three-dimensional Data

GATE Galileo Telescope Environment • GARP Atlantic Tropical Experiment • General Access to Transport Extension • Generalized Algebraic Translator Extended • Generic Automatic Test Equipment • Global Atmospheric Tropical Experiment • Graphical Analysis Tool Environment

GATEC Government Acquisition Through Electronic Commerce

GateD Gateway Daemon

GATES Global Access Telehealth and Educational System • Global Access Telehealth and Education System

GATF Graphics Art Technical Foundation

GATI Grupo de Análisis y Tratamiento de la Información

GATIQ Groupe pour l'Avancement Technologique et Industriel du Québec (Canada-QC)

GATP Ground Acceptance Test Procedure • Ground Article Test Procedure

GATS General Agreement on Trade in Services • Global Automotive Telematics Standard • Ground-Air Telerobotic System • Guidance Acceptance Test Set

GATES Global Air Transportation Execution System

GATP Ground/Air Transportation Plan

GATR Ground-Air Transmitter-Receiver

GATSS Global Air Transportation Systems and Services

GATT Gate-Assisted Turnoff Thyristor • General Agreement on Tariffs and Trade • Government Acceptance Test Team

GATU Geophysical Automatic Tracker Unit

GAU Gauhati (India) (location identifier) • General Accounting Unit • Größerer Anzunehmender Unfall • Guwahati (India) (location identifier)

GAUCHO General Architecture for Unified Communication in Heterogeneous OSI environments

GAUS Geoscience Australia

GAUSS Galactic Ultrawide-angle Schmidt System • Galaktisches Ultraweitwinkel-Schmidt-System

GAV Gag Island (Indonesia) (location identifier) • Garantie Accident de la Vie • Garde A Vue • Gaveto • Gross Annual Value • Gruppo Astrofili Vesuviano (Italy) • Gruppo Astrofili Vicentini (Italy)

GAVO German Astrophysical Virtual Observatory

GAVRT Goldstone Apple Valley Radio Telescope (USA-CA)

GAW Gangaw (Myanmar) (location identifier) • Generally Associated With • Global Atmosphere Watch (*) • Global Atmospheric Watch • Guaranteed Annual Wage

GAWA German-American Writers' Association

GAWH Gruppo Astrofili "William Herschel" (Italy)

GAWV Gunstig Antwoord Wordt Verwacht

GAX Gamba (Gabon) (location identifier)

GAY Gaya (India) (location identifier) • Good As You

GAYA Great Austral Young Association

Gaz Gazetta • Gazette • Gazetteer

GAZ Guasopa (Papua New Guinea) (location identifier)

GAZS Gesamtverzeichnis ausländischer Zeitschriften und Serien

GazT Gazodinamika i Teploobmen (periodical code)

Gb Gigabit • Gilbert

G&B Gordon & Breach

G/B Gateway/Bridge • Gearbox

GB Gain Bandwidth • Gamma-ray Burst • Gauss-Bonnet • Gaussian Beam • Gebiedsbestuur • General Background • General Budget • Génie Biologique • Geologische Bundesanstalt (Austria) • Giant Branch • Gigabyte • Good Bye • Gould Belt • Governing Board • Grand Bazar • Grand Bibliothécaire • Grande Bretagne • Great Britain • Great Burst • Green Bank • Green Belly • Groot-Brittannië • Ground-Based • Group Buffer • Grundbuch • Guanabara (Brazil) • Guide Book • Royaume-Uni (code ISO de pays) • United Kingdom (international vehicle registration code) • United Kingdom (ISO country code)

GBA Alderney (international vehicle registration code) • Big Bay (Vanuatu) (location identifier) • GAS Bridge Assembly • Global Biodiversity Assessment • Global Block Adjustment • Grand Bibliothécaire Anonyme • Grundbuchamt

GBAT Government Baggage Ticket

GBB Gara Djelibet (Algeria) (location identifier) • Generic Blackboard

GBC Gasuke (Papua New Guinea) (location identifier) • Ground-Based Computer

GBCL Gravitationally-Bound Current Loop

GBCS Ground-Based Common Sensor

GBD Gamma Burst Detector • Generalized Brans-Dicke • Geometric Database • Global Burst Detector • Government Business Division • Great Bend (USA-KS) (location identifier)

GBDF Generalized Backward Differentiation Formula

GBDL Ground-Based Data Link

GB&E Gravitational Biology & Ecology

GBE Gaborone (Botswana) (location identifier) • Gödel - Bach - Escher (trilogy) • Gravitational Biology and Ecology • Ground-Based Element • Gun-Barrel Equation

GBERL Gulf Breeze Environmental Research Laboratory

GBEV Ground-Based Experimental Version

GBf Guterbahnhof

GBF Geographic Base File • Gesellschaft für Biotechnologische Forschung (Germany) • Gravitational Biology Facility • Negarbo (Papua New Guinea) (location identifier)

GBFC Ground-Based Facilities Committee (UK)

GBFEL Ground-Based Free Electron Laser

Gbg Garbage

GBG Galesburg (USA-IL) (location identifier) • Geschlossene Benutzergruppe • Guernsey (international vehicle registration code)

GBH Galactic Black Hole • Galbraith Lake (USA-AK) (location identifier) • Group Busy Hour

GBHC Galactic Black Hole Candidate

GBHRIL Ground-Based High-Resolution Imaging Laboratory

GBI Goodwin Biotechnology Inc. • Grand Bahama (Bahamas) (location identifier) • Green Bank Interferometer • Green Biomass Index • Ground-Based Interceptor

GBIC Gigabit Interface Connector • Gigabit Interface Converter (*)

GBIF Global Biodiversity Information Facility

GBII Ground-Based Infrared Instrumentation

GBIP GAP-Based Image Processor • General-Purpose Interface Bus

gBIS Graphical Issue Based Information System

GBIS Geo-Based Information System • Geographic Base Information System

GBIIS Ground-Based Infrared Instrumentation System

GBJ Jersey (international vehicle registration code) • Marie Galante (Guadeloupe) (location identifier)

GBK Gbangbatok (Sierra Leone) (location identifier)

gbl Global (file extension)

GBl Gesetzblatt

GBL Generation Breakdown List • Goulburn Island (Australia-NT) (location identifier) • Government Bill of Lading • Great Belt Link • Ground-Based Laboratory • Ground-Based Laser

GBLOC Government Bill of Lading Office Code

GBM Gamma-ray Burst Monitor • Garbaharey (Somalia) (location identifier) • GLAST Burst Monitor • Global Book Market • Isle of Man (international vehicle registration code)

GBML Genetic-Based Machine Learning • Genetics-Based Machine Learning

GBN Global Business Network • Graphite Boron Nitride

Gbno Gobierno

GBNP Global Block Numbering Plan

Gbo Grabador

GBO Baltimore Greenbelt (USA-MD) (location identifier) • Goods in Bad Order • Green Bank Observatory (USA-WV) • Ground-Based Observatory • Grundbuchordnung

Gbor Gobernador

GBP Gamboola (Australia-QLD) (location identifier) • Global Business Partnership • Group Buffer Pool • Pound Sterling (UK) (ISO currency code)

GBPC Ground-Based Programme Committee

GBPP Green Bank - Berkeley Pulsar

GBPPP Green Bank - Berkeley Pulsar Processor

G/bps Gigabits per Second

Gbps Gigabits per Second

GBR Glass Bead Rating • Governed Burst Rate • Great Barrier Reef • Great Barrington (USA-MA) (location identifier) • Green - Blue - Red • Ground-Based Radar • Ground - Balloon - Rocket • Royaume-Uni (code ISO de pays) • United Kingdom (ISO country code)

GBRF Ground-Based Radio Frequency

GBRS Ground-Based Radar System

GBRT Ground-Based Radar Terminal

Gb/s Gigabits per Second

GBS Gamma Burst Source • Geographic Base System • Global Broadcast Satellite • Global Broadcast Service • Global Broadcast System • Goal-Based Scenario • Government Bureau of Standards • Ground-Based Scanner • Ground-Based Sensor • Ground-Based Software • Ground-Based System • Ground-Beacon System • Ground-Burst Simulator • Port Fitzroy (New Zealand) (location identifier)

GBSN Ground-Based Support Navigation

GBSRN Global Baseline Surface Radiation Network

GBSS Ground-Based Surveillance System

GBT Generalized Brillouin Theorem • Green Bank Telescope (USA-WV) • Gridded Brightness Temperature • Ground-Based Telemetry

GBTS Ground-Based Training System

GBU Khashm El Girba (Sudan) (location identifier)

GBUV Geostationary Backscatter Ultraviolet

GBV Gate Breakdown Voltage • Gemeinsamer Bibliothekverband (Germany) • Gibb River (Australia-WA) (location identifier) • Graphischer Bildungsverband

GBW Guild of Book Workers (USA)

GBX Glovebox

Gbyte Gigabyte

GBZ Gibraltar (international vehicle registration code) • Great Barrier Island (New Zealand) (location identifier)

GBzG Gerlands Beiträge zur Geophysik (periodical code)

GB2 Green Bank Second Survey

Gc Gigacycle

G&C Guidance & Control

GC Gain Control • Gaining Command • Galactic Center • Galactic Cluster • Garbage Collect • Garbage Collection • Garbage Collector • Gas Chromatograph • Gas Chromatographic • Gas Chromatography • Gaseous Chromatography • General Catalog • General Comment • General Contractor • General Corpus • General Council • Generic Code • Genetic Computing • Geneva Convention • Gigacycle • Gimbal Case • Global Catalog • Global Change • Globular Cluster • Gold Coast • Good Condition • Good Conduct • Governing Council • Government Center • Government Contractor • Government Contribution • Grande Coupole • Graphics Context • Great Circle • Green Bank C • Ground Computer • Ground Contract • Ground Control • Ground Controller • Ground Cover • Groupe de Coordination • Guidance and Control • Guidance Computer • Guidance Control • Gyro Compass • Gyro Control • Gyrodine Complex

GCA Gain Control Amplifier • Galactic Center Arm • Gas Control Assembly • Geophysics Corporation of America • Gesellschaft für Computeranwendungen • Global Change Atlas • Global Coalition for Africa • Government Contracting Activity • Graphics Communication Association (USA) • Ground Control Approach • Ground-Controlled Approach • Group Capacity Assessment • Growth Curve Analysis • Guacamaya (Colombia) (location identifier) • Guatemala (international vehicle registration code) • Guidance and Control Assembly • Guidance Control Assembly • Guidance - Control - Avionics • Guidance Coupler Assembly • Gyro Control Assembly

GCAC Generic Connection Admission Control

GCAG Grand Challenge Application Group

GCAL Gemini Calibration (unit)

GCAM Government Cost Account Manager

GCAP Generalized Circuit Analysis Program • Graduate Cost Analysis Program

GCAPL Grand and Cooperative Agreement Policy Layer

GCAS Generic Configuration Accounting System • Gram-Charlier A-Series • Ground Collision Avoidance System

GCAT Gestion Commerciale des Activités Techniques

GCB German Conference on Bioinformatics • Graphitized Carbon Black

GCC Gaining Command Code • GAME Command Chain • General Compilation Catalogue • General Conference Control • Generic Command Center • Gillette (USA-WY) (location identifier) • Global Change Category • Global Change Coalition • Global Change Consortium • Global Climate Coalition • Global Climatic Change • Global Control Center • Globular Cluster Candidate • Gnu C Compiler • Goddard Communications Center • Goddard Computing Center • Governmental Coordinating Committee • GPS Command Center • Gross Cycle Control • Ground Command and Control • Ground Communications Controller • Ground Communications Coordinator • Ground Component Command • Ground Component Commander • Ground Computer Controller • Ground Control Center • Guidance Checkout Computer • Guidance and Control Computer • Gulf Cooperation Council

GCCA Generic Command Center Architecture • Graphic Communications Computer Association

GCCB Gaining Command Coordination Briefing

GCCC Ground Control Computer Center

GCCD Global Climate Change Digest

GCCE Graphics Correctness Criteria Editor

GCCF Generic Cost Comparison Form

GCCHK Global Cross-Check

GCCO Ground Control Checkout

GCCS Global Command and Control System

GCCU Guidance and Control Coding Unit

GCD Gain Control Driver • Galactic Chemodynamics • Gaze-Contingent Display • General Catalog of Documents • Geometric Correction Data • Gnu C Debugger • Great Circle Distance • Greatest Common Denominator • Greatest Common Divisor • Ground Controlled Descent

GCDA Geocentric Datum of Australia

GCDB Geographic Coordinate Database

GCDC Global Change Data Center • Ground Checkout Display and Control

GCDE Galactic Central Deep Exposure

GCDIS Global Change Data and Information System

GCDMWG Global Change Data Management Working Group

GCDP Global Change Database Project

GCDU Graphical Control Display Unit

GCE Galactic Chemical Enrichment • Galactic Chemical Evolution • General Certificate of Education • Gestion Commerciale d'Entreprises • Global Change Encyclopaedia • Government Cost Estimate • GPS Cargo Element • Grand Contournement Est • Ground-based Cloud Experiment • Ground Checkout Equipment • Ground Communications Electronics • Ground Communications Equipment • Ground Control Equipment • Groupe de Coordination Espace

GCEM Geographic Correlator Exploration Machine • Goddard Cumulus Ensemble Model

G&CEP Guidance & Control Equipment Performance

GCEP Global Change Education Program • Guidance and Control Equipment Performance

GCERF Governor's Council on Education Reform and Funding

GCEV Grupo Canario de Estrellas Variables (Spain)

GCF General Citedness Factor • Gestion Comptable et Financière • Globular Cluster Feature • Greatest Common Factor • Ground Calibration File • Ground Command Facility • Ground Communications Facility

GCFAP Guidance and Control Flight Analysis Program

GCFP Globular Cluster Fundamental Plane

GCG Galactocentric Chemical Gradient • Generalized Conjugate Gradient • Gravity-Controlled Gyro • Ground Command Guidance • Guidance Control Group

GCGS Gravity-Controlled Gyro System

GCGSS General Catalogue of Galactic S Stars

GCH Generalized Continuum Hypothesis

GCHC Gulf Coast Hydroscience Center

Gche Gauche

GCHQ Government Communications Headquarters (UK)

GCHX Ground Cooling Heat Exchanger

G&CI Guidance & Control Integration

GCI General Circuit Interface • Generalized Communications Interface • Geocentric Inertial • Gestion de Collections Informatisées • Ground Control Intercept • Ground-Controlled Intercept • Ground-Controlled Interception • Group Complexity Index • Guernsey - La Villiaze (Channel Islands) (location identifier) • Guest Computational Investigator • Guidance and Control Integration

GCID Ground Computer Interface Device

GCIL Ground Command Interface Logic • Ground Control Interface Logic

GCILC Ground Command Interface Logic Controller

GCILU Ground Command Interface Logic Unit

GCIP GEWEX Continental International Project

GCIR Goodwin Institute for Cancer Research

GCJ Johannesburg - Grand Central (South Africa) (location identifier)

GCK Garden City (USA-KS) (location identifier)

gcl Graphing Calculator (file entension)

GCl Globular Cluster

GCL Galactic Center Lobe • Gestion de Configuration Logicielle • Global Change Laboratory • Global Commerce Link • Globular Cluster • Graphics Card Language • Graphics Command Language • Ground Control Landing • Ground Coolant Loop

GCLF Globular-Cluster Luminosity Function

GCM General Circulation Model (*) • General Collective Model • General Control Module • Geodetic Coordinates Manual • Global Circulation Model • Global Classification Model • Global Climate Model • Globally Coupled Map • Grand Cayman (Cayman Islands) (location identifier) • Graph-Cache Manager • Greatest Common Measure • Greatest Common Multiple • Ground Control Message • Groupe de Concertation de Mécanique • Groupe de Contrôle Modifié • Guidance and Control Module

GCMA Government Contract Management Association (USA)

GCMAM Global Climate/Middle Atmosphere Model

GCMC Grand Canonical Monte Carlo

GCMD Gaining Command • Global Change Master Directory

GCMF Globular Cluster Mass Function

GCMR Ground Control Message Request

GCMRC Grand Canyon Monitoring and Research Center (USA-AZ)

GC/MS Gas Chromatograph/Mass Spectrometer

GCMS Gas Chromatograph and Mass Spectrometer • Gas Chromatograph/Mass Spectrometer • Gas Chromatograph Mass Spectrometry

GCN Gage Control Number • Galactic Center Newsletter • GIRD Change Notice • Global Communications Network • Government Computer New • Grand Canyon National Park (USA-AZ) (location identifier) • Grand Contournement Nord • GRB Circular Network • GRB Coordinates Network • Ground Communications Network • Ground Control Network

GCNMS Gas Chromatograph and Neutral Mass Spectrometer

GCNS Glasgow College of Nautical Studies (UK-Scotland)

GCO Gemeenschappelijk Centrum voor Onderzoek • General Counsel Office • Generation Certification Official • Government Concept of Operations • Grand Contournement Ouest • Ground Checkout • Ground Control Officer • Grove Creek Observatory (Australia-NSW)

GCOM Global Change Observation Mission

GCOP Ground Common Operating Picture • Ground Common Operational Picture

GCOS General Comprehensive Operating System • General Computer Operating System • General Computer Operational System • Global Change Observing System • Global Climate Observation System • Global Climate Observing System (*) • Ground Computer Operating System

GCP Generalized Computer Program • Geodetic Control Point • Gestion du Cycle de Projet • Global Change Program • GLO/CRYOHP Payload • Glow and Cryogenic Payload • Good Clinical Practices • Graph Coloring Problem • Ground Command Post • Ground Control Planning • Ground Control Point • Ground Control Position • Ground Control Processor • Group Command Post • Group Control Point • Guidance Control Package

GCPA General Catalogue of stars with the system of Photoelectric Astrolabes

GCPB General Catalogue of Printed Books

GCPD General Catalogue of Photometric Data

GCPO Gemini Canadian Project Office

GCpS Gigacycles per Second

GCPS Gigacycles Per Second

GCQMS Groupe Conseil pour la Qualité des Matériaux Spatiaux

GCR Galactic Cosmic Radiation • Galactic Cosmic Ray • Gamma Cosmic Ray • Gas-Cooled Reactor • General Component Reference • Grand Cercle de Référence • Gravitation et Cosmologie Relativistes • Gray Component Replacement • Ground-Controlled Radar • Group-Coded Recording • Group-Coded Request • Group-Code Recording • Group-Code Request

GCRA Generic Cell Rate Algorithm

GCRI General Collective Routing Indicator

GCRIO Global Change Research Information Office

GCRN Galactic Cosmic-Ray Nucleosynthesis

GCRP Galactic Cosmic Ray Particle • Global Change Research Plan • Global Change Research Programme

GCRS Galactic Cosmic-Ray Source • Geocentric Celestial Reference Frame

GCRV General Catalog of Radial Velocities • Generalized Conjugate Residual

Gc/s Gigacycles Per Second

GCS Galactic Clusters Survey • Gas-Controlled Switch • Gate-Controlled Switch • Gate Controller Switch • General Classification Scheme • General Communication System • Generator-Controlled Switch • Geo-Common System • Geostationary Communications Satellite • Global Conceptual Scheme • Globular-Cluster System • Grand Contournement Sud • Ground Central Site • Ground Command System • Ground Communications System • Ground Computer System • Ground Control Station • Ground Control Surveillance • Group Clock Selector • Group Control System • Guidance and Control System • Guidance Control Section • Guidance Cutoff Signal • Gyroless Control System

GCSA Genetic Configuration Similarity Algorithm

GCSAS Generic Configuration Status Accounting System

GCSC Guidance Control and Sequencing Computer

GCSD Computer Graphics System Development

GCSE General Certificate of Secondary Education (UK) • Generative and Component-based Software Engineering (symposium)

Gc/sec Gigacycles per Second

GCSP Grand Canyon Star Party (USA)

GCSS GEWEX Cloud System Study • Global Cloud System Study • Global Communications Satellite System • Ground-Controlled Space System • Ground Control Station Simulator

GCT General Classification Test • Geocentric Coordinated Time • Government Competitive Test • Greenwich Civil Time (*) • Greenwich Conservatory Time • Gregory-Coudé Telescope • Ground Checkout and Test • Guidance Command Test • Guidance Computer Test

GCTA Ground-Commanded Television Assembly

GCTC Gagarin Cosmonaut Training Center (Russia)

GCTE Global Change and Terrestrial Ecosystems (*) • Global Change in Terrestrial Ecosystems • Guidance Computer Test Equipment

GCTM Global Chemical Transport Model

GCTP General Catalogue of Trigonometric Parallaxes • General Coordinate Transformation Package • Geographic Coordinate Transformation Package

GCTS Gas Component Test Stand • Ground Communications Tracking System

GCTSP General Catalogue of Trigonometric Stellar Parallaxes

GCU General Control Unit • Generator Control Unit • Glasgow Caledonian University (UK-Scotland) • GNC Computer Unit • Grand Canyon University (USA-AZ) • Ground Checkout Unit • Ground Control Unit • Ground Cooling Unit • Guidance and Control Unit • Guidance Coupler Unit • Gyro Coupling Unit

GCUpS Giga Connection Updates per Second

GCUPS Giga Connection Updates Per Second

GCV Generalized Cross-Validation • Global Change Video • Gross Calorific Value

GCVS General Catalogue of Variable Stars

GCW Galactic Center Workshop (USA-HI) • Gross Combination Weight • Gross Combined Weight

GCWD Globular Cluster White Dwarf (star)

GCX Galactic Center X-ray • Gas Chromatograph Transmitter

GCXSE GSFC Cosmic X-ray Spectroscopy Experiment

GCY Greenville (USA-TN) (location identifier)

GC2 Glass Capillary Gaseous Chromatography

GC3 Grand Challenge Cosmology Consortium • Great Circle Cell Count

GC3A Global Command, Control, and Communications Assessment

GC4A Global Command, Control, Communications, and Computers Assessment

gd Guard • Scots Gaelic (ISO language code)

Gd Gadolinium • Good • Grand • Ground • Guard

GD Gap Detector • Garbage Disposal • Gate Driver • Gauche-Droite • General Delivery • General Design • General Development • General Diagram • General Discharge • General Duties • General Duty • General Dynamics • Generic Decryption • Geologic Division • Geospatial Data • Giclas Dwarf (star) • Glow Discharge • Good Delivery • Gradient Descent • Grand Design • Grand Duchess • Grand Duke • Graph Drawing • Graphic Display • Graphics Digitization • Graphics Disable • Gravimetric Density • Grenada (ISO country code) • Grenade (code ISO de pays) • Ground Demonstration • Ground Detector • Guard • Guarda • Guidance

GDA Galvo-Drive Amplifier • GEBCO Digital Atlas • Geographic Data Analysis • Gimbal Drive Actuator • Gimbal Drive Assembly • Gimballed Dish Antenna • Global Data Area • Global Directory Agent • Gounda (Central African Rep.) (location identifier) • Groupement Départemental d'Apprentissage

GDAAC Goddard Distributed Active Archive Center

GDAAS GAIA Database Access and Analysis Study

GDAM General Data Access Method

GDAO GSFC Data Assimilation Office

GDAP GEOS Data Adjustment Program

GDAR Generalized Data-sset Attribute Record

GDAS Global Data Assimilation System • Global Deployment Analysis System • Ground Data Acquisition System

GDAU General Data Acquisition Unit

GDB Genome Database • Geodetic Data Beacon • Gesellschaft für Dokumentation und Bibliographie (Austria) • Global Database • Gnu Debugger • Graphical Database

GDBMS Generalized Database Management System • Global Database Management System

GDBS Generalized Database System • Grenada Bureau of Standards

GDC General Design Criteria • General Digital Computer • General Dynamics Corporation • Geocentric Dust Cloud • Geodetic Data Center • Geographic Data Council • Geological Data Center • Geomagnetic-Data Center • Geophysical Data Center • Gesellschaft für Datensysteme und Computer • Global Data Centre • Global Data Control • Glow Discharge Cleaning • Graphic Display Console • Greenville - Donaldson Center (USA-SC) (location identifier) • Gross Dependable Capacity • Guarded Definition Clause • Guidance Data Converter • Gyro Display Coupler

GDCC Guarded Definite Clause with Constraints

GDCD Glow Discharge Cleaning Device

GDCF Generic Data Capture Facility

GDCh Gesellschaft Deutscher Chemiker (Germany)

GDCLS General Dynamics Commercial Launch Services

GDCP Geostationary Data Collection Platform

GDCS Government Documents Catalog Service

GDD Gesellschaft für Datenschutz und Datensicherung (Hermany) • Global Data Dictionary • Global Data Directory • Global Data Disable • Gordon Downs (Australia-WA) (location identifier) • Graphics Data Display

GDDL Graphical Data Definition Language

GDDM Graphical Data Display Manager • Graphics Data Display Manager

GDDS Gemini Deep Deep Survey • Global Data Dictionary Services

GDDSYN Geodesic Distribution Binary Synthesis

Gde Gourde (currency) • Grande

GDE Generalized Data Entry • Geospatial Data Exploitation • Gimbal Drive Electronics • Gode (Ethiopia) (location identifier) • Ground Data Equipment • Group for Development and Environment

GDEL Grand Dictionnaire Encyclopédique Larousse

GDEM Generalized Digital Elevation Model • Global Digital Elevation Model

GDEW Global Delivery Express Worldwide

GDF Gaz De France • General Data File • Generalized Data Format • Geographic Data File • Geophysical Data Facility • Grammar Description File • Group Distribution Frame

GDFCF Gross Domestic Fixed Capital Formation

GDG Generation Data Group • Generation Data-set Group • Magdagachi (Russia) (location identifier)

GDH Gerasimov - Drell - Hearn (sum rule) • Golden Horn (USA-AK) (location identifier) • Ground Data Handling

GDHS Ground Data Handling System

GDI Gasoline Direct Injection • General Dissemination of Information • General Drafting Interface • Gordil (Central African Rep.) (location identifier) • Graphical Device Interface • Graphics Device Interface (*) • Grupo de Desarollo Instrumental

GdIG Gadolinium Iron Garnet

GDJ Gandajika (Zaire) (location identifier)

Gdk Gdansk (Poland)

GDL Gas Dynamic Laser • Gas Dynamics Laboratory • Global Data Link • Glow Discharge Lamp • Grammar Description Language • Grupo de Descontaminación Luminica • Guadalajara - Miguel Hidalgo (Mexico) (location identifier)

GDLC Generic Data Link Control

GDLI Global Distance Learning Initiative

GDM Gardner (USA-MA) (location identifier) • General Development Model • General Display Manager • Generalization Data Model • Geodetic Distance Measurement • Geodynamical Model • Geospatial Data Management • Gesellschaft für Didaktik der Mathematik (Germany)

GDMO General Definition of Managed Objects • Global Definition for Managed Objects • Guidelines for the Definition of Managed Objects

GDMP Grid Data Mirroring Package

GDMS Generalized Data Management System • Generic Data Management System • Global Database Management System • Global Data Management Service • Global Data Management System • Glow Discharge Mass Spectroscopy • Ground Data Management System

GDMT Global Desired Merge Total

Gdn Garden • Guardian

GDN Garden • Gdansk (Poland) (location identifier) • Goal Dependency Network • Government Data Network

GDNÄ Gesellschaft Deutscher Naturforscher und Ärzte (Germany)

Gdnc Guidance

GDNC Guidance

Gdns Gardens

GDNS Gardens

GDNT Global Desired Non-merge Total

GDO Global Disk Oscillation • Guasdualito (Venezuela) (location identifier)

GDOP Geometric Dilution Of Precision

Gdor Gobernador

GDP Gás de Portugal • General Deployment Plan • General Development Plan • Generalized Direct Processor • Generalized Distributor Program • Generalized Documentation Processor • Generalized Drawing Primitive • Gestion de Données Produit • Global Data Process • Global Daily Pay • GOME Data Processor • Government Distribution Point • Grammar Description Programme • Graphic Display Processor • Graphic Draw Primitive • Gross Domestic Product • Ground Data Processing • Ground Demonstration Program • Guadalupe (Brazil-PI) (location identifier)

GDPL Gordhandas Desai Pvt. Ltd. (India)

GDPS Geographically-Dispersed Parallax Sysplex • Global Data Processing System • Government Document Publishing Service • Ground-based Data Processing System

GDQ Gondar (Ethiopia) (location identifier)

GdR Groupe de Recherche • Groupement de Recherche

GD&R Grinning, Ducking & Running

GDR Gas-to-Dust Ratio • General Designer's Review • Generalized Data Retrieval • Geodetic Data Reduction • Geophysical Data Record • German Democratic Republic (now merged into FRG) • Giant Dipole Resonance • Global Depository Receipt • Grinning, Ducking and Running • Groupement De Recherche

GDRC Gyro Drift Rate Compensation

GDRQ General Design Requirement

Gds Goods

GDS Gated Digital Sender • General Data Stream • General Declassification Schedule • Generation Data Set • Geodetic Data Site • Geographic Data Service • Global Directory Service • Global Distribution System • Goldstone • Graphical Data Set • Graphical Data Syntax • Graphical Data System • Graphic Data Set • Graphic Data Syntax • Graphic Data System • Graphic Display System • Great Dark Spot (Neptune) • Great Dust Storm (Mars) • GRETACODER Data Systems AG (Switzerland) (*) • Gretag Data Systems AG (Switzerland) • Ground Data System • Ground Development System • Ground Display System • Guarded Discrete Stochastic • Guide and Directory Services

GDSD Ground Data Systems Division

GDSIDB Global Digital Sea-Ice Data Bank

GDSM Ground Data Systems Manager

GDSMQ General Department for Standardization Metrology and Quality Control

GDSN Global Digital Seismic Network • Global Digital Seismograph Network

GDSO Ground Data Systems Officer

GDSS General Dynamics Space Systems • Geospatial Decision Support System • Global Decision Support System • Global Distress Safety System • Group Decision Support System

GdT Gemeinschaftsausschuß der Technik

GD&T Geometric Dimensioning & Tolerancing

GDT Geometric Dimensioning and Tolerancing • Gestion de Données Techniques • Global Descriptor Table • Grand Turk (Turks and Caicos Islands) (location identifier) • Graphic Display Terminal • Graphics Development Tool • Graphics Development Toolkit • Ground Data Terminal

GDTA Groupement pour le Développement de la Télédétection Aérospatiale

GDU Graphic Display Unit

GDV Gesellschaft für Datenverarbeitung • Glendive (USA-MT) (location identifier)

GDW Gladwin (USA-MI) (location identifier) • Global Display Workstation • Graphics Display Workstation

GDX Magadan (Russia) (location identifier)

GDZ Gelendzik (Russia) (location identifier)

GDZS Gesamtverzeichnis deutschsprachiger Zeitschriften und Serien

ge Gilt Edges

gE Giant Elliptical (galaxy)

Ge Germanium

G/E Graphite/Epoxy

GE Gas Ejection • Gas Exchange • Gaussian Elimination • Gebühreneinheit • General Education • General Election • General Electric Co. • General Equipment • Geocentric Gravitational Constant • Geographic Equivalent • Georgia (ISO country code) • Géorgie (code ISO de pays) • Geoscience Electronics • Germany • Gewerbeeinheit • Gigabit-Ethernet • Grande École • Grande Entreprise • Graphics Enable • Graphite Epoxy • Greater than or Equal • Greatest Elongation • Greedy Execution • Ground Equipment • Group of Experts • Gyro Error

GEA Generalized Extensions Agreement (FITS) • Geometric Element Analysis • Gestion des Entreprises et des Administrations • Gigabit Ethernet Alliance • Gimbal Electronics Assembly • Global Environment Assessment • Grup d'Estudis Astronomics (Spain) • Grupo de Estudos de Astronomia (Brazil) • Nouméa Magenta (New Caledonia) (location identifier)

Ge&Ae Geomagnetizm i Aeronomiia (periodical code)

GeAer Geodeziia i Aerofotos (periodical code)

GEANT Gigabit European Academic Network

GEAP Grupo de Estudios Astronómicos de Puertollano (Spain)

GEAR General Environment for ATM Research

GEAUDE Gestion Automatisée Décentralisée

geb geboren • gebunden

Geb Gebergte • Gebirge

GEB Gebe (Indonesia) (location identifier) • Gödel - Escher - Bach (trilogy) • Gemeentelijk Energiebedrijf

GEBA Gebäudeautomation • Global Energy Balance Archive • Government Excess Baggage Authorization

GEBCO General Bathymetric Chart of the Oceans

GEBECOMA Groupement Belge des Constructeurs de Matériel Aérospatial (Belgium)

Gebr Gebroeders • Gebrüder

GEBR Galactic and Extragalactic Background Radiation

GEC Gecitkale (Cyprus) (location identifier) • General Education Curriculum • General Electric Co. • Genetic and Evolutionary Computation • Genetic and Evolutionary Computing • Geospace Electrodynamics Connections (mission) • Global Electrodynamic Monitor • Global Environmental Change • Ground Environment Complex

GeCAS Geochimica et Cosmochimica Acta Supplement (periodical code)

GECAS General Electric Capital Aviation Services

GECCO Genetic and Evolutionary Computation Conference

GECG Grandes Écoles de Commerce et de Gestion

GECHS Global Environmental Change and Human Security

GECJR GEC Journal Research (periodical code)

GECMA Global Electrodynamics and Coupling through the Middle Atmosphere

GeCoA Geochimica et Cosmochimica Acta (periodical code)

GECOM Generalized Compiler

GECOS General Electric Comprehensive Operating System

GECR Global Environment Change Report

GECS Geostationary Earth Climate Sensor

ged Graphics Editor (file extension)

GeD Germanium Detector

GED Gasoline Engine Driver • General Educational Development (*) • General Education Development • General Equivalency Degree • General Equivalency Diploma • Georgetown - Sussex County (USA-DE) (location identifier) • Gestión Electrónica de Documentos • Gestion Électronique de la Documentation • Gestion Électronique de Documents • Global Electrodynamics • Graduate Equivalent Diploma • Groupe Étude Dosimétrie

GEDAE Graphical Entry Distributed Application Environment

GEDEMON Geometrics, Derivatives, Moments and Noise

GEDEONEX Générateur de Dossiers d'Environnement d'Océanographie Numérique Expérimental

GEDES Groupement Étudiant de l'Enseignement Supérieur

GEDEX Geodetic Demonstration Experiment • Greenhouse Effect Detection Experiment

GEDI Groupement Europages de Diffusion d'Information • Group on Electronic Document Interchange

GEDIS Geological, Exploration and Development Information System

GEDP General Education Development Program

GEDR Global Emissivity Data Record

GEDT General Educational Development Test

GEE Galaxy Evolution Explorer • Generalized Estimating Equation (*) • Generalized Estimation Equation • George Town (Australia-TAS) (location identifier) • Greatest Eastern Elongation • Ground Exploration System • Group of Economic Experts

GEEP Group of Experts on Environmental Pollutants

GEERS Groupe d'Études Européennes pour la Recherche Spatiale (now defunct)

GEF Gain Equalization Factor • Gauss Error Function • Gestion Électronique des Factures • Geva (Solomon Islands) (location identifier) • Global Environmental Facility • Global Environment Facility (*) • Global Environment Fund • Ground Equipment Failure • Groupement d'Exercice Fonctionnel

GEFD Geophysical and Environmental Fluid Dynamics

GEFE Grupo de Estudios de Formación Estelar

GEFES Grupo Especializado de Física del Estado Solido (Spain)

GEFIE Groupe d'Enseignants de Formation et d'Intervention Éducative

gefl gefälligst

GEG Spokane (USA-WA) (location identifier)

GeGe Geophysica et Geodaetica (periodical code)

gegr gegründet

geh geheftet

GEH Gasetagenheizung

GEHN Groupe d'Experts de Haut Niveau

GEHR Giant Extragalactic HII Region

GEI Gesellschaft für Elektronische Informationsverarbeitung mbH • Gestion Électronique de l'Information • Grand Établissement Industriel • Graphics Engine Interface • Green Islands (Papua New Guinea) (location identifier)

GEIA Global Emission Inventory Activity

GEID Gestion Électronique de l'Information et de Documents

GEIDE Gestion Électronique de l'Information et des Documents pour l'Entreprise (forum)

GEIE Groupement Européen d'Intérêt Économique

GEII Génie Électrique et Informatique Industrielle

GEIP Greenhouse Education and Information Programme

GEIS Gebäudeinformationssystem • General Edited Information Set • General Electric Information Service • General Electric Information System • General Environmental Impact Statement • Generic Edited Information Set • Geographical Environmental Information System • German Encyclopaedic Internet Service • Groupe d'Étude sur l'Information Scientifique

GEISA Gestion et Études des Informations Spectroscopiques Atmosphériques

GEISCO General Electric Information Services Co.

GEISHA Geodetic Inertial Survey and Horizontal Alignment • Gestion, Extraction, Interrogation d'une Structure Hiérarchisée Arborescente • Groningen Exportable Infrared Survey High-Resolution Analysis

GEITA Government Electronics and Intelligence Technology Association (USA)

GEK Ganes Creek (USA-AK) (location identifier) • Geomagnetic Electrokinetograph • Lari (Georgia) (ISO currency code)

GeKaA Geodeziia Kartografiia i Aerofotos (periodical code)

GeKar Geodeziia i Kartografiia (periodical code)

GEL Santo Angelo - Sepe Tiaraju (Brazil-RS) (location identifier)

Gelio Geliotekhnika (periodical code)

GELIS Ground Emitter Locations and Identification System

GELOC Geographic Location (code)

gem gemäß

Gem Gemeinde • Gemeinschaft • Gemini-Geminorum (constellation)

GEM Galactic Emission Mapping (*) • Galactic Events per Megayear • Galaxy Emission Mapping • Galileo Europa Mission • Gamma - Electron - Muon (collaboration) • Gas Electron Multiplier • Gaseous Electron Multiplier • Gateway to Education Materials (consortium) • Geminids (meteor stream) • General Energy Management • General Epitaxial Monolith • Generalized Emulation Microcircuit • Generalized Energy Management • Generalized Ensemble Method • Generated Energy Management • Generic Electronics Module • Generic Engineering Model • Geometric Element Modeling • Geospace Environmental Modeling • Geospace Environment Modeling (*) • German Equatorial Mount • Giant Earth Mover • Giotto Extended Mission • Global Electrodynamics Monitor • Global Enterprise Management • Global Enterprise Manager • Global Environmental Multiscale (model) • Global Evaluation Model • Goddard Earth Model (*) • Goddard Environment Model • Gonsalves Error Metrics • Graphical Environment Manager • Graphical Execution Manager • Graphic Engine Monitor • Graphics Environment Manager • Graphite Epoxy Motor • Gravity Earth Model • Ground Effect Machine • Ground Electronics Maintenance • Ground Environmental Model • Ground Equipment Model • Guiding Error Minimizer

GEMCOS Generalized Message Control System

Gemin GEMINI Newsletter (periodical code)

GEMINI General Excitation Mechanisms in Nightglow • Global Emergency Management Information Network Initiatie (USA)

GEMM General Electronics Maintenance Model

GEMMS Geophysical Exploration Manned Mobile Submersible

GEMO Ground Electronic Maintenance Officer

GEMPAK General Meteorology Package

GEMPS Groupe d'Études en Médecine et Physiologie Spatiales

GEMS Galaxy Evolution Morphology and SEDs • Gas Exchange Monitoring System • General Education Management System • General Electric Manufacturing Simulator • General Electric Multispectral System • General Energy Management System • Generalized Energy Management Steering • Generalized Energy Management System • Generalized Evaluation Model Simulator • Geographic and Environmental Modeling System • Geographic Monitoring System • Geological Exploration and Mapping System • Geostationary European Meteorological Satellite • GITS Electronic Mail Steering • Glass-Embedded Metals Sulfides • Glasses with Embedded Metals and Sulfides • Global Environmental Monitoring System (*) • Global Environment Monitoring System • Goodyear Electronic Mapping System • GPS Engineering Management System • Graphical Exposure Modeling System • Graphics Engineering and Mapping System • Great Explorations in Math and Science

GEMU Global Electronic Multimedia University

GemVO Gemeinnützigkeitsverordnung

gen genannt • Generator

Gen General • Generally • Generator • Generic • Genitive

Gén Général • Género

Gen Generaal • General • Generate • Generated • Generation • Generator • Genesis • Geneva (Switzerland) • Genève (Switzerland) • Gennaio • Genossenschaft • Genova (Italy)

GEN General • Generation • Generator • Generic • Global Engineering Network • Global European Network • Grande Entreprise Nationale (France) • Oslo Gardermoen (Norway) (location identifier)

Gend Gendarme

GENE GPS Enhanced Navigation System

Geneal Genealogía

GENESIS Generalized Environment for the Simulation of Integrated System • Global Ecosystem Numerical Estimation Using Satellite Imaging System

GenG Genossenschaftsgesetz

GENIAC Genius Almost Automatic Computer

GENIE General Electric Network for Information Exchange • Global Environmental Network for Information Exchange • Ground-based European Nulling Interferometer Experiment

GENISIS General Imaging Spectrometry Interpretation System

Génit Génitif

Genit Genitive • Genitivo

GENIUS Germanium in liquid Nitrogen Underground Setup • Global Environment Network and Information User System

Genl General

GENLY Generally

Gen.Mgr. General Manager

Genn Gennaio

GENREP General Report

GENSER General Security • General Services

GENSUP General Supply • General Support

Gent Gentleman

GENTEXT General Text

Geo Geographic • Geology • Geology (periodical code) • George

GEO Geocentric • Georgetown Timehri (Guyana) (location identifier) • Georgia (ISO country code) • Géorgie (code ISO de pays) • Geostationary Earth Observatory • Geostationary Earth Orbit • Geostationary Earth Orbiter • Geosynchronous Earth Orbit • German-England Observatory • Global Environment Outlook • Gravitational European Observatory • Grupo Especial de Operaciones

Geoch Geochemistry (periodical code)

GeocJ Geochemical Journal (periodical code)

Géod Géodésie • Géodésique

Geod Geodesia • Geodesy • Geodetic

GEODAS Geophysical Data System

GEODE Geodynamics and Ore Deposit Evolution

GEODES Geographic Decision Expert System

GEODESIC Geo-electrodynamics and Electro-optical Detection of Electron and Suprathermal Ion Currents (project)

GEODIS Geographic Design and Implementation System

GEODOM Geographic Data, Overlay and Manipulation

GEODS Geosynchronous Electro-Optical Data System

GEODSS Ground-based Electro-Optical Deep-Space Space Surveillance • Ground-based Electro-Optical Deep-Space Surveillance System (USAF)

Geofís Geofísica

Geog Geography

Géogr Géographie • Géographique

Geogr Geografía • Geographer • Geographic • Geographical • Geography

GeoIn Geocarto International (periodical code)

GeoIs Geomagnitnye Issledovaniia (periodical code)

GeoJ Geophysical Journal (periodical code)

GeoJI Geophysical Journal International (periodical code)

Geokh Geokhimiia (periodical code)

Geol Geological • Geology

Géol Géologie • Géologique

Geol Geología • Geologic • Geological • Geologist • Geology

GEOLAN Geological Local Area Network

GEOLOC Geographical Location

Geom Geometría • Geometric • Geometry

GeoM Geological Magazine (periodical code)

Géomag Géomagnétique • Géomagnétisme

Geomag Geomagnetic • Geomagnetism

GEOMER Geostationary Multispectral Electroscanning Radiometer

Geomorf Geomorfología

GEOMPAK Geometric Manipulation Package

GEOMULDAT Geochemical Multielement Database

GEON Gyro-Erected Optical Navigation

GEONESS Geostationary Earth Orbiting Nadir Etalon Sounding Spectrometer

GeoNr Geophysica Norvegica (periodical code)

GEONS Geomagnetic Event Observation Network by Students

GeoOM Georgetown Observatory Monograph (periodical code)

Geoph Geophysica (periodical code)

Geophys Geophysical • Geophysics

Géophys Géophysique

Geopol Geopolitics

GeoRE Georgetown Observatory Reprints (periodical code)

GEOREF Geographical Reference • Geological Reference

GEOREFS Geographical Reference System

GeoRL Geophysical Research Letters (periodical code)

GEOS Geodetic Earth-Orbiting Satellite • Geodetic Earth-Observation Satellite • Geodetic Satellite • Geodynamics Experimental Ocean Satellite • Geostationary Earth-Orbiting Satellite • Geosynchronous Earth Observation System • Goddard Earth Observing System • Graphical Environment Object System • Graphic Environment Operating System • Groupement Européen d'Observation Stellaire • Grupo Europeo de Observaciones Estelares • Gruppo Europeo di Osservazione Stellare

GEOSAR Geologic Synthetic Aperture Radar • Geosynchronous Synthetic Aperture Radar

GEOSAT Geodesy Satellite • Geodetic Satellite

GeoSb Geofizicheskii Sbornik (periodical code)

GEOSC Groupement Européen d'Observation Stellaire - Circulars (periodical code)

GeoSCIA Geostationary Scanning Imaging Absorption Spectrometer

GEOSEPS Geosynchronous Solar Electric Propulsion Stage

GeoSJ Geological Society Journal (periodical code)

GEOSN Groupement Européen d'Observation Stellaire - Note Circulaire (periodical code)

GeoSu Geophysical Surveys (periodical code)

Geot Geotimes (periodical code)

Geoth Geothermics (periodical code)

GeoZh Geofizicheskii Zhurnal (periodical code)

GEP Generalized Eigenvalue Problem • Goddard Experiment Package • Good Engineering Practice • Gross Energy Product • Ground Entry Point

GEPAC General Electric Process Automation Computer • General Electric Programmable Automatic Computer

GEPAN Groupe d'Études sur les Phénomènes Aérospatiaux Non-identifiés (France)

GEPATA Groupe d'Étude des Problèmes de l'Automatisation des Travaux Administratifs (France)

GEPP Gaussian Elimination with Partial Pivoting

GEPS Geostationary Earth Processes Spectrometer

GEQI Groupe d'Expertise de la Qualité Image

Ger German • Germany • Gerundio • Gerundive

GER Gerência • Global Environmental Research • Gross Energy Requirement • Groupement d'Étude et de Réflexion • Nueva Gerona (Cuba) (location identifier)

GERAILP Groupe d'Étude et de Recherches pour les Applications Industrielles des Lasers de Puissance (France)

GERB Geostationary Earth Radiation Budget

GERD Graduate Education and Research Development • Gross Expenditure on Research and Development

GERDIEN General European Road Data Information Exchange Network

GERDSM Groupe d'Études et de Recherches en Détection Sous-Marine

GERE Génie Électrique et Radioélectricité

GEREMI Groupe de Réflexion sur la Maintenance et l'Instrumentation

GERES Geometric Rectification Expert System

GERG Global Environment Research Group

GERIS Geophysical and Environmental Research Imaging Spectrometer • Group on Evaluation of Research in Information Science

Germ Germanía • Germanic • Germánico

GERMEA Groupe d'Entraînement et de Recherche pour les Méthodes d'Éducation Active (France)

GERS Guaranteed Error Reduction Strategy

GERSIS General Electric Range Safety Instrumentation System

GERSULP Groupe d'Étude et de Recherche sur les Sciences de l'Université Louis Pasteur (France)

GERT Giant Equatorial Radiotelescope • Graphical Evaluation and Review Technique

GERTS General Electric Radio Tracking System • General Electric Range Tracking System • General Electric Remote Terminal System

GERV General Electric Reentry Vehicle

Ges Gesamt- • Gesamtschule (Germany) • Gesellschaft • Gesetz

GES Gateway Earth Station • Gaz à Effet de Serre • General Santos - Buayan (Philippines) (location identifier) •

Generic External Shell • Geographic Entry System • Gesellschaft für Elektronische Systemforschung (Germany) • Gift and Exchange Section • Global Enterprise Services • Government Electronic System • Ground Earth Station • Ground Electronic System • Ground Engineering System • Ground Equipment System • Ground Exploitation System

GESAMP Group of Experts on Scientific Aspects of Marine Pollution

Gesch Geschichte

Geschw Geschwindigkeit • Geschwister

Gesell Gesellschaft

GESF General Equation Specification File

ges.gesch. gesetzlich geschützt

GESMAR Geodetic Survey Marks Register

GesmbH Gesellschaft mit beschränkter Hafting

GESNS Galileo European Satellite Navigation System

GESOC General Electric Satellite Orbit Control

GESPER Gestion du Personnel

GESPL Generalized Edit System Programming Language

GESS Geostationary Environmental Satellite System • Goddard Experiment Support System • Grating Element Support Structure

GESSY Groupe d'Études de Signaux et Systèmes

gest gestorben

GEST Galactic Exoplanet Survey Telescope • Goddard Earth Sciences and Technology Center (USA)

Get Getekend

GET Generalized Einstein Theory • Geraldton (Australia-WA) (location identifier) • Get - Execute - Trigger • Ground Elapsed Time • Groupe d'Enseignement Technologique • Groupe d'Études Techniques • Group Execute Trigger

GETA Government Employees Training Act (USA) • Groupe d'Études pour la Traduction Automatique (France)

GETF Global Environment Trust Fund

GETG General Environmental Test Guidelines

GETI Ground Elapsed Time of Ignition • Groupe d'Études Techniques et Informatiques (France)

GETL Ground Elapsed Time of Landing

GETMS Geostationary Experimental Temperature and Moisture Sounder

GETO Ground Equipment Turnoff

GETOL Ground Effect Takeoff and Landing

GETS General Environmental Test Specification • Government Emergency Telecommunications Service • Ground Equipment Test Set

GETSCO GE Technical Services Co.

GEU GAS Electronics Unit • Grism Exchange Unit • Gyroscope Electronics Unit

GeV Giga Electronvolt

GEV Gällivare (Sweden) (location identifier) • Gallivare (Sweden) (location identifier)

GEVS General Environmental Verification Specification

GEW Gewoya (Papua New Guinea) (location identifier)

GewB Gewerkschaftsbibliothek

GEWEX Global Energy and Water-cycle Experiment (*) • Global Energy Water-cycle Experiment

GewO Gewerbeordnung

GewStDV Gewerbesteuer-Durchführungsverordnung

GewStG Gewerbesteuergesetz

GewStR Gewerbesteuer-Richtlinien

GEx Gas Exchange

GEX Gas Exchange • Gas Exchange Experiment • Gaussian Extraction • Geelong (Australia-VIC) (location identifier)

GEY Greybull - South Big Horn County (USA-WY) (location identifier)

gez gezeichnet

GF French Guyana (ISO country code) • Gain Factor • Galactic Field • Galactic Fountain • Gap Filler • Garfield Factor • Gauge Factor • General Form • General Format • General Formula • Generator Field • Generic Failure • Generic Function • Geschoßfläche • Gigafranc • Girlfriend • Glass Fiber • Globular Filament • Government Form • Government-Furnished • Gradient Freezing • Green Flash • Green Form • Green Function • Ground Fog • Ground Force • Group Framing • Growth Factor • Guinean Franc (currency) • Guyana (French_) (ISO country code) • Guyane Française (code ISO de pays)

GFA Good Fair Average • Great Falls - Malmstrom (USA-MT) (location identifier) • Groupe Fermé d'Abonnés

GFAAS Graphite Furnace Atomic Absorption Spectroscopy

GFAC Ground Forward Air Controller • Groupement Français pour l'Analyse des Contraintes (France)

GFAE Government-Furnished Aeronautical Equipment • Government-Furnished Aerospace Equipment • Government-Furnished Aircraft Equipment • Government-Furnished Avionics Equipment

GFAI Gesellschaft für Förderung der Angewandten Informationsforschung (Germany)

GFB Go For Broke • Government-Furnished Baseline • Togiak Fish (USA-AK) (location identifier)

GFBL Government-Furnished Baseline

GFC Gas Filter Correlation • Gel Filtration Chromatography • Generic Flow Control • Gestion Financière et Comptable • Green Flare Cluster

GFCC Ground Facilities Control Center

GFCF Gross Fixed Capital Formation

GFCI Ground-Fault Circuit Interruptor

GFCM General Fisheries Council for the Mediterranean

GFCP Generic Front-end Communications Processor

GFCR Gas Filter Correlation Radiometer

GfD Gesellschaft für Datenverarbeitung

GFD General Forecasting Division • General Functional Description • Geophysical Fluid Dynamics • Gone For the Day • Government-Furnished Data • Government-Furnished Documentation • Greenfield Pope (USA-IN) (location identifier) • Ground Fault Detector

GFDL Geophysical Fluid Dynamics Laboratory (NOAA)

GFE Galaxy Formation Efficiency • Government-Furnished Equipment • Graphical Front End • Grenfell (Australia-NSW) (location identifier) • Gross Feasibility Estimator

GFEL Government-Furnished Equipment List

GFE&M Government-Furnished Equipment & Materiel

GFEM Generalized Finite Element Method • Government-Furnished Equipment and Materiel

GFERD Government-Furnished Equipment Requirement Description

GFF Government-Furnished Facility • Gradient Freeze Furnace • Griffith (Australia-NSW) (location identifier)

GFFC Geophysical Fluid Flow Cell

GFFIL Groupement Français des Fournisseurs d'Information en Ligne (France)

GFH Global Failure Handler

GFI Gas Flow Indicator • General Format Identification • General Format Identifier • Government Final Inspection • Government Free Issue • Government-Furnished Information • Government-Furnished Item • Ground Fault Indicator • Ground Fault Interrupt • Ground Fault Interruptor

GFID General Format Identifier

GFII Groupement Français de l'Industrie de l'Information (France) • Groupement Français de l'Industrie de l'Informatique (France)

GFIM Government-Furnished Information Material

GFIT Glass Fiber Insulation Tubing

GfK Gesellschaft für Klassifikation (Germany)

GFK Grand Forks (USA-ND) (location identifier)

Gfl Gewerbefläche

GFL Gas Film Levitation • Gas Film Levitator • Glens Falls - Warren County (USA-NY) (location identifier)

GFlops Billion Floating-Point Operations per Second

GFLOPS Billion Floating-Point Operations Per Second • Giga Floating-Point Operations Per Second

GFM Generic Function Manager • Global Forecast Model • Government-Furnished Material

GFML Gluten Free Meal

GFMS Gaseous Flow Measuring System • Generalized File Management System • Generic Function Management System

GFN Grafton (Australia-NSW) (location identifier)

GFO Bartica (Guyana) (location identifier) • General Fiber Optics • GEOSAT Follow-On • GPS Field Office

GFOES Groupement Français pour l'Observation et l'Étude du Soleil (France)

GFOS Ground Facilities and Operations Section • Ground Facilities Operations Section

GF&P Gases, Fluids & Propellants

GFP General Forecasting Program • Generalized Filter Program • Global Functional Plan • Government Full Period • Government-Funded Procurement • Government-Funded Program • Government-Furnished Program • Government-Furnished Property • Graphical Form Painter • Green Fluorescent Protein

GFPBBD Groupement Français des Producteurs de Bases et Banques de Données (France)

GFPI Goddard Fabry-Pérot Imager

GFPL Government-Furnished Property List

GFPT Green-Function Perturbation Technique

GFQ Gradient Facility with Quenching • Gradient Furnace with Quenching

GFR Galactic Formation Rate • Gap Filler Radar • General Functional Requirement • German Federal Republic • Glass Fiber Reinforced • Government Facilities Request • Government Flight Representative • Granville (France) (location identifier) • Grim File Reaper • Guaranteed Frame Rate

GFRC Graphite-Fiber-Reinforced Composite

GFRP Glass-Fiber-Reinforced Plastic (*) • Graphite-Fiber-Reinforced Plastic

GFRS Global Foundation for Research and Scholarship (Japan)

GFS Gas Flow Shielded • Generalized Finite Sequence • Government-Furnished Service • Government-Furnished Software • Grandfather - Father - Son

GFSK Gaussian Filtered Shift Keying • Gaussian Frequency Shift Keying

GFSR General Functional System Requirement

GFT General Flying Test

GFTM Generic Fluids Transfer Model

GFTU General Federation of Trade Unions

GFU Government-Furnished Unit • Groupe Fermé d'Usagers • Groupe Fermé d'Utilisateurs • Group Framing Unit

GFURD Ground Facilities User Requirements Document

GFV Gestion des Forces de Vente

GfW Gesellschaft für Weltraumforschung

GFW Ground Fault Warning

GFY Government Fiscal Year • Grootfontein (Namibia) (location identifier)

GFZ Geodätisches Forschungszentrum • Geoforschungszentrum • Geschossflächenzahl

G&G Geodetic & Geophysical • Geology & Geophysics

G-G Ground to Ground

G/G Gas Generator • Ground to Ground

GG Galileo Galilei (1564-1642) • Gas Generator • Golden Gate • Governor General • Grande Gueule • Gravity Gradient • Gravity Gradiometer • Great Gross • Ground Guidance • Grundgesetz

GGA Groß-Gemeinschafts-Antenne

GGAO Goddard Geodetic and Astronomy Observatory (formerly GORF) (USA-MD) • Goddard Geophysical and Astronomical Observatory (USA-MD)

GGAV Gefahrgutausnahmeverordnung

GGB Golden Gate Bridge (USA-CA)

GGC Galactic Globular Cluster • Gemeenschappelijke Geautomatiseerde Catalogssystemen • Generalized Genetic Code • Ground Guidance Computer • Lumbala (Angola) (location identifier)

GGCL Government-to-Government Communications Link

GGCM Geospace General Circulation Model

GGD Gemeentelijke Geneeskundige Dienst • Gregory Downs (Australia-QLD) (location identifier) • Gyulbudaghian - Glushkov - Denisyuk (catalog)

GGDPAC Government Geoscience Database Policy Advisory Committee

GGE General Graphical Editing • Georgetown (USA-SC) (location identifier) • Ground Guidance Equipment

ggf gegebenfalls

GGF Global Geophysical Fluid • Global Grid Forum • Guidance and Guarantee Fund

GGFC Global Geophysical Fluids Center

GGFM Generalized Green's Function Method

GGFR Geosynchronous Gas Filter Radiometer

GGG Gadolinium Gallium Garnet • Glacial Geology and Geomorphology • Kilgore (USA-TX) (location identifier) • Longview - Gregg County (USA-TX) (location identifier)

GGGS Galileo Geometry and Graphics Software

GGI Generic Graphical Interface • Global Geospatial Information • Global Grid Initiative • GPS Geoscience Instrument • Greenhouse Gas Index

GGI&S Global Geospatial Information & Services

GGIS Global Geographic Information and Services • Global Geospatial Information and Services

GGL Gilgal (Colombia) (location identifier)

GGM Gravity Gradient Mode • Gravity Gradiometer Mission

GGN Gagnoa (Ivory Coast) (location identifier)

GGNT Gallium-Germanium Neutrino Telescope

GGO Guiglo (Ivory Coast) (location identifier)

GGP Gateway-to-Gateway Protocol • General Global Parameter • Generic Gateway Protocol • Generic Genetic Programming • Global Geodynamics Project • GPS Guidance Package • Gross Global Product

GGPD General Global Parameter Definition • Generalized Gaussian-Poisson Distribution

GGPP Giant Gaseous Protoplanet

G. gr. Great Gross

GGR Garoe (Somalia) (location identifier)

GGS Geometry and Graphics Software • Global Geoscience • Global Geospace • Global Geospace Science • Global Geospace System • Gobernador Gregores (Argentina-SC) (location identifier) • Goddard GRO Simulator • Graphic Generator System • Gravity Gradient Satellite • Gravity Gradient Sensor • Ground Guidance Station • Ground Guidance System

GGSE Gravity Gradient Stabilization Experiment

GGSN Gateway GPRS Support Node

GGSS Geodetic and Geophysical Sites Subcommission • Gravity Gradiometer Survey System • Grid Giant Star Survey

GGSSS Global Geospace Surveillance Satellite System

GGT George Town - Exuma (Bahamas) (location identifier) • Global Geoscience Transects • Gravity Gradient Torque

GGTS Gravity Gradient Test Satellite

GGTTS Group on GPS Time Transfer Standards

GGVM Gas Generator Valve Module

GGVS Gefahrgutverordnung Straße

GGW Glasgow (USA-MT) (location identifier)

GGWAG Goddard Gravitational Wave Astrophysics Group (NASA)

GH Gaseous Hydrogen • Gas Heater • Ghana (international vehicle registration code) • Ghana (ISO country code) • Giclas Hyades • Global Header • Good Health • Graphics Hardware • Ground Handling • Gaseous Hydrogen

GHA General Housekeeping Area • Ghana (ISO country code) • Ghardaia (Algeria) (location identifier) • Greenwich Hour Angle • Gyro Header Assembly

GHAAS Greater Hazleton Area Astronomical Society (USA-PA)

GHAO Guillermo Haro Astrophysical Observatory (Mexico)

GHASP Gassendi H-Alpha survey of Spirals

GHAT Ground Handling And Transportation

GHB Gesamthochschulbibliothek • Governor's Harbour (Bahamas) (location identifier) • GSFC Handbook

GHC Cedi (Ghana) (ISO currency code) • Gating Half-Cycle • Generalized Hyperbolic Class • Giant Hydrogen Cloud • Great Harbour Cay (Bahamas) (location identifier) • Ground Half-Coupling • Group Handling Council • Guarded Horn Clause

GHCC Global Hydrology and Climate Center (USA)

GHCD Growth Hormone Concentration and Distribution

GHCN Global Historical Climatology Network

GHDBS Generic Hierarchical Database System

GHe Gaseous Helium

GHE Garachine (Panama) (location identifier) • Gauss Hypergeometric Equation • Gibbs-Helmholtz Equation • General Hierarchical Editor • Ground Handling Equipment

GHF Gauss Hypergeometric Function • Generalized Hartree-Fock (method) • Gradient Heating Facility • Gradient Heating Furnace

GHG Greenhouse Gas

GHII Giant HII (region)

GHIIR Giant HII Region

GHIS GOES High-resolution Interferometer Sounder

GHK Gush Katif (Israel) (location identifier)

GHL Garden Hill (Canada) (location identifier)

GHM Centerville (USA-TN) (location identifier) • Generalized Hall Model • Global History Manager • Golla - Hughes - McTavish (method) • Gravitohydrodynamics

GHMA Greek Hellenic Mapping Agency

GHMI Ground Human Machine Interface

GHN Get Hold Next • Guanghan (China-PRC) (location identifier)

GHO Guillermo Haro Observatory (Mexico-SON) • Gunn - Hoessel - Oke (catalog)

GHOST Global Horizontal Sounding Technique

G-HOU Global Hands On Universe (program)

GHOU Global Hands On Universe (program)

GHP Gas High Pressure • Geroch - Held - Penrose (formalism) • Grand Handicapé Physique

GHPU Ground Hydraulic Power Unit

GHQ General Headquarters

GHR Giant HII Region

GHRIL Ground-based High-Resolution Imaging Laboratory

GHRS Goddard High-Resolution Spectrograph

GHS Gas Handling System • Geohazards System • Ground Handling System

GHSG Guest Housing

GHSM Groupe Hydraulique Servomoteur

GHSOM Growing Hierarchical Self-Organizing Map

GHT Gas Holding Tank • Generalized Hough Transform • Ghat (Libya) (location identifier)

GHU Get Hold Unique • Gualeguaychu (Argentina-ER) (location identifier)

GHW George Herbert Wells (1866-1946) • Ground Handling Wheel • Guaranteed Hourly Wage

GHX Ground Heat Exchanger

GHz Gigahertz

GHZ Galactic Habitable Zone • Greenberger - Horne - Zeilinger (experiment)

GH2 Gaseous Hydrogen

gi Gill

Gi Gilbert

G/I Ground Idle

GI Galilean Invariance • Galvanized Iron • Gated Integrator • Gateway Integration • General Index • General Information • General Inspection • General Investigator • General Issue • Generalized Instance • Generic Issue • Geodesic Isotensoid • Geographic Information • Geological Institute • Geometric Identification • Geophysical Institute • Gesellschaft für Informatik (Germany) • Gibraltar (ISO country code) • Governmental and Industrial • Government-Initiated • Government Issue • Gradient Index • Grammatical Inference • Gravitational Instability • Grazing Incidence • Guerre Informatique • Guest Investigator • Guidance Initialization

GIA Draconids (meteor stream) • General-purpose Interface Adapter • Geographical Information Analysis • Geographic Information and Analysis • Gestionnaire Informatique d'Applications • Glacial Isostatic Adjustment • Gruppo Italiano Astrometristi (Italy)

GIAC Geomatics Industry Association of Canada • Graphic Input Aggregate Control • Graphic Interface Aggregate Controller

GIADA Grain Impact Analyzer and Dust Accumulator (Rosetta)

GIAG Government-Industry Advisory Group

GIAL Geographic Information and Analysis Laboratory

GIAM Global Implications of Applied Microbiology

GIANS German Integrated Air Navigation System

GIANT Genealogical Information and Name Tabulating (system) • General Information and Analysis Tool • General Integrated Analytical Triangulation

GIAO Gauge-Invariant Atomic Orbital

GIAPP Geological Image Analysis Program Package

GIAS Global Integration And Synthesis • GPS Integration Alternate Study

gib Graph-in-the-Box Chart (file extension)

Gib Gibraltar

GIB General Instruction Book • General Interface Board • Generic Interface Board • Generierung Intelligent Bedienoberflächen • Geographic Information Board • Gibraltar (ISO country code) • Gibraltar (location identifier) • Guy In the Back

GIBIS GAIA Image and Basic Instrument Simulator • Graphical Issue-Based Information System

GIBS Générateur d'Interface Bord-Sol

GIC Boigu Island (Australia-QLD) (location identifier) • Gabarit International de Chargement • Generic Instrument Class • Geocarto International Centre • Geomagnetically-Induced Current • GPC Interface Console • Grande Installation de Combustion • Graphite Intercalation Compound • Ground-Induced Current

GICC Geographic Information Coordinating Committee

GICEL Groupement des Industries de la Construction Électrique (France)

GICi Gruppo Italiano RV Tauri - Circolare (periodical code)

GICLA Grupo de Implantação do Centro de Lançamento de Alcântara (Brazil-MA)

GICOD Good Idea Cutoff Date

GICR Government/Industry Cooperative Research

GICS Geocoded Image Correction System • Geographic Identification Code Scheme • Grants Information and Control System

GID Gesellschaft für Information und Dokumentation (Germany) • Gestion de l'Information des Décisions • Gitega (Burundi) (location identifier) • Grazing Incidence Diffraction • Group Identification • Group Identificator

GIDAP Guidance Inertial Data Analysis Program

GIDB Geographic Information Database

GIDD GSE Interface Definition Document

GIDEP Government and Industry Data Exchange Program • Government-Industry Data Exchange Program (*) • Government Inter-agency Data Exchange Program

GIDPS Ground Image Data Processing Software

Gidro Gidromekhanika (periodical code)

GIDS General-purpose Interactive Display System • Generic Intelligent Driver Support • Generic Intelligent Driver System

GIDSC GLAST Instrument Development Steering Committee

GIDW General Inverse-Distance Weighted (model)

GIE Global Information Environment • Government-Installed Equipment • Greatest Illuminated Extent • Groupe d'Intérêt Économique • Groupement d'Intérêt Économique • Ground Instrumentation Equipment

GIEC Groupe Intergouvernemental sur l'Évolution du Climat

GIEE Groupe d'Intérêt Économique Européen • Groupement d'Intérêt Économique Européen

GIERS General Industrial Equipment Reserve

GIEWS Global Information and Early Warning System

gif Graphics Interchange Format (file extension)

GIF GAME Interface • Gamma Irradiation Facility • General Image Format • General Information File • German-Israeli Foundation • Gradient Index Fiber • Graphical Interchange Format • Graphic Information File • Graphics Interchange Format • Gravito-Inertial Force • Ground Interface Facility • Guidance Integrated Fuse • Gulf It to Fortran • Winter Haven - Gilbert (USA-FL) (location identifier)

GIFAD Graphical Interface For Astronomical Data • Graphical Interface For Astronomical Databases

GIFAS Groupement des Industries Françaises Aéronautiques et Spatiales (France)

GIFF Graphical Interchange File Format

GIFO Groupement des Industries Françaises de l'Optique (France)

GIFOP Groupe Interprofessionnel de Formation, d'Orientation et de Perfectionnement

GIFOV Ground Instantaneous Field Of View

GIFP Groupe Interprofessionnel de Formation et de Perfectionnement

GIFS Generalized Interrelated Flow Simulation

GIFT Gas-Insulated Flow Tube • General Internal FORTRAN Translator • General Internet File Transfer • Grupo Interuniversitario de Física Teórica (Spain)

GIG Generalized Inverse Gaussian (distribution) • Geometric Interpolation Grid • Global Information Grid • GPS IERS and Geodynamics (experiment) • GPS Integration Guide • Group Infrastructure Grants • Rio de Janeiro (Brazil-RJ) (location identifier)

GIGA Grupo de Informatica Gráfica Avanzada

GiGi Gidroprivod i Gidropnevmoavtomatika (periodical code)

GIGO Garbage In - Garbage Out

GIGOT Garbage In - Garbage Out Theorem

GIGP Generalized Inverse Gaussian-Poisson (distribution)

GIGT Groupe d'Instrumentation des Grands Télescopes (France)

GIGUE Groupement pour l'Informatique de Gestion des Universités et des Établissements

GIH Global Information Highway

GIHS Generalized Information Handling System

GII Global Information Infrastructure • Government Information Infrastructure • Graphics Interactive Interface • Siguiri (Guinea) (location identifier)

GIIC Global Information Infrastructure Commission

GIIS Groupement International d'Informatique de Santé

GIIT Groupe Institutionnel de Terminologie et de Documentation

GIITS General Imagery Intelligence Training System

GIIV Gated-Image Intensifier Viewer

GIK Glucose - Insulin - Potassium

GIL Gaseous Ion Laser • Gilgit (Pakistan) (location identifier) • Graphical Intermediate Language • Green Indicating Lamp

GILC Global Internet Liberty Campaign

GILD Gas-Immersed Laser Doping

GILDAS Grenoble Image and Line Data Analysis Software

GILS Global Information Locator Service • Government Information Locator Service

GIM Generalized Information Management • Génie Industriel et Maintenance • Geomagnetically-induced Image Motion • Geomagnetically Induced Motion • Geomagnetic Image Motion • Global Ionospheric Map • Miele Mimbale (Gabon) (location identifier)

GIMADS Generic Integrated Maintenance and Diagnostic System • Generic Integrated Maintenance Diagnostic System (*)

GIMEX Greenland Ice Margin Experiment

GIMF Galactic Initial Mass Function

GIMI Gestion Intégrée de Maintenance Informatisée • Global Imaging Monitor of the Ionosphere

GIMM Grand Interféromètre Millimétrique

GIMMS Geographic Information Mapping and Management System • Global Inventory Modeling and Monitoring

Study • Global Inventory Monitoring and Modeling Study • Global Inventory Mounting Modeling and Simulation

GIMP Geomagnetically-induced Image Motion Problem (*) • Geomagnetically-Induced Motion Problem • Geomagnetic Image Motion Problem • GNU Image Manipulator Program

GIMS Generalized Information Management System • Geographic Information and Modeling System • Ground Identification of Missions in Space • Group on Instrument and Measurement Science

GIN Geomagnetic Information Node • Global Information Network • Governmental In Nature • Graphics Input • Green Index Number • Guinea (ISO country code) • Guinée (code ISO de pays)

GINA GPS Inertial Navigation Assembly

GINAS GPS Integrated Navigation and Attitude System

GINO Graphical Input Output

GINOF Graphical Input and Output in Fortran

Gio Giovedi

GIO Gamma Interoperability • Government Insurance Office (Australia-NSW) • Graphics Input/Output • Great Island Observatory (USA-RI) • Gurushikhar Infrared Observatory (India)

GIOC Generalized Input/Output Controller

GIOD Globally Interconnected Object Databases (project)

GIOP General Inter-ORB Protocol

GIOSP Groupement International d'Observation des Surfaces Planétaires

GIP Gaussian Image Point • General Information Program • Genetic Improvement Program • Gileppe (Belgium) • Glazed Imitation Parchment • Global Internet Project • Global Inventory Project • Gross Internal Product • Groupement d'Intérêt Public • Guest Investigator Program • Guidance Improvement Program • Pound (Gibraltar) (ISO currency code)

GIPB General-Purpose Instrument Bus

GIPC Global Information Security Partnership Conference

GIPE Generation of Interactive Programming Environments

GIPME Global Investigation of Pollution in the Marine Environment

GIPR Gigabit Internet Protocol Router

GIpS Giga-Instructions per Second

GIPS Generalized Image Processing System • Geographical Information Processing System • Giga-Instructions Per Second • Ground Information Processing System

GIPSY General Image Processing System • General Information Processing System • Generalized Information Processing System • Graphic Information Presentation System • Groningen Image Processing System

GIR Generalized Information Retrieval • Girardot (Colombia) (location identifier) • Groupe Informatique Régional

GIRAFFE Graphic Interface For Finite Elements • Grating Instrument for Radiation Analysis with a Fibre-Fed Échelle

GIRAS Geographic Information Retrieval and Analysis System

GIRD General Incentive for Research and Development (Canada) • Ground Integration Requirements Document • Group for the Study of Reaction Motion

GIREP Groupe International de Recherches en Éducation de la Physique

GIRL Generalized Information Retrieval Language • German Infrared Laboratory • Graph Information Retrieval Language

GIRLS General Indexing in Reciprocal Lattice Space • Generalized Information Retrieval and Listing System • Global Interrogation Recording and Location System

GIRMOS Gemini Infrared Multiobject Spectrograph

GIROS Graphic Interactive Raster-Oriented System

GIRS General Information Retrieval System • Graph Information Retrieval System

GIRSS General Information Retrieval System Simulation

GIRTA Gestion Intégrée des Rechanges par Traitement Informatisé

GIRTS Generic Infrared Training System

GI&S Geospatial Information & Services

GIS Gas Imaging Spectrometer • Gas-scintillation Imaging Spectrometer • Gemini Interlock System • General Information System • Generalized Information System • Generalized Inquiry System • Generalized Instance Set • Geografisch Informatie Systeem • Geographical Information System • Geographic Information System • Geohydrologic Information Service • Geological Information Service • Geological Information Society • Geoscience Information Society • Geospatial Information and Services • Geospatial Information Services • Geospectral Imaging Services (Australia-WA) • Gisborne (New Zealand) (location identifier) • Global Indexing System • Global Information Society • Global Ionospheric Studies • Global Iterative Solution • Globally Integrated Storm • Government Information Services (USA) • Government Information System • Grant Information System • Graphic Information System • Graphic Input System • Grazing Incidence Spectrometer • Groupement d'Intérêt Scientifique • Guidance Information System

GISA Geochemical Interactive Systems Analysis

GiSan Gigiena i Sanitariia (periodical code)

GISAT Ground Identification of Satellites

GISB Gas Industry Standards Board

GISC Government Information Services Committee • Guest Investigator Selection Committee

GISE Generalized Integrated Square Error

GISIG Geographic Information System Interest Group • Geographic Information Systems International Group

GISMO General Interpretative System for Matrix Operations

GISMOS Global Ionospheric Simultaneous Measurements Of Substorms

GISO General Industry Safety Order

GISP General Information System for Planning • Geophysical In Situ Probe • Greenland Ice Sheet Project • Group Independent Study Project

GISS Galileo Interim Support Structure (ESA) • Globally Integrated Substorm • Goddard Institute for Space Studies (USA-NY)

GISSAT Geographic Information System and Spatial Analysis Technologies

GISSEL Galaxy Isochrone Synthesis Spectral Evolution Library

GISSMO Gas Ionization Solar Spectral Monitor

GIST Geographic Information System Toolkit • Geographic Information System for Transportation • Geographic Information System Tutorial • Global Information System Technology • Global Information System Test • Government Information Systems Technology • Graphic Intelligence Support Terminal • Greeting - Inquiring - Suggesting - Thanking • Groupe Informatique Scientifique et Technique

GISTDA Geoinformatics and Space Technology Development Agency (Thailand)

GISW Generalized Interacting Stellar Wind

GIT Gastrointestinal Tract • Geita (Tanzania) (location identifier) • General Information Test • Georgia Institute of Technology (USA-GA) • Graph Isomorphism Tester

GITA Geospatial Information and Technology Association • Ground Instructional Training Aircraft • Groupe d'Informatique et de Traitement Automatique

GITAC General Intelligence Training Advisory Committee

GITC General Intelligence Training Council

GITCAD Global Ionosphere-Thermosphere Coupling And Dynamics • Global Ionospheric Thermospheric Coupling And Dynamics

GITEC General Information Technology • Government Information Technology Executive Council (USA)

GITG Ground Interface Technical Group

GITI Group for Integration of Technical Information

GITIS Government Integrated Technical Information System

GITP Groupement des Industries de Télécommunication et d'Électronique Professionnelle (France) • Ground Integration Test Program

GITS Geosynchronous Imaging Fourier Transform Spectrometer • Government Information Technology Services (USA)

Giu Giugno

GIU Graphics Interface Unit • Guidance Integration Unit

GIUA Geophysical Institute of the University of Alaska (USA-AK)

GIUB Geodetic Institute of the University of Bonn (Germany)

GIUK Greenland - Iceland - United Kingdom

GIUSA Government Industry University Software Alliance (USA)

GIV Geological Information Visualization

GIVN Geodetic International VLBI Network

GIWG Ground Interface Working Group

GIWTLAF Globally Integrated Workbench for Teaching and Learning Accounting and Finance

GIX Global Information Exchange • Global Internet Exchange • Government Information Exchange

GIXS Grazing-Incidence X-ray Scattering

GIY Giyani (South Africa) (location identifier)

GIZ Gizan (Saudi Arabia) (location identifier)

GIZC Gestión Integrada de Zonas Costeras

GI2T Grand Interféromètre à Deux Télescopes

GI3T Grand Interferometer of Three Telescopes • Grand Interféromètre à Trois Télescopes

GJ Gigajoule • Gliese-Jahreiß (catalog) • Grapefruit Juice • Grown Junction

GJA Guanaja (Honduras) (location identifier)

GJI Geophysical Journal International

GJL Jijel (Algeria) (location identifier)

GJM Guajará-Mirim (Brazil-RO) (location identifier)

GJP Galactic Jupiter Probe • Good Job Partner • Graphic Job Processor

GJR Gjogur (Iceland) (location identifier)

GJRO German-Japanese (Earth) Rotation Observations

GJT Grand Junction (USA-CO) (location identifier)

Gk Greek

GK Grundkurs

GKA Goroka (Papua New Guinea) (location identifier)

GKD Gemeinsame Körperschaftsdatei • Gordon - Kendall - Davison (notation)

GKE Geilenkirchen (Germany) (location identifier)

GKG Gerichtskostengesetz

GKH Gorkha (Nepal) (location identifier)

GKK Gantmacher - Krein - Kotelyanski (matrix)

GKL Gacs - Kurdyumov - Levin (rule) • Great Keppel Island (Australia-QLD) (location identifier)

GKN Gulkana (USA-AK) (location identifier)

GKO Kongoboumba (Gabon) (location identifier)

GKP Global Knowledge Partnership

GKS Graphical Kernel System (*) • Graphics Kernel System

g/l Grams per Liter

gl Galician (ISO language code) • Gill • Greenland

Gl Glaciar • Glacier • Glas • Gleichung • Gleis • Gloss • Glucinum • Gold

G/L General Ledger

GL Garage Level • Gas Laser • Gauge Length • General Ledger • General Lexicon • General Linear • Generic List • Geophysical Laboratory • Geophysics Laboratory • Geosciences Laboratory • Giant Lobe • Gliese (catalog) • Gothic Letter • Government Laboratory • Grade Level • Grande Ligne • Graphical Link • Graphics Language • Graphics Library • Gravitational Lens • Gravitational Lensing • Gravitationally-Lensed • Gray Level • Greatest Length • Greenland (ISO country code) • Green Light • Groenland (code ISO de pays) • Ground Level

GLA Gamma-Linolenic Acid • General Ledger Account • Giant Luminous Arc • Glasgow (Starlink internal code) • Glasgow (UK-Scotland) (location identifier) • Greater London Authority (UK-England) • Ground Long Activity • GSFC Laboratory for Atmospheres

GLAC General Ledger Accounting Code

GLACS Ground Loop Attitude Control System

GLADIS Global Astronomical Data and Information Server

GLADYS Geophysics Laboratory Array Detector Spectrometer

GLAG General Ledger Accounting Guide • Ginzburg - Landau - Abrikosov - Gorkov (formulation)

GLAI Green Leaf Area Index

Glam Glamorgan (UK-Wales) • Glamorganshire (UK-Wales)

GLAN Global Local Area Network

GLARE Glass-Reinforced

GLARS Geoscience Laser Altimetry Ranging System

Glas Glasgow (UK-Scotland)

GLAS Geodesic Laser Altimeter System • Geoscience Laser Altimeter System • Goddard Laboratory for Atmospheric Sciences • Global Land AVHRR Sampled

GLASOD Global Assessment of Soil Degradation

GLASS Generalized Language and Speech System • Germanium-Lithium-Argon Scanning System

GLAST Gamma Large-Area Silicon Telescope • Gamma-ray Large-Area Space Telescope (*)

GLAT Geomagnetic Latitude • Groupe de Linguistique Appliquée des Télécommunications

GLB Glas Lubricated Bearing • Global Location Broker • Grease Lubricated Bearing • Greatest Lower Bound • Ground Location Beacon

GLBC Great Lakes Basin Commission

GLC Gas-Liquid Chromatography • Geladi (Ethiopia) (location identifier) • Generator Line Contractor • Global Linear Collider • Gray-Level Cooccurrence • Greater London Council • Ground-Level Concentration

GLCF Galaxy Luminosity Correlation Function

GLCM Gray-Level Cooccurrence Matrix

Gld Guilder • Gulden

GLD Gas Leak Detector • Generalized Levenshtein Distance • Goodland (USA-KS) (location identifier) • Gross Logical Design • Ground Laser Designator • Groupe de Légitime Défense

GLDH Gray Level Difference Histogram

GLDT Gas Laser Discharge Tube

GLDV Gesellschaft für Linguistische Datenverarbeitung (Germany) • Gray Level Difference Vector

GLE Gainesville (USA-TX) (location identifier) • Generalized Langevin Equation • Generalized Luminosity Evolution • Government Loaded Equipment • Ground Level Effect • Ground Level Enhancement • Ground Level Event

GLEAM Graphic Layout and Engineering Aid Method • Ground-based Lidar Experiment for Atmospheric Measurement

GLEP Group for Lunar Exploration and Planning

GLERL Great Lakes Environmental Research Laboratory

GLF Galaxy Luminosity Function • Gaussian Lens Formula • Generalized Lambda Family • Golfito (Costa Rica) (location identifier)

GLG Glengyle (Australia-QLD) (location identifier)

GLH Greenville (USA-MS) (location identifier)

GLI Glen Innes (Australia-NSW) (location identifier) • Global Imager

GLIDS Grey Level Indicating Display System

GLIM Generalized Linear Interactive Modeling • Generalized Linear Model

GLIMPCE Great Lakes International Multidisciplinary Program on Crustal Evolution

GLIMPSE Galactic Legacy Infrared Mid-Plane Survey Extraordinaire (SIRTF) • Global Implicit Search

GLIN Global Legal Information Network • Grijze Literatuur in Nederland

GLINT Gated Laser Intensifier • GEO Light Imaging National Testbed • Global Intelligence • Ground-based Lidar Investigation Transportable

GLIPAR Guidelines for Investigation, Planning and Research

GLIS GEAC Library Information System • Geographic and Land Information Society • Global Land Information System

GLK Galcaio (Somalia) (location identifier)

GLL Gol Klanten (Norway) (location identifier)

GLLD Ground Laser Location Designator • Ground Locator Laser Designator

GLLS Generalized Linear Least Squares (program)

GLM Generalized Linear Model • General Land Office (USA) • General Linear Model • Gigabaud Link Module • Gigabit Link Module • Glenormiston (Australia-QLD) (location identifier) • Government-Loaned Material • Graduated-Length Method • Grand Livre du Mois

GLN Glen • Goulimime (Morocco) (location identifier)

GLNS Glens

GLO Galaxy-Like Object • Get the Lead Out • Gloucester (UK-England) (location identifier) • Goddard Launch Operations • Ground Liaison Officer • Ground Logistics Operation

GLOB Galaxy-Like Object

GLOBE Global Backscatter Experiment • Global Land One-kilometer Base Elevation • Global Observations to Benefit the Environment • Global Learning and Observations to Benefit the Environment (*) • Global Learning by Observations to Benefit the Environment

GLOBEC Global Ocean Ecosystem

GLOBECOM Global Communications

GLOBUS Global Budget of the Stratosphere

GLOCHANT Global Change Antarctica

GloED Global Emissions Database

GLOM Gross Lift-Off Mass

GLOMAC Global Modeling of Atmospheric Constituents • Global Modeling of Atmospheric Chemistry • Global Modeling of the Atmosphere Chemistry

GLOMEX Global Meteorological Experiment (now GARP)

GLOMO Global Mobile

GLOMR Global Low-Orbiting Message Relay • Global Low-Orbit Message Relay (*)

GLONASS Global Navigation Satellite System

GLOPAC Gyroscopic Low-Power Attitude Control

GLORe Goethe Link Observatory Reprints (periodical code)

GLORI Global and Local Radar Imager

GLORIA Geological Long-Range Inclined Asdic

Glos Glossary • Gloucestershire (UK-England)

GLOS Gaseous Luminosity on Optical Surface

Gloss Glossary

GLOSS Global Observation Surveillance System • Global Ocean Surveillance System • Global Sea Level Observing System • Global Sounding System

GLOTRACK Global Tracking (system)

GLOUP Global Undersea Pressure

GLOW Gross Lift-Off Weight (*) • Ground Lift-Off Weight • GSFC Lidar Observatory for Winds

GLP Gas Licuado de Petróleo • General Layout Plan • General Linguistic Processor • Good Laboratory Practice • Government Loaned Property • Gross Lawyer Product • Guadeloupe (code ISO de pays) • Gulgubip (Papua New Guinea) (location identifier)

GLPA Great Lakes Planetarium Association (USA)

Glpg Glowplug

GLPS Genetic Logic Programming System

GLQ Glennallen (USA-AK) (location identifier)

GLR Gaylord - Otsego County (USA-MI) (location identifier) • Government Land Register • Graphic Level Recorder

GLRAL Great Lakes Region of the Astronomical League (USA)

GLRC Gas-Liquid Radiochromatography • Great Lakes Research Center

GLRD Great Lakes Research Division

GLRS Geodynamics Laser Ranging System • Geoscience Laser Ranging System

GLS Galveston Scholes (USA-TX) (location identifier) • Generalized Least Squares • Generic Learning Support • Glide Slope • Global Learning Scheme • Global Logical Schema • Green Laser System • Ground Launch Sequencer

GLSA General Ledger Subsidiary Account

GLSB Giant-Low-Surface-Brightness (galaxy) • Global Load Server Balancing

Glsgw Glasgow (UK-Scotland)

GLSR Geodynamics Laser Ranging System

Glsry Glossary

GLSVLSI Great Lakes Symposium on Very-Large-Scale Integration

GLT Gas Laser Tube • Gladstone (Australia-QLD) (location identifier) • Ground Laser Tracking

GLTN Goddard Laser Tracking Network (USA-MD)

GLTP General Language Transformation Process

GLU General Linking Utility • Générateur de Liens Uniformes (*) • Générateur de Liens Universels • Global Land Use • Group Logic Unit

GLUE Global User Exit • Goldstone Lunar Ultra-high Energy (neutrino experiment)

GlV Globe Valve

GLV Ginzburg-Landau Vortex • Golovin (USA-AK) (location identifier)

GLW Glasgow (USA-KY) (location identifier)

GLX Galela (Indonesia) (location identifier)

gly Glossary (file extension)

GLY Glossary • Goldsworthy (Australia-WA) (location identifier)

Gly Glycine

Glyc Glycerite

Glycn Glycerine

GLZ Breda Gilze-Rijen (Netherlands) (location identifier)

gm Gigameter • Gram

Gm Gigameter

G&M Geography & Map

GM Gambia (ISO country code) • Gambie (code ISO de pays) • Gaseous Mixture • Gas Module • Gauss-Markov • Geiger-Müller (counter) • Generalized Method • General Maintenance • General Manager • General Meeting • General Member • General Merchandise • General Merit • General Mess • General Mortgage • General Motors Corporation • Genetically Modified • Geometric Mean • Grand Master • Gravitational Mass • Gravitational Microlensing • Gravitomagnetic • Greenwich Meridian • Grid Modulation • Ground Malfunction • Ground Map • Ground Mobile • Ground Mode • Group Mark • Group Modem • Group Mobile • Guam • Guidance Manager

GM&A General Management & Administration

GMA Gefahren-Meldeanlage • Gemena (Zaire) (location identifier) • General Management and Administration • General Mental Ability • Giant Molecular Association • Government Modification Authorization • Graphe des Marquages Accessibles • Gruppo Marsicano Astrofili (Italy) • Guest Observer

GMAB Give Me A Break

GMAG Geophysics Magnetometer

GMAN Global Microlensing Alert Network

GMAO Gestion de Maintenance Assistée par Ordinateur

GMAP General Micro-Assembler Processor • General Micro-Assembler Program • Geometric Modeling Applications Programme • Geometric Modeling Applications Project • Geometric Modeling Applications Protocol

GMAPS General Map Analysis Planning System

GMAS General Management Assessment System

GMAT General Management Admission Test • Graduate Management Admission Test • Graduate Management Aptitude Test • Greenwich Mean Astronomical Time

GMATS Government Metropolitan Area Telephone Service (USA)

GMAW Gas Metal Arc Welding

GMB Gambela (Ethiopia) (location identifier) • Gambia (ISO country code) • Gambie (code ISO de pays) • Gran Mancha Blanca (Saturno)

GmbH Gesellschaft mit beschränkter Hafting

Gmc Germanic

GMC Generalized Magnetic Coordinates • General Management Committee • Giant Molecular Cloud • Glinka Memorial Collection • Gross Maximum Capacity • Ground Mission Control • Guerima (Colombia) (location identifier)

GMCA Genetic Maximum Clique Algorithm

GMCC General Monte Carlo Code • Geophysical Monitoring for Climate Change • Ground Mobile Control Centr

GMCF Galaxy Mass Correlation Function • Goddard Mission Control Facility

GMCL Ground Measurements Command List

GMCP Ground Mobile Command Post

GMCS Giant Molecular Cloud Structure

GMD Dalasi (Gambia) (ISO currency code) • Gel Microdrop • General Management Directive • Geometric Mean Distance • Gesellschaft für Mathematik und Datenverarbeitung (Germany) • Gravitational Measurements Division (SAI) • Ground Meteorological Detector • Ground Meteorological Device • Group Multiplexer/Demultiplexer

GMDA Group Method of Determining Arguments

GMDB Grid Meta-Database

GMDES Gridded Meteorological Data Extraction System

GMDH Group Method of Data Handling

GMDL GEC Marconi Dynamics Ltd.

GMDSS Global Maritime Distress and Safety System

GME Generic Model Editor • Gestion et Management des Entreprises • Gimbal Module Electronics • Goddard Medium Energy (experiment) • Gomel (Belarus) (location identifier) • Grupo de Mecánica Espacial

GMES Global Monitoring for Environment and Security

GMF General Mail Facility • Global Market Forecast • Government Master File • Ground Module Force • Ground Monitor Facility

GMG Gross Maximum Generation

GMH General Mail Handling

GMHS General Mail Handling System

GMI Gasmata (Papua New Guinea) (location identifier) • Génie Mathématique et Informatique • Goddard Management Instruction • Government Manufacturing Group • GSFC Management Instruction • Grupo de medios Impresos

GMIC Glass Microwave Integrated Circuit

GMIP Government Mandatory Inspection Point

GMIR Global Merged Interaction Region

GMIRS Global Merged Interaction Regions

GMIS Generalized Management Information System • GSSC Management Information System

GMITS Guidelines for the Management of Information Technology Security

GMK Gyromagnetic Kompass

GML Generalized Markup Language • Generalized Maximum Likelihood • General Markup Language • General Measurement Loop • General Modeling Language • Geographic Markup Language • Giant Magnetic Loop • Graphic Machine Language • Gravitational Microlensing

GMLCC Ground Mobile Launch Control Center

GMLS Geometric Method of Least Squares

GMM Gamboma (Congo) (location identifier) • Gaussian Mixture Model • Generalized Mixed Model • Generic Mars Mission • Geometric Mathematical Model

GMMSS Global Multimedia Mobile Satellite System

GMMWJ Gesellschaft für angewandte Mathematik und Mechanik - Jahrestagung (periodical code)

GMN Greymouth (New Zealand) (location identifier)

GMO Garantie Multioptions • Genetically-Manipulated Organism • Genetically-Modified Organism • Group of Meteorological Observations

GMOS Gemini Multiobject Spectrograph

GMP General catalog of Proper Motions • Genetically Modified Plant • Global Mobile Professional • Good Manufacturing Practice • Graph-Matching Problem • Guaranteed Minimum Pension

GMPCS Global Mobile Personal Communications by Satellite • Global Mobile Personal Communications System (*)

GMPI General and Mechanical Patents Index

GMPL Global Managed Private Line

GMPLS Generalized Multiprotocol Label Switching • Global Multimode Precision Landing System

GMPR Geostationary Microwave Precipitation Radiometer

GMQ Good Merchantable Quality

GMQS Gruppo Milanese Quadranti Solari (Italy)

GMR Galactic Molecular Ring • Gambier Island (location identifier) • Geometric Mean Radius • Giant Magnetoresistance • Giant Magnetoresistive • GOES-Meteosat Relay • Ground Mapping Radar

GMRA Government Management Reform Act (USA)

GMRES Generalized Minimal Residual • Generalized Minimum Residual

GMRF Gaussian Markov Random Field

GMROC Ground Mobile Regional Operations Center

GMRP GARP Multicast Registration Protocol

GMRT Giant Meterwave Radio Telescope (India)

GM&S General, Medical & Surgical

GMS Gas Management System • General Mail System • General Maintenance System • General Military Science • Geophysical Monitoring Satellite • Geophysical Monograph Series (periodical code) • Geostationary Meteorological Satellite • Global Management System • Global Mapping Satellite • Global Messaging Service • Global Messaging System • Grant-Maintained Status • Grants Management System • Gravitational Mass Sensor • Gravity Measuring System • GRB Monitoring System • Ground Maintenance Support • Ground Mobile Segment • Guidance Monitor Set • Guimarães (Brazil-MA) (location identifier)

GMSC Geologic Map Standards Committee

GMSD Gas Microstructure Detector

GMSK Gaussian-filtered Minimum-Shift Keying • Gaussian Minimum-Shift Keying

GMSP Great Manitou Star Party (Canada-ON)

GMSS Geostationary Meteorological Satellite System • Global Meteorological Satellite System

GMST Greenwich Mean Sidereal Time

GMSW&D Guaranteed Maximum Shipping Weight & Dimensions

GMSWD Guaranteed Maximum Shipping Weight and Dimensions

GMT Geiger-Mueller Tube • General Machine Test • Generic Mapping Tool • Geomarine Technology • Giant Modular Telescope • Granite Mountain (USA-AK) (location identifier) • Greenwich Mean Time

GMTA Great Minds' Think Alike

GMTF Geometric Modulation Transfer Function

Gmtry Geometry

GMU George Mason University (USA-VA) • GPS Monitoring Unit • Greenville Downtown (USA-SC) (location identifier)

GMuD Gesellschaft Mathematik und Datenverarbeitung - Multigrid Methods Special Topics Applications (periodical code)

GMV Gram-Molecular Volume • Grupo de Mecánica de Vuelos SA (Spain) • Guaranteed Minimum Value

GMW Gram Molecular Weight • Große Magellansche Wolke

GMY Rheindahlen (Germany) (location identifier)

GMZ Grundmietzeit

gn Green • Guarani (ISO language code) • Guinea

G&N Guidance & Navigation

G/N General Notes

GN Gauss-Newton • General Notes • Graduate Nurse • Ground Network • Guidance and Navigation • Guinea (ISO country code) • Guinée (code ISO de pays)

GNA Gaseous Nitric Acid • Global Network Adressing • Globewide Network Academy (USA-TX) • Grodna (Belarus) (location identifier)

GNAA Gridless Narrow-Angle Astrometry

GNAAC Global Network Associated Analysis Center

GNAO Gemini North Adaptive Optics

GNAS General National Airspace System

GNAT General Numerical Analysis of Transport • Global Network of Astronomical Telescopes • Global Network of Automatic Telescopes

GNATS General Non-linear Analysis of Two-Dimensional Structures

GNB Grenoble Saint-Geoirs (France) (location identifier) • Guinea-Bissau (ISO country code) • Guinée-Bissau (code ISO de pays)

GNBS Guyana National Bureau of Standards

GN&C Guidance, Navigation & Control

GNC Geologic Names Committee • Global Navigation Chart • Goddard Network Control • Graduated Non-Convexity • Graphical Numerical Control • Graphic Numerical Control • Groupe National de Contrôle • Guidance and Navigation Computer • Guidance and Navigation Control • Guidance, Navigation and Control (*)

GNCC Guidance Navigation Control Computer

GNCIS Guidance, Navigation and Control Integration Simulator

GN&CS Guidance, Navigation & Control System

GNCS Guidance and Navigation Control System • Guidance, Navigation and Control System

Gnd Ground

GND Grenada (Windward Islands) (location identifier) • Ground

GNDÄ Gesellschaft Deutscher Naturforscher und Ärzte (Germany)

GNDFG Ground Fog

GNE Gent (Belgium) (location identifier) • Greatest Northern Elevation • Guidance and Navigational Electronics • Guidance and Navigation Electronics • Guidance and Navigation Equipment

GNF Franc (Guinea) (ISO currency code)

GNFMS Gaseous Nitrogen Flow Measuring System

GNFS General Number Field Sieve

GNG Gaussian Noise Generator • Gooding (USA-ID) (location identifier)

GNI Get Net Info • Global Network Infrastructure • Google News Index • Green Island (China-ROC) (location identifier) • Gross National Income • Gross National Investment

GNIR Get Net Info Reply

GNIRS Gemini Near-Infrared Spectrograph

GNIS Geographic Names Information System

GNL Gaz Naturel Liquéfié • Greatest Northern Latitude

GNLS GPS Navigation and Landing System

Gnltd Granulated

GNM Grand Nuage de Magellan • Guanambi (Brazil-BA) (location identifier)

GNMA Global Network Management Architecture

GNMP Government Network Management Profile • Government Network Management Protocol

GNMS Gaseous Nitrogen Measuring System

GNN Ghinnir (Ethiopia) (location identifier) • Global Network Navigator

GNO Gallium Neutrino Observatory • Gamma Normids (meteor stream) • Gross Product Originating

GNOL Good Neighbor Lighting

GNOLE Good Neighbor Lighting Experiment

Gnom Gnomónica

GNOME GNU Network Object Model Environment

GNOMES Generalized NRAO Observe, Multitelescope Experiment Scheduler

GNOS Goddard Network Operations Support

GNOSC Global Network Operations and Security Center

GNOSIS Gesellschaft für Normierte Organisations-Software und Informationsstandards

GNP Graphics Nesting Program • Gross National Product • Guidance Navigation and Processor

GNPC Global Navigation and Planning Chart

GNQ Equatorial Guinea (ISO country code) • Guinea (Equatorial_) (ISO country code) • Guinée Équatoriale (code ISO de pays)

GNR Gaseous Nuclear Rocket • General Roca (Argentina-RN) (location identifier) • Genetics - Nanotechnology - Robotics • Geographical Names Register

Gnrl General • Général

GNRT Generalized Near-Real Time

Gnrx Généraux

gns Guineas • Guinées

G&NS Guidance & Navigation System

GNS Franc (Guinea) (ISO currency code) • Get Nearest Server • Global Navigation System • Global Network Service • GOMR Nadir Sounder • Graphics Name Service • Guidance and Navigation Simulator • Guidance and Navigation System • Guidance Navigation System • Gunungsitoli (Indonesia) (location identifier)

GNSO Goddard Network Support Operations

GNSS Global Navigation Satellite System • Ground Network Scheduling System

GNT Grants Milan (USA-NM) (location identifier)

GNTO General Nathan Twining Observatory (USA-NM)

GNU General Neural Unit • Geologic Names Unit • GNU is Not Unix • Goodnews Bay (USA-AK) (location identifier)

GNULEX Geologic Names Unit Lexicon

GNUT Goddard-Navy Universal Time

GNV Gainesville (USA-FL) (location identifier) • Gaz Naturel pour Véhicule

GNVQ General National Vocational Qualification

GNW Global Nuclear War

GNY Granby - Grand County (USA-CO) (location identifier)

GNZ Ghanzi (Botswana) (location identifier)

GN2 Gaseous Nitrogen

G&O Gain & Offset

GO Garantie d'Origine • Gas Operated • Gaussian Orbital • Gebruikerondersteuning • General Observer • General Office • General Officer • General Order • General Organization • Generalized Operation • Geodetic Observatory • Geometric Optics • Geometry-Optimized • Global Optimization • Goiás (Brazil) • Government Office • Government-Operated • Government-Owned • Grandes Ondes • Grant Officer • Graphitic Oxide • Gros Oeuvre • Guest Observer

GOA General Operating Agency • Generic Open Architecture • Gênes (Italy) (location identifier) • Genoa (Italy) (location identifier) • Geologisch - Oceanografisch - Atmosferisch • Gyro-Optics Assembly • Gyro Output Amplifier

GOAA Greater Orlando Airport Authority (USA-FL)

GOADO Global One-Armed Disk Oscillations (model)

GOAL Galactic Origin and the Acceleration Limit • General Orientation Analysis Language • Ground Operations Aerospace Language

GOALI Grant Opportunities for Academic Liaison with Industry

GOALS Global Ocean-Atmosphere-Land System

GOAS Guidance Optical Alignment Shelter

GOAT Gerber Oscillogram Amplitude Translator • Goes Over All Terrain

Gob Gobernación

GOB Goba (Ethiopia) (location identifier)

GOBELIN GOME Breadboard Model Experimental Link

GOBEP Generalized One-Boson Exchange Potential

GOBI General Officer Bright Idea

GOBILS Government Bill of Lading System (USA)

GOC General Optical Council • Global Operations Control • Gora (Papua New Guinea) (location identifier) • Government Of Canada • Ground Operations Coordinator • Ground Operations Center

GOCA Ground Operations Control Area

GOCAP Graphic Output Circuit Analysis Program

GOCC GATE Operational Control Center • Geodetic Operations Control Center • Global Operations Control Center

GOCE Gravity and Ocean Circulation Explorer

GOCI General Operator-Computer Interaction

GOCO Government-Owned - Contractor-Operated

GOCP Groupe des Observateurs du Ciel Profond (France)

GOD Giant Optical Device

GODAE Global Ocean Data Assimilation Experiment

GODAS Graphically-Oriented Design and Analysis System

GODC Geosciences Operational Data Center

GODIVA GOME Data Interpretation, Validation and Application

GODSEP Guidance and Orbit Determination for Solar Electric Propulsion

GOE Gaussian Orthogonal Ensemble • General Operating Expense • Gonalia (Papua New Guinea) (location identifier) • Government-Owned Equipment • Ground Operating Equipment • Ground Operational Equipment • Ground Outfitting Equipment

GOED Global Oceans Ecosystems Dynamics

GOES Geostationary Operational Environmental Satellite (*) • Geostationary Orbiting Environmental Satellite • Geosynchronous Operational Environmental Satellite • Geosynchronous Orbiting Environmental Satellite • Global Omnibus Environmental Survey

GOEZS Global Ocean Euphotic Zone Study

GOF Global Ocean Flux • Goodness Of Fit • Government-Owned Facility • Group Of Frames • Guest Observer Facility • San Angelo - Goodfellow (USA-TX) (location identifier)

GOFC Global Observations of Forest Cover (project)

GOFS Global Ocean Flux Study

GOGO Government-Owned - Government-Operated

GOH Guest Of Honor • Nuuk (Greenland) (location identifier)

GOHREM Geosynchronous Orbit High-Resolution Earth Monitoring

GOHSS Galileo OH Subtracted Spectrometer

GOI Geodynamisch Onderzoek Instituut (Netherlands) • Geostationary Operational Imager • Goa Dabolim (India) (location identifier) • Government Of India • Government-Owned Installation

GOIN Global Observation Information Network

GOJ Nizhnij Novgorod (Russia) (location identifier)

GOK Guthrie (USA-OK) (location identifier)

GOL General Operating Language • Gold Beach State (USA-OR) (location identifier)

GOLD General Online Display • General Organization of Large Databases • Geophysical Online Data • Global Online Data • Global Online Directory

GOLDIS Geophysics Online Data and Information System

GOLDS General Online Display Station • General Online Display System

GOLEM Großpeicher Orientierte Listenorganisierte Ermittlungsmethode

GOLF Global Oscillations Low-Frequency (SOHO)

GOLISS Galileo Online Search System

GOLL Graphic Online Language

GOLS German Oceanic Lidar System

GOM Global Ocean Monitoring • Goma (Zaire) (location identifier) • Government-Owned Material • Grand Old Man • Ground Operations Manager • Groupe d'Observations Météorologiques

GOME Global Ozone Monitoring Experiment

GOMI Global Ozone Monitoring Instrument

GOMM General Organizational Maintenance Manual

GOMMS Ground Operations and Material Management System

GOMOS Global Ozone Monitoring by Occultation of Stars

GOMPS Global Ocean Monitoring Payload Study

GOMR Global Ozone Monitoring Radiometer

GOMS Geostationary Orbiting Meteorological Satellite • Global Ocean Monitoring Satellite • Global Ozone Monitoring Satellite • Goals - Operators - Methods - Selection • Ground Operations Management System • Geostationary Operational Meteorological Satellite

GOMY Get Out of My Way!

GON Groton (USA-CT) (location identifier) • New London (USA-CT) (location identifier)

GONG Global Oscillations Network Group

GOO Generalized Overhauser Orbital • Goondiwindi (Australia-QLD) (location identifier) • Ground Observer Organization • Ground Operation Order

GOOB Going Out Of Business

GOODS Great Observatories Origins Deep Survey

GOOS Global Ocean Observing System (*) • Global Ozone Observing System

GO3OS Global Ozone Observing System

GO&P General Operations & Plans

GOP General Operating Procedures • General Operational Plot • General Operations and Plans • General Outpost • Generated Options Plan • Geographical Observatories Programme • Good Old Paper • Gorakhpur (India) (location identifier) • Government-Owned Property • Grand Old Party (USA) • Ground Operations Panel • Ground Operations Plan • Group Of Pictures • Guest Observer Program

GOPE Government-Owned Plant Equipment

GOPEX Galileo Optical Experiment

GOPG Ground Operations Planning Group

GOpS Gigabyte Operations per Second • Giga Operations per Second

GOPS Gigabyte Operations Per Second • Giga Operations Per Second

GOQ Golmud (China-PRC) (location identifier)

GOQS General Online Query System

GOR Gained Output Ratio • General Operating Room • General Operational Requirements • Gore (Ethiopia) (location identifier) • Grant Officer Representative • Ground Operations Review

GORBI Graphical Object Retrieval and Browsing Interface

GORD Ground Operations Requirements Document

GORDES Groupe Opérationnel de Recherche, de Documentation et d'Étude sur la Science (Belgium)

GORF Goddard Optical Research Facility (now GGAO)

GORMP Global Ozone Research and Monitoring Program • Global Ozone Research and Monitoring Project

GORP Goal-Oriented Re-Planner • Ground Operational Requirements Plan • Ground Operations Requirements Plan • Ground Operations Review Panel

GORR Ground Operations Readiness Review

GORS General Online Retrieval System • General Organization Of Remote Sensing

GOS Gemenebest van Onafhenkelijke Staten • Generalized Oscillator Strength • Geodetic Optical System • Geomagnetic Observing System • Geostationary Operational Sounder • Global Observing System • Global Ozone Sensor • Gosford (Australia-NSW) (location identifier) • Grade Of Service • Gross Operating Surplus • Ground Operations System • Ground Optical Station • Guest Operating System

GOSAM Gelatin of Sols: Applied Microgravity Research

GOSB General Observer Support Branch

GOSC Global Operations and Security Center

GOSEAC Group of Specialists on Environmental Affairs and Conservation

GOSIP Government Open Systems Interconnection Profile

GOSP Goldendale Observatory State Park (USA-WA) • Ground Operational Support Plan

GOSRP Ground Operational Support Requirements Plan

GOSS GOES Operational Support System • Ground Operational Support System • Ground Operations Support System • Groupe des Opérateurs de Stations SPOT

GOSSP Global Observing Systems Space Panel

GOSST Global Ocean Sea Surface Temperature

GOST Gravito-Optical Surface Trap

Gót Gótico

GOT General Observing Time • Göteborg Landvetter (Sweden) (location identifier) • Gothenburg Landvetter (Sweden) (location identifier) • Government-Owned Terminal • Ground Operations Team

GOTF Goddard Optical Test Facility

Goth Gothic

GOTS Governmental Off-The-Shelf • Government Off-The-Shelf • Gravity-Oriented Test Satellite

Gou Gourde (currency)

GOU Garoua (Cameroon) (location identifier)

Gouv Gouvernement • Gouvernemental

Gov Government • Governmental • Governor

GOV Generator Output Voltage • Gove Nhulunbuy (Australia-NT) (location identifier) • Government • Governor

Gov. Gen. Governor General

Govt Government

GOVT Government

GOW Garrow Lake (Canada-NT) (location identifier)

GOWG General Officer Working Group • Ground Operations Working Group

GOWI Get On With It

GOX Gaseous Oxygen

GOY Gal Oya - Amparai (Sri Lanka) (location identifier)

GOYA Get Off Your After-end

GOZ Gorna Orjahovica (Bulgaria) (location identifier) • Groupement Opérationnel de Zone

GO2 Geaseous Oxygen

GO3OS Global Ozone Observing System

g/p Giro Postal

gp Galley Proofs • Group

Gp Group

GP Galactic Plane • Galactic Probe • Galaxy Population • Gallup Poll • Gamma-Poisson (process) • Gas-Permeable • Gas Plasma • Gas Pressure • General Paralysis • General Pause • General Practitioner • General Processing • General Processor • General Product • General Protection • General Provision • General Publication • General Purpose • Generalized Programming • Genetic Programming • Geographical Pole • Geographical Position • Geographic Point • Geometric Pitch • Geometric Progression • Geschichte der Physik • Giant Planet • Glidepath • Global Positioning • Global Power • Gloria Patri • Government Personnel • Government Property • Graduated Pension • Grand Prix • Grand Public • Graphic Production • Graphics Processor • Gratuity Passenger • Gravity Probe • Great Primer • Ground Processor • Ground Protective • Group Permit • Group Processor • Guadeloupe (code ISO de pays) • Guard period • Guidance Package • Guidance Processor • Guided Projectile • Guinea Pig • Gunn-Peterson (effect) • Gun Position • Gyro Package

GPa Gigapascal

GPA Galactic Plane A (survey) • Gas Path Analysis • General-Purpose Amplifier • General-Purpose Analysis • Gestion Partagée des Approvisionnements • GPRA Performance Assessment (USA) • Grade Point Average • Graphical PERT Analog • Gravity Probe A • Great Problem of Astrophysics • Grosse Pénétration Active • Ground Plane Antenna • Grounded Plate Amplifier • Groupe Patrimoine Astronomique (France) • Guidance Platform Assembly • Guidance Positioning Assembly • Patras Araxos (Greece) (location identifier)

GPAA General Physics Advanced Abstracts

GPAC General-Purpose Analog Computer

gpad Gallons per Acre per Day

GPADS Guided Parafoil Airborne Aerial Delivery System • Guided Precision Aerial Delivery System

GPALS Global Positioning And Location System

GPAMS General-Purpose Automated Maintenance System • Ground Processing Automated Maintenance System

GPAO Gestion de Production Assistée par Ordinateur

GPAPD Gallons Per Acre Per Day

GPARS Generic Phased Array Radar Simulation

GPAS General Performance Appraisal System • General-Purpose Airborne Simulator • Global Print Accounting System

GPATS General Purpose Automatic Test System

GPB General-Purpose Buffer • Gravity Probe B • Group Parameter Block • Guarapuava - Tancredo Thomaz Faria (Brazil-PR) (location identifier)

GPBIM General-Purpose Buffer Interface Module

GP&C Global Positioning & Communication

GPC Gauge Pressure Control • Gel Permeation Chromatograph • Gel Permeation Chromatography • General Precision Connector • General-Purpose Committee • General-Purpose Computer • General-Purpose Container • General-Purpose Counter • Global Planetary Change (periodical code) • Global-Processing Center • Government Procurement Code • Granville-Phillips Co. • Graphic Performance Characterization • Graphics Performance Committee • Groupe de Pilotage de Circonscription • Gyro Pulse Count

GPCC Global Precipitation Climatology Centre

gpcd Gallons per Capita per Day

GPCDU General-Purpose Control and Display Unit

GPCE Generative Programming and Component Engineering

GPCI General-Purpose Channel Interface

GPCIPL General-Purpose Computer Initial Program Load

GPCL General-Purpose Closed Loop

GPCOE General-Purpose Checkout Equipment

GPCP Generalized Process Control Programming • Global Precipitation Climatology Project

GPCPD Gallons Per Capita Per Day

GPCR G-Protein Coupled Receptor

GPCSC General-Purpose Computer Support Center

gpd Gallons per Day

GPD Gallons Per Day • Gaseous Protoplanetary Disk • Gas Pressure Dominated • Generalized Probabilistic Descent • General Presentation Descriptor • General Publishing Division • General-Purpose Data • Gibbs Probability Distribution • Gimbal Position Display • Graphics Products Division • Groupe de Pilotage Départemental

GPDC General-Purpose Digital Computer

GPDL Graphical Picture Drawing Language

GPDS General-Purpose Display System

GPE Galactic Plane Enhancement • Gas Power Exchange • General-Purpose Equipment • Gestion Prévisionnelle des Emplois

GPEA Government Paperwork Elimination Act (USA)

GPEC Gestion Prévisionnelle des Emplois et des Compétences

GPEE General Purpose Encryption Equipment

GPEF Generic Package of Elementary Function

GPEM Groupement Permanent d'Étude des Marchés

GPEN Ground Penetration

GPES Generalized Polyelastic Sampling

GPETE General-Purpose Electronic Test Equipment

GPF Gas Proof • General Protection Fault • General-Purpose Force • Geospatial Prototype Facility • Ground Processing Facility

GPFS General-Purpose File System • General-Purpose Flight Simulator

GPFU Gas Particulate Filter Unit

GPG Grant Proposal Guide • Graphic Product Generation

GPGL General-Purpose Graphic Language

gph Gallons per Hour • Graph (file extension)

Gph Graphite

GPH Gallons Per Hour

GPHIS Geoplatform High-Resolution Interferometer Sounder

GPHS General-Purpose Heat Source

GPI General Physics Institute • General-Purpose Interface • General-Purpose Inverter • Genuine Progress Indicator • Gimbal Position Indicator • Glide Path Indicator • Global Precipitation Index • GOES Precipitation Index • Government Preliminary Inspection • Grand Projet Innovant • Graphics Programming Interface • Ground Point of Interception • Ground Position Indicator • Guapi (Colombia) (location identifier)

GPIA General-Purpose Interface Adapter • General-Purpose Interface Assembly

GPIB General-Purpose Information Bus • General-Purpose Interface Bus

GPID Goals, Processes and Indicators of Development

GPIO General-Purpose Input/Output

GPIP General-Purpose Image Processor

GPIRS GSFC Photographic Image Retrieval System

GPKD General-Purpose Keyboard and Display

GPL Gaz de Pétrole Liquéfié • General Public Licence • General-Purpose Laboratory • General-Purpose Language • General-Purpose Link • Generalized Programming Language • Geometric Processing Library • Gestion de Produits Listés • Global Package Link • GNU Public License • Graphic Programming Language • Growing-Point Language • Guapiles (Costa Rica) (location identifier)

GPLA General-Purpose Line Adapter

GPLD Government Property Lost or Damaged • Government Property Lost or Destroyed

GPLI Groupement Permanent de Lutte contre l'Illettrisme (France)

GPLP General-Purpose Linear Programming

GPLR Government-Purpose License Rights

gpm Gallons per Minute

GpM Gallons per Mile • Groups per Minute • Groups per Message

GPM Gallons Per Mile • Gallons Per Minute • Gaussian-Peak Method • General Payload Model • General Processing Module • General-Purpose Macrogenerator • Genotype-Phonetype Mapping • Gestalt Photomapper • Global Precipitation Measurement • Global Precipitation Mission • Gradient Particle MHD • Graduated Payment Mortgage • Grant Policy Manual • Ground Potential Model • Ground Precautionary Message • Groups Per Minute • Groups Per Message

GPMC General-Purpose Mail Container

GPME General-Purpose Mission Equipment

GPMP Government Property Management Plan

GPMS Galileo Probe Mass Spectrometer • General-Purpose Microprogram Simulator • Global Position Management System

GPN Garden Point (Australia-NT) (location identifier) • General Performance Number

GPNDB Galactic Planetary Nebulae Database

GPNDS Global Positioning and Nuclear Detection System

Gpo Grupo

GPO Gemini Project Office • General Pico (Argentina-LP) (location identifier) • General Post Office • General-Purpose Oscilloscope • General-Purpose Output • German Patent Office (Germany) • Government Printing Office • Grand Prisme-Objectif

GPOE General-Purpose Operating Environment

GPOS Government Printing Office Style (manual)

GPP Gamma-Poisson Process • General Plant Projects • General-Purpose Processor • Generic Packetized Protocol • Gestion Prévisionnelle des Personnels • GLOW/PADSE Payload • Green Patch Panel • Gross Primary Production • Gyro Pitch Position

GPPA Great Plains Planetarium Association (USA)

GPPB General Power Plant Bulletin

GPPE General Purpose Processing Equipment

GPPF Generic Package of Primitive Functions • Giant Planet Probe Facility • Gravitational Plant Physiology Facility

GPPL General-Purpose Programming Language

gppm Graphics Pages per Minute

GPR General-Purpose Radiometer • General-Purpose Receiver • General-Purpose Register • General-Purpose Relay • General-Purpose Representative • Genio Populi Romani • Geometric Propagation • Government Plan Representative • Government Property Register • Government Purpose Right • Graphical Patient Record • Greenwich Photoheliographic Result • Ground-Penetrating Radar

GPRA Government Performance and Results Act (USA)

GPRC Geophysical and Polar Research Center

GPRDR Global Power Reflection Data Record

GPRN GOAL Procedure Release Notice

GP& RP Government Production & Research Property

GPRP Government Production and Research Property • Government Production Research Property

GPRR General-Purpose Radio Receiver

GPRS General Packet Radio Service • General-Purpose Radio Service

GPRT General-Purpose Radio Transmitter

gps Gallons per Second

GPS Galactic Plane Scan • Galactic Plane Survey • Galapagos Island (Ecuador) (location identifier) • Gauge Pressure Switch • Generality and Problem Solving • Generalized Parabolic Sheave • General Problem Solver • General Processing System • General-Purpose Segment • General-Purpose Shelter • General-Purpose Simulator • Generic Presentation Service • Genetic Programming System • Geophysical Processor System • Geophysical Products System • Géopositionnement Par Satellite • GHz-Peaked Spectrum • Gigahertz Peaked Spectrum • Global Positioning Satellite • Global Positioning System (*) • Global Position Sensor • Global Position System • Global Protection System • Global Pulse of Science • Government Paper Specification • Government Paper Standard • Graduated Pension Scheme • Ground-based Planetary Spectroscopy • Ground Playback Station • Ground Positioning System • Ground Power Supply • Ground Processing Simulation • Ground Processing System • Groupe de Physique des Solides • Groups of Pulses per Second • Guidance Power Supply • Gyroscope Parameter Shift

GPSC General-Purpose Synchronous Communication • GLIMPSE Point Source Catalog

GPSCO Global Position System Consortium

GPSCS General-Purpose Satellite Communications System

GPSDIC General-Purpose Scientific Document Image Code

GPSDR Global Positioning System Demonstration Receiver

GPSDW General-Purpose Scientific Document Writer

GPSE General-Purpose Simulation Environment

GPSI General-Purpose Serial Interface

GPSIC Global Positioning System Information Center

GPSIM Global Position System Integrity Monitoring

GPSP General-Purpose Signal Processor • General-Purpose Software Program • Great Plains Star Party (USA)

GPSR Global Positioning System Receiver

GPSS General-Purpose Simulation System • General-Purpose Simulator System • General-Purpose Systems Simulator

GPSSP General-Purpose Simulation System Package

GPST Global Positioning System Time

GPSU Ground Power Supply Unit

GPSW Global Positioning System World (periodical code)

Gpt Groupement

GPT Biloxi (USA-MS) (location identifier) • Galileo Project Team • General-Purpose Technology • General-Purpose Tool • Geometric and Positional Tolerance • Grand Programme Technique • Guidance Position Tracking • Gulfport (USA-MS) (location identifier)

GPTA Guangdong Post and Telecommunications Administration (China-PRC)

GPTE General-Purpose Test Equipment

GPTED General-Purpose Telegraphy Encryption Device

GPTS Geomagnetic Polarity Time Scale

GPU Gas Power Unit • General Postal Union • General Public Utilities • General-Purpose Unit • General-Purpose User • Generating Power Unit • Graphics Processing Unit • Ground Power Unit

GPV General-Public Virus • General-Purpose Vehicle • Grand Projet de Ville • Gyroscope Pickoff Voltage

GPVI Graphic Processor Video Interface

GPVS General-Purpose Visual Simulator

GPW Global Point Warning

GPWS General-Purpose Work System • Ground Proximity Warning System

GPX Galactic-Plane X-ray • Generalized Programming Extended • General Programming Extended

GPZ Grand Rapids (USA-MN) (location identifier)

GQ Equatorial Guinea (ISO country code) • Galactic Quadrant • General Quarters • Generation Qualifier • Geologic Quadrangle • Guinea (Equatorial_) (ISO country code) • Guinée Équatoriale (code ISO de pays)

GQA Get Quick Answer • Give Quick Answer • Government Quality Assurance

GQCS General Queuing Collection System

GQED Gravitational Quasi-Equilibrium Distribution

GQG Grand Quartier Général

GQJ Machrihanish (UK) (location identifier)

GQL Graphical Query Language

GQM Global Query Manager • Goal - Question - Metric

GQQ Galion (USA-OH) (location identifier)

GQR Gauss Quadrature Rule • Giant Quadrupole Resonance

GQSA Global Quantitative Sensitivity Analysis

gr Grain • Gram • Greece • gross • Grote

Gr Gear • Graad • Graben • Grade • Grain • Grammar • Grand • Grashof Number for Mass Transfer • Gravity • Great • Greece • Greek • Grey • Griego • Groschen (currency) • Ground • Group • Grove • Gründung

GR Game Reserve • Gamma Ray • Gas Retention • Géante Rouge • Gear • General Reconnaissance • General Register • General Relativistic • General Relativity • General Research • General Reserve • Generic Relation • Generic Requirement • Giclas Red • Glide Ratio • Global Reach • Government Regulation • Government Report • Government Representative • Government-Required • Graduated Response • Grande Randonnée • Grande Remise • Grand Répulseur • Graphics • Gravitational Radiation • Great Repulsor • Grèce (code ISO de pays) • Greece (international vehicle registration code) • Greece (ISO country code) • Greece (postal country code) • Greedy Refinement (algorithm) • Greek • Grid Reference • Gross Requirement • Ground Rule • Groupe de Recherche • Growth Rate • Grundfläche

gra Graph (file extension)

GRA Gamarra (Colombia) (location identifier) • Gamma-Ray Amplification • Gamma-Ray Astronomy • Gamma Resonance Absorption • Graduate Research Assistant • Graduate Research Assistantship • Graduate Research Associate • Gyro Reference Assembly

GRAAA Grand Rapids Amateur Astronomical Association (USA-MI)

GRAAL Graph Algorithmic Language • Groupe de Recherche en Astronomie et Astrophysique du Languedoc (France)

Grab Grabado

GRAB Galactic Radiation And Background • Gamma-Ray Alerted Binoculars (project) • Gamma-Ray Binocular

GRACE Gamma-Ray Astrophysics Coordinated Experiments • Gamma-Ray (burst) Afterglow Collaboration at ESO • Generic Reusable Ada Components for Engineering • Graphic Arts Composing Equipment • Gravity Recovery And Climate Experiment • Ground Radar Calibrating Equipment

Grad Gradient • Grading • Gradual • Graduate • Graduated

GRAD Gamma-Ray Advanced Detector • Generalized Remote-Access Database • General Recursive Algebra and Differentiation • Gradient • Graduate Resume Accumulation and Distribution

GRADB Generalized Remote Access Database

GRADD Graphics Adapter Device Driver

GRADE Graphical Airspace Design Environment

GrADS Grid Analysis And Data System • Grid Applications Development Software (project)

GRADS Generalized Remote Access Database System • Grid Analysis And Data System • Grid Analysis And Display System

GrAeH Grumman Aerospace Horizons (periodical code)

GRAF Graphic Addition to Fortran

GRAFH Groupe de Radioastronomie Franco-Helvétique

GRAFIC Groupe de Recherche et d'Application en Formation Initiale et Continue

GRAFSS General Relativistic Astrophysical Flow and Shock Solver

GR&AI Government Reports & Announcements Index

GRAI Government Reports and Announcements Index

GRAID Graphical Aid

GRAIL General Real-time AIDA Interface Language

GRAIN Graphics-oriented Relational Algebraic Interpreter

Gral General

Gram Gramática • Grammar • Grammarian • Grammatical

GRAM Generalized Rank Annihilation Method • General Reference Atmospheric Model • Generic Radio Access Network • GPS Receiver Applications Module • Grid Resource Access and Management • Ground Resource Access and Management • Grupo de Rádio-Astronomia Mackenzie (Brazil-SP) (now CRAAM)

GRAMME Groupement pour la Recherche et les Applications de la Micropesanteur aux Matériaux de l'Espace

GRAMPA General Analytical Model for Process Analysis

GRAMS Ground Receiving And Monitoring System

Gran Granada (Spain) • Granite • Granular • Granulated

GRAND Gamma-Ray Astrophysics at Notre-Dame

GRANDE Gamma-Ray And Neutrino Detection Experiment • Gamma-Ray And Neutrino Detector

GRANTECAN Gran Telescopio de Canarias

GRAPE Gamma-Ray Attenuation Porosity Evaluator (*) • Gamma-Ray Polarimeter Experiment • Graphical Analysis of Program Execution • Gravitational Pipe • Gravitational Pipeline • Gravity Pipe (*) • Gravity Pipeline

GRAPES Group for the Advancement of Positional and Evolutionary Studies

GRAPHDEN Graphic Data Entry

GRAPHIC Group And Phase Ionosphere Calibration

GRAPWG Gamma-Ray Astronomy Program Working Group (NASA)

GRARD Goddard Range And Range Data

GRARR Goddard Range And Range Rate

GRAS Generally Recognized As Safe • Generally Regarded As Safe • General Relativistic Atomic Structure (program) • George R. Agassiz Station (USA-AZ) • Glonass Receiver for Atmospheric Sounding • GNSS Receiver for Atmospheric Sounding

GRASER Gamma-Ray Amplification by Simulated Emission of Radiation

GRASP Gamma-Ray Astronomy with Spectroscopy and Positioning • Generalized Reentry Application Simulation Program • Generalized Retrieval and Storage Program • General-purpose Relativistic Atomic Structure Programme • General Resource Allocation and Selection Program • General Retrieval And Sort Processor • Generic Retrieve and Archive Services Protocol • GONG Reduction and Analysis Software Package • Government Relations And Science Policy • Graphical System for Presentation • Graphic Animation System for Professionals • Graphic Service Program • Gravitational Radiation Analysis and Simulation Package

GRASS Geographical Resources Analysis Support System

GRATE Generic Rules and Agent model Testbed Environment

GRATIS Gamma-Ray Arcminute Telescope Imaging System • Generation, Reduction And Training Input System

Grav Gravitational • Gravity • Gravure

GRAV Groupe de Recherche d'Art Visuel

GRAVES Grands Réseaux Appliqués à la Veille Spatiale

GRAZAR Gamma-Ray BLAZAR

GRB Gamma-Ray Burst • Gamma-Ray Burster • Green Bay (USA-WI) (location identifier) • Grid Resource Broker

GRBM Gamma-Ray Burst Monitor

Gr. Br. Great Britain

Gr. Brit. Great Britain

Grbx Gearbox

GRBX Gearbox

GRC General Research Corporation • Generation of Reproducible Copy • Gestion de la Relation Client • Glassfibre Reinforced Concrete • Glenn Research Center (NASA) (USA-OK) • Global Reference Code • Grand Cess (Liberia) (location identifier) • Grèce (code ISO de pays) • Greece (ISO country code) • Ground Radio Communications • Ground Resolution Cell

GRCA Ground Reference Coverage Area

Gr. Cap. Grande Capitale

GRCDA Governmental Refuse Collection and Disposal Association

Grd Grand • Grind • Guard

GRD Drachma (Greece) (ISO currency code) • General Requirements Document • Genetic Reconfiguration of DSPs • Graphics Display • Greatest Response Data • Greenwood (USA-SC) (location identifier) • Grenada (ISO country code) • Grenade (code ISO de pays) • Ground • Ground-Resolved Distance • Guard

GRDB Geoscientific Resource Database

GRDC Global Runoff Data Centre

Grde Grande

GRDEN Garden

Grdes Grandes

GRDF Georgia Research and Development Foundation (USA-GA)

Grdgr Grundstücksgröße

GRDL Geodetic Research and Development Laboratory • Geosciences Research and Development Laboratory • Gradual

GRDN Garden

GRDNS Gardens

grds grundsätzlich

Grds Grands

GRDS Generic Radar Display System

Grdst Grundstück

Grdtn Graduation

Gre Greek

GrE Graphite Epoxy

GRE Galactic Ridge Emission • Gamma-Ray Experiment • Gamma-Ray Explorer • Generic Routing Encapsulation • Giotto Radio-science Experiment • Graduate Record Examination • Grant-Related Expenditure • Graphics Engine • Greenville (USA-IL) (location identifier) • Ground Read-out Equipment • Ground Reconstruction Electronics • Ground Resolution Element

GREAS Generic Resource Event and Activity Scheduler

GREAT German Receiver for Astronomy at Terahertz (frequencies)

GREB Galactic Radiation Experiment Background

GREC Graphics Recognition • Groupement de Recherche sur les Émetteurs de Chaleur

GRECE Groupement de Recherches et d'Études sur la Civilisation Européenne

GRECO Groupement de Recherches Coordonnées

Grecolat Grecolatino

GRED Generalized Random Extract Device • General Relativistic Electrodynamics

GREDB Gamma-Ray Event Database

GREDR Global Reflectivity Data Record

GREE Groupe de Recherche sur l'Éducation et l'Emploi

GREEN Global Data Resource for Earth and the Environment

GREENS Global Redevelopment with Energy Environment Sustainability

GREG Groupe de Relativité Générale

GReGr General Relativity and Gravitation (periodical code)

GREM Geopotential Research Explorer Mission

GREOS Groupe de Recherches et d'Études en Océanographie Spatiale • Groupement pour la Recherche En Océanographie Spatiale

GREP Generalized Regular Expression Parser • Global Regular Expression Print • Groupe de Recherche sur l'Éducation Permanente

GRES Groupe de Recherche sur l'Enseignement Supérieur

GRESIC Groupe de Recherche Expérimentale sur les Systèmes Informatisés de Communication

GRESS Gradient Enhanced Software System

GrEStG Grunderwerbsteuergesetz

GRETA Groupe d'Établissements

GREX Gamma-Ray Experiment

grf Graph • Graphic Data (file extension)

GRF Gamma Radiation Facility • Gaussian Random Field • Gibbs Random Field • Gradual Rise and Fall • Graduate Research Fellowship • Graph • Gravity Research Foundation • Großer Roter Fleck (Jupiter) • Ground Reaction Force • Ground Relay Facility • Tacoma Gray (USA-WA) (location identifier)

GRFL Groundwater Remediation Field Laboratory

GRG Gardez (Afghanistan) (location identifier) • General Relativity and Gravitation • Giant Radio Galaxy • Gross Reserve Generation

GR/GP Global Reach/Global Power

GRGS Groupe de Recherches de Géodésie Spatiale

GRH Garuahi (Papua New Guinea) (location identifier) • Gestion des Ressources Humaines • Gramm - Rudman - Hollings (act) (USA)

GRI Generic Recce Interface • Government Reports Index • Grand Island (USA-NE) (location identifier) • Groupe Régional d'Intervention (France) • Group Repetition Interval

GRIB Gridded Binary

GRID Gamma-Ray Imaging Detector • Gamma-Ray Imaging Device • Globalisation des Ressources Informatiques et des Données • Global Resources Information Database • Gradient Imaging Display Software • Graphic Interactive Display • Graphic Remote Integrated Display • Graphic Retrieval and Information Display

GRIDS Geographic Resources Information and Data System • Gridded Resource Inventory Data System • Grid-Referenced Information Display System • Grid Research Integration Development and Support

GRIMP Groupement de Recherches et d'Intervention en Milieux Périlleux

GRIN Graded Index • Gradient Index • Gradient Radial Index • Graphical Input

GRIND Graphic Relation Index

GRINS General Retrieval Inquiry Negotiation Structure

GriO Griffith Observer (periodical code)

GrIP Gravis Interface Protocol

GRIP Gamma-Ray Imaging Payload • General Retrieval of Information Program • Greenland Icecore Project

GriPhyN Grid Physics Network

GRIPS GCM-Reality Intercomparison for SPARC (project) • General Relation-Based Information Processing System • Government Raster Image Processing Software • Groupe Informel de Problèmes Stratégiques

GRIS Gamma-Ray Imaging Spectrometer • Global Resources Information System • Grating Infrared Spectrometer

GRISM Grazing Incidence Prism

GRIST Grazing Incidence Solar Telescope

GRITS Gamma-Ray Imaging Telescope System • Goddard Range Instrumentation Tracking System

GRJ George (South Africa) (location identifier)

GRK Graduierten Kolleg (Germany) • Killeen Gray (USA-TX) (location identifier)

GRL Gamma-Ray Line • Garasa (Papua New Guinea) (location identifier) • Geophysics Research Letters (periodical code) • Greenland (ISO country code) • Groenland (code ISO de pays) • Gross Requirement List

GRLD Graphic Remote Interface Display

GRM Geopotential Research Mission • Grand Marais - Devils Track (USA-MN) (location identifier) • Growing Rigour Method • Guidance Rate Measurement

gRMF Grating Response Matrix File

GRMHD General Relativistic Magnetohydrodynamics

GRMS Generalized Rate Monotonic Scheduling

Grn Green

GRN Good Receipt Note • Gordon (USA-NE) (location identifier) • Green • Ground

GRNC Group Not Counted

Grnd Ground

GRND Ground

GRNN General Regression Neural Network

GRNP General Remote Network Processor

GRNS Greens

gro gross

Gro Gros • Grove

GRO Gamma-Ray Observatory • General Register Office • Gerona (Spain) (location identifier) • Government Research Organization • Gravitational Radiation Observatory • Guerrero (Mexico)

GROCSE Gammay-Ray Optical Counterpart Search Experiment

GROG Groupe Régional d'Observation des Galaxies

GroR Kapteyn Astronomical Laboratory Groningen Report (periodical code)

GROSSC GRO Science Support Center (USA-MD)

GROT Grota

GROV Grove

grp Group • Group Definition (file extension)

Grp Group

GRP Gamma-Ray Projector • Gaussian Random Process • Geophysical Research Program • Glass-Reinforced Plastic (*) • Glass-Reinforced Polyester • Government-Required Property • Ground Reference Point • Ground Relay Panel • Group • Groupement pour la Recherche en Productique • Groupement Régional Permanent • Guide de Randonnée Pédestre • Gurupi (Brazil-TO) (location identifier)

GRPA Government Performance and Results Act (USA)

GRPBF Group Buffer

GRPFR Group Framing

Grph Graphic

GRPN Group Name

GRPOT Group Output Controller

GRQ Groningen (Netherlands) (location identifier)

GRQI Grassman-Rayleigh Quotient Iteration

GFR Galaxy Remaining Rate • Grand Rapids (USA-MI) (location identifier) • Graphical Rewrite Rules • Guidance Reference Release

GRRH Gestion des Ressources et des Relations Humaines

grs gross

Grs Grains • Grease • Gros • Gross

GRS GAIA Reference System • Galactic Radio Sources • Galactic Ring Survey • Gamma-Radiation Source • Gamma-Radiation Spectrometer • Gamma-Ray Sensor • Gamma-Ray Spectrometer • Gamma-Ray Spectrum • Generalized Retrieval System • General Records Schedule • General Retrieval System • Generic Retrieval System • Geocentric Reference System • Geodetic Reference System • Giant Radio Source • Global Receiving System • Global Reference System • Global Resource Serialization • Global Retirement System • Graph-Rewriting System • Gravity Reference Signal • Gravity Reference System • Great Red Spot (Jupiter) • Grid Reference System • Grille de Référence SPOT • Gross • Grosseto (Italy) (location identifier) • Ground Receiving Station • Gyro Reference System

GRSE Gamma-Ray Spectrometric Equipment

GRSFE Geologic Remote Sensing Field Experiment (*) • Global Remote Sensing Field Experiment

GRSG Geological Remote Sensing Group

GRSP General Range Safety Plan

GRSS Geoscience and Remote Sensing Society

GRSST Gamma-Ray Silicon Space Telescope

GrStG Grundsteuergesetz

GrStR Grundsteuer-Richtlinien

GRSU Geography Remote Sensing Unit

Grt Gracht • Great

GRT Gamma-Ray Telescope • Gamma-Ray Tube • Generalized Ray Theory • General Relativity Theory • Generic Run Time • Germanium Resistance Thermometer • Gestionnaire du Réseau de Transport • Government Rate Tender • Great • Gross Registered Tonnage • Gross Registered Tonnes • Ground-based Radio Telescope • Ground Radio Telescope • Ground-Received Time • Ground Receive Terminal • Gujrat (Pakistan) (location identifier)

GRTAD Ground-based Real-Time Attitude Determination

Grtg Grating

GRTLKS Great Lakes

GRTLS Glide Return To Launch Site

GrTOn Gravitatsiia i Teoriia Otnositel nosti (periodical code)

Grtr Greater

GRTR Groupement de Recherche en Télédétection et Radiométrie

GRTS General Remote Terminal System • Goddard Real-Time System

Gru Grus-Gruis (constellation)

GRU Ground Replaceable Unit • Guidance Regulator Unit • Gyroscope Reference Unit • São Paulo - Guarulhos (Brazil-SP) (location identifier)

Grv Groove • Grove

GRV Grove • Groznyj (Russia) (location identifier)

Grvd Grooved

Grvl Gravel

GRVS Groves

GRW Galactic Radio Wave • Graciosa Island (Azores) (location identifier)

gr. wt. Gross Weight

GRWT Gross Weight

GRX Granada (Spain) (location identifier)

GRXE Galactic Ridge X-ray Emission

GRY Grímsey (Iceland) (location identifier) • Gross Redemption Yield

GRZ Graz (Austria) (location identifier)

g/s Grant Supported

gs Ghostscript • Groundspeed • Ground State

Gs Gauss

G/S Glide Slope • Ground Segment • Ground Speed • Ground Station

GS Galactic Shell • Galvanized Steel • Gas Servicer • Generalized Scidar • Generalized System • General Schedule • General Search • General Secretary • General Service • General Staff • General Station • General Support • General Surveillance • General System • Genetic Society • Geographical Survey • Geographic Scope • Geological Survey • Geomagnetic Storm • Georgia and South Sandwich Islands (South_) (ISO country code) • Geprüfte Sicherheit • Giant Shell • Giclas Shell • Gigasamples • Glide Slope • Global Symmetry • Ghostscript • Grammar System • Grande Section • Grand Schmidt (telescope) • Gray Scale • Grigg-Skjellerup (comet) • Ground Segment • Ground Speed • Ground Stabilized • Ground State • Ground Station • Ground System • Groupement Scientifique • Groupe Spécialisé • Group Separator • Grundstück • Guidance Station • Guidance System • Guide Star • Heliocentric Gravitational Constant • Sandwich Islands (South Georgia and South_) (ISO country code) • South Georgia and South Sandwich Islands (ISO country code) • South Sandwich Islands (South Georgia &_) (ISO country code)

G&SA Goods & Service Allowance

GSA Gaseous Sulfuric Acid • Gas Storage Area • Gemini Science Archive • General Services Administration • Generalized Semi-Analytical • Generalized Similarity Analysis • General Services Administration (USA) • Geological Society of Africa • Geological Society of America (USA-CO) • Geological Society of Australia • Geologic Spatial Analysis • GNC Simulation Assembly • Gnomonicae Societas Austriaca (Austria) • Global Security Architecture • Goods and Service Allowance • Government Service Administration • Graduate Student Association • Ground Support Agency • Guidance System Astrometer • Long Pasia (Malaysia) (location identifier)

GSABC General Services Administration Board of Contracting

GSAC Graduate Student Advisory Committee • Greenhouse Science Advisory Committee • Ground Systems Associate Contractor • Groupement pour la Sécurité de l'Aviation Civile (France)

GSAI Garny Space Astronomy Institute (Armenia)

GSAM Generalized Sequential Access Method • Generalized Standard Addition Method

GSAMB Geological Society of America Bulletin (periodical code)

GSAMm Geological Society of America Memoir (periodical code)

GSAO Gemini South Adaptive Optics

GSAOI Gemini South Adaptive Optics Imager

GSAR Generalized Synthetic Aperture Radar • Guide Star Availability Request

GSARP Global Scale Atmospheric Research Program

GSARRTS Generator, Starter, Alternator, Regulator, Rectifier Test Stand

GSARS General Services Acquisition Regulation Supplement

GSAS General Sequence Action Schedule • Geostationary Satellite Archive System • Graduate School in Arts and Sciences

GSAT Geosynchronous Satellite • Ground System Acceptance Test

GSB Gamma-Strahl-Blitze • Gateway Signaling Buffer • Goldsboro - Seymour Johnson (USA-NC) (location identifier) • Guandong Seismological Bureau (China-PRC)

GSBCA General Services Board of Contract Appeals

GS&C Global Surveillance & Communications

GSC Galileo Science Catalog • Gascoyne Junction (Australia-WA) (location identifier) • Gas Slit Camera • Gas-Solid Chromatography • Gas-Star Complex • Gemini Science Committee • Generalized Self-Calibration • General Staff Council • Geological Survey of Canada • Global Standards Collaboration • Global Surveillance and Communications • Government Security Committee • Graduate Studies Committee • Ground Safety Console • Ground Switching Center • Guidance System Console • Guide Star Catalog (*) • Guide Star Coordinate

GSCR Generic Switching Center Requirement • Ground Service Change Request • Ground Service Contract Request

GSCU Ground Service Cooling Unit • Ground Support Cooling Unit

GSD General Supply Depot • General Support Division • General System Description • Geodetic Survey Division • Geographic Situation Display • Geological Survey Division • Geometric Spot Diagram • Globally Significant Difference • Graphical Situation Display • Greelay Storage Division • Ground Sample Distance • Ground Slant Distance • Guide Star Data • Guide Star Determination

GSDC Gas Scintillation Drift Chamber

GSDF Global Sustainable Development Facility

GSDI Gesellschaft zum Schutz (privater) Daten in Informations- und Komunikationsdiensten (Germany) • Global Spatial Data Infrastructure • Global Spatial Data Initiative

GSDL Ground Software Development Laboratory

GSDM Global Spatial Data Model

GSDR Global Slope Data Record

GSDS Geospatial Decision Support • Global Spatial Data System

GSDSF Ground Station Detailed Schedule File

GSDSM Global Circulation Dust and Smoke Model

GSE General Supply Equipment • General System Engineering • Geocentric Solar Ecliptic (coordinates) • Gesellschaft für Software-Engineering • Global Security Environment • Government Sponsored Enterprise • Government Support Equipment • Göteborg Saeve (Sweden) (location identifier) • Gothenburg Saeve (Sweden) (location identifier) • Grad-Shafranov Equation • Graduate School of Education • Graphic Screen Editor • Ground Segment Equipment • Ground Service Equipment • Ground Servicing Equipment • Ground Station Equipment • Ground Support Equipment • Ground Systems Engineering

GSED Ground Systems Engineering Department

GSE&I General Systems Engineering & Integration

GSEI General Systems Engineering and Integration

GSEL Ground Support Equipment List

GSEnv Ground Support Environment

GSEQ Geocentric Solar Equatorial (coordinates)

GSERD Ground Support Equipment Recommendation Data • Ground Support Equipment Requirement Data • Ground System Engineering Research and Development

GSESP Graduate in Science and Engineering Survey Programme

GSF General Survey of Finland • Gesellschaft für Strahlen- und Umweltforschung mbH (Germany) • Global Star Formation • Global Status Flag • Goldreich - Schubert - Fricke (instability) • Ghostscript Font • Graduate Student Council • Gyro Scale Factor

GSFC Goddard Space Flight Center (NASA) (USA-MD)

GSG General Support Group • Gesundheitsstrukturgesetz • Governance and Science Group

GSGAC GSFC Space Geodesy Analysis Center (USA-MD)

GSGE Computational Science Graduate Fellowship

GSGP Global Sedimentary Geology Program

GSH Global Schoolhouse • Goshen (USA-IN) (location identifier)

GSHAP Global Seismic Hazard Assessment Program (*) • Global Seismic Hazard Assessment Project

GSI Galileo Sensitive Information • Generalized Scale Invariance • General Server Interface • Génie des Systèmes Industriels • Geographical Survey Institute (Japan) • German SOFIA Institute • Gesellschaft für Schwerionenforschung (Germany) • Giga-Scale Integration • Glide Slope Indicator • Government Secure Intranet • Government Source Inspection • Grand Scale Integration • Grid Security Infrastructure • Ground Segment Infrastructure • Ground Support Item • Gruppo Sole Italia (Italy) • Guadalcanal (Solomon Islands) (location identifier) • Guide Star Index

GSIG Ground Systems Integration Group

GSII Government Services Information Infrastructure

GSIM Global Satellite Image Mapping

GSIP Glideslope Intercept Point

GSIR Guide Star Inventory Request

GSIS Geographic Snow Information System • Graphics Standard Interface Standard • Ground Safety Information System

GSIU Ground Standard Interface Unit

GSIWG Ground System Integration Working Group

GSJ Geological Survey of Japan

GSK General Semantic Knowledge

GSKM Galactic Structure and Kinematic Model

GSL Generalized Simulation Language • Geological Society of London (UK-England) • Global Sea Level • Global Service Logic • Graphics Software Laboratories • Greatest Southern Latitude • Guaranteed Student Loan

GSLIS Graduate School of Library and Information Science (UCLA)

GSLMP Global Sea Level Monitoring Programme

GSLS Ground Station Link Survivability

GSLV Geostationary Launch Vehicle • Geostationary Satellite Launch Vehicle • Geostationary Space Launch Vehicle • Geosynchronous Satellite Launch Vehicle

gsm Grams per Square Meter

GSM Generalized Seeing Monitor • General Sales Manager • General Support Maintenance • General Systems Model • Generic Simulation Manager • Geocentric Solar Magnetospheric (coordinates) • Global Satellite Messaging • Global Shared Memory • Global Sky Model • Global Spectral Model • Global Standard for Mobiles • Global State Manager • Global System for Mobiles • Grating Scale Monitor • Ground Safety Manager • Ground Segment Manager • Ground Station Module • Ground Support Module • Groupe Spécial Mobile

GSMC Geospatial Standards Management Committee • Global System for Mobile Communications

GSME Ground Support Maintenance Equipment

GSMN Gerencia del Servicio Meteorológico Nacional

GSMNP Great Smoky Mountains National Park

GSMO Ground Systems Management Office

GSMP Generalized Semi-Markov Process • General Switch Management Protocol

GSMQ Global Services Management Queue

GSMS Ground System Management System

GSMT Giant Segmented Mirror Telescope

GSN Gigabyte System Network • Global Seismic Network • Global Seismographic Network (*) • Global Subscriber Number • GPRS Support Node • Mount Gunson (Australia-SA) (location identifier)

GSNF Growing Season Net Flux

GSNP Group on Statistical and Non-linear Physics

GSNW Gateway Service for NetWare

GS&O Ground Systems & Operations

GSO General Service Officer • General Spin Orbital • General Staff Officer • Geostationary Orbit • Geostationary Satellite Orbit • Geosynchronous Orbit • Global Seismic Oscillation • Global Software Outsourcing • Graduate School of Oceanography • Greensboro (USA-NC) (location identifier) • Ground Safety Office • Ground Safety Officer • Ground Staff Office • Ground Support Office • Ground Support Operations • Ground Systems and Operations • Ground Systems Operations • GSFC Security Office • Guang Sheng Optical (China-ROC) • Gyro Storage Oven • High Point (USA-NC) (location identifier) • Winston-Salem (USA-NC) (location identifier)

GSOC German Space Operations Center

GSOIA General Security Of Information Agreement

GS&OP Ground Systems & Operations Project

GSOP Ground Segment Operation • Ground Systems and Operations Project • Ground Systems Operations Project • Guidance Systems Operation Plan

GSOPS Ground Segment Operations

GSOR General Staff Operational Requirements

GSORTS Global Status Of Resources and Training System

GSOS GPS Surface Observing System

GSOV Ground Segment Overall Verification

GSOWG Ground Systems Operations Working Group

GSOWM Global Spectral Ocean Wave Model

GSP Generalized Scheme of Preferences • Generalized System of Preferences • General Simulation Program • General System of Preferences • Geodetic Satellite Program • Geological Survey of Pakistan • Glassfibre-Strengthened Polyester • Global Standards Program • Good Scientometric Practice • Government-Specified Purchase • Graphics System Processor • Graphic Subroutine Package • Greenland Sea Project • Greenville (USA-SC) (location identifier) • Greer (USA-SC) (location identifier) • Gross Social Product • Gross State Product • Ground Safety Plan • Guidance Signal Processor • Spartanburg (USA-SC) (location identifier)

GSPB Geodetic Satellite Policy Board

GSPC Gas Scintillation Proportional Counter • Guide Star Photometric Catalogue

GSPD Ground Speed

GSPDC Geostationary Satellite Precipitation Data Centre

GSPE Generic Sample Polyelasticity

GSPF Ground Software Production Facility

GSPH Gridded Smooth Particle Hydrodynamics

GSPL Global Services Priority List

GSPM Guelph Scanning Proton Microprobe (Canada)

GSPN Generalized Stochastic Petri Net

GSPO Ground Systems Project Office

GSPR Groupement Spécial de Protection Rapprochée • Guidance Signal Processor-Repeater • Guide Star Parallel Request

GSQ Generalized Sinusoidal Quantity

GSQA Government Source Quality Assurance

GSQL Gateway to Structured Query Language

GSR Galactic Standard of Rest • Galvanic Skin Reflex • Galvanic Skin Response • Gardo (Somalia) (location identifier) • General Staff Requirements • General Support Reinforcing • Générateur Solaire Rigide • Generic Record Syntax • Geostationary Ring • Glide Slope Receiver • Global Self-Regulation • Ground Surveillance Radar

GSRC Geological Survey Research Committee

GSRF Ground Software Reference Facility

GSRN Global Surface Radiation Network

GSRP Graduate Student Research Program

GSRR Ground Segment Requirements Review

GSRS General Support Rocket System

GS/s Gigasamples per Second

GSS General Security Service • General Security Support • General Service School • General Staff Support • General Supply Support • General Support System • Generic Security Service • Generic Security Standard • Geodetic Survey Squadron • Geographic Support System • Geostationary Satellite • Geosynchronous Space Station • Global Space Station • Global Surveillance System • Government Statistical Service • Government Systems Sector • Grad - Schlüter - Shafranov (equation) • Graphics State Stack • Graphic Support System • Groth Survey Strip • Ground Station System • Ground Support Software • Ground Support System • Grouped Sweeping Scheduling • Group Support System • Growth Space Station • Grumman Sensor Systems Inc. • Grumman Space Systems Inc. • Guidance System Simulator • Guided Self-Scheduling • Guide Star Selection • Sabi Sabi (South Africa) (location identifier)

GSSA General Support Supply Activity • General Support Supply Area • Ground Support Systems Activation

GSSAP Generic Security Service Applications Programming

GSSAPI Generic Security Service Application Program Interface

GSSB General Supply Support Base

GSSC General Systems Simulation Center • Ground Support Simulation Computer • Ground Support Systems Contractor

GSSD General Support Systems Division

GSSDR Global Surface Slope Data Record

GSSF General Services Stock Fund

GSSG General Schedule Supervisory Guide

GSSI Ground Support System Integration

GSSI&C Ground Support Systems Installation & Checkout

GSSIC Ground Support Systems Installation and Checkout

GSSP Galaxy Simple Stellar Population • Generally-accepted System Security Principle • Global Stratotype Section and Point

GSSR Goldstone Solar System Radar • Ground Support System Review • Guide Star Selection Request

GSSS Ground Support System Specification • Guide Star Selection System • Guide Star Support System

GSST Global Sea Surface Temperature

GST General Systems Theory • Geographical Specialist Team • Goods and Services Tax • Government Standard Translator • GPS Science and Technology • Greenwich Sidereal Time • Greenwich Standard Time • Ground Satellite Terminal • Ground Sensor Terminal • Ground Station Terminal • Ground Support Terminal • Ground Systems Test • Groupe de Servitude Table • Groupe de Soutien Technique • Gust • Gustavus (USA-AK) (location identifier)

GSTA Ground Surveillance and Target Acquisition

GSTDN Ground Spacecraft Tracking and Data Network • Ground Spaceflight Tracking and Data Network

GSTE Geometric Similarity as Topological Equivalence • Guidance System Test Equipment

GSTF Ground Systems Test Flow

GSTN General Switched Telephone Network • Global Switched Telephone Network

GSTP General Supporting Technology Programme (*) • General Support Technology Programme • Ground System Test Procedure

GSTS Ground-based Surveillance and Tracking System • Groupe Scientifique de Télédétection Spatiale • Gusts

GSTU Guidance System Test Unit

GSTV Ground System Thermal Vacuum

GSTY Gusty

GSU Gas Servicer Unit • Gedaref (Sudan) (location identifier) • General Service Unit • General Support Unit • Geographically-Separated Unit • Georgia Southern University (USA-GA) • Georgia State University (USA-GA) • Graduate Students' Union (Ireland)

GSUF Gesellschaft für Strahlen- und Umweltforschung mbH (Germany)

GSV Geostationary Service Vehicle • Glove Stop Valve • Guided Space Vehicle

GSVD Generalized Singular Value Decomposition

GSVOO Ground Station Visibility of Orbiting Objects

GSVP Ground Support Verification Plan

GSVS Global Secure Voice System

gsw GraphShow Worksheet (file extension)

GSW Ground Saucer Watch

GSWM Global-Scale Wave Model

GSWR Galvanized Steel Wire Rope

Gsy Guernsey (Channel Islands)

GSY Grimsby Binbrook (UK-England) (location identifier)

gt Gilt • Gross Ton • Guatemala

Gt Gent (Belgium) • Gigatonne • Great • Groot

G/T Facteur de Mérite • Figure of Merit • Gain/Temperature

GT Galilean Telescope • Galilean Transformation • Game Theory • Gamow-Teller • Gas Tank • Gas Turbine • Gauge Theory • General Tool • Generic Task • Geographic Term • Geometry and Topology • Gigaton • Government Technical • Government Test • Gravitational Telescope • Greater Than • Greenwich Time • Gross Ton • Gross Tonnage • Ground Team • Ground Temperature • Ground Test • Ground Transmit • Ground Transmitter • Ground Truth • Groupe de Travail • Group Technology • Growth Test • Guaranteed Time • Guatemala (ISO country code) • Guidance Trajectory • Guidance Transmitter • Gyro Torque

GT&A Ground Test & Acceptance

GTA Gas-Tungsten Arc • Gatokae (Solomon Islands) (location identifier) • Gestionnaire des Titres d'Accès • Global Title Address • Graduate Teaching Assistantship • Graduate Teaching Associate • Graphic Training Aid • Greater Toronto Area (Canada-ON) • Ground Test Accelerator • Ground Test Access • Ground Test and Acceptance • Ground Test Article • Ground Test Assembly • Ground Torquing Assembly • Groupe de Télédétection Aquatique • Groupe de Travail Académique • Guaranteed Traffic Agreement

GTAC GMRT Time Allocation Committee

GTAS Grand Traverse Astronomical Society (USA-MI)

GTAW Gas Tungsten Arc Welding

GTB Genting (Malaysia) (location identifier) • Gestion Technique du Bâtiment

Gt. Br. Great Britain

Gt. Brit. Great Britain

GTC Gain Time Constant • Gain Time Control • General Telephone Company • General Transformation Corporation • Global Theme Conference • Glycol Trim Console • Good Till Cancelled • Grand Tour Cluster • Gran Telescopio de Canarias • Green Turtle (Bahamas) (location identifier) • Ground Test Conductor

GTCC Gas Turbine Combined Cycle • Group Technology Characterization Code • Guilford Technical Community College (USA-NC)

GTCE Global Tropospheric Chemistry Experiment

GTCP Global Tropospheric Chemistry Program

GTCU Ground Thermal Conditioning Unit

GTC3 Global Terrestrial Critical Control Circuit

GTC3S Global Terrestrial Critical Control Circuit System

Gtd Guaranteed

GTD Generalized Tracking Data • General Telephone Digital • Geometric Theory of Diffraction • Geothermodynamic Data • Graphic Tablet Display • Groupe de Travail Disciplinaire • Groupe Technique Disciplinaire

GTDAR GEC-Thompson-DASA Airborne Radar

GTDM Generic Time Division Multiplex

GTDR Global Topographic Data Record

GTDS Goddard Trajectory Determination System

GTE General Telephone and Electric • General Telephone Electronics • Génie Thermique et Énergie • Global Troposphere Experiment • Global Tropospheric Experiment • Government Technical Evaluator • Greater Than or Equal • Groote Eylandt - Alyangula (Australia-NT) (location identifier) • Ground Test Equipment • Ground Transportation Equipment • Guidance Test Equipment

GTED Gas Turbine Engine Driven (generator)

Gtee Guarantee

GTF Generalized Transformation Function • Gestion Transaction Financement • Global Temporary File • Government Test Facility • Great Falls (USA-MT) (location identifier) • Ground Test Facility

GTG Got To Go • Grantsburg (USA-WI) (location identifier) • Ground-To-Ground • Groupe de Tomographie Géophysique

GTGS Gas Turbine Generator Set

GTHRES Gravity Threshold

GTI Gesellschaft für Technische Informationssysteme • Global Taxonomy Initiative • Good Time Interval • Grand Turk Island • Graphiques - Textes - Images • Grote Technologische Instituten • Ground Test Instrumentation • Guettin (Germany) (location identifier)

GTIC Groupe Technique Interministériel des Composants

GTIM Groupe de Travail Inter-Ministériel

GTIS Geographical Technical Information System • Government Technology Information System • Government Telecommunications and Information Services

GTK Gross Tonne Kilometer • Sungei Tekai (Malaysia) (location identifier)

GTL Gas To Liquid • Gross Task List • Gunning Transceiver Logic

gTLD Generic Top Level Domain

GTLD Generic Top Level Domain

GTLV GFE Target Launch Vehicle (*) • Government-furnished Target Launch Vehicle

GTM Generative Topographic Mapping • Geometry Technology Module • Glass To Metal • Global Telecom Market • Global Track Manager • Global Trade Manager • Global Transaction Manager • Global Transport Model • Good This Month • Gran Telescopio Milimétrico • Ground Team Manager • Ground Test Mission • Ground Test Motor • Guatemala (ISO country code)

GTMS Graphical Text Management System

GTN Glasgow Telecolleges Network (UK-Scotland) • GLAST Telescope Network • Global Terrestrial Network • Global Transportation Network • Mount Cook - Glentanner (New Zealand) (location identifier)

Gto Guanajuato (Mexico)

GTO Gate Turnoff • Gaussian-Type Orbital • Generic Technical Order • Geostationary Transfer Orbit • Geosynchronous Transfer Orbit • Gorontalo (Indonesia) (location identifier) • Guanajuato (Mexico) • Guaranteed Time Observation • Guaranteed-Time Observer • Guide To Operations

Gtos Gastos

GTOS Global Terrestrial Observing System

GTOT Grand Total

GTOW Gross Takeoff Weight

GTOWG Guaranteed Time Observer Working Group

GTP General Test Plan • Geophysical Turbulence Program • Glan-Thompson Prism • Global Theme Program • Go to Plain (text) • Government Telecommunications Program • GPRS Tunneling Protocol

GTPA Gestion Technologique des Passagers Aériens

GTPZ Gigiena Truda i Professional nye Zabolevaniia (periodical code)

GTQ Quetzal (Guatemala) (ISO currency code)

Gtr Greater

GTR Columbus - Golden Triangle (USA-MS) (location identifier) • Garantie de Temps de Réparation • General Theory of Relativity • Génie des Télécommunications et des Réseaux • Geoid/Topography Ratio • Global Test Range • Government Transportation Request • Grand Trunk Railway • Ground Truth Radiometer • Starkville (USA-MS) (location identifier) • West Point (USA-MS) (location identifier)

GTRE Gas Turbine Research Establishment

GTREP Georgia Tech Regional Engineering Program (USA-GA)

GTRF GPS Terrestrial Reference Frame

GTRI Georgia Tech Research Institute (USA-GA)

GTS Gas Turbine Ship • Gas Turbine Starter • General - Technical - Security • General Technical Services • General Technology Systems • General Test Support • Geostationary Technology Satellite • Global Telecommunications Service • Global Telecommunications Society • Global Telecommunications System • Global Translation System • Goldstone Tracking Station (NASA) • Go To Standby • Government Telecommunications System (USA) • Granites (Australia-NT) (location identifier) • Graphics Terminal Scheduler • Graphics Terminal System • Great Trigonometric Survey • Greenwich Time Signal • Ground Telemetry System • Ground Terminal System • Ground Test Station • Ground Tracking Station • Guam Tracking Station (NASA) • Guidance Test Set • Guidance Test Station • Gyro Tilt Signal

GTSF Guidance Test and Simulation Facility

GTSN Global Telemetered Seismograph Network

GTSS General Time-Sharing System

gtt Guttae

GTT Geiger Tube Telescope • Georgetown (Australia-QLD) (location identifier) • Global Title Transmission • Government Technical Test • Government Technical Testing • GPS Telemetry Transmitter

GTTS Gyro Transfer Table System

GTU Global Telecommunications University • Ground Test Unit • Guidance Test Unit

GTUB Geographic Tabulation Unit Base

GTV Gate Trigger Valve • Ground Test Vehicle • Ground Transport Vehicle • Guidance Test Vehicle • Guided Test Vehicle

GTW Gesellschaft für Terminologie und Wissenstransfer • Good This Week • Gross Takeoff Weight • Zlin Holesov (Czech Rep.) (location identifier)

GTWAY Gateway

GTWY Gateway

GTY Gettysburg (USA-PA) (location identifier)

GTZ Gesellschaft für Technische Zusammenarbeit (Germany) • Guidance Time Zero

gu Guam • Guinea • Gujarati (ISO language code)

Gu Guest • Guinea

GU Georgetown University (USA-DC) • Glasgow University (UK-Scotland) • Government Unit • Guam (ISO country code) • Guardia Urbana • Guatemala • Guidance Unit • Guinea

GUA Gestionnaire Utilisateur d'Applications • Grupo Universitario de Astronomía (Spain) • Guatemala City - La Aurora (Guatemala) (location identifier) • Guidance Unit Assembly

GUACAMOLE Guiding, Acquisition and Calibration Module

Guad Guadalajara (Spain)

Guadal Guadalajara (Spain)

GUAM Georgia - Ukraine - Azerbaijan - Moldova

Guar Guarantee • Guaranteed

Guat Guatemala

Guay Guyaquil

GUB Generalized Upper Bounding • Greatest Upper Bound • Guerrero Negro (Mexico) (location identifier)

GUC Crested Butte (USA-CO) (location identifier) • Gunnison (USA-CO) (location identifier)

GUCP Ground Umbilical Carrier Plate

GUD Goundam (Mali) (location identifier)

GUE Gaussian Unitary Ensemble • Guriaso (Papua New Guinea) (location identifier)

GUESS General Purpose Expert System Shell

GUF French Guiana (ISO country code) • General University Funds • Guiana (French_) (ISO country code) • Guyane Française (code ISO de pays) • Gulf Shores - Edwards (USA-AL) (location identifier)

GUG Guari (Papua New Guinea) (location identifier)

GUGK Glavnoye Upravileniie Geodesii i Kartografii (Russia)

GUH Gunnedah (Australia-NSW) (location identifier)

Gui Guiana

GUI Give Us Incentives • Graphical User Interface • Graphic User Interface • Guidance • Guiria (Venezuela) (location identifier)

GUIB Graphical User Interface for the Blind

GUID Globally Unique Identifier • Global Unique Identifier • Global Universal Identifier • Group User Identifier • Guidance

GUIDE Glasgow University Initiative in Distance Education • Graphical User Interface Designed for Education • Graphical User Interface for a Data Environment • Grating User Interactive Data Extension

GUIDO Guidance Officer

Guil Guilder (currency)

GUILD Government, University, Industry, and Laboratory Development

Guin Guinea

Guip Guipúzcoa (Spain)

GUIPS Glasgow University Inverse Problem Software

GUIRR Government-University-Industry Research Roundtable

GUISDAP Grand Unified Incoherent Scatter Design and Analysis Package • Grand Unified Incoherent Scatter Design and Analysis Program (*)

Guj Gujarat

GUJ Guaratinguetá (Brazil-SP) (location identifier)

GUL Geschichte und Lichtwechsel der Veränderlichen Sterne (periodical code) • Glasgow University Library (UK-Scotland) • Goulburn (Australia-NSW) (location identifier) • GSE Utilization List

GULP General Utility Language Processor • General Utility Library Program • Global Upgrade LAN Product • Grotesque Large Useless Project

GULS Generic Upper Layer Security

GUM Gamma Ursa Majoris • Glówny Urzad Miar (Poland) • Guam Agana (Guam) (location identifier) • Guatemala (ISO country code)

GUN General Usage Network • Montgomery Gunter (USA-AL) (location identifier)

GUO Gauhati University Observatory (India) • Gualaco (Honduras) (location identifier)

GÜP Gebäude-Übergabepunkt

GUP Gallup (USA-NM) (location identifier) • Gas Under Pressure

GUPS Grand Unified Photon Spectrum

GUPT Grand Unified Phase Transition

GUQ Guanare (Venezuela) (location identifier)

GUR Alotau (Papua New Guinea) (location identifier)

GUREN Groupe Universitaire de Recherche sur l'Environnement

GURME GAW Urban Research Meteorological Environment

GUS Gemeinschaft der Unabhängigen Staten • Peru Grissom (USA-IN) (location identifier)

GUSP Gateway to the Universe Star Party (Canada-ON)

GUSTO Guidance Using Stable Tuning Oscillations

GUT Grand Unified Theory (*) • Grand Unified Transition • Great Unification Theory • Groupe francophone des Utilisateurs de TEX • Gütersloh (Germany) (location identifier)

GUU Grundarfjördur (Iceland) (location identifier)

GUUG Gross Unit Unavailable Generation

GUUS Greatest Useful Unfounded Set

guv gerecht und vollkommen

GUV Mougulu (Papua New Guinea) (location identifier)

GUVI Global Ultraviolet Imager (TIMED)

GUW Guryev (Kazakhstan) (location identifier)

GUX Guna (India) (location identifier)

GUY Guyana (international vehicle registration code) • Guyana (ISO country code) • Guyane (code ISO de pays) • Guymon (USA-OK) (location identifier)

G-V Gravity-Velocity

G&V Graphics & Visualization

GV Gebäudeverteiler • Generalversammlung • Giant Void • Gigavolt • Goed Voor • Grande Visite • Grande Vitesse • Gran Velocidad • Graphics and Visualization • Gravimetric Volume • Grid Variation • Gross Valuation • Ground Validation • Ground Visibility

GvA Gesellschaft für volkstümliche Astronomie (Germany)

GVA Geneva Cointrin (Switzerland) (location identifier) • Genève Cointrin (Switzerland) (location identifier) • Gyroscope Vibration Absorber

GVB Generalized Valence Bond

GVC Global Value Chain • Grand Visiteur Cosmique

GVD Group Velocity Dispersion • Group View Display

GVDS Global Vegetation Data Set • Global Virtual Data Structure

Gve Genève (Switzerland)

GVE Gordonsville (USA-VA) (location identifier)

GVF Goodness of Variance Fit

GVG Gerichtsverfassungsgesetz

GVH Graphical Virtual Hand

GVHRR Geosynchronous Very-High-Resolution Radiometer

GVI Global Vegetation Index (*) • Green Vegetation Index • Green River (Papua New Guinea) (location identifier)

GVIB Gearbox Vibration

GVIS Geographic Visualization

Gvl Gravel

GVL Gainesville - Lee Gilmer (USA-GA) (location identifier) • Graphics and Visualization Laboratory • Graphics Visualization Laboratory • Ground Vibration Laboratory

GVLS Global Verification and Location System

GVM Global Virtual Manufacturing • Gross Vehicle Mass

GVMK Gibridnye Vychislitel nye Mashiny i Kompleksy (periodical code)

GVN Global Volcanism Network

GVO Global Virtual Observatory • Global Virtual Organization

GVP Global Volcanism Program (Smithsonian Institution) • Greenvale (Australia-QLD) (location identifier) • Gross Value of Production

GVQ Generalized Vector Quantization

GVR Governador Valadares (Brazil-MG) (location identifier)

GVRP GARP VLAN Registration Protocol

GVS Ground Vehicle System • Ground Vibration Study

GVSC General-purpose VLSI Spaceborne Computer • Generic VHSIC Spaceborne Computer

GVSM Generalized Vector Space Model

GVSP Goddard Visiting Scientist Program

GVSU Grand Valley State University (USA-MI)

Gvt Government

GVT Global Virtual Team (*) • Global Virtual Time • Greenville Majors (USA-TX) (location identifier) • Ground Vibration Test

GVTA Ground Vibration Test Article

GVTF Ground Vibration Transfer Function

Gvtl Governmental

GVU Graphics, Visualization and Usability

GVV Gewählte Virtuelle Verbindung

GVW Grandview - Richards Gebaur (USA-MO) (location identifier) • Great Virtual Wall • Gross Vehicle Weight • Gross Vehicular Weight

GVWG Global VLBI Working Group

GVX Gävle (Sweden) (location identifier) • Gavle (Sweden) (location identifier)

Gw Gigawatt

G/W Gateway • Gross Weight

GW Galactic Wind • Galton-Watson (process) • Gas Well • Gateway • Gee Whiz • Generalized Waring (distribution) • General Warning • Gesture Workshop • Gierer-Witz • Gigawatt • Global Warming • Graphics Workstation • Gravitational Wave • Gravity Wave • Gross Weight • Ground Water • Guinea Bissau (ISO country code) • Guinée Bissau (code ISO de pays)

GWA General Work Area • Global Work Area • Gwa (Myanmar) (location identifier)

GWAAT Global Wide-Area Applications Testbed

GWAC Government Wide Agency Contract

GWADIR Gravitational-Wave Antenna Detector Interferometer

GWAI German Workshop on Artificial Intelligence

Gway Gateway

GWB Gravitational Wave Background

GWBB Generalized Wind Blown Bubble (model)

GWC Generic WIYN Client • Global Weather Center • Graduate Work in Chemistry

GWD Gwadar (Pakistan) (location identifier)

GWDG Gesellschaft für Wissenschaftliche Datenverarbeitung Göttingen (Germany)

GWE Global Weather Experiment • Gravitational-Wave Experiment • Greatest Western Elongation • Gweru (Zimbabwe) (location identifier)

GWEC Global Wireless Education Continuum

GWEN Ground Wave Emergency Network

GWG Gesellschaft für Wirtschaftsförderung Göttingen (Germany)

GWh Gigawatt-Hour

GWHIS Global-Wide Help and Information System

GWI Gravitational Waves Institute

GWIB Gravity-Wave-Induced Bias

GWIC Gravitational Waves International Committee

GWIGWO Good Will In - Good Will Out

GWL Gateway Link • Gwalior (India) (location identifier)

GWM Generic Window Manager • Global Wind Measurement • Graphics Windows Manager • Guaranteed Weekly Minimum

GWN Gnarowein (Papua New Guinea) (location identifier)

GWO General Watch Officer • Greenwood Leflore (USA-MS) (location identifier)

GWP Generalized Waring Process • Gloval Warning Potential • Global Water Partnership • Government White Paper • Graduate Work in Physics • Greenhouse Warming Potential • Gross World Product • Peso (Guinea-Bissau) (ISO currency code)

GWPESD German Workshop on Performance Engineering within Software Development

GWPI Guelph-Waterloo Physics Institute (Canada-ON)

GWP2 Guelph-Waterloo Programme for Graduate Work in Physics

GWR Gravitational Wave Radiation

GWS Gemini Wavefront Sensor • General Working Scheme • Glashow - Weinberg - Salam (theory) • Glenwood Springs (USA-CO) (location identifier) • Global Wave Statistics • Graphical Windowing System • Graphics Workstation • Great White Spot (Saturn)

GWSI Ground Water Site Inventory

GWT Griffith-Woodley Technique • Gross Weight • Ground Winds Tower • Guided Wave Theory • Westerland (Germany) (location identifier)

GWTR Ground Water Treatment Rule

GWU Gardner-Webb University (USA-NC) • George Washington University (USA-DC)

GWUI Ground Water Under the Influence

GWUP Gesellschaft zur Wissenschaftlichen Untersuchung von Parawissenschaften eV (Germany)

GWV Glendale (USA-WV) (location identifier)

GWW Berlin Gatow (Germany) (location identifier) • Gauss-Weierstraß Wavelet • Guaranteed Weekly Wage

GWY Galway (Ireland) (location identifier)

GX Galactic X-source • Gateway Accelerated

GXF Seiyun (Yemen) (location identifier)

GXG Negage (Angola) (location identifier)

GXH Mildenhall (UK-England) (location identifier)

GXQ Coyhaique (Chile) (location identifier)

GXS Global Exchange Service

GXUVDC German XUV Data Centre

GXX Yagoua (Cameroon) (location identifier)

Gxy Galaxy

GXY Greeley (USA-CO) (location identifier)

Gy Gray • Gymnasium

GY Good Year • Greene-Young (catalog) • Guernsey (Channel Islands) • Guyana (ISO country code) • Guyane (code ISO de pays) • Gypsy

GYA Guayaramerin (Bolivia) (location identifier)

GYD Dollar (Guyana) (ISO currency code) • Guide

GYE Guayaquil - Simon Bolivar (Ecuador) (location identifier)

GYI Gisenyi (Rwanda) (location identifier)

GYL Argyle (Australia-WA) (location identifier)

GYM Guaymas (Mexico) (location identifier)

GYN Goiânia - Santa Genoveva (Brazil-GO) (location identifier)

GYO Get Your Own

GYP Gympie (Australia-QLD) (location identifier) • Gyro Yaw Position

Gyr Gigayear • Gyroscope

GYR Goodyear Litchfield (USA-AZ) (location identifier)

Gyro Gyroscope

Gyrs Gigayears

GYY Chicago Gary (USA-IN) (location identifier)

gz GZIP Compression (file extension)

G-Z Giacobini-Zinner (comet)

GZ Giacobini-Zinner (comet) • Goldene Zahl • Ground Zero

GZI Ghazni (Afghanistan) (location identifier)

GZK Greisen - Zatsepin - Kuzmin (cutoff)

GZM Gozo (Malta) (location identifier)

GZO Gizo (Solomon Islands) (location identifier)

GZS Gesamtverzeichnis der Zeitschriften und Serien

GZT Gaziantep (Turkey) (location identifier) • Greenwich Zone Time

G2RA Groupement Régional de Recherche en Acoustique

G3C Great Circle Cell Count

G7 Groupe des Sept • Group of Seven Industrial Nations (now G8) • Grupo de los Siete

G8 Groupe des Huit • Group of Eight Industrial Nations (formerly G7) • Grupo de los Ocho

H

h Altitude • Elevation • Enthalpy • Habitant • Habitante • Hacia • Harbour • Hard • Hardness • Hart • Header (file extension) • Heat • Heat Transfer Coefficient • Heavy Sea • Hecto • Height • Heiß • Hekto • Hell • Help • Hence • Heure • High • Hinten • Hoch • Höhe • Homme • Hoogte • Hora • Horizontal • Horse • Hot • Hour • House • Hundred • Husband • Hybrid • Miller Index • Planck Constant • Specific Enthalpy • Step Size • Thickness

H Absolute Helligkeit • Altitude • Angular Impulse • Boltzmann Function • Dose Equivalent • Dynamic Amplification Factor • Eccentric Anomaly of Hyperbolic Orbit • Elevation • Enthalpy • Haben • Hafen • Haft • Haftung • Hail • Haler • Half • Halt • Haltestelle • Hamiltonian • Handel • Handschrift • Harbour • Hard • Hardness • Haupt • Haus • Hauteur • Hazardous • Haze • Head • Header • Heat Capacity • Heater • Heating Value of the Fuel • Heavy • Heck • Heft • Height • Heilige • Heim • Heimat • Heiß • Heizung • Helligkeit • Help • Hengst • Henry • Hermano • Hersteller • Hessen (Germany) • High • Highest • High-Pressure Center • Hill • Hinfahrt • Hired • Hitze • Hoch • Hochdruck • Hochschule • Höhe • Hoheit • Hôpital • Hora • Horizontal Component of the Magnetic Vector • Horn • Hospital • Hot • Hotel (international alphabet) • Hour • Hour Angle • Hub • Hubble Constant • Hund • Hundert • Hungary (international vehicle registration code) • Hungary (postal country code) • Hydrant • Hydrogen • Hydrogène • Hydrogenium • Light Exposure • Magnetic Field Strength • Magnetic Intensity • Momentum • Thickness of Lithosphere • Two Hundred (Roman numeral) • Wasserstoff • 200 (Roman numeral)

ha hauptamtlich • Hausa (ISO language code) • Hectare • Hektar • High Altitude • Hoc Anno • Huius Anni

hA herrschende Ansicht

Ha Hafen • Hahnium • Haiti • Handel • Hartmann Number • Hawaii

H&A Health & Accident

H/A Hazardous Area

HA Apogee Altitude • Haftanstalt • Haiti • Half-Adder • Handausgabe • Handicapé Auditif • Harvard Annals (periodical code) • Hauptabteilung • Hauptanschluß • Hauptamt • Hautes Alpes (France) • Hawaii • Hazard Analysis • Hazard Assessment • Hazardous Area • Health Authority • Height Adjust • Height of Apogee • Hexanoic Acid • High Altitude • High Angle • High Aperture • High Availability • Higher Authority • Highlights of Astronomy • Hit Avoidance • Hochamt • Holding Area • Home Agent • Homogeneous Arc • Host Agent • Host Astronomer • Hour Angle • Humanitarian Assistance • Hungary (international aircraft registration code) • Hydrologic Atlas

HAA Haloacetic Acid • Hamilton Amateur Astronomers (Canada-ON) • Hasvík (Norway) (location identifier) • Hawkesbury Astronomical Association (Australia-NSW) • Height Above Airport • High-Altitude Abort • Hungarian Astronomical Association (Hungary)

HAADS High-Altitude Attitude Determination System

HAAP High-Altitude Aerial Photograph

HAARP High-frequency Active Auroral Research Program

HAARS High-Altitude Attitude Reference System

HAAS High Altitude Air Sampler

hab habilitieren • habilitiert

Hab Habilitand • Habilitation • Habitant • Habitat • Habitation

HAB Habitação • Habitation • Hamilton - Marion County (USA-AL) (location identifier) • Hammelburg (Germany) • Harmful Algal Bloom • High-Altitude Burst • Horizontal Assembly Building • Horizontal Axis Bearing

HABAS Hastings and Battle Astronomical Society (UK-England)

HABCIT High-Altitude Balloon Circumstellar Imaging Telescope

Hab.Corp. Habeas Corpus

HABE High-Altitude Balloon Experiment

habil Habilitatus

Habil Habilitand • Habilitation

HABP Hypersonic Arbitrary Body Program

Habt Habeat

HAC Hachijojima (Japan) (location identifier) • Haleakala Atmospheric Characterization • Half Alignment Chart • Harbor and Approach Chart • Heading Alignment Circle • Heading Alignment Cone • Heading Alignment Cylinder • Head of Astronomical Computing • Helicopter Active Control • Hereinafter Described • Hilltopper Astronomy Club (USA-KY) • House Appropriations Committee (USA) • Hughes Aircraft Co. • Hydrogenated Amorphous Carbon

HACBSS Homestead And Community Broadcasting Satellite Service

HACC Humanitarian Assistance Coordination Center

HACCP Hazard Analysis and Critical Control Point • Hazard Analysis Critical Control Point (*)

HACD Home Area Customer Dialing

HACE Hubble Advanced Camera for Exploration

HACMP High-Availability Cluster Multiprocessing

HACP HTML Astronomy Classroom Project

HACR High-Accuracy Cryogenic Radiometer

HACT Helicopter Active Control Technology

HAD Historical Astronomy Division (AAS) • Half-Amplitude Duration • Halmstad (Sweden) (location identifier) • Hauptabteilungsdirektor • Health Assessment Document • Heat Actuated Device • Helium Abundance Detector • High-Accuracy Data • Historical Astronomy Division • Hospitalisation À Domicile

HADD Hardness Assurance Design Documentation

HADES High-Altitude Debris Environmental System

HADS Hierarchical Applications Data Structure • High-Accuracy Digital Sensor • High-Amplitude Delta Scuti (star) • Hypersonic Air Data Sensor

HADTS High-Accuracy Data Transmission System

HAdW Heidelberger Akademie der Wissenschaften (Germany)

HAE Havasupai (USA-AZ) (location identifier) • Height Above Ellipsoid • Helicity-induced Alfvén Eigenmode • High-Altitude Endurance • High-Amplification Event

HAeBe Herbig Ae/Be (star)

HAEBE Herbig Ae/Be (star)

HAEC Hughes Aerospace and Electronics Co.

HAEMP High-Altitude Electromagnetic Pulse

Haf Hafen

HAF Half Moon (USA-CA) (location identifier) • Height Above Field • High-Altitude Fluorescence

HAFB Holloman Air Force Base (USA-NM) (formerly AAFB)

HaFraBa Hamburg-Frankfurt-Basel (highway)

Haft Haftung

Haftb haftbar

Haftg Haftung

HAG Den Haag (Netherlands) (location identifier) • Hampshire Astronomical Group (UK-England) • Hardware Advisory Group • Heimarbeitsgesetz • Help Authoring Guide • Hochschularbeitsgemeinschaft

HAGE Human Activity and Global Environment

HAH Moroni Hahaya (Comoros) (location identifier)

HAHO High-Altitude High-Opening

HAI Helicopter Association International • Hellenic Aerospace Industry (Greece) • Three Rivers - Dr. Haines (USA-MI) (location identifier)

HAIC Hetero-Atom In Context

HAIL High Altitude Inversion Layer

HAINS High-Accuracy Inertial Navigation System

HAIR High-Accuracy Instrumentation Radar

HAIRDS High-Altitude Infrared Detecting Set

HAIRS High-Accuracy Intersatellite Ranging System

HAIS Hughes Applied Information Systems

HAIZY Hamburger Isochron-Zyklotron (Germany)

HAJ Hanover Langenhagen (Germany) (location identifier)

HAK Haikou (China-PRC) (location identifier) • HETG Analysis Kit • Horizontal Access Kit

Hal Halogen • Hôpital • Hospital

HAL Halali (Namibia) (location identifier) • Hard Array Logic • Hardware Abstraction Layer • Hardware Access Layer • Hardware Adaptation Layer • Hardware Array Logic • Height Above Landing • Heuristically Programmed Algorithmic (2001) • Highly Automated Logic • High-order Articulated Language • High-order Assembly Language • High-temperature Acoustic Levitator • Hindustan Aeronautics Ltd. (India) • House-programmed Array Logic • Houston Aerospace Language • Hyperspace Analog to Language (algorithm)

halbj halbjährig

Halbj Halbjahr

halbjhg halbjährig

Halbs Halbsatz

HALCA Highly Advanced Laboratory for Communications and Astronomy (formerly VSOP)

HALDB High-Availability Large Database

HALE High-Altitude Long-Endurance

HALLPASS Hadamard Transform Laser Long-Path Absorption Spectrometer System

HALO High-Altitude Large Optics • High-Altitude Learjet Observation • High-Altitude Learjet Observatory • High-Altitude Lift Off • High-Altitude Long Operation • High-Altitude Low-Observable • High-Altitude Low-Opening • Hughes Automated Lunar Observer

HALOE Halogen Occultation Experiment

HalRe Hale Observatories Reprints (periodical code)

HALS Hydrographic Airborne Laser Sounder • High Approach Landing System

HALSS High-Altitude Lidar Sensing Station

HALT High-Altitude Laser Tracking

Ham Hamburg (Germany) • Hameau

HAM Hamburg Fuhlsbüttel (Germany) (location identifier) • Hardware Associative Memory • Heavy - Apperance - Magnetic (meteorite) • Height Adjustment Maneuver • Hierarchical Access Method • Hierarchical Archive Manager • Hold And Modify • Hopfield Associative Memory • Horizontal Access Module • Hydrological Angular Momentum • Hypertext Abstract Machine

HAMI Host-Automated Message Interface

HAMM Hypertext Abstract Machine Management

HAMOTS High-Accuracy Multiple-Object Tracking System

HAMP Handling And Moving Procedure • High-Altitude Measurement Probe

HamS Hamburger Sternwarte Sonderdrücke (periodical code)

HAMS Half-Age Main Sequence • Hardness Assurance, Maintenance and Surveillance

HAMT Human-Aided Machine Translation

Han Hanover (Germany)

HAN Hanoi (Viet Nam) (location identifier) • Health Alert Notice • Home Area Network • Home ATM Network

HANA Haiti National Airlines

HANAP Hardness Assessment and Network Analysis Program

HAND Have A Nice Day

Handb Handbook • Handbuch

handgeschr handgeschrieben

handl handlich

Handl Handling • Handlung

HANDS High-Altitude Nuclear Detection Studies

handschr handschriftlich

Handschr Handschreiben • Handschrift

Handw Handwerk

Handwb Handwörterbuch

HANE High-Altitude Nuclear Environment • High-Altitude Nuclear Explosion

HANG Hawaiian Air National Guard

HANOMAG Hannoverische Maschinenbau-Aktiengesellschaft (Germany)

HANOVA High-dimensional Analysis Of Variance

HANS Head And Neck Support • High-Altitude Navigation System

Hants Hampshire (UK-England)

HANUL High-energy Astrophysics Neutrino Laboratory (Korea-ROK)

HAO Hamilton (USA-OH) (location identifier) • High-Altitude Observatory (USA-CO) • Hydrogenated Anthracene Oil

HAP Half-Arcsecond Programme • Handelsabgabepreis • Hardware Allocation Panel • Hazard Abatement Program • Hazardous Air Pollutant • High Accident Potential • High-Altitude Platform • High-Altitude Probe • High-speed Arithmetic Processor • Honeywell Array Processor • Horizontal Axis Pivot • Host Access Protocol • Hydrocarbure Aromatique Polycyclique • Hydroxyl Ammonium Perchlorate • Kontiki Resort - Whitsunday - One Hundred (Australia-QLD) (location identifier) • Long Island (Australia-QLD) (location identifier)

HAPAG Hamburg-Amerikanische Packetfahrt-Aktiengesellschaft

HAPDAR Hard Point Demonstration Array Radar

HAPE High-Altitude Pulmonary Endema

HAPEX Hydrological Atmospheric Pilot Experiment

HAPHEEP Hadronic Physics with High-Energy Electromagnetic Probes

HAPI Harris Application Program Interface • High-Altitude Plasma Instrument

HAPL High Average Power Laser

HAPP High-Altitude Powered Platform

HAPPE High-Altitude Particle Program Experiment

HAPPEX Hall A Proton Parity Experiment

HAPRS High-Accuracy Positioning Reference System

HAPS High-Altitude Platform Station • High-Altitude Platform System • Huddersfield Astronomical and Philosophical Society (UK) • Hybrid Approximated Solution • Hydrazine Auxiliary Propulsion System

HAQ Hanimaadhoo (Maldives) (location identifier)

Har Harbour • Harmonização • Harmonizada • Harmonisation • Harmonisé • Harmonization • Harmonized

HAR Harrisburg (USA-PA) (location identifier) • Hazard Analysis Report • Heterodyne Array Receiver • High-Altitude Research • High-Angular Resolution • H-transformed Autoregression (solution)

HARA High-Altitude Radar Altimeter

HarAC Harvard College Observatory Announcement Card (periodical code)

HARAC High-Altitude Resonance Absorption Calculation

HarCi Harvard College Observatory Circular (periodical code)

HARB Harbor • Harbour

HARC Houston Advanced Research Centre (USA-TX)

HARD High-Accuracy Reflector Development

HARDI High Angular Resolution Deployable Interferometer

HARDS High-Altitude Radiation Detection System

HARDTS High-Accuracy Radar Data Transmission System

HAREM Halieutic Radar Experiment in the Mediterranean

harm Harmonisch

Harm Harmonic • Harmonie • Harmonique • Harmonium • Harmony

HARM Hazard Assessment Rating Methodology

harml Harmlos

HARN High-Accuracy Reference Network

HARNET Hong-Kong Academic and Research Network

HaRP Hawaiian Rainband Project

HARP Hazardous Assessment of Rocket Propellants • Hermite Accelerator Pipeline • Heterodyne Array Receiver Program • High-Altitude Reconnaissance Platform • High-Altitude Release Point • High-Altitude Research Project • High-gain Avalanche Rushing Photoconductor • HST Aberration Recovery Program

HARPI Height Azimuth Range Position Indicator

HARPS High-Accuracy Radial-velocity Planetary Search • High-Accuracy Radial-velocity Planet Searcher

HarRe Harvard College Observatory Reprints (periodical code)

HARS Heading And Reference System • Heading/Attitude Reference System • High-Altitude Route System

HART Harmonic Aerodynamic Rotor Test • Hayden Analysis and Reporting Tool • Heartland Astronomical Research Team (USA-KS) • Height Area Rain Threshold • Hypermedia Authoring Research Toolkit

HartRAO Hartebeesthoek Radio Astronomy Observatory (South Africa)

HARTRAO Hartebeesthoek Radio Astronomy Observatory (South Africa)

HARTS Hydrometeorological Automatic Recording and Telemetering System

Harv Harvard

HARV High-Alpha Research Vehicle • High-Angle-of-attack Research Vehicle • High-Aspect Ratio Vessel • Hybrid Augmented Reusable Vehicle

HarZi Harthaer Beobachtungs-Zirkular (periodical code)

HAS Hail (Saudi Arabia) (location identifier) • Halmstads Astronomiska Sällskap (Sweden) • Hamilton Astronomical Society (New Zealand) • Harrogate Astronomical Society (UK) • Hawaiian Astronomical Society (USA-HI) • Hazard Analysis Summary • He Atom Scattering • Helicopter Air Service • High-Altitude Sensor • Highlands Alkali Suite • Holddown Alignment Support • Horizontal Air Shower • Houston Astronomical Society (USA-TX) • Hungarian Academy of Sciences • Hydraulic Actuation System • Hydrogen Actuation System

HASAWA Health And Safety At Work Act (USA)

HASC High-Altitude Surveillance Concept • House Appropriations Subcommittee (USA)

HASE High-Assurance Systems Engineering

HASI Huygens Atmospheric Structure Instrument

HASL Health And Safety Laboratory • Hot Air Solder Levelling

HASP Health And Safety Plan • High-Accuracy Satellite Position • High-Altitude Sampling Program • High-Altitude Sounding Project • High-Altitude Space Platform • Houston Automatic Spooling Priority • Hub And Spoke Program • Hub And Spoke Project

HASPA High-Altitude Superpowered Powered Aerostat

HASS H-Alpha Sky Survey

HAST High-Altitude Supersonic Target • High-Altitude Supersonic Test • Horizontal Automated Solar Telescope • Humanitarian Assistance Survey Team

HASTA H-Alfa Solar Telescope for Argentina

HASVR High-Altitude Space Velocity Radar

HASYLAB Hamburg Synchrotron Laboratory (Germany)

HAT Handling and Archiving Tool • Hardened Antenna Technology • Hatfield (Starlink internal code) • Heathlands (Australia-QLD) (location identifier) • Height Above Touchdown • Hierarchical Alias Table • High-Altitude Test • High-Altitude Testing • Hot-Anode Thruster • Hungarian Automatic Telescope

HATII Humanities Advanced Technology and Information Institute (UK-Scotland)

HATM Home Automatic Teller Machine

HATR Hazardous Air Traffic Report

HATS Heuristic Automated Transportation System

HATT Hattinger Astronomie- und Trödeltag (Germany)

HATVAD Hatfield (Starlink internal code)

HAU Haugesund (Norway) (location identifier)

Haub Haubitze

haupts hauptsächlich

haus hausieren

Hausbes Hausbesetzer • Hausbesitzer • Hausbesorger • Hausbesuch

Hausgeh Hausgehilfe

Haush Haushalt • Haushaltung • Hausherr

Hausm Hausmacht • Hausmädchen • Hausmarke • Hausmeister

Hausr Hausrat • Hausrecht

haust Haustus

Hausverw Hausverwalter • Hausverwaltung

hausw hauswirtschaftlich

Hausw Hauswirtschaft

Hav Havana

HAV La Habana - José Marti (Cuba) (location identifier)

HAVAGO Horizontal Adjustment and Vertical Adjustment of Geodetic Observations

HAVI Home Audio Video Interoperability

HAVO Hoger Algemeen Voortgezet Onderwijs

Haw Hawaii (USA)

HAW Haverfordwest (UK-England) (location identifier) • Hawaii (STDN) • Hazardous Analysis Worksheet • Heidelberger Akademie der Wissenschaften (Germany)

HAWAII HgCdTe Astronomical Wide-Area Infrared Imager

HAWC High-resolution Airborne Wideband Camera

HAWIE Hamburg-Wechsler-Intelligenztest für Erwachsene

HAWIK Hamburg-Wechsler-Intelligenztest für Kinder

HAWK Have Alimony - Will Keep

HawRe University of Hawaii Reprints (periodical code)

HAWS Hawaiian Area Wideband System

HAWT H-Alpha and White-light Telescope • Horizontal-Axis Wind Turbine

HAX Muskogee Hatbox (USA-OK) (location identifier)

HAY Haycock (USA-AK) (location identifier)

Haz Hazard • Hazardous

HAZ Hatzfeldthaven (Papua New Guinea) (location identifier) • Hazardous • Heat-Affected Zone • Heiße Abfall-Zellenanlage • High-latitude Activity Zone

HAZCOM Hazardous Communications

HAZCON Hazardous Consideration • Hazardous Condition

HAZMAT Hazardous Material • Hazardous Materiel

HAZOP Hazard and Operability (study)

HAZTOX Hazardous and Toxic

HAZU Hrvatska Akademija Znanosti i Umjetnosti (Croatia)

HAZW Hazardous Waste

HAZWOPER Hazardous Waste Operations and Emergency Response

h/b Hardbound

hb Halb • Heilbar • Hemoglobin

Hb Handbedienung • Handbetrieb • Handbuch • Hausbriefkasten • Hemoglobin • Herberge

H-B Hexadecimal to Binary

H/B Half Board • Hyperbaric

HB Half-Breadth • Handbibliothek • Handbook • Handbücherei • Handelsbetrieb • Handelsbilanz • Hardback • Hard Black • Harvard Bulletin (periodical code) • Hauptbibliothek • Hauptbuch • Hauptbüro • Heat Budget • Heavy Boat • Height Balanced • Helvetisches Bekenntnis • Herringbone • Heure Budgétaire • Hexachlorobutadien • Hierarchical Basis • High Bay • High Burst • Hochbau • Hochschulbibliothek • Hofbibliothek • Hollow Beam • Homogeneous Band • Horizontal Baffle • Horizontal Branch • Horizontal Bridgman • Human Being • Hydrogen Burning • Hydrostatic Balance • Hyperbaric • Liechtenstein (international aircraft registration code) • Switzerland (international aircraft registration code)

HBA Health Benefits Advisor • Hobart (Australia-TAS) (location identifier) • Hochbauamt • Host Bus Adapter

HBAPP Hamburg Block Adjustment Program Package

HBB Handbook Bulletin • Hobbs - Industrial Park (USA-NM) (location identifier) • Horizontal Blue Branch • Hot Bottom Burning • Hot Big Bang

HBBC Hot Big Bang Cosmology

HBBM Hot Big Bang Model

HBBW Half-Beam Bandwidth

HBC Handlebar Control • Herbig-Bell Catalog • Hierarchical Bayesian Clustering • High Breaking Capacity • House Budget Committee • Hydrogen Bubble Chamber

HBCSE Homi Bhabha Centre for Science Education (India)

HBCU Historically Black Colleges and Universities (USA)

Hbd Halbband • Hochbedarf

HBD Had Been Drinking • Has Been Drinking

HBDE Hirota Bilinear Difference Equation

HBDS Hierarchical Binary Digital Search

HBE First-overtone Blue Edge • Hélicoptère Bombardier d'Eau

HBEO Human Beyond Earth Orbit (project)

HBES Human Behavior and Evolution Society

Hbf Hauptbahnhof • Heimatbahnhof

HBF Horizon Break File

HBFG Hochschulbauförderungsgesetz • Host Behavior Functional Group

hbfl hauptberuflich

Hbg Hamburg (Germany)

HBG Hattiesburg (USA-MS) (location identifier)

HBH Hauptbuchhalter • Hobart Bay (USA-AK) (location identifier)

Hbhf Hauptbahnhof

HBI Harbour Island (Bahamas) (location identifier)

Hbj Halbjahr

HBJ Harcourt Brace Jovanovich

Hbl Halbblut

HBl Handelsblatt • Hauptblatt

HBL Bebelegi (South Africa) (location identifier) • Health-Based Level • High-energy-peaked BL Lacertae (object) (*) • High-frequency BL Lacertae (object) • Hydrogen Burning Limit

HBLL Hydrogen Burning Limiting Layer

HBLR Hidden Broad-Line Region

HBM Halo Beaming Model • High Boost Mode • Hydrologic Benchmark

HBML Hydrogen-Burning Mass Limit (*) • Hydrogen-Burning Minimum Luminosity

HBMM Hydrogen-Burning Minimum Mass

HBMN Hydrologic Benchmark Network

HBMS Hyperbase Management System

HBN High-Brightness Network • Phu-bon (Viet Nam) (location identifier)

HBO Hoger Beroeps Onderwijs • Home Box Office • Horizontal-Branch Oscillation • Humboldt (USA-NE) (location identifier) • Hydrogen Burning Object

HBOSC Howard B. Owens Science Center (USA-MD)

HBP High Blood Pressure • Hydrogen Burning Phase

HBPIC High Blood Pressure Information Center

HBQS Homogeneous Bright QSO Survey • Homogeneous Bright Quasar Survey

Hbr Harbour

HBr Hydrogen Bromide

HBR Harbor • Harvard Business Review • Has Been Reviewed • High Bit Rate • High Burst Rate • Hobart (USA-OK) (location identifier) • Hydrogen Burning Region

HBRA High Background Radiation Area

HBRACW Has Been Reviewed And Concurred With

HBRS Harbors

HBS Hardware Breakdown Structure • Harvard Business School (USA-MA) • Hogere Burger School • Hydrogen Burning Sequence • Hydrogen Burning Shell • Hydrostatic Bearing System

Hbschr Hubschrauber

HBSN Humboldt Bay Seismic Network

HBSS High-Band SIGINT System • High-Band Subsystem

hbstl halbstündlich

Hbstr Herbiester (Belgium)

Hbt Hobart (Australia-TAS)

HBT Hafr Albatin (Saudi Arabia) (location identifier) • Hanbury-Brown Twiss • Heflex Bioengineering Test • Heterojunction Bipolar Transistor

HBTD High Bit Density Tape

HBTM Hybrid Bispectral Threshold Method

HBV Hamburg-Bergedorf Variable (star)

HBW Half Bandwidth • Heated Bridge Wire • Hot Bridge Wire

HBX Hubli (India) (location identifier)

Hby Hereby

HBZ Hochschulbibliothekszentrum • Hydrogen Burning Zone

h/c Held Covered

hc Honoris Causa

H/C Hand Carry • Head Count • Health Check • Held Covered • Hydrogen/Carbon

H&C Hot & Cold

HC Ecuador (international aircraft registration code) • (The) Hague Convention • Hand Control • Hand Controller • Handling Capacity • Hard Color • Hard Copy • Hard Core • Harvard Circular (periodical code) • Hazardous Cargo • Hazardous Code • Head Count • Header Compression • Health Certificate • Heat Content • Heater Controller • Helical Coil • Heterogeneous Computing • Heuristic Concept • Hierarchical Clustering • High Capacity • High Carbon • High Court • High Current • High-speed CMOS • Hipparcos Catalogue • Historical Commission • Holding Coil • Hollow Cathode • Hollow Charge • Hôpital Civil • Horizontal Cable • Hors Catégorie • Hors Commerce • Hors Concours • Hospices Civils • Host Country • Hot Component • Hot Core • House of Commons • Human Civilization • Humidity Control • Hybrid Computer • Hydrocarbon • Hydrocarbure • Hydrogen Chloride

HCA Big Spring - Howard County (USA-TX) (location identifier) • Half-Cone Angle • Heading Crossing Angle • Head of Contracting Activity • Head of Contracting Agency • Hierarchical Cluster Analysis • High-Capacity Ambulance • Hill County Astronomers (USA-TX) • Humanitarian Assistance • Hungarian Communications Authority (Hungary)

HCAA Hellenic Civil Aviation Authority (Greece)

Hcap Handicap

HCAS Harford County Astronomical Society (USA-MD)

HCB Hexachlorobenzene • House of Commons Bill • Shoal Cove (USA-AK) (location identifier)

HCBP Hexachlorobiphenol

HCC Han Characters Collection • Haut Comité Scientifique • Hexachlorocyclohexan • Hibbing Community College (USA-MN) • High-Context Culture • Highly Competitive Carrier • Hollow Cathode Contactor • Honeycomb-Chain-Channel (model) • Hot Compact Core • Hudson - Columbia County (USA-NY) (location identifier) • Hydraulic Cement Concrete

HCCB Hardware Configuration Control Board

HCCCHO Propynal

HCCD Historical Canadian Climate Database • High-performance Charge-Coupled Device

HCCI Haut Conseil pour la Coopération Internationale • Homogeneous Charge Compression Ignition

HCCUP Health Care Cost and Utilization Project

HCD Hardware Configuration Definition • Hardware Configuration Dialog • High-Current Density • High-Current Diode • High-speed Copper Drop • Hollow Cathode Discharge • Hot-Carrier Diode

HCDB Historical Cost Database

HCDD Hexachlorodibenzodioxin

HCDM Hot Cold Dark Matter • Hybrid Compending Delta Modulation

HCDN Hydroclimate Data Network

HCE Hardware Configuration Engineer • Heater Control Electronics • Hic Conditus Est • High-Cycle Fatigue • Hollow-Cathode Effect • Human-Caused Error

HCED Hand Controller Engage Driver

HCF Halo Compression Formula • Halt and Catch Fire • Heat Control Filter • Height Correction Factor • High-Calorific Fuel • High-Carbon Ferrochrome • High-Cycle Fatigue • Highest Common Factor • Highest Confidence First • Honeycomb Foundation • Hundred Cubic Feet

HCFA HARPS Cassegrain Fibre Adapter • Health Care Financing Administration

HCFC Hydrochlorofluorocarbon

HCFO Hispanic Community Fund Office

HCFP Haut Comité à la Formation Professionnelle

HCG Hickson Compact Group (of galaxies) • Hierarchical Concept Graph • Horizontal (location of) Center of Gravity

HCH Hechingen (Germany) • Hexachlorocyclohexane

HCHO Formaldehyde

HCHy Hovering Craft and Hydrofoil (periodical code)

HCI Hardness-Critical Item • Hierarchically Classified Index • High-Current Inductor • High-Current Injector • Highly Charged Ion • Horizontal Contracting Integration • Host Controller Interface • Hughes Communications Inc. • Human-Computer Interaction • Human-Computer Interface • Human-Computer Interfacing

HCIL Hague Conference on International Law • Human-Computer Interaction Laboratory

HCIMD Human Computer Interaction with Mobile Devices (conference)

HCIRV Human Computer Interface Research Centre

HCIS Human-Centered Information System

HCISIMS Highly-Charged Ion-Secondary Ion Mas Spectrometry

HCIT High-Contrast Imaging Testbed

HCITE Horizontal Cargo Integration Test Equipment

HCJ High Court of Justice

HCl Hydrogen Chloride

HCL Hardware Closed Loop • Hardware Compatibility List • Hardware Configuration List • Harvard College Library (USA-MA) • Helium Cadmium Laser • High - Common - Low • High Cost of Living • Hollow Cathode Lamp • Horizontal Center Line • Horizontal Component of Lift • Hummingbird Communications Ltd. (Canada) • Hydrochloric Acid • Hydrochloride • Hyperchannel Communications Link

HCLF Haut Comité à la Langue Française

HCLOS High-Capacity Line Of Sight

HCLP Hierarchical Constraint Logic Programming • Hierarchically-Constrained Logic Programming

HCM Eil (Somalia) (location identifier) • Hard Copy Module • Heat Capacity Mapper • Human Capital and Mobility • Hybrid Coupled Model

HCMC Ho Chi Minh City (Viet Nam)

HCMM Heat Capacity Mapping Mission

HCMOS High-performance Complementary Metal Oxyde Semiconductor

HCMP Hardware Configuration and Maintenance Plan

HCMR Heat Capacity Mapping Radiometer

HCMT HOSS Communications Model Toolkit

HCMTS High-Capacity Mobile Telephone System

HCN Historical Climatology Network • Host Country National • Hydrocyanic Acid • Hydrogen Cyanide • Hyperswitch Communications Network

HCO Formyl Radical • Harper College Observatory (USA-IL) • Harvard College Observatory (USA-MA) • Hastings College Observatory (USA-NH) • Health Care Organization

HCOA Harvard College Observatory Archives

HCON Hour-Confirmed

HCOOH Formic Acid

Hcp Handicap

HCP Habitat Conservation Plan • Hardcopy Printer • Hardness Critical Process • Hazardous Critical Procedure • Health Care Provider • HET Completion Project • Hexachlorocyclopentadiene • Hexachlorophene • Hexagonal Close-Packed • High-Calcium Pyroxene • Highly Cited Paper • House of Commons Paper • Hybrid Combustion Process • Hydraulic Circulation Pump

HCPG Hard Copy Processor Group

HCPI Highly Cited Paper Index

Hcptr Helicopter

HCQ Halls Creek (Australia-WA) (location identifier)

HCR High Cross-Range • Highway Contract Route • Holy Cross (USA-AK) (location identifier) • Horizon Crossing Sensor

HCRB Hardware Configuration Review Board

HCRC Human Communications Research Centre

HCRD Human-Carrying Rotation Device

HCRF Hipparcos Celestial Reference Frame

HCRO Hat Creek Radio Observatory (USA-CA) • High Cross-Range Orbiter

HCRS Heritage Conservation and Recreation Service (USA)

HCRT Hanna City Robotic Telescope

HCRW Hot and Cold Running Water

HCS Hard-Clad Silica (fiber) • Hazard Communication Standard • Header Checksum • Heliographic Coordinate System • Heliospheric Current Sheet • High-Carbon Steel • Homogeneous Computer System • Johannesburg Randburg (South Africa) (location identifier)

HCSAS Hardware Configuration Status Accounting System

HCSD High-Speed Circuit Switched Data

HCSDS High-Capacity Satellite Digital Service

HCSHT High-Carbon Steel Heat-Treated

HSCI Hughes Communications Services Inc.

HCSL Hybrid Computation and Simulation Laboratory

HCSS Herschel Common Science System • High-Capacity Storage System

hct Hora Cum Tempore

HCT Hardware Compatibility Test • HASP Communication Table • HgCdTe • High Critical Temperature • Hoffer Sternfreunde-Treffen (Germany) • Hollow Cathode Tube • Home Communications Terminal • Hypertext Compiling Tools • Mercury Cadmium Telluride

HCTDS High-Capacity Terrestrial Digital Service

HCTR High-Capacity Trunk Radio

HCTS Hydraulic Component Test Stand

HCU Handheld Computer Unit • Handheld Control Unit • Heater Control Unit • Helium Charging Unit • High-capacity Computer Unit • High-Capacity Unit • Host Connection Unit • Hydraulic Charging Unit • Hydraulic Control Unit • Hydraulic Cycling Unit

HCUP Health Cost and Utilization Project

HCV Hawaii Center for Volcanology (USA-HI) • Helium Check Valve • High-Capacity Voice • Hydrogen Check Valve

HCVM Human-Centered Virtual Machine

HCW Cheraw (USA-SC) (location identifier) • Heterogeneous Computing Workshop • Horizontal Component of Weight

HCX High Current Experiment

HCZ Hydrogen Convection Zone

hd Hand • Hard • Head • Hora Decubitus • Hundert

Hd Hand • Handel • Hard • Head • Hemd • Hund • Hundert

H-D Hexadecimal to Decimal

H&D Hardened & Dispersed

HD Hafendienst • Half Duplex • Hand-Drawn • Hard Disk • Hard Drive • Hardened and Dispersed • Harmonic Distortion • Harmonization Document • Hauptdeck • Hausdorff Distance • Haut Débit • Head Diameter • Heavy Distillate • Heavy Duty • Heißdampf • Helium Dewar • Henry Draper (catalog) • Hierarchical Diagnosis • High Density • High Definition • High Dose • High Dynamics • Higher Dimension • Highly Desirable • Hochdecker • Hochdruck • Holddown • Honorable Discharge • Horizontal Drain • Horizontal Drive • Hurter-Driffield • Hydrodynamic • Hydrodynamics • Hydrogen Deuteride • Hydrogen-Deuterium • Hydrogen Drain • Hydrographic Department

Hda Hacienda

HDA Handbuch der Astrophysik (periodical code) • Hard-Disk Attack • Hardware Disposition Area • Head Disk Assembly • Hexanediamine • Hexamethylenediamine • Hidden Falls (USA-AK) (location identifier) • High-Density Amorphous • High-Duty Alloy • Host Data Adapter • Housekeeping Data Acquisition • HST Data Archive • Hubble Data Archive (HST) • Hybrid Detector Assembly

HDAL Hexadecenal • Hexadecimal

HDAM Hierarchical Direct-Access Method

HDAS Hinckley and District Astronomical Society (UK) • Hydrographic Data Acquisition System

HDATZ High-Density Air-Traffic Zone

Hdb Handbuch

HdB Hydrogen-deficient Binary (star)

HDB Handbook • Haut Débit Binaire • HEASARC Database • Heidelberg (Germany) (location identifier) • Hierarchical Database • High-Density Bipolar • Housing and Development Board • Housing Development Board • Hypersonic Database

HDBC High-Density Bipolar Code

Hdbk Handbook

HDBK Handbook

hdbr handbreit

Hdbr Handbremse

HDBV Host Database View

HdC Hydrogen-deficient Carbon-rich

HDC Hard Disk Controller • Hasselblad Data Camera • Hawaii DUMAND Center (USA-HI) • Henry Draper Catalogue • Hierarchical Distributed Control • Higher Degrees Committee • High-Definition Contrast • Highly Developed Country • Hungarian Data Center • Hybrid Device Controller • Hydrodynamic Chromatography

HDCC High-Dependability Computing Consortium

HDCD High-Definition Compatible Digital

HDCL High-level Data Link Control

Hdcp Handicap

HDCP High-Bandwidth Digital Content Protection • High-Dependability Computing Program

HDCS HIPPARCOS Dedicated Computer System

HDD Hard-Disk Drive • Hardware Design Document • Head-Down Display • High-Definition Display • High-Density Data • High-Density Digital • High-Density Disk • Horizontal Directional Drilling • Hyderabad (Pakistan) (location identifier)

HDDA Hexanediol Diacrylate

HDDBMS Heterogeneous Distributed Database Management System

HDDI Host Document Data Interchange

HDDL High-Definition Distance Learning

HDDMC High-Definition Display Manufacturing Consortium

HDDR High-Density Digital Recorder • High-Density Digital Recording

HDDS High-Density Data System

HDDT High-Density Digital Tape

HDDTR HHDT Recorder

HDE Hauptverband des Deutschen Einzelhandels (Germany) • Head-Drop-Einheit • Henry Draper Extension (catalog) • Holdrege Brewster (USA-NE) (location identifier) • Homogeneous Differential Equation

HDEC Henry Draper Extension Chart

HDEP High-Density Electronic Packaging

hdf Hierarchical Data (file extension)

HDF Hardware Display Function • Hartmann Dispersion Formula • Hierarchical Data Format • High-frequency Direction Finder • High-frequency Direction Finding • Historical Data File • Hubble Deep Field

HDF-N Hubble Deep Field North

HDFN Hubble Deep Field North

HDF-S Hubble Deep Field South

HDFS Hubble Deep Field South

Hdg Heading

HDG Heading

HDGCP Human Dimension of Global Change Programme • Human Dimension of Global Change Project

HDGEC Human Dimensions of Global Environmental Change • Human Dimensions of Global Environment Change

hdgest handgesteuert

hdgew handgewebt

hdgm handgemacht

hdgr handgroß

Hdgr Handgriff

HDGRS High-Density Ground Recording System

HDH HDLC Distant Host • Oahu Dillingham (USA-HI) (location identifier)

Hdhbg Handhabung

HDHN High-Degree Helioseismology Network

HDI Haftpflichtverband der Deutschen Industrie • Header Information • Head-to-Disk Interface • Head-to-Disk Interference • High-Definition Imaging • High-Density Inclusion • High-Density Interconnect • Human Development Index • Human Development Institute

HDIC Helium Double Ionization Cell

HDIS High-Density Instrumentation System • High Dynamic Instrument Set

HdJ Haus der Jugend

HDJ Hydrographic Department of Japan

HdK Handkasse • Haus der Kultur

Hdkf Handkerchief

hdl handlich

Hdl Handel • Handle • Handlung

HDL Hardware Description Language • Hardware Design Language • High-Density Lipoprotein • High-Density Liquid

Hdlbg Heidelberg (Germany)

HDLC Higher-level Data Link Control • High-level Data Link Control (*) • High-speed Data Link Control

HDLD Headland

Hdlg Handling • Handlung

HDM Hamadan (Iran) (location identifier) • Hardware Device Module • Hierarchical Development Methodology • High-Duty Metal • Hot Dark Matter • Hypermedia Design Model • Hypermedia Document Modeling • Hypertext Design Model • Hypertext Document Model

HDMAC High-Definition Multiplexed Analogue Component

HDMI High-Density Multichip Interconnect

HDML Handheld Device Markup Language

HDMR High-Density Multitrack Recording

HDMS Heidelberg Dark Matter Search • Heritage Documentation Management System • Hermes Documentation Management System • High-Density Memory Set • High-Density Memory System

Hdn Harden

HDN Hayden - Yampa Valley (USA-CO) (location identifier) • Steamboat Springs (USA-CO) (location identifier)

Hdns Hardness

HDO Deuterated Water • Eau Lourde • Heavy Water

HDOP Horizontal Dilution Of Precision

HDOS Hughes Danbury Optical Systems (USA-CT)

HDP Hardware Development Plan • High-Definition Progressive • High-Density Polyethylene • Holddown Post • Horizontal Data Processing • Human-Dimension Programme • Hydroxydimethylpyrimidine

HDPE High-Density Polyethylene

HDPLX Half Duplex

HDPS High-Density Power Supply

HDQ Headquarters

Hdqrs Headquarters

Hdqtrs Headquarters

hdr Header (file extension)

Hdr Header

HDR Header • Health Data Recorder • Heißdampfreaktor • High Data Rate • High Data Register • High-Density Recorder • High-Density Region • High-Density Rule • High-Dose Rate • Hold Down and Release • Hot Dry Rock • Humanitarian Daily Ration

HDRA High Data Rate Assembly

HDRF High-Density Random Fiber • Honolulu Data Reduction Facility (USA-HI)

HDRI High-Dynamic-Range Imaging

HDRL High-Data-Rate Laser • High-Dose Reference Laboratory

HDRM High Data Rate Multiplexer

HDRR High Data Rate Recorder

HDRRM High-Density Radiation-Resistant Microelectronics

HDRS High-Data-Rate System

HDRSAT High-Data Rate Satellite Terminal

HDRSS High-Data-Rate Storage System

HDRT High Data Rate Terminal • High Dynamic Range Telescope

HDS Hachette Distribution Services (France) • Hardware Description Sheet • Hawaii Deep Survey • Hierarchical Data Structure • Hierarchical Data System • High-Density Sample • High-Dispersion Spectrograph • High-resolution Data Stream • Hitachi Data Systems • Homogeneous Data Set • Hydrogen Disposal System

HDSB Heavy Dry Support Bridge

HDSBUV High-Density SBUV

HDSC High-Density Signal Carrier

hdschr handschriftlich

Hdschr Handschreiben • Handschrift

Hdshk Handshake

HDSL High-bit-rate Digital Subscriber Line (*) • High-data-rate Digital Subscriber Line • High-data-rate Digital Subscriber Link • High-data-rate Digital Subscriber Loop • High-rate Digital Subscriber Line • High-rate Digital Subscriber Link

HDSS Holographic Data Storage System

Hdst Headset

HDST Headset

hdt hundert

Hdt Hundert

HdT Haus der Technik

HDT Hard-Disk Toolkit • Head-Down Tilt • Head-Down Tilted • Heat Distortion Temperature • High-Density Tape • High Driver Task • Hochdruckturbine • Hospitalisation à la Demande d'un Tiers • Host Data Terminal • Host Digital Terminal

HDTOMS High-Density TOMS

HDTP Handheld Device Transport Protocol • Hoofddirectie Telecommunicatie en Post (Netherlands) • Hyperdata Transfer Protocol

HDTR High-Density Tape Recorder

HDTS High-Density Transcription System

HDTV High-Definition Television

HDU Hard Disk Unit • Head-Drop-Unit • Header and Data Unit • Header Data Unit (*) • Helmet Display Unit

HDUE High Dynamic User Equipment

HDUX Half-Duplex

HdV Heure de Vérité

HDV High Dollar Value

HDVC High-Density Voice Compressor

HDVD High-Definition Video Disk • High-Definition Video Display • High-Definition Volumetric Display

Hdw Handwagen • Handwerk • Hardware

Hdwb Handwörterbuch

HDW Hardware

Hdwd Hardwood • Headword

HDWDM High-Density Wavelength Division Multiplexer

Hdwe Hardware

HDWE Hardware

Hdwr Hardware

HDx Half Duplex

HDX Half Duplex

HDY Hat Yai (Thailand) (location identifier) • Heavy Duty

Hdz Handzeichen

HDZ Hochdruckzylinder

he Hic Est • Hoc Est

He Helium • Hélium

H/E Heat Exchanger

HE Half Effect • Hamburg-ESO (survey) • Hardware Evaluation • Hardware Evaluator • Hauptentschädigung • Hautes Études • Header Extension • Health Expenditure • Heat Engine • Heat Exchange • Heat Exchanger • Heavy Equipment • Hebrew • Henize • Her Eminence • Her Excellency • High Eccentricity • High Efficiency • High End • High Energy • Higher Education • High Explosive • Highly Explosive • His Eminence • His Excellency • Höheneinheit • Horas Extra • Horizontal Equivalent • Host Environment • Hub End • Human Engineering • Hydroelectric • Hydrogen Embrittlement • Hydrostatic Equilibrium • Hyperbolic Equation

HEA Health Effects Assessment • Herat (Afghanistan) (location identifier) • High-Energy Approximation • High-Energy Astrophysics • Higher Education Act (USA) • Higher Education Authority

HEAB High-Energy Astrophysics Branch

HEAD AAS High-Energy Astrophysics Division Newsletter (periodical code) • Handheld Encryption Authentication Device • High-Energy Astrophysics Division

HEADS High-Energy Astrophysics Database Service

HEAL Human Exposure Assessment Location

HEAMOWG High-Energy Astrophysics Management Operations Working Group

HEAMWG High-Energy Astrophysics Management Working Group

HEANet Higher Education Authority Network

HEAO High-Energy Astronomical Observatory • High-Energy Astronomy Observatory • High-Energy Astrophysical Observatory (*) • High-Energy Astrophysics Observatory • Hoger Economisch en Administratief Onderwijs

HEAP High-Energy Aim Point • High-Energy Astrophysics • High-Energy Astrophysics Program

HeaPh Health Physics (periodical code)

HEAR High-Efficiency Anti-Reflecting • High-Efficiency Anti-Reflection • Human-Error Action Report

HEART Hardcore Entry And Routing Technique • Health Evaluation And Risk Tabulation

HEASARC High-Energy Astrophysics Science Archive Research Center

HEASRC High-Energy Astrophysics Science Research Center

HEAT Headquarters Effectiveness Assessment Tool • High-Energy Antimatter Telescope • High-Enthalpy Ablation Test • High-lift Engine Aeronautics Technology

Heb Hebrew

HEB Handheld Electronic Book • Henzada (Myanmar) (location identifier) • Hersbruck (Germany) • High-Energy Buncher • High-Excitation Blob • Hollow Electron Beam • Hot Electron Bolometer

Hebd Hebdomadaire

HEBIS Hessisches Bibliotheks-Informations-System

HEBM Hot Electron Bolometer Mixer

Hebr Hebrew • Hebrides

HEC Hardened Electronic Component • Hasselblad Electric Camera • Hautes Études Commerciales • Header Error Check • Header Error Control • Header Error Correction • Heliocentric Ecliptic Coordinates • Heure de l'Europe Centrale • High-Emission Cathode • High-Energy Chemistry • Higher Education Coordinating (board) • Higher Education Coordination • Hollerith Electronic Computer • Hydroelectricity Commission • Hydrological Engineering Center

HECaR High-sensitivity Electrically Calibrated Radiometer

HECC Higher Education Coordinating Council • High-End Computing and Computation

HECK Header Error Check • High-Energy Chemical Kernel

HECP Harbor Entrance Control Post • Herbaceous Energy Crops Program

HECR High-Energy Cosmic-Ray

HECRE High-Energy Cosmic-Ray Experiment

HECSE Higher Education Consortium for Special Education (USA)

HeCV Helium Cataclysmic Variable (star)

HECV Helium Check Valve

HECZ He-flash-driven Convective Zone

HED Herendeen (USA-AK) (location identifier) • High-Energy Density • High-Energy Detector • Horizontal Electrical Dipole • Howardite - Eucrite - Diogenite (meteorites) • Hypohidrotic Ectodermal Dysplasia

HEDC Hasselblad Electric Data Camera • HESSI European Data Center • HESSI Experimental Data Center (*)

HEDCC Human-Error Data Control Center

HEDCO Higher Education for Development Cooperation

HEDM High-Energy Density Matter

HEDR Human Engineering Domain Report

HEDs Howardites - Eucrites - Diogenites (meteorites)

HEDS Human Exploration and Development of Space (office) (NASA)

HEE Helena - Thompson-Robbins (USA-AR) (location identifier) • Higher Education Exchange • Human Energy Expenditure

HEEC Heure d'Été de l'Europe Centrale

HEED Health and Environmental Effects Document • High-Energy Electron Diffraction

HEEEL High-Energy Electronically Excited Laser

HEEF High-Energy Electron Fluxmeter

HEEN Human Evolution Education Network

HEEO Highly Elliptical Earth Orbit • Highly Elliptical Equatorial Orbit

HEEP Health and Environmental Effects Profile

HEES Hypertext Evaluation Expert System

HEF High Efficiency • Hochschulkonferenz für Erziehungswissenschaft und Fachdidaktik

HEFA Higher Education Facilities Act (USA)

HEFC Higher Education Funding Council

HEFCE Higher Education Funding Council for England (UK)

HEFCS Higher Education Funding Council for Scotland (UK)

HEFCW Higher Education Funding Council for Wales (UK)

HEFIN Hoge-Energiefysica Instituut Nijmegen (Netherlands)

HEFOE Hydraulic Electrical Fuel Oxygen Engine

HEFT High-Energy Focusing Telescope

HeG Helium Gauge

HEG Helium Gauge • High-Energy Gamma-ray • High-Energy Grating • High-Enthalpy (facility at) Göttingen (Germany) • High-Excitation Galaxy

HEGCS High-Energy Gas Čerenkov Spectrometer

HEGR High-Energy Gamma Ray

HEGRA High-Energy Gamma-Ray Array (*) • High-Energy Gamma-Ray Astronomy

HeGV Helium Gauge Valve

HEGV Helium Gauge Valve

HEH Heho (Myanmar) (location identifier)

HEHI High-Energy Heavy Ion

HEI Hautes Études Industrielles • Hautes Études Internationales • Heide Büsum (Germany) (location identifier) • Higher Education Institute • Higher Education Institution

HEIAC Hydraulic Engineering Information Analysis Center

HEIC Hubble ESA Information Centre

HEIDI High-Efficiency Intermediate-Dispersion Instrument • High-Energy Imaging Device

HEIG Hautes Études d'Informatique et de Gestion

Heim Heimat

heimatl heimatlich • heimatlos

HeimG Heimgesetz

HeiTS Heidelberg Transport System

HEK Hanseatische Ersatzkasse • Heihe (China-PRC) (location identifier)

Hel Helicopter

HEL Hardware Emulation Layer • Header Extension Length • Heliocentric • Helsinki (Finland) (location identifier) • High-Energy Laboratory • High-Energy Laser • Human Engineering Laboratory

HELA High-Energy Laser Astrophysics

HELAS Hellenic Astronomical Society (Greece) • High-Energy Large-Area Survey

HELB High-Energy Laser Beam

HELCUSS High-Energy Laser Computerized Unmanned Space Synthesis

HELD Health Effects Laboratory Division

HELF High-Energy Laser Fluence

Helg Helgoland

Heli Heliport

Hélico Hélicoptère

Heliogr Heliograph • Heliographie

HELIO Heliocentric

HELIOS Handicapped European Living In Open Society

hell hellenisch

HELLO High-Energy Laser Light Opportunity

HELMET High-Energy Laser Meteorology

HELO High Elliptical Lunar Orbit

HelOb Helwan Institute of Astronomy and Geophysics Bulletins (periodical code)

HELOS Highly Eccentric Lunar Occultation Satellite

HELP Highly Extendable Language Processor

HelR Helsinki Observatory and Astrophysics Laboratory Reports (periodical code)

HELS High-Energy Laser System

HELSTF High-Energy Laser System Test Facility

HELTA High-Endurance Lighter-Than-Air (project)

helv helvetisch

Helv Helvetia • Helvetica

HEM Harmonization of Environmental Measurement • Heat Exchanger Method • Heavy Equipment Maintenance • Helsinki Malmi (Finland) (location identifier) • High-Energy Monitor • Higher Education Manager • Hitchhike Experiment Module • Hybrid Electromagnetic Wave

HEMA High-Energy Neutral Atom

HEMCO Heavy Equipment Maintenance Company

HemD Hemel en Dampkring (periodical code)

HEMI Hautes Études de Management International • Hemispherical

HEMIS Harmonization of Environmental Measurement Information System

HEMM Heavy Earth-Moving Machinery

HEMP HET Echo Mapping Project • High-altitude Electromagnetic Pulse • High Energy Magnetic Pulse

HEMS Helicopter Emergency Medical Service • High-level Entity Management System

HEMT High Electron Mobility Transistor (*) • High Electron Movability Transistor • High Energy Microwave Transistor

HeMV Helium Manual Valve

HEMV Helium Manual Valve

HEN Hendon (UK-England) (location identifier)

HENA High-Energy Neutral Analyzer • High-Energy Neutral Atom

HENAP High-Energy Neutrino Astrophysics Panel (IUPAP)

HEND High-Energy Neutron Detector

HENP High Energy and Nuclear Physics

HENSA Higher Education National Software Archive • Higher Education National Software Authorities

HEO Haelogo (Papua New Guinea) (location identifier) • High Earth Orbit • High Earth Orbiter • High-Energy Orbit • High Eccentric Orbit • High Elliptical Orbit • Higher Executive Officer • Highly Eccentric Orbit • Highly Elliptical Orbit

HEOD Honeywell Electrical Optical Division

HEOO Honeywell Electro-Optical Operations

HEOR Historical Event Operation Record

HEOS High-Eccentricity Earth-Orbiting Satellite • Highly Eccentric Orbit Satellite (*) • Highly Elliptical Orbit Satellite

HEP Heliospheric Energetic Particle • Heterogeneous Element Processor • High-Energy Particle • High-Energy Physics • High-Energy Pulse • Human Error Probability • Hydroelectric Power

HEPA High-Efficiency Particle Absolute • High-Efficiency Particle Accumulator • High-Efficiency Particulate Absorber • High-Efficiency Particulate Air (filter) (*) • High-Efficiency Particulate Atmosphere

HEPAD High-Energy Proton and Alpha Detector

HEPAP High-Energy Physics Advisory Panel (USA)

HEPC High-Energy Proportional Counter

HEPI High Earch Processes Imager

HEPIC High-Energy Physics Information Center

HEPL High-Energy Physics Laboratory

HEPNet High-Energy Physics Network

HEPP High Energy and Particle Physics

HEPS High-Energy Particle Spectrometer

HEQ Heliocentric Equatorial (coordinates)

HEQC Higher Education Quality Council (UK)

Her Herald • Hercules-Herculis (constellation) • Hertog

HER Heraklion (Greece) (location identifier) • High-Energy Resolution • High Erection Rate • Human-Error Rate

HeRA Hermes Robot Arm • Hermes Robotic Arm

HERA Hadron Electron Research Accelerator • Hadron Electron Ring Accelerator (*) • Hadron Elektron Ring Anlage • Hermes Robot Arm • Hermes Robotic Arm

HERAP Human-Error Research and Analysis Program

HERAS Hull and East Riding Astronomical Society (UK)

HERB High-Energy Resolution Burst

HERC Hercules • High-Energy Cosmic Ray

HERCMA Hellenic European Research on Computer Mathematics and its Applications (conference)

HERCULES Hand-held, Earth-oriented, Real-time, Cooperative, User-friendly, Location-targeting and Environmental System • High Efficiency and Resolution Canterbury University Large Echelle Spectrograph

HERD Herdade • High Education and Research Development

HERE Hazards of Electromagnetic Radiation to Equipment

Héréd Héréditaire

Hered Hereditary

HERF Hazards of Electromagnetic Radiation to Fuel • High-Energy Radiated Field • High-Energy Radio Frequency

herg hergestellt

Herg Hergabe • Hergang

hergest hergestellt

HERL Health Effects Research Laboratory

HERMES High-Energy Radiation Megavolt Electron Source • High-resolution Evaluation of Radiances from Meteorological Satellites

HERMS Human Engineering, Reliability, Maintainability, and Safety

HERO Hazards of Electromagnetic Radiation to Ordnance • High-Energy Replicated Optics • Hyperextremely Red Object

HEROS Heidelberg Extended Range Optical Spectrograph

HERP Hazards of Electromagnetic Radiation to Personnel • Hydrological Emscher Radar Project

HERS High-Energy Range Spectrometer • Higher Education Resource Services

HERSCP Hazardous Exposure Reduction and Safety Criteria Plan

Herst Hersteller • Herstellung

HERT Headquarters Emergency Relocation Team

Herts Hertfordshire (UK-England)

HERU Higher Education Research Unit

HES Hamburg/ESO Survey • Hautes Études Sociales • Hermiston State (USA-OR) (location identifier) • Hypertext Editing System

HESAC Hermes Safety Advisory Committee

HESF Hermes Engineering Support Facility

HESP High-Energy Solar Physics

HESPE High-Energy Solar-Physics Explorer

hess hessisch

Hess Hessen (Germany)

HESS High-Energy Squib Simulator • High-Energy Stereoscopic System • Hydrology and Earth System Sciences (journal)

HESSE Hard Embedded Software Support

HESSI High-Energy Solar Spectroscopic Imager

HESYRL Hefei National Synchrotron Radiation Laboratory (China-PRC)

Het Heterodyne

HET Hall Effect Thruster • Heavy-Equipment Transporter • High-Energy Telescope • High-Energy Transfer • Hobby-Eberly Telescope • Hohhot (China-PRC) (location identifier) • Hot Electron Bolometer • Human Engineering Test

HETAF Heated Exhaust Toxic Area Forecast

HETC High-Energy Transport Code

HETCP Hobby-Eberly Telescope Completion Project

HETE High-Energy Transient Experiment (*) • High-Energy Transient Explorer

HETG High-Energy Transmission Grating

HETGS High-Energy Transmission Grating Spectrometer

HETM Hybrid Engineering Test Model

HETP Healthcare Engineering and Technology Policy • Height Equivalent per Theoretical Plate • Height Equivalent to a Theoretical Plate • Hexaethyl Tetraphosphate

HETS Heavy-Equipment Transporter System • High-Energy Telescope System • Hispanic Educational Telecommunications System (USA)

HEU Highly Enriched Uranium

Heur Heuristic

HEUS High-Energy Upper Stage

HEV Heavy • Hybrid Electric Vehicle

HEW Half-Energy Width • Hamburgische Elektricitätswerke • Health - Education - Welfare

HeWD Helium White Dwarf (star)

HEWG Human Engineering Working Group

HEWS Humanitarian Early Warning System

Hex Hexadecimal • Hexagon • Hexagonal

HEX Heat Exchanger • Hexadecimal • Hexagon • High-Energy X-ray • High-frequency Exciter • Santo Domingo - Herrera (Dominican Rep.) (location identifier)

HEXAGONE High-Energy X- And Gamma-ray Observatory of Nuclear Emission

HEXE High-Energy X-Ray Experiment

HEXIS High-Energy X-ray Imaging Spectrometer

HEXOS Humidity Exchange Over the Sea

HEXTE High-Energy X-ray Timing Experiment

HEY Ozark Hanchey (USA-AL) (location identifier)

HEZ Natchez - Hardy-Anders (USA-MS) (location identifier)

HEZG Hinterbliebenenrenten- und Erziehungszeitengesetz

hf Half

Hf Hafen • Hafnium • Handelsform • Hauptform • Hypothèse Faible

H/F Homme/Femme • Human Factor

HF Halbe Fahrt • Hard Failure • Hard Firm • Hartree-Fock • Hauptfunktion • Haute Fréquence • Haze Filter • Height Finder • Heure de Formation • Hickey-Frieden • High Fidelity • High Frequency • Hilfsfunktion • Hochfrequenz • Horizontal Flight • Hors Fonction • Hot Firing • Human Factor • Hydrofluoric Acid • Hydrogen Fill • Hydrogen Fluoride • Hyperfiltration • Hypothèse Forte

Hfa Haifa (Israel)

HFA Haifa (Israel) (location identifier) • Hartree-Fock Approximation • High-Frequency Accelerometer • Housing Finance Agency • Hydrofluoroalkane

HFAARP High-Frequency Active Auroral Research Program

HFAARS High-Frequency Adaptive Antenna Receiving System

HFAC House Foreign Affairs Committee • Human Factors

HFAS High-Frequency Acoustic Sounder • High-Frequency Antenna System

HFB Hartree - Fock - Bogoliubov (theory) • Heptafluorobutyrate

HFBA Heptafluorobutyric Acid

HFBC High-Frequency Broadcasting

hf. bd. Half Binding • Half Bound

Hfbf Hafenbahnhof

HFBI Heptafluorobutyrylimidazole

HfBK Hochschule für Bildende Künste

HFBR High-Flux Beam Reactor

HFBS High-Fluc Backscattering Spectrometer • High-Fluc Backscattering Spectrometry

HFC Heat Flow and Convection • Helsinki Fair Centre (Finland) • High-Frequency Correction • High-Frequency Current • Hybrid Fiber and Coax • Hybrid Fiber Coax • Hybrid Fiber Coaxial • Hydraulic Flight Control • Hydrofluorocarbon • Hydrofluorocarbure

HFCA Health Care Financing Administration

HFCC Hermes Flight Control Centre • High Frequency Communications Central

HFCL Hydraulic Flight Control Laboratory

HFCT Hydraulic Flight Control Test

HFCV Helium Flow Control Valve

HFCW Hydrogen Fluoride Continuous Wave

HfD Hauptanschluß für Direktruf

HFD Hartford Brainard (USA-CT) (location identifier) • Heat Flow Density • High-Frequency Domain • Hrvatsko Fizikalno Drustvo (Croatia)

HFDF High-Frequency Direction Finder • High-Frequency Direction Finding • High-Frequency Distribution Frame

HFDL High-Frequency Data Link • Host Forms Description Language

HFDP Human Factors Development Plan

HFDS Head-up Flight Display System • Hydrogen Fluid Distribution System

HFDU High-Frequency Data Unit

HFE Heavy Front Engine • Heavy Fuel Engine • Hefei (China-PRC) (location identifier) • High-Frequency Excess • High-Frequency Executive • Historia de Formación Estelar • Human Factors and Ergonomics • Human-Factors Engineering

HFEA Human-Factors Engineering Analysis • Human-Factors Engineering Assessment

HFEI Host Front End Interface

HFEN High Frequency Emergency Network

HFEP Host Front End Processor

HFES Human Factor and Ergonomics Society

HFET Hellmann-Feynman Electrostatic Theorem • HEM Field-Effect Transistor • Heterojunction Field-Effect Transistor

HFF Hellmann-Feynman Force • High-Frequency Filter • High-Frequency Furnace • Hochschule für Fernsehen und Film • Hoffman Mackall (USA-NC) (location identifier) • Hubble Flanking Field

HfG Hochschule für Gestaltung

HFG Harmonic Frequency Generator • High-Frequency Generator • Hochfrequenzgesetz

HFH High-Frequency Hopping

HFI High-Frequency Instrument (Planck)

HFIP Hexafluoroisopropanol • Hexafluoroisopropyl • High-Frequency Improvement Program

HFIR High Flux Isotope Reactor

HFIS History and Foundations of Information Science

HFL Dutch Guilder (currency) • Hollandse Florijn (currency)

HFM Hand and Foot Monitor • Hazardous Fluids Module • Health and Fault Management • Hollow Fiber Membrane • Human-Factor Modeling • Human Factors and Medicine • Human Figure Model

HFML High Fibre Meal • High Field Magnet Laboratory

HFMP Human Factors Management Plan

HFMRF High Field Magnetic Resonance Facility (USA-WA)

HFMS High Fidelity Mechanical Simulator • High-Frequency Monitoring System

HFMU High-Fidelity Mock-Up • High-Frequency Mock-Up

Hfn Hafen • Hafner

HFN High-Frequency Noise • Höfn (Iceland) (location identifier) • Hornafjördur (Iceland) (location identifier)

HFO Heavy Fuel Oil • High-Frequency Oscillation • High-Frequency Oscillator

HFOF Hierarchical Friends Of Friends

HFOSC Hanle Faint Object Spectrograph Camera (India)

HFOVR Hemispherical Field-Of-View Radiometer

HFP Hardened Focal Plane • Hexafluoropropylene • High-Frequency Peak • High-Frequency Peaked • High-Frequency Peaker • Host Front Processor

HFPL HF-induced Plasma Line

HFPM Human Figure Performance Model

HFPO Hexafluoropropylene Oxide

HFPS Helium Fill and Pressurization System

HFR Hartree-Fock Relativistic (method) • Height-Finding Radar • High-Flux Reactor • High Focal Ratio • High-Frequency Range • High-Frequency Receiver • High-Frequency Resistor • Hochflußreaktor • Human and Family Resources

HFRA High-Frequency Recovery Antenna

HFRB High-Frequency Radio Broadcast • High-Frequency Regional Broadcast

HFRG High-Frequency Radio Group

HFRT High-Frequency Radio Transmitter

HFRW High-Frequency Resistance Welding

HFS Hagfors (Sweden) (location identifier) • Heliogeophysical Forecasting Service • Hierarchical File System (*) • Hierarchical Filing System • High-Fidelity Simulator • High-Fidelity Sound • High-Field Side • Horizontal Flight Simulator • Hunters - Floaters - Sinkers • Hyperfine Structure • Hypothetical Future Sample

HFSD Heating and Fueling Systems Division

HFSE High Field Strength Element

HFSJG Hochalpine Forschungsstationen Jungfraujoch und Gornergrat (Switzerland)

HFSN High-Frequency Seismic Noise

HFSP Hierarchical and Functional Software Process • Human Frontier Science Program

HSFRQ High-energy-peaked Flat-Spectrum Radio Quasar

HFSS High-Frequency Structure Simulation • Hochschule für Sozialpädagogik und Sozialökonomie

HFSWR High-Frequency Surface Wave Radar

Hft Haftung

HFT Hammerfest (Norway) (location identifier) • Harbor File Transfer • Heating Facility for Turbines • High-Flux Telescope • High Function Terminal • Horizontal Flight Test • Human Factors in Telecommunications • Human Factors Test • Hyperbolic Flux Tube

HFT&E Human Factors Test & Engineering • Human Factors Test & Evaluation

HFTE Human Factors Test and Engineering • Human Factors Test and Evaluation

HFTF Horizontal Flight Test Facility

Hftg Haftung

HFTS Horizontal Flight Test Simulator

HFTTY High-Frequency Teletype

HFTUAAM High-Frequency Tuning Unit Antenna Automatic Matching

HFU Heat-Flow Unit

HFW Henschel Flugzeug-Werke

HFWA High-Frequency Wave Analyzer

HFX High-Frequency Transceiver • High-Frequency Transducer

hg Hectogram • herausgegeben

Hg Hang • Herausgeber • Hergang • Hofgebäude • Hügel • Mercure • Mercury • Quecksilber

HG Handelsgenossenschaft • Handelsgericht • Handelsgesellschaft • Hand Generator • Harmonic Generator • Haute-Garonne (France) • Henyey-Greenstein • Her Grace • Hierarchical Group • High Gain • High German • His Grace • Hot Gas • Hyades Giclas • Hydride Generation • Hydrogen Generator

HGA Hardware Graphics Accelerator • Hargeisa (Somalia) (location identifier) • Hercules Graphics Adapter • High-Gain Antenna • Human-Genome Analysis • Hybrid Genetic Algorithm

HGAC High-Gain Antenna Controller

HGAS High-Gain Antenna System

hgb herausgegeben

Hgb Herausgeber

HGB Handelsgesetzbuch

Hgbf Hauptgüterbahnhof

HGC Hercules Graphics Card

HGCC Hot Gas Clean Up

HgCdTe Mercury-Cadmium Telluride

HGCP Hercules Graphics Card Plus

HGD Hand-Grip Dynamometer • Hughenden (Australia-QLD) (location identifier)

HGDS Hazardous Gas Detection System

HGE Handling Ground Equipment

HGF Hermann-von-Helmholtz-Gemeinschaft deutscher Forschungszentren (Germany) • Hierarchical Galaxy Formation

hgg herausgegeben

HGG Hot Gas Generator

hGH Human Growth Hormone

HGH Hangzhou (China-PRC) (location identifier) • Human Growth Hormone

hgi HP Font Information (file extension)

HGI High Gain Initiative

hgl HP Graphics Language (file extension)

Hgl Hügel

HGL Helgoland (Germany) (location identifier) • High-Gain Link • High Galactic Latitude

hgm hausgemacht

HGM Hot-Gas Manifold

HGMF High-Gradient Magnetic Filtration

HGMP Human Genome Mapping Project

HGMS High-Gradient Magnetic Separation

HGN Mae Hong Son (Thailand) (location identifier)

HGO Haleakala Gamma Observatory (USA-HI) • Henry Grieb Observatory (USA-CA) • Hermes Global Orbiter • Hidalgo (Mexico) • Korhogo (Ivory Coast) (location identifier)

HGP Human Genome Project

h. g. pt. Hard Gloss Paint

Hgr Hangar • Hanger

HGR Hagerstown (USA-MD) (location identifier) • Hangar • High Growth Rate

HGrG Haushaltsgrundsätzegesetz

HGS Freetown Hastings (Sierra Leone) (location identifier) • Head-up Guidance System • Human Genome Sciences • Hydrogen Gas Saver • Hydrogenous Gas System

HGSC Hubble Guide Star Catalogue

Hgsk Hogskole

HGSt Hauptgeschäfstelle

Hgt Height

HGT Height • High Group Transmit • Jolon Hunter (USA-CA) (location identifier)

Hgts Heights

HGU Mount Hagen (Papua New Guinea) (location identifier)

HGV Heavy Goods Vehicle • Hydrogen Gas Valve • Hypersonic Glide Vehicle

HGVT Horizontal Ground Vibration Test

HGW Herbert George Wells • Higher Gross Weight • Hochgeschwindigkeitswolken

Hgwy Highway

Hgy Highway

HGZ Habitable Galactic Zone • Handels- und Gaststättenzählung • Hogatza (USA-AK) (location identifier)

hh Hands

HH Double Hard • Half Height • Haiti (international aircraft registration code) • Handelshochschule • Handheld • Handhold • Hansen-Hurwitz • Hare Hoogheid • Health Hazard • Heavy Hydrogen • Herbig-Haro (object) • Heren • Her Highness • Hermanos • Herr • Hinterhaus • His Highness • His Holiness • Hitchhiker • Hochhaus • Hochschulen • Horizontal-Horizontal • Hour • Hydrometeor Heating

HHA Health Hazard Analysis • Health Hazard assessment • Huanghua (China-PRC) (location identifier)

HHAR Health Hazard Analysis Report

HHB Hot Horizontal Branch

HHC Hammer Head Crane • Handheld Computer • Handheld Controller • Higher Harmonic Control • Hydrohalocarbon

HHCU Handheld Control Unit

HHD Handheld Display • Hypertensive Heart Disease

HHE Hachinohe (Japan) (location identifier)

H-HEAD Handheld Electronic Authentication Device

HHEAD Handheld Electronic Authentication Device

HHeF Hot Helium Flasher

HHF Höchstfrequenz • Hyper-High Frequency

HHFA Housing and House Finance Agency

HHFW High-Harmonic Fast Wave

HHG High Harmonic Generation • Household Goods

HHH Hilton Head Island (USA-SC) (location identifier) • Treble Hard

HHI Heinrich-Hertz-Institut (Germany) • Wahiawa Wheeler (USA-HI) (location identifier)

HHIP Handheld Information Processor

HHL Handheld Laser

HHLL Handheld Laser Locator

HHM Handheld Monitor

HHMI Howard Hughes Medical Institute (USA)

HHMS Hydrostatic Head Monitoring Station

HHMMSS Hours - Minutes - Seconds

H-HMMWV Heavy Highly Mobile Multipurpose Wheeled Vehicle

HHMMWV Heavy Highly Mobile Multipurpose Wheeled Vehicle

HHMU Handheld Maneuvering Unit

HHN Hahn (Germany) (location identifier)

HHO Herbig-Haro Object • Hypertext Hands-On

HHOK Ha Ha Only Kidding

HHOS Ha Ha Only Serious

HHP Hong Kong (China-PRC) (location identifier)

HHPA Hexahydrophthalic Anhydride

HHPC Handheld Programmable Calculator

HHQ Höchste Quantität • Hua Hin (Thailand) (location identifier)

HHR Habitat Holding Rack • Handheld Radar • Hawthorne (USA-CA) (location identifier)

HHRD Hipparcos Hertzsprung-Russell Diagram

HHRR Handheld Reflectance Radiometer • Handheld Ratioing Radiometer

HHS Handelshochschule • Health and Human Services • Heterodyned Holographic Spectrometer

HHSC Health Hazard Severity Category

HHT Handheld Terminal • Handheld Tester • Handheld Tranceiver • Heinrich Hertz Telescope (USA-AZ) • High-Temperature Tetragonal

HHTV Handheld Terminal Viewer

HHV Higher Heating Value • High Heating Value

HHVIT Handheld Video Imagery Terminal

HHW Höchstes Hochwasser

Hhz Hochdruckdampfheizung

HHZ Hikueru (French Polynesia) (location identifier)

hi Hindi (ISO language code)

Hi Hindi

H&I Harassing & Interdicting

HI Atomic Hydrogen • Dominican Rep. (international aircraft registration code) • Halbinsel • Handelsinstitut • Handling Instruction • Hawaii (USA) • Hawaiian Islands (USA) • Health Insurance • Hearing Impaired • Heat Index • Height Indicator • Height of Instrument • Hic Iacet • High • High Intensity • High Ionization • Honeywell Inc. • Horizontal Interval • Host Integrator • Host Interface • Human Interaction • Human Interface • Humidity Index • Hydriodic Acid • Hydrodynamic Interaction • Hydrogen Iodide • Hydrographisches Institut • Neutral Hydrogen

HiA Highlights in Astronomy (periodical code)

HIA Herzberg Institute of Astrophysics (Canada) • Hot Ion Analyzer

HIAA Hydroxyindoleacetic Acid

Hi.Ac. High Accuracy

HIAD Handbook of Information for Aircraft Designers

HIAL High-Intensity Approach Light • High-Intensity Approach Lighting

HIALS High-Intensity Approach Light System

HIAPER High-performance Instrumented Airborne Platform for Environmental Research

HIAS Human Intellect Augmentation System

HI/AYH Hostelling International/American Youth Hostels

HIB Chisholm (USA-MN) (location identifier) • Heavy-Ion Beam • Hibbing-Chisholm (USA-MN) (location identifier)

HIBAL High-altitude Balloon • High-Ionization Broad Absorption Line

HIBE Hildesheimer Borde Experiment

HIBEX High Acceleration Booster Experiment

HiBFiN High-Brightness Filamentary Network

HIBP Heavy Ion Beam Probe

HIBREL High-Brightness Relay

HIC Habitat International Coalition • Hardware Integration Contractor • Headend Interface Converter • Heavy Ion Counter • HI Cloud • Highest Incoming Channel • High-Integrity Container • High-Intensity Conflict • HIPPARCOS Input Catalogue • Human In Control • Human Interaction Capability • Human Interaction Component • Humidity Indicator Controller • Hybrid Integrated Circuit • Neutral Hydrogen Cloud • Pretoria Iscor (South Africa) (location identifier)

HICA HI Continuum Absorption • Neutral Hydrogen Continuum Absorption

HICAMP Highly Calibrated Airborne Measurements Program

HICANS High-speed Collision Avoidance and Navigation System

HICAP Histogram Cluster Analysis System

HICAS High-velocity Impacts of Composite Aircraft Structures

HICAT High-Altitude Clear Air Turbulence

HICCS Hardware Inventory Configuration Control System

HICDRP Hybrid Incoherent Coherent Dispersion Removal Processor

HICDS Highly Integrated Control and Data System

HICIP Hybrid Image Classification Instruction Package

HICOG High Commissioner for Germany

HICOM High Commission for Germany

HICON Higher Control

HICP Harmonized Index of Consumer Prices

HICSS Hawaii International Conference on System Science

HID Hardness-Intensity Diagram • Hardware Interface Device • Heavy-Ion Discontinuity • High-Intensity Discharge • Horn Island (Australia-QLD) (location identifier)

HIDACZ High-Density Airspace Control Zone

HIDAD High-Density Array Development

HIDAM Hierarchical Indexed Direct-Access Method

HiDAS High Desert Astronomical Society (USA-CA)

HIDAS Helicopter Integrated Defensive Aids System

HiDEC High Density Electronic Center (USA) • Highly-integrated Digital Electronic Control

HIDEC Highly-Integrated Digital Electronic Control

HIDES High-Dispersion Echelle Spectrograph

HIDM High Information Delta Modulation

Hidrául Hidráulica

Hidrom Hidrometría

HIDSS Helmet Interface Display Sighting System • Helmet-Integrated Display Sighting System

HIE Helicopter-Installed Equipment • HI Emission • Neutral Hydrogen Emission • Whitefield (USA-NH) (location identifier)

HIER Higher

HIES Hydrographic Image Exploitation System

HIF Hand-edited Information File • Horizontal Internetwork Field • Host Interface • Human-Initiated Failure • Hydrologic Instrumentation Facility • Hyperfile Interface Language • Hyper-G Interchange Format • Ogden Hill (USA-UT) (location identifier)

HIFAM High-Fidelity Amplitude Modulation

HIFD High-Density Floppy Disk

HIFE Height Interpolation by Finite Elements

Hi-Fi High Fidelity

HIFI Hawaii Imaging Fabry-Pérot Interferometer • Heterodyne Instrument for FIRST • High Fidelity • Hypertext Interface For Information

HIFINS High-Fidelity Network Simulation

HIFLUGCS Highest (X-ray) Flux Galaxy Cluster Sample

HIFOGS High-efficiency Infrared Faint-Object Grating Spectrometer (*) • High-resolution Faint-Object Grating Spectrometer

HIFR Helicopter-In-Flight Refueling • Hover-In-Flight Refueling

HIFU High-Intensity Focused Ultrasound

HIFX High-Energy Flash X-ray

Hig Higiene

HIG Hawaii Institute of Geophysics (USA-HI) • Highbury (Australia-QLD) (location identifier) • High-Integrating Gyroscope • Honeywell Integrating Gyro • Hyperluminous Infrared Galaxy

HIGP Hawaii Institute of Geophysics and Planetology (USA-HI)

HIGED Handbook of Instructions for Ground Equipment Designers

HIGFET Heteroinsulating Gate Field Effect Transistor

HIGM Hot Intergalactic Medium

HIGS High-Intensity Gamma-ray Source

HIH Her Imperial Highness • His Imperial Highness • Hook Island (Australia-QLD) (location identifier)

HIHO High Insertion - High Opening

HII High Input Impedance • Human-Information Interaction • Ionized Hydrogen • Lake Havasu City (USA-AZ) (location identifier)

HIIP Harvard Information Infrastructure Project

HIIPS Home Institute Image Processing System • Home Institution Image Processing Subsystem (*) • Home Institution Image Processing System • HUD-Integrated Information Processing Services

HIIR HII Region • Ionized Hydrogen Region

HIIS Heavy Ions In Space

HIJ Hiroshima (Japan) (location identifier)

HIJASS HI Jodrell All-Sky Survey

HIK Honolulu Hickam (USA-HI) (location identifier)

HIL Hardware In the Loop • Hazardous Immiscible Liquid • High-Ionization Line • High-Intensity Light • High-Intensity Lighting • Human Interface Link • Human In the Loop • Shillavo (Ethiopia) (location identifier)

HILAC Heavy-Ion Linear Accelerator

HILAT High-Latitude

HILDCAA High-Intensity Long-Duration Continuous Auroral Activity

HILEX High-Level Exercise

HILI Higher Level Interface

HILO High Insertion - Low Opening

HILS Hardware-In-the-Loop Simulation

HILT Heavy Ion Large Telescope

HILV High Level

HIM Hardware Interface Module • Highly Ionized Medium • Hot Ionized Material • Hot Ionized Medium (*) • Hot Intercloud Medium • Hypertext - Information Retrieval - Multimedia

HIMAC Heavy-Ion Medical Accelerator

HIMAGV High Mobility and Agility Ground Vehicle

HIMARS History Management And Retrieval System

HiMAT Highly Maneuverable Aircraft Technology

HIMAT Instituto Colombiano de Hidrología, Meteorología y Adecuación de Tierras (Colombia)

HiMaTE High Mach Turbine Engine

HIMEM High Memory

HIMES Highly Maneuverable Experimental Space-vehicle

HIMF HI Mass Function • Mass Function of Neutral Hydrogen

HIMO High Mobility

HIMOV High Mobility Vehicle

HIMS Hot Interstellar Medium Spectrometer • Housing Information Management System • Human Information Management System

HIMSIS Hymalayan Snowcover Information System

HIMSPEC High- and Mid-latitude Speciation

HIMSS Healthcare Information and Management System Society (USA) • High-Resolution Microwave Spectrometer Sounder

Hin Hindi

HIN Chinju (Korea-ROK) (location identifier) • Health Information Network • Hors Instituts Nationaux • Hybrid Integrated Network

Hind Hindi • Hindu • Hindustan • Hindustani

HINDU Histogram Inspired Neighborhood Discerning Unsupervised

HINE Heterogeneous Inhouse Networking Experiment (project)

HINER High-Ionization Nuclear Emission-line Region

HINIL High-Noise-Immunity Logic

HINLA HI Narrow-Line Absorption

HINS Helicopter Integrated Navigation System

hint hinten • hinter

HInt High Intensity

HINT Hanoi Institute of Nuclear Technology (Viet Nam)

Hinw Hinweg • Hinweis • Hinwendung

HIO Hillsboro Portland (USA-OR) (location identifier) • Horizontal Interoperability

HIORP HST Independent Optical Review Panel

HIOS High-Intensity Optical Source • High-performance Input/Output System

HIP Hardening Improvement Program • Hardness Improvement Program • Hardware Interface Panel • Hardware Interface Processor • Hardware Interface Program • Headingly (Australia-QLD) (location identifier) • Health Information Provider • Health Insurance Plan • Hierarchical Information Processor • High-Impact Paper • High-Impact Pressure • High Interface Processor • Highly Ionized Plasma • Hilpoltstein (Germany) • Hipparcos (catalog) • Host Interface Port • Host Interface Protocol • Hot Isostatically Pressed • Hot Isostatic Press • Hot Isostatic Pressed • Hot Isostatic Pressing • Hot Isostatic Processing

HIPA Health Information Portability Act (USA)

HIPAA Health Insurance Portability and Accountability Act (USA)

HIPAR High-Power Acquisition Radar

HIPAS Hunting Image Processing and Analyzing System

HIPASS HI Parkes All Sky Survey

HiPC High-Performance Computing

H-IPC Heterogeneous Interprocess Communications

HIPC Heavily Indebted Poor Country • Heterogeneous Interprocess Communications • High Chamber Pressure • Highly Indebted Poor Country

H-IPCS Heterogeneous Interprocess Communications Software

HIPCS Heterogeneous Interprocess Communications Software

HIPDA Hydroxyisophthalyl Dihydroxamic Acid

HIPERLAN High-Performance Local Area Network

HIPERNAS High-Performance Navigation System

HIPO Hierarchical Input-Process-Output • Hierarchy plus Input-Process-Output • High-speed Imaging Photometer for Occultations

HiPot High Potential

HIPOTT High Potential Test

Hipp Hippocrates

HIPPARCOS High-Precision Parallax Collecting Satellite

HIPPI High-Performance Parallel Interface

HIPPO Hawaii Imaging Polarimeter for Proto Objects

HIPPS Highly Integrated Pluto Payload System

HIPR Hypermedia Image Processing Reference

HIPS Harwell Image Processing System • High-Impact Polystyrene • Highly Interactive Problem Solver

HIPSS Hub Integrated Power Switching System

HIPWAC Heterodyne Instrument for Planetary Wind and Composition

HIQ High Intelligence Quotient

HIR Harmonized Interoperability Representation • Higher • Honiara (Solomon Islands) (location identifier)

HiRAAS High-Resolution Airglow/Aurora Spectroscopy

HIRAC High-Range Radiation Counter • High-Resolution Adaptive Camera

HIRAM High-Resolution Infrared Auroral Measurement

HiRAN High-Range Navigation

HiRAP High-Resolution Accelerometer Package

HiRDES High-Resolution Double Echelle Spectrograph

HIRDES High-Resolution Double Echelle Spectrograph

HiRDLS High-Resolution Dynamics Limb Sounder

HIRDLS High-Resolution Dynamics Limb Sounder

HiREFS High-Resolution Erect Field Spectrometer

HiREGS High-Resolution Gamma-ray Spectrometer

HiRel High Reliability

HiRELS High-Resolution Emission-Line Spectrometer

HiRENS High-Resolution Narrow Swath

HiRes High Resolution

HiRES Highly Integrated Raster-based Exploration System

HIRES High-Resolution Echelle Spectrograph

HIRF High-Intensity Radiated Field • High-Intensity Radiation Field • High-Intensity Radio Field • High-Intensity Radio Frequency

HIRG Hyperluminous Infrared Galaxy

HIRIS High-Resolution Imaging Spectrometer

HIRISE High-Resolution Imaging Science Experiment

HIRL High-Intensity Runway Light • High-Intensity Runway Lighting

HIRLAM High-Resolution Limited Area Model (*) • High-Resolution Local Area Model

HIRMS Hypothetical Information Resource Management System

HIRO HiScat International Radio Observatory

HIRRLS High-Resolution Research Limb Sounder

HIRS High-Impulse Retrorocket System • High-resolution Infrared Radiation Sounder (*) • High-resolution Infrared Sounder • Holographic Information Retrieval System

His Histoire • Historia

HIS Hauptinformationsstelle • Hayman Island (Australia-QLD) (location identifier) • Heavy-Ion Source • Hic Iacet Sepultus • Hic Iacet Sepulta • Hierarchical Intensive Search • High-Intensity Spectrometer • High-resolution Infrared Sounder • High-resolution Interferometer Sounder • Hochschul-Informations-System • Home Information System • Host Integration Server • House Information System • Hue - Intensity - Saturation • Hybrid Infrared Source • Hydrologic Information System • Hypermedia Indexing Scheme • Hypertext Information System

HISA HI Self-Absorption • Neutral Hydrogen Self-Absorption

HISAM Hardware-Initiated Standalone Memory • Hierarchical Indexed Sequential Access Method

HISAR Hughes Integrated Synthetic Aperture Radar

HISARS Hydrologic Information Storage and Retrieval System

HISDAM Hierarchical Indexed Sequential Direct Access Method

HISDM Hybrid Intelligent Systems for Data Mining (workshop)

HISI Healthcare Information System Infrastructure

HISL High-Intensity Strobe Light • High-Intensity Strobe Lighting

Hisp Hispanic • Hispánico

HISP Hybrid Image and Signal Processing

HISPARS High-Speed Pattern Analysis and Representation System

HISS Heavy Ion Superconducting Spectrometer • Heritage Information Statistical System • Holographic Ice Surveying System

HISSR Hydrogeological Information System with Spatial Reference

HiSSS High Speed Strike System

Hist Histoire • Histology • Historia • Historian • Historical • Historique • History

HIST High-Sensitivity Telescope

HISTEP High-speed Integrated Space Transportation Evaluation Program

HISTORIA Heraldic Images Storing Applications

HIT Hardware Integration Test • Heavy Ion Telescope • Hierarchical Incremental Testing • High-Intensity Training • High-Interest Telecommunications • High-Isolation

Transformer • High Time (resolution) • Holographic Interferometry Technique • Human-Interface Technology

HITC Hughes Information Tecnology Co.

HiTec High Technology (periodical code)

HITF Hypervelocity Impact Test Facility (JSC)

HITL Hardware In The Loop • Human Interface Technology Laboratory • Human In The Loop • Hydroacoustic Impact Timing System

HITLS Hardware In The Loop Simulation

HiTOP High Integration of Technology, Organization and People

HITS Hierarchical Integrated Test Simulator • High-speed Integrated Test System • Humans, Interaction, Technology and Strategy (conference) • Hyperlink Induced Topic Search

Hitt Hittite

HITT Holland Institute for Traffic Technology (Netherlands)

HIU Headset Interface Unit • Host Interface Unit • Human Intelligence Unit • Hydrologic Information Unit

HIUS Hispanic Institute of the United States

HIV Helium Isolation Valve

HIVE Highly-parallel Integrated Virtual Environment

HIVOS High-Vacuum Orbital Simulator

HIWAS Hazardous In-flight Weather Advisory Service

HIWY Highway

HIX Hiva Oa (French Polynesia) (location identifier) • Holiday Inn Express

HizEC High-Redshift Elliptical Candidate

HIZSS HI ZOA Shallow Survey • Neutral Hydrogen ZOA Shallow Survey

HiZSST High Z Supernova Search Team

hj halbjährig • halbjährlich • Hic Jacet

Hj Haji • Halbjahr • Hilfsjahr

H-J Hamilton-Jacobi (equation)

HJ Halt and Jump • Hamilton-Jacobi (equation) • Hic Jacet

HJB Halbjahresbeitrag • Hamilton - Jacobi - Bellman (equation)

HJBI Hamilton - Jacobi - Bellman - Isaacs (equation)

HJBT Heterojunction Bipolar Transistor

HjD Halbjahresdurchschnitt

HJD Heliocentric Julian Date • Heliocentric Julian Day

HJR House Joint Resolution • Khajuraho (India) (location identifier)

hjs Hic Jacet Sepultus • Hic Jacet Sepulta

HJS Hic Jacet Sepultus • Hic Jacet Sepulta

HJT Khujirt (Mongolia) (location identifier)

h/k Housekeeping

hk Hoek • Hong Kong (China-PRC) • Housekeeping

Hk Hauptkabel • Heck

H&K Homing & Kill

H/K Housekeeping

HK Colombia (international aircraft registration code) • Handelskammer • Hauptkasse • Hauptkommission • Hefner-Kerze • Heilkosten • Heizkosten • Hochspannungskabel • Hohe Kommission • Hoger Kommissar • Homing and Kill • Hong Kong (international vehicle registration code) (obsolete) • Hong Kong (ISO country code) (obsolete) • Hook • Housekeeping

HKA Blytheville (USA-AR) (location identifier)

HKAAS Kong Kong Amateur Astronomical Society

HKAS Kong Kong Astronomical Society

HKB Healy Lake (USA-AK) (location identifier)

HKCS Hong Kong Computer Society (China-PRC)

HKD Dollar (Hong Kong) (ISO currency code) (obsolete) • Hakodate (Japan) (location identifier) • Housekeeping Data

HKDB Housekeeping Database

HKE Housekeeping Equipment

Hkf Handkerchief

HKF Housekeeping File

HKFRD Hoher Kommissar der Französischen Republik in Deutschland

HKG Hong Kong (ISO country code) (obsolete) • Hong Kong - Kai Tak (China-PRC) (location identifier)

HKH Hare Koninklijke Hoogheid

HKIEd Hong Kong Institute of Education (China-PRC)

HKIGS Hong Kong Internet and Gateway Services (China-PRC)

HKIX Hong Kong Internet Exchange (China-PRC)

HKJ Jordan (international vehicle registration code)

HKK Hokitika (New Zealand) (location identifier)

HKl Handelsklasse

HKMS Hong Kong Mathematical Society (China-PRC)

HKN Hoskins (Papua New Guinea) (location identifier)

HKO Hong Kong Observatory (China-PRC) (formerly RHKO)

HKP Housekeeping • Hubble Key Program

HKPG Hubble Key Program Group

HKPU Hong Kong Polytechnic University (China-PRC)

HKS Jackson Hawkins (USA-MS) (location identifier)

HKSAR Hong Kong Special Administration Region (China-PRC)

Hksw Hookswitch

HKT Hollow Kathode Tube • Hot Kathode Tube • Phuket (Thailand) (location identifier)

HKTA Hong Kong Tourism Association (China-PRC)

HKTDC Hong Kong Trade Development Council (China-PRC)

HKU Hogeschool voor de Kunsten Utrecht (Netherlands) • Hong Kong University (China-PRC)

HKUST Hong Kong University of Science and Technology (China-PRC)

HKV Heizkostenvorschuß • Hit-to-Kill Vehicle • Kaskovo (Bulgaria) (location identifier)

HKW Heizkraftwerk

HKY Hickory (USA-NC) (location identifier)

Hkz Hauptanschlußkennzeichen

HKZ Heizkostenzuschluß

hl Haul • Hectoliter • Heilig • Hell • Hoc Loco

hL herrschende Lehre

Hl Heilige • Heiliger

H/L Hard Line

HL Half Life • Hard Lander • Hard Line • Harmonic Line • Haute-Loire (France) • Headline • Heel Line • Higher Loop • High Level • Hill • Hinge Line • Hochleistung • Hora Local • Horizontal Length • Hotline • House of Lords • Hydrogen Line • Korea (ROK) (international aircraft registration code) • ROK (international aircraft registration code)

HLA High-Level Analog • High-Level Architecture • High-Level Assembler • Home Location Area • Human Leucocyte Antigen • Lanseria (South Africa) (location identifier)

HLAL High-Level Assembler Language

H-LAM/T Human using Language, Artifacts and Methodology in which he is Trained

HLAMT Human using Language, Artifacts and Methodology in which he is Trained

HLARTI High-Level Architecture Runtime Infrastructure

HLASM High-Level Assembler

hlb Help Library (file extension)

HLB Batesville Hillenbrand (USA-IN) (location identifier) • Harvard Library Bulletin • Hydrophilic-Lipophilic Balance

HLC Heavy-Lift Capability • High-Latitude Cloud • High-Level Committee • High-Level Compatibility • Hill City (USA-KS) (location identifier) • Hybrid Load Controller

HLCF Hardened Launch Control Facility

Hlcl Helical

HLCL High-Level Command Language • High-Level Concurrent Language

HLCM High-Level Control Module

HLCO Hard Labor Creek Observatory (USA-GA) • High - Low - Close - Open

Hlcptr Helicopter

HLD Hailar (China-PRC) (location identifier) • Head-Level Display • High-Level Discrete • Hold

HLDC High-Level Data Control

Hldg Holding

Hldn Holddown

Hldr Holder

HLDS Hydrogen Leak Detection System

HLE Heure Limite d'Enregistrement • High-Level Exposure • Homogeneous Linear Estimator • Hydrogen Line Emission

HLF Half • High-Latitude Field • High-Level Formatting • Horizontal Laminar Flow • Hultsfred (Sweden) (location identifier)

HLFC Hybrid Laminar Flow Control

HLFD High-Level Functional Description

hlg heilig • hochliegend

Hlg Halogen • Heilige

HLG Helpline Group • High-Level Group • Wheeling - Ohio County (USA-WV) (location identifier)

HLH Heavy-Lift Helicopter • High-L Helioseismometer • Ulanhot (China-PRC) (location identifier)

HLHS Hidden-Line Hidden-Surface

HLHSR Hidden-Line Hidden-Surface Removal

HLI Higher-Layer Information • High-Level Interface • High-speed LAN Interface • Hollister (USA-CA) (location identifier) • Host Language Interface

HLIG Hyperluminous Infrared Galaxy

HLIPS High-Level Image Processing System

HLIV High-Level Input Voltage

Hlk Heilkunde

HLK Handlochkarte

Hll Heilige

HLL Halbleiterlabor • High-Level Language • Hill • Hillside (Australia-WA) (location identifier)

HLLAPI High-Level Language Application Programming Interface

HLLS Hills

HLLV Heavy-Lift Launch Vehicle

HLLW High-Level Language Worm • Hollow

Hlm Hefner-Lumen

HLM Habitation à Loyer Modéré • Heterogeneous LAN Management • High-Level Meeting • Holland Park Township (USA-MI) (location identifier)

HLMT Helmet

HLN Helena (USA-MT) (location identifier)

HLNW High-Level Nuclear Waste

HLO Hoher List Observatory (Germany)

HLOS Horizontal Line Of Sight

HLOV High-Level Output Voltage

hlp Help (file extension)

HLP Hardy - Littlewood - Pólya (inequality) • Help • Hertzsprung-Luyten Diagram • Jakarta - Halim Perdana Kusuma (Indonesia) (location identifier)

Hlpr Helper

HLPS Heavy-Lift Preposition Ship • Hot Liquid Process Simulator • Human Language Processing Server

HLQ High-Level Qualifier • High-Luminosity Quasar • Highly Luminous Quasar (*)

HLR High-Level Radiation • High-Level Radioactivity • High-Lift Rotor • Home Location Register • Home Location Regulator • Human-Language Resource • Killeen - Fort Hood (USA-TX) (location identifier)

HLRKA Höchstleistungsrechenzentrum Karlsruhe (Germany)

HLRM High-Level Radio Modulator

HLRS Höchstleistungsrechenzentrum Stuttgart (Germany)

HLRW High-Level Radioactive Waste

HLRZ Höchstleistungsrechenzentrum

HLS Halt Line State • Harvard Law School (USA-MA) • Hefei Light Source (China-PRC) • Hills • Hogere Landbouwschool • Hue - Luminance - Saturation • Saint Helens (Australia-TAS) (location identifier)

HLSC Helicopter Logistic Support Center

HLSP Heitler - London - Slater - Pauling • High-Level Science Product

HLSTO Hailstones

HLSUA Honeywell Large Systems Users Association

HLSW High-Level Software

HLT Halt • Hamilton (Australia-VIC) (location identifier) • Helium II Telescope • Human-Languages Technologies (conference) • Human-Language Technology

HLTA Halt Acknowledge

Hlth Health

HLTL High-Level Test Language • High-Level Transistor Logic

HLTP Hardline Test Point • Hilltop

HLTTL High-Level Transistor-Transistor Logic (*) • High-Level Transistor-Translator Logic

HLU High-Level Unit • Houailou (New Caledonia) (location identifier)

HLUE Homogeneous Linear Unbiased Estimator

HLV Heavy-Lift Vehicle • Helenvale (Australia-QLD) (location identifier)

HLW High-Level (radioactive) Waste • Hluhluwe (South Africa) (location identifier)

HLWG High-Level Working Group

Hlx Hefner-Lux

HLX Hyperluminous X-ray (source)

HLY Holyhead (UK-England) (location identifier)

HLZ Hamilton (New Zealand) (location identifier) • Helicopter Landing Zone

hm Hallmark • Hectometer • Hectometric • Hoc Mense • Huius Mensis • Hydroxymethyl

hM Harmonisches Mittel • herrschende Meinung

H/m Henry per Meter

Hm Halbmesser • Hausmeister • Heckium • Heimat • Tiempo Solar Medio

HM Hallmark • Handelsmarine • Handelsministerium • Handicapé Moteur • Hardness Maintenance • Harmonic Mean • Haute-Marne (France) • Hazardous Material • Headmaster • Head Motion • Health Management • Heard and McDonald Islands (ISO country code) • Heavy Maintenance • Heavy Metal • Hertzsprung Method • Hierarchical Merging • High Metal • Hinge Mount • Hochleistungsmotor • Homogenization Medium • Hungarian Method • Hydrogen-line Magnetometer • Hydrogen Maser • Hydrogen Monitor • Îles Heard et McDonald • McDonald Islands (Heard and_) (ISO country code)

HMA High-Memory Area • Hub Management Architecture • Hybrid Mirror Antenna • Hypergolic Maintenance Area • Hypergol Maintenance Area • Malmö City (Sweden) (location identifier)

HMAC Hashed Message Authentication Code

HMAF Half Metallic Antiferromagnet

Hmb Hamburg (Germany)

HMB Happy Monkey Blunder • Hard Mobile Basing • Hexamethylbenzene

HMBA Hydroxymethylbenzanthracene

HMBP High-Mass Binary Pulsar

Hmc Harmonic

HMC Halley Multicolor Camera (Giotto) • Hardware Management Console • Harvey Mudd College (USA-CA) • Heterogeneous Multicast • Horizontal Meridian Circle • Horizontal Microcode • Hot Molecular Core • Hybrid Microcircuit • Hydrogen Maser Clock • Hypergol Maintenance and Checkout

HMCA Hardware Management Console Application

HMCC Harbor Master Communications Center • Houston Mission Control Center (NASA) • Hypergolic Maintenance and Checkout Cell • Hypergol Maintenance and Checkout Cell

HMCF Hypergol Maintenance and Checkout Facility

HMCK Hummock

HMCRI Hazardous Materials Control Research Institute

HMCS Helmet-Mounted Cuing System

HMCT Highway Movement Control Team

HMCTAT Hazardous Material Cost Tradeoff Analysis Tool

Hmd Humid • Humidity

HMD Handbuch der Maschinelle Datenverarbeitung • Hardware Maintenance Depot • Head-Mounted Device • Head-Mounted Display • Heard and McDonald Islands (ISO country code) • Helmet-Mounted Display • Heterodyne Matrix Detector • Hydraulic Mean Depth • Hydrostatic Motor Driven • Îles Heard et McDonald • McDonald Islands (Heard and_) (ISO country code)

HMDA Hitchhiker Motorized Door Assembly

HMDE Hanging Mercury Drop Electrode

HMDS Hawaii Medium Deep Survey

Hme Homme

HME Handheld Microgravity Experiment • Hassi Messaoud (Algeria) (location identifier) • High Magnification Event • Home-Made Explosive • Hough Mode Extension

HMF Health Maintenance Facility • Heavy-Metal Fluoride • Heliospheric Magnetic Field • High Mach Flow • High Magnetic Field • Horizontal Mating Facility • Hypergolic Maintenance Facility • Hypergol Maintenance Facility

HMG Hermannsburg (Australia-NT) (location identifier) • Hydraulic Motor Generator

HMGF High Modulus Glass Fiber

HMI Hahn-Meitner Institute (Germany) • Hami (China-PRC) (location identifier) • Handbook of Maintenance Instructions • Hardware Monitor Interface • Helioseismic and Magnetogram Imager • Host Message Interface • Hub Management Interface • Human-Machine Interface

HMIHY How May I Help You?

HMINET Hahn-Meitner-Institut Netz (Germany)

HMIS Hazardous Materials Information Sheet • Hazardous Materials Information System

HMJ Khmelnitskii (Ukraine) (location identifier)

HML Hardened Mobile Launcher • Hard Mobile Launcher • Heure Moyenne Locale • Hierarchical Model Learning • Hot Mockup Laboratory • Human-Machine Language

HMLE Hypermedia-based Learning Environment

HMM Half-Molecule Model • Hidden Markov Model • Hidden Markov Modeling • Hierarchical Merging Model

HMMP Hazardous Material Management Plan • Hypermedia Management Protocol

HMMR High-resolution Multifrequency Microwave Radiometer

HMMS Hypermedia Management Schema • Hypermedia Management Scheme

HMMU Hardware Memory Management Unit

HMMWV High-Mobility Multipurpose Wheeled Vehicle • High-Mobility Multiwheeled Vehicle • Humvee

HMN Alamogordo Holloman (USA-NM) (location identifier) • Halo de Matière Noire

HMNAO Her Majesty's Nautical Almanac Office (UK)

HMNS Houston Museum of Natural Sciences (USA-TX)

HMO Health Maintenance Organization • Hermosillo (Mexico) (location identifier) • High-Minded Outsider • Honolulu Magnetic Observatory (USA-HI) • Hueckel Molecular Orbital

HMOM Hypermedia Object Manager

HMOS High-density Metal Oxide Semiconductor • High-performance Metal Oxide Semiconductor • High-speed Metal Oxide Semiconductor

HMP Hand-Made Paper • Haughton-Mars Project (NASA) • Hazardous Material Pharmacy • Hexasodium Metaphosphate • High Melting Point • Hoc Monumentum Posuit • Host Management Protocol • Host Monitoring Protocol • Host Monitor Protocol • Humidity Monitoring Panel • Hydraulic Maintenance Panel • Hydrazine Monopropellant • Hydrozene Monopropellant • Latvian Meteorological Agency

HMPA Hexamethylphosphoramide

HMPAC Helmet-mounted Multipurpose Applications Control

HMPC Hazardous Maintenance Procedure Code

HMPO High-Mass Protostellar Object

HMPT Hexamethylphosphorous Triamide • Human factors - Manpower - Personnel - Training

HMQAM Hamming Quadrature Amplitude Modulation

Hmr Hammer

HMR Half-Mass Radius • Hamar (Norway) (location identifier) • Hermes Resource Module

HMRPP Hazardous Material Reduction Prioritization Process

HMRS High-Mobility Rocket System

HMRT Harmonic Mean Response Time

HMS Hardened Memory System • Harmonic Multiplier Source • Hazard Monitor System • Hazards Monitoring System • Health Management System • Health Monitoring System • Heavy Materiel Supply • Heavy Media Separation • Helmet-Mounted Sight • Hermes • Hessen - Mittelrhein - Saar (Germany) • Hierarchical Memory Storage • High-Mass Star • High-Momentum Spectrometer • History Memory System • Holonic Manufacturing System • Home Media Server • Homeshore (USA-AK) (location identifier) • Hours - Minutes - Seconds • Huge Magnetic Field • Human Safety Modification • Hungarian Meteorological Service • Hypermedia Management System

HMSO Her Majesty Stationary Office • His Majesty Stationary Office • Honolulu Magnetic and Seismological Observatory (USA-HI)

Hmstd Homestead

HMSU High-resolution Microwave Sounding Unit

HMT Handheld Multimedia Terminal • Hazardous Materials Technology • Hemet Ryan (USA-CA) (location identifier) • Hexamethylenetetramine • High-Mobility Trailer • High-Mobility Traveller • High-performance Management Team

HMTA Hazardous Materials Transportation Act (USA)

HMTF Hydrodynamic Modulation Transfer Function

HMTUSA Hazardous Materials Transportation Uniform Safety Act (USA)

HMU Hardware Mockup • Health Monitoring Unit • Helmet-Mounted Unit • Hydraulic Mockup • Hydromechanical Unit

HMUX Hybrid Multiplexer

HMV Heavy Maintenance Visit • High Magnification Viewer • His Master's Voice • Hydrodynamically Modulated Voltammetry • Hydrogen Manual Valve

HMVS Helmet-Mounted Vision System

HMW Health - Morale - Welfare • High Molecular Weight

HMWS Hawaii Medium Wide Survey

HMX Cyclotetramethylene Tetranitramine

HMXB High-Mass X-ray Binary

HMXRB High-Mass X-Ray Binary

hn Hac Nocte • Honduras

Hn Horn

HN Blister Agent • Home Net • Honduras (ISO country code) • Horn • Host Nation • Host to Network • Hotel Night • Hypernova

Hna Hermana

HNA High Nickel Alloy • Host Nation Agreement • Host Nation Approval • Morioka (Japan) (location identifier)

Hnas Hermanas

HNB Huntingburg (USA-IN) (location identifier)

HNBGU Hemvati Nandan Bhauguna Garhwal University (India)

HNC Hatteras (USA-NC) (location identifier) • Heavy Nucleus Collector • Higher Education Certificate • Hybrid Newton-Raphson • Hydrogen Isocyanide • Isocyanure d'Hydrogène

HNCO High-precision Numerical-Controlled Oscillator • Isocyanic Acid

HNCS Isothiocyanic Acid

HND Higher National Diploma • Honduras (ISO country code) • Hundred • Tokyo Haneda (Japan) (location identifier)

Hndbk Handbook

HNDLER Handler

HNDLG Handling

Hndrl Handrail

Hndst Handset

Hndwl Handwheel

HNE Heavy Nuclei Experiment • Heure Normale de l'Est • Tahneta Pass Lodge (USA-AK) (location identifier)

HNEI Hawaii Natural Energy Institute (USA-HI)

HNEO Hazardous Near-Earth Object

HNF Hydrazinium Nitroformate

Hng Hinge

HNG Hanging • Hienghene (New Caledonia) (location identifier) • Hinge

HNGL Helium Neon Gas Laser

HNGR Hangar • Hanger

HNH Hoonah (USA-AK) (location identifier)

HNI Heiweni (Papua New Guinea) (location identifier)

HNIL High-Noise-Immunity Logic

HNK Hinchinbrooke Island (Australia-QLD) (location identifier)

HNL Helium Neon Laser • Honolulu (USA-HI) (location identifier) • Lempira (Honduras) (ISO currency code)

HNM Hana (USA-HI) (location identifier)

HNML Hindu Meal

HNMS Hellenic National Meteorological Service (Greece)

HNN Honinabi (Papua New Guinea) (location identifier)

HNNC Hecht-Nielsen Neurocomputer Corporation

Hno Hermano

HNO Hercegnovi (Yugoslavia-Montenegro) (location identifier) • Nitroxyl Radical

Hnos Hermanos

HNO3 Nitric Acid

HNPA Home Numbering Plan Area

HNR Henry Norris Russell (1877-1957) • Hypernova Remnant

HNRC HOMS National Reference Centre

Hnrs Honors

HNS Haines (USA-AK) (location identifier) • Hexanitrostilbene • High Nitrogen Steel • Host Nation Support • Hughes Network Systems • Hybrid Neural System

HNSC House National Security Committee (USA)

Hntg Hunting

HNX Hanna (USA-WY) (location identifier) • Heavy Nuclei Explorer

HNY Happy New Year • Hengyang (China-PRC) (location identifier)

Ho Holmium • Hotel • House

H/O Handover

HO Habitual Offender • Hale Observatories (USA-CA) • Handelsorganisation • Handover • Harmonic Oscillator • Haystack Observatory (USA-MA) • Head Office • High Output • History Office • Hochofen • Hoger Onderwijs • Hold Over • Home Office • Honduras • Hopi • Hours of Operation • House Officer • Hvar Observatory (Croatia) • Hydrogen-Oxygen • Hydrographic Office • Hydroxyl Radical

HOA Hands-Off Automatic • Hands-On Astrophysics (project) • Head Of Agency • Heavy Observation Aircraft • Higher-Order Approximation • High-Order Algebra • Hola (Kenya) (location identifier)

HOAO High-Order Adaptive Optics

HoB Header of Bus

HOB Harvest Object Browser • Height Of Burst • High-Order Bit • High-Order Byte • Hobbs - Lea County (USA-NM) (location identifier)

HOBIS Hotel Billing Information System

HOBL High-Order Balmer Line

HOBU Hoger Onderwijs Buiten de Universiteit

HoC House of Commons

HOC Halogenated Organic Compound • Hands-On Component • Head Of Component • Heavy Organic Chemical • Held On Charge • Higher-Order Crossing • High Output Current • Hourly Operating Cost • Humanitarian Operations Center • Hydrophobic Organic Compound • Komako (Papua New Guinea) (location identifier)

hochd hochdeutsch

Hochf Hochformat

Hochsch Hochschule

Hochschwarzw Hochschwarzwald (Germany)

höchst höchstens

hochw hochwertig • hochwürdig

Hochw Hochwasser

HOCl Hypochlorous Acid

HOCUS Hand Or Computer Universal Simulation

HoD Head of Delegation • Head of Department

HÖD Handbuch des öffentlichen Dienstes

HOD Halo Occupation Distribution • Head Of Department • Hodeidah (Yemen) (location identifier)

HODAC Head Of Departments And Centres

HOE Height Of Erection • Holographic Optical Element • Homing Optical Experiment • Homing Overlay Experiment • Houeisay (Laos) (location identifier) • Hydraulically Operated Equipment

HOEAS Heart of England Astronomical Society (UK)

HOED Hypermedia Online Educational Database

HOEM Heliosynchronous Out-of-Ecliptic Mission

HOF Hall Of Fame • High-Order Field • Hofuf (Saudi Arabia) (location identifier) • Home Office Facility

H. of C. House of Commons

HOFCO Horizontal Functional Checkout

HOFD Heterogeneous Opposed Flow Diffusion

H. of L. House of Lords

höfl höflich

Höfl Höflichkeit

H. of R. House of Representatives

HOFS Hydrogen-Oxygen Fuel System

HOFT High-Order Formal Tool

HOG Hermann-Oberth Gesellschaft (Germany) • High-Order Godunov • High-Order Gravity • Hofgeismar (Germany) • Holguin (Cuba) (location identifier)

HOGE Hover Out of Ground Effect

Hoh Hoheit

HOH Hash Of Hashes • Hofheim (Germany) • Hohenems (Austria) (location identifier) • Water

Hohenz Hohenzollern

HOHI High Order Harmonic Interferometer

HOI Hao Island (Tuamotu Islands) (location identifier) • Headquarters Operating Instruction • Homogeneous Object Identifier

HOK Hooker Creek (Australia-NT) (location identifier)

Hol Holandés • Holiday • Holland • Hollandais • Hollande • Holocene

HOL Head Of Line • Higher Order Language • High-Order Language • Holikachu (USA-AK) (location identifier)

HOLAB Holographic Laboratory Brassboard

HOLC High-Order Language Computer

HOLIS Harvard Online Information Server

Holl Holland • Hollandais • Hollande

HOLM Higher Order Language Machine

HOLOP Holographic Optics • Holographische Optiklabor

Holst Holstein (Germany)

HOLW Hollow

HOLWG High-Order Language Working Group

HOM Hectometric (radiation) • Homer (USA-AK) (location identifier) • Homogeneity • Hotine Oblique Mercator

HOME Highly Optimized Microscope Experiment • Hydrology, Oceanography, Meteorology for the Environment (project)

HOMES Huron - Ontario - Michigan - Erie - Superior (lakes)

HOMO Highest Occupied Molecular Orbit • Highest Occupied Molecular Orbital

HOMS Hydrological Operational Multipurpose Subprogramme • Hydrological Operational Multipurpose System (*)

Hon Honor • Honorable • Honorably • Honorary

HON Handover Number • Honorary • Huron (USA-SD) (location identifier)

Hond Honduras

HONN Higher-Order Neural Network

Hono Honolulu (USA-HI)

Hons Honors

Hon.Sec. Honorary Secretary

HOO Quanduc - Nhon Co (Viet Nam) (location identifier)

HOOD Hierarchical Object-Oriented Design

HOOP Handbook Of Operating Procedures • Hoger Onderwijs- en Onderzoekplan • Hydrogen Out-Of-Plane

HOOPE Hierarchical Object-Oriented Parallel Environment

HOOPS Hierarchical Object-Oriented Picture System

HOP Hands-Off Point • Harvest Object Protocol • Hierarchical Objective Procedure • Hierarchical Observing Protocol • High-Order Partition • High Output Paraffin • High-Oxygen Pressure • Hopkinsville Campbell (USA-KY) (location identifier) • Hourly Operating Plan • Hybrid Operating Program

HOPE Halley Optical Probe Experiment • H-II Orbiting Plane Experiment • Hydrogen-Oxygen Primary Extraterrestrial • H-Z Orbital Plane

HOPFI High-Order Fabry-Pérot Interferometer

HOPG Highly-Oriented Pyrolytical Graphite

HOPI High-speed Occultation Photometer and Imager

HOPL History Of Programming Languages

HOPLA Harmonized Optical Programmable Logic Array

HOPOS History Of Philosophy Of Science (working group)

HOPR Hubble Observation Problem Report

HOPS Haystack Observatory Postprocessing Software

HOPU Head Of Planning Unit

HOQ Hof (Germany) (location identifier)

Hor Horizon • Horizontal • Horologium-Horologii (constellation) • Horology

HOR Heliocentric Orbit Rendezvous • Home Of Record • Horb (Germany) • Horizontal • Horta (Azores) (location identifier)

hor. decub. Hora Decubitus

HoReCa Hotels - Restaurants - Cafés

Horiz Horizon • Horizontal

HORIZ Horizon • Horizontal

HORM Human and Organizational Risk Management • Hybrid Orbital Rehybridization Method

Horol Horological • Horology

HORSE Horizon Science Experiment

HORU Head Of Resource Unit

Horz Horizontal

HÖS Höchstadt (Germany)

HOS Chos Malal (Argentina-NE) (location identifier) • Hardware Operating System • Harvest Object System • Heckscher - Ohlin - Samuelson (model) • High-level Observing Software • High-level Operating Software • High-level Operation Software • High-Order Software • Hybrid Operating System

HOSC Huntsville Operations Support Center (USA-AL)

HOSM Host Operations Systems Monitor

Hosp Hospices • Hospital

HOSS Halo Orbit Space Station • Hamilton Outer Space Services

Host Hostería

HOST Hardened Optical Sensor Testbed • Helicopter Overall Simulation Tool • Hubble Orbital System Test

Hot Hotel

HOT Hands-On Technology • Hands-On Training • Health-Oriented Telecommunications • Highly Optimized Tolerance • Holographic On-Tube • Holographic Optical Trapping • Home Order Television • Hot Springs (USA-AR) (location identifier)

HOTAS Hands On Throttle And Stick

HOTAY Hands-On-Throttles-And-Yokes

HOTLIPS Home and Office Techniques for a Local Image Processing System

HOTOL Horizontal Takeoff and Landing

HOTS Holographic Optical Telescope and Scanner

HOTT Hot Of The Tree

Hotzenw Hotzenwald (Germany)

HOU Hands On Universe (program) • Houston Hobby (USA-TX) (location identifier)

HOV High-Occupancy Vehicle • Homogeneity Of Variance • Hover • Ørsta-Volda (Norway) (location identifier)

HOW Fort Kobbe - Howard (Panama) (location identifier) • Handover Word • Handover Work

HOx Odd Hydrogen

HOX Homalin (Myanmar) (location identifier)

HOY Hoy Island (UK) (location identifier)

Hp Haltepunkt • Hilfspunkt

H/P Heliport

HP Halbpension • Half Pay • Half Plate • Haltepunkt • Handicapped Person • Handling and Propulsion • Handling Procedure • Hard Point • Harmonic Progression • Harvard Pulsar • Hauptpreis • Hauptpunkte • Haute Pression • Haute Provence • Hautes Pyrénées (France) • Haut-Parleur • Haverah Park (UK-England) • Head-on Penetration • Health Physicist • Health Physics • Heated Pitot • Heating Plant • Heat Pipe • Height of Perigee • Heliopause • Herbstprüfung • Heure Poste • Hewlett-Packard • Higher Power • Highly Polarized • High Pass • High Performance • High Power • High Pressure • High Priority • Himachal Pradesh (India) • Hiper Pool • Hire Purchase • Hit Point • Hochparterre • Holding Pattern • Holiday Pass • Homepage • Horizontally Polarized • Horizontal Parallax • Horizontal Parallelism • Horizontal Polarization • Horsepower • Hot Penetration • Hot Pressed • Hot Prop • Human Performance • Hydrazinium Perchlorate • Hydrogen Purge • Hydrophone • Hydrostatic Pressure • Panama (international aircraft registration code)

hPa Hecto-Pascal

HPA Ha'Apai (Tonga) (location identifier) • Handling and Positioning Aid • Head of Processing Activity • Head of Procuring Activity • Heavy Penetration Assistance • Heuristically-Programmed Algorithm • Hidrocarburo Policíclico Aromático • High-Performance Absorber • High-Power Amplifier • High-Power Antenna • High-Powered Amplifier • Holding and Positioning Aid • Hot Plasma Analyzer • Hydrocarbure Polycyclique Aromatique

HPAC Hazard Prediction Analysis Code • Hazard Prediction Assessment Capability • High-Pressure Air Compressor

HPAD Host Packet Assembler/Disassembler

HPAG High-Power Amplifier Group

HPAL House-Programmed Array Logic

HPAU High-Power Amplifier Unit

HPB Hooper Bay (USA-AK) (location identifier)

Hpbf Hauptpersonenbahnhof

HPBRM High Payload Baseline Reference Mission

HPBW Half-Power Bandwidth • Half-Power Beam Width

HPC Handheld Personal Computer • High-density Projection Chamber • High-Performance Computer • High-Performance Computing • High-Performance Controller • High-Pressure Compressor • High Probability of Completion • Hollerith Punched Card • Hydrometeorological Prediction Center

HPCA High-Performance Computer Architecture

HPCAT High-Pressure Collaborative Access Team

HPCB High-Performance Computing Branch

HPCC High-Performance Computer and Communications • High-Performance Computing and Communications (*) • High-Performance Computing Center • High-Performance Control Center

HPCCC High-Performance Computing and Communications Center

HPCCD High-Performance Charge-Coupled Device

HPCCIT High-Performance Computing, Communications and Information Technology

HPCCP High-Performance Computing and Communications Program

HPCC&S High-Performance Computing, Communications & Simulation

HPCCS High-Performance Computing, Communications and Simulation

HPCCV High-Performance Computing, Communications and Visualization

HPCE High-Performance Capillary Electrophoresis

HPCF High-Performance Computing Facility

HPCG Handheld Protein Crystal Growth

HPCI High-Performance Computing Initiative • High-Performance Controller Interface

HPCJ High-Performance Compiler for Java

HPCMP High-Performance Computing Modernization Program

HPCN High-Performance Computing and Networking

HPCR High-Pressure Compressor Rotor

HPCRC High-Performance Computing Research Center

HPCS High-Performance Computing System

HPCU High-Performance Computing Unit

HPCWG High-Performance Computing Working Group

HPC2 High-Performance Computing Center

HPD Half-Power Diameter • Highest Posterior Density • High-Performance Demodulator • High-Power Density • High-Power Destruction • High Progressivity/Density • Horizontally Polarized Dipole • Hourly Precipitation Data • Hybrid Phase Discriminator • Hybrid Photocathode Detector • Hybrid Photodetector • Hybrid Photon Detector • Hydrometeorological Processes Division

HPDC High-Performance Distributed Computing

HPDD High-Performance Device Driver • High-Performance Disk Driver

HPDF High-Performance Demonstration Facility

HPDJ Hewlett-Packard DeskJet (printer)

HPDM High-Performance Data Mover

HPDT High-Pressure Drift Tube

HPDU Hydraulic Power Drive Unit

HPE Heat Pipe Experiment • High-Power Effect • Hope Vale (Australia-QLD) (location identifier)

HPEP High Pipe Experiment Package

HPER Health - Physical Education - Recreation

HPES Hierarchical Polyelastic Sampling

hpf HP Font (file extension)

HPF Hazardous Processing Facility • High-level Processing Facility • High-Pass Filter • High-Performance Fortran • High-Power Field • Highest Portable Frequency • Highest Possible Frequency • Highest Probable Frequency • Hochpass-Filter • Horizontal Processing Facility

Hpfl Halbpflicht

HpflG Halbpflichtgesetz

HPFS High-Performance File System • High-Pressure Fire Service

HPFT High-Pressure Fuel Turbopump

HPFTP High-Pressure Fuel Turbopump

hpg HPGL Graphics (file extension)

HPG Hard (copy) Processor Group • Hewlett-Packard Graphics • High-Performance Generator • High-Performance Group • High-Power Generator • High-Power Group • High-Pressure Gas • Homopolar Generator

HPGe High-Purity Germanium

HPGL Hewlett-Packard Graphics Language

HPGN High-Precision Geodetic Network

HPGS High-Pressure Gas System

HPGSPC High-Pressure Gas Scintillation Proportional Counter

HPH Haiphong Catbi (Viet Nam) (location identifier) • Horsepower-Hour

HPHD High-Pressure High-Density

HPHS Hartford Public High School (USA-CT)

HPI Hazard Potential Index • High-Performance Insulation • High-Powered Illuminator • Host Processor Interface • Horizontal Production Integration

HPIB Hewlett-Packard Interface Bus

HPICS Heat Pipe Instrumentation and Control System

HPIL Hewlett-Packard Interface Loop

HPJ High-Performance Java

HPKB High-Performance Knowledge Base

HPL Hewlett-Packard Language • High Polar Latitude • High-Power Laser • High-Priority List • High-Productivity Language • Hybrid Programming Language • Hypergolic Plume Impingement

HPLC High-Performance Liquid Chromatograph (*) • High-Performance Liquid Chromatography (*) • High-Precision Liquid Chromatograph • High-Precision Liquid Chromatography • High-Pressure Liquid Chromatograph • High-Pressure Liquid Chromatography

HPLJ Hewlett-Packard LaserJet (printer)

HPLMN Home Public Land Mobile Network

HPLV High Priority Low Volume

HPM Harvard Project Manager • Haut-Plateau Marécageux • Hierarchical Particle Mesh • High-Pass Mode • High-Performance Motor • High-Powered Microwave • High-Power Microwave • High Proper Motion • Holdup Processing Mode • Holdup Process Mode • Hydro Particle Mesh • Hyperpage Mode

HPMA High-Power Microwave Assembly

HPME High-Pressure Melt Ejection

HPMSK High-Priority Mission Support Kit

HPMT HOSS Process Model Toolkit

HPN High-Pass Network • Westchester County (USA-NY) (location identifier) • White Plains (USA-NY) (location identifier)

HPNS High-Pressure Nervous Syndrome

HPN2 High-Purity Nitrogen

HPO Head Post Office • High-Performance Option • High-Pressure Oxygen • Highway Post Office • Horizontal Parallax Only • Hourly Post Flight • Huntsville Program Office (USRA)

HPOC Huygens Payload Operations Centre • Huygens Probe Operations Center (*)

HPOL Halfwave Polarimeter • High-Power Optics Laboratory

HPOP High-Precision Orbit Propagator • High-Pressure Oxidizer Pump

HPot Helical Potentiometer

HPOT Heliopotentiometer • High-Pressure Oxidizer Turbopump

HPOTP High Pressure Oxidizer Turbopump

HPP Half Page Printer • Heat Pipe Performance • High-Precision Photogrammetry • High-Pressure Polyethylene • High-Profile Promotion • Holding under Promise of Payment • Homogeneous Poisson Process

HPPA Hewlett-Packard Precision Architecture • High-Power Power Accelerator

HPPF Horizontal Payloads Processing Facility

HPPI High-Performance Parallel Interface • High-Performance Peripheral Interface • High-Precision Parallel Interface

HPPN Hydrogen-Poor Planetary Nebula

HPPP High-Performance Parallel Processor

HPQ Highly Polarized Quasar (*) • High-Polarization Quasar

HPR Half-Power Radius • Hardware Problem Record • Hardware Problem Report • High-Performance Rate • High-Performance Router • High-Performance Routing • High-Power Receiver • High-Pressure Regulator • Horizontal Pressure Regulator • Human Performance Reliability • Hydrogen Pressure Regulator • Pretoria Central (South Africa) (location identifier)

HPRF High Power Radio Frequency • High Pulse Recurrence Frequency • High Pulse Repetition Frequency

HPRV Hydrogen Pressure Relief Valve

HPS Health Physics Society (USA-MD) • Hexapod Pointing System • High Performance System • High Performance Systems Inc. (USA-NH) • High-Precision Subset • High-Pressure Sodium • High-Pressure Steam • High-Protein Supplement • Hybrid Propulsion System • Hydraulic Power System • Hygiène - Prévention - Secourisme

HPSC High-Performance Scientific Computing • High-Performance Supercomputer

HPSEQT Hiper Pool Sequential Threshold

HPSIP Hazard Prediction System Integration Program

HPSIR Hewlett-Packard Serial Infrared

HPSK Hybrid Phase Shift Keying

HPSLT High Power Semiconductor Laser Technology

HPSN High-Performance Scalable Networking

HPSP High-Precision Scanning Platform • High-Precision Scan Platform

HPSS Helmet Position Sensing System • High-Performance Storage System

HPT Hampton (USA-IA) (location identifier) • Hazard Prediction Tool • Hexapod-Teleskop • High-Payoff Target • High-Potential Target • High-Potential Test • High-Power Transmission • High-Precision Thermostat • High-Pressure Test • High-Pressure Tin • High-Pressure Turbine • High-Priority Target • High-Productivity Tool • Horizontal Plot Table

Hptb Hauptband • Hauptberuf • Hauptbuch

Hptbd Hauptband

HPTE High-Precision Tracking Experiment

HPTET High-Performance Turbine Engine Technology

HPTF High-Pressure Test Facility

Hptl Hauptlehrer

HPTLC High-Performance Thin-Layer Chromatography

Hptm Hauptmann

Hpts Hauptsache

HPTS High-Performance Third Stage • High-Performance Transaction System

Hptsch Hauptschule

Hptst Hauptstadt • Hauptstation

hptw hauptwörtlich

Hptw Hauptwache • Hauptwerk • Hauptwort

HPU High-Power Unit • High-Pressure Unit • Hydraulic-Power Unit • Hydraulic-Propulsion Unit • Hydraulic-Pumping Unit

HPUX Hewlett-Packard Unix

HPV Helium Pressure Vessel • High-Performance Video • High-Pressure Valve • Human-Powered Vehicle • Kauai Island - Princeville (USA-HI) (location identifier)

HPVL High-Performance Visualization Laboratory

HPVM High-Precision Visibility Mode

HPW Half Power Width • High-Performance Workstation

HPWM HP Window Manager

HPWREN High-Performance Wireless Research and Education Network

HPY Baytown (USA-TX) (location identifier)

HPZ Helicopter Protection Zone

hq Hoc Quaere

Hq Headquarters

HQ Hardware Qualification • Headquarters • Hydroquinone • Highly Qualified • High Quality

HQCS Heraldic Quality Control System

HQDM Headquarters Data Manager

HQE Haute Qualité Environnementale

HQF Homogeneous Quadratic Form

HQL High-Quality Line

HQM Aberdeen (USA-WA) (location identifier) • Hamburg Quasar Monitoring • Hoquiam Bowerman (USA-WA) (location identifier)

HQMD Headquarters Management Directive

HQO Headquarters Originator

HQ&OPS Headquarters & Operations

HQP Highly Qualified Personnel • Hybrid Query Processor

HQR Handling Quality Rating

HQS Hamburg Quasar Survey • High-Quality Screening • High-Quality Stuff

hr Croatia • Croatian (ISO language code) • Hear • Here • Hour

Hr Heer • Herr • Hier • Horn • Hour

H-R Hertzsprung-Russell (diagram)

H/R High Rate

HR Croatia (ISO country code) • Croatie (code ISO de pays) • Hail and Rain • Half Rate • Half Resolution • Handelsrat • Handelsrechnung • Handelsrecht • Handelsregister • Handling Room • Hand Receipt • Hardness Ratio • Harvard Revised • Hardware Reliability • Hauptredaktion • Hauptreferent • Hazard Report • Heart Rate • Heat-Resistant • Henry Russell • Hertzsprung-Russell (diagram) • High Range • High Rate • High Reduction • High Resolution • Historical Record • Hit Rate • Holiday Rambler • Honduras (international aircraft registration code) • Horizontal Rule • Hour • House of Representatives • House Resolution • Human Relation • Human Reliability • Human Resource • Humidité Relative • Hydrogen Relief

HRA Handelsregister Abteilung • Haute Résolution Angulaire • Health Resource Administration • Hemispherical Reflective Antenna • High-Rate Acquisition • High-Resolution Apodizer • Horizontal Resolution Advisory • Human Resources Administration • Human Rights Act (UK) • Mansehra (Pakistan) (location identifier)

HRAA High-Rate Acquisition Assembly

HRAM Hard Random Access Memory

HRAO Hartebeesthoek Radio Astronomy Observatory (South Africa)

HRAP Hydrologic Rainfall Analysis Project

HRAS Harvard Radio Astronomy Station

HRAV High-Resolution Airborne Video

HRB Handelsregisterblatt • Harbin (China-PRC) (location identifier) • Hazard Review Board • Highway Research Board

HRC Hasselblad Reflex Camera • Herbig-Rao Catalog • Highly Reflective Cloud • High-Resolution Camera • High-Resolution CCD • High-Resolution Channel • High-Resolution Chromatography • High-Resolution Converter • Horizontal Redundancy Check • Human Resources Center • Human Resources Committee • Human Rights Commission • Hybrid Receiver Circuit • Hybrid Ring Control

HRCam High-Resolution Camera

HRCD Handheld Radio Control Device • Handheld Remote Control Device

HRCGC High-Resolution Capillary Gas Chromatography

HRCQ Highway Route Controlled Quality

HRCS Hazard Response Computer System • High-Resolution Camera System

HRCT High-Resolution Computed Tomography • High-Resolution Computerized Tomography

HRD Hertzsprung-Russell Diagram • High-Rate Densitometer • High-Rate Detector • High-Rate Discharge • High-Resolution Data • High-Resolution Diagnostic • High-Resolution Diskette • High-Resolution Display • Human Resources Data • Human Resources Department • Human Resources Development • Hurricane Research Division (NOAA)

HRDA High-Rate Data Assembly

HRDB High-Resolution Database

HRDC Human Resources Development Canada • Human Resources Development Council

HRDD High-Rate Data Delivery • Human Resources Development Division

HRDI High-Resolution Doppler Imager • High-Resolution Doppler Instrument • High-Resolution Doppler Interferometer

HRDL High-Resolution Doppler Lidar

HRDM High-Rate Demultiplexer

HRDP Hypothetical Reference Digital Path

HRDR High-Rate Data Radio • High-Rate Digital Recorder

HRDS High-Rate Data Station • High-Resolution Display System

HRDTM High-Resolution Digital Terrain Model

HRE Harare (Zimbabwe) (location identifier) • Hardware Reuse Evaluation • Hardware Routing Engine • Heavy Rear Engine • Highly Relativistic Electron • Holy Roman Emperor • Holy Roman Empire • Homogeneous Reactor Experiment • Human Research and Engineering • Hydrazine Rocket Engine

HR&ED Human Research & Engineering Directorate

HRED Human Research and Engineering Directorate

HREE Heavy Rare Earth Element

HREELS High-Resolution Electron Energy Loss Spectrometer • High-Resolution Electron Energy Loss Spectroscopy

HREF Hypertext Reference

HReg Handelsregister

HREM High-Resolution Electron Microscope • High-Resolution Electron Microscopy

HREP Highly Relativistic Electron Precipitation

HRES High-Resolution Electronic System

HRESC High-Resolution Still Camera

HRF Height Ranger Finder • High-Resolution Facsimile • Human Research Facility

HRFD High-Resolution Fourier Diffractometer

HRFG Human Resources Framework Group

HRFM High-Rate Frame Multiplexer

HRFPI High-Resolution Fabry-Pérot Imaging

HRFR High-Resolution Film Recorder

HRFRS High-Resolution Film Recorder System

hrg herausgegeben

Hrg Hearing • Herausgeber

HRG Haute Résolution Géométrique • Hemispherical Resonating Gyro • Hemispherical Resonator Gyro (*) • High Ratio Gas • High-Resolution Geometric • High-Resolution Graphics • Hochschulrahmengesetz (Germany) • Hurghada (Egypt) (location identifier)

HRGC High-Resolution Gas Chromatography

HRGM High-Resolution Ground Map

HRH Her Royal Highness • His Royal Highness

HRHR High-Resolution Hamilton Retrofit

HRI Harish Research Institute (India) • High-Resolution Image • High-Resolution Imager • High-Resolution Imaging • High-Resolution Interferometer • Horizontal Requirements Integration • Human Readable Interpretation

HRII High-Resolution Imaging Interferometer

HRIIS High-Resolution Infrared Imaging Spectrometer

HRIN Human Resource Information Network

HRIP Hic Requiescit In Pace

HRIR High-Resolution Infrared Radiometer

HRIRS High-Resolution Infrared Radiation Sounder

HRIS High-Resolution Imaging Spectrometer • High-Resolution Infrared Sounder • Human-Resource Information System

HRISI High-Resolution Imaging Spectrometer Industrial

HRIT High-Rate Information Transmission

HRJ Chaurjhari (Nepal) (location identifier)

HRK Hochschulrektorenkonferenz (Germany) • Kharkov (Ukraine) (location identifier) • Kuna (Croatia) (ISO currency code)

HRL Harlingen (USA-TX) (location identifier) • High-Refraction Layer • High-Repetition Laser • High Return Loss • Horizontal Reference Line • Hughes Research Laboratories • Hypermedia Research Laboratory (USA-TX) • San Padre Island (USA-TX) (location identifier)

HRLD High-Resolution Laser Diffraction

HRM Handheld Radiation Meter • Hassi R'Mel - Tilrempt (Algeria) (location identifier) • High-Rate Multiplexer • High-speed Road Monitor • HOMS Reference Manual • Hopwise Reliable Multicast • Hudson River Museum (USA-NY) • Human Resource Management

HRMA High-Resolution Mirror Assembly

HRMAS High-Resolution Magic-Angle Spectroscopy

HRMI High-Resolution Microwave Imager

HRMM Human Resources Management Model

HRMP Harvard Radio Meteor Project • Haui-Recurrent Markov Process

HRMS High-Resolution Mass Spectrometry • High-Resolution Mass Spectroscopy • High-Resolution Microwave Survey • High-Resolution Multispectral Scanner • Human-Resources Management System

HRMSC H.R. MacMillan Space Centre (Canada-BC)

HRMSI High-Resolution Multispectral Stereo Imager

HMSS Human-Resources Management Support System

HRMTI High-Resolution Multispectral Thermal Imager

HRMVS High-Resolution Multispectral Video System

Hrn Herren • Herrn

HRN Heron Island (Australia-QLD) (location identifier) • Human Resources Need

HRNTWT High Reynolds Number Transonic Wind Tunnel

HRO Harmonic Reference Oscillator • Harrison (USA-AR) (location identifier) • Housing Referral Office

HROI High-Resolution Optical Imager • High-Resolution Optical Instrument

HROM High-Resolution Ozone Mapping

HROS High-Resolution Optical Spectrograph

HRP High-Resolution Profile • High Resolving Power • High Risk Personnel • Highway Regulating Point • Horizontal Radiation Pattern • Human Reliability Program

HRPI High-Resolution Pointable Imager

HRPS Hazard Reduction Precedence Sequence • Human Reliability Program Status

HRPT High-Resolution Picture Transmission

HRR Herrera (Colombia) (location identifier) • Higher Reduced Rate • High-Range Resolution • High-Resolution Radar • High-Resolution Radiometer • Höchstrichterliche Rechtsprechung

HRRC Human Resources Research Center

HRRDA Human Resources Regional Development Assistant

HRRE Hydrogen Recombination Radiation Experiment

HRRI Human Resources Research Institute

HRRL Human Resources Research Laboratory

HRRO Human Resources Research Office

HRRR High-Range Resolution Recognition

HRRS High-Range Resolution Signature

Hrs Heures • Horas • Hours

HRS Harrismith (South Africa) (location identifier) • Haute Résolution Spatiale • Haute Résolution Spectrale • Haute Résolution Stéréoscopique • Hazard Ranking System • High-Rate Science • High-Rate Switch • High-Resolution Scenario • High-Resolution Spectrograph • High-Resolution Spectrum • High-Resolution System • Horizon Reference System • Hours • Housing Referral service • Human Restraint System

HRSC High-Resolution Shock Capturing (method) • High-Resolution Stereo Camera • Hiroshima Synchrotron Radiation Center (Japan) • Human Resources Steering Committee

HRSCMR High-Resolution Surface Composition Mapping Radiometer

hrsg herausgegeben

Hrsg Herausgeber

HRSG Heat Recovery Steam Generator

HRSGS High-Resolution Shuttle Glow Spectroscopy

HRSI High-temperature Reusable Surface Insulation

HRSMR High-Resolution Soil Moisture Radiometer

HRSO High-Resolution Solar Observatory

HRSP HERCULES Reduction Software Package • Hurricane Ridge Star Party (USA-WA)

HRSR High-Resolution Scanning Radiometer

HRSS High-Resolution Simulator System • High-Resolution Stereo Camera

Hrst Hersteller • Herstellung

HTST History of Recent Science and Technology (project)

HR&T Heat Rejection & Transport

HRT Harrogate Linton-on-Ouse (UK-England) (location identifier) • Heat Rejection and Transport • Helmholtz Reciprocal Theorem • High-Rate Telemetry • High-Resolution Telescope • High-Resolution Thermometer • High-

Resolution Tracker • High Res Technologies (USA-NY) • Highway Resolution Team

HRTEM High-Resolution Transmission Electron Microscope • High-Resolution Transmission Electron Microscopy

HRTF Head-Related Transfer Function • Huygens Recovery Task Force (ESA/NASA)

HRTI High-Resolution Thermal Imager

HRTIR High-Resolution Thermal Infrared • High-Resolution Thermal Infrared Radiometer

HRTM High-Resolution Tiltmeter

HRTR High-Rate Tape Recorder

HRTS High-Rate Telemetry System • High-Resolution Telescope and Spectrograph (*) • High-Resolution Telescope Spectrograph • High-Resolution Transition Region

HRU High-Risk Unit

HRUS High-Rate User Station

HRV Croatia (ISO country code) • Croatie (code ISO de pays) • Handelsregisterverfügung • Haute Résolution dans le Visible • Haute Résolution Visible • Heart Rate Variability (*) • Heat Rate Variability • High Resolution in the Visible • High-Resolution Video • High-Resolution Visible • Home Rekey Variable

HRVIR Haute Résolution Visible-Infrarouge • High-Resolution Visible-Infrared

HRVO Hausratsverordnung

HRW Human Rights Watch

HRWFS High-Resolution Wavefront Sensor

HRWSR High-Resolution Weather Satellite Receiver

HRX Hypothetical Reference Connection

HRY Henbury (Australia-NT) (location identifier)

Hrzg Herzog

Hrzgt Herzogtum

Hrzn Horizon

HRZN Horizon

hs handschriftlich • Hic Situs • Hoc Sensu • Hora Sommi

Hs Handschrift • Haus • Hours

H/S Health and Safety • Heat Shield • High Speed

H-S Hamilton Standard • Horizon Scanner • Horizon Sensor

H&S Headquarters & Service • Health & Safety • Health & Status

HS Habitability System • Half Subtractor • Hail and Snow • Haltestelle • Hamburg Schmidt (survey) • Hamilton Standard • Handschrift • Handset • Hard Silo • Hardstanding • Harmonized System • Haute Saône (France) • Haute Sécurité • Hawker Siddeley • Header Size • Headquarters and Service • Health Service • Heat Shield • Heat Stop • Heat Suppression • Heavy Side • Heck-Sause (comet) • Helical Scan • Heliosphere • Heliospheric Shock • Heure Supplémentaire • Hic Sepulta • Hic Sepultus • Hic Situs • Hierarchically Structured • Hierarchical Signature • Highest Score • High School • High Session • High Speed • High State • History of Science • Hochschule • Home Secretary • Home Station • Hora Sidérea • Horizon Scanner • Horizon Sensor • Hors Service • Host Scientist • Hot Short • Hot Spot • Human Services • Hybrid System • Hydrogen Sulfide • Hypersensibilisation • Hypersensibilisierung • Hypersensitization • Hyperstructural • Hyperstructure • Thailand (international aircraft registration code)

HSA Hardware System Area • Headspace Analysis • Health Service Area • Heat Shield Abort • Heure Supplémentaire-Année • High-Speed Access • High-Speed Arithmetic • Hispanic Society of America (USA) • Historic Sites Act (USA) • Hören - Sprechen - Aufnehmen • Hollandse Signaalapparaten BV (Netherlands) • Horizon Sensor Assembly • HRMA Shutter Assembly

HSAC High-Speed Analog Computer

HSAM Hierarchical Sequential Access Method

HSAP Honeycomb Sandwich Aluminium Panel

HSAR Holographic Synthetic Aperture Radar

HSAS Homeland Security Advisory System (USA)

H-Sat Heavy Satellite

HSAT Heavy Satellite

Hsb Hefner-Stilb

HSB Harrisburg Raleigh (USA-IL) (location identifier) • Heat-Shield Boost • High-Speed Boat • High-Speed Bus • High Surface Brightness • Hochleistungsschnellbahn • Hochschulbibliothek • Hue - Saturation - Brightness • Hungarian Space Board

HSBG High Surface Brightness Galaxy

HSBIC High-Speed Balanced Interface Card

HSBIP Hsinchu Science-Based Industrial Park (Taiwan)

HSC Hardware/Software Coordination • Hardware System Command • Harmon Science Center (USA-OK) • Headwaters Science Center (USA-MN) • Herschel Signal Chain • Hierarchical Storage Controller • Higher School Certificate • Higher State of Consciousness • High-Speed Channel • High-Speed Craft • Honeycomb Strip Chamber • Host Software Component • House Spacecraft • Human Systems Center • Shaoguan (China-PRC) (location identifier)

HSCA Harvard-Smithsonian Center for Astrophysics (USA-MA)

HSCaRS Hydrology, Soil Climatology and Remote Sensing

HSCC Hybrid Systems: Computation and Control (workshop)

HSCDM High-Speed Cable Driver Modem

Hsch Hochschule

hschr handschriftlich

Hschr Handschreiben • Handschrift

HSCL High-Speed Command Link

HSCP High-Speed Card Punch • High-Speed Command Punch

HSCR High-Speed Card Reader • High-Speed Command Reader

HSCS High-Speed Communications System

HSCSD High-Speed Circuit-Switched Data

HSCT High-Speed Civil Transport (*) • High-Speed Commercial Transport • High-Speed Compound Terminal • Hypersonic Civil Transport

HSCU Hydraulic Supply and Checkout Unit

HSD Hamilton Standard Division • Health and Safety Directive • Heat Storage Device • High Salt Diluent • High-Speed Data • High-Speed Detector • High-Speed Device • High-Speed Diesel • High-Speed Displacement • Historical Schedule Database • Honestly Significant Difference • Horizontal Situation Display • Human and Social Dynamics (NSF) • Human Systems Division

HSDA High-Speed Data Acquisition • High-Speed Data Assembly

HSDARS High-Speed Data Acquisition and Reduction System

HSDB High-Speed Data Bus

HSDF High-Speed Digital Filter

HSDG Hamburg-Südamerikanische Dampfschiffahrts-Gesellschaft • High School Diploma Graduate

HSDL High-Speed Data Line • High-Speed Data Link • High-Speed Digital Line • Hypertext Structure Description Language

HSDM High-Speed Data Monitor • High-Speed Data Mover

HSDMC High-Speed Direct Multiplex Control

HSDS High-Speed Detector Survivability

HSDT Hypersonic Small-Disturbance Theory

Hse House

H&SE Health & Safety Executive

HSE Halley-centered Solar Ecliptic • Headquarters Support Element • Health and Safety Executive • Heat Shield Entry • Heure Supplémentaire d'Enseignement • Heure Supplémentaire Effective • Hic Sepulta Est • Hic Sepultus Est • Hic Situs Est • High-Speed Encryption • High-speed Switching Element • Hydrostatic Equilibrium

Hsekpr Housekeeper

HSEP High-Speed Electrostatic Printer

HSES High-Speed Experimentation and Simulation • Hughes Satellite Earth Station

HSF HERA Simulator Facility • Hot Section Factor • Human Space Flight • Hypergolic Servicing Facility • Hypergol Servicing Facility

HSFC High-Speed Framing Camera

HSFD High-Speed Flight Demonstration • High-Speed Flight Demonstrator

HSFEC High-Speed Forward Error Correction

HSF-P HERA Simulator Facility Prototype

HSFP HERA Simulator Facility Prototype

HSFS High Sierra File System • Hobart Space Frontier Society (Australia-TAS)

HSFU Hard Summary Fault Unit

Hsg Housing

HSG Hierarchical Symbolic Grouping • High School Graduate • High Sierra Group • High Speed Gear • High Sustained Gs (acceleration) • Hochschulgemeinschaft • Horizontal Spectrograph

HSGC Hawaii Space Grant Consortium (USA-HI)

HSGT High-Speed Ground Transportation

Hsh Haushalt • Haushaltung

HSH Her Serene Highness • His Serene Highness • Las Vegas - Henderson (USA-NV) (location identifier)

HSHC High-Speed H-alpha Camera

Hshld Household

HSHT Helen Sawyer Hogg Telescope

HSI Habitat Suitability Index • Handling Speed Intelligently • Hardware/Software Integration • Hastings (USA-NE) (location identifier) • Heat Stress Index • High-resolution Spectroscopic Imager • High-Sensitivity Intensity • High Solar Intensity • High-Speed Imager • High-Speed Interface • High-Speed Interferometer • High-speed Serial Interface • Hispanic Serving Institution (USA) • Horizontal-Situation Indicator • Host Speed Interface • Host System Interface • Hot Section Inspection • Hot System Interface • Hue - Saturation - Intensity • Human-Systems Integration • Human-Systems Interaction • Human-Systems Interface • Hyperspectral Imager • Hyperspectral Imagery • Hyperspectral Instrument

HSIAC Human Systems Information Analysis Center

HSIC High-Speed Integrated Circuit

HSID High-Speed Interface Device • Historical Solar Image Database

HSIF Hardware/Software Integration Facility

HSIG Hermetically Sealed Integrating Gyroscope

HSIP Human System Integration Plan

H/SIR Hardware/Software Integration Review

HSIR Hardware and Software Integration Review • High-Speed Information Retrieval

HSIT Hardware/Software Integration Test

HSJ Honeycombed Sandwich Joint

HSK Horizontal Sling Kit

HSKP Housekeeping

hsl häuslich

HSL Hardware Simulation Laboratory • Hazardous Substance List • Health Service Laboratory • High-Speed Launch • High-Speed Logic • Hora Sidérea Local • Hubble Science Legacy • Hue - Saturation - Lightness • Hue - Saturation - Luminance • Huslia (USA-AK) (location identifier)

HSLA High-Speed Line Adapter • High Strength - Low Alloy

HSLAN High-Speed Local Area Network

HSLCM High-Speed Line Control Module

HSLN High-Speed Local Network

HSLS Health Service Logistics Support

HSLT High-Speed Telemetry Link

HSM Hardware Specific Module • Helical Scam Magnetic • Hermetically Sealed Microcircuit • Hierarchical Storage Management • Hierarchical Storage Manager • High-Speed Memory • Horsham (Australia-VIC) (location identifier) • Host Security Module • Hyperbase Session Manager

HSMA Hotel Sales and Marketing Association

HSMB Hybrid Superconductor Magnet Bearing

HSMD High-Speed Multiplexer/Demultiplexer

HSMI Health Services Management Inspector

HSMIMP High-Speed Modular Interface Message Processor

HSMS Hazardous Substance Management System • Health Status and Monitoring System • High-Speed Microwave Switch

HSMT Helical Scan Magnetic Tape

HSN High-Speed Network • Home Shopping Network

HSNG Housing

H/S/O Hardware/Software/Operations

HSO Habitation and Space Operations • Health and Safety Officer • Heliosynchronous Orbit • Herschel Space Observatory (formerly FIRST) • Hoger Secundair Onderwijs • Hungarian Space Office

HSOM Hierarchical Self-Organizing Map

HSOS Huairou Solar Observing Station (China-PRC)

Hsp Hörspiel • Hospital

HSP Haleakala Stokes Polarimeter • Health Stabilization Program • Hermes Spaceplane • High-Speed Photometer • High-Speed Printer • High-Speed Processor • High-Speed Protocol • High-Speed Pulse • Hot Springs (USA-VA) (location identifier) • Hot Subdwarf Primary • Huronia Star Party (Canada-ON) • Hydrological Simulation Programme • Hygiène et Sécurité Publique

HSPA High-Speed Parallel Adder

HSPD Hybrid Superconducting Pixel Detector

HSPF High-Speed Passenger Ferry

HSPR High-Speed Pulse Restorer

HSPTAL High-Speed Paper Tape Absolute Loader

HSPTP High-Speed Paper Tape Punch

HSPTR High-Speed Paper Tape Reader

HSQ Heat Shield Qualification

HSR Hardware Specification Review • Hardware Status Register • Harvest Server Registry • Health Service Region • Heat Shield Recovery • High-Speed Reader • High-Speed Research • History Study Reports (periodical code) • Hochschulring

HSRA Harvard-Smithsonian Reference Atmosphere • High-Speed Regeneration Assembly

HSRC Human Sciences Research Council (South Africa)

HSRD Hermes System Requirements Document

HSRG Hochschulrahmengesetz

HSRL High Spectral Resolution Lidar

HSRM High Spectral Resolution Mode

HSRMWR High Spatial Resolution Millimeter Wave Radiometer

HSRN High Spectral Resolution Sounder

HSRO High-Speed Repetitive Operation

HSRP High-Speed Research Program • High-Speed Rotary Prism • Hot Standby Router Protocol • Hot Standby Routing Protocol

HSRS High Spectral Resolution Sounder

HSRV Heat Source Reentry Vehicle

Hss Handschriften • Hessen

HSS Hamburg/SAO Survey • Health Service Support • Hic Sepulti Sunt • Hic Siti Sunt • High-School Student • High-Speed Scientific • High-Speed Service • High-Speed Steel • High-Speed Storage • High-Speed Stream • High-Speed System • High Stress Strain • Hissar (India) (location identifier) • History of Science Society (USA) • Hot Spot Simulator • Huairou Solar Station (China-PRC) • Hybrid Simulation System • Hydraulic System Simulator • Hyperspectral System

HSSA Health Service Support Area

HSSDB High-Speed Serial Data Buffer

HSSDC High-Speed Serial Data Connection

HSSDS High-Speed Switched Digital Service

HSSFC Humanities and Social Sciences Federation of Canada

HSSG High-Speed Study Group

HSSI High-Speed Serial Interface

HSSM Hardened Space Sensor Model • High-Speed Sorting Machine • Hydrocarbon Spill Screening Model

HSSMF High School Science and Math Faculty

HSSN Header Station Serial Number

HSSP High-Speed Switch Port

HSSS Highly Structured Stochastic System

HSSU High-Speed Storage Unit • Housing and Supporting Structure Unit

HSSWS High-Speed Solar Wind Stream

hst History (file extension) • Hora Sine Tempore • Host (file extension)

Hst Haltestelle • Handsteuerung • Hauptstadt • Hauptstelle • Hilfstelle

HST Harmonized Sales Tax • Hawaiian Standard Time • Health Sciences and Technology • Highest Spring Time • High-Speed Taxiway • High-Speed Technology • High-Speed Telemetry • High-Speed Train • High-Speed Transit • High-Speed Transport • History • Hoge-Snelheids Tunnel • Hoist • Homestead (USA-FL) (location identifier) • Horizontal Shock Tube • Horizontal Solar Telescope • Host • Hubble Space Telescope • Hygiène et Sécurité du Travail • Hypersonic Transport

HSTCT Hubble Space Telescope Comet Team

HSTCXO High-Stability Temperature-Compensated Crystal Oscillator

HSTDD Home Station Departure Date

HSTF Hypergolic System Test Facility

HSTJ HST Superconducting Tunnel Junction

HSTP Hard Stop • Hardware and Software Turnover Plan

HSTR High-Speed Token Ring

HSTRA High-Speed Token Ring Alliance

Hstrx Hestreux (Belgium)

HSTS Heure Supplémentaire à Taux Spécifique • High-Speed Text System • Horizontal Solar Telescope with Spectrograph

HSTSP Hardware and Software Turnover Support Plan

HSTTL High-Speed Transistor-Transistor Logic

HSTTY High-Speed Teletypewriter

HSTV High-Survivability Test Vehicle

HSU Heater Switching Unit • Helium Service Unit • Helium Servicing Unit • Humboldt State University (USA-CA) • Hygiène Scolaire et Universitaire

HSV Decatur (USA-AL) (location identifier) • Hermes Space Vehicle • Hue - Saturation - Value • Huntsville - Madison County (USA-AL) (location identifier) • Hypersonic Vehicle

HSW Hauptschlagwort • Health and Safet at Work (act) • High Temperature Winkler

HSWA Hazardous and Solid Waste Act (USA) • Hazardous and Solid Waste Amendment

HSWG Helioseismology Working Group

HSWO Hierarchical Stepwise Optimization

HSWS Hartmann-Schack Wavefront Sensor

HSWT Huygens Science Working Team

HSX Helical Stellarator Experiment

HSYNC Horizontal Synchronization

ht Hoc Tempore • Hoc Titulo • Hyperterminal (file extension)

Ht Haut • Heat • Height

H&T Handling & Transportation

H/T Heat Threat

HT Hadamard Transform • Haiti (ISO country code) • Halftime • Half Title • Halftone • Halley-Type (comet) • Halt and Transfer • Handling Time • Hankel Transform • Hatch Team • Haute Tension • Hawaiian Time • Head-Tail • Head-to-Tail • Heat Threat • Heat Transfer • Heat Treatment • Heavy Tank • Heavy Terminal • Height • High Technology • High Temperature • High Tensile • High Tension • High Tide • Hilbert Transformation • History Tape • Hochtemperatur • Hoc Tempore • Hoc Titulo • Holding Time • Horizontal Tab • Horizontal Tabulation • Horizontal Technology • Horsepower Tonnage • Hors Taxes • Hors Texte • Hough Transform • Hubble Type • Hypertext

Hta Huerta

HTA Chita (Russia) (location identifier) • Haute Tension Aérienne • Hazardous Test Area • Heavier Than Air • Hungarian Trademark Association • Hypertension Artérielle • Hypertextual Application

HTAR Highly Turbulent Accretion Region • High-Temperature Accretion Region

HTB Haupttagebuch • High-Temperature Branch • High-Tension Battery • Terre de Bas (Guadeloupe) (location identifier)

HTC Halley-Type Comet • Heat Transfer Coefficient • Hermes Training Complex • High-Tech Curve • High-Technology Center • High-Temperature Carbon • Hybrid Technology Computer • Hydrographic and Topographic Center • Hydrothermical Coefficient

Htd Heated

HTD Highway Traffic Division

HTDP Horizontal Time-Dependent Positioning

Hte Haute

HTE Hand-Triggered Erection • Hazardous Top-level Event • High Temperature Elongation • Horwitz-Thompson Estimator • Hypertext Engine

H. Tens. Haute Tension

Hter Hinter

HTF Hatfield (UK-England) (location identifier) • Helium Transfer Facility • Horizontal Test Facility • Hyper-G Text Format • Hypertext Functionality

HTFA Host Transfer File Access

Htg Heating

HTG Gourde (Haiti) (ISO currency code) • Hatanga (Russia) (location identifier) • Heating

HTGCR High-Temperature Gas-Cooled Reactor

HTGI How To Get It

HTGR High-Temperature Gas-cooled Reactor • High-Temperature Gas Reactor

HTH Hawthorne (USA-NV) (location identifier) • Helix - Turn - Helix • High Test Hypochlorite • Hope This Helps

HTHL Horizontal Takeoff and Horizontal Landing

HTHP High Temperatures and High Pressures (periodical code)

HTHY Hope This Helps You

HTI Haiti (ISO country code) • Hamilton Island (Australia-QLD) (location identifier) • Handwerkstechnisches Institut • Heure à Taux Spécifique • High Temperature Incendiary • Horizons Technology Inc. • Horizontal Technology Insertion • Horizontal Technology Integration

HTIO Horizontal Technology Integration Office

HTIP Horizontal Technology Integration Plan

HTIS Hadamard Transform Imaging Spectroscopy

HTIWG Horizontal Technology Integration Working Group

HTJR Heat Transfer Japanese Research (periodical code)

HTK HMM Toolkit

Htl Hotel

HTL High Threshold Logic • Höhere Technische Lehranstalt • Houghton - Roscommon County (USA-MI) (location identifier)

HTLL High Test Level Language

HTLRS Hydrographic Transportable Laser Ranging System

htm Hypertext (file extension)

HTM Hard To Match • Hard Tube Modulator • Heat Transfer Medium • Hierarchical Triangular Mesh • High-Temperature Material • High-Throughput Mission • Hypotheses Testing Model

HTMA High-Throughput Mirror Assembly

HTMIAC High-Temperature Materials Information Analysis Center

html Hypertext (file extension)

HTML Hypertext Markup Language

HTMPL High-Temperature Materials Processing Laboratory

HTMS High-Temperature Mass Spectrometry • High-Temperature Measurement System

HTN Hotan (China-PRC) (location identifier)

HTO East Hampton (USA-NY) (location identifier) • High-Temperature Oxidation • Horizontal Takeoff • Hospital Transfer order • Hydrous Titanium Oxide • Hypertext Optimization • Tritiated Water

HTOC High-Throughput Organic Chemistry

HTOHL Horizontal Takeoff - Horizontal Landing

HTOL Horizontal Takeoff and Landing

HTOVL Horizontal Takeoff - Vertical Landing

Htp Haltepunkt

HTP Hardness Test Plan • HEMT Total Power • Hermes Technology Programme • High Temperature-Pressure • High Test Peroxide • Highway Traffic Point • Hochtemperaturpyrolyse • Hubble Treasury Program • Hydroxyl Terminated Propellant

HTPB Highly Terminated Polybutadiene • Hydroxyl-Terminated Polybutadiene (*)

HTPE Hybrid Temporal Processing Element

HTQL Hypertext Query Language

Htr Hauteur • Heater

HTR Hateruma (Japan) (location identifier) • Heater • High-Temperature Reactor • High-Temperature Region • High-Temperature Resistor • High Time Resolution

HTRA High Time Resolution Astrophysics (conference)

HTrEn Heat Transfer Engineering (periodical code)

HTRM High Time Resolution Mode

HTRP High Time Resolution Processor

HTRQ Hypertext Request

HTRS Heaters

Hts Heights

HTS Ashland (USA-KY) (location identifier) • Harm Targeting System • Hawaii Tracking Station (USA-HI) • Hazard Tracking System • Heat Transfer Section • Heat Transfer System • Heat Transport Section • Heat Transport System • Heights • Hierarchical Time Steps (technique) • Higher Technical School • High-Technology School • High-Technology System • High-Temperature Superconducting • High-Temperature Superconductivity • High-Temperature Superconductor • High-Tensile Strength • High-Throughput Screening • High-Throughput Sensing • Hogere Technische School • How To Support • Huntington (USA-WV) (location identifier) • Hybrid Test Set • Hydrodynamic Test System • Hypertext System

HTSA Host-Tenant Support Agreement

HTSC High-Temperature Superconducting • High-Temperature Superconductivity • High-Temperature Superconductor

HTSF Hard Target Smart Fuze

HTSR Heat Transfer Soviet Research (periodical code)

HTemS High Temperature Science (periodical code)

HTrEn Heat Transfer Engineering (periodical code)

HTSUS Harmonized Tariff Schedule of the United States

Htt Hütte

HTT Hofer Teleskoptreffen (Germany)

HTTB High-Technology Testbed

HTTC Higher Teacher Training College

HTTL High-Power Transistor-Transistor Logic

HTTP Hypertext Transfer Protocol (*) • Hypertext Transmission Protocol • Hypertext Transport Protocol

HTTPD Hypertext Transport Protocol Daemon

HTTPS Hypertext Transport Protocol Server

HTTR High-Temperature Test Reactor

HTU Handheld Terminal Unit • Heat Transfer Unit • Height of Transfer Unit • Hopetoun (Australia-VIC) (location identifier)

HTV H-II Transfer Vehicle • Hope Transfer Vehicle • Hover Test Vehicle • Huntsville (USA-TX) (location identifier)

HTVA Hors Taxe sur la Valeur Ajoutée

HTVL Horizontal Takeoff - Vertical Landing

HTW Chesapeake - Huntington County (USA-OH) (location identifier) • Hazardous Toxic Waste

htx Hypertext (file extension)

HTX HTML Extension

HTXS High-Throughput X-ray Spectroscopy

HTZ Hato Corozal (Colombia) (location identifier) • Helicopter Taxi Zone

hu Hungarian (ISO language code) • Hungary

HU Harvard University (USA-MA) • Hauptuntersuchung • High Up • High Usage • Histogram Updating • Hongrie (code ISO de pays) • Hospital Unit • Humboldt-Universität • Hungary (ISO country code)

HUA Huntsville Redstone (USA-AL) (location identifier) • Hydrologic Unit Area

HUAC House Unamerican Activities Committee

HUB Hold-Up Battery • Housing and Urban Development • Humbert River (Australia-NT) (location identifier) • Humboldt-Universitätsbibliothek (Germany) • Humboldt-Universität zu Berlin (Germany)

HUBE Hopkins Ultraviolet Background Explorer • Hot Universe Background Explorer

Hubschr Hubschrauber

HUC Handheld and Ubiquitous Computing (symposium) • Health Unit Coordinator • Human Use Committee • Hydrologic Unit Code

HUCR Harvard University Character Recognition • Harvard University Character Recognizer • Highest Useful Compression Ratio • Humacao (Puerto Rico) (location identifier)

HUD Header and Data Unit • Head-Up Display • Housing and Urban Development • Humboldt (USA-IA) (location identifier)

HUDE Head-Up Display Electronics

HUDOM Heads of University Departments of Mathematics

HUE Halo Ultraviolet Explorer • Humera (Ethiopia) (location identifier)

HUF Forint (Hungary) (ISO currency code) • Terre Haute (USA-IN) (location identifier)

HUFEB Hokkaido University Faculty Engineering Bulletin (periodical code)

HUFEM Hokkaido University Faculty Engineering Memoirs (periodical code)

HUFS Histogram Updating with Full Sorting

HUG HEASARC Users Group • HOL Users Group • Huehuetenango (Guatemala) (location identifier)

HUGO Highly Unusual Geophysical Operation • Human Genome Organization

HUH Huahine Island (Society Islands) (location identifier)

HUI Hue - Phu Bai (Viet Nam) (location identifier)

HUJ Hebrew University of Jerusalem (Israel) • Hugo (USA-OK) (location identifier)

HUK Haftpflicht-, Unfall- und Kraftverkehrsversicherung • Hukuntsi (Botswana) (location identifier)

HUL Hardware Utilization List • Harvard University Libraries (USA-MA) • Houlton (USA-ME) (location identifier)

HULL Hardware Utilization and Location List

HULLAC Hebrew University/Lawrence-Livermore Atomic Code

hum Humilis • Humor (file extension)

Hum Human • Humane • Humanities • Humanity • Humble • Humidity • Humorous

HUM Highly Unstable Mother • Hilbert Uniqueness Method • Houma Terrebonne (USA-LA) (location identifier) • Humanities • Humidity

HUMINT Human Intelligence

HUMS Health and Usage Monitoring System

HÜN Hünfeld (Germany)

Hun Hungarian

HUN Hongrie (code ISO de pays) • Hualien (China-ROC) (location identifier) • Hungary (ISO country code)

Hund Hundred

Hung Hungarian • Hungary

HÜP Hausübergabepunkt

HUP Harvard University Press (USA-MA) • Horizon Ultraviolet Program • Hot Uniaxially Pressing

HUPD Harvard University Police Department

HUQ Houn (Libya) (location identifier)

Hur Hurricane

HUR Hardware Usage Report • Heat Up Rate

HURA Hexagonal Uniformly Redundant Array

HURCN Hurricane

HURD Heads Up Range Device

HUREVAC Hurricane Evacuation

Huricn Hurricane

HURT High-Utility Research Telescope

HUS HAMOTS Upgrade System • Hardened Unique Storage • Hughes (USA-AK) (location identifier) • Husum (Germany)

Husb Husbandry

HUSK Hardened Unique Storage Key

HUT Hard Upper Torso • Helsinki University of Technology (Finland) • Hopkins Ultraviolet Telescope • Hutchinson (USA-KS) (location identifier)

HUTE Hard User Terminal Element

HUTSPA Hopkins Ultraviolet Telescope Spartan

HUU Huanuco (Peru) (location identifier)

HUV Hudiksvall (Sweden) (location identifier)

HUW Hurricane Watch

HUX Huatulco (Mexico) (location identifier) • Santa Cruz - Huatulco (Mexico) (location identifier)

HUY Humberside (UK-England) (location identifier)

HUZ Huizhou (China-PRC) (location identifier)

hv Hoc Verbum

H/V Height/Velocity • Horizontal/Vertical

H&V Heating & Ventilation

HV Handelsverbindung • Handelsverkehr • Handelsvertrag • Handelsvertreter • Handelsvertretung • Hand Valve • Hardware Virtualizer • Harvard Variable (star) • Haulage Vehicle • Hauptverteiler • Hauptvortrag • Hausverwalter • Hausverwaltung • Haute-Volta (code ISO de pays) • High Vacuum • High Value • High Velocity • High Voltage • High Volume • Hydrogen Vent • Hypervelocity • Hyperventilation • Upper Volta (ISO country code) • Vatican (international aircraft registration code)

HVA Analalava (Madagascar) (location identifier) • Hauptverwaltung Aufklärung • High Value Unit • High-Velocity Absorption • High-Voltage Amplifier • High-Voltage Assembly

HVAC Heating - Ventilating - Air Conditioning • High Vacuum • High-Voltage Actuator • High-Voltage Alternating Current

HVAG Halls Valley Astronomical Group (USA-CA)

HvaOB Hvar Observatory Bulletin (periodical code)

HVAR High-Velocity Aircraft Rocket

HVASC High-Value Asset Security Cage • High-Value Asset Security Container

HVB Hervey Bay (Australia-QLD) (location identifier) • High-Voltage Bias

HVC Hardened Voice Channel • Hardened Voice Circuit • High-Value Compound • High-Velocity Cloud • High-Velocity Complex • High-Vision Camera • High-Voltage Connector • High-Voltage Control • Hovercraft

HVD High-Voltage Differential • Humanistischer Verband Deutschlands (Germany) • Khovd (Mongolia) (location identifier)

HVDC High-Voltage Direct Current

HVDF High and Very (high frequency) Direction Finder

HVDS Hazardous Vapor Detection System • Hypergol Vapor Detection System

HVE Hanksville (USA-UT) (location identifier) • High-Vacuum Environment • High-Vacuum Evaporation • High-Voltage Electronics

HVEC High Vacuum Equipment Corporation

HVEM High-Voltage Electron Microscope • High-Voltage Electron Microscopy

HVES High-Vacuum Evaporation System

HVF Harmonically Varying Field • High-Velocity Flow • High-Viscosity Fluid • Hi-res Variable Frequency

HVG High-Velocity Gas • High-Voltage Gradient • Honningsvåg (Norway) (location identifier)

HVHMD Holographic Visor Helmet Mounted Display

HVI High-Value Item • High-Velocity Impact • High-Volume Impactor

HVISS High-Value Item Security System

HVIT High Volume Information Transfer

HVK Hólmavík (Iceland) (location identifier)

HVL Half-Value Layer • Hora Verdadera Local • Hypervelocity Launcher

HVM Helium Vector Magnetometer • High-Velocity Maser • Huairou Vector Magnetograph (China-PRC) • Hvammstangi (Iceland) (location identifier) • Hydraulic Valve Module

HVMI Helioseismic and Vector Magnetic Imager

HVN Haven • New Haven (USA-CT) (location identifier)

HVNS High-Velocity Neutron Star

HVO Hawaiian Volcano Observatory (USA-HI)

HVOA High-Voltage Operational Amplifier

HVoDS Hunstville Voice Distribution System (USA-AL)

HVOF High-Velocity Oxygen Fuel

HVP Half-Value Period • High-Vacuum Pump • High-Velocity Pulsar • High Video Pass • High-Voltage Pump • Horizontal and Vertical Position • Horizontal and Vertical Projection

HVPA High-Voltage Protection Assembly

HVPF Hauptverwaltung für das Post- und Fernmeldewesen

HVPS High-Voltage Power Supply • High-Volume Printing System

HVPVE High-Voltage Photovoltaic Effect

HVR Havre (USA-MT) (location identifier) • Heavier • High-Vacuum Rectifier • High-Voltage Rectifier • High-Voltage Regulator • High-Voltage Relay • High-Voltage Resistor • Home Video Recorder • Hover

HVS Hartsville (USA-SC) (location identifier) • High-Velocity System • High-Voltage Supply • High-Voltage Switch • High-Volume Sampler • Hubble Volume Simulation • Human Visual System

HVSA High-Voltage Solar Array

HVSE High-Voltage Solar Experiment

HVSF Honeywell Verification Simulation Facility

HVSL Holidays - Vacation - Sick Leave

HVSP High-Voltage Solar Panel • Hudson Valley Star Party (USA-NY)

HVSS High-Velocity Separation System

HVSt Hauptvermittlungsstelle

HVST High-Voltage Switching Transistor

HVt Hauptverteiler

HVT Half-Value Thickness • High-Value Target • Hypersonic Vehicle Technology

HVTA High-Value Target Acquisition

HVTIP High Volume Transaction Interface Package

HVTL High Value Target List

HVTU High Voltage Transformer Unit

HvU Hogeschool van Utrecht (Netherlands)

HVU High-Value Unit • High-Voltage Unit

HVV Hausverteil-Verstärker • Hypervelocity Vehicle

HVW High-Voltage Wire • Hochschule für Verwaltungswissenschaften

Hvy Heavy

HVY Heavy

HVYR Heavier

h/w Herewith • Husband and Wife

Hw Handwagen • Hauptwagen • Hauptwert • Hauswart

H/W Hardware • Head Wind

HW Half Wave • Half Width • Half Word • Halle/Westfalen (Germany) • Handelsware • Handelswert • Handelswissenschaft • Handwerk • Handwörterbuch • Hardware • Hardwood • Hazardous Waste • Head Wind • Heavy Weight • High Water • Hochwasser • Hot Water • Hot Wire

HWA Hawabango (Papua New Guinea) (location identifier)

HWAI Hand-Written Address Interpretation

HWAY Highway

Hwb Handwörterbuch

HWB Halbwertbreite

HWC Hazardous Waste Coordinator

HWCA Human Waste Collection Assembly

HWCI Hardware Configuration Item • Hardware Critical Item

HWCP Hardware Code Page

hwd Hollywood (file extension)

HWD Hayward (USA-CA) (location identifier) • Height - Width - Depth • Horizontal Weather Depiction

HWDMS Hazardous Waste Data Management System

Hwg Handwagen • Hochwohlgeboren

HWG Haustürwiderrufungsgesetz

HWH Hochwasserhöhe

HWHM Half Width Half Maximum

HWI Hawk Inlet (USA-AK) (location identifier) • Hochwasserintervall

H/WIL Hardware In Loop

HWIL Hardware In Loop

HWK Handwerkskammer • Hawker - Wilpena Pound (Australia-SA) (location identifier)

HWL Hazard Warning Label • High-Water Line

HWM High-Water Mark • Horizontal-Wind Model

HWMC Harware Management Console

HWMCA Harware Management Console Application

HWMP Hazardous Waste Management Plan

HWMS Handbook of Writing for the Mathematical Sciences

HWN Hwange National Park (Zimbabwe) (location identifier)

HwO Handwerksordnung

HWO Hazardous Waste Operations • Hollywood - North Perry (USA-FL) (location identifier)

hwp Hangul Word Processor (file extension)

HWP Half-Wave Plate • Hazardous Waste Program • Hochschule für Wirtschaft und Politik

Hwr However

HWR Heavy-Water Reactor

HWRD Hardware Requirements Document

HWRP Hydrology and Water Resources Programme

HWS Hawaii Wide Survey • Heat Wedge Spectrometer • Hurricane Warning System

HWSFC Hazardous Waste Superfund Collection

HWSFD Hazardous Waste Superfund Database

HWSTD High Water Speed Technology Demonstrator

HWT Heavyweight Tank • Hypersonic Wind Tunnel

HWU Hazardous Waste Unit • Heriot-Watt University (UK-Scotland)

HWVG Handwerkerversicherungsgesetz

Hwvr However

HWW Handheld Wearable Wireless (device) • High-Wind Warning • Hot-and-Warm Worked

HWWI Halley Worldwide Investigations

HWWRS Hyperfiltration Wash Water Recovery System

HWWS Hyperfiltration Wash Water (recovery) System

Hwy Highway

HWY Highway

HWZ Hochwasserzeit • Hochwasserzeit

HWZI Half-Width at Zero Intensity

Hx Homunculus

H/X Heat Exchanger

HX Hard X-ray • Heat Exchanger

HXD Hard X-ray Detector

HXDA HRMA X-ray Detection Assembly

HXDS HRMA X-ray Detection System (*) • HRMA X-ray Detector System

HXIS Hard X-ray Imaging Spectrometer (*) • Hard X-ray Imaging Spectrometry

HXR Hard X-Ray

HXRB Hard X-Ray Background

HXRBS Hard X-Ray Burst Spectrometer

HXRS Hard X-Ray Spectrometer

HXRT Hard X-Ray Telescope

HXS Hard X-ray Spectrometer

HXT Hard X-ray Telescope • High-energy X-ray Telescope

HXTS High Throughput X-ray Spectroscopy

hxwxl Height by Width by Length

HXX Hay (Australia-NSW) (location identifier)

hy Armenian (ISO language code)

Hy Heavy • Highway • Hydraulique • Hydrazine

HY Heavy • High Yield

Hya Hydra-Hydrae (constellation)

HYA Hyannis (USA-MA) (location identifier)

HYADES Hydroxyl Atmospheric Detection System

Hyb Hybrid

HYBALL Hybrid Analog Logic Language

Hybr Hybrid

HYC High Wycombe (UK-England) (location identifier)

HYCOL Hybrid Computer Link

HYCOS Hydrological Cycle Observing System

HYCOTRAN Hybrid Computer Translator

Hyd Hydrant • Hydrate • Hydraulic • Hydraulics • Hydraulique • Hydrographic • Hydrostatic • Hydrostatics

HYD Hyderabad (India) (location identifier) • Hydraulic • Hydraulics • Hydrology • Sigma Hydrids (meteor stream)

HYDAC Hybrid Digital-Analog Computer

HYDAPT Hybrid Digital-Analog and Pulse Time

Hydm Hydrometer

Hydr Hydrant • Hydrate • Hydraulic • Hydraulics • Hydraulique • Hydrazine • Hydrographic • Hydrographique • Hydrography • Hydrostatic • Hydrostatics

HYDR Hydraulic • Hydraulics

HYDRA Hierarchical Yest Dynamic Radio Architecture • Hierarchical Yest Dynamic Reprogrammable Architecture

Hydraul Hydraulic • Hydraulics

Hydrelc Hydroelectric

Hydrelcty Hydroelectricity

Hydrogr Hydrographic • Hydrographique • Hydrography

Hydros Hydrostatic • Hydrostatics

Hydt Hydrant

HYEX Hydraulic Excavator

HYF Hayfields (Papua New Guinea) (location identifier)

HYFA Hydrometeor Free Atmosphere

HyFaCC Hypergravity Facility for Cell Culture

HYFLEX Hypersonic Flight Experiment

Hyg Hygiène • Hygiene • Hygiènique

HYG Hydaburg (USA-AK) (location identifier) • Hypergol • Hypergolic

HYGMA Hygromagmaphile Mantle Array

Hygr Hygromètre • Hygrométrie

Hyi Hydrus-Hydri (constellation)

HYI High-Yield Investment

HyInt Hyperfine Interactions (periodical code)

HYL Hollis (USA-AK) (location identifier)

HyLIG Hyperluminous Infrared Galaxy

HyLIRG Hyperluminous Infrared Galaxy

HYMORS Hybrid Morphology Radio Source

HYMOSS Hybrid Mosaic On a Silicon Stack

HYN Huangyan (China-PRC) (location identifier)

Hyp Hypodermic • Hypophyse • Hypotenuse • Hypothèque • Hypothèse • Hypothesis • Hypothetical

HYP Hyphenation

HYPACE Hybrid Programmable Attitude Control Electronics

Hyper Hyperbole • Hyperbolic • Hyperbolique

HyperTIES Hypertext based on The Interactive Encyclopaedia System

Hypn Hypnose • Hypnotic • Hypnotique

Hypot Hypotenuse

Hypoth Hypothèse • Hypothesis • Hypothetical • Hypothétique

HYR Hayward (USA-WI) (location identifier) • Higher

HYS Hays (USA-KS) (location identifier)

HYSAS Hydrographic Source Assessment System

HYSAT Hybrid Satellite • Hybrid Sensor Experimental Satellite • Hyperspectral Satellite

HySTP Hypersonic System Technology Program • Hypervelocity System Technology Program

HYTEA Hypertext Authoring

HyTech Hypersonic Technology

HYV Hyvinkää (Finland) (location identifier)

Hz Handzeichen • Heizung • Hertz • Hochzeit

HZ Dust Haze • Habitable Zone • Haftungszeit • Harrison-Zel'dovich • Haze • Hertz • Humason-Zwicky (survey) • Saudi Arabia (international aircraft registration code)

hzb heizbar

HZB Hazebrouck Merville-Calonne (France) (location identifier)

HZD Hydrated Zirconium Dioxide

HZE High Atomic Number and Energy

Hzg Heizung • Herzog

HZG Hanzhong (China-PRC) (location identifier)

Hzgin Herzogin

Hzgt Herzogtum

Hzgtm Herzogtum

HZK Hessischer Zentralkatalog • Húsavík (Iceland) (location identifier)

HZL Hazleton (USA-PA) (location identifier)

HZMP Horizontal Impulse

HZN Horizon

HZO Hydrated Zirconium Oxide

HzRG High-Redshift Radio Galaxy

HZRG High-Redshift Radio Galaxy

HZSS Habitable Zone of the Solar System • High-Z Supernova Search

Hzt Herzogtum

Hztm Herzogtum

HZW Hazardous Waste

Hzy Hazy

H0 Hubble Constant

H2 Hydrogen

H2O Water

H2O2 Hydrogen Peroxide

H2S Hydrogen Sulfide

H4 Solomon Islands (international aircraft registration code)

I

i Bahnneigung • Ice • Im • Imaginary Number • Impulse • In • Inclination • Indicate • Information • Innen • Innerhalb • Innerlich • Instantaneous Current • Interest • Intermediate (file extension) • Intransitive • Island • Isle • Orbital Inclination • Phase Angle • Pitch Angle • Tourist Information • Unit Coordinate Vector • van't Hoff Factor

I Current • Electric Current • Idle • Île • Ilha • Immediate • Imperator • Imperium • Important • Impulse • Inch • Inclination • Independent • India (international alphabet) • Indian • Indicator • Individual • Inductive Effect • Inductometric Effect • Industrie • Industry • Inertia • Information • Injected • Input • Insel • Inspection • Inspector • Institute • Instruction • Instructor • Instrumentation • Intelligence • Intensity • Intermediate • International • Interpreter • Interrupt • Interstate • Inti (currency) • Invar • Iode • Iodine • Ionic Strength • Ionization Potential • Ireland • Irish • Irradiated • Isla • Island • Isle • Isola • Isolation • Isolement • Isoleucine • Isospin Quantum Number • Isotopic Spin • Issue • Italian • Italy (international aircraft registration code) • Italy (international vehicle registration code) • Italy (postal country code) • Jod • Luminous Intensity • Moment of Inertia • Momentum • Nuclear Spin Quantum Number • One (Roman numeral) • Particular Affirmative • Point-Spread Function • Radiant Intensity • Sound Intensity • Supergiant Star • Unit Matrix • 1 (Roman numeral)

ia im allgemeinen • In Absentia • Inter Alia • Interlingua (ISO language code)

iA Im Allgemeinen • Im Aufbau • Im Auftrag • In Abwesenheit • In Ausbildung

Ia Iowa (USA) • Supergiant Star of High Luminosity • Supersupergiant (star)

I/A Inspection/Acceptance

I&A Identification & Authentication • Identification & Authorization • Identify & Authenticate • Indexing & Abstracting • Integration & Assembly

IA Ice Age • Illuminated Atmosphere • Imagery Analyst • Im Allgemeinen • Immediate Action • Immediately Available • Impedance Angle • Implementation Agency • Implementation Agreement • Implementing Agency • Impulse Analysis • In Absentia • Incentive Award • Incremental Analysis • Indicated Airspeed • Indicated Altitude • Indirect Access • Indirect Addressing • Industry Application • Information Architect • Information Architecture • Information Asset • Information Assurance • Ingénieur de l'Armement • Inherent Availability • Initial Allowance • Initial Appearance • Initial Assessment • Input Axis • Inspecteur d'Académie • Inspection Académique • Inspection Authority • Inspection Authorization • Institute of Astronomy • Instrumentation Amplifier • Intel Architecture • Intelligence Artificielle • Intelligence Assessment • Intelligent Agent • Intelligenzalter • Interactive • Interactive Analysis • Interagency Agreement • Inter-American • Intercept Angle • Interface Agreement • Interface Amplifier • Interim Action • Interim Archive • Internal Audit • Internal Auditor • International Agreement • International Alphabet • Internet Access • Intersite Architecture • Inverter Assembly • Iowa (USA) • Iron Age • Isochron Age • Issuing Agency • Issuing Authority

IA&A Ingenieria Aeronautica y Astronautica (periodical code)

IAA Igarka (Russia) (location identifier) • Immediate Action Authority • Institut d'Astronomie et d'Astrophysique • Institute of Applied Astronomy • Instituto Antarctica Argentina • Instituto de Astrofísica de Andalucía (Spain) • Instrumental Activation Analysis • Interactive Achievement Award • Interconnected Accredited Automated (system) • Interim Access Authorization • Interim ASOC Automation • International Academy of Astronautics • International Advertising Association • International Aerospace Abstracts • International Astronautical Academy • Internationales Arbeitsamt • Irish Astronomical Association (Ireland) • Irish Aviation Authority (Ireland) • Ithaca Amateur Astronomers (USA-NY)

IAAA I Am An Accountant • International Association for Astronomical Arts (USA-CA) • International Association for Astronomical Artists (USA-CA) (*)

IAAB Indefinite Appointment Advisory Board • Interamerican Association of Broadcasters

IAAC Interamerican Accreditation Cooperation • International Association for Analog Computing

IAAE Institute of Automotive and Aeronautical Engineers

IAAER International Association for the Advancement of Educational Research

IAAF International Amateur Athletic Federation

IAAFA Inter-American Air Forces Academy (USA)

IAAG Institut d'Astronomie, d'Astrophysique et de Géophysique • ISEE Archiving Advisory Group

IAAI Innovative Applications of Artificial Intelligence (conference)

IAAIL International Association for Artificial Intelligence and Law

IAAL I Am A Lawyer

IAAMBA I Am A Master in Business Administration

IAAP Institute of Astrophysics and Atmospheric Physics • International Association of Applied Psychology

IAARC International Administrative Aeronautical Radio Conference • International Association for Automation and Robotics in Construction

IAASM International Academy of Aviation and Space Medicine

IAASS Institute of Astronomy And Space Sciences

IAAT Institut für Astronomie und Astrophysik Tübingen (Germany)

IAB ICSU Abstracting Board • Independent Accountability Board • Industrial Advisory Board • Industrial Arbitration Board • Industry Advisory Board • Industry Arbitration Board • Information Access Branch • Inter-American Bank • Interessengemeinschaft Astrofotografie Bochum (Germany) • International Abstracting Board • International Arrivals Building • International Association of Broadcasting • Internet Activities Board (*) • Internet Architecture Board • IUS Assembly Building • Wichita McConnell (USA-KS) (location identifier)

IABA International Aircraft Broker's Association • International Amateur Boxing Association

IABBE International Association for Better Basic Education

IABC International Association of Business Communicators

IABE Ibero-American Bureau of Education

IABEM International association for Boundary Element Methods

IABG Industrie Anlagen-Betriebs Gesellschaft mbH (Germany)

IABLA Inter-American Bibliographical and Library Association

IABO International Association for Biological Oceanography

IABP International Arctic Buoy Programme

IABS Integrated Apogee Booster System • International Association for Business and Society • International Association of Biological Standardization

IABSE International Association for Bridge and Structural Engineering

IABV Integriertes Automatisches Besteuerungsverfahren

IAC Image Analysis Center • Import Advisory Committee • Improved Anode Catalyst • In Any Case • Independent Accountability Class • Industrial Advisory Council • Industrial Application Center • Industrial Assessment Center • Industry Advisory Commission • Industry Advisory Committee • Information Analysis Center • Infrared Astrometric Catalog • Ingénieurs, Assimilés et Cadres • Instituto de Astrofísica de Canarias (Spain) • Instruction Atomicity Controller • Instrument Approach Chart • Instrument Array Cable • Instrument-on-A-Card • Integrated Analysis Center • Integrated Avionics Computer • Integrating Assembly Contractor • Integrating Associate Contractor • Integration, Assembly and Checkout • Intelligence Analysis Center • Interacademy Council • Interactive Courseware • Interagency Assessment Committee • Interagency Committee • Inter-American Conference • Inter-Application Communicationis • Inter-Array Communications • Interconnection Access Charge • Interim Acceptance Criteria • Interim Archive Catalogue • International Access Code • International Activities Committee • International Advisory Committee • International Agricultural Centre (Netherlands) • International Air Convention • International Algebraic Compiler • International Analysis Code • International Astronautical Congress • Internet Application Component • Internet As Control • Interstate Aviation Committee • Investment Advisory Committee • Item Activity Code

IACA Independent Air Carriers' Association • Inter-American Cultural Association • International Air Carriers' Association • International Air Charters' Association

IACB International Advisory Committee on Bibliography

IACBDT International Advisory Committee on Bibliography, Documentation and Terminology

IACC Industrial Analysis and Control Council • Integrating Assembly and Checkout Contractor • Interagency Cartographic Committee • International Advanced Correlator Consortium • International Air Cargo Corporation

IACCA Inter-Agency Committee on Climate Change

IACCP Inter-American Council of Commerce and Production • International Association for Cross-Cultural Psychology

IACCR Industrial Activity Capability and Capacity Response

IACDLA International Advisory Committee on Documentation, Libraries, and Archives

IACE International Air Cadet Exchange

IACED Inter-Agency Committee on Environment and Development

IACET International Association for Continuing Education and Training

IACG Interagency Committee on Geomatics • Interagency Consultative Group

IACGEC Interagency Committee on Global Environmental Change

IACHEI International Association of Consultants in Higher Education Institutes

IACHR Inter-American Commission on Human Rights

IACI Inter-American Copyright Institute • International Association of Conference Interpreters

IACIS Intelligent Area Communication and Information System

IACIT International Association of Conference Interpreters and Translators

IACKL Interrupt Acknowledgment Latency

IACL International association of Constitutional Law

IACM International Association for Computational Mechanics

IA&CO Installation, Assembly & Checkout

IACO Installation, Assembly, and Checkout

IACODLA International Advisory Committee on Documentation, Libraries and Archives

IACOMS International Advisory Committee on Marine Sciences

IACP International Aerosol Climatology Project • International Association of Computer Programmers

IACR International Agreement Competitive Restriction • International Association for Cryptologic Research

IACRS Inter-Agency Committee on Remote Sensing

IACS Indian Association for the Cultivation of Science (India) • Inertial Attitude Control System • Institute for Astronomy and Computational Sciences • Institute for Astrophysics and Computational Sciences (CUA) • Integrated Access and Crossconnect System • Integrated Acoustic Communications System • Integrated Avionics Control System • Interactive Computer System • International Association of Classification Societies

IACSE Interagency Advisory Committee on Security Equipment

IACSS Inter-American Conference of Social Security

IACT Imaging Atmospheric Čerenkov Telescope

IACWD Interagency Advisory Committee on Water Data

IaD Imad-ad-Dean (USA-MD)

IAD Immediate Action Directive • Indemnité Annuelle de Départ • Information And Documentation • Information Assurance Document • Inspecteur Adjoint au Directeur • Installation, Assembly or Detail • Institut d'Assistance au Développement • Integrated Access Device • Integrated Antenna Detector • Integrated Automatic Documentation • Intelligence Artificielle Distribuée • Interactive Debug • Interface Adapter Device • Interface Agreement Document • Interface Analysis Document • Interim Acceptance Demonstration • Intermediate Astrometric Data • International Astrophysical Decade • Internationaler Austauschdienst • Internet Addiction Disorder • Inventory Adjustment Document • Ion-Assisted Deposition • Washington Dulles (USA-DC) (location identifier)

IADB Inter-American Development Bank

IADC Inter-Agency (space) Debris Coordination • Inter-Departmental Advisory and Development Committee

IADD Index to American Doctoral Dissertations

IadEn Iadernaia Energiia (periodical code)

IADIC Integration Analog-to-Digital Converter

IADIWU International Association for the Development of International and World Universities

IADL Instrument Activity Deviation List • International Association of Democratic Lawyers • Ion-Acoustic Double Layer

IADLA International Association for the Development of Libraries in Africa

IADP INTELSAT Assistance and Development Programme

IADPC Interagency Data Processing Committee

IADPG Intelligence Automatic Data Processing Group

IADS Interactive Authoring and Display System

IADT Integrated Automatic Detection and Tracking

IAE Impuesto sobre Actividades Económicas (Spain) • In Any Event • Inflatable Antenna Experiment • Information Analysis and Evaluation • Infrared Auroral Emission • Institut d'Administration d'Entreprise • Institut d'Administration et d'Économie • Institute for the Advancement of Engineering • Institute of Atomic Energy •

Instituto de Aeronáutica e Espaço (Brazil) • Integral Absolute Error • Integral of Absolute Error • Intermediate-Age Elliptical (galaxy) • International Aero Engines • ISDN-Anschluß-Einheit

IAEA Inter-American Education Association • International Atomic Energy Agency (Austria) (*) • International Atomic Energy Authority

IAEAA International Antitrust Enforcement Assistance Act (USA)

IAEB Institute of Atomic Energy of Beijing (China-PRC)

IAEC International Atomic Energy Committee

IAECOSOC Inter-American Economic and Social Council

IAEE International Association for Earthquake Engineering • International Association of Energy Economists

IAEEA International Association for the Evaluation of Educational Achievement

IAEEL International Association for Energy Efficient Lighting

IAEG International Association for Engineering Geology • International Association of Engineering Geologists • International Association of Engineering Geology (*)

IAEN Inspecteur de l'Administration de l'Éducation Nationale (France)

IAEO International Atomic Energy Organization • Internationale Atomenergie-Organisation

IAEP International Assessment of Educational Progress

IAES Intelligent Array Expansion System • International Association of Exchange Students

IAESM IEEE Aerospace Electronic Systems Magazine (periodical code)

IAESR Institute of Applied Economic and Social Research

IAESTE International Association for the Exchange of Students for Technical Experience

IAET International Association of Educational Television

IAEVG International Association for Education and Vocational Guidance

IAEVI International Association for Educational and Vocational Information

IAEWG International Acquisition Education Working Group

IAF Image Access Facility • Image Analysis Facility • Imageur d'Aurores pour FREJA • In Another File • Indian Air Force (India) • Initial Approach Fix • Interactive Applications Facility • Interactive Facility • Inter-American Foundation • International Accreditation Forum • International Aeronautical Federation • International Astronautical Federation • International Automobile Federation • Interoperability Assessment Facility • Istituto de Análisi Numerica

IAFA Internet Anonymous FTP Archive

IAFC Instantaneous Automatic Frequency Control

IAFE Instituto de Astronomía y Física del Espacio (Argentina)

IAFET Instituto de Astronomía y Física del Espacio de Buenos Aires (periodical code)

IAFIS Integrated Automated Fingerprint Identification System

IAG Information Assurance Group • Instituto Astronômico e Geofísico • Instituto de Astronomía y Geodesia • Instruction Address Generation • Instrument Advisory Group • Interactive Application Generation • Interagency Advisory Group • Interagency Group • International Association of Geodesy (*) • International Association of Geology • International Association of Geomorphologists (*) • Niagara Falls (USA-NY) (location identifier)

IAGA International Association of Geomagnetism and Aeronomy

IAGB Industrieanlagen Betriebsgesellschaft

IAGBN International Absolute Gravity Base Network

IAGC Instantaneous Automatic Gain Control • International Association on Geochemistry and Cosmochemistry • International Association of Geophysical Contractors

IAGIS International Association of Geodesy Information Service

IAGL Institut d'Astrophysique et de Géophysique de Liège (Belgium) (formerly IAL)

IAGO Interaction Atmosphère Glace Océan

IAGOD International Association on the Genesis of Ore Deposits

IAGRG Indian Association for General Relativity and Gravitation (India)

IAGS Inter-American Geodetic Survey

IAH Houston (USA-TX) (location identifier) • Interface Attachment Hardware • International Association of Hydrogeologists (*) • International Association of Hydrogeology

IAHB International Association of Human Biologists

IAHC International Ad Hoc Committee • Internet Ad Hoc Committee

IAHF International Aerospace Hall of Fame

IAHR International Association for History of Religion • International Association for Hydraulic Research

IAHS International Academy of History of Science (obsolete) • International Association of Hydrological Sciences

IAI Initial Acceptability Item • Institut Africain d'Informatique • Instituto de Automática Industrial • International Acquisition and Interpretation • International African Institute • International Alliance for Interoperability • International Anthropological Institute • International Association for Identification • International Automotive Institute • Internet Application Integration • Ion-Acoustic Instability

IAIA International Association for Impact Assessment

IaIaF Iakutsk Iakutskii Filial SO AN SSSR (periodical code)

IAIPS Intelligence Analyst Integrated Processing System

IAIR Institute for Advanced Interdisciplinary Research (USA)

IAIS Industrial Aerodynamics Information Service • Institute for Advanced International Studies

IAIWC International Association of Independent Web Consultants

IAJ Irish Astronomical Journal

IAK Information - Anchor - Knowledge • Internationale Arbeitskonferenz • Internet Access Kit • Internet Administrator Kit

IAKV Interessengemeinschaft Autoren, Komponisten und Musikverleger (Austria)

IAL Immediate Action Letter • Information Access Laboratory • Immediate Action Letter • Immediate Action Limit • In Another Life • Incoming Async Link • Institut d'Astrophysique de Liège (Belgium) (now IAGL) • Instrument Activity List • Instrument Analysis Laboratory • Instrument Approach and Landing • International Algebraic Language • International Algorithmic Language • International Association of Limnologists • International Association of Linguistics • International Astronomical Landscape • Investment Analysis Language • Irish Academy of Letters (Ireland)

IALA International Association of Lighthouse Authorities • International Auxiliary Language Association

IALBA Inter-American Biographical and Library Association

IALD International Association of Lighting Designers

IALE Individual Annual Loss Expectancy

IALEIA International Association of Law Enforcement Intelligence Analysts

IALHI International Association of Labour History Institutions

IALL International Association for Learning Laboratories

i. allg. im allgemeinen

IALPG Interprétation et Adaptation Locale des Prévisions Générales

IALS International Association of Legal Science

IAL1K Institut d'Astronomie de Lausanne (periodical code)

IAM Iberian Atlantic Margin • ILA Associative Memory • Image Analysis Mode • Improved Acoustic Modeling • Impuls-Amplituden Modulation • In Amenas (Algeria) (location identifier) • Information Asset Management • Initial Acquisition Mode • Initial Address Message • Institute for Applied Mechanics • Institute of Administrative Management • Institute for Advanced Materials • Integrated Assessment Model • Interactive Algebraic Manipulation • International Academy of Management • International Association of Machinists • International Association of Meteorology • Internationale Arbeitsgemeinschaft für Müllforschung

IAMAP International Association of Meteorology and Atmospheric Physics (now IAMAS)

IAMAS International Association of Meteorology and Atmospheric Sciences (formerly IAMAP)

IAMAW International Association of Machinists and Aerospace Workers

IAMCR International Association for Mass Communication Research

IAMG International Association for Mathematical Geology

IAMM Interoperability Assessment Management Model

IAMOT International association for the Management Of Technology

IAMP Imagery Acquisition and Management Plan • Industrial Applied Mathematics Program

IAMR Intermediate-Age Merger Remnant

IAMS Initial Attack Management System • Institute for Atomic and Molecular Sciences (China-ROC)

IAMT International Association for Machine Translation

IAN Instant Astronomical News • Institutional Area Network • International Astronomical Network • Internationale Anatomische Nomenklatur • Kiana (USA-AK) (location identifier)

IANA Internet Assigned Numbers Authority

IANAA I Am Not An Accountant

IANAL I Am Not A Lawyer

IANAMBA I Am Not A Master in Business Administration

IANC International Airline Navigators Council • International Anatomical Nomenclature Committee

IANDS International Association for Near-Death Studies

IANEC Inter-American Nuclear Energy Commission

IANI Intelligent Access to Nordic Information

IANOR Institut Algérien de Normalisation (Algeria)

IANS Institute of Air Navigation Services

IANT International Advances in Non-destructive Testing (periodical code)

IANW International Academic Networkshop

IAO In And Out of Cloud • Indexation Assistée par Ordinateur • Indian Astronomical Observatory (India) • Information Activities Office • Information Activities Officer • Ingénierie Assistée par Ordinateur • Institut für Arbeitswirtschaft und Organisation (Germany) • Interdepartementaal Antarctica Overleg (Netherlands) • Interim Approval to Operate • Internal Automation Operation • International Amateur Observatory • Internationale Arbeitsorganisation

i. ao. D. im außerordentlichen Dienst

IAOE International Arctic Ocean Expedition

IAOG International ADMD Operators Group

IAOO International Association of Assessing Officers

IAOPA International Council of Aircraft Owners and Pilot Associations

IAOS International Association for Official Statistics

IAP Image Analysis Processor • Image And Pattern • Imagery Architecture Plan • Imaging Atom Probe • Improved Accuracy Program • Independent Activities Period • Independent Assessment Plan • Independent Assessment Program • Indiscrétions Astronomiques Publiables • Information Architecture Plan • Information Assurance Panel • Information Assurance Posture • Initial Aiming Point • Initial Approval Plan • Institut d'Astrophysique de Paris (France) • Institut für Atmosphärenphysik (Germany) • Institute for Atmospheric Physics • Institute of Applied Physics • Institute of Atmospheric Physics • Institute of Atomic Physics • Instrument Approach Plate • Instrument Approach Procedure • Integrated Action Plan • Integrated Architecture Panel • Integrated Avionics Package • Inter-Academy Panel • Interactive Application Processor • Inter-Array Processor • International Academy of Pathology • International Association of Photoplatemakers • International Association of Planetology • International Association of Psychotechnics • Internet Access Provider • Internet Application Platform

IAPA Instrument Approach Procedure Automation • Inter-American Press Association • International Airline Passengers Association

IAPCM Institute of Applied Physics and Computational Mathematics (China-PRC)

IAPCO International Association of Professional Congress Organizers

IAPG Institute of Astronomical and Physical Geodesy • Institut für Astronomische und Physikalische Geodäsie (Germany) • International Association of Physical Geography

IAPH International Association of Ports and Harbours

IAPI International Application Programming Interface

IAPIP International Association for the Protection of Industrial Property

IAPM IEEE Antennas Propagation Magazine (periodical code)

IAPO International Association of Physical Oceanography

IApp Inspecteur de l'Apprentissage

IAPP Inter-Access Point Protocol • International Association of Plant Physiology • International ATOVS Processing Package

IAPPP International Amateur-Professional Photoelectric Photometry (USA-TN)

IAPR International Association for Pattern Recognition

IAPS Integrated Apogee Boost System • Integrated Avionics Processing System • Interim Antenna Pointing System • International Academy of Political Science • International Association for the Properties of Steam • International Association of Physics Students

IAPSO International Association for the Physical Sciences of the Ocean

IAPT International Association for Plant Taxonomy

IAPU Improved Auxiliary Power Unit

IAPW In A Perfect World

IAQ Bahregan (Iraq) (location identifier) • Indoor Air Quality

IAR Independent Annual Review • Independent Architecture Review • Independent Assessment Report • Inertial Angular Reference • Institute for Aerospace Research • Institute for Astrophysical Research • Institut für Atmosphärische Radioaktivität • Instituto Argentino de Radioastronomía (Argentina-BA) • Institut pour les Applications de la Recherche • Instruction Address Register

• Intelligence Assessment Report • Intelligence Automation Requirements • Internal Assessment Review • Internal Audit Report • Interoperability Assessment Report • Yaroslavl (Russia) (location identifier)

IARA Inter-Allied Reparations Agency

IARAS Institute of Astronomy of the Russian Academy of Sciences

IARB Inspection Analysis Review Board • Interagency Audiovisual Review Board • Inter-American Review of Bibliography

IARC International Agency for Research on Cancer • International Arctic Research Center (USA-AK)

IARD Incendie - Accidents - Risques Divers (insurance) • Information Analysis and Retrieval Division

IARDS Interim Automated Reduction and Display System

IARES Illustrateur Autonome de Robotique pour l'Exploration Spatiale

IARIW International Association for Research in Income and Wealth

IARR Internal Average Relative Reflectance

IARS Independent Air Revitalization System • Integrated Airborne Reconnaissance Strategy

IART Initial Access Response Time

IARU International Amateur Radio Union

IARUS International Association for Regional and Urban Statistics

iAS Internet Application Server

IAS Iasi (Romania) (location identifier) • Ideal Adsorbed Solution • Image Acquisitions Systems • Image Analysis System • Image Assessment System • Immediate Access Storage • Immediate Access Store • Impact Assessment Sheet • Impact Assessment Study • Indiana Astronomical Society (USA-IN) • Indian Academy of Sciences (India) • Indicated Airspeed • Individual Article Supply • Inertial Angle Sensor • Information Access Service • Information Acquisition System • Information Adaptive System • Information Assurance Solution • Infrared Absorbed Spectroscopy • Infrastructure Architecture Solution • Institut d'Astrophysique Spatiale (France) • Institute for Advanced Studies • Institute for Advanced Study (USA-NJ) • Institute for Atmospheric Sciences • Institute of Aeronautical Sciences • Institute of Applied Systems • Institute of the Aerospace Sciences • Institut für Angewandte Systemtechnik • Instrument Activity Specification • Instrument Adaptor System • Instrument Approach System • Instrumentation Acquisition System • Integrated Avionics Suite • Integrated Avionics System • Intelligence Analysis System • Intelligent Array Subsystem • Intelligent Autonomous System • Interactive Analysis System • Interactive Application System • Interface Applications Software • International Accounting Standard • International Accounting System • International Aircraft Standards • International Association of Sedimentologists • International Association of Seismology • Internationale Amateur-Sternwarte eV (Germany) • Internationale Arbeitsgemeinschaft Ski • Internationaler Antwortschein • Internet Authentication Service • Ion Acoustic Structure • Ion-beam Acoustic Structure • Iowa Academy of Science (USA-IA) • Irish Astronomical Society (Ireland) • Istituto di Astrofisica Spaziale (Italy)

IASA Improving America's Schools Act (USA) • International Academy of Sciences and Arts • International Air Safety Association • International Association of Sound Archives • International Aviation Safety Assessment

IASAI Image Analysis Systems and Artificial Intelligence

IASAP Indian Academy of Sciences - Aryabhata Proj (periodical code)

IASB Institut d'Aéronomie Spatiale de Belgique (Belgium) • International Accounting Standards Board • International Aircraft Standards Bureau

IASBS Institute for Advanced Studies in Basic Sciences (India)

IASc Indian Academy of Sciences (India)

IASC Indexing and Abstracting Society of Canada • Institut Aéronautique et Spatial Canadien • Interagency Standing Committee • Interagency Steering Committee • International Arctic Science Committee • International Association for Statistical Computing (Netherlands)

IASCC Irradiation-Assisted Stress Corrosion Cracking

IASE Information Assurance Support Environment • International Association for Statistics Education

IASF Istituto di Astrofisica Spaziale di Frascati (Italy)

IASFC Istituto di Astrofisica Spaziale e Fisica Cosmica (Italy)

IASG Internetwork Address Subgroup • Internetworking Address Subgroup

IASH International Association of Scientific Hydrology

IASI Improved Atmospheric Sounder in the Infrared • Improved Atmospheric Sounding Interferometer • Infrared Atmospheric Sounding Instrument • Infrared Atmospheric Sounding Interferometer (*) • Integral Applied Studies Institute (Moldova) • Inter-American Statistical Institute

IASIW Institute for the Advanced Study of Information Warfare

IASL Institute of Air and Space Law • Inter-American School of Librarianship (Venezuela) • International Association of School Librarianship

IASLS Institute for Advanced Studies in Life Support

IASM Institute of Aerospace Safety and Management

IASN Intelligence Analysis Support Network

IASO Industrial Analysis Support Office

IASOR Ice And Snow On Runway

IASP Integrated Acquisition Strategy Panel • Integrated Acquisition Strategy Process • Interactive Automatic System for Photointerpretation • International Association for Social Progress • International Association of Scholarly Publishers

IASPEI International Association of Seismology and Physics of the Earth's Interior

IASPM Infrared Atmospheric and Signature Prediction Model

IASPS International Association for Statistics in Physical Sciences

IASR Intermediate Altitude Sounding Rocket

IASS Institute of Astronomy and Space Sciences • International Association of Survey Statisticians • Inverter/ATCS Support Structure

IASSIST International Association for the Social Sciences Information Services and Technology

IASSP IEEE ASSP Magazine (periodical code)

IASTE International Association for the Exchange of Students for Technical Experience

IASTED International Association of Science and Technology for Development

IASTG International Association of Structural/Tectonic Geologists

IASTP Inter-Agency Solar-Terrestrial Physics

IASY International Active Sun Year

IA&T Integration, Acceptance & Test • Integration, Assembly & Test

I&AT Integration & Acceptance Test

IAT Image Annotation Tape • Import Address Table • Individual Acceptance Test • Information Assessment Team • Informazioni e Assistenza Turistica (Italy) • Initial Acceptance Test • Inside Air Temperature • Installation and Acceptance Test • Integrated Assembly Test • Integrated Avionics Test • Integrated Avionics Trainer • Integration,

Acceptance and Test • Integration and Acceptance Test • Integration, Assembly and Test • Intelligence Analysis Team • International Alias Table • International Atomic Time • Inverse Abel Transform • Ion Acoustic Turbulence

IATA Intelligent Agents for Telecommunication Applications (workshop) • International Air Transport Association • Is Amended To Add

IATAC Information Assurance Technology Analysis Center

IATAFI International Association for Technology Assessment and Forecasting Institutions

IATCA International Auditor and Training Certification Association

IATD Is Amended To Delete

IATE Intermediate Automatic Test Equipment

IATEFL International Association of Teachers of English as a Foreign Language

IATM International Association for Testing Materials

IATME International Association of Terrestrial Magnetism and Electricity

IATO Independent Acceptance Test Organization • Interim Approval To Operate

IATOS Ingénieurs, Administratifs, Techniques et Ouvriers de Service • Ingénieurs, Administratifs, Techniques, Ouvriers et de Service (*)

IATR Is Amended To Read

IATU Inter-American Telecommunications Union

IATUL International Association of Technical University Libraries • International Association of Technological University Libraries (*)

IATUR International Association for Time Use Research

IATV Interactive Alphanumeric Television

IAU IAU AG URL • Iaura (Papua New Guinea) (location identifier) • Image Alignment Unit • Incremental Analysis Update • Infrastructure Accounting Unit • Instituto Antártico Uruguayo (Uruguay) • Interface Adapter Unit • Internal Alignment Unit • International Academic Union • International Association of Universities • International Astronomical Union

IAUC International Astronomical Union Circular (periodical code)

IAUCo International Astronomical Union Colloquium (periodical code)

iAufR im Auftrage und für Rechnung

IAUFWG International Astronomical Union FITS Working Group

iAuiV im Auftrag und in Vertretung

IAUL IDL Astronomy User's Library

IAUNAM Instituto de Astronomía de la Universidad Nacional Autónoma de México (Mexico)

IAUP International Association of University Presidents • Internet User Account Provider

IAUPL International Association of University Professors and Lecturers

IAUS International Astronomical Union Symposium (periodical code)

IAUT Transactions of the International Astronomical Union Symposium (periodical code)

IAV International Association of Vulcanology • Internationales Aisschreibungsverfahren • Inventory Adjustment Voucher

IAVC Instantaneous Automatic Video Control • Instantaneous Automatic Volume Control

IAVCEI International Association of Volcanology and Chemistry of the Earth's Interior

IAVG International Association for Vocational Guidance

IAVSD International Association for Vehicle System Dynamics

iaW in anderen Worten

IAW Imagery Analysis Workpoint • In Accordance With • In Accord With • In Agreement With • In Another World • International AGN Watch • Isotopic Atomic Weight

IAWA International Association of Wood Anatomists

IAWG Infrastructure Analysis Working Group • International Affiliation of Writer's Guild

IAWPRC International Association on Water Pollution Research and Control (now IAWQ)

IAWQ International Association on Water Quality (formerly IAWPRC)

IAWR Internationale Arbeitsgemeinschaft im Rheineinzugsgebiet

IAX Image Application Executor

IAYC International Astronomical Youth Camp

ib Ibidem • im besonderem

iB im Bau • im Bayern • im Betrieb • in Buchstaben

Ib Ibidem • Supergiant Star of Low Luminosity

I/B Inboard

IB Inboard • In Bond • Inbound • Incentive-Based • Indice Brut • Indirect Budget • Industrial Business • Industriebank • Industriebau • Inert Building • Information Bottleneck (method) • Information Bulletin • Information Bureau • Informationsbank • Informationsblatt • Informationsbüro • Ingenieurbau • Ingenieurbüro • Inlet Busy • Inner Button • Inspection Bulletin • Instruction Book • Instruction Buffer • Interacting Binary (star) • Interdisciplinary Body • Interface Bus • Internal Bremsstrahlung • International Baccalaureate • International Bank • International Board • International Boundary • Internationales Büro • Interrogation Bus • Inverted Barometer • Invoice Book • Ionization-Bounded • Isobar • Issue Book

ibA in besonderem Auftrag

IBA Ibadan (Nigeria) (location identifier) • Igniter Booster Assembly • Independent Broadcasting Authority • Industrial Base assessment • Inner Buffer Assembly • Inspection By Attribute • INTEGRAL Burst Alert • Interest-Bearing Account • International Bar Association • International Biographical Archive • International Book Association • Internationale Buchausstellung • Internationales Biographisches Archiv • Ion Beam Analysis

IBAC Identification-Based Access Control • International Business Aviation Council

IBAD Ion-Beam-Assisted Deposition

IBAMA Instituto Brasileiro de Meio Ambiente (Brazil)

IBAME International Board on the Applications of the Mössbauer Effect

IBB International Broadcasting Bureau

IBBD Instituto Brasileiro de Bibliografia e Documentação (Brazil)

IBBH Internationaler Bund der Bau- und Holzarbeiter

IBBN Inhomogeneous Big Bang Nucleosynthesis (*) • Inhomogeneous Black Body Nucleosynthesis

IBBR Interbank Bid Rate

IBBY Internationale Board on Books for the Youth

IBC Impurity Band Conduction • Impurity Band Conductor • Industrial Base capability • Information-Based Complexity • Inside Back Cover • Instrument Bus Computer • Integrated Broadband Communications • Interferometric Beam Combiner • Intermediate Bulk Container • International Biographical Centre (UK) • International Boundary Commission • International Brightness Coefficient • International Broadcasting Corporation • International Business Co.

IBCA International Book Collectors Association

IBCC International Building Classification Committee • International Bureau of Chambers of Commerce

IBCCA International Bathymetric Chart of the Caribbean

IBCCEA International Bathymetric Chart of the Central Eastern Atlantic

IBCE International Bureau for Cultural Exchange

IBCM International Bathymetric Chart of the Mediterranean

IBCN Integrated Broadband Communications Network • Integrated Broad Channel Network

IBCO Innermost Bound Circular Orbit

IBCS Inflight Blood Collection System • Intel Binary Compatibility Specification • Intel Binary Compatibility Standard

IBCWIO International Bathymetric Chart of the Western Indian Ocean

ibd Ibidem

IBD Intermediate Block Diagram • International Business Development • Ion Beam Deposition • Iterative Blind Deconvolution

IBE Ibague (Colombia) (location identifier) • Inner Back End • Institute of British Engineers • International Bureau for Education (*) • International Bureau of Education

IBEC International Bank for Economic Cooperation • Irish Business and Employers Confederation

IBECO Inboard Booster Engine Cutoff

IBEP Instituto Brasileiro de Edições Pedagógicas (Brazil) • International Border Environmental Plan

Ibér Ibérico

i. bes. im besonderes

i. bes. A. im besonderen Auftrag

IBEW International Brotherhood of Electrical Workers

IBEX International Burst Explorer • International Business Exchange • Iodine, Betas and X-rays

IBF Internally Blown Flap • International Book Festival

IBFF Impulse Base Flow Facility

IBFM Interacting Boson-Fermion Model

IBFN Integriertes Breitbandfernmeldenetz

IBG Institute of British Geographers • Interblock Gap • Intermediate Blue Galaxy • Internationale Bodenkundliche Gesellschaft

IBGE Instituto Brasileiro de Geografia e Estatística (Brazil)

IBGIS Image-Based Geographic Information System

IBGM Internationales Büro für Gewichte und Maße

IBH Instituto Bibliográfico Hispánico (Spain)

IBHB Intermediate Blue Horizontal Branch

IBHE Illinois Board of Higher Education (USA-IL)

IBI I Believe It • Iboki (Papua New Guinea) (location identifier) • Impuesto de Bienes Inmuebles (Spain) • Information-Based Index • International Bureau for Informatics • Internationales Burgenforschungsinstitut • Invoice Book Inwards

IBIA Institut de Bureautique et d'Informatique Appliquée

ibid Ibidem

IBICT Instituto Brasileiro de Informação em Ciência e Tecnologia (Brazil)

IBID International Bibliographical Description

IBIP Information-Based Index Program • International Bureau of Intellectual Property

IBIR Internal Band Irradiance Response

IBIS Image-Based Information System • Image on-Board INTEGRAL Spacecraft • Imager on-Board INTEGRAL Spacecraft (*) • Instant Business Information System • Instrument de Biologie Spatiale • Instrument for Biological Investigations in Space • Integrated Bibliographic Information System • Integrated Business Information System • Interagency Border Inspection System • Interaktives Bildinterpretationssystem • Interferometer-Based Information System • Interferometric Bidimensional Spectrometer • Intranet Business Information System • Issue-Based Information System

IBIT Initiated Built-In Test • Intergovernmental Bureau for Information Technology • Intrusive Built-In Test

IBJ Internationales Buchjahr

IBK Internationale Beleuchtungskommission

IBL Inner Bipolar Lobe • Inter-Bibliothecair Leenverkeer • Intermediate BL Lacertae (object) • Internal Boundary Layer • Intraband Luminescence

IBM Inferior But Marketable • Insidious Black Magic • Integrated Background Monitoring • Interacting Boson Model • International Business Machines Corporation • It's Better Manually

IBMAL Ion Beam Modification and Analysis Laboratory (UNT)

IBMJ International Business Machines Journal (periodical code)

IBMP Institute for Biomedical Problems

IBMBR Interbank Market Bid Rate

IBMC Industrial Base Maintenance Contract • Institut de Biologie Moléculaire et Cellulaire

IBMG Internationales Büro für Maße und Gewichte

IBMP Institut de Biologie Moléculaire des Plantes • Institute for Biomedical Problems

IBMR International Bureau for Mechanical Reproduction

IBMS Integriertes Breitbandiges Mobilkommunikationssystem • Ion Beam Mass Spectrometer

IBMtr Inboard Motor

IBN Identification Beacon • Indexed By Name • Institut Belge de Normalisation (Belgium) • Integrated Business Network • Integrative Biology and Neuroscience • International Biosciences Network

Ibnd Inbound

IBNORCA Instituto Boliviano de Normalización y Calidad (Bolivia)

IBO Ibo (Mozambique) (location identifier) • Information Bearing Object • International Baccalaureate Organization • International Broadcasting Organization • Invoice Book Outwards

IBOC In-Band On-Channel

IBOL Interactive Business-Oriented Language

IBOM Installation Bill of Materiels

IBOP International Balance Of Payment

IBOY International Biodiversity Observation Year

IBP Iberia (Peru) (location identifier) • Industrial Base Planning • Initial Boiling Point • Institut für Bauphysik (Germany) • Institute of British Photographers • Integrated Business Package • International Balance of Payment • International Biological Programme • International Boundary Point • Ion Beam Projector • Itty-Bitty Processor

IBPA International Book Printers Association • Irish Book Publishers Association

IBPGR International Board for Plant Genetic Resources

IBPIO International Buoy Programme for the Indian Ocean

IBPM Industrial Base Program Manager

i. Br. im Breisgau

IBR Implementation Baseline Review • Incorporate By Reference • Integrated Baseline Review • Intermediate Bit Rate • Investment Balance Review

IBRD International Bank for Reconstruction and Development

IBRO International Brain Research Organization

IBRS International Business Reply Service • Intra-Base Radio System

IBS Imaginary Bulk Syndrome • Impulse Balance System • Information Brokering System • Integrated Booking System • Integrated Broadcast Service • Integrated Broadcast System • INTELSAT Business Service • Interafrican Bureau of Soils • International Business Service • Intervened Billing System • Intrabeam Scattering • Ion Beam Scanning • Ion Beam Spectrometer • Ion Beam Sputtering • Ion Beam Synthesis

IBSAT Indexing By Statistical Analysis Techniques

IBSE Initial Blood Storage Equipment

IBSH Informational Bulletin of the Southern Hemisphere (periodical code)

IBSL Internationales Berufssekretariat für Lehrer

IBSO Installation Branch Safety Office

IBSP Inverse boundary Spectral Problem

IBSR Interactive Bibliographic Search and Retrieval

IBSRAM International Board for Soil Research And Management

IBSS Information Business Systems Solution • Infrared Background Signature Survey • Institute of Biology of the Southern Seas

IBT Income Before Tax • International Brotherhood of Teamsters • International Bureau of Time • Internet-Based Training • Ion Beam Technology

IBTE International Bureau for Technical Education

IBTN Integrated Broadband Telecommunications Network

IBTT International Bureau for Technical Training

IBTTA International Bridge, Tunnel and Turnpike Association

IBU Interference Blanker Unit • International Broadcasting Union • International Business Unit

IBVPA Inverse Boundary-Value Problems in Aerodynamics

IBVPH Inverse Boundary-Value Problems in Hydrodynamics

IBVS Information Bulletin on Variable Stars (periodical code)

IBW Impulse Bandwidth • Instantaneous Bandwidth • Intelligent Bibliographic Workstation • Ion Bernstein Wave

IBWC International Boundary Water Commission

IBWD Interacting Binary White Dwarf (star)

IBWM Intelligent Bandwidth Manager • International Boundary and Water Commission • International Bureau of Weights and Measures

IBWZ Internationale Bank für Wirtschaftliche Zusammenarbeit

IBY International Book Year (1972)

IBZ Ibiza (Spain) (location identifier) • Internationales Begegnungszentrum

i/c Icing Condition • In Charge of

I&C Installation & Checkout • Instrument & Control • Instrumentation & Communications • Interoperability & Commonality • Issues & Criteria

Ic Icarus

I/C Implementation/Conversion • In Command • Intercept • Intercom • Interoperability/Compatibility

I&C Installation & Checkout • Instrumentation & Communication • Integration & Checkout

IC Ice Crystal • Iceland • Identification Card • Identification Code • Identity Card • Image Capture • Image Compression • Imperial College (UK) • Implementation and Conversion • Implementation Convention • Important Customer • Impulse Conductor • In Charge • Incident Commander • Incomplete Contract • Incremental Cost • Indenture Code • Independent Carrier • Index Catalog • Index Correction • Indicator and Control • Inductance-Capacitance • Inductive Coupling • Industrial Court • Industrialized Country • Industry Commission • Inertial Component • Informal Communication • Information and Communication • Information Carrier • Information Center • Information Circular • Information Content • Infrastructure Committee • Ingénieur Commercial • Ingénieur Conseil • Ingénieur en Chef • Ingénieurs et Cadres • Initial Contact • Initial Condition • Inner Cluster • Inorganic Carbon • Input Circuit • Input Context • Input Controller • Input Current • Inspection Card • Inspection Committee • Inspection General • Inspector General • Instabilité de Courant • Installation and Checkout • Installed Capacity • Instruction Card • Instruction Code • Instruction Counter • Instrumentation and Communication • Instrument Controller • Instrument Correction • Insulated Conductor • Integrated Chip • Integrated Circuit • Integrating Command • Integration and Checkout • Integration Cell • Integration Contractor • Intellectual Capital • Intelligence Center • Intelligence Collection • Intelligence Community • Intelligent Civilization • Intensive Care • Interaction Coefficient • InterCity • Intercept • Intercom • Intercommutation • Intercomparison • Intercomputer • Intercomputer Communication • Interexchange Carrier • Interface Certification • Interface Control • Interface Controller • Interface Coordinator • Interference Canceler • Intergovernmental Council • Interim Change • Interior Communications • Intermittent Combustion • Internal Combustion • Internal Communication • Internal Connection • Internal Conversion • International Conference • International Congress • International Cooperation • Interrupt Controller • Intra-Cluster • Inverse Check • Inverse Compton (scattering) • Inversion Code • Ion Chromatography • Ionization Chamber • Iron Column • Itinerário Complementar • Ivory Coast

ICA Icabaru (Venezuela) (location identifier) • Ignition Control Additive • Immediate Constituent Analyzer • Independent Component Analysis • Independent Computing Architecture • Independent Console Architecture • Independent Cost Analysis • Independent Cost Assessment • Industrial Communications Association • Instrument Coordination Analyzer • Integrated Chameleon Architecture • Integrated Circuit Array • Integrated Communications Adapter • Integrated Control Architecture • Intelligent Communication Agent • Intelligent Console Architecture • Interapplication Communications Area • Interface Control Agreement • Interim Change Analysis • International Cartographic Association • International Commission on Acoustics • International Communication Agency • International Communication Association • International Congress on Acoustics • International Cooperation Administration • International Cooperative Alliance • International Council on Archives • International Court of Arbitration • Intra-application Communication Area • Intracavity Assembly • Invalid Care Allowance • Ion Composition Analyser • Item Change Analysis • Item Control Area

ICAA Instituto de Cinematografía y Artes Audiovisuales (Spain) (*) • Instituto de las Ciencias y los Artes Audiovisuales (Spain)

ICAAD Interactive Computer-Aided Analysis and Design • International Civil Airports Association

ICAAP International Consortium for Alternative Academic Publications

ICAC Independent Commission Against Corruption • Information Centre Advisory Committee • Instituto de Contabilidad y Auditoría de Cuentas

ICACGP International Commission on Atmospheric Chemistry and Global Pollution

ICACS International Conference on Atomic Collisions and

Solids

ICAD Individual Chemical Agent Detector • Individual Concern And Deficiency • Integrated Computer-Aided Design • Integrated Control And Display • International Committee for Accessible Document Design • International Community for Auditory Display

ICADD International Committee for Accessible Document Design

ICADE Instituto Católico de Administración y Dirección de Empresas (Spain) • Interactive Computer-Aided Design Evaluation

ICADS Integrated Correlation And Display System

ICAE Integrated Computer-Aided Engineering • International Commission on Agricultural Engineering • International Commission on Atmospheric Electricity • International Conference of Agricultural Economists • International Council for Adult Education

ICAENA Ion Composition Analyser and Energetic Neutral Atom (imager)

ICAES International Congress of Anthropological and Ethnological Sciences

ICAG International Comparison of Absolute Gravimeters

ICAI Image Classification Artificial Intelligence • Institute for Computer-Assisted Information • Intelligent Computer-Assisted Instruction • Interactive Computer-Aided Instruction • International Commission for Aid to Intellectuals

ICAIL International Conference on Artificial Intelligence and Law

ICAIR International Centre for Antarctic Information and Research

ICAIS ICSC Conference on Autonomous Intelligent Systems

ICAITI Instituto Centroamericano de Investigación y Tecnología Industrial

ICAL Image and Chemical Analysis Laboratory • Internal Calibration

ICALP International Colloquium on Automata, Languages and Programming

ICALPE International Centre for Alpine Environment

ICAM Improved Chemical Agent Monitor • Institut Catholique des Arts et Métiers • Integrated Communications Access Method • Integrated Computer-Aided Manufacturing • Integrated Computer-Assisted Manufacturing • Interactive Cartographic Automated Mapping

ICAMD International Commission on Antigens and Molecular Diagnostics

ICAME International Conference on the Applications of the Mössbauer Effect

ICAMER International Center for Astronomical, Medical and Ecological Research (Ukraine)

ICAMP Incident Cost-Analysis Modeling Project

ICAN Individual Circuit Analysis • Integrated Communications Advanced Network • International Commission for Air Navigation

ICANN International Conference on Artificial Neural Networks • Internet Corporation for Assigned Names and Numbers

ICAO Institute of Chartered Accountants of Ontario (Canada-ON) • International Civil Aviation Organization (Canada-QC)

ICAOB International Civil Aviation Organization Bulletin (periodical code)

ICAOC Interim Combined Air Operations Center

ICAP Improved Capability • Improved Capability Program • Increased Capacity • Integrated Communications Access Package • Integrated Contractor Assessment Program • International Conference on Atomic Physics • International Crashworthiness Analysis Program • Internet Calendar Access Protocol • Internet Content Adaptation Protocol

ICAPA Integrated Circuit Active Phased Array

ICAPC Increased Capability Adaptive Payload Carrier

ICAPP Interim Configuration Management and Procurement System

ICAPR International Conference on Advances in Pattern Recognition

ICAPS Interactions in Cosmic and Atmospheric Particle Systems • Interactive Computer-Aided Provisioning System • Internal Control Audit Planning Summary

Icar Icarus (periodical code)

ICAR Illinois Consortium for Accelerator Research (USA-IL) • Indian Council of Agricultural Research (India) • Initiated Corrective Action Report • Integrated Camera And Radiometer • Interface Control Action Request • Investigation and Correction Action Report • Investigation and Corrective Action Report (*)

ICARDA International Centre for Agricultural Research in Dry Areas

ICARE Image Cartography Expert • International Center for the Advancement of Research and Education

ICARUS Image Correction And Registration Utility System • Imaging Cosmic and Rare Underground Signal

ICAS Image Capture and Analysis System • Instrument Crew Alerting System • Integrated Computer-Assisted Surveillance • Integrated Condition Assessment System • Integrated Configuration Accounting System • Integrated Cost And Schedule • Intel Communicating Applications Specification • Interdepartmental Committee for Atmospheric Sciences • Interdepartmental Committee on Aerial Surveys • Intermittent Commercial and Amateur Service • Internal Compliance and Assessment System • International Committee for Atmospheric Sciences • International Computer Access System • International Council of Aeronautical Sciences (*) • International Council of Aerospace Sciences

ICASA Institute for Complex Additive Systems Analysis (USA-NM)

ICASE Institute for Computer Applications in Science and Engineering • Integrated Computer-Aided Software Engineering • Integrated Computer-Aided Software Environment • Integrated Computer-Aided Sustaining Engineering • Integrated Computer-Aided Systems Engineering • International Council of Associations for Science Education

ICASL International Center for the Advancement of Scientific Literacy (USA-IL)

ICASS International Cooperative Administrative Support Service

ICASSP International Conference on Acoustics, Speech and Signal Processing

ICAST International Conference on Adaptive Structures and Technologies

ICASVR International Committee on Atmosphere-Soil-Vegetation Relations

ICAT Institut Canadien d'Astrophysique Théorique (Canada-ON)

ICATI Instituto Centroamericano de Tecnología Industrial

ICATPN International Conference on Application and Theory of Petri Nets

ICATU International Council of Arab Trade Unions

ICB Information Collection Budget • Inner Core Boundary • Instituto Catalán de Bibliografía • Instrument Control Bus • Integrated Circuit Board • Interface Change Board • Interface Control Board • Interim Change Bulletin • Interior Control Board • International Competitive Bid •

International Competitive Bidding • International Cooperation Board • Internet Citizen's Band • Interrupt Control Block • Inventions and Contributions Board

ICBC IMAX Cargo Bay Camera • International Cataloguing and Bibliographic Control

ICBL International Committee of Ball Lightning

ICBM Intercontinental Ballistic Missile

ICBN Initiative Canadienne sur les Bibliothèques Numériques (Canada) • International Code of Botanical Nomenclature

ICBP International Committee for Bird Preservation (now BirdLife)

ICBT Improved Common Bridge Transporter • Interactive Computer-Based Training

ICC Ice Crystal Cloud • Image Converter Camera • Import Commodity Clearance • Incident Command Center • Independent Communications Center • Industrial Communications Council • Information and Computer Center • Information and Coordination Center • Information and Coordination Central • Information Center Complex • Information Coding Classification • Information Control Center • Information Coordination Center • Information Coordination Central • Information Coordination Control • INSAT Coordination Committee • Inspectorate Coordinating Committee • Installation, Calibration and Checkout • Installation Calibration Coordinator • Instructor Control Console • Instrumentation Control Center • Instrument Competency Check • Instrument Control Center • Instrument Control Computer • Instrumentation Checkout Complex • Instrumentation Control Center • Integrated Cargo Carrier • Integrated Chip Circuit • Integrated Circuit Card • Integrated Circuit Chip • Integrated Communications Center • Integrated Communications Control • Integrated Composites Center • Intelligence - Command - Control • Intelligence Coordination Center • Intelligent Call Center • Intelligent Channel Controller • Intercompany Correspondence • Intercomputer Channel • Intercomputer Communications Corporation • Interchannel Communicator • Intercomputer Channel • Intercomputer Communication • Interface Control Chart • Interface Control Contractor • Intergovernmental Consultative Committee • Intergovernmental Copyright Committee (UNESCO) • Interim Consultative Committee • International Cartographic Conference • International Chamber of Commerce • International Climatological Commission • International Color Consortium • International Communications Committee • International Communications Conference • International Computation Centre • International Computing Centre • International Conference on Communications • International Congregational Council • International Congress Center • International Consultative Committee • International Control Commission • International Convention Centre • International Coordinating Committee • International Copyright Convention • International Copyright Convention • International Cultural Center • Interstate Commerce Commission (USA) • Inventory Control Center • Item Category Code

ICCA Interagency Coordinating Committee for Astronomy • International Conference on Computer Applications • International Congress and Convention Association • International Council of Chemical Associations

ICCAD International Conference on Computer-Aided Design

ICCAIA International Coordination Council of Aerospace Industries Associations

ICCAL International Conference on Computer-Assisted Learning • International Conference on Computers And Learning

ICCB Integrated Change Control Board • Interface Configuration Control Board • Intergovernmental Consultation and Coordination Board • Internet Configuration Control Board • Interoperability Configuration Control Board

ICCBR International Conference on Case-Based Reasoning

ICCC Interclient Communication Convention • Intercouncil Coordination Committee • International Computers Communications Conference • International Conference on Computers and Communication • International Council for Computer Communications

ICCCM Inter-Client Communication Conventions Management • Inter-Client Communication Conventions Manual

ICCD Intensified Charge-Coupled Device • International Conference on Computer Design

ICCE International Commission on Continental Erosion • International Conference on Computers in Education • International Council of Commerce Employers

ICCEE International Classification Commission for Electrical Engineering

ICCF I Can Count on Failure • Interactive Control and Computing Facility • Interpolated Cross-Correlation Function

ICCG Incomplete Choleski Conjugate Gradient (method) • Interagency Coordinating Committee for Geophysics

ICCHP International Conference on Computers for Handicapped Persons

ICCI International Conference on Computing and Information

ICCIP Intercenter Council on Information Processing

ICCIT International Conference on Computing and Information Technologies

ICCITS Intercenter Council on Information Technology Security

ICCL Interface Control Configuration List • International Commission on Climate • International Committee on Computational Linguistics • International Conference on Computer Languages

ICCM International Conference on Computational Mathematics • Internet Channel Conflict Management

ICCN Intercenter Council on Computer Networking • Intercenter Council on Networking

ICCP Improved Computer Control Panel • Indemnité Compensatrice des Congés Payés • Institute for Certification of Computer Professionals (*) • Institute of Certified Computer Professionals • Institute of Certified Computing Professionals • Interclient Communications Protocol • Interface Coordination and Control Procedure • International Cloud Climatology Program • International Commission on Cloud Physics • International Commission on Clouds and Precipitation • International Conference on Cataloguing Principles • International Conference on Computational Physics

ICCPR International Covenant on Civil and Political Rights

ICCR Institute for Cosmic-Ray Research (Japan)

ICCS Ice Centre Communications System • Improved Command and Control System • Integrated Command and Control System • Integrated Communications Cabling System • Intercomputer Communication System • International Commission for Control and Supervision • International Conference in Computer Science • International Conference on Conceptual Structures • International Council for Canadian Studies

ICCSIP International Command and Control Systems Interoperability Program

ICCSTI Interdepartmental Coordinating Committee on Scientific and Technical Information (UK)

ICCTA International Committee for the Coordination of Terminological Activities • International Consultative Council of Travel Agents

ICCTI Instituto de Cooperação Científica et Tecnológica Internacional (Portugal)

ICCTL International Conference on College Teaching and Learning

ICCU Intercomputer Control Unit • Istituto Centrale per il Catalogo Unico (Italy)

ICCV International Conference on Computer Vision

ICCWG Interoperability Configuration Control Working Group

ICD Image Component Data • Industry Council for Development • Input Character Dispatching • Installable Client Driver • Installation Completion Date • Instrumentation Control Document • Interactive Call Distribution • Interface Connecting Device • Interface Control Diagram • Interface Control Document • Interface Control Documentation • Interface Control Drawing • International Center for Documentation • International Classification of Diseases • International Climatic Decade • International Code Designation • International Code Designator • International Cooperative Division • Interoperability Control Document

ICDA Integrated Cached Disk Array

ICDAR International Conference on Document Analysis and Recognition

ICDB Ice Core Database • Integrated Communications Database • Integrated Configuration Database

ICDD International Centre for Diffraction Data (USA-PA)

ICDE International Conference on Data Engineering • International Conference on Distance Education

ICDG Institute of Chemistry and Dynamics of the Geosphere

ICDL Integrated Circuit Design Language • International Centre for Distance Learning

ICDH Intelligence Collection Data Handling

ICDM Interactive Communications Data Management • Interclass Distance Measure • International Conference on Data Mining (IEEE)

ICDO International Civil Defense Organization

ICDP Intercontinental Drilling Project • International Conference on Distributed Platforms

ICDR Incremental Critical Design Review • Initial Critical Design Review • Internal Critical Design Review

ICDS Ice Coring and Drilling Services • Industrial and Commercial Development Service • Integrated Control and Display System • Interface Control Document Schedule

ICDT International Conference on Database Theory

ICDU Inertial Coupling Data Unit • Inertial Coupling Display Unit • Integrated Control and Display Unit

Ice Iceland • Icelandic

ICE Immediate Cable Equalizer • Improved Cost Estimate • In-Circuit Emulation • In-Circuit Emulator • Independent Collection Equipment • Independent Cost Estimate • Individual Compass Error • Information Catalog Environment • Information Center for the Environment • Information Center on Education • Information - Communication - Education • Initial Cooling Experiment • Initial Cost Estimate • Innovative Control Effector • Input-Checking Equipment • Input Control Element • Institut du Cinéma Éducatif (France) • Institution of Civil Engineers • Instituto de Ciencias de la Educación (Spain) • Instrumentation/Communication Equipment • Instrument Checkout Equipment • Instrument/Communication Equipment • Integrated Communications Environment • Integrated Conceptual Environment • Integrated Cooling for Electronics • Integration with Controlled Error • Intelligent Concept Extraction • Interactive Computer Ephemeris • Interactive Conference Environment • Interactive Control Experience • InterCity Express • Interdiction Control Element • Interface Configuration Equipment • Interface Configuration Experiment • Interference Cancelling Equipment • Interferometry Center of Excellence • Internal Combustion Engine • Internal Communications Element • Internal Communications Exchange • International Cirrus Experiment • International Cometary Explorer (formerly ISEE-3) • International Corpus of English • International Cultural Exchange • Intrusion Countermeasure Electronics • Inventory Control Effectiveness • IRAF Control Environment • Isolated and Confined Environment

ICEAM International Committee on Economic and Applied Microbiology

ICEAR International Centre for Equatorial Atmospheric Research

ICEC Internet Computing and Electronic Commerce (conference)

ICECC Information Centre for European Culture Collections

ICED International Council for Educational Development • Intraprofessional Council on Environmental Design • Ionospheric Conductivity and Electron Density

ICEDIS International Committee on Electronic Data Interchange for Serials

ICEF International Children's Emergency Fund • International Council for Educational Film

ICEHL International Conference on English Historical Linguistics

ICEI Institution of Civil Engineers of Ireland • Instituto Complutense de Estudios Internacionales (Spain)

ICEIS International Conference on Enterprise Information Systems

Icel Iceland • Icelandic

ICEM Incremental Cost Effectiveness Model • Independent Cluster Emission Model • Institut Coopératif de l'École Moderne (France) • Integrated Coastal and Estuarine Management • Integrated Computer-aided Engineering and Manufacturing • Interdisziplinäres Centrum für Elektronenmikroskopie und Mikroanalyse (Germany) • Intergovernmental Committee for European Migration • International Conference on Electronic Materials • International Council for Educational Media

ICEMAP Interagency Committee for Extramural Mathematics Program

ICEO Imperial College Environment Office (UK-England)

ICEP Information Certification Evaluation Plan

ICEPRO Icelandic Procedures

ICER Information Centre of the European Railways

ICES Initiative pour la Coopération dans l'Europe du Sud-Est • Interference-Causing Equipment Standard • Internal-Conversion Electron Spectroscopy • International Committee on Erosion and Sedimentation • International Conference on Environment Systems • International Conference on Evolvable Systems • International Council for the Exploration of the Seas • International Council of Environmental Sciences

ICESA International Commission for Earth Sciences in Africa

ICESat Ice, Cloud and (land) Evaluation Satellite

ICESCR International Covenant on Economic, Social and Cultural Rights

ICESD Integrated Circuit and Embedded System Design

ICET International Centre for Earth Tides • International Centre of Earth Tides (*) • International Council for Engineering and Technology • International Council on Education for Teaching

ICETT Industrial Council for Educational and Training Technology

ICEUM International Conference on Exploration and Utilization of the Moon

ICEV Internal Combustion Engine Vehicle

ICEX Ice and Climate Experiment • Instituto de Comercio Exterio (Spain)

ICF Inertial Confinement Fission • Inertial Confinement Fusion (*) • Information and Content Exchange • Information Conversion Function • Initial Confinement Fission • Innovation Capital Funds • Installation Confinement Facility • Institut Cinématographique de France • Instrument Control Facility • Integrated Citation File • Integrated Catalog Facility • Integrated Coupling Facility • Integration and Checkout Facility • Interact Communication Facility • Interconnect Facility • Interface Control Function • International Computer Forum • International Congress on Fracture • Intracellular Fluid • Intrinsic Correlation Function • Ionization Correction Factor • Isothermal Casting Furnace

ICFA Institut Consulaire de Formation Alternée (France) • International Committee for Future Accelerators

ICFAC Inertial Confinement Fusion Advisory Committee

ICFES Instituto Colombiano para el Fomento de la Educación Superior (Colombia)

ICFG International Commission on Fungal Genetics • Internetwork Coordination Functional Group

ICFM International Commission on Food Mycology

ICFMH International Committee on Food Microbiology and Hygiene

ICFP International Conference on Functional Programming

ICFPM International Centre for Fundamental Physics in Moscow (Russia)

ICFT Inverse Chip Fourier Transform

ICFTU International Confederation of Free Trade Unions

ICG Icing • In-flight Coverall Garment • Institut de Contrôle de Gestion • Intellectual Capital Group • Interactive Computer Graphics • Interactive Computerized Graphics • International Commission on Glass • International Commission on the Geoid • International Coordinating Group • International Crisis Group • Internet Commerce Group • Intersector Coordination Group • Interunion Commission on Geodynamics

ICGA IEEE Computer Graphics Applications (periodical code) • International Conference on Genetic Algorithms

ICGC International Conference on Gravitation and Cosmology

ICGEB International Centre for Genetic Engineering and Biotechnology

ICGI International Colloquium on Grammatical Inference • International Conference on Grammar Inference

ICGRG International Commission on General Relativity and Gravitation

ICGS Interactive Computer-Graphics System

ICGTMP International Colloquium on Group Theoretical Methods in Physics

ICGW International Commission on Ground Water

ICH Institut de Culture Humaine • Instructor Contact Hours • Interchanger • International Conference on Harmonization • I/O Controller Hub • Ion Cyclotron Heating

ICHA Integrated Cargo Hazard Assessment • Inter-union Commission for History of Astronomy (IA-IUHPS)

ICHB International Comet Handbook

ICHEO Inter-University Council for Higher Education Overseas

ICHEP International Conference on High-Energy Physics

ICHM Intelligent Component Health Monitor • International Commission on History of Meteorology • International Commission on the History of Mathematics

ICHMT International Centre for Heat and Mass Transfer • International Communications in Heat and Mass Transfer (periodical code)

ICHS International Congress of History of Science

ICI Cicia (Fiji) (location identifier) • Image Component Information • Imagery/Communication Interface • Imperial Chemical Industries • Improved Control Interval • Incoming Call Identification • Information and Communication Infrastructure • Institut de Créativité Industrielle • Instituto de Cooperación Iberoamericana • Instituto de Cooperación Internacional • Intelligent Communication Interface • Interactive Communications Interface • Intercarrier Interface • Interface Control Information • Interim Configuration Inspection • International Collaborative Index • International Commission on Illumination • International Copyright Institute • Internet Computer Index • Ion Composition Instrument

ICIA International Communications Industries Association

ICIAM International Conference on Industrial and Applied Mathematics

ICIAP International Conference on Image Analysis and Processing

ICIC Intercalibrations-Intercomparisons • International Copyright Information Center

ICICIS International Conference on Interoperability in Cooperative Information Systems

ICICS International Conference on Information and Communication Security

ICID Intensified Charge Injection Device • International Commission of Irrigation and Drainage

ICIDR International Centers for Infectious Diseases Research

ICIE International Center for Information Ethics

ICIL Institut de Chimie Industrielle de Lyon (France)

ICIMOD International Centre for Integrated Mountain Development (Nepal)

ICIO Interim Cargo Integration Operations

ICIP Intercarrier Interface Protocol • International Conference on Image Processing

ICIPE International Centre for Insect Physiology and Ecology

ICIRM Interagency Commission on Information Resources Management

ICIS Integrated Cockpit Information System • Integrated Contract Imaging System • Inter-Campus Information System • Interdepartmental Committee on Internal Security • International Conference on Information Systems

ICISC International Conference on Information Security and Cryptology

ICiSM IEEE Circuits Systems Magazine (periodical code)

ICIST Institut Canadien de l'Information Scientifique et Technique (Canada-ON)

ICIT Industry Coalition on Information Technology • Instituto Cubano de Investigaciones Tecnológicas (Cuba)

ICITV Institut de la Carte Internationale du Tapis Végétal

ICIV Institut de la Carte Internationale de la Végétation

ICJ International Court of Justice

ICK Nieuw Nickerie (Suriname) (location identifier)

icl Icon Library (file extension)

ICL Clarinda (USA-IA) (location identifier) • Incoming Line • Initial Configuration Load • Initial Contact Link • Innovative Computing Laboratory • Institut de Chimie de Lille (France) • Instrumentation Connection List • Instrument Calibration Laboratory • Integrated Cloud Liquid • Interface Clear • Interface Control Layout • International Computers Ltd. • International Confederation of Labour • Interpretive Coding Language • Interunion Commission on the Lithosphere • Intracluster Light

ICLA International Commission on the Low Atmosphere

ICLAS International Council on Laboratory Animal Science

ICLC International Cognitive Linguistic Conference

ICLD International Commission on Large Dams

ICLI Institute for Logic, Cognition, Language and Information • Integrated Call Level Interface

ICLID Incoming-Caller Identification • Incoming-Call Line Identification

Iclnd Iceland

ICLP International Conference on Logic Programming

ICLS Intercavity Laser Spectroscopy • Intracavity Laser Spectroscopy • Irish Central Library for Students

icm Image Color Matching (file extension)

ICM Image Color Matching • Incoming Communications Manager • Increased Coherence Model • Ingénieur Civil des Mines • Installable Compression Manager • Instantaneous Center of Motion • Institute of Computer Management • Instrumentation and Communication Monitor • Integrated Call Management • Integrated Circuit Mask • Integrated Cognitive Modeling • Integrated Cost Model • Integrated Cost Module • Integrated Coverage Measurement • Intelligent Call Management • Interactive Condition Mode • Interbus Caching Module • Intercloud Medium • Intercluster Medium • Intercommunication • Interface Coordination Meeting • Interface Coordination Memorandum • Interference Control Monitor • Intergovernmental Committee for Migrations • Interim Control Module • Intermediate Coupled Model • International Conference on Magnetism • International Conference on Materials • International Congress of Mathematicians • Interoperability Configuration Manager • Intra-Cluster Medium • Inventory Control Manager • Iterated Conditional Mode • Iterated Conditioned Mode • Iterative Conditional Mode

ICMA International Card Manufacturers Association • International Commission on the Middle Atmosphere

ICMAS International Conference on Multiagent Systems

ICMC International Cryogenic Materials Conference

ICMCTF International Conference on Metallurgical Coatings and Thin Films

ICME Institute for Computational Mathematics and Engineering • International Conference on Mathematical Education • International Congress on Mathematical Education • Interplanetary Coronal Mass Ejection

ICMES International Cooperation on Marine Engineering Systems

ICMF Initial Cluster Mass Function • Integrated Coupling Migration Facility

ICMG International Commission for Microbial Genetics

ICMH Institut Canadien de Microreproductions Historiques (Canada)

ICMI International Commission for Mathematical Instruction • International Conference on Multimodel Interfaces

ICMID International Committee for Microbiological and Immunological Documentation

ICMIS Integrated Computerized Management Information System

ICML Instrument Control Markup Language • International Conference on Machine Learning

ICMO Indirect Cost Monitoring Office • Integrated Configuration Management Office

ICMOD International Conference on Management of Data

ICMOVPE International Conference on Metal Organic Vapor Phase Epitaxy

ICMP Intelligence Collection Management Process • Interchannel Master Pulse • Internet Control and Message Protocol • Internet Control Message Protocol (*)

ICMPP Interim Configuration Management and Procurement Program

ICMR Instrument Calibration and Maintenance Record

ICMS Indirect-Cost Management System • Integrated Circuit and Message Switch • International Centre for Mathematical Sciences

ICMSF International Commission on Microbiological Specifications for Foods

ICMSSR Interagency Committee on Meteorological Services and Supporting Research

ICMST International Conference on Machine Searching and Translation

ICMT Intercontract Material Transfer • International Commission on Mycotoxicology

ICMUA International Commission on Meteorology of the Upper Atmosphere

icn Icon (file extension)

ICN Ice Condensation Nucleus • Idle Channel Noise • Implementation Change Notice • Institut Commercial de Nancy (France) • Instrumentation and Calibration Network • Integrated Computer Network • Inter-Computer Network • Interface Change Notice • Internal Control Number • International Communications Network • International Conference on Networking • Iodine Cyanide • Item Control Number

ICNAF International Council for Northwest Atlantic Fisheries

ICNI Integrated Communications Navigation and Identification

ICNIA Integrated Communications, Navigation, Identification Analysis

ICNIRP International Commission on Non-Ionising Radiation Protection

ICNN International Conference on Neural Networks

ICNP International Committee for Nuclear Physics

ICNPAR International Conference on Non-linear Problems in Aviation and Aerospace

ICNS Integrated Communications and Navigation System

ico Icon (file extension)

I&C/O Integration & Checkout

ICO Icon • Idle Cutoff • In Case Of • Input Current Offset • Instituto de Crédito Oficial (Spain) • Instrumentation Control Officer • Instrumentation Control Operator • Integrated Checkout • Integrated-Communication Officer • Integration and Checkout • Intensity-Controlled Oscillator • Intermediate Circular Orbit • International Carbohydrate Organization • International Commission for Optics • International Crew Office • Internet Connectivity Option • Sicogon Island (Philippines) (location identifier)

ICOA International Conference on Oriental Astronomy • Internet Change Of Address

ICOD International Centre for Ocean Development • International Conference On Databases

ICODP International Conference on Open Distributed Processing

ICOHTEC International Committee for Cooperation in the History of Technology

ICOL International Conference on the Origin of Life

ICOLC International Coalition Of Library Consortia

ICOLD International Commission On Large Dams

ICOLE Indiana Council on Outdoor Lighting Education (USA-IN)

ICOLP Industry Cooperative for Ozone Layer Protection

ICOM Imbedded Communications • Input-Control-Output Mechanism • Integrated Communications • International Council Of Museums

ICOME International Commission on Microbial Ecology

ICOMIDC International Council On Mathematics In Developing Countries

IComM IEEE Communications Magazine (periodical code)

ICOMOS International Council Of Monuments and Sites

iCOMP Intel Comparative Microprocessor

ICON Imagery Communications and Operations Node • Integration Contract • Intermediate Coordination Node

ICONE Index Comparatif Entre Normes Nationales et Européennes

ICONIP International Conference on Neural Information Processing

ICONO International Conference on Coherent and Nonlinear Optics

ICONTEC Instituto Colombiano de Normas Técnicas y Certificación (Colombia)

ICOR Intergovernmental Conference on Oceanic Research

iCORE Informatics Centre of Research Excellence (Canada-AB)

ICORP International Cooperative Research Project

ICOS Improved Crew Optical Sight • Integrated Checkout System

ICoSM IEEE Control Systems Magazine (periodical code)

ICOT Isothermal Corrosion Oxidation Test

ICOTA International Conference on Optimization: Techniques and Applications

ICOTS International Conference On the Teaching of Statistics

ICOU Inertial Coupling Data Unit

ICP Ignition Control Programmer • IMA Control Protocol • Image Coprocessor • Imperial College Press (UK) • Incident Command Post • Incident Control Point • Incremental Change Package • Independent Content Provider • Indicated Contribution Profile • Indicator Control Panel • Inductively-Coupled Plasma • Information Computer Technology • Infrastructure à Clés Publiques • Inherently Conducting Polymer • Initial Connection Protocol • Initial Contract Period • Institut de la Communication Parlée • Institute of Chemistry Physics (Russia) • Instituto das Comunicações de Portugal • Instrument Calibration Procedure • Instrument Command Parameter • Integrated Channel Processor • Integrated Circuit Package • Integrated Communications Provider • Integrated Contingency Plan • Intensive Course Programme • Intercomputer Processor • Interface Change Proposal • Interface Control Panel • Interim Change Package • Internal Communications Processor • International Classification of Patents • International Commission on Photobiology • International Comparison Programme • International Computer Programs • International Congress of Publishers • International Control Plan • International Cooperative Program • Internet Cache Protocol • Internet Caching Protocol • Internet Content Provider • Intertask Communication Protocol • Intra-Cluster Pattern • Intracranial Pressure • Inventory Communications Processor • Inventory Control Point • Item Control Point

ICPA International Cooperative Program Activity

ICPAE International Commission on Planetary Atmospheres and their Evolution

ICPAES Inductively Coupled Plasma Atomic Emission Spectrometer

ICPAM International Centre for Pure and Applied Mathematics

ICPAQGP International Conference on Physics and Astrophysics of Quark-Gluon • Plasma

ICPD International Conference on Population and Development

ICPE International Commission on Physics Education

ICPEAC International Conference on Photomic, Electronic and Atomic Collisions

ICPI Institut de Chimie et de Physique Industrielle

ICPIA International Conference on Parallel Image Analysis

ICPIC International Cleaner Production Information Clearinghouse

ICPL Istituto Centrale per la Patologia del Libro (Italy)

ICPM International Commission for Polar Meteorology

ICPMC Intermediate Contract Performance Management Course

ICP/MS Inductive Coupled Plasma/Mass Spectrometry

ICPMS Inductively Coupled Plasma Mass Spectrometer

ICPN Intracluster Planetary Nebula

ICPNe Intracluster Planetary Nebulae

ICPO International Criminal Police Organization (Interpol)

ICPP Interactive Computer Presentation Panel • International Conference on Parallel Processing • International Conference on Plasma Physics • International Congress on Plasma Physics

ICPR Information Center for Physics Research (Korea-ROK) • International Conference on Pattern Recognition

ICPS Image Collecting and Processing System • Informatique et Calcul Parallèle de Strasbourg • Integrated Command Post System • International Conference of Physics Students • International Conference on Physics of Semiconductors

ICPSR Inter-university Consortium for Political and Social Research

ICQ International Comet Quarterly • I Seek You

ICR Ice Crystal Replicator • In-Circuit Reconfigurable • Indirect Collective Referencing • Indirect Cost Recovery • Individually Carried Record • Inductance - Capacitance - Resistance • Information Collection and Report • Information Collection Request • Initial Cell Rate • Input and Compare Register • Institute for Cooperative Research • Institute for Creation Research • Institute of Cancer Research • Instruction Change Request • Intelligence Collection Requirements • Intelligent Character Reader • Intelligent Character Recognition • Interdepartementaale Commissie Ruimtevaart (Netherlands) • Interface Compatibility Record • International Commission on Rheology • Interrupt Control Register • Inverse Compton Radiation • Ion Cyclotron Resonance • Ionic Cyclotronic Resonance • Nicaro (Cuba) (location identifier)

ICRA International Center for Relativistic Astrophysics (Italy)

ICRAD Interactive Color Radar Display

ICRAF International Council for Research in Agroforestry

ICRB Interface Change Review Board

ICRC Industry Cooperative Research Center • Interagency Classification Review Committee (USA) • Internal Calibration Repair Center • International Committee of the Red Cross • International Cosmic-Ray Conference

ICRCCM Intercomparison of Radiation Codes in Climate Models

ICRCM International Centre for Recent Crustal Movements • International Centre on Recent Crustal Movements

ICR&D International Cooperative on Research & Development

ICRD International Cooperative on Research and Development

ICRF IERS Celestial Reference Frame • International Celestial Reference Frame • Ion Cyclotron Radio Frequency • Ion Cyclotron Range of Frequencies

ICRG Impact Cratering Research Group

ICRH Ion Cyclotron Resonance Heating

ICRI Interswitch Collective Routing Indicator

ICRIS Integrated Customer Record Information System

ICRISAT International Crops Research Institute for Semi-Arid Tropics

ICRM International Committee on Radionuclide Meteorology

ICRO International Cell Research Organization

ICRP International Climatic Research Program • International Commission on Radiation Protection • International Commission on Radiological Protection (*)

ICRR Institute for Cosmic-Ray Research (Japan)

ICRS Imprimatur Common Reference Set • Improved Command Receiver Set • Improved Command Receiver System • International Celestial Reference System

ICRSDT International Committee on Remote Sensing and Data Transmission

ICRSE International Committee for Remote Sensing of the Environment

ICRT International Consumer Research and Testing

ICRU International Commission on Radiation Units

ICS IBM Cabling System • Identification Code Sort • Image Creation Studio • Image Creation System • Improved Composite Structure • Incident Command Structure • Incident Command System • Independent Cost Study • Industrial Chemistry Society • Industrial Computing Society • Inertial Control System • Information and Computer Science • Information and Computer Systems • Information Control System • Infrared Calibration System • Infrared Camera System • Infrared Communications System • Input Contactor Switch • Insert Card Section • Institute of Computer Sciences • Instrumentation Checkout Station • Instrumentation and Communications System • Instrumentation and Control System • Instrumentation Control Software • Instrumentation Control Supervisor • Instrument Command Sequence • Instrument Control Software • Instrument Control System • Integrated Checkout System • Integrated Command System • Integrated Communications System • Integrated Control Storage • Integrated Control System • Integration Control Sequencer • Integration Control System • Interactive Chatting Service • Interagency Communications System • Intercommunication System • Interdepartmental Committee on Space • Interface and Control Systems Inc. (USA) • Interface Control Specification • Interferometer Control Software • Interim Contractor Support • Interim Contract Support • Internal Communications System • Internal Customer Support • International Centre for Science • International Chamber of Shipping • International Classification of Standards • International Commission on Stratigraphy • International Conference on Supercomputing • International Consultancy Service • Internet Client Station • Internet Connection Sharing • Internet Connector Service • Interphone Control Station • Interpretive Computer Simulator • Intuitive Command Structure • Inventory Control System • Inverse Compton Scattering • Istituto Centrale di Statistica (Italy)

ICSA International Computer Security Association

ICSAR Interagency Committee on Search And Rescue

ICSB International Committee on Systematic Bacteriology

ICSC International Civil Service Commission • International Committee for Science Communication • International Computer Science Conference

ICSD International Continental Scientific Drilling

ICSE Impact of Computing in Science and Engineering (periodical code) • Initial Common Support Equipment • International Certificate of Secondary Education • International Conference on Software Engineering

ICSECA International Contributions for Scientific, Educational and Cultural Activities

ICSED Interamerican Center for Sustainable Ecosystems Development

ICSEI International Cooperation in the Search and the Examination of Inventions

ICSEKE International Conference on Software Engineering and Knowledge Engineering

ICSG International Civil Service Commission

ICSH Interstitial Cell Stimulating Hormone

ICSI Interdepartmental Committee on Security and Intelligence • International Commission on Snow and Ice • International Computer Science Institute (USA-CA) • International Conference on Scientific Information

ICSID International Center for the Settlement of Investment Disputes • International Council of Societies of Industrial Design

ICSL Interactive Continuous Simulation Language • Internet Communication Search Language

ICSLS International Convention for Safety of Life at Sea

ICSM IEEE Conference on Software Maintenance • Intergovernmental Committee on Surveying and Mapping • International Committee on Scientific Management • International Conference on Software Maintenance • International Conference on Synthetic Materials

ICSN Institut de Chimie des Substances Naturelles (France)

ICSO International Conference on Space Optics

ICSOC International Conventions and Scientific Organizations

ICSOS International Conference on Structure of Surfaces

ICSPRO Intersecretariat Committee on Scientific Programmes Relating to Oceanography

ICSPS International Council for Science Policy Studies

ICSR International Conference on Software Reuse

ICSRD Interdepartmental Committee on Scientific Research and Development

ICSS IBM Continuous Speech Series • Initial Conventional Support System • Integrated Communications Switching System • Intelligent Customer Support System • Interim Contractor Supply Support • International Communications Switching Systems • Internet Connection Secure Server

ICSSEA International Conference on Software and System Engineering and their Applications

ICST Imperial College of Science and Technology (UK) • Information and Communication Science and Technology • Institute for Computer Sciences and Technology • Integrated Combined System Test • International Centre for Science and Technology (Russia) • International Cooperation in Science and Technology

ICSTI International Council for Scientific and Technical Information

ICSTIS Independent Committee for the Supervision of Standards of Telephone Services

ICSTM Imperial College of Science, Technology and Medicine (UK)

ICSU International Council of Scientific Unions

ICSW International Commission on Surface Water

I&CT Information & Communication Technology

IC&T Installation, Checkout & Test

ICT Image Converter Tube • Individual and Collective Training • Influence Coefficient Test • Información Científica y Técnica • Information and Communications Technology • Information and Communication Technologies • Information Centre for Technology • Ingénieur Civil Technico-commercial • Input Character Transfer • Input Character Type • Installation, Checkout and Test • Institute for Chemical Technology • Institute for Computer Typesetting • Institute of Computer Technology • Instrument Command Template • Instrument Control Terminal • Integer Cosine Transform • Integrated Circuit Tester • Integrated Computer Telemetry • Integrated Concept Team • Interactive Command Test • Interactive Coordinate Transformation • Interface Compatibility Test • Interface Control Tooling • Interface Coordination Team • International Call for Tenders • International Cooperative Team • Wichita Mid-Continent (USA-KS) (location identifier)

ICTA International Centre for Tropical Agriculture • International Council of Travel Agents

ICTAF Interdisciplinary Center for Technological Analysis and Forecasting

ICTAP Interagency Career Transition Assistance Program

ICTB Individual and Collective Training Branch • International Customs Tariffs Bureau

ICTC Inertial Component Temperature Controller

ICTD Interchannel Time Displacement

ICTDR International Center for Tropical Disease Research

ICTE Inertial Component Test Equipment • Information and Communication Technologies in Europe

ICTF International Commission on the Taxonomy of Fungi • Inverse Collection Term Frequency

ICTL International Conference on Temporal Logic

ICTM Institut Centralien de la Technologie et du Management (France)

ICTO Interkerkelijk Comité voor Tweezijdige Ontwapening

ICTP Individual and Collective Training Plan • International Centre for Theoretical Physics (Italy)

ICTS Innovative Concepts and Theory of Stellarators (conference) • Integrated Circuit Test Set

ICTV International Commission on the Taxonomy of Viruses

ICU Image Capture Unit • Image Control Unit • Indicator Control Unit • In-flight Calibration Unit • Information and Communications University (Korea-ROK) • Institut pour la Coopération Universitaire • Instruction Cache Unit • Instruction Control Unit • Instrumentation Control Unit • Instrument Controller Unit • Instrument Control Unit • Intelligent Control Unit • Intensive Care Unit • Interactive Chart Utility • Interface Control Unit • Interface Converter Unit • International Code Use • ISA Configuration Utility

IC&V Intelligent Collaboration & Visualization

ICV Integrity Check Value • Intelligent Collaboration and Visualization • Internal Combustion Vehicle

ICVA International Council of Voluntary Agencies

ICVAD Imperial College of Science, Technology and Medicine (Starlink internal code)

ICVS IRAS Catalogue of Variable Sources • IRAS Catalogue of Variable Stars

ICW In Compliance With • In Conjunction With • In Connection With • In Coordination With • Initialization Command Word • Input Control Word • Integrated Cooperative Workspace • Interactive Courseware • Interrupted Continuous Wave • Ion-Cyclotron Wave

ICWA Institute of Current World Affairs

ICWE International Conference on Water and the Environment

ICWED Informal Conference on Western European Directors

ICWG Interface Control Working Group • International Cooperation Working Group • International Coordination Working Group

ICWM International Committee on Weights and Measures

ICWQ International Commission on Water Quality

ICWRS International Commission on Water Resources Systems

ICY Icy Bay (USA-AK) (location identifier)

ICYYLM International Commission on Yeasts and Yeast-Like Microorganisms

ICZ Initial Contact Zone • International Climate Zone • Intertropical Convergence Zone

ICZN International Commission on Zoological Nomenclature

IC4I Integration for Command, Control, Communication, Computers, and Intelligence

id Disk Identification (file extension) • Idem • in dem • in den • in der • Indonesia

Id Idaho (USA) • Idem • Identification • Identity

I/D Instruction/Data

I&D Information & Documentation • Investigação & Desenvolvimento

I+D Investigación + Desarollo

ID Idaho (USA) • Identification • Identification Data • Identification Dissector • Identified • Identifier • Identify • Identity Document • Image Dissector • Immediate Delivery • Incremental Delivery • Indefinite Delivery • Identification • Identity • Index Diagram • Index of Discrimination • Index of Dissimilarity • Indicating Device • Individual Development • Indonesia (ISO country code) • Indonésie (code ISO de pays) • Industrial Democracy • Infectious Desease • Information Deficiency • Information Department • Information Documentaire • Informationsdienst • Initial Distribution • Inner Detector • Insertion Device • Inside Diameter • Inside Dimensions • Installation Data • Institut für Datenverarbeitung • Instrumentation Directorate • Instrument-on-a-Disk • Integrated Demonstration • Integrated Diagnostics • Integrated Dictionary • Intelligence Department • Intelligent Design • Interactive Debugging • Interactive Discrimination • Interdigit • Interdigital • Interface Device • Interface Diagnostics • Interface Diagram • Interface Document • Interférométrie Différentielle • Intermediate Description • Internal Development • Internal Diameter • Internet Draft • Intrusion Detection • Iraqi Dinar (currency) • Isotropic Diffuse • Issue Date • Item Description • Item Documentation • Iterative Deconvolution • Izquierda-Derecha

Ida Idaho (USA)

IDA Idaho Falls (USA-ID) (location identifier) • Image Dissector Assembly • Indemnisation Directe des Assurés • In-Depth Analysis • Industrial Development Agency • Information - Decision - Action • Informatique Distribuée et Applications • Infrared Detection Array • Initial Data Analysis • Initial Denial Authority • Initial Design Activity • Input Data Assembler • Installation Design Authority • Institut de l'Audiovisuel • Institute for Defense Analyses • Integrated Data Analyzer • Integrated Digital Access • Integrated Digital Avionics • Integrated Diode Array • Integrated Dust Analyzer • Integro-Differential Analyzer • Intelligent Data Analysis • Intelligent Disk Array • Intelligent Drive Array • Interactive Data Archive • Interactive Differential Analyzer • Interactive Digital Archive • Interchange of Data between Administrations • Intercommunication Data Area • Interface Definition Agreement • Interface Display Assembly • Intermediate Data Array • International Dark-sky Association (USA-AZ) • International Data-casting Association • International Deployment of Accelerometers • International Development Agency • International Development Association • International Documentary Association • Internet Developers Association • Ionospheric Data Acquisition • Ionospheric Dispersion Analysis • ISO Data Archive • Isotope Dilution Analysis

IDAAS Integrated Data Acquisition and Analysis Station • International Directory of Astronomical Associations and Societies (merged with IDPAI into ASpScROW) (now StarGuides)

IDAC Instant Data Access Control • Integrated Data Acquisition and Control • Interconnecting Digital-Analog Converter

IDACA Integrated Data Acquisition and Control Algorithm

IDACS Integrated Digital Audio Control System

IDAM Image Data Access Method • Indexed Direct Access Method

IDAMAP Intelligent Data Analysis in Medicine and Pharmacology

IDAMIS Integrated Data Analysis and Management Information System

IDAMS Image Display And Manipulation System • Integrated Data Analysis and Management System • Internationally developed Data Analysis and Management Software

IDAPI Independent Database Application Programming Interface • Integrated Database Application Programming Interface • Integrated Data Application Programming Interface

IDAPS Image Data Processing System

IDAR Iterim Data Assessment Review

IDAS Ilkeston and District Astronomical Society (UK) • Image Data Analysis System • Image Digitization and Analysis System • Image Display and Access Software • Integrated Data Acquisition System • Integrated Data Analysis System • Integrated Design Automation System • Integrated Development Approval System • Integrated Digital Avionics System • International Database Access Service

IDASS Integrated Data Assimilation and Sounding System • Intelligence Data Analysis System for Spacecraft

IDAST Interpolated Data and Speech Transmission

IDATE Institut de l'Audiovisuel et des Télécommunications en Europe

IDATS International Data Acquisition and Transmission System

IDAW Indirect-Data-Address Word

IDB Idre (Sweden) (location identifier) • Image Database • Incremental Data Box • Industrial Development Board • Industrial Development Bond • Inertial Data Box • Information Database • Instant Database • Instrument Database • In-suit Drink Bag • Integrated Database • Integrating Double Buffer • Intentional Database • Inter-american Development Bank • Internal Database • Internal Data Bus • International Database • Interoperability Database • Islamic Development Bank

IDBL Initial Database Load

IDBM Intelligent Database Management

IDBMS Image Database Management System • Integrated Database Management System

IDBS Interoperability Database System

IDBTF Integrated Database Transaction Format

idc In Due Course

ID&C Integrated Display & Control

IDC IBM Data Connector • Image Dissector Camera • Indefinite Delivery Contract • Indice de Disparité de la Consommation • Indirect Cost • In Due Course • Industrial Development Certificate • Information and Demonstration Center • Information and Documentation Center • Information Dominance Center • Infrared Detector Cryostat • Inner Drift Chamber • Input Display Console • Inspection Data Card • Insulation Displacement Connection • Insulation Displacement Connector • Integrated Database Connector • Integrated Desktop Connector • Integrated Disk Channel • Integrated Display and Control • Integrated Display Assembly • Intelligent Design Creationism • Inter-Departmental Committee • Interface Document Control • International Data Center • International Data Corporation • Internet Database Connector • Internet Data Center • Interactive Devices Control • Interface Document Control • Inverse Dual Converter • ISOPHOT Data Center • Item Design Change

ID&CA Inverter Distribution & Control Assembly

IDCA International Development Cooperation Agency • Inverter Distribution and Control Assembly

IDCC International Data Communications Centre • International Data Coordination Centre

IDCCC Interior Department Cartographic Coordinating Committee (USA)

IDCE Institut pour le Développement du Conseil d'Entreprise

IDCMA Independent Data Communications Manufacturers Association

IDCNS Inter-Divisional Committee on Nomenclature and Symbols (IUPAC)

IDCP Interdivision Change Proposal

IDCS Image Dissector Camera System • Integrated Data Coding System • Intelligent Data Collection System • International Data Collecting System • International Data Collection System • ISO Dedicated Computer System

IDCT Initial Detection and Classification Trainer • Inverse Discrete Cosine Transform

IDCU Ice Detector Control Unit • Improved Digital Controller Unit • Integrated Data Communications Utility

idd Instrument Definition (file extension)

ID&D Inspection Damage & Discrepancy

IDD Image Definition Device • Indirect by Direct • Information and Data Dissemination • Information Delivery Device • Initial Delivery Date • Inspection Damage and Discrepancy • Instrument Deployment Device • Instrument Description Document • Integrated Data Denial • Integrated Data Dictionary • Interface Definition Document • Interface Description Document • Interface Design Description • Interface Design Document • Interface Development Document • International Direct Dial • International Direct Dialing • Internet Data Distribution

ID-DA Indefinite Delivery - Definite Quantity

IDDA Indefinite Delivery - Definite Quantity

IDDD International Direct Distance Dialing

IDDE Integrated Development and Debugging Environment • Integrated Development and Debugging Equipment • Interactive Development and Debugging Environment

IDDF Intermediate Data Distribution Facility

IDDL Integrated Data Description Language

IDDP Interface Device-Dependent Port

IDDS Improved Data Display System • Information Distribution and Display System • Instrumentation Data Distribution System • Integrated Data Display System • Integrated Data Distribution System • Integrated Display Development Station

ide Integrated Development Environment (file extension)

I&DE Investigação & Desenvolvimento Experimental

IDE Imbedded Drive Electronics • Implicit Differential Equation • Independent Developmental Evaluator • Information - Documentation - Evaluation • Initial Design Evaluation • Inspecteur Départemental de l'Éducation (France) • Institute of Developing Economies • Instructional Design Environment • Integrated Data Environment • Integrated Design Environment • Integrated Development Environment • Integrated Device Electronics • Integrated Digital Environment • Integrated Disk Electronics • Integrated Drive Electronics (*) • Intelligent Drive Electronics • Interactive Design and Engineering • Interactive Development Environment • Interface Design Enhancement • Interim Data Element • Internal Display Element • Interplanetary Dust Experiment • Investissement Direct Étranger

IDeA Intelligent Design Aid

IDEA Improved Data Effectiveness and Availability • Inductive Data Exploration and Analysis • Initiative to Develop Education through Astronomy • Innovation Development and Engineering Applications • Instituto Internacional de Estudios Avanzados • Instrumentation Development for Environmental Activities • Instrument Driver for Experimental Applications • Integrated Database Environment for Assessment • Integrated Data for Enforcement Analysis • Intelligence - Diagnostic - Expertise - Administration • Interact Decision Evolution Aid • Interactive Data Entry Access • Interactive Data Exploration and Analysis • Interactive Deconvolution with Error Analysis • Interface and Display Electronics Assembly • International Data Encryption Algorithm • International Digital Electronic Access Library

IDEAL Initiating - Diagnosing - Establishing - Acting - Leveraging • Intelligent Data Engineering and Automated Learning (conference) • Interactive Databased Environment for Accelerated Learning • Interactive Database Editor And Linker • Interactive Dialogues for Explanation And Learning • International Decade for the East African Lakes • International Digital Electronic Access Library

IDEALS Information Directory of European Automated Library Services • Intelligent Direct European Access to Library Systems

IDEAM Instituto de Estudios Ambientales • Instituto de Estudios del Ambiente

IDEAS Information Dissemination, Editing and Searching System • Initial Design and Evaluation Analysis System • Initiative to Develop Education through Astronomy and Space Science • Integrated Design and Engineering Automated System • Intelligence Data Elements Authorized Standards • Interactive Design and Evaluation of Advanced Spacecraft • Interactive Document Easy Access System • Interior Department's Electronic Acquisition System • International Database Engineering and Applications Symposium

IDECM Integrated Defensive Electronic Countermeasures

IDECS Image Discrimination, Enhancement and Combination System

IDEE Institut de Documentation et d'Études Européennes (France)

IDEEA Information and Data Exchange Experimental Activity

IDEF Integrated Data Exchange Facility • Integrated Definition

IDEI Institut de Développement des Entreprises Industrielles (France)

iDEN Integrated Digital Enhanced Network

Iden Identification • Identify

IDEN Inspecteur Départemental de l'Éducation Nationale (France) • Interactive Data Entry Network

Ident Identical • Identification • Identified • Identifier • Identify

IDEO Insertion - Dynamisation - Évaluation - Orientation

IDEP Industry Data Exchange Program • Institut pour le Développement de l'Éducation Permanente (France) • Interagency Data Exchange Program • International Data Exchange Program • Interservice Data Exchange Program

IDER Integrated Data Extraction and Reduction

IDES Image Detail Enhancement System • Image Dissector Echelle Spectrograph • Information and Data Exchange System • Interactive Data Entry System • Interactive Doppler Editing Software • International Diffuse EUV Spectrometer

IDEX Image-Data Exchange • Image Data Exploitation

idF in der Fassung

IDF Idiofa (Zaire) (location identifier) • IDL Data Format • Image Description File • Image Display Format • Industrial Diesel Fuel • Information Design Facility • Instantaneous Direction Finder • Institut für Diagnostikforschung • Integrated Data File • Intel Developer's Forum • Interactive Dialogue Facility • Interactive Distant Learning • Interface Definition File • Intermediate Data File • Intermediate Distributing Frame (*) • Intermediate Distribution Facility • Intermediate Distribution Frame • Internal Delay Factor • International Dairy Federation • International Development Fund • International DOI Foundation (USA) • Inverse Document Frequency (*) • Inverted Document Frequency • Isotropic Distribution Function

IDFA International Duty Free Association

IDFOP Institut pour le Développement de la Formation Permanente (France)

IDFT Inverse Discrete Fourier Transform

idFv in der Fassung von

IDG Ida Grove (USA-IA) (location identifier) • Institute of Dynamics of Geospheres • Integrated Drive Generator • Inter-Dialog Gap • International Data Group (USA-MA)

IDGC Institutional Dimensions of Global Change

IDGE Isothermal Dendritic Growth Experiment

IDGS Isotope-Dilution Gamma Spectrometry

IDH Interface Design Handbook

IDHA Images Distribuées Hétérogènes pour l'Astronomie (project)

IDHAF One-Dimensional Heisenberg Antiferromagnet

IDHCP Identification Heat Conduction Problem

IDHERA Institut Des Hautes Études et de Recherches Appliquées (France)

IDHS Information Data Handling System • Integrated Data Handling System • Intelligent Data Handling System

IDHT Instrument Data Handling and Transmission

IDHTS Instrument Data Handling and Transmission System

I+D+I Investigación + Desarollo + Innovación

IDI Image Display Interface • Imaging Doppler Interferometer • Improved Data Interchange • Indiana (USA-PA) (location identifier) • Information Dynamics Inc. (USA-DC) • Initial Domain Identifier • Institut de Droit International • Instrumentation Data Item • International Development Institute • Investigación - Desarollo - Innovación • Ion Dipole Interaction • Investissement Direct International

IDIAS Ice Data Integration and Analysis System

IDICT Instituto de Documentación e Información Científica y Técnica

IDID Integrated Dual Image Detector

IDIDAS Interactive Digital Image Display and Analysis System

IDIMS Interactive Data Integration and Management System • Interactive Digital Image Manipulation System

IDIOT Ice Depth Instrument Operator Transportable • Instrumentation Digital Online Transcriber

ID/IQ Indefinite Delivery/Indefinite Quantity

ID-IQ Indefinite Delivery - Indefinite Quantity

IDIQ Indefinite Delivery - Indefinite Quantity

IDIS Integrated Data and Information System • Integrated Data Information System • International Directory Inquiry System

IDIV Integer Divide

IDK I Don't Know • Indulkana (Australia-SA) (location identifier)

IDL Image Display Language • Indentured Drawing List • Information Description Language • Instruction Definition Language • Intelligent Database Language • Interactive Data Language • Interface Definition Language (*) • Interface Description Language • Interface Design Language • Interim Data Link • Intermediate-Density Lipoprotein • International Date Line • International Data Link • Interoperable Data Link

IDLC Integrated Digital Logic Circuit

IDLE Integrated Distributed Learning Environment

IDLH Immediately Dangerous to Life and Health

IDLI International Development Law Institute

IDLT Identification Light

IDM Identification of Dark Matter • Ignition Diagnostic Module • Impact Delay Module • Improved Data Modem

• Induced Dipole Moment • Information and Data Management • Information Dissemination Management • Information Distribution Management • Instrument Demonstration Module • Intelligent Database Machine • Intelligent Data Management • Intelligent Document Management • Interactive Dynamic Map • Inter-Departmental Meeting • Interim Data Modem • Ion Drift Meter

IDMEP Intrusion Detection Message Exchange Protocol

IDMM Intermediate and Depot Maintenance Manual

IDMP International Daylight Measurement Program

IDMRD Intermediate and Depot Requirements Document

IDMRS Intermediate and Depot Requirements Specification

IDMS Image Database Management System • Information and Data Management System • Ingest Debris Monitoring System • Integrated Database Management System • Integrated Data Management System • Intelligent Data Management System • Interactive Distributed Multimedia Systems (workshop) • Interim Deployable Maintenance System • Isotope Dilution Mass Spectrometry

IDMT Intelligence Digital Message Terminal

IDMTF Interim Data Management and Transfer Facility

IDN Indagen (Papua New Guinea) (location identifier) • In Dei Nomine • Indonesia (ISO country code) • Indonésie (code ISO de pays) • Initial Distribution Number • Integrated Data Network • Integrated Delivery Network • Integrated Digital Network • Intelligent Data Network • Interface Definition Notation • International Directory Network • Internetwork Datagram Protocol

IDNDR International Decade for Natural Disaster Reduction (UNO)

IDNX Integrated Digital Network Exchange

IDO Integrated Development Operation • Interdivisional Operation • Interface Definition Object • Santa Isabel do Morro (Brazil-TO) (location identifier)

IDOC Inner Diameter of Outer Conductor

IDOE International Decade for Ocean Exploration

IDOH Instrument Data Operations Handbook

IDOS Integrated Display Operating System

IDOT Instrument Digital Online Transcriber

IDP Image Data Processing • Image Display Paradigm • IMU Data Processing • IMU Data Processor • Incremental Data Processor • Incremental Development Package • Independence (USA-KS) (location identifier) • Individual Development Plan • Individual Development Program • Industrial Data Processing • Inertial Data Processor • Information Data Processing • Initial Delay Position • Initial Domain Part • Input Data Processor • Institute of Data Processing • Instrument Development Program • Integrated Data Package • Integrated Data Presentation • Integrated Data Processing • Integrated Data Processor • Integrated Development Programme • Integrated Display Processor • Integration Decision Paper • Intelligent Data Processing • Intelligent Data Product • Intensified Deep Penetration • Interactive Display Panel • Interface Demonstration Plan • Interface Development Plan • Interface Digital Processor • Interim Digital Processor • International Driving Permit • Internet Datagram Protocol • Internetwork Datagram Protocol • Interplanetary Dust Particle (*) • Interstellar Dust Particle

IDPAI International Directory of Professional Astronomical Institutions (merged with IDAAS into ASpScROW) (now StarGuides)

IDPC Integrated Data Processing Center • Integrated Data Protocol Controller

IDPD ISO Data Product Document

IDPF Integrated Digital Photogrammetric Facility

IDPGF Institutional Data Products Generation Facility

IDPI Institut de Développement de Projets Industriels (France) • Isophoromdiisocyanate

IDPM Institute of Data Processing Management (UK)

IDPMA International Data Processing Management Association

IDPMX Interim Digital Processor with Multiple Execution

IDPO Indefinite Delivery Purchase Order

IDPR Illinois Department of Professional Regulation (USA-IL) • Interdomain Policy Routing

IDPRS Interactive Data Processing and Research System

IDPS Image Data Production System • Instrument Data Processing System • INTEGRAL Data Processing System • Integrated Data Processing System • Integrated Design and Production System • Integrated Distributed Power System • Interim Deployable Process Station

IDPT Image Dissector Photomultiplier Tube

IDPU Information, Documentation and Publication Unit • Instrument Data Processing Unit • Integrated Data Processing Unit

idR in der Regel

IDR Image Data Resampler • Incremental Data Release • Incremental Design Review • Indore (India) (location identifier) • Industrial Data Reduction • Information Dissemination and Retrieval • Initial Design Review • Input Data Rate • Input Data Request • Inspection Discrepancy Report • Installation Data Record • Installation Design Review • Instantaneous Data Rate • Institute for Dream Research • Instrumental Data Release • Instrumentation Development Request • Intelligent Document Recognition • Interface Design Requirement • Interface Design Review • Interim Data Review • Interim Design Review • Interim Discrepancy Report • Intermediate Data Rate • Intermediate Data Record • Intermediate Design Review • Internal Design Review • Internal Development Report • International Drawing Rights • Internet Daily Report • Rupiah (Indonesia) (ISO currency code)

IDRA Interdivision Revision Authorization • Istituto per la Didattica e la Ricerca Applicata

IDRAC Institut De Recherche et d'Action Commerciale (France)

IDRC Improved Data Recording Capability • Indefinite Delivery Requirements Contract • International Development Research Centre

IDRD Information Definition Requirements Document • Internal Data Requirements Document

IDRF Integrated Diagnostic Research Facility

IDRIS Institut de Développement et des Ressources en Informatique Scientifique

IDRL Integrated Data Requirements List

IDRM Income-Driven Resource Model

IDRP Interdomain Routing Protocol

IDRS Integrated Data Retrieval System

IDR2 Second Instrumental Data Release

idS in diesem Sinne

IDS Image Display System • Image Dissection System • Image Dissector Scanner • Improved Data Set • Income Data Services • Index of Double Stars • Inertial Data System • Inertial Doppler System • Information and Documentation Service • Information and Decision System • Information Data System • Information Display System • Information Dissemination Server • Information Dissemination Service • Information Dissemination System • Information Distribution System • Informix Dynamic Server • Infrared Detection Set • Infrared Detection System • Infrared Discrimination System • Initiative de Défense Stratégique • Input-Data Strobe • Institut de Documentation Scientifique • Institute of Development Studies • Instructional Design Strategy • Instrument Data System • Instrument Development Section •

Integrated Database Storage • Integrated Data Storage • Integrated Data Store • Integrated Data System • Integrated Decision Support • Integrated Demonstration System • Integrated Design Specification • Integrated Design Support • Integrated Design System • Integrated Diagnostic System • Integrated Display Set • Integrated Display System • Integrated Dynamic System • Interactive Data System • Interactive Development System • Interactive Display System • Interdepartmental Committee on Space • Interdisciplinary Science • Interdisciplinary Scientist • Interdisciplinary Study • Interface Data Sheet • Interface Design Specification • Interface Design Standard • Interim Decay Storage • Intermediate Direct Support • Intermediate-Dispersion Spectrograph • Internal Directory System • International Data Service • International Development Strategy • International Digitization Strategy • International DORIS Service • Internet Database Service • Internet Distribution Service • Interstellar Dust System • Intrusion Detection System • Item Description Sheet

IDSA Interactive Digital Software Association (USA)

IDSC Information and Decision Support Centre

IDSCS International Decade of Solar Cycle Studies

IDSD Individual Data Storage Device • Institutional Data System Division (JSC) • Integrated Diagnostics Support Demonstration

IDSET Institut pour le Développement de la Science, de l'Éducation et de la Technologie (France)

IDSIPS Interactive Digital Satellite Image Processing System

IDSL ISDN Digital Subscriber Line

IDSM Intermediate Direct Support Maintenance

IDSN Integrated Decision Support Network

IDSO Interdivisional Sales Order

IDSR Initial Design Status Review • Inspecteur Départemental de Sécurité Routière • Interdivisional Support Request

IDSS Image Data System Simulation • Inquiry Delivery Schedule Status • Integrated Data Sounding System • Integrated Decision Support System • Integrated Design Support System • Integrated Diagnostic Support System • Intelligent Decision Support System • Interoperability Decision Support System

IDST Integrated Digital Services Terminal

IDT Idêntica • Identification and Tracking • Image Dissector Telescope • Image Dissector Tube • Indefinite-Delivery Type (contract) • Independent Development Tester • Industrial Design Technology • Information - Documentation - Transfert • Ingénieur Divisionnaire des Travaux • Instrument Dedicated Team • Instrument Definition Team • Instrument Development Team • Integrated Device Technology • Interactive Development Tool • Interactive Display Terminal • Interdigital Transducer • Interdisciplinary Team • Interface Design Tool • Interférométrie Différentielle des Tavelures • International Direct Telecommunications • International Discount Telecommunications • Interrupt Descriptor Table • Investigação e Desenvolvimento Tecnológico • Investigation Definition Team • Isodeclination Tile • Isodensitracer

IDTA Interdivisional Technical Agreement

IDTC IEEE Design Test Computers (periodical code) • Indefinite-Delivery-Type Contract

IDTL Independent Detector Testing Laboratory

IDTN Interim Data Transmission Network

IDTP Integrated Data Transmittal Package

IDTS Improved Doppler Tracking System • Instrumentation Data Test Station • Instrumentation Data Transmission System • Instrument Data Telemetry System • Integrated Data Transport System • Integrated Development Test Schedule • Integrated Digital Telecommunications Switch • Integrated Digital Telecommunications System

IDTV Improved Definition Television

IDU Industrial Development Unit • Industrial Documentation Unit • Information Development Unit • Instrument Development Unit • Integrated Display Unit • Interactive Display Unit • Interface Data Unit • Interface Demonstration Unit • Interim Demonstration Unit

IDUA Integrated Directory User Agent

IDUM ISO Data User's Manual

IDUP Independent Data Unit Protection

i. Durchschn. im Durchschnitt

IDV Indirect View • Individualism • Individualverkehr • Institut für Datenverarbeitung • Integrating Digital Voltmeter • Integrierte Datenverarbeitung • Interlibrational Derived Vehicle • Intra-Day Variability • Intra-Day Variable (object)

IDVC Integrated Data/Voice Channel

IDVS Integriertes Datenverarbeitungssystem

IDW Informations- und Dokumentationswissenschaft • Institut für Dokumentationswesen • Investigation-Derived Waste • IRAF Developers' Workshop

IDWA Interdivisional Work Authorization

IDWG Inter-Departmental Working Group • IRAF Developers' Workshop

idx Index (file extension)

Idx Index

IDX Index

IDY Identify • Île d'Yeu (France) (location identifier)

IDYL Image et Dynamique de la Litosphère • Imagerie et Dynamique de la Litosphère (*)

IDZ Informations- und Dokumentationszentrum • Inner Detection Zone

ID3 International Direct Distance Dialing

ie Id Est • im einzelnen • Interlingue (ISO language code) • Ireland

I&E Information & Education • Integration & Evaluation

IE Imaging Element • Independent Evaluation • Independent Evaluator • Index Error • Individual Equipment • Indo-European • Industrial Engineer • Industrial Engineering • In Excess • Information Element • Information Engineer • Information Engineering • Information Environment • Information Exchange • Information Expert • Information Extraction • Ingénieur d'Études • Initial Equipment • Inside Edge • Inspection and Enforcement • Instrumental Engineering • Instrumentation Engineer • Integrated Environment • Intelligence Économique • Interactive Engagement • Interface Engineer • Integrated Experiment • Integration Exercise • Intermediate Equipment • Intermediate Erection • International Exchange • Internet Explorer (Microsoft) • Ion Exchange • Ionization Energy • Ionospheric Explorer • Ireland (ISO country code) • Irlande (code ISO de pays)

IEA Image Enhancement Algorithm • Implicit Engineering Assumption • Industrial and Engineering Applications (conference) • Industrial Editors Association • Industrial Engineering Activity • Information Exchange Agreement • Information Exchange Annex • Institute of Energy Analysis • Institution of Engineers - Australia • Instituto de Estudos Avançados • Instytut Energii Atomowej (Poland) • Integrated Electronics Assembly • International Energy Agency • International Ergonomics Association • Ionization Equilibrium and Abundances

IEAEE Institution of Engineers, Australia - Electrical Engineering Transactions (periodical code)

IEAS Ice Elevation Altimeter System

IEAUST Institution of Engineers, Australia

IEB Independent Evaluation Briefing • Initial Evaluation Briefing • Instituto de Estudos Brasileiros (Brazil) • Instruction Engineering Bit

IEBS Institute of Environmental and Biological Sciences

IEC Imagery Exploitation Cell • Industrial Energy Conservation • Information - Education - Communication • Information Exchange Center • Institut d'Estudis Catalans (Spain) • Integrated Electronic Component • Integrated Electronic Control • Integrated Environmental Control • Integrated Equipment Component • Integration and Evaluation Center • Intelligence Économique et Concurrentielle • Intelligence - Electronics - Communications • Interexchange Carrier • Interim Executive Committee • Interlayer Exchange Coupling • International Electrotechnical Commission (*) • International Electrotechnical Committee • International Engineering Consortium • International Executive Committee • Ion Exchange Chromatography • Ionospheric Electron Content • Item Entry Control • Item Essentiality Code

IECA Independent Election Corporation of America

IECC International Express Carriers Conference

IECG Io - Europa - Ganymede - Callisto

IECM Induced Environment Contamination Monitor

IECO Inboard Engine Cutoff

IECP Interface Engineering Change Proposal

IECQ International Electrotechnical Common Quality

IECS Igloo Environment Control System • Institut d'Études Commerciales de Strasbourg (France) • Integrated Environmental Control System

IECTC International Electrotechnical Commission Technical Committee

IECU Improved Environmental Control Unit

IED Information Engineering Directorate • Infrastructure Engineering Division • Institut Européen de Design • Institution of Engineering Designers • Interacting Equipment Document • Intercambio Electrónico de Datos • Internet Engineering Design • Intrinsic Event Discrimination • Ionospheric Electron Density

IEDAL International Electronic Digital Access Library

IEDI Interned-based Electronic Data Interchange

IEDK Individual Equipment Decontamination Kit

IEDL IEEE Electron Device Letters (periodical code)

IEDM International Electron Devices Meeting • International Electronic Devices Meeting

IEDRM Interactive Electronic Diagnostic and Repair Manual

IEDS International Environment and Development Service

IEE Implicit Equivalence Principle • Induced Electrical Effect • Induced Electron Emission • Initial Environmental Evaluation • Institution of Electrical Engineers (UK) • Instituto de Electrotécnica e Energia

IEEE Institute of Electrical and Electronic Engineering • Institute of Electrical and Electronics Engineers (*)

IEEEN Institute of Electrical and Electronics Engineers Network (periodical code)

IEEEP Institute of Electrical and Electronics Engineers Proceedings (periodical code)

IEEES Institute of Electrical and Electronics Engineers Spectrum (periodical code)

IEEP Incapacitated Emergency Egress Practice • Institute for European Environment Policy • Institution of Electrical Engineers Proceedings (periodical code) • International Environment Education Program

IEERv Institution of Electrical Engineers Reviews (periodical code)

ief Inventory Exchange Format (file extension)

IEF Impressed Electron Field • Information Engineering Facility • Informatique Électronique Française (France) • Infrared Emission Feature • International Environment Facility • International Environment Forum • Inventory Exchange Format • Isoelectric Focusing

IEFP Institut Européen pour la Formation Professionnelle

IEFT Integrated Experiment Functional Test • Internet Engineering Task Force

IEG Independent Evaluation Group • Information Exchange Group • Institut d'Électronique de Grenoble (France) • Zielona Gora (Poland) (location identifier)

IEGI Institut Européen pour la Gestion de l'Information

IEGP Information Exchange Group Program

IEH International EUV/FUV Hitchhiker

IEHT Intensifier Electrical High Tension

IEI Industrial Engineering Institute • Institution of Engineers of Ireland

IEIC International Education Information Center

IEICE Institute of Electronics, Information and Communication Engineers • Institute of Electronics, Information and Communication Engineers Transactions (periodical code)

IEIWG International Element Interface Working Group

IEJ Iejima (Japan) (location identifier)

IELS Isotope Exciter Light Source

IELV Intermediate Expandable Launch Vehicle

IEM Impulsion Électromagnétique • Initial Evaluation Memorandum • Installation Equipment Management • Installation Equipment Manager • Institute of Experimental Meteorology • Instituto de Enseñanza Media • Integral Equation Model • Integrated Electronics Module • Interférence Électromagnétique • Interim Examination and Maintenance • Institution of Engineers - Malaysia • Invalid End of Message

IEMATS Improved Emergency Message Automatic Transmission System

IEMC Intrasystem Electromagnetic Compatibility

IEMCCAP Intrasystem Electromagnetic Compatibility Computer Analysis Program

IEMG International Experiment Management Group

IEMP Induced Electromagnetic Pulse • Intermittent Electromagnetic Pulse • Internal Electromagnetic Pulse • International Experiment Management Program

IEMS Installation Equipment Management System • Integrated Emergency Management System

IEMSI Interactive Electronic Mail Standard Identification

IEN Individualized Electronic Newspaper • Inspection de l'Éducation Nationale • Integrated Enterprise Network • Integrated Enterprise Node • Interactive Employment Network • Internet Engineering Note • Internet Experimental Note • Internet Experiment Note • Istituto Elettrotecnico Nazionale (Italy)

IENM Ingénieur de l'École Nationale de la Météorologie

IENR Investigation of the Earth Natural Resources

IEO Incoherent Electronic Oscillator • Independent Evaluation Office • Inner Earth Orbit • Instituto Español de Oceanografía (Spain) • Interior to Earth's Orbit (object) • International Exchange Office

IEOS International Earth Observing System

IEOSC International Earth Observation Satellite Committee

IEP Independent Electric Plant • Independent Evaluation Plan • Individual Educational Plan • Individualized Educational Programme • Information Exchange Plan • Information Exchange Program • Information Exchange Project • Inner Equipment Platform • Instantaneous Effective Photocathode • Institut de l'Éducation Permanente • Institut d'Études Politiques • Instituto Electrotécnico Português (Portugal) • Integrated Engineering Program • Intergovernmental Enterprise Panel • International Executive Program • Inverse Eigenvalue Problem • Isoelectric Point • Punt (Ireland) (ISO currency code)

IEPA Independent Electron Pair Approximation

IEPC Institut d'Études Préparatoires Commerciales • Interagency Emergency Planning Commission • ISDN Exchange Power Controller

IEPG Independent European Programme Group • Intercontinental Engineering Planning Group • Internet Engineering Planning Group

IEPRC International Electronic Publishing Research Centre

IEPS International Emergency Preference Scheme

IEQ Indoor Environment Quality

IER Import/Export Ratio • Independent Evaluation Report • Independent Evaluation Review • Individual Evaluation Report • Industrial Equipment Reserve • Information Exchange Requirement • Initial Evaluation Report • Installation Enhancement Release • Institute of Engineering Research • Institute of Environmental Research • Instrument Engineering Request • Interface Evaluation Report • Internationale Elektronische Rundschau (periodical code) • Interoperability Evaluation Report

IERE Institution of Electronic and Radio Engineers

IEREJ Institution of Electronic and Radio Engineers Journal (periodical code)

IERI Interagency Education Research Initiative

IERS International Earth Reference System • International Earth Rotation Service (*) • International Ellipsoid Reference System

ieS im engeren Sinne

I&ES Information & Electronics System

IES Illuminating Engineering Society • Illustrative Evaluation Scenario • Image Enhancement System • Imagery Exploitation Segment • Imagery Exploration System • Imaging Electron Spectrometer (Cluster) • Industrial Engineering Standard • Information and Electronics System • Information Exchange System • Information Extraction System • Initiation Économique et Sociale • Institute for Environmental Studies • Institute of Environmental Sciences • Instituto de Educación Secundaria • Instituto de Estudios Superiores (Spain) • Integral Energy Spectrum • Integral Error Squared • Intelligent Energy Saver • Interactive Exploration System • International Exchange Service • Intrinsic Electrical Strength • Inverted Echo Sounder • Ion and Electron Sensor • Ion and Electron Spectrometer • Ion Electron Spectrometer • Ion Energy Selector • Ion Engine Simulator • Ion Engine System • IRAS Explanatory Supplement • Isolated Emission Source

IEASANZ Illuminating Engineering Society of Australia and New Zealand

IESC Information Exchange Steering Committee

IESD Instrumentation and Electronic Systems Division

IESE Instituto de Estudios Superiores de la Empresa (Spain)

IESG Internet Engineering Steering Group

IESI Institut Européen des Sciences Informatiques

IESJ Institute of Electrostatics Japan

IESM Institut d'Études Supérieures du Marketing (France)

IESNA Illuminating Engineering Society of North America (USA)

IESP Informatique et Électronique en Sciences Physiques • Integrated Electronic Signal Processor

IESS Imagery Exploitation Support System • Internet Entrepreneurs Support Server

IESSD Institute for Environmental Studies for Sustainable Development

IEST Institut Européen des Standards de Télécommunications

IESTO Institut d'Études Supérieures des Techniques d'Organisation

IET Impact Excited Transmitter • Independent Evaluation Team • Initial Engine Test • Initial Entry Training • Inspecteur de l'Enseignement Technique • Institut d'Électrotechnique de Toulouse (France) • Integrated Engineering Team • Interactive Evaluation Tool • Interest Equalization Tax • International Evaluation Team • Irregular Erection Time

IETA Ingénieur des Études et Techniques d'Armement

IETE Institution of Electronics Telecommunication Engineers

IETF Indian Engineering Trade Fair (India) • Internet Engineering Task Force

IETM Interactive Electronic Technical Manual

IETS Inelastic Electron Tunneling Spectroscopy

IEU Input Expansion Unit • Instrument Electronics Unit • Interface Electronics Unit • Ion Exchange Unit

IEV Integrated Experimental Verion • International Electrotechnical Vocabulary • Kiev Zhulyany (Ukraine) (location identifier)

IEVD Integrated Electronic Vertical Display

I&EW Intelligence & Electronic Warfare

I/EW Intelligence/Electronic Warfare

IEW In Eén Woord • Information Engineering Workbench • Intelligence and Electronic Warfare • International Education Network • International Energy Workshop

I&EWCS Intelligence & Electronic Warfare Control System

I/EWCS Intelligence and Electronic Warfare Control System

IEWCS Intelligence and Electronic Warfare Common System • Intelligence and Electronic Warfare Control System

IEWG Information Environment Working Group

IEWS IEE Electromagnetic Waves Series (periodical code)

IEX Inter-Exchange

IExp Institute of Electrical and Electronics Engineers Expert (periodical code)

IEY International Education Year (1970)

IEZ Intense Erection Zone

if Ipse Fecit

I/F Interface • Intermediate Frequency

IF Image Fusion • Impact Factor • Importance Factor • Incomplete Factorization • Individual Focus • Industrial Fund • Inertial Fusion • Information Feedback • Information Fetch • Information Filtering • Information Flow • Inhibiting Factor • Initialization File • Inlet Frame • Inside Frosted • Institut de France • Instrument Flying • Integration Facility • Interactive Facsimile • Interface • Interference Filter • Intermediate Fix • Intermediate Focus • Intermediate Forward • Intermediate Frequency • Internal Fault • Interrupt Flag • Inverted File • Investissement Formation • Ionization Front

IfA Institute for Astronomy

IFA Igniter-Fuel Assembly • Immunofluorescence Assay • Independent factor Analysis • Independent Financial Adviser • Industry Film Association • In-Flight Abort • In-Flight Analysis • In-Flight Anomaly • Information Flow Analysis • Information Flow in Astronomy (diagram) • Institut de Formation à l'Animation • Institute For Astronomy • Institute for Atomic Physics • Institut for Fysik og Astronomi • Instrumented Fuel Assembly • Integrated Feed Antenna • Integrated File Adapter • Integrated Financial Adviser • Intelligent Filtering Agent • Intensive Flux Array • Interagency Funding Agreement • Interface Functional Analysis • Intermediate Frequency Amplifier • Intermediate Frequency Attenuator • International Financial Advisor • Ionization Front Agent • Iowa Falls (USA-IA) (location identifier) • Istituto di Fisica dell'Atmosfera

IFAA International Federation of Advertising Agencies

IFAC Institut de Formation d'Animateur de Collectivités (France) • International Federation of Automatic Control

IFAD International Fund for Agricultural Development

IFAE Interamerican Federation for Adult Education

IFAEPE Institut Fédéral pour l'Aménagement, l'Épuration et la Protection des Eaux (Switzerland)

IfAG Institut für Angewandte Geodäsie

IFAG Institut de Formation d'Assistant de Gestion (France)

IFAID Institut de Formation et d'Appui aux Initiatives de Développement (France)

IFALPA International Federation of Airline Pilots' Associations

IFAM Inverted File Access Method

IFAORS Institute for Atmospheric Optics and Remote Sensing

IFAPA International Foundation of Airline Passengers Associations

IFARE Institut Franco-Allemand de Recherches sur l'Environnement

IFAS Idaho Falls Astronomical Society (USA-ID) • Industrial Fund Accounting System • Institute of Food and Agricultural Sciences

IFASC Integrated Functions Assessment Steering Committee

IFAST Integral Field Array Spectrography in (real) Time • Integrated Facility for Avionic Systems Test • Integration Facility for Avionic Systems Testing

IFATCA International Federation of Air Traffic Controllers Associations

IFATT Integrated Feature And Terrain Triangulation (algorithm)

IFB Institut de Formation des Bibliothécaires • Invitation For Bids

IFBel Institut Fiziki Nauchnaia Sessiia Minsk Belarussian SSR Zhurnal (periodical code)

IFC Industry Foundation Class • In-Flight Calibration • Inside Front Cover • Institut de Formation Continue • Instrument Flight Center • Intelligent Forms Language • Intensive Field Campaign • Interface Clear • Interface Controller • International Finance Corporation • Internet Foundation Class • Item For Collection

IFCA Instituto de Física de Cantabria (Spain) • Internal Functional Configuration Audit • International Committee on Future Accelerators

IFCAI International Federation of Commercial Arbitration Institutions • Istituto di Fisica Cosmica ed Applicazioni dell'Informatica (Italy)

IFCE In-Flight Contamination Experiment • Institut de Formation au Commerce Extérieur

IFCES Institut de Formation aux Carrières Éducatives et Sociales

IFCF Intermediate Frequency Crystal Filter

IFCG Interface Control Group

IFCIS International Foundation on Cooperative Information Systems

IFCL Interfacility Communication Link

IFCM Integrated Flight Control Module

IFCN Interfacility Communication Network

IFCO Inspecteur de la Formation Continue et de l'Orientation

IFCP Interfacility Communications Processor

IFCPAR Indo-French Centre for the Promotion of Advanced Research

IFCPLV Institut Français de la Communication et de la Publicité sur le Lieu de Vente (France)

IFCS In-Flight Checking System • International Federation for Computer Sciences • International Federation of Classification Societies • Integrated Fire Control System • Integrated Flight Control System • Irish Federation of Computer Societies

IFCTR Istituto per Ricerche in Fisica Cosmica e Tecnologie Relative (Italy)

IFCU Interface Control Unit • International Federation of Catholic Universities

IFCV Institut de Formation du Cadre de Vie

IfD Institut für Dokumentation • Institut für Dokumentationswesen

IFD Image File Directory • Industrial Facilities Discharge • Inflatable Floating Device • In-Flight Deployment • Information Flow Diagram • Initial Fill Date • Instantaneous Frequency Discriminator • Integrated Feasibility Demonstration • Interim Feasibility Demonstration • Invariant Factor Decomposition

IFDA International Federation for Data Processing

IFDAPS Integrated Flight Data Processing System

IFDC Improved Field Data Collector • International Fertilizer Development Center

IFDL Interfacility Data Link

IFDM Integrated Finite-Difference Method

IFDO International Federation for Data Organizations

IFDR Incremental Final Design Review

IFDS Inertial Flight Data System • Interfacility Data System

IfE Institut für Erdmessung (Germany)

IFE Industrially-Furnished Equipment • Industry-Furnished Equipment (*) • Inertial Fusion Energy • In-Flight Emergency • In-Flight Entertainment • Initial Funding Estimate • Inner Front End • Institut für Fernerkundung (Germany) • Instituto Federal Electoral • Instrument Flight Examiner • Intelligent Front End • Introduced Foreign Element • Isoelectric Focusing Experiment

IFEA In-Flight Emergency Assistance • Initial Front-End Analysis • Integrated Furnace Experiment Assembly

IFEC Institut Français d'Enseignement par Correspondance (France)

IFEL Induction Free-Electron Laser

IFEM Integrated Force and Execution Management

IfEN Institut für Erdmessung und Navigation (Germany)

IFEN Institut Français de l'Environnement (France) • Intercompany File Exchange Network

IFEOS International Forum for Earth Observations from Space

IFEP Institut de Formation et d'Éducation Permanente

IFEPP Institut de Formation et d'Études Psychosociologiques et Pédagogique (France)

IFERS International Flat Earth Research Society

IFEU Institut für Energie- unf Umweltforschung GmbH (Germany)

iff If and Only If • Image File Format (file extension) • Interchange File Format (file extension)

IFF Identification Friend or Foe • Iffley (Australia-QLD) (location identifier) • Image File Format • Independent Fourier Frequency • Induced Fluid Flow • Institute For the Future • Interchangeable File Format • Interchange File Format • Intergovernmental Forum on Forests • Ionized Flow Field

IFFA Interactive Flash Food Analyzer • International Federation of Film Archives

IFFC Integrated Fire and Flight Control

IFFET Institut Français du Film Éducatif et Technique (France)

IFFN Identification Friend-Foe-Neutral

IFFP Institut Français de Formation Permanente (France)

IFFRO International Federation of Reproduction Rights Organizations

IFFS International Federation of Film Societies

IFFT Inverse Fast Fourier Transform

IfG Institut für Gewerbezentren

IFG Incoming Fax Gateway • Information File Generator • Intellectual Freedom Group • Interframe Gap • Investment For Growth

IFGL Initial File Generation Language

IFGU In-Flight Guidance Update

IfH Institut für Handelsforschung

IFH Image File Header • In-Flight Helium

IFI Independent Fourier Interval • International Film Institute • International Fund for Ireland

IFIA International Federation of Inventors' Associations

IFIAS International Federation of Institutes for Advanced Study (Canada) • International Federation of Institutes of Advanced Studies

IFICS Inflight Interceptor Communications System

IFID International Federation for Information and Documentation

IFIG Intellectual Freedom Interest Group

IFIM International Flight Information Manual

IFIN Institutul de Fizica si Inginerie Nucleara (Romania)

IFIP Inofficieller Mitarbeiter • Institut de Formation et d'Information Permanente • International Federation for Information Processing (*) • International Federation of Information Processors • Inverse First Ionization Potential

IFIPS International Federation of Information Processing Societies (now IFIP)

IFIR Instituto de Física de Rosario (Argentina-SF)

IFIRS Integral Field Infrared Spectrograph • Integral Field Infrared Spectrometer (*)

IFIT International Forest Investigation Team

IFJ International Federation of Journalists • Isafjördur (Iceland) (location identifier)

IFK Integral of First Kind

IFL Implementation of Functional Languages (workshop) • In a Former Life • Innisfail (Australia-QLD) (location identifier) • Integrated Facility for Linux • Intelligent Forms Language • International Frequency List

IFLA International Federation of Library Associations

IFLC Initial Force Level Control • International Frequency List Committee

IFLCC Initial Force Level Control Capability

IFLCS Initial Force Level Control System

IFLOT Intermediate Focal Length Optical Tracker

IFLOTS Intermediate Focal Length Optical Tracker System

IFLOWS Integrated Flood Observation and Warning System

IFLTT Intermediate Focal Length Tracking Telescope

IFM Improved Frequency Modulation • Incidental Frequency Modulation • In-Flight Maintenance • Instantaneous Frequency Measurement • Institut de Formation des Maîtres • Integrated Financial Management • Integrated Force Management • Integrated Formal Methods (conference) • Intelligent Flow Management • Intermediate Frame Memory • Intermediate Frequency Modulation • Io Flow Model

IFMA Information Functional Mission Area • Institut Français de Mécanique Avancée (France)

IFMBE International Federation for Medical and Biological Engineering

IFMF Initial-Final Mass Function

IFMIS Integrated Facilities Management Information System

IFMMS Integrated Facility Maintenance Management System

IFMO Institut de Formation à la Maîtrise d'Ouvrage

IFMP IP Flow Management Protocol • Ipsilon Flow Management Protocol

IFMR Initial-Final Mass Relationship

IFMS Impact Force Measuring System • In-Flight Management System • Integrated Financial Management System • Integrated Flexible Manufacturing System

IFMU Integrated Flight Management Unit

IFN Ice-Forming Nucleus • Ice-Freezing Nucleus • Info-Fuzzy Network • Institut Français de Navigation (France) • Interferon • Isfahan (Iran) (location identifier) • Item For Negotiation

IFNA International Federation of Non-linear Analysts

IFNP International Federation of Newspaper Publishers

IFO Identified Flying Object • Improved Fiber Optics • Information For Officers • Intensive Field Observation • Intensive Field Operation • International Field Office • Ivano-Frankovsk (Ukraine) (location identifier)

IFOAM International Federation of Organic Agriculture Movements

IFOBS International Forum on Open Bibliographic Systems

IFOD Ingénierie Financière des Produits Dérivés

IFODA Institut Français d'Organisation Documentaire et Administrative (France)

IFOG Interferometric Fiberoptic Gyro

IFOP Institut Français d'Opinion Publique (France) • Instrument Flight Operation Plan • Instrument Flight Operation Procedure

IFOR Implementation Force • Interactive Fortran

IFORS International Federation of Operational Research Societies

IFOT In-Flight Operations and Training

IFOU Instrument Flight Operations Understanding

IFOV Individual Field Of View • Instantaneous Field Of View • Instrument Field Of View

IfP Institut für Planetologie (Germany)

IFP Bullhead City (USA-AZ) (location identifier) • In-Flight Performance • Information Flow Process • Institut de Formation Pédagogique • Institut Français du Pétrole (France) • Instruction Fetch Pipeline • Intensive Field Program • Interface Processor • Interféromètre Fabry-Pérot • Interferometry Panel • Intermediate Frequency Patch • Invitation For Proposals • Laughlin Bullhead (USA-AZ) (location identifier)

IFPA Industrial Film Producers Association (USA) • Information Film Producers of America • Institute for Foreign Policy Analysis (USA) • Instituto de Física de Porto Alegre (Brazil-RS) • Integrated Focal Plane Assembly

IFPC International Fund for the Promotion of Culture (UNESCO)

IFPI Imaging Fabry-Pérot Interferometer • International Federation of the Phonographic Industry

IFPM In-Flight Performance Monitor • In-Flight Performance Monitoring

I&FPP International & Foreign Parcel Post

IFPP Industrial Facilities Protection Program • Information For Proposal Preparation • Institut de Formation Pédagogique et Psychosociologique • Instruction For Proposal Preparation • International and Foreign Parcel Post • International Federation of the Periodical Press

IFPS Imaging Fabry-Pérot Spectrograph • Imaging Fabry-Pérot Spectrometer • International Federation of Palynological Societies • Intra-Formation Positioning System

IFPSI Institut Français de Presse et des Sciences de l'Information (France)

IFPU Integrated Flight Processing Unit

IFPUG International Function Point Users Group

IFQ Invitation For Quotations • Invitation For Quotes

IFQT Informal Qualification Test • Initial Formal Qualification Test

IFR Image-to-Frame Ratio • Increased Frequency Reporting • Increasing Failure Rate • In-Flight Refueling • Infrared Filter Radiometer • Initial Flight Requirement • Instantaneous Frequency-measuring Receiver • Institut Fédératif de Recherche • Instrument Flight Recovery • Instrument Flight Rules • Instrument Flying Regulation • Integral Fast Reactor • Interface Repository

IFRB Intermediate Frequency Review Board • International Frequency Registration Board

IFRC International Federation of Red Cross • International Fusion Research Council

IFREMER Institut Français de Recherche pour l'Exploitation de la Mer (France) (*) • Institut Français de Recherche pour l'Exploitation des Mers (France)

IFRI Institut Français des Relations Internationales (France)

IFRR Institut für Raumflug- und Reaktortechnik (Germany)

IFRRO International Federation of Reproduction Rights Organizations

IFRT Intellectual Freedom Round Table

IFRTP Institut Français pour la Recherche et la Technologie Polaires (France)

IFRU In-Flight Replaceable Unit • In-Flight Replacement Unit • Interference Frequency Rejection Unit

IfS Institut für Sozialdienste

IFS Impact and Flotation System • In-Flight Safety • Information Flow Structure • Information Flow System • Installable File System • Institute for Fusion Studies (USA-TX) • Institute for Future Studies • Institute of Fundamental Studies (Sri Lanka) • Integral Field Spectrograph • Integral Field Spectrography • Integral Field Spectroscopic • Integral Field Spectroscopy • Integrated Facilities System • Integrated Flight Simulator • Integrated Forecasting System • Integrated Forest Study • Interchange File Separator • Interface Specification • Interface System • Interframe Space • Integrated Field Spectrograph • Integrated Flight Simulation • Integrated Flight Simulator • Integrated Forecasting System • Internal Field Separator • International Federation of Surveyors • International Foundation for Science • Ionospheric Forward Scatter • Iterated Function System

IFSA International Fuzzy Systems Association

IfSAR Interferometric Synthetic Aperture Radar

IFSARE Interferometric Synthetic Aperture Radar for Elevation

IFSC Instituto de Física de São Carlos (Brazil-SP)

IFSD In-Flight Shutdown

IFSE International Federation of Science Editors (*) • International Federation of Scientific Editors

IFSEA International Federation of Scientific Editors' Association

IFSEG International Federation of Societies of Economic Geologists

IFSEM International Federation of Societies for Electron Microscopy

IFSG Industrial Forecasting Support Group

IFSI Istituto di Fisica dello Spazio Interplanetario (Italy)

IFSM Integrated Facilities System Modernization

IFSMgr Installable File System Manager

IFSR International Federation for Systems Research

IFSS Index of Federal Specifications and Standards (USA) • Inspection Field Support System • Instrumentation and Flight Safety System • International Flight Service Station • IRASS Faint Source Survey

IFSTD Interim Fund for Science and Technology for Development

IFT In-Flight Test • Ingénieur Divisionnaire des Travaux • Input Frequency Tolerance • Instantaneous Fourier Transform • Instituto de Física Teórica • Interactive Flight Technology • Interface Tool • Intermediate Flight Time • Intermediate Frequency Transformer • International Federation of Translators • International Frequency Table • Inverse Fourier Transform • Io Flux Tube • Ion Focusing Technique • Irregular Flares of Temper

IFTA International Federation of Teachers' Association • International Federation of Television Archives • International Federation of Travel Agencies • International Free Trade Area

IFTAD Initial and Final Terminal Arrival Date

IFTC International Film and Television Council

IFTD Interface Terminal Distributor

IFTE Integrated Family of Test Equipment • Intermediate Forward Test Equipment

IFTF Institute For The Future

IFTM In-Flight Test and Maintenance • In-Flight Test Maintenance

IFToMM International Federation for the Theory of Machines and Mechanisms

IFTPP International Federation of the Technical and Periodical Press

IFTR Ion Flow Tube Reactor

IFTS Imaging Fourier Transform Spectrograph • Imaging Fourier Transform Spectrometer • Indemnités Forfaitaires pour Travaux Supplémentaires • In-Flight Test System • Institut de Formations Technologiques Supérieures

IFTU International Federation of Teachers' Unions • International Federation of Trade Unions

IFTV Integrated Flight Test Vehicle

I/FU Interface Unit

IFU Imprimé Fiscal Unique • Institut Für Umweltforschung • Institut Für Umweltwissenschaften (Germany) • Integral Field Unit • Integrated Field Unit • Interface Unit

IFUG Instituto de Física de la Universidad de Guanajuato (Mexico)

IFV Institut de Force de Vente

IFVV Internationale Feste Virtuelle Verbindung

IFW Information Framework

IFWEA International Federation of Workers' Education Association

IFWG Interface Working Group

IFYCSEM Integrated First-Year Curriculum in Science, Engineering, and Mathematics

IFYGL International Field Year for the Great Lakes

IFYS International Forum of Young Scientists

Ig Ignition • Ignitor

IG Igloo • Ignitor • Image Generator • Indo-Germanic • Industrial Group • Industriegewerkschaft • Inertial Guidance • Information Gateway • Infrastructure Group • Ingénieur Général • Inlet Group • Inner Gimbal • Inspecteur Général • Inspection Générale • Inspector General • Institute of Geophysics • Instruction Guide • Instructor Guide • Instrumentation Group • Instrument Ground • Intelligence Group • Interacting Galaxies • Interagency Group • Interessengemeinschaft • Intergalactic • Internal Guidance • Inverse Gain • Ionization Gauge

IGA Inagua (Bahamas) (location identifier) • Infrared Galaxy Atlas • Inner Gimbal Angle • Inspecteur Général de l'Administration • Instituto de Geofísica y Astronomía • Integrated Graphics Array • Integrating Gyro Accelerometer • Interdivisional Group on Accelerators • Intergimbal Assembly • Inter-Governmental Agreement • International Geographic Association • International Geothermal Association • Ion Gun Assembly

IGAC Instituto Geografico Augustin Codazzi (Colombia) • Interessengemeinschaft Astronomie Crimmitschau (Germany) • International Global Atmospheric Chemistry (program) (*)

IGACEM Inspection Générale de l'Aviation Civile et de la Météorologie

IGACS Integrated Guidance And Control System

IGACSM Inter-Governmental Advisory Committee on Surveying and Mapping

IGAD Intergovernmental Authority on Development

IGADD Inter-Governmental Authority on Drought and Development

IGAE Inter-Governmental Agreement on the Environment

IGAEN Inspecteur Général de l'Administration de l'Éducation Nationale

IGAFS Issledovaniia Geomagnetizmu Aeronomii i Fizike Solntsa (periodical code)

IGAM Institut für Geophysik, Astrophysik und Meteorologie

IGAP International Global Aerosol Program

IGARSS International Geoscience and Remote Sensing Society • International Geoscience and Remote Sensing Symposium

IGAX Inner Gimbal Axis

IGB Ingeniero Jacobacci (Argentina-RN) (location identifier) • Internationaler Genossenschaftsbund • International Gravity Bureau

IGBMC Institut de Génétique et de Biologie Moléculaire et Cellulaire

IGBP International Geosphere-Biosphere Program

IGBT Insulated Gate Bipolar Transistor

IGC Independent Government Cost • Institut du Génie Chimique • Institute for Global Communications (USA-CA) • Institutional Grants Committee • Integrated Graphical Controller • Integrated Graphics Controller • Intelligent Graphics Controller • Intergovernmental Committee • Intergovernmental Conference • Intergovernmental Council • International Geological Congress • International Geophysical Committee • International Geophysical Cooperation • International Gravity Commission • Internet Governance Committee • Inverse Gas Chromatography • Ion Gun Collector

IGCC Integrated Gasification Combined Cycle • Intergovernmental Copyright Committee

IGCE Independent Government Cost Estimate

IGCP Instituto Geografico e Cadastral do Portugal • International Geological Correlation Program

IGCRE Institute for Global Change, Research and Education (USA)

IGCS Integrated Guidance and Control System

IGCY International Geophysical Cooperation Year (1958-1959)

IGD Institut für Graphische Datenverarbeitung • Integrated Genomic Database • Interaction Graphics Display • Interactive Graphical Document • Interactive Graphical Documentation • Interessengemeinschaft Dokumentation • Inverse Gaussian Distribution

IGDB Instrument Group Database

IGDC Interagency Geographic Data Committee

IGDR Interim Geophysical Data Record • Intermediate Geophysical Data Record

IGDS Image Generation and Display System • Interactive Graphics Design System • Iodine Generating and Dispensing System

IGE Iguela (Gambia) (location identifier) • Image Generating Equipment • In-Ground Effect • Instrumentation Graphics Environment • Instrumentation Ground Equipment • International Geophysical Extension

IGEB Interagency GPS Executive Board

IGeC International Geoid Commission

IGEC International Gravitational Event Collaboration

IGEN Inspecteur Général de l'Éducation Nationale (France)

IGERT Integrative Graduate Education and Research Training • Interdisciplinary Graduate Education and Research Training (program)

IGeS International Geoid Service

IGES Initial Geometric Exchange Specification • Initial Graphical Exchange Specification • Initial Graphics Exchange Specification (*) • Initial Graphics Exchange Standard • Institute for Geography and Earth Studies • Institute of Global Environment and Society (USA-MD) • Interagency Government Executive Board • Intermediate Graphics Exchange Standard • International Graphics Exchange System

IGEX International GLONASS Experiment

IGF Inert Gas Fusion • Inset Graphics Format • Insulin-like Growth Factor • International Graphical Federation

IGFA International Group of Funding Agencies

IGFES Interactive Graphics Finite Element System

IGFET Insulated-Gate Field-Effect Transistor (*) • Isolated Gate Field-Effect Transistor

IGFM Internal Gamma Flux Monitor

IGFN Italian GPS Fiducial Network

IGFOV Instantaneous Geometric Field Of View

IGG Igiugig (USA-AK) (location identifier)

IGGOS Integrated Global Geodetic Observing System

IGH Ingham (Australia-QLD) (location identifier) • Intelligent Gas Handling • Intervening Galaxies Hypothesis

IGHC International Group Handling Council

IGI Imaging Grism Instrument • Informatique de Gestion pour l'Imprimerie • Instrument Ground Instructor • Interactive Graphics Interpreter • International Gemnological Institute (Belgium)

IGIA Industrial Guest Investigator Agreement

IGIF International Geographic Information Foundation

IGIP International Gesellschaft für Ingenieurpädagogik (Austria)

IGIPAC Integrated GPS Inertial Position and Attitude Control

IGIPAS Interagency Group on International Programs in Atmospheric Sciences

IGIS Integrated Geographical Information System • Intelligent Geographical Information System

IGL Information Grouping Logic • Integrated Galaxy Light • Interactive Graphics Library • Ionized Gas Laser • Izmir Cigli (Turkey) (location identifier)

IGLB Interactive Gravitational Lensing Bibliography

IGLD International Great Lakes Datum

IGLO Informal Group of Liaison Offices

IGM Inertial Guidance Mode • Inspection Générale de la Métrologie (Belgium) • Instituto Geográfico Militar • Interactive Guidance Mode • Intergalactic Medium • Inter-Governmental Meeting • Intragroup Medium • Iterative Guidance Mode • Kingman (USA-AZ) (location identifier)

IGMA Instituto Geográfico Militar de Argentina

IGMF Intergalactic Magnetic Field

IGMIRS Inspector General Management Information and Reporting System

IGMP Internet Group Management Protocol (*) • Internet Group Membership Protocol • Internet Group Message Protocol • Internet Group Multicast Protocol

ign Ignotus

Ign Ignite • Ignition • Ignotus

IGN IBM Global Network • Igniter • Ignition • Iligan - Maria Cristina (Philippines) (location identifier) • Institut Géographique National • Instituto Geográfico Nacional

IGN&C Integrated Guidance, Navigation & Control

IGNC Integrated Guidance, Navigation and Control

IGNOU Indira Gandhi National Open University (India)

IGNRQ Ignition Request

IGNS Institute of Geological and Nuclear Sciences Inc. (New Zealand)

IGO Chigorodo (Colombia) (location identifier) • Inter-Governmental Organization • International Gamsberg Observatory (Namibia) • International Gemini Observatory • International Gemini Office • Stichting voor Isotopen-Geologisch Onderzoek

IGOE Inputs - Guides - Outputs - Enablers

IGOR Intelligent Graphics Object Recognition • Intercept Ground Optical Recorder

IGORE Integrated Gamma-Ray Observatory Reduction Environment • Interactive GRO/OSSE (data) Reduction Environment

IGORTT Intercept Ground Optical Recorder Tracking Telescope

IGOS Inspecteur Général de l'Organisation Scolaire (France) • Integrated Global Observing Strategy • Interactive Graphic Orbit System

IGOSS Industry/Government Open Systems Specification • Integrated Global Ocean Services System • Integrated Global Ocean Station System (*)

IGP Igneous and Geothermal Processes • Inertial Guidance Package • Inertial Guidance Platform • Information Generation and Publishing • Institut für Geodäsie und Photogrammetrie • Institute of General Physics • Instituto Geofísico del Perú • Intelligent Gateway Processor • Interior Gateway Protocol • International Gemini Project • International Geodynamics Project • Internet Gate Protocol • Internet Gateway Protocol • Internet Governance Project (ACM) • Ion Getter Pumping

IGPA Índice General de Precio de las Acciones

IGPCF Inert Gas Phase Condensation Facility

IGPD Interdivisional Group on Physics for Development

IGPF Internationale Gesellschaft für Photogrammetrie und Fernerkundung

IGPO International Gemini Project Office • International GEWEX Project Office

IGPP Institute of Geophysics and Planetary Physics (USA-CA) • Interactive Graphics Packaging System

IGPRAD Inter-Governmental Panel on Radioactive waste

IGPS Ignition Power System • Intelligent Gateway Processing System • International Galactic Plane Survey

IGQMS Ion-Guide Quadrupole Mass Separation (technique) • Ion-Guide Quadrupole Mass Spectrometer

IGR Igreja • Iguassu Falls (Argentina-MI) (location identifier) • Iguazu Cataratas (Argentina-MI) (location identifier) • Index to Government Regulation (USA) • Institute for Gravitational Research

IGRAC International Groundwater Assessment Centre

IGRAP Institut Gassendi pour la Recherche Astronomique en Provence (France)

IGRF International Geomagnetic Reference Field (*) • International Geomagnetic Reference File

IGRP Interior Gateway Routing Protocol (*) • Internet Gateway Routing Protocol

IGRSS International Geoscience and Remote Sensing Society

IGS IBM Global Servies • Improved Gray Scale • Individual Generation Site • Inertial Guidance System • Inspection Générale des Services • Instrumentation Ground System • Instrument Guidance System • Integrated Graphics System • Integrated Ground Segment • Interactive Graphics System • Interaktives Graphisches System • Interchange Group Separator • Interconnection Ground Subnetwork • Intergovernmental Support • Intermediate General Support • International Geodynamics Service • International Geophysics Series (periodical code) • International Geosynthetics Society • International Glaciological Society • International GPS Service • International Ground Station • Internet Go Server

IGSC International Group for Scientific Coordination

IGSCB International GPS Service Central Bureau

IGSE Instrument Ground Support Equipment

IGSM Interim Ground Station Module

IGSMA Inertial Guidance System Maintenance Area

IGSN International Gravity Standardization Network

IGSPC Imaging Gas Scintillation Proportional Counter

IG StaR Interessen Gemeinschaft Sternfreunde an der Ruhr (Germany)

Igt Ingot

IGT Initial Glider Training • Institute of Gas Technology • Integrated Government Testing • Integrated Ground Test • Interactive Graphics Terminal

IGTES ISOCAM Guaranteed Time Extragalactic Survey

IGU Iguassu Falls (Brazil-PR) (location identifier) • International Gas Union • International Geographical Union • Saltos do Iguaçu (Brazil) (location identifier)

IGV Inlet Guide Vane

IGVS Inspecteur Général de la Vie Scolaire (France)

IGW Increased Gross Weight • Integral Gateway • Integrated Gateway • Internal Gateway • Internal Gravity Wave

IGWMC International Ground Water Modeling Center

IGY International Geophysical Year (1957-1958)

Igzu Iguazu (Argentina-MI)

I/H In House

IH Iacet Hic • Information Handling • Information Hub • Insect Habitat • Installation Handbook • Integrated Hardware • Intercept Heading • Interferometric Hartmann • Internet Hub • Interrupt Handler

iha In Het Algemeen

IHA Indian Housing Authority (India) • Information Handling in Astronomy • Institut Herzberg d'Astrophysique (Canada) • Interface Hazard Analysis • Interim Housing Allowance • Niihama (Japan) (location identifier)

IHADSS Integrated Helmet and Display Sight System

IHAP Image Handling And Processing

IHAS Integrated Hazard Avoidance System • Integrated Helmet Assembly System

ihb In Het Bijzonder

IHB In-House Bid • Intermediate Horizontal Branxh • International Hydrographic Bureau

IHC Information Handling Committee • Intellectually Handicapped Child • Ionic Heated Cathode

IHCI Internet Hardware Configuration Item • Internetworking Hardware Communications Interface

IHCL International Conference on Historical Linguistics

IHCP Inverse Heat Conduction Problem

IHD Integrated Help Desk • International Hydrological Decade • Ischaemic Heart Disease

IHDA Illinois Housing Department Authority (USA-IL)

IHDP International Human Dimension Program

IHDS Infrared High-Dispersion Spectrograph

IHE Infrastructural - Hydraulic - Environmental • Institut pour l'Hygiène et l'Épidémiologie • Instituut voor Hygiëne en Epidemiologie • Intermediate High Energy • Internal Hydrogen Embrittlement

IHEA Irish Higher Education Authority

IHEDN Institut des Hautes Études de Défense Nationale (France)

IHEG Institute of Hydrogeology and Engineering Geology

IHEP Institute of High-Energy Physics

IHERv Ishikawajima Harima Engineering Review (periodical code)

IHES Institut des Hautes Études Scientifiques (France)

IHEU International Humanist and Ethical Union

IHF Improved High Frequency • Inverse Hyperbolic Function • Isothermal Heating Facility

IHFAS Integrated High-Frequency Antenna System

IHFF Inhibit Halt Flip-Flop

IHFR Improved High Frequency Radio • Institut Hydrométéorologique de Formation et de Recherche (Algeria)

IHFT Institut für Hochfrequenztechnik (Germany)

IHHA Initial Health Hazard Assessment

IHHAR Initial Health Hazard Assessment Report

IHI Improved Holographic Image • Ishikawajima-Harima Industries

IHIL Internal Hardware-In-the-Loop

IHIS Interoperability of Heterogeneous Information Systems

IHK Industrie- und Handelskammer

ihl In Hoc Loco

IHL Internet Header Length • IP Header Length

IHM Inspection History Maintenance • Integrated Health Monitoring • Interface Homme-Machine

IHMC Institute for Human and Machine Cognition (USA-FL)

IHMS Integrated Health Monitoring System

IHN Qishn (Yemen) (location identifier)

IHO Ihosy (Madagascar) (location identifier) • In Honor Of • International Hydrographic Organization

IHP Indicated Horsepower • Information Handling Project • Institut Henri Poincaré (France) • Institutional Homepage • Internationaal Hydrologisch Programma • International Hydrological Programme

IHPAn Annales de l'Institut Henri Poincaré (periodical code)

IHPEG Information Highway Parental Empowerment Group

IHPH Indicated Horsepower-Hour

IHPI Improved High Power Illumination • Improved High Power Illuminator

IHPP Institute of High-Pressure Physics

IHPRPT Integrated High Payoff Rocket Propulsion Technology

IHPS Intelligent Host Processing System

IHPST International History and Philosophy of Science Teaching

IHPTET Integrated High-Performance Turbine Engine Technology

IHR Infrared Heterodyne Radiometer • Inhomogeneous Reionization • Institute of Historical Research • Integrate - Hold - Reset

IHRWG International Halon Replacement Working Group

IHS Information Handling Service • Information Handling System • Integrated Headgear System • Integrated Hospital Support • Intensity - Hue - Saturation • International Heliospheric Study • Internet Host Server • Isotope Heat Source

IHSA Interrupt Handler Save Area

IHSDS Interconnections within High-Speed Digital Systems

IHSL Institute for the Historical Study of Language

IHSM Institute of Health Services Management

IH/SR Integrated Harware/Software Review

IHSR Integrated Harware - Software Review

IHT Inheritance Tax • Internationales Heide-Teleskoptreffen (Germany) • International Herald Tribune • International Hydrology Today (periodical code)

IHTEP Integrated High-Performance Turbine Engine Technology

IHTP Institut d'Histoire du Temps Présent (France)

IHTS Intermediate Heat Transport System

IHTV Interim Hypersonic Test Vehicle

IHU Ihu (Papua New Guinea) (location identifier) • Image Handling Utility

iHv in Höhe von

IHV Independent Hardware Vendor

IHX Interloop Heat Exchanger

IHW International Halley Watch

IHwy Info Highway • Information Highway

I&I Interoperability & Interchangeability

I/I Inboard/Inboard

II Bright Giant Star • Igniter Initiator • Image Intensification • Image Intensifier • Imagery Interpretation • Immediacy Index • Implementing Instruction • Information for Industry • Information Index • Initial Issue • Input Impedance • Instituto de Informática • Interdisciplinary Investigator • Ion Implantation • Item Identification • Luminous Giant Star

IIa Bright Giant Star

IIA Image Intensifier Assembly • Indian Institute of Astrophysics (India) • Indicator of Industrial Activities • Inertial Instrument Assembly • Information Industry Association (USA) • Information Interchange Architecture • Inishmaan (Ireland) (location identifier) • Institute of Industrial Administration • Instituto de Investigaciones Antropológicas (Mexico) • Instrument Interface Agreement • Intelligent Industrial Automation • Intelligent Information Agent

IIab Bright Giant Star

IIACE Instituto de Investigaciones Aplicadas de Ciencias Espaciales (Argentina)

IIAF Idaho Institute Analysis Facility (USA-ID)

IIAL International Institute of Arts and Letters

IIAP Institut International d'Administration Publique

IIApN Indian Institute of Astrophysics Newsletter (periodical code)

IIAS Interactive Image Analysis System • International Institute for Administrative Sciences • International Institute for Advanced Studies

IIASA International Institute for Applied Systems Analysis

IIATP Impact Ionization Avalanche Transit Point

IIb Bright Giant Star

IIB Institut International de la Bibliographie • Institut International des Brevets • Interagency Information Bureau • International Investment Bank

IIBR Inheritance Inference-Based Refinement

IIC Identifiable Incremental Cost • Ignitor Initiator Cartridge • Ignitor Initiator Container • Image Interpretation Cell • Independent Inspection Co. • Interactive Information Centre • Investigation In Charge • Ion-Ion Collision • Istituto Italiano di Cultura • Item Identification Code

IICC Improved Information and Coordination Center

IICCD Image-Itensified Charge Coupled Device

IICE Information Integration for Concurrent Engineering • Interactive Image Client Environment

IICF Interdisciplinary Investigator Computing Facility

IICIS Institute for International Commerce and Information Sciences

IICO Ion Implantation Corporation

IICP Integrated Interoperability Controlled Phase

IICS International Interoperability Catalog System

IICT Indian Institute of Chemical Technology (India) • Instituto de Investigação Científica et Tropical (Portugal)

iid Independently and Identically Distributed

IID Image Intensifier Device • Impact Ionization Diode • Independently and Identically Distributed • Independently Identically Distributed • Instantiation Identifier • Instrument Interface Document • Integrated Information Display • Integrated Interoperability Demonstration • Interface Identifier • Ion Implantation Doping • Ionospheric Ion Density

IIDB Integrated Intelligence Database

IIDP Integrated Instrument Development Program

IIDS Integrated Instrumentation Display System • Integrated Instrument Display System

IIDT Ion Implantation Doping Technique

IIE Initial Ion Event • Institute for International Education • Institute of Industrial Engineers • Institute of International Education

IIED International Institute for Environment and Development

IIEE Ion-Induced Electron Emission

IIEF Institut International d'Études Françaises

IIELR International Institute for Electronic Library Research (UK)

IIEP Institute of International Educational Programmes (Russia) • International Institute of Educational Planning

IIF Image Information File • Image Interchange Facility • Imagery Interpretation Facility • Immediate Interface • Information Item File • Institute of International Finance (USA-DC) • Instituto de Investigaciones Filosóficas (Mexico) • Instituto Internacional de Finanzas • Intelligent Information Fusion • International Impact Factor

IIFGS ISO-IRAS Faint Galaxy Survey

IIFS Intelligent Information Fusion System

IIG Independent Investigation Group • Indian Institute of Geomagnetism (India) • Instantaneous Inverse Gain • Interrange Instrumentation Group

IIGDT Interactive Intelligent Image Graphics Display Terminal

IIGM Indian Institute of Geomagnetism (India)

IIGS Initial Image Generating System

IIH Instituto de Investigaciones Históricas (Mexico) • Integration Information Handler • Integrated Interactive Home

IIHE Institut Universitaire des Hautes Énergies (Belgium)

III Giant Star • Improved Initialization Inversion • Incapacitating Illness or Injury • Institute for Information Industry (China-ROC) • Intelligent Information Interface • Interesting - Informative - Impressive

IIIa Giant Star

IIIA Internet Information Infrastructure Architecture

IIIab Giant Star

IIIb Giant Star

IIID International Institute of Information Design

IIIT Indian Institute of Information Technology (India)

IIJ Instituto de Investigaciones Jurídicas (Mexico)

IIK Imagery Interpretation Key

IIL Induction Ion Laser • Institute of International Law • Integrated Injection Logic • Intermediate Inversion Layer

IILC Information Industry Liaison Committee

IILR Inner Inner Lindblad Resonance

IILS Internet Information Locator Service

IiM Integration in Manufacturing

IIM Implicit Integral Method • Institute for Industrial Mathematics (Israel) • Institute for Information Management • Institution of Industrial Managers

IIMAS Instituto de Investigaciones en Matemáticas Aplicadas y en Sistemas

IIMI International Irrigation Management Institute (Sri Lanka)

IIMP Integrated Information Management Protocol

IIMS Imaging Ion Mass Spectrometer • Integrated Information Management System • Intensive Item Management System

IIMSES Initial Implementation of Mode S Enhanced Surveillance

IIN Inlet Identification Number • Instituto Isaac Newton (Chile) • Integrated Information Network • Item Identification Number • Nishinoomote (Japan) (location identifier)

IINA Istituto Italiano di Navigazioni - Atti (periodical code)

IInfSc Institute of Information Scientists

IINREN Interagency Interim National Research and Education Network

IINS Integrated Inertial Navigation System

IIO Information Item Only • Inspecteur de l'Information et de l'Orientation

IIOC Intelligent Input-Output Controller

IIOE International Indian Ocean Expedition

IIOP Internet Inter-Operability Protocol • Internet Inter-ORB Protocol (*) • Interoperable Inter-ORB Protocol • Intra/Internet Object Protocol

IIOS Integrated Information Operation Strategy

I/IP Installation/Implementation Plan

I&IP Industrial & International Programs

IIP Implementation and Installation Plan • Implementation Information Plan • Industrial and International Programs • Information and Installment Plan • Installation and Implementation Plan • Instantaneous Impact Point • Instantaneous Impact Prediction • Institute for Image Processing • Institute of Intellectual Property • Institut International de la Presse • Instrument Information Profile • Instrument Interface Proposal • Intelligence Interface Processor • Intelligent Information Processing (conference) • Intended Impact Point • International Informatics Programme • Interoperability Improvement Plan • Interoperability Improvement Program

IIPA International Intellectual Property Alliance

IIPAH International Institute for Problems of the Asteroid Hazard

IIPE Investigation of the Ionospheric Plasma Escape

IIPH ISO Instantaneous Pointing History

IIPL Interactive Image Processing Laboratory

IIPS Interactive Image Processing System • Interactive Information and Processing System

IIPT Integrating Integrated Process Team • Integrating Integrated Product Team

IIPTI International Intellectual Property Training Institute

IIQ Initial Issue Quality • Initial Issue Quantity

IIR Imagery Interpretation Report • Imaging Infrared • Immediate Impulse Response • Indefinite-duration Impulse Response • Infinite Impulse Response • Initial Integration Review • Inspectorate for Ionizing Radiation (Czech Rep.) • Institute for International Research • Instrument Interface Review • Integrated Instrumentation Radar • Intelligence Information Report • Intelligent Information

Retrieval • Intergalactic Infrared • Intermediate Information Requirement • Intermediate Infrared • International Institute of Refrigeration

IIRA Integrated Inertial Reference Assembly

IIRD International Interdependent Research and Development

IIRF Intergalactic Infrared Radiation Field

IIRG Istituto Internazionale per le Ricerche Geotermiche (Italy)

IIRK Interarea Interswitch Rekeying Key

IIRM Information Interchange Reference Model

IIRN Inactive Item Review Notification

IIRP Improved Item Replacement Program • Individual Item Replacement Program

IIRS Indian Institute of Remote Sensing (India) • Institute for Industrial Research and Standards • Integrated Inventory Reduction System • Intelligent Information Retrieval System

IIRTC International Informatics Research and Training Centre (Ukraine)

IIRV Improved Interrange Vector

IIRW Intelligent Information Retrieval Workstation

IIS Image Intensified System • Imagery Interpretation Segment • Implant Ion Sensor • Industry Information Security • Information and Intelligent Systems • Information Improvement Support • Information-Interaction Space • Infrared Image Scanner • Infrared Imager Spectrograph • Infrared Imaging Spectrometer • Infrared Imaging System • Infrared Information Symposium • Infrared Instrumentation System • Infrared Intelligent Spectroradiometer • Infrared Interferometer Spectrometer • Infrared Intruder Sensor • Infrastructure and Information Systems • Inspection Item Sheet • Institute of Information Scientists • Institut für Integrierten Schaltungen (Germany) • Institut International de la Soudure • Institut International de Statistique • Integrated Information System • Integrated Instrument Simulator • Integrated Instrument System • Intelligent Information Subfile • Intelligent Information System • Interactive Instructional System • Interface Identification Sheet • Interface to Information Sources • Internal Integral Starter • International Information System • Internet Information Server • Internet Information Services • Internetted Information System • Intervehicular Information System • Ion Implantation Study • Nissan Island (Papua New Guinea) (location identifier)

IISA Institut International des Sciences Administratives

IISc Indian Institute of Science (India)

IISC Intelligence Information Systems Committee

IISE Intelligence Information Services Enhancement

IIS/G Internal Integral Starter/Generator

IISG Interferometric Imaging Science Group • Interferometric Instrumentation Science Group • Internal Integral Starter/Generator

IISI International Iron and Steel Institute

IISL International Institute of Space Law (France)

IISME Industry Initiatives for Science and Math Education

IISN Institut Interuniversitaire des Sciences Nucléaires

IISP Information Infrastructure Standards Panel • Interim Interface Switch Protocol • Interim Inter-Switch Protocol • Interim Inter-Switch Signaling Protocol

IISR Institute of Ionospheric and Space Research

IISS Image Inventory Search and Summary • Integrated Information Support System • International Institute for Strategic Studies (UK)

IIST International Institute for Software Technology

IISU ISRO Inertial System Unit

II&T Instrument Integration & Test • Instrument Integration & Testing

IIT Illinois Institute of Technology (USA-IL) • Image Intensification Tube • Image Intensifier Tube • Indian Institute of Technology (India) • Industrial Information Transfer • Institute for Information Technology • Instituto de Investigación Tecnológica • Instrument Integration and Test • Instrument Integration and Testing • Integrated Idea Team • Integrated Information Technology • Internal Information Transfer • Israel Institute of Technology

IITA Image Intensifier Tube Assembly • Information Infrastructure Technologies and Applications (*) • Information Infrastructure Technology Applications • Institute of Information Technology Assessment (Korea-ROK) • International Institute of Tropical Agriculture (Nigeria)

IITAP International Institute of Theoretical and Applied Physics

IITB Indian Institute of Technology, Bombay (India)

IITCS Igloo Internal Thermal Control Section

IITD Indian Institute of Technology, Delhi (India)

IITE Institute for Information Technologies in Education (UNESCO)

IITF Information Infrastructure Task Force

IITK Indian Institute of Technology, Kampur (India)

IITKa Indian Institute of Technology, Kanpur (India)

IITKh Indian Institute of Technology, Kharagpur (India)

IITM Indian Institute of Technology, Madras (India) • Indian Institute of Tropical Meteorology (India)

IITND Indian Institute of Technology, New Delhi (India)

IITS Igniter Initiator Test Set

IITV Image-Intensified Television • Image-Intensifying Television

IIU Input Interface Unit • International Islamic University

IIV Image Intensifier Viewer • Initial Input Variable

IIVC Institut National de l'Ingénierie de la Vision et du Contrôle (France)

IIVD Image Intensifier Viewing Device

IIVRS International Institute for Vital Registration and Statistics

IIW International Institute of Welding

IIWG Interdisciplinary Investigator Working Group

IIWPA International Information/Word Processing Association

iJ im Jahre

IJ Identité Judiciaire • IJssel • Indemnité Journalière

IJA International Journal of Astrobiology (periodical code)

IJAA International Journal of Adhesion and Adhesives (periodical code)

IJACS International Journal of Adaptive Control and Signal Processing (periodical code)

IJAEM International Journal of Analytical and Experimental Model Analysis (periodical code)

IJAF Instrumentcenter for Jordbaseret Astronomisk Forskning (Denmark)

IJAIS International Journal of Accounting Information Systems

IJAP International Journal of Aviation Psychology (periodical code)

IJBC International Journal on Bifurcation and Chaos (periodical code)

IJBS Integrated Joint Broadband System

IJC International Joint Commission • International Journal of Control (periodical code)

IJCAI International Joint Conference on Artificial Intelligence

IJCAII International Joint Conferences on Artificial Intelligence Inc.

IJCAR International Joint Conference on Automated Reasoning

IJCM International Journal of Computer Mathematics (periodical code)

IJCMS International Journal of Computer Mathematics - Section B (periodical code)

IJCPR International Joint Conference on Pattern Recognition

IJCS Integrated Joint Communications System

IJDAR International Journal on Document Analysis and Recognition

IJDM International Journal of Damage Mechanics (periodical code)

IJE International Journal of Electronics (periodical code) • Inverse Joule Effect

IJEFM International Journal of Engineering and Fluid Mechanics (periodical code)

IJEM International Journal of Electronic Markets

IJER International Journal of Energy Research (periodical code)

IJES International Journal of Engineering Science (periodical code)

IJET International Journal of Educational Telecommunications (AACE)

IJFa International Journal of Fatigue (periodical code)

IJFE International Journal of Fusion Energy (periodical code)

IJFr International Journal of Fracture (periodical code)

IJGIS International Journal of Geographical Information Systems

IJHE International Journal of Hydrogen Energy (periodical code)

IJHFF International Journal of Heat and Fluid Flow (periodical code)

IJHM International Journal of Hybrid Microelectronics (periodical code)

IJHMT International Journal of Heat and Mass Transfer (periodical code)

IJICS International Journal of Intelligent Control and Systems

IJIE International Journal of Impact Engineering (periodical code)

IJIMW International Journal of Infrared Millimeter Waves (periodical code)

IJIS International Journal of Information Security (periodical code)

IJIST International Journal of Imaging Systems Technology (periodical code)

IJK Izhevsk (Russia) (location identifier)

IJLT IEEE Journal of Lightwave Technology (periodical code)

IJMeS International Journal of Mechanical Sciences (periodical code)

IJMF International Journal of Multiphase Flow (periodical code)

IJMHG Indian Journal of Meteorology, Hydrology, and Geophysics (periodical code)

IJMM International Journal of Mini and Microcomputers (periodical code)

IJMoS International Journal of Modeling and Simulation (periodical code)

IJMPA International Journal of Modern Physics A (periodical code)

IJMPB International Journal of Modern Physics B (periodical code)

IJMPC International Journal of Modern Physics C (periodical code)

IJMPD International Journal of Modern Physics D (periodical code)

IJMPE International Journal of Modern Physics D (periodical code)

IJMSI International Journal of Mass Spectrometry and Ion Processes (periodical code)

IJNA IMA Journal on Numerical Analysis (periodical code)

IJNLM International Journal of Non-Linear Mechanics (periodical code)

IJNME International Journal for Numerical Methods in Engineering (periodical code)

IJNMF International Journal for Numerical Methods in Fluids (periodical code)

IJO Individual Job Order

IJODL International Journal On Digital Libraries (periodical code)

IJOE IEEE Journal of Oceanic Engineering (periodical code)

IJP Ink Jet Printer • Ink Jet Printing • International Journal of Plasticity (periodical code)

IJPS Initial Joint Polar System

IJQC International Journal of Quantum Chemistry (periodical code)

IJQE IEEE Journal of Quantum Electronics (periodical code)

IJRA IEEE Journal of Robotics Automation (periodical code)

IJRAI International Journal of Radiation Applications and Instrumentation (periodical code)

IJRR International Journal of Robotics Research (periodical code)

IJRS International Journal of Remote Sensing (periodical code)

IJRSP Indian Journal of Radio and Space Physics (periodical code)

IJSA International Journal of Supercomputer Applications (periodical code)

IJSAC IEEE Journal on Selected Areas in Communications (periodical code)

IJSC International Journal of Satellite Communications (periodical code)

IJScA International Journal of Supercomputer Applications (periodical code)

IJSEKE International Journal of Software Engineering and Knowledge Engineering

IJSS International Journal of Solids and Structures (periodical code)

IJSSC IEEE Journal of Solid-State Circuits (periodical code)

IJSSR Indemnités Journalières de Sujétion Spéciale de Remplacement

IJTJE International Journal of Turbo Jet Engines (periodical code)

IJTP International Journal of Theoretical Physics (periodical code)

IJU Ijuí - João Batista Bos Filho (Brazil-RS) (location identifier)

IJVTOP International Jupiter Voyager Telescope Observations Program • International Jupiter Voyager Telescope Observing Program

IJW International Jupiter Watch • It Just Works

IJX Jacksonville (USA-IL) (location identifier)

ik Inupiak (ISO language code)

IK Image Kernel • Indemnité Kilométrique • Inner Keel • Installation Kit • Institut für Kernphysik • Interchange Key

IKAT Interactive Keyboard And Terminal

IKB International Klein Blue • Wilkesboro - Wilkes County (USA-NC) (location identifier)

IKBS Intelligent Knowledge-Based System

IKE Integrated Knowledge Environment • Internet Key Exchange

IKEA Ingvar Kamprad, Elmtaryd, Agunnaryd

IKES Ion Kinetic Energy Spectrometry

IKET Individual Knowledge Evaluation Test

IKF Institut für Kosmoforschung

IKFIA Institute of Cosmophysical Research and Aeronomy (Russia)

IKH Instabilité de Kelvin-Helmholtz

IKI Iki (Japan) (location identifier) • Institut Kosmicheskij Isledovanja (Russia) • Space Research Institute (Russia)

IKIP Institut Keguruan dan Ilmu Pendidikan (Indonesia)

IKK Kankakee - Greater Kankakee (USA-IL) (location identifier)

IKL Ikela (Zaire) (location identifier)

IKMP Internet Key Management Protocol

IKO Nikolski (USA-AK) (location identifier)

IKON Interkerkelijke Omroep Nederland

IKP Inkerman (Australia-QLD) (location identifier) • Instructor and Key Personnel • Internet Keyed Payment

I&KPT Instructor & Key Personnel Training

IKPT Instructor and Key Personnel Training

IKR Isokinetic Relationship

IKS Image Kernel System • Integrated Key Set • Inverse Kinetics Simulator • Tiksi (Russia) (location identifier)

I&KT Informations- & Kommunikationstechnologien

IKT Informations- und Kommunikationstechnologien • Irkutsk (Russia) (location identifier)

IKU Interface Keying Unit

IKV Interkerkelijk Vredesberaad

IKZ Impulskennzeichen • Internationales Kongresszentrum

il Icon Library (file extension) • Israel

Il Illustrated • Illustration

I/L Import Licence

I&L Installation & Logistics

IL Illinois (USA) • Ilyushin • Imaging Library • Important Lady • Important Launch • Incoming Letter • Inertial Lab • Informative Language • Insertion Loss • Inside Leg • Instruction Leaflet • Instrumentation Laboratory • Instrument Landing • Interaction List • Interactive Lyndenmaier • Intermediary Letter • Intermediate Language • Intermediate Loop • Internal Letter • International Light (USA-MA) • International Logistics • Interpretive Language • Inversion Line • Investigation Leader • Ionization Loss • Israel (international vehicle registration code) • Israel (ISO country code)

ILA Illaga (Indonesia) (location identifier) • Image Light Amplifier • Independent Logistic Assessment • Individual Learning Account • Induction Linear Accelerator • Initial Line Address • Instruction Look Ahead • Instrument Landing Approach • International Law Association • Internet Learning Agent

ILAC International Laboratory Accreditation Cooperation

ILAFA Instituto Latinoamericano del Fierro y el Acero

ILAMS Infrared Laser Atmospheric Monitoring System

ILAN Implementation of Local Area Networks

ILANA Inertial Land Navigation

ILAP Integrated Local Area Planning • Integrated Logistics Analysis Program • Interactive Lively Application Project

ILAS Improved Limb Atmospheric Spectrometer • Infrared Limb Atmospheric Spectrometer • Instrument Landing Approach System • International Linear Algebra Society • Interrelated Logic Accumulating Scanner

Ilat Ilativo

ILB Interactive Linear Browser

ILBM Image Lithologic Boundary Manifestation

ILBS Image Library and Browse System

ILC Idle Load Compensator • Industrial Liaison Committee • Inhibit Launch Command • Initial Launch Capability • Instruction Length Code • Integrated Launch Complex • Integrated Logic Circuit • Interim Library Catalog • Interim Load Cycle • Intermediate Load Cycle • International Linear Collider • International Logistics Center • Irrevocable Letter of Credit

ILCA International Livestock Centre for Africa

ILCBT Instructor-Led Computer-Based Training

ILCC Initial Launch Capability Complex

ILCCS Integrated Launch Control and Checkout System

ILCE Integrated Logistics and Concurrent Engineering

ILCM Information Life Cycle Management

ILCO Instantaneous Launch Control Officer • International Logistics Control Office • International Logistics Control Organization

iLCS Integrated Laser Control System

ILCS International Liquid Crystal Society • International Logistics Communication System

ILCSM IEEE LCS Magazine (periodical code)

ILCSO Illinois Libraries Computer System Organization

ILD Initial Launch Data • Injection Laser Diode • Integrated LAN Driver • Integrated Logistics Data • Intensity Level Difference • Interaural Level Difference • Interlayer Dieletric • Internal Locking Device • International Lunar Dateline

ILDA Intensified Linear Diode Array • International Laser Display Association (USA-FL)

ILDC Intern Leadership Development Course

ILDE Initiative Locale de Développement et d'Emploi

ILDF Integrated Logistics Data File

ILDP Instrument Library Developer Program

ILDS Integrated Logistics Data System

Ile Ilustre

ILE Fort Hood (USA-TX) (location identifier) • Initiative Locale pour l'Emploi • Installations - Logistics - Environment • Institute for Laser Engineering (Japan) • Institution of Lighting Engineers (UK) • Interactive Learning Environment • Interbase Local Engine • Intermediate-Level Education • Killeen (USA-TX) (location identifier)

ILEC Incumbent Local-Exchange Carrier • Independent Local-Exchange Carrier (*)

ILECS Incumbent Local-Exchange Carriers

ILES Indian Landsat Earth Station (India)

ILEWG International Lunar Exploration Working Group

ILF Ilford (Canada-MB) (location identifier) • Inductive Loss Factor • Infra Low Frequency • Institut für Lehrerfortbildung (Germany) • Integrated Luminosity Function • Intermediate Latitude Field • Internal Logical File • Internal Luminosity Function

ILFC International Lease Financial Corporation

ILFO International Logistics Field Office

ILFOSS International Lunar Far-side Observatory and Science Station

ILG International Loading Gauge • Wilmington (USA-DE) (location identifier)

ILH Illisheim (Germany) (location identifier)

ILHP Integrated Liquid Hydrogen Plant

ILI Iliamna (USA-AK) (location identifier) • Instant Lunar Ionosphere • Institute for Land Information • Instructional Laboratory Improvement • Instrumentation and Laboratory Improvement • Intelligent Line Interface • Intelligent Link Interface • International Law Institute

ILIAD Internet Library Information Assembly Database

ILID International Logistic Identification Data • International Logistic Identification Directory

ILIERS Integrated Library Information Education and Retrieval System

ILIF International Logistic Identification File

ILIM Institute for Land Information Management

ILIP In-Line Instrument Package

ILIR Independent Laboratory Inhouse Research • Independent Laboratory Innovative Research

ILISS Integrated Logistics Information Support System

ILJS ISO Loading Jack System

ILK Ilaka (Madagascar) (location identifier)

ill Illustrissimus

Ill Illinois (USA) • Illustrate • Illustrated • Illustration • Illustrator • Illustré • Illustrissimus

ILL Impact Limit Line • Input Logic Level • Institut (Max von) Laue - (Paul) Langevin (France) • Interlibrary Loan • Willmar (USA-MN) (location identifier)

ILLIAC Illinois Integrator and Automatic Computer

IlliGAL Illinois Genetic Algorithms Laboratory (USA-IL)

ILLINET Illinois Library and Information Network (USA-IL)

Illit Illiterate

IllOB Bulletin of the Astronomical Observatory of the University of Illinois (periodical code)

Ills Illustrations

ILLS Interlibrary Loan System

ILLSS Identification List of Lines in Stellar Spectra

Illum Illuminate • Illuminated • Illumination

ILLUM Illuminate • Illuminated • Illumination

Illus Illustrate • Illustrated • Illustration • Illustrator

Illust Illustrate • Illustrated • Illustration • Illustrator

ILM Immeuble à Loyer Moyen • Independent Landing Monitor • Information Logic Machine • Integrated Layer Management Interface • Interface Logiciel-Matériel • Intermediate-Level Maintenance • Wilmington (USA-NC) (location identifier)

ILMF Input Language Media Format • Intermediate-Level Maintenance Facility

ILMI Integrated Local Management Interface • Interim Local Management Interface

Ilma Ilustrísima

Ilmo Ilustrísimo

ILMP Integrated Logistics Management Program

ILMR International Laboratory of Marine Radioactivity

ILMRS Integrated Logistics Support Milestone Reporting System

ILMS Improved Launcher Mechanical System • Integrated Library Management System

ILMT Integrated Logistics Management Team • International Liquid Mirror Telescope

ILN Immeuble à Loyer Normal • Intermediate Logic Network • Wilmington Clinton (USA-OH) (location identifier)

ILNP Interim Local Number Portability

ILO Iloilo (Philippines) (location identifier) • Injection-Locked Oscillator • In Lieu Of • Integrated Logistics Operations • Integrated Logistics Overhaul • International Labour Office • International Labour Organization (*) • International Logistic Office

ILOC Intermediate Location • International Lunar Occultation Centre

ILOGS Integrated Logistics System

ILOM International Latitude Observatory of Mizusawa (Japan)

ILOSS International Logistics Overseas Support System

ILP Île des Pins (New Caledonia) (location identifier) • Incapacidad Laboral Permanente • Inductive Logic Programming • In-Line Printer • Instruction Level Parallelism • Integrated Layer Processing • Integrated Logistics Panel • Integrated Logistics Plan • Intermediate Language Processor • International Lithosphere Program • International Logistics Program

ILPC Intermediate Long-Period Comet

ILPES Instituto Latinoamericano de Planificación Económica y Social

ILPF Ideal Low-Pass Filter • Internet Law and Policy Forum

ILPO International Lunar Planning Office

ILPS International Logic Programming Symposium

ILPT Interlayer Pair Tunneling

ILQ Ilo (Peru) (location identifier) • Intermediate Level Qualifier • International Lunar Quinquennium

ILR Ilorin (Nigeria) (location identifier) • Ingoing Longwave Radiation • Inner Lindblad Resonance • Ingenieurbüro für Luft- und Raumfahrt • Integrated Logistics Report • Intermediate Line Region • Intranet Location Register

ILRAD International Laboratory for Research on Animal Diseases

ILRI Institute for Land Reclamation and Improvement (Netherlands)

ILRO Industrial Labor Relations Officer

ILRS International Laser Ranging Service

ILRV Integral Launch and Recovery Vehicle • Integral Launch and Reentry Vehicle • Integrated Launch and Recovery Vehicle • Integrated Launch and Reentry Vehicle

ILRVS Integral Launch and Recovery Vehicle System • Integral Launch and Reentry Vehicle System • Integrated Launch and Recovery Vehicle System • Integrated Launch and Reentry Vehicle System

ILS Idle Line State • Illumination System • Inertial Location System • Instrument Landing System • Instrument Line of Sight • Integrated Light Spectrum • Integrated Linearization System • Integrated Logistics Support • Integrated Logistics System • Interactive Laboratory System • Interferometric Laser Source • International Language Support • International Latitude Service • International Launch Services • International Lunar Society • Internet Locator Server • Internet Locator Service • Intracavity Laser Spectroscopy • Inventory Locator System • Inverse Least Squares • Ionization Loss Spectroscopy • Iterative Least-Squares • Shekel (Israel) (ISO currency code)

ILSA Improved Lightweight Satellite Antenna • Improved Lightweight Solar Array • Integrated Logistics Support Analysis

ILSC Industrial Logistics Systems Center • Integrated Launch Services Contractor • Integrated Launch Support Center • Integrated Logistics Support Concept

ILSCM Integrated Logistics Support Control Manual

ILSCP Integrated Logistics Support Concept Plan • International Land Surface Climatology Project

ILSDF Integrated Logistics Support Data File

ILSDP Integrated Logistics Supply Delivery Plan • Integrated Logistics Support Delivery Plan

IL&SE Integrated Logistics & Specialty Engineering

ILSE Integrated Logistics and Specialty Engineering

ILSES Integrated Library and Survey Extraction Service

ILSF Integrated Logistics Support Facility

ILSI International Life Sciences Institute

ILSL Initial Logistics Supprt Letter

ILSM Integrated Logistics Support Manager

ILSMIS Integrated Logistics Support Management Information System

ILSMP Integrated Logistics Support Management Plan

ILSMR Integrated Logistics Support Management Review • Integrated Logistics Support Milestone Reporting

ILSMS Integrated Logistics Support Management System

ILSMT Integrated Logistics Support Management Team

ILSO Integrated Logistics Support Office • Integrated Logistics Support Officer • Integrated Logistics Support Organization

ILSP Institute for Language and Speech Processing (Greece) • Integrated Logistics Support Plan • Integrated Logistics Support Planning • Island Lake Star Party (USA-MI)

ILSPM Integrated Logistics Support Program Manager

ILSS Integrated Life Support System • Integrated Logistics Support Standards • Integrated Logistics Support System

ILSSE Integrated Life Science Shuttle Experiment

ILST Indonesian Low-Speed Tunnel

ILST&E Integrated Logistics Support Test & Evaluation

ILSTE Integrated Logistics Support Test and Evaluation

ILSW Interrupt Level Status Word

ILSWG Integrated Logistics Support Working Group

ILT Imprecisely Located Target • Incapacidad Laboral Transitoria • Independent Laboratory Testing • Installation Lead Time • Institute for Learning and Teaching (UK) • Institute for Learning Technologies • Institut für Lasertechnik (Germany) • Instructor-Led Training

ILTF Internet Law Task Force

Iltma Ilustrísima

Iltmo Ilustrísimo

ILTP Integrated Long-Term Programme

Iltre Ilustre

ILU I Love You • Independent Logical Unit • Interlanguage Unification • Kilaguni (Kenya) (location identifier)

ILUG Innovative Low Users Group

ILV Industrial Launch Vehicle • Internationaler Leihverkehr

ILW Immediate-Level Waste • Intermediate-Level Waste (*) • International Leonid Watch

ILWAS Integrated Lake-Watershed Acidification Study

ILWIS Integrated Land and Water Information System • Integrated Land and Watershed Information System

ILWS International Living With a Star (project)

ILX Ileg (Papua New Guinea) (location identifier)

ILY I Love You • Islay (UK) (location identifier)

ILZ Zilina (Slovakia) (location identifier)

iM im Mittel • im Monat

I&M Improvement & Modernization • Inspection & Maintenance • Installation & Maintenance

I/M Intermediate Maintenance

IM Image Motion • Imaging Mode • Impulse Modulation • Indice Majoré • Information Memory • In Maintenance • Industrial Manager • Informal Memorandum • Information Management • Information Manager • Information Model • Information Modeling • Infrastructure Model • Infrastructure Module • Ingénieur de la Météorologie • Initialization Mode • Inner Marker • Inspection Manual • Inspection Memorandum • Installation and Maintenance • Instant Messager • Instant Messaging • Instruction Management • Instruction Manual • Instrumentation Manager • Instrumentation and Measurement • Instrument Module • Integration Management • Integration Manager • Intensity Modulation • Interest Manager • Interface Management • Interface Module • Interim Memorandum • Intermediate Maintenance • Intermediate Model • Intermediate Modulation • Intermodulation • Internal Memorandum • Internal Message • Interpersonal Mode • Intra Muros • Intra-Muscular • Inventory Management • Inventory Manager • Inverted Microscope • Inverted Microscopy • Ionospheric Mapper • Irregular Marking • Isle of Man • Item Management • Item Manager • Item Mark

IMA Iamalele (Papua New Guinea) (location identifier) • Ice Motion Algorithm • Impedance Matching Attenuator • Independent Medical Assessor • Information Management and Archives • Information Management Architecture • Information Management Area • Information Médicale Automatisée • Information Mission Area • Institut de Mathématique Appliquée • Institute of Management Accountants • Institute of Mathematics and its Applications • Integrated Modular Avionics • Intelligent Maintenance Aid • Interactive Measurement and Analysis • Interactive Multimedia Association (USA) • Intermediate Maintenance Activity • Intermodulationsabstand • International Mineralogical Association • Inverse Multiplexing for ATM • Ion Microprobe Analyzer

IMAC Integrated Microwave Amplifier Converter • Interim Message Automated Capability • Intermediate Maintenance Area Coordinator • Isochronous Media Access Control • Isochronous Media Access Controller

IMACS Image Management And Communication System • Inamori Magellan Areal Camera and Spectrograph • Integrated Manufacturing and Accounting System • Integrated Manufacturing Control System • International association for Mathematics And Computers in Simulation

IMAD Instructional Materials and Assessment Development

IMADS Information Management Assimilation and Delivery System

IMAFF Instituto de Matemáticas y Física Fundamental

Imag Imaginary • Imagination • Imagine

IMAG Information Management Advisory Group • Informatique et Mathématiques Appliquées de Grenoble (France) • Institut des Mathématiques Appliquées de Grenoble (France) (*)

IMAGE Imager for Magnetopause-to-Aurora Global Exploration • Integrated Mapping And Geographic Encoding • Integrated Model for the Assessment of the Greenhouse Effect • Integrated Model to Assess the Greenhouse Effect • International Monitor for Auroral Geomagnetic Effects • Investigative Materials About Global Environment

IMAGES Integrated Mapping And Geographic Encoding System • International Marine Global Change Study

IMAGINE Integrated Multiagent Interactive Environment

IMALS Integrated Maintenance of Assets and Logistics Support

IMAO Ingénierie Mécanique Assistée par Ordinateur

IMAP Imager for the Magnetosphere, Aurora and Plasmasphere • Indian Middle Atmosphere Programme • Input - Management - Analysis - Presentation • Interactive Mail Access Protocol • Internet Message Access Protocol (*) • Internet Messaging Access Protocol

IMAPI Interactive Multimedia Application Program Interface

IMAPP International MODIS/AIRS Processing Package

IMAPS Interactive Microcomputer Array Processing System • Interstellar Medium Absorption Profile Spectrograph

IMAR Institute for Manufacturing and Automation Research

IMAS Institutional Management Accounting System • Integrated Mass Announcement System • Intelligent Mobile Autonomous System

IMASP Instantiable MASP • Instantiated MASP

IMASS Intelligent Multiple Access Spectrum Sharing

IMAT Interactive Multisensor Analysis Training

IMATI Istituto per la Matematica Aplicata e Tecnologie Informatiche (CNR)

IMAU Instituut voor Marien en Atmosferisch Onderzoek Utrecht (Netherlands)

IMaX Imaging Magnetograph Experiment

IMAX Image Maximum • Isotope Matter Antimatter Experiment

IMB Imbaimadai (Guyana) (location identifier) • Indice Moyen Budgétaire • Information Management Branch • Input Memory Buffer • Interface Management Board • Investigator of Micro-Biosphere • Irvine - Michigan - Brookhaven (collaboration)

IMBH Intermediate-Mass Black Hole (*) • Intermediate Massive Black Hole

IMBLMS Integrated Medical Behavioral Laboratory Measurement System

IMBIBER Impact By Incident Baryon Excitation Rate

IMBP Intelligent Manufacturing and Business Process • Intermediate Mass Binary Pulsar

IMC Ice Mass Content • Image Motion Compensation • Image Motion Compensator • Image Motion Configuration • Information Management Center • Information Management Conference • Information Modeling by Composition • Information Modeling Convergence • Índice de Masa Corporal • Indice de Masse Corporelle • Isochronous Maintenance Channel • Institute of Management Consultants • Institute of Measurement and Control • Instructional Media Center • Instrument Meteorological Conditions • Integrated Maintenance Chart • Integrated Maintenance Complex • Integrated Maintenance Concept (*) • Integrated Management Center • Integrated Maneuver Controller • Integrated Microelectronic Circuitry • Integrated Microwave Circuit • Integrated Monolithic Circuit • Interactive Markov Chain • Interim Message Change • Intermediate-source Mars-Crosser • Internal Management Control • International Machine Consortium • International Maintenance Control • International Management Council • International Maritime Committee • International Meteor Conference • International Meteorological Centre • International Meteorological Committee • Internet Measurement Conference • Internet Microcode • Item Management Code • Item Marking Code

IMCC Image Motion Compensation and Calibration • Integrated Mission Control Center • Item Management Control Code

IMCCAS International Conference on Instrumentation, Measurement, Control, Circuits and Systems

IMCCE Institut de Mécanique Céleste et de Calcul des Éphémérides (France)

IMCD Input Marginal Checking and Distribution

IMCE Independent Management and Cost Evaluation • Inter-Ministerial Committee for Environment • ISS Management and Cost Evaluation

IMCO Imager-Coronagraph • Information Management Control Officer • Intergovernmental Maritime Consultative Organization (now IMO)

IMCoS International Map Collectors Society

IMCP Information Management Career Program • Integrated Monitor and Control Panel

IMCR Instrument Mode Change Request • Integrated Master Control Record • Internal Management Control Review • Internet Measurement Condensed Report

IMCS Integrated Machinery Control System • Integrated Maritime Communications System • Integrated Monitoring and Control System

IMCT Intermodule Compatibility Testing

IM&D Information Management & Display • Information Management & Distribution

IMD Imonda (Papua New Guinea) (location identifier) • Implementation Management Document • India Meteorological Department • Informational Macrodynamics • Information Management and Display • Information Management and Distribution • Information Management Department • Information Management Directorate • Information Management Display • Information Media Department • Initial Mass Distribution • Institute for Management Development (*) • Institute for Managing Development • Integrated Map Display • Interactive Multimedia Document • Intermediate • Intermittent Motion Driver • Intermodulation Distortion

IMDA International Map Dealers Association

IMDACS Integrated Multivariate Data Analysis and Classification System

IMDAS Interactive Multispectral Data Analysis System

IMDB Instruction Management Database • Integrated Maintenance Database

IMDC Instructional Media Development Center • Internal Message Distribution Center

IM/DD Intensity Modulation with Direct Detection

IMDD Intensity Modulation with Direct Detection

IMDE Integrated Model Development Environment

IMDG International Maritime Dangerous Goods

IMDPS INSAT Meteorological Data Processing System

IMDS Image Data Stream • Integrated Maintenance Data System • Integrated Monitoring and Diagnostics System • International MD Service

IMDT Intermediate

IME Indirect Manufacturing Expense • Information Management and Engineering • Input Method Editor • Institution of Mechanical Engineers • Institut Médico-Éducatif • Institut Monétaire Européen • Intelligent Mapper Engine • International Magnetospheric Explorer • International Materiel Evaluation • Interplanetary Meteoroid Experiment

IMEC Interuniversitair Micro-Electronica Centrum • Interuniversity Micro-Electronics Centre

IMechE Institute of Mechanical Engineers

IMEI International Mobile Equipment Identity

IMEKO International Measurement Confederation

IMEO Initial Mass in Earth Orbit

IMEP Independent Master Evaluation Plan • Indicated Mean Effective Pressure • Institution of Mechanical Engineers - Proceedings (periodical code) • International Measurement Evaluation Programme

IMERSE Indonesian Marine Environment Remote Sensing Experiment

IMeRv International Metals Review (periodical code)

IMETS Integrated Meteorological System

IMEWG International Mars Exploration Working Group

IMEX Inner Magnetosphere Explorer

IMF Ice Mass Flux • Image Furnace • Imphal (India) (location identifier) • Improved Message Facility • Incoming Mail Facility • Individual Master File • Initial Mass Function • Instituteur - Maître - Formateur • Intense Magnetic Field • Intermediate Maintenance Facility • Intermediate Mass Fragment • Internal Magnetic Focus • International Monetary Fund • Interplanetary Magnetic Field • Interstellar Magnetic Field • Inventory Master File

IMFAA Information Management Functional Area Assessment

IMFG Information Management Functional Group

IMFK Integrated Multifunction Key

IMFL Institut de Mécanique des Fluides de Lille (France) • Intensive Management Force List

IMFM Institut de Mécanique des Fluides de Marseille (France)

IMFP Inelastic Mean Free Path • Interaction Mean Free Path

IMFR Investigative Memorandum For Record

IMFT Institut de Mécanique des Fluides de Toulouse (France) • Interferometric Moving Target Focusing

img Image (file extension)

IMG Image • Imager • Industrial Management Group • Information Management Group • Inhaminga (Mozambique) (location identifier) • Institut de Mécanique de Grenoble (France) • Institut für Meteorologie und Geophysik • Interferometric Monitor for Greenhouse (gases) • Inverted Magnetron Gauge

IMGC Istituto di Metrologia G. Colonnetti (Italy)

IMGG Institute of Meteorology and Geophysics of Graz (Austria)

IMGWL IEEE Microwave and Guided Wave Letters (periodical code)

IMH Institute of Meteorology and Hydrology • Interactive Message Handler • Interactive Message Handling

IMHD Ideal Magnetohydrodynamics • Incompressible Magnetohydrodynamics

IMHE Institutional Management of Higher Education

IMHO In My Humble Opinion

IMHS Integrated Mail Handling System

IMI Industrial Maintenance Inspection • Ine Island (Marshall Islands) (location identifier) • Inner Magnetosphere Imager • Inner Magnetosphere Imaging • Installation and Maintenance Instruction • Institut des Matériaux Industriels (France) • Interactive Menu Interface • Interactive Multimedia Instruction • Intermediate Maintenance Instruction • Intermodulation Interference • Internal Multilayer Insulation • Internal Multiscreen Insulator • International Meteorological Institute (Sweden) • Ion Mass Imager

IMIA International Medical Informatics Association

IMIC Integrated Management Information Center • Integrated Management Information Computer

IMICS Integrated Management Information Control System

IMINT Imagery Intelligence

IMIO Image Input-Output

IMIP Industrial Modernization Incentive Program

IMIQ Instituto Méxicano de Ingenieros Químicos (Mexico)

IMIS Integrated Maintenance Information System • Integrated Management Information System • Ion Mass Imaging Spectrograph

Imit Imitate • Imitation • Imitative

IMK Institut für Meteorologie und Klimaforschung • Simikot (Nepal) (location identifier)

IML Imperial (USA-NE) (location identifier) • Initial Measurement List • Initial Microcode List • Inner Mold Line • Inside Mold Line • Institut Médico-Légal • Instrument Markup Language • Interactive Maintenance Language • Intermediate Maintenance Level • Internationale Mietleitungen • International Microgravity Laboratory • Internet Microcode Load

IMLI International Maritime Law Institute

IMLR Iron Mass to Light Ratio

IMLT Improved Mixing Length Theory

Imm Immeuble

IMM Immokalee (USA-FL) (location identifier) • Inbound Message Manager • Independent Manned Manipulator • Information Management Model • Input Message Manual • Institute of Mining and Metallurgy • Institute of Molecular Medicine • Integrated Maintenance Management • Integrated Maintenance Manual • Integrated Materiel Management • Integrated Materiel Manager • Integrated Memory Module • Intelligent Memory Manager • Interactive Multimedia • Intermediate Maintenance Manual • International Mail Manual • International Mars Mission • International Monetary Market • International Money Market

IMMA Ion Microprobe Mass Analyzer

IMMC Integrated Materiel Management Center

IMME Isobaric Multiplet Mass Equation

Immed Immediate

IMMI Integrated Man-Machine Interface

IMMP Information Management Master Plan • Information Mission Management Plan • Integrated Modification Management Plan

IMMPS Intelligent Multimedia Presentation System

IMMR Institut des Matériaux et Mesures de Référence (Belgium)

IMMS Integrated Maintenance Management System

Immun Immunity • Immunization

IMN Imane (Papua New Guinea) (location identifier) • Indicated Mach Number • Intermediate Node

IMNHO In My Not (so) Humble Opinion

IMNSHO In My Not So Humble Opinion

IMO Information Management Office • Information Management Officer • Information Market Observatory • Initiatives de Microprocesseurs Ouverts • In My Opinion • Installation Maintenance Officer • Installation Mobility Officer • Integration Management Office • Interband Magneto-Optic • Interferometer Module • Internationale Meteorologische Organisatie • International MARC Office • International Maritime Organization • International Mathematical Olympiad • International Meteorological Organization (now WMO) • International Meteor Organization • Zemio (Central African Rep.) (location identifier)

IMOAS Information Management and Office Automation System

IMOP Instruments and Methods of Observation Programme

IMOS Integrated Modeling of Optical Systems • Ion-implanted Metal Oxide Semiconductor

IMOX Implanted Oxide

Imp Impact • Impasse • Impedance • Imperative • Imperator • Imperatrix • Imperfect • Imperial • Impersonal • Implement • Implementation • Implement • Implemented • Important • Import • Important • Imported • Importer • Imprenta • Impression • Imprimatur • Imprimé • Imprimeur • Imprint • Improper • Improve • Improved • Improvement • Impulse

IMP Ice Mass Path • Imager for Mars Pathfinder • Impasse • Imperatriz (Brazil-MA) (location identifier) • Imperial College of Science, Technology and Medicine (Starlink internal code) • Implement • Implementation • Implementation and Monitoring Plan • Implementation Management Plan • Importance • Improve • Improvement and Modernization Plan • Important • Improvement Maintenance Program • Improving Management Performance • Inclinaison de Meilleure Pénétration • Indeterminate Mass Particle • Inflatable Micrometeoroid Paraglider • Information Management Plan • Information Management Program • Information Message Processing • Information Message Processor • Infrastructure Modernization Program • Initial Memory Protection • Injection into Microwave Plasma • Input Message Processing • Input Message Processor • Institut Médico-Pédagogique • Integrated

Maintenance Plan • Integrated Maintenance Procedure • Integrated Management Plan • Integrated Manpower Program • Integrated Master Plan • Integrated Memory Processor • Integrated Microwave Products • Integrated Monitoring Panel • Integrated Multiprotocol Processor • Intelligence Message Processor • Interface Management Plan • Interface Message Processor • Interim Monitoring Program • Internal Macro-instruction Procedure • International Mars Patrol • International Match Point • Interplanetary and Magnetospheric Physics • Interplanetary Magnetometer Probe • Interplanetary Measurement Probe • Interplanetary Monitoring Platform • Interplanetary Monitoring Probe • Interprocess Message Parsing • Interprocess Message Passing • Intrinsic Multiprocessing • Inventory Management Plan • Item Management Plan

IMPA Intelligent Multiport Adapter • International Master Printers' Association

IMPAC International Merchant Purchase Agreement Card • International Merchant Purchase Authorization Card

IMPACT Illinois Microarchitecture Project utilizing Advanced Computer Technology • Implementation, Planning, And Control Technique • Improved Management Procurement and Contracting Technique • Information Market Action Programme • Innovations in Mobile, Pervasive and Agile Computing Technologies • Innovative Microelectronic Pixellated and Advanced CCD Technology • Integrated Manufacturing, Planning and Control Technology • Integrated Material Productivity Action • International Monitoring Programs for Asteroid and Comet Threat • Inventory Management Program And Control Technique • Investigations of Magnetospheric Particle Acceleration and Turbulence

IMPACTS Integrated Manpower, Personnel, And Comprehensive Training plus Safety

IMPAE Institute of Meteorology and Physics of the Atmospheric Environment

IMPATT Impact Avalanche and Transit Time (diode)

IMPC Independent Mail Processing Center • Inventory Management Processing Code

Impce Importance

Impd Impedance

IMPD Impedance

IMPE Information Management Processing Equipment

IMPEG Information Management Program Evaluation Group

Impér Impératif

Imper Imperative • Imperativo

Impérat Impératif

Imperat Imperative • Imperativo

Imperf Imperfect

Impers Impersonal

IMPEX Import/Export

Impf Imperfect

IMPF Image Processing Facility • International Microgravity Plasma Facility

Impft Imperfect

IMPI Internal Microprogrammed Interface

Impl Implement • Implementation • Implemented

IMPLAN Implementation Plan • Improvement Plan

IMPLS Impulse

IMPO Industrial Maintenance and Plant Operation

IMPOL Imaging Polarimeter

Imposs Impossible

IMPOV In My Point Of View

IMPP Imaging Photometer and Polarimeter

IMPPACT Integrated Modeling of Products and Processes using Advanced Computer Technologies

Impr Imprenta • Impreso

IMPR Improving

IMPReSS Image Perimeters of Sky Surveys

IMPRIMATUR Intellectual Multimedia Property Rights Model And Terminology for Universal Reference

IMPRINT Imagery Prototype Initiative • Improved Performance Research Integration Tool

Imprl Imperial

Improv Improvement • Improvisation

IMPRS International Max-Plack Research School (Germany)

Imprsn Impression

IMPS Institut de Médecine et Physiologie Spatiales • Integrated Master Programming and Scheduling • Integrated Meteorological Processing System • Interaction Measurement Payloads • IRAS Minor Planet Survey

Impt Important

IMPT Important

Imptr Importer

IMPTS Improved Programmer Test Station

Impv Imperative

IMPVD Improved

IMQ Item Manager Quantity

IMR Imaging Microwave Radiometer • Indice Moyen Réel • Individual Medical Report • Informal Memorandum Report • Input Message Report • Insect Monitoring Radar • Intergrain Magnetoresistance • Intermetallic Reaction • Internal Management Review • Internet Measurement Report • Internet Monthly Report • Inventory Management Review • Ion-Molecule Reaction • Item Management Review

IMRADS Information Management, Retrieval And Dissemination System

IMRB Independent Modification Review Board • Interim Mission Reconfiguration Nase • Item Management Review Board

IMRC Imaging Media Research Center (KITST) • Innovative Manufacturing Research Centre (UK) • International Mission Review Committee

IMRD Initial Mass Ratio Distribution

IMRMS Ion-Molecule Reaction Mass Spectrometry

IMRO Inspection, Modification, and Repair Order

IMRR Inspection Maintenance and Repair Record • Institute for Reference Materials and Measurements

IMRT Imagerie Médicale et Radiologie Thérapeutique • Indo-Mauritian Radiotelescope • Intensity-Modulated Radiation Therapy

IMRv International Materials Review (periodical code)

IMS Image Manipulation System • Image Motion Simulator • Imaging Spectrometer • Impact Measuring System • Indirect Measuring System • Industrial Mathematics Society (USA-MI) • Industrial - Medical - Scientific • Inertial Measuring Set • In-flight Management System • Information Management Service • Information Management System • Information Manufacturing System • Infrared Mapping Spectrometer • Institute for Molecular Science • Institute of Management Services • Institute of Mathematical Statistics • Institute of Museum Services • Institut Suisse de Météorologie (Switzerland) • Instructional Management System • Instructional Media Service • Integrated Maintenance System • Integrated Management Schedule • Integrated Management System • Integrated Mapping System • Integrated Master Schedule • Integrated Meteorological System • Integrated Mission System • Integrated Multiplex System • Intelligent Manufacturing System • Intelligent Microimage Scanner • Interim Meteorological Satellite • Interlock Management System • Intermediate Maintenance Standard • Intermediate Map

Scale • Intermediate-Mass Star • International Magnetosphere Study • International Magnetospheric Study (*) • International Metallographic Society • International Metric System • International Monitoring System • Internet Map Server • Interoperability in Multidatabase Systems (workshop) • Interplanetary Mission Support • Interprocess Message Switch • Inventory Management Simulator • Inventory Management Specialist • Inventory Management System • Ion Mass Spectrometer • Ion Mobility Spectrometer • Ion Mobility Spectrometry • Ion Mobility Spectroscopy • Ionization and Momentum Sensor • Irradiance Measuring System • Item Management Specialist

IMSA Illinois Mathematics and Science Academy (USA-IL) • Installation Medical Support Activity

IMSAT Imaging Satellite

IMSc Institute of Mathematical Sciences

IMSC Information Management Support Council • Installation Management Steering Committee • Installation Management Support Council • Integrated Media Systems Center • International Mobile Satellite Conference

IMSCIE International Magnetosphere Study Central Information Exchange

IMSD Information Management and Services Division • Integrated Mission for Sustainable Development

IMSE Improved Mobile Subscriber Equipment • Integrated Mean Square Error • Integrated Methods in Science and Engineering (conference) • Integrated Modeling Support Environment

IMSI Information Management System Interface • Intelligent Multisensor Integration • Inter-MAN System Interface • International Mobile Subscriber Identity

IMSL Integrated Mathematics and Statistics Library • International Mathematical and Statistical Library • International Mathematical Subroutine Library

IMSO Installation Materiel Support Office • Intensive Management Staff Office • Intensive Management Staff Officer • International Maritime Satellite Organization

IMSP Institut de Mathématiques et de Sciences Physiques • Intense Morning Sector Precipitation • Internet Message Support Protocol

IMSR In-flight Medical Support System • Interplanetary Mission Support Requirement

IMSRR Interim Modeling and Simulation Resource Repository

IMSS In-flight Medical Supply System • Integrated Mission Support System • Integrated Multisensor System • Istituto e Museo di Storia della Scienza (Italy)

IMST Integrated Mission Simulation Test • Integrated Mission Simulator Test

IM&T Information Management & Telecommunications

IMT Immediately • Inductive Modeling Technology • Industrial and Materials Technologies • Industrial Matters and Technology • Information Management and Telecommunications • Information Management Technology • Instantiated Mission Timeline • Institutional Maintenance Trainer • Integrated Modeling and Test • Interactive Modeling Technology • Interface Maintenance Testing • Intermachine Trunk • International Mobile Telecommunications (ITU) • International Mobile Telephony (conference) • Internet Media Type • Inverted Magneton Transducer • Iron Mountain (USA-MI) (location identifier)

IMTA Intermediate Mail Transfer Agent • International Map Trade Association

IMTC Interactive Media Technology Center (USA-GA) • International Multimedia Teleconferencing Consortium

IMTDS Introduction to the Modern Theory of Dynamical Systems

IMTF Integrated Manufacturing Test Facility

IMTI Instrumentation Moving Target Indicator

IMTR Institute for Meteorological Training and Research (Kenya)

IMTS Improved Mobile Telephone Service • Improved Mobile Telephone System • Initial Main Test Set • Interactive Modular Training System

IMTV Interactive Multimedia Television

IMU Impedance Matching Unit • Income Maintenance Unit • Inertial Measurement Unit • Inertial Measuring Unit • Interface Master Unit • Intermatrix Unit • Internal Measurement Unit • Internal Measuring Unit • International Mathematical Union • International Measurement Unit

IMUIMG ISDN Memorandum of Understanding Implementation Management Group

IMUL Integer Multiply

IMUPLM Inertial Measurement Unit Prelaunch Monitor

IMUX Inverse Multiplexer • Inverse Multiplexing

IMV Imaging Motion Magnetograph • Intermodule Ventilation

IMVDS Item Management Vehicle Data System

IMVP Institute of Metrology for Time and Space (Russia)

IMW Instituut voor Milieu Wetenschappen • Intelligent Machining Workstation • Intelligent Manufacturing Workstation • Interactive Management Workshop • Intermediate Molecular Weight • Intermediate Wave • International Map of the World • Internet Measurement Workshop

IMWG International Mission Working Group

IMX International Monetary Exchange

IMXB Intermediate-Mass X-ray Binary

IMYCS International Meeting of Young Computer Scientists

IMZ Nimroz (Afghanistan) (location identifier)

in Inch • India • Indonesian (ISO language code) • Intortus

In India • Indian • Indium • Inertia • Inertial • Inlet • Inner

I/N Interference/Noise

IN Ice Nucleus • Impulse Noise • Inca • Inde (code ISO de pays) • India (ISO country code) • Indiana (USA) • Individual Network • Inexact Newton (algorithm) • Information Need • Inlet • Input • Institut National • Instructor • Instrumentation Notice • Instrument Note • Intelligence • Intelligence Network • Intelligent Network • Intelligent Networking • Interference-to-Noise (ratio) • Internal Note • International Notice • Intranetwork

INA Information National Association • Information Network Architecture • Information Networking Alliance • Information Networking Architecture • Institute for Numerical Analysis (USA-CA) • Institut National Agronomique (France) • Institut National de l'Audiovisuel (France) • Inta (Russia) (location identifier) • Integrated Network Analysis • Interactive Network Adapter • Interessengemeinschaft Niederrheinischer Astronomievereine (Germany) • International Normal Atmosphere • Interoperability Network Activity • Iron Nickel Alloy

INAA Instrumental Neutron Activation Analysis • Integrated Neutron Activation Analysis

INADS Initialization and Administration System

INAE Indian National Academy of Engineering (India)

INAF Instituto Nazionale di Astrofisica (Italy) • International Name Authority File

INAG Institut National d'Astronomie et de Géophysique (France) (now INSU)

INALCO Institut National des Langues et Civilisations Orientales

INAMHI Instituto Nacional de Meteorología e Hidrología (Ecuador)

INAMI Institut National d'Assurance Maladie-Invalidité (Belgium)

INAOE Instituto Nacional de Astrofísica, Optica y Electrónica (Mexico)

INAP Integrated Neutron Activation Prediction • Intelligent Network Application Part

INAPI Institut National Algérien de la Propriété Industrielle (Algeria)

InARP Inverse Address Resolution Protocol

INAS Indexing and Abstracting Service • Institutional Need Analysis System • Institut National d'Administration Scolaire • Integrated Network Analysis System • International Near-Earth Asteroid Survey

INASAN Institut Astronomii Akademii Nauk (Russia)

INASI Institut National d'Assurance Sociale pour Travailleurs Indépendants (Belgium)

InASP Indian Academy of Sciences Proceedings (periodical code)

INASP International Network for the Availability of Scientific Publications

InAsSb Indium - Arsenide - Antimonide

Inaug Inaugurate • Inaugurated

INAV Inertial Navigation

INB Independence (Belize) (location identifier) • Internal Network Bus

INBAD Institut National du Baccalauréat à Distance (France)

Inbd Inboard • Inbound

INBD Inbound

Inbegr Inbegriffen

INBIT Input Bit

inc Incisus • Include (file extension) • Incus

Inc Include • Included • Including • Inclusive • Income • Incomplete • Incorporate • Incorporated • Incorporation • Income • Incoming • Incomplete • Inconclusive • Incorporated • Increase • Increased • Increasing • Increment • Incumbent

INC In Cloud • Incomplete • Incorporated • Increase • Increment • Indice National de la Consommation • Inertial Navigation Computer • Inhibit Command • Installation Notice Card • Installation Notification Certificate • Institut National de Cartographie • Interface Network Card • Interface Network Controller • Intergovernmental Negotiating Committee • International Carrier • International Negotiating Committee • Intranuclear Cascade • Investigations Numériques en Cosmologie • Yinchuan (China-PRC) (location identifier)

INCA Indian National Cartographic Association (India) • Input Catalogue • Integrated Nuclear Communications Assessment • Intelligence Communications Architecture • Ion and Neutral Camera (Cassini)

Incand Incandescent

INCB International Narcotics Control Board

Ince Insurance

INCE Instituto Nacional de Calidad y Evaluación (Spain)

InCh Inorganic Chemistry (periodical code)

INCH Intelligent Chart

INCI Internetworking Configuration Item

Incid Incidental

Incin Incinerator

Incl Inclination • Incline • Inclined • Include • Included • Including • Inclusion • Inclusive • Incoming • Incomplete • Inconclusive

INCL Inclination • Include • Including

INCM Intelligent Network Conceptional Model

INCNR Increment Number

INCO Initialize and Checkout • Install and Checkout • Instrumentation Communications Officer • Integrated Communications Officer

Incoat Incoativo

Incog Incognito

Incoh Incoherent

Incompl Incomplete

INCOMPLT Incomplete

Incor Incorporated • Incorporation • Incorrect

INCOR Incorrect

Incorp Incorporated • Incorporation

Incorr Incorrect

INCOSE International Council on System Engineering

INCOSPAR Indian National Committee for Space Research (India)

Incr Increase • Increased • Increasing • Increment

INCR Increase • Interrupt Control Register

Incrt Increment

INCS Institut National de Cinématographie Scientifique

in d. In Dies

ind Index (file extension) • In Nomine Dei

Ind Independence • Independent • Index • India • Indian • Indiana (USA) • Indicate • Indicated • Indication • Indicative • Indicator • Indices • Indies • Indigo • Indirect • Indirectly • Individual • Indonesia • Indonesian • Inductance • Induction • Indus-Indi (constellation) • Industrial • Industrie • Industriel • Industry

IND India (international vehicle registration code) • Inde (code ISO de pays) • India (ISO country code) • Indianapolis (USA-IN) (location identifier) • Indication • Indicator • Indústria • In Nomine Dei • Investigative New Drug

INDAC Inductor Accelerator

INDAE Instituto Nacional de Derecho Aeronáutico y Espacial (Argentina)

INDAT Incoming Data

INDB Integrated Network Database

INDC Indicate • International Nuclear Data Committee (IAEA)

indd In-Design Document (file extension)

INDD Intranodal Data Distribution

IndE Industrial Engineer

INDECOPI Instituto Nacional de Defensa de la Competencia y de la Protección de la Propiedad Intelectual (Peru)

Indéf Indéfini

Indef Indefinido • Indefinite

INDEF Indefinite • Indefinitely

INDEFLY Indefinitely

Indep Independent

INDES Integrated Nuclear Data Evaluation System

Indét Indéterminé

Indet Indeterminado

INDFT Indefinite

INDI Ion-Neutral Dynamics Investigation

Indic Indicating • Indication • Indicative • Indicativo • Indicator

INDIGO Interservice Data Integration for Geodetic Operations

Indiv Individual

Individ Individual

Indl Industrial

Indn Induction

INDO Intermediate Neglect of Differential Overlap

Indo-Eur Indo-European

INDOEX Indian Ocean Experiment

Indo-Ger Indo-German • Indo-Germanic

Indon Indonesia

INDOTEC Instituto Dominicano de Tecnologia Industrial (Dominican Rep.)

INDREG Inductance Regulator

indt In-Design Template (file extension)

Indt Indent

Indtry Industry

Induc Induction

Indum Indumentaria

Indus Industrial • Industrious • Industry

Indust Industrial • Industrious • Industry

Indv Individual

INE Chinde (Mozambique) (location identifier) • Inertial Navigation Equipment • Inline Network Encryptor • Instituto Nacional de Empresas (Spain) • Instituto Nacional de Estadística (Spain)

INEAS International Near-Earth Asteroid Survey

INEC Institut National de l'Eau et de la Chimie (France)

Ined Ineditus

INEEL Idaho National Engineering and Environmental Laboratory (USA-ID)

INED Institut National d'Études Démographiques (France) • International Network for Educational (information)

INEGI Instituto de Engenharia Mecânica e Gestão Industrial (Portugal) • Instituto Nacional de Estadística, Geografía e Informatica (Mexico)

INEL Idaho National Engineering Laboratory (USA-ID) • International Network of Emerging Libraries

INEM Instituto Nacional de Empleo (Spain)

INEMET Instituto Nacional de Meteorologia (Brazil-DF)

INEN Instituto Ecuatoriano de Normalización (Ecuador)

INEP Institut National d'Éducation Populaire (France)

InEPS Indian Earth and Planetary Sciences (periodical code)

INER Institute of Nuclear Energy Research

INERIS Institut National de l'Environnement Industriel et des Risques (France)

InES Indian Academy of Sciences Proceedings: Section C Engineering Sciences (periodical code)

INES Innovative Navigation European System • International Network of Engineers and Scientists • International Nuclear Event Scale • IUE Newly Extracted Spectra

INESC Instituto Nacional de Engenharia de Sistemas e Computadores (Portugal)

INET Institute of Nuclear Energy Technology

INETER Instituto Nicaragüense de Estudios Territoriales (Nicaragua)

INETI Instituto Nacional de Engenharia e Tecnologia Industrial (Portugal)

INETOP Institut National d'Étude du Travail et d'Orientation Professionnelle (France)

INEWS Integrated Electronic Warfare System

in ex. In Extenso

in f. In Fine

inf Infinum • Information (file extension) • Infunde

Inf Inferior • Infinite • Infinitive • Infinity • Infinum • Infirmier • Infirmière • Influence • Inform • Informal • Information • Informed • Infra

INF In a Next Life • Infinite • Information • In Guezzam (Algeria) (location identifier) • Interactive Network Facility • Interface • Interference

INFA Infanta • Institut National de Formation des Adultes (France)

InfCo Information Control (periodical code)

INFCO Information Committee

InfD Information Display (periodical code)

INFE Infante

INFEP Institut National de Formation à l'Enseignement et la Pédagogique (France)

Infin Infinitive

Infinit Infinitivo

INFIS International Federation of Information Services

InFiZ Inzhenerno Fizicheskii Zhurnal (periodical code)

INFJ Introverted Intuitive Feeling Judging

Infl Inflammable • Inflated • Influence • Influenced • Influencia • Influido

INFLIBNET Information and Library Network (India)

Infm Information

I/NFM Intergraph Network File Manager

INFM Istituto Nazionale di Fisica della Materia (Italy)

INFN Istituto Nazionale di Fisica Nucleare (Italy)

INFNET Instituto Nazionale di Fisica Nucleare Network (Italy)

Info Inform • Information • Informed

INFO Information

In-Fo In-Folio

InFo In-Folio

INFO Information Network and File Organization • Information Network For Operations • Integrated Network Fiber Optics

INFOCEN Information Central

INFOL Information-Oriented Language

Inform Informática • Information

INFORMS Institute For Operations Research and the Management Sciences (merging of ORSA and TIMS) • Institute For Operations Research and the Management Society

INFONET Information Network

INFOSEC Information Security

INFP Introverted Intuitive Feeling Perceiving

InfPh Infrared Physics (periodical code)

INFRA Infrastructure

INFRAL Information Retrieval Automatic Language

INFREP Institut National de Formation et de Recherche sur l'Éducation Permanente (France)

INFSWG Information Services Working Group

INFU Institut für Umweltforschung (Germany)

InFuJ Institute of Fuel Journal (periodical code)

Ing Ingeniero • Ingenieur • Ingénieur • Inglaterra • Inglés

ING Incremented Number Generator • Inertial Navigation and Guidance • Inertial Navigation Gyro • Intense Neutron Generator • Isaac Newton Group (Canary Islands) • Istituto Nazionale di Geofisica (Italy) • Lago Argentino (Argentina-SC) (location identifier)

InGaAs Indium Gallium Arsenide

IngAr Ingenieur Archiv (periodical code)

INGE Intelligent Generator

Ingeg Ingegneria (periodical code)

Ingen Ingeniería

InGeo International Geophysics (periodical code)

Ingl Inghilterra

INGO International Non-Governmental Organization

INGRES Institut für Gewerblichen Rechtschutz (Switzerland) • Interactive Graphics and Retrieval System

INGRID Isaac Newton Group Red Imaging Device

Inh Inhaber • Inhabitant • Inhalt

INH Inhambane (Mozambique) (location identifier) • Inhibit

Inhab Inhabitant

INHCI Internetworking Hardware Configuration Item

Inher Inheritance

ini Initialization (file extension)

INI Information Network Interface • Initialize • Isaac Newton Institute (Chile) • Nis (Yugoslavia-Serbia) (location identifier)

INIB Internationale Normeninformationsbank

INIC Instituto Nacional de Investigação Científica (Portugal)

INID Intercept Network Inward Dialing

INIE Innovations in Nuclear Infrastructure and Education (program)

INIS International Nuclear Information System (IAEA)

InISA Indian Institute of Science Journal of Aeronomical Society of India (periodical code)

INISI Instituto de Derecho y Sociedad de la Información (Spain)

InISJ Indian Institute of Science Journal (periodical code)

INISSE International Institute of Space Sciences and Electronics

INIST Institut National de l'Information Scientifique et Technique (France)

Init Initial • Initialization • Initialize • Initially • Initio

INIT Initial • Initialization • Initialize • Initialized • Initiate

INITCO Initialization and Checkout

INIU Interim Network Interface Unit

Inj Injection • Injury

INJ Injune (Australia-QLD) (location identifier) • Institut National de la Jeunesse (France)

InJAE Indian Journal of Aerospace Engineering (periodical code)

inject Injectio

Inject Injection

INJEP Institut National de la Jeunesse et de l'Éducation Populaire (France)

InJET Indian Journal of Electronics Telecommunication (periodical code)

InJME Indoan Journal of Mechanical Engineering (periodical code)

InJP Indian Journal of Physics (periodical code)

InJPS Indian Journal of Physics - Section B (periodical code)

InJRSP Indian Journal of Radio and Space Physics (periodical code)

InJTP Indian Journal of Theoretical Physics (periodical code)

INK Information Network of Kansas (USA-KS) • Information - Networks - Knowledge • Wink (USA-TX) (location identifier)

inkl Inklusive

Inl Inlet

INL Integral Non-Linearity • Internal Noise Level • International Falls (USA-MN) (location identifier) • Internode Link

INLD Inland

in lim. In Limine

in loc. In Loco

in loc. cit. In Loco Citato

InLoP Instytut Lotnicza Prace (periodical code)

INLR Item No Longer Required

INLT Inlet

INM Indice Nouveau Majoré • Innamincka (Australia-SA) (location identifier) • Institut National de Métrologie • Instituto Nacional de Meteorología (Spain) • Integrated Network Management • Integrated Noise Model • International Nautical Mile

INMARSAT International Maritime Satellite Organization

in mem. In Memoriam

INMET Instituto Nacional de Meteorologia (Brazil)

INMETRO Instituto Nacional de Metrologica, Normalização e Qualidade Industrial

INMG Instituto Nacional de Meteorologia e Geofísica (Portugal)

INMH Instituto Nacional de Meteorología e Hydrología (Ecuador)

InMP Instytut Maszyn Przeplywowych (periodical code)

InMPP Instytut Maszyn Przeplywowych Prace (periodical code)

InMS Indian Academy of Sciences Proceedings Mathematical Sciences (periodical code)

INMS Integrated Network Management System • Ion and Neutral Mass Spectrometer

Inn Inning

INN Innsbruck Kranebitten (Austria) (location identifier) • Instituto Nacional de Normalización (Chile) • Intermediate Network Node • International Negotiation Network • Internet News • Internode Network

INNERTAP Information Network on New and Renewable Energy Technologies for Asia and the Pacific

INNORPI Institut National de la Normalisation et de la Propriété Industrielle (Tunisia)

INNS International Neural Network Society

INO Inflow - No Outflow • Inongo (Zaire) (location identifier) • Institute for Naval Oceanography • Iraqi National Observatory (Iraq) • Istituto Nazionale di Ottica (Italy) • Iterative Natural Orbital

INOBA Intelligent Node for Basic Access

INOC Internal Network Operations Center • Internet Network Operations Center

INode Index Node

Inop Inoperative

INOPV Inoperative

INORAC Inquiry - Order - Accounting (system)

Inorg Inorganic

INOSHAC Indian Ocean and Southern Hemisphere Analysis Centre

Inp Input

InP Indium Phosphide

INP If Not Possible • Initial Program Load • Input • Institute of Nuclear Physics • Institut National Polytechnique • Interdepartementaal Natuurbeleids Plan • Interim Number Portability • Intranode Policing

INPA Institute for Nuclear and Particle Astrophysics • Instituto Nacional de Pesquisas da Amazonia (Brazil)

INPAC Institute for Nuclear and Particle Astrophysics and Cosmology (USA-CA)

INPACK Input Acknowledge

INPADOC International Patent Documentation Center

INPC International Nuclear Physics Conference

INPE Instituto Nacional de Pesquisas Espaciais (Brazil)

INPG Institut National Polytechnique de Grenoble (France)

INPHO Information network for Public Health Officials

INPI Institut National de la Propriété Industrielle (France) • Ion and Neutral Particle Imager

INPL Institut de Physique Nucléaire de Lyon (France) • Institut National Polytechnique de Lorraine (France) • Israel National Physical Laboratory

INPO Information Network Program Office • Institute of Nuclear Power Operations

in pr. In Principio

INPRES Instituto Nacional de Prevención Sísmica (Argentina)

In Pro In Proportion

INPROC In Processing

INPS Institut National Pour la Simplification • Intense Pulsed Neutron Source

INPT Institut National Polytechnique de Toulouse (France)

INPUT Induced Pulse Transient

Inq Inquiry • Inquisition

INQ Inisheer (Ireland) (location identifier)

INQUA International Union for Quaternary Research

Inr Inner

INR Impôt de Non-Résident • Inference-to-Noise Ratio • Initial Nuclear Radiation • Institute for Nuclear Research • Intercept Recorder • Rupee (India) (ISO currency code) • Sault Sainte Marie - Kincheloe (USA-MI) (location identifier)

INRA Institute for Natural Resources in Africa • Institut National de Recherche Agronomique (France)

INRC Identique - Négative - Réciproque - Corrélative (opération)

INRDP Institut National de Recherche et de Documentation Pédagogique (France)

INRE Institute of Natural Resources and Environment

INREQ Information Request

INRENA Instituto Nacional de Recursos Naturales (Peru)

INRETS Institut National de Recherche sur les Transports et leur Sécurité (France)

INRIA Institut National de Recherche en Informatique et en Automatique (France)

INRNE Institute for Nuclear Research and Nuclear Energy (Bulgaria)

INRP Institut National de Recherche Pédagogique (France)

INRPVC Institut National de Recherche sur la Prévention du Vieillissement Cérébral (France)

INRS Institut National de la Recherche Scientifique • Institut National de Recherche et de Securité

in s. In Situ

ins Inches • Installation (file extension)

Ins Inscribe • Inscribed • Inscription • Insert • Inserted • Insertion • Inserto • Inside • Inspector • Insular • Insulate • Insulated • Insulation • Insurance • Insure

INS Immigration and Naturalization Service (USA) • Improved Navigational Satellite • Indian Springs (USA-NV) (location identifier) • Inelastic Neutron Scattering • Inertial Navigational System • Inertial Navigation Sensor • Inertial Navigation System (*) • Information Network System • Input String • Insensitive • Insert • Institut National de la Statistique • Instruction • Instrumentation • Insulation • Integrated Navigation Set • Integrated Navigation System • Integrated Network Server • Integrated Network System • International Navigation System • International Network Service • Ion-Neutralization Spectroscopy • Isolated Neutron Star

INSA Indian National Science Academy (India) • Institut National des Sciences Appliquées (France)

INSAG International Nuclear Safety Advisory Group

INSAP Indian National Science Academy Proceedings (periodical code) • Inspiration of Astronomical Phenomena

InSAR Interferometry using Synthetic Aperture Radar

INSAR Interferometric Synthetic Aperture Radar

INSAT Indian National Satellite

insb insbesondere

InSb Indium Antimonide

Insc Inscribe • Inscribed • Inscripción • Inscription

INSC Interoperable Networks for Secure Communications

Insce Insurance

Inscr Inscribe • Inscription

INSDOC Indian National Scientific Documentation Centre (India)

INSEC Internal Security

INSEE Institut National de la Statistique et des Études Économiques (France)

Insep Inseparable

Insép Inséparable

INSERM Institut National de la Santé et de la Recherche Médicale (France) (*)

INSIC Information Storage Industry Consortium (USA)

INSIDE Interior Structure and Internal Dynamical Evolution

InSIGHT Investigative Stimuli for Intuitive Growth using High Technology

INSITE Institutional Space Inventory Technique • Integrated Sensor Interpretation Technique • Interactive Nonlinear Systems Investigative Toolkit for Everyone

INSIVUMEH Instituto Nacional de Sismología, Vulcanología, Meteorología e Hidrología (Guatemala)

INSM Integrated Network and System Management

Insol Insoluble

Insolv Insolvent

Insp Inspect • Inspected • Inspection • Inspector

INSP Inspection • Integrated National Space Programme

INSPEC Information Services for the Physics and Engineering Communities

INSPIRE Integrated Spatial Information Research and Engineering • Interactive NASA Space Physics Ionosphere Radio Experiment

InSSR Akademiia Nauk SSSR Investiia Mekhanika Zhidkosti i Gaza (periodical code)

InSST Indian Institute of Science Space Science Technology Applications (periodical code)

Inst Install • Installation • Installed • Instance • Instant • Instantaneous • Institución • Institut • Institute • Institution • Institutional • Institutionnel • Instituto • Instruct • Instruction • Instructional • Instructor • Instrument • Instrumental • Instrumentation • Instrumented

INST Information Standard • Installation • Institute • Instruction • Instrument • Instrumentation

INSTA Internordisk Standardisering

INSTAAR Institute for Arctic and Alpine Research

INSTAR Inertialess Scanning, Tracking And Ranging

INSTARS Information Storage And Retrieval System

INSTBY Instability

Instd Instead

IN-STEP In-Space Technology Experiment Program

INSTEP In-Space Technology Experiment Program

Instl Install • Installation

Instld Installed

INSTLR Installer

Instm Instrumentation

Instn Institution

INSTN Institut National des Sciences et Techniques Nucléaires

Instr Instruction • Instructor • Instrument • Instrumentation • Instrumental

INSTR Instruct • Instrument • Instrumentation

Instrum Instrumental • Instrumental • Instrumentation • Instrumented

INSU Institut National des Sciences de l'Univers (France)

Insuf Insufficient

Insul Insulate • Insulation

Insuld Insulated

INSURV Inspection and Survey

int Interface (file extension)

Int Integer • Integral • Integrated • Intelligence • Intendente • Intensity • Interest • Intérêt • Interim • Interior • Intérieur • Interjection • Intermediate • Intermittent • Internal • International • Interned • Interpret • Interpretation • Interpreter • Interrogate • Interrupt • Intersection • Interval • Interview • Intransitive

INT Institut National des Télécommunications (France) • Institut National du Travail (France) • Integer • Integrate • Integrated • Integration • Intelligence • Intendente • Intensity • Intercept • Interior • Intermittent • Internal • International • Interrupt • Isaac Newton Telescope (Canary Islands) • Winston-Salem (USA-NC) (location identifier)

INTA Instituto Nacional de Técnica Aerospacial (Spain) • Instituto Nacional de Tecnologia Agropecuaria (Argentina) • International Trademark Association (formerly USTA) • Interrupt Acknowledge

IntaC INTA CONIE (periodical code)

INTACS Integrated Tasking and Control System

int. al. Inter Alia

INTAP Interoperability Technology Association for Processing

INTASAT Instituto National de Ténica Aerospatial Satellite

Intchg Interchangeable

INTCHG Interchange

INTCM Intercom

Intcp Intercept

INTCP Intercept

INTCTR Integrating Center

INTD Institut National des Techniques de la Documentation (France)

INTEC Institut National des Techniques Économiques et Comptables (France)

INTECH Institute for New Technologies

INTECO Instituto de Normas Técnicas de Costa Rica

INTECOL International Association for Ecology

Integ Integral • Integrate • Integration

INTEG Integrate • Integrated • Integration

INTEGR Integrate • Integration

INTEGRAL International Gamma-Ray Astrophysics Laboratory

Integrg Integrating

Intel Intelligence • Intelligent

INTEL Intelligence

Intell Intelligence • Intelligent

INTELSAT International Telecommunication Satellite Consortium

Intens Intensifier • Intensify • Intensity • Intensive • Intensivo

INTENS Intensity

Inter Interavia (periodical code) • Interim • Interior • Intermediate • International • Interrogation • Interrogative

INTER Intermediate • Intermittent

Intercomm Intercommunication

Interj Interjección • Interjection • Interjectiva

INTERMAG International Magnetics (conference)

INTERMAPS Interactive Multimedia Access Publishing services

Intern International

Internac Internacional

Internat International

InterNIC Internet Network Information Center

Interop Interoperability

INTEROP Interoperability • Interoperational

Interp Interpreter

Interplanet Interplanetary

Interrog Interrogate • Interrogation • Interrogative • Interrogativo

INTERROG Interrogate

INTERTACOS Interpretation, Treatment, Analysis and Correlation Of Spectra

INTESA Iniziative Telematiche per Servizi Applicativi

INTF Interface • Interim National Test Facility (USA)

Intg Interrogation

INTG Integration

Intgr Integrate

INTI Instituto Nacional de Tecnologia Industrial

INTIPS Integrated Information Processing System

INTJ Introverted Intuitive Thinking Judging

Intk Intake

Int'l International

Intl Internal • International

INTL International

Intlk Interlock

INTLK Interlock

Intlz Initialize

Intmd Intermediate

INTMED Internal Medicine

INTMK Itogi Nauki i Tekhniki Seriia Meteorologiia i Klimatologiia (periodical code)

Intmt Intermittent

INTMT Intermittent

INTN Instituto Nacional de Tecnología y Normalización

INTNET Intelligence Network

INTO Interrupt if Overflow

INTOP International Operation

INTOR International Tokamak Reactor

INTP Introverted Intuitive Thinking Perceiving

INTPHTR Interphase Transformer

Intpr Interpret

Intr Interior • Intransitif • Intransitive • Intransitivo • Introduce • Introduced • Introducing • Introduction • Introductory

INTR Interior

INTRA Intelligent Tracker

INTRAN Input Translator

Intrans Intransitive

In Trans In Transit

INTREP Intelligence Report

Int. Rev. Internal Revenue

INTREX Information Transfer Experiment

Intrf Interference

Intrg Interrogate

INTRG Interrogate

INTRLVR Interleaver

Intro Introduce • Introduced • Introduction • Introduire • Introduit

Introd Introducción • Introduce • Introduced • Introduction • Introduire • Introduit

Intrpl Interpolate

Intrpt Interrupt

INTRPT Interrupt

INTS Improved Night Targeting System • Institut National des Techniciens Supérieurs (France) • Intense

INTSA Itogi Nauki i Tekhniki Seriia Astronomiia (periodical code)

Intsct Intersect

INTSF Intensify • Intensifying • Itogi Nauki i Tekhniki Seriia Fizika Plazmy (periodical code)

INTSFY Intensify

INTSI Itogi Nauki i Tekhniki Seriia Issledovanie Kosmicheskogo Prostranstva (periodical code)

IntSM Interavia Space Markets (periodical code)

INTSO Itogi Nauki i Tekhniki Seriia Okeanologiia (periodical code)

INTSR Itogi Nauki i Tekhniki Seriia Radiotekhnika (periodical code)

INTSS Itogi Nauki i Tekhniki Seriia Sovremennye Problemy Matematiki (periodical code)

INTST Intensity

Intstg Interstage

INTSUM Intelligence Summary

INTSV Itogi Nauki i Tekhniki Seriia Vozdushnyi Transport (periodical code)

INTUG International Telecommunications Users Group

Intvl Interval

INTVL Interval

INTVLM Intervalometer

INTX Interrupt and Timing Extension

INU Inertial Navigation Unit • Integration Unit • Ion Nucleation • Nauru Island (location identifier)

INUP Intelligent Network User Part

Inus Inusitado • Inusité

inv Invenit

Inv Invenit • Invent • Invented • Inverter • Invention • Inventor • Inventory • Inverse • Inversed • Inversion • Invert • Inverted • Inverter • Investigate • Investigated • Investigation • Investment • Invitation • Invite • Invited • Invoice • Invoiced

INV Inductive Null Voltage • Institut National du Verre (Belgium) • Invalid • Inventory • Inverness (UK-Scotland) (location identifier) • Inverter

Invar Invariable

Invest Investigate • Investigated • Investigation • Investigational

inv. et del. Invenit et Delineavit

InvHG Investitionshilfegesetz

INVOF In Vicinity Of

INVOL Involuntary

InvPr Inverse Problems (periodical code)

INVRN Inversion

Invs Inverse

Invt Inventory • Invert

Invtr Inverter

InvZulG Investitionszulagengesetz

INW Institut für Nachrichtentechnik und Wellenausbreitung • Inward • Winslow (USA-AZ) (location identifier)

INWATS Inward Wide Area Telephone Service

INWG International Network Working Group • Internetwork Working Group

inx Index (file extension)

INX Inanwatan (Indonesia) (location identifier) • Information Exchange

INXS Internet Exchange Service • Internet Exchange System

INY Inyati (South Africa) (location identifier)

INZ In Salah (Algeria) (location identifier)

IN2 Square Inches

IN2P3 Institut National de Physique Nucléaire et de Physique des Particules (France)

In2S3 Indium Sulfide

IN3 Cubic Inches • Internet Interdisciplinary Institute

Io Iowa (USA)

I/O Inboard/Outboard • In/Out • Input/Output • Inspecting Order

I&O Inlet & Outlet

IO British Territory of Indian Ocean (ISO country code) • Idea Organizer • Identification Officer • Image Orthicon • Indian Ocean (British Territory of_) • Indirect Object • Industrial Object • Industrial Operations • Inflow-Outflow • Information Objective • Information Officer • Information Operation • Information-Orientation • Information Outlook • Inlet and Outlet • In Order • Information Officer • Initial Only • Input-Output • Inspection Order • Inspection Outline • Instruction Officielle • Integrated Optics • Intelligence Officer • Intelligence Oversight • Interior Orientation • Intermediate Objective • International Officer • International Organization • Interoperability • Interpretive Operation • Inventory Objective • Investigating Officer • Inward Operator • Ion Engine • Ionospheric Observatory • Issue Order • Iterative Operation • Territoire Britannique de l'Océan Indien (code ISO de pays)

IoA Institute of Astronomy

IOA Independent Operational Assessment • Initial Operational Assessment • Input-Output Adapter • Input-Output Address • Input-Output Assembly • Inquiry - Ordering - Accounting • Instrumentation Operating Area • Instrument Operating Assembly • Internal Operations Assessment • Ioannina (Greece) (location identifier)

IOAA Independent Office Appropriations Act (USA)

IOAT Indicated Outside Air Temperature • In-Orbit Automation Technology

IOAU Input-Output Access Unit

I/OB Input/Output Bus

IOB Ink On Bitmap • Input-Output Block • Input-Output Box • Input-Output Buffer • Input-Output Bus • Installation Operating Budget • Intelligence Oversight Board • Inventory Operation Branch

IOBC International Organization for Biological Control

IOBM In Ocean By Mistake • Input-Output Bitmap

IOBPS Input-Output Box and Peripheral Simulator

I/OC Input/Output Controller

IOC Image Orthicon Camera • Image Orthicon Control • Inception Of Contract • Indirect Operating Cost • Information Operations Center • Initial Operating Capability • Initial Operating Configuration • Initial Operational Capability • Initial Operations Capability • Initiation Of Contract • Inner Oort Cloud • Innovation and Organizational Change • In-Orbit Calibration • In-Orbit Checkout • In-Orbit Communication • Inorganic Chemical • In-Out Converter • Input Offset Current • Input-Output Card • Input-Output Channel • Input-Output Comparator • Input-Output Controller • Input-Output Converter • Integrated Operations Configuration • Integrated Optical Circuit • INTELSAT Operations Center • Intergovernmental Oceanographic Commission • Interim Operational Capability • International Observations Campaign • International Oceanographic Commission • International Olympic Committee • International Operating Center • International Operating Center • International Ozone Commission • Interoffice Channel • Interorbit Communication • Investigations Operations Center • ISDN Ordering Code • Item Of Concern • Iterative Orbit Calculator

IOCC Information Operations Coordination Cell • Input-Output Channel Converter • Input-Output Control Center • Input-Output Control Command • Input-Output Controller Card • Input-Output Controller Chip • Integrated Operations Control Center

IOCD Initial Operational Capability Date • International Organization for Chemistry in Development

IOCDS Input-Output Configuration Data Set

IOCG International Organization on Crystal Growth

IOCM Interim Operational Contamination Monitor

IOCRD In-Orbit Calibration Requirements Document

IOCS In-Office Cost System • Input-Output Computer Services • Input-Output Control System

IOCTL Input-Output Control

IOCU Input-Output Control Unit • International Organization of Consumers Unions

IOD Identified Outward Dialing • Images On Demand • Information Object Definition • Information On Demand • Initial Operational Date • Initial Orbit Determination • Input-Output Daemon • Input-Output Device • Integrated Optical Density • Integration and Operational Demonstration

IODA Input-Output Drive Assembly

IODB Instrument Operations Database • Integrated Object Database

IODC Indian Ocean Data Coverage • Input-Output Data Channel • Input-Output Data Controller • Input-Output Delay Counter

IODE International Oceanographic Data and Exchange • International Oceanographic Data Exchange

IODP Input-Output Digital Processor

IODS Improved Optical Detection System

IoE Institute Of the Environment

IOE Illogical Operation Error • Independent Operational Evaluation • Independent Operational Evaluator • Initial Operating Experience • Initial Operational Evaluation • Input-Output Error • Input-Output Expander • Institute Of the Environment • Internal Operating Expenditure

IOEM Integrated Opto-Electronics Module

IOEP Independent Operational Evaluation Plan

IOER Independent Operational Evaluation Report

IOF Initial Operational Flight • Input-Output Facility • International Oceanographic Foundation

IofM Isle of Man (UK)

IOFPS Integrated Operations Facility-Platform-System

IOG Input-Output Gate

IOGE Input-Output Operational Ground Equipment

IOH Integrated Open Hypermedia

IOHA Integrated Operating Hazard Analysis

IOHR Imagerie Optique à Haute Résolution

IOI Imaging Optical Interferometer • In-Orbit Infrastructure • Interactive Observing Infrastructure • Interim Operating Instruction • Internal Operating Instruction • International Ocean Institute • International Olympiad in Informatics • International Ozone Institute

IOIC Input-Output Interrupt Code

IOIS Integrated Office Information System • Inter-Organization Information System

IOJ International Organization of Journalists

IOK Iokea (Papua New Guinea) (location identifier)

IOL Initial Outfitting List • Instantaneous Overload • Intermediate Objective Lens • Interoffice Letter • Interorbit Link • Intraocular Lens • Ireland On Line

IoM Institute of Meteoritics

IOM Input-Output Module • Input-Output Multiplexer • Installation - Operation - Maintenance • Install - Operate - Maintain • Institute Of Meteoritics • Instituto de Optica de Madrid (Spain) • International Organization for Migration • Interoffice Memorandum • Isle of Man (UK) (location identifier)

IOMAC Indian Ocean Marine Affairs Cooperation

IOMAS Ion Mass Spectrometer

I/OMI Integration, Operations, and Maintenance Instruction

IOMP International Organization for Medical Physics

Ion Ionian

IoN Institute of Navigation

ION Impfondo (Congo) (location identifier) • Input-Output Node • Institute Of Navigation • Interlending OSI Network • International Ocean Network

IONA Input-Output Network Adapter

IONCAP Ionospheric Communications Analysis Prediction

IONDS Integrated Operational Nuclear Detection System • Integrated Operational Nuclear Detonation System

IONIC Integrated Optics Near-infrared Interferometric Camera

IonIs Ionosfernye Issledovaniia (periodical code)

Iono Ionosphere (periodical code)

Ionos Ionosphere • Ionospheric

IONL Internal Organization of the Network Layers

IONO Ionospheric

IONPRO Ionospheric Propagation

IOOFC Integrated Optics and Optical Fiber Communication

IoP Institute of Physics

I/OP Inboard/Outboard Profile

I&OP In and Out Processing

IOP Increment Operation Plan • Independently Observable Parameter • Indian Ocean Programme • Initial Operational Period • In-Orbit Plane • Input-Output Package • Input-Output Parallelism • Input-Output Part • Input-Output Processor • Input-Output Pulse • Installation Operating Program • Institute of Ocean Sciences • Institute Of Physics • Instituto de la Opinión Física (Spain) • Integrated Office Package • Integrate On Pad • Intensive Observation Period • Intensive Observing Period • Intensive Operational Period • Interface Operating Procedure • Interface Operational Procedure • Interim Operating Procedure • Internal Ocular Pressure • Internal Operating Procedure • International Opacity Project • Interoperability • Intraocular Pressure • Ioma (Papua New Guinea) (location identifier)

IOPAB International Organization for Pure and Applied Biophysics

IOPB Input-Output Parameter Buffer

IOPC International Oil Pollution Compensation

IOPG Institut et Observatoire de Physique du Globe

IOPL Input-Output Privilege Level • Integrated Open Problem List

IoPP Institute of Physics Publishing

IOPP Institute Of Physics Publishing

IOPS Input-Output (operations) Per Second • Input-Output Programming System • Interface Operating Procedures

IOPT Information Operations Planning Tool • Input-Output Parameter Table

IOQE Input-Output Queue Element

IOR Immediate Operational Requirement • Index Of Refraction • Indian Ocean Region • Industrially-Oriented Research • Inishmore (Ireland) (location identifier) • Input-Output Register • Interoperability Requirement • Interoperable Object Reference • Interorder Overlap Region • Iterative Orthophoto Refinements

IORD Input-Output Read • Integrated Operational Requirements Document

IORL Input-Output Requirements Language

IORTD Input-Output Relationships and Timing Diagram

IOS Ilhéus - Eduardo Gomes (Brazil-BA) (location identifier) • Image Optical Scanner • Image Orthicon System • Índice de Oscilação Sul • Initial Operational System • In-Orbit Servicing • Input-Output Selector • Input-Output Sense • Input-Output Skip • Input-Output System • Input-Output Supervision • Input-Output Supervisor • Input-Output Switch • Inspection Operation Sheet • Inspection Operation System • Institute of Oceanographic Sciences • Instructor-Operator Station • Instrumentation Operation Station • INTEGRAL Overall Simulator • Integrated Observation System • Integrated Observing System • Integrated Office System • Integrated Operations Schedule • Interim Operational Supplement • Interim Operational System • Internetworking Operating System • Interoperability System • Interorganizational System • Intranet Operating System • Inventory and Order System

IOSB Input-Output Standard Block

IOSC Integrated Operations Support Center • Integrated Operations Support Complex

IOSF In-Orbit Simulation Facility

IOSGA Input-Output Support Gate Array

IOSH Institution of Occupational Safety and Health

IOSPO Information Operations and Special Project Office

IOSS Input-Output Subsystem • Interagency Operations Security Support

IOSSS Interagency Operating Security Support Staff

IOST ISEMSI Operations Support Team

IOSTE International Organization for Science and Technology Education

I/OT Input/Output Test

IOT British Territory of Indian Ocean (ISO country code) • Independent Operational Test • Indian Ocean Territory • Initial Officer Training • Initial Orbit Time • Initial Operational Test • In-Orbit Test • In-Orbit Testing • In-Out_thru • Input-Output Table • Input-Output Termination • Input-Output Transfer • Instrument Operations Team • Integral Operator Trainer • Intelligent Optical Terminal • Territoire Britannique de l'Océan Indien (code ISO de pays)

IOTA Infrared and Optical Telescope Array • Infrared-Optical Telescope Array (*) • Institut d'Optique Théorique et Appliquée (France) • Interfaces Océan-Terre-Atmosphère • International Occultation Timing Association

IOTA-ES International Occultation Timing Association - European Section

IOTAES International Occultation Timing Association - European Section

IOTD Image Of The Day

IOT&E Independent Operational Test & Evaluation • Initial Operational Test & Evaluation • Installation Operational Test & Evaluation • Integration Operational Test & Evaluation • Internal Operational Test & Evaluation

IOTA Inbound/Outbound Traffic Analysis • Information Overload Testing Analysis

IOTC Information Operations Technical Center

IOT&E Initial Operational Test & Evaluation

IOTE Initial Operational Test and Evaluation • Integration Operational Test and Evaluation

IOTG Input-Output Task Group

IOTM Image Of The Month

IOTR Item Operation Trouble Report

IOTW Image Of The Week

IOTY Image Of The Year

I/OU Input/Output Unit

IOU Île Ouen (New Caledonia) (location identifier) • Input-Output Unit • Input-Output Utility • I Owe You

IOUBC International Office for Universal Bibliographic Control

IOV Input Offset Voltage

IOVF Independent Overflow

IOW Inflow-Outflow Wind • In Other Words • Iowa City (USA-IA) (location identifier) • Isle Of Wight (UK)

IOWR Input-Output Write

IOX Input-Output Exchange • Input-Output Expansion • Input-Output Transfer (unit)

I/P Current to Pressure Ratio • Identification of Position • Impact Predictor • In Progress • Input • Insurance/Pension

I&P Inserting & Preheating • Intelligence & Planning

IP Ice Pellet • Identification Point • Identification of Position • Igloo Pallet • Image Point • Image Processing • Imaginary Part • Impact Point • Impact Prediction • Impact Predictor • Impingement Point • Implementation Plan • Implementation Procedure • Important Paper • Important Passenger • Important Person • Improvement Program • Incentive Pay • Increment Plan • Indexed and Paged • Index of Performance • Indicator Panel • Indice de Productividad • Indirect Proof • Individual Permit • Induced Polarization • Industrial Preparedness • Industry Program • Inertial Platform • Inertial Processing • Informational Program • Information Processing • Information Processor • Information Product • Information Protection • Information Provider • Ingénieur Principal • Initial Phase • Initial Plan • Initial Point • Initial Position • Initial Production • Initial Provisioning • Inner Pulse • In Plant • Input Power • Input Primary • Input Processor • Installation Plan • Installation Procedure • Installment Plan • Institut Polytechnique • Instruction Pamphlet • Instruction Pointer • Instructor Personnel • Instructor Pilot • Instrumental Polarization • Instrumental Profile • Instrumentation Payload • Instrumentation Plan • Instrumentation Power • Instrument Payload • Instrument Processor • Integer Program • Integer Programming • Integrated Process • Integrated Project • Intellectual Property • Intelligence Processing • Intelligent Peripheral • Intense Penetration • Interacting Processing • Interaction Point • Intercept Point • Interested Parties • Interface Processor • Interfacing Processor • Interior Point • Intermediate Pallet • Intermediate Point • Intermediate Polar • Intermediate Pressure • Intermediate Processor • Internal Protocol (*) • Internal Provider • International Partners • Internet Protocol • Internetwork Protocol • Interpedia • Interpersonal • Interplanetary • Interrupt Process • Interservice Publication • Intervention

Point • Interworking Port • Invalidez Permanente • Invalidité Permanente • Invited Paper • Invited Presentation • Ionic Propulsion • Ionization Potential • Ion Pairing • Iron Pipe • IRON Project • Irregular Penetration • Issue Paper • Issuing Point • Item Processing • Itinerário Principal

IPA Image Product Archive • Imagery Product Archive • Independent Public Accountant • Indice Ponctuel d'Abondance • Industrial Preparedness Action • Infinitesimal Perturbation Analysis • Information Process Analysis • Information Processing Activity • Information Processing Association • Initial Performance Assessment • Institut de Pédagogie Active • Institute for the Physics of the Atmosphere • Institute of Public Affairs • Institut für Physik der Atmosphäre • Integrated Photodetection Assembly • Integrated Program Assessment • Intergovernmental Personnel Act (USA) • Intergovernmental Personnel Agreement • Intermediate-Power Amplifier • International Paleontological Association • International Permafrost Association • International Pharmaceutical Abstracts • International Phonetic Alphabet (*) • International Phonetic Association • International Press Association • International Priority Airmail • International Publishers Association • Interproject Agreement • Intrinsic Polarization Angle • Invalidez Permanente Absoluta • Invalidité Permanente Absolue • Ipota (Vanuatu) (location identifier) • Isopropyl Alcohol

IPAA Instrumental Photon Activation Analysis

IPAC Industrial Programme Advisory Committee (PPARC) • Information Processing And Control • Infrared Processing and Analysis Center (USA-CA) (*) • Integrated Phase and Amplitude Controller • IRAS Processing and Analysis Center

IPACS Integrated Power and Attitude Control System

IPAD Integrated Program for Aerospace Design

IPADS Improved Processor And Display System • Interactive Processing And Display System

IPAG Information Planning and Analysis Group • Institut de Préparation à l'Administration Générale

IPAL Image Processing Algorithms Library

IPAM Incremental Probabilistic Action Modeling • Institute for Pure and Applied Mathematics

IPAP Industrial Preparedness Action Plan • Information Protection Assessment Program • Interagency Placement Assistance Program

IPAR Improved Pulse Acquisition Radar • Initial Product Assessment Report • Instantiated Parameterized Action Representation • Intrapulse Polarization Agile Radar • Irregular Parts Authorization Request

IPAS Image Processing and Analyzing System • Integrated Preparation and Analysis System • Integrated Problem Assessment System

IPAT Impact - Population - Affluence - Technology

IPAUC INTERKOSMOS Prague Astronomicky Ustav Ceskoslovenské Akademie Ved (periodical code)

IPB Illuminated Push Button • Illustrated Parts Book • Illustrated Parts Breakdown • Industrial Products Branch • Industrial Products Bulletin • Inert Processing Building • Information Policy Board • Installation Parts Breakdown • Installation Property Book • Institut de Physique Biologique • Institut für Pflanzenernährung und Bodenkunde (Germany) • Interoperability Policy Board • Ion Pileup Boundary

IPBS Installation Property Book System

IpC Instructions per Clock • Instructions per Cycle

IPC Easter Island (Chile) (location identifier) • Illustrated Parts Catalogue • Image Processing Center • Imaging Proportional Counter • Impact Programme Committee • Indentured Product Code • Indicative Planning Council • Índice de Precios al Consumidor • Índice de Precios al Consumo (Spain) • Industrial Policy Committee • Information and Publicity Committee • Information Policy Committee • Information Processing Center • Information Processing Code • Information Product Characteristics • Infrastructure Protection Center • Initial Planning Conference • Inner Tracking Chamber • Institute for Interconnecting and Packaging • Instructions Per Clock • Instructions Per Cycle • Instrument Processing Center • Integrated Pollution Control • Integrated Printed Circuit • Integrated Procedures Control • Integrated Processor Controller • Integrated Procurement Control • Integrated Protocol Converter • Intellectual Property Council • Intermediate Processing Center • Intermittent Positive Control • International Patent Classification • International Polar Commission • International Privacy Coalition • International Program Committee • International Pyrheliometer Comparison • Internet Proxy Cache • Interplanetary Communication • Interplanetary Plasma Cloud • Interprocess Communication (*) • Interprocessor Communication • Isla de Pascua (Chile) (location identifier)

IPCA Internal Physical Configuration Audit

IPCB ISO Parts Coordination Board

IPCC Intergovernmental Panel on Climate Change • International Property Cooperation Center • International Publishers Copyright Council • Internet Protocol Contact Center (Cisco)

IPCD Image Photon Counting Detector

IPCDN Internet Protocol over Cable Data Network

IPCDT IEE Proceedings on Computers and Digital Techniques (periodical code)

IPCE Independent Parametric Cost estimate • Industrial Pedagogical College-Enterprise (Russia) • International Phone Card Exchange

IPCES Intelligent Process Control by Expert Systems

IPCL Instrumentation Program and Component List • Instrument Program and Component List

IPCME Interplanetary Coronal Mass Ejection

IPCMS Institut de Physique et Chimie des Matériaux de Strasbourg (France)

IPCO Integer Programming and Combinatorial Optimization (conference)

IPCP Internet Protocol Control Protocol

IPCRP IEE Proceedings on Communications and Radar Processing (periodical code)

IPCRS IEE Proceedings on Communications, Radar and Signal Processing (periodical code)

IPCS Image Photon Counting System • Image Production Control System • Integrated Propulsion Control System • Intelligent Process-Control System • Interactive Problem Control System • International Package Consignment Service • International Programme on Chemical Safety

IPCSV IEE Proceedings on Communications, Speech, and Vision (periodical code)

IPCT Integrated Parametric Current Transformer

IPCTA IEE Proceedings on Control Theory and Applications (periodical code)

IPCU Interface Power Control Unit

IPD Image Plane Detector • Imaging Photon Detector • IMU Performance Data • Individual Package Delivery • Individual Potential Demand • Information Processing Division • Initial Production Delivery • Insertion Phase Delay • Institute for Professional Development • Instructional Program Development • Integrated Pin Diode • Integrated Powerhead Demonstration • Integrated Product Development • Integrating Product Division • Intelligent Product Data • International Program Directive • Interplanetary Dust • Inverse Power Distance • Isotope-Powered Device • Issue Priority Designator • Iterated Prisoner's Dilemna

IPDA Intrapulse Demodulation Analysis

IPDB Integrated Product Database

IPDC Integrated Product Development Center • International Programme for the Development of Communication • Internet Protocol Device Control

IPDD Initial Project Design Description • Integrated Product Design and Development

IPDE Institut Postdoctoral Européen • Integrated Product Data Environment • Integrated Product Development Environment

IPDF Indexed Portable Document Format • Integrated Processing and Distribution Facility

IPDM Isolated Power and Data Module

IPDN International Paleoclimatic Data Network

IPDP Integrated Product Development Process • Interval of Pulsations of Diminishing Period • Irregular Pulsation of Decreasing Period

IPDPG Image Processing Data Products Group

IPDPS International Parallel and Distributed Processing Symposium

IPDR Incremental Preliminary Design Review • Initial Program Design Review • Internal Preliminary Design Review • Internet Protocol Detail Record

IPDS IBM Personal Dictation System • Image Processing and Dissemination System • Improved Position Determining System • Integrated Product Design System • Intelligent Printer Data Stream

IPDT Integrated Product Development Team • Integrated Product Development Test

IPDU Instantaneous Panoramic Display Unit • Internet Protocol Data Unit

IPE Image Pac Extension • Image Processing Engine • Image Processing Equipment • Immeuble du Parlement Européen • Improved Performance Engine • Incentive PERT Event • Increased Performance Engine • Individual Protective Equipment • Industrial Plant Equipment • Information Processing Equipment • Initial Production Evaluation • Integrated Platform Electronics • Integrated Programming Environment • Interplanetary Ejection • Interpret Parity Error • Inverse Photoelectric Effect • Ipil (Philippines) (location identifier)

IPEA International Preliminary Examining Authority • Internet Protocol Encapsulated Addressing

IPEC IVS Proposal Evaluation Committee

IPECS IEE Proceedings on Electronics, Circuits, and Systems (periodical code) • Integrated Power and Environmental Control System

IPEE Inclination of a Plane to the Earth's Equator • Introduction Progressive d'Éléments Étrangers

IPEEFI Identification de Projets Énergétiques Éligibles à des Financements Internationaux

IPEFE Institut de Préparation à l'Exercice de Fonctions d'Encadrement

IPEI Ionospheric Plasma and Electrodynamics Instrument

IPELS Interrelationship between Plasma Experiments in Laboratory and Space

IPEM Institute of Physics and Engineering in Medicine

IPEMB Institute of Physics and Engineering in Medicine and Biology

IPEN Instituto de Pesquisas Energéticas e Nucleares (Brazil-SP)

IPENZ Institution of Professional Engineers of New Zealand

IPEPA IEE Proceedings on Electric Power and Applications (periodical code)

IPERS International Peer Review System

IPES Improved Proposed Encryption Standard • Inorganic Properties Estimation System • Intermediate Polyelastic Sampling • Inverse Photoemission Study

IPET Inspecteur Principal de l'Enseignement Technique

IPEV Institut Paul-Émile Victor (France)

IPEX Interferometry Program Experiment

IPF Image Processing Facility • Imagery Processing Facility • Immediate Parameter File • Imposición a Plazo Fijo • Industrial Physics Forum (AIP) • Information Presentation Facility • Information Processing Facility • Initial Production Facility • Initial Production Funds • Institut Polytechnique Féminin • Instrument Pointing File • Integrated Processing Facility • Interactive Processing Facility • Intergovernmental Panel on Forests • International Partner Facility • International Publication Frequency • Isochrone Probability Function

IPFC Indo-Pacific Fisheries Council • Information Presentation Facility Compiler

IPFM Integral Pulse Frequency Modulation

IPFR Institute of Plasma and Fusion Research

IPG Image Processing Group • Implementation Planning Group • Information Power Grid (program) • Institut de Physique du Globe • Institut für Physikalische Geodäsie (Germany) • Institut für Planetarische Geodäsie (Germany) • Instrument Projects Group • Interference Protection Group • International Planning Group • Interpacket Gap • Ipiranga (Brazil-AM) (location identifier) • Isotope Power Generator • Issue Priority Group

IPGE Institut de Préparation aux Grandes Écoles

IPGP Institut de Physique du Globe de Paris (France)

IPGR Iterative Parallel Region Growing

IPGS Institut de Physique du Globe de Strasbourg (France) • International Partner Ground System

IPGTD IEE Proceedings on Generation, Transmission, and Distribution (periodical code)

iph Impressions per Hour • Inches per Hour

IPH Ipoh (Malaysia) (location identifier) • Instantaneous Pointing History • Interplanetary Hydrogen

IPHAC Ion Produced Hydrogenated Amorphous Carbon

IPHIR Interplanetary Helioseismology by Irradiance

IPhO International Physics Olympiad

IPHO International Physics Olympiad

IPHT Institut für Physikalische Hochtechnologie (Germany)

IPI Image Processing and Interchange • Image Processing Interchange • Índice de Producción Industrial (Spain) • Industrial Process Improvement • Inner Port Injection • In-Process Inspection • Input Package Interface • Institute of Production Innovation • Instrument Principal Investigator • Intelligent Peripheral Interface • Internal Process Improvement • International Patent Institute • International Press Institute • International Public Information • Internet Process Improvement • Interpulse Interval • Inventory of Principal Item • Ipiales (Colombia) (location identifier)

IPIC Intellectual Property of Integrated Circuits • ISS Payload Integration Contract

IPICC International Partners' Instrument Control Center

IPII Intermediate Population II

IPIP Information Processing Improvement Program

IPIPS Interactive Planetary Image Processing System

IPIR Initial Photographic Interpretation Report • Initial Programmed Interpretation Report • Internal Peak Irradiance Response

IPIX Intelligent Pixel-processing

IPK Internationale Patentklassifikation

IPL El Centro - Imperial County (USA-CA) (location identifier) • Identified Parts List • Illustrated Parts List • Image Processing Laboratory • Image Processing Log • Image Product Library • Imperial County (USA-CA) (location identifier) • Indentured Parts List • Information Processing Laboratory • Information Processing Language • Information Programming Language • Initial Program Load • Initial Program Loader • Instrumentation Program List • Integrated Priority List • Internet Public Library • Interrupt Priority Level • Ion Projection Lithography • Isotope Products Laboratories (USA-CA)

IPLAN Integrated Planning and Analysis

IPLB Internet Protocol Load Balancing

IPLC Initial Program Load Complete • International Private Leased Channel

IPLI Internet Private Line Interface

IPLL Instituto Português do Livro e da Leitura (Portugal) • International Private Leased Line

IPLR Interplanetary Laser Ranging

IPLS Initial Program Load Start • Institut pour la Promotion du Lien Social (France)

i. Pl. v. In Plaats Van

IPLV Intermediate Payload Launch Vehicle

ipm Inches per Minute

IpM Impulses per Minute • Interruptions per Minute

IP&M Information Processing & Management

IPM Illustrated Parts Manual • Implementation Program Manager • Impulses Per Minute • Inches Per Minute • Incidental Phase Modulation • Incidental Pulse Modulation • Incremental Phase Modulator • Industrial Preparedness Measure • Information Processing and Management • Inspection Procedure Manual • Institute for Practical Mathematics • Institut für Physikalische Meßtechnik • Instrument Performance Monitoring • Integrated Process Management • Integrated Product Management • Integration Program Manager • Intelligent Processing of Materials • Interactive Performance Monitor • Interference Prediction Model • Internal Polarization Modulation • International Prototype Meter • Internet Policy Management • Inter-Personal Message • Inter-Personal Messaging • Interplanetary Medium • Interruptions Per Minute • Inventory Projection Model • Italian Panoramic Monochromator (SOHO)

IPMAF IRAS Post Mission Analysis Facility

IPMAP IEE Proceedings on Microwaves, Antennas, and Propagation (periodical code)

IPMCT Institute of Precision Mechanics and Computer Technology

IPME Institut de Pédagogie Moderne et d'Enseignement

IPMG Instrument Physical Modeling Group

IPMI Information Processing in Medical Imaging (conference) • Intelligent Platform Management Interface • Internet Provider Multicast Initiative

IPMO Implementation Program Management Office • Isolated Planetary-Mass Object

IPMOA IEE Proceedings on Microwaves, Optics, and Antennas (periodical code)

IPMOC International Partners' Mission Operations Center

IPMP Investigation into Polymer Membranes Processing

IPMS Impact Predictor Monitor Set • Information Processing Management System • Initial Performance Measurement System • Institution of Professionals, Managers and Specialists • Integrated Product Management System • International Polar Motion Service (now IERS) • Interpersonal Message Service • Interpersonal Message System • Interpersonal Messaging System • Impact Predictor Monitor Set

IPMU Information Processing and Management of Uncertainty (conference)

IPN Institut de Physique Nucléaire • Instituto Politécnico Nacional (Mexico) • Institut Pédagogique National • Intelligence Processing Network • International Product News • Interplanetary Network • Interprocessor Node • Intracluster Planetary Nebula • Ipatinga Usiminas (Brazil-MG) (location identifier)

IPNE Institut de Physique Nucléaire Expérimentale

IPng Internet Protocol Next Generation

IPnG Internet Protocol Next Generation

IPNG Internet Protocol Next Generation

IPNNI Integrated Private Network-to-Network Interface

IPNO Institut de Physique Nucléaire d'Orsay (France)

IPNS Intense Pulsed Neutron Source

IPN1 First Interplanetary Network

IPN2 Second Interplanetary Network

IPN3 Third Interplanetary Network

IPO IEEE/IEI Publication On-disk • Industrial preparedness Operations • Information Protection Office • Initial Public Offering • Input - Process - Output • Instantaneous Power Output • Integrated Program Office • Intellectual Property Owners Inc. • Interactive Planning Optimization • International Planning Office • International Post Organization • International Procurement Office • International Programs Office

IPoA Internet Protocol over ATM

IPoATM Internet Protocol over Asynchronous Transfer Mode

iPOC Interim Policy Oversight Committee

IPOC International Partner Operations Center

IPOMS International Polar Orbiting Meteorological Satellite

IPOP Integrated Point Of presence • International Polar Orbiting Platform

IPOpt IEE Proceedings on Optoelectronics (periodical code)

IPoS Internet Protocol over SDH

iPP Isostatic Polypropylene

IPP Imaging Photopolarimeter • Imaging Polarimeter and Photopolarimeter • Impact Point Prediction • Incapacidad Permanente Partial • Incapacité Permanente Partielle • Índice de Precios a la Producción • Individual Program Plan • Industrial Partnering Program • Industrial Preparedness Plan • Industrial Preparedness Planning • Industrial Preparedness Program • Information Processing Profile • Information Processing Protocol • Information Production Process • Inhomogeneous Poisson Process • Initiative for Proliferation Prevention • Input Processor Package • Institute of Physical Problems • Institut für Plasmaphysik (MPG) • Institute of Plasma Physics • Instrument Programmable Parameter • Integrated Plotting Package • Integrated Product Plan • Intense Progressive Penetration • Intermittent Positive Pressure • Internet Presence Provider • Internet Printing Protocol • Inverse Polarity Protection • Ionospheric Propagation Path • Irregular Parcels and Pieces

IPPAP Institute for Particle Physics and Astrophysics

IPPB Illustrated Provisioning Parts Breakdown

IPPC International Policies and Plans Committee

IPPD Integrated Process and Product Development • Integrated Product and Process Development • Integrated Product Process Design

IPPDT Integrated Product and Process Development Team

IPPE Institute for International Politics and Economics (Yugoslavia) • Institute for Physics and Power Engineering (Russia)

IPPEX Internet Plasma Physics Education Experience

IPPJ Institute of Plasma Physics of Japan

IPPL Industrial Preparedness Planning List

IPPM Industrial Preparedness Planning Measure • Integrated Product and Process Management • Internet Provi-

IPPNW International Physicians for the Prevention of Nuclear War

IPPP Industrial Preparedness Planning Program

IPPR Internal Peak Power Response

IPPS Interactive Picture Processing System • International Parallel Processing Symposium

IPPSM IEE Proceedings on Physical Science and Measurements (periodical code)

IPPT Integrated Process and Product Team • Integrated Product and Process Team

IPPV Intermittent Positive Pressure Ventilation

IPQ Information Processing Quotient • Information Processing Quality • Information Product Quality • Instituto Português da Qualidade (Portugal)

IPQC In-Process Quality Control

IPQM In-Process Quality Metrics

ipr Inches per Revolution

IPR Ice-Point Reference • Implementation Planning Record • Incremental Performance Review • Individual Pay Record • Industrial Plant Representative • Injector Pressure Regulator • In-Plant Review • In-Process Report • In-Process Review • In-Progress Review • Informal Progress Report • Inspecteur Pédagogique Régional • Institute for Plasma Research • Institute of Plasma Research • Institute of Public Relations • Instrumentation Performance Report • Instrument Performance Report • Integrated Process Review • Intellectual Property Right • Intellectual Proprietary Right • Interim Problem Report • Interim Program Review • Interim Progress Report • Interim Progress Review • Internal Progress Report • Internet Protocol Router • Ion Production Rate • Irish Publishing Record

IPRA International Public Relations Association • Internet PCA Registration Authority • Internet Policy Registration Authority

IPRC International Pacific Research Center (USA-HI)

IPRF Impage Processing Research Facility • Impuesto sobre la Renta de Personas Físicas

IPRG Institute for Petroleum Research and Geophysics (Israel)

IPRM Indium Phosphide and Related Materials (conference)

IPRO International Patent Research Office

IPRS Information and Public Relations Service • International Postgraduate Research Scholarship

IPRSP IEE Proceedings on Radar and Signal Processing (periodical code)

IPRT Institute for Physical Research and Technology

IPRWG Intellectual Property Rights Working Group

ips Inches per Second • Instructions per Second • Interconnections per Second • Interruptions per Second

Ips Ipswich (UK)

I/PS Insurance/Pension Specialist

IPS Ice Protection System • Illustrative Planning Scenario • Image Processing Software • Image Processing Specialist • Image Processing System • Impact Predictor System • Improved Processing System • Inches Per Second • Indirect Pricing System • Inertial Pointing System • Inertial Positioning System • Inertial Propulsion System • Information and Photographic Service • Information Processing Standard • Information Processing System • Inlet Particle Separator • Installation Performance Specification • Institute of Polar Studies • Institut Paul Scherrer (Switzerland) • Instructions Per Second • Instrumentation Power Supply • Instrumentation Power System • Instrument Pointing System • Insurance and Pension Specialist • Integral Propulsion System • Integrated Personnel Support • Integrated Photonic System • Integrated Power Source • Integrated Power System • Integrated Procurement System • Integrated Product Service • Integrated Product Summary • Integrated Program Strategy • Integrated Program Study • Integrated Program Summary • Integrated Program System • Integrated Propulsion System • Integrated Publishing System • Integration Plan and Schedule • Intelligent Page Store • Intelligent Power System • Intensity Period Search • Interconnections Per Second • Interface and Power System • Interface Power System • Interface Problem Sheet • Interim Policy Statement • International Pipe Standard • International Planetarium Society (USA-UT) • Internet Protocol Storage • Internet Protocol Suite • Interoperability Planning System • Interplanetary Scattering • Interplanetary Scintillation • Interpretative Programming System • Interruptions Per Second • Interstellar Propulsion Society (USA-CA) • Inverter Power Supply • Invoice Processing System • Inward Primary Sorting • Ionospheric Prediction Service (Australia) • Ion Propulsion System • Israel Physical Society • Italian Physical Society (Italy) • Item Processing System

IPSA Institut Polytechnique des Sciences Appliquées • Interplanetary Space Agency (UK) • Irish Physics Students' Association

IPSC Ice Protection System Controller • Information Processing Standards for Computers • International Program for Scientific Cooperation • IRAS Point Source Catalog

IPSD Image Processing Systems Division

IPSE Integrated Programming Support Environment • Integrated Project Support Environment • Intelligent Program Support Environment • Internship in Public Science Education (NSF) • Italian Package for Science Experiments (MSR)

IPSec Internet Protocol Security

IPSEC Internet Protocol Security

IPSG International Programs Steering Group

IPSGSM Instituto Politécnico Superior General San Martín (Argentina-SF)

IPSI Institut pour la Promotion des Sciences de l'Ingénieur (France) • Integrated Publication Systems Institute

IPSIM Interactive Postal Simulator

IPSJ Information Processing Society of Japan

IPSL Institut Pierre-Simon Laplace (France) • Interface Problem Status Log

IPSN Institut de Protection et de Sûreté Nucléaire

IPSO Information Processing Service Organization • Institut pour la Promotion des Systèmes Ouverts • Internet Protocol Security Option

IPSP Initialize Programmable Signal Processor • Integrated Program Schedule Plan • Intelligence Priorities for Strategic Planning • Internet Protocol Security Protocol

IPSRU Inertial Pseudo Star Reference Unit

IPSS Image Processing Subsystem • Information Processing System Simulator • Initial Preplanned Supply Support • Institute of Planetary and Space Science • Integrated Parts Statusing System • Integrated Planning and Scheduling System • Integrated Project Scheduling System • International Packet Switched Service • IRAS Point Source Survey

IPSSE IEE Proceedings on Solid-State Electron Devices (periodical code)

IPST In-Process Self Test • Institute for Physical Science and Technology • Institut Professionnel des Sciences et Technologies

IPT Image Perspective Transform • Image Perspective Transformation • Immersive Projection Technology • Improved Programming Technology • Incapacidad Permanente Total • Incapacité Permanente Totale • Incremental Proof Testing • Indexed - Paged - Titled • Informatique Pour Tous (France) • Initial Production Test • In-Plant Team • In-Plant Test • In-Plant Transporter • In-Process Test

• Instituto de Pesquisas Tecnológicas • Instrument Procedures Trainer • Insurance Premium Tax • Integrated Process Team • Integrated Product Team • Integrated Program Team • Integrated Project Team • Intelligent Printhead Technology • Intermediate Pressure Turbine • Internal Parameter Table • International Pipe Thread • Interplanetary Travel • Interprocess Communications • Inverse Planck Transform • Io Plasma Torus • Williamsport (USA-PA) (location identifier)

IPTC International Press Telecommunications Council

IPTCS Igloo Passive Thermal Control System

IPTEL Internet Protocol Telephony

IPTF Infrastructure Protection Task Force

IPTL IEEE Photonics Technology Letters (periodical code)

IPTO Independent Power Takeoff • Information Processing Techniques Office

IPTR In-Process Technical Review

IPTS Institute for Prospective Technological Studies • Institut für Praxis und Theorie der Schule (Germany) • International Practical Temperature Scale

IPU Image Processing Unit • Inert Prefill Unit • Information Processing Utility • Information Provider Unit • Instruction Processing Unit • Integrated Plot Utility • Integrated Power Unit • Intelligent Processing Unit • Interface and Priority Unit • Interface Processor Unit • Inter-Parliamentary Union • Interphase Unit • Ipiaú (Brazil-BA) (location identifier) • Isotope Power Unit

IPUT In-Plant User Test

IPV Improve • Individual Pressure Vessel • Industrial Pressure Vessel • In Plaats Van • Interpercentile Velocity • Intrapixel Variation • Intraplate Volcanism • Isentropic Potential Velocity • Isentropic Potential Vorticity

IPVC International Permanent Virtual Connection

IPVS Ion Pump Vacuum System

IPW Ipswich (UK-England) (location identifier)

IPWG Integrated Project Working Group • Internet Privacy Working Group

IPX Integrated Packet Exchange • Internet Packet Exchange • Internet Protocol Exchange • Internetwork Packet Exchange (*)

IPXCP Internet Packet Exchange Control Protocol

ipy Inches per Year

IPY International Polar Year • Ion Pair Yield

IPZ Intense Penetration Zone

IP2D Integrated Product and Process Development

iq Idem Quod

I-Q In-Quadrature

I/Q Information/Query • In-Quadrature

IQ Image Quality • Image Query • Indefinite Quantity • Information Quality • In Quadrature • Integrity Quotient • Intelligence Quotient • Intelligent Queuing • Interactive Query • International Quota • Intrinsic Quality • Iraq (ISO country code)

IQA Institute for Quality Assurance (UK) • Irish Quality Association

IQAS International Quality Assurance System

IQC Indefinite Quantity Contract

IQCS In-service Quality Control System

IQD Dinar (Irak) (ISO currency code) • Image Quality Detector • International Quiet Day • Interquartile Difference

IQE Image Quality Estimator • In-plant Quality Evaluation • Interactive Query Expansion

IQEC International Quantum Electronics Conference

IQF Interactive Query Facility • Intrinsic Quality Factor

IQI Image Quality Indicator • Institute for Quantum Information (USA-CA)

IQL Information Query Language • Intelligent Query Language • Intelligent Query Linker • Interactive Query Language • Intermediate Query Language

IQLA International Quantitative Linguistics Association

IQM Institute for Quality and Management • Qiemo (China-PRC) (location identifier)

IQMSS International Quality Management System Standard

IQN Inner Quantum Number • Qingyang (China-PRC) (location identifier)

IQP Intelligent Query Processor • Interactive Qualifying Project

IQQ Iquique - D. Aracena (Chile) (location identifier)

IQR Inquiry

IQRP Interactive Query and Report Processor

IQS Intelligent Query Service • International Quality Study

IQSA International Quantum Structures Association

IQSY International Quiet Sun Year (1964-1965)

IQT Independent Quality Team • Informal Qualification Test • Initial Qualification Training • Intelligent Query Tree • Iquitos (Peru) (location identifier)

IQUE In-plant Quality Evaluation

IQV Ion-Quiver Velocity

ir Infrared • Iran

iR im Ruhestand

Ir Ingenieur • Ingénieur • Ireland • Iridium • Irish • Irregular (galaxy)

I&R Inspect & Review • Interchangeability & Replaceability

I/R Inverter/Regulator

IR Ice on Runway • Imaging Radiometer • Immediate Release • Immediate Reserve • Implementation Requirement • Impulse Response • Incidence Rate • Incident Record • Incident Report • Index Register • Individual Rational • Individual Reinforcement • In-flight Refueling • Informal Report • Information Rate • Information Ratio • Information Report • Information Request • Information Requirement • Information Resource • Information Retrieval • Infrared • Infrared Radiation • Infra-Rot • Infra-Rouge • Ingénieur de Recherche • Initial Release • Inland Revenue • Inner Roll • Inside Radius • Inspect and Review • Installation Report • Institutional Representative • Instruction Register • Instrumentation Report • Instrument Rating • Instrument Reading • Instrument Reference • Insulation Resistance • Intelligence Ratio • Intelligence Requirement • Interchangeability and Replaceability • Interim Report • Intermediate Rear • Intermediate Review • Internal Reliability • Internal Report • Internal Resistance • Internal Revenue • Internal Review • International Registration • International Rendezvous • Internet Registry • Interpretation Report • Interrogation Report • Interrogator • Interrogator-Responder • Interval Rate • Investigation Request • Ionizing Radiation • Iran (international vehicle registration code) • Iran (ISO country code) • Iranian Rial (currency) • Ireland • Iroquois • Irreductible Representation • Isoprene Rubber

IRA Independent Reporting Activity • Independent Research Agency • Individual Retirement Account • Infrared Astronomy (periodical code) • Infrared Radiometer • Institut de Radiophysique Appliquée (Switzerland) • Institute of Radio Astronomy • Institutional Research Award • Institut Régional d'Administration • Input Reference Axis • Intelligence Reuse Advisor • Interface Requirement Agreement • Interim Remedial Action • Istituto di Radioastronomia (Italy) • Kirakira (Solomon Islands) (location identifier)

IRAA Indoor Radon Abatement Act (USA)

IRAC Infrared Array Camera • Information Resources Administration Council • Interdepartmental Radio Advisory

Committee • Interim Rapid Action Change • Internal Review and Audit Compliance • International Radio Advisory Committee (*) • International Relations Advisory Committee • International Requirements And Criteria

IRACE Infrared Array Control Electronics

IRAD Independent Research And Development • Infrared Radiometer

IRADS Infrared Acquisition and Detection System

IRAF Image Reduction and Analysis Facility

IRAIT Italian Robotic Antarctic Infrared Telescope

IrAJ Irish Astronomical Journal (periodical code)

IRAL Information Retrieval with Asian Languages (workshop)

IRAM Implicitly-Restarted Arnoldi Method • Improved Reliability And Maintainability • Improved Repairable Asset Management • Indexed Random Access Method • Input Random Access Memory • Institut de Radioastronomie Millimétrique • Instituto de Racionalización de Materiales (Argentina) • Instituto de Radioastronomía Milimetrica

IRAMMP Infrared Analysis Measurement and Modeling Program

IRAMS Inertial Reference Alignment Monitor System

Iran Iranian

IRAN Inspect and Repair As Necessary • Inspection and Repair As Necessary

IRAND Interactive RAND

IRAO Itapetinga Radio Astronomy Observatory (Brazil)

IRAP Industrial Research Assistance Programme (Canada) • ISLSCP Retrospective Analysis Project (now IDS-LSC)

IRAPS Image Recording And Processing System

IRAR Infrared Airborne Radar

IRAS Infrared Absorption Spectroscopy • Infrared Astronomical Satellite • Istituto Spezzino Ricerche Astronomiche (Italy)

IRASI Internal Review And System Improvement

IRATS Infrared Ranging And Tracking System

IRAW International Registration of Audiovisual Works

IRB Individual Records Brief • Industrial Relations Bureau • Industrial Revenue Bond • Infrared Background • Infrastructure Review Board • Inspection Review Board • Institutional Review Board • Instruction Recoder Buffer • Instruction Recorder Buffer • Instruction Reorder Buffer (*) • International Review Board • Inter-Repeater Bus • Iraan (USA-TX) (location identifier)

IRBC Information-Rich Bar Code

IRBM Intermediate Range Ballistic Missile

IRBS Infrared Background Sensor • Intermediate-Range Booster System

IRc Infrared Compact

IRC Circle (USA-AK) (location identifier) • Imaging Research Center • Individual Request for Classification • Information Research Center • Information Resource Catalogue • Information Resources Centre • Information Retrieval Center • Information-Rich Commerce • Infrared Camera • Infrared Catalogue • Infrared Companion • Initial Response Call • Initial Response Cell • Innovation Relay Centre • Inspection Record Card • Instrument Remote Control • Interactive Rule Checking • Interagency Radio Committee • Interdisciplinary Research Center • Interdisciplinary Research Collaboration • Internal Revenue Code • International Radiation Commission • International Record Carrier • International Red Cross • International Reference Centre • International Reply Coupon • International Rescue Committee • International Research Council (now ICSU) • International Response Coupon • Internet Relay Chat • Interregion Communication • Interrupt Request Controller • Ionosphere Research Committee • Ion Recombination Chamber • Item Responsibility Code • IUCAA Reference Centre

IRCA International Research Course on Atmospheres

IRCAM Infrared Camera • Institut de Recherche et Coordination Acoustique-Musique

IRCANTEC Institut de Retraite Complémentaire des Agents Non Titulaires de l'État et des Collectivités (locales) (France)

IRCAS Information Requirements Control Automated System

IRCC Indian River Community College (USA-FL) • International Radio Consultative Committee

IRCCD Infrared Charge Coupled Device

IRCCM Infrared Counter-Countermeasure

IRCD International Research Centers Directory

IRCEP International Centre for Experimental Physics (QUB)

IRCFE Infrared Communications Flight Experiment

IRCI Infrared Cloud Imager

IRCM Infrared Countermeasures

IRCN Interagency Report Control Number

IRCS Inertial Reference and Control System • Infrared Camera and Spectrograph • Institute for Research in Cognitive Science • Integrated Radiocommunications System • Integrated Resource Control System

IRCSET Irish Research Council for Sciences, Engineering and Technology (Ireland)

IRCTR International Research Centre for Telecommunications-transmission and Radar

IRCTTR International Research Centre for Telecommunications, Transmission and Radar

IR&D Independent Research & Development • Internal Research & Development • International Research & Development

IRD Independent Research and Development • Information Requirements Description • Information Requirements Document • Information Research Department • Information Resource Dictionary • Information Resources Department • Information Resources Division • Infrared Data • Infrared Detector • Infrared Display • Infrared Divergence • Initial Requirement Determination • Initiating Reference Document • Institut de Recherche pour le Développement (France) (formerly ORSTOM) • Integrated Receiver-Decoder • Integrated Receiver-Descrambler • Integrated Record Data System • Integration Requirements Document • Interface Requirements Document • Internal Research and Development • International Relations Division • International Research and Development • Internet Resource Database • Interoperability Requirements Document • Ishurdi (Bangladesh) (location identifier)

IrDA Infrared Data Association

IRDA Industrial Research and Development Authority • Infrared Data Association • Infrared Detection Array • Infrared Developers Association • Infrared Device Association

IRDAC Industrial Research and Development Advisory Committee (EU)

IRDB Information Retrieval Data Bank • Integrated Regional Database

IRDC Infrared Dark Cloud • Intelligence Research and Development Council

IRDE Instruments Research and Development Establishment (India)

IRDG Interrange Documentation Group

IRDL Information Retrieval and Display Language

IRDM International Rendezvous and Docking Mission

IRDNET Ice Reconnaissance Data Network

IRDP ICMP Router Discovery Protocol

IRD&S International Research, Development & Standardization

IRDS Information Resource Dictionary Standard • Information Resource Dictionary System • Infrared Detection Set • Infrared Detection System • Infrastructure Resource Dictionary System • International Research, Development, and Standardization

IRDSS Infrared Defeating Smoke System

IRDT Inflatable Reentry and Descent Technology

IRDU Internet Research and Development Unit

Ire Ireland

IRe Infrared Extended

IRE Immediate Ready Element • Information Resource Engineering • Infrared Emission • Initial Resources Estimate • Institut de Recherche et d'Évaluation • Institut des Radioéléments (Belgium) • Institute for Radioelectronics and Engineering (Russia) • Institute of Radio Engineers • Institute of Radioengineering and Electronics • Institut Régional d'Éducation • Irecê (Brazil-BA) (location identifier)

IREB Intense Relativistic Electron Beam

IREBL Infrared Extragalactic Background Light

iRED Interim Resistive Exercise Device

IRED Infrared Emitting Diode

IREE Institute for Radio Engineering and Electronics • Institution of Radio and Electronics Engineers (Russia)

IREG International Roaming Expert Group

IREM Institut de Recherche et d'Enseignement des Mathématiques

IRENA Instituto Nicaragüense de Recursos Naturales y del Ambiente (Nicaragua)

IREO Infrared Electro-Optical • Institut Régional d'Éducation et d'Orientation

IREP Integrated Reliability Evaluation Program • Interim Reliability Evaluation Program • Interdisciplinary Research Equipment Program

IREPA Institut Régional de Promotion de la Recherche Appliquée

IREPS Integrated Refractive Effect prediction System

IReS Institut de Recherches Subatomiques

IRES Imagery Reporting and Exploitation System • Infrared Earth Sensor

IRESTE Institut de Recherche et d'Enseignement Supérieur aux Techniques de l'Électronique (France)

IRET Interrupt Return

IREX International Research and Exchange • International Researcher Exchange

IRF Imagery Readiness Facility • Information Retrieval Facility • Inheritance Rights Filter • Inherited Rights Filter • Instantaneous Radiative Flux • Institut de Recherche Fondamentale • Institutet för Rymdfysik (Sweden) • Integrated Research Facility • Intermediate Routing Function • Internal Raster File • Internal Raster Format • International Road Federation

IRFA Initial Regulatory Flexibility Analysis • Institut Régional de Formation pour Adultes

IRFC Infrared Flux Collector

IRFDA Infrared Focal-Plane Assembly

IRFED Institut de Recherche et de Formation pour l'Éducation et le Développement

IRFEST Institut de Recherche sur les Fondements et les Enjeux des Sciences et des Techniques

IRFITS Infrared Fault Isolation Test System

IRFK Institutet för Rymdfysik, Kiruna (Sweden)

IRFM Infrared Flux Method

IRFN Inhibited Red Fuming Nitric (acid)

IRFNA Inhibited Red Fuming Nitric Acid

IRFPA Infrared Focal Plane Array

IRFREP Institut Régional de Formation et de Recherche sur l'Éducation Permanente

IRFU Institutet för Rymdfysik, Uppsala (Sweden)

IRFUm Institutet för Rymdfysik, Umeå (Sweden)

IRG Inertial Rate Gyro • Information Research Group • Infrared Galaxy • Infrared Generator • Inner Roll Gimbal • Interdepartmental Regional Group • Internet Research Group • Interrecord Gap • Lockhart River (Australia-QLD) (location identifier)

IRGA Infrared Gas Analyzer

IRH Informatique des Ressources Humaines • Initial Radiohologram

IRHC Ingénieur de Recherche Hors Classe

IRHDF Infrared Hubble Deep Field

IRHS Infrared Heterodyne Spectrometer • Infrared Heterodyne Spectroscopy • Infrared High-Dispersion Spectrograph

Iri Irish

IRI Industrial Research Institute • Infrared Imagery • Infrared Instrumentation • Interagency Research Internet • International Reference Ionosphere • Ionospheric Research Instrument • Iringa (Tanzania) (location identifier) • Istituto per la Recostruzione Industriale (Italy)

IRIA Infrared Information Analysis

IRIC Infrared Image Converter

IRICC Interagency Resource Information Coordination Council

IRICE Instituto Rosario de Investigación en Ciencias de la Educación (Argentina-SF)

IRID Infrared Image Display

IRIE Infrared Imaging Equipment • Infrared Imaging Experiment • Infrared Information Exchange

IRIG ILL Protocol Implementation Group • Inertial Rate Integrating Gyroscope • Inertial Reference Integrating Gyroscope • Integrated Range Instrumentation Group • Interrange Instrumentation Group

IRIM Infrared Imager

IRIMO Islamic Rep. of Iran Meteorological Organization

IRIS Image Reading Instrument System • Image Retrieval and Information System • Images, Reflets, Initiation Scientifique (France) • Incorporated Research Institutions for Seismology • Increased Readiness Information System • Inferential Retrieval Indexing System • Information, Robotics and Intelligent Systems • Infrared Imaging Seeker • Infrared Imaging Spectrometer • Infrared Imaging Surveyor • Infrared Imaging System • Infrared Instrumentation System • Infrared Interferometer Spectrometer • Infrared Radiation Interferometer Spectrometer • Initiative for Research in Informatics applied to Society • Instant Response Information System • Institute for Research in Information and Scholarships • Institute for Robotics and Intelligent Systems • Integrated Radar Imaging System • Integrated Radio Information System • Integrated Raster Imaging System • Integrated Remote Imaging System • Integrated Resource Information System • Integrated Risk Information System • Integriertes Real-time Informationssystem • Interactive Radar Information System • Interactive Research Imaging System • Interactive Retrieval of Information System • Interconexión de Recursos Informáticos • International Radiation Investigation Satellite • International Radio Interferometric Surveying (*) • International Radio Interferometry Survey • International Relations Information Service • International Research and Information Service • International Research on the Interior of the Sun • Interrogation Requirement Information System • Italian Research Interim Stage

IRISA Institut de Recherche en Informatique et Systèmes Aléatoires (France)

IRISI Inter-Regional Information Society Initiative

IRISP Infrared Imaging and Spectropolarimeter

IRIT Institut de Recherche en Informatique de Toulouse (France)

IRJ La Rioja (Argentina-LR) (location identifier)

IRJE Internet Remote Job Entry

IrJST Iranian Journal of Science Technology (periodical code)

IRK Implicit Runge-Kutta • Institut für Radiumforschung und Kernphysik (Austria) • Interswitch Rekeying Key • Kirksville (USA-MO) (location identifier)

Irl Irlandés

IRL Indemnité Représentative de Logement • Indexed Retrieval Language • Industrial Research Ltd. (New Zealand) • Industrial Robot Language • Information Retrieval Language • Infrared Lamp • Infrared Light • Infrared Lens • Initiating Reference Letter • Initiating Reference List • In Real Life • Institute for Research on Learning • Interactive Reader Language • Interface Requirements List • Interrepeater Link • Ireland (international vehicle registration code) • Ireland (ISO country code) • Irlande (code ISO de pays) • Irish Pound (currency)

IRLAN Infrared Local Access Network

IrLAP Infrared Link Access Protocol

IRLAP Infrared Link Access Protocol

IRLDA Infrared Linear Detector Array

IRLED Infrared Light Emitting Diode

IRLM IMS Resource Lock Manager • Internal Resource Lock Manager

IrLMP Infrared Link Management Protocol

IRLMP Infrared Link Management Protocol

IRLS Infrared Line Scanner • Integrated Remote Imaging System • Interrogation, Recording and Location System • Interrogation Recording Location System • Iteratively Reweighted Least Squares

IRM IERS Reference Meridian • Image Refresh Memory • Image Rejection Mixer • Imagerie par Résonnance Magnétique • Induced Remanent Magnetization • Information and Records Management • Information Resources Management • Information Resources Manager • Information Resources Monitor • Infrared Mapper • Infrared Measurement • Infrared Montgolfier • Inherent Rights Mask • Initial Release Memorandum • Institut Royal Météorologique • Integrated Range Mission • Interconnect Resource Manager • Interim Research Memorandum • Intermediate Range Monitor • International Relations Manager • Ion Release Module • Isothermal Remanent Magnetization

IRMA Information Régionale Météorologique Automatisée • Information Resources Management Association (USA) • Information Revision and Manuscript Assembly • Infrared Michelson Array • Institut de Recherche Mathématique Avancée

IRMAC Information Resource Management Advisory Committee

IRMB Institut Royal Météorologique de Belgique (Belgium)

IRMBP Institut Royal Météorologique de Belgique - Publications (periodical code)

IRMC Information Resource Management College

IRMD Information Resource Management Department

IRME Initiator Resistance Measuring Equipment

IRML Interactive Media Research Laboratory

IR&MM Infrared & Millimeter

IRMM Institute for Reference Materials and Measurements (Belgium) • Isothermal Remanent Magnetization

IR/MMW Infrared/Millimeter Wave

IRMO Information Resource Management Office • Information Resource Management Officer • Information Resource Manager Office

IRMOS Infrared Multiobject Spectrograph • Infrared Multiobject Spectrometer

IRMP Information Resources Management Plan • Infrared Measurement Program • Integrated Risk Management Process

IRMS Infrared Mapping System • Information Resources Management Service • Information Resources Management Software • Information Resources Management System • Information Retrieval and Management System • Integrated Radio Management System • Isotope Ratio Mass Spectrometry

IRN Independent Radio News • Interface Revision Note • Interface Revision Notice • Interim Revision Notice • Intermediate Routing Node • Internal Reference Number • Invoice Register Number • Iran (ISO country code) • Iriona (Honduras) (location identifier)

IRO Birao (Central African Rep.) (location identifier) • Immediate Response Option • Independent Review Officer • Industrial Relations Office • Industrial Relations Officer • Inland Revenue Office • Institut für Rationalisierung und Organisation • Internal Representation Optimization • Internal Review Office • Internal Review Officer • International Refugee Organization • International Relief Organization • Interrange Operations • Iowa Robotic Observatory (USA-IA)

IROAN Inspect and Repair Only As Necessary • Inspect and Repair Only As Needed

IROBEX Infrared Object Exchange (protocol)

IROD Instantaneous Readout Detector

IROG Interrange Operations Group

IROL Imagery Requirements Objective Listing

IROMM International Register of Microform Masters

Irón Irónico

IRON Interrange Operations Number

IROPS Institut Régional d'Orientation Professionnelle et Scolaire

IROR Internal Rate Of Return

IROS Improved Reliability Operational System • Increase Reliability of Operational Systems • Instant Order Response System

IROSS Infrared Overlay Scene Simulation

IRP IERS Reference Pole • Image Restoration Project • Independent Research Project • Individualized Reading Program • Individual Reinforcement Plan • Inertial Reference Package • Information Requirements Plan • Infrared Passive • Initial Receiving Point • Innovative Research Program • Installation Restoration Program • Institut de Recherches et d'Études Politiques (France) • Instrument Resource Profile • Integrated Resource Planning • Interest Rate Parity • Intermediate Rated Power • Invited Review Paper • I/O Request Packet • Isiro (Zaire) (location identifier) • Issue Resolution Plan

IRPA Intensification of Research in Priority Areas • International Radiation Protection Agency • International Radiation Protection Association (*)

IrPh International Conference on Infrared Physics (periodical code)

IRPP Impôt sur les Revenus des Personnes Physiques (France) • Industrial Readiness Planning Program • Installation Resource Protection Plan • Ionizing Radiation Protection Program

IRPPS Information Resource Planning and Projection System

IRPS Image Retrieval and Processing System • Infrared Photometer Spectrometer • Infrared Photometer System

IRPTC International Register of Potentially Toxic Chemicals

IRQ Interrupt Request • Iraq (international vehicle registration code) • Iraq (ISO country code)

IRQC Infrared Quantum Counter

IRQZ International Radio Quiet Zone

Irr Irregular

IRR Independent Readiness Review • Information Research and Retrieval • Infrared Radiation • Infrared Radiometer • Infrared Receiver • Infrared Reflectography • Initial Requirements Review • Inspection Rejection Report • Installation and Removal Record • Integral Rocket Tamjet • Integrated Radio Room • Integrated Readiness Report • Integrated Readiness Review • Integration Readiness Review • Interface Requirements Review • Interim Readiness Review • Interim Release Request • Interim Requirements Review • Internal Rate of Return • Introduction Rapide Répétitive • Iterative Residual Rescaling (algorithm) • Rial (Iran) (ISO currency code)

IR-RAD Infrared Radiometer

IRRAD Infrared Radiometer

IRRB Integrated Requirements Review Board

IRRC Inter-regional Research Collaboration

IRRD International Road Research Documentation

Irreg Irregular • Irregularly

Irrglr Irregular

IRRI International Rice Research Institute

IRRL Information Retrieval Research Laboratory

IRRM Information Report Requirement Manager

IRRMP Information Report Requirement Management Program • Infrared Radar Measurement Program

IrrP Irregular Penetration

IRRS Infrared Reflection Spectroscopy

IRRV Institute of Revenues, Rating, and Valuation

I&RS Information & Referral Service • Instrumentation & Range Safety

IRS Indian Remote-sensing Satellite • Indian Resources Satellite • Inertial Reference System • Information and Referral Service • Information and Research Service • Information Recovery System • Information Reporting System • Information Requirements Study • Information Retransmission System • Information Retrieval Service • Information Retrieval System • Informationsrecherchesystem • Infrared Scanner • Infrared Sounder • Infrared Source • Infrared Spectrograph • Infrared Spectrometer • Infrared Spectroscopy • Inquiry and Reporting System • Installation Readiness System • Institut de Recherche Scientifique • Institut für Raumfahrtsysteme • Instrumentation and Range Safety • Instrumentation Radar System • Integrated Rate System • Integrated Record System • Intensified Reticon Scanner • Interchange Record Separator • Interactive Retrieval Software • Interface Requirements Specification • Intermediate Reference Structure • Intermediate Resolution Spectrograph • Internal Reflection Spectroscopy • Internal Revenue Service (USA) • International Radiation Symposium • International Radio Silence • International Reference Stars • International Referral System • Ionizing Radiation Standard • Ionospheric Radio Signal • Romanian Institute of Normalization • Sturgis Kirsch (USA-MI) (location identifier)

IRSA Infrared Science Archive • Institute for Remote Sensing Applications • Irisg Research Scientists' Association

IRSAC Institut de Recherche Scientifique en Afrique Centrale (now defunct)

IRSAP Interactive Remote Sensing Applications Package

IRSB Infrared Surface Brightness

IRSBF Infrared Surface Brightness Fluctuation

IRSC Institute for Reconfigurable Smart Components (USA-AL) • International Relations Subcommittee • Iranian Remote Sensing Center

IRSCS Interregional Subject Covering Scheme

IRSD Information Retrieval, Storage and Dissemination

IRSF Infrared Survey Facility

IRSG Information Retrieval Specialist Group • Internet Research Steering Group

IRSHELL Infrared Echelle Spectrometer

IRSI Infrared Space Interferometer • Infrared Space Interferometry

IRSIA Institut pour l'encouragement de la Recherche Scientifique dans l'Industrie et l'Agriculture (Belgium)

IRSN Institut de Radioprotection et de Sûreté Nucléaire (France) • Institut Royal des Sciences Naturelles (Belgium)

IRSNB Institut Royal des Sciences Naturelles de Belgique (Belgium)

IRSOL Istituto Ricerche Solari Locarno (Switzerland)

IRSp Informationsrecherchesprache

IRSP Infrared Scene Projector

IRSPEC Infrared Spectrometer

IRSR Infrared Scanning Radiometer

IRSS Indian Remote Sensing Satellite • Inertial Reference Stabilization System • Infrared Sensor System • Instant Recall Signal Storage • Instrumentation and Range Safety System

IRSSP Interactive Remote Sensing Software Package

IRS&T Infrared Search & Track

IRST Infrared Search and Track • Infrared Search and Tracking • Institut de Recherches Scientifiques et Techniques

IRSTS Infrared Search and Track System

IRSU Infrared Radio Scientific Union

IRSV Infrared Survey of Valinhos

IRT Icing Research Tunnel • Incident Response Team • Independent Review Team • Individual Reliability Test • Information Retrieval Technique • Informationsrecherchetechnologie • Infrared Radiation Thermometer • Infrared Telescope • Infrared Thermometer • Infrared Tracker • Infrared Triplet • Infrared Tube • Initialize Reset Tape • In Reference To • In Regard To • In Reply To • In Response To • Instantaneous Radiative Transfer • Institut für Rundfunkteknik • Integrated Radar Target • Integrated Range Tracker • Integrated Readiness Testing • Integrated Retail Terminal • Interborough Rapid Transit (USA-NY) • Interrogator - Responder - Transponder • IUE Review Team

IRTAD International Road and Traffic Accident Database

IRTC Information Technology Research Centre

IRTCM Integrated Real-Time Contamination Monitor

IRT&E Independent Review, Test & Evaluation

IRTE Independent Review, Test and Evaluation • Integrated Road Telematics Environment

IRTF Infrared Telescope Facility (USA-HI) • Infrared Tully-Fisher (relation) • Internet Research Task Force • Iowa Robotic Telescope Facilities (USA-IA)

IRTFRD Internet Research Task Force on Resource Discovery

IRTM Infrared Thermal Mapper • Infrared Thermal Mapping • Infrared Transparent Material

IRTOS Intelligent Real-Time Operating System

IRTP Integrated Reliability Test Program • Internet Reliable Transaction Protocol

I&RTS Integration & Runtime Specification

IRTS Incident Reporting and Tracking System • Infrared Telescope in Space • Infrared Temperature Sounder • Integration and Runtime Specification

IRU Indefatigable Right of Use • Industrial Rehabilitation Unit • Inertial Reference Unit • Information Retrieval Unit • Instrument Reference Unit • Integrated Recovery Utility • Intermediate Registration Unit

IRUS Irvine Research Unit in Software (USA-CA)

IR/UV Infrared/Ultraviolet

IRUVDR Infrared Ultraviolet Double Resonance

iRv im Rahmen von

IRV Improved Recovery Vehicle • Insulated Reentry Vehicle • Interrange Vector • Irregular Variable (star) • Isoradial Velocity • Isotope Reentry Vehicle

IRVAT Infrared Video Automatic Tracker

IRVB Infrared Video Bolometer

IR/VIS Infrared/Visible

IRVR Instrument Runway Visual Range

IRW Indirect Reference Word • Infrared Window

IRWC International Registry of World Citizens

IRWG Image Reconstruction Working Group • Image Restoration Working Group

IRWL Interchangeability and Replaceability Wording List

IRX Information Retrieval Experiment

IRZ Informations- und Recherchezentrum

is Iceland • Icelandic (ISO language code)

iS im Sinn

Is Îles • Îlets • Ilhas • Island • Islands • Islas • Isle • Isles • Islets • Isolas • Israel • Israeli

I&S Intelligence & Surveillance • Interchangeability & Substitutability • Interchangeability & Substitution

I/S Installation of Systems

IS Iceland (international vehicle registration code) • Iceland (ISO country code) • Iceland (postal country code) • Image Selection • Image Sharpening • Image Subtraction • Incomplete Sequence • Indexing in Source • Indian Standard (India) • Industrial Society • Industrial Specialist • Infinite Server • Information Science • Information Security • Information Separator • Information Series • Information Service • Informationsstelle • Informationssystem • Information Superiority • Information System • Initialize • Initial Synchronization • Inner Shell • Input Secondary • Inquiry Server • Installation Sanitaire • Installation Support • Instituteur Spécialisé • Instruction Sheet • Instructor Station • Instrumentation Summary • Instrumentation System • Instrument Scientist • Instrument Section • Instrument Station • Integrated Satellite • Integrated Service • Integrated Software • Integrated Starlight • Integrative System • Intelligence Specialist • Interchangeability and Substitutability • Interconnecting Station • Interface Specification • Interference Suppressor • Interim Standard • Intermediate School • Intermediate State • Intermediate System • Internal Security • Internal Shield • International Staff • International Standard • Internet Standard • Interplanetary Scintillation • Interrupt Set • Interrupt Status • Interstellaire • Interstellar • Interswitch • Interval Signal • Intrinsically Safe • Intrinsic Safety • Invited Speaker • Island • Islande (code ISO de pays) • Isomer Shift • Isotropic Shift • Israel

ISA Ignition and Separation Assembly • Imprimé Sans Adresse • Independent Schedule Assessment • Individual Savings Account • Industrial Standard Architecture • Industry Standard Architecture (*) • Inertial Sensor Assembly • Information Science Abstracts • Information Science and Applications (conference) • Information Security Architecture • Information Superiority Assessment • Information System Architecture • Initial Sample Address • Initial Sun Acquisition • Installation and Service Activity • Institute of Space Aeronomy • Institut Supérieur des Affaires • Instructional Systems Approach • Instruction Set Architecture • Instrumentation Society of America (*) • Instrument Society of America • Instruments SA Inc. • Instrument Subassembly • Insulated Structure Assembly • Integrated Services Architecture • Integrated Support Area • Intelligent Software Agent • Intelligent Software Assistant • Intelligent Systems and Applications (conference) • Interactive Services Association • Interface Switching Assembly • Interim Stowage Assembly • Intermediate Specific Activity • Internal Security Act (USA) • International Search Authority • International Searching Authority • International Security Affairs • International Security Assistance • International Sociological Association • International Standard Atmosphere • International Standardization Agreement • International Symposium on Algorithms • Internet Security Association • Interservice Agreement • Interservice Support Agreement • Interstage Adapter • Israel Space Agency • Mount Isa (Australia-QLD) (location identifier)

ISAA Israel Society of Aeronautics and Astronautics

ISAAC Infrared Spectrograph And Array Camera • Infrared Spectrometer And Array Camera (*) • International Society for Archaeoastronomy and Astronomy in Culture • International Symposium on Algorithms And Computation • Internet Security, Applications, Authentication and Cryptography

ISAAG Imaging Spectrometer Science Advisory Group

ISAB International Scientific Advisory Board • International Society for Adaptive Behavior

ISABP International South Atlantic Buoy Programme

ISAC Information Society Activity Center • INTELSAT Solar Array Coupon • Interferometry Science Advisory Committee • Isotope Separator Accelerator • Isotope Separator and Accelerator • ISRO Satellite Centre (India) • Issue Screening and Analysis Committee

ISACA International Symposium on Applications of Computer Algebra

ISAD Information Society And Development (conference)

ISADPM International Society for the Abolition of Data Processing Machines

ISADS International Symposium on Autonomous Decentralized Systems

ISAGEX International Satellite Geodesy Experiment

ISAI International Symposium on Artificial Intelligence

ISAIA Interoperable Systems for Archival Information Access

ISAKMP Internet Security Association and Key Management Protocol

ISAL Information Service Access Line • Information System Access Line • International Surface Air Lift • Investigation of STS Atmospheric Luminosities

ISAM Indexed Sequential-Access Management • Indexed Sequential-Access Method • Integrated Switching And Multiplexing • Intermediate Software Acquisition Management

ISAMC Intermediate Software Acquisition Management Course

ISAMS Improved Stratosphere And Mesosphere Sounder • Improved Stratospheric And Mesospheric Sounder (*)

ISAN International Standard Agreement Number • International Standard Astronomical Number • International Standard Audiovisual Number • Isomer Spectroscopy of Activated Nuclei

ISAO Istituto di Scienze dell'Atmosfera e dell'Oceano (Italy)

ISAP Index to South African Periodicals • Individual System Automation Plan • Information Sort And Predict • Information System Architecture Plan • Integrated System for Agriculture Productivity • ISA Proceedings (periodical code) • ISO Spectral Analysis Package

ISAPI Information Server Application Programming Interface • Internet Server Application Programming Interface

(*) • Internet Service Application Programming Interface

ISAR Information Storage And Retrieval • Instrument Support Activity Request • Intelligent Synthetic Aperture Radar • Interferometric Synthetic Aperture Radar • International Symposium on Augmented Reality • Inverse Synthetic Aperture Radar

ISARA Interferometric Synthetic Aperture Radar Altimeter

ISARD Integrated Support for Aeronautical Research and Development

ISAS Illinois State Academy of Science (USA-IL) • Imager Spectrometer and Atmospheric Sounder • Institute for Space and Astronautical Science (Japan) (now JAXA) • Institute of Space and Atmospheric Studies • INTEGRAL Scientific Analysis System • Integrated Situational Awareness System • International School for Advanced Studies (Italy) • Ipswich School Astronomical Society (UK)

ISAT Information Science And Technology • ISA Transactions (periodical code)

ISATS Integrated Space (vehicle) Automation Test System • Integration Spacecraft Automated Test System

ISAV Instituto de Sistemas Audiovisuales

ISB Independent Sideband • Industrial Security Bulletin • Instantaneous Starburst • Instrument Support Branch • Intermediate Staging Base • Intermediate Supply Base • Internal Serial Bus • International Scientific Brotherhood • International Society of Biometeorology • Interstellar Bubble • Islamabad (Pakistan) (location identifier)

ISBA Information Systems Business Area • Interactions Surface-Biosphère-Atmosphère • International Seabed Authority • International Society for Bayesian Analysis

ISBC International Smal Business Congress

ISBD International Standard Bibliographic Description

ISBF Interactive Search of Bibliographic Files

ISBL Information System Base Language • Institution Sans But Lucratif

ISBN Integrated Satellite Business Network • International Standard Book Number

ISBRv International Space Business Review (periodical code)

IS&C Information Security & Cryptography

ISC Icing System Controller • Incremental Support Capability • Index of Status Characteristics • Indice de Satisfaction de la Clientèle (*) • Indice de Satisfaction du Client • Individual Service Contract • Industrial Support Contractor • Inertial Start Command • Information and Software Control • Information Security and Cryptography • Information Service Center • Information Society of Canada • Information Systems Center • Information Systems Committee • Initial Software Configuration • Institute for Scientific Computation • Institut Supérieur du Commerce (France) • Instruction Set Computer • Integrated Satellite Control • Integrated Storage Control • Integrated System Check • Integrated System Control • Intelligent Subject Code • Intercommunications Set Control • Interdisciplinary Scientific Commission • Interferometry Science Center • Interim Steering Committee • Intermediate Switching Center • International Seismological Centre • International Signal and Control • International Space Corporation • International Statistical Classification • International Statistical Congress • International Steering Committee • International Student Conference • International Student Council • International Switching Center • Inter-Software Communications • Inter-State Commission • Interstellar Cloud • Interstellar Communication • Inter-Systems Communications • Isles of Scilly - Saint Mary's (UK) (location identifier) • Item Status Code • Iterim Storage Cask • IVS Steering Committee

ISCA Information Systems Consultants Association • Integrated Systems Control Architecture • Intelligent Synchronous Communication Adapter • International Speech Communication Association • International Symposium on Computer Architecture • Ionization Spectroscopy for Chemical Analysis

ISCAN Inertialess Steerable Communications Antenna

ISCAP Institute for Strings, Cosmology and Astroparticle Physics (USA-NY)

IScB International Scientific Brotherhood

ISCC Information and Software Control Center • International Standard Commodity Classification

ISCCB Internal Software Change Control Board

ISCCO Information Systems Command Contracting Office

ISCCP International Satellite Cloud Climatology Project

ISCD Interface Specification Control Document

ISCDR Incremental Software Critical Design Review

ISCE Information Systems Cost Estimation • Integrated Scientific Computing Environment • International Staff Committee Europe (ESO)

ISCED International Standard Classification of Education

ISCGM International Steering Committee for Global Mapping

ISCJS ISRO Satellite Centre Journal of Spacecraft Technology (periodical code)

ISCM Initial Software Configuration Map

ISCMS Integrated Supply Chain Management System

ISCN International Shared Cost Number • International Software Consulting Network • International Standard Cartographic Number

ISCNET Inter-Systems Communications Network

ISCO Initial Systems Checkout • Innermost Stable Circular Orbit • International Soil Conservation Organisation (India) • International Standard Classification of Occupations

ISCON Integrated Systems Control

ISCOPE International Scientific Computing in Object-oriented Parallel Environments (conference)

ISCP Integrated Services Control Point

ISCR Institute for Scientific Computing Research (LLNL)

ISCRO Industrial Security Clearance Review Office

ISCS Information Service Computer System • Integrated Satellite Control System • Intermediate Science Curriculum Study • International Solar Cycle Studies

ISCT Institute for the Study of College Teaching (RSCNJ)

ISCZ Intershell Convective Zone

iSd im Sinne der • im Sinne des

ISD Image Section Descriptor • Independent School District • Induction System Deposit • Information Systems Development • Information Systems Directive • Information Systems Division • Inner Stellar Disk • In-Service Date • In-Source Decay • Inspection Supervision Disposition • Installation Start Date • Instructional Systems Design • Instructional Systems Development • Instructional Systems Device • Instruction Sequence Diagram • Integrated Service Delivery • Integrated Spatial Database • Integrated Symbolic Debugger • Integrated Systems Development • Integrated Systems Dictionary • Intelligent Spooling Device • International Standard Depth • International Subscriber Dialing • International Subscriber Dialog • Interoperability Standardization Directorate • Interstellar Dust • Iscuande (Colombia) (location identifier)

ISDA Imaging Spectrometer Data Analyzer • INTEGRAL Spectrometer Data Analysis Group • Intelligent Systems Design and Applications (conference) • Interactive Software Developers Association • Invariant Subspace Decomposition Algorithm

ISDB Information Systems Development Branch • Integrated Services Digital Broadcasting • Integrated Support Database • IRF Scientific Database

ISDC INTEGRAL Science Data Centre • International Space Development Conference

ISDE Integrated Software Development Environment

ISDEIV International Symposium on Discharges and Electrical Insulation in Vacuum

ISDF Institut de Sûreté De Fonctionnement

ISDL International School on Digital Libraries

ISDN Integrated Services Data Network • Integrated Services Digital Network (*) • Integrated Switch Data Network • Integrated Switch Digital Network • Integrated Switched Data Network • Integrated Switched Digital Network • Integrated Switching Data Network • Integrated Switching Digital Network • International Standard Decimal Number

ISDNX Integrated Services Digital Network Exchange

ISDR International Strategy for Disaster Reduction

ISDS Illini Space Development Society (USA-IL) • Image Switching and Distribution System • Inadvertent Separation Destruct System • International Serials Data System

ISDSS International Society for Decision Support Systems

ISDT Integrated Systems Development Tool

ISDU Interim Secure Data Unit

ISDWG Information Systems Development Working Group

I&SE Installation & Service Engineering

ISE Image Symbol Editor • Improved Spectrum Efficiency • Informal Science Education • Information and Software Engineering • Information Science Education • Information Systems Engineering • Information Systems Executive • Initial Support Element • In-Service Engineer • Installation and Service Engineering • Installation Support and Evaluation • Institute for Software Engineering (USA) • Institutional Self-Evaluation • Institut Supérieur d'Électronique • Instrumentation Suitability Evaluation • Integral Squared Error • Integrated Sensing Experiment • Integrated Space Experiment • Integrated Storage Elements • Integrated Systems Evaluation • Intelligence Support Element • Interact Simulation Environment • Intercept System Environment • International Space Enterprises (USA-CA) • International Space Explorer • International Stock Exchange • Interpret Sign Error • Ion Selective Electrode

ISEA In-Service Engineering Activity • In-Service Engineering Agent • International Symposium on Electronic Art • International Symposium on Equatorial Aeronomy

ISEB Information Systems Executive Board • International Symposium on Electronic Business

ISEC Information Security Evaluation Criteria • Information Systems Electronics Command • Information Systems Engineering Command • International Symposium on Electronic Commerce

ISECCo International Space Exploration and Colonization Company (USA-AK)

ISED Integrated Spectral Energy Distribution • Ionizing Spectral Energy Distribution

ISEE Integrated Software Engineering Environment • Integrated System Engineering Environment • International Society for Environmental Ethics • International Sun-Earth Explorer

ISEF International Science and Engineering Fair

ISEIC Information Systems Engineering and Integration Center

ISELA International Space Enterprises - Lavochkin Associates

ISEM Information Systems, Engineering and Management • Integrated Space Environmental Model • International Society for Ecological Modeling

ISEMP Integrated Systems Evaluation Management Plan

ISEMS Improved Spectrum Efficiency Modeling and Simulation

ISEMSI Isolation Study for the European Manned Space Infrastructure

ISEN Interactive Satellite Education Network • International Standard Exploration Number

ISENA Imaging Spectrometer for Energetic Neutral Atom

ISEO In-Service Engineering Office

ISEP Institute of Solid-Earth Physics (Norway) • Instructional Scientific Equipment Program • Isolated/Stabilized Exercise Platform

ISEPS International Sun-Earth Physics Satellite

ISER Integral Systems Experimental Requirement

ISES Information Sciences Experiment System • Information-Society-Energy-System • Integrated Solid Earth System • International Solar Energy Society • International Space Environment Service

ISESA Information Systems Engineering Support Activity

ISESE International Symposium on Empirical Software Engineering

ISET Independent Software Evaluation Test

ISF Implied Shortage Factor • Impôt de Solidarité sur la Fortune (France) • Improved Support Facility • Industrial Space Facility • Industrial Support Fund • Information Science Function • Information Systems Factory • Infrasonic Frequency • Installation de Simulation Fonctionnelle • Integral Shaped Filament • Integrated Space Factory • Integrated Subject File • Integrated Support Facility • Integrated Systems Facility • Integration Support Facility • International Science Foundation • Invalid Security Field • Iowa Science Foundation (USA-IA) • Israel Science Foundation

ISFA International Scientific Film Association

ISFET Ion-Sensitive Field Effect Transistor

ISFJ Introverted Sensing Feeling Judging

ISFL International Scientific Film Library

ISFMS Indexed Sequential File Management System

ISFN International Standard File Number

ISFP Introverted Sensing Feeling Perceiving

ISFRA Institut Supérieur de Formation et de Recherche Appliquée

ISFUG Integrated Software Federal User Group

ISG Image Synthesis Group • Industry Steering Group • Information Systems Group • Infrastructure Support Group • Instructional Systems Group • Integrated Survey Grid • Intelligence Support Group • Interoperable Systems Group • Interstage • Interstellar Gas • Ishigaki (Japan) (location identifier)

ISGA International Society for Genetic Algorithms (now ISGEC)

ISGE Instrument Software Systems Engineer

ISGEC International Society for Genetic and Evolutionary Computation

ISGGN International Space Geodetic and Gravimetric Network

ISGI International Service of Geomagnetic Indices

Isgn Insignia

ISGN International Space Geodetic and Gravimetric Network • International Space Geodetic Network • International Standard Graphical Number

ISGRG International Society on General Relativity and Gravitation

ISGRI INTEGRAL Soft Gramma-Ray Imager

ISGS INTEGRAL Science Ground Segment

ISH Information Superhighway • Ischia (Italy) (location identifier)

ISHA Integrated System Hazard Analysis

ISHAM International Society for Human and Animal Mycology

ISHC International Society of Heterocyclic Chemistry

ISHM International Society for Hybrid Microelectronics

ISHPC International Symposium on High-Performance Computing

ISHS International Society of Horticultural Science

I&SI Information & Software Integration

ISI Image System Interface • Indian Standards Institution (India) • Indian Statistical Institute (India) • Induced Spatial Incoherence • Informatics Society of Iran • Information and Software Integration • Information Sciences Institute • Information Services International • Information Society Initiative • Information Systems Inventory • Infrared Spatial Interferometer • Infrared Spectroscopic Imaging • Initial Support Increment • Initial Support Item • Initial Systems Installation • In-Service Inspection • Institute for Scientific Information • Institute of Scientific Instrument (Czech Rep.) • Institut Supérieur de l'Ingénieur • Institut Supérieur d'Informatique • Instrumentation Support Instruction • Integrated Solutions Inc. • Intelligent Serial Interface • Interdisciplinary Science Investigation • Interim Support Item • Internally Specified Index • International Statistical Information • International Statistical Institute • International Systems Institute • Ion Source Injector • Isisford (Australia-QLD) (location identifier)

ISIC Information and Software Integration Contractor • Information Seeking In Context • Integrated Software Infrastructure Center • International Standard Industrial Classification • International Student Identity Card

ISID International Symposium on Inverse Problems

ISIDE Interactive System for Image Data Elaboration

ISIF Intermediate Standard Transformation Format

ISIL Interim Support Items List

ISIM Instrument Simulator • Integrated Science Instrument Module

ISIMM International Society for the Interaction of Mechanics and Mathematics

ISIN Information Systems Internetting • International Standard Image Number

ISIP International Science and Its Partners

ISIR Improved SGM Imaging Radiometer • Infrared Spectral Imaging Radiometer • International Space Industry Report

ISIRA International Science Initiative in the Russian Arctic

ISIRI Institute of Standards and Industrial Research of Iran

ISiS Indiana Silicon Sphere (experiment)

I&SIS Interoperability & Standard Information System

IS-IS Intermediate System to Intermediate System

ISIS Imaging Spectrograph for Interstellar Shocks • Improved Speech Intelligibility System • Inflatable Sunshield In Space • Institute for Science and International Security (USA-DC) • Institute for Software Integrated Systems (USA-TN) • Institute for Systems, Informatics and Safety • Integrated Safety Information System • Integrated Scientific Information System • Integrated Ship Information System • Integrated Student Information System • Integrated Software for Image Spectrometry • Integrated Standards Information System • Integrated Systems and Information Services • Integrated Systems for Installation Support • Interactive Sharable Image System • Interactive Spectral Interpretation Software • Interferometric Stellar Imaging System • Intermediate-dispersion Spectrograph and Imaging System • Intermediate System to Intermediate System • Internally Switched Interface System • International Satellite for Ionospheric Studies • International Space Information Service • Interoperability and Standard Information System • Interoperability and Standardization of Information System • Investigative Support Information System • ISRIC Soil Information System • Item Standardization Information System

ISITE Integrated Self-Instructive Test Equipment • Intensified Silicon Intensifier Target

ISITEM Institut des Sciences de l'Ingénieur en Thermique-Énergétique des Matériaux (France)

ISJ Information Systems Journal • Isla Mujeres (Mexico) (location identifier)

IsJAP Istanbul University Faculty of Science Journal of Astronomy Physics (periodical code)

IsJT Israel Journal of Technology (periodical code)

ISK Icelandic Króna (currency) • Insert Storage Key • Internet Starter Kit • Ion Source Kit • Krona (Iceland) (ISO currency code) • Nasik (India) (location identifier)

ISKN International Standard Key Number

ISKO International Society for Knowledge Organization

ISKZ Issledovaniia Solntsa i Krasnykh Zvezd (periodical code)

Isl Island • Isle

I&SL Interchangeability & Substitutability Logistics

ISL Iceland (ISO country code) • Industrial Security Letter • Inertial Systems Laboratory • Information Search Language • Information System Language • Initial Support List • Initial System Loading • Institute of Space Law • Instructional Systems Language • Instrument Standards Laboratory • Integrated Schottky Logic • Integrated Stock List • Interactive Simulation Language • Interactive System Language • Interchangeability and Substitutability Logistics • Internetworking Switch Link • Intersatellite Link • Interswitching Link • Interswitch Link • Inverse Square Law • Ionen-Strahl-Labor • Isabel Pass (USA-AK) (location identifier) • Island • Islande (code ISO de pays) • Item Selection List • Item Study Listing

ISLA Information System for Los Angeles (USA-CA) • International Sign Linguistics Association • International Stratospheric Laboratory for Astrophysics

ISLAN Integrated Services Local Area Network

ISLC International Semiconductor Laser Conference

Isld Island

Islds Islands

ISLE Individual Single Loss Expectancy • Information Systems for Law Enforcement

ISLI Institute of System Level Integration

ISLM Integrated Services Line Module • Integration Shop and Lab Manager

Isln Isolation

ISLN Integrated Services Local Network • Isolation

ISLPED International Symposium on Low-Power Electronic Design

ISLR Integrated Sidelobe Ratio

Isls Islands

ISLS Islands

ISLSCP International Satellite Land Surface Climate Project • International Satellite Land Surface Climatology Program • International Satellite Land Surface Climatology Project (*)

ISLSSPWG International Space Life Sciences Strategic Planning Working Group

ISLU Integrated Services Line Unit

ISLV Institut Supérieur des Langues Vivantes

ISM Image Sensor Module • Image Steering Mirror • Image Storage Module • Image Support Module • Industrial - Scientific - Medical • Industrial Security Manual • Information Sources Map • Information Systems Management • Information Systems Manager • Information Systems for Management • Infrared Sky Monitor • Infrared Spectrometer • Input Signal Management • Installation Support Module • Institute of Scientific Modeling • Institute of Statistical Mathematics • Institut Supérieur des Métiers (France) • Institut Suisse de Météorologie (Switzerland) • Institut Svizzer da Meteorologia (Switzerland) • Integrated Sight Module • Integrated Structures Model • Integrated Sustainment Maintenance • Integrated System Management • Interactive Surface Modeling • Intermediate Service Module • International Safety Management • International Solar Month • International Staff Member • Internet Service Manager • Interpretive Structural Modeling • Interstellar Matter • Interstellar Media (USA-CA) • Interstellar Medium • ISDN Standards Management • Istituto Svizzero di Meteorologie (Switzerland) • Kissimmee (USA-FL) (location identifier)

ISMA Improved Satellite Multichannel Antenna • Information Systems Management Activity • Information Systems Management Agency • International Satellite Monitoring Agency • Internet Streaming Media Alliance

ISMAR International Symposium on Mixed and Augmented Reality

ISMARSC Information Systems Mapping And Remote Sensing Consultancy

ISMB Information Systems Management Board • Intelligent Systems for Molecular Biology

ISMC Inflight Services Management Council • Integrated Sustainment Maintenance Concept • International Seminars on Mathematical Cosmology

ISMCC Integrated Systems Management Control Center

ISMCS International Station Meteorological Climate Summary

ISMDA International Symposium on Medical Data Analysis

ISMDWG Infrared, Submillimeter, and Millimeter Detector Working Group

ISME Instrument Software Maintenance Environment • International Subject Matter Expert

ISMEBC International Society for Molecular Electronics and Biocomputing

ISMED Institute of Sciences and Mathematics Education and Development

ISMF Interactive Storage Management Facility

ISMIS International Symposium on Methodologies for Intelligent Systems

ISML Intermediate System Mockup Loop

ISMLS Interim Standard Microwave Landing System

ISMM INTEGRAL Software Management Meeting • International Symposium on Memory Management

ISMN International Standard Music Number

ISMO Information Systems Management Order • Information Systems Mission Order

ISMRB Information Systems Management Review Board

ISMS Image Store Management System • Integrated Security Management System • Interactive Supply Management System • International Society for Mine Surveying • International Society for Mushroom Science

IsMTD Issledovaniia Mekhanike i Teploobmenu Dvukhfaznykh Sred (periodical code)

ISMV International Symposium on Multiple Valued Logic

IS&N Intelligence in Services & Networks (conference)

ISN Information Systems Network • Initial Sequence Number • Immediate Short Negative • Institut des Sciences Nucléaires • Integrated Systems Network • Intelligence in Services and Networks (conference) • Internal Sequence Number • Internal Statement Number • International Standard Nomenclature • International Supernovae Network • Internet Shopping Network • Williston (USA-ND) (location identifier)

ISNAR International Service for National Agricultural Research (Netherlands)

ISND IGN Seismic Network Data Bank

ISNS Inside Science News Service (AIP) • International Satellite Navigation System • Internet Storage Name Service

ISNSA Independent Software Nuclear Security Analysis

ISNU Illinois State Normal University (USA-IL)

ISO Imaging Spectrometer Observatory • Imaging Spectrometric Observatory (*) • Imaging Spectroscopic Observatory • Incentive Stock Option • Indemnité de Suivi et d'Orientation • Independent System Operator • Individual System Operation • Information Security Office • Information Systems Office • Information Systems Officer • Infrared Space Observatory • In Search Of • Installation Supply Officer • Institute for Space Observations (USA-NY) • Institut Supérieur d'Optique • Integration Staff Office • Integration Staff Officer • International School's Observatory (project) • International Science Organization • International Standardization Office • International Standardization Organization • International Standards Organization • Interoperable System Office • Interseasonal Oscillation • Isolation • ISRO Safety Office • Kinston (USA-MO) (location identifier)

ISOC INTEGRAL Science Operations Centre • Integrated Science Operations Center • Internet Society • IVS Science Operations

ISOCAM Infrared Satellite Observatory Camera

ISOCLS Interactive Self-Organizing Cluster Program

ISODE International Standards Organization Development Environment

ISOE Integrated Sequence Of Events

ISO-IP International Standards Organization Internetworking Protocol

Isol Isolate • Isolated • Isolation

ISOL Isolated • Isolation • Isotope Separator Outline

ISOLAN Isolated Local Area Network

ISOLD Isolated

ISOM Industrial Security Operating Manual • Information Systems and Operations Management • International Symposium on Optical Memory

ISOMAX Isotope Magnet Experiment

ISON International Standard Organization Number

ISONET ISO Information Network

ISOO Information Security Oversight Office

ISOOMS International Symposium on Object-Oriented Methodologies and Systems

ISOON Improved Solar Observing Optical Network

ISOP Integrated Spacecraft Operations Plan • International Seismological Observation Period

ISOPAR Improved Symbolic Optimizing Assembly Routine

ISOR Industrial security Operating Regulation • Initial Statement Of Requirement

ISORC International Symposium on Object-oriented Realtime Computing

ISOS International Southern Ocean Studies

ISOSDC Infrared Space Observatory Spectrometer Data Center

ISO-SP International Standards Organization Session Protocol

ISOSS Infrared Space Observatory Serendipity Survey

ISOSWS Infrared Space Observatory Short Wavelength Spectrometer

ISOT International Scientific Oversight Team

ISOTAS International Symposium on Object Technologies for Advanced Software

ISOTOF Isochronous Time-Of-Flight

ISO-TP International Standards Organization Transport Protocol

ISOWAKE International Standardized Oil Wake Experiment

ISOWG Information Systems and Operations Working Group

Isp Specific Impulse

ISP Idaho Star Party (USA-ID) • Identifier Son Potentiel • Image Synthesis Processor • Immediate Short Positive • Implementation Support Plan • Independent Study Program • Independent Study Project • Indexed Sequential Processor • Index Sequential Processing • Index Sequential Processor • Industrial Security Plan • Informatie Stimuleringsplan • Information Security Program • Information Services Provider • Information Sharing Protocol • Information Systems Plan • Information Systems Planning • Information Systems Program • Initial Set of Parameters • Initial Specific Impulse • In-Space Propulsion • Institute of Solid-state Physics • Institut Supérieur de la Pédagogie • Institut Supérieur de Préparation • Instructional Staff Presentation • Instructional System Package • Instruction Set Processor • Instrumentation Support Plan • In-System Programmable • Integrated Service Panel • Integrated Space Plan • Integrated Steel Plant • Integrated Studies Program • Integrated Support Plan • Intelligence Support Plan • Intelligent Services Peripheral • Intented Support Plan • Interferometric SAR Processor • Interim Support Period • Interim Support Plan • Intermediate Service Part • Internally Stored Program • International Scientific Product • International Scientific Publication • International Society for Photogrammetry • International Standardization Profile • International Standardized Profile • International Standard Profile • International Study Program • Internet Service Provider • Interrupt Stack Pointer • Interrupt Status Port • Interstellar Particle • Interstellar Polarization • Interstellar Probe • Inverse Sampling Procedure • Inverse Scattering Problem • Ion-Sensitive Probe • Islip - Long Island (USA-NY) (location identifier) • Isotope Separation Power • Italian Society of Physics • Long Island - Macarthur (USA-NY) (location identifier)

ISPA International Society of Parametric Analysts • International Software Producers Association • Internet Service Providers Association • Inverted Socket Process Architecture

ISPAE Institute for Space Physics, Astrophysics, and Education (USA-TN)

ISPBX Integrated Services Private Branch Exchange

ISPC Interim System Planning Computer • International Statistical Program Center • ISI Studies and Publishing Center

ISPCS Italian Solar Physics Community Server

ISPD International Symposium on Physical Design

ISPDR Integrated System Preliminary Design Review

ISPE Improved Sonar Processing Equipment • Interim Software Progress Emulation

ISPEC International Space Physics Educational Consortium (USA-CA)

ISPF Information System Project File • Interactive System Processing Facility • Interactive System Productivity Facility • Interactive System Programming Facility

ISPG Institutional Support Planning Group

ISPI Information Systems Processing Installation • Infrared Side Port Imager • Intermediate-Scale Product Inventory • Internationales System für Patentinformation

ISPL Incremental System Processing Language

ISPLED International Symposium on Low Power Electronics and Design

ISPM IEEE Signal Processing Magazine (periodical code) • Informal Security Policy Model • Information Security Program Manager • International Solar Polar Mission (now Ulysses)

ISPN International Standard Picture Number • International Standard Program Number • International Standard Protocol Number

ISPO Information Society Project Office

ISPP Industry Standard Plotting Package • Information System for Policy Planning • Information Systems Program Plan • Institute for Solid-State Physics • Institut Supérieur de Préparation Professionnelle

ISPR Institute for Survey and Policy Research • International Society for Paranormal Research • International Society for Psychotherapy Research • International Standard Payload Rack • Inter Services Public Relations

ISPRS International Society for Photogrammetry and Remote Sensing

ISPS Instruction Set Processor Specification • International Standard Paper Size

ISPT Inverse Scattering Perturbation Theory

ISPTC Information Systems Performance Technical Center

ISPU Information Services Planning Unit

ISPWG International Software Process Working Group

ISQ In Status Quo • Manistique - Schoolcraft County (USA-MI) (location identifier)

ISQAP International Standards and Quality Assurance Programme

ISQL Interactive Structured Query Language

ISQP Information Systems Quality Program

Isr Israel • Israeli

IS&R Information Storage & Retrieval

ISR Ikeda Sum Rule • Image Storage and Retrieval • Implementation Support Requirement • Incoherent Scatter Radar • Independent Science Review • Independent Sufficiency Review • Index to Scientific Reviews • Individual Schedule Request • Individual School Requirement • Industrial Security Regulation • Information Source Repository • Information Storage and Retrieval • Information Systems Research • Informatique de Systèmes et Réseaux • Infrared Scanning Radiometer • Infrared, Submillimeter and Radio • Ingoing Shortwave Radiation • Initial System Release • Installation Status Report • Institute for Space Research • Institute of Social Research • Instrument Science Report • Integrated Safety Review • Integrated Switch Router • Integrated Systems Review • Intelligence - Surveillance - Reconnaissance • Interim Scientific Report • Interim Shop Release • Interim System Review • Intermediate Session Routing • Internal Scientific Report • International Simple Resale • International Society for Radiology • Interrupt Service Routine • Interrupt Status Register • Intersecting Storage Ring • Intersecting Storage Road • Inventory Status Report • Investissement Socialement Responsable • Israel (ISO country code)

ISRA Intelligente Systeme Roboter und Automatisierung • International Symposium on Robotics and Automation

ISRAD Information Storage, Retrieval And Dissemination

ISRAM International Symposium on Robotics and Automated Manufacturing

ISRAMAP Information System for Regional Automated Mapping Analysis and Planning

ISRB Integrated Systems Requirements Board • Integrated Systems Review Board

ISRC Information Services Readiness Center • International Standard Recording Code

ISR&D Information Storage, Retrieval & Dissemination

ISRD Information Storage, Retrieval and Dissemination • Information Systems Requirement Document

ISRF Interstellar Radiation Field

ISRG Independent Space Research Group • Inferred Stochastic Regular Grammar

ISRI Information Science Research Institute • Information Storage and Retrieval Interface • Institute for Software Research International (USA)

ISRIC International Soil Reference and Information Centre (Netherlands)

ISRM Information System Resource Manager • International Society of Rock Mechanics

ISRN International Standard Recording Number • International Standard Reporting Number • International Standard Report Number

ISRO Indian Space Research Organization (India) • Institute for Space Research Observatory

IS&RP Initial Spares & Repair Parts

ISRP Initial Spares and Repair Parts

ISRS Indian Society of Remote Sensing (India) • Inflatable Space-Rigidized Structure • Information Search and Report System • Information Search and Retrieval System • Information Storage and Retrieval System • Information System Resource Software • Integrated Status Reporting System

ISRU In-Situ Resource Utilization

ISRv Interdisciplinary Science Reviews (periodical code)

IsRvC Istanbul University Faculty Science Review - Serie C (periodical code)

Iss Issue

ISS Ice Status System • Ice Suppression System • Ideal Solidus Structure • Ignition Shielding System • Image Sensor System • Image Stabilization System • Imaging Science System • Indemnité de Sujétion Spéciale • Industry Standard Specification • Inertial Sensor System • Inertial Subsystem • Information and Software Support • Information Search System • Information Seeking Situation • Information Sharing System • Information Storage System • Information Subsystem • Information Support System • Information Systems Security • Initial Space Station • Input - State - Stability • Input Subsystem • In Seam Seismic • In-Service Support • Installation Support Services • Institute for Space Sciences • Institute for Space Studies • Institute for Systems Sciences • Institute of Soil Science • Instruction Set Simulator • Instruction Summary Sheet • Instrumentation Support Service • Instrument Servo System • Instrument Software System • Instrument Support Structure • Instrument Support System • Integrated Satellite System • Integrated Scheduling System • Integrated Sensor System • Integrated Sounding System • Integrated Storage System • Integrated Sunlight Spectrometer • Integrated Support Station • Integrated Switching System • Integrated Systems and Services • Integrated System Schematic • Integrated Systems Support • Integration Support Service • Intelligent Support System • Interface Subset • Interface Subsystem • Interferometer Supervisor Software (VLTI) • Interim Safety Supplement • Interim Stowage Shelf • Interim Supply Support • Internal Secondary Storage • Internal Switching System • International Scientific Secretariat • International Space Station • International Summer School • International Switching System • Intersatellite Service • Interstellar Scattering • Interstellar Scintillation • Ionosphere Sounding Satellite • Ion-Scattering Spectrometer • Ion-Scattering Spectrometry • Ion-Scattering Spectroscopy • Iron and Steel Society • Islands • Isotope Separation System • Issue • Istituto Superiore di Sanitá (Italy) • Wiscasset (USA-ME) (location identifier)

ISSA Information Systems Security Association (USA) • Infrared Sky Survey Atlas • Institute for Satellite and Software Applications (South Africa) • Institute of Space Sciences and Astronomy • Integrated Software Safety Analysis • International Space Station Alpha • Interservice Support Agreement • IRAS Sky Survey Atlas

ISSAA Information Systems Selection and Acquisition Agency

ISSAC International Symposium on Symbolic and Algebraic Computation

ISSAE ISDN Supplementary Services Application Entity

ISSAG Imaging Spectrometer Science Advisory Group

ISSAS Integrated Space Segment Analysis and Simulation

ISSAT Institut des Sciences Spatiales et Applications de Toulouse (France) • International Space Station Amateur Telescope

ISSB Information Systems Standards Board

ISSC Information and Software Support Center • Information Service Support Center • Information Software Support Center • Information Systems Security Committee • Information Systems Service Center • Information Systems Software Center • Information Systems Steering Committee • Information Systems Support Center • Integrated Systems Solution Corporation • International SKA Steering Committee • International Social Science Council

ISSCC International Solid-State Circuit Conference • International Space Station Coordination Committee

ISSCL Integrated Services with Specific Link Layers (IETF)

ISSCO Integrated Software Systems Corporation

ISSE Information Security Solutions Europe (conference) • Information Systems Security Engineering • International Secondary School in Eindhoven (Netherlands)

ISSEP Integrated Systems and Software Engineering Process • International Soros Science Educational Programme

ISSES Institute for Space Sciences and Engineering Systems (USA)

ISSG Information Strategy Steering Group • Information Systems Support Group

ISSI International Society of Scientometrics and Infometrics • International Software Systems Inc. • International Space Science Institute (Switzerland)

ISSL Initial Spares Support List • Initial Supplies Support List

ISSLL Integrated Services over Specific Link Layer Architecture

ISSM Inertial Surveying System Mining • Information System Security Manager

ISSMFE International Society of Soil Mechanics and Foundation Engineering

ISSMO International Society for Structural and Multidisciplinary Optimization

ISSN Integrated Special Services Network • International Standard Serial Number

ISSO Information Services Support Office • Information Services Support Officer • Information Special Security Officer • Information Systems Security Office • Information Systems Security Officer • Information Systems Security Organization • Information Systems Staff Officer • International Small Satellite Organization • Interplanetary Solar Stereoscopic Observatory

ISSOF International Solar System Observer's Fund

ISSOL International Society for the Study of the Origin of Life

ISSP ICSU Solar-System Panal • Incoming Secondary Sort Program • Information Systems Security Program • Information Systems Standardization Program • Information Systems Strategic Plan • International Solar Systems Program

ISSPM Information Systems Security Program Manager • Information Systems Security Project Manager

ISSPP Information Systems Strategic Planning Project

ISSR Incremental Software Specification Review • Information Storage, Selection and Retrieval • Information Systems Security Representative • Internal Software Specification Review

ISSRE International Symposium on Software Reliability Engineering

IsSRT Israel Space Research and Technology Information Bulletin (periodical code)

ISSRU Information Science and Scientometrics Research Unit (Hungary)

ISSS IBM Speech Server Series • Information Society Standardization System (CEN) • Initial Sector Suite System • Installation Service Supply Support • Intelligent Signals, Sensors and Surveillance • International School of Space Simulations • International Society for the Systems Sciences • International Society of Soil Science • International Student Space Simulations • International Symposium on System Synthesis

ISSSR International Symposium on Spectral Sensing Research

ISST Institute for Space Science and Technology (USA-FL)

ISSTA International Symposium on Software Testing and Analysis

ISSTI Information System for Scientific and Technical Information

IssUP Issledovaniia Uprugosti i Plastichnosti (periodical code)

ISSW Instrument Specific Software

ISSWG International ScaRab Science Working Group

IssZK Issledovanie Zemli iz Kosmosa (periodical code)

Ist Istituto

IS&T Information Science & Technology • Information Systems & Technologies • Information Systems & Technology • Innovative Science & Technology • Invite, Slow & Test

IST Icelandic Standard • Imagery Support Terminal • Implementation Systems Test • Indian Standard Time (India) • Information Science and Technology • Information Sciences Technology • Information Scientifique et Technique • Information Security Technology • Information Set Trailer • Information Society Technologies • Information Systems and Technologies • Information Systems and Technology • Information Systems Technology • Initial Service Test • Initial Support Team • Initial System Test • Innovative Science and Technology • In-service School Teacher • Installed System Test • Installed System Testing • Institut des Sciences et Techniques • Institute for Simulation and Training • Institute of Science and Technology • Institute of Science Technology • Institute of Software Technology • Institut für Software-Technologie (Germany) • Instituto Superior Técnico • Institut Supérieur de Technologie • Instrumentation Support Team • Instrument Science Team • Instrument Support Terminal • Instrument Support Toolkit • Integrated Science and Technology • Integrated Surveillance Team • Integrated Systems Test • Integrated Systems Testing • Integrierte Systemtest • Interactive Speech Technology • International Space Technology Inc. • International Standard Thread • International Standard Time • Interstellar Travel • Interswitch Trunk • Invite - Show - Test • ISO Science Team • Issues in Science and Technology (periodical code) • Istanbul Atatürk (Turkey) (location identifier)

ISTA International Seed Testing Association • Intertank Structural Test Assembly

ISTAG Information Society Technologies Advisory Group (EU)

ISTAP Information Systems Technology Application Program • International Space Technology Assessment Program

ISTAR Imagerie Stéréo Appliquée au Relief • Image Storage Translation And Reproduction • Information Storage Translation And Reproduction • Intelligence, Surveillance, Target Acquisition and Reconnaissance

ISTB Information Sciences and Technology Branch • Integrated Sensing Testbed • Integrated System Testbed

ISTC Incunabula Short Title Catalogue • Institute of Scientific and Technical Communicators • Institut Supérieur des Techniques de Communication (France) • Integrated System Test Capability • International Science and Technology Centre • International Student Travel Confederation

ISTD In Situ Thermal Desorption • Integrated Space Technology Demonstration

ISTDT Integrated Science and Technology Definition Team

ISTE Information Scientifique, Technique et Économique • Integrated Service Terminal Equipment • International Society for Technology in Education

ISTEA Initial Screening Training Effectiveness Analysis

ISTeC International Small Telescope Cooperative (formerly NASTeC)

ISTEC Information Systems Test and Evaluation Center • International Superconductivity Technology Center (Japan)

ISTEEM Institut des Sciences de la Terre, de l'Eau et de l'Espace de Montpellier (France)

ISTEI Institute for Scientific, Technical and Economic Information

ISTF Information Strategy Task Force • Installed System Test Facility • Integrated System Test Flow • Internet Societal Task Force

ISTG Institut des Sciences et Techniques de Grenoble (France)

ISTGG Interferometric Spectrometer for Temperature and Greenhouse Gases

Isth Isthmus

ISTI Institut des Sciences et Techniques de l'Information • Institute of Scientific and Technical Information • International Space Technology Inc. • Istituto di Scienza e Tecnologie dell'Informazione (Italy)

ISTIC Institute of Scientific and Technical Information of China

ISTJ Introverted Sensing Thinking Judging

ISTL Issues in Science and Technology Librarianship

ISTM Information System Management Team

ISTN Institute Sains and Technology Nasional (Indonesia)

ISTO Institut des Sciences de la Terre d'Orléans (France)

ISTP Index of Scientific and Technical Publications • Index to Scientific and Technical Proceedings • Information Systems Tasking Plan • Institut Supérieur des Techniques Productiques (France) • Integrated Space Transportation Plan • Integrated Systems Test Procedure • International Solar-Terrestrial Physics (*) • International Solar-Terrestrial Program • Introverted Sensing Thinking Perceiving

ISTR Indexed Sequential Table Retrieval • Interoperability Standards Technical Report

ISTRA Inter-Satellite Tracking

ISTRAC ISRO Telemetry, Tracking And Command

ISTS Institute for Space and Terrestrial Science (Canada-ON) (now CRESTech) • International Symposium on Space Technology and Science • Intersite Transmission System

ISTTT International Symposium on Transportation and Traffic Theory

ISTV Integrated System Test Vehicle

ISU Idaho State University (USA-ID) • Independent Site Unit • Indiana State University (USA-IN) • Inertial Sensing Unit • Infrared Sounding Unit • Instrument Support Unit • Integrated Service Unit • Integrated Sight Unit • Interface Switching Unit • International Scientific Union • International Space University • International System of Units • Iowa State University (USA-IA) • Irkutsk State University (Russia) • Isochronous Service Unit

ISUP Institut de Statistique de l'Université de Paris (France) • Integrated Services User Part • ISDN User Part

ISURSL Indiana State University Remote Sensing Laboratory

ISUS Imaging Stratospheric Ultraviolet Spectrometer • Information Services and User Support

ISV Independent Software Vendor • Intelligence Secure Voice • Input Signal Voltage • International Scientific Vocabulary

ISVA International Satellite Verification Agency

ISVCS Improved Secure Voice Conferencing System

ISVL Internet Software Visualization Laboratory

ISVN Interim Secure Voice Network

ISVR Institute of Sound and Vibration Research • Instrument Science Verification Review

ISVS International Switched Voice Service • International Switched Voice System

ISVT Integrated Secure Voice Terminal

ISVTB Integrated Systems Verification Testbed

ISVTOP International Saturn Voyager Telescope Observing Program

ISW Industrial SPARC Workstation • Information Security Workshop • Integrated Sachs-Wolfe (effect) • Interacting Stellar Wind • Wisconsin Rapids - Alexander (USA-WI) (location identifier)

ISWA International Science Writers Association (USA-MA)

ISWC International Semantic Web Conference • International Standard Work Code • International Symposium on Wearable Computers

ISWE Integrated Sachs-Wolfe Effect

ISWG Imperial Standard Wire Gauge • Industry Standard Working Group • Information Systems Working Group • Integrated Support Working Group • Intelligence Support Working Group • Interim Science Working Group

ISWN International Standard Work Number

ISWT INTEGRAL Science Working Team

ISX Impurity Studies Experiment

ISY International Space Year (1992)

ISYA International School for Young Astronomers

ISZ Isskustvenij Sputnik Zemli

IS2 Information System Integration Support

IS4ALL Information Society for All (project)

it Italian (ISO language code) • Italy

It Italia • Italian • Italiano • Italie • Italien • Italy

I&T Inspection & Test • Installation & Test • Integration & Test • Integration & Testing

I/T Intertank

IT Identification Transponder • Ignition Temperature • Important Traveller • Importation Temporaire • Inactivity Test • Inclusive Tour • Income Tax • Index Term • Individual Training • Industrial Technology • Informatie Technologie • Information Technology • Information Theory • Information Transport • Ingénieur des Travaux • Injection Tank • Inlet Terminal • Innovative Test • Input Translator • Inspecteur du Travail • Inspection Tag • Installation Test • Installation Thermique • Instruction Technique • Instrument Team • Instrument Test • Instrument Tree • Insulating Transformer • Integrated Technology • Integration and Test • Integration Test • Integration Testing • Intelligent Terminal • Interline Transfer • Internal Thread • International Table • Interoperability Test • Interoperability Testing • Interrogator-Transponder • Intersection Table • Interval Timer • In Transit • Inverse Telescope • Ion Thruster • Italie (code ISO de pays) • Italy (ISO country code) • Item Transfer

Ita Italian

IT&A Initial Test & Acceptance

ITA Independent Television Authority • Industrial and Technical Assistance • Industrial Telecommunication Association (USA) • Industry and Trade Administration • Industry Technical Agreement • Ingénieurs - Techniciens - Administratifs • Initial Teaching Alphabet • Initial Test and Acceptance • Institut du Transport Aérien • Institute for Theoretical Astronomy • Institut für Theoretische Astrophysik • Instituto Tecnológico de Aragón (Spain) • Instrumentation Technology Associates • Integrated Technology Architecture • Integrated Test Area • Integrated Test Article • Integrated Thruster Assembly • Integrated Truss Assembly • Intelligent Task Automation • Interface Test Adapter • International Telegraph Alphabet • International Trade Administration • International Typographic Alphabet • Ionization Test Apparatus • IRAS-Texas Associations (catalog) • Itacoatiara (Brazil-AM) (location identifier) • Italie (code ISO de pays) • Italy (ISO country code) • Iterative Transfer Algorithm

ITAA Information Technology Association of America (USA)

ITAB Institut Teoreticheskoi Astronomii Biulleten (periodical code)

ITABBS Information Technology Acquisition Bulletin Board System

ITAC IEEE Transactions on Automatic Control (periodical code) • Information Technology Acquisition Center • Information Technology Association of Canada • International Target Audience Code • International Time Allocation Committee

ITAE Integrated Time and Absolute Error

ITAES IEEE Transactions on Aerospace Electronic Systems (periodical code)

ITAG Inertial Terrain-Aided Guidance

Ital Italian • Italic • Italicized • Italique • Italy

ITALY I Trust And Love You

ITAM Instituto Tecnológico Autónomo de México • Integrated Training Area Management

ITAMP Institute for Theoretical Atomic and Molecular Physics (USA-MA)

IT&AP Inspection Test & Analysis Plan

ITAP IEEE Transactions on Antennas and Propagation (periodical code) • Information Technology Acquisition Paper • Integrated Technical Assessment Panel • Inspection Test and Analysis Plan

ITAR International Traffic in Arms Regulations

ITARF Ingénieurs, Techniciens et Administratifs de Recherche et de Formation

ITAS IEEE Transactions on Applied Superconductivity (periodical code) • Institute for Technology Assessment and Systems Analysis (Germany) • Integrated Target Acquisition System

ITASE International Transantarctic Scientific Expedition

ITASS IEEE Transactions on Acoustics, Speech, and Signal Processing (periodical code)

ITAV Independent Test and Verification Plan

ITB IEEE Transactions on Broadcasting (periodical code) • Industry Training Board • Information Technology Branch • Institut Teknologi Bandung (Indonesia) • Integral Terminal Block • Integrated Testbed • Intermediate Text Block • Intermediate Transmission Block • Internal Thermal Barrier • Internal Transport Barrier • International Time Bureau • Invitation To Bid • Ion Thruster Beam • Itaituba (Brazil-PA) (location identifier)

ITBD Intraplate Tectonics and Basin Dynamics

ITBE IEEE Transactions on Biomedical Engineering (periodical code)

ITC Igloo Thermal Control • Incoming Trunk Connection • Independent Telephone Co. • Independent Television Commission • Independent Television Committee • Industrial Training Council • Information Technology Center •

Information Technology Committee • Ingénieur Technico-Commercial • Inlet Terminal Circuit • Installation Time and Cost • Instructional Telecommunications Council • Instructor Training Course • Integrated Telemetry Complex • Integrated Test Center • Integrated Test Configuration • Integrated Trajectory Computation • Integration Time Calculator • Interface Compatibility Test • International Telecommunications Convention • International Telephone Code • International Teletraffic Congress • International Television Commission (UK) • International Television Committee • International Test Conference • International Toki Conference (Japan) • International Trade Center • International Trade Commission • International Traffic Congress • International Training Center • International Translation Centre • International Typeface Corporation • Intertask Communication • Intertropical Confluence • In-Trail Climb • Investment Tax Credit • Ionic Thermoconductivity

ITCA International Topographic Composition Association

ITCAD IEEE Transactions on Computer-Aided Design (periodical code)

ITCAN Inspect, Test and Correct As Necessary

ITCC Intelligence Technical Coordinating Committee

ITCF Installateur en Télécommunications et Courants Faibles

ITCG Information Technology Coordinating Group

ITCHM IEEE Transactions on Components, Hybrids and Manufacturing Techniques (periodical code)

ITCJ ITC Journal (periodical code)

ITCmp IEEE Transactions on Computers (periodical code)

ITCom IEEE Transactions on Communications (periodical code)

ITCS IEEE Transactions on Circuits and Systems (periodical code) • Information Technology Customer Satisfaction • Integrated Target Control System • Internal Thermal Control System

ITCTO Interim Time Compliance Technical Order

ITCZ Intertropical Convergence Zone

ITD Ibrahim's Time Domain • Inception To Date • Individual Training Directorate • Informal Test Description • Information Technology Directorate • Information Technology Division • Inner-Truncated Disk • Input Transition Detection • Institute for Technology Development • Institute of Theoretical Dynamics • Instrumentation Technology Development • Integrated Technology Demonstration • Integrated Technology Development • Integrated Test Director • Integrated Test Document • Integrated Test Documentation • Integration Test and Demonstration • Interaural Time Difference • Interim Terrain Data • Intertropical Discontinuity • In-Trail Descent

ITDB Inorganic Thermochemistry Database

ITDC Instructional Technology Development Consortium

ITDE Interchannel Time Displacement Error

ITDI Industrial Technology Development Institute

ITDM Intelligent Time-Division Multiplexer

ITDO Input Transaction Data Object

ITDP International Telecom Development Corporation

ITDR Integrated Technology and Design Review • Integrated Technology Demonstration Review • Integrated Technology Design Review • Interim Technical Design Review

ITDS Improved Technical Data System • Integrated Technical Data System • International Trade Data System

IT&E Independent Test & Evaluation • Information Technology & Engineering • Integration, Test & Evaluation

ITE Independent Test and Evaluation • Independent Trading Exchange • Information Technology and Engineering • Information Technology Equipment • Informationstechnische Einrichtung • Input Test Equipment • Institute of Teacher Education • Institute of Terrestrial Ecology • Institute of Transportation Engineers • Instrumentation Test Equipment • Instrument Test Equipment • Instrument Test Engineer • Integral Throat Entrance • Integrated Technology Experiment • Integration, Test and Evaluation • Integration Test Equipment • Intensity Transport Equation • International Transmission Exchange • International Tunguska Expedition • Intersite Transportation Equipment • Inverse Time Element • Ituberá (Brazil-BA) (location identifier)

ITEA International Test and Evaluation Association

ITEC IEEE Transactions on Electromagnetic Compatibility (periodical code) • Information Technology Electronic Commerce • Information Technology, Electronics and Communications • Information Technology Executive Committee • International Training Equipment Conference

ITECS Integrated Terminal Emulation Software

ITED IEEE Transactions on Electronic Devices (periodical code)

ITEdu IEEE Transactions on Education (periodical code)

ITEE Information Technology in Environmental Engineering (conference)

ITEI IEEE Transactions on Electrical Insulation (periodical code)

ITEIC IEEE Transactions on Electromagnetic Compatibility (periodical code)

ITEM IEEE Transactions on Engineering Management (periodical code) • Information Technology in Educational Management

ITEME Institution of Technician Engineers in Mechanical Engineering

ITEnC IEEE Transactions on Energy Conversion (periodical code)

ITEP Independent Technical Evaluation Plan • Individual Training and Evaluation Program • Institute for Theoretical and Experimental Physics • Integrated Test and Evaluation Program • International Test and Evaluation Procedure • International Traineeship Exchange Program

ITEPC Inter-Mars Tissue Equivalent Proportional Counter

ITER Independent Technical Evaluation Report • Individual Tracking for Excellence Report • International Thermonuclear Experimental Reactor • International Tokamak Engineering Reactor • International Toroidal Experimental Reactor

Iterat Iterativo

ITeSRE Istituto di Tecnologie e Studio delle Radiazioni Extraterrestri (Italy)

ITEU Information Technology Education Unit

ITF Implementation Task Force • Impulse Transfer Function • Information Technology Facility • Integrated Test Facility • Integrated Thermal Flux • Integration Task Force • Integration Test Facility • Integration Test Folder • Intelligence Task Force • Intensity Transfer Function • Interactive Terminal Facility • Interactive Test Facility • International Training Forum • International Transfer Format • Intertropical Front • Inverse Trigonometric Function

ITFA In The Final Analysis

ITFE Inductive Terrain Feature Extraction

ITFS IERS Terrestrial Reference Frame • Instructional Television Fixed Service

ITG Independent Test Group • Informationstechnische Gesellschaft (Germany) • Information Technology Group • Integrated Terminal Guidance • Interactive Test Generator • International Technology Group • International Travel Grant • Internet Translation Gateway • Ion Temperature Gradient

ITGA Isothermogravimetric Analysis

ITGB Interféromètre à Très Grande Base

ITGE IEEE Transactions on Geoscience Electronics (periodical code) • Instituto Tecnológico Geominero de España (Spain)

ITGRS IEEE Transactions on Geosciences and Remote Sensing (periodical code)

ITGS Intelligent Traffic Guidance System

ITH Ithaca (USA-NY) (location identifier)

ITHB Imagerie à Très Haute Dynamique

ITI Indian Telephone Industries (India) • Industrial Technology Institute • Information Technology Industry • Information Technology Integration • Information Technology Interfaces (conference) • Inspection and Test Instruction • Inspection/Test Instruction • Integrated Technical Information • Interactive Terminal Interface • International Technologies Inc. • International Thermal Instrument Co. (USA-CA) • International Training Institute • Itambacuri (Brazil-MG) (location identifier)

ITIA IEEE Transactions on Industry Applications (periodical code)

ITIC Information Technology Innovation Center • International Tsunami Information Centre • Internet Telephony Interoperability Consortium

ITiCSE Innovation and Technologies in Computer Science Education (conference)

ITIE IEEE Transactions on Industrial Electronics (periodical code)

ITIEC IEEE Transactions on Industrial Electronics and Control (periodical code)

ITIL Information Technology Infrastructure Library

ITIM IEEE Transactions on Instrumentation Measurement (periodical code)

ITIMP Integrated Technical Item Management and Procurement

ITIMS In-service Transmission Impairment Measurement Set • Integrated Technical Information Management Set

Itin Itinerary

ITINS Integrated Triangular Irregular Network Service

ITIP IEEE Transactions on Image Processing (periodical code)

ITIR Infrared Thermal Imaging Radiometer • Intermediate Thermal Infrared Radiometer

ITIS Industrial Technical Information Service • Interagency Taxonomic Information System • Internal Translation Information System

ITIT IEEE Transactions on Information Theory (periodical code)

ITIWG International Test Integration Working Group

ITJSE Indian Institute of Technology Journal on Section Engineering Technology (periodical code)

ITK Infrastructure Toolkit • Instrument Time Key • Itokama (Papua New Guinea) (location identifier)

Itl Italian

ITL Independent Test Laboratory • Information Technology Liaison • Integrated Teaching and Learning • Integrated Test Line • Integrated Transport Launcher • Integrate - Transfer - Launch • Integration, Test and Launch • Intent To Launch • Inverse Time Limit • Lira (Italy) (ISO currency code)

ITLB Instruction Translation Lookaside Buffer

ITLD International Top-Level Domain

ITLRS Italian Transportable Laser Ranging System

ITLSA Integrated Torso Limb Suit Assembly

ITM IEEE Transactions on Magnetics (periodical code) • Inductive Turing Machine • Information Technology Measurement • Information Tools and Methods • Ingénieur des Travaux de la Météorologie • Installation Team Meeting • Interim Technical Memorandum • Intertask Message • Ionospheric - Thermospheric - Mesospheric • Osaka Itami (Japan) (location identifier)

ITMA Institute of Trade Mark Agents

ITMag IEEE Transactions on Magnetics (periodical code)

ITMC Information Technology Management Council

ITMG Integrated Thermal Micrometeoroid Garment

ITMO Inspection du Travail et de la Main d'Oeuvre

ITMP Integrated Test Management Plan • Integrated Training Management Plan

ITMRA Information Technology Management Reform Act (USA)

ITMS Intelligent Text Management System • Ion Trap Mass Spectrometry

ITMSC Information Technology Management Strategy Council

ITMSP Information Technology Management Strategic Plan

ITMTT IEEE Transactions on Microwave Theory Techniques (periodical code)

ITN Identification Tasking and Networking • Independent Television News • Individual Telephony Number • Information Transfer Node • Installation Transport Network • Instituto Tecnológico e Nuclear (Portugal) • Interim Technical Note • Itabuna (Brazil-BA) (location identifier)

ITNN IEEE Transactions on Neural Networks (periodical code)

ITNS IEEE Transactions on Nuclear Science (periodical code)

ITO Hilo (USA-HI) (location identifier) • Identified Technical Order • Impulse Transfer Orbit • Independent Test Organization • Indium Tin Oxide • Individual Travel Order • Information Technology Office • Initial Timeout • Instrument Takeoff • Integrated Tasking Order • Integrated Task Order • Integrated Technical Order • Integration and Test Order • Interface Technology Operation • Interim Technical Order • International Test Organization • International Trade Organization • Invitational Travel Order

IToBu Instituto de Tonantzintla Boletín (periodical code)

ITOCP International Target-Oriented Complex Project

ITOF Isochronous Time-Of-Flight

ITOG Instrument Test and Operations Group

ITOP International Test Operations Procedure

ITOpS Information Technology Operation Services

ITOPS Interim Terminal Overseas Processing System • International Test Operation Procedures

ITOS Improved TIROS Operational Satellite • Improved TIROS Operational System • Interactive Terminal Operating System

ITOY International Tropospheric Ozone Year

ITP Impact Time Prediction • Impuesto sobre Transmisiones Patrimoniales • Incidental Take Permit • Independent Test Plan • Individual Training Plan • Individual Training Proposal • Informal Test Plan • Informal Test Procedure • Informatic Technology Programme • Information Technology Professional • Information Transformation Point • Input Translator Program • Inspection Test Procedure • Installation Test Program • Institute of Theoretical Physics • Institut Technique Professionnel • Instruction To Proceed • Integral Tile Part • Integrated Terminal Program • Integrated Test Plan • Integrated Test Program • Integrated Training Plan • Intention To Proceed • Intent To Purchase • International Thomson Publishing • International Time Programme • International Time Project • International Training Programme • Internet Transaction Processing • Interoperability Test Panel

ITPA Institute for Terrestrial and Planetary Atmospheres • Institute for Theoretical Physics and Astronomy

ITPAM IEEE Transactions on Pattern Analysis and Machine Intelligence (periodical code)

ITPAMI IEEE Transactions on Pattern Analysis and Machine Intelligence (periodical code)

ITPAS IEEE Transactions on Power Apparatus Systems (periodical code)

ITPB Information Technology Policy Board • Integrated Test Program Board

ITPE IEEE Transactions on Power Electronics (periodical code)

ITPO International TOGA Project Office

ITPHP IEEE Transactions on Parts, Hybrids, and Packaging (periodical code)

ITPP International TOVS Processing Package • Interoperability Test Program Plan

ITPR Infrared Temperature Profile Radiometer

ITPS IEEE Transactions on Plasma Science (periodical code) • Integrated Teleprocessing System • Integrated Test Program Schedule

ITQ Individual Transferable Quota • Itaqui (Brazil-RS) (location identifier)

ITR IEEE Transactions on Reliability (periodical code) • Incremental Tape Recorder • Individual Training Record • Infinite Time Response • Informal Test Result • Information Technology Research • Initial Training Requirement • Input To Register • Inspection Test Report • Institute for Telecommunications Research (Australia-SA) • Instrumentation Tape Recorder • Instrumented Test Range • Instrument Test Rig • Integrated Technology Review • Integrated Testing Regimen • Integrated Test Range • Integrated Test Requirement • Intense Thermal Radiation • Interfacial Thermal Resistance • Interim Technical Report • Interim Test Report • Interim Test Review • Internal Technical Report • Internal Throughput Rate • Internet Talk Radio • In-Transit Rendezvous • Inverse Time Relay

ITRA IEEE Transactions on Robotics Automation (periodical code) • Intercomparison of Transmittance and Radiation Algorithms • International Telescope for Radio Astronomy

ITRB Integrated Technical Review Board • Initial Test Review Board

ITRC Information Technology Research Center

ITRD Integrated Test Requirements Document

ITRDB International Tree-Ring Data Bank

ITRF IERS Terrestrial Reference Frame • Ingénieurs et Techniciens de Recherche et de Formation • International Terrestrial Reference Frame

ITRI Industrial Technology Research Institute (USA-OH) • International Technology Research Institute (USA-VA)

ITRM Information Technology Resources Management • Institutional Training Resource Model

ITRO Interservice Training Review Organization

ITRR Integrated Technology Requirements Review • Internal Test Readiness Review • Internal Throughput Rate Ratio

ITRS International Terrestrial Reference System

IT&S Information Technologies & Systems

ITS Ignition Test Simulator • Impactor Targeting System • Improved Tandem Switch • Incompatible Time-sharing System • Individual Time Step • Individual Training Standard • Individual Training Strategy • Individual Training System • Industrial Technology Services • Industrial Television Services Inc. • Infinite Time Span • Information and Technology Services • Informationstekniska Standardiseringen • Information Technologies and Systems • Information Technology Security • Information Technology Service • Information Technology Standards • Information Technology System • Information Transfer Satellite • Information Transfer System • Information Transmission Satellite • Information Transmission System • Inner Tracking System • Instantaneous Terrestrial System • Institute for Telecommunications Sciences • Institute of Telecommunication Science • Institute of Transportation Studies (USA) • Instrumentation Telemetry Station • Instrumentation Television System • Instrumentation Terrestre et Spatiale • Insulation Test Specification • In-Tank Solidification • Integrated Test Schedule • Integrated Training Strategy • Integrated Training System • Integrated Trajectory System • Integrated Transfer System • Integrated Truss Segment • Integrated Truss Structure • Intelligent Terminal Segment • Intelligent Transaction System • Intelligent Transportation System • Intelligent Transport System • Intelligent Tutoring System • Intelligent Tutor System • Interactive Terminal System • Interactive Tutoring System • Intercept Trajectory Simulation • Interim Teleprinter System • Interim Trainer Support • Intermarket Trading System • Intermediate Tape Store • International Television System • International Temperature Scale • Internet Telephony Server • Internet Terminal Server • Intersectional Transportation Service • Inventory Tracking System • Ion Thrust System

ITSA Institute for Telecommunications Sciences and Aeronomy

ITSB Information Technology Systems Bridge • Information Technology Systems Budget

ITSC Information Technology Strategy Committee • Intelligent Transportation System Council (IEEE) • International Telecommunications Satellite Consortium (now INTELSAT) • International TOVS Study Conference

ITSCG Information Technology Strategy Cooperation Group

ITSCM Information Technology Service Continuity Management

ITSE IEEE Transactions on Software Engineering (periodical code) • International Society for Technology in Education

ITSEC Information Technology Security • Information Technology Security Evaluation Criteria

ITSEF Information Technology Security Evaluation Facility

ITSEM Information Technology Security Evaluation Methodology

ITSG Information Technology Standard Guidance

ITSI Information Technology Standard Information

ITSIE Intelligent Training System in Industrial Environments

ITSIG Information Technology Strategies Implementation Group

ITSM Instrument Team Software Manager

ITSMC IEEE Transactions on Systems, Man, and Cybernetics (periodical code)

ITSN Italian Telemetered Seismic Network

ITSO International Telecommunications Satellite Organization

ITSOP Integrated Telecommunications Systems Operational Planning

ITSP IEEE Transactions on Signal Processing (periodical code) • Internet Telephony Service Provider (*) • Internet Telephony System Provider

ITSPO Interoperability Testing Standards Program Office

ITSQC Information Technology Services Qualification Center (USA)

ITs Individual Training Standards

ITSS Incompatible Time-Sharing Systems • Information Technology and Scientific Services • Integrated Tactical Surveillance System

ITST International Tsunami Survey Team

ITSTC Information Technology Steering Committee

ITSTEC Integrated Transmission Switching and Technical Control

ITSU IEEE Transactions on Sonics and Ultrasonics (periodical code) • Information Technology Standard Unit

itt In Tegenstelling Tot

I&TT Integration & Test Team

IT&T Information Technologies & Telecommunications

ITT Imagery Training Tool • Image Transformation Table • Information Technologies and Telecommunications • Integrated Test Team • Integration and Test Team • Intensity Transfer Table • Intensity Transformation Table • Interactive Training Tool • Interface Task Team • Interfacial - Transport - Thermodynamics • Internationales Teleskoptreffen (Germany) • International Telephone and Telegraph • Interstage Turbine Temperature • Inter-Turbine Temperature • Invitation To Tender • Invitation To Transmit

ITTA Information Technology Training Association

ITTAC International Telephone and Telegraph Advisory Committee

ITT&C Instrumentation Telemetry Tracking & Command

ITTC Information and Telecommunication Technology Center (USA-KS) • Instrumentation Telemetry Tracking and Command

ITTCC International Telephone and Telegraph Consultative Committee

ITTF Information Technologies Task Force (*) • Information Technology Task Force

ITTI Initiation à la Télédétection et au Traitement de l'Image

ITTO International Tropical Timber Organization

ITTP Individual Task Training Package

ITTS Interactive Text Transmission System

ITU Image Transfer Unit • Information Transport Utility • Infrared Telescope Unit • Institute for Transuranium • Integrated Terrain Unit • Intensive Therapy Unit • International Technological University • International Telecommunication Union (formerly CCITT) • International Telegraphic Union • International Typographical Union • Interrogator Translator Unit • Istanbul Technical University (Turkey)

ITUFF IEEE Transactions on Ultrasonics, Ferroelectrics, and Frequency Control (periodical code)

ITUG International Telecommunications User Group

ITUTJ International Telecommunication Union Telecommunication Journal (periodical code)

IT&V Integrated Test & Verification

iTV Interactive Television

ITV Independent Television • Industrial Television • Inspección Técnica de Vehículos (Spain) • Inspection Technique des Véhicules • Instructional Television • Instrumented Test Vehicle • Integrated Technology Validation • Integrated Test and Verification • Integrated Test Vehicle • Integration - Test - Verification • Interactive Television • Internationales Teleskoptreffen am Vogelsberg (Germany) • In-Transit Visibility

ITVE Integrated Technology Validation Experiment

ITVT IEEE Transactions on Vehicular Technology (periodical code)

ITW Integrated Test Wafer • International TOAD Watch

ITWG In-plant Test Working Group • Interface Technical Working Group

ITWS Integrated Terminal Weather System • International Tsunami Warning System

ITX Information Technology Exchange • Intermediate Text

IT2P International Technology Transfer Panel

iü im übrigen

I/U Instrumentation Unit

IU Image Understanding • Immunizing Unit • Impedance Unit • Indiana University (USA-IN) • Information Unit • Initial User • Instrument Unit • Integer Unit • Intendant Universitaire • Interface Unit • Interface Utilisateur • Interference Unit • Intermatrix Unit • Internal User • International Unit • Internet User • Interrogation Unit • Inter-Union • Intonation Unit

IUA Image Understanding Architecture • Inertial Unit Assembly • Interface Unit Adapter • International Union of Academies • International University of America • Ionosphere and Upper Atmosphere

IUAA International Union of Amateur Astronomers

IUAES International Union of Anthropological and Ethnological Sciences

IUAP Internet User Account Provider

IUAPPA International Union of Air Pollution Prevention Associations

IUB International Union of Biochemistry • International Universities Bureau

IUBMB International Union of Biochemistry and Molecular Biology

IUBS International Union of Biological Sciences

IUC Independent User Center • Independent User Conference • Information Unit for Conventions • Information Users Committee • International Union of Crystallography • International Users' Committee • Inter-University Centre (India)

IUCAA Inter-University Centre for Astronomy and Astrophysics (India)

IUCAF Interunion Commission on Frequency Allocations for Radio Astronomy and Space Science

IUCC Information Unit on Climate Change • Interuniversity Committee on Computing

IUCDAEF Inter-University Consortium for the Department of Atomic Energy Facilities (India)

IUCF Indiana University Cyclotron Facility (USA-IN)

IUCH International Union for Circumpolar Health

IUCI Interunion Committee on the Ionosphere

IUCN International Union for the Conservation of Nature

IUCr International Union of Crystallography

IUCRC Industry-University Cooperation Research Centre

IUCRM Interunion Commission on Radio Meteorology

IUCS Instrumentation Update Command System • Interunion Commission on Spectroscopy

IUCSTP Interunion Commission on Solar-Terrestrial Physics

IUCV Intra User Communication Vehicle

IuD Information und Dokumentation

IUD Implementation Use Descriptor • Intrauterine Device

IUDS Integrated Utilities Distribution System

IUDT Interface Unit Data Terminal

IUE Image Understanding Environment • Institut Universitaire Européen • Interface Unit Error • International Ultraviolet Explorer • Niue (location identifier)

IUEAC International Ultraviolet Explorer Allocation Committee

IUEAIMS International Ultraviolet Explorer Automated Information Management System

IUED Institut Universitaire d'Études du Développement

IUEDAC International Ultraviolet Explorer Data Analysis Center

IUEEN International Ultraviolet Explorer ESA Newsletter (periodical code)

IUEF International University Exchange Fund

IUEFA International Ultraviolet Explorer Final Archive

IUEFPA International Ultraviolet Explorer FES Photometric Archive

IUELDA International Ultraviolet Explorer Low-Dispersion Atlas

IUEML International Ultraviolet Explorer Merged Log

IUENN International Ultraviolet Explorer NASA Newsletter (periodical code)

IUEOCC International Ultraviolet Explorer Operations Control Center

IUESIPS International Ultraviolet Explorer Spectral Image Processing System • International Ultraviolet Explorer Standard Image Processing Software

IUESOC International Ultraviolet Explorer Science Operations Center

IUF Institut Universitaire de France

IUFM Institut Universitaire de Formation des Maîtres (France)

IUFOMC International Unidentified Flying Object Monitoring Center

IUFoST International Union of Food Science and Technology

IUFRO International Union of Forestry Research Organizations

IUGG International Union of Geodesy and Geophysics

IUGS International Union of Geological Sciences

IUH Instructor Utilization Handbook

IUHEI Institut Universitaire des Hautes Études Internationales (Switzerland)

IUHPS International Union of the History and Philosophy of Science

iUHR Inner Ultra-Harmonic Resonance

IUHS International Union for History of Science (now IUHPS)

IUI Integrated User Interface • Intelligent User Interface • Interactive User Interface • Interim Use Item • International Union of Interpreters

IUIC Interuniversity Information Committee

IUID Internal Unit Identification

IUIS International Union of Immunological Societies

IuK Informations- und Kommunikationssystem

IUL Ilu (Indonesia) (location identifier)

IULA International Union of Local Authorities

IULC Interuniversity Library Council

IUM Interim Use Material • Summit Lake (USA-CA) (location identifier)

IUMRS International Union of Materials Research Societies

IUMA Interim Use Material Authorization

IUMJ Indiana University Mathematics Journal (periodical code)

IUMP International Upper Mantle Program

IUMS International Union of Microbiological Sciences • International Union of Microbiological Societies (*)

IUNS International Union of Nutritional Sciences

IUNT Instituto Uruguayo de Normas Técnicas (Uruguay)

IUO Interunit Order

IUP Indiana University of Pennsylvania (USA-PA) • Indiana University Press (USA-IN) • Installed User Program • Institut für Umweltphysik (Germany) • Instituto Universitario de Posgrado (Spain) • Institut Universitaire Professionalisé (France) • Institut Universitaire Professionnel • Internet Usage Policy • Irish University Press • IRS Utilization Programme • Israel Universities Press

IUPAB International Union of Pure and Applied Biophysics

IUPAC International Union of Pure and Applied Chemistry

IUPAP International Union of Pure and Applied Physics • International User Resource Allocation Panel

IUPESM International Union for Physical and Engineering Sciences in Medicine

IUPHAR International Union of Pharmacology

IUPHY Inter-University Programme in Hydrology

IUPP Introductory University Physics Project

IUPS International Union for Philosophy of Science (now IUHPS) • International Union of Physiological Sciences

IUPsyS International Union of Psychological Sciences

IUPUI Indiana University - Purdue University at Indianapolis (USA-IN)

IUR Interuser Reliability

IUS Inertial Upper Stage • Initial Upper State • Interchange Unit Separator • Interim Upper Stage • Interim Use Sheet • International Union of Students • Inus (Papua New Guinea) (location identifier)

IUSB Indiana University in South Bend (USA-IN)

IUSM International Union for Surveys and Mapping

IUSS International Undergraduate Summer School • Integrated Unit Simulation System

IUT Implementation Under Test • Institut Universitaire de Technologie (France) • Item Under Test

IUTAM International Union of Theoretical and Applied Mechanics

IUTE Industry University Technology Experiment

IUTOX International Union of Toxicology

IUV Interlibrational Utility Vehicle

IUVSTA International Union for Vacuum Science, Technique and Applications

IUWDS International URSIGRAM and World Days Service

iV in Vertretung • im Vorjahre • in Verbindung • in Vorbereitung

I/V Ida/Vuelta • Inlet Valve

IV Implementation Visit • Increased Value • Informationsvermittlung • Information Visualization • Initialization Variable • Initialization Vector • Initial Velocity • Inner Vessel • Input Voltage • Insignificant Vector • Integrated Vehicle • Intensifier Vidicon • Interactive Video • Interface Volume • Intermediate Valency • Intermediate Voltage • Intravehicular • Intravenous • Intravenously • Inventory Variance • In Vocabulary • Invoice Value • Subgiant Star

IVa Subgiant Star

IVA Ambanja (Madagascar) (location identifier) • Immediate Visual Assessment • Impuesto al Valor Agregado • Impuesto sobre el Valor Añadido • Individual Voluntary Arrangement • Ingenjörsvetenskapsakademien (Sweden) • Insertion dans la Vie Active • Instrument de Vision Artificielle • International Voice Association • Intravehicular Activity

IVab Subgiant Star

iVAD Integrated Voice And Data

IVAR Insertion Velocity Adjust Routine • Internal Variable

IVb Subgiant Star

IVB Intermediate Vector Boson

IVBC Integrated Vehicle Baseline Characteristics • Integrated Vehicle Baseline Configuration

I&VC Integration Verification & Validation

IVC Inner Vacuum Center • Institute Visiting Committee • Integrated Vacuum Circuit • Integration Verification and Validation • Intermediate-Velocity Cloud • International Visiting Committee • Intervehicular Communications • Invercargill (New Zealand) (location identifier)

IVCE Internal Voice Communications Equipment • International Video and Communications Exhibition

IVCM Integration Vegetation-Complex Map

IVCS Inner Vapor Cooled Shield • Internal Voice Communications System

IVD Information Viewing Device • Integrated Voice and Data • Integrated Voice/Data • Interactive Video Disk • Interactive Visual Display • Ion-Vapor-Deposited

IVDA Inland Valley Development Authority (USA-CA)

iVDGL International Virtual Data Grid Laboratory

IVDLAN Integrated Voice Data Local Area Network

IVDS Integrated Voice Distribution System

IVDT Integrated Voice/Data Terminal

IVDW Integrated Voice/Data Workstation

IVE Interface Verification Equipment • Interval-Valued Epistemic

IVF International Video Federation • In Vitro Fertilization

IVG Intermediate-Velocity Gas • Ivangrad (Yugoslavia-Montenegro) (location identifier)

IVH Ivishak (USA-AK) (location identifier)

IVHS Intelligent Vehicle Highway System

IVHSA Intelligent Vehicle-Highway System Act (USA)

IVI Incremental Velocity Indicator • Interactive Video Interface • Interchangeable Virtual Instrument • Ion Velocity Instrument

IVIA Interactive Video Industry Association (USA)

IVIC Instituto Venezolano de Investigaciones Científicas (Venezuela)

IVICS Interactive Visual Image Classification System

IVIS Interactive Vehicular Information System • Interactive Velocity Interpretation System • Interactive Video Information System • Intervehicular Information System

IVL Independent vendor League • Individually Validated License • Institutet för Vatten och Luftvård (Sweden) • Intel Verification Laboratory • Ivalo (Finland) (location identifier)

IVLN Integrated Vehicle Logic Network

ivm In Verband Met

IvM Instituut voor Milieuvraagstukken (Netherlands)

iVm in Verbindung mit

IVM Imaging Vector Magnetograph • Informationsverarbeitungsmaschine • Instituut Voor Milieuvraagstukken (Netherlands) • Interactive Volume Modeling

IVME Interpreters, Virtual Machines and Emulators (conference)

IVMMS Integrated Vehicle Mission Management System

IVMU Inertial Velocity Measurement Unit

IVN Integrated Voice Network • Interactive Video Network

IVO Chivolo (Colombia) (location identifier) • Individueel Voortgezet Onderwijs • International Virtual Observatory • International Visitors Office • In View Of • Inward Voucher Office

IVOA International Virtual Observatory Alliance

IVOD Integrated Video On Demand

IVoIP Improved Voice over Internet Protocol

IVP Initial-Value Problem • Installation Verification Procedure • Installation Verification Process • Installation Verification Program • Integrity Verification Procedure • Interface Verification Plan • Ion Vacuum Pump

IVPA Initial Value Problem Approach • International Virtual Pilots Association

IVPN International Virtual Private Network

IVPS Integrated Vacuum Processing System

IVR Inner Vertical Resonance • Integrated Voice Response • Integrated Voltage Regulator • Interactive Voice Responder • Interactive Voice Response • International Vehicle Registration • Intramolecular Vibrational Relaxation • Inverell (Australia-NSW) (location identifier)

IVRCU Intravehicular Remote Control Unit

IVRS Interactive Voice Response System • Interim Voice Response System

IVS IBM-Verkablungssystem • Informationsverarbeitungssystem • Informationsvermittlungsstelle • Informationsvermittlungsystem • Input Voltage Supply • Interactive Videodisk System • Interactive Vsualization System • Interface Verification Summary • International VLBI Satellite • International VLBI Service (*)

IVSI Instantaneous Vertical Speed Indicator

IVSN Initial Voice Switching Network

IVT Initial Vocational Training • Integrated Verification Test • Integration - Verification - Training • Interface Verification Test • Interface Verification Testing • Interrupt Vector Table • Intervehicular Transfer

IVTS Interactive Video Training System • International Video Teleconferencing Service

IVU Interface Verification Unit • Intravehicular Umbilical

IVUS Intravascular Ultrasound

IV&V Independent Validation and Verification • Independent Verification & Validation

IVV Independent Validation and Verification • Independent Verification and Validation • Informations-Vermittlung und Verarbeitung • Instantaneous Vertical Velocity • Internationale Vereinigung der Vermessungsingenieure

IVVI Instantaneous Vertical Velocity Indicator

IVVS Instantaneous Vertical Velocity Sensor

IVVV Internationaal Verbond van Vrije Vakverenigingen

IVW Interferometrically Verified Window • International Venus Watch • Inverway (Australia-NT) (location identifier)

iw Hebrew (ISO language code)

I/W In Work

I&W Indications & Warnings • Intelligence & Warning

IW Ice Water • Indirect Waste • Influence Weight • Information Warfare • Inside Width • Inside Wiring • Internet World • Interstellar Wind • Intrawing • In Work • Isle of Wight (UK) • Isolated Word • Isotopic Weight

IWA IAYC Workshop for Astronomy • Institute of World Affairs • International Workshop Agreement • Ivanova (Russia) (location identifier)

IWAA Interdisziplinäres Workshop Astronomie und Astrophysik

IWAI International Workshop on Affective Interactions

IWALC International Water Associations Liaison Committee

IWAN International Workshop on Active Networks

IWANN International Workshop on Artificial and Natural (neural) Networks

I-WAY Information Highway

IWAY Information Highway • Information Wide Area Year

IWBNI It Would Be Nice If

IWBS Indirect Work Breakdown Structure

IWC Ice Water Column (*) • Ice Water Content • Inertia Weight Class • International Whaling Commission • Iowa Wesleyan College (USA-IA)

IWCA International Workshop on Cometary Astronomy

IWCASE International Workshop on Computer-Aided Software Engineering

IMCO International World Commission on the Oceans

IWCS Integrated Wideband Communications System

IWCV International Workshop on Climate Variability

IWD Inland Waters Directorate (Canada) • Integral Wind Defect • Ironwood (USA-MI) (location identifier)

IWDM International Workshop on Database Machines

IWDS Integrated Warning and Display System • International World Days Service • Interwork Distributed System

IWEDA Integrated Weather Effects Decision Act (USA)

IWEM Institute of Water and Environmental Management

IWEX Internal Wave Experiment

IWF Institut für den Wissenschaftlichen Film (Germany) • Institut für Weltraumforschung (Austria) • Integrated Warning Function • Internationaler Währungsfonds • Internetworking Function • Interworking Function

IWFIPT International Workshop on Future Information Processing Technologies

IWFS Integrated Waste Fluid System

IWG Impact Working Group • Implementation Working Group • Instrumentation Working Group • Integration Working Group • Intelligence Working Group • Interagency Working Group • Intercommission Working Group • Interdisciplinary Writing Program • Interface Working Group • Interoperability Working Group • Intersessional Working Group • Intersystem Working Group • Investigating Working Group • Investigator Working Group

IWGATWR International Working Group on Advanced Technologies for Water-Cooled Reactors

IWGDMCG Interagency Working Group on Data Management for Global Change

IWGFPT International Working Group on Fuel Performance and Technology

IWGFR International Working Group on Fast Reactors

IWGGCR International Working Group on Gas-Cooled Reactors

IWGGE Interdepartmental Working Group on the Greenhouse Effect

IWGMFS International Working Group on Magnetic Field Satellite

IWGMP Intergovernmental Working Group on Marine Pollution

IWI Innovatie Wetenschappelijke Informatie

IWIA International Workshop on Information Assurance

IWIS International Water Information System

IWJ Iwami (Japan) (location identifier)

IWLCS International Workshop on Learning Classifier Systems

IWLP Institution-Wide Language Programme

IWLS International Workshop On Logic and Synthesis

IWMM International Workshop on Memory Management

IWMSC Internetworking Mobile Switching Center

IWNLG International Workshop on Natural Language Generation

IWO Inter-White-Oval (region) (Jupiter) • Iwo Jima (Japan) (location identifier)

IWP Ice Water Path • Independent Weather Productions • Indirect Work Package • In-service Work Plan • Intergovernmental WOCE Panel • Interim Working Party • Internal Working Paper

IWPA International Word Processing Association

IWPC Institute of Water Pollution Control

IWQoS International Workshop on Quality of Service

IWR Information World Review • Interdisziplinäres Wissenschaftliches Rechnen • Internet Weather Report • Internet Weekly Report

IWRA International Water Resources Association

iwS im weiteren Sinne

I&WS Indications & Warning System

IWS Indications and Warning System • Instrument Workstation • Integrated Water System • Integrated Wiring System • Integrated Work Statement • Intelligent Workstation • West Houston (USA-TX) (location identifier)

IWSA International Water Supply Association

IWSAPF International Workshop on Software Architectures for Product Families

IWSAS International Workshop on Self-Adaptive Software

IWSM International Workshop on Software Management

IWT Industrial Waste Treatment • Inland Waterway Transport • Integral Wavelet Transform • Inverse Wavelet Transform

IWTC International World Trade Center

IWTO International Wool Textile Organization

IWU Illinois Wesleyan University (USA-IL) • Intermediate Working Unit • Interworking Unit

IWV Impulswahlverfahren

IWVF International Workshop on Visual Form

IWW Inland Waterway • Institut für Wirtschaftspolitik und Wirtschaftsforschung (Germany)

IWWERT International Workshop on Word Equations and Related Topics

IWWG Internetwork Working Group

IW3C1 First International WWW Conference

IW3C2 Second International WWW Conference • International WWW Conference Committee

IW3C3 Third International WWW Conference

IW3C4 Fourth International WWW Conference

ix Index (file extension)

IX Interactive Executive • Inverted Index • Ion Exchange

IXA Agartala (India) (location identifier) • Internet Exchange Architecture

IXAE Indian X-ray Astronomy Experiment

IXAS Indian X-ray Astronomy Satellite

IXB Bagdogra (India) (location identifier) • Intelligent X Board

IXC Chandigarh (India) (location identifier) • Interchange Carrier • Interexchange Carrier

IXD Allahabad (India) (location identifier)

IXE Internet Exchange Extension • Mangalore (India) (location identifier)

IXF Integrated Exchange Format

IXG Belgaum (India) (location identifier)

IXH Kailashahar (India) (location identifier)

IXI Interim X.25 Infrastructure • International X.25 Infrastructure • International X.25 Interconnect • Lilabari (India) (location identifier)

IXJ Jammu (India) (location identifier)

IXK Keshod (India) (location identifier)

IXL Leh (India) (location identifier)

IXM Inter-Exchange Mileage • Madurai (India) (location identifier)

IXMA Imaging X-ray Microcalorimeter for Astrophysics

IXMP Information Standards Management Panel

IXN Khowai (India) (location identifier)

IXO Intermediate-luminosity X-ray Object

IXP Information Exchange Plan • Pathankot (India) (location identifier)

IXQ Kamalpur (India) (location identifier)

IXR Intelligent Transfer Restore • Ranchi (India) (location identifier)

IXS Inelastic X-ray Scattering • Silchar (India) (location identifier)

IXT Ideas Exchange Traffic • Pasighat (India) (location identifier)

IXU Aurangabad (India) (location identifier)

IXV Along (India) (location identifier)

IXW Jamshedpur (India) (location identifier)

IXY Kandla (India) (location identifier)

IXZ Port Blair (Andaman Islands) (location identifier)

IYAS International Years of the Active Sun

IYDK If You Didn't Know

IYF International Year of Freshwater (2003)

IYFEG Insert Your Favorite Ethnic Group

IYGTFUI If You Get The Font, Use It

IYH Inside Your Head • In Your Head

IYIS Internal Photopic Illuminance Response

IYK Inyokern (USA-CA) (location identifier) • Ridgecrest (USA-CA) (location identifier)

IYL International Youth Library • In Your Life

IYM Investment Year Method

IYO International Year of the Ocean • In Your Opinion

IYOR International Year Of the Reef

IYQS International Year of the Quiet Sun (1964-1965)

IYR Internet Yearly Report

IYSWIM If You See What I Mean

IYV International Year of Volunteers (2001)

IZ Informationszentrum

IzAlm Izvestiia Astrofizicheskogo Instituta Alma-Ata (periodical code)

IzAne Izvestiia na Fizicheskiia Institut's Aneb (periodical code)

IzArm Izvestiia Akademiia Nauk Armyanskoi (periodical code)

IzAsh Izvestiia Akademii Nauk Turkmenskoi SSR Ashkhabad (periodical code)

IzAvT Izvestiia VUZ Aviatsionnaia Tekhnika (periodical code)

IzAze Akademiia Nauk Azerbaidzhanskoi SSR Izvestiia Seriia Fiziko Tekhnicheskikh i Matematicheskikh Nauk (periodical code)

IZB Independent Zone Boundary

Izda Izquierda

Izdo Izquierdo

IzDus Dushanbe Izdatel Donish (periodical code)

IzFru Frunze Izdatel Ilim (periodical code)

IzGeo Tbilisi Georgian SSR Izdatel Metsniereba (periodical code)

IzGla Izvestiia Glavnoi Rossiiskoi Astronomicheskoi Observatorii (periodical code)

IzKaz Izvestiia Astronomicheskoi Engel'gardt obskoi Observatorii Kazan (periodical code)

IzKie Izvestiia Glavnoi Astronomicheskoi Observatorii Kiev (periodical code)

IzKry Izvestiia Ordena Trudovogo Krasnogo Znameni Krymskoi Astrofizicheskoi Observatorii (periodical code)

IzLat Akademiia Nauk Latviiskoi SSR Izvestiia Seriia Fizicheskikh i Tekhnicheskikh Nauk (periodical code)

IzLen Leningrad, Izdatel'stvo Nauka (periodical code)

iZm im Zusammenhang mit

IZM Institut für Zuverlässigkeit und Mikrointegration (Germany) • Izmir (Turkey) (location identifier)

IzMin Minsk Izdatel BGU (periodical code)

IZMIRAN Institute of Terrestrial Magnetism, Ionosphere and Radio Wave Propagation (Russia)

IzMol Akademiia Nauk Moldavskoi SSR Izvestiia Seriia Fiziko Tekhnicheskikh i Matematicheskikh Nauk (periodical code)

IzmP Ege University Izmir Publications of the Department of Astronomy (periodical code)

IzNic Izvestiia Nikolaevskoi Glavnoi Astronomicheskoi Observatorii (periodical code)

IZO Izumo (Japan) (location identifier)

IzOde Odesskij Gosudarstvennyi Universitet im. I. I. Mechnikova Izvestiia Astronomicheskoi Observatorii (periodical code)

IzPul Izvestiia Glavnoi Astronomicheskoi Observatorii v Pulkove (periodical code)

Izq Izquierda • Izquierdo

Izqda Izquierda

Izqdo Izquierdo

IzRad Izvestiia VUZ Radiofizika (periodical code)

IzRos Rossijskaia Akademiia Nauk Izvestiia Seriia Fizicheskaia (periodical code)

IZS Informationszentrum Socialwissenschaften • International Zurich Seminar

IzSib Akademiia Nauk SSSR Sibirskoe Otdelenie Izvestiia (periodical code)

IzSSR Akademiia Nauk SSSR Izvestiia Seriia Fizicheskaia (periodical code)

IZT Ixtepec (Mexico) (location identifier)

IzTas Tashkent Izdatel Fan (periodical code)

IzTif Tiflis Izdatel Tbilisskogo Universiteta (periodical code)

IzTur Akademiia Nauk Turkmenskoi SSR Izvestiia Seriia Fiziko Tekhnicheskikh Khimicheskikh i Geologicheskikh Nauk (periodical code)

IzUkr Akademiia Nauk Ukrainian SSSR Izvestiia Seriia Fizicheskaia (periodical code)

IzUzb Akademiia Nauk Uzbekskoi SSR Izvestiia Seriia Fiziko Matematicheskikh Nauk (periodical code)

IzVil Vilnius Izdatel Mokslas (periodical code)

IzYak Vsesoiuznaia Konferentsiia Kosmicheskim Lucham Yakutsk USSR Akademiia Nauk SSSR (periodical code)

IzYer Vsesoiuznaia Konferentsiia Kosmicheskim Lucham Yerevan Armenian SSR Akademiia Nauk SSSR Izvestiia Seriia Fizicheska (periodical code)

I2 Image Intensification • Image Intensifying

I2CS Innovative Internet Computing Systems (conference)

I2I Individual to Individual

I2ICE Integrated Instrumentation and In-Circuit Emulation

I2L Integrated Injection Logic

I2M Integrated Internet Marketing

I2O Intelligent Input-Output

I2R Imaging Infrared

I2RG Imaging Infrared Guidance

I2S Infrared Imaging System

I2S2 Integrated Intelligence Support System • Intelligence Information Subsystem

I2T Interférométrie à 2 Télescopes

i2u Italian Unix Systems User Group

I2UI Intelligent Integrated User Interface

I2V2 Independent Integration, Verification, and Validation

I2WD Intelligence and Information Warfare Directorate

I3 Intelligent Information Interface • Intelligence Integration of Information

I3A Informatique et Astronomie pour Astronomes Amateurs (France) • International Imaging Industry Association

I3L Isoplanar Integrated Injection Logic
I3R Intelligent Interface for Information Retrieval
I3WG Interservice Interoperability Implementation Working Group
I4 International Information Integrity Institute
I4T Interférométrie à Quatre Télescopes

J

j Current Density • Imaginary Number • Jährlich • Jemand • Jetzt • Jour • Jung • Unit Coordinate Vector

J Action Integral • Advance Ratio • Angular Momentum • Coupling Constant • Current Density • Electric Current Density • Exchange Integral • Flux Density • Jack • Jacobian Determinant • Jahr • Jahrgang • Jamming • January • Japan (international vehicle registration code) • Jet • Job • Jod • Joule • Journal • Judge • Jugend • Jugendlich • Juliet (international alphabet) • July • Junction • June • Junge • Jupiter • Jurassic • Justice • Magnetic Polarization • Massieu Function • Moment of Inertia • Nuclear Spin Quantum Number • One (Roman numeral) • Performance Criterion • Quantum Number of Total Angular Momentum • Rotational Quantum Number • Sound Intensity • Total Angular Momentum Quantum Number • 1 (Roman numeral)

ja Japanese (ISO language code)

Ja January

J/A Joint Account

J&A Justification & Approval • Justification & Authorization

JA Jahresabschluß • Jahresausgleich • Jamaica (international vehicle registration code) • January • Janvier • Japan (international aircraft registration code) • Job Aid • Joint Account • Joint Affairs • Joint Agent • Judge Advocate • Jugendamt • Jump Address • Jump if Above • Junior Achievement

JA&A Journal of Astrophysics & Astronomy

JAA Jackson Astronomers Association (USA-GA) • Jackson Astronomical Association (USA-MS) • Jalalabad (Afghanistan) (location identifier) • Joint Airworthiness Authorities • Joint Aviation Authorities • Jornada de Astronomía Amateur

JAAD Justification, Approval, and Acquisition Documentation

JAAR Joint After-Action Report

Jaarb Jaarboek

JAARS Joint After-Action Reporting System

JAAS Java Authentication and Authorization Service • Jewish Academy of Arts and Sciences

JAB Jabiru (Australia-NT) (location identifier) • Jahresabschlußbericht • Joint Advisory Board

JAC Jackson Hole (USA-WY) (location identifier) • Joint Advisory Committee • Joint Analysis Center • Joint Analytical Center • Joint Astronomy Centre (USA-HI) • Jongerenadviescentrum • Junta Aeronáutica Civil

JACARA Joint Australian Centre for Astrophysical Research in Antarctica

JACEE Japanese-American Cooperative Emulsion Experiment

JACERS Japanese Committee of Environmental Remote Sensing

JACH Joint Astronomy Centre in Hawaii

JAChS American Chemical Society Journal (periodical code)

JACKPHY Japanese - Arabic - Chinese - Korean - Persian - Hebrew - Yiddish

JACM Journal of the Association for Computing Machinery (periodical code)

JAco Journal d'Acoustique (periodical code)

JACO Joint Actions Control Office

JACoW Joint Accelerator Conference Website

JACS Jersey Astronomical Conference and Showcase (USA-NJ) • Jet Attitude Control System • Journal Article Copying Service • Journal of the American Ceramic Society (periodical code) • Journal of the American Chemistry Society

JAD Jandakot (Australia-WA) (location identifier) • Joint Analysis Directorate • Joint Analysis Division • Joint Application Design • Joint Application Development • Joint Astrophysics Division • Journal of Astronomical Data (periodical code) • Julian Astronomical Day

JADC Japan Aircraft Development Co.

JADE Japanese Dynamic Earth (observing) • Joint Advanced Demonstration Environment

JADL JIAC Agent Description Language

JADS Joint Advanced Distributed Simulation

JADT Journée d'Analyse des Données Textuelles

JAE Atlanta - Technology Park (USA-GA) (location identifier) • Java Application Environment • Journal of Acoustic Emission (periodical code) • Jump if Above or Equal

JAEC Japan Aero Engine Corporation • Japan Atomic Energy Commission • Joint Atomic Energy Committee

JAEI Japan Aviation Electronics Industry Ltd.

JAEIC Joint Atomic Energy Intelligence Committee

JAEL JSC Avionics Evolution Laboratory

JAEMA Journal of Analytical and Experimental Modal Analysis (periodical code)

JAerE Journal of Aerospace Engineering (periodical code)

JAERI Japan Atomic Energy Research Institute

JAerP Journal of Aerospace Power (periodical code)

JAerS Journal of Aerosol Science (periodical code)

JAES Journal of Audio Engineering Society (periodical code)

JAF Jaffna Kankesanturai (Sri Lanka) (location identifier) • Java (Beans) Activation Framework • Joint Architecture Framework • Journal des Astronomes Français (periodical code)

JAFCO Just Another Foolish Console Operator

JAfES Journal of African Earth Sciences (periodical code)

JAFoE Japan-America Frontiers of Engineering (symposium)

JAFOOS Joint Australian Facility for Ocean Observing Systems

JAFSA Japan Foundation for Shipbuilding Advancement

JAG Jacobabad (Pakistan) (location identifier) • Jaggies Are Gone • JKT Acquisition and Guiding (unit) • Joint Analyses Group • Joint Association of Geophysics (now BGA) • Judge Advocate General • Jugendarbeitsgemeinschaft

JAH Aubagne Agora (France) (location identifier)

Jahrb Jahrbuch

Jahresber Jahresbericht

Jahrg Jahrgang

Jahrh Jahrhundert

Jahrt Jahrtausend

Jahrtsd Jahrtausend

JAH2 Journal of Astronomical History amd Heritage

JAI Jaipur (India) (location identifier) • Japanese Aerospace Industry • Java Advanced Imaging • Joint Administrative Instruction

JAICI Japan Association for International Chemical Information

JAIED Journal of Artificial Intelligence in Education (AA-CE)

JAIEG Joint Atomic Information Exchange Group

JaIMJ Japan Institute of Metals Journal (periodical code)

JAIN Java AIP for Integrated Network

JAIPA Japan Association for Information Processing in Astronomy

JAir Journal of Aircraft (periodical code)

JAIR Journal of Artificial Intelligence Research (periodical)

JAJ Atlanta - Perimeter Mall (USA-GA) (location identifier)

JaJAP Japanese Journal of Applied Physics (periodical code)

JAK Jacmel (Haiti) (location identifier)

JAKIS Japan Keyword Indexing Simulator

Jal Jalisco (Mexico)

JAL Jalapa (Mexico) (location identifier) • Jalisco (Mexico) • Jet Approach and Landing • Journal of Academic Librarianship

JAllC Journal of Alloys and Compounds (periodical code)

JALN Journal of Asynchronous Learning Networks

JALPG Joint Automatic Language Processing Group

Jam Jamaica • Jamaican

JAM Jamaica (ISO country code) • Jamaique (code ISO de pays) • Jambol (Bulgaria) (location identifier) • Job Analysis Memorandum • Job Assignment Memorandum • Journal of Applied Meteorology • Jugend-Astrocamp Mühlhausen (Germany) • JYACC Application Manager

JAMA Japan Automobile Manufacturers Association • Journal of the American Medical Association

JAMAC Job Analysis Memorandum Activity Chart

JAMAS Japan Medical Abstracts Society

JAMES Joint Automated Message Editing Software • Joint Automated Message Editing System

JAMH Joint Automated Message Handling

JAMIC Japanese Microgravity Center

JAMIE Joint Analogue Microsystems Initiative of Europe

JAMP Joint Automated Mapping Project

JAMPS Joint Automated Message Preparation System

JAMR Joint Application Materiel Request

JAMS Joint Acquisition Management System

JAMSS Joint Applications Management Standard System

JAMSTEC Japan Marine Science and Technology Center

Jan Janeiro • Janitor • Januar • January

JAN Jackson Thompson (USA-MS) (location identifier) • Janeiro • January • Janvier • Joint Army-Navy

JANA Jamahiriya Arab News Agency (Libya)

JANAP Joint Army-Navy-Air Force Publication

JANet Joint Academic Network (UK)

JANIP Joint Army Navy Instrumentation Program

JANNAF Joint Army, Navy, NASA, and Air Force

JANOP Joint Automatic Network Operating Procedure

JAnSc Journal of Astronautical Sciences (periodical code)

JANUS Joint Academic Network Using Satellites

Janv Janvier

JANZOS Japan-Australia-New-Zealand Observation of Supernovae

JAO Atlanta - Beaver Ruin (USA-GA) (location identifier) • Joint ALMA Office

JAOS Junior Amateur Observers' Society (USA-NY)

JAOT Journal of Atmospheric and Oceanic Technology

Jap Japan • Japanese • Japon • Japonais • Japonés

JAP Jahresarbeitsprogramm • Joint Active Penetration • Joint Astronomy Programme (IISc) • Journal of Applied Physics (periodical code) • Jupiter Atmospheric Probe

JApA Journal of Astrophysics and Astronomy (periodical code)

JAPACS Japanese Pacific Climate Studies

JAPATIC Japan Patent Information Center

JAPE Journal of Applied Photographic Engineering (periodical code)

JApEl Journal of Applied Electrochemistry (periodical code)

JAPh Journal of Applied Physiology (periodical code)

JAPHAR Joint Air-breathing Propulsion for Hypersonic Applications Research

JAPIO Japan Patent Information Organization

JApMa Journal of Applied Mathematics (periodical code)

JApMe Journal of Applied Meteorology (periodical code)

JApMw Journal of Applied Metalworking (periodical code)

Japn Japan • Japanese

JAPOA Japan Amateur Photoelectric Observers Association

JAPSS Joint Automated Planning Support System

JAQ Jacquinot Bay (Papua New Guinea) (location identifier) • Job Activities Questionnaire

Jar Jargon

JAR Arcadia - Civic Center (USA-CA) (location identifier) • Java Archive • Joint Airworthiness Regulations • Joint Airworthiness Requirement • Joint Aviation Requirement • JOSO Annual Report • Jovian Atmospheric Reference • Jupe Arrière

JARAC Joint Architecture Review and Analysis Council

JArbSchG Jugendarbeitsschutzgesetz

JARE Japanese Antarctic Research Expedition

JARECT Japan Annual Reviews in Electronic Computers and Telecommunications

JARG Joint Agency Requirements Group

JARI Journal of Applied Radiation and Isotopes (periodical code)

JARING Joint Advanced Research Integrated Networking

JAROS Japan Resources Observation System

JARP Joint Analysis Review Panel

JARS Java Applet Rating Service

JAS Japanese Awareness Service • Jasper County (USA-TX) (location identifier) • Java Alternate Synchronizer • Job Accounting System • Job Analysis System • Job Applicant System • Joint Administration Services • Joint Astrophysical Seminar • Jordanian Astronomical Society (Jordan) • Journal of Aerospace Science • Journal of the Atmospheric Sciences • July - August - September • Junior Astronomical Society (UK)

JASA Japan Astronomical Study Association • Joint Accreditation Support Activity

JASAC Japan Astronomical Study Association Circulars (periodical code)

JAsAs Journal of Astrophysics and Astronomy (periodical code)

JASC Joint Actions Steering Committee

JASCE Journal of the American Society of Civil Engineers (periodical code)

JASchG Jugendarbeitsschutzgesetz

JASCOME Japan Society for Computational Methods and Engineering

JASE Journal of Applied Science and Engineering (periodical code)

JASEg Journal of the Astronomical Society of Egypt (periodical code)

JASG Joint Advanced Study Group

JASHOPS Joint Atlantic Seminar on the History Of the Physical Sciences

JASIN Joint Air-Sea Interaction Experiment

JASIS Journal of the American Society for Information Science

JASJa Journal of the Acoustical Society of Japan (periodical code)

JASMINE Japanese Astrometry Satellite Mission for Infrared Exploration

JASORS Joint Advanced Special Operations Radio System

JASPIC Joint American-Soviet Intercalibration Campaign

JASPP Joint Acquisition Systems Protection Program

JASRAC Japanese Society of Rights of Authors and Composers

JASS Journal of Applied Systems Studies • Journal of Astronomy and Space Sciences (periodical code)

JASTIS Japanese Science and Technology Information System

JASTP Journal of Atmospheric and Solar-Terrestrial Physics (periodical code)

JASV Journal of the Astronomical Society of Victoria Melbourne (periodical code)

JAT Jabat (Marshall Islands) (location identifier) • Japanese Alias Table

JAtC Journal of Atmospheric Chemistry (periodical code)

JATC Joint Atomic Time Commission

JATCC Joint Air Traffic Control Center

Jato Jahrestonne

JATO Jet-Assisted Takeoff

JAtOT Journal of Atmospheric and Oceanic Technology (periodical code)

JATP Journal of Atmospheric and Terrestrial Physics (periodical code)

JAtS Journal of Atmospheric Sciences (periodical code)

JATS Jamming Analysis and Transmission Selection

JAU Jauja (Peru) (location identifier)

JAuGG Journal of Australian Geology and Geophysics (periodical code)

jav Java (file extension)

Jav Java • Javanese

JAV Jahresarbeitsverdienst • Jakobshavn (Greenland) (location identifier) • Jupe Avant

java Java (file extension)

JAVA James A. Van Allen (1914-)

JAVIC Japan Audio-Visual Information Center

JAVSO Journal of the American Association of Variable Star Observers (periodical code)

JAWG Joint Airmiss Working Group

JAWOP Joint Automated Weather Observing Program

JAWS Joint Airport Weather Study • Junk Acronyms When Speaking

JAX Jacksonville (USA-FL) (location identifier)

JAXA Japan Aerospace Exploration Agency (merging of ISAS, NAL and NASDA)

JAZ Jahresarbeitszeit

Jb Jahrbuch • Jahresbericht

JB Jahresbeitrag • Jardin Botanique • Joint Board • Jugendbewegung • Jugendbund • Jukebox • Jump if Below • Junction Box

JBAA Journal of the British Astronomical Association (periodical code)

JBAF Java Beans Activation Framework

JBall Journal of Ballistics (periodical code)

JBAn Jodrell Bank Annals (periodical code)

JBAS Journal of the British Astronomical Society (periodical code)

JBASp Jodrell Bank Ann Ser Pt (periodical code)

Jbb Jahrbücher

Jber Jahresbericht

Jberr Jahresberichte

JBC Boston City (USA-MA) (location identifier) • Japan Broadcasting Corporation • Joint Board on Communication

JBD Jordan - Brans - Dicke (theory) • Joint Board of Directors

JBDC Japanese Beam Physics Club

JBE Joint Blind Equalization • Jump if Below or Equal

JBEM Junction Box Environmental Module

JBEMC Junction Box Environmental Module Controller

Jber Jahresbericht

JBES Jodrell Bank Experimental Station (UK)

JBF Jump Back-Forwards

JBIG Joint Bilevel Image-experts Group • Joint Bilevel Image Group • Joint Binary Image Group

JBiom Journal of Biomechanics (periodical code)

JBIS Journal of the British Interplanetary Society (periodical code)

JBK Berkeley (USA-CA) (location identifier)

JBl Justizblatt

JBM Judgment Base Module

JBO Jodrell Bank Observatory (UK-England) • Journal of Biomedical Optics (periodical code) • June Bootids (meteor stream)

JBOD Just a Bunch Of Disks

JBOSC Joint Base Operations Support Contract

JBOX Junction Box

JBP Los Angeles - Commerce Business Plaza (USA-CA) (location identifier)

JBPA Japan Book Publishers Association

JBPDS Joint Biological Point Detection System

JBR Jonesboro (USA-AR) (location identifier)

JBRe Jodrell Bank Reprints (periodical code)

JBS Jamaica Bureau of Standards • Jamaican Bureau of Standards • Joint Broadcast Service • Joint Broadcast System • Joly Black Screen • Pleasanton - Hacienda Business Park (USA-CA) (location identifier)

JBSPO John Bryan State Park Observatory (USA-OH)

JBT Bethel City Landing (USA-AK) (location identifier)

JBU John Brown University (USA)

JBVAD Jodrell Bank (Starlink internal code)

JBW Jugendbildungswerk

JBY Just Be Yourself

J&C Justification & Certification

JC Job Card • Joint Committee • Joint Compound • Joule Cycle • Julius Caesar • Jump if Carry • Jump on Condition • Junior College

JCA Cannes Croisette (France) (location identifier) • Jettison Controller Assembly • Joint Center for Astrophysics (USA-MD) • Journal Co-citation Analysis • Jupiter Closest Approach

JCAB Japanese Civil Aviation Bureau

JCAC Joint Civil Affairs Committee

JCAE Joint Committee on Atomic Energy

JCALS Joint Computer-aided Acquisition and Logistics Support • Joint Computer-aided Acquisition and Logistics

System • Joint Continuous Acquisition and Lifecycle Support

JCAM Journal of Climate and Applied Meteorology (periodical code)

JCAMP Joint Committee on Atomic and Molecular Physics

JCAP Journal of Cosmology and Astroparticle Physics (periodical code)

JCAR Joint Conference on Automated Reasoning

JCAST Joint Committee on the Archives of Science and Technology

JCAT Joint Crisis Action Team

JCATS Joint Cartographic Analysis Tool Set

JCB Job Control Block • Jocaba (Brazil-SC) (location identifier) • Joint Communication Board • Juris Canonici Baccalaureatus • Juris Civilis Baccalaureatus

JCC Java Competence Center • Job Control Card • Joint Communications Center • Joint Computer Conference • Joint Consultative Committee • Joint Consultative Council • Joint Coordination Center • San Francisco - China Basin (USA-CA) (location identifier)

JCCB Joint Configuration Control Board

JCCC Joint Command Control Center • Joint Communications Control Center

JCCDS Joint Committee on Cartographic Data Standardization (USA)

JCCE Journal of Computing in Childhood Education (AACE)

JCCG Joint Combined Coordination Group

JCCOMNET Joint Coordination Center Communications Network

JCCP Joint Casualty Collection Point

JCCRG Joint Command and Control Requirements Group

JCCRS Joint Contingency Construction Requirements System

JCCS JIVE Correlator Control Software • Job Card Control System

JCCSA Joint Communication Contingency Station Assets

JCCSD Joint Communication Command System Designator

JCD Jeune Cadre Dynamique • Juri Canonici Doctor • Juri Civilis Doctor • Sainte Croix Island (US Virgin Islands) (location identifier) • Ste. Croix Island (US Virgin Islands) (location identifier)

JCDCG Japanese Conference on Discrete and Computational Geometry

JCDE Java Certified Developer Examination

JCDL Joint Conference on Digital Libraries (IEEE)

JCE Java Cryptographic Extension • Jeune Chambre Économique • Juriste-Conseil d'Entreprise • Oakland - Convention Center (USA-CA) (location identifier)

JCEE Joint Council on Economic Education

JCEF Jeune Chambre Économique Française (France)

JCEOI Joint Communications-Electronics Operating Instruction

JCER Japan Center for Economic Research

JCET Joint Center for Earth Technology • Joint Council on Educational Telecommunications • Joint Council on Educational Television

JCEWS Joint Command, Control, and Electronic Warfare System

JCF Joint Citedness Factor • Joint Communications Facility • Joint Control Facility

JCG Joint Consultative Group • Joint Coordinating Group

JCh Journal of Chromatography (periodical code)

JCH Christianshåb (Greenland) (location identifier)

JChPh Journal of Chemical Physics (periodical code)

JChS Journal of Chromatographic Science (periodical code)

JCI Joint Communications Instruction • Kansas City - Johnson Industrial Estate (USA-MO) (location identifier)

JCIC Joint Compliance and Inspection Committee

JCIP Joint Combined Interface Procedure

JCIS Joint Command Information System • Joint Communications Information Synchronization • Joint Communications Infrastructure Synchronization • Journal of Colloid and Interface Science (periodical code)

JCIST Japan Information Center of Science and Technology

JCJ Cheju - Chu Ja (Korea-ROK) (location identifier)

JCK Jacobus Cornelius Kapteyn (1851-1922) • Java Compatibility Kit • Julia Creek (Australia-QLD) (location identifier)

JCKBSE Joint Conference on Knowledge-Based Software Engineering

JCL Jeans' Color Law • Job Command Language • Job Control Language (*) • Joint Consolidated List

JCLA Journal of Current Laser Abstracts

JCLC Joint Committee on the Library of Congress

JCLDS Joint Committee for Library and Documentation Standardization

JCLE Joint Committee on Library Education

JCli Journal of Climate (periodical code)

JClim Journal of Climatology (periodical code)

JCLL Joint Center for Lessons Learned

JCLO Joint Communication Leasing Office

JCLOT Joint Closed-Loop Operations Test

JClP Journal of Clinical Pharmacology (periodical code)

JCLWC Joint Committee on Library Works as a Career

JCM Jacobina (Brazil-BA) (location identifier) • Joint Commission Meeting • Joint Committee on Microcards • Joint Consultative Machinery • Journal of Computational Mathematics (periodical code)

JCMC Joint Crisis Management Capability

JCME Joint Collection Management Element

JCMS Johnson County Middle School (USA-KY)

JCMST Journal of Computers in Mathematics and Science Teaching (AACE)

JCMT James Clerk Maxwell Telescope (USA-HI) • Joint Collection Management Tool

JCMTB James Clerk Maxwell Telescope Board

JCN Incheon (Korea-ROK) (location identifier) • Job Change Notice • Job Control Number • Joint Communications Network • Junction

JCO Comino (Malta) (location identifier) • Joint Certification Office • Jupiter-Crossing Object

JCoAM Journal of Computational and Applied Mathematics (periodical code)

JCOC Joint Combined Operations Center

JCoCh Journal of Computational Chemistry (periodical code)

JCOE Joint Common Operating Environment

JCom Journal of Communications (periodical code)

JCoM Journal of Computational Mathematics (periodical code)

JCoMa Journal of Composite Materials (periodical code)

JCOMM Joint Commission for Oceanography and Marine Meteorology

JCoPh Journal of Computational Physics (periodical code)

JCORE Joint Cost-Oriented Resource Estimating

JCOT Joint Collection Observation Team

JCP Java Community Process • Joint Committee on Printing • Joint Cost Position • Journal of Chemical Physics (periodical code)

JCPCUS Joint Committee on Printing of the Congress of the United States

JCPDS Joint Committee on Powder Diffraction Standards

JCPES Joint Contingency Planning and Execution Support

JCPI Jpont Committee on Paleontologic Information

JCPMIS Joint Communications Planning Management Information System

JCPMS Joint Communications Planning and Management System

JCR Jacareacanga (Brazil-PA) (location identifier) • Joint Casualty Resolution • Journal Citation Reports

JCRF Joint Climate Research Fund

JCrGr Journal of Crystal Growth (periodical code)

JCRLCMP Joint Computer Resource Life Cycle Management Program

JCRP Joint Command Readiness Program

JCRT Joint Concept Review Team

JCS Job Code Schedule • Joint Chiefs of Staff • Journal Citation Score • Journal of the Chemical Society

JCSAT Japan Communication Satellite (now merged with SAJAC into JSS)

JCSC Joint Communications Satellite Center • Joint Communications Support Command • Joint Computer Security Conference

JCSI Joint Command System Integration

JCSS Joint Communications Support Squadron • Journal of the Chinese Silicate Society (periodical code)

JCSTC Joint Council for Scientific and Technical Information

Jct Junction

JCT Japanese Cordless Telephone • Job Control Table • Joint Central Team • Joint Committee on Technology • Junction • Junction - Kimble County (USA-TX) (location identifier)

JCTEA Joint Cost and Training Effectiveness Analysis

JCTI Joint Council for Technology Insertion

JCTION Junction

Jctn Junction

JCTN Joint Composite Tracking Network • Junction

JCTNS Junctions

JCTR Journal of Composites and Technology Research (periodical code)

JCTS Junctions

JCU James Cook University (Australia-QLD) • John Caroll University (USA-OH)

JCUDI Japan Computer Usage Development Institute

JCULS Joint Committee on the Union List of Serials

JCV Joule-Clausius Velocity

JCWS Java Card web Servlet

JCX Los Angeles - Citicorp Plaza (USA-CA) (location identifier)

JCY Johnson (USA-TX) (location identifier)

JC2 Journées CNES des Jeunes Chercheurs (France)

Jd Joined

J-D Jacobi-Davidson (method)

JD Doctor of Jurisprudence • Doctor of Laws • Jacobi-Davidson (method) • Jahresdurchschnitt • Joint Discussion • Jordan Dinar (currency) • Julian Date • Julian Day • Julianisches Datum • Juris Doctor • Justice Department

JDA Japanese Dark-sky Association • Japanese Defense Agency • Japan's Defense Agency • John Day (USA-OR) (location identifier) • Joint Development Agreement • Joint Duty Assignment

JDAL Joint Duty Assignment List

JDAM Joint Direct Attack Munition

JDB Dallas Downtown (USA-TX) (location identifier) • Java Debugger

JDBC Java Database Connect • Java Database Connectivity (*)

JDBE Joint Database Element

JDC Japan Documentation Center • Japanese Digital Cellular • Jet Drift Chamber • Job Description Card • Joint Deployment Community

JDCP Joint Distributed Collaborative Planning

JDD JAICI Document Delivery • Job Data Documentation

JDE Journal of Differential Equations (periodical code)

JDF Juiz de Fora - Francisco de Assis (Brazil-MG) (location identifier)

JDH Jodhpur (India) (location identifier)

JDI Just Do It

JDIMP Joint Data and Information Management Panel

JDK Java Development Kit

JDL Job Description Language • Job Description Library • Job Descriptor Language • Joint Data Library • Joint Directors of Laboratories

JDM Miami Downtown (USA-FL) (location identifier)

JDMAG Joint Depot Maintenance Analysis Group

JDMAPI Java Data Mining Application Programming Interface

JDN Jordan (USA-MT) (location identifier) • Julian Day Number

JDO Java Data Object • Juàzeiro do Norte - Cariri (Brazil-CE) (location identifier)

JDOC Joint Desired Operational Capability

JDOP Joint Doppler Operational Project

JDP Héliport de Paris (France) (location identifier) • Joint Departmental Publication • Joint Development Plan • Joint Development Project

JDPA Japan Design Protection Association

JDS Japan Documentation Society • Job Data Set • Job Data Sheet • Joint Decision Statement • Joint Deployment System

JDSCR Joint Deployment System Change Request

JDSIP Joint Deployment System Interface Processor

JDSIR Joint Deployment System Incident Report

JDSQP Joint Deployment System Query Processor

JDSS Joint Decision Support System

JDSUP Joint Deployment System Update Processor

JDT Joint Development Team • Joint Development Test • Julian Date Time • Minneapolis Downtown (USA-MN) (location identifier)

JDT&E Joint Development Test & Evaluation

JDTE Joint Development Test and Evaluation

JDUP Joint Deployment Update Processor

JDUPO Joint Dual Use Project Office

JDX Houston - Central Business District (USA-TX) (location identifier)

JDY Downey (USA-CA) (location identifier)

JDZ Jingdezhen (China-PRC) (location identifier)

Je June

JE Japanese Ephemeris • Job Enrichment • Job Evaluation • Joshi Effect • Joule Effect • Jump if Equal • Junta Electoral

JEA Japan Environmental Agency • Joint effectiveness Analysis • Joint Endeavor Agreement • Journal of Experimental Algorithmics

JEAS Journal of Engineering and Applied Sciences (periodical code)

Jeb Jebel

JEB Joint Executive Board

JEC Joint Economic Committee • Joint European Conference • Joint Evaluation Committee • Junta Electoral Central

JECC Japan Electronic Computer Corporation

JECCS Joint Enhanced Core Communications System

JECF Java Electronic Commerce Framework

JECFA Joint Expert Committee on Food Additives

JECG Joint Exercise Control Group

JECL Job Entry Control Language

JECP Joint Electronic Commerce Program • Joint Experimentation Campaign Plan

JECPO Joint Electronic Commerce Program Office

JECS Journal of Electronic Circuits and Systems (periodical code)

JED Jeddah - King Abdul Aziz (Saudi Arabia) (location identifier) • Joint Exercise Division • Journal of Electronic Defense • Julian Ephemeris Data • Julian Ephemeris Date (*)

JEDC Joint Environmental Data Center

JEDEC Joint Electronic Devices Engineering Committee • Joint Electronic Devices Engineering Council (*)

JEdI Joint Education Initiative

JEDI Joint Education Initiative • Joint Electronic Data Interchange • Joint Exercise Driver for Intelligence

JEDMICS Joint Engineering Data Management Information and Control System • Joint engineering Drawing Management and Control System

JEE Jeremie (Haiti) (location identifier) • Journal of Earthquake Engineering

JEEEA Journal of Electrical and Electronic Engineering - Australia (periodical code)

JEEC Joint ETSI ECMA Committee

JEEM Journal of Environmental and Economic Management (periodical code)

JEEP Joint Emergency Evacuation Plan

JEF Jacobi Elliptical Function • Jefferson City (USA-MO) (location identifier) • Jeunesses Européennes Fédéralistes

JEG Egedesminde (Greenland) (location identifier) • Joint Evaluation Group

JEHLT Jet Engine of High Limited Thrust

JEI Joint Education Initiative (formerly JEdI) • Journal of Electronic Imaging (periodical code) • Junta Electoral Institucional

JEIA Joint Electronic Information Agency • Joint Export-Import Agency

JEIDA Japan Electronics Industry Department Association • Japan Electronics Industry Development Association • Japanese Electronics Industry Development Association (*)

JEIM Jet Engine Intermediate Maintenance

JEIO Joint Effectiveness and Interoperability Office

JEIPAC JICST Electronic Information Processing Automatic Computer

JEJ Jeh (Marshall Islands) (location identifier)

JEL Joint Electronic Library

JElas Journal of Elasticity (periodical code)

JElCo Electronics Communications of Japan (periodical code)

JElEn Electrical Engineering of Japan (periodical code)

JElS Journal of the Electrochemical Society (periodical code)

jem jemals • jemand

JEM Emeryville (USA-CA) (location identifier) • Japanese Experimental Module • Japanese Experiment Module (*) • Japan Experimental Module • Japan Experiment Module • JET Engine Modulation • Joint European Monitor (INTEGRAL) • Joint Exercise Manual

JEMat Journal of Electronic Materials (periodical code)

JEMD Japan English Magazine Directory

JEMH Journal of Educational Multimedia and Hypermedia (AACE)

JEMIC Jet Engine Mishap Investigation Course

JEMP Japanese Ephemerides of Minor Planets • Joint Exercise Management Package

JEMRMS Japanese Experimental Module Remote Manipulator System

JEMT Journal of Electron Microscopy Technique (periodical code)

JEN Junta Electoral Nacional

JENAM Joint European and National Astronomy Meeting

JENC Joint European Networking Conference

JEner Journal of Energy (periodical code)

JEngM Journal of Engineering and Mechanics (periodical code)

JEngS Journal of Engineering Sciences (periodical code)

JEnMa Journal of Engineering and Mathematics (periodical code)

JenRv Jena Review (periodical code)

jens jenseitig • jenseits

JEnvS Journal of Environmental Sciences (periodical code)

JEOC Japanese Earth Observing Center

JEOCG Joint Exercise Operations Control Group

JEOCN Joint European Operational Communications Network

JEOS Janus Earth Observation Satellite • Japanese Earth Observation System • Japanese Earth Observing Satellite • Japanese Earth Observing System • Journal of the European Optical Society

JEP Joint European Project • Joint Experiment Programme • Jupiter Entry Probe

JEPES Joint Engineer Planning and Execution System

JEPI Joint Electronic Payments Initiative

JEPS Joint European Printer Server

JEQ Jequié (Brazil-BA) (location identifier)

Jer Jersey (Channel Islands)

JER Jersey (Channel Islands) (location identifier) • Joint Ethics Regulation

JERC Joint Electronic Research Committee

JERI Joint Ecology Research Initiative

JERIC Journal on Educational Resources in Computing (ACM) (periodical code)

JERICO Joint European Robotics Interactive and Calibration Operations

JERS Japanese Earth Resources Satellite • Japan's Earth Resources Satellite (*)

JES Japanese Engineering Standard • Jet Expansion Spectroscopy • Job Entry System • Joint Environment Simulator

JESA Japan Engineering Standards Association

JESAUG Journal of the Environmental Satellite Amateur Users' Group

JESC Japan Engineering Standards Committee

JESG Job Entry System Group

JESCT Job Entry System Communication Table

JESRP Journal of Electron Spectroscopy and Related Phenomena (periodical code)

JESS Joint Exercise Simulation System • Joint exercise Support System

JESSE Journal of Earth System Science Education

JESSI Joint European Submicron Silicon Initiative • Joint European Submicron Silicon (*)

JEST Joint Entrance Screening Test

JET Japan Exchange and Teaching • Jet-Ending Penetration • Job Element Text • Job Euro Transfer • Joint Estimate Team • Joint European Telescope • Joint European Torus • Joint European Transport • Joint Evaluation Team • Jury d'Éthique Publicitaire

JETDAS Joint Electronic Type Designated Automatic System

JETDLAG Joint European development of Tunable Diode Laser Absorption spectroscopy for measurement of Atmospheric trace Gases

JETDS Joint Electronics Type Designation System

JETEC Joint Expendable Turbine Engine Concept

JETh Journal of Engineering and Thermophysics (periodical code)

JETP Journal of Experimental and Theoretical Physics • Joint European Torus Project

JETRO Japanese External Trade Organization

JETS Joint Electronic Type System

JETh Journal of Engineering and Thermophysics (periodical code)

JETT Java Engagement for Teacher Training (ACM) • Jettison

JETTA Joint Environment Testing, Training and Analysis

Jeu Jeudi

JEV Évry (France) (location identifier) • Jever • Joint European Venture

jew jeweilig • jeweils

JEWA Journal of Electromagnetic Waves and Applications (periodical code)

JEWM Joint Expert Working Meeting

JEX Jupiter Express

JEXAM Japanese Experiment on Asia Monsoon

JEXM Joint European X-ray Monitor

J/F Journal Folio

JF Junior Fiction • Jupiter Family

JFACT Joint Flight Acceptance Composite Test

JFAA Journal of Fourier Analysis and Applications

JFAP Joint Faculty Appointment Program • Joint Frequency Allocation Panel

JFAS Jet Fuel Accounting System

JFAST Joint Flow and Analysis System for Transportation

JFC Java Foundation Class • Journal of Fluid Control (periodical code) • Jupiter Family Comet

JFCB Job File Control Block

JFCC Joint frequency Coordination Committee

JFCL Jump if Flag set and Clear it

JFD Japanese Flight Demonstration • Japanese Flight Demonstrator

JFET Junction Field Effect Transistor

JFF Journal of Fire and Flammability (periodical code) • Just For Fun

JFI James Frank Institute (USA-IL)

JFIF JPEG File Interchange Format

JFIP Japanese Facilities Improvement Project

JFIS Journal to Field Impact Score

JFIT Joint Framework for Information Technology

JFK John Fitzgerald Kennedy (USA-NY) • New York - John Fitzgerald Kennedy (USA-NY) (location identifier)

JFKL John Fitzgerald Kennedy Library

JFL Joint Frequency List

JFM Fremantle (Australia-WA) (location identifier) • January - February - March • Job Function Manual • Journal of Fluid Mechanics (periodical code) • Jupiter Flyby Mission

JFMIP Joint Financial Management Improvement Program

JFMSES Joint Frequency Management and Spectrum Engineering System

JFN Jefferson Ashtabula (USA-OH) (location identifier)

JFOC Joint Future Operational Capability

JFP Joint Frequency Panel

JFPCC Joint Fusion Power Coordination Committee

JFR Frederikshåb (Greenland) (location identifier) • Joint Functional Requirement

JFRC Joint Functional Requirement Council

JFRDT Joint Functional Requirement Determination Team • Joint Functional Requirement Development Team

JFROM Joint Force Reception and Onward Movement

JFS Joint Feasibility Study • Journaled File System • Journal of Fluid Structures (periodical code)

JFT Job File Table • Joint Firing Test • Joint Functional Team

JFTDS Joint Electronic Type Designation System

JFTOT Jet Fuel Thermal Oxidation Tester

JFTR Joint Federal Travel Regulation

JFTX Joint Field Training Exercise

JFUB Joint Facilities Utilization Board

JFV Jupiter Flyby Vehicle

jg jung

Jg Jaargang • Jahrgang

J/G Joules/Gram

JG Job Guide • Joint Group • Journal of Geology (periodical code) • Jugendgericht • Jump if Greater • Junior Grade • Junta General

JGA Jamnagar (India) (location identifier) • Jovian Gravitational Assist • Jovian Gravitational Assistance

JGB Jagdalpur (India) (location identifier)

JG&C Joint Guidance & Control

JGC Grand Canyon (USA-AZ) (location identifier)

JGCD Journal of Guidance Control and Dynamics (periodical code)

Jgd Jugend

JGD Japanese Geodetic Datum • JOPES Global Dictionary

Jgdl Jugendlich

JGE Geoje (Korea-ROK) (location identifier) • Jump if Greater or Equal

JGeo Journal of Geophysics (periodical code)

Jgg Jahrgänge

JGG Journal of Geomagnetism and Geoelectricity (periodical code) • Jugendgerichtsgesetz

JGGS Journal of Geomagnetism and Geoelectricity - Supplements (periodical code)

JGI Jolly Good Idea

JGL Atlanta Galleria (USA-GA) (location identifier)

JGla Journal of Glaciology (periodical code)

JGM Joint Gravity Model

JGN Japan Gigabit Network • Japan Global Network • Jiayuguan (China-PRC) (location identifier)

JGO Godhavn (Greenland) (location identifier)

JGOFS Joint Global Ocean Flux Study • Joint Global Ocean Forecasting System

JGP Houston - Greenway Plaza (USA-TX) (location identifier) • Journal of Geometry and Physics (periodical code)

JGQ Houston - Transco Tower Galleria (USA-TX) (location identifier)

JGR Gronnedal (Greenland) (location identifier) • Journal of Geophysical Research (periodical code)

JGRS Journal of Geophysical Research - Supplements (periodical code)

JGSE Joint Group on System Engineering

JGuC Journal of Guidance Control (periodical code)

JGX Glendale (USA-CA) (location identifier)

JGZG Journal of Geophysics - Zeitschrift für Geophysik (periodical code)

Jh Jahresheft • Jahrhundert

JH Jägerhaus

JHA Japan Hydrographic Association • Job Hazard Analysis • Journal for the History of Astronomy (periodical code) • Justice and Home Affairs

JHATD Johns Hopkins APL Technical Digest (periodical code)

Jhb Jahrbuch

JHB Johannesburg (South Africa • Johor Bahru (Malaysia) (location identifier)

JHC Garden City (USA-NY) (location identifier)

JHD Japan Hydrographic Department (Japan) (now JHOD)

Jhdt Jahrhundert

JHE Hälsingborg (Sweden) (location identifier) • Helsingborg (Sweden) (location identifier)

JHENP Journal of High-Energy and Nuclear Physics (periodical code)

JHEP Journal of High-Energy Physics (periodical code)

Jhg Jahrgang

JHG Java Harmonic Generator • Jinghong Gasa (China-PRC) (location identifier) • Joule Heat Gradient

Jhgg Jahrgänge

Jhh Jahreshefte

J. Hist. A. Journal for the History of Astronomy

JHM Kapalua (USA-HI) (location identifier)

JHMCS Joint Helmet-Mounted Cueing System

JHO Jeremiah Horrocks Observatory (UK-England)

JHOD Japan Hydrographic and Oceanographic Department (Japan) (formerly JHD)

JhPCh Journal of Physical Chemistry (periodical code)

JHQ Shute Harbour (Australia-QLD) (location identifier)

Jhrb Jahrbuch

Jhrg Jahrgang

JHS Holsteinborg (Greenland) (location identifier) • Junior High School

Jhtsd Jahrtausend

Jhtsde Jahrtausende

JHU Johns Hopkins University (USA-MD)

JHV Jahreshauptversammlung

JHW Jamestown (USA-NY) (location identifier) • Jugendhilfswerk

Jhy Jalhay (Belgium)

JHY Cambridge - Hyatt Regency (USA-MA) (location identifier)

JHyd Journal of Hydronautics (periodical code)

ji Yiddish (ISO language code)

JI Io (satellite of Jupiter) • Job Instruction • Joint Implementation • Joint Inspection • Junction Isolation • Jupiter Inlet

JIA Juiná (Brazil-MT) (location identifier)

JIAC Java Intelligent Agent Componentware

JIAFS Joint Institute for the Advancement of Flight Sciences

JIAPS Journal of the International Association of Physics Students

JIATF Joint Interagency Task Force

JIAWG Joint Integrated Avionics Working Group

JIB Djibouti (location identifier) • Job Information Block • Job Information Bureau

JIBSG Joint Integrated Broadcast Steering Group

JIC Jet Interaction Control • Job Instruction Card • Joint Ice Center • Joint Implementation Committee • Joint Industrial Council • Joint Information Center • Joint Intelligence Center • Joint Intelligence Committee • Joint Intelligence Curriculum • Joint International Conference • Journal of Internet Cataloguing • Just In Case

JICA Japan International Cooperation Agency • Joint Intelligence Collecting Agency

JICC Japanese Instrument Control Center • Java and the Internet in the Computing Curriculum (conference) • Joint Intelligence Coordination Center

JICO Joint Interface Control Officer

JICST Japan Information Centre of Science and Technology

JID Joint Intrusion Detection • Los Angeles - City of Industry (USA-CA) (location identifier)

JIDC Joint Interrogation and Debriefing Center

JIDM Joint Interdomain Management

JIDRP Joint Intel Dissemination Requirements Panel

JIE Journal of Industrial Ecology

JIEB Joint Intelligence Executive Board

JIECE Journal of the Institute of Electronic Communication Engineers (periodical code)

JIEO Joint Interoperability and Engineering Office • Joint Interoperability and Engineering Organization • Joint Interoperability Evaluation Organization

JIEP Joint Intelligence Estimate for Planning

JIEq Journal of Integral Equations (periodical code)

JIES Joint Interoperability Evaluation System

JIETE Journal of the Institution of Electronics and Telecommunications (periodical code)

jif JPEG Image Format (file extension)

JIF Joint Infrastructure Fund (UK) • Joint Intelligence Facility • Joint Interrogation Facility • Journal Impact Factor

JIFFI Job Instruction File Format for Industry

JIFT Joint Institute for Fusion Theory (Japan)

JIFTS Joint In-Flight Transmission Service

JIG Jane's Information Group • Joint Implementation Guide • Joint Integration Group

JIGI Java Interface for Geospatial Information

JIHOP J.I. Holcomb Observatory and Planetarium (USA-IN)

JII Europa (satellite of Jupiter) • Joint Integration Interface

JIIB Joint Intelligence Interoperability Group

JIII Ganymedes (satellite of Jupiter) • Japanese Institute of Invention and Innovation (Japan)

JIIM Journal of Information and Image Management

JIIS Journal on Intelligent Information Systems (periodical code)

JIK Ikaria Island (Greece) (location identifier)

JIL Jilin (China-PRC) (location identifier)

JILA Joint Institute for Laboratory Astrophysics (USA-CO)

JILE Joint Intelligence Liaison Element

JILO Joint information Liaison Office

JILR Journal of Interactive Learning Research

JILSIPT Joint Integrated Logistics Support Integrated Product Team

JILSMT Joint Integrated Logistics Support Management Team

JILSP Joint Integrated Logistics Support Plan

JIM Jimma (Ethiopia) (location identifier) • Job Information Memorandum • Job Instruction Manual • Jovian Ionosphere Model • Jovian Ionospheric Model

JIMA Joint Institute for Millimeter Astronomy (now IRAM)

JIMAR Joint Institute for Marine and Atmospheric Research

JIMC Joint Information Management Center

JIMIA Journal of the Institute of Mathematics and Its Applications (periodical code)

JIMMT JIM Materia;s Transactions (periodical code)

JIMO Journal of the International Meteor Organization (periodical code) • Jupiter Icy Moon Orbiter

JIMPP Joint Industrial Mobilization Planning Process

JIMS Joint information Management System • Jukebox Interface Management System

JIMSS Journal of Intelligent Material Systems and Structures (periodical code)

JIMW Journal of Infrared and Millimeter Wave (periodical code)

JIN Jinja (Uganda) (location identifier) • Joint Implementation Network

JINI Java Intelligent Network Infrastructure

JINR Joint Institute for Nuclear Research (Russia)

JINS Japan Institute of Nuclear Safety

JIO Joint Information Office • Joint Intelligence Organization • Joint Interoperability • Ontario (USA-CA) (location identifier)

JION Joint Interoperable Operator Network

JIOP Joint Integrated Operating Procedure • Joint Interface Operational Procedure

JIOS Joint Interoperability System

JIP Jipijapa (Ecuador) (location identifier) • Job Improvement Plan • Job Installation Plan • Joint Identifier Program • Joint Input Processing • Joint interface Program

JIPA Java-based Image Processing Application • Java Image Preview Application

JIPC Joint Imagery Production Complex

JIPDEC Japan Information Processing Development Centre

JIPID Japanese International Protein Information Database

JIPNET Japan Information Processing Network

JIPS JANET Internet Protocol service

JIPT Joint integrated Process Team

JIR Jiri (Nepal) (location identifier) • Job Improvement Request • Joint Intelligence Room • JOPES Information Requirement • Journal of Irreproducible Results

JIRA Japanese Robotic Industries Association

JIRAD Joint Independent Research And Development

JIRC Journal of International Research Communications

JIRONET Joint Intelligence Research Office Network

JIRS Joint Infrared Standard

JIRSG Joint Interservice Regional Support Group

JIRW Journal of Indexing and Reference Work

JIS Jamaica Information Service • Japanese Industrial Standard • Japan Industrial Standard • Japan Institute of Standards • Joint Integrated Simulation • Joint Integration Simulation • Joint Interoperability System • Journal of Information Science

JISAO Joint Institute for the Study of Atmosphere and Ocean (USA-WA)

JISC Japanese Industrial Standards Commission • Japanese Industrial Standards Committee (*) • Joint Information Services Committee • Joint Information Systems Committee (UK)

JISP Joint Integrated Simulation Procedure

JISR Joint Information Search and Retrieval • Joint Intelligence, Surveillance and Reconnaissance

J/IST JSF/Integrated Subsystem Technology

JIST Journal of Imaging Science and Technology (periodical code) • JSF Integrated Subsystem Technology

JISWG Joint Integrated Simulation Working Group • Joint IRAS Science Working Group

JISYA Japan ISY Association

JIT Job Information Test • Joint Interest Test • Just Interoperability Tool • Just In Time

JITC Joint Interoperability Technology Center • Joint Interoperability Test Center

JITF Joint Integration Test Facility • Joint Intelligence Test Facility • Joint Interface Task Force • Joint Interface Test Force

JITP Joint Interface Test Plan

JITR Japan Industry Technical Reports

JITS Joint interface Test System • Joint Interoperability Test System • Just-In-Time System

JiTT Just-in-Time Teaching (project)

JIU Jiujiang (China-PRC) (location identifier) • Joint Inspection Unit

JIV Callisto (satellite of Jupiter)

JIVA Joint intelligence Virtual Intelligence

JIVE Joint Institute for VLBI in Europe

JIW Jiwani (Pakistan) (location identifier)

JIWG Joint Imagery Working Group • Joint Industry Working Group

JIWP Joint Interim Working Party

JIX Job (queue) Index • Sinope (satellite of Jupiter)

jJ jedes Jahr

J/J Januar/Juli

JJ Josephson Junction • Jour Julien • Judges • Justices

JJA Jour Julien Astronomique • June - July - August

JJAPL Japanese Journal of Applied Physics - Letters (periodical code)

JJAPR Japanese Journal of Applied Physics - Regular Papers, Short Notes and Reviews (periodical code)

JJAPS Japanese Journal of Applied Physics - Supplement (periodical code)

JJC Joliet Junior College (USA-IL) • Journées Jeunes Chercheurs (SF2A)

JJH Jour Julien Héliocentrique

JJI Juanjui (Peru) (location identifier)

JJM Jour Julien Modifié

JJMO John J. McCarthy Observatory (USA-CT)

JJN Jinjiang (China-PRC) (location identifier)

JJPPAA Jornadas de Puertas Abiertas

JJT Josephson Junction Transistor • Jour Julien Tronqué • Jumbo Jet Transport

JJU Julianehåb (Greenland) (location identifier)

J/K Joule per Kelvin

Jk Jack

JK Jugendkreis • Julianischer Kalender

JKAS Journal of Korean Astronomical Society (periodical code)

JKC Johnson - Kron - Cousins (system)

JKG Jönköping (Sweden) (location identifier) • Jönköping (Sweden) (location identifier)

JKH Chios (Greece) (location identifier)

JKM Joint Key Management • Just Keep Moving

JKMIWG Joint Key Management Infrastructure Working Group

JKMP Joint Key Management Plan

JKMS Joint Key Management System

JKPS Journal of the Korean Physical Society

JKR Janakpur (Nepal) (location identifier)

Jkt Jacket

JKT Jacobus Kapteyn Telescope (Canary Islands) • Jakarta (Indonesia) (location identifier) • Job Knowledge Test

JKU Johannes-Kepler-Universität (Austria)

JKV Jacksonville (USA-TX) (location identifier)

jl Jongsleden

Jl Journal • July

JL Jaidee-Lyngå (survey) • Joint Logistics • Joule's Law • Juillet • July • Jump if Less

JLA Cooper Lodge - Quartz Creek (USA-AK) (location identifier) • Jamaica Library Association • Japan Library Association

JLB Long Beach (USA-CA) (location identifier)

JLC Japan Linear Collider • Joint Logistics Command

JLCM Journal of Less Common Metals (periodical code)

JLD Hälsingborg Landskrona (Sweden) (location identifier) • Helsingborg Landskrona (Sweden) (location identifier)

JLE Jump if Less than or Equal

JLF J-band Luminosity Function

JLH Arlington Heights (USA-IL) (location identifier)

JLIT Joint Logistics Integrated Team

JLMIS Joint Logistics Management Information System

Jln Jalan

JLN Joplin (USA-MO) (location identifier)

JLNCHREP Joint Launch Report

JLNT Japanese Large National Telescope

JLO Jesolo (Italy) (location identifier)

JLOG Joint Logistics

JLOTS Joint Logistics Over-The-Shore

JLP Juan-les-Pins (France) (location identifier)

JLPPG Joint Log and Personnel Policy and Guidance

JLR Jabalpur (India) (location identifier)

JLRA Joint Laboratories of Radio Astronomy

JLRB Joint Logistics Review Board

JLRC Joint Logistics Readiness Center

JLREID Joint Long-Range Estimate Intelligence Document

JLRSA Joint Long-Range Strategic Appraisal • Joint Long-Range Strategic Assessment

JLRSE Joint Long-Range Strategic Estimate

JLRSS Joint Long-Range Strategic Study

JLS Jales (Brazil-SP) (location identifier) • Jamaica Library Service • Joint Least Squares

JLSC Joint Logistic Service Committee • Joint Logistic System Center

JLSP Joint Logistics Support Plan • Joint Logistics Support Program

JLTP Journal of Low-Temperature Physics (periodical code)

JLwT Journal of Lightwave Technology (periodical code)

JLX Los Angeles - Union (USA-CA) (location identifier)

Jly July

jm Jamaica

jM jeder Monat

J/M Jettison Motor

JM Jahresmiete • Jamaica (ISO country code) • Jamaique (code ISO de pays) • Jettison Motor • Joint Mission • Joint Movement • Jolly Mad • Justizminister

JMA Houston Marriott (USA-TX) (location identifier) • Japan Management Association • Japan Meteorological Agency • Joint Mission Analysis • Joint Mission Application • Joint Mission Area

JMAA Joint Mission Area Assessment • Journal of Mathematical Analysis Applications (periodical code)

JMAARS Joint Model After-Action Review System

JMAC Japan Market Awareness Committee

JMagR Journal of Magnetic Resonance (periodical code)

JMAPI Java Management Application Programming Interface

JMAPS Joint Message Analysis and Processing System • Joint Message And Processing System

JMAS Joint Manpower Automation System • Joint Mission Application Software

JMASS Joint Modeling And Simulation System

JMAT Joint Mobility Assistance Team

JMatR Journal of Materials Research (periodical code)

JMatS Journal of Materials Science (periodical code)

JMB Jamba (Angola) (location identifier) • Joint Meteorological Board

JMBl Justizministrialblatt

JMBM Journal of Mechanical Behavior and Materials (periodical code)

JMC Japan Map Center • Joint Maritime Course • Joint Movement Center • Sausalito - Marin County (USA-CA) (location identifier)

JMCAP Joint Maritime Crisis Action Planning

JMCC Joint Movement Control Center

JMCIS Joint Maritime Command and Information System

jmd jemand

JMD Dallas - Market Center (USA-TX) (location identifier) • Dollar (Jamaica) (ISO currency code) • JOPES Management Division

JMDC Joint Message Distribution Center

JMDS Joint Manpower Data System

JME Jones Matrix Eigenanalysis

JMec Journal de Mécanique (periodical code)

JMEC Joint Materiel Exploitation Center

JMecA Journal de Mécanique Appliquée (periodical code)

JMecE Journal of Mechanical and Engineering Science (periodical code)

JMecT Journal de Mécanique Théorique et Appliquée (periodical code)

JMED Jungle Message Encoder-Decoder

JMemS Journal of Microelectromechanical Systems (periodical code)

JMENS Joint Mission Element Needs Statement

JMeOp Jemna Mechanika Optika (periodical code)

JMES Journal of Materials and Energy Systems (periodical code)

JMet Journal of Metals (periodical code)

JMET Joint Mission Essential Task

JMETL Joint Mission Essential Task List

JMEX Jupiter Magnetospheric Explorer

JMF Java Media Framework

JMFU Joint Meteorological Forecast Unit

JMG Joint Meteorological Group

JMH Joint Mission Hardware • Schaumburg Marriott (USA-IL) (location identifier)

JMHC Joint Mission Hardware Contractor

JMI Java Metadata Interface • Jim Mobile Inc. (USA-CO) • Journal of Medical Imaging (periodical code)

JMIE Joint Maritime Information Element

JMiMi Journal of Micromechanics and Microengineering (periodical code)

JMIS Journal of Management Information Systems

JMK Mykonos (Greece) (location identifier)

JML Job Master Listing • Job Materiel List

JMLC Joint Modular Languages Conference

JMLR Journal of Machine Learning Research

JMLS John Menzies Library Service (UK)

JMM Jacobi Matrix Method • Malmö Harbour (Sweden) (location identifier)

JMMM Journal of Magnetism and Magnetic Materials (periodical code)

JMN Mankato (USA-MN) (location identifier)

JMNA Joint Mission Need Analysis

JMNS Joint Mission Need Statement

JMO Joint Maritime Operations • Jomsom (Nepal) (location identifier)

JMOA Joint Memorandum Of Agreement • Journal of Microwave, Optics, and Acoustics (periodical code)

JMOC Jupiter Microwave Observing Campaign

JMoEl Journal of Molecular Electronics (periodical code)

JMolE Journal of Molecular Evolution (periodical code)

JMOp Journal of Modern Optics (periodical code)

JMoSp Journal of Molecular Spectroscopy (periodical code)

JMoSt Journal of Molecular Structure (periodical code)

JMP Joint Management Plan • Joint Manpower Program • Joint Mission Processor • Joint Monitoring Programme • Joint Monitor Protocol • Journal of Mathematical Physics (periodical code) • Jump

JMPAB Joint Materiel Priorities and Allocation Board

JMPC Joint Mapping and Photography Committee (USA)

JMPE Joint Mission Processing Environment • Joint Mission Processing Equipment

JMPMS Journal of Materials, Processing, and Manufacturing Science (periodical code)

JMPS Journal of Mathematical and Physical Sciences (periodical code)

JMPSo Journal of Mechanics, Physics, and Solids (periodical code)

JMQ Java Message Queuing

JMR Job Management Record • Journal of Magnetic Resonance (periodical code)

JMRC Joint Mobile Relay Center

JMRO Joint Medical Regulating Office

JMRR Joint Monthly Readiness Report • Joint Monthly Readiness Review

Jms Jamais

JM&S Joint Modeling & Simulation

JMS Jamestown (USA-ND) (location identifier) • Java Message Service • Java Messaging Service • Joint Modeling and Simulation • Journal of Molecular Spectroscopy (periodical code)

JMSAC Joint Meteorological Satellite Advisory Committee

JMSEC Journal of Mathematical Systems, Estimation, and Control

JMSJa Journal of the Meteorological Society of Japan (periodical code)

JMSL Journal of Materials Sciences - Letters (periodical code)

JMSME Journal of Materials Science and Materials Electronics (periodical code)

JMSNS Joint Mission Systems Need Statement

JMSW Joint Mission Software System

JMSWG Joint Message Standard Working Group

JMT Joint Management Team

JMTAS Journal de Mécanique Théorique et Appliquée - Suppléments (periodical code)

JMTK Joint Mapping Tool Kit

JMTP Joint Master Test Plan

JMTSS Joint Multichannel Trunking and Switching System

JMU James Madison University (USA-VA) • Jiamusi (China-PRC) (location identifier) • John Moores University (UK)

JMV Jahresmitgliederversammlung

JMW Journal of Marginal Worthiness

JMX Java Management Extension

JMY Freetown - Mammy Yoko (Sierra Leone) (location identifier)

Jn Join • Junction • Junior

JN Jahresnachweis • Jet Navigation • Journal Name • Juin • June

JNA Januária (Brazil-MG) (location identifier) • Journée Nationale de l'Astronomie • Jump if Not Above

JNACC Joint Nuclear Accident Coordinating Center

JNAE Jump if Not Above or Equal

JNAF Jefferson National Accelerator Facility (USA-VA)

JNav Journal of Navigation (periodical code)

JNB Johannesburg - Jan Smuts (South Africa) (location identifier) • Jump if Not Below

JNBE Jump if Not Below or Equal

Jnc Junction

JNC Japan Nuclear Cycle • Jet Navigational Chart • Joint Negotiating Committee

JNCA Journal of Network and Computer Applications

JNCASR Jawaharlal Nehru Centre for Advanced Scientific Research (India)

JNCS Journal of Non-Crystalline Solids (periodical code)

JND Just Noticeable Difference

JNDI Java Naming and Directory Interface • Java Naming Directory Interface • Java Network Directory Interface

Jne Jaune • June

JNE Journal of Non-destructive Evaluation (periodical code)

JNET Journal of Non-Equilibrium Thermodynamics (periodical code)

JNF Junior Non-Fiction

JNG Jining (China-PRC) (location identifier) • Jump if Not Greater

JNGE Jump if Not Greater or Equal

JNH Dallas - North Park Inn (USA-TX) (location identifier)

JNI Java Native Interface • Junin (Argentina-BA) (location identifier)

JNICT Junta Nacional de Investigação Científica e Tecnológica (Portugal)

JNIRE Journal of the National Institute for Resources and Environment (periodical code)

JNK Junk

Jnl Journal

JNL Journal • Jump if Not Less

JNLE Jump if Not Less or Equal

JNLM Journal of Non-Linear Mechanics (periodical code)

Jnls Journals

JNLS Journals

Jnlst Journalist

JNLT Japanese National Large Telescope • Japan National Large Telescope

JNM Joint Network Management

JNMCCB Joint Network Management Configuration Control Board

JNMWG Joint Network Management Working Group

JNN Nanortalik (Greenland) (location identifier)

JNND Just Not Noticeable Difference

JNNIE Joint NSF-NASA Initiative in Evaluation

JNNS Japanese Neural Network Society (Japan)

JNO Jump if No Overflow

JNP Jump if No Parity • Newport Beach (USA-CA) (location identifier)

JNPEA Japan Newspaper Publishers and Editors Association

Jnr Junior

JNr Jahresnummer

JNR Japan National Railways • Junior

JNS Jump if No Sign • Just Noticeable Shift • Narsaq (Greenland) (location identifier)

Jnt Joint

JNT Java Network Technology • Joint Network Team

JNTF Joint National Testing Facility

JNU Jawaharlal Nehru University (India) • Juneau (USA-AK) (location identifier)

JNuM Journal of Nuclear Materials (periodical code)

JNWPU Joint Numerical Weather Prediction Unit

JNX Naxos Island (Greece) (location identifier)

JNZ Jump if Not Zero

JO Jagdordnung • Job Order • Jordan (ISO country code) • Jordanie (code ISO de pays) • Journal des Observateurs (periodical code) • Journal Officiel • Journal of Optics (periodical code) • Junior Officer • Junta Ordinaria • Jupiter Orbiter

JOA Joint Occupancy Agreement • Joint Operating Agreement • Joint Operation Area • Joint Operational Architecture • Joint Operations Agency

JOAN Job Order Account Number

JOB Joint Oversight Board

JOBRESA Joint Board of Remote Sensing Activities

Jo'burg Johannesburg (South Africa)

JOBS Job Opportunity in Basic Skills

JOC Job Order Contracting • Joint Operations Center • Joint Operations Community • Joint Organizing Committee • Joint Oversight Committee • Journal of Optical Communications (periodical code) • Journal of Organic Chemistry (periodical code) • Santa Ana (USA-CA) (location identifier)

JOCAR Joint Communications Allocation Requirement

JOCAS Job Order Cost Accounting System

JOCC Joint Operations Control Center

JOCE Journal Officiel des Communautés Européennes

JOCF Joint Operations Capability File

JOCR Joint Observatory for Cometary Research (USA-NM)

JOCS Joint Operational Climatological Support • Joint Operations Command System

JOD Dinar (Jordan) (ISO currency code) • Jodrell Bank (Starlink internal code) • Joint Occupancy Date • Joint Occupancy Data • Joint Operations Division • Journal Of Development

JODC Japan Ocean Data Center • Japan Oceanographic Data Center

JODL Journal Of Digital Libraries

JOE Java Objects Everywhere • Job Output Element • Joensuu (Finland) (location identifier)

JOEG Joint Operations Evaluation Group

JOF Joint Operating Facility

JOG Joint Operations Graphic • Joint Oversight Group • Yogyakarta (Indonesia) (location identifier)

JOGS Joint Operations Graphics System

JOH Port Saint Johns (South Africa) (location identifier)

Johan Johannesburg (South Africa)

JOI Joint Oceanographic Institutions • Joinville Cubatão (Brazil-SC) (location identifier) • Jupiter Orbital Insertion • Jupiter Orbit Insertion

JOIDES Joint Oceanographic Institutions for Deep Earth Sampling

JOIN Joint Optimal Information System

JOINT-net Japanese Open Integration Network

JOINTnet Japanese Open Integration Network

JOIP Joint Operations Interface Plan • Joint Operations Interface Procedure

JOIS JICST Online Information System • Job Order Information System

JOISTS Joint Operational Interface Simulation Training System

JOK Joshkar-Ola (Russia) (location identifier)

JOL Job Organisation Language • Joint Operations Laboratory • Jolo (Philippines) (location identifier)

JOLA Journal Of Library Automation • Just Overlapping Line Approximation

JOLIS Joint Library and Information System

JOM Japanese Optical Module • Job Operation Manual • Journal of Metals (periodical code) • Njombe (Tanzania) (location identifier)

JOMACHEM Journal of Mathematical Chemistry

JOMAR Joint Office for Mapping and Research (USGS)

JOMOQ Joint Monitoring of Optical Quasars

JON Job Order Number • Johnston Island (US Minor Outlying Islands) (location identifier)

JONSWAP Joint North Sea Wave Analysis Project

JOOD Junior Officer Of the Deck

JOOW Junior Officer Of the Watch

JOp Journal d'Optique (periodical code)

JOP Job Operating Procedure • Joint Observing Programme • Joint Operating Plan • Joint Operating Procedure • Joint Operation Procedure • Josephstaal (Papua New Guinea) (location identifier) • Jupiter Orbiter Probe (now Galileo)

JOPAL Journal of Patent-Associated Literature

JOPES Joint Operational Planning and Execution System • Joint Operations Planning and Execution System

JOPESIR Joint Operations Planning and Execution System Incident Reporting

JOPREP Joint Operational Reporting System

JOPS Joint Operational Planning System • Joint Operations Planning System

JOpt Journal of Optics (periodical code)

JOptT Journal of Optical Technology (periodical code)

JOR Job Operational Requirement • Job Operations Report • Job Order Request • Joint Operational Requirement • Jordan (ISO country code) • Jordanie (code ISO de pays) • Orange (USA-CA) (location identifier)

JORAM Java Open Reliable Asynchronous Messaging

Jord Jordan • Jordanian

JORD Joint Operational Requirement Document

JORS Joint Operational Requirements Statement

JOS Job Order Supplement • Jos Heipang (Nigeria) (location identifier)

JOSE Jindalee Ocean Surveillance System • Joint Observatories Seeing Evaluation (programme) • Jose Observatory Site Evaluation

JOSO Joint Organization for Solar Observations

JOSS Joint Object Services Specification • Joint Office for Science Support • Joint Overseas Switchboard Switch • Joint Overseas Switching System

JOT Joint Observer Team • Joint Operational Test • Joliet (USA-IL) (location identifier)

JOT&E Joint Operational Test & Evaluation

JOTE Joint Operational Test and Evaluation

JOTA Journal of Optimization Theory and Applications (periodical code)

JOTS Joint Operational Telephone System

JOULE Joint Opportunities for Unconventional Long-term Energy • Joint Opportunities for Unconventional or Long-term Energy supply (*)

Jour Journal • Journalism • Journalist • Journey • Journeyman

Journ Journal • Journalism • Journalist • Journey • Journeyman

JOV Joint Orbital Verification • Joint Orbit Verification

JOVE Joint Venture

JOVIAL Joule's Own Version of the International Algebraic Language • Jule's Own Version of the International Algorithmic Language

JOWIP Joint Ocean Wave Investigation Project

JOY Job Opportunity for Youth

jp Japan

jP juristische Person

JP Jahresprämie • Japan (ISO country code) • Japon (code ISO de pays) • Jet Pilot • Jet Propellant • Jet-Propelled • Jet Propulsion • Joint Publication • Julian Period • Justice de Paix • Justice of the Peace

JPA Jamaica Press Association • Japan Partnering Award • Japan Patent Association • Japan Printers Association • Japan Procurement Agency • João Pessoa - Castro Pinto (Brazil-PB) (location identifier) • Job Package Authorization • Job Performance Aids • Johnstone Plasma Analyzer • Joint Participation Agreement • Joint Powers Authority

JPal Journal of Paleontology (periodical code)

JPALS Joint Precision Approach/Landing System

JPAM Joint Program Assessment Memorandum

JPAO Joint Public Affairs Office

JPApS Journal of Pure and Applied Sciences (periodical code)

JPAPS Job Performance Aids Production System

JPAR Joint Potential Assessment Report

JPAS Journal of Practical Applications in Space (periodical code)

JPASO Joint Pacific Area Scheduling Office

JPAT Joint Process Action Team

JPATS Joint Primary Aircraft Training System

JPB Jahrprüfungsbericht • Joint Planning Board • Joint Policy Board • Joint Procurement Board • Joint Program Board

JPBM Joint Policy Board for Mathematics

JPC Japan Patent Center • Jet Penetration Chamber • Joint Planning Committee • Joint Planning Council • Joint Power Conditioner • Joint Power Conditions • Joint Production Council • Joint Propulsion Conference • Journal of Physical Chemistry (periodical code)

JPCC Joint Propulsion Coordinating Committee

JPCG Joint Policy Coordinating Group

JPCM Journal of the Physics of Condensed Matter (periodical code)

JPCRD Journal of Physical and Chemical Reference Data (periodical code)

JPCS Journal of Physics and Chemistry of Solids (periodical code)

JPD Joint Planning Document • Joint Potential Designator • Joint Program Directive • Pasadena (USA-CA) (location identifier)

JPDC Journal of Parallel Distributed Computing (periodical code)

JPDRD Joint Programme Definition and Requirements Document

jpe JPEG Format (file extension)

JPE Journal of Political Economy (periodical code) • Jump if Parity Even • Pegasids (meteor stream)

JPEC Joint Planning and Execution Committee • Joint Planning and Execution Course

JPEDB Joint Planning and Execution Database

jpeg Joint Photographic Experts Group (file extension)

JPEG Joint Photographic Experts Group (*) • Joint Photographic-image Experts Group • Joint Picture Experts Group • Joint Planning and Execution Group

JPEP Joint Planning and Execution Process

JPER Journal of Physics Education Research

JPESC Joint Planning and Execution Steering Committee

JPet Journal of Petrology (periodical code)

J-PEX Joint Plasmadynamic Experiment

JPF Job Planning Form • Jovian Planet Finder

jpg Joint Photographic Experts Group (file extension)

JPG Joint Planning Group • Joint Planning Guidance

JPGM Joint Planning Graphics Module • J. Paul Getty Museum (USA-CA)

JPhA Journal of Physics A Mathematical & General Physics (periodical code)

JPhB Journal of Physics B Atomic & Molecular Physics (periodical code)

JPhC Journal of Physics C Solid-State Physics (periodical code)

JPhCh Journal of Physical Chemistry (periodical code)

JPhD Journal of Physics D Applied Physics (periodical code)

JPhE Journal of Physics E Scientific Instruments (periodical code)

JPhF Journal of Physics F Metal Physics (periodical code)

JPhG Journal of Physics G Nuclear Physics (periodical code)

JPhGS Journal of Physics G Nuclear Physics Supplements (periodical code)

JPhys Journal de Physique (periodical code)

JPI Jensen and Partners International • Joint Packaging Instrumentation

JPIC Joint Programme Integration Committee

JPITL Joint Prioritized Integrated Target List

JPJ Paterson (USA-NJ) (location identifier)

JPL Jet Propulsion Laboratory (USA-CA)

JPlPh Journal of Plasma Physics (periodical code)

JPM Job Performance Management • Job Performance Measure • Joint Program Management • Joint Program Manager

JPMO Joint Product Management Office • Joint Program Management Office • Joint Project Management Office

JPMR Joint Projected Manpower Requirements

Jpn Japan • Japanese

JPN Japan (ISO country code) • Japon (code ISO de pays) • Joint Planning Network • Washington Pentagon (USA-DC) (location identifier)

JPO Japanese Patent Office • Joint Program Office • Joint Program Officer • Joint Project Office • Joint Project Officer • Joint Prototyping Office • Joint Prototyping Officer • Journal of Physical Oceanography (periodical code) • Journée Portes Ouvertes • Jump if Parity Odd • Junior Professional Officer • Jupiter Polar Orbiter

JPOC Joint Planning Orientation Course • JSC Payload Operations Center

JPOES Joint Planning Operations and Execution System

JPOP Japanese Polar Orbiting Platform • Jupiter Polar Orbiter with Probes (mission)

JPoSc Journal of Polymer Science (periodical code)

JPOTF Joint Psychological Operations Task Force

JPOTG Joint Psychological Operations Task Group

JPP Joint Planning Process • Joint Plenary Presentation • Joint Program Plan • Journal of Propulsion Power (periodical code)

JPR Jiparaná (Brazil-RO) (location identifier) • Joint Program Review • Journal of Philosophical Research (periodical code)

JPRD Joint Program Requirements Document

JPRS Joint Publications Research Service • Journal of Photogrammetry and Remote Sensing (periodical code)

JPS Jan Paagman Sterrenwacht (Netherlands) • Japan Physical Society • Jet-Propulsion System • JICST Photoduplication Service • Joint Planning Staff • Joint Polar System • Journal of Power Sources (periodical code)

JPSG Japanese Phrase Structure Grammar

JPT Houston - Park Ten (USA-TX) (location identifier) • Jet Pipe Temperature • Journal of Propulsion Technology (periodical code)

JPTM Joint Procedures Training Manual

JPTO Jet-Propelled Takeoff

JPTS Jet Propellant Thermally Stable

JPU Job Processing Unit • Paris - La Défense (France) (location identifier)

JPVRC Japan Pressure Vessel Research Council

JPW Job Processing Word

JPWG Joint Project Working Group • Joint Protocol Working Group

JPY Yen (Japan) (ISO currency code)

JP2AB Joint Pollution Prevention Advisory Board

JP2PP Java Phase Two Proposal Preparing (tool)

JQ Job Questionnaire

JQE Jaque (Panama) (location identifier) • Job Queue Element • Job Queue Entry

JQG Joint Question Group

JQQ Just a Quick Question

JQR Job Qualification Requirement

JQS Job Qualification Standard

JQSRT Journal of Quantitative Spectroscopy and Radiative Transfer (periodical code)

JQT Joint Question Team • Journal of Quality Technology (periodical code)

Jr Jahr • Januar • Jour • Journal • Junior

JR Jahresrechnung • Joanneum Research • Job-Related • Job-Routed • Joint Resolution • Joint Review

JRA Joint Rear Area • New York - West 30th Street (USA-NY) (location identifier)

JRAC Joint Rear Area Coordinator • Journal of Radioanalytical Chemistry (periodical code)

JRADS Joint Resource Assessment Database System

JRAMS Joint Readiness Automated Management System

JRASA Journal of the Royal Aeronautical Society of London (periodical code)

JRASC Journal of the Royal Astronomical Society of Canada (periodical code)

J. RAS Canada Journal of the Royal Astronomical Society of Canada

JRASN Journal of the Royal Astronomical Society of New Zealand (periodical code)

JRATA Joint Research And Test Activity

JRB Joint Review Board • New York - Manhattan (USA-NY) (location identifier)

JRC Jet Reaction Control • Joint Reception Centre • Joint Reporting Centre • Joint Research Centre • Rochester (USA-MN) (location identifier)

JRCC Joint Reception Coordination Center • Joint Rescue Coordination Center

JRD Jardim • Justification Review Document • Riverside (USA-CA) (location identifier)

JRDC Japan Research Development Corporation

JRDOD Joint Research and Development Objectives Document

JRE Java Runtime Environment • Joint Readiness Exercise • New York - East 60th Street (USA-NY) (location identifier)

JReAt Journal de Recherches Atmosphériques (periodical code)

JRECIN Japan Research Career Information Network

JREF James Randi Educational Foundation (USA-FL)

JREI Joint Research Equipment Initiative (UK)

JRes Journal of Research (periodical code)

JResB Journal of Research B Mathematical Sciences (periodical code)

JRFL Joint Radio Frequency List • Joint Restricted Frequency List

JRH Jorhat (India) (location identifier)

JRheo Journal of Rheology (periodical code)

JRI Journaliste Reporter d'Images

JRIM Joint Requirements Integration Manager

JRK Arsuk (Greenland) (location identifier)

JRMB Joint Requirements and Management Board • Joint Resource Management Board • Joint Review Management Board

JRMET Joint Reliability and Maintainability Evaluation Team

JRMI Java Remote Method Invocation

JRMP Java Remote Method Protocol

JRMS Japanese Remote Manipulator System

JRN Juruena (Brazil-MT) (location identifier)

JRNBS Journal of Research of the National Bureau of Standards (periodical code)

Jrnl Journal

Jrnx Journaux

JRO Jicamarca Radio Observatory (Peru) • Kilimanjaro (Tanzania) (location identifier)

JROC Joint Required Operational Capability • Joint Requirements Oversight Committee • Joint Requirements Oversight Council

JROCA Joint Requirements Oversight Council Assessment

JROD Jet Remote Opening Device

JROP Joint Range Operating Procedure

JRoS Journal of Robotic Systems (periodical code)

JRP Joint Requirement Panel • Joint Requirement Planning

JRPA Joint Readiness, Planning and Assessment

JRPC Journal of Reinforced Plastics and Composites (periodical code)

JRPO Joint Research Project Office

JRRC Japanese Reprographic Rights Center

Jrs Jours

JRS Jerusalem (location identifier) • Joint Reporting Service • Joint Reporting Structure • Joint Reporting System

JRSp Journal of Raman Spectroscopy (periodical code)

JRSS Journal of the Royal Statistical Society (periodical code)

JRT Joint Rapporteur Team

JRTC Joint Readiness Training Center

JRWG Joint Requirements Working Group

JRX Joint Readiness Exercise

js Java Script (file extension)

J/S Jammer/Signal • Jamming/Signal

JS Java Script • Job Specification • Joint Services • Joint Staff • Jornada Semanal • Jump if Sign

JSA Jaisalmer (India) (location identifier) • Japanese Standards Association • Job Safety Analysis • Joint Security Area • Joint Service Agreement • Joint System Architecture

JSAI Japan Society for Artificial Intelligence

JSAM Joint Security Assistance Memorandum

JSAP Japan Society of Applied Physics • Joint Security Assistance Planning

JSAPI Java Speech Application Programming Interface

JSAPJ Japan Society of Air Pollution Journal (periodical code)

JSAPS Japanese Journal of Applied Physics - Supplement (periodical code)

JSAR Journal of Space Astronomy Research (periodical code)

JSARO Joint Service and Agency Representative Office

JSASJ Japan Society of Aeronautical Space Sciences Journal (periodical code)

JSASL Journal of Southern Academic and Special Librarianship

JSASS Japan Society for Aeronautical and Space Sciences

JSAST Japan Society of Aeronautical Space Sciences Transactions (periodical code)

JSAT Joint Security Assistance Training • Joint Structural Assessment Team • Joint System Acceptance Test • Joint System Acceptance Testing

JSBC Japan Satellite Broadcasting Co.

JSBD Japanese Standard Bibliographic Description

JSBSS Japan Society for Biological Sciences in Space

JSC Japanese Science Council • Japan Scientific Committee • Johnson Space Center (USA-TX) • Joint Scientific Committee • Joint Spectrum Center • Joint Steering Committee

JSCAS Johnson Space Center Astronomical Society (USA-TX)

JSCB Job Step Control Block

JSCC Joint Services Coordination Committee

JSCE Joint Services Communications Element

JSCM Japanese Science Council Manual

JSCMJ Japan Society of Composite Materials Journal (periodical code)

JSCNOET Joint Standing Committee on Nuclear and Other Energy Technologies

JSCom Journal of Scientific Computing (periodical code)

JSCP Joint Service Cost Position

JSD Jackson System Development • Job Status Document • Stratford Sikorsky (USA-CT) (location identifier)

JSDAP Jovian System Data Analysis Program • Jupiter System Data Analysis Program

JSDF Japanese Self-Defence Forces

JSDK Java Software Development Kit

JSDT Java Shared Data Toolkit

JSE Jamieson Research and Engineering (USA-CA) (now VRI) • Johannesburg Stock Exchange (South Africa) • Journal of Scientific Exploration • Journal of Statistics Education • Juneau (USA-AK) (location identifier)

JSEC Joint Scientific and Educational Center

JSEP Job Skills Education Program • Joint Services Electronics Program

JSET Jonction Supraconductrice à Effet Tunnel

Jsey Jersey (Channel Islands)

JSF Japan Science Foundation • Japan Space Forum • Joint Strike Fighter

JSG Joint Service Guidance • Joint Standardization Group • Journal of Structural Geology (periodical code) • San Rafael (USA-CA) (location identifier)

JSGA Japan Spaceguard Association

JSGCC Joint Service Guidance and Control Committee

JSGeo Journal of Series Geophysics (periodical code)

JSH Sitia (Greece) (location identifier)

JShR Journal of Ship Research (periodical code)

JSHS Joint Science and Humanities Symposium • Junior Science and Humanities Symposium

JSI Joint Service Interface • Jozef Stefan Institute (Slovenia) • Skiathos (Greece) (location identifier)

JSIA Japan Software Industry Association

JSIAM Japan Society for Industrial and Applied Mathematics

JSIC Joint Space Intelligence Center

JSIG Joint Service Integration Group

JSIK Japan Society of Information Knowledge

JSIMS Joint Simulations

JSIPS Joint Services Imagery Processing System

JSIR Joint Spectrum Interference Resolution

JSIS Japan Science Information Service

JSIT Journal of Systems and Information Technology

JSK Saint Cloud (USA-MN) (location identifier)

JSL Atlantic City - Steel Pier (USA-NJ) (location identifier) • Jet Select Logic • Job Specification Language

JSLA Japan Special Library Association

JSLC Jinquan Satellite Launch Facility (China-PRC) • Joint Standing Committee on Library Cooperation

JSLEJ Japan Society of Lubrication Engineers Journal (periodical code)

JSLR Journal of Soviet Laser Research (periodical code)

JSLS Japan Society of Library Science

JSLWG Joint Spacelab Working Group

JSM José de San Martin (Argentina-CB) (location identifier) • Journal of Structural Mechanics (periodical code)

JSMAJ Japan Society of Microgravity Applications Journal (periodical code)

JSMB Joint Space Management Board • Joint Surface Movements Board

JSME Japan Society of Mechanical Engineers

JSMEB Japan Society of Mechanical Engineers Bulletin (periodical code)

JSMET Japan Society of Mechanical Engineers Transactions (periodical code)

JSMG Joint Service Materiel Group

JSMS Joint Spectrum Management System

JSMSJ Japan Society of Materials Science Journal (periodical code)

JSN Junction Switch Number • Los Angeles - Sherman Oaks (USA-CA) (location identifier)

JSNPE Joint Strategic Nuclear Planning Element

JSO Joint Service Office • Joint Service Officer • Joint Specialty Office • Joint Specialty Officer • Joint Staff Office • Joint Staff Officer • Sodertalje (Sweden) (location identifier) • Södertälje (Sweden) (location identifier)

JSOA Joint Special Operations Agency • Joint Special Operations Area

JSOC Joint Science Operations Centre • Joint Service Operational Center • Joint Service Operational Concept • Joint Service Operations Command

JSOF Joint Special Operations Forces

JSOFI Joint Special Operations Forces Institute

JSOLE Joint Special Operations Liaison Element

JSOP Joint Strategic Objectives Plan

JSOR Joint Service Operational Requirement • Joint Statement Of Requirement • Joint Systems Operational Requirements

JSORD Joint Systems Operational Requirements Document

JSOW Joint Stand Off Weapon

JSP Cheju Sogwipo (Korea-ROK) (location identifier) • Jackson Structured Programming • Java Server Page • Joint Service Plan • Joint Strategic Planning • Joint Seismic Program • Joint Service Program • Joint Support Plan • Journal of Statistical Physics (periodical code)

JSPA Job Separator Page Area

JSPC Joint Seismic Program Center

JSPD Joint Strategic Planning Document

JSPO Joint Strategic Planning Office

JSPP Joint Service Program Plan

JSpRo Journal of Spacecraft and Rockets (periodical code)

JSPRS Japan Society of Photogrammetry and Remote Sensing

JSPS Japan Society for the Promotion of Science • Joint Strategic Planning System

JSpT Journal of Spacecraft Technology (periodical code)

JSR Jessore (Bangladesh) (location identifier) • Joint Strategic Review • Joint Status Review

JSRA Joint Sponsored Research Agreement

JSRB Joint Software Review Board

JSRC Joint Search and Rescue Center • Joint Services Review Committee

JSRCC Joint Search and Rescue Coordination Center

JSS Japan Satellite Systems (merger of JCSAT & SAJAC) • JavaScript Style Sheet • Joint Service Software • Joint Services Standard • Joint Shared Server • Joint Surveillance System • Spetsai Island (Greece) (location identifier)

JSSA Joint Subcommittee on Space Astronomy

JSSAG Joint Strategic Space Advisory Group

JSSC Joint Strategic Survey Committee

JSSCh Journal of Solid-State Chemistry (periodical code)

JSSE Java Secure Sockets Extension

JSSED Journal of Solid-State Electron Device (periodical code)

JSSEE Joint Service Software Engineering Environment

JSSMF Joint Summer Science and Mathematics Facility

JSSPP Job Scheduling Strategies for Parallel Processing (workshop)

JSST Joint Space Support Team

JST Japanese Standard Time • Japan Science and Technology (corporation) • Japan Society of Translators • Japan Standard Time • Job Site Training • Job Standard • Johnstown (USA-PA) (location identifier) • Joint Systems Test • Justice

JStA Journal of Strain Analysis (periodical code)

JSTA Joint Science and Technology Alliance

JSTARS Joint Surveillance Target Attack Radar System

JSTC Joint Scientific and Technical Committee

JSTF Joint Society Task Force

JStG Jahressteuergesetz

JSTP Joint System Test Program

JSTPS Joint Strategic Targeting Planning Staff

JSU Jackson State University (USA-MS) • Sukkertoppen (Greenland) (location identifier)

JSup Journal of Superconductivity (periodical code)

JSUS Joint Science Utilization Study

JSV Journal of Sound Vibration (periodical code)

JSWG Joint Science Working Group

JSWT Joint Science Working Team

JSY Syros Island (Greece) (location identifier)

JSZ Saint-Tropez (France) (location identifier)

Jt Jahrestonne • Jahrtausend • Joint

J-T Joule-Thompson (effect)

JT Jahn-Teller (effect) • Joint Test • Joule-Thompson (effect) • Journal Télévisé

J&TA Job & Task Analysis

JTA Japan Trademark Association • Java Transaction API • Job Table Area • Job and Task Analysis • Job Task Analysis • Joint Table of Allowances • Joint Technical Architecture • Joint Test Assembly • Joint Transfer Agreement

JTAB Joint Technical Advisory Board

JTAC Joint Tactical Area Communications • Joint Technical Advisory Committee

JTACS Japanese Total Access Communication System

JTAG Joint Task Action Group • Joint Test Action Group

JTAGG Joint Turbine Advanced Gas Generator

JTAPI Java Telephony Application Programming Interface

JTASC Joint Training, Analysis and Simulation Center

JTASG Joint Target Automation Steering Group

JTATE Journal of Technology and Teacher Education (AACE)

JTAV Joint Total Asset Visibility

JTAWG Joint Technical Architecture Working Group

JTB Joint Target Board • Joint Targeting Board • Joint Transportation Board

JTC Joint Tactical Centre • Joint Technical Committee • Joint Technology Center • Joint Trade Committee • Joint Training Center • Joint Training Configuration • Joint Transfer Correlation • Joint Transfer Correlator • Joint Tripartite Committee • Joule-Thomson Coefficient • Joule-Thomson Cooler

JTCB Joint Targeting Coordination Board

JTCCG Joint Technical Configuration Control Group

JTCG Joint Technical Coordinating Group • Joint Technical Coordination Group

JTC3A Joint Tactical Command, Control, and Communications Agency

JTD Job Task Data • Joint de Traversée Dalle • Joint Table of Distribution • Joint Technology Demonstration • Joint Test Director

JTDAS Joint Type Designation Automation System

JTDE Joint Technology Demonstrator Engine

JTDL Joint Tactical Data Link

JTDP Joint Training Development Plan

JT&E Joint Test & Evaluation

JTE Joint Test and Evaluation • Joint Test Element • Joule-Thomson Effect • Journal of Technology Education

JTech Journal of Technology (periodical code)

JTeEv Journal of Testing Evaluation (periodical code)

JTEMP Joint Test and Evaluation Master Plan

JTEP Joint Technology Exchange Program

JTePh Journal of Technical Physics (periodical code)

JTF Joint Task Force • Joint Test Facility • Joint Test Force • Joule-Thomson Flow

JTFA Joint Time-Frequency Analysis

JTFEX Joint Task Force Exercise

Jtg Jahrestag

JTG Joint Task Group • Joint Technical Group • Joint Training Group

JTGCM Jovian Thermospheric General Circulation Model

JThCM Journal of Thermoplastic Composite Materials (periodical code)

JThSt Journal of Thermal Stresses (periodical code)

JTHT Journal of Thermophysics and Heat Transfer (periodical code)

JTI Jatai (Brazil-GO) (location identifier) • Job Task Inventory

JTIDS Joint Tactical Information Distribution System (*) • Joint Tactical Integrated Display System • Joint Technical Information Distribution System

JTIS Japanese Technical Information Service

JTJ Josephson Tunnel Junction

JTL Joint Target List

Jtly Jointly

JTM Job Transfer and Management • Job Transfer and Manipulation • Job Transfer Manipulation • Joint Technical Manual • Joint Training Manual

JTMD Joint Terminology Master Database

JTMP Job Transfer and Manipulation Protocol • Joint Training Master Plan

JTMS Joint Technical Manual System • Joint Training Master Schedule • Justification-based Truth Maintenance System

JTN Joint Targeting Network

JTO Joint Technical Operations • Jump Takeoff • Thousand Oaks (USA-CA) (location identifier)

JTP Job Training Package • Joint Technical Panel • Joint Training Plan

JTPA Job Training Partnership Act (USA)

JTPC Joint Technical Planning Committee • Joint Technical Programming Committee

JTPO Joint Terminal Program Office • Joint Test Program Outline

JTPT Job Task Performance Test

JTPWG Joint Test Planning Working Group

JTR Joint Training Readiness • Joint Travel Regulation • Thira (Greece) (location identifier)

JTRB Joint Telecommunications Resources Board

JTS Java-based Tuplespace • Java Transaction Service • Job Training Standards • Joint Training System

JTSAST Joint Training System Automated Support Tools

JTSC Joint Test Support Center

Jtsd Jahrtausend

JTSSG Joint Telecommunications Standards Steering Group

JTSST Joint Training System Support Team

JTST Jet Stream • Journal of Thermal Spray Technology (periodical code)

JTSTR Jet Stream

JTT Joint Training Team

JTTAS Japan Technology Transfer Association

JTTP Joint Tactics, Techniques, and Procedures

JTTPRG Joint Tactics, Techniques, and Procedures Review Group

JttU Journey through the Universe (program)

JTU Jackson Turbidity Unit

JTWC Joint Typhoon Warning Center

JTX Joint Training Exercise

JTY Astypalaia Island (Greece) (location identifier)

Ju June

JU Joint Use • Junge Unternehmer

JUA Joint Usage Agreement • Juara (Brazil-MT) (location identifier)

Jub Jubilar • Jubiläum

JUB Juba (Sudan) (location identifier)

JUC Los Angeles - Universal City (USA-CA) (location identifier)

JUCS Journal of Universal Computer Science

Jud Judge • Judgment • Judicial • Judiciary

JUD Juris Utriusque Doctor

Judgt Judgment

JUDI Joint Universal Data Interpreter

Juev Jueves

JUF Joint Users File

JUG Java User Group • Joint User Group

JUGFET Junction-Gate Field-Effect Transistor

JUGL JANET User Group for Libraries

JUH Joint User Handbook

JUI Juist (Germany) (location identifier)

JUIC Joint Unit Identification Code

Juil Juillet

Juill Juillet

JUJ Jujuy - El Cadillal (Argentina-PJ) (location identifier)

Jul Julho • Julio • July

JÜL Jülich (Germany)

JUL Julho • Juliaca (Peru) (location identifier) • July

JUM Jumla (Nepal) (location identifier)

Jun June • Junio • Junior • Junta

JUN Jundah (Australia-QLD) (location identifier) • June • Junho

Junc Junction

JUNCO Japan University Consortium

JUNet Japanese University Network • Japanese Unix Network • Japanese UUCP Network

JUNET Japan Unix Network

JUNIC Joint United Nations Information Committee

JUNO Joint Ultraviolet Night-sky Observer

Junr Junior

JUO Jurado (Colombia) (location identifier)

Jup Jupiter

JUP Upland Cable (USA-CA) (location identifier)

JUR Jurien Bay (Australia-WA) (location identifier)

JURG Joint Users Requirements Group

Juris Jurisprudence

JURIS Justice Retrieval and Enquiry System

Jurisp Jurisprudence

JUS Japanese UNIX Society

JUSE Japanese Union of Scientists and Engineers

Just Justice • Justicia • Justification

JUST Japan Universal System Transport

JUSTIS Joint Uniform Service Technical Information System

JUSTSAP Japan - United States Science, Technology and Space Applications Program

JUT Juticalpa (Honduras) (location identifier)

JUTCPS Joint Uniform Telephone Communications Precedence System

Juv Juvenile

JUV Upernavik (Greenland) (location identifier)

Juw Juwelen • Juwelier

Jux Juxtaposition

JUZ Juzhou (China-PRC) (location identifier)

Jv Java • Javanese

JV Amalthea (satellite of Jupiter) • Jahresversammlung • Joint Venture • Joint Vision • Journal Voucher • Jugendverband • Jugendvertreter • Jugendvertretung • Jules Verne (1828-1905)

JVA Ankavandra (Madagascar) (location identifier) • Jahresverbrauchsabrechnung

JVAP Joint Venture Analysis Plan

JVAS Jodrell Bank - VLA Astrometric Survey

JVB Joint Visitors Bureau

JVCIR Journal of Visual Communication and Image Representation (periodical code)

JVD Joint Venture Directorate

JVDN Joint Voice and Data Network

JVE Joint Verification Experiment • Journal of Vibration Engineering (periodical code)

JVEL Joint Virtual Environment Laboratory

JVerz Jahresverzeichnis

JVF Joint Venture Funding

JVGR Journal of Volcanology and Geothermal Research (periodical code)

JVH Joint Vertical Handling

JVI Himalia (satellite of Jupiter) • Joint Visual Inspection • Manville Kupper (USA-NJ) (location identifier)

JVIA Joint Visual Information Activity

JVIDS Joint Visually Integrated Display System

JVII Elara (satellite of Jupiter)

JVIII Pasiphae (satellite of Jupiter)

JVIRD Joint Verification and Integration Requirements Document

JVIS Joint Visual Information Services

JVISDA Joint Visual Information Services Distribution Activity

JVJ Joint Virtual Journal

JVL Janesville (USA-WI) (location identifier) • Joint Virtual Laboratory

JVM Java Virtual Machine

JVMF Joint Variable Message Format

JvNC John von Neumann Center

JvNCC John von Neumann Computer Center

JVO Japanese Virtual Observatory (Japan)

JVOQL Japanese Virtual Observatory Query Language

JVP Joint Verification Plan

JvS Jongeren Vereniging voor Sterrenkunde (Belgium)

JVS Jeudi - Vendredi - Samedi

JVST Journal of Vacuum Science and Technology (periodical code)

JVT Joint Venture Team

jw Javanese (ISO language code)

JW Jahreswagen • Jewish

JWA Japan Weather Association • Jwaneng (Botswana) (location identifier)

JWARS Joint Warning And Reporting System

JWasA Journal of the Washington Academy of Sciences (periodical code)

JWC Joint Wet Case • Los Angeles - Warner Center Business Plaza (USA-CA) (location identifier)

JWF Job Work Folder

JWG Joint Working Group

JWH Houston - Westchase Hilton (USA-TX) (location identifier)

JWICS Joint Worldwide Information Communications System

JWL Houston Woodlawns (USA-TX) (location identifier)

Jwlr Jeweler

JWO Job Work Order • Jupiter's White Oval

JWS Java Workshop

JWST James Webb Space Telescope (formerly NGST)

JWT Job Wait Time

JX Lysithea (satellite of Jupiter)

JXCG Joint Exercise Control Group

JXI Carme (satellite of Jupiter)

JXII Ananke (satellite of Jupiter)

JXIII Leda (satellite of Jupiter)

JXIV Thebe (satellite of Jupiter)

JXN Jackson (USA-MI) (location identifier)

JXST Journal of X-ray Science and Technology (periodical code)

JXV Adrastea (satellite of Jupiter)

JXVI Metis (satellite of Jupiter)

Jy Jansky • July

JY Jersey (Channel Islands) • Jordan (international aircraft registration code)

JYP John Young Planetarium (USA-FL)

JYV Jyväskylä (Finland) (location identifier)

JZ Jahreszahl • Jodzahl • Joint Zone • Jump if Zero

J2 Djibouti (international aircraft registration code)

J2EE Java 2 Enterprise Edition

J2ME Java 2 Micro Editor

J3 Grenada (international aircraft registration code)

J5 Guinea Bissau (international aircraft registration code)

J6 Saint Lucia (international aircraft registration code)

J7 Dominica (international aircraft registration code)

J8 Saint Vincent and the Grenadines (international aircraft registration code)

K

k Absorption Coefficient • Boltzmann Constant • Capacity • Circular Wave Number • Circular Wave Vector • Cold • Constant • Coupling Coefficient • Curvature • Effective Thermal Conductivity • Extinction Coefficient • Gaussian Gravitational Constant • Grid Index • Heat Conductivity • Height Index • Kaiserlich • Kalt • Karat • Katholisch • Kill • Kilo • Kilogram • Kindergarten • King • Knight • Knot • Kopeck (currency) • Mass Transfer Coefficient • Miller Index • Molecular Conductivity • Multiplication Factor • Phase Factor • Propagation Vector • Radius of Gyration • Rate Coefficient • Rate Constant • Stiffness • Thermal Conductivity • Unit Coordinate Vector • von Kármán Constant

K Aberration Constant • Bulk Modulus • Cambodia (international vehicle registration code) • Capacity • Carat • Cathode • Compression Modulus • Contract • Cretaceous • Eddy Viscosity • Equilibrium Constant • Heat Transfer Coefficient • Inertia • Kaap • Kabel • Kaiser • Kalium • Kalorie • Kalt • Kälte • Kammer • Kanal • Kaon • Kap • Kapitel • Kapp • Karat • Kärnten (Austria) • Kasse • Kathode • Kaution • Kelvin • Kerma • Kette • Kill • Kilo (international alphabet) • Kilocycle • Kilogram • Kina (currency) • Kindergarten • Kinetic Energy • King • Kip (currency) • Kirche • Kiste • Knight • Knot • Köchel • Kommandant • Kommissar • König • Konsonant • Konsul • Konto • Konzentration • Koordinaten • Kopf • Korps • Kote • Krona • Krone (currency) • Kwacha (currency) • Kyat (currency) • Lift-Dependent Drag • Luminous Efficacy • Moment of Inertia • One Thousand • Permeability • Potassium • Quantum Number of Component of Angular Momentum along Axis • Rotational Quantum Number • Solar Absorption Index • Solar Constant • Spectral Type • Stiffness • Thousand • Zero Point

ka Georgian (ISO language code)

kA Kiloampere

Kä Kärnten (Austria)

Ka Auroral Absorption Index • Thousands of Years

KA Karelia • Kassenanwalt • Keep Alive • Kernel Algorithm • Kill Assessment • Kilo-Ampere • Kleine Anfrage • Klimaanlage • Knotenamt • Knowledge Acquisition • Kommunikationsanschluß • Kontoauszug • Kulturamt • Kunstakademie • Kurzarbeit

KAA Kasama (Zambia) (location identifier)

KAAS Kharkov Amateur Astronomers' Society (Ukraine)

KAAU King Abdul Aziz University (Saudi Arabia)

KAAUSO King Abdul Aziz University Solar Observatory (Saudi Arabia)

Kab Kabel • Kabine • Kabinett

KAB Kariba (Zimbabwe) (location identifier)

KABLE KSC Atmospheric Boundary Layer Experiment

KAC Kameshli (Syria) (location identifier) • Krasnoyarsk Astronomical Club (Russia)

KACST King Abdulaziz City for Science and Technology (Saudi Arabia)

Kad Kadenz • Kader • Kadett

KAD Kaduna (Nigeria) (location identifier)

KADC Korean Astronomical Data Center (Korea-ROK)

KaDeWe Kaufhaus des Westens (Berlin)

KADC Korean Aircraft Development Consortium

KADOC Katholiek Archief en Documentatiecentrum

KADS Knowledge Acquisition and Design Support • Knowledge Acquisition and Documentation System

KAE Kake (USA-AK) (location identifier) • Key Agile Encryption

KAERI Korean Atomic Energy Research Institute

KAF Karato (Papua New Guinea) (location identifier) • Karlskrona Astronomi Förening (Sweden) • Københavns Astronomiske Forening (Denmark) • Kost - Assekuranz - Fracht

KAFB Kirtland Air Force Base (USA-NM)

KAG Kangnung (Korea-ROK) (location identifier) • Kapitalanlagegesellschaft • Kreisarbeitsgemeinschaft • Künstlerische Arbeitsgruppe

Kah Kahoolawe (USA-HI)

KAH Melbourne City (Australia-VIC) (location identifier)

KAI Kaieteur (Guyana) (location identifier)

KAIS Korea Advanced Institute of Science (Korea-ROK)

KAIST Korea Advanced Institute of Science and Technology (Korea-ROK)

KAIT Katzmann Automatic Imaging Telescope (USA-CA)

Kaj Kajüte

KAJ Kajaani (Finland) (location identifier)

KAK Kar (Papua New Guinea) (location identifier) • Key - Auto - Key

kal Grammkalorie (now cal.)

Kal Kalendae • Kalendas • Kalender • Kaliber • Kilokalorie (now Kcal.)

KAL Kaltag (USA-AK) (location identifier) • Keep-Alive Loop • Key Assets List

Kalf Kalfactor

KALI Korea Academy of Library and Information Science (Korea-ROK)

Kam Kamerad

KAM Kamaran Island (Yemen) (location identifier) • Key Alive Memory • Kinematic Analysis Method • Knowledge Acquired by Media • Knowledge Acquisition Model • Knowledge Area Module • Kolmogorov - Arnold - Moser (theorem)

KAMA Korea Automobile Manufacturers Association (ROK)

KAMBIS KARMAC Modulair Bibliotheek Informatie Systeem

KaMI Kaons at the Main Injector

KAMLAND Kamioka Liquid (scintillator) Anti-Neutrino Detector

KAMS Kawasaki Automatic Machining System

KAMSC Kalamazoo Area Math and Science Center (USA-MI)

Kan Kanaal • Kanada • Kanadier • Kanal • Kamister • Kansas (USA)

KAN Kano (Nigeria) (location identifier) • Kansas (USA) • Knowledge And Naming • Kriminalaktennachweis

Kand Kandidat

KANDIDATS Kansas Digital Image Data System

Kans Kansas (USA)

KANSA Kansas Association for Networked Supercomputer Applications (USA-KS)

kant kantonal

Kant Kanton

KAO Kansas Astronomical Observers (USA-KS) • Korea Astronomy Observatory (Korea-ROK) • Kuiper Airborne Observatory (USA-CA) • Kuusamo (Finland) (location identifier)

KAON Kaon, Antiproton, Other hadron and Neutrino

KAOS Kansas Astrophotographers and Observers Society (USA-KS) • Kashima Automatic Operating System • Kiepenheuer Adaptive Optics System • Kiepenheuer-Institut Adaptives Optiksystem • KIS Adaptive Optics System • Kilo-Aperture Optical Spectrograph

Kap Kapazität • Kapelle • Kapillare • Kapital • Kapitän • Kapitel

KAP Kapanga (Zaire) (location identifier) • Kluwer Academic Publishers • Kommunikativer Arbeitsplatz

KapErG Kapitalerhöhungsgesetz

KAPIS Kluwer Academic Publishers Information Service

KAPL KSC Approved Parts List

KAPN KSC Automated Payloads Notice

KAPP Key Assets Protection Program

KAPPA Kernel Application Package

KAPPS KSC Automated Payloads Project Specification

KAPSE Kernel Ada Programming Support Environment

Kapt Kapitän

KAPTUR Knowledge Acquisition for Preservation of Tradeoffs and Underlying Rationales

KapVStG Kapitalverkehrssteuergesetz

KAQ Kamulai (Papua New Guinea) (location identifier)

Kar Karabiner • Karachi (Pakistan) • Karaffe • Karat • Karosse

KAR Kamarang (Guyana) (location identifier) • Karlstadt (Germany) • Kodak Automated Retrieval

KARA Karachentseva (catalog)

KARC Kansai Advanced Research Center (Japan)

Kard Kardinal • Kardiogramm

Karel Karelia • Karelian

KARI Korea Aerospace Research Institute (Korea-ROK) (*) • Korean Aerospace Research Institute (Korea-ROK)

KARMA Knowledge-based Augmented Reality for Maintenance Assistance

KARMEN Karlsruhe Rutherford Intermediate Energy Neutrino • Karlsruhe Rutherford Medium-Energy Neutrino (experiment) (*)

KARP Knowledge Acquisition, Representation and Processing

KARRN Kyushu Area Research Regional Network

KARS Kansas Applied Remote Sensing Program • Kilometer-square Area Radio Synthesis Telescope

KARST Kilometer-square Array Radio Synthesis Telescope

kart kartoniert

Kart Kartell • Kartoffel • Kartograph • Karton

KAS Kalamazoo Astronomical Society (USA-MI) • Kansas Academy of Sciences (USA-KS) • Karasburg (Namibia) (location identifier) • Kentucky Academy of Sciences (USA-KY) • Kern Astronomical Society (USA-CA) • Knowledge Availability System • Korean Astronomical Society (Korea-ROK)

KASC Knowledge Availability System Center

KASCADE Karlsruhe Shower Core and Array Detector

Kash Kashmir

KASL Kansas Association of School Librarians (USA-KS)

KASLIB Kanagawa Association of Special Libraries and Information Bureaus

Kass Kassation • Kasserolle • Kassette • Kassier

KASSPR Knowledge Augmented Severe Storms Predictor

Kat Katalog • Katalysator • Katamaran • Katastrophe • Kategorie

KAT Kaitaia (New Zealand) (location identifier) • Key Alias Table • Knowledge Acquisition Team • Koordinierungsausschuß Terminologie

KATE Knowledge-based Autonomous Test Engineer

Kath Kathedral

KATRIN Karlsruhe Tritium Neutrino (experiment)

KATS Kennedy Avionics Test Set • Korean Agency for Technology and Standards (Korea-ROK)

KAU Kauhava (Finland) (location identifier) • Keystone Adapter Unit • Kilo Accounting Units

KAUDYTE Knowledge Acquisition and Use in Dynamic Task Environments

Kaufh Kaufhalle • Kaufhaus

Kaufm Kaufmann

KAUS Knowledge Acquisition and Utilization System

KAV Kavanayen (Venezuela) (location identifier) • Kindergeldauszahlungs-Verordnung

KAVAS Knowledge Acquisition, Visualization and Assessment Study

KAvB Koninklijke Academie van Belgien (Belgium)

KAW Kawthaung (Myanmar) (location identifier) • Kinetic Alfvén Wave

KAWD Contract Award Date

KAX Kalbarri (Australia-WA) (location identifier)

KAY Wakaya Island (Fiji) (location identifier)

KAZ Kau (Indonesia) (location identifier) • Kazakhstan (ISO country code) • Kazayan (catalog)

Kazakh Kazakhstan

KAZMENST Committee for Standardization, Metrology and Certifications of Kazakhstan

KazOB Astronomicheskoi Observatorii Kazan Byulleten (periodical code)

kb Kilobar • Kilobit

K&B Kitchen & Bathroom

KB Kantonsbibliothek • Kartenbeilage • Kassenbericht • Kaushītaki Brāhmana • Key Based • Keyboard • Keyboard Button • Kilobyte • Knowledge Base • Knowledge-Based • Knowledge Building • Kommissionsberatung • Kommissionsbericht • Kongelige Bibliotek (Denmark) • Königliche Bibliothek • Koninklijk Besluit • Koninklijke Bibliotheek • Koninkrijk België • Kontaktbereich • Kostenbereich • Kriegsbeschädigte • Kuiper Belt • Kulturbund • Kungliga Biblioteket • Kurbestimmungen • Kursbuch • Kurzbericht

KBA Kabala (Sierra Leone) (location identifier) • Knowledge-Based Accounting • Knowledge-Based Analysis

KBAC Kennedy Booster Assembly Contractor

KBAD Knowledge-Based Analysis and Design

KBAIS Knowledge-Based Action Information System

kbar Kilobar

KBB Kirkimbie (Australia-NT) (location identifier) • Knowledge-Based Bibliography

KBBE Knowledge-Based Business Engineering

KBC Birch Creek (USA-AK) (location identifier) • Kapton Beta Cloth

KBCA Knowledge-Based Circuit Analysis

KBCD Knowledge-Based Circuit Design

KBCE Knowledge-Based Control Engineering

KBCID Knowledge-Based Concurrent Integral Design

Kbcm Kubikzentimeter

KBCS Knowledge-Based Computer System

KBCSD Knowledge-Based Control System Design

KBD Keyboard • Kimberley Downs (Australia-WA) (location identifier) • Knowledge-Based Database • Knowledge-Based Design • Knowledge-Based Detection • Knowledge-Based Development • Knowledge-Based Device • Knowledge-Based Diagnosis • Knowledge-Based Directory • Knowledge-Based Dispatching • Knowledge-Based Drafting

KBDA Knowledge-Based Design and Analysis • Knowledge-Based Distribution and Assignment

KBDAT Knowledge-Based Design And Test

KBDB Knowledge-Based Database

KBDBMS Knowledge-Based Database Management System

KBDCE Knowledge-Based Distributed Computing Environment

KBDD Knowledge-Based Design and Drafting • Knowledge-Based Drafting and Design

KBDDS Knowledge-Based Design Drafting System

KBDE Knowledge-Based Design and Evaluation • Knowledge-Based Design Engineering • Knowledge-Based Document Engineering

KBDET Knowledge-Based Design and Electrical Test • Knowledge-Based Design Experiment Translator

KBDIC Knowledge-Based Design of Electrical Circuits

KBDIS Knowledge-Based Design Information System

Kbdm Kubikdezimeter

KBDMP Knowledge-Based Data Management Procedure

KBDMS Knowledge-Based Database Management System

KBDS Knowledge-Based Design System • Knowledge-Based Dispatching System

KBDSS Knowledge-Based Decision Support System

KBDT Knowledge-Based Doctrine Tool

KBE Bell Island - Hot Springs (USA-AK) (location identifier) • Keyboard Encoder • Keyboard Entry • Knight Commander of the British Empire • Knowledge-Based Education • Knowledge-Based Engineering • Knowledge-Based Entry • Knowledge-Based Environment • Knowledge-Based Estimating

KBEDE Knowledge-Based Engineering and Design for Electronics

KBEG Knowledge-Based Engineering Graphics

KBES Knowledge-Based Execution System • Knowledge-Based Expert System (*)

KBET Knowledge-Based Electrical Test

KBF Karubaga (Indonesia) (location identifier) • Koordinationsstelle für Forschungsprojekten

KBFE Knowledge-Based Film Editor • Knowledge-Based Front End

KBFF Knowledge-Based Feature Finder

KBFIT Knowledge-Based Fault Isolation Test

KBFS Knowledge-Based File System

KBG Kabalega Falls (Uganda) (location identifier) • Knowledge-Based Gauging • Knowledge-Based Graphics

KBGD Knowledge-Based Geometric Design

KBGIS Knowledge-Based Geographical Information System

KBH Kalat (Pakistan) (location identifier) • Kerr Black Hole

KBHD Knowledge-Based Hypermedia Development

KBHDE Knowledge-Based Hypermedia Development Environment

KBHDM Knowledge-Based Hypermedia Design Method

Kbhvn København (Denmark)

KBI Knowledge-Based Instruction • Kreisbrandinspektor • Kribi (Cameroon) (location identifier)

KBIC Key Base In Context

KBID Knowledge-Based Imaging Device

KBIM Keyboard Interface Module

KBIN Koninklijk Belgisch Instituut voor Natuurwetenschappen (Belgium)

Kbit Kilobit

KBJ Kings Canyon (Australia-NT) (location identifier)

KBK Klag Bay (USA-AK) (location identifier)

KBL Kabul (Afghanistan) (location identifier) • Knowledge-Based Language • Knowledge-Based Learning • Knowledge-Based Library • Knowledge-Based Listing

KBLA Knowledge-Based Loads Analysis

KBLCE Knowledge-Based Life Cycle Analysis

Kblg Kabellänge

KBLL Knowledge-Based Learning Analysis

KBLPS Knowledge-Based Logistics Planning Shell

KBLS Knowledge-Based Logistics and Support • Knowledge-Based Logistics Support • Knowledge-Based Logistics System

Kbm Kubikmeter

KBM Kabwum (Papua New Guinea) (location identifier) • Keyboard Monitor • Kinetic Ballooning Mode • Knowledge-Based Maintenance • Knowledge-Based Management • Knowledge-Based Manufacture • Knowledge-Based Manufacturing • Knowledge-Based Mapping

KBMAC Knowledge-Based Measurement And Control

KBMAT Knowledge-Based Design, Manufacture, And Test • Knowledge-Based Design, Manufacturing, And Testing

KBMCS Knowledge-Based Maintenance and Control System

KBMD Knowledge-Based Mechanical Drawing

KBMM Knowledge-Based Maintenance Management • Knowledge-Based Manufacturing Management

KBMP Knowledge-Based Management Planning

KBMS Knowledge-Based Management System • Knowledge-Based Manipulation System • Knowledge Base Management System

KBMU Knowledge-Based Makeup

KBN Kabinda (Zaire) (location identifier) • Knowledge-Based Navigation • Knowledge-Based Navigator • Knowledge-Based Numerics

KBO Kabalo (Zaire) (location identifier) • Kuiper Belt Object

KBOS Knowledge-Based Observation System • Knowledge-Based Operating System

KBP Keyboard Process • Kiev Borispol (Ukraine) (location identifier) • Knowledge-Based Pattern • Knowledge-Based Planning • Knowledge-Based Printing • Knowledge-Based Production • Knowledge-Based Programming • Knowledge-Based Publishing

kBpI Kilobits per Inch

KBPL Knowledge-Based Programming Language

KBPP Knowledge-Based Process Planning • Knowledge-Based Production Planning

kbps Kilobits per Second

Kbps Kilobits per Second

KBps Kilobytes Per Second

KBPS Kilobytes Per Second • Knowledge-Based Programming and Systems • Knowledge-Based Programming System

KBQA Knowledge-Based Quality Assurance

KBR Knowledge-Based Retrieval • Knowledge Base Repository • Kota Bharu (Malaysia) (location identifier)

KBRA Knowledge-Based Research Analysis

KBRD Knowledge-Based Research and Development

KBRIS Knowledge-Based Resource Information System

KBRP Knowledge Base Research Program

KBRS Knowledge-Based Reference Service • Knowledge-Based Routing System

kb/s Kilobits per Second

KBS Bo (Sierra Leone) (location identifier) • Ka-Band System • Knowledge-Based Software • Knowledge-Based Styling • Knowledge-Based System • Knowledge Base of Stars • Knowledge-Base Server • Korean Bibliographical Society (Korea-ROK) • Korean Bureau of Standards (Korea-ROK)

KBSA Kenya Booksellers ans Stationers Association • Knowledge-Based Software Assistant • Knowledge-Based System Analysis

KBSD Knowledge-Based Software Design • Knowledge-Based Structural Design • Knowledge-Based System Design

KBSE Knowledge-Based Support Equipment • Knowledge-Based Systems Engineering • Knowledge-Based Systems Evaluation

KBSF Knowledge-Based Sensor Fusion

KBST Knowledge-Based Software Testing • Knowledge-Based Structural Technology

KBT Kaben (Marshall Islands) (location identifier) • Knowledge-Based Technology • Knowledge-Based Tomography • Knowledge-Based Trading • Knowledge-Based Transcription • Knowledge-Based Typesetting

KBTE Knowledge-Based Test Equipment

KBTM Knowledge-Based Testing and Maintenance

KBTP Knowledge-Based Typesetting Process

KBTS Knowledge-Based Telemetry System • Knowledge-Based Test Shop • Knowledge-Based Test System • Knowledge-Based Trading System • Knowledge-Based Troubleshooting

KBU Keyboard Unit • Kotabaru (Indonesia) (location identifier)

KBUI Knowledge-Based User Interface

kbV keine besonderen Vorkommnisse

KBV Kleinbetragsverordnung • Knowledge-Based Verification • Knowledge-Based View

KBW Chignik Bay (USA-AK) (location identifier) • Knowledge-Based Warfare

KBX Kambuaya (Indonesia) (location identifier)

KBY Streaky Bay (Australia-SA) (location identifier)

Kbyte Kilobyte

KBZ Kaiboura (New Zealand) (location identifier)

kc Kilocurie • Kilocycle

KC Kansas City (USA) • Key Change • King's Counsel • Kyle Classification

KCA Ku-band Communications Adapter • Kuqa (China-PRC) (location identifier)

kcal Große Kalorie • Kilocalorie • Kylocycle

KCAO Kumamoto Civil Astronomical Observatory (Japan)

KCAS Knots Calibrated Airspeed

KCC Coffman Cove (USA-AK) (location identifier) • Keyboard Common Contact • Knowledge Consistency Checker

KCDC Korea Commercial Aircraft Development Consortium

KCE Collinsville (Australia-QLD) (location identifier)

KCG Chignik Fisheries (USA-AK) (location identifier) • Kappa Cygnids (meteor stream)

KCH Kuching (Malaysia) (location identifier)

kCh/s Kilochannels per Second

kCi Kilocurie

KCI Kenya Computer Institute • Kon (Indonesia) (location identifier)

KCK Kansas City - Fairfax (USA-KS) (location identifier)

KCl Potassium Chloride

KCL Chignik Lagoon (USA-AK) (location identifier) • Kirchoff Current Law • Kyoto Common LISP

KCLN Kansas City Library Network

KCMS Kodak Color Management System

KCMSD Kansas City Missouri School District (USA-MO)

KCN Chernofski (USA-AK) (location identifier)

KCNA Korean Central News Agency (Korea-ROK)

KCO Kappler Crystal Optics (USA-MA) • Keystone College Observatory (USA-PA)

KCP Kamenets-Podolskiy (Ukraine) (location identifier) • Kheda Communications Project

KCPG Karachentsev Catalog of (isolated) Pairs of Galaxies

KCPT Kinematic Carrier Phase Tracking

KCQ Chignik Lake (USA-AK) (location identifier)

KCR Colorado Creek (USA-AK) (location identifier) • KSC Change Request

KCRT Keyboard Cathode Ray Tube

kc/s Kilocycles per Second

kcs Kilocycles per Second

Kčs Koruna (currency)

KCS Keyboard Configuration Studies • Key Configuration Study • Kings Creek Station (Australia-NT) (location identifier) • Korean Chemical Society (Korea-ROK)

KCSC Kansas Cosmosphere and Space Center (USA-KS)

kc/sec Kilocycles per Second

KCU Masindi (Uganda) (location identifier)

KCV KKV Carrier Vehicle

KCZ Kochi (Japan) (location identifier)

KD Keyboard Display • Key Distribution • Knock Down • Knocked Down • Known Distance • Kongeriget Danmark • Kontrolldienst • Kreisdirektor • Kundendienst • Kursor Down • Kuwaiti Dinar (currency)

KDA Kolda (Senegal) (location identifier)

KDAAS Kettering and District Amateur Astronomers Society (UK) (now NAA)

KDAD Knowledge Discovery from Advanced Databases

KDB Kambalda (Australia-WA) (location identifier) • Kernel Debugger • Kinematically Decoupled Bulge • Koordinaten-Datenbanksystem • Küche - Diele - Bad

KDC Kandi (Benin) (location identifier) • Key Distribution Center • Kinematically Decoupled Core • Kinematically Distinct Core • Knocked-Down Condition • Korean Decimal Classification

KDCI Key Display Call Indicator

KDD Key Distribution Device • Khuzdar (Pakistan) (location identifier) • Knowledge Discovery in Data (conference) • Knowledge Discovery in Databases • Kokusai Denshin Denwa Co.

Kde Kunde

KDE K Desktop Environment • Kernel Density Estimator • Koroba (Papua New Guinea) (location identifier)

KdF Kraft durch Freude

KDF Keck Deep Field • Key Distribution Facility • Knocked Down Flat • Kouba (Algeria) (location identifier)

Kdg Kündigung

KDG Kardjali (Bulgaria) (location identifier)

Kdgg Kündigung

KDH Kandahar (Afghanistan) (location identifier) • Key Distribution Host

KDI Kendari (Indonesia) (location identifier) • Knowledge and Distributed Intelligence • Korea Development Institute (Korea-ROK)

KDJ N'Djole (Gambia) (location identifier)

KDK Kodiak (USA-AK) (location identifier)

Kdm Kingdom

KDM Kaadedhoo (Maldives) (location identifier) • KDE Desktop Manager • Kongelige Danske Marine (Denmark)

KDMS Kennedy Data Management System

KDN Kinetically Designed Nozzle • N'Dende (Gabon) (location identifier)

Kdo Kommando

KDO Kadhdhoo (Maldives) (location identifier)

KDP Kandep (Papua New Guinea) (location identifier) • Potassium Dihydrogen Phosphate

KDPN KSC DOD Payloads Notice

KDPSN KSC DOD Payloads Project Specification

Kdr Kalender • Kinder • Kommandeur

KDR Kandrian (Papua New Guinea) (location identifier) • Keyboard Data Recorder

Kdre Kommodore

KDS Kamaran Downs (Australia-QLD) (location identifier) • Kathode Dark Space • Key Display System • Knowledge Data System • Knowledge Dictionary System • Knowledge Discovery Service • KSC-Developed Software

KDSI Thousand of Delivered Source Instructions

Kdt Kommandant

KDT Keyboard and Display Test • Keyboard Display Terminal • Key Definition Table • Kinetic Desintegration Technology • Knowledge Discovery from Text • Knowledge Discovery Tool

KDU Keyboard and Display Unit • Keyboard Display Unit • Skardu (Pakistan) (location identifier)

KDV Kandavu (Fiji) (location identifier) • KKV Deployment Vehicle

KDVG Kriegsdienstverweigerungsgesetz

KDVNG Kriegsdienstverweigerungsneuordnungsgesetz

KDVZ Kommunale Datenverarbeitungszentrale

KDWG Karachentsev Dwarf Galaxy (catalog)

Ke Kessel • Kette

KE Karteneinheit • Kenya (ISO country code) • Keypad End • Kinetic Energy • Kinetic Equation • Knowledge Engineering • Knowledge Exchange

KEA Keisah (Indonesia) (location identifier) • Key Exchange Algorithm

KEALSE Key-search Algorithm Selection

KEAS Kanzelhöhe-Elektronische-Arkiv-System • Knots Equivalent Airspeed • Knots Estimated Airspeed

KEB English Bay (USA-AK) (location identifier)

KEBS Kenya Bureau of Standards

KEC Kasenga (Zaire) (location identifier) • Key Encryption Component

KECNET Kentucky Educational Computer Network

KECU Kilo European Currency Unit • Thousand European Currency Units

KED Kaedi (Mauritania) (location identifier) • Kinetic Energy Dominated

KEDDS Kansas Educational Dissemination/Diffusion System

KEE Keele (Starlink internal code) • Kelle (Congo) (location identifier) • Knowledge Engineering Environment

KEEP Kestrel Educational Exchange Program

KEEPS Kodak Ektaprint Electronic Publishing System

KEF Reykjavík Keflavík (Iceland) (location identifier)

KEFIR Key Findings Reporter

KEG Keglsugl (Papua New Guinea) (location identifier) • Knowledge Engineering Group • Kommission der Europäischen Gemeinschaft

KEGS KAO Echelle Grating Spectrometer • Kuiper Echelle Grating Spectrometer

KEH Kenmore (USA-WA) (location identifier)

KEI Kepi (Indonesia) (location identifier) • Key Enabling Investment

KEIS Kentucky Economic Information System

KEJ Kemerovo (Russia) (location identifier)

KEK Ekwok (USA-AK) (location identifier) • Key Encryption Key • Kommission fßr Erschliessung and Katalogmanagement • Konferenz der Europäischen Kirchen • Koo Energy Ken

KEL Kehl (Germany) • Kiel (Germany) (location identifier)

KELOG Kiso Emission-Line Objects and Galaxies

KELV KKV/ESM Launch Vehicle

KEM Kemi Tornio (Finland) (location identifier) • Kemnath (Germany)

KEMAR Knowles Electronics Manakin for Acoustic Research

Ken Kensington • Kentucky (USA) • Kenya

KEN Kenema (Sierra Leone) (location identifier) • Kent (Starlink internal code) • Kenya (ISO country code)

Kennz Kennzahl • Kennzeichen • Kennziffer

KENVAD Kent (Starlink internal code)

KEO Odienne (Ivory Coast) (location identifier)

KEOF Kalman Earth Orientation Filter

Kep Kepulauan

KEP Nepalganj (Nepal) (location identifier)

KEQ Kebar (Indonesia) (location identifier)

KER Kerman (Iran) (location identifier) • Kinetic Energy Release

KERA Kentucky Education Reform Act (USA-KY)

KERMIT Kl-10 Error-free Reciprocal Micro Interconnect Over Tty lines

KernR Kernforschungsanlage Report Results Solid-State Nuclear Physics (periodical code)

KES Kesley (Canada) (location identifier) • Key Escrow System • Knowledge-based Engineering System • Knowledge Engineering Software • Knowledge Engineering System • Shilling (Kenya) (ISO currency code)

KET Kentucky Educational Television (USA-KY) • Kiel Electron Telescope • Keng Tung (Myanmar) (location identifier)

keV Kiloelectronvolt

KEV Kinetic-Energy Vehicle • Kuorevesi Halli (Finland) (location identifier)

KEX Kanabea (Papua New Guinea) (location identifier)

KexT Kexue Tongbao (periodical code)

KEY Keep Extending Yourself • Kericho (Kenya) (location identifier)

KEYB Keyboard

KEYBBE Keyboard for Belgium

KEYBBR Keyboard for Brazil

KEYBCF Keyboard for Canada-France

KEYBCZ Keyboard for Czech Republic

KEYBDK Keyboard for Denmark

KEYBFR Keyboard for France

KEYBGR Keyboard for Germany

KEYBHU Keyboard for Hungary

KEYBIT Keyboard for Italy

KEYBLA Keyboard for Latin America

KEYBNL Keyboard for Netherlands

KEYBNO Keyboard for Norway

KEYBPL Keyboard for Poland

KEYBPO Keyboard for Portugal

KEYBSF Keyboard for Switzerland-France

KEYBSG Keyboard for Switzerland-Germany

KEYBSL Keyboard for Slovakia

KEYBSP Keyboard for Spain

KEYBSU Keyboard for Finland

KEYBSV Keyboard for Sweden

KEYBUK Keyboard for United Kingdom

KEYBUS Keyboard for United States

KEYBYU Keyboard for Yugoslavia

kF Kilofrancs

Kf Kauf • Kurzfassung

KF Kalman Filter • Kaltfront • Kemisk Forening (Denmark) • Kilo Francs • Kleine Fahrt • Knowledge Finder • Kümmerly and Frey AG

KfA Kommission für Astronomie

KFA Kernforschungsanlage • Kiffa (Mauritania) (location identifier) • Kosten - Fracht - Assekuranz

KFAS Keyed File Access System • Kuwait Foundation for the Advancement of Science

KFDD Kronecker Functional Decision Diagram

Kffr Kauffrau

KFG Kalkurang (Australia-NT) (location identifier)

Kfh Kaufhaus

KfH Kammer für Handelssachen

KFIT Kinetic Flight Integration Test

KFK Kernforschungszentrum Karlsruhe (Germany)

KFL Key Facilities List • Kinderferienlager

Kfm Kaufmann • Kaufmännisch

KFNT Kinematika i Fisika Nebesbykh Tel. (periodical code)

KFP False Pass (USA-AK) (location identifier) • Konferenz der Fachbereiche Physik (Germany)

KFS Kabelfernsehen

KFT Kilofeet

Kfz Kraftfahrzeug • Kubisch-flächenzentrierte

KfZ Kraftfahrzeug

kg Kilogram

kG Kilogauss

Kg Kampong • Kegel • Kenntnis genommen • König

KG Kammergericht • Kantonsgericht • Kassationsgericht • Kaufmannsgericht • Kellergeschoß • Key Generator • Keying Generator • Kilogram • Kindergarten • Klein-Gordon (equation) • Kommanditgesellschaft • Korte Golf • Kreisgericht • Kyrgyzstan (ISO country code)

KGA Kananga (Zaire) (location identifier)

KGaA Kommanditgesellschaft auf Aktien

KGB Komitet Gosudarstvennoi Bezopasnosti (USSR) (obsolete) • Konge (Papua New Guinea) (location identifier)

KGC Kingscote (Australia-SA) (location identifier) • Kurt Gödel Colloquium

KGD Kaliningrad (Russia) (location identifier) • Known Good Die

Kgdm Kingdom

KGer Kammergericht

kgf Kilogram-Force

Kgf Kriegsgefangene

KGF Karaganda (Kazakhstan) (location identifier) • Konstruktionsgrundfläche

KGG Kedougou (Senegal) (location identifier)

KGH Yongai (Papua New Guinea) (location identifier)

KGI Kalgoorlie (Australia-WA) (location identifier)

KGJ Karonga (Malawi) (location identifier)

KGK New Koliganek (USA-AK) (location identifier)

kgl königlich

KGL Kigali (Rwanda) (location identifier)

kgm Kilogram-Meter

KGM Known Good Module • Kungum (Papua New Guinea) (location identifier)

Kgn Königin

KGN Kasongo Lunda (Zaire) (location identifier)

KGO Kirovograd (Ukraine) (location identifier)

kgps Kilograms per Second

KGPS Kilograms Per Second • Kinematic Global Positioning System

Kgr Kongress • Königreich

KGR Kammergerichtsrat • Kulgera (Australia-NT) (location identifier)

KGRMHD Kerr General Relativistic Magnetohydrodynamics (code)

kgs Kilograms

KGS Kings • Known Geological Structure • Kos (Greece) (location identifier) • Kurt Gödel Society (Austria) • Som (Kyrgyzstan) (ISO currency code)

KGU Keningau (Malaysia) (location identifier)

KGUG Kontoguthabenumstellungsgesetz

KGV K Giant Variable (star) • Kleinste Gemene Veelvoud

KGW Kagi (Papua New Guinea) (location identifier)

KGX Grayling (USA-AK) (location identifier)

KGY Kingaroy (Australia-QLD) (location identifier)

KGZ Glacier Creek (USA-AK) (location identifier) • Kyrgyzstan (ISO country code)

Kh Krankenhaus

K-H Kelvin-Helmholtz (instability)

KH Cambodge (code ISO de pays) • Cambodia (ISO country code) • Kelvin-Helmholtz (instability) • Keyhole • Khmer Republic (ISO country code) • Kinderheim • Know How • Kulturhaus

KHA Khaneh (Iran) (location identifier)

KharI Kharkov Izdanie IRE AN USSR (periodical code)

KHB KSC Handbook

KHBV Krankenhaus-Buchführungsverordnung

KHC Kerch (Ukraine) (location identifier)

KHD Katastrophenhilfsdienst • Klöckner-Humboldt-Deutz AG (Germany)

Khdw Kunsthandwerk

KHE Kherson (Ukraine) (location identifier)

Khf Kirchhof

KHG Kashi (China-PRC) (location identifier)

KHH Kaohsiung (China-ROC) (location identifier)

KHI Karachi - Quaid-e-Azam (Pakistan) (location identifier) • Kawasaki Heavy Industries • Kelvin-Helmholtz Instability

KHIL KKV Hardware In the Loop

KHILS KKV Hardware In the Loop Simulator

KHIT KKV Hovering Interceptor Test

KHJ Kauhajoki (Finland) (location identifier)

KHK Khark Island (Iran) (location identifier)

KHL Khulna (Bangladesh) (location identifier)

KHM Cambodge (code ISO de pays) • Cambodia (ISO country code) • Khamti (Myanmar) (location identifier)

KHN Nanchang (China-PRC) (location identifier)

KHO Khoka Moya (South Africa) (location identifier)

KhPl Khimiia Plazmy (periodical code)

KHR Kharkhorin (Mongolia) (location identifier) • Riel (Cambodia) (ISO currency code)

KHS Khasab (Oman) (location identifier) • Knight of the Holy Sepulchre

KHT Khost (Afghanistan) (location identifier)

KHTT Know-How Transfer and Training

KHU Konferenz der Hessischen Universitäten • Kremenchug (Ukraine) (location identifier) • Kyung Hee University (ROK)

KHV Khabarovsk (Russia) (location identifier)

KHW Khwai River Lodge (Botswana) (location identifier)

kHz Kilohertz

Ki Kilo • Kind • Kinder • Kirche

KI Keck Interferometer • Kinetic Inhibition • Kiribati (ISO country code) • Kiwanis International • Knowledge Information • Known Icing • Kosmicheskie Issledovaniia • Kreisinspektor • Künstliche Intelligenz

KIA Kaiapit (Papua New Guinea) (location identifier) • Keck Interferometric Array

KIAM Keldysh Institute of Applied Mathematics (Russia)

KIAS Knots Indicated Airspeed • Korea Institute for Advanced Studies (Korea-ROK)

KIB Ivanof Bay (USA-AK) (location identifier) • Kiso Information Bulletin

KIC King City - Mesa del Rey (USA-CA) (location identifier)

KICH Kyoto International Conference Hall (Japan)

KID Kristianstad (Sweden) (location identifier)

KIDS Kestrel Interactive Development System • Knowledge-Based Integrated Design System

KIDU Knowledge - Information - Data - Utilities

KIE Kieta (Papua New Guinea) (location identifier) • Knowledge Integration Environment

KiePr Kievskii Ordena Lenina Gosudarstvennyi Universitet Shevchenko Astronomicheskaia Observatoriia - Preprints (periodical code)

KIER Korea Institute of Energy and Resources (Korea-ROK)

KIET Korea Institution for Economics and Technology (Korea-ROK)

KIF Kingfisher Lake (Canada-ON) (location identifier) • Knowledge Interchange Format

KIFU Koordinationsstelle Für Umweltforschung

KIG Koingnaas (South Africa) (location identifier)

KIH Kish Island (Iran) (location identifier)

KIIA Keck Imaging Interferometric Array

KiIND Kiev Izdatel Naukova Dumka (periodical code)

KiInF Kiev Institut Fiziki AN USSR (periodical code)

KiInM Kiev Institut Matematiki AN USSR (periodical code)

KiIzT Kiev Izdatel Tekhnika (periodical code)

KIJ Niigata (Japan) (location identifier)

KIK Kirkuk (Iraq) (location identifier)

Kil Kilometer

KIL Kilwa (Zaire) (location identifier)

Kild Kildare (Ireland)

Kilk Kilkenny

KIM Kalev Iteration Method • Keyboard Input Matrix • Kharkov Institute of Metrology (Ukraine) • Kimberley (South Africa) (location identifier) • Kinetic Isotope Method • Knowledge-based Information Management

KIMMJ Korean Institute of Metals and Materials Journal (periodical code)

KIMS Kodak Image Management System • Korean Invisible Mass Search • KSC Inventory Management System

KIN Kingston - Norman Manley (Jamaica) (location identifier)

Kind Kindergarten

Kinderg Kindergarten

King Kingdom

Kingd Kingdom

KINITI Korea Institute of Industry and Technology Information

KIO Key Interoperability • Keyword Interface Object • Kili Island (Marshall Islands) (location identifier)

KIP Keyboard Input Processor • Key Information Provider • Key Intelligence Position • Knowledge Information Processing • Thousand Pounds • Wichita Falls - Kickapoo (USA-TX) (location identifier)

KIPME Kiev Institut Problem Modelirovaniia Energetike AN USSR (periodical code)

KIPO Keyboard Input Printout • Korea Industrial Property Office (Korea-ROK)

KIpS Thousands of Instructions per Second

KIPS KAIST Image Processing System • Knowledge Information Processing System • Thousands of Instructions Per Second

KIQ Kira (Papua New Guinea) (location identifier)

KIR Kerry County (Ireland) (location identifier) • Kiribati (ISO country code) • Kyocera Image Refinement

KIRDI Kenya Industrial Research and Development Institute

KIS Keyboard Input Simulation • Kiepenheuer-Institut für Sonnenphysik (Germany) • Kisumu (Kenya) (location identifier) • Kitting Instruction Sheet • Knowbot Information Service • Knowledge-based Information System • Kodak Infrared Scope • Ku-band Interface System

KISMIF Keep It Simple, Make It Fun

KISR Kuwait Institute for Scientific Research

KISS Keep It Short and Simple • Keep It Simple, Sir • Keep It Simple, Stupid • Keep It Straight and Simple • Keyed Indexed Sequential Search • Kinematic Surveying System • Knowledge-based Interactive Signal-monitoring System • Korea Information Science Society • KPNO International Spectroscopic Survey • Kurs Information Service System

KIST Korea Institute of Science and Technology (Korea-ROK)

KISTI Korean Institute of Science and Technology Information (Korea-ROK)

KIT Kernel for Intelligent Terminals • Key Issue Tracking • Kithira (Greece) (location identifier) • KWIC Interactive Tracker

KITE Kuiper Infrared Technology Experiment

KITP Kavli Institute for Theoretical Physics (USA-CA)

KITSAT Korea Institute of Technology Satellite

KIU Kiunga (Kenya) (location identifier)

KIUSN KSC Inertial Upper Stage Notice

KIUPSN KSC Inertial Upper Stage Project Specification

KIV Kishinev (Moldova) (location identifier)

KIW Kitwe (Zambia) (location identifier)

KIX Osaka Kansai (Japan) (location identifier)

KIY Kilwa (Tanzania) (location identifier)

KIZ Karlsruhe Isochron Zyklotron (Germany) • Kikinonda (Papua New Guinea) (location identifier)

KIzIM Kiev Izdanie Instituta Matematiki AN USSR (periodical code)

KIzKU Kazan Izdatel Kazanskogo Universiteta (periodical code)

KIzS Kishinev Izdatel Shtiintsa (periodical code)

KIzVS Kiev Izdatel Vishcha Shkola (periodical code)

kJ Kilojoule • Kommenden Jahres • Künftigen Jahres

Kj Kalenderjahr • Konjunktion • Konjunktiv

KJA Krasnoyarsk (Russia) (location identifier)

KJK Kortrijk (Belgium) (location identifier)

KJP Kerama (Japan) (location identifier)

KJS Kenya Journal of Sciences

KJU Kamiraba (Papua New Guinea) (location identifier)

kk Kazakh (ISO language code)

KK Kabushiki Kaisha • Kaiserlich-Königlich • Kaluza-Klein • Kempen-Krefeld (Germany) • Kernkraft • Kleinkaliber • Kompaßkurs • Kontaktkreis • Kreiskarte

KKA Koyuk (USA-AK) (location identifier)

KKB Kitoi Bay (USA-AK) (location identifier)

KKC Khon Kaen (Thailand) (location identifier)

KKD Kokoda (Papua New Guinea) (location identifier)

KKE Bay of Islands (New Zealand) (location identifier) • Kerikeri (New Zealand) (location identifier)

KKG Konawaruk (Guyana) (location identifier) • Kundenkreditgemeinschaft

KKH Kongiganak (USA-AK) (location identifier)

KKI Akiachak (USA-AK) (location identifier)

KKJ Kita Kyushu - Kokura (Japan) (location identifier)

KKK Kalakaket (USA-AK) (location identifier) • Kreiskontrollkommission

KKL Kanäle kleiner Leistung • Karluk Lake (USA-AK) (location identifier) • Kultur- und Kongresszentrum Luzern (Switzerland)

KKM Lop Buri (Thailand) (location identifier)

KKN Kirkenes (Norway) (location identifier)

KKO Kaikohe (New Zealand) (location identifier)

KKP Koolburra (Australia-QLD) (location identifier)

KKR Kagaku Kogaku Ronbunshu (periodical code) • Kaukura Atoll (Tuamotu Islands) (location identifier) • Korringa - Kohn - Rostoker (method)

KKT Karush - Kuhn - Tucker (theorem) • Kentland (USA-IN) (location identifier)

KKU Ekuk (USA-AK) (location identifier) • Khon Kean University (Thailand)

KKW Kernkraftwerk • Kikwit (Zaire) (location identifier)

KKX Kikaiga Shima (Japan) (location identifier)

KKY Kilkenny (Ireland) (location identifier)

KKZ Koh Kong (Cambodia) (location identifier)

kl Greenlandic (ISO language code) • Kiloliter • klagen • klar • klären • klassisch • kleben • kleiden • klein • kleine • klingen • klingeln • klug

Kl Klage • Klang • Klappe • Klasse • Klassifikation • Klassifizierung • Klassik • Klavier • Kleid • Kleider • Kleidung • Klein • Kleine • Kleiner • Klemme • Klient • Klinge • Klingel • Klinik • Klinker • Klippe • Kloster • Klub

KL Kabellänge • Kalmyk • Karhunen-Loeve (transform) • Kartenleser • Keller's Language • Kelvin Law • Kernel Language • Key Length • Kleinmann-Low (nebula) • Knowledge Level • Kondo Lattice • Kontrollampe • Kuala Lumpur (Malaysia) • Kullback-Leibler (distance) • Kursor Left

KLA Kansas Library Association • Karachi Library Association • Kentucky Library Association • Kenya Library Association • Korean Library Association

klass klassisch

Klass Klassifikation • Klassifizierung • Klassiker

Klassif Klassifikation • Klassifizierung

K'lautern Kaiserslautern

Klautern Kaiserslautern

Klav Klavier

Klb Thousands of Pounds

KLB Kalabo (Zambia) (location identifier)

KlBer Kleinheubacher Berichte (periodical code)

Klbf Kleinbahnhof

KlC Kleincomputer

KLC Kaolack (Senegal) (location identifier) • Karl Lambrecht Corporation (USA-IL)

KLD Kalinin (Russia) (location identifier)

KLE Kaele (Cameroon) (location identifier)

Kleinb Kleinbahn

Kleinbf Kleinbahnhof

klf klassifiziert

KLF Kaluga (Russia) (location identifier) • K-band Luminosity Function

KLG Kalskag (USA-AK) (location identifier) • Kindererziehungsleistungsgesetz • Ks Luminosity Function

KLH Keyhole Limpet • Kolhapur (India) (location identifier)

KLI Kota Koli (Zaire) (location identifier)

KLIC KBS Life Cycle • Key-Letter-In-Context

KLJ Klaipeda (Lithuania) (location identifier)

KLK Kalokol (Kenya) (location identifier)

KLL Levelock (USA-AK) (location identifier)

KLM Kerr-Lens Mode • Kilometer • Knowledge-Level Model

KLN Larsen Bay (USA-AK) (location identifier)

KLNN Kullback-Leibler Nearest Neighbor

KLO Kalibo (Philippines) (location identifier)

KLOC Kilolines Of Code • Thousand Lines Of Code

KLOE k Long Experiment

KLP Kalp Bay (USA-AK) (location identifier)

KLQ Keluang (Indonesia) (location identifier)

KLR Kalmar (Sweden) (location identifier)

Kls Kloster

KLS Kelso Longview (USA-WA) (location identifier)

KLSS Korean Library Science Society

KLT Karhunen-Loeve Transform

KLTL Keeping Line Time Law

KLu Koninklijke Luchtmacht (Netherlands)

KLU Klagenfurt (Austria) (location identifier)

KLUN Kinematics of the Local Universe

KLUT Knowledge Look-Up Table

KLV Karlovy Vary (Czech Rep.) (location identifier)

KLVAD Keele (Starlink internal code)

KLW Klawock (USA-AK) (location identifier)

KLX Kalamata (Greece) (location identifier)

KLY Kalima (Zaire) (location identifier)

KLZ Kleinzee (South Africa) (location identifier) • Kopier- und Lieferzentrale

km Cambodian (ISO language code) • Kilometer • Kilometric • Kilómetro

Km Kingdom

kM Künftigen Monats

K-M Kaplan-Meier (estimator)

K&M Klystron & Modulator (system)

KM Comores (code ISO de pays) • Comoros (ISO country code) • Kaltmiete • Kaplan-Meier (estimator) • Kinematic Misalignment • Knowledge Management • Knowledge Manager • Koninklijke Marine (Netherlands)

KMA Kerema (Papua New Guinea) (location identifier) • Koninklijke Militaire Academie • Korean Meteorological Administration • Koronaler Massenauswurf (*) • Koronaler Materieauswurf • Kreuzmodulations-Abstand

KMASE Key Management Application Service Element

KMB Koinambe (Papua New Guinea) (location identifier)

KMBAA Knowledge and Model-Based Algorithm Adoption

kMc Kilomegacycle

KMC Key Management Center • King Model Cluster • McCord (USA-AK) (location identifier)

KMCB Knowledge Management Certification Board

KMD Kenya Meteorological Department • Key Management Device • Mandji (Gabon) (location identifier)

KME Kamembe (Rwanda) (location identifier)

KMF Franc (Comoros) (ISO currency code) • Kamina (Papua New Guinea) (location identifier)

KMG Kunming (China-PRC) (location identifier)

km/h Kilometers per Hour

KMH Kilometers per Hour • Kosmischer Mikrowellenhintergrund • Kuruman (South Africa) (location identifier) • Speed

KMI Key Management Infrastructure • Kleine en Middelgrote Industrieën • Knowledge Media Institute (UK) • Koninklijk Meteorologisch Instituut • KSC Management Instruction • Miyazaki (Japan) (location identifier)

KMIB Koninklijk Meteorologisch Instituut van België (Belgium)

KMID Key Management Identification Number • Key Material Identifier

KMIO Kaliningrad Magnetic Ionospheric Observatory (Russia)

KMIPS KAIST Map and Image Processing Station

KMIT King Mongkut's Institute of Technology (Thailand)

KMJ Kumamoto (Japan) (location identifier)

KMK Makabana (Congo) (location identifier)

KML Kamileroi (Australia-QLD) (location identifier)

KMM Kimam (Indonesia) (location identifier)

KMN Kamina (Zaire) (location identifier) • Knowledge Media Networking

KMO Kleine en Middelgrote Ondernemingen • Manokotak (USA-AK) (location identifier)

KMODC Key Material Ordering and Distribution Center

KMOS Cryogenic Multiobject Spectrometer

KMP Keetmanshoop (Namibia) (location identifier) • Key Management Plan • Key Management Protocol • Key Measurement Point • Knuth - Morris - Pratt (algorithm)

KMPA Kenya Master Printers Association

KMPDU Key Management Protocol Data Unit

kmph Kilometers per Hour

kmps Kilometers per Second

KMPS Kernel Multiple-Processing System

KMQ Komatsu (Japan) (location identifier)

KMR Karimui (Papua New Guinea) (location identifier) • Key Milestone Review • Knowledge Metadata Repository

KMRP Knowledge Management Research and Practice (periodical code)

km/s Kilometers per Second

KMS Kabelmanagement-System • Key Management System • Knowledge Management System • Koninklijke Militaire School (Belgium) • Korean Mathematical Society (Korea-ROK) • Kubo - Martin - Schwinger (condition) • Kumasi (Ghana) (location identifier)

KMSA Key Management System Agent

km/sec Kilometers per Second

KMSS Key Management Software Support

km/st Stundenkilometer

km/t Kilometer/Time

KMT Kampot (Cambodia) (location identifier) • Kreiss Matrix Theorem

km/u Kilometer per Uur

KMU Kismayu (Somalia) (location identifier) • Kleine und Mittlere Unternehmen

KMUA Key Management User Agent

KMUL Karl-Marx-Universität Leipzig (former GDR)

KMV Kalemyo (Myanmar) (location identifier)

KMW Kleine Magellansche Wolke • Kostroma (Russia) (location identifier)

KMX Khamis Mushait (Saudi Arabia) (location identifier)

KMY Moser Bay (USA-AK) (location identifier)

KMZ Kaoma (Zambia) (location identifier)

kn Kannada (ISO language code) • Krona (currency) • Krone (currency)

Kn Croatian Kuna (currency) • Knot • Knudsen Number

KN Îles Saint Kitts et Nevis (code ISO de pays) • Kaon-Neon • Kip (currency) • Knot • Knowledge Networking • Kongeriket Norge • KSC Notice • Nevis (Saint Kitts and_) (ISO country code) • Saint Kitts and Nevis (ISO country code)

KNA Kenya News Agency • Saint Kitts and Nevis (ISO country code) • Viña de Mar (Chile) (location identifier)

KNAAAS Kenya National Academy for Advancement of Arts and Sciences

KNAC Keck Northeast Astronomy Consortium • Koninklijke Nederlandse Automobielclub (Netherlands)

KNAS Kenya National Academy of Sciences

KNAW Koninklijke Nederlandse Akademie van Wetenschappen (Netherlands)

KNB Kanab (USA-UT) (location identifier)

KNC Ji'An (China-PRC) (location identifier)

KNCSS Thousand of Non-Comment Source Statements

KND Kindu (Zaire) (location identifier) • Kirchlicher Nachrichtendienst (Germany)

KNE Kanainj (Papua New Guinea) (location identifier)

KNET Kangaroo Network

KNF Kings Lynn - Marham (UK-England) (location identifier) • Klein-Nishina Formula

KNG Kaimana (Indonesia) (location identifier)

KNH Kinmen (China-ROC) (location identifier)

KNI Katanning (Australia-WA) (location identifier)

KNIFE Kashima Nobeyama Interferometer

KNITQ Korean National Institute of Technology and Quality (Korea-ROK)

KNJ Kindamba (Congo) (location identifier)

KNJK Koninklijke Nederlandse Journalistenkring (Netherlands)

KNK Kakhonak (USA-AK) (location identifier)

KNL Kelanoa (Papua New Guinea) (location identifier) • Knoll

KNLS Knolls

KNM Kaniama (Zaire) (location identifier) • Kongelige Norske Marine (Norway)

KNMI Koninklijk Nederlands Meteorologisch Instituut (Netherlands)

k-NN k Nearest Neighbors

kNN k Nearest Neighbors

KNN Kankan (Guinea) (location identifier)

KNO Knokke - Het Zoute (Belgium) (location identifier)

KNOB Knowledge Object

KNOBS Knowledge-Based System

KNP Capanda (Angola) (location identifier)

KNQ Kone (New Caledonia) (location identifier)

KNS Kerridge Knowledge System • King Island (Australia-TAS) (location identifier)

Knt Knight

KNT Kennett (USA-MO) (location identifier) • Knight

Knts Knots

KnU Knowledge Utility

KNU Kanpur (India) (location identifier) • Kyiv National University (Ukraine) • Kyungpook National University (Korea-ROK)

KNUB Koninklijke Nederlandse Uitgeversbond (Netherlands)

KNV Knights Inlet (Canada-BC) (location identifier)

KNVD Koninklijk Nederlands Verbond van Drukkerijen (Netherlands)

KNVNG Knappschaftsrentenversicherungs-Anpassungsgesetz

KNVvL Koninklijke Nederlandse Vereniging voor Luchtvaart (Netherlands)

KNW New Stuyahok (USA-AK) (location identifier)

KNX Kununurra (Australia-WA) (location identifier)

KNY Kinoosao (Canada-SK) (location identifier)

KNZ Kenieba (Mali) (location identifier)

ko Knockout • Korean (ISO language code)

Ko Kilo-octet

KO Contracting Officer • Keep Off • Keep Out • Kilo-Octet • Knockout • Knowledge Organization • Knoxville Observers (USA-TN) • Konkursordnung

KOA Kailua (USA-HI) (location identifier) • Kampgrounds Of America • Kona - Ke-Ahole (USA-HI) (location identifier)

KOB Koutaba (Cameroon) (location identifier)

KOC Key Operational Facility • Koumac (New Caledonia) (location identifier)

KOD Karaoke On Demand • Kick-Off Drift • Knowledge-Oriented Design • Kotabangun (Indonesia) (location identifier)

KodOB Kodaikanal Observatory Bulletin (periodical code)

KodRe Kodaikanal Observatory Reprints (periodical code)

KOE Kupang (Indonesia) (location identifier)

KOG Khong (Laos) (location identifier)

KOH Koolatah (Australia-QLD) (location identifier) • Potassium Hydroxide

kohm Kilo-Ohm

KOI Kirkwall (UK-Scotland) (location identifier) • KSC Operation Instruction • Orkney Island (UK) (location identifier)

KOJ Kagoshima (Japan) (location identifier)

KOK Kokkola-Pietarsaari-Kruunupyy (Finland) (location identifier)

KOL Kamerlingh Onnes Laboratorium (Netherlands) • Koumala (Central African Rep.) (location identifier)

Kolloq Kolloquium

KOM Komo-Manda (Papua New Guinea) (location identifier) • KSC Organizational Manual

KOMA Koordinaten-Meßapparat

KomMe Komety i Meteory (periodical code)

Komm Kommentar • Kommission

Komp Kompanie

Kon Koninklijk

KON Kontum (Viet Nam) (location identifier)

Konf Konferenz

Koni Kochnische

KONIC Kiso Observatory Near-Infrared Camera

KOO Kongolo (Zaire) (location identifier)

kop Kopeck

KOP Nakhon Phanom (Thailand) (location identifier)

KOpS Thousands of Operations per Second

KOPS Thousands of Operations Per Second (*)

Kor Korea • Korean

KOR Corée du Sud (code ISO de pays) • Knowledge Of Results • Kokoro (Papua New Guinea) (location identifier) • Korea (ROK) (ISO country code) • ROK (ISO country code)

KORDI Korea Ocean Research and Development Institute

KORG Knowledge in Organizations Research Group

KOS Kompongsom (Cambodia) (location identifier) • Sihanoukville (Cambodia) (location identifier)

KOSC KSC Operations Support Center

KOSEF Korea Science and Engineering Foundation (Korea-ROK)

KOSI Kaiser Optical Systems Inc. • Knowledge-Oriented Statistical Inference • Kodak Optical Storage Interface • Kometen-Simulations (experiment)

KosIs Kosmicheskie Issledovaniia (periodical code)

KosIU Kosmicheskie Issledovaniia Ukraine (periodical code)

KosLu Kosmicheskie Luchi (periodical code)

KOSMA Kölner Observatorium für Submillimeter Astronomie (Switzerland) • Kõlner Observatorium für Submillimeter und Millimeter Astronomie (Switzerland)

KosNT Kosmicheskaia Nauka i Tekhnika (periodical code)

KOSS Kirschner - Oemler - Schechter - Schechtman (void)

KostO Kostenordnung

KOT Kotlik (USA-AK) (location identifier)

KOU Koulamoutou (Gabon) (location identifier)

KOV Kokchetav (Kazakhstan) (location identifier)

KOW Ganzhou (China-PRC) (location identifier)

KOX Kokonao (Indonesia) (location identifier)

KOY Olga Bay (USA-AK) (location identifier)

Kozmo Kozmos (periodical code)

KOZ Ouzinkie (USA-AK) (location identifier)

kp Kilopond

KP Corée du Nord (code ISO de pays) • DPRK (ISO country code) • Kaufpreis • Keypad • Key Parameter • Key Person • Key Personnel • Key Point • Key Processor • Key Programme • Key Project • Key Pulse • Key Punch • Kitt Peak (USA-AZ) • Klein Paradox • Kommunikationsprotokoll • Korea (North_) (ISO country code) • Kukarkin-Parenago • Kurie Plot

kPa Kilopascal

KPA Kenya Publishers Association • Key Penetration Access • Key Personnel Area • Key Process Area • Key Process Attribute • Klystron Power Amplifier • Kopiago (Papua New Guinea) (location identifier) • Kraft Paper Association

KPB Point Baker (USA-AK) (location identifier)

kpc Kiloparsec

KPC Kakuda Propulsion Center (Japan) • Keyboard-Printer Control • Keypunch Cabinet • Port Clarence (USA-AK) (location identifier)

KPCB Kinematics and Physics of Celestial Bodies (periodical code)

KPCM Key Privilege and Certificate Management

KPCMS Kodak Precision Colour Management System

KPD King of Prussia (USA-PA) (location identifier) • Kitt Peak - Downes (survey) • KSC Program Directive

KPE Yapsiei (Papua New Guinea) (location identifier)

KPES K-best Polyelastic Sampling

KPG Kurupung (Guyana) (location identifier)

KPGO Kokee Park Geophysical Observatory (USA-HI)

kph Kilometers per Hour

KPH Kilometers Per Hour • Pauloff Harbor (USA-AK) (location identifier)

KPI Kapit (Malaysia) (location identifier) • Key Performance Indicator • Kernel Programming Interface • Korean Periodical Index

KPIC Key Phrase In Context

KPK Key Production Key • Parks (USA-AK) (location identifier)

KPL Key Personnel Locator • Kit Parts List

KPM Kathode Pulse Modulation • Kometen - Planetoiden - Meteore • Kompiam (Papua New Guinea) (location identifier)

KPN Kipnuk (USA-AK) (location identifier)

KPNO Kitt Peak National Observatory (USA-AZ)

KPO Keypunch Operator • P'Ohang (Korea-ROK) (location identifier)

KPP Kalpowar (Australia-QLD) (location identifier) • Key Performance Parameter • Key Programme Publication • Kolmogorov - Petrovsky - Piskunov (equation) • Korean Pacific Press

kpps Kilopulses per Second

Kpr Keeper

KpR Knots per Revolution

KPR Knots Per Revolution • Kodak Photo Resist • Port Williams (USA-AK) (location identifier)

KPRD KSC Program Requirements Document

KPRNA Konza Prairie Research Natural Area (USA-KS)

KpS Kilometers per Second

KPS Kempsey (Australia-NSW) (location identifier) • Kilometers Per Second • Klystron Power Supply • Knowledge Pool System • Knowledge Processing System • Koch Process Systems Inc. • Korean Physical Society (Korea-ROK)

kpsf KIPs per Square Foot

kpsi KIPs per Square Inch • One Thousand Pounds per Square Inch

KPSM Klystron Power Supply Modulator

KPT Jackpot (USA-NV) (location identifier) • Kai's Power Tool

KPTC Kenya Post and Telecommunications Corporation

KPU Keyboard Printer Unit

KPUP Key Personnel Update System

KPV Perryville (USA-AK) (location identifier)

KPVT Kitt Peak Vacuum Telescope (USA-CA)

KPW Won (Korea-DPRK) (ISO currency code)

KPY Port Bailey (USA-AK) (location identifier)

KPZ Kardar - Parisi - Zhang (equation)

KQA Akutan (USA-AK) (location identifier)

KQB Koonibba (Australia-SA) (location identifier)

KQL Kol (Papua New Guinea) (location identifier)

KQML Knowledge Query and Manipulation Language

kr Korea (South_) • Kreutzer • Króna (currency) • Krona (currency) • Krone (currency) • Kroner (currency) • Kronor (currency) • Krónur (currency) • Kursor Right • ROK

Kr Contractor • Kreis • Krona (currency) • Króna (currency) • Krone (currency) • Kroner (currency) • Kronor (currency) • Krónur (currency) • Krypton

K&R Kernighan & Ritchie

KR Contractor • Corée du Sud (code ISO de pays) • Knowledge Representation • Kommunikationsrechner • Korea (South_) (ISO country code) • ROK (ISO country code)

KRA Kerang (Australia-VIC) (location identifier) • Key Recovery Agent • Key Recovery Alliance

KraftStG Kraftfahrzeugsteuergesetz

KRB Karumba (Australia-QLD) (location identifier)

KRC Kerinci (Indonesia) (location identifier) • Key Recovery Center • Kodak Reflex Camera

KRD Kurundi (Australia-NT) (location identifier)

KRDL Kent Ridge Digital Laboratories (Singapore)

KRE Key Recovery Encryption • Kirundo (Burundi) (location identifier)

KREEP Potassium, Rare-Earth Elements and Phosphorus

KREMU Kenya Rangeland Ecological Monitoring Unit

KRF Key Recovery Field • Korea Research Foundation (Korea-ROK) • Kramfors (Sweden) (location identifier)

KRG Kantonale und Regionale Gesellschaften • Karasabai (Guyana) (location identifier)

KRI Kikori (Papua New Guinea) (location identifier)

KRIMO Kenya Rift Monitoring

KRIS Korea Research Institute of Standards (Korea-ROK)

KRISS Korea Research Institute of Standards and Science (Korea-ROK) (formerly KSRI)

KRJ Karawari (Papua New Guinea) (location identifier)

KRK Cracow (Poland) (location identifier) • Kraków (Poland) (location identifier)

KRL Key Revocation List • Khan Research Laboratories (Pakistan) • Kirchoff Radiation Law • Knowledge Representation Language • Korla (China-PRC) (location identifier)

KRM Karanambo (Guyana) (location identifier) • Kiruna Rymd- och Miljöcampus (Sweden) • Klein-Rydberg Method

KRMS Ka-band Radiometric Mapping System

KRN Kiruna (Sweden) (location identifier)

KRO Kathode Ray Oscilloscope • Kurgan (Russia) (location identifier)

KRP Karup (Denmark) (location identifier) • Kevlar-Reinforced Plastic • Known Reference Point • Kodak Relief Plate

KRQ Kramatorsk (Ukraine) (location identifier)

KRR Knowledge Representation and Reasoning • Krasnodar (Russia) (location identifier)

krs Kurus

Krs Kreis

KRS Knowledge Retrieval System • Knowledge-Based Replanning System • Kristiansand (Norway) (location identifier)

Krsn Kerosene

KRT Kathode Ray Tube • Key Renewal Theorem • Keyword Review Team • Khartoum (Sudan) (location identifier)

KRTO Kathode Ray Tube Oscillograph

KRTS Kathode Ray Tube Shield

KRTT Kathode Ray Tube Tester

KRU Kerau (Papua New Guinea) (location identifier)

KRV Kerio Valley (Kenya) (location identifier)

KRW Krasnovodsk (Turkmenia) (location identifier) • Won (Korea-ROK) (ISO currency code)

KRX Kar Kar (Papua New Guinea) (location identifier)

KRY Karamay (China-PRC) (location identifier)

Krz Kubisch-raumzentrierte

KRZ Kiri (Zaire) (location identifier)

ks Kashmiri (ISO language code)

K-S Kolmogorov-Smirnov (test)

K/S Kick Stage

KS Kansas (USA) • Kantowski-Sachs • Katroptic System • Keypad Start • Kick Stage • Knowledge Sharing • Knowledge Source • Knowledge Space • Kodak Standard • Kolmogorov-Sinai (entropy) • Kolmogorov-Smirnov (test) • Konungariket Sverige • Koordinatensystem • Kosmische Strahlung • Kuramoto-Sivashinsky (equation) • Kustaanheimo-Stiefel (regularization) • Kvant Source (catalog)

KSA K-band Single Access • Key Search Algorithm • Kingdom of Saudi Arabia • Knowledge - Skills - Abilities • Knowledge - Skills - Attitudes • Kosrae (Caroline Islands) (location identifier) • Ku-band Single Access

KSAs Knowledge - Skills - Abilities • Knowledge - Skills - Attitudes

KSAM Keyed Sequential Access Mode

KSB Kasanombe (Papua New Guinea) (location identifier) • Koninklijk Sterrenwacht van België (Belgium)

KSC Kagoshima Space Center (Japan) • Kearney State College (USA-NE) (now UNK) • Kennedy Space Center (USA-FL) • Kosice (Slovakia) (location identifier) • Krasnoyarsk Scientific Center (Russia)

KSCAP Kennedy Space Center Area Permit

KSchG Kündigungsschutzgesetz

KSCVC Kennedy Space Center Visitor Complex (USA-FL)

KSD Karlstad (Sweden) (location identifier) • Key Storage Device

KSE Kasese (Uganda) (location identifier)

KSEC Kopernik Space Education Center (USA-NY)

KSF Kassel Calden (Germany) (location identifier)

KSG Keplerian Self-Gravitating • Keplerian Self-Gravity • Kisengan (Papua New Guinea) (location identifier) • Kolen- en Staalgemeenschap

KSGA Koninklijk Sterrenkundig Genootschap van Antwerpen (Belgium)

KSGC Kamisaibara Spaceguard Center (Japan) • Kentucky Space Grant Consortium (USA-KY)

KSh Kenya Shilling (currency)

KSH Bakhtaran (Iran) (location identifier) • Kermanshah (Iran) (location identifier) • Korn Shell • Kugelsternhaufen

KSI Kissidougou (Guinea) (location identifier) • Knowledge Source Index • Knowledge Systems Inc.

KSIE Kinetic Solvent Isotope Effect

KSJ Kasos Island (Greece) (location identifier)

KSK Karlskoga (Sweden) (location identifier) • Kommando Spezialkräfte

KSL Kassala (Sudan) (location identifier)

KSLOC Thousand Source Lines Of Code

KSM Kungliga Svenska Marinen (Sweden) • Saint Mary's (USA-AK) (location identifier)

KSME Korean Society of Mechanical Engineering (Korea-ROK)

KSMP Kosher Meal

KSN Kit Serial Number • Kustanay (Kazakhstan) (location identifier)

KSNM Thousand of Square Nautical Miles

KSNPS Kuo-Sheng Nuclear Power Station (China-ROC)

KSO Kanzelhöhe Sonnenobservatorium (Austria) • Karl-Schwarzschild-Observatorium (Germany) • Kastoria (Greece) (location identifier)

KSOC Key Symbol Out of Context

KSOS Kernalized Secure Operating System

KSP Key Science Programme • Key Science Project • Keystone Project • Key Supplier Process • Kinderspielplatz • Kisatchie Star Party (USA-LA) • Kodak Special Plate • Kosipe (Papua New Guinea) (location identifier)

KSPEC K-band Spectrograph

KSpH Keystrokes per Hour

KSPH Keystrokes Per Hour

KSQ Karshi (Uzbekistan) (location identifier)

KSR Keyboard Send-Receive • Kopf-Schwanz-Radioquelle • Sandy River (USA-AK) (location identifier)

KSRC Kashima Space Research Centre (Japan)

KSRI Korea Standards Research Institute (Korea-ROK) (now KRISS)

KSS Kandy School of Sciences (India) • Kenya Soil Survey • Knowledge Support System • Sikasso (Mali) (location identifier)

KSSS Korean Space Science Society (Korea-ROK) • KSC Station Set Specification

KST Kiruna - Sodankylä - Tromsø (EISCAT) • Knowledge Space Theory • Korean Standard Time • Kosti (Sudan) (location identifier)

KSU Kansas State University (USA-KS) • Kennesaw State University (USA-GA) • Kent State University (USA-OH) • Key Service Unit • Key System Unit • King Saud University (Saudi Arabia) • Kristiansund (Norway) (location identifier) • Kyoto Sangyo University (Japan)

KSV Katalog Sluzhbij Vremeni • Springvale (Australia-WA) (location identifier)

KSVG Künstlersozialversicherungsgesetz

KSVH Kungl. Svenska Vetenskapsakademiens Handlingar (periodical code)

KSW Kapteyn Sterrenwacht Werkgroep (Netherlands) • Kiryat Shmona (Israel) (location identifier) • Kuffner-Sternwarte Wien (Austria)

KSX Yasuru (Papua New Guinea) (location identifier)

KSY Kars (Turkey) (location identifier)

KSZ Kinematic Sunyaev-Zel'dovich (effect) • Kotlas (Russia) (location identifier)

KSZE Konferenz über Sicherheit und Zusammenarbeit in Europa

kt Karat • Kilotonne • Knot

Kt Kanton • Kaution • Knight

K/T Cretaceous/Tertiary

KT Contract • Cretaceous-Tertiary • Kanalteiler • Kayser-Threde GmbH (Germany) • Kiloton • Kirchoff Theory • Kit • Knot • Kohler Tube • Kuhn-Tucker • Kuzmin-Toomre (disk)

KTA Karratha (Australia-WA) (location identifier)

KTAS Knots True Airspeed

KTB Cretaceous-Tertiary Boundary • Kosterlitz - Thouless - Berezinski (transition) • Thorne Bay (USA-AK) (location identifier)

KTC Katiola (Ivory Coast) (location identifier)

KTD Kitadaito (Japan) (location identifier)

KTDS Key To Disc Software

kte Korte

KTE Kerteh (Malaysia) (location identifier)

KTeV Kaons at the Tevatron

KTF Kaua'i Test Facility • Takaka (New Zealand) (location identifier)

KTFR Kodak Thin-Film Resist

KTG Kerntechnische Gesellschaft (Germany) • Ketapang (Indonesia) (location identifier)

KTH Kungliga Tekniska Hoegskolan (Sweden) • Tikchik (USA-AK) (location identifier)

KTHNY Kosterlitz - Thouless - Halperin - Nelson - Young (theory)

KTI Kratie (Cambodia) (location identifier)

KTK Kanua (Papua New Guinea) (location identifier)

KTL Keck Task Library • Kitale (Kenya) (location identifier)

KTM Kathmandu Tribhuvan (Nepal) (location identifier) • Keyword Translation Module • Kineto-Tracking Mount

KTN Ketchikan (USA-AK) (location identifier)

KTNF Kodak Timing Negative Film

Kto Konto

KTO Kato (Guyana) (location identifier) • Kenting Observatory (China-ROC)

KTP Kingston Tinson (Jamaica) (location identifier) • Potassium Titanyl Phosphate

KTQ Kitee (Finland) (location identifier)

Ktr Contractor

KTR Contractor • Katherine Tindal (Australia-NT) (location identifier)

kts Knots

KTS Kelvin Temperature Scale • Key Telephone Set • Key Telephone System • Knots • Komitee für Terminologie und Sprachfragen • Teller Mission (USA-AK) (location identifier)

KTT Kittilä (Finland) (location identifier) • Kodaikanal Tower Telescope (India)

KTTM Khimiia i Tekhnologiia Topliv i Masel (periodical code)

KTU Key Telephone Unit • Kota (India) (location identifier)

KTV Kamarata (Venezuela) (location identifier) • Kongress für Textverarbeitung

KTW Katowice (Poland) (location identifier)

KTX Koutiala (Mali) (location identifier)

KTY Terror Bay (USA-AK) (location identifier)

ku Kurdish (ISO language code)

Kü Küche

Ku Kurchatovium

KU Keyboard Unit • Knowledge Unit • Kursor Up • Kuwait

KUA Kapoliskian University of Athens (Greece) • Kuantan (Malaysia) (location identifier)

KUB Katholieke Universiteit Brabant • Kidneys - Ureters - Bladder • Kommission für Umweltbeobachtung • Kommunal- und Unternehmensberatung

KUC Kuria (Kiribati) (location identifier)

KUD Kudat (Malaysia) (location identifier)

KUE Kukundu (Solomon Islands) (location identifier)

KUED Kodak Unitized Engineering Data

KUF Samara (Russia) (location identifier)

KUFEM Kyushu University Faculty Engineering Memoirs (periodical code)

KüFu Küstenfunk

KUG Katholieke Universiteit Brabant (Netherlands) • Kiso UV-excess Galaxies (catalog) • Kubin Island (Australia-QLD) (location identifier)

KUH Kushiro (Japan) (location identifier)

KUI Kawau Island (New Zealand) (location identifier) • Keyboard User Interface

KUJ Kushimoto (Japan) (location identifier)

KUK Kasigluk (USA-AK) (location identifier)

KUL Katholieke Universiteit Leuven (Belgium) • Kuala Lumpur (Malaysia) (location identifier) • Kuwait University Library

KUM Yakushima (Japan) (location identifier)

KUN Katholieke Universiteit Nijmegen (Netherlands) • Kaunas (Lithuania) (location identifier)

KündFG Kündigungsfristengesetz

KUO Kuopio (Finland) (location identifier)

KUP Kupiano (Papua New Guinea) (location identifier)

KUQ Kuri (Papua New Guinea) (location identifier)

Kur Kurdish

KUR Kuran-o-Munjan (Afghanistan) (location identifier)

KURR Kyoto University Research Reactor (Japan)

KUS Kulusuk Island (Greenland) (location identifier)

KUSP Ku-band Signal Processor (*) • Ku-band Single Processor

KUT Kutaisi (Georgia) (location identifier)

KUTD Keep Up To Date

KUTGW Keep Up the Good Work

KUU Kulu (India) (location identifier)

KUV Kiso UV-excess (object) • Kunsan (Korea-ROK) (location identifier)

Kuw Kuwait

KUY Uyak (USA-AK) (location identifier)

KUZ Kusan (Korea-ROK) (location identifier)

kV Kilovolt

K&V Korteweg-de Vries

KV Kabelverteiler • Kennis Verrijkt • Kerr Vector • Kilovolt • Köchel Verzeichnis

kVA Kilovolt-Ampere

KVA Kavala (Greece) (location identifier) • Kungliga Vetenskaps Akademien (Sweden)

KVACC KHILS Vacuum Cold Chamber

kVAhm Kilovolt-Ampere Hour Meter

KvanE Kvantovaia Elektronika Moscow (periodical code)

kVAr Kilovar

KVAS Kanawha Valley Astronomical Society (USA-WV)

KVB Skövde (Sweden) (location identifier) • Skovde (Sweden) (location identifier)

KVC King Cove (USA-AK) (location identifier)

KVCC Kalamazoo Valley Community College (USA-MI)

KVdc Kilovolt Direct Current

KVDT Keyboard Video Display Terminal

KVDU Keyboard Visual Display Unit

KVE Key Vision Event • Kitava (Papua New Guinea) (location identifier)

KVeBB Kleine Veröffentlichungen der Universitätssternwarte zu Berlin (periodical code)

KVEG Kostendämpfungs-Ergänzungsgesetz

KVG Kapitalverkehrssteuergesetz • Kavieng (Papua New Guinea) (location identifier) • Key Variable Generator

KVHAA Kunliga Vitterhets Historie och Antikvitets Akademien (Sweden)

KVHS Kreisvolkshochschule (Germany)

KVI Kernfysisch Versnellerinstituut (Netherlands)

KVIV Koninklijk Vlaamse Ingenieurs Vereniging (Belgium)

KVK Karlsruhe Virtual Katalog • Kirovsk (Russia) (location identifier)

KVL Kirchoff's Voltage Law • Kivalina (USA-AK) (location identifier)

KVM Kalamazoo Valley Museum (USA-MI) • Keyboard - Video - Mouse • Kilobyte Virtual Machine

KVnT Kibernetika i Vychislitel naia Tekhnika (periodical code)

KVO Keyword Visual Organizer • Korean Virtual Observatory (Korea-ROK) • Kraftverkehrsordnung

kVp Kilovolts at Peak

KVP Kodak Vacuum Probe • Kodak Versamat Processor • Kontinuierlicher Verbesserungsprozess

KVQ Key Value Question

KVSt Knotenvermittlungsstelle

KVStG Kraftverkehrsteuergesetz

KVT Kanzelhöhe Vacuum Telescope (Austria)

KVU Korolevu (Fiji) (location identifier)

KVX Kirov (Russia) (location identifier)

KVz Knotenverzweiger

kw Kuwait

kW Kilowatt

KW Keyword • Koweit (code ISO de pays) • Kuwait (ISO country code)

KWA Kwajalein (Marshall Islands) (location identifier)

KWAC Keyword And Context

KWADE Keyword As a Dictionary Entry

KWC Kentucky Wesleyan College (USA-KY)

KWD Dinar (Kuwait) (ISO currency code) • Kawadjia (Central African Rep.) (location identifier)

KWE Guiyang (China-PRC) (location identifier)

KWEST Keyword Extracted as a String of Terms

KWF Keyword File • Waterfall (USA-AK) (location identifier)

KWG Kaiser-Wilhelm Gesellschaft (Germany) (now MPG) • Krivoy Rog (Ukraine) (location identifier)

kWh Kilowatt-Hour

KWH Khwahan (Afghanistan) (location identifier)

kWhm Kilowatt - Hour - Meter

kWhr Kilowatt-Hour

KWI Kuwait (location identifier)

KWIC KAO Wide-field Infrared Camera • Keyword In Context • Kuiper Wide-field Infrared Camera

KWICK Knowledge Workers Intelligently Collecting, Coordinating, Creating Consulting Knowledge

KWIDR Keyword In Defined Rotation

KWIK Cross-Wind Integrated Concentration

KWIM Know What I Mean

KWIOC Keyword In/Out of Context

KWIP Keyword In Permutation

KWIT Keyword In Title

KWJ Kwangju (Korea-ROK) (location identifier)

KWK Kwigillingok (USA-AK) (location identifier)

KWL Guilin (China-PRC) (location identifier)

KWLF Kodak Wratten Light Filter

KWM Kowanyama (Australia-QLD) (location identifier)

KWN Quinhagak (USA-AK) (location identifier)

KWNX Keyword Index

KWO Kawito (Papua New Guinea) (location identifier)

KWOC Keyword Out of Context

KWOT Keyword Out of Title • Kilometer-Wave Orbiting Telescope

KWP West Point (USA-AK) (location identifier)

KWPSI Keyword Phrase Subject Index

KWPT Keyword Plus Title

KWR Kwai Harbour (Solomon Islands) (location identifier)

KWS Kwailabesi (Solomon Islands) (location identifier)

KWT Koweit (code ISO de pays) • Kuwait (international vehicle registration code) • Kuwait (ISO country code) • Kwethluk (USA-AK) (location identifier)

KWU Kansas Wesleyan University (USA-KS) • Mansion House (New Zealand) (location identifier)

KWUC Keyword and UDC out of Context

KWV Kurwina (Papua New Guinea) (location identifier)

KWX Kiwai Island (Papua New Guinea) (location identifier)

kwy Keyway

KWY Kiwayu (Kenya) (location identifier)

KWZ Kolwezi (Zaire) (location identifier)

KX Quit

KXA Kasaan (USA-AK) (location identifier)

KXE Klerksdorp (South Africa) (location identifier)

KXF Kodak X-ray Film • Koro Island (Fiji) (location identifier)

KXK Komsomolsk na Amure (Russia) (location identifier)

KXR Karoola (Papua New Guinea) (location identifier)

ky Kirghiz (ISO language code) • Kyat (currency) • Thousand Years

Ky Kentucky (USA)

KY Cayman Islands (ISO country code) • Îles Cayman (code ISO de pays) • Kentucky (USA) • Kenya • Key

KYA Konya (Turkey) (location identifier)

KYB Yangoonabie (Australia) (location identifier)

Kybd Keyboard

KYBD Keyboard

KYC Know Your Customer

KYD Dollar (Cayman Islands) (ISO currency code) • Orchid Island (China-ROC) (location identifier)

KYE Tripoli (Lebanon) (location identifier)

KYF Yeelirrie (Australia-WA) (location identifier)

Kyffh Kyffhäuser

KYI Yalata Mission (Australia-SA) (location identifier)

KyITB Kyushu Institute Technology Bulletin (periodical code)

KYK Karluk (USA-AK) (location identifier)

KYL Key Largo - Port Largo (USA-FL) (location identifier)

KYN Milton Keynes (UK) (location identifier)

KYO Tampa - Top of Tampa (USA-FL) (location identifier)

KyoMe Kyoto University Faculty Engineering Memoirs (periodical code)

KYP Kyaukpyu (Myanmar) (location identifier)

Kypd Keypad

kyr Thousands of Years

Kyrgyz Kyrgyzstan

KYS Kayes (Mali) (location identifier) • Keys

KYT Kyaukpyu (Myanmar) (location identifier)

KYU Koyukuk (USA-AK) (location identifier)

KyUAM Kyushu University Research Institute Applied Mechanics Reports (periodical code)

KyUSR Kyushu University Research Institute Industrial Science Reports (periodical code)

KyUTR Kyushu University Technology Reports (periodical code)

KYX Yalumet (Papua New Guinea) (location identifier)

KYZ Kyzyl (Russia) (location identifier)

kz Duststorm • Kazakhstan • Sandstorm

K-Z Kilble-Zureck (mechanism)

KZ Kazakhstan (ISO country code) • Kurzzug

KZB Zachar Bay (USA-AK) (location identifier)

KZC Kompong-Chhna (Cambodia) (location identifier)

KZD Krakor (Cambodia) (location identifier)

KZF Kaintiba (Papua New Guinea) (location identifier)

KZH Kizhuyak (USA-AK) (location identifier)

KZI Kozani (Greece) (location identifier)

KZK Kompong Thom (Cambodia) (location identifier)

KZN Kazan (Russia) (location identifier)

KZP Katalog Zvezd Zapodozrennykh v Peremennosti

KZS Kastelorizo (Greece) (location identifier)

KzSib Akademiia Nauk SSSR Sibirskoi Otdelenie Izvestiia Seriia Tekhnicheskikh Nauk (periodical code)

KZT Tenge (Kazakhstan) (ISO currency code)

KZU Kennzeichenumsetzer

K2K KEK to Kamioka

K7 Cassette

L

l Direct Supply • Ecliptic Longitude • Galactic Longitude • Geometric Path Length • Heliocentric Longitude • Lake • Lambda • Land • Lang • Large • Late • Lateral • Latitude • Laut • Law • Leaf • League • Left • Legal • Legitimate • Leicht • Length • Length Scale • Lepton Number • Leu (currency) • Lies! • Light • Lightning • Line • Link • Links • Lira • Line • Links • Liquid • Liter • Literate • Litre • Little • Loch • Long • Lost • Low • Mean Anomaly of the Moon • Mean Free Path • Miller Index • Mixing Length • Orbital Angular Momentum Quantum Number • Pendulum Length • Pound (currency) • Quantum Number of Rotation • Quantum Number of Vibrational Angular Momentum • Reduced Pendulum Length • Relaxation Length • Thickness

L Angular Momentum • Avogadro Constant • Characteristic Length • Conservation Equation • Convective Operator • Diffusion Length • Fifty (Roman numeral) • Geographical Longitude • Inductance • Inductor • Lac • Ladung • Lady • Lage • Lago • Lagóa • Lagoon • Lagrangian • Laguna • Lahn-Dill-Kreis (Germany) • Lake • Lambert • Land • Language • Large • Largeur • Last • Latent Heat • Latin • Latitude • Launch • Lautsprecher • Law • League • Learner • Left • Lei (currency) • Leistung • Lempira (currency) • Length • Leu (currency) • Lev (currency) • Leva (currency) • Level • Lex • Ley • Liber • Liberal • Libra • Licentiate • Licht • Lied • Lieferung • Lift • Liga • Light • Lima (international alphabet) • Limited • Line • Linear • Linear Energy Transfer • Linear Integral Operator • Link • Linking Number • Lira (currency) • Lire (currency) • Liste • Liter • Litre • Littéraire • Live • Lobby • Local • Locator • Loch • Locus • Lodge • Loi • London (UK-England) • Long • Longitude • Longueur • Lord • Lorentz Coefficient • Lough • Low • Lowest • Low-Pressure Center • Lumen • Luminance • Luminosity • Luxembourg (international vehicle registration code) • Luxembourg (postal country code) • Luyten (catalog) • Kinetic Potential • Mean Longitude of a Planet • Orbital Angular Momentum Quantum Number • Pound (currency) • Radiance • Self-Inductance • Solar Luminosity • Sound Intensity • 50 (Roman numeral)

l/a Letter of Authority

la Lacunosus • Latin (ISO language code) • Lege Artis

lA laut Akte • laut Angabe • laut Auftrag

La Lane • Lanthan • Lanthane • Lanthanum • Louisiana (USA)

LA Laboratoire d'Aérologie • Land Access • Landesamt • Landesarchiv (Germany) • Lane • Language Availability • Lao PDR (ISO country code) • Laos (code ISO de pays) • Large Aircraft • Large Aperture • Large Array • Latin • Latin America • Latin American • Launch Abort • Launch Aft • Launch and Ascent • Launch Area • Launch Azimuth • Law Agent • Lead Agent • Lead Angle • Leading Arm (galaxies) • Learning Agreement • Leave Allowance • Legal Adviser • Legislative Affairs • Legislative Assembly • Legislative Assistant • Leseranalyse • Letter of Activation • Letter of Authority • Library Administration • Library Administrator • Library Association (UK) • Library Automation • Licensing Act (USA) • Licensing Authority • Lightning Arrester • Limited Area • Line Amplifier • Linear Accelerator • Link Assurance • List Attributes • Load Address • Loan Agreement • Local Agent • Local Archive • Local Authority • Location Area • Log Analyzer • Logistics Assistance • Long Arc • Longitudinal Acoustic • Los Angeles (USA-CA) • Louisiana (USA) • Low Altitude • Low Attitude • Lycée Agricole (France)

LAA Lamar (USA-CO) (location identifier) • Landesarbeitsamt • Laser Attenuator Assembly • Launch Area Antenna • Lepton Asymmetry Analyzer • Library Association of Australia • Limited Access Authorization • Local Access Authorization • Local Airport Advisory • Locally-Administered Address • Los Alamos Astrophysics (USA-NM) • Louisville Amateur Astronomers (USA-KY)

LAAB Laboratoire d'Astronomie et d'Astrophysique de Bordeaux (France)

LAAFB Los Angeles Air Force Base (USA-CA)

LAAM Laboratoire d'Astrophysique, Atomes et Molécules

LAAME Large Antenna Assembly and Measurement Experiment

LAAOL Los Alamos Astrophysical Opacity Library

LAAPD Large-Area Avalanche Photodiode

LAAR Liquid-Air Accumulator Rocket

LAAS Laboratoire d'Analyse et d'Architecture des Systèmes (France) • Laboratoire d'Automatique et d'Analyse des Systèmes (France) • Local Area Augmentation System • Los Angeles Astronomical Society (USA-CA) • Low-Altitude Alerting System • Low-Altitude Alert System

Lab Label • Labial • Laborant • Laboratoire • Laboratorium • Laboratory • Labour • Labourer • Labrador

LAB Lablab (Papua New Guinea) (location identifier) • Laboratory • Latency Adjustment Buffer • Latvijas Astronomijas Biedriba (Latvia) • League of American Bicyclists (USA) • Legal Advisory Board • Legal Aid Board • Linear Alkyl Benzene • Lithosphere-Asthenosphere Boundary • Local Activities Board • Logic Array Block

LABCEDE Laboratory Cryogenic Electron-Dependent Emissions

LABOCA Large Bolometer Camera

lac Locus Ante Citatus

Lac Lacerta-Lacertae (constellation)

LAC Laboratório Asociado de Computação • Lake City (Papua New Guinea) (location identifier) • Large Area Counter • Laser Amplifier Chain • Last Acquisition Cost • Latin America and the Caribbean • Latin American Country • LEISA Atmospheric Corrector • Levelized Annual Cost • Library Advisory Committee • Library Advisory Council • Limited Agency Check • Linear Absorption Coefficient • Linear Accelerator Center • Linear Amplifier Circuit • Link Access Control • Local Agency Check • Local Area Communication • Local Area Controller • Local Area Coordinator • Local Area Coverage • Long-run Average Cost • Loop Around Circuit • Loop Assignment Center • Lunar Astronautical Chart

LACAC Latin American Civil Aviation Conference

LACATE Lower Atmosphere Composition And Temperature Experiment

LACB Landing Aids Control Building

LACC Los Angeles City College (USA-CA)

LACCD Large-Area Charge-Coupled Device

LACE Laser Atmospheric Compensation Experiment • Liquid-Air Cycle Engine • Limited Area of Central Europe • Local Automatic Circuit Exchange • Los Angeles Contemporary Exhibition • Low-power Atmospheric Compensation Experiment (*) • Lunar Atmospheric Composition Experiment

LACEF Los Alamos Critical Experiments Facility (USA-NM)

LACES Low-Altitude Conical Earth Sensor

LACF Lead Analyst Control Function

LACH Lightweight Amphibious Container Handler

LASI Laser Autostigmatic Cube Interferometer

LACIE Large-Area Crop Inventory Experiment • Laser Altimeter Camera Irim Experiment

LACIP Large-Area Crop Inventory Program

LACL Logical Aspects of Computational Linguistics (conference)

LACMA Los Angeles County Museum of Arts (USA-CA)

LACN Local Area Communications Network

LACONIQ Laboratory Computer Online Inquiry

LACOS Large-scale Correct System

LACS Laboratory Acquisition and Control System • Light Aircraft Camera System • Locatable Address Conversion System

LACT Large Atmospheric Čerenkov Telescope

LACV Lighter Air Cushioned Vehicle

LACVD Laser-Activated Chemical Vapor Deposition

lad laden

Lad Ladung

LAD Laboratory of Advanced Design • Ladeira • Language Acquisition Device • Large Aircraft Division (EADS) • Large-Area Detector • Large-Area Display • Latest Arrival Date • Least Absolute Deviation • Least Absolute Difference • Least (sum of) Absolute Deviations • Lecture Automatique de Documents • Liquid Agent Detector • Location Aid Device • Logical Analysis Device • Logical Analysis of Data • Logical Aptitude Device • Logistics Anchor Desk • Logistics Approval Data • Logistics Assistance Division • Logistics Automation Directorate • Low-Accuracy Data • Low-Accuracy Designation • Low-Altitude Dispenser • Luanda (Angola) (location identifier) • Lunar Atmosphere Detector

LADA Langage de Description d'Architectures • Local Area Data Access

LADAR Laser Detection And Ranging • Laser Radar • Light Amplification for Detection And Ranging

LADD Lens Antenna Deployment Demonstration

LADDR Layered Device Driver

LADECO Linea Aérea Del Cobre

Ladest Ladestelle

LADGPS Local Area Differential Global Positioning System

LADIES Lapan Digital Image Evaluation System • Los Alamos Digital Image Enhancement System

LADIMAS Latitudinal Middle-Atmosphere Structure

LADISIS Laboratoire des Dimensions Sociales et Incidences Subjectives du Langage

LADL Laboratoire d'Automatique Documentaire et Linguistique (France)

LADN Local-Area Distribution Network

Ladp Ladyship

LADS Laser Airborne Depth Sounder • Local-Area Data Service • Logistics Automated Data System • Logistics Automated Documentation System • Low-Amplitude Delta Scuti (star)

LadSchlG Ladenschlußgesetz

LADT Local Access Data Transport

LAE Laboratoire d'Astrophysique Expérimentale • Lae (Papua New Guinea) (location identifier) • Left Arithmetic Element • Légion Astronomique Étrangère • Linear Algebraic Equation • Liquid Apogee Engine • Load Address Extended

LAEDP Large-Area Electronic Display panel

LAEFF Laboratorio de Astrofísica Espacial y Física Fundamental (Spain)

LAEMW Large-Amplitude Electromagnetic Wave

LaEO Laser ElektroOptik (periodical code)

LAEO Low-Altitude Electro-Optical

LAER Lowest Achievable Emission Rate

LAERTE Laboratoire des Écoulements Réactifs et de leurs Techniques d'Études

LAES Latin American Economic System • Letters in Applied and Engineering Sciences (periodical code)

LAET Limiting Actual Exposure Time

Laev Laevus

LaF Louisiana French

LAF Lafayette (USA-IN) (location identifier) • Leading Arm Feature • Local Adaptive Filter • Lower Atmospheric Facility

LAFM Limited Area Fine Mesh

LaFoc Laser Focus (periodical code)

LAFORIA Laboratoire des Formes et d'Intelligence Artificielle

LAFTA Latin-American Free Trade Association

Lag Lager • Lagerung • Lagging • Lagoon • Laguna • Lagune

LAG Laboratoire d'Astrophysique et de Géophysique • Laboratoire d'Automatique de Grenoble (France) • Lagrangian • La Guaria (Venezuela) (location identifier) • Landesarbeitsgericht (Germany) • Lastenausgleichgesetz • Launch and Analysis Group • Library Action Group • Linzer Astronomische Gesellschaft (Austria) • Load and Go Assembler • Logical Application Group • Lovers of Active Galaxies (collaboration)

LAGB Late Asymptotic Giant Branch

LAGCM Loa Alamos General Circulation Model

LAGEOS Laser Geodetic Satellite • Laser Geodynamic Earth-Orbiting Satellite • Laser Geodynamics Satellite (*)

LAGG Laboratory of Aero-Gas Dynamics

LAGOS Laser Gravitational-wave Observatory in Space • Laser-interferometer Gravitational-wave Observatory in Space

LAGS Laser-Activated Geodetic Satellite • Launch Abort Guide Simulation

Lah Lahore (Pakistan)

LAH Labuha (Indonesia) (location identifier) • Local Analyzer of Hypothesis

LAHS Low Altitude - High Speed

LAHSO Land And Hold Short Operations

LAI Laboratory for Astronomical Imaging • Land - Atmosphere - Ice • Lannion (France) (location identifier) • Leaf Area Index • Library Association of Ireland • Location Area Identity • Logistics Artificial Intelligence

LAIA Latin American Integration Association

LAICA Laboratorio di Informatica e Calcolo • Large Area Imager for Calar alto

LAID Land-Atmosphere Interaction Dynamics

LAIOO Landelijk AiO- en OiO Overleg (Netherlands)

LAIR Laboratory for Astronomical Image Reduction

LAIRE Laboratory for Astronomical Instrumentation, Research and Education (RIT)

LAIRS Land/Air Integrated Reduction System • Lockheed Arizona Infrared Spectrometer (USA-AZ)

LAIRT Large Aperture Infrared Telescope

LAIRTS Large Aperture Infrared Telescope System

LAIS Local Automatic Intercept System

LAIT Logistics Assistance and Instruction Team

LAIU Launch Abort Interface Unit

LaJ Lapidary Journal (periodical code)

LAJ Lages (Brazil-SC) (location identifier)

LAK Aklavik (Canada-NT) (location identifier) • Kip (Laos) (ISO currency code)

LAL Laboratoire de l'Accélérateur Linéaire (France) • Lakeland (USA-FL) (location identifier)

LALA Large Area Lyman Alpha (survey)

LALD Laboratory of Astronomical Light Detectors

LALL Longest Allowed Lobe Length

LALP Laboratoire d'Application des Lasers de Puissance (France)

LALPV Large-Amplitude Long-Period Variable (star)

LALR Look-Ahead Left-Recursive

LALSR Low-Altitude Large-Scale Reconnaissance

Lam Laminated

LAM Laboratoire d'Astrophysique de Marseille (France) • Laboratoire d'Astrophysique de Meudon (France) • Lameira • Lane Adaptor Module • Liberalium Artium Magister • Liquid Apogee Motor • Limited Area Model • Line Access Module • Liquid Apogee Motor • Lobe Access Module • Lobe Attachment Module • Local Area Model • Local Area Multicomputer • Lock Assembly Machine • Logistics Attrition Model • Long-Axis Mode • Los Alamos (USA-NM) (location identifier) • Low-Altitude Mission • Lyman Alpha Monitor • Master of Liberal Arts

LAMA Laboratoire de la Montagne Alpine (France) • Large Active Mirror for Astronomy • Large Active Mirror in Aluminium • Large Aperture Mirror Array • Large Area Mass Analyzer • Large Astronomical Mercury-mirror Array • Laser Machine • Library Administration and Management Association • Local Automatic Message Accounting

LAMAR Large Area Modular Array of Reflectors

LAMARS Large-Amplitude Multimode Aerospace Research Simulator

LAMAS Laboratoire de Médecine Aérospatiale

LAMBADA Large-scale Atmospheric Moisture Balance of Amazonia using Data Assimilation

LAMBDA Laboratory Measurements and Borehole Data Analysis • Legacy Archive for Microwave Background Data Analysis • Location And Movement Analysis System

LAME Lick Automatic Measuring Engine

LAMF Layered Aspect Moderator Framework

LAMM Late Arrival Managed Mail

LAMMA Laser Microprobe Mass Analyzer

LAMMR Large Antenna Multichannel Microwave Radiometer • Large Antenna Multifrequency Microwave Radiometer

LAMMS Local-Area Maintenance Management System

LAMOD Laser Optical Motion Detector

LAMOST Large-Area Multiobject Spectroscopic Telescope

LAMP Labor And Management Partner • Laboratoire Associé de Météorologie Physique • Large Active Mirror Program • Large Advanced Mirror Program • Laser, Atomic and Molecular Physics • Local Area Multiprocessor • LODE Advanced Mirror Program • Logistical Assessment of Modifications Programme • Logistics Automation Master Plan • Low-Altitude Manned Penetration • Low-Altitude Mapping Photogrammetry • Lunar Analysis and Mapping Program

LAMPE Langage Algorithmique de Mise en Page et d'Édition

LAMPF Los Alamos Meson Physics Facility (USA-NM)

LAMPS Laser-scan Automated Map Production System • Light Airborne Multipurpose System • Limited Area Mesoscale Prediction System • Linear and Mathematical Programming System

LAMRDT Large Aperture Microwave Radiometer Development Test

LAMS Large-Area Mosaicking System • Large-Area Mosaic System • Launch Acoustic Measuring System • Light Aircraft Maintenance Schedule

Lan Langley

LAN Inland • Landau/Isar (Germany) • Lansing (USA-MI) (location identifier) • Lembaga Administrasi Negara (Indonesia) • Lineas Aéreas Nacionales • Local Apparent Noon • Local Area Network • Local Asymptotic Normality • Longitude of the Ascending Node

LANA Local Area Network Adapter

LANACS Local Area Network Asynchronous Connection Server

Lanc Lancaster (UK-England)

LANCC Local Area Network Communications Controller • Local Area Network Control Center

Lancs Lancashire (UK-England)

LANCSE Los Alamos Neutron Science Center (USA-NM)

LAND Large-Area Neutrino Detector • Lead Astrophysical Neutrino Detector

Landk Landkarte

Landkr Landkreis

LANDP Local Area Network Distributed Platform

Landr Landrat

LANDS Land Analysis and Decision Support

LANDSAT Land Satellite

LANDSCAPE Large Astrophysical Network Detecting Showers from Cosmic Accelerated Particles to the Earth

Landsch Landschaft • Landschutz

Landstr Landstraße

LANE Local Area Network Emulation • Local Area Network Extension

Lang Language • Languedoc (France)

LANHIP Local Area Network Host Interface Protocol

LANICA Lineas Aéreas de Nicaragua

LANIU Local Area Network Interface Unit

LANL Los Alamos National Laboratory (USA-NM)

LANNET Large Artificial Neural Network

LANRES Local Area Network Resource Extension and Services

LANS Local Area Network System

LANSCE Los Alamos Neutron Science Center (USA-NM)

LANSEC Local Area Network Security (workshop)

LANSM Local Area Network Security Monitor

LANTIRN Low-Altitude Navigation and Targeting Infrared (system) for Night

LANTR LOX-Augmented Nuclear Thermal Rocket

Lao Laothian

LAO Laoag (Philippines) (location identifier) • Laos (international vehicle registration code) • Laos (code ISO de pays) • Large Assembly Order • Legal Assistance Officer • Logistic Assistance Office

LAOB Laboratoire d'Astrophysique de l'Observatoire de Besançon (France)

LAOF Lower Atmospheric Observing Facility

LAOG Laboratoire d'Astrophysique de l'Observatoire de Grenoble (France)

LAOL Los Alamos Opacity Library

LAOTEC Laboratory for Advanced Optical Technology (China-ROC)

Lap Lapland • Lappish

LAP Laboratório Asociado de Plasma • Langmuir Probe (Rosetta) • Language Analysis Package • Language Assistance Plan • La Paz (Mexico) (location identifier) • Large Area Panel • L-Arginine Phosphate • Leadership Assessment Program • Least Action Principle • Least Activation Path • Lesson Assembly Program • Library Access Program • Linear Accelerometer Package • Linear Array of

Processors • Link Access Procedure • Link Access Protocol • List Assembly Programming • Load - Assembly - Pack • Loading - Assembling - Packing • Local Access Point • Logistics Assistance Program • Low-Altitude Photography

LaPaB Laser Particle Beams (periodical code)

LAPACK Linear Algebra Package

LAPAN National Institute of Aeronautics and Space (Indonesia)

LAPB Link Access Procedure Balanced (*) • Link Access Protocol Balanced

LAPD Laser-Aided Plasma Diagnostics • Limited Axial Power Distribution • Link Access Procedure D channel • Link Access Procedure Direct • Link Access Protocol Digital

LAPE Local-Area Predictor Evaluation • Low-Altitude Parachute Extraction (system)

LAPERS Labor and Production Effectiveness Reporting System

LAPES Low-Altitude Parachute Extraction System

LAPH Laboratoire d'Annecy-le-Vieux de Physique Théorique (France)

LAPI Low-Altitude Plasma Instrument • Low-level Applications Programming Interface

LAPM Link Access Procedure for Modems • Link Access Protocol Modem

LAPO Lake Afton Public Observatory (USA-KS)

LAPP Laboratoire d'Annecy-le-Vieux de Physique des Particules (France)

LAPR Linear Array Pushbroom Radiometer

LAPRIDI Laboratorio de Procesamiento de Imágenes Digitalizadas

LAPS LAN Adapter and Protocol Support • LAN Adapter and Protocol Support • Left Aft Propulsion System • Limited-Area Prediction System • Local Analysis and Prediction System

LAPSS Large Area Pulsed Solar Simulator

LAPST Lunar And Planetary Sample Team

LAPW Linear Augmented Plane Wave • Linearized Augmented Plane Wave

LAPX Link Access Procedure Extended

LAQ Beida (Libya) (location identifier)

LAR La Alternativa Racional (periodical) • Laramie (USA-WY) (location identifier) • Large Adaptive Reflector • Launch Acceptability Region • Leaf Area Ratio • Ley de Arrendamientos Rústicos • Libya (international vehicle registration code) • Ligne À Retard • Limit Address Register • Linear Associative Retrieval • Load Access Right • Local Acquisition Radar • Logistics Area Representative • Logistics Assessment Review • Logistics Assistance Representative • Low-Angle Reentry

LARA Laminar Flow Research Action • Launch Area Risk Analysis

LARAM Line Addressable Random Access Memory

LaRC Langley Research Center (NASA)

LARC Langley Research Center (NASA) • Large Automatic Research Computer • Livermore Automatic Research Computer • Local Accessibility Rank Condition • Los Alamos Radiometry Code

LARCA Laboratoire de Réalisation de Composants pour l'Astronomie

LARCON Launch Area Risk Contour

LARCS Las Campanas/AAT Rich Cluster Survey

LARE Local Asymptotic Relative Efficiency

LAREG Laboratoire de Recherche en Géodésie

Larg Largamente • Largeur • Largo

LARGE Laboratoire de Recherche en Gestion

LARGOS Laser-Activated Reflecting Geodetic Optical Satellite

LARI Los Alamos Radiometric Instrument

LARIE Laser-Activated Reactive Ion Etching

LAROE Large Area Operational Experiment

LARP Live Action Role Play • Local And Remote Printing • Local Approvals Review Program • Locus Address Resolution Protocol • Low-Accretion Rate Polar

LARPS Local And Remote Printing Station

LARS Laboratory for Applications of Remote Sensing • Laminar Angular Rate Sensor • Laser Angular Rate Sensor • Launch And Recovery System • Load Alleviation and Ride Smoothing • Lower Airspace Radar-advisory Service • Lower-Atmosphere Research Satellite

LARST Land Applications of Remote Sensing Techniques • Local Applications of Remote Sensing Techniques

LARV Large-Amplitude Red Variable (star)

LAS Laboratoire d'Astronomie Spatiale • Laboratório de Sensores e Materiais • LAGEOS Apogee Stage • Land Analysis Software • Land Analysis System • Landsat Assessment System • Large Angular Size • Large Area Smoke • Large Area Survey • Large Astronomical Satellite • Large Atmosphere Structure • Laser Absorption Spectrometer • Las Vegas - McCarran (USA-NV) (location identifier) • Latvian Academy of Sciences (Latvia) • Latvian Astronomical Society (Latvia) • Launch Area Supervisor • Launch Auxiliary System • League of Arab States • Lebanese Astronomical Society (Lebanon) • Leeds Astronomical Society (UK) • Leicester Astronomical Society (UK) • Lethbridge Astronomy Society (UK) • Lettre Autographe Signée • Level of Archive Services • Leven Astronomical Society (UK) • Library Access System • Library Association of Singapore • Light-Activated Switch • Ligue of Arab States • Limb Atmospheric Spectrometer • Linear Array Scanner • Liverpool Astronomical Society (UK) • Local Area Screening • Local Area System • Longmont Astronomical Society (USA-CO) • Loughton Astronomical Society (UK-England) • Louisville Astronomical Society (USA-KY) • Low-Altitude Satellite • Low-Altitude Sensor • Low-Angle Scattering • Lunar Attitude System

LASA Landesagentur für Struktur und Arbeit (Germany) • Large Aperture Seismic Array • Large Area Solar Array • Laser Atmospheric Sounder and Altimeter • Lidar Atmospheric Sounder and Altimeter

LASC Library Automation Steering Committee • Lockheed Aeronautical Systems Co

LASCAR Large-Scale Reprocessing

LASCO Large-Angle Spectrometric Coronagraph

LASCOM Laser Communication

LASCR Light-Activated Silicon-Controlled Rectifier

LASD Large Angle Silicon Detector

LASE Large Aperture Seismic Experiment • Large Aperture Speckle Experiment • Large-Area Scintillation (counter) Experiment • Lidar Acquisition and Sizing Experiment • Lidar Atmospheric Sensing Experiment • Lidar Atmospheric Sounder Experiment • Logistic Asset Support Estimate

LASER Laboratory for Atmospheric Sciences and Environmental Research • Light Amplification by Stimulated Emission of Radiation • Lincoln Advanced Science and Engineering Reinforcement • London And Southeastern Region (UK-England)

LASIC Laboratoire Associé des Sciences de l'Information et de la Communication (France)

LASII Laser Stabilized Imaging Interferometer

LasJ Laser Journal (periodical code)

LASL Los Alamos Scientific Laboratory

LASMAS Laboratoire d'Analyse Secondaire et des Méthodes Appliquées à la Sociologie (France)

LASP Laboratory for Astronomy and Solar Physics • Laboratory for Atmospheric and Solar Physics • Laboratory for Atmospheric and Space Physics • Leaflet of the Astronomical Society of the Pacific (periodical code) • Low-Altitude Space Platform

LASR Laboratory for Astrophysics and Space Research • Laboratory of Automation of Scientific Research • Library Access, Search and Retrieval • Low-level Altitude Surveillance Radar

LASRE Linear Aerospike SR-71 Experiment

LASS Land Application Satellite System • Large-Amplitude Space Simulator • Large Area Sky Survey • Laser Aim Scoring System • Launch Area Support Ship • Library Automated Service System • Local Area Signaling Service • Local Area Sounding System • Logistics Activities Status System • Low-Altitude Surveillance System

LASSII Low-Altitude Satellite Study of Ionospheric Irregularities

LASSO Laramie Astronomical Society and Space Observers (USA-WY) • Laser Search and Secure Observer • Laser Synchronization from Satellite Orbit • Laser Synchronization from Stationary Orbit (*) • Logistics Automation Systems Support Office

LASSP Laboratory of Atomic and Solid-State Physics (Cornell)

LAST Local Area Storage Transport

LASTIR Laser Sensor Thermal Infrared

LASTP Linhas Aéreas de São Tome e Principe

LASTport Local Area Storage Transport

LAstr L'Astronomie (periodical code)

LASTREE Laboratoire de Sociologie du Travail, de l'Éducation et de l'Emploi (France)

LASV Low-Altitude Supersonic Vehicle

Lat Latent • Lateral • Latin • Latín • Latitude • Latus • Latvia • Latvian

LAT Laboratoire d'Astrophysique de Toulouse (France) • Laboratoire d'Astrophysique Théorique • Land Application Tool • Language Actions Table • Large Amateur Telescope • Large Angle Torque • Large-Area Telescope • Large-Area Triangulation • Large Atacama Telescope (project) • Latch • Lateral • Latitude • La Uribe (Colombia) (location identifier) • Lauterbach (Germany) • Local Access Terminal • Local Address Table • Local Apparent Time • Local Area Terminal • Local Area Transport • Logistics Asset Tracking • Long Alias Table • Los Angeles Times (periodical code) • Lot Acceptance Test

LATA Local Access and Transport Area • Local Access Transport Area

LATAS Laser True Airspeed System

LATCC London Air Traffic Control Centre (UK-England)

LATI Landsat Advanced Technology Instrument

LATIS Loop Activity Tracking Information System

Latl Lateral

LATM Local Asynchronous Transfer Mode • Local Asynchronous Transport Mode

LatME Latviiskii Matematicheskii Ezhegodnik (periodical code)

Latr Latrine

LATRS Laser Acquisition and Tracking System

LATTE Lesson and Technical Training Environment

Latv Latvia

Latvn Latvian

Lau Laundry

LAU Lamu (Kenya) (location identifier) • Launcher Unit • Ley de Autonomía Universitaria (Spain) • Linear Amplifier Unit

LAUC Librarians Association of the University of California (USA-CA)

Lauenb Lauenburg (Germany)

Laun Launch • Launched

LAUNOP Launch Opportunity

Lav Lavatory

LAV Lalomalava (Samoa) (location identifier) • Law of Averted Vision • Load Average • Local As View

LAVA Linear Amplifier for Various Applications • Look-Ahead Vector Analysis

LAVC Local Area VAX Cluster • Los Angeles Valley College (USA-CA)

LAVE Laser-Activated Vacuum Evaporation

Lavg Average Level

LAVIP Land-Atmosphere-Vegetation Interaction Project

LAVO Lager Algemeen Voortgezet Onderwijs

LAVS Lavatories

Law Lawine • Lawyer

LAW Fort Sill (USA-OK) (location identifier) • Lawton (USA-OK) (location identifier) • Local Authority Workstation • Low-Altitude Warning

LAWA Los Angeles World Airport (USA-CA)

LAWD Luyten Atlas of White Dwarfs (stars)

LAWN Local Area Wireless Network

LawOB Laws Observatory Bulletin (periodical code)

LAWP Lower Atmosphere Wind Profiler

LAWRS Limited Aviation Weather Reporting System

LAWS Laser Atmospheric Wind Sounder • Low-Altitude Warning System • Low-Altitude Wind Shear

LAX Los Angeles (USA-CA) (location identifier)

LAXPC Large-Area Xenon-filled Proportional Counter

LAXRV Large-Amplitude X-Ray Variability

LAXS Large Area X-ray Spectroscopy

LAXSM Large Area X-ray Spectroscopy Mission

LAXT Large-Area X-ray Telescope

LAY Ladysmith (South Africa) (location identifier)

Laz Lazarett

LAZ Bom Jesús da Lapa (Brazil-BA) (location identifier) • Launch Azimuth • Lebensarbeitszeit

lb Binary Logarithm • Libra • Liebe • Lieber • Pound

lB laut Bericht

Lb Leben • Lehrbuch • Leib • Libelle • Lichtbild • Luftbild

LB Label Boundary • Landbund • Landesbezirk • Landesbibliothek • Langmuir-Blodgett • Large Business • Launch Boost • Launch Booster • Launch Bus • Leaky Bucket • Lebanon (ISO country code) • Lectori Benevolo • Lecture Bottle • Left Button • Lesebuch • Letter Ballot • Letter Box • Liban (code ISO de pays) • Liberia (international vehicle registration code) • Libya • Light Box • Light Bridge • Limb Brightening • Line Block • Line Buffer • Load Bank • Local Battery • Local Board • Local Bubble • Logbook • Long Baseline • Longbow • Low Bay • Lower Brace • Luyten Blue (survey) • Pound

LBA Large-scale Biosphere-Atmosphere • Learning By Analogy • Leeds Bradford (UK-England) (location identifier) • Linear Block Array • Logical Block Address • Logical Block Addressing • Long-Baseline Array • Luftfahrt-Bundesamt

lb. ap. Pound (Apothecaries)

lb. av. Pound (Avoirdupois)

LBB Leak Before Break • Little Brown Bird • Lubbock (USA-TX) (location identifier)

LBBY Lobby

Lbc Lübeck (Germany)

LBC Laboratory of Biophysical Chemistry • Liaison Banque-Client • Lübeck (Germany) (location identifier)

LBCG Luminous Blue Compact Galaxy

LBCI Loss of Bit Count Integrity

LBCR Logbook Change Request

lbd lebend

LBD Laser Beam Deflection • Laser Beam Detector • Laser Beam Director • Learning By Demonstration • Light Beam Deflection

lbdg lebendig

LBDS Leiden-Berkeley Deep Survey

LBE Latrobe (USA-PA) (location identifier) • Lattice Boltzmann Equation • Learning By Example • Ley Básica de Empleo (Spain) • Locally-owned Business Enterprise • Location-Based Entertainment • Low-Bearing Equipment • Low Business Ethics

LBET Large Beam Expander Technology

lbf Pound-Force

LBF Labor für Betriebsfestigkeit • North Platte (USA-NE) (location identifier)

LBFL Low-power Buffered FET Logic

LBG Lyman Break Galaxy • Paris - Le Bourget (France) (location identifier)

lf-ft Pound-Foot

LBH Length - Breadth - Height • Linear Black Hole • Lyman - Birge - Hopfield (band) • Sydney - Palm Beach (Australia-NSW) (location identifier)

LBHL Lyman-Birge-Hopfield Long (band)

LBHS Long Beach High School (USA-CA)

LBI Albi (France) (location identifier) • Large Block Interface • Legally-Binding Instrument • Long-Baseline Interferometer • Long-Baseline Interferometry

LBIR Laser Beam Image Recorder • Long-Band Infrared • Low-Background Infrared • Low-Background Infrared Radiometer

LBJ Labuan Bajo (Indonesia) (location identifier) • Lyndon Baines Johnson (1908-1973)

LBJL Lyndon B. Johnson Library (USA)

LBK Left Bank • Liboi (Kenya) (location identifier)

lbl Label (file extension)

Lbl Label

LBL Label • Laminar Boundary Effect • Land Between the Lakes (USA-KY) • Lawrence Berkeley Laboratory (USA-CA) • Left Buttock Line • Liberal (USA-KS) (location identifier) • Line By Line • Low-energy-peaked BL Lacertae (object)

LBLRR Lawrence Berkeley Laboratory Research Review (periodical code)

LBLRTM Line-By-Line Radiative Transfer Model

LBLs Low-frequency-peaked BL Lacertae (objects)

LBLS Laminar Boundary-Layer Separation

lbm Pounds-Mass

LBM Lattice Boltzmann Method • Liquid Boost Module • Load Buffer Memory • Locator Back Marker • Logic Bus Monitor • Long Burn Motor • Luabo (Mozambique) (location identifier)

LBMR L-Band Microwave Radiometer

LBN Lake Baringo (Kenya) (location identifier) • Latvia (ISO country code) • Lettonie (code ISO de pays) • Logical Bibliographic Network • Logical Block Number • Lynds Bright Nebulae (catalog)

LBNL (E.O.) Lawrence Berkeley National Laboratory (USA-CA)

LBNP Lower-Body Negative Pressure

LBNPD Lower-Body Negative Pressure Device

LBO Las Brisas Observatory (USA-CO) • Laser Blow-Off • Lean Blow Out • Leveraged Buyout • Line Buildout • Lusambo (Zaire) (location identifier)

LBP Length Between Perpendiculars • Logarithmic Barrier Problem • Long Banga (Malaysia) (location identifier) • Lower Boundary of the Plate • Pound (Lebanon) (ISO currency code)

LBQ Lambarene (Gabon) (location identifier)

LBQS Large Bright QSO Survey • Large Bright Quasar Survey

Lbr Labour • Lumber

LBR Lábrea (Brazil-AM) (location identifier) • Laser Beam Recorder • Laser Beam Recording • Laser Beam Rider • L-Band Radiometer • Liberia (ISO country code) • Librarian • Library • Low Bit Rate

Lbry Library

lbs Pounds

LBS Labasa (Fiji) (location identifier) • LAN Bridge Server • L-Band System • Lectori Benevolo Salutem! • Line Block Simulator • Line Buffer System • Link Buffer Storage • Load Balance System • Load Balancing Service • Locaal Bibliotheek System • Location-Based Service • London Business School • Low-Brightness Source • Pounds

LBSAAS Life and Biomedical Sciences and Applications Advisory Subcommittee

LBSAD Life and Biomedical Sciences and Applications Division

LBSD Low-Barrier Schottky Diode

LBSS Low-Band SIGINT System • Low-Band Subsystem

lb. t. Pound (Troy)

LBT Large Binocular Telescope • Listen Before Talk • Long-Based Transponder • Long-Baseline Transponder • Low Bit Test • Lumberton (USA-NC) (location identifier)

LBTB Large Binocular Telescope Beteiligungsgesellschaft (Germany)

LBTC Large Binocular Telescope Corporation

LBTI Large Binocular Telescope Interferometer

LBTM Layer Bispectral Threshold Method

LBTO Large Binocular Telescope Observation

LBTP Launch Base Test Plan • Launch Base Test Procedure

LBTS Land-Based Test Site

LBU Labuan (Malaysia) (location identifier) • Launcher Booster Unit

LBV Leuchkräftige Blaue Veränderliche (Stern) • Libreville (Gabon) (location identifier) • Local Bus Video • Low Bandwidth Video • Luminous Blue Variable (star)

LBVDS Lightweight Broadband Variable Depth Sonar

LBVN Luminous Blue Variable Nebula

LBW Live Body Weight • Long Bawan (Indonesia) (location identifier)

LBX Laser Beam Expander • Local Bandwidth X • Local Bus Accelerator • Lubang (Philippines) (location identifier)

LBY La Baule - Montoir (France) (location identifier) • Libya (ISO country code) • Libye (code ISO de pays) • Line Busy

l/c Letter of Credit

lc Loco Citato

L/C Land Cover • Letter of Credit • Load Crew

L&C Laboratory & Checkout

LC Inductance-Capacitance • Label Cover • Laboratory and Checkout • Land Cover • Landing Chart • Launch Center • Launch Complex • Launch Conductor • Launch Cost • Launch Countdown • Land Cover • Large Coronagraph • Launch Complex • Launch Conference • Launch Control • Launch Coordinator • Launch Count • Launch Countdown • Launch Critical • Lead Covered • Left Center • Left Circular • Legal Currency • Letter Contract • Letter of Credit • Lettres et Cartes • Level Control • Level Crossing • Library of Congress • Light Curve • Lightweight Computer • Ligne Centrale • Limit Circle • Linear Chain • Linear Collider • Linear Communication • Linear Contact • Line Carrying • Line Connector • Line Contractor • Line of Communication • Line of Contact • Link-Cell • Link Control • Liquid Chromatography • Liquid Crystal • Load Carrier • Load Center • Load Compensating • Local Call • Local Channel • Local Chebychev • Local Committee • Local Control • Local Controller • Logistics Command • Low Carbon • Low Carbon • Lower California • Lower Canada • Lower Cassette • Luminosity Class • Lumped Constant • Saint Lucia (ISO country code)

LCA Laboratory for Computational Astrophysics • Land Cover Analysis • Larnaca (Cyprus) (location identifier) • Last Common Ancestor • Launch Control Amplifier • Launch Control Analyst • Launcher Control Area • Life Cycle Assessment • Light-Curve Analysis • Line Check Airman • Load Control Assembly • Load Controller Assembly • Local Call Area • Local Communications Area • Local Component Analysis • Local Context Analysis • Logistic Control Activity • London City Airport (UK-England) • Longitudinal Chromatic Aberration • Loopback Command Audio • Low-Cost Automation • Lower Case Alphabet • Lufthansa Cargo Airlines • Nicosia Larnaca (Cyprus) (location identifier) • Saint Lucia (ISO country code)

LCAAS La Crosse Area Astronomical Society (USA-WI)

LCAJ Low-Cost Anti-Jam (system)

LCAM Laboratoire des Collisions Atomiques et Moléculaires (France)

LCAMIMS Logistics Capability Assessment Models Information Management System

LCAMOS Loop Cable Maintenance Operation System

LCAO Linear Combination of Atomic Orbitals

LCAP Loosely Coupled Array of Processors

LCAR Launch Complex Assessment Report

LCAS Lake County Astronomical Society (USA-IL)

LCATS Large Capacity Automated Telecommunications System • Low-Cost Advanced Target Sensor

LCB Launch Control Building • Least Common Bit • Limited Commitment Basis • Line Control Block • Line of Constant Bearing • Local Control Box • Low-Cost Booster • Lower Crustal Block

LCC Lambert Conformal Conical (projection) • Land Cover Change • Landing Control Center • Landing Command and Control • Langage Commun de Commande • Langage de Commande de Communications • Langley Complex Coordination • Language for Conversational Computing • Launch Command and Control • Launch Commit Criteria • Launch Control Car • Launch Control Center • Launch Control Complex • Launch Control Console • Leadless Chip Carrier • Lecce Galatina (Italy) (location identifier) • Leeward Community College (USA-HI) • Levelized Compiled Code • Library Coordination Committee • Library of Congress Classification • Life-Cycle Cost • Life-Cycle Costing • Lighterage Control Center • Limited Capability Configuration • Link Communications Circuit • Liquid Crystal Cell • Load-Carrying Capability • Load Command Center • Local Communications Complex • Local Communications Console • Local Control Center • Local Control Common • Lockheed-Martin California Co. • Logic and Computational Complexity (workshop) • Logic Control Cell • Logistic Control Code • Logistics Control Center • Logistics Coordination Cell • Logistics Coordination Center • Loran C navigational Chart • Lost Calls Cleared • Low-Context Culture • Lower Centaurus-Crux (association)

LCCA Laboratório de Computação Científica Avançada • Life Cycle Cost Analysis • Logical Configuration Communication Area

LCCB Local Change Control Board • Local Configuration Control Board

LCCBA Life Cycle Cost Benefits Analysis

LCCC Launch Control Communications Center

LCCD Launch Commit Criteria Document

LCCE Life Cycle Cost Estimate

LCCF Local Cross-Correlation Function

LCCIS Local Common Channel Interoffice Signaling

LCCMP Life Cycle Cost Management Plan

LCCMS Launch Control Center Measuring Station

LCCN Library of Congress Card Number

LC&CO Launch Control & Checkout

LCCO Launch Control and Checkout • Life Cycle Cost of Ownership

LCCP Light Communication Configured Package

LCCS Large Capacity Core Storage • Life Cycle Contractor Support • Life Cycle Contract Support

LCD Laboratoire de Combustion et de Détonique (France) • Laboratory for Computational Dynamics • Language for Computer Design • Launch Control Design • Launch Countdown • Learner-Centered Design • Least Common Denominator • Liquid Crystal Diode • Liquid Crystal Display (*) • List of Chosen Descriptors • Load Classification Group • Local Change Directory • Local Climatological Data • Local Consignment Deport • Local Curvature Diminishing • Logistics Communication Division • Loopback Command Digital • Loss-of-Cell Delineation • Lost Calls Delayed • Louis Trichardt (South Africa) (location identifier) • Low Collated Damage • Lowest Common Denominator (*)

LCDA Level Converting Data Adapter • Local Committee Development Assistant

LCDB Land Cover Database

LCDCS Las Campanas Distant Cluster Survey

LCDIA Low-Cost Data Image Analyzer

LCDL Large Components Test Loop • Low-Cost Data Link

LCDM Logical-density Cold Dark Matter (model) • Low-density Cold Dark Matter (model) (*)

LCDN Last Called Directory Number

LCDP Lateral Control Departure Parameter

LCDRS Las Campanas Deep Redshift Survey

LCE La Ceiba (Honduras) (location identifier) • La Commande Électronique (France) • Latest Cost Estimate • Launch Complex Engineer • Launch Complex Equipment • Launch Control Equipment • Launch Countdown Exercise • Lettres et Civilisations Étrangères • Ley de Contratos del Estado (Spain) • Liaison Coordination Element • Limited Career Extension • Line Conditioning Equipment • Load Carrying Equipment • Local Communications Environment • Logistics Capability Estimator • Lyapunov Characteristic Exponent

LCEB Launch Control Equipment Building

LCEC Local Committee Exchange Controller • Luxembourg Commission European Communities (periodical code)

LCED Life Cycle Environmental Document

LCEF Launch Centered Earth-Fixed

LCELSS Lunar CELSS

LCER Lewis Center for Educational Research (USA-CA)

LCES Line Conditioning Equipment Scanner

LCET Logistics Cost Estimating Tool

LCF Language Central Facility • Language Control Facility • Large Cooling Flow • Launch Control Facility • Least Common Factor • Level Control Function • Liquid Crystal Filter • Local Control Facility • Local Cycle Fatigue • Log Control Function • Logical for Computable Functions • Low-Cost Fiber • Low-Cycle Fatigue • Lowest Common Factor

LCFH Large Capacity Field Heater

LCFL Last Closed Field Line

LCFS Last Closed Flux Surface • Last Come - First Served

LCG La Coruña (Spain) (location identifier) • LHC Computing Grid • Linear Congruence Generator • Linear Congruential Generator (*) • Liquid Cooled Garment • Liquid Cooling Garment • Load Classification Group • Longitudinal Centre of Gravity • Low-Cost Guidance • Luminous Compact Galaxy

LCGN Logical Channel and Group Number

LCGT Large Cryogenic Gravitational-wave Telescope

LCH Lake Charles (USA-LA) (location identifier) • Latch • Launch

LCHTF Low Cycle High Temperature Fatigue

Lchr Launcher

LCHR Launcher

LCHS Large Component Handling System

LCH4 Liquid Methane

LCI Laconia (USA-NH) (location identifier) • Langage Commun de l'Information • Launch-Centered Inertial • Launcher Control Indicator • Liquid Crystal Institute (USA-OH) • Local Craft Interface • Logical Channel Identifier • Lyapunov Characteristic Indicator

LCIE Laboratoire Central des Industries Électriques (France)

LCIIO Life Cycle Information Integration Office

LCIS Logistics Communications Information System

LCK Columbus Rickenbacker (USA-OH) (location identifier) • Library Construction Kit • Lock

LCKS Locks

Lcl Local

LCl Lycée Climatique (France)

LCL La Coloma (Cuba) (location identifier) • Laser Centerline Localizer • Latching Current Limiter • Less than Car Load • Lifting Condensation Level • Linkage Control Language • Local • Lower Confidence Limit • Lower Control Limit • Lunar Cargo Lander

LCLLY Locally

LCLO Low-Cost Low Observable

LCLOC Line Card Location

LCLS Linac Coherent Light Source • Low-Cost Launch System

LCLV Liquid-Crystal Light Value • Liquid-Crystal Light Valve • Liquid-Crystal Light Wave • Low-Cost Launch Vehicle

LCLY Locally

LCM Laboratório Central de Metrologia • Laboratory Contract Manager • La Cumbre (Argentina-CD) (location identifier) • Large-Core Memory • Launch Control Monitor • Launch Crew Member • LCD Clock Module • Least Common Multiple • Least Common Multipler • Life Cycle Management • Life Cycle Manager • Life Cycle Model • Lightweight Communications Module • Line Center Magnetogram (technique) • Line Control Module • Liquid Crystal Meter • Local Classification Model • Local Committee Meeting • Local Content Manager • Local Customer Manager • Logistics Capability Model • Logistics Composite Model • Longitude of Central Meridian • Lowest Common Multiple

LCMCS Launch Control Monitor and Checkout System

LCMI Laboratoire des Champs Magnétiques Intenses

LCML Low Calorie Meal

LCMM Life Cycle Management Model

LCMN Lauthanum Cerium Magnesium Nitrate

LCMP Life Cycle Management Plan

LC-MS Liquid Chromatography - Mass Spectrometry

LCMS Last Closed Magnetic Surface • Launch Control and Monitoring System • Life Cycle Management System • Liquid Chromatography - Mass Spectrometry • Logistics Capability Measurement System • Low-Cost Modular Spacecraft

LCMT Lfe-Cycle Mutation Testing

LCMWG Life Cycle Management Working Group

LCN Balcanoona (Australia-SA) (location identifier) • Load Classification Number • Local Civil Noon • Local Communications Network • Local Computer Network • Logical Channel Number • Logic Change Notice • Logistics Control Number • Loosely Coupled Network • Lyapunov Characteristic Number

LCNC Leiden Center for Natural Computing (Netherlands)

LCNO Linear Combination of Orbitals

LCNP Long Capitalized Noun Phrases

LCNVG Low-Cost Night Vision Goggle

LCO Laboratoire de Chimie Océanographique • Lague (Congo) (location identifier) • Las Campanas Observatory (Chile) • Launch Control Office • Launch Control Officer • Launch Control Operation • Launching Control Officer • Limit-Cycle Oscillation • Limited Condition of Operability • Limiting Condition for Operation • Liquid Crystal Device • Loan Control Officer • Local Control Organization • Loopback Command Off • Lower Critical Orbit

LCOG Local Committee Operations Guide

LCOM Lack of Cohesion in Methods • Local Committee Operations Manual • Logic Control Output Module

Lconc Local Concavity

LCOSE Launch Complex Operational Support Equipment

LCP Language Conversion Program • Large Combustion Plant • Last Complete Program • Launch Control Panel • Least-Cost Planning • Left Circularly-Polarized • Left Circular Polarization • Light Charged Particle • Light Communications Package • Linear Complementarity Problem (*) • Linear Complementary Problem • Link Configuration Protocol • Link Control Protocol • Liquid-Crystal Polymer • Local Calibration Procedure • Local Committee President • Low-Calcium Pyroxene • Low-Cost Production • Lower Cost Processor

LCPC Laboratoire de Contrôle des Ponts et Chaussées (France) • Languages and Compilers for Parallel Computing (workshop)

LCPE Laboratoire de Chimie et Physique de l'Environnement

LCPL Left Circularly Polarized Light

LCPM Local Committee Presidents' Meeting

LCQ Light-Cone Quantization • Logical Channel Queue

l/cr Lettre de Crédit

LCR Inductance - Capacitance - Resistance • Laboratoire Central de Recherches • La Chorrera (Colombia) (location identifier) • Laser Clean Room • Least Cost Router • Least Cost Routing • Line Control Register • Listed Character Reference • Local Cosmic Ray • Log Count Rate • Low Cross-Range

LCRS Las Campanas Redshift Survey

LCRSSS Libyan Centre for Remote Sensing and Space Science (Libya)

LCS Laboratory for Computer Science • Landsat Customer Services • Large Capacity Storage • Large Cassegrain Spectrograph • Large Complex System • Large Core Storage • Las Cañas (Costa Rica) (location identifier) • Laser

Cross-link System • Laser Cross-Section • Launch Control Sequence • Launch Control Simulation • Launch Control Station • Launch Control System • LCU Common Software • Learning Classifier System • Ley de Contrato de Seguro • Library Circulation System • Library Computer System • Library Control System • Life Cycle Support • Lightcurve Shape • Limiting Control Setting • Line Coding Storage • Liquid-Crystal Shutter • List of Commands Signal • Literature Classification System • Load Carrying Structure • Local Common Software • Local Conceptual Schema • Local Control System • Longest Common Subsequence • Loop Clock Selector • Loop Control System • Lotus Communication Server • Loudness-Contour Selector

LCSA Life Cycle Support Agent • Load Carrying Structure Assembly

LCSB Low Central Surface Brightness

LCSDE Life Cycle Software Development and Engineering

LCSDEC Life Cycle Software Development and Engineering Center

LCS&E Laboratory for Computational Science & Engineering

LCSE Laboratory for Computational Science and Engineering • Laser Communication Satellite Experiment • Life Cycle Software Engineering • Life Cycle Support Environment

LCSEC Life Cycle Software Engineering Center

LCSH Library of Congress Subject Heading

LCSI Launch Critical Support Item

LCSMM Life Cycle System Management Model

LCSO Launch Complex Safety Officer • Launch Control Safety Officer

LCSOP Launch Complex Safety operating Procedure

LCSP Launch Complex Safety Plan • Life Cycle Support Plan • Liquid Crystal-based Stokes Polarimeter • Local Call Service Position

LCSQA Laboratoire Central de Surveillance de la Qualité de l'Air

LCSRM Loop Current Step Response Authority

LCSS Launch Control and Sequencer System • Life Cycle Software Support

LCSSC Life Cycle Software Support Center

LCSSE Life Cycle Software Support Environment

LCST Laboratoire des Composants et Systèmes pour Télécommunications • Launch Combined System Test

LCT Laboratoire Central de Télécommunications • Laplace-Carson Transform • Ley de Contrato de Trabajo (Spain) • Line Characteristic Table • Link Confidence Test • Local Civil Time • Local Correction Tracking • Local Correlation Technique • Local Correlation Tracking • Local Craft Terminal • Low-Cost Terminal • Low Critical Temperature

Lctd Located

LCTES Languages, Compilers, and Tools for Embedded Systems (workshop)

LCTF Liquid Crystal Tunable Filter

LCTGM Library of Congress Thesaurus for Graphic Materials

LCTI Large Components Test Installation

L/CTL Launch Control

LCTMP Little Change in Temperature

Lctr Locator

LCTS Laboratory Computer Test Set

LCTV Liquid-Crystal Television

LCTY Locality

LCU Laptop Computer Unit • Large Close-Up • Last Cluster Used • Launch Control Unit • Launch Correlation Unit • Library Control Unit • Lightweight Computer Unit • Line Control Unit • Line Coupling Unit • Load Control Unit • Local Control Unit

LCUG Liquid Cooling Undergarment

LCUP Least-Cost Utility Planning

LCV Light Commercial Vehicle • Light Contingency Vehicle • Logistics Cargo Vehicle • Loopback Command Video • Lucca (Italy) (location identifier)

LCVD Least Voltage Coincidence Detection

LCVG Liquid Cooling and Ventilation Garment • Liquid Cooling/Ventilation Garment

LCVIP License Contractor Vendor Inspection Report

LCVR Liquid-Crystal Variable Retarder

LCW Linear Collider Workshop • Low-Conductivity Water

LCX Logistics Coordination Exercise

LCY London City (UK-England) (location identifier)

ld ledig • leitend • leuchtend

LC3 Logistics Command, Control, and Communications

lD Lichter Durchmesser

Ld Ladung • Land • Landung • Lead • Leading • Limited • Load • Lord

L/D Length to Diameter • Letter of Deposit • Lift to Drag

L&D Landing & Deceleration • Loss & Damage

LD Landau Damping • Landau-Darrieus (instability) • Landesdienst • Landing and Deceleration • Large Diffusion • Laser Designator • Laser Desorption • Laser Diode • Lateral Direction • Launch Dependability • Launch Director • Lead Detector • Learning Disability • Learning-Disabled • Lethal Dose • Letter Directive • Level Discriminator • Libyan Dinar (currency) • Light Discipline • Limb-Darkened • Limb Darkening • Linear Dimension • Line Drawing • Line Driver • Line of Departure • Line of Duty • Linguistics in Documentation • List Dump • Loading Dock • Lobe-Dominated • Local Disk • Local Distribution • Logic Driver • Logistics Demonstration • Long Distance • Low Delay • Low Density • Low Dutch

Lda Licenciada • Limitada

LDA Landing Directional Aid • Landing Distance Available • Laser Diffraction Analyzer • Laser Doppler Anemometer • Laser Doppler Anemometry • Latissimus Dorsi Anterior • Leading Dimension of the Array • Linear Diode Array • Linear Discriminant Analysis • Line Driving Amplifier • Local Data Acquisition • Local Database Agent • Local Density Approximation • Localizer Directional Aid • Locate Drum Address • Logical Device Address • Logistics Design Appraisal • Low-Density Amorphous • Malda (India) (location identifier)

LDAC Leiden Data Analysis Center (Netherlands)

LDAIS Learning in Distributed Artificial Intelligence Systems (workshop)

LDAP Leader Development Action Plan • Lightweight Directory Access Panel • Lightweight Directory Access Protocol (*)

LDAR Lightning Detection and Ranging System

LDAS Land Data Access Subsystem • Letchworth and District Astronomical Society (UK) • Local Data Access System

LDB Large Database • Launch Data Bus • Lexical Database • Light Distribution Box • Linear Decision Boundary • Lithium Depletion Boundary • Limited Data Block • Local Data Block • Logistics Data Bank • Londrina (Brazil-PR) (location identifier) • Long-Duration Balloon • Long-Duration Ballooning

LDBD Logical Database Design

LDBF Long-Duration Balloon Flight

LDBS Land Database System

LDC Labor Distribution Code • Latitude Data Computer • Least-Developed Country • Less-Developed Country •

Level Decision Circuit • Light Direction Center • Lightweight Deployable Communications • Lindeman Island (Australia-QLD) (location identifier) • Line Drop Compensator • Linguistic Data Consortium • Local Data Center • Local Distribution Center • Local Distribution Company • Logistics Data Center • Logistics Data Communications • Long-Distance Call • Long-Distance Carrier • Long-Distance Communication • Lot Date Code • Lotus Development Corporation • Low-Developed Country

LDCC Large Diameter Component Cask • Leaded Chip Carrier

LDCE Limited Duration Candidate Exposure

LD-CELP Low-Delay Code Excited Linear Prediction

LDCELP Low-Delay Code Excited Linear Prediction

LDCS Long-Distance Control System

LDD La Définition des Données • Lasting Deep Penetration • Lightly Doped Drain • Logical Database Design • Long-Duration Disease • Loss - Damage - Destruction • Low-Density Data • Lunar Dust Detector

LDDE Luminosity-Dependent Density Evolution

LDDI Local Distributed Data Interface • Logical Distributed Data Interface

LDDL Logical Data Definition Language

LDDMWG LGSOWG Data Distribution and Marketing Working Group

LDDN Leader Development Decision Network

LDDS Langage de Description des Données Statistiques • Long-Distance Discount Service • Low-Density Data System

LDE Lagrangian Differential Equation • Laminar Defect Examination • Linear Differential Equation • Liquid Discharge Eject • Long-Decay Event • Long-Delayed Echo • Long-Duration Event • Lourdes Tarbes Ossun (France) (location identifier) • Low-Density Environment

LDEC Lunar Docking Events Controller

LDEF Land Data Entry Form • Long-Duration Exposure Facility

LDEO Lamont-Doherty Earth Observatory (USA-NY)

Lderry Londonderry (Northern Ireland)

LDEV Logical Device

LDF Laboratoire David Florida (Canada-ON) • Large Deployable Reflector • Lateral Density Function • Lateral Distribution Function • Line Depression Function • Liquid Drop Facility • Lightweight Digital Facsimile • Linear Display File • Log-normal Distribution Function • Long-Duration Flare

LDFD Logical Data Flow Diagram

LDFOCCS Local Distribution Fiber Optics Cable Communications System

Ldg Ladung • Landing • Landung • Leading • Loading • Lodging

LDG Laboratoire de Détection et de Géophysique • Landing • Landing Gear • Leading • Light Decontamination Group • Lodge • Logistics Data Gateway

LDGE Lodge

LDGO Lamont-Doherty Geological Observatory

LDGPS Landing Gear Position • Local-area Differential Global Positioning System

LDH Large Dynamic Hyperbase • Lord Howe Island (Australia-NSW) (location identifier)

LDI Lactate Dehydrogenase • Landing Direction Indicator • Landsat Digital Image Analysis System • Large Diffusion Internationale • Lean Direct Injection • Life Detection Instrument • Lindi (Tanzania) (location identifier) • Linear Differential Inclusion • Linear Displacement Indicator • Line Driver Interface • Long Dwell Imagery

LDIF LDAP Data Interchange Format

LDIM Lunar Digital Image Model

LDINT Line Driver Interface

LDIS Land Data Ingest System • Lightning Data and Information System • Low-Dynamics Instrumentation Set

LDIU Launch Data Interface Unit

LDJ Linden (USA-NJ) (location identifier)

Ldk Landkarte

LDK Lidköping (Sweden) (location identifier) • Lower Deck

LDL Language Description Language • Large Diffusion Locale • Logical Data Language • Logic-based Data Language • Long Delay Line • Low-Density Lipoprotein • Low-Density Liquid • Lower Deviation Level

LDLE Luminosity-Dependent Luminosity Evolution

LDLP Low-Density Lipoprotein

LDM Last Day of the Month • Lens Data Manager • Library Data Management • Limited-Distance Modem • Linear Delta Modulation • Liquid-Drop Model • Local Data Manager • Local Domain Manager • Logical Data Model • Logistics Data Manager • Logistics Decision Model • Long Data Message • Long-Distance Modem • Ludington - Mason County (USA-MI) (location identifier)

LDMA Laboratoire de Dynamique Moléculaire et Atomique

LDMD Localized Dark Mantle Deposit

Ldmk Landmark

LDMK Landmark

Ldmn Lendemain

LDMP Library Data Management Project

LDMS Laboratory Data Management System • Laser Distance Measuring System • Laser Desorption Mass Spectrometer • Laser Desorption Mass Spectrometry • Long-Duration Meteor Stream • Lunar Distance Measuring System

LDMX Local Digital Message Exchange

Ldn London

LDN Lamidanda (Nepal) (location identifier) • Large Diffusion Nationale • Less-Developed Nation • Level Day/Night • Local Data Network • Lynds Dark Nebulae (catalog)

LDNG Landing

LDNS Lightweight Doppler Navigation System

Ldo Licenciado

LDO Ladouanie (Suriname) (location identifier) • Laser Designation Operator • Leader Development Office • Letter District Office • Limited Duty Officer • Linear Distribution of Objects • Logically Deleted Object • Lower Domain Ontology • Lunar Discovery Orbiter

LDOCE Longmans Dictionary of Contemporary English

LDOCF Long-Distance Operational Control Facility

L/DOS Launch/Deploy Operations Segment

LDOS Landsat Data and Operations System • Local Density of Occupied States • Local Density Of States

Ldp Ladyship • Lordship

LDP Label Distribution Protocol • Language Data Processing • Language Designated Position • Large Deviation Principle • Leader Development Program • Library Development Program • Linux Documentation Project • Local Data Package • Local Data Processor • Logistics Data Package • Long-Distance Path • Long-Duration Penetration • Lorentz Doppler Profile • Low-Density Polyethylene

LDPE Low-Density Polyethylene

LDQ Lobe-Dominant QSO • Lobe-Dominated QSO (*) • Lobe-Dominant Quasar • Lobe-Dominated Quasar (*)

Ldr Ladder • Leader • Ledger • Lodger

LDR Large Deployable Reflector • Large Diffusion Régionale • Leader • Lethal Dose of Radiation • Level Distribution Recorder • Light-Dependent Resistor • Light-to-Dark Ratio • Linear Depolarization Ratio • Linear Dynamic Range

• Line-Dependent Resistor • Liquid-Droplet Radiator • Lodar (Yemen) (location identifier) • Lorentz Double Refraction • Low Data Rate • Low Data Register • Low-Density Region • Low Dose Rate

LDRC Logic Design Rules Checker

LDRD Laboratory Directed Research and Development

LDRI Low Data Rate Input

LDRLQ Lobe-Dominated Radio-Loud Quasar

LDRM Low Data Rate Multiplexer

LDRP Logistics Depot Repair Procedure

Ldry Laundry

lds Laus Deo Semper

Lds Loads

LDS Landing/Deceleration System • Landing - Deservicing - Safing • Large Disk Storage • Laser Detecting Set • Laser Docking Sensor • Latter Day Saints • Launch Data System • Launch Detection Satellite • Layered Defense System • Leadership Development Seminar • Leicester Database System • Leiden-Dwingeloo Survey • Lightweight Decontamination System • Linear Decision Surface • Linear Dynamic System • Local Distribution System • Logical Data Structure • Logistics Data System • Long-Distance Service • Low-Density Sample • Low-Dispersion Spectrograph • Luyten Double Stars

LDSA Local Directory System Agent

LDSL LBR Digital Subscriber Line

LDSS Launch Decision Support System • Leader Development Support System • Lunar Deep Seismic Sounding • Low-Dispersion Spectroscopic Survey • Low-Dispersion Survey Spectrograph

LDSVR Letter Design Specification Validation Report

LDT Laser Display Technology • Level Delay Time • Library Development Team • Linear Differential Transformer • Linear Displacement Transducer • Local Descriptor Table • Logic Design Translator • Long-Distance Transmission

LDTP Laboratoire de Dynamique Terrestre et Planétaire (OMP)

LDTS Laser Designator Tracker System

LDU Lahad Datu (Malaysia) (location identifier) • Level Distribution Unit • Light Distribution Unit • Line Driver Unit

LDUUV Large-Diameter Unmanned Undersea Vehicle

LDV Laser Doppler Velocimeter • Laser Doppler Velocimetry • Laser Doppler Velocity • Laser Doppler Vibrometer • Leonardo Da Vinci • Linear Differential Vector • Linguistische Datenverarbeitung • Lock Data View • Low-Data Voice

LDVE Linear Differential Vector Equation

LDVR Letter Design Validation Report

LDVS Logistics Data Validation System

LDW Lansdowne (Australia-WA) (location identifier) • Long-Distance Wireless • Loran Drop Windsonde

LDWG Large Display Working Group

LDX Long-Distance Extender • Long-Distance Xerography • Saint-Laurent-du-Maroni (French Guiana) (location identifier)

LDY Londonderry (Northern Ireland) (location identifier)

LDZ Lithium-Depleted Zone • Londolozi (South Africa) (location identifier)

Le Lebanese • Lebanon • Leone (currency) • Lewis Number

L/E Left Engine

LE Egyptian Pound (currency) • LAN Emulation • Language Engineering • Language Environment • Langues Étrangères • Large Enterprise • Launch Effectiveness • Launch Equipment • Launch Escape • Launching Equipment • Leading Edge • Left End • Left Engine • Left Eye • Lesotho • Less or Equal • Less than or Equal to • Library Edition • Light Equipment • Lightning Effect • Limited Edition • Limit or Error • Linear Encoder • Linear Equalizer • Lobster Eye (geometry) • Long Element • Lorien Engine • Low End • Low Energy • Low Excitation • Luminosity Evaluation • Lyapunov Exponent

Lea League • Leave

LEA Laboratoire Européen Associé • Laboratorium voor Elektromagnetisme en Acustica • Laboratory for Experimental Astrophysics • Laboratory of Ephemeris Astronomy • Langues Étrangères Appliquées • Langues Étrangères Approfondies • Late Embryonic Abundance • Launch Escape Assembly • Law Enforcement Agency • Learmonth (Australia-WA) (location identifier) • Light-Emitting Array • Ligue des États Arabes • Limited Exclusion Area • Load Effective Address • Local Education Agency • Local Education Authority • Local Examination Authority • Logistics Engineering Analysis • Logistics Evaluation Agency • Lycée d'Enseignement Adapté (France)

LEAD Leadership Education And Development • Leadership for Environment And Development • Logistic Engineering Advanced Demonstration • Low-cost Encryption and Authentication Device

LEADS Law Enforcement Automated Data System

LEAF Land Ecosystem Atmosphere Feedback • Large Effective Area Fiber • Large European Acoustic Facility • Laser Environmental Active Fluorosensor • Laser Excited Automatic Fluorometer • Law Enforcement Access Field • Logistics Education Assistance Fund

LEAFS Laser Excited Atomic Fluorescence Spectrometry

LEAM Lunae Ejecta And Meteoroids

LEANDRE Lidar embarqué pour l'Étude des Aérosols, des Nuages, de la Dynamique, du Rayonnement et des Espèces Minoritaires

LEAO Lager Economisch en Administratief Onderwijs

LEAP Lasers and Electro-optic Applications Program • Leading Edge Advanced Package • Lightweight Exoatmospheric Projectile • Low-Energy Antiproton • Lunar Environment Assessment Package

LEAPS Low Electron Attachment Potential Species

LEAR Logistics Evaluation and Review Technique • Low-Energy Antiproton Ring

LEARP LAN Emulation Address Resolution Protocol

LEAS Lease Electronic Accounting System

LEAST Learning Systems Standardization

LEAU Law Enforcement Assistance Unit

Leb Lebanese • Lebanon

LEB Hanover (USA-NH) (location identifier) • Label External Boundary • Launch Equipment Building • Lebanon (USA-NH) (location identifier) • Low-Energy Booster • Low-Energy Buncher • Lower Equipment Bay • White River (USA-VT) (location identifier)

LeBAN Leningrad Biblioteka Akademii Nauk SSSR (periodical code)

LEBES Laboratory Electron Beam Exposure System

LEBS Low-Emission Boiler System

LEC LAN Emulation Client • Launch Equipment Console • Launch Escape Control • Light-emitting Electrochemical Cell • Link Edit Control • Liquid Encapsulated Czochralski • Local Exchange Carrier

LECF Lateral Energy Correction Function

LECID LAN Emulation Client Identifier

LECIS Laboratory Equipment Central Interface Specification

LECP Logistic Engineering Change Proposal • Low-Energy Charged Particle

LECR Low-Energy Cosmic Ray

LECRe Low-Energy Cosmic-Ray Electron

LEC/S Low-Energy Concentrator/Spectrometer

LECS Land Evaluation Computer System • LAN Emulation Configuration Server • Launching Equipment Checkout Set • Liquid Effluent Collection System • Low-Energy Concentration Spectrometer • Low-Energy Concentrator-Spectrometer (*)

Lect Lectio • Lecture • Lecturer

LECT Liquid Effluent Collection Tank

Lectr Lecturer

Led Ledig

LED Language Editor • La Silla Engineering Department (ESO) • Lateral Effect Diode • Law Enforcement Desk • Light-Emitting Diode • Low-Energy Detector • Low-Energy Diffraction • Low-Energy Dislocation • Saint Petersburg - Pulkovo (Russia) (location identifier)

LEDA Library of Efficient Data and Algorithms • Light and Easily Deployable by Air • Local Employment Development Action • Lunar European Demonstration Approach • Lyon-Meudon Extragalactic Database • Online Earthnet Data Availability

LEDAS Leicester Database and Archive Service • Leicester Database Archive Service

LEDC Low-Energy Detonating Charge

LEDCOP Light Element, Detailed Configuration Opacity (code)

LEDD Light-Emitting Diode Display

LEDR Laboratory for Environmental Data Research

LEDSS Laboratoire d'Études Dynamiques et Structurales de la Sélectivité (France)

LEDT Limited-Entry Decision Table

LEE Leeds (Starlink internal code) • Leesburg (USA-FL) (location identifier) • Low-Energy Electron

LEED Low-Energy Electron Diffraction

LEEM Low-Energy Electron Microscope • Low-Energy Electron Microscopy

LeEne Leningrad Energoizdat (periodical code)

LEER Low-Energy Electron Reflection

LEET Limiting Equivalent Exposure Time

LEF Large Energy Fluctuation • Leading Edge Flap • Lebakeng (Lesotho) (location identifier) • Light-Emitting Film • Linear Electric Field • Low-frequency Electric Field

LEFI Local Electric Field Instrument

LEFM Linear Elasticity Fracture Mechanics

leg Legacy Document (file extension)

Leg Legajo • Legal • Legislation • Legislative • Legislature

LEG Aleg (Mauritania) (location identifier) • Landesentwicklungsgesellschaft (Germany) • Légua • Low-Energy Gamma-ray • Low-Energy Germanium • Loe-Excitation Galaxy

LEGG Longevity Evaluation Government Group

LEGI Laboratoire des Écoulements Géophysiques et Industrielles (France)

LeGid Leningrad Gidrometeoizdat (periodical code)

Legis Legislation • Legislative • Legislature

LEGM Low-Energy Gamma Monitor

LEGOS Laboratoire d'Études en Géophysique et Océanographie Spatiales (OMP)

LEGR Low-Energy Gamma-Ray

LEGRI Low-Energy Gamma-Ray Imager

LEGS Liquid Encapsulated Gallium-arsenide Solidification

LEGSPC Low-Energy Gas Scintillation Proportional Counter

LEGT Lycée d'Enseignement Général et Technologique (France)

LEGTA Lycée d'Enseignement Général et Technologique Agricole (France)

LEGTI Lycée d'Enseignement Général et Technologique Industriel (France)

LEGTP Lycée d'Enseignement Général, Technologique et Professionnel (France)

LEGTPI Lycée d'Enseignement Général, Technologique, Professionnel et Industriel (France)

LEH Launch/Entry Helmet • Le Havre - Octeville (France) (location identifier)

LEI Almería (Spain) (location identifier) • Laser-Enhanced Ionization • Leicester (Starlink internal code) • Legal Education Institute • Local Employment Initiative

LEICA Low-Energy Ion Composition Analyzer

Leics Leicestershire (UK-England)

LEID Low-Energy Ion Detector

LEIDS Logistics Electronic Information Delivery System

Leip Leipzig (Germany)

LEIR Low-Energy Ion Ring

LEISA Linear Etalon Imaging Spectral Array

LEIT Low-Energy Imaging Telescope

LeIzE Leningrad Izdatel Energiia (periodical code)

LeIzG Leningrad Izdatel Gidrometeoizdat (periodical code)

LeIzK Leningrad Izdatel Klimiia (periodical code)

LeIzM Leningrad Izdatel Mashinostroenie (periodical code)

LeIzN Leningrad Izdatel Nauka (periodical code)

LeIzS Leningrad Izdatel Sudostroenie (periodical code)

LeIzU Leningrad Izdatel Leningradskogo Universiteta (periodical code)

LEJ Leipzig (Germany) (location identifier)

LEK Labe (Guinea) (location identifier) • Liquid Encapsulated Kyropoulos

LEL Laser à Électrons Libres • Lake Evella (Australia-NT) (location identifier) • Linear Energy Loss • Link-Edit Language • Link, Embed and Launch-to-edit • Lower Explosive Limit • Lower Exposure Limit

LELE Low-Extinction Level Event

LeLen Leningrad Lenizdat (periodical code)

LELS Low-Energy Laser System

LELU Launch Enable Logic Unit

LEM Laboratoire d'Études des Microstructures • Laboratory Environment Model • Landform Elevation Model • Language Extension Module • Languages in Educational Mobility • Launch Escape Motor • Lecteur Électronique de Microfiches • Lemmon (USA-SD) (location identifier) • Light Equipment Maintenance • Link Error Module • Link Error Monitor • Logistic Element Manager • Long-Elements Method • Low-Energy Monitor • Lunar Excursion Module • Lunar Exploratory Module

LEMA Long-term Ecological Modeling Activity

LEMAC Leading Edge Mean Aerodynamic Chord

LEMD Laboratoire d'Électrostatique et de Matériaux Diélectriques (France)

LEMDE Lunar Excursion Module Descent Engine

LEMMS Low-Energy Magnetospheric Measurement System

LEMO Laboratoire d'Électromagnétisme Microondes et Optoélectronique (France)

LEMP Lightning Electromagnetic Pulse

LEMPA Low-Energy Magnetospheric Particle Analyzer

LEMS Lunar Excursion Module Simulator

LEMT Low-Energy Matrix Telescope

LEMTG Laboratoire d'Étude des Mécanismes de Transfert en Géologie (OMP)

LEMZ Liquid Encapsulated Melt Zone

len Lenticaris

LEN Large Extension Node • Length • León (Mexico) (location identifier) • Line Equipment Number • Local Electrical Neutrality • Low Entry Networking

LENA Low-Energy Neutral Astrophysics • Low-Energy Neutral Atom (imager) (*)

LEND Large External Network Database

LENS Large Extension Node Switch • Low-Energy Neutrino Spectrometer • Low-Energy Neutrino Spectroscopy

LENSC Lake Erie Nature and Science Center (USA-OH)

LENST Large Energy National Shock Tunnel

LENTIC Laboratoire d'Étude des Nouvelles Technologies de l'Information et de la Communication (Belgium)

Leo Leo-Leonis (constellation)

LEO Leconi (Gabon) (location identifier) • Leonids (meteor stream) • Liaison Engineering Order • Librating Equidistant Observer • Linear Electro-Optic • Link Everything Online • Logistics Engineering Office • Low Earth Orbit • Low Earth Orbiter

LEOA Launch Early Orbit Analysis

LEOMA Laser and Electro-Optics Manufacturers Association

Leon Leonés

LEONESS Low Earth Orbiting Nadir Etalon Sounding Spectrometer

LEOP Launch and Early Operations Phase • Launch and Early Orbit Phase (*) • Low Earth Operational (orbit) • Low Earth Orbit Phase

LEOPR Laboratoire d'Électrochimie Organique et Photochimie Redox (France)

LEOS Laboratory for Electro-Optic Sensors • Low Earth Orbiting Satellite

LEOSAR Low Earth Orbit Synthetic Aperture Radar

LEOSAT Low Earth Orbit Satellite

Lep Lepus-Leporis (constellation)

LEP Laboratoire d'Électronique Philips • Laboratory for Extraterrestrial Physics • Large Electron-Positron (collider) • Large External Pressurized (tank) • Laser-Electron Photon • Layered Electrical Product • Least Energy Principle • Leopoldina (Brazil-MG) (location identifier) • Light-Emitting Polymer • Limited English Proficiency • Linear Error Probable • Localized Erection Phenomenon • Low-Energy Particle • Lower End Plug • Lowest Evaluated Price • Lycée d'Enseignement Professionnel (France)

LEPA Lycée d'Enseignement Professionnel Agricole (France)

LEPC Local Emergency Planning Committee • Low-Energy Proportional Counter

LEPEDEA Low-Energy Proton and Electron Differential Energy Analyzer

LEPES Laboratorie d'Études des Propriétés Électroniques des Solides (France)

LEPI Laboratoire d'Évaluation et de Prospective Internationales

LEPL Locally Essential Particle List

LEPMI Laboratoire d'Électrochimie et de Physicochimie des Matériaux et Interfaces (France)

LEPOR Long-term and Expanded Program of Oceanic Research

LEPORE Long-term and Expanded Program of Oceanic Research and Exploration

LEPP Laboratory for Elementary Particle Physics (Cornell)

LEPS Laser-Electron Photon (facility) at SPring-8 (Japan) • Launcher Environmental Protective System • Launch Escape Propulsion System • Lidar Echo Processing System

LEPSF Long-Exposure Point Spread Function

LEPSI Laboratoire d'Électronique et de Physique des Systèmes Instrumentaux

LEPT Low-Energy Particule Telescope

LEQ Lands End (UK-England) (location identifier) • Liaison École-Quartier

LER Label Edge Router • Launch Equipment Room • Launcher Equipment Room • Lebensgestaltung - Ethik - Religion • Leinster (Australia-WA) (location identifier) • Licensee Event Report • Light-Emitting Resistor • Line-Emitting Region • Link Error Rate • Long Eye Relief (optics) • Loss Exchange Ratio • Lunar Eyepiece Receptacle

LERAM Littoral Ecosystem Risk Assessment Model

LeRC Lewis Research Center (NASA) (now John H. Glenn Research Center)

LERCSI Laboratoire d'Étude de la Réactivité Catalytique des Surfaces et Interfaces (France)

LERec L'Écho des Recherches (periodical code)

LERECO Laboratoire d'Études et de Recherches Économiques (France)

LERS Learning from Examples based on Rough Sets • Lewis Research Center (NASA)

LERTS Laboratoire d'Études et de Recherches en Télédétection Spatiale (France)

LERU League of European Research Universities

LERX Leading Edge Root Extension

LES Land Earth Station • LAN Emulation Server • LAN Emulation Service • Large Eddy Simulation • Large Eddy Simulator • Large Extension Switch • Launch and Entry Suit • Launch Enabling System • Launch/Entry Suit • Launch Environmental Simulator • Launch Equipment Shop • Launch Escape System • Leading Edge Simulation • Leading Edge System • Lesobeng (Lesotho) (location identifier) • Licensing Executives Society • Lincoln Experimental Satellite • Loi sur l'Enseignement Supérieur (France) • Lunar Escape System

LESA Land Evaluation and Site Assessment • Legal Expert Systems Association • Local Environment System Architecture

LESC Life and Environmental Sciences • Life and Environmental Sciences Committee • Life and Environment Standing Committee • Lockheed Engineering and Sciences Co

LESD Logistics Exercise and Simulation Directorate

LESIA Laboratoire d'Études Spatiales et d'Instrumentation en Astrophysique

LESIP Little Experimental System for Image Processing

LESOC Lincoln Experimental Satellite Operations Center

LESR Limited Early Site Review

LESS Lake Effect Snow Studies • Laser Excited Sholskiij Spectroscopy • Leading Edge Structural System • Leading Edge Structure System • Leading Edge Subsystem • Leascost Estimating and Scheduling System (IBM) • Least Squares Solution

L. ès Sc. Licencié ès Sciences

LEST Laboratoire d'Économie et de Sociologie du Travail (France) • Large Earth Survey Telescope • Large Earth-based Solar Telescope (*) • Large Earth-borne Solar Telescope • Large Earthbound Solar Telescope • Large Earth Survey Telescope • Large European Solar Telescope

LESU Laboratoire d'Études Supérieures Urbaines (France)

Let Lettish

LET Laboratoire d'Essais des Télécommunications (France) • Launch Escape Tower • Leonhard Euler Telescope • Leticia (Colombia) (location identifier) • Letter • Light

Equipment Transport • Limited Environment Test • Limiter Experiment Testband • Linear Energy Transfer • Linearer Energietransfer • Link Evaluation Terminal • Local Essential Tree • Long Erection Time • Low-Energy Telescope • Lycée d'Enseignement Technique (France)

LETAM Low End Telecommunications Access Method

LETF Launch Equipment Test Facility

LETG Low-Energy Transmission Grating

LETGS Low-Energy Transmission Grating Spectrometer

LETI Laboratoire d'Électronique et des Techniques de l'Informatique • Laboratoire d'Électronique et de Technologie de l'Informatique

LETS Liquid Engine Transient Simulation • Local Employment Trading System • Local Exchange Trading System • Low-Energy Telescope System

Lett Letters • Lettres

LEU Large External Unpressurized (tank) • Launch Enable Unit • Launcher Electronic Unit • Low-Enriched Uranium • Seo de Urgel (Spain) (location identifier)

LEUWS Low End Unix Workstation

Lev Levante

LEV Bureta Levuka (Fiji) (location identifier) • Launch Escape Vehicle • Levuka (Fiji) (location identifier) • Luminous Eruptive Variable (star) • Lunar Excursion Vehicle

LEW Auburn (USA-ME) (location identifier) • Lewiston Auburn (USA-ME) (location identifier)

LEWEX Labrador Extreme-Wave Experiment

LEWP Line Echo Wave Pattern

LEWS Lightweight Extreme Weather Shelter

Lex Lexical • Lexicon • Lexikon

LEX Leading Edge Extension • Lexicon • Lexington - Blue Grass (USA-KY) (location identifier) • Local Exchange

LEXAS Laboratory for Experimental Analysis (WUStL)

LExEn Life in Extreme Environments

LEXS Laser-driven Electron X-ray Source

LExSWG Lunar Exploration Science Working Group (NASA)

LEZ La Esperanza (Honduras) (location identifier) • Lunar Equatorial Zone

L/F Load Factor

LF Franc (Luxembourg) (currency) • Laboratory Frame • Large Field • Large Format • Late Finish • Launch Facility • Launch Forward • Lead Facilitator • Left Forward • Leuchtkraftfunktion • Life Finder • Light Face • Linear Feet • Linear File • Line Feed • Line Finder • Line Fitting • Load Factor • Loaf • Local Fluff • Locator Field • Log File • Logical File • Loss of Frames • Low Frequency • Luminosity Function

LfA Laboratory for Astrophysics • Linux for Astronomy • Low-frequency Active

LFA Lead Federal Agency (USA) • Locally Flat Approximation • Logistics Functional Analysis • Long-Focus Astrograph • Low-Frequency Active • Low-Frequency Adjunct • Low-Frequency Astronomy

LFAF Low-Frequency Accelerometer Flutter

LFAM Low-Frequency Accelerometer Mode

LFAP Low-Frequency Accelerometer Pogo

LFAS Low-Frequency Active Sonar

LFC Large-Format Camera • Level of Free Convection • Limit of Free Convection • Lit Fluidisé Circulant • Local File Check • Loss Free Counting • Low-Frequency Control • Low-Frequency Current

LFCS Logical Foundations of Computer Science (symposium)

LFCTR Laboratorio di Fisica Cosmica e Tecnologie Relative (CNR)

lfd laufend

LFD Lasting Flexible Penetration • Lietuvos Fiziku Draugija (Lithuania) • Linear Finite Difference • Live Functional Diagram • Local Fractional Derivative • Low-Fat Diet • Low-Frequency Domain

LFDF Low-Frequency Direction Finder

LFDLTS Low-Frequency Deep-Level Transient Spectroscopy

lfd. M. laufenden Monats

LFDN Launch Facility Down

LFE Laboratory For Electronics • Leyes Fundamentales del Estado

LFED Lightweight Forward Entry Device

LFEEP Ligue Française de l'Enseignement et de l'Éducation Permanente (France)

LFER Linear Free Energy Relationship

LFES Launch Facility Environmental Shelter

LFF Logical Feature Filter • Logistics Factors File • Lower Forward Fuselage

LFFT Long Form Functional Test

Lfg Lieferung

LFG Lexical Functional Grammar

LFGM Low-Frequency Gravitomagnetic

LFH Local Failure Handler • Lunar Far Horizon

LFI Hampton Langley (USA-VA) (location identifier) • Laser Fantasy International (USA-WA) • Last File Indicator • Limited Field Investigation • Link Fragmentation and Interleaving • Low-Frequency Imager • Low-Frequency Instrument (*) • Lost and Found International

LFIE Live Flight Integration Exercise

LFIRG Luminous Far-Infrared Galaxy

LFK Log File Key • Lufkin - Angelina County (USA-TX) (location identifier)

LFL Linear Field Line • Long Focal Length • Lower Flammability Limit

LFM Lateral Force Microscope • Lateral Force Microscopy • Lateral Force Mode • Limited-area Fine Mesh • Limited Fine Mesh • Limited Fine-mesh Model • Linear Fields per Minute • Linear Frequency Modulated • Linear Frequency Modulation • Local Focus Mechanism • Long Focus Mechanism • Lower Flammability Limit • Low-Frequency Modulation • Low-powered Fan Marker • Lyon - Fedder - Mobarry (code)

LFML Low Fat Meal

LFMR Low-Frequency Microwave Radiometer

LFN Long File Name • Louisburg Franklin (USA-NC) (location identifier) • Low-Frequency Noise

LFO Low-Frequency Oscillator • Kelafo Callaf (Ethiopia) (location identifier)

LFOE Large Format Optical Enlarger

LFOP Landing and Ferry Operations Panel

LFOS Launch and Flight Operations System • Launch Facility Out of Sync

LFOV Large Field Of View

LFP Lakefield (Australia-QLD) (location identifier) • Large-Format Printing • Logistics Funding Profile

LFPCA Large-Format Photon Counting Array

LFPL Lewis Flight Propulsion Laboratory (NASA)

LFpM Linear Fields per Minute

LFPM Linear Fields Per Minute

LFPRL Lewis Flight Propulsion Research Laboratory (NASA)

LFR Franc Luxembourgeois (currency) • La Fría (Venezuela) (location identifier) • Launch and Flight Reliability • Less

Favoured Region • Liechtenstein Förderkreis für Raumfahrt • Lietuvos Fizikos Rinkinys (periodical code) • Low-Frequency Range • Low-Frequency Receiver • Luxembourg Franc (currency)

LFRA Low-Frequency Radio Astronomy

LFRD Low-Frequency Reliability Definition

LFRIG Luminous Far-Infrared Galaxy

LFRR Low-Frequency Radio Range

LFS Labour Force Survey • Laser Fluorescent Spectroscopy • Laser Fluorosensor • Laser-induced Fluorescent Spectroscopy • Life From the Stratosphere • Live From the Stratosphere (*) • Logic Fault Simulator • Loop Feedback Signal

LFSA Logical Force Structure Assessment

LFSAR Low-Frequency Synthetic Aperture Radar

LFSO Lunar Far Side Observatory

LFSR Linear Feedback Shift Register

LFSRQ Low-energy-peaked Flat-spectrum Radio Quasar

LFSS Launch Facility Security System

LFST Largest Feasible Steerable Telescope

Lft Leaflet

LFT Lafayette (USA-LA) (location identifier) • Least Frequent Trigram • Lift • Long Flight Time • Low-Function Terminal • Luyten Five Tenths • New Iberia - Lafayette (USA-LA) (location identifier)

LfU Landesamt für Umweltschutz

LFU Laser Firing Unit • Least Frequently Used • Lunar Flying Unit

LFV Low-Frequency Variability • Low-Frequency Variable (object) • Low-Frequency Vibration • Lunar Flying Vehicle

LFVN Flow-Frequency VLBI Network

LFW Lome (Togo) (location identifier)

LFZ Laminar Flow Zone • Line-Free Zone

lg Common Logarithm • lang • long

Lg Lagoon • Lange • Large • Length • Liège (Belgium) • Long

LG Label Generator • Landgericht (Germany) • Landing Gear • Landing Ground • Largo • Lecture Guide • Length • Ley General • Line Generator • Linguistic Geometry • Local Government • Local Group • Locomotive Generator • Logistics Group • Low Gain • Low German • Lycée Général (France)

LGA Lattice Gas Automaton • Local Government Agency • Local Government Area • Local Government Authority • Low-Gain Antenna • New York - La Guardia (USA-NY) (location identifier)

LGAI Laboratori General d'Assaigs i Investigacions (Spain)

LGB Laser-Guided Bomb • Lateral Geniculate Body • Long Beach (USA-CA) (location identifier)

Lgc Logic

LGC La Grange - Calloway (USA-GA) (location identifier) • Large Gas Chromatograph (PV) • Launch Guidance Computer • Local Group Census • Loop Generation Cask • Lunar Gas Chromatograph • Lunar Geological Camera

LGCA Laboratoire de Géodynamique des Chaînes Alpines (France)

LGCAI Latin - Greek - Cyrillic - African - IPA

LGCP Lexical-Graphical Composer Printer

LGCR Laboratoire de Gravitation et Cosmologie Relativistes (France)

LGCU Landing Gear Control Unit

LGD La Grande (USA-OR) (location identifier) • Large Group Display

LGDT Load Global Descriptor Table • Local Global Descriptor Table • Local Government Development Program

lge large

Lge Liège (Belgium)

LGE Logic Gate Expander • Lunar Geological Equipment

LGEIES Local Government Environmental Information Exchange Scheme

LGF Yuma Laguna (USA-AZ) (location identifier)

LGG Liège Bierset (Belgium) (location identifier)

LGGE Laboratoire de Glaciologie et de Géophysique de l'Environnement (France)

LGGST Laboratoire de Géologie-Géomorphologie Structurale et de Télédétection

LGH Leigh Creek (Australia-SA) (location identifier)

LGI Deadmans Cay (Bahamas) (location identifier) • Long Island (Bahamas) (location identifier)

LGIO Local Government Information Office

LGIS Local Government Information Service

LGIT Laboratoire de Géophysique Interne et Tectonophysique (France)

LGIU Laser Gyro Interface Unit

LGI2P Laboratoire de Génie Informatique et de l'Ingénierie de Production (France)

LGK Langkawi (Malaysia) (location identifier)

LGL Legal • Long Lellang (Malaysia) (location identifier) • Low Galactic Latitude

LGM Laboratoire de Géologie et Mécanique (France) • Laiagam (Papua New Guinea) (location identifier) • Last Glacial Maximum • Link Gateway Module • Little Green Man • Loop Group Multiplexer

LGN Lagoon • Landelijke Grondgebruiksclassificatie van Nederland • Landsat Ground Network • Line Gate Number • Linga Linga (Papua New Guinea) (location identifier) • Logical Group Node • Logical Group Number

LGO Langeoog (Germany) (location identifier) • Low Gravity Orbit

LGP Laboratoire de Géographie Physique • Legaspi (Philippines) (location identifier) • Length Growth Period • Ley General Presupuestaria (Spain) • Limit Generation Program

LGPL Lesser General Public License • Library General Public License

LGQ Lago Agrio (Ecuador) (location identifier)

Lgr Longueur

LGR Cochrane (Chile) (location identifier)

LGRNG Long Range

Lgs Longs

LGS Landing Gear System • Laser Guide Star • Local Group Suspected (member) • Local Group System • Malargue (Argentina-MD) (location identifier)

LGSCV Local Group of Santa Clarita Valley (USA-CA)

L/GSE Launch/Ground Support Equipment

LGSE Launch and Ground Support Equipment

LGSF Laser Guide Star Facility

Lgsltd Legislated

Lgsltr Legislature

LGSM Light Ground Station Module

LGSOWG Landsat Ground Station Operation Working Group • Landsat Ground Station Operators Working Group

LGSR Local Galactic Standard of Rest • Local Group Standard of Rest (*)

LGST Logistic Group Service Test

Lgstcs Logistics

LGT Las Gaviotas (Colombia) (location identifier) • Ley General del Trabajo • Ley General Tributaria (Spain) • Light • Liquid-Gauging Technology • Local Geomagnetic

Time • Low Group Transmit • Lycée Général et Technologique (France)

LGTA Landing Gear Test Article

Lgtd Lighted

Lgth Length

Lgtps Longtemps

Lgts Lights

LGTS Lights

LGU Local Government Unit • Logan Cache (USA-UT) (location identifier) • London Guildhall University (UK-England)

lgue longue

LGUPM Leningradskii Gosudarstvennyi Universitet Problemy Matematicheskoi Fiziki (periodical code)

LGV Ligne à Grande Vitesse • Long Goods Vehicle

LGW London Gatwick (UK-England) (location identifier) • Low Gross Weight

LGWD Large Group Wall Display

LGWR Log Writer

LGX Light Guide Crossconnect • Lugh Ganana (Somalia) (location identifier)

LGY Lagunillas (Venezuela) (location identifier)

LGZ Leguizamo (Colombia) (location identifier)

lH lichte Hohe

L/H Local/Horizontal

LH Labour Hour • Left Half • Left Hand • Left Handed • Lick H-alpha (catalog) • Light Helicopter • Light-Hour • Line Hunting • Link Header • Liquid Hydrogen • Load High • Local/Horizontal • Long Haul • Love-Hate • Lower Half • Lower Hybrid • Lucke-Hodge

LHA Laboratoire de l'Horloge Atomique (France) • Lahr (Germany) (location identifier) • Lamont-Hussey H-Alfa • Landshauptarchiv • Local Hour Angle (*) • Lower Hour Angle

LHATS Local Hour Angle of True Sun

LHB Landes- und Hochschulbibliothek • Late Heavy Bombardment • Line History Block • Local Hot Bubble • Lokale Heiße Blase • Lost Harbor (USA-AK) (location identifier)

LHBl La Houille Blanche (periodical code)

LHc Liquid Hydrocarbon

LHC Large Hadron Collider • Left-Hand Circular • Logistics Headquarters Center • Lower Hatch Closed

LHCb Large Hadron Collider beauty (experiment)

LHCC Large Hadron Collider Committee

LHCD Logical Hierarchical Connectivity Data • Lower Hybrid Current Drive

LHCN Long-Haul Communications Network

LHCP Left-Hand Circularly Polarized • Left-Hand Circular Polarization • Left-Hand Circular Polarized

LHD Doctor of Human Letters • Landing Helicopter Dock • Large Helical Device • Left-Hand Drive • Litterarum Humaniorum Doctor

LHDI Lower Hybrid Drift Instability

LHDLC Logical High-level Data Lead Control

LHDr Left-Hand Drive

LHe Liquid Helium

LHE Lahore (Pakistan) (location identifier)

LHEA Laboratory for High-Energy Astrophysics

LHe4 Liquid Helium

LHFEB Left-Hand Forward Equipment Bay

LHG Lightning Ridge (Australia-NSW) (location identifier)

LHGR Linear Heat Generation Rate

LHGXA Lightweight High-Gain X-band Antenna

LHI Lereh (Indonesia) (location identifier)

LHIA Logistics Horizontal Integration Analysis

LHITA Long-Haul Information Transfer Architecture

LHK Guanghua (China-PRC) (location identifier)

LHM Laser-Hardened Material • Learning in Humans and Machines • Left-Handed Material

LHMEL Laser Hardening Materials Evaluation Laboratory

LHMS Lithuanian Hydrometeorological Service

LHMT Letters on Heat Mass Transfer (periodical code)

LHN Long-Haul Network • Lishan (China-ROC) (location identifier)

LHNO Lager Huishoud- en Nijverheidsonderwijs

LHO Lower Hatch Open

LHOP Laser-Hardened Optical Payload

LHP Left-Hand Panel • Lehu (Papua New Guinea) (location identifier) • Loop Heat Pipe

lhr Lumen-Hour

LHR Left Hand Rule • London Heathrow (UK-England) (location identifier) • Love-Hate Relationship • Lower Hybrid Resonance

LHRCS Left-Hand Reaction Control System

lhs Literate Haskell Source (file extension)

LHS Laser Heterodyne Spectrometer • Las Heras (Argentina-SC) (location identifier) • Lawrence Hall of Science (USA-CA) • Layered Half Space • Left-Hand Side • Library of Health Sciences • Liquid Hydrogen Storage • Lunar Horizon Sensor • Luyten Half-Second (catalog)

LHSC Left-Hand Side Console • Liquid Hydrogen System Complex

LHSS Linear Hashing Superimposed Signature

LHSSC Left-Hand Side Storage Container

LHSTF Liquid Hydrogen Structural Test Facility

LHSV Liquid Hourly Space Velocity

LHT Layman's Hypertext • Left Hand Turn • Line Haul Tractor • Local Heaven Time

LHU Lake Havasu (USA-AZ) (location identifier) • Local Health Unit

LHUP Lock Haven University of Pennsylvania (USA-PA)

LHV Lock Haven - W.T. Piper (USA-PA) (location identifier) • Lower Heating Value

LHW Langzhou (China-PRC) (location identifier)

lhx LHA Compression (file extension)

LHX Liquid Hydrogen

LHY Liquid Hydrogen

LH2 Liquid Hydrogen

Li Link • Linker • Lithium

L&I Launch & Impact

LI Language-Independent • LAN Interface • Laser Illuminator • Laser Ionization • Lead Investigator • Leeward Islands • Left Intermediate • Length Indicator • Letter of Intent • Letter of Introduction • Level Indicator • Less Important • Liberia • Liechtenstein (ISO country code) • Lifted Index • Linearized Image • Line Interface • Line Item • List • List Item • Lithuania • Load Immediate • Local Interconnect • Long Island (USA-NY) • Longitudinal Interval • Lorentz Invariance • Lower Intercept • Low Intensity • Low Ionization • Lyons Instruments Ltd. (UK)

LIA Laboratoire d'Informatique d'Avignon (France) • Laboratory of Image Analysis • Land Information and Analysis • Laser Industry Association • Laser Institute of America (USA-FL) • Legacy Interface Adapter • Liangping (China-PRC) (location identifier) • Limiting Interval Availability • Linear-Induction Accelerator • Little Ice Age •

Logistics Impact Analysis • Logistics Integration Agency • Lorien Iterative Algorithm • LWS Interactive Analysis (package)

LIAC Liège International Astrophysical Colloquium

LIACC Laboratorio de Inteligência Artificial e Ciência de Computadores

LIADA Liga Ibero-Americana de Astronomía

LIAI Leningrad Institute of Aviation Instrumentation (Russia)

LIAISE Local Interactive In-school Science and Engineering

LIAM Local Infrastructure and Access Maintenance

LIAS Library Information Access System

lib Liber • Library (file extension)

Lib Liber • Liberal • Liberia • Liberty • Libra-Librae (constellation) • Librarian • Library • Libretto

LIB Left Inboard • Liberdade • Library • Light-Ion Beam • Limbunya (Australia-NT) (location identifier) • Logistics Information Board

LIBAL Low-Ionization Broad Absorption Line

LIBER Ligue des Bibliothèques Européennes de Recherche

LIBG Luminous Infrared Bright Galaxy

LIBHE Laboratoire Interuniversitaire Belge des Hautes Énergies (Belgium)

LIBI/O Library Input/Output

LIBIO Library Input/Output

LIBIS Leuvens Integraal Bibliotheek- en Informatiesysteem

Libr Libration

LIBRA Logiciel Intégré pour les Bibliothèques en Réseau Automatisé

LIBRIME Library and Information Management in Europe

LIBRIS Land Image-Based Resource Information System

LIBS Laser-Induced Breakdown Spectroscopy • Library Information and Bibliographic System

lic License (file extension)

Lic Licenciado • License • Licensed • Licenciado • Licentiate

LIC Land Information Center • Land Information Council • Language Identifier Code • LAN Interface Card • Licensed Internal Code • Ligne Inférieure de Contrôle • Limon (USA-CO) (location identifier) • Linear Integrated Circuit • Line Integral Convolution • Line Integrity Check • Line Interface Card • Line Interface Coupler • Load Interface Circuit • Local Interstellar Cloud • Logistics Indicator Code • Loop Insertion Cell • Low-Intensity Conflict

LICA Low-energy Ion Composition Analyzer

Licdo Licenciado

LICIL Long Island Consortium for Interconnected Learning (USA)

LICK Lightweight Communications Kit • Local Interstellar Cloud Kernel

LicOB Lick Observatory Bulletin (periodical code)

LICQ Laboratório Industrial de Controle da Qualidade (Portugal) • Linear Independence Constraint Qualification

LiCr Liquid Crystals (periodical code)

LICS Library Information Control System • Laser-Induced Continuum Structure • Logic In Computer Science (symposium) • Lotus International Character Set

LICU Laparoscopic Intracorporeal Ultrasound

LID Lander Interface Document • Land Information Division • Laser-Induced Damage • Laser-Induced Desorption • Laser Interference Diffractometer • Leadless Inverted Device • Leiden Valkenburg (Netherlands) (location identifier) • Library Issue Document • Lift Improvement Device • Light-Induced Damage • Light-Induced Drift • Like Item Distribution • Line Interface Device • Line Item Description • Linear Imaging Device • Local Injection and Detection • Local Island Diversity • Local Island Divertor • Locked-In Device • Logical Input Device • Logic In Databases (workshop) • Logistics Identification Document • Luminescent Indicia Detector

LIDAR Laser Image Detection And Ranging • Laser Imaging Detector And Ranger • Light Detection And Ranging

LIDB Line Information Database

LIDIA Liquid Dispensating Agency

LIDOS Long-slit Imaging Dual Order Spectrograph

LIDQA Landsat Image Data Quality Analysis • Landsat Image Data Quality and Analysis (*)

LIDS Laser Illumination Detection System • Logistics Integrated Data System • Loral Information Display System

LIDSE Laboratoire Interdisciplinaire de Didactique des Sciences Expérimentales (France)

LIDT Load Interrupt Descriptor Table

LIE Left Inboard Elevon • Libenge (Zaire) (location identifier) • Liechtenstein (ISO country code) • Lightning-Induced Electron • Limited Information Estimation • Linear Incremental Encoder • Localized Induction Equation • Lower Ionization Energy

Liech Liechtenstein

LIEF Launch Information Exchange Facility

LIEP Large Internet Exchange Packet

LIESIB Laboratoire d'Informatique, d'Électronique, du Signal, et de l'Imagerie de Bourgogne (France)

LIET Local Inclusive Electron Trigger

LiF Laser-induzierte Fluoreszenz • Lithium Fluoride

LIF Large Isothermal Furnace • Laser-Induced Fluorescence • Laser Interference Filter • Layaway Industrial Facility • Lifou (Loyalty Islands) (location identifier) • Light Interference Filter • Linking Interface • Logical Interchange Format • Low Insertion Force

LIFD Laser-Induced Fluorescence Detection

LiFE Learning is a Family Experience

LIFE Laboratory for International Fuzzy Engineering • Language Industry For Europe • Laser-Induced Fluorescence Emission • Laser Infrared Flyout Experiment

LIFES Laser-Induced Fluorescence and Environmental Sensing

LIFFE London International Finance Futures Exchange

LIFIA Laboratoire d'Informatique Fondamentale et d'Intelligence Artificielle

LIFO Last In - First Out

LIFS Laser-Induced Fluorescence Spectroscopy

LIFT Laser-Induced Fluorescence of Trees • Laser-Induced Forward Transfer • Life science Technology • Linking Innovation, Finance and Technology (EU) • Logically Integrated Fortran Translator • Logistics Improvement Facility Technology • Long Term - Full Time

Lig Liguria

LIG Limoges Bellegarde (France) (location identifier) • Luminous Infrared Galaxy

LIGHT Life-cycle Global Hypertext

LIGM Luminous Infrared Galaxy Merger

LIGMALS Landsat Interactive Grey Map And Level Slice

LIGO Laser Interferometer Gravitational-wave Observatory

LIH Kauai Island (USA-HI) (location identifier) • Laser Interferometric Holography • Light Intensity High • Lihue (USA-HI) (location identifier)

LiI Lithium Iodide

LII Mulia (Indonesia) (location identifier)

LIIP Laser-Induced Infrared Photochemistry

LIJ Leaf-Initiated Join • Long Island (USA-AK) (location identifier)

LIJP Leaf-Initiated Joint Parameter

LIK Likiep Island (Marshall Islands) (location identifier)

LIL Large Immersion Lens • Law of Iterated Logarithm • Library Interconnection Language • Light Intensity Low • Lille Lesquin (France) (location identifier) • Linear Injector for LEP (CERN) • Local Interaction Language • Low-Ionization Line

LILE Large-Ion Lithophile Element

LILIA LTE Inversion based on Lorien Iterative Algorithm

LILO Last In - Last Out • Linux Loader

LILT Low-Intensity Low-Temperature

Lim Limerick (Ireland) • Limit • Limitation • Limité • Limited • Limiting

LIM Land Identification Method • Land Information Management • Land Inventory and Management • Laser Intensity Modulation • Light Intensity Modulation • Lima (Peru) (location identifier) • Limit • Limiter Impurity • Linear Induction Motor • Line Interface Module • Local Information Manager • Losing Inventory Manager • Lotus - Intel - Microsoft • Low- and Intermediate-Mass (star)

LIMA Laboratoire d'Instrumentation Météorologique Aéroportée (France) • Laser Ionization Mass Analyzer • Laser Ionization Mass Analysis • Lotus - Intel - Microsoft - AST • Low-Inertia Magnetic Accretor

LIMAC Large Integrated Monolithic Array Computer

LIMAP Landsat Imagery Analysis Package

LIMARS Laser Imaging and Ranging System

LIMD Large-scale In-Memory Database

LimDis Limited Dissemination • Limited Distribution

LIMDO Light Intensity Modulation Direct Overwrite

LIME Laboratorie d'Instrumentation en Micro-informatique et Électronique (France)

LimFac Limiting Factor

LIMGR Local Information Manager

LIMIS Liquid Metal Ion Source

LIMM Laser Intensity Modulation Method • Light Intensity Modulation Method

LIMOS Laser Intensity Modulation System

LIMP Lunar-anchored Interplanetary Monitoring Platform

LIMRC LRU Identification and Maintenance Requirements Catalog

LIMS Laboratory Information Management Software • Laboratory Information Management System • Land Information Management System • Landsat Image Management System • Laser Ionization Mass Spectroscopy • LeRC Information Management System • Library Information Management System • Lightning Mapper Sensor • Limb Infrared Monitoring System • Limb Infrared Monitor of Stratosphere • Limb Inversion Monitoring of the Stratosphere • Logistics Information Management System • Logistics Inventory Management System

LIMSI Laboratoire d'Informatique pour la Mécanique et les Sciences de l'Ingénieur (France)

LIMSS Logistics Information Management Support System

Lin Lineal • Linear

LIN Less Industrialized Nation • Linear • Line Item Number • Liquid Nitrogen • Logical Infostructure Name • Milano - Forlanini Linate (Italy) (location identifier)

LINAC Linear Accelerator

LINC Laboratory Instrument Computer • LBT Interferometric Camera • Logic and Information Network Compiler

LINCOTT Liaison, Interface, Coupling, Technology Transfer

Lincs Lincolnshire (UK-England)

LINDA Langmuir Interferometry and Density for Astrid 2 • Line Drawing Analyzer

LINE Library Information Network Exchange

LINEAR Lincoln Near-Earth Asteroid Research

LINER Low-Ionization Nuclear Emission-line Region • Low-Ionization Nuclear Emission Region

Ling Lingüística

LINJET Liquid Injection Electric Thruster

LINK Live and Interactive Network of Knowledge

lin-log Linear-Logarithmic

LINLOG Linear-Logarithmic

LINPAC Linear Algebra Package

LINS Laser Inertial Navigation System

LINUS Laser-Induced Nuclear Simulation

LINX London Internet Exchange

LINZ Land Information of New Zealand

LIO Land Information Office • Limon (Costa Rica) (location identifier) • Longitudinal Interoperability

LIOC Launch and In-Orbit Checkout

LIOCS Logical Input/Output Control System

LiOH Lithium Hydroxide

LIOMAS Large-Ion Mass Spectrometer

LIOME Learning, Interaction, and Organization in Multiagent Environments (workshop)

LION Lunar International Observer Network

LIP Laboratoire d'Information du Parallélisme • Laboratory for Image Processing • Land Impact Program • Large Igneous Province • Large Installation Program • Large Internet Packet • Laser-Induced Plasma • Launch In Process • Lightning Instrument Package • Limited Installation Program • Lins (Brazil-SP) (location identifier) • Local Image Processing • Local Interested Party • Loop Initialization Primitive • Loop Initialization Process • Lunar Impact Probe

LIPAS Lidar Performance Analysis Simulator

LIPI Indonesian Institute of Sciences • Lembaga Ilmu Pengetahuan Indonesia

LIPL Linear Information Processing Language

LIpS Logical Instructions per Second • Logical Interferences per Second

LIPS Laboratory Information Processing System • Laser Image Processing Scanner • Library of Image Processing Software • Logical Instructions Per Second • Logical Interferences Per Second • Logistics Information Processing System

LIPSAS Lidar Performance Analysis Simulator

LIPX Large Internetwork Package Exchange

Liq Liquid • Liquor

LIQ Liquid • Lisala (Zaire) (location identifier)

LIR Large Infrared Radiometer (PV) • Liberia (Costa Rica) (location identifier) • Line Integral Refractometer

LIRA Library and Information Retrieval Application • Linear Recognition and Analysis • Lunar Infrared Array

LIRACS Library Information Resources Automated Control System

LIRAD Lidar-Radiometer

LIRES Literature Retrieval System

LIRF Low-Intensity Reciprocity Failure

LIRG Luminous Infrared Galaxy

LIRIS Long-slit Intermediate Resolution Infrared Spectrograph

LIRS Laser Imaging Radar Sensor • Library Information Retrieval Service • Library Information Retrieval System • Limb Infrared Sounder • Logical Information Retrieval System

Lis Lisboa (Portugal)

L&IS Library & Information Science

LIS Laboratory Information System • Land Information System • Landsat Information System • Language-Independent Specification • Language-Independent Standard • Laser Interferometer System • Laser Isotope Separation • Launch Instant Selector • Law of Inevitable Shrinkage • Learning and Intelligent System • Letter Information Sample • Letter Information System • Ley del Impuesto sobre Sociedades (Spain) • Library and Information Science • Library and Information Service • Library and Instructional Services • Library Information System • Lightning Imaging Sensor • Lightning Imaging Spectrometer • Line Item Summary • Lisboa Portela (Portugal) (location identifier) • Lisbon Portela (Portugal) (location identifier) • Lithium Ion Storage • Logical IP Subnetwork • Logistics and Inventory System • Logistics Information System • Loop Input Signal

LISA Laboratoire Interuniversitaire des Systèmes Atmosphériques (France) • Land Information System for Agriculture • Large Interferometer for Submillimeter Astronomy (*) • Large Installation System's Administration • Large Installation System's Administrator • Large Isotope Spectrometer for Astromag • Laser Interferometer Space Antenna • Laser Interferometric Space Antenna (*) • Library and Information Science Abstracts • Library and Information Services in Astronomy • Library Systems Analysis • Lichthinder Informatie Site Associatie (Netherlands) • Limb Imaging Spectrograph for Airglow • Linked Indexed Sequential Method • Local Integrated Software Architecture • Logistics Information Systems Analysis

LISAA Low-cost Imaging System for Advanced Applications

LISAG Laser Interferometry Satellite Array for Gravimetry

LISAN Libraries on the Information Superhighway Advocacy Network

LISAR LaRC Image Scanning, Archival and Retrieval

LISB Lithium Ion Storage Battery

LISC Land Information System Committee • Lockheed-Martin Integrated Solutions Co.

LISD Logistics Information Systems Division

LISE Laboratoire d'Interférométrie Stellaire et Exoplanétaire (France)

LISI Laser-Induced Surface Improvement • Level of Information Systems Interoperability

LISITT Laboratorio Integrado de Sistemas Inteligentes y Tecnologias en Trafico

LISKA Lithium Ion Source Karlsruhe (Germany)

LISM Local Interstellar Medium

LISN Line Impedance Stabilization Network

LiSOCl2 Lithium-Thionyl Chloride

LISP List Processing • List Processor

LISPB Lithospheric Seismic Profile of Britain (UK)

LISPEC LLNL Immersion Spectrograph

LISR Line Information Storage and Retrieval

LISRF Local Interstellar Radiation Field

LISS Linear Imaging Self-Scanner • Linear Imaging Self-Scanning • Linear Imaging Self-scanning Sensor • Lockheed Intermediate Sun Sensor

LISSA Life Support Systems Analysis

LISSP Laboratory for Image Signal and Statistical Processing

LISSR Line Item Schedule Status Record

LIST Library Index Search and Transcribe • LISA International Science Team • Lunar Interferometry Study Team

LISTED Library Integrated System for Telematics-based Education

Lit Literal • Literalmente • Literally • Literary • Literatur • Literatura • Literature • Lithuanian • Litre • Litter • Little

LIt Lira Italiana (currency)

LIT Italian Lira (currency) • Laboratório de Integração e Testes • Large Image Terminal • Lead-In Training • Linear Ion Trap • Line Identifier Tag • Lira Italiana (currency) • Literature Interpretation Theory • Little Rock (USA-AR) (location identifier) • Local Implementation Team • Local Information Transfer • Low-Impedance Transmission

LITA Library and Information Technology Association (*) • Library Information and Technology Association • Local Information Transfer Architecture

LITAS Low-Intensity Two-Colour Approach System

LITC Library Information Technology Centre

LITE Large Imagery Telescope • Large Imaging Telescope • Laser In-space Technology Experiment • Legal Information Through Electronics • Lidar In-space Technology Experiment • Light Inquiry Through Experiments (project) • Light-Time Effect • Lightweight Imaging Technology Experiment • Limited Interoperability Test and Evaluation • Links and Interfaces for Tool-data Exchange • Lower Ionosphere Thermosphere Explorer

Lith Lithograph • Lithographer • Lithographic • Lithography • Lithuania • Lithuanian

Litho Lithos (periodical code)

Lithol Lithology

LITINT Literature Intelligence

LITS Lights In The Sky • Linear Ion Trap Standard • Local Information Transfer System • Logistics Information Technology Strategy

Litt Littéraire • Littera • Littéral • Littérature

LITU Life In The Universe

LITVC Liquid Injection Thrust Vector Control

LIU LAN Interface Unit • Line Interface Unit • Long Island University (USA-NY)

LIUAPV Laboratoire d'Informatique de l'Université d'Avignon et des Pays de Vaucluse (France)

LIUM Laboratoire d'Informatique de l'Université de Mons (Belgium)

Liv Liverpool (UK-England) • Livonian • Livre

LIV Line Item Value • Livengood (USA-AK) (location identifier) • Low-Input Voltage • Lunar and Interplanetary Vehicle

LIVA Liga Venezolana de Astronomía (Venezuela) (formerly LIVAA)

LIVAA Liga Venezolana de Aficionados a la Astronomía (Venezuela) (now LIVA)

LIVC Low-Input Voltage Converter

LIVCR Low-Input Voltage Conversion and Regulation

LIVE Large Image Visualization Environment

LIVES Landsat Image Verification and Extraction System

LIVR Low-Input Voltage Regulation

LIVS Lvov Izdatel Vishcha Shkola (periodical code)

LIW Loikaw (Myanmar) (location identifier) • Long Instruction Word

LIWC Linguistic Inquiry and Word Count (program)

LIY Hinesville Wright (USA-GA) (location identifier)

LIZ Limestone Loring (USA-ME) (location identifier)

LIzNe Leningrad Izdatel Nedra (periodical code)

lJ laufenden Jahres

Lj Lichtjahr

LJ Loja

LJA Lodja (Zaire) (location identifier)

LJC Louisville (USA-KY) (location identifier)

LJJ Long Josephson Junction

LJLV Little Joe Launch Vehicle

LJM Liverpool (Starlink internal code)

LJMU Liverpool John Moores University (UK-England)

LJN Lake Jackson - Brazoria County (USA-TX) (location identifier)

LJO Learjet Observatory • Licksele Ionospheric Observatory (Sweden)

LJU Ljubljana - Brniki (Slovenia) (location identifier)

LK Lake • Leak • Lochkarte • Lock • Sri Lanka (ISO country code)

LKA Landeskriminalamt (Germany) • Larantuka (Indonesia) (location identifier) • Sri Lanka (ISO country code)

LKB Lakeba (Fiji) (location identifier) • Large Knowledge Base • Lochkartenanlage

LKBS Leids Kerkhoven Bosscha Fonds (Indonesia)

LKC Lekana (Congo) (location identifier)

Lkd Locked

LKD Lakeland Downs (Australia-QLD) (location identifier)

LKE Seattle - Lake Union (USA-WA) (location identifier)

LKEI Lockheed Martin - Khrunichov - Energia International

LKF Leuchtkraftfunktion

Lkg Locking

LKG Loop Key Generator

Lkge Linkage

LKGI Loop Key Generator Interface

LKHA Lick H-Alpha

LKI Duluth Lakeside (USA-MN) (location identifier)

LKK Kulik Lake (USA-AK) (location identifier)

LKL Lakselv (Norway) (location identifier) • Lochkartenleser

LKLY Likely

LKM Local Key Manager • Lochkartenmaschine

LKN Leknes (Norway) (location identifier) • Lembaga Kimia Nasional (Indonesia)

LKO Lichtklimatisches Observatorium (Switzerland) • Lucknow (India) (location identifier)

LKP Lake Placid (USA-NY) (location identifier) • Last Known Position

Lkr Locker

LKR Las Khoreh (Somalia) (location identifier) • Lieferkoordinierungsrichtlinie • Rupee (Sri Lanka) (ISO currency code)

LKS Lakes • Lakeside (USA-TX) (location identifier)

LKT Lakota (Ivory Coast) (location identifier)

LKU Lake Rudolf (Kenya) (location identifier)

LKV Lakeview - Lake County (USA-OR) (location identifier) • Landes- und Kommunalverwaltung • Lochkartenverfahren

LKW Lastkraftwagen

LKY Lake Manyara (Tanzania) (location identifier)

LKZ Brandon Lakenheath (UK-England) (location identifier) • Landeskennzahl

ll Loco Laudado

L-L Line-to-Line

L/L Land Line • Latitude/Longitude

L&L Launch & Landing • Legislative Liaison

LL Laatstleden • Landau-Lifshitz • Landline • Late Latin • Latent Lethality • Launch and Landing • Launch Left • Lave-Linge • Laws • Leased Line • Lebanese Pound (currency) • Legislative Liaison • Lessons Learned • Lever Lock • Light Line • Ligne Louée • Limiting Level • Limited Liability • Lincoln Laboratory (MIT) • Line-by-Line • Linear-Linear • Line Length • Lines • Link by Link • Link Layer • Liquid Limit • Little Link • Live Load • Lobby Level • Lobe Length • Local Language • Local Loop • Logical Link • Long Lead • Long Line • Loose Leaf • Lords • Lower Left • Lower Level • Lower Limit • Lower Lobby • Lower Loop • Low Lead • Low Level • Low Luminosity • Lunar Landing

LLA Latitude - Longitude - Altitude • Linklist Lookaside • Logical Layered Architecture • Luleå (Sweden) (location identifier)

LLabJ Lincoln Laboratory Journal (periodical code)

LLAC Long Line Accounting Classification

LLAGN Low-Luminosity Active Galactic Nucleus

LLAM Logical Link Accepted Message

LLAP Local Link Access Protocol

LLAPI Low-Level Air Picture Integration • Low-Level Air Picture Interface

LLB Bachelor of Laws • Laboratoire Léon Brillouin (France) • Large Liquid Booster • Legum Baccalaureus • Line Loopback

LLBF LENS Low-Background Facility

LLBL Low-Latitude Boundary Layer

LLC Limited Liable Company • Line Length Code • Link Layer Control • Liquid-Liquid Chromatography • Limited Liability Corporation • Local Learning Center • Logical Layer Control • Logical Link Control • Long Life Cooler • Lower-Layer Compatibility • Lower Left Centre • Low Layer Compatibility

LLCC Low Life Cycle Cost

LLCF Launch and Landing Computational Facilities

LLCP Logical Link Control Protocol

LLCS Length of Longest Common Subsequence

LLCSC Low-Level Computer Software Component

LLD Doctor of Laws • Lamp Lumen Depreciation • Legum Doctor • Location de Longue Durée • Logic Level Driver • Lower Level Discriminator • Lower Level of Detection • Lower Limit of Detection • Low-Level Detector • Low-Level Discrete

LLDC Least Developed Country

LLDM Logical Link Denied Message

LLDPE Linear Low-Density Polyethylene

LLDT Load Local Descriptor Table

LLE Laboratory for Laser Energetics • Liquid-Liquid Extraction • Load List Element • Loss of Life Expectancy • Low-Level Exposure • Low-Luminosity Elliptical • Malelane (South Africa) (location identifier)

LLEBG Local Law Enforcement Block Grant (USA)

LLEIS Lower-Level End Item Subdivision

LLF Large Low-speed Facility • Local Luminosity Function • Low-Level Format • Low-Level Formatting

LLFBB Large Liquid Fast Burn Booster

LLFM Low-Level Flux Monitor

LLFS Lyman-Limit Flat Spectrum

LLG Chillagoe (Australia-QLD) (location identifier) • Low-Luminosity Galaxy

LLGCX Low-Luminosity Globular-Cluster X-ray (source)

LLH Las Limas (Honduras) (location identifier)

LLI Lalibela (Ethiopia) (location identifier) • Latitude and Longitude Indicator • Limited-Life Item • Local Lorentz Invariance • Logical Link Identifier • Long-Lead Item

LLIA Local Language Inference Algorithm

LLIC Low Level Interaction Component

LLID Loopback Location Identification

LLIL Long Lead Item List • Long Leadtime Items List

LLIV Low-Latitude Intermediate Velocity (arch) • Low-Level Input Voltage

LLL Later Life Learner • Lawrence Livermore Laboratory (now LLNL) • Learning Language in Logic (workshop) • Left Lower Lobe • Limited-Life List • Lissadell (Australia-WA) (location identifier) • Logistics Lesson Learned • Lowest Landau Level • Low-Level Laboratory • Low-Level Logic • Low Light Level

LLLLLL Laboratories Low-Level Linked List Language

LLLTV Low-Light-Level Television

LLLW Liquid Low-Level Waste

LLM L-band Land Mobile • Long Lama (Malaysia) (location identifier) • Lunar Landing Mission • Lunar Landing Module

LLMC Lower-Latitude Meridian Circle (*) • Low-Latitude Meridian Circle

LLMR Low-Level Meteorological Rocket

LLMS Little LEO Messaging System

LlN Louvain-la-Neuve (Belgium)

LLN Kelila (Indonesia) (location identifier) • Line Link Network • Lire La Nature (Canada-QC) • Louisiana Library Network (USA-LA)

LLNL Lawrence Livermore National Laboratory (USA-CA)

LLNW Low-Level Nuclear Waste

LLO Legislative Liaison Officer • Low Lunar Orbit

LLOS Landmark Line Of Sight

LLOV Low-Level Output Voltage

LLOX Lunar Liquid Oxygen

LLO2 Lunar Liquid Oxygen

LLP Launch and Landing Project • Liaison Logique Permanente • Lightning Location and Prediction • Lightweight Laser Printer • Limited Liability Payout • Linda Downs (Australia-QLD) (location identifier) • Line Link Pulsing • Link Layer Protocol • Link Level Protocol • Long Lead Procurement • Long Lead Production • Long Line Program • Lower-Layer Protocol • Lower-Level Protocol • Lunar Landing Program • Lyman Line Pumping

LLPI Link Level Protocol Interpreter

LLPN Lumped, Linear, Parametric Network

LLPO Launch and Landing Project Office

LLPR Lattimer, Lamb, Pethick and Ravenhall

LLQ Left Lower Quadrant • Low Latency Queuing • Low-Level Qualifier

LLR Lessons Learned Report • Load-Limiting Resistor • Lunar Laser Ranging

LLRA LapLink Remote Access

LLRC Length Longitudinal Redundancy Check • Library Learning Resource Center (USA) • Luneberg Lens Rapid Commutator

LLRES Load-Emitting Resistor

LLRF Lunar Landing Research Facility

LLRS Library and Learning Resource System • Luminous Linear Radio Source

LLRV Lunar Landing Research Vehicle

LLRW Low-Level Radioactive Waste

L&LS Launch & Landing Site

LLS Laser Light Scattering • Laser Light Source • Laser Line Scan • Laser Line Scanner • Las Lomitas (Argentina-FO) (location identifier) • Latitudinal Layered Structure • Launch and Landing Site • Lightning Location System • Lunar Landing Simulator • Lunar Logistics System • Lyman Limit System

LLSA Literary Legacy of Space Age

LLSF Linear Least Squares Fit

LLSI Laser Light Scattering Instrument

LLSN Lawrence Livermore Seismic Network

LLST Linear Least Squares Transformation

LLSV Lunar Logistics System Vehicle

LLT Lander Local Time • Laser Launch Telescope • Long Lead Time • Low-Level Terminal

LLTDS Launch Landing Tracking Data System

LLTI Long Lead-Time Item

LLTIL Long Lead Time Item List

LLTM Long Lead Time Materiel

LLTR Low-Level Transit Route

LLTV Low Lightlevel Television • Low-Light Television • Lunar Landing Training Vehicle

LLU Lending Library Unit

LLV Lockheed-Martin Launch Vehicle • Long-Life Vehicle • Lunar Landing Vehicle • Lunar Logistics Vehicle

LLVI Low-Level Voice Intercept

LLVPG Large Launch Vehicle Planning Group

LLW Language Learning Workbench • Lilongwe Kamuzu (Malawi) (location identifier) • Low-Level (radioactive) Waste

LLWAS Low-Level Windshear Alert System

LLWS Low-Level Windshear

LLWSAS Low-Level Wind Shear Alert System

LLX Lyndonville (USA-VT) (location identifier)

LLY Mount Holly - Burlington County (USA-NJ) (location identifier)

Llz Localizer

l/m Lines per Minutes

lm Lumen

lM laufenden Monats

Lm Maltese Lira (currency)

L/M Leadership/Management • Lines per Minute • Link Margin • List of Materials

L-M Levenberg-Marquardt (algorithm)

L&M Light & Medium

LM Laboratory Module • Labor Month • Land Mobile • Langage de Machine • Language Modeling • Langue Maternelle • LAN Manager • Large Model • Laser Module • Launch Mode • Launch Mount • Leadership-Management • Left Mid • Leistungsmerkmal • Lettre de Motivation • Levenberg-Marquardt (algorithm) • Library Management • Light and Medium • Light Microscope • Light Microscopy • Light-Minute • Lightning Mapper • Limiting Magnitude • Line Monitor • Linguistic Model • Liquid Metal • Liquid Mirror • List of Materials • Livello del Mare • Load Management • Load Master • Load Multiple • Local Manufacture • Logic Module • Logistics Management • Logistics Manager • Logistics Manual • London Museum • Long Module • Loop Modem • Lord Mayor • Low Mass • Low Metal • Lunar Module

LMA Laboratory for Millimeter Astronomy • Laboratory for Millimeter-wave Astronomy • Lake Minchumina (USA-AK) (location identifier) • Large Mixing Angle (solution) • Laser Microprobe Analyser • Leave Me Alone • Lockheed Martin Astronautics • Lunar Meteoroid Analyzer • Lunar Module Adapter

LMAC Leonid Multiinstrument Airborne Campaign • Lockheed Martin Astronautics Corp. (USA)

LMAE Lunar Module Ascent Engine

LMAN Last-Minute Astrometry Network

LMARS Library Management and Retrieval System

LMATC Lockheed Martin Advanced Technology Center

LMB Left Mouse Button • Le Micro-Bulletin (CNRS) • Local Message Box • Low-Mass Binary • Salima (Malawi) (location identifier)

LMBCS Lotus Multibyte Character Set

LMBP Low-Mass Binary Pulsar

L&MC Launch & Mission Control

L/MC Launch/Mission Control

LMC Laboratoire de Modélisation et Calcul (France) • La Macarena (Colombia) (location identifier) • Large Magellanic Cloud • Laser Mirror Coating • Late Mid-Course • Line Management Code • Line Manager Code • Link Monitor and Control • Liquid-Metal Coolant • Local Management Committee • Low-Mass Companion

LMCA Land Management Cooperative Agreement • Logistics Materiel Control Activity

LMCG Low-Master Cluster Galaxy

L/MCP Launch/Mission Control Processor

LMCS Local Multipoint Communications System

LMD Laboratoire de Météorologie Dynamique • Laboratory of Meteorological Dynamics • La Manipulation des Données • Large Mirror Development • Let Me Do • Local Management Device • Local Medical Doctor • Logistics Maintenance Demonstration • Logistics Maintenance Director • Logistics Maintenance Directorate • Loop Multiplexer/Demultiplexer • Los Menucos (Argentina-RN) (location identifier) • Lunar Meteoroid Detector

LMDA Lunar Meteoroid Detector-Analyzer

LMDC Leadership and Management Development Center

LMDE Lunar Module Descent Engine

LMDS Local Multipoint Distribution Service (*) • Local Multipoint Distribution System • Logistics Management Data System

LME Launch Monitor Equipment • Layer Management Entity • Layer Management Entry • Le Mans - Arnage (France) (location identifier) • Lewis-Michaels Engineering (USA-NY) • Lightweight Maintenance Enclosure • Liquid Motion Experiment • Locally-Manufactured Equipment • Lower Mid Fuselage • Lunar Module Engine

LMER Land Margins Ecosystem Research

LMES Laboratory for Meteorology and Earth Sciences

L/MF Low/Medium Frequency

LMF Language Media Format • Light Manufacturing Facility • Limited-Match Filter • Linear Matched Filter • Load Module Format • Low and Medium Frequency • Lower Mid-Fuselage • Low/Medium Frequency

LMFBR Liquid-Metal-cooled Fast Breeder Reactor

LMFC Limited Field Moisture Capacity

LMG Lamassa (Papua New Guinea) (location identifier) • Left Main Gear • Let Me Go • Longitudinal Magnetograph

LMGC Lunar Module Guidance Computer

LMH Limón (Honduras) (location identifier)

lm-hr Lumen-Hour

LMi Leo Minor - Leonis Minoris (constellation)

LMI Layer Management Interface • Les Marques Internationales • Linear Matrix Inequality • Link Management Interface • Local Management Interface • Lockheed-Martin Intersputnik • Logistics Management Information • Logistics Management Institute • Lumi (Papua New Guinea) (location identifier)

LMIC Land Management Information Center

LMIF Lower Mechanical Integration Frame

LMIS Liquid-Metal Ion Source • Logistics Management Information System

LMJ Laser Megajoule

LMK Landmark • Let Me Know

LML Lae Island (Marshall Islands) (location identifier) • Let Me Look • Lotus Messaging Link

LMLE Local Maximum Likelihood Estimate

LMLG Left Main Landing Gear

LMLM Letter Mail Labelling Machine • Local Mixing-Length Model

LMLR Load Memory Lockout Register

LMLV Lockheed Martin Launch Vehicle

L&MM Logistics & Materiel Management

LMM Le Matériel Magnétique • Length-Measuring Machine • Linear Multistep Method • Locator Middle Marker • Logical Memory Manager • Los Mochis (Mexico) (location identifier) • Lunar Mapping Module • Lundi - Mardi - Mercredi

LMMF Local Maintenance and Management of Facilities

LMMS Laser Microprobe Mass Spectrometry • Local Message Metering System • Lockheed Martin Missiles & Space • Logistics Maintenance Management System

LMN Library Management Network • Limbang (Malaysia) (location identifier)

Lmnt Lentement

LMO Learning Material Object • Lens-Modulated Oscillator • Lossiemouth (UK-Scotland) (location identifier) • Low Mars Orbit • Low-Mass Object • Lucile Miller Observatory (USA-NC) • Lunar Magma Ocean

LMOF Local Media Output Format

LMOS Line Maintenance Operating System • Line Maintenance Operations System

LMP Lampedusa (Italy) (location identifier) • Land Management Planning • LAN Management Protocol • Library Management Protocol • Link Management Protocol • List of Measurement Points • Locally Most Powerful • Lunar Module Pilot

LMPP Lockheed Martin Power Products

LMPRT Locally Most Powerful Rank Test

LMQ Marsa Brega (Libya) (location identifier)

LMR Land Mobile Radio • Large Microwave Radiometer • Laser Magnetic Resonance • Launch Mission Rule • Launch Monitor Room • Lime Acres (South Africa) (location identifier) • Limite Maximale de Résidus • Liquid-Metal Reactor • Long Message Recovery • Lowest Maximum Range

LMRA Land Mobile Radio Architecture

LMRD Launch Mission Rules Document

LMREP Lower Mississippi River Environmental Program

LMRI Laboratoire de Métrologie des Rayonnements Ionisants (France) (now LPRI) • Letter Mail Readibility Index

LMRT Logistics Management Responsibility Transfer

LMS Land Mobile Service • Landsat Mapping System • Large-scale Mapping System • Laser Mass Spectrometer • Laser Mass Spectrometry • Lead Mission Scientist • Learning Management System • Least Mean Square • Least Median of Squares • Legg Middle School (USA-MI) • Let Me Speak • Library Management Software • Library Management System • Life and Microgravity Sciences • Life and Microgravity Spacelab • Lightning Mapper Sensor • Linear Measuring Set • Load Measurement System • Local Management Services • Local Monitoring Station • Local Monitoring System • Logistics Master Schedule • Logistics Management Specialist • Logistics Management Support • Logistics Management System • Loi de Modernisation Sociale (France) • London Mathematical Society (UK-England) • Loss of Memory Syndrome • Lotus Messaging Switch • Louisville - Winston County (USA-MS) (location identifier) • Low-Mass Star • Lunar Mass Spectrometer • Lunar Measuring System • Lunar Module Simulator

LMSA Large Millimeter and Submillimeter Array (Japan) (now merged into ALMA)

LMSAAC Life and Microgravity Sciences and Applications Advisory Committee

LMSAL Lockheed-Martin Solar and Astrophysics Laboratory (USA-CA)

LMSC LAN and MAN Standards Committee • Local and Metropolitan Standards Committee • Lockheed-Martin Missiles and Space Co. • Logistics Management Systems Center

LMScAAC Life and Microgravity Sciences and Applications Advisory Committee

LMSF Low-Mass Star Formation

LMSI Local Mobile Subscriber Identity

LMSS Land Mobile Satellite Service • Land Mobile Satellite System • Lunar Mapping and Survey System

LMST Lightweight Multiband Satellite Terminal

LMSW Load Machine Status Word

LMT Klamath Falls (USA-OR) (location identifier) • Large Millimeter Telescope • Large Millimeter-wave Telescope • Launch Motor Test • Length - Mass - Time • Let Me Think • Limit • Liquid-Mirror Telescope • Load Module Table • Local Mean Time • Local Minima Table • Logistics Management Transfer

LMTA Library and Media Technical Assistant

LMTAS Lockheed Martin Tactical Aircraft Systems

LMTBCF Launch Mean Time Between Critical Failure

LMTBDE Launch Mean Time Between Downing Event

LMTE Liquid Metal Thermal Experiment

LMTL Long-term Mission Timeline

LMTO Linear Muffin-Thin Orbital (method)

Lmtr Limiter

LMTT Low-Mass T Tauri (star)

LMTTRF Launch Mean Time To Restore Function

LMTTRS Launch Mean Time To Restore System

LMU LAN Management Utilities • LAN Manager for Unix • Lightning Mockup • Lincoln Memorial University (USA) • Line Modem Unit • Line Monitor Unit • Logical Memory Unit • Loyola Marymount University (USA) • Ludwig-Maximilians-Universität (Germany)

LMV Laboratoriet för Mekaniska Vågor (Sweden)

lm/W Luman/Watt

LMW Low Molecular Weight

LM/X LAN Manager for Unix

LMX LAN Manager for Unix • Lopez de Micay (Colombia) (location identifier)

LMXB Low-Mass X-ray Binary

LMXRB Low-Mass X-Ray Binary

LMY Lake Murray (Papua New Guinea) (location identifier)

LMYSO Low-Mass Young Stellar Object

LMZ Palma (Mozambique) (location identifier)

ln Lingala (ISO language code) • Natural Logarithm

Ln Laan • Lane • Link

LN Lane • Large Nephelometer (PV) • League of Nations • Line • Link • Liquid Nitrogen • Load Number • Localization • Local National • Logarithm (Natural_) • Logistics Needs • Log-Normal (distribution) • Luna Nuova • Natural Logarithm • Norway (international aircraft registration code)

LNA Laboratório Nacional de Astrofísica • Launch Numerical Aperture • Layered Network Algorithm • Licensed National Agent • Linear Non-Adiabatic • Local Network Adapter • Local Network Attachment • Logistics Net Assessment • Low-Noise Amplifier • Low-Noise Antenna • Low-Voice Amplifier • West Palm Beach - Palm Beach County (USA-FL) (location identifier)

LNAAC Local Network Associate Analysis Center

LNAI Lecture Notes on Artificial Intelligence (Springer)

LNB Lamen Bay (Vanuatu) (location identifier) • Linear Noise Block • Load Noise Buffer • Low-Noise Block

LNC Lengbati (Papua New Guinea) (location identifier) • Liquid-Nitrogen Cooling • Loch Ness Centre (UK-Scotland) • Low-Noise Cable • Low-Noise Converter

LNCC Ligue Nationale Contre le Cancer (France)

LNCH Launch

LNCS Lecture Notes in Computer Science (Springer)

LND Lander Hunt (USA-WY) (location identifier) • Limiting Noise Dive

LNDF Log Normal Distribution Function

Lndg Landing

LNDG Landing

LN:DI Lotus Notes Document Imaging

LNDI Lotus Notes Document Imaging

LNDNG Landing

LNE Laboratoire National d'Essais (France) • Late Network Entry • Lonorore (Vanuatu) (location identifier)

LNEn Lecture Notes in Engineering (Springer) (periodical code)

LNES Lecture Notes in Earth Sciences (Springer) (periodical code)

LNF Laboratori Nazionali di Frascati (Italy) • Low-Noise Feed • Munbil (Papua New Guinea) (location identifier)

LNEC Laboratório Nacional de Engenharia Civil (Portugal)

LNEMS Lecture Notes in Economic and Mathematical Systems (Springer)

LNEn Lecture Notes in Engineering (Springer) (periodical code)

LNETI Laboratório Nacional de Engenharia e Tecnologia Industrial (Portugal)

Lng Landing

LNG Lese (Papua New Guinea) (location identifier) • Liquefied Natural Gas • Liquid Natural Gas

LNGS Laboratori Nazionali del Gran Sasso (Italy)

Lnh Lunch

LNH Lake Nash (Australia-NT) (location identifier) • Large Numbers Hypothesis • Lunar Near Horizon

LNHM Lafayette Natural History Museum (USA-LA)

LNI Local Network Interconnect • Lonely Dew (USA-AK) (location identifier)

LNK Lincoln (USA-NE) (location identifier) • Link

LNM Langimar (Papua New Guinea) (location identifier) • LAN Network Manager • Lecture Notes in Mathematics (Springer) (periodical code)

LNMRI Laboratório Nacional de Metrologia das Radiaçoes Ionizantes (Brazil)

LNMS Large Neutral Mass Spectrometer (PV)

LNN Linear Neural Network • Willoughby - Lost Nation (USA-OH) (location identifier)

LNNI LAN Network Node Interface • LAN Network-to-Network Interface

LNO Leonora (Australia-WA) (location identifier) • Liaison Office • Liaison Officer • Limited Nuclear Option • Line Number of Origin

LNP Lecture Notes in Physics (Springer) (periodical code) • Liquid Nitrogen Processing • Local Number Portability • Loch Ness Productions (USA-MA) • Local Number Portability • Lomb Normalized Periodogram • Loss of Normal Power • Lunar Neutron Probe • Wise (USA-VA) (location identifier)

LNPA Low-Noise Preamplifier

LNPAM Lecture Notes in Pure and Applied Mathematics (Springer) (periodical code)

LNPE Lunar Neutron Probe Experiment

LnPT Level n Page Table

LNQ Loani (Papua New Guinea) (location identifier)

LNR Line Number • Lone Rock - Tri County (USA-WI) (location identifier) • Low-Noise Receiver

LNRF Locally Non-Rotating Frame

Lnrty Linearity

LNS Laboratory for Nuclear Science • Lancaster (USA-PA) (location identifier) • Land Navigation System • Leader Node Switch • Lines

LNT Linear No-Threshold (theory) • Liquid-Nitrogen Temperature

LNTB Lateral Nucleus of the Trapezoidal Body

LNTS Liquid-Nitrogen Transfer System

LNTWA Low-Noise Traveling Wave Amplifier

LNU Last Name Unknown

LNUG Library Network User Group

LNV Landbouw, Natuurbeheer en Visserij (Netherlands) • Londolovit (Papua New Guinea) (location identifier)

LNX Smolensk (Russia) (location identifier)

LNY Lanai City (USA-HI) (location identifier)

LNZ Linz (Austria) (location identifier)

LN2 Liquid Nitrogen

lo Laothian (ISO language code)

Lo Lodge • Lord

L/O Lift Off • Limited Operation

LO Lager Onderwijs • Launch Operations • Launch Operator • Law and Order • Leading Order • Learning Objective • Letter of Offer • Letter Orders • Level Off • Ley Orgánica (Spain) • Liaison Office • Liaison Officer • Lichtenknecker Optics (Belgium) • Lift-Off • Linear Optimization • Line Occupancy • Local Oscillator • Located Object • Lock-On • Lock-Out • Logical Operation • Loi Organique • Longitudinal Optical • Low • Low Observable • Lowest Order • Low Observable • Low Orbit • Lubricating Oil • Lunar Observer • Lunar Occultation • Lunar Orbiter • Lund Observatory (Sweden)

LOA Laboratoire d'Optique Appliquée • Laboratoire d'Optique Atmosphérique • Landing Operations Area • Launch On Assessment • Launch Operations Agency • Launch Operations Area • Lead Operational Authority • Leave Of Absence • Length Overall • Letter Of Acceptance • Letter Of Agreement • Letter Of Assist • Letter Of Authorization • Letter of Offer and Acceptance • Light Observational Aircraft • Line Of Assurance • Local Ocean Area • Lorraine (Australia-QLD) (location identifier)

LOAA Letter Of Agreement and Acceptance

LOAC Likelihood to Overseas Access Co-authorship

LOAD Letter of Offer and Acceptance Data

LOADT Loi d'Orientation pour l'Aménagement et le Développement du Territoire (France)

LOAL Lock-On After Launch

LOAMP Logarithmic Amplifier

LOAN Library Online Automated Network • Low Observable Axisymmetric Nozzle

LOAO Low-Order Adaptive Optics

LOAP List Of Applicable Publications

LOAS Lift-Off Acquisition System

LOB Laboratory Office Building • Large Object • Launch Operations Branch • Launch Operations Building • Left Outboard • Lick Observatory Bulletin • Line Of Balance • Line Of Bearing • Line Of Business • Low-Order Bit • Low-Order Byte

LOBES Low-Budget Extraterrestrial-intelligence Search

LOBL Lock-On Before Launch

LOBO Low-frequency Background Observatory

LOBSTER Large Angle Observatory for Burst, Supernova and Transient Event Research

Loc Local • Locality • Localizer • Locate • Located • Locating • Location • Locative • Locator • Locución

LoC Library of Congress

LOC Language-Oriented Computer • LAN Operations Center • Large Optical Cavity • Launch Operations Center • Launch Operations Complex • Launch Operations Contractor • Launch Operations Control • Launch Operator's Console • Letter Of Comment • Letter Of Consent • Letter Of Credit • Level Of Conflict • Library Of Congress (USA) • Limited Operational Capability • Limit Of Convection • Line Of Code • Line Of Communication • Line Of Contact • Local • Localidade • Localizer • Locally • Locally Optimally-emitting Cloud • Local Operating Company • Local Operations Console • Local Optimally-emitting Cloud • Local Organizing Committee • Location • Lock (Australia-SA) (location identifier) • Logic Optimizing Compiler • Logistics Operations Center • Loop Online Control • Loss Of Cell • Loss Of Control • Lowest Outgoing Channel

LOCA Loss Of Coolant Accident

LOCAAS Low-Cost Autonomous Attack System

LOCAL Laboratory Of Computer-Assisted Learning

LOCAP Low Capacitance • Low Capacity

LOCB Launch Operations Control Building

LOCC Launch Operations Control Center • Logistical Operations Control Center

LOCF Last Observation Carried Forward

LOCI Logarithmic Computing Instrument

loc. cit. Loco Citato

LOCE Ley Orgánica de Calidad de la Educación (Spain)

LOCID Location Identifier

LOCIS Library Of Congress Information System

LOCK Logical Co-processing Kernel • Logical Co-processor Kernel

LockH Lockheed Horizons (periodical code)

loc. laud. Loco Laudato

Locn Location

LOCPOD Low-Cost Powered Dispenser

loc. primo cit. Loco Primo Citato

LOCREP Location Report

LOCS Land-Ocean Contamination Study • Librascope Operations Control System • Local Optical Correction System • Logic and Control Simulator

LOCSS Land-Ocean Climate Satellite System

LOD Launch Operations Directive • Launch Operations Director • Launch Operations Directorate • Launch Operations Division • Launch Order Document • Leading Ones Detector • Legion Of Doom • Length Of Day • Letter Of Delegation • Level Of Detail • Limit Of Detection • Line Of Demarcation • Line Of Departure • Line Of Duty • Locally One-Dimensional • Longana (Vanuatu) (location identifier) • Loss Of Data

LODA Large Object Database

LODCS Lunar Orbiter Data Conversion System

LODE Large Optics Demonstration Equipment • Large Optics Demonstration Experiment • Ley Orgánica del Derecho a la Educación (Spain)

LODESMP Logistics Data Element Standardization and Management Process

LODESTAR Logically Organized Data, Entry, Storage And Recording

LODF Line Opacity Distribution Function

LODI List Of Deleted Items

LODIA Low-Cost Digital Image Analyzer

LODP Lunar Orbiter Data Printer

LODR Large Optics Demonstration Radar • Length Of Day - Regularized

LODSB Load String Byte

LODSC Logistics Operations Decision Support Center

LODTM Large Optics Diamond Turning Machine

LODUS Low Data Rate UHF Satellite

LODYC Laboratoire d'Océanographie Dynamique et de Climatologie (*) • Laboratoire d'Océanographie et de Dynamique du Climat

LOE Left Outboard Elevon • Letter Of Evaluation • Letter Of Execution • Level Of Effort • Limited Objective Experiment • Limit Of Exploitation • Line Of Effort • Line Of Engagement • Loei (Thailand) (location identifier)

LOEAT Lowest Temperature Equaled for All Time

LOEFM Lowest Temperature Equaled For the Month

LOEle L'Onde Électrique (periodical code)

LOEP Limited Objective Experiment Plan

LOEPS Land-Ocean Evolutionary Perspective Study

LOESE Lowest Temperature Equaled So Early

LOF Lift-Off • Limitation Of Fund • List Of Figures • Local Oscillator Frequency • Lock Off-line • Loen (Marshall Islands) (location identifier) • Loss Of Flow • Loss Of Frame • Loss Of Function • Lowest Operating Frequency

LOFAR Low-Frequency Acquisition And Ranging • Low-Frequency Array

LofC Lines of Communication

Lo-Fi Low Fidelity

LoFi Low Fidelity

LoFLYTE Low-Observable Flight Test Experiment

LOFO Low-Frequency Oscillation

LOFP Launch Operations and Flight Plan

LOFPI Low-Order Fabry-Pérot Interferometer

LOFS Loopback File System

LOFT Launch Operations Flight Test • Line-Oriented Flight Training • Low-Frequency Radio Telescope

log Log (file extension) • Logarithm

Lóg Lógica

Log Logic • Lógica • Logical • Logistical • Logistics

LoG Laplacian of Gaussian

L&OG Logistics & Operations Group

LOG Large Optical Generator • Linear Quadratic Gaussian • Logarithm • Login • Longview (USA-WA) (location identifier) • Lumicon Off-axis Guider

LOGACS Low-G Accelerometer Calibration System

LOGAI Logical Addressing And Interoperability

LOGAIS Logistics Automated Information System

LOGAL Logical Algorithmic Language

LOGALGOL Logical Algorithmic Language

LOGAM Logistics Analysis Model

LOGAMP Logistics and Acquisition Management Program

LOGC Logistics Center

LOGCA Logistics and Command Assessment • Logistics Civil Augmentation (program)

LOGCAB Logistics Center Advisory Board

LOGCAP Logistical Civil Augmentation Program • Logistics and Command Assessment of Projects

LOGCCIS Logistics Command and Control Information System

LOGCEN Logistics Center

LOGDEMO Logistics Demonstration

LOGDET Logistics Detail

LOGDIS Logistics Data Integration System

LOGDSS Logistics Decision Support System

LOGEL Logic Generating Language

LOGEX Logistical Exercise

LOGFAC Logistics Feasibility Analysis Capability

LOGGEN Logistics Generator

LOGICA Logistics Operations Graphical Integration Capability Assessment

LOGIPAC Logical Processor And Computer

LOGIT Logical Inference Tester

LOGLAN Logical Language

LOGMDS Logistics Management Data System

LOGMIS Logistics Management Information System

LOGMOD Logic Model • Logistics Module

LOGNET Logistics Network

LOGO Limitation Of Government Obligation • Limit Of Government Obligation

LOGOIS Logistics Operation Information System

LOGPAC Logistics Package

LOGPARS Logistic Planning and Requirement System

LOGPLAN Logistics Planning

LOGR Logistical Ratio • Logradouro

LOGRAM Logistics Readiness Assessment Model

LOGREP Logistics Replenishment • Logistics Report • Logistics Representative • Logistics Requirement

LOGRUN Logistics Remote Users Network

LOGS Logistics Supportability

LOGSA Logistics Support Activity

LOGSAC Logistics Structure And Composition

LOGSACS Logistics Structure And Composition System

LOGSAFE Logistics Sustainment And Feasibility Estimator

LOGSE Ley de Ordenación General del Sistema Educativo (Spain)

LOGSME Logistics Subject Matter Expert

LOGSTAR Logistics Status Report

LOGSTAT Logistics Status

LOGSUM Logistics Summary

LOGSWG Landsat Operations Ground Segment Working Group

LOGTAB Logical tables

LOGTIM Logistics Technical Interchange Meeting

LOH Loja (Ecuador) (location identifier)

LOHAE Low-Observable High-Altitude Endurance

LoHCo Local Helioseismology Comparisons (workshop)

LoI Letter of Instruction • Letter of Intent • Letter of Interest • List of Items

LOI Letter Of Instruction • Letter Of Intent • Letter Of Interest • List Of Items • Loss Of Ignition • Loss Of Input • Loss On Ignition • Low-resolution Oscillation Imager • Luminosity Oscillation Imager • Lunar Orbital Insertion • Lunar Orbit Insertion

LOIC Logistics Organization Integration Committee

LOICZ Land-Ocean Interactions in the Coastal Zone

LOID Location Identifier

LOIS Land Oceans Interactions Study • Library Orders Information System • Logistics Information Simulator • Lowell Observatory Instrumentation System

LOK Lodwar (Kenya) (location identifier)

LOL Laughing Out Loud • Length Of Lead • Limited Operating Life • Limit Of Liability • Loss Of Learning • Lovelock Derby (USA-NV) (location identifier) • Low-Order Language

LOLA Library Online Acquisition • Lunar Orbit and Lunar Approach • Lunar Orbit Landing Approach

LOLI Limited-Operational-Life Item

LOLITA Language for the Online Investigation and Transformation of Abstractions • Library Online Information and Text Access

LO/LO Lift-On/Lift-Off

LOLP Loss Of Load Probability

LOM Lagos de Moreno (Mexico) (location identifier) • Laser Optical Modulator • Launch Operations Manager • Learning Object Metadata (working group) • Life Of Mission • List Of Materiel • List Of Modifications • Locator at Outer Marker • Low-Order Memory • Lunar Orbital Mission

LOMA Letters Of Map Amendment • Life Office Management Association

LOMAPS Logical and Operational Methods in the Analysis of Programs and Systems (workshop)

Lomb Lombard • Lombardy

LOMC Letters Of Map Correction

LOME Low-rate Minimum Energy

LOMICS Low-energy Magnetospheric Ion Composition Sensor

LOMMCA Logistics Operations Manpower and Materiel Cost Analysis

LOMR Letters Of Map Revision

LOMUX Low-frequency Multiplexing

Lon Longitude

LON LAN Outer Network • Launch On Need • Letter Of Notification • Line Of Nodes • Local Operating Network • Lock Online • London Metropolitan (UK-England) (location identifier) • Longitude

Lond London (UK-England) • Londonderry (Northern Ireland)

LONEOS Lowell Observatory's Near-Earth Object Search (*) • Lowell Observatory's Near-Earth Object Survey

LONEX Laboratory Office Network experiment

Long Longitude • Longitudinal • Longueur

LONG Longitude

Longl Longitudinal

LonGSp Long-slit Gornergrat Spectrometer

LONGSTOP Long-term Gravitational Stability Test for the Outer Planets

LONS Laboratory Office Network System • Local Online Network System

LOO Laghouat L'Mekrareg (Algeria) (location identifier) • Launch Operations Order • Letter Of Offer • Log Of Operation

LOOM Laboratoire d'Optique de l'Observatoire de Marseille (France) • Low Overhead Object Management

LOOMS Laser Optics Operations and Maintenance Support

LOOP Library-Operated Outreach Program • Low Overhead Object Protocol

LOOPE Loop while Equal

LOOPNE Loop while Not Equal

LOOPNZ Loop while Not Zero

LOOPS LISP Object-Oriented Programming System

LOOPZ Loop while Zero

LOP Last Operation Completed • Launch Operations Panel • Launch Operations Phase • Launch Operator Panel • Letter Of Procurement • Letter Of Proposal • Life Of Program • Limitation Of Payment • Line Of Position • Loss Of Payload • Loss Of Pointer • Low-Oil Pressure • Low-Order Partition

LOPAR Low-Power Acquisition Radar

LOPARC Lowell Observatory Public Astronomical Research Center (USA-AZ)

LOPC Lunar Orbit Plane Change

LOPLAS Letters on Programming Language and Systems

LOPP Lunar Orbiter Photographic Project

LOPS Launch Operations • Lunar Orbiting Photographic System

LOPSTR Logic Program Synthesis and Transformation (workshop)

loq Loquitur

LOQ Lobatse (Botswana) (location identifier)

Lor Lorong

LOR Large Optical Reflector • Letter Of Request • Level Of Repair • Light Output Ratio • Line Outage Recorder • Loss Of Reference • Lunar Orbital Rendezvous • Lunar Orbit Rendezvous • Ozark Lowe (USA-AL) (location identifier)

LORA Lecturer-Oriented Response Analysis • Level-Of-Repair Analysis

LORAC Long-Range Accuracy

LORAD Long-Range Active Detection

LORAN Long-Range Aid to Navigation • Long-Range Navigation (*)

LORCS League of the Red Cross and Red Crescent Societies

LORD Launch Operations Requirements Document • Local Optical Reference Device

LORE Langley Online Research Explorer (NASA) • Line-Oriented Editor

LOREORS Long-Range Electro-Optical Reconnaissance System

LoRes Low Resolution

LORIS Long-Range Infrared System

LORI Logistics Operation Readiness Inspection

LORL Large Orbital Research Laboratory

LO/RO Lift-On/Roll-Off

LORO Lobe On Receive Only

LOROP Long-Range Oblique Photography

LOROPS Long-Range Oblique Photography System

LORV Low Observable Reentry Vehicle

LOS Lagos (Nigeria) (location identifier) • Land Observation Satellite • Large Optics Segment • Launch On Schedule • Launch Operations System • Length Of Service • Length Of Stay • Liaison Office Support • Lift-Off Simulator • Line Of Sight • Line Of Supply • Line Operational Simulation • Liquid Oxygen Storage • Loop Output Signal • Loss Of Sight • Loss Of Signal • Loss Of Synchronization • Lunar Orbiting Satellite

LOSF Lunar Orbit Staging Facility

LOSGEN Line Of Sight Generator

LOSHEA Line-Of-Sight Hard Earth Angle

LOSIC Logiciel de Simulation de Couverture

LOSIM Low Observable Simulation

LOSM Launch Operations Simulation Model

LOSP Landing Operations Support Panel • Loss of Off-Site Power

LOSPMA Line Of Sight Pointing Mirror Assembly

LOSR Line-Of-Sight Radiance • Line-Of-Sight Range

LOSREP Loss Report

LOSS Large Object Salvage System • Leuschner Observatory Supernova Search • Lick Observatory Supernova Search • Light Observatory Supernova Search • Local Oscillator Subsystem • Lunar Orbital Space Station • Lunar Orbital Survey System • Lunar Orbit Space Station

LOSSAM Line Of Sight Sky Absorption Monitor

LOSTA Line Of Sight Tangent Altitude

LOSV Line-Of-Sight Velocity

LOSVD Line-Of-Sight Velocity Distribution

Lot Lotion

LOT Large Optical Telescope • Large Orbital Telescope • Large Orbiting Telescope • Ley de Ordenación de las Telecomunicaciones (Spain) • Ley de Ordenación del Territorio (Spain) • Lift-Off Time • Limited Operational Test • Linear Onset Time • Line Of Travel • List Of Tables • Lockport Lewis (USA-IL) (location identifier) • Low-Observable Technology • Lulin One-meter Telescope (China-ROC)

LOTA Low-Observable Technology Application

LOTAWS Laser Obstacle Terrain Avoidance Warning System

LOTEAM Loteamento

LOTEX Life of Type Extension

LOTIS Livermore Optical Transient Imaging System • Logics - Timing - Sequencing

LOTMOP Lick Observatory Torus Monitoring Program

LOTOS Language Of Temporal Ordering Specification • Long-Term Observation Of Stars • Long-Term Ozone Simulation

LOTREX Longitudinal Traverse Experiment

LOTS Low-Overhead Time-Sharing System

LOTV Launch Operations and Test Vehicle

Lou Louisiana (USA)

LOU Letter Of Understanding • Ley Orgánica de las Universidades (Spain) • Limited Official Use • Louisville Bowman (USA-KY) (location identifier)

LOUISA Lunar-Optical-Ultraviolet-Infrared Synthesis Array

LoV Line of Variation

LOV Limit Of Visibility • Line Of Variation • Loss Of Vehicle • Loss Of Video • Loss Of View • Loss Of Visibility • Loss Of Vision • Loss Of Visual • Monclova (Mexico) (location identifier)

LOVLA Lunar Optical Very Large Array

LoW Line of Work

LOW Line Of Work • Link Orderwire

LOWESS Locally Weighted Least Squares

LOWG Landing Operations Working Group

LowOB Lowell Observatory Bulletin (periodical code)

LOWR Lower

LOX Liquid Oxygen

LOXT Large Orbiting X-ray Telescope

LOY Loyengalani (Kenya) (location identifier)

LOZ Corbin London (USA-KY) (location identifier)

LO2 Liquid Oxygen

Lp Ladyship • Lordship

L-P Low Pressure

LP Laboratoire Propre • Landing Point • Land Process • Langmuir Probe • Language Planning • La Pampa (Argentina) • Lapp • Large Paper • Large Particle • Large Pénétration • Large Photometer • Large Post • Laser Printer • Lasting Penetration • Last Post • Latent Period • Launch Pad • Launch Platform • Launch Point • Launch Position • Launch Preparation • Launch Processing • Layer Protocol • Learning Prototype • Legacy Project • Legal Procurator • Lesson Plan • Ley del Patrimonio (Spain) • Life Policy • Light Pollution • Limited Procurement • Limited Production • Limit Point • Linearly Polarized • Linear Phase • Linear Polarization • Linear Polarizer • Linear Potential • Linear Prediction • Linear Program • Linear Programming • Line Printer • Linguistic Processor • Linkage Programme • Linking Protection • Liquid Petroleum • Liquid Propellant • Listening Post • List Processing • List Processor • Local Purchase • Local Purchaser • Logic Programming • Log Periodic (antenna) • Lonely Planet • Long Pass • Long Penetration • Long Period • Long Play • Long Player • Long Playing • Long Primer • Loop • Lower Panel • Low Pass • Low Point • Low Power • Low Pressure • Lubricated Penetration • Luna Piena • Lunar Prospector • Luyten Palomar (catalog) • Lycée Professionnel (France)

LPA Gran Canaria (Canary Islands) (location identifier) • Launch Phase Analyst • Launch Preparation Analysis • Least Perturbative Action • Ley de Proceso Administrativo (Spain) • Light Penetration Assistance • Light Pollution Abatement • Light Pulser Array • Linear Power Amplifier • Link Pack Area • Local Potential Approximation • Log Periodic Antenna • Log Periodic Array • Low-Power Amplifier

LPAA Log Periodic Array Antenna

LPAC Launching Programme Advisory Committee

LPAD Learning Potential Assessment Device

LPAI Logic Programming and Artificial Intelligence

LPAO La Plata Astronomical Observatory (Argentina)

LPAP Laboratoire de Physique Atmosphérique et Planétaire • Light Pollution Abatement Program (RASC)

LPAPS Low-Power Amplifier and Power Supply

LPAR Large Phased Array Radar • Logical Partition • Logical Partitioning • Logic Programming and Automated Reasoning (conference)

LPARL Lockheed-Martin Palo Alto Research Laboratory (USA-CA)

LPATS Lightning Position And Tracking System

LPB La Paz - El Alto (Bolivia) (location identifier)

LPBBA Log Periodic Broadband Antenna

LPC Label Print Center • Label Printing Center • Laboratoire de Physique Corpusculaire • Laboratory Precision Connector • Laser Pressure Catapulting • Laser Processing Consortium • Launch Pad Container • Launch Processing Contractor • Leading Particle Cascade • Linear Plate Constant • Linear Prediction Coding • Linear Prediction Coefficient • Linear Predictive Coding • Line Printer Controller • Link Priority Change • Local Procedure Call • Lompoc (USA-CA) (location identifier) • Longitudinal Parity Check • Long-Period Comet (*) • Long Periodic Comet • Low-Power Compact (source) • Low-Pressure Chamber

LPCA Lunar Pyrotechnic Control Assembly

LPCDF Low-Profile Combined Distributing Frame

LPCE Laboratoire de Physico-Chimie de l'Environnement • Laboratoire de Physique et Chimie de l'Environnement (France)

LPCLK Loop Clock

LPCM Laboratoire de Physico-Chimie de Matériaux • Linear Pulse Code Modulation

LPCP Lightweight Phone Control Protocol

LPCVD Low-Pressure Chemical Vapor Deposition

LPD Landing Platform Deck • Landing Platform Dock • Landing Point Designator • Language Processing and Debugging • La Pedrera (Colombia) (location identifier) • Laser Polarization Detector • Launch Point Determination • Launch Preparation Document • Launch Procedure Document • Linear Power Density • Line Printer Daemon

• List of Physical Dimensions • Local Parcel Depot • Log Periodic Dipole • Low Power Density • Low-Power Device • Low-Power Difference • Low-Pressure Difference • Low Probability of Detection • Luyten Palomar Double

LPDA Laboratory Program Development Activity • Laser-Phased Doppler Anemometer • Linear Photodiode Array • Link Problem Determination Aid • Log Periodic Dipole Array

LPDGWY Line Printer Daemon Gateway

LPDM Lagrange Particle Dispersion Model • Lagrangian Particle Dispersion Model • List of Physical Dimensions

LPDS Laboratory of Parallel and Distributed Systems (Hungary)

LPDT Low-Power Distress Transmitter

LPDU LLC Protocol Data Unit

LPE Laboratoire de Physique de l'Environnement • Laboratoire de Physique pour l'Enseignement • Lambda Point Experiment • La Primavera (Colombia) (location identifier) • Lasting Pulsating Erection • Launch Point Estimate • Launch Preparation Equipment • Liquid-Phase Epitaxial • Liquid-Phase Epitaxy • Loi sur la Protection de l'Environnement (France) • Look Point Equation • Loop Preparation Equipment • Low Probability of Exploitation • Lunar and Planetary Ephemerides

LPEC Launch Preparation Equipment Compartment

LPEE Liquid-Phase Electro-Epitaxial

LPEO Local Public Employment Office

LPES Launch Preparation Equipment Set • Localized Polyelastic Sampling

LPETS Lunar Prospector Electrical Test Set

LPEV Launch Preparation Equipment Vault

LPF Lambda Point Fridge • Launch Pad Facility • League for Programming Freedom • Library Processing Facility • Low-Pass Filter • Low-Power Field

LPFT Low-Pressure Fuel Turbopump

LPFTP Low-Pressure Fuel Turbopump

LPG Langage de Programmation Numérique • La Plata (Argentina-BA) (location identifier) • Last Page Generator • Launch Preparations Group • Liquefied-Petroleum Gas • Liquid-Petroleum Gas • Liquid Plume Generator • Liquid-Propane Gas • Liquid-Propellant Gun • Long-Path Gas • Low-Pressure Gas

LPGE LEM Partial Guidance Equipment • Ley de Presupuestos Generales del Estado (Spain)

LPGN Lexicon of Greek Personal Names

LPGP Laboratoire de Physique des Gaz et des Plasmas

LpH Lines per Hour

LPH Lines Per Hour • Lochgilphead (UK-Scotland) (location identifier)

LPHA Laboratoire de Physique des Hautes Atmosphères

LPHD Local Parton-Hadron Duality

LPHS Lunar and Planetary Horizon Scanner

lpi Lines per Inch

LPI Language Proficiency Interview • Large Pénétration Internationale • Latent Photographic Image • Launching Position Indicator • Lines Per Inch • Linköping (Sweden) (location identifier) • Linux Professional Institute (Canada-ON) • Lithium Pellet Injector • Local Position Invariance • Lotus Programming Language • Low-Power Illuminator • Low-Pressure Impactor • Low Probability of Intercept • Low Probability of Interception • Low Probability of Interference • Lunar and Planetary Institute (USA-TX) • Lunar and Planetary Science Conference (periodical code)

LPIB Laser Pulse Induced Breakdown • Lunar and Planetary Information Bulletin (*) • Lunar and Planetary Institute Bulletin

LPIC Lycée Professionnel Industriel et Commercial (France)

LPID Logical Page Identifier

LPIE Liquid Plume Imagery Experiment

LPIF Launch Package Integration Facility

LPIX Liquid Propellant Identification Experiment

LPJ Pijiguaos (Venezuela) (location identifier)

LPK Lipetsk (Russia) (location identifier)

LPKR Logic Programming and Knowledge Representation (workshop)

LPL Laboratoire de Physique des Lasers (France) • Large Pénétration Locale • Laser - Pumped-Laser • Ley de Procedimiento Laboral (Spain) • Linear Programming Language • List Processing Language • List Programming Language • Liverpool (UK-England) (location identifier) • Long-Path Laser • Long-Pulse Laser • Lotus Programming Language • Low Polar Latitude • Lunar and Planetary Laboratory • Lunar Personnel Lander • Lunar Projects Laboratory

LPlaC Observatorio Astronomico de La Plata Circular (periodical code)

LPlaS Observatorio Astronomico de La Plata Separata Astronomica (periodical code)

LPLC Low-Pressure Liquid Chromatography

LPLS Lockheed Payload Launch System

LPLWS Launch Pad Lightning Warning System

lpm Latidos por Minuto • Lines per Millimeter • Lines per Minute • Liters Per Minute

LPM Label-Producing Machine • Laboratorie de Physique Mathématique • Lamap (Vanuatu) (location identifier) • Landau - Pomeranchurk - Migdal (effect) • Landsat Program Management • Licensing Project Manager • Linear Piezo Motor • Line Processing Module • Linkable Protection Module • Link Power Monitor • Load Point Marker • Logical Process Manager • Loss Projection Model • Lunar Payload Module • Lunar Portable Magnetometer • Luyten Proper Motion

LPMA Laboratoire de Physique Moléculaire Atmosphérique • Laboratoire de Physique Moléculaire et Applications (France) • Limb Profile Monitor of the Atmosphere • Loose Parts Monitor Assembly

LpMin Lines per Minute

LPML Low Protein Meal

lpmm Lines per Millimeter

LPN Large Pénétration Nationale • Leron Plains (Papua New Guinea) (location identifier) • Licensed Practical Nurse • Logical Page Number • Lomb Periodogram - Normalized • Low-Pass Network

LPNC Logical Private Network Capability

LPNHE Laboratoire de Physique Nucléaire des Hautes Énergies (France)

LPNMR Logic Programming and Non-Monotonic Reasoning (conference)

LPO La Palma Observatory (Canary Islands) • La Plata Observatory (Argentina-BA) • Laporte (USA-IN) (location identifier) • Lattice Preferred Orientation • Low-Power Output • Lunar Parking Orbit • Lunar Polar Orbiter • Lycée Polyvalent (France)

L'pool Liverpool (UK-England)

LPOP Low-Pressure Oxidizer Pump

LPOR Lycée Polyvalent Régional (France)

LPOT Low-Pressure Oxidizer Turbopump

LPOTP Low-Pressure Oxidizer Turbopump

LPOX Low-Pressure Oxygen

LPP Laboratoire de Physique des Polymères (France) • Lappeenranta (Finland) (location identifier) • Launch Point

Prediction • Lean Premixed Prevaporized • Learning Prototype Phase • Leightweight Presentation Protocol • Licensed Program Product • Lightweight Presentation Protocol • Lignes Par Pouce • Line Printer/Plotter • Long Periodic Perturbation • Low-Power Physics • Low-Profile Pallet • Lunar Precepts Positioner

LPPM Laboratoire de Photophysique Moléculaire

LPQ Low-Polarization Quasar • Luang Prabang (Laos) (location identifier)

LPR Large Pénétration Régionale • Large Pulkovo Radiotelescope • Leadership Potential Rating • Least-Positive Residue • Light Penetration Resistance • Light Pollution Reduction • Light Pollution Rejection • Line Printer • Liquid Propellant Rocket • Local Payment Receipt • Local Polynomial Regression • Low-cost Packet Radio

LPRB Loaded Program Request Block

LPRC Launch Pitch Rate Control

LPRD Light Pollution Reduction Device

LPRI Laboratoire Primaire des Rayonnements Ionisants (France) (formerly LMRI)

LPRL Lockheed-Martin Palo Alto Research Laboratory (USA-CA)

LPRQ Low-optically Polarized Radio-loud Quasar

LPRT Low-Power Radio Transmitter

lps Liters per Second • Loops

LPS Label Printing System • Laboratory for Physical Sciences • Landsat Processing System • Large Poincaré System • Large Pointing System • Large Processing Station • Laser Particulate Spectrometer • Laser Power Supply • Latvian Physical Society (Latvia) • Launch Phase Simulator • Launch Processing System • Launch Protection System • Limited Protection System • Linear Programming System • Liquid-Phase Sintering • Liquid Propulsion Stage • Lithuanian Physical Society • Local Processor Service • Local Production System • Logistics Planning Support • Loop Prominence System • Loops • Lopez Island (USA-WA) (location identifier) • Low-Power Schottky • Low-Power Sodium • Low-Pressure Sodium (*) • Lunar Penetrometer System • Lunar Pilotage System

LPSA Lunar Periodic Scattering Array

LPSC Liquid Propulsion System Centre (ISRO) • Lunar and Planetary Science Conference

LPSE Launch Platform Support Equipment

LPSH Laboratoire de Physique Solaire et de l'Héliosphère

LPSIRS Linear Potential-Sweep Infrared Reflectance Spectroscopy

LPSP Laboratoire de Physique Stellaire et Planétaire • La Palma Stokes Polarimeter

LPSS Launch Processing Software System • Local Property and Supply System • Logic Programming Summer School

LPST Latest Possible Start Time

LPT Lampang (Thailand) (location identifier) • Laplacian Pyramid Transform • Large Petal Telescope • Lewthwaite - Pickover - Truchet (tiling) • Limited Private Type • Limited Production Test • Line Printer • Local Printer Terminal • Long-Term Prediction • Low-Power Test • Low-Power Transmitter • Low-Pressure Test • Low-Pressure Turbine • Lunar Propellant Tanker

LPTA Lowest Price Technically Acceptable

LPTF Laboratoire Primaire du Temps et des Fréquences (France) • Low-Power Test Facility

LPTHE Laboratoire de Physique Théorique et des Hautes Énergies

LPTL Laboratoire de Physique Théorique des Liquides (France)

LPTO Long-Term Programme Office

LPTTL Low-Pressure Transistor-Transistor Logic

LPTV Low-Power Television

LPT1 First Parallel Printer Port

LPT2 Second Parallel Printer Port

LPT3 Third Parallel Printer Port

LPU Limited Production Urgent • Line Printer Unit • Line Processing Unit • Local Programming Unit • Logical Program Unit • Long Apung (Indonesia) (location identifier) • Low Pay Unit

LPUU Linear Programming Under Uncertainty

LPV Laser Protective Visor • Launching Point Vertical • Line Profile Variability • Line Profile Variable • Line Profile Variation • Long-Period Variable (star) • Lunar Piloted Vehicle

LPVAD Preston (Starlink internal code)

lpW Lumens per Watt

LPW Linear Polarized Wave • Little Port Walter (USA-AK) (location identifier) • Lumens Per Watt

LPWA Lucent Personalized Web Assistant

LPWG Logistics Planning Working Group

LPX Liepaya (Latvia) (location identifier) • Liquid Plume Experiment • Lunar and Planetary Exploration

LPY Le Puy (France) (location identifier)

Lpz Leipzig (Germany)

LPZ Low Population Zone

lq Lege Quaeso

LQ Argentina (international aircraft registration code) • Laser Quality • Lege Quaeso • Leptoquark • Letter Quality • Life Quality • Linear Quadratic • Lower Quartile

LQA Line Quality Analysis • Link Quality Analysis

LQC Laboratory Quality Control

LQE Linear Quadratic Equation

LQG Large Quantity Generator • Large Quasar Group • Linear Quadratic Gaussian • Local Quasar Group

LQI Laboratory Quality Initiative

LQIP Laboratory Quality Improvement Program

LQK Pickens (USA-SC) (location identifier)

LQM Link Quality Monitor • Link Quality Monitoring

LQN Qala-Nau (Afghanistan) (location identifier)

LQP Letter-Quality Printer

LQR Linear Quadratic Regulator • Link Quality Report

LQS Local Query Server

LQSS Loral Qualcomm Satellite Services Inc.

lr Laar

lR laufende Rechnung

Lr Lawrencium • Lower

L/R Launch and Recovery • Locus of Radius

LR Labeled Release • Laboratory Report • Laboratory Request • Landing Radar • Landing Report • Language Resource • Lapse Rate • La Rioja (Argentina) • Laser Reflectometer • Latent Root • Launch Reliability • Launch Right • Lead Radial • Left Rear • Left-Right • Left Rudder • Letter Report • Letter Requirement • Lettre Recommandée • Liaison Report • Liaison Request • Liberia (ISO country code) • Library Representation • Ligne Réseau • Limited Recoverable • Linear Representation • Line Receiver • Line Request • Link Register • Liquid Rocket • Living Room • Load Ratio • Load Register • Local Reference • Local Register • Logistics Regression • Log Run • Long Range • Long Run • Loop Resistance • Lorentz Resonance • Loss of Retranslation • Lower Right • Lucy-Richardson • Luft und Raumfahrt (periodical code) • Luminance Ratio

LRA Laboratoire de Radioastronomie • Large Reflector Assembly • Larisa (Greece) (location identifier) • Laser Reflector Array • Laser Retroreflector Array • Local Register Allocation • Local Reproduction Authorized • Long-Range Aircraft • Long-Range Aviation

LRAD Logistics Readiness Analysis Division

LRAE Long-Range Acquisition Estimate

LRAF Low-Radiative (efficiency) Accretion Flow

LRAM Livermore Risk Analysis Model

LRAP Long-Range Action Plan • Long Route Analysis Program

LRAU Ley Reguladora de la Actividad Urbanística (Spain)

LRB Laboratoire René Bernas (France) • Leribe (Lesotho) (location identifier) • Level Reference Base • Liquid Rocket Booster

LRBA Laboratoire de Recherches Ballistiques et Aérodynamiques (France)

LRC Laarbruch (Germany) (location identifier) • Langley Research Center (NASA) • Learning Resources Center • Lesser Regional Conflict • Lesser Regional Contingency • Lewis Research Center (NASA) • Ley del Registro Civil (Spain) • Library Research Committee • Library Resource Center • Lighting Research Center (USA-NY) • Limited Regional Conflict • Load Ratio Control • Local Register Cache • Logical Redundancy Check • Logistics and Readiness Center • Longitudinal Redundancy Character • Longitudinal Redundancy Check • Longitudinal Redundancy Checking • Long-Range Cargo • Longitudinal Redundancy Character • Longitudinal Redundancy Check • Luneberg Rapid Commutator

LRCCP Laboratoire de Recherche et de Contrôle du Caoutchouc et des Plastiques (France)

LRCO Limited Remote Communications Outlet

LRCP Long-Range Construction Program

LRCS Long-Range Communications System • Long-Range Computer Study • Low Radar Cross-Section

LRCSS Logistics Resource Control and Support System

LRD Dollar (Liberia) (ISO currency code) • Laredo (USA-TX) (location identifier) • Launch Readiness Date • Launch Recovery Division • Launch Requirements Document • Liberian Dollar (ISO currency code) • Lightning and Radio-emission Detector • Long-Range Dependence • Long-Range Detection • Low-Resolution Data

LRDA Land Resources Data Analysis

LRDB Low-Resolution Database

LRDC Land Resources Development Centre

LRDG Learning Resource Development Group

LRDM Low-Rate Data Multiplexer

LRDP Long-Range Development Plan

LRDTF Low-Rate Data Transcription Facility

LRE Language Research and Engineering • Lasting Rigid Erection • Latest Revised Estimate • Launch and Recovery Element • Launch and Recovery Equipment • Linguistic Research and Engineering • Linguistic Research Engineering • Liquid Rocket Engine • Longreach (Australia-QLD) (location identifier) • Low Risk Enhancement • Lunar Retrograde Engine

LREC Language Resources and Evaluation Conference • Local Research Ethics Committee (UK)

LRECL Logical Record Length

LREE Light Rare-Earth Element

LRES Liquid Rocket Engine Stability

LRF Jacksonville - Little Rock (USA-AR) (location identifier) • Large-scale Research Facility • Linear Ramp Filter • Liquid Rocket Fuel • Local Rest Frame • Long-Range Fuel

LRFS Logistic Requirement and Funding Summary

Lrg Large

LRG Label Raster-Generator • Laser Ring Galaxy • Library Research Group • Lithium-Rich Giant (star) • Logistics Review Group • Long Range • Lora Lai (Pakistan) (location identifier) • Low-Ratio Gas • Luminous Red Galaxy

LRH La Rochelle (France) (location identifier)

LRI Least Recently Loaded • Limited Range of Intercept • Line Replaceable Item • Low-Resolution Image • Low-Resolution Imager • Low-Resolution Imaging • Lorica (Colombia) (location identifier)

LRIA Level Removable Instrument Assembly

LRIC Long Run Incremental Cost

LRIDP Land Resource Inventory Demonstration Project

LRIP Limited-Rate Initial Production • Low-Rate Initial Production

LRIR Limb Radiance Inversion Radiometer • Low-Resolution Infrared Radiometer

LRIS Land Resources Information System • Low-Resolution and Imaging Spectrograph • Low-Resolution Imaging Spectrograph (*) • Low-Resolution Imaging Spectrometer

LRIT Low-Rate Image Transmission • Low-Rate Information Transmission

LRITF Land Resources Inventory Task Force

LRJ Lemars (USA-IA) (location identifier)

LRK Lincoln Rock (USA-AK) (location identifier)

LRL Lama Kara (Togo) (location identifier) • Lawrence Radiation Laboratory • Learning Resource Laboratory • Least Recently Loaded • Lectronic Research Laboratories (USA-NJ) • Lunar Receiving Laboratory • Niamtougou (Togo) (location identifier)

LRLF Local Radio Luminosity Function

LRLS Long Range Lineup System

LRM Casa de Campo (Dominican Rep.) (location identifier) • Laboratório Regional de Metrologia • Language Reference Manual • La Romana (Dominican Rep.) (location identifier) • Laser Raman Microprobe • Licensed Radioactive Material • Likelihood Ratio Method • Limited Register Machine • Linear Referencing Method • Line - Reflect - Match • Line-Replaceable Module • Line-Replacement Module • Liquid Rocket Motor • Logistics Reassignment Monitor • Long Radius Meridian • Low-Rate Multiplexer • Lunar Reconnaissance Mission • Lunar Reconnaissance Module

LRMF Liquid Rocket Motor Facility

LRN Local Reference Number • Local Resource Name • Location Routing Number • Moses Lake - Larson (USA-WA) (location identifier)

LRNAV Long-Range Navigation

LRNOD Long-Range Night Observation Device

LRNR Low-Resolution Non-scanning Radiometer

LRO Large Radio Observatory • Lathrop Sharpe (USA-CA) (location identifier) • Local and Remote Operations • Logistics Readiness Officer • Long-Range Objective • Long-Range Order

LRONIS Low-Resolution Optical/Near-Infrared Spectrophotometer

LROR Low-Resolution Omnidirectional Radiometer

LROSS Long-Range Optical System Study

LRP Lander Robotic Payload • Lasting Rigid Penetration • Launching Reference Point • Load and Roll Pallet • Logistics Recovery Program • Logistics Release Point • Long-Range Path • Long-Range Plan • Long-Range Planning • Low-Rate Production • Low-Resolution Profile • Low-Resolving Power • Lunar Radiation Principle

LRPA Liferaft Portside Aft • Long-Range Patrol Aircraft

LRPDS Long-Range Position Determination System

LRPF Liferaft Portside Forward

LRPG Long-Range Planning Group

LRPM Liferaft Portside Middeck

LRPP Long-Range Planning Panel

LRPS Long-Range Precision Strike

LRPT Large Repair Parts Transporter • Low-Rate Picture Transmission • Low-Resolution Picture Transmission

LRQ Laurie River (Canada-MB) (location identifier)

LRR Laser Radiation Receiver • Laser Retro-Reflector • Launch Readiness Report • Launch Readiness Review • Long-Range Radar

LRRDA Long-Range Research and Development Acquisition

LRRDAP Long-Range Research, Development, and Acquisition Plan

LRRR Laser Ranging Retro-Reflectometer

LRRS Library and Learning Resource Service

L&RS Launch & Recovery System

LRS Laboratoire de Recherches Spatiales (Observatoire de Paris-Meudon) • Lander Radio System • Language Resource • Language Resource System • Laser Raman Spectroscopy • Laser Ranging Station • Launch and Recovery Shelter • Launch and Recovery Site • Launch and Recovery System • Leros (Greece) (location identifier) • Linear Referencing System • Line Repeater Station • Linguistic Research System • Locally Rotationally Symmetric • Low-Rate Science • Low-Rate Switch • Low-Resolution Scenario • Low-Resolution Spectrograph • Low-Resolution Spectrometer • Low-Resolution Spectrum

LRSA Land Remote Sensing Assembly • Liferaft Starboard Aft

LRSF Liferaft Starboard Forward

LRSI Low-temperature Reusable Surface Insulation

LRSM Liferaft Starboard Middeck • Long-Range Seismograph Measurement

LRSN Log Record Sequence Number

LRSP Long-Range Security Program • Low-Rate Science Processor

LRSR Long-Range Storage Requirement

LRSS Logistics Readiness Simulation System • Long-Range Survey System

LRS&TP Long-Range Science & Technology Plan

LRSTP Long-Range Science and Technology Plan

LRSU Local Remote Switching Unit • Long-Range Sensor Unit

LRT Label Reader-Translator • Launch, Recovery and Transport • Lehrstuhl für Raumfahrttechnik • Light Rail Train • Light Rail Transport • Likelihood Radio Test • Local Rapid Transport • Loi de Réglementation des Télécommunications (France) • London Regional Transport (UK-England) • Long-Range Transport • Lorient Lann-Bihoué (France) (location identifier)

LRTAP Long-Range Transport of Air Pollutants

LRTF Linear Radial Transmission Filter • Long Range Technical Forecast

LRTGT Last-Resort Target

LRTM Long-Range Training Mission

LRTS Launch Range Technology Study

LRU Las Cruces (USA-NM) (location identifier) • Last Recently Used • Last Register Used • Least Recently Used • Ley de Reforma Universitaria (Spain) • Linear Replaceable Unit • Line Replaceable Unit • Line Replacement Unit • Logistical Replacement Unit • Loi de Réforme Universitaire • Lowest Replaceable Unit

LRUS Low-Rate User Station

LRV Launch Readiness Verification • Lifting Reentry Vehicle • Light Rail Vehicle • Long-Range Video • Los Roques (Venezuela) (location identifier) • Lunar Rover Vehicle • Lunar Roving Vehicle (*)

LRVT Launch Readiness Verification Test

LRW Left Rendezvous Window

LRZ Leibniz-Rechenzentrum (Germany)

ls Locus Sigilli

L/S Landing Site • Load System • Low Speed

L&S Land & Sea • Lange & Springer • Launch & Steering • Logistics & Support

LS Labor Service • Land and Sea • Landing Site • Language Specification • Late Start • Launch and Steering • Launching Station • Launching System • Launch Safety • Launch Sequence • Launch Service • Launch Simulator • Launch Site • Launch Station • Launch System • Leading Spot • Lead Sheet • Learning Step • Least Significant • Least Squares • Lectori Salutem • Left Side • Legacy Survey • Legal Seal • Lesotho (international vehicle registration code) • Lesotho (ISO country code) • Letter Service • Letter Signed • Lettre Signée • Ley del Suelo • Liaison Scientist • Liaison Spécialisée • Library Science • Library Search • Licensed Surveyor • Licensing Server • Lichtsäule • Life Science • Life Support • Life System • Lighting System • Light Second • Light Shield • Light Source • Ligne Spécialisée • Limit Switch • Lindsey-Smith • Line Scanner • Liquid Scintillator • List Status • Livret Suiveur • Load System • Local Sensor • Local Source • Local Store • Local Sunset • Local Symmetry • Locus Sigilli • Logistics and Support • Logistics Summary • Logistics Support • Lomb-Scargle (periodogram) • Long Scarf • Long Shot • Long Sight • Loop System • Loudspeaker • Low Session • Low Smoke • Low Speed • Low State • Luminous Stars (survey) • Lump Sum • Lunar Surface • Syrian Pound (currency)

LSA Labor Service Agency • Labor Surplus Area • LAN and SCSI Adapter • Language Students Abroad • LAN System Administrator • Large Scale Area • Large-scale Sounding Array • Large Science Aperture • Large Southern Array (now merged into ALMA) • Large Space Antenna • Large Spherical Array • Laser with Saturable Absorber • Latent Semantic Analysis • Lateral Spherical Aberration • Launch Services Agreement • Lead Standardization Activity • Library Services Act (USA) • Libre de Service d'Affranchissement • Linear Stability Analysis • Line Sharing Adapter • Linguistic Society of America • Link State Advertisement • Link State Algorithm • Local Security Authority • Logistics Support Agency • Logistics Support Analysis • Logistics Support Area • Logistics Systems Application • Logistic Sustainability Analysis • Longitudinal Spherical Aberration • Loss of Situation awareness • Losuia (Papua New Guinea) (location identifier) • Low-cost Solar Array • Low-Speed Access • Low Specific Activity

LSAA Library Services Authority Act (USA)

LSAC Life Sciences Advisory Committee

LSACN Logistics Support Analysis Control Number

LSAD Logistic Support Analysis Documentation

LSAH Launch Site Accommodation Handbook

LSALT Lowest Safe Altitude

LSAMP Lewis Stokes Alliance for Minority Participation

LSAN LWS Automatic Analysis Result

LSAP Launch Sequence Applications Program • Layer Service Access Point • Linear Systems Analysis Program • Link Service Access Point • Logical Service Access Point • Logistics Support Analysis Plan • Logistics Support Analysis Process • Logistics Support Analysis Program • Low Sun Angle Photography

LSAPI License Service Applications Programme Interface

LSAPT Lunar Sample Analysis Planning Team

LSAR Logistics Support Analysis Receipt • Logistics Support Analysis Record • Logistics Support Analysis Report • Logistics Support Analysis Requirement

LSAS Logistics Support Analysis Strategy

LSAST Least Squares Ambiguity Search Technique

LSAT Large Scale Applications Test • Laser Satellite • Logistics Support Analysis Task

lsb Least Significant Bit

LSB Launcher Support Building • Launch Service Building • Launch Support Building • Least Significant Bit • Least Significant Byte • Line Segment Block • Linux Standard Base • Logistics Support Base • Lordsburg (USA-NM) (location identifier) • Lower Sideband • Low-Speed Breaker • Low Surface Brightness • Lunar Surface Base

LSBD Low Surface Brightness Dwarf (star)

LSBF Least Square Best Fit

LSBG Low Surface Brightness Galaxy

LSBY Least Significant Byte

lsc Loco Supra Citato

LSc Licentiate in Sciences

LSC Land Surface Climatology • Large-Scale Computer • La Serena (Chile) (location identifier) • Laser Spectral Control • Launch Sequence Control • Launch Services Contract • Launch Services Contractor • Launch Services Corporation • Launch Support Center • Launch Support Contractor • Launch Support Controller • Least Significant Character • Least-Square Circle • Least-Squares Collocation • Lecture Series Committee • Lens Sign Convention • Liberty Science Center (USA-NJ) • Life Support Cost • Lightspeed c • Ligne Supérieure de Contrôle • LIGO Science Collaboration • LIGO Scientific Collaboration • Linear Sequential Circuit • Linear-Shaped Charge • Line Scan Camera • Line State Control • Liquid Scintillation Counter • Liquid Scintillation Counting • Liquid-Solid Chromatography • LISA Science Centre • Live Sequence Chart • Load Standardization Crew • Local Staff Committee • Local Supercluster • Local Systemic Change • Loco Supra Citato • Logistics Support Cadre • Logistics Support Charge • Logistics Support Concept • Logistics Support Cost • Louisville Science Center (USA-KY)

LSCA Library Services Construction Act (USA) • Limited Security Control ACID • Logistics Support Cost Analysis

LSCAD Lightweight Standoff Chemical Agent Detector

LSCB Launching Section Control Box

LSCBO Launching Section Control Box Operator

LSCC Line-Sequential Color Composite

LSCD Large Screen Color Display • Laser Standoff Chemical Detector

LSCDM Low-Speed Cable Driver Modem

LSCE Launch Sequence and Control Equipment

LSCF Least Squares Curve Fitting

LSCM Laser Scan Confocal Microscope • Laser Scan Confocal Microscopy • Least-Squares Collocation Method • Least-Squares Conform Map

LSCO Lanthanum Strontium Copper Oxide

LSCOP Launch Complex Safety Operating Procedure

LSCP Low-Speed Card Punch

LSCT Laser Spectral Control Technique

lsd Librae Solidi Denarii

LSd Sudanese Pound (currency)

LSD Lagrangian Simulation and Droplets • Lake Shore Drive • Landing Ship Dock • Landing Site Determination • Land Surface Datum • Language for Systems Development • Large-Scale Display • Large-Scale Domain • Large Screen Display • Launch Support Data • Launch Support Division • Launch Systems Data • Law of Selective Declination • Least Separation Distance • Least Significant Difference • Least Significant Digit (*) • Least-Squares Deconvolution • Lexington Creech (USA-KY) (location identifier) • Librae - Solidi - Denarii • Local Spin Density • Logarithmic Series Distribution • Low-Speed Data • Lunar Surface Drill

LSDA Life Sciences Data Archive • Local Spin-Dependent Approximation

LSDB Launch Support Database

LSDE Launch Support Equipment

LSDS Large-Screen Display System • Less Structured Data Set • Low-Speed Digital System

LSDU Link Service Data Unit

LSDV Link Segment Delay Value

LSE Laboratory Support Equipment • La Crosse (USA-WI) (location identifier) • Language Sensitive Editor • Large-Scale Enterprise • Launch Sequencer Equipment • Launch Support Equipment • Learning Support Equipment • Least Squares Estimation • Life Support Equipment • Limited Signed Edition • Liquid-Solid Extraction • Liquid-Solvent Extraction • Local Security Environment • Local Stellar Environment • Local Subscriber Environment • Logistics Support Element • Logistics Support Expenses • London School of Economics (UK-England) • London Stock Exchange (UK-England) • Luminous Stars Extension • Lunar Surface Experiment

LSEA Lifetime Support Engineering Agent

LSEAT Launch Systems Evaluation Advisory Team

LSEC Logistical Support Element Communications

LSEET Laboratoire de Sondages Électromagnétiques de l'Environnement Terrestre (France)

LSEP Large Solar Energetic Particle • Lunar Surface Experiment Package

LSEQ Lockout Sequential

LSER Labor Standards Enforcement Report

LSET Logistics Supportability Evaluation Team • Logistics Support Evaluation Team

LSEV Lunar Surface Exploration Vehicle

LSF Columbus Lawson (USA-GA) (location identifier) • Fort Benning - Lawson (USA-GA) (location identifier) • Laboratory Simulation Facility • Laser Support Facility • Launch Support Facility • Leaf Setup Request • Least Squares Filter • Least Squares Fit • Linear Swept Frequency • Line Spectral-pair Frequency • Line Spread Function (*) • Liquid Structure Facility • Load Sharing Facility • Logistics Support Facility • Log-Structured File • Luminosity Spread Function • Lunar Scientific Facility

LSFA Logistic System Feasibility Analysis

LSFE Life Sciences Flight Experiment

LSFN List of Selected File Numbers

LSFR Large Solar Flux Radiometer (PV) • Launch Site Flow Review

LSG Law of Selective Gravitation • Liaison Scientists Group • Linear Sequence Generator • Local Sponsor Group • Local Studies Group • Logistics Support Group • Low Specific Gravity • Lunar Surface Gravimeter

LSGER Laboratory of Space Geodesy and Earth Rotation (Russia)

LSH Laboratoire de Spectroscopie Hertzienne • Lashio (Myanmar) (location identifier)

LSHI Large-Scale Hybrid Integration

LSHTIPMP Laughing So Hard That I Pee My Pants

LSI Laboratoire de Systèmes Intégrés • Laboratório de Sistemas Integráveis • Laboratory for Scientific Interrogation (USA-AZ) • Large-Scale Integrated • Large-Scale Integration • Latent Semantic Indexing • Launch Success Indicator • Lead System Integrator • Life Space Interviewing • Linear Shift Invariant • Line Status Indication • Liquid Silicon Infiltration • Lunar Surface Instrument • Shetland Islands - Sumburgh (UK) (location identifier)

LSIC Large-Scale Integrated Circuit • Launch Support Integration Contractor • Launch System Integration Contractor

LSID Launch Sequence and Interlock Document • Lunar Sequence and Interlock Document

LSIEP Least Squares Inverse Eigenvalue Problem

LSIIT Laboratoire des Sciences, de l'Image, de l'Informatique et de la Télédétection (France)

LSIN Logistics System Information Network

LSJ Long Island (Papua New Guinea) (location identifier)

LSK Large-Scale Kinetic (simulation) • Lochstreifenkarte • Lusk (USA-WY) (location identifier)

LSKBS Link-Structured Knowledge-Based System

LSKS Large-Scale Kinetic Simulation

LSL Landing Ship Logistic • Larch Shared Language • Link Support Layer • Load Segment Limit • Los Chiles (Costa Rica) (location identifier) • Loti (Lesotho) (ISO currency code) • Lower Specification Limit • Low-Speed Logic

LSLE Life Sciences Laboratory Equipment

LSLF Large Short-Lived Fluctuation

lsm Litera Scripta Manet

LSM Laboratoire Souterrain de Modane (France) • Land Surface Model • Large Solid Motor • Laser Scan Microscope • Laser Scan Microscopy • Laser Scanning Microscope • Laser Scanning Microscopy • Launcher Status Multiplexer • Launching System Module • Layered Synthetic Microstructure • Least Squares Matching • Least Squares Mean • Least Squares Method • Letter Sorting Machine • Level Set Method • Life Science Module • Linear Synchronous Motor • Link Services Manager • Linux Software Map • Loader Select Modification • Local State Manager • Local System Management • Local System Manager • Logistics Support Manager • Longitudinal Solar Magnetograph • Long Semado (Malaysia) (location identifier) • Lunar Surface Magnetometer

LSMC Life Sciences Master Catalog

LSMDB Logistics Simulation Models and Databases

LSMF Large-Scale Magnetic Field

LSMI Logistics Support Management Information

LSMIS Logistics Support Management Information System

LSML Low Salt Meal

LSMM Lower Semi-Midmean

LSMQ Local Services Management Queue

LSMS LNP Service Management System

LSMU Launcher and Sensor Mockup

LSN Large-Scale Network • Large-Scale Networking • Linear Scaling Number • Line Stabilization Network • Local Stock Number • Log Sequence Number • Los Banos (USA-CA) (location identifier) • Luminous Stars Northern

LSND Liquid Scintillator Neutrino Detector

LSO Lager Secundair Onderwijs • Landing Signal Officer • Landing Support Officer • Large-sized Symmetric Object • Large Solar Observatory • Large Solar Orbiter • La Silla Observatory (Chile) • Launch Safety Officer • Launch Support Operations • Law of Selective Observation • Lesotho (ISO country code) • Les Sables - Talmont (France) (location identifier) • Life Systems Officer • Logistics Support Office

LSOC Launch Support Operations Contractor • Logistical Support Operations Center

LSOCE Linear Stochastic Optimal Control and Estimation

LSOP Lunar Surface Operations Planning

LSP Labelled Switch Path • Label Switched Path • Laboratoire de Spectrométrie Physique (France) • Laboratoire des Systèmes Photoniques • Large-Scale Phenomenon • Large Space Photography • Las Piedras (Venezuela) (location identifier) • Lasting Soft Penetration • Launcher Status Panel • Least Significant Portion • Least-Squares Polynomial • Library Software Package • Lightest Superparticle • Lightest Supersymmetric Particle (*) • Lightest Supersymmetric Partner • Light Scattering Photometer • Lightspeed Pascal • Linear-Spectrum Pair • Linguistic String Processor • Link State Packet • Link State Protocol • Localized Surface Plasmon • Logical Support Program • Logistical Support Plan • Logistics Support Plan • Logistics Support Priority • Lomb-Scargle Periodogram • Long Secondary Period • Long/Stable Penetration • Low-Speed Printer • Lunar Solar Power • Lunar Spectral Photometrics • Lunar Surface Probe • Pounds - Shillings - Pence

LSPC Local Serial Port Controller • Logistic System Policy Committee • Long-term Space Policy Committee • Lunar Solar Power Coalition

LSPD Large Screen Plasma Display

LSPDF Life Science Payload Development Facility

LSPE Lunar Seismic Profiling Experiment

LSPES Logistics System Planning and Execution System

LSPET Lunar Sample Preliminary Examination Team

LSPL Local Services Priority List

LSPO Large-Scale Parameter Optimization • Lunar Surface Project Office

LSPP Large-Scale Printer Plotter

LSPPS Logistics Support Plan for Preoperational Support

LSPR Low-Speed Pulse Restorer

LSPS Landelijk Samenwerkende Publiekssterrenwachten (Netherlands)

LSPTP Low-Speed Paper Tape Punch

LSQ Least Squares • Line Squall • Loi de Sécurité Quotidienne (France) • Los Angeles (Chile) (location identifier)

LSQA Local System Queue Area

LSQE Language-Sensitive Quality Editor

LSR Label Switched Router • Label Switching Router • Land-Sea Rescue • Lanthanide Shift Reagent • Laser Scanning Radar • Launch Signal Responder • Launch Site Recovery • Launch Support Request • Launch Support Requirement • Leaf Setup Request • Leased Service Request • Level of System Robustness • Licensed Storage Capacity • Load Shifting Resistor • Locally Shared Resource • Local Service Request • Local Shared Resource • Local Space Rectangular • Local Standard of Rest • Local Sunrise • Location Server Request • Location Server Router • Location Service Router • Logically Shared Resource • Logiciel, Systèmes et Réseaux • Logistics Shop Release • Logistics Success Rate • Logistics Support Requirement • Logistics Support Review • Lost River (USA-AK) (location identifier) • Low Storage Requirement

LSRA Landing Systems Research Aircraft

LSRC Logistics Support Review Committee

LSRF Logistics Support Ressource Funds

LSRG Lineat Shift Register Generator

LSRM Life Science Research Module

LSRO Life Sciences Research Office

LSRP Laser Printer

LSRPS Line-Scan Receiving and Processing System

LSRR Launch Site Readiness Review • Location Server Request Router • Loose Source and Record Route

LSRRI Leningrad Scientific Research Radiotechnical Institute (Russia)

LSRV Lunar Surface Rover Vehicle • Lunar Surface Roving Vehicle

LSS Landing and Separation Simulator • Landing Sensor System • Langley Software Server (NASA) • Large-Scale Structure • Large-Scale Survey • Large-Scale System • Large Sky Survey • Large Space Simulator • Large Space Structure • Large Space System • Large Steep Spectrum • Last Scattering Surface • Launch Sequence Simulator • Launch Site Support • Launch Status Summarizer • Launch Support Section • Launch Support Services • Launch Support Structure • Launch Support System • Licensing Support System • Life-Saving Station • Life-Support

System • Limited Storage Site • Link Server System • Link Status Signal • Literature Search Service • Lithuanian Scientific Society (Lithuania) • Load Sensing System • Local Speed of Sound • Local Subscriber Switch • Logical Source Statement • Logistics Support Section • Logistics Support Services • Logistics Support Station • Logistics Support System • Logistics Systems Support • Long-Slit Spectrograph • Long-Slit Spectrography • Long-Slit Spectroscopy • Loop Switching System • Low-Speed Stream • Luminous Stars Southern • Lunar Surveying System • Terre-de-Haut (Guadeloupe) (location identifier)

LSSA Licence de Sciences Sociales Appliquées • Logistics System Support Activity

LSSC Large Space Simulation Chamber • Launch Services Support Contract • Launch Services Support Contractor • Launch Site Support Contract • Launch Site Support Contractor • Launch Support Services Contract • Launch Support Services Contractor • Lincoln Space Surveillance Complex (USA-MA)

LSSD Level-Sensitive Scan Design • Level-Sensitive Scan Detection • Lunar Surface Sampling Device

LSSE Laser Sensor Support Equipment

LSSF Life Sciences Support Facility • Limited Service Storage Facility

LSSG Logistics Studies Steering Group • Lunar Study Steering Group

LSSI Ley sobre los Servicios de la Sociedad de la Información (Spain)

LSSK Local Site Spares Kit

LSSL Life Sciences Space Laboratory

LSSM Launch Site Support Manager • Local Scientific Survey Module

LSSO Large-Size Symmetric Object

LSSP Launch Site Support Plan • Lunar Surveying System Program

LSSS Large-Scale Structure Survey • Limiting Safety System Setting

LSST Large-aperture Synoptic Survey Telescope • Large-Scale Seismic Test • Large Space System Technology • Launch Site Support Team

LSSTO Large-aperture Synoptic Survey Telescope Observatory

LSSU Lake Superior State University (USA-MI) • Link State Signal Unit

LSSW Launch Site Software

lst List (file extension)

Lst Ladestelle

LST Lage-Snelheids Tunnel • Land Surface Temperature • Laplace-Stieltjes Transform • Large Simple Trial • Large Space Telescope • Large Stellar Telescope • Laser Scattering Tomograph • Launceston (Australia-TAS) (location identifier) • Launch Support Team • Light Scattering Temperature • Lightweight Satellite Terminal • Limited Streamer Tube • Liquid Storage Tank • List • Lithuanian Standardization Office • Local Sidereal Time • Local Solar Time • Local Standard Time • Lunar Surface Telescope • Lunar Surface Transponder • Lyman-Spitzer Telescope

L*STAR Lockheed-Martin Satellite Telemetry Analysis in Real Time

LSTC Large Sensor Test Chamber • Laser Systems Test Center • Link Security Transit Case • Livermore Software Technology Corp. (USA-CA)

LSTD Laser System Technology Demonstration • Lunar Satellite Tracking Data

LSTDM Low-Speed Time Division Multiplexing

LStDV Lohnsteuer-Durchführungsverordnung

LSTE Launch Site Transportation Equipment

LSTH Liu - Siebahn - Truhlow - Horowitz

LSTM Lander Static Test Model

LSTP Logistics System Training Program

LStR Lohnsteuer-Richtlinien

LSTS Launching Station Test Set • Launch Site Test Set • Lunar Surface Thermal Simulator

LSTSRFA Launch Station Test Set Radio Frequency Adapter

LSTTY Low-Speed Teletypewriter

LSTWA List of Serial Title Word Abbreviations

LSU La Silla University (ESO) • Launcher Selector Unit • Library Storage Unit • Life Support Umbilical • Life Support Unit • Line State Unknown • Local Source Unit • Lochschriftübersetzer • Long Sukang (Malaysia) (location identifier) • Louisiana State University (USA-LA)

LSUR Launch Site Utilization Report

LSUS Louisiana State University in Shreveport (USA-LA)

LSV Landelijk Samenwerkende Volkssterrenwachten (Netherlands) • Las Vegas - Nellis (USA-NV) (location identifier) • Launch Support Van • Law of Selective Vegetation • Letztschichtversagen • Line Status Verifier • Logistics Support Vessel • Low-Speed Vehicle • Lunar Shuttle Vehicle • Lunar Surface Vehicle

LSVD Large-Screen Visual Display

LSVE Large-Scale Virtual Environment

LSVI Large-Size Visual Interface

LSVT Lead Site Verification Test

LSW Landessternwarte • Lhok Seumawe (Indonesia) (location identifier)

LSWD Large-Screen Wall Display

LSWG Life Science Working Group

LSVT Large Solar Vacuum Telescope

LSX Lhok Sukon (Indonesia) (location identifier)

LSY Lismore (Australia-NSW) (location identifier)

LSZ Limited-Speed Zone • Low-Speed Zone • Mali Losinj (Croatia) (location identifier)

LS3 Life Saving - Life Sustaining - Life Supporting

lt Light • Lithuania • Lithuanian (ISO language code)

Lt Laut • Light • Long Ton • Low Temperature • Low Tension • Luitenant

L&T Laboratory & Test

L/T Load Test • Long Ton

LT Laboratory and Test • Landing Team • Landing Time • Language Translation • Laplace Transform • Laser Transmitter • Launch Telescope • Lead Time • Leak Test • Left • Left Turn • Less Than • Letter of Transmittal • Light • Light Terminal • Linear Triangulation • Line Telecommunications • Line Telegraphy • Line Terminal • Line Termination • Line Traffic • Link Terminal • List Translation • Lithuania (international vehicle registration code) • Lithuania (ISO country code) • Lithuanie (code ISO de pays) • Liverpool Telescope (JMU) • Live Theatre • Load Testing • Local Terminal • Local Time • Local Timing • Locum Tenens • Logic Theory • Logic Tree • London Transport (UK-England) • Long Ton • Long Tour • Lorentz Transformation • Lote • Lot Traceability • Lovell Telescope (JBO) • Lower Tester • Low Technology • Low Temperature • Low Tension • Lycée Technique (France) • Turkish Lira (currency)

LTA Lead Time Analysis • Library Technical Assistant • Lighter Than Air • Liquid Ternary Aerosol • Local Training Area • Long-Term Archive • Lower Torso Assembly • Low-Temperature Adsorber • Tzaneen (South Africa) (location identifier)

LTAB Line Test Access Bus

LTAS Laboratoire des Techniques Aéronautiques et Spatiales • Launch Trajectory Acquisition System

LTB Latrobe - Westmoreland County (Australia-TAS) (location identifier) • Lemaître - Tolman - Bond (metrics) • Line Term Buffer • Low-Temperature Branch • Low-Tension Battery

LTBR Long-Term Bedrest

LTBT Limited Test Ban Treaty

LTC Lai (Chad) (location identifier) • Language Translation Computer • Library Technology Center • Light Transfer Characteristics • Line Traffic Coordinator • Live Time Correction • Local Test Cabinet • Local Time Clock • Local Trainer Center • Longitudinal Time Code • Long-Term Contract • Long-Term Costing • Long Time Coming • Long Time Constant • Low-Temperature Carbonization • Lunar Terrain Camera • Lunar Topographic Camera

LTCC Low-Temperature Cofired Ceramics

LTCM Long-Term Capital Management

LTCS Laser Traffic Control System • Lower Thermosphere Coupling Study • Lower Thermospheric Coupling Study

Ltd Limited

LTD Ghadames (Libya) (location identifier) • LaTEX Tree Drawer • Launch Test Directive • Lift-To-Drag • Limited • Live Test Demonstration • Local Test Desk • Long-Term Depression • Long-Term Disability • Low-Temperature Demagnetization • Low-Temperature Detector

Ltda Limitada

LTDS Launch Tracking Data System • Launch Trajectory Data System • Linear Time Delay System

LTDT Langley Transonic Dynamics Tunnel (NASA)

LTE Laboratoire du Temps et de l'Espace • Large Table Electroplotter • Less Than or Equal • Letter To the Editor • Limited-Term Employee • Limited Test Equipment • Linear Threshold Element • Line Terminating Equipment • Link Terminating Equipment • Local Thermal Equilibrium • Local Thermodynamic Equilibrium (*) • Long-Term Effect • Low-Thrust Engine • Lycée Technique d'État (France)

Ltée Limitée

LTEF Long-Term Events File

LTER Long-Term Ecological Research

LTerm La Termotécnica (periodical code)

LTF Leitre (Papua New Guinea) (location identifier) • Lightwave Terminating Frame • Line Trunk Frame • Local Temporary File

LTFD Logic and Test Function Drawer

LTFT Long-Term Full Time

Ltg Lettering • Lighting

LTG Langtang (Nepal) (location identifier) • Lightning • Linear Tangent Guidance • Line Trunk Group • Local Timing Generator • Logistics Task Group • Low-Temperature Grown • Low-Temperature Growth • Lunar Traverse Gravimeter • Lund Telescope Group

LTH Lathrop Wells (USA-NV) (location identifier) • Length • Lethality and Target Hardening

LTHG Laboratoire d'Étude des Transferts en Hydrologie et Environnement (France)

LTHIS Laboratory Test Hardware Inventory Status

LtHo Light House

Lthr Leather

LtHse Light House

LTI Altai (Mongolia) (location identifier) • Laser Thermal Imaging • Linear Time-Invariant • Logistics Technical Information • Long-Term Instrument Plan

LTIRF Laboratoire de Traitement d'Images et Reconnaissance de Formes (France)

LTIV Lunar Trajectory Injection Vehicle

LTK Latakia (Syria) (location identifier)

LTL Lastourville (Gabon) (location identifier) • Law of Temporary Loss • Less Than Lethal • Linear Temporal Logic • Line-To-Line • Lit (Lithuania) (ISO currency code) • Little • Little Template Library • Long-term Timeline

LTLCG Little Change

LTLCS Long-Term Life Cycle Support

LTM LAN Traffic Monitor • Lateral Translation Mode • Lead Time Matrix • Lethem (Guyana) (location identifier) • Life Test Model • Linear Tangent Model • Line Terminal Multiplexer • Load Transformation Matrix • Long-Term Maintenance • Long-Term Memory • Long-Term Monitoring

LTMH Largeur Totale à Mi-Hauteur

LTMS Local Time Management System • Lunar Terrain Measuring System

LTN London Luton (UK-England) (location identifier)

LTNG Lightning

LTNR Local Thermonuclear Runaway

LTO Lease To Ownership • Linear Tape Open • Little Thompson Observatory (USA-CO) • Logistics Task Order • Logistics Technology Office • Long-Term Orientation • Loreto (Mexico) (location identifier) • Low-Temperature Orthorhombic • Lunar Transfer Orbit

LTOC Lowest Total Overall Cost

LTOE Living Table of Organization and Equipment

LTON Long Ton

LTOP Lease To Ownership Plan • Long-Term Operating Plan • Long-Term Operation Plan

LTP Laboratory for Terrestrial Physics • Late Thermal Pulse • Leader Training Plan • Leader Training Program • Learning Technologies Project • Learning Theory and Practice (conference) • Library Technology Project • LISA Technology Package • Long-Term Planning • Long-Term Potentiation • Long-Term Prediction • Long Trace Profile • Low-Temperature Phosphorescence • Low-Temperature Phosphorimetry • Low-Temperature Physics • Low-Temperature Platform • Lunar Transient Phenomenon • Lyndhurst (Australia-QLD) (location identifier)

LTPG Long-Term-Planning Group

LTPMS Lidar Temperature/Pressure/Moisture Sounder

LTPO Lease To Purchase Options • Long-Term Program Office

LTPP Ley de Tasas y Precios Públicos (Spain)

LTPS Low-Temperature Polysilicon

LTPV Long-Term Photometric Variables • Long-Term Photometry of Variables

LTQ Le Touquet (France) (location identifier)

Ltr Letter • Lighter

LTR Laboratory Test Report • Lander Trajectory Reconstruction • Large Tuning Range • Later • Left To Right • Letter • Letterkenny (Ireland) (location identifier) • Load Task Register • Long-Term Relationship • Long-Term Research • Loop Transfer Recovery • Low-Temperature Region

LTRE Life Table Response Experiment

LTREB Long-Term Research in Environmental Biology

LTRF LLR Terrestrial Reference France

LTRMP Long-Term Resource Monitoring Program

LTRR Laboratory for Tree-Ring Research

LTRRS Long-Term Regional Research Site

LTRS Langley Technical Report Server (NASA) • Letters • Letter Shift

Lts Lights

LTS Altus (USA-OK) (location identifier) • Labelled Transition System • Lagere Technische School • Language Translation System • Large Technical System • Large Truss Structure • Laser Tracking System • Launcher Test Set

• Launch Telemetry Station • Launch Telemetry System • Launch Tracking Station • Launch Tracking System • Lexicon/Thesaurus System • Library Technical Service • Lights • Light Tactical Satellite • Linearity Test Set • Long-Term Schedule • Long-Term Scheduler • Long-Term Scheduling • Long-Term Stability • Long-Term Storage • Loss Test Set • Low-Temperature Solidification • Low-Temperature Superconducting • Low-Temperature Superconductivity • Low-Temperature Superconductor

LTSA Learning Technology Systems Architecture • Long-Term Space Astrophysics (program)

LTSAB Long-Term Stability

LTSC Learning Technology Standardization Committee (IEEE) • Learning Technology Standards Committee

LTSEM Low Temperature Scanning Electron Microscope

LTSG Laser-Triggered Spark Gap

LTSP Long-Term Science Plan • Long-Term Space Plan

LTSPC Long-Term Space Policy Committee

LTST Local True Solar Time • Long-Term Scheduling Tool

LTSV Long Time Scale Variability

LTT Language Teacher and Tester • Latent Trait Theory • Learning - Teaching - Technology • Light-Travel Time • Long-Term Training • Low-Temperature Test • Lunar Test Table • Lunar Transfer Trajectory • Lunar Transit Telescope • Luyten Two Tenths (catalog) • Saint-Tropez (France) (location identifier)

LTTE Link Terminal Terminating Equipment

LTTL Long-Term Timeline • Low-power Transitor-Transistor Logic

LTTM Long-Term Terrain Model

LTTR Long-Term Technology Review

LTTS Laboratory Transponder Test Set • Lead Time Tracking System

LTU Linear Threshold Unit • Line Terminating Unit • Line Termination Unit • Line Test Unit • Lithuania (ISO country code) • Lithuanie (code ISO de pays) • Luleå Tekniska Universitet (Sweden)

LTUM Line Termination Unit Module

LTV Large Test Vessel • Launch Test Vehicle • Logistics Transfer Vehicle • Long Time (scale) Variability • Long Tube Vertical • Lotusvale (Australia-QLD) (location identifier) • Lunar Transfer Vehicle

LTVAD Leicester (Starlink internal code)

LTVC Linear Terminal Velocity Constraint

LTVGF Low-Temperature Vapor Growth Facility

LTW Leonardtown - Saint Marys County (USA-MD) (location identifier) • Localized Traveling Wave

LTWG Landsat Technical Working Group • Large Telescope Working Group • Launch Test Working Group • Logistics Test Working Group

LTWT Lightweight

LTWTA Linearized Traveling-Wave Tube Amplifier

LTX Lane Training exercise

LTXDB Leicester (Starlink internal code)

lt-yr Light-Year

LT2 Learning Through Technology (project)

lu Luxembourg

Lu Lucerne (Switzerland) • Luglio • Luglio • Lutécium • Lutetium • Luzern (Switzerland)

L/U Land Use

LU Land Unit • Leicester University (UK-England) • Light Unit • Link Unavailable • Litterae Uraniae • Logical Unit • Logistical Unit • Loudness Unit • Luminosity Uniformity • Luxembourg (code ISO de pays) • Luxembourg (ISO country code)

LUA Logical Unit Application • Lukla (Nepal) (location identifier)

lub Least Upper Bound

Lub Lubricant • Lubricate • Lubricating • Lubrication

LUB Least Upper Bound • Logical Unit Block • Lubricate • Lumid Pau (Guyana) (location identifier) • Lunds Universitetsbibliotek (Sweden)

LubEn Lubrication Engineering (periodical code)

Lubt Lubricant

LUC Convertible Luxembourg Franc (ISO currency code) • Labor Utilization Code • Laucala Island (Fiji) (location identifier) • Limburgs Universitair Centrum (Belgium) • Loyola University Chicago (USA-IL)

LUCA Last Unicellular Common Ancestor • Last Universal Common Ancestor

LUCC Land Use and Cover Change

LUCI Leading Us in Commerce Industry

LUCID Language for Utility Checkout and Instrumentation Development

LUCIL Lutte Contre l'Illettrisme (France)

LUCS Library Updating and Compiling System

LUD Lower-Upper Decomposition • Luderitz (Namibia) (location identifier)

LUDA Land Use Data Analysis • Land Use Development Assistance

LUE Linear Unbiased Estimator • Lucenec (Slovakia) (location identifier)

LUEMA Land Use and Environmental Management Authority

LUF Franc (Luxembourg) (ISO currency code) • Laboratoire Univers Froid • Lowest Usable Frequency • Phoenix Luke (USA-AZ) (location identifier)

Lug Luggage • Luglio

LUG Linux Users Group • Local Users Group • Lugano (Switzerland) (location identifier) • Lugar

LUGAM Lunar Gamma Spectrometer

LUH Light Utility Helicopter • Ludhiana (India) (location identifier)

LUHF Lowest usable High Frequency

LUHG Lowest Useful High Frequency

LUI La Unión (Honduras) (location identifier) • Life Usage Index • Linux Utility for Installation • Load Upper Immediate • Local User Input

LUIS Landes-Umwelt-Informations-System • Library User Information Service

LUJ Lusikisiki (South Africa) (location identifier)

LUK Cincinnati (USA-OH) (location identifier)

LUKAS Lowell-UK Asteroid Survey

LUL Financial Luxembourg Franc (ISO currency code) • Laboratoire Univers Lointain • Laurel (USA-MS) (location identifier) • Left Upper Lobe

LULI Laboratoire d'Utilisation des Lasers Intenses (France)

Lum Luminosity • Luminous

LuM Luchtmacht

LUM Living Utility Module • Luxi Mangshi (China-PRC) (location identifier) • Lynx - Ursa Major

LUMAS Lunar Mapping System

LUME Light Utilization More Efficient

LUMI Lunar Microwave Instrument

LUMIS Land Use Management Information System

LUMO Lowest Unoccupied Molecular Orbital

LUMS Lahore University of Management Sciences (Pakistan) • Land Use Management System

Lun Lundi • Lunedi • Lunes

LUN Logical Unit Name • Logical Unit Number (*) • Logic Unit Number • Lusaka (Zambia) (location identifier)

LUNA Laboratory for Underground Nuclear Astrophysics (Italy)

LUNI LAN Emulation User Network Interface

LUNIS Loyola University Nuclear Information System

LUNO Loyola University New Orleans (USA-LA)

LUNOX Lunar Oxygen

LUNR Land Use and Natural Resources

LUO Luena (Angola) (location identifier)

Lup Lupus-Lupi (constellation)

LUP Kalaupapa (USA-HI) (location identifier) • Location Update

LUPF Linear Utility Prediction Function

LUPI Laser Unequal-Path Interferometer

LUPIN Land Use Planning Information Network

LUPIS Land Use Planning Information System

LUPS Logistics Unit Productivity Study • Logistics Unit Productivity System

LUPWT Langley Unitary Plan Wind Tunnel (NASA)

LUQ Left Upper Quadrant • San Luis (Argentina-SL) (location identifier)

LUR Cape Lisburne (USA-AK) (location identifier) • Labor Utilization Report

LuRaQ Luft und Raumfahrt Quarter (periodical code)

LURE Laboratoire pour l'Utilisation du Rayonnement Électromagnétique (France) • Lunar Ranging Experiment

LUS Liquid Upper Stage • Lusanga (Zaire) (location identifier)

LUSC Lightweight Undersea Sensor Component

LUSE Land Use Suitability Evaluation

LUSEX Lunar Surface Explorer

LUSI Lunar Surface Inspection

LUSIS Land Use and Suitability Information System

LUSS Lunar Subsatellite

LUST Leaking Underground Storage Tank

LUSTEC Lunar Stereo Camera

LUT Land Utilization Type • Launch Umbilical Tower • Laura Station (Australia-QLD) (location identifier) • Limited User Test • Limited User Training • Local User Terminal • Look-Up Table

LUTE Limited User Test and Evaluation • Lunar Ultraviolet Telescope Experiment

LUTH Laboratoire Univers et Théorie

LUTIOA Limited User Test Initial Operational assessment

LUTT Launcher Umbilical Tower Transporter

LUU Laura (Australia-QLD) (location identifier)

LUV Langgur (Indonesia) (location identifier) • Lunar Utility Vehicle

LUVIMS Lunar UV-Visible-Infrared Mapping Spectrometer

LUVIT Lunar Ultraviolet Imaging Telescope

LUVO Lunar Ultraviolet Observatory

LUW Luwuk (Indonesia) (location identifier)

Lux Luxembourg • Luxembourg

LUX Luxembourg (code ISO de pays) • Luxembourg (location identifier) • Luxembourg (ISO country code)

Luxmbrg Luxembourg

LuxF Luxembourg Franc (currency)

LUY Lushoto (Tanzania) (location identifier)

LUZ Lushan (China-PRC) (location identifier)

lv Latvia • Latvian (ISO language code) • Leave • Lettish (ISO language code)

Lv Leave • Lev (currency) • Livre

L/V Launch Vehicle • Lethality/Vulnerability

L&V Lethality & Vulnerability

LV Argentina (international aircraft registration code) • Landesverfügung • Langue Vivante • Large Volume • Lasting Value • Latching Valve • Latvia (international vehicle registration code) • Latvia (ISO country code) • Launch Vehicle • Lave-Vaisselle • Leave • Leaving • Lebensversicherung • Leistungsverhältnis • Lettonie (code ISO de pays) • Lev (currency) • Leva (currency) • Levenberg-Marquardt • Lift Vector • Light and Variable • Limit Value • Linear Velocity • Loading Valve • Load Vertical • Local Vertical • Local Void • Local Volume • Logical Volume • Logiciel Vol • Long Vector • Loss Variance • Low Velocity • Low Voltage • Luncheon Voucher

LVA Latvia (ISO country code) • Launch Vehicle Adaptor • Launch Vehicle Availability • Laval Entrammes (France) (location identifier) • Lettonie (code ISO de pays) • Local Velocity Anomaly

LVAAS Lehigh Valley Amateur Astronomical Society (USA-PA)

LVAD Low-Velocity Air Drop

LVAR Launch Vehicle Assessment Report • Launch Vehicle Assessment Review

LVAS Las Vegas Astronomical Society (USA-NV)

LVB Livramento - Dos Galpões (Brazil-RS) (location identifier) • Luchtverkeersbeveiliging

L/V/C Live/Virtual/Constructive

LVC Launch Vehicle Contractor • Low-Velocity Cloud

LVCD Least Voltage Coincidence Detection

LVCM Launch Vehicle Cost Model

LVCSR Large-Vocabulary Continuous-Speech Recognition

LVD Large Volume Detector • Laser Video Disk • Lime Village (USA-AK) (location identifier) • Low-Velocity Drop • Low Voltage Differential

LVDA Launch Vehicle Data Adapter • Launch Vehicle Deployment Assembly

LVDC Launch Vehicle Data Center • Launch Vehicle Digital Computer

LVDE Large Volume Data Exchange • Low-Vibration Drive Electronics

LVDS Low Voltage Differential Signal (*) • Low Voltage Differential Signaling • Low Voltage Differential Swing

LVDT Linear Variable Differential Transformer • Linear Variable Differential Transducer (*) • Linear Velocity Displacement Transducer • Linear Voltage Differential Transformer • Linear Voltage Displacement Transducer

LVE Linear Variable Etalon • Linear Variable Equation • Low-Voltage Electronics

LVES Low-Vision Enhancement System

LVET Left Ventricular Ejection Time

LVF Linear Variable Filter • Linear Vector Function

LVFC Launch Vehicle Flight Control

LVFCS Launch Vehicle Flight Control System

LVFRR Launch Vehicle Flight Readiness Review

LVG Large Velocity Gradient • Low-Vapor(-pressure) Gas • Low-Velocity Gas • Low-Viscosity Gyro

LVGC Launch Vehicle Guidance Computer

LVGSE Launch Vehicle Ground Support Equipment

LVHGM Low-Velocity High-Gain Maser

LVHV Low Volume High Velocity

LVI Low Voltage Initiator • Livingstone (Zambia) (location identifier) • Low Viscosity Index

LVIP Launch Vehicle Integration Plan

LVIRD Launch Vehicle Interface Requirements Document

LVIS Launch Vehicle Instrumentation System • Live Video Insertion System • Logistics Verification Information System

LVK Livermore (USA-CA) (location identifier)

Lvl Level

LVL Lat (Latvia) (ISO currency code) • Lawrenceville (USA-VA) (location identifier) • Level

LVLASO Low-Visibility Landing And Surface Operations

LVLH Local Vertical - Local Horizontal

LVM Liceo Víctor Mercante (Argentina-BA) • Livingston - Mission Field (USA-MT) (location identifier) • Logical Volume Management • Logical Volume Manager

LVMD Limb Volume Measurement Device

LVMI Local VAN Management Interface

LVN Library Video Network • Licensed Vocational Nurse

LVNL Luchtverkeersleiding Nederland (Netherlands)

LVO Label-Value Object • Large Virtual Organization • Launch Vehicle Operation • Laverton (Australia-WA) (location identifier)

LVOF Low-Velocity Overflow

LvoTs L'vovskij Ordena Lenina Gosudarstvennyi Universitet Tsirkulyar (periodical code)

LVP Lagrangian Variational Principle • Lavan (Iran) (location identifier) • Linear Variational Problem • Low-Voltage Pressure • Low-Voltage Protection

LVPA Lease Versus Purchase Analysis

LVPD Launch Vehicle Pressure Display

LVPG Launch vehicle Planning Group

LVPP Launch Vehicle Propulsion Program

LVPS Low-Voltage Power Supply

LVQ Learning Vector Quantification • Learning Vector Quantization (*) • Linear Vector Quantization

LVR Large Vocabulary Recognition • Laser Videodisk Recorder • Local Vertical Reference • Low-Voltage Release

LVRS Lightweight Video Reconnaissance System

LVS Large Volume Sampling • Las Vegas (USA-NM) (location identifier) • Latvian Standard • Launch Vehicle Simulator • Launch Vehicle System • Limburgse Volkssterrenwacht (Belgium) • Logistics Vehicle System • Low-Voltage Supply • Low-Volume Sampling

LVSEM Low-Voltage Scanning Electron Microscope • Low-Voltage Scanning Electron Microscopy

LVSG Launch Vehicle Study Group

LVSSTS Launch Vehicle Safety System Test Set

LVST Large Vacuum Solar Telescope (Russia) • Lead Verification Site Test

LVT Low Volume Terminal

LVTC Launch Vehicle Test Conductor

LVTD Linear Variable Differential Transformer

LVTV Launch Vehicle Transport Vessel

LVU Liaison Virtuelle Unitaire • Low-Voltage Unit

LVZ Low-Velocity Zone

l/W Luman/Watt

lW lichte Weite

Lw Lawrencium (now Lr)

L/W Lightweight

LW Launch Window • Lazy Write • Leeway • Left Wing • Leitungswähler • Light Weight • Line Weight • Liquid Water • Lockman Window • Longwave • Longwavelength • Low Water • Low Weight • Low Wing

LWA Land Withdrawal Act (USA) • Limited Work Authorization • Local Wavefront Analyzer • Long-Wavelength Array • Long Wire Antenna • Lwbak (Philippines) (location identifier)

LWB Lewisburg (USA-WV) (location identifier)

LWC Lawrence (USA-KS) (location identifier) • Lightweight Coated Paper • Liquid-Water Concentration • Liquid-Water Content • Long Wavelength Channel

LWCF Land and Water Conservation Fund

LWCI Leaf Water Content Index

LWD Launch Window Display • Left Wing Down • Logistics Working Group • Loomis-Wood Diagram • Luyten White Dwarf (star)

LWDS Local weather Dissemination System

LWE Lewoleba (Indonesia) (location identifier)

LWEST Low-Water Equinoctial Spring Tide

LWF Long Wavelength Frequency

LWFA Laser Wakefield Acceleration • Laser Wakefield Accelerator

LWFPA List of Wide-Field Plate Archives

LWG Landelijke Werkgemeenschap

LWH Lawn Hill (Australia-QLD) (location identifier)

LWHS Lightweight Headset

LWI Laser Weather Identifier • Lasing Without Inversion • Lowai (Papua New Guinea) (location identifier)

LWIC Long-Wavelength Infrared Camera

LWIM Low-power Wireless Integrated Microsensor

LWIR Long-Wave Infrared • Long-Wavelength Infrared

LWIRC Long-Wavelength Infrared Camera

LWIRHSS Long-Wave Infrared Hyperspectral Sensor System

LWK Shetland Islands - Lerwick Tingwall (UK) (location identifier)

LwL Lichtwellenleiter

LWL Lichtwellenleiter • Wells Harriet (USA-NV) (location identifier)

LWLA Long Wavelength Large Aperture

LWLC Low-Weight Low-Cost

LWLD Lightweight Laser Designator

LWM Lawrence (USA-MA) (location identifier) • Local Window Manager • Low-Water Mark

LWN Leninakan (Armenia) (location identifier)

LWO L'viv (Ukraine) (location identifier) • L'vov (Ukraine) (location identifier)

LWOP Lease With Option to Purchase • Leave Without Pay

lwp Lotus Word Pro (file extension)

LWP Leave With Pay • Lightweight Process • Liquid Water Path • Long-Wavelength Primary (camera) • Long-Wavelength Prime (camera)

Lwr Lower

LWR Laser Warning Receiver • Leeuwarden (Netherlands) (location identifier) • Level Wage Rate • Light-Water Reactor • Locally Weighted Regression • Local Wage Rate • Long-Wavelength Radar • Long-Wavelength Radiation • Long-Wavelength Redundant (camera) • Lower

LWRHU Lightweight Radioisotope Heater Unit

LWRM Longwave Radiation Model

LWS Laser Warning System • Lewiston (USA-ID) (location identifier) • Lightning Warning System • Linear White Space • Living With a Star (initiative) • Loathy Worm of Sendmail • Long-Wavelength Spectrograph • Long-Wavelength Spectrometer

LWSA Long Wavelength Small Aperture

LWSD Large Wall-Screen Display

LWT Legal-Weight Truck • Lewistown (USA-MT) (location identifier) • Lightweight Tank • Listen While Talk • Long-Wavelength Transmission

LWV Lawrenceville (USA-IL) (location identifier)

LWW Launch Window Width

LWX LAN/WAN Exchange

LWY Lawas (Malaysia) (location identifier)

Lwyr Lawyer

lx Lux • Meter-Candle

LX Low Index • Luxembourg (international aircraft registration code)

LXA Lhasa (China-PRC) (location identifier)

LXe Liquid Xenon

LXE Lightguide Express Entry

LXeGRIT Liquid Xenon Gamma-Ray Imaging Telescope

LXeTPC Liquid Xenon Time Projection Chamber

LXG Luang Namtha (Laos) (location identifier)

LXI Linxi (China-PRC) (location identifier)

Lxmbrg Luxembourg

LXN Lexington (USA-NE) (location identifier)

LXR Luxor (Egypt) (location identifier)

LXS Lemnos (Greece) (location identifier)

LXT Local Exchange Terminal

LXU Lukulu (Zambia) (location identifier)

LXV Leadville (USA-CO) (location identifier)

ly Light Year

Ly Lyman

LY Libya (ISO country code) • Libye (code ISO de pays) • Light Year • Lithuania (international aircraft registration code)

LYA Luoyang (China-PRC) (location identifier)

LYB Little Cayman (Cayman Islands) (location identifier)

LyC Lyman Continuum

LYC Lycksele (Sweden) (location identifier)

LYD Dinar (Libya) (ISO currency code)

LYE Lyneham (UK-England) (location identifier)

LYG Lianyun'Gang (China-PRC) (location identifier)

LYH Lynchburg (USA-VA) (location identifier)

LYI Linyi (China-PRC) (location identifier)

LYK Lunyuk (Indonesia) (location identifier)

Lyn Lynx-Lyncis (constellation)

LYN Lyon Bron (France) (location identifier)

LYO Lift Your Own • Lyons - Rice County (USA-KS) (location identifier)

LyoRe Reprints de l'Observatoire de Lyon (periodical code)

LYP Faisalabad (Pakistan) (location identifier) • Lycée Polyvalent (France)

Lyr Layer • Lyra-Lyrae (constellation)

LYR Layer • Layered • Longyearbyen (Norway) (location identifier) • Lyrids (meteor stream)

LYRA Lowestoft and Great Yarmouth Regional Astronomers (UK)

LYRD Layered

LYRIC Language for Your Remote Instruction by Computer

LYS Lyon Saint-Exupéry (France) (location identifier) • Lyon Satolas (France) (location identifier)

Lyt Layout

LYT Lady Elliot Island (Australia-QLD) (location identifier)

LYU Ely (USA-MN) (location identifier)

LYX Lydd (UK-England) (location identifier)

LZ Bulgaria (international aircraft registration code) • Landau-Zener (approximation) • Landing Zone • Langzug • Lazy Write • Lempel-Ziv • Level Zero • Luftschiff Zeppelin

LZA Landau-Zener Approximation • Luiza (Zaire) (location identifier)

LZB Landeszentralbank (Germany)

LZC Lazaro Cardenas (Mexico) (location identifier)

LZCO Landing Zone Control Officer

LZD Lanzhou (China-PRC) (location identifier) • Launch Zone Display

LZEEBE Long-term Zonal Earth Energy Budget Experiment

LZF Launch Zone Flag

LZGF Lewis Zero Gravity Facility (NASA)

LZH Lempel - Ziv - Huffmann (algorithm) • Liuzhou (China-PRC) (location identifier)

LZI Luozi (Zaire) (location identifier)

LZO Launch Zone Override • Luzhou (China-PRC) (location identifier)

LZP Level Zero Processing

LZS Lempel - Ziv - Stacker (algorithm)

LZR Lizard Island (Australia-QLD) (location identifier)

LZT Large Zenith Telescope • Lead Zirconate-Titanate • Local Zone Time

LZW Lempel - Ziv - Welch (algorithm)

L1 First Language • Lagrange Point # 1

L1PT Level 1 Page Table

L2 Lagrange Point # 2 • Second Language

L2B Linienzugbeeinflussung

L2CAP Logical Link and Control Adaptation Protocol

L2F Layer 2 Forwarding

L2FET Logic Level Field Effect Transistor

L2M Laboratoire de Microstructures et de Microélectronique

L2PT Level 2 Page Table

L2SS Level 2 System Simulator

L2TP Layer 2 Tunneling Protocol

L3 Lagrange Point # 3 • Latitude/Longitude Locator • Third Language

L3AB Laboratoire d'Astrodynamique, d'Astrophysique et d'Aéronomie de Bordeaux (France)

L3C Laboratoire Capteurs-Caractérisation-Contrôle

L3D Life-Long Learning and Design

L3E Laboratoire d'Études Expérimentales en Électromagnétisme (France)

L3G Language de Troisième Génération

L3TV Low-Light-Level Television

L4 Lagrange Point # 4

L4G Language de Quatrième Génération

L5 Lagrange Point # 5

L6 Laboratories Low-Level Linked List Language

M

m Apparent Magnitude • Electromagnetic Moment • Electron Mass • Inertial Constant • Mach Number • Magnetic Pole Strength • Magnetic Quantum Number • Magnetic Moment • Magnitude • Magnitude Apparente • Mailer • Male • Manipulus • Mare • Mark (currency) • Masculine • Masculino • Mass • Mass of Electron • Matured • Medieval • Medium • Member • Membre • Memorandum • Meridian • Meridies • Meridional • Meta • Meter • Mètre • Metric • Midday • Middle • Mile • Mill (currency) • Mille • Milli • Million • Minim • Minor • Minute • Misce • Mist • Mittlere • Moderate • Modification • Modulus • Mois • Molar • Molarity • Molecular Mass • Mole Mass • Month • Moon • Morning • Mort • Mountain • Mouth • Noon • Number of Phases • Particle Mass • Phase Number • Scheinbare Helligkeit

M Absolute Helligkeit • Absolute Magnitude • Angular Momentum • Anomalie Moyenne • Em • General Precession in Right Ascension • Loti (currency) • Luminous Exitance • Mach Number • Macintosh • Magister • Magnetic • Magnetic Moment • Magnetic Polarization • Magnetic Quantum Number • Magnetization • Magnitude • Magnitude Absolue • Maintainability • Maintenance • Major • Male • Maloti (currency) • Malta (international vehicle registration code) • Mandatory • Manitoba (Canada) • Manoeuver • March • Marginal • Mark (currency) • Marker • Markka (currency) • Marquis • Masculine • Mass • Master • May • Me • Mean • Mean Anomaly • Mean Molar Mass • Measure • Mechanized • Medal • Medicine • Medieval • Medium • Meer • Mega • Member • Mer • Meridian • Meridies • Mesomeric Effect • Messier (catalog) • Metal • Meter • Metropolitan • Mezzo • Middle • Migration Length • Mike (international alphabet) • Mile • Million • Minim • Ministère • Ministerio • Minus • Mitte • Mittlere Anomalie • Modal • Modifier • Molar Mass • Moment of Force • Monday • Monsieur • Monte • Monthly • Mother • Motorway • Mountain • Moyen • Moyenne • Mud • Multi • Mutual Inductance • Noon • One Thousand (Roman numeral) • Radiant Exitance • Sea Mile • Spectral Type • Thermal Insulance • Thermal Insulation Coefficient • Thousand (Roman numeral) • Wind Speed • 1000 (Roman numeral)

m/a My Account

ma Myria

mA Milliampere

Ma Mach Number • Mars • Master • Million Years

M&A Maintenance & Assembly • Management & Administration • Management & Analysis • Monitor & Alarm

M/A Memorandum of Association • Monitoring/Accounting

MA Macedonian • Machine Accountant • Magnetic Amplifier • Magneto-Acoustics • Mai • Maintenance • Maintenance Ability • Maintenance Action • Maintenance Advisor • Maintenance Administrator • Maintenance Agency • Maintenance Analysis • Maintenance Area • Maître Assistant • Maître Auxiliaire • Malaysia • Management Accounting • Management Adviser • Management Architecture • Manager Agent • Mandan • Manufacturing Assembly • Map Analysis • Mapper Application Interface • Maranhão (Brazil) • Maritime Administration • Marketing Assistance • Maroc (code ISO de pays) • Marquesas Islands • Massachusetts (USA) • Master • Master Alarm • Master of Arts • Material Authorization • May • Mean Annual • Mechanical Alloying • Mediation Agent • Medium Adapter • Medium Angle • Megaampere • Memory Address • Mental Age • Mercury-Atlas • Merging Analysis • Message Assembler • Metabolic Analyzer • Meter Amplifier • Middeck Act (USA) • Middle Age • Middle Amazonian • Mike Amplifier • Millennium Assessment • Milliarcsecond Astrometry • Mirror Assembly • Missed Approach • Mission Abort • Mission Analysis • Mission Area • Mission Authority • Mobile Agent • Mobility Allowance • Model Age • Modified Achromat • Modulation d'Amplitude • Monitor and Alarm • Monitoring Agency • Morocco (international vehicle registration code) • Morocco (ISO country code) • Moving Average • Moyen-Age • Multiple Access • Multiplex Advantage

MAA Madras Meenambakkam (India) (location identifier) • Madras Astronomical Association (India) • Martz Astronomical Association (USA-NY) • Materiel Access Area • Mathematical Association of America (USA) • Maximum Authorized Altitude • Mechanical Arm Assembly • Mémoire à Accès Aléatoire • Message Authenticator Algorithm • Metropolitan Area Acquisition • Mission Area Analysis • Mission Area Assessment • Moist Adiabatic Adjustment • Monitoring Angle of Attack • Mutual Assistance Agreement

MAAA Mankato Area Astronomy Association (USA-MN)

MAAB Materials Application Advisory Board

MAACS Multiaddress Asynchronous Communication System

MAADP Mission Area Analysis Deficiency Program • Mission Area Analysis Development Plan

MAAF Mishap Analysis and Animation Facility

MAAP Maintenance And Administration Panel

MAAPR Materiel Acceptance and Accounts Payable Report

Maar Maarschalk

MAAR Mandatory Annual Audit Requirement • Monthly Associate Administrator's Review

MAARC Magnetic Annular Arc

MAAS Manpower Allocation and Accounting System • Mission Area Analysis System

MAATAG Mission Area Analysis Test Advisory Group

MAAMAW Modeling Autonomous Agents in a Multiagent World (workshop)

MAB Magnetic-Amplifier Bridge • Main Asteroid Belt • Malfunction Analysis Branch • Management Advisory Business • Man And the Biosphere • Marabá (Brazil-PA) (location identifier) • Maschinelles Austauschformat für Bibliotheken • Master Acquisition Bus • Materials Applications Board • Materials Advisory Board • Mechanical Automation Breadboard • Membership Activities Board • Mise À Bord

MABAS Mutual Aid Box Alarm System

MABIE Mauna Loa Aerosol Backscatter Intercomparison Experiment

MABL Marine Atmospheric Boundary Layer • Mass Addition Boundary Layer

MABM Multilayer Absorbing Bottom Layer

MABP Mean Arterial Blood Pressure

MABS Multiagent-Based System

MABV Makle- und Bauträgerverordnung

Mac Macintosh

MAc Master of Accountancy

MAC Macau (ISO country code) • Machine-Aided Cognition • Macon Smart (USA-GA) (location identifier) • Macro • Magnetic Attitude Control • Magnetic Automatic Calculator • Maintenance Action Code • Maintenance Administration Center • Maintenance Advisory Committee • Maintenance Aid Computer • Maintenance Allocation Chart • Maintenance Analysis Center • Major Activity Center • Management Action Center • Management Advisory Committee • Man And Computer •

Mandatory Access Control • Maneuver Analysis and Command • Mapping Advisory Committee • Mapping Applications Center • Mass Absorption Coefficient • Massive Algebraic Computation • Material Availability Commitment • Maximum Allowable Concentration • Maximum Allowable Cost • Maximum Atmospheric Concentration • Maxwellian Average Capture • Mean Aerodynamic Chord • Mechanical Analog Computer • Media Access Control • Medium Access Control • Meeting of Aid Coordinators • Memory Access Controller • Memory-Assisted Cognition • Meridian Axial Circle • Message Authentication Code • Message Authenticity Check • Method of Asymptotic Collocations • Microgravity Advisory Committee • Middle-Atmosphere Cooperation • Midland's Astronomy Club (USA-SC) • Millimeter Autocorrelator • Millstream Astronomy Club (USA-OH) • Mineralogical Association of Canada • Miniature Advanced Coronagraph • Mirror-Advanced Coronagraph • Mission d'Action Culturelle • Money Access Center • Monitoring and Assessment Centre • Months After Contract • Motion Analysis Camera • Mouse-Activated Computer • Multiaccess Computer • Multiinstrument Airborne Campaign • Multiinstrument Aircraft Campaign • Multimission Application Concept • Multiphase Atmospheric Chemistry • Multiple-Access Computer • Multiple-Access Computing • Multiple-Access Control • Multiple-Access Controller • Multiple-Address Code • Multiplex Analog Component • Multiplexed Analog Component • Multiplicator/Accumulator • Multiplier And Accumulator • Multiply/Accumulate • Multisensory Airborne Campaign • Multiservice Access Concentrator

MACA Mars Atmospheric Chemistry and Astrobiology (conference) • Master of Arts in Communication Arts • Midair Collision Avoidance • Modular Automatic Conferencing Arranger • Months After Contract Award

MACABRE Material Ablation with Chemically Active Boundary Layer in Reentry

MACAO Multiple Application Concept for Adaptive Optics (ESO) • Multiple Application Curvature Adaptive Optics

MACAS Magnetic Capability and Safety System

MACBDI Materiel Chemical-Biological Defense Initiative

MACBM Multiaccess Cable Billing System

MACCS Magnetometer Array for Cusp and Cleft Studies

MACDIF Map And Chart Data Interchange Format • Mapping And Charting Data Interchange Format

MACDS Monitor And Control Display System

MACE Machine-Aided Composition and Editing • Macintosh Audio Compression and Expansion • Maintenance Analysis Checkout Equipment • Management Applications in Computer Environment • Material Ablation Conduction Erosion • Mechanical Antenna Control Electronics • Meeting on Asteroids and Comets in Europe • Middeck Active Control Experiment • Minimum Average Correlation Energy

Maced Macedonia • Macedonian

MACF Multiple Application Control Function

Mach Machine • Machinery • Machinist

MACH Machinable • Machine • Maximum Average Correlation Height • Mesothermal Auroral Charging Code • Multilayer Actuator Head

MACHO Massive Astrophysical Compact Halo Object (*) • Massive Compact Halo Object • Massive Condensed Halo Object

MACI Management Agent Configuration Item • Monitor Access and Control Interface • Multimedia Advanced Computational Infrastructure

MACIS Management and Contracts Information Service

MACMATC Mid-Atlantic Consortium for Mathematics and its Applications Throughout the Curriculum

MACMIS Maintenance And Construction Management Information System

MACO Major Assembly Checkout • Mars Atmospheric Constellation Observatory

MaCom Mathematics of Computation (periodical code)

MACON Matrix Connector

MACOS Modeling and Analysis for Controlled Optical Systems

MACP Mission Analysis Computer Program

MACPOL Management and Access Control Policy

MACR Materiel Acquisition Control Record

MACRES Malaysia Centre for Remote Sensing • Malaysian Centre for Remote Sensing

MACRIIM Marine And Coastal Resources Integrated Information and Management

MACRO Merge And Correlate Recorded Output • Monopoles, Astrophysics and Cosmic-Ray Observatory • Muon And Cosmic-Ray Observatory

MACS Magellanic Catalogue of Stars • Maintenance Assistance Capability for Software • Management And Control System • Massive Cluster Survey • Medium-Altitude Communications Satellite • Membership And Circulation System (AIP) • Microwave Attitude Control Sensor • Modular Attitude Control System • Modulator Attitude Control System • MPI Array Correlator System • Multiaccess Conversional System • Multiaircraft Cockpit Simulator • Multiaperture Coronal Spectrometer • Multifunction Automated Crawling System • Multiple Access Communication System • Multiproject Automated Control System • Mutually Agreeable Commercial Software

MACSAT Multiple-Access Communications Satellite

MACSE Minority Access to Careers in Science and Engineering • Module Accélérateur à Cavités Supraconductrices pour Electrons • MPI Array Correlator System for Effelsberg

MACSS Medium-Altitude Communications Satellite System

MACSYMA Massive Algebraic Computation Symbolic Manipulation • Massive Algebraic Computation Symbolic Manipulator

MACWS Machine Area Cooling Water System

Mad Madame • Madeira (Azores)

MA&D Mission Analysis & Design

MAD Dirham (Morocco) (ISO currency code) • Machine Analysis Display • Madrid Barajas (Spain) (location identifier) • Magnetic Ambipolar Diffusion • Magnetic-Anomaly Detection • Magnetic-Anomaly Detector • Magnetic-Anomaly Director • Magnetic-Azimuth Detector • Main Assembly Drawing • Maintenance Analysis Data • Maintenance Analysis Document • Maintenance - Assembly - Disassembly • Management Accounting Division • Management Analysis Division • Manufacturing Assembly Drawing • Material Analysis Data • Material Availability Date • Materiel Acquisition and Delivery • Maximum Aspheric Deviation • Mean Absolute Deviation • Mean Absolute Difference • Median Absolute Deviation • Message Address Directory • Metal Abundance Deficiency • Metallicity-Age Dependent • Méthode d'Analyse de Données • Michigan Algorithmic Decoder • Mis À Disposition • Mission Analysis Division • Mission Area Deficiency • Mission Area Director • Momentum Accumulation and Dumping • Monitor Address Data • Multiaccess Device • Motor Assembly and Disassembly • Multiaperture Device • Multiconjugate AO Demonstrator • Multiple-Access Device • Multiple-Aperture Device • Multiply And Add • Multispectral Analyzer Display • Multiwavelength Anomalous Diffraction • Mutual Assured Destruction

MADA Mutiple Access Demand Assignment • Multiple-Access Discrete Address

Madag Madagascar

MADAM Mean And Dispersion Additive Model • Moderately Advanced Data Management • Multipurpose Automatic Data Analysis Machine

MADAR Malfunction Analysis Detection And Recording

MADARS Maintenance Analysis, Detection and Reporting System • Malfunction Analysis Detection And Recording System

MADC Machine-Assisted Detection and Classification • Multiplexer Analog to Digital Converter

MADCAP Mammoth Decimal Arithmetic Program • Mosaic Array Data Compression And Processing • Multicast Address Dynamic Client Allocation Protocol

MADCAPM Mosaic Array Data Compression And Processing Module

MADCS Modular Attitude Determination and Control System

MADD Mirror Added Dual Doppler • Module for Automatic Dock and Detumble • Mothers Against Drunk Driving

MADDAM Macromodule and Digital Differential Analyzer Machine

MADE Manufacturing and Automated Design Engineering • Manufacturing Automation and Design Engineering • Microalloy Diffused Electrode • Micron Accuracy Deployment Experiment • Multichannel Analog-to-digital Data Encoder

MaDes Machine Design (periodical code)

MADES Menu-Assisted Data Entry System

MADEX Magnetic-Anomaly Detection Exercise

MADGE Microwave Aircraft Digital Guidance Equipment

MADIC Multidisciplinary Analysis and Design Industrial Consortium

MADL Maryland Advanced Development Laboratory (USA-MD)

Madm Madame

MADM Maintenance Automated Data Management • Multiattributive Decision Making • Multiple-Attributes Decision Making

MADN Multiple Access Directory Number

MADO Mulliken Approximation for Differential Overlap

MADP Materiel Acquisition Decision Process • Mission Area Development Plan

Madr Madras (India) • Madrid (Spain)

MADR Maximum Allowable Defect Rate

MADRAF Midwestern Astronomical Data Reduction and Analysis Facility

MADRE Magnetic Drum Receiving Equipment • Manufacturing Data Retrieval

MADREC Malfunction and Detection Recording

MADS Management And Distribution System • Mars Atmosphere Density Sensor • Meteorological Airborne Data System • Mission Approach Detector System • Mission Area Deficiency Statement • Modular Attitude Control System • Modular Auxiliary Data System • Multiple Area Data Set

MADT Mean Administrative Delay Time • Microalloy Diffused-Base Transistor • Modified Anomalous Diffraction Theory

MADYMO Mathematical Dynamic Model • Mathematical Dynamic Modeling

MaE Master in Engineering

MAE Machine Arithmétique Électronique • Macintosh Application Environment • Madera (USA-CA) (location identifier) • Make An Ending • Management Application Entity • Master of Aeronautical Engineering • Matrix Arithmetic Expression • Mean Absolute Error • Mean Area of Effectiveness • Medium Altitude Endurance • Météorologie Appliquée et Environnement • Metropolitan Area Exchange • Migration Assessment Editor • Ministère des Affaires Étrangères • Mission Accomplishment Estimate • Mission Analysis and Engineering • Multiaccess Enclosure • Musée de l'Air et de l'Espace (France)

MAEB Macintosh Application Environment

MAECI Ministère des Affaires Étrangères et du Commerce International

MAED Middle Atmosphere Electrodynamics

MAEDS Meteosat Argos Extended Dissemination Service • Multisatellite Applications Extended Dissemination Service

MAEG Multiple Asynchronous Exponential Growth

MAEIF Mean Absolute Error Improvement Factor

MAEL Mobile Aeronautics Education Laboratory

MAEO Medium Altitude Electro-Optical

MAEP Minimum Autoland Entry Point

MAER Multiple Assets Effectiveness Report

MAES Magnetized Accretion-Ejection Structure • Management Assistant Expert System • Mexican American Engineers and Scientists

MAESA Measurements for Assessing the Effects of Stratospheric Aerosols

Maest Maestro

MAESTRO Machine-Assisted Educational System for Teaching by Remote Operation • Measurements of Aerosol Extinction in the Stratosphere and Troposphere by Occultation • Méthodes d'Administration, d'Exploitation de Systèmes et de Télédistribution en Réseau Ouvert • Mission Analysis Evaluation and Space Trajectory Operations • Multiple Airborne Experiments Towards Radar Observation

MAET Microwave Amplifier Electron Tube

maf Microsoft Access Form (file extension)

MAF Magnetic-Anisotropy Field • Main Assembly Fixture • Maintenance Action Form • Maintenance and Administration Facility • Major Academic Field • Management Application Function • Manpower Authorization File • Master Address File • Maximum Amine Fuel • Maximum Amplitude Filter • Michoud Assembly Facility (NASA) • Midland (USA-TX) (location identifier) • Million Acre Feet • Mission À la Formation • Mission Analysis File • Mixed Amine Fuel • Multiple-Access Facility • Odessa (USA-TX) (location identifier)

MAFB McClellan Air Force Base (USA)

MAFD Minimum Acquisition Flux Density

MAFET Microwave and Analog Front-End Technology

MAFF Ministry of Agriculture, Fisheries and Food

MAFG Mission Applications Functional Group

MAFIS Mobile Area Field Instrumentation System

MaFiz Matematicheskaia Fizika (periodical code)

MAFPEN Mission Académique de la Formation des Personnels de l'Éducation Nationale (France)

MAFR Merged Accountability Fund Reporting

mag Magnetic • Magnitude

Mag Magasin • Magazine • Maggio • Magnesium • Magnet • Magnetic • Magnetism • Magneto • Magnetometer • Magnification • Magnitude • Magnus • Magyar

MAG Madang (Papua New Guinea) (location identifier) • Magazine • Magnetic • Magneto-Acoustic-Gravity (wave) • Magnetometer • Magnitude • Metropolitan Area Gateway • Mission Advisory Group • MPP Advisory Group

MAGA Medium-Accuracy Gyro Assembly

MAGAMP Magnetic Amplifier

MAGE Map Authoring and Generalization Expert • Marine Aerosol and Gas Exchange • Mechanical Aerospace Ground Equipment

MAGELLAN Magasin Électronique d'Archivage Numérique

MAG/ER Magnetometer/Electron Reflectometer (MO)

MAGER Magnetometer/Electron Reflectometer (MO)

MagFE Magnetic Fusion Energy (periodical code)

Magg Maggio • Maggiore

MAGG Modular Alphanumeric Graphics Generator

MAGGE Medium-Altitude Gravity Gradient Experiment

MagGi Magnitnaia Gidrodinamika (periodical code)

MAGGS Modular Advanced Graphics Generation System

MAGIA Méthodes Avancées de Gestion de l'Information en Astronomie

MagIC Magellan Instant Camera

MAGIC Machine for Automatic Graphics Interface to a Computer • Magnetic And Germanium Integer Calculator • Major Atmospheric Gamma-ray Imaging Čerenkov (telescope) • Manipulator And Gantry Intelligent Control • Matrix Algebra General Interpretive Coding • Matrix Analysis via Generative and Interpretive Computations • Model of Acidification of Groundwater In Catchment • Modern Analytical Generator of Improved Circuits • Modified Action Generated Input Control • Modular Architecture for Graphics Image Control • MPI for Astronomy General-purpose Infrared Camera • Multidimensional Applications and Gigabit Internetwork • Multimission Advanced Ground Intelligent Control Consortium • Multipurpose And Generalized Interface to COBOL

MAGICA Multimedia Agent-based Interactive Catalogues

MAGICS Meteorological Application Graphics Integrated Colour System

MagIs Magnitosfernye Issledovaniia (periodical code)

MAGLEV Magnetically-Levitated • Magnetic Levitation

MAGLOC Magnetic Logic Computer

MAGMA Magnetic fields near Mars

MAGMOD Magnetic Modulator

Magn Magnet • Magnetic • Magnetismo • Magnetron

MAGN Mildly Active Galactic Nucleus

MAGNUM Multicolor Active Galactic Nuclei Monitoring

MAGO Martian Atmospheric Grain Observer

MAGPIE Machine Automatically Generating Production Inventory Evaluation • Mega Ampere Generator for Plasma Implosion Experiments

MAGPIES Magnetospheric Plasma Imaging Explorer Satellite

MAGR Miniature Advanced GPS Receiver • Miniature Airborne GPS Receiver • Miniaturized Airborne GPS Receiver

MAGRET Modèle Agrométéorologique d'Évaporation et de Température

MAGSAT Magnetic-field Satellite • Magnetometer Satellite

MAGSS Medium-Altitude Global Satellite System

MAGT Magnetic Tape

Magtd Magnitude

MAGW Maximum Alternate Gross Weight

mAH Milliampere-Hour

Mah Maharashtra (India) • Mahogany

MAH Mass Aggregation History • Menorca (Baleares) (location identifier) • Multilingual Aligned Hypertext

MAHC Maximum Allowable Housing Cost

MAHFIL Midday Astronomy Hour For Interaction and Lunch (IUCAA)

MAHIS Meridian Automatic Horizontal Instrument

MAHLOVS Middle And High Latitudes Ocean Variability Study

MAHO Mobile-Assisted Handover

MAHRSI Middle-Atmosphere High-Resolution Spectrograph Instrument • Middle-Atmosphere High-Resolution Spectrograph Investigation (*)

MAHT Machine-Assisted Human Translation

Mai Maio

MAI Machine-Aided Indexing • Mail • Management Action Indicator • Mangochi (Malawi) (location identifier) • Manufacturer Access Interface • Mean Annual Increment • Melnikov-Arnold Integral • Morphological Age Index • Multiapplication Interface • Multilateral Agreement on Investments • Multiple-Access Interface • Multiple-Address Instruction • Myanmar Airways International

MAIAA Member of the American Institute of Aeronautics and Astronautics

MAIB Marine Accident Investigation Branch (UK)

MAICO Manufacturing - Assembly - Integration - Checkout

MAID Maintenance Automatic Integration Director • Market Analysis and Information Database • Merger Acquisition Improved Decision • Multimedia Assets for Industrial Design • Multipurpose Automatic Inspection and Diagnostic • Münchener Arbeitskreis für Information und Dokumentation

MAIDS Machine-Aided Information and Dissemination System • Management Automated Information Display System • Multipurpose Automatic Inspection and Diagnostic System

MAIL Mockup And Integration Laboratory

MAILS Maintenance Activity Information Logistics System

MAIMT Associazione Italiana di Meccanica Teorica ed Applicata, Milano (periodical code)

MAIN Maintenance • Material Automated Inventory Network • Molecular gas in Active and Inactive Nuclei • Multiple-Access Internal Network

MAINE Maintaining Availability in Europe

MAINS Modular Automated Integration System

MAINSITE Modular Automated Integration System Interoperability Test and Evaluation

Maint Maintain • Maintained • Maintenance

MAINT Maintenance

MAIP Matrix Algebra Interpretive Program • Member of the Australian Institute of Physics

MAIPL Mid-America Image Processing Laboratory (USA)

MAIQ Multiple Award Indefinity Quantity

MAIR Manufacturing And Inspection Record • Multisensor Autonomous Image Registration

MAIS Major Automated Information System • Materiel Acquisition Integrated System • Mobile Automated Information System • Module Airborne Imaging Spectrometer • MOTIF Advanced Imaging System • Multicolor Airglow Imaging System • Multicolor All-sky Imaging System • Multiresource Analysis and Information System

MAISAP Major Automated Information System Acquisition Program

MAISON Multiwavelength Astronomical Image Service Online

MAISRC Major Automated Information System Review Council • Materiel Acquisition Integrated System Review Committee • Materiel Acquisition Integrated System Review Council

MAIT Maintenance Assistance and Instruction Team • Management Automation Information Technology • Manufacturers' Association for Information Technology

Maj Majoor • Major • Majority

MAJ Major • Majuro (Marshall Islands) (location identifier) • Mise É Jour • Mission Astronomie Jeunesse

MAJA Mise À Jour Annuelle • Mise À Jour Automatique

MAJC Microarchitecture for Java Computing

MAJH Mise À Jour Hebdomadaire

MAJM Mise À Jour Manuelle • Mise À Jour Mensuelle

MAJQ Mise À Jour Quotidienne

mak Makefile (file extension)

MAK Malakal (Sudan) (location identifier) • Mariestads Astronomiska Klubb (Sweden)

MAKE Most Admired Knowledge Enterprises (program)

Mál Málaga (Spain)

Mal Malay • Malayan • Malaysia • Malaysian • Malta • Maréchal

MAL Macro-Assembly Language • Malaysia (international vehicle registration code) • Malfunction • Material Allowance List • Materiel Allowance List • Mangole (Indonesia) (location identifier) • Mathematics and Astronomy Library • Member-At-Large • Meta-Assembly Language • Mobile Airlock • Model Assembly List

MALA Manpower And Logistics Analysis

MALC Management of Acquisition Logistics Course

MALCD Matrix-Addressed Liquid Crystal Display

MALD Master of Arts in Law and Diplomacy • Miniature Air-Launched Decoy

MALDI Matrix-Assisted Laser Desorption Ionization

MALE Medium Altitude - Long Endurance • Member-At-Large Entry • Multiaperture Logic Element

Malf Malfunction

MALF Malfunction • Mobile Aerobee Launch Facility

MALI Mémoire Adaptée aux Langages Indéterministes • Multilevel Accelerated Lambda Iteration

MALIGN Masses And Luminosities of Infrared Galaxies Nearby (survey)

MALITAS Mali Tinbouctou Air Service

MALL Malleable

MALS Master of Arts in Liberal Studies • Master of Arts in Library Science • Medium-intensity Approach Lighting System

MALT Mnemonic Assembly Language Translator • Model Assembly List Transmittal

MALTED Mesosphere and Lower Thermosphere Equatorial Dynamics

mam Microsoft Access Macro (file extension)

MAM Maintenance Assistance Module • Management Analysis Model • Management and Administration Manual • March - April - May • Matamoros (Mexico) (location identifier) • Materiel Acquisition Management • Measurement of Atoms and Molecules • Memory Allocation Map • Microarcsecond Metrology • Middle Atmosphere Model • Mission Area Manager • Mirror Attenuation Mosaic • Modular Assembly Model • Mot-À-Mot • Multifunctional Autocorrelation Method • Multiple-Access Memory • Multiple Allocation Memory • Musée des Arts et Métiers

MAMA Machine à Mesurer Automatique • Machine Automatisée à Mesurer pour l'Astronomie (*) • Machine Automatique à Mesurer pour l'Astronomie • Man-Machine • Manual-Automatic Multipoint Apparatus • Mobile Automated Metabolic Analyzer • Multianode Microchannel Array • Multianode Multiarray

MAMBO Max-Planck Millimeter Bolometer • Mediterranean Association for Marine Biology and Oceanography

MAMDT Mean Active Maintenance Downtime

MaMeA Sankt Peterburgskii Universitet Vestnik Seriia Matematika Mekhanika Astronomiia (periodical code)

MAMI Machine-Aided Manufacturing Information • Machine-Assisted Manufacturing Information

MAMIE Minimum Automatic Machine for Interpolation and Extrapolation

MAMMAX Machine-Made and Machine-Aided Index

MOMMOS Magnetic Amplifying Magneto-Optical System

MAMOS Marine Automatic Meteorological Observing Station

MAMP Mission Area Materiel Plan

MAMS Maintenance Activity Management System • Materiel Acquisition Management System • Medical Administration Management System • Model Attitude Measuring System • Multispectral Atmospheric Mapping Sensor • Multiple Access to Memory System

MAMT Mean Active Maintenance Time

mAn Meiner Ansicht nach

Man Management • Manager • Manchester (UK-England) • Manila (Philippines) • Manitoba (Canada) • Manual • Manually • Manuel • Manufacture • Manufactured • Manufacturer • Manufacturing

MAN Manchester (Starlink internal code) • Manchester (UK-England) (location identifier) • Mandatory • Maneuver • Manitoba (Canada) • Manual • Maschinenfabrik Augsburg-Nürnberg (Germany) • Medium Area Network • Metropolitan Access Network • Metropolitan Area Network (*) • Microwave Aerospace Navigation

Manch Manchester (UK-England) • Manchuria

Man. Dir. Managing Director

MANDOP Multiple Analytical Doppler

Man. Ed. Managing Editor

Manf Manifold • Manufacturer

MANF Manifold • May - August - November - February

ManGe Manuscripta Geodaetica (periodical code)

MANIA Multichannel Analysis of Nanosecond Intensity Alterations (*) • Multichannel Analyzer of Nanosecond Intensity Alterations

MANIAC Mathematical Analyzer, Numerical Integrator, and Computer • Mechanical And Numerical Integrator And Computer • Mid- And Near-Infrared Array Camera

MANIP Manual Input

Manit Manitoba (Canada)

MANOLIA Matière Noire et Laboratoires Interactifs

Man.Op. Manually Operated

MANOVA Multivariate Analysis Of Variance

Manp Manpower

MANPAGE Manual Page

MANPER Manpower/Personnel

MANPLAN Management Plan

man. pr. Mane Primo

Mans Mansion

MANS Mansarda • Microcosm Autonomous Navigation System

MANSAT Manned Satellite

MANTECH Manufacturing Technology

MANTLE Mapping Night-Time Light Emissions

MANTRA Middle Atmosphere Nitrogen Trend Assessment

Manuf Manufacture • Manufactured • Manufacturer • Manufacturing

Manufac Manufacture • Manufactured • Manufacturer • Manufacturing

Manz Manzana

MAO Magnetic-Amplifier Output • Mailing Address Only • Main Astronomical Observatory (Ukraine) • Maintenance Assistée par Ordinateur • Major Access Option • Manaus - Eduardo Gomes (Brazil-AM) (location identifier) • Maragha Astronomical Observatory (Japan) • Modulateur Acousto-Optique • Moerser Astronomische Organisation (Germany) • Moletai Astronomical Observatory (Lithuania) • Musique Assistée par Ordinateur

MAOA Meteorological Aspects of Ocean Affairs

MAOPR Minimum Acceptable Operational Performance Requirement

MAOS Magnetic-Amplifier Output Stage

MAOT Maximum Allowable Operating Time

MA&P Maintenance Analysis & Planning

Map Mapping

MAP Macro Arithmetic Processor • Macroassembly Program • Maintenance Analysis and Planning • Maintenance Analysis Procedure • Maintenance Analysis Program • Maintenance and Administration Position • Maintenance Automation Program • Major Air Pollutant • Mamai (Papua New Guinea) (location identifier) • Managed Application Provider • Management Access Protocol • Management Action Plan • Management Analysis Program • Management Assessment Program • Maneuverable Atmosphere Probe • Manifold Absolute Pressure • Manpower Analysis Paper • Manufacturing Applications Protocol • Manufacturing Automation Protocol • Map Analysis Package • Margins of Active Plants • Mars Atmosphere Probe • Massive Assisted Penetration • Master Action Plan • Master Activity Programming • Master Analysis Package • Master Automation Plan • Materiel Acquisition Plan • Materiel Acquisition Process • Mathematical Analysis without Programming • Maximum A Posteriori • Maximum Average Price • Mean Arterial Pressure • Mean Average Precision • Measurement Allocation Plan • Media Access Project • Medical Aid Post • Mediterranean Action Plan • Medium Altitude Probe • Mesoscale Alpine Programme • Message Acceptable Pulse • Message Acceptance Pulse • Metric Analysis Plan • Microabrasion Package • Microgravity Application Promotion • Microwave Anisotropy Probe • Middle-Atmosphere Program • Minimum Ascending Path • Missed Approach Point • Missed Approach Procedure • Mission Area Plan • Mission Area Planning • Mission Application Program • Mission Automation Plan • Mobile Access Protocol • Mobile Antenna Platform • Mobile Application Part • Model Accreditation Plan • Modern Aids to Planning • Modified Access Pricing • Modified American Plan • Modular Application System • Mosaic Archive Pipeline • Multiannual Programme • Multichannel Aperture Photometer • Multichannel Astrometric Photometer (*) • Multifoil Abrasion Package • Multinational Account Program • Multiple Aim Point • Multiple-Allocation Procedure • Multiple-Array Processor • Multiple-Avoidance Program • Mutual Assistance Program • Mutually Agreed Position

MAPAQ Ministère de l'Agriculture, des Pêcheries et de l'Alimentation du Québec (Canada-QC)

MAPCHE Mobile Automatic Programmed Checkout Equipment

MAPCO Map Code

MAPDU Management Application Protocol Data Unit

MAPE Mean Absolute Percentage Error • Mean Absolute Percent Error

MAPI Mail Application Program Interface • Mail Application Programming Interface • Message Application Programming Interface • Messaging Application Program Interface • Messaging Application Programming Interface (*) • Microsoft Application Programming Interface (*)

MAPICS Manufacturing, Accounting, and Production Information Control System • Morphological Analysis with Parameters Information Coding Scheme

MAPID Machine-Aided Preparation of Instruction Data

MAPIPS MacAulay Automated Photogrammetric and Interactive Processing System

MAPK Mitogen-Activating Protein Kinase

MAPL Manufacturing Assembly Parts List • Multiple Array Programmable Logic

MAPM Minimum-cost Augmented Path Method

MAPMIS Manpower And Personnel Management Information System

MAPMT Multianode Photomultiplier Tube

MAPOS Multiple Access Protocol Over SONET

MAPP Manpower And Personnel Plan • Master Acquisition Planning Program • Master Acquisition Program Plan • Mission Analysis and Performance Program

MAPPER Maintaining, Preparing and Processing Executive Reports

MAPPIT Masked Aperture-Plane Interference Telescope

MAppSc Master of Applied Science

MaPr Mathematical Proceedings (periodical code)

MAPR Marine Airborne Patrol Radar • Miniaturization and Automation of Personnel Record

MAPRP Mesoscale Atmospheric Processes Research Program

MAPS Madrid Agreement and Protocol System • Mail Abuse Prevention System • Maintenance Analysis and Procedures System • Malabar Advanced Photometric System • Management Analysis and Planning System • Manpower And Personnel System • Map Analysis Package System • Material Acquisition Processing System • Measurement of Air Pollution from Satellite • Measurement of Atmospheric Pollution from Satellites • Mesoscale Analysis and Prediction System • Meteorological Application and Presentation Software • Meteorological Application and Presentation System • Microcomputer-based Automated Projection System • Microwave Active and Passive Spectrometer • Middle Atlantic Planetarium Society (USA-PA) • Miniature Autonomous Power System • Minnesota Automated Plate Scanner • Mitsubishi Array Processing System • Mobile Atmospheric Profiling System • Mobility Analysis and Planning System • Modular Azimuth Positioning System • Monopropellant Accessory Power Supply • Moteur Atomique de Propulsion Spatiale • Multichannel Astrometric Photometer with Spectrograph • Multicolor Automatic Projection System • Multidimensional Analysis Processing Scheme • Multiple-Address Processing System • Multiple Aircraft Positioning System • Multisatellite Attitude Program System • Multiservice Access Platforms • Multispectral Active and Passive System • Multispectral Active/Passive Scanner • Multivariate Analysis and Prediction of Schedules • Multivariate Analysis, Participation and Structure

MAPSAC Machine-Aided Programming, Scheduling And Control

MAPSAT Mapping Satellite

MAPSE Minimum Ada Programming Support Environment • Mission Analysis Program for Solar Electric Propulsion (periodical code)

MAPSEP Mission Analysis Program for Solar Electric Propulsion

MAPSS Maintenance and Analysis Plan for Special Services

MAPSTAR Middle Atmosphere Periodic Structured Associated Radiance

MAPT Maintenance Activation Planning Team

MAPTIP Marine Aerosol Properties and Thermal Imager Performance

MAPTIS Materials And Processes Technical Information System

MAPU Multiple Address Processing Unit

MAPUG Meade Advanced Products User Group

MAP3S Multistate Atmospheric Power Production Pollution Study

maq Microsoft Access Query (file extension)

MAQ Mae Sot (Thailand) (location identifier) • Maximum Acceptance Quantity

mar Microsoft Access Report (file extension)

Mar March • Maréchal • Marina • Marine • Maritime • Martedi • Marzo

MAR Maatschappelijke Advies Raad • Magnetic-Amplifier Relay • Maintenance Action Report • Maintenance And Refurbishment • Major Assembly Release • Management Assessment Report • Management Assessment Review • Maracaibo - La Chinita (Venezuela) (location identifier) • March • Março • Maroc (code ISO de pays) • Mars • Marzo • Material Availability Report • Mean Annual Runoff • Memory Address Register • Microprogram Address Register • Mid-Atlantic Ridge • Middeck Accommodations Rack • Minimal Angle Resolution • Minimum Acceptable Reliability • Mission Analysis Representative • Missing At Random • Molecular-Activated Recombination • Molecular-Assisted Recombination • Monthly Activity Report • Morocco (ISO country code) • Multiple-Access Relay

MARA Multimode Airborne Radar Altimeter

MARB Materiel Acquisition Review Board

MARBI Machine-Readable Bibliographic Information

MaRBu Materials Research Bulletin (periodical code)

Marc Marcato

MARC Machine-Readable Cataloguing (*) • Machine-Readable Catalogue • Manpower Allocation Requirements Criteria • Manpower Authorization Requirements Criteria • Materiel Acquisition Review Committee • Microgravity Advanced Research Centre • Mid-America Remote Sensing Center (USA-KY) • Middle Atmosphere Response to Change • Modern Asia Research Centre • Monitoring and Assessment Research Centre • Mean Annual Relative Citedness • Multiaccess Research Catalog

MARCEP Maintainability And Reliability Cos-Effectiveness Program

March Maréchal

MARCI Mars Color Imager

MARCIA Mathematical Analysis of Requirements for Career Information Appraisal

MARCS Multispectral Airborne Reference-Aided Calibrated Scanner

marcsec Milliarcsecond

MARDAC Manpower Research and Data Analysis Center

MARDI Malaysian Agricultural Research and Development Institute (Malaysia) • Mars Descent Imager

MARE Marine Activities, Resources and Education

MARECS Maritime Communications Satellite

MARES Muscular Atrophy Research and Exercise System

MARF Meteorological Archive and Retrieval Facility (EUMETSAT)

MARFE Multifaceted Asymmetric Radiation From the Edge

Marg Margin • Marginal

MarGe Marine Geodesy (periodical code)

MARGIE Minute-of-Arc Resolution Gamma-ray Imaging Experiment

MARGO Mars Aerosol and Radiative Gas Observer

MarGR Marine Geophysical Researches (periodical code)

MARIA Methods for Assessing the Radiological Impact of Accidents

MARIAN Multiple Access and Retrieval of Information with Annotations

MARICA Mariage de l'Image et de la Carte

MARIE Martian Radiation Environment (experiment) • Mobile Autonomous Robot in an Industrial Environment

MARIS Mississippi Automated Resource Information System

Marit Maritime

Markgräflerl Markgräflerland (Germany)

MARNR Ministerio del Ambiente y de los Recursos Naturales Renovables (Venezuela)

MAROSIM Multiarm Robot Simulator

MAROTS Maritime Orbital Test Satellite

MARP Manpower Requirements Plan • Markov Analysis and Reliability Prediction • Multiple Address Resolution Protocol

MarPG Marine Petroleum Geology (periodical code)

Marq Marquês • Marquess • Marquis

Marqa Marquesa

MARR Manpower Authorization Requirements Review

MarRQ Marconi Review Quarter (periodical code)

MARRS Mission Analysis and Reconfiguration Requirements System

MARS Machine À Réduction Symbolique • Machine À Résolution Symbolique • Machine Retrieval System • Maintenance Activities and Resources Simulation • Malmö Astronomi- & Rymdfarts Sälskap (Sweden) • Management Analysis Reporting Service • Management Analysis Reporting System • Manned Aerodynamic Reusable Spaceship • Manned Astronautical Research Station • Maritime Access and Retrieval System • Marketing and Advertising Reference Service • Master Attitude Reference System • Materials Analysis Research Station • Mechanically Actuated Reconfigurable Slit • Meeting of Astronomy Research Students • Membrane with Active Rapid Segment • Memory Address Register Storage • Message Automatic Reporting System • Meteorological and Aeronautical Retrieval System • Meteorological Automatic Reporting Station • Microcomputer-Assisted Rating System • Microdynamics Archiving and Retrieval System • Microgravity Advanced Research and Support • Microwave Atmospheric Remote Sensing • Mid-Air Recovery System • Mid-Air Retrieval System • Miniature Attitude Reference System • Mirror Alignment Recovery System • Mission Analysis and Reporting System • Mobile Agricultural Radar Scatterometer • Mobile Access and Retrieval System • Mobile Access for Resources on Space • Mobile Atmospheric Remote Sensing • Modular Airborne Recorder System • Monitoring Agriculture with Remote Sensing • Monograph Acquisitions and Record System • Mountain Astronomical Research Section (AL) • Multiaccess Radio System • Multiactivity and Resources Scheduling • Multiaperture Reluctance Switch • Multicast Address Resolution Server • Multiinterface Advection and Reconstruction Solver • Multiple-Access Ranging System • Multiple-Access Retrieval System • Multiuser Archival and Retrieval System • Multivariate Adaptive Regression Splines • Multivariate Analysis, Retrieval and Storage

MARSIS Marine Remote Sensing Information System

MARSNET Mars Network

MARSS Maintenance Activity Reporting and Scheduling System • Maintenance And Repair Support System • Microwave Airborne Radiometer and Scanning System • Multiactivity and Resources Scheduling System

MARSSS Meteorological and Range Safety Support System • Modular Airborne Remote Sampling and Sensing System

MARSYAS Marshall System for Aerospace Simulation (NASA)

MART Maintenance Analysis Review Technique • Mean Active Repair Time • Modular Automated Remote Terminal

MARTINI Multiple Aperture Real-Time Image Normalization Instrument

MARV Maneuverable Reentry Vehicle • Medium-Amplitude Red Variable (star)

MARVEL Machine-Assisted Realization of the Virtual Electronic Library

MARVS Materiel Acquisition Requirements Validation System

mas Milliarcsecond

mAS Milliampere-Second

Mas Masculine

MAS Macintosh Application Services • Madison Astronomical Society (USA-WI) • Malta Astronomical Society • Management Advisory Service • Manchester Astronomical Society (UK-England) • Manipulateur Agile Spécialisé • Manuscrit Autographe Signé • Manus Island (Papua New Guinea) (location identifier) • Marquette Astronomical Society (USA-MI) • Mars Approach Sensor • Masculinity • Mass Announcement System • Master Analysis Scheme • Master of Accounting Science • Master of Advanced Studies • Master of Applied Science • Mattatuck Astronomical Society (USA-CT) • Measurement Approval Sheet • Media Arts and Sciences • Medium-Altitude Sensor • Memory and Auxiliary Storage • Metal - Alumina - Silicon • Metastable Atomic State • Meteorological Aids Service • Michiana Astronomical Society (USA-IN) • Microwave Active Spectrometer • Microwave Active System • Microwave Atmospheric Sounder • Millimeter-wave Atmospheric Sounder • Milwaukee Astronomical Society (USA-WI) • Minnesota Academy of Science (USA-MN) • Minnesota Astronomical Society (USA-MN) • Mission Analysis Summary • Mississippi Academy of Sciences (USA-MS) • Mobile Access Structure • Mobile Astronomical Society (USA-AL) • Modélisation Aléatoire et Statistique • MODIS Airborne Simulator • Mountain Astronomical Station • Multiaccess Spool • Multiagent System • Multiamplitude Signaling • Multiple Access System • Multiple Arm System • Multiple Award Schedule • Multiplex Applications Software • Multiprotocol Access Service • Murdoch Astronomical Society (Australia) • Muskegon Astronomical Society (USA-MI)

MASA Massachusetts Astronomy and Space Association (USA-MA) • Member of the Astronomical Society of Australia • Modular Avionics Systems Architecture

MASAR Mid-Atlantic Slope And Rise

Masc Masculine

MASc Master of Applied Science

MASC Methylaluminium Sesquichloride • Modeling, Analysis, and Simulation Center • Mountain Administration Support Center • Multiple Award Schedule Contract • Multiscale Alternating Shadow Collimator

MASCAM Masking-pattern Adaptive Subband Coding And Multiplexing

MASCO Massive-Star-sized Compact Object

MASCON Mass Concentration

MASCOT Multiaccess System Control Terminal

MASCOTTE Montage Autonome Simplifié pour la Cryocombustion dans l'Oxygène et Toutes Techniques Expérimentales

MA&SE Mission Analysis & System Engineering

MASE Mechanical Ground Support Equipment • Message Administration Service Element • Mission Analysis and System Engineering

MASER Material Science Experiment Rocket • Microwave Amplification by Stimulated Emission of Radiation • Multispectral Airborne System for Environmental Research

MASERATI Middle Atmosphere Spectrometric Experiment on Rockets for Analysis of Trace-gas Influences • Middle Atmosphere Spectroscopic Experiment on Rockets for Analysis of Trace-gas Influences

MASEX Massive Air-Sea Transfer Experiment • Mesoscale Air-Sea Exchange (*)

MASH Multimedia Architecture that Scales across Heterogeneity

Mashi Mashinostroenie (periodical code)

MASIIS Maintenance Analysis Structural Integrity Information System

MASIS Management And Scientific Information System

MASIV Microarcsecond Scintillation-Induced Variability (survey)

MASL Material Article Service List • Multimission Analysis Subroutine Library

MASM Macro Assembler • Motorized Antenna Switching Matrix

MASN Moving-Actor Subnetwork

MASP Mid-Atlantic Star Party • Model for Assisted Software Process • Modular Atmosphere Simulation Program

MASPLAS Mid-Atlantic Student workshop on Programming Languages and Systems

MASPnet Multiple Access-type Satellite Packet Network

MASPS Minimum Aircraft System Performance Standard

MASR Microwave Atmospheric Sounding Radiometer

Mass Massachusetts (USA)

MASS Magnetically Active Stellar System • Manned Activity Scheduling System • Mathématiques Appliquées aux Sciences Sociales • Mathématiques Appliquées et Sciences Sociales • Matter Antimatter Superconducting Spectrometer • Maximum Availability and Support System • Mesoscale Analysis and Space Sensor • Mesoscale Atmospheric Simulation Sensor • Michigan Automatic Scanning System • Miniature Acquisition Sun Sensor • Mobility Analysis Support System • Model Applications and System Software • Model Atmospheres and Spectrum Synthesis • Monitor And Assembly System • Multiaccess Storage System • Multiaircraft Simplified Simulator • Multiaperture Scintillation Sensor • Multiline Accuracy Support System • Multiple-Access Sequential Selection

Mass. Ave. Massachusetts Avenue

MASSDAR Modular Analysis, Speedup, Sampling and Data Reduction

MASSIF Masses and Stellar Systems with Interferometry (program)

MASSPO Manned Space Flight Support Project Office

MaST Maneuvering Systems Technology

MAST Maintenance Standardization Team • Marine Action in Science and Technology • Marine Atmosphere Science and Technology • Marine Science and Technology • Marine Sciences and Technologies • Mass Spectrometer Telescope • Measurement And Stimulus • Mega-Amp Spherical Tokamak • Model Assembly Sterilizer for Testing • Multiapplication Sonar Trainer • Multimission Archive at STScI • Multiple-Applications Storage Tube

MASTER Manufacturing And Structures Technology for Efficient Rotorcraft • Mars and Asteroid (mission) • Matching Available Student Time to Educational Resources • Meteoroid And Space Debris Terrestrial Environment Reference • Millimeter-wave Acquisitions for Stratosphere/Troposphere Exchange Research • Multiple-Access Shared-Time Executive Routine

MASTEX Mediterranean Aircraft and Ship Transmission Experiment

MASTIC Malaysian Science and Technology Information Center

MASTIF Multiaxis Space Test Inertia Facility (*) • Multiaxis Spin Test Inertia Facility

MASTIFF Middle Atmosphere Studies of Instabilities, Fluctuations and Fields

MASTIR Microfilmed Abstract System for Technical Information Retrieval

MASTR Multiple Assets Status Report

MASW Multiple Application Software

mat Microsoft Access Table (file extension)

Mat Matemáticas • Material • Matériel • Materiel • Matière • Matinée • Matrix • Maturity

MA&T Manufacturing Assembly & Test

MAT Machine-Aided Translation • Machine-Assisted Translation • Machine Available Time • Maintainability Analysis Tool • Maintenance Access Terminal • Management Advisory Team • Manufacture And Test • Manufacturing Assembly and Test • Marconi Applied Technologies • Marine Air Temperature • Marine - Aviation - Transport • Master Alias Table • Master Archive Tape • Master of Arts in Teaching • Matadi (Zaire) (location identifier) • Mathematical Astronomical Text • Mathematical Automata Theory • Mathematics • Matrix • Measurement of Atmospheric Turbulence • Mechanical Aptitude Test • Medial Axis Transform • Meteorological Atmospheric Turbulence • Microalloy Transistor • Microwave Antenna Tower • Migration Assessment Tool • Mission Area Team • Mobile Aeronautical Telemetry • Mobile Anisotropy Telescope • Module Acceptance Test • Molecular Analysis Team • Mount Alignment Tool • Multilingual Aligned Text • Multiple-Access Test • Multiple Actuator Test • Multiple-Address Telegram • Multiple-Attribute Tree

MATA Mobile Agents for Telecommunication Applications (workshop) • Multiple Answering Teaching Aid

MATADOR Mars Atmosphere And Dust in the Optical and Radio

MATC Mathematics and their Applications Throughout the Curriculum • Middle Atmosphere and Tropospheric Climate

MATCH Mesoscale Atmospheric Transport and Chemistry

MATCHES MAPS-Assisted Translation and Classification Help for Examiners System

MATCO Materials Analysis, Tracking and Control

MATCON Microwave Aerospace Terminal Control

MATE Mainframe Automatic Test Equipment • Mars Array Technology Experiment • Measuring And Test Equipment • Ministère de l'Aménagement du Territoire et de l'Environnement (France) • Mission Analysis Technique for Experiments • Mobile Analysis and Training Extension • Modular Approach to Technology Education • Modular Automated Test Equipment • Modular Automatic Test Equipment • Multiaircraft Training Environment • Multiband Automatic Test Equipment • Mutiple-Access Time-Division Equipment • Multipurpose Automatic Test Equipment • Multisystem Automatic Test Equipment

MaTeJ ManTech Journal (periodical code)

Mater Materialprüfung (periodical code)

MATES Manufacturing And Tooling Expert System • Multiband Antiship Tactical EW System

MatEv Materials Evaluation (periodical code)

MatG Mathematical Geology (periodical code)

Math Mathematical • Mathematically • Mathematician • Mathematics • Mathematika (periodical code) • Mathématique • Mathématiques

MathML Mathematical Markup Language

Maths Mathematics

MATIAS Mediciones Atmosféricas, Telecomunicaciones e Ingeniería y Aplicaciones de los Satélites

MATIC Multiple-Area Technical Information Center

MATICO Machine Applications to Technical Information Center Operations

MATIF Marché À Terme International de France

MATIP Mapping of Airline Traffic over Internet Protocol

MatIs Matematicheskie Issledovaniia (periodical code)

MATISSE Modèle Avancé de la Terre pour l'Imagerie et la Simulation des Scènes et de leur Environnement

Matl Material

MatL Materials Letters (periodical code)

MATL Material

MATLAN Matrix Language

MATLS Materials

MatM Mathematical Modeling (periodical code)

MATM Multiple Assets Tasking Message

MATMO Medical Advanced Technology Management Office

MATOC Mobility Air Terminal Operations Center

MATOGW Maximum Allowable Takeoff Gross Weight

MAtom Moscow Atomizdat (periodical code)

MATPS Machine-Aided Technical Processing System

MatTr Materials Transactions JIM (periodical code)

MATR MOPITT Airborne Test Radiometer

MATRIS Manpower And Training Research Information System

MATS Maintenance Analysis Test Set • Major Assembly Tracking System • Manufacturing And Test Schedule • Mesoscale Atmospheric Transport Study • Milestone Automated Tracking System • Mission Analysis and Trajectory Simulation • Monitoring And Test System • MOTIF Acquisition Telescope System • Museum Advisory and Training Service

MATSS Mobile Aerial Target Support System

MatST Materials Science Technology (periodical code)

MATSYM Material Symbol

MATT Mapping Analysis Tools for Transportation • Message Analysis Test Tool • Multimission Advanced Tactical Terminal

MatTe Matériaux et Techniques (periodical code)

MATU Micro-Accelerometer Test Unit

MATV Master Antenna Television • Multiaxis Thrust Vectoring

MatVe Matematichki Vesnik (periodical code)

MAU Maintenance Analysis Unit • Math Acceleration Unit • Mathematical Acceleration Unit • Maupiti Island (Society Islands) (location identifier) • Media Access Unit • Media Attachment Unit • Medium Access Unit • Medium Attachment Unit • Mega Accounting Units • Million Accounting Units • Modular Avionics Unit • Mount Allison University (Canada-NB) • Multiattribute Utility • Multipoint Access Unit • Multistation Access Unit

MAUDE Morse Automatic Decoder

Maur Mauritius

Mau Re Mauritian Rupee (currency)

MAUS Materialwissenschaften Autonome (Experiment) Unter Schwerelosigkeit • Multiscale Adaptive Upwind Scheme

MAUT Multiattribute Utility Theory

MAUTO Miranda - Ariel - Umbriel - Titania - Oberon • Mispronunciation Afflicts Uranus Too Often • My Angel Uriel Takes Opium

MAUTx Monographs in Astronomy - University of Texas (periodical code)

mA/V Milliamperes per Volt

MAV Maloelap Island (Marshall Island) (location identifier) • Mars Ascendant Vehicle • Mars Ascent Vehicle • Mathematical Association of Victoria (Canada-BC) • Maturitätsanerkennungs-Verordnung • Mean Absolute Value • Micro Air Vehicle • Minimum Acceptable Value

MAVAD Manchester (Starlink internal code)

MAVDM Multiple Application Virtual DOS Machine

MAVES Mars And Venus Exploration Studies

MAVIN Machine-Assisted Vendor Information Network

MAVIS Multiaxis Vibration System

MAVIX Multimedia Audiovisual Information Exchange

MAVO Middelbaar Algemeen Voortgezet Onderwijs

MAVS Mean Absolute Value Slope • Multispectral Airborne Video System

MAVSS Monostatic Acoustic Vortex Sensing System

MAVT Multiattribute Value Theory

maW mit anderen Worten

MAW Magneto-Acoustic Wave • Malden (USA-MO) (location identifier) • Mathematics Awareness Week • Met Andere Woorden • Metropolitan Area Window • Microsoft At Work • Ministry of Agriculture and Water • Minor Assisting Work • Mission Adaptive Wiring • Modular Abstandwaffe

MAWD Mars Atmospheric Water Detection • Mars Atmospheric Water Detector (*)

MAWG Maintenance Analyzer Working Group

MAWP Maximum Allowable Working Pressure

Max Maximum

MAX Matam (Senegal) (location identifier) • Maximum • Message Automatic Exchange • Metropolitan Area Exchange • Metropolitan Area Express • Microwave Anisotropy Experiment • Mid-infrared Array Expandible • Millimeter-wave Anisotropy Experiment • Millimeter Anisotropy Experiment • Mountain Air Express

MAXAT Maximum Aperture Telescope

MAXENT Maximum Entropy

MAXI Modular Architecture for Exchanging Information • Modular Architecture for Exchanging Intelligence • Monitor of All-sky X-ray Image

MAXIE Magnetospheric Atmospheric X-ray Imaging Experiment

MAXIM Micro-Arcsecond X-ray Imaging Mission • Micro-Arcsecond X-ray Interferometer Mission • Micro-Arcsecond X-ray Interferometric Mission (*)

MAXIMA Millimeter Anisotropy Experiment Imaging Array

MAXLIMA Maximum-Likelihood Mass

MAY Mangrove Cay (Bahamas) (location identifier) • Mayo

MAZ Magnetische Aufzeichnung • Main Activity Zone • Mayaguez - Eugenio M. de Hostos (Puerto Rico) (location identifier)

mb Millibar

Mb Megabit • Millibar

M&B Mild & Bitter

MB Maandelijks Bulletin • Magellanic Bridge • Magnetband • Magnetic Bearing • Magnetic Break • Magnetic Breaking • Main Base • Main Battery • Main Bus • Maintenance Busy • Management Baseline • Management Board • Manitoba (Canada) • Manned Base • Marine Board • Marker Beacon • Master Beacon • Matter-Bounded • Mbundu • Measurement Base • Megabaud • Megabyte • Memory Buffer • Mid-Body • Middle Button • Millibar • Monthly Bulletin • Moored Buoy • Multiband • Multibeam • Musterbedingung

MBa Mombasa (Kenya)

MBA Main Baffle Assembly • Main-Belt Asteroid • Marge Brute d'Autofinancement • Massbus Adapter • Master in Business and Administration • Master of Business Administration (*) • Material Balance Area • Memory Bank Assignment • Mentoring Business Agreement • Minimum Burst Altitude • Mombasa (Kenya) (location identifier) • Multibeam Antenna • Multiple Beam Antenna • Multiple Berthing Adapter

MBAC Marshall Booster Assembly Contractor • Master of the British Association of Chemists • Measurement-Based Admission Control

MBAL Minimum Biologically Acceptable Level

mbar Millibar

MBAR Marshall Baseline Accounting and Reporting (system)

MBARI Monterey Bay Aquarium Research Institute (USA-CA)

MBAS Methylene Blue Active Substance

MBASIC Microsoft BASIC • Model-Based Analysis Synthesis Image Coding

MBB Make-Before-Break • Marble Bar (Australia-WA) (location identifier) • Messerschmitt-Bölkow-Blohm GmbH (Germany) • Modular Building Block • Multiband Beacon

MBBL Thousands of Barrels

MBC Manual Battery Control • Maximum Breathing Capacity • M'Bigou (Gabon) (location identifier) • Memory Bus Controller • Meteor Burst Communication • Methyl Benzimidazolecarbamate • Metropolitan Borough Council • Modular Building Concept • Multiballoon Canister

MBCP Mission Budget Capture Program • Multibyte Code Page

MBCS Meteor Burst Communication System • Motion-Base Crew Station • Multibyte Character Set

mbd Million Barrels per Day

MBD Magnetic Bubble Device • Mail Bag Depository • Mission Baseline Description • Mmabatho (South Africa) (location identifier) • More - Better - Different • Multibody Dynamics

MBDB Microsecond Burst Database

MBDS Minimum Basic Data Set

MBE Massive Brutal Erection • Minority Business Enterprise • Minority-owned Business Enterprise • Modified Bernouilli Equation • Molecular Beam Epitaxy • Monbetsu (Japan) (location identifier)

MBF Molecular Beam Facility • Mount Buffalo (Australia-VIC) (location identifier)

MBFP Manufacturing Build and Flow Plan

MBG Mobridge (USA-SD) (location identifier) • Montréal Blue Galaxy

MBGP Multiprotocol Border Gateway Protocol

mbH mit beschränkter Haftung

MBH Maryborough (Australia-QLD) (location identifier) • Massive Black Hole

MBI Major Budget Issue • Management Buy-In • Manpower Build-up Index • Mathematical Biosciences Institute (USA-OH) • May Be Issued • Mbeya (Tanzania) (location identifier) • Molecular Biology Institute • Multiband Imager • Multibus Interface

Mbit Megabit

MBIT Maintenance Built-In Test

MBIU Multibus Interface Unit

MBJ Montego Bay (Jamaica) (location identifier)

MBK Matupa (Brazil-MT) (location identifier) • Medications and Bandage Kit

Mbl Mobile

MBL Manistee (USA-MI) (location identifier) • Marine Boundary Layer • Martian Boundary Layer • Mechanical Boundary Layer • Memory-Based Learning • Microcomputer-Based Laboratory • Mobile

MBLCC Marine Boundary Layer Convective Complex

MBLV Multiple-Beam Laser Velocimeter

MBM Magnani - Blitz - Mundy (catalog) • Magnetic Bubble Memory • Metal - Barrier - Metal • Mkambati (South Africa) (location identifier) • Model Buffer Manager • Motherboard Monitor

MBMFR Multiband Multifunctional Radio

MBMM Massively Parallel Mesoscale Model

MBMMR Multiband Multimode Radio

MBMR Multiband Multirole

MBMS Model Base Management System • Molecular-Beam Mass Spectrometry

MBN Mount Barnett (Australia-WA) (location identifier)

MBO Main Belt Object • Mamburao (Philippines) (location identifier) • Management Buy-Out • Management By Objectives • Middelbaar Beroeps-Onderwijs • Modine-Benstead Observatory (USA-WI) • Monostable Blocking Oscillator • Motor Burn Out

MBOH Minimum Break Off Height

MBOL Motor Burn Out Locking

MBONE Multicast Backbone

MBOR Management By Objectives and Results

MBox Mail Box

MBP Magnetic Binary Problem • Magnetic Bright Point • Management By Participation • Mean Blood Pressure • Medium-Band Photometer • Moyobamba (Peru) (location identifier) • Multiband Prism • Multibody Platform

mbpd Million Barrels per Day

MBPL Minor Body Priority List

MBPN Multilayer Back Propagation Network

MBPR Memory-Based Program Response

MBPRE Multitype Branching Process in a Random Environment

Mbps Megabits per Second • Millions Bits per Second

MBps Megabytes per Second

MbpS Megabits per Second

MBPS Main Bus Power Supply • Megabytes Per Second (*) • Million Bits Per Second

MBPT Many-Body Perturbation Theory

MBQ Mbarara (Uganda) (location identifier)

MBQS Medium-Bright QSO Sample • Medium-Bright QSO Survey (*) • Medium-Bright Quasar Sample • Medium-Bright Quasar Survey (*)

Mbr Member

MBR Mail-Based Replication • Management By Results • Marker Beacon Receiver • Mars Balloon Relay • Master Boot Record • Materiel Balance Report • Mbout (Mauritania) (location identifier) • Mechanical Buffer Register • Medium Bit Rate • Memory-Based Reasoning • Memory Buffer Register • Microwave Background Radiation • Minimum Bounding Rectangle • Mission Briefing Room • Model-Based Reasoning

MBRA Multibeam Radiometer Antenna

MBRR Magnetic Bubble Recorder-Reproducer

MBRV Maneuverable Ballistic Reentry Vehicle

Mb/s Megabits per Second

MBS Bay City (USA-MI) (location identifier) • Magnetron Beam Switching • Malta Board of Standards (now MSA) • Master Boot Sector • Maximum Burst Size • Midland (USA-MI) (location identifier) • Minimum Bias Set • Mission Bit Stream • Mjølner BETA System • Mobile Base Support • Mobile Broadband Service • Mobile Broadband System • Mobile Broadcast Service • Mobile Broadcast System • Mortgage-Backed Securities • MRS Base System • Multibody System • Multimedia Broadcast Service • Saginaw (USA-MI) (location identifier)

MBSc Master of Business Science

MBSL Multiple Bubble-Sonoluminescence

MBST Multiple-Switching Tube

MBSU Main Bus Switching Unit

mbt Met Betrekking Tot

MBT Main Body Temperature • Masbate (Philippines) (location identifier) • Maximum Burst Tolerance • Mean Body Temperature • Mechanical Bathythermograph • Metal-Base Transistor • Meyer Binocular Telescope • Motor Burning Time

MBTA Massachusetts Bay Transportation Authority (USA-MA)

MBTI Myers-Briggs Temperament Indicator • Myers-Briggs Type Indicator (*)

MBTT Migration-Based Travel Time

MBU Mbambanakira (Solomon Islands) (location identifier) • Modern Burner Unit • Multiple Bit Upset

mbv Met Behulp Van

MBV Main Base Visit • Masa (Papua New Guinea) (location identifier) • Model-Based Vision

MBW Mid Blue Wing • Moorabbin (Australia-VIC) (location identifier)

MBWA Management By Wandering Around

MBWO Microwave Backward-Wave Oscillator

MBX Mailbox • Maribor (Slovenia) (location identifier)

MBY Moberly (USA-MO) (location identifier)

Mbyte Megabyte

MBZ Mandatory Broadcast Zone • Maués (Brazil-AM) (location identifier) • Must Be Zero

MBZS Maximum Bandwidth Zero Suppression

m/c Mi Cuenta • Motorcycle

mc Megacycle • Millicurie

mC Millicoulomb

Mc Megacurie • Megacycle

M&C Maintenance & Calibration • Maintenance & Checkout • Management & Control • Monitor & Control • Monitoring & Control

M/C Main Chamber • Manchester (UK-England) • Mission Control • Modules/Circuits

MC Machine Check • Machine Console • Magellanic Cloud • Magnetic Cloud • Magnetic Clutch • Magnetic Core • Magnetic Course • Magnetospheric Constellation • Main Chamber • Main Chute • Maintenance Console • Maintenance Control • Maître de Conférence • Major Component • Management Center • Management Committee • Maneuver Control • Mantle Cavity • Manual Control • Manufacturing Cost • Mapping Camera • Mapping Center • Mare Crisium (Moon) • Marginal Check • Marginal Cost • Maritime Commission • Markov Chain • Mars Chart • Mastercard • Master Clock • Master Control • Master Controller • Master of Ceremonies • Material Change • Material Code • Material Control • Materiel Change • Materiel Control • Maximum Clique • Medical Center • Medical Certificate • Medical Corps • Medium Capacity • Medium Coeli • Megacycle • Member of Congress • Member of Council • Memorandum of Conditions • Memory Chip • Memory Clear • Memory Configuration • Memory Consumption • Memory Control • Mention Complémentaire • Méridien Central • Message Center • Message Code • Message Composer • Meter-Candle • Metric Camera • Metropolitan Counties (UK) • Microchannel • Micronesia • Midcourse • Midcourse Correction • Mills Cross • Mini-Cartridge • Mini-Calorimeter • Mínimos Cuadrados • Minimum Conventionnel • Mission Capability • Mission Capable • Mission Commander • Mission Completion • Mission Computer • Mission Continuation • Mission Control • Mission Controller • Mission-Critical • Mnemonic Code • Mobility Center • Mobility Control • Mobility Corridor • Mode Change • Model Coordinates • Modular Computer • Modulation Collimator • Moindres Carrés • Mois Courant • Molded Component • Molecular Cloud • Molecular Cluster • Molecular Computing • Molonglo Cross • Momentary Contact • Monaco (international vehicle registration code) • Monaco (ISO country code) • Monaco (postal country code) • Monitoring Center • Monte Carlo • Monte-Charge • Montréal-Cambridge (survey) • Morse Code • Movement Control • Moyens Communs • Multicarrier • Multichannel • Multichip • Multicoated • Multiconfiguration • Multiple Choice • Multiple Contact • Multiple Copy • Multiple Core • Multiply-Convolve • Mushroom-shaped Current

MCA Macenta (Guinea) (location identifier) • Machine à Calculer Analogique • Machine à Calculer Arithmétique • Magneto-Centrifugal Acceleration • Main Console Assembly • Main Control Assembly • Major Commercial Account • Management Control Activity • Management Control Authority • Maneuver Control Agency • Maneuver Control Application • Caster calling Area • Master Clock Assembly • Master Control Assembly • Master Cooperative Agreement • Materiel Control and Accountability • Maximum Calling Area • Maximum Crossing Altitude • Median Case Analysis • Median Cut Algorithm • Message Channel Agent • Microchannel Adapter • Microchannel Architecture • Microwave Control Assembly • Minimum Controllable Airspeed • Minimum Crossing Altitude • Mise en Configuration d'Arrêt • Mission Concept Approval • Monetary Compensatory Amount • Monitoring and Control Assembly • Monitor Controller Assembly • Motor Control Assembly • Motoring and Control Assembly • Movement Control Agency • Multichannel Analyzer • Multichip Assembly • Multicriteria Analysis • Multiple Channel Analyzer • Multiple-Classification Analysis • Multiple Correspondence Analysis

MCAA Minority Career Advancement Award

MCAC McLennan County Astronomy Club (USA-TX)

MCACES Microcomputer-Aided Cost-Estimating Support

MCAD Mechanical Computer-Aided Design

MCAE Mechanical Computer-Aided Engineering • Module Collectif d'Accès à l'Emploi

MCAI Maximum Calling Area Indicator • Microcomputer-Assisted Instruction

MCAIERS Microcomputer Aerial Image Enhancement Reconnaissance System

MCAO Mount Cuba Astronomical Observatory (USA-DE) • Mt. Cuba Astronomical Observatory (USA-DE) • Multiconjugate Adaptive Optics

MCAP Maximum Calling Area Precedence • Maximum Calling Area Procedure • Multiple-Channel Analysis Program

MCAR Missing Completely At Random • Multichannel Acoustic Relay

MCARM Multichannel Airborne Radar Measurement

MCAS Marine Corps Air Station • Medina County Astronomical Society (USA-OH)

MCAT Manufacturing Cell Action Team • Medical College Admission Test

MCATS Medium-Capacity Automated Telecommunications System

MCATU Master Code Acquisition and Tracking Unit

MCAV Modified Constant Angular velocity

MCB Massive Close Binary • McComb (USA-MS) (location identifier) • Memory Control Block • Message Control Block • Miniature Circuit Breaker • Module Control Block • Molecular and Cellular Biosciences • Multilateral Coordination Board

MCBF Mean Cycle Between Failures

MCBIAS Microcomputer Based Image Analysis System

MCBP Multiconfiguration Breit-Pauli (method)

McC McCormick

MC&C Measurement, Command & Control

MCC Main Combustion Chamber • Main Communications Center • Main Control Circuit • Main Control Console • Maintenance Control Center • Maintenance Control Circuit • Management - Command - Control • Management Control Center • Man-Computer Communication • Maneuver Control Center • Maneuver Control Command • Master Control Center • Master Control Computer • Master Control Console • Material Control Code • Measurement, Command and Control • Mechanical-Chemical Code • Medical Care Centre • Mesoscale Cellular Convection • Mesoscale Convection Complex • Mesoscale Convective Complex • Message Control Center • Micro-Climate Cooling • Micro Climatic Cooling • Micro Compact Car • Microelectronics and Computer Corporation • Midcourse Correction • Middlesex Community College • Ministère de la Culture et de la Communication (France) • Mission Control Cell • Mission Control Center • Mission Control Complex • Mission Control Console • Mobile Command Center • Mobile Communications Center • Mobile Country Code • Mobility Control Center • Mode Coherence Coefficient • Molecular Cooling Catastrophe • Mosaic Communications Corporation (USA-IL) • Motor Control Center • Movement Control Center • Multichannel Communications Controller • Multiple-Chip Carrier • Multiple-Communications Control • Multiple-Computer Complex • Museo de la Ciencia y del Cosmos (Spain) • Sacramento McClellan (USA-CA) (location identifier)

MCCA Manual Cloud Cover Assessment • Marché Commun Centre-Américain • Media Conversion Computer Assembly

MCCB Modification Configuration Control Board • Multinational Configuration Control Board

MCCC Maragone Convection on Closed Containers (USML1) • Mission Control and Computing Center

MCCD Mees Charge-Coupled Device • Monthly Cold Cloud Duration • Mosaic Charge-Coupled Device

MCCEM Multichamber Concentration and Exposure Model

MCCF Modernized Cross-Correlation Function

MCCG Microgravity Combustion Coordination Group • Mission Command and Control Group

MCCI Monte Carlo Configuration Interaction

MCCL Maintainability Criteria Checklist • Maintenance Criteria Checklist • Microcomputer Compiler Language

MCCP Main Communications Control Panel • Maintenance Control Console Panel • Mission Contamination Control Plan • Mission Control Computer Program • Multichannel Colour Photometer

MCCR Mean Citing Country Rate • Mechanically-Cooled Cryogenic Radiometer • Medical Care Cost Recovery • Mission-Critical Computer Resource

MCCRFP Mission-Critical Computer Resources Focal Point

MCCS Mission Control Center Simulation • Mission Control Center System • Mission-Critical Computer System

MCCT Materials Center for Combustion Turbines • Mission Control Operator Trainer

MCCU Multiple-Communications Control Unit

McD McDonald

M&CD Metals & Ceramics Division (ORNL)

MCD Mackinac Island (USA-MI) (location identifier) • Magnetic Charge Distribution • Magnetic Circular Dichroism • Major Civil Division • Mars Climate database • Masse Cachée dans le Disque • Massimo Comune Divisore • Master Clerical Data • Master Countdown • Mathematics and Computer Division • Maximum-likelihood Convolutional Decoder • Mean Cell Diameter • Mean Corpuscular Diameter • Memory Control Data • Migration Control Data • Migration Control Data-set • Minimum Cost Design • Minor Civil Division • Mission Capability Document • Mission Control Directorate • Modèle Conceptuel de Données • Monitor Criteria Data • Months for Cyclical Dominance • Multichannel Deconvolution • Multichannel Detector • Multicolor Disk • Multiconcentration Diffusion • Multimedia Cartridge Drive

MCDCC Master Classified Document Control Center

MCDF Multiconfiguration Dirac-Fock (model)

MCDM Multicriteria Decision Making • Multiple Criteria Decision Making

MCDN Micro Cellular Digital Network

MCDP Microprogrammed Communication Data Processor • Mode Control Display processor

MCDR Mission Critical Design Review

MCDS Maintenance Control and Display System • Malfunction CRT Display System • Management Communication and Data System • Migration Control Data Set • Mission Control and Display System • Mission Control Data Set • Modular Cargo Delivery System • Monitor and Control Display System • Motor Control Development Station • Multicommand Data System • Multifunction CRT Display System

MCDU Multifunction Control Display Unit • Multifunction CRT Display Unit • Multipurpose Computer Display Unit

MCDV Maximum Cell Delay Variance

MCDW Monthly Climatic Data for the World (*) • Monthly Climatic Data of the World

MCE Machine à Calculer Électronique • Magnetic Cumulative Explosion • Management Center Europe • Manpower and Cost Estimate • Marché Commun Européen • Marginal Cost Efficiency • Master of Civil Engineering • Maximum Creditable Event • Mean Chance Expectation • Media Conversion Equipment • Mercato Comune Europeo • Merced (USA-CA) (location identifier) • Mini-Creation Event • Mission Control Element • Mission Control Equipment • Model Control Element • Modular Camera Electronics • Modular Control Element • Modular Control Equipment • Monitor and Control Electronics • Multiple Coulomb Excitation

MCELS Magellanic Cloud Emission-Line Survey

MCEP Mean Celestial Ephemeris Pole

MCES Modular Command and Control Evaluation Structure

MCET Management and Cost Evaluation Team • Massachusetts Corporation for Educational Telecommunications (USA-MA)

Mcf Thousands Cubic Feet

MCF Magnetic Confinement Fusion • Main Checkout Facility • Maintenance and Checkout Facility • Maintenance Control Facility • Marie Curie Fellowship • Master Calibration File • Master Code File • Master Contract File • Master Control Facility • Master Control File • Mean Cumulative Function • Metacontent Format • Metacontent Framework • Metadata Configuration File • Methane Conversion Factor • Mission Control Facility • Mission Control Forecast • Mission-Critical Function • Moderate Cooling Flow • Modular Combustion Facility • Modulation Contrast Function • Monolithic Control Filter • Monolithic Crystal Filter • Multicomplementary Filter • Multiple-Cassegrain Feed • Multiple-Code Field • Multiple-Cost Factor • Mutual Coherence Function • Tampa - Mac Dill (USA-FL) (location identifier) • Thousands Cubic Feet

MCFA Marie Curie Fellowship Association (EU) • Mission-Critical Function Analysis

MCFC Molten Carbonate Fuel Cell

Mcfd Thousands Cubic Feet per Day

MCFM Mission-Critical Failure Mode

MC&G Mapping, Charting & Geodesy

MCG Magnetic-Cumulative Generator • Magnetocardiography • Man-Computer Graphics • Mapping, Charting and Geodesy • Master Control Group • McGrath (USA-AK) (location identifier) • Mid-Course Guidance • Modèle de Circulation Générale • Monitor Control Group • Morphological Catalogue of Galaxies • Movement Control Group • Multimedia Communications Group

MCGA Modèle de Circulation Générale de l'Atmosphère • Monochrome/Color Graphics Adapter • Multicolor Graphics Adapter • Multicolor Graphics Array (*) • Multicompartmented Gateway Agent

MCG&I Mapping, Charting, Graphics & Imagery

MCGI Mapping, Charting, Graphics and Imagery

m/cgo Mi Cargo

McGU McGill University (Canada-QC)

MCH Machala (Ecuador) (location identifier) • Maintenance Channel • Methylcyclohexane • Momentum Coupling Hypothesis • Monitor and Command Handler • Monitor and Control Handler

mchan Multichannel

MCHB Maintenance Channel Buffer

MChE Master of Chemical Engineering

MCHF Multiconfiguration Hartree-Fock (method)

MCHFR Minimum Critical Heat Flux Rate

MCHO Mobile-Controlled Handover

Mcht Merchant

Mchry Machinery

Mchy Machinery

mCi Millicurie

MC&I Master Control & Interface

MCI Kansas City (USA-MO) (location identifier) • Malicious Call Identification • Management Configuration Item • Massive Cluster Ionization • Master Configuration Index • Master Control and Interface • Matsushita Communication Industries (Japan) • Media Command Interface • Media Control Interface • Meridian Control Integrator • Microwave Communications Inc. • Midcourse Interceptor • Ministère du Commerce et de l'Industrie • Mission Capability Inspection • Modular Common Inlet • Monte Carlo Inversion • Multiport Communications Interface

MCIA Microchannel Interface Adapter

MCIAS Multichannel Intelligent Announcement System • Multichannel Intercept Announcement System

MCIC Metals and Ceramic Information Center

MCID Modular Common Inlet Device

McIDAS Man-computer Interactive Data Access System (*) • Map-computer Interactive Data Access System

MCIDAS Man-Computer Interactive Data Access System

MCIF Miniature Card Implementers Forum • Molecular Contamination Investigation Facility

MCIM Minnesota Center for Industrial Mathematics (USA-MN)

MCIR Materiel Change Information Report • Mission Configuration and Implementation Requirements

MCIS Maintenance Control Information System • Management Control Information System • Master of Communication and Information Studies • Material Control and Information System

MCIT Master of Computer and Information Technology

MCIU Manipulator Controller Interface Unit • Master Control and Interface Unit • Mission Control and Interface Unit • Master Control Interface Unit

MCJ Maicao (Colombia) (location identifier)

MCK McCook (USA-NE) (location identifier)

MCL Denali National Park (USA-AK) (location identifier) • Maison de la Culture et des Loisirs • Markov Cluster (algorithm) • Mass Cargo Lander • Mass Change Log • Master Configuration List • Maximum Contaminant Level • Maximum Contamination Level • Message Control Language • Microfax Command Language • Microprogram Control Logic • Microsoft Compatibility Laboratories • MIDAS Command Language • Middle Compass Locator • Minimal Computer Load • Mount McKinley (USA-AK) (location identifier) • Mt. McKinley (USA-AK) (location identifier) • Multicolor Laser

MCLA Massachusetts College of Liberal Arts (USA-MA)

MCLB Maximum Cell Loss Ratio

MCld Molecular Cloud

MCLD Molecular Cloud • Multicolor Laser Display

MCLG Maximum Contaminant Level Goal

MCLR Maximum Cell Loss Ratio

MCLS Multicolor Lightcurve Shape (method)

MCLT Maximum Cruise Level Thrust

mcm Minimo Comune Multiplo

MCM MADS Control Module • Magnetic-Core Memory • Magnetospheric Constellation Mission • Main Control Module • Manned Circumlunar Mission • Martian Climate Model • Materiel Change Management • Mathematical Contest in Modeling • Maximum-Correlation Method • Media and Communication Management • Megawatt Cassegrain Monopulse • Memorandum of the Chairman • Memory Controller Module • Meson Cloud Model • Microcircuit Module • Microcontaminometer • Million Cubic Meters • Ministerial Council Meeting • Monitor and Control Module • Montants Compensatoires Monétaires • Monte Carlo (Monaco) (location identifier) • Monte Carlo Method • Movement Control Module • Multicarrier Modulation • Multicarrier Modulator • Multichip Module • Multicloud Model • Multicommodity Machine • Multiconference Management • Multiple Comparison Method • Multiply - Convolve - Multiply

MCMC Markov Chain Monte Carlo (algorithm)

MCMG Man-Carrying Motion Generator

MCMM Memory Controller Module with Memory • Multichip Memory Module

MCMN Marché Commun de la Mer Noire

MCMS Multichannel Memory System

MCMT Mean Corrective Maintenance Time

MCMU Mass Core Memory Unit

MCN Macon - Lewis B. Wilson (USA-GA) (location identifier) • Maintenance Change Notice • Maintenance Control Number • Management Change Notice • Management Control Number • Manual Control Number • Manufacturing Change Notice • Marégraphe Côtier Numérique • Master Change Notice • Master Control Number • Material Change Notice • Mathieu Characteristic Number • Micro Cellular Network • Minimum-Cost Network

MCNC Microelectronics Center of North Carolina

MCNE Master Certified Novell Engineer

MCNF Minimum Cost Network Flow

MCNP Monte Carlo Neutron Photon

MCNS Multimedia Cable Network System

Mco Morocco

M&CO Maintenance & Checkout

MCO Maintenance-Capable Office • Maintien en Condition Opérationnelle • Managed Care Organization • Maneuver Control Office • Maneuver Control Officer • Mars Climate Orbiter • Miscellaneous Charges Order • Mission Control Operation • Monaco (ISO country code) • Multiple-Channel Oscilloscope • Multicanister Overpack (system) • Orlando (USA-FL) (location identifier)

MCOMIB Mars Climate Orbiter Mishap Investigation Board (NASA)

MComM Mathematical and Computer Modeling (periodical code)

MCOO Modified Combined Obstacle Overlay

MCOP Mission Control Operations Panel

MCOPS Millions of Complex Operations Per Second

MCOR Member of Committee of Region

MCOS Multichannel Objective System

MCOV Main Chamber Oxidizer Valve

MCP Macapá (Brazil-AP) (location identifier) • Magnetic Chemically Peculiar (star) • Main Control Panel • Maintenance Collection Point • Maintenance Control Point • Management and Control of Provisioning • Management Control Plan • Maneuver and Constraints Program • Manual Control Panel • Massively Concurrent Processing • Master Change Package • Master Change Proposal • Master Computer Program • Master Controller Process • Master Control Program • Master of City Planning • Materials Control Plan • Materiel Change Plan • Materiel Change Program • Mathematics and Computer Physics (conference) • Maximum Call Precedence • Maximum Clique Problem • Maximum Continuous Power • Measurements Control Procedure • Medical Continuation Pay • Memory-Centered Processing • Memory-Centered Processor • Merrillville Community Planetarium (USA-IN) • Message Control Program • Metal-Coated Particle • Meteorological Communications Package • Microchannel Plate • Microsoft Certified Professional • Minimum Convex Polygon • Minor Cycle Period • Mission Capability Package • Mission Communications Payload • Mission Concept Paper • Mission Control Program • Mission Control Programmer • Mission Coordinating Paper • Mode Control Panel • Molecular Cloud Phase • Monitor and Control Protocol • Monitoring and Control Panel • MOO Client Protocol • Mount Control Processor • Multibeam Communications Package • Multichannel Communications Program • Multichannel Plate • Multicomponent Plasma • Multiple-Chip Package • Multiple-Comparison Procedure • Multiprotocol Communications Processor

MCPA Memory Clock Pulse Amplifier

MCpB Multiple Channels per Burst

MCPB Multiple Channels Per Burst

MCpC Multiple Channels per Carrier

MCPC Manipulator Controller Power Conditioner • Multiple Channels Per Carrier

MCPE Modular Collective Protection Equipment

MCPF Modular Containerless Processing Facility

MCpH Maintenance Cost per Hour

MCPH Maintenance Cost Per Hour

MCPN Magellanic Cloud Planetary Nebula

MCPO Master Chief Petty Officer

MCPP Mission Configuration Product Plan • Multicircuit Patch Panel

MCPPM Materiel Control Policies and Procedure Manual

MCP-PMT Microchannel Plate Photomultiplier

MCPPMT Microchannel Plate Photomultiplier

MCPR Maximum Critical Power Ratio

MCPS Major Cost Proposal System • Mechanical Copyright Protection Society (UK) • Microchannel Plates • Microsoft Certified Product Specialist

MCPT Microchannel Plate Tube

MCQ Miskolc (Hungary) (location identifier)

MCR Magnetic Card Reader • Magnetic Cartridge Recorder • Magnetic Character Reader • Magnetic Character Reading • Magnetic Character Recognition • Main Conference Room • Main Control Room • Maintenance Control Report • Manpower Change Request • Manpower Control Report • Manual Change Release • Manufacturing Capability Requirement • Manufacturing Change Request • Master Change Record • Master Change Request • Master Clock Receiver • Master Control Record • Master Control Room • Master Control Routine • Maximum Continuous Rating • Mean Citation Rate • Mechanically-cooled Cryogenic Radiometer • Melchor de Menco (Guatemala) (location identifier) • Message Completion Rate • Middle Common Room • Minimum Cell Rate • Minor Clarification Request • Mission Concept Review • Mission Control Room • Mission Control Routine • Mobile Cellular Radio • Mobile Communications Radio • Mobile Control Room • Modem Control Register • Modular Cellular Radio • Multicast Radio • Multichannel Radiometer • Multichannel Receiver • Multispectral Cloud Radiometer

MCRC Master Control and Reporting Center

MCRD Master Cross-Reference Data

MCRDF Mission Configuration Requirements Data File

MCRF Mission Configuration Requirements File

MCRG Mission Configuration Requirements Group

MCRL Master Cross-Reference Library • Master Cross-Reference List

MCRR Machine Check Recording and Recovery • Movement Control and Readiness Reporting

MCRS Maintenance Computing and Recording System • Malaysian Centre for Remote Sensing

MCRT Monte Carlo Radiative Transfer • Monte Carlo Ray Tracing (algorithm) • Multichannel Rotary Transformer

MCS Maintenance and Checkout Station • Maintenance and Control System • Maintenance Control System • Maintenance Cost System • Major Component Schedule • Management Control System • Maneuver Control System • Manned Core Station • Manufacturing Control System • Mapping Camera System • Maritime Communication System • Master Circuit System • Master Control Station • Master Control System • Mathematical Code System • Mathematics and Computer in Simulation • Matrix Common Substructure • Matrix Common Structure • Mean Crew Size • Measurements Calibration System • Media Compensation System • Medical Computer Services • Mercalli - Cancani - Sieberg (scale) • Meridian Control Signal • Mesoscale Cloud System • Mesoscale Convective System • Message Control System • Method of Constant Stimuli • Michigan Curtis Schmidt • Microcomputer System • Microcryogenic System • Microsoft Consulting Services • Microwave Carrier Supply • Microwave Communication System • Mid-Course Sensor • Mid-Course System • Minimum Chi-Square • Miscellaneous Command Ship • Mission Control Station • Mission Control Supervisor • Mission Control System • Mobile Communications System • Mobile Control Station • Model-Controlled System • Modular Causeway System • Modular Charge System • Modular Computer System • Module Computer System • Monitor and Control System • Monitor Control System • Monitoring and Control System • Monte Carlo Simulation • Monte Caseros (Argentina-CR) (location identifier) • Monthly Cost Summary • Motor Circuit Switch • Mount Control System • Multicast Server • Multichannel Scaler • Multichannel Scaling • Multichannel Survey • Multimedia Communications Station • Multimedia Computer System • Multiple Character Set • Multiple Chemical Sensitivity • Mutiple Classifier System • Multiple-Computer System • Multiple-Console Support • Multipoint Communications Service • Multiprogrammed Computer System • Multivendor Customer Service

MCSA Maneuver Control System Application • Microsoft Certified Systems Administrator • Mir Cooperative Solar Array • Multichannel Spectrum Analyser

MCSC Microcathod Strip Chamber

MCSCF Multiconfiguration Self-Consistent Field

MCSD Maneuver Control System Directorate • Microsoft Certified Solution Developer

MCSE Microsoft Certified Systems Engineer

MCSEE Member of the Canadian Society of Electrical Engineers

MCSET Microsoft Certified Systems Engineer Training

MSCF Mission Critical Subfunction

MCSP Mission Completion Success Probability • Multichannel Spectrometer

MCSR Mean Citing Subfield Rate • Minimum Commercial Security Requirement

MCSS Mask Cassette Support Structure • Mathematics of Control, Signals, and Systems (periodical code) • Midcourse Sensor Study • Monitor and Control Subsystem • Multiformat Communications Signal Simulator

MCSSCC Multicolor Spin-Scan Cloud Camera

MCSSP Multicomponent Spatial Search Program

MCSST Multichannel Sea Surface Temperature

MCST Materials Information for Science and Technology • Mission Culture Scientifique et Technique

MCSV Mars Crew Sortie Vehicle

Mct Mean Corrective Time

M&CT Mathematics & Computing Technology

MCT Magnetic Character Typewriter • Magnetic Charge Topology • Mainstream Corporation Tax • Master Control Terminal • Mathematical Control Theory • Mathematics and Computing Technology • Maximum Climb Thrust • Maximum Continuous Thrust • Maximum Corrective Time • Mean Corrective Time • Mean Correct Time • Mechanical Comprehensive Test • Memory Cycle Time • Mercury Cadmium Telluride • Metabolic Control Test • Methodology Coordination Team • Microwave Ceramic Triode • Minimum Competency Test • Minimum Connecting Time • Minimum Connection Time • Ministério de Ciência e Tecnologia (Brazil) • Mission Control Table • Mission Control Team • Mobile Communications Terminal • Module Confidence Test • Montréal - Cambridge - Tololo (survey) • MOS-Controlled Thyristor • MOS-Controlled Transistor • Movement Control Team • Multipurpose Community Telecenter • Multistrip Cesium Thruster • Muscat Seeb (Oman) (location identifier)

m/cta Mi Cuenta

MCTD Maximum Cell Transfer Delay

MCTEX Marine Continental Thunderstorm Experiment

MCTF Microcomputing Task Force

MCTR Message Center

MCTRAP Mechanized Customer Trouble Report Analysis Plan

MCTS Marie Curie Training Site (EU) • Master Central Timing System

MCTT Metal-Ceramic Transmitting Tube

MCTX Merge Count Threshold Extended

M&CU Monitor & Control Unit

MCU Machine Control Unit • Machines, Communications, and Universality (conference) • Main Control Unit • Maintenance Communications Unit • Maître de Conférences des Universités • Management Control Unit • Manual Control Unit • Master Clock Unit • Master Control Unit • Mechanism Control Unit • Medium Close-Up • Memory Control Unit • Message Construction Unit • Message Control Unit • Microcomputer Unit • Microcontroller Unit • Micro Control Unit • Microprogram Control Unit • Microprogrammed Control Unit • Minimum Coded Unit • Mission Computer Upgrade • Mission Control Unit • Mobile Control Unit • Modem Control Unit • Modular Concept Unit • Montluçon (France) (location identifier) • Multicast Unit • Multichip Unit • Multipoint Conferencing Unit • Multipoint Control Unit

MCUIS Master Control and User Interface Software

MCUOX Multicomponent Uniform Order-based Crossover

mCV Magnetic Cataclysmic Variable (star)

MCV Magnetic Cataclysmic Variable (star) • Magnetic Cushion Vehicle • Main Control Valve • McArthur River (Australia-NT) (location identifier) • Mean Cell Volume • Microbial Check Valve • Minimum Carrier Vehicle • Mission-Capable Vehicle • Mission Carrier Vehicle • Movable Closure Valve

MCVD Modified Chemical Vapor Deposition

MCVF Multichannel Voice Frequency

MCVM Multicommodity Vending Machine

MCVP Materials Control and Verification Program

MCVT Modification Change Verification Test

mcw Microsoft Word (file extension)

MC&W Master Caution & Warning

MCW Mason City (USA-IA) (location identifier) • Master Caution and Warning • Merge Count Weight • Modulated Carrier Wave • Modulated Continuous Wave

MCWS Magnetically Confined Wind Shock

MCX Makhachkala (Russia) (location identifier) • Movement Coordination Exercise

MCXO Microcomputer Compensated Crystal Oscillator

MCY Maroochydore (Australia-QLD) (location identifier) • Sunshine Coast - Maroochydore (Australia-QLD) (location identifier)

MCyT Ministerio de Ciencia y Tecnología (Spain)

MCYT Ministerio de Ciencia Y Tecnología (Spain)

MCZ Maceio Palmares (Brazil-AL) (location identifier)

MC1 Molonglo Catalogue 1

MC2 Master Catalogue of Magellanic Clouds • Measurement - Computing - Communications • Molonglo Catalogue 2 • Multiwavelength Catalogue towards the Magellanic Clouds

MC2R Mobile Computing and Communications Review

MC3 Molonglo Catalogue 3 • Multichannel Cryptocontroller

MC4 Molonglo Catalogue 4

MC5 Molonglo Catalogue 5

m/d Months after Date

Md Maryland (USA) • Mendelevium • Mendélévium • Milliarde

M/D Man/Day • Monitor/Decoder

M&D Maintenance & Diagnostics • Modeling & Design

MD Doctor of Medicine • Machine Direction • Madagascar • Magnetic Deflection • Magnetic Disk • Magnetic Drum • Main Droite • Main Drum • Maintenance Diagnostic • Maintenance Drop • Make Directory • Malfunction Detection • Malicious Damage • Management Data • Management Directive • Management Domain • Managerial Development • Managing Director • Manual Data • Manual Direct • Manual Disconnect • Manufacturing Development • Map Distance • Market Day • Marketing Direct • Maryland (USA) • Mass Donor • Massive Disk • Master Diagram • Master Dimension • Master Directory • Materiel Developer • Materiel Development • Materiel Directorate • Matter-Dominated • Maturity Difficulty • Mean Defect • Mean Deviation • Measured Discard • Mechanical Dummy • Mediation device • Medical Department • Medicinae Doctor • Medium Dynamic • Memorandum of Deposit • Mendoza (Argentina) • Mentally Deficient • Message Data • Message Development • Message Digest • Metadata • Metallicity Distribution • Meteorological Data • Meteorology Department • Microdot • Middeck • Middle Dutch • Millennium Dome • Millidiopter • Mini Disk • Miscellaneous Document • Mission Day • Mission Directive • Mission Director • Mobile Disk • Modèle Dynamique • Modular Design • Modular Dewar • Modulation Depth • Moldavie (code ISO de pays) • Moldova (ISO country code) • Molecular Dynamics • Monitor Display • Monochrome Display • Monotically Decreasing • Monsoon Depression • Months after Date • Motor Drive • Movement Directive • Multidirectional • Multidomain • Multinomial Distribution

MDA MacDonald, Dettwiler and Associates Ltd. (Canada) • Magnetic Deflection Amplifier • Magnetic Disk Assembly • Mail Delivery Agent • Main Distribution Assembly • Maintainability Design Approach • Maintenance Data Analysis • Major Decision Authority • Management Domain Architecture • Manufacturing Design Analysis • Mars Dust Analyzer • Materials Dispersion Apparatus • McDonnell Douglas Aerospace • Measurement, Decision and Actuation • Mechanically Despun Antenna • Medical Digital Assistant • Metal Deactivator • Meteoroid Detector-Analyzer • Methodology of Decision Analysis • Middeck Assembly • Milestone Decision Authority • Minimum Descent Altitude • Minimum Detectable Activity • Minimum Detectable Amount • Mission Data Analysis • Mobile Data Access • Model-Driven Architecture • Modular Digital Architecture • Moldavie (code ISO de pays) • Moldova (ISO country code) • Monochrome Display Adapter • Motor Drive Amplifier • Motorized Door Assembly • Multidimensional Access • Multidimensional Analysis • Multidocking Adapter • Multiple Diode Array • Multiple-Discriminant Analysis • Multiple-Docking Adapter • Multivariate Data Analysis • Munich Dust Analyzer • San Antonio - Martindale (USA-TX) (location identifier)

MDAC McDonnell Douglas Astronautics Co. • Multichannel Data Acquisition Computer • Multiplying Digital-to-Analog Converter

MDACS Modular Digital Access Control System

MDAIS McDonnell Douglas Aerospace Information Services

MDAM Multidimensional Access Memory

MDAP Mutual Defense Assistance Programme

MDAR Malfunction Detection, Analysis, and Recording

MDARS Mobile Detection Assessment Response System

MDAS Magnetic Drum Auxiliary Sender • Meteorological Data Acquisition System • Mission Data Acquisition System • Mobile Data Access System • Mount Diablo Astronomical Society (USA-CA) • Multiple Database Access System • Multispectral Data Analysis System

MDASER Multispectral Data Analysis System for Earth Resources

MDAT Meteorological Data Acquisition Terminal

MDAU Maintenance Data Acquisition Unit

MdB Mitglied des Bundestages

MDB Massive Disk and Bulge • Master Database • Melinda (Belize) (location identifier) • Memory Data Bank • Meteorological Database • Mission Database • Mission Data Book • Mission Display Board • Movie Database • Multidimensional Database • Multilateral Development Bank • Multiple-Device Boot • Multiple-Drive Block • Multiplexed Data Bus • Museum Data Bank

MDBG Master Database Generator

MDBL Maintainability Data Baseline • Maintainability Design Baseline

MDBMS Multimedia Database Management System • Multiuser Database Management System

MDBS Multidatabase System

MDBST Multidimensional Binary Search Tree • Multidimensional Binary Spanning Tree

MDC Magnetic Disk Cartridge • Main Display Console • Maintenance Data Collection • Maintenance Dependency Chart • Major Distribution Center • Manado (Indonesia) (location identifier) • Manager Designator Code • Manager Development Course • Manipulation Detection Code • Marker Distributor Control • Mars Dust Counter • Master DAMA Controller • Master Data Center • Master Direction Center • Material Distribution Center • Materiel Development Cell • Maximum Dependable Capacity • McDonnell Douglas Corporation • Meridian Digital Centrex • Message Display Console • Message Distribution Center • Meteorological Data Collection • Metropolitan District Commission • Metropolitan District Council • Microform Data Center • Midcourse Data Center • Middle Developed Country • Minimum Detectable Concentration • Minimum Detectable Contrast • Minimum Distance Classification • Mission Director Center • Mission Duty Cycle • MMA Development Consortium • Mobile Data Center • Modification and Design Control • Momentum Distribution Curve • More Developed Country • Moscow Data Center (Russia) • Motor Driver Chassis • Movement Designator Code • Multidevice Controller • Munich Dust Counter (HITEN)

MDCA Main Distribution Control Assembly

MDCC Master Data Control Console

MDCD Mobile Data Collection Device

MDCS Maintenance Data Collection System • Malfunction Display and Control System • Master Digital Command System • Material Data Collection System • Mission Data Collection Sheet

MDD Machine-Dependent Data • Maintenance Due Date • Mate/Demate Device • Maximum Drive Distance • Meteorological Data Dissemination • Meteorological Data Distribution • Microsoft Disk Defragmenter • Midland (USA-TX) (location identifier) • Minimum Detectable Dose • Mission Data Display • Model-Dependent Data • Mole Deployment Device • Multidimensional Database

MDDB Multidimensional Database

MDDBMS Multidimensional Database Management System

MDDS Material Directory Data Sheet • Meteorological Data Dissemination Service • Mobile Data Display System • Modular Design Display System • Multimedia Digital Distribution System

Mddx Middlesex (UK-England)

MDE Magnetic Decision Element • Main Distribution Frame • Maîtrise de la Demande d'Électricité • Materials Degradation Experiment • Matrix Difference Equation • Matter-Dominated Era • Mechanical Design Engineer • Mechanical Drive Electronics • Mechanism Drive Electronics • Medellín - José Maria Cordova (Colombia) (location identifier) • Medellín - Rio Negro (Colombia) (location identifier) • Message Distribution Element • Meteoroid Detection Experiment • Mission-Dependent Element • Mission-Dependent Equipment • Mission-Dependent Experiment • Mobile Datenerfassung • Modular Design Environment • Modular Design of Electronics • Modular Display Electronics • Most Durable Erection • Multimedia Document Environment

MDEA Master Data Exchange Agreement

MDES Multiple Data Entry System

MDF Macroscopic Degree of Freedom • Magnetic Direction Finding • Main Distributing Frame • Main Distribution Frame (*) • Manipulator Development Facility • Manpower Data File • Mask Definition File • Master Document File • Mate/Demate Facility • Mating/Demating Facility • Maxwellian Distribution Function • Mean Daily Frequency • Mean Distribution Function • Medford (USA-WI) (location identifier) • Median Destructive Field • Medium-Density Fiberboard • Medium-Density Fibre • Menu Definition File • Metallicity Distribution Function • Microcomputer Development Facility • Mild Detonating Fuse • Minimum-Dose Focusing • Minimum Duration Flight • Mission Data File • Mittwoch - Donnerstag - Freitag • Mutual Data File

MDFOS Minimum Design Factor Of Safety

MDFP MAST Data Format Processing • Mission Data Format Processing

MDFRR Mission Directors Flight Readiness Review

MDG Madagascar (ISO country code) • Mission Data Generator • Mission Definition Group • Mudanjiang (China-PRC) (location identifier) • Multidisciplinary Group • Multiplier Decoder Gate • Multipurpose Display Group

MDGC Multidimensional Gas Chromatography

MDGD Mercury-Doped Germanium Detector

MDH Carbondale - Southern Illinois (USA-IL) (location identifier) • Magnetic Drum Head • Mechanical Data Handling • Minimum Descent Height • Multimedia Data Handling

MDHMS McDonnell Douglas Human Modeling System

MDHS Meteorological Data Handling System

MDI Magnetic Direction Indicator • Magnetic Doppler Imaging • Makurdi (Nigeria) (location identifier) • Malfunction Detection Indicator • Management Data and Information • Manual Data Input • Manufacturing Development Initiative • Market Data Input • Master Direction Indicator • McDonnell Douglas Inc. • Medium-Dependent Interface • Memory Display Interface • Metered Dose Inhaler • Michelson-Doppler Imager • Michelson-Doppler Interferometer • Micro Design International • Minimum Discrimination Information • Miss Distance Indication • Miss Distance Indicator • Mission Data Input • Mobile Data Interface • Model Deployment Initiative • Multidocument Interface • Multiple-Display Indicator • Multiple-Document Interface

MDIC Manchester Decoder and Interface Chip • Microwave Dielectric Integrated Circuit

MDIE Mother-Daughter Ionosphere Experiment

MDIF Manual Data Input Function • Map Data Interchange Format

MDIG Multiple Display Indicator Group

MDIM Mars Digital Image Map • Mars Digital Image Mosaic

MDIMS Material Distribution and Inventory Management System

MDIO Multidimensional Input/Output

MDIS Manual Data Input Section • Medical Diagnostic Imaging Support • Medical Diagnostic Imaging System • Metadata Interchange Specification • Miss Distance Indicator System

Mdise Merchandise

MDIU Manned Data Insertion Unit • Manual Data Input Unit

MDJ Madras (USA-OR) (location identifier)

MDK Mbandaka (Zaire) (location identifier) • Multimedia Developers Kit • Multimedia Development Kit

Mdl Middle • Model • Module

MdL Mitglied des Landtages

MDL Lei (Moldova) (ISO currency code) • Machine-Dependent Library • Macro Description Language • Magnetic Delay Line • Magnetic Double Layer • Maintenance Diagnostic Logic • Management Data List • Mandalay (Myanmar) (location identifier) • Maquette de Déploiement Lanceur • Master Data Library • Microdevices Laboratory (JPL) • Middeck Locker • Minimum Description Length • Minimum Detectable Level • Minimum Detectable Limit • Mirrored Data Link • Mission Data Link • Mission Data Load • Model Descriptor List • Modular Design Language • Modular Dummy Load • Module Design Language • Momentum Dump Logic

MDLA Multiple Damped Lyman Alpha (system)

MDLING Modeling

Mdlle Mademoiselle

MDLP Minidisk Long Play • Minimum Description Length Principle • Mobile Data Link Protocol

MDLS Master's Degree in Library Science

MDLT Mobile Data Link Terminal

Mdm Madam • Medium

MDM Magnetic Disc Memory • Manipulator Deployment Mechanism • Materials Database Management • Materials Division Manual • Maximum Detectable Momentum • Maximum Design Meter • Maximum Disk Model • Media Device Manager • Medical Decision Making • Medical Monitor • Mega DM • Metadata Manager • Metal - Dielectric - Metal • Metric Distance Minimization • Michigan - Dartmouth - MIT (observatory) • Micro-Deformable Mirror • Micro-machined Deformable Mirror • Midnight Density Maximum • Minimum Distance Method • Mirror Demonstration Model • Mirror Development Model • Miss Distance Measurement • Mission Data Message • Mixed Dark Matter • Mobile Data Management • Mobile Depot Maintenance • Mobility Diagnostic Module • Module Design Methodology • Molecular Dynamics Module • Monolithic Diode Matrix • Montag - Dienstag -

Mittwoch • Mount Diablo Meridian • Moving Domain Memories • Multimode Deformable Mirror • Multiplexer-Demultiplexer • Munduku (Papua New Guinea) (location identifier)

MDMA Multidimensional Multiple Access

Mdme Madam

MDMI Monodisperse Microparticulate Injection

MDMO Michigan-Dartmouth-MIT Observatory (USA)

MDMPI Monodisperse Microparticulate Injection

MDMS Maintenance Data Management System • Multiple Database Management System

Mdn Median

MDN Madison Jefferson (USA-IN) (location identifier) • Mobile Directory Number

MDNS Managed Data Network Service

Mdnt Midnight

Mdnz Modernize

Mdnzd Modernized

M&DO Mission & Data Operations

MDO Massive Dark Object • Middleton Island (USA-AK) (location identifier) • Mobile District Office • Multidisciplinary Design Optimization (*) • Multidisciplinary Optimization

MDOA Mount Diablo Observatory Association (USA-CA)

M&DOD Mission & Data Operations Directorate

MDOD Mission and Data Operations Directorate

MDOP Maximum Design Operating Pressure

MDP Main Display Panel • Maintenance Data Panel • Maintenance Display Panel • Malfunction Detection Package • Management Development Programme • Management Directives and Practices • Markovian Decision Process • Master Data Processor • Master Development Plan • Master Display Panel • Meteorid Debris Protection • Meteorological Datum Plane • Mindiptana (Indonesia) (location identifier) • Minimum Deceptive Problem • Minimum Distance Probability • Mission Data Pressure • Mission Data Processing • Mission Data Processor • Mission-Dependent Processor • Modified Dynamic Programming • Modular Display Processor • Mound Prominence • Multidesignation Protocol • Multidestination Protocol • Multidivisional Programme

MDPG Magnetic Digital-Pulse Generator

MDPH Maquette de Déploiement Partie Haute

MDPPS Microfilm Digital Printer/Plotter System

MDPS Metric Data Processing System • Micrometeorid and Debris Protection Shield • Micrometeorid Debris Protection Shield • Mission Data Planning System • Mission Data Preparation System • Mission Data Processing System • Mobile Data Processing System • MOPS Data Processing System

MDPSK Modified Differential Phas-Shift Keying

MDQ Mar del Plata (Argentina-BA) (location identifier) • Minimum Deliverable Quantity • Minimum Detectable Quantity

MDQL Multidimensional Query Language

MDQS Management Data Query System

MDR Magnetic Dipole Radiation • Magnetic Disc Recorder • Magnetic Document Reader • Magnetic Drum Recorder • Maintenance Data Record • Maintenance Demand Rate • Maintenance Design Requirement • Major Design Review • Manually Digitized Radar • Manufacturing Design Review • Master Data Record • Master Discrepancy Report • Materials Disposition Review • Materiel Deficiency Report • Materiel Deficiency Reporting • Maximum Desired Result • Maximum Detectable Rigidity • Mechanical Design Review • Medfra (USA-AK) (location identifier) • Medium Data Rate • Memory Data Register • Metric Data Reduction • Mid-term Definition Review • Mid-term Design Review • Milestone Decision Review • Minimum Daily Requirement • Minimum Data Recorder • Minimum Design Requirement • Minimum Detectable Radiance • Minor Discrepancy Repair • Missing Data Report • Mission Data Reduction • Mission Definition Review • Mission Director Representative • Mock-up Discrepancy Report • Module Description Resource • Monthly Demand Rate • Monthly Director's Review • Morphology-Density Relation • Motion Detection Radar • Multichannel Data Recorder • Multidisk Reader • Multidrug-Resistant • Multifunction Dual-band Radar

MDRA Materiel Deficiency Report Analysis

MDRAM Multiband Dynamic Random Access Memory

MDRC McDonnell Douglas Research Center • Mean Daily Relative Citedness

MDRD Mission Data Requirements Document

MDRE Mass Driver Reaction Engine

MDRI Mission Des Relations Internationales • Multipurpose Display Repeater Indicator

MDRS Management Data Reporting System • Manufacturing Data Retrieval System • Mission Data Retrieval System

MDRU Mineral Deposit Research Unit

M&DS Maintenance & Diagnostics Software

MDS Macintosh Development System • Magnetic Disk System • Magnetic Drum System • Mail Distribution Scheme • Maintenance and Diagnostics Software • Maintenance Data System • Maintenance Documentation System • Malfunction Detection System • Management Data System • Manpower Data System • Mass Digital Storage • Massive Disk Star • Master Delivery Schedule • Master Development Schedule • Mécanique Des Structures • Mechanized Documentation System • Medium-Deep Survey • Memory Disk System • Message Distributing System • Message Distribution System • Metadata Standards • Metadata System • Metadirectory Services • Meteoroid Detection Satellite • Meteorological Data System • Microprocessor Development System • Microsoft Directory Services • Microwave Design System • Middle Caicos (Turks & Caicos Islands) (location identifier) • Minimum Detectable Signal • Minimum Discernible Signal • Minimum Discernible System • Mission Data Set • Mission Demonstration Satellite • Mission Design Specification • Mission Development Simulator • Mission Display System • Mobile Data Service • Mobile Display System • Modern Data System • Modular Data System • Modular Decontamination System • Modular Display System • Modular Dissemination System • Modulate-Demodulate System • Molecular Dynamic Simulation • Monitoring and Discovery Services • Motion Detection System • Motivational Development Seminar • Multichannel Distribution Service • Multidimensional Scaling • Multidisciplinary Simulation • Multiple-Data-set System • Multiple Data Source • Multiple-input Delivery Service • Multipoint Distribution Service • Multipoint Distribution System

MDSC Management Data Service Center • Modular Digital Scan Converter

MDSCC Madrid Deep Space Communications Complex (Spain)

MDSD Magnetic Disk Storage Device • Mate/Demate Stiffleg Derrick • Monodigitizer/Stereodigitizer

MDSDS Mass Data Storage and Delivery System

Mdse Merchandise

MDSE Manufacturing and Development Support Equipment • Message Delivery Service Element

MdsF Milliards de Francs

MDSF Mass Data Storage Facility

MDSL MBR Digital Subscriber Line

MDSOC Multidimensional Separation Of Concerns

MDSP Mason-Dixon Star Party (USA-PA) • Multichannel Subtractive Double Pass

MDSRS Morphology-Dependent Stimulated Raman Scattering

MDSS Magnetic Drum Storage System • Masonry Damage Diagnostic System • Mass Digital Storage System • Meteorological Data Sounding System • Mission Data Staging Service • Mission Data Staging System (*) • Mission Data Support Service (*) • Mission Data Support System • Multiple Distribution Switching System

MDSSC McDonnell Douglas Space Systems Corporation

MDST Mission Director Support Team

MDSU Mobile Diving and Salvage Unit

MDSV Manned Deep Space Vehicle

MDT Harrisburg (USA-PA) (location identifier) • Maintenance Data Terminal • Maintenance Demand Time • Mal Des Transports • Management Discovery Tool • Mean Detonating Time • Mean Downtime • Measurement Descriptor Table • Message Distribution Terminal • Middleton - Olmsted State (USA-PA) (location identifier) • Minimum Detectable Target • Minimum Detectable Temperature • Mission Data Table • Mission Data Terminal • Mobile Data Terminal • Moderate • Modified Data Tag • Modified Distance Transform • Modular Differentiation Technique • Monitored Drift Tube • Mountain Daylight Time

MDTA Manpower Development and Training Act (USA)

MDTF Mixing Due to Tidal Forces

MDTG Multiple Digital Transmission Group • Multiplexed Digital Transmission Group

MDTRS Mobile Digital Trunked Radio System

MDTS Megabit Digital Troposcatter System • Mission Data Transfer System • Mobile Doppler Tracking Station • Modular Data Transport System

MDTSCO McDonnell Douglas Technical Services Co.

MDTSP Materiel Developer's Training Support Package

MDTUM Mean Downtime for Unscheduled Maintenance

MDT2 Multiservice Distributed Training Testbed

MDT2B Multiservice Distributed Training Testbed

MDU Maharishi Daganand University (India) • Marker Decoder Unit • Market Development Unit • Master Drive Unit • Mendi (Papua New Guinea) (location identifier) • Message Decoder Unit • Missing Data Unit • Mission Data Unit • Mobile Development Unit • Multidimensional Unfolding • Multidisplay Unit • Multifunction Display Unit • Multiplexer and Distribution Unit • Multiplexer/Demultiplexer Unit • Multiplexer Decoder Unit

MdUC Milliards d'Unités de Compte

MDUC Meteorological Data Utilization Centre

MDUE Missing Data Unit Entry

MDUS Machine Drawing Understanding System • Medium-Scale Data Utilization Station • Meteorological Data Utilization Station

MDV Maldives (code ISO de pays) • Map and Data Viewer • Medouneu (Gabon) (location identifier)

MDW Chicago Midway (USA-IL) (location identifier) • Mars Departure Window • Meadow

MDWG Midwing

MDWS Meadows

mdx Multiindex (file extension)

MDX Maintenance and Diagnostic Extended • Maintenance Diagnostic Extended • Manpower Data Extract • Mercedes (Argentina-CR) (location identifier) • Modular Digital Exchange • Multidimensional Expression

MDY Midway Island (US Minor Outlying Islands) (location identifier) • Month - Day - Year

MDZ Mendoza - El Plumerillo (Argentina-MD) (location identifier)

me Multiengine • Read-Me Text (file extension)

mE meines Erachtens

Me Maine (USA) • Maître • Mercury • Methyl

M&E Materiel & Equipment • Mechanical & Electrical • Monitoring & Enforcement

ME Magnetic Energy • Magnification Evaluation • Magnitude Estimation • Maine (USA) • Main Engine • Main Entry • Maintenance Error • Management Engineering • Managing Editor • Maneuver Enhancement (mode) • Manpower Estimate • Manufacturing Engineer • Manufacturing Engineering • Marine Engineer • Marine Engineering • Mark End • Maser Emission • Mass Effectiveness • Master of Education • Master of Engineering • Matrix Element • Maximum Effort • Maximum Energy • Maximum Entropy • Mean Error • Mechanical Efficiency • Mechanical/Electrical • Mechanical Energy • Mechanical Engineer • Mechanical Engineering • Medical Examiner • Medio Evo • Medium Energy • Melanesia • Metabolizable Energy • Metal Evaporated • Metallurgical Engineer • Methodist Episcopal • Mexico • Microelectronics • Middeleeuwen • Middle East • Middle Eastern • Middle English • Mieteinnahme • Milne-Eddington (model) • Minimum Energy • Minimum Entropy • Mining Engineer • Mining Engineering • Miscellaneous Equipment • Mission d'Étude • Mission Equipment • Mission Envelope • Mission Evaluation • Mission Executive • Miteigentum • Mobiele Eenheid • Mobile Equipment • Mobile Express • Mobility Equipment • Modulation Efficiency • Molecular Electronics • Moneda Extranjera • Monitoring and Enforcement • Monoscopic Extraction • Mössbauer Effect • Most Excellent • Motion Estimator • Mouvement Européen • Movimiento Europeo • Multiengine • Multiple Element • Multiprotocol Encapsulation

MEA Macaé (Brazil-RJ) (location identifier) • Main Electronics Assembly • Main Error Amplifier • Maintainability Engineering Analysis • Maintenance Engineering Analysis • Major Emergency Actions • Management Engineering Activity • Materials Experiment Assembly • Mean En-route Altitude • Medium-Energy Accelerator • Member of the European Assembly • Minimum Energy Absorbed • Minimum En-route Altitude • Miteigentumsanteil • Monoethanolamine • More Electric Aircraft • Multimode Error Analysis

MeAar Meddelser fra Ole Romer Observatoriet Aarhus (periodical code)

MEAD Maintenance Engineering Analysis Data

MEADS Maintenance Engineering Analysis Data System • Medium Extended Air Defense System

MEAL Master Equipment Allowance List • Master Equipment Authorization List

MEAO Middelbaar Economisch en Administratief Onderwijs

MEAR Maintenance Engineering Analysis Record • Maintenance Engineering Analysis Request • Manufacturing Engineering Analysis Report

MeARB Mémoires de l'Académie Royale de Belgique (periodical code)

Meas Measurable • Measure • Measurement • Measurements (periodical code)

MEAS Measure • Measurement • Measuring

MEASAT Malaysia East Asia Satellite

MeAut Meres es Automatika (periodical code)

MEB Main Electronics Box • Master Electronics Board • Maximum Entropy Bubble • Medium-Energy Burst • Melbourne Essendon (Australia-VIC) (location identifier) • Memory Expansion Board • Microscope Électronique à Balayage • Microscopie Électronique à Balayage • Mission Element Board

MEBD Medical Evaluation Board

MEBO Main Engine Burnout

MEBU Minimum Essential Backup • Mission Essential Backup

Méc Mécanique

Mec Mecánica

MEC Main Element Contractor • Main Engine Console • Main Engine Controller • Manta (Ecuador) (location identifier) • Manual Emergency Control • Master Event Controller • Master Executive Council • Master of Engineering Chemistry • Maximum Endurable Concentration • Medgar Evers College (USA-NY) • Member Education Cycle • Member of the Executive Council • Meson Exchange Current • Microbiological Education Committee • Microelectronics Center • Ministerio de Educação e Cultura (Brazil) • Ministerio de Educación y Ciencia (Spain) • Mission Events Controller • Mobile Equipment Console • Modular Electronics Concept • Multiplex Exchange Channel

MECA Main Engine Controller Assembly • Maintainable Electronics Component Assembly • Malfunctioned Equipment Corrective Action • Mars Environmental Compatibility Assessment • Medical Electronic Customer Assistance • Mercury Evaporation and Condensation Analysis • Metal Casting • Multielement Component Array

MECAS Mission Equipment Configuration Accounting System

MECB Missão Espacial Completa Brasileira

Mecc Meccanica (periodical code)

MECC Minnesota Educational Computing Consortium (USA-MN)

MECCA Master Electrical Common Connector Assembly • Mechanized Catalog • Model Evaluation Consortium for Climate Assessment

MECD Ministerio de Educación, Cultura y Deportes (Spain)

MECDL Mission Equipment Control Data Link

MECE Mutually Exclusive and Collectively Exhaustive

MecEn Mechanical Engineering (periodical code)

MECF Main Engine Computational Facility

Mech Mechanic • Mechanical • Mechanically • Mechanics • Mechanism • Mechanist • Mechanized

MECH Mechanical • Mechanism • Mechanized

MECI Mission Essential Contingency Item

Meckl Mecklenburg (Germany)

MECL Minimum Essential Circuit List • Motorola Emitter Coupled Logic

MECLSI Mission Equipment Configuration Launch Site Installation

MECO Main Engine Cutoff • Manual Equipment Checkout • Muon to Electron Conversion

MECR Maintenance Engineering Change Request • Mean Expected Citation Rate

MeCS Mercury Cornerstone

MEC/S Medium-Energy Concentrator/Spectrometer

MECS Mechanical Elevator Control System • Media Equipment Control System • Medium-Energy Concentrator/Spectrometer • Minuteman Entry Control System • Modular Embedded Computer Software

MECSLSI Mission Equipment Cargo Support Launch Site Installation

MECT Mission Endurance Cycle Test

MECU Million European Currency Units

med Mediocris

Méd Médico

Med Median • Medical • Médical • Medicina • Médecine • Medicine • Medieval • Mediterranean • Medium

MEd Master of Education

ME&D Manufacturing Engineering & Development

MED Madinah - Prince Mohammad Bin Abdulaziz (Saudi Arabia) (location identifier) • Manual Electron Device • Manual Entry Device • Manufacturing Engineering and Development • Maximum Equivalent Dose • Mechanical Equipment Design • Median • Medical • Medicine • Medina (Saudi Arabia) (location identifier) • Mediterranean • Medium • Message Element Dictionary • Message Entry Device • Message Exchange Device • Microelectrical Device • Microelectronic Device • Minimum Effective Dose • Molecular Electronic Device • Momentum Exchange Device • Monitor Execution and Dump

MEDA Multiplex Electronic Doppler Analyzer

MEDAC Meteor Echo Detection And Collection

MEDAD Mean Effective Depth of the Atmosphere Duct

MEDAL Micromechanical Engineering Data for Automated Logistics

MEDALUS Mediterranean Desertification And Land Use

MEDAS Meteorological Data Acquisition System

MEDAX Message Data Exchange

MEDC Mössbauer Effect Data Center

MEDCEN Medical Center

Med Mededeling

MEDE Message Entry and Distribution Equipment

MEDEA Materialwissenschaftliche Experimente Doppelrack für Einzelexperimente und Anlagen • Measurements of Earth Data for Environmental Analysis • Microelectronics Development for European Applications • Multidiscipline Engineering Design, Evaluation and Analysis

MEDEVAC Medical Evacuation

MEDF Maximum Energy Distribution Function

MEDH Maintainability Design Handbook

MEDI Marine Environmental Data Information • Mössbauer Effect Data Index

MEDIA Modèle Eulérien de Dispersion Atmosphérique

MEDICARE Medical Care

MEDIS Modular Electronic Document Information System

MEDICS Medical Information Computer System

Medit Mediterranean

MEDL Mission Equipment Development Laboratory

MEDLAS Mediterranean Laser (campaign)

MEDLL Multipath Estimation Delay Lock Loop

MEDO Multipole Expansion of Diatomic Overlap

MeDOC Multimedia Electronic Documents

MEDOC Multiexperiment Data and Operations Centre (SOHO) • Multiexperiment Data Operations Centre

MEDP Mission Essential Data Processing

MedPh Medical Physics (periodical code)

MEDREG Medical Regulation

MEDRES Medium Resolution

MEDS Maintenance Engineering Data System • Marine Environmental Data Service • Mechanical Engineering Design System • Medical Evaluation Data System • Meteorological Data System • Minimum Essential Data Set • Multifunctional Electronic Display System (*) • Multifunction Electronic Display System

MEDSTAT Medical Status

MEDUSA Miniaturized Electrostatic Dual-top-hat Spherical Analyzer

MEDY Mesure Dynamique

MEE Maintenance Engineering Evaluation • Mare (Loyalty Islands) (location identifier) • Mass Energy Equivalent • Master of Electrical Engineering • Mechanical Evaluation Equipment • Metal-Enriched Environment • Minimum Essential Equipment • Mission-Essential Equipment

• Multimedia Enabled Ethernet • Multiple Effect Evaporator

MEEC Mars Experiment on Electrostatic Charging

MEECN Minimum Essential Emergency Communications Network

MEED Medium-Energy Electron Diffraction • Microbial Ecology Evaluation Device

MEEGA Mars-Earth-Earth Gravity Assist

MEEL Mission Equipment Essentiality List

MEEP Management and Equipment Evaluation Program • Mir Environmental Effects Payload

MEEPS Mir Environmental Effects Payloads

MEER Mechanical-Electrical Equipment Room

MeeRe Kenneth Mees Observatory Reprints (periodical code)

MEES Multipurpose Electromagnetic Environment Simulator

MEET Minimum Essential Equipment for Training

MEETAT Maximum Improvement in Electronics Through Advanced Techniques

MEF MAD Environmental Filter • Major Equipment File • Master E-mail Facility • Melfi (Chad) (location identifier) • Mémoire Fixée • Microwave Emission Fluctuation • Microwave Emission Flux • Minimum Essential Facilities • Minimum Essential Force • Minimum Essential Functions • Mission Économique et Financière • Mission Essential Functions • Module Élémentaire de Formation • Multiextension FITS (*) • Multiextension File • Multiextension Format • Multiple Extension FITS

MEFE Mission Essential Function Equipment

MEFI Ministère de l'Économie, des Finances et de l'Industrie (France)

MEFISTO Mise En Forme de l'Information Scientifique et Technique

MEFOS Meudon-ESO Fibre Optics Spectrograph

MEFS Microelectrofluidic System

MEFT Minimum Essential Functional Task

MEFTA Middle-East Free-Trade Agreement

MEG Magnetoencephalogramme • Magnetoencephalography • Malange (Angola) (location identifier) • Medium-Energy Grating • Megabyte • Message Entry Generator • Miniature Electrostatic Gyro

MEGA Mars-Earth Gravity Assist • Meridian of Galaxy • Microlensing Exploration of the Galaxy and Andromeda (project)

MEGASTAR Meaning of Energy Growth: an Assessment of Systems, Technologies And Requirements

MEGG Merging

MeGen Mededelingen Universiteit te Gent Sterrenkundig Instituut (periodical code)

MeGid Meteorologiia i Gidrologiia (periodical code)

MEGO My Eyes Glaze Over

MEGRIN Multipurpose European Ground-Related Information Network

MEH Mehamm (Norway) (location identifier)

MEHALICE Methane and Halocarbons Intercalibration Experiment

MEHQ Monomethyl Ether Hydroquinone

Mehrw.St. Mehrwertsteuer

MEHT Minimum Eye Height over Threshold

MEI Main Economic Indicator • Main Engine Ignition • Maintenance Effectiveness Inspection • Maintenance Engineering Inspection • Maintenance Evaluation Inspection • Major End Item • Management Effectiveness Inspection • Management Engineering Inspection • Manual of Engineering Instructions • Master End Item • Mathematical Engineering Industry (periodical code) • Meridian (USA-MS) (location identifier) • Message Exchange Interface • Ministry of Electronic Industries • Modified - Exclusive - Invalid • Multivariate ENSO Index

MEIC Member of the Engineering Institute of Canada • Multichannel Electronic Image Classifier

MEIFU Million Element Integral Field Unit

MEIG Main Engine Ignition

MEILU Million Element Integral Field Unit

MEIMN Multiend Item Modification Notice

MEIN Medium-Energy Intense Neutron

MEIP Market Economic Investor Principle

MEIR Minimum Essential Information Requirement

MEIRS Machinery and Electric Information Retrieval Service

MEIS Medium Energy Ion Scattering • Mineral Exploration Information System • Multispectral Electro-optical Imaging Spectrometer

MEISR Minimum Essential Improvement in System Reliability

MEIT Multiple Element Integration Test

MEITS Mission Effective Information Transmission System • Mission Essential Information Transmission System

MEIU Main Engine Interface Unit • Management Education Information Unit

Méj Méjico

Mej Mejuffrouw

MEJ Meadville (USA-PA) (location identifier)

Mejic Mejicano

MEK Meknes (Morocco) (location identifier) • Methyl Ethyl Ketone • Mission Equipment Kit

MEKASPEK Mehrkanal-Spektral-Photometer

MEKC Micellar Electrokinetic Chromatography

MEKF Mission Equipment Kit Facility

MekGS Mekhanika Giroskopicheskikh Sistem (periodical code)

MekKM Mekhanika Kompozitnykh Materialov (periodical code)

MekP Mekhanika Polimerov (periodical code)

MekTT Mekhanika Tverdogo Tela (periodical code)

Mel Melotte (catalog)

MEL Maintenance Expenditure Limit • Manufacturing Engineering Laboratory • Many-Element Laser • Master Environment Library • Master Equipment List • Master Events List • Material Engineering Laboratory • Mechanical Engineering Laboratory • Medium-Energy Laser • Medium Expandable Launcher • Melbourne Tullamarine (Australia-VIC) (location identifier) • Messy Astronomer Logics • Minimum Equipment List • Mobile Electronics Laboratory • Modular Electromagnetic Levitator

Melan Melanesia • Melanesian

Melb Melbourne (Australia-VIC)

MELCO Mitsubishi Electric Corporation

MELE Mid-Extinction Level Event

MELEC Microelectronics

MELEM Microelement

MeLeu Mededelingen van het Astronomisch Instituut van de Katholieke Universiteit Leuven (periodical code)

MELI Master Equipment List Index • Minimum Equipment List Index

MELIOS Miniature Eye-safe Laser Infrared Observation Set

MELiSSA Micro-Ecological Life-Support System Alternative

MELISSA Micro-Ecological Life-Support System Alternative

MELP Mixed-Excitation Linear Prediction • Mixed-Excitation Linear Predictive

MELT Mantle Electromagnetic and Tomography (experiment)

MeLu1 Meddelanden fran Lunds Astronomiska Observatorium Serie I (periodical code)

MeLu2 Meddelanden fran Lunds Astronomiska Observatorium Serie II (periodical code)

MELV Medium Expendable Launch Vehicle • Medium-light Expendable Launch Vehicle • Medium-lite Expendable Launch Vehicle

Mem Member • Memento • Memoir • Memorandum • Memorial • Memory

M-E-M Model-Exercise-Model

MEM Main Engine Maneuver • Main Engine Mode • Mars Excursion Module • Maximum Entropy Method • Membach (Belgium) • Memory • Memory Module • Memphis (USA-TN) (location identifier) • Meteoroid Exposure Module • Méthode d'Entraînement Mental • Microelectromechanical • Microelectromechanical Mirror • Microerosion Meter • Micrometeorite Exposure Module • Middeck Electronics Module • Minimum Essential Medium • Ministry for Energy and Mines • Mission Effectiveness Model • Model - Exercise - Model • Model - Experiment - Model • Module Exchange Mechanism • Multimode Equivalent Network

MEMA Microelectronic Modular Assembly

MeMaT Mechanism and Machine Theory (periodical code)

Memb Member

MEMDT Message Execution Matrix Display Task

MEME Multitasking Extensible Messaging Environment

MEMEX Memory Extended

MEMI Master Equipment Management Index

MEMIC Medical Microbiology Interdisciplinary Committee

Mem'l Memorial

Meml Memorial

Memo Memorandum

MEMO Mission Essential Maintenance Only • Mission Essential Maintenance Options

MEMS Microbial Ecological Monitoring System • Micro Electrical-Mechanical System • Micro-Electronics Mechanical System • Microelectromechanical Scale • Microelectromechanical System (*) • Micro-fabricated Electromechanical System

MEMSAC Mega Earth Mobile Seismic Array Consortium

MEMU Manned Extravehicular Manipulating Unit

MEMV Model - Exercise - Model - Validate

Men Mensa-Mensae (constellation)

MEN Management Event Notification • Mende Brenoux (France) (location identifier) • Menu • Ministère de l'Éducation Nationale • Multiple-Event Network

MENA Medium-Energy Neutral Atom • Mission Element Need Analysis

MEner Moscow Energoizdat (periodical code)

MENESR Ministère de l'Éducation Nationale, de l'Enseignement Supérieur et de la Recherche (France)

MENESTRIP Ministère de l'Éducation Nationale, de l'Enseignement Supérieur, de la Recherche et de l'Insertion Professionnelle (France)

MEng Master of Engineering

MENJS Ministère de l'Éducation Nationale, de la Jeunesse et des Sports

MENO Modified Essentially Non-Oscillatory

MENRT Ministère de l'Éducation Nationale, de la Recherche et de la Technologie (France)

Mens Mensile • Mensual • Mensuel

MENS Mission Element Needs Statement

Mensur Mensuration

Ment Mention • Mentioned

Mentd Mentioned

MENTOR Meteor Echoes - No Transmitter - Only Receivers

MEO Maintenance Engineering Order • Major Engine Overhaul • Manned Earth Orbit • Manned Extravehicular Operation • Manteo - Dare County (USA-NC) (location identifier) • Manufacturing Engineering Order • Mars Environmental Observer • Mass in Earth Orbit • Mean Opinion Score • Medium Earth Orbit • Medium Earth Orbiter • Message Exchange Occurrence • Mid-Earth Orbit • Most Efficient Organization • Multiple-Entry Object • Municipal Environment Officer

MEOC Moreton/EIT Observing Campaign

MEOM Manned Earth Orbit Mission

MEOP Maximum Expected Operating Pressure • Maximum Expected Operational Pressure

MEOS Multidisciplinary Earth Observation Satellite

MEOSS Monocular Electro-Optical Stereo Scanner

MEOTBF Mean Engine Operating Time Between Failures

MEOW Mono-Extraction Orthophotographic Workstation

MEP Mail Exit Point • Main Engine Propellant • Main Entry Point • Main Equipment Platform • Maintenance and Enhancement Plan • Maintenance Engineering Program • Maintenance Enhancement Program • Maison de l'Éducation Permanente • Management Engineering Plan • Management Engineering Program • Management Evaluation Program • Manipulator Engineering Panel • Manufacturing Extension Partnership • Mars Entry Probe • Mars Exploration Program • Mars Explorer Program • Massive Erection Phenomenon • Master Evaluation Plan • Master Experimental Plan • Master of Engineering Physics • Mean Effective Pressure • Mean Escape Probability • Membre of the European Parliament • Mersing (Malaysia) (location identifier) • Meteorological Package • Meteosat Exploitation Project • Microcircuit Emulation Program • Microfilm Enlarger Printer • Minimum Entry Point • Mission de l'Éducation Permanente • Mission Effects Projector • Mission Equipment Package • Mobile Electric Power • Multiengine Piston • Multiple Equipment Package

MEPA Master of Engineering and Public Administration • Meteorological and Environmental Protection Administration • Meteorology and Environmental Protection Administration (Saudi Arabia)

MEPAG Mars Exploration Program Analysis Group • Mars Exploration Program Assessment Group

MEPC Maritime Environment Protection Committee

MEPCO Meeting of European Planetary and Cometary Observers (*) • Meeting of European Planetary and Comet Observers

MEPED Medium Energy Proton and Electron Detector

MEPGS Mobile Electric Power Generating Source

MEPHISTO Matériel pour l'Étude des Phénomènes Intéressant la Solidification sur Terre et en Orbite

MEPI Medium-Energy Particle Instrument

MEPIS Management Engineering Program Information System

MEPP Management Efficiency Pilot Program • Maximum Entropy Production Principle • Middle East Peace Process

MEPPS Multiple Entry Point Payment System

MEPS Medical Expenditure Panel Survey • Medium-Energy Particle Spectrometer • Message Editing and Preparation System • Modification Engineering Program Status • Modular Electrical Power Station

meq Milli Equivalent

MEQ Meulaboh (Indonesia) (location identifier) • Minimum Essential Quality • Mission Essential Quality • Modelo de Engenharia-Qualificação

MEQPT Major Equipment

Mer Mercantile • Merchandise • Mercoledi • Mercredi • Mercure • Mercury • Meridian • Meridional • Merionethshire (UK-Wales)

MeR Mise en Route

M/ER Mechanical/Electrical Room

MER Main Equipment Room • Maintenance Engineering Report • Maintenance Evaluation Report • Manned Earth Reconnaissance • Manpower Estimate Report • Manpower Estimating Relationship • Mars Exploration Rover • Mass Energy Relationship • Material Específico de Riesgo • Maximum Emission Region • Mechanical Equipment Room • Medium-Energy Resolution • Merced Castle (USA-CA) (location identifier) • Meridian • Message Error Rate • Metal Evaporated Resistor • Minimum Energy Requirement • Minimum Essential Requirement • Mise En Route • Mission des Études et de la Recherche • Mission Evaluation Room • Mitigated Emergency Receiver • Multiechelon Reporting • Multielement Radar • Multiple-Ejector Rack

MERASEX Mesoscale Rain and Snowfall Experiment

MeRAu Mesures Régulation Automatisme (periodical code)

Merc Mercado • Mercantile • Mercoledi • Mercredi • Mercure • Mercurial • Mercury

MERC Media Equipment Resource Center • Middle East Regional Cooperation

MERCATOR Miniaturized European Carrier for Applications in Telescience in Orbit

MERCI Montage d'Étude du Refroidissement par Convection Interne

MERCOSUR Mercado Comun del Cono Sur

Mercu Mercury (periodical code)

MeReC Mechanics Research Communications (periodical code)

MERES Mineral and Energy Resources Exploration Satellite

MERF Magnetic Engineering Research Facility (USA) • Mobile En-route Radar Facility • Monthly Estimated Rainfall

MERG Merger

MERGE Model for Evaluating Regional and Global Effects

MERI Media Equipment Resource Index • Moon-Earth Radio Interferometer

MERID Meteosat Real-time Image Dissemination

MERIS Medium-Resolution Imaging Spectroradiometer

MERIT Measure Earth Rotation and Intercompare Techniques • Monitor Earth Rotation and Intercompare Techniques • Monitoring Earth Rotation and Intercomparison of Techniques • Monitoring of Earth Rotation and Intercomparison of Techniques (*)

MERL Materials Engineering Research Laboratory • Materials Equipment Requirement List • Mitsubishi Electric Research Laboratories

MERLIN Mailing Evaluation Readability Lookup Instrument • Managing Effective Research in the Library Information Network • Multielement Radio-Linked Interferometer Network

MERMAIDS Mediterranean Eddy Resolving Modeling and Interdisciplinary Studies

MEROD Message Entry and Readout Device

MERP Marine Ecosystem Response Project

MerRe Instituto Venezolano de Astronomía Mérida (periodical code)

MERS Message and Error Reporting System • Mobility Environmental Research Study • Most Economical Route Selection

MERSAT Meteorology and Earth Observation Satellite

MERTU Medical Entomology Research and Training Unit

MERU Mechanical Engineering Research Unit • Milliearth Rate Unit

MERWS Modular Extendable Rigid Wall Shelter

M&ES Magnetic & Electromagnetic Silencing

MES Magnetic and Electromagnetic Silencing • Mail Equipment Shop • Main-Engine Start • Main Equipment Supplier • Manchester Echelle Spectrograph • Manchester Echelle Spectrometer (*) • Manufacturing Enterprise System • Manufacturing Execution System • Master Earth Station • Mated Elements Simulator • Mated Events Simulator • Medan Polonia (Indonesia) (location identifier) • Medium-Energy Source • Member and Education Services • Message Entry System • Message Exchange System • Microsoft Education Services • Mission Events Sequence • Mobile Earth Station • Molecular Emission Spectrometry • Mössbauer Emission Spectroscopy

MESA Manned Environmental Systems Assessment • Materials - Engineering - Sensors - Actuators • Mathematics - Engineering - Science Achievement • Maximum-Entropy Spectrum Analysis • Microsystems and Engineering Science Applications (USA-NM) • Miniature Electrostatic Accelerometer • Modular Equipment Stowage Area • Modularized Equipment Storage Area • Modularized Equipment Storage Assembly • Modularized Equipment Stowage Area • Modularized Equipment Stowage Assembly • Molecular Engineering, Sensors and Actuators • Multirole Electronically Scanned Array

MESAR Microwave Electronically Scanned Adaptive Radar • Minimum Essential Security Assistance Requirement • Multifunctional Electronically-Scanned Adaptive Radar • Multifunction Electronically-Scanned Adaptive Radar

MESc Master of Engineering Science

MESC Master Event Sequence Controller • Member of the Economic and Social Committee • Ministry of Education, Science and Culture • Mission Events Sequence Controller

MESCAM Microwave Emission and Scattering Model

MESCO Moteur d'Étage Supérieur Cryogénique Optimalisé

MeScT Measurement Science and Technology (periodical code)

MESE Maximum Entropy Spectral Emission

MESEEC Methodologies to Estimate Social, Environmental, and Economic Consequences

MESFET Metal-Semiconductor Field Effect Transistor

MESG Maximum Experimental Safe Gap • Microelectrostatically-suspended Gyro • Microelectrostatic Gyro

MESGA Microelectrostatic Gyro-Accelerometer

MeSH Medical Subject Headings

MESH Management, Environment, Safety and Health

MESI Modified - Exclusive - Shared - Invalid (protocol)

MeSJJ Journal of the Meteorological Society of Japan (periodical code)

MESL Minimum Essential System List • Mission Essential Systems List • Museum Educational Site Licensing

MESM Message Memory • Mission Essential Systems Matrix • Multiechelon Supply Model

MESO Mesoscale • Microsat Experiment for Sounding Oxygen Densities

MESOE Méditerranée Enseignement Secondaire Observation et Environnement (France)

MESP Management Engineering Scheduling Program

MESPREP Message Preparation

MESR Ministère de l'Enseignement Supérieur et de la Recherche

MeSRM Mémoires Scientifiques de la Revue de Métallurgie (periodical code)

MESS Market Expert Surveillance System • Misalignment Estimation Software System • Monitor Event Simulation System

MESSAGE Modular Electronic Solid-State Aerospace Ground Equipment

MESSENGER Mercury Surface, Space Environment, Geochemistry and Ranging (mission) (NASA)

MeSSR Akademiia Nauk SSSR Mekhanika Zhidkosti i Gaza (periodical code)

MESSR Multispectral Electronic Self-Scanning Radiometer (*) • Multispectrum Electronic Self-Scanning Radiometer

Messrs Messieurs

MEST Maintenance Engineering Support Team • Mobile Earth Station

MESUR Mars Environmental Survey

MESZ Mitteleuropäische Sommerszeit

met Metafile Format (file extension)

Met Metal • Metallurgical • Metallurgy • Metaphor • Metaphysics • Meteor • Meteoritic • Meteoritics • Meteorological • Meteorology • Metropolitan

Mét Métal • Métallurgie • Métallurgique • Météorologie • Météorologique

MET Main Entry Title • Management Engineering Team • Management Entity Task • Manpower Evaluation Team • Marshall Engineering Thermosphere • Master Events Timer • Mean Elapsed Time • Mécanisme Européen de Technologie • Memory Enhancement Technology • Metafile • Meteorological • Microscope Électronique à Transmission • Midden Europese Tijd • Minimum Energy Trajectory • Minimum Exposure Time • Mission Elapsed Time • Mission Entry Time • Mission Environment Tape • Mission Events Timer • Mission Event Time • Mobile Earth Terminal • Mobile Education Team • Mobile Equipment Technology • Modified Expansion Tube • Modular Equipment Transporter • Modularized Equipment Transporter • Moreton (Australia-QLD) (location identifier) • Multi-element Test • Multi-image Exploitation Tool

MetA Meteorological Aid Service

META Megachannel Extraterrestrial Array • Megachannel Extraterrestrial Assay (*) • Method for Evaluating Technology Alternatives • Methodology for Extracting Text Automatically • Methods for Extracting Text Automatically • Million-channel Extraterrestrial Assay (*)

METAR Meteorological Aviation Report

Metab Metabolic • Metabolism

Métab Métabolique • Métabolisme

METAC Meteorological Advisory Committee

MetAP Meteorology and Atmospheric Physics (periodical code)

Metal Metallofizika (periodical code) • Metallurgical • Metallurgy • Metalurgia

Metall Metallurgical • Metallurgy

MetAP Meteorology Atmospheric Physics (periodical code)

METAP Mediterranean Environmental Technical Assistance Programme

Metaph Metaphysical • Metaphysics

Metaphys Metaphysics

Metapl Metaplasmo

Metát Metátesis

METCAL Metrology and Calibration

METCAM Metric Camera

METDAS Meteorological Data Analysis System

METDRTHIP Mimas - Enceladus - Tethys - Dione - Rhea - Titan - Hyperion - Iapetus - Phoebe • Miriam's Enchiladas Taste Divine Recently. Tell Her I'm Proud

MetE Metallurgical Engineer

Meteo La Météorologie (periodical code)

Meteor Meteorología • Meteorological • Meteorology

Météor Météorologie • Météorologique

Meteorol Meteorological • Meteorology

Météorol Météorologie • Météorologique

METEOR Multiple Experiment Transporter to Earth Orbit and Return

METEOSAT Meteorological Satellite

MeTeS Mechanika Teoretyczna i Stosowana (periodical code)

METG Medium-Energy Transmission Grating

Meth Methodist

MetHy Meteorology Hydrology JPRS (periodical code)

METI Meteorological Information • Ministry of Economy, Trade and Industry (Japan)

Metic Meteoritics (periodical code)

Metik Meteoritika (periodical code)

MetIs Meteornye Issledovaniia (periodical code)

METIS Meteorological Information System

Metit Meteorites (periodical code)

M-et-L Maine-et-Loire (France)

METL Mission Essential Task List

M-et-M Meurthe-et-Moselle (France)

MetMa Meteorological Magazine (periodical code)

m. et n. Mane et Nocte

METOC Météo-Océanographie • Meteorological and Oceanographic • Meteorology and Oceanography

METOP Meteorological Operational Platform • Meteorological Operational Polar (satellite) • Meteorological Operations

METP Materiel Equipment Test Plan

Métr Métrica

Metr Metrología

METRAN Managed European Transmission Network

METRIC Multiechelon Technique Recoverable Item Control

Metro Metrologia (periodical code)

METRO Meridional Transport of Ozone

Metrol Metrological • Metrology

Metrop Metropolitan

Metropol Metropolitan

MetRR Meteornoe Rasprostranenie Radiovoln (periodical code)

MetRu Meteorologische Rundschau (periodical code)

METS Meteorological Environment Test Support • Meteorological System • Microwave Energy Transmission System • Mobile Electronic Test Set • Modular Engine Test Stand • Multienvironment Test System

METSAT Meteorological Satellite

METSIS Meteorological Satellite Information System

METU Middle East Technical University (Turkey)

METV Main Engine Thrust Vector

METVC Main Engine Thrust Vector Control

MetVy Metody Vychislenii (periodical code)

METWATCH Meteorological Watch

MetZe Meteorologische Zeitschrift (periodical code)

MetZp Meteorologicke Zpravy (periodical code)

MEU Main Electronics Unit • Mask Exchange Unit • Medical Education Unit • Medium-Enriched Uranium • Memory Expansion Unit • Message Encoder Unit • Mission Essential Unit • Monte Dourado (Brazil-PA) (location identifier) • Multiplexer Encoder Unit

MeUpp Meddelanden fran Astronomiska Observatorium Uppsala (periodical code)

MEUR Mega Euro • Millions Euros

Mev Mevrouw

MeV Mega Electronvolts • Million Electronvolts

MEV Manned Entry Vehicle • Mass Excursion Vehicle • Minden - Douglas County (USA-NV) (location identifier)

Mevr Mevrouw

MEVT VMars: Evolution of Volcanism, Tectonics and Volatiles (experiment)

MEVVA Metal Vacuum Vapor Arc

MEW Mweka (Zaire) (location identifier)

MEWG Maintenance Engineering Working Group • Master Evaluation Working Group

MEWS Mono-Extraction Workstation

MEWT Matrix Electrostatic Writing Technique

Mex Mexican • Mexico

MEx Mars Express (mission)

MEX Estado de México (Mexico) • Mars Express (mission) • Mercury Express (mission) • Mexico (international vehicle registration code) • Mexico (ISO country code) • México City - Benito Juarez (Mexico) (location identifier) • Mexique (code ISO de pays) • Microelectronics • Modèle d'Expérimentation • Mondays Excepted

MExE Mobile Execution Environment

MEXH Multiple-Energy X-ray Holography

Mex. Sp. Mexican Spanish

MEXT Ministry of Education, Culture, Sports, Science and Technology (Japan)

MEY Maximum Economic Yield • Meghauli (Nepal) (location identifier)

Mez Mezzo • Mezza • Mezzanine

MEZ Maritime Exclusion Zone • Messina (South Africa) (location identifier) • Mitteleuropäische Zeit

MEZT Midden Europese Zomer Tijd

m/f Mi Favor

mf Medium Frequency • Metafont (file extension) • Microfilm • Minor Frame

mF Millifarad • mittlerer Fehler

Mf Massif

M/F Mainframe • Maintenance Factor • Male or Female

M&F Manufacturing & Facilities • Materials & Facilities

MF Magnetic Field • Magnetic Focus • Main Feed • Main Force • Main Frame • Maintenance Factor • Maintenance Function • Major Frame • Major Function • Management Facility • Mandatory Frequency • Mass Fraction • Mass Function • Master File • Master Frame • Mate and Ferry • Mated Flight • Mean Field • Measurement Facility • Mediation Function • Medium Frequency • Mégafranc • Membrane Filter • Metal Factor • Metallic Film • Microfiche • Microfilm • Middeck Forward • Middle French • Middle Frequency • Midfuselage • Millions de Francs • Minkowski Functional • Mobile Facility • Moderately Fast • Modulated Frequency • Modulation de Fréquence • Modulazione de Frequenza • More Fragments • Motor Field • Move Fragment • Moyenne Fréquence • Multifield • Multifiltration • Multifractal • Multifrequency • Multifunction • Multifunctional • Multiple Frequency • Multiplicity Factor • Multiplying Factor

MFA Mafia (Tanzania) (location identifier) • Main Frame Access • Major Facilities Access • Management Functional Area • Maneuver Functional Area • Manned Flight Awareness • Marconi-Franklin Antenna • Master of Fine Arts • Mean Field Annealing • Mean Field Approximation • Minister of Foreign Affairs • Ministry of Foreign Affairs • Mission Functional Analysis • Mission Functional Area • Monofluoroacetate • Moveable Finite Automata • Multifunction Antenna • Multifunction Attachment • Multifurnace Assembly • Multiple Frequency Analysis • Museum of Fine Arts

MFAASAR Multifrequency Antenna Array Synthetic Aperture Radar

MFAC Master of Fine Arts in Computing

MFAP Manned Flight Awareness Program

MFAS Maneuver Functional Area System • Multifibre Area Spectrograph

MFB Mixed Functional Block • Monfort (Colombia) (location identifier) • Multifunction Buffer

MFBM Modified Fuzzy Bog Model

MFBP Manufacturing Flow and Building Plan • Manufacturing Flow and Build Plan

MFC Mafeteng (Lesotho) (location identifier) • Main Fuel Control • Manual Flight Control • Manual Frequency Control • Map Feature Coding • Mass Flow Controller • Master File Copy • Maximum Fault Coverage • Message Formatting Committee • Microfilm Frame Card • Microfunctional Circuit • Microgravity Facilities for Columbus • Microsoft Foundation Class • Motor Fuel Consumption • Multifrequency Code • Multifunction Computer • Multifunction Console • Multinational Force Compatibility • Multiple-Flight Computer • Multiple-Flight Controller

MFCC Multifunction Common Console

MFCD Multifunction Color Display

MFCL Multifrequency Cavity Laser

MFCM Multifunction Card Machine

MFCO Manual Fuel Cutoff

MFCQ Mangasarian-Fromovitz Constraint Qualification

MFCS Magnetic-Field Calibration System • Manual Flight Control System • Materiel Finance Control System • Mathematical Foundations of Computer Science (conference) • Mechanical Flight Control System • Model Following Control System

MFCSIT Mathematical Foundation of Computer Science and Information Technology (conference)

Mfd Manufactured

MFD Magnetic Frequency Detector • Main Feed • Malfunction Detection • Malfunction Display • Manipulator Flight Demonstration (*) • Manipulator Flight Development • Mansfield Lahm (USA-OH) (location identifier) • Master File Directory • Maximum Frequency Difference • Mean Field Dynamo • Medium-Frequency Domain • Metal Friction Device • Minimum Fatal Dose • Mode Field Diameter • Moving-Finite Difference • Multifunctional Display • Multifunction Display

MFDBS Mathematical Fundamentals of Database Systems (symposium)

MFDD Message Format Definition Database

MFDP Mesoscale Frontal Dynamics Project

MFDSUL Multifunction Data Set Utility Language

MFDT Magnetic Field Directional Turning • Mean Fractional Dead Time

MFE Magnetic-Field Energy • Magnetic-Field Experiment • Magnetic-Field Explorer • Magnetic Fusion Energy • Manpower Force Element • McAllen (USA-TX) (location identifier) • Mean Field Electrodynamics • Mid-Frequency Executive • Minimum Free Energy • Mission (USA-TX) (location identifier) • Most Frequent Erection • Moving-Finite Element

MFECC Magnetic Fusion Energy Computer Center

MFED Manned Flight Engineering Division

MfEE Ministry for Education and Employment (UK)

MFENet Magnetic-Fusion Energy Network

MFENET Magnetic-Fusion Energy Network

MFEX Microrover Flight Experiment

mff MIDI File Format (file extension)

MFF Magnetic Field Function • Magnetic Flip-Flop • Millions de Francs Français • Millions of French Francs • Mixed Fleet Flying • Moanda (Gabon) (location identifier)

MFFE Multifrequency Frontend

MFFS Microsoft Flash File System

MFFT Multifunction Focal Tester

Mfg Manufacturing

MFG Major Functional Group • Manufacture • Manufacturing • Master Frequency Generator • Math Font Group • Multifunction Gateway • Muzaffarabad (Pakistan) (location identifier)

MFH Mehrfamilienhaus • Mesquite (USA-NV) (location identifier) • Multifrequency Heterodyne

MFHBF Mean Flight Hours Between Failures

MFHBUMA Mean Flight Hours Between Unscheduled Maintenance Action

MFHBVF Mean Flight Hours Between Verified Failures

M&FI Maintenance & Fault Isolation

MFI Magnetic-Field Indicator • Magnetic-Field Intensity • Magnetic-Field Investigation • Maintenance and Fault Isolation • Major Force Issue • Marshfield (USA-WI) (location identifier) • Medium/Format Identifier • Metafile Input • Método del Flujo Infrarojo • Multifunction Interpreter

MFIDS Multifunction Information Distribution System

MFIE Magnetic-Field Integral Equation

MFIMR Multifrequency Imaging Microwave Radiometer

MFiNM Matematicheskaia Fizika i Nelineinaia Mekhanika (periodical code)

MFIR Medium-Far Infrared • Moderate Field-of-view Imaging Radiometer

MFJ Moala (Fiji) (location identifier) • Modification of Final Judgment • Modified Final Judgment

MFKP Multifrequency Key Pulsing

MFL Magnetic-Field Line • Major Force List • Mount Full Stop (Australia-QLD) (location identifier) • Multifunction Keyboard • Multiple Folded Laser

MFLOP Mega-Floating-point Operation

MFLOpS Million of Floating Point Operations per Second

MFLOPS MegaFLOPS • Million of Floating Point Operations Per Second (*)

MFLT Modified Fast Lapped Transform

MFLTU Multifrequency Line Termination Unit

MFM Magnetic-Field Modeling • Magnetic-Field Modulation • Magnetic-Field Monitor • Magnetic Force Microscope • Magnetic Force Microscopy • Master File Maintenance • Modified Frequency Modulated • Modified Frequency Modulation • Multilevel Flow Modeling

MFMA Mean Field Mean Annealing • Monolithic Ferrite Memory Array

MFMMIC Multifunction Monolithic Microwave Integrated Circuit

MFMR Multifrequency Microwave Radiometer

MFMT Microwave Frequency Modulation Transmitter

MFN Milford Sound (New Zealand) (location identifier) • Most Favored Nation

MFNS Most Favored Nation Status

MFO Major Function Overlay • Master Frequency Oscillator • Metafile Output • Meteorologie en Fysische Oceanografie • Mixed-Function Oxidase • Multiple-Facilities Organization • Ras Nasrani (Egypt) (location identifier)

MFOC Multifunction Operator Console

MFOD Manned Flight Operations Directive

MFOI Major Force Oriented Issue

MFOS Multifunctional Optical System

MFOV Medium Field Of View • Multifield Of View • Multiple Field Of View

MFP Magnetic-Field Perturbation • Magnetic Flow Parameter • Main Free Path • Major Force Program • Management Framework Plan • Manners Creek (Australia-NT) (location identifier) • Mass Flow Parameter • Massive Flexible Penetration • Master File Program • Matched Filter Performance • Maximum Freezing Point • Mean Free Path • Middle Free Path • Mid-Frequency Preference • Mission Flight Plan • Molecular Force Probe • Molecular Free Path • Monofluorophosphate • Mosaic Focal Plane • Multifactor Productivity • Multifunction Peripheral • Multifunction Product • Mutualité de la Fonction Publique (France)

MFPA Magnetic-Field Position Angle • Mosaic Focal Plane Array

MFPE Mission From Planet Earth

MFPI Multifunction Peripheral Interface

MFPL Message Format Processing Language

MFPS Mathematical Foundations of Programming Semantics (conference)

MFPT Machine Failure Prevention Technology

MFQ Maradi (Niger) (location identifier) • Mouvement Français pour la Qualité (France)

Mfr Manufacture • Manufacturer

MFr Middle French

MFR Major Frame Ratio • Manipulator Foot Restraint • Maximum Flight Rate • Medford (USA-OR) (location identifier) • Medium Frequency Radar • Medium Frequency Range • Memorandum For Record • Minimum Funding Requirement • Multifilter Radiometer • Multifractal • Multifrequency Radar • Multifrequency Receiver • Multifunctional Receiver • Multifunctional Review • Multilink Frame Relay

m/fra Mi factura

Mfre Manufacture

MFRM Mobile Foot Restraint Manipulator

Mfrs Manufacturers

MFRS Major Frame Ratio Signal

MFRSR Multifilter Rotating Shadowband Radiometer

MFS Macintosh File System • Magnetic-Field Strength • Magnetic-tape Field Search • Manned Flight Simulator • Manned Flying System • Master Flexible System • Master of Foreign Service • Master of Foreign Study • Memory-based File System • Memory File System • Miraflores (Colombia) (location identifier) • Model Formulation System • Modified Filing System • Multichannel Flare Spectrograph • Multifrequency Synthesis • Multifunction Sensor • Multifunction Switch • Multifunction System • Multiple-Frequency Synthetizer

MFSE Maintenance Flying Scheduling Effectiveness

MFSK Multifrequency Shift Keying • Multilevel Frequency Shift Keying • Multiple Frequency Shift Keying

MFSP Macrofunction Signal Processor

MFSPP Mediterranean Forecasting System Pilot Project

MFSR Memory Fault Status Register

MFSS Multifunction Sensor Suite

MFT Machu Pichu (Peru) (location identifier) • Master File Table • Master of Foreign Trade • Mean Flight Time •

Metal Film Resistor • Metallic Facility Terminal • Meter Fix Time • Mobilfunktechnologien • Module Fixture Transporter • Multiprogramming Fixed Task • Multiprogramming with a Fixed number of Tasks

MFTAD Master Flight Test Assignment Document

MFTIO Minimum Flyout Time for Orbital Interception

MFTP Multicast File Transfer Protocol

MFTL My Favorite Toy Language

MFTP Multicast File Transfer Protocol

MFTRS Magnetic Flight Test Recording System

MFTS Multilateral File Transfer System

MFU Mfuwe (Zambia) (location identifier) • Most Frequently Used

MFUA Medical Follow-Up Agency

MFV Magnetic-Field Vector • Main Fuel Valve • Mars Flyby Vehicle • Mehrfrequenzwahlverfahren • Melfa - Accomack County (USA-VA) (location identifier) • Microfilm Viewer • Motor Fleet Vessel

MFVT Mixed-Flow Vectored Thrust

MFWD Mobile Firewall Daemon

MFWS Multifunction Workstation

Mfy Manufactory

MFY Mayfa'ah (Yemen) (location identifier)

mg Malagasy (ISO language code) • Milligram

mG Milligauss

Mg Magnesium • Magnésium • Megagram

M&G Mapping & Geodesy

MG Machien Gun • Madagascar (ISO country code) • Main Gauche • Major General • Malgasia (ISO country code) • Management • Management Group • Marchandises Générales • Médecin Généraliste • Media Gateway • Medium Gain • Megagauss • Membership Grade • Meridional Gradient • Message Gateway • Message Generator • Meteorological Group • Meteorology Group • Microgravity • Middle Gimbal • Minas Gerais (Brazil) • Minimum Garanti • Mission Generation • MIT-Greenbank (survey) • Mobile Generator • Modified Gaussian (model) • Monguor • Motor-Generated • Motor Generator • Multigage • Multigrid

Mga Maga

MGA Managua - Augusto Cesar Sandino (Nicaragua) (location identifier) • Matrix Graphics Architecture • Media Graphics Architecture • Medium-Gain Antenna • Middle Gimbal Angle • Monochrome Graphics Adapter • Multigraph Architecture • Multimedia Graphics Architecture • Multiple Group Analysis

MGAA Medium-Gain Autotrack Antenna

MgAL MgII Absorption Line

MGAS Middle Georgia Astronomical Society (USA-GA)

MGAWD Make Good All Works Disturbed

MGB Metropolitan Green Belt • Mount Gambier (Australia-SA) (location identifier)

MGBX Middeck Glovebox

MGC Main Giant Clump • Manual Gain Control • Media Gateway Controller • Memoryless Gaussian Channel • Michigan City (USA-IN) (location identifier) • Middle Georgia College (USA-GA) • Millennium Galaxy Catalog • Morphological Catalogue of Galaxies • Moyens Généraux Communs

MGCD Microgravity Compatible Drive

MGCM Mars General Circulation Model (*) • Mars Global Circulation Model

MGCO Mars Geoscience/Climatology Orbiter

MGCP Media Gateway Control Protocol • Multiple Gateway Control Protocol

MGCS Meteosat Ground Computer System • Mobile Ground Control System

MGD Magdalena (Bolivia) (location identifier) • Maximally Generalized Description • Mean Gain Deviation • Mercury Germanium Detector • Millions of Gallons per Day

MGDB Microgravity Database

MGDR Merged Geophysical Data Record

MGDS Multimission Ground Data System

MGE Maintenance Ground Equipment • Marietta Dobbins (USA-GA) (location identifier) • Modular GIS Environment • Multigaussian Expansion

MGEN Mutuelle Générale de l'Éducation Nationale (France)

MGES Maintenance Ground Equipment Section

mget Multiple Get

Mgf Magnify

Mgfd Magnified

MgF Magnesium Fluoride

MGF Franc (Madagascar) (ISO currency code) • Magnetic Field • Maringá (Brazil-PR) (location identifier)

MGFEL Master Government-Furnished Equipment List

MGFLD Multigroup Flux-Limited Diffusion

MGG Margarima (Papua New Guinea) (location identifier) • Marine Geology and Geophysics • Memory Gate Generator

MGGE Mongolian Geophysical and Geological Exploration

MGGF Monitoring Global Geophysical Fluids

MGGI Morphic Generator Grammatical Inference

MGH Margate (South Africa) (location identifier) • Mathematische Gesellschaft in Hamburg (Germany)

MG/I Major Gateway/Internet

MGI Management Group Identifier • Mandatory Government Inspection • Matagorda Island (USA-TX) (location identifier) • Mission Générale d'Insertion

MGIM Micro-Gravity Isolation Mount

MGIMO Moskovskii Gosudarstvennyi Universitet Institut Mekhaniki Otchet (periodical code)

MGIO Mount Graham International Observatory (USA-AZ)

MGIS Micro Geographic Information System

MGk Medieval Greek

MGK Mong Ton (Myanmar) (location identifier) • Multigraph Kernel

Mgl Mitglied • Mogul

MGL Matrix Generator Language

MGLH Multivariate General Linear Hypothesis

Mglia Mongolia

MGM Master Group Multiplexer • Materiel Group Manager • Mechanics of Granular Materials • Metro-Goldwyn-Mayer • Modified Gaussian Model • Montgomery Dannelly (USA-AL) (location identifier)

Mgmt Management

MGMT Management

Mgn Magneto • Margin

MGN Magangue Baracoa (Colombia) (location identifier) • Magellan

Mgna Montagna

Mgne Montagne

Mgng Managing

MGO Main Geophysical Observatory (Russia) • Manega (Gabon) (location identifier)

mgp Mètre Géopotentiel

MGP Magnetic Granulometry Plot • Manga (Papua New Guinea) (location identifier) • Manufactured-Gas Plant • Multiannual Guidance Programme

MGpD Millions of Gallons per Day

MGPS Molonglo Galactic Plane Survey

MGQ Maximum Guaranteed Quantity • Mogadishu (Somalia) (location identifier)

Mgr Manager • Monseigneur • Monsignor

MGr Medieval Greek

MGR Manager • Miniature GPS Receiver • Modified General Relativity • Modular Gas-cooled Reactor • Moultrie Thomasville (USA-GA) (location identifier) • Multiple GPS Receiver

MGRA Master Geographical Reference Area

Mgrs Managers • Monseigneurs • Monsignors

MGRW Matrix Generator and Report Writer

mgs Meter - Gram - Second

MgS Magnesium Suite

MGS Mangaia Island (Cook Islands) (location identifier) • Marine Geophysical Survey • Mars Geoscience Surveyor • Mars Global Surveyor (*) • Medium Ground Station • Mercury Gamma-ray Spectrometer • Message Gateway Server • Message Generation System • Microcomputer Graphic System • Microgravity Science • Mission Ground Station • Mobile Ground Station • Mobile Ground System • Moment Gyro System • Multilayer Grating Spectrometer

MGSE Maintenance Ground Support Equipment • Mechanical Ground Support Equipment

MGSM Medium Ground Station Module

MGSO Multimission Ground Systems Office

Mgt Management

MGT Magnetic Tape • Major Ground Test • Management • Measured Gas Temperature • Millingimbi (Australia-NT) (location identifier) • Mobile Global Title • Mobile Ground Terminal

MGTDR Magellan Global Topography Data Record

MGTP Model Generation Theorem Prover

MGTVM Moskovskii Gosudarstvennyi Tekhnicheskii Universitet Vestnik Seriia Mashinostroenie (periodical code)

MGTVP Moskovskii Gosudarstvennyi Tekhnicheskii Universitet Vestnik Seriia Priborostroenie (periodical code)

MGU Manaung (Myanmar) (location identifier) • Midcourse Guidance Unit

MGUkr Akademiia Nauk Ukrainskoi SSR Morskie Gidrofizicheskie Issledovaniia (periodical code)

MGUMN Moskovskii Gosudarstvennyi Universitet Institut Mekhaniki Nauchnye Trudy (periodical code)

MGV Margaret River (Australia-WA) (location identifier)

MGVI MERIS Global Vegetation Index

MGVT Mated Ground Vibration Test

MGW Maximum Gross Weight (*) • Measured Gross Weight • Morgantown (USA-WV) (location identifier)

MGWR Morgan-Good Wolf-Rayet

MGX Moabi (Gabon) (location identifier)

MGY Dayton - Montgomery County (USA-OH) (location identifier)

MGZ Mergui (Myanmar) (location identifier) • Myeik (Myanmar) (location identifier)

mH Millihenry • Millihour

MH Îles Marshall (code ISO de pays) • Magnetic Head • Magnetic Heading • Magnetic Hole • Mail Handler • Mail Header • Main Hatch • Man-Hour • Marital History • Marshall Islands (ISO country code) • Massive Halo • Medal of Honor • Medium Haul • Meierhof • Mental Health • Mesh Heading • Message Handler • Message Handling • Metal Halide • Mexican Hat • Mobile Host • Modified Huffmann • Mount Wilson H-alpha • Mt. Wilson H-alpha

Mha Million Hectares

MHA Mahdia (Guyana) (location identifier) • Maintenance Hazard Analysis • Maximum Hypothetical Accident • Merrill H-Alpha • Minimum Hold Altitude • Minimum Holding Altitude • Mission Hydrographique de l'Atlantique • Modified Handling Authorized • Modular Hypermedia Application • Multiple Headset Adapter

MHAA Mid-Hudson Astronomy Association (USA-NY)

MHB Auckland - Mechanics Bay (New Zealand) (location identifier) • Maintenance Handbook

MHBM Minimum Hydrogen Burning Mass

MHC Macmahon - Camp 4 (Australia-SA) (location identifier) • Manipulator Hand Controller • Materiel Handling Crane • Mile-High Club • Modified Huffmann Code • Most Helpful Cycle • Mount Holyoke College

MHCC Mount Hook Community College (USA-OR)

MHCI Management Hardware Configuration Item

Mhd Masthead

MHD Magnetohydrodynamics • Magnétohydrodynamique • Mashad (Iran) (location identifier) • Met Hartelijke Dank • Michigan Henry Draper (catalog) • Mihalas - Hummer - Däppen (equation of state) • Minimum Hamming Distance • Moving Head Disk • Multihead Disk

MHDG Magnetic Heading

MHDR MNG Header

MHDS Mosaic Data Handling System

MHE Material Handling Equipment • Materiel Handling Equipment • Message Handling Element • Mitchell (USA-SD) (location identifier) • Molecular Hydrogen Emission

MHEG Multimedia and Hypermedia Experts Group

MHEO Multiregional HEO

MHF Massive Hydraulic Fracture • Massive Hydraulic Fracturing • Medium High Frequency • Mexican Hat Function • Ministerium für Hoch- und Fachschulwesen • Mixed Hydrazine Fuel • Morichal (Colombia) (location identifier)

MHFEP Multihost Front End Processor

MHG Mannheim (Germany) (location identifier) • Meteorological Hydrogen Generator • Middle High German • Miethöhegesetz

MHGS Mikrowellen-Hintergrundstrahlung

MHH Marsh Harbour (Bahamas) (location identifier)

MHI Mitsubishi Heavy Industries • Musha (Djibouti) (location identifier)

MHIR Millstone Hill Ionospheric Radar (USA-MA)

MHITR Mitsubishi Heavy Industries Technical Review (periodical code)

MHK Manhattan (USA-KS) (location identifier)

MHL Îles Marshall (code ISO de pays) • Marshall (USA-MO) (location identifier) • Marshall Islands (ISO country code) • Metastable Helium Level • Microprocessor Host Loader • Minimum Helium Loss

MHM Minchumina (USA-AK) (location identifier)

MHMT Multihead-Multitail

MHN Mullen (USA-NE) (location identifier)

MHO Massive Halo Object • Meteorology - Hydrology - Oceanography • Millstone Hill Observatory (USA-MA) • Mount Hopkins Observatory (USA-AZ) • Mt. Hopkins Observatory (USA-AZ) • Mount House (Australia-WA) (location identifier) • My Humble Opinion

MHOA Mount Hook Observatory Association (USA-OR)

M. Hon. Most Honorable

MHP Mécanique de Haute Précision • Mobile Host Protocol • Modification Hardware Package • Mostly Harmless Planet • Move and Handling Procedure

MHPCC Maui High-Performance Computing Center (USA-HI)

MHQ Mariehamn (Finland) (location identifier)

MHR Man-Hours Required • Member of the House of Representatives • Microwave Hologram Radar • Millstone Hill Radar (USA-MA) • Mission Hydrographique de Recherche • Modular Heliotron Reactor • Modular Helium Reactor • Sacramento Mather (USA-CA) (location identifier)

MHRS Man-Hours • Message Handling and Retrieval System • Modified Hazard Ranking System

MHRSR Man-Hours Required

MHS Magnetohydrostatistics • Mail Handling System • Many Hilbert Spaces • Marseille H-alpha Survey • Medical History Sheet • Mercury Horizon Sensor • Message Handling Service • Message Handling System (*) • Microwave Humidity Sounder • Mount Shasta (USA-CA) (location identifier) • Multiaircraft Horizontally Separated

MHSA Mars Horizon Sensor Assembly (MO)

MHSC Manipulator Handset Controller

MHSTR Millstone Hill Satellite Tracking Radar (USA-MA)

MHT Manchester (USA-NH) (location identifier) • Message Handling Table • Mild Heat Treatment • Mutiple Hypothesis Tracking

MHTC Most Helpful Total Cut

MHTCC Most Helpful Total Cut Cancelling

MHTGR Modular High-Temperature Gas-cooled Reactor • Modular High Temperature Gas Reactor

MHU Mount Hotham (Australia-VIC) (location identifier)

MHV Message Header Validation • Miniature Homing Vehicle • Mojave - Kern County (USA-CA) (location identifier)

MHVR Multimedia, Hypermedia, and Virtual Reality (conference)

MHW Mean High Water • Mexican Hat Wavelet • Multi-hundred-Watt

MHWCI Management Hardware Configuration Item

MHX Manihiki Island (Cook Islands) (location identifier)

MHY Morehead (Papua New Guinea) (location identifier)

mHz Millihertz

MHz Megahertz

MHZ Mildenhall (UK-England) (location identifier)

mi Maori (ISO language code) • Mile • Mill (currency) • Miscelleanous (file extension)

Mi Mile • Minor • Mittwochs

M/I Minimum Impulse

M&I Management & Integration • Modernization & Improvement • Municipal & Industrial

MI Magnetoimpedance • Main Impulse • Maintenance Industrielle • Maintenance Instruction • Maître d'Internat • Major Issue • Malleable Iron • Management and Integration • Management Indicator • Management Information • Management Instruction • Management Interface • Management of Information • Manual Input • Mare Imbrium (Moon) • Market Investigation • Marshall Islands (international aircraft registration code) • Master Index • Master Item • Materiel Improvement • Maturity Index • Medical Imaging • Medium Intensity • Message Indicator • Michigan (USA) • Micro-Image • Middle Initial • Mijns Inziens • Mineral-Insulated • Minimum Impulse • Ministry of Information • Minority Institution • Mischgebiet • Misiones (Argentina) • Mission Independent • Mode Indicate • Modèle d'Identification • Modification Instruction • Modulation Index • Moment of Inertia • Monitoring Information • Monotically Increasing • Movement Instruction • Multiple Invocation • Municipal and Industrial • Mutual Inductance • Mutual Information • Mutual Interaction • Mutual Interference • Phobos (satellite of Mars)

MIA Management Information Analysis • Master of International Affairs • Meeting of International Authorities • Metal Interface Amplifier • Miami (USA-FL) (location identifier) • Michigan Array • Minimum IFR Altitude • Missing In Action • Mission-Independent Area • Mobile Internet Access • Model-Independent Algorithm • Multiple-Interface Adapter • Multiplexer Interface Adapter • Multiplex Interface Adapter • Multivariate Image Analysis • Multivendor Integration Architecture

MIAC Material Identification and Accounting Code • Meteorites and Impacts Advisory Committee (Canada)

MIADS Map Information Assembly and Display System

MIAeE Member of the Institute of Aeronautical Engineers

MIAGE Maîtrise d'Informatique Appliquée à la Gestion des Entreprises

MIALS Medium Intensity Approach Light System

MIAM McGill Institute for Advanced Materials (Canada-QC) • Member of the Institute of Administrative Management

MIAMM Multidimensional Information Access using Multiple Modalities

MIAOU Mathématiques et Informatique de l'Automatique et de l'Optimisation pour l'Utilisateur

MIAP Member of the Institution of Analysts and Programmers

MIAPR Medical Image Analysis and Pattern Recognition

MIAR Message Initiate Alarm Reset

MiARB Mitteilungen Astronomisches Rechen-Institut Serie B (periodical code)

MIARG Management Information Analysis and Report Generator

MiARI Mitteilungen Astronomisches Rechen Institut Serie A (periodical code)

MIARS Maintenance Information Automated Retrieval System

MIAS Management Information and Accounting System • Marine Information Advisory Service • Marine Information and Advisory Service • Mathématique, Informatique et Applications aux Sciences • Multipoint Interactive Audiovisual System • Multispectral Image Analysis System

MI&B Ministry of Information & Broadcasting

MIB Management Information Base • Management Information Block • Manual Input Buffer • Master Instruction Book • Master Interconnect Board • Message Information Base • Metric Interface Buffer • Minimum-Impulse Bit • Minot (USA-ND) (location identifier) • Mission Information Base • Module Interface Board • Multilayer Interconnection Board

MiBas Mitteilungen der Astronomisch-Meteorologischen Anstalt der Universität Basel Astronomische Reihe (periodical code)

MIBK Methyl Isobutyl Ketone

MIBOM Major Item Bill Of Material

MiBon Mitteilungen der Astronomischen Institute der Universität Bonn (periodical code)

MiBud Mitteilungen der Sternwarte der Ungarischen Akademie der Wissenschaften Budapest (periodical code)

mic Microsoft Image Composer (file composer)

Mic Micrometer • Microphone • Microscopium-Microscopii (constellation)

MIC Macro Interpretive Command • Magnetosphere-Ionosphere Coupling • Maintenance Inventory Center • Major Item Center • Management Indicator Code • Management Information Center • Management Information Class • Management Information Control • Management

Integration Consortium • Mathematics Information Center • Media Interface Connector • Medium Interface Connector • Memory In Cassette • Memory Interface Connection • Memory Interface Connector • Memory Interface Control • Message Identification Course • Message Integrity Check • Message Integrity Code • Metastable Interstitial Composite • Meteorological Information Committee • Micrometer • Microphone • Microsoft Internet Chat • Microwave Integrated Circuit • Microwave Integrated Circuitry • Microwave Interference Coordination • Minimum-Impulse Controller • Minimum Ignition Current • Ministère de l'Industrie et du Commerce • Ministry of Information and Communications • Minneapolis Crystal (USA-MN) (location identifier) • Mission Catalogue • Mission Information Center • Model-Integrated Computing • Modulation d'Impulsions Codées • Modulation par Impulsions Codées (*) • Module Interface Card • Monitoring, Identification and Correlation • Monolithic Integrated Circuit • Multichannel Image Camera • Multiinstrument Chamber • Multimedia Information Center • Mutual Interference Chart

MICA Macroinstruction Compiler Assembler • Macroinstruction Computer Assembler • Mirror Coronagraph for Argentina • Modem ISDN Channel Aggregation • Multiyear Interactive Computer Almanac

MICAI Mexican International Conference on Artificial Intelligence

MICADO Michelson Interferometer for Coordinated Auroral Doppler Observations

MICAP Mission Capability • Mission Capable • Monte Carlo Ionization Chamber Analysis Package

MICAS Miniature Integrated Camera Spectrometer

MICB Mission Integration Change Board

MICC Master Information Coordination Central

MICCAI Medical Image Computing and Computer-Assisted Intervention (conference)

MICCG Modified ICCG

MICE Map Image Congruency Evaluation • Member of the Institution of Civil Engineers • Modular Integrated Communications Environment • Multiinterface Computer Equipment

MicEc Microbial Ecology (periodical code)

MICEFA Mission Interuniversitaire de Coordination des Échanges Franco-Américains

MICEI Member of the Institution of Civil Engineers of Ireland

MICELEM Microphone Element

MICG Mercury Iodide Crystal Growth

MicgQ Microgravity Quarterly (periodical code)

Mich Michigan (USA) • Michoacan (Mexico)

MICH Michoacan (Mexico)

MICIS Material Inventory Control and Information System • Mechanized Information Control Inventory System • Mechanized Inventory Control Information System

MICNIA Modular Integrated Communications, Navigation and Identification Avionics

MICNS Modular Integrated Communication and Navigation System

MICO Management Information Control Officer

MICON Mission Concept

MICOS Multifunctional Infrared Coherent Optical Sensor

Micr Microscope • Microscopy

MICR Magnetic Ink Character Recognition

MICRA Motion, Imaging Contouring and Roughness Amplitude

MicRe University of Michigan Observatory Reprints (periodical code)

MICREX Microgravity Research Experiment

Micro Micronesia • Microscope • Microscopist • Microscopy

MICRO Multiple Indexing with Console Retrieval Operations

Microbiol Microbiología

Micros Microscope • Microscopist • Microscopy

MICROSAT Microwave Satellite

MICROSCOPE Microsatellite à traînée Compensée pour l'Observation du Principe d'Équivalence

MICRS Magnetic Ink Character Recognition System • Main Instrument Console and Readout Station

MICS Macro Interpretive Commands • Magnetospheric Ion Composition Spectrometer • Management Information and Control System • Management Information Control System • Management Integrated Control System • Manned Interactive Control Station • Mathematical, Information and Computational Science • Mid-Infrared Camera and Spectrometer • Mobile Integrated Communications System

MICSIS Micro Spatial Information System

MicWa Microwaves (periodical code)

mid MIDI Data (file extension)

Mid Middle • Midlands • Midnight • Midshipman

MID Magnetically Insulated Diode • Management Information Division • Master of Industrial Design • Materiel for Interactive Document • Mérida (Mexico) (location identifier) • Message Identification • Message Identifier • Message Input Device • Middle • Mobile Information Device • Multiple-Ion Detection

MIDAC Management Information for Decision And Control

MIDAN Microprocessor Data Analyzer

MIDAP Multichannel Image Display and Analysis Program

MIDAR Microwave Detection And Ranging

MIDAS Mail International Dispatch and Accountability System • Maintenance Information Data Automation System • Maintenance Integrated Data Access System • Maklumat Industri dan Sains (Malaysia) • Management Information, Data and Accounting System • Management Integrated Data Accumulating System • Management Issues in a Distributed Ada Support • Manufacturing Information Distribution and Acquisition System • Map Information and Data Analysis System • Measurement Information and Data Analysis System • Measurement Information Data Acquisition System • Memory-Implemented Data Acquisition System • Meteorological Interactive Data Access System • Metrology Information And Documentation System • Microcomputer-Interfaced Data Acquisition System • Microimaged Data Addition System • Micro Imaging Dust Analysis System (Rosetta) • Miniature Data Acquisition System • Miniature Inertial Digital Attitude System • Mission Integrated Design Analysis Simulator • Mobile Integrated Digital Automatic System • Modified Integration Digital Analog Simulator • Modular Interactive Data Acquisition System • Modulator Isolation Diagnostic Analysis System • MOTIF Integrated Data Applications System • Multidiscipline Interactive Display and Analysis System • Multimedia Industry Development for Academic Software • Multimedia Information Demonstration And Support • Multiple Index Direct Access System • Multiple-Input Data Acquisition System • Multiple-Input Digital Autocorrelation Spectrometer • Multiproject Investigation During the Alignment of Spacecraft • Multisensor Image Database System • Multitask Ice Data Analysis System • Multivariate Interactive Digital Analysis System • München Image Data Analysis System (ESO) • Munich Image Data Analysis System (ESO)

MIDB Major Item Database • Modernized Integrated Database

MIDDLE Microprogram Design Description Language

MIDDS Meteorological Interactive Data Display System

Middx Middlesex (UK)

MidEx Medium-class Explorer (*) • Medium Explorer

MIDEX Medium-class Explorer (*) • Medium Explorer

MIDF Major Item Data File • Multiple Input Describing Function

MIDI Multiple Instrument Digital Interface • Musical Instrument Digital Interface

MIDILAH Mail Distribution Instructions and Labelling Handbook

MIDIST Mission Interministérielle de l'Information Scientifique et Technique (France)

Midl Midlands

MIDL Miniaturized Interoperable Data Link • Modular Interoperable Data Link

MIDLA Media Independent Data Link Architecture

MIDMS Machine-Independent Data Management System

Midn Midshipman

MIDN Midnight

MIDO Manufacturing Inspection District Office

MIDOT Multiple Interferometer Determination Of Trajectories

MIDP Major Item Distribution Plan • Mars Instrument Development Program

MIDPM Member of the Institute of Data Processing Management

MIDR Mosaicked Image Data Record

MIDS Management Information and Data System • Management Information Display System • Matrix Information and Directory Service • Matrix Information Directory Service • Mechanized Information and Documentation System • Miniaturized Interoperable Data System • Movement Information Distribution Station • Multifunctional Information Distribution System • Multifunction Information Distribution System (*) • Multimode Information Distribution System • Multinational Information Distribution System

MIDSIM Midcourse Simulation

M&IE Meals & Incidental Expenses

MIE Master of Industrial Engineering • Meals and Incidental Expenses • Meteor Ionizing Efficiency • Mission-Independent Equipment • Model Institution for Excellence • Most Impressive Erection • Most Intense Earthquake • Muncie (USA-IN) (location identifier)

MIEA Maître Itinérant d'École d'Application • Master Information Exchange Arrangement

MIEBA Moscow Izdatel Energiia Biblioteka Avtomatike (periodical code)

MIEBR Moscow Izdatel Energiia Biblioteka Radioelektronike (periodical code)

MIEC Meteorological Information Extraction Centre • Mission Item Essentiality Code

MIEE Member of the Institution of Electrical Engineers

MIEEE Member of the Institute of Electrical and Electronics Engineers

MIEG Mission à Intérêt Économique Général

MIEP Multiplicative Inverse Eigenvalue Problem

MIER Master Interoperability Evaluation Register

MIES Modernized Imagery Exploitation System • Modernized Integrated Exploitation System

MIESR Matrix Isolation Electron Spin Resonance

MIEWF Master Integrated Engineering Work File

MIF Main Image File • Maker Interchange Format • Management Information File • Management Information Format • Master Index File • Master Item File • Mean Information Field • Mechanical Integration Frame • Minimum Internetworking Functionality • Mode Indication Function • Modernization and Implementation Facility • Module Integration Facility • Module Interconnect Framework • Module Interface Formalism • Monahans - Roy Hurd (USA-TX) (location identifier) • Monopulse Interference Filter • Monthly Impact Factor • Multiscale Image Fusion

MIFDT Multifunction Information Data Terminal

MIFF Management Information Format File

MIFIR Microwave Instantaneous Frequency Indication Receiver

MIFR Master International Frequency Register • Monitored International Frequency Register

MIFS Multiplex Interferometric Fourier Spectroscopy

MIG MACH Interface Generator • Magnetized Ionized Gas • Major Installation Grant • Metal-Inert Gas • Metal-In-Gap • Migration • Miniature Integrating Gyroscope • München Neubiberg (Germany) (location identifier)

MIGA Message-based Interface Gate Array • Mid-Infrared Galaxy Atlas

MIGMAS Microgravity Mass Spectrometer

MiGoe Astronomische Mitteilungen der Universitäts-Sternwarte zu Göttingen (periodical code)

MiGra Mitteilungen der Universitäts-Sternwarte Graz (periodical code)

MIGRACION Milieu Interstellaire Galactique, Rayons Cosmiques et Ionisation dans les Nuages

MIH Major Industrial Hazard • Mitchell Plateau (Australia-WA) (location identifier) • Move In Housing Allowance

MiHam Mitteilungen der Hamburger Sternwarte (periodical code)

MiHar Mitteilungen der Bruno-H.-Bürgel-Sternwarte Hartha (periodical code)

MIHDAS Mid-Infrared High-Dispersion AO-fed Spectrograph

MII Deimos (satellite of Mars) • Marília - Dr. Gastão Vidigal (Brazil-SP) (location identifier) • Master Inspection Item • Mathematics In Industry • Media-Independent Interface • Microsoft - IBM - Intel • Ministry of Information Industry (China-PRC)

MIIC Major Item Information Center

MIICS Master Item Identification Control System

MIIE Member of the Institution of Industrial Engineers

MIIF Master Item Intelligence File

MIIL Master Item Intelligence List

MiInn Mitteilungen der Sternwarte Innsbruck (periodical code)

MiINT Minsk Izdatel Nauka i Tekhnika (periodical code)

MIIP Maneuver Interoperability Implementation Plan

MIIPS Multipurpose Interactive Image Processing System

MIIRS Modular Imagery Interpretation and Reporting System

MIIS Mid-Infrared Imager Spectrometer • Miscellaneous Inputs Information System

Mij Maatschappij

MIJ Mili Island (Marshall Islands) (location identifier)

MIJA Multiple Impact Jet Apparatus

MiJen Mitteilungen der Universitäts-Sternwarte zu Jena (periodical code)

MiJo Microwave Journal (periodical code)

MIK Mikkeli (Finland) (location identifier)

MIKE Mass-analyzed Ion Kinetic Energy

MIKES Mass-analyzed Ion Kinetic Energy Spectrometry

MiKon Mitteilungen der Konkoly Sternwarte Budapest-Szabadsaghegy (periodical code)

Mikro Mikroelektronika (periodical code)

MIKSAM Multiple Indexed Keyed Sequential Access Method

MikSb Mikroelektronika Sbornik (periodical code)

Mil Milano (Italy) • Mileage • Million

MIL Machine Interface Layer • Magnetic Indicator Loop • Malfunction Investigation Laboratory • Man In the Loop • Master Incident List • Master Index List • Master Instrumentation List • Micro-Implementation Language • Milano (Italy) (location identifier) • Module Interconnection Language • One Million

MILA Merritt Island Launch Area (NASA)

MILAP Maintenance Information Logically Analyzed and Presented

MILC Modified Intermediate Load Cycle

MILES Magnetic Intrusion Line Sensor • Monotically Integrated Large Eddy Simulation

MILIC Microwave Insular Line Integrated Circuit

Mill Million

MILLIPOL Millimeter Polarimeter

MILO Magnetically Insulated Line Oscillator

MILP Mixed-Integer Linear Program • Mixed-Integer Linear Programming

MILS Microwave Instrument Landing System • Minicomputer Library System

Milw Milwaukee (USA-WI)

MIM Magnetic Interaction Mechanism • Maintenance Instructions Manual • Management Information Model • Maquette d'Interrogation Multilingue • Market Information Machine • Master in International Management • Merimbula (Australia-NSW) (location identifier) • Metal Injection Moulding • Metal - Insulator - Metal • Microgravity Isolation Mount • Minicomputer Interface Module • Minimum Information Method • Misappropriation - Interference - Misrepresentation • Mobile Instant Messaging • Modem Interface Module • Modified Index Method • Multiple Instrument Mount

MIMA Mighty Image Analyzer

MIMC Multiple Indicator - Multiple Cause

MIMD Multiple Instruction - Multiple Data

mime Multipurpose Internet Mail Extension (file extension)

MIME Media Internet Message Exchange • Multimedia Internet Mail Extension • Multipurpose Internet Mail Extension (*)

MIMEX Major Item Materiel Excess

MIMI Magnetospheric Imaging Instrument (Cassini) • Mesospheric Imaging Michelson Interferometer • Millimetric Ice Mapping Instrument

MIMIC Micro-Moulding In Capillaries • Microwave Monolithic Integrated Circuit • Millimeter-wave Monolithic Integrated Circuit

MIMIS Major Item Management Information System

MIMLT Magnetospheric, Ionospheric, Mesospheric and Lower Thermospheric

MIMO Man In - Machine Out • Mini Mosaic (imager) • Multiple Input - Multiple Output

MIMOSA Micro-Measurement Of Satellite Acceleration • Mission Modes and Space Analysis

MIMR Magnetic Ink Mark Recognition • Microwave Imaging Radiometer • Monthly Integration Management Review • Multiband Imaging Microwave Radiometer • Multifrequency Imaging Microwave Radiometer

MIMS Major Item Management System • Major Item Movement System • Material Information Management System • Medical Information Management System • Meteosat Infrared and Microwave Sounder • Multiitem Multisource

MIMSE Microwave Infrared Mesospheric Stratospheric Experiment

MIMU Miniature Inertial Measurement Unit

MiMue Mitteilungen des Astronomischen Instituts der Universität Münster (periodical code)

MiMun Mitteilungen der Sternwarte München (periodical code)

MIMUSA Matrix Iteration Method of Unfolding Spectra

Min Mineral • Minéral • Mineralogical • Mineralogy • Minéralogie • Minería • Minim • Minimum • Mining • Minister • Ministère • Ministerial • Ministerium • Ministre • Ministry • Minor • Minority • Minuscule • Minute

MIN Minimum • Minister • Minnipa (Australia-SA) (location identifier) • Minute • Mobile Identification Number • Mobile Identity Number • Multistage Interconnection Network

MINAC Miniature Navigation Airborne Computer

MINAR Multimedia Information Analysis and Retrieval (workshop)

MINC Minimum Curvature

MINCV Moscow Izdatel Nauka AN SSSR Chteniia imeni Vernadskogo (periodical code)

MIND Magnetic Integration Neutron Duplicator

MinDe Mineralium Deposita (periodical code)

MIndE Master of Industrial Engineering

MINDM Miniature Dynametric Model

MINEFI Ministère de l'Économie, des Finances et de l'Industrie (France)

Mineral Mineralogía • Mineralogical • Mineralogy

Minéral Minéralogie • Minéralogique

MINFI Moscow Izdatel Nauka AN SSR Fizicheskii Institut Trudy (periodical code)

MINGI Moscow Izdatel Nauka AN SSSR Geologicheskoi Institut Trudy (periodical code)

MINIS Multipurpose Interactive NASA Information System

Minist Ministerial • Ministry

MINK Missouri - Iowa - Nebraska - Kansas

MINLP Mixed-Integer Non-Linear Program • Mixed-Integer Non-Linear Programming

MinM Mineralogical Magazine (periodical code)

MINMI Moscow Izdatel Nauka AN SSSR Matematicheskii Institut Trudy (periodical code)

Minn Minnesota (USA)

MINOS Main Injector Neutrino Oscillation Search • Marks Information Optically Stored • Minimum Norm Solution

MINQE Minimum Invariant Quadratic Estimate

Min. Res. Minister Resident • Minister Residentiary

MINS Minor In Need of Supervision • Multisource Integrated Notification System

MINT Magnetic Information Technology • Management Information Network for Training • Managing the Integration of New Technology • Marée Interne • Micromagnetics and Information Technologies • Millimeter Interferometer • Millimeter-wave Interferometry • Mutual Interference • Mutual Interference Test

MINTERM Miniature Terminal

MINTF Moscow Izdatel Nauka Teoreticheskaia Fizika (periodical code)

Mintr Miniature

MINUET Minnesota Internet Users Essential Tool

MINVS Moscow Izdatel Nauka Vychislitel naia Seismologiia (periodical code)

MINW Master Interface Network

Min. Wt. Minimum Weight

MINX Multimedia Information Network Exchange

Mio Million • Miocene

MIO Maidanak International Observatory (Uzbekistan) • Maintained Interoperability • Management Information Office • Management Integration Office • Man In Office • Map Information Office • Meteoritic Impact Origin • Metric Information Office • Miami (USA-OK) (location identifier) • Mobile Ionospheric Observatory • Modular Input/Output • Multipurpose Input/Output

MIONP Microwave-Induced Optical Nuclear Polarization

MIOP Mission Instruction and Operation Plan • Multilateral Interoperability Program • Multiple Input-Output Processor • Multiplexing Input-Output Processor • Musée de l'Instrumentation Optique (France)

MIOS Multispectral Imaging Optical Sensor

MiOTL Microwave and Optical Technology Letters (periodical code)

MIP Machine Instruction Processor • Mainframe Internetting Project • Maintenance Improvement Program • Maintenance Index Page • Maison d'Information Professionnelle • Management Improvement Plan • Management Improvement Program • Management Incentive Program • Management Information Plan • Mandatory Inspection Point • Man Instrument Panel • Manual Input Processing • Manual Input Program • Mars In-situ Propellant (experiment) • Materiel Improvement Program • Materiel Improvement Project • Matrix Inversion Program • Maximum Intensity Pixel • Maximum Investment Plan • Mean Indicated Pressure • Measurement Integration Panel • Medium Interface Point • Mercury Intrusion Porosimetry • Message Input Processing • Message Interface Processor • Meteorological Improvement Program • Methods Improvement Program • Microwave-Improved Plasma • Microwave-Induced Plasma • Microwave Interference Protection • Minimum Impulse Pulse • Minimum Ionising Particle • Mission Implementation Plan • Mission Integration Panel • Mitspeh Ramon (Israel) (location identifier) • Mixed Integer Programming • Mobility Improvement Program • Model Implementation Plan • Model Improvement Plan • Model Improvement Program • Model Installation Program • Modification Instruction Package • Molecularly-Imprinted Polymer • Monthly Investment Plan • Most Important Person • Multiband Imaging Photometer • Multilateral Interoperability Program • Multiple Injection Program

MIPAS Management Information Planning and Accountancy Service • Michelson Interferometer for Passive Atmospheric Sounding • Michelson Interferometric Passive Atmospheric Sounder

MIPB Materiel Improvement Project Board

MIPC Management Information Processing Center • Metropolitan Information Processing Conference • Multimission Image Processing Laboratory

MIPE Modular Information Processing Equipment

MIPEA Multiple Item Progressive Electronic Auctions

MIPG Master Index Pulse Generator

MIPI Mission Interministérielle pour la Promotion Informatique (France)

MIPL Multimission Image Processing Laboratory (*) • Multimission Image Processing Library

MIPO Multiple-Item Purchase Order

MiPot Mitteilungen des Astrophysikalischen Observatoriums Potsdam (periodical code)

MIPP Main Injector Particle Production

MIPR Monthly Interim Progress Report

MIPRB Materiel Improvement Program Review Board

MIpS Mega Instructions per Second • Million Instructions per Second

MIPS Management Information Processing System • Map and Image Processing System • Master Integrated Program Schedule • Mega Instructions Per Second (*) • Metric Integrated Processing System • Microcomputer Image Processing System • Microwave Pulse Storage System • Million Instructions Per Second • Mini-Image Processing System • Minimum Integration Protection System • Mir Interface to Payload Systems • Mission and Information and Planning System • Mission and Information Planning System • Mission Information and Planning System (*) • Mission Information Planning System • Modified Integrated Program Summary • MSS Image Processing System • Multiband Imaging Photometer for SIRTF • Multiimage Processing System • Multimedia Information Presentation System • Multimedia Information Processing System • Multimission Image Processing System • Multipurpose Image Processing System

MiPul Mitteilungen der Nikolai-Hauptsternwarte zu Pulkowo (periodical code)

MIQ Machine Intelligence Quotient • Omaha Millard (USA-NE) (location identifier)

MIr Middle Irish

MIR Magnetic Ink Reader • Malfunction Investigation Report • Management Information Report • Management Information Repository • Management Information Retrieval • Market Investigation Report • Master Inventory Record • Maximum Information Rate • Memory Information Register • Memory Input Register • Merged Interaction Region • Method of Iterated Resolvents • Micro-Instruction Register • Microwave Imaging Radiometer • Middle-Infrared • Mid-Infrared (*) • Millimeter-wave Imaging Radiometer • Mirror • Mission Integration Review • Monastir (Tunisia) (location identifier) • Montgolfière Infrarouge • Multilingual Information Retrieval • Multimedia Information Retrieval • Multiple Internal Reflection • Multiple Isomorphous Replacement • Music Information Retrieval

MIRA Mars International Reference Atmosphere • Medium-wave Infrared Array • Mitaka Infrared Array (Japan) • Monterey Institute for Research in Astronomy (USA-CA) • Multifunctional Inertial Reference Assembly

MIRABU Mid-Infrared Array for Boston University

MIRAC Management Information Research Assistance Center • Mid-Infrared Array Camera

MIRACL Mid-Infrared Advanced Chemical Laser • Missouri Research Consortium of Libraries (USA-MO)

MIRACLE Magnetometer Ionospheric Radars All-sky Camera Large Experiment • Mid-Infrared Array Camera of Least Effort • Multidisciplinary Integrated Research Activities in Complex Laboratory Environments

MIRAGE Multimedia Information Retrieval and Gathering Environment

MIRAM Multiple Indexed Random Access Method

MIRAS Mid-Infrared Array Spectrometer • Mid-Infrared Array Spectropolarimeter • Mortgage Interest Relief At Source

MIRC Manufacturing Information Resource Center • Mosaic Infrared Camera

MIRCEN Microbial Resource Center

MIRD Medical Internal Radiation Dose • Mission Implementation Requirements Document

MIRDC Metals Industry Research and Development Center

MiRe Microelectronics Reliability (periodical code)

MIRESOL Measurements and Identifications of active Regions on the Solar Limb

MIRF Multiple Instantaneous Response File

MIRFAC Mathematics In Recognizable Form Automatically Compiled

MIRFI Mid-Infrared Fabry-Pérot Interferometer

MIRFS Multifunction Integrated Radio Frequency System

MIRI Mid-Infrared Imager • Mid-Infrared Instrument • Multichannel Infrared Interferometer

MIRIAD Multichannel Image Reconstruction, Image Analysis and Display (*) • Multichannel Image Reconstruction with Interactive Analysis and Display

MIRIS Michigan Inventory and Resource Information System • Modified Infrared Interferometer Spectrometer

MIRO Microwave Instrument for the Rosetta Orbiter

MIROS Microwave Remote Sensor for the Ocean Surface • Modulation Inducing Retrodirective Optical System

MIRP Manipulated Information Rate Processor • Minnesota Infrared Polarimeter

MIRPS Multiple-Information Retrieval by Parallel Selection

MIRR Materiel Inspection and Receiving Report

MIRROR Management Information Reporting and Review of Operational Resources System

MIRROS Modulation Inducing Reactive Retrodirective Optical System

MIRS Mail Item Retrieval System • Maintenance Information Retrieval System • Management Information and Reporting System • Manpower Information and Retrieval System • Merged Interaction Regions • Mid-Infrared Spectrometer • Miniaturized Imagery Receive System • MOTS Information Retrieval System • Multiple Internal Reflection Spectroscopy

MIRSE Multipurpose Imaging Radiometer Spectrometer Equipment

MIRSI Mid-Infrared Spectrometer and Imager

MIRSL Microwave Remote Sensing Laboratory

MIRSP Mobile Infrared Scene Projector

MIRST Multiple Infrared Scattered Light Recorder

MIRT Mission-Integrated Readiness Testing • Mission Integration and Readiness Testing • Mission Integration Readiness Test • Molecular Infrared Track

MIRTL Model of Information Relevant to Terminological Logic • Multimedia Information Retrieval Terminological Logic

MIRV Major Item Requisition Validation • Multiple Independently-targeted Reentry Vehicle • Multiple Independent Reentry Vehicle

Mis Mississippian

MIS Magnetic Isolation System • Maintenance Information System • Managed Internet Service • Management Information Science • Management Information Server • Management Information Service • Management Information Specialist • Management Information System • Management of Information Systems • Man In Space • Manufacturing Information System • Map Information System • Master Index Schedule • Master Integrated Schedule • Master of Information Systems • Master of Science in Information Systems • Mathematik Informationssystem • Matière Interstellaire • Message Input Segment • Metal-Insulated Semiconductor • Metal - Insulator - Semiconductor • Meteorological Impact Statement • Meteorological Information System • Metrology Information Service • Microcomputer Information Service • Microencapsulation In Space • Milieu Interstellaire • Misima Island (Papua New Guinea) (location identifier) • Mission Information System • MONS Information System • Moyens Informatiques et Services • Multimedia Information Source • Multimedia Information System • Multispectral Imaging Spectrometer • Multispectral Imaging System

MISAM Multiple Index Sequential Access Method

MISAR Miniature Information Storage and Retrieval

Misc Miscellaneous • Miscellany

MISC Management Information Services Committee • Micro Image Scanner • Mid-Infrared Steering Committee • Miscellaneous

MISCAP Mission Capability

MISE Mean Integrated Squared Error

MISEP Mutual Information System of Employment Policies

MISER Management Information System for Expenditure Reporting

MISETA Multiinstrumented Studies of Equatorial Thermospheric Aeronomy

MISCF Miscellaneous Frame

MISCHA Magnetic fields in Interplanetary Space during Comet Halley's Approach

MISD Multiple Instruction - Single Data

MISDC Multicharged Ion Spectra Data Center (Russia)

MISDE Mir Structural Dynamics Experiment

MISFET Metal-Insulator-Semiconductor Field-Effect Transistor

MISG Mathematics-in-Industry Study Group • Missing

MISHA Magnetic fields in Interplanetary Space during Halley's Approach

MISI Mathematics and Information Sciences for Industry • Modular Interactive Simulation

MISICC Management Information System Input to Command and Control

MISL Malfunction Investigation Support Laboratory • Management Information System Laboratory

MISM Metal-Insulator-Semiconductor Metal

MISMA Micro-Structured Magnetic Atmosphere • Micro Structures Magnetic Atmospheres (project)

MISMAC Matrix-Isolation Spectroscopy of Mass-selected Clusters

MISO Management Information Systems Office • Manufacturing Inspection Satellite Office • Multiple Input - Single Output

MISOE Management Information System for Occupational Education

MISolFA Moniteur d'Images Solaires Franco-Algérien

MiSon Mitteilungen der Sternwarte Sonneberg (periodical code)

MISP Management Information System Plan • Minnesota Infrared Spectropolarimeter

MISQ Management Information Systems Quarterly

MISr Muy Ilustre Señor

MISR Major Item Status Report • Mars ISRU Sample Return • Multiangle Imaging Spectroradiometer (*) • Multidirectional Imaging Spectroradiometer

MISREP Mission Report

MISRI Moscow Izdatel Sovietskoe Radio Ionosfernie Issledovaniia (periodical code)

Miss Mission • Missionary • Mississippi (USA)

MISS Management Information Subsystem • Man In Space Simulator • Man In Space Soonest • Manpower Information Systems Support • Mechanical Interruption Statistical Summary • Meteorological Interactive Software System • Minicomputer Interfacing Support System • Missing • Mobile Integrated Support System • Multiband Image Scanning System • Multiband Infrared Source Solution • Multichannel Image Solar Spectrograph • Multiitem Single Source

MISSI Multilevel Information System Security Initiative

MISSIL Management Information System Symbolic Interpretive Language

MisSp Missiles and Spacecraft (periodical code)

MISST Moscow Izdatel Sviaz Statisticheskaia Teoriia sviazi (periodical code)

mist Mistura

MiST Microgravity Science and Technology (periodical code)

MIST Magnetosphere, Ionosphere and Solar-Terrestrial (physics) • Maintenance International Standardization System

• Meteorological Information Self-briefing Terminal • Microburst and Severe Thunderstorm (program) • Mission de l'Information Scientifique et Technique • Modular Interoperable Surface Terminal • Mosaic Infrared Sensor Technology • Multiinput Standard Tape • Multispectral Image Storage Tape Format

MISTE Miniaturized Integrated Satellite Terminal Equipment

MISTER Model-based Intelligent Star Target Exploitation and Recognition

MISTr Matematicheskii Institut imeni Steklova Trudy (periodical code)

MISTR Management of Item Scheduled for Repair • Management of Item Subject to Repair

Mistrans Mistranslation

MISVE Management Information System for Vocational Education

MISWG Multinational Industrial Security Working Group

Mit Mitología • Mittwoch

MIT Management Information Tree • Massachusetts Institute of Technology (USA-MA) • Master Instruction Tape • Master of International Trade • Memorandum for Information Technology • Mercury Ion Transfer • Miles In Trail • Minimum Individual Training • Mission Integration Team • Module Information Table • Module Integration Test • Most Important Test • Most Intense Tornado • Schaffer - Kern County (USA-CA) (location identifier)

MITA Microcomputer Industry Trade Association • Modified Iterative Transform Algorithm

MitAe Max-Planck-Institut für Aeronomie - Mitteilungen (periodical code)

MitAG Mitteilungen der Astronomischen Gesellschaft (periodical code)

MITAS Martin Marietta Interactive Thermal Analysis System

MITASK Mission Tasking

MiTau Mitteilungen des Karl-Schwarzschild-Observatoriums Tautenburg (periodical code)

MITC Microfilm and Information Technology Center • Mortgage Investment Tax Credit

MITCH Mars Investigation of Total Climatological Hazards

MitE Miteigentum

MITE Magnetic Insulation Test Experiment • Master Instrumentation Timing Equipment • Megapixel Imaging Technology Experiment • Microelectronic Integrated Test Equipment • Microelectronics Test and Evaluation • Multiple Input Terminal Equipment

MITEC Machine Intelligent Technical Controller • Microwave Technology Centre

MITES Miniature Interferometer Terminal for Earth Survey • Minority Introduction to Engineering and Science • Model-driven Iterative Texture Segmentation

MITF Mailing Industry Task Force

Mitgl Mitglied

MITH Milli-Kelvin Thermostat

MITI Ministry of International Trade and Industry

MITL Magnetically-Insulated Transmission Line • Man In The Loop • Matsushita Information Technology Laboratory (USA-NJ)

MITLL Massachusetts Institute of Technology Lincoln Laboratory (USA-MA)

MITM Magnetosphere - Ionosphere - Thermosphere - Mesosphere • Magnetospheric - Ionospheric - Thermospheric - Mesospheric

MITN Michigan Interferometer Technology Network

MITO Millimeter and Infrared Testagrigia Observatory (Italy)

MITOC Multiple Intercommunications Technical Operations Communications

MITOL Machine-Independent Technology-Oriented Language

MITP Management - Integration - Transition - Plans • Master Intern Training Plan

MITR Monthly Inventory Transaction Report

MITRE Miniature Individual Transmitter-Receiver Equipment

MITREX Microwave Time and Ranging Experiment

MITS Maintenance Institutional Training System • Management Information and Text System • Microcomputer Interactive Test System • Microfiche Image Transmission System • Module Initialization, Test and Software

Mitt Mitte • Mitteilung

Mittelfr Mittelfranken (Germany)

Mittelholst Mittelholstein (Germany)

MITTS Minimalist Tutorial and Tools for SmallTalk

MITTSE Mobile Integrated Tracking Telescope System

Mittw Mittwochs

MiTue Mitteilungen des Astronomischen Instituts der Universität Tübingen (periodical code)

MitVS Mitteilungen über Veränderliche Sterne (periodical code)

MIU Maiduguri (Nigeria) (location identifier) • Malfunction Insertion Unit • Marycrest International University (USA-IA) • Message Interception Unit • Multiplex Interface Unit • Multistation Interface Unit

MIUR Ministero dell'Istruzione, dell'Università e della Ricerca (Italy)

MIUS Modular Integrated Video System • Modulated Integrated Video System

miv Met Ingang Van

MIV Main Instrumentation Van • Millville (USA-NJ) (location identifier) • Mobile Instrumentation Van • Motorisierter Individualverkehr

MIVIS Multispectral Infrared and Visible Imaging Spectrometer

MIVP Mission Interface Verification Plan

MIW Management Information on the Web • Marshalltown (USA-IA) (location identifier)

MiWie Mitteilungen der Universitäts-Sternwarte Wien (periodical code)

Mix Mixture

MIX Magnetic Ionization Experiment • Member Information Exchange • Metropolitan Information Exchange • Miriti (Colombia) (location identifier) • Modular Interface Extension

Mixt Mixture

MIY Mittiebah (Australia-NT) (location identifier)

MIZ Mainoru (Australia-NT) (location identifier) • Management-Informationszentrum • Marginal Ice Zone

MIZEX Marginal Ice Zone Experiment

MIzLI Moscow Izdatel Legkaia Industriia (periodical code)

MIzMa Moscow Izdatel Mashinostroenie (periodical code)

MIzMe Moscow Izdatel Metallurgiia (periodical code)

MIzMU Moscow Izdatel Moskovskovo Universiteta Pt (periodical code)

MIzNe Moscow Izdatel Nedra (periodical code)

MIZNZ Moscow Izdatel Znanie Novoe Zhizni Nauke Tekhnike Seriia Fizika (periodical code)

MIzRS Moscow Izdatel Radio Sviaz (periodical code)

MIzSR Moscow Izdatel Sovietskoe Radio (periodical code)

MIzVS Moscow Izdatel Vysshaia Shkola (periodical code)

mJ Millijoule

MJ Mari • Master of Journalism • Mechanical Joint • Megajoule • Modular Jack

MJA Manja (Madagascar) (location identifier)

MJB Mejit Island (Marshall Islands) (location identifier)

MJC Maison des Jeunes et de la Culture • Man (Ivory Coast) (location identifier)

MJCA Mid-body Jettison Control Assembly

MJCOM Major Command

MJD Management Job Description • Modified Julian Date • Modified Julian Day (*) • Mohenjodaro (Pakistan) (location identifier)

MJE Majkin (Marshall Islands) (location identifier)

MJEMS Modular Jet Engine Management Simulator

MJF Mosjøen (Norway) (location identifier)

MJG Mayajigua (Cuba) (location identifier)

MJH Majma (Saudi Arabia) (location identifier)

MJI Magneto-Jeans Injection • Maji (Ethiopia) (location identifier) • Monojet Injection

MJJ May - June - July • Moki (Papua New Guinea) (location identifier)

MJK Monkey Mia (Australia-WA) (location identifier)

MJL Mouila (Gabon) (location identifier)

MJM Mbuji-Mayi (Zaire) (location identifier)

MJN Majunga (Madagascar) (location identifier)

MJNPE Mobile Joint Nuclear Planning Element

MJO Madden-Julian Oscillation • Mariner Jupiter Orbit • Master Job Order • Mount Etjo Lodge (Namibia) (location identifier) • Mt. Etjo Lodge (Namibia) (location identifier)

MJP Manjimup (Australia-WA) (location identifier)

MJQ Jackson (USA-MN) (location identifier)

MJR Maintenance Job Request • Major Job Requirement • Management Job Review • Mariner Jupiter-Saturn • Miramar (Argentina-BA) (location identifier)

MJS Maganja da Costa (Mozambique) (location identifier) • Manipulator Jettison System • Mariner Jupiter-Saturn (spacecraft) • Member of the Japan Society

MJSD March - June - September - December

MJT Mytilene (Greece) (location identifier)

MJTS Maintenance Job Tracking System

MJU Mamuju (Indonesia) (location identifier) • Mariner Jupiter-Uranus • Multijunction Unit

MJUO Mount John University Observatory (New Zealand) • Mt. John University Observatory (New Zealand)

MJV Mercredi - Jeudi - Vendredi • Murcia (Spain) (location identifier)

MJW Move - Jump - Walk

MJWG Maintenance Joint Working Group

MJX Toms River - Robert J. Miller (USA-NJ) (location identifier)

mJy Millijansky

MJy Megajansky

MJY Mangunjaya (Indonesia) (location identifier)

MJZ Mirnyj (Russia) (location identifier)

mk Macedonian (ISO language code) • Mark (currency) • Markka (currency)

MK Macédoine (code ISO de pays) • Macedonia (ISO country code) • Malawi Kwatcha (currency) • Manual Keying • Mark • Markarian (catalog) • Mauna Kea (USA-HI) • Metaknowledge • Modification Kit • Morgan-Keenan

MKA Marianske Lazne (Czech Rep.) (location identifier)

MKAS Mauna Kea Astronomical Society (USA-HI) • Milton Keynes Astronomical Society (UK-England)

MKB Mekambo (Gabon) (location identifier) • Metaknowledge Base • Methods Knowledge Base

MKBEEM Multilingual Knowledge-Based European Electronic Market

MKC Kansas City (USA-MO) (location identifier)

Mkd Marked

MKD Chagni (Ethiopia) (location identifier) • Macédoine (code ISO de pays) • Macedonia (ISO country code) • Mehrkörperdynamik • Metekel (Ethiopia) (location identifier)

mkdir Make Directory

MKDIR Make Directory

MKE Milwaukee - General Mitchell (USA-WI) (location identifier) • Molecular Kinetic Energy

MKF Columbus McKenna (USA-GA) (location identifier)

MKG Muskegon (USA-MI) (location identifier)

MKH Mokhotlong (Lesotho) (location identifier)

MKI M'Boki (Central African Rep.) (location identifier)

MKIR Maina Kea Infrared (instrument)

MKJ Makoua (Congo) (location identifier)

MKK Hoolehua (USA-HI) (location identifier) • Morgan - Keenan - Kellman (atlas)

MKL Jackson (USA-TN) (location identifier)

MKM Mukah (Malaysia) (location identifier)

MKMP Modular Key Management Protocol

MKN Malekolon (Papua New Guinea) (location identifier)

MKO Mauna Kea Observatory (USA-HI) • Muskogee Davis (USA-OK) (location identifier)

MKP Makemo (Tuamotu Islands) (location identifier)

MKQ Merauke (Indonesia) (location identifier)

Mkr Marker

MKR Marker • Meekatharra (Australia-WA) (location identifier)

mks Marks • Metre - Kilogram - Second

MKS Manual Keying Station • Mehrkörpersysteme • Mekane Selam (Ethiopia) (location identifier) • Meter - Kilogram - Second • Microwave Keying Switch

MKSA Meter - Kilogram - Second - Ampere

MKSP Mount Kobau Star Party (Canada-BC)

MKSS Microwave Keying Switch Station

MKST Multidimensional k-ary Search Tree

Mkt Market • Markt

MKT Mankato (USA-MN) (location identifier) • Market

Mktg Marketing

MKTG Marketing

MKTI Mission Kit Technical Instruction

MKU Madurai Kamaraj University (India) • Makokou (Gabon) (location identifier)

MKV Master Key Vector • Mount Cavenagh (Australia-NT) (location identifier)

MKW Manokwari (Indonesia) (location identifier) • Morgan - Kayser - White (catalog)

MKX Mukalla (Yemen) (location identifier)

MKY MacKay (Australia-QLD) (location identifier)

MKZ Malacca (Malaysia) (location identifier)

ml Mail • Malayalam (ISO language code) • Milliliter

mL Millilambert

M/L Maintenance Loop • Manual/Local • Mass/Light • Mass/Luminosity

M&L Maintenance & Logistics

M-L Mass-Luminosity • Metallic-Longitudinal

ML Machine Language • Machine Learning • Macro Language • Magnetisches Loch • Magnitude Locale • Mail List • Main Lobe • Maintenance and Logistics • Maintenance Level • Major Lobe • Mali (ISO country code) • Management Level • Management List • Mass-Luminosity • Master Library • Master of Law • Master of Letters • Materiel Laboratory • Maximum Likelihood • Mean Level • Medieval Latin • Medium Level • Memory Load • Mesa Laboratory (NCAR) • Message Line • Metalanguage • Method of Limits • Microlensing • Microprogramming Language • Middeck Left • Middle Latin • Mill • Mini Laboratory • Mirroir Liquide • Mission Life • Mission Load • Mission Locale • Mixed Layer • Mobile Launcher • Modeling Library • Mold Line • Monolayer • Motor Launch • Multilayer • Multilingual • Multiple Lobe • Multiplexed Local

MLA Magnetic Lens Assembly • Mail List Agent • Malta (location identifier) • MDM Launch Aft • Meander Line Antenna • Medical Library Association • Member of Legislative Assembly • Microlens Array • Microprocessor Language Assembler • Microwave Linear Accelerator • Mission Load Allowance • Mixing-Length Approximation • Modern Language Association • Monochrome Lens Assembly • Multilevel Analysis • Multiline Addressing • Multilinear Array • Multiplex Line Adapter • Multispectral Linear Array

MLAM Mars Limited Area Model

MLAN Mobile Local Area Network

MLAP Multilingual Action Plan

MLAPI Multilingual Application Programming Interface

MLAT Multilevel Adaptive Technique

MLB Main Lobe Blanking • Mass-Loading Boundary • Melbourne (USA-FL) (location identifier) • Message Link Buffer • Metallic Link Belt • Multiband Layer • Multilayer Board

MLBR Medium Low Bit Rate

MLBSR Multilingual Bank Statistics Reporting

M&LC Mission & Launch Control

MLC Machine Level Control • Maison des Loisirs et de la Culture • Maneuver Load Control • Master Labor Contract • Maximum Likelihood Classification • McAlester (USA-OK) (location identifier) • Member of the Legislative Council • Memory Load Command • Micellar Liquid Chromatography • Microelectric Logic Circuit • Midland Lutheran College (USA-NE) • Minimum Level Cataloguing • Mission and Launch Control • Mobile Launcher Computer • Mobile Liaison Cell • MOL Launch Complex • Multilayer Ceramic • Multilayer Circuit • Multileaf Collimator • Multilens Camera • Multilevel Cell • Multiline Communications • Multiline Control • Multiline Controller

MLCB Multilayer Circuit Board

MLCC Multilook Cross-Correlation

MLCD Maximum-Likelihood Convolutional Decoder

MLCG Medical Logistics Control Group

MLCH Major Logistics Control Headquarters

MLCI Management Logistics Communications Interface • Management Logistics Configuration Item

MLCP Mission and Launch Control Process • Mission and Launch Control Processor • Mobile Land Command Post

MLCS Multicolor Lightcurve Shape (method)

Mld Mould • Moulded

MLD Malad City (USA-ID) (location identifier) • Maximum-Likelihood Detection • Median Lethal Dose • Message Loading Device • Method of Least Distortion • Mild • Minimal Lethal Dose • Minimum Lethal Dose • Mixed Layer Depth • Multicast Listener Discovery • Multilevel Decoding • Multiload Design

MLDAS Meteorological and Lighting Data Acquisition System

MLDC Museum Digital Licensing Collective

Mldg Moulding

MLDI Meter List Display Interval

MLDM Machine Learning and Data Mining (workshop)

MLDR Maximum Likelihood Decision Rule

MLDT Mean Logistics Delay Time

MLE Male (Maldives) (location identifier) • Manned Lunar Exploration • Maximum-Likelihood Estimate • Maximum-Likelihood Estimation • Maximum-Likelihood Estimator • Maximum Likely Earthquake • Mesoscale Lightning Experiment • Microprocessor Language Editor • Minimum Level of Effort • Multiline Editor

MLELV Medium-Light Expendable Launch Vehicle

MLEP Manned Lunar Exploration Program

MLES Multiple Line Encryption System

MLF Mechanical Luminosity Function • Milford (USA-UT) (location identifier) • Mobile Launcher Facility • MOL Launch Facility

MLFA Machine-Learned Fragment Analysis

MLFAT MOL Launch Facility Acceptance Team

MLFG Multiplicative Lagged-Bibonacci Generator

MLFM Multilayer Feedforward Network

MLFN Multiple-Layer Feedforward Network

Mlg Malagasy

MLG Main Landing Gear • Malang (Indonesia) (location identifier) • Maximum-Likelihood Gaussian • Middle Low German • Modular Logic Grammar

MLGD Main Landing Gear Door

MLGS Microwave Landing Guidance System

MLH Bâle Mulhouse (Switzerland) (location identifier) • Basel Mulhouse (Switzerland) (location identifier) • Basle Mulhouse (Switzerland) (location identifier) • Medium-Lift Helicopter • Mulhouse Bâle (France) (location identifier) • Mulhouse Basel (France) (location identifier) • Mulhouse Basle (France) (location identifier) • My Left Hand

MLHR Maximum-Likelihood Ratio • Maximum-Likelihood Reduction

MLI Machine Language Instruction • Machine Learning and Inference • Magnetic Level Indicator • Mali (ISO country code) • Master Listing Index • Matrix Lightcurve Inversion • Mid-Life Improvement • Mission Locale pour l'Insertion • Mixing-Length Theory • Moline (USA-IL) (location identifier) • Multilayered Insulation • Multilayered Insulator • Multilayer Insulation • Multilayer Insulator • Multiple Link Interface

MLIA Multilingual Information Access

MLib Master of Library Science

MLibSc Master of Library Science

MLID Monitor Location Identifier • Multilink Interface Driver • Multiple Link Interface Driver

MLIR Maximum-Likelihood Image Restoration • Monolingual Information Retrieval • Multilingual Information Retrieval

M&LIS Mapping & Land Information Service

MLIS Molecular Laser Isotope Separation • Multilevel Information System • Multilingual Information Society • Multiple-Level Indexing Scheme

MLISP Meta List Processor • Meta List Processor • Ministry of Land, Infrastructure and Transport (Japan)

MLJ Machine Learning Journal • Milledgeville - Baldwin County (USA-GA) (location identifier)

MLK Malta (USA-MT) (location identifier) • Maschinenlochkarte

MLL Manned Lunar Landing • Marshall (USA-AK) (location identifier) • Mars Logistics Lander • Maximum Lobe Length • Multilayer Logic

MLLA Modified Leading Logarithm Approximation

Mlle Mademoiselle

Mlles Mademoiselles

MLLP Manned Lunar Landing Program

MLM Mailing List Manager • Manufacturing - Logistics - Management • Maximum-Likelihood Method • Mid-Level Manager • Mixed Lambda Model • Mixed Level Matrix • Morelia (Mexico) (location identifier) • Multilayer Metallization

MLN Melilla (Spain) (location identifier) • Message Ledger Number • Metropolitan Library Network

MLO Manned Lunar Orbiter • Master Layout • Master Local Oscillator • Mauna Loa Observatory (USA-HI) • Mechanized Letters Office • Milos (Greece) (location identifier) • Model Lighting Ordinance • Mount Laguna Observatory (USA-CA)

MLOC Million Lines Of Code

MLOCR Multiline Optical Character Reader

MLOF Mount Lemmon Observing Facility (USA-AZ) • Mt. Lemmon Observing Facility (USA-AZ)

MLOPEX Mauna Loa Observatory Photochemistry Experiment

MLOWG Model Lighting Ordinance Working Group (IDA)

MLP Machine Language Program • Malabang (Philippines) (location identifier) • Master Logistics Plan • Maximum-Likelihood Poisson • Maximum-Likelihood Programme • Message Load Plan • Microsoft Licence Pack • Mid-Level Processor • Mirror Link Protocol • Mobile Launcher Platform • Mobile Launch Platform (*) • Modern Laplacian Theory • Moon Literature Project • Multilayer Perception • Multilayer Perceptron (*) • Multilayer Protocol • Multilevel Parallelism • Multilevel Precedence • Multilevel Process • Multilevel Programmer • Multilink Procedure • Multiple Line Printer

MLPC Multilayer Printed Circuit

MLPCB Machine Language Printed Circuit Board • Multilayer Printed Circuit Board

MLPCO Mechanized Letters and Parcels Concentration Office

MLPF Miniature Low Pass Filter

MLPP Multilevel Precedence and Preemption • Multilevel Priority and Preemption

MLPPP Multilink Point-to-Point Protocol

MLPRF Modular Low Power Radio Frequency

MLPS Multilink Processing System

MLQ Malalaua (Papua New Guinea) (location identifier)

M/LR Maintenance Loop Recorder

MLR Magnetic Line Ratio • Maintenance Loop Recorder • Mass - Luminosity - Radius • Mass-Luminosity Relation • Medium Lift Replacement • Memory Lockout Register • Millicent (Australia-SA) (location identifier) • Minimum Lending Rate • Monodisperse Latex Reactor • Monotone Likelihood Ratio • Multilevel Relational • Multiple Linear Regression

MLRA Major Land Resource Area

MLRD Mass - Luminosity - Radius - Distance

MLRF Minilaser Rangefinder

MLRO Matera Laser Ranging Observatory (Italy)

MLRS McDonald Laser Ranging Station (USA-AZ) (*) • McDonald Laser Ranging System • Multilingual Reference System • Multiple-Launch Rocket System

MLRT Multilevel Regression Testing

MLS Machine Literature Searching • Manned Lunar Surface • Master Line State • Master of Library Science • Master of Library Service • Microwave Landing System • Microwave Limb Scanner • Microwave Limb Sounder • Mid-Latitude Summer • Miles City (USA-MT) (location identifier) • Mills • Mixed Language System • MOL Launch Site • Molonglo Quasar Sample • Multilanguage System • Multilevel Secure • Multilevel Security • Multilevel Staging • Multiline Selection • Multimedia Learning Station • Multiple Listings Service

MLSA Mid-Level Service Area

MLSc Master of Library Science

MLSC Measured Logistics Support Cost

MLSE Mechanical Launch Support Equipment

MLSG Mobile Logistics Support Group

MLSO Mauna Loa Solar Observatory (USA-HI)

MLSP Multiple-Link Satellite Program

MLSPH Moving Least Squares Particle Hydrodynamics

MLSR Maximum-Length Shift Register

MLSS Multilevel Secure Server

MLSWG Multilevel Security Working Group

MLT Machine Language Translation • Machine Learning Toolbox • Magnetic Local Time • Malta (ISO country code) • Malte (code ISO de pays) • Manned Lunar Test • Manufacturing Lead Time • Master Library Tape • Mean Length of Turn • Mean Life Time • Mean Logistical Time • Mechanized Loop Testing • Média de Longo Tempo • Mesosphere - Lower Thermosphere • Millinocket (USA-ME) (location identifier) • Mixing-Length Theory • Mobile Launch Tower • Mobile Launch Transport • Modulated Lapped Transform • Multilevel Transmission • Multiple Logical Terminal

MLTA Multiple Line Terminal Adapter

MLTCS Mesosphere Lower-Thermosphere Coupling Study

MLTI Mesosphere-Lower Thermosphere-Ionosphere

MLU Mean Length of Utterance • Memory Lock-Up • Memory Logic Unit • Mid-Life Update • Mid-Life Upgrade • Module Lighting Unit • Monitor and Logic Unit • Monroe (USA-LA) (location identifier) • Multiple Logical Unit • Multiplicity Logic Unit

MLV Medium Launch Vehicle • Medium-Lift Vehicle • Memory-Loader Verifier • Merluna (Australia-QLD) (location identifier)

MLVP Manned Lunar Vehicle Program

MLVQ Modified Learning Vector Quantization

MLW Maximum Landing Weight • Mean Low Water • Mid-Latitude Winter • Monrovia Sprigg-Payne (Liberia) (location identifier)

MLWFA Multilingual Weather Forecast Assistant System

MLX Malatya (Turkey) (location identifier) • Multilingual Extended

MLY Manley Hot Springs (USA-AK) (location identifier)

MLZ Melo (Uruguay) (location identifier)

m/m Même Mois • Moi-Même

mm Mass Mailer • Millia • Millimeter • Millimetric • Moi-Même • Mutatis Mutandis

mM Millimolare • Millimole

M/M Maximum/Minimum • Multimission

M&M Materials & Maintenance

MM Magnetic Monopole • Magnitude Method • Main Memory • Main Module • Maintenance Manual • Maintenance Mission • Major Mode • Managed Mail • Management Manual • Management Meeting • Man-Month • Manned Module • Manual Morse (code) • Manufacturing Manual • Manufacturing Method • Mariner Mark • Mass Memory • Master Monitor • Material Management • Material Manager • Material Manual • Materials Measurement • Materiel Management • Materiel Manager • Mathematical Model • Mathematical Morphology • Math Model •

Mechanics and Mathematics • Memory Module • Memory Multiplexer • Mesoscale Model • Messier Marathon • Messieurs • Metadata Management • Metadata Manager • Method of Moment • Methodology Management • Methodology Manager • Michelson-Morley (experiment) • Mickey Mouse • Middle Marker • Midmean • Millimeter • Minor Merger • Minutes • Mirror Model • Mirror Module • Mission Management • Mission Manager • Mixed Morphology • Mobility Management • Modèle Mathématique • Modification Management • Modification Manager • Modified Mercalli • Molecular Magnet • Monatsmiete • Month • Motor Module • Mount Model • Multimedia • Mutatis Mutandis • Myanmar, (ISO country code) • Two Thousand

MMA Main Mission Antenna • Malmö Metropolitan (Sweden) (location identifier) • Martin Marietta Astronautics • Masse Maximum Autorisée • Mathematical Modeling and Analysis (conference) • Memory-to-Memory Adapter • Metropolitan Museum of Art • Michigan Muon Array (USA-MI) • Microcomputer Managers Association • Microgravity Measurement Assembly • Microgravity Measuring Assembly • Micro Mirror Array • Millimeter Array (USA) (now merged into ALMA) • Ministerio del Medio Ambiente • Moving Mechanical Assembly • Multimedia Assistant • Multimeter Array • Multimission Antenna • Multiple-Module Access • Multiplexed Matrix Array

MMAC Materiel Management Aggregation Code • Multimedia Mathematics Across the Curriculum

MMACE Millimeter-wave Advanced Computer Environment

MMACS Maintenance Management And Control System

MMACTEE Mathematical Methods And Computational Techniques in Electrical Engineering (conference)

MMAD Mass Median Aerodynamic Diameter

MMAOC Millimeter Array Oversight Committee

MMAP Multimedia Access Profile

MmARB Mémoires de l'Académie Royale de Belgique (periodical code)

MMART Multimass Adaptive Refinement Tree

MMAS Martin Marietta Astro Space • Material Management Accountability System • Material Management Accounting System • Material Management and Accounting System • Material Management and Accounting System • Mathematical Methods in the Applied Sciences (periodical code) • Mechanized Material Accountability System

MmASI Memoirs of the Astronomical Society of India (periodical code)

MMath Master of Mathematics

MMAU Master Multiattribute Utility

MMB Maintenance Management Branch • Mast Mounting Bracket • Memanbetsu (Japan) (location identifier) • Middle Mouse Button • Mirror Maintenance Building • Mixed-Mode Bending

MmBBA Memoirs of the British Astronomical Association (periodical code)

MMBF Mean Miles Between Failures

MMBH Medium-Mass Black Hole • Modest-Mass Black Hole

MMBUMA Mean Miles Between Unscheduled Maintenance Actions

MMC Ciudad Mante (Mexico) (location identifier) • Magnesium Methyl Carbonate • Maintenance Monitoring Console • Man-Machine Communication • Manpower Management Computer • Manual Mode Command • Martin Marietta Corporation • Master Microprocessor Control • Master Mission Computer • Matched Memory Cycle • Materiel Management Center • Materiel Management Code • Maximum Metal Condition • Medical Management Center • Meet-Me Call • Meet-Me Conference • Memory Management Controller • Metal Matrix Composite • Microcomputer Marketing Council • Microsoft Management Console • Midcourse Measurement Correction • Midmotor Controller • Minicomputer Maintenance Center • Mission Management Center • Mission Management Computer • Mobile Management Control • Mobile Mission Coordinator • Mobile-to-Mobile Call • Modular Mission Computer • Monopolies and Mergers Commission • Movements Monitoring Center • Multimedia Card • Multimedia Command

MMCA Mid-body Motor Control Assembly • Midmotor Controller Assembly

MMCBT Multimedia Computer-Based Training

MMCC Mission Management and Control Centre • Mission Management Control Centre • Modèle de Markov Caché Continu • Multimedia Conference Control

MMCCB Multimission Change Control Board

MMCD Multimedia Compact Disk

MMCF Multimedia Communications Forum

MMCIAC Metal Matrix Composite Information Analysis Center

MMCIP Media Manager Control Interface Protocol

MMCL Master Measurement and Control List

MMCLP Multimode Communications Link Processor

MMCM Mars Mesoscale Circulation Model

MMCN Multimedia Computing and Networking (conference)

MMCR Millimeter Cloud Radar

MMCS Multimedia Communications System • Multimedia Computer System • Multiple-Mission Command System

MMCX Multimedia Communications Exchange

MMCXS Multimedia Communications Exchange Server

MMD Masse Maximum au Décollage • Mass Median Diameter • Master Monitor Display • Material, Maintenance and Distribution • Materiel Management Decision • Materiel Management Directorate • Mean Mass Density • Mean Mission Duration • Memory Module Directory • Minami Daito (Japan) (location identifier) • Minimization of Maximum Deviation • Minimum Metric Distance • Mixed-Mode Disk • Mobile Maintenance Depot • Modified Metropolis Dynamics • Multimode Digital • Multimode Display

MMDA Money Market Deposit Account

MMDB Main-Memory Database • Mass Memory Database • Master Measurement Database • Mirror Metrology Database

MMDC Master Message Display Console

MMDD Matra Marconi - DASA - Dornier

MMDF Maintenance Master Data File • Mission Model Data File • Multichannel Memorandum Distribution Facility

MMDI Monthly Mean Percentage Increase

MMDR Multimode Digital Radio

MMDS Maintenance Management Data System • Multichannel Microwave Distribution System • Multichannel Multipoint Distribution Service • Multichannel Multipoint Distribution System • Multimegabit Data Service • Multipoint Microwave Distribution System • Multipoint Multichannel Distribution System • Multipoint Multimedia Distribution System

Mme Madame

MME Manufacturing Machines and Equipment • Master Mode Entry • Master of Mechanical Engineering • Master of Mining Engineering • Master Multiplexer Encoder • Michaelis-Menten Enzyme • Micrometeoritic Erosion • Microsoft Multimedia Extension • Mobile Management Entity • Mobile Meßeinheit • Multimedia Extension • Tees-Side (UK-England) (location identifier)

MMEB Martin Marietta Enhanced Black (coating)

MMED Mariner Mars Spectral Editor

MMEL Master Minimum Equipment List

Mmes Mesdames

MMES Martin Marietta Energy Systems

mmf Microsoft Mail (file extension)

MMF Magnetomotive Force • Mamfe (Cameroon) (location identifier) • Master Message File • Mean Magnetic Field • Missing Minor Frame • Mission Management Facility • Mobile Magnetic Field • Mobile Maintenance Facility • Module Maintenance Facility • Moving Magnetic Feature • Multimedia Forum • Multimode Fiber • Multimode Fiberoptic (cable)

MMFMP Matematicheskie Metody i Fiziko Mekhanicheskie Polia (periodical code)

MMFS Manufacturing Message Format Standard • Mean Magnetic Field of the Sun

MMG Motor-Motor Generator • Mount Magnet (Australia-WA) (location identifier)

MMgtE Master of Management Engineering

MMH Maintenance Man-Hour • Mammoth Lakes (USA-CA) (location identifier) • Monomethyl Hydrazine

mmHg Millimeter of Mercury

MMHH Mijne Heren

MMI Athens - McMinn County (USA-TN) (location identifier) • Maintenance Micro-Informatique • Management and Maintenance Inspection • Man-Machine Interface • Manpower Management Information • Micro-Mechanical Inertial • Multimode Interference • Mutual Mortgage Insurance • 2001

MMIA Methods and Models for Integrated Assessment • Monolithic Microwave Integrated Amplifier

MMIC Microwave Monolithic Integrated Circuit • Millimeter-wave Monolithic Integrated Circuit • Monolithic Microwave Integrated Circuit

MMICS Maintenance Management Information and Control System • Maintenance Management Information and Control System

MMIDS Man-Machine Interface Development Station

MMII Mariner Mark II

MMIL Multimodal Interface Language

MMIMU Micromechanical Inertial Measuring Unit

MMIO Multimedia Input/Output

MMIP Mass Memory Integration Plan • Media Manager Interchange Protocol

MMIPS Man/Machine Interactive Processing System

MMIR Multimedia Information Retrieval

M-MIS Multimissions Integrating System

MMIS Maintenance Management Information System • Materials Manager Information System • Materiel Manager Information System • Minimum Mean-square Image Spread • Multimedia Information System • Multimissions Integrating System • Multimissions Integration System • Multimodel Information System

MMIT Man-Machine Integration Technique

MMITG Man-Machine Interaction Transaction Group

MMITS Modular Multifunction Information Transfer System • Modular Multifunction Information Transmission System • Modular Multifunction Information Transport System

MMJ Mardi - Mercredi - Jeudi • Matsumoto (Japan) (location identifier) • Modified Modular Jack • Modular Molded Jack

MMK Kyat (Myanmar) (ISO currency code) • Mensch - Maschine - Kommunikation • Murmansk (Russia) (location identifier)

MmKMO Kakioka Magnetic Observatory Memoirs (periodical code)

MmKyo Memoirs of the College of Sciences of the University of Kyoto (periodical code)

MML Man-Machine Language • Marker Mark Language • Marshall (USA-MN) (location identifier) • Master Materials List • Master Measurement List • Multimedia Laboratory

MMLE Modified Maximum Likelihood Estimate

MMLP Microsoft Multilicenses Pack

MMLQ Multiple Maximum Likelihood Quantization • Multiple Model Linear Quadratic

MMLS Martin Marietta Launch Services • Mobile Microwave Landing System • Model - Modes - Loads - Stresses

MMM Magnetospheric Mapping Mission • Magnetospheric Multiscale Mission • Maintenance and Materiel Management • Maintenance Management Manual • Manned Mars Mission • Mars Mesoscale Model • Mars Mission Module • Mask Manufacturing Machine • Mathematical Methods and Models (workshop) • Meaningful Measures of Merit • Measuring Monitoring Module • Medical Materiel Manager • Mesoscale and Microscale Meteorology • Middlemount (Australia-QLD) (location identifier) • Minimum - Maximum - Mean • Mobile Media Mode • Mode Management Module • Multihead Multitail Mosaic • Multimission Module

MMMC Medical Materiel Management Center • Modified Modular Mission Computer

MMMIS Maintenance and Material Management Information System

MMMOS Mobile Micrometeorological Observation System

MMMS Maintenance and Material Management System

MmMtS Memoirs of the Mount Stromlo Observatory (periodical code)

MMMW Monolithic Microwave Millimeter Wave

MMN Mismatch Negativity • Stow - Minute Man (USA-MA) (location identifier)

MMNA Mathematical Modeling and Numerical Analysis (periodical code)

MmNap Memorie del Osservatorio di Capodimonte in Napoli (periodical code)

MMNS Management of Multimedia Networks and Services (conference)

MMO Main Meteorological Office • Maintenance Management Order • Maio (Cape Verde) (location identifier) • Maria Mitchell Observatory (USA-MA) • Mercury Magnetospheric Orbiter • Middelbaar Middenstandonderwijs • Mission Management Organization

MMOC Minicomputer Maintenance Operations Center

MMOD Micromodule

MMOG Monmouth Mobile Observer's Group (USA-NJ)

mmol Millimole

MMOP Modeling Measures Of Performance

MMOS Multimode Optical Sensor

MMP Maintenance Management Plan • Maintenance Management Program • Maintenance Monitoring Panel • Maintenance Monitor Panel • Managed Mail Program • Media Management Protocol • Microphysical Measurement Package • Millennium Mathematics Project • Millimeter-wave Polarimeter • Mission Management Plan • Mixture Melting Point • Modular Mission Payload • Momentum Management Program • Mompos (Colombia) (location identifier) • Monthly Maintenance Plan • Multiple Multipole Program • Multiplexed Message Processor • Multiplex Message Processor

MMPA Marine Mammal Protection Act (USA)

MMPI Minnesota Multiphasic Personality Inventory

MMPM Multimedia Presentation Manager

MMPO Mission Management and Planning Office

MMPP Markov Modulated Poisson Process

MMPS Multimicroprocessing System

MMPSE Multiuse Mission Payload Support Equipment

MMPT Man-Machine Partnership Translation

MMPU Memory Manager and Protect Unit

MMQ Mbala (Zambia) (location identifier) • Minimum Manufacturing Quantity

MMR Austin Maybry (USA-TX) (location identifier) • Magnetic Memory Record • Maintenance Management Report • Maintenance Management Review • Mass-Magnitude Relationship • Mass Memory Ready • Maximal Marginal Relevance • McMath-Pierce Telescope (KPNO) • Mean Motion Resonance • Milton Multiband Radiometer • Minimum Mandatory Requirement • Minimum Marginal Return • Minor Merger Remnants • Modified Modified Read • Modular Multiband Radiometer • Monthly Management Review • Multiband Multimode Radio • Multimode Radar • Multimode Radiometer • Multimode Receiver • Myanmar (ISO country code)

MMRALS Moveable Microwave Remote Area Landing System

MmRAS Memoirs of the Royal Astronomical Society (periodical code)

MMRB Materiel Release Review Board

MMRC Mars Mission Research Center • Mean Monthly Relative Citedness

MMRD Maximum Magnitude versus Rate of Decline

MMRI Mandatory Mission Rule Item • Multimode Radar Input

MMRS Mission Maintenance and Reliability Simulation • Multimikrorechnersystem • Multinational Meeting on Remote Sensing

Mms Manuscritos

MMS Magnetospheric Multiscale (mission) • Mail Management System • Main Memory Status • Maintenance Mode System • Maintenance Monitoring System • Malaysian Meteorological Service • Manifest Mailing System • Man-Machine System • Manned Maneuvering System • Manpower Management System • Manufacturing Message Specification • Manufacturing Message Standard • Mardi Mercredi Samedi • Marine Meteorological Service • Maritime Mobile Service • Marks Selfs (USA-MS) (location identifier) • Mars Mai Septembre • Massive Meteor Stream • Mass Memory Store • Mass Memory System • Mast-Mounted Sight • Materiel Management System • Mathematical Modeling of Systems • Matin Midi Soir • Matra Marconi Space • Mauritius Meteorological Services • Media Management System • Media Messaging Services • Memory Management System • Mes Meilleurs Souvenirs • Meteorological Measurement System • Meteorological Measuring Set • Meteorological Measuring System • Michigan Multispectral Scanner • Microfiche Management System • Micrometeorological Measurement System • Microsoft Metadirectory Services • Middelbare Meisjesschool • Millimeter Source • Minerals Management Service • Mission Modular Spacecraft • Mission Module System • Modem Management System • Modern Men of Science (book) • Modification Management System • Modified Monitoring State • Modular Measurement System • Modular Multiband Scanner • Modular Multispectral Scanner • Modulation Measuring System • Module Management System • Monolithic Miniature Spectrometer • Movement Monitoring System • Moving Media Statistics • Multimedia Messaging Service • Multimedia Survey • Multimedia System • Multimission Modular Spacecraft • Multimission Spacecraft • Multimission Spectrum • Multimission System • Multimode Seeker • Multiplex Modulation System • Multiscale Modeling and Simulation

MmSAI Memorie della Società Astronomica Italiana (periodical code)

MMSc Master of Medical Science

MMSC Micro Miniature Signal Conditioner

MMSD Madison Metropolitan School District (USA-WI) • Metric Mapping Support Data

MMSE Minimum Mean Squared Error • Minimum Mean Square Error • Minimum Mean Square Estimation • Mission Module Simulation Equipment • Multimission Support Equipment • Multiuse Mission Support Equipment

MMSI MMS Interface

MMSL Microgravity Materials Science Laboratory

MMSP Mount Magazine Star Party (USA-AR)

MMSR Multimission Support Recording

MMSS Manned Maneuverable Space System • Manual Mode Space Simulator • Marine Meteorological Services System • Maritime Mobile Satellite Service • Martin Marietta Space Systems • Mast Mounted Sight System • Material Management Standard System • Multimode Space Station

MMST Microelectronics Manufacturing Science and Technology

MMSW Multimedia Software

MM&T Manufacturing Methods & Technology

MMT Columbia McEntire (USA-SC) (location identifier) • Machine à Mesurer Tridimensionnelle • Main Mantle Thrust • Manufacturing Methods and Technology • Mass Memory Test • Master Message Terminal • Materiel Management Team • Methylcyclopentadienyl Manganese Tricarbonyl • Miniature Moving Target • Mission Management Team • Monolithic Mirror Telescope • Monthly Mean Temperature • Multimedia Terminal • Multiple-Mirror Telescope (USA-AZ) • Multiple-Mission Telemetry • Multiresolution Median Transform

MMTET Matematicheskoe Modelirovanie i Teoriia Elektricheskikh Tsepei (periodical code)

MMTF Maryland-Magellan Tunable Filter

MMTO Multiple-Mirror Telescope Observatory (USA-AZ)

MMTPS Maintenance Materiel Transaction Processing System

MM&TS Master Manufacturing & Test Schedule

MMTS Master Manufacturing and Test Schedule • Maximum-Minimum Temperature System • Multiple-Mission Telemetry System

MMTT Multipurpose Multiple Telescope Testbed

MMU Main Memory Unit • Manchester Metropolitan University (UK-England) • Manned Maneuvering Unit • Mask Manufacturing Unit • Mass Memory Unit • Memory Management Unit • Memory Mass Unit • Midcourse Maneuvering Unit • Midcourse Measurement Unit • Million Monetary Units • Mobile Meteorological Unit • Mobile Monitoring Unit • Modular Maneuvering Unit • Morristown (USA-NJ) (location identifier)

MMUAV Multimission Unmanned Air Vehicle

MMUI Multimachine User Interface • Multimedia User Interface

MMUS Mass Memory Unit Simulator

mmv Met Medewerking Van

MMV Mal (Papua New Guinea) (location identifier) • Mobile Maintenance Van • Mobile Maintenance Vehicle • Multimedia Video

MMVP Model - Map - View - Praxis

MMW Major Midwinter Warming • Millimeter Wave • Millimeter Wavelength • Moma (Mozambique) (location identifier) • Multimega Watt

MMWR Millimeter Wave Radar • Millimeter Wave Radio

MMX Malmö Sturup (Sweden) (location identifier) • Matrix Manipulation Extension • Memory Multiplexer • Multimedia Extension

MMY Miyako Jima (Japan) (location identifier)

MMZ Maimana (Afghanistan) (location identifier)

mm3 Cubic Millimeter

m/n Moneda Nacional

mn Met Name • Moneda Nacional • Mongolian (ISO language code) • Mutato Nomine

Mn Main • Mangan • Manganese • Manganèse

MN Magnetic North • Main • Management Network • Mande • Manitoba (Canada) • Mantle Nerve • Materiel Need • Météorologie Nationale • Midnight • Mine • Minnesota (USA) • Mobile Node • Moneda Nacional • Mongolia (ISO country code) • Mongolie (code ISO de pays) • Month • Monthly Notices

MNA Melangguane (Indonesia) (location identifier) • Member of National Assembly • Methylnitroaniline • Mission Needs Analysis • Modèle Numérique d'Altitude

MNAS Member of the National Academy of Sciences

MNASSA Monthly Notes of the Astronomical Society of Southern Africa

MNB Moanda (Zaire) (location identifier)

MNBA Mono-Normal-Butylamine

MNBDF Metanitrobenzenediazonium

MNBLE Modified Nearly Best Linear Estimator

MNC Minimum Necessary Change • Mobile Network Code • Multinational Centre • Multinational Company • Multinational Corporation • Nacala (Mozambique) (location identifier)

MnCAR Minnesota Center for Automotive Research (USA-MN)

MNCIS Management Numerical Control Information System

MNCP Network Management Communication Protocol

MNCPC Mission Nationale de Contrôle des Précurseurs Chimiques (France)

Mncpl Municipal

MNCS Master Net Control Station • Multipoint Network Control System

MND Medina (Colombia) (location identifier) • Mission Need Determination • Mission Needs Document • Multinational Division

MNDP Multinational Data Processing

MNDV Monocular Night Vision Device

MNE Master of Nuclear Engineering • Mnemonic • Modèle Numérique d'Élévation • Multinational Enterprise • Mungeranie (Australia-SA) (location identifier)

MNEE Mission Non-Essential Equipment

MNEF Mutuelle Nationale des Étudiants de France

MNF Mana Island (Fiji) (location identifier) • Minimum Noise Fraction

Mnfrm Mainframe

MNFS Minor Frame Synchronization

Mng Managing

MNG Maningrida (Australia-NT) (location identifier) • Microwave Negative Grid • Mongolia (ISO country code) • Mongolie (code ISO de pays) • Multinet Gateway • Multiple Network Graphics

MNGE Mission Nationale de la Gestion des Échanges (France)

Mngmnt Management

Mngmt Management

Mngr Manager • Monsignor

MNH Minneriya (Sri Lanka) (location identifier)

MNHN Muséum National d'Histoire Naturelle

MNI Micro-Mechanical Inertial • Minimum Number of Individuals • Montserrat (Leeward Islands) (location identifier)

MNIR Mid and Near Infrared

MNISM Mongolian National Institute for Standardisation and Metrology

MNJ Manajary (Madagascar) (location identifier) • Midnight Job

MNK Maiana (Kiribati) (location identifier)

Mnl Manila (Philippines) • Manual

MNL Manila - Ninoy Aquino (Philippines) (location identifier)

MNLD Mainland

MNM Menominee (USA-MI) (location identifier) • Minimum

MNN Marion (USA-OH) (location identifier)

MNO Manono (Zaire) (location identifier)

MNODE Major Node

MNOS Metal Nitride Oxide Semiconductor

MNP Îles Mariannes du Nord (code ISO de pays) • Magnetic North Pole • Mariana Islands (Northern_) (ISO country code) • Maron (Papua New Guinea) (location identifier) • Master Navigation Plan • Microcom Networking Protocol • Microcom Network Protocol (*) • Microsoft License Pack • Mobile Number Portability • Mobile Number Portability • Northern Mariana Islands (ISO country code)

MNPS Minimum Navigation Performance Specification

MNPSA Minimum Navigation Performance Specifications Airspace

MNPSD Mean Normal Point System Delay

MNQ Monto (Australia-QLD) (location identifier)

Mnr Manor • Mijnheer

MNR Manor • Ministry of Natural Resources • Mongu (Zambia) (location identifier)

MNRA Ministerio de Recursos Naturales y del Ambiente (Nicaragua)

MNRAS Monthly Notices of the Royal Astronomical Society (periodical code)

Mnrl Mineral

MNRS Manors

Mns Moins

MNS Management Network Services • Mansa (Zambia) (location identifier) • Mean Number of States • Mercury Neutron Spectrometer • Meta-Nitride Semiconductor • Minutes • Mirror Neurons System • Mission Needs Statement

MnSCU Minnesota State Colleges and Universities (USA-MN)

MNSF Median Normalized Spatial Filter

MNSSA Monthly Notes of the Astronomical Society of South Africa (periodical code)

MNSSP Multilevel Network System Security Program

Mnstb Monostable

MNT Minto (USA-AK) (location identifier) • Modèle Numérique de Terrain • Modern Network Theory • Mononitrotoluene • Mount • Tugrik (Mongolia) (ISO currency code)

MNTB Medial Nucleus of the Trapezoidal Body • Multinode Testbed

MNTF Multinational Task Force

MNTN Mountain

MNTNS Mountains

Mntrl Montréal (Canada-QC)

MNTY Maintainability

MNu Mare Nubium (Moon)

MNU Mathematischer und Naturwissenschaftler Unterricht • Maulmyine (Myanmar) (location identifier) • Menu • Moulmein (Myanmar) (location identifier)

MNV Mountain Valley (Australia-NT) (location identifier)

MNVD Monocular Night Vision Device

MNVR Maneuver

MNW MacDonald Downs (Australia-NT) (location identifier)

MNX Manicoré (Brazil-AM) (location identifier)

MNY Mono (Solomon Islands) (location identifier)

MNZ Manassas (USA-VA) (location identifier)

m/o Mi Orden

mo Mail Order • Modus Operandi • Moldavian (ISO language code)

Mo Mégaoctet • Missouri (USA) • Molybdän • Molybdène • Molybdenum • Moment • Monday • Montags • Month • Mouth

M/O Meteorological/Oceanographic

M&O Machinery & Optics • Maintenance & Operations • Maintenance & Overhaul • Management & Operating • Management & Organization • Manpower & Organization • Maps & Overlays

MO Macau (ISO country code) • Machine Operation • Machine-Outil • Maître Ouvrier • Managing Organization • Modus Operandi • Magneto-Optic • Magneto-Optical • Mail Order • Maintenance Officer • Maintien de l'Ordre • Major Objective • Managed Object • Management Office • Manila Observatory (Philippines) • Manned Orbiter • Manner of Operating • Manpower Office • Manually Operated • Manual Orientation • Manual Output • Manufacturing Order • Maps and Overlays • Mark Off • Mars Observer • Mars Orbiter • Massive Object • Mass Observation • Master Oscillator • McKee-Ostriker (model) • Medical Officer • Member Organization • Memory Operation • Meteorological Observatory • Meteorological Office • Meteorological Officer • Meteorology Officer • Method of Operation • Middeck Overhead • Middelbaar Onderwijs • Miscellaneous Operation • Mission of Opportunity • Mission Operations • Mission Order • Mission Oriented • Missouri (USA) • Mobile-Originated • Model Orientation • Mode of Operation • Modus Operandi • Molecular Orbital • Mondays Only • Money Order • Mongolia • Monitor Output • Month • Mordvinian • Motor Operated • Move • Movement Order • Multicriteria Optimization • Municipal Officer

MoA Memorandum of Agreement

MOA MACHO Observations in Astrophysics • Made On Assembly • Make On Arrival • Matrix Output Amplifier • Medium Observation Aircraft • Memorandum Of Agreement • Microlensing Observations in Astrophysics (collaboration) • Minute Of Angle • Moa (Cuba) (location identifier) • Module d'Orientation Approfondie

MOAD Mother Of All Demos

MOAS Museum Of Arts and Sciences

MOAT Mechanical/Optical Alignment Test

Mob Mobile

MOB Main Operating Base • Main Operations Buffer • Make Or Buy • Medisch-Opvoedkundig Bureau • Memory-Order Buffer • Mobile • Mobile (USA-AL) (location identifier) • Mobility • Memory Order Buffer • Pascagoula (USA-AL) (location identifier)

MOBDP Mission Onboard Data Processor

MOBIS Management-Oriented Budget Information System • Mobile Information Station

MOBL Macro-Oriented Business Language

MOBLAS Mobile Laser System

MOBO Machinenorientierte Bibliotheksorganisation • Motherboard

MOBS Multiple Observation Sequence

MoC Memorandum of Consultations

MO&C Mission Operations & Control

MOC Magnetic Optic Converter • Maintenance Operational Check • Maintenance Operations Center • Managed-Object Class • Management and Operating Contractor • Management Of Change • Manual Operations Control • Marine Operation Center • Mars Observer Camera • Mars Orbiter Camera (*) • Master Operational Controller • Master Operations Center • Master Operations Console • Master Operations Control • Mathematical Operations Computer • Memorandum Of Conditions • Memorandum Of Cooperation • Memory Operating Characteristic • Meteorological Operational Capability • Method Of Characteristics • Microfilm Output Computer • Minimum Operating Capability • Mission Operations and Control • Mission Operations Center • Mission Operations Complex • Mission Operations Computer • Mission Operations Contractor • Mobile-Originated Call • Modular Operations Center • Montes Claros (Brazil-MG) (location identifier) • Monthly Operating Cost

MoCA Movie Content Analysis

MOCA Minimum Obstacle Clearance Altitude • Minimum Obstruction Clearance Altitude • Module d'Orientation Collective Approfondie • Museum Of Contemporary Arts

MOCAM Mosaic Camera

MOCC Meteosat Operations Control Center • Miniature Operations Control Center • Mission Operations Control Center • Molecular-Orbital Close-Coupling (method)

MOCCA Mobile Operations Center for Communications and Analysis

MOCCT Method of Characteristics-Constrained Transport

MOCD Magneto-Optical Compact Disk • Mission Operations Concept Document

MOCF Mission Operations Computational Facility

MOCIC Molecular Orbital Constraint of Interaction Coordinates

MOCN Machine-Outil à Commande Numérique

MOCOCO Movement Coordinating Committee

MOCP Mission Operational Computer Program

MOCR Mean Observed Citation Rate • Mission Operations Control Room

MOCS Managed Objects Conformance Statement • Master Operations Control System • Multichannel Ocean Color Sensor • Multiorder Citation Score

MOCSW Monitor and Operations Control Software

MOCV Manual Oxygen Control Valve

MOCVD Metallo-Organic Chemical Vapor Deposition (*) • Metal Organic Chemical Vapor Deposition

Mod Model • Moderate • Moderator • Modern • Modernization • Modernize • Moderno • Modification • Modified • Modifier • Modify • Modulate • Modulated • Modulation • Modulator • Module • Modulo • Modulus

MoD Ministry of Defense

MOD Magnetic Optical Display • Magneto-Optical Disk • Magneto-Optical Drive • Mail-Order Delivery • Mail-Order Department • Management Of Data • Masters Of Deception • Masters Of Destruction • Microfilm-Output Device • Microwave Oscillating Diode • Milliseconds of Day • Ministry of Defence • Miscellaneous Obligation Document • Mission Objectives Document • Mission Operating Directive • Mission Operations Department • Mission Operations Director • Mission Operations Directorate • Mission Operations Division • Model • Modem • Moderate • Modernization • Modesto (USA-CA) (location identifier) • Modification • Modified • Modify • Modular • Modulation • Modulator • Module • Modulus • Movies On Demand

MO&DA Mission Operations & Data Analysis

MODA Magneto-Optical Doppler Analyzer • Mission Operations and Data Analysis • Motion Detector and Alarm

MOD/AO Modular Adaptive Optics

MODAP Multiple-Operational Data Acquisition Program

MODAPS Modal Data Acquisition And Processing System

MODAPTS Modular Arrangement of Predetermined Time Standards

MODAS Maintenance and Operational Data Access System • Mobile Open Data Access Standard • Modular Data Acquisition System • Modular Ocean Data Assimilation System

MODB Mission Operational Database • Mission Operations Database

MODCA Mixed Object Document Content Architecture

MODCPS Multiple-Output Direct-Current Power Supply

mod. dict. Modo Dicto

MODE Middeck 0-Gravity Dynamics Experiment • Mid-Ocean Dynamics Experiment

MODEM Modulate/Demodulate • Modulation/Demodulation • Modulator/Demodulator

MODEST Modeling Dense Stellar (systems) (project)

MODFET Modulation-Doped Field-Effect Transistor

ModGe Modern Geology (periodical code)

ModGk Modern Greek

Modif Modification

MODIL Manufacturing Operations, Development and Integration Laboratory

MODIS Moderate-resolution Imaging Spectrometer • Moderate-resolution Imaging Spectroradiometer (*)

MODM Magneto-Optical Display Memory • Multiobjective Decision Making

MODOK Modulare Dokumentation

mod. praes. Modo Praescripto

MODS Major Operation Data System • Management Operating Data System • Manned Orbital Development Station • Manned Orbital Development System • Mission Operations Data System • Modifications • Modular-Oriented Direct Support • Multiobject Double Spectrograph

MO&DSD Mission Operations & Data Systems Division

MODSD Mission Operations and Data Systems Division

MODTRAN Moderate-resolution Transmission

MoE Measure of Effectiveness

MOE Major Organization Entity • Mars Orbit Ejection • Measure Of Effectiveness • Ministry Of Education • Ministry Of the Environment • Mission Operations Engineer • Mission-Oriented Equipment • Modèle d'Objets d'Entreprise • Momeik (Myanmar) (location identifier)

MOEA Multiobjective Evolutionary Algorithm

MoEBA Moscow Energoatomizdat Biblioteka Automatike (periodical code)

MOEBIUS Mobile Experimental Broadcast Interconnection Using Satellites

MoEIT Moskovskii Energeticheskii Institut Trudy (periodical code)

MOEM Micro-Opto-Electro-Mechanical

MOEMS Micro-Opto-Electro-Mechanical System

MOEP Meteorological and Oceanographic Equipment Program

MOES Minimal Optical Echelle Spectrometer • Multiorder Etalon Sounder

MOF Magneto-Optical Filter • Managed Object Format • Manned Orbital Flight • Mass Octave Function • Maumere (Indonesia) (location identifier) • Maximum Observed Frequency • Maximum Operating Frequency • Metal-Oxide Film • Metaobject Facility • Method Of Frames • Michoud Operations Facility (NASA) • Mini-Orange Filter • Mission Operations Facility

MOFA Mole Fraction of A

MOFAT Multiphase Flow And Transport

MOFDN Method Of Frames Denoising

MOFE Memorandum Of Final Evaluation

MOFFIT Moon Free-Flyer Interferometer Trade-off

MoFi Mondfinsternis

MOFS Multiobject Fibre Spectrograph

MOFT Maneuver-Oriented Flight Training

MOFWG Magneto-Optical Filter Working Group

MOG Micro-Ordinateur de Guichet • Minicomputer Operations Group • Mong Hsat (Myanmar) (location identifier) • Monthly Orbit Geometry

MOGA Microwave and Optical Generation and Amplification

MOGAS Motor Gasoline

MoGid Moscow Gidrometeoizdat (periodical code)

MOGR Moderate or Greater

MOGS Mission Operation Ground System

MOGUNTIA Model of the Global Universal Tracer In the Atmosphere

moh Meter ober Hav

MOH Maximum Operating Hours • Medal Of Honor • Meteorologisches Observatorium Hamburg (Germany) • Mohanbari (India) (location identifier) • Music On Hold

MOHAWC Model Of Human Actions in Work Context

Mohd Mohammad • Mohammed

MOHLL Machine-Oriented High-Level Language

MOHp Meteorologisches Observatorium Hohenpeissenberg (Germany)

MOI Magneto-Optical Imaging • Maintenance Operating Instruction • Mars Orbital Insertion • Mars Orbit Insertion • Mars Orbiter Insertion • Maximum Obtainable Irradiance • Memorandum Of Instruction • Mercury Orbital Insertion • Mercury Orbit Insertion • Method Of Instruction • Ministry Of Information • Ministry Of Interior • Ministry Of the Interior • Mission-Oriented Item • Mitiaro Island (Cook Islands) (location identifier) • Moment Of Inertia • Moon Orbital Insertion • Moon Orbit Insertion

MOIA Mission-Oriented Item Activity

MOID Minimum Orbital Intersection Distance

MoIKI Moscow Institut Kosmicheskikh Issledovanii AN (periodical code)

MoINA Moscow Izdatel Nauka AN SSSR Fizicheskii Institut Trudy (periodical code)

MoINI Moscow Izdatel Nauka Issledovaniia Geomagnetizmu Aeronomii i Fizike Solntsa (periodical code)

MoIPM Moscow Institut Prikladnoi Matematiki AN SSSR (periodical code)

MOIRCS Multiobject Infrared Camera and Spectrograph

MOIS Mission Operations Intercommunication System

MOISE Mission-Oriented Information System Engineering

MoIzA Moscow Izdatel Atomizdat (periodical code)

MoIzE Moscow Izdatel Energiia (periodical code)

MoIzK Moscow Izdatel Khimiia (periodical code)

MoIzM Moscow Izdatel Moskovskogo Universiteta (periodical code)

MoIzN Moscow Izdatel Nauka (periodical code)

MoIzS Moscow Izdatel Sviaz (periodical code)

MoIzT Moscow Izdatel Transport (periodical code)

MoIxV Moscow Izdatel VINITI (periodical code)

MoIzZ Moscow Izdatel Znanie (periodical code)

MOJ Moengo (Suriname) (location identifier)

MOK Mankono (Ivory Coast) (location identifier)

MOKA Mitaka-Okayama-Kiso Archive

MOKE Magneto-Optic Kerr Effect

mol Mole

Mol Molecular • Molecule • Molokai (USA-HI)

MOL Machine-Oriented Language • Mac-On-Linux • Mailing Online • Mail-Order Library • Manned Orbital Laboratory • Manned Orbiting Laboratory • Maximum Operating Level • Maximum Order Limitation • Maximum Output Level • Metallo-Organic Laser • Meteorologisches Observatorium Lindenberg (Germany) • Method Of Lines • Microsoft Open License • Minimum Operating Level • Ministry Of Labour • Model Lighting Ordinance • Molde (Norway) (location identifier) • Molecular

MOLA Mars Observer Laser Altimeter • Mars Orbiter Laser Altimeter (*) • Multiplicative Optimally Localized Average

MOLAP Multidimensional Online Analytical Processing

Mold Moldavia • Moldavian

MOLD Model Of Light Diode

MOLDS Management Online Data System • Multiple Online Debugging System

Moldv Moldavia • Moldavian

MOLE Molecular Optics Laser Examiner • Multichannel Online Evaluator • Multichannel Operational Line Evaluator

Molec Molecular • Molecule

Molek Molekulare

MOLI Microsoft Online Institute

MOLL Metallo-Organic Liquid Laser

MOLLE Modular Lightweight Load Equipment

MoLLP Moshchnye Lazery i Lazernaia Plazma (periodical code)

MolPh Molecular Physics (periodical code)

MOLS Magnetic-Operated Limit Switch • Mutually Orthogonal Latin Squares

MOLTOL Manned Orbiting Laboratory Test-Oriented Language

mol. wt. Molecular Weight

Mom Moment • Momentary • Momentum

MoM Method of Moments

MOM Maintenance Operations Management • Manager Of Managers • Manned Orbiting Mission • Measure Of Merit • Message-Oriented Middleware • Metal-Oxide Metal • Microsoft Office Management • Microsoft Office Manager (*) • Middle Of Mission • Middle Of Month • Minutes Of Meeting • Mission Océanographique de la Méditerranée • Mission Operations Manager • Modular Ocean Model • Moment • Monthly Orbit Milestones • Moudjeria (Mauritania) (location identifier)

MOMA Message-Oriented Middleware Association • Museum Of Modern Art

MOMBE Metal Organic Molecular Beam Epitaxy • Metal Oxide Molecular Beam Epitaxy

MOMF Multiobject Multifilter

MOMFOS Multiobject Multifiber Optical Spectrograph

MoMGK Moscow Mezhduvedomstvennyi Geofizicheskii Komitet (periodical code)

MOMI Manchester Occulting Mask Imager • Museum Of Moving Images (USA-CA)

MOMO MOS Management and Operation

MOMS Map, Operator and Maintenance Station • Mobile Optical Measurement System • Modular Optical Multispectral Scanner • Modular Optoelectronic Multispectral Scanner • Monospectral Optical Monitoring System • Multiple Orbit - Multiple Satellite • Multispectral Optical Monitoring System

MoMTN Moscow Mezhdunarodnyi Tsentr Nauchnoi i Tekhnicheskoi Informatsii (periodical code)

MOMV Manned Orbital Maneuvering Vehicle

Mon Maison • Monaco • Monasterio • Monastery • Monat • Monatlich • Monday • Monetary • Mongolian • Monitor • Monmouthshire (UK-Wales) • Monoceros-Monocerotis (constellation) • Monsignor • Monsoon • Montag • Month • Monthly • Monument

MON Memorandum Of Notification • Mixed Oxides of Nitrogen • Monday • Monitor • Monocerotids (meteor stream) • Motor Octane Number • Mount Cook (New Zealand) (location identifier) • Mt. Cook (New Zealand) (location identifier)

MONA Manager Of Network Activity

MONA LISA Modeling Natural Images for Synthesis and Animation

MonAP Communications du Département d'Astrophysique de la Faculté des Sciences de Mons (periodical code)

MOND Modified Newtonian Dynamics

MONET Meeting On the Network • Mobile Network • Monitoring Network of Telescopes • Multiwavelength Optical Networking

MONEX Monsoon Experiment

Mong Mongolia • Mongolian

Mongol Mongolia • Mongolian

MONICA Monitoring of Neutral and Ionized Atoms Clouds • Montréal Infrared Camera

MONISAR Monitoring with Synthetic Aperture Radar

Mono Monophonic

Monogr Monograph • Monographie

MONOLITH Massive Observatory for Neutrino Oscillations or Limits on Their existence

Mons Monsenhor • Monseñor • Monsieur

MONS Measuring Oscillations in Nearby Stars (project)

MONSEE Monitoring the Sun-Earth Environment

Mont Montags • Montana (USA) • Montería • Montgomeryshire (UK-Wales) • Monument

MONTBLEX Monsoon Through Boundary Layer Experiment

Montep Montepio

Montgom Montgomeryshire (UK-Wales)

Montr Montréal (Canada-QC)

MONTR Monitor

MOO Measure Of Outcome • Money Order Office • Moomba (Australia-SA) (location identifier) • MUD Object-Oriented • Multiuser Object-Oriented

MOODS Master Ocean Observation Data Set

MOONLIT Motif OpenLook Intrinsics Toolkit

MoonP Moon and Planets (periodical code)

MOOSE Man Orbital Operations Safety Equipment • Man Out Of Space Easiest

MOOSSE Manned Orbital Oceanographic Survey System Experiment

MOP Maintenance Operational Procedure • Maintenance Operations Protocol • Mandatory and Optional Parameters • Manned Orbital Platform • Manual Operations Panel • Manual Override Panel • Manuscript On Paper • Master Operations Plan • Matrix Operations Programming • Maximum Operating Pressure • Measure Of Performance • Memorandum Of Policy • Memorandum Of Procedures • Message Output Processing • Message Output Processor • Meta-Object Protocol • Meteorologisches Observatorium Potsdam (Germany) • Meteosat Operational Programme • Méthode de l'Onde de Pression • Microwave Observing Project • Ministerio de Obras Públicas (Spain) • Mission Océanographique du Pacifique • Mission Operations Plan • Mission Operations Processor •

Mobility Operating Procedure • Model Operational Plan • Modular Operating Procedure • Modulation On Pulse • Monthly Obligation Plan • Monthly Observation Plan • Monthly Observing Plan • Monthly Operating Plan • Mount Pleasant (USA-MI) (location identifier) • Multiple Online Programming • Pataca (Macau) (ISO currency code)

MOPA Master Oscillator Power Amplifier

MOPAR My Only Problems Are Repairs

MOPC Mission Operations Planning Center

MOPCA May Our Paths Cross Again

MOPE Multiple Object Parameter Estimation

MOPI Maximum Rate Output Initiator

MOPIC Motion Picture

MOPITT Measurements Of Pollution In The Troposphere

MOPL Machine-Oriented Program Language

MOPP Mission-Oriented Protective Posture

MOPR Manner Of Performing Rating • Mission Operations Planning Review • Mission Operations Planning Room

MOPS Machine-Oriented Programming System • Man-Operated Propulsion System • Maneuver Operations Program System • Marine Oil Pollution System • McDonald Observatory Planetary Search • Mega Operations Per Second • Message Output Processing System • Million Operations Per Second • Minimum Operational Performance Specifications • Minimum Operational Performance Standards • Minimum Operating Performance Standard • Mission Operations • Mission Operations Planning System

MOPSY Multiprogramming Operating System

MOpt Master of Optometry

MOPTMA Ministerio de Obras Públicas, Transportes y Media Ambiente (Spain)

MOPU Ministerio de Obras Públicas y Urbanismo (Spain)

MOPWG Meteosat Operational Programme Working Group

MOQ Minimum Order Quantity • Morondava (Madagascar) (location identifier)

m. or. Mismo Origen

Mor Moroccan • Morocco

MOR Magneto-Optical Rotation • Management Operating Ratio • Manufacturing Operations Record • Manufacturing Order Record • Mars Orbital Rendezvous • Memorandum Of Record • Memory Output Register • Middle Of the Road • Mission Objectives and Requirements • Mission Operations Report • Mission Operations Review • Mission Operations Room • Modulus Of Rupture • Monthly Operating Report • Morelia (Mexico) • Morristown - Moore-Murrell (USA-TN) (location identifier)

MORA Minimum Off Route Altitude

MORACC Modeling Radiation, Clouds, and Convection

Morad Moradia

MORAN Management of Resources, Accesses and Networks

MORANE Mobile Radio for Railway Networks in Europe

MORAS Mobile Radiation Station

MORB Mid-Ocean Rich Basalt • Mid-Ocean Ridge Basalt (*) • Mid-Ocean Rigid Basalt

MORD Miscellaneous Obligation Reimbursement Document • Mission Operations Requirements Document

mor. dict. More Dicto

MORDS Manned Orbital Research and Development System

MORE MARC Optical Recognition • Multimedia Object Retrieval Environment • Multimedia-Oriented Repository Environment

MORECS Meteorological Office Rainfall and Evapotranspiration Calculation System

MORENet Missouri Research and Education Network

MORF Manned Orbital Research Facility

Morg Morgado

MorGI Morskie Gidrofizicheskie Issledovaniia (periodical code)

MORIS Moderate Resolution Imaging Spectrometer

MORL Manned Orbital Research Laboratory • Manned Orbiting Research Laboratory

Morn Morning

MORO Moon Orbiting Observatory

MORP Meteorite Observation and Recovery Programme • Meteorite Observation and Recovery Project

Morph Morphological • Morphology

Morphol Morphological • Morphology

MORPHS Minicomputer-Operated Retrieval Partially Heuristic System

MORSE Metal Organic Research for Semiconductor Epitaxy

mor. sol. More Solito

Mort Mortal • Mortality • Mortgage • Mortuary

MORT Management Oversight and Risk Tree • Ministry Of Research and Technology

MORTE Multispectral Open Air Test Environment

Mos Months • Mosaic • Moscow (Russia) • Moselle

MoS Ministry of Supply

MÖS Mössbauer Spectrometer

MOS Machinenorientierte Sprache • Machinenorientiertes System • Magneto-Optical Storage • Magneto-Optical System • Maintenance Operating System • Major Operating System • Management Operating System • Manned Orbital Station • Manual Override Switch • Manufacturing Operating System • Mapping - Ordering - Searching • Marine Observation Satellite • Maritime Observation Satellite • Mass Online Storage • Master Operating Schedule • Mean Opinion Score • Measure Of Success • Measure Of Suitability • Media Operating Software • Memorandum Of Support • Metal-Oxide Semiconductor • Metal - Oxide - Silicon • Method Of Support • Minimum Operating Strip • Mission Operations Strategy • Mission Operations System • Mobile Object System • Model Output Statistics • Modular Operating System • Modular Opto-electronic Scanner • Modular Opto-electronic System • Monitor-Only Station • Monolithic Oxide Silicon • Moses Point (USA-AK) (location identifier) • Mössbauer Spectrometer • Multibeam Optical Sensor • Multiobject Setting • Multiobject Slit • Multiobject Spectrograph • Multiobject Spectrometer • Multiobject Spectroscopy • Multiple Object System • Multiprogramming Operating System • Multiple Optical Switch • Multispectral Opto-electronical Scanner • Multiuser Operating System

MOSAIC Macro Operation Symbolic Assembler and Information Compiler • MDM-Ohio State Active Infrared Camera • Metal Oxide Semiconductor Advanced Integrated Circuit • Meteosat Operational System for Acquisition and Interchange • Model-Oriented Software Automatic Interface Converter • Modular Open System Architecture for Industrial Control • Mosaicking of Satellite (images) using Automated Image Correlation • Multiuser Online System for Automated Information Communication

MOSAICS Multiobservation Satellite Image Correction System

MOSAIQUE Modélisation des Systèmes Aéroportés par Informatique

MOSAR Modulation Scan Array Radar

MOSART Moderate Spectral Atmospheric Radiance and Transmittance

MOSC Management Operations Support Center • Manned Orbital Systems Concept • Mission Operations Support Center • Mixed Open Systems Consortium

MOSCA Multiobjekt Spectrograph für den Calar Alto

MOSES Massive Open System Environment Standard • Mission Operations and Systems Engineering Software (*) • Mission Operations and Systems Engineering Support • Missions Operations Support Engineering System • Molecules in Outer Space and the Earth's Stratosphere • Multiorder Solar EUV Spectrograph

MOSFET Metal-Oxide Semiconductor Field-Effect Transistor

MOSHED Multiplanar Organic Scintillator High-Energy Detector

MoSI Museum of Science and Industry

MOSIE Menu of Opportunities for Scientists in Education (project)

MOSIS MOS Implementation System

MosIZ Moscow IZMIRAN (periodical code)

MOSLM Magneto-Optic Spatial Light Modulator

MOSM Metal-Oxide Semi-Metal • Mission Operations System Manager

MOSN Magnetic and Optical Storage on Network

MOSO Mission Operations Support Office • Multimission Operations Systems Office

MOSP Mission Operations Support Program • Multimission Optronic Stabilized Payload

MOSPF Multicast Open Shortest Path First

MOSPO Mobile Satellite Photometric Observatory

MOSR Mission and Observation Scheduling Room • Mission Operations Support Requirement

MOSS Maintenance and Operator Subsystem • Maintenance Operations Support Set • Manned Orbital Space Station • Map Overlay and Statistical System • Map Overlay and Statistics System • Meteorological Observing System for Ships • MIME Object Security Service • Miniaturized Optimized Smart Sensor • Mission Operations Support Services • Modular Star Sensor • Monitor Output Signal Strength • Moving Object Scheduling System • Moving Object Support System

MOSST Ministry of State for Science and Technology

MOST Management Of Social Transformations (UNESCO) • Management Operation System Technique • Manned Operational System Test • Manned Orbital Solar Telescope • Metal-Oxide-Semiconductor Transistor • Metal-Oxide-Silicon Transistor • Microvariability and Oscillations of Stars • Microwave Observation of Stratosphere and Troposphere • Ministry Of Science and Technology • Mission-Oriented Simulator Training • Mission-Oriented System Tape • Mobile Station • Molonglo Observatory Synthesis Telescope (Australia-NSW) • Multimodal Occupation Strategy • Multiple Operating System Tool • Museum Of Science Technology

MOSTAC Mosaic Sensor Test And Calibration

MOSTE Ministry Of Science, Technology and Environment (Malaysia)

MOSTL Metal-Oxide Semiconductor Transistor Logic

MOSTT Mosaic Optical Sensor Technology Testbed

MosVI Moscow Voennoe Izdatel (periodical code)

MosVo Moscow Voenizdat (periodical code)

Mot Motor • Motorized

MOT Managed Object Toolkit • Management Of Technology • Magneto-Optical Trap • Main Optical Telescope • Manned Orbital Telescope • Maturity Operational Test • Maximum Operating Time • Maximum Overhaul Time • Mean Operating Time • Means Of Test • Mechanical Operability Test • Member Of the Tribe • Method Of Test • Ministry of Transport • Minot (USA-ND) (location identifier) • Mission Operations Team • Molecular-Orbital Theory • Month Of Travel • Motor • Motor Operating Time

MOTA Materiel Open Test Assembly

MOTAR Modular Thermal Analyzer Routine

MOTAS Mission-Oriented Terminal Area Simulator

MOTD Message Of The Day

MOT&E Multiservice Operational Test & Evaluation

MOTE Multiservice Operational Test and Evaluation

MOTESS Moving Object and Transient Event Search System

MOTF Manganese-Oxide Thin Film

MOTHS Mirror Over-The-Horizon Simulation

MOTI Management Of Technical Innovation

MOTIF Maui Optical Tracking and Identification Facility (USA-HI) • Message-Oriented Transfer Integrated File

MOTIS Message-Oriented Text Interchange System

MOTIVE Modular Timing Verifier • Motion Video Engine

MOTNE Meteorological Operational Telecommunications Network Europe

MOTOR Mobile Oriented Triangulation Of Reentry

MOTR Multiple-Object Tracking Radar

MOTS Minitrack Optical Tracking Station • Minitrack Optical Tracking System • Mobile Optical Tracking System • Module Test Set

MOTSC Maintenance Overhaul and Technical Service Center

MOTU Mobile Optical Tracking Unit

MOTV Manned Orbital Transfer Vehicle

MotZe Motortechnische Zeitschrift (periodical code)

MoU Memorandum of Understanding

MOU Matching Optics Unit • Memorandum Of Understanding • Mountain Village (USA-AK) (location identifier)

MO&UCB Multilateral Operations & Utilization Control Board

MOUCB Multilateral Operations and Utilization Control Board

MOUDI Micro-Orifice Uniform Deposit Impactor

MOUS Maîtrise d'Oeuvre Urbaine et Sociale • Microsoft Office User Specialist

MOUSE Minimum Orbital Unmanned Satellite of the Earth

MOUT Message Out

mov Movie (file extension)

Mov Movement • Movimiento

MOV Main Oxidizer Valve • Manned Orbiting Vehicle • Materiel Obligation Validation • Materiel Order Validation • Measure Of Vulnerability • Metal-Oxide Varistor • Moranbah (Australia-QLD) (location identifier) • Motor-Operated Valve • Move • Movement • Moving

MOVE Manipulator Operation using Value Encoding • Microsoft Overlay Virtual Environment

MOVG Moving

MoVIN Moscow VINITI (periodical code)

MoVNT Moscow Vychislitel nyi Tsentr AN SSSR (periodical code)

MOVPE Metallo-Organic Vapor Phase Epitaxy • Metal Organic Vapor Phase Epitaxy

MOVS Move String

Movt Movement

MoVyS Moscow Vysshaia Shkola (periodical code)

MOW Maintenance Orderwire • Ministerie van Openbare Werken • Mission Operations Wing • Moscow (Russia) (location identifier)

MOWG Management and Operations Working Group • Management Operations Working Group • Mission Operations Working Group

MOWS Make Only What's Sold

MOX Mars Oxidant Experiment • Mixed Oxide • Morris (USA-MN) (location identifier)

MOXE Monitoring X-ray Experiment

MOY Monterrey (Colombia) (location identifier)

MOYRDA Month - Year - Day

moz Mozilla (file extension)

Moz Mozambique

MOZ Mittlere Ortszeit • Moorea (Society Islands) (location identifier) • Mozambique (code ISO de pays)

Mozár Mozárabe

mp Modo Praescripto

M/P Main Parachute • Mass Properties • Memorandum of Partnership • Monitor/Programmer

M&P Materials & Processes • Moon and Planets (periodical code)

MP Îles Mariannes du Nord (code ISO de pays) • Madhya Pradesh (India) • Magnetic Particle • Magnetic Point • Magnetic Pressure • Magnetopause • Main Phase • Main Program • Main Pulse • Mains Propres • Maintainability Plan • Maintenance Panel • Maintenance Period • Maintenance Plan • Maintenance Planning • Maintenance Point • Maintenance Print • Maintenance Procedure • Maintenance Program • Maître Principal • Major Program • Maladie Professionnelle • Management Package • Management Plan • Mandat-Poste • Maneuver Program • Manifold Pressure • Manpack • Manpower • Manpower and Personnel • Manpower Plan • Manufacturing Plan • Manufacturing Process • Maquette de Pilotage • Mariana Islands (Northern_) (ISO country code) • Maritime Patrol • Massively Parallel • Massive Penetration • Mass Properties • Master Pattern • Master Plan • Matching Pursuit • Materiel Professional • Mathematical Programming • Mathematics-Physics • Matsya Purāna • Maximum Principle • Measuring Point • Medium Pressure • Meeting Point • Melting Point • Melting Pot • Member of Parliament • Membre du Parlement • Mercator's Projection • Merchant Profiling • Message Parser • Message Processor • Metal-Piercing • Metal-Poor • Meteorological Package • Meteorology Panel • Metropolitan Police • Mezzo Piano • Michoud Plant (NASA) • Microcode Processor • Microgravity Pressure • Microprint • Microprocessor • Middle Point • Mid-Phase • Mile Post • Milestone Programming • Minimum Phase • Minister Plenipotentiary • Miscellaneous Paper • Miscellaneous Publication • Mission Payload • Mission Planner • Mission Planning • Mission Processing • Mission Profile • Modèle Ponctuel • Modernization Plan • Modus Ponens • Molonglo Pulsar • Monitoring Plan • Monitor Panel • Monitor Printer • Months after Payment • Mounted Police • Mouvement Perpétuel • Movement Planner • Moyenne Pression • Multilayer Perceptron • Multilink PPP • Multiple Penetration • Multiple Processors • Multiplexing Procedure • Multiplier Phototube • Multipole • Multiprocessing • Multiprocessor • Multipurpose • Northern Mariana Islands (ISO country code)

MPa Megapascal

MPA Magazine Publishers' Association (USA) • Magnetospheric Plasma Analyzer • Mail Processing Area • Main Processing Area • Manpower Authorization • Man-Powered Aircraft • Manufacturing Process Audit • Manufacturing Producibility Analysis • Maritime Patrol Aircraft • Master of Professional Accounting • Master of Public Administration • Master of Public Affairs • Master Public Archive • Materials Preparing Activity • Mathematics - Physics - Astronomy • Max-Planck-Institut für Astrophysik (Germany) • Master of Public Administration • Microgravity Pressure - Ambient • Microprocessor Application • Microsoft Product Activation • Microwave Power Amplifier • Miniature Pendulum Accelerometer • Mission and Payload Assessment • Mission/Payload Assessment • Mission Performance Assessment • Mission Phase Analysis • Mission Planning Agent • Mission Profile Analysis • Modification Proposal Analysis • Modification Proposal and Analysis • Mpacha (Namibia) (location identifier) • Multiple Parameter Analysis • Multiple-Period Average • Multiprecision Arithmetic • Multiuse Planning Area

MPAA Motion Picture Association of America (USA) • Multibeam Phased-Array Antenna • Multifunctional Phased-Array Antenna • Multimode Processing Arrays and Architecture

MPAC Materials and Parts Availability Control • Mission Planning and Analysis Center • Multipurpose Application Console

MPACS Management Planning And Control System

MPAD Manpower Personnel Assignment Document • Mission Planning and Analysis Division (JSC)

MPAe Max-Planck-Institut für Aeronomie (Germany)

Mpal Municipal

MPAM Maritime Polar Air Mass

MPAPW Minor Planet Amateur-Professional Workshop

MPAR Maintainability Problem and Assessment Report • Maintenance Program Analysis Report

MPASS Modular Processing And Support System

MPB Magnetic Particle Brake • Magnetic Pileup Boundary • Maintenance Parts Breakdown • Mathématiques et Physique de Base • Matrix Program Board • Miami (USA-FL) (location identifier) • Minor Planet Bulletin • Missing Persons Bureau • Molecular Particle Beam

MPBB Maximum Permissible Body Burden

MPBCS Mail Processing Bar Code Sorter

MPBG Manpower Program Budget Guidance

MPBRS Maintenance, Production and Backlog Reporting System

MPBS Main Power Bus Supply

Mpc Megaparsec

MP&C Maintenance Planning & Control

MPC Machine Punch Card • Magnetic Particle Clutch • Mail Processing Center • Mail Processing Characteristic • Maintenance Parts Catalog • Maintenance Procedure Chart • Manpower and Personnel Center • Manpower Planning Council • Manpower Priorities Committee • Manual Pointing Controller • Manufacturing, Planning and Control • Manufacturing - Planning - Control • Marine Prediction Center • Mars Polar Camera • Massively Parallel Computer • Material Procurement Cost • Materials Property Council • Materiel Program Code • Mathematics of Program Construction (conference) • Mathematics - Physics - Chemistry • Mathématiques - Physique - Chimie • Maximum Permissible Concentration • Mechanical Positioning Control • Mediterranean Partner Country • Memory Protection Check • Message-Passing Coprocessor • Message Processing Center • Meteorological Prediction Center • Midbody Pyro Controller • Mid-Power Controller • Mid-term Planning Conference • Minor Planet Center (USA-MA) • Minor Planet Circular • Mission Processing Center • Mitsubishi Precision Co. • Manpower Personnel Center • Materiel Program Code • Mission Phase Code • Mission Processing Center • Mobile Processing Center • Monetary Policy Committee • Monitoring Proportional Counter • Monitor Proportional Counter • Monterey Peninsula College (USA-CA) • Most Positive Cut • Most Probable Cost • Motion Picture Camera • Muko-Muko (Indonesia) (location identifier) • Multichannel Pulse Count • Multimedia Personal Computer • Multiparty Control • Multipath Channel • Multiple-Process Chart • Multiple-Purpose Communications • Multiplicateur de Puissance Crête • Multiprobe Carrier • Multiprocessor Computer • Multiprogram Control • Multiproject Chip • Multiprotocol Client • Multipurpose Canister • Multipurpose Carrier • Multipurpose Center • Multipurpose Concept • Multipurpose Console • Multipurpose Container

MPC&A Materials, Protection, Control & Accounting

MPCA Manufacturing Process Capability Assessment • Materials, Protection, Control and Accounting • Mid-Power Controller Assembly • Multiprotocol Communications Adapter

MPCAC Minor Planet Center Advisory Committee

MPCC Motion Picture Control Panel • Multiprocessor Computer Complex • Multiprotocol Communications Controller • Multipurpose Coma Corrector

MPCD Manufacturing Process Control Document • Minimum Perceptible Color Difference • Multipurpose Color Display

MPCG Modified Preconditioned Conjugate Gradient

MPCH Main Parallel Channel

MPCMP Mass Properties Control and Management Plan

MPCORB Minor Planet Center Orbit (database)

MPCP Modified Plant Clearance Program

MPCRL Massively Parallel Computing Research Laboratory

MP&CS Mission Planning & Control System

MPCS Management Performance Control System • Mission Planning and Control Station • Mission Planning and Control System • Multiprocessing Control System

MPCT Modular Parametric Current Transformer

MPCTS Modular Portable Consolidated Test System

MPCU Main Power Control Unit • Message Processing Control Unit

MPCWG Multimedia Personal Computer Working Group

MPD Magnetoplasma Dynamic • Magnetoplasma Dynamics • Magnification Probability Distribution • Main Power Distributor • Maintenance Policy Document • Management Policy and Directives • Management Position Description • Manufacturing Process Difficulty • Master Power Distribution • Materials Property Data • Maximum Permissible Dosage • Maximum Permissible Dose (*) • Minimum Permissible Dose • Mini Port Driver • Mirpur Khas (Pakistan) (location identifier) • Mission Planning Document • Multiple Personality Disorder • Multipurpose Display

MPDA Magnetoplasmadynamic Arcjet

MPDB Mission Planning Database

MPDD Meteorological Penetration Detection Development

MPDI Microwave Polarization Difference Index

MPDM Maintenance Planning Data Manual

MPDN Matching Pursuit Denoising • Materials Property Data Network

MPDS Maintenance Planning Data Support • Message Processing and Distribution System • Micrometeorid/Debris Protection Shield

MPDT Message Processing Data Terminal • Microwave Polarization Difference Temperature

M-PDU Mutiplexed Protocol Data Unit

MPDU MAC Protocol Data Unit • Master Power Distribution Unit • Mutiplexing Protocol Data Unit

mpe Motion Picture Experts Group (file extension)

MP&E Maintenance Planning & Execution

MPE Madison (USA-CT) (location identifier) • Magnetic Properties Experiment • Maintenance Planning and Execution • Manufacturing and Production Engineering • Map Publishing Environment • Mathematics - Physics - Engineering • Maximum Permissible Exposure • Max-Planck-Institut für Extraterrestrische Physik (Germany) • Memory Protection Electronics • Minimum Polynomial Extrapolation • Minimum Potential Energy • Mission-Peculiar Equipment • Mission Performance Envelope • Mission Planning Element • Mission to Planet Earth • Mobile Platform Equipment • Molecular Probe Ellipsometry • Monthly Project Evaluation • Most Probable Earthquake • Most Prominent Erection • Multiple Phase Ejector • Multiple Programming Executive • Multipoint Extension • Multipurpose Equipment

MPEC Mathematical Program with Equilibrium Constraint • Minor Planet Electronic Circular (periodical code) • Multipurpose Experiment Canister

MPECC Multiprocessor Experimental Computer Complex

MPEF Meteorological Products Extraction Facility

mpeg Motion Picture Experts Group (file extension)

MPEG Motion Picture Experts Group (*) • Moving Picture Experts Group

MPEL Maximum Permissible Exposure Level

MPEP Manual of Patent Examining Procedures

MPES Mean Polyelastic Sample

MPESS Mission Peculiar Equipment Support Structure • Mission Peculiar Experiment Support Structure • Multipurpose Equipment Support Structure

MPF Mail Processing Facility • Mapoda (Papua New Guinea) (location identifier) • Mars Pathfinder • MasPar Fortran • Master Personnel File • Materials Processing Facility • Message Parameter File • Mindless Pop Fluff • Mission Plan File • Mission Planning Facility • Mission Planning Forecast • Momentary Power Failure • Multiple Primary Feed • Multipurpose Facility • Multipurpose Fotometer • Multispectral Photographic Facility

MPFE Modified Poole-Frenkel Effect

MPFM Multipurpose Fork Mount

MPFS Multiple Primary Feed System • Multipupil Fiber Spectrograph • Multipupil Field Spectrograph

MPFSM Multiposition Flats Sorting Machine

mpg Miles per Gallon

MPG Magnetoplasma Generator • Makini (Papua New Guinea) (location identifier) • Matched Power Gain • Max-Planck-Gesellschaft (Germany) • Meteor Physics Group • Miles Per Gallon • Mountain Programming Group • Multiple Point Ground • Multipoint Grounding • Multiprogramme Groundstation

MPGC Micro-Pattern Gas Chamber • Micro-Pixel Gas Chamber • Multipatch with Geometrical Constraints • Multiphoto Geometrically Constrained (*) • Multiprotocol Geoinformation Client

MPGHM Mobile Payload Ground Handling Mechanism

MPGS Multipurpose Graphic Software

mph Miles per Hour

MPh Master of Philosophy

MPH Caticlan Malay (Philippines) (location identifier) • Malay (Philippines) (location identifier) • Master of Public Health • Miles Per Hour

MPHCC Maui High Performance Computing Centre (USA-HI)

M/PHE Material or Personnel Handling Equipment

MPHE Material and Personnel Handling Equipment

MPHF Minimal Perfect Hash Function

MPI Machine Performance Integration • Magnetic Particle Inspection • Malaysian Press Institute • Mamitupo (Panama) (location identifier) • Martin-Puplett Interferometer • Master Patient Index • Max-Planck-Institut (Germany) • Mean Point of Impact • Message-Passing Interface • Microprecision Interferometer • Mission Payload Integration • Molecular Parameter Index • Multiphoton Ionization • Multiple Power Input • Multiple Protocol Interface • Multiprecision Integer • Multiprotocol Interface

MPIA Max-Planck-Institut für Astronomie (Germany)

MPIAT Mars Program Independent Assessment Team (NASA)

MPIC Message Processing Interrupt Count

MPICD Mission-Peculiar Interface Control Document

MPID Multiphoton Ionization and Dissociation

MPIE Mixed Potential Integral Equation

MPIfR Max-Planck-Institut für Radioastronomie (Germany)

MPIIN Modification Procurement Instrument Identification Number

MPIK Max-Planck-Institut für Kernphysik (Germany)

MPIL Magnetic Polarity Inversion Line

MPIO Mission and Payload Integration Office • Mission Payload Integration Office

MPIP Maintenance Posture Improvement Program • Max-Planck-Institut für Plasmaphysik (Germany)

MPIQO Max-Planck Institut für Quantum Optik (Germany)

MPIR Multispectral Pushbroom Imaging Radiometer

MPIS Multipurpose Integration Stand

MPJ Message Passing in Java • Morrilton - Petit Jean (USA-AR) (location identifier)

mpk Manpack

MPK Manpack • Message Passing Kernel • Mokpo (Korea-ROK) (location identifier)

mpkg Metapackage (file extension)

MPL Machine-Processable Language • Maintenance Parts List • Mandatory Parts List • Manipulation Positioning Latch • Marine Physical Laboratory (USA) • Mars Polar Lander (*) • Mars Probe Lander • MasPar Language • Master Procedure List • Master Procedure Listing • Master Program Library • Material and Process List • Mathematical Programming Language • Maximum Penalized Likelihood • Maximum Permissible Level • Maximum Potential Loss • Mechanical Parts List • Mechanical Properties Loop • Megaprogramming Language • Microphysics Laboratory • Microprocessor Programming Language • Microprogramming Language • Micropulse Lidar • Minimum Power Level • Mission Planning Laboratory • Montpellier Fréjorgues (France) (location identifier) • Montpellier Méditerranée (France) (location identifier) • Mozilla Public License • Multiplatform Launcher • Multiple Payload Launcher • Multiple-Purpose Language • Multiprogramming Level • Multipulse Laser

MPLA Modern Physics Letters A (periodical code) • Multipoint Link Aggregation

MPLB Modern Physics Letters B (periodical code)

MPLC Mass - Period - Luminosity - Color

MPLCC Most Probable Life Cycle Cost

MPLE Maximum Penalized Likelihood Estimate

MPLIS Multipurpose Land Information System

MPLM Main Pressurized Logistics Module • Mini Pressurized Logistics Module • Multipurpose Logistics Module (*) • Multipurpose Pressurized Logistics Module

MPLN Maintenance Planning

MPLS Maximal Principle Least Squares • Multiprotocol Label Swapping • Multiprotocol Label Switch • Multiprotocol Label Switching (*) • Multiprotocol Lambda Switching

MPLSM Multiposition Letter Sorting Machine

mpm Meters per Minute

MPM Magnetic Phase Modulator • Main Propulsion Motor • Maintenance Planning Manual • Maintenance Program Management • Major Program Memorandum • Major Project Manager • Manipulator Positioning Mechanism • Manpower Planning Model • Maputo (Mozambique) (location identifier) • Marginal Probability Maximum • Materials Processing and Manufacturing • Maximizer of Posterior Marginals • Medical Planning Module • Medium Power Microwave • Message Passing Library • Meters Per Minute • Metra-Potential Method • Microframe Project Manager • Microphysical Model • Microwave Power Module • Milestone Planning Meeting • Millimeter-wave Propagation Model • Milwaukee Public Museum (USA-WI) • Monolithic Piezoelectric Mirror • Movement Planning Module • Multiple-Processor Module

MPMC Multinational Program Management Course

MPMD Multiple Program Multiple Data

MPMI Micro-Phase Measuring Interferometer

MPML Minor Planet Mailing List

MPMP Mass Properties Management Plan • Master Program Management Plan • Microwave Point-to-Multipoint • Modification Program Management Plan

MPMS Magnetic Property Measurement System

MPMSE Multiuse Payload and Mission Support Equipment

MPMT Mean Preventive Maintenance Time

MPN Message Packet Network • Mobility Position Number • Most Probable Number • Mout Pleasant (Falkland Islands) (location identifier)

MPNE Microgravity Plant Nutrient Experiment

mpnt Mac Paint (file extension)

MPO Magnetospheric Plasma Oscillation • Main Post Office • Main Production Order • Management and Personnel Office • Manufacturing Production Order • Maximum Power Output • Mercury Planetary Orbiter • Message-Passing Object • Minor-Planet Observer • Minor-Planet Orbit • Mobile Post Office • Modular Payload Operator • Mount Pokono (USA-PA) (location identifier) • Multiproduct Operation

MPOA Multiprotocol Over ATM

MPOC Mission Payload and Operations Center • Multipurpose Operators Console

MPOD Mean Planned Outage Duration • Multimode Proximity Operations Device

MPOI Master Program Of Instruction

MPOM Maintenance Program Operations Management

MPOO Manager of Post Office Operations

MPOR Master Power-On Reset

MPOW Multiple Purpose Operator Workstation

M&PP Materials & Plant Protection

MPP Maintainability Program Plan • Major Program Proposal • Manufacturing Process Procedure • Markov Point Process • Mars Polar Path Finder • Martens Polarization Photometer • Massively Parallel Platform • Massively Parallel Processing • Massively Parallel Processor • Material Processing Procedure • Materials and Plant Protection • Maximum Power Point • Max-Planck-Institut für Physik (Germany) • Mechanical Parts Production • Medical Principles and Practice • Medical Provisioning Point • Member of Provincial Parliament • Message Posting Protocol • Message Processing Program • Methodology - Pedagogy - Philosophy • Microfilm Printer-Plotter • Microprogrammable Processor • Microprogrammable Processor Part • Mission Planning Phase • Mission Preparation Processor • Modernization Planning Process • Most Probable Position • Moyens Provisoires Pourvus • Mulatupo (Panama) (location identifier) • Multiphase Pinned • Multiphase Pinning • Multipinned Phase • Multipin Phase • Multiple Particle Plasma • Multiple Penetration Project

MPPD Multipurpose Peripheral Device

MPPE Massive Parallel Portal Engine • Microsoft Point-to-Point Encryption

MPPL Multipurpose Processing Language • Multipurpose Programming Language

MPPM Mission Prediction and Performance Module

MPPP Multilink Point-to-Point Protocol

MPPRC Materiel Procurement Priorities Review Committee

MPPS Master Program Planning Schedule • Million Pulses Per Second

MPPSE Multipurpose Payload Support Equipment

MPPT Maximum Power Point Tracker • Maximum Power Point Tracking • Message Preparation Processing Task

MPPU Multipurpose Processor Unit

MPQ Maan (Jordan) (location identifier) • Max-Planck Institut für Quantenoptik (Germany) • Most Probable Quantity

MPQP Multiprotocol Quad Port

MPR Maintainability Problem Report • Major Project Report • Malfunction Problem Record • Management Program Review • Manpack Radio • Manpower • Mars Pressurized Rover • Mass Properties Report • Master Personnel Records • McPherson (USA-KS) (location identifier) • Mean Planetary Radius • Medium Power Radar • Message Processing Region • Microburst Prediction Radar • Microwave Precipitation Radiometer • Mission Planning Room • Mockup Purchase Request • Monthly Program Review • Monthly Progress Report • Monthly Project Report • Multiband Profile Radiometer • Multipart Repeater • Multiple Protective Shelter • Multiple Protective Structure • Multiport Repeater • Multiprotocol Router • Multipurpose Recorder

MPRAS Multipath Redundant Avionics Suite

MPRC Manpower and Personnel Readiness Center • Multipurpose Range Complex

MPRD Modified Proportionnal Radial Displacement

MPRF Medium Pulse Repetition Frequency

MPRG Modification Planning and Review Group

MPRI Multiphoton Resonance Ionization

MPRL Massively Parallel Richardson-Lucy • Minimum Proficiency Requirement Loading

MPRP Mean Prioritized Replacement Position

MPRS Mission Planning and Rehearsal System

MPRSA Marine Protection, Research and Sanctuaries Act (USA)

mps Meters per Second

M&PS Meteoritics and Planetary Science (periodical code)

MP&S Manufacturing, Planning & Scheduling • Materials, Processes & Structures

MPS Magnetic Pole Strength • Mail Preference Service • Main Power Switch • Main Propulsion System • Maintenance Problem Summary • Maintenance Processor System • Management of Parts Shortage • Management Policy Statement • Maneuvering Propulsion System • Manpower Planning System • Manpower System • Manual Phase Shifter • Manufacturer's Part Specification • Manufacturing, Planning, and Scheduling • Manufacturing Process Specification • Maschinenprogrammsprache • Maschinenprogrammsystem • Mask Preparation Software • Massive Passive Service • Master Production Schedule • Master Program Schedule • Material Processing in Space • Material Processing Science • Material Processing Specification • Material Processing System • Materials, Processes and Structures • Materiel Planning Study • Mathematical and Physical Sciences • Mathematical Processing System • Mathematical Programming Society (USA) • Mechanical Phase Shifter • Memory Processor Switch • Message Passing Service • Message Passing System • Message Processing System • Meters Per Second • Microdrop Particle Search • Microlensing Planet Search (collaboration) • Microwave Pressure Sounder • Milwaukee Public Schools (USA-WI) • Minimum Performance Standards • Mission Payload System • Mission Performance Standard • Mission Planning Simulator • Mission Planning Station • Mission Planning System • Mission Preparation Sheet • Mission Preparation Software • Mission Preparation System • Mission Profile Simulator • Mobile Position Search • Mobile Profiler System • Moment Period Search • Moteur à Propergol Solide • Motor Pump System • Mount Pleasant (USA-TX) (location identifier) • Moving Platform Simulator • Multiple Port Sharing • Multiple Protection Shelter • Multiple Protective Shelter • Multiple Protective Structure • Multiprocessor Specification • Multiprogramming System • Multiprotocol Server • Multipurpose Presentation System • Multipurpose Ship

MPSA Multiprocessor Parallel Server Architecture • Multiprocessor System Architecture

MPSAC Mathematical and Physical Sciences Advisory Committee

MPSCL Mathematical Programming System Control Language

MPSE Mexican Payload Specialist Experiment

MPSET Mars Program Systems Engineering Team

MPSF Multipupil Fiber Spectrograph • Multipupil Field Spectrograph

MPSI Magnetic Plage Strength Index • Mathématiques - Physique - Sciences de L'Ingénieur • Mees Photometric Spot Index

MPSK Multilevel Phase Shift Keying

MPSL Materials and Processes Specification List

MPSO McMath Pierce Solar Observatory (USA)

MPSOC Multipurpose Satellite Operations Center

MPS&P Material Processing Specifications & Procedures

MPSP Material Processing Specifications and Procedures • Mathematical Problem-Solving Project (NSF)

MPSR Mission Profile Storage and Retrieval • Multipurpose Support Room

MPSS Mission Payload System Segment • Multipurpose Sampling System

MPST Message Processing Subscriber Terminal • Mission pour la Parité en Science et Technologie (France) • Multipurpose Support Team

MPSX Mathematical Programming System Extended

mpt Microsoft Project Template (file extension)

M. Pt. Melting Point

MPT Main Propulsion Test • Maintenance Panel Trainer • Maison Pour Tous (France) • Male Pipe Thread • Maliana (Indonesia) (location identifier) • Maneuver Planning Table • Manpower - Personnel - Training • Mean Pulse Time • Mechanical Power Transmission • Melting Point • Memory Point Track • Message Preparation Terminal • Message Processing Time • Microprogramming Technique • Mid-Point of Test • Minimum Process Time • Ministry of Post and Telecommunications • Ministry of Posts and Telecommunications • Mission Planning Table • Mission Planning Team • Mission Planning Terminal • Multiport Transceiver • Multipurpose Telescope • Multipurpose Tray

MPTA Main Propulsion Test Article • Manpower Personnel and Training Assessment

MPTE Mission Planning and Training Environment • Mission Preparation and Training Environment • Mission Preparation and Training Equipment • Multipurpose Test Equipment

MPTF Main Propulsion Test Facility • Maintenance Patch and Test Facility • Mission Planning Task Force

MPTI Moscow Physical-Technical Institute (Russia)

MPTN Multiprotocol Transport Network • Multiprotocol Transport Networking

MPTO Method and Procedure Technical Order

MPTP Main Propulsion Test Program

MPTQM Mail Preparation Total Quality Management

MPTS Manpower - Personnel - Training - Safety • Manpower - Personnel - Training - Simulation • Mobile Photographic Tracking Station • Multiprotocol Transport Services • Multipurpose Test Set • Multipurpose Tool Set • Multipurpose Tracking System

MPTWT Medium Power Traveling Wave Tube

MPU Main Processing Unit • Main Propulsion Unit • Manpack Unit • Mapua (Papua New Guinea) (location identifier) • Médecine Préventive Universitaire • Message Processing Unit • Metadata Populating Utility • Microprocessing Unit • Microprocessor Unit • Mission Planning Unit • Mission Programming Unit • Mobility Processing Unit • Modem Processor Unit • Monitor Printing Unit • Motorola Processor Unit • Motor Pressurization Unit

mput Multiple Put

MPV Barre (USA-VT) (location identifier) • Magnetic Polarization Vector • Man-Powered Vehicle • Mars Piloted Vehicle • Methane-Powered Vehicle • Montpelier (USA-VT) (location identifier) • Multipurpose Variant • Multipurpose Vehicle

MPVA Main Propellant Valve Actuator

MPVU Multiplot Variable Updating

MPW Macintosh Programmer's Workshop • Mac Path Way • Mariupol (Ukraine) (location identifier) • Maximum Permitted Weight • Microsoft Programmer's Workshop • Modified Plane Wave • Multipole Wiggler • Multiproject Wafer

MPWCS Mail Processing Work Credit System

MPWG Maintenance Planning Working Group

mpx Microsoft Project Export (file extension)

MPX Miyanmin (Papua New Guinea) (location identifier) • Mutiplex • Mutiplexer • Multiprocessor Extension • Multiprogramming Executive

MPY Maripasoula (French Guiana) (location identifier)

MPZ Mount Pleasant (USA-IA) (location identifier)

MQ Marketing Questionnaire • Martinique (code ISO de pays) • Merit Quotient • Message Queuing • Metro Quadrato • Modèle de Qualification

MQA Mail Quality Analysis • Mandora (Australia-WA) (location identifier)

MQAM Message Queue Access Method

MQB Macomb (USA-IL) (location identifier)

MQBO Mesospheric Quasibiennial Oscillation

MQC Main Quality Control • Many-Quark Cluster • Miquelon (Saint Pierre & Miquelon) (location identifier)

MQCC Maintenance Quality Control Check

MQD Maquinchao (Argentina-RN) (location identifier) • Monolithic Quad Device

MQDP Méthodologie et Qualité de Développement de Projets

MQDT Multichannel Quantum Defect Theory

Mqe Martinique

MQE Marqua (Australia-NT) (location identifier) • Massive and Quick Erection

MQF Magnitogorsk (Russia) (location identifier) • Mobile Quarantine Facility

MQH Memory Queue Handler • Minacu (Brazil) (location identifier)

MQI Message Queue Interface • Message Queuing Interface • Quincy (USA-MA) (location identifier)

MQIL Miniature Quartz Incandescent Lamp

MQL Mildura (Australia-VIC) (location identifier) • Miniature Quartz Lamp

MQM Matching Quadrupole Magnet • Message Queuing Middleware

MQN Magnetic Quantum Number • Mo I Rana (Norway) (location identifier)

MQO Measurement Quality Objective

MQP Major Qualifying Project • Motor Qualification Program

MQQ Moundou (Chad) (location identifier) • Multiple Quick Query

MQR Modified Quick Rank • Mosquera (Colombia) (location identifier) • Multiply Quotient Register

MQS Modified Q-Switching • Molonglo Quasar Sample • Mustique (Grenadine Islands) (location identifier)

M-QSPR Multivariate Quantitative Structure-Property Relationship

MQSPR Multivariate Quantitative Structure-Property Relationship

MQT Macroscopic Quantum Tunneling • Marquette County (USA-MI) (location identifier) • Mission Qualification Training • Model Qualification Test • Motor Qualification Test

MQU Mariquita (Colombia) (location identifier) • Multiplier Quotient Unit

MQW McRae Telfair-Wheeler (USA-GA) (location identifier) • Multiple Quantum Well (*) • Multiquantum Well

MQX Makale (Ethiopia) (location identifier)

MQY Smyrna (USA-TN) (location identifier)

m/r Mi Remesa

mr Marathi (ISO language code) • Milliradian • Milliroentgen

mR Milliroentgen

Mr Meester • Mister • Monsieur

M-R Mass-Radius (relation)

M/R Mixture Ratio

M&R Maintenance & Refurbishment • Maintenance & Repair

MR Macchia Rossa • Machine Record • Magnetic Reconnection • Magnetic Recorder • Magnetic Resonance • Magnetic Rotation • Magneto-Resistance • Magneto-Resistive • Main Resonance • Main Ring • Maintenance and Refurbishment • Maintenance Ratio • Maintenance Report • Maintenance Review • Malfunction Report • Management Requirement • Management Reserve • Management Review • Manpower Requirement • Manufacturer's Representative • Manufacturing Risk • Many Rows • Map Reference • March • Marginal Return • Marker Ranger • Mars • Master Reset • Material Request • Material Review • Matériau de Référence • Materiel Release • Materiel Review • Mathematical Reviews (periodical code) • Mauritania (ISO country code) • Mauritanie (code ISO de pays) • Medium Range • Medium Resolution • Memorandum for Record • Memorandum Receipt • Memorandum Report • Memory Recall • Memory Register • Merger Remnants • Message Register • Message Repeat • Message Router • Metabolic Rate • Metal-Rich • Methane Recovery • Microwave Radiometer • Milestone Review • Milliradian • Minister Residentiary • Minnesota Pulsar • Minor Repair • Miscellaneous Receipt • Miscellaneous Report • Mission Ready • Mission Reliability • Mission Report • Mixed Reality • Mixture Ratio • Model Reduction • Modem Ready • Modification Request • Modification Requirement • Modified Read • Moisture-Resistant • Molecular Replacement • Monitor Recorder • Monthly Report • Monthly Review • Morocco • Motivational Research • Motivation Research • Mover Ring • Multirole

MRA Machine Records Activity • Machine Reference Architecture • Macintosh Reference Architecture • Main Ring Assembly • Management and Reconfiguration Assembly • Manufacturing Requirements Analysis • Maximum Rendezvous Altitude • Mechanical Readiness Assessment • Message Release Authority • Message Review Authority • Metric Reduction Analysis • Milestone Review and Approval • Minimum Reception Altitude • Mission Ready and Available • Mission Requirements Analysis • Mission Requirements and Allocation • Misurata (Libya) (location identifier) • Mobile Robotic Actor • Multimode Radar Altimeter • Multiresolution Analysis • Multistaged Radial Axial • Mutual Recognition Agreement

mrad Milliradian

MRAD Mass Random Access Disk

MRADS Mass Random Access Data Storage

MRAFS Multiresident Andrews File System

MRAM Magneto-resistive Random Access Memory

MRAMS Mars Regional Atmospheric Modeling System

MRAO Mullard Radio Astronomy Observatory (UK)

MRAR Mean Reciprocal Answer Rank

MRAS Member of the Royal Astronomical Society

MRASS Modular Radar Analysis Software System

MRB Magnetic Recording Borescope • Magnetospheric Radio Burst • Maintenance Requirements Branch • Maintenance Review Board • Malfunction Review Board • Management Review Board • Martinsburg (USA-WV) (location identifier) • Material Review Board • Mission Reconfiguration Board • Mission Review Board • Modification Requirements Board • Modification Review Board • Multinational Review Board • Multireference Bit

MRBAB Musées Royaux des Beaux-Arts de Belgique (Belgium)

MRBB Master Reference Blackbody

MRBSR Material Review Board Standard Repair

MRC Columbia - Maury County (USA-TN) (location identifier) • Machine-Readable Character • Machine-Readable Code • Magnetic Research Corporation • Mail Recovery Center • Maintenance and Repair Cycle • Maintenance Requirements Card • Maître de Recherche Contractuel • Major Regional Conflict • Major Regional Contingency • Manpower Requirements Change • Marmara Research Centre (Turkey) • Matériau de Référence Certifié • Materiel Release Confirmation • Mathematics Research Center • Message-based Reliable Channel • Measurement Requirements Committee • Medical Research Council (UK) • Meteorological Research Committee • Mineralogy Research Centre • Minimum Resolvable Contrast • Mission Requirements Change • Mobile Radio Communications • Modified Read Code • Molonglo Reference Catalogue • Monthly Recurring Charge • Monthly Relative Citedness • Multiple Register Counter • Multiple Regression/Correlation • Multiresolution CLEAN

MRCAS Medical Robotics and Computer-Assisted Surgery (conference)

MRCAT Materials Research Collaborative Access Team

MRCC Maritime Rescue Coordination Center

MRCF Microsoft Real-time Compression Format

MRCI Microsoft Real-time Compression Interface

MRCR Measurement Requirement Change Request

MRCS Mobile Range Control System • Mobile Reporting and Control System

mrd Millirutherford

Mrd Milliarde

MRD Machine-Readable Data • Machine-Readable Dictionary • Mandatory Retirement Date • Material Review Disposition • Materiel Requirements Document • Measuring and Reference Device • Mechanized Range Documentation • Memory Raster Display • Memory Reduction Technology • Mérida (Venezuela) (location identifier) • Milestone Review Documentation • Minimum Reflectance Difference • Mission Rehearsal device • Mission Requirements Definition • Mission Requirements Document • Most Reliable Data

MRDA Marketing Regional Development Assistant • Mission Requirements and Definition Analysis • Mission Requirements Definition and Analysis

MRDB Mapping Requirements Database • Mission Requirements Database • Multiple Representative Database

MRDE Multireference Double Excitation

MRDF Machine-Readable Data File • Mail Run Data File

MRDM Model Requirements and Development Manual

MRDOS Mapped Real-time Disk Operating System

MRDR Material Review Disposition Record

MRDS Maintenance Requirements Data System

MRDT Mission Readiness Demonstration Test

MRe Mauritian Rupee (currency)

MRE Major Research Equipment • Mara Lodges (Kenya) (location identifier) • Material Review Engineering • Meal Ready-to-Eat • Mean Radial Error • Microbiological Research Establishment • Microrocket Engine • Mid-Range Estimate • Mineral Resources Engineering • Ministère de la Recherche et de l'Espace • Multiple Representation Environment • Multiple-Response Enable

M/ref Mi Referencia

MREF Major Research Equipment and Facilities (NSF)

MREFC Major Research Equipment and Facilities Construction (NSF)

mrem Millirem

MREN Metropolitan Research and Education Network • Mission des Relations avec les Entreprises

MRES Ministère de la Recherche et de l'Enseignement Supérieur

M&RF Maintenance and Refurbishing Facility

MRF Magnetic Resonance Facility • Magnetorheological Finishing • Maintenance and Refurbishment Facility • Maintenance Repair Facility • Maintenance Replacement Factor • Maintenance Reset Function • Maintenance Responsibility File • Marfa (USA-TX) (location identifier) • Marine Research Facilities • Markov Random Field • Maximum Retarding Force • Measurements Request Form • Medium-Range Forecast • Metagalactic Radiation Field • Meteorological Research Facility • Meteorological Rocket Facility • Metering Research Facility • Milestone Reference File • Mission Reliability Factor • Multipath Reduction Factor

MRFM Magnetic Resonance Force Microscopy

MRFPI Moderate-Resolution Fabry-Pérot Imaging

MRFS Multipurpose Radio Frequency Surveillance

MRG Magnetic Radiation Generator • Mareeba (Australia-QLD) (location identifier) • Mission Requirements Group • Mission Routing Guide • Mission Rules Guideline

MRGCP Metal-Rich Globular Cluster Population

MRGFUS Magnetic Resonance - Guided Focused Ultrasound Surgery

MRGL Marginal

MRGS Member of the Royal Geographical Society

MRGT Multirouter Graphing Tool

MRH Magnetic Recording Head • May River (Papua New Guinea) (location identifier) • Mechanical Recording Head • Mission-Related Hardware • Module de Ressources Hermès • My Right Hand

MRHD Magneto-Radiation-Hydrodynamics

MRHS Master Requisition History and Status • Materiel Request History Status

MRI Anchorage Merrill (USA-AK) (location identifier) • Machine Records Installation • Magnetic Resonance Image • Magnetic Resonance Imager • Magnetic Resonance Imaging • Magnetic Resonance Institute • Magnetic Rotation Instability • Magneto-Rotational Instability • Major Research Instrumentation • Master Record Index • Materiel Receiving Instruction • Mathematical Research Institute • Measurement Requirements and Interface • Medium-Resolution Imager • Mehta Research Institute (India) • Metadata Reference Information • Meteorological Research Institute (Japan) • Molecular Research Institute • Monopulse Resolution Improvement

MRINDO Modified Rydberg Intermediate Neglect of Differential Overlap

MRIP Mobile Range Instrumentation Platform

MRIR Medium-Resolution Imaging Radiometer • Medium-Resolution Infrared Radiometer

MRIS Machine-Readable Identification System • Moderate Resolution Imaging Spectrometer

MRJ Marcala (Honduras) (location identifier)

mrk Markup (file extension)

Mrk Markarian

MRK Manual Remote Keying • Marco Island (USA-FL) (location identifier) • Menschenrechtskonventionen

Mrkr Marker

Mrktg Marketing

Mrkting Marketing

MRL Machine Representational Language • Main Reference Layer • Main Ring Length • Maintenance Repair Level • Maintenance Requirements List • Manipulator Retention Latch • Manipulator Retention Lock • Master Report List • Master Requirements List • Material Requirements List • Materiel Requirements List • Maximum Recording Level • Maximum-Residual Likelihood • Maximum Residue Limit • Measurement Requirements List • Media Response Lines • Medical Research Laboratory • Miners Lake (Australia-QLD) (location identifier) • Minimum-Risk Level • Mobile Rocket Launcher • Multiple Rocket Launch • Multiple Rocket Launcher • Multiple Ruby Laser

MRLS Microgravity Research and Life Sciences

MRLU Map Reproduction and Liaison Unit

MRM Maintenance Reporting and Management • Management Reform Memorandum • Management Responsibility Matrix • Manare (Papua New Guinea) (location identifier) • Manufacturing Resource Management • Materiel Readiness and Modernization • Metabolic Rate Monitor • Meter Reply Mail • Mid-term Review Meeting • Mission Reconfiguration Master • Mission Requirements Matrix • Most Recently-used Master • Motif Resource Manager

MRMS Maintenance Reporting and Management Service • Maritime Resource Management Service • Mission Reconfiguration Management System • Mobile Remote Manipulator System

MRMU Mobile Radiological Measuring Unit

MRN Magnetic Reynolds Number • Maritime Radionavigation • Meteorological Rocket Network • Minimum Rejection Number • Morganton Lenoir (USA-NC) (location identifier)

mRNA Messenger Ribonucleic Acid

Mrng Morning

MRNG Morning

MRNS Maritime Radionavigation Service

MRNSS Maritime Radionavigation Satellite Service

MRO Machine-Readable Output • Magdalena Ridge Observatory (USA-NM) • Maintenance - Repair - Operation • Maintenance - Repair - Overhaul • Management Review Officer • Manastash Ridge Observatory (USA-WA) • Mars Reconnaissance Orbiter • Masterton (New Zealand) (location identifier) • Material Release Order • Medical Regulating Office • Medical Regulating Officer • Message Releasing Office • Message Releasing Officer • Metsähovi Radio Observatory (Finland) • Mid-Range Objective • Movement Report Office • Multiregion Operation • Ouguiya (Mauritania) (ISO currency code)

MROA Magnetic Raman Optical Activity

MROC Miniature Ruggedized Optical Correlator • Multicommand Required Operational Capability

MROF Maintenance, Repair and Operation of Facility

MROL Minimum Resolvable Object Length

MROVG Munich R. Oldenburg Verlag GmbH (periodical code)

MRP Machine-Readable Passport • Manned Reusable Payload • Manpower Requirements and Personnel • Manpower Requirements Program • Manual Repair Point • Manufacturers' Recommended Price • Manufacturing Requirements Planning • Manufacturing Resource Planning • Marla (Australia-SA) (location identifier) • Massive Repeated Penetration • Master in Regional Planning • Master of Regional Planning • Material Requirement Planning • Material Resource Planning • Material Review Process • Materials and Resources Planning • Materiel Release Point • Materiel Release Process • Materiel Release Program • Materiel Requirements Planning • Mathematics Resources Project (NSF) • Maximum Resolving Power • Medium-Range Plan • Medium-Range Planning • Microfiche Reader/Printer • Microwave Radiometric Profiler • Mid-Range Plan • Mission Reference Plane • Mobile Repair Party • Mobile Reporting Post • Monitoring and Response Program • Monthly Report of Progress • Multiple Rate Processor • Multisecond Radio Pulsar

MRPA Modified Random Phase Approximation

MRPC Multicost Remote Control Procedure • Multigap Resistive Plate Chamber

MRPF Maintenance of Real Property Facilities

MRPG Minority Research Planning Grant

MRPL Main Ring Path Length

MRQ Marinduque (Philippines) (location identifier) • Maximum Release Quantity

MRR Macara (Ecuador) (location identifier) • Maintenance Repair Rate • Maintenance Replacement Rate • Manufacturing Readiness Review • Mars Return Rate • Mass-Radius Relationship • Mass Return Rate (*) • Master Release Record • Material Rejection Report • Material Reliability Report • Materiel Readiness Report • Materiel Release Point • Mechanical Reliability Report • Microfilm Reader Recorder • Milestone Readiness Review • Minimum Reporting Requirement • Minimum-Risk Route • Mission Readiness Review • Mission Reconfiguration Request • Mission Requirements Review • Molecular Rotational Resonance • Monthly Review Report • Multirole Radio

MRRB Maintenance Requirements Review Board • Materiel Release Review Board

MRRC Maintenance Requirements Review Committee

MRRG Melderechtsrahmengesetz

MRRL Materiel Repair Requirements List • Midlands Regional Research Laboratory (UK)

MRRS Materiel Readiness Reporting System • Mobile Remote Receiving System

MRRT Maintenance Requirements Review Team

MRRU Maximum Received Reconstructed Unit

Mrs Messieurs • Mistress • Missus

MRS Magnetic Resonance Spectroscopy • Magnetic Resonance Spectrum • Maintenance, Repair and Service • Maintenance Reporting System • Management and Reporting System • Management Review System • Marseille - Marignane (France) (location identifier) • Marseille - Provence (France) (location identifier) • Master Repair Schedule • Material Request Summary • Material Requirement Summary • Material Routing Slip • Materials Research Society (USA-PA) • Mathematics Research Center • Mean Root Square • Mechanized Range Scheduling • Media Recognition System • Medical Readiness System • Medium-Range Search • Medium-Range Sensor • Medium-Range Surveillance • Medium-Resolution Spectrograph • Medium-Resolution Spectrometer • Memorandum Routing Slip • Merchandise Return Service • Microcontroller Readable Sector • Microwave Radiometer Spacecraft • Microwave Ranging System • Mission Radio System • Mission Rehearsal System • Mission-Related Software • Mobile Radar Search • Mobile Remote Servicer • Mobility Requirement Study • Moderate Relevance Scheme • Modification Record Sheet • Monitored Retrievable Storage • Movement Report System • Multicommand Requirements Submission • Multidiscipline Refurbishable Satellite • Multiresonance Spectrum • Multiresonance Structure • Multirole System • Multispectral Resource Sampler

MRSA Major River Systems Analysis • Materiel Readiness Support Activity • Materiel Readiness Support Agency • Microwave Radiometer/Scatterometer and Altimeter

MRSBu MRS Bulletin (periodical code)

MRSC Manitoba Remote Sensing Centre (Canada-MT) • Maritime Remote Sensing Committee • Member of the Royal Society of Chemistry • Mississippi Remote Sensing Center (USA-MS)

MRSD Maximum Rated Standard Deviation • Mission Requirements on System Design

MRSE Message Retrieval Service Element • Microwave Remote Sensing Experiment

MRSEC Materials Research Science and Engineering Center (USA)

MRSH Marsh

MRSP Marginalized Rural Sectors and Peoples • Microwave Remote Sensing Program • Mobility Readiness Spares Package • Multicast Router-Switch Protocol • Multiresponse Sequence Procedure • Münster Redshift Survey Project

MRSPA Monthly Register of the Society for Practical Astronomy (defunct)

MRSR Mars Rover and Sample Return (*) • Mars Rover Sample Return

MRSRM Mars Rover and Sample Return Mission (*) • Mars Rover Sample Return Mission

MRSS Mechanized Range Scheduling System • Mobile Range Safety System

MRST Mobile Subscriber Radio Terminal

MRSV Maneuverable Recoverable Space Vehicle

MRT Machine-Readable Tape • Magnetic Resonance Tomograph • Magnetic Resonance Tomography • Magnetresonanz-Tomograph • Magnetresonanz-Tomographie • Mail Relay Tester • Maintenance Recovery Team • Maintenance Review Team • Management Review Team • Manpack Radiotelephone • Mass Rail Transit • Mass Rapid Transit • Mass Rapid Transport • Mauritania (ISO country code) • Mauritanie (code ISO de pays) • Mauritius Radio Telescope • Mean Radiative Transfer • Mean Repair Time • Mean Residence Time • Mean Response Time • Memorandum Report of Test • Message Routing • Microbeam Radiation Therapy • Miniature Receiver Terminal • Miniature Receive Terminal • Minimum Reaction Time • Minimum Resolvable Temperature • Ministère de la Recherche et de la Technologie • Ministry of Research and Technology • Mission Readiness Test • Mission Recovery Team • Mobile Radio Terminal • Mobile Repair Team • Morehead Radio Telescope (USA-NC) • Moroak (Australia-NT) (location identifier) • Movement Regulating Team • Multipôle Régional Technologique • Multirole Radiotelephone

MRTA Mechanical Response Tissue Analyzer

MRTC Multiple Real-Time Command

MRTD Minimum Resolvable Temperature Difference

MRTF Major Range and Test Facility

MRTFB Major Range and Test Facility Base

MRTG Multirouter Traffic Grapher

MRTM Maritime

MRTP Mission Readiness Training Program

MRTS Material Requirements Tracking System • Meteorological Real-Time System • Mobile Real-Time System • Mobile Remote Tracking System • Modification and Repair Tracking System

MRTT Modular Record Traffic Terminal • Multirole Tanker Transport

MRTU Multiplexed Remote Terminal Unit • Multiplex Remote Terminal Unit

MRU Machine Record Unit • Main Regulator Unit • Maintenance Recorder Unit • Management and Regulation Unit • Manpower Research Unit • Material Recovery Unit • Materiel Recovery Unit • Mauritius (location identifier) • Maximum Receive Unit • Message Retransmission Unit • Meteorological Research Unit • Microwave Relay Unit • Minimum Replacement Unit • Mobile Radio Unit • Mobile Receiving Unit • Mobile Repair Unit • Monitoring and Reconfiguration Unit • Most Recently Used • Mountain Rescue Unit • Multicampus Research Unit • Port Louis (Mauritius) (location identifier)

MRV Machine Readable Visa • Mineralnie Vody (Russia) (location identifier) • Mission Roel Vehicle • Mitte-Rand-Variation • Multiple Reentry Vehicle • Multiple Rotational Variant

MRVC Multiple Rate Voice Card

MRVI Medium Resolution Visible Imager

MRVS Meet-, Registratie- en Verwerkingssysteem

MRVT Materiel Release Verification Test

MRW Maribo (Denmark) (location identifier) • Maximum Ramp Weight • Mid Red Wing

MRWC Multiple Reading and Writing with Computing • Multiple Read-Writing Compute

MRWG Mission Requirements Working Group

MRWS Manned Remote Workstation • Monoscopic Revision Workstation

MRX Magnetic Reconnection Experiment • Magneto-Resistive Extended • Message Routing Exchange

MRY Carmel (USA-CA) (location identifier) • Monterey Peninsula (USA-CA) (location identifier)

MRZ Meteorologisches Rechenzentrum • Moree (Australia-NSW) (location identifier)

m/s Meters per Second

ms Malay (ISO language code) • Manuscript • Manuscrit • Millisecond

M/s Meters per Second

Ms Miss • Mistress

M&S Maintenance & Security • Maintenance & Supply • Mapping & Surveying • Marred & Scarred • Materials & Structures • Modeling & Simulation • Models & Simulations

M/S Mail Stop • Main Stage • Measurement Stimulus • Meters per Second • Milestone • Modeling/Simulation • Months after Sight • Motorship

MS Machine Screw • Machine Selection • Machine Steel • Macromodular System • Magellanic Storm • Magellanic Stream • Magnetic South • Magnetic Storage • Magnetic Susceptibility • Magnetosheath • Mail Steamer • Main Sequence • Main Stage • Main Station • Main Store • Main Structure • Main Switch • Maintenance and Service • Maintenance Shelter • Maintenance Software • Maintenance Standard • Maintenance State • Maîtrise Spécialisée • Management Science • Manned System • Mano Sinistra • Manufacturing Specification • Manufacturing Standard • Manufacturing Status • Manufacturing Support • Manufacturing System • Manuscript • Manuscrit • Manuskript • Mare Serenitatis (Moon) • Margin of Safety • Marine Stratocumulus • Market Survey • Mark Start • Massenspektrometrie • Massive Star • Mass Spectrometer • Mass Spectrometry • Mass Spectroscopy • Mass Spectrum • Mass Storage • Mastère Spécialisé • Master of Science • Master Schedule • Master Scheduler • Master Station • Master Switch • Material Specification • Material Standard • Material Support • Materiel Specification • Mating Sequence • Mato Grosso do Sul (Brazil) • Maturity Status • Mauritius (international vehicle registration code) • Maximum Stress • Mean Seasonal • Mean Square • Measurement Set • Measuring Set • Measuring System • Mechanical System • Media Status • Medical Service • Medical Staff • Medium Sensitivity (survey) • Medium Shot • Medium Solid • Memoriae Sacrum •

Memory System • Message Store • Message Switch • Message Switching • Meteoritical Society • Meteoroid Shield • Metric System • Microsecond • Microsoft Corporation • Microstrip • Milchstraße • Milestone • Millisecond • Minisymposium • Minus • Mirror Shell • Mission Schedule • Mission Scientist • Mission Sequence • Mission Simulator • Mission Specialist • Mission Station • Mission Station • Mission Support • Mississippi (USA) • Mobile Service • Mobile Station • Mobile Subscriber • Modalité de Service • Modèle Standard • Modular SmallTalk • Molecular Sieve • Monitor Station • Monochromatic Scanner • Montserrat (ISO country code) • Moonshine • Mössbauer Spectrometer • Mössbauer Spectroscopy • Motor Ship • Moving Surface • Multiple Sclerosis • Multiservice • Multispectral • Multispectral Sensor • Multistream

MS&A Modeling, Simulation & Assessment

MSA Main Solar Array • Maintenance Support Activity • Maintenance Support Agreement • Major Subcontract Administrator • Major Statistical Area • Major System Acquisition • Malta Standardization Authority (formerly MBS) • Management System Analysis • Master Schedule Authorization • Master Science Archive • Material Service Area • Materiel Surveillance Assembly • Mechanical Signature Analysis • Medical Services Account • Medium-Speed Access • Meteorological Support Activity • Methane Sulfonic Acid • Metropolitan Statistical Area • Metropolitan Survey Area • Microgravity Science and Applications • Millennium Star Atlas • Mineralogical Society of America (USA) • Minimum Safe Altitude • Minimum Sector Altitude • Minimum Surface Area • Mirror Support Adapter • Mission Support Area • Mixer Subassembly • Mobile Subscriber Access • Modeling, Simulation and Assessment • Modern Structured Analysis • Monitor and Switching Assembly • Moral Supprt Activity • Multiple Statistical Analysis • Multisource Agreement • Muskrat Dam (Canada-ON) (location identifier)

MSAA Microsoft Active Accessibility

MSAAS Microgravity Science and Application Advisory Subcommittee

MSAC Medical Situational Awareness and Control • Microgravity Sciences Advisory Committee • Multiservice Access Concentrator

MSACM Microsoft Audio Compression Manager

MSACP Microgravity Science and Applications Cooperative Program

MSACSS Mount Stromlo Abell Cluster Supernova Search

MSAD Materials Summary Acceptance Document • Microgravity Science and Applications Division • Multisatellite Attitude Determination • Multisensor Algorithm Development

MSAE Master of Science in Aeronautical Engineering

MSAG Multifunction Self-Aligned Gate

MSAI Memorie della Società Astronomica Italiana

MSAIQ Memorie della Società Astronomica Italiana - Quarterly (periodical code)

MSAIS Memorie della Società Astronomica Italiana - Supplement (periodical code)

MSAM Master of Science in Applied Mechanics • Medium-Scale Anisotropy Measurement • Multisequential Access Method

MSAP Message Store Access Protocol • Modeling and Simulation Asset Plan • Multisatellite Attitude Prediction

MSAR Manpower Standards Application Routine • Microwave Spectrometer And Radiometer • Miniature Synthetic Aperture Radar • Monthly Significant Activity Report

MSART Multifunction Synthetic Aperture Radar Technology

MSAS Manpower Standards Application System • Mansfield and Sutton Astronomical Society (UK) • Mexborough and Swinton Astronomical Society (UK) • Mountain Skies Astronomical Society (USA-CA) • MSAT Satellite-based Augmentation System

MSASM Multiselection Automatic Sorting Machine

MSAT Minnesota Scholastic Aptitude Test • Mobile Satellite • Monitoring of Selected Astrophysical Targets (program) • Multisensor Acquisition and Targeting • Multisensor Aided Targeting

MSATT Mars' Surface and Atmosphere Through Time • Martian Surface and Atmosphere Through Time

MSAU Multistation Access Unit

MSAV Microsoft Antivirus

MSAW Minimum Safe Altitude Warning

MSAWS Minimum Safe Altitude Warning System

msb Minor Significant Breakthrough

MSB Magnetschwebebahn • Maintenance Standard Book • Major Significant Breakthrough • Manufacturing Studies Board • Material Support Branch • Medium Solid Booster • Minimally Schedulable Block • Mission Security Briefing • Mission Simulator Building • Mission Status Board • Mitteilungen der Sternfreunde der VHS Buchloe (Germany) • Molecular Superbubble • Most Significant Bit • Most Significant Byte • Saint-Martin Marigot (Guadeloupe) (location identifier)

MSBA Master of Science in Business Administration

MSBC Master of Science in Building Construction

MSBF Mean of Swaps Between Failures

MSBL Magnetosheath Boundary Layer

MSBLS Microwave Scanning Beam Land Station • Microwave Scanning Beam Landing System

MSBM Maîtrise en Sciences Biologiques et Médicales

MSBN Mobile Satellite Business Network

MSBP Median Surface Brightness Profile • Multiseed Backpropagation

Msc Miscellaneous

MSc Marine Stratocumulus • Master of Science

MSC Macro Selection Compiler • Magnetic Surface Current • Main Storage Control • Main System Controller • Maintenance Support Center • Major Support Command • Major System Command • Management Sectional Center • Manned Spacecraft Center (USA-TX) (now JSC) • Mapping Science Committee • Marginal Stability Criterion • Marine Science Council • Maritime Safety Committee (IMO) • Maryland Science Center (USA-MD) • Mass Spin Coupling • Mass Storage Control • Mass Storage Controller • Master Schedule Chart • Master Sequence Controller • Materials Service Center • Materiel Status Committee • Materiel Support Command • Materiel Support Concept • Mathematics Study Center • Memory Storage Control • Mesa Falcon (USA-AZ) (location identifier) • Message Sequence Chart • Message Switching Center • Meteorological Satellite Center • Meteorological Service of Canada • Meteorological Synthesising Center • Michigan Spectral Catalogue • Milestone Schedule Chart • Minimal Sufficient Condition • Ministerio de Sanidad y Consumo (Spain) • Minnesota Supercomputer Center (USA-MN) • Minnesota Supercomputing Center • Mirror Sign Convention • Mirror Spacecraft (XEUS) • Miscellaneous • Mission Support Calibration • Mission Support Center • Mission Support Component • Mobile Satellite Communication • Mobile Service Carrier • Mobile Service Center • Mobile Servicing Center • Mobile Switching Center • Modal Spiral Case • Model Selection Criterion • Mode Selector Controller • Mode Sequencing and Control • Moding - Sequencing - Control • Most Significant Character • Moved - Seconded - Carried • Moving Switching Centre • Multiple Scan Correlator • Multiple Star Catalog • Multiple Systems Coupling • Multiplicative Signal Correction • Multiservice Center

MSCADC Miniature Standard Central Air Data Computer

MSCB Manufacturing Status Control Board

MSCC Manned Space Laboratories Control Centre

MScD Master of Medical Science

MSCD Metropolitan State College of Denver (USA-CO) • Mirror Scan Correction Data

MSCDEX Microsoft CD-ROM Extension • Microsoft Compact Disk Extension

MSCE Main Storage Control Element • Master of Science in Civil Engineering

MSCF Master Surveillance Control Facility

MSCFO Master Surveillance Control Facility Operator

MSChE Master of Science in Chemical Engineering

MSCG Micro Strip Gas Chamber

M/SCI Mission/Safety Critical Item

MSCI Minnesota Supercomputer Center Inc. (USA-MN) • Mission Safety Critical Item

MSCIT Medium-Speed Communications Interface Terminal

MSCL Measurement Standards and Calibration Laboratory

MScMed Master of Medical Science

MSCP Management Systems Control Program • Massive Star Cluster Population • Mass Storage Control Processor • Mass Storage Control Protocol

MSCR Materiel System Computer Resources • Measurements/Stimuli Change Request

MSCS Microsoft Cluster Server • Mobile Service Center System • Multiservice Communications System

MSCTN Meteorological Satellite Center Technical Note (periodical code)

MSCU Modulated Shot and Continuum Uniform • Multisystem Communications Unit

MSD Doctor of Medical Science • Magnetic Storage Drum • Main-Sequence Detached (binary) • Major Seismic Disturbance • Management Support Division • Management Systems Designer • Manual Simulations Directive • Mapping Support Data • Marginal Support Date • Mass Storage Device • Material Support Data • Materiel Support Date • Materials and Structures Division • Mean Solar Day • Mean Squared Deviation • Mean Squared Difference • Mean Square Deviation • Mean Square Difference • Mean Square Displacement • Measurement in a Small Diaphragm • Medium Screen Display • Message Sequence Diagram • Meteorological Satellite Downlink • Method of Steepest Descent • Microsoft Diagnostics • Microsoft System Diagnostics • Minimization of the Sum of Deviations • Minimum Safe Distance • Mission Sequence Document • Mission Support Directorate • Mission Systems Data • Mobile Security Division • Modem Sharing Device • Most Significant Digit • Mount Pleasant (USA-UT) (location identifier) • Multiband Selection Device • Multiple Site Damage • Multisensor Display • Multispark Discharge • Museum of Scientific Discovery (USA-PA)

MSDA Multichip Systems Design Advisor

MSDB Main Storage Database • Mission Schedule Database

MSDC Mass Spectrometry Data Centre (RSC) • Midwest Space Development Corporation (USA)

MSDE Microsoft Desktop Engine

MSDL Microsoft Download Service

MSDN Microsoft Developer Network • Microsoft Developer's Network

MSDO Mapping Support District Office

MS-DOS Microsoft Disk Operating System

MSDOS Mass Storage Digital Operating System • Microsoft Disk Operating System

MSDP Maintenance Systems Development Program • Mission System Distribution Panel • Multicast Source Discovery Protocol • Multichannel Subtractive Double Pass (spectrograph)

MSDR Maintenance Signal Data Recorder • Master Sensor Data Record • Multiplexed Streaming Data Request

MSDRS Master Signal Data Recording Set

MSDS Magnetic Storage Drum System • Manpower Standards Development System • Material Safety Data Sheet • Materiel Safety Data Sheet • Microsoft Developer Support • Mission-Specific Data Set • Multispectral Scanner and Data System

MSDSS Microsoft Directory Synchronization Services

MSDT Maintenance Strategy Diagramming Technique • Multiple Shuttle Data Sheet

MSDU MAC Service Data Unit

MS&E Materials Science & Engineering

MSE Maintenance Support Equipment • Manned-Spaceflight Engineer • Manston (UK) (location identifier) • Mass Storage Editor • Master of Science in Education • Master of Science in Engineering • Master of Software Engineering • Materiel Status Evaluation • Mean Squared Error • Mean Square Error • Measuring and Stimuli Equipment • Mechanical Support Equipment • Medical Support Equipment • Mercury Surface Element • Microelectronics Systems Education • Microsoft Scripting Engine • Milton S. Eisenhower • Mission Staff Engineer • Mission Support Element • Mission Support Equipment • Mobile Subscriber Equipment • Modified Semi-Empirical (method) • Motional Stark Effect • Multicultural Science Education • Multimedia Software Engineering (symposium) • Multiple Site Event • Multiscale Entropy

MSEA Member of the Society of Engineering Associates • Modeling and Simulation Executive Agent

MSEB Mathematical Sciences Education Board

m/sec Meters per Second

msec Millisecond

MSec Millisecond

MSEC Master Separation Events Controller • Millisecond

MSECV Mean Square Error of Cross-Validation

MSEd Master of Science in Education

MSED Messaging Systems Engineering Directorate

MSEDS Maryland Students for the Exploration and Development of Space (USA-MD)

MSEE Master of Science in Electrical Engineering • Mean Square Error Efficiency • Mean Square Error of Estimation

MSEG Multispectral Environment Generator

MSEG&C Multispectral Environment Generator & Chamber

MSEGC Multispectral Environment Generator and Chamber

MSEI Mean Square Error Inefficiency

MSEIF Mean Square Error Improvement Factor

MSEL Master Scenario Events List • Master Sequence of Events List

MSELC Mean Square Error of Leverage Correction

MSEM Master of Science in Engineering Mechanics • Mission Status and Evaluation Module

MSEng Materials Science Engineering (periodical code)

MSEO Modified Simulated Evolutionary Optimization

MSEP Maintenance Standardization and Evaluation Program • Mean Square Error of Prediction • Multiple Systems Evaluation Program • Multisensor Electronic Package

MSEPS Modular Space Electrical Power Station

MSER Management System Evaluation Review

MSET Macvicar School of Education and Technology • Maintenance Standardization and Evaluation Team • Mathematics, Science, Engineering, and Technology

MSETI Microwave Search for Extraterrestrial Intelligence

MSEU Mass Storage Expansion Unit

MSF Maintenance Source File • Maintenance Support Facility • Manned Space Flight • Manufacturing, Science and Finance • Masse Sombre Froide • Massive Star Formation • Master Security File • Master Source File • Médecins Sans Frontières • Medium Standard Frequency • Message Scenario File • Message Security Filter • Meta-Scoring Function • Microsoft Solutions Framework • Million Square Feet • Mission Simulator Facility • Mission Support Facility • Mission Support Force • Month - Second - Foot • Mount Swan (Australia-NT) (location identifier) • Multiplex Signal Format • Multisensor Fusion

MSFBB Medium Solid Fast Burn Booster

MSFC Marshall Space Flight Center (NASA)

MSFDPS Manned Space Flight Data Processing System

MSFEB Manned Space Flight Experiments Board

MSFM Magnetospheric Specification and Forecast Model

MSFN Manned Space Flight Network

MSFNOC Manned Space Flight Network Operations Center

MSFO Manned Space Flight Operations

MSFP Manned Space Flight Program

MSFR Massive Star Formation Rate • Minimum Security Function Requirement

MSFS Mail Slot File System • Maintenance Shop Floor System • Microsoft Flight Simulator

MSFSG Manned Space Flight Support Group

MSFSRD Manned Space Flight Support Requirements Documentation

msg Message (file extension)

Msg Message

MSG Maintenance Steering Group • Maintenance Study Group • Maintenance Support Group • Maison des Sciences de l'Homme • Maîtrise en Sciences de Gestion • Management Supervisory Group • Marine Security Guard • Materiel System Group • Matsaile (Lesotho) (location identifier) • Mechanical System Group • Message • Meteosat Second Generation • Microgravity Science Glovebox • Miscellaneous Simulation Generator • Mission Support Group • Monosodium Glutamate • Multisource Group

MSGC Maryland Space Grant Consortium (USA-MD)

MSGCEN Message Center

MSGCM Mars Simple General Circulation Model

MSGID Message Identifier

MSGM Master of Science in Government Management

MSGMgt Master of Science in Government Management

Msgr Messenger • Monseigneur • Monsignor

MSGR Mobile Support Group

MSH Masirah (Oman) (location identifier) • Melanocyte-Stimulating Hormone • Mills - Slee - Hill (catalog)

MSHA Mine Safety and Health Administration • Mine Safety and Health Agency

MSHE Master of Science in Home Economics

Mshl Marshal

MSI Magnetic Shearing Instability • Maintenance Significant Item • Maintenance Status Interrogation • Maintenance Support Item • Management System Indicator • Manned Satellite Inspector • Masalembo (Indonesia) (location identifier) • Mass Sharing Index • Mathematical Science Institute • Medina Software Inc. (USA-FL) • Medium-Scale Integrated (circuit) • Medium-Scale Integration • Minnesota Supercomputer Institute (USA-MN) • Minority-Serving Institution (USA) • Mission Success Incentive • Moisture Stress Index • Moon Sphere of Influence • Multisensor Imagery • Multisensor Integration • Multisource Integration • Multispectral Imager • Multispectral Imagery • Museum of Science and Industry

MSIA Multispectral Image Analyzer

MSIAC Modeling and Simulation Information Analysis Center

MSIAS Multispectral Image Analysis System

MSIC Medical Support and Integration Contract

MSID Mass Spectrometric Isotope Detection • Master Stimulus Identification Document • Measurement and Stimulation Identification • Measurement Stimulation Identification • Measurement Stimulus Identification • Medium-Scale Integration Device • Minimization of the Sum of Interior Distances • Mnemonic String Identifier • Mobile Station Identification Number

MSIE Master of Science in Industrial Engineering • Microsoft Internet Explorer

MSIEP Modeling and Simulation Implementation and Execution Plan

MSIF Multisystem Integration Facility

MSIFO Marine Stratocumulus Intensive Field Observation

MSII Matra-SEP Image Informatique

MSIL Microsoft Intermediate Language

MSIM Mirror Simulator

MSIMD Mutiple Single Instruction - Multiple Data

MSIN Mobile Subscriber Identification Number

MSIO Mass Storage Input-Output

MSIP Modeling and Simulation Investment Plan • Multinational Staged Improvement Program • Multiple Source Interface Processor • Multispectral Image Processor • Multispectral Imagery Processor • Multistage Improvement Plan • Multistage Improvement Program

MSIS Management Science and Information Systems • Manned Satellite Inspection System • Man-Systems Integration Standards • Mass Spectrometer and Incoherent Scatter • Mass Spectrometer Incoherent Scatter • Mass Spectrometer Incoherent Scanner • Microgravity Science Information System • Modeling and Simulation Information System • Model State Information System • Multisensor Integrated System

MSISE Mass Spectrometer and Incoherent Scatter Extended

MSIT Major Switch Interoperability Test

MSIX Metered Services Information Exchange

MSJ Magnetic Society of Japan • Master of Science in Journalism • Misawa (Japan) (location identifier) • Multimedia Systems Journal (ACM)

MSK Manual-Selection Keyboard • Mastic Point (Bahamas) (location identifier) • Minimum Shift Key • Minimum Shift Keying • Minimum Spares Kit • Mission Support Kit

Mskg Masking

Mskr Manuskript

MSL Florence (USA-AL) (location identifier) • Machine Specification Language • Main Sea Level • Maintenance Supply Liaison • Map Specification Library • Marvel Strategy Language • Master of Science in Linguistics • Master Station Log • Master Support List • Master System Library • Materials Science Laboratory • Material Support List • Maximum Segment Lifetime • Mean Sea Level • Measurement Standards Laboratory • Mechanical Systems Laboratory • Meteorological Satellite Laboratory • Microgravity Science Laboratory • Microsoft Software Library • Mirrored Server Link • Mode Select Logic • Multilevel Secure • Multistrategy Learning • Muscle Shoals (USA-AL) (location identifier) • Sheffield (USA-AL) (location identifier)

MSLAP Modular Spectral Line Analysis Program

MSLCC Manned Space Laboratory Control Centre

MSLD Mass Spectrometer Leak Detector

MSLF Main Sequence Luminosity Function

MSLIP Multistage Stochastic Linear Programming System

MSLP Mean Sea Level Pressure • Medium-Speed Line Printer

MSLS Maneuverable Satellite Landing System • Master of Science in Library Science • Multiservice Launch System

MSLT Multiple Sleep Latency Test

Msly Mostly

MSM Magnetosphere Specification Model • Magnetospheric Specification Model • Manned Spaceflight and Microgravity (ESA Directorate) • Manned Support Module • Manufacturing Standards Manual • Mars Surface Module • Masi Manimba (Zaire) (location identifier) • Master Structure Management • Master Structure Manager • Materials Science and Mechanics • Mechanics Structures Machines (periodical code) • Media Support Module • Medical Sourcing Model • Meritorious Service Medal • Metal - Semiconductor - Metal • Mission Simulation Model • Mission Support Model • Mode Selection Mechanism • Mode Select Mechanism • Modified Source Manipulation • Multilayer Switch Module • Multiple Shooting Method • Multispectral Mapper • Multisystem Manager

MSMA Maintenance des Systèmes Mécaniques Automatisés

MSMC Master Schedule and Milestone Chart

MSME Master of Science in Mechanical Engineering

MSMFR Seminario Matematico e Fisico, Rendiconti, Milano (periodical code)

MSMIS Master of Science in Management Information Systems

MSMLC Medium-Speed Multiline Controller

MSMP Manned Support Module Program • Modeling and Simulation Master Plan • Multispectral Measurements Program

MSMQ Microsoft Message Queuing System

MSMR Multichannel Scanning Microwave Radiometer • Multispectral Multiresolution

MSMS Magnetic Signature Measurement System • Meteorological Systems Management Section

MSMSE Modeling Simul. Mater. Sci. Eng (periodical code)

MSMU Moscow State Mining University (Russia)

MSMW Magnetically Suspended Momentum Wheel

Msn Mission

MSN Madison (USA-WI) (location identifier) • Manned Space Network • Material Supply Notice • Message Switching Network • Microsoft Network • Mission • Mission Support Network • Monitoring Sequence Number • Multiple Subscriber Number • Multiple Systems Network

MSNAP Microwave Steerable Null Antenna Processor

MSNet Microsoft Network

MSNF Multiple System Networking Facility • Multisystem Network Facility

Msngr The Messenger (periodical code)

Msnry Masonry

MSO Main Scientific Organizer • Maintenance Standard Order • Management Science Office • Manned Solar Observatory • Manned Spacecraft Operation • Manufacturing Safety Office • Manufacturing Sequence Outline • Mars Surface Operation • Mass Spectrometer Outgasing • Master Shop Order • Materiel Status Office • Medial Superior Olive • Medium-size Symmetric Object • Medium Symmetric Object • Mees Solar Observatory (USA-HI) • Missoula (USA-MT) (location identifier) • Model for Spares Optimization • Mountain Skies Observatory (USA-WY) • Multiple-Systems Operator • Multiservice Operational

MSOB Manned Spacecraft Operations Building

MSOCC Multisatellite Operations Control Center

MSOE Milwaukee School Of Engineering (USA-WI) • Multiband Spectral Observation Equipment

MSOGS Molecular Sieve Oxygen Generating System

MSOR Modification Statement Of Requirement • Multiple Source Of Repair

MSOS Mass Storage Operating System

MSOSA Modeling and Simulation Operational Support Activity

MSOT Multiservice Operational Test

msp Manuscript Page • Microsoft Paint (file extension)

MSP Magnetic Scalar Potential • Magnetic South Pole • Maintenance Support Plan • Maintenance Support Program • Maintenance Surveillance Procedure • Management Service Provider • Management System Planning • Manpower Support Plan • Manuscript Page • Massive Slow Penetration • Mass Storage Processor • Master Simulator Program • Master Support Plan • Math and Science Partnership • Medical Supply Point • Medium-Speed Printer • Medium-Stressed Platform • Memory System Performance • Message Security Protocol • Message System Primitive • Microgravity Sensor Package • Microsoft Paint • Mid-level Service Provider • Millisecond Pulsar • Minimum Sustaining Power • Ministerio de Salud Pública • Minneapolis (USA-MN) (location identifier) • Mission Security Panel • Mission Sequence Plan • Mission Support Plan • Modular SAR Processor • Modular System Program • Mosaic Sensor Program • Motion-Sensing Platform • Multiband Sensor Package • Multiscale Profiler • Multisector Planning • Multiservice Platform • Multispectral Photography • Multispectral Sensor Package • Mutual Support Program • Saint Paul (USA-MN) (location identifier)

MSPB Merit Systems Protection Board

MSPC Microstrip Proportional Counter

MSPD Million Subscriber-Page-Days

MSPDS Microcomputer Suggestion Program Data System

MSPE Maintenance Safety and Protection Equipment

MSPF Multispectral Photographic Facility

MSPG Materiel Support Planning Guidance

MSPH Master of Science in Public Health

MSPhar Master of Science in Pharmacy

MSPHE Master of Science in Public Health Engineering

MSPLT Master Source Program Library Tape

MSPM Materiel Safety Program Manager

MSPO Middelbaar Sociaal-Pedagogische Opleiding

MSPR Maneuvering System Program Review

MSPRB Meteorological Satellite Program Review Board

MSPS Modular Self-Protection System • Modular Space Power Station

MSQ Managing Service Quality • Minsk (Belarus) (location identifier)

MSQL Multilevel Structured Query Language

MSR Machine-Specific Register • Machine State Register • Magnetic Shift Register • Magnetic Storage Ring • Magnetic Superresolution • Main Supply Route • Maintenance Status Reply • Management Status Report • Manufacturing Specification Request • Mark Sheet Reader • Mars Sample Return • Master Schedule Report • Material Status Report • Materiel Status Report • Mean Spring Rise • Measurement Stimuli Request • Measurement Support Request • Mechanized Storage and Retrieval • Meteorological Sounding Rocket • Meteorological Survey Radar • Microwave Scanning Radiometer • Milestone Review • Miniature Satellite Receiver • Minimum Security Requirement • Minimum Spatial Representation • Mission Support Recording • Mission Support Request • Mission Support Room • Model-Specific Register • Modification Status Report • Modified Stepwise Regression • Module

Support Rack • Monthly Status Report • Monthly Status Review • Montserrat (ISO country code) • Muon Spin Rotation • Mus (Turkey) (location identifier)

MSRAL Mid-States Region of the Astronomical League (USA)

MSRC Major Shared Resource Center • Marine Sciences Research Center (USA-NY)

MSRD Modeling and Simulation Requirements Document

MSRE Mobile Subscriber Radio Equipment • Mobile Subscriber Remote Element • Molten Salt Reactor Experiment • Moon Signal Rejection Equipment

MSRF Microwave Space Research Facility

MSRI Mathematical Sciences Research Institute (USA-CA)

MSRL Modeling and Simulation Reuse Library

MSRM Mars Sample Return Mission • Mass Storage Reference Model

MSRN Mobile Station Roaming Number

MSRO Mars Sample Return Orbiter

MSRP Manufacturer's Suggested Retail Price • Meteorological Sounding Rocket Program • Mission Support Real Property • Multichannel Solar Radio Polarimeter

MSRR Mission and System Requirements Review • Model and Simulation Resource Repository

MSRS Madurai Solar Radio Spectrograph • Materiel System Requirements Specification • Meteoroid Shield Release System

MSRSL Mémoires de la Société Royale des Sciences de Liège (periodical code)

MSRT Meter-wave aperture Synthesis Radiotelescope • Miyun Synthesis Radiotelescope (China-PRC) • Mobile Subscriber Radio Terminal • Mobile Subscriber Receiver-Transmitter

MSRTP Micelle-Stabilized Room Temperature Phosphorescence

MSRV Message Switch Rekeying Variable

mss Manuscripts • Manuscrits

Mss Manuskripte

MsS Magnetostrictive Sensor

MSS Magnetic Sensing System • Magnetic Storm Satellite • Magnetostrictive Sensor • Main Stellar Spectrograph • Main Support Structure • Maintenance Status System • Management Subsystem • Management Support Schedule • Management Support System • Manned Space Station • Manpower Staffing Standards • MAN Switching System • Manual Safety Switch • Manufacturers Standardization Society (USA) • Manuscripts • Manuscrits • Marconi Space Systems • Maritime Surveillance System • Massena (USA-NY) (location identifier) • Mass Storage System • Master of Social Sciences • Master of Social Service • Master of Space Studies • Maximum Segment Size • MBS Base System • Mechanical Support System • Medium-deep Spectroscopic Survey • Medium-sensitivity Sky Survey • Medium Satellite Servicing • Meeting Support System • Message Source Selector • Message Support System • Message Switching Station • Meteorological Satellite Section • Meteorological Service Singapore • Meteorological Sounding Station • Meteorological Sounding System • Metropolitan Switching System • Michigan Spectral Survey • Microwave Switching Station • Midlands Spaceflight Society (UK) • Miniature Surveillance System • Minimum Size of Shipment • Mission Safety System • Mission Simulation System • Mission Simulator System • Mission Specialist Station • Mission Status Summary • Mission Support Software • Mission Support System • Mobile Satellite Service • Mobile Service Structure • Mobile Servicing System • Mobile Subscriber System • Mobile Support Station • Mobile Support System • Modular Space Station • Multichannel Spectrum Scanner • Multiple Satellite System • Multiprotocol Switched Services • Multisatellite System • Multisensor Support Summary • Multishaker System • Multispectral Scanner • Multispectral Scanner System • Multistation Solution • Multisystem Software

MSSA Maîtrise en Sciences Sociales Appliquées

MSSAR Multispectral Synthetic Aperture Radar

MSSc Master of Social Sciences

MSSC Mass Storage System Communications • Maui Space Surveillance Complex (USA-HI) • Missouri Southern State College (USA-MO)

MSSCB Mass Storage and Scientific Computing Branch

MSSCC Multicolor Spin-Scan Cloud Camera

MSSCS Manned Space Station Communications System

MSSE Master of Science in Software Engineering • Message Submission Service Element

MSSI Multilevel System Security Initiative

MSSL Mullard Space Science Laboratory (UK)

MSSM Mars Spinning Support Module • Medium-Speed Sorting Machine • Minimal Supersymmetric Standard Model

MSSN Mission

MSSP Manpower Scheduling and Staffing Program • Mechanical Systems and Signal Processing (periodical code)

MSSQL Microsoft Structured Query Language

MSSR Mars Soil Sample Return • Mars Surface Sample Return • Mission and Science Scheduling Room • Monopulse Secondary Surveillance Radar

MSSRV Mars Surface-Sample Return Vehicle

MSSS Malin Space Science Systems (USA-CA) • Manned Space Station Simulator • Manned Static Space Simulator • Mass Spectral Search System • Maui Space Surveillance Site (USA-HI) • Maui Space Surveillance System • Multiple-Start Systematic Sampling • Multisatellite Support System

MSSSO Mount Stromlo and Siding Spring Observatories (Australia)

MSST Mean Sea Surface Temperature • Ministry of State for Science and Technology • Multiside Spectroscopic Telescope

MSSTA Multispectral Solar Telescope Array

MSSU Message Switching System Upgrade

MSSW Mission Support Software

MS&T Manufacturing Science & Technology

MST Maastricht (Netherlands) (location identifier) • Magnetospheric - Stratospheric - Tropospheric • Main Spacecraft Thruster • Maintenance Support Team • Maîtrise de Sciences et Techniques • Management Survey Team • Manufacturing Science and Technology • Mass Spectrometer Tube • Master of Science in Teaching • Mauritius Standard Time • MCCC Telemetry System • Mean Sidereal Time • Mean Solar Time • Mean Survival Time • Measurement Status Table • Mechanical Support Team • Mechanics Support Team • Mesosphere - Stratosphere - Troposphere • Microprocessor Simulation Technology • Miniature Satellite Transceiver • Minimal Spanning Tree • Ministry of Science and Technology • Mission Scheduling Test • Mission Science Team • Mission Selection Team • Mission Sequence Test • Mission Support Team • Mission Support Timeline • Mission System Trainer • Mobile Service Tower • MODIS Science Team • Module Service Tool • Monolithic Systems Technology • Mountain Standard Time • Multimission Software Transmission • Multisensor Tracking

M-STAR Merced Society for Telescope and Astronomy Recreation (USA-CA)

MSTAR Merced Society for Telescope and Astronomy Recreation (USA-CA)

MSTB Mission Simulator and Training Building

MSTC Manned Spacecraft Test Center • Multisource Track Control

MST&E Multiservice Test & Evaluation

MSTE Mechanical Special Test Equipment • Multiservice Test and Evaluation

MSTG Mass Storage Task Group • Materiel Safety Task Group

MSTI Miniature Seeker Technology Initiative • Miniature Seeker Technology Integration • Miniature Sensor Technology Integration (program) (*) • Moving/Stationary Target Indicator • Multispectral Thermal Imaging

MSTIRC Multiservice Test Investment Review Committee

MSTLY Mostly

MSTM Multisource Track Management

MSTO Main-Sequence Turnoff

MSTP Manned Space Transportation Programme

Mstr Master • Moisture

MSTR Master • Moisture • Multispectral Transient Radiometer

MSTS Multisource Tactical System • Multisubscriber Time-Sharing • Multisubscriber Time-sharing System

MSTTS Multiservice Terminal Test Set

MSTV Microsoft Television

MSU Main Storage Unit • Management System Unit • Mankato State University (USA-MN) • Maseru Moshoeshoe (Lesotho) (location identifier) • Mass Storage Unit • Mathematical Study Unit • Measuring Stimuli Unit • Medical Service Unit • Memory Service Unit • Memphis State University (USA-TN) • Message Service Unit • Message Signal Unit • Message Switching Unit • Message System Upgrade • Metropolitan State University (USA-MN) • Michigan State University (USA-MI) • Microwave Sounder Unit • Microwave Sounding Unit (*) • Million of Service Units • Minot State University (USA-ND) • Mississippi State University (USA-MS) • Mode Selector Unit • Modular Sorting Unit • Montana State University (USA-MT) • Montclair State University (USA-NJ) • Moorhead State University (USA-MN) • Morgan State University (USA-MD) • Moscow State University (Russia) • Murray State University (USA-KY)

MSUM Minnesota State University at Moorhead (USA-MN)

mSv Millisievert

MSV Maintenance Support Vessel • Major Software Vendor • Manned Space Vehicle • Martian Surface Vehicle • Mean Square Velocity • Mechanisms of Solar Variability • Mission Support Vehicle • Model and Scenario Visualization • Monitored Sine Vibration • Monticello (USA-NY) (location identifier) • Multispectral Video

MSVC Mass Storage Volume Control • Multispectral Visible Camera

MSVD Message Status Visual Display

MSVDP Message Status Visual Display Panel

MSVIAS Multispectral Video Imaging and Analysis System

MSVP Master Shuttle Verification Plan

MSW Machine Status Word • Magnetic Surface Wave • Mammals Species of the World • Massawa (Ethiopia) (location identifier) • Master in Social Work • Master of Social Welfare • Master of Social Work • Microswitch • Mikheyev - Smirnov - Wolfenstein (effect) • Multiple Shock Wave • Municipal Solid Waste

MSWD Mean Square Weighted Deviation

MSWG Megaprime Survey Working Group • Message Standards Working Group • Modeling and Simulation Working Group • Multilateral Science Working Group

MSWI Municipal Solid Waste Incinerator

MSWM Microsoft Windows Manager

MSWWF Ministerium für Schule, Weiterbildung, Wissenschaft und Forschung (Germany)

MSX Midcourse Space Experiment • Mossendjo (Congo) (location identifier)

MSY Maximum Sustainable Yield • New Orleans (USA-LA) (location identifier)

MSZ Namibe (Angola) (location identifier)

MSZT Magyar Szabványügyi Testület (Hungary)

MS2I Matra SEP Imagerie et Informatique

mt Maltese (ISO language code) • Megaton • Metric Ton

Mt Measurement • Megaton • Metical (currency) • Mont • Mount • Mountain

M/T Empty • Mail Transfer • Measurement Tons

MT Empty • Machine Translation • Maddalena Thaddeus • Mean Time • Magnetic Tape • Magnetic Test • Magnetic Torquer • Magnetic Tube • Magnetotelluric • Mail Transfer • Main Tank • Main Telescope • Maintenance Team • Maintenance Technician • Maintenance Test • Maintenance Time • Malta (ISO country code) • Malte (code ISO de pays) • Management Threshold • Mandated Territory • Man-Tended • Manual Test • Manufacturing Technology • Many Thanks • Mare Tranquilitatis (Moon) • Maritime Tropic • Mass-Temperature (relation) • Mass Transport • Master Timer • Master Tool • Materiel Test • Mato Grosso (Brazil) • Maximum Torque • Mean Tide • Mean Time • Measured Time • Measurement Ton • Mechanical Technician • Mechanical Time • Mechanical Translation • Mechanical Transport • Mechanized Terminal • Médecine du Travail • Media Table • MediaTheque • Medio-Temporal • Medium Technology • Medium Terminal • Megaton • Member Traceability • Même Temps • Message Transfer • Message Type • Metatheory • Metric Ton • Microptic Theodolite • Microwave Transducer • Minitutorial • Mismo Tiempo • Mission Time • Mission Timer • Mission Trainer • Mission Trajectory • Mi-Temps • Mittagshöhe • Mobile Target • Mobile Terminal • Mobile Terminated • Mobile Transfer • Mobile Transporter • Module Test • Modus Tolens • Monitor Telescope • Monotype • Monotype Times • Montana (USA) • Monsoon Through • More Than • Motor Transport • Mount • Mountain • Mountain Time • Movement Tracker • Moving Target • Multifrequency Transducer • Multiple Transfer • Multithreading

Mta Meseta

M&TA Monitoring & Trends Analysis

MTA Macintosh Telephony Architecture • Magnetic Torque Assembly • Magyar Tudományos Akadémia (Hungary) • Mail Transfer Agent • Mail Transport Agent • Maintenance Task Analysis • Maintenance Training Aid • Major Test Article • Major Trading Area • Major Training Area • Management Technology Analysis • Mass Thermal Analysis • Master Time Assembly • Materials Transfer Agreement • Matamata (New Zealand) (location identifier) • Mathematics Teaching Assistant • Message Traffic Analysis • Message Transfer Agent • Message Transport Agent • Metering Truss Assembly • Metropolitan Transit Authority • Metropolitan Transportation Authority • Micro-Thermal Array • Minimum Terms Agreement • Modification and Test Agreement • Mojave Test Area (USA-CA) • Monitoring and Trends Analysis • Motion-Time Analysis • Multiple Terminal Access • Multiterminal Adapter • Multithread Architecture • Multithreaded Architecture

MTAC Mailer's Technical Advisory Committee

MTACCS Marine Tactical Command and Control System

MTAE Message Transfer Agent Entity

MTAG Moving Target Advisory Group

Mtagne Montagne

MTAM Maritime Tropical Air Mass

MTAP Multitask Analysis Program

MTAS Middle Tennessee Astronomical Society (USA-TN) • Modular Target Acquisition System • Multisensor Target Acquisition System

MTAT Mean Turnaround Time

MTB Magnetic-Torquer Bar • Main Terminal Board • Market-To-Book (value) • Mass Transfer Binary • Materials Testing Branch • Mobility Testbed • Monte Libano (Colombia) (location identifier)

MTBB Mean Time Between Breakdowns

MTBCF Mean Time Between Critical Failures

MTBCM Mean Time Between Corrective Maintenances

MTBCMA Mean Time Between Corrective Maintenance Actions

MTBD Mean Time Between Demands • Mean Time Between Discrepancies

MTBDE Mean Time Between Downing Events

MTBDL Mean Time Between Data Loss

MTBE Mean Time Between Erections • Mean Time Between Errors • Mean Time Between Events

MTBEMA Mean Time Between Essential Maintenance Actions

MTBERA Mean Time Between Essential Replacement Actions

MTBF Mean Time Before Failure • Mean Time Between Failures (*) • Meson Test Beam Facility

MTBFA Mean Time Between False Alarms

MTBFF Mean Time Between First Failures

MTBFL Mean Time Between Function Losses

MTBFRO Mean Time Between Failures Requiring Overhaul

MTBHCF Mean Time Between Hardware Critical Failures

MTBI Mathematical and Theoretical Biology Institute (Cornell) • Mean Time Between Incidents

MTBJ Mean Time Between Jams

MTBM Mean Time Between Maintenance • Mean Time Between Modifications

MTBMA Mean Time Between Maintenance Actions

MTBMCF Mean Time Between Mission-Critical Failures

MTBMF Mean Time Between Mission Failures

MTBNDR Mean Time Between No Defect Removals

MTBO Mean Time Between Outages • Mean Time Between Overhauls

MTBOMF Mean Time Between Operational Mission Failures

MTBOUMA Mean Time Between Operational Unscheduled Maintenance Actions

MTBPM Mean Time Between Preventive Maintenance

MTBR Mean Time Between Removals • Mean Time Between Repairs • Mean Time Between Replacements

MTBS Mean Time Between Services

MTBSD Mean Time Between Supply Demands

MTBSF Mean Time Between Significant Failures • Mean Time Between System Failures

MTBSM Mean Time Between Scheduled Maintenance

MTBUM Mean Time Between Unscheduled Maintenance

MTBUMA Mean Time Between Unscheduled Maintenance Actions

MTBUR Mean Time Between Unscheduled Removals • Mean Time Between Unscheduled Replacements

MTC Magnetic Tape Control • Magnetic Tape Controller • Maintenance Task Cycle • Maintenance Time Constraint • Man-Tended Capability • Manual Thrust Controller • Manual Traffic Control • Manufacturing Technology Center • Massachusetts Technology Collaborative • Master Tape Control • Master Thrust Control • Master Thrust Controller • Master Timing Center • Mathematics Throughout the Curriculum • Maximum Track Capacity • Measurement Technology Center • Measurement Type Code • Members - Team - Committee • Memory Test Computer • Message Type Code • Meteorological Training Center • Metropolitan Technology Center (SJSU) • MIDI Time Code • Midlands Technical College (USA-SC) • Mission and Test Computer • Mission and Traffic Control • Mitsui Trading Co. (Japan) • Mobile Terminating Call • Moderator Temperature Coefficient • Module de Transfert de Charge • Monitor Test Control • Mount Clemens - Selfridge (USA-MI) (location identifier) • Movement To Contact • Moving Trajectory Coordinates

MTCA Minimum Terrain Clearance Altitude • Monitor and Test Control Area • Multiple-Terminal Communication Adapter

MTCAC Midlands Technical College Astronomy Club (USA-SC)

MTCC Master Timing and Control Circuit

MTCD Medium Term Conflict Detection

MTCE Million Tonnes of Carbon Equivalent

MTCF Mean Time to Catastrophic Failure

Mtchd Matched

MTCIP Manual Technical Control Improvement Program

MTCN Management Task Control Number

MTCP Modern Technology Cockpit Program

MTCR Million Tonnes of Coal Equivalent • Mission Technology Control Regime

MTCS Meteor Trail Communications System • Minimal Terminal Communications System • Mobile Telemetry Computing System • Model for Time-Critical Systems • Modular Terminal Communications System

MTCU Magnetic Tape Control Unit

Mtd Mounted

MTD Machine-Traceable Dictionary • Maintenance Training Device • Master Tape Data • Master Time Display • Master Tracking Data • Material Technology Division • Maximum Tolerated Dose • Mean Temperature Difference • Mechanical Thermal Dummy • Microelectronics Technology Division • Minimal Toxic Dose • Mobility Technology Demonstration • Mounted • Mount Sandford (Australia-NT) (location identifier) • Moving Target Detection • Multimedia Transport Protocol • Muon Tracking Detector

MTDA Mean Time to Data Availability • Mesocyclone Tornado Detection Algorithm • Modification Table of Distribution and Allowances

MTDC Modified Total Direct Cost

MTDD Master Time Division Demultiplexer • Mean Time between Disclosure and Diagnosis

MTDE Multimedia Title Development Environment

MTDEC Magnetic Tape Dubbing and Evaluation Center

MTDF Master Tracking Data File

MTDL Mean Time to Data Loss

MTDM Master Time Division Multiplexer

mtDNA mitochondrial DNA

MTDP Multisensor Trajectory Data Package

MTDR Mean Time between Diagnosis and Repair

MTDS Manufacturing Test Data System • Masuda Tracking and Data Station (Japan)

MTDSK Magnetic Tape Disk

Mte Monte

M&TE Measuring & Test Equipment

MTE Magnetic-Tape Emulator • Magnetic-Torquer Electronics • Magnetosphere Thermosphere Explorer • Mail Transport Equipment • Maintenance Test Equipment • Maintenance Training Equipment • Maximum Tracking

Error • Mean Transverse Energy • Measuring and Test Equipment • Modular Toolkit Environment • Monte Alegre (Brazil-PA) (location identifier) • Moving Target Exploitation • Multisystem Test Equipment

MTEC Mail Transport Equipment Center • Maintenance Test Equipment Catalog

MTEE Mission Time Extreme Environment

MTEFP Ministère du Travail, de l'Emploi et de la Formation Professionnelle

MTEL Manning Table and Equipment List

MTEM Maintenance Test Equipment Module

MTEP Maritime Technology Enhancement Program • Mission Training Evaluation Plan

Mtes Montes

MTESC Mail Transport Equipment Service Center

MTF Maintenance Test Facility • Maßtensorfunktion • Master Test Frame • Materials Test Facility • Mean Time to Failure • Mechanical Time Fuze • Medical Treatment Facility • Message Text Format • Message Transfer Facility • Microsoft Tape Format • Mid-Tandem Fan • Mizan Teferi (Ethiopia) (location identifier) • Mobile Test Facility • Modulated Transfer Function • Modulation Transfer Frequency • Modulation Transfer Function (*) • Move To Front

MTFD Mean Time between Failure and Disclosure • Minimum Tracking Flux Density

MTFE Mid-Thrust Family Engine • Multifrequency Front End

MTFF Man-Tended Free Flyer • Mean Time to First Failure

MTFFCC MTFF Control Center

MTFHE Mean Time to First Human Error

MTFO Modular Training Field Option

MTFR Mean Time between Failure and Repair

mTFS Multichannel Fourier Transform Spectrometer

MTFS Magnetic Tape Field Search • Medical Treatment Facility System

Mtg Meeting • Mortgage • Mounting

MTG Master Timing Generator • Mato Grosso (Brazil-MT) (location identifier) • Meeting • Mounting

MTGCM Mars Thermospheric General Circulation Model

MTGCT Master Timing Generator Controller

MTGDR Master Timing Generator Driver

Mtge Montagne • Mortgage

Mtgee Mortgagee

Mtgne Montagne

Mtgor Mortgagor

MTGP Monitor Tables Generator Program

MTGW Maximum Total Gross Weight

Mth Month

MTH Magnetic Tape Handler • Magnetic Tape Holder • Marathon (USA-FL) (location identifier)

Mthbd Motherboard

Mthd Method

MTHE Mean Time Between Human Errors

MTHLY Monthly

MTHM Metric Ton of Heavy Metal

MTHR Moderate To High Resolution

Mths Months

Mti Monti

MTI Mass Transfer Instability • Méthode et Traitement de l'Information • Morton Thiokol Inc. (USA) • Mosteiros (Cape Verde) (location identifier) • Moving-Target Identification • Moving-Target Identificator • Moving-Target Imaging • Moving-Target Indication • Moving-Target Indicator • Multilayer Thermal Insulation • Multispectral Thermal Imager

MTIAC Manufacturing Technology Information Analysis Center

MTIGCM Mesosphere/Thermosphere/Ionosphere General Circulation Model

MTIE Microthrust Ion Engine

MTIM Mass Transfer Instability Model

M-TIS Micro Test Item Simulator

M/TIS Management/Technical Information System

MTIS Maintenance Task Information System • Micro Test Item Simulator

MTJ Magnetic Tunnel Junction • Montrose (USA-CO) (location identifier)

MTK Makin Island (Kiribati) (location identifier) • Mechanical Time Keeping • Mechanized Time Keeping

Mtl Material • Monatlich

MTL Lira (Malta) (ISO currency code) • Main Transfer Line • Maitland (Australia-NSW) (location identifier) • Master Tape Loading • Master Target List • Master Task Listing • Master Timeline • Material • Mean Tide Level • Medial Temporal Lobe • Merged-Transistor Logic • Message Transfer Layer • Minimum Time Limit • Mission Target List • Mission Tasking Letter • Mission Time Life • Mission Timeline • Moving Target Locator • Multiconductor Transmission Line

MTLP Master Tape Loading Program • Monitor Table Listing Program

MTLRS Modular Transportable Laser Ranging System

M-T-M Model-Test-Model

MTM Maintenance Test Module • Mean Time Measurement • Mechanical Test Model • Method of Task Measurement • Methods-Time Measurement • Metlakatla (USA-AK) (location identifier) • Midnight Temperature Mission • Million - Ton - Miles • Mission Table Management • Mission Tasking Message • Mission To Mars • Model - Test - Model • Modification Transmittal Memorandum • Modified Trimmed Mean (filter) • Monsieur Tout le Monde • Multiple Terminal Manager • Multitaper Method • Multitarget Model • Multitask Monitor • Multiterminal Monitor

MTMB Master Target Model Book

MTMM Multitrait Multimethod • Multitrait Multimethod Model

MTMO Mated Test Management Office

MTMS Magnetic Tape Messaging System • Methyl Tri-Methoxy Silane

MTMT Master Target Model Tape

Mtn Motion • Mountain

MTN Baltimore - Glenn L. Martin (USA-MD) (location identifier) • Main Telecommunications Network • Méthodes et Techniques Nouvelles • Mobile Telephone Network • Mountain • Multilateral Trade Negotiations • Multilevel Tree Network • Multinational Trade Negotiations

MTNS Mathematical Theory of Networks and Systems • Mountains

Mto Monumento

MTO Made To Order • Magnetic Tape Operator • Master Timing Oscillator • Mattoon (USA-IL) (location identifier) • Maximum Timeout • Message To Observer • Ministry of Transport of Ontario (Canada-ON) • Mission - Task - Objectives • Mission Type Order • Mississippi Test Operations • Mobile Technical Office • Modification Task Outline • More Than One • Motor Transport Officer • Mount Tamborine Observatory (Australia-QLD) • Mt. Tamborine Observatory (Australia-QLD) • Multilateral Trade Organization

MTOB Manned Test Operations Board

MTObM Metallovedenie i Termicheskaia Obrabotka Metallov (periodical code)

MTOE Million Tonnes of Oil Equivalent • Modification Table of Organization and Equipment

MTOF Mass-determining Time Of Flight

MTOGW Maximum Takeoff Gross Weight

MTON Measurement Ton • Metric Ton

MTOpS Million Theoretical Operations per Second

MTOPS Million Theoretical Operations Per Second (*)

MTOS Magnetic Tape Operations System • Metal-Thick Oxide Silicon

MTOW Maximum Takeoff Weight

MTP Magnetic Tape Processor • Mail Transfer Protocol • Maintenance Test Package • Maintenance Test Pilot • Maintenance Test Plan • Manufacturing Technical Procedure • Master Test Plan • Master Test Procedure • Master Timing Pulse • Master Training Plan • Materiel Transfer Plan • Materiel Transportability Plan • Mathematical Tables Project • Mechanical Thermal Pulse • Mediterranean Targeted Project • Medium-Term Perspective • Medium-Term Plan • Medium-Term Prediction • Message Transfer Part • Meteosat Transitional Programme • Meteosat Transition Programme (*) • Microwave Temperature Profile • Microwave Temperature Profiler • Mission Tasking Package • Mission Test Plan • Mission Training Plan • Montauk (USA-NY) (location identifier) • Multicast Transport Protocol • Multiple-Task Performance

MTPE Mission To Planet Earth

MTPF Maximum Total Peaking Factor • Minimal Total Processing Time

MTPMM Metallurgical Transactions Physical Metallurgy Materials Science (periodical code)

MTPN Multiprotocol Transport Network

MTPS Magnetic Tape Plotting System • Magnetic Tape Programming System • Multitransport Protocol Service

MTQ Magnetic Torquer • Martinique (code ISO de pays) • Mitchell (Australia-QLD) (location identifier)

Mtr Meter

MTR Magnetic Tape Reader • Magnetic Tape Recorder • Magnetic Tape Recording • Materiel Testing Report • Maximum Tracking Range • Mean Time to Removal • Mean Time to Repair • Mean Time to Replacement • Mean Time to Restore • Mechanized Time Reporting • Medium-Term Research • Meter • Minimum Time Rate • Minitemperature Recorder • Mobile Telemetry Receiver • Mobile Test Rig • Modernization Transition Plan • Modular Tree Representation • Module Test and Repair • Monteria (Colombia) (location identifier) • Motor • Multiple Tracking Range • Multiple Track Radar • Multisensor Trajectory Report

MTRAP Magnetic Transition region Probe

MTRC Mobile Teacher Resource Center

MTRD Master Test Requirement Document

MTRE Magnetic-Tape Recorder End

MTRF Mean Time to Reduced Functionality

Mtrg Metering

MTRMA Meteor Forschungsergebnisse Reihe B Meteorologie und Aeronomie (periodical code)

MTRR Magnetic-Tape Recorder-Reproducer

MTRS Magnetic-Tape Recorder Start • Magnetic-Tape Reformatting System • Mobile Telemetry Repeater System

Mts Monts • Mountains • Mounts

MTS Magnetic-Tape Station • Magnetic-Tape System • Magnetic Torquing System • Maintenance Training Set • Main Thruster System • Managerial Training System • Manzini (Swaziland) (location identifier) • Master Test Schedule • Master Test Station • Master Timing System • Material Test Specification • Medium-Term Schedule • Medium-Term Scheduler • Medium-Term Scheduling • Member of the Technical Staff • Memory Test System • Message Telecommunications Service • Message Telephone Service • Message Toll Service • Message Transfer Service • Message Transfer System • Meteoroid Technology Satellite • Metering Truss Structure • Meter - Ton - Second • Metric Time System • Microsoft Transaction Server • Microwave Temperature Sounder • Microwave Tracking System • Middelbare Technische School • Millimeter-wave Temperature Sounder • Mission Telemetry System • Mission test Specification • Mobile Telephone Service • Mobile Tracking Station • Mobile Training Set • Modernization Through Spares • Modular Training System • Modulation Transfer Service • Modulation Transfer System • Module Test Set • Motor-operated Transfer Switch • Motor Turbine Ship • Mountains • Movement Tracking System • Moving Target Simulator • Moving Time Series • MSX Tracking Station • Multichannel Television Sound • Multiple Time Scale • Multithreaded Server

MTSA Multitaper Spectral Analysis

MTSAT Multifunction Transport Satellite

MT/SC Magnetic Tape Selective Composer

MTSC Magnetic Tape Selective Composer • Maintenance Technical Support Center

MTSE Magnetic-Tape Splicing Equipment • Message Transfer Service Element

MTSL Microelectronics Technology Support Laboratory

MTSO Mobile Telephone Switching Office

MtSOM Mount Stromlo Observatory Mimeographs (periodical code)

MTSP Maintenance Test Support Package • Multiple Traveling Salesmen Problem

MTSR Magnetic Tape Selective Recorder • Mean Time to Service Repair • Mean Time to Service Restoral • Mid-Term Status Report

MtSeM Mount Stromlo Observatory Reprints (periodical code)

MTSS Magnetic-Tape Storage System • Maintenance and Test Simulation System • Ministerio de Trabajo y Seguridad Social (Spain)

MTST Medium-Term Scheduling Tool

MTSU Magnetic-Tape Search Unit • Middle Tennessee State University (USA-TN)

MTSV Medium Timescale Variability

MTT Magnetic-Tape Terminal • Magnetic-Tape Transport • Maintenance Training Team • Maximum Touch Temperature • Mean Transit Time • Migration Tracking Tool • Minatitlan (Mexico) (location identifier) • Mi-Temps Thérapeutique • Mobile Test Trailer • Mobile Training Team • Multiple Target Tracking • Multiple Telescope Telescope • Multitelescope Telescope • Multitransaction Timer

MTTA Mean Time To Accomplish • Mean Time To Assemble

MTTB Mean Time To Bench

MTTC Magnetic Tape Transport Controller

MTTD Mean Time To Defect • Mean Time To Diagnose • Mean Time To Diagnosis

MTTE Magnetic-Tape Terminal Equipment • Mean Time To Exchange

MTTF Mean Time To Failure • Mean Time To Fix

MTTFF Mean Time To First Failure

MTTL Mean Time To Load

MTTM Mean Time To Maintain

MTTPO Mean Time To Planned Outage

MTTR Mean Time To Recover • Mean Time To Recovery • Mean Time To Removal • Mean Time To Repair • Mean

Time To Replacement • Mean Time To Restore • Mean Time To Return • Medium-Term Technology Review • Moyenne des Temps Totaux de Réparation

MTTRes Mean Time To Restore

MTTRF Mean Time To Restore Functions

MTTRS Mean Time To Restore Service • Mean Time To Restore System

MTTS Magnetic Tape Terminal Station • Microwave Theory and Technical Society • Multiple Target Tracking System • Multitask Training System

MTTU Modular Timing Terminal Unit

MTTUO Mean Time To Unplanned Outage

MTTY Maintenance Module Teletype

MTU Magnetic-Tape Unit • Maintenance Termination Unit • Master Terminal Unit • Master Time Unit • Master Timing Unit • Maximum Transfer Unit • Maximum Transmission Unit • Message Transfer Unit • Michigan Technological University (USA-MI) • Mission Timer Unit • Mobile Test Unit • Mobile Training Unit • Momentum Transfer Unit • Montepuez (Mozambique) (location identifier) • Motoren- und Turbinen-Union • Multiplexer and Terminal Unit

MTV Mars Transfer Vehicle • Mean Transformed Value • MEA Test Vehicle • Medium Timescale Variability • Message Translation and Validation • Mostly Traditional Value • Mota Lava (Vanuatu) (location identifier) • Motoren und Turbinen Verband • Music Television

MTVC Manual Thrust Vector Control

MTVS Mission Test and Video System

MTW Manitowoc (USA-WI) (location identifier) • Maximum Taxi Weight • Mechanical and Thermal Works • Monday - Tuesday - Wednesday • Mountain Waves

MTWA Maximum Total Weight Authorized

MtWAR Mount Wilson and Palomar Observatory Annual Report (periodical code)

MtWRe Mount Wilson and Palomar Observatory Reprints (periodical code)

MTWRF Monday - Tuesday - Wednesday - Thursday - Friday

MTWY Motorway

Mtx Matrix

MTX Fairbanks Metro (USA-AK) (location identifier) • Mobile Telephone Exchange

MTY Empty • Monterrey - General Mariano Escobedo (Mexico) (location identifier)

MTZ Magnetic Transition Zone • Massada (Israel) (location identifier) • Miller - Tucker - Zemlin (formulation)

m/u Make Up

M/U Mockup

MU Île Maurice (code ISO de pays) • Machine Unit • Macquarie University (Australia) • Mail Unit • Maintenance Unit • Make Up • Maneuvering Unit • Master Unit • Math Unit • Mauritania • Mauritius (ISO country code) • Measurement Unit • Memory Unit • Message Unit • Microwave Unit • Middle and Upper • Minority University • Mobile Unit • Mobile User • Mockup • Model Updating • Monetary Unit • Multiple Unit • Multiplexing Unit

MUA Mail User Agent • Materials Usage Agreement • Maximum Usable Altitude • Memorandum of Understanding and Agreement • Message User Agent • Munda (Solomon Islands) (location identifier) • Mutual Usage Agreement

MUAO Mars Upper Atmosphere Orbiter

MUAP Massive Unassisted Penetration

MUB Maun (Botswana) (location identifier)

MUBAR Messed Up Beyond All Recognition

MUBEE Moscow University Balloon Emulsion Experiment

MUC McDonald (Observatory) Users Committee • Mesoscale Uncinus Complex • Message Understanding Conference • Million d'Unités de Compte • Mount Union College (USA-OH) • München - Franz-Josef Strauss (Germany) (location identifier) • Munich - Franz-Josef Strauss (Germany) (location identifier)

MUCELS Multiple Communications Emitter Locating System

MUCOOL Muon Cooling (experiment)

MUCS Modulated Uniform and Continuum Shot • Multiuser Compliant Scheduling

MUD Memory Unit Drum • Miscellaneous User Documents • Mueda (Mozambique) (location identifier) • Multiple User Dialogue • Multiple User Dimension • Multiple User Dungeon • Multiuser Dialog • Multiuser Dimension • Multiuser Domain • Multiuser Dungeon

MUDAS Modular Universal Data Acquisition System

MUDOF Metalearning Using Document Features

MUDPAC Modular Unit Deployable Package

MUDS Man-hour Use Data System

MUDSS Mobile Underwater Debris Survey System

MUE Kamuela (USA-HI) (location identifier) • Mission Unique Electronics • Mission Unique Equipment • Modification Unique Equipment

MUET Mission Universitaire et Éducative de Tunisie (Tunisia)

MuF Multiplicity Factor

MUF Material Unaccounted For • Materiel Unaccounted For • Maximum Urgency First • Maximum Usable Frequency • Muting (Indonesia) (location identifier)

MUFA Microwave Ultrafast Absorption

MUFFIN Multichannel Finite-impulse-response Filter and Integrator

MUFON Mutual Unidentified Flying Object Network

MUFP Minority Undergraduate Fellows Program

MÜG Mietenüberleitungsgesetz

MUG Maximum Usable Gain • McIDAS Users Group • Microelectronics Users Group • Mulege (Mexico) (location identifier)

MUH Mersa Matruh (Egypt) (location identifier)

MUI Fort Indiantown - Muit (USA-PA) (location identifier) • Management User Interface • Multimedia User Interface

MUID Map Unit Identification Data

MUIR Map Unit Interpretation Record

MUJ Mui (Ethiopia) (location identifier)

MUK Mauke Island (Cook Islands) (location identifier)

MUKLIMO Mikroskaliges Urbanes Klima-Modell

MUL Master Unit List • Master Urgency List • Material Usage List • Mission-Unique List • Mobile User Link • Moultrie Spence (USA-GA) (location identifier) • Multiply

MULC Monthly Usage License Charge

MULDEM Multiplexer-Demultiplexer

MULE Manned-Unmanned Lunar Explorer • Modular Universal Laser Equipment

MULSAM Multispectral Auger Microscope • Multispectral Auger Microscopy

MULSORANNS Multiple Linked Self-Organized Artificial Neural Networks

Mult Multiple • Multiplexer • Multiplication

MULT Multiple • Multiply

MULTDOP Multiple Doppler Radar Analysis Program

MULTICS Multiplexed Information and Computing Service • Multiplex Information Computer Service

MULTIFADS Multispectral Fast Area Digitizing Scanner

Multr Multiplier

MUM Maximum Useful Magnification • Mesoscale Unified Model • Method of Unweighted Means • Miszipped/Unzipped Mail • Mopra User's Manual • Multiple-Unit Message • Multiuse Manuscript • Multiuser Minicomputer • Mumias (Kenya) (location identifier)

MUMS Mobile Utility Module System • Multifrontal Massively Parallel Solver

Mun Municipal • Municipality

MUN Maturin (Venezuela) (location identifier) • Memorial University of Newfoundland (Canada-NF)

Muni Municipal

Munic Municipal • Municipality

MUNICS Munich Near-Infrared Cluster Survey

MUO Mountain Home (USA-ID) (location identifier)

MUOD Mean Unplanned Outage Duration

MUP Massive Unassisted Penetration • Master of Urban Planning • Metrics Utilization Plan • Mobile User Part • Mulga Park (Australia-NT) (location identifier) • Multiple UNC Provider

MUPPET Maryland University Project in Physics and Educational Technology

MUPPS Mission-Unique Planning and Processing Schedule

MUPS Mobile Uninterruptible Power System • Momentum Unloading Propulsion System • Multiple-Utility Peripheral System

MUQ Main Users Quarters • Muccan (Australia-WA) (location identifier)

MUR Manpower Utilization Report • Marudi (Malaysia) (location identifier) • Maximum Uncertainty Requirement • Rupee (Mauritius) (ISO currency code)

MURA Midwestern Universities Research Association (USA)

Murc Murcia (Spain)

MURC Minority University Research Center (USA)

MURED Minority University Research and Education Division (NASA)

MURF Materiel Utilization Reference File

MURG Macquarie University Research Grant

MURI Multidisciplinary University Research Initiative (USA)

MURS Management Update and Retrieval System

MURST Ministero dell'Università, della Ricerca Scientifica e Tecnologica

Mús Música

mus Music (file extension)

Mus Musca-Muscae (constellation) • Museum • Music • Musical • Musician

MUS Île Maurice (code ISO de pays) • Marcus Island (Japan) (location identifier) • Mauritius (ISO country code) • Mission-Unique Software • Multilateral Utilization Study • Multiuser System • Multiutility System

MUSA Multiple-Unit Steerable Antenna

MUSAP Multisatellite Augmentation Program

MUSC Microgravity User Support Center

MUSCLE Multiple Scattering Lidar Experiment

MUSE Mapping Utility Software Environment • Measurements of the Unresolved Spectrum of the Earth • Monitor of Ultraviolet Solar Energy • Multidimensional User-oriented Synthetic Environment • Multielement Ultraviolet Spectroscopic Experiment • Multimedia-based System Engineering • Multiple Subnyquist Encoding • Multiunit Spectroscopic Explorer • Multiuser Science Environment • Multiuser Shared Environment • Multiuser Simulated Environment

MUSES Materials Use: Science, Engineering and Society • Mu Space Engineering Satellite (*) • Mu Space Engineering Spacecraft

MUSH Multiuser Shared Hallucination

MUSIC Management - User interface - System interface - Information - Communication • Metrics for Usability Standards In Computing • Multifunctional Stereo Imaging Camera • Multiple Signal Classification • Multispectral Infrared Camera • Multiwavelength Sampling of Interacting Clusters

MUSICOS Multisite Continuous Spectroscopy

MUSIP Multisensor Image Processing • Multisensor Image Processor

MUSS Module Utility Support Structure • München Universitäts Schriften Serie (periodical code)

MUSSL Marine Upwelled Spectral Signature Laboratory

MUSSWAS Multisensor Surface Wind Acquisition System

MUST Mbarara University of Science and Technology (Uganda)

MUSTARD Multiunit Space Transport And Recovery Device

MUSTIP Multilateral Strategic and Tactical Investigation Process

Mut Mutilated • Mutual

MUT Manpack UHF Terminal • Mean Up Time • Measurements Unit Table • Model Updating Technique • Module Under Test • Monitor Under Test • Muscatine (USA-IA) (location identifier)

MUTA Multiple Unit Training Assembly

MUTE Multiplex Unit for Transmission Elimination

MUTTS Multiple-Unit Terminal Test Set

MUU Main User Unit • Mount Union (USA-PA) (location identifier)

muv Met Uitzondering Van

MUV Mechanized Utility Vehicle • Middle Ultraviolet • Mid-Ultraviolet • Philadelphia Mustin (USA-PA) (location identifier)

MUVE Multiuser Virtual Environment

MUVI Multispectral Ultraviolet Imager

MUVSK Moskovskii Universitet Vestnik Seriia Khimiia (periodical code)

MÜW Monopolübertragungswege

MUW Main User Workshop • Mascara (Algeria) (location identifier)

MUX Multan (Pakistan) (location identifier) • Multiplex • Multiplexer • Multiplexing

MUXing Multiplexing

MUY Mouyondzi (Congo) (location identifier)

MUZ Main User Zone • Musoma (Tanzania) (location identifier)

mv Move • Visual Magnitude

mV Millivolt

Mv Absolute Visual Magnitude • Mendelevium • Move

M/V Motor Vessel

MV Main Verb • Maldives (ISO country code) • Manned Vehicle • Manpower Voucher • Manual Valve • Manufacturing Verification • Mare Vaporum (Moon) • Market Value • Mass-Velocity • Mean Value • Mean Variation • Measured Value • Medium Voltage • Meervoud • Megavolt • Merchant Vessel • Mercury Vapor • Messagerie Vocale • Mietverhältnis • Mietvertrag • Miniature Vehicle • Mitgliedversammlung • Mixed Valence • Mobile VLBI • Motion Video • Motor Vessel • Multivibrator

MVA Machine Vision Applications (workshop) • Main Valve Actuator • Manufacturing Value Added • Mean Vertical Acceleration • Megavolt-Ampere • Migration Velocity Analysis • Minimum Vectoring Altitude • Motion Video Accelerator • Motor Vehicle Accident • Müllverbrennungsanlage • Multidomain Vertical Alignment • Multivendor

Architecture • Mývatn Reykiahlid (Iceland) (location identifier)

MVAAC Martha's Vineyard Amateur Astronomy Club (USA-MA)

MVAC Mission Valley Astronomy Club (USA-MT)

MVACS Mars Volatiles And Climate Surveyor

MVAR Magnetic Variation

MVAS Magic Valley Astronomical Society (USA-ID) • Mahoning Valley Astronomical Society (USA-OH) • Miami Valley Astronomical Society (USA-OH) • Mohawk Valley Astronomical Society (USA-NY) • Multipurpose Ventricular Actuating System

MVB Franceville (Gabon) (location identifier) • Minimum Variance Bound • Multimedia Viewer Book

Mvbl Movable

MVBL Magneto-Viscous Boundary Layer

MVC Manual Volume Control • Master Volume Control • Maximum Value Composite • Medium-Velocity Cloud • Model View Controller • Monroeville - Monroe County (USA-AL) (location identifier) • Multimedia Viewer Compiler • Multivolume Cartridge

MVCO Mahoning Valley Cortese Observatory (USA-OH)

MVCP Management Visibility and Control of Provisioning

MVCU Multivariable Control Unit

MVD Map and Visual Display • Mission Verification Data • Montevideo Carrasco (Uruguay) (location identifier) • Multivalent Document

MVDA Multivariate Discriminant Analysis

MVDM Multiple Virtual DOS Machine

MVDS Microwave Video Distribution System

MVE Minimum-Volume Ellipsoid • Montevideo (USA-MN) (location identifier) • Multivalued Dependency • Multivariate Exponential (distribution)

MVEd Master of Vocational Education

MVEMJSUNP Mein Vater erklärt mir jeden Sonntag unsere niedlichen Planeten • My Very Educated Mother Just Showed Us Nine Planets • Mercury - Venus - Earth - Mars - Jupiter - Saturn - Uranus - Neptune - Pluto (*)

MVF Manned Vertical Flight • Mossoró - Dixsept Rosado (Brazil-RN) (location identifier)

MVFR Marginal Visual Flight Rules

MVFV Manned Venus Flyby Vehicle

Mvg Moving

MVG Mevang (Gabon) (location identifier) • Moving

MVGA Monochrome Video Graphics Array

MVGVT Mated Vertical Ground Vibration Test

MVH Macksville (Australia-NSW) (location identifier)

MVI Medium Value Item • Menatai (Papua New Guinea) (location identifier) • Microgravity Vestibular Investigation • Multivendor Interaction • Multizonal Visual Information

MVIEP Multivariate Inverse Eigenvalue Problem

MVIP Multivendor Integration Protocol

MVJ Mandeville Marlboro (Jamaica) (location identifier)

MVK Mulka (Australia-SA) (location identifier)

MVL Main Vertical List • Morrisville Stowe (USA-VT) (location identifier) • Multiple Valued Logic

MVLP Microsoft Variable Licence Pack

MVLT Multivalued Logic Tree

MVLU Minimum Variance Linear Unbiased

MVM Kayenta (USA-AZ) (location identifier) • Mariner Venus/Mercury • Medium-Voltage Mode • Metal - Vacuum - Metal • Mission and Vehicle Management • Modèle de Vision Multiéchelle • Multiple Virtual Machine • Multiscale Vision Model

Mvmt Movement

MVMT Movement

MVN Mount Vernon (USA-IL) (location identifier) • Mt. Vernon (USA-IL) (location identifier) • Multivariate Normal

MVNO Mobile Virtual Network Operator

MVO Mongo (Chad) (location identifier) • Motor Vehicle Operator

MVOI Multivariate Optimum Interpolation

MVOS Mobile Video Optical System

MVP Magnetic Vector Potential • Maintenance Verification Plan • Massive Violent Penetration • Master Verification Plan • Mechanical Vacuum Pump • Member Value Plus (ACM) • Method of Variation of Parameters • Microsoft Valued Professional • Mitu (Colombia) (location identifier) • Modular Voice Processor • Most Valuable Player • MOS Verification Programme • Multiline Variety Package • Multimedia Value Point • Multimedia Video Processor • Multimission Vicar Planner

MVPG Move Page

MVPS Mechanical Vacuum Pump System

MVQ Model-based Vector Quantization • Mogilev (Belarus) (location identifier) • Motion Vector Quantization

Mvr Maneuver

MVR Rufiyaa (Maldives) (ISO currency code) • Maneuver • Maroua (Cameroon) (location identifier) • Measurement Velocity Radial • Mechanical Vapor Recompression • Mesure de Vitesse Radiale

MVRF Mobile Vehicle Repair Facility

MVS Message Validation System • Middle-Value Select • Mid-Value Selected • Minimum Variance Simulation • Minimum Visible Signal • Mission Video System • Mixed-Variable Symplectic • Modularized Vehicle Simulation • Monthly Visits Statistics • Motor Vehicle Service • Mucuri (Brazil-BA) (location identifier) • Multiaircraft Vertically Stacked • Multiple Virtual Storage • Multiple Virtual System

MVSB Motor Vehicle Storage Building

MVSFA Moskovskii Universitet Vestnik Seriia Fizika Astronomiia (periodical code)

MVSMM Moskovskii Universitet Vestnik Seriia Matematika Mekhanik (periodical code)

MVSPC Multivariate Statistical Process Control

MVSRF Man-Vehicle Systems Research Facility

MVSS Motor Vehicle Storage Shed

Mvt Movement

MVT Mataiva (Tuamotu Islands) (location identifier) • Module Verification testing • Multiprogramming Variable Tasks

MVU Minimum Variance Unbiased • Musgrave (Australia-QLD) (location identifier)

MVUE Man/Vehicular User Equipment • Minimum Variance Unbiased Estimate • Minimum Variance Unbiased Estimator

MVULE Minimum Variance Unbiased Linear Estimator

MVV Mean Vertical Velocity • Megève (France) (location identifier)

MVW Mount Vernon - Skagit (USA-WA) (location identifier) • Mt. Vernon - Skagit (USA-WA) (location identifier)

MVWT Mated Vehicle Wind Tunnel

MVX Minvoul (Gabon) (location identifier)

MVY Martha's Vineyard (USA-MA) (location identifier)

MVZ Masvingo (Zimbabwe) (location identifier)

mw MacWrite (file extension)

mW meines Wissens • Milliwatt

Mw Megaword • Mevrouw

M/W Make Work • Microwave

MW Magellansche Wolken • Malawi (international vehicle registration code) • Malawi (ISO country code) • Man-Week • Master Warrant • Medium Wavelength • Medium Wave • Megawatt • Message Waiting • Microwave • Mid-west • Milky Way • Mittelwasser • Mixed Waste • Module Width • Molecular Weight • Momentum Wheel • Multi-wavelength

MWA Marion (USA-IL) (location identifier) • Midwestern Astronomers (USA-KY) • Momentum Wheel Assembly

MWAGN Multiwavelength Active Galactic Nuclei (survey)

MWAP Marshall W. Alworth Planetarium (USA-MN)

MWB Master Workbook • Maxwell-Wien Bridge • Microwave Background • Microwave Burst • Monitoring Workbench • Morawa (Australia-WA) (location identifier)

MWC Mary Washington College (USA-VA) • Milwaukee - Lawrence J. Timmerman (USA-WI) (location identifier) • Mount Wilson Catalog

MWCN Mobile and Wireless Communication Networks (conference)

MWCS Mini-World Coordinate System

MWD Magnetic White Dwarf (star) • Material Withdrawal Document • Measurement While Drilling • Mianwali (Pakistan) (location identifier) • Microwell Detector • Millimeter Wave Device • Molecular Weight Distribution

MWDI Master Water Data Index

MWDP Master Warning Display Panel

MWE Merowe (Sudan) (location identifier) • Multiword Expression

MWF Maewo (Vanuatu) (location identifier) • Ministerium für Wissenschaft und Forschung • Monday - Wednesday - Friday

MWG Magnetospheric Working Group • Maintainability Working Group • Management Working Group • Marawaka (Papua New Guinea) (location identifier) • Meteorological Working Group • Milky Way Galaxy

MWGB Mid-Wing Gearbox

MWGBI Mannheim West Germany Bibliographisches Institut AG (periodical code)

MWGC Milky Way Globular Cluster

MWh Megawatt Hour

MWH Moses Lake - Grant County (USA-WA) (location identifier)

MWHG Minimum Weighted Hamiltonian Graph

MWHGS Mikrowellen-Hintergrundstrahlung

MWI Malawi (ISO country code) • Many-Worlds Interpretation • Maramuni (Papua New Guinea) (location identifier) • Message Waiting Indication • Message Waiting Indicator • Mount Wilson Institute (USA-CA) • Multi-mode Waveguide Interferometer

MWIC Millimeter-Wave Integrated Circuit

MWIR Medium-Wave Infrared • Medium-Wavelength Infrared • Mid-Wave Infrared

MWIS Millimeter-Wave Imaging System

MWJ Matthews Ridge (Guyana) (location identifier)

MWK Kwacha (Malawi) (ISO currency code) • Matak (Indonesia) (location identifier) • Ministerium für Wissenschaft und Kunst (Germany)

mWL Milliwatt Logic

MWL Mean Water Level • Mineral Wells (USA-TX) (location identifier)

MWLCO Mt. Wilson and Las Campanas Observatories

MWLS Microwave Line Scanner

MWLT Mess White Light Telescope (USA-HI)

MWM Maxwell-Wagner Mechanism • Motif Window Manager • Windon (USA-MN) (location identifier)

MWMS Maintenance Workload Management System

mwN mit weiteren Nennungen • mit weiteren Nachweisen

MWN Message Waiting Notification • More Work Needed • Mwadui (Tanzania) (location identifier)

MWNT Miniwall Nanotube

MWO Maintenance Work Order • Manufacturing Work Order • Meteorological Watch Office • Meyer-Womble Observatory (USA-CO) • Middletown Hook (USA-OH) (location identifier) • Millimeter Wavelength Oscillator • Millimeter Wave Observatory • Modification Work Order • Mt. Wilson Observatory (USA-CA)

MWOA Mt. Wilson Observatory Association (USA-CA)

MWODF Mount Wilson Observatory Director's Files

MWOR Maintenance Work Order Request

MWP Maneuvering Work Platform • Manufacturer's Whole-sale Price • Maximum Working Pressure • Meteorologist Weather Processor • Millimeter Wave Propagation

MWPC Multiwire Proportional Counter

MWPR Monthly Work Package Report

MWPS Millimeter Wave Power Source

MWQ Magwe (Myanmar) (location identifier)

MWR Maintenance Work Request • Maintenance Work Requirement • Mean Width Ratio • Meteor Wind Radar • Method of Weighted Residuals • Microwave Radiometer • Mikrowellen-Wasserdampf-Radiometrie • Motswari (South Africa) (location identifier) • Multiwavelength Radiometer

MWRC Mean Weekly Relative Citedness

MWRS Millimeter-Wave Radio System

MWRv Monthly Weather Review (periodical code)

MWS Maintenance Work Specification • Major Weapon System • Management Workstation • Maximum Wind Speed • Microwave Station • Microwave Wind Scatterometer • Millimeter-Wave Sounder • Minimum Wheel Speed • Modular Workstation • Mount Wilson (USA-CA) (location identifier) • Mt. Wilson (USA-CA) (location identifier) • Multimedia Wireless System • Multimedia Workgroup System • Multiworkstation • Multiworkstation Systen

MWSC Microwave Scatterometer

MWSI Mount Wilson Sunspot Index

MWSR Magnetic Wire Shift Register • Microwave Scanning Radiometer

MWSSHG Multiwaveband Spectra Spectroheliograph

MwSt Mehrwertsteuer

MWT Microwave Tube • Millimeter Wave Tube • Moolawatana (Australia-SA) (location identifier)

MWTD Metal-Weak Thick Disk

MWU Mussau (Papua New Guinea) (location identifier)

MWV Maximum Working Voltage

MWWG Millimeter-Wave Working Group (IUCAF)

MWX Multiple Way Examination

MWY Miranda Downs (Australia-QLD) (location identifier)

MWZ Mwanza (Tanzania) (location identifier)

mx Mexico

Mx Maxwell • Middlesex (UK-England) • Multiplexer

MX Mail Exchange • Mail Exchanger • Massachusetts X-ray • Merged Extracted (image) • Mexico (ISO country code) • Mexique (code ISO de pays) • Mixed • Mondays Excepted • Multiplex • Multiplexer

MXA Manila (USA-AR) (location identifier)

MXB Masamba (Indonesia) (location identifier) • Massachusetts X-ray Burster (catalog) • Medium X-ray Band

MXC Matrix Controller • Monticello - San Juan County (USA-UT) (location identifier)

MXCC Module X-bus Cache Controller

Mxd Mixed

MXD Marion Downs (Australia-QLD) (location identifier) • Mixed

MXDMX Multiplexer/Demultiplexer

MXE Massive X-ray Experiment • Maxton (USA-NC) (location identifier)

MXF Message Transfer Facility • Montgomery Maxwell (USA-AL) (location identifier)

Mxg Mixing

MXG Marlborough (USA-MA) (location identifier) • Mixing

MXH Maximum X-ray Handler • Maximum X-ray Handling • Moro (Papua New Guinea) (location identifier)

MXI Mati (Philippines) (location identifier) • Multisystem Extension Interface

MXJ Main Crossing Junction • Minna (Nigeria) (location identifier)

MXK Mindik (Papua New Guinea) (location identifier)

MXL Maximum Likelihood • Mexicali (Mexico) (location identifier)

MXLO Mixed Extracted Low-dispersion (IUE image)

MXM Morombe (Madagascar) (location identifier)

MXN Morlaix (France) (location identifier)

MXO Manually-Extended Operation • Monticello (USA-IA) (location identifier)

MXP Milano Malpensa (Italy) (location identifier) • Peso (Mexico) (ISO currency code)

MXQ Mitchell River (Australia-QLD) (location identifier)

Mxr Mixer

MXR Mail Exchange Record • Mirgorod (Ukraine) (location identifier) • Multiplexer

MXRB Massive X-Ray Binary

MXS Maota - Savail Island (Samoa) (location identifier) • Mercury X-ray Spectrometer • Microsoft Exchange Server

Mxt Mixture

MXT Maintirano (Madagascar) (location identifier) • Memory Expansion Technology

MXU Mask Exchange Unit • Mullewa (Australia-WA) (location identifier) • Multiplexer Unit

MXUVI Multiple Extreme Ultraviolet Imager

MXV Moron (Mongolia) (location identifier)

MXW Mandalgobi (Mongolia) (location identifier)

MXX Mora (Sweden) (location identifier)

MXY McCarthy (USA-AK) (location identifier)

MXZ Meixian (China-PRC) (location identifier)

my Burmese (ISO language code) • Malaysia

MY Malaysia (ISO country code) • Malaysie (code ISO de pays) • Man-Year • Mayan • Mid Year • Million Years • Minyong • Multiyear

MYA Mid-Year Analysis • Moruya (Australia-NSW) (location identifier)

MYB Mayoumba (Gabon) (location identifier) • Multiyear Budget

MYC Maracay (Venezuela) (location identifier) • Massive Young Cluster • Multiyear Contract • Multiyear Contracting

MYD Malindi (Kenya) (location identifier) • Multiyear Directive

MYDL Middle Years Digital Library (project)

MYE Man-Year Equivalent • Miyake Jima (Japan) (location identifier) • Multiyear Extension

MYF San Diego - Montgomery (USA-CA) (location identifier) • Multiyear Financing

MYG Mayaguana (Bahamas) (location identifier) • Multiyear Guidance

MYH Marble Canyon (USA-AZ) (location identifier) • Multiyear Hosting

MYI Multiyear Investment • Murray Island (Australia-QLD) (location identifier)

MYJ Main Yield Junction • Matsuyama (Japan) (location identifier)

MYK May Creek (USA-AK) (location identifier)

MYL McCall (USA-ID) (location identifier) • Multiyear License

mym Myriameter

MYM Managing Your Money • Monkey Mountain (Guyana) (location identifier) • Multiyear Management

MYN Mareb (Yemen) (location identifier)

MYO Multiyear Offer • Myroodah (Australia-WA) (location identifier)

MYOB Mind Your Own Business

MYP Mary (Turkmenia) (location identifier) • Multiyear Procurement • Multiyear Project

MYPB Multiple Year Packaged Buy

MYQ Multiyear Qualification • Mysore (India) (location identifier)

Myr Megayear • Millions of Years

MYR Ringgit (Malaysia) (ISO currency code) • Mid-Year Review • Myrtle Beach (USA-SC) (location identifier)

MYRA Multiyear Rescheduling Agreement

MYRC Mean Yearly Relative Citedness

MYS Malaysia (ISO country code) • Malaysie (code ISO de pays) • Moyale (Ethiopia) (location identifier)

MYSO Massive Young Stellar Object

Myst Mystery

MYT Mayotte (code ISO de pays) • Mid-Year Term • Myitkyina (Myanmar) (location identifier)

Myth Mythological • Mythology

Mythol Mythological • Mythology

MYU Mekoryuk (USA-AK) (location identifier) • Multiyear Usage

MYV Marysville Yuba (USA-CA) (location identifier) • Multiyear Validity

MYW Mtwara (Tanzania) (location identifier) • Multiyear Warranty

MYX Menyamya (Papua New Guinea) (location identifier) • Mid-Year Exchange

MYY Miri (Malaysia) (location identifier)

MYZ Monkey Bay (Malawi) (location identifier)

mZ mangels Zahlung

Mz Mehrzahl • Mesozoic

MZ Mach-Zehnder • Mozambique (ISO country code) • Multizone

MZA Multizone Agreement • Muzaffarnagar (India) (location identifier)

MZB Mocimboa Praia (Mozambique) (location identifier) • Multizone Budget

MZC Mitaka Zenith Catalogue • Mitzic (Gabon) (location identifier)

MZD Mendez (Ecuador) (location identifier) • Multizone Directive

MZE Manatee (Belize) (location identifier) • Multizone Experiment

MZF Manganese Zinc Ferrite • Mzamba (South Africa) (location identifier)

MZFW Maximum Zero Fuel Weight

MZG Makung (China-ROC) (location identifier)

MZH Merzifon (Turkey) (location identifier) • Mid-Zone Halting

MZI Mopti (Mali) (location identifier) • Mid-Zone Inversion

MZJ Marana (USA-AZ) (location identifier) • Mid-Zone Junction

MZK Marakei (Kiribati) (location identifier)

MZL Manizales (Colombia) (location identifier) • Multizone License

MZM Metical (Mozambique) (ISO currency code) • Metz (France) (location identifier)

MZN Minj (Papua New Guinea) (location identifier) • Multizone Network

MZO Manzanillo (Cuba) (location identifier) • Multizone Operations

MZP Motueka (New Zealand) (location identifier) • Multizone Project

MZQ Mkuze (South Africa) (location identifier)

MZR Mazar-I-Sharif (Afghanistan) (location identifier) • Multizone Recording

MZS Multizone System • Mostyn (Malaysia) (location identifier)

MZT Mazatlán (Mexico) (location identifier) • Mercury Zinc Telluride

MZU Multizone Usage • Muzaffarpur (India) (location identifier)

MZV Multizone Validity • Mulu (Malaysia) (location identifier)

MZW Multizone Width

MZX Mena (Ethiopia) (location identifier) • Multizone Extension

MZY Mossel Bay (South Africa) (location identifier)

MZZ Marion (USA-IN) (location identifier)

M0DE Middeck O-gravity Dynamics Experiment

M1 Primary Mirror

M2 Medium Two

M2A Multimedia Asia (satellite)

M2P2 Multimission Production Plan

M2S Modular Multiband Scanner • Modular Multispectral Scanner • Multimission Seeker

M2TS Mirror Tip/Tilt System

M3 Multimission Module

M3IC Monolithic Microwave and Millimeter-wave Integrated Circuit

M4 Milky-Way Magnetic-field Mapping Mission

N

n/ Near • Nuestro

n Amount of Substance • And • Electron Number Density • Indefinite Number • Index of Refraction • Load Factor • Mean Daily Angular Planet Motion • Molecule Number Density • Nach • Nacional • Name • Nano • Natus • Né • Near • Negative • Nephew • Net • Neuter • Neutro • Neutron • Neutron Number Density • New • Night • Nominative • Noon • Nordlich • Normal • Nota • Notre • Noun • Nous • Nuestro • Nuevo • Number • Number Density • Number Density of Molecules • Number of Degrees of Freedom • Number of Stages • Order of Reflexion • Particle Density • Principal Quantum Number • Refractive Index • Rotational Frequency

N And • Aperture • Asparagine • Avogadro Number • Azote • Loudness • Molecule Number • Naira (currency) • National • Navy • Nebula • Neptune • Neutral • Neutron Number • New • Newton • Ngultrum (currency) • Night • Ninety (Roman numeral) • Nitrogen • No • Noise • Noon • Nord • Norden • Nordlich • Normal • Norte • North • Northern • Norway (international vehicle registration code) • Norway (postal country code) • Note • Notice • Nova • November (international alphabet) • Nuclear • Nucleon • Number of Molecules • Number of Objects per Square Degree • Number of Particles • Number of Particles per Cell • Number of Turns • Nutation Constant • Quantum Number of Total Angular Momentum • Particule Number • Precession in Declination • Refractivity • Spectral Type • Stickstoff • Telescope Aperture • Turn Number • United States of America (international aircraft registration code) • USA (international aircraft registration code) • 90 (Roman numeral)

n/a No Account • Not Applicable • Not Available

na Nauru (ISO language code)

nA Nanoampere

Na Fast Nova • Natrium • Nuestra • Sodium

N/A New Account • New Address • Next Assembly • No Action • No Advice • Non Acceptance • Non Applicable • Not Applicable • Not Authorized • Not Available

NA Namibia (ISO country code) • Namibie (code ISO de pays) • Nanai • Narrow Angle • National Academician • National Archives • National Atlas • Nautical Almanac • Navigation Assistant • Nébuleuse Annulaire • Netherlands Antilles (international vehicle registration code) • Network Administrator • Network Analysis • Network Appliance • Network Archivist • Network Auditor • Neutral Atom • Neutral Axis • New Account • New Address • Next Action • Next Address • Night Assistant • No Account • Nodal Architect • Non-African • North America • North American • Not Add • Not Applicable • Not Assigned • Not Authorized • Not Available • Now Available • Numerical Analysis • Numerical Aperture • Numérotation Abrégée

NAA Naperville Astronomical Association (USA-IL) • Narrabri (Australia-NSW) (location identifier) • Narrow-Angle Astrometry • National Aeronautic Association (USA) • National Airports Authority (USA) • National Airworthiness Authority • Network Analysis Area • Neutron Activation Analysis • North American Aircraft • North American Aviation • North Atlantic Assembly (NATO) • Northants Amateur Astronomers (UK) (formerly KDAAS) • Nürnberger Astronomische Arbeitsgemeinschaft (Germany)

NAAA Northern Arizona Astronomy Association (USA-AL)

NAAAC National astronomy and Astrophysics Advisory Committee (USA)

NAAC National Assessment and Accreditation Council (India) • National Aviation Associations Coalition (USA)

NAACL North American Association of Computational Linguistics

NAACP National Association for the Advancement of Colored People (USA)

NAAG National Association of Attorneys General (USA)

NaAIJ Nanjing Aeronautical Institute Journal (periodical code)

NAAIL NSSDC Application of Artificial Intelligence Laboratory

NAAL Network of Alabama Academic Libraries (USA-AL) • North American Aerodynamic Laboratory

NAAP Nederlands Antarctisch Onderzoek Programma

NAAPO North American Astrophysical Observatory (USA-OH)

NAAQ National Ambient Air Quality

NAAQS National Ambient Air Quality Standards

NAAR Night Air-to-Air Refueling

NAARS National Automated Accounting Research Service (USA)

NAAS Newbury Amateur Astronomical Society (UK-England) • North American Automated Systems Co. (USA) • Northeast Arkansas Astronomical Society (USA-KS)

NAB Albany (USA-GA) (location identifier) • National Association of Broadcasters (USA) • Nordic Advisory Board • Northern Arctic Band

NABA North American Broadcasters Association

NABD Normenausschuß Bibliotheks- und Dokumentationswesen

NABE National Association of Book Editors (USA) • North Atlantic Bloom Experiment

NABI NASA Astrobiology Institute

NABP National Association of Book Publishers (USA)

NABS Nomenclature des Activités-Budgets de la Science (France)

NABST National Advisory Board on Science and Technology (Canada)

NABT National Association of Biology Teachers (USA)

NABTS North American Broadcast Teletext Standard • North American Broadcast Television System

NABUCO Nouvelle Approche Budgétaire et Fiscale

Nac Nacelle • Nacional

NAC Nacelle • Naracoorte (Australia-SA) (location identifier) • Narrow Absorption Component • Narrow-Angle Camera • NASA Advisory Council • National Accelerator Centre (South Africa) • National Aerospace Centre • National Advisory Council • National Aerospace Centre • National Archives Center • National Archives Council • Naval Avionics Center • Near Ambient Chamber • Nederlandse Astronomenclub (Netherlands) • Nepean Astronomy Centre (Australia-NSW) • Network Access Center • Network Access Component • Network Access Controller • Network Adapter Card • Network Administration Center • Network Advisory Committee • North American Craton • North Atlantic Council • NSO Aladdin Camera • Null Attachment Concentrator

NACA National Advisory Committee for Aeronautics (USA) (now NASA)

NACAA National Australian Convention of Amateur Astronomers

NACAC National Association for College Admission Counseling (USA)

NACC National Air Communications Council • Necessary Ad-hoc Coordinating Committee • North American Calibration Cooperation • North Atlantic Cooperation Council

NACCB National Accreditation Council for Certification Bodies (USA)

NACCH Network Access and Connectivity Channel

NACCIRN North American Coordinating Committee for Intercontinental Research Networking (USA)

NACD National Association of Computer Dealers (USA) • North Atlantic Climatological Data-set

NACDF National Area Coverage Data File

NACE National Association of Corrosion Engineers (USA) • Nomenclature d'Activités dans la Communauté Européenne • North American Commission on the Environment

NacEl Nachrichten Elektronik (periodical code)

NACEPT National Advisory Committee for Environmental Policy and Technology

NACES Navy Aircrew Escape System

Nachf Nachfolger

Nachm Nachmittags

Nachr Nachricht

NachwG Nachweisgesetz

NACIS North American Cartographic Information Society

NACK Negative Acknowledgment

Nacl Nacional

NaCl Sodium Chloride

NACL National Advisory Committee on Libraries (USA)

NACLIS National Advisory Commission on Libraries and Information Science (USA) • National Advisory Council for Library and Information Science (South Africa)

NACM Network Authentication Control Module • New Area Calculation Method

NACME National Action Council for Minorities in Engineering (USA)

NACo National Association of Counties (USA)

NACO Name Authority Cooperative Project • NAOS + CONICA

NACS National Advisory Committee on Semiconductors (USA) • National Association of College Stores (USA) • National Association of Computer Stores (USA) • National Association of County Surveyors (USA) • NetWare Asynchronous Communications Service • Network-Assisted Coordinated Science • Network Asynchronous Communications Services • Neutral Atmosphere Composition Spectrometer

NACSEM National Communications Security Emanations Memorandum

NACSI National Communications Security Instruction

NACSIM National Communications Security Instruction Memorandum

NACSIS National Centre for Science Information System (Japan)

NACSM National Advisory Committee on Surveys and Mapping (Canada)

NACT Network Account Table

NACU National Association of Colleges and Universities (USA)

NacZe Nachrichtentechnische Zeitschrift (periodical code)

NAD Macanal (Colombia) (location identifier) • Nansen Arctic Drilling • National Astronomy Day • Network Access Delay • Network Access Device • Network Adaptor Device • Nicotinamide Adenine Dinucleotide • No Apparent Defect • No Appreciable Difference • Noise Amplitude Distribution • North American Datum • Not on Active Duty • NRAO ALMA Division

NADAC National Damage Assessment Center

NADAMS National Agricultural Drought Assessment and Monitoring System (India)

NADAR Nadir Altimeter for Rain

NADB National Air Data Branch • Network Accessible Database

NADC National Alumni Data Center • Naval Air Development Center (USA) • North American Digital Cellular • North American Dual-mode Cellular

NADDIS Narcotics and Dangerous Drugs Information System

NADE National Association for Development Education (USA)

NADEP Naval Aviation Depot

NADF North American Directory Forum

NADGE NATO Air Defense Ground Environment

NADIN National Airspace Data Interchange Network (USA)

NADIR Network Anomaly Detection Intrusion Reporter

NADIS Nachrichtendienstliches Informationssystem • National Aerometric Data Information System (USA)

NADP NASA Ames Development Plan • National Atmospheric Deposition Program • Nicotinamide Adenine Dinucleotide Phosphate

NADUC Nimbus/ATS Data Utilization Center

NADW North Atlantic Deep Water

NAE National Academy of Engineering (USA) • National Aeronautical Establishment (USA) • Natitingou (Benin) (location identifier) • Not Above or Equal • Nuclear Astrophysics Explorer • Núcleo de Assuntos Europeus

NAEB National Association of Educational Broadcasters (USA)

NAEC National Atomic Energy Commission (USA) • Naval Air Engineering Center

NAEP National Assessment of Educational Progress (USA) • National Association of Environmental Professionals (USA)

NAEYC National Association for the Education of Young Children (USA)

NAf Netherlands Antillean Guilder (international vehicle registration code)

NAF Banaina (Indonesia) (location identifier) • National Analytical Facility • Negation As Failure • Nernst Approximation Formula • Network Access Facility • Nomenclature des Activités Françaises (France) • Non-Appropriated Funds • Norsk Astronautisk Forening (Norway)

NAFB Norton Air Force Base (USA-CA)

NAFC National Average Fuel Consumption

NAFD Non-Associated Forbush Decrease

NAFEO National Association For Equal Opportunity (USA)

NAFF Numerical Analysis of Fundamental Frequencies

NAFI National Association of Flight Instructors (USA) • Non-Appropriated Fund Instrumentality

NAFIP National Foreign Intelligence Program

NAFIPS North American Fuzzy Information Processing Society (USA)

NAFIS NASA Accounting and Financial Information System

NAFISS Non-Appropriated Funds Information Standard System

NAFMIS Non-Appropriated Funds Management Information System

NAFRMS Non-Appropriated Funds Resource Management System

NAFS Name Authority File Service • Novell Advanced File System

NAFSMA National Association of Flood and Stormwater Management Agencies (USA)

NAFT Network for Analysis of Fireball Trajectories

NAFTA New Zealand and Australia Free Trade Agreement • North American Free Trade Agreement (*) • North American Free Trade Area

Nag Nagasaki (Japan)

NAG Nagpur (India) (location identifier) • National Advisory Group • Net Annual Gain • Netherlands Aerospace Group • Network Address Group • Network Advisory Group • Network Architecture Group • Noyau Actif de Galaxie • Numerical Algorithms Group

NAGAP National Acid Precipitation Assessment Program

NAGARA National Association of Government Archives and Records Administration (USA)

NAGC National Association of Government Communicators (USA)

NaGe National Geographic (periodical code)

NAGE National Association of Government Employees (USA)

NAGG Non-Aliased Gravity Grid

NAGPS National Association of Graduate-Professional Students (USA)

NAGT National Association of Geoscience Teachers (USA)

NAGUA Numerical Algorithms Group Users Association

NAH Naha (Indonesia) (location identifier) • Network Administration Host • Not At Home

NAHSTE Navigational Aids for the History of Science, Technology and the Environment

NAI Annai (Guyana) (location identifier) • Named Area of Interest • NASA Astrobiology Institute (USA) • National Archives of Ireland • National Association of Interpretation (USA) • Net Attributable Income • Neutral Atom Imager • No Action Indicated • No-Address Instruction • Non-Accidental Injury

NAIADE Normalized Automatic Interpreter in Ada for Data Exchange

NAIC National Address Information Center (USA) • National Air Intelligence Center (USA) • National Astronomy and Ionosphere Center (USA) • Network Applications and Information Center (USA-CA) • Newly Agro-Industrialized Country

NAICC National Association of Independent Computer Companies (USA) • Nuclear Accident Incident Control Center

NAICS North American Industry Classification System

NAICU National Association of Independent Colleges and Universities (USA)

NAID Note Added In Proof

NAIF Nanjing Astronomical Instruments Factory (China-PRC) (now NAIRC) • Navigation Ancillary Information Facility

NAIIU Not Authorized If Issued Under

NAILS National Airspace Integrated Logistics Support • NCAR Airborne Infrared Lidar System

NAIR Narrow Absorption Infrared • Network Action Item Report

NAIRC Nanjing Astronomical Instruments Research Facility (China-PRC) (formerly NAIF)

NaiRe Pradesh State Observatory Naini Tal Reprints (periodical code)

NAIS NASA Acquisition Internet Service • National Atlas Information Service (Canada)

NAISS New Abstracting and Indexing Support System

NAIST Nara Institute of Science and Technology (Japan)

NAJ National Association of Journalists (UK)

NAK Nakhon Ratchasima (Thailand) (location identifier) • Negative Acknowledge • Negative Acknowledgment • Non-Acknowledged • Non-Acknowledgment • Normenausschuß Klassifikation • Norrköpings Astronomiska Klubb (Sweden)

NAL Nalchik (Russia) (location identifier) • Narrow Absorption Line • National Acoustics Laboratory • National Aeronautical Laboratory • National Aerospace Laboratory • NetWare Application Launcher • Novell Application Launcher • Numerical Analysis Laboratory

NALC National Association of Letter Carriers (USA)

NALIN National Library Information Network

NALM Non-linear Amplifying Loop Mirror

NALMS North American Lake Management Society

NALPS North-American Presentation-Level Protocol Syntax

NALTA North American Large-area Time-coincidence Array • North American Large-scale Time-coincidence Array

N. Am. North America • North American

NAM Name and Address Module • Namibia (ISO country code) • Namibie (code ISO de pays) • Namlea (Indonesia) (location identifier) • National Account Manager • National Association of Manufacturers (USA) • Nautical Air Mile • Navy Aerosol Model • Nederlandse Aardolie Maatschappij (Netherlands) • NetWare Application Manager • Network Access Machine • Network Access Method • Network Access Mode • Network Analysis Model • Network Assessment Model • Non-Aligned Movement • Normalized Axial Moment • Northern (hemisphere) Annular Mode • Nuclear Assessment Methodology • Number Assignment Module • Numerical Action Method

NAMAC North American Market Awareness Committee

NAMAS National Measurement Accreditation Service

NAME Naval Architecture and Marine Engineering

NAMEDA National Association of Minority Program Administrations (USA)

NAMG Narrow-Angle Mars Gate

NAML National Association of Marine Laboratories (USA) • Naval Aircraft Materials Laboratory

NAMN North American Meteor Network (USA-SC)

NAMP National Association of Magazine Publishers (USA)

NAMPS Narrow-band Advanced Mobile Phone Service

NAMRIA National Mapping and Resources Information Authority (Philippines)

NAMS Network Administrative Management System

NAMT National Advanced Manufacturing Testbed (NIST)

NaN Not a Number

NAN Nadi (Fiji) (location identifier) • National Academy of Neuropsychology (USA) • No Action Necessary • Numerical Assignment Number

NANBA North American National Broadcasters Association

NAND Not And

NA-NET Numerical Analysis Network

NANET Numerical Analysis Network

NANOFET Nanometer-scale Field Effect Transistor

NANOG North American Network Operators Group

Nanot Nanotechnology (periodical code)

NANP North American Numbering Plan

NANU Notice Advisory to NAVSTAR Users

NAO Nanchong (China-PRC) (location identifier) • National Astronomical Observatory • National Audit Office (UK) • Nautical Almanac Office (USNO) • Navigation Assistée par Ordinateur • Nikolaev Astronomical Observatory (Ukraine) • North American Operations • North Atlantic Oscillation • Notation Assistée par Ordinateur

NAOC National Astronomical Observatories of China

NAOCO New Advanced One-unit Correlator

NAOE Naval Architecture and Ocean Engineering

NAOGE National Association of Government Engineers (USA)

NAOJ National Astronomical Observatory of Japan

NAOM National Astronomical Observatory - Misuzawa (Japan)

NAOMI Nasmyth Adaptive Optics for Multiple Instrumentation • Natural (guide star) Adaptive Optics (system) for Multipurpose Instrumentation

NAOMZ National Astronomical Observatory - Misuzawa (Japan)

NAOOS Not Assigned or Out Of Service

NAOPIA New Architectures for Optical Processing in Industrial Applications

NAOS NASA Aircrew Oxygen System • Nasmyth Adaptive Optics System • North Atlantic Ocean Stations

NAOSMM National Association Of Scientific Material Managers (USA)

NAOT National Astronomical Observatory - Tokyo (Japan)

Nap Napoli (Italy)

NAP Naples Capodochino (Italy) (location identifier) • Napoli Capodochino (Italy) (location identifier) • National Academy Press (USA-DC) • National Action Plan • National Action Programme • National Aim Point • National Association of Publishers (USA) • National Awareness Partner • Navigation Analysis Program • Network Access Point • Network Access Protocol • Network Access Provider • Network Applications Platform • Noise Abatement Procedure • Nomenclature des Activités Professionnelles • Non-linear Assignment Problem • Northern Affairs Program • Not A Priori • Not At Present • Not Authorized Positioning • Nuclear-Active Particle • Nuclear Auxiliary Power

NAPA National Academy of Public Administration (USA)

NAPAG National Academies Policy Advisory Group

NAPAMA National Association of Performing Artists Managers and Agents (USA)

NAPAP National Acid Precipitation Assessment Program (USA)

NAPCAE National Association for Public and Continuing Education (USA)

NapCo Osservatorio Astronomico di Capodimonte Napoli Contributi Astronomici (periodical code)

NAPDD Non-Acquisition Program Definition Document

NAPE National Association for Primary Education (UK)

NAPF North Atlantic Polar Front

Naph Naphtha

NAPIS National Agricultural Pest Information System

NAPL National Airphoto Library (Canada)

NAPLPS North American Presentation Level Protocol Syntax

NAPM National Association of Photographic Manufacturers (USA) • National Association of Purchasing Managers (USA)

NAPNOC No Acceptable Price - No Contract

NAPO National Association of Planetary Observers (USA)

NAPP National Aerial Photography Program (USA)

NAPRS National Airspace Performance Reporting System

NAPS North American Project Scientist (ALMA) • Numerical Analysis and Prediction System

NAPSS Numerical Analysis Problem Solving System

NAPU Nuclear Auxiliary Power Unit

NAPUS Nuclear Auxiliary Power Unit System

NAPWPT National Association of Professional Word Processing Technicians (USA)

NAQ National Agency Questionnaire • Qanaq (Greenland) (location identifier)

NAQP National Association of Quick Printers (USA)

Nar Narrow

NAR Nare (Colombia) (location identifier) • Narrow • Net Assimilation Rate • No Action Required • No Answer Required • Noise-Adding Radiometer • Non-Adiabatic Reversible • Non-Advocate Review • Non-linear Autoregression • North American Route • North Atlantic Route • Not According to Routine • Nuclear Acoustic Resonance • Numerical Analysis Research

NARA National Archives and Records Administration (USA)

NARC National Association of Regional Councils (USA) • National Atmospheric Radiation Centre (USA) • Non-Automatic Relay Center • Non-Automatic Relay Code

NARCOM North Atlantic Relay Communication

NARDAC Navy Regional Data Automation Center (USN)

NARE North Atlantic Regional Experiment

NARESA Natural Resources, Energy and Sciences Authority (Sri Lanka)

NARFA National Allied Radio Frequency Agency

NARFE National association of Retired Federal Employees (USA)

NARL National Arctic Research Laboratory • National Air and Radiation Laboratory • Naval Arctic Research Laboratory

NARMA Non-linear Auto-Regressive Moving Average

NARPA National Advanced Research Projects Agency

NARS National Accident Reporting System • National Administration for Remote Sensing (Egypt) • National Archives and Records Services (USA) • Netwerk voor Autonoom Registrerende Seismografen • Network of Autonomously Recording Stations

NARSA National Advanced Remote Sensing Application

NARSAP National Advanced Remote Sensing Application Program

NARSIM NLR ATC Research Simulator

NARSTO North American Research Strategy for Tropospheric Ozone

NARTS National Association of Rocketry Technical Services (USA)

Nas Nasmyth

NAS Name and Addressing Service • NASA Advanced Supercomputing (division) • Nassau (Bahamas) (location identifier) • National Academy of Sciences • National Advanced Systems • National Aerodynamic Simulation • National Aerospace System • National Aircraft Standard • National Airspace Study • National Airspace System • Naval Air Station • Netware Access Server • Network Access Server • Network Access Switch • Network Analysis System • Network Application Support • Network-Attached Storage • Newcastle Astronomical Society (Australia-NSW) • Newcastle-upon-Tyne Astronomical Society (UK) • Newly Associated tate • Nigerian Academy of Sciences • Noise Abatement System • Non-Avionics Simulator • Non-linear Antenna System • Norfolk Astronomical Society (USA-VA) • Norwich Astronomical Society (UK) • Nottingham Astronomical Society (UK) • Numerical Aerodynamic Simulation • Numerical Aerodynamic Simulator • Numerical Analysis Subroutine • Numerical And Atmospheric Sciences

NASA National Aeronautics and Space Act (USA) • National Aeronautics and Space Administration (USA)

NASAC National Aeronautics and Space Administration Conference Publication (periodical code)

NASAL National Aeronautics and Space Administration Circular Letter (periodical code)

NASAP Network Analysis for Systems Application Program

NASAR National Aeronautics and Space Administration Reference Publication (periodical code) • National Association for Search And Rescue (USA)

NASAS National Aeronautics and Space Administration Special Publication (periodical code)

NASC National Aeronautics and Space Council (USA) • National Aircraft Standards Committee • North American Stratigraphic Code

NASCA Nagoya Air Shower Counter Array (Japan)

NASCAP NASA/Air Force Spacecraft Charging Analyzer Program

NASCOM NASA Communications (network)

NASCOP NASA Communications Operating Procedures

NASD National Association of Securities Dealers (USA) • Network-Attached Secure Disk • Network-Attached Storage-Device

NASDA National Space Development Agency (Japan) (now JAXA)

NASDAQ National Association of Securities Dealers Automatic Quotation

NASENI National Agency for Science and Engineering Infrastructure (Nigeria)

Nash Nashville (USA-TN)

NASI NetWare Asynchronous Interface • NetWare Asynchronous Services Interface

NASIG North American Serials Interest Group

NASIP NASA Airframe Structural Integrity Program

NASIRC NASA Automated Systems Incident Response Capability (*) • NASA Automated Systems Internet Response Capability

NASIRE National Association of State Information Resources Executives (USA)

NASIS NASA Aerospace Safety Information System • Northeast Australian Satellite Imagery System

NASL National Association of State Libraries (USA)

NASM National Air and Space Museum (USA-DC) • National Association for School Magazines (USA)

NASNET National Airspace System Network

NASO North American Space Operations

NASP National Aero-Space Plane • National Airspace System Plan • National Atmospheric Sciences Program • Northern Arizona Star Party (USA-AZ)

NaSPA Network and Systems Professionals Association (USA-WI)

NASPA National Association of Student Personnel Administrators (USA)

NASPAC National Airspace System Performance Analysis Capability

NASPR NASA Procurement Regulation

NASRDA National Space Research and Development Agency (Nigeria)

NASREC National Soil Reference Collection

NASRS Non-Available Status Reporting System

Nass Nassau (Bahamas)

NASS National Agricultural Statistics Service • National Air and Space Society (USA-DC) • North American Sundial Society (USA-CT)

NASSCOM National Association of Software and Service Companies (USA)

NASSE National Association of Space Simulating Educators (USA-OH)

NASSL National Association of Spanish Speaking Librarians (USA) • New Activation Supply Support List

NAST National Academy of Science and Technology (USA)

NASTeC North-American Small Telescope Cooperative (now ISTeC)

NASTI NPOES Atmospheric Sounder Tested Interferometer

NASU National Academy of Sciences of Ukraine • National Association of State Universities (USA)

NASW National Association of Science Writers (USA)

NASWSC North American Society for Water and Soil Conservation

nat naturwissenschaftlich

Nat Natal (South Africa) • National • Nationalist • Nationality • Native • Natur • Natural • Nature • Naturel • Natus

NAT Narrow Angle Tail • Narrow Angle Tailed • Narrow Angle Tracker • Natal - Augusto Severo (Brazil-RN) (location identifier) • National • Natural Unit of Information • Network Address Translation (*) • Network Address Translator • Network Analysis Team • New Alias Table • Nitric Acid Trihydrate • No Action Taken • Nodes-Attached Table • Non-conformance Action Team • Normal Allowed Time • Normenausschuß Terminologie • North Atlantic Traffic • Not Air Transportable • Not Air Transportation

NATA National Air Transportation Association (USA)

NATAS NOAA AVHRR Transcription and Archive System

NATCA National Air Traffic Controllers Association (USA)

NATCO Northwest Aerospace Training Corporation

NATF Naval Air Test Facility

NATFB National Archives Trust Fund Board (USA)

NatH Natural History (periodical code)

NATi Nordisk Astronomisk Tidsskrift (periodical code)

NATICH National Air Toxics Information Clearinghouse (USA)

Natl National • Nationality

NATL National

NATMA Non-linear Analysis Theory Methods and Applications (periodical code)

NATMAP National Mapping

NATMC NATO Air Traffic Management Committee

NATO Narrow-Angle Target of Opportunity • North Atlantic Treaty Organization

NATOTS North Atlantic Organized Track System

NATS National Air Traffic Services (UK) • Naval Air Transport service • North American Terrestrial System

Nat.Sc. Natural Science

Nat.Sci. Natural Science

NATSF Naval Air Technical Services Facility

NATT Northern Australian Tropical Transect

Natur Nature (periodical code)

Naturwiss Naturwissenschaft • Naturwissenschaftlich

Natuurkd Natuurkunde

NAU Napuka Island (Tuamotu Islands) (location identifier) • Network Access Unit • Network Addressable Unit • Network Address Unit • NGAS Archiving Unit • Northern Arizona University (USA-AZ)

NAUCA Nomenclatura Arancelaria Uniforme Centroamericana

NAUI National Association for Underwater Instructors (USA)

NAUN Nearest Active Upstream Neighbor • Next Active Upstream Neighbor

Náut Náutica

Naut Nautical • Nautique

nav Naar Aanleiding van

Nav Naval • Navarra (Spain) • Navigable • Navigate • Navigation • Navigational • Navigator

NAV Naval • Navigate • Navigation • Net Asset Value • Norton Anti-Virus

NAVA National Audiovisual Association (UK)

NAVAID Navigational Aid

NAVAIDS Navigational Airborne Integrated Data Systems

NavAu Navigation Australia (periodical code)

NAVCAM Navigation Camera

NAVCOM Naval Communications (USN)

NAVCON Navigation Control

NAVD North American Vertical Datum

NAVDAC Naval Data Automation Command (USN)

NAVDAD Navigational Derived Air Data • Navigationally-Derived Air Data

NAVEX Navigation Experiment • Navigation Expert • NAVNET Experimental

Navig Navigation • Navigation (periodical code) • Navigator

NAVMN Navigation Monitor and Control

NAVNET Naval Network • Navy VLBI Network

NAVO Noordatlantische Verdragsorganisatie

NavPa Navigation Paris (periodical code)

NAVSAT Navigational Satellite (*) • Navigation Satellite

NAVSATCOM Naval Satellite Communications

NAVSOC Naval Satellite Operations Center

NAVSPASUR Naval Space Surveillance Radar

NAVSSI Navigation Sensor System Interface

NAVSTAR Navigational System Timing And Ranging (*) • Navigational System Tracking And Ranging (*) • Navigation Satellite Timing And Ranging (*) • Navigation System with Time And Ranging

NAVSWC Naval Surface Weapons Center (USN)

NAVT NASA Altimeter Verification Telemetry

NAVY Never Again Volunteer Yourself

N/AW Night/Adverse Weather

NAW Narathiwat (Thailand) (location identifier) • National Astronomy Week • Negri-Adler-Wetzel (technique)

NAWAS National Warning System

NAWAU National Aviation Weather Advisory Unit

NAWDEX National Water Data Exchange

NAWPF National Aviation Weather Processing Facility

NAWPS National Association of Word Processing Specialists (USA)

NAWQUA National Water Quality Assessment

NAWS Negotiate About Window Size

NAWT Negri-Adler-Wetzel Technique

NAX Barbers Point (USA-HI) (location identifier)

NAY Beijing Nanyuan (China-PRC) (location identifier) • Nayarit (Mexico)

Naz Nazionale

NAZ Nachrichtenblatt der Astronomischen Zentralstelle (periodical code) • Normal Analytical Zone

Nazl Nazionale

n/b Noir/Blanc

Nb Niob • Niobium • Slow Nova

N/B Narrow Band • Noir/Blanc

NB Naine Blanche • Naive Bayes (method) • Narrow Band • Narrow Beam • Nationalbibliothek • National Body • Navigation Base • Negative Binomial (distribution) • Neubau • Neutral Beam • New Brunswick (Canada) • Nitrogen Base • No Bias • No Bid • No Bidder • Nonthermal Bremsstrahlung • Normal Branch • Northbound • Northern Boundary • Nota Bene • Notebook • Nouveau Brunswick (Canada)

NBA Nambaiyufa (Papua New Guinea) (location identifier) • Narrow-Band Adapter • Narrow-Band Amplifier • Narrow-Band Analyzer • National Basketball Association (USA) • National Book Award

NBAA National Business Aircraft Association (USA)

NBAANB New Brunswick Astronomy - Astronomie Nouveau Brunswick Inc. (Canada-NB)

NBB Barrancominas (Colombia) (location identifier) • Number of Bytes of Binary

NBBI Nederlands Bureau voor Bibliotheekwezen en Informatieverzorging (Netherlands)

NBC Naberevnye Chelnye (Russia) (location identifier) • Narrow-Band Conducted • National Broadcasting Corporation (USA) • Noise Balancing Circuit • Nuclear - Bacteriological - Chemical • Nuclear - Biological - Chemical • Number Base Conversion

NBCC Nuclear, Biological, and Chemical Contamination

NBCCS Nuclear, Biological, and Chemical Contamination Survivability

NBCD Natural Binary-Coded Decimal • Network-Based Collaborative Development • Nuclear, Biological, and Chemical Defense

NBCDCE Nuclear, Biological, and Chemical Defense Control Element

NBCE Nuclear, Biological, and Chemical Element

NBCMAA Nuclear, Biological, and Chemical Mission Area Analysis

NBCS Northern Boundary Containment System

NBCV Narrow-Band Coherent Video

NBCWRS Nuclear, Biological, and Chemical Warning and Reporting System

NBD Narrow-Band Data • Narrow-Band Detector • National Book Day • Nederlands Bibliotheek Dienst (Netherlands) • Negative Binomial Distribution • Neutral Beam Divider • No Big Deal

NBDL Narrow-Band Data Line

NBDM Non-Baryonic Dark Matter

NBDP Narrow-Band Direct Printing

NBDS Nuclear Burst Detection System

NBDVS Narrow-Band Digital Voice System

NbE North by East

NBE Not Below or Equal

Nber Number

NBER National Bureau of Economic Research (USA)

Nbers Numbers

Nbf Nebenbahnhof

NBF Narrow-Band Filter • Neutral Buoyancy Facility

NBFCP NetBIOS Frames Control Protocol

NBFM Narrow-Band Frequency Modulation

NBG New Orleans - Alvin Callendar (USA-LA) (location identifier) • Nobelis Best Glory

NBH Nambucca Heads (Australia-NSW) (location identifier) • Network Busy Hour

NBHF Narrow-Band High Frequency

NBI Neutral-Beam Injection • Niels Bohr Institute (Denmark) • Nothing But Initials • Nouvelle Bonification Indiciaire • Nuclear Burst Indicator

NBII National Biological Information Infrastructure

NBIOME Northern Biosphere Observation and Modeling Experiment

NBIOS Network Basic Input Output System

NBIS Neutral Beam Injection System • Northern Biosphere Information System

NBIT New Bedford Institute of Technology (USA-MA)

NBL Neue Bundesländer (Germany) • Neutral Beam Line • Neutral Buoyancy Laboratory (USA-TX) • San Blas (Panama) (location identifier)

NBLC Nederlands Bibliotheek en Lektuur Centrum (Netherlands)

NBM National Book Month • National Bureau of Metrology • Network Business Management • Nothing By Mouth

NBMA Non-Broadcast Multiaccess • Non-Broadcast Multiple Access

NBMS National Bulk Mail System

NBN Narrow-Band Network • Narrow-Band Noise • National Bibliography Number • National Book Number

NBNC Noted But Not Corrected

NBND Northbound

NBNS NetBIOS Name Server

NBO Nairobi - Jomo Kenyatta (Kenya) (location identifier) • Network Build Out • Networked Business Object • Normal Branch Oscillation

NBOI Nederlands Bureau voor Onderzoek Informatie (Netherlands)

NBOSS New-Business Operations Support System

NBP Name Binding Protocol • Near-real-time Batch Processing • Network Bridge Processor • Network Bright Point • New York - Battery Park City (USA-NY) (location identifier) • Normal Boiling Point

NBpC Number of Bits per Column

NBPC Neutral Beam Power Conversion • Number of Bits Per Column

NBPM Narrow-Band Phase Modulation

NBpP Number of Bits per Pixel

NBPP Navy/Berkeley Pulsar Processor • Number of Bits Per Pixel

NBpR Number of Bits per Row

NBPR Number of Bits Per Row

NBQ Nuclear - Bacteriológica - Química

n. Br. nördliche Breite

Nbr Nombreux • Number

NBR Nambour (Australia-QLD) (location identifier) • Narrow-Band Radiated • National Business Review (periodical code) • Norges Byggstandardiseringsråd (Norway)

Nbre Noviembre

NBRF National Biomedical Research Foundation • Neutral Buoyancy Research Facility

NBRI National Botanical Research Institute

NBRVF Narrow-Band Radio Voice Frequency

Nbrx Nombreux

NBS National Biological Survey (USA) • National Board of Survey • National Budget System • National Bureau of Standards (USA) (now NIST) • Netzwerkbetriebssystem • Neuroscience and Biomedical Systems • Neutral Buoyancy Simulation • Neutral Buoyancy Simulator • New British Standard • Notice Board System • Numeric Backspace

NBSAMRF National Bureau of Standards Advanced Manufacturing Research Facility

NBSM National Bureau of Surveying and Mapping

NBSS Narrowband Switching System

NBSS&LUP National Bureau of Soil Survey & Land Use Planning

NBSSLUP National Bureau of Soil Survey and Land Use Planning

NBSSP National Bureau of Standards Special Publication (periodical code)

NBST Narrowband Secure Terminal (*) • Narrowband Subscriber Terminal • National Board of Science and Technology (Ireland) • National Board for Science and Technology (Indonesia)

NBSTN National Bureau of Standards Technical Note (periodical code)

NBSV Narrowband Secure Voice

NBT National Book Trust (India) • Neutral Buoyancy Trainer • Next Big Telescope • Null Balance Transmission

NBTDR Narrowband Time Domain Reflectometry

NBTF Neutral Beam Test Facility

NBTR Narrow-Band Tape Recorder

NBTS Neutral Beam Test Stand

NBTU Narrowband Trunking Unit

NBU Glenview (USA-IL) (location identifier) • National Bibliographical Unit • Neil Brown Unit • NGAS Buffering Unit

NBV Cana Brava (Brazil-MG) (location identifier) • Net Book Value

NBVCXO Narrow-Band Voltage-Controlled Crystal Oscillator

NBVM Narrow Band Voice Modulation

NBVS Neutral Beam Vacuum System

NBVST Narrow Band Voice Secure Terminal

NbW North by West

NBW Guantanamo (Cuba) (location identifier) • National Book Week • Noise Bandwidth

NBX Nabire (Indonesia) (location identifier)

NBY National Book Year

NBZ Neutral Buoyancy Zone

n/c No Charge • Nuestra Casa • Nuestra Cuenta • Nuestro Cargo

nc No Change

Nc Very Slow Nova

N-C Nitrogen-Carbon

N/C No Change • No Charge • No Connect • No Coverage • Normally Closed • Nose Cone • Not Critical • Numerical Control • Numerically Controlled

NC Narrowbeam Coverage • Narrow Coverage • National Certificate • National Coarse • National Committee • National Conference • National Congress • National Coordinator • National Council • National Course • National Curriculum • Navigation Computer • Navigation Console • Neanderthal Conservative • Network Computer • Network Connect • Network Connection • Network Congestion • Network Control • Network Controller • Network Countdown • Neurocomputing • Neutral Current • Neutralizing Capacitance • Neutralizing Coil • New Caledonia (ISO country code) • Newtonian Collimator • Nickel-Cadmium • Nitrocellulosis • Nitrogen Compound • No Carrier • No Carry • No Change • No Charge • No Circuit • No Coil • No Comment • No Connection • No Contact • No Correction • No Cost • Nodal Controller • Node Center • Noise Criterion • Nomenclature Combinée • Nominal Correction • Nominating Committee • Non-Catalogued • Non Communiqué • Non Corrosive • Non Coupé • Non-recurring Cost • Nordic Council • Normal Confinement • Normal Corrective • Normally Closed • North Carolina (USA) • Nose Cone • Not Carried • Not Changing • Not Coded • Not Collected • Not Connected • Not Critical • Notification Control • Nouvelle-Calédonie (code ISO de pays) • Number of Columns • Numerical Control • Numerical Controller • Numerically Controlled • Numerus Clausus

NCA National Capital Astronomers (USA-DC) • National Command Authority • National Committee for Astronomy • National Committee of Astronomy • National Communication Association (USA) • National Computer Association • National Cooperative Account • Nederlandse Comité Astronomie (Netherlands) • Network Communications Adapter • Network Computing Architecture • Network Control Analysis • Nickel Copper Alloy • No Control Access • North Caicos (Turks & Caicos Islands) (location identifier) • Northern Control Area • Nuclear and Chemical Agency

NCAA National Collegiate Athletic Association (USA) • Nuclear Consumable Annual Analysis

NCAC National Copyright Administration of China (China-PRC) • Nicolaus Copernicus Astronomy Center (Poland)

NCAER National Council of Applied Economic Research (USA)

NCAHF National Council Against Health Fraud (USA)

NCAI National Committee on Atomic Information (USA)

NCAICM National Center for Advanced Information Components Manufacturing (USA)

NCAIR National Center for Automated Information Retrieval (USA)

NCAL New Caledonia • Nouvelle-Calédonie

NCAM National Center for Accessible Media (USA) • National Consortium for Aviation Mobility (USA-VA) • Network Communications Access Method

NCAP Nature Center And Planetarium • New Car Assessment Program • Northwest Coalition for Alternatives to Pesticides (USA)

NCAPI Netscape Client Application Programming Interface

NCAR National Center for Atmospheric Research (USA-CO) • National Conference on the Advancement of Research (USA) • Non-conformance and Corrective Action Reporting

NCARAI Navy Center for Applied Research in Artificial Intelligence (USA)

NCARC National Civil Aviation Review Commission (USA)

NCAS Network and Customer Administration System • Northern Colorado Astronomical Society (USA-CO)

NCASI National Council on Air and Stream Improvement (USA)

NCASR National Conformity Assessment System Recognition (USA)

NCAST National Center for Remote Sensing in Transportation (USA)

NCAT National Center for Advanced Technologies (USA)

NCATE National Council for Accreditation of Teacher Education (USA)

NCAW Neutralized Current Acid Waste

NCB National Codification Bureau • Near-Contact Binary • NetBIOS Control Block • Network Control Block • Nickel-Cadmium Battery • Non-spherical Collapse Boundary

NC-C Network Computer - Client

NCC NASA Class Code • NASA Communications Control • National Climate Center • National Climatic Center (NOAA) • National Climatological Center • National Color Code • National Command Center • National Computer Center • National Computer Conference (USA) • National Computing Centre • National Consultative Committee • National Consultative Council • National Consumer Council • National Coordinating Center • National Coordination Centre • National Coordination Committee • Navigation Computer Control • Navigation Control Console • NAVSTAR Control Center • Nederlandse Centrale Catalogus (Netherlands) • Nederlandse Classificatie Commissie (Netherlands) • Negotiated Contract Cost • Netware Console Commander • Network Communications Center • Network Communications Central • Network Communications Control • Network Connection and Control • Network Control Center • Network Control Code • Network Coordination Center • Neural Correlate of Consciousness • New Common Carrier • New Computational Challenge • Noise Control Committee • Noise Criterion Curve • Nominal Corrective Combination • Non-Cancellable Commitment • Non-Commercial Constituency • Non-Commutative Collision • Non-Commuting Collision • Non-Cooling Cluster • NORAD Command Center • Normal Corrective Combination • North Celestial Cap • North Cusp Cap • Nouveau Centre de Calcul • Numerical Control Code

NCCA National Council for Curriculum and Assessment

NCCAM National Center for Complementary and Alternate Medicine (USA)

NCCAS National Center of Communication Arts and Sciences (USA)

NCCB National Configuration Control Board • Network Configuration Control Board • NITF Configuration Control Board

NCCC North Cusp Cap Collar

NCCDS Network Control Center Data System

NCCE Novell Collaborative Computing Environment

NCCF Network Communications Control Facility • NOAA Central Computer Facility

NCCI Network Control Connection Identifier

NCCN National Committee on Computer Networks

NCCP Navigation Control Console Panel

NCCR National Center of Competence in Research (Switzerland)

NCCS NASA Center for Computational Sciences • National Computer Crime Squad • Navigation Command and Control System • Nodal Control Circuit Switch

NCCU North Carolina Central University (USA-NC)

NCCUSL National Conference of Commissioners on Uniform State Laws (USA)

NCD National Capability Demonstrator • Network Communications Division • Network Computing Device • Network Computing Devices Inc. (USA-CA) • Network Control Device • No Can Do • Node Connectivity Diagram • Non-City Delivery • Normalized Cumulative Deviation • Null Cell Discard

NCDB National Cartographic Database

NCDC National Climate Data Center • National Climatic Data Center (USA-NC) (*)

NCDCDS National Committee on Digital Cartographic Data Standards (USA)

NCDCS Narrowband Coherent Data Collection System

NCDD No Change in the Due Date

NCDM National Center for Data Mining (USA-IL) • Numerically-Controlled Drafting Machine

NCDNHC Non-Commercial Domain Name Holders Constituency

NCDP Navigation Control/Display Panel

NCDS NASA Climate Data System • National Cartographic Data Standard (USA) • National Centre for Development Studies (USA) • NPOESS Climate Data System

NCDSB National Cartographic Data Standard Base (USA)

NCDT Non-Chargeable Downtime

NCDU Navigation Control and Display Unit

NCDW Network and Control Development Network

NCE National Committee for the Environment • Navigation and Command Equipment • Net Carbon Emission • Network Centre of Excellence • Network Control Element • Network Control Engine • New Chemical Entity • Nice -

Côte d'Azur (France) (location identifier) • No Change in Estimates • Noise Control Engineering (periodical code) • Nomadic Computing Environment • Nouveau Contrat pour l'École • Nouvelle Compagnie Européenne

NCEAS National Center for Ecological Analysis and Synthesis (USA-CA)

NCEC National Committee Exchange Controller

NCED National Center for Employee Development (USA)

NCEE National Commission on Excellence in Education

NCEM National Center for Electron Microscopy (USA-CA)

NCEP National Center for Environmental Prediction (USA)

NCER National Center for Earthquake Research (USGS)

NCERQA National Center for Environmental Research and Quality Assurance (USA)

NCERT National Council for Educational Research and Training (India)

NCES National Center for Education Statistics (USA)

NCET National Council for Educational Technology (UK)

NCEX Neutral Current Excitation

NCF National Computing Facility • National Control Facility • Nationale Computerfaciliteiten (Netherlands) • Near-Caustic Field • Net Control Facility • NetWare Command File • Network Command Facility • Network Control Facility • Newton-Cotes Formula • Next Century of Flight (AW&ST) • No Clean Flux • Non-Cold Front • NSSDC Computer Facility • Nucleonics Calibration Facility

NCFC National Computer Facility of China

NCFLGT National Computing Facility for Lattice Gauge Theory (Australia)

NCFR National Council on Family Relations (USA)

NCFSK Non-Coherent Frequency Shift Keying

NCG National Control Group • Non-Condensible Gas • Nuclear Criteria Group • Nuevas Casas Grandes (Mexico) (location identifier) • Numerical Control Graphics

NCGA National Computer Graphics Association (USA)

NCGIA National Center for Geographic Information and Analysis (USA)

n/cgo Nuestro Cargo

NC&GS National Charting & Geodetic Survey (USA)

NCH Nachingwea (Tanzania) (location identifier) • National Clearinghouse

NCHARAN New Characteristic Analysis

NCHC National Center for High-performance Computing

Nchf Nachfolger

NCHMX Nine-Channel Multiplexer

NCHP Nickel-Chromium Honeycomb Panel

NCHPC National Consortium for High-Performance Computing (USA)

n. Chr. Nach Christus

NCHRP National Cooperative Highway Research Program (USA)

NCHS National Center for High-performance Supercomputing (China-ROC)

nCi Nanocurie

NCI National Cancer Institute (USA) • National Computing Industry • Navigation Control Indicator • Necocli (Colombia) (location identifier) • Negative Chemical Ionization • Network Computing Infrastructure • Network Control Interface • New Community Instrument • No Common Interest • No-Cost Item • Nomenclature Color Index • Non-Coded Information • Non-Commercial Item • Nuclear Cities Initiative • Nuclear Control Institute (USA-DC)

NCIA Native Client Interface Architecture

NCIC National Career Information Center (USA) • National Cartographic Information Center (USGS) • National Crime Information Center (USA) • Non-Confidential Information Center

NCIIA National Collegiate Inventors and Innovators Alliance (USA)

NCimB Nuovo Cimento B (periodical code)

NCimC Nuovo Cimento C (periodical code)

NCimD Nuovo Cimento D (periodical code)

NCimL Nuovo Cimento Lettere (periodical code)

NCimR Nuovo Cimento Rivista (periodical code)

NCimS Nuovo Cimento Serie (periodical code)

NCIMS Negative Chemical Ionization Mass Spectra • Numerical Control Information Management System

NCimZ Nuovo Cimento A (periodical code)

NCIR National Conference on Industrial Research (USA)

NCIS Nadir Climate Interferometer/Spectrometer • National Chemical Information System (USA)

NCIU Network Common Interface Unit • NEXRAD Communications Interface Unit

NCK Neck

NCKU National Cheng Kung University (China-ROC)

NCL Nacelle • National Central Library • National Chemical Laboratory • National Consumers League • Net Capacity Load • Network Control Language • New Caledonia (ISO country code) • Newcastle (UK-England) (location identifier) • Nouvelle-Calédonie (code ISO de pays)

NCLANG Nacelle Angle

NCLB No Child Left Behind (act) (USA)

NCLIS National Commission on Libraries and Information Science (USA)

NCLP Numerically Controlled Line Plotter

NCM National Centre of Metrology • Network Configuration Management • Network Control and Management • Network Control Manager • Non-Corrosive Metal • Non-Crew Member • Non-dissipative Closure Model • Nozzle Closure Mechanism • Núcleo de Cooperação Multilateral • Numerical Controlled Machine

NCMA National Contract Management Association (USA)

NCMC National Critical Materials Council (USA) • NORAD Cheyenne Mountain Complex (USA-WY)

NCMD Nine-Channel Multiplexer/Demultiplexer

NCMLS North Carolina Museum of Life and Science (USA-NC)

NCMN National Crustal Motion Network

NCMP Network Configuration Management Plan • Non-constituency Member of Parliament

NCMR National Center for Microgravity Research (USA) • Non-Conforming Material Report

NCMRWF National Centre for Medium-Range Weather Forecasting

NCMS National Center for Manufacturing Sciences (USA) • National Classification Management Society (USA) • Network Connection Management Subprotocol

NCMSACS Nodal Control Mobile Subscriber Access Circuit Switch

NCMT Numerical Control for Machine Tools

NCMU Network Control Master Unit

NCN New Chenega (USA-AK) (location identifier)

NCNR NIST Center for Neutron Research (USA)

NCO National Coordination Office • National Credit Office • Net Control Operator • Non-Commissioned Officer • Non-Constrained Optimization • Non-linearly-Constrained Optimization • Northern Cross Observatory (USA-MI) • Number-Controlled Oscillator • Numerically-Controlled Oscillator • Quonset Point (USA-RI) (location identifier)

NCOA National Change Of Address • Non-Commissioned Officer Academy • Non-Commissioned Officers Association

NCOIC Non-Commissioned Officer In Charge

NCOP National Civic Outreach Program

NCOQ Non-Cost Of Quality

NCOS National Commission On Space • Network Computer Operating System

NCOSE National Council On Systems Engineering (USA)

NCP Luzon Island - Cubi Point (Philippines) (location identifier) • National Climate Program • National Committee for Physics • National Committee President • National Computer Policy • National Contact Point • National Contingency Plan • Natural Current Profile • Net Circular Polarization • Net Control Procedure • NetWare Core Protocol • Network Control Point • Network Control Processor • Network Control Program • Network Control Protocol • Network Core Protocol • No-Copy Paper • Noctilucent Cloud Particle • Non-Consultative Party • Non-linear Complementarity Problem • Normal Circular Pitch • North Celestial Pole • Not Copy Protected • Nouveau Code Pénal (France)

NCPAS National Computer Program Abstract Service

NCPCO National Climate Program Coordinating Office

NCPDI National Coastal Pollutant Discharge Inventory

NCPI National Computer Program Index (UK)

NCPO National Climate Program Office (NOAA)

NCPS National Center for Patient Safety (USA) • NetWare Cross-Platform Services • Non-Contributory Pension Scheme • Nuclear Contingency Planning System

NCPTWA National Clearinghouse for Periodical Title Word Abbreviations (USA)

NCQ Marietta Atlanta (USA-GA) (location identifier)

NCQLP National Council on Qualifications for the Lighting Professionals (USA)

NCR National Capital Region • National Cash Register Co. • National Council for Research • National Cryptologic Representative • Nippon Cataloguing Rules • Network Change Request • No Calibration Required • No Carbon Required • Non-Compliance Report • Non-Computer Readable • Non-Conformance Record • Non-Conformance Report • Notification of Change Report • San Carlos (Nicaragua) (location identifier)

NCRA National Center for Radio Astrophysics (India) • National Cooperative Research Act (USA)

NCRC Nickel-Cadmium Rechargeable Cell

NCRD National Council for Research and Development

NCREMP North Central Region Educational Materials Project

NCREN North Carolina Research and Education Network

NCRHI National Council for Reliable Health Information (USA)

NCRP National Climatic Research Program • National Council on Radiation Protection • Non-Compulsory Reporting Point

NCRRT National Centre for Radiation Research Technology (Egypt)

NCRS Non-Conformance Report Supplement

NCRT National Research Council of Thailand

NC-S Network Computer - Server

NCS National Cartographic Standards • National Communications System • National Cryptologic System • NATO Committee for Standardization • Natural Color System • Navigational Computer Set • Navigation Control Simulator • Nearline Control Solution • Negative Central Shear • Net Control Station • Net Control System • NetWare Connect Services • Network Computing System • Network Control Series • Network Control Station • Network Control System • Network Coordination Station • Neural Computer Science • Neutron Compton Scattering • Newcastle (South Africa) (location identifier) • New Classification Society (UK) • NICMOS Cooling System • Nodal Central Switch • Nodal Control System • Node Center Switch • Node Control System • Non-Critical Sensitive • Normal Color Sequence • Numerical Category Scaling • Nutation Control System

NCSA National Center for Supercomputing Applications (USA-IL) • National Computer Security Agency • National Computer Security Association (USA) • NATO Consultation, Command and Control Agency • North Central Study Area

NCSC National Communications Security Committee • National Computer Security Center (USA) • National Computer Security Classification • National Computer Security Committee • National Computer Security Council • National Customer Support Center (USA) • North Carolina State College (USA-NC)

NCSCI National Center for Standards and Certification Information (USA)

NCSD National Communications System Directive • Natrona County School District (USA-WY)

NCSE National Center for Science Education (USA-CA) • NIST Communications Support Element

NCSES Nuclear Cross-Section Evaluation System

NCSF Northern Cross Science Foundation (USA-WI)

NCSG National Center Sponsor Group

NCSI National Centre for Science Information (India) • Network Communications Services Interface

NCSL National Computer Systems Laboratory (NIST) • National Conference of Standards Laboratories • National Conference of State Legislatures (USA)

NCSM Nagoya City Science Museum (Japan) • National Council of Science Museums (India)

NCSO National Central Sorting Office (USA)

NCSR National Centre for Scientific Research • National Committee for Space Research • National Committee on Space Research • National Council for Scientific Research

NCSS National Council for the Social Studies (USA) • Net-based Customer Service System

NCSSM North Carolina School of Science and Mathematics (USA-NC)

NCS&T National Council for Science & Technology

NCST National Center for Science and Technology (Saudi Arabia) (formerly SANCST) • National Committee on Science and Technology • National Council for Science and Technology • Naval Center for Space Technology (USA)

NCSTC National Council for Science and Technology Communications

NCSTRL Networked Computer Science Technical Reports Library

NCSU North Carolina State University (USA-NC)

NCSX National Compact Stellarator Experiment (USA)

NCT Nanosatellite Constellation Trailblazer • NATO Comparative Test • Network Control Terminal • Nicoya Guanacaste (Costa Rica) (location identifier) • Non-Cooperative Target • North Central Texas (USA) • Nuclear Compton Telescope

n/cta Nuestra Cuenta

NCTA National Cable Television Association (USA)

NCTAMS Naval Computer and Telecommunications Area Master Station

NCTC NRAO Charlottesville Technology Center

NCTD National Committee for Technological Development • Non-Consecutive Triple Difference

NCTE National Council of Teachers of English • Network Channel-Terminating Equipment

NCTI Non-Cooperative Target Identification

NCTL National Computer and Telecommunications Laboratory

NCTM National Council of Teachers of Mathematics (USA)

NCTR National Council on Teacher Retirement (USA) • Non-Cooperative Target Recognition • Northern California Technical Representative (USA-CA)

NCTRS Network Command and Telemetry Router System • Network Control and Telemetry Router System

NCTS Network Configuration and Test Section • Network Configuration and Test Station • Not Clear To Send

NCTU National Chiao Tung University (China-ROC)

NCU National Central University (China-ROC) • Navigation Computer Unit • Network Configuration Unit • Network Control Unit • NGAS Control Unit • Nitrogen Control Unit • Notebook Computer Unit • Nozzle Control Unit • Nukus (Uzbekistan) (location identifier)

NCUA National Credit Union Administration (USA)

NCUF National Computer Users Forum

NCUP No Commission Until Paid

NCUR National Conference on Undergraduate Research (USA)

NCURA National Council of University Research Administrators (USA)

NCURE National Center for Ultra-Reliability Engineering

NCV Narrowband Compressed Voice • Net Calorific Value • Net Contract Value • No Commercial Value • Noise Control Vehicle • Normalized Critical View

NCVQ National Council for Vocational Qualification (UK)

NCWM National Conference on Weights and Measures (USA)

NCWTD Nationaal Centrum voor Wetenschappelijke en Technische Dokumentatie (Belgium)

NCX Non Charge Exchange

NCXC National Committee Exchange Controller

NCY Annecy (France) (location identifier)

NC3O NATO Consultation, Command and Control Organization

n/d No Date

nD n-Dimensional (*) • n-Directional

Nd Néodyme • Neodymium

N-D N-Dimensional

N/D N-Dimensional • Need Date

ND NASA Document • National Debt • Navigation Display • Navy Department • N-Dimensional • Need Date • Negative Declaration • Neighbor Discovery • Neutral Density • Neutron Detector • Neutron Diffraction • Next Day • No Data • No Date • No Decision • No Deed • No Delay • No Demand • No Detect • No Drawing • Non Delayed • Non-Delivery • Non-Detectable • Non-Dimensional • Normalized Difference • North Dakota (USA) • Nose Down • Not Dated • Not Deeded • Not Detected • Not Determined • Not Done • Not Drawn • Nothing Doing • Nutzungsdauer

NdA Non-disclosure Agreement

NDA Bandanaira (Indonesia) (location identifier) • Nançay Decametric Array (France) • National Distribution Authority • Neutral Detector Assembly • No Detectable Activity • Non-Destructive Analysis • Non-Destructive Assay • Non-Dimensional Analysis • Non-Disclosure Agreement • Northern Delta Aquarids (meteor stream)

NDAA National Defense Authorization Act (USA)

NDAB Numerical Data Advisory Board

NDAC Network Design and Analysis Center • Northern Data Analysis Consortium (Hipparcos) • Not Data Accepted

NDADS NASA Data Archive and Distribution Service (*) • NASA Data Archive and Distribution System • NSSDC Data Archive and Distribution Service • NSSDC Data Archive and Distribution System

NDAG Notre Dame Astrophysics Group (USA-IN)

N. Dak. North Dakota (USA)

NDAP Neptune Data Analysis Program

NDAR National Directory of Australian Resources

NDAS North Devon Astronomical Society (UK-England)

NDATS Neutral Density And Temperature Spectrometer

NDB N-Dibutylamine • Neue deutsche Biographie • Non-Directional Beacon • Nouadhibou (Mauritania) (location identifier) • Numerical Database • Numeric Database

NDBC National Data Buoy Center (USA)

NDBO NOAA Data Buoy Office • Non-Directional Beacon Observation

NDC Nanded (India) (location identifier) • National Data Center • National Data Communications • National destination Code • National Documentation Center • Network Data Center • Night Display Component • Nippon Decimal Classification • Non-Dairy Cattle • Normalized Device Coordinates • Notre Dame College (USA-IN) • NRSA Data Centre • Nuclear Detector Circuit

NDCBU Neighborhood Delivery and Collection Box Unit

NDCDB National Digital Cartographic Database (USGS)

NDCC National Disaster Coordinating Council (Philippines) • Network Data Collection Center

NDCS National Digital Cartographic Standard • Network Data Control System

NDD Natural Diamond Detector • Network Delivery Device • Neutron Drip Density • Norton Disk Doctor • Sumbe (Angola) (location identifier)

NDDK Network Device Development Kit

NDDL Neural Data Definition Language

NDDO New Drug Development Office

NDDS Network Delivery Data Service • Networking Delivery Data Service • Nuclear Detonation Detection System

NdE Note de l'Éditeur

NDE Mandera (Kenya) (location identifier) • Near-Death Experience • News Development Environment • No Delay Expected • Non-Destructive Evaluation • Non-Destructive Examination • Non-Linear Differential Equation

NDEA National Defense Education Act (USA)

NDEF Not to be Defined

NDF Narrative Data File • N-dimensional Data Format • Neutral Density Filter • NTT Deep Field

NdFeB Neodymium-Iron-Boron (magnet)

NDG National Distribution Guide • No Date Given • Qiqihar (China-PRC) (location identifier)

NDGL Neodymium-Doped Glass Laser

NDGPS Nationwide Differential Global Positioning System

NDH Network Data-set Handler • No Damage History

NDI Namudi (Papua New Guinea) (location identifier) • No Deception Indicated • Non-Destructive Inspection • Non-Developmental Item • Non-Development Item • Numerical Designation Index

NDIA National Defense Industrial Association (USA)

NDIC National Defence Industries Council (UK) • Nuclear Data Information Center (USA)

NDIF Non-Destructive Inspection Facility

NDIR Non-Dispersive Infrared • Non-Dispersive Infrared Radiation

NDIS Network Device Interface Specification • Network Device Interface System • Network Driver Interface Specification • Network Driver Interface Standard • Non-Dimensional Impact Stress

NDIT Non-Destructive Inspection and Testing

NDIU Network Data Interface Unit

NDJ Ndjamena (Chad) (location identifier) • November - December - January

NDK Namorik Island (Marshall Islands) (location identifier)

NDL National Digital Library • Ndele (Central African Rep.) (location identifier) • Network Database Language • Network Data Language • Network Definition Language • Numerical Drawing List

NdlR Note de la Rédaction

NDLR Note de la Rédaction

NDLTD National Digital Library of Theses and Dissertations • Networked Digital Library of Theses and Dissertations (*)

NDM Mendi (Ethiopia) (location identifier) • Negative Differential Mobility • Negative Dispersion Mirror • Network Data Mover • Neutron Dose Monitor • Newtonian Dark Matter • No Dark Matter • Normal Disconnect Mode

NDMP Network Data Management Protocol

NDMS NetWare Distributed Management Server • NetWare Distributed Management Services

NDN Nadunumu (Papua New Guinea) (location identifier) • Nodal Data Network • Non-Delivery Notification • Nuclear Data Network

NDO Network Development Office

NDOS New Disk Operating System

NDP National Defense Plan • National Disclosure Policy • Net Domestic Product • Neutron Depth Profiling • Nom De Plume • Non-Directly (observable) Parameter • Normal Diametral Pitch • Normal Diametric Pitch • Numerical Data Package • Numerical Data Processor • Numeric Data Processor • Pensacola Ellyson (USA-FL) (location identifier)

NDPC National Data Processing Centre

NDPD National Data Processing Division • Numeric Data Processing Division

NDPF NASA Data Processing Facility • National Data Processing Facility

NDPS National Data Processing Service • National Disclosure Policy System • Novell Distributed Print Services

NDQ NASA Delta Quotation • No Difficult Question

Ndr Nieder

NdR Note du Rédacteur

NDR Nador (Morocco) (location identifier) • Negative Differential Resistance • Network Data Reduction • Network Data Representation • New Document Reference • Non-Destructive Read • Norddeutscher Rundfunk (Germany)

NDRC National Defense Research Committee (USA)

NDRE Norwegian Defence Research Establishment (Norway)

NDRO Non-Destructive Readout

NDRW Non-Destructive Read/Write

NDS Nationalverband Deutscher Schriftsteller (Germany) • Natural Decision System • Navigation Display System • NetWare Directory Services • Network Directing Station • Network Distribution Service • Non-Deliverable Software • Non-Developmental Software • Non-parametric Detection Scheme • Novell Directory Services • Nuclear Data Section • Nuclear Detection System • Nuclear Detonation System • Sandstone (Australia-WA) (location identifier)

NDSC Network for the Detection of Stratospheric Change

NDSE Non-Deliverable Support Equipment

NDSF Non-Dispersion Shifted Fiber

NDSI Normalized Difference Snow Index

NDSIC National Documentation and Scientific Information Centre

NDSL National Direct Student Loan • Non-Domestic Substances List

NDSOS National Deep-Sky Observers Society (USA-KY)

NDSS National Directory Support System

NDST Non-Dimensional Special Tool

NDSU North Dakota State University (USA-ND)

NDSW National Dark Sky Week (USA)

NdT Note du Traducteur

NDT Net Data Throughout • Newfoundland Daylight Time • Nil-Ductility Transition • Non-Destructive Test • Non-Destructive Testing • Non-deterministic Decision Table • Non-Distributive Trade

NDTA Night Driving Training Aid

NDTE Non-Destructive Test Equipment

NDTF Non-Destructive Test Facility

NDTI Non-Destructive Testing and Inspection

NDTIB Non-Destructive Test and Inspection Building

NDTL Non-Destructive Test Laboratory

NDTT Nil-Ductility Transition Temperature

NDU National Defense University • Navigation Data Unit • Navigation Distribution Unit • Network Data Unit • Nuclear Data Unit • Rundu (Namibia) (location identifier)

NDUS North Dakota University System (USA-ND)

NDV Anacostia (USA-DC) (location identifier) • NASP Derived Vehicle

NDVI Normalized Difference Vegetation Index

NDW Norton Desktop for Windows • Notgemeinschaft der Deutschen Wissenschaften (Germany) (now DFG)

NDWFS NOAO Deep Wide-Field Survey

ndx Index (file extension)

NDX Index

NDY Sanday (UK) (location identifier)

NDZ Nordholz-Spieka (Germany) (location identifier)

n/e New Edition • No Effects

ne Nepali (ISO language code) • Nueva Época

Ne Neon • Néon • Nepal • Nepalese • Netherlands

N/E New Edition • Not Entered • Not Exceeding

NE Named Entity • Name Error • National Emergency • National Engineer • National Estate • National Executive • Natural Erection • Naval Engineer • Near Encounter • Nebraska (USA) • Neghidhian • Nepal • Network Element • Network Engineering • Neuguen (Argentina) • Neutral Excitation • New Edition • New England • New Equipment • Niger (ISO country code) • Nine • No Effects • No Evolution • Noise Equivalent • Non-Emptiness • Non Essential • Nord-Est • Northeast • Northeastern • Not Equal • Not Essential • Not Exceeding • Notice of Exception • Nuclear Emission • Nuclear Energy • Nuclear Explosion

NEA Glynco (USA-GA) (location identifier) • National Education Association (USA) • National Endowment for the Arts (USA) • National Energy Authority (Iceland) • Near-Earth Asteroid • Negative Electron Affinity • Network Element Architecture • Network Equivalent Analysis • Noise Equivalent Angle • Non-Explosive Actuator • Northeastern African • Nuclear Energy Agency • Null Error Amplifier

NEAB Northern Examinations and Assessment Board (UK)

NEAD Negative Electron Affinity Detector

NEAF Northeast Astronomy Forum (USA)

NEAM Non-volatile Electrically Alterable Memory

NEAP National Environmental Action Plan (USA) • Near-Earth Asteroid Prospector

NEAR National Electronic Accounting and Reporting (system) • Near-Earth Asteroid Rendezvous

NEARnet New England Academic and Research Network

NEARS Near Earth Asteroid Returned Sample

NEARSS Northeast Area Remote Sensing System

NEARTIP Near-Term Improvement Program

NEASC New England Association of Schools and Colleges (USA)

NEAT Natural-source Electromagnetic Array Technique • Near-Earth Asteroid Tracking • Novell Easy Administration Tool

NEATM Near-Earth Asteroid Thermal Model

NEATT Northeast Association of Technical Teachers (USA)

neb Nebulosus

Neb Nebraska (USA) • Nebulous

NEB National Environmental Board • Net Environmental Benefit • Noise-Equivalent Bandwidth • North Equatorial Belt (jupiter) • Nudged Elastic Band

NEBA NASA Employees Benefit Association • North East Bolivian Airways

NEBD North Equatorial Band Disturbance

NEbE Northeast by East

NEbN Northeast by North

NEBn North Equatorial Belt - Northern Component (Jupiter)

Nebr Nebraska (USA)

NEBs North Equatorial Belt - Southern Component (Jupiter)

NEBS Network Equipment Building System • New Equipment Building System

NEBz North Equatorial Belt - Internal Zone

Nec Necessary • Necessity

NEC National Economic Council • National Electrical Code (*) • National Electric Code • National Engineering Centre • National Executive Committee • National Exhibition Center • Necochea (Argentina-BA) (location identifier) • Network Equipment Controller • New Eruptive Cosmogony • Nippon Electric Co. (Japan) • Noise Electron Count • North Equator Crossing • North Equatorial Current • Not Elsewhere Classified • Null Energy Condition

NECAP NASA Energy-Cost Analysis Program • Navigation Equipment Capability Analysis

NECOS Network Coordination Station

Necr Necrosis

Necrol Necrology

NECS National Electrical Code Standard • Nationwide Educational Computer Service • Normal Environmental Control System

NECSTS Network of European Centers in Science and Technology Studies

NECTAR Network of European Communication and Transport Activities Research • Network of European Communications and Transport Activity Research (*)

NECUSE New England Consortium for Undergraduate Science Education

NECZ Near-Equatorial Convergence Zone

NED NASA Extragalactic Database • Navigation Error Data • Neutron Detector • New English Dictionary • No Evidence of Desease • No Expiration Date • Normal Equivalent Deviation • North - East - Down

NEDA National Economic and Development Agency • National Economic Development Authority • National Electronics Distributors Association (USA)

NEDC National Economic Development Council

NEDR Noise Equivalent Dalta Reflectivity

NEDRES National Environmental Data Referral System (USA)

NEDS National Emission Data System

NEDT Noise Equivalent Delta Temperature

NEE Noise-Equivalent Energy • Non-linear Evolution Equation

NEED Non-Maxwellian Electron Energy Distribution

NEEDS NASA End-to-End Data System • National Engineering Education Delivery System • Nikkei Economic Electronic Data System

NEEP Nuclear Engineering and Engineering Physics

Néérl Néérlandais

Neerl Neerlandés

NEES Network for Earthquake Engineering Simulation

NEESDI Netherlands Environmental Earth Systems Dynamics Initiative

NEET Nuclear Excitation by Electronic Transition

nef nicht entflammbar

NEF NASA Extended Fortran • National Extra Fine • Neftekamsk (Russia) (location identifier) • Network Element Function • Noise-Equivalent Flux • Noise Exposure Forecast • Nuit des Étoiles Filantes

NEFAS Northeast Florida Astronomical Society (USA-FL)

NEFD Noise-Equivalent Flux Density

NEFF Non-Essential-Function Failure

NEFT Negative Extended Faceted Taxonomy

Neg Negate • Negación • Negation • Negative • Negatively • Neglect • Negligence • Negligible • Negotiate

NEG Negate • Negative • Neglect • Negligible • Negril (Jamaica) (location identifier) • Non-Evaporable Getter • Numerical Experimentation Group

Negat Negativo

NEH National Endowment for the Humanities (USA)

NEHA National Environmental Health Association (USA)

NEHRP National Earthquake Hazards Reduction Program (USA)

NEI National Eye Institute (USA) • New Equipment Introduction • Noise-Equivalent Input • Noise-Equivalent Intensity • Noise-Equivalent Irradiance • Non-Equilibrium Ionization • Non-Equilibrium Ionized (plasma) • Non Est Inventus • Non-Explosive Initiation • Northern Engineering Industries • Not Elsewhere Included (*) • Not Elsewhere Indicated • Nouvel Espace Indiciaire • Nouvel État Indépendant

NEIA National Energy Information Administration (USA)

NEIC National Earthquake Information Center (USA) • National Energy Information Center (USA) • National Enforcement Investigations Center (USA)

NEIF Near-Earth Instrumentation Facility

NEIR Narrative End Item Report

NEIS National Earthquake Information Service (USGS) • National Energy Information System (USA) • National Environmental Information System

NEIT New Equipment Introductory Team

NEIU Northeastern Illinois University (USA-IL)

NEJ Nejjo (Ethiopia) (location identifier)

NEJM New England Journal of Medicine (periodical code)

NEK Nekemt (Ethiopia) (location identifier) • Norsk Elektroteknisk Komite (Norway)

NEKAAL Northeast Kansas Amateur Astronomers League (USA-KS)

NEL Lakehurst Naec (USA-NJ) (location identifier) • Narrow Emission Line • National Exploitation Laboratory • Network Element Layer • Network Event Level • Nuclear Enterprises Ltd.

NELAS North East London Astronomical Society (UK)

NELG Narrow-Emission-Line Galaxy

NELH Natural-Energy Laboratory of Hawaii (USA-HI)

NELLCO New England Law Library Consortium (USA)

NELPAG New England Light Pollution Advisory Group (USA-MA)

NELR Narrow Emission Line Region

NELS NASA Electronic Library System • Northern European LORAN System • Nuclear Effects Link Simulator

NELY Northeasterly

NEM Nanoelectromechanical • Negative Energy Mode • Noise-Equivalent Magnitude • Non-Electronic Maintenance • Not Elsewhere Mentioned

NEMA National Electrical Manufacturers Association (USA)

NEMAC Normal Error Model Analysis Chart

NEMAG Negative Effective Mass Amplifiers and Generators

nem. con. Nemine Contradicente

nem. dis. Nemine Dissentiente

NEMO Neutral Extinction of Massive Objects • Neutrinoless Experiment on Molybdenum • Neutrino Mediterranean Observatory (Italy) • Non-Empirical Molecular Orbital • Not Everybody Must Observe • Nouvelle Expérience Majeure de l'Observatoire

NEMP Network Error Management Protocol • Nuclear Electromagnetic Pulse

NEMPS National Environmental Monitoring and Prediction System

NEMS NASA Equipment Management System • National Emergency Management System • National Energy Modeling System • Near-Earth Magnetospheric Satellite • New England Meteoritical Service (USA-MA) • Nimbus E Microwave Spectrometer

NEMT National Emergency Management Team

NEN Whitehouse (USA-FL) (location identifier)

N. Eng. New England • Northern England

NENL Near-Earth Neutral Line

NEO Near-Earth Object • Near-Earth Orbit • Network Enterprise Objects • Northeastern Operations Office

NEOC National Earth Observations Center (USA)

NEOCD Near-Earth Object Control Page

NEOCP Near-Earth Object Confirmation Page

NEOCS Near-Earth Object Coordination System

NEODyS Near-Earth Object Dynamics Site

NEOF No Evidence Of Failure

Neol Neologism • Neologisme • Neologismo

NEOME New Electroactive Organic Materials for Electronics

NEON National Ecological Observatory Network (USA) • Network of European Observatories in the North

NEONET Netherlands Earth Observation Network

NEOPAT Near-Earth Object Patrol

NEOS National Earth Orientation Service (USA-DC)

NEOSC Near-Earth Object Search Committee

NEOWG Near-Earth Object Working Group

Nep Nepal • Nepalese • Neptune

NEP National Extension Program • Near-Earth Phase • Nearest Equivalent Product • Negative Equally Probable • Net Ecosystem Production • Network Entry Point • New Economic Policy • New Edition Pending • Noise Equal to Power • Noise-Equivalent Power • Nominal Entry Point • Non-Electronic Part • Non-Elutable Polar • Normal Entry Point • North Ecliptic Pole • Northern Equatorial Pacific • Nuclear Electric Propulsion • Nuclear Emulsion Plate • Nuclear Energy Power • Nuclear Environment Protection

NEPA National Electric Power Authority • National Energy Policy Plan • National Environmental Policy Act (USA) • National Environmental Protection Agency (USA) • National Environment Protection Authority

NEPAE Núcleo de Estudos e Pesquisas Aeroespaciais (Brazil-RS)

NEPAL Nouvelles Expériences de Physique de l'Accélérateur du LAL

NEPC National Environment Protection Council

NEPD Noise-Equivalent Power Density

NEPM National Environment Protection Measure

NEPN Near-Earth Phase Network

NEPP National Environment Policy Plan • National Environment Protection Plan

NEPR North Ecliptic Pole Region

NEPRF Naval Environmental Prediction and Research Facility (USA)

NEPS Novell Enterprise Print Services

Nept Neptune

NEPTUNE New European Programme for Technology Utilization in Education

NEQ Não Equivalente • Non-Equivalence • Non-Équivalence • Normal Equation

NER Neryungri (Russia) (location identifier) • Never Exceed Redline • Niger (ISO country code) • Noise-Equivalent Radiance • Non-conformance Event Record

NERA Northern (extremity of) Earth's Rotation Axis

NERAL Northeast Region of the Astronomical League (USA)

NERC Named Entity Recognition • National Environmental Research Center (USA) • Natural Environment Research Council (UK)

NERCS National Energy Research Supercomputer Center

NERDAS NASA Earth Resources Data Annotation System

NERF Naval Electromagnetic Radiation Facility

NERI National Environmental Research Institute

NERIST Northeastern Regional Institute of Science and Technology (USA)

NERL Northeast Research Libraries

NERN Northeastern

NERO Near-Earth Rescue and Operations

NEROC Northeast Radio Observatory Corporation (USA-MA)

NERP National Environmental Research Park (USA) • National Environmental Research Policy (USA)

NERS Natural Environmental Radioactivity Survey

NERSC National Energy Research Scientific Center (USA-CA)

NERVA Nuclear Energy for Rocket Vehicle Application • Nuclear Engine for Rocket Vehicle Application (*)

NES National Education Supercomputer • National Election Studies • National Encryption System • National Energy Strategy • National Environment Secretariat • Network Enabled Server • Network Encryption System • New York - East 34th Street (USA-NY) (location identifier) • Noise Equivalent Signal • Non-Erasable Storage • Not Elsewhere Specified • Nuclear Elastic Scattering

NESA Near East and South Asia

NESB Non-English-Speaking Background

NESC National Electrical Safety Code • National Energy Software Center • National Energy Software Corporation • National Environmental Satellite Center • National Environmental Supercomputing Center • New England Science Center (USA-MA)

NeSDaC Netherlands SAX Data Centre

NESDD NOAA Earth System Data Directory

NESDIS National Environmental Satellite Data and Information Service (USA-NC) (*) • National Environmental Satellite Data Information Service

NESDP National Economic and Social Development Plan

NE/SE Network Engineering/System Engineering

NESE Network Engineering - System Engineering • New and Renewable Source of Energy

NESFD Noise Equivalent Spectral Flux Density

NESHAP National Emission Standards for Hazardous Air Pollutants

NESLI National Electronic Site License Initiative

NESPUT National Educational Simulations Project Utilizing Telecommunications

NESR Natural Environment Support Room • NGST External Review Board • Noise Equivalent Spectral Radiance

NESS National Earth Satellite Service (NOAA) (now merged with EDIS into NESDIS) • National Environmental Satellite Service (NOAA) • New England Skeptical Society (USA)

NEST New and Emerging Science and Technologies • Novell Embedded Systems Technology • Nuclear Effects Support Team • Nuclear Emergency Search Team • Nuclear Energy Search Team

NESTA National Earth Science Teachers Association (USA) • National Endowment for Science, Technology and the Arts (UK)

NESTOR Netherlands Educational and Scientific Titles for Online Retrieval • Neutrino Experimental Submarine Telescope with Oceanographic Research • Numérisation et Sériation pour Traitement des Informations Radar

net Network (file extension)

Net Network

NET National Educational Television • National Emergency Technology • National Emerging Technologies • Near-Equatorial Through • Net Energy Thrust • Net Equivalent Temperature • Network • Network Entity Title • Network Equipment Technologies • Network of Expanded Telecommunications • New England Telephone (USA-MA) • New Equipment Training • Newton Emission Theory • Next European Torus • Nimbus Experiment Team • Noise-Equivalent Target • Noise-Equivalent Temperature • Normal Environment Temperature • Norme Européenne des Télécommunications • Not Earlier Than • Nuclear Emergency Team • Nuclear Emission Type

NETA No Electron Theft Act (USA)

NetBEUI NetBIOS Extended User Interface

NETBEUI NetBIOS Extended User Interface

NetBIOS Network Basic Input/Output System

NETBIOS Network Basic Input/Output System

NetBlT Network Block Transfer

NETBLT Network Block Transfer

NETC National Emergency Transportation Center (USA)

NETCAP Network Capability

netCDF Network Common Data Form

NETCOM Network Communications

NETD Noise Equivalent Temperature Difference

NetDDE Network Dynamic Data Exchange

NETDDE Network Dynamic Data Exchange

NETDS Near-Earth Tracking and Data System

NETEV Network Experimental Version

NETFD Non-Equilibrium Thermo-Field Dynamics

Neth Netherlands

Neth. Ant. Netherlands Antilles

NETID Network Identification • Network Identifier

NETIPC Network Interprocess Communications

NETMAN Network Management

NETMGR Network Manager

NETMON Network Monitor

NETP New Equipment Training Plan • New Equipment Training Program

NETR NATO Electronic Technical Recommendation

NETS National Educational Technology Standardsi (ITSE) • National Emergency Telecommunications Service (USA) • Nationwide Emergency Telecommunications Service • Nationwide Emergency Telecommunications System • Network of Sensors

NETSEC Network Security

NETSET Network Synthesis and Evaluation Technique

NETSIM Network Simulation

NETSP New Equipment Training Support Package

Nett Nettoyage

NETT Network for Environmental Technology Transfer • Network for Environment Technology Transfer • New Equipment Training Team

NETTSP New Equipment Training Test Support Package

NETV National Educational Television

NETWX Network Weather

NEU Network Encryption Unit • Northeastern University (USA-MA) • Sam Neua (Laos) (location identifier)

Neurol Neurological • Neurology

Neut Neuter • Neutral • Neutralize • Neutralized

Nev Nevada (USA)

NEV Network Experimental Version • Nevada (USA) • Nevis (Saint Kitts & Nevis) (location identifier)

NEVADA Net Energy Verification And Determination Analyzer

NEVEC NOVA-ESO VLTI Expertise Centre

NEVS NASA Equipment Visibility System

New Newton • Newtonian

NEW NASA Education Workshop • Net Economic Welfare • New Orleans - Lakefront (USA-LA) (location identifier)

NewA New Astronomy (periodical code)

NewAR New Astronomy Review (periodical code)

NEWEST NASA Educational Workshop for Elementary School Teachers

Newf Newfoundland (Canada)

NEWFIRM NOAO Extremely Wide-Field Infrared Imager

New M New Mexico (USA)

NEWMAST NASA Educational Workshop for Math, Science and Technology

NEWPIMS Neutral Environment With Plasma Interactions Monitoring System

NeWS Network Windows System

NEWS National Early Warning System • NetWare Early Warning System • Network Extensible Windows System • New Engineering Work Schedule • New Engineering Work System • NOAA's Earth Watch Service • Non-linear Elastic Wave Spectroscopy • Novell Electronic Webcasting Service

NewSc New Scientist (periodical code)

NEWSIPS New Spectral Image Processing System

Newsl Newsletter

NEWSTAR Northeast Wisconsin Star Gazers (USA-WI)

NEXAFS Near-Edge Absorption Fine Structure

NEXRAD Next-generation Radar

NExT New Experiences in Teaching

NEXT NASA Evolutionary Xenon Thruster • Near-End Crosstalk • NLR Engineering Expert Toolkit

n/f No Funds • Nuestro Favor

n.F. Neue Fassung

N/F No Funds

NF Île Norfolk (code ISO de pays) • Nanofarad • National Fine • National Formulary • Natural Flood • Navigation Filter • Near Face • Neue Folge • Neuro-Fuzzy • Neutral Fraction • Newfoundland (Canada) • Niederfrequenz • No Feedback • No Fool • No Form • No Funds • Noise Factor • Noise Figure • Noise Frequency • Non-Ferrous • Non-Fiction • Non-Filtered • Non-Flickering • Non-Fundable • Norfolk Island (ISO country code) • Norges Forskningsråd (Norway) • Normal Formula • Norman French • Norme Française (France) • Nose Fairing • Not Found • Number of Fields • Nutzfläche

NFA Network File Access • Nitrogen Filling Assembly • No-Fire Area • No Further Action • Null Filling Antenna

NFAC National Foreign Assessment Center

NFAIS National Federation of Abstracting and Indexing Services (USA) (*) • National Federation of Abstracting and Information Services • National Federation of American Information Services

NFAM Network File Access Method

NFANP North Florida Association of Networking Professionals (USA-FL)

NFB Mount Clemens - Detroit (USA-MI) (location identifier) • National Film Board • Negative Feedback • No Feedback • Nordisk Forskningsbiblioteken

NFBC National Film Board of Canada

NFBO Normal-Flaring Branch Oscillation

NFC Near-Field Communication • Negative Factor Counting • Negative Feedback Circuit • Neuro-Fuzzy Classifier • No Further Consequences • Not Favorably Considered

NFCT Non-Federal Control Tower

Nfd Newfoundland (Canada)

NFD Net Filter Discrimination • Neutron Flux Density • Newfoundland (Canada) • Nodal Fault Diagnostics • No Fixed Date • Nürnberger Flugdienst

NFDC National Flight Data Center

NFE Network Front End • Non-Fundable Expenses • Normalized Fuzzy Entropy • Not Fully Equipped

NFER National Foundation for Educational Research (UK)

NFF No Fault Found • Non-Formula Funded

NFFE National Federation of Federal Employees (USA)

NFG Nefteyugansk (Russia) (location identifier) • Non-Functional Goodie

NFGS Nearby Field Galaxy Survey

NFI Narrowband Filter Imager • Narrow-Field Imaging • Narrow-Field Instrument (*) • Narrow-FOV Instrument • Nationale Faciliteit Informatica (Netherlands) • Neutron Flux Integrator • No-Frame Inconditional • Noise Figure Indicator • Nouvelles Formations d'Ingénieurs

NFIC National Fraud Information Center

NFIP National Flood Insurance Program • National Foreign Intelligence Program

NFK Île Norfolk (code ISO de pays) • Norfolk Island (ISO country code)

Nfl Nutzfläche

NFL Fallon (USA-NV) (location identifier) • National Film Library • Nerve Fiber Layer • New Foreign Launch • No-Fire Line • Non-Fermi Liquid

NLFC National Foreign Language Center (USA-DC)

Nfld Newfoundland (Canada)

NFLD Newfoundland (Canada)

NFLNA Noise Figure Low-Noise Amplifier

NFM Narrow-band Frequency Modulation • Network Facilities Manual • Network File Manager • Noise Figure Meter • Non-Ferrous Metal

NFMA National Forest Management Act (USA)

NFMC Not Fully Mission Capable

NFMS Nitrogen Flow Measuring System • Noise Figure Meter System

NFN No Form Necessary • No Further Need

nfo Information (file extension)

NFO Near-Field Optical • Near-Field Optics • Niuafo'Ou (Tonga) (location identifier)

NFOEC National Fiber Optic Engineer Conference (USA)

NFOM Near-Field Optical Microscope

NFOV Narrow Field Of View (*) • Near Field Of View

NFOVR Narrow Field Of View Radiometer

NFP National Focal Point • National Forschungsprogramme • National Fusion Program (Canada) • New Framework Programme • NOAO Fundamental Plane (survey) • Normal Failure Period • Not For Profit

NFPA National Fire Protection Association (USA)

NFPL Near Field Perfect Lens

NFPM Nuclear Flight Propulsion Module

NFPO Non For Profit Organization

NFPPE Nuclear Fore Policy, Planning and Execution

NFPS NOAO Fundamental Plane Survey • Nuclear Flight Propulsion System

NFR Naturvetenskapliga Forskningsrådet (Sweden) • Near Field Recording • Near Final Return • Negative Flux Rate • Net Flux Radiometer • Network Flight Recorder • No Further Requirements • Non-Functional Requirement • Not For Report • Nothing Further to Report • Nuclear Fission Reactor

n/fra Nuestra Factura

NFRA Netherlands Foundation for Radio Astronomy • Netherlands Foundation for Research in Astronomy (now ASTRON)

NFRE Net Flux Radiometer Experiment

NFS NASA FAR Supplement • National and Foreign Stations • Network File Server • Network File Service • Network File Structure • Network File System (*) • Neuro-Fuzzy System • Neutron Flux Spectrum • Nitrogen Flow System • Noise Frequency Spectrum • Not For Sale • Not on Flying Status • Nozzle Flow Sensor • Number Field Sieve (method)

NFSI National Foundation for Science and Industry (USA)

NFSP No Frills Star Party

NFSS New Font Selection Scheme

NFSU Non-Flying Support Unit

NFT National Fly-by-wire Testbed • Network File Transfer • Non-oscillatory Forward-in-Time • Normal-Field Telescope

NFTC National Flight Test Center

N/FTS Near-/Far-Term System

NFU Not For Us • Not Frequently Used

NFW Navarro - Frenk - White (model) • Non-Fuel-Wasting

NFWO Nationaal Fonds voor Wetenschappelijk Onderzoek (Belgium) (see now FWO)

NFZ No-Fly Zone • Nuclear Free Zone

NF2 Non-First Normal Form

N&G Navigation & Guidance

ng Nanogram • Norton Guide (file extension)

Ng Norwegian

N&G Navigation & Guidance

NG Narrow Gauge • National Gallery • National Geographic (magazine) • National Guard (USA) • Natural Gas • Navigation and Guidance • Nebulizer Gas • Network Gateway • New Growth • New Guinea • Next Generation • Niger • Nigeria (ISO country code) • Nitrogen Gauge • Nitroglycerine • Nitroglycerol • No Go • No Good • Non-Graduate • Nose Gear • Not Given • Not Good

NGA National Geographic Association • National Governors' Association • Next-Generation Array • Nigeria (ISO country code) • Young (Australia-NSW) (location identifier)

NGAF Non-GPRS Alert Flag

NGAMS Next Generation Archive Management System

NGAO North Georgia Astronomical Observatory (USA-GA)

NGAS Next Generation Archive System

NGASS Next Generation of Australian Scientists Society

NGAST Next Generation Archive System Technologies (project)

NGAT New Generation Astronomical Telescope

NGB Next Generation Browse • Ningbo (China-PRC) (location identifier)

NGBN National Gravity Base Net

NGBT Next Generation Binocular Telescope

NGC Grand Canyon - North Rim (USA-AZ) (location identifier) • National Geophysical Committee • National Grid Co. plc (UK) • Navigation - Guidance - Control • Networked Group Communication (workshop) • New General Catalogue • Next Generation Computer • North Galactic Cap • North Georgia College (USA-GA) • Nozzle Gap Control

NGCAD Next Generation Computer-Aided Design

NGCFHT Next-Generation Canada-France-Hawaii Telescope

NGCIA National Centre for Geographic Information and Analysis

NGCP Navigation, Guidance and Control Processor

NGCR Next Generation Computer Resource • Next Generation Computing Resources

NGCSU North Georgia College and State University (USA-GA)

NGD Anegada (British Virgin Island) (location identifier)

NGDB National Geo-Database • National Geodetic Database

NGDC National Geophysical Data Center (USA-CO)

NGE Navigation Guidance Equipment • N'Gaoundere (Cameroon) (location identifier) • Not Greater or Equal

NGeb Nebengebäude

NGEDMS Next Generation Engineering Data Management System

NGEM Next Generation Extension Mode

NGEO Non-Government Educational Organization

nGF Normalized Garfield Factor

NGF Nerve Growth Factor • Nervous Growth Factor • Nettogrundfläche • Next Generation Focuser • Non-Gravitational Force

NGFAW Nachrichtentechnische Gesellschaft Fachtagung über Antennen (periodical code)

NGfL National Grid for Learning (UK)

NGG Negative Grid Generator

NGGRB Next-Generation Gamma-Ray Burst (mission)

NGH NASA Grant Handbook • Natural Gas Hydrate

NGHEG Next-Generation High-Energy Gamma-ray (mission)

NGI Nationaal Geografisch Instituut (Belgium) • National Geographic Institute • National Geographic Interactive • Nederlands Genootschap voor Informatica (Netherlands) • Next Generation Internet • Ngau Island (Fiji) (location identifier) • Non-Governmental Institution • Norwegian Geotechnical Institute (Norway)

NGII Next-Generation Information Infrastructure

NGIMS Neutral Gas and Ion Mass Spectrometer

NGIO Next Generation Input/Output

NGIS National Geographic Information System • Next Generation Information System

NGISE National Geographic Information System on Environment

NGIT Next-Generation Internet Technology

NGITS Next Generation Information Technologies and Systems (workshop)

NGk New Greek

NGL Natural-Gas Liquid • Neodymium Glass Laser • Neon Glow Lamp • Ngala (South Africa) (location identifier)

NGld Nederlandse Gulden

NGLR Neodymium Glass Laser Rod

NGLT Next Generation Lowell Telescope

NGM Guam Agana (Guam) (location identifier) • Nested Grid Model • Nouvelle Gestion des Moyens

NGN Naira (Nigeria) (ISO currency code) • Nargana (Panama) (location identifier) • National Geodetic Network • Next-Generation Network

NGNP Nominal Gross National Product • Non-Geographic Number Portability

NGO Nagoya (Japan) (location identifier) • National Gemini Office • Non-Governmental Organization

NGOS Next Generation Optical Spectrograph

NGP Corpus Christi (USA-TX) (location identifier) • Nearest Grid Point • North Galactic Pole

ngr Nanogram

NGR New General Relativity • Ningerum (Papua New Guinea) (location identifier)

NGRI National Geophysical Research Institute (India)

NGRS National Geodetic Reference System (USA)

N&GS Navigation & Guidance System

NGS Nagasaki (Japan) (location identifier) • National Geodetic Society • National Geodetic Survey • National Geographic Society (USA-DC) • Natural Guide Star • Network Gateway Server • Non-Government Specification • Non-Government Standard • NSSDC Graphics System • Nuclear Generating Station

NGSB Non-Government Standard Body

NGSC NOAO Gemini Science Center

NGSDC National Geophysical and STP Data Center

NGSIC National Geodetic Survey Information Center (NOAA)

NGSO Non-Geostationary Orbit • Non-Geostationary Satellite Orbit • Non-Geosynchronous Orbit

NGSOC National Geodetic Survey Operations Center (NOAA)

NGSP National Geodetic Satellite Program

NGSS Nearby Galaxies Supernova Search • Next-Generation Sky Survey

NGSSC National Graduate School of Scientific Computing

NGST New-Generation Space Telescope (now JWST) • Next-Generation Space Telescope (*)

NGSTDC National Geophysical and Solar-Terrestrial Data Center (NOAA)

NGT NASA Ground Terminal • New-Generation Telescope • New-Generation Trainer • Newton Gravitation Theory • Next-Generation Telescope (*) • Night • Noise Generator Tube

NGTS New-Generation Training System

NGU Norfolk Chambers (USA-VA) (location identifier)

NGV Natural-Gas for Vehicles • Natural-Gas Vehicle • Navire à Grande Vitesse • Ngiva (Angola) (location identifier)

NGVCC Natural Gas Vehicle Composite Cylinders

NGVD National Geodetic Vertical Datum

NGVHET Next-Generation Very-High-Energy Telescope

NGW Corpus Christi - Cabaniss (USA-TX) (location identifier)

NGWIC National Ground Water Information Center

NGX Manang (Nepal) (location identifier)

NGXO Next Generation X-ray Observatory

NGZ Alameda (USA-CA) (location identifier)

nH Nanohenry

NH National Heritage (UK) • National Highway • National Hot • New Hampshire (USA) • Nitrogen Hydride • Nominal Height • Non-Hydroscope • Normal Height • Northern Hemisphere • Nuclear Head • Nuclear-Hardened

NHA Next Higher Assembly • Next Higher Authority • Nha Trang (Viet Nam) (location identifier) • Non-Hydrogen Atom

NHAB Nürnberger Hobbyastronomenbund (Germany)

NHAC North Houston Astronomy Club (USA-TX)

NHAO Nishi-Harima Astronomical Observatory (Japan)

NHAP National High-altitude Aerial Photography • National High-Altitude Photography

NHAPP National High-altitude Aerial Photographic Program

NHar Não Harmonizada

NHB NASA Handbook • Near-Hydrogen Burning (region) • Nickel-Hydrogen Battery

NHBU Networking Hardware Business Unit

NHC National Hurricane Center (NOAA) • Next Hop Client

NHCWP North Hollywood California Western Periodicals (periodical code)

NHD Minhad Ab (UAE) (location identifier)

NHDR Network Header

NHE National Heat Exchange • No High Energy • Normal Hydrogen Electrode

N. Heb. New Hebrides

N. Héb. Nouvelles Hébrides

NHEML National Hurricane Experimental Meteorological Laboratory

NHF New Halfa (Sudan) (location identifier)

NHG New High German

NHGDS Non-Hydrogenous Gas Delivery System

NHI National Health Insurance • Northwestern Hawaiian Islands (USA-HI)

NHK Patuxent River (USA-MD) (location identifier)

NHL National Hockey League (USA) • Normal Hydrogen Line

NHLBI National Heart, Lung and Blood Institute (USA-MD)

n.h.M. nach herrschender Meinung

NHM National History Museum (UK) • Natural History Museum (UK) • No Hot Metal • Nuclear Hyperfine Magnetic

NHMF Non-polluting Hypergolic Miscible Fuel

NHMFL National High-Magnetic-Field Laboratory (USA-FL)

NHML Northern Hemisphere Mid-Latitude

NH&MRC National Health & Medical Research Council

NHMRC National Health and Medical Research Council

NHMT Nuclear-Hardened Mosaic Technology

NHMTF Natural Hazards Mitigation Trust Fund

NHO Northern Hemisphere Observatory (Canary Islands)

NHP Nitrogen High Pressure • Nominal Horsepower

NHPN National Highway Planning Network (USA)

NHPP Non-Homogeneous Poisson Process

NHQ Nuclear Hyperfine Quadrupolar

NHQP Number of High-Quality Papers

NHR Natural Handwriting Recognition • Non-Hierarchical Routing

NHRA Next Higher Recoverable Assembly • Next Higher Repairable Assembly

NHRC National Hydrology Research Centre

NHRE National Hail Research Experiment

NHRI National Health and Retirement Insurance • National Hydrology Research Institute

NHRP Next-Hop Resolution Protocol • Next-Hop Routing Protocol

NH&S Nuclear Hardening & Survivability

NHS National Health Service • National Highway System (USA) • National Hydrological Service • Next-Hop Server • Nuclear Hardening and Survivability • Nushki (Pakistan) (location identifier)

NHSC NASA Herschel Science Center (USA-CA) • National Home Study Council (USA)

NHSE National HPCC Software Exchange

NHSG Non-High-School Graduate

NHST Next-generation Hubble Space Telescope

NHT Northolt (UK-England) (location identifier)

NHTA Numerical Heat Transfer Applications (periodical code)

NHTCU National High-Tech Crime Unit

NHTF National Hover Test Facility (USA)

NHTSA National Highway Traffic Safety Administration (USA)

NHU National Hispanic University (USA-CA)

NHV Nuku Hiva (Marquesas Islands) (location identifier)

NHX Foley Barin (USA-AL) (location identifier)

NHZ Brunswick (USA-ME) (location identifier) • Non-Habitable Zone

NH3 Amonia

NH4 Hydrazine

ni Nicaragua

Ni Nickel

N/I Non-Interlaced

NI National Identification • National Instruments • Natural Intelligence • Navigation Initialization • Network Indicator • Network Interface • New Impression • Nicaragua (ISO country code) • Nigeria • No Information • Noise Index • Non-Inductive • Non-Interchangeable • Non-Interlaced • Non-Intervention • Norfolk Island • Northern Ireland • North Island (New Zealand) • Note d'Information • Noted Item • Not Interested • Not Issued • Numerical Index • Número de Investigadores • Triton (satellite of Neptune)

NIA National Information Account • National Institute of Aerospace (USA-VA) • National Institute on Aging (USA) • National Intelligence Authority • National Irrigation Administration • Negative Impedance Amplifier • Network Integration and Alignment • Network Interface Adapter • Network Interoperability Alliance • Newspaper Institute of America • Newspaper Institute on Aging • Next Instruction Address • Nickel-Iron Alloy • Nimba (Liberia) (location identifier) • No Information Available • No Input Acknowledge • Northern Iota Aquarids (meteor stream)

NIAA National Institute of Aerospace Associates (USA-VA)

NIAAA National Institute on Alcohol Abuse and Alcoholism

NIAC NASA Industrial Application Center • NASA Institute for Advanced Concepts • National Information Accounts Clearinghouse

NIAE National Institute of Adult Education (UK)

NIAG NATO Industrial Advisory Group • Netherlands Institute of Applied Geoscience • Northern Indiana Astronomical Group (USA-IN)

NIAID National Institute of Allergies and Infectious Diseases (USA)

NIAM Nijssen Information Analysis Methodology • Nijssen Information Analysis Model

NIAMS National Institute of Arthritis and Musculoskeletal and Skin Diseases

NIAR National Institute of Atmospheric Research

NIAS National Institute of Advanced Studies • National Institute of Aeronautical Sciences • Novell Internet Access Server

NIB National Information Bureau • Negative Impedance Booster • Negative Ion Beam • Negative Ion Blemish • Nikolai (USA-AK) (location identifier) • Non-Interference Basis • Nordic Investment Bank

NIBIB National Institute of Biomedical Imaging and Bioengineering (USA)

NIBS Nuffield Interactive Book System

Nic Nicaragua • Nicaraguan

NIC Córdoba (Nicaragua) (ISO currency code) • National Ice Center (USA) • National Identity Card • National Industrial Council • National Informatics Centre (India) • National Information Centre • National Insurance Contribution • Near-Infrared Camera • Nearly Instantaneous Coding • Negative Impedance Converter • Negative Ion Chamber • Networked Improvement Community • Networked Information Center • Networked Information Community • Network Imaging Corporation • Network-Independent Clocking • Network Information Center • Network Information Community • Network Interface Card • Network Interface Control • Network Interface Controller • Neumann Institute for Computing (Germany) • Newly Industrialized Country (*) • Newly Industrializing Country • Nicaragua (international vehicle registration code) • Nicaragua (ISO country code) • Nicosia (Cyprus) (location identifier) • Not In Contact • Not In Contract • Nouvel Instrument Communautaire • Nuclei In the Cosmos (conference) • Numeric Intensive Computing

Nica Nicaragua • Nicaraguan

NICA Non-Interactive Computer Application

NiCad Nickel-Cadmium

NICAD Nickel-Cadmium

NICADD Northern Illinois Center for Accelerator and Detector Development (USA-IL)

NICAM Near-Instantaneous Commanded Audio Multiplex

NICAP National Investigation Committee on Aerial Phenomena (USA)

Nicar Nicaragua • Nicaraguan

NICC National Inventory Control Center

Ni/Cd Nickel-Cadmium

NiCd Nickel-Cadmium

NICE Narrative, Immersive, Constructionist/Collaborative Environment • NASA Science Internet Interactive Computer Environment • Near-Infrared Color Excess • Network for Industrial Catalysis in Europe • Network Information and Control Exchange

NICHD National Institute on Child Health and Human Development

NICI Near-Infrared Coronagraphic Imager • Near-Infrared Coronagraph/Imager • Nijmeegs Instituut voor Cognitieonderzoek en Informatietechnologie (Netherlands)

NICK Nature's Initial Cosmic Kickstart

NICMOS Near-Infrared Camera and Multiobject Spectrometer • Near-Infrared Camera Multiobject Spectrograph (*)

NICNET National Informatics Centre Network

NICO Near-Infrared Chromosphere Observatory

NICOLAS Network Information Center Online Aid System

NICOLE Non-LTE Inversion Code based on Lorien Engine

NICP National Inventory Control Point • Nuclear Incident Control Plan

NICS National Inter-facility Communications System • NATO-Integrated Communications System • Near-Infrared Camera Spectrograph • Newly Industrialized Countries • Novell Internet Caching System

NID National Identity • National Interest Determination • Network In-Dial • Network Information Database • Network Interactive Display • Network Interface Device • Network Intrusion Detector • Network Inward Dialing • New Interactive Display • Next Identification • Non-equilibrium Ionospheric Disturbance • Non-Interactive Display

NIDA National Institute on Drug Abuse (USA) • Numerically-Integrated Differential Analyzer

NIDC National Information and Documentation Center

NIDCD National Institute of Deafness and Other Communications Disorders

NIDL National Information Display Laboratory (USA) • Network Interface Definition Language

NIDOO Network Inward Dial Operator Outward

NIDR Networked Information Discovery and Retrieval

NIDRR National Institute on Disability and Rehabilitation Research • National Institute on Disability Rehabilitation and Research

NIDS National Institute for Discovery Science (USA-NV) • Network Interface Data System • Network Intrusion Detection System • NEXRAD Information Display System • NEXRAD Information Dissemination Service

NIE National Institute for the Environment • National Institute of Education • National Intelligence Estimate • Negative Ion Erosion • Networking Infrastructure for Education • Networking Infrastructure in Education • Neutron Ionization Effect • Newly Industrializing Economy • Niblack (USA-AK) (location identifier) • Non-Interference Equipment • Non-Ionization Equilibrium • Not Included Elsewhere

NIED National Institute for Earth and Disasters

Niederbay Niederbayern (Germany)

Niedersachs Niedersachsen (Germany)

NIEHS National Institute of Environmental Health Sciences

NIEI NASA-Industry Education Initiative

NIEL Non-Ionizing Energy Level

NIEM National Industrial Engineering Mission • Network for Industrial Environmental Management

NIEO New International Economic Order

NIER National Industrial Equipment Reserve

NIES National Institute for Environmental Studies • National Institute of Environmental Science

NIEX No-notice Interoperability Exercise

NIF National Ignition Facility (USA) • National Ignition Furnace • National Issue Forum • Neighborhood Information Frame • Neighbor Information Frame • Network Information File • Noise Improvement Factor • Non-Industrial Facility • Nordisk Industrifond • Normalized Impact Factor • Note Issuance Facility • Not In File • Nuclear Information File • Número de Identificación Fiscal (Spain)

NIFO Next In - First Out

NIFOE Naval Observatory - Indiana University Fiber Optics Echelle

NIFTE Neon-Indicating Functional Test Equipment

NIFTP Network Independent File Transfer Protocol

NIFTR Near Isothermal Flowing Test Rig

NIFS National Institute for Fusion Science (Japan) • Near-infrared Integral-Field Spectrograph

Nig Nigeria • Nigerian

NIG Neues Institutsgebäude • Nikunau (Kiribati) (location identifier) • Non-Isolated Gate • Normal Inverse Gauss

NIGA Neutron-Induced Gamma Activity

NIGEC National Institute for Global Environmental Change

NIGMS National Institute of General Medical Sciences (USA)

NIH National Institute of Health (USA) • Not Invented Here

NIHL Noise-Induced Hearing Loss

NII National Information Infrastructure (USA) • Nereid (satellite of Neptune) • Nuclear Installation Inspectorate (UK)

NIIAC National Information Infrastructure Advisory Council

NIICP No Increase In Contract Price

NIII Naiad (satellite of Neptune) • National Information Infrastructure Initiative (USA)

NIIIP National Industrial Information Infrastructure Protocol

NIIM National Integrated Information Management

NIIN National Information Interchange Network • National Item Identification Number (USA)

NIIR Non-Imaging Infrared

NIIRS National Imagery Interpretation Rating Scale • National Imagery Interpretation Resolution Standards

NIIT National Information Infrastructure Testbed (USA) • National Institute for Information Technology (USA) • National Institute of Instructional Technology (USA)

NIJ National Institute of Justice (USA)

NIK Niokolo Koba (Senegal) (location identifier)

NIKHEF Nationaal Instituut voor Kernfysica en Hoge-Energiefysica (Netherlands)

NIL Non-monotonic and Inductive Logic (workshop)

nILF Normalized Integrated Luminosity Function

NILF Normalized Integrated Luminosity Function

NILPRP National Institute for Laser, Plasma and Radiation Physics (Romania)

NILU Norwegian Institutt for Luftforskning (Norway)

NIM National Information Market • National Institute of Metrology • Near Infrared Magnetograph • Network Installation Management • Network Installation Manager • Network Install Manager • Network Interface Machine • Network Interface Module • Niamey (Niger) (location identifier) • Non-Interrupt Mode • Normal Integration Mode • Nothing In Mind • Nuclear Instrumentation Module

NIMA National Imagery and Mapping Agency (USA)

NIMBY Not In My Backyard

NIMD National Institute of Management and Development

NiMH Nickel-Metal-Hydride

NIMH National Institute of Mental Health (USA)

NIMHANS National Institute of Mental Health And Neuro Sciences

NIMI National Internet Measurement Infrastructure

NIMIS NRSA Integrated Management Information System

n. imp. New Impression

NIMPA Nuclear Instruments and Methods in Physics Research A (periodical code)

NIMPB Nuclear Instruments and Methods in Physics Research B (periodical code)

NIMPH Network Interface Message Processing Host

NIMPHE Nuclear Isotope Monopropellant Hydrazine Engine

NIMPOL N-band Imaging Polarimeter

NIMPR Nuclear Instruments and Methods in Physics Research (periodical code)

NIMROD Northern Illinois Meteorological Research on Downbursts

NIMS Near-Infrared Mapping Spectrometer (Galileo) • NSSDC Information Management System

NIMSC Non-consumable Item Materiel Support Code

NIMSR Non-consumable Item Materiel Support Request

NIMT National Institute for Management Technology

NIN Name Index Number • National Information Network • Ninilchik (USA-AK) (location identifier)

NINCH National Initiative for a Networked Cultural Heritage (USA)

NINDS National Institute on Neurological Disorders and Stroke (USA)

Ninfo Nauchnye Informatsii (periodical code)

NINJA Numerically Intensive Java

NINO No Inflow - No Outflow • No Inspector - No Operator • Nothing In - Nothing Out

NINT National Institute for Nanotechnology (Canada-AB)

NIO National Institute of Oceanography • National Intelligence Officer • Native Input/Output • New Initiatives Office (AURA) • Nioki (Zaire) (location identifier) • Noeud d'Interconnexion d'Opérateurs • No Inflow - Outflow • Nominal Interoperability

NIOBE Numerical Integration Of the Boltzmann Equation

NIOD Network In-Out Dial • Network Input-Output Dialing • Network Inward-Outward Dialing

NIOF Neutron Interferometer and Optics Facility (USA)

NIOSH National Institute of Occupational Safety and Health (USA)

NIOZ Nederlands Instituut voor Onderzoek der Zee (Netherlands)

Nip Nippon • Nipponese

NIP Jacksonville (USA-FL) (location identifier) • National Indicative Programme • National Information Point • National Information Policy • National Institute of Physics (Philippines) • National Inventory Programme • Negative Ion Plasma • Network Implementation Plan • Network Input Processor • Network Integrated Processing • Network Interface Processor • New Investigator Program • Normal Incidence Pyrheliometer • Nuclear Initialization Program • Nucleus Initialization Program • Numeric Indicator Performance

NIPALS Non-linear Iterative Partial Least Squares

NIPC Nacelle Ice Protector Controller • National Infrastructure Protection Center (USA)

NIPD Not In the Public Domain

NIPDE National Initiative for Product Data Exchange

NIPDWS National Interim Primary Drinking Water Standard

NIPI Near-Infrared PICNIC Imager

NIPO Nederlands Instituut voor de Publieke Opinie (Netherlands) • Negative Input - Positive Output

NIPR National Industrial Plant Reserve • National Institute of Polar Research (Japan)

ni. pri. Nisi Prius

NIpS Network I/Os per Second

NIPS National Information Processing System • Network I/Os Per Second • Neural Information Processing System • News in Physiological Sciences • Nottingham Image Processing System

NIPSS Near-Infrared Photographic Sky Survey

N. Ir. Northern Ireland

NIR Beeville Chase (USA-TX) (location identifier) • Near Infrared • Networked Information Resource • Networked Information Retrieval • Network Information Retrieval • Night-time Infrared • No Individual Requirement • Non-Ionizing Radiation

NIRA National Industrial Recovery Act (USA) • National Institute for Research Advancement (Japan) • Near Infrared Analysis • Near Infrared Reflectance Analysis

NIRC Near-Infrared Camera • Negative Ion Recombination Chamber

NIRDIF Near-Infrared Deployable Integral-Field (spectrograph)

N. Ire. Northern Ireland

NIRE National Institute for Resources and Environment (Japan)

NIrES Near Infrared Echelle Spectrograph

NIRI Near-Infrared Imager

NIRIM National Intitute for Research in Inorganic Materials (Japan) • Near-Infrared Imager • Near-Infrared Imaging (camera)

NIRIS Near-Infrared Imaging Spectrograph

NIRMOS Near-Infrared Multiobject Spectrograph

NIRP National Institute for Radiation Protection (Sweden)

NIRRL Northen Ireland Regional Research Laboratory

NIRS Near-Infrared Spectrometer • Northern International Reference Stars • Nuclear Information and Resource Service

NIRSP Near-Infrared Spectropolarimeter

NIRST Near Infrared Solar Telescope

NIRVANA Near-Infrared and Visible Adaptive Interferometer for Astronomy

NIS Israeli Shekel (currency) • Name Information Service • Naming Information Service • National Information Service • National Information System • National Input Segment • National Institute for Standards • National Institute of Science • National Intelligence Survey • Near-Infrared Spectrograph • Near-Infrared Spectrometer • Negative Ion Source • Net Identification Sign • Network Information Service • Network Information System • Neutral and Ion Spectrometer • Neutron Inelastic Scattering • Neutron Instrumentation System • New Independent State • Newly Independent State • Nickel-Iron System • Night Illumination System • Noise Information System • Non-Imaging Sensor • Non-proliferation and International Security • Normal Incidence Spectrometer • Not In Stock • Not In System

NISAC National Industrial Security Advisory Committee • National Information Security Assessment Center • National Infrastructure Simulation and Analysis Center (USA)

NISARC National Information Storage And Retrieval Center

NISC NAS Implementation Support Contract • National Independent Study Center • National Industrial Safety Committee • National Industrial Space Committee • National Information Service Corporation (USA) • National Information System for Chemistry

NISDC National Information Systems Development Center

NISDDT NASCOM Interface Standards for Digital Data Transmission

N-ISDN Narrowband Integrated Service Digital Network

NISDN Narrowband Integrated Service Digital Network

NISE National Institute for Science Education (USA) • Naval In-Service Engineering • Normalized Integral Squared Error

NISH National Industries for the Severely Handicapped • National Institute for the Severely Handicapped (USA)

NISM National Information System for Mathematics

NISN NASA Integrated Services Network

NISO National Information Standards Organization (USA-MD) • Netherlands Industrial Space Organisation

NISP National Industrial Security Program • National Information System for Physics • Networked Information Services Project

NISS NASA Institutional Support System • National Information on Software and Services • National Information Services and Systems (UK) • National Information Society Strategy • National Institute of Statistical Sciences (USA) • National Integrated Scheduling System • Nuclear Information Sharing System

NISSAT National Information System for Science and Technology (India)

NISSBB National Information on Software and Services Bulletin Board

NISSC National Information Systems Support Center

NISSPAC National Information on Software and Services Public Access Collections

NIST National Information System for Science and Technology • National Institute of Science and Technology • National Institute of Standards and Technology (USA-CO) • National Intelligence Support Team

NISTADS National Institute of Science Technology and Development Studies

NISTED National Institute of Science and Technology Policy (Japan)

NISTJ National Institute of Standards and Technology Journal (periodical code)

NISZ Nabliudeniia Iskusstvennykh Sputnikov Zemli (periodical code)

NIT National Imagery Transmission • National Institute of Technology • Negative Income Tax • Network Integration Test • Network Integration Testbed • Network Integrity Test • Network Interface Tag • Niort (France) (location identifier) • Node Information Table

NITAAC National Information Technology Acquisition and Assessment Center

NITB Neoclassical Internal Transport Barrier

NITC National Information Technology Center • National Information Transfer Center • Northern Ireland Technology Centre

NITE Nanoinfiltration and Transient Eutectic-phase

NITEC NASA Information and Technology Exchange Council • National Imaging Technology in Education Conference (USA)

NITF National Imagery Transmission Format • National Intelligence Transmission Format

NITFS National Imagery Transmission Format Standard

NITFTP National Imagery Transmission Format Transfer Protocol

NITG Nederlands Instituut voor Toegepaste Geowetenschappen (Netherlands)

NITI National Institute of Technical Information

NITR National Institute for Telecommunications Research

NITRD Networking and Information Technology Research and Development

NITS National Institute of Test Standards • Near-term Information Transport Security • Network Independent Transport Service

NITSC National Information Technology Steering Committee

NIU Nacelle Interface Unit • NASCOM Interface Unit • National Information Unit • Network Interface Unit • New Internet User • Niue (ISO country code) • Northern Illinois University (USA-IL)

NIUA National Institute of Urban Affairs (India)

NIUF North-American ISDN Users' Forum

NIUNA Napoli Istituto Universitario Navale - Annali (periodical code)

NIV Thalassa (satellite of Neptune)

NIVR Nederlands Instituut voor Vliegtuigontwikkeling en Ruimtevaart (Netherlands)

NIWAR National Institute of Water and Atmospheric Research (New Zealand)

NIWF Nationaal Instituut voor de Wetenschappelijke Film (Belgium)

NIX NASA Image Exchange • Nioro (Mali) (location identifier)

NIXT Normal Incidence X-ray Telescope

NIZ Nationales Informationszentrum

NizRe Nizamiah Observatory Reprints (periodical code)

nJ nächsten Jahres • Nanojoule

NJ Nächsten Jahres • New Jersey (USA) • Non-Juxtaposed

NJA Atsugi (Japan) (location identifier)

NJAA New Jersey Astronomical Association (USA-NJ)

NJAC National Joint Advisory Council

NJAG North Jersey Astronomical Group (USA-NJ)

NJB New Jersey Bell Co. (USA-NJ)

NJC National Joint Council • Nizhnevartovsk (Russia) (location identifier)

NJCC National Joint Computer Committee • National Joint Consultative Committee

NJCL Network Job Control Language

NJE Network Job Entry

NJFA NATO Joint Frequency Agreement

NJGL New Jersey Good Lighting (forum) (USA-NJ)

NJH Network Job Header

NJIT New Jersey Institute of Technology (USA-NJ)

NJJ Nombre de Jours Juliens

NJK El Centro (USA-CA) (location identifier)

NJMM Neues Jahrbuch Mineralogie Monatshefte (periodical code)

NJNC National Joint Negotiating Committee

NJP Network Job Processing • New Journal of Physics • Normalized Journal Position

NJPh New Journal of Physics (periodical code)

NJR Non Job Routed

NJRS Nederlandse Jeugdvereniging voor Ruimtevaart en Sterrenkunde (Netherlands)

NJV Norske Justervesen (Norway)

NJW Neue Juristische Wochenschrift (periodical code)

NK Nebenkosten • Not Known

NKA Nkan (Gabon) (location identifier) • No Known Allergies

NKAOS New Kashima Automatic Operating System

NKB Noonkanbah (Australia-WA) (location identifier)

NKC Nouakchott (Mauritania) (location identifier)

NKG Nanjing (China-PRC) (location identifier) • Nishimura - Kamata - Greisen (function) • Non-linear Klein-Gordon (equation)

NKI Naukiti (USA-AK) (location identifier)

Nkl Nickel

NKL Nkolo (Zaire) (location identifier)

NKN Nankina (Papua New Guinea) (location identifier)

NKO Nederlandse Kalibratie Organisatie (Netherlands)

nKOM Narrow-band Kilometric • Narrow-band Kilometric Radiation

NKOS Network Knowledge Organization Source • Network Knowledge Organization System

NKr Norwegian Krone (currency)

NKR Netzknotenrechner • Norwegian Krone (currency)

NKRE Non-Key Recovery Encryption

NKS Nesis Kolmogorov Spectrum • Netzkontrollserver • Nkongsamba (Cameroon) (location identifier)

NKÜ Netzknoten der Übertragungstechnik

NKU Nkaus (Lesotho) (location identifier) • Northern Kentucky University (USA-KY)

NKV Nichen Cove (USA-AK) (location identifier)

NKX San Diego - Miramar (USA-CA) (location identifier)

NKY Nkayi (Congo) (location identifier)

NKz New Kwanza (currency)

n/l Nuestra Letra

nl Dutch (ISO language code) • nachleuchtend • Namelijk • Netherlands • New Line • Non Licet • Non Liquet

Nl Namelijk • National • Nova-Like

N/L Navigation/Localizer • No Limit

NL National League • National Library • Natural Language • Negative Large • Neon Lamp • Netherlands (international vehicle registration code) • Netherlands (ISO country code) • Netherlands (postal country code) • Network Layer • Neutral Line • New Latin • New Line • Newsletter • Next Line • Niederlausitz (Germany) • Night Letter • Noise Level • No Label • No Liability • No Limit • Non-Labelled • Non-Linear • Non-Listed • Non-Local • North Latitude • Not Listed • Not Located • Nouvelle Lune • Nova-Like • Nuevo León (Mexico) • Number of Layers • Pays-Bas (code ISO de pays)

NLA National Land Agency (Japan) • National Libraries Authority (UK) • National Library of Australia • Native Language Assistant • Ndola (Zambia) (location identifier) • New Large Aircraft • Newspaper Licensing Agency (UK) • Next-Layer Aggregation • Next-Level Aggregation • Next-Level Aggregator • Next Lower Assembly • Nine-Letter Acronym • No Longer Affiliated • No Longer an Aircraft • Non-Linear Amplifier • Notice of Last Action • Numerical Linear Algebra

NLAB National Library Advisory Board

NLAC National Library Advisory Council

NLAF Northern Lights Astronomy Festival (USA-WI)

NLANR National Laboratory for Applied Network Research (USA)

NLAPS National Landsat Archive Production System (USGS)

N. Lat. North Latitude

NLB Network Load Balancing • Number of Lines of Binary

NLC Lemoore Reeves (USA-CA) (location identifier) • Narrow-Line Component • National Laboratory Consortium • National Libraries Committee (UK) • National Library of Canada • New Line Character • Next Linear Collider • Noctilucent Cloud

NLCTA Next Linear Collider Test Accelerator

NLD NASA Launch Director • Netherlands (ISO country code) • Network Logistics Depot • Nuevo Laredo (Mexico) (location identifier) • Pays-Bas (code ISO de pays)

NLDA National Library Development Act (USA)

NLDB Natural Language Database • Natural Language to Databases (conference)

NLDC National Legal Data Center

NLDM Network Logical Data Manager

NLDN National Lighting Data Network • National Lighting Detection Network

NLDP National Launch Development Program

NLDS National Leadership Development Seminar

NLE National Library for the Environment • Niles - Jerry Tyler (USA-MI) (location identifier) • Non-Linear Element • Not Less or Equal • Nuclear Lighthouse Effect

NLEA Nutrition Labelling and Education Act (USA)

NLECTC National Law Enforcement and Corrections Technology Center (USA-CO)

NLEV National Low-Emission Vehicle (USA)

NLF Darnley Island (Australia-QLD) (location identifier) • National Laser Facility • Natural Laminar Flow • Nearest Landing Field

NLFF Non-Linear Force-Free

NLFM Noise-Level Frequency Monitor

NLFSR Non-Linear Feedback Shift Register

NLG Guilder (Netherlands) (ISO currency code) • Natural-Language Generation • Nelson Lagoon (USA-AK) (location identifier) • Nose Landing Gear • Null Line Gap

NLGC Noise Level Gain Control

NLGD Nose Landing Gear Door

NLI National Library of Ireland • Natural Language Inc. • Natural Language Interface • Neodymium Laser Illuminator • Noise Limit Indicator • Non-Linear Interpolating

NLII National Learning Infrastructure Initiative (USA)

NLIN Non-Linear

NLIS National Land Information Service (UK)

NLK Norfolk Island (location identifier)

NLKF Non-Linear Kalman Filter

NLL Negative Logic Level • Next-to-Leading Logarithmic • Night-Low Level • Nullagine (Australia-WA) (location identifier)

NLLR Non-Linear Lorentz Resonance

NLLST National Lending Library for Science and Technology

NLLU National Lending Library Unit (UK)

nlm NetWare Loadable Module (file extension)

NLM National Library of Medicine (USA) • Nautical Mile • NetWare Loadable Module • Network Loadable Module • Non-Linear Mapping • Non-Linear Material

NLMC National Library of Medicine Classification • Non-Linear Monte Carlo

NLML No Lactose Meal

NLN National Learning Network • National Library Network (USA) • No Longer Needed

NLnr Non-Linear

NLO Kinshasa N'Dolo (Zaire) (location identifier) • National Liaison Officer • Next-to-Leading Order • Non-Linear Optical • Non-Linear Optics • Norman Lockyer Observatory (UK-England)

NLOC Non-commented Line of Code

N-LOS Non-Line Of Sight

NLOS Natural Language Operating System • Non-Line Of Sight

NLP Name Looking Protocol • Natural Language Processing • Neglected Language Program • Nelspruit (South Africa) (location identifier) • Network Layer Protocol • Neurolinguistic Programming • Non-Linear Potential • Non-Linear Programming • Normal Lead Point • Normal Link Pulse

NLPAC National Light Pollution Abatement (RASC)

NLPID Network Layer Protocol Identifier

NLPIP National Lighting Products Information Program (USA-NY)

NLQ Near-Letter Quality

NLQP Natural Language Query Processor

NLQSO Narrow-Line Quasar

NLR Narrow-Line Region • Nationaal Lucht- en Ruimtevaartlaboratorium (Netherlands) • NEAR Laser Rangerfinder • Nuclear Lindblad Resonance

NLRB National Labor Relations Board (USA) • National Level Bibliographic Record

NLRG Narrow-Line Radio Galaxy

NLRGC Nationaal Lucht- en Ruimtevaartgeneeskundig Centrum (Netherlands)

NLRO North Liberty Radio Observatory (USA-NM)

NLRRN Non-Linear Random Resistor Network

NLRS National Laser Ranging System • NATMAP Laser Ranging System

NLS On-Line System • Na-Layer Spot • Narrow-Line Seyfert (galaxy) • NASA Launch Services • National Land Survey • National Language Support • National Launch System • National Library Service • Native Language Support • Natural Language System • Negative Lens System • Neodymium Laser System • Nestor Learning System • Network Library System • Network License Server • Nicholson (Australia-WA) (location identifier) • Niet-Lineaire Systemen • Noise Line State • No-Line-of-Sight • Non-Linear Smoothing • Non-Linear System • Northern Loop System • Online System

NLSE Non-Linear Schrödinger Equation

NLSFUNC National Language Support Function

NLSM National Launch System Model • Non-Linear Sigma Model

NLSP NetWare Link Services Protocol • NetWare Link State Protocol • Network Layer Security Protocol

NLST Name List

NLSUS National Launch System Upper Stage

NLSy Narrow-Line Seyfert (galaxy)

NLSYM National Launch System Model

NLT Natural Language Technology • Not Later Than • Not Less Than

NLTE Non-Local Thermal Equilibrium • Non-Local Thermodynamic Equilibrium (*)

NLTS Near Launch Tracking System

NLTSG Natural Language Translation Specialist Group

NLTT Natural Language Theory and Technology • New Luyten Two-Tenth (catalog)

NLU Mexico City - Santa Lucia (Mexico) (location identifier) • National Louis University (USA-IL) • Natural Language Understanding

NLUUG Netherlands Local Unix systems User Group

NLV National Language Version • Nikolaev (Ukraine) (location identifier)

NLVD Natural Language Voice Document

NLW National Library Week

NLXG Narrow(-emission)-Line X-ray Galaxy

Nly Northerly

nm Nachmittags • Namiddag • Nanometer • Nautical Mile • Nocte et Mane • Nutmeg

nM Nächsten Monats • Nanomole

N/m No Mark

Nm Nachmittag • Next Matter

NM Nächsten Monats • Namibia • Nanomanufacturing • Nautical Mile • Nedler-Mead • Network Management • Network Manager • Network Module • Neutron Monitor • New Mexico (USA) • New Moon • Newtonian Mechanics

• Newton-Meter • Nieuwe Maan • Night and Morning • Night Message • Nissan Motors (Japan) • Noise Meter • No Mark • No Message • No Mission • Non-Metallic • Not Marked • Not Measured • Noun Modifier • Nuclear Magnetic • Nuclear Magnetron

NMA Namangan (Uzbekistan) (location identifier) • National Management Association • Network Management and Administration • Network Management Architecture • New Mexico Array • Nobeyama Millimeter Array (Japan) • Normal Method of Acquisition • Norwegian Mapping Authority

NMAB National Materials Advisory Board (USA)

NMAC Near Mid-Air Collision

NMag Non-Magnetic

NMAH National Museum of American History (USA-DC)

NMAP Network Management Application Program

NMAS National Map Accuracy Standard • Network Management Alert System • Novell Modular Authentication Services

NMAT Non-Mathematical Astronomical Text

NMB Daman (India) (location identifier) • National Mapping Bureau • National Mediation Board • National Metric Board • Nucleaire Magnetische Resonantie

NMBMMR New Mexico Bureau of Mines and Mineral Resources (USA-NM)

NMBPC Neural Model Based Predictive Control

NMBR Number

NMBS Nationale Maatschappij van de Belgische Spoorwegen (Belgium)

NMC NASA's Master Catalog • Natal Midlands Centre (ASSA) • National Map Collection • National Mathematical Centre • National Meteorological Center • National Metrological Center • Network Management and Control • Network Management Center • Network Management Console • Network Monitoring and Control • Network Monitoring Center • New Media Center • No More Credit • Non-Mission Capable • Non-Modifying Construction • Norman's Cay (Bahamas) (location identifier) • Northwestern Michigan College (USA-MI) • Not Mission Capable • Nozzle Closure Mechanism • NSSDC Master Catalog • Numerical Meteorological Centre

NMCC NASCOM Management Control Center • Network Management Control Center

NMCCR Normalized Mean Citing Country Rate

NMCE Network Monitoring and Control Equipment

NMCI Network Management Communications Interface • Network Management Configuration Item

NMCM Numerical Methods and Computational Methods (conference)

NMCR Normalized Mean Citation Rate

NMCS Network Monitoring and Control System • Not Mission Capable Supply

NMCSR Normalized Mean Citing Subfield Rate

NMCV Non-Magnetic Cataclysmic Variable (star)

NMD NASA Management Delegation • NASA Master Directory • National Mapping Division • Network Measurements Division • Nuclear Molecular Disk

NMDA N-Methyl D-Asparate (receptor)

NMDB Network Management Database

NMDG Network Management Development Group

NMDR Nuclear Magnetic Double Resonance

NME Network Management Entity • Network Monitoring Experiment • Nightmute (USA-AK) (location identifier) • Noise-Measuring Equipment • Non-linear Mesoscopic Elastic • Non-linear Mesoscopic Elasticity

NMEE Numerische Mathematik Electronic Edition

NMELP Non-Monotonic Extensions of Logic Programming (workshop)

NMERI New México Engineering Research Institute (USA-NM)

N. Mex. New Mexico (USA)

NMF Nación Más Favorizada • Network Management Facility • Network Management Forum • Network Management Function • Node Management Facility

NMFF National Magnetic Fusion Facility (USA)

NMGECC National Magnet Fusion Energy Computer Center

NMFS National Marine Fisheries Service (USA)

NMG San Miguel (Panama) (location identifier)

NMH Nickel-Metal-Hydride

NMHC Non-Methane Hydrocarbon

NMHS National Meteorological and Hydrological Service

NMHU New Mexico Highlands University (USA-NM)

nmi Nautical Mile

NMi Nederlands Meetinstituut (Netherlands)

NMI NASA Management Instruction • National Metrology Institute • National Multimedia Institute • Nautical Mile • Netherlands Measurement Institute • New Millennium Interferometer • No Middle Initial • Non-Major Item • Non-Maskable Interrupt • Non-Masking Interrupt

NMIB Network Management Information Base • New Materiel Introductory Briefing

NMIBT New Materiel Introductory Briefing Team

NMICS National Maintenance Information and Control System

NMIL New Materiel Introductory Letter

NMIMT New Mexico Institute of Mining and Technology (USA-NM)

NMIP Network Management Integration Plan

NMIS Network Management Integration System • Nuclear Materials Information System

NMIT New Material Introductory Team

NMKW Nichtmethankohlenwasserstoffe

Nml Normal

NML National Measurement Laboratory • National Media Laboratory • Network Management Layer • Neugebauer - Martz - Leighton (star)

NMLA New Mexico Library Alliance

Nmlz Normalize

NMM NASA Management Manual • NetWare Management Map • Network Management Module • Neutron Magnetic Moment • Non-hydrostatic Mesoscale Model

NMMBA National Museum of Marine Biology and Aquarium

NMMP Network Management Manager Process

NMMW Near-Millimeter Wave

NMN Niedersächsischer Monographiennachweis • Nomane (Papua New Guinea) (location identifier) • No Middle Name

NMNH National Museum of Natural History

Nmns Néanmoins

NMNS National Museum of Natural Science

NMO National Member Organization • Network Management Option • Non-Machinable Outside • Normal Manual Operation • Normal Mode Operation • Normal Moveout

NMOC National Meteorological Operations Centre • Network Management Operations Center

NMOG Non-Methane Organic Gas

NMOS Negative-channel Metal-Oxide Semiconductor • Negative-type Metal-Oxide Semiconductor • Negatively-doped

Metal-Oxide Semiconductor • Network and Mission Operations Support

NMP National Maintenance Point • National Mapping Program • Network Management Processor • Network Management Protocol • New Millennium Program (NASA) • New Moon (Australia-QLD) (location identifier) • Nominated Member of Parliament

NMPA Normal Materiel Production Allowance

NMPC National Maintenance Publications Center

NMPDN National Materials Property Data Network

NMPE Network Management Protocol Entry

NMPF Network Management Productivity Facility • Normal Magnitude Probability Function

NMPH Nautical Miles Per Hour

NMPP National Monitoring and Promising Practices (USA) • Nouvelles Messageries de la Presse Parisienne (France)

NMpS Nautical Miles per Second

NMPS Nautical Miles Per Second

Nmr Namur (Belgium)

NMR Nappa Merry (Australia-QLD) (location identifier) • Nautical Mile Radius • News Media Representative • No Maintenance Required • No Maintenance Requirement • Normal Mode Rejection • Nuclear Magnetic Relaxation • Nuclear Magnetic Resonance (*)

NMRA Nor Mission Ready and Available

NMRAS Nuclear Material Report and Analysis System

NMRF National MST Radar Facility (India)

NMRI Nuclear Magnetic Resonance Imaging

NMRON Nuclear Magnetic Resonance of Oriented Nuclei

NMRP Nuclear Magnetic Resonance Program

NMRQ Non-Master Request

NMRS National Magnetic Resonance Society (India) • NSSDC-Maintained Record Separator • NSSDC Maintenance Record Separator • Numbers

NMRT Nimbus Meteorological Radiation Tape

NMS Namsang (Myanmar) (location identifier) • National Master Specification • National Measurement Service • National Meteorological Service • National Metrology Service • Naval Meteorological Service • Navigational Management System • Navigation Management System • Navire Météorologique Stationnaire • NetWare Management System • Network Management Service • Network Management Station • Network Management System • Network Measurement System • Network Monitoring Station • Neutral Mass Spectrometer • New Management Strategy • New Manning System • NGST Mission Simulation • NGST Mission Simulator • Nippon Meteor Society (Japan) • Numerical Methods Symposium

NMSA National Meteorological Services Agency

NMSAE Netherlands Ministry for Social Affairs and Employment

NMSC National Merit Scholarship Corporation (USA)

NMSD Next Most Significant Digit • NGST Mirror System Demonstrator

NMSE Normalized Mean Squared Error

NMSI National Mobile Subscriber Identity • National Museum of Science and Industry (UK)

NMSL NetWare Mirrored Server Link

NMSO Nuclear Medicine Science Officer

NMSQT National Merit Scholarship Qualifying Test

NMSR Non-consumable Materiel Support Request

NMSS Non-Minimum State Space

NMST New Materiel System Test

NMSU New Mexico State University (USA-NM)

NMT Namtu (Myanmar) (location identifier) • Network Management Terminal • Network Management Tool • New Mexico Tech (USA-NM) • Non-Mechanized Terminal • Nordic Mobile Telephone • Nordic Mobile Telephony • Notification of Master Tool • Not More Than • Nuclear Materials Technology

NMTA National Metal Trades Association (USA)

NMTBA National machine Tool Builders' Association (USA)

NMTRO New Mexico Tech Robotic Observatory (USA-NM)

NMTS Next Manned Transportation System • Noise Measurement Test Set

NMTU Net Manufacturing Time per Unit

NMU Namu (Marshall Islands) (location identifier) • NGAS Master Unit • Northern Michigan University (USA-MI)

NMUP Network Management User Process

NMV Neubaumietenverordnung • Nitrogen Manual Valve

NMVT Network Management Vector Table • Network Management Vector Transport

NMW National Minimum Wage

NMWC New Mexico Western College (USA-NM)

NMZ Nationales Metrologisches Zentrum

nn Names • Notes

N/N Noise-to-Noise • Not to be Noted

NN Far North • Far Northern • NASA Notice • Nearest Neighbor • Network Node • Neural Network • Neutralization Number • Next Node • Nomen Nescio • No Name • No Need • No News • Normalnull • Nucleon-Nucleon

NNA Kenitra (Morocco) (location identifier) • Neural Network Analysis • Neural Network and Applications (conference)

NNAC Northeast Nebraska Astronomy Club (USA-NE)

NNAPS Night Navigation And Pilotage System

NNB Santa Ana Island (Solomon Islands) (location identifier)

NNC Neural Network Classifier • Neural Network Council • Non-Noun Compound • Normes Nationales du Canada

NNCS NICS Network Control System

NND Nangade (Mozambique) (location identifier)

NNDC National Nuclear Data Center

NNE Network Node Equipment • Non-Nuclear Energy • Nord-Nord-Est • North-Northeast • North-Northeastern

NNEO Natural Near-Earth Object

NNERN North-Northeastern

NNF Narrowband Nebular Filter

NNG Nanning (China-PRC) (location identifier) • National Numbering Group

NNHS Northamptonshire Natural History Society (UK-England)

NNI Namutoni (Namibia) (location identifier) • National Nanotechnology Initiative (USA) • Nederlands Normalisatie-Instituut (Netherlands) • Network-Network Interface • Network Node Interface • Network-to-Network Interface • Neural Network Interpolation • Neural Network Interpolator • Noise and Number Index • Noise Nuisance Index • Nucleon-Nucleon Interaction

NNK Naknek (USA-AK) (location identifier) • Non-Nuclear Kill

NNL Neural-Network Language • Nondalton (USA-AK) (location identifier)

NNLL Next-to-Next-to-Leading Logarithmic

NNM Naryan-Mar (Russia) (location identifier) • Network Node Manager • Non-linear Normal Mode

NNMF Nearest Neighbor Median Filter

NNMI Non-linear Normal Mode Initialization

NNMRTP Northern New Mexico Rural Telemedicine Project

NNNDO Neglect of Non-Neighbor Differential Overlap

NNNTB North North North Temperate Belt

NNO Noordnoordoost • Nord-Nord-Ouest

NNP Negative Node Point • Net National Product

NNR National Nature Reserve • No-Net-Rotation • Spiddal (Ireland) (location identifier)

NNRIC National Natural Resources Information Centre

NNRMS National Natural Resources Management System

NNRP Network News Reader Protocol

NNRT Not-so-Near Real Time

NNS National Numbering Scheme • Navy Navigation Satellite • Nearest-Neighbor Spacing • Nearest-Neighbors Statistics • NetWare Name Service • Neural Network Simulator • Novell Name Service • Nucleon-Nucleon Scattering

NNSA National Nuclear Safety Administration (USA)

NNSC National Network Service Center

NNSF National Natural Science Foundation (China-PRC)

NNSFC National Natural Science Foundation of China (PRC)

NNSG NASCOM Network Scheduling Group

NNSN No National Stock Number

NNSS National Navigational Satellite System • Naval Navigational Satellite System • Navy Navigational Satellite System (*)

NNT Nan (Thailand) (location identifier) • Non-Nodal Terminal • Not Necessarily True • Nouvelle Numérotation Téléphonique

NNTB North North Temperate Belt (Jupiter)

NNTP National News Transfer Protocol • Net News Transfer Protocol • Network News Transfer Protocol (*) • Network News Transmission Protocol • Network News Transport Protocol • News Network Transfer Protocol

NNTT National New Technology Telescope

NNTZ North North Temperate Zone

NNU Nanuque (Brazil-MG) (location identifier)

NNV Nederlandse Natuurkundige Vereniging (Netherlands)

NNW Noordnoordwest • North-Northwest • North-Northwestern

NNWRN North-Northwestern

NNWS Non-Nuclear Weapon State

NNWWD North-Northwestward

NNX Network Numbering Exchange • Nunukan (Indonesia) (location identifier)

NNY Nanyang (China-PRC) (location identifier)

n/o Notre Ordre • Nuestra Orden

no Norway • Norwegian (ISO language code) • Number • Número

No Nobelium • Nobélium • North • Northern • Norway • Norwegian • Nuestro • Number • Numero • Numéro

N/O No Orders • Normally Open

NO Neptune Orbiter • New Orleans (USA-LA) • Nitric Oxide • Nitrous Oxide • Nitrogen Monoxide • Non-Official • Non-Original • No Office • No Opinion • Noordoost • Nordosten • Nord-Ouest • Normally Open • Norvège (code ISO de pays) • Norway (ISO country code) • Not Operational • Not Or • Not Out • November • Novembre • Nuclear Orientation • Nuffield Observatory (UK-England) • Number

N-O-A Not-Or-And

NOA National Observatory of Athens (Greece) • National Optometric Association • National Ordering Agreement • Nature Of Action • Nature Of Address • New Obligation Authority • Notice Of Availability • Not - Or - And • Not Otherwise Authorized • Nowra (Australia-NSW) (location identifier) • NSWC Office Automation

NOAA National Oceanic and Atmospheric Administration (USA-DC)

NOAASIS National Oceanic and Atmospheric Administration Satellite Information System

NOAC Narrow Optical Absorption Component • National Operational Advisory Committee (USA)

NOADN National Oceanic and Atmospheric Data Network

NOAEL No Observed Adverse Effect Level

NOAO National Oceanographic and Atmospheric Organization • National Optical Astronomy Observatories (USA-AZ)

NOARL Naval Oceanographic and Atmospheric Research Laboratory (USA)

Nob Nobis • Noble

NOB No Other Business • Nosara Beach (Costa Rica) (location identifier) • Northern Operating Base • Not Observed

NOBIM Norsk forening for Bildebehandling og Mønstergjenkjenning (Norway)

NOC Connaught (Ireland) (location identifier) • National Organizing Committee • Network Operating Center • Network Operations Center • Network Operations Control • Notation Of Content • Notice Of Change • Notice Of Completion • Notice Of Contents • Not Otherwise Classified • Not Otherwise Coded • Nuclear Optical Core • Number Of Children • Numerical Optimization Center

NOCA Notice Of Contract Action

NOCC Network Operations Control Center

NOCLIMP Nordic Climate Modeling Project

NOCMIS Network Operations Center Management Information System

No.Co. Northern Countries

NOCR No Carriage Return

NOCS Network Operations Center System

Noct Nocte

NOD Network Operations Directive • Network Outward Dialing • News On Demand • Night Observation Device • Non-Offensive Defense • Norden (Germany) (location identifier) • Notice Of Discrepancy

NODC National Oceanographic Data Center (USA-DC)

NODDES Navy Oceanographic Data Distribution Expansion System

NODDS NOAA Oceanographic Data Distribution System

NODEID Node Identification

NODIS NASA Online Data and Information Service • NASA Online Directory and Information System • No Distribution • NSSDC Online Data and Information Services (*)

NODLR Night Observation Device - Long Range

NODO NASA Orbital Debris Observatory

NODS NASA Ocean Data System

NoE Need of Expertise • Network of Excellence • Number of Employees

NOE Nap Of the Earth • Network Of Excellence • Neutrino Oscillation Experiment • Norddeich (Germany) (location identifier) • Not Otherwise Enumerated • Notice Of Exception • Notice Of Execution • Nuclear Overhauser Enhancement • Number Of Employees • Number Of Errors

NOEL No-Observed-Effect Level • Number Of Errors Left

NOES National Operational Environmental Satellite

NOESS National Operational Environmental Satellite System

NOESY Nuclear Overhauser Effect Spectroscopy

NOF National Optical Font • Network Operations and Facilities • Not On File

NOFC Notification Of Future Change

NOFORN Not Releasable to Foreign Nationals

NOFS (US) Naval Observatory Flagstaff Station (USA-AZ)

NOFT Notice Of Foreign Travel

NOG Nearby Optical Galaxy • Nederlandse Onderzoeks Data-bank (Netherlands) • Net Operating Goal • Nogales (Mexico) (location identifier)

NOGAPS Navy Operational Global Atmospheric Prediction System (USA)

NOH Chicago (USA-IL) (location identifier) • Number Of Horrors

NOHIC Nizamiah JAPAL Rangapur Observatories Hyderabad India Contributions (periodical code)

NOHP Not Otherwise Herein Provided

NOHRSC National Operational Hydrological Remote Sensing Center (USA)

NOI National Optics Institute (Canada) • Neptune Orbital Insertion • Neptune Orbit Insertion • Notice Of Inquiry • Notice Of Intent • Not Otherwise Identified • Not Otherwise Indexed • Novorossijsk (Russia) (location identifier)

NOIBN Not Otherwise Indexed By Name • Not Otherwise Indicated By Name

NoIGG Novosibirsk Institut Geologii i Geofiziki (periodical code)

NOIP National Office of Industrial Property

NoIzN Novosibirsk Izdatel Nauka (periodical code)

NOJ Nationally-Oriented Journal • Nojabrxsk (Russia) (location identifier)

NOK Krone (Norway) (ISO currency code) • Next Of Kin

NOL Naval Ordnance Laboratory (USN) • Normal Operating Loss

NOLAN NASCOM Operational Local Area Network

nol. con. Nolo Contendere

NOLOC No Location

nol. pros. Nolo Prosequi

Nom Nomenclature • Nominal • Nominate • Nominated • Nomination • Nominative • Nominee

NOM Network Object Management • Network Object Manager • Network Operations Manager • Network Output Multiplexer • Nomad River (Papua New Guinea) (location identifier) • Nomenclature • Nominal • Normal Operative Mode • Normas Oficiales de México

NOMAC Noise Modulation And Correlation

NOMAD Navy Oceanographic and Meteorological Automatic Device • Neutrino Oscillation Magnetic Detector • Non-Mass Dependent

Nom. Cap. Nominal Capital

NOMEK Nordisk Meteorolog Efteruddannelses Kursus

Nomen Nomenclature

NOMEN Nomenclature

NOMHICE Non-Methane Hydrocarbon Intercomparison Experiment

Nomin Nominatif • Nomination • Nominative

Nominat Nominativo

nom. nov. Nomen Novum

NOMS Network Operations Management System • Nuclear Operations Monitoring System

NOMSS National Operational Meteorological Satellite System • Naval Oceanographic and Meteorological Support System

Non Nonas

NON Nonouti (Kiribati) (location identifier)

NONA Notice Of Non-Availability • NASA Science Internet Online Network Aid

noncit Non-Citizen

Noncom. Non-Commissioned

Noncon Non-Conformist

non cul. Non Culpabilis

Nonli Non-linearity (periodical code)

non obs. Non Obstante

non pros. Non Prosequitur

non rep. Non Repetatur

non repetat. Non Repetatur

non seq. Non Sequitur

Nonstand Non-Standard

Nonstd Non-Standard

Nonsyn Non-Synchronous

NOO Naoro (Papua New Guinea) (location identifier) • National Oceanographic Office • Naval Oceanographic Office (USA)

no-op. No Operation

NOOP No Operation

NOOS Nuclear Orbit-to-Orbit Shuttle

NOOU Not One Of Us

NOP Mactan Island - Nab (Philippines) (location identifier) • National Opinion Poll • Navigation Operating Procedure • Network Operations Procedure • Nightly Observing Program • No Operation • Normal Operating Procedure • Not Otherwise Provided • Not Our Publication

NOPA National Office Products Association • Network Operations Performance Analysis

NOPAC North Pacific

NOPAR Non-Orbiter Parts Authorization Request

NOPC NOAA Ocean Products Center

NOPEX Northern Hemisphere Processes Experiment

NOPPA National Ocean Pollution Protection Act (USA)

NoPT No Procedure Turn

Nor Norma-Normae (constellation) • Normal • Norman • North • Northern • Norway • Norwegian • Norwich (UK-England)

NOR Net Operating Result • Nordfjördur (Iceland) (location identifier) • Norvège (code ISO de pays) • Norway (ISO country code) • Notice Of Revision • Not Or

NORAD North American Aerospace Defence Command (now USSC) • North American Air Defense Command

NORADS Naval Operations Regional Atmospheric Prediction System (model)

NORAS Northern ROSAT All-Sky (survey)

NORC National Opinion Research Center (USA-IL)

NORCSEX Norwegian Continental Shelf Experiment

Nórd Nórdico

Nord Nordic

NORDA Naval Ocean Research and Development Activity

Nordb Nordbahn

NORDITA Nordic Institute for Theoretical Physics (Denmark) • Nordisk Institut for Teoretisk Atomfysik (Denmark)

NORDO No Radio

NORDUNET Nordic University Network

NOREK Nordiske Elektrotekniske Komiteer

NORESS Norwegian Regional Seismic Array System (Norway)

Norf Norfolk

NORGEN Network Operations Report Generator

NoRH Nobeyama Radio Heliograph (Japan)

NORI Nobeyama Radio Interferometer

Norm Normal • Normalize • Normalized • Normalizer • Norman

NORM Naturally Occurring Radioactive Material • Non-Operational Ready Maintenance • Normal • Not-Operationally-Ready Maintenance

NORMA No Remote Memory Access

NORP Nobeyama Radio Polarimeter

NORPEX North Pacific Experiment

NORS Nobeyama Radiospectrograph • Not-Operationally-Ready Supply

NORSE Nuclear Optical and Radar Systems Effects

NORSEX Norwegian Remote Sensing Experiment

NORSTAR Northern Solar Terrestrial Array

NORT Network of Oriental Robotic Telescopes

Northants Northamptonshire (UK-England)

Northd Northumberland (UK-England)

Northld Northumberland (UK-England)

Northumb Northumberland (UK-England)

NORTIM Nordiska Föreningen För Tillämpad och Industriell Mathematik

Norw Norway • Norwegian

Nos Numbers • Numéros

NOS National Occupational Standards (UK) • National Oceanic Service • National Ocean Service • National Ocean Survey • Nature Of Science • Network Operating System • Night Observation System • Nodal Operating System • Non-Oriented Satellite • Nossi-Be (Madagascar) (location identifier) • Not Otherwise Specified • Not Otherwise Stated • Nuclear Optical Source • Numbers

NOSB National Organic Standards Board (USA)

NOSC Naval Ocean Systems Center (USA-CA) • Network Operations and Security Center

NOSes Network Operating Systems

NOSF Network Operation System Function

NOSFERATU New Object-oriented System For Effective Retrieval From Texts with Uncertainty

NOSHDB National Ocean Service Hydrographic Database

NOSIE NODC Ocean Science Information Exchange • NODC SPAN Information Exchange

NOSIG No Significant (change)

NOSL Night/Day Optical Survey of Lightning

NOSP NASA Orbital Support Plan • Near-Object Scanning Probe • Network Operating Support Plan • Network Operations Support Plan • Network Operation Support Program

NOSS National Oceanic Satellite System (USA) • National Orbiting Space Station

NOSSA Notice Of Source Selection Action

NOSSDAV Network and Operating System Support for Digital Audio and Video (workshop)

NOST NASA/OSSA Office of Standards and Technology

Not Notación • Notary • Notationes • Note • Notes • Notice • Notices • Notizi

NOT New Organization Training • Nordic Optical Telescope • Nottingham (Starlink internal code) • Novato (USA-CA) (location identifier)

NOTA None Of The Above

NOTAL Not To All

NOTAM Notice to Airmen

NOTAR No Tail Rotor

NOTAS Naval Observatory Time Alternate Station

NOTIPG Nordic Optical Telescope Image/Photo Gallery

NOTIS Network Operator Trouble Information System

NOTMAR Notice to Mariners

NOTS Naval Ordnance Test Station

NOTSA Nordic Optical Telescope Scientific Association

Nott Nottingham (UK-England)

Nottm Nottingham (UK-England)

Notts Nottinghamshire (UK-England)

NOTU Naval Operational Training Unit

Notwg Notwithstanding

NOU Nouméa Tontouta (New Caledonia) (location identifier)

Nov Novel • Novelist • November • Novembre • Novembro • Noviembre • Novice

NOV Huambo (Angola) (location identifier) • Nature Of Violation • Notice Of Availability • November • Novembre • Novembro

NOVA Nederlandse Onderzoekschool Voor Astronomie (Netherlands) • Network Omnivision Array • North Valley Astronomers (USA-CA) • Northwest Ohio Visual Astronomers (USA-OH)

NOVAC Northern Virginia Astronomy Club (USA-VA)

NOVAM Naval Oceanic Vertical Aerosol Model

NOVAS Naval Observatory Vector Astrometry Software (*) • Naval Observatory Vector Astrometry Subroutines

NOVIB Nederlandse Organisatie voor Internationale Bijstand (Netherlands)

NovIG Novosibirsk Institut Gidrodinamiki Dinamika Sploshnoi Sredy (periodical code)

Nouv Nouveau

NOVRAM Non-Volatile Random Access Memory

NOW Negotiable Order of Withdrawal • Neogene of the Old World • Network Of Workstations • NOAA Ocean Wave

NOWESP Northwest European Shelf Programme

NOx Nitrogen Oxide

NOX Nitrogen-Oxygen

NOY Not Out Yet

NOYAA National Outstanding Young Astronomer Award

Noz Nozzle

NOZ Novokuznetsk (Russia) (location identifier)

NOZE National Ozone Expedition

NO2 Nitrogen Dioxide

NO3 Nitrate Radical

np Nisi Prius • Nombre Proprio

Np Neper • Neptunium

N/P Not Provided

NP National Park • National Pipe • National Priority • Natural Park • Naval Publication • Navigation Payload • Near Point • Nébuleuse Planétaire • Nebulosa Planetaria • Negative Polarization • Nepal (ISO country code) • Net Proceeds • Network Planner • Network Planning • Network Processor • Network Program • Network Protocol • Neuropsychiatry • Neutral Porosity • Newman-Penrose • New Paragraph • New Pattern • New Pence (currency) • New Physics • New Providence (Bahamas) • Nitronium Perchlorate • Nobel Prize • Nombre Proprio • Non-deterministic Polynomial • Non-Participating • Non-Polar • Non-Polynomial • Non-Print • Non-Proliferation • No Pagination • No Parking • No Printer • No Protest • No Publisher • Normal Pitch • Normal Point • Normal Pressure • Norma Portuguesa • North Pole • Notary Public • Not Paginated • Not Provided • Noun Phrase • Nouvelle Physique • Number of Publications • Number Portability

nPa Nanopascal

NpA Neutrons per Absorption

NPA Network Performance Analyzer • Network Principal Astronomer • Network Printer Alliance • Network Provider

Agent • Neutral Particle Analyzer • Neutrons Per Absorption • Newspaper Publishers' Association • Non-Precision Approach • Non-Principal Axis • No Power Alarm • Nordic Planetarium Association (Sweden) • Notice of Proposed Amendment • Nulle Part Ailleurs • Numbering Plan Area • Number Plan Area • Numerical Production Analysis • Pensacola (USA-FL) (location identifier)

NPAC National Program for Acquisitions and Cataloguing • National Provisioning Administration Center • Network Principal Astronomer's Computer • Number Portability Administration Center

NPACI National Partnership for Advanced Computational Infrastructure (NSF)

NPAI N'habite Pas à l'Adresse Indiquée

NPAM Non-Professional Astronomy Magazine

NPAP Network Printing Alliance Protocol

NPB Negative Polarization Branch • Neutral Particle Beam • Neutron Particle Beam

NPBG Nébuleuse Planétaire du Bulbe Galactique

NPBS Non-Primary Basic Subject

NPBSV Neutral Particle Beam Space Vehicle

NPC NASA Procurement Circular • NASA Publication Control • National Patent Council (USA) • National Photocolor Corporation (USA-NY) • National Point of Contact • National Postgraduate Committee (UK) • National Project Coordinator • Needle Punch Card • Network Parameter Control • Network Personal Computer • Network Protocol Control • Non-Patent Citation • Non-Player Character • Non-Printing Character • Normal Plane Change • North Permanent Cap • North Polar Cap

NPCI National Programme of Community Interest

NPCR National Point of Contact Representative • No Periodic Calibration Required • No Programmed Calibration Required

NPCSD Normal Point Calibration Standard Deviation

NPCV Non-Polar Cataclysmic Variable (star)

NPD Named Protection Domain • NASA Policy Directive • Ne Pas Déranger • Network Products Division • New Product Development • Nitrogen-Phosphorous Detector • Noise Power Density • Nominal Percent Defective • No Place or Date • North Polar Distance • North Pole Distance

NPDA Network Problem Determination Aid • Network Problem Determination Application • Network Program Determination Application

NPDB Number Portability Database

NPDES National Pollutant Discharge Elimination System • National Pollution Discharge Elimination System

NPDF N-dimensional Probability Density Function • Normal Probability Distribution Function

NPDN Nordic Public Data Network

NPDS Nuclear Particle Detection System

NPDSA National Public Domain Software Archive

NPDU Network Protocol Data Unit

NPDWR National Primary Drinking Water Regulation

NPE Napier Hastings (New Zealand) (location identifier) • Network Peer Exchange • Network Planning and Engineering • New Preliminary Evaluation • Non-Potential Energy • Nuclear Photographic Emulsion

NPEA National Printing Equipment Association (USA)

NPEF New Product Evaluation Form

NPER National Professional Engineers Register

NPES Normalized Polyelastic Sampling • Nuclear Planning and Execution System

NpF Neutrons per Fission

NPF National Park Foundation (USA) • National Postal Forum (USA) • Negative Preflare • Neutrons Per Fission • Newtonian Potential Function • North Polar Feature • Not Provided For • Nuclear Power Facility

NPFF Normal Probability Frequency Function

NPFL Near-Field Perfect Lens

NPFR Northern Polarity Flare Ribbon

NPFS Named Piped File System

NPG NASA Procedures and Guidelines • Naval Proving Ground • Net Participant Group • Network Participating Group • Nipa (Papua New Guinea) (location identifier) • Nuclear Planning Group

NPGS Naval Postgraduate School

NPH Nephi (USA-UT) (location identifier) • Non-Processed Header • Non-Proton Hydrate • Normal Paraffin Hydrocarbon

NPhS Nature Physical Science (periodical code)

NPI Navy Prototype Interferometer • Network Printer Interface • Neutral Particle Imager • New Product Information • Non-Precision Instrument • Non-Priced Item • Non-static Presentation Information • Nouveau Pays Industrialisé • Nuclear Physics Institute (Russia)

NPIC National Photographic Interpretation Center (USA) • National Photo Interpretation Center

NPIRS National Pesticide Information Retrieval System

NPIS National Physics Information System (USA)

NPISH Non-Profit Institution Serving Households

NPK Nitrogen - Phosphorus - Potassium

Npl Nameplate

NPL National Physical Laboratory • National Priority List • National Project Leader • Nepal (ISO country code) • New Plymouth (New Zealand) (location identifier) • New Product Line • New Programming Language (now PL/1) • Noise Pollution Level • Non-Patent Literature • Non-Personal Liability • Non-Procedural Language • Normal Power Level • Nuclear Physics Laboratory

NPLI National Physical Laboratory of India

Np/M Neper per Meter

NPM Network Performance Monitor • Non-Print Media • Northern Proper Motion (survey)

NPMAS Neville Public Museum Astronomical Society (USA-WI)

NPMF Negative-Polarity Magnetic Field

NPMG Nordic Planetarium Management Group

NPMHU National Postal Mail Handlers Union

NPMLE Non-Parametric Maximum Likelihood Estimate • Non-Parametric Maximum Likelihood Estimator

N-P-N Negative-Positive-Negative

NPN NASA Part Number • Negative - Positive - Negative • New Public network • Non-Protein Nitrogen

NP/ND Not Published - No Date

NPND Not Published - No Date

npo Ne per Oris

NPO Nangapinoh (Indonesia) (location identifier) • National Program Office • National Project Office • Non-Profit Organization • Non-Purchase Offer • North Pines Observatory (USA-AZ) • Nuclear Propulsion Office

NPOC National Point Of Contact • Naval Polar Oceanography Center

NPOE Numerical Predictions of Orbital Elements

NPOES National Polar-orbiting Operational Environmental Satellite

NPOESP National Polar-orbiting Operational Environmental Satellite Program

NPOESS National Polar-orbiting Operational Environmental Satellite System (USA)

NPOI Navy Prototype Optical Interferometer

NPOP NASA Polar Orbiting Platform

NPOST Non-Perturbative Open-Shell Theory

NPP Napperby (Australia-NT) (location identifier) • National Policy Paper • Net Primary Production • Net Primary Productivity • New Policy Proposal • No Passed Proof • No Prior Publication • NPOESS Preparatory Program • NPOESS Preparatory Project • Nuclear Power Plant • Numéro de Patient Permanent

NPPO NASA Parts Project Office

NPQ Not Physically Qualified

NPR National Partnership for Reinventing (government) (USA) • National Performance Review • National Public Radio (USA) • Noise Power Ratio • Noise Prediction and Reduction • Non-Patent Reference • Non-Persistent Reference • Non-Processor Request • Non-Penetrating Radiation • North Polar Region (Jupiter) • Nuclear Pulse Rocket • Rupee (Nepal) (ISO currency code)

NPRD Non-electronic Parts Reliability Data

NPREC Normalized Precision

NPRF Noise Power Ratio Floor

NPRL National Physical Research Laboratory

NPRM Notice of Proposed Rule-Making

Nprn Neoprene

NPRS NASA Procurement Regulation Supplement • National Programme for Remote Sensing • Non-Persistent

NPRT National Performance Review Team • New Policy Review Team

NPRV Nitrogen Pressure Relief Valve

NPS NASA Planning Study • National Park Service (USA) • National Periodicals System (USA) • Naval Postgraduate School (USA-CA) • Netware Protocol Stack • Network Print Server • Network Product Server • Non-Point Source • Non-Propulsive Stage • No Prior Service • North Polar Sequence • North Polar Spot • North Polar Spur • Nuclear Power Source • Nuclear Power Station • Nuclear Power System • Numerical Plotting System • Numerical Products Service

NPSD Non-Programmed Suspense Date • Normalized Power Spectrum Density

NPSG Nuclear Physics Support Group

NPSH Net Positive Suction Head

NPSI NCP Packet Switching Interface • Network Packet Switching Interface

NPSNET NASCOM Packet Switched Network

NPSO Non-Paired Spatial Orbital

NPSP Net Positive Static Pressure • Net Positive Suction Pressure

NPSS NASA Packet Switched System • NASA Packet Switching System • Numerical Propulsion System Simulation

NPT National Pipe Thread • Navy Pointer and Tracker • Network Planning Technique • Network Planning Terminal • Network Planning Tool • New Planetary Telescope • Newport (USA-RI) (location identifier) • Noise Producing Test • Non-Proliferation Treaty • Normal Pressure and Temperature

NPTF National Pipe Thread - Female

NPTM National Pipe Thread - Male

NPTN National Pesticide Telecommunications Network • National Public Telecomputing Network

npu Ne Plus Ultra

NPU Natural Processing Unit • Nitrogen Purge Unit • San Pedro Uraba (Colombia) (location identifier)

NPUJ Northwestern Polytechnical University Journal (periodical code)

NPV Net Present Value (*) • New Present Value • Nitrogen Pressure Valve • No Par Value • Not Present Value

NPX Numeric Processor Extension

NP0 Negative Power Zero

NQ Neural Quantum • No Quota • Nouvelle Qualification

NQA Memphis (USA-TN) (location identifier) • National Quality Assurance • Niveau de Qualité Acceptable • No Questions Asked

NQAM Normalized Quadratic Axial Moment

NQAS Not Quite A Smile

NQC NASA Quality Control • National Quality Control

NQCC Network Quality Control Center

NQD Notice of Quality Discrepancy

NQE Network Querying Environment

NQF National Qualifications Framework (UK)

NQI Kingsville (USA-TX) (location identifier) • National Quality Index • Nuclear Quadrupole Interaction

NQL Network Query Language • Niquelândia (Brazil-GO) (location identifier)

NQN Neuquen (Argentina-NE) (location identifier)

NQOC Not Quite Our Class

NQR Nuclear Quadrupole Resonance

NQS National Quality Standard • Network Queue Scheduler • Network Queue System • Network Queuing System

NQT Neutral Qualification Test

NQU Nuqui (Colombia) (location identifier)

NQX Key West (USA-FL) (location identifier)

NQY Newquay (UK-England) (location identifier)

n/r Nuestra Remesa

nr Non-Registered • Number

Nr Near • Numero • Numéro • Nummer • Recurrent Nova

N/R Not Required

N-R Newton-Raphson (method) • Normal-Reversed

NR National Range • National Record • Natural Resources • Nauru (ISO country code) • Negative Resistance • Newton-Raphson (method) • Nitrat-Reduktase • Noise Rating • Noise Ratio • Non-Rated • Non-Reactive • Non-Recoverable • Non-Reflective • Non-Regional • Non-Registered • No Ranking • No Release • No Report • No Requirement • No Risk • Normalisatieraad (Norway) • Normalized Rank • Normal Range • Not Rated • Not Reported • Not Required • Nova Remnant • Nuclear Reactor • Nuclear Ring • Number • Number of Records • Number of Reports • Number of Rows • Nyberg-Rueppel (algorithm)

NRA Narrandera (Australia-NSW) (location identifier) • NASA Research Announcement • National Reconstruction Act (USA) • National Recovery Administration • National Recreation Area • National Rivers Authority (UK) • Natural Resources Authority • Network Resolution Area • Never Refuse Anything • Non-Registered Accountable • Nuclear-Reaction Analysis

NRA&A Nouvelle Revue d'Aéronautique et d'Astronautique (periodical code)

NRAC National Radio Astrophysics Centre • National Resources Analysis Cener • Naval Research Advisory Council

NRAL National Radio Astronomy Laboratory • Nuffield Radio Astronomy Laboratories (UK)

NRAO National Radio Astronomy Observatory (USA-VA)

NRAOR National Radio Astronomy Observatory Reprints (periodical code)

NRAP Network Risk Analysis Plan

NRAT Network Risk Analysis Team

NRB Mayport (USA-FL) (location identifier)

NRBC Nuclear - Radiological - Bacteriological - Chemical

NRC Crows Landing (USA-CA) (location identifier) • National Regulatory Commission (USA) • National Research Council • National Response Center • Natural Resources

Canada • Network Reliability Council • Neutron Radiation Capture • Non-Recurring Charge • Non-Recurring Cost • Nuclear Regulatory Commission (USA) • Número de Referencia Completo

NRCA Non-conformance Reporting and Corrective Action

NRCAN National Resources of Canada

NRCC National Range Control Center (USA) • National Research Council of Canada

NRCE Non-Redundant Common Equipment

NRCI National Research Council of Iran

NRCP National Research Council of the Philippines

NRCS National Radio Communications System • Natural Resources Conservation Service • Normalized Radar Cross-Section

NRCSE National Resource Centre for Satellites in Education

NRCT National Research Council of Thailand

NR/D Non-Relevant Document

NR&D Naval Research & Development

NRD National Range Division • Naval Research and Development • Negative Resistance Diode • Noeud Régional Distribué • Non-Relevant Document • Norderney (Germany) (location identifier) • Nuclear Radiation Detector

NRDC National Research and Development Corporation • Natural Resources Defense Council (USA-DC) • Naval Research and Development Command

NRDD National Resources Data Directory • Natural Resources Data Directory

NRDF Nuclear Reaction Data File

NRDN NOAA Radar Data Network

NRDP National Research and Development Program

NRDS Nuclear Rocket Development Station

NRE Namrole (Indonesia) (location identifier) • Negative Resistance Effect • Non-Recurring Effort • Non-Recurring Engineering • Non-Recurring Expense • Non-Relativistic Electron • Non-Rotating Earth • Nuclear Radiation Effect • Nuclear Rocket Engine

NREC National Resource Evaluation Center (USA) • National Robotics Engineering Continuum

NRED Natural Resources and Energy Division

n/ref Nuestra Referencia

NREL National Renewable Energy Laboratory (USA)

NREM National Research and Engineering Network • Non-Rapid Eye Motion • Non-Rapid Eye Movement

NREN National Research and Education Network (USA)

NRER Non-Rejected Earth Radiance

NRF National Research Foundation (South Africa) (formerly FRD) • Nouvelle Revue Française

NRFA Navigation Radio Frequency Assembly • Netherlands Foundation for Radio Astronomy

NRFD Not Ready For Data

NRFI Not Ready For Issue

NRG Énergie • Energy • Narrogin (Australia-WA) (location identifier) • Notional Requirements Generator

NRH Nançay Radioheliograph (France) • Nobeyama Radio Heliograph (Japan)

NRI Natural Resources Institute • Network Radio Interface • Non-Recurrent Investment • Shangri-La (USA-OK) (location identifier)

NRIA National Research Investment Act (USA)

NRIAG National Research Institute for Astrophysics and Geophysics (Egypt)

NRIC National Resource Information Centre • Network Reliability and Interoperability Council (USA-DC)

NRIHVE National Research Institute for Higher and Vocational Education

NRIM National Research Institute for Metals • National Research Institute for Materials (Japan) (*) • National Research Institute of Metals (Japan)

NRIS Natural Resource Information System • Network Resource Information System

NRK Norrköping (Sweden) (location identifier)

NRL National Radiation Laboratory • National Reference Library • National Research Laboratory • Natural Rubber Latex • Naval Research Laboratory (USA-DC) • Normal Response Level • North Ronaldsay (UK) (location identifier)

NRLA Non-Redundant Linear Array

NRLM National Research Laboratory of Metrology (Japan)

NRM Nara (Mali) (location identifier) • National Resources Management • Natural Remanent Magnetism • Natural Remanent Magnetization • Network Resource Manager • Non-Recurring Maintenance • Non-Redundant Masking • Normal Response Mode

NRMC National Resources Management Centre

NRMDP Natural Resource Management and Development Project

Nrml Normal

NRMP Natural Remanent Polarization • Non-Resident Matching Program

NRMS National Resource Management System

NRN National Research Network • Network Routing Number • Neutron-Rich Nucleus • Non-Receipt Notification • No Reply Necessary • Northern • Novell Remote Network

NRO National Reconnaissance Office (USA) • National Reconnaissance Organization • National Research Office • Negative Resistance Oscillator • Network Routing Order • Nichtregierungsorganisation • Nobeyama Radio Observatory (Japan) • Non-Radial Oscillation • Non-Rotating Origin

NROMM Netherlands Register Of Microform masters

NROp Nouvelle Revue d'Optique (periodical code)

NROSS Navy Remote Ocean Sensing System (USA)

N-rotation Numérotation

NRP NASA Research Park (USA-CA) • National Research Plan • National Research Program • National Resource Protection • National Route Program • Navigation Reference Point • NEA Review Panel (IAU) • Net Radio Protocol • Network Radio Protocol • Non-Radial Pulsation • Non-Registered Publication • Normal Rated Power • Notice of Research Project • Nuclear Reprocessing Plant

NRPA National Recreation and Park Association (USA)

NRPB National Radiological Protection Board (UK) • National Resources Planning Board

NRPGC National Research Program for Global Change

NRPIF Nordic Regional Planetary Image Facility (Finland)

NRPS Nuclear Resonance Photon Scattering

NRQ N'Riquinha (Angola) (location identifier)

NRQCD Non-Relativistic Quantum Chromodynamics

NRQZ National Radio Quiet Zone (USA-VA)

NRR Net Reproduction Rate • Noise Rejection Ratio • No Reply Required • Nuclear Rocket Reactor • Roosevelt Roads (Puerto Rico) (location identifier)

NRRC Nuclear Risk Reduction Center

NRRF Naval Radio Receiving Facility

NRRv Naval Research Reviews (periodical code)

NRs Nepalese Rupee (currency)

NRS Imperial Beach (USA-CA) (location identifier) • Name Registration Scheme • Name Repository Server • National Reference System • Naval Radio Station • Next Receive

Sequence • Nitrogen Recharge Station • Non-conformance Reporting System • Nuclear Radiation Shield • Nuclear Rocket Shuttle

NRSA National Remote Sensing Agency (India) • National Remote Service Award (USA)

NRSC Nansen Remote Sensing Centre (Norway) • National Remote Sensing Centre (UK) • Network Reliability Steering Committee

NRSCM National Remote Sensing Centre for Mauritius

NRSF National Remote Sensing Facility (Zimbabwe)

NRSP National Remote Sensing Programme

NRSQA National Registration Scheme for Quality Assessors

NRT Nançay Radiotelescope • National Response Team • Near Real Time • Net Registered Tonnage • Network Readiness Test • News Retrieval Tool • Newtonian Reflector Telescope • NIVR Ruimtetechnologie • Non-Real Time • Tokyo Narita (Japan) (location identifier)

NRTA Near Real-Time Analysis

NRTD Near Real-Time Display • Near Real-Time Dissemination

NRTF National RCS Test Facility (USA)

NRTL Nationally Recognized Testing Laboratory

NRtn Non-Return

NRTS National Reliable Transfer Server • Near Real-Time System • Network Resources and Training Site • NITF Reliable Transfer Server • Non Repairable at This Station

NRTU Non-Real-Time Unit

NRTVNR Non-Real-Time Variable Bit Rate

nrtVFR Non-Real-Time Variable Frame Rate

NRU National Research Universal (reactor) (Canada) • Nauru (ISO country code)

NRUS Non-linear Resonant Ultrasound Spectroscopy

NRV Guam - Uscg Shore (Guam) (location identifier) • Natural Resource Valuation • Net Realizable Value • Non-Return Valve

NRvsbl Non-Reversible

Nrw Norwegian

NRW Narrow • Nebular Rossby Wave • Nordrhein-Westphalen (Germany)

NRWS Narrows

NRY Newry (Australia-NT) (location identifier)

NRZ Non-Return to Zero

NRZI Non-Return to Zero IBM • Non-Return to Zero Indicated • Non-Return to Zero Inverse (*) • Non-Return to Zero Inverted

NRZL Non-Return to Zero Level

NRZM Non-Return to Zero Mark

NRZS Non-Return to Zero Space

n/s Non-Smoker • Not Sufficient

ns Nanosecond • Non Significatif • Non-Standard

Ns Nimbostratus • Nous

N-S Navier-Stokes (equation)

N/S Nord-Sud • North-South

NS Nachschrift • Naked Singularity • Name Server • Nanosecond • Nassi-Schneidermann • National Security • National Service • National Society • Natural Science • Natural State • Navier-Stokes (equation) • Near Side • Near Space • Nederlandse Spoorwegen (Netherlands) • Negative Small • Netscape • Network Server • Network Service • Network Supervisor • Neutron Spectrometer • Neutron Star • New Series • New South Wales (Australia) • New Style • New System • Next System • Nickel Steel • Nimbostratus • Nitrogen Sulfide Radical • Nitrogen Supply • Nitrogen System • Node Switch • Non-Scheduled • Non-Serviceable • Non-Significatif • Non-Smoker • Non-Specified • Non-Standard • Non-Stop • Non-Sufficient • Nord-Sud • Norse • Norsk Standard (Norway) • North-South • No Smoking • No Sort • Nossa • No Swap • Notification Service • Not Satisfactory • Not Significant • Not Specified • Not Sufficient • Not Suppressed • Not Switchable • Nova Scotia (Canada) • Nuclear Science • Nuclear Ship • Nuclear Shuttle • Nuclear Survivability • Nuclear System • Number of Stations

NSA Nançay Surveillance Antenna (France) • National Security Agency (USA) • National Seismological Archive (UK) • National Shipping Authority • National Sound Archive • National Standards Association • National Student Association • Negotiated Services Agreement • Network Supported Account • Next Station Addressing • NOAO Science Archive • Noise Suppressor Assembly • Non-Standard Analysis • Non-Sterling Area • Noosa (Australia-QLD) (location identifier) • Norma Spiral Arm • Northwest Suburban Astronomers (USA-IL) • Nosewheel Steering Amplifier • No Suitable Applicant • Not Seasonally Adjusted • Not Strongly Absorbing

NSAAC North Shore Amateur Astronomy Club (USA-MA)

NSAC NRSA Scientific Advisory Council • Nuclear Safety Information Center (USA) • Nuclear Science Advisory Committee

NSAD Network Security Architecture and Design

NSAI National Standards Authority of Ireland

NSAP National Service Assurance Plan • Network Service Access Point

NSAPI Netscape Application Programming Interface • Netscape Server Application Programming Interface

NSAS Non-spinning Sun and Aspect Sensor • Northern Skies Astronomical Society (USA-ND) • Northern Sydney Astronomical Society (Australia-NSW)

NSAU National Space Agency of Ukraine

NSB Bimini North (Bahamas) (location identifier) • National Science Board • National Standards Board • Naval Studies Board • Near Surface Burst • Neue Schweizer Biographie • Neutron Star Binary • Night Sky Background • Night Sky Brightness

NSBBN Non-Standard Big Bang Nucleosynthesis

NSBD Narrow Spectral Band Detection • Normal Stars and Brown Dwarfs (stars) (project)

NSBE National Society of Black Engineers

NSBF National Scientific Balloon Facility (USA-NM)

NSBRI National Space Biomedical Research Institute (USA)

NSC National Safety Council • National Science Center • National Science Council • National Security Council (USA) • National Seismic Center • National Service Code • National Simulation Capability • National Simulations Center • National Standards of Canada • National Steering Committee • Navigation and Sensor Computer • Navigational Star Catalogue • Navigation Star Catalogue • Nehru Science Centre (India) • Network Service Center • Network Service Class • Network Support Center • Network Support Committee • Network Switching Center • Network Systems Corporation • Neutron Source Cask • Neutron Star Candidate • Nordic Satellite Co. • Norwegian Space Centre (Norway) • No Significant Cloud • Notice of Schedule Change • Nuclear Science Centre (India) • Numerical Sequence Code

NSCAT NASA Scatterometer

NSCCA Nuclear Safety Cross-Check Analysis

NSCD National Spatial Cadastral Database

NSCG Neue Schweizerische Chemische Gesellschaft (Switzerland)

NSCI Neuroscience Citation Index (ISI)

NSCL National Supercomputing Cyclotron Laboratory (USA-MI)

NSCORT NASA Specialized Center Of Research and Training

NSCP National Scalable Cluster Project

NSCS National Security Council System • Network Service Center System • New Swiss Chemical Society (Switzerland)

NSC1 Network Systems Cypher One

NSD National Security Decision • National Security Directive • Network Security Device • Network Security Directive • Network Service Definition • Network Statistics Database • Network Status Display • Night-Side Darker

NSDC National Serials Data Centre (UK) • Natural Satellites Data Center

NSDD National Security Decision Directive • Nuclear Structure and Decay Data

NSDF NSSDC Supplementary Data File

NSDI National Spatial Data Infrastructure (USA) • National Spatial Data Initiative • Non-Sequential Double Ionization

NSDIR National Software Data and Information Repository

NSDL National Science Digital Library

NSDM National Security Decision Memorandum

NSDO National Solar Digital Observatory

NSDS Neutron Spectrometer Digital System

NSDSSO NASA Science Data Systems Standards Office

NSDU Network Service Data Unit

NS&E Nanoscale Science & Engineering • Natural Sciences & Engineering

NSE Milton Whiting (USA-FL) (location identifier) • National Software Exchange • National Space Enterprise • National Stock Exchange • National Support Element • Natural Sciences and Engineering • Natural Space Environment • Navier-Stokes Equation • Navigation Support Equipment • Network Software Environment • Non-Scientist/Engineer • Noordwijk Space Expo (Netherlands) • Nuclear Science Engineering (periodical code) • Nuclear Statistical Equilibrium

NSEAP Nuclear Safety Evaluation and Assistance Program

nsec Nanosecond

NSEC Network Switching Engineering Center

NSEP National Security and Emergency Preparedness

NSERC Natural Sciences and Engineering Research Council (Canada)

NSES National Science Education Standards (USA)

NSESCC NASA Space and Earth Sciences Computing Center

N/S/F Not Sufficient Funds

NSF Camp Springs - Andrews (USA-MD) (location identifier) • Nationale Simulatie Faciliteit (Netherlands) • National Science Foundation (USA-DC) • National Simulation Facility • Nomenclature des Spécialités de Formation • Non-Spin-Flip • Norges Standardiseringsforbund (Norway) • Norwegian Standards Association (Norway) • No Star Formation • Not Sufficient Funds • Nuclear Structure Facility

NSFC National Science Foundation of China • Natural Science Foundation of China

NSFL National Science Film Library (Canada)

NSFLEX Navier-Stokes solver using characteristic Flux Extrapolation

NSFNET National Science Foundation Network

NSFS Newcastle Space Frontier Society (Australia)

NSG Naturschutzgebiet • Nippon Silica Glass • Non-Self-Gravitating • Non-Statutory Guidelines • North-Seeking Gyroscope • No Self-Gravity • Not So Good • Nuclear Supplier Group

NSGC Numeric, Symbolic and Geometric Computation

NSGP National Space Grant Consortium (USA)

NSGT Non-Self-Governing Territory

NSGW NASCOM Service Gateway

NSH No Special Handling • Now Shahr (Iran) (location identifier)

NSHA Neutron Source Holder Assembly

NSHC North Sea Hydrographic Commission

NSHS Nosewheel Steering Hydraulic System

NSI NASA Science Internet • NASA Standard Initiator • National Science Indicator • National Security Information • National Soil Institution • National Space Institute • National System of Innovations • Network and Systems Integration • Network Support Item • Nitrogen Solubility Index • Noise Source Instrumentation • Non-Standard Item • Not Seriously Injured • Nouveaux Systèmes d'Information • Nouveaux Systèmes Informatiques • Nuclear Systems Initiative (NASA) • Numeric System Identifier • Yaounde Nsimalen (Cameroon) (location identifier)

NSIA National Security Industrial Association (USA)

nSIC Normalized Specific Impact Contribution

NSIC National Storage Industry Consortium • Nuclear Safety Information Center (ORNL)

NSIDC National Snow and Ice Data Center (USA)

NSIE Network Security Information Exchange

NSIF Near-Space Instrumentation Facility

NSII Narrabri Stellar Intensity Interferometer

NSINIC NASA Science Internet Network Information Center

NSIOD National Science Indicators On Diskette (ISI)

NSIP NASA Student Involvement Program • Network Security Improvement Program

NSIPO NASA Science Internet Project Office

NSIPP NASA Seasonal to Interannual Prediction Project

NSIS Norske Statens Institutt for Strälehygiene (Norway)

NSIUWG NASA Science Internet Users Working Group

NSIX Network Security Information Exchange

NSJI North-South Joint Implementation

NSK Neutron Star Kick • Non-Stop Kernel • Noril'sk (Russia) (location identifier)

NSKS Navier - Stokes - Kuramoto - Sivashinsky (equation)

NSL National Science Laboratories • National Science Library • National Storage Laboratory • National Suppliers List • Network Service Layer • Network Software Laboratory • Nominal Scanning Law • Non-Standard Logics • Non-Stockage List • Nuclear Safety Line

NSLA National Science Library of Australia

NSLDS National Student Loan Data System

NSLI National Service Life Insurance

NSLIN Non-Standard Line Item Number

NSLR NASA Satellite Laser Ranging • Nuclear Spin-Lattice Relaxation

NSLRSA National Satellite Land Remote Sensing Archives

NSLRSDA National Satellite Land Remote Sensing Data Archives

NSLS National Synchrotron Light Source (BNL)

NSM Natural Semantic Metalanguage • Negative-Stiffness Mechanism • Network and Systems Management • Networked System Management • Network Security Management • Network Security Manager • Network Status Monitor • Network Systems Management • Neutron Star Magnetosphere • Neutron Star Merger • New Standard Model • Niet-Stationaire Meetmethode • Non-Standard Model

• Norseman (Australia-WA) (location identifier) • North-Seeking Module

NSMGC Neutron Star Merger Grand Challenge

NSMP National Security Management Plan • National Semiconductor Metrology Program

NSN NASA Science Network • NASA Space Network • National Satellite Network • National Stock Number • Nelson (New Zealand) (location identifier) • No Stock Number • No Such Number • Novell Script for NetWare

NSNA No Stock Number Assigned

NSNF Non-Strategic Nuclear Force

NSO NASA Safety Officer • NASA Support Operation • National School Observatories • National Security Office • National Security Officer • National Solar Observatory (USA-AZ) • National Statistical Office • National Statistical Officer • Network Security Office • Network Security Officer • Network Support Office • Network Support Officer • Night Sky Observers (USA-KY) • Nuclear Safety Office • Nuclear Safety Officer • Numerical Stock Objective • Scone (Australia-NSW) (location identifier)

NSOC National Security Operations Center

NSOM Near-field Scanning Optical Microscope

NSOR Nederlandse Stichting voor Operations Research (Netherlands)

NSP Name Service Protocol • NASA Support Plan • National Service Provider • National Space Program • Native Signal Processing • Nebraska Star Party (USA-NE) • Network Service Provider • Network Services Processor • Network Services Protocol • Network Signal Processing • Network Signal Processor • Network Support Plan • Non-Sequenced Poll • Non-Sequential Programming • Non-volatile Serially Programmable • Normalized Stokes Parameter • Normal Safety Precaution • Not Separately Priced • Nuclear Spin Relaxation • Stanley Point (Philippines) (location identifier)

NSPAC Nova Scotia Planetarium Advisory Committee (Canada-NS)

NSPAR Non-Standard Parts Approval Request

NSpC National Space Committee • National Space Council

NSPD National Space Policy Directive

NSPE National Society of Professional Engineers (USA)

NSPF Not Specifically Provided For

NSPMP Network Switching Performance Evaluation Plan

NSPO National Space Program Office

NSPR Netscape Portable Runtime

NSPS National Society of Professional Surveyors (USA) • New Source Performance Standard • Nutrient Supply Predicting Submodel

NSPSR New Source Performance Standard Review

NSPW New Security Paradigms Workshop

NSR National Scientific Register (USA) • National Security Review • Network Service Request • Network Surveillance and Reconfiguration • Neutron Source Reactor • Night Sky Radiation • NISN Service Request • Nominal Slow Rate • Non-linear Scaling Relationship • Non-Shared Resource • Non-Source Routed • Non-Synchronous Rotation • Normal Slow Rate • Northern Starburst Region • Nosso Senhor • No Staff Responsibility • Nuclear Spin Relaxation • Nutrient Supply Response

NSRA National Remote Sensing Agency (India) • Nossa Senhora

NSRB National Security Resources Board

NSRC National Science Research Council • National Science Resources Center (USA) • National Solar Radiation Center (PAGASA) • National Synchrotron Research Center (Thailand) • Network Startup Resource Center

NSRD National Software Reuse Directory

NSRDC National Standard Reference Data Center (USA)

NSRDP National Standard Reference Data Program (USA)

NSRDS National Standard Reference Data Series • National Standard Reference Data System (USA)

NSRF Nutrient Supply Response Function

NSRL Natural Science Research Laboratory

NSRM New Solid Rocket Motor

NSRP Neutral Seat Reference Point

NSRS National Spatial Reference System

NSRT Nassau Station Robotic Telescope (USA-OH)

NSS National Seismic Station • National Space Society (USA-DC) • National Stream Survey • Navigational Safety System • NERC Scientific Services • Network Simulation System • Network Subscription Services • Network Subsystem • Neutron Spectrometer System • Network Synthesis System • Nitrogen Supply System • Nodal Switching System • Non-Sea Salt • Non-Self-Sustaining • Non-Sequential Surface • Nordic Statistical Secretariat • Normal Saline Solution • North Star Systems (USA-NY) • NSSDC Storage System

NSSA National Space Society of Australia

NSSB National Skills Standard Board (USA)

NSSC NASA Safety Standards Committee • NASA Standard Spacecraft Computer • National Space Science Centre (UK) • Natural Science Societies of Canada • NORAD Space Surveillance Center • Nouvelle Société Suisse de Chimie (Switzerland)

NSSDC NASA Space Science Data Center • National Space Science Data Center (USA-MD) (*)

NSSE National Survey of Student Engagement (USA)

NSSFC National Severe Storms Forecast Center (USA)

NSSK North-South Station Keeping

NSSL National Severe Storms Laboratory (USA)

NSSN National Standard System Network

NSSO National Solar Space Observatory

NSSOG National Security Standards Oversight Group

NSSS National Space Surveillance System

NSST New Swedish Solar Telescope

NSSTC National Space Science and Technology Center (USA)

NSSTS Near-Surface Salinity and Temperature System

NSt Nebenstelle

NST Nakhon Si Thammarat (Thailand) (location identifier) • Network Support Team • Newfoundland Standard Time • New Serial Titles • Not Sooner Than

NStA Nebenstellenanlage

NSTA National Safe Transit Association • National Science Teachers Association (USA)

NSTAC National Security Telecommunications Advisory Committee (USA)

NSTAG National Science and Technology Analysis Group (USA)

NSTAR NASA SEP Technology Application Readiness

NSTB National Science and Technology Board (Singapore)

NSTC National Science and Technology Centre • National Science and Technology Council • Nova Scotia Technical College (Canada-NS)

Nstd Non-Standard

NSTD Non-System Training Device

NSTDA National Science and Technological Development Agency

NSTDC National Scientific and Technical Documentation Center

NSTIC National Scientific and Technical Information Center

NSTL National Software Testing Laboratory (USA) • National Space Technology Laboratory • National Study Team Leader

NSTR NASA Standard Tape Recorder • National Solar-Terrestrial Research • Nordic Sea Test Range • Nothing Significant To Report

NSTDR Non-System Training Device Requirement

NSTIS National Security Telecommunications and Information System (USA)

NSTO New Systems Training Office

NSTP National Security Telecommunications Policy

NSTS National Secure Telephone System • National Space Transportation System

NSTSPO National Space Transportation System Project Office

NSTSS National Space Transportation and Support Study • National Space Transportation System

NSTX National Spherical Tokamak Experiment

NSU Network Service Unit • Network Synchronization Unit • Nitrogen Supply Unit • Node Switching Unit • Norfolk State University

NSV Negative Supply Voltage • New Suspected Variable (star) • Node Support Vehicle • Non-Spinning Vehicle • Noosaville (Australia-QLD) (location identifier)

NSVN NATO Secure Voice Network

NSVS NATO Secure Voice System

NSW Nebenschlagwort • New South Wales (Australia) • Non-Secure Warning • No Significant Weather • NSP Status Word

NSWC Naval Surface Warfare Center • Naval Surface Weapons Center (USA)

NSWP National Space Weather Program (USA)

NSWR Nuclear Salt Water Rocket

NSWS National Surface Water Survey

NSWT Non-Secure Warning Tone

NSWTI Nationales System für die Wissenschaftliche und Technische Information

NSX North Sound (British Virgin Islands) (location identifier)

NSXB Neutron Star X-ray Binary

NSY Sigonella (Italy) (location identifier)

nT Nanotesla

Nt Net • Newton • Nit • Niton

NT Nachrichtentyp • Narrower Term • National Treatment • Navigation Technology • Near Term • Neighbor Table • Net Terms • Net Tonnage • Network Terminal • Network Termination • Network Terminator • Neutral Zone between Iraq and Saudi Arabia (ISO country code) • Newfoundland Time • New Technology • New Testament • Nodal Terminal • Non-Telluric • Non-Thermal • Normal Threat • Northern Territory (Australia) • Northwest Territories (Canada) • No Test • No Torque • Not Tested • Not Titled • Nouvelle Technologie

Nta Nuestra

NTA Natadola (Fiji) (location identifier) • National Technology Agency • National Telecommunications Administration • Net Technical Assessment • Network Termination Analog • News Transfer Agent • Nippon Total Access Communication System • Nitrogen Tank Assembly • Northern Taurids (meteor stream) • Nouvelles Technologies Appliquées

NTAC National Test Administration Center (USA) • National Time Allocation Committee (Gemini)

NTAD National Transportation Atlas Data

NTAM Notice To Airmen

NTAP Notice To Airmen Publication

NTAPS Near-Term Active Protection System

NTAS New Technology Advanced Server

NTB National Technical Board • National Testbed • Network Testbed • NITF Technical Board • Non Tariff Barrier • Non-Thermal Bremsstrahlung • North Temperate Belt (Jupiter) • North Tidal Bridge • Notodden (Norway) (location identifier)

NTBA Network Termination of Basic Access

NTBIC National Testbed Integrating Contractor

NTBn North Temperate Belt - Northern Component (Jupiter)

NTBR Not To Be Resuscitated

NTBs North Temperate Belt - Southern Component (Jupiter)

NTBT Nuclear Test Ban Treaty

Ntc Notice

NTC National Technological Capability • National Telecommunications Conference • National Training Center • Negative Temperature Coefficient • Negative Torque Control • Network and Training Center • Network Terminal Computer • Network Terminal Concentrator • Normal Test Condition • Noshiro Testing Center (Japan) • Nouvelles Technologies de Communication • Santa Carolina (Mozambique) (location identifier)

NTCB Network Trusted Computing Base

NTCDA Naphthalenetetracarboxylic Dianhydride

NTCF NATO Technical Control Facility

NTCI Nouvelles Technologies de Communication et d'Information

NTCN National Telecommunications Coordinating Network

NTCo Non-destructive Testing Communications (periodical code)

NTCS National Television Committee System

NTD NASA Test Director • National Topographic Database • Neural Tube Defect • Neutron Transmutation-Doped (*) • Neutron Transmuted-Doped • Nova Technology Development Pte. Ltd. (Singapore) • Port Hueneme - Point Mugu (USA-CA) (location identifier)

NTDB National Topographic Database • National Trade Data Bank

NTDR Near-Term Data Radio • Near-Term Digital Radio

Nte Norte

NTE Nantes Atlantique (France) (location identifier) • NASA Technology Experiment • National Teacher Examination • Naturally-Triggered Erection • Navigation Telemetry Encoder • Neutron Transient Effect • Not To Exceed • Nouvelles Technologies d'Enseignement • Nouvelles Technologies Éducatives

NTEC Network Technical Equipment Center

NTER National Engineering Technologists Register • Normalized Transmission Energy Requirement

NTEU National Treasury Employees Union (Australia)

NTESS NASA Technical Experience for Select Students

NTeZ North Temperate Zone

ntf Lotus Notes Templates (file extension)

NTF National Test Facility • National Transfer Format • Neutron Therapy Facility • Non-Thermal Filament • No Trouble Found • Nouvelles Technologies de Formation • Nuclear Task Force • Nuclear Test Facility

NTFS New Technology File System

Ntfy Notify

NTG Nantong (China-PRC) (location identifier) • Narrower Term - Generic • Nederlandse TeXgebruikers Groep (Netherlands) • Newtonian Theory of Gravitation • NGST Task Group • Nitroglycerine • Non-Technical Generator • Nordic Telescope Group • Nuclear Theory Group

NTGDS Non-Tritium Gas Delivery System

Nth North

NTH Norges Tekniske Hogskole (Norway)

NTHG Nothing

Nthn Northern

NTHU National Tsing-Hua University (China-ROC)

NTI Bintuni (Indonesia) (location identifier) • Nanyang Technology Institute (Singapore) • National Technology Initiative • Near-Term Improvement • Near-Term Initiative • Network Terminating Interface • Noise Transmission Impairment • No Travel Involved • Nouvelles Technologies de l'Information • New Technology Inc.

NTIA National Telecommunications and Information Administration (*) • National Telecommunications and Information Agency

NTIAC Non-destructive Testing Information Analysis Center

NTIC National Technical Information Center • Non-destructive Testing Information Center (USA) • Nouvelles Technologies de l'Information et de la Communication

NTIM Not That It Matters

NTIMM Not That It Matters Much

NTIPS National Total Image Processing System

NTIS National Technical Information Service (USA-VA)

NTISS National Telecommunications and Information System Security

NTJ Manti Ephraim (USA-UT) (location identifier)

NTK Need To Know • Net Tonne Kilometer

NTL National • Natural Thermoluminescence • Newcastle Williamtown (Australia-NSW) (location identifier) • Non-Threshold Logic • No Time Lost • Nuclear Thermionics Laboratory

NTLAM New Technology Lunar Astronomy Mission

NTLM New Technology LAN Manager

NTM Miracema do Norte (Brazil-TO) (location identifier) • National Technical Means • National Telecommunications Management • Network Terminal Management • Network Terminal Manager • Nombre de Transparents par Minute • Non-Tariff Measure • No Target Model • Notice To Mariners • Number of Transparencies per Minute

NTMA National Telecommunications Management Architecture

NTMS National Telecommunications Management Structure • National Telecommunications Management System • National Traffic Management System

NTN National Trends Network • Network Terminal Number • Neutralized Twisted Nematic • Neutral Tree Network • New Technology News (periodical code) • Normanton (Australia-QLD) (location identifier)

NTNU National Taiwan Normal University (China-ROC) • Norges Teknisk-Naturwetenskapelige Universitet (Norway)

Nto Nuestro

NTO Network Terminal Option • Nitrogen Tetroxide • Not Taken Out • Nuevas Tecnologías Observacionales (Spain) • Santo Antao (Cape Verde) (location identifier)

NTON National Transparent Optical Network

NTOPS New Technology for Open and Portable Systems

NTOT New Technology Orbiting Telescope

NTP Narrower Term - Partitive • National Technology Programme • National Transportation Policy • Near-Term Penetrator • Network Termination Point • Network Test Panel • Network Time Protocol • Network Timing Protocol • Network Transport Protocol • New Tape Product • New Technology Prototype • Normal Temperature and Pressure • Notice To Proceed • No Title Page • Nuclear Test Plant • Nuclear Thermal Propulsion • Numerical Tape Punch

NTpM Number of Transparencies per Minute

NTPM Number of Transparencies Per Minute

NTPS National Transportation Planning System

NTQO Nanango TIE QUT Observatory (Australia-QLD)

NTR Monterrey Norte (Mexico) (location identifier) • National Tape Repository • Nernst-Thomson Rule • Net Time Reference • Neutron Test Reactor • New Technology Report • Nightly Total Rate • Nothing To Report • Nouvelles Technologies de Reproduction • Nuclear Test Reactor • Nuclear Thermal Rocket • Nuclear Thermal Rocketry

Ntra Nuestra

NTRAS New Technology Remote Access Services

NTrB North Tropical Belt

NTRI NCP/Token Ring Interconnection

Ntro Nuestro

NTRS NASA Technical Report Server

NTrZ North Tropical Zone (Jupiter)

NTRZ North Tropical Zone (Jupiter)

NTS NASA Test Support • National Topographic System (USA) • National Traffic System • Naval Telecommunications System • Navigation Technology Satellite • Near-Term Schedule • Negative Torque Sensing • Negative Torque Signal • Negative Torque System • Network Task Scheduler • Network Technical Support • Network Test System • Network Time Service • Network Time System • Nevada Test Site (USA-NV) • New Typesetting System • Night-Targeting System • Non-avionics Test Station • Non-Tariff Size • Non-Technical Skill • Not To Scale

NTSA National Training Systems Association (USA) • NetWare Telephony Services Architecture

NTSB National Transportation Safety Board (USA)

NTSC National Television Standards Committee (USA) • National Television Systems Committee • Never The Same Color • Never Twice Same Color (*)

NTSI Non-linear Thin Shell Instability

NTSO NASA Test Support Office

NTSS Nitrogen Transfer and Storage System

NTT Naked T Tauri • New Technology Telescope • Nippon Telegraph and Telephone (Japan) • Niuatoputapu (Tonga) (location identifier) • Numbered Test Trunk

NTTAA National Technology Transfer and Advancement Act (USA)

NTTC National Technology Transfer Center (USA-WV)

NTTDF New Technology Telescope Deep Field

NTTF NASA Test and Training Facility • Networking and Telecommunications Task Force • Network Test and Training Facility • Networks Technical Training Facility

NTTN National Technical Transfer Network

NTTRv NTT Review (periodical code)

NTTS Naked T Tauri Star

NTTUP New Technology Telescope Upgrade Plan

NTU National Taiwan University (China-ROC) • National Technological University • Navigation Training Unit • Network Terminating Unit • Network Termination Unit • Normalized Turbidity Unit • Northern Territory University (Australia-NT) • Oceana (USA-VA) (location identifier)

NTUA National Technical University of Athens (Greece)

NTUHEP National Taiwan University High-Energy Physics (group)

NTV Nader Te Verdelen • NASA Television • Nippon Television • Number To View

Ntwk Network

NTWK Network

nt. wt. Net Weight

NTX Natuna Ranai (Indonesia) (location identifier) • Neutralized Transport Experiment

NTY Sun City (South Africa) (location identifier)

NTZ North Temperate Zone (jupiter) • North Tropical Zone • No Transgression Zone

Nu Ngultrum (currency) • Nusselt Number

NTZ Neutral Zone between Iraq and Saudi Arabia (ISO country code)

NÜ Netzübergang

NU Naciones Unidas • Name Unknown • National Union • Nations Unies • Natural Uranium • Nebraska University (USA-NE) • Network Utility • Niue (ISO country code) • Northern Union • Northwestern University (USA-IL) • Nose Up • Number Uniformity • Number Unobtainable • Numerical Universe

NUA Network User Address • News User Agent

NUAAJ Nanjing University Aeronautics and Astronautics Journal (periodical code)

NUAC Nuage Acide

NUAN New User Action Notation

NUB Nederlandse Uitgeversbond (Netherlands) • Numbulwar (Australia-NT) (location identifier)

NUBIA Nubus Interface Adaptor

Nuc Nuclear

NUC National Union Catalogue • Non-Uniformity Compensation

NUCATE Núcleo de Ciência, Aplicações e Tecnologia Espaciais (Brazil)

NucFS Nuclear Fusion Supplement (periodical code)

NucFu Nuclear Fusion (periodical code)

NucIM Nuclear Instruments and Methods (periodical code)

Nucl Nuclear • Nucléaire • Núcleo • Nucleus

NucPh Nuclear Physics (periodical code)

NucTe Nuclear Technology (periodical code)

NucTF Nuclear Technology Fusion (periodical code)

NUD En Nahud (Sudan) (location identifier) • Not Updated

NUDET Nuclear Detonation

NUDETS Nuclear Detection System • Nuclear Detonation System

NUDIAS National University Digital Image Analysis System

NUDTJ National University of Defense Technology Journal (periodical code)

NUE New Universe Exploration • Nürnberg (Germany) (location identifier)

NuEnD Nuclear Engineering and Design (periodical code)

NUFEM Nagoya University Faculty of Engineering Memoirs (periodical code)

NUFFIC Netherlands Universities Foundation For International Cooperation

NUG Nuguria (Papua New Guinea) (location identifier)

NUGA Nuclei of Galaxies (project)

NuGeo Nuclear Geophysics (periodical code)

NUGRO Nucleation and Growth

NUH Nunchia (Colombia) (location identifier)

NUI National University of Ireland • Network User Identification • Network User Identifier • Network User Interface • Notebook User Interface • Nuiqsut (USA-AK) (location identifier) • Numéro Unique d'Identification

NUK Nukutavake (Tuamotu Islands) (location identifier)

Nukl Nuklear

NUL Non-Utilization List • Nulato (USA-AK) (location identifier) • Null

NULO NASA Unmanned Launch Operations • Negative Unliquidated Obligation

NULOS Negative Unliquidated Obligations

Núm Número

Num Number • Numeral • Numeric • Numerical • Numérique

NUM Northern University of Malaysia • Numeric • Número

NUMA Non-Uniform Memory Access (*) • Non-Uniform Memory Architecture

NUMAPS Networks for Use in Math and Physical Science

NuMat Numerische Mathematik (periodical code)

Numb Number • Numbered

NumHT Numerical Heat Transfer (periodical code)

NuMI Neutrinos at the Main Injector (project)

NUMI Number of Images

Núms Números

NUN Pensacola Saufley (USA-FL) (location identifier)

NUP Nunapitchuk (USA-AK) (location identifier)

NUPACU Nuclear Powered Air Conditioning Unit

NuPECC Nuclear Physics European Collaboration Committee

NuPhA Nuclear Physics A (periodical code)

NuPhB Nuclear Physics B (periodical code)

NuPhS Nuclear Physics Supplements (periodical code)

NUQ Mountain View - Moffett (USA-CA) (location identifier)

NUR Nullarbor (Australia-SA) (location identifier)

NURBS Non-Uniform Rational B-Spline

NURC National University Research Council

NURE National Uranium Resource Evaluation (USA)

NUREC Network on Urban Research in th European Community

NURO National Undergraduate Research Observatory (USA)

NURWEC Netherlands UK Radar Wavebuoy Experimental Comparison

NUS National Union of Students (UK) • National University of Singapore • New Upper Stage • Norsup (Vanuatu) (location identifier)

NUSAC Nuclear Science Advisory Committee

NUSC Naval Underwater Systems Center (USA)

NuSEA Nucleonic Sea (collaboration)

NUSEL National Underground Science and Engineering Laboratory (USA)

NUSEM Nagoya University School of Engineering Memoirs (periodical code)

NUSL National Underground Science Laboratory (USA-SD)

NUSP National User Support Programme

NUSS Nuclear Safety Standards

NUSSAG Nuclear Safety Standards Advisory Group

NUSSE Non-Uniform Simple Surface Evaporation

N-u-T Newcastle-upon-Tyne (UK-England)

NuT Newcastle-upon-Tyne (UK-England)

NUT National Union of Teachers • Network Utility Transfer • Nutuve (Papua New Guinea) (location identifier)

NUTEC Norwegian Underwater Technology Centre (Norway)

NuTeV Neutrinos at the Tevatron

NUTI NASCOM User Traffic Interface

NUTIS NERC Unit for Thematic Information Systems • Numerical and Textual Information System

Nutr Nutrition

NUTS Nomenclature des Unités Territoriales Statistiques

NutzEV Nutzungsentgeltverordnung

NUU Nakuru (Kenya) (location identifier) • New University of Ulster

NuUB National- und Universitätsbibliothek

NUV Near Ultraviolet

NUVEL Northwestern University Velocity (model)

NUVI Near Ultraviolet Imager

NUVIEWS Narrow-band Ultraviolet Imaging Experiment for Wide-field Surveys

NUVO Near Ultraviolet/Optical

NUW Whidbey Island - Ault (USA-WA) (location identifier)

NUWIG NCC User Working Interface Group

NUX Novii Urengoi (Russia) (location identifier)

Nv Nanovolt

N/V No Value

NV Despina (satellite of Neptune) • Naamloze Vennootschap • Naked Vision • Nationale Vereinigung • Navajo • Near Vertical • Nevada (USA) • New Version • Night Vision • Node Visit • No Overflow • Nuclear Vehicle

NVA Negative Vorticity Advection • Neiva (Colombia) (location identifier) • Norske Vitenskapakademi (Norway)

NVAFB North Vandenberg Air Force Base (USA-CA)

NVAL Not Available

NVAO National Virtual Aeronomical Observatory (USA)

NVB Navigational Base

NVC National Visiting Committee • Non-Verbal Communication • Non-Verbal Conversation

NVCASE National Voluntary Conformity Assessment Systems Evaluation (USA)

NVCN National Vertical Control Network (USA)

NVD Nevada (USA-MO) (location identifier) • Night-Vision Device

NVDET Network Virtual Data Entry Terminal

NVDS National Virtual Data System

NVE Constant Volume-Energy • Network Visible Entry • Night Vision Equipment • Non-Visual Eyepiece • Non-Volatile Electronics

Nveau Nouveau

Nvel Nouvel

Nvelle Nouvelle

Nvelles Nouvelles

NVEO Night Vision Electro-Optics

NVEOC Night Vision Electro-Optics Center

NVEOL Night Vision Electro-Optics Laboratory

NVES Night Vision Electronic Sensor

Nvx Nouveaux

NVFEL National Vehicle and Fuel Emissions Laboratory

NVFRAM Non-Volatile Ferroelectric Random Access Memory

NVG Night Vision Goggles • Nueva Guinea (Nicaragua) (location identifier)

NVGC Night Vision Goggle Compatibility

NVI Galatea (satellite of Neptune) • Non-Value Indicator • Normalized Vegetation Index

NVII Larissa (satellite of Neptune)

NVIII Proteus (satellite of Neptune)

NVIS Night Vision Imaging Spectrometer • Night Vision Imaging System

NVK Narvík (Norway) (location identifier)

NVLAP National Voluntary Laboratory Accreditation Program (USA)

NVLP Nederlandse Vereniging van Lucht- en Ruimtevaart Publicisten (Netherlands)

NVLSPM Night Vision Laboratories Static Performance Model

NVM Non-Volatile Matter • Non-Volatile Memory

NVMS Nose and Vibration Monitoring System

NVO National Virtual Observatory (USA)

nVoD Near Video on Demand

NVOD Near Video On Demand

NVP Network Voice Protocol • Night Visibility Plan • Nominal Velocity of Propagation • Novo Aripuanã (Brazil-AM) (location identifier)

NVPAS Night Vision Pilot Aids System

NVPOWG NASA/VAFB Payload Operations Working Group

NVPP Night Vision Pilotage Package

NVPS Night Vision Pilotage System

NVQ National Vocational Qualification

NVR Naval Vessel Register • Nederlandse Vereniging voor Ruimtevaart (Netherlands) • Non-Volatile Residue • No Verification Required • Novgorod (Russia) (location identifier) • No Voltage Release

NVRAM Non-Volatile Random Access Memory

NVRP Northern Virginia Recreational Parks (USA-VA)

NVS Nachrichtenblatt der Vereinigung der Sternfreunde (periodical code) • Nevers (France) (location identifier) • Night Vision System • Non-Verbal Signal • Non-Volatile Storage

NVSIMM Non-Volatile Single Inline Memory Module

nvSRAM Non-Volatile Static Random Access Memory

NVSS NRAO VLA Sky Survey

NVSSP North Valley Star Party (USA-CA)

NVT Constant Volume-Temperature • Navegantes (Brazil-SC) (location identifier) • Netware Virtual Terminal • Network Virtual Terminal • Novell Virtual Terminal

NVTM Northern Virginia Telescope Meet (USA-VA)

NVTS Network Virtual Terminal Service

NVVSR Nederlandse Vereniging voor Weer en Sterrenkunde Werkgroep Veranderlijke Sterren Reports (periodical code)

NVWS Nederlandse Vereniging voor Weer- en Sterrenkunde (Netherlands)

NVY Neyveli (India) (location identifier)

nw nordwestlich

nW Nanowatt

N/W Network

NW Nachwort • NASA Waiver • Naturwissenschaften (periodical code) • Net Weight • Network • Net Worth • Niedrigwasser • Nominal Width • Non-Weighting • Noordwest • Nordrhein-Westfalen (Germany) • Nordwest • North Wales • Northwest • Northwestern • Nose Wheel • Not Waiverable • No Wind

NWA Moheli (Comoros) (location identifier) • National Water Alliance • National Weather Association • Neighborhood Watch Area

NWACC Northwest Academic Computing Consortium (USA)

NWAG North West Astronomy Group (USA-OR)

NWB New Work Brief • Normal Wideband • Not Weight Bearing

NWBLTU Normal Wideband Line Termination Unit

NWbW Northwest by West

NWC NetWare Connect

NWCI Network Client Item • Network Configuration Item

NWCS NetWare Workstation Compatible Services

NWDA Northwest Deformation Array

NWDB National Wasteland Development Board

nwe Nieuwe

NWE Nuclear Weapon Effect

NWF Nuclear Waste Fund • Numerical Weather Facility

NWfl Nettowohnfläche

Nwfld Newfoundland (Canada)

NWFS Netware File System

NWG Natural Work Group • Network Working Group

NWGAS North West Group of Astronomical Societies (UK)

NWH Newport Parlin (USA-NH) (location identifier) • Normal Working Hours • Northwestern Half

NWI Network Interface • Nordic Web Index • Norwich (UK-England) (location identifier)

NWL Natural Wavelength

NWLTU Normal Wideband Line Termination Unit

NWLY Northwesterly

NWMP National Water Management Programme • National Weapons Monitor Plan

NWMS Non-linear Wave Modulation Spectroscopy

NWN Non-White Noise

NWO NASA Washington Office • Nederlandse Organisatie voor Wetenschappelijk Onderzoek (Netherlands)

NWOC New Way Of Computing

NWP Argentia (Canada-NF) (location identifier) • National Weather Prediction • North-Western Province (India) • Northwest Passage • Numerical Weather Prediction

NWPA Nuclear Waste Policy Act (USA)

NWPM Numerical Weather Prediction Model

NWR National Workhour Report • Next Work Request

NWRC National Weather Records Center

NWRI National Water Research Institute

NWRN Northwestern

NWRS National Workhour Reporting System

NWS National Weather Service (USA) • Network Weather Service • New York - Pier 11 - Wall Street (USA-NY) (location identifier) • Nimbus Weather Satellite • Normal Water Surface • Nose Wheel Steering • Novell Web Server • Nuclear Weapon State

NWSA Nose Wheel Steering Actuator • Nose Wheel Steering Amplifier

NWSC National Weather Service Center

NWSFO National Weather Service Forecast Office

NWSMC National Weather Service Modernization Committee (USA)

NWSRFS National Weather Service River Forecast System

NWSS National Weather Satellite System

n. wt. Net Weight

NWT Non-Watertight • Normal Wideband Terminal • Northwest Territories (Canada) • Nowata (Papua New Guinea) (location identifier)

NWTC National Wind Tunnel Complex

NWTRB Nuclear Waste Technical Review Board (USA)

NWU Bermuda (location identifier) • Nebraska Wesleyan University (USA-NE) • Northwestern University (USA-IL) • North - West - Up

NWV Numerische Wettervorhersage

NWW Nose Wheel Well

NWWA National Water Well Association

NWWS NOAA Weather Wire Service

NX Normal Crossover • Normal to X-axis • Not Exceeding • Not Expendable

NXB Non-X-ray Background

nXD Non X-ray Dominant

NXLG Narrow-emission-line X-ray-Luminous Galaxy

NXX Willow Grove (USA-PA) (location identifier)

NY New Year • New York (USA) • Normal to Y-axis • Number of Years

NYA National Youth Administration • Not Yet Answered

NYAA National Young Astronomer Award (AL) • North York Astronomical Association (Canada-ON)

NYAM New York Academy of Medicine (USA-NY)

NYAS New York Academy of Sciences (USA-NY)

NYASA New York Academy of Sciences Annals (periodical code)

NYC New York (USA-NY) (location identifier) • New York City (USA-NY)

NYCRL New York Comprehensive Research Libraries (USA-NY)

NYD Not Yet Diagnosed

NYE Nyeri (Kenya) (location identifier)

NYG Quantico (USA-VA) (location identifier)

NYHoS New York Hall of Science (USA-NY)

NYHS New York Hall of Science (USA-NY)

NYI Not Yet Impacted • Sunyani (Ghana) (location identifier)

NYK Nanyuki (Kenya) (location identifier)

Nyl Nylon

NYM Anonymous • Nadym (Russia) (location identifier)

NYN Nyngan (Australia-NSW) (location identifier)

NYNEX New York - New England Exchange

NYO Nyköping (Sweden) (location identifier)

NYP Not Yet Published

NYPL New York Public Library

NYPTC New York Patent, Trademark and Copyright Inc. (USA-NY)

NYR Not Yet Reported • Not Yet Required • Not Yet Returned

NYS New York State (USA)

NYSCATE New York State Committee for the Advancement of Technology Education (USA-NY)

NYSDOL New York State Department Of Labor

NYSE New York Stock Exchange (USA-NY)

NYSERNet New York State Educational and Research Network

NYSTEC New York State Technology Enterprise Corporation (USA-NY)

NYT New York Times (periodical code)

NYU New York University (USA-NY) • Nyaung-u (Myanmar) (location identifier)

nz New Zealand

NZ Neutrality Zone • Neutral Zone • New Zealand (international vehicle registration code) • New Zealand (ISO country code) • Normal Acceleration • Northern Zodiacal • Nouvelle-Zélande (code ISO de pays)

NZA Niobium Zinc Alloy

NZAP New Zealand Antarctic Programme • New Zealand Associated Press

NZATE New Zealand Association for the Teaching of English

NZB Non Zero Binary

NZBC New Zealand Broadcasting Corporation

NZBN New Zealand Bibliographic Network

NZC Jacksonville Cecil (USA-FL) (location identifier)

NZCS New Zealand Computer Society

NZCSICOP New Zealand Committee for the Scientific Investigation of Claims of the Paranormal

NZD Dollar (New Zealand) (ISO currency code)

NZDF Non-Zero Dispersion Fiber

NZDL New Zealand Digital Library

NZDS New Zealand Daylight Saving (time) • Non-Zero Dispersion Shifted

NZDSF Non-Zero Dispersion Shifted Fiber

NZDSIR New Zealand Department of Scientific and Industrial Research

NZDT New Zealand Daylight Time

NZE Nzerekore (Guinea) (location identifier)

N. Zeal. New Zealand

NZF Near Zero Field

NZG Near Zero Gravity

NZHRM Noord- en Zuidhollandse Reddingsmaatschappij (Netherlands)

NZIIR New Zealand Institute for Industrial Research

NZIP New Zealand Institute of Physics

NZJS New Zealand Journal of Science (periodical code)

NZL New Zealand (ISO country code) • Nouvelle-Zélande (code ISO de pays)

NZMG New Zealand Map Grid

NZMS New Zealand Mathematical Society • New Zealand Meteorological Service

NZN Niedersächsischer Zeitschriftennachweis

NZO Nzoia (Kenya) (location identifier)

NZOI New Zealand Oceanographic Institute

NZP National Zoological Park

NZPA New Zealand Press Agency • New Zealand Publishers' Association

NZSA New Zealand Spaceflight Association

NZSI New Zealand Standards Institute (now SANZ)

NZSLO New Zealand Scientific Liaison Office

NZST New Zealand Standard Time

NZV Netzzugangsverordnung

NZW South Weymouth (USA-MA) (location identifier)

NZY San Diego - North Island (USA-CA) (location identifier)

NZZ Neue Zürcher Zeitung

N1NF Non-First Normal Form

N2 Nitrogen

N2HO4 Nitrogen Peroxide

N2H2 Hydrazine

N2H4 Hydrazine

N2O Nitrogen Oxide • Nitrous Oxide

N2O4 Nitrogen Tetroxide

N2O5 Nitric Anhydride • Nitrogen pentoxide

N30 Normal System Epoch 1930

N° Numéro

O

o/ Orden

o Obra • Occasional • Octarius • Octavo • Off • Old • Only • Optimus • Orden • Order • Ortho • ostlich • Over • Overcast

O/ Orden

O Eleven (Roman numeral) • Oberösterreich (Austria) • Object • Observer • Ocean • Octarius • Octavo • October • Oculus • Oeste • Office • Officer • Oficina • Ohio (USA) • Old • Ontario (Canada) • Oost • Open • Operational • Operator • Orden • Order • Order of Magnitude • Ordinary • Ordonnance • Oregon (USA) • Organization • Oscar (international alphabet) • Ost • Osten • Ostlich • Ostrov • Ottava • Ouest • Output • Ouvert • Over • Ovest • Owner • Oxydizer • Oxygen • Quantum Number of Orbital Angular Momentum due to Molecular Rotation • Sauerstoff • Spectral Type • Zero • 11 (Roman numeral)

o/a On Account • On or About

oä oben ähnliches

oa oben aufgeführt • oben angegeben • Onder Andere • Onder Anderen • Overall

O/a On Account of

O/A On Account of • On or About • Operations/Administration • Orbit/Attitude • Overall

O&A Offer & Acceptance • Operations & Analysis • Over & Above

OA Objective Analysis • Objective Area • Obligation Authority • Observatoire de l'Aigle (Canada-QC) • Observing Assistant • Off-Axis • Office Address • Office Automation • Office of Administration • Office of Aeronautics • Office of Applications • Office of the Administrator • Officier d'Académie • Offline Analysis • Old Account • Omnidirectional Antenna • Onboard Acquisition • Onder Andere • On or About • Open Account • Open Architecture • Operating Agency • Operating Assembly • Operating Authorization • Operational Aft • Operational Analysis • Operational Architecture • Operational Archive • Operational Area • Operational Assessment • Operational Availability • Operationally Available • Operations Adviser • Operations Analysis • Operations Analyst • Operations Area • Operations Astronomer • Optical Afterglow • Optical Amplifier • Optique Adaptative • Opto-Acoustic • Orbit Adjust • Orbital Accuracy • Orbital Assembly • Orbit Analyst • Orbiter Assembly • Other Articles • Output Amplitude • Output Axis • Overall

ØAA Østjyske Amateur Astronomers (Denmark)

OAA Observatório Astronômico Antares (Brazil-BA) • Oglethorpe Astronomical Association (USA-GA) • Ontario Association of Archivists (Canada-ON) • Open Agent Architecture • Optical Acquisition Aid • Orbiter Access Arm • Orbiter Alternate Airfield • Organisation pour l'Alimentation et l'Agriculture • Orient Airlines Association • Osservatorio Astronomico di Arcetri (Italy)

oab Outlook Address Book (file extension)

ÖAB Öffentliche Allgemeinbildende Bibliothek

OAB Ocean Affairs Board • One-to-All Broadcast • Operational Advisory Broadcast • Ordnance and Assembly Building • Osservatorio Astronomico di Brera (Italy)

OABNN Orthogonal Activation Based Neural Network

oac Obra Antes Citada

OAC Oakland Astronomy Club (USA-MI) • Objective Assessment Chart • Observatoire de la Côte d'Azur (France) • Observatorio Astronómico de Córdoba (Argentina-CO) • Observatory Advisory Committee • Office of Academic Computing • Oficina de Atención al Ciudadano • On-Line Analysis Computer • On-orbit Activation and Calibration • Ontario Academic Credit • Operating Agency Code • Operations Analysis Center • Operator Access Console • Optimal Automatic Control • Osservatorio Astrofisico di Catania (Italy) • Osservatorio Astronomico di Cagliari (Italy) • Osservatorio Astronomico di Capodimonte (Italy) • Outer Arm Cloud

OACC Oceanic Area Control Centre

OACCB Office Automation Computer Coordination Board • Office Automation Configuration Control Board

OACD Observatoire de l'Association Culturelle de Dax (France)

OACDF Osservatorio di Capodimonte Deep Field

OACH Office of the Assistant Chief Hydrologist

OACI Organisation de l'Aviation Civile Internationale • Organización de Aviación Civil Internacional

OACIS Over and Above Centralized Information System

OACL Osservatorio Astronomico Colle Leone (Italy)

OACS Observatorio Astronómico Centroamericano de Suyapa (Honduras) • Optical Absorption Coefficient Spectroscopy

OACT Office of Advanced Concepts and Technology • Organisation Africaine de Cartographie et Télédétection • Osservatorio Astronomico di Colluriana-Teramo (Italy)

OACUL Ontario Association of College and University Libraries (Canada-ON)

OAD Object Management and Definitions • Open Architecture Driver • Operational Availability Data • Operational Availability Date • Orbit and Attitude Division • Orbit Attitude Division • Orbiter Atmospheric Drag (experiment) (PV) • Overall Depth

OADCES On-Line Aerotriangulation Data Collection and Editing System

OADM Optical Add/Drop Multiplexer

OADR Originating Agency's Determination Required

OADSS Office Assignment Decision Support System

ÖAdW Österreichische Akademie der Wissenschaften (Austria)

OAE Operational Area Evaluation • Optical Alignment Equipment • Organisatie van Afrikaanse Eenheid

OAEM Observatorio Astronômico da Escuela de Minas (Brazil-MG)

OAET Office of Aeronautics, Exploration and Technology

OAF Open Architecture Framework • Optical Alignment Facility • Optimum Array Filter

OAFA Observatorio Astronómico Felix Aguilar (Argentina-SJ)

OAFD Orbiter Aft Flight Deck

OAFTO Orbiter Atmospheric Flight Test Office

OAG Off-Axis Guider • Open Applications Group • Operational Advisory Group • Orange Springhill (Australia-NSW) (location identifier) • Osservatorio Astronomico di Genova (Italy)

OAGH Observatorio Astrofísico Guillermo Haro (Mexico)

OAGM Off-Axis Generating Machine • Österreichische Arbeitsgemeinschaft für Mustererkennung (Austria)

OAH Organization of American Historians • Overall Height

OAHE Observatório Astronômico Herschel-Einstein (Brazil)

OAHTS Orbiter Avionics Hardware Tracking System

OAI Office of Automation Integration • Oficina de Analisis de Información • Ohio Aerospace Institute (USA-OH) • OK As Is • Open Applications Interface • Open Archives Initiative

OAIA Organisation des Agences d'Information d'Asie

OAII Ocean-Atmosphere-Ice Interaction

OAIS Open Archival Information System

OAJ Jacksonville (USA-NC) (location identifier)

OAJCG Operation Alliance Joint Control Group

OAK Oakland (USA-CA) (location identifier) • Object Application Kernel

OAL Cacoal (Brazil-TO) (location identifier) • Observatoire Astronomique de Lyon (France) • Observatório Astronômico de Lisboa (Portugal) • ODF Access Layer • Office of Arts and Libraries (UK) • Open Access Layer • Operations Approval Letter • Overall Level

OALA Ohio Academic Libraries Association (USA-OH)

OALC Ogden Air Logistics Center (USA-UT)

OALD Oxford Advanced Learner's Dictionary

OALM Observatorio Astronomico "Los Molinos" (Uruguay)

OALP Observatorio Astronómico de La Plata (Argentina-BA)

OALT Ontario Association of Library Technicians • Operational Acceptable Level of Traffic • Operational Acceptance Level of Traffic

o. alt. hor. Omnibus Alternis Horis

OA&M Operations, Administration & Maintenance

OAM Oamaru (New Zealand) (location identifier) • Observatoire des Alpes Maritimes (France) • Observatorio Astronómico de Mallorca (Baleares) • Observatorio Astronómico Móvil (Argentina) • Observatório Astronômico Monoceros (Brazil-MG) • Oceanic Angular Momentum • Office of Administration and Management • Office of Aerospace Medicine • Office of Alternative Medicine (USA) • Office of Automation and Manpower • Operations - Administration - Maintenance • Operations - Administration - Management • Operations And Maintenance • Operator Assistance Menu • Oracle Access Manager • Orbit Adjust Module • Orbital Assembly Module • Orbiting Adaptation Method

OAMC Operations, Administration and Maintenance Centre • Optique Adaptative Multiconjugée

OAMCE Optical Alignment, Monitoring and Calibration Equipment

OAME Orbital Attitude and Maneuvering Electronics

OAMEX Ocean-Atmosphere Materials Exchange

OAMF Observatorio Astronómico Municipal de Funes (Argentina)

OAML Oceanographic and Atmospheric Master Library

OAMM Observatoire Astronomique du Mont Mégantic (Canada-QC) • Observatorio Astronómico Municipal de Mercedes (Argentina) • Office of Acquisition and Material Management

OAM&P Operations, Administration, Maintenance & Provisioning

OAMP Operations, Administration, Maintenance and Provisioning • Optical Airborne Measurements Program • Optical Analog Matrix Processing

OAMR Observatorio Astronómico Municipal de Rosario (Argentina-SF)

OAMS Onboard Alignment Measurement System • Orbital Attitude and Maneuvering System • Organic and Atmospheric Mass Spectrometer

OAN Observatório Astronômico Nacional • Observatorio Astronómico Nacional • Olanchito (Honduras) (location identifier) • Optical Access Network

OANA Organization of Asian News Agencies

OAO Off And On • Office of Aircraft Operations (NOAA) • Okayama Astrophysical Observatory (Japan) • One And Only • Orbital Astronomical Observatory • Orbited Assembly Operation • Orbiting Astronomical Observatory • Over And Out

OAOG Ottawa Valley Astronomy and Observers Group (Canada-ON)

OAOI On And Off Instrument

OAP Observatoire Astronomique des Pises (France) • Observatorio Astronómico de Patacamaya (Bolivia) • Observatório Astronômico de Piedade (Brazil-MG) • Odessa Astronomical Publications (periodical code) • Off-Axis Paraboloid • Offset Aiming Point • Old-Age Pension • Old-Age Pensioner • Orbit Activity Plan • Osservatorio Astronomico di Padova (Italy) • Osservatorio Astronomico di Palermo (Italy) • Outline Acquisition Plan • Oxygen at Atmospheric Pressure

OAPEC Organization of Arab Petroleum-Exporting Countries

OAPI OPUS Applications Programming Interface • Organisation Africaine de la Propriété Intellectuelle

OAPM Optimal Amplitude and Phase Modulation

OAPN Optimal Augmented Proportional Navigation

OAPS Orbit Adjust Propulsion System

OAQ Observatorio Astronómico de Quito (Ecuador)

OAQPS Office of Air Quality Planning and Standards

OAQPSTTN Office of Air Quality Planning and Standards Technology Transfer Network

OAR Monterey Fritzsche (USA-CA) (location identifier) • Oceanic and Atmospheric Research • Off-Axis Rejection • Office of Aerospace Research • Office of Air and Radiation • Office of Analysis and Review • Office of Atmospheric Research • Open Air Range • Open Architecture for Reasoning • Operational Analysis and Research • Operations Analysis Report • Optical Amplified Receiver • Optical Automatic Ranging • Osservatorio Astronomico di Roma (Italy) • Overtime Authorization Request

OARE Orbital Acceleration Research Experiment (*) • Orbiter Acceleration Research Experiment

OARL Ontario Academic Research Libraries (Canada-ON)

OARM Office of Administration and Resources Management

OARME Oxygen Atom Reaction Materials Experiment

OARNET Ohio Academic Resources Network (USA-OH)

OARS On-Line Automated Reference Service • Ontario Association for Remote Sensing • Open Architecture Retrieval System • Orbiter Avionics Radiation Specification

OARSA Optical Aberration Reduction and System Analysis

ÖAS Östergötlands Astronomiska Sällskap (Sweden)

OAS Observatoire Astronomique de Strasbourg (France) • Office Automation System • Office of Aerospace Study • Ogden Astronomical Society (USA-UT) • Oklahoma Academy of Sciences (USA-OK) • Olympic Astronomical Society (USA-WA) • Omaha Astronomical Society (USA-NE) • On Active Service • One-to-All Scatter • Open Accounting System • Operations Automation System • Optical Absorption Spectroscopy • Optical Access Switch • Optical Alignment Sight • Optical Amplifier Section • Optical Assembly Simulator • Optically-Activated Semiconductor • Orbit Adjustment System • Orbiter Aeroflight Simulator • Orbiter Atmospheric Simulator • Orbiter Atmospheric Sounder • Orbiter Avionics Simulator • Orbiter Avionics System • Organisation de l'Armée Secrète • Organization of American States • Orpington Astronomical Society (UK)

OASC Office Automation Support Center

OASCB Orbiter Avionics Software Control Board

OASD Office of the Assistant Secretary of Defense

OASES Outpost for Analysis and Spectroscopy of Extrasolar Systems

OASF Office Automation System Facility • Orbiting Astronomical Support Facility

OASI Old Age and Survivors Disability Insurance • Orwell Astronomical Society Ipswich (UK) • Osservatorio Antarctica Submillimetrico e Infrarosso

OASIS Oceanic Analysis and Surveying Information System • Oceanic and Atmospheric Satellite Imaging System • Oceanic and Atmospheric Scientific Information System • OEX Autonomous Supporting Instrumentation System • Okayama Astrophysical System for Infrared (imaging and) Spectroscopy • Online Archive Science Information Services • Online Archive Science Information System • Open Access Strategic Information System • Open Access Symbiotic Information System • Open Architecture Scientific Information System • Open Architecture for Simulation Systems • Operational Airport Safety Information System • Operational Analysis Strategic Interaction Simulator • Operational Automatic Scheduling Information System • Operations And Science Instrument Support • Operations And Science Instrument System • Operations Audit Sunnary Information System • Optically Adaptive System for Imaging Spectrography • Optically Adaptive System for Imaging Spectroscopy • Orbit Avionics Software Integration Study • Orbiter Autonomous Supporting Instrumentation System • Organization for the Advancement of Space Industrialization and Settlement • Organization for the Advancement of Structured Information Standards • Outil d'Aide à la Spécification des Informations-Systèmes

OAsk Ortsanschlußkabel

OAsl Ortsanschlußleitung

OASM Office of Aerospace Medicine • Operational Analysis Support Model

OASP On-Array Signal Processor

OASPL Overall Sound Pressure Level

OASSIS Operations And Support for Science Instrument Systems

OASSO Operational Applications of Satellite Snowcover Operations

OAST Office of Aeronautics and Space Technology

OASV Orbital Assembly Support Vehicle

OASYS Obstacle Avoidance System • Office Automation System • Operability Assessment System

OAT Obligation Assimilable du Trésor • Ocean Acoustic Tomography • Office Advanced Technology • Operating Acceptance Test • Operating Ambient Temperature • Operational Acceptance Test • Operational Acceptance Testing • Optical Adaptive Technique • Optically-Amplified Transmitter • Optometry Admission Test • Original Alias Table • Osservatorio Astronomico di Trieste (Italy) • Outside Air Temperature • Overall Test

OATA Optical Acquisition and Tracking Aid

OATH Optical Athermalization

OATM Orbiter Antenna Test Model

OATo Osservatorio Astronomico di Torino (Italy)

OATP Operational Acceptance Test Procedure

OATS Office Automation Technology and Services • On-Aircraft Test Software • Optical Attenuation Test Set • Orbit Adjust Thrust System • Orbiter Avionics Tracking System • Organización Andina de Telecomunicaciones por Satélite • Original Article Tear Sheet • Original Article Text Service • Overall Test Set

OATUU Organization of African Trade Union Unity

OATV Open Architecture Television

OAU Observatorio Astronômico de Uberlândia (Brazil) • Optical Alignment Unit • Organization of African Unity (now AU)

OAUV Observatorio Astronomico de la Universidad de Valencia (Spain)

OAV Operational Aerospace Vehicle

ÖAW Österreichische Akademie der Wissenschaften (Austria)

OAW Off-Airway Weather • On Another World • Optically-Assisted Winchester

OAWF Objective Acquisition Workforce

OAWMN Österreichische Akademie Wissenschaften - Mathematisch-Naturwissenschaftliche Klasse Sitzungsberichte (periodical code)

OAWP Operations Analysis Working Paper

OAX Oaxaca (Mexico) • Oaxaca Xoxocotlan (Mexico) (location identifier)

OAXTC Ocean-Atmosphere Exchange of Trace Compounds

ob ober • Obiit • Obiter • Obra • Observation • Obsolete

oB Ohne Befund

Ob Ober • Obligatory

O-B Octal to Binary

O/B Octal to Binary • On Board • Operational Base • Ordered Back • Outboard

O&B Out & Back

OB Oberbegriff • Observation Block • Observed Bearing • Observing Block • Off Broadway • Official Board • Onboard • Openbare Bibliothek • Opening of Books • Operating Budget • Operational Base • Optical Bench • Orange Belly • Ordered Back • Ordinary Building • Ordinary Business • Oscillator Bath • Outband • Outboard • Outlet Busy • Output Buffer • Outside Broadcast • Outside Broadcasting • Outside Broadcasting • Outward Bound • Peru (international aircraft registration code)

OBA Oban (Australia-QLD) (location identifier) • Object Behavior Analysis • Oferta de Bucle de Abonado • Olimpiada Brasileira de Astronomia (Brazil) • Omnibase Administrator • Onboard Allowance • Operating Budget Authority • Optical Base Assembly • Optical Bench Assembly • Outer-Belt Asteroid • Outer Buffer Assembly • Oxygen Breathing Apparatus

OBAC Organization-Based Access Control

OBAD Operating Budget Authority Document

OBAFGKMRNS O Be A Fine Girl, Kiss Me Right Now Sweatheart (spectral-type mnemonic sequence) (*) • Oh Backward Astronomer, Forget Geocentricity: Kepler's Motions Reveal Nature's Simplicity (spectral-type mnemonic sequence)

OBAN Onboard Autonomous Navigation • Operating Budget Account Number

obb Obbligato

ÖBB Österreichische Bundesbahnen (Austria)

OBC Obock (Djibouti) (location identifier) • Officer Basic Course • Onboard Checkout • Onboard Computer • Optical Bar Camera

OBCE Onboard Checkout Equipment • Operational Baseline Cost Estimate

OBCCS Onboard Computer Calibration Source

OBCM Object-Based Conceptual Modeling

OBCO Onboard Checkout

OBCP Onboard Computer Program

OBCS Onboard Calibration Source • Onboard Checkout System

OBCTE Onboard Computer Test Equipment

obd Microsoft Office Binder (file extension)

OBD Obano (Indonesia) (location identifier) • Onboard Diagnostics • Optimal Brain Damage • Overall Block Diagram

OBDD Oblivious Binary Decision Diagram • Ordered Binary Decision Diagram • Ordered Boolean Decision Design

OBDH Onboard Data Handler • Onboard Data Handling

OBDHTE Onboard Data Handling Test Equipment

OBDI Onboard Decoder Interface

OBDM Onboard Data Management

OBDP On-Board Data Processing • On-Board Data Processor

OBDS Onboard Decoder Simulator

Obdt Obedient

OBDU Onboard Diagnostic Unit

OBE Observation Equipment • Okeechobee County (USA-FL) (location identifier) • Onboard Equipment • Operating Basis Earthquake • Operating Basis Event • Optical Bench Enclosure • Outcome-Based Education • Out-of-Body Experience • Overcome By Events • Overtaken By Events

OBECO Outboard Engine Cutoff

Oberbay Oberbayern (Germany)

Oberfr Oberfranken (Germany)

Oberpf Oberpfalz (Germany)

OBES Orthonormal Basis of an Error Space

OBETC Optical Bench Enclosure Thermal Control

OBEX Object Exchange

OBF Oberpfaffenhofen (Germany) (location identifier)

OBFCS Onboard Fire Control System

OBFS Offshore Bulk Fuel System

OBG Observatoire du Bénévolat et de la Générosité (France)

OBH Ocean Bottom Hydrophone (survey)

OBI Obidos (Brazil-PA) (location identifier) • Observation Interval • Omnibearing Indicator • On-Line Book Initiative • Open Business Interchange • Open Buying on the Internet

OBIC Optical Beam Induced Current

OBIDS Onboard Integrated Diagnostics System

OBIGGS Onboard Inert-Gas-Generating System

OBIGS Onboard Inert-Gas-generating System

OBIOS Open Basic Input/Output System

OBIPS Optical Bench Imager and Photometer System

OBIS Object-Based Information System

Obit Obituary

obj Object (file extension)

Obj Object • Objection • Objective • Objekt

OBJ Object • Objective

Objv Objective

OBK Northbrook (USA-IL) (location identifier)

Obl Obligated • Obligation • Oblige • Oblique • Oblong

OBL Obligated • Ocean Bill of Lading • Operating Budget Ledger • Operational Base Launch • Operations Base Launch • Outer Bipolar Lobe

Oblg Obligate

OBLSS Operational Base Launch Safety System

OBM Morobe (Papua New Guinea) (location identifier) • Object Manager

OBMC Outbound Midcourse Correction

OBMM Onboard Mission Manager

OBMS Onboard Management System

OBMU Onboard Management Unit

OBN Oban Connel (UK-Scotland) (location identifier) • Openbaar Bibliotheek Netwerk (Netherlands)

OBNI Objet Brilliant Non Identifié

OBO Obihiro (Japan) (location identifier) • O'Brien Observatory (USA) • Official Business Only • Or Best Offer • Output Back-Off

OBOGS Onboard Oxygen Generating System

OBP Observation Package • Odorant Binding Protein • Onboard Processing • Onboard Processor

OBPC Optical Bar Panoramic Camera

OBPDC Object-Based Parallel and Distributed Computation (workshop)

OBPR Office of Biological and Physical Research (NASA)

OBRA Omnibus Budget Reconciliation Act (USA)

OBRC Operating Budget Review Committee

Obre Octobre

OBRG Onboard Range Compression

OBRM Observatorio del Roque de los Muchachos (Canary Islands)

OBRP Object Representation • Onboard Repair Part

OBRQ Observation Request

Obs Obscure • Observações • Observateur • Observation • Observatoire • Observatorio • Observatório • Observatory • The Observatory (periodical code) • Observe • Observed • Observer • Observing • Obsolete • The Observatory (periodical code)

OBS Aubenas Vals-Lanas (France) (location identifier) • Observation • Observatorium Bad Salzschlirf (Germany) • Observatory • Observatory Scientist • Observe • Observer • Ocean Bottom Seismometer • Omnibearing Selector • Onboard Software • Operational Bioinstrumentation System • Operational Biomedical System • Optimal Brain Surgery • Organic Brain Syndrome • Organizational Breakdown Structure • Oscillating Blue Straggler (star) • Out-of-Band Signaling

Obsc Obscure • Obscured

OBSC Obscure • Obscured

OBSCD Obscured

Obser Observation pi (periodical code)

Observ Observateur • Observatoire • Observatorio • Observatório • Observatory • Observe • Observed • Observer • Observing

OBSET Observations Set

OBSM Onboard Software Maintenance

Obsn Observation

Obsol Obsolete

obsp Obiit Sine Prole

OBSS One-Bundle Sliding Shelf

Obst Obstacle • Obstruction

OBST Obstacle

Obstn Obstruction

Obsv Observation

OBSW Onboard Software

Obt Obedient • Obtained

OBT Oakland-Coliseum (USA-MD) (location identifier) • Observation de la Terre • Onboard Time • Onboard Training • Organically-Bound Tritium

Obtd Obtained

OBTR Observation Block Tracking and Reporting

OBU Kobuk (USA-AK) (location identifier) • Offshore Banking Unit • Oklahoma Baptist University (USA-OK) • Optical Bench Unit

Obv Observe • Observed

OBV Observation vehicle • Oxidizer Bleed Valve

ObVyS Obzory Vysokotemperaturnoi Sverkhprovodimosti (periodical code)

OBX Obo (Papua New Guinea) (location identifier)

OBY Operating Budget Year • Scoresbysund (Greenland) (location identifier)

o/c Overcharge

oc Obra Citada • Occitan (ISO language code) • Ocean • On Center • Opere Citato

Oc Ocean

O/C Observer/Collector • Observer/Controller • On Completion • Operations/Critical • Overcharge

O-C Observed-Computed

O&C Operations & Checkout • Operations & Control

OC Object Calculus • Object Calibration • Object Class • Object Classification • Observational Catalogue • Observatory Council • Observer Computer • Observer/Controller • Oceania • October • Octobre • Of Course • Office Copy • Office of the Comptroller • Officer Candidate • Officer Commanding • Official Classification • Official Collaborator • On Center • On Condition • On Course • Ondes Courtes • Oort Cloud • Oort Cluster • Open Circuit • Open Cluster • Open Collector • Open Connect • Open Contract • Operating Center • Operating Characteristic • Operating Curve • Operational Capability • Operational Check • Operational Computer • Operational Concept • Operational Condition • Operations Center • Operations-Central • Operations Communications • Operations Control • Operations Coordinator • Operator Centralization • Operator Centre • Opere Citato • Opposite Circular • Optical Carrier • Optics Control • Optics Controller • Optimized Configuration • Oral Contribution • Orbital Check • Ordinary Chondrite • Original Cover • Outer Cluster • Outflow Connected • Output Computer • Output Controller • Outside Circumference • Over Center • Overcontact (binary star) • Overcurrent

OCA Obligation Cost Authority • Observatoire de la Côte d'Azur (France) • Observatorio Cerro Armazones (Chile) • Obstacle Clearance Altitude • Ocean and Coastal Areas • Oceanic Control Area • Ocean Reef (USA-FL) (location identifier) • Office of Consumer Advocate (USA) • Open Communications Architecture • Operating Cost Authority • Operational Contracting Authority • Operational Control Authority • Optical Corporation of America (USA) • Orbiter Capability Analysis • Orbiter Capability Assessment • Organisme Collecteur Agréé • Original Classification Authority • Oxygen Control Assembly

OCAAS Observatory Control and Astronomical Analysis System

OCAC Operations Control and Analysis Center • Orbiter Cabin Air Cleaner

OCAL On-Line Cryptanalytic Aid Language

OCAM Optimal Control Applications and Methods (periodical code) • Organisation de la Communauté Africaine et Malgache

OCAN Observatoire de la Côte d'Azur, Nice (France)

OCARINA Ocean, Atmosphere, Research and Investigation with Acoustics

OCAS On-Line Cryptanalytic Aid System • Organization of Central American States

OCAST Oklahoma Center for the Advancement of Science and Technology (USA-OK)

OCAT Observation Catalog • Observing Catalog

OCB Office Central des Bibliothèques (France) • Operational Configuration Baseline • Operations Control Board • Operations Coordinating Board • Outgoing Call Blocked • Output Current Booster • Override Control Bits

OCBB Operating Cost-Based Budget

Occ Occasion • Occasional • Occasionally • Occident • Occidental • Occidente • Occupation • Occupé • Occupied • Occupy • Occur • Occurred • Occurrence

OCC Coca (Ecuador) (location identifier) • Observation Control Center • Occultation • Office of Contract Compliance • Open Cooperative Computing • Operational Computer Complex • Operational Control Center • Operational Control Console • Operations and Control Centre • Operations Control Centre • Operations Control Console • Operator Control Command • Optical Crossconnect • Optically Coupled Common • Organic Carbon Cycle • Originating Center Control • Other Common Carrier

OCCA Ocean Cargo Clearance Authority • Open Cooperative Computing Architecture

OCCAM Ocean Circulation Climate Advanced Modeling

Occas Occasion • Occasional • Occasionally

OCCB Operational Configuration Control Board

OCCC Operations and Communications Control Center

OCCE Office Central de Coopération à l'École (France)

OCCG Operations Center Coordinator Group

OCCI Optical Coincidence Coordinate Indexing

OCCIS Operations Command and Control Information System

Occit Occitan • Occitano

OCCL On-Line Computer Library Center

OCCM On-Condition Cyclic Maintenance

Occn Occasion

OCCP Orbit Command Computer Program

OCCR Organization Cost Center Record

OCCS Operational Command and Control System • Operator Control Console System

OCCULT Orser Complete Conversational User-Language Translator

OCD Obsessive Compulsive Disorder • Ocean Chemistry Division • Octal Communications Device • On-Line Communications Driver • Operational Capabilities Document • Operational Capability Date • Operational Capability Development • Operational Concept Demonstration • Operational Concept Description • Operational Concept Document • Operations Concept Document • Orderwire Clock Distributor • Out-of-Cell Delineation

OCDE Organisation de Coopération et de Développement Économiques (*) • Organisation pour la Coopération et le Développement Économiques • Organización para la Cooperación y el Desarollo Económico

OCDF Operations Control and Display Facility

OCDM Office of Civil Defense and Mobilization • Open Cold Dark Matter (model) • Optical Code Division Multiplexing

OCDR Orbiter Critical Design Review

OCDU Optics Coupling Data Unit • Optics Coupling Display Unit

OCE Ocean • Ocean City (USA-MD) (location identifier) • Ocean Color Experiment • Office of the Chief Engineer • Ontario Center of Excellence (Canada-ON) • Open Collaboration Environment • Open Collaborative Environment (*) • Open Computing Environment • Open Continuing Education • Optical Control Electronics • Oscillating Current Element • Other Common Equipment

OCEA Office de Culture et d'Éducation Audiovisuelle

Ocean Oceanus Summer (periodical code)

OCEAN Ocean Colour Environment Archive Network • Ocean Colour European Archive Network • Open Corrosion Expertise Accesses Network

OCEANIC Ocean Network Information Center

Oceanogr Oceanografía

OcEng Ocean Engineering (periodical code)

OCenW Onderwijs - Cultuur - Wetenschappen

OCEP Office Central d'Édition et de Publicité (France)

OCEPA Office of Communications, Education and Public Affairs

OCETA Ontario Centre for Technology Advancement (Canada-ON)

OCF Object Component Framework • Object Correlation File • Object Correlation Frame • Observer Comments File • Ocala Taylor (USA-FL) (location identifier) • Onboard Computational Facility • Open Channel Flow • Operational Check Flight • Optical Coating Facility • Orbiter Computational Facilities

OCFNT Occluded Front

OCG Ocean Catalog Group • Operations Control Group • Optimal Code Generation

ÖCG Österreichischen Computergesellschaft (Austria)

OCGA Office of Congressional and Government Affairs

OCGF Organic Crystal Growth Facility

OCH Nacogdoches (USA-TX) (location identifier) • Obstacle Clearance Height • Office of the Chief Hydrologist • Orbiter Common Hardware

OCHA Office for the Coordination of Humanitarian Affairs (UNO) • Office of Coordination for the Humanitarian Assistance (UNO)

OCI Object Code Insertion • Ocean Color Imager • Oceanic (USA-AK) (location identifier) • Office of Complaint Investigation • Office of Computer Information • Office of Corollary Interest • Oficina Central de Información • Open Communications Infrastructure • Optical Coherence Imaging • Organic Chemistry Institute • Organizational Conflict of Interest • Out of City Indicator • Ouvrages - Cartes - Instruments

OCIA Onizuka Center for International Astronomy (USA-HI)

OCIAM Oxford Centre for Industrial and Applied Mathematics (UK-England)

OCIE Organizational Clothing and Individual Equipment

OCIF Out Card In File

OCIMF Oil Companies International Maritime Forum

OCIP Ottawa Carleton Institute of physics (Canada-ON)

OCIR Optical Coincidence Information Retrieval

OCIS Organized Crime Information System

OCIW Observatories of the Carnegie Institution of Washington

OCJ Ocho Rios (Jamaica) (location identifier)

OCK Optics Cleaning Kit

OCl Open Cluster

OCL Object Connecting Language • Object-Constraint Language (*) • Obstacle Clearance Limit • Occlude • Open Cluster • Operational Check List • Operational Control Level • Operation Control Language • Operations Checklist • Operator Control Language

OCLC Ohio College Library Center (USA-OH) • On-Line Computer Library Catalog • On-Line Computer Library Center

OCld Oil-Cooled

Ocle Oncle

OCLD Occlude

OCLI Optical Coating Laboratory Inc. (USA-CA)

OCLIP Optical Cellular Logic Image Processor

OCLN Occlusion

OCLSC Ontario Centre for Large-Scale Computations (Canada-ON)

OCM Object Class Model • Ocean Color Mission • Ocean Color Monitor • Ocean Color Monitoring • Office de Contrôle des Mutualités (Belgium) • On-Channel Map • On-Condition Maintenance • Optical Countermeasure • Orbit Correction Mode • Orbiter Celestial Mechanics (experiment) (PV) • Organic Content Monitor • Organisation Commune des Marchés (EU) • Organisation du Commerce Mondial • Organización Común de Mercados (EU) • Organization Computer Manager • Outgoing Communications Manager

OCMG Oxygen Controlled Melt Growth

OCMR Ontario Centre for Materials Research (Canada-ON)

OCMS Organisme Certificateur des Métiers et des Services (France)

OCMW Openbaar Centrum voor Maatschappelijk Welzijn

OCN Cyanate • Occasion • Oceanside (USA-FL) (location identifier) • Open Computer Network • Operation Completion Notice • Operation Control Number • Order Change Number • Order Control Number • Organization Change Notice

OCNL Occasional • Occasionally

OCNLY Occasionally

OCO Object Code Only • Offload Control Officer • Open - Close - Open • Operational Capability Objective • Operational Checkout • Operational Contracting Officer • Operations Control Office • Operations Control Operator • Optimiseur de Code Objet • Orbiting Carbon Observatory

OCOE Overall Checkout Equipment

OCOG Offset Centre Of Gravity

OCONUS Outside Continental United States

OCOP Outline Contingency Operation Plan

OCP Office of Civilian Personnel • Office of Commercial Programs • One-Component Plasma • Open-Cluster Pattern • Operating Control Procedure • Operational Capability Plan • Operational Checkout Procedure • Operational Computer Program • Operational Configuration Processing • Operational Control Panel • Operational Control Process • Operations Checkout Procedure • Operations Control Plan • Operations Control Procedure • Optical Continuum Peak • Orbital Combustion Process • Orbital Correction Program • Organic Change Proposal • Organic Conducting Polymer • Organizational Concept Plan • Output Control Pulse

OCPA Optically-Controlled Phased-array Antenna • Overseas Chinese Physics Association

OCPC Office of Compensation and Position Classification

OCPI Office Central de la Presse Illustrée (France)

OCPO Office of Civilian Personnel Operations

OCPP Orbital Cloud Photopolarimeter • Orbiter Cloud Photopolarimeter (PV)

OCPREP Object Correlation Preparation

OCPS Orbiter Camera Payload System • Oxygen Cabin Pressurization Section

OCPSF Organic Chemicals, Plastics, and Synthetic Fibers

OCR Observatorio Cristo Rey (Argentina) • Observer Citation Rate • Occur • Ocean Colour Radiometer • Office for Civil Rights • Office of Coordinating Responsibility • Open Cluster Remnant • Operational Capability Readiness • Operational Capability Release • Operational Capability Requirements • Operational Change Report • Operational Change Request • Operational Concept Review • Operational Control Room • Optical Character Reader • Optical Character Recognition (*) • Optical Character Recognizer • Optimal Character Recognition • Optimum Charge Regulator • Organization Change Request • Overcurrent Relay • Overhaul Component Requirement

OCRA Office of Classification and Rates Administration • One-Centimeter Receiver Array • Overseas Companies Registration Agents

OCRE Optical Character Recognition Element • Optical Character Recognition Equipment

OCRIT Optical Character Recognizing Intelligent Terminal

OCRM Orbiter Crash and Rescue Manual

OCRS Ontario Centre for Remote Sensing • Operational Change Reporting System • Optical Character Recognition System

OCRUA Optical Character Recognition Users Association

OCRWM Office of Civilian Radioactive Waste Management (USA)

OCS Carbonyl Sulfide • Observation Control System • Observatory Control System • Ocean Colour Scanner • Office of Cadastral Survey • Office of the Chief Scientist • Officer Candidate School • Onboard Checkout System • On-Card Sequencer • One Chip System • Operational Control Segment • Operational Control System • Operations Control Supervisor • Operations Control System • Operator Control Station • Optical Character Scanner • Optical Character Scanning • Optical Control System • Optimum Class Size • Outer Continental Shelf

OCSE Operational Concept for Security Equipment • Optical Calibration Sphere Experiment • Organizzazione di Cooperazione e Sviluppo Economico

OCSG Open Computing Security Group

Ocsnl Occasional

OCSP On-Line Certificate Status Protocol

Ocst Overcast

OCST Office of Commercial Space Transportation

Oct Octagon • Octans-Octantis (constellation) • Octavo • October • Octobre • Octubre

OC&T Office of Certification & Training

OCT Observation Control Table • Observatório das Ciências e das Tecnologias (Portugal) • Ocean Color and Temperature • Octal • Octave • October • Octobre • Office of Certification and Training • Office of Communications and Technology • Officer Candidate Test • Operational Climatic Test • Operations Control Team • Optical Coherence Tomography • Optimal Control Theory • Orbital Circularization Technique • Overseas Countries and Territories

OCTC Operators Console Transfer Channel

Octl Octal

OCTL Open-Circuited Transmission Line

Octn Octane

OCTOPUS Ocean Colour Techniques for Observation, Processing and • Utilization Systems

OCTS Ocean Color and Temperature Scanner

OCTV Open-Circuit Television

OCTW Optical Communication Through the Shuttle Window

OCU Office Channel Unit • Oklahoma City University (USA-OK) • Operational Capability Update • Operational Control Unit • Operational Conversion Unit • Operations Control Unit • Operator Control Unit • Optical Conversion Unit • Orderwire Control Unit • Ordnance Control Unit • Organización de Consumidores y Usuarios (Spain) • Output Control Unit

OCUFA Ontario Confederation of University Faculty Associations (Canada-ON)

OCUL Ontario Council of University Libraries

OCULA Ontario College and University Library Association (Canada-ON)

OCV Ocaña (Colombia) (location identifier) • Open-Circuit Voltage

OCVCXO Oven Controlled-Voltage Controlled Crystal Oscillator

OCVD Open-Circuit Voltage Decay

OCW Operation Command Word • Washington Warren (USA-NC) (location identifier)

OCWG Ocean Colour Working Group

ocx Object Linking and Embedding (file extension)

OCX OLE Control • OLE Custom Control

OCXO Oven-Controlled Crystal Oscillator

OCYT Oficina de Ciencia y Tecnología

OCZ Outer Convective Zone

OCzAS Ondrejov Czechoslovak Academy Sciences Astronomical Institute Publications (periodical code)

o/d On Demand • Overdose • Overdraft

od Oculus Dexter

O/d On Demand • Overdraft

Od Oder

O&D Organization & Deployment • Origin & Destination

O/D On Demand • On Dock • Outside Diameter • Overdraft

OD Absorbance • Doctor of Optometry • Lebanon (international aircraft registration code) • Object Description • Object Descriptor • Object Draw • Observable Difference • Observational Data • Observatory Director • Observed Drift • Oculus Dexter • Office of the Director • Officer of the Day • Old Dutch • On Demand • On Duty • Operational Detachment • Operational Downlink • Operations Directive • Operations Director • Operations Division • Operations Document • Optical Density • Optical Disk • Orbit Determination • Ordinateur Domestique • Ordnance Department • Organization Development • Original Design • Original Dirac • Origin-Destination • Other Denominations • Outer Detector • Outer Diameter • Output Data • Outside Diameter • Outside Dimensions • Overdoctoral • Overdose • Overdraft • Overdrawn • Oxygen Drain

oda Open Document Architecture (file extension)

ODA Octal Debugging Aide • Office D'Annonces • Office Document Architecture • Official Development Assistance • Omnidirectional Antenna • Open Document Architecture • Operational Data Analysis • Operational Design and Analysis • Optical Data Associates • Optical Diagnostics Assembly • Optical Diffraction Analysis • Original Design Activity • Ouadda (Central African Rep.) (location identifier) • Overseas Development Administration • Overseas Development Agency (UK) • Oxydianiline

ODAC ODA Consortium • Office Départemental d'Action Culturelle (France) • Operations Distribution Administration Center

ODAL Omnidirectional Approach Lighting System

ODAP Officer Development Action Plan

ODAPI Open Database Application Programming Interface

ODAPS Oceanic Display And Planning System • Oceanic Display And Processing Station

ODAR Omnidirectional Airborne Radar

ODAS OCA-DLR Asteroid Survey (now ADAS) • Office décentralisé de l'Action Sociale (France) • Officer Distribution and Assignment System • ONERA-DLR Aerospace Symposium • Ocean Data Acquisition System • Ocean Data Station • Oceanic Data Assimilation System • Optical Disk Archive System

OdB Oeil-de-Boeuf

ODB Córdoba (Spain) (location identifier) • Object-oriented Database • Operational Database • Operational Data Book • Operations Database • Orbit Determination Beacon • Output to Display Buffer

ODBC Object-oriented Database Connectivity • Open Database Connectivity (*)

ODBMS Object-oriented Database Management System

ODBS Optical Disk-Based System

ODC Open Design Center • Operational Data Center • Operational Data Corrector • Operational Design Criteria • Operational Document Control • Optical Disk Cartridge • Orbital Data Collector • ORFEUS Data Centre • Oscilloscope Digital Control • Other Data Centre • Other

Direct Costs • Outil de Développement Cognitif • Ozone-Depleting Chemical • Ozone-Depleting Containment

ÖDD Österreichischer Dokumentationsdienst (Austria)

ODD Objective Digital Division • Oodnadatta (Australia-SA) (location identifier) • Optical Data Digitizer • Optical Data Disk • Optical Digital Disk • Orbiting Debris Device • Ordered Decision Design • Overdetermined Dual-Doppler

ODDL Onboard Digital Data Load

ODDP Object-Dependent Data Processor

ODDR&E Office of the Director of Defense Research & Engineering

ODDS Operational Data Delivery Services

ODDSC Optical Disk Data Storage Capability

Ode Oude

OD&E Operational Demonstration & Evaluation

ODE Odense (Denmark) (location identifier) • Office de Documentation Économique • Office Developer Edition • On-Line Data Entry • Only During Eclipse • Open Development Environment • Operational Demonstration and Evaluation • Ordinary Differential Equation

ODECA Organización de Estados Centroamericanos

Odenw Odenwald (Germany)

ODERACS Orbital Debris Radar Calibration Spheres

ODES Ocean Data Evaluation System • Optical Discrimination Evaluation Study • Optimal Design of Engineered Structures

ODESSA Oceanographic Database and Environmental Satellite System • Ordinary Differential Equation Solver with Sensitivity Analysis

ODETT Organization for Data Exchange by Teletransmission

ODEX Optical Dynamics Experiment

ODF Observation Data File • Observation Definition File • One-Dimensional Flow • Opacity Distribution Function • Optimal Decision Function • Orbit Determination Facility • Ordering and Distribution Facility • Orientation Distribution Function • Original Data File • Output Data Formatter

OdG Ordine del Giorno

ODG Office of the Director General • Operational Data Group • Operational Design Group • Orbit Data Generator

ODGSE Operational Development Ground Support Equipment

ODH Odiham (UK-England) (location identifier) • Oxygen Deficiency Hazard

ODI Office de Documentation Industrielle • Office Document Index • On Demand Information • One-Degree Imager • Open Data Interface • Open Data-link Interface • Open Device Interconnect • Open Driver Interface • Optical Digital Imagery • Optical Digital Image • Optical-Disk Image

ODID Operational Display and Input Development

ODIF Office Data Interchange Format • Office Document Interchange Format • Open Document Interchange Format

ODIN Oosterhouts Digitaal Informatie Netwerk (Netherlands) • Operational Digital Network • Orbital Design Integration (system) • Outsourcing Desktop Initiative for NASA

ODIS Origin-Destination Information System

ODISS Optical Digital Image Storage System

ODIT Optical Digital Imaging Technology

OdJ Ordre du Jour

ODJ Ouanda Djalle (Central African Rep.) (location identifier)

ODK One-Dimensional Kinetics

ODL Cordillo Downs (Australia-SA) (location identifier) • Object Definition Language • Object Description Language • Observation Data Log • Oceanic Data Link • Office Document Language • Open and Distance Learning • Open Document Language • Optical Data Link • Optical Delay Line

ODLI Optical Data Link Interface

ODM Oakland (USA-MD) (location identifier) • Object Data Manager • Office of Defense Mobilization (USA) • Oil Debris Monitoring • One-Day Mission • Open Drakmatter Model • Operating Data Manual • Operational Data Management • Operational Data Message • Operational Development Memorandum • Operational Development Model • Operational Direction Message • Operation Data Management • Operation Data Message • Operation Development Memorandum • Optical Demonstration Model • Optical Display Memory • Orbital Determination Module • Orbit Deflection Maneuver • Orbiter Deflection Maneuver • Orbiter Disconnect Mechanism • Organizational Domain Modeling • Original Design Manufacturer

ODMA Open Document Management API

ODMG Object Database Management Group • Object Data Management Group

ODMR Optical Detection of Magnetic Resonance • Optically Detected Magnetic Resonance

ODMS Object Data Management System

ODN Long Seridan (Malaysia) (location identifier) • Open Data Network • Optical Distribution Network • Outdial Notification • Own Doppler Nullifier

ODNAO Orbital Dynamics of Natural and Artificial Objects

ODNMR Optical Detection of Nuclear Magnetic Resonance

ODO Orbital Debris Object • Other Disbursing Office

ODP Object-Dependent Processing • Object-Dependent Processor • Ocean Drilling Program (*) • Ocean Drilling Project • Officer Distribution Plan • Official Development Planning • Open Data Path • Open Directory Project • Open Distributed Processing • Open Distributing Processing • Open-Door Policy • Operational Development Plan • Operational Development Program • Operational Display Procedure • Optical Data Processing • Orbit Determination Program • Original Document Processing • Outline Development Plan • Output Data Processing • Overall Development Planning • Ozone-Depleting Potential

ODPCS Oceanographic Data Processing and Control System

ODPI Office of the Director of Public Information

ODPS Optical Data Processing System

ODR Omnidirection Range • Orbital Data Record • Ord River (Australia-WA) (location identifier) • Original Data Record • Output Data Redundancy

ODRAM Optical Dynamic Random Access Memory

ODRAN Operational Drawing Revision Advance Notice

ODRC Orbiter Data Reduction Center

ODRN Orbiting Data Relay Network

ODRS Orbiting Data Relay System

ODRSS Orbiting Data Relay Satellite System

ODS Object Description Skeleton • Observation Data-file System • Observation Data Set • Observation Data System • Odessa (Ukraine) (location identifier) • Online Document System • Open Data Services • Operational Data Server • Operational Data Store • Operational Data Summary • Operator and Display System • Operator Display System • Optical Data System • Optical Detector System • Optical Disk Storage • Optical Disk System • Optical Displacement Sensor • Optical Docking System • Orbital Disconnect Support • Orbit Determination System • Orbiter Docking System • Orbiter Dynamic Simulator

• Output Data Strobe • Overhead Data Stream • Oxide Dispersion Strengthening • Oxygen Depletion Sensor • Ozone Depleting Substance

ODSC Optical Disk Storage Capability

ODSI Open Directory Services Interface • Optical Domain Service Interconnect

ODST Observation Data Set Template

ODT Octal Debugging Technique • Omnidirectional Transmission • Omnidirection Transmission • On-Line Debugging Technique • Open Desktop • Operational Data Team • Operational Demonstration Test • Optical Data Transmission • Optical Detector Team • Optical Digital Technology • Order-Disorder Transition • Overseas Deployment Training • Oxford-Dartmouth Thirty-degree (survey)

ODTC Office of Defense Trade Controls

ODTF Operational Development Test Facility

ODTM Orbiter Dynamic Test Model

ODTP Optical Discrimination Technology Program

ODTS Operational Development Test Site • Optical Data Transmission System • Optical Discrimination and Tracking System

ODU Old Dominion University (USA-VA) • Optical Display Unit • Organizational Documentation Update • Output Display Unit

ODUS Ozone Dynamics Ultraviolet Spectrometer

ODV Omnidirectional Vehicle

ODVAR Orbit Determination and Vehicle Attitude Reference

Odw Odenwald (Germany)

ODW Oak Harbor (USA-WA) (location identifier)

ODWE Oblique Detonation Wave Engine

ODY Oudomxay (Laos) (location identifier)

ODZ Outer Defense Zone

OD3 Operational Data Dictionary/Directory

oe Omissions Excepted

Oe Oersted

O/E Optical to Electrical

ÖE Öleinheit

OE Austria (international aircraft registration code) • Observatoire de l'Énergie (France) • Office of Enforcement • Old English • Omissions Excepted • Open Edition • Open End • Operating Engineer • Operating Environment • Operational Effectiveness • Operational Element • Operational Environment • Operations Engineer • Operations Engineering • Opposition Effect • Optical Element • Optical Engineer • Optical Estimation • Order Entry • Ordinary Erection • Organizational Effectiveness • Original Error • Outboard Engine • Output Enable • Ouvrier d'État • Oxygen-Enriched Air

OEA Observatoire Européen de l'Audiovisuel • Office for External Activities • Office of Economic Adjustment • Operator Error Analysis • Organización de los Estados Americanos • Organisation des États Américains • Organizational Expense Account • Ouvrier d'Entretien et d'Accueil • Vincennes Oneal (USA-IN) (location identifier)

OEAS Orbital Emergency Arresting System

OEB Office Européen des Brevets

OEC Ocussi (Indonesia) (location identifier) • Office of Energy Conservation • Office of Environmental Compliance • Operational Employment Concept • Operations Engineering Center • Orbital Electron Capture • Orbiting Experimental Capsule • Other Early Capability • Oxygen-Enriched Combustion • Oxygen-Evolving Complex

OECD Organization for Economic Cooperation and Development (formerly OEEC)

OECE Organisation Européenne de Coopération Économique • Organizzazione Europea per la Cooperazione Economica

OECO Outboard Engine Cutoff

OECS Organization of Eastern Caribbean States

OEC2 Opposed Entry Command and Control

OED Offer Expiration Date • Office of Emergency Preparedness (USA) • On-Line Event Display • Operational Effectiveness Demonstration • OSIS Evolutionary Development • Oxford English Dictionary (OUP)

OEDG Opportunities for Enhancing Diversity in the Geosciences

OEDIP Outil Expert de Diagnostic Interactif de Panne

OEDIPUS Observations of Electric-field Distributions in the Ionosphere Plasma - a Unique Strategy (project)

OEE Observation of Earth and its Environment • Open Enterprise Environment • Overall Equipment Effectiveness

OEEC Organization for European Economic Cooperation (now OECD)

OEEPE Organisation Européenne d'Études Photogrammétriques Expérimentales

OEF Olimpiada Española de Física (Spain) • Optical Evaluation Facility

OEFD Orbiter Electric Field Detector (PV)

OeFOSC Österreich Faint Object Spectrograph and Camera

OEG Office of Environmental Guidance • Operation Exposure Guide • Organization and Equipment Guide

OeGA Österreichische Gesellschaft für Akustik (Austria)

OeGAI Österreichische Gesellschaft für Artificial Intelligence (Austria)

OEH Order Entry Handling

OEI Oficina de Educación Ibero-americana (Spain) • One-Engine Inoperative • Optical-Electronic Interface • Organisation Européenne d'Information • Organización de Estados Iberoamericanos • Overall Efficiency Index

OEIC Open-Ended Investment Company • Opto-Electronic Integrated Circuit

OEID Origin and Evolution of Interplanetary Dust

OEIS Orbiter Electrical Interface Simulator

OEL Occupational Exposure Limit • Optional Endorsement Line • Orel (Russia) (location identifier) • Organic Electroluminescence • Organization Equipment List

OEM Object Exchange Model • Office of Emergency Management • Office of Executive Management • Onde Électromagnétique • Optimal Estimation Method • Original Engine Manufacturer • Original Equipment Manufacturer (*) • Other Equipment Manufacturer • Own Equipment Material

OEMA Optional Equipment Maintenance Allowance • Organización Europea de Medio Ambiente

OEMCP Optical Effects Module Electronic Controller and Processor

OEMI Office Equipment Manufacturer's Institute (USA)

OEMM Octree-based External Memory Mesh • Operational Experiment on Mesoscale Meteorology

OEMT Operational Emergency Management Team

OEN Operational Evaluation Network

O. en W. Onderwijs en Wetenschappen

OEO Office of Economic Opportunity (USA) • Office of Equal Opportunity • Optical and Electro-Optical • Optical - Electronic - Optical • Osceola (USA-WI) (location identifier)

OEOB Old Executive Office Building

OEOF Oeuvre d'Expression Originale Française

OEOP Office of Equal Opportunity Programs

OEP Office of Economic Planning • Office of Economic Preparedness • Office of Emergency Planning • Office of Emergency Preparedness • Office of Environmental Policy • Oficina Europea de Patentes • Operand Execution Pipeline • Operational Employment Plan

OEPE Optoelectronic Processing Element

OEPM Oficina Española de Patentes y Marcas (Spain)

OEPS Operationally Efficient Propulsion System

OEQ Organisation Européenne pour la Qualité

OER Odd-Even Rule • Office of Energy Research • Office of Environmental Restoration • Officer Effectiveness Report • Officer Efficiency Report • Officer Evaluation Report • Operational Effectiveness Rate • Operational Equipment Requirement • Operations Engineering Report • Organization for European Research • Original Equipment Request • Örnsköldsvik (Sweden) (location identifier) • Ornskoldsvik (Sweden) (location identifier) • Overhead Expenditure Request • Oxygen Enhancement Ratio

OERC Optimum Earth Reentry Corridor

OERD Office of Energy Research and Development

OERI Office of Educational Research and Improvement (USA)

OERR Office of Emergency and Remedial Response

OERS Officer Evaluation Reporting System • Operational Earth Resources System

OES Object Encapsulation Software • Ocean Energy System • Office Evaluation System • Office of Earth Science (NASA) • Office of Economic Stabilization • Office of Emergency Services • Officer Education System • Officer Evaluation System • Open Electrode Structuring • Optical Emission Spectrometer • Optical Emission Spectroscopy • Optische und Elektronische System GmbH (Germany) • Orbital-Escape System • Orbital Extravehicular Suit • Orbiter Emergency Site • Organization of European States • Other Evaluation System • San Antonio Oeste (Argentina-RN) (location identifier)

OESA Office of Earth Science Applications

OESC Organisation pour l'Éducation et la Science (UNESCO)

OESD Organic Electronic Spectral Data

OESO Organisatie voor Economische Samenwerking en Ontwikkeling

OESS Operating Environment Segment Specification • Orbiter/ET Separation System • Organizational Effectiveness Survey System

OESZ Ostereuropäische Sommerszeit

OET Object Engineering Technique • Objective End Time • Office of Emergency Transportation • Orbiter Engineering and Test • Orbiter External Tank

OETF Observing Efficiency Task Force

OETP Orbiter Electron Temperature Probe (PV)

OEW Operating Empty Weight

OEWG Operational Evaluation Working Group

OEX Orbiter Experiment

OEXP Office of Exploration

of Officio

Of Oficina

O/F Orbital Flight • Oxidizer/Fuel

OF Observation File • Officer • Old Face • Old French • Open File • Operational Forward • Operational Functionality • Operation File • Optical Fiber • Optical Frequency • Optically Faint • Optional Form • Orbital Flight • Orbit File • Order Form • Organisme de Formation • Out File • Outflow • Output Factor • Output Format • Output Formatter • Outside Face • Overflow • Overflow Flag • Oxidizer-to-Fuel (ratio) • Oxidizing Flame • Oxygen Fill

OFA Object-Free Area • Observatoire Fédéral d'Astrophysique (Canada-BC) • Office Fédéral de l'Air (Switzerland) • Office Franco-Allemand • Office of Fiscal Administration • Off-Line Archive • Optical-Fiber Amplifier • Organisme de Formation Agréé

OFAB Oil Filter Actual Bypass

OFAC Office Fédéral de l'Aviation Civile (Switzerland)

OFACIL Office Français d'Action Culturelle Internationale et de Langues (France)

OFAD Optical Field Angle Distortion

OFAP Observational Facilities Advisory Panel

OFAT One Factor At a Time

OFB Operational Facilities Branch • Output Feedback

Ofc Office

OFC Office • On-Line File Corrector • Open Financial Exchange • Operational Flight Control • Operations Flight Control • Optical Fiber - Conductive • Optical Fiber Control • Optical Frequency Conversion • Oxygen-Free Copper

OFCC Office of Federal Contract Compliance (USA)

OFCCP Office of Federal Contract Compliance Programs (USA)

OFCCS Optical Fiber Communications and Control System

Ofcl Official

OFCO Office of the Federal Coordinating Officer (USA)

Ofcr Officer

OFCR Officer

OFCS Operational Flight Control System

OFD Oberfinanzdirektion • One-Function Diagram

OFDA Office of Foreign Disaster Assistance (USA)

OFDD Ordered Functional Decision Design

OFDM Optical Frequency Division Multiplexing • Orthogonal Frequency Division Multiplex • Orthogonal Frequency Division Multiplexing

OFDPS Offshore Flight Data Processing System

OFDS Orbiter Flight Dynamics Simulator • Oxygen Fluid Distribution System

OFE Office • Office Français d'Édition (France) • Open Finance Exchange • Operative Functional Element

OFEC Office of Federal Employees' Compensation (USA)

OFEFP Office Fédéral de l'Environnement, des Forêts et des Paysages (Switzerland)

OFELIA Optical Fibres for Electrical Industry Applications

OFES Office Fédéral de l'Éducation et de la Science (Switzerland) • Office of Fusion Energy Sciences

Off Offer • Offered • Office • Office • Officer • Official • Officiate • Officiel • Officinal

OFF Object Feasibility Filtering • Office • Omaha Offutt (USA-NE) (location identifier) • Optical Fabrication Facility

OFFCDT Officer Cadet

Offcl Official • Officinal

OFFM Office of Federal Financial Management

Offn Offnung

Offr Officer

OFG Optical Frequency Generator

OFGR Objective Force Gross Requirement

OFHC Oxygen-Free High-conductive Copper • Oxygen Free High Conductivity

OFI Operational Flight Instrumentation • Ouango Fitini (Ivory Coast) (location identifier)

OFIA Office of Federal and International Affairs (USA)

OFIB Oil Filter Impending Bypass

OFID Optical Free Induction Decay

OFIL Observatoire Français des Industries de la Langue (France)

OFJ Ólafsfjördur (Iceland) (location identifier)

OFK Official Flight Kit • Optical Flight Kit • Norfolk (USA-NE) (location identifier)

Ofl Offline

OFLI Other Force-Level Information

OFLS Off-Line System

OFltr Oil Filter

OFM Office of Fiscal Management • Office of Flight Missions • Online File Maintenance • Operational Forecast Model

OFMET Office Fédéral de Métrologie (Switzerland)

OFML Open Forms Markup Language

OFMS Overseas Financial Management System

OFMT Output Format

OFN Optical Fiber - Non-conductive

OFNI Objet Flottant Non Identifié • Objet Fumant Non Identifié

OFNPS Outstate Facility Network Planning System

OFO Office of Flight Operations • Orbiting Frog Otolith

OFP Occluded Frontal Passage • On-the-Fly Processing • Operating Force Plan • Operational Flight Profile • Operational Flight Program • Operational Format Program • Orbiter Flight Program • Original Flight Path • Ozonforschungsprogramm

OFPP Office of Federal Procurement Policy (USA)

OFPS Office of Field Project Support

OFQJ Office Franco-Québécois de la Jeunesse

OFr Old French

OFR Observatoire Fédéral de Radioastrophysique (Canada-BC) • Off-Frequency Rejection • On-Frequency Repeater • Open Failure Report • Open File Report • Organization Form Registry • Original Format Recognition • Output From Register • Overfrequency Relay

OFRATEME Office Français des Techniques Modernes d'Éducation (France)

OFRP On-the-Fly Reprocessing

OFRS Optically Faint Radio Source

O&FS Operations & Flight Support

OFS Object File System • Office of the Foreign Secretary • Officer Foundation Standard • Officer Foundation System • Operations and Flight Support • Optical-Fiber System • Optics-Free Spectrometer • Orange Free State (South Africa) • Orbital Flight System • Orbiter Flight Simulator • Orbiter Flight System • Orbiter Functional Simulator • Order of Free State • Output Field Separator

OFSD Operating Flight Strength Diagram

OFSHR Offshore

OFSS Old Font Selection Scheme • Open-File Services Section

Oft Outfit

OFT Office of Fair Trading • Operational Feasibility Test • Operational Flight Test • Operational Flight Trainer • Optical Fiber Thermometry • Optical Fourier Transform • Optimal Foraging Theory • Orbital Flight Test • Orbital Flight Training • Outer Fix Time

OFTA Office For Telecommunications Authority • On-The-Fly Adder

OFTDA Office of Flight Tracking and Data Acquisition

OFTDS Orbital Flight Test Data System

OFTE Oscillatory Thermocapillary Flow Experiment

OFTEL Office of Telecommunications (UK)

OFTM On-orbit Flight Technique Meeting

OFTP Iff-Line Transaction Processing

OFTR Orbital Flight Test Requirement

OFU Observation and Fixation Unit • Ofu (American Samoa) (location identifier)

OFUP Office Universitaire de Presse (France)

OFVR Out-of-Field-of-View Rejection

OFX Open Financial Exchange

OFWG OGIP FITS Working Group

OFXB Observatoire François-Xavier Bagnoud (Switzerland)

OFZ Obstacle-Free Zone

ÖFZS Österreichisches Forschungszentrum Seibersdorf (Austria)

og oben genannt • Ogee

OG Obergeschoß • Observation Group • Ogee • Onde de Gravité • Open Group • Operational Group • Operations Group • Organisme Gestionnaire • Organizational Genetics • Or Gate • Original Gum • Ostrogoth • Outer Gap • Outer Gimbal • Outer Gradient • Outgoing • Outlet Group • Overground • Oxygen Gauge

OGA Ogallala Searle (USA-NE) (location identifier) • Open GIS Architecture • Opérateur de Gestion Aéronautique • Option Generation Aid • Other Government Agency • Outer Gimbal Angle • Outer Gimbal Axis

ÖGAA Österreichische Gesellschaft für Astronomie und Astrophysik (Austria)

OGAR On-Ground Attitude Reconstruction

OGAW Organismus für gemeinsame Anlagen in Wertpapieren

OGB Olbers Gesellschaft Bremen eV (Germany) • Orangeburg (USA-SC) (location identifier)

OGBD Orbiter Gamma Burst Detector (PV)

OGC Office of General Counsel • Old Globular Cluster • Open GIS Consortium

OGCM Ocean General Circulation Model

OGD Ogden (USA-UT) (location identifier)

OGDC Office of Geographic Data Coordination

OGE Office of Government Ethics (USA) • Ogeranang (Papua New Guinea) (location identifier) • Open GIS Environment • Operating Ground Equipment • Operational Ground Equipment • Optical Gyro Encoder • Out of Ground Effect

OGF Offspring Generation Function • Operational Green's Function

OGG Kahului - Maui Island (USA-HI) (location identifier)

OGI Oculogyral Illusion • Oregon Graduate Institute (USA-OR) • Output Graphic Interface

OGIP Office of Guest Investigator Programs

OGIRS Oklahoma Geographic Information Retrieval System

OGIS Open Goodata Interoperability Specification

OGIST Oregon Graduate Institute of Science and Technology (USA-OR)

OGL Ogle (Guyana) (location identifier) • Open General Licence • Overall Grade Level • Overlay Generation Language

OGLE Optical Gravitational Lensing Experiment

OGM On-Going Merger • Optimum Gradient Method • Ordinary General Meeting • Organisme Génétiquement Modifié

OGMS Open Gradient Magnetic Separation

OGMT Orbiter Greenwich Mean Time

OGN Okayama Gigabit Network (Japan) • Yonaguni Jima (Japan) (location identifier)

OGO Abengourou (Ivory Coast) (location identifier) • Orbiting Geophysical Observatory

OGOT Oficina General de Operaciones Técnicas

OGP Office of Global Programme • Operational Ground Program • Organisation et Gestion de Production

OGPE Orbiter Gas Plasma Environment (PV)

OGPSH Országos Global Positioning System Hálózat (Hungary)

OGR Bongor (Chad) (location identifier) • Outgoing Repeater

OGRAM Oscillogram

OGRC On-Ground Range Compression

OGRIP Ohio Geographically Referenced Information Program

O&GS Operations & Ground Systems

OGS Objective Grating Spectrometer • Offset Guidance Sensor • Ogdensburg (USA-NY) (location identifier) • Operation Ground Segment • Operator to General Support • Optical Ground Station • Optical Guidance System • Optical Guidance Systems (USA-PA) • Outer Galaxy Survey • Outer Glideslope • Overseas Ground Station • Oxide Glass-like Semiconductor • Oxide Glassy Semiconductor • Oxygen Generation System

OGSA Open Grid Services Architecture

OGSE Operational Ground Support Equipment • Optical Ground Support Equipment

OGSEL Operational Ground Support Equipment List

OGT Observatoire Géodésique de Tahiti • Opérateur Géomètre Topographe • Outgoing Trunk

OGU Outgoing Unit

OGV Outlet Guide Vane • Oxygen Gauge Valve

OGW Office of Ground Water

OGWP Office of Ground Water Protection

OGX Ouargla (Algeria) (location identifier)

OGY Ott - Grebogi - Yorke (method)

OGZ Vladikavkaz (Russia) (location identifier)

oh Omni Hora

O-H Octal to Hexadecimal

O/H On Hand • Overhead

OH Finland (international aircraft registration code) • Hydroxyl Radical • Ofenheizung • Off Hook • Office Hours • Ohio (USA) • Ohmic Heating • Old Halo • On Hand • Open House • Operational Hardware • Optical Head • Overhaul • Overhead

OHA Ohakea (New Zealand) (location identifier) • Operating Hazard Analysis • Operational Hazard Analysis • Orbital Height Adjustment • Oscillator Housing Assembly • Outside Helix Angle • Oxidizer Holdoing Area

OHAN Optical Hybrid Access Network

OHANA Optical Hawaiian Array for Nanoradian Astronomy (USA-HI)

OHB Orbital- und Hydrotechnologie Bremen (Germany)

OHC Northeast Cape (USA-AK) (location identifier) • Onboard Hard Copier • Optics Hand Controller • Overhead Camshaft • Oxygen Hole Center

OHCI Open Host Controller Interface

OHD Ohrid (Macedonia) (location identifier) • Over-the-Horizon Detection

OHEA Office of Health and Environmental Assessment

OHER Office of Health and Environmental Research

OHF Occupational Health Facility • Off Hook Factor • On-Line Help Facility

oHG offene Handelsgesellschaft

OHG Offene Handelsgesellschaft • Old High German • Operators Harmonization Group

OHGC Old Halo Globular Cluster

OHGVT Orbital Horizontal Ground Vibration Test

OHI Organisation Hydrographique Internationale • Oshakati (Namibia) (location identifier)

OHIM Office for Harmonization in the Internal Market (EU)

OHIO Open Host Interface Object

OHL Observatorium Hoher List (Germany)

OHM Hydroxyl Megamaser • Occupational Health Management • Office of Hazardous Materials • OH Megamaser

OHMIS Occupational Health Management Information System

OHMS Oil and Hazardous Materials

OHO Ohotsk (Russia) (location identifier)

OHP Oban (UK-Scotland) (location identifier) • Observatoire de Haute-Provence (France) • Open Hypermedia Protocol • Operational Hydrological Program • Operational Hydrology Programme • Organizational Homepage • Overhead Projector

OHQ Originating Headquarters • Ouvrier Hautement Qualifié

OHR Office of Human Resources • Operational Hydrology Report • Wyk auf Föhr (Germany) (location identifier)

OHRE Office of Human Resources and Education

OH&S Occupational Health & Safety

OHS Hydroxyl Radical Suppression • Hydroxyl Radical Suppressor • Observation Handling System • Occupational Health Services • Old Halo Star • Open Hyperdocument System • Open Hypermedia System (*) • Open Hypertext System

OHSPC Oil and Hazardous Substance Pollution Contingency

OHSU Oregon Health Sciences University (USA-OR)

OHSWG Open Hypermedia System Working Group

OHT Kohat (Pakistan) (location identifier) • Orbital History Table

OHTEX Ocean Heat Transport Experiment

OHV Off-Highway Vehicle • Overhead Valve

OHW Office of Hazardous Wastes

OHWM Ordinary High Water Mark

oi Onzes Inziens

O/I Operations/Intelligence • Outboard/Inboard

O&I Operations & Intelligence • Organization & Intermediate

OI Office of Information • Office Instruction • Onzes Inziens • Operating Income • Operating Instruction • Operational Increment • Operational Instrumentation • Operational Issue • Operations Interface • Optical Imager • Optically Improved • Optimal Interpolation • Optimum Interpolation • Orbital Insertion • Orbit Insertion • Orbiter Insertion • Orbiter Instrumentation • Ordinateur Individuel • Organizational Integrator • Organization Integration • Other Investigator • Output Impedance • Overstemming Index

OIA Office Information Architecture • Office of Integrative Activities • Office of Internal Affairs • Office of International Activities • Office of International Administration • Office of International Affairs • Operator Information Area • Orbiter Interface Adapter

OIAC Organisation Internationale de l'Aviation Civile

OIB Oceanic Island Basalt • Ocean Island Basalt • Office International de Bibliographie • Official Information Base • Operating Improvement Board • Operations Integration Branch • Optische Interferenz Bauteile GmbH (Germany) • Orbiter Interface Box

OIBD Office International de Bibliographie et de Documentation

OIbF Olimpiada Iberoamericana de Física

OIC Norwich Eaton (USA-NY) (location identifier) • Office of Information Control • Office of Instructional Consultation • Officer In Charge • Oh I See • Operational Issue and Criteria • Operations Instrumentation Coordinator • Operator's Instruction Chart • Optical Integrated Circuit • Orbital Integrated Checkout • Organisation Internationale du Commerce • Orientation - Information - Conseil

OICC Office in Charge of Construction • Officer in Charge of Construction • Operations Interface Control Chart

OICETS Optical Inter-orbit Communications Engineering Test Satellite

OICO Office of Integration and Checkout

OICP Office for International Cooperative Program • Office of International Communications Policy

OICR Operation, Implementation, and Capabilities Report

OICTP Outile of Individual and Collective Training Programs

OICV Organisation Internationale de la Commission de Valeurs

oid Of Iets Dergelijks

OID Object Identification • Object Identifier • Object Interaction Diagram • Octal Identifier • Oficina de Información Diplomática • Optical Image Database • Organisation Intégrée de Développement

OIDD Orbiter Internal Density Distribution (experiment) (PV)

OIDL Object Interface Definition Language

OIE Office International de l'Eau • Operational Independent Evaluator • Orbiter Integration Engineering • Organisation Internationale des Employeurs

OIEA Office of Integrated Environmental Analysis • Organización Internacional para la Energía Atómica

OIEC Organisation Internationale des Échanges Culturels • Organization for International Economic Cooperation

OIES Office for Interdisciplinary Earth Studies

OIF Object Infrastructure Framework • Observation Index File • Old IRAF Format • Optical Interface • Optical Internetworking Forum • Optimum Index Factor • Original IRAF Format (*)

OIG Office of Inspector General • Office of International Geology • Office of the Inspector General • Operational Improvement Program • Operations Interface Group • Optical Imager • Optically Isolated Gate • Organisation Intergouvernementale

OIH Office of International Hydrology

OII Office of Invention and Innovation (USA) • Open Information Interchange • Operations Integration Instruction

OIJ Organisation Internationale des Journalistes

OIL Oil City - Splane (USA-PA) (location identifier) • Optical Imaging Instrument • Orbital International Laboratory

OILM Organizational and Intermediate Level Maintenance

OILR Outer Inner Lindblad Resonance

OILSAR Ocean-Ice-Land Synthetic Aperture Radar

OIM Optical Interference Microscope • Orbit Insertion Maneuver • Organización Internacional para las Migraciones • Organizational Intermediate Maintenance • Oshima (Japan) (location identifier) • OSI Internet Management • Oxygen Interaction with Materials

OIML Organisation Internationale de Métrologie Légale

OIMS Orbiter Ion Mass Spectrometer (PV)

OIN Office International de Normalisation

OING Organisation Internationale Non-Gouvernementale

OiO Onderzoeker in Opleiding

OIO Onderzoeker-In-Opleiding • Onderzoek-In-Opleiding • Operational Interoperability • Operations Integration Officer

OIP Office for International Programmes • Open Internet Platform • Operational Improvement Plan • Operational Improvement Program • Operations Interface Procedure • Operator Interface Process • Operator Interface Program • Optical Image Processor • Optronics Instruments and Products • Orbital Improvement Program

OIPC Organisation Internationale de Police Criminelle (Interpol)

OIPN Office International pour la Protection de la Nature

OIPP Office of Intellectual Property Policy

OIPS Optical Image Processing System

OIr Old Irish

O/IR Optical/Infrared

OIR Object Interface Repository • Office of Industrial Relations • Okushiri (Japan) (location identifier) • Online Insertion and Removal • Open Item Report • Open Item Review • Operational and Information Requirements • Operations Integration Review • Optical and Infrared • Optical Infrared • Orbiter Infrared Radiometer (PV) • Organisation Internationale de Radiodiffusion

OIRA Office of Industrial Resource Administration • Office of Information and Regulatory Affairs

OIRE Optical Infrared Equipment

OIRG Operational Interoperability Requirements Group

OIRM Office of Information Resources Management

OIRO Orbiting Infrared Observatory

OIRT Organisation Internationale de Radio et Télévision

OIS Observation Implementation Scientist • Office Information System • Office of Information Security • Open Information Society • Operating Information System • Operational Information Service • Operational Instrumentation System • Operational Integration Site • Operational Intercommunication System • Operations Information System • Optical Imager Study • Optical Image Sensor • Optical Image Subtraction (method) • Orbiter Insertion System • Orbiter Instrumentation System • Orbit Injection System • Orbit Inserting System • Orbit Insertion System • Other Information Source

OISAR Optical Inverse Synthetic Aperture Radar

OISCA Organization for Industrial, Spiritual and Cultural Advancement

OISD Outsourced Information Sciences Development

OISE Office Information System Equipment

OISNP Oxford Pergamon Press International Series on Natural Philosophy (periodical code)

OISR Open Item Status Report

OISS Open Item Status System • Operational Intelligence Support System

OISSP Office of Interim Space Station Program

OISTP Oxygen-Iodine Supersonic Technology Program

OISTS Oxford Pergamon Press International Tables Selected Constants (periodical code)

OIt Old Italian

OIT Object Identification Test • Office of Information Technology • Oita (Japan) (location identifier) • Operations - Information - Technology • Orbiter Integrated Test • Orbiter Integration Test • Orbiter Interface Test • Organisation Internationale du Travail

OITDA Optoelectronic Industry and Technology Development Association

OITP Office for Information Technology Policy

OIU Observer Interface Unit • Office Interface Unit • Operator Interface Unit

OIV Oxidizer Isolation Valve

OIVO Observatory-Independent Visibility Object

OIVS Orbiter Interface Verification Set

OIW OSE Implementation Workshop

OIWFS Onboard Instrument Wavefront Sensor • On-Instrument Wavefront Sensor (*)

oJ ohne Jahr

OJ Official Journal • Orange Juice • Ordre du Jour • Outer Join

ÖJAB Österreichische Jungarbeiterbewegung (Austria)

OJAE On-Line Journal of Astronomy Education

OJC Kansas City - Johnson Executive (USA-MO) (location identifier)

OJCCT On-Line Journal of Current Clinical Trends

OJD Official Journal Date • Oficina para la Justificación de la Difusión (Spain)

OJE On-the-Job Education • On-the-Job Experience

OJEC Official Journal of the European Communities

OJL Optoelectronics Joint Laboratory

OJPS On-Line Journal Publishing Service (AIP)

OJS On-Line Journal Service

OJSA On-Line Journals Subscription Agreement

OJT Office de Justification des Tirages (France) • On-the-Job Training

OJTA Officer Job Task Analysis

oK ohne Kosten

OK All Correct • All Right • Czech Republic (international aircraft registration code) • Oklahoma (USA) • Oklahoma-Kansas • Operatiekamer • Oxidation Kinetics • Slovakia (international aircraft registration code)

OKA Okinawa Naha (Japan) (location identifier) • Otherwise Known As

Okean Okeanologiia (periodical code)

OKAPI Open Kernel for Access to Protected Interoperable (services)

OKB Orchid Beach - Fraser Island (Australia-QLD) (location identifier) • Ordered Knowledge Base

OKC Oklahoma City - Will Rogers (USA-OK) (location identifier)

OKCAC Oklahoma City Astronomy Club (USA-OK)

OKD Sapporo Okadama (Japan) (location identifier)

OKE Okino Erabu (Japan) (location identifier) • Optical Kerr Effect

OKF Okaukuejo (Namibia) (location identifier)

OKFDD Ordered Kronecker Functional Decision Diagram

OKG Okoyo (Congo) (location identifier)

ÖKG Österreichische Klimatechnik Gesellschaft (Austria)

OKH Oakham Cottesmor (UK-England) (location identifier)

OKI Oki Island (Japan) (location identifier)

OKJ Okayama (Japan) (location identifier)

OKK Kokomo (USA-IN) (location identifier)

OKL Oksibil (Indonesia) (location identifier)

Okla Oklahoma (USA)

OklRe University of Oklahoma Observatory Reprints (periodical code)

OKM Okmulgee (USA-OK) (location identifier)

OKN Okondja (Gabon) (location identifier) • Optokinetic Nystagmus

OKO Tokyo Yokota (Japan) (location identifier)

OKP Oksapmin (Papua New Guinea) (location identifier)

OKPSK Offset Keyed Phased-Shift Keying

OKQ Okaba (Indonesia) (location identifier)

OKQAM Offset Keyed Quadrature Amplitude Modulation

OKR Yorke Island (Australia-SA) (location identifier)

OKS Operator Keying Station • Oshkosh (USA-NE) (location identifier)

Okt Oktober

OKT Oktiabrskij (Russia) (location identifier)

OKU Mokuti Lodge (Namibia) (location identifier) • Osaka Kyoiku University (Japan)

OKY Oakey (Australia-QLD) (location identifier)

ol Oculus Laevus

Ol Oil • Olympiade

O/L Open Loop • Overload

OL Oberlausitz (Germany) • Occupational Level • Oculus Laevus • Offered Load • Old Latin • One-Lobed • Online • Oosterlengte • Open Loop • Operating Level • Operating Licence • Operating Location • Operating Log • Operational Lifetime • Opposite Linear • Optimal Level • Orbital Launch • Ordered List • Organizational Learning • Organization List • Orienting Line • Or Less • Orthogonal Linear • Outlet • Overflow Level • Overlap • Overlap Level • Overlay • Overload

ÖL Östliche Länge

OLA Off-Line Analysis • Ontario Library Association (Canada-ON) • Optimal Localized Average • Optimally Localized Average • Optimally Localized Averaging (method) • Orbital Lock Assembly • Orland (Norway) (location identifier) • Overwhelmingly Large Array

OLAA Open Learning Agency of Australia

OLAC Off-Line Adaptive Computer • Online Archive Client

OLADE Organización Latinoamericana de Energía

OLAE Opération Locale Agri-Environnement

OLAF Office de Lutte Anti-Fraude (UE) • On-Line Archive Facility

OLAFS Orbiting and Launch Approach Flight Simulator

OLAP On-Line Analytical Processing • On-Line Analytical Processor

OLAS On-Line Archive System

OLB Olbia (Italy) (location identifier)

OLBI Optical Long Baseline Interferometry

OLBIN Optical Long Baseline Interferometry Newsletter

OLBW Open-Loop Bandwidth

OLC Office of Library Coordination (Canada) • On-Line Computer • Operating Location Clerk • Optical Loop Carrier • Outdoor Lighting Code (USA)

OLCA On-Line Circuit Analysis

OLCOP On-Line Control and Operational Program

OLCP On-Line Complex Processing

old Obsolete (file extension)

OLD Off-Load Device • Old Town (USA-ME) (location identifier) • On-Line Diagnostics

Oldb Oldenburg (Germany)

OLDB On-Line Data Bank • On-Line Database

OLDC On-Line Data Collection

OLDECC Ordinary Linear Differential Equation with Constant Coefficients

OLDI On-Line Data Interchange

OLDP On-Line Data Processing

OLDS On-Line Display System • On-Line Log Data Set • Ozone Layer Depleting Substance

OLDT Ordered selection strategy with Linear resolution for Definite classes with Tabling

OLE Object Linking and Embedding • Olean (USA-NY) (location identifier) • Opto and Laser Europe (UK) • Organizational Learning in Enterprises

OLEB Origins of Life and Evolution of the Biosphere (periodical code)

OLED Organic Light-Emitting Diode

OLEG Omnidirectional Low-Energy Geoplasma

OLERD Orbiter Landing Emergency Rescue Data

OLES Office of Law Enforcement Standards (USA)

OLEW Open Learning Experimental Workshop

OLF Observational Luminosity Function • Optical Luminosity Function • Orbital Launch Facility • Orbiter Landing Facility • Orbiter Lifting Frame • Outlying Field • Wolf Point (USA-MT) (location identifier)

OLG Oberlandesgericht • Objective Lens Group • Observatory Lustbühel Graz (Austria) • Office of Local Government • Old Low German • Open-Loop Gain

OLGA Organic Land Grid Array

OLH Old Harbor (USA-AK) (location identifier)

Oli Oligocene

OLI Ólafsvík (Iceland) (location identifier) • Optical Line Inlet • Optical Line Interface • Optronic Laboratories Inc. (USA-FL)

OLIC Official Language Implementation Committee

OLIF Orbiter Landing Instrumentation Facility

OLIP On-Line Instrument Package

OLIPaC Office of Land Information Policy and Coordination

OLIS Oxford Library Integrated System

OLIVE Optically-Linked VLBI Experiment

OLJ Olpoi (Vanuatu) (location identifier)

OLL Organic Liquid Laser • Output Logic Level

OLLRC Ontario Laser and Lightware Research Centre (Canada-ON)

OLM Olympia (USA-WA) (location identifier) • One-Leg Method • On-Line Monitor

OLMC Output Logic Macrocell

OLMF Output Language Media Format

OLMOS Onboard Life Monitoring System

OLMSA Office of Life and Microgravity Sciences and Applications (NASA)

OLN Colonia Sarmiento (Argentina-CB) (location identifier)

OLO Olomouc (Czech Rep.) (location identifier) • On-Line Operation • Operations Launch Order • Optical Local Oscillator • Optical Line Outlet • Orbital Launch Operation

OLOM Orbiter Lift-Off Mass

OLOS On-Line Order Status • Operational Land Observation System • Out of Line Of Sight

OLOW Orbiter Lift-Off Weight

OLP Off-Line Processing • Olympic Dam (Australia-SA) (location identifier) • On-Line Programming

OLPAC Ohio Light Pollution Advisory Council (USA-OH)

OLPARS On-Line Pattern Analysis and Recognition System

OLPR Office for Library Personnel Resources

OLPS On-Line Programming System

OLQ Olsobip (Papua New Guinea) (location identifier) • Online Query

OLR Objective Loudness Rating • Off-Line Reader • Off-Line Recovery • Online Reference • On-Load Reactor • Open-Loop Receiver • Open-Loop Receiver • Outer Lindblad Resonance • Outgoing Longwave Radiation

OLRP On-Line Report Processor

OLRT On-Line Real Time

OLS Nogales (USA-AZ) (location identifier) • Oceanographic Lidar System • Omnilink System • On-Line Library System • On-Line Search • On-Line Service • On-Line System • Operational Launch Station • Operational License Stage • Operational Line Scanner • Operational Linescan System • Optical Landing System • Optical Line Scanner • Optical Line System • Orbiting Lunar Station • Ordinary Least Squares • Over Larger Scale

OLSAP On-Line Support Application Processor

OLSC On-Line Scientific Computer • On-Line Scientific Computing

OLSCA Orientation Linkage for a Solar Cell Array

OLSE Ordinary Least Squares Estimators

OLSMA Office of Life Science and Microgravity Applications (now BPR) (NASA)

OLSP On-Line Service Provider • Operational Logistics Support Plan • Orbiter Logistics Support Plan

OLSS Off-Line Systems List • On-Line Software System • Operational Logistics Support Summary

OLT Observatory Level Test • On-Line Test • Optical Line Termination • Oskar-Lühning-Teleskop

OLTC Open-Loop Tracking Complex

OLTE Organizational Level Test Equipment

OLTEP On-Line Test Executive Program

OLTIP On-Line Technical Information Package

OLTM Optical Line Terminating Multiplexer

OLTP On-Line Transaction Processing • On-Line Transaction Processor

OLTS On-Line Test System • On-Line Time Share

OLU Columbus (USA-NE) (location identifier) • Optical Light Unit

OLUIT Object-oriented Librarian User Interface Tool

OLUQ Optimum Learning Vector Quantization

OLUS On-Line Users Service

OLUT Output Lookup Table

OLV Olive Branch (USA-MS) (location identifier) • Online Validation • Open Low Voltage • Orbital Launch Vehicle

OLVG Open-Loop Voltage Gain

OLVIMS Online Vehicle Interactive Management System

OLVM Open Look Window Manager

OLWM Ordinary Low Water Mark

OLY Olney Noble (USA-IL) (location identifier)

om Omni Mane • Oromo (ISO language code)

Om Ostmark (currency) (now obsolete)

O/M Operations and Maintenance • Optical Modem

O&M Operations & Maintenance • Operator & Maintainer • Organization & Management • Organization & Members • Organization & Methods • Overhaul & Modification

Ö/M Öffentlichkeitsarbeit/Marketing

OM Object Manager • Object Model • Observation Manager • Observatoire de Marseille (France) • Office Management • Office Manager • Old Measurement • Oman (ISO country code) • Openbare Ministerie • Open Market • Operate Mode • Operating Memorandum • Operating Memory • Operational Management • Operational Manager • Operations and Maintenance • Operations Maintenance • Operations Manager • Operations Manual • Operations Module • Operator's Manual • Optical Marker • Optical Master • Optical Microscope • Optical Microscopy • Optical Model • Optical Module • Optical Monitor • Optical Multimode • Optimus Maximus • Orbital Maneuvering • Orden Ministerial • Ordinary Member • Ordre de Mission • Organic Matter • Organisation and Management • Organizational Maintenance • Organizational Memory • Our Message • Outer Marker • Outlet Module • Output Magnitude

OMA Object Management Architecture • Observatório Municipal de Americana (Brazil-SP) • Office of Multidisciplinary Activities (NSF) (*) • Office of Multidisciplinary Affairs (NSF) • Omaha Eppley (USA-NE) (location identifier) • Open Management Architecture • Operational Maintenance Activity • Operations and Maintenance Appropriation • Operations Maintenance Area • Operations Management Application • Optical-Mechanical Assembly • Optical Multichannel Analyzer • Optical Multichannel Array • Optomechanical Assembly • Orbiter Maintenance Area • Orderly Marketing Agreement • Organisme de Mutualisation Agréé • Organizational Maintenance Activity

OMAC Open Modular Architecture Controller

OMAG Orbiter Magnetometer (PV)

OMAP Open Multimedia Application Platform • Operations, Maintenance, and Administration Part • Operations Maintenance Application Part

OMAPI Office Malgache de la Propriété Industrielle (Madagascar)

OMAS Official Mail Accounting System • Operations Management Application Software

OMAT Optimal Maturity Parameter

OMB Office of Management and Budget (USA) • Omboue (Gabon) (location identifier) • Originating Message Buffer • Outer Marker Beacon

OMBC Office of Management and Budget Circular

OMBUU Orbiter Mid-Body Umbilical Unit

OMC Object Modeling Card • Observatório Municipal de Campinas (Brazil-SP) • Occupational Measurement Center • Operation and Maintenance Center • Optical Monitoring Camera (INTEGRAL) • Organisation Mondiale du Commerce • Operating and Maintenance Cost • Operation and Maintenance Center • Operation and Maintenance Cost • Operations Monitoring Computer • Optical Memory Card • Optical Monitor Camera • Optical Monitoring Camera • Orbiter Maintenance and Checkout • Organic Matrix Composite • Organic Molecular Crystal • Organisación Mundial del Comercio • Organometallic Compound • Orion Molecular Cloud • Ormoc (Philippines) (location identifier)

OMCF Operations and Maintenance Control File • Orbiter Maintenance and Checkout Facility

OMCO Official Mail Control Officer

OMCS Orbiter and Maintenance Communications System • Ozone Monitor Comparison System

OMCSR Object Model Content Standards Repository

OMCVD Organometallic Chemical Vapor Deposition

OMD Operations and Maintenance Document • Operations and Maintenance Documentation • Operations Maintenance Division • Operator and Maintenance Decision • Optical Measurement Device • Oranjemund (Namibia) (location identifier) • Orbiter Mating Device

OMDD Object Model Data Dictionary

OMDDS Object Model Data Dictionary System

OMDM Optomechanical Display Module

OMDP Orbiter Maintenance Down Period

OMDR Operations and Maintainability Data Record • Operations and Maintenance Data Record • Optical Memory Disk Recorder

OMDT Object Model Development Tool

OME Nome (USA-AK) (location identifier) • Open Messaging Environment • Opérateur Mathématique Électronique • Operational Mission Environment • Orbital Maneuvering Engine • Orbiter Main Engine

OMED Ocean Minerals and Energy Division

OMEGA Orbiting Medium Explorer for Gravitational Astrophysics

OMEI Other Major End Item • Other Major Equipment Item

OMET Orbiter Mission Elapsed Time

OMEWG Orbiter Maintenance Engineering Working Group

OMEX Ocean Margin Exchange • Ocean Margin Experiment

OMF Mafraq - King Hussein (Jordan) (location identifier) • Object Management Facility • Object Management Framework • Object Management Function • Object Module Format • Office Master File • Oil Mist Filter • Open Media Framework • Open Message Format • Open Modeling Form • Operational Mission Failaure • Operation and Maintenance of Facilities • Optical Matched Filter • Organization Master File • Oscillatory Magnetic Field

OMFBAA Operation and Maintenance of Facilities Budget Activity Account

OMFCA Operation and Maintenance of Facilities Cost Account

OMFCU Outboard Message Format Conversion Unit

OMFSCA Operation and Maintenance of Facilities Summary Cost Account

OMFT Optical Matched Filter Technique

OMG Object Management Group • Observation Manager • Omega (Namibia) (location identifier) • Operational Maneuver Group

OMGA Operations Management Ground Application

OMGS Orbite, Manoeuvre et Gestion du Système

OMH Officier Marinier Hydrographe • Országos Méréstügyi Hivatal (Hungary) • Urmieh (Iran) (location identifier)

OMI Observatoire du Marché de l'Information • Office des Migrations Internationales • Open Messaging Interface • Open Microprocessor Initiative • Operational Maintenance Instruction • Operation, Maintenance and Instruction • Operations and Maintenance Instruction • Operator/Machine Interface • Operator Message Input • Optical Mapping Instrument • Optical Media International • Optical Modulation Index • Organisation Maritime Internationale • Organisation Météorologique Internationale (now OMM) • Ozone Monitoring Instrument

OMIJC Operations and Maintenance Instruction Job Card

OMIM On-Line Mendelian Inheritance in Man

OMIMO On-Line Multimedia Information for Maintenance and Operation

OMIPE Office Mondial d'Information sur les Problèmes d'Environnement

OMIS Office Management Information System • Operational Maintenance Information System • Operational Management Information System (*) • Operations and Management Information Systems • Operations Management and Information Systems • Operations Management Information System • Organisation Mondiale de l'Image et du Son

OMISS Operation and Maintenance Instruction Summary Sheet

OMJ Omura (Japan) (location identifier)

OMK Omak (USA-WA) (location identifier) • Orbiter Mission Kit

OMKM Office of Mauna Kea Management (USA-HI)

ömL öffentlicher mobiler Landfunk

OML Object Manipulation Language • Object Model Library • Omkalai (Papua New Guinea) (location identifier) • Orbital Motion Limited (theory) • Orbiter Mold Line • Organizational Maintenance Level • Outer Mold Line • Outside Mold Line

OMLA Organizational Maintenance Level Activity

OMLB Originating Message Line Block

OMM Marmul (Oman) (location identifier) • Observatoire du Mont-Mégantic (Canada-QC) • Operational Maintenance Manual • Operation and Maintenance Manual • Orbiter Major Modification • Organisation Météorologique

Mondiale • Outbound Message Manager • Output Message Manual

OMMH Optical Multimark Hypercube • Orbiter Maintenance Man-Hour

OMMSQA Office of Modeling, Monitoring Systems and Quality Assurance

OMN Oman (ISO country code) • Osmanabad (India) (location identifier)

OMNI Organizing Medical Networked Information • Origins Mission for Nuclei Investigation

OMNIS Observatory for Multiflavor Neutrino Interactions from Supernovae (project)

OMNS Open Network Management System

OMO Mostar (Bosnia-Hercegovina) (location identifier) • Occupational Medicine Office • One-Man-Operated • One-Man-Operation • One-Man-Operator • Operator Message Output • Other Mapping Organization

OMP Observation Management Project • Observatoire des Matières Premières (France) • Observatoire Magnétique Planétaire • Observatoire Midi-Pyrénées (France) • Observatório Municipal de Piracicaba (Brazil-SP) • Ocean Margins Program • Office of Microelectronics Programs (NIST) • Offline Mission Processing • Operating Maintenance Plan • Operations and Maintenance Plan • Operations and Management Platform • Ordre de Mission Permanent • Organometallic Polymer (coating) • Orthogonal Matching Pursuit • Osculatrice de Meilleure Pénétration

OMPA Octamethyl Pyrophosphoramide

OMPE Organisation Mondiale de la Profession Enseignante

OMPF Operation and Maintenance Processor Frame

OMPI Organisation Mondiale de la Propriété Intellectuelle

OMPP Organisation Mondiale de la Presse Périodique

OMPR Operational Maintainability Problem Reporting • Optical Mark Page Reader • Optimal Multiple Point Reassignment

OMPRA One-Man Propulsion Research Apparatus

OMPS Orbit Maneuvering Propulsion System • Outboard Mission Planning System

OMPT Observed Man Point Trajectory • Observed Mass Point Trajectory (*)

OMPTE Office of Mission To Planet Earth

OMR Old Merger Remnant • Operation Management Room • Operations and Maintenance Requirement • Operations Management Room • Operations Manager's Report • Optical Mark Reader • Optical Mark Reading • Optical Mark Recognition • Oradea (Romania) (location identifier) • Orbiter Management Review • Organic Magnetic Resonance • Output Message Report • Rial (Oman) (ISO currency code)

OMRB Object Modeling Resource Base • Operating Material Review Board

OMRC Operational Maintenance Requirements Catalog

OMRF Orbiter Maintenance and Refurbishment Facility • Orbiter Modification and Refurbishment Facility

OMRP Operations and Maintenance Requirements Plan

OMRS Operations and Maintenance Requirements and Specifications • Operations and Maintenance Requirements Specification • Operations and Maintenance Requirements System • Optical Mark Reader System

OMRSCB Operations and Maintenance Requirements and Specifications Control Board

OMRSD Operations and Maintenance Requirements and Specifications Document • Operations and Maintenance Requirements Specifications Documentation

OMRV Operational Maneuvering Reentry Vehicle

OMS Object Management System • Observatory Monitoring System • Omsk (Russia) (location identifier) • On-board Maintenance System • Opcode's MIDI System • Operational Message Standard • Operational Meteorological Satellite • Operational Mode Standard • Operational Mode Summary • Operational Mode System • Operational Monitoring System • Operations and Maintenance System • Operations Management System • Operator Maintenance System • Optical Maser System • Optical Modulation System • Optimization Methods and Software • Orbital Maneuvering System • Orbital Mapping System • Orbital Multifunction Satellite • Orbiter Maneuvering System • Organic Mass Spectroscopy • Organisation Mondiale de la Santé • Organizational Maintenance Shop • Origin of the Modern State • Oscillating Motion Support • Output per Man Shift

OMSC Object Management Standards Category

OMSCJ Office Municipal des Sports, de la Culture et des Sports

OMSEC Office Municipal Socio-Éducatif et Culturel

OMSF Office of Manned Space Flights

OMSI Oregon Museum of Science and Industry (USA-OR)

OMSP Operational Maintenance Support Plan

OMSS Organizational Management Staffing System

OMSV Organizational Maintenance Storage Van

OMT Object-Modeling Technique • Object Model Technique • Object Model Template • Operational Maintenance Test • Operational Maintenance Testing • Operations Management Team • Optical Modeling Tool • Organisation Mondiale du Tourisme • Organización Mundial del Turismo • Organizational Maintenance Technician • Organizational Maintenance Test • Organizational Maintenance Trainer • Orthogonal Mode Transducer • Orthomode Transducer

OMTP Operational Maintenance Test Plan

OMTPE Office of Mission To Planet Earth (NASA)

OMTS Operational Maintenance Test Station • Organizational Maintenance Test Set • Organizational Maintenance Test Station

OMU Operational Measuring Unit • Operational Mock-Up • Optical Measuring Unit • Optical Multiplex Unit • Other Minority University (USA)

OMV Orbital Maneuvering Vehicle (*) • Orbiter Maneuvering Vehicle • Oxygen Manual Valve

OMVA Optimum Magnified Visual Angle

OMVS Open Multiple Virtual System

OMW Object Management Workbench

OMY Oddor Meanche (Cambodia) (location identifier)

on Omni Nocte

ÖN Österreichisches Normungsinstitut (Austria)

ON Observatório Nacional • Observatorio Nacional • Occultation Newsletter • Octane Number • Ontario (Canada) • Operation Notice • Optical Nerve • Optical Nucleus • Organizational Notary • Original Negative • Ortsnetz

ONA Office of National Assessments • On-Line Archive • Open Network Architecture • Winona (USA-MN) (location identifier)

ONAC Operations Network Administration Center

ONAFTS Office National d'Allocations Familiales pour Travailleurs Salariés (Belgium)

ONAL Off Network Access Line

ONAP Orbit Navigation Analysis Program

ONB Ononge (Papua New Guinea) (location identifier) • Orthonormal Basis

ONBA Observatorio Naval de Buenos Aires (Argentina-BA)

ONBD On Board

ONBI Office National des Brevets d'Inventions (France)

ONBT Orbiter Neutral Buoyancy Trainer

ONC Open Network Computing • Open Networking Computing • Operational Navigational Chart • Operational

Navigation Chart • Ordinary National Certificate • Orion Nebula Cluster

ONCE On-Line Network for Continuing Education

OND October - November - December • Ondangwa (Namibia) (location identifier) • Operational Need Date • Operator Need Date • Order Need Date • Ordinary National Diploma • Overnight Delivery

ONDCP Office of National Drug Control Policy (USA)

ONDINES Ondes Internes

ONDS Open Network Distribution Services

ONE Onepusu (Solomon Islands) (location identifier) • OPAC Network in Europe • Open Network Environment • Organization Number of Employees

ONEB Office of National Environmental Board

ONEM Office National de l'Emploi (Belgium)

ONER Oceanic Navigational Error Report

ONERA Office National d'Études et de Recherches Aérospatiales (France)

ONERN Oficina Nacional de Evaluación de Recursos Naturales (Peru)

ONET Output Network

ONF Office National des Forêts (France) • Old North French

ONFCSM Observatorio Nacional de Física Cosmica - San Miguel (Argentina)

ONFOC Office National de Formation Continue

ONFr Old North French

ONG Mornington Island (Australia-QLD) (location identifier) • Organisation Non-Gouvernementale • Organización No-Gubernamental

ONGC Oil and Natural Gas Commission

ONH Oneonta (USA-NY) (location identifier)

ONI Moanamani (Indonesia) (location identifier) • Office of Naval Intelligence • Oficina Nacional de Información • Operator Number Identification • Optical Network Interface • Organismo Nacional de Información • OSI Network Interface

ONIEP Office National d'Information sur les Enseignements et les Professions (France)

ONIITEM Oficina Nacional de Invenciones, Información Técnica y Marcas (Cuba)

ONIM Oh No It's Monday

ONIOP Office National d'Information pour l'Orientation Pédagogique (France)

ONIRIS OSU-NOAO Infrared Imaging Spectrometer

ONIS OSU-NOAO Infrared Spectrometer

ONISEP Office National d'Information sur les Enseignements et les Professions (France)

ONIST Office National d'Information Scientifique et Technique

ONKz Ortsnetzkennzahl

ONL O'Neill (USA-NE) (location identifier) • On-Line • Optique Non Linéaire

ONLAS Online Archive System

ONLFD Online Fault Detection

ONLS Online System

ONM Office National Météorologique • Socorro (USA-NM) (location identifier)

ONMS Open Network Management System • Optical Network Management System • Orbiter Neutral Mass Spectrometer (PV)

ONN Onion Bay (USA-AK) (location identifier)

ONNUM Oficina Nacional de Normas y Unidades de Medida (Costa Rica)

ONO Ontario (USA-OR) (location identifier) • Or Nearest Offer • Or Near Offer • Ouest-Nord-Ouest

ONP Newport (USA-OR) (location identifier) • Office National des Pensions (Belgium) • Open Network Provision

ONPI Office National de la Propriété Industrielle (Algeria) • Office of News and Public Information

ONPT Office National des Postes et Télécommunications

ONQ Ouvrier Non Qualifié

ONR Monkira (Australia-QLD) (location identifier) • Office of Naval Research (USA)

ONRAS Occasional Notes of the Royal Astronomical Society (periodical code)

ONRJ Observatório Nacional de Rio do Janeiro (Brazil-RJ)

ONRS Organisme National de la Recherche Scientifique (Algeria)

ONS Office for National Statistics (UK) • On-Line News Service • Onslow (Australia-WA) (location identifier) • Operational Need Statement • Oversampled Noise-Shaping

ONSHR Onshore

ONSS Office National de la Sécurité Sociale (Belgium)

ONSSAPL Office National de la Sécurité Sociale des Administrations Provinciales et Locales (Belgium)

Ont Ontario (Canada)

ONT Office of Naval Technology • Ontario (USA-CA) (location identifier) • Optical Network Termination

ONTC Optical Networks Technology Consortium

ONU Ohio Northern University (USA-OH) • Olivet Nazarene University (USA-IL) • Ono i Lau (Fiji) (location identifier) • Optical Networking Unit • Optical Network Unit • Organisation des Nations Unies

ONUDI Organisations des Nations Unies pour le Développement Industriel

ONUEF Office National des Universités et des Écoles Françaises (France)

ONUESC Organisme des Nations Unies pour l'Éducation, les Sciences et la Culture

ONVA Office National des Vacances Annuelles (Belgium)

ONWI Office of Nuclear Waste Insulation

ONX Colón (Panama) (location identifier)

ONY Olney (USA-TX) (location identifier)

Onz Onzijdig

o/o On Order • Order of

oo On Order • Order of

oO ohne Ort

O/o Offers Over

O-O Object-Oriented

O/O On/Off • Operations Order • Optical to Optical • Ordering Officer • Outboard/Outboard

O&O Operational & Organizational • Organizations & Operations

OO Belgium (international aircraft registration code) • Object-Orientation • Object-Oriented • Once Over • On Order • Operation Order • Organisateur d'Ordre

OOA Object-Oriented Analysis • Optimum Orbital Altitude • Oskaloosa (USA-IA) (location identifier) • Out Of Action • Out Of Area

OOA&D Object-Oriented Analysis & Design

OOAD Object-Oriented Analysis and Design • Object-Oriented Application Design

OOB Object-Oriented Bibliography • Out Of Band

OOBE Object-Oriented Business Engineering • Out-Of-Box Experience

OOBI Objet Orbitant Bien Identifié

OOC On-Off Control • Outer Oort Cloud • Out Of Character • Out Of Core • Out Of Cycle

OOD Object-Oriented Design • Object-Oriented Development • Object-Oriented Directory • Officer Of the Day • Officer Of the Deck • Operational Oceanography Center • Orbiter On Dock

OODA Object-Oriented Design with Assemblies • Observation - Orientation - Décision - Action • Observe - Orient - Decide - Act

OODB Object-Oriented Database

OODBMS Object-Oriented Database Management System

OODBS Object-Oriented Database System

OODCE Object-Oriented Distributed Computing Environment

OODMS Object-Oriented Database Management System

OODP Out-Of-Detent Pitch

OODR Out-Of-Detent Roll

OODT Object-Oriented Data Technology (*) • Object-Oriented Digital Technology

OODV Orbit-On-Demand Vehicle

OoE Out of Eclipse

OOE Object-Oriented Environment • Only Outside of Eclipse • Out-Of-Ecliptic (later ISPM, now Ulysses)

OOEF Object-Oriented Enterprise Framework

OOER Object-Oriented and Entity-Relationship (modeling)

OOERM Object-Oriented Entity-Relationship Model

OOF Out Of Frame

OOFDM Object-Oriented Functional Data Model

OOFF Object-Oriented Feature Finder

OOFS Object-Oriented File System

OOG Object-Oriented Graphics

OOGL Object-Oriented Graphics Language

OOGP Object-Oriented Genetic Programming

OOH Orbital Operations Handbook

OOHDM Object-Oriented Hypermedia Design Method • Object-Oriented Hypermedia Design Model (*)

OOI Object-Oriented Interface • Other On-Line Issue • Other Outstanding Item

OOIS Object-Oriented Information System

OOK On-Off Keying • Toksook Bay (USA-AK) (location identifier)

OOL Gold Coast - Coolangatta (Australia-QLD) (location identifier) • Object-Oriented Language • Object-Oriented Library • Object-Oriented Listing • Operator-Oriented Language • Optimized Optical Link • Out Of Limits • Out Of Lock

OOLR Overall Objective Loudness Rating

OOM Cooma (Australia-NSW) (location identifier) • Object-Oriented Method • Observable Operator Model • On-Orbit Maintenance

OOMM Organizational Operations and Maintenance Manual

OOMP Object-Oriented Management Planning

OOMPS Open Ocean Monitoring Payload Studies

OOMR Oscillating Oblique Magnetic Rotator

OOMS On-Orbit Maintenance/Servicing • Open Ocean Monitoring Studies

OON Object-Oriented Numerics

OONI Objet Orbitant Non Identifié

OOO Old Orchard Observatory (USA-MO) • Out Of Order • Ozone

OOOI Out - Off - On - In

OOOS Object-Oriented Operating System

O&OP Operational & Organizational Plan

OOP Object-Oriented Pattern • Object-Oriented Programming • Office for Official Publications • Operational and Organizational Plan • Out Of Plane • Out Of Position • Out Of Print • Overseas Order Point

OOPC Ocean Observations Panel for Climate

OOPEC Office for Official Publications of the European Communities

OOPIC Object-Oriented Programmable Integrated Circuit

OOPL Object-Oriented Programming Language

OO.PP. Obras Públicas

OOPS Fundamental'nye Osnovy Opticheskoi Pamiati i Sredy (periodical code) • Object-Oriented Programming and Systems • Object-Oriented Programming System • Off-Line Operating Simulator • Okayama Optical Polarimeter and Spectrometer • Okayama Optical Polarimetry and Spectroscopy • Operational Ozone Processing System • Operational Ozone Product System

OOPSLA Object-Oriented Programming Systems Language and Applications (conference)

OOR Mooraberree (Australia-QLD) (location identifier) • On-Orbit Review

OORA Object-Oriented Requirement Analysis

OORALD Object-Oriented Requirements Analysis and Logical Design

OORDB Object-Oriented Relational Database

OORTSF Object-Oriented Real-Time Software Foundation

OOS Object-Oriented Software • Object-Oriented System • Office for Outer Space • Off-line Operating Simulator • On-Orbit Station • On-Orbit Support • Optical Oscillator Strength • Orbit-to-Orbit Shuttle • Orbit-to-Orbit Stage • Out Of Service • Out Of Synchronization

OOSA Object-Oriented Systems Analysis • Office for Outer Space Affairs (UNO)

OOSD Object-Oriented Software Development • Object-Oriented Systems Development • On-Orbit Support Document • Orbital Operations Support Document

OOSDP Oceans Observation System Development Panel • On-Orbit Station Distribution Panel

OOSE Object-Oriented Software Engineering • Object-Oriented Systems Engineering

OOSS On-Orbit Support System

OOT Object-Oriented Technology • Onotoa (Kiribati) (location identifier) • Out Of Tolerance • Out Of Town

OOTM Object-Oriented Testing and Maintenance

OOTW Operations Other Than War

OOUI Object-Oriented User Interface

oOuJ ohne Ort und Jahr

OOV Object Of Verification • Object-Oriented View • Out Of Vocabulary

OOVRO Observations of Owens Valley Radio Observatory (periodical code)

OOVT Out Of Vocabulary Term (rate)

OOVW Out Of Vocabulary Word (rate)

OOWS Object-Oriented Web Service

OOZE Object-Oriented Z Environment

o/p Out of Print

op Opacus • Opaque • Opera • Operate • Operated • Operation • Operational • Operator • Opinion • Opposite • Optical • Optimus • Opus

oP Ordentlicher Professor

O/p Orden de Pago

Op Operation • Operational • Operator • Optical • Opus • Output

O/P Out of Print • Output • Ownership/Purpose

O&P Operations & Procedures • Outsourcing & Privatization

OP Observation Post • Observation Procedure • Observatoire de Paris (France) • Observing Programme • Occupation Probability • Office of Personnel • Office of Policy • Office Power • Oficina Principal • Old Persian • Ooty Pulsar • Opacity Project • Open • Open Policy • Operate • Operating Plan • Operating Procedure • Operation • Operational • Operational Performance • Operational Phase • Operational Policy • Operational Priority • Operational Procedure • Operational Program • Operational Project • Operational Publication • Operations • Operations Plan • Operator • Opposite Polarization • Optical • Optical Parameter • Optical Probe • Optional Procedure • Orbital Period • Orbital Probe • Ordering Possibility • Ordinary Penetration • Ordinateur Personnel • Ordnance Pamphlet • Orientation Professionnelle • Orthogonal Polarization • Osmotic Pressure • Other People • Other Procurement • Outer Panel • Outer Planet • Out of Print • Output • Output Primary • Output Processor • Outside Plant • Ouvrier Professionnel • Over Pressure • Overproof • Oxide Precipitate • Oxygen Purge

Opa Opaque

OPA Kópasker (Iceland) (location identifier) • Océan Parallélisé • Oferta Pública de Acciones • Oferta Pública de Adquisición • Office of Price Administration • Office of Public Affairs • Officer Personnel Act (USA) • Offre Publique d'Achat • On-Line Privacy Alliance (USA) • Opaque • Operational Amplifier • Operations Planning Analysis • Operations Procurement Authority • Optical Parametric Amplification • Orbiter Plasma Analyzer • Organisme Paritaire Agréé • Other Procurement Authority • Output Plate Assembly • Overarching Partnering Agreement • Overseas Publishers Association

OPAAL Optimized Portable Algorithms and Application Libraries

OPAC Office Public d'Aménagement et de Construction (France) • On-Line Payment And Collection (system) • On-Line Public Access Catalog (*) • Open Public Access Catalog • Operations Planning Advisory Committee • Organisme Paritaire Collecteur Agréé (France)

OPAEP Organisation des Pays Arabes Exportateurs de Pétrole

OPAG Open Programme Area Group

OPAL Object-Process-Actor Language • Omnipurpose Apparatus for LEP • Open Programmable Architecture Language • Operational Performance Analysis Language • Optical Platform Alignment Linkage • Optimized Portable Application Library

OPALS Officer Projection Aggregate Level System

OPANAL Organismo para la Proscripción de las Armas Nucleares en la America Latina

OPAQUE Optical Atmospheric Quantities in Europe

OPARS Optimum Path Aircraft Routing System

OPAS Operational Public Address System • Overpass

OpAt Optika Atmosfery (periodical code)

OpAtO Optika Atmosfery i Okeana (periodical code)

OPAX Opération Pilote Aristote X500

OPB Observation Planning Branch • Open Bay (Papua New Guinea) (location identifier) • Oxidizer Preburner

OPBOV Oxidizer Preburner Oxidizer Valve

OPBS Orbiter Payload Bay Simulator

OPC Observing Programme Committee • Ocean Products Center • Office Products Center • OLE for Process Control • On-Line Public Catalogue • OpenGL Performance Characterization • Operations Planning Center • Optical Particle Counter • Optical Phase Conjugation • Optical Photoconductor • Optical Proximity Correction • Organic Photoconducting Cartridge • Organic Photoconductor • Organization Postcode • Originating Point Code

OPCA Organisme Collecteur Paritaire Agréé (France)

OPCC Oficina para la Protección de la Calidad de los Cielos (Chile)

OPCGA Observable Protein Crystal Growth Apparatus

op. cit. Opera Citata • Opere Citato

OPCODE Operational Code • Operations Code

OPCOM Operating Command • Operational Command • Operations Communications

OPCON Operational Control • Optimizing Control

OPCONCTR Operations Control Center

OPCS Office of Population Censuses and Surveys (UK)

OPCT Operational Products Control Table • Operations Planning and Control Team

OPCVM Organisme de Placements Collectifs en Valeurs Mobilières (France)

OPD Object-Process Diagram • Observatorio do Pico dos Dias (Brazil) • Office of Program Development • Operand • Operations and Processing Division • Operations Process Document • Operations Department • Optical Particle Detector • Optical Path Delay • Optical Path Difference • Optical Proximity Detector • Orbiting Propellant Depot • Output Driver

OPDAR Optical Detection And Ranging

OPDD Operational Plan Data Document

OPDEP Operations Deputy

OPDIR Operational Directive

OPDOC Operational Documentation

OPDS Officer Professional Development Seminar • Off-shore Petroleum Discharge System

OPDU Operations Protocol Data Unit

OpE Once per Exposure

OPE Offre Publique d'Échange • Once Per Exposure • Operational Planning Estimate • Operational Processing Equipment • Operations Project Engineer • Operator Product Expansion • Optical Pointing Error • Optical Probe Experiment • Orbiting Primate Experiment • Other Plant Equipment • Other Project Element • Outer Planets Explorer

OPEC Organization of Petroleum-Exporting Countries

OPECST Office Parlementaire d'Évaluation des Choix Scientifiques et Technologiques (France)

OPED Other Pay Entry Date

OPEEPE Organisation Européenne d'Études Photogrammétriques Expérimentales

OPEF Overall Plume Enhancement Factor

OPEI Office of Public Education and Information

OPEIU Office and Professional Employees International Union

OPELINT Operational Electronic Intelligence

OPEN Object-oriented Process, Environment and Notation • Open Protocol Enhanced Network • Optical Pan European Network • Origin of Plasma in Earth's Neighborhood (project)

OPEOS Outside Plant Engineering and Operations System

OPEP Orbital-Plane Experiment Package • Organisation des Pays Exportateurs de Pétrole

Oper Operate • Operating

OPER Operate • Operation • Operations • Operator

OPERA Onde di Plasma e Radiazioni Aurorali • Operational Programme for the Exchange of (weather) Radar (information) • Operations for ALIS • Optimisation des Performances d'Ariane • Oscillation Project with Emulsion-Tracking Apparatus

OPERREQ Operator Requests

OPERRESP Operator Response

OPES Ordinary Polyelastic Sample

OPEST Oil Protection of Emission System Test

OPET Organization for the Promotion of Energy Technologies

OPEVAL Operational Evaluation

OPEX Olympus Propagation Experiment • Olympus Propagation Experimenter • Operating Expenses

OPF Miami - Opa Locka (USA-FL) (location identifier) • Official Personnel File • Official Personnel Folder • Optical Propagation Facility • Orbiter Processing Facility

OPFAC Operating Facility • Operational Facility • Operations Facility

OPFC Orbiter Preflight Checklist

OPFL Observer Personal Feed Line

ÖPG Österreichische Physikalische Gesellschaft (Austria)

OPG Operations Planners Group • Operations Planning Group • Optical Parametric Generation • Optical Precipitation Gauge

OPGC Observatoire de Physique du Globe de Clermont-Ferrand (France)

OPGEN Operations Plan Generation

OPGT Outer Planets Grand Tour

OPGUID Optimum Guidance

Oph Ophiucus-Ophiuchi (constellation)

OPH Observatoire Planétarium de la Hague (France) • Operational Propellant Handling

OPHF Orbital Polarized Hartree-Fock

OPHLM Office Public d'Habitations à Loyer Modéré (France)

OPHS Operational Propellant Handling System

Ophtal Ophtalmologiste • Ophtalmologie

Ophth Ophthalmologist • Ophtalmology

OP&I Office of Patents & Inventions

OPI Oenpelli (Australia-NT) (location identifier) • Office de la Propriété Industrielle • Office of Primary Interest • Office of Public Information • Open Prepress Interface • Operations Procedure Instruction • Operator Interface • Opposite Polarity Intrusion • Optional Procedure Indicator • Oral Proficiency Interview • Orbital Position Indicator • Orbiter Payload Interrogator • Organismo Público de Investigación • Outer Port Injection

OPIDF Operational Planning Identification File

OPIE One-time Password In Everything

OPIG Octopole Ion-beam Guide

OPIM Order Processing and Inventory Monitoring

OPIS Orbiter Prime Item Specification • Organismo Público de Investigación (Spain)

OPIT Opération Pilote Interministérielle de Télédétection (France) • Oxide Powder in Tube

OPJ Officier de Police Judiciaire

Opl Operational

OPL Observatoire Populaire de Laval (France) • On-Line Programming Language • Opelousas - Saint Landry Parish (USA-LA) (location identifier) • Open Power Law • Open Problem List • Operational • Optical Path Length • Orbiter Parts List • Organiser Programming Language

OPLA Optical Programmable Logic Array

OpLAN Operational Local Area Network

OPLAN Operational Plan • Operations Plan

op. laud. Opere Laudato

OPLE Omega Position Location Experiment • Optical Path Length Equalizer

OPLED Organic Polymer Light-Emitting Device

OPLF Orbiter Processing and Landing Facility

OPLIN Ohio Public Library Information Network

OPLL Optical Phase-Locked Loop

OPLOC Operating Location • Operations Location

OPLSM Operational Phase Logistic Support Manager

OPLSS Optimized Portable Life-Support System

OPLT Operational Probable Level of Traffic

OpM Operations per Minute • Output per Man

OPM Object-Process Methodology • Object Protocol Model • Observatoire de Paris-Meudon (France) • Office of Personnel Management • Office of Prime Minister • Office of Procurement and Materiel • Office of Product Management • Office of Program Management • Office of the Product Manager • Office of the Program Manager • Office of the Project Manager • Operations Message • Operations Per Minute • Operations Procedure Manual • Operator Programming Method • Optical Power Meter • Optical Properties Monitor • Orbital Propulsion Module • Other People's Money • Outer Planet Mission • Output Per Man • Outside Plant Module

OpMeP Optiko Mekhanicheskaia Promyshlennost (periodical code)

OPMODEL Operations Model

OpMS Output per Man Shift

OPMS Output Per Man Shift

OPMT Observatoires du Pic du Midi et de Toulouse (France) (now OMP)

Opn Operation • Opinion • Option

OPN Ondes dans les Plasmas Naturels • Open • Operation • Orate Pro Nobis

OPNAV Optical Navigation

OPNET Operational Network Evaluation Tool

OPNL Operational

OPNOTE Operational Note

Opnr Opener

Opns Operations

OPNS Operations

OPO Office of Public Outreach • Offre à Prix Ouvert • One-Person Operation • One-Person Operator • Operational Performance Objective • Operations Order • Operations Project Office • Oporto - Dr. Francisco Sa Carneiro (Portugal) (location identifier) • Optical Parametric Oscillator • Orbiter Project Office • Orbiting Planetary Observatory • Oscillation Paramétrique Optique • Porto - Dr. Francisco Sa Carneiro (Portugal) (location identifier)

OPOCE Office des Publications Officielles des Communautés Européennes

OPOL Optimization-Oriented Language

OPORD Operations Order

OPOV Oxidizer Preburner Oxidizer Valve

opp Opera • Opuses

Opp Opera • Opportunity • Oppose • Opposed • Opposite • Opposition • Opuses

OPP Octal Print Punch • Office of Pesticide Programs • Office of Polar Programs (NSF) • Oficina de Planeamiento y Presupuesto (Uruguay) • Operational Positional Procedure • Opportunity • Opposite • Optical Precipitate Profiles • Order Processing Point • Other Personal Property • Other Physical Principles • Other Physical Properties • Other Programme Participation • Outer Planet Program • Outer Planet Project • Out of Print at Present • Ouvrier Professionnel Principal • Oxygen Partial Pressure

OPPA Operation Plan Package Appraisal

OPPAR Orbiter Project Parts Authorization Request

OPPE Office of Policy, Planning, and Evaluation • Operations Planning Project Engineer

OPPL Orbiter Project Parts List

OPPM Outside the Principal Period of Maintenance

OPPS Operations Planning and Project Support

OPPT Office of Pollution Prevention and Toxics

OPPTS Office of Prevention, Pesticides and Toxic Substances

OPPTY Opportunity

OpPuT Optoelektronika i Poluprovodnikovaia Tekhnika (periodical code)

Oppy Opportunity

OPQ Occupational Personality Questionnaire • Ouvrier Professionnel Qualifié

OPQCM Office Professionnel de Qualification des Conseils en Management

OPQF Office Professionnel de Qualification de la Formation

Opr Operate • Operator

OP&R Operations Planning & Requirements

OPR Office of Planning and Research • Office of Primary Responsibility • Officer Performance Report • Official of Primary Responsibility • Offsite Procurement Request • Operational Project Requirement • Operations Planning Report • Operations Planning Review • Operations Problem Report • Operator • Optical Page Reader • Optical Page Reading • Optical Pattern Recognition • Orbiter Payload Recorder • Organisme Public de Recherche • Organization of Primary Responsibility • Ortho/Para Ratio • Output Product Request • Overall Pressure Ratio • Oxygen Pressure Regulator

OPRA Occupational Pension Regulatory Authority

OPRD Organisme Public de Recherche et Développement

OPRDY Operationally Ready

Oprg Operating

OPRI Office de Protection contre les Rayonnements Ionisants (France)

OPROM Optical Programmable Read-Only Memory

OPRQ Operation Request

OPRV Oxygen Pressure Relief Valve

Ops Operates • Operations • Operations per Second

OpS Operations per Second

OPS Objective Prism Survey • Observation Preparation System • Observatory Project Scientist • Oferta Pública de Suscripción • Office of Price Stabilization • Off-Premises Station • Offre Publique de Souscription • Oil Pressure Switch • Onboard Payload Support • Online Process Synthesizer • Open Profile Standard • Open Profiling Standard • Operational Paging System • Operational Power Supply • Operational Protection System • Operational Sequence • Operations • Operations Per Second • Operations Processing System • Operations Sequence • Opération Ponctuelle de Symétrie • Optical Processing System • Optical Propagation Study • Optical Sensor • Oracle Parallel Server • Orbiter Project Schedule • Orbiting Primate Spacecraft • Oxygen Purge System • Sinop (Brazil-MT) (location identifier)

OPSA Origin and evolution of Planetary and Satellite Atmospheres

OPSAT Operational Satellite

OPSB Orbiter Processing Support Building • Output Product Status Block

OpSCN Optical Sciences Center Newsletter (periodical code)

OPSCON Operations Concept • Operations Control

OPScr Opération Ponctuelle de Symétrie Cristallographique

OPSDB Operations Database

OPSDBA Operations Database Administrator

OPSDEP Operations Deputy

OPSE Optical Science and Engineering

OPSEC Open Platform for Secure Enterprise Connectivity • Operations Security

OPSF Orbital Propellant Storage Facility

OPSG Operations Planning Steering Group

OPSGP Operations Group

OPSL Optically-Pumped Semiconductor Laser

OPSN Output Product Status Number

OPSNET Operations Network (ESA)

OPSNOTE Operations Note

OPSORD Operations Order

OpSp Optika i Spektroskopiia (periodical code)

OPSPEC Operational Specification

OPSR Operations Supervisor • Output Product Status Request

OPSS Office of Public Service and Science • Orbital Propellant Storage System

OPST Outer Planets Study Team

OPSTAT Operational Status

OPSUM Operational Summary • Operations Summary

OPSW Operational Switch

OPSWG Outer Planets Science Working Group

Ópt Óptica

Opt Optical • Optics • Optimal • Optimum • Option • Optional • Optique

OPT Observatory - Planetarium - Theater • Open Protocol Technology • Operational Planning Team • Operational Pressure Transducer • Operations Planning Team • Operations Planning Tool • Optical • Optical Point Transfer • Optics • Optimum • Option • Optional • Orbiting Planetary Telescope • Orientation Phase Transition • Outline Test Plan • Overhead Projection Transparency • Ozone Processing Team

OPTA Optimal Performance Theoretically Attainable

OPTADS Operations Tactical Data System

OPTAR Operating Target

OPTB Operational Program Time Base

OptCo Optics Communications (periodical code)

OptD Doctor of Optometry

OPTE Operator Proficiency Training Equipment • Organisme pour la Promotion des Technologies Énergétiques

OPTEC Operational Test and Evaluation Center • Operational Test and Evaluation Command

OptEn Optical Engineering (periodical code)

OPTIC Optical Procedural Task Instruction Compiler

OPTICAL Optics and Topography Involving the Cornea And Lens

OPTICON Optical and Infrared Coordination Network

OPTID Optical Identification

OPTINT Optical Intelligence

Optl Optional

OptL Optics Letters (periodical code)

OptLE Optics and Lasers in Engineering (periodical code)

OptLT Optics and Laser Technology (periodical code)

OptMa Optical Materials (periodical code)

OptN Optics News (periodical code)

OPTN Option

OPTOL Optimized Test-Oriented Language

OPTOMA Ocean Prediction Through Observation, Modeling and Analysis

OptPN Optics & Photonics News (periodical code)

OptQE Optical Quantum Electronics (periodical code)

OPTR Optical Punched-Tape Reader • Orifice Pulse Tube Refrigerator

OptSp Optical Spectra (periodical code) • Optics and Spectroscopy (periodical code)

Opty Opportunity

OPU Balimo (Papua New Guinea) (location identifier) • Open University (Starlink internal code) • Operations Priority Unit • Overlay Processor Unit

OPUS Observation/Postobservation Unified System • Obvious Password Utility System • Octal Program Updating System • Office of Public Understanding of Science • Olympus Payload Utilization Secretariat • Offshore Permanent Upwelling Structure • Operational Pipeline Unified System • Organization of Persistant Upwelling Structure • OSS and PODPS Unified System • Outstation Production Unified System

OPUS3 Observing Platform for Ultraviolet Solar-System Science

OpV Ohms per Volt

OPV Oferta Pública de Venta • Offre Publique de Vente • Ohms Per Volt • Optical Pressure Vessel • Optics Pressure Vessel

OPVR Operational Performance Validation Report

OPW Office of Public Works • Operating Weight • Opuwa (Namibia) (location identifier) • Orthogonalized Plan Wave

OPWS Orbital Payload Workstation

OPX Off-Premises Exchange • Off-Premises Extension

OPZ Operational Pointing Zone

OPZONE Operation Zone

OQ Optical Quality • Ouvrier Qualifié

OQA Operator's Questions and Answers

OQB Oscillation Quasi-Biennale

OQC Office of Quality Control

OQD Optical Quantum Detector

OQE Objective Quality Evidence

OQF Operator's Questions Forum

OQL Object Query Language • Observed Quality Level • On-Line Query Language

OQP Operator's Questions Procedure • Optimum Qualification Procedure

OQPSK Offset Quadratic Phase Shift Keying

OQQ Ongoing Questions and Queries • Optically Quiet QSO • Optically Quiet Quasar

OQR Officer's Qualification Record • Operator's Questions and Responses

OQS Object Query Service

OQT Operator's Questions Time

o/r On Request

or Oriya (ISO language code)

Or Orient • Oriental • Oriente • Origen • Original • Other

O/R On Request • Operational Requirement • Originator/Recipient • Outside Radius • Outside Relief • Overrange • Override • Owner's Risks • Oxygen Relief

O&R Overhaul & Repair

O-R Object-Relational

OR Object-Relational • Object Relationship • Object Reuse • Observation Request • Observatoire de Rouen (France) • Occurrence Report • Odds Ratio • Official Receiver • Old Roman • Ondernemingsraad • On Return • Onward Routing • Operating Reactor • Operating Resources • Operating Room • Operationally Ready • Operationally Risky • Operational Readiness • Operational Report • Operational Requirement • Operational Research • Operational Right • Operation Record • Operations Requirement • Operations Research • Operations Review • Operations Room • Optical Reader • Optical Repeater • Orderly Room • Oregon (USA) • Organic-Refractory • Originating Register • Other Rank • Outer Roll • Out of Range • Output Register • Output Relevance • Outside Radius • Overhaul and Repair • Overload Relay • Owner's Risk • Own Risk • Oxygen Relief

OR&A Operational Research & Analysis • Operations Research & Analysis

ORA Observatoire Régional de l'Air (France) • Occultation Radiometer • Office of Research and Applications • Operational Readiness Assessment • Operational Requirements Analysis • Operations Research Analysis • Optical Reference Axis • Optical Research Associates • Optimum Repair Analysis • Orán (Argentina-SA) (location identifier) • O'Reilly and Associates • Organizational Registration Authority • Organization Registration Authority • Output Reference Axis • Output Register Address

ORAC Observatory Reduction and Acquisition Control • Oxygen Radical Absorbance Capacity

ORACLE Oak Ridge Automatic Computer and Logical Engine • Office Régional d'Action Culturelle de Liaisons et d'Échanges • On-Line Retrieval and Computational Language for Economists • Optimized Reliability And Component Life Estimate • Optional Reception of Announcements by Coded Line Electronics • Ozone Research with Advanced Cooperative Lidar Experiments (project)

ORACT Operational Readiness and Confidence Test

ORAD Orbiter Radar (mapper) (PV)

ORAM Output Random Access Memory

ORAN Orbital Analysis • Organisation Régionale Africaine de Normalisation

ORANGE Optronique et Radar de Nouvelle Génération

ORAP Organisation Associative Parallèle

ORAS Oil Region Astronomical Society (USA-PA)

ORASA Operational Research And Systems Analysis

Orat Oration • Orator • Oratorical • Oratorically • Oratorio • Oratory

ORATOS Orbit Attitude Operations System

ORAU Oak Ridge Associated Universities (USA-TN)

ORAVEP Observatoire des Ressources Audiovisuelles pour la Formation Permanente (France)

Orb Orbit • Orbital • Orbiter • Orbiting

ORB Object Request Broker • Observatoire Royal de Belgique (Belgium) • Officer Record Brief • Omnidirectional Radio Beacon • Operations Record Book • Operations Review Board • Optimum Radiation Baffle • Örebro (Sweden) (location identifier) • Orbit • Orbital • Orbiter • Orebro (Sweden) (location identifier) • Orthogonal Recursive Bisection • Outer Radiation Belt

ORBIS Orbiting Radio Beacon Ionospheric Satellite

ORBIT Oracle Binary Internal Translator

ORBL Open Relay Black List

ORBS Open Relay Behavior System • Orbital Rendezvous Base System • Oregon Rule-Based System

ORBZ Open Relay Blackhole Zone

ORC Offer Release Code • Online Reactivity Computer • Operational Readiness Check • Operations Request Card • Operations Research Center • Operations Resource Control • Optical Rotation Curve • Orbital Research Centrifuge • Ordinance Rocket Center • Organic Ranking Cycle • Orocue (Colombia) (location identifier)

ORCA Open Reliable Communication Architecture • Operational Requirement-based Casualty Assessment • Operations Requirements Continuity Assessment

ORCC Operations Resource Control Center

ORCH Orchard

ORCHIS Oak Ridge Computerized Hierarchical Information System

ORCHRD Orchard

ORCLAC Oficina Regional para la Cultura en Latina America y el Caribe

ORCON Originator Control

ORCP Output Reference Central Pin

ORCS Online Remote Compile System

ORCUS Orbital Recording Capability and Update System

ORCYT Oficina Regional de Ciencia Y Tecnologia

Ord Orden • Order • Ordinaire • Ordinary • Ordinal • Ordinance • Ordinary • Ordnance

ORD Chicago O'Hare (USA-IL) (location identifier) • Object Relation Diagram • Office of Research and Development • Operational Rate Distortion • Operational Readiness Data • Operational Readiness Date • Operational Readiness Demonstration • Operational Readiness Document • Operational Ready Data • Operational Ready Date • Operational Requirements Document • Operations Requirements Document • Optical Reference Device • Optical Rotary Dispersion • Orbital Requirements Document • Ordinary • Ordnance • Organizational Requirements Document

ORDB Object Relational Database • Occupational Research Database • Open Relay Database

ORDBMS Object Relational Database Management System

ORDC Orbiter Data Reduction Center

ORDEAL Orbit Rate Display - Earth And Lunar

ORDEM Orbital Debris Engineering Model

ORDENG Ordnance Engineering

ORDIS Optical Reading Direct Input System

Ordn Ordnance

ORDVAC Ordnance Variable Automatic Computer

Ordza Ordenanza

Ore Oregon (USA)

ORE Occupational Radiation Exposure • On-orbit Repair Experiment • Operational Readiness Evaluation • Operational Readiness Exercise • Operations Requirements Extract • Oregon (USA) • Orléans (France) (location identifier) • Output Request Element

OREF Observatoire Régional pour l'Emploi et la Formation (France)

Oreg Oregon (USA)

ORELA Oak Ridge Electron Linear Accelerator

OREO Orbiting Radio Emitting Observatory

OREP Office Régional d'Éducation Permanente (France)

OREX Orbital Reentry Experiment • Orbiting Reentry Experiment

Orf Orifice

OR/F Orbiter Refrigerator/Freezer

ORF Norfolk (USA-VA) (location identifier) • Open Reading Frame • Operational Readiness Float • Orbiter Refrigerator/Freezer • Orifice • Österreichische Rundfunk (Austria) • Portsmouth (USA-VA) (location identifier) • Virginia Beach (USA-VA) (location identifier)

ORFA Oxygen Resonance Fluorescence and Absorption

Orfebr Orfebrería

ORFEUS Orbiting and Retrievable Far and Extreme Ultraviolet Spectrometer

ORFI Optical-RF Interface

ORFM Outlet Region Feature Model

Org Organic • Organism • Organization • Organizational • Organize • Organized • Organizer • Origin • Original

ORG Operations Research Group • Optical Rain Gauge • Organization • Organizational • Origin • Paramaribo - Zorg en Hoop (Suriname) (location identifier)

Organ Organic • Organization

ORGD Organization Directorate

ORGDEV Organization Developer

OrGeo Organic Geochemistry (periodical code)

ORGPHC Orographic

Orgzn Organization

ORH Office of the Regional Hydrologist • Operations Requirement Handbook • Worcester (USA-MA) (location identifier)

ori Original (file extension)

Ori Origin • Original • Orion-Orionis (constellation)

ORI Occurrence of Reinforcing Information • Ocean Research Institute (Japan) • Office of Research and Information • Office of Research Integrity • Ongoing-Research Information • Online Retrieval Interface • Operational Readiness Inspection • Operations Research Inc. • Orbital Replacement Instrument (*) • Orbit Replaceable Instrument • Original • Orionids (meteor stream) • Port Lions (USA-AK) (location identifier)

ORID Orthorectified Image Data

ORIDE Override

ORIE Orientation

ORIEN Orientation

Orig Origin • Original • Originally • Originate • Originated

ORIG Origin • Original

ORION Oak Ridge Isochronous Observation Network (USA-TN) • Optically Recorded Information On-Line Network

ORISE Oak Ridge Institute for Science Education (USA-TN)

ORIT Operational Readiness Inspection Team • Operational Readiness Inspection Test

ORJ Orinduik (Guyana) (location identifier)

ORJETS Online Remote Job Entry Terminal System

Ork Orkney Islands

ORK Cork (Ireland) (location identifier) • Optimizable Recursive Keyword

Orkn Orkney Islands

ORL Optical Return-Loss • Orbital Research Laboratory • Orlando (USA-FL) (location identifier)

ORLA Optimum Repair Level Analysis

ORLaND Oak Ridge Large Neutrino Detector

OrLi Origins of Life (periodical code)

ORLL Operational Reports - Lessons Learned

ORLY Overload Relay

ORM Northampton (UK-England) (location identifier) • Object Role Modeling • Observatorio del Roque de los Muchachos (Canary Islands) • Ole Rømer Museet (Denmark) • Operational Rules Model • Operations Reference Model • Optical Reference Manual • Optical Reference Measurement • Optical Remote Module • Organizational Record Manager

ORMAS Operational Resource Management Assessment System

ORMP Oblique Rotating Magnetic Pulsator

ORMS Operating Resource Management System • Operations Resource Management System

ORMSS Ohio River Main Stem Study

ORMU Orbital Remote Maneuvering Unit

Orn Orange • Ornitología

ORN Oran - Es Senia (Algeria) (location identifier)

OrNav Ortung und Navigation (periodical code)

Ornge Orange

ORNI Objet Roulant Non Identifié

ORNL Oak Ridge National Laboratory (USA-TN)

ORO Office of Regional Operations • Old Royal Observatory (UK-England) • On Request Only • Orbiter Radio Occultation (PV) • Yoro (Honduras) (location identifier)

OROM Optical Read-Only Memory

OROS Optical Read-Only Storage

ORP Optical Reflecting Projector • Optional Retirement Program • Orapa (Botswana) (location identifier) • Orbital Rendezvous Procedure • Overhaul Recovery Percent • Oxidation-Reduction Potential

ORPA Office of Research and Project Administration • Orbiter Retarding Potential Analyzer (PV)

ORPC Object Remote Call Procedure

ORPICS Orbital Rendezvous Positioning, Indexing and Coupling System

ORQ Norwalk (USA-CT) (location identifier)

ORR Operational Radar Replacement • Operational Readiness Review • Operational Ready Rate • Operations Requirements Review • Orbital Readiness Review • Orbital Rendezvous Radar • Orbit Readiness Review • Overhaul Replacement Rate • Yorketown (Australia-SA) (location identifier)

ORRAS Optical Research Radiometrical Analysis System

ORRH Office of the Regional Research Hydrologist

ORROE Occasional Reports of the Royal Observatory Edinburgh (periodical code)

ORRT Operational Readiness and Reliability Test

Ors Others

ORS Obligation Réglementaire de Service • Office Reporting System • Old Red Sandstone • On-Line Retrieval Service • On-Line Retrieval System • Operational Reports Section • Operational Requirements Statement • Operational Research Society (UK) • Optical Redshift Survey • Orbital Refueling System • Orbital Reservicing System • Orbital Resupply System • Orbiter Refueling System • Orbiter Relay Simulator • Orbiting Research Satellite • Order Routing System • Organization Rating Scale • Orpheus Island Resort (Australia-QLD) (location identifier) • Oscillographic Recording System • Other Research Sponsors • Output Record Separator • Overseas Research Student • Overseas Research Studentship (program)

ÖRS Österreichische Raumfahrt- und Systemtechnik GesmbH (Austria)

ORSA Oceanic Remote Sensing Assembly • Ogive Recovery System Assembly • Operations Research Society of America (now merged with TIMS into INFORMS) • Operations Research Systems Analyst • Organizzazione Ricerche e Studi d'Astronomia (Italy)

ORSAC Oak Ridge Systems Analysis Code

ORSDI Oak Ridge Selective Dissemination of Information (system)

ORSER Office for Remote Sensing of Earth Resources • Office of Remote Sensing for Earth Resources

ORSO Ottawa River Solar Observatory (Canada-ON)

ORSS Operational Research Society of Singapore

ORSTOM Office de la Recherche Scientifique et Technologique d'Outre-Mer (France) (now IRD)

ORT Northway (USA-AK) (location identifier) • Oblique Rotator Theory • Ooty Radiotelescope (India) • Operational Readiness Test • Operational Readiness Trainer • Operational Readiness Training • Operational Ready Time • Optical Ranging Telescope • Optical Relay Tube • Optimum Resolution Technique • Orbital Rendezvous Technique • Orbiting Radio Telescope • Orbit Readiness Test • Organisation - Reconstruction - Travail (school) • Oriental Robotic Telescopes (project) • Overland Radar Technology

ORTA Office of Research and Technology Applications

ORTF Office de Radiodiffusion-Télévision Française (France) (now Radio France)

Orth Orthodox

ORTHOG Orthogonal

ORTN Oriental Robotic Telescope Network

Ortogr Ortografía

ORTP Operational Readiness Training Program

ORTS Operational Readiness Test System

ORTT Operational Readiness Training Test

ORU Operational Readiness Unit • Operational Replaceable Unit • Opération de Renouvellement Urbain • Operator Radio Unit • Optical Reference Unit • Orbital Replaceable Unit • Orbital Replacement Unit (*) • Orbit Replaceable Unit • Orbit Replacement Unit • Organized Research Unit • Oruro (Bolivia) (location identifier)

ORUS Orbital Replacement Units

ORV Noorvik (USA-AK) (location identifier) • Off-Road Vehicle • Orbital Rescue Vehicle • Orbital Research Vehicle

ORVO Optical Radial-Velocity Orbit

ORW Ormara (Pakistan) (location identifier)

ORWG Operational Requirements Working Group

ORX Oriximiná (Brazil-PA) (location identifier)

ORY Paris Orly (France) (location identifier)

ORZ Orange Walk (Belize) (location identifier) • Outer Radiation Zone

o/s Old Style • On Sale • Out of Service • Out of Stock • Outsize • Outstanding

os Oculus Sinister • Old Series • On Station • Outside

O/s Out of Stock • Outstanding

Os Osmium

O&S Operating & Support • Operations & Support • Operations & Sustainment

O/S Old Style • Operating System • Operations and Support

OS Object Services • Object Space • Observation Software • Observation Summary • Observatory Scientist • Observing Software • Observing Station • Observing System • Oculus Sinister • Odd Symmetric • Old Saxon • Old Series • Old Style • On Sale • On Sample • On Schedule • On Side • Opacity Sampling • Open System • Operating Schedule • Operating Software • Operating Strength • Operating System • Operational Sequence • Operational Sheet • Operational Specification • Operational Suitability • Operator Service • Opposite Sense • Opposition Spike • Optical Scanning • Optical Site • Optical Station • Optogalvanic Signal • Optospinography • Ordered Subsets • Ordinary Seaman • Ordnance Survey (UK) • Ordre de Service • Organizational Source • Organizational Support • Oscan • Ostyak • Other Sources • Otherwise Specified • Outer Shell • Outer Space • Out of Stock • Output Secondary • Output Specification • Outsize • Outstate • Ouvrier Spécialisé • Overseas • Overshoot • Sulfur Monoxide

OSA Objective Switching Architecture • Object System Adapter • Offshore Acquisition • Old Style Antique • Open Scripting Architecture • Open Service Architecture • Open Systems Adapter • Open Systems Approach • Open Systems Architecture (*) • Operational Support Aircraft • Operational Support Airlift • Operational Support Area • Operations Support Area • Optical Society of America (USA-DC) • Optimal Scheduling Algorithm • Organizations and Strategies in Astronomy (book series) • Osaka (Japan) (location identifier) • Outer Soundings Array • Overseas Sterling Area

OSAC Oxford System of Automated Cartography

OSAD Outer Space Affairs Division

OSAJ Optical Society of America Journal (periodical code)

OSAJB Optical Society of America Journal B - Optical Physics (periodical code)

OSAJI Optical Society of America Journal - Optics and Image Science (periodical code)

OSAM Overflow Sequential Access Method

OSAMM Optimum Support And Maintenance Model

OSAR Optical Storage And Retrieval

OSAS Observation du Soleil Au Sol • Oxford Space and Astronomical Society (UK-England) (see now OUSAS)

OSAT Office of Space Access and Technology (NASA) • On-Site Acceptance Test • Optical Sensor And Tracker • Out-of-Service Analog Test

OSATES Ocean Space Advanced Technologies European Show

OSB Oberer Sonderkanal-Bereich • Ocean Studies Board • Operational Stations Book • Operations Software Branch • Operations Support Building • Optimal Search Behavior • Osage Beach (USA-MO) (location identifier)

Osc Oscillate • Oscillation • Oscillator

OSC Objective Supply Capability • Oblate Spherical Coordinates • Observatorio San Calixto (Bolivia) • Observatory Science Council • Ocean Science Committee • Office of Scientific Computing • Office of Special Counsel • Off-line Satellite Calibration • Ohio Supercomputer Center (USA-OH) • On-Scene Coordinator • On-site Safety Committee • Ontario Science Centre (Canada-ON) • Open Star Cluster • Operating System Control • Operating Systems and Compilers • Operational Simulator Console • Operational Summary Console • Operational Switching Cabinet • Operational System Control • Operations Sequence Chart • Operation Subcommittee • Operations Support Call • Operations Support Cell • Operations Support Center • Operations Support Contract • Optical Sciences Center • Optical Signature Code • Optical Supervisory Channel • Orbital Sciences Corporation (USA-VA) • Orbit-Spin Coupling • Organic Sulfur Compound • Organizational Supply Code • Organization Science (periodical code) • Oscillator • Orlando Science Center (USA-FL) • Oscoda Wurtsmith (USA-MI) (location identifier) • Outer Space Committee • Outsourcing Service Centre

OSCA Optical Sensors Collaborative Association • Optimized Stellar Coronagraph for AO

OSCAR On-Line Satellite Catalog And Request (system) • On-Screen Configuration and Activity Reporting • Operating Sequence Control Array • Operations, Scheduling, Control And Reporting • Optically Scanned Character Automatic Reader • Optical Systems Characterization and Analysis Research • Optimum Systems Covariance Analysis Results • Orbiting Satellite Carrying Amateur Radio • Outside Cable Rehabilitation

OSCC Open Skies Consultative Commission

OSCE Objective Structured Clinical Examination • Office Statistique des Communautés Européennes • Organization for Security and Cooperation in Europe

OSCF Operations Support Computing Facility

OSCI Organisation de Séjours Culturels Internationaux

OSCIR Observatory Spectrometer and Camera for the Infrared

OSCL Operating System Control Language

OSCN Orbiter Specification Change Notice

OSCP On-Site Computer Programmer

OSCR Ocean Surface Current Radar • Operating and Support Cost Reduction • Orbiter Software Change Request

OSCRL Operating System Command and Response Language • Operating System Command Response Language

OSCRS Orbital Spacecraft Consumables Resupply System

OS&D Over, Short & Damaged

ÖSD Österreichische Staatsdrückerei (Austria)

OSD Object State Diagram • Object Structure Definition • Office of the Secretary of Defense (USA) • On-Screen Display • Open Software Description • Open Software Distribution • Operating Strength Deviation • Operational Scenarios Document • Operational Sequence Diagram • Operational Systems Development • Optical Scanning Device • Organization Subject Detail • OSI Session Driver • Östersund (Sweden) (location identifier) • Ostersund (Sweden) (location identifier) • Overseas Surveys Directorate (UK) • Over, Short and Damaged

OSDA Optical Shallow-Defect Analyzer

OSDB Operations Support Database

OSDBU Office of Small and Disadvantaged Business Utilization

OSDH Orbiter System Definition Handbook

OSDI Operating Systems Design and Implementation (conference)

OSDIT Office of Software Development and Information Technology

OSDM Optical Space Division Multiplexing

OSDN Open Source Development Network

OSDP Onboard Signal and Data Processing • Onboard Signal and Data Processor • On-Site Data Processing • On-Site Data Processor • Operational System Development Program

OSDPD Office of Satellite Data Processing and Distribution • Office of Space Data Processing and Distribution

OSDR Over, Short, and Damaged Report

OSDS Oceanic System Development and Support • Operating System for Distributed Switching

OSDU Output Signal Distribution Unit

OS/E Operating System/Environment

OS&E Optical Science & Engineering

OSE Observing System Experiment • Office Server Extension • Omora (Papua New Guinea) (location identifier) • On-Scene Endurance • Open Systems Environment • Operating Support Equipment • Operational Security Environment • Operational Security Evaluation • Operational Status Equipment • Operational Support Equipment • Orbital Support Equipment • Orbiter Support Equipment • Other Support Equipment

OSEA Ontario Society of Education through Art (Canada-ON)

OSEC Operations Security

OSEDA Office of Social and Economic Data Analysis

OSEF Office Scolaire d'Étude par le Film

OSEM Opto-electronic Sensor Electromagnetic Motor

OSEP Office of Scientific and Engineering Personnel

OSes Operating Systems

OSES Operation Support Equipment Software

OSETI Optical Search for Extraterrestrial Intelligence

OSF Observation Status Form • Observation Summary Form • Office of Space Flights (NASA) • Oldenburger Sternfreunde eV (Germany) • Open Software Foundation • Open Systems Foundation • Operational Support Facility • Operations Support Facility • Operation System Function • Optronique Secteur Frontal • Orbiter Simulation Facility • Order Statistic Filter • Ordnance Storage Facility

OSFM Office of Spacecraft and Flight Missions

OSG Observatory for Satellite Geodesy • Observatory Support Group • Open Systems Group • Operability Service Group • Operand Select Gate • Operations Support Group • Optical Systems Group • Ossima (Papua New Guinea) (location identifier)

OSGB Ordinance Survey of Great Britain

OSGC Ohio Space Grant Consortium (USA-OH)

OSH Occupational Safety and Health • On-Line Shorthand • Organisations Sociales et Humanitaires • Oshkosh (USA-WI) (location identifier)

O&SHA Operating & Support Hazard Analysis

OSHA Occupational Safety and Health Act (USA) • Occupational Safety and Health Administration (USA) • Operating and Support Hazard Analysis

OSI Objective Systems Integration • Office of Science Information • Office of Scientific Information • Office of Scientific Integrity • Office of Scientific Intelligence • Office of Special Investigations • Office of Strategic Information • On-Site Inspection • Open-Source Initiative • Open Standards Interconnection • Open Switching Interval • Open Systems Integration • Open Systems Interconnect • Open Systems Interconnection • Operating Space Item • Operational Status Indicator • Operational Status Interrogation • Operational System Interface • Optical Society of India • Optical Stellar Interferometer • Optical Storage International • Orbiting Stellar Interferometer • Organisation de Solidarité Internationale • Organisation du Système d'Information • Osijek (Croatia) (location identifier)

OSIA On-Site Inspection Agency

OSIC Optimization of Subcarrier Information Capacity

OSID Objective System Interface Drawing • Operating System Interface Definition

OSIE Operational Support Integration Engineering

OSIFA Operational Sequence and Information Flow Analysis

OSINET Open Systems Interconnection Network

OSIP Operational and Safety Improvement Program • Operational Safety and Improvement Program • Operational Satellite Improvement Program

OSIR Office of Scientific Investigation and Research

OSIRIS Ohio State Infrared Imager and Spectrometer • Ohio State Infrared Imaging Spectrograph • Ohio State Infrared Imaging System • OH-Suppressing Infrared Imaging Spectrograph • Optical Simulation for Imaging Reconnaissance and Intelligence Sensors • Optical Spectrograph and Infrared Imaging System • Optical, Spectroscopic and Infrared Remote Imaging System (Rosetta) • Optical System for Imaging and low-Resolution Integrated Spectroscopy • Oxide Superalloy Improved by Resolidification In Space

OSIRM Open Systems Interconnection Reference Model

OSIS Office of Science Information Services • Open Shop for Information Services • Open Source Information System • Operating Systems Installation Support • Optionally Stabilized Imager and Spectrometer

OSJ Objet Stellaire Jeune • Observatorio San José (Argentina)

OSJaJ Oceanographical Society of Japan Journal (periodical code)

OSK Oskarshamn (Sweden) (location identifier)

Osl Oslo (Norway)

OSL Object Script Language • Old Style Latin • On-Site Laboratory • Operating System Language • Optically Stimulated Luminescence • Optical Science Laboratory • Orbiting Solar Laboratory • Orbiting Space Laboratory • Oslo (Norway) (location identifier)

OSLM Operations Shop or Lab Manager

OSLR Ocean Sciences and Living Resources

OS&M Operational Support & Maintenance

OSM Mosul (Iraq) (location identifier) • Office of Special Mining • Office Security Manager • Office Security Monitor • Off-line Services Manager • Off-Screen Menu • Off-Screen Model • On Station Mode • Open Service Market • Open Storage Manager • Operating System Manual • Operating System Module • Operational Status Monitoring • Operational Support and Maintenance • Operations Status Monitoring • Operations Systems Manager • Operator's Service Manual • Optical Strength Meter • Optical Support Measure • Optical Surveillance Measure • Orbital Service Module • Orbital Support Model • Orbiter Service Module • Oscillating Secondary Mirror

OSMA Office of Safety and Mission Assurance (NASA)

OSMAE Open System Management Application Entity

OSMAP Open System Management Application Process

OSMC Operational Software Maintenance Complex

OSME Open Systems Message Exchange

OSMIS Operating and Support Management Information System

OSMOS Operating System Managed Object Specifications

OSMP Operational Support Maintenance Plan

OSMV One-Shot Multivibrator

OSN Observatorio de Sierra Nevada (Spain) • Open Storage Networking • Osan (Korea-ROK) (location identifier)

OSNI Ordnance Survey of Northern Ireland

OSNR Optical Signal to Noise Ratio

OSNS Open Systems Network Session • Open Systems Network Support

OSO Objective and Strategies Overview • Ocean Systems Operations • Office of Space Operations • Onsala Space Observatory (Sweden) • Operational Support Office • Orbiting Solar Observatory • Originating Signaling Office • Ouest-Sud-Ouest

OSOA Optimal Statistical Objective Analysis

OSOD On-Site Orbit Determination

OSOP Orbiter Systems Operating Procedure

osp Obiit Sine Prole

OSp Old Spanish

OSP Objective and Strategies Plan • Observatoire de la Santé Publique • Observer Selection Program • Ocean Surveillance Product • Office of Special Projects • Office of State Planning • Office of Strategic Phenomena (NRL) • Online Service Provider • On-Screen Programming • On-Site Programmer • Operating System Principle • Operations Support Plan • Optical Storage Processor • Optical Surveillance Platform • Optimal Scanning Path • Orbital Space Plane • Orbital Support Plan • Oregon Star Party (USA-OR) • Outer Space Place • Outside Plant • Slupsk (Poland) (location identifier)

OSPA Office of Special Political Affairs • Open Signal Coprocessing Architecture

OSPAT Operations and Spacecraft Performance Assessment Team

OSPE Organizational Spare Parts and Equipment

OSPF Open Shortest Path First

OSPFIGP Open Shortest-Path First Internet Gateway Protocol

OSPI Office of the Superintendent of Public Instruction

OSPREY Open Structure for Protecting Rights Electronically

OSPU Odessa State Polytechnic University (Ukraine)

OSQAM Operating Standards and Quality Assurance Manual

OSQL Object Structured Query Language

OSR Occupational Survey Report • OEM Service Release • OEM System Release • Office of Science and Research • Office of Scientific Research • Office of Secondary Responsibility • On-Site Representative • Operational Scanning Recognition • Operational Status Reply • Operational Status Response • Operational Support Readiness • Operational Support Requirement • Operations Safety Requirement • Operations Support Room • Optical Scanning Recognition • Optical Signature Report • Optical Solar Reflector • Optical Surface Reflector • Organization for Scientific Research • Ostrava (Czech Rep.) (location identifier) • Outgoing Shortwave Radiation • Output Shift Register

OSRA Observatorium für Solare Radioastronomie (Germany)

OSRAP Optimum Storage Requirements Analysis Program

OSRBD Operational Software Reliability Benchmark Development

OSRD Office of Scientific Research and Development

OSRI Originating Station Routing Indicator

OSRMS Ocean Surface Roughness Measurement System

OSRP Optical Switching and Routing Protocol

OSRS Oceanic Satellites and Remote Sensing • Operational Status Recording System • Overhead Status Reporting System

OSRT Ooty Synthesis Radio Telescope (India)

Oss Osservatorio

OSS Observation Support Software • Observation Support System • Observatoire du Sahara et du Sahel • Observatory Support System • Observer Support Software • Ocean Sensor System • Ocean Surveillance Ship • Ocean Surveillance System • Office of Space Science • Office of Space Sciences • Office of Space Station • Office of Special Services (USA) • Office of Strategic Services (now CIA) (USA) • Office Subsystem • Office Support System • Omnidirectional SETI System • One-Stop Shop • Online Scheduling System • Open Storage Solution • Operating Support System • Operating Subsystem • Operational Storage Site • Operational Support System • Operations Support System • Operator Support System • Optical Scoring System • Optical Subsystem • Optical Support Structure • Optical Surveillance System • Optics Subsystem • Optics Support Structure • Orbital Space Station • Orbital Stabilization System • Orbiting Space Station • Osh (Kyrgyzstan) (location identifier) • Outer Solar System • Oxygen Storage System

OSSA Office of Space Science and Applications (USA)

OSSE Observing-System Simulation Experiment • Office of Space Science Education • On-Site System Evaluation • Oriented Scintillation Spectrometer Experiment (*) • Oriented Scintillation Spectroscopy Experiment

OSSEC Organisation de Séjours Sportifs, Éducatifs et Culturels

OSSI Ohio State Spectroscopic Institute (USA-OH) • Open Storage System Interconnection • Optronic Sensor for Stereographic Imagery

OSSIU Orbiter-Space Station Interface Unit

OSSL Operating System Simulation Language • Organisation Scolaire de Séjours Linguistiques

OSSN Originating Station Sequence Number • Originating Station Serial Number

OSSO Observation Science Support Office • Observing Simulations and Simulating Observations (panel)

OSSOM Office de Sécurité Sociale d'Outre-Mer (Belgium)

OSSP Organizational Standard Software Process • Organization Standard Software Process • Outer Solar System Probe

OSSRH Orbiter Subsystem Requirements Handbook

OSSS Optical Space Surveillance System • Orbital Space Station Study • Orbital Space Station System

OSSTF Ontario Secondary School Teachers Federation (Canada-ON)

OSSWG Operating Systems Standards Working Group

OST Objectif Scientifique et Technique • Objective Start Time • Objectives, Strategy, and Tactics • Observational Sequence Table • Observation Sequence Table • Observation Summary Table • Observatoire des Sciences et des Techniques (France) • Ocean Surface Temperature • Office of Science and Technology • Office of the Secretary of Transportation (USA) • Office Scientifique et Technique • On-Shift Test • Operational System Test • Operationnelle Systemtest • Operations Support Team • Operations Support Timeline • Operation Stress Test • Optical Star Tracker • Optic Support Table • Orbiter Support Trolley • Orbit Stay Time • Order and Shipping Time • Organisation Scientifique du Travail • Oostende (Belgium) (location identifier) • Organisation Scientifique du Travail • Ostdeutsche Sterntagung (Germany)

OSTA Office of Space and Terrestrial Applications • Office of Space Technology and Applications (USA) • Optical Storage Technology Association

OSTC Office of Science, Technology and Culture • Optical Science and Technology Center

OSTDS Office of Space Tracking and Data Systems

OSTEO Osteoporosis Experiment in Orbit

Osterzgeb Osterzgebirge (Germany)

OSTF Off-Site Test Facility • Operational System Test Facility • Ordnance Survey Transfer Format

OSTI Office of Science and Technology Infrastructure • Office of Scientific and Technical Information • Organization for Social and Technological Innovation

OSTO Office of Space Transportation Operations

OSTP Office of Science and Technology Policy (USA) • Office of Science Technology Planning • On-Site Test Procedure • Orbiting System Test Plan

OSTR On-Site Technical Representative

OSTS Office of Space Transportation Systems

OSU Columbus OSU (USA-OH) (location identifier) • Observatoire des Sciences de l'Univers • Odessa State University (Ukraine) • Ohio State University (USA-OH) • Oklahoma State University (USA-OK) • Operational Switching Unit • Optical Simulation Unit • Oregon State University (USA-OR)

OSUMMPMRSFC Ohio State University Multimode Prism Monochromator, Reflectomator and Spectral Flux Calibrator

OSURO Ohio State University Radio Observatory (USA-OH)

OSUT On-Site User Test • On-Site User Testing • On-Station Unit Testing

OsUTR Osaka University Technology Reports (periodical code)

OSV Ocean Station Vessel • Orbital Support Vehicle • Oriented Space Vehicle

OSVA Off-Site Vital Area

OSVS Orbiter Space Vision Systems

OSW Office of Solid Waste • Office of Surface Water • Orsk (Russia) (location identifier)

OSWER Office of Solid Waste and Emergency Response

OSWS Operating System Workstation

OSX Kosciusko - Attala County (USA-MS) (location identifier)

OSY Namsos (Norway) (location identifier)

OSZ Koszalin (Poland) (location identifier)

OSZE Organisation für Sicherheit und Zusammenarbeit in Europa

OS/2 Operating System Two

o/t Other Times

O/T Other Times

OT Oberer Tubus • Object Technology • Object Theory • Observation Tool • Observatorio del Teide (Canary Islands) • Observer-Target • Observing Tool • Occupational Therapist • Occupational Therapy • Off Time • Old Testament • On Target • Ontario (Canada) • On Time • On Track • Open Technology • Open Transport • Operating Temperature • Operating Time • Operational Task • Operational Test • Operational Tester • Operational Testing • Operational Trajectory • Operations Team • Operator Task • Operator Terminal • Opposite Term • Optical Tracker • Optical Tracking • Optical Transceiver • Optical Transient • Optimality Theory • Optimal Trajectory • Orientation Tour • Original Title

• Ortsteil • Other Than • Other Time • Other Transaction • Ouest • Outlet Terminal • Output Term • Out Temperature • Overall Test • Overtime • Overtone

OT&A Operational Test & Acceptance

OTA Mota (Ethiopia) (location identifier) • Observatory for Time-domain Astronomy • Office of Technical Assistance • Office of Technology Assessment (USA-DC) (now defunct) • Open Test Assembly • Operational Test Activity • Operational Test Agency • Operational Test Agent • Operational Test and Acceptance • Operational Trajectory Analysis • Operational Transconductance Amplifier • Operation-Triggered Architecture • Optical Telescope Assembly • Optical Tracking Aid • Optical Tube Assembly • Orbital Thruster Assembly • Organic Trace Analysis • Other Transactions Authority • Over The Air

OTAA Ohio Turnpike Astronomers Association (USA-OH)

OTAC Observatory Time Allocation Committee • Observing Time Allocation Committee

OTAD Overseas Terminal Arrival Date

OTADA Office of Tracking And Data Acquisition

OTAF Office of Technology Assessment and Forecast

OTAG Operations and Training Analysis Group

OTAI Over-The-Air Initialization

OTAM Over-The-Air Management

OTAN Organisation du Traité de l'Atlantique Nord

OTAR Over-The-Air Rekeying

OTASE Organisation du Traité de l'Asie du Sud-Est

OTASIM Optical Telescope Assembly Simulator

OTAT Over-The-Air Transfer

OTAU Optical Test Access Unit • Over-The-Air Updating

OTB Object Transaction Broker • Occupation Thérapeutique Bénévole • Off-Track Betting • Orbiting Tanker Base • Oxide Titanium Bronze

Otbd Outboard

OTBD Outboard

OTBE Overtaken By Events

OTC Bol (Chad) (location identifier) • Objective, Time and Cost • Office of Telecommunications • Onsite Transfer Cask • Operational Test Center • Operational Test Command • Operational Test Coordinator • Operational Text Classification • Operational Training Capability • Operational Training Center • Operational Training Course • Optical Test Chamber • Optical Transient Camera • Orbiter Test Conductor • Orbiting Trajectory Computation • Organization for Trade Cooperation • Outgoing Trunk Connection • Outlet Terminal Circuit • Over The Counter

OTCC Operator Test Control Console

OTCCC Open-Type Control Circuit Contact

OTCCD Orthogonal-Transfer Charged-Couple Device

OTCM Orbiter Thermal Control Model

OTCS Operational Teletype Communications System

OTD Contadora (Panama) (location identifier) • Office of Technology Development • Offset Tilted Dipole • One Time Developed • Operational Technical Documentation • Operational Test Director • Operator Training Device • Optical Tracking Device • Optical Transient Detector • Orbital Transfer Device • Orbiter Test Director

OTDA Office of Tracking and Data Acquisition

OTDB Operations Tasking Database

OTDC Operational Test Design Concept

OTDM Optical Time-Division Multiplexing

OTDO Output Transaction Data Object

OTDR Optical Test Data Receiver • Optical Time Domain Reflectometer • Optical Time Domain Reflectometry

Ote Oriente

OT&E Operational Test & Evaluation • Operational Testing & Evaluation • Operator Training & Evaluation • Optical Test & Evaluation

OTE Observatoire des Technologies Éducatives • Observing Time Estimator • Open Toolkit Environment • Open Transaction Environment • Operational Test and Evaluation • Operational Test Environment • Operational Test Equipment • Operational Testing and Evaluation • Operator Training and Evaluation • Operator Training Equipment • Optical Telescope Element • Optical Test and Evaluation • Optical Tracking Electronics • Optically Transparent Electrode • Orbital Transfer Equipment • Oriente

OTEA Operational Test and Evaluation Agency

OTEAS Organisation Tunisienne d'Éducation Aéronautique et Spatiale (Tunisia)

OTEC Ocean Thermal-Energy Conversion • Operational Test and Evaluation Center

OTECC Operational Test and Evaluation Coordinating Committee

OTECI Office Technique d'Études et de Coopération Internationale

OTEMP Overtemperature

OTES Orbiter Thermal Effects Simulator

otf OpenType Font (file extension)

OTF Observation Test File • Off The Film • On-Target Flag • On-The-Fly • Open Token Foundation • Operational Time Fraction • Optical Transfer Function • Orbital Test Flight

OTFC On-The-Fly-Calibration

OTFR On-The-Fly-Reprocessing

OTFT Organic Thin Film Transistor

OTG Objective Transmission Grating • Oracle Transport Gateway • Worthington (USA-MN) (location identifier)

OTH North Bend (USA-OR) (location identifier) • Over the Horizon

OTHB Over the Horizon Backscatter

OTHQ Ouvrier Très Hautement Qualifié

OTHT Over the Horizon Targeting

OTI Morotai Island (Indonesia) (location identifier) • Office of Technical Information • Office of Technical Integration • Open Topology Interface • Ordered Tree Inclusion • Organización de la Televisión Iberoamericana • Original Transmitter Identification

OTIA Office of Technical Information Agency

OTIM Operational Test Interchange Meeting

OTIS Occupational Training Information System • Office of Telecommunications and Information Systems • Oklahoma Telecommunications Interlibrary System (USA-OK)

OTIU Overseas Technical Information Unit

OTJ Off-The-Job • On-The-Job

OTK Ocean Toolkit • Oostelik Transvaalse Kooperasie (South Africa) • Oxidizer Tank

OTL Boutilimit (Mauritania) (location identifier) • Ooty Lunar Occultations • Operating Temperature Limit • Operating Time Log • Ordnance Test Laboratory • Out To Lunch • Overtime Loss

OTLAMR Ocean Temperature Large Antenna Microwave Radiometer

OTLC Orbiter Timeline Constraint

OTLF One-Time License Fee

OTLK Outlook

OTLPF Odd Tile Length/Low Pass First

OTLT One Time, Last Time

OTM Object Transaction Manager • Observatory Technique Magazine • Occupational Training Manager • Office of Telecommunications Management • Online Training Module • On The Move • On-Time Marker • Operational Test Model • Optical Transient Monitor • Orbit Transfer Maneuver • Orbit Trim Maneuver • Ottumwa (USA-IA) (location identifier) • Output Transformation Matrix

OTMOP Operational Test Measures Of Performance

OTMS Operational Technical Managerial System

OTMX Optical Terminal Multiplexer

OTN Oaktown Green (USA-IN) (location identifier) • Octal Track Number • Optical Transport Network • Operational Teletype Network • Optisches Transportnetz • Oracle Technology Network

OTO One-Time Only • Operational Test Organization • Optical Tracker Operator • Otto Vor (USA-NM) (location identifier)

OtObI Otbor i Obrabotka Informatsii (periodical code)

OTOE Objective Table of Organization and Equipment

OTOH On The Other Hand

OTOS Orbit-To-Orbit Stage

OTP Bucharest Otopeni (Romania) (location identifier) • Bucureşti Otopeni (Romania) (location identifier) • Office of Telecommunications Policy • On-Time Pad • On-Time Password • On-Time Programmable • On Top • Open Trading Protocol • Operational Targeting Program • Operational Test Plan • Operational Test Procedure • Operational Test Program • Operations Test Plan • Operations Turnaround Plan • Optically-Thin Plasma • Orbital Test Plan • Outline Test Plan

OTPC Oficina Técnica para la Protección del Cielo (IAC)

OtPI Otbor i Peredacha Informatsii (periodical code)

OTPI On Top Position Indicator

OTPO Operational Test Project Officer

OTPP Operational Transfer Point Plan

OTPS Operational Test Program Set

OTR Coto 47 (Costa Rica) (location identifier) • Observed Temperature Rise • Oceanic Transition Route • On The Rag • Ooty Radio Telescope (India) • Operating Temperature Range • Operating Time Record • Operational Test Readiness • Operational Test Report • Operational Time Record • Operational Turn-Round • Optical Tracking • Optical Transition Radiation • Optical Transmit • Other • Outer • Outer Tracking • Output Routine • Over The Road

OTRC Optoelectronics Technology Research Corporation

OTRI Oficina de Transferencia de Resultados de Investigación (IAC)

OTRR Operational Test Readiness Review

OTRRWG Operational Test Readiness Review Working Group

OTRS Operational Test Readiness Statement • Operational Training Readiness Statement • Outline Test Readiness Statement

OTS Anacortes (USA-WA) (location identifier) • Object Transaction Service • Observatoire des Technologies Stratégiques (France) • Office of Technical Services • Office of Technical Support • Office of Thrift Supervision • Office of Toxic Substances • Officers' Training School • Off-The-Shelf • Open Trading System • Operational Test Site • Operational Time Synchronization • Operations Technical Support • Opportunities To See • Optical Technology Satellite • Optical Tracking System • Optical Transient Source • Optical Transmission Section • Optical Transport System • Optimum Telemetry System • Orbital Test Satellite • Orbital Test System • Orbital Transfer Stage • Orbital Transfer System • Orbital Transport Services • Orbital Transport System • Orbiter Transfer System • Orbiting Test Satellite • Organized Track System

OTSDF Optimal Trade-off Spatial Distance Filter

OTSI Office du Tourisme et Syndicat d'Initiative • Oscillating Two-Stream Instability

OTSR Optimum Track Ship Routing

OTSS Operational Telecommunications Switching System • Operational Teletype Switching System • Operational Test Support System

Ott Ottava • Ottawa (Canada-ON) • Ottobre • Ottoman

OTT Oficina de Transferencia de Tecnología • One-Time Tape • Open Tower Telescope (now DOT) • Operational Test Team • Operational Training Test • Operations Team Trainer • Optional Team Targeting • Over The Top

OTTB Optically-Thin Thermal Bremsstrahlung

OTT&E Operational Test Training & Evaluation

OTTE Operational Testing, Training and Evaluation

OTTER Oregon Transect Ecosystem Research

OTTLE Optically Transparent Thin-Layer Electrode

OTTOMH Off The Top Of My Head

OTU One-Time Use • Operational Taxonomic Unit • Operational Test Unit • Operational Training Unit • Optical Translator Unit • Organisme pour le Tourisme Universitaire • Otu (Colombia) (location identifier)

OTV Operational Television • Orbital Test Vehicle • Orbital Transfer Vehicle (*) • Orbiter Transfer Vehicle • Orbit Test Vehicle • Orbit Transfer Vehicle

OTVS Onboard TV System

OTW Off The Wall • Over The Wall

OTWG Operational Test Working Group

OTY Oria (Papua New Guinea) (location identifier)

OTZ Kotzebue (USA-AK) (location identifier)

ou Oculi Uterque • Oculus Uterque

OU Oculi Uterque • Oculus Uterque • Open Under • Open University (UK) • Operations Unit • Order of Use • Organizational Unit • Organization Unit • Oxford University (UK-England)

OUA Organisation de l'Unité Africaine (now UA) • Organizational User Agent • Ouagadougou (Burkina Faso) (location identifier)

OUAS Oxford University Astronomical Society (UK-England)

OUAT Olsztyn University of Agriculture and Technology (Poland)

OUCF Office Universitaire et Culturel Français

OUCS Oxford University Computing Services

OUD Operational Use Data • Oujda - Les Angads (Morocco) (location identifier)

OUE Operational Utility Evaluation • Orbital Uncertainty Estimate • Ouesso (Congo) (location identifier)

OUG Ouahigouya (Burkina Faso) (location identifier)

OUH Oudtshoorn (South Africa) (location identifier)

OUI Ban Houei (Laos) (location identifier) • Open University of Israel • Organizationally Unique Identifier

OUIC Operational Unit Identification Code

OUK Outer Skerries (UK) (location identifier)

OUL Oulu (Finland) (location identifier)

OULIPO Ouvroir de Littérature Potentielle

OUM Oum Hadjer (Chad) (location identifier)

OUN Norman - Max Westheimer (USA-OK) (location identifier)

ouO ohne unser Obligo

OUO Official Use Only

OUP Oxford University Press

OUR Batouri (Cameroon) (location identifier)

OURS Open Users Recommended Solution • Orbiting Unification Ring Satellite

OUS Oscillateur Ultra-Stable • Ourinhos (Brazil-SP) (location identifier) • Outer User System

OUSAS Oxford University Space and Astronomical Society (UK-England)

OUST Office of Underground Storage Tanks

out Output (file extension)

Out Outlet • Output

OUT Bousso (Chad) (location identifier) • Operational Unit Transportable • Orbiter Utility Tray • Outlet • Outline • Output • Outsize • Outubro • Outward

Outbd Outboard

OUTBD Outboard

Outg Outgoing

OUTR Outeiro

OUTRAN Output Translator

OUTS Operational Unit Transportable System • Output String

OUTWATS Outward Wide Area Telephone Service

OUU Ouanga (Gabon) (location identifier)

OUV Optical-Ultraviolet

OUVS Orbiter Ultraviolet Spectrometer (PV)

OUZ Zouerate (Mauritania) (location identifier)

oV ohne Verfasser

Ov Over

O/V Overvoltage

OV Object-Verb • Observatory Verification • Office Vision • Oklahoma Variable (star) • Operational Value • Operational Verification • Operational View • Operativer Vorgang • Optical Verification • Orbital Vehicle • Orbital Verification • Orbiter Vehicle • Orientation Visit • Ostriker-Vishniac (effect) • Outer Vessel • Output Voltage • Overflow • Overshooting • Overvoltage • Oxygen Vent

Ova Ostrova

OVA Bekily (Madagascar) (location identifier) • Optical Visual Analysis

OVAL Object-based Virtual Application Language • Object-based Visual Application Language

OVAM Orbital Vehicle Assembly Mode

OVAOG Ottawa Valley Astronomy and Observers Group (Canada-ON)

OVAS Ocular Vergence and Accommodation Sensor • Ohio Valley Astronomical Society (USA-KY) • Ohio Valley Astronomical Society (USA-WV)

OVB Novosibirsk (Russia) (location identifier)

Ovbd Overboard

OVBD Overboard • Overboard Dumping Assembly

OVC Ocean Variability and Climates • Optical Velocities Catalogue • Other Valuable Consideration • Outer Vacuum Chamber • Overcast

OVCO Operational Voice Communication Office

OVCP Orbiting Vehicle Checkout Procedure

OVCS Operational Voice Communication System

OVD Asturias (Spain) (location identifier) • Optical Video Disk • Outer Vapor Deposition • Outside Vapor Deposition • Overdue • Oviedo (Spain) (location identifier)

OVDLM Open View Data Line Monitor

OVDS Oxidizer Vent Drain System

ÖVE Österreichischer Verband für Elektrotechnik (Austria)

OVE Observatoire de la Vie Étudiante (France) • On-Vehicle Equipment • Operator's Vehicle Equipment • Organisation for Vedvarende Energi (Denmark) • Oroville (USA-CA) (location identifier)

OVEI Orbiter Vehicle End Item

OVF Organe (de calcul) en Virgule Flottante • Overfill • Overflow

Ovfl Overflow

OVFL Overflow

OVG Oberverwaltungsgericht

OVH Overhead

Ovhd Overhead

OVHD Overhead

Ovhl Overhaul

OVHL Overhaul

OVHT Overheat

OVI Operational Validation Inspection • Oxygen Ventilation Network

OVIP Orbital Verification Implementation Plan

OVIS Optical Video Integrated System • Optical Visual Information System

OVIT Orbital Verification Integrated Timeline

OVIWG Orbital Verification Implementation Working Group

ovl Overlay (file extension)

OVL Ortsverbindungsleitung • Oval • Ovalle (Chile) (location identifier) • Overlay

OVLA Optical Very Large Array

OVLAHT Optical Very Large Array Hypertelescope

OVLBI Orbiting Very Large Baseline Interferometry

Ovld Overload

OVLD Overload

Ovlp Overlap

Ovly Overlay

OVMA Owens Valley Millimeter Array

OVN Optical - Visible - Near-infrared

OVNI Objet Volant Non-Identifié

OVNO Or Very Near Offer

OVO Outward Vouching Office

OVOP Orbital Verification Operating Procedure

OVOT Orbital Verification Operating Timeline

OvP Oval Point

OVP Office of the Vice President • OSHA Voluntary Program • Over Voltage Protection

Ovpd Overpaid

OVPF Office of the Vice President for Finances

OVPN Optical Virtual Private Network

OVPR Office of the Vice President of Research

OVPT Orbital Verification Performance Test

ovr Overlay (file extension)

OVR Orbiting Vehicle Requirement • Outer Vertical Resonance • Over

Ovrd Override

OVRD Override

Ovrn Overrun

OVRNG Overrunning

OVRO Owens Valley Radio Observatory (USA-CA)

OVS Observation of Variable Stars (periodical code) • Operational Voice System • Orbiting Vehicle System • Overshooting

OVSA Owens Valley Solar Array (USA-CA)

OVSB Oberbayerische Volkssternwarte Berg (Germany) • Orbiting Vehicle Support Building

OVSF Orthogonal Variable Spreading Factor

Ovsht Overshoot

Ovshtg Overshooting

OVst Ortsvermittlungsstelle

ÖVSV Österreichischer Versuchssenderverband (Austria)

OVSW Orbital Verification Software

Ovsz Oversize

OVT Object au Voisinage de la Terre • Operational Validation Test • Operational Verification Test

OVTQ Overtorque

Ovtr Overtravel

OVTR Operational Video Tape Recorder

OVU Optical Visualization Unit

Ovv Overvoltage

OVV Optically Violently Variable • Optically Violent Variable (star) • Optically Violent Variation • Optical Violent Variable (star) (*) • Optical Violent Variation • Overvoltage

OVW Open View Windows

O/W Order Wire

O&W Onderwijs & Wetenschappen • Onderzoek & Wetenschappen

OW Obstacle Warning • Office of Water • Oil Well • One Way • Openbare Werken • Open Workstation • Optical Warning • Optical Window • Order Wire • Orientation Workshop • Outer Wing • Over-Write

OWA Open-World Assumption • Ordered Weighting Averaging • Owatonna (USA-MN) (location identifier)

OWB Owensboro (USA-KY) (location identifier)

OWC Open Waste Container

OWCP Office of Workers Compensation Program

OWD Norwood (USA-MA) (location identifier) • One-Way Doppler

OWDC Office of Water Data Coordination

OWDE One-Way Doppler Extraction

OWE Operating Weight Empty • Optimum Working Efficiency • Owendo (Gabon) (location identifier)

OWF Optimum Working Frequency

OWG Operational Working Group • Operations Working Group

OWI Office Work Instruction • Open Widget Interface

OWIS Open When In Sync

OWK Norridgewock - Central Maine (USA-ME) (location identifier)

OWL Object Windows Library • Object Wrapper Layer • Office Workstation Ltd. • On-line Weather Looper • Ontology Web Language • Open Windows Library • Optical Warning Laser • Orbiting (array of) Wide-angle Light-collectors • Overwhelmingly Large (telescope)

OWLT One-Way Light Time

OWM Office of Weights and Measures

OWO On Work Order

OWPE Office of Waste Programs Enforcement

OWR Optical Warning Receiver

OWRR Office of Water Resources and Research

OWRT Office of Water Research and Technology

OWS Ocean Weather Service • Ocean Weather Ship • Ocean Weather Station • Odyssey Worldwide Services • Oil-Water Separator • Operational Weather Support • Operator Workstation • Orbital Workshop • Orbiting Workshop • Oxford WFC Survey

OWSE Operational WWW System Evaluation

OWT Orthogonal Wavelet Transform

OWU Ohio Wesleyan University (USA-OH)

OWWS Operational Windshear Warning System

Ox Oxford (UK-England) • Oxidator Mass

OX Order Crossover • Oxidizer

OXB Bissau (Guinea Bissau) (location identifier)

OXC Optical Crossconnect • Waterbury Oxford (USA-CT) (location identifier)

Oxd Oxidized

OXD Oxford - Miami University (USA-OH) (location identifier) • Oxide • Oxidizer

OXDZR Oxidizer

Oxf Oxford (UK-England) • Oxfordshire (UK-England)

OXF Oxford (Starlink internal code)

OXID Oxidizer

OXO Orbiting X-ray Observatory • Orientos (Australia-QLD) (location identifier)

Oxon Oxoniensis • Oxfordshire (UK-England)

OXR Oxnard Ventura (USA-CA) (location identifier)

OXS Objective Crystal Spectrometer • Objective X-ray Spectrometer

OXVAD Oxford (Starlink internal code)

OXY Morney (Australia-QLD) (location identifier) • Oxygen

OY Denmark (international aircraft registration code) • Option Year • Out Year

OYA Goya (Argentina-CR) (location identifier)

OYC Out Year Costs

OYE Oyem (Gabon) (location identifier)

OYG Moyo (Uganda) (location identifier)

OYK Oiapoque (Brazil-AP) (location identifier)

OYL Moyale (Kenya) (location identifier)

OYN Ouyen (Australia-VIC) (location identifier)

OYO On Your Own • Tres Arroyos (Argentina-BA) (location identifier)

OYP Saint-Georges-de-l'Oyapock (French Guiana) (location identifier)

OYRA Outstanding Young Researcher Award (OCPA)

OYS Yosemite National Park (USA-CA) (location identifier)

oz Ounce

Oz Ozero

OZ Outer Zone

OZA Ozona (USA-TX) (location identifier)

oz. ap. Ounce (Apothecaries)

oz. av. Ounce (Avoirdupois)

OZC Ozamis City - Labo (Philippines) (location identifier)

OZE Outer Zone Electron

OZEP Outer Zone Electron Precipitation

OZH Zaporozhye (Ukraine) (location identifier)

OZP Morón (Spain) (location identifier)

OZR Ozark Cairns (USA-AL) (location identifier)

ozs Ounces

oz. t. Ounce (Troy)

OZZ Ouarzazate (Morocco) (location identifier)

O2 Object-Oriented • Oxygen

O2A Object-Oriented Analysis

O2AD Object-Oriented Analysis and Design

O2B Object-Oriented Bibliography

O2C Organisation de la Culture Commerciale

O2D Object-Oriented Design • Object-Oriented Development • Object-Oriented Directory

O2DB Object-Oriented Database

O2DBMS Object-Oriented Database Management System

O2DCE Object-Oriented Distributed Computing Environment

O2E Object-Oriented Environment

O2ERM Object-Oriented Entity-Relationship Model

O2FS Object-Oriented File System

O2G Object-Oriented Graphics

O2GL Object-Oriented Graphics Language

O2I Object-Oriented Interface

O2L Object-Oriented Language • Object-Oriented Library • Object-Oriented Listing

O2MP Object-Oriented Management Planning

O2N Object-Oriented Numerics

O2P Object-Oriented Pattern • Object-Oriented Programming

O2PL Object-Oriented Programming Language

O2PS Object-Oriented Programming and System • Object-Oriented Programming System

O2RA Object-Oriented Requirement Analysis

O2RALD Object-Oriented Requirements Analysis and Logical Design

O2S Object-Oriented Software • Object-Oriented System

O2SA Object-Oriented Systems Analysis

O2SE Object-Oriented Software Engineering

O2T Object-Oriented Technology

O2UI Object-Oriented User Interface

O2V Object-Oriented View

O2ZE Object-Oriented Z Environment

O3 Ozone

P

p Electric Dipole Moment • Electron Spectral Index • Generalized Momentum • Hole Number Density • Molecular Momentum • Momentum • Page • Página • Pamphlet • Para • Paragraph • Parallax • Part • Participio • Particle • Partim • Pascal (file extension) • Pass • Passed • Past • Pater • Peak • Peculiar • Pence • Penny • Per • Perch • Permanent Dipole Moment of a Molecule • Person • Peseta (currency) • Peso (currency) • Phosphate • Piano • Piastre (currency) • Pico • Pied • Pint • Pipe • Pius • Plasma Pressure • Polar • Pole • Pole Pairs Number • Population • Por • Port • Post • Pouce • Pour • President • Pressure • Primus • Pro • Probability Density Function • Production • Professional • Proton • Resonance Escape Probability • Shower • Sound Pressure • Static Pressure

P/ Periodic Comet

P Dielectric Polarization • DPRK (international aircraft registration code) • Electric Polarization • Four Hundred (Roman numeral) • Gauss Vector • Korea (DPRK) (international aircraft registration code) • Pad • Papa (international alphabet) • Paragraph • Parental • Parity • Park • Parking • Parkplatz • Paseo • Pass • Passing • Pater • Pedestrian Crossing • Pence • Penny • Père • Perimeter • Period • Permian • Personnel • Peseta (currency) • Peso (currency) • Peta • Pharmacy • Phase • Phosphate • Phosphor • Phosphore • Phosphorus • Pic • Pico • Pin • Piso • Pitch • Plate • Pluto • Poids • Poise • Polar • Polynomial • Poor • Populus • Portugal (international vehicle registration code) • Portugal (postal country code) • Position Angle • Positive • Post • Postage • Poster • Posterior • Post Office • Power • Precedence • Precision • Predicate • Pregunta • Preliminary • President • Pressure • Priest • Prince • Priority • Probability • Probability Density • Probability Distribution • Progressive • Prohibited • Promotable • Promoted • Proposed • Proprietary • Proton • Public • Pula (currency) • Pulau • Pulsar • Pump • Punta • Pursuit • Quantum Number of Component of Angular Momentum Along Symmetry Axis • Radiant Flux • Radiant Power • Sidereal Period • Sound Energy Flux • Sound Power • Space Inversion • Weight • 400 (Roman numeral)

p/a Personal Account

pa Panama • Per Adres • Per Anno • Per Annum • Por Ausencia • Punjabi (ISO language code)

pA per Adresse • Picoampere

Pa Pascal • Pennsylvania (USA) • Protactinium

P/A Particular Average • Physics/Astronomy • Power Amplifier • Preamplifier • Pressure-Actuated • Private Account • Problem Analysis • Propulsion/Avionics

P&A Personnel & Administration • Photometry & Astrometry • Physics & Astronomy • Planning & Assessment • Price & Availability • Priorities & Allocations • Programming & Analysis

PA Pad Abort • Paging and Area • Paired Array • Palmes Académiques • Pamphlet • Panama (international vehicle registration code) • Panama (ISO country code) • Pará (Brazil) • Paraguay • Parametric Amplifier • Par Avion • Parents' Association • Parent Atom • Participial Adjective • Particular Average • Passenger Agent • Pattern Analysis • Paying Agent • Payload Accommodation • Pedagogische Academie • Pending Availability • Pendulous Axis • Pénétration Assistée • Pennsylvania (USA) • Per Adres • Per Annum • Performance Analysis • Performance Analyzer • Performance Assessment • Performance Assurance • Permanent Address • Permanent Appointment • Personal Accident • Personal Account • Personal Allowance • Personal Assistant • Perturbation After • Petites Annonces • Phased Array • Photoacoustic • Photo Area • Photodiode Amplifier • Physical Address • Physician's Assistant • Pilotless Aircraft • Pitch Angle • Pitch Axis • Plasma Analyzer • Plasma Armature • Plateau Age • Play Aid • Point of Aim • Polar Angle • Polarization Analyzer • Popular Astronomy (periodical code) • Por Autorización • Porous Aggregata • Port A • Position Angle • Posta Aerea • Power Alarm • Power Amplifier • Power of Attorney • Preamplifier • Prearbitrated • Precision Angle • Predictive Analyzer • Predictive Annotation • Predictive Avoidance • Preliminary Acceptance • Preliminary Assessment • Preparing Activity • Presidential Appointment • Press Agent • Press Association • Press Attaché • Pressure Altitude • Pressure Angle • Primary Access • Primary Amplifier • Principal Astronomer • Principal Axes • Privacy Act (USA) • Private Account • Probability of Acceptance • Probability of Arrival • Problem Analysis • Procedimiento Administrativo • Process Algebra • Processing Activities • Processing Application • Procurement Appropriation • Procurement Authorization • Procuring Activity • Product Analysis • Product Assurance • Professeur Adjoint • Professeur Agrégé • Professional Association • Professional Astronomer • Program Address • Program Administrator • Program Analysis • Program Assessment • Program Assistant • Program Authority • Program Authorization • Prohibited Area • Project Analyst • Propagation Algorithm • Property Administrator • Proponent Agency • Protected Area • Proton Affinity • Provider Agent • Provisional Acceptance • Provisioning Activity • Public Accountant • Public Address • Public Administration • Public Affairs • Public Archive • Publicity Agent • Publishers Association (UK) • Pulsating Arc • Pulse Amplifier • Pulsed Annealing • Purchasing Agent

PAA Pa-an (Myanmar) (location identifier) • Pattern Analysis and Applications (periodical code) • Payload Assembly Area • Peer Access Approval • Phase Array Antenna • Phased-Array Antenna • Photon Activation Analysis • Piedmont Amateur Astronomers (USA-SC) • Planar Array Antenna • Policy Approving Authority • Ports Application Analysis • Power Amplifier Assembly • Priority Abatement Area • Problem Assessment Area • Program Associate Administrator • Prueba de Aptitud Académica

PAABS Panamerican Association of Biochemical Societies

PAAC Program Analysis Adaptable Control

PAACE Pennsylvania Association for Adult Continuing Education (USA-PA)

PAAD Program Acronyms, Abbreviations and Definitions • Program Assistance Approval Document

PAAM Phased Array Antenna Module

PAAP Planetary Astronomy and Atmospheres Program

PAARL Philippines Association of Academic and Research Libraries

PAAS Pakistan Association for the Advancement of Science • Phased Array Antenna System • Publications of the American Astronomical Society (periodical code)

PAASCU Philippine Accrediting Association of Schools, Colleges and Universities (now AACCUP)

PAAT Passive Acoustics Analysis Trainer

PAAW Policy Approving Authority Workstation

pab Personal Address Book (file extension)

Pab Pabellón

PAB Balboa (Panama) (ISO currency code) • Bilaspur (India) (location identifier) • Panamean Balboa (ISO currency code) • Payload Assembly Building • Periodic Area Balancing • Personal Address Book • Policy Advisory Board • Program Advisory Board • Project Astro Boston

• Public Affairs Bureau • Publications of the Astrophysics Branch (periodical code)

PABD Precise Access Block Diagram

PABei Progress in Astronomy (periodical code)

PABF Precision Air Bearing Facility

PABLA Problem Analysis By Logical Approach

PABMB Pan American Association for Biochemistry and Molecular Biology

PABO Pedagogische Academie voor het Basisonderwijs

PABST Primary Adhesive Bonded Structure Technology

PABX Private Automated Branch Exchange • Private Automatic Branch Exchange

Pac Pacific

PaC Pénétration à Chaud

PAC Pacific • Pacific Accredition Cooperation • Package Attitude Control • Packaged Assembly Circuit • Packet • Paging Area Controller • Panama City - Paitilla (Panama) (location identifier) • Parametric Amplifier Converter • Parent - Adult - Child • Particle Accelerator Conference (USA) • Passive Air Cycle • Pedagogic Automatic Computer • Performance Acceptance Certified • Performance Analysis and Control • Performance Appraisal Committee • Personal Analog Computer • Personal Authenticator Card • Personnel and Administration Center • Perturbed Angular Correlation • Photo Annotation/Classification • Physics Action Council • Planetarium Association of Canada • Planetary Astronomy Committee • Planning Advisory Committee • Player access Control • Pneumatic Analog Computer • Pneumatic Auxiliary Console • Polar Atmospheric Chemistry • Policy Advisory Committee • Political Action Committee • Politique Agricole Commune • Popular Astronomy Club Inc. (USA-IL) • Positional Astronomy Centre (India) • Postattenuation Correction • Postaward Conference • Powdered Activated Carbon • Power Assurance Check • PPP Access Concentrator • Prairie Astronomy Club (USA-NE) • Prescott Astronomy Club (USA-AZ) • Preservation And Conservation • Pressure Alpha Center • Price Adjustment Code • Primary Address Code • Prime Associate Contractor • Principal Associate Contractor • Principal Astronomer's Computer • Privilege Attribute Certificate • Probably Approximately Correct • Probe Aerodynamic Center • Problem Action Center • Processing and Archiving Centre • Processor Augmentation Card • Product Acceptance Criteria • Productibility Assurance Center • Production Acceptance Certification • Program Activity Center • Program Advisory Committee • Program Analyzer Console • Program Acquisition Cost • Programme d'Actions Communautaires • Promotion et Assistance Commerciale • Public Access Catalog • Publications Advisory Committee • Pure and Applied Chemistry

PACA Payload Accommodation Configuration Assessment • Phased Array Communications Antenna • Physics And Chemistry of the Atmosphere • Priority Access and Channel Assessment • Provence - Alpes - Côte d'Azur (France)

PACAS Personnel Access Control Accountability System

PACB Programme Atmosphère et Cycle Biogéochimique

PACC PERT-Associated Cost Control • Problem Action Control Center • Product Assessment and Certification Center

PACCOM Pacific Computer Communications

PACCS Postattack Command Control System

PACCT PERT And Cost Correlation Technique

PACE Pacific-Arizona Crustal Experiment • Parallel Automated Coding Expert • Passive Attitude Control Experimental • Performance And Cost Evaluation • Phased-Array Control Electronics • Physics And Chemistry Experiments • Pilot and Aircraft Courtesy Evaluation • Plant Acquisition and Construction Equipment • Plasma-Assisted Chemical Etching • Plymouth Atmospheric Corrosion Experiment • Postappointment and Continuing Education • Power And Control Electronics • Power And Control Equipment • Precision Analog Computing Equipment • Preflight Acceptance Checkout Equipment • Preflight Automatic Checkout Equipment • Prelaunch Automatic Checkout Equipment • Priority Access Control Enabled • Producible Alternative to Cadmium-Telluride Epitaxy • Product Acceptance Configuration Element • Professional and Administrative Career Examination • Program Acquisition Cost Estimate • Program for Advanced Continuing Education • Programmable Aerospace Control Equipment • Programming Analysis Consulting Education • Projects to Advance Creativity in Education • Property Advisor to Common Estate • Proving and Adjustment for Communications Efficiency

PACED Program for Advanced Concepts in Electronic Design

PACEM Propagation en Air Clair et Météorologie

PACER Part And Component Evaluation Report • Post-analysis Critique and Exercise Review • Process Assembly Case Evaluator Routine • Program-Assisted Console Evaluation and Review • Program for the Acceleration of Commercial Energy Research • Programmed Automatic Communication Equipment • Public Access to Count Electronic Records

PACES Pan American Center for Earth and Environmental Studies • Political Action Committee for Engineers and Scientists

PACI Partnership for Advanced Computational Infrastructure • Payload and Attitude Control Interface

Pacif Pacific

PACIMS Passive Chemical Ionization Mass Spectrometry

PACIR Propulsion, Aerodynamics, Control Integration Research

PACIS Pacific Asia Conference on Information Systems • Pulsed Arc Cluster Ion Source

PACK Payload Active Cooling Kit

PACL Power Assurance Check Left

PACLING Pacific Association for Computational Linguistics

PACM Passive Access Control Module • Passive-Active Collective Memory • Program in Applied and Computational Mathematics • Pulse Amplitude Code Modulation

PACMIG Pressurized Air-Cooled Metal Inert Gas

PACNET Pacific Network

PACO Parts Control • Prelaunch Automatic Checkout

PACOS Palomar Abell Cluster Optical Survey

PACOSS Passive and Active Controls of Space Structures

PACP Propulsion Auxiliary Control Panel

PACR Power Assurance Check Right

PACS Pacific Area Communications System • Palomar Asteroid and Comet Survey • Passenger Automated Check-in System • Personal Access Communications System • Photoconducting Array Camera and Spectrometer • Photoconductor Array Camera and Spectrometer • Photodetector Array Camera and Spectrometer (*) • Physical Abstract Classification Scheme • Physics and Astronomy Classification Scheme • Picture Archiving and Communication System • Planet Asteroid and Comet Survey • Pointing and Attitude Control System • Polar Acquisition and Command System • Programming And Computation Service • Public Access Computer System

PaCT Parallel Computing Technologies (conference)

PACT Parallel Architectures and Compilation Techniques (conference) • Parchmarki Array of Čerenkov Telescopes (India) • Palo Alto Collaborative Testbed • Pay Actual Computer Time • PCTE Added Common Tools • Power-Augmented Catalytic Thruster • Practical Application of Constraint Technology • Prefix Access Code Translator •

Primary Access to Chemical Titles • Production Analysis Control Technique • Program for Automatic Coding Techniques • Programmed Analysis Computer Transfer • Programmed Automatic Circuit Tester • Project for the Advancement of Coding Techniques

PACU Philippines Association of Colleges and Universities

PACVD Plasma-Activated Chemically Vapor Deposited

PACX Private Automatic Computer Exchange

PAD Packet Assembler-Disassembler • Packet Assembly-Disassembly • Packetizer-Depacketizer • Padding • Paderborn (Germany) (location identifier) • Parts Approval Document • Payable After Death • Penetration Aid Device • Pitch Angle Distribution • Pitch Axis Definition • Plan d'Action Discrétion • Planning Analysis Document • Planning and Analysis Division • Polar and Auroral Dynamics • Polyaperture Device • Poor Acquisition Data • Portable Attachment Device • Positive Additive Distribution • Power Amplifier Device • Power Amplifier Driver • Precise Access Diagram • Preferred Arrival Date • Preliminary Advisory Data • Prêt À Diffuser • Product Assurance Directorate • Production - Archiving - Distribution • Production Assurance Directorate • Program Action Directive • Program Analysis for Documentation • Program Approval Document • Programmation Automatique des Données • Project Approval Document • Propellant Acquisition Device • Propellant-Actuated Device • Proposal Administrative Data • Protoplanetary Accretion Disk • Provisional Acceptance Date • Public Affairs Division • Pulse Amplitude Detector • Pulse Amplitude Discriminator

PADAR Passive Detection And Ranging • Photoacoustic Detection And Ranging • Program Approved Disposal and Redistribution

PADC Planetary Atmospheric Dispersion Corrector

PadCR Osservatorio Astronomico di Padova - Comunicazioni e Rassegne (periodical code)

PADD Plan d'Aménagement et de Développement Durable

PADDS Procurement Automated Data Documentation System

PADEM Projet Amélioration des Enquêtes de Ménage

PADI Preserving Access to Digital Information

PADISNet Pan African Development Information System Network

PADL Part and Assembly Description Language • Practical Aspects of Declarative Languages (workshop)

PADO Proposed Advanced Development Objective

PADOC Pay Adjustment Document

PADP Production Automated Design Process

p. Adr. Per Adresse

PADR Preferential Arrival and Departure Route • Production Administration Delinquency Report

PADRE Portable Automatic Data Recording Equipment

p. ae. Partes Aequales

PADS Parallel And Distributed Simulation • Passive-Active Data Simulation • PDS Archive Delivery System • Pen Application Development System • Performance Analysis and Design Synthesis • Performing Arts Data Service • Position Azimuth Determining System • Price and Availability Data Sheet • Problem Action Data System • Program Access to Data Set

PADSIM Precision Attitude Determination Simulation

PADT Postalloy-Diffused Transistor

PA&E Plan Analysis & Evaluation • Product Acceptance & Evaluation • Product Assurance & Engineering • Program Analysis & Evaluation

PAE Everett - Snohomish County (USA-WA) (location identifier) • Parc d'Activités Économiques • Peer Access Enforcement • Performance Assurance Engineer • Phase Angle Error • Pilote Automatique Embarqué • Plan d'Action pour l'Environnement • Postaccident Environment • Power-Added Efficiency • Preliminary Acceptance Europe (VLTI) • Presses Académiques Européennes (Belgium) • Preventive Action Engineer • Problem Assessment Engineer • Problem Assessment Engineering • Program Analysis and Evaluation • Projet d'Action Éducative

PAEC Pakistan Atomic Energy Commission • Parc d'Activités Économiques et Commerciales (France) • Planned and Archival Exposures Catalog (HST)

PAECT Pollution Abatement and Environmental Control Technology

PA&ED Program Analysis & Evaluation Director

PAED Program Analysis and Evaluation Director • Program Analysis and Evaluation Directorate

PAEM Program Analysis and Evaluation Model

PAEMST Presidential Award for Excellence in Mathematics and Science Teaching (USA)

PAEO Public Affairs and Educational Outreach

PAESMEM Presidential Awards for Excellence in Science, Mathematics and Engineering Mentoring (USA)

PAET Planetary Atmosphere Entry Test (*) • Planetary Atmosphere Experimental Test

PAEW Personnel And Equipment Working

PaF Pénétration à Froid

PAF Pakuba (Uganda) (location identifier) • Parcours d'Approfondissement de Formation • Particle and Field Package • Payload Attach Fitting • Payload Attachment Fitting • Paysage Audiovisuel Français • Peak Annual Funding • Performance Accelerator Feature • Peripheral Address Field • Personnel Authorization File • Plan Académique de Formation (France) • Police de l'Air et des Frontières (France) • Postal Address File • Postcode Address File • Processing and Archiving Facility • Processing Archiving Facility • Project Approval Form • Protocol Assessment Facility • Puissance Au Frein

PAFB Patrick Air Force Base (USA)

PAFS Predictive Analysis Flagging System

Pág Página

PAG Pagadian (Philippines) (location identifier) • Phenylacetyl Glutamine • Plena Analiza Gramatiko • Poole Astronomical Society (UK-England) • Precision Alignment Gyrocompass • Precursor Active Galaxy • Preliminary Analysis Group • Project Advisory Group • Protective Action Guide • Public Affairs Guidance

PAGASA Philippine Atmospheric, Geophysical and Astronomical Service Administration

PAGB Postasymptotic Giant Branch

PAGE Periodicals Accuracy, Grading and Evaluation • PERT Automated Graphical Extension • Piston Arrestment Gas Entrapment • Polyacrylamide Gel Electrophoresis • Program for the Advancement of Geoscience Education • Programmable Aerospace Ground Equipment

PAGEOS Passive Geodetic Earth Orbiting Satellite

PAGES Past Global Changes

PAGIS Pastoral and Agricultural Geographic Information System • Planning And Geographic Information System • Programmable Agents for Group Interaction Systems

PAGODA Profile Alignment Group on Open Document Architecture

Págs Páginas

PAGSI Programme d'Action Gouvernementale pour la Société de l'Information (France)

PAH Paducah (USA-KY) (location identifier) • Payload Accommodations Handbook • Physical Activity and Health • Polycyclic Aromatic Hydrocarbon • Push And Hold

PAHO Pan American Health Organization

PAHT Power-Augmented Hydrazine Thruster

PA&I Planning, Analysis & Integration • Plan Analysis & Integration • Program Analysis & Integration

PAI Pailin (Cambodia) (location identifier) • Paysage Astronomique International • Point d'Accueil International • Portable Auroral Imager • Probabilistic Affinity Index • Programa de Actuación Integrada • Programme d'Action Intégrée • Programmer Appraisal Instrument • Projet d'Accueil Individualisé • Projet d'Action Innovante • Protocol Address Information • Provisional Acceptance Inspection

PAIB Publications of the Astronomy Institute of Bonn (periodical code)

PAIC Public Address Intercom System • Publications of the Astronomical Institutes of Czechoslovakia (periodical code)

PAICz Publications of the Astronomy Institute of the Czechoslovak Academy of Sciences (periodical code)

PA&ID Plan Analysis & Integration Director • Plan Analysis & Integration Directorate • Program Analysis & Integration Director • Program Analysis & Integration Directorate

PAID Plan Analysis and Integration Director • Plan Analysis and Integration Directorate • Procedures - Applications - Infrastructures - Data • Program Acquisition Information Database • Program Analysis and Integration Director • Program Analysis and Integration Directorate

PAIDAI Planetary Astronomy InSb Detector Array Infrastructure

PAIGH Pan American Institute of Geography and History

PAIH Public-Access Internet Host

PAILS Publication Automated Information Locator System

PAIO Permanence d'Accueil, d'Information et d'Orientation

PAIP Performance Assurance Implementation Plan

PAIR Partnership Award for the Integration of Research • Polarization-Agile Instrumentation Radar

PAIS Portable Ada Interface Set • Prestigious Applications of Intelligent Systems (conference) • Process Appraisal Information System • Product Assurance Information System • Project Analysis Information System • Public Access Internet Site • Public Affairs Information Service

PAIV Particle-Acceleration-Induced Viscosity • Power as An Independent Variable

PAJ Para Chinar (Pakistan) (location identifier)

Pak Pakistan • Pakistani

PAK Hanapepe - Port Allen (USA-HI) (location identifier) • Packed • Pakistan (international vehicle registration code) • Pakistan (ISO country code) • Polyaromatische Kohlenwasserstoffgemische • Program Attention Key

PAKD Practical Application of Knowledge Discovery

PAKISTAN Pendjab - Afghania - Kashmir - Iran - Sind - Turkmenistan

PAKM Practical Aspects of Knowledge Management

pal Paintbrush Palette (file extension)

Pal Palace • Palacio • Palais • Palencia (Spain) • Paleocene • Paleography • Paleontology • Palestine • Palomar

PAL Palacio • Palanquero (Colombia) (location identifier) • Pallet • Paradox Application Language • Parallel Aligned Liquid-crystal • Parc d'Activités Logistiques • Parcel Airlift • Parts Authorization List • Pedagogic Algorithmic Language • Performance Assessment Logic • Peripheral Availability List • Permanent Artificial Lighting • Permissive Action Link • Personnel Allowance List • Phase-Alternate Line • Phase-Alternating Line • Phase-Alternation Line • Pohang Accelerator Laboratory (ROK) • Portable - Accurate - Lightweight • Postal Answer Line • Preliminary Allowance List • Present Atmospheric Level • Price and Availability List • Prime Agricultural Land • Privileged-Architecture Library • Process Assembler Language • Process Assembly Language • Process Asset Library • Process Automation Language • Program Assembler Language • Programmable Array Logic • Programmed Array Logic • Programmed Application Library • Programmer Assistance and Liaison • Programming Assembly Language • Projected Average Loss • Propulseur Auxiliaire Liquide • Propulseur d'Appoint Liquide • Propulsion Auxiliaire Liquide • Prototype Application Loop • Psychoacoustic Laboratory • Public Ada Library • Publication Applicability List • Push And Latch

PALAO Palomar Adaptive Optics

PALC Plasma-Addressed Liquid Crystal (display) • Point Arguello Launch Complex (USA-CA) • Professional-Amateur Liaison Committee

PALCD Plasma-Addressed Liquid Crystal Display

PALCI Pennsylvania Academic Library Connection Initiative (USA-PA)

PALD Phase Alteration Line Delay

PALE Paleoclimates of Arctic Lakes and Estuaries

Paleo Paleocene • Paleography • Paleontology

Paleog Paleography

Paleogr Paleografía

Paleon Paleontology

Paleont Paleontología

Paleontol Paleontology

Palest Palestine • Palestinian

PALI Procurement Annex Line Item

PALILA Photon Acquisition at Low Intensity Level for Astronomy

PALIS Polarized Airborne Laser Imaging Sensor • Ponds And Lakes Information System

PALL Pallet

PAllO Publications of the Allegheny Observatory (periodical code)

PALME Procédure d'Acquisition de Logiciels Microsoft pour l'Enseignement

PALMS Particle Analysis by Laser Mass Spectrometry • Partnership Advancing the Learning of Math and Science

PALOP Países de Língua Oficial Portuguesa

PALR Permissive Action Link Report

PALS Permissive Action Link System • Photo Area and Location System • Point Arguello Launch Site (USA-CA) • Prague Asterix Laser System • Precision Approach and Landing System • Principle of the Alphabet Literacy System • Project for Automated Library System • Public Access Library System

PALSD Production Assurance Logistics Support Division

PALSG Personnel And Logistics Systems Group

PALT Pressure Altitude • Procurement Acquisition Lead Time • Procurement Administrative Lead Time

Pam Pamphlet

P/AM Propulsion/Avionics Module

PAM Panama City - Tyndall (USA-FL) (location identifier) • Parallel Associative Memory • Parameter Adjusting Mechanism • Parameter Assignment Map • Payload Assist Module • Periodic Absorption Modulation • Periodic Anderson Model • Peripheral Adapter Module • Permanent Archive Manager • Personal Applications Manager • Phase Amplitude Modulation • Phased Array Module • Physics - Astronomy - Mathematics • Planetary Aerosol Monitor • Plataforma Antiruido de Mejorada • Pluggable Authentication Module • Point Access Method • Point d'Accès Métropolitain • Pole Amplitude Modulation • Portable Automated Mesonet • Postaccident Monitoring • Power Amplifier Module • Predictive Aircraft Maintenance • Preliminary Aerosol Monitor • Primary Access Method • Priorities and Allocations Manual • Procurement and Acquisition Management • Professional Astronomy Magazine • Programa de Alimentación Mundial

(ONU) • Program Analysis Memorandum • Programme Alimentaire Mondial • Propulsion and Avionics Module • Propulsion Avionics Module • Pulse-Address Modem • Pulse-Amplification Modulation • Pulse-Amplifier Modulation • Pulse-Amplitude Modulation (*) • Pupil Adjust Mechanism • Pupil Alignment Mechanism (*)

PAMA Preassigned Multiple Access • Professional Aviation Maintenance Association • Pulse-Address Multiple Access

PAMC Provisional Acceptable Means of Compliance

PAMCS Panel on Airspace Management and Control System

PAM-D Payload Assist Module - Delta class

PAMD Price And Management Data • Public Access Machine-Readable Document

PAME Projet d'Atelier de Modélisation Électromagnétique

PAMELA Payload for Antimatter Matter Exploration and Light-nuclei Astrophysics • Phased Array Mirror Extensible Large Aperture • Process Abstraction Method for Embedded Large Applications

PAMINA Palatinat - Mittlerer Oberrhein - Nord Alsace (Germany/France)

PAMIPS Parallel Algorithm for Mixed Integer Programming in Scheduling

PAMIR Passive and Active Microwave and Infrared Radiometer

PAMIRASAT Passive Microwave Radiometer Satellite

PAMIS Processing And Manufacturing In Space

PAMM Permanent Archive and Media Management • Physical Access and Media Management

PAMOS Programme Atmosphère Météorologique et Océan Superficiel

PAMOY Programme Atmosphère Moyenne

PAMPER Practical Application of Mid-Points for Exponential Regression

PAMPL Program Approved Materials and Processes Selection List

PAMR Private Access Mobile Radio

PAMRI Peripheral Adaptor Module Replacement Item

PAMRS Parameter Adaptive Model Reference System

PAMS Pad Abort Measuring System • Partitioned Access Method System • Passive Aerodynamically-stabilized Magnetically-damped Satellite • Patent Automated Management System • Pentax Analytical Measurement System • Perceptual Analysis and Measurement System • Photogrammetric Analytical Measurement System • Portable Automated Mesonet Station • Predictive Aircraft Maintenance System • Program Acquisition Management System

PAMSCO Program Article Master Schedule Commitment

pan Pannus

Pan Panamá • Panchromatic • Panoramic

PAN Panama (ISO country code) • Passive/Active Neutron • Pattani (Thailand) (location identifier) • Peroxyacetylnitrat • Peroxyacetyl Nitrate • Personal Access Number • Personal Account Number • Personal Area Network • Personennamen der Antike • Plan D'Aménagement National • Polska Akademia Nauk (Poland) • Polyacrylonitrile • Primary Alerting Network • Probabilistic Analysis Network • Production Approval Number • Public Access Network • Project voor Automatische Namenverwerting

PANAFTEL Pan-African Telecommunications Network

PANAGIC Particle And Nuclear Astrophysics in Gravitation International Committee

PANAMSAT Pan American satellite

PANASH Paleoclimates of the Northern And Southern Hemispheres

Pan. Can. Panama Canal

PANDA Pacific Neotectonic Deformation Array • Panel Design Analysis • Personnel And Administration • Planetary Atmospheres Node Data Analysis

PANES Program for Analysis of Non-linear Equilibrium and Stability

PANGIS Pan-African Network for a Geological Information System

PANH Polycyclic Aromatic Nitrogen Heterocyclic

PANI Phénomène Aérospatial Non Identifié

PANIC Particles and Nuclei International Conference • Persson's Auxiliary Nasmyth Infrared Camera • Potential And Needs - Investments - Capabilities • Prototype Adaptive-optics Near-Infrared Camera • PtSi Array Near-Infrared Camera • PtSi Astronomical Near-Infrared Camera

PANS Pretty Advanced New Stuff • Procedures for Air Navigation Services

PAO Palo Alto (USA-CA) (location identifier) • Period of Autonomous Operations • Physical Abstract Object • Pierre Auger Observatory • Polissage Assisté par Ordinateur • Polyalpha Olefin • Postacademiaal Onderwijs • Primary Action Office • Privacy Act Officer • Production Assistée par Ordinateur • Program Action Officer • Programmation Assistée par Ordinateur • Project Action Officer • Project Administration Officer • Property Accounting Office • Public Affairs Office • Public Affairs Officer • Publication Assistée par Ordinateur • Pulse Avalanche-diode Oscillator • Pyongyang Astronomical Observatory (Korea-DPRK)

PAOC Postaward Orientation Conference

PAONET Public Astronomical Observatory Network

PAOS Program in Atmospheric and Oceanic Sciences (Univ. Colorado)

PAOT Persons At One Time

Pap Papel • Paper • Papier • Papua • Papuan

PAP Packet Authentication Protocol • Packet-level Procedure • Password Authentication Protocol • Part Acceptance Probability • Payload Accommodation Plan • Payload Adapter Plate • Peak Advertised Performance • Performance Assurance Plan • Personnel Assistance Point • Personnel Assurance Program • Photo-Addressable Polymer • Physics and Astronomy Program • Plug And Play • Port au Prince (Haiti) (location identifier) • Practical Application of PROLOG • Prêt À Poster • Primary Aerosol Particle • Printer Access Protocol • Producibility Assessment Procedure • Product Assurance Plan • Program Action Plan • Projet d'Action Personnalisé • Propulseur d'Appoint à Poudre • Public Access Profile • Pulmonary Arterial Desease

PAPA Photon Analog Precision Address • Precision Analog Photon Address • Programmer And Probability Analyzer • Public Affairs Proponency Activity

PAPAO Photometry and Astrometry Programme for Adaptive Optics

PAPCO Panel Plot Composer

PAPE Photoactive Pigment Electrophotography

PApGe Pure and Applied Geophysics (periodical code)

PAPI Personal And Performance Information • Precision Approach Path Indicator

PAPL Preliminary Allowance Part List • Program Approved Parts List • Project Approved Parts List

PAPO Papers Acquisition Program Office

PApOp Pure and Applied Optics (periodical code)

PAPP Product Assurance Program Plan

PAPS Pooled Analytical Plotter System • Portable Ada Programming System • Propulsion-Associated Parameter Set

PAQ Palmer (USA-AK) (location identifier) • Pending Action Queue • Perceived Air Quality • Plan d'Amélioration de la Qualité • Point Association Question • Position Analysis Questionnaire • Process Average Quality • Processus d'Amélioration de la Qualité

Pár Párrafo

Par Parágrafo • Paragraph • Paragraphe • Paraguay • Parallax • Parallel • Parallèle • Parameter • Paramètre • Parenthèse • Parenthesis • Parish

P/AR Peak-to-Average Ratio

PAR Pacific Antarctic Ridge • Parallel • Parameter • Parametrized Action Representation • Parcel • Paris Metropolitan (France) (location identifier) • Parity • Partial Acceptance Review • Participatory Action Research • Parts Approved Request • Patient Accounting and Reporting • Payload Adapter Ring • Peak Accelerometer Recorder • Peak-to-Average Ratio • Performance Analysis and Review • Performance Appraisal Report • Performance Assessment Report • Performance Assurance Requirements • Perimeter Acquisition Radar • Personal Animation Recorder • Personnel-At-Risk • PERT Analysis Report • Phased-Array Radar • Photoexcitation by Accidental Resonance • Photosynthetically Active Radiation • Photosynthetically Available Radiation • Planned All Around • Planning Action Request • Planning Activity Report • Population-At-Risk • Positive Acknowledgment and Retransmission • Positive Acknowledgment Retransmission • Power Analyzer and Recorder • Precision Approach Radar • Preferential Arrival Route • Preliminary Analysis Review • Price Analysis Report • Priority Action Report • Priority Action Request • Probabilistic Analysis of Risk • Problem Accountability Record • Problem Action Record • Problem Action Request • Problem Analysis Report • Problem Analysis and Resolution • Product Acceptance Review • Production Analysis Report • Program Acquisition Request • Program Activity Register • Program-Aid Routine • Program Analysis Review • Program Appraisal and Review • Program Assessment Report • Program Assessment Review • Programmable Access Router • Programmed Annual Rate • Project Appraisal Report • Project Assurance Requirements • Project Authorization Request • Projected Automation Requirement • Projet d'Acte Règlementaire • Proposal Analysis Report • Provisional Acceptance Review • Pulse Acquisition Radar • Purchase Approval Request • Purchasing Approval Request • Push And Release

Para Paragraph • Paraguay

PARA Parabolic • Paragraph • Problem Analysis and Recommended Action

Parab Parabola

PARABOLA Portable Apparatus for Rapid Acquisition of Bidirectional Observations of Land and Atmosphere

PARADE Passive-Active Ranging And Determination

PARADISE Piloting A Researchers Directory Service for Europe • Piloting A Researchers Directory Service in Europe

PARAGON Processing and Archiving of Radar And Gauge data Off-line and in Near-real-time

Paragr Paragraph • Paragraphe

PARAM Parameter

PARAMP Parametric Amplifier

PARASYN Parametric Synthesis

PARB Problem Assessment Review Board

Parc Parcella

ParC Parallel Computing (periodical code)

PARC Palo Alto Research Center (XEROX) • Pennsylvania Astronomical Research Consortium (USA-PA) • Polarimetric Active Radar Calibrator • Public Access Resource Center

PARCA Pan American Railway Congress Association • Parallel Channel Adapter

PARCC Precision - Accuracy - Representativeness - Completeness - Correctness

Parch Parchemin • Parchment

ParCo Parallel Computing

PARD Parts Application Reliability Data • Periodic And Random Deviation • Personnel Actions and Records Directorate • Phased Array Radar Detection • Pilotless Aircraft Research Division • Precision Annotated Retrieval Display • Preliminary Automation Requirements Document

PARDP Perimeter Acquisition Radar Data Processor

PARE Plan d'Aide au Retour à l'Emploi (France) • Physiological and Anatomical Rodent Experiment

PaReL Pattern Recognition Letters (periodical code)

Paren Parenthesis

Parens Parentheses

PARES Preprocessing of Airborne Remote Sensing

PARG Perth Astronomy Research Group (Australia-WA)

PARI Pisgah Astronomical Research Institute (USA-NC)

PARINFO Programme d'Aide à la Recherche en Informatique

PARIS Parallel Instruction Set • Passive Reflectometry and Interferometry System • Pulse Analysis-Recording Information System

Parl Parallel • Parliament • Parliamentary

PARM Participating Manager • Program Acquisition Resource Manager • Program Analysis for Resource Management

PARMA Program for Analysis, Reporting and Maintenance

PARMIS Planning And Resource Management Information System

PARNet Palermo Academic and Research Network

PARR Procurement Authorization and Receiving Report • Program Analysis and Resources Review • Program Assessment Review Report

PARRSSIS Program for Advanced Research into the Regulation, Steering and Stability of the International System

PARS Performance Analysis Reporting System • Postal Automated Redirection System • Procurement Accounting and Reporting System • Property Accountability Record System • Property Accounting and Reporting System

PARSCA Paranal Seeing Campaign (ESO)

PARSE Phased Array Radar Simulation Environment

Parsec Parallax-Second • Parser and Extensible Compiler

PARSEC Paranal Artificial Reference Source for Extended Coverage (ESO)

PARSECS Program for Astronomical Research and Scientific Experiments Concerning Space

PARSIFAL Programming And Real-time System Interface For ALIS

PARSYN Parametric Synthesis

Part Parte • Partial • Partícula • Particular • Partie • Participate • Participial • Particle • Partícula • Particular • Particule • Partition • Partner • Partnership • Party

PART Particular • Planning for Activities with Resources and Time Constraints • Planning of Activities, Resources and Technology • Program Assessment Rating Tool • Projection and Angular and Radial Transformation (method)

part. aeq. Partes Aequales

PartGG Partnerschaftsgesellschaftgesetz

PARTNER Proof of Analog Results Through a Numerical Equivalent Routine

PARTS Parts Analysis and Review Technique for Spares • Parts Assembly and Reuse Tool Set • Parts Automated Repairable Tracking System • Price Analysis and Review Technique for Support

part. vic. Partibus Vicibus

PARU Photographic And Reproduction Unit

Parv Parvis

pas Pascal (file extension)

Pas Pasarela • Passage • Passive

PA&S Product Assurance & Safety

PAS Pacific Aerial Survey • Pan American Satellite • Parameter Adaptive System • Parametric Amplifier System • Paros (Greece) (location identifier) • PASCAL • Payload Accommodations Study • Payload Adaptor System • Peninsula Astronomical Society (USA-CA) • Peoria Astronomical Society (USA-IL) • Perigee-Apogee Satellite • Perigee-Apogee Stage • Perigee-Apogee System • Personal Amplification System • Personnel Accounting Symbol • Personnel Accounting System • Personnel Automation Section • Phase Address System • Phase Alteration Standard • Phase Array System • Philippine Astronomical Society (Philippines) • Phoenix Astronomical Society (USA-AZ) • Photo-Acoustic Spectroscopy • Physics Academic Software (USA-NC) • Pilot Alert System • Plasma-Activated Source • Planning and Analysis Section • Planning And Scheduling • Plasma Analyzer Spectrometer • Platform Analysis System • Plymouth Astronomical Society (UK-England) • Point Antisolaire • Polynomial Approximation Scheme • Pontchartrain Astronomy Society (USA-LA) • Positron Annihilated Spectroscopy • Postage Assessment System • Postal Addressing Standards • Power-Assisted Steering • Preaward Survey • Precision Acquisition System • Prêt à l'Accession Sociale • Primary Alerting System • Primary Alert System • Primary Ascent System • Principal Assistant Secretary • Principal Axis System • Print Accounting System • Privacy Act Statement • Probabilistic Argumentation System • Problem Assessment System • Processed Array Signal • Producibility - Affordability - Supportability • Product Acceptance Specification • Product Assurance and Safety • Productivity - Affordability - Supportability • Program Address Storage • Programmable Audio Synthesizer • Programme d'Ajustement Structurel • Programme d'Application SPOT • Program of Advanced Studies • Public Access System • Public Address System • Public Announcement Service • Publicly Available Specification • Publicly Available Standard • Publicly Available Submitter • Pulse Analysis System

PASA Participating Agency Service Agreement • Publications of the Astronomical Society of Australia (periodical code)

PASAM Polish Academy of Sciences - Arch. Mech (periodical code)

PASAS Polish Academy of Sciences - Artificial Satellites (periodical code)

PASAu Proceedings of the Astronomical Society of Australia (periodical code)

PASC Pacific Area Standards Congress • Pan-American Standards Commission • Particle Accelerator Society of China • Planning Advisory Subcommittee • Polar Atmospheric and Snow Chemistry • Proposed/Actual Support Cost

PASCA Pachón All-Sky Camera

PASCAL Philips Automatic Sequence Calculator

PASDE Photogrammetric Appendage Structural Dynamics Experiment

PASE Polar Air-Snow Experiment • Port Area Support Element • Power-Assisted Storage Equipment

PASEG Parliamentary Astronomy and Space Environment Group (UK)

PASEP Passed Separately

PASH Polynuclear Aromatic Sulphur Heterocyclic

PASI Preliminary Assessment and Site Inspection • Pragmatic Approach to Subject Indexing

PASIG Polish Academy of Sciences - Institute of Geophysics Publications (periodical code)

PASIPS Parallel Signal and Image Processing System

PASJ Publications of the Astronomical Society of Japan (periodical code)

PASK Publications of the Academy of Sciences of Kazakhstan (periodical code)

PASKA Polarized Atomic Beam Source Karlsruhe (Germany)

PASL Primary Area Switch Locator

PASM Preaward Survey Manager

PASO Particle Acceleration Solar Orbiter • Plateau d'Architecture des Systèmes Orbitaux • Preaward Support Office

PASP Photovoltaic Array Space Power • Publications of the Astronomical Society of the Pacific (periodical code)

PASPS Publications of the Astronomical Society of the Pacific Supplement (periodical code)

PASR Problem Analysis Status Report

pass Passim

Pass Passage • Passenger • Passim • Passive

PASS Passegem • Passager • Passenger • Payload Assist Subsystem • Personal Access Satellite System • Photogrammetric Archival Storage System • Physics and Astronomy Student Society (Canada-BC) • Planar Array of Superheated Superconductors • Planetarium Activities for Student Success • Planning And Scheduling System • POCC Application Software Support (*) • POCC Application Support Software • Ports Application Software System • Preaward Survey System • Precision Autocollimating Solar Sensor • Prediction of Activity Spectra for Substances • Primary Avionics Software System • Prince Albert Satellite Station (Canada) • Procurement Aging and Staging System • Procurement Automated Source System • Production Automated Scheduling System • Professional Airways Systems Specialist • Program Aid Software System • Program Alternative Simulation System • Prometheus All Sky Survey

PASSCAL Program for Array Seismic Studies of the Continental Lithosphere

PaßG Paßgesetz

PASSIM Presidential Advisory Staff on Scientific Information Management (USA)

PASSION Parallel And Scalable Software for Input-Output • Program for Algebraic Sequences Specifically of Input-Output Nature

passwd Password

PASSWORD Piloting Authentication and Security Services Within OSI for the Research and Development

PASSYS Passive Solar System

Past Pasteur • Pastoor

PAST Planetary Astronomy (program) • Professeur Associé Temporaire

PASTA Polish Academy of Sciences - Theory of Aggregation (periodical code)

PASTE Program Analysis for Software Tools and Engineering (conference)

PASTEC Passager Technologique

PASTEL Passager SPOT de Télécommunications par Laser • Passeport pour Téléphoner

PASTIC Pakistan Science and Technology Information Centre

PASTIS Projet d'Analyse de Scènes et de Traitement d'Images Symboliques

PASV Passive

pat Pattern (file extension)

Pat Patent • Patente • Patented • Pattern

P. At. Peso Atomico • Poids Atomique

PA&T Product Assurance & Test

PAT Parametric Artificial Talker • Passive Angle Tracking • Patch • Patent • Patna (India) (location identifier) • Pattern Analysis Test • Personalized Array Translator • Physics Achievement Test • Physics and Advanced Technologies • Point After Touchdown • Polygon Attribute Table • Polyvalent Alias Table • Port and Address Translation • Port Assistance Team • Positron Annihilation Technique • Post Acoustic Test • Power Alarm Test • Power Ascension Testing • Preauthorized Automatic Transfer • Precedence Access Threshold • Prediction Analysis Technique • Preliminary Acceptance Test • Preliminary Acceptance Testing • Preliminary Acceptance Trial • Primary Acceptance Test • Primary Access Terminal • Prime d'Aménagement du Territoire (France) • Priority Action Table • Priority Air Travel • Problem Action Team • Process Action Team • Process-Activated Training • Process Analysis Team • Product Acceptance Test • Product Assurance Test • Production Acceptance Test • Production Assessment Test • Program Activation Team • Program Analysis Team • Program Attitude Test • Programme in Advanced Technologies • Pseudo-Adder Tree • Pseudo-Automatic Translation • Public Affairs Team

Pata Patagonia

PATA Pneumatic All-Terrain Vehicle

PATAC Prévision Améliorée - Techniques d'Affinement de la Climatologie

PATAT Practice And Theory of Automated Timetabling (conference)

PATC Peninsula Advanced Technology Center (USA-VA) • Professional - Administrative - Technical - Clerical

Patd Patented

PAT&E Product Acceptance, Test & Evaluation • Production Acceptance, Test & Evaluation

PATE Performance Analysis and Technical Evaluation • Performance Analysis, Test, and Evaluation • Product Acceptance, Test and Evaluation • Product Assurance Test and Evaluation • Production Acceptance, Test and Evaluation • Programmed Automatic Test Equipment

PATF Program Activation Task Force

PatG Patentgesetz

PATG Productivity Assessment Task Group

Path Pathological • Pathology

PAth Publications of the Laboratory of Astronomy - University of Athens (periodical code)

PATH Performance Analysis and Test History • Port Authority Trans-Hudson (USA-NY) • Postflight Attitude and Trajectory History

Pathol Pathological • Pathology

PATHS Precursor Above-The-Horizon Sensor

PATi Popular Astronomisk Tidsskrift (periodical code)

PATM Planetary Atmospheres (program)

PATMOS Power and Timing Modeling, Optimization and Simulation (workshop) • Power And Timing Modeling System

Pat. Off. Patent Office

PATOM Programme Atmosphère et Océan à Moyenne échelle

PATOME Programme Atmosphère et Océan à Moyenne Échelle

PATOS Personnel Administratif, Technique, Ouvrier et de Service

PATP Parallel Application Technology Program • Preliminary Authority To Proceed

Pat. Pend. Patent Pending

PATRACK Package for Tracking Anomalous Changes of Knowledge

PATRICIA Particle Tracking by (microscopical) Imaging and Correlation for In-situ Analysis • Practical Algorithm To Retrieve Information Coded In Alphanumeric • Proving And Testing for Reliability Improvement of Complex Integrated Architectures

PATRIOT Provide Appropriate Tools Required to Intercept and Obstruct Terrorism (act) (USA)

PATROL Program for Administrative Traffic Reports On-Line

PATS Parametrized Abstract Test Suite • Payload Avionics Test Station • Plant Account Tracking System • Precedence Access Threshold Setting • Precision Automated Tracking System • Predicasts Abstract Terminal System • Program for Analysis of Time Series • Programmatic And Technical Support • Propulsion Analysis Trajectory Simulation • Prototyped Automated Telecommunications System

PATSY Parametric Test Synthesis

Patt Pattern

PATT Panel for the Allocation of Telescope Time • Partial Automatic Translation Technique • Pattern • Project for the Analysis of Technology Transfer

PATU Pan-African Telecommunications Union

PATWAS Pilots Automatic Telephone Weather Answering Service

PAU Pan American Union • Pattern Articulation Unit • Pauk (Myanmar) (location identifier) • Pilotless Aircraft Unit • Piscis Austrinids (meteor stream) • Platform Acceptance Unit • Present Address Unknown • Probe Aerodynamic Upper • Programa de Actuación Urbana (Spain) • Programs Analysis Unit • Pupil Alignment Unit

PAUC Program Acquisition Unit Cost

PAUSD Palo Alto Unified School District (USA-CA)

PAUTx Publications in Astronomy of the University of Texas (periodical code)

Pav Pavillon • Pavo-Pavonis (constellation)

PAV Parallel Access Volume • Paulo Afonso (Brazil-BA) (location identifier) • Pavilhão • Personnel Allotment Voucher • Phase Angle Voltmeter • Pool Adjacent Violators (method) • Position And Velocity • Pressure-Actuated Valve • Price and Availability System • Primary Access Volume • Progress Assistance Visit

PAVE Presort Accuracy Validation and Evaluation • Principles and Applications of Value Engineering

PAVM Phase Angle Voltmeter

PAVT Position And Velocity Tracking

PAW Pambwa (Papua New Guinea) (location identifier) • Peachtree Accounting for Windows • Photo-Absorptionswirkungsquerschnitte • Physics Analysis Workstation • Plasma Arc Welding • Powered All the Way • Producibility Assessment Worksheet

PAWE Program for Analysis of the World Ecosystem

PAWES Performance Assessment and Workload Evaluation System

PAGW Public Affairs Working Group

PAWOS Portable Automatic Weather Observing Station

PAWS Parallel Application Workstation • Phased-Array Warning System • Physics Analysis Workstation • Polar Automatic Weather Station • Portable Analyst Workstation • Portable Automatic Weather Station • Protection Against Wrapped Sequence • Prototype Analyst Workstation • Provision Analyst Workstation

PAX Passenger • Payload Accelerometer Experiment (Olympus) • Portable Archive Exchange • Port de Paix (Haiti) (location identifier) • Private Automatic Exchange • PSDE Accelerometer Experiment (Olympus)

PAY Pamol (Malaysia) (location identifier)

PAYDIRT Processing Architecture Yielding Deductions In Real Time

PAYE Pay As You Earn • Pay As You Enter

Paymr Paymaster

Payt Payment

PAYV Pay As You View

PAZ Partial Annealing Zone • Poza Rica - Tajin (Mexico) (location identifier)

PAZh Pis'ma Astronomicheskii Zhurnal (periodical code)

p/b Paperbound

Pb Blei • Lead • Plomb • Problem • Problème

P&B Planning & Budgeting

P/B Piggyback • Playback • Preburner • Pushbutton

P-B Pays-Bas

PB Palomar-Berger (survey) • Pañcaviṁśa Brāhmana • Paraíba (Brazil) • Particle Beam • Parts Breakdown • Pay Back • Peripheral Buffer • Personal Best • Perturbation Before • Petabyte • Pharmacopoeia Britannica • Phonetically Balanced • Photovoltaic Battery • Pipe Break • Plage Brightness • Plasma Boundary • Playback • Plug Board • Poisson-Boltzmann (equation) • Polybutylene • Port B • Postal Bulletin • Prayer Book • Preburner • Precision Bus • President's Budget • Pressure-Balanced • Pressure Breathing • Pressure Broadening • Primary Buffer • Primordial Binary • Printer Background • Probability of Blocking • Processor Board • Product Baseline • Product Bulletin • Program Baseline • Program Board • Program Budget • Project Book • Prominence Brightness • Property Book • Proton Beam • Publications Board • Publications Branch • Publishing Bulletin • Pulsating Band • Pulse Beacon • Pulsed Beam • Purple Belly • Pushbutton

PBA Barrow (USA-AK) (location identifier) • Pencil Beam Antenna • Penn Bolometer Array • Pill Box Antenna • Point Barrow (USA-AK) (location identifier) • Pool Boiling Apparatus • Production Base Analysis • Programming By Analogy • Public Buildings Administration

PBAB Please Bring A Bottle

PBAC Program Budget Advisory Committee

PBAN Polybutadiene-Acrylonitrile

PBAO Publications of the Beijing Astronomical Society (periodical code)

PBAS Program Budget Accounting System

PBB Parallel By Bit • Paranaíba (Brazil-MS) (location identifier) • Polybrominated Biphenyl

PBBE Performance-Based Business Environment

PBBS Physical Baseline Build Standard

PBC Parallel By Character • Performance-Based Contract • Performance-Based Contracting • Period Batch Control • Peripheral Bus Computer • Peripheral Bus Controller • Postal Business Center • Pressure Beta Center • Price Break Calculator • Processor Bus Controller • Program and Budget Committee • Program Budget Committee • Public Benefit Conveyance • Puebla (Mexico) (location identifier)

PBCA Physically-Based Computer Animation

PBCR Portable Bar Code Recorder

PBCS Package Bar Code Sorter • Postboost Control System

PBCT Proposed Boundary Crossing Time

PBD Payload Bay Door • Photothermal Beam Deflection • Playback Drive • Porbandar (India) (location identifier) • Precise Block Diagram • Pressboard • Production Buy Decision • Program Begin Date • Program Budget Decision • Program Budget Directive • Programmer Brain Damage • Programming By Demonstration

PBDF Payload Bay Door Forward

PBDG Pushbutton Data Generator

PBDI Position Bearing and Distance Indicator

PBDM Payload Bay Door Mechanism

Pbdo Poblado

PBDS Photothermal Beam Deflection Spectroscopy

PBE Paschen-Back Effect • Piggyback Experiment • Pool Boiling Experiment • Programming By Example • Prompt By Example • Protective Breathing Equipment • Proton Balance Equation • Proton Binding Energy • Puerto Berrio (Colombia) (location identifier)

PBefG Personenbeförderungsgesetz

PBeiO Publications of the Beijing Astronomical Observatory (periodical code)

PBEO Programme Board Earth Observation

PBET Photonics Basics for Engineers and Technicians

Pbf Personenbahnhof

PBF Pine Bluff - Grider (USA-AR) (location identifier)

PBFA Particle Beam Fusion Accelerator

PBG Photonic Bandgap • Plattsburgh (USA-NY) (location identifier) • Program and Budget Guidance • Program Budget Guidance

PBGA Plastic Ball Grid Array

PBH Paro (Bhutan) (location identifier) • Physikzentrum Bad Honnef (Germany) • Primordial Black Hole

PBHP Pounds per Brake Horsepower

PBHT Poisson Brackets of Hydrodynamics Type • Polybutadiène Hydroxy-Téléchélique • Postbond Heat Treatment

PBI Pacific Bell Internet • Parallel Binary Integration • Partial Background Investigation • Person-Based Interface • Playback Interface • Polybenimidezole • Postburst Increase • Primeval Baryon Isocurvature • Producto Bruto Interno • Pushbutton Indicator • West Palm Beach (USA-FL) (location identifier)

PBIC Programmable Buffer Interface Card

PBIM Programmable Buffer Interface Module

PBIMR Push Broom Imaging Microwave Radiometer

PBIO Portable Binary Input/Output

PBIS Peachtree Business Internet Suite

PBIT Parity Bit • Periodic Built-In Test

PBJ Paama (Vanuatu) (location identifier)

pbk Phonebook (file extension)

Pbk Paperback • Phonebook

PBK Pack Creek (USA-AK) (location identifier) • Payload Bay Kit • Phonebook

PBL Payload Bay Liner • Performance-Based Language • Planetary Boundary Layer • Polarized Broad Line • Probable • Problem-Based Learning • Product Baseline • Product Baseline Library • Puerto Cabello (Venezuela) (location identifier)

Pbld Published

Pblg Publishing

Pblr Publisher

pbm Portable Bitmap (file extension)

PBM Paramaribo - John Adolf Pengel (Suriname) (location identifier) • Paramaribo Zanderli (Suriname) (location identifier) • Peribacteroid Membrane • Permanent Bench Mark • Physically-Based Modeling • Play By Mail • Portable Bitmap • Postal Booklet Machine • Prenegotiation Briefing Memorandum • Probability-Based Matching • Program Budget Manager • Program Business Management

PBMA Production Base Modernization Activity

PBMR Pebble Bed Modular Reactor • Push Broom Microwave Radiometer

PBMS Pacific Bell Mobile Services

PBN Porto Amboim (Angola) (location identifier) • Production - Bremsstrahlung - Nuclear (interactions) • Pyrolytic Boron Nitride

PBNCO Property Book Non-Commissioned Officer

PBNI Pacific Bell Network Integration

pbo Profiler Binary Output (file extension)

PBO Paraburdoo (Australia-WA) (location identifier) • Performance-Based Organization • Photon Bubble Oscillation • Pine Bluff Observatory (USA-WI) • Profiler Binary Output • Project-Based Organization • Property Book Officer

PBOM Priced Bill of Materiel

PBosO Publications of Bosccha Observatory (periodical code)

PBP Pellin-Broca Prism • Performance-Based Pay • Point By Point • Postboost Phase • Preburner Pump • Program and Budget Planning • Program Board Panel

PBPF Propeller Blade Passing Frequency

PBPL Particle Beam Physics Laboratory

PBPS Postboost Propulsion System

PBq Petabecquerel

PBQ Pimenta Bueno (Brazil-RO) (location identifier)

PBQS Palomar Bright Quasar Survey

PBR Packed Bed Reactor • Paschen-Back Regime • Payment By Results • Policy-Based Routing • Program Budgetary Review • Puerto Barrios (Guatemala) (location identifier)

PBRC Program Budget Review Committee

PBRF Parallel Blind Relevance Feedback • Pilot Briefing

PBrn Publications of the Astronomical Institute - University of Borneo (periodical code)

Pbs Problems • Problèmes

PbS Lead Sulfide

PBS Payload Berthing System • Personal Bibliographic System • Personal Business Services • Polarized Beam Structure • Polarizing Beam Splitter • Portable Base Station • Portable Batch System • Power Bus Simulator • Powered Breaking System • Pressure Balance Structure • Primary Basic Subject • Production Base Support • Profit Before Tax • Program Breakdown Structure • Programming By Selection • Project Baseline Summary • Project Breakdown Structure • Protective Breathing System • Public Buildings Service • Public Broadcasting Service • Public Broadcasting System (*)

PBSC Performance-Based Service Contracting

PBSM Postal Booklet and Stamp Machine

PBSRAM Pipelined Burst Static Random Access Memory

PbSt Lead Stearate

pbt Profiler Binary Table (file extension)

PBT Payback Time • Performance-Based Training • Permeable Base Transistor • Piggyback Tape • Preliminary Breath Test • Profiler Binary Table • Public Benefit Transfer

PBTP Polybutylentherephthalat

PBTS Payload Bay Transfer System

PBU Peribacteroid Unit • Putao (Myanmar) (location identifier)

PBV Postboost Vehicle

PBW Palibelo (Indonesia) (location identifier) • Parts By Weight • Personal Bandwidth • Power-By-Wire • Proportional Bandwidth

PBWA Plasma Beat-Wave Accelerator

PBWS Performance-Based Work Statement

PBWU Projektgruppe Bayern zur Erforschung der Wirkungen von Umweltschadstoffen

PBX Playback Transmitted • Princeton Beta Experiment • Private Branch Exchange

PBX-M Princeton Beta Experiment - Modified

PBXM Princeton Beta Experiment - Modified

PBZ Plettenberg Bay (South Africa) (location identifier)

p/c Pour Compte • Price Current

pc Parsec • Partially Cloudy • Percent • Percentage • Pica • Picocurie • Piece • Post Cibos • Pour Cent • Pour Condoléance • Price

pC Picocoulomb • Picocurie • Post Christo

Pc Parc

P/C Pay Code • Petty Cash • Physico-Chemical • Pitch Control • Price Current

P-C Performance-Cost • Period-Colour • Pluto-Charon (system) • Polar to Cartesian • Price Current • Primitive-Composite

P&C Procurement & Contracts • Propulsion & Control • Purchasing & Contracting

PC Package Control • Package Controller • Pad Character • Pad Coordinator • Palomar Confirmed • Panama Canal • Panel Coordinator • Panoramic Camera • Paper Chromatography • Paper Copy • Parliamentary Commissioner • Partial Coverage • Partially Cloudy • Partially Contained (event) • Parts Catalogue • Patent Citation • Patent Country • Path Control • Peace Corps • Percent • Percentage • Peripheral Computer • Peripheral Control • Personal Code • Personal Computer • Personal Correction • Personality Conflict • Personnel Code • Petty Cash • Phase Coherent • Phase Conjugation • Photocell • Photoconductive • Photoconductor • Photon Counter • Physical Conditioning • Physical Contact • Physico-Chemical • Physique-Chimie • Pilot Carrier • Pilot-in-Command • Pitch Circle • Pitch Control • Plane Change • Planetary Camera • Plenipotentiary Conference • Pocket Calculator • Point of Curve • Polar Cap • Polar Crane • Police Constable • Political Correctness • Politically Correct • Polycarbonate • Port C • Port Call • Position Control • Postal and Courier • Postal Card • Postcard • Post Cibum • Post Commander • Post Computer • Poste de Commandement • Potentially Correct • Power Control • Power Conversion • Power Converter • Praça • Predictor-Corrector • Preparatory Commission • Preparatory Committee • Press Club • Press Conference • Pressure Chamber • Pressure Controller • Prêt Conventionné • Price Current • Primary Center • Primary Circuit • Primary Control • Prime Contractor • Prime Cost • Prince Consort • Principal Component • Principals Committee • Printed Circuit • Printer Control • Priority Control • Privileged Character • Privy Council • Probabilistic Computing • Process Cluster • Process Control • Process Controller • Process Coordinator • Processor Control • Processor Controller • Procurement Cycle • Product Center • Production Control • Production Cost • Professional Computer • Professional Corporation • Program Change • Program Commission • Program Computer • Program Control • Program Coordination • Program Coordinator • Program Counter • Programmable Controller • Programma Commissie • Programmatic Compression • Programme Committee • Programmed Check • Programming Committee • Project Control • Project Coordinator • Proportional Counter • Proposal Coordinator • Proposed Change • Propositional Calculus • Propulsive Coefficient • Propulsive Consumable • Propylene Carbonate • Protocol Control • Protocol Converter • Provisional Cost • Pseudocode • Pseudocumene • Public Convenience • Pulsating Current • Pulse Command • Pulse Compression • Pulse Counter • Pulverized Coal • Punched Card • Purchasing and Contracting

Pça Praça

P/CA Price/Cost Analysis

PCA Pacific Communications Area • Parallel Channel Adapter • Particle-Cluster Agglomeration • Parts Control Area • Patient Care Assistant • Performance and Coverage Analyzer • Performance Coverage Analyzer • Permanent

Change of Assignment • Permanent Court of Arbitration • Personal Channel Agent • Personal Communication Agent • Perturbative Convolution Approximation • Photon-Counting Array • Physical Configuration Assessment • Physical Configuration Audit • Pitch Control Assembly • Plane Circular Aperture • Planet-Crossing Asteroid • Plasma-Covered Antenna • Pneumatic Control Assembly • Point of Closest Approach • Polar-Cap Absorption • Polar-Cap Anomaly • Policy Certification Authority • Policy Creation Authority • Portage Creek (USA-AK) (location identifier) • Positive Control Area • Positively-Controlled Airspace • Power Calibration Area • Power Challenge Array • Power Conditioning Assembly • Power Control Assembly • Power Controller Assembly • Precontractual Authorization • Pressure Control Assembly • Prime Contractor Address • Principal-Component Analysis • Printed Card Assembly • Printed Circuit Assembly • Production Configuration Audit • Program Change Analysis • Program Commitment Agreement • Program Control Analyst • Project Change Agreement • Proportional Counter Array • Propulsion-Controlled Aircraft • Pyrotechnic Control Assembly

PCAA Passaic County Astronomical Association (USA-NJ)

PCAC Partially Conserved Axial Current • Porterville College Astronomy Club (USA-CA) • Primary Control and Analysis Center

PCAD Pointing Control and Aspect Determination • Pointing, Control and Attitude Determination

PCAF Perfect Core-Annular Flow

PCAL Position Calibration

PCAM Protein Crystallization Apparatus for Microgravity • Punched Card Accounting Machine

PCAOD Public Company Accounting Oversight Board

PCAR Platform Core Activity Request

PCARSS Plant Clearance Automated Reutilization Screening System

PCAS Persistent Chemical Agent Stimulant • Personnel Cost Accounting System • Planet-Crossing Asteroid Survey • Primary Central Alarm Station • Punched Card Accounting System

PCASP Passive Cavity Aerosol Spectrometer

PCASS Parts Control Automated Support System • Program Compliance Assurance and Status System

PCAST President's Council of Advisors on Science and Technology (USA)

PCASTRD Philippine Council for Advanced Science and Technology Research and Development

PCat Pubblicazione dell'Osservatorio Astrofisico di Catania (periodical code)

PCAT Personal Computer Access Tool • Precedence Category

PCAW Policy Creation Authority Workstation

PCB Parts Control Board • Petty Cash Book • Physique - Chimie - Biologie • Polychlorinated Biphenyl • Pondok Cabe (Indonesia) (location identifier) • Power Circuit Breaker • Power Control Board • Power Control Box • Precataclysmic Binary (system) • Printed-Circuit Board • Process Control Block • Processor Control Block • Product Configuration Baseline • Produit Criminel Brut • Program Change Board • Program Control Block • Project Change Board • Project Control Block • Project Coordination Board • Propulsion Control Box • Protocol Control Box

PCBA Printed Circuit Board Assembly

PCBC Plain Cipher Block Chaining

PCBMS Portable Computer-Based Maintenance System

PCBS Programme on Capacity Building in Science

PCC Pad Control Center • Panama Canal Commission • Panel on Climate Change • Partial Crystal Control • Partnership Coordination Cell • Pasadena City College (USA-CA) • Payload Control and Checkout • PC Cluster Controller • Per Copia Conforme • Permanent Cool Container • Personal Communications Computer • Personnel Contingency Cell • Photon Counting Camera • Physics and Chemistry of Comets • Pilot Command Center • Plastic Chip Carrier • Point of Compound Curve • Polarity Coincidence Correlator • Policy Coordinating Committee • Portable C Compiler • Postal Concentration Center • Postal Customer Council • Post- Collapse Core (*) • Postcore Collapse • Postcore Collapsed (cluster) • Poste de Commande Centralisé • Pour Copie Conforme • Power Control Console • Precommand Course • Press Control Center • Primary Command Center • Primary Control Center • Printed-Circuit Card • Process Control Computer • Production Control Center • Production Control Clerk • Professional Competency Credit • Program Change Control • Program Controlled Computer • Program Coordinating Centre • Program for Cooperative Cataloguing • Protective Ceramic Coating • Protocluster Cloud • Protoglobular Cluster Cloud • Provisioning Control Code • Puerto Rico (Colombia) (location identifier)

PCCA Physical Configuration Communications Area • Portable Computing and Communication Association

PCCB Program Change Control Board • Program Configuration Control Board

PCCD Peristaltic Charged Couple Device • Portable Charged Couple Device

PCCE Payload Common Communication Equipment • Product Center Civil Engineering

PCCIE Power Conditioning and Continuation Interfacing Equipment

PCCIP President's Commission on Critical Infrastructure Protection (USA)

PCCL Precontract Cost Letter

PCCM Parallel Community Climate Model • Primary Care Case Management • Princeton Center for Complex Materials • Program Change Control Management • Program Change Control Manager

PCCN Port Call Control Number • Provisioning Contract Control Number

PCCP Preliminary Contract Change Proposal

PCCS Personnel Command and Control System • Photographic Camera Control System • Program Change Control System

PCCTB Pilot Command Center Testbed

PCCU Power Conditioning and Control Unit • Punched Card Control Unit

PCD Patchy Corona-Disk (model) • Payload Correction Data • Performance Criteria Document • Photo Compact Disk • Photoconductive Decay • Photon Counting Detector • Pitch Circle Diameter • Plataforma de Coleta de Dados • Platform Compatibility Demonstration • Polycrystalline Diamond • Power Control and Distribution • Power Control Device • Prairie du Chien (USA-WI) (location identifier) • Precision Course Direction • Prime Contractor Department • Procedural Change Directive • Process Control Division • Procurement Control Document • Procurements and Contracts Division • Product Confidence Data • Product Configuration Demonstration • Production Control Desk • Program Change Decision • Program Control Document • Pseudo-Coherent Detection

PCDA Personal Communicating Digital Assistant • Program Controlled Data Acquisition

PCDB Project Control Database

PCDD Polychlorodibenzo-p-Dioxine

PCDF Polychlorodibenzofuranne

PCDMA Personal Computer Direct Marketers Association • Personal Computer Direct Marketing Association

PCDN Plan Communal de Développement de la Nature (Belgium)

PCDP Pilot Control Display Panel • Project Control Database Program • Punched Card Data Processing

PCDS Pilot Climate Data System (now NCDS) • Power Conditioning and Distribution System • Power Converter and Distribution System • Power Conversion and Distribution System

PCDU Payload Command Decoder Unit • Power Control and Distribution Unit

Pce Princesse

PCE Painter Creek (USA-AK) (location identifier) • Parametric Cost Estimate • Peripheral Control Element • Physics and Chemistry of the Earth (journal) • Postal Computing Environment • Power Conditioning Electronics • Power Conditioning Equipment • Power Control Electronics • Power Conversion Efficiency • Power Conversion Equipment • Preliminary Closure Estimate • Pressure to Clutch Engage • Prêt à la Création d'Entreprise (France) • Principle of Computational Equivalence • Process-Centered Environment • Processor Control Element • Professional Continuing Education • Program Cost Estimate • Programmable Crypto Engine • Punch Card Equipment

PCEA Pointing Control Electronics Assembly

PCEB Postcommon Envelope Binary

PCEI Prime Contract End Item

PCEM Propulsion Contamination Effects Module

PCER Per-Comparison Error Rate

PCES Physics - Chemistry - Earth Sciences • Postal Career Executive Service

PCESP Program Communications Electronic Support Program

PCET Professeur de Collège d'Enseignement Technique

PCETF Power Conversion Equipment Test Facility

pcf Pounds per Cubic Foot

PCF Packet Control Facility • Payload Control Facility • Personal Computing Facility • Personnel Control Facility • Phone Confusability Factor • Piecewise Continuous Function • Plastic Cladding Fiber • Point Coordination Function • Polar Crown Filament • Portable Compiled Format • Portable Complied Format • Possible Cooling Flow • Potentially Critical Failure • Pounds per Cubic Foot • Primary Calibration File • Probability of Consequence Factor • Process Control File • Processed Citation File • Program Change Factor • Programmable Command Format • Protein Crystallization Facility

PCFC Polytechnics and Colleges Funding Council (UK)

PCFE Prime Contractor Furnished Equipment

PCFG Probabilistic Context-Free Grammar

PCFOV Partially Coded Field Of View

PCFP Predicted Comparative Failure Probability

PCFS Personal Computer File System

PCG Parts Control Group • Paso Caballos (Guatemala) (location identifier) • Phonocardiogram • Planetário Calouste Gulbenkian (Portugal) • Planning and Control Guide • Power Conditioning Group • Power Converter Group • Preconditioned Conjugate Gradient • Printed Circuit Generator • Production Control Group • Program Coordination Group • Project Coordination Group • Protein Crystal Growth

PCGA Principios Contables Generalmente Aceptados

PCGC Pulverized Coal Gasification and Combustion

PCGD Pollution Control Guidance Document • Post-, Cheque- en Girodienst

PCGN Permanent Committee on Geographic Names

PC&H Packaging, Crating & Handling

PCH Pacific Coastal Highway • Packaging, Crating and Handling • Paging Channel • Palacios (Honduras) (location identifier) • Parallel Channel • Port C High • Professional Competency Hour

PCHE Phénomènes Cosmiques de Haute Énergie

Pchs Purchase

Pchsg Purchasing

Pcht Parchment

PCH&T Packaging, Crating, Handling & Transportation

PCHT Packaging, Crating, Handling and Transportation

pci Pounds per Cubic Inch

P/CI Payload/Cargo Integration

PCI PABX Computer Interface • Panel Call Indicator • Patent Citation Index • Pattern Correspondence Index • Payload/Cargo Integration • Peripheral Command Indicator • Peripheral Communications Interface • Peripheral Component Interconnect • Peripheral Component Interconnect • Peripheral Component Interface • Peripheral Controller Interface • Personal Computer Interface • Phase Contrast Interferometer • Physical Configuration Identification • Postal Commodity Machine • Postcollision Interaction • Postcorrelator Integrator • Poste Cartographique Interactif • Probability of Chance Improvement • Procedure Change Instruction • Procurement Contracting Manual • Product Configuration Identification • Production Count Indicator • Program Check Interruption • Program Control Input • Program-Controlled Input • Programmable Communication Interface • Proof of Concept Initiative • Protocol Control Information

PCIA Personal Communications Industry Association • Principal Component Image Analysis

PCIC Personal Computer Card Interrupt Controller

PCID Personal Computer Interface Device • Physically-Constrained Iterative Deconvolution

PCIE Permis de Conduire Informatique Européen • President's Council on Integrity and Efficiency (USA)

PCIERD Philippine Council for Industry and Energy Research and Development

PCIES Protective Clothing and Individual Equipment System

PCIL Pilot-Controlled Instrument Landing

PCIM Policy Core Information Model

PCIN Program Change Identification Number • Program Change Integration

PCinO Publications of the Cincinnati Observatory (periodical code)

PCIPI Permanent Committee on Industrial Property Information

PCIPS Personal Computer Image Processing System

PC-I/O Program Controlled Input/Output

PCIO Program Controlled Input/Output

PCIRD Personal Computer Independent Research and Development

PCIS Personal Computer Information Service • Portable Common Interface Set • Prototype Communications Interface System

PCIU Programmable Communications Interface Unit • Programmable Computer Interface Unit

PCK Payload Command Kit • Porcupine Creek (USA-AK) (location identifier)

PCKB Printed Circuit Keyboard

Pcl Parcel • Pencil

PCL Pallet Coolant Loop • Parallel Communications Link • Parcel • Peltier Current Lead • Personal Clearance • Personnel Security Clearance • Pilot-Controlled Lighting • Planning Class Library • Portable Common Loop • Port C Low • Positive Control Launch • Postconference List • Power Control List • Predesignated Conventional Load • Primary Coolant Line • Primary Coolant Loop • Printer Command Language • Printer Control Language • Procedural Control Language • Process Control Language •

Programming Checklist • Pucallpa (Peru) (location identifier) • Purple Crow Lidar

PCLA Power Control Linkage Assembly

PCLC Personal Computer Lifecycle Team

PCLDI Prototype Closed-Loop Development Installation

PCLP PC LAN Program

PCLS Portable Chopped Light Source • Prototype Closed-Loop System • Public Contract Law Section

PCLSAR Personal Computer Logistic Support Analysis Record

PCLT Prototype Closed Loop Test

pcm Per Calendar Month

PCM Package Command Manager • Parabolic Collimator Mirror • Parallel Climate Model • Parity Check Matrix • Parts Control Management • Personal Computer Manufacturer • Phase Calibration Module • Phase Change Material • Phase Code Modulation • Phase Comparison Monopulse • Phase Contrast Microscope • Phase Contrast Microscopy • Photo-Chemical Machinery • Photoelectric Cell Module • Physical Connection Management • Physics and Chemistry of Minerals (periodical code) • Pitch Control Motor • Plane Change Mechanism • Planning and Control Memorandum • Playa del Carmen (Mexico) (location identifier) • Plug-Compatible Mainframe • Plug-Compatible Manufacturer • Plug-Compatible Module • Polyatomic Ceramic Material • Power Control Mission • Power Control Module • Power Converter Module • Price Competition Memorandum • Printed-Circuit Module • Printer Cartridge Metric • Process Control Module • Production Cost Model • Profiling Current Meter • Program Configuration Manager • Program Coordination Meeting • Program Cost Management • Project Cycle Management • Propose - Critique - Modify • Protein-Calorie Malnutrition • Pulse Coded Modulation • Pulse Code Modulated • Pulse Code Modulation (*) • Pulse Code Modulator • Pulse Counting Mode • Pulse-Count Modulation • Punch Card Machine • Punched Card Machine (*)

PCMC PCI Cache Memory Controller • Precision Coherent Monopulse C-band

PCMCIA Personal Computer Memory Cards Industries Association • Personal Computer Memory Cards International Association (*)

pCMD Pixel-by-pixel Color Magnitude Diagram

PCMD Pulse Code Modulation Digital

PCMDI Program for Climate Model Diagnosis and Intercomparison

PCME Pulse-Code Modulation Event

PCMGI Physico-Chimie des Molécules et des Grains Interstellaires • Physico-Chimie des Molécules et Grains Interstellaires

PCMI Parks City Mathematics Institute (USA-NJ) • Photochromic Microimage • Physico-Chimie des Molécules Interstellaires • Physique et Chimie du Milieu Interstellaire (*) • President's Council on Management Improvement

PCMIM Personal Computer Media Interface Module

PCMM People Capability Maturity Model • Personnel Capability Maturity Model

PCMMU Pulse Code Modulation Master Unit

PCMOS Positive Channel Metal Oxide Semiconductor

PCMP Parts Control Management Program

PCMR Probability of Correct Message Receipt

PCMS Process and Configuration Management Software • Process Configuration Management Software • Product Configuration Management System • Project by Contract Management System • Pulse Code Modulation Shared • Punched Card Machine System • Purchase Card Management System

PCMTE Pulse Code Modulation and Timing Equipment

PCMTEA Pulse Code Modulation and Timing Electronics Assembly

PCMU Programmable Control and Monitor Unit

PCN Package Control Number • Page Change Notice • Part Control Number • Pavement Classification Number • Personal Communications Network • Personal Computer Network • Picton (New Zealand) (location identifier) • Pitcairn (ISO country code) • Planning Change Notice • Polyester-Coated Nylon • Procedure Change Notice • Processing Control Number • Procurement Control Number • Product Control Number • Production Control Number • Program Change Notice • Program Change Number • Program Control Number • Project Control Number • Proposal Control Number • Provisioning Control Number • Publication Change Notice

PCNB Pentachloronitrobenzene

PCNFS Personal Computer Network File System

PCNIU Personal Computer Network Interface Unit

PCNS Polar Coordinates Navigation System

Pco Pico

PCO Parcels Concentration Office • Point of Control and Observation • Polar Cap Observatory • Postcheckout • Postcheckout Operations • Primary Control Officer • Principal Contracting Office • Principal Contracting Officer • Procedure Change Order • Procurement Contracting Office • Procurement Contracting Officer • Procuring Contracting Office • Procuring Contracting Officer • Professional Conference Organizer • Program Change Order • Program Contracting Office • Program Contracting Officer • Program-Controlled Output • Program Control Office • Program Control Officer • Program Coordination Office • Program Coordination Officer • Project Control Office • Project Control Officer • Publications Coordination Office • Publications Coordination Officer • Punta Colorada (Mexico) (location identifier) • Purchase Change Order

PCON Proof of Concept

PCooO Publications of Cook Observatory (periodical code)

PCopO Publikationer og Mindre Meddeler fra Kobenhavns Observatorium (periodical code)

PCOS Primary Communications-Oriented System • Proportional Chamber Operating System

PCOT Payload Center Operations Team

PCP Packet Core Protocol • Parallel Cascade Processor • Parallel Circular Plate • Parallel Coordinate Plot • Parts Control Program • Passenger Control Point • Perfect Cosmological Principle • Peripheral Control Program • Peripheral Control Pulse • Personnel Control Point • Photon-Coupled Pair • Planning Command Pool • Poisson Cluster Process • Postconstruction Permit • Potential Contractors Program • Power Control Panel • Preliminary Cost Proposal • Pressurization Control Panel • Primary Control Program • Prime Commercial Paper • Principe Island (location identifier) • Prise de Culot Pneumatique • Private Carrier Paging • Process Control Processor • Processor Control Program • Program Change Proposal • Program Control Plan • Programmable Data Processor • Programme de Coopération Postgraduée • Project Change Proposal • Project Control Plan • Project Cost Plan • Punched Card Punch

PCPA Polar Cap Phase Anomaly

PCPC Personal Computers Peripheral Corporation

PCPD Project Control and Planning Division

PCPG Primary Clock Pulse Generator

PCPGA Printed-Circuit Pin-Grid Array

PCPI Permanent Committee on Patent Information

PCPL Proposed Change Point Line

PC&PM Program Control & Performance Management

PCPM PERT Cost Performance Measurement • Program Control and Performance Management

PCPN Precipitation

PCPP Parallel Computing using the Prefix Problem • Parts Control Program Plan

PCPS Portable Collective Protection System • Pulse-Coded Processing System

PCPV Prestressed Concrete Pressure Vessel

pcq Parce que

PCR Packet Completion Rate • Payload Changeout Room • Payload Checkout Room • Payload Clean Room • Peak Cell Rate • Peer Code Review • Pen Character Recognition • Phase-Conjugate Reflectivity • Photoconductive Relay • Photoexcitation by Continuum Resonance • Pitch Circle Radius • Phase Change Rewritable • Planning Change Request • Planning Check Request • Polar-Cap Radar • Polymerase Chain Reaction • Power Change Request • Preventive Cyclic Retransmission • Primary Cosmic Radiation • Primary Cosmic Ray • Prime Contractor Region • Principal-Component Regression • Print Contrast Ratio • Problem Change Report • Procedure Change Request • Procurement Center Representative • Production Center Response • Production Control Record • Production Cost Requirement • Program Change Report • Program Change Request • Program Clock Reference • Program Control Register • Program Control Report • Programme Cadre de Recherche • Project Control Room • Project Cost Record • Project Cost Report • Projected Communications Requirement • Publication Change Request • Puerto Carreno (Colombia) (location identifier) • Punched Card Reader • Punched Card Requisition

PCRB Program Change Review Board • Project Change Review Board • Property Control Record Book • Proposal Commitment Review Board

PCRCW Parallel Computer Routing and Communication Workshop

PCRD Programme Cadre de Recherche et de Développement (France)

PCRDT Programme Cadre de Recherche et de Développement Technologique (France)

PCRK Permanent Cable Repair Kit

PCRS Payload Cover Removal System • Performance and Cosr Report System • Pointing Control Reference Sensor

Pcs Pieces • Prices

P&CS Pointing & Control System

PCS Panoramic Camera System • Particle Counting System • Partition Communication System • Patchable Control Store • Patent Classification Service • Pattern Code Security • Payload Calibration System • Payload Checkout System • Payload Control Supervisor • Payload Control System • Payload Correction System • Peltier Current Source • Penrose Compton Scattering • Periodical Control System • Permanent Change of Station • Performance Command System • Performance Control System • Permanent Change of Station • Permit Compliance System • Persistent Current Switch • Personal Communications Service • Personal Communications System • Personal Conferencing Specification • Personnel Communications System • Phase Control System • Photo Correlation Spectroscopy • Photo Correlation System • Photon Correlation Spectroscopy (*) • Photon Counting Spectrometer • Photon Counting System • Physical Coding Sublayer • Physical Control Space • Physical Control System • Physically-Controlled Space • Picos (Brazil-PI) (location identifier) • Pieces • Pilot Control System • Pitch Control System • Planning and Control System • Planning Control Sheet • Plasma Current Switch • Plastic-Clad Silica • Pointing and Control System • Pointing Control System • Polar Coordinate System • Polymer Cladding Silica • Portable Computer System • Portable Control Station • Position - Course - Speed • Postcamp-Station • Power Conversion System • Precision Clock System • Precision Correction System • Preliminary Component Specification • Pressure Control System • Primary Calibration System • Primary Control Ship • Primary Control Station • Primary Coolant System • Principal Component Substitution • Principal Coordinating Scientist • Print Contrast Scale • Print Contrast Signal • PRISMA Computer System • Print Contrast System • Probability of Crew Safety • Procedure Completion Sheet • Process Control System • Processor Controlled Strapping • Procured Component Software • Product Control Service • Program Control System • Program Counter Store • Programmable Controller System • Project Control System • Project Coordination Staff • Project Control System • Protected Cable System • Proxy Cache Server • Punched Card System

PCSA Personal Computer System Architecture • Personal Computing System Architecture • Property Control System Analysis

PCSB Pulse-Coded Scanning Beam

PCSC Payload and Crew Support Computer • Power Conditioning, Switching and Control

PCSD Philippine Council for Sustainable Development • Proposed Contract Summary Document

PCSIU Pointing Control System Interface Unit

PCSN Private Circuit Switching Network

PCSO Presidential Communications Support Office (USA)

PCSP Polar Continental Shelf Project

PCSR Parallel Channels Signaling Rate • Planck Calibrated Sky Radiometer

PCST Public Communication of Science and Technology (network)

Pct Percent

PCT Pacific Crest Trail (USA) • Parallel Change Task • Pass Control • Patent Cooperation Treaty • Percent • Performance Correlation Technique • Personal Constructs Theory • Photometric Calibration Target • Photometric Calibration Test • Photon-Coupled Transistor • Physico-Chimie de la Troposphère • Picture • Picture Contour Table • Plan à Court Terme • Planning and Control Technique • Polychlorinated Triphenyl • Power Control Test • Praceta • Primary Care Trust • Prime Contract Termination • Princeton (USA-NJ) (location identifier) • Principal Component Transform • Private Communication Technology • Process Control Table • Processor Control Table • Product Consistency Test • Program Coordinator Team • Pulse Compression Test

PCTA Personal Computer Terminal Adapter

PCTC Payload Crew Training Complex

PCTE Portable Commercial Test Equipment • Portable Common Tools Environment

PCTF Passive Cooler Test Facility • Plant Component Test Facility

PCTJ Parallel Connected Twin Junction

PCTL Percentile

PCtM Pulse Count Modulation

PCTM Pulse Count Modulation

PCTN Pacific Consolidated Telecommunications Network

PCTO Payload Cost Tradeoff Optimization

PCTR Pad Connection Terminal Room • Pipeline Control • Product Center • Prominence-Corona Transition Region

PCTS POSIX Conformance Test Suite

PCTSS Provisioning, Cataloguing and Technical Support System

PCU Packet Control Unit • Passenger Car Unit • Passenger Control Unit • Payload Checkout Unit • Payload Control Unit • Picayune (USA-MS) (location identifier) • Plant Control Unit • Platinum Cobalt Unit • Portable Computer Unit • Portable Control Unit • Power Conditioning Unit • Power Control Unit • Power Conversion Unit • Power Convertor Unit • Pressure Control Unit • Pressurization Control Unit • Primary Conditioning Unit • Print Control Unit • Process Control Unit • Programmable Conditioner Unit • Progress Control Unit

• Propeller Control Unit • Proportional Counter Unit • Punched Card Utility

PCUA Power Controller Unit Assembly

PCV À Percevoir • Passenger-Carrying Vehicle • Passenger-Controlled Vehicle • Payable Chez Vous • Performance, Calibration and Verification • Polar Cataclysmic Variable (star) • Pollution-Controlled Valve • Pollution Control Valve • Positive Crankcase Ventilation • Potential Cataclysmic Variable (star) • Precheck Verification • Preliminary Concept Verification • Premeasurement - Calibration - Verification • Pressure Containment Vessel • Pressure Control Valve • Primordial Cataclysmic Variable (star) • Punta Chivato (Mexico) (location identifier) • Purge Control Valve

PCVB Pyro Continuity Verification Box • Pyrotechnic Continuity Verification Box

PCVD Plasma-Activated Chemical Vapor Deposition

PCVL Pilot-Controlled Visual Landing

PCW Personal Computer World • Postconsumer Waste • Postconsumption Waste • Previously Complied With • Pulsed Continuous Wave

PCWG Personal Conferencing Work Group • Personal Conferencing Working Group

PCWQM Process Control Water Quality Mirror

PCWUIS Personal Computer Work Unit Information System

pcx Picture Image (file extension)

PCX Pellet Charge Exchange • Picture Image • Plasma Confinement Experiment • Private Call Exchange

PCZ Panama Canal Zone • Physical Control Zone • Physically-Controlled Zone • Positive Control Zone • Protected Communications Zone

PCZA Pseudo-Circumzenithal Arc

PCZNN Politechnika Czestochowska Zeszyty Naukowe Nauki Techniczne Mechanika (periodical code)

PCZST Panama Canal Zone Standard Time

PC1 First Principal Component

PC2 Second Principal Component

PC3 Third Principal Component

PC4 Fourth Principal Component

PC5 Fifth Principal Component

p/d Postdated

pd Per Day • Per Diem • Period • Postage Due • Postdated

P/d Porte Debido

Pd Paid • Palladium • Passed • Probability of Detection

P&D Panels & Displays • Pay & Display • Production & Deployment • Production & Development

P/D Price-Divident

PD Doctor of Pharmacy • Packet Decimal • Parameter Data • Parent Distribution • Partial Discharge • Participatory Design • Penetration Duration • Per Diem • Performance Demonstration • Peripheral Decoder • Peripheral Device • Phase Discriminator • Phase Diversity • Phenomenology Document • Photodiode • Photoelectron Diffraction • Photon Drag • Phrase Dictionary • Physical Delivery • Physics Department • Pictorial Display • Pitch Diameter • Pitch Down • Planned Derating • Planning Directive • Planning Document • Plasma Desorption • Point of Delivery • Point of Departure • Polar Distance • Police Department • Policy Deployment • Policy Directive • Policy Division • Port Dû • Port Dues • Port of Delivery • Position Description • Positive Displacement • Postage Due • Postage District • Postal District • Postdata • Postdoctoral • Potential Difference • Potsdam Durchmusterung • Power Distribution • Precision Digitizing • Predigital • Preliminary Design • Presidential Directive • Price-Divident • Prime Driver • Printer Driver • Priority Designator • Priority Directive • Prism Diopter • Privatdozent • Probability of Damage • Probability of Destruction • Probability of Detection • Problem Description • Problem Domain • Procedure Description • Process Data • Processed Data • Procurement Data • Procurement Directive • Procurement Division • Procurement Document • Procurement Drawing • Product Definition • Product Delivery • Product Directorate • Production and Deployment • Production Department • Program Development • Program Directive • Program Director • Program Document • Project Definition • Project Directive • Project Director • Project Document • Project Documentation • Projected Date • Projected Decision • Propagation Delay • Propellant Dispersal • Property Damage • Proportional Differential • Propulsion Directorate • Protective Device • Protocol Draft • Public Domain • Pulse Doppler • Pulse Duration • Punch Driver

PDA Packet Data Analysis • Packet Data Analyzer • Parallel Data Adapter • Parametric Design Analysis • Partial Discriminant Analysis • Payload Disconnect Assembly • Percent Defect Allowed • Per Diem Allowance • Performance Data Archive • Personal Data Assistant • Personal Digital Assistant (*) • Phased Doppler Anemometer • Photodiode Array • Photon Detector Assembly • Planning and Decision Aid • Polydiacethylene • Portable Data Access • Portable Digital Assistant • Postdesign Analysis • Pour Dire Adieu • Power Distribution Assembly • Precision Drive Axis • Predelivery Acceptance • Predicted Drift Angle • Preliminary Design Approval • Preliminary Design Audit • Preliminary Design Authorization • Present Duty Assignment • Principal Decision Authority • Principal Design Activity • Principal Developing Agency • Principal Development Activity • Probability Distribution Analyzer • Processor and Distribution Assembly • Production Data Analysis • Program Development Activity • Project Definition and Approval • Propellant Drain Area • Property Disposal Agent • Public Display of Affection • Public Domain Algorithm • Puerto Inirida (Colombia) (location identifier) • Pump Drive Assembly • Push Down Automaton

PDAA Publications of the Department of Astronomy in Austin (periodical code)

PDAC Paris Data Analysis Center (France)

PDAD Proposed Draft Addendum

PDAF Probabilistic Data Acquisition Filter • Probabilistic Data Association Filter

PDAG Process Definition Advisory Group

PDAM Proposed Draft Amendment

PDAO Publications of the Dominion Astrophysical Observatory (periodical code)

PDAP Portable Data Acquisition Package

PDAR Preferential Departure and Arrival Route • Program Description And Requirements

PDAS Photo Data Analysis System

PDAU Physical Delivery Access Unit

PDAUC Publications of the Department of Astronomy - University of Chile (periodical code)

pdb Palm Database (file extension)

PdB Bachelor of Pedagogy • Plateau de Bure (France)

PDB Parachute Database • Pedro Bay (USA-AK) (location identifier) • Performance Data Book • Personal Database • Planar-Doped Barrier • Postdeployment Built • Power Distribution Box • President's Daily Brief • Primary Database • Project Database • Project Development Brochure • Protein Data Bank

PdBI Plateau de Bure Interferometer (IRAM)

PDBS Personalized Database System

P&DC Printing & Distribution Center • Processing & Distribution Center

PDC Mueo (New Caledonia) (location identifier) • Parallel Data Centre • Parallel Data Controller • Partial Distribution Curve • Participatory Design Conference • Pattern

Description Compiler • Performance Data Computer • Personal Digital Cellular • Personnel Dispatch Centre • Personnel Dispersal Centre • Phenomenology Data Center • Plumes Data Center • Policy Development Committee • Postal Data Center • Power Disk Cartridge • Power Distribution and Control • Power Distribution Control • Predefined Command • Predeparture Clearance • Predeployment Checkout • Predefined Command • Premission Documentation Change • Primary Development Contract • Primary Digital Carrier • Primary Domain Controller • Printing and Distribution Center • Priority Designator Code • Problem Domain Component • Processing and Distribution Center • Procurement Document Change • Product Directorate Code • Professional Development Center • Proficiency Data Card • Program Data Coordinator • Program Delivery Control • Program Designator Code • Program Development Center • Programming - Design - Construction • Project Data Card • Project Data Control • Pulse-Discharge Cleaning

PDCA Plan - Do - Check - Act

PDCF Protein Crystallization Diagnostics Facility

PDCM Product Data Control Model • Product Data Control Module

PDCO Property Disposal Contracting Officer

PDCP Packet Data Convergence Protocol

PDCR Project Data Compliance Report • Proprietary Data Control Record

PD&CS Power Distribution & Control Sequencer

PDCS Palomar Deep Cluster Survey (*) • Palomar Distant Cluster Survey • Parallel and Distributed Computer Systems (conference) • Parallel Digital Computing System • Performance Data Computer System • Power Distribution and Control Sequencer • Power Distribution and Control System • Predictably Dependable Computing System • Processing Distribution and Control System

PDCU Plotting Display Control Unit • Power Distribution and Control Unit

pdd Photo Deluxe Document (file extension)

PdD Doctor of Pedagogy • Palladium Deuteride

PDD Package Designation and Description • Past Due Date • Payload Description Document • Personal Detection Device • Physical Device Driver • Portable Delivery Device • Postdial Delay • Preliminary Design and Development • Presidential Decision Directive • Priority Delivery Date • Probability Density Distribution • Probability Distribution Density • Product Definition Data • Product Delivery Device • Program Description Document • Program Definition Document • Program Design Data • Program Design Description • Program Design Document • Program Directive Document • Project Disclosure Document • Proposal Due Date • Prospective Decision Date • Provisioning Description Data • Pulsating Delayed Detonation

PDDA Performance Problem Diagnostic Assistance • Power Driven Decontamination Apparatus

PDDB Peripheral Data Definition Block

PDDF Preliminary Data Description File

PDDI Product Data Definition Interface • Product Definition Data Interchange • Product Definition Data Interface

PDDO Publications of David Dunlap Observatory (periodical code)

PDDR Preliminary Design Data Review • Programme de Développement des Données Radar • Projet de Développement de Données Radar

PDDS Product Design Data System • Program Definition Data Sheet

PDDU Power Distribution and Drive Unit

PDE Page Directory Entry • Pandie Pandie (Australia-SA) (location identifier) • Partial Differential Equation • Particle Dispersion Experiment • Parzen Density Estimator • Personal Digital Entertainment • Preliminary Data Element • Process Directory Entry • Product Data Exchange • Product Design Engineering • Programa de Divulgación Escolar • Projected Operational Environment • Promenade • Prospective Data Element • Pulsed Detonation Engine • Pulse Detonation Engine (*) • Pure Density Evolution

P-de-C Pas-de-Calais (France)

P-de-D Puy-de-Dôme (France)

PDEF Protean Data Exchange Format

PDEFO Partial Differential Equation of First Order

PDEIS Preliminary Draft Environmental Impact Statement

PDEL Partial Differential Equation Language

PDES Preliminary Draft Environmental Statement • Product Data Evolvment System • Product Data Exchange Specification • Product Data Exchange Standard • Product Data Exchange using STEP • Product Description Exchange Standard

PDET Probability of Detection

pdf Portable Document Format (file extension) • Probability Density Function

Pdf Densely Fracture Plain

PDF Package Definition File • Parameter Definition File • Particle Distribution Function • Parton Distribution Function • Patient Data Form • Peak Distribution Function • Planar Deformation Feature • Plan Départemental de Formation (France) • Plant Design Factor • Platform Data File • Portable Data File • Portable Document File • Portable Document Format (*) • Postdoctoral Fellow • Powder Diffraction Data • Power Density Function • Power Distribution Frame • Prado (Brazil-BA) (location identifier) • Precision Direction Finder • Printer Description File • Probability Density Fluctuation • Probability Density Function • Probability Distribution Function • Procedure Definition File • Processing and Distribution Facility • Processor-Defined Function • Program Data File • Program Data Form • Program Development Facility • Programmable Data Formatter • Pulsar Data File

PDFA Praseodymium-Doped Fiber Amplifier

PDFF Pseudo-Derivative Feed-Forward

PDFI Probability of Detected Fault Isolation

PDFRR Program Directors Flight Readiness Review

PDG Padang (Indonesia) (location identifier) • Particle Data Group • Patent Documentation Group • Payload Data Generator • Paymaster Director-General • Polarization-Dependent Gain • Président Directeur Général • Priority Designator Grouping • Program Development Group • Proposal Development Group

PDGF Power and Data Grapple Fixture (*) • Power Data Grapple Fixture

PDH Payload Deployment Hardware • Planned Derated Hours • Plesiochronous Digital Hierarchy • Pretty Damn High • Professional Development Hour

PDHE Payload Data Handling Electronics

PDHID Pulsed Discharge Helium Ionization Detector

PDI Palmer Drought Index • Partial Differential Inequality • Patent Documentation and Information • Payload Data Interleaver • Persistent Document Identifier • Phase-Dispersion Introscopy • Pindiu (Papua New Guinea) (location identifier) • Plan Départemental d'Insertion • Planet-Disk Interaction • Point-Diffraction Interferometer • Postdetection Integration • Power and Data Interface • Power Distance Index • Powered Descent Initiation • Predelivery Inspection • Predesignated Item • Preservation Description Information • Privately Developed Item • Product Data Interchange • Program Development Identifier • Programmed Device Introduction • Programme de Développement Intellectuel • Project Data Index

PDIA Process-Driven Information Automation

PDIAL Public Dialup Internet Access List

PDIEP Partially Described Inverse Eigenvalue Problem

PDII Pusat Dokumentasi Ilmiah dan Informasi (Indonesia)

PDIO Photodiode

PDIP Program Decision Increment Package • Program Development Increment Package

PD&IS Power Distribution & Interface Support

PDIS Parallel and Distributed Information System • Power Distribution and Interface Support • Preparation for Dissemination

PDIU Propulsion Discrete Interface Unit

PDK Atlanta - De Kalb - Peachtree (USA-GA) (location identifier) • Planare Driftkammer • Processing Declarative Knowledge (workshop) • Publishing Development Kit

PDKD Parallel and Distributed Knowledge Discovery

pdl Poundal

P&DL Program & Data Library

PDL Page Description Language • Pattern Description Language • Permanent Duty Location • Picture Description Language • Polarization-Dependent Loss • Polarization Diversity Lidar • Polarized Diversity Lidar • Ponta Delgada (Azores) (location identifier) • Poverty Datum Line • Preliminary Design Language • Procedure Definition Language • Procedure Distribution List • Process Design Language • Procurement Data List • Product Development Laboratory • Program Definition Language • Program Description Language • Program Design Language • Programmed Digital Logic • Project Document List • Propositional Dynamic Logic • Public-Domain Library • Pulsed Dye Laser • Pushed Down List

PDLA Proximate Damped Lyman Alpha (system)

PDLAC Preliminary Dynamic Loads Analysis Cycle

PDLC Preliminary Design Load Cycle

PDLM Periodic Depot-Level Maintenance

PdM Master of Pedagogy

PDM Partial Descriptive Method • Payload Deployment Module • Percent Deviation from the Median • Periodic Depot Maintenance • Phase Difference Modulation • Phase Dispersion Minimization • Physical Data Model • Physical Device Manager • Physical Distribution Management • Planned Depot Maintenance • Please Detect Me • Power Distribution Module • Precedence Diagramming Method • Precision Doppler Measurement • Primary Deployment Mechanism • Primary Drive Mechanism • Primary Drive Motor • Probabilistic Dilution Model • Probe Data Model • Processor Data Monitor • Product Data Management • Product Data Modeling • Program Data Manager • Program Decision Meeting • Program Decision Memorandum • Programmed Depot-level Maintenance • Project Data Manager • Project Design Memorandum • Property Driven Model • Propellant Distribution Module • Pulse Delta Modulation • Pulse Division Multiplex • Pulse-Duration Modulation • Push Down Memory

PDMA Polarization Division Mutiple Access

PDMC Parallel and Distributed Model Checking (workshop)

P-DME Precision Distance Measuring Equipment

PDME Precision Distance Measuring Equipment

PDMF Predisaster Mitigation Fund • Present-Day Mass Function

PDMFM Pulse-Duration Modulation Frequency Modulation

PDML Procedural Data Manipulation Language

PDMM Push Down Memory Modem

PDMP Program Data Management Plan • Project Data Management Plan

PDMS Phase Difference Measurement System • Plasma Desorption Mass Spectroscopy • Polidimethylsiloxane • Program Definition and Management System

PDMSS Programmed Depot Maintenance Scheduling System

PDMT Predominant

PDMU Passive Data Memory Unit

Pdn Production

PDN Packet Data Network • Parndana (Australia-SA) (location identifier) • Position-Dependent Normalization • Power Dividing Network • Project Deviation Notice • Public Data Network

Pdo Pasado

PDO Pacific Decadal Oscillation • Packed Data Optimized • Parcel Delivery Office • Parcel District Office • Pendopo (Indonesia) (location identifier) • Physical Data Object • Planning and Development Office • Portable Distributed Object • Postdetection Observation • Postman's Delivery Office • Postman's District Office • Procedure Description Overview • Program Directive for Operations • Property Disposal Office • Pseudo-Differential Operator • Publications Distribution Office • Publications of Dominion Observatory (periodical code)

PDOP Positional Dilution Of Precision • Position Dilution Of Precision

Pdor Parador

PD&OS Production Deployment & Operational Report

PDOS Parallel and Distributed Operating Systems • Partial Density Of States • Production Deployment and Operational Report • Publishing Distribution Office System

PDP Parallel Distributed Processing • Parkinson Disease Patient • Passive Dosimeter Packet • Payload Data Pack • Personal Development Planning • Phase Doppler Pneumometer • Piecewise Deterministic Process • Plasma Diagnostics Package • Plasma Display Panel • Policy Decision Point • Postinsertion Deorbit Preparation • Power Distribution Panel • Preliminary Definition Plan • Preliminary Design Phase • Preliminary Design Proposal • Procurement Data Package • Production Data Package • Professional Development Plan • Professional Development Program • Program Decision Package • Program Definition Phase • Program Development Plan • Programmable Data Processor • Programmed Data Processor • Programmed Digital Processor • Project Data Package • Project Definition Phase • Project Development Plan • Punta del Este (Uruguay) (location identifier)

PDPA Program Decision Principal Advisor

PDPF Project Data Processing Facility

PDPS Parts Data Processing System • Planning and Data Processing System • Program Data Processing Section

PDPTA Parallel and Distributed Processing Techniques and Applications (conference)

PDQ Panel on Data Quality • Parallel Data Query • Partnership for Developing Quality • Peachtree Data Query • Pertinent Data Quest • PODPS Data Quality • Position Description Questionnaire • Preroute Delay Qualifier • Pretty Damn Quick • Price and Delivery Quotation • Procedural Data Quality • Proposal Data Query • Protocol Data Query

PDQS Portable Document Query System

Pdr Powder

PDR Partially Dissociated Region • People's Democratic Republic • Phase Data Recorder • Phase Delay Rectifier • Phase-Derived Range • Photodissociated Region • Photodissociation Region (*) • Photon-Dominated Region • Pitch-to-Diameter Ratio • Power Directional Relay • Precision Depth Recorder • Preferential Departure Route • Preliminary Data Report • Preliminary Data Requirements • Preliminary Design and Review • Preliminary Design Review (*) • Preliminary Document Review • Previous Document Reference • Priority Data Reduction • Processed Data Recorder • Processing Data Rate • Procurement Data Reference • Program Design and Reviewer • Program Design Review • Program Deviation Report

• Program Director's Review • Program Discrepancy Report • Program Document Requirement • Program Drum Recording • Programmable Digital Radio • Project Definition Report • Project Design Review • Publications Data Request

PDRA Postdoctoral Research Assistant • Postdoctoral Research Associate

PDRB Preliminary Design Review Baseline

PDRC Professional Development and Recognition Committee • Program Development Review Committee

PDRD Procurement Data Requirements Description • Procurement Data Requirements Document • Program Data Requirements Document • Program Definition and Requirements Document

PDRDP Phased-Derived Range Data Package

PDreO Publications of Dresden Observatory (periodical code)

PDREP Product Deficiency Reporting and Evaluation Program

PDRF Postdoctoral Research Fellow

PDRL Preliminary Design Requirements List • Procurement Data Requirements List • Program-Derived Requirements List

PDRM Payload Deployment and Retrieval Mechanism

PDRP Program Data Requirement Plan

PD&RR Program Definition & Risk Reduction

PDRR Payload Design Requirements Review • Preliminary Design Requirements Review • Program Definition and Risk Reduction (*)

PD&RS Payload Deployment & Retrieval System

PDRS Payload Data and Retrieval System • Payload Deployment and Retrieval System • Primary Data Receiving Station • Probe Data Relay System

PDRSS Payload Deployment and Retrieval System Simulation • Payload Deployment and Retrieval System Simulator

PDRSTA Payload Deployment and Retrieval System Test Article

PDRT Photodissociation Region Toolbox

PDRW Postdoctoral Research Worker

PDS Package Data System • Packet Driver Specification • Page Data Set • Page Display System • Parachute Design System • Paradynamic Sketch • Parallel and Distributed Systems (conference) • Partitioned Data Set • Passenger Documentation System • Passive Detection System • Passive Dosimeter System • Pathfinder Demonstration Satellite • Payload Data Segment • Peripheral Device Status • Persistent Data Segment • Personal Data Server • Personal Dictation System • Personal Display System • Personnel Daily Summary • Personnel Data System • Phase Differential Scattering • Phase-Diverse Speckle (imaging) • Phased Development Shuttle • Phoswich (crystal) Detector System • Photodischarge Spectroscopy • Photometric Data Systems • Photothermal Deflection Spectroscopy • Physical Delivery System • Picos dos Dias Survey • Piedras Negras (Mexico) (location identifier) • Pilot Data System • Pilot Distributed System • Planetary Data System • Planning Data Sheet • Platform Data System • Pneumatic Distribution System • Polydimethylsiloxane • Portable Data Store • Portable Data System • Portable Document Software • Positional Data System • Postal Delivery System • Postdesign Support • Posture Dorsale de Sauvegarde • Power Demand Signal • Power Density Spectrum • Power Distribution System • Practice Dangerous to Security • Premises Distribution System • Previously Developed Software • Problem Data System • Procedures Development Simulator • Processed Data Stream • Processing and Data System • Processor Direct Slot • Procurement Data Sheet • Product Design Standard • Product Distribution System • Production Data Set • Production Data Sheet • Production Data Specialist • Professional Development Seminar • Program Data Sheet • Program Data Source • Program Design Specification • Program Development Section • Program Distribution System • Programming Documentation Standards • Project Definition Statement • Project Definition Study • Project Designation Study • Project Design Standard • Projected Distribution System • Protected Distribution System • Publishing Distribution System • Pyrotechnic Device Simulator

PDSA Plan - Do - Study - Act

PDSAR Public Document Status of Assessment Report

PDSE Partitioned Data Set Extension

PDSF Physics Detector Simulation Facility

PdSi Palladium Silicide

PDSI Palmer Drought Severity Index • Phase-Diverse Speckle Imaging

PDSL Power Demand Signal Left • Powerline Digital Subscriber Line • Process Data Sensitivity Level • Public Domain Software Library

PDSM Product Data System Modernization

PDSO Packet Data Security Overlay

PDSP Peripheral Data Storage Processor • Position Determination on Schmidt Plates

PDSR Power Demand Signal Right • Principal Director of Scientific Research

PDSS Payload Data Services System • PIMOS Development Support System • Postdeployment Software Support • Postdevelopment and Software Support

PDSSBN Planetary Data System Small Bodies Node

PDST Pacific Daylight Saving Time

PDSU Power Distribution and Switching Unit

PD&T Professional Development & Training

PDT Pacific Daylight Time • Parallel Data Transmission • Payload Data Tape • Pendleton (USA-OR) (location identifier) • Performance Demonstration Test • Performance Diagnostic Tool • Personal Development Technology • Photodynamic Therapy • Plasma Display Terminal • Precision Digitizing Table • Predictor Display Technique • Product Development Team • Printer Definition Table • Product Development Team • Professional Development and Training • Programmable Data Terminal • Programmable Drive Table • Propagation Delay Time

PDTA Power and Data Transfer Assembly

Pdte Presidente

PDTS Program Development Tracking System

PDU Packet Data Unit • Paysandú (Uruguay) (location identifier) • Physical Data Unit • Pilot Display Unit • Plan de Déplacements Urbains • Plug Distribution Unit • Position Display Unit • Power Distribution Unit • Power Drive Unit • Premises Distribution Upgrade • Pressure Distribution Unit • Process Development Unit • Professional Development Unit • Programme de Développement axé sur les Utilisateurs • Protocol Data Unit • Pulse Detection Unit

PDUS Primary Data User Station (*) • Primary Data Utilization Site • Principal Data User Station

PdV Point de Vue • Pressure-driven Volume

PDV Path Delay Value • Personaldatenverarbeitung • Plovdiv (Bulgaria) (location identifier) • Pressure Disconnect Valve

PDVS Patentdatenverarbeitungssystem

PDW Preliminary Design Walkthrough • Priority Delayed Weather

PDWE Pulsed Detonation Wave Engine

PDWS Passive Doppler Wind Sensor • Precision Digitizing Workstation

PDX Paradox • Phone Data Exchange • Poloidal Divertor Experiment • Portland (USA-OR) (location identifier) • Princeton Divertor Experiment

PDZ Pedernales (Venezuela) (location identifier)

PDZAMF Present-Day Zero-Age Mass Function

Pdzo Pasadizo

p/e Por Ejemplar • Price Earnings Ratio

pe Par Exemple • Per Esempio • Perlucidus • Peru • Photoelectric • Photoelectron • Por Ejemplo • Probable Error

Pe Péclet Number

P/E Port of Embarkation • Precipitation/Evaporation • Price-to-Earnings • Profit-to-Earnings

P-E Perkin-Elmer Corporation

P&E Plant & Equipment • Pricing & Estimating • Propellant & Explosives

PE Padre • Parabolic Equation • Parity Error • Parity Even • Parlamento Europeo • Parlement Européen • Payload Element • Pellets • Peripheral Equipment • Perkin-Elmer • Permanent Echo • Permanent Erection • Permissible Error • Pernambuco (Brazil) • Pérou (code ISO de pays) • Persian • Personal Effects • Peru (international vehicle registration code) • Peru (ISO country code) • Phase Encode • Phase-Encoded • Phase Encoder • Phase Encoding • Photoelectric • Photoelectron • Photoexcitation • Photogrammetric Engineer • Physical Education • Physical Entity • Physical Extent • Planetary Explorer • Planning Estimate • Plant Equipment • Plasma Effect • Plasma Emission • Plastic Explosive • Pluto Express • Pocket Edition • Polyethylene • Port Elizabeth - H.F. Verwoerd (South Africa) • Port of Embarkation • Position Error • Postencounter • Post Engineer • Potential Energy • Potential Evapotranspiration • Potentiel Évoqué • Power Enrichment • Practical Exercise • Precipitation/Evaporation • Precision Echo • Precision Electronics • Prediction Efficiency • Preequilibrium • Preliminary Estimator • Preliminary Evaluation • Presiding Elder • Primary Eclipse • Prime Equipment • Primitive Equation • Prince Edward Island (Canada) • Principal Engineer • Printer's Error • Probability of Error • Probable Error • Processeur Élémentaire • Processing Element • Processor Element • Procurement Executive • Product Engineering • Production Engineering • Professeur des Écoles • Professional Engineer • Professorat des Écoles • Program Element • Program Engineer • Program Exception • Progressive Erection • Project Engineer • Project Engineering • Prolific Editor • Protect Enable • Protective Entrance • Pulsating Erection • Purchased Equipment

PEA Pattern Error Analysis • Payload Electronics Assembly • Payload Enclosure Assembly • Paysage Électronique Astronomique • Penneshaw (Australia-SA) (location identifier) • Permanent Electronic Address • Phillips Exeter Academy (USA-NH) • Pitch Error Amplifier • Plan d'Épargne en Actions (France) • Platform Electronics Assembly • Pocket Ethernet Adapter • Portable Electrical Appliance • Preliminary Engineering Authorization • Preliminary Environmental Analysis • Primary Expense Account • Programmatic Environment Analysis • Programme d'Étude Amont • Proposal Expansion Award • Push-Effective Address

PEACE Plasma Electron And Current Experiment (Cluster) • Plasma Energy Angle and Composition Experiment • Production Expansion/Acceleration Capability Enhancement • Protection of the Environment for Assuring a Cleaner Earth

PEAD Presidential Emergency Action Document (USA)

PEAG Procurement Evaluation and Analysis Group

PEAGB Postearly-Asymptotic Giant Branch

PEAM Personal Electronic Aid for Maintenance • Portable Electronic Aid for Maintenance

PEAN Pan European ATM Network

PEAP Pad Emergency Air Pack • Personal Egress Air Pack

PEAR Polyether Amide Resin • Program Explanation Analysis Required

PEARLS Physics Electronic Academic Reference Library for Simulation

PEARLSS Particulate Effects Analysis on Response Levels of Space-borne Sensors

PEAT Programme Élargi d'Assistance Technique

PEB Payload Equipment Bay • PCM Expansion Bus • Pebane (Mozambique) (location identifier) • Performance Evaluation Board • Physical Evaluation Board • Prêt Entre Bibliothèques • Program Element Breakdown • Propulsion Examining Board • Pulsed Electron Beam

PEBB Power Electronic Building Block

PEbd Polyéthylène Basse Densité

PEBD Pay Entry Basic Date

PEBES Personal Earnings and Benefits Statement

PEBI Perpendicular Bisector (grid)

PEBS Pulsed Electron Beam Source

Pec Peculiar

PEC Passive Equipment Cabinet • Peak Electrode Current • Pelican (USA-AK) (location identifier) • Periodic-Error Correction • Perkin-Elmer Corporation • Photoelectric Cell • Photoelectrochemical Cell • Physics - Engineering - Chemistry • Planetary Entry Capsule • Potentiel Évoqué Cognitif • Preexisting Condition • Prêt à Être Changé • Previous Element Coding • Primary Environment Care • Production Engineering Control • Production Equipment Code • Program Element Code • Program Environment Control • Program Evaluation Committee • Program Executive Command • Programme Européen de Coopération • Projet Européen Commun • Proposal Evaluation Criteria • Propulsion Environmental Chamber • Pulsed Eddy Current • Pulse-height to Energy Conversion

PECASE Presidential Early Career Awards for Scientists and Engineers (USA)

PECC Product Engineering Control Center

PECD Photoelectric Conversion Device

PECE Proposed Engineering Change Estimate

PECF Profile of the Effective Cloud Fraction

PECH Polyepichlorohydrine

PECI Preliminary Equipment Component Index • Preliminary Equipment Component Inventory • Productivity Enhancement Capital Investment

PECK Patient Evacuation Contingency Kit

PECL Positive Emitter-Coupled Logic

PECO Pays d'Europe Centrale et Orientale

PECOD Program Element Code

PECOS Perspectives on Cooperative Systems • Picture Enhancement Computer Operating System • Program Environment Checkout System

PECP Preliminary Engineering Change Proposal

PECR Program Error Correction Report

PECS Portable Environmental Control System • Powered Elevator Control System

PECU Powered Elevator Control Unit

PECVD Plasma-Enhanced Chemical-Vapor Deposition

Ped Pedal • Pedestal • Pedestrian • Pediatrics

PEd Proposal Editor

PED Packet Encryptor/Decryptor • Pays En Développement • Pedal • Pedestal • Pedestrian • Period End Date • Personal Electronic Device • Personnel Equipment Data • Photoemission Diode • Photon-Emitting Defect • Plasma Emission Detector • Pôle Européen de Développement • Portable Electronic Device • Prime d'Encadrement Doctoral • Procurement Early Development • Production Eligibility Date • Production Engineering Division • Program Editor • Program Element Directive • Program Entry Device • Program Execution Directive • Programmable

Encryption Device • Promotion Eligibility Date • Proposal Editor • Proton-Enhanced Diffusion • Pulsed Electrochemical Detection • Purchase Early Development

Pedag Pedagogía

PEDAS Panel on Environmentally Detrimental Space Activities • Potentially Environmentally Detrimental Activities in Space (COSPAR)

PEDB Payload Engineering Database • Planning and Execution Database

PedD Doctor of Pedagogy

PEDD Portable Electronic Display Device

PEDM Preliminary Exploratory Development Model

PEDN Planned Event Discrepancy Notification

PEdO Physical Education Officer

PEDRO Pneumatic Energy Detector with Remote Optics

PEDS Pedestrians • Program Element Descriptive Summary • Protective Equipment Decontamination Section

PEE Perm (Russia) (location identifier) • Photoelectron Emission • Photoemission Effect • Pôle d'Études Économiques • Polychromatic Encircled Energy • Poste d'Expansion Économique • Program Estimating Equation

PEEC Personnel Emergency Estimator Capability

PEEK Polyether Ether Ketone

PEEL Programmable Electrically Erasable Logic

PEELS Primary Electrical Equipment Life Support

PEEM Panel of Experts on Environmental Management • Photoelectron Emission Spectroscopy • Photoemission Electron Spectroscopy

PEEP Parallel Evaluation and Experimentation Platform • Pilot's Electronic Eye-Level Presentation • Planetary Event and Eclipse Predictor • Porous Electrode Electrostatic Precipitation

PEERS Physics Encyclopaedia of Email Records

PEF Packet over Ethernet over Fiber • Payload Encapsulation Facility • Penetration with Excessive Friction • Personal Effects Floater • Physical Electronics Facility • Polyethylene Foam • Prediction Error Filter • Program Estimating Factor • Proposal Evaluation Form • Pulsed Electric Field

PEFC Pan-European Forest Certification

PEFO Payload Effects Follow-On

PEFOS Payload Effects Follow-On Study

PEFPN Physica Energiae Fortis et Physica Nuclearis (periodical code)

PEFT Positive Extended Faceted Taxonomy

PEFU Payload Effects Follow-Up

Peg Pegasus-Pegasi (constellation)

PEG Passive Elliptical Galaxy • Performance Evaluation Group • Perugia - Sant Egidio (Italy) (location identifier) • Physics Education Group • Polyethylene Glycol • Port Entry Guide • Powered Explicit Guidance • Program Element Group • Program Evaluation Group • Program Execution Group

PEGA Program in Extragalactic Astronomy

PEGASUS Preliminary Expert Ground Analysis Scheduling User System

PEGC Professeur d'Enseignement Général des Collèges

PEGE Pôle Européen de Gestion et d'Économie

PEGI Programme Environnement de la Géosphère Intertropicale

PeGRS Penetrator Gamma-Ray Spectrometer

PEH Pehuajo (Argentina-BA) (location identifier)

PEhd Polyéthylène Haute Densité

PEHD Polyéthylène Haute Densité

PEHG Payload Ethernet Hub Gateway

PEI Pereira (Colombia) (location identifier) • Phase-Encoded Information • Plan d'Épargne Interentreprise (France) • Preliminary Engineering Inspection • Prince Edward Island (Canada) • Principal End Item • Professional Engineering Institution • Profile for Enterprise Integration • Projet Éducatif Individuel • Publication Effective Index • Publication Effectiveness Index

PEIC Periodic Error Integrating Controller

PEID Program Element Identification

PEIDA Photométrie et Étude d'Images Destinées à l'Astrophysique

PEIR Project Equipment Inspection Record

PEIS Programmatic Environmental Impact Statement

p. ej. Por Ejemplo

Pek Peking (China-PRC)

PEK Beijing Capital (China-PRC) (location identifier) • Peking Capital (China-PRC) (location identifier) • Phase-Exchange Keying

Pel Picture Element

PEL Pelaneng (Lesotho) (location identifier) • Permissible Exposure Level • Permissible Exposure Limit • Photoelectron Layer • Physics and Engineering Laboratory • Physics Express Letters • Picture Element • Precision Elastic Limit • Protocol Engineering Laboratory

PElec Photoelectric

PEM Pacific Exploratory Mission • Parametric Earth Model • Particle Environment Monitor • Payload Ejection Mechanism • Payload Electronics Module • Pelletier Effect Module • Phased Equipment Modernization • Photoelastic Modulator • Photoelectromagnetic • Piezoelastic Modulator • Plant Engineering and Maintenance • Plastic-Encapsulated Microcircuit • Plastic-Encapsulated Microelectronics • Polymer Electrolyte Membrane • Positive Energy Mode • Postal Electronic Mail • Preemptive Multitasking • Privacy-Enhanced Mail (*) • Privacy-Enhanced Messaging • Probable Error of Measurement • Processing Element Memory • Product Effectiveness Manual • Product Error Message • Production Engineering Measure • Program Element Manager • Program Element Monitor • Program Engineering Manager • Programmer's Environment Manager • Project Engineering Memorandum • Proton-Exchange Membrane • Puerto Maldonado (Peru) (location identifier)

Pembs Pembrokeshire (UK)

PEMD Personalized Educational Materials Development

PEMF Pulsating Electromagnetic Field

PEMFC Proton Exchange Membrane Fuel Cell

PEMO Production Engineering and Manufacturing Organization • Protective Equipment Maintenance Organization

PEMP Photoelastically Modulated Polarimeter

PEMRC Program Executive Management Review Committee

PEMS Portable Environmental Measuring System

Pen Penal • Penetration • Península • Peninsular • Peninsule • Penisola • Pennsylvanian

PEN Nuevo Sol (Peru) (ISO currency code) • Particle Entity Notation • Penang (Malaysia) (location identifier) • Peninsula • Plan Energético Nacional (Spain) • Poets - Essayists - Novelists • Professeur d'École Normale • Program Element Number • Program Error Note • Public Education Network • Public Electronic Network

PENAID Penetration Aid

PENAIDS Penetration Aids

PENCIL Pictorial Encoding Language

P.Eng. Professional Engineer

PENIS Polyelastic Normal Integration of Samples

Penn Pennsylvania (USA)

Penna Pennsylvania (USA)

Pens Pensamiento

Pent Pentagon • Pentecost

PEO Patrol Emergency Officer • Polyethene • Polythylene • Postemergency Organisation • Procédure d'Évaluation et d'Orientation • Product Engineering Office • Professional Employment Organization • Professional Engineers of Ontario (Canada-ON) • Program Evaluation Office • Program Executive Office • Program Executive Officer • Project Engineering Office • Project Executive Office • Project Executive Officer

PEOC Presidential Emergency Operations Center (USA)

PEOPLE Precollege Enrichment Opportunity Program for Learning Excellence

PEOS Propulsion and Electrical Operating System

PEP Packet Exchange Protocol • Packetized Ensemble Protocol • Pan-European Paging • Paperless Environment Project • Partitioned Emulation Program • Partitioned Emulator Program • Pauli Exclusion Principle • Payload Exploitation Plan • Peak Energy Product • Peak Envelope Power • Peer Evaluation Program • Peppimenarti (Australia-NT) (location identifier) • Perceptions in Emerging Physics (IUCAA) • Performance Evaluation Plan • Performance Evaluation Program • Peripheral Event Processor • Perkin-Elmer Processor • Personal Equity Plan • Personnel Exchange Program • Phased Execution Plan • Phase-Encoded Part • Photoelectric Photometer • Photoelectric Photometry • Photoelectric Potential • Photoemission Polarimeter • Physics and Engineering Physics • Piper External Power • Planar Epitaxial Passivated (transistor) • Planetary Ephemeris Program • Planetary Exploration Plan • Planning and Evaluation Processor • Planning Event Processor • Plant Engineering Procedure • Plant Equipment Package • Plasma Exhaust Process • Platform Electronic Package • Platform Evaluation Program • Plume Exposure Pathway • Polar Energetic Particle • Pole - Equator - Pole • Policy Enforcement Point • Political and Economic Planning • Positron Electron Project (*) • Positron - Electron - Proton • POSIX Environment Platform • Power Entry Panel • Power Evaluation Program • Power Extension Package • Precipitation Enhancement Program • Predictability Engineering Planning • Preejection Period • Preliminary Evaluation Plan • Producibility - Engineering - Planning • Productivity Enhancement Program • Productivity Enhancement Project • Program Evaluation Procedure • Project Element Plan • Promotion Evaluation Pattern • Proposal Entry Processing • Proposal Entry Processor (*) • Proposal Entry Proposer • Proposal Evaluation Panel • Proposal Evaluation Plan • Protocol Extension Protocol • Proton - Electron - Proton

PEPAT Protocol on Environment Protection of the Antarctic Treaty

PEPC Polynomial Error Protection Code

PEPDB Proposal Entry Processing Database

PEPE Parallel Element Processing Ensemble • Persistant Elevated Pollution Episode • Plasma Experiment for Planetary Exploration

PEPG Piezoelectric Power Generation

PEPGB Plasma Exhaust Pumping Glovebox

PEPH Penumbra-Photosphere

PEPI Physics of the Earth and Planetary Interiors (periodical code)

PEPL Photo-Electron-enhanced Plasma Line • Plasmadynamics and Electric Propulsion Laboratory (USA-MI)

PEPP Planetary Entry Parachute Program • Princeton Earth Physics Project • Producibility Engineering and Production Planning

PEPR Precision Encoding and Pattern Recognition

PEPS Preliminary Evaluation Programme for SPOT • Programme d'Étude Probatoire de SPOT • Programme d'Évaluation Préliminaire SPOT

PEPSI Potsdam Echelle Polarimetric and Spectroscopic Instrument • Proposal Entry Processor System

PEPSOC Portable ESOC Package for Synchronous Orbit Control

PEPSY Precision Earth-Pointing System • Proposal Entry Processor System

PEPT Professeur d'Enseignement Professionnel Technique

PEQ Pecos City (USA-TX) (location identifier)

PEQUOD Pacific Equatorial Ocean Dynamics

Per Percentile • Period • Periodic • Periodical • Perigee • Perseus-Persei (constellation) • Persia • Persian • Person • Personnel

Pér Période • Périodique

PER Packet Error Rate • Parametric Estimating Relationship • Parity Error Rate • Perform • Performance • Performance Evaluation Report • Perigee • Period • Pérou (code ISO de pays) • Perseids (meteor stream) • Personnel • PERT Event Report • Perth (Australia-WA) (location identifier) • Peru (ISO country code) • Physics Education Research • Preejection Rigidity • Preenvironmental Review • Preliminary Engineering Report • Price Earning Rate • Price-Earnings Ratio • Professional Employment Register • Program Event Recording • Program for Ecological Research • Proposal Evaluation Report • Public Environment Report

PERA Planning and Engineering for Repair Activity

per an Per Annum

Perc Percussion

PERC Portable Executive for Reliable Control

PerCo Perkins Observatory Contributions (periodical code)

per con Per Contra

PERCOS Performance Coding System

PERD Panel on Energy Research and Development

PERDM Program Execution Review Decision Memorandum

PERE Person Eligible to Receive Effects

Perf Perfect • Perfection • Perfecto • Perforated • Perforation • Performance • Performed

PERF Perform • Performance

PerFECT Peroxide in Fuel Estimation and Concentration Test

PERG Production Equipment Redistribution Group

PERGO Project Evaluation and Review with Graphic Output

PERGS Portable Earth Resources Ground Station

Perh Perhaps

Peri Perigee

PERI Production Equipment Redistribution Inventory

PERIDOT Prévision à Échéance Rapprochée Intégrant des Données Observées et Télédétectées

Périph Périphérique

Period Periodic • Periodical

Péris Péristyle

PERL Pathologically Eclectic Rubbish Lister • Practical Extraction and Report Language • Prepositioned Equipment Requirements List • Process Enactment Resource Library

Perm Permanent • Permanently • Permission

PERM Parallele Elektronische Rechenmaschine München (Germany) • Permanent • Program Evaluation for Repetitive Manufacturing

PERMANENT Projects to Employ Resources of the Moon and Asteroids Near Earth in the Near Term

PERMARS Personnel Management Reports System

PERMAS Personnel Management Assistance System

Permb Permeability

PERMIS Personnel Management Information System

PERMIT Polar Orbiter Effective Rainfall Monitoring Integrative Technique

PerMS Perceptual Motor Skills (periodical code)

PERMS Personnel Electronic Records Management System • Planning Execution Resources Monitoring System

Perp Perpendicular • Perpetual

PERP Parts and Equipment Reliability Program

per pro Per Procurationem

per proc Per Procurationem

PERRMS Personnel Resource Management System

Pers Persia • Persian • Person • Persona • Personal • Personalize • Personalized • Personally • Personen • Personnel • Perspective

PE&RS Photogrammetric Engineering & Remote Sensing (journal)

PERS Performance Evaluation Reporting System • Person

PERSEC Personal Security

Persp Perspectiva • Perspective

PERSREP Personnel Report

PERSTAT Personnel Status

PERSTATR Personnel Status Report

Pert Pertain • Pertaining

PERT Performance Evaluation and Review Technique • Program Evaluation and Research Task • Program Evaluation and Review Technique (*) • Program Evaluation Research Task • Program Evaluation Review Technique • Project Evaluation and Review Technique

PERTCO Program Evaluation and Review Technique with Cost

PERTEC Pôle Européen de Recherches en Turbulence et Convection

PERU Production Equipment Records Unit

Peruv Peruvian

PES Nuevo Sol (Peru) (ISO currency code) • Packetized Elementary Stream • Packet over Ethernet over Sonet • Parallel Enterprise Server • Patent Examining System • Personal Earth Station • Petrozavodsk (Russia) (location identifier) • Photoelectric Scanner • Photoelectron Spectroscopy • Planning and Evaluation Service • Pointing Error Sensor • Polyelastic Sampling • Polyethersulfone • Positioning Error Signal • Positive Enable System • Postenumeration Survey • Potential-Energy Surface • Potentiel Évoqué Somesthésique • Preparedness Evaluation System • Processor Enhancement Socket • Production Engineering Specification • Professional Engineering Services • Program Element Summary • Programmable Electronic System • Proposed Encryption Standard • Public Enquiry System

PESA Particle Elastic Scattering Analysis

PESC PCM Encoder and Signal Conditioning • Physical and Engineering Sciences • Physical and Engineering Sciences Committee • Politique Étrangère et de Sécurité Commune • Politique Extérieure et de Sécurité Commune • Public Expenditure Survey Committee

PESD Personnel Electrostatic Discharge • Program Execution Subdirective

PESDS Program Element Summary Data Sheet

Pesh Peshawar (Pakistan)

PESHE Programmatic Environment, Safety, and Health Evaluation

PESM Photoelectron Spectromicroscope • Photoelectron Spectromicroscopy

PESO Product Engineering Services Offer

PESR Planning Element System Report

PEST Parameter Estimation by Sequential Testing • Political - Environmental - Social - Technological

PESTO Photosynthesis Experiment and System Testing Operations • Public Engagement and Science/Technology Options

Pet Petroleum • Petrological • Petrologist • Petrology

PET Pan-European Technology • Particle Electrostatic Thruster • Patterned Epitaxial Technology • Pelotas (Brazil-RS) (location identifier) • Performance/Endurance Tradeoff • Performance Enhancement Team • Performance Evaluation Test • Periodic Environmental Test • Periodic Evaluation Test • Peripheral Equipment Tester • Personal Electronic Transactor • Phase Elapsed Time • Photoemission Tube • Phototropic Energy Transfer • Photovoltaic Engineering Testbed • Physical Equipment Table • Pilotline Experiment Technology • Pixel Extraction Table • Plasma Electro-Thermal • Polyéthylène Téréphthatale • Polyethylene Terephtalate • Portable Earth Terminal • Portable Electronic Telephone • Portable Electronic Translator • Portable Executive Telephone • Position - Event - Time • Positron Emission Tomography • Postendurance Test • Potentially Exempt Transfer • Preliminary Evaluation Team • Preliminary Examination Team • Print Enhancement Technology • Privacy Enhancement Technology • Probe Ephemeris Tape • Production Environmental Test • Production Environmental Testing • Production Evaluation Test • Program Elapsed Time • Program Evaluator and Tester • Proposal Evaluation Tool • Propulsion Engineering Trainer • Proton/Electron Telescope • Prototype Evaluation Testing

PETA Performance Evaluation and Trend Analysis • PetaFLOPS Enabling Technologies and Applications

PetE Petroleum Engineer

PETE Priority End To End

PETFE Polyethylene Tetrafluoroethylene

Petn Petition

PETN Pentaerythritoltetranitrate

PETP Polyethylenterephthalat

Petro Petroleum

Petrog Petrography

Petrol Petrology

PETS Parametrized Executable Test Suite • Payload Environmental Transportation System • Performance Endurance Tradeoff Study • Performance Evaluation and Test System • POCC Experiments Timeline System • Posting and Enquiry Terminal System • Prior to Expiration of Term of Service • Programmed Extended Time Sharing

PETSC Portable Extensible Toolkit for Scientific Computation

P. et T. Postes et Télécommunications

PEU Peripheral Electronics Unit • Pod Electronics Unit • Puerto Lempira (Honduras) (location identifier)

PEUL Probabiliste Eulérien Lagrangien

PEUM Penumbra-Umbra

PeV Petaelectronvolt

PEV Potentiel Évoqué Visuel • Pyroelectric Vidicon

PEW Packet over Ethernet over WDM • Peshawar (Pakistan) (location identifier)

p. ex. Par Exemple • Por Extensión • Por Extenso

PEX Pechora (Russia) (location identifier) • PHIGS Extensions to X • Physics Expectation • Pipeline Executive • Plasmadynamic Experiment • PLuto Express (mission) • Prestation Exceptionnelle • Protocol Extensions to X-Windows

PEY Penong (Australia-SA) (location identifier)

PEZ Penza (Russia) (location identifier)

PE3 Protocol Engineering Environment in Estelle

pf Per Favore • Perfect • Pfarrer • Pfennig • Power Factor • Preferred • Pro Forma • Proof

pF Picofarad

Pf Pfeiler • Pfennig • Postfach • Probability of Failure

P/F Poll/Final

PF French Polynesia (ISO country code) • Packet Format • Page Footing • Page Formatter • Panchromatic Film • Parachute Facility • Parallel Fold • Parallel Fortran • Pathfinder • Payload Function • Payload Forward • Perforated Layer • Performance Factor • Permeability Factor • Pérot-Fabry • Pilot Flying • Plan de Formation • Pluto Flyby • Point-like Feature • Poloidal Field • Polynesia (French_) (ISO country code) • Polynésie Française (code ISO de pays) • Powered Flight • Power Factor • Prefetch • Preflight • Preliminary Failure • Primary Feed • Primary Focus • Prime Focus • Prime Function • Priming Fluid • Probability of Failure • Production Function • Program Function • Programmable Format • Programmed Function • Project File • Pseudo-Format • Pseudo-Frequency • Publication Frequency • Public Funding • Pulse Frequency • Pulverized Fuel

PFA Paf Warren (USA-AK) (location identifier) • Parallel Finite Automaton • Participating Field Activity • Perfluoralkoxy (polymer) • Personnel Functional Assessment • Planet Finder Array • Popular Flying Association (UK) • Postflight Analysis • Predictive Failure Analysis • Primary Faculty Advisor • Prime Factor Algorithm • Principal Factor Analysis • Printer Font ASCII • Probabilité de Fausse Alarme • Probability of False Alarm • Pulverized Fuel Ash

PFAM Programmed Frequency Amplitude Modulation

PFAR Power Fail Automatic Restart • Preliminary Failure Analysis Report

PFAT Prefirst Article Test • Preflight Acceptance Test

PFB Parallel Filter Bank • Passo Fundo (Brazil-RS) (location identifier) • Payload Feedback • Payload Forward Bus • Position Feedback • Postscript Font Binary (*) • Pressure-Fed Booster • Pressurized Fluidized Bed • Printer Font Binary

PFBC Pressurized Fluidized Bed Combustion

PFBIT Preflight Built-In Test

PFBR Prototype Fast Breeder Reactor

PFC Pacific City (USA-OR) (location identifier) • Passenger Facility Charge • Perfluorocarbon Compound • Performance Flight Certification • Photographic Field Center • Physics Frontiers Center • Plasma-Facing Component • Plasma Fusion Center • Porous Friction Course • Portable Four Channel • Postflight Checklist • Power Factor Corrector • Preflight Console • Preliminary Flight Certification • Preliminary Flight Configuration • Preparing For Cutover • Pricing - Finance - Claim • Primary Flight Control • Primary Focus Camera • Prime-Focus Camera • Prime-Focus Corrector • Priority Foreign Country • Private First Class • Prototype Flight Cooler

PFCB Perfluorocyclobutyl

PFCD Primary Flight Control Display

Pfce Performance

PFCF Payload Flight Control Facility

PF&CO Publications of Flower & Cook Astronomical Observatory (periodical code)

PFCP Program Funding Change Proposal

PFCS Primary Flight Control System

Pfd Pfund • Preferred

PF&D Personnel, Fatigue & Delay

PFD Particle Flux Density • Peak Flux Density • Personal Floatation Device • Personnel, Fatigue, and Delay • Phase Front Disturbance • Pinch-Force Dynamometer • Port Frederick (USA-AK) (location identifier) • Position-Fixing Device • Power Flux Density • Poynting-Flux-Dominated (jet) • Preliminary Functional Description • Primary Flight Display • Primary Function Display • Probability of Fault Detection • Procurement Fraud Division • Proton Flux Density

PFDA Postflight Data Analysis • Precision Frequency Distribution Amplifier

pfdr Pathfinder

PFDR Postflight Data Reduction

Pfd. Sp. Preferred Spelling

PFE Période de Formation en Entreprise • Photometric Front End • Plan de Formation en Entreprise • Poole-Frenkel Effect • Postflight Evaluation • Professional File Editor • Professional Flight Engineer • Programmer's File Editor • Promotion Fitness Examination

PFEP Programmable Front-End Processor

PFER Per-Family Error Rate

PFES Portable Field Emission Spectrometer • Prime Focus Echelle Spectrometer • Proposed Final Environment Statement

PFET Power Field-Effect Transistor

PFEU Phase and Frequency Estimation Unit

PFF Page Fault Frequency • Physical Fitness Facility • Planning Factors File • Pluto Fast Flyby • Postfactory Functional • Presidential Faculty Fellow • Primary Focus Feed

Pfg Pfennig

PFG Primary Frequency Generator • Product Feature Guide

PFGS Potential Fields Graphics System

PFH Per Flight Hour

PFHE Plate-Fin Heat Exchanger

PfHSN Protocol for High-Speed Network

PFI Pack File Indexing • Partnership For Innovation • Port-Fuel Injection • Power Failure Indicator • Pricing For Information • Printing For Information • Private Finance Initiative • Pro Forma Invoice • Provisional Full Internet

PFIAB President's Foreign Intelligence Advisory Board

PFIC Procurement Fraud Irregularities Coordinator

PFIP Primary Focus Instrument Package • Prime Focus Instrument Package • Prime Focus Instrument Platform

PFIRCAM Prime Focus Infrared Camera

PFIRES Policy for Integrating Risk in E-business Security

PFIS Prime Focus Imaging Spectrograph • Property Fabric Information System

PFJ Patreksfjördur (Iceland) (location identifier) • Proposed Final Judgment

PFK Payload Function Key • Program Function Keyboard • Programmed Function Keyboard

PFL Phase Fresnel Lens • Plant Fuel Load • Postflare Loop • Primary Freon Loop • Propulsion Field Laboratory

PflegeVG Pflegeversicherungsgesetz

PFLP Point-Feature Label Placement

pfm PostScript Font Metrics (file extension)

PFM Peak Fitting Module • Plan For Maintenance • Plasma Facing Material • Platform • Postal Furnished Materiel • Potential Field Model • Power Factor Meter • Precision Flight Mode • Predictor Frame Memory • Preflight Module • Pressure Flow Meter • Printer Font Metrics • Program Financial Management • Protoflight Model • Psion Freeware Movement • Pulse-Frequency Modulation

PFMP Plan de Formation en Milieu Professionnel

PFMR Program Fund Management Record • Project Fund Management Record • Project Fund Manager Record

PFMT Personal Financial Management Training

PFN Panama City - Bay County (USA-FL) (location identifier) • Point For Negotiation • Pulse Forming Network

PFNA Pulsed Fast Neutron Analysis

PFNS Position Fixing Navigation System

PFO Paphos (Cyprus) (location identifier) • Particulate Fallout • Piñon Flat Observatory (USA-CA) • Postal Finance Officer

PFOB Perfluorooctylbromide

Pfor Professor

PFOV Per-pixel Field Of View (*) • Pixel Field Of View

P&FP Program & Financial Plan

PFP Parallel Function Processor • Personal Financial Planning • Powered Flight Program • Prime Focal Point • Prime Focus Platform • Probability of Performance Failure • Program Financial Plan • Program Forecast Period • Programmable Field Processor • Programmable Function Panel

PFPD Pulse-Flame Photometric Detector

PFPU Processor Frame Power Unit

PFQ Preflight Qualification

Pfr Fractured and Ridged Plain

PFR Ilebo (Zaire) (location identifier) • Parts Failure Rate • Personal Financial Record • Plug Flow Reactor • Portable Font Resource • Portable Foot Restraint • Positive Feedback Region • Power-Fail Restart • Präzisions-Filterradiometer • Precision Filter Radiometer • Preflight Review • Prime Focal Ratio • Problem/Failure Report • Program Financial Review • Prototype Fast Reactor

PFRA Prairie Farm Rehabilitation Administration

PFRC Pacific Forest Research Centre

Pfrs Parfois

PFRS Portable Field Recording System • Portable Field Reflectance Spectrometer • Power Failure Recovery System

PFRT Preliminary Flight Rating Test • Preliminary Flight Readiness Test

P&FS Particles & Fields Satellite

PFS Parallel File System • Particles and Fields Satellite • Percent Full Scale • Perfect-Forward Secrecy • Performance Funding System • Periodical Fine Structure • Physical File System • Picture Frustration Study • Planetary Fourier Spectrometer • Plug-in Finder Service • Precision Frequency Source • Prefeasibility Study • Primary Flight Software • Primary Flight System • Product Function Specification • Programmable Frequency Standard • Propellant Feed System

PFSA Pour Faire Ses Adieux

PFSMETE Postdoctoral Fellowship in Science, Mathematics, Engineering and Technology Education

PFSS Potential Field Source Surface

PFT Parallel Fourier Transform • Perfluorocarbon Tracer • Periodic Flare of Temper • Postflight Testing • Prefiltering Technique • Pressed For Time • Production Flight Test • Projective Field Theory • Pulse Fourier Transform

PFTA Payload Flight Test Article

PFTB Preflight Test Bus

PFTI Plateforme Technique pour l'Indemnisation

PFTM Preliminary Flight Test Memo

PFTR Postflight Trajectory Reconstruction • Preliminary Flight Test Report

PFTS Permanent Field Training Site

PFTU Portable Fuel Transfer Unit

PFU Particle Flux Unit • Peoples Friendship University (Russia) • Please Follow Up • Prime Focus Unit • Protoflight Unit

PFV Pour Faire Visite • Probability of Failure of Vehicle

PFW Progressive Free Wave

Pfx Prefix

PFY Postal Fiscal Year • Prior Fiscal Year

pg Page • Página • Persona Grata • Picogram • Photographic

Pg Page • Palaeogene • Petagram • Photogrammetria (periodical code) • Portugal • Portuguese

P&G Penny & Giles

PG Page • Palomar-Green (survey) • Papouasie Nouvelle-Guinée (code ISO de pays) • Papua New Guinea (ISO country code) • Parental Guidance • Pasquill-Gifford • Past Grand • Paying Guest • Peer Group • Penumbral Grain • Percolation Group • Permanent Grade • Persona Grata • Planetary Geophysics • Planning Group • Planning Guidance • Plasmagrenze • Plasma Grid • Portugal • Postgraduate • Power Gain • Power Generation • Pregnant • Pressure Gauge • Primeval Galaxy • Processing Gain • Procurator General • Program Group • Program Guidance • Project Group • Protogalaxy • Proving Ground • Pulse Generator • Pyrolytic Graphite • Pyrotechnic Gyro

PG/A Programmable Gain Amplifier

PGA Page (USA-AZ) (location identifier) • Parallel Genetic Algorithm • Peak Ground Acceleration • Pendulous Gyro Accelerometer • Pin Grid Array • Power Gain Antenna • Power Generating Assembly • Pressure Garment Assembly • Professional Graphics Adapter • Programmable Gain Amplifier • Programmable Gate Array • Proposal Generation Aid

PGADS Personnel Guided Aerial Delivery System

PGAOI Pulkovo Glavnaia Astronomicheskaia Observatoriia Izvestiia (periodical code)

PGB Pangoa (Papua New Guinea) (location identifier) • Pico Gravity Box

pgc Per Gyrocompass

PGC Pacific Geosciences Centre (Canada) • Petersburg - Grant County (USA-WV) (location identifier) • Plan Général de Coordination • Plant Growth Chamber • Policía Guardia Civil (Spain) • Policy Guidance Council • Poorly Graphiticized Carbon • Predicted Group Crossing • Prediction of Group Crossings • Pressure Gamma Center • Principal Galaxies Catalogue • Produit de Grande Consommation • Program Group Control • Programmed Gain Control • Promotion Graphics Council • Propeller Gear Case • Protogalactic Cloud (*) • Protogalactic Clump

PGCC Protoglobular Cluster Cloud

PGCD Plus Grand Commun Dénominateur • Plus Grand Commun Diviseur

PGCE Postgraduate Certificate of Education

PGCS Photogrammetric Graphics Compiling System

PGD Planetary Geosciences Division • Pulse Generator Display • Punta Gorda - Charlotte County (USA-FL) (location identifier)

PGDA Power Generation and Distribution Assembly

PGD&CS Power Generation, Distribution, & Control System

PGDCS Power Generation, Distribution, and Control System

PGDF Product Generation and Dissemination Facility

PgDn Page Down

PGDN Page Down

PGDS Pioneer Ground Data System • Pulse Generator Display System

PgE Photogrammetric Engineering (periodical code)

PGE Platinum Group Element • Presupuestos Generales del Estado • Product Generation Engine • Product Generation Executable • Purge • Yegepa (Papua New Guinea) (location identifier)

PGen Publications de l'Observatoire de Genève (periodical code)

PGenA Publications de l'Observatoire de Genève - Series A (periodical code)

PGenB Publications de l'Observatoire de Genève - Series B (periodical code)

PgERS Photogrammetric Engineering and Remote Sensing (periodical code)

PGF Perpignan Rivesaltes (France) (location identifier) • Peter Gruber Foundation • Plant Growth Facility • Presentation Graphics Facility • Pressure Gradient Force • Probability Generating Function • Protein Growth Facility

PGG Project Galactic Guide

PGGA Permanent Geodetic GPS Array • Permanent GPS Geodetic Array

PGH Pantnagar (India) (location identifier) • Programme du Génome Humain

PGHM Payload Ground Handling Mechanism

PGI Chitato (Angola) (location identifier) • Plan de Gestion Informatique • Progiciel de Gestion Intégré • Programme Général d'Information

PGID Process Group Identification • Process Group Identificator • Process Group Identifier

PGIP Predicted Ground Impact Point

PGK Kina (Papua New Guinea) (ISO currency code) • Pangkalpinang (Indonesia) (location identifier)

PGL Pascagoula - Jackson County (USA-MS) (location identifier) • Peer Group Leader • Planetary and Geology Laboratory • Portable Gas Laser • Professional Graphics Language • Pulsed Gas Laser

PGLO Publications of Goethe Link Observatory (periodical code)

pgm Portable Gray Map (file extension)

Pgm Program

PGM Permanent Ground Monument • Planning Guidance Memorandum • Portable Greymap • Port Graham (USA-AK) (location identifier) • Pragmatic Group Multicast • Precision-Guided Munition • Privacy Graphic Mail • Processing Graph Method • Product Generation Module • Product Group Manager • Program • Program Group Manager • Program Guidance Memorandum • Portable Gray Map

PGML Precision Graphics Markup Language

Pgmt Pigment

PGN Pangia (Papua New Guinea) (location identifier) • Pôle Galactique Nord • Positive Gauss-Newton

PGNCS Primary Guidance, Navigation, and Control System

PGO Pettinger-Guilay Observatory (USA-WA) • Principal Guest Observer • Productivity of Global Ocean

PGOC Payload Ground Operations Contract • Payload Ground Operations Contractor

PGooO Publications of Goodsell Observatory (periodical code)

PGOR Payload Ground Operation Requirements

PGORS Payload Ground Operation Requirements Study

PGOU Plan General de Ordenación Urbana (Spain)

PGP Physique des Gaz et des Plasmas • Porto Alegre (Portugal) (location identifier) • Pretty Good Privacy • Primary Geodetic Parameter • Prism - Grating - Prism • Program Parameter

PGPS Parallel Genetic Programming System • Pulses Generated Per Second

PGR Paragould (USA-AR) (location identifier) • Parental Guidance Recommended • Planning and Ground Rule • Planning Ground Rule • Population Growth Rate • Postglacial Rebound • Precision Graphic Recorder • Procuradoría General de la República • Psychogalvanic Reflex

PGRC Program Guidance and Review Committee

PGRo Publications of the Kapteyn Astronomical Observatory in Groningen (periodical code)

PGRWG Payload Ground Requirements Working Group

PGS Passive Geodetic Satellite • Passive Gravity Stabilization • Peach Springs (USA-AZ) (location identifier) • Penetration Guiding System • Personal Handyphone System • Plane Grating Spectrograph • Plasma Generator System • Pôle Galactique Sud • Portable Ground Station • Power Generation Section • Power Generation System • Precision Gamma-ray Spectrometer • Primary Ground Station • Product Generation Segment • Product Generation Service • Product Generation System • Program Generation System • Proposal Generation System

PGSC Payload and General Support Computer • Payload General Support Computer

PGSE Payload Ground Support Equipment • Peculiar Generalized Structure Element • Peculiar Ground Support Equipment • Programming Graph Simulation Environment • Pulsed Gradient Spin Echo (technique)

PGSTK Product Generation System Toolkit

PGT Peace and Global Transformation • Per Gross Ton • Pistol Grip Tool • Proposal Generation Tool

PGTL Planning Group for Training and Logistics

PGTV Pulse-Gated Television

PGU Package Generator Utility • Plant Growth Unit • Power Generating Unit

PgUp Page Up

PGUP Page Up

PGUS Plan General de Usos del Suelo

PGV Greenville (USA-NC) (location identifier)

PGVF Plant Genome Venture Fund

PGWG Particles and Gases Working Group

PGWS Primary Groupware Server

PGX Périgueux (France) (location identifier)

PGZ Ponta Grossa - Sant'Ana (Brazil-PR) (location identifier)

PGZNM Politechnika Gdanska Zeszyty Naukowe Mechanika (periodical code)

ph per Hour • Phase • Philippines

pH Degree of Acidity or Alkalinity • Potential of Hydrogen Ions

Ph Phase • Phenyl • Philosophy • Phone • Phosphorus • Phot

PH Netherlands (international aircraft registration code) • Packet Handler • Packet Header • Pädagogische Hochschule • Page Heading • Paralaje Horizontal • Parallaxe Horizontale • Parity High • Penthouse • Period Hours • Personal Hygiene • Petroleum Hydrocarbon • Phase • Philippines (ISO country code) • Phrase Hypothesis • Pinhole • Power House • Precipitation Hardening • Preliminary Hazard • Presentation Header • Prohibited (area) • Proposal Handling • Proton Hole • Proton Hydrate • Provincial Highway • Public Health • Pulse Height

PHA Phan Rang (Viet Nam) (location identifier) • Photoelectric Astrolabe • Potentially Hazardous Asteroid • Preliminary Hazard Analysis • Process Hazards Analysis • Public Health Act (USA) • Public Housing Administration • Public Housing Authority • Pulse Height Amplitude • Pulse Height Analysis • Pulse Height Analyzer

PHAB Photon Application Build

PHAK Pilot's Handbook of Aeronautical Knowledge

PHAMOS Primate Hemodynamics and Metabolism on an Orbiting Satellite

Phar Pharmaceutical • Pharmacology • Pharmacopoeia • Pharmacy

PHARAO Projet d'Horloge Atomique à Refroidissement d'Atomes en Orbite

PharB Bachelor of Pharmacy

PharD Doctor of Pharmacy

Pharm Pharmaceutical • Pharmacology • Pharmacopoeia • Pharmacy

PharM Master of Pharmacy

PharmB Bachelor of Pharmacy

PharmD Doctor of Pharmacy

PharmM Master of Pharmacy

PHARO Palomar High-Angular-Resolution Observer

PHAROS Phased-Array Radar Operational Simulation • Proposal for a High Angular Resolution Optical Survey

PHARS Phased Array SAR

PHARUS Phased Array Universal SAR

Phas Phase

PhASp Physics and Astrophysics from Spacelab (periodical code)

PHAST Physics-Astronomy

PHAZ Poleward-moving High-latitude Activity Zone

PhB Bachelor of Philosophy • Physics Bulletin (periodical code)

PHB Parnaíba - Santos Dumont (Brazil-PI) (location identifier) • Per-Hop Behavior • Photographic Bulletin

PhC Pharmaceutical Chemist

PHC Pacific Hurricane Center • Photographic Change • Port Harcourt (Nigeria) (location identifier) • Primary Health Care • Probable Hydrologic Consequence • Probe Heater Controller • Public Health Committee

PHCA Proyecto Hidrometeorológico Centroamericano

PhChE Physics and Chemistry of the Earth (periodical code)

PhChH Physico-Chemical Hydrodynamics (periodical code)

PhD Doctor of Philosophy • Philosophiae Doctor

PHD New Philadelphia - Harry Clever (USA-OH) (location identifier) • Payload Handling Device • Phase Shift Driver • Photohydrodynamic • Project History Document • Pulse Height Discrimination • Pulse Height Distribution (*)

PhDEd Doctor of Philosophy in Education

PHDR Preliminary Hardware Design Review

PHDS Parallel Hypertext Data Structure

PhDT PhD Thesis

PHDU Primary Header Data Unit

Phe Phoenix-Phoenicis (constellation) • Photoelectric

PHE July Phoenicids (meteor stream) • Port Hedland (Australia-WA) (location identifier) • Preflight Heat Exchanger • Public Health Engineer

PHEDRE Partenariat Hermes de Développement Régional

PHEL Patton Hawksley Education Ltd. (UK)

PHEMT Pseudo-morphic High Electron Mobility Transistor

PHEMU Phénomènes Mutuels

Phen Phenolic

PHEN Phenomenon

PHENIX Pioneering High-Energy New Ion Experiment • Pioneering High-Energy Nuclear Interaction Experiment

PHF Hampton (USA-VA) (location identifier) • Newport News (USA-VA) (location identifier) • Parts History File • Payload Handling Fixture • Perfect Hash Function • Personal History Form • Personal Hygiene Facility • Plug Handling Fixture • Williamsburg (USA-VA) (location identifier)

PhFl Physics of Fluids (periodical code)

PhFlB Physics of Fluids B (periodical code)

PhG Graduate in Pharmacy

PHH Phan Thiet (Viet Nam) (location identifier)

PHHC Programmable Hand-Held Calculator

PHI Packet Handler Interface • Permanent Health Insurance • Photonic Hit Indicator • Pinheiro (Brazil-MA) (location identifier) • Procedural Hierarchy of Issues • Public Health Inspector

PHIAT Passive Hybrid Imaging on the AMOS Telescope

PHIC Portable Human Computer Interface

PHIGS Programmer's Hierarchical Interactive Graphics Standard • Programmer's Hierarchical Interactive Graphics System (*) • Programmer's Hierarchical Interface to Graphics Standard

PhiJR Philips Journal Research (periodical code)

Phil Philadelphia (USA-PA) • Philippines • Philosopher • Philosophical • Philosophy

PHIL Personal Hot Interest List

Phila Philadelphia (USA-PA)

Phil. I. Philippines Islands

Philipp Philippines

Phil. Is. Philippines Islands

Philol Philological • Philology

Philos Philosopher • Philosophical • Philosophy

Phil. Trans. Philosophical Transactions

PhInt Photo Interpretation (periodical code)

PHIP Photo-Injecteur Prototype

PhiRR Philips Research Reports (periodical code)

PHIS Population and Housing Information System

PhiTR Philips Technical Review (periodical code)

PHIVOLCS Philippines Institute of Volcanology and Seismology

PHJ Port Hunter (Australia-NS) (location identifier)

PHK Pahokee - Palm Beach County (USA-FL) (location identifier) • Personal Hygiene Kit

PhL Licentiate in Philosophy • Licentiate of Philosophy • Physics Letters (periodical code)

PHL Palomar - Haro - Luyten (survey) • Peculiar Hydrogen Line • Philadelphia (USA-PA) (location identifier) • Philippines (ISO country code) • Preliminary Hazards List • Presse Hebdomadaire Locale

PhLA Physics Letters A (periodical code)

PhLB Physics Letters B (periodical code)

PhLC Physics Letters C (periodical code)

PhLF Physics Laser Fusion (periodical code)

PHLIS Public Health Laboratory Information System

PHLS Public Health Laboratory Service (UK)

PhM Master of Philosophy • Phase Modulation

PHM Patterson-Harker Method • Per Hundred Million • Phase Meter • Phase Modulation • Power Hybrid Microcircuit • Prognostic Health Monitoring • Prognostics and Health Management • Pump Head Module • Pyrheliometer

PHMC Probe Heater Motor Controller

PHN Plan Hidrológico Nacional (Spain) • Port Huron - Saint Claire County (USA-MI) (location identifier) • Presse Hebdomadaire Nationale

PHNEO Potentially Hazardous Near-Earth Object

PHO December Phoenicids (meteor stream) • Phone • Photometric • Point Hope (USA-AK) (location identifier) •

Potentially Harzardous Object • Public Health Observatory

PHOEBE Photochemistry for Exobiology and Exoplanets (code)

Phoen Phoenix - Mitteilungsblatt für Veränderlichenbeobachter (periodical code)

PHOKA Photosphärenkamera

Phon Phonetics

Phonet Phonetics

Phono Phonograph

Phonol Phonology

Phos Phosphate

PHOSI Preliminary Handbook of Operations and Service Instructions

PhoSp Photonics Spectra (periodical code)

Phot Photograph • Photographer • Photographic • Photographical • Photographie • Photographique • Photography • Photometric

PHOTINT Photographic Intelligence

Photo Photograph • Photographer • Photographic • Photographical • Photographie • Photographique • Photography • Photometric

Photog Photograph • Photographer • Photographic • Photographical • Photographie • Photographique • Photography

Photogr Photograph • Photographer • Photographic • Photographical • Photographie • Photographique • Photography

Photom Photometric • Photometrical • Photométrie • Photométrique • Photometry

PHOXA Photochemical Oxidant and Acid (deposition)

php Pounds per Horsepower • Pump Horsepower

PHP Pacific Home Phone (system) • Page Hypertext Processor • Payload Handling Panel • Personal Handy Phone • Personal Homepage • Peso (Philippines) (ISO currency code) • Philip (USA-SD) (location identifier) • Planetary Horizon Platform • Planetary Horizontal Platform • Pounds per Horsepower • Preferred Hiring Program

PHPA Professional Helicopter Pilots Association

PHPL Parallel Hardware Processing Language

PHQ Paper of High Quality • Postal Headquarters • Proportion of High Quality

PHQP Proportion of High-Quality Papers

Phr Phrase

PhR Physical Reports (periodical code) • Physical Reviews (periodical code)

PHR Pacific Harbour (Fiji) (location identifier) • Payload Hazardous Report • Pound-force per Hour • Power-to-Heat Ratio • Presse Hebdomadaire Régionale • Pulse-Height Resolution

PHRI Public Health Research Institute

PHRIR Pointable High Resolution Imaging Radiometer

PHRK Power and Heat Rejection Kit

Phrm Pharmacy

PHRM Personnel and Human Resources Management

PHRR Payload High-Rate Recorder

PHRS Portable Heat Rejection System • Proposal Handling and Reporting System

PhRv Physical Review (periodical code)

PhRvA Physical Review A (periodical code)

PhRvB Physical Review B (periodical code)

PhRvC Physical Review C (periodical code)

PhRvD Physical Review D (periodical code)

PhRvE Physical Review E (periodical code)

PhRvL Physical Review Letters (periodical code)

PHS Packaging - Handling - Storage • Payload Handling Station • Payload Handling System • Personal Handyphone Service • Personal Handyphone System • Personal Hygiene System • Phitsanulok (Thailand) (location identifier) • Photohydrography System • Primary Heavy Shelter • Proposal Handling System • Public Health Service

PHSA Public Health Service Act (USA)

PHSC Physical Sciences Planning Unit

PHSF Payload Hazardous Servicing Facility

PhSFK Phase-Shift Keying

PHSIG Phenomenology Special Interest Group

PHSL Program Hardware Software List

PHSPS Preservation, Handling, Storage, Packaging and Shipping

PhST Physica Scripta Volume T (periodical code)

PHS&T Packaging, Handling, Shipping & Transportation • Packaging, Handling, Storage & Transportation

PHST Packaging, Handling, Shipping, and Transportation • Packaging, Handling, Storage, and Transportation

PHST&D Packaging, Handling, Storage, Transportation & Disposal

PHST Packaging, Handling, Storage, Transportation, and Disposal

PhT Physics Today (periodical code)

PH&T Packaging, Handling & Transportation

PHT Packaging, Handling and Transportation • Paris - Henry County (USA-TN) (location identifier) • Plötzlicher Herztod • Port Hold Time

PhTec Physics Technology (periodical code)

PHTR Packaging, Handling, and Transportation Record

PHT&S Packaging, Handling, Transportation & Storage

PHTS Packaging, Handling, Transportation, and Storage

PHU Phu Vinh (Viet Nam) (location identifier) • Primary Header Unit

PHVPS Primary High-Voltage Power Supply

PHW Phalaborwa (South Africa) (location identifier)

PHWR Pressurized Heavy-Water Reactor

Phx Phoenix (USA-AZ)

PHX Phoenix (USA-AZ) (location identifier)

Phy Physica (periodical code) • Physics • Physik • Physique

PHY Physical • Physics

PhyA Physica A - Statistical Mechanics and its Applications (periodical code)

PhyB Physica B - Condensed Matter (periodical code)

PhyC Physica C - Superconductivity (periodical code)

PhyD Physica D - Non-linear Phenomena (periodical code)

PhyE Physica E - Low-Dimensional Systems and Nanostructures (periodical code)

PhyEd Physics Education (periodical code)

PhyEs Physics Essays (periodical code)

PhyL Physics Letters (periodical code)

PhylS Physiologist Supplement (periodical code)

Phys Physica (periodical code) • Physical • Physically • Physician • Physicist • Physics • Physiological • Physiologist • Physiology • Physique

PhyS Physica Scripta (periodical code)

PhysB Physica B (periodical code)

PhySB Physica Scripta B (periodical code)

PhysC Physica C (periodical code)

PhysD Physica D (periodical code)

PHYSEC Physical Security

Physiol Physiological • Physiologist • Physiology

PHYSnet High-Energy Physics Network

Phys. Rev. Physical Review

PhysS Physica Section (periodical code)

Phys. Sc. Physical Science

PHz Petahertz

PHZ Pannenhilfezentrale • Phi Phi Island (Thailand) (location identifier)

PH3 Phosphine

pi Par Intérim

Pi Piaster (currency) • Piastre (currency)

P&I Performance & Interface • Planning & Implementation • Privileges & Immunities

PI Charon (satellite of Pluto) • Parallel Input • Par Intérim • Patent Information • Path Integral • Payload Interrogator • Pen and Ink • Pending Interrupt • Performance and Interface • Performance Improvement • Performance Index • Performance Indicator • Periodic Inspection • Peripheral Interface • Personal Identification • Personal Injury • Personal Investigador • Philippine Islands • Photographic Intelligence • Photographic Interpretation • Photographic Interpreter • Photointerpretation • Photointerpreter • Photoionization • Physik Instrumente (Germany) • Piauí (Brazil) • Planetary Imager • Planet Imager • Plan Information • Point Initiating • Point Insulating • Point of Impact • Point of Interception • Point of Intersection • Politically Incorrect • Polynomial Interpolation • Position Indicator • Position Invariant • Power Injection • Precedence Indicator • Power Input • Precipitation Index • Precision Instrument • Predicted Impact • Pré-Inscription • Preliminary Impulse • Preliminary Inspection • Preliminary Investigation • Preliminary Issue • Pressure Indicator • Préstamo Interbibliotecario • Primary Impulse • Primary In • Primary Input • Prime Item • Principal Investigator • Priority Index • Priority Interrupt • Pristine Ice • Private Investigator • Probability of Incapacitation • Procedural Item • Procedure Interface • Process Instruction • Process Intermittent • Process of Induction • Procurement Instruction • Procurement Item • Product Improvement • Productivity Index • Program Indicator • Program Instruction • Program Integrator • Program Interrupt • Program Interrupter • Program Interruption • Program Introduction • Programmed Information • Programmed Instruction • Programmed Introduction • Program of Instrumentation • Progress Indicator • Project Instruction • Proportional Integral • Propriété Intellectuelle • Protocol Interpreter • Pseudo-Intelligence • Pubblica Istruzione • Public Information • Pulse Invariance • Pulse Invariant

PI&A Personnel Identification & Authentication

PIA Particle Impact Analyser • Payload Interface Adapter • Peoria (USA-IL) (location identifier) • Peripheral Interface Adapter • Personnel Identification and Authentication • Personal Investment Authority • Photometers for Imaging the Aurora • PHOT Interactive Analysis • Plug-In Administrator • Plug-In Amplifier • Portable Image Analizer • Positive Ion Accelerator • Poste d'Indexation Automatique • Prácticas de Investigación en Astronomía (project) • Preinspection Acceptance • Preinstallation Acceptance • Pressure-Induced Amorphization • Processor Interface Adapter • Project Impact Analysis • Public Information Access • Public Information Act (USA)

PIAB Particle In A Box

PIACH Proactive Interactive Adaptive Computer Help

PIAF Placement Interactif et Automatique de Formes

PIAG Publications de l'Institut d'Astronomie et de Géophysique de Louvain (periodical code)

PIAMA Photo-Induced Activation of mm-wave Absorption

PIANC Permanent International Association of Navigation Congresses

PIAR Project Impact Analysis Report

PIARC Permanent International Association of Road Congresses

Pias Piaster (currency)

PIB Laurel Pine Belt (USA-MS) (location identifier) • Particle In a Box • Payload Integration Bay • Polar Ionospheric Beacon • Policy Information Base • Poste d'Indexation de Bureau • Preliminary Instruction Book • Production Intérieure Brute • Producto Interior Bruto • Produit Intérieur Brut (*) • Program Information Briefing • Propellant Inspection Building • Pyrotechnic Installation Building

PIBE Process Integration in Biochemical Engineering

PIBOR Paris Interbank Offered Rate

PIBS Polar Ionospheric Beacon Satellite

pic Picture (file extension)

Pic Pictor-Pictoris (constellation) • Pictorial • Picture

P-I-C Particle-In-Cell

PIC Parent Indicator Code • Particle-In-Cell • Particulate Inorganic Carbon • PASSCAL Instrument Center • Payload Integration Center • Payload Integration Committee • Payload Integration Contractor • Performance Indicator Chart • Peripheral Interface Card • Peripheral Interface Control • Personal Identification Code • Personal Information Carrier • Personal Intelligent Communicator • Personal Interest Checklist • Personnel Investigations Center • Phase Tracking Circuit • Photographic Industry Contract • Photointerpretation Console • Photonic Integrated Circuit • Pilot In Command • Pine Cay (Turks and Caicos Islands) (location identifier) • Plastic-Insulated Cable • Polyethylene-Insulated Cable • Position Independent Code • Positive Ion Chamber • Potential Impact Category • Precision Industrial Components Corporation (USA-CT) • Preferred Interexchange Carrier • Preinstallation Calibration • Preinstallation Checkout • Preliminary Interface Certification • Pressure Indicator Controller • Primary Independent Carrier • Primary Interexchange Carrier • Primordial Isocurvature • Principal Interexchange Carrier • Prior Informed Consent • Priority Interrupt Controller • Product Information Catalog • Product of Incomplete Combustion • Professional Interest Committee • Program Identification Code • Program In Computing • Program Information Center • Program Interrupt Control • Program Interrupt Controller • Programmable Integrated Circuit • Programmable Interconnect Component • Programmable Interrupt Controller • Programmable Interval Clock • Programme d'Initiative Communautaire • Programme Interuniversitaire de Coopération • Proton-Initiated Cascade • Public Information Center • Pulsed Ionization Chamber • Pyro Ignition Control • Pyro Initiator Capacitor • Pyro Initiator Controller • Pyrotechnic Ignition Control • Pyrotechnic Initiator Capacitor • Pyrotechnic Initiator Circuit • Pyrotechnic Initiator Controller

PICA Performance-enhanced I/O and CPU Architecture • Phenolic Impregnated Carbon Ablator • Picosecond Imaging Circuit Analysis • Platform-Independent Cryptography API • Preintegrated Columbus Assembly • Preintegrated Columbus ATM • Preliminary Inventory Control Activity • Press Independence and Critical Ability • Primary Inventory Control Activity • Primary Inventory Control Agency • Primary Item Control Activity • Program Issues and Criteria Annex • Project for Integrated Catalogue Automation

PICAD Program Integrator's Contractor Assessment Data

PICAO Provisional International Civil Aviation Organization (now ICAO)

PICAPI Platform-Independent Cryptography Application Programming Interface

PICASSO Pathfinder Instruments for Cloud and Aerosol Spaceborne Observation

Picc Piccolo

PICC Payload Interface Control Computer • Peripheral Interface Controller Card

PICCA Positive Ion Cluster Composition Analyzer

PICCG Parallelized Incomplete Choleski Conjugate Gradient (method)

PICD Platform Interface Control Document

PICE Programmable Integrated Control Equipment

PICES Programmed Interactive Cost Estimating System

PICF Payload Integration and Checkout Facility • Principal Investigator Computing Facility

PICG Programme International de Corrélation Géologique

PICMA Program for International Cooperation in Mathematics and its Applications

PiCO Portable Interactive Computing Object

PICO Pic du Midi Coronagraph

PICOE Program - Initiations - Commitments - Obligations - Expenditures

PICP Program Interface Control Plan

PICRS Program Information Control and Retrieval System • Program Information Coordination and Review Service

PICS Parts Interpretive Catalog System • Perpetual Inventory Control System • Personnel Information Communication System • Personnel Information Control System • Photographic Integrated Control System • Photo Index and Cataloguing System • Photo Indexing and Cataloguing System • Pioneer Image Converter System • Planetary Image Cartography System • Planetary Integrated Camera-Spectrometer • Platform for Internet Content Selection • Plug-in Inventory Control System • Predefined Input Control Sequence • Procurement and Inventory Control System • Procurement Information Control System • Production Information and Control System • Production Inventory Control System • Programme International de Coopération Scientifique • Protocol Implementation Conformance Statement • Protocol Information Conformance Statement

PICsIT Pixellated Imaging Cesium Iodide Telescope

pict Picture (file extension)

Pict Picture

PICT Picture • Programme in Information and Communication Technologies • Pune Institute of Computer Technology (India)

PICU Pediatric Intensive Care Unit • Priority Interrupt Control Unit

PiD Proportionnel Intégral Dérivé

P&ID Piping & Instrumentation Diagram • Piping & Instrumentation Drawing • Process & Instrumentation Diagram

PID Nassau - Paradise Island (Bahamas) (location identifier) • Parameter Identification • Passive Identification Device • Patentinformationsdienst • Payload Insertion Device • Permanent Station Identifier • Persistent Identifier • Personal Identification Device • Photoionization Detector • Pictorial Information Digitizer • Position Identifier • Position Integral Derivative • Prestation Invalidité Décès • Prime Item Development • Privater Informationsdienst • Process Identification • Process Identifier • Procurement Identification Description • Procurement Item Description • Program Identifier • Program Implementation Directive • Program Information Document • Program Introduction Document • Programmable Interface Device • Project Initiation Document • Proportional - Integral - Derivative • Proportional Integral Deviation • Proportional - Integral - Differential • Proportional - Integral - Differentiated • Proportional Integrated-Derivative • Protocol Identification • Protocol Identifier • Prototype International Directory • Public Information Division

PIDA Payload Installation and Deployment Aid • Photonics Industry and technology Development Association

PIDAS Portable Instant Display and Analysis Spectrometer • Portable Intelligent Display and Analysis Spectrometer

PIDDP Planetary Instrument Definition and Development Program

PIDP Portable Image Display System • Product Initialization Data Pack • Programmable Indicator Data Processor

PIDS Parameter Inventory Display System • Pattern Identification and Detection System • Polarization Intensity Differential Scattering • Portable Image Display System • Prime Item Development Specification

PIDT Photometric Instrument Dedicated Team

PIE Clearwater (USA-FL) (location identifier) • Parallel Instruction Execution • Payload Integration Equipment • Payload Interface Engineer • Paysage Internet Européen • Personal Integrated Electronics • Personal Interactive Electronics • Physics Information Exchange • Plug-In Electronics • Polar Ice Extent • Positive-Ion Emission • Postirradiation Examination • Process Improvement Experiment • Process-Integrated Environment • Product Integrity Engineering • Program Interrupt Entry • Programmable Interface Electronics • Programme International de l'Énergie • Prompt Initial Enrichment • Protoindoeuropean • Pulse Interference Emitting • Saint Petersburg (USA-FL) (location identifier)

PIECP Preliminary Impact Engineering Change Proposal

PIEMD Pseudo-Isothermal-Elliptical Mass Distribution

PIEP Pseudo-Isothermal Elliptic Potential

PIER Project Information and Evaluation Report

PIES Planning-Implementation-Evaluation System • Program Information and Evaluation System

pif Picture Interchange Format (file extension)

PIF Paid In Full • Payload Integration Facility • Paysage Internet Français • Personal Identification Feature • Photo Interpretation Facility • Picture Interchange Format • Pilot Information File • Pingtung (Taiwan-ROC) (location identifier) • Plan Individual de Formation • Postirradiation Facility • Process Information File • Productivity Improvement Fund • Productivity Investment Funding • Program Information File • Provision of Industrial Facilities • Pulsed Ionospheric Flow

PIFOV Planet In Field Of View

PIFP Pan-Institutional Formation Program • Personal Investigador en Formación y Perfeccionamiento • Programme d'Intégration et de Formation Professionnelle • Public Interest Fellowship Program

PIFS Palomar Integral Field Spectrograph • Process Interchange Format Standard

PIFT Package for Information Formatting and Transformation

PIG Partially Ionized Globule • Pendulous Integrating Gyro • Pitinga (Brazil-AM) (location identifier) • Postburst Infrared Galaxy • Primary Image Guider • Process Ink Gamut • Product Information Guide • Projet d'Intérêt Général

PIGA Pendulous Integrating Gyro Accelerometer

PIGB Programme International Geosphère-Biosphère

PIGE Proton-Induced Gamma-ray Emission

PIGMI Position Indicating General Measuring Instrument

Pigmt Pigment

PIGS Portable Inertial Guidance System

PIGU Pendulus Integrating Gyro Unit

PIGWAD Planetary Interactive GIS-on-the-Web Analyzable Database

PIH Pocatello (USA-ID) (location identifier) • Public and Indian Housing

PIHC Plate-forme d'Insertion pour l'Humanitaire et la Coopération

PII Fairbanks Phillips (USA-AK) (location identifier) • Paysage Internet International • Persatuan Insinyur Indonesia • Personally Identifiable Information • Positive Immittance Inverter • Procurement Instrument Identification • Programa Intergubernamental de Informatica • Program Integrated Information • Publisher Item Identification • Publisher Item Identifier (*)

PIIC Politique de l'Information, de l'Informatique et des Communications • Proyecto de Investigación Industrial Concertada (Spain)

PIII Plasma Immersion Ion Resonance

PIIN Procurement Instrument Identification Number • Procurement Item Identification Number

PIIT Programa Interamericano de Información Popular

PIJR Program Improvement Joint Review

PIK Glasgow Prestwick (UK-Scotland) (location identifier) • Payment In Kind • Potsdamer Institut für Klimaforschung (Germany)

PIKS Programmer's Imaging Kernel System

pil Pileus

PIL Passenger Information List • Payment In Lieu • Personal Interest List • Pilar (Paraguay) (location identifier) • Pitt Interpretive Language • Played Item List • Polarity Inversion Line • Preferred Item List • Processing Information List • Procurement Instruction Letter • Programme d'Insertion Locale

PILL Programmed Instruction Language Learning

PILO Public Information Liaison Officer

PILOT Phase-Integrated Laser Optics Technology • Program Integration and Loading Tool • Programmed Inquiry, Learning Or Teaching

PILP Parametric Integer Linear Program

PILS Payload Integration Library System • Payload Interactive Library System

PIM Parallel Inference Machine • Parametrized Ionosphere Model • Parametrized Ionospheric Model • Passive Intermodulation • Payload Integration Manager • Paysage Internet Mondial • Peripheral Interface Module • Personal Information Management • Personal Information Manager • Pine Mountain - Garden Harris County (USA-GA) (location identifier) • Planning Interchange Meeting • Plug-In Module • Point of Intended Movement • Position and Intended Movement • Precision Indicator of the Meridian • Precision Instrument Mount • Presentation and Interaction Module • Prestation Interministérielle • Pretrained Individual Manpower • Previously Intended Movement • Primary Interface Module • Processor-In-Memory • Product Information Management • Program Integration Manual • Programmatic Interchange Meeting • Programme Intégré Méditerranéen • Protocol-Independent Multicast • Pulse Intensity Modulation • Pulse Interval Modulation

PIMA Photographic and Imaging Manufacturers Association • Plug-In Module Adapter • Plug-In Module Assembly • Portable Interactive Maintenance Aid

PiMF Paar in Metall-Folie

PIMOS Parallel Inference Machine Operating System • Personal Information Manager Operating System

PIMMS Portable Interactive Multimission Simulator • Project to Increase Mastery of Mathematics and Science

PIMR Polar Ice Mapping Radiometer • Program Integration Management Review • Pushbroom Imaging Microwave Radiometer

PIMRC Personal, Indoor and Mobile Radio Communications (symposium)

PIMS Pacific Institute for the Mathematical Sciences • Payload Information Management System • Personnel Information Management Service • Photoionization Mass Spectroscopy • Procurement Information Management System • Profit Impact of Market Strategy • Project Information Management System

PIMSL Personal Information Manager System Language

PIN Pacific Island Network • Page and Item Number • Parcel Identification Number • Parintins (Brazil-AM) (location identifier) • Particle-Induced Noise • Password Identification Number • Patent Information Network • Personal Identification Number • Personal Increment Number • Personal Item Number • Pesticide Information Network • Piece Identification Number • Plan Identification Number • Positive Intrinsic Negative (diode) • Preinvitation Notice • Primary Interactive Network • Process Identification Number • Processor-Independent NetWare • Product Identification Number • Production Identification Number • Produit Intérieur Net • Program Identification Number • Programme Indicatif National • Proposal Identification Number

PINC Printer's Industry of Northern California (USA-CA) • Programme Indicatif Nucléaire pour la Communauté

PINCH Payload Integration Nominal Cost Hardware

PIND Particle Impact and Noise Detection • Particle Impact Noise Detection

PINE Pine Is Not Elm

PINET Physics Information Network

PING Packet Internet Groper

PINO Positive Input - Negative Output

PINOCCHIO Pinpointing Otbit-Crossing Collapsed Hierarchical Objects

PINR Peak Intensity Response

PINS Person In Need of Supervision • Personnel Information System • Pipeline Inspection Notification System • Point In Space • Precise Integrated Navigation System • Proliferation Information Network System

Pint Pintura

PIO Parallel Input/Output • Pilot-Induced Oscillation • Pisco (Peru) (location identifier) • Policy Implementation Office • Position Iterative Operation • Press Information Officer • Prime Interoperability • Processor Input/Output • Production Intégrée par Ordinateur • Program Integration Office • Programmable Input/Output • Programmed Input/Output • Project Implementation Order • Provisioned Item Order • Public Information and Outreach • Public Information Office • Public Information Officer • Public Inquiries Office

PIOCS Physical Input-Output Control System

PIOL Periodic Input/Output Link

PION Physics In Our Neighborhood

PIONIER Persoonsgerichte Impuls voor Onderzoekgroepen met Nieuwe Ideeën voor Excellente Research

PIOP Payload Integration Operations Plan

PIOS Planning Information Overlay System

PIOT Plan Insular de Ordenación del Territorio (Canary Islands)

PIOU Parallel Input/Output Unit

PIOUS Parallel Input/Output System

PIP Packet Interface Port • Packet Interface Processor • Parallel Image Processor • Participant Instrumentation Package • Payload Integration Plan • Payload Interface Plan • PCM Interface Port • Performance Improvement Program • Peripheral Interchange Program • Persistent Internal Polarization • Personal Incentive Program • Personal Identification Project • Personal Interview Program • Personal Investigador de Plantilla • Petite Introduction Préliminaire • Photo Interpretive Program • Picture In Picture • Pilot Point (USA-AK) (location identifier) • Planning Integration Process • Plant Instrumentation Program • Policy Improvement Program • Polymer Impregnation and Pyrolysis (process) • Portable Instrument Pedestal • Position-Indicating Probe • Power Input Panel

• Precipitation Intercomparison Project • Predicted Impact Point • Preliminary Impact Point • Preliminary Information Pamphlet • Presence Information Protocol • Pretty Important Person • Primary Indicating Position • Primary Injection Point • Priority Information Programme • Probabilistic Information Processing • Probability Information Processing • Problem Input Preparation • Problem Isolation Procedure • Procedural Information Pamphlet • Process Improvement Plan • Product Improvement Phase • Product Improvement Plan • Product Improvement Program • Product Improvement Proposal • Product Information Program • Product Integration Program • Production Instrumentation Package • Productivity Investment Program • Professional Interest Profile • Program Implementation Plan • Program Integration Plan • Program Introduction Plan • Program In Process • Programmable Integrated Processor • Programmable Interconnect Point • Programme d'Information Prioritaire • Programmed Instructional Package • Project Implementation Plan • Project on Information Processing • Probability Information Processing • Project Information Package • Project Initiation Period • Proof In Print • Proportional - Integral - Plus • Proposal Implementation Plan • Proposal Information Package • Proposal Instruction Package • Proposer's Information Package • Public Information Package • Pulsed Integrating Pendulum • Pulse Interval Processor • Pyro Interception Plug

PIPA Pulse-Integrating Pendulum Accelerometer • Pulse-Integrating Pendulum Assembly

PIPAC Philippines Institute for Pure and Applied Chemistry

PIPD Planning Input for Program Development

PIPE Parallel Image Processing Engine • Product Integrity and Production Engineering • Pulsed Irradiance of a Particle Element

PIPO Parallel In - Parallel Out

PIPOR Programme for International Polar-Ocean Research

PIPPS Publication Information Processing and Printing Standard • Publication Information Processing and Printing System

PIPPY Person Inheriting Parent's Property

PIPS Passivated Implanted Planar Silicon • Passively Ion-Planted Silicon (detector) • Personal Image Processing System • Plans Integration Partition System • Portable Interactive Planning System • Postinjection Propulsion System

PIPSS PPARC Industrial Programme Support Scheme

PIPT Program Integrated Product Team • Program-level Integrated Product Team

PIQ Person In Question • Pipillipai (Guyana) (location identifier)

PIR Page Information Request • Peak Intensity Ratio • Peak Irradiance Response • Phase Increment Register • Pierre (USA-SD) (location identifier) • Precision Infrared Radiometer • Pressure Ignition Rocket • Primary Image Record • Primary Intelligence Requirements • Priority Information Requirement • Priority Intelligence Report • Priority Intelligence Requirement • Problem Identification Report • Problem Information Report • Program Integration Review • Programme Interdisciplinaire de Recherche • Programming Information Requirement • Project Independence Report • Protein Identification Resource • Protocol-Independent Routing • Publications and Information Resources

PIRAMHYD Passive Infrared Atmospheric Measurement of Hydroxyl

PIRAQ PC Integrated Radar Acquisition

PIRC Pensions Investment Research Council (UK)

Pirchnx Pironchêneux (Belgium)

PIRCS Payload Interactive Requirements Collection System • Project to Intercompare Regional Climate Simulations

PIRD Program Instrumentation Requirements Document • Project Instrumentation Requirements Document

PIRE Puissance Isotrope Rayonnée Équivalente

PIREN Programme Interdisciplinaire de Recherche sur l'Environnement

PIRENEA Piège à Ions pour la Recherche et l'Étude de Nouvelles Espèces Astrochimiques

PIREP Pilot Report

PIRG Public Interest Research Group

PIRL Patent Information Retrieval Language • Planetary Image Research Laboratory

PIRMAT Programme Interdisciplinaire de Recherche sur les Matériaux

PIRN Preliminary Interface Revision Notice • Proposed Interface Revision Notice

PIROCEA Programme Interdisciplinaire de Recherche Océanographique

PIROG Pointed Infrared Observing Gondola

PIROLA Pilotage Robuste des Lanceurs

Pironch Pironchêneux (Belgium)

PIRR Parts Installation and Removal Record • Problem Investigation and Repair Record

PIRS Personal Information Retrieval System • Photo-Image Retrieval System • Productivity Information Reporting System • Program Information Retrieval System • Project Information Retrieval System

PIRSEM Programme Interdisciplinaire de Recherche sur les Sciences pour l'Énergie et les Matières Premières

PIRMAT Programme Interdisciplinaire de Recherche sur les Matériaux

PIRT Precision Infrared Tracking • Precision Infrared Triangulation • Projet d'Inventaire des Ressources Terrestres

PIRTTEM Programme Interdisciplinaire de Recherche sur la Technologie, le Travail, l'Emploi et les Modes

PIRU Public Information Reference Unit

PIS Patent Information Service • Personal Information System • Personalinformationssystem • Pointable Imaging Spectrometer • Poitiers Briard (France) (location identifier) • Positive Ion Source • Poste d'Indexation Simplifié • Process Instrument Sheet • Product Information Specialist • Prospero Interface Server • Provisional International Standard • Pulse Integration System • Purchase Information System

PISA Position, Intensity and Shape Analysis • Programme for International Student Assessment (OECD) • Prototype Image Spectrum Analyzer

PISCO Polarimeter with Instrumental and Sky Compensation • Polarization with Instrumental and Sky Compensation

PISEMA Polarization Inversion Spin Exchange at the Magic Angle

PISH Program Instrumentation Summary Handbook

PISN Pair-Instability Supernova • Private Integrated Services Network

PISO Parallel In - Serial Out

PISP Polar Ice Sheet Program

PIST Plasma Initiation Separated from the Projectile

PIstO Publications of Istanbul University Observatory (periodical code)

PISW Process Interrupt Status Word

PIT Page Information Transfer • Passive Intermediate Tracker • Paul Ion Trap • People - Ideas - Tools • Performance Improvement Test • Peripheral Input Tape • Personal Income Tax • Phase Inversion Temperature • Phase I Tool • Photographic Interpretation Technique • Photointerpretation Tool • Pilot Instructor Training •

Pittsburgh (USA-PA) (location identifier) • Portable Input Tracer • Postinstallation Test • Preinstallation Test • Preintegrated Truss • Principal - Interests - Taxes • Prioritized Image Transmission • Process Improvement Team • Processing Index Terms • Processor Interface Table • Product Improvement Test • Program Instruction Tape • Product Integration Team • Programmable Interval Timer • Programmed Instruction Test • Progressive Image Transmission • Projected Inactive Time • Project Interface Team • Proposal Implementation Team • Pulsed Inductive Thruster

PITAC President's Information Technology Advisory Committee (USA)

PITCOM Parliamentary Information Technology Committee

PITE Plug-In Test Equipment

PITG Payload Integration Task Group

PITMAS Paul Ion Trap Mass Spectrometer

PITN Pain In The Neck

PITS Payload Integration Test Set • Payload Interface Test System

PITT PGS Integration and Test Team

Pitts Pittsburgh (USA-PA)

PIU Path Information Unit • Payload Interface Unit • Photographic Interpretation Unit • Piura (Peru) (location identifier) • Plasma Interface Unit • Plug-In Unit • Power Interface Unit • Power Isolation Unit • Process Input Unit • Processor Interface Unit • Programmable Interface Unit • Project Implementation Unit • Pyro Initiator Unit • Pyrotechnic Initiator Unit

PIUP Planetary Instrument Update Program

PIUS Process-Inherent Ultimate Safety

PIV Partial Image Velocimeter • Partial Image Velocimetry • Particle Image Velocimeter • Particle Image Velocimetry • Particle Imaging Velocimetry (*) • Payload Integration and Verification • Payload Interface Verification • Peak Inverse Voltage • Personal Independent Vehicle • Philippines Institute of Volcanology • Planet In View • Plan International de Validation • Plena Ilustrita Vortaro • Product Improvement Verification

PIVEX Platform-Independent VLBI Exchange (format)

PIVR Payload Interface Verification Requirement

PIVS Payload Interface Verification Summary • Payload Integration and Verification Simulator

PIVT Product Insurance Validation Test

PIV&V Processing Independent Verification & Validation

PIVV Processing Independent Verification and Validation

PIW Person In Water • Pikwitonei (Canada-MB) (location identifier)

PIWG Performance Interface Working Group • Performance Issue Working Group • Permanent Interoperability Working Group • Platform Integration Working Group • Product Improvement Working Group • Propulsion Instrumentation Working Group

PIX Pico Island (Azores) (location identifier)

PIXE Particle-Induced X-ray Emission (*) • Proton-Induced X-ray Emission

PIXEL Picture Element

PIXIE Polar Ionospheric X-ray Imaging Experiment

PIXIT Protocol Implementation Extra Information for Testing

PIXS Preaward Information Exchange System

PIZ Partially Ionized Zone • Point Lay (USA-AK) (location identifier)

PIZA Patch Interchange Zel'dovich Approximation

PJ Netherlands Antilles (international aircraft registration code) • Pajama • Perijove • Période Julienne • Petajoule • Pièce Jointe • Poder Judicial • Police Judiciaire • Project • Provincia Jujuy (Argentina) • Pulsating Jet

PJA Proceedings of Japan Academy (periodical code) • Pulsating Jet Accretion

PJB Payson (USA-AZ) (location identifier)

PJC Pedro Juan Caballero (Paraguay) (location identifier) • Pensacola Junior College (USA-FL)

PJCE Plasma Jet Chemical Etching

Pje Pasaje

PJE Program for Joint Education

PJF Photogrammetric Journal of Finland • Pulsating Jet Force

PJG Panjgur (Pakistan) (location identifier)

PJI Pulsating Jet Impact

PJM Puerto Jimenez (Costa Rica) (location identifier)

PJML Pulsating Jet Mass Loss

PJNAO Publications of the Japan National Astronomical Observatory (periodical code)

PJNF Project Joint Normal Form

PJO Pioneer Jupiter Orbit

PJOP Preliminary Joint Operation Procedure

PJS Port San Juan (USA-AK) (location identifier) • Pulsating Jet Spreading

Pjtr Projector

PJX Pulsating Jet Exchange

PJZ Puerto Juarez (Mexico) (location identifier)

Pk Pack • Package • Park • Peak • Peck

P/K Pluto/Kuiper

PK Indonesia (international aircraft registration code) • Pakistan (international vehicle registration code) • Pakistan (ISO country code) • Park • Peak • Perek-Kohoutek (catalog) • Personal Knowledge • Postkeplerian • Professional Knowledge • Protokollkonverter • Psychokinesis

PKA Napaskiak (USA-AK) (location identifier) • Public Key Algorithm

PKAF Public Key Authentication Framework

PKAS Publication of Korean Astronomical Society (periodical code)

PKB Marietta (USA-OH) (location identifier) • Parkersburg Marietta (USA-WV) (location identifier) • Pluto-Kuiper Belt (mission) • Portable Keyboard

PKBO Plutonian Kuiper Belt Object

PKC Petropavlovsk-Kamchatsky (Russia) (location identifier) • Phase Knife Coronagraph • Public Key Cryptography

PKCat Publ. House Czechoslovak Academy of Sciences (periodical code)

PKCS Public Key Cryptographic Standard • Public Key Cryptography Standard (*)

PKD Parallel k-d Tree • Parallel Knowledge Discovery • Park Rapids (USA-MN) (location identifier) • Programmable Keyboard and Display

PKE Parkes (Australia-NSW) (location identifier) • Plasmakristall-Experiment • Pluto-Kuiper Express • Private Key Encryption • Public Key Encryption

PKF Park Falls (USA-WI) (location identifier)

Pkg Package • Packing

PKG Package • Packaged • Pangkor (Malaysia) (location identifier)

Pkge Package

PKH Porto Kheli - Alexion (Greece) (location identifier)

PKI Philips Kommunikations Industrie AG (Germany) • Private Key Infrastructure • Public Key Infrastructure (*)

PKirO Publications of Kirkwood Observatory (periodical code)

PKIS Public Key Infrastructure Service

PKK Pakokku (Myanmar) (location identifier)

PKL Pakatoa Island (New Zealand) (location identifier)

PKM Perigee Kick Motor • Port Kaituma (Guyana) (location identifier)

PKN Pangkalanbun (Indonesia) (location identifier) • Polski Komitet Normalizacyjny (Poland)

PKNM Polski Komitet Normalizacji Miar

PKO Parakou (Benin) (location identifier) • Peacekeeping Operations

PKoFi Problemy Kosmicheskoi Fiziki (periodical code)

PKP Public Key Partners • Puka Puka (Tuamotu Islands) (location identifier)

Pk-Pk Peak to Peak

PKR Pokhara (Nepal) (location identifier) • Rupee (Pakistan) (ISO currency code)

Pks Parks • Peaks

PKS Paksane (Laos) (location identifier) • Parkes Catalogue • Peaks • Public Key System

PKSD Programmable Key Storage Device

Pkt Packet • Paket • Pocket • Punkt

PKT Packet • Phase Keying Technique • Pocket • Port Keats (Australia-NT) (location identifier)

PKU Pekanbaru (Indonesia) (location identifier)

PKV Pskov (Russia) (location identifier)

PKW Personenkraftwagen • Selebi-Phikwe (Botswana) (location identifier)

Pkwy Parkway

PKWY Parkway

PKWYS Parkways

PKY Palangkaraya (Indonesia) (location identifier) • Parkway

PKZ Pakse (Laos) (location identifier)

pl Perl (file extension) • Pipeline • Place • Poland • Polish (ISO language code)

Pl Plaats • Place • Plain • Planche • Plate • Platz • Plaza • Plug • Plural • Pole

P/L Partial Loss • Parts List • Payload • Postlanding • Profit and Loss • Purchased Labor

P&L Production & Logistics • Profit & Loss

P-L Period-Luminosity (relation)

PL Packing List • Padding Length • Parity Low • Partial Loss • Parting Line • Parts List • Passenger Liability • Payload • Pénétration Locale • Period-Luminosity (relation) • Permanent Link • Personal Librarian • Phase Line • Phillips Laboratory • Photoluminescence • Physical Layer • Physical Link • Piecewise Linear • Piping Load • Place • Plain Language • Plan • Plastic Limit • Plate • Pleine Lune • Plug • Plus • Poids Lourd • Point de Lagrange • Poland (international vehicle registration code) • Poland (ISO country code) • Poland (postal country code) • Pologne (code ISO de pays) • Positive Large • Postlanding • Powered Lift • Power Law • Power Level • Prelaunch • Price List • Private Line • Probability of Leakage • Production Language • Product Liability • Product Licence • Programmed Learning • Programming Language • Project Leader • Propagation Loss • Propellant Loading • Property List • Public Law • Public Library • Purchased Labor

Pla Península • Plaza

PLA Panchromatic Linear Array • Parachute Location Aid • Patent License Agreement • Patent Licensing Agreement • Payload Adapter • Perceptron Learning Algorithm • Perturbative Least Action • Phase Locked Array • Physiological Learning Aptitude • Physique Largement Avantageux • Plain Language Address • Plain Language Addressee • Planadas (Colombia) (location identifier) • Planned Labor Application • Planned Landing Area • Plan of Launch Azimuth • Power Lever Angle • Practice Low Approach • Procurement Liaison Auditor • Programmable Logic Array • Programmed Logic Array • Proton Linear Accelerator • Public Library Association

PLAC Postlaunch Analysis of Compliance

PLACE Position Location Aircraft Communications Equipment • Position Location And Communications Experiment • Post-Landsat Advanced Concept Evaluation • Programming Language for Automatic Checkout Equipment

PLACID Payload Aboard - Caution In Descent

PLAD Plain Language Address

PLADS Plain Language Address System • Pulsed Light Airborne Depth Sounder

PLAID Position Location And Identification • Precision Location And Identification (system)

PLAIL Public Libraries And Independent Learners

PLAIT Pupils' Learning and Access to Information Technology

PLAM Price-Level Adjusted Mortgage

Plan Planet • Planetarium

PLAN Payload Activity Network • Program Language Analyzer • Public Library Access Network • Public Library Automation Network

Planet Planetary

PLANET Packetized Lightwave Architecture Network • Personal Laconic Access Network/Ethereal Transmission • Private Local Area Network • Probing Lensing Anomalies Network

PLANIT Programming Language for Interactive Teaching

PLANS Positioning, Location and Navigation Symposium • Programming Language for Allocation and Network Scheduling

PLAO Poste de Lecture Assistée par Ordinateur

PLAR Private-Line Automatic Ring-down

Plas Plaster • Plastic

PLAS Performance Labor Accounting System • Professional Library Automation System

PLASMAE Plasmas Magnétisés de l'Environnement

PlaSpaSci Planetary and Space Sciences

PlAst Planetary Astronomy (periodical code)

PLAST Propellant Loading and All Systems Test

Plat Plateau • Platform • Platinum

PLAT Pilot Landing Aid Television • Plateau • Platform • Postlaunch Analysis of Telemetry • Prelaunch Acceptance Test

Platf Platform

PLATF Plain Language Terminal Forecast

PLATO Planetarium Learning And Teaching Opportunity • Planetary Analysis Tools • Programmed Learning for Automated Teaching Operations • Programmed Logic for Automatic Teaching Operations (*) • Programmed Logic for Automatic Training Operations

PLATS Postlaunch Analysis of Telemetry Systems

PLATT Page Level Availability Time Test

PLB Payload Bay • Perioden-Leuchtkraft-Beziehung • Personal Locator Beacon • Picture Level Benchmark • Platform Bay • Plattsburgh (USA-NY) (location identifier)

PLBD Payload Bay Door

PLBK Playback

plc Public Ltd. Co.

P-L-C Period-Luminosity-Colour (relation)

PLC Pacific Lumber Co. (USA-CA) • Payload Carrier • Payload Commander • Period - Luminosity - Colour (relation) • Planar Lightwave Circuit • Planeta Rica (Colombia) (location identifier) • Power-Line Carrier • Power-Line Communications • Power-Line Conditioner • Preliminary Load Cycle • Pressurized Logistics Carrier • Probe Launch Complex • Product Life Cycle • Professorat des Lycées et des Collèges • Programmable Logic Control • Programmable Logic Controller • Programmation Logique avec Contraintes • Programming Language Committee • Propellant Loading Console • Public Limited Company

PLCA Product Life Cycle Assessment

PLCB Pseudoline Control Block

PLCC Payload Launch Control Center • Plastic Leaded Chip Carrier (*) • Plastic Leadless Chip Carrier • Provisioning List Category Code

PLCCB Product Level Configuration Control Board

PLCCE Program Life-Cycle Cost Estimate

PLCI Physical Link Connection Identifier • Pressurized Logistics Carrier Interface

PLCL Plant Clearance

PLCM Propellant Loading Control Monitor

PLCO Plant Clearance Office • Plant Clearance Officer

PLCON Palomar - Las Campanas Observatory - NOAO (survey)

PLCOP Prelaunch Checkout Plan

PLCP Physical Layer and Convergence Protocol • Physical Layer Convergence Protocol • Planning Command Pool

PLCR Period-Luminosity-Colour Relation

PLCS Propellant Loading Control System

PLCU Propellant Level Control Unit

Plcy policy

PLCZR Period-Luminosity-Color-Metallicity Relation

Pld Payload

PLD Payload • Phase-Lock Demodulator • Plafond Légal de Densité • Playa Samara (Costa Rica) (location identifier) • Polar Layered Deposit • Postal Logistics Directory • Potentially Lethal Damage • Program Listing Document • Program Load Disk • Programmable Logic Device • Programming Language Design • Public Libraries Division • Pulsed Laser Deposition • Pulse-Length Discriminator

PLDC Preliminary List of Design Changes • Primary Leadership Development Course

PLDF Payload Data File

PLDI Payload Data Interleaver • Programming Language Design and Implementation (conference)

PLDIS Public Libraries Development Incentive Scheme

PLDM Payload Management

PLDMI Precise Laser Distance Measuring Instrument

PLDP Photon-noise Limited Doppler Precision

PLDS Payload Support • Pilot Land Data System

PLDTS Propellant Loading Data Transmission System

ple Pleistocene

P.le Piazzale

Ple Passerelle

PLE Paiela (Papua New Guinea) (location identifier) • Partial Lunar Eclipse • Phased Loading Entry • Phase Locked Epitaxy • Photoluminescence Excitation • Preliminary Logistics Evaluation • Programmable Logic Element • Public Local Exchange • Pulsed Laser Experiment • Pure Luminosity Evolution

PLED Polymeric Light-Emitting Diode

PLEGPAY Plasma Electron Gun Payload

Plen Plenipotentiary

PLEVAL Payload Evaluation

PLEXAS Polarimeter for Low-Energy X-ray Astrophysical Sources

plf Pounds per Linear Foot

Plf Plaintiff

PLF Pala (Chad) (location identifier) • Payload Fairing • Phase Lock Frequency • Plasma and Fluids Dynamics • Point Location File • Progression Level Factor

PLFA Phospholipid Fatty Acid • Primary Level Field Activity

Plff Plaintiff

PLG La Plagne (France) (location identifier) • Public Libraries Group • Pulsed Light Generator

PLGR Precise Lightweight GPS Receiver • Precision Lightweight GPS Receiver • Precision Low-cost GPS Receiver

PLGS Polychromatic Laser Guide Star

PLGSS Payload Ground Support System

PLH Payload Handling • Plymouth (UK-England) (location identifier)

PLHX Payload Heat Exchanger

Pli Pliocene

Pl&I Plasmas and Ions (periodical code)

PLI Palm Island (Saint Vincent & The Grenadines) (location identifier) • Payload Interrogator • Plan Local d'Insertion • Position Location Information • Power Level Indicator • Power Loss Indicator • Principles, Logics and Implementations (conference) • Private Line Interface • Procurement Line Item • Programming Language Interface • Programming Language One • Propellant-Level Indicator • Pulsed Laser Interferometry

PLIANT Procedural Language Implementing Analog Technique

PLIAR Postlaunch Instrumentation Accuracy Report

PLIC Payload Intelligent Controller • Piecewise Linear Interface Calculation • Public Libraries and Information Council

PLicO Publications of Lick Observatory (periodical code)

PLIE Plan Local d'Insertion Économique

PLIF Planar Laser-Induced Fluorescence

PLILP Programming Language Implementation and Logic Programming (conference) • Programming Languages - Implementations - Logics - Programs

PLIM Payload Installation Mechanism • Postlaunch Information Message • Propellant Linear Insulation Matrix

PLIMHR Payload Integration Management Hardware Review

PLIMR Payload Integration Management Review

PLINS Payload Integration Review

PLIS Plannungs-Informationssystem

PLISN Provisioning Line Item Sequence Number • Provisioning List Item Sequence Number

PLJ Placencia (Belize) (location identifier)

PLK Branson (USA-MO) (location identifier) • Poincaré - Lighthill - Kuo (method) • Point Lookout - M. Graham Clark (USA-MO) (location identifier)

Pll Pallet

PLL Phased Lock Loop • Phase-Locked Loop (*) • Phase Lock Loop • Ponta Pelada (Brazil-AM) (location identifier) • Positive Logic Level • Preliminary Load List • Prelinked Library • Prescribed Load List • Programming Language for Linguistics

PLLR Phase Lock Loop Receiver

plm Por lo Mejor

PL/M Programming Language for Microcomputers • Programming Language for Micros

PLM Palembang (Indonesia) (location identifier) • Passive Lunar Marker • Payload Management • Payload Module • Payload Monitoring • Performance Level Modeling • Phase Length Modulation • Polarized Light Microscopy • Power Line Modulation • Prelaunch Monitor • Product Level Manager • Programming Language for Microcomputers • Programming Language for Micros • Programming Logic Manual • Pulse-Length Modulation

Plmb Plumber • Plumbing

PLMD Payload Mating Dolly

Plmg Plumbing

PLMN Public Land Mobile Network

PLMP Program Logistics Management Plan

Plmr Plumber

PLMR Postlaunch Memorandum Report

PLMS Program Logistics Master Schedule

Pln Plane • Plein

PLN Pellston (USA-MI) (location identifier) • Perusahaan Listrik Negara (Indonesia) • Plain • Plan • Private Line Number • Program Line Number • Program Logic Network

PLNAG Public Libraries Networking Advisory Group

PLNASG Public Libraries Networking Awareness Steering Group

Plng Planning

PLNG Planning

Plnm Plenum

Plnr Planar

PLNS Plains • Plans

Plnt Planet

Plntrm Planetarium

Plnty Planetary

PLO Pacific Launch Operation • Passenger Liaison Office • Perform Locked Operations • Phase-Locked Oscillator • Please Leave On • Port Lincoln (Australia-SA) (location identifier) • Primary Learning Objective • Proton Liberation Organization • Publications of the Lick Observatory (periodical code) • Public Liaison Officer • Pulsed Laser Oscillator • Pulsed Locked Oscillator

PLOAM Physical Layer Operation, Administration and Maintenance

PLOD Periodic List of Data

PLOO Pacific Launch Operations Office

PLOP Planetary Landing Observation Package

PLOS Primary Line of Sight • Probe Line of Sight

PLOT Porous Layer Open Tubular • Probability of Launch On Time

PLP La Palma (Panama) (location identifier) • Packet Layer Protocol • Packet Level Procedure • Packet Level Protocol • Parallel Logic Programming • Pattern Learning Parser • Pitch Lock Pin • Postlaunch Phase • Power Law Process • Power Law Profile • Presentation Level Protocol • Procedural Language Processor • Product Launch Point • Pulsed Laser Photolysis

PLPA Pageable Link Pack Area

PlPh Plasma Physics (periodical code)

PLPla Publications of La Plata Observatory (periodical code)

Plprt Plupart

PLPS Propellant Loading and Pressurization System

PLPV Periodic Line Profile Variation

Plq Plaque

PLQ Palanga (Lithuania) (location identifier)

Plr Pillar • Plier • Puller

PLR Pell City - Saint Claire County (USA-AL) (location identifier) • Period-Luminosity Relation • Planar Lightwave Circuit • Power Line Radiation • Prelayout Review • Public Lending Right • Pulse Link Repeater

PLRD Payload Requirements Document

PLRG Public Libraries Research Group

P&LRP Productivity & Long-Range Planning

PlRp Planetary Report (periodical code)

PLRP Productivity and Long-Range Planning

Plrs Pelorus

PLRS Position Locating and Reporting System • Position Location Reporting System

Plrty Polarity

PLRV Payload Launch Readiness Verification

Pls Plates • Please • Plus • Pulse

PLS Palletized Loading System • Palomar-Leiden Survey • Partial Least Squares • Payload Simulator • Payload System • Personal Launch System • Personal Librarian System • Personal Library Software • Personal Library System • Personal Location System • Personnel Launch System • Personnel Locator System • Phone Lattice Scanning • Physical Layer Signaling • Plasma • Plasma Light Source • Plasma Science • Pohang Light Source (ROK) • Point-Like Structure (Sun) • Position Latérale de Sécurité • Postlanding and Safing • Prelaunch Survivability • Preliminary Landing Site • Primary Landing Site • Primary Link Station • Probability of Launch Success • Producibility and Logistics Support • Product Logistic Support • Professional Land Surveyor • Programmable Logic Sequencer • Propellant Loading Sequencer • Propellant Loading System • Providenciales (Turks & Caicos Islands) (location identifier) • Publishers' Licensing Society (UK) • Pulsed Laser System • Pulsed Light Source

PLSA Partial Length Shield Assembly • Probabilistic Latent Semantic Analysis • Public Library Services Act (USA)

PLSC Plasma Science Committee

PLSDU Physical Layer Service Data Unit

PLSF Phononic Lattice Solid with Fluids

PLSI Phononic Lattice Solid by Interpolation • Probabilistic Latent Semantic Indexing

PLSL Propellants and Life Support Laboratory

PSLM Participating Logistics Support Manager

PLSO Phonetic Letters Spelled Out

PLSP Payload Signal Processor • Prelaunch Survival Probability

PLSR Partial Least Squares Regression • Postlaunch Service Review

Plsrs Plusieurs

PLSS Personal Life Support System • Portable Life Support System • Position Location Sensor System • Postlanding Survival System • Prelaunch Status Simulator • Primary Life Support System • Public Land Survey System (USA)

Plstc Plastic

PLSV Polar Launch Satellite Vehicle

PLSVP Preliminary SV Planning

plt Plotter Drawing (file extension)

Plt Pallet • Pilot • Plant

PLT Pallet • Payload Type • Pilot • Pilot Lead Time • Pipeline Time • Plan à Long Terme • Plato (Colombia) (location identifier) • Power Line Termination • Power Line Transient • Precision Live Timing • Princeton Large Tokamak • Princeton Large Torus • Procurement Lead Time • Production Lead Time • Product Line Testing • Program Library Tape • Program Lead Time • Program List Table • Program Lock Token • Pulsed Light Theodolite

Plta Plazoleta

Pltc Political

Pltf Plaintiff • Platform

PLTFM Platform

Pltg Planting • Plating

PLTS Precision Laser Tracking System

Plu Plural • Pluto • Pluton

PLU Belo Horizonte - Pamphula (Brazil-MG) (location identifier) • Pacific Lutheran University (USA-WA) • People Like Us • Periodic Location Update • Plage de Lumière Uniforme • Plan Local d'Urbanisme • Position Location Uncertainty • Price Lookup • Primary Logical Unit • Program Load Unit • Programmable Load Unit • Propellant Loading and Utilization

PLUM Payload Umbilical Mast • Planning Land Use Model • Polymerization with Light Under Microgravity

Plur Plural • Plurality

PLUS Platform Utilization Study • Pragmatic Language Understanding System • Program Library Update System • Programmed Learning Under Supervision • Programming Language for UNIVAC Systems • Project Literary United States (USA)

Plut Pluto • Pluton

PLUTO Pipe Line Under the Ocean

PLV Payload Launch Vehicle • Payload Vehicle • Poltava (Ukraine) (location identifier) • Postlanding Valve • Postlanding Vent • Postlanding Ventilation • Powered-Lift Vehicle • Probe Launch Vehicle • Production Level Video • Pseudo-Local Vibration • Publicité sur le Lieu de Vente

PLVC Postlanding Vent Control

PLVT Physical Layer Voice Telephony

PLW Palau (ISO country code) • Palu (Indonesia) (location identifier) • Plowed • Plowed

PLX Parallax • Propellant Loading Exercise • Semipalatinsk (Kazakhstan) (location identifier)

Ply Plymouth (UK)

PLY Play • Plymouth (USA-IN) (location identifier)

Plywd Plywood

PL/Z Programming Language Zilog

Plz Plaza • Please

PLZ Plaza • Port Elizabeth (South Africa) (location identifier) • Postleitzahl • Programming Language for the Zilog • Złoty (Poland) (ISO currency code)

Plza Plaza

Plzd Polarized

PL/1 Programming Language One

pm Permanent Magnet • Picometer • Post Meridiem

Pm Premium • Promethium

P&M Processes & Materials • Procure & Manufacture • Productibility & Manufacturing • Production & Maintenance

PM Miquelon (Saint Pierre et_) (code ISO de pays) • Pad Mechanic • Panama • Panel Meter • Parabolic Mirror • Paramagnetic • Partial Match • Particle Mesh • Particulate Matter • Past Master • Payload Management • Payload Midbody • Payload Module • Paymaster • Pénétration Massive • Penetration Measurement • Pénétration Moyenne • Per Message • Per Month • Performance Management • Performance Monitor • Periodic Maintenance • Peripheral Module • Permanent Magnet • Permanent Memory • Permanent Motion • Permanent Movement • Permittivity Meter • Personality Module • Perturbation Method • Phase Modulated • Phase Modulation • Phase Monitor • Photomultiplier • Photosynthetic Measurement • Physical Medium • Planetary Mission • Plan de Matière • Planning Module • Plant Management • Plasma Mantle • Plus-Minus • Point Mass • Point-to-Multipoint • Polarization Modulation • Polar Motion • Police Magistrate • Policy Manual • Polymer Morphology • Portable Magnetometer • Postmaster • Post Meridiem • Post Mortem • Potentiometer • Pounds per Minute • Pour Mémoire • Power Module • Predictive Maintenance • Premier Ministre • Presentation Manager • Pressurized Module • Preventive Maintenance • Preventive Medicine • Primary Memory • Primary Mirror • Prime Minister • Principle of Multiplying • Problem Management • Problem Mode • Procedure Memory • Procedures Manual • Processes and Materials • Process Manager • Process Manual • Process Model • Procurement Management • Procurement Manager • Procurement Manual • Product Management • Product Manager • Programa Marco • Program Management • Program Manager • Program Memorandum • Program Milestone • Program Monitoring • Project Management • Project Manager • Propellant Management • Proper Motion • Proposal Management • Proposal Manager • Propulsion Module • Provost Marshal • Pulse Modulated • Pulse Modulation • Saint Pierre et Miquelon (code ISO de pays)

PMA Parts Manufacturer Approval • Parts Manufacturing Approval • Pays Moins Avancé • Pemba (Tanzania) (location identifier) • Performance Monitor Annunciator • Permanent Magnet Alternator • Permanent Magnet Array • Permanent Magnet Assembly • Permanent Mailing Address • Personal Mower Assistant • Personennamen des Mittelalters • Physical Medium Attachment • Policy Management Architecture • Portable Maintenance Aid • Poste Médical Avancé • Postmission Archive • Preliminary Mission Analysis • President's Management Agenda (USA) • Pressurized Mating Adapter • Primary Mirror Assembly • Procréation Médicalement Assistée • Program Management Agreement • Program Management Assistant • Program Management Authority • Program Memory Area

PMAA Parallel Matrix Algorithms and Applications (conference)

PMAC Packet Media Access Controller • Parallel Memory Address Counter • Performance Management Advisory Committee • Peripheral Module Access Controller • Personnel Management Action Committee • Portable Multipurpose Applications Console • Program Management Allocation Criterion • Programmable Multiple-Axis Controller

PMad Publicaciones de la Faculdad de Ciencias de Madrid (periodical code)

PMAD Performance Monitor Annunciation Driver • Performance Monitor Annunciator Driver • Personnel Management Authorization Document • Portable Maintenance Aid Device • Power Management And Distribution • Professional Map Analysis Package

PMAG Photographic Magnitude • Program Management Advisory Group

PMALS Preliminary Miniature Air-Launched System

PManO Publications of Manila Observatory (periodical code)

PMAP Performance Monitor Annunciation Panel • Port Mapper • Professional Map Analysis Package

PMAR Preliminary Maintenance Analysis Report • Program Manager's Assessment Report

PMAS Patient Management and Analysis System • Performance Management Analysis System • Performance Management Application Service • Phase-coupled Multiaperture Spectrograph • Potsdamer Multiapertur Spektrophotometer • Potsdam Multiaperture Spectrophotometer • Programmed Manpower Authorizations System • Property Management Accounting Software

PMAT Page Map Address Table • Portable Maintenance Access Terminal • Program Management Assistant Team

PMATS Performance Monitor And Tracking System • Producibility Methodology and Associated Tool Set • Project Management Automated Tool Set

PMB Parkes Multiband • Payload Module Bus • Pembina (USA-ND) (location identifier) • Performance Management Baseline • Performance Measurement Baseline • Physics Medicine and Biology (periodical code) • Pitch Momentum Bias • Port and Maritime Board • Primordial Main Belt • Private Mail Box • Produit Matériel Brut • Program Management Board • Programme Management and Budgeting • Programme Management Board

PMBS Property Management and Budgeting System

PMBX Private Manual Branch Exchange

PM&C Plant Maintenance & Construction

PMC Parallel Multiunit Circle • Partially Mission Capable • Partial Mission Capable • Payload Module Computer • Payload Monitoring and Control • Payload Monitoring Control • Perceptron Multicouche • Permanently Manned Capability (*) • Permanently Manned Configuration • Permanent Manned Capability • Permanent Manned Configuration • Personal Mobile Communication • Photoelectric Meridian Circle • Pitman's Measure of Closeness • Plan de Mesure et de Commande • Planning Ministers Council • Plant Maintenance and Construction • Polar Mesospheric Cloud • Pollen Mother Cell • Port Management Center • Postal and Mailing Center • Postmaintenance Check • Postmanufacturing Checkout • Postministerial Conference • Preliminary Maintenance Concept • President's Management Council • Pressure-Modulator Cell • Primary Management Contractor • Private Medical Communication • Procurement Method Code • Programmable Machine Controller • Program-Managed Contract • Program Management Center • Program Management Charter • Program Management Committee • Program Management Council • Program Management Course • Program Modification Card • Programme Management Council • Project Management Charter • Project Management Committee • Propellant Monitor and Control • Pseudo Machine Code • Puerto Montt (Chile) (location identifier)

PMCB Preventive Maintenance Checks and Balance

PMCC Platforms Mission Control Center • Port Management Control Center

PMcCO Publications of Leander McCormick Observatory (periodical code)

PM&CE Plant Maintenance & Construction Engineer

PMCE Plant Maintenance and Construction Engineer • Program Manager's Cost Estimate

PMCM Programmable Mobile Communications Modem

PMCN Program in Mathematical and Computational Neuroscience

PMCS Power Management and Control System • Preventive Maintenance Checks and Services • Primary Mirror Control System • Programmable Modular Communications System • Program Management Control System • Project Managers Control System

PMC2 Program Manager Command and Control

P&MD Productivity & Manufacturing Development

PMD Packet Mode Data • Pakistan Meteorological Department • Palmdale (USA-CA) (location identifier) • Particulate and Multiphase Processes • Physical-layer Medium-Dependent • Physical-Media Dependent • Physical-Medium Dependent • Polarization Measurement Device • Polarization Measuring Device • Polarization Mode Dispersion • Power Management and Distribution • Productivity and Manufacturing Development • Product Marketing Division • Program for Management Development • Program Management Data • Program Management Department • Program Management Directive • Program Management Division • Program Management Document • Program Management Documentation • Programmed Multiple Development • Program Monitoring and Diagnosis • Projected Map Display • Propellant Management Device

PMDAR Promedios Móviles Determinísticos Autoregresivos

PMDB Photographic Meteor Database (IMO) • Program Management Decision Brief • Program Management Development Directorate • Project Master Database • Proposal Management Database

PMDE Policy and Management Development and Environment

PMDF PASCAL Memorandum Distribution Facility • Project Master Data File

PMDM Payload Multiplexer/Demultiplexer

PMDP Project Management Development Plan • Project Managers Development Program • Prototype Model Design Plan

PMDR Parametric Monotone Decreasing Ratio • Programmed Monthly Demand Ratte • Program Monthly Demand Ratte

PME Paramagnetic Meissner Effect • Payload Mechanical Engineer • Performance Monitoring Equipment • Petites et Moyennes Entreprises • Photomagnetoelectric • Porte-Monnaie Électronique • Portsmouth (UK-England) (location identifier) • Precision Measurement Equipment • Precision Measuring Equipment • Prime Mission Equipment • Process and Manufacturing Engineering • Product Model Enhancement • Program Management for Executives • Projected Mars Estimator • Protective Multiple Earthing • Pyramid Maximum Entropy

PMEC Photomultiplier Exposure Counter

PMEE Prime Mission Electronic Equipment

PMEI Petits et Moyens Établissements Industriels

PMEL Pacific Marine Environmental Laboratory • Pacific Marine Environment Laboratory • Precision Measurement Equipment Laboratory

PMEO Performance Modeling, Evaluation and Optimization

PMES Physical, Mathematical, and Engineering Sciences

PMEST Personality - Matter - Energy - Space - Time (formula)

PMetR Papers in Meteorological Research (periodical code)

PMF Parameter Management Frame • Parma (Italy) (location identifier) • Passive Matched Filter • Penetration with Minimal Friction • Performance Monitor Function • Performance Monitoring Function • Perigee Motor Firing • Permanent Magnetic Field • Personnel Master File • Planetary Mass Function • Polarization-Maintaining Filter • Poloidal Magnetic Field • Potential of Mean Force • Precision Measurement Facility • Present Mass Function • Principal Management Facility • Probability Mass Function • Probable Maximum Flood • Professional Medical Film • Program Management Facility • Provisioning Master File • Pulsating Magnetic Field

PMFA Permanent Magnet Focus Assembly • Permanent Magnetic-Focus Assembly

PMFBI Pardon Me For Butting In

PMFJI Pardon Me For Jumping In

Pmflt Pamphlet

PMFS Pulsed Magnetic Field System

PMFVS Protective Mask Fit Validation System

PM&G Planetary Materials & Geochemistry

PMG Papers on Meteorology and Geophysics (periodical code) • Parts Management Group • Paymaster General • Permanent Magnet Generator • Phase Modulation Generator • Plasma Motor Generator • Ponta Porã (Brazil-MS) (location identifier) • Postmaster General • Process Manager • Program Maintenance Group • Program Management Group

PMGM Program Manager Guidance Memorandum

PMGT Power Management

PMGW Primary Mission Gross Weight

pmh Per Man-Hour

PMH Portsmouth (USA-OH) (location identifier) • Previous Medical History • Probable Maximum Hurricane

PMHEMT Pseudo-Morphic High Electron Mobility Transistor

PMHL Preferred Measurement Hardware List

PMHS Physical Material Handling Segment

pmi Photographic Microimage

PMI Palma de Mallorca (Baleares) (location identifier) • Passive Microwave Imager • Patient Movement Item • Permission Management Infrastructure • Personnel Management Information • Petites et Moyennes Industries • Physical Medium Interface • Physical Model Interface • Point of Maximum Intensity • Power Margin Indicator • Primary Maintenance Inspection • Preventive Maintenance Inspection • Preventive Maintenance Instruction • Primary Measurement Instrument • Principal Maintenance Inspector • Private Medical Insurance • Privilege Management Infrastructure • Programmable Machine Interface • Programmable Memory Interface • Programmable Multispectral Imager • Program Management Instruction • Programme du Marché Intérieur • Project Management Institute

PMIA Parallel Multiplexer Interface Adapter

PMIC Periodic Maintenance Information Card • Personnel Management Information Center • Portfolio Management for Intellectual Capital • Precious Metal Indicator Code • Programa Mundial de Investigaciones Climáticas

PMIG Programmer's Minimal Interface to Graphics

PMIP Paleoclimate Modeling Intercomparison Project • Program Management Improvement Project

PMIR Pressure Modulator Infrared Radiometer • Program Manager's Integration Review

PMIRR Pressure Modulated Infrared Radiometer • Pressure Modulation Infrared Radiometer (*) • Pressure Modulator Infrared Radiometer

PMIS Passive Microwave Imaging System • Personnel Management Information System • Printing Management Information System • Production Management Information System • Project Management Information System

PMISP Personnel Management Information System Plan

PMIT Program Management Integration Test • Program Manager Integration Test

PMJEG Performance Measurement Joint Executive Group

PMJI Pardon My Jumping In

Pmk Postmark

PMK Palm Island (Australia-QLD) (location identifier)

PML Perfectly Matched Layer • Physics and Mathematics Lyceum • Port Moller (USA-AK) (location identifier) • Preliminary Materials List • Prins Maurits Laboratorium (TNO) • Probable Maximum Loss • Process Modeling Language • Programmable Macro Logic • Program Management Line

PMLC Performance Metrics of Link Characteristics

PMLCD Passive-Matrix LCD

PMM Personnel Management Manual • Phanom Sarakham (Thailand) (location identifier) • Planetary Mission Module • Power Management Mode • Precision Measuring Machine • Precision Measuring Microdensitometer • Precision Metrology Mount • Probability Matching Method • Program Memory Module • Project Manager Meeting • Property Management Manual • Pulse-Mode Multiplex

PMMA Polymethylmethacrylate • Polyméthylméthacrylate

PMMD Programmable Magnetic Momentum Dump

PMME Product Management Mission Element

PMMEL Proposed Master Minimum Equipment List

PMML Predictive Model Markup Language

PMMP Preventive Maintenance Management Program

PMMR Project Manager Monthly Review

PMMRS Passive Microwave Mapping and Retrieval Simulator

P&MMSC Procurement & Materiel Management Service Center

PMMSC Procurement and Materiel Management Service Center

PMMU Paged Memory Management Unit

PMMW Passive Millimeter Wavelength (technology)

PMN Parkes - MIT - NRAO (survey) • Premanufacture Notification • Presse Mensuelle Nationale • Produit Matériel Net • Program Management Network • Pumani (Papua New Guinea) (location identifier)

PMO Palermo - Punta Raisi (Italy) (location identifier) • Personnel Management Officer • Perturbation Molecular Orbital • Pine Mountain Observatory (USA-GA) • Planetary-Mass Object • Present Method of Operation • Prime Minister's Office • Principal Meteorological Office • Production Management Officer • Product Managers Office • Program Management Office • Project Management Office • Project Management Organization • Project Managers Office • Purple Mountain Observatory (China-PRC)

PMOC Payload Mission Operations Center • Pioneer Mission Operations Center (NASA) • Prototype Mission Operations Center

PMOCCO Preconditioning Methods for Optimal Control and Constrained Optimization (conference)

PMOD Physikalisch-Meteorologisches Observatorium Davos (Switzerland)

P. Mol. Peso Moleculare • Poids Moléculaire

PMOM Performance Management Operating Manual • Performance Management Operations Manual

PMON Performance Management Operations Network • Process Monitor

PMOS Polarimetric Multiobject Spectroscopy • Positive (channel) Metal Oxide Semiconductor • Positively-doped Metal Oxide Semiconductor

PM&P Parts, Materials & Processes

PMP Parallel Microprogrammed Processing • Parallel Microprogrammed Processing • Parallel Microprogrammed Processor • Participative Management Plan • Part Management Program • Parts, Materials and Packaging • Parts, Materials and Processes • Passive Measurement Program • Performance Management Package • Performance Monitoring Program • Phased Maintenance Program • Pimaga (Papua New Guinea) (location identifier) • Point to Multipoint • Polymethylpentene • Pontryagin Maximum Principle • Prelaunch Monitor Program • Preliminary Mission Plan • Preliminary Mission Profile • Premodulation Processor • Preventive Maintenance Plan • Prime Mission Product • Prism - Mirror - Prism • Probable Maximum Precipitation • Program Management Plan • Program Master Plan • Program Monitor Panel • Project Management Plan • Project Management Professional • Project Master Plan • Pump

PMPBE Program Manager for Budget and Performance Evaluation

PMPC Priority Mail Processing Center

PMPCB Parts, Materials and Processes Change Board • Parts, Materials and Processes Control Board

PMPD Portable Mission Preview Device

PMPE Program Materials and Processes Engineer

PMPFR Program Manager's Preflight Review

PMPL Preferred Mechanical Parts List

PMPPC Priority Mail Postal Processing Center

PMPR Project Manager Program Report

PMPS Program Management Performance System

PMPSL Parts, Materials and Processes Selection List

PMQ Perito Moreno (Argentina-SC) (location identifier)

Pmr Paymaster

PMR Palmerston North (New Zealand) (location identifier) • Parabolic Microwave Reflector • Parts Material Request • Passive Microwave Radiometer • Patient Movement Request • Payload Manifest Review • Payload Mass Ratio • Payload Mixing Review • Performance Measurement Report • Performance Monitoring Receiver • Planned Maintenance Requirement • Portable Microfiche Reader • Portable Microfilm Reader • Power Monitor Relay • Precious Metals Recovery • Precision Microwave Radiometer • Preliminary Materials Review • Presse Mensuelle Régionale • Pressure-Modulated Radiometer • Pressure Modulation Radiometer • Pressure Modulator Radiometer (*) • Primary Mission Readiness • Private Mobile Radio • Problem Management Report • Procurement Management Review • Professional Mobile Radio • Program Management Report • Program Management Responsibility • Program Management Review • Program Manager's Report • Program Manager's Review • Programmed Monthly Rate • Program Modification Request • Program Monthly Review • Project Management Report • Project Management Review • Proton Magnetic Resonance • Provisioning Master Record • Public Mobile Radio

PMRB Primary Material Review Board

PMRC Programme Mondial de Recherche sur le Climat

PMRD Preliminary Mission Requirements Document

PMRF Precision Machining Research Facility (NIST)

PMRM Periodic Maintenance Requirements Manual

PMRO Producibility - Maintainability - Reliability - Operations

PMRP Precious Metal Recovery Program

PMRR Paramedic Rapid Response • Premanufacturing Readiness Review • Premate Readiness Review

PMRS Performance Management and Recognition System

PMRT Program Management Responsibility Transfer

PMRTP Program Management Responsibility Transfer Plan

PMS Palmyra (Syria) (location identifier) • Pantone Matching System • Payload Monitoring System • Payload Mounting Structure • People Moving System • Performance Management System • Performance Measurement System • Performance Monitoring System • Periodical Management System • Personal Mail Service • Phase Measurement System • Physiological Monitoring System • Pipeline Management System • Planar Magnetic Structure • Planned Maintenance Support • Planned Maintenance System • Plant Monitoring System • Platform Management System • Policy Management System • Pollution Monitoring Satellite • Portable Monitoring Set • Post-Main-Sequence • Power Management Software • Power Management System • Precision Metrology Station • Premain Sequence (star) • Preventive Maintenance System • Probability of Mission Success • Probable Maximum Surge • Processor - Memory - Switch • Production Management Segment • Production Management Specialist • Production Management Support • Product Monitoring System • Program Management Support • Program Management System • Program Master Schedule • Project Management System • Property Management System • Propulsion Module System • Protection Management System • Psycho-Médico-Social • Pulsating Meteor Stream

PMSA Primary Metropolitan Statistical Area • Project Management System Assessment

PMSC Programme Management Support and Coordination • Purchasing and Materials Service Center

PMSD Program Management and Services Directorate

PMSE Percentage Mean Squared Error • Polar Mesosphere Summer Echo • Polar Mesospheric Summer Echo (*) • Predictive Mean Square Error

PMSEIC Prime Minister's Science, Engineering and Innovation Council (Australia)

PMSFN Planetary Manned Space Flight Network

PMSO Program Manager Support Office

PMSP Photon-counting Microspectrophotometer • Preliminary Message Security Protocol • Premessage Security Protocol

PMSR Performance Measurement Status Report

PMSS Physique Mathématique - Modélisation - Simulation • Pre-main-Sequence Star • Program Management Support Services • Program Management Support System • Project Manager's Support System • Proposal Management System for the Subaru (telescope)

PMSWG Platform Management System Working Group

Pmt Payment

P&MT Production & Manufacturing Technology

PMT Paramakotoi (Guyana) (location identifier) • Performance Measurement Training • Phase-Modulated Transmission • Photoelectric Magnetograph Telescope • Photomechanical Transfer • Photomultiplier Tube • Physical Modeling Tool • Plan à Moyen Terme • Planning/Management Team • Portable Magnetic Tape • Positive Media Transfer • Prepare Master Tape • Preventive Maintenance Time • Production Materiel Technician • Production Monitoring Test • Program Management Team • Program Master Tape • Project Management Team • Prominent • Pyramidal Median Transform

PMTF Payload Module Transfer Platform • Zhurnal Prikladnoi Mekhaniki i Tekhnicheskoi Fiziki (periodical code)

PMTP Pay Me Transfer Protocol • Program Management Transfer Point • Prototype Model Test Plan

PMTPP Program Management Transfer Point Plan

PMTR Program Manager's Technical Representative

PMTS Personnel Management Training System • Predetermined Motion Time System • Processor Module Test Set

PMtv Publicaciones de la Faculdad de Humanidades y Ciencias de Montevideo (periodical code)

PMU Paimiut (USA-AK) (location identifier) • Payload Monitoring Unit • Performance Monitor Unit • Personal Maneuvering Unit • Physical Mock-Up • Plaster Mock-Up • Polarization Modulator Unit • Portable Memory Unit • Power Management Unit • Precision Measurement Unit • Pressure Measuring Unit • Programmable Master Unit • Pulse Modulation Unit

PMunO Publications of Munich Observatory (periodical code)

PMV Panneau à Message Variable • Permanently-Manned Vehicle • Porlamar del Caribe (Venezuela) (location identifier) • Predictive Mean Vote • Private Motor Vehicle

PMW Palmas (Brazil-TO) (location identifier) • Passive Microwave • Pitch Maneuver Wheel • Project Manager Workbench

PMWG Producibility and Manufacturing Working Group

PMWO Preventive Maintenance Work Order

PMWS Project Manager Workstation

PMX Packet Multiplexer • Palmer (USA-MA) (location identifier) • Partially Matched Crossover • Private Manual Exchange

PMY Puerto Madryn - El Tehuelche (Argentina-CB) (location identifier)

PMZ Palmar (Costa Rica) (location identifier)

PMZF Programmable Multizone Furnace

pn Per Night

Pn Prochain

P-N Positive-Negative

P/N Part Number • Promissory Note

PN Parc National • Parque Nacional • Part Number • Patrimonio Nacional (Spain) • Pénétration Nationale • Performance Number • Personal Name • Personal Number • Peso Neto • Petri Net • Phosphorus Nitride • Physical Network • Physique Nucléaire • Pitcairn (ISO country code) • Planetarischer Nebel • Planetary Nebula • Please Note • Poids Net • Pôle Nord • Position Navigation • Positive-Negative • Postnewtonian • Preliminary Notification • Previous Node • Primary Node • Primordial Nucleosynthesis • Prior Notice • Processing Node • Professional Network • Programme National • Program Note • Project Number • Promissory Note • Proper Noun • Proton Number • Pseudo-Noise • Public Network • Pulse Network

Pna Panama

PNA Pacific - North American • Pamplona (Spain) (location identifier) • Parallel Network Architecture • Peptide Nucleic Acid • Phoneline Networking Alliance • Postnewtonian Approximation • Programmable Network Access

PNAA Prompt-Neutron Activation Analysis

PNAEE Programme National d'Amélioration de l'Efficacité Énergétique (France)

PNAF Potential Network Access Facility

PNAOJ Publications of the National Astronomical Observatory of Japan (periodical code)

PNAS Proceedings of the National Academy of Sciences (periodical code)

PNB Pôrto Nacional (Brazil-TO) (location identifier) • Producto Nacional Bruto • Produit National Brut

PNC Paging Network Controller • Parity Non Conserving • Personnel Navigant Commercial (*) • Personnel Navigant de Cabine • Physique Nucléaire et Corpusculaire • Ponca City (USA-OK) (location identifier) • Programmed Numerical Control • Programme National de Cosmologie (France)

PNCA Programme National de Chimie Atmosphérique (France)

PNCGC Programme National de Cartographie Géoscientifique du Canada

PNCI Plan Nacional de Calidad Industrial (Spain)

PNCP Peripheral Node Control Point

Pnd Pond

PND Passive Nutation Damper • Personennamendatei • Position/Navigation Device • Present Next Digit • Presse-Nachrichtendienst (Austria) • Probabilité de Non-Détection • Punta Gorda (Belize) (location identifier)

PNdB Perceived Noise Decibel

PNDC Parallel Network Digital Computer

Pndg Pending

PNDR Programme National sur le Déterminisme du Recrutement (France)

PNDS Ponds

PNe Planetary Nebulae

PNE North Philadelphia (USA-PA) (location identifier) • Pacific National Exhibition • Peaceful Nuclear Explosion • Pine

PNEA Programma Nazionale di Ricerche in Antàrtide

PNEAT Programme National d'Efflorescences Algales Toxiques (France)

PNEDC Programme National d'Étude de la Dynamique du Climat (France)

PNEK Postnuclear Event Key

PNER Programme de Numérisation pour l'Enseignement et la Recherche (France)

PNES Pines

PNETS Public Non-Exclusive Telecom Service

Pneu Pneumatic

Pneum Pneumatic

PNF Peterson's Point (USA-AK) (location identifier) • Plan National de Formation (France)

PNFE Prototype Network Front End

png Portable Network Graphics (file extension)

PNG Papouasie Nouvelle Guinée (code ISO de pays) • Papua New Guinea (international vehicle registration code) • Papua New Guinea (ISO country code) • Paranaguá (Brazil-PR) (location identifier) • Persona Non Grata • Pilot Non Flying • Portable Network Graphics • Programme National Galaxies (France)

PNGCS Primary Navigation, Guidance, and Control System

PnGid Pnevmatika i Gidravlika (periodical code)

PNGV Partnership for a New Generation of Vehicles (USA)

PNH Phnom-Penh (Cambodia) (location identifier)

PNHRA Programme National de Haute Résolution Angulaire

PNHRAA Programme National de Haute Résolution Angulaire en Astronomie • (France)

PNI Permit Next Increase • Pharmaceutical News Index • Pictorial Navigation Indicator • Pohnpei (Caroline Islands) (location identifier) • Ponape (Caroline Islands) (location identifier) • Prix Nobel Ignoré

PNIC Programme National d'Intérêt Communautaire • Prototype Near-Infrared Camera

PNID Precedence Network In Dialing

PNIE Programa Nacional de Investigación Espacial (Spain)

PNJ Polar Night Jet

PNK Pontianak (Indonesia) (location identifier) • Proper Noun Keyword

Pnl Panel

PNL Pacific Northwest Laboratory (USA) • Panel • Pantelleria (Italy) (location identifier) • Prescribed Nuclear Load • Programmable Netlist Library • Programmation Neuro-Linguistique

PNLA Pseudo Next-Level Aggregator

Pnlbd Panelboard

Pnld Paneled

PNLF Planetary Nebulae Luminosity Function

Pnlg Paneling

PNM Petit Nuage de Magellan • Pressão ao Nível do Mar • Pression au Niveau de la Mer • Price Negotiation Memorandum • Pulse Number Modulation

PNMC Packet Network Management Center • Packet Network Management Console

PNMS Permanent Noise Monitoring System

PNN Planetary Nebula Nucleus • Princeton (USA-ME) (location identifier) • Probabilistic Neural Network • Producto Nacional Neto • Profesor No Numerario

PNNi Planetary Nebula Nuclei

PNNI Private Network-Network Interface • Private Network-to-Network Interface (*)

PNNL Pacific Northwest National Laboratory (USA-WA)

Pnnt Pennant

PNNV Planetary-Nebula Nucleus Variable (star)

PNO Pinotepa (Mexico) (location identifier) • Public Network Operator

PNOC Programme National d'Océanographie Côtière (France) • Proposed Notice Of Change

PNOK Primary Next Of Kin

pnp Positive - Negative - Positive

PnP Plug and Play

PNP Personnel Non-Permanent • Plug and Play • Popondetta (Papua New Guinea) (location identifier) • Positive - Negative - Positive • Prenegotiation Position • Private Non-Profit (sector) • Private Numbering Plan • Programme National de Planétologie (France)

PNPI Petersburg Nuclear Research Institute (Russia)

PNPN Positive - Negative - Positive - Negative

PNPS Programme National de Prospective Scientifique (France) • Public Network Professional Services

PNQ Poona (India) (location identifier)

PNQR Pure Nuclear Quadrupole Resource

Pnr Pioneer

PNR Parc Naturel Régional • Passenger Name Record • Passive Noise Reduction • Point de Non-Retour • Pointe Noire (Congo) (location identifier) • Point of No Return • Power Network Repeater • Prior Notice Required • Programme National de Recherche • Prompt Nuclear Reaction

PNRA Programma Nazionale di Ricerche in Antartide (Italy)

PNRCO Programme National des Récifs Coraliens (France)

PNRH Programme National de Recherche en Hydrologie (France)

PNRI Philippines Nuclear Research Institute

PNRN Programme National sur les Risques Naturels (France)

PNRP Pollution and Natural Resources Programme

PNRS Preliminary Natural Resources Survey

PNS Parabolized Navier-Stokes • Pensacola (USA-FL) (location identifier) • Peripheral Nervous System • Philippine National Standard • Pilot Navigation System • Population Number Synthesis • Postnormal Science • Precision Navigation System • Prescribed Nuclear Storage • Primary Name Server • Projeto de Novos Satélites

PNSC Palomar-Norris Sky Catalog (*) • Palomar Northern Sky Catalog

PNSS Philippine National Science Society • Pseudo-Noise Spread Spectrum

Pnt Paint

PNT Personnel Navigant Technique • Positioning, Navigation, and Training • Private Network Termination • Processeur Numérique Transparent • Puerto Natales - Teniente J. Gallardo (Chile) (location identifier)

PNTD Personnel Neutron Threshold Detector • Plastic Nuclear Track Detector

Pntgn Pentagon

PNTI Ported Number Translation Indicator

PNTPM Physique Nucléaire Théorique et Physique Mathématique

PNTS Programme Nationale de Télédétection Spatiale (France)

PNU Panguitch (USA-UT) (location identifier) • Philippine Normal University (Philippines) • Physics News Update • Pli Non Urgent • Power Network Unit • Pusan National University (Korea-ROK)

PNUD Programa de las Naciones Unidas para el Desarollo (ONU) • Programme des Nations-Unies pour le Développement (ONU)

PNUE Programme des Nations-Unies pour l'Environnement (ONU)

PNUMA Programa de Naciones Unidas para el Medio Ambiente

PNV Potential Network Value

PNVAL Previously Not Available

PNVD Passive Night Vision Device

PNVG Panoramic Night Vision Goggle

PNW Pacific Northwest • Personal NetWare • Program Notes and Waivers

PNX Sherman-Denison - Grayson County (USA-TX) (location identifier)

pnxt Pinxit

PNY Pondicherry (India) (location identifier)

PNZ Petrolina (Brazil-PE) (location identifier)

p/o Part of

po Par Ordre • Per Ommegaande • Per Order • Per Os • Por Orden

Po Paseo • Paso • Polonium

P&O Peninsular & Oriental Steam Navigation Co. • Plans & Operations

P/O Part Of • Preservation by Operation • Provisioning Order

PO Parallel Output • Parametric Oscillation • Parity Odd • Parking Orbit • Par Ordre • Partitioned Organization • Passport Office • Patent Office • Payload Office • Peaks-Only (model) • Permanence Opérationnelle • Persistent Object • Personnel Office • Personnel Officer • Petites Ondes • Petty Officer • Phoenician • Photogrammetric Operations • Physical Optics • Planetary Orbit • Planning Objective • Planning Office • Planning Order • Plans and Operations • Point d'Observation • Poland • Polar Orbiter • Polynesia • Por Orden • Postal Order • Poste d'Observation • Post Office • Post Orbit • Power On • Power-Operated • Power Oscillator • Power Output • Preoperational • Previous Order • Primary Out • Primary Output • Printout • Prisme-Objectif • Privately Owned • Process-Oriented • Procurement Order • Production Offset • Product Order • Programmation Objects • Programmed Oscillator • Programme Opérationnel • Program Objective • Program Office • Program Officer • Program Order • Program Originator • Project Office • Project Officer • Project Order • Proportional Output • Province of Ontario (Canada) • Publication Officielle • Public Office • Pulkovo Observatory (Russia) • Purchase Order • Purchasing Office • Pyrénées-Orientales (France)

POA Persistent Object Adapter • Plan d'Option sur Actions • Plan Of Action • Polaris Observatory Association (USA-CA) • Portable Object Adapter • Porto Alegre - Salgado Filho (Brazil-RS) (location identifier) • Port Of Arrival • Post Office Agent • Postobservational Archives • Postoperational Archives (*) • Postoxidation Annealing • Power Open Association • Primary Optical Area • Provisional Operating Authorization

POAA Planetary Operations Analysis Area

POA&M Plan Of Actions & Milestones

POAM Plan Of Actions and Milestones • Polar Ozone Aerosol Monitor • Polar Ozone and Aerosol Measurement

PoAn Postepy Astronautyki (periodical code)

POAN Publications of the National Astronomical Observatory (periodical code)

POAS Psychological Operations Automated System

PoAst Postepy Astronomii Kraków (periodical code)

POAT Psychological Operations Assessment Team

POB Fayetteville Pope (USA-NC) (location identifier) • Peripheral Order Buffer • Persons On Board • Place Of Birth • Point Of Beginning • Point Of Burst • Post Office Box

POBeo Publi. Obs. Astron. Beograd (periodical code)

Pobl Población

POBol Pubblicazione dell'Osservatorio Astronomico di Bologna (periodical code)

POBor Publications de l'Observatoire de Bordeaux (periodical code)

POBY Prior Operating Budget Year

POC La Verne - Brackett (USA-CA) (location identifier) • Parallel Optical Computer • Particulate Organic Carbon

• Payload Operations Center • Physical Optics Corporation • Plan d'Opérations Combinées • Point Of Contact • Point Of Control • Policy Oversight Committee • Portable Occultation CCD • Port Of Call • Posted Out of Course • Precision Oscillator Crystal • Preliminary Operating Capability • Preliminary Operational Concept • Privately-Owned Conveyance • Probe Operations Centre • Process Operator Console • Product Of Combustion • Program Organizing Committee • Proof Of Concept • Public Call Office • Purchase Order Closeout • Purchase Order Contract

PoCa Polar Camera

POCA Point Of Closest Approach

POCC Payload Operations Control Center • Platform Operations Control Center • Project Operations Control Center

POCCNET Payload Operations Control Center Network

POCE Pantone Open Color Environment • Proof-Of-Concept Experiment

POCIBO Polar Circling Balloon Observatory

POClo Publications de l'Observatoire de Cluj (periodical code)

POCM Parallel Ocean-Climate Model

POCN Purchase Order Change Notice

POCO Physiology Of Chimpanzees in Orbit • Power On - Clock On • Purchase Order Change Order

PoCom Polymer Composites (periodical code)

POCP Program Objectives Change Proposal

POCR Program Objectives Change Request

POCS Passice Ocean Colour System • Patent Office Classification System (USA) • Projection Onto Convex Sets • Proof Of Concept Switch

POCSAG Post Office Code Standardization Advisory Group

POCT Passive Optical Component Technology • Point Of Care Testing

pocul Poculum

POD Parameter Override Display • Payload Operations Director • Payload Operations Division • Payment On Delivery • Pay On Death • Pay On Delivery • Perl Online Documentation • Piece Of Data • Place Of Discharge • Planning and Observational Data • Plan Of the Day • Podor (Senegal) (location identifier) • Point Of Delivery • Port Of Delivery • Port Of Disembarkation • Post Office Department • Post Office Directory • Precise Orbit Determination • Precision Orbit Determination • Preliminary Orbit Determination • Pressurized Optical Detector • Printing On Demand • Print On Demand • Probability Of Detection • Production Operations Directive • Program Operation Description • Program to Output Data • Project Operations Director • Project Organization Demographic • Proof Of Delivery • Proposed Operational Dates • Protective Oceanic Device • Publishing On Demand • Pulse Omission Detector • Purchase Order Deviation

PODAAC Physical Oceanography Distributed Active Archive Center

PODAF Postoperation Data Analysis Facility

PODAPS Portable Data Processing System

PODAS Portable Data Acquisition System

PODB Product Order Database

PODC Principles Of Distributed Computing (symposium)

PODDP Principles Of Digital Document Processing (workshop)

PODE Publications de l'Observatoire Dun Echt (periodical code)

PODIUM Presentation Overlay Display for Interactive Users of Media

PODM Preliminary Orbit Determination Method

PODP Principles Of Document Processing (workshop) • Public Office Dialing Plan

PODPS Postobservation Data Processing System

PODS Passive Orbital Disconnect Strut • Pilot Ocean Data System • Postobservation Data Processing System • Precise Orbit Determination System • Precision Orbit Determination System

PODTG Precision Orbit Determination Task Group

POE Fort Polk (USA-LA) (location identifier) • Pilot Operational Equipment • Point Of Embarkation • Point Of Entry • Port Of Embarkation • Port Of Entry • Power Open Environment • Precise Orbit Ephemeris • Precision Orbit and Ephemeris • Print Out Effect • Probability Of Error • Probing Origins of EGBR (network) • Program Office Estimate • Projected Operational Environment • Property Owning Expenses • Proximity Optical Device

POEAS Planetary Orbiter Error Analysis Study

POED Program Organization for Evaluation and Decision

POEM Photonics and Optoelectronic Material • Physical Oceanography of the Eastern Mediterranean • Polar Orbit Earth-observing Mission • Polar-Orbiting Earth Mission • Polar-Orbiting Earth Monitoring • Polar-Orbiting Environmental Mission • Principle Of Equivalence Measurement

POEMS Positron Electron Magnetic Spectrometer • Positron Electron Magnet Spectrometer (*)

POES Polar Operational Environmental Satellite • Polar-Orbiting Environmental Satellite • Polar-Orbiting Environmental System

POESY Pulsar Observation with Effelsberg Systems (MPIfR)

Poet Poetic • Poetical • Poetry

POET Primed Oscillator Expandable Transponder

POETRY Public Outreach, Education, Teaching and Reaching Youth (project)

P/OF Pinhole/Occulter Facility

POF Payload Operations Facility • Physics Of Failure • Pinhole/Occulter Facility • Planned Observation File • Planned Outage Factor • Plastic Optical Fibre • Polymer Optical Fiber • Poplar Bluff - Earl (USA-MO) (location identifier) • Programmable Operator Facility

POFS Private Operational Fixed Service

POG Peace - Order - Government • Psychological Operations Group • Port Gentil (Gabon) (location identifier) • Pressure Omega Center • Proposers' Observatory Guide • Propulsion Operating Guide

POGASIS Planetary Observation Geometry And Science Instrument Sequence

POGO Polar-Orbiting Geophysical Observatory • Programmer-Oriented Graphics Operation

POGS Polar-Orbiting Geomagnetic Survey

POGSI Policy Group on Scientific Information

POH Path Overhead • Pilot's Operating Handbook • Planned Outage Hours • Pocahontas (USA-IA) (location identifier) • Power-On Hours

POHel Publications de l'Observatoire de Helsinki (periodical code)

POHM Page-Oriented Holographic Memory

POHP Publications de l'Observatoire de Haute-Provence (periodical code)

POI Panoramic Optical Imager • Parking Orbit Injection • Path Overhead Identifier • Période d'Observation Intensive • Period Of Interest • Plan d'Opérations Intercalées • Plan Of Instruction • Pluto Orbital Insertion • Pluto Orbit Insertion • P-mode Oscillation Instrument • Point Of Information • Point Of Interconnection • Point Of Interest • Point Of Interface • Potosí (Bolivia) (location identifier) • Power-On/Off Interrupt • Principal Operations Inspector • Probability Of Intercept • Product Of Inertia • Program Of Instruction • Purchase Order Item

POIC Payload Operations and Integration Center • Payload Operations Integration Center

POID Power-On/Off Interrupt Delay

POIL Power Density Imbalance Limit

POINTER Particle Orientation Interferometer

POINTS Precision Optical Interferometer in Space (*) • Precision Optical Interferometry in Space

POIROT Passively-cooled Orbiting Infrared Observatory Telescope

POIS Processing Operations Information Center • Prototype Online Instrument System • Purchase Order Information System

POISE Panel On In-flight Scientific Experiments • Photosynthetic Oxygenation Illuminated by Solar Energy • Preoperational Inspection Services Engineering

POIst Publications de l'Observatoire d'Istanbul (periodical code)

Pol Poland • Polar • Polarity • Polarization • Polarize • Polarized • Polarizer • Police • Policía • Policy • Poligono • Polish • Political • Politics • Polygone

POL Pacífico Oriental Latinoamericano • Pemba (Mozambique) (location identifier) • Pénétration d'Orifice Lubrifié • Petroleum - Oil - Lubricants • Physics Online Library • Poland (ISO country code) • Polar • Pologne (code ISO de pays) • Port Of Loading • Post Online • Problem-Oriented Language • Procedure-Oriented Language • Process-Oriented Language • Program-Oriented Language • Proudman Oceanographic Laboratory (UK) • Provisional Operating Language

POLA Pointable Optical Linear Array

POLAD Political Adviser

POLAR Polarization Observations of Large Angular Regions (experiment) • Polarization Of Large Angular Regions • Potentials Of Large objects in the Auroral Region • Production Order Location And Record • Production Order Location And Record • Production Order Location And Reports

POLARIS Polar motion Analysis by Radio Interferometric Studies

POLARSCAT Polarimetric Scatterometer

PolAt Pollution Atmosphérique (periodical code)

POLC Pennsylvania Outdoor Lighting Council (USA-PA)

POLCA Pennsylvanians for an Outdoor Lighting Control Act (USA-PA)

POLDER Polarization and Directionality of Earth Radiation • Polarization and Directionality of Earth Reflectances (*)

POLDIRAD Polarimetric Diversity Radar

POLDPS Pioneer Off-Line Data Processing System

POLE Polar Ozone Lidar Experiment

POLEX Polar Experiment

Polg Poligono

POLGEN Problem-Oriented Language Generator

POLID Personnes - Organismes - Langues - Implantations - Dates

Polig Poligono

POLIS Parliamentary Online Information System • Polarimetric Littrow Spectrometer • Promoting Operational Links with Integrated Services

Polít Política

Polit Political • Politician

POLITE Plasmaspheric Observations of Light Ions in the Topside Exosphere

Poll Pollution

POLMF Preferred Output Language Media Format

POLO Problem-Oriented Language Organizer

PolPP Poluprovodnikovye Pribory i ikh Primenenie (periodical code)

PolRe Polar Research Memoirs (periodical code)

POLS Provisioning Online System

PolSi Poliarnye Siianiia (periodical code)

Polstr Polystyrene

Polthn Polyethylene

PolTM Poluprovodnikovaia Tekhnika i Mikroelektronika (periodical code)

Poly Polynesia • Polynesian • Polytechnic

Polyg Polygono

Polyn Polynesia • Polynesian

POLyo Publications de l'Observatoire de Lyon (periodical code)

PoM Plan of Management

POM Pallet-Only Mode • Particulate Organic Matter • Période Opérationnelle Moyenne • Persistent Object Manager • Petite Opération Millimétrique • Phase Of the Moon • Pilot's Operating Manual • Polycyclic Organic Matter • Polyoxymethylene • Port Moresby (Papua New Guinea) (location identifier) • Postal Operations Manual • Prefix-Omission Method • Printer Output Microfilm • Program Objective Memorandum • Program Operation Mode • Project Office Memorandum • Project Operations and Milestones • Proximity Operation Mode

POMA Pointable Optical Mosaic Array

PomAK Pomiary Automatyka Kontrola (periodical code)

POMC Period of Minimum Change • Principles Of Mobile Computing (conference)

POMDP Partially Observable Markov Decision Process

POME Programs - Objectives - Methods - Evaluation

POMI Preliminary Operating and Maintenance Instruction

POMic Publications of Michigan Observatory (periodical code)

POMil Pubblicazioni dell'Osservatorio Astronomico di Milano-Merate (periodical code)

POMin Publications of the Astronomical Observatory of the University of Minnesota (periodical code)

POMO Production-Oriented Maintenance Organization

POMOLA Poor Man's Optical Landing System

POMS Panel on Operational Meteorological Satellites • Parallel Observation Matching System • Polar-Orbiting Meteorological Satellite • Postmain-Sequence • Probability Of Mission Success

POMSEE Performance, Operating and Maintenance Standards for Electronic Equipment

POMSS Prepositioned Operational Materiel Storage Site

POMT Planning Operations Management Team

POMWG Program Objective Memorandum Working Group

PON Particulate Organic Nitrogen • Passive Optical Network • Poptun (Guatemala) (location identifier) • Project Order Number • Project Originator Number

PONAM Polar North Atlantic Margin

PONS Picture of NAG Splinefit

PONZ Precise Object Near Zenith

POO Poços de Caldas (Brazil-MG) (location identifier) • Post Office Order • Precision Optics Organization (EKC) • Principles Of Operations • Priority Operational Objective • Programmation Orientée Objet

POOC Posted Out Of Course

POOF Peripheral Online Oriented Function

POOL Parallel Object-Oriented Language • Probabilistic Object-Oriented Logic

POOMA Parallel Object-Oriented Methods and Applications

POOZ Permanently Open Ocean Zone

Pop Popular • Popularly • Populate • Populated • Population

PoP Period of Performance • Point of Presence

POP Parallel Observing Program • Parallel Ocean Program • Payload Operations Plan • Payload Optimized Program • Performance-Oriented Packaging • Period Of Performance • Perpendicular-to-Orbit Plane • Persistent Organic Pollutant • Physics Optical Propagation • Physics Outreach Program • Picture Out of Picture • Plug-in Optronic Payload • Point Of Presence • Point Of Purchase • Polar-Orbiting Platform • Polar Outflow Probe • Postobject Programming • Post Office Preferred • Post Office Protocol • Preburner Oxidizer Pump • Preflight Operations Procedure • Prelaunch Operations Plan • Principal Oscillation Pattern • Printing-Out Paper • Printout Paper • Programa de Orientación Plurianual • Programmed Operators and Primitives • Programme d'Orientation Pluriannuel • Program Operating Plan • Project Objective Plan • Project Operating Plan • Proof Of Principle • Proof Of Purchase • Provisioning Operational Plan • Puerto Plata (Dominican Rep.) (location identifier)

POPA Panel On Public Affairs • Payload Ordnance Processing Area • Programme d'Orientation Pluriannuel

POPad Pubblicazione dell Osservatorio di Padova (periodical code)

POPAE Protons On Protons And Electrons

POPal Pubblicazioni dell'Osservatorio Astronomico di Palermo (periodical code)

POPar Publication de l'Observatoire de Paris - Notes et Informations (periodical code)

POPB Palais Omnisports de Paris-Bercy (France)

PoPC Pile of Personal Computers

POPC Payload Operations Planning Center

POPE Polar-Orbiting Platform Element

POPIN Population Information Network

POPL People-Oriented Processing Language • Principles Of Programming Languages • Problem-Oriented Programming Language

POPO Push-On - Pull-Off

POPOS Post Office Point Of Sale

POPot Publikationen des Astrophysikalischen Observatoriums zu Potsdam (periodical code)

POPs Persistent Organic Pollutants • Principal Oscillation Patterns

POPS Pantograph Optical Projection System • Paperless Order Placement Service • Paperless Order Placement System • Paperless Order Processing Service • Paperless Order Processing System • Participatory Orientated Planetarium in the Schools • Participatory-Oriented Planetarium Show • People-Oriented Parsing Service • People-Oriented Processing System • Port Operational Performance Simulator • Positioning Orbital Propulsion System • Principal Oscillation Pattern System • Program for Operator Scheduling • Pyrotechnic Optical Plume Simulator

POPSAT Precise Orbit Positioning Satellite

POPSI Postulate-based Permuted Subject Indexing

POPU Push Over/Pull Up

POQ Polk Inlet (USA-AK) (location identifier)

Por Porosity • Porous • Portion • Portuguese

POR Pacific Ocean Region • Packet Overlay Retrofit • Parking Orbit Rendezvous • Pay On Receipt • Pay On Return • Payable On Receipt • Periodic Operation Report • Point Of Reference • Point Of Resolution • Pori (Finland) (location identifier) • Port Of Refuge • Power On Reset • Price On Request • Problem-Oriented Routine • Project Officer Report • Purchase Order Request

PORB Production Operations Review Board

PORCN Production Order Records Change Notice

POrCo Progress in Organic Coatings (periodical code)

PORD Performance and Operations Requirements Document

PORE Procurement-Oriented Requirements Engineering

PORM Plus Or Minus

PorMe Poroshkovaia Metallurgiia (periodical code)

POROTECS Potsdam Robotic Telescope Control System

PORR Preliminary Operations Requirements Review

PORS Periodiek Overleg Remote Sensing

PORSEC Pacific Ocean Remote Sensing Conference

Port Portable • Portrait • Portuario • Portugal • Portugués • Portuguese

PORT Photo-Optical Recorder Tracker • Portable • Postweb Open Resource Toolkit

Porta Portada

PORTAL Polar Orbiter Remapping and Transformation Application Library • Process-Oriented Real-Time Algorithmic Language

PORTS POCC Real-Time Support • Portable Remote Telecommunications System • Preliminary Operations Requirements and Test Support

PORV Pilot-Operated Relief Valve • Pressure-Operated Relief Valve

Pos Position • Positional • Positive • Possession • Possessive

PoS Point of Sale

POS Pacific Ocean Ship • Packet Over Sonet • Part Of Speech • Payload Operations Schedule • Payload Operations Simulation • Pénétration d'Orifice Sec • Period Of Service • Persistent Object Service • Personal Operating Space • Photo Optic System • Physics On Stage (project) • Plan d'Occupation des Sols • Plane Of the Sky • Point Of Sale (*) • Point Of Scale • Portable Oxygen System • Port of Spain - Piarco (Trinidad & Tobago) (location identifier) • Port Of Support • Position • Positive • Preferred Observation Schedule • Preferred Observation Sequence • Preoperational Support • Primary Operating Stocks • Primary Operating System • Primary Oxygen System • Probability Of Success • Probability Of Sufficiency • Probably Old System • Professional Operating System • Programmable Object Select • Programmable Option Select • Program Order Sequence • Purchase Order Supplement • Trinidad (Trinidad & Tobago) (location identifier)

POSA Preliminary Operating Safety Analysis

POSANG Position Angle

POSC Personnel Occupational Specialty Code • Primary Operations System Code

POSD Pointed Observation by Survey Detectors • Program for Optical System Design • Project Operation Support Division

POSDCORB Planning, Organizing, Staffing, Directing, Coordinating, Reporting and Budgeting

POSE Picture-Oriented Software Engineering • Portable Operating System Environment

POSEIDON Positioning Ocean Solid Earth Ice Dynamics Orbiting Navigator • Projet d'Observatoire de Surveillance et d'Études Intégrées de la Dynamique des Océans

POSH Parents for Objective Science and History (USA)

POSIT position • Profile for Open System Internetworking Technology

POSIX Portable Operating System for Interactive Executive • Portable Operating System Interface Exchange • Portable Operating System Interface for Unix (*)

Posn Position

POSN Position

POSNAV Position and Navigation • Positioning and Navigation • Position Navigation

POSO People Over Sixty Online

Poss Possession • Possessive • Possibility • Possible • Possibly

POSS Palomar Observatory Sky Survey • Passive Optical Satellite Surveillance • Prototype Optical Satellite Surveillance • Prototype Optical Surveillance System

POSSE Pluto and Outer Solar System Explorer

Post Posterior • Postulación

PoST Position-Sensing TES

POST Pacific Ocean Science and Technology • Parliamentary Office of Science and Technology • Part-Of-Speech Tagger • Passive Optical Scan Tracker • Passive Optical Seeker Technique • Passive Optical Seeker Technology • Payload Operations Support Team • Piezoelectric-Oscillator Self-Tuned • Point Of Sale Terminal • Polar Observatory Stratospheric Telescope • Polar Stratospheric Telescope • Portable Optical Sensor Tester • Portable Optical Sensor Tracking • Portable Submillimeter Telescope • Positive • Power-On Self-Test • Predevelopment Operational System Test • Production-Oriented Scheduling Technique • Programmable Optical Sensor Tractor • Programmer Operating Standards Technique • Program to Optimize Shuttle Trajectories • Program to Optimize Simulated Trajectories

POSTCODE Polar Stratospheric Clouds and Ozone Depletion

Postf Postfach

PostGebO Postgebührenordnung

Posth Posthume • Posthumous • Posthumously

Posthum Posthumous • Posthumously

PoSTI Policies for Sustainable Technological Innovation

POSTNET Postal Numeric Encoding Technique

Postop Postoperative • Postoperatively

POStr Publications de l'Observatoire de Strasbourg (periodical code)

Pot Potassium • Potential • Potentiometer

POT Picture Object Table • Pitch-Orthogonal Thrust • Point Of Termination • Port Antonio (Jamaica) (location identifier) • Potable • Potentiometer • Preplanned Observation Tape • Preproduction Operational Test • Proof Of Technology • Proximity Operations Trainer

POTC PERT Orientation and Training Center

POTG Psychological Operations Task Group

POTO Plans, Operations, and Training Officer

POTor Pubblicazioni Varie Fuori Serie dell'Osservatorio Astronomico di Torino (periodical code)

POTou Publications de l'Observatoire de Toulouse (periodical code)

POTP Power On Time Pad

POTS Pan-Only Triangulation System • Plain Old Telephone Service • Plain Old Telephone System • Plain Ordinary Telephone Service • Portable Optical Tracking Station • Precision Optical Tracking System • Purchase Of Telecommunications Services • Purchase Of Telephone and Services

POTUS President Of The United States

POTV Personnel Orbit Transfer Vehicle

POTW Publicly-Operated Treatment Work • Publicly-Owned Treatment Work

PoU Point of Use

POU Point Of Use • Poughkeepsie (USA-NY) (location identifier)

POUS Pousada

pov Persistence of Vision (file extension)

P-ov Poluostrov

POV Peak Operating Voltage • Perfect Operational Volatility • Persistence Of Vision • Pneumatically-Operated Valve • Pneumatic-Operated Valve • Point Of View • Póvoa • Presov (Slovakia) (location identifier) • Privately-Owned Vehicle • Program-Oriented View • Proximity Operations Vehicle

POVRO Publications of the Owens Valley Radio Observatory (periodical code)

POW Please Oblige With • Portoroz (Slovenia) (location identifier) • Prisoner Of War • Program Of Work

Powd Powder

POWER Performance Optimization With Enhanced RISC • Performance Optimized With Enhanced RISC • Portable Workstation for Education in Europe

POWG Payload Operations Working Group • Payload Orbital Working Group

POWR Predictive Optical Word Recognition

POWS Project Operating Work Statement

POX Point Of Exit • Pontoise (France) (location identifier) • Pressure Oxidation • Prism Objective X

POxf Publications University of Oxford Department of Astrophysics (periodical code)

POY Powell (USA-WY) (location identifier)

POZ Poznań (Poland) (location identifier)

P° Paseo • Paso

pp Pages • Páginas • Par Procuration • Past Participle • Per Person • Perpetuo • Per Procura • Per Procurationem • Postpaid • Prepaid

Pp Pappband • Porte Pagado

P-P Peak-to-Peak • Pitch-Pitch • Probability-Probability • Push-Pull

P&P Planning & Programming • Policies & Procedures • Postage & Packing • Principles & Practices • Procurement & Production

P/P Patch Panel • Point-to-Point • Prelaunch/Postlanding • Preplanned • Printer/Plotter • Publication/Printing

PP Brazil (international aircraft registration code) • Pacco Postale • Pacific Plate • Page Printer • Pair Production • Panel Point • Parallel Port • Parcel Post • Parish Priest • Parliamentary Paper • Partial Pressure • Partial Program • Participatory Program • Particle-Particle • Partitioned Parallelism • Passage Point • Past Participle • Patch Panel • Pay Period • Paysan Pingre • Peak-to-Peak • People's Photometer • Perfect Program • Peripheral Processor • Permanent Party • Permanent Pass • Persistent Penetration • Personal Property • Petite Pénétration • Physical Profile • Piecewise Polynomial • Pillar Point • Pilot Project • Pipeline Processing • Pisces-Perseus • Pitch-Pitch • Planetary Program • Planning Package • Planning Purpose • Plasma Processing • Plasma Propulsion • Play or Pay • Point-to-Point • Polypropylene • Port Payé • Posa Piano • Positive Polarization • Possibly Present • Postage Paid • Postpaid • Post Pay • Postponed • Postprocess • Pouvoirs Publics • Power Package • Power Plant • Power Pole • Precise Prediction • Precise Predictor • Preference Profile • Premodulation Processor • Prepaid • Preparatory Programme • Prepositional Phrase • Preprinted • Preprocessor • Preproduction • Present Position • Pressure Proof • Primary Point • Principal Point • Principal Polarization • Principles and Practices • Printer Page • Printer-Plotter • Print/Punch • Priority Precedence • Prior Permission • Privately Printed • Probability-Probability • Process Planning • Procurement Plan • Professeur Principal • Program Package • Program Performance • Program Plan • Programming Plan • Program Performance • Program Plan • Progressive Penetration • Progress Payment • Projected Proponency • Projection Pursuit • Project Proposal • Proof Plot • Proof Plotter • Propeller Pitch • Propulsion Power • Protection Plan • Proton-Proton •

Protoplanet • Pruppacher-Pitter • Public Policy • Public Property • Pulsar Processor • Pulsating Penetration • Pulse Polarography • Pushchino Pulsar • Push-Pull

ppa PowerPoint Add-in (file extension)

PP&A Prepay & Add

PPA Pacific Planetarium Association (USA-CA) • Pampa - Perry Lefors (USA-TX) (location identifier) • Parallel Performance Analysis • Parallel Plate Analyzer • Parité de Pouvoir d'Achat • Payload Performance Analysis • Periodical Publishers' Association (UK) • Photo Peak Analysis • Physical Page Address • Pilots Personal Assistant • Pitch Precession Amplifier • Pollution Prevention Act (USA) • Power Pitch-Around • Precision Pulsar Astrophysics • Preplanned Attack • Prepay and Add • Prime Provisioning Activity • Product Performance Agreement • Professional Programmers Association • Program Problem Area • Program, Project and Activity • Program, Project or Activity • Prompt Payment Act (USA) • Proposal Preparation Assistant • Pulsed Power Amplifier • Pulse Plasma Accelerator • Pupil Plane Assembly

PPAC Parallel Plate Avalanche Counter

PPAD Program Planning and Analysis Department

PPAG Proposed Public Affairs Guidance

PPAP Programme Personnalisé d'Aide et de Progrès

PPAR Program/Project Approval Review • Purpose, Priority, Allocation and Restrictions

PPARC Particle Physics and Astronomy Research Council (UK)

Ppart Plupart

PPAS Parallel Poll Active State • Photometric Periods of Artificial Satellites • Piezoelectric Photoacoustic Spectroscopy

ppb Parts per Billion

PPB Paper - Printing - Binding • Parts Per Billion • Planning - Programming - Budgeting • Plans - Program - Budget • Polar Patrol Balloon • Positive Polarization Branch • Positive Pressure Breathing • Power Plant Bulletin • Presidente Prudente A. de Barros (Brazil-SP) (location identifier) • Primary Propulsion Branch • Private Posting Box • Program Performance Baseline • Provisioning Parts Breakdown • Public Policy Brief (AAS)

PPBAS Planning-Programming-Budgeting-Accounting System • Program Planning-Budgeting-Accounting System

PPBEE Planning - Programming - Budgeting - Enactment - Execution

PPBERS Program Performance and Budget Execution Review System

PPBES Planning-Programming-Budgeting-Evaluation System • Planning-Programming-Budgeting-Execution System • Program Planning-Budgeting-Evaluation System

PPBG Preliminary Program and Budget Guidance

PPBIT Preperiodic Built-In Test

ppbm Parts per Billion by Mass

PPBM Pulse Polarization Binary Modulation

PPBS Planning, Programming and Budgeting System • Positive Pressure Breathing System

ppbv Parts per Billion per Volume

PP&C Production Planning & Control • Program Planning & Control • Project Planning & Control

PPC Parallel Poll Configure • Payload Preparation Complex • Pay Per Channel • Phased Provisioning Code • Platform Position Computer • Pointed Proportional Counter • Positive Peer Control • Pour Prendre Congé • Power Pack Charger • Power Plant Change • Preprocessing Center • Preprotostellar Core • Primary Power Control • Procurement Policy Committee • Production Planning and Control • Programmable Protocol Converter • Program Planning and Control • Program-to-Program Communication • Project Planning and Control • Prospect Creek (USA-AK) (location identifier) • Provisional Performance Criterion • Public Policy Committee • Pulsed Power Circuit

PPCA Preliminary Physical Configuration Audit • Probabilistic Principal Component Analysis

PPCC Past Performance Coordinating Committee • Past Performance Coordinating Council • Peaks Pike Community College (USA-CO)

PpCF Pounds per Cubic Foot

PPCF Plasma Physics and Controlled Fusion (periodical code) • Pounds Per Cubic Foot

PpCI Pounds per Cubic Inch

PPCI Pounds Per Cubic Inch

PPclC Preprotocluster Core

PPCM Plus Petit Commun Multiple

ppcn Per Person per Night

PPCO Provisioning Principal Contracting Officer • Provisioning Procuring Contracting Officer

ppCO2 Partial Pressure of Carbon Dioxide

PPCP Postprocessing Computer Program • Principles and Practice of Constraint Programming (workshop)

PPCS Person-to-Person Collect Special • Precision Pointing Control System • Production Planning and Control System

PPCU Partial Payload Checkout Unit

PPC4I Power Projection Command, Control, Communications, and Computers and Information • Power Projection Command, Control, Communications, and Computers and Infrastructure

ppd Postscript Printer Description (file extension)

Ppd Postpaid • Postponed • Prepaid

PPD Humacao - Palmas del Mar Hotel (Puerto Rico) (location identifier) • Partial Packet Discard • Particle Physics Department • Particle Physics Division • Payload Position Data • Peripheral Pulse Distributor • Plan-Position Display • Point Position Data • Postscript Printer Definition • Postscript Printer Description • Postponed • Power Pitch-Down • Predefined Process Diagram • Primary Point Digitization • Printer Page Description • Proficiency Pay Designator • Program for Persons with Disabilities • Program Package Document • Program Planning Directive • Program Planning Document • Project Planning Directive • Propulsion and Power Division • Provinciale Planologische Dienst • Provisioning Procurement Data

PPDB Point Positioning Database

PPDD Preliminary Project Design Description

PPDF Poisson Probability Distribution Function

PPDG Particle Physics Data Grid

p. pdo. Próximo Pasado

PPDO Per Person - Double Occupancy

PPDP Preliminary Project Development Plan • Preprogram Definition Phase • Principles and Practice of Declarative Programming (conference)

PPDR Prepreliminary Design Review

PPDS Particle Physics Data System • Personal Printer Data Stream • Physical Property Data Service • Pilot Planetary Data System

PPDU Payload Power Distribution Unit • Presentation Protocol Data Unit

P&PE Principles & Practices of Engineering

PPE People - Place - Events • Personal Protection Equipment • Personal Protective Equipment (*) • Personnel Protection Equipment • Phase Partitioning Experiment • Philosophy - Politics - Economics • Plan - Prepare - Execute • Plants, Property and Equipment • Platform Position Equipment • Politics - Philosophy - Economics

• Portable Purge Equipment • Power Plant Electric • Predicted Period of Effect • Premodulation Processing Equipment • Preproduction Evaluation • Principles and Practices of Engineering • Problem Program Efficiency • Problem Program Evaluation • Programme Prioritaire pour l'Environnement • Program Planning and Evaluation • Prototype Production Evaluation • Puerto Penasco (Mexico) (location identifier) • Purchasing Power Equivalent

PPEC Permanent Periodic Error Correction

PPECO Preprogrammed EVA Controlled Operations

PPEDC Printing and Publishing Economic Development Committee

PPEE Periodica Polytechnica on Electrical Engineering (periodical code)

PPEP Particles Physics Exhibits Project • Pen Plotter Emulation Program • Plasma Physics and Environment Perturbation

PPES Photopyroelectric Spectroscopy

PPEUC Particle Physics and Early Universe Conference

PPF Paging Proceed Flag • Paper Publication Frequency • Parsons (USA-KS) (location identifier) • Payload Processing Facility • Personal Property Floater • Phase Pushing Factor • Polar Platform • Preprints in Particles and Fields (SLAC) • Primary Processing Facility • Provision of Production Facilities

PPFA Page Printer Formatting Aid

PPFD Photosynthetic Photon Flux Density

PPFF Poisson Probability Frequency Function

PPFP Programme Pluriannuel de Finances Publiques (France)

PPFS Portable Parallel File System • Product Performance Feedback System

PPG Pago Pago (American Samoa) (location identifier) • Parallel Programming Group • Personnel Priority Group • Personnel Processing Group • Piezoelectric Power Generation • Planning and Programming Guidance • Planning Policy Guidance • Plasma Power Generator • Plasma Pulse Generator • Program Planning Group • Program Planning Guidance • Project Planning Group • Propulsion and Power Generation • Pulses Per Group

PPGA Plastic Pin Grid Array

PPGM Planning and Programming Guidance Memorandum

pph Parts per Hundred

Pph Pamphlet

PpH Pounds per Hour • Pulses per Hour

PPH Paid Personal Holidays • Parts Per Hundred • Peraitepuy (Venezuela) (location identifier) • Pounds Per Hour • Primeval Particle Horizon • Pulses Per Hour

PPHA Peak Pulse Height Analysis

pphm Parts per Hundred Million

PPHM Parts Per Hundred Million

PpHP Pounds per Horsepower

PPHP Pounds Per Horsepower

PpI Particles Per Inch • Pixels per Inch

PPI Packing, Postage and Insurance • Parcel Post Insured • Particles Per Inch • Patient Package Insert • Pay Per Incident • Pixels Per Inch • Planetary Plasma Interaction • Planned Position Indicator • Plan Particulier d'Intervention • Plan Position Indicator • Policy Proof of Interest • Port Pirie (Australia-SA) (location identifier) • Preplanned Product Improvement • Preprocessing Instruction • Printed Postage Impression • Processor-to-Processor Interface • Producer Price Index • Programmable Peripheral Interface • Programmation Pluriannuelle d'Investissements • Programme Prévisionnel des Investissements • Program Package for Improvement • Project Procurement Instruction • Proposal Preparation Instruction • Public Policy Institute • Public-Private Interface • Pulse Position Indicator

PPIB Program Protection Implementation Plan

PPIC Pollution Prevention Information Clearinghouse

PPID Parent Process Identification • Parent Process Identificator • Parent Process Identifier

PPIE Pollution Prevention Information Exchange

PPIF Photographic Processing and Interpretation Facility • Photo Processing Interpretation Facility

PPII Protostars and Planets II (conference)

PPIII Protostars and Planets III (conference)

PPIL Parts Provisioning Index List • Priced Provisioned Item List

PPIMS Past Performance Information Management System • Personal Property Information Management System

PPIO Priced Provisioned Item Order

PPIP Physics Postdoctoral Information Pool

PPITB Printing and Publishing Industry Training Board (UK)

PPIV Protostars and Planets IV (conference)

PPJ Production au Plus Juste • Pulau Panjang (Indonesia) (location identifier)

PPK Personal Preference Kit • Petropavlovsk (Kazakhstan) (location identifier)

PPKDD Principles and Practice of Knowledge Discovery in Databases (conference)

PPKTP Periodic Poled Potassium Titanyl Phosphate

Ppl Participle

P/PL Primary Payload

PPL Packet-by-Packet Layer • Parts Provisioning List • People • Phaplu (Nepal) (location identifier) • Planning Parts List • Plasma Physics Laboratory • Plasma Propulsion Laboratory • Polymorphic Programming Language • Populated Place • Predictive Period Laser • Preferred Parts List • Preferred Products List • Preliminary Parts List • Priced Parts List • Primary Payload • Private Pilot License • Process Programming Language • Production Payload • Program Parts List • Program Production Library • Provisional Parts List • Provisioning Parts List • Purge Parameter List • Pyrotechnically-Pumped Laser

PPLAN Programming Plan • Program Plan

Pple Participle

PpLF Pounds per Linear Foot

PPLF Pounds Per Linear Foot

PPLI Precise Participant Location and Identification • Precise Participant Location Identification • Precise Participant Location Information • Precise Position Location Information • Provisioning Parts Lists Index

PPLL Programmable Phase-Locked Loop

PP&LS Production Planning & Logistics Support

PPLS Production Planning and Logistics Support • Propellant and Pressurant Loading System

PPLUS Polar Platform Utilization Study

ppm Pages per Minute • Parts per Million • Parts per Minute • Portable Pixelmap (file extension) • Pulsaciones por Minuto • Pulses per Minute

PPM Pages Per Minute • Pair Production in Matter • Parametric Potential Method • Parts Per Million • Payload Project Manager • Peak Power Meter • Peak Programme Meter • Perfect Prog Method • Periodic Permanent Magnet • Period of Performance and Maintenance • Perl Package Manager • Piecewise Parabolic Method • Planned Preventive Maintenance • Pompano Beach (USA-FL) (location identifier) • Portable Pixmap • Position and Pay Management • Positions and Proper Motions • Pounds

Per Minute • Principal Period of Maintenance • Priority Placement Program • Production Prototype Model • Programmable Pointing Mode • Program Project Management • Project Profile Manual • Proton Precession Magnetometer • Pulse Phase Modulation • Pulse-Position Modulation • Pulse Power Module • Pulses Per Minute • Punch Press Machine

PPMA Printing Paper Manufacturers Association

PPMDE Piecewise Parabolic Method - Direct Eulerian

PPME Pacific Plate Motion Experiment • Periodica Polytechnica on Mechanical Engineering (periodical code)

PPMF Positive-Polarity Magnetic Field

PPMG Professional Publishers Marketing Group

PPMN Pulse-Position Modulation of Natural (type)

PPMS Photogrammetry Production Monitoring System • Physical Property Measurement System • Program and Project Management System • Program Performance Measurement System • Program Planning and Management System • Project Performance Measurement System

PPMtO Publications of the Purple Mountain Observatory (periodical code)

ppmv Parts per Millions per Volume

PPN Parameterized-Postnewtonian (model) • Popayan (Colombia) (location identifier) • Preplanetary Nebula • Procurement Program Number • Project Programmer Number • Protoplanetary Nebula (*) • Public Packet Network

PPNe Preplanetary Nebulae • Protoplanetary Nebulae (*)

PPNL Pacific Northwest National Laboratory

PP&O Plans, Policies & Operations

PPO Pédagogie Par Objectifs • Petit Prisme-Objectif (OHP) • Photographic Program Office • Pikes Peak Observatory (USA-CO) • Planetary Protection Officer • Plans, Policies and Operations • Powell Point (Bahamas) (location identifier) • Power Plant Operating • Preferred-Provider Organization • Prior Permission Only • Procurement Planning Officer • Program Planning Office • Projected Program Objective • Publications and Printing Office • Push-Pull Output

PPOC Procurement Point Of Contact

PPOP Principles and Practices of Parallel Programming (conference)

PPOV Pilot Point Of View

ppO2 Partial Pressure of Oxygen

ppp Pixels par Pouce • Points par Pouce

PPP Package Processing Point • Parallel Pattern Processor • Parcours Pédagogique Personnalisé • Payload Patch Panel • Peak Pulse Power • Penrose Pair Production • Pensée Positive Permanente • Permanent Power Pack • Personal Pension Plan • Personal Performance Profile • Personnel Performance Profile • Personne Physique Payée • Phased Project Planning • Planning Purpose Proposal • Plasma Physics Package • Point Protocol Problem • Points Par Pouce • Point-to-Point Protocol • Polluter Pays Principle • Pople - Parr - Pariser (method) • Power Projection Platform • Préfecture de Police de Paris (France) • Preliminary-Phase Pilot • Preliminary Program Plan • Préparation de Pénétration Profonde • Pretty Poor Planning • Primary Patch Panel • Priority Placement Program • Program Protection Plan • Program Protection Planning • Proserpine (Australia-QLD) (location identifier) • Psychology - Philosophy - Physiology • Public Private Partnership • Publishing and Print Planning • Purchasing Power Parity

PPPI Preplanned Product Improvement

PPPJ Principles and Practice of Programming in Java (conference)

PPPL Princeton Plasma Physics Laboratory (USA-NJ) • Printed Planning Parts List

PPPM Particle - Particle - Particle - Mesh • Preservation - Packaging - Packing - Marking

PPPoE Point-to-Point Protocol over Ethernet

PPPP Proposed Partial Package Program

PPPS Postpostpostscriptum

P&PQ Productivity & Product Quality

PPQ Paraparaumu (New Zealand) (location identifier) • Possible Parliamentary Question • Productivity and Product Quality

P&PQI Productivity & Product Quality Improvement

PPQI Productivity and Product Quality Improvement

P&PQIC Productivity & Product Quality Improvement Council

PPQIC Productivity and Product Quality Improvement Council

PPQT Preproduction Prototype Qualification Test • Preproduction Qualification Test

PPQT&E Preproduction Qualification Test & Evaluation

PPQTE Preproduction Qualification Test and Evaluation

Ppr Paper • Proper

PPR Pasir Pangarayan (Indonesia) (location identifier) • Payload Preparation Room • Peak Power Response • Periodic Performance Report • Periodic Personnel Report • Permanent Pay Record • Photopolarimeter Radiometer • Physical Progress Review • Plan de Prévention des Risques • Portable Pulsar Receiver • Prepublication Review • Present Participle • Printed Paper Rate • Prior Permission Required • Programme Prioritaire • Program Planning Review • Program Progress Review • Projection Pursuit Regression • Proprietary Procurement Request • Purchase Parts Request

PPRAG Past Performance Risk Assessment Group

PPRB Preliminary Proposal Review Board

PPREPT Periodic Personnel Report

PPRI Plasma Physics Research Institute (USA-CA)

PPRN Plan de Prévention des Risques Naturels • Preliminary Publication Revision Notice • Purchased Parts Requirement Notice

PPRNG Parallel Pseudo-Random Number Generator

p. pro. Per Procurationem

PPRR Preliminary Program Requirements Review • Program Prereleased Review

PPRS Promotion and Placement Referral System

Pprt Plupart

pps Packets per Second • Post Postscriptum • PowerPoint Slideshow (file extension) • Pulses per Second

PpS Packets per Second • Pounds per Second • Pulses per Second

PPS Additional Postscript • Packets Per Second • Page Printing System • Parallel Pattern Search • Parallel Processing System • Paraphenylene Sulfide • Partial-Pressure Suit • Patchboard Programming System • Payload Pointing System • Payload Positioning System • Payload Power Switch • Peripheral Processing System • Photographic Processing System • Photopolarimeter Spectrometer • Photopolarimeter System • Photovoltaic Power Supply • Piece Part Specification • Pipeline Processing System • Pisces-Perseus Supercluster • Plant Protection System • Plasma Power Supply • Pneumatic Power System • Postpostscriptum • Postproduction Support • Pounds Per Second • Power Personal Systems • Precise Positioning Service • Precise Positioning System • Precision Pointing System • Precision Power Supply • Precision Processing System • Préparation de Pénétration Soutenue • Primary Power Standard • Primary Power Supply • Primary Power System • Primary Propulsion System • Princeton Planetary Society (USA-NJ) • Printer/Plotter System • Production Planning Software • Product Performance Survey

• Product Performance System • Programmable Patch System • Programmable Power Supply • Program Performance Schedule • Program Performance Specification • Program Protection Survey • Program-Provided Summary • Projektplannungsgesellschaft • Proposal Pricing System • Provisioning Performance Schedule • Public Products Service • Puerto Princesa (Philippines) (location identifier) • Pulsed Power Supply • Pulse Power Supply • Pulses Per Second • Purchasing Power Standard • Second Postscript

PPS&C Programming Practices, Standards & Conventions

PPSC Physical Profile Serial Code • Programming Practices, Standards and Conventions

PpSF Pounds per Square Foot

PPSF Pounds Per Square Foot

PpSI Pounds per Square Inch

PPSI Pounds Per Square Inch • Prompt Photometric Sunspot Index

PPSL Program Parts Selection List

PPSN Parallel Problem Solving from Nature (conference) • Public Packet Switched Network • Public Packet Switching Network

PPSO Personal Property Shipping Office

PPSP Postproduction Support Plan

PPSR Progress Payment System Review

PPSS Postproduction Software Support • Precise Positioning Service Security

PPST Photothermal and Photoacoustic Science and Technology

ppt Parts per Thousand • Parts per Trillion • Powerpoint Presentation (file extension)

Ppt Plupart • Precipitate

PPT Papeete (Tahiti) (location identifier) • Parts Per Thousand • Parts Per Trillion • Peak Power Tracker • Peak Power Transfer • Performance Price Tradeoff • Periodic Programs Termination • Petite Propriété Terrienne • Physical Prototyping Team • Picture Page Table • Picture Planet Table • Pitch Precision Torquer • Plasma Phase Transition • Port Processing Time • Positive Permanent Thinking • Postproduction Test • Prelaunch Package Test • Preproduction Test • Program Performance Test • Program Properties Table • Project Planning Technique • Pulsed-Plasma Thruster • Pulse Pair Timing • Pulse Plasma Thruster • Punched Paper Tape

Pptd Precipitated

PPTD Products/Process Technology Division

PPTE Pays Pauvre Très Endetté • Periodica Polytechnica on Transportation Engineering (periodical code) • Polyphenyleneterephthalamide

PPtF Passivated Platinum Film

Pptg Precipitating

Pptn Precipitation

PPTP Point-to-Point Tunneling Protocol • Program Performance Test Procedure

PPTR Punched Paper Tape Reader

pptv Parts per Trillion per Volume

Ppty Property

PPU Papun (Myanmar) (location identifier) • Parallel Poll Unconfigure • Payload Processing Unit • Peripheral Processing Unit • Pi Puppids (meteor stream) • Pixel Processing Unit • Power Processor Unit • Power Providing Unit • Preprototype Unit • Prime Power Unit • Program-Produced Underlining • Protocol Processing Unit

PpV Pay per View

PPV Pay Per View • Plus Proche Voisin • Polyphényle de Vinyle • Polyphenylene Vinylene • Port Protection (USA-AK) (location identifier) • Position-Position-Velocity • Postproduction Verification • Preproduction Verification

PPVP Production Vacuum Pumping System

PPW Papa Westray (UK) (location identifier) • Plane-Polarized Wave

PP&WC Process Planning & Work Control

PPWC Process Planning and Work Control

ppx PagePlus Template (file extension)

PPX Param (Papua New Guinea) (location identifier) • Point-to-Point Switching • Private Packet Exchange

PPY Pouso Alegre (Brazil-MG) (location identifier)

PPZ Proton Polar Zone • Puerto Paez (Venezuela) (location identifier)

pp2pp Proton-Proton to Proton-Proton (experiment)

P&Q Productivity & Quality

PQ Parliamentary Question • Parque • Personality Quotient • Planetary Quarantine • Postal Quarter • Preceding Question • Premier Quartier • Previous Question • Primo Quarto • Priority Queue • Province de Québec (Canada) • Québec (Canada)

PQA Palm Query Application • Preliminary Quantitative Analysis • President's Quality Award (USA) • Procurement Quality Assurance • Production and Quality Assurance • Professionally Qualified Amateur • Protected Queue Area

PQAP Planned Quality Assurance Program • Product Quality Assurance Plan

PQBD Projet Qualité des Bases de Données

PQC Phu Quoc - Duong Dang (Viet Nam) (location identifier) • Precision Quartz Crystal • Procurement Quality Control • Production Quality Control • Product Quality Council

PQCD Perturbative Quantum Chromodynamics

PQD Percentage Quartile Deviation • Prequalification Document • Pyroelectric Quad Detector

PQDMB Percentage Quartile Deviation Median Bias

PQDR Product Quality Deficiency Report

Pqe Parque

PQE Path Quality Evaluation • Procurement Quality Engineering • Product Quality Evaluation • Progress in Quantum Electronics (periodical code)

PQEP Product Quality Evaluation Plan

PQET Print Quality Enhancement Technology

PQFP Plastic Quad Flat Pack • Plastic Quad Flat Package

PQG Plasma of Quarks and Gluons

PQGS Propellant Quantity Gauging System

PQI Presque Isle (USA-ME) (location identifier) • Professional Qualification Index • Propellant Quantity Indicator

PQIC Productivity and Quality Improvement Council • Product Quality Improvement Council

PQJ Parkes-Quarter-Jansky (sample)

PQL Picture Query Language • Presse Quotidienne Locale

PQLI Physical Quality of Life Index

PQM Palenque (Mexico) (location identifier) • Preliminary Qualification Motor • Product Quality Management • Pulse Quaternary Modulation

PQMD Propellant Quantity Measuring Device

PQN Presse Quotidienne Nationale • Principal Quantum Number

PQP Planetary Quarantine Plan • Prequalification Prototype

PQPF Probabilistic Quantitative Precipitation Forecasting

PQQ Port Macquarie - Wauchope (Australia-NSW) (location identifier)

PQQPRI Provisional Qualitative and Quantitative Personnel Requirements Information

PQR Presse Quotidienne Régionale • Project Quality Responsible

PQRVU Programmazione Quadratica Ricorsiva con Vincoli di Uguaglianza

PQS Parallel Query Server • Personnel Qualification Standard • Pilot Station (USA-AK) (location identifier) • Promotion Qualification Score

PQT Preliminary Qualification Test • Preliminary Qualification Testing • Production Qualification Test • Production Qualification Testing • Production Quality Test • Production Quality Testing • Professional Quick Take • Prototype Qualification Test • Prototype Qualification Testing

pr Payroll • Per

Pr Pair • Paper • Power • Praça • Prandtl Number • Praseodym • Praséodyme • Praseodymium • Preference • Preferred • Present • Pressure • Price • Priest • Priester • Prime • Prince • Prins • Print • Printed • Printer • Printing • Private • Professeur • Pronoun • Proper • Prove • Proved • Provençal • Provence (France) • Provincial

P+R Park + Ride

P-R Period-Radius • Pitch-Roll • Poynting-Robertson • Precision-Recall

P/R Payroll • Planner/Replanner • Productivity/Respiration

P&R Performance & Resources • Plans & Requirements • Plans & Resources • Price & Review

PR Packet Radio • Pair • Parallax and Refraction • Paraná (Brazil) • Parcel Receipt • Parliamentary Report • Partially Reusable • Partial Response • Path Replacement • Pattern Recognition • Payroll • Pearson-Readhead • Peer Review • Pénétration Réduite • Pénétration Régionale • Pénétration Répétitive • People's Republic • Percentage Rate • Percentile Rank • Performance and Resources • Performance Ratio • Performance Report • Performance Requirement • Period-Radius • Personal Representative • Photographic Reconnaissance • Photo Reconnaissance • Photoreflectance • Photo-Refractive • Photo Revision • Phrase Recognition • Physical Review • Pitch Ratio • Pitch-Roll • Planetary Radar • Planning Reference • Please Return • Polarization Ratio • Populus Romanus • Portaria • Porto Rico (code ISO de pays) • Position Record • Postal Regulations • Poste Restante • Potential Requirement • Power Reactor • Poynting-Robertson • Prairie • Precipitation Radar • Precipitation Rate • Precision-Recall • Préfecture • Preliminary Report • Preliminary Review • Premise Router • Press Release • Press Representative • Pressure • Pressure Ratio • Pressure Regulator • Primary • Primary Responsibility • Priorities • Priority Regulation • Privé • Prize Ring • Probabilistic Reasoning • Problem Report • Procurement Regulation • Procurement Request • Procurement Requisition • Production Rate • Production Requirement • Profit Rate • Program Register • Program Reliability • Program Requirement • Program Resource • Program Return • Progress Report • Project Report • Proportional Representation • Proposed Regulation • Proposed Request • Pro Rata • Pseudo-Random • Public Relations • Puerto Rican • Puerto Rico (ISO country code) • Pulse Rate • Purchase Request • Purchase Requirement • Pure Radiation • Pyrolytic Release

pra Praecipitatio

PRA Paperwork Reduction Act (USA) • Parabolic Reflector Antenna • Parana (Argentina-ER) (location identifier) • Partly Resolved Array • Peak Recording Accelerograph • Pendulous Reference Axis • Physical Review A (periodical code) • Physical Review Abstracts (periodical code) • Planetary Radio Astronomy • Preamble • Precession Axis • Precision Radar Approach • Primary Rate Access • Primary Review Authority • Print Alphanumerically • Probabilistic Relational Algebra • Probabilistic Risk Analysis • Probabilistic Risk Assessment • Probability Risk Assessment • Production Reader Assembly • Programme de Recherches Avancées • Program Reader Assembly • Projected Requisition Authority • Public Relations Adviser • Public Roads Administration

PrAA Progress in Astronautics and Aeronautics (periodical code)

PraAc Prague Academia (periodical code)

PRAB Program Resource Allocation Board

PRAC Program Resource Advisory Committee • Public Relations Advisory Committee

PRACA Problem Reporting, Analysis and Corrective Action • Problem Reporting And Corrective Action

PRACDA Polarimetric Radar Control and Data Acquisition

PRACL Page-Replacement Algorithm and Control Logic

PRACSA Public Remote Access Computer Standards Association (USA)

Prad Pradesh

PRAD Pitch Ratio Adjust Device

PrAeS Progress in Aerospace Sciences (periodical code)

Prag Pragmatic • Pragmatism

PRAG Performance Risk Analysis Group • Performance Risk Assessment Group • Procurement Risk Analysis Group • Professeur Agrégé

PrAiA Problemy Arktiki i Antarktiki (periodical code)

Pral Principal

PRAM Parallel Random Access Machine • Parallel Random Access Memory • Parameter Random Access Memory • Pipelined Random Access Memory • Productivity - Reliability - Availability - Maintainability • Programmable Random Access Memory • Program Requirements Analysis Method

Prama Pramana (periodical code)

PRAME Pôle de Recherches Aquitain sur les Matériaux de l'Espace (France)

PRAMS Passenger Reservation And Manifesting System

prand Prandium

PRAO Présentation Assistée par Ordinateur • Pushchino Radioastronomy Observatory (Russia)

PraP Publications of the Astronomical Institute of the Charles University (periodical code)

PRAR Postal Revenue Activity Report

PRARE Positioning Range And Rate Experiment • Precise Range And Range-rate Equipment (*) • Precise Range And Range-rate Experiment • Precise Range And Rate Equipment

PRAREE Precise Range And Range-rate Equipment Extended

PRARS Pitch, Roll, Azimuth Reference System

PRAS Passenger Revenue Accounting System • Polynomial Randomized Approximation Scheme

PRAT Prediction Resolution Analysis Tool • Production Reliability Acceptance Test

PrAtO Problemy Atmosfernoi Optiki (periodical code)

PRB Panel Review Board • Parachute Refurbishment Building • Paso Robles (USA-CA) (location identifier) • Performance Review Board • Personnel Records Branch • Physical Review B (periodical code) • Planning and Resource Board • Polar Research Board • Postretirement Benefit • Preliminary Review Board • Procurement Review Board • Program Request Block • Program Requirement Baseline • Program Review Board • Project Review Board

PRBAC Partition Rule-Based Access Control

PRBE Precision/Recall Break-Even (point)

PRBG Pseudo-Random Bit Generator

PrBio Problemy Bioniki (periodical code)

Prblc Parabolic

PRBLTY Probability

PRBS Pseudo-Random Binary Sequence • Pseudo-Random Bit Sequence

PRBSG Pseudo-Random Binary Sequence Generator

prc Palm Resource Code (file extension)

PRC Palm Resource • Partial Response Coding • Passenger Reservation Center • People's Republic of China • Personnel Readiness Center • Personnel Reporting Code • Photoreactive Crystal • Physical Review C (periodical code) • Physics Resource Center (AIP) • Planning Research Corporation (USA) • Point of Reverse Curve • Polar-Ring (galaxy) Catalog • Policy Review Committee • Populace and Resources Control • Portable Recompression Chamber • Postal Rate Commission • Post Roman Conditam • Power Reflection Coefficient • Precipitation • Prescott (USA-AZ) (location identifier) • Price Regulation Committee • Primary Rate Card • Procedure Review Committee • Programme de Recherche Concerté • Programme de Recherches Coordonnées • Program Review Committee • Project Review Committee • Propulsion and Reaction Control • Provisioning Class • Public Research Centre

PRCA Pitch and Roll Channel Assembly • Problem Reporting and Corrective Action • Professional Rodeo Cowboys Association

PRCB Program Requirements Change Board • Program Requirements Control Board • Program Review Control Board

PRCBD Program Requirements Control Board Directive • Program Review Control Board Directive

PRCE Prime Régional à la Création d'Entreprise (France)

prcht Parachute

Prchte Parachute

PRCO Publications of the Riverview College Observatory (periodical code)

PRCP Power Remote Control Panel

Prcs Process

PRCS Premature Retirement Compensation Scheme • Primary Reaction Control System • Propulsion and Reaction Control System

Prcsr Processor

PRCSR Processor

Prcst Precast

PRCU Power Regulating and Control Unit

PRCV Partially Reusable Core Vehicle

prd Printer Driver (file extension)

PRD Pardoo (Australia-WA) (location identifier) • Partial Redistribution • Payload Requirements Document • Period • Personal Radiation Dosimeter • Personnel Readiness Date • Personnel Readiness Division • Personnel Records Division • Personnel Requirements Document • Physical Resources Department • Physical Review D (periodical code) • Piezoelectric Resonating Device • Pilot Requirements Document • Power Range Detector • Power Requirement Data • Preliminary Requirements Document • Presidential Review Directive • Printer Driver • Print Reflectance Difference • Procurement Regulation Directive • Procurement Requirements Document • Product Requirements Document • Program Requirements Data • Program Requirements Document • Projected Rotation Date • Project Requirements Data • Project Requirements Document

PRDA Parada • Program Research and Development Activity • Program Research and Development Announcement

PRDDO Partial Retention of Diatomic Differential Overlap

PRDE Program Requirements Document Extract

PRDMD Private Directory Management Domain

Prdn Production

Prdr Producer

PRDR Production Reliability Design Review

PrDRV Problemy Difraktsii i Rasprostraneniia Voln (periodical code)

PRDT Process Rights Description Table

PRE Partial Reflection Experiment • Personal Rescue Enclosure • Plan de Régulation de l'Énergie • Planetary Radio Emission • Pore (Colombia) (location identifier) • Power Regulation Equipment • Preamp/Readout Electronics • Program Requirements Estimate • Preliminary Research Equipment • Preston (Starlink internal code) • Prime Régional à l'Emploi (France) • Processor Recovery Element • Pulse Radiation Effect

PREAMP Preamplifier

PréAO Présentation Assistée par Ordinateur

Prec Precede • Preceded • Precedence • Preceding • Precision

Préc Précédent • Précéder • Précision

Precdg Preceding

Precip Precipitation

PRECIS Preserved Context Index System

PRECISE Precision Estimation and Control In Scientific and Engineering (computing)

Precp Precipitation

PrECS Progress in Energy and Combustion Science (periodical code)

Pred Predicate • Predicative • Predicted • Predicting • Prediction

Préd Prédicat • Prédiction • Prédire • Prédit

PRED Predicted • Prediction

Predic Predicate • Predicative

Préf Préface • Préfecture • Préférence • Préférer • Préfixe

Pref Preface • Prefaced • Prefatory • Prefazione • Prefecture • Prefer • Preferably • Preference • Preferred • Prefijo • Prefix • Prefixed

PREF Preference • Preferential • Propulsion Research Environmental Facility

Prefab Prefabricated

Prefect Prefecture

PREFLT Preflight

PREIS Prototype Engineering Information System

Prej Prejudice

Prel Preliminary

PREL Preliminary

Prél Préliminaire

Prelim Preliminary

Prélim Préliminaire

Prem Premium

PrEM Probabilistic Engineering Mechanics (periodical code)

PREM Preliminary Reference Earth Model

PREMO Presentation Environment for Multimedia Object

prEN Project of European Norm

Preop Preoperative • Preoperatively

PREOS Performance and Range of Electro-Optical Systems

Prep Preparation • Preparatory • Prepare • Prepared • Preparing • Preposición • Preposicional • Preposition

Prép Préparation • Préparer • Préposition

PReP PowerPC Reference Platform

PREP Preparation • Prepare • Programmable Electronics Performance • Programmed Educational Package

Prepd Prepared

Prepg Preparing

PREPHA Programme de Recherche sur la Propulsion Hypersonique Avancée

Prepn Preparation

PREPNet Pennsylvania Research and Economic Partnership Network (USA-PA)

Prepr Preprint

PreR Precambrian Research (periodical code)

Pres Presbyterian • Presence • Present • Presente • Presentation • Preserve • Preserved • Presidency • President • Presidential • Pressure • Presumed • Presumptive

Prés Présence • Présent • Préservé • Préserver • Présidence • Président

PRES Present • Presented

PRESAC Photographic Reconnaissance System Analysis by Computer

Presb Presbyterian

Presbyt Presbyterian

Prescr Prescription

Press Pression • Pressure • Pressuriser • Pressurize

PRESS Pressurant • Pressure

PREST Policy Research in Engineering, Science and Technology • Politique de la Recherche Scientifique et Technique

PRESTI Pattern Recognition Technique for ISOCAM

PRESTO Precursory Research for Embryonic Science and Technology • Program for Rapid Earth-to-Space Trajectory Optimization • Programmable Etalon Spectrometer for Twilight Observations • Program Reporting and Evaluation System for Total Operation • Project to Reengineer Space Telescope Observing

Presv Preservation • Preserve

Pret Preterit

Prév Préventif • Prévention

Prev Prevention • Preventive • Previous • Previously

PREV Previous

PREVLV Prevalve

PrEx Propellants and Explosives (periodical code)

PrExP Propellants and Explosives Pyrotechnics (periodical code)

Prf Proof

PRF Parachute Refurbishment Facility • Partial Recursive Function • Personal Resource Form • Personnel Readiness File • Personnel Resources File • Pixel Response Function • Plan Régional de Formation • Point Response Function • Port Johnson (USA-AK) (location identifier) • Potential Requirements File • Potentiel de Répartition Futur • Preference • Processing and Reference Facility • Programme de Recherche Fédérateur • Promotion Recommendation Form • Pseudo-Relevance Feedback • Publication Reference File • Pulse Rate Frequency • Pulse Recurrence Frequency • Pulse Repetition Frequency

PrFA Problemy Fiziki Atmosfery (periodical code)

PRFC Plastique Renforcé de Fibres de Carbone

PRFE Polar Reflection Faraday Effect

PrFKL Problemy Iadernoi Fiziki i Kosmicheskikh Luchei (periodical code)

Prfm Performance

PRFM Pseudo-Random Frequency Modulated

Prfnl Professional

Prfrd Proofread

Prfrdg Proofreading

Prfrdr Proofreader

Prfs Parfois

Prft Parfait

PRFVT Plastiques Renforcés et Fibres de Verre Textile (periodical code)

Prfx Prefix

prg Program (file extension)

Prg Pergamino

PRG Patent Resources Group • Peculiar Red Giant (star) • Periodic Rate Group • Polar-Ring Galaxy • Powerful Radio Galaxy • Prague Ruzyne (Czech Rep.) (location identifier) • Praha Ruzyne (Czech Rep.) (location identifier) • Priority Review Group • Program • Program Review Group • Pseudo-Random (number) Generator • Pulsating Red Giant

Prgm Program

PRGM Program

Prgmg Programming

Prgmr Programmer

PRH Phrae (Thailand) (location identifier) • Probability Ranking Hypothesis • Programme de Recherche en Hydrologie • Program Requirements Handbook

PRHB Postretirement Health Benefit

Prhps Perhaps

PRHR Passive Residual Heat Removal

Pri Primary • Primer • Priority • Private

PRI Phase-Referenced Imaging • Photographic Reconnaissance and Interpretation • Porto Rico (code ISO de pays) • Praslin Island (Seychelles) (location identifier) • Primary • Primary-Rate Interface • Princeton Research Instruments (USA-NJ) • Priority • Programme de Recherche en Infométrie • Progressive Routing Indicator • Provisional Retrieval Interface • Provisioning Instance • Public Radio International • Puerto Rico (ISO country code) • Pulse Rate Indicator • Pulse Recurrence Interval • Pulse Repetition Interval • Pupil Research Initiative

PRIA Publicación en Revista Internacional con Arbitros • Public Relations Institute of Australia

PRIAM Plasmas Réactifs en Interaction Avec les Matériaux

PRIAMM Programme pour la Recherche et l'Innovation dans l'Audiovisuel et le Multimédia (France) (now RIAM)

Prib Priborostroenie (periodical code)

PRICAI Pacific Rim International Conference on Artificial Intelligence

PRICE Parametric Review of Information for Costing and Evaluation • Programmed Review of Information for Costing and Evaluation

PRID Provisioning Instance Identifier

PRIDE Performance Readiness Initiative Decision Evaluation (system) • Programmed Reliability In Design • Provisioning Relational Integrated and Distributed Electronics • Puerto Rico Dust Experiment

Prign Prignitz (Germany)

Prim Primaire • Primary • Primer • Primitive

PriM Prikladnaia Mekhanika (periodical code)

PRIM Primary • Primordial

PRIMA Phase-Referenced Imaging and Microarcsecond Astrometry

PRIMAS Precipitation Radar Image Management System

PRIMATE Personal Retrieval of Information by Microcomputer And Terminal Ensemble

PRIME Physics of Regolith Impacts in Microgravity Experiment • Precision Integrator for Meteorological Echoes • Precision Recovery Including Maneuvering Entry • Prime Explorer • Priority Management Effort • Product Information for Manufacturing Excellence • Program Independence, Modularity, Economy • Programmed Instruction from Management Education • Program of Research in

International Management and Economy • Promoting Reform In Mathematics Education

PRIMER Product Improvement Management Information Report

PriMM Prikladnaia Matematika i Mekhanika (periodical code)

PriMP Prikladnaia Matematika i Programmirovanie (periodical code)

Prin Principal • Principally • Principle

PRINC Principe

PRINCE Parts Reliability Information Center • Projects in Controlled Environments

PRIND Present Indications

Print Printing

PRINTF Print with Formatting

PRIOR Program for In-Orbital Rendezvous

PRIP Pattern Recognition and Image Processing

Prir Priroda (periodical code)

PRIS Payload Remote Integration System • Personal Resource Impact Simulation • Power Reaction Information System • Power Reactor Information System (*) • Program Resource Information System • Propeller Revolution Indicator System

PRISE Page Reader Input System with Editing

Prism Prismatic

PRISM Perkins Reimaging System • Photorefractive Information Storage Material • Pôle Régional et International des Sciences de la Mesure • Portable, Reusable, Integrated Software Module • Power Reactor Inherently Safe Module • Programmed Integrated System Maintenance • Program Reliability Information System for Management

PRISMA Parallel Inference and Storage Machine • Probing Rotation and Interior of Stars: Microvariability and Activity

PRISME Programmation des Robots Intelligents et des Systèmes Manipulateurs Évolués

PrISMS Program Information Systems Mission Services (NASA/CSC)

PRISMS Plan of Research for Integrated Soil Moisture Studies

PRISTE Programme de Recherches Interdisciplinaires et Techniques de l'Environnement (CNRS)

PRIT Pôle de Ressources et d'Innovation Technologique

PriTE Pribory i Tekhnika Eksperimenta (periodical code)

Priv Private • Privative • Privativo • Privé • Privilege • Privilège • Privilegio

Priv. Comm. Private Communication

prj Project (file extension)

PRJ Capri (Italy) (location identifier) • Paper in a Refereed Journal

PRJMP Pressure Jump

PRK Corée du Nord (code ISO de pays) • DPRK (ISO country code) • Korea (DPRK) (ISO country code) • Park • Prieska (South Africa) (location identifier)

Prl Parallel

PRL Page Revision Log • Pattern Recognition Logic • Payload Retention Latch • Photonics Research Laboratory • Physical Research Laboratory (India) • Physical Review Letters (periodical code) • Plasma Research Laboratory • Port Oceanic (USA-AK) (location identifier) • Priority Rate Limiting • Program Requirements Logic • Project Research Laboratory • Propulsion Research Laboratory

PRLA Payload Retention Latch Actuator • Payload Retention Latch Assembly

PRLP Planetary Rocket Launcher Platform

PRLS Portable Reconfigurable Line Sensor • Pulsed Ruby Laser System

Prm Premium

PRM Partially Redundant Mask • Partially Reflecting Mirror • Partially Regulated Module • Partial Response Method • Payload Retention Mechanism • Performance and Reliability Module • Perijove Raise Maneuver • Personal Radiation Monitor • Personnel Readiness Manager • Pocket Radiation Meter • Portable Radio Modem • Portimão (Portugal) (location identifier) • Posigrade Rocket Motor • Power Range Monitor • Precision Runway Monitor • Preliminary Requirements Model • Presidential Review memorandum • Private Mobile Radio • Production Representative Model • Programme Review Meeting • Programmer's Reference Manual • Programming and Resources Management • Program Resource Methodology • Project Review Meeting • Prospect Relationship Management • Protocol Reference Model • Pulse Rate Modulation • Pulse Repetitions per Minute • Pyrradiometer

PrMas Problemy Mashinostroeniia (periodical code)

PRMC Pacific Regional Monitoring Center • Partial-Response Maximum Likelihood

PrMD Private Management Domain

PRMD Private Management Domain

PRMFL Permanent File

PRMG Performance Risk Management Group

PRMIS Printing Resources Management Information System

PRML Low Purin Meal • Partial-Response Maximum Likelihood

PRMO Productibility - Reliability - Maintainability - Operability

PRMP Programme de Restitution des Mots de Passe

PRMPA Presidential Recordings and Materials Preservation Act (USA)

PrMS Progress in Materials Science (periodical code)

Prmtr Parameter

prn Printer Output (file extension) • Pro Re Nata

PRN Pôle de Recherche National • Postal Routed Network • Practitioners Relief Network • Presse Quotidienne Régionale • Printer • Pristina (Yugoslavia-Serbia) (location identifier) • Program Release Notice • Program Revision Number • Pseudo-Random Noise • Pseudo-Random Number • Pulse Range Navigation • Pulse Ranging Navigation • Pulse Ranging Network

PRNET Packet Radio Network

PRNG Pseudo-Random Number Generator

Prnl Pronominal

Prntd Printed

Prntg Printing

Pro Procedure • Proceed • Professional • Project • Prophylactic

Pr. O. Press Officer

PRO Perry (USA-IA) (location identifier) • Planning Resident Order • Plant Representative Office • PLL Reference Oscillator • Precision RISC Organization • Preplanned Response Option • Proceed • Profile • Programme • Publication Requirement Officer • Public Records Office (UK) • Public Records Officer • Public Relations Office • Public Relations Officer • RAL (Astrophysics cluster) (Starlink internal code)

Prob Probabilité • Probability • Probable • Probably • Probate • Probation • Problem

PROB Probabilities • Problem

PROBA Project for Onboard Autonomy

PROBAR Programmable Band Radiometer

ProbE Probabilistic Estimation

PROBE Pilot Radiation Observation Experiment • Portable Remote Observation of the Environment • Program Optimization and Budget Evaluation

Probl Problem

PROBS Prototype RVD Onboard Software

Proc Procedure • Procédure • Proceed • Proceedings • Process • Processing • Processor • Proclamation • Proctor • Procure • Profession • Professional

PROC Procedure • Process • Processor • Procurement • Proposed Required Operational Capability

PROCAS Process-Oriented Contract Administration Services

PROCEDE Process Engineering Design Environment

PROCOM Projection Compositor

PROCURE Procurement

Prod Prodigy • Produce • Produced • Producer • Product • Production • Productor • Produire • Produit

PROD Product

PRODAC Programmed Digital Automatic Control

PRODEX Programme for the Development of Experiments (ESA)

PROE Publications of the Royal Observatory of Edinburgh (periodical code)

Prof Profesor • Professeur • Profession • Professional • Professionnel • Professor • Proficiency • Profile

PROFDA Processed Frame Database

PROFILE Passive Radio Frequency Interference Location Experiment

PROFIT Programmed Reviewing, Ordering and Forecasting Inventory Technique • Propulsion Flight Control Integration Technology

PROFOEG Programa de Fotometría y Estructura Galáctica

Profrship Professorship

PROFS Professional Office System (IBM) • Program for Regional Observing and Forecasting Services • Prototype for Regional Observing and Forecasting Services

Prog Prognosis • Program • Programmable • Programmer • Programming • Progress • Progressive

PROG Prognosis • Prognostic • Program • Programmer • Progress

PROGERT Projet de Recherche pour l'Observation et la Gestion des Ressources Terrestres

Progr Program • Programmable • Programmer • Programming • Progress • Progressive

PROI Public Records Office of Ireland

Proj Project • Projected • Projectile • Projection • Projector

PROJ Project • Projectile

PROJTA Projectada

PROJTO Projectado

Pról Prólogo

Prol Prologue

PROLA Physical Review Online Archive

PROLOG Programming in Logic

Prolong Prolongación • Prolongamento

Prolyd Protoplanetary Disk

Prom Prominent • Promontory • Promote • Promoter • Promotion • Promulgate

PROM Pockels Readout Optical Modulator • Programmable Read-Only Memory • Prominence Research: Observation and Modeling

PROMATS Programmable Magnetic Tape System

PROMENADES Projet de Mesures en Astrométrie de Distances d'Étoiles avec un Schmidt

PROMIS Polar Regions and Outer Magnetosphere International Study • Polar Regions Outer Magnetosphere International Study • Program for an Operational Meteorological Information System • Project-Oriented Management Information System

PROMISE Process-Oriented Machine Independent Simulation Environment

PROMPT Production, Reviewing, Organizing and Monitoring of Performance Techniques • Program Monitoring and Planning Technique • Project Management and Production Team • Project Reporting Organization and Management Planning Technique

PROMS Protocols for Multimedia Systems (conference)

PROMYS Program in Mathematics for Young Scientists

Pron Pronombre • Pronominal • Pronoun • Pronounce • Pronounceable • Pronounced • Pronouncement • Pronouncing • Pronunciation

PRON Procurement Request Order Number

PRONAOS Programme National d'Observation Submillimétrique (France)

PRONTO Program for Numerical Tool Operation

Pronun Pronunciación

Pronunc Pronunciation

Prop Propagate • Propagation • Propellant • Propeller • Propelling • Proper • Properly • Property • Proportion • Proportional • Proposal • Propose • Proposed • Proposer • Proposition • Proprietary • Proprietor • Propulse • Propulsion

PROP Performance Review for Operating Programs • Planetary Rocket Ocean Platform • Propagation • Propellant • Property • Proportioning • Proposal • Propulsion

Propag Propagate • Propagation

PROPIN Proprietary Information

Propl Proportional

PROPLT Propellant

PROPLYD Protoplanetary Disk

PROPLYDs Protoplanetary Disks

Propn Proportion

Propr Proprietor

Propt Proportionnel

PRORISC Programme for Research on Integrated Systems and Circuits

ProRR Production Readiness Review

Pros Prosodia • Prospectus

PROS Parts and Repair Ordering System • Postreduction Off-line Software • Prevention, Repair and Operational Services

PROSE Problem Solution Engineering

ProSEDS Propulsive Small Expandable Deployer System

Prot Protective • Protectorate

PROT Protect • Protected • Protection • Protective

ProTe Promyshlennaia Teplotekhnika (periodical code)

PROTEL Procedure-Oriented Type Enforcing Language • Proton Telescope

pro tem Pro Tempore

PROTEUS Profile Telemetry of Upper Ocean Currents • Prototype Real-time Operational Test, Evaluation and User Simulation

PROTO Prototype

Prov Provençal • Provence (France) • Proverb • Proverbial • Proverbially • Provide • Provided • Provider • Province • Provincia • Provincial • Provinz • Provision • Provisional • Provost

PROV Provision • Provisioning

Provis Provision • Provisional

prox Proximo

Prox Proximité

PROX Proximity

prox. acc. Proximo Accessit

prox. luc. Proxima Luce

Prp Purpose

PRP Peak Radiated Power • Performance Recovery Program • Performance-Related Pay • Performance - Requirements - Practices • Personnel Reliability Program • Personnel Requirements Report • Pipelined Resampling Processor • Potentially Responsible Party • Precision Registration Processor • Prévention des Risques Professionnels • Prise Rapide de Parole • Probability Ranking Principle • Production Readiness Plan • Proficiency Recognition Profile • Profit-Related Pay • Program Redefinition Proposal • Program Review Panel • Program Review Proposal • Proposal Review Panel • Propriano (France) (location identifier) • Pseudo-Random Pulse • Pulse Recurrent Period • Pulse Repetition Period

PRPDC Pacific Regional Planetary Data Center

PRPF Planar Radial Peeking Factor

PrPh Progress of Physics/Fortschritte der Physik (periodical code)

Prphl Peripheral

PrPNP Progress in Particle and Nuclear Physics (periodical code)

pr. pr. Praeter Propter

PrPro Problemy Prochnosti (periodical code)

PRPS Programming Requirements Process Specification

Prpsl Proposal

PRQ Prerequisite • Presidente Roque Saenz Peña (Argentina-CH) (location identifier)

PRQA Plan Régional pour la Qualité de l'Air (France)

PR&R Project Report & Review

PRR Paruima (Guyana) (location identifier) • Parts Replacement Request • Parts Rework or Replacement • Peak Radiance Response • Personnel Requirements Report • Pilot Record Request • Preflight Readiness Review • Preliminary Report of Results • Preliminary Requirements Review • Production Readiness Review • Production Revision Record • Production Revision Request • Program Requirements Review • Program Revision Report • Project Requirements Review • Projet de Recherche en Réseau • Proton Relaxation Rate • Prototyping Results Review • Pulse Recurrence Rate • Pulse Repetition Rate

PRRB Payload Requirements Review Board

PRRC Petroleum Recovery Research Station (USA-NM) • Procurement Requirements Review Committee

PRRD Proprietary Rights to Registration Data

PRRDA Public Relations Regional Development Assistant

PRRG Prerelease Review Group

PRRM Program Review and Resources Management

PRRS Positioning Reporting Recording System

prs Printer Output (file extension)

Prs Pairs

PRs Pakistan Rupee (currency)

PRS Pacific Rocket Society (USA-CA) • Packet Routing Switch • Parallel Resource Scheduler • Parasi (Solomon Islands) (location identifier) • Passive Ranging Subsystem • Pattern Recognition Society (USA) • Pattern Recognition System • Payload Retention System • Peculiar Red Star • Performing Rights Society (UK) • Personal Recording System • Personal Rescue System • Personal Response System • Personnel Readiness System • Personnel Rescue Service • Personnel Rescue Sphere • Personnel Rescue System • Pointing Reference System • Position Reference System • Power Reactant System • Power Relay Satellite • Precision Ranging System • Preliminary Requirements Specification • Preliminary Resource Schedule • Premium Rate Service • Primary Recovery Ship • Primary Recovery Site • Primary Reference Source • Primary Rescue Site • Process Radiation Sampler • Production Readiness Strategy • Production Reporting System • Program Requirements Summary • Provisioning Requirements Statement • Pseudo-Random Sequence • Public Relations Section • Public Research System

PRSA Power Reactant Storage Assembly

Prsd Pressed

PRSD Power Reactant Storage and Distribution • Power Reactant Supply and Distribution

PRSDS Power Reactant Storage and Distribution System

PRSFR Pressure Falling Rapidly

prsmn Pressman

PRSN Provisional Relative Sunspot Number

PRSNT Present

PrSP Problemy Sluchainogo Poiska (periodical code)

PRSP Prineville Reservoir Star Party (USA-OR)

PRSRR Pressure Rising Rapidly

Prsrz Pressurize

PrST Prairie Standard Time

PRST Persist • Probability Ratio Sequential Test

PRSUG Puerto Rico Spectrum Users Group

Prsvn Preservation

prt Printer Output (file extension)

Prt Print

P&RT Physical & Recreational Training

PRT Participating Research Team • Pattern Recognition Technique • Personal Rapid Transit • Personnel Research Team • Personnel Research Test • Physical Readiness Training • Platinum Resistance Thermometer • Poincaré Recurrence Theorem • Point Retreat (USA-AK) (location identifier) • Polarized Radiation Transfer • Port • Portable Remote Terminal • Porteiro • Portugal (ISO country code) • Power Ratchet Tool • Power Recovery Train • Precision Radiation Thermometer • Preliminary Reference Trajectory • Prestation Technologique Réseau • Primary Ranging Terminal • Print • Printer • Procurement Review Team • Program Reference Table • Program Requirements Team • Provisioning Review Team • Pulse Recurrence Time • Pulse Repetition Time • Pulsed Radar Transmitter

PrTE Problemy Tekhnicheskoi Elektrodinamiki (periodical code)

Prtg Printing

PrTGE Problemy Teorii Gravitatsii i Elementarnykh Chasitits (periodical code)

Prtl Partial

PRTLS Powered Return to Launch Site

PRTM Printing Response-Time Monitor

Prtng Printing

PRTOT Prototype Real-Time Optical Tracker

PrTPT Problemy Teploenergetiki i Prikladnoi Teplofiziki (periodical code)

Prtr Printer

PRTS Ports

PRTSC Print Screen

PRTU Packet Remote Terminal Unit

PRTV Peterlee Relational Test Vehicle • Production Representative Test vehicle

PRU Pattern Recognition Unit • Performance Reference Unit • Personnel Readiness Unit • Photographic Reconnaissance Unit • Power Regulation Unit • Primary Replacement Unit • Program Research Unit • Prome (Myanmar) (location identifier) • Pyro Regulation Unit • Pyrotechnic Regulation Unit

PRUG Plan Rector de Uso y Gestión

PrUkr Akademiia Nauk Ukrains'koi RSR Dopovidi Matematika Prirodoznavstvo Tekhnichni Nauki (periodical code)

Prus Prussia • Prussian

Pruss Prussia • Prussian

PRV Peak Reverse Voltage • Perfect Rotational Variant • Pour Rendre Visite • Precise Radial Velocity • Precision Radial Velocity • Prerov (Czech Rep.) (location identifier) • Pressure-Reducing Valve • Pressure Reduction Valve • Pressure Regulator Valve • Pressure Relief Valve • Prüfungsrichtlinienverordnung • Pseudo-Relative Velocity

PrVP Proceedings of Vibration Science (periodical code)

PRVS Penetration Room Ventilation System

Prvw Preview

PRW Peachtree Report Writer • Planetary Rossby Wave • Prentice (USA-WI) (location identifier)

PRWD Priority Regular World Day

Prx Prefix

PRX Paris (USA-TX) (location identifier)

PRY Paraguay (ISO country code) • Pretoria Wonderboom (South Africa) (location identifier)

PRZ Prineville (USA-OR) (location identifier)

PRZM Pesticide Root Zone Model

ps Pashto (ISO language code) • Picosecond • Pieces • Postscript (file extension) • Pseudonym • Pushto (ISO language code)

Ps Positronium

P&S Personnel & Security • Planning & Scheduling • Purchase & Sales

P/S Parallel to Serial • Payload Specialist • Planning/Scheduling • Power Supply • Principal/Secondary

PS Packaging Specialist • Packet Switch • Packet Switching • Palomar Schmidt • Parachute System • Parity Switch • Parity Symmetry • Parliamentary Secretary • Partenaire Social • Participating Scientist • Particle Separator • Particle Size • Passenger Steamer • Payload Specialist • Payload Station • Payload Support • Performance Standard • Permanent Secretary • Permanent Signal • Personal Station • Personal System • Personnel Services • Petite Section • Pferdestärke • Phase Shift • Philippine Standard • Photographic Service • Photometric System • Phrase Structure • Physical Security • Physical Sequential • Physics Specification • Picosecond • Planetary Society • Planetary System • Planning and Scheduling • Planning Study • Plasma Sheath • Plasma Switch • Plate Scanner • Plotting System • Plus • Pointing System • Point of Symmetry • Point Sémaphore • Point Source • Polar Stereographic • Pôle Sud • Polystyrene • Polystyrène • Population Synthesis • Porous Silicon • Position-Switched (mode) • Positive Small • Postal Schedule • Postal Service • Poste Supplémentaire • Postscript • Post Scriptum • Postskript • Postsleep Activity • Power Section • Power Source • Power Spectrum • Power Steering • Power Supply • Power System • Preliminary Study • Preparation Software • Presleep Activity • Press-Schechter (theory) • Press Secretary • Pressure Switch • Pressurized Section • Primary Structure • Prime Select • Principal Shock • Prior Service • Private Secretary • Privy Seal • Probability of Success • Probability of Survival • Problem State • Problem Statement • Processing System • Processor Sharing • Process Specification • Process Status • Procurement Specification • Production Sheet • Production System • Product Standard • Programmed Symbol • Programming System • Program Specification • Program Start • Program Store • Program Summary • Project Scientist • Project Stock • Promotion Sociale • Proof Stress • Propellant Supply • Propellant System • Proportional Spacing • Proposed Standard • Protective Security • Protective System • Proton Switch • Proton Synchrotron (CERN) • Public School • Public Service • Publication Standard • Publishing Service • Pull Switch • Pulsating Surface

Psa Presa

PsA Piscis Australis - Piscis Australis (constellation) • Piscis Austrinus - Piscis Austrini (constellation)

PSA Pacific Science Association • Pacific Sociological Association • Pacific-South American • Pacific Southwest Airlines • Panoramic Spectral Analyzer • Particle Size Analyzer • Particle Surface Area • Path of Steepest Ascent • Payload Service Area • Payload Support Avionics • Peacekeeping Services Agreement • Pedestal Stand Assembly • Personal Satellite Assistant (NASA) • Phase-Shift Analysis • Philosophy of Science Association • Photographic Society of America • Physical Sector Address • Pisa (Italy) (location identifier) • Plataforma Solar de Almería (Spain) • Port Support Activity • Postshakedown Availability • Postsleep Activity • Power and Servo Assembly • Power Servo Amplifier • Power Servo Assembly • Power Spectral Analysis • Power Supply Assembly • Power Switching Assembly • Preferred Storage Area • Prefixed Save Area • Prefixed Storage Area • Presleep Activity • Pressure Suit Assembly • Pressure Swing Adsorption • Pressure Switch Assembly • Prices Surveillance Authority • Primary Science Aperture • Principal Staff Assistant • Probabilistic Safety Analysis • Probabilistic Safety Assessment • Problem Statement Analyzer • Processor Save Area • Procurement System Analyst • Productibility - Supportability - Affordability • Program on Space Applications • Program Study Authorization • Program Support Agreement • Project Support Agreement • Prostatic Specific Agent • Provisional Site Acceptance • Provisions Stowage Assembly • Public Service Announcement • Pulse Shape Analysis

PSAB Personnel Security Appeal Board

PSAC Parent Science Advisory Council • Passive Satellite Attitude Control • Presidential Scientific Advisory Committee (USA) • President's Science Advisory Committee (USA)

PSACI Plasma Science Advanced Computation Initiative

PSACR Power Sum Attenuation Crosstalk Ratio

PSAD Predicted Site Acquisition Data • Prediction - Simulation - Adaptation - Decision

PSAG Phenomenology Science and Analysis Group

PSAIL Pubblicazioni della Stazione Astronomica Internazionale di Latitudine (periodical code)

PSAL Programming System Activity Log

PSAM Partitioned Sequence Access Method

PSAO Primary Staff Action Officer

PSAP Presentation Service Access Point • Prospective Special Access Program • Public Safety Answering Point

PSAR Preliminary Safety Analysis Report • Programmable Synchronous/Asynchronous Receiver

PSAS Phenytoin for Space Adaptation Syndrome • Physical-Space Assimilation Analysis System • Physical-space Statistical Analysis System

PSAT Predicted Site Acquisition Table • Preliminary Scholastic Aptitude Test • Preliminary Scholastic Assessment Test • Programmable Synchronous/Asynchronous Transmitter

PSB Bellefonte (USA-PA) (location identifier) • Clearfield (USA-PA) (location identifier) • Panel on Scientific Ballooning • Philipsburg Mid-State (USA-PA) (location identifier) • Poststarburst • Prepress Service Bureau • Productivity and Standards Board (Singapore) • Program

Specification Block • Program Status Book • Pyrotechnic Storage Building

PSBL Planetary Sheet Boundary Layer • Plasma Sheet Boundary Layer • Possible • Public Sector Bidder List

PSBLS Permanent Space-Based Logistics System

PSBW Program Specification Block Workpool

Psc Pisces-Piscium (constellation)

PSC Pacific Science Center • Pacific Science Council • Packaging Services Contract • Parallel Subspace Correction • Parallel Switch Control • Parallel to Serial Converter • Parliamentary Space Committee (UK) • Pasco (USA-WA) (location identifier) • Payload Support Contractor • Peachtree Support Center • Pembroke State College (USA) • Performance Seeking Control • Periodic Surveillance Committee • Personal Service Contractor • Personal Supercomputer • Personnel Security Clearance • Personnel Service Center • Personnel Status Change • Phase-Sensitive Converter • Photosensitive Cell • Physical Sciences Center • Physical Sciences Committee • Pittsburgh Supercomputer Center (USA-PA) • Pittsburgh Supercomputing Center (USA-PA) • Planetary Spacecraft • Planning - Scheduling - Control • Platform Support Center • Point Source Catalogue • Polar Science Center • Polar Stratospheric Cloud • Port Security Company • Postal Service Center • Poststorage Checkout • Potomac State College (USA-MD) • Power Supply Calibrator • Pressure Suit Circuit • Pressure System Control • Prestellar Core • Price Signal Code • Prime Service Contractor • Principle of System Compatibility • Print Server Command • Problem-Solving Context • Processing and Spectral Control • Processor Support Circuit • Procurement Source Code • Production Surveillance Category • Product Service Center • Professional Support Centre • Profile Steering Committee • Program Schedule Chart • Program Science Center • Program Status Chart • Program Structure Code • Program Support Center • Program Support Communication • Program Support Concept • Program Support Contract • Program Support Contractor • Progress in Scientific Culture • Project Steering Committee • Project Systems Control • Protosolar Cloud • Protostar Cloud • Provide Station Control • Public Service Commission • Public Service Company

PS&CA Power Supply & Control Assembly

PSCA PASS SMS Composition Aids • Power Supply and Control Assembly • Pressure Suit Conditioning Assembly

PSCB Padded Sample Collection Bag • Protected Step Control Block

PSCCE Production Satellite Configuration Control Element

PSCDS Publications Spéciales du Centre de Données de Strasbourg (periodical code)

PSCF Personnel Security Clearance File

PSci Pour la Science (periodical code)

PSCL Programmed Sequential Control Language • Propellants System Cleaning Laboratory • Propellants System Components Laboratory

PSCM Problem Space Computational Model • Progression - Southerly - Cyclonicity - Meridionality

PSCN Permanent System Control Number • Preliminary Specification Change Notice • Production Schedule Completion Notice • Program Support and Communications Network • Program Support Communications Network • Project Support Communications Network • Proposed Specification Change Notice

PSCNET Pittsburgh Supercomputing Center Network

PSCNi Program Support Communications Network Internet

PSCO Personnel Survey Control Officer

PSCR Production Schedule Completion Report

PSCRD Program Support Communications Requirement Document

PSCS Pacific Space Centre Society (Canada-BC) • Program Support Control System

PSCU Power Supply and Control Unit • Power Supply Control Unit • Propellant Storage and Control Unit

PSCz Point Source Catalog Redshift (survey)

psd Photoshop (file extension)

P&SD Procurement & Supply Department

PSD Page Setup Dialog • Panel on Satellite Dynamics (COSPAR) • Particle Size Distribution • Patent Search Documentation • Pattern Standard Deviation • Performance and Security Document • Personnel Service Division • Phase-Sensitive Demodulator • Phase-Sensitive Detector • Phase - Space - Density (parameter) • Philadelphia School District (USA-PA) • Photometric Sunspot Deficit • Photon-Sensitive Device • Photon-Stimulated Desorption • Pitch Servo Drive • Pits - Spots - Domes • Planning Systems Division • Platform-Specific Driver • Poorly Significant Difference • Port Said (Egypt) (location identifier) • Position-Sensitive Detector • Position-Sensitive Device • Positive Semi-Definite • Post Source Decay • Potsdamer Spektraldurchmusterung • Power Spectral Density • Power Spectrum Density • Power System Demonstrator • Power Systems Division • Prestation Spécifique Dépendance • Prevention of Significant Deterioration • Procedural Support Data • Processing Status Display • Procurement and Supply Department • Product System Demonstration • Program Status Data • Program Status Document • Program Summary Document • Program Support Document • Program System Description • Projects and Systems Division • Propellant Storage Depot • Proportional Scintillation Detector • Pseudo-Single Domain • Pulse Shape Discrimination • Pulse Shape Discriminator

PSDC Postal Service Data Center • Public Switched Digital Capability

PSDD Preliminary Software Design Document • Preliminary System Design Description

PSDE Payload and Spacecraft Demonstration and Experiment • Payload and Spacecraft Development and Experimentation

PSDF Power Spectral Density Function • Principal's Strategic Development Fund • Printer Snap Definition File • Propulsion System Development Facility

PSDN Packet Switched Data Network • Partially Stochastic Petri Net • Public Switched Data Network

PSDP Payload Station Distribution Panel • Payload System Distribution Panel • Preliminary System Design Plan

PSDPC Polar Satellite Data Processing Centre

PSDR Planning and Scheduling Document Record • Preliminary System Design Review • Property Set Definition Requirements • Public Sector Debt Requirement

PSDS Packet Switched Data Service • Postal Source Data System • Public Switched Data Service • Public Switched Digital Service

PSDU Power and Signal Distribution Unit • Power Switching Distribution Unit • Presentation Service Data Unit

PSE Pacific Stock Exchange • Packet-Switch Exchange • Packet-Switching Exchange (*) • Parallel Software Environment • Partial Solar Eclipse • Passive Seismic Experiment • Path Switching Element • Payload Service Equipment • Payload Servicing Equipment • Payload Support Engineer • Payload Support Equipment • Peculiar Support Equipment • Periodensystem der Elemente • Periodic System of Elements • Personal and Social Education • Personal Service Environment • Personnel System Element • Photosensitive Element • Physical Sciences and Engineering • Physical Security Equipment • Physiological Systems Experiment • Phytochemical Society of Europe • Pidgin Sign English • Pitch Steering Error • Planned Special Exposure • Platform Support Equipment • Polystyrène Expansé • Polystyrène Expansible • Ponce Mercedita (Puerto Rico) (location identifier)

• Power System Engineering • Precision Shooting Equipment • Principal Staff Element • Priority Standardization Effort • Probe Support Equipment • Problem-Solving Environment • Process Systems Enterprise • Product Subsidy Equivalent • Professional Server Expert • Professional Software Engineering • Professional Sound Equipment • Programming Support Environment • Program Support Engineer • Program Support Environment • Program Support Equipment • Program Support Estimate • Program System Evolution • Project Support Environment • Propulsion Support Electronics • Psychological Stress Evaluator • Pulsar-Signal-Entzerrer

PSEA Pointing and Safemode Electronics Assembly • Pointing/Safing Electronics Assembly

PSEAG Physical Security Equipment Action Group

psec Picosecond

PSEG Populations Stellaires et Évolution des Galaxies

PSEL Priced Support Equipment List

PSEMO Physical Security Equipment Management Office

PSEP Passive Seismic Experiments Package • Professional Student Exchange Program

PS&ER Production Support & Equipment Replacement

PSER Personnel Status and Effectiveness Report • Production Support and Equipment Replacement

PSERE Population Sans Emploi Recherchant un Emploi

PSERVER Print Server

PSESWG Programming Support Environment Software Working Group

PSET Program Support Engineering Test

Pseud Pseudonym • Pseudonymous

psf Pounds per Square Foot

PSF Pakistan Science Foundation • Passive Solar Foundation • Payload Structure Fuel • Permanent Swap File • Phase Structure Function • Pittsfield (USA-MA) (location identifier) • Planetary Studies Foundation (USA-IL) • Planning Skeleton File • Point Source Function • Point Spread Function (*) • Prestación Social Sustitoria (Spain) • Print Services Facility • Probability of Stress Failure • Processing and Staging Facility • Processing and Storage Facility • Program for the Study of the Future • Programme Star File • Progressive Space Forum • Projection Specification Field • Propellant Servicing Facility • Punktbild-Streufunktion

PSFAM Parameter Sensitive Frequency Assignment Method

PSFB Plan Spécialisé de Fortes Précipitations

PSFC Permanent Service on the Fluctuation of Glaciers • Power Supply Fuel Cell

PSFG Permanent Service on Fluctuation of Glaciers

PSG Passage • Passing • Personnel Support Group • Petersburg (USA-AK) (location identifier) • Phase Screen Generator • Phrase-Structure Grammar • Power Subsystem Group • Preliminary Study Group • Prioritization Steering Group • Programmable Sound Generator • Program Security Guard • Program Steering Group • Project Science Group

PSGD Position Sensitive Germanium Detector

PSGE Passage

Psgr Passenger

PSGS Palomar Scan Grism Survey

PSH Permanent Superhumper • Program Support Handbook • Push • Sankt Peter (Germany) (location identifier)

PSHS Philippine Science High School

PSHTI Packet Switch Host Terminal Interface • Packet Switch Host/Trunk Interface

psi Pounds per Square Inch

PSI Packetnet System Interface • Packet-Switched Interface • Pakistan Standards Institution • Partial Shipment Indicator • Pasni (Pakistan) (location identifier) • Paul Scherrer Institute (Switzerland) • Personalized System of Instruction • Personnel Security Investigation • Perspectives of System Informatics (conference) • Photometric Sunspot Index • Physique - Sciences de l'Ingénieur • PI-class Science Instrument (SOFIA) • Planetary Science Institute (USA-AZ) • Plane Speed Indicator • Plasma Source Instrument • Pollutant Standards Index • Postscript Interpreter • Pounds per Square Inch • Power per Square Inch • Power Static Inverter • Present Serviceability Index • Preshipment Inspection • Princeton Synergetics Inc. (USA-NJ) • Principal-investigator-class Science Instrument (SOFIA) • Principle of System Incompatibility • Process System Index • Process to Support Interoperability • Program-Specific Information • Program Status Information • Project Starlight International • Protocol Service Information

psia Pounds per Square Inch - Absolute

Psic Psicólogo

PSIC Planning and Systems Integration Center • Provisioning System Identifier Code

PSICP Program Support Inventory and Control Point

psid Pounds per Square Inch - Differential

PSID Postscript Image Data • Preliminary Safety Information Document

PSIF Packet and Science Instrument File

psig Pounds per Square Inch - Gauge

PSIG Propulsion Systems Integration Group

PSII Plasma Source Ion Implementation

PSIM Power System Instrumentation and Measurement

PSIN Postal System Identification Number

psis Pounds per Square Inch - Scaled

PSIS Programming Systems Information System

PSIT Pilotage des Systèmes d'Information et de Télécommunications

PSIU Packet Switch Interface Unit • Power and Signal Interface Unit

PSIWA Penn State In-service Workshops in Astronomy (USA-PA)

PSJ Physical Society of Japan • Poso (Indonesia) (location identifier)

PSJaJ Physical Society of Japan Journal (periodical code)

PSK Dublin - New River Valley (USA-VA) (location identifier) • Phase Shift Key • Phase Shift Keyed • Phase Shift Keying (*) • Power Supply Kit • Program Selection Key

PSKM Phase-Shift Keying Modem

PSKMP Photuris Session Key Management Protocol

PSL Personal Launch System • Perth (UK-Scotland) (location identifier) • Photographic Science Laboratory • Physical Sciences Laboratory • Pipeline Scripting Language • Polystyrene Latex • Portable Standard LISP • Potential Source List • Power Source Logic • Pressure Seal • Prestabilized Laser • Primary Standards Laboratory • Princeton Science Library • Private Sector Liquidity • Problem Specification Language • Problem Statement Language • Process Specification Language • Process Support Language • Programming Support Language • Programming Systems Laboratory • Program Statement Language • Program Supply Logic • Program Support Letter • Program Support Library • Program Support Logistics • Propulsion Systems Laboratory • Protected Services List • Protocol Specification Language • Public Sector Loan

PSLI Pilot Site Licence Initiative

PSLIC Pseudo-Stray Light Isochronal Contour

PSLLS Pulse Solid-State Laser Light Source

PSLR Portable Satellite Laser Ranging

PSLS Parallel Symbolic Languages and Systems (workshop) • Programming Support Library System

PSLT Programme Spatial à Long Terme

PSLU Power Supply Logic Unit

PSLV Polar Satellite Launch Vehicle • Polar Space Launch Vehicle

Psm Prism

P&SM Procurement & Subcontract Management • Procurement & Systems Management

PSM Packet Service Module • Parcel Sorting Machine • Payload Support Module • Peak Selector Memory • Permanent Storage Memory • Persistent Stored Module • Personalized Service Manager • Personnel Status Monitor • Personnel System Manager • Phase-Sensitive Modulation • Phase-Sensitive Modulator • Phase Shifter Module • Pinhole Solar Monitor • Please See Me • Point Scattering Model • Point-Source Method • Portsmouth Pease (USA-NH) (location identifier) • Position-Switched Mode • Position Switching Module • Postal Service Manual • Power Storage Module • Power System Module • Practical Software Measurement • Prévision et Suivi des Moyens • Printing Systems Manager • Problem-Space-level Model • Process Safety Management • Process Support Module • Procurement and Subcontract Management • Product Sales Manager • Product Support Manual • Professional Staff Member • Program Support Management • Project Support Manager • Propellant Storage Module • Pyro Substitute Monitor • Pyrotechnic Substitute Monitor

PSMA Prototype Secondary Mirror Assembly

PSMAT Provisioning Screening Master Address Table

PSME Philippine Society for Mechanical Engineers

PSMG Parallel Superconvergent Multigrid

PSML Processor System Modeling Language

PSMN Pôle Scientifique de Modélisation Numérique

PSMOA Program-Specific Memorandum Of Agreement

PSMOS Planetary Scale Mesopause Observing System

PSMR Parts Specification Management for Reliability

PSMSL Permanent Service for Mean Sea Level

Psn Position

PSN Packet Switched Network • Packet Switched Node • Packet Switching Network • Packet Switching Node • Packet Switch Node • Palestine (USA-TX) (location identifier) • Pasifik Satelit Nusantara (Indonesia) • Piano Spaziale Nazionale (Italy) • Position • Position Switched Network • Postal System Number • Private Switching Network • Processor Serial Number • Process Serial Number • Program Schedule Network • Program Summary Network • Public Switched Network

PSND Partial Sequence Numbers PDU

Psnl Personal

PSNR Power Signal-to-Noise Ratio

Pso Paso

PSO Pad Safety Officer • Paranal Science Operations (ESO) • Passeio • Pasto (Colombia) (location identifier) • Pauli Spin Operator • Payload Support Office • Pending Service Order • Permanent Solar Orientation • Persistent SQL Object • Personal Staff Officer • Pilot Systems Operator • Planet Sensor Output • Point Symmetry Operation • Polar Stratospheric Ozone • Postman's Sorting Office • Post Security Officer • Pressure Shut Off • Primary Standardization Office • Principal Scientific Officer • Procurement Service Office • Program Secretarial Office • Program Security Office • Program Security Officer • Program Staff Office • Program Staff Officer • Project Science Office • Protective Service Operations • Provisions Supply Office • Public Service Obligation • Public Service Office • Public Service Officer

PSOC Physics Survey Overview Committee (USA) • Preliminary System Operational Concept

PSOLA Pitch-Synchronous Overlap and Add

PSOP Payload Systems Operating Procedure • Peak Side On Pressure • Power System Optimization Program

P/SOS Payload/Spacecraft Operations Simulation

PSP Packet Switching Processor • Packet Switch Process • Palm Springs (USA-CA) (location identifier) • Payload Signal Processor • Payload Specialist Panel • Payload Support Plan • Peak Sideband Power • Peinture Sensible à la Pression • Perforated Steel Planking • Performance Shaping Parameter • Performance Standards Program • Personal Software Process • Personal Software Products • Personal Success Program • Personal Survival Pack • Phased Support Plan • Pierced Steel Planking (*) • Pierced Steel Plate • Piezo-Sensitive Paint • Planetary Scan Platform • Planetary Scan Probe • Planet Scan Platform • Planned Standard Programming • Platform Sensor Package • Portable Service Processor • Positive Static Pressure • Postage Stamp Pipeline • Power Support Platform • Power System Planning • Precision Solar Pyranometer • Precision Spectral Pyranometer • Pré-Séquence Principale • Pressure Sensitive Paint • Primary Supply Point • Private Sector Partnership • Product Support Program • Programmable Scanner Position • Programmable Signal Processor • Program Security Plan • Program Segment Prefix • Program Support Plan • Project Schedule Plan • Project Support Plan • Public Star Party

PSPA Pilotage de Systèmes de Production Automatisé • Pressure Static Probe Assembly

PSPC Permanent South Pacific Commission • Position Sensitive Proportional Counter

PSPCT Programmable Signal Processor Configuration Table

PSPDC Polar satellite Precipitation Data Centre

PSPDN Packet Switched Public Data Network

PSPF Potential Single-Point Failure

PSPFLI Pulsed Single-Photon Fluorescence Lifetime Instrumentation

PSPKR Pusat Standardisasi dan Penelitian Keselamatan Radiasi (Indonesia)

PSPL Picosecond Pulse Labs Inc. (USA-CO) • Priced Spare Parts List

PSPM Penn State Pulsar Machine

PSPMT Position-Sensitive Photomultiplier Tube

PSPOS Payload Signal Processor Output Simulator

PSPP Proposed System Package Plan

PSprO Publications of the Sproul Observatory (periodical code)

PSPS Passive Surface Pressure Sounder • Planar Silicon Photoswitch • Program Support Plan Summary

PSPT Planar Silicon Power Transistor • Position Sensitive Phototube • Precision Solar Photometric Telescope

PSPU Physical Sciences Planning Unit

PSQ Personnel Security Questionnaire • Philadelphia (USA-PA) (location identifier)

PSQA Pageable System Queue Area

PSQM Perceptive Speech Quality Measurement

PSR Packed Snow on Runway • Pad Safety Report • Parachute Status Report • Patient Summary Report • Payload Safety Review • Performance Status Report • Performance Summary Report • Personal Service Radio • Personnel Status Report • Pescara (Italy) (location identifier) • Phase Shift Register • Phrase-Structure Rule • Polarimeter Sensitive Reflector • Precision Segmented Reflector • Preship Review • Previous Slot Received • Price-Sales Ratio • Problem Status Report • Processor State Register • Procurement Status Report • Production Safety Review • Production Student Record • Product Specific

Realization • Programming Status Report • Programming Support Representative • Program Status Report • Program Status Review • Program Study Request • Program Summary Record • Program Support Requirement • Progress Summary Report • Project Safety Review • Project Scan Record • Project Status Report • Project Status Review • Project Summary Report • Project Support Room • Proton Storage Ring • Public Sector Research • Pulsar

PSRA Program Summary Record Address

P-SRAM Pseudo-Static Random Access Memory

PSRAM Pseudo-Static Random Access Memory

PSRB Physical security Review Board • President's Scientific Research Board (USA)

PSRC Physical Sciences Resource Center

PSRD Planetary Science Research Discovery • Program Support Requirements Document

PSRE Propulsion System Rocket Engine

PSRG Planetary Science Research Group • Privacy and Security Research Group

PSRI Planetary Sciences Research Institute

PSRO Passenger Standing Route Order • Professional Standards Review Organization

PSRR Preliminary Systems Requirements Review • Primary-Shadow RTO Replication

PSRS Patient Safety Reporting System • Plant Service Recall System • Portable Seismic Recording System • Preliminary Software Requirements Specification

PSRT Passive Satellite Research Terminal

P&SS Planetary and Space Science (periodical code) • Planning & Scheduling System

PSS Packet-Switched Service • Packet Switching Service • Packet Switching Stream • Packet Switching System • Packet Switch Stream • Pad Safety Supervisor • Palomar Sky Survey • Panel Support Scientist • Passive Sensor System • Pauli Spin Susceptibility • Payload Safing Sequence • Payload Specialist Station • Payload Status System • Payload Support System • Performance Summary System • Performance Support System • Personnel Service Support • Personnel Support Summary • Personnel Support System • Photostationary State • Physical Source Statement • Pipeline Subsystem • Plafond de Sécurité Sociale • Planetary and Space Science • Planetary Scan System • Planning and Scheduling System • Portable Satellite Simulator • Portable Security System • Portable Survival Shelter • Posadas (Argentina-MI) (location identifier) • Postal Supply Schedule • Postscripts • Power Subsystem • Power Supply Section • Power Supply System • Power System Simulator • Power System Stabilizer • Premature Separation System • Printing and Stationery Service • Private Signaling System • Procedures, Specifications and Standards • Production System Support • Professional Server Specialist • Professional Services Section • Programme des Sciences Spatiales • Program Support Specialist • Project Summary Sheet • Propellant Supply System • Proprietary Software System • Proprietary Support System • Propulsion Subsystem • Propulsion System Structure • Propulsion Support System • Protective Security Service

PSSA Planned Software Schedule Adequacy

PSSAR Physica Status Solidi Applied Research (periodical code)

PSSBR Physica Status Solidi B Basic Research (periodical code)

PSSC Physical Science Study Committee • Point Source Sensitivity - Continuum

PSSG Peach State Star Gaze (USA-GA) • POSIX Standards Style Guide • Protocol Standards Steering Group

PSSI Pacific/Sulfur/Stratus Investigation

PSSL Point Source Sensitivity - Lines

PSSM Passive Solid-State Magnetic (sensor) • Photographic Systems Simulation Model

PSSNN Poliarnye Siianiia i Svechenie Nochnogo Neba (periodical code)

PSSO Permanent Solar-Stellar Orientation

PSSP Payload Specialist Station Panel

PSSR Performance Status Summary Report • Performance Status Summary Review • Preliminary System Software Review • Problem Status and Summary Report

PSSRI Planetary Space Sciences Research Institute (OU)

PSSS Packet Scan Status Stack • Palomar Schmidt Sky Survey

PSSST Project on the Social Studies of Science and Technology

PS&SSU Power Supply & System Selector Unit

PSSSU Power Supply and System Selector Unit

PSST Periodic Significant Scheduled Task • Planet Search Survey Telescope • Plasma Sources Science Technology (periodical code)

PSSU Patch Survey and Switching Unit

PSSWG Planetary System Science Working Group

Pst Paste

P&ST Physics & Space Technology

PST Pacific Standard Time • Pacific Summer Time • Paired Selected Ternary • Parallel/Serial Technology • Paulucci Space Theatre (USA-MN) • Performance Specification Tree • Periodic Self-Test • Photometric Solar Telescope • Photometrisches Sonnenteleskop • Phototronics Solartechnik GmbH (Germany) • Physics and Space Technology • Pied-Support de Terrace • Planning and Scheduling Team • Point of Spiral Tangent • Point Source Transmission • Point Source Transmittance • Polar Star Tracker • Polar Stratospheric Telescope • Polished Surface Technique • Preliminary Service Testing • Preliminary Systems Test • Preservice School Teacher • Preston (Cuba) (location identifier) • Primary School Teacher • Product Support Team • Professional, Scientific and Technical • Program Support Team • Project Science Team • Project Support Team • Proposal Submission Tool • Propulsion Science and Technology • Provincial Sales Tax

PSTA Packaging Science and Technology Abstracts

PSTAT Precipitation Static

PSTC Public Service Training Council

PSTE Personnel System Test and Evaluation • Production Special Test Equipment

PSTF Payload Spin Test Facility

PSTJ Provence Sciences Techniques Jeunesse (France)

Pstl Pistol • Postal

PSTL Pressure Statis and Transient Launch

Pstn Piston

PSTN Public Switched Telephone Network

Psto Puesto

PSTOL PORTS System Test and Operations Language • Pseudo STOL

PSTP Protocol Standards Technical Panel

PSTT Photometric System Test Target

PSU Package Splitter Utility • Packet Switching Unit • Passenger Service Unit • Penn State University (USA-PA) • Pennsylvania State University (USA-PA) • Power Supply Unit • Power Switching Unit • Primary Sampling Unit • Processor Speed Up • Process Support Unit • Program Storage Unit • Publications Services Unit • Public Service Union • Putussibau (Indonesia) (location identifier) • Pyrotechnic Switching Unit

PSV Papa Stour (UK) (location identifier) • Photographic Spatial Volume • Pilotage Sans Visibilité • Planetary Space Vehicle • Preliminary Site Visit • Pressure Sequencing

Valve • Production System Verification • Pseudo-Synthetic Video • Public Service Vehicle

PSVD Partial Singular Value Decomposition

PSVP Pilot Secure Voice Project

PSVT Pilot Secure Voice Terminal

Psvtn Preservation

PSW Passos (Brazil-MG) (location identifier) • Processor Status Word • Processus Status Word • Program Status Word

PSWG Personnel Security Working Group

PSWM Prestage Watermark

PSWO Picture and Sound World Organization

PSX Palacios (USA-TX) (location identifier)

PSY Port Stanley (Falkland Islands) (location identifier) • Professional Staff Year • Psychiatry

Psych Psychic • Psychical • Psychological • Psychologist • Psychology

Psychol Psychological • Psychologist • Psychology

PSZ Partially Stabilized Zirconia • Puerto Suarez (Bolivia) (location identifier)

PSZNM Politechnika Slaska Zeszyty Naukowe Mechanika (periodical code)

p/t Part-Time

pt Pint • Point • Portugal • Portuguese (ISO language code) • Pro Tempore

pT Picotesla

Pt Part • Patient • Payment • Petit • Platin • Platine • Platinum • Point • Pont • Poort • Port • President • Président • Puerto • Punt

P/T Part-Time • Power Turbine • Prototype

P&T Packaging & Transportation • Personnel & Training • Pipeline & Terminal

PT Brazil (international aircraft registration code) • Pacific Time • Package Table • Packaging and Transportation • Page Table • Paper Tape • Part • Part Task • Part Time • Patient • Payload Type • Penetrant Testing • Penetration Test • Personal Telecommunications • Personnel Training • Perte Totale • Perturbation Theory • Phase Transition • Photothermal • Physical Therapy • Physical Training • Physics Today (periodical) • Pie Town (USA-NM) • Pint • Pitch Trim • Plain Test • Plain Text • Plein Traitement • Pneumatic Tube • Point • Point of Tangency • Point of Turn • Point of Turning • Point to Point • Porta • Portugal (ISO country code) • Positive Test • Postal Telegraph • Post Town • Potential Transformer • Power Transfer • Power Turbine • Preferential Treatment • Pressure Transducer • Previously Treated • Primary Target • Primary Term • Printer Terminal • Private Type • Procedure Turn • Production Test • Productivity Team • Product Team • Professeur Technique • Proficiency Testing • Program Tab • Program Timer • Project Team • Propulsion Test • Protected Text • Protocol Translation • Protocol Translator • Prototype • Publication Type • Pulsar Time • Pulse Time • Punched Tape • Pupil Teacher • Purchase Tax

Pta Peseta (currency) • Ponta • Pretoria (South Africa) • Puerta • Punta

PTA Parent-Teacher Association • Personnel Telecommunications Assistant • Peseta (currency) • Phototransistor Amplifier • Physical Terminal Address • Physique Totalement Avantageux • Pie Town Antenna (NRAO) • Planar Turbulence Amplifier • Pointing and Tracking Assembly • Point of Total Assumption • Polskie Towarzystwo Astronomiczne (Poland) • Port Alsworth (USA-AK) (location identifier) • Posts and Telecommunications Administration • Posttest Analysis • Potential Toxic Area • Preferential Trade Area • Prepaid Ticket Advice • Primary Target Area • Process Technology Advancement • Procurement Technical Assistance • Production Test Array • Professeur Technique Adjoint • Programme de Travail Académique • Program Time Analyzer • Propellant Tank Assembly • Propellant Transfer Area • Property Transfer Authorization • Proposed Technical Approach • Propulsion Test Article • Pulsar Timing Array • Pulse Torquing Assembly

PTAC Poids Total Autorisé en Charge • Procurement Technical Assistance Center

PTADB Planning Terrain Analysis Database

PTarO Publications of Tartu Observatory (periodical code)

Ptas Pesetas

PTAS Performance Trend Analysis System

PTB Pass-Through Bus • Payload Timing Buffer • Permian-Triassic Boundary • Petersburg (USA-VA) (location identifier) • Photometric Testbench • Photometry Testbench • Photonics Tech Briefs • Physikalisch-Technische Bundesanstalt (Germany) • Pounds per Thousand Barrels • Producto Territorial Bruto • Program Time Base • Project Testbed

Ptbl Portable • Potable

PTBR Punched Tape Block Reader

PTC Pacific Telecommunications Council • Packet Transmission Controller • Parent-Teacher Consultation • Passive Thermal Conditioning • Passive Thermal Control • Personnel Transfer Capsule • Phase Tracking Circuit • Photographic Type Composition • Pointing Telescope Computer • Portable Temperature Controller • Port Alice (USA-AK) (location identifier) • Positive Temperature Coefficient • Positive Train Control • Positive Transaction Control • Posttest Critique • Power Transfer Coefficient • Presentation Technique Component • Pretaxi Clearance • Primary Technical Control • Primary Traffic Channel • Primary Training Centre • Programmable Test Control • Program Transmission Control • Proof Test Capsule • Propellant Tanking Console • Propulsion Test Complex • Pulse Time Code

PTCC Priorities and Tasking Control Committee • Propulsion Test Control Center

PTCDA Perylenetetracarboxylic Dianhydride

PTCE Pilot Tropical Cirrus Experiment

PTCHY Patchy

PTCR Pad Terminal Connection Room

PTCS Passive Thermal Control Section • Passive Thermal Control System • Payload Test and Checkout System • Planning, Training, and Checkout System • Pointing, Tracking, and Control System • Precision Telescope Control System • Propellant Tanking Computer System

PTCT Professeur Technique Chef de Travaux

Ptd Painted • Printed

PTD Parafoil Technology Demonstration • Parallel Transfer Disk • Penetration Test Device • Physical Theory of Diffraction • Port Alexander (USA-AK) (location identifier) • Preliminary Technical Directive • Programmable Threshold Detector • Provisioning Technical Documentation

PTDD Parallel Transfer Disk Drive

PTDLDP Pas Tombé De La Dernière Pluie

PTDOM Pays, Territoires et Départements d'Outre-Mer

PTDS Preliminary Training Device Study

Pte Plate • Pointe • Poniente • Ponte • Porte • Private • Puente

P&TE Packaging & Transport Engineering

PTE Escudo (Portugal) (ISO currency code) • Packaging and Transport Engineering • Page Table Entry • Peculiar Test Equipment • Personal Tax Edge • Poniente • Ponte • Portable Test Equipment • Port Stephens (Australia-NSW) (location identifier) • Portuguese Escudo (ISO currency code) • Posttest Examination • Preliminary Test Estimator • Pressure-Tolerant Electronics • Private Trading Exchange

PTEA Preliminary Training Effectiveness Analysis

PTEL Prequalified Training Equipment List

PTEP Prequalified Training Equipment Program • Professeur Technique d'Enseignement Professionnel

PTER Processed Term Error Rate

P&TF Prototype & Test Facility

PTF Malololailai (Fiji) (location identifier) • Patch and Test Facility • Payload Test Facility • Phase Transfer Function • Pilot Training Facility • Plasma Turbulent Reactor • Probabilistic Term Frequency • Problem Temporary Fix • Problem Trouble Fix • Productividad Total de los Factores • Program Temporary Fix • Program Trouble Fix • Propellant Tank Flow • Prototype and Test Facility

PTFE Polytetrafluorethylene

Ptg Pleistogene • Portugal • Portuguese • Printing

PTG Personal Travel Guide • Pietersburg (South Africa) (location identifier)

PTGS Palomar Transit Grism Survey • Portable Telemetry Ground Station

PTGT Primary Target • Privilege Ticket Granting Ticket

PTH Plated-Through Hole • Port Heiden (USA-AK) (location identifier)

PThPh Progress in Theoretical Physics (periodical code)

PThPS Progress in Theoretical Physics Supplement (periodical code)

PTI Packet Type Identifier • Palomar Testbed Interferometer • Parkes-Tidbindilla Interferometer (Australia) • Path Table Information • Payload Type Identifier • Payload Type Indicator • Personal TEX Inc. (USA-CA) • Personnel Transaction Identifier • Physical Training Instructor • Physical Trait Identification • Physico-Technical Institute • Port Douglas (Australia-QLD) (location identifier) • Preliminary Test Information • Press Trust of India • Programmed Test Input • Public Tool Interface

PTIA Perte Totale et Irréversible d'Autonomie

PTIC Program Track Interface Computer

PTIP Production Test and Inspection Plan • Prototype Transition and Implementation Plan

PTIS Programmed Test Input System • Program Test Input System • Propulsion Test Instrumentation System • Public Travel Information Service

PTIT Planetary Target Implementation Team

PTJ Portland (Australia-VIC) (location identifier)

PTK Passport To Knowledge • Pontiac (USA-MI) (location identifier)

Ptl Patrol

PTL Payload Test Laboratory • Phase Tracking Loop • Photographic Technology Laboratory • Physical Transport Layer • Plasma Turbulence Layer • Port Armstrong (USA-AK) (location identifier) • Potential Termination Liability • Primary Task List • Probabilistic Terminological Logic • Process and Test Language • Professeur Technique de Lycée • Protecting Technical Lead

PTLA Pseudo Top-Level Aggregator

PTLS Partial Total Least Squares

PTLXAU Public Telex Access Unit

Ptly Partly

PTLY Partly

PTM Packet Transfer Mode • Packet Transport Mode • Palmarito (Venezuela) (location identifier) • Parent-Teacher Meeting • Pattern Transformation Memory • Pays Tiers Méditerranéens • Phase Time Modulation • Photographic Transit Method • Pneumatic Telescope Mast • Point To Multipoint • Preliminary Technical Manual • Program Timing and Miscellaneous • Proof Test Model • Pulse Time Modulation • Pulse Time Multiplex • Pulse Transmission Mode • Pulse Type Modulation

PTMLE Preliminary Test Maximum-Likelihood Estimator

PTMP Point-To-Multipoint

PTMPT Point-To-Multipoint

PTMS Pattern Transformation Memory System • Precision Torque Measuring System

PTMU Power and Temperature Monitor Unit

Ptn Partition • Pattern • Portion

PTN Morgan City (USA-LA) (location identifier) • Partly • Patterson Williams (USA-LA) (location identifier) • Physique et Techniques Nucléaires • Plant Test Number • Private Telecommunications Network • Procedure Turn • Public Telephone Network • Public Transportation Network

Ptnr Partner

PTNX Private Telecommunications Network Exchange

Pto Portillo • Porto • Puerto • Puesto

PTO Parent-Teacher Organization • Participating Test Organization • Part-Time Operation • Patent and Trademark Office (USA) • Pátio • Pato Branco (Brazil-PR) (location identifier) • Permit To Operate • Personal Time Off • Please Turn Over • Post and Telephone Operator • Power Takeoff • Power Test Operation • Preliminary Technical Order • Professeur Technique et Ouvrier • Proof Test Orbiter • Propellant Transfer Operation • Public Telecommunications Operator • Public Telephone Operator • Pulse Tube Refrigerator

PTOC Part Time On Call

PTOE Plan Table of Organization and Equipment

PTOM Pays et Territoires d'Outre-Mer

P&TOP Pipeline & Terminal Operating

PTOP Program Test and Operations Plan

PTOR Power Turn On Reset

PTOS Paper Tape-Oriented Operating System

PTP Packet Translator Process • Paper Tape Perforator • Paper Tape Printer • Paper Tape Punch • Patch and Test Panel • Peak-To-Peak • Piezoelectric Tilting Platform • Pointe-à-Pitre (Guadeloupe) (location identifier) • Point-To-Point • Postgraduate Training Partnership • Preliminary Task Plan • Preoperational Test Procedure • Prepaid Transportation Program • Primary Target Point • Probability Thresholding Principle • Probability To Penetrate • Programmable Text Processor • Program Task Planning • Program Test Plan

PTPH Projected Total Piece Handling

PTPR Program Test Procedures and Reports

PTPS Package Test Power Supply • Propellant Transfer Pressurization System

pt/pt Point to Point

PTPU Program Tape Preparation Unit

PTQ Pôrto de Moz (Brazil-PA) (location identifier) • Put Together Quickly

Ptr Printer

PTR Pacific Test Range • Paper Tape Reader • Perforated-Tape Reader • Photoelectric Tape Reader • Photo-Thermo-Refractive • Physikalisch-Technische Reichsanstalt (Germany) (now PTB) • Pleasant Harbor (USA-AK) (location identifier) • Pointer • Pole Tip Recession • Portable Tape Recorder • Posttraumatic Recurrence • Power Transformer • Precision Transmitter Receiver • Preliminary Technical Report • Preliminary Technical Review • Preliminary Test Report • Prestation Technologique de Réseau • Printer • Problem Trouble Record • Problem Trouble Report • Processor Tape Read • Programmer Trouble Report • Program Technical Review • Program Test Review • Program Trouble Report • Protocol de Terminal Réseau • Proton Transfer Reaction • Punched-Tape Reader

PTRA Physics Teaching Resource Agent

PTRK Probability of Track

pTRM Partial Thermoremanence

PTRM Partial Thermoremanent Magnetization

PTRMS Proton Transfer Reaction Mass Spectrometry

PTRR Preliminary Test Readiness Review

PTRSL Philosophical Transactions of the Royal Society of London (periodical code)

Pts Parts • Payments • Pesetas (currency) • Pints • Points • Ports • Portsmouth (UK-England)

P&TS Program & Technical Support

PTS Palermo Technical Scale • Parallel Transaction Server • Partículas Totales en Suspensión • Payload Test Set • Payload Transportation System • Permanent Threshold Shift • Photothermal Spectroscopy • Phototypesetting • Pittsburgh (USA-KS) (location identifier) • Plasma Technology Section • Plasmaturbulenz-Schicht • Pneumatic Test Set • Point de Transfert Sémaphore • Points • Polar Track Structure • Polished Thin Section • Portas • Positive Torque Signal • Post Telephone Switch • Posttraumatic Stress • Power Transfer Switch • Practical Test Standard • Precision Timing System • Presentation Time Stamp • Pressurized Thermal Shock • Prime Target System • Principal Test Stand • Program Test System • Propellant Transfer System • Pure Time Sharing

PTSA Parent, Teacher, Student Association

PTSD Posttraumatic Stress Disorder

PTSE PNNI Topology Sate Element

PTSF Payload Trunnion Support Fixture

PTSG Plasma Theory and Simulation Group

pts/hr Parts per Hour

PtSi Platinum Silicide

PTSI Physique - Technologie - Sciences de l'Ingénieur • Politique des Télécommunications et des Services de l'Information

Ptsmth Portsmouth (UK-England)

PTSP PNNI Topology State Packet

PTSR Preliminary Technical Survey Report • Preliminary Technical System Review

PTSS Precision Time Synchronization Service

Ptt Partout • Petit • Pourtant

P&TT Pointing & Target Tracking

PTT Pacific Telephone and Telegraph Co. • Part Task Trainer • Performances and Phonograms Treaty • Platform Terminal Transmitter • Pointing and Target Tracking • Posterijen - Telegrafie - Telefonie • Postes - Télécommunications - Télédiffusion • Postes - Télégraphes - Téléphones • Post - Telegraph - Telephone • Post T Tauri (star) • Pratt (USA-KS) (location identifier) • Prenegociated Target Time • Press To Transmit • Program Test Tape • Projet de Transfert Technologique • Protocol Test Tool • Public Telephone and Telegraph • Push To Talk

PTTC Paper Tape Transmission Code

Ptte Petite

Pttes Petites

PTTI Precise Time and Time Interface • Precise Time and Time Interval (*)

pt-tm Part Time

pttm Part Time

Ptts Petits

PTTS Platform Test and Training System • Portable Transponder Test Set • Post T Tauri Star

PTTX Phase Transfer Transmitter

PTU Parallel Transmission Unit • Plan and Tilt Unit • Plan de Transporte Urbano • Platinum (USA-AK) (location identifier) • Portable Test Unit • Power Terminal Unit • Pression - Température - Humidité • Pressure - Temperature - Humidity • Probe/Transducer Unit • Process Technology Utilization

PTUC Participating Test Unit Coordinator

PTUV Pression - Température - Humidité - Vent

PTV Particle Tracking Velocimetry • Passenger Transport Vehicle • Pathfinder Test Vehicle • Peak To Valley • Penetration Test Vehicle • Pitch Thrust Vector • Porterville (USA-CA) (location identifier) • Predetermined Time Valve • Programmierte Textverarbeitung • Projection Television • Propulsion Test Vehicle • Prototype Test Vehicle • Public Television • Punched Tape Verifier

PTVA Propulsion Test Vehicle Assembly

PTVC Pitch Thrust Vector Control

PTVE Propulsion Test Vehicle Engineering

PTW Per Thousand Words • Playing To Win • Pottstown Limerick (USA-PA) (location identifier) • Primary Translation Word

PTWC Pacific Tsunami Warning Center (USA-HI)

PTX Pitalito (Colombia) (location identifier) • Practical Training Exercise • Preliminary Training Exercise • Pressure Transducer

Pty Party • Planetary • Principality • Property • Proprietary

PTY Panama City - Balboa - Omar Torrijos (Panama) (location identifier) • Panama City - Tocumen (Panama) (location identifier) • Party • Proprietary

PtyS Planetary Sciences

PTZ Pastaza (Ecuador) (location identifier)

p/u Pick up

Pu Plutonium • Purple

P/U Power Unit

PU Paid Up • Participating Unit • Penumbra • Peripheral Unit • Physical Unit • Picked Up • Pick Up • Planning Unit • Plant Unit • Polyurethane • Portable Unit • Potchefstroom University (South Africa) • Potins d'Uranie • Power Unit • Pressure Unit • Princeton University (USA-NJ) • Printer Unit • Prix Unitaire • Processing Unit • Production Unit • Professeur des Universités • Propellant Unit • Propellant Utilization • Propulsion Unit • Prototype Unit • Public • Public Utility • Pueblo

PUA Planning Unit Administrator • Primary Using Audience • Profiling User Agent • Puas (Papua New Guinea) (location identifier)

PUAE Peaceful Uses of Atomic Energy

PUAMA Poznań Uniwersytet im Adama Mickiewicza Seria Akustika (periodical code)

PUAMF Poznań Uniwersytet im Adama Mickiewicza Seria Fizika (periodical code)

pub Publication (file extension)

Pub Public • Publication • Publicity • Publish • Published • Publisher • Publishing

PUB Physical Unit Block • Public • Publication • Publish • Published • Publisher • Publishing • Pueblo (USA-CO) (location identifier)

Pubbl Pubblicazione

Pubd Published

Pub. Doc. Public Document

Publ Public • Publication • Publicity • Published • Publisher • Publishers • Publishing

PUBLICA Public Libraries Concerted Action

Pubn Publication

Pubr Publisher

PUBS Publications

Pub. Wks. Public Works

PUC Pacific Union College (USA-CA) • Paper Under Consideration • Peripheral Unit Controller • Permanent Unit Code • Personal Universal Controller • Pick Up Car • Pontificia Universidad Católica (Chile) • Presidential Unit

Citation • Presse Universitaire Canadienne (Canada) • Price (USA-UT) (location identifier) • Processing Unit Cabinet • Procurement Unit Cost • Program Unit Code • Public Utilities Commission • Public Utility Commission (USA)

PUCCh Pontificia Universidad Católica de Chile

PUCE Pattern - Unit - Component - Evaluation • Pour l'Utilisation de Composants Électroniques

PUCH Precursor of Ultracompact HII (region)

PUCS Propellant Utilization Control System

PUCU Propellant Utilization Control Unit

PUD Physical Unit Directory • Pick Up and Delivery • Planned Unit Development • Pop-Up Detector • Puerto Deseado (Argentina-SC) (location identifier)

PUDLV Postuniversaldienstleistungsverordnung

PUDT Propellant Utilization Data Translator

Pue Puebla (Mexico)

PUE Puebla (Mexico) • Puerto Obaldia (Panama) (location identifier) • Puesta

PUEO Probing the Universe with Enhanced Optics

PUF Pau Pyrénées (France) (location identifier) • Paysage Universitaire Français • Planning Units File • Presses Universitaires de France (*) • Program Unit Funding

PUFG Prototype UDS File Generator

PUFI Processed Under Federal Inspection

PUFir Pubblicazioni della Università degli studi di Firenze (periodical code)

PUFS Programmer's Utility Filing System

PUFT Projective Unified Field Theory

PUG Pascal Users' Group • Port Augusta (Australia-SA) (location identifier) • Prime Users Group • PROS Users' Guide

PUGS Portable Uplink Ground Station • Propellant Utilization and Gauging System

PUH Pochutla (Mexico) (location identifier)

PUHK Polytechnic University of Hong Kong (China-PRC)

PUI Pen User Interface • Perceptive User Interface • Perceptual User Interface • Pick-Up Ion • Pureni (Papua New Guinea) (location identifier)

PUID Personal User Identification • Project Unique Identifier

PUJ Punta Cana (Dominica) (location identifier)

PUK Pack Up Kit • Personal Unlocking Key • Pukarua (Tuamotu Islands) (location identifier)

PUL Parent Unit Level • Plan d'Urbanisme Local • Poulsbo (USA-WA) (location identifier) • Preses Universitaires de Lyon • Program Update Library • Propellant Utilization and Loading

PULL Plan d'Urbanisme Local Libéralisé

PULS Persoonsgericht Universitair Postdoc Stimulering • Propellant Utilization and Loading System

Pulsar Pulsating Star

PULSAUR Pulsating Aurora

pulv Pulvis

PUM Pomala (Indonesia) (location identifier) • Presses Universitaires de Montréal (Canada-QC) • Primitive Upper Mantle

PuMa Pulsar Machine

PUMA Portable University Model of the Atmosphere • Pour l'Utilisation de Matériaux Avancés • Preparation for Use of MSG in Africa • Punching Machine

PUMAS Practical Uses of Maths And Science

PUMPA Parallel Universal Message Passing Architecture

PUMS Physical Unit Management Services • Public Use Microdata Samples

PUN Physical Unit Number • Punia (Zaire) (location identifier)

Punc Punctuation

Punj Punjab (India)

PUO Prudhoe Bay (USA-AK) (location identifier)

PUOW Proposed Unit Of Work

Pup Puppis (constellation)

PUP PARC Universal Packet • PARC Universal Protocol • Performance Update Program • Peripheral Unit Processor • Po (Burkina Faso) (location identifier) • Polytechnic University of the Philippines • Polyurethane Paint • Pop-Up Point • Portable Uninterruptible Power • Power Upgrade Program • Princeton University Press • Principal User Processor • Product Update Program • Puppids/Velids (meteor stream)

PUPD Poona University Physics Department (India)

PUPO Pull Up - Push Over

PUPS Portable Uninterruptible Power Supply

PUQ Punta Arenas - Presidente Ibañez (Chile) (location identifier)

Pur Purchase • Purchaser • Purple • Pursuant • Pursuit

PUR Payload Under Review • Pileup Rejection • Program of University Research • Puerto Rico (Bolivia) (location identifier) • Purchasing

Purch Purchase • Purchaser

PUREX Plutonium and Uranium Extraction

PURL Persistent Uniform Resource Locator

Purp Purple • Purpose

PUS Packet Utilization Standard • Parliamentary Undersecretary • Performance Upgrade Socket • Permanently Unfit for Service • Permanent Undersecretary • Processor Upgrade Socket • Propellant Utilization System • Public Understanding of Science • Pusan (Korea-ROK) (location identifier)

PUSAS Proposed United States of America Standard

PUSH Public Understanding of Science and Humanities

PUSK Publikationen der Königlichen Sternwarte in Kiel (periodical code)

PUSNO Publications of the US Naval Observatory - Second Series (periodical code)

PUSPIPTEK Pusat Penelitian Ilmu Pengetahuan dan Teknologi (Indonesia)

PUSS Pallet Utility Support Structure

PUSSYCATS Pedestrian Urban Safety System and Comfort at Traffic Signal

PUST Public Understanding of Science and Technology

PUSV Packet Utilization Standard Validation

PUT Primary User Terminal • Programmable Unijunction Transistor • Puttaparthi (India) (location identifier)

PUTT Propellant Utilization Time Trace

PUU Puerto Asis (Colombia) (location identifier)

PUV Poum (New Caledonia) (location identifier) • Propellant Utilization Valve

PUW Planetarium und Uraniasternwarte Wien (Austria) • Pullman Moscow (USA-WA) (location identifier)

PUY Pula (Croatia) (location identifier)

PUZ Puerto Cabezas (Nicaragua) (location identifier)

pv Cauchy Principal Value • Per Value

P/V Photovoltaic • Portable/Vehicle

PV Packetized Voice • Paketvermittlung • Parallel Vector • Parity Violation • Path Verification • Peak to Valley • Peculiar Velocity • Pequeña Velocidad • Performance Verification • Personalvertretung • Petite Visite • Petite Vitesse • Photovoltaic • Physical Volume • Piccola Velocità • Pioneer Venus • Pixel Value • Planetary Vehicle •

Plan View • Plena Vortaro • Position-Velocity • Positive Volume • Potential Velocity • Potential Vorticity • Power Voltage • Present Value • Pressure Vessel • Pressurization Valve • Prevalve • Prime Vendor • Printed Version • Private Venture • Privileged Information • Process Verification • Procès-Verbal • Production Validation • Program Visualization • Promesse de Vente • Pyro Valve • Pyrotechnic Valve

PVA Perigee Velocity Augmentation • Pivalic Acid • Polyvinyl Acetate • Polyvinyl Alcohol • Population Viability Analysis • Portable Voice Amplifier • Positive Vorticity Advection • Preburner Valve Actuator • Propellant Valve Actuator • Providencia (Colombia) (location identifier)

PVAA Pomona Valley Amateur Astronomers (USA-CA)

PVAO Planned Versus Actual Obligation • Platte Valley Astronomical Observers (USA-NE)

PVAS Primary Voice Alerting System

PVASI Pulsating Visual Approach Slope Indicator

PVasO Publications of the Vassar College Observatory (periodical code)

PVATS Portable Volume Automated Test System

PVB Polyvinyl Butyral

PV&C Performance Verification & Calibration

PVC Performance Verification and Calibration • Permanent Virtual Circuit • Permanent Virtual Connection • Pigment Volume Concentration • Polychlorure de Vinyle • Polyvinylchloride • Position and Velocity Computer • Postal Vending Console • Potential Volume Change • Pressure Vacuum Chamber • Pressure Vacuum Compensator • Pressure-Volume Compensated • Provincetown (USA-MA) (location identifier)

PVCC Permanent Virtual Channel Connection

PVCS Popular Version Control System

PV&D Purge, Vent & Drain

PVD Pairwise Velocity Dispersion • Paquet avec Valeur Déclarée • Passive Vibration Damping • Pays en Voie de Développement • Peculiar Velocity Distribution • Physical Vapor Deposition • Physique Vraiment Désavantageux • Plan View Display • Plasma Vapor Deposition • Position-Velocity Diagram • Primary Volume Descriptor • Prise de Vues Directe • Providence (USA-RI) (location identifier) • Pulsed Voltammetric Detection • Purge - Vent - Drain

PVDF Polyvinyl Difluoride • Polyvinylidene Fluoride

PVDG Petascale Virtual Data Grid

PVDS Propellant Vent and Drain System

PVDST Perennisation et Valorisation des Données Scientifiques et Techniques (conference)

P&VE Propulsion & Vehicle Engineering

PVE El Porvenir (Panama) (location identifier) • Photovoltaic Effect • Point de Vente Électronique • Process Validation Enterprise • Propulsion and Vehicle Engineering

PVED Propulsion and Vehicle Engineering Division

PVF Peripheral Visual Field • Placerville (USA-CA) (location identifier) • Polyvinyl Fluoride

PVFS Parallel Virtual File System

PVG Personalvertretungsgesetz • Perspective View Generator

PVGC Pioneer Venus Gas Chromatograph

PVH Pôrto Velho - Belmonte (Brazil-RO) (location identifier) • Premier Vol Habité

PVI Paranavaí (Brazil-PR) (location identifier) • Perpendicular Vegetation Index • Pilot Vehicle Interface • Point of Vertical Intersection • Postage Validation Imprinter • Postal Validation Imprinter • Principe de Vision Insuffisante

PVK Polyvinylcarbazole • Preveza Lefkas (Greece) (location identifier)

PVL Parameter Value Language • Physical Volume Library • Prevail

PVLA Purchase Versus Lease Analysis

PVLAN Protocol Virtual Local Area Network

PVM Parallel Virtual Machine • Particle Volume Monitor • Pass-through Virtual Machine • Potentiometric Voltmeter • Product Verification Manager • Projection Video Monitor

PVN Pleven (Bulgaria) (location identifier) • Private Virtual Network

Pvnt Prevent

Pvntv Preventive

PVO Pioneer Venus Orbiter • Portoviejo (Ecuador) (location identifier) • Private Voluntary Organization • Private Volunteer Organization

PVOC Polar Volatile Organic Compound

PVOR Precision VHF Omnirange

PVOUVS Pioneer Venus Orbiter Ultraviolet Spectrometer

PVP Packet Video Processor • Packet Video Protocol • Parallel Vector Platform • Parallel Vector Processing • Parallel Vector Processor • Performance Verification Plan • Permanent Virtual Path • Pipelined Vector Supercomputer • Planned Value Profile • Plasma Vaporization Process • Precio de Venta al Público • Prix de Vente Public • Product Verification Program

PVPC Permanent Virtual Path Connection

PVPP Polyvinylpolypyrrolidon

PV/PS Pressure Vessel and Pressure System

PVPS Pressure Vessel and Pressure System

PVR Personal Video Recorder • Photovoltaic Radiator • Physical Volume Repository • Precision Voltage Reference • Procedure Validation Report • Production Verification Review • Puerto Vallarta (Mexico) (location identifier)

PVRA Pioneer Venus Radar Archive

PVRD Purge - Vent - Repressurize - Drain

PVS Parallel Visualization Server • Parameter Variation Space • Performance and Verification System • Performance Verification System • Performance Visualization System • Periscope View Simulator • Personal Vision System • Physical Vapor Synthesis • Physical Volume Repository • Postal Vehicle Service • Pressure Vacuum System • Prime Vendor Support • Program Validation Service • Project Visualization System • Propellant Venting System • Provideniia (Russia) (location identifier)

PVSA Prime Visual Signal Area

PVSD Programmable Variable Storage Device

PVSG Penobscot Valley Star Gazers (USA-ME)

PVSM Periodic Volume Stimulus Method

PVSS Publications of Variable Star Section RASNZ (periodical code)

PVST Prime Verification System Test

Pvt Private

PVT Page View Terminal • Par Voie Télégraphique • Passive Visible Tracker • Performance Verification Test • Permanent Virtual Terminal • Physical Vapor Transport • Point of Vertical Tangent • Polyvinyltoluene • Position - Velocity - Time • Position - Vitesse - Temps • Preflight Verification Test • Pressure - Volume - Temperature • Previous Value Table • Private • Probe Velocity Transducer • Production Validation Test • Production Verification Test • Product Validation Test • Product Verification Test • Product Verification Testing • Projet de Validation Technologique • Prototype Version Test • Pyrotechnic Verification Test

PVTCS Photovoltaic Thermal Control System

PVTOS Physical Vapor Transport of Organic Solids

PVTR Portable Video Tape Recorder

PVU Potential Vorticity Unit • Provo (USA-UT) (location identifier)

PVV Pressure, Vent and Vacuum

PVVO Publications of the Van Vleck Observatory (periodical code)

PVVP Pioneer Venus and Venera Probe

PVW Plainview - Hale County (USA-TX) (location identifier) • Principle of Virtual Work

PVWA Planned Value of Work Accomplished

PVWS Planned Value of Work Scheduled

PVX Provedenia (Russia) (location identifier)

PVY Pope Vanoy (USA-AK) (location identifier)

PVZ Painesville Casement (USA-OH) (location identifier)

pw Per Week

pW Picowatt

P&W Pratt & Whitney

P/W Password • Plasma/Wave

PW Palau (ISO country code) • Password • Pawnee • Per Week • Phase-Weighted • Phonon Wing • Plain Washer • Planetary Wave • Plane Wave • Plan Washer • Plan Window • Positionswinkel • Potable Water • Practical Work • Precipitable Water • Precipitation Water • Printed Wiring • Production Work • Project Work • Public Works • Publieke Werken • Pulse Width

PWA Oklahoma City - Wiley Post (USA-OK) (location identifier) • Partial Word Aligner • Permittivity and Wave Analyzer • Pratt and Whitney Aircraft • Printed Wire Assembly • Printed Wiring Assembly • Private Write Area • Probabilistic Web Algebra • Product Work Authorization • Program-Wide Activities • Project Work Authorization • Public Works Administration

PWasO Publications of the Washburn Observatory (periodical code)

PWB Printed Wire Board • Printed Wired Board • Printed Wiring Board • Programmer's Workbench • Pulsar Wind Bubble

PWBA Printed Wiring Board Assembly

PWBS Preliminary Work Breakdown Structure • Program Work Breakdown Structure • Project Work Breakdown Structure

PWC Pacific Weather Centre • Performing Work Center • Physical Work Capacity • Proportional Wire Chamber • Proportional Wired Chamber • Public Works Center • Pulse-Width Coded

pwd Password • PocketWord Document (file extension)

PWD Participatory Work Design • Plentywood Sherwood (USA-MT) (location identifier) • Power Distribution • Prewhite Dwarf (star) • Print Working Directory • Proceed With Design • Procurement Work Directive • Program Work Directive • Proximity Warning Device • Public Works Department • Pulsating White Dwarf (star) • Pulse Width Discriminator

PWDN Power Down

PWDS Protected Wire Distribution System • Protected Wireline Distribution System

PWE Pauli-Weisskopf Equation • Pevek (Russia) (location identifier) • Pulse-Width Encoder

PWEA Printed Wiring and Electronic Assemblies

PWEAR Preliminary WBS Element Audit Review

PWEN Physics World Electronic News

PWF Personnel Working File • Photoelectric Work Function • Poleward-Facing • Present Worth Factor • Propellant Weight Fraction • Pulse Wave Form

PWFA Plasma Wakefield Acceleration • Plasma Wakefield Accelerator

PWFS Peripheral Wavefront Sensor

PWG Parallels Working Group • Payload Working Group • Permanent Working Group • Planar Waveguide

PWGS Pacific Weather Graphics System

PWGSC Public Works and Government Services of Canada

PWHT Postweld Heat Treatment

PWI Pawi (Ethiopia) (location identifier) • Pilot Warning Instrument • Plasma Waves Instrument • Plasma Waves Investigation • Proximity Warning Instrument • Public Windows Interface

PWIA Plane Wave Impulse Approximation

PWK Chicago Palwaukee (USA-IL) (location identifier)

PWL Password List • Piecewise Linear • Purwokerto (Indonesia) (location identifier)

PWLA Portable Work Light Assembly

PWLL Profile Word Logic List

PWM Plattform Wissensmanagement • Portland (USA-ME) (location identifier) • Probability-Weighted Moment • Pulse-Width Modulated • Pulse-Width Modulation • Pulse-Width Modulator • Pulse-Width Multiplier

PWMD Printed Wiring Master Drawing

PWMR Pulse Width Modulated Regulator

PWN Pitts Town (Bahamas) (location identifier) • Pulsar Wind Nebula

PWNe Pulsar Wind Nebulae

PWO Pweto (Zaire) (location identifier)

PWP Planning Work Package • Plasticized White Phosphorus • Portable Work Platform • Price When Perfect

PWQ Pavlodar (Kazakhstan) (location identifier)

Pwr Plain with Wrinkle Ridges • Power

PWR Port Walter (USA-AK) (location identifier) • Power • Pressurized-Water Reactor • Program Work Requirement • Publication Work Request • Pulse-Width Regulation

PWRI Public Works Research Institute (Japan)

Pwr. Sup. Power Supply

PWS Paley-Wiener Space • Parallel Working System • Parks and Wildlife Service • Peer Web Services • Performance Work Statement • Plane-Wave Spectrum • Plasma and Wave System • Plasma Wave System • Possible World Semantics • Potable Water Storage • Predictive Wind Shear • Preliminary Work Statement • Pricing Work Statement • Programmable Workstation • Program Work Statement • Proximity Warning System • Public Water System • Public Weather Services

PWSCS Programmable Workstation Communication Services

PW&SO Publications of the Warner & Swasey Observatory (periodical code)

pwt Pennyweight

PWT Bremerton (USA-WA) (location identifier) • Participating Work Team • Progressive Wave Tube • Propulsion Wind Tunnel

PWTN Power Train

Pwtr Pewter

PWV Precipitable Water Vapor • Pretoria-Witwatersrand Vereeniging (South Africa) • Pulse Wave Velocity

PWW Project West Wing

pwz PowerPoint Wizard (file extension)

PX Penetration Exercise • Physical Examination • Pixel • Please Exchange • Post Exchange • Primary Index • Private Exchange

PXD Powder X-ray Diffraction

PXE Phenyl-o-Xylylethane

pxl Pocket Excel (file extension)

PXL Polacca (USA-AZ) (location identifier) • Pulsed Xenon Laser

PXLS Pulsed Xenon Light Source

PXLSS Pulsed Xenon Light Source System

PXM Puerto Escondido (Mexico) (location identifier)

PXO Porto Santo (Madeira Islands) (location identifier)

PXP Packet Exchange Protocol

pxr Pixar Image (file extension)

PXS Pulsed Xenon System

PXSS Pulsed Xenon Solar Simulator

pxt Pinxit

PXU Pleiku (Viet Nam) (location identifier) • Portable X-ray Unit

py Paraguay • Pyridine

P-Y Pitch-Yaw

P/Y Pitch/Yaw

PY Paraguay (international vehicle registration code) • Paraguay (ISO country code) • Pygmy • Pitch-Yaw • Planning Year • Prior Year • Program Year • Publication Year • Pyrometer

PYA Puerto Boyaca (Colombia) (location identifier)

PYB Jeypore (India) (location identifier)

PyC Pyrolytic Carbon

PYC Playon Chico (Panama) (location identifier)

PYD Put Yourself Down

PYE Penrhyn Island (Cook Islands) (location identifier)

PYerO Publications of Yerkes Observatory (periodical code)

PYF French Polynesia (ISO country code) • Personal Yearly Funding • Polynesia (French_) (ISO country code) • Polynésie Française (code ISO de pays)

PYG Guarani (Paraguay) (ISO currency code)

PYH Puerto Ayacucho (Venezuela) (location identifier)

PYJ Polyarnyj (Russia) (location identifier)

PYL Perry Island (USA-AK) (location identifier)

PYM Plymouth (USA-MA) (location identifier)

PYME Pequeña Y Mediana Empresa

PYMES Pequeñas Y Medianas Empresas

Pymt Payment

PYN Payan (Colombia) (location identifier)

PYO Pick Your Own • Putumayo (Ecuador) (location identifier)

PYP Photoactive Yellow Protein • Photoreactive Yellow Protein

Pyr Pyramid

P-Y-R Pitch-Yaw-Roll

PYR Pitch - Yaw - Roll • Prior Year Report • Pyrgos (Greece) (location identifier) • Pyro Release • Pyrotechnic Release

PYREX Pyrénées Experiment

Pyro Pyrotechnics

Pyrotech Pyrotechnics

PYS Partial Yield Spectroscopy • Probably Young System

PYT Pretty Young Thing

PYTC Parent-Year Tutor Consultation

PYunO Publications of Yunnan Observatory (periodical code)

PYV Yaviza (Panama) (location identifier)

Pyx Pyxis-Pyxidis (constellation)

PYX Pattaya (Thailand) (location identifier)

Pz Palaeozoic

PZ Phase Zero • Peremennye Zvezdy (periodical code) • Pickup Zone • Precautionary Zone • Suriname (international aircraft registration code)

Pza Piazza • Plaza

PZA Paz de Ariporo (Colombia) (location identifier) • Postzeitungsamt

PZAA Polarized Zeeman Atomic Absorption

PZB Pädagogische Zentralbibliothek • Palast und Zentralbibliothek • Pietermaritzburg (South Africa) (location identifier) • Planning - Zoning - Building

PZC Pickup Zone Coordinator • Point of Zero Charge

PZE Penzance (UK-England) (location identifier)

PZETF Pis ma Zhurnal Eksperimental noi i Teoreticheskoi Fiziki (periodical code)

PZH Zhob (Pakistan) (location identifier)

PZhTF Pis ma Zhurnal Tekhnicheskoi Fiziki (periodical code)

PZJA Protected Zone Joint Authority

PZL Phinda (South Africa) (location identifier)

Pzla Plazuela

PZO Puerto Ordaz (Venezuela) (location identifier)

PZP Peremennye Zvezdy Prilozhenie (periodical code) • Phase Zero Program • Phase Zone Plate

PZT Lead Zirconate Titanate • Photographic Zenithal Telescope • Photographic Zenithal Tube • Photographic Zenith Telescope • Photographic Zenith Tube • Photographische Zentralteleskop • Piezoelectric Transducer • Piezo Translator

PZU Port Sudan (Sudan) (location identifier)

PZY Piestany (Slovakia) (location identifier)

P2 Papua New Guinea (international aircraft registration code) • Pollution Prevention

P2C Population II Cepheid

P2C2G Process Too Complicated To Explain

P2C4I Power Projection Command, Control, Communications, Computers, and Intelligence

P2I Preplanned Productivity Increase

P2K Passport to Knowledge

P2M2 Pseudoparticle Multipole Method

P2N Physique Neutronique et Nucléaire

P2P Peer to Peer

P2PDT Peer-to-Peer Data Transfer

P2PP Phase Two Proposal Preparation

P3 Parallel Porting and Productizing • Perception and Performance Prototyper • Platform for Privacy Preferences • Portable Plotting Package • Préparation de Pénétration Profonde

P3I Preplanned Product Improvement • Preplanned Production Improvement

P3IP Preplanned Product Improvement Plan • Preplanned Product Improvement Program

P3M Particle-Particle/Particle-Mesh • Personnel Policy Projection Model

P3P Platform for Privacy Preferences Project

P3T Parallel Porting and Productizing Team

P4 Aruba (international aircraft registration code)

P4I Preplanned Product and Process Improvement

Q

q Charge of Particle • Circular Wave Vector • Constant of Proportionality • Deceleration Constant • Density of Heat Flow Rate • Dynamic Pressure • Electric Charge • Generalized Coordinate • Heat Current Density • Heat Flow Rate • Heat Flux • Local Speed • Nodal Displacement • Partition Function • Perihelion Distance • Pitch Angular Velocity • Propagation Vector • Quadrans • Quadrat • Quaestio • Quark • Quart • Quarter • Quarterly • Quarto • Quasi • Que • Query • Question • Quick • Quintal • Quire • Quit • Relative Quantum Efficiency • Slowing Down Density • Solar Forcing Ratio • Specific Humidity • Squall

Q Aphelion Distance • Damping Factor • Desintegration Energy • Dynamic Pressure • Electric Charge • Electricity Quantity • Ensemble des Nombres Rationnels • Five Hundred (Roman numeral) • Gauss Vector • Heat • Heat Addition • Heating Rate • Heat Quantity • Injection Rate of Relativistic Electrons • Integral Heat Load • Light Quantity • Nuclear Quadrupole Moment • Partition Function • Pitch Rate • Pulsational Constant • Quadrature • Quadrupole Moment • Quai • Qualification • Qualified • Qualität • Qualité • Quality • Quality Factor • Quantität • Quantité • Quantity • Quantity of Electricity • Quantity of Heat • Quantity of Light • Quart • Quartal • Quarter • Quarterly • Quarternion • Quartier • Quarto • Quarz • Quaternary • Québec (international alphabet) • Québec (Canada) • Queen • Queensland (Australia) • Query • Question • Quetzal (currency) • Queue • Quintal • Quintus • Quit • Quitter • Radiant Energy • Reaction Energy • Reactive Power • Set of Rational Numbers • Solution Vector • Squall • Throughput • Torque • 500 (Roman numeral)

qa Quick Assembly

Q&A Qualification & Acceptance • Question & Answer

Q/A Qualification Approval • Quality/Accounting • Question/Answer • Questioning/Answering

QA Qatar (ISO country code) • Qualification Approval • Quality Analysis • Quality Assessment • Quality Assurance • Quality Audit • Quasi-Asymmetric • Query Analysis • Query Analyzer • Question and Answer • Question Answering • Queue-Arbitrated • Quiescent Aerial • Quiescent Antenna

QAA Quality Assurance Agency (UK) • Quality Assurance Assistant

QAA/P Quality Assurance Assessment and Plan

QAAP Quality Assurance Assessment and Plan

QAB Quality Assurance Board

QABP Quantum Aspects of Beam Physics (workshop)

QAC Qatar (ISO country code) • Quality and Accounting Capsule • Quality Assurance Chart • Quality Assurance Checklist • Quality Assurance Code

QACAD Quality Assurance Corrective Action Document

QACC Quality Assurance Consultative Committee

QAD Quality Assurance Directorate • Quality Assurance Division

QADS Quality-Assurance Data System

QAE Quality Assurance Engineer • Quality Assurance Evaluation • Quality Assurance Evaluator • Quality Assurance Executive

QAEN Quality Assurance and Enhancement Network

QAF Quality Achievement factor • Query Absolute Frequency • Queue-Arbitrated Function • Quick Action Firing

QAFM Quality Assurance Forms Manual

QAHN Quality Assurance Hold Notice

QAI Quality Assurance Inspection • Quality Assurance Institute • Quality Assurance Instruction

QAID Quick Access Identification

QAIP Quality Assurance Inspection Procedure

QAL QSO Absorption Line • Quality Activities List • Quasar Absorption Line

QALC Quasar Absorption Line Cloud

QALI Quality Assurance Letter of Instruction

QAM Quadraface Amplitude Modulation • Quadratic Amplitude Modulation • Quadratur-Amplitudenmodulation • Quadrature Amplitude Modulation (*) • Quadripartite Agreed Materiel • Quality Assurance Manager • Quality Assurance Manual • Quality Assurance Monitor • Queued Access Method

QAMR Quadripartite Agreed Materiel Requirement

QAMS Quality Assurance Management Staff

QANGO Quasi-Autonomous Non-Government Organisation

QANTAS Queensland And Northern Territory Airline System

QA&O Quality Assurance & Operations

QAO Quality Assurance Office

QAOP Quality Assurance Operating Procedure

QAP Quadratic Assignment Procedure • Quadripartite Advisory Publication • Qualification Acceptance Program • Qualification and Acceptance Program • Quality Assurance Plan • Quality Assurance Procedure • Quality Assurance Product • Quality Assurance Program • Quality Assurance Provisions • Quantum Action Principle

QAPE Quadripartite Afreed Plan of Engineering

QAPI Quality Assurance Program Index

QApMa Quarterly of Applied Mathematics (periodical code)

QAPP Quality Assurance Program Plan • Quality Assurance Project Plan

QAPSK Quadrature Amplitude Phase Shift Keying

QAQC Quality Assurance - Quality Control

QA/R Quality Assurance/Reliability

QA&R Quality Assurance & Revalidation

QAR Quality Acceptance Report • Quality Assessment Report • Quality Assurance and Revalidation • Quality Assurance/Reliability • Quality Assurance Report • Quality Assurance Representative • Quality Assurance Requirement • Quality Assurance Responsible • Quasi-Adiabatic Representation • Quick-Access Record • Quick-Access Recorder • Quick-Access Recording • Rial (Qatar) (ISO currency code)

QA/RM Quality Assurance/Risk Management

QARP Question-and-Answer Reinforcement Package

QAS Quality Assurance Specialist • Quasi-Asymmetric • Quasi-Asymmetry • Quaternary Ammonium Sulphanilamide • Question Answering System • Question/Answer System

QASAR Quality and Safety Achievement Recognition (NASA award)

QASL Quality Assurance Systems List • Quebec Association of School Librarians (Canada-QC)

QASP Quality Assurance Support Plan • Quality Assurance Surveillance Plan

QA&T Quality Assurance & Test

QAT Qatar (ISO country code) • Quality Action Team • Quality Assurance and Test • Quality Assurance Team • Query Alias Table

QATDP Quality Assurance Technical Development Program

QATS Quality Assurance Test Suite • Quality Assurance Test System

QATT Qualification for Acceptance Thermal Testing

QAU Quality Assurance Unit

QAVP Quad Analog Voice Port

QAVT Qualification Acceptance Vibration Test • Qualification for Acceptance Vibration Testing

QAW Anniston - Fort McLellan (USA-AL) (location identifier)

QB Quality Board • Quick Basic • Quick Break

QBA Quasi-Boson Approximation

qbb QuickBooks Backup (file extension)

Qbc Québec (Canada)

QBC Bella Coola (Canada-BC) (location identifier) • Query By Committee

QBD Quasi Birth and Death (process) • Query By Diagram

QBE Query By Example

QBF Query By Form

QBFJOTF Quick Brown Fox Jumped Over The Fence

QBFJOTLD Quick Brown Fox Jumped Over The Lazy Dog

QBE Query By Example

QBF Brack (Libya) (location identifier) • Query By Form

qBH Quiescent Black Hole

QBIC Query By Image Content

QBL Qualified Bidders List

QBM Quantum Brownian Motion

QBN Query By Navigation

QBO Quasi-Biannual Oscillation • Quasi-Biennial Oscillation (*)

QBR Quarterly Billing Report • Quarterly Business Review

QBS Quality-Based Selection • Quick Blue Survey

QBSA Quarterly Bulletin on Solar Activity

QBT Quadrax Biaxial Tape • Query By Table

qbw QuickBooks Windows (file extension)

QBX Quick Basic Extended

Q/C Quality Control • Quality Controller • Quality/Cost

QC Qualification Course • Quality Circle • Quality Conference • Quality Control • Quality Controller • Quality/Cost • Quantifiable Criteria • Quantum Cascade • Quantum Cryptography • Quantum Cycle • Quasi-Continuous • Québec (Canada) • Queen's Counsel • Query Case • Query Command • Query Component • Query Coordinator • Quick Change • Quit Claim

QCA Qualification and Curriculum Authority (UK) • Quality Control Analysis • Quick Change Assembly

QCAB Quote Commerce Access Broker

QCAC Queen's College Astronomers' Club (Hong Kong)

QCAL Qualified Contractor Access List

QCAM Queued Communications Access Method

QCAS Quad Cities Astronomical Society (USA-IA) • Quality Control and Applied Statistics

QCB Quality Control Board • Queue Control Block

QCC Qualitative Construction Capability • Quality Control Committee • Quantum Computing and Communications • Quenched Carbonaceous Carbon • Quenched Carbonaceous Composite

QCCOD QMW - Cambridge - CalTech - Oxford - Durham (collaboration)

QCD Quality Control Data • Quantenchromodynamik • Quantum Chromodynamics

QCDM Quintescence Cold Dark Matter (model)

QCDP Quality Color Dithering Process

QCDR Quality Control Deficiency Report

QCE Quality Control and Evaluation • Quality Control Engineering • Quality Control Expert

QCF Quasi-Continuous Fragmentation

QCFI Quality Conformance Final Inspection

QCG Quality Control Group

QCGD Quality Circle Gesellschaft Deutschland

QCI Quality Conformance Inspection • Quality Control Information • Quality Control Inspection

QCIC Quality Control, Assurance and Inspection

QCIF Quarter Common Interchange Format • Quarter Common Intermediate Format

QC Is Queen Charlotte Islands (Canada)

QCL Quality Control Level • Quantum Cascade Laser

qcm Quadratzentimeter

QCM Quality Control Manager • Quality Control Manual • Quantitative Computer Management • Quark Cluster Model • Quartz Crystal Microbalance • Quartz Crystal Mirror • Quartz Crystal Monitor • Question à Choix Multiple • Questionnaire à Choix Multiple

QCO Quality Control Officer • Quality Control Organization • Quartz Crystal Oscillator

Qconque Quiconque

QCOP Quality Control Operating Procedure

QCP Quality Control Plan • Quality Control Procedure • Quality Control Program • Quantum Critical Point

QCPI Quality Conformance Preliminary Inspection

QCPSK Quadrature Coherent Phase Shift Keying

QCQC Quantum Computing and Quantum Communications (conference)

QCQP Quadratically Constrained Quadratic Program

QCR Qualitative Construction Requirement • Quality Control Reliability • Quality Control Report • Quality Control Representative • Quality Control Review • Quality Control Room

QCRT Quick Change in Real Time • Quick Change Real Time

QCS Qualified Calibration Source • Quality Control Standard • Quality Control Synopsis • Quality Control System • Quality Customer Service • Query Control Station

QCT Quality Control Technology • Quality Control Test • Quantitative Computed Tomography • Quasi-Classical Trajectory

QCTR Quality Control Test Report

QCU Quartz Crystal Unit

qd Quaque Die • Quasi Dicat • Quasi Dictum

Qd Quand

Q&D Quick & Dirty

Q-D Quantity Distance

QD Quadruple Distortion • Quadrupolar Distortion • Quality Distance • Quantity Discount • Quantum Dot • Quantum Dynamics • Quaque Die • Query Decomposition • Queuing Delay • Quick and Dirty • Quick Disconnect

Qda Quebrada

QDA Quadratic Discriminant Analysis • Quantitative Descriptive Analysis • Quantity Discount Agreement

QDAS Quick Data Analysis System • Quick Distributed Analysis Software

QDBMS Query for Database Management System

qdbv Quod Deus Bene Vertat

QDC Charge to Digital Converter

QDE Quality Data Evaluation • Quantitative Differential Equation • Quantum Detection Efficiency

QDHT Quasi-Discrete Hankel Transform

Qdisc Quick Disconnect

qdk Quarterdeck (file extension)

QDLL Quick Data Definition Language

qdm Quadratdezimeter

QDM Dammam (Saudi Arabia) (location identifier) • Quadrature Discretization Method • Quantitative Debris Monitor • Quasi-Diffusion Method

QDMBPT Quasi-Degenerate Many-Body Perturbation Theory

QDN Query Direct Number

QDO Quadripartite Development Objective • Quantum Defect orbital

QDOS Quick and Dirty Operating System

QDOT Queen Mary and Westfield - Durham - Oxford - Toronto

QDP Qualified Data Package • Quality Deficiency Report • Quantity Discount Procedure • Quick and Dandy Plotter • Quickly Dissipated Packet

QDR Quad Data Rate • Quadrennial Defense Review • Qualification Design Review • Quality Deficiency Report

QDRI Qualitative Development Requirement Information

Qdrnt Quadrant

Qdrtr Quadrature

qds Quater in Die Sumendus

QDS Quality Data System • Quick and Dirty Spectrograph • Quick-look Data Set

QDTA Quantitative Differential Thermal Analysis

QDU Düsseldorf Hauptbahnhof (Germany) (location identifier) • Düsseldorf Hbf. (Germany) (location identifier) • Düsseldorf Main Station (Germany) (location identifier) • Quantized Distortion Unit

QD3D Quickdraw Tridimensional

qe Quod Est

QE Quadranteinheit • Quadrant Elevation • Quality Engineer • Quality Engineering • Quality Evaluation • Quantum Efficiency • Quantum Electronics • Quasi-Electrostatic • Query Expansion • Question Écrite • Quotient Émotionnel • Quotient Intellectuel

QEA Quantum Electronics and Applications

QEAM Quick Erect Antenna Mast

QEAMS Quick Erect Antenna Mast System

QEB Quick Engine Bulletin

QEC Quadrantal Error Correction • Quick Engine Change

qed Quod Erat Demonstrandum

QED Quality Education Data • Quantenelektrodynamik • Quantitative Evaluation of Deficiencies • Quantum Effect Design • Quantum Efficiency Detector • Quantum Electrodynamics • Quantum Electronics Division • Quasar Energy Distribution • Query - Explore - Demonstrate • Quick Easy Disk • Quick Epicenter Determination • Quick-Erection Dome • Quite Evidently Derived • Quod Erat Demonstrandum

QEDR Quick-look Experiment Data Record

QEER Quick Erect Expandable Mast

qef Quod Erat Faciendum

QEF Quod Erat Faciendum

QEH Quantum Efficiency Hysteresis

qei Quod Erat Inveniendum

QEII Queen Elizabeth II

QEL Quality Evaluation Laboratory • Queue Element

QELS Quality Evaluation of Literature System • Quantum Electronics and Laser Science (conference)

QEM Quality Education for Minorities (network) (USA)

QEMM Quarterdeck Expanded Memory Manager • Quarterdeck Extended Memory Manager

QEOD Quantum Electronics and Optics Division (EPS)

QEP Quadratic Eigenvalue Problem • Quality Enhancement Program • Quality Evaluation Program • Query Execution Plan

QEPCAD Quantifier Elimination by Partial Cylindrical Algebraic Decomposition

QEPD Que En Paz Descanse

QER Qualitative Equipment Requirement

QERD Quick-look Experiment Data Record

QEST Quality Evaluation System Test

QET Quantum Effect Transistor • Query Execution Tree

QEX Questions Extended

QF Quadruple Fraction • Qualitätsfaktor • Quality Factor • Quality Form • Quick Fix • Quotient Familial

QFA Quick File Access

QFAA Quartz Furnace Atomic Absorption

QFC Quantum Flow Control

QFD Quality Functional Deployment • Quality Function Deployment • Quality Function Development • Quantum Flavordynamics

QFF Quadrupole Flip-Flop

QFH Frederikshavn (Denmark) (location identifier)

QFI Qualified Flight Instructor

QFIRC Quick Fix Interference Reduction Capability

QFL Qualified Products List

QFM Qualification Flight Model • Quantized Frequency Modulation

QFP Quad Flat Pack • Quad Flat Package • Question Fréquemment Posée

QFPL Qualified Film Producer List

QFS Quartz Flexure Suspension

QFT Quantum Field Theory

QG Qualitätsgruppe • Quantum Gravity • Quartier Général • Quasigeostrophic • Query Generator

QGA Quaternary Geological Atlas

QGAR Querying Graphics through Analysis and Recognition

QGP Quark-Gluon Plasma

QGS Quantity Gauging System • Quark-Gluon String

QGSM Quark-Gluon String Model

QGV Frankfurt - Neu-Isenburg (Germany) (location identifier)

qh Quaque Hora

QH Quadruple Hydrant • Quartz-Halogen (bulb)

QHD Quantum Hydrodynamics

QHE Quantum Hall Effect

QHI Qualified Helicopter Instructor

QHJ Hjørring (Denmark) (location identifier)

QHO Quantum Harmonic Oscillator

QHR Quality History Record • Quantized Hall Resistance • Quantum Hall Resistance (*)

QHS Quartier de Haute Sécurité

QHz Quantum Hertz

Q/I Question/Issue

QI Quality Improvement • Quality Increase • Quality Index • Quality Indicator • Quality Information • Quantity Index • Quartier d'Isolement • Quartz Iodine • Quasi-Inertial • Quasi-Invariant • Quotient d'Intelligence • Quotient Intellectuel

QIC Quality Information Control • Quality Information using Cycle • Quantum Information and Computation (journal) • Quarter-Inch Cartridge

QICC Quarterly Index to Current Contents

qid Quater in Die

QIEC Quasi-Isoenergetic Complex

QIF Quicken Interchange File • Quicken Interchange Format

QINMS Quadrupole Ion Neutral Mass Spectrometer

QIO Qualified Interoperability • Queue Input/Output

QIP Quad Inline Package • Quality Improvement Paradigm • Quality Improvement Program • Quality Improvement Prototype • Quality Indicator Program • Quartz Iodine Polarized • Quiescat In Pace

QIRA Quantitative Infrared Analysis

QIS Quality Information System • Quellenorientiertes Informationssystem

QISAM Queued Indexed Sequential Access Method

QISM Quantum Inverse Scattering Method

QIT Quadrupole Ion Trap • Quality Improvement Team • Quality Information and Test • Quantum Information Theory • Queensland Institute of Technology (Australia-QLD)

QITF Quarterly Inspection of Inactive Facilities

QIV Qualité Image Végétation

QIZ Quellenorientiertes Informationszentrum

QJ Quarterly Journal

QJLC Quarterly Journal of the Library of Congress

QJM Quantified Judgment Model • Quantitative Judgment Method

QJMAM Quarterly Journal of Mechanics and Applied Mathematics (periodical code)

QJRAS Quarterly Journal of the Royal Astronomical Society (periodical code)

QJRMS Quarterly Journal of the Royal Meteorological Society (periodical code)

QJS Quick Jump Sideways

Qk Quick

QK Quantum Kinematics

QKD Quantum Key Distribution

QKL Cologne Main Station (Germany) (location identifier) • Köln Hauptbahnhof (Germany) (location identifier) • Köln Hbf. (Germany) (location identifier)

QK-LK Quick Look

QKLK Quick Look

qkm Quadratkilometer

QKM Quadratic Kohn Method

QKT Quietly Killing Time

QKU Cologne-Deutz (Germany) (location identifier) • Köln-Deutz (Germany) (location identifier)

ql Quantum Libet • Quintal

Q/L Quick Look

QL Quantitative Literacy • Quasi-Longitudinal • Queensland (Australia) • Querleitung • Query Language • Quick Look

QLA Quebec Library Association (Canada-QC) • Quick-Look Analysis • Quick-Look Assessment

QLAP Quick-Look Analysis Program

qlb Quick Library (file extension)

QLB Quasi-Lower Bound

QLBB Quick-Look Bulletin Board

QLCI Quick-Look Capabilities Index

Qlconque Quelconque

Qld Queensland (Australia)

QLD Queensland (Australia)

QLDA Quick-Look Data Analysis • Quick-Look Display Area

QLDAC Quick-Look Data Analysis Center

QLDS Quick-Look Data Station

QLF QSO Luminosity Function • Quasar Luminosity Function • Quick-Look Facility

QLI Quality-of-Life Index • Query Language Interpreter • Quick-Look Image

QLIP Quick-Look Image Producer

QLIRS Quick-Look Image Recording System

QLIS Quality of Life in Information Society • Quick-Look Image Server

QLIT Quick-Look Intermediate Tape

QLLC Qualified Logical Logical Link

QLM Qualified Left Monotonicity • Quasi Littrow Mode

qLMXB Quiescent Low-Mass X-ray Binary

QLN Quantitative Links Network

QLOP Quarterly Limitation On Payment

QLOPR Quick-Look Product

QLP Quadrilateral Protocol • Qualified Parts List • Query Language Processor

QLPF Quick-Look Processing Facility

QLPLDF Quasilinear Plateau Distribution Function

QLPS Quick-Look Processing System

QLR Quick-Look Report

QLS Quick-Look Station • Quiet Line State

QLSD Quick-Look Selection and Display

QLT Quasi-Linear Theory

Qlty Quality

QLV Quiet Little Village

QLX Lauterach (Austria) (location identifier)

Qly Quarterly

qm Quadratmeter • Quaque Mane • Quomodo • Quoque Matutino

Qm Quintal Métrico

QM Qualification Method • Qualification Model • Qualification Motor • Quality Management • Quality Manual • Quality Memorandum • Quality Monitoring • Quantenmechanik • Quantum Mechanical • Quantum Mechanics • Quark Model • Quarter Master • Query Manager

QMA Qualitative Materiel Approach

QMB Quality Management Board

QMBT Quantum Many-Body Theory

QMC Quantum Monte Carlo • Queen Mary College (UK-England) • Quick Modification Concept

QMCMV Queen Mary and Westfield College (Starlink internal code)

QMCS Quantitative Methods and Computer Science

QMD Quantum Molecular Dynamics • Quartz-Monzodiorite

QMDE Quantum Mechanical Differential Equation

QMDO Qualitative Materiel Development Objective

QMF Quadrature Mirror Filter • Quadrupole Mass Filter • Query Management Facility

QMG Qualité Maximale Garantie • Quarter Master General • Quenched Melt Growth

QMI Quality Management Institute • Quotidien des Marchés de l'Informatique

QMIPS Quantitative Methods In Parallel Systems

QMIT Quasi-Monolithic Integration Technology

QML Qualified Manufacturers List • Qualified Manufacturing Line • Qualified Manufacturing Listing

qmm Quadratmillimeter

QMM Marina di Massa (Italy) (location identifier)

QMO Qualitative Materiel Objective

QMOCC Quantum Molecular-Orbital Close-Coupling (method)

QMP Qualitative Management Program • Quality Measurement Plan

QMQ Murzuq (Libya) (location identifier) • Quantitative Management of Quality

QMR Qualitative Material Requirement • Quarterly Management Review • Quasi-Minimal Residual

QMRS Quasi-Minimal Residual Squared

QMS Quadrupole Mass Spectroscopy • Qualification Model Spacecraft • Quality Management System • Quarter Master Sergeant • Queue Mess System

QMU Quarters - Maintenance - Utilities

QMUL Queen Mary University of London (UK-England)

QMV Montvale (USA-NY) (location identifier) • Qualified Majority Voting • Quantum Mechanics Violation

QMW Queen Mary and Westfield • Queen Mary and Westfield College (Starlink internal code)

QMWC Queen Mary and Westfield College (Starlink internal code)

qn Quaque Nocte

Qn Queen • Quelqu'un • Question • Quotation

QN Quantum Number • Queuing Network • Quiescent Nova • Quinoline

QNA Qatar News Agency • Quarterly National Accounts

QND Quantum Non-Demolition

QNM Queuing Network Model

qNS Quiescent Neutron Star

Qns Questions

QNS Quantity Not Sufficient

Qnsld Queensland (Australia)

Qnty Quantity

QNY New York - Marine (USA-NY) (location identifier)

QO Quality Office • Quasi-Optical

QOD Quality Of Data • Quick-Opening Device

QOF Kiev Borispol (Ukraine) (location identifier)

QoFIS Quality of Future Internet Services (workshop)

QOH Kiev Hotel Rus (Ukraine) (location identifier)

QOL Quality Of Life

QoR Query on Release

QOR Qualitative Operational Requirement • Queue Owning Region

QORGS Quasi-Optimal Rendezvous Guidance System

QoS Quality of Service

QOS Quality Of Service • Query Optimization System

QOT Quadratic Onset Time • Qualification Of Tooling • Quasi-Optical Technique

QOT&E Qualification Operational Test & Evaluation

QOTE Qualification Operational Test and Evaluation

QOV Query Out of Vocabulary

QOVT Query Out of Vocabulary Term (rate)

QOVW Query Out of Vocabulary Word (rate)

qp Quantum Placet

QP Quadratic Programming • Qualification Pay • Qualification Proposal • Quality Procedures • Quantitative Process • Quantum Physics • Quasi-Periodic • Query Processing • Query Processor • Quick-look (image) Producer • Quick-look (image) Production • Quick Propagation • Quiescent Prominence

QPA Quantity Per Application • Quantity Per Assembly

QPAF Quarter Panel Alignment Fixture

QPC Query Processor Coordinator

QPD Quality Planning Document

QPE Query by Pictorial Example

QPEI Quantity Per End Item

QPES Quadratic Polyelastic Sampling

QPF Quadruple Point Failure

QPG Quantum Phase Gate • Singapore - Paya Lebar (Singapore) (location identifier)

QPI Quality Performance Indicator • Quality Program Instruction • Quantitative Phase Imaging

QPIG Quadrupole Ion Guide

q. pl. Quantum Placet

QPL Qualified Parts List • Qualified Products List • Quality Parts List

QPLT Quiet Propulsion Lift Technology

QPM Quality Program Manager • Quasi-Phase Matched • Quick Panel Manager

QPMIS Quantitative Process Management Information System

QPO Quasi-Periodic Oscillation

QPOE Quasi-Periodic Oscillation Extraction • Quick Position-Ordered Event (*) • Quick Position-Oriented Event

QPP Quality Program Plan

QPPM Quantized Pulse Position Modulation

QPQ Quid Pro Quo

QPR Qualitative Performance Requirement • Quality Project Requirement • Quarterly Progress Report • Quarterly Progress Review

QPRD Quality Planning Requirements Document • Quality Program Requirements Document

QPRH Quality Planning Requirements Handbook

QPS Quality Planning Specification • Quantity Planning Specification • Quantitative Physical Science • Quasi-Poloidally Symmetric • Quasi-Poloidal Stellarator

QPSD Quality and Productivity Systems Division

QPSK Quadrature-Phase Shift Key • Quadrature-Phase Shift Keying (*) • Quadri-Phase Shift Keying • Quaternary-Phase Shift Key • Quaternary-Phase Shift Keying

QPSX Queued Packet and Synchronous Exchange • Queued Packet Switch Exchange

QPT Quad Port Transceiver • Quadrant Power Tilt • Quantum Phase Transition

qq Questions

Qq Quartos • Quelques

Q-Q Quantile-Quantile

QQ Quantile-Quantile • Quick Query

QQA Quick Questions and Answers

Qqch Quelque Chose

Qqe Quelque

QQE Quick - Quiet - Effective

Qqes Quelques

Qqf Quelquefois

Qqfois Quelquefois

Qqfs Quelquefois

QQO Quasi-Quadrennial Oscillation

QQP Quick Query Program

QQPR Qualitative and Quantitative Personnel Requirements

QQPRI Qualitative and Quantitative Personnel Requirements Information

Qqprt Quelque Part

QQR Quinquennial Review

Qque Quelque

Qques Quelques

Qqu'un Quelqu'un

qq. v. Quae Vide • Quod Vide

qr Quadrans • Quarter • Quarterly • Quire

Qr Quarter

Q&R Quality & Reliability

Q/R Query/Response

QR Qatar Riyal (currency) • Quadrupole Resonance • Quality and Reliability • Quality-Rated • Quality Rating • Quality Record • Quality Review • Quantitative Requirement • Quarterly Review • Quick Rank • Quick Response

QR&A Quality & Reliability Assurance

QRA Johannesburg Randgermiston (South Africa) (location identifier) • Quality and Reliability Assurance • Quarterly Review and Analysis • Quaternary Research Association • Quick Reaction Alert • Quick Reaction Assessment

QRAL Quality and Reliability Assurance Laboratory

QRB Qualification Review Board • Quality Review Board

QRBM Quasi-Random Band Model

QRC Qualitative Construction Capability • Quantum Research Corporation (USA-MD) • Quick Reaction Capability • Quick Reaction Communications

QRCG Quasi-Random Code Generator

QED Query Regarding Decision

QRE Quick-Reaction Element • Quick-Reaction Estimate • Quick-Reaction Event

QREI Quality Reliability Engineering International (periodical code)

QRF Query Relative Frequency • Quick-Reaction File • Quick-Reaction Force

QRFL Queries-R-Focused Link

QRH Quick Reference Handbook • Rotterdam - Central Station (Netherlands) (location identifier)

QRI Qualitative Requirements Information • Quick-Reaction Integration

QRIA Quick-Reaction Integration Activity

QRIP Quick Return on Investment Program

QRL Queries-R-Linked • Quick Reference List

QRM Qualified Right Monotonicity • Quick Response Modification

QRMP Quick-Reaction Multicolor Printer • Quick-Reaction Multipurpose Printer • Quick-Response Multicolor Printer • Quick-Response Multipurpose Printer

QRO Querétaro (Mexico)

QROC Questionnaire à Réponse Ouverte et Courte

QROO Quintana Roo (Mexico)

QRP Qualified Recycling Program • Quality Review Program • Quick-Reaction Package • Quick-Reaction Program • Quick-Reference Page • Quick-Response Posture • Quick-Response Program • Quick-Response Proposal

QRPA Quasi-particle Random-Phase Approximation • Quasi-Random Phase Approximation

QRPS Quick-Reaction Procurement System

QRR Qualification Results Review • Qualitative Research Requirement • Quarterly Requisition Report

Qrs Quarters

QRS Qualification Review Sheet • Quantum Readout System • Quartz Rate Sensor • Query Response System • Quick-Reaction Sortie

QRSA Quick-Reaction Satellite Antenna

QRSIN Queensland Remote Sensing Information Note

QRSL Quick Reaction Space Laboratory

QRSR Quarterly Risk Status Report

QRSS Quasi Random Signal Source

QRT Quarto • Quick Ray Tracer • Quick-Response Team

QRTLY Quarterly

qry Query (file extension)

Qry Quarry • Query

QRY Query

qs Quantum Satis • Quantum Sufficit

QS Quadraphonic-Stereophonic • Quadratic Splines • Qualitative State • Quantity Surveyor • Quarter Section • Quasi-Stellar • Query Summary • Question Statement • Quick Sort • Quiet Sun • Quota Source

QSA Quadripartite Standardization Agreement • Qualification Site Approval

QSAE Quality and Standards Authority of Ethiopia

QSAM Quadrature Sideband Amplitude Modulation • Quasi-Static Acceleration Measurement • Quasi-Steady Acceleration Measurement • Queued Sequential Access Method

QSAR Quality System Assessment Recognition • Quantitative Structure-Activity Relation • Quantitative Structure-Activity Relationship

QSB Quarter Swing Baffle

QSC Quadrilaterized Spherical Cube • Quick-Setting Concrete

QSCB Quality System Certification Body

QSCV Quasi-Stationary Convection Vortex

QSD Quad Synchronous Data • Quantum Spin Dynamics • Quantum Structural Diagram • Quasi-Sudden Discharge

QSDAM Queued Sequential Data Access Method

QSDC Quantitative Structural Design Criteria

QSDG Quality of Service Development Group

QSE Qualified Scientist and Engineer • Qualitätssicherungseinheit • Quantum Size Effect • Queue Shared Element

QSF Quasi-Stationary Floculus

QSG Quasi-Steady Glide • Quasi-Stellar Galaxy • Quick-Setting Glue

QSGVT Quarter Scale Ground Vibration Test

QSH Quick Set Hard

QShAR Quantitative Structure-Shape Relation

QSHS Quantitative Spectroscopy of Hot Stars

QSI Quality Service Indicator • Quality Step Increase

QSIF Quarter Source Input Format

QSIM Queuing Simulation

QSJ Que Sais-Je?

QSL Q-Switch Laser • Qualification Status List • Qualified Source List • Qualified Supplier List • Quarter Scan Line • Quasi Separatrix Layer

QSM Quad Surface Mount • Quantum-Statistical Model • Quarter Scale Model • Queued Serial Module

QSMVT Quarter Scale Model Vibration Testing

QSN Quantum Shot Noise

QSNO Relative Signal/Noise Ratio

QSO Quasi-Stellar Object • Queued Service Observing • Queue Service Observing

QSOAL Quasi-Stellar Object Absorption Line

QSOALS Quasi-Stellar Object Absorption Line System

QSOP Quadripartite Standing Operating Procedure

QSPR Quantitative Structure-Property Relationship

QSQ Query/Subquery

QSR Qualified Status Report • Quality System Review • Quarterly Statistical Report • Quasar • Quasi-Stellar Radiosource • Query and Reporting System

QSRS Quasar • Quasi-Stellar Radiosource

QSS Quadrant Star Sensor • Quality-Securing System • Quality Surveillance System • Quasi-Static System • Quasi-Steady State • Quasi-Stellar Source • Quick Start Service • Quick Supply Store

QSSC Quasi-Steady-State Cosmology

QSSR Quarterly Specialized Safety Report

QSSS Quasi-Stationary Spiral Structure

QSTAG Quadripartite Standardization Agreement

Qstn Question

QSTNRY Quasistationary

QSTOL Quiet Short Takeoff and Landing

QSTR Quick Short Test Report

QSY Quiet Sun Year

qt QuickTime (file extension)

Qt Quant • Quantity • Quart • Quiet

QT Quad Tree • Qualification Test • Quantum Theory • Quarteirão • Quasi-Transverse • Query Tool • Queuing Theory • Quick Time • Quick Turn • Quiet • Quiet Trader

Qta Quinta

QTA Queue Terminal Access • Quick-Turn Area • Quinta

QTAI Quick Turnaround Interconnect

QTAL Quintal

QTAM Queued Telecommunications Access Method • Queued Teleprocessing Access Method • Queued Transmission Access Method

QTAS Quintas

QTB Quarterly Training Brief

QTC Charge-to-Time Converter • Quick-Time Conferencing • Quick Transfer Centre

QTD Quadrupole Terminal Digit • Quasi-Two-Dimensional

QTDAM Queued Telecommunications Data Access Method

QT&E Qualification Test & Evaluation

QTE Qualification Test and Evaluation

QTEL Quartel

QTERS Quarters

QTG Quantum Theory of Gravitation • Quarterly Training Guidance

QTGG Quantum Theories and Gravitation Group

QTH Quartz Tungsten Halogen

QTIG Queensland Telecommunications Industry Group

QTL Quantitative Trait Locus

Qtly Quarterly

qtm QuickTime Movie (file extension)

QTM Quadratic Texture Map • Qualification Test Model • Quaternary Triangular Mesh • Quick-Time Movie • Quick To Market

QTNHA Quintinha

Qto Quarto

QTO Quasi-Triennial Oscillation

QTOL Quiet Takeoff and Landing

QTP Quadrant Tool Piezoelectric • Qualification Test Plan • Qualification Test Procedure • Qualification Test Program • Qualification Training Package • Quality Test Plan • Quality Through Partnership • Quantum Test Programme

Qtr Quarter • Quarterly

QTR Qualification Test Report • Quarter • Quarterly Training Report

QTRLY Quarterly

QTRS Quarters

Qts Quarts

QTS Quartz Temperature Sensor

QTU Qualification Test Unit

QTT Quasi-particle Trapping Transistor • Quatratran

Qtté Quantité

Qtty Quantity

QTV Qualification Test Vehicle

QTW Quick-Time for Windows

Qty Quantity

QTY Quantity

qu Quasi • Quechua (ISO language code)

Qu Quart • Quartal • Quarter • Quarterly • Quelle • Query • Question

QU Québec (Canada)

QUA Quarantids (meteor stream)

Quad Quadrangle • Quadrant • Quadratic • Quadrilateral • Quadruple

QUAD Quadrangle • Quadrant • Quadratic • Quadrilateral • Quadruple

quadr quadratisch

Quadr Quadragesima • Quadrant • Quadratic • Quadratur • Quadrilateral • Quadrille • Quadrupel • Quadruple

QUADS Quadrants • Quality Achievement Data System • Quality Assurance Data System

qual qualifizieren • Qualifiziert • Qualitativ

Qual Qualification • Qualified • Qualifikation • Qualify • Qualitative • Quality

QUAL Qualification • Qualified • Qualifikation • Qualify • Qualitative • Quality

Qualn Qualification

QUAM Quadrature Amplitude Modulation

QUANGO Quasi-Autonomous Non-Governmental Organization

Quant Quantifier • Quantität • Quantitative • Quantity

QUANTOS Quantum-Optical Spectrometer

Quar Quarantine • Quartal • Quarter • Quarterly

QUARG Queen's University Astronomy Research Group (Canada-ON)

Quart Quarter • Quarterly

QUASAR Quasi-Stellar Radiosource

QUASAT Quasar Satellite

QUASRT Qualified Software Review Team

Quat Quattuor

QUAT Quaternion

QUB Queen's University of Belfast (Northern Ireland) • Ubari (Libya) (location identifier)

QUBIC Quantum and Biologically Inspired Computing

QudV Qualität durch Verantwortung

Que Québec (Canada)

QUE Québec (Canada)

QUEEN Quaternary Environment for the Eurasian North

QUEL Quelha • Query Language

QUELD Queen's University Experiment in Liquid Diffusion

QUELH Quelhas

Ques Question

QUES Quality Evaluation System

Quest Quest (periodical code)

QUEST Quality Electrical Systems Test • Quasar Equatorial Survey Team • Quasi-Electrostatic • Quasi-Electrostatic Test • Quasi-Electrostatic Trap • Question

QUESTA Querying Systems: Theory and Applications

QUESTER Quick and Effective System to Enhance Retrieval

Questn Questionnaire

QUESTOL Quiet Experimental Short Takeoff and Landing

Quí Química

QUIC Quadrupole Ioffe Configuration • Quality Information and Control • Quick Infrared Camera

QUIDS Quick Interactive Documentation System

QUILT Query User Interface with Light Translations

Quím Química

Quin Quintuple

Quinq Quinque

Quint Quintuple • Quintuplicate • Quintus

QUIP Query Interactive Processor • Questionnaire Interpreter Programme

QUIRC Quick Infrared Camera

QUIS Queen's University Information System • Questionnaire for User Interaction Satisfaction • Questionnaire for User Interface Satisfaction

QUIST Quick Infrared Survey Telescope

Quitt Quittung

QUK Queen's University Kingston (Canada-ON)

QUL Ulm Bus Station (Germany) (location identifier)

QUJM Quincy University Journal of Mathematics (periodical code)

QUnJ Qinghua University Journal (periodical code)

Quor Quorum

Quot Quotation • Quote • Quotient

QUOTA Query Optimization Transformation Approach

QUOTE Quality and User Oriented Typesetting Editor

Quotid Quotidie • Quotidien

QUP Quality Unit Pack • Quantity per Unit Pack

QUS Quantitative Ultrasound

QuSOp Quantum and Semi-classical Optics (periodical code)

QUT Queensland University of Technology (Australia-QLD)

QUVAD Belfast (Starlink internal code)

qv Quantum Vis • Quantum Vult • Quod Vide

QV Quantum Vector • Quick Visual

QVANTOS Quantum-Optical Spectrometer

QVC Quartz Vibrating Crystal

QVD Qualified Vehicle Driver

QVE Quasi-Vibrational Energy

QVI Quality Verification Inspection

QVLBI Quasi-Very-Long-Baseline Interferometry

QVP Quality Vendor Program • Quality Verification Plan

QVPL Qualified Videotape Producer List

QVR Quality Verification Report

QVT Qualified Verification Testing • Quality Verification Testing

QVVT Qualified Verification Vibration Testing

QW Quality of Water • Quantum Well

QWA Quarter-Wave Antenna

QWD Quarterly World Day

QWE Quality Week Europe

QWG Charlotte Wilgrove (USA-NC) (location identifier) • Quadripartite Working Group • Quarkonium Working Group

QWI Quantum Well Intermixing (process)

QWICPIC QWIP Wide-field Imaging Color Prime Focus Camera

QWIKTRAN Quick Fortran

QWIP Quantum Well Infrared Photodetector

QWIPM Quantum Well Infrared Photomultiplier

QWIPS Quantum Well Infrared Photodetectors

QWL Quality of Working Life • Quality of Work Life • Quick Weight Loss

QWP Quarter Waveplate • Quite Well Presented

qxd QuarkXPress Document (file extension)

QXFF Quartz Crystal Fabrication Facility

QXI Queue Executive Interface

qxl QuarkXPress Library (file extension)

qxt QuarkXPress Template (file extension)

Qy Query

Qz Quartz

QZ Qualitätszirkel • Quiet Zone

QZG Qualitätszielgruppe

R

r Distance • Distance from the Sun • Distance to Earth Center of Gravity • Heliocentric Distance of Perihelion • Internal Resistance • Mass Fraction of Individual Stage • Mixture Ratio • Molecular Position • Mole Ratio of Solution • Particle Position Vector • Polar Coordinate • Position • Radius • Railroad • Railway • Rain • Range • Rare • Rate • Ratio • Ray • Rayon • Read • Received • Rechts • Recipe • Recto • Red • Reflection Coefficient • Replacing • Reply • Reserve • Residence • Resistance • Response • Retirado • Retired • Revolution • Right • Río • River • Rivière • Road • Rod • Roentgen • Rouble (currency) • Royal • Rubber • Ruble (currency) • Rue • Ruled • Run • Rupee (currency)

R Cosmic Scale Factor • Distance from the Galactic Center • Eighty (Roman numeral) • Gas Constant • Gauss Vector • Lattice Vector • Linear Range • Mass Ratio • Molar Gas Constant • Multiple Correlation Coefficient • Nuclear Radius • Particle Position Vector • Radar • Radial • Radical • Radiancy • Radiometer • Radiosonde • Radius • Railroad • Railway • Rain • Ram • Rand (currency) • Range • Rankine • Rapido • Rate • Ratio • Rayleigh • Reaction Force • Reaction Rate • Read • Real • Rear • Réaumur • Recall • Receive • Recht • Recipe • Recommandation • Recommandé • Recommendation • Rector • Red • Redactor • Redistribution • Redundant • Reflectance • Refraction • Regina • Registered • Registro • Regular • Regulate • Reine • Reinforcing • Relative • Release • Reliability • Reluctance • Replace • Reply • Réponse • Report • Republic • Republican • Required • Reserve • Resistance • Resolution • Resonance Effect • Response • Respuesta • Restricted • Return • Reverse • Revolution • Rex • Ria • Rial • Right • Ring • Río • River • Rivière • Road • Rod • Roentgen • Roi • Roll • Roman • Romania (international vehicle registration code) • Romania (postal country code) • Romeo (international alphabet) • Rotated • Rouble (currency) • Rough • Route • Routine • Royal • Rua • Ruble (currency) • Rúbrica • Rue • Ruin • Run • Rupee (currency) • Rydberg Constant • Set of Real Numbers • Sound Reduction Index • Spectral Type • Thermal Resistance • Total Mass Ratio • Wolf Number • 80 (Roman numeral)

ra Radiatus • RealAudio (file extension)

Ra Radium • Range • Rayleigh Number

R/A Radar Altimeter • Return to Author

R&A Reliability & Availability • Report & Analysis • Research & Analysis • Responsibility & Action • Review & Analysis • Review & Approval

RA Argentina (international vehicle registration code) • Radar Altimeter • Radar Altimetry • Radio Antenna • Radio Astronomy • Radiocommunication Assembly • Radiocommunications Agency • Range Area • Readiness Assessment • Real Audio • Real-time Analysis • Rear Admiral • Rain • Random Access • Raster • Receipt Acknowledgment • Rechtsanwalt • Reduction of Area • Reference Architecture • Regional Association • Registration Act (USA) • Registration Authority • Relational Algebra • Relative Address • Release Authorization • Remedial Action • Remote Agent • Remote Authentication • Rental Agreement • Repeat to Address • República Argentina • Requesting Agency • Requirements Analysis • Research Announcement • Research Assistant • Resident Astronomer • Resolution Advisory • Resource Adviser • Resource Allocation • Return Authorization • Ricerche Astronomiche Specola Vaticana (periodical code) • Right Adjusted • Right Aft • Right Ascension • Ring Acknowledge • Risk Analysis • Risk Assessment • Rocket-Assisted • Roll Axis • Routing Area • Royal Academician • Royal Academy • Russia (international aircraft registration code)

raa Reduction Ad Absurdum

RAA Rainwater Astronomical Association (USA-MS) • Rakanda (Papua New Guinea) (location identifier) • Rapid Access Archive • Regional Airline Association • Required Asset Availability • Resource Apportionment Aid • Responsibility - Authority - Accountability • Risk Acceptance Authority • Rockford Amateur Astronomers (USA-IL) • Root Addressable Area • Royal Academy of Arts (UK)

RAAA Reliability Allocations, Assessment and Analysis

RAAB Remote Amplifier and Adaptation Box • Remote Amplifier and Advisory Box

RAAC Remote Active Archive Center

RAAD Rapid Architect Application Development

RAAF Royal Australian Air Force

RAAGE Regionale Astronomische Arbeitsgemeinschaft Euskirchen (Germany)

RAAM Recursive Auto-Association Memory

RAAN Right Ascension of Ascending Node

RAAO Remote Access Automatic Observatory

RAAP Remote Access Astronomy Project

RAAG Real Algebraic and Analytical Geometry

RAASP Reusable Ada Avionics Software Package

RAAT Remote Access Automatic Telescope

RAAWS Radar Altimeter and Altitude Warning System

Rab Rabat (Morroco)

RAB Rabaul (Papua New Guinea) (location identifier) • Radiometer Box • Rapid Application Builder • Readiness Assessment Board • Registrar Accreditation Board • Registration Accreditation Board • Restoration Advisory Board

RABAL Radiosonde Balloon

RABAR Radiosonde Balloon Release

RABATS Rapid Analytical Block Aerial Triangulation System

RABF Radioastronomy Bandstop Filter

RABIN Raad van Advies voor Bibliothekwezen en Informatieverzorging (Netherlands)

RABMN Remote Area Business and Messaging Network • Remote Area Business Management Network • Remote Area Business Message Network

RAC Racine Horlick (USA-WI) (location identifier) • Radio Astronomy Centre • Radio Astronomy Controller • Radiometric Area Correlator • Raleigh Astronomy Club (USA-NC) • Rapid Action Change • Read Address Counter • Receiver - Absorber - Converter • Recruitment and Admissions Committee • Redshifted Absorption Component • Reflective Array Compressor • Reliability Action Center • Reliability Analysis Center • Remedial Action Contract • Remote Access Computer • Remote Access Concentrator • Remote Analysis Computer • Repair, Alignment and Calibration • Request Authority to Conclude • Remote Assessment Commission • Request for Authority to Complete • Request for Authority to Conclude • Request for Authority to Contract • Risk Assessment Code • Robotic Arm Camera • Rochester Astrophysics Consortium (USA-NY) • Rockland Astronomy Club (USA-NY)

RACAL Radar Calibration

RACAR Remote Access Computer-Assisted Retrieval

RACC Repairable Asset Control Center • Reporting Activity Control Card

RACCH Random Access Channel

RACCS Remote Automatic Checkout and Calibration System

RACE Random Access Card Equipment • Random Access Computer Equipment • Random Access Control Equipment • Rapid Automatic Checkout Equipment • R&D in Advanced Communications for Europe • Research in Advanced Communications Environment • Research on Advanced Communications in Europe • Research on Automatic Computation Electronics

RACEP Random Access and Correlation for Extended Performance

RACES Remote Acoustic Electronic System

RACF Remote Access Control Facility • Resource Access and Control Facility • Resource Access Control Facility (*)

RACH Random Access Channel

RACI Royal Australian Chemical Institute

RACIA Random Access Charge Injection Device

RACIP Research And Communications In Physics

RACMO Regional Atmospheric Climate Model

RACON Radar Beacon

RACS Radar Acquisition Control System • Random Access Communications System • Random Closed Set • Redundant Attitude Control System • Remote-Access Calibration System • Remote-Access Computing System • Remote Automatic Calibration System • Remote Automatic Control System • Request for Approval of Contractual Support • Rotating Antenna C-band Scatterometer

RACT Reasonably Available Control Technology • Remote Access Computer Technique

RACU Remote Acquisition and Control Unit • Russian-to-American Converter Unit

rad Radian

Rad Radian • Radiatif • Radiation • Radiative • Radical • Radio • Radium • Radius • Radnorshire (UK-Wales)

R&AD Research & Advanced Development

RAD Radar • Radial • Radian • Radiation • Radiation Absorbed Dose • Radiation Accumulated Dose • Radiator • Radical • Radio • Radius • Random Access Data • Random Access Device • Random Access Disk • Rapid Access Disk • Rapid Access Drive • Rapid Application Design • Rapid Application Development • Rapid Automatic Drill • Recommendation and Approval Document • Regional Accountable Depot • Remote Antenna Driver • Request for Authority to Develop • Required Availability Data • Required Availability Date • Requirements Action Document • Research Alert Direct • Research and Advanced Development • Research And Development • Retiree Activity Day • Review and Approval Document • Roentgen Absorbed Dosage • Tortola (British Virgin Islands) (location identifier)

RADA Random-Access Discrete Address

RADAC Radar Analog Data And Control • Radar Data Acquisition Converter • Rapid Digital Automatic Computing

RADALT Radar Altitude

RADAR Radio Detection and Ranging • Response And Distribution Alternative Relationship • Rights - Availability - Distribution - Analysis - Reporting

RADARSAT Radar Satellite

RADAS Real-time And Distributed Ada Services

RADB Routing Arbiter Database

RADC Reliability Analysis Data Center

RADCAD Radar Computer-Aided Design

RADCAL Radar Calibration • Radiation Calibration

RADCC Radiation Control Center

RADCOM Radar and Communications

RADCON Radiological Control

RADEC Radiation Detection Capability

RadEf Radiation Effects (periodical code)

Radel Radioelektronika (periodical code)

RADEM Random Access Delta Modulation • Rocket Ascent Demonstrator Mission

RADEO Rapid Design Exploration and Optimization

RADEX Rapid Deployment Exercise

RADFAC Radiating Facility

RadFET Radiation-sensitive Field-Effect Transistor

RADFET Radiation-sensing Field-Effect Transistor

RADH Radiation Dose Rate per Hour

RADHAZ Radiation Hazard

RADI Radial Injector

RADIAC Radiation - Detection - Indication - Computation

Radiad Radiatif • Radiation • Radiative

RADIC Rapidly Deployable Integrated Control (system)

RADIG Radar Investigation Group

RADINT Radar Intelligence

RADIR Random Access Document Indexing and Retrieval

RADIST Radar Distance (indicator)

RADIT Radio Teletype

RaDiUS Research and Development in the United States

RADIUS Remote Authentication Dial-In User Server • Remote Authentication Dial-In User Service • Research And Development for Image-Understanding Systems • Research Association for the Development of the Industrial Use of Space • Residue Arithmetic Digital Image Understanding System

Radl Radiological • Radiology

RADL Radial

RADLAC Radio-Frequency Linear Accelerator

RADLFO Radiological Fallout

RADLMON Radiological Monitoring

RADLOPS Radiological Operations

RADLSAFE Radiological Safety

RADLSO Radiological Survey Officer

RADLSV Radiological Survey

RADM Regional Acid Deposition Model

RADMO Radio Mode

Radn Radiation

RADN Radiation

RADNOTE Radio Note

RADOME Radar Dome • Réseau d'Acquisition de Données d'Observations Météorologiques Étendu

RADOP Radar Operator

RADOT Real-time Automatic Digital Optical Tracker • Recording Automatic Digital Optical Tracker

RadRe Radcliffe Observatory Reprints (periodical code)

RADREL Radio Relay

RADS Radar Acquisition and Data System • Radar Acquisition Data System • Radiation And Dosimetry Services • Rapid Area Distribution Support • Requirements Analysis and Definition Summary • Risk Analysis Decision Support

RADSL Rate-adaptive Asymmetrical Digital Subscriber Line • Rate-Adaptive Digital Subscriber Line (*)

RADSO Radiological Survey Officer

RADT RPC-ASO Data Transfer

RADVS Radar Altimeter and Doppler Velocity Sensor

RADWASS Radioactive Waste Safety Standards

RAE Arar (Saudi Arabia) (location identifier) • Radio Astronomy Explorer • Range - Azimuth - Elevation • Real Academia Española (Spain) • Research Aircraft Experiment • Research Assessment Exercise • RGS Analogue Electronics • Right Ascension Encoder • Right of Assistance Entry • Royal Aerospace Establishment (UK) • Royal Aircraft Establishment

RaEE Radio and Electronic Engineer (periodical code)

RaEl Radiotekhnika i Elektronika (periodical code)

RAEL Renewable and Appropriate Energy Laboratory

RAEN Radio Amateur Emergency Network

RAEng Royal Academy of Engineering (UK)

RAeS Royal Aeronautical Society

RAES Remote Access Editing System

RAET Range - Azimuth - Elevation - Time

RaF Radiofizika (periodical code)

RAF Ras an Naqb (Egypt) (location identifier) • Régime Administratif et Financier • Requirements Analysis Form • Research Aviation Facility • Resource Allocation Frame • Royal Air Force (UK)

RAFAS Report of Accounting and Finance Activities System

RAFC Regional Area Forecast Centre

RAFCAP Radio Astronomy Frequency Committee in the Asia-Pacific Region

RAFI Rural Advancement Foundation International

RAFP Repartição de Aministração Financeira e Patrimonial

RAFS Regional Analysis and Forecast System

RAFT Radially-Adjustable Facility Tube

RAFUS Radio Frequency Usage Reporting System

RAG Radio Astronomy Group • Radiocommunications Advisory Group • Raglan (New Zealand) (location identifier) • Rama Asintótica de Gigantes • Readiness Analysis Group • Requirements Assessment Group • Responsable Assurance Qualité • Reusable Agenda • Row Address Generator

RAGG Residual Aliased Gravity Grid

RAH Rafha (Saudi Arabia) (location identifier)

RAHF Research Animal Holding Facility

RAI Praia (Cape Verde) (location identifier) • Radar Altimeter Indicator • Radioactive Interference • Radio Audizioni Italia • Rainfall Anomaly Index • Random Access and Inquiry • Remote Air Intake • Remote Alarm Indication • Review Action Item • Roll Attitude Indicator • Runway Alignment Indicator

RAID Recent Advances in Intrusion Detection (workshop) • Recherche - Assistance - Intervention - Dissuasion • Recherche Automatique de l'Information Disponible • Redundant Array of Independent Disks • Redundant Array of Independent Drives • Redundant Array of Inexpensive Disks (*) • Remote Access Interactive Debugger • Research Amplifying Imaging Detector

RAIDS Rapid Availability of Information and Data for Safety • Rapid Information Dissemination System • Remote Atmospheric and Ionospheric Detection System

RAIE Random Access Index Edit

RAIF Relative Annual Impact Factor

RAIL Remote Array of Inexpensive Libraries • Runway Alignment Indicator Lights

RAILE Retro-Assisted Imaging Laser Experiment

RAILS Remote Area Instrument Landing Sensor

RAIM Receiver Autonomous Integrity Monitoring

RAIN Reduction of Airframe and Installation Noise (project) • Relation Algebraic Interpreter • Réseau d'Accès à Internet • Rochester Astronomical Information Network

RAIP Requester's Approval In Principle

RAIPT Resource Allocation Integrated Process Team

RAIR Random Access Information Retrieval • Remote-Access Immediate Response

RAIRO Revue d'Automatique, Informatique et Recherche Opérationnelle

RAIRS Reflection-Absorption Infrared Spectroscopy

RAIS Range Automated Information System • Redundant Array of Inexpensive System • Research Assistant Industrial Secondment

RAISE Rigorous Approach to Industrial Software Engineering

RAIST Réseau Africain d'Institutions Scientifiques et Techniques

RAIT Redundant Array of Inexpensive Tapes

RAIUB Radioastronomisches Institut der Universität Bonn (Germany)

Raj Rajasthan (India)

RAJ Rajkot (India) (location identifier)

RAJPO Range Application Joint Program Office

RAK Marrakech (Morocco) (location identifier) • Regelwerk für die Alphabetische Katalogisierung

RAKA Rossiiskoe Aviaciono-Kosmicheskoe Agentsvo (Russia) (formerly RKA)

RAL Radio Astronomy Laboratory • RAL (Astrophysics cluster) (Starlink internal code) • Random Assortive Law • Rapid Access Loop • Registered Asset List • Responsibility Assignment List • Revue d'Aptitude au Lancement • Ringer Approximate Loading • Riverside (USA-CA) (location identifier) • Rule-extended Algorithmic Language • Rutherford Appleton Laboratory (UK-England)

RALI Radar Lidar

RALPH Reduction and Acquisition of Lunar Pulse Heights

RALS Random Access Logical Subfile • Remote Augmented Lift System

RALSA Random Access Line Scanned Array

RALSDC RAL Space Data Centre

RALT Radar Altimeter

RALU Register Arithmetic Logic Unit

ram RealAudio (file extension)

R/A/M Reliability, Availability & Maintainability

RA&M Reliability, Availability & Maintainability

RAM Radar-Absorbent Material • Radar-Absorbing Material • Radar-Absorption Material • Radio Attenuation Measurement • Raised Angle Marker • Ramal • Ramingining (Australia-NT) (location identifier) • Random Access Machine • Random Access Measurement • Random Access Memory (*) • Random-Axis Model • Rank Annihilation Method • Rapid Area Maintenance • Read Active Memory • Readiness Assessment Module • Rechargeable Alkaline Manganesedioxide • Reconnection and Microscale Probe • Red Academicia de Mexico • Reliability Analysis Methodology • Reliability Assessment Model • Reliability - Availability - Maintainability • Remote-Access Modem • Remote-Access Monitor • Requirements Allocation Matrix • Research Allocation Methodology • Research and Applications Module • Resident Access Method • Resource Allocation Manager • Resource Allocation Model • Resource Allocation Module • Resource Assignment Module • Responsibility Assignment Matrix • Reverse Annuity Mortgage • Rocket-Assisted Motor

RAMA Reliability - Availability - Maintainability - Analysis • Remote Access to Museum Archives

RAMAC Random Access Method of Accounting and Control

RAMADCS Reliability, Availability, Maintainability Automated Data Collection System

RAMAS Reliability, Availability, Maintainability Analysis System

RAMB Random Access Memory Buffer

RAMBO Robust Association of Massive Baryonic Objects

RAMC Regional Administrative Management Center • Regional Assessment Management Committee

RAMCAD Reliability And Maintainability in Computer-Aided Design

RAMCS Remote Automatic Measuring and Calibration System

RAMD Random Access Memory Device • Receiving Agency Materiel Division • Reliability - Availability - Maintainability - Dependability • Reliability - Availability - Maintainability - Durability

RAMDAC Random Access Memory Digital-to-Analog Converter

RAMDAS Reliability And Maintainability Data Access System

RAMEAU Répertoire d'Autorité-Matières Encyclopédique et Alphabétique Unifié

RAMEP Reliability, Availability, Maintainability, and Enhanced Performance

RAMES Reliability, Availability, Maintainability Engineering System

RAMIS Rapid Access Management Information System • Reliability And Maintainability Information System

RAMM Random Access Memory Module • Rapid Agile Metrology for Manufacturing • Regional And Mesoscale Meteorology

RAMMS Radiating Millimeter Source • Regional Atmospheric Mesoscale Modeling System • Responsive Automated Materiel Management

RAMOS Radiation Monitoring in Space • Remote Automatic Meteorological Observation System • Remote Automatic Meteorological Observing Station

RAMP Rapid Acquisition of Manufactured Parts • Rating Maintenance Phase • Remote Access Maintenance Protocol • Remote Atmospheric Measurements Program • Remote Atmospheric Monitoring Project • Replenishment Activity Management Plan • Report And Message Processor • Requirement And Management Plan • Results, Applications and Modeling Phase • Rural Abandoned Mine Program

RAMPS Resource Allocation and Multiproject Scheduling • Resource Analysis and Mission Planning System

RAMRR Reliability, Availability, and Maintainability Rationale Report

RAMS Random Access Measurement System • Range Measurement System • Regional Agricultural Management System • Regional Atmospheric Modeling System • Reliability - Availability - Maintainability - Safety • Remote Atmospheric Measurements System • Reproduction Administrative Management System • Research Aircraft Measurement System • Results Analysis and Management System

RAMSES Radar Aéroporté Multispectral d'Étude des Signatures • Recherche Appliquée aux Méthodes de Séparation en Électrophorèse Spatiale • Recherche Appliquée sur les Méthodes de Séparation en Électrophorèse Spatiale (*)

RAMSQ Reliability, Availability, Maintainability, Safety and Quality

RAMTIP Reliability And Maintainability Technology Insertion Program

RAMU Removable Auxiliary Memory Unit

RAMUS Remote Access Multiuser System

RAN Radio Access Network • Radio Area Network • Random Number • Ravenna - La Spreta (Italy) (location identifier) • Remote Access Node • Request for Authority to Negotiate • Royal Australian Navy

RANC Radar Absorption Noise and Clutter

RANCID Radio Astronomy Numerical Computation and Imaging Device • Real And Not Corrected Input Data

RAND Radio Array Neutrino Detection • Radio Array Neutrino Detector • Random • Request Activity and Name Directory • Research And Development • Rural Area Network Design

RANDAM Random Access Non-Destructive Advanced Memory

RANDID Rapid Alphanumeric Digital Indicating Device

RangKoM Rangkaian Komputer Malaysia

RANLP Recent Advances in Natural Language Processing (conference)

RANN Research Applied to National Needs

RANS Reynolds-Averaged Navier-Stokes (method)

RAO Radio Astronomical Observatory • Recherche Assistée par Ordinateur • Regional Acquisition Office • Research Announcement of Opportunity • Revenue Accounting Office • Ribeirão Preto - Leite Lopes (Brazil-SP) (location identifier) • Rothney Astronomical Observatory (Canada-AB)

RAOB Radio Atmospheric Observation Balloon • Radiosonde Observation

RAOC Regional Air Operations Center

RAOOS Radio Astronomy Observatory Operating System

RAORE RPC-ASO Oneshot Reply

RAORQ RPC-ASO Oneshot Request

RAOU Recherches Astronomiques de l'Observatoire d'Utrecht (periodical code)

Rap Rapids

RAP Radar-Absorbent Paint • Radiation and Atmospheric Pollutants • Radio Access Point • Radio Astronomy Preprint • Radiological Assistance Program • Random Access Point • Random Access Program • Rapid Acquisition Package • Rapid Application Prototyping • Rapid Assessment Procedure • Rapid City (USA-SD) (location identifier) • Ready-Assembled Price • Real Application Performance • Reconnaissance Automatique de la Parole • Redundancy Adjustment of Probability • Regional Access Point • Regional Application Program • Regression Analysis Program • Relational Associative Processor • Relative Ability of Prediction • Reliability Assurance Program • Reliable Acoustic Path • Remedial Action Programme • Remedial Action Project • Remote Advisory Panel • Rencontres Astronomiques du Pilat (France) • Research Applications Program • Research Assessment Panel • Research Associateship Program • Reserve Auxiliary Payload • Resource Allocation Process • Resource Allocation Program • Responsibility Accounting Program • Retail Analysis Program • Retirement Annuity Plan • Revised Approved Program • Right-Angle Program • Risk Aversion Plan • Rocket-Assisted Projectile • Root Anchor Point • Route Access Protocol • RSVP Admission Policy

RAPA Radio Access Point Antenna

RAPAT Radiation Protection Advisory Team

RaPC Radiation Physics and Chemistry (periodical code)

RAPCO Radar Approach Control

RAPCOE Random Access Programming and Checkout Equipment

RAPCON Radar Approach and Control • Radar Approach Control

RAPD Regional Accountable Paper Depository • Response Amplitude Probability Data

RAPDS Radar Parameters Data Set

RAPI Royal Australian Planning Institute

RAPID Random-Access Personnel Information Disseminator • Random-Access Photographic Index and Display • Relative Address Programming Implementation Device •

Remote-Access Planning for Institutional Development • Remote Atmospheric Probing Information Display • Research in Automatic Photocomposition and Information Dissemination • Research with Adaptive Particle Imaging Detectors (Cluster) • Responsive Acquisition Processing for Improved Delivery • Retrieval through Automated Publication and Information Digest • Retrieval And Processing Information for Display • Reusable Ada Product for Information Development • Riparian Aerial Photographic Inventory of Disturbance

RAPIDS Random Access Personnel Information Dissemination System • Real-time Acquisition and Processing Integrated Data System • Real-time Automated Personnel Identification System

RAPIV Radiographic Particle Image Velocimeter

RAPOE Regional Aerial Port Of Embarkation

Rapp Rapport

RAPP Registered Air Parcel Post • Resource And Planning Program

RAPPI Random Access Plan-Position Indicator

RaPro Raschety Prochnost (periodical code)

RAPS Radio Planning System • Remote Area Power System • Requisition Automated Processing System • Retrieval Analysis and Presentation System • Review of Automated Personnel System • Right Aft Propulsion System • Risk Approval of Program System • Rural Adjustment Processing System

RAPSE Report on Active and Planned Spacecraft and Experiments

RAPTOR Rapid Telescopes for Optical Response

RAQ Raha (Indonesia) (location identifier) • Réseau des Archives du Québec (Canada-QC)

RAR Radar Arrival Route • Rapid Access Recording • Rarotonga (Cook Islands) (location identifier) • Real-Aperture Radar • Reciprocal Answer Rank • Regulatory Affairs Review • Resource Accommodation Report • Return Address Register • Risk Assessment Report

RARA Retractable Advanced Rigid Array

RARB Risk Assessment and Review Board

RARC Regional Administrative Radio Conference • Ritter Astrophysical Research Center (USA-OH)

RARD Royal Armaments Research and Development Establishment (UK)

RaRe Radiation Research (periodical code)

RARE Réseaux Associés pour la Recherche Européenne

RAREP Radar (weather) Report

RaRLJ Radio Research Laboratories Journal (periodical code)

RaRLR Radio Research Laboratories Review (periodical code)

RARP Reverse Address Resolution Protocol

RARS Reusable Ada Repository System

ras Raster Graphics (file extension)

Ras Raspado • Raspadura

RAS Racine Astronomical Society (USA-WI) • Radar Absorbent Structure • Radar Advisory Service • Radar Altimeter Set • Radar Altimeter System • Radar à Antenne Synthétique • Radio Astronomy Service • Random-Access Storage • Rapid-Access Storage • Rasht (Iran) (location identifier) • Raw Address Strobe • Reacquisition Sensor • Reader Admission System • Reading Astronomical Society (UK-England) • Rectified Airspeed • Redémarrage Automatique du Serveur • Refractive Aberrated Simulator • Région Administrative Spéciale • Registration - Admission - Status • Relay Antenna System • Reliability - Availability - Serviceability • Remote Access Server • Remote Access Service • Remote Access Set • Remote Analysis Station • Remote Analysis System • Requirements Accounting System • Requirements Allocation Sheet • Requirements Analysis Sheet • Réseau d'Aides Spécialisées • Resource Allocation Software • Reuse All Systems • Richland Astronomical Society (USA-OH) • Richmond Astronomical Society (USA-VA) • Rien À Signaler • Riverside Astronomical Society (USA-CA) • Rotating Anode Source • Row Address Select • Row Address Strobe • Royal Academy of Sciences (UK) • Royal Astronomical Society (UK) • Runway Alert System • Rural Adjustment Scheme • Russian Academy of Sciences • Ryerson Astronomical Society (USA-IL)

RASA Russian Aviation and Space Agency (Russia) (formerly RSA)

RASAB Royal Academies for Science and Arts of Belgium

RASAPI Remote Access Service Application Programming Interface

RaSc Radio Science (periodical code)

RASC Radio Atmospheric Science Center • Radio Atmospheric Study Centre • Random Access Storage Controller • Residence Account Service Center • Royal Army Service Corps • Royal Astronomical Society of Canada

RASCAL Rotorcraft Aircrew Systems Concepts Airborne Laboratory

RASCAS Recherches sur l'Aérodynamique, la Structure et la Commande de l'Avion Souple

RASCIBA Radar - SCIDAR - Ballon

RASCOM Regional African Satellite Communication System

RaScS Radio Science Supplement (periodical code)

RASD Radio and Atmospheric Science Division

RASER Radio-frequency Amplification by Stimulated Emission of Radiation

RASG Radar Antenna System Group

RASH Rain Shower

RASM Remote Analog Submultiplexer • Romanian Astronomical Society for Meteors (Romania)

RASME Reconfigurable Automated Software Metrics Environment

RASN Rain and Snow

RASNZ Royal Astronomical Society of New Zealand

RASO Rear Airfield Supply Organisation

RaSol Radioizluchenie Solntsa (periodical code)

RASP Random Access Stored Program • Rapid Acquisition Spectrum Processor • Regional Assimilation and Prediction System • Remote Antenna Signal Processor • Retrieval And Sort Processor

R/ASR Revise As Required

RASS Radar Altimeter Simulation System • Radar Altimeter Speedometer Sounder • Radar Altimeter System Simulator • Radar Analysis and Support System • Radar and Acoustic Sensing System • Radio Acoustic Sounding System • ROSAT All-Sky Survey

RASSBSC ROSAT All-Sky Survey Bright Source Catalog

RASSCALS ROSAT All-Sky Survey Center-for-Astrophysics Loose Systems

RASSFSC ROSAT All-Sky Survey Faint Source Catalog

RASSR Reliable Advanced Solid-State Radar

RAST Radio Standardization • Recovery Assistance and Traversing System • Reliability and System Test

RASTAC Random Access Storage And Control

RASTAD Random Access Storage And Display

RASTRO Rama de Astronomía

RASU Random Access Storage Unit

Rat Rateable • Rating • Ratio

RaT Radiotekhnika (periodical code)

R&AT Research & Advanced Technology

RAT Radar Altimeter Set • Radiation Air Thermometer • Radioactive Tracer • Radioastronomical Telescope • Radiological Assistance Team • Raduzhnyi (Russia) (location identifier) • Ram Air Turbine • Rapport d'Activité Territoriale • Rattaché • Rattachement • Regional Assessment Team • Register Alias Table • Reliability Assessment Test • Reliability Assurance Test • Remote Access Terminal • Reserve Auxiliary Transformer • Resource Assignment Toll • Riverside Astronomical Telescope (USA-CA) • Robust Audio Tool • Rock Abrasion Tool • Routing Analysis Tool

RATA Required Application Technology Advancement

RATAC Radar Target Acquisition

RATAN Radar and Television Aid to Navigation

RATCC Radar Air Traffic Control Center

RATCF Radar Air Traffic Control Facility

RATE Remote Automatic Telemetry Equipment

RATEL Radiotelephone

RATELO Radiotelephone Operator

RATER Response Analysis Tester

RATG Radiotelegram • Radiotelegraph • Radiotelegraphic • Radiotelegraphy

RatK Radiotekhnika Kharkov (periodical code)

RATO Rocket-Assisted Takeoff

RATP Régie Autonome des Transports Parisiens (France) (*)

RATS Radio Amateur Telecommunications Society • RAID Array Tape System • Random-Accessible Text System

RATT Radio Teletypewriter • Runaway T Tauri (star)

RATTS Runaway T Tauri Star

RAU Radio Access Unit • Rand Afrikaans University (South Africa) • Rangpur (Bangladesh) (location identifier) • Red Académica Uruguaya • Remote Access Unit • Remote Acquisition Unit • République Arabe Unie

RAUIS Remote Acquisition Unit Interconnecting Station

RaUkr Vsesoiuznyi Simpozium Difraktsii i Rasprostraneniiu Voln Lvov Ukrainian SSR (periodical code)

RAV Cravo Norte (Colombia) (location identifier) • Revue d'Aptitude au Vol • Risk-Adjusted Value

RAVE Radial Velocity Experiment • Random-Access Video Editing • Relevance Assessment Vehicle • Rendering Acceleration Virtual Engine

RAVEN Radar Analysis and Visualization Environment

RAVI Représentation Audiovisuelle Interactive

RAVIR Radar Video Recording

RAVU Radiosonde Analysis and Verification Unit

raw Raw Image (file extension)

RAW Arawa (Papua New Guinea) (location identifier) • Read After Write • Read And Write • Reconfigurable Architectures Workshop • Reduction of Administrative Workload

RAWB Raad van Advies voor het Wetenschapsbeleid (Netherlands)

RAWD Read-After-Write Dependency

RAWG Regional Association Working Group

RAWS Radar Altimeter Warning System • Radar Wind Sounder • Remote Automatic Weather Station

RAWT Remote-Access Wind Tunnel

RAX Oram (Papua New Guinea) (location identifier) • Remote Access • Remote Agent Experiment • Rural Automatic Exchange

RAY Rothesay (UK-Scotland) (location identifier)

RAZ Rawala Kot (Pakistan) (location identifier) • Remise À Zéro

RAZA Radio Astronomy Zoning Act (USA-WV)

RAZEL Range - Azimuth - Elevation

Rb Rubidium

R/B Rate Buffered • Red/Black

R&B Rehabilitation & Betterment

RB Botswana (international vehicle registration code) • Radar Beacon • Radial Bearing • Radiation Belt • Radio Beacon • Radiocommunication Bureau • Rapid Burster • Read Buffer • Record Byte • Red Belly • Reduction Block • Reentry Body • Reference Burst • Regular Budget • Rehabilitation and Betterment • Relative Bearing • Representative Body • República Boliviana • Request Block • Requirements Board • Response Bus • Return to Bias • Review Body • Ridge Belt • Right Back • Right Button • Rigid Body • Rood-Baum (catalog) • Rubber Band • Rubber Base • Rule Based

RBA Rabat (Morocco) (location identifier) • Radar Beacon Antenna • Rapid Biodiversity Assessment • Recovery Beacon Antenna • Reimbursable Budget Authority • Relative Byte Address • Rigid Bottom Approximation • Risk-Benefit Analysis

RBAC Role-Based Access Control

RBB Borba (Brazil-AM) (location identifier) • Residential Broadband

RBBS Remote Bulletin Board System

RBC Radial Bay Camera • Radio Beam Communications • Red Blood Cell (*) • Red Blood Count • Redundant Battery Charger • Relative Bus Cycle • Relocatable Binary Code • Remote Balance Control • Request for Business Clearance • Return Beam Camera • Robinvale (Australia-VIC) (location identifier) • Rotating Beam Ceilometer

RBCC Rocket-Based Combined Cycle

RbCP Retrieval by Joint Probability

RBCS Remote Bar Coding System

RBD Dallas Redbird (USA-TX) (location identifier) • Randomized Block Design • Relative Band Depth • Reliability Block Diagram

RBDC Read Back Data Clock

RBDE Radar Bright Display Equipment • Read Back Databus Enable

RBDNRQ Received But Did Not Returned Questionnaire

RBDP Rocket Booster Development Program

RBDPE Radar Beacon Data Processing Equipment

RBDS Radio Broadcast Data System

RBDST Richard B. Dunn Solar Telescope (USA-NM) (formerly VTT)

RBE Radiation Belt Environment • Radiation Biological Effectiveness • Ratanankiri (Cambodia) (location identifier) • Record-Breaking Erection • Red Brown Earth • Region of Bizarre Emptiness • Relative Biological Effectiveness • Replacement Battery Equipment • Reply By Endorsement • Résultat Brut d'Exploitation • Rule-Based Engine

RBerG Rechtsberatungsgesetz

RBES Rule-Based Expert System

RBF Big Bear (USA-CA) (location identifier) • Radial Basis Function • Remote Batch Facility • Rockefeller Brothers Fund

RBG Red Blue Green • Réseau de Boutiques de Gestion • Roseburg (USA-OR) (location identifier)

RbGLI Retrieval by General Logic Imaging

RBH Brooks Lodge (USA-AK) (location identifier)

RBHC Regional Bell Holding Co.

RBHF Relativistic Brueckner-Hartree-Fock (approximation)

Rbí Recibí

RBI Rabi (Fiji) (location identifier) • Relative Bearing Indicator • Remote Bus Interface • Require Better Information • Resource-Based Industry • Response Bus Interface

• Ripple-Blanking Input • River Basin Investigation • Runs Batted In

RBJ Rebun (Japan) (location identifier)

RbJP Retrieval by Joint Probability

RBK Right Bank

Rbl Rouble (currency)

RBL Radio BL Lacertae (object) • Radio-loud BL Lacertae (object) (*) • Radio-selected BL Lacertae (object) • Readiness-Based Level • Real-time Blackhole List • Red Bluff (USA-CA) (location identifier) • Reliability-Based Logistics • Right Buttock Line

Rbla Rambla

RBM Radiation Belt Mapper • Radiation Burst Monitor • Random-Band Matrix (model) • Ray Bundle Method • Readiness-Based Maintenance • Real-time Batch Monitor • Running-Box Mean • Rutherford Backscattering Method • Straubing (Germany) (location identifier)

RBMS Resource Base Monitoring System

RBMT Retrospective Bibliographies on Magnetic Tape

RBN Radio Beacon • Ribbon • Ring Back Non-secure

RBO Red Buttes Observatory (USA-WY) • Ripple-Blanking Output • Robore (Bolivia) (location identifier)

RBOC Regional Bell Operating Co.

RBOR Request Basic Output Report

RBP Rabaraba (Papua New Guinea) (location identifier) • Radio Breaking Point • Rat Bundesdeutscher Planetarien (Germany) • Record-Breaking Penetration

RBPE Red de Bibliotecas Públicas del Estado (Spain)

RBQ Really Big Question • Rurrenabaque (Bolivia) (location identifier)

Rbr Rubber

RBR Receiver Buffer Register • Relevance-Based Ranking • Reserve Book Room • Rio Branco - Presidente Medici (Brazil-AC) (location identifier)

RBRC Riken-BNL Research Center

RBrFi Revista Brasileira de Fisica (periodical code)

RBS Orbost (Australia-VIC) (location identifier) • Radar Beacon System • Radio Base Station • Ranked Bubble Sort • Rare Book Section • Reasy Business System • Real Big Scope • Reconstructed Bit Stream (*) • Reduction Block Scheduler • Reliable Broadcast System • Remote Batch Station • Repository Brokerage Services • Reproduced Bit Stream • Richardson-Brealy Spectrograph • Ring Back Secure • Robbed-Bit Signaling • ROSAT Bright Survey • Rule-Based System • Rutherford Backscattering • Rutherford Backscattering Spectroscopy

RBSE Repository-Based Software Engineering (*) • Repository-Base for Software Engineering • Research-Based Science Education

RBSS Recoverable Booster Support System

Rbt Roundabout

RBT Marsabit (Kenya) (location identifier) • Random Boundary Tolerance • Random Breadth Testing • Raumbildtechnik GmbH (Germany) • Reality-Based Trading • Rede Brasileira de Teleinformatica (Brazil) • Registration By Telephone • Remote Batch Terminal • Requirements-Based Testing • Review of Business Taxation

RBTF Reusable Booster Turnaround Facility

RBU Response Bus Unit • Roebourne (Australia-WA) (location identifier) • Répertoire Bibliographique Universel

RBV Radiative Bulk Viscosity • Return Beam Vidicon

RBW Relativistic Blast Wave • Resolution Bandwidth • Walterboro (USA-SC) (location identifier)

Rbx Roubaix (France)

RbXO Rubidium-Crystal Oscillator

RBY Ruby (USA-AK) (location identifier)

rc Run Commands Module (file extension)

R-C Radiative-Convective

R/C Radio Command • Radio Control • Range Clearance • Rate of Climb • Recredited • Remote Control

R&C Read & Compute • Research & Consultancy

RC China (international vehicle registration code) • PRC (international vehicle registration code) • Radiative-Convective • Radio Code • Radio Command • Radio Continuum • Radio Control • Radio Controlled • Range Clearance • Range Command • Range Contractor • Range Control • Range Correction • Rapid Change • Rate of Change • Rate Command • Rate of Climb • Ray Control • Reaction Control • Read Cursor • Reader Code • Real Consejo • Receive Clock • Recent Change • Recognition Context • Recoverability Code • Recovery Controller • Recurring Cost • Red Clump • Red Core • Red Cross • Reference Catalogue • Reflection Coefficient • Regional Center • Regional Coordinator • Regional Council • Registre de Commerce • Registro Civil • Reimbursable Code • Reinforced Concrete • Rekencentrum • Relational Chart • Relative Citedness • Relay Computer • Release Candidate • Release Clause • Remote Component • Remote Control • Reproductive Capacity • Request Counter • Rés-do-Chão • Research Centre • Reverse Course • Research Corporation • Resin Coated • Resistance-Capacitance (*) • Resistance-Conductance • Resistor-Capacitor • Resonance Circuit • Resource Center • Response Cell • Responsibility Center • Ressources Centrales • Reticular Complex • Return Code • Reverse Current • Review Cycle • Revised Catalog • Right Centre • Right Circular • Ring Cell • Riser Cable • Risk Criterion • Ritchey-Chrétien • River Current • Rotary Combustion • Rotational Control • Rotation Center • Rotation Control • Rotation Curve • Route Control • Taiwan (China-ROC) (international vehicle registration code)

Rca Rocca

RCA Central African Republic (international vehicle registration code) • Radar-Controlled Approach • Radiative-Convective Atmospheric • Radio Corporation of America • Radiological Control Area • Rapid City - Ellsworth (USA-SD) (location identifier) • Rate Change Authorization • Reaction Control Assembly • Receiver Collision Avoidance • Regional Cooperative Agreement • Regulatory Contracting Approval • Related Component Analysis • Remote Control Amplifier • République Centrafricaine • Research Center for Astronomy • Residual Capabilities Assessment • Reuse Cost Avoidance • Revealed Comparative Advantage • Revenue Cost Analysis • Right Coronary Artery • Rose City Astronomers (USA-OR) • Royal Canadian Academy

RCAC Remote Computer Access Communications

RCAEn RCA Engineer (periodical code)

RCAF Royal Canadian Air Force

RCAG Remote Center Air/Ground • Remote Communications Air/Ground • Remote Control Air/Ground

RCARv RCA Review (periodical code)

RCAS Reserve Component Automation System • Residual Capability Assessment Survey

RCAT Radio-Code Aptitude Test

RCB Congo (international vehicle registration code) • Radiation Control Board • Randomized Complete Block • R Coronae Borealis • Region Control Block • Remote Circuit Breaker • Richards Bay (South Africa) (location identifier) • Right Centre Back

RCBG Reference Catalogue of Bright Galaxies

RCBR Renegociated Constant Bit Rate

RC/C Reinforced Carbon-Carbon

R&CC Recorder & Communication Control

RCC Radio Common Carrier • Radio Communications Coordinator • Radio Control Complex • Range Control Center • Range Control Complex • Rates and Classification

Center • Real-time Current Control • Receive Control Character • Recorder and Communication Control • Recovery Control Centre • Reductive Citrate Cycle • Redundancy Connection Control • Regional Census Center • Regional Commonwealth of Communications • Regional Computer Center • Regional Consignment Center • Regional Control Center • Regional Coordinating Center • Reinforced Carbon-Carbon • Relocation Coordinating Center • Remote Center Compliance • Remote Cluster Controller • Remote Communications Center • Remote Communications Console • Remote Communications Control • Remote Control Center • Report Control Code • Request for Contract Clearance • Rescue Coordination Centre • Reservations Control Center • Reset Control Circuit • Resource Control Center • Resource Control Console • Reusable Carbon-Carbon • Reverse Command Channel • Ritchey-Chrétien-Coudé (telescope) • Riverside Community College (USA-CA) • Robotic Command Center • Rough Combustion Cutoff • Routing Control Center

RCCA Rough Combustion Cutoff Assembly

RCCB Remote Central Circuit Breaker • Remote Control Circuit Breaker • Remotely-Controlled Circuit Breaker

RCCBPL Resource-Constrained Capabilities-Based Program List

RCCC Range Communications Control Center • Regional Communications Control Center

RCCL Robot C Control Language • Royal Caribbean Cruise Line (now RCI)

RCCN Research Center for Cosmic Neutrinos (Japan)

RCCO Reverse Current Cutout

RCCP Recorder and Communications Control Panel

RCCS Rate Command Control System • Rotary Cell Culture System

Rcd Received

RCD Radar Cloud Detection • Radial Cumulative Distribution • Read Configuration Data • Record • Remote Change Directory • Repetitive Countdown • Residual Current Device • Retirement Computation Date • Revue Critique de Définition

RCDAMP Research Center for Dielectric and Advanced Matter Physics (Korea-ROK)

RCDATA Replaceable Character Data

RCDC Radiation Chemistry Data Center

RCDD Remote Colour Display Driver

RCDI Rate of Climb and Descent Indicator

RCDP Record Parallel

Rcdr Recorder

RCDR Recorder

RCDS Range Countdown and Display System • Range Countdown System

RCE Reaction Control Equipment • Redundant Common Equipment • Regional Coordinating Element • Remote-Control Equipment • Réseau Climatologique d'État • Réseau et Communication d'Entreprise • Resonant-Cavity Enhanced • Restricted Coulomb Energy • Roche Harbor (USA-WA) (location identifier) • Rotary Combustion Engine

RCEA Route Centre Europe - Atlantique

RCEC Repayment of Capital-Excess Compensation

RCEI Range Communications Electronics Instructions

RCEM Regional Contingency Engineering Management • Research Center for Energetic Materials (USA-NM)

RCEP Research Center for Earthquake Prediction (Japan)

RCF Radio Control Function • Registration Confirm • Relative Centrifugal Force • Remote Call Forwarding • Remote Communications Facility • Remote Collection Facility • Remote Concentration Facility • Research Computing Facility

RCFP Reduced Coefficient of Fractional Parentage

RCFT Randomized Controlled Field Trial • Rational Conformal Field Theory

RCfTA Research Centre for Theoretical Astrophysics

RCG Radiation Concentration Guide • Radiocommand-Guided • Rayonnement Cosmique Galactique • Red Clump Giant (star) • Regulatory Coordinating Group • Relativity and Cosmology Group • Rose Compact Group

RCGS Raster Color Graphics System • Royal Canadian Geographic Society

RCH Chile (international vehicle registration code) • Reach • Riohacha (Colombia) (location identifier) • Route Characteristics File

RChA Red Chilena de Astronomía (Chile)

RCHCS Regenerable CO and Humidity Control System

RCHG Reaching

RCI Radar Coverage Indicator • Range Communications Instructions • Registre de Contrôle Individuel • Relative Citation Impact • Relative Citation Index • Relativistic Configuration Interaction • Remote Control Interface • Request for Contract Investigation • Research Career Initiative • Resource Critical Information • Resource Critical Item • Rodent Cage Interface • Royal Canadian Institute • Royal Caribbean International (formerly RCCL)

RCIC Regional Coastal Information Center

RCII Relative Citation Impact Index

RCIP Replacement Cost Integration Program

RCIRS Recoverable Consumption Item Requirements System

RCIS Remotely-Controlled Intelligent System • Resource Center Information System

RCIU Remote Control and Interface Unit

RCJ Reaction Control Jet • Royal Court of Justice

RCK Radial Conformal Killing • Rockdale Coffield (USA-TX) (location identifier)

RCKY Rocky

Rcl Recall

RCL Radio Communications Link • Radio Control Link • Receive Control Logic • Recoilless • Reconnecting Current Layer • Redcliffe (Vanuatu) (location identifier) • Reliability Control Level • Resource Class Library • Reusable Components Library • Robotics Control Library • Rotate Carry Left • Ruling Case Law • Runway Centerline

RCLEP Resonant-Cavity Light Emitting Diode

RCLF Robust Control Lyapunov Function

RCLIS Research Center for Library and Information Science

Rclm Reclaim

RCLM Runway Centerline Marking

RCLO Reports Liaison Control Officer

RCLP Russian Conference on Logic Programming

RCLS Runway Centerline Light System

RCM Radiative Convective Model (*) • Radioactive Convective Model • Radiomètre Cryogénique à refroidissement Mécanique • Random-Choice Method • Range Cell Migration • Recent Crustal Movement • Recessed Contact Mechanism • Reliability-Centered Maintenance • Reliability Corporate Memory • Remote Carrier Module • Requirements Correlation Matrix • Responsibility Center Manager • Reverse Cuthill-McKee (algorithm) • Rice Convection Model • Richmond (Australia-QLD) (location identifier)

RCMA Relativity for Celestial Mechanics and Astrometry (working group)

RCMAM Relativity for Celestial Mechanics, Astrometry and Metrology (IAU WG)

RCMP Royal Canadian Mounted Police (Canada)

RCMU Remote Control and Monitoring Unit

Rcn Recreation

RCN American River (Australia-SA) (location identifier) • Reactor Centrum Nederland (Netherlands) • Receipt of Change Notice • Record Control Number • Recovery Communications Network • Registry Control Number • Report Change Notice • Report Control Number • Requirements Change Notice • Resource Control Number • Royal Canadian Navy

RCNC Range Communications Network Controller

Rcndt Recondition

RCNS Range Communications Network Supervisor

RCO Radiation Control Office • Radiation Control Officer • Range Control Office • Range Control Officer • Record Communications Operator • Regional Compliance Office • Regional Compliance Officer • Regional Coordinating Office • Regional Coordinating Officer • Registration Confirmation Packet • Remote Communications Outlet • Remote-Control Office • Remote-Control Oscillator • Rendezvous Compatible Orbit • Reports Control Office • Reports Control Officer • Requirements Control Office • Requirements Control Officer • Resident Contracting Officer • Rochefort Saint-Agnan (France) (location identifier)

RCOC Regional Communications Operations Center

rcp Remote Copy

RCP Radiation Constraints Panel • Radon Contractor Proficiency • Raster Color Printer • Real Continuous Pairwise • Recherche Coopérative sur Programme • Recording Control Panel • Recovered Clock Path • Rectangular Coordinate Plotter • Recurring Content Pattern • Regional Cooperation Project • Registration Control Point • Relevant Common Picture • Reliability Critical Problem • Remote Communications Processor • Remote-Control Panel • Remote Copy • Remote Copy Program • Remote Copy Protocol • Request Change Proposal • Restore Cursor Position • Resynchronization Control Panel • Retention Control Point • Right Circularly-Polarized • Right Circular Polarization • Right Circular Polarizer • Robust Communications Protocol

RCPA Ratio of Cited to Published Articles

RCPL Right Circularly Polarized Light

RCPSP Resource Constrained Project Scheduling Project

RCPSPDCF Resource Constrained Project Scheduling Project with Discounted Cash Flows

Rcpt Receipt • Reception

RCPT Receipt • Receptacle • Recipient

Rcptn Reception

RCQ Reconquista (Argentina-SF) (location identifier)

RCR Radio Communications Report • Relative Citation Rate • Relative Cost of Reuse • Remote Computer Reader • Requirements Change Request • Retrofit Configuration Report • Reusable Component Recovery • Reverse Current Relay • Rochester - Fulton County (USA-IN) (location identifier) • Rotate Carry Right • Runway Condition Rating • Runway Condition Reading

RCRA Resource Conservation and Recovery Act (USA)

RCRE Receive Corrected Reference Equivalent • Required Collaborative Research Experience

RCRM Radiomètre Cryogénique à Refroidissement Mécanique

RCRS Regenerative Carbon-dioxide Removal System

RCRU Radio Communications Research Unit

RCS Radar Cross-Section • Radio Communications System • Range Control Supervisor • Rate Control System • Rayonnement Cosmique Solaire • Reaction Control System • Readiness Control System • Reconnecting Current Sheet • Record Creation System • Red-sequence Cluster Survey • Reentry Control System • Reference Color Space • Regional Control Station • Remote Communications System • Remote Computing Service • Remote Connection Service • Remote Control System • Remote Core Sampler • Reply Card Scanning • Reports Control System • Requirements Collection System • Requirements Computational System • Research and Consultancy Services • Research Camera System • Resource Control System • Retention Category Summary • Retrofit Configuration System • Return Channel System • Revision Control System • Rochester (UK-England) (location identifier) • Rochester Connectionnist Simulator • Rudder Control System • Rural Counseling Service

RCSC Rates and Classification Service Center • Reaction Control System Controller • Remote Spooling Communications System • Requirement Computation Suppression Code

RCSDE Reaction Control System Drive Electronics

RCSP Remote Call Service Position

RCSR Radar Cross-section Reduction

RCSS Random Communications Satellite System • Records Communications Switching System

RCSSMRS Regional Centre for Services in Surveying, Mapping and Remote Sensing

Rct Receipt

RCT Radiometric Calibration Target • Randomized Clinical Trial • Randomized Control Trial • Reaction Control Thruster • Reality Capture Technologies Inc. (USA-CA) • Receive Cipher Text • Reed City - Miller (USA-MI) (location identifier) • Region Control Table • Region Control Task • Reliability Confidence Testing • Remote Control Transmitter • Remotely-Controlled Telescope • Repair Cycle Time • Rescue Coordination Team • Resolve Control Transformer • Resource Control Table • Rework Completion Tag • Ritchey-Chrétien Telescope • Robotically-Controlled Telescope

RCTC Receive Cipher Text Clock • Reversible Cooke Transit Circle

RCTF Réunion des Chimistes et Théoriciens Francophones (conference)

RCTL Resistor-Capacitor-Transistor Logic

RCTLR Receiver Controller

Rcto Recinto

RCTS Remote Commercial Television Service

RCTT Regional Center for Technology Center

RCU Radar Control Unit • Radio Channel Unit • Rate Changes Unit • Recovery and Control Unit • Reference Control Unit • Regional Coordination Unit • Rekeying Control Unit • Relay Control Unit • Remote-Control Unit • Retail Consolidation Unit • Rio Cuarto (Argentina-CD) (location identifier)

RCUA Responsable de la Charge Utile d'Ariane

RCUH Research Corporation of the University of Hawaii (USA-HI)

Rcv Receive

RC/V Recent Change/Verify

RCV Reaction Control Valve • Receive • Recent Change-Verify • Remotely-Controlled Vehicle

Rcvd Received

Rcvg Receiving

Rcvr Receiver

RCVR Receiver

Rcvs Receives

RCVS Remote Control Video Switch • Remote-extension of Concurrent Versions System

RCVY Recovery

RC&W Read, Compute & Write

RCW Rapidly Changing World • Reaction Control Wheel • Rodgers - Campbell - Whiteoak (catalog)

RCWA Rigorous Coupled Wave Analysis

RCWP Rapidly Changing World Policy

RCWR Relative Cost of Writing for Reuse

RCY Recovery • Remaining Cycles • Rum Cay (Bahamas) (location identifier)

RCZ Radiation Control Zone

RC2 Second Reference Catalogue • Second Revised Catalogue

RC3 Third Reference Catalogue • Third Revised Catalogue

rd Rund

Rd Radium (now Ra) • Read • Rendered • Road • Rod • Rond • Round • Rutherford

R-D Reaction-Diffusion

R&D Read & Destroy • Recherche & Développement • Research & Development

R/D Rate of Descent • Refer to Drawer • Relevant Document • Research and Development • Resolver to Digital

RD Radiation Detection • Radiation-Dominated • Radiation Dose • Radiation-Driven • Radio Determination • Random Drift • Range Development • Raster Device • Rate of Detection • Read • Read Data • Readiness Data • Readiness Date • Real Decreto • Real Drift • Rear Door • Receive Data • Received Data • Red Dwarf (star) • Reference Designation • Reference Designator • Reference Directory • Reference Document • Regional Director • Register Drive • Règlement Différé • Relative Density • Released Drawing • Relevance Description • Reliability Development • Remedial Design • Remove Directory • Remove Duplicate • República Dominicana • Request for Deviation • Required Data • Required Date • Requirements Document • Research and Development • Research Department • Responsibility/Distribution • Restricted Data • Restricted Distribution • Rettungsdienst • Ring Data • Ring Diagram • Ring Down • Rive Droite • Road • Root Diameter • Rotation Dominated • Rotational Discontinuity • Round • Route Départementale • Route Destination • Routing Domain • Running Days • Rural Delivery

Rda Redonda • Ronda

R&DA Research & Data Analysis

RD&A Research, Development & Acquisition

RDA Read Data Available • Receive Data • Recommended Daily Allowance • Recommended Dietary Allowance • Regional Development Assistant • Relational Distributed Architecture • Reliability Design Analysis • Remote Data Access • Remote Database Access • Remote Dial-in Analysis • Representational Difference Analysis • République Démocratique Allemande (now merged into RFA) • Research and Data Analysis • Research, Development and Acquisition • Resident Data Area • Rockhampton Downs (Australia-NT) (location identifier) • ROSAT Data Archive • Rotary Drive Actuator

RDAAC Remote Data Analysis and Archive Center

RD&AC Research, Development & Acquisition Committee • Research, Development & Acquisition Council

RDAC Research, Development and Acquisition Committee • Research, Development and Acquisition Council

RD&AP Research, Development & Acquisition Plan

RDAP Remote Data Access and Protocol • Research, Development and Acquisition Plan

RDAS Requirements Definition and Analysis Summary

R&DAT Research and Development Acceptance Test • Research and Development Accepted Testing

RDAT Research and Development Acceptance Test • Research and Development Accepted Testing • Rotary-Head Digital Audio Tape

RDAU Remote Data Acquisition Unit

RDAF Regional Data Analysis Facility

RDAP Research and Development Applied to Projects

RDATA Resource Data

RDAV Remote Data Acquisition Van

RDB Raw Data Bus • Receive Data Buffer • Red Dog (USA-AK) (location identifier) • Reference Database • Relational Database • Requirements Data Bank • Research and Development Board

RDBI Relational Database Interface

RDBM Relational Database Management

RDBMS Relational Database Management System

RDBS Relational Database System

RdC Rez-de-Chaussée

RDC Range Data Controller • Read Device Characteristics • Real-time Data Cell • Redenção (Brazil-PA) (location identifier) • Reference Designation Code • Reference Designator Code • Regional Data Center • Regional Dissemination Center • Regional Documentation Center • Reliability Data Center • Remote Data Change • Remote Detector Chamber • Reproduction and Distribution Center • Request for Design Change • Request for Document Change • Request for Drawing Changes • Research and Documentation Center • Review Draft Copy • Running Down Clause • Rural District Council

RDCLK Received Clock

RDCO Research Development Coordination Office

RDCOMP Round Complete

RDCP Remote Display Control Panel

RDCS Reconfiguration Data Collection System • Regional Data Collecting System • ROSAT Deep Cluster Survey • ROSAT Distant Cluster Survey

RD&D Research, Development & Demonstration

RDD Radar Detector Detector • Random Digital Dialing • Redding (USA-CA) (location identifier) • Replaceable Database Driver • Required Delivery Date • Requirements Definition Developer • Requirements Definition Document • Requirements Description Document • Requirements-Driven Database • Requirements-Driven Design • Requirements-Driven Development • Required Delivery Date • Research - Development - Demonstration • Residual Dark-current Distribution • Residual Dark Distribution • Residual of Dark Distribution • Responsibility-Driven Design • Ruggedized Disk Drive

RDDBS Redundant Distributed Database System

RDDF Ruggedized Disk Drive Fixed

RDDP Radar Data Development Programme

RDDR Ruggedized Disk Drive Removable

RDDS Range Data Distribution System • Recoil-Distance Doppler-Shift • Research and Development Descriptive Summary

RDDT&E Research, Development, Demonstration, Testing & Evaluation

RDDTE Research - Development - Demonstration - Testing - Evaluation

RD&E Research, Development & Engineering

RDE Merdey (Indonesia) (location identifier) • Radial Defect Examination • Radiation Detection Experiment • Radiation-Dominated Era • Remote Data Entry • Research and Development Establishment • Research and Development Expenditure • Research - Development - Engineering • RGS Digital Electronics • Rotating-Disc Electrode • Route Determination Entity

RdeC Rez-de-Chaussée

RD&EC Research, Development & Engineering Center

RDEC Requirement Data Exchange Card • Research, Development and Engineering Center

R.-de-Ch. Rez-de-Chaussée

RdeCh Rez-de-Chaussée

RDEMS Regional Disaster Emergency Management System

RD&ES Requirements Determination & Evaluation System

RDES Remote Data Entry System • Requirements Determination and Evaluation System • Requirements Determination and Execution System

RDEX Readiness Exercise

RDF Radio Direction Finder • Radio Direction Finding • Range Data Facility • Rate Decrease Factor • Rationalized Data Format • Real Data Fine • Record Definition Field • Regional Development Fund • Request Denied Frame • Resource Data File • Resource Denied Frame • Resource Description Format • Resource Description Framework • Revised Delivery Forecast

RDFD Requirement Decomposition Functional Definition

RDFOC Rapidly-Deployable Fiber-Optic cable

RDFR Restricted Direct Fuzzy Reasoning

RDFS Resource Description Framework Schema

RDFT Real Discrete Fourier Transform

RDG Random Data Generator • Reading (USA-PA) (location identifier) • Research Development Grant • Ridge

RDGE Ridge

RDGS Reconfiguration Display Generation System • Ridges

RDGT Reliability Development and Growth Test

RDI Range-Doppler Interferometer • Range-Doppler Interferometry • Raw Data Interface • Reconnaissance - Detection - Identification • Released Data Index • Remote Data Input • Remote Defect Indication • Remote Defect Indicator • Research and Development Institute • Research, Development and Industrialization • Research - Development - Innovation

RDIF Relative Daily Impact Factor

RDIS Remediation Design and Implementation Schedule • Research and Development Information System

RDIT Replication - Distribution - Installation - Training

RDIU Radiation Device Interface Unit

RDJFCB Read Job File Control Block

RDL Radio Data Link • Rapid Deployment Launcher • Reconfigurable Design Language • Record Description Language • Remote-Display Link • Requirements Document Library

RDLAP Radio Data Link Access Protocol

RDLO Research and Development Liaison Office

RdM Résistance des Matériaux

RD&M Research, Development & Marketing

RDM Bend (USA-OR) (location identifier) • Random • Rate Dampping Mode • Real-time Data Manager • Recoil Distance Method • Recording Demand Meter • Redmond (USA-OR) (location identifier) • Reduced-Density Matrix • Relational Data Mining • Relational Data Model • Relational Document Manager • Report Display Manager • Repulsive Dark Matter • Research, Development and Marketing • Rotterdamsche Droogdok Mij. BV

RDMD Regional Dark Mantle Deposit

RDMF Rapid Deployable Medical Facility

RDMS Range Data Management System • Range Data Measurement System • Relational Database Management System

RDMSR Read Machine Specific Register

RDMSS Remotely Deployable Mission Support System

RDMTR Radiometer

RDN Radio Data Network • Relative Distinguished Name • Resources Discovery Network • Routing Domain Name

rDNA Recombinant DNA

RDNH Reverse Domain Name Hijacking

Rdnr Randnummer

Rdnrn Randnummern

Rdnt Redundant

Rdo Radio • Recuerdo

R&DO Research & Development Operations

RDO Radio • Redistribution Order • Relational Database Operator • Remote Data Object • Request for Deployment Order • Research and Development Objective • Research and Development Operations • Resource Definition Online

RDOC Reference Designation Overflow Code

RDOME Radiatively-Driven Orbital Mass Ejection

RDOP Relative Dilution Of Precision

RDOS Real-time Disk Operating System

RDP Radar Data Processing • Radar Development Project • Radial-Distance Profile • Rapid Deployment Planning • Rat Deutscher Planetarien (Germany) • Readiness Directive Plan • Receiver and Data Processor • Reconstruction and Development Programme • Recursive Dynamic Programming • Reliable Datagram Protocol • Reliable Data Protocol • Remote Data Processor • Remote Desktop Protocol • Remote Display Processor • Report Documentation Page • Requirements Data Plan • Requirements Definition Program • Requirements Development Plan • Research and Development Personnel • Research and Development Project • Research Data Program • Resource Distribution Pattern • Ribosomal Data Project

RDPC Radar Data Processing Center • Regional Data Processing Center

RDPI Remote Data Processing Installation • Revue du Droit de la Propriété Intellectuelle (periodical)

RDPL Laos (international aircraft registration code)

R&DPP Research & Development Program Planning

RDPS Radar Data Processing System

RdPt Rond Point • Rond Punt

RDQ Really Dumb Question

Rdr Radar

RDR Radar • Radar Departure Route • Radio Downstream Receiver • Raw Data Record • Raw Data Recorder • Reddening-Distance Relation • Red River - Grand Forks (USA-ND) (location identifier) • Reduced Data Record • Reduced Data Requirements • Release Design Review • Requirements Definition Review • Research and Development Report • Rudder

RDRAM Rambus Dynamic Random Access Memory

RDRO Robinson/Downs Rooftop Observatories (USA-CA)

RDS Radio Data System • Radio Determination Service • Radio Digital System • Radio Distribution Search • Rapid Dissemination System • Rat Deutscher Sternwarten (Germany) (formerly RWS) • Real Data Smooth • Relational Data System • Remboursement de la Dette Sociale • Remote Data Service • Remote Device System • Remote Diagnostic Support • Remote Display System • Replaceable Disk Store • Report Distribution Software • Requirements Data Sheet • Rincón de los Sauces (Argentina-NE) (location identifier) • Roads • ROSAT Deep Survey • Rossiter Double Star • Running Digital Sum

RDSC Research and Development Standard Contract

RDSES Random Dynamics in Spatially Extended Systems

RDSI Red Digital de Servicios Integrados

RDSR Receiver Data Service Request

RDSS Radio Determination and Satellite System • Radio Determination Satellite Service • Regional Development

Simulation System • Research and Data Support System • Research and Development Streamlined Solicitation

RDT Radio Digital Terminal • Radio Downstream Transmitter • Rapid Distorsion Theory • Recherche et Développement Technologique • Reference Data Table • Reliability Development Test • Remote Data Transmission • Remote Data Transmitter • Remote Destination Table • Remote Display Terminal • Réseau de Développement Technologique • Réseau de Diffusion Technologique • Richard Toll (Senegal) (location identifier) • Rights Description Table • Rise Time Discriminator • Rudimentary Debugging Tool

RDT&E Research, Development, Technology & Evaluation • Research, Development, Test & Engineering • Research, Development, Test & Evaluation • Research, Development, Testing & Engineering • Research, Development, Training & Evaluation

RDTE Remote Data Terminal Entry • Research, Development, Technology and Evaluation • Research, Development, Test and Engineering • Research, Development, Test and Evaluation • Research, Development, Testing and Engineering • Research, Development, Training and Evaluation

RDTL Resistor Diode Transistor Logic

RDTM Rated Distribution Training Management

RDTO Receive Data Transfer Offset

RDTR Radiator • Research Division Technical Report

RDTS Radar Data Transfer System • Research and Development Tracking System

RDU Durham (USA-NC) (location identifier) • Raleigh (USA-NC) (location identifier) • Remote Data Unit • Remote Display Unit • Resource Development Unit

RDV Red Devil (USA-AK) (location identifier) • Relevance Description Vector • Rendezvous

RDVT Reliability Design Verification Test

RDVU Rendezvous

RDW Rapid Development Workshop • Record Descriptor Word • Request for Deviation/Waiver • Response Data Word • Return Data Word

RDX Cyclotrimethylene Trinitramine • Rapid Detonating Explosive • Realistic Display Mixer • Road Crossing

RDY Ready

RDYLD Ready to Load • Ready to Load Data

Rdz Randziffer

RDZ Radiation Danger Zone • Rodez (France) (location identifier)

Rdzvs Rendezvous

re in Regards to • rechter

Re Reference • Regarding • Reinsurance • Reply • Respecting • Reynolds Number • Rhenium • Rhénium • Rupee (currency)

R&E Recovery & Evacuation • Research & Education • Research & Engineering • Research & Enterprise • Research & Experimentation • Review & Evaluation

R/E Reentry • Right-hand Engine

RE Île de la Réunion (code ISO de pays) • Radial Error • Radiated Emission • Radiative Equilibrium • Range Equipment • Rare Earth • Rate Effect • Rate of Exchange • Real Estate • Recall • Receive Element • Rechtsentscheid • Record End • Recovery-Enabled • Red Edge • Reference • Regular Editor • Regular Exposure • Relative Eccentricity • Relative External (use) • Relatively External • Renewable Energy • Repeated Erection • Répondeur-Enregistreur • Requirement Engineering • Research Engineering • Research Establishment • Resolution Element • Resonant Excitation • Responsible Engineer • Retirado • Reunion • Réunion Island (ISO country code) • Reverse Engineering • Revised Edition • Riccati Equation • Richards Equation • Right End • Right Excellent • Risk Exposure • Robust Estimation

REA Raad van Economische Aangelegenheden • Range Error Average • Real Estate Agent • Reao (Tuamotu Islands) (location identifier) • Reentry Angle • Request for Engineering Assistance • Request for Equitable Adjustment • Resource - Event - Agent • Responsible Engineering Authority • Rocket Engine Assembly • Rural Electrification Administration (USA)

ReAA Recherches Aéronautiques et Aérospatiales (periodical code)

ReAeB La Recherche Aérospatiale - Bimonthly Bulletin ESA (periodical code)

Reac Reactor

REAC Radiation Emergency Assistance Center

REACH Remote Access to Circular History • Remote External Access to Circular History

REACQ Reacquisition

REACT Rain Echo Attenuation Compensation Technique • Rapid Evaluation And Correction Team • Rapid Evaluation And Corrective Team • Research and Education Automatically Controlled Telescope

REACTS Radiation Emergency Assistance Center Training Site

READ Reader for Event Analysis and Detection • Readiness • Real-time Electronic Access and Display • Relative Element Address Destinate • Remote Electronic Alphanumeric Display

READI Rights for Electronic Access to Delivery of Information (*) • Rights to Electronic Access and to Delivery of Information

readme Read-Me Text (file extension)

READS Re-Entry Air Data System

ReAer La Recherche Aérospatiale (periodical code)

REAF Revised Engineering Active Force

REAFAMS Rapid Easy Accurate Finderscope Alignment Mounting System

Realsch Realschule

REAM Rocket Engine Assembly Module

REAMS Resource Evaluation and Management System

REAP Reengineering Application Project • Regional Environment Assessment Program • Research and Engineering Apprenticeship Program • Resource Evaluation Aerial Photography • Rural Education Access Program

REAPS Reportable Excess Automated Property System

REAR Reliability Engineering Analysis Report

REARS Redundant Attitude Reference System

REAUMUR Réseau Aquitain des Utilisateurs des Milieux Universitaires et de Recherche (France)

REB Relativistic Electron Beam

REBA Radiometer Electronics Box Assembly

REBIUN Red de Bibliotecas Universitarias (Spain)

REBOOT Reuse-Based Object-Oriented Technology • Reuse By Object-Oriented Techniques

REBS Radiation Energy Balance System

rec Recorder (file extension)

Réc Réception

Rec Receipt • Receive • Recensement • Recent • Recepción • Reception • Recipe • Recommend • Recommendation • Recommended • Record • Recorded • Recorder • Recording • Recreation • Recreational • Rectification

REC Radiative Electron Capture • Recado • Receive • Receiver • Recife Guararapes (Brazil-PE) (location identifier) • Reconstituable Enduring Communication • Reconstruction • Record • Recorder • Record of Environmental

Consideration • Recreation • Rectifier • Regional Environment Center • Regional Evaluation Center • Request for Engineering Change • Research - Evaluation - Communications

RECA Residual Capability Assessment

Recap Recapitulate • Recapitulation

RECAP Reliability Evaluation Corrective Action Program • Review and Command Assessment of Projects

RECAS Residual Capability Assessment System

RECAT Residual Capabilities Assessment Team

RECC Regional Emergency Communications Coordinator • Remote Element Control Center

Recd Received

RECD Received

Recep Reception

Recept Receptacle

RECERT Recertification

RECFM Record Format

Rech La Recherche (periodical code) • Recherche

RECHG Recharge

Recíp Recíproco

Recip Reciprocal • Reciprocating • Reciprocity

RECIP Reciprocate

Recirc Recirculate

RECIRC Recirculate • Recirculating

RECITE Regions and Cities for Europe

Reclam Reclamation

Recm Recommend

Recmd Recommended

Recncln Reconciliation

RECODIS Recherche Collaborative à Distance

Recog Recognition • Recognize

RECOG Recognition

Recogd Recognized

Recon Reconnaissance

RECON Reconnaissance • Remote Console • Rendezvous with a Comet Nucleus • Retrospective Conversion

RECONFIG Reconfiguration

RECONN Reconnaissance

RECONS Research Consortium on Nearby Stars

RECORD Russia-Europe Cooperation on Rocket engine Demonstration • Russian-European Cooperation on Rocket Demonstration (*)

RECOV Recovery

RECP Receptacle • Record Engineering Change Proposal • Regional Emergency Communications Planner • Request for Engineering Change Proposal

Recpt Receipt

RECRC Recirculate

RECREP Reconnaissance Report

RECS Range Embedded Computer System • Representative Environmental Control System • Reseller Electronic Communication System

RECSTA Receiving Station

Rect Receipt • Rectangle • Rectangular • Rectifier • Rectify

RECT Rectaguarda • Rectifier • Rectify

RECTAS Regional Centre for Training in Aerospace Surveys

Récup Récupération

RECV Receive

RECVR Receiver

Red Redeemable • Reduce • Reduced • Reduction • Redundant

RE&D Research, Engineering & Development

RED Lewiston - Mifflin County (USA-PA) (location identifier) • Random Early Detect • Random Early Detection • Range Error Detector • Rapid Excess Disposal • Reduction • Reedsville - Mifflin County (USA-PA) (location identifier) • Relais Électroniques Deutsch • Remisión Electrónica de Documentos • Research - Engineering - Development • Revision Effective Date • Risk Estimate Distance

REDAP Reentrant Data Processing

REDAR Radiation and Environmental Data Acquisition and Recording • Range Engineering and Data Acquisition and Reduction

REDARS Reference Engineering Data Automated Retrieval System

REDCAP Real-time Digitally Controlled Analyzer and Processor

REDCOM Readiness Command

REDCON Readiness Condition

REDG Réseau Européen de Documentation Géologique

REDI Relational Database for Images

REDMAC Readiness Management Center

REDOPS Readiness Operations

REDOSI Répertoire de la Documentation Scientifique et Informelle

REDS Radar Error Determination System

REDSTAR Resource Data Storage And Retrieval

REDTRAIN Readiness Training

REDUN Redundancy

REDVLPG Redeveloping

REE Lubbock Reese (USA-TX) (location identifier) • Rare-Earth Element • Régimen Especial de Estudiantes • Remote Exploration and Experimentation (NASA) • Requirements Engineering Environment • Responsible Equipment Engineer

REEP Regional Environment Employment Program • Regression Estimation of Event Probabilities

REES Reactive Electronic Equipment Simulator

Réf Référence

Ref Refer • Referee • Reference • Referencia • Referred • Refining • Reform • Reformed • Refrán • Refresher • Refugio • Refund • Refunding

REF Range Error Function • Refer (to) • Reference • Reference File • Refresh • Refrigeration • Refurbish • Refurbishment • Relative Efficiency Factor • Répartiteur Électronique de Freinage • Répartition Électronique de Freinage

REFCE Régimen Especial de Funcionaros Civiles del Estado (Spain)

Refd Referred • Refund

REFIR Radiation Explorer in the Far Infrared

Refl Reflect • Reflection • Reflective • Reflex • Reflexive

REFLEX Remote Fluvial Experiment • ROSAT ESO Flux Limited X-ray (sample)

Refrig Refrigerate • Refrigeration • Refrigerator

REFRIG Refrigerate

REFRMG Reforming

REFSPEC Reflectance Spectrometer

REFURB Refurbish

reg Register • Registration (file extension)

Reg Region • Register • Registered • Registration • Registre • Registro • Registry • Regla • Regulación • Regular • Regularly • Regulate • Regulated • Regulating • Regulation • Regulator • Regulatory

Rég Région

REG Reggio Calabria (Italy) (location identifier) • Region • Register • Registration • Regulate • Regulation • Regulator

REGAIN Recombinateur du Grand Interféromètre

REGAL Range and Elevation Guidance for Approach and Landing • Rigid Epoxy Glass Acrylic Laminate

Reg.-Bez. Regierungsbezirk

Reg.Bez. Regierungsbezirk

Regd Registered

Regelsch Regelschule

REGEN Regenerate

REGIS Regional Geographical Information System • Remote Graphics Instruction Set

Regr Registrar

Regres Regresivo

Regt Regent • Regiment

REH Rehoboth Beach (USA-DE) (location identifier)

Rehab Rehabilitate • Rehabilitation

REHE Relativistic Effects in Heavy Elements

REI Range from Entry Interface • Reactor Experiments Inc. (USA-CA) • Réduction pour Embauche et Investissement • Regina (French Guiana) (location identifier) • Research - Education - Innovation • Research-Engineering Interface • Rigid External Insulation • Runway End Identification

REIC Rare Earth Information Center (USA-IA) • Running Edge Image Compression

REIL Runway End Identification Lights

REILS Runway End Identification Light System

Reimp Reimpresión

REIMP Regional Environmental Information Management Project

Reinf Reinforce

REINF Reinforce

Reinfd Reinforced

Reinfmt Reinforcement

REINIT Reinitialize

REINS Requirements Electronic Input System

REIO Regional Economic Integration Organization

Reit Reiteration

REIT Real-Estate Investment Trust

REJ Radiative and Expanding Jet • Reject

REK Reykjavík (Iceland) (location identifier)

REKLIP Regio-Klima-Projekt

Rej Reject

REJ Reject

REJASE Reusing Junk As Something Else

Rel Relate • Related • Relating • Relation • Relative • Relatively • Release • Released • Relié • Relief • Relieve • Relieved

REL Rapidly Extensible Language • Recommended Exposure Limit • Recovered Energy Logic • Relative • Relativistic • Relayer • Release • Released • Reliability • Reliable • Renaissance Experimental Laboratory • Restricted Energy Loss • Rights Expression Language • Ripper Excavator Loader • Trelew (Argentina-CB) (location identifier)

RELAB Reflectance Experiment Laboratory

RELAFI Red Latina-Americana de Física

Relat Relatif • Relation • Relative • Relativität

RelComm Reliable Communications

Relctd Relocated

RELDMS Rocket Engine Leak Detector Mass Spectrometer

RELI Reliability

RELISH Responses Evaluated by Listing Information in SEED Headers

RELMAT Relative Matrix

Reloj Relojería

RELOS Regional Locator System • Relative radiometer for Oscillations

RELP Random Exited Linear Predictive Coding • Random Excitor Linear Prediction

RELSAT Reliability for Satellite

RELSECT Relative Sector

Rem Remainder • Remaining • Remark • Remarque • Remit • Remittance • Remote • Remove

REM Radiation Environment Monitor • Radiation Experiments and Monitors Ltd. (UK) • Random Entry Memory • Rapid Eye Mount (telescope) • Rapid Eye Movement • Rare-Earth Metal • Rat Enclosure Module • Reaction Engine Module • Recognition-Extended Memory • Reflection Electron Microscope • Relabeling Exchange Method • Release Engage Mechanism • Release Engagement Monitor • Release Engine Module • Release Escape Mechanism • Reliability Engineering Model • Remaining • Remark • Remote • Remote Equipment Module • Remote Ethernet Management • Resource and Environmental Management • Ring Error Monitor • Ripple Effect Matrix • Rocket Engine Module • Röntgen Equivalent in Man • Router Engine Module

REMAP Regional Environmental Mapping and Analysis Process • Research Maximization And Prioritization (NASA task force)

REMAPP Remote Sensing Array Processing Procedures

Remb Remboursable • Remboursé • Remboursement

REMCO Reference Materials Committee (ISO)

REMD Red Medium Dispersion

REMIDAS Resource Evaluation and Monitoring Integrated Data Analysis System

REMIS Reliability and Maintainability Information System

Remitt Remittance

REML Radiological Environmental Monitoring Laboratory • Residual Maximum Likelihood Estimation

REMOB Remote Observation

REMOBS Remote Observation System

REMOS Rapid Eye Movement Observing Software

REMP Relative Electromagnetic Particle

REMPEC Regional Maritime Pollution Emergency Centre

REMPI Resonantly-Enhanced Multiphoton Ionization (*) • Resonance-Enhanced Multiphoton Ionization

REMS Registered Equipment Management System • Remotely Employed Sensor • Remote Sensor • Rocket Engine Modules

REMSAT Real-time Emergency Management via Satellite

REMT Radiological Emergency Medical Team • Regional Emergency Management Team

REN Orenburg (Russia) (location identifier) • Range Extension Node • Reduced Exterior Noise • Remote Enable • Rename • Ring Equivalence Number • Ringer Equivalency Number

RENASS Réseau National de Surveillance Sismique (France)

RENATA Recherche Nationale en Aéroélasticité de Turbomachines Aéronautiques (France)

RENATER Réseau National de Télécommunications pour la Technologie, l'Enseignement et la Recherche (France) (*)

RENC Regional Electronic Navigation Center • Regional Electronic Navigation Chart

REND Rendezvous

ReNEv Research in Non-destructive Evaluation (periodical code)

Renf Renfrewshire (UK-Scotland)

RENFE Red Nacional de Ferrocarriles Españoles (Spain)

Renor Renormalization (periodical code)

R/EO Radar/Electro-Optical

REO Radar Electro-Optical • Rare-Earth Oxide • Readiness Engineering Office • Removable - Erasable - Optical • Responsible Engineering Office • Rome State (USA-OR) (location identifier)

REOPE Reflexive Evaluation Of Predictive Expertises (method)

REOSC Recherches et Études d'Optique et de Sciences Connexes (France)

REOT Right-End-Of-Tape

rep Repetatur • Report (file extension)

Rep Repair • Repeat • Repetition • Report • Reported • Reporter • Represent • Representative • Represented • Representing • Reproducción • Reprint • Republic • República • Republican • Republik

Rép Réponse • Républicain • République

REP Range Error Probable • Rapport d'Étude Professionnelle • Réacteur à Eau Pressurisée • Réacteur à Eau sous Pression • Regional Energy-saving Plan • Regional Environmental Plan • Regional Exchange Point • Remote Experimental Package • Rendezvous Evaluation Pad • Repeat • Replacement • Replica • Reply • Reporting Point • Representative • Reproducing Program • Research Enhancement Program • Resource Enhancement Program • Resource Enhancement Project • Resource Estimating Procedure • Robots Exclusion Protocol • Roentgen Equivalent Physical • Siem Reap (Cambodia) (location identifier)

REPART Repartição

REPE Repeat while Equal

REPIN Reply If Negative

Repl Replace • Replaced • Replacement

REPL Replace

REPLEX Redundancy Planning Expert

REPLI Radar-coupled Emitter/Passive Location and Identification

REPNE Repeat while Not Equal

REPNZ Repeat while Not Zero

REPP Relative Evaluation and Policy Project (ANU)

Repr Représentant • Represent • Representation • Représentation • Representative • Represented • Représenter • Representing • Reprint • Reprinted • Reproduce • Reproductif • Reproduction • Reproductive • Reproduire

REPR Repressurization

Repres Represent • Representation • Representative

Représ Représentant • Représentation • Représenter

Repro Reproduce • Reproduced • Reproduction

REPRO Regional Program Office • Reproduction

REPS Replicas

REPSHIP Report of Shipment

Rept Receipt • Report

REPTI Resonance-Enhanced Two-Photon Ionization

Repub Republic • Republican

REPZ Repeat while Zero

Req Request • Require • Required • Requirement • Requisition

REQ Request • Require • Required

REQCONF Request Confirmation

Reqd Required

REQD Required

REQMT Requirement

REQMTS Requirements

REQS Requirements • Requisitions

Reqt Requirement

REQUID Required Identification

REQVAL Requisition Validation

REQVER Requirements Verification

RER Renewable Energy Resource • Réseau Express Régional • Resource Evaluation Report • Reusable - Expendable - Reusable • Reverse Error Reporting • Rough Endoplasmic Reticulum

RERC Regional Educational Resource Center

RERE Radiotekhnika i Elektronika Radio Engineering (periodical code)

REREX Remote Readout Experiment

RERF Radiation Effects Research Foundation

RERIC Renewable Energy Resources Information Center

RERV Radar Estimated Rain Volume

res Resource

Res Rescue • Research • Researcher • Reservation • Reserve • Reserved • Reservoir • Residence • Resident • Residue • Resigned • Resistance • Resistor • Resolution • Resource • Resume

Rés Réserve • Réservé • Réserver • Résidence

RES Rachat d'Entreprise par les Salariés • Radar-Enhanced System • Radiation-Enhanced Sublimation • Radiation Exposure State • Radiation Exposure Status • Radio Emission Surveillance • Radio Equipment and Systems • Range Extension System • Record Evaluation System • Regional Earth Station • Regional Economic Study • Regional Evaporation Study • Remote Emitter System • Remote Evaluation System • Remote Execution Service • Renewable Energy Source • Renewable Energy Systems Ltd. • Request for Engineering Services • Réseau d'Échange de Savoir • Reserve • Reserved • Reservoir • Reset • Resistencia (Argentina-CH) (location identifier) • Resistor • Resolution • Resource • Restriction • Resume Message • Risk Estimate of Situation

RESA Research, Evaluation, and System Analysis

RESAC Regional Earth Science Applications Center

RESACENT Remote Sensing Applications Centre

RESAR Reference Safety Analysis Report

Resc Rescind

RESCEU Research Center for the Early Universe (Japan)

RESCU Remote Emergency Satellite Cellular Unit

Resd Residência • Residêncial

RESDA Remote Sensing Development Agency

RESEDA Remote Sensor Data Analysis

RESEED Retirees Enhancing Science Education through Experiment and Demonstration

resel Resolution Element

Resgnd Resigned

Resid Residual

RESID Residencial • Residential • Résidentiel • Residual

Resig Resignation

RESINS Redundant Strap-down Inertial Navigation System

RESIS Remote Sensing Information Sensing • Resistance

Resln Resolution

ReSLT Reusable Space Launch Technology

RESMO Respiratory Monitoring

Resn Resonant

RESOLV Resolver

RESOMS Resource Scheduling and Operational Management System

RESORS Remote Sensing Online Retrieval Service • Remote Sensing Online Retrieval System (*)

Resp Respectif • Respective • Respectively • Respectivement • Respektive • Respondent • Response • Respublika

ReSp La Recherche Spatiale (periodical code)

RESP Response • Responsibility

RESPAS Remote Sensing Preprocessing and Archiving System • Remote Sensing Processing and Archiving System

RESPECT Rapid Electromagnetic Scattering Predictor for Extremely Complex Targets

Ress Ressource

RESSAC Radar pour l'Étude du Spectre des Surfaces par Analyse Circulaire

RESSYS Regional Ecosystem Simulation System

Rest Restaurado • Restauré • Restoration • Restored • Restrict • Restricted • Restriction

REST Reader Evaluation and Selection Technique • Restart

RESTEC Remote Sensing Technology Center (Japan)

Restr Restaurant • Restorer • Restrict • Restricted • Restriction

RESVR Reservoir

Ret Retail • Retain • Retained • Reticulum-Reticuli (constellation) • Retire • Retired • Retórica • Return • Returnable • Returned

RE&T Research Engineering & Test

RET Reliability Evaluation Test • Renewable Energy Technology • Research Effort/Test • Research Experiment for Teachers (project) • Resolution Enhancement Technology • Retire • Retired • Retract • Return • Røst (Norway) (location identifier)

RETA Regional Environmental Technical Assistance • Retrieval of Enriched Textual Abstracts • Revised Estimated Time of Arrival

RETAIN Remote Technical Assistance and Information Network

RETC Regional Emergency Transportation Center

RETCO Regional Emergency Transportation Coordinator

Retd Retained • Retired • Returned

RETD Red Especial de Transmisión de Datos (Spain)

RETE Research on Electrodynamic Tether Effects

RETEN Retention

RETF Radiation Effects Test Facility

RETIM Réseau Télé-Informatique de la Météorologie

RETINA Réseau Téléinformatique National (France)

RETM Rare-Earth Transition Metal

Retn Retain

Retnr Retainer

ReToh Scientific Reports of Tohoku University (periodical code)

RETOOLE Remote Tools for EUVE

RETP Research Experience in Teacher Preparation

Retr Retract • Retractable • Retrieval • Retrieve • Retrieved

RETR Retrieval • Retrieve

Retrans Retransmission

Retro Retroactive • Retrofit • Retrograde

RETRO Retrofire • Retroreflector

RETRV Retrieve

RETS Reconfigurable Electrical Test Stand • Regional Emergency Transportation System

RETV Retrieval

REU Île de la Réunion (code ISO de pays) • Remote Electronics Unit • Remote Emitter Unit • Research Experience for Undergraduates • Research Evaluation Unit • Réunion Island (ISO country code) • Reus (Spain) (location identifier)

REUN Réseau Européen Unifié de Nivellement

REUNIR Réseaux des Universités et de la Recherche

Rév Révérend • Révision • Révolution

Rev Revelation • Revenue • Reverend • Reverse • Reversed • Review • Reviewed • Revisado • Revise • Revised • Revision • Revisión • Revolution • Revolve • Revolving • Revue

REV Reentry Vehicle • Relative Esteem Value • Reverendo • Reverse • Review • Revision • Revolution

REVA Recommended Vehicle Adjustment

REVAD Royal Observatory Edinburgh (Starlink internal code)

REVCON Revolutionary Concept

REVP Research Expansion Verification Program • Research Experimental Verification Program

REVS Requirements Engineering and Validation System

REVTOLC Retrieval Experiment - Virginia Tech Online Catalogue

Rew Reward • Rewind

REW Rest (frame) Equivalent Width • Rewa (India) (location identifier) • Rewind

REX Radiation Experiment • Radio-Emitting X-ray (source) • Radio (science) Experiment • Readiness Exercise • Relocatable Executable • Retour d'Expérience • Reynosa (Mexico) (location identifier) • RIU Exerciser

rexec Remote Execute

REXEC Remote Executable • Remote Execution

REXMIT Retransmit • Retransmitted

REXRB Radio-Emitting X-Ray Binary

REXX Restructured Extended Executor

REY Reyes (Bolivia) (location identifier)

rf Radio Frequency

Rf Reef • Rufiyaa (currency) • Rutherfordium

RF Radio Facility • Radio Fréquence • Radio Frequency • Range Finder • Rating Factor • Raumfahrtforschung (periodical code) • Read Forward • Receiver Function • Reception Fair • Redistribution Function • Reducing Flame • Reef • Relationship Factor • Relative Flow • Released Frontier • Relevance Feedback • Replacement Factor • Reporting Frequency • Representative Fraction • Reproduction Fidelity • Reprographic Facility • République Française • Research Foundation • Responsable de Formation • Response Force • Retention Factor • Return Flux • Reusable Function • Right Forward • Risk Factor • Roll Film • Rwanda Franc (currency)

RFA Rdaio Frequency Authorization • Rafai (Central African Rep.) (location identifier) • Rate of False Alarms • Relieved From Assigned • Remote File Access • République Fédérale d'Allemagne • Request For Action • Request For Algorithm • Request For Application • Request For Approval • Request For Assistance • Request for Final Approval

RFAA Relieved From Attached and Assigned

RFAD Relative Frequency in All Documents

RFB Ready For Baseline • Recording For the Blind • Remote Frame Buffer • Request For Bid

RFB&D Recording For the Blind & Dislexic

RFBD Recording For the Blind and Dislexic

RFBR Russian Foundation for Basic Research

RFC Radio Frequency Cabinet • Radio Frequency Chart • Radio Frequency Choke • Radio Frequency Compatibility • Recommended Flight Clearance • Regenerative Fuel Cell • Remote Function Call • Request For Change • Request For Comments • Request For Connection • RGS Focal Camera • Royal Flying Corps

RFCA Radiator and Flow Control Assembly • Radiator Flow Control Assembly

RFCM Radio Frequency Countermeasure

RFCP Radio Frequency Compatibility Program • Request For Computer Program

RFCT Radio-Frequency Compatibility Test

RFD Radio-Frequency Device • Range Focused Doppler • Ready For Data • Ready For Delivery • Ready For Deviation • Ready For Discussion • Reporting For Duty • Request For Delivery • Request For Deviation • Request For Discussion • Request For Documentation • Requirement Formulation Document • Revue Finale de Définition • Rockford - Greater Rockford (USA-IL) (location identifier) • Rural Free Delivery

RFDA Radio Frequency Deployed Assembly

RFDD Range Focused Doppler Display

RFDIU Radio Frequency Digital Interface Unit

RFDL Radio Frequency Data Link

RFDP Rotating Full Disk Photometer

RFDS Royal Doctor Flying Service (Australia)

RFDU Radio Frequency and Digital Unit • Radio Frequency Distribution Unit

RFE Radar Front End • Regional Finite Element • Request For Enhancement • Request For Estimate • Request For Exception • Requisition For Expenditure

RFEA Request For Equitable Adjustment

RFECP Request For Engineering Change Proposal

RFEEI Radio Frequency Engineers Education Initiative (UK)

RFEI Request For Engineering Information

RFF Remote File Facility • Request For File • Request For Funds • Resources For the Future

RFG Radar Field Gradient • Radio Frequency Gateway • Refugio Rooke (USA-TX) (location identifier) • Rentenreformgesetz

RFGC Revised Flat Galaxy Catalogue

Rfgt Refrigerant

RFH Remote Frame Handler

RFHA Reusable Flight Hardware Authorization

RFI Radio Frequency Interface • Radio Frequency Interference • Radio Frequency Interferometer • RCRA Facility Investigation • Ready For Installation • Ready For Issue • Relative Fluorescence Intensity • Remote Facility Inquiry • Remote File Inquiry • Representative of a Foreign Interest • Request For Information • Request For Inspection • Request For Intelligence

RFIC Radio-Frequency Integrated Circuit

RFID Radio-Frequency Identification

RFIS Radio Frequency Instrument System • Remote Frequency Indicator Selector • Remote Frequency Interface System

RFIT Radio Frequency Interference Test

RFK Anguilla Rollang (USA-MS) (location identifier)

RFL Reflect

Rflg Refueling

RFLP Restriction Fragment Length Polymorphism

Rflx Reflex

RFM Reactive Factor Meter • Receiver Function Method • Reconfigurable Furnace Module • Reconnection Flare Model • Refocusing Mechanism • Reliable Flow Manager • Remote Fault Multiplexer • Repair Forecasting Model • Replay File Management

RFML Radio Frequency Measurements Laboratory

RFMS Radio Frequency Management System • Remote File Management System • Research Flight Management System • Risk Feedback Management Strategy

RFMSS Range Facility Management Support System

RFMV Radio Frequency Measurements Van

RFN Radio Frequency Noise • Raufarhöfn (Iceland) (location identifier) • Reflection Nebula

RFNA Red Fuming Nitric Acid

RFNM Ready For Next Message • Request For Next Message

RFO Reason For Outage • Request For Offer • Request For Order • Request For Outage

RFOG Resonator Fiber Optic Gyro

RFOI RF-Optical Interface

RFOV Resolution Field Of View

RFP Raiatea Island (Society Islands) (location identifier) • Raumflugplanetarium (Halle, Germany) • Request For Payment • Request For Procurement • Request For Product • Request For Program • Request For Proposal • Requirements and Formulation Phase • Requirements Formulation Phase • Reversed-Field Pinch • Rutgers Fabry-Pérot (interferometer)

RFPA Request For Proposal Authorization

RFPC Radio-Frequency Pulse Compression

RFPITS Radio Frequency Portable Input Terminal

RFPSO Request For Proposal Support Office • Request For Proposal Support Organization

RFPU Request For Proposal Update

RFQ Radio-Frequency Quadrupole • Request For Quotation • Request For Quote

RFR Radio Frequency Radiation • Rate of False Resolutions • Refrigerator Freezer Rack • Réseau Formation-Recherche • Retract Feed Rate • Rio Frio (Costa Rica) (location identifier)

Rfrd Referred

RFREQ Receive Frequency

RFrM Revue Française de Mécanique (periodical code)

RFrMQ Revue Française de Mécanique Quarterly (periodical code)

RFS Radio Frequency Sensor • Radio Frequency Shift • Radio Frequency Surveillance • Radio Frequency System • Ready For Sea • Règle Fondamentale de Sûreté • Remote File Service • Remote File Sharing • Remote File System (*) • Renormalized Forward Scattering • Request For Service • Request For Shipping • Request For Specification • Research Flight Simulator • Response Feedback System

RFSC Rubberized Friction Seal Coat

RFSE Radio Frequency Shielded Enclosure

RFSH Refresh

RFSM Radio Frequency Spectrum Management

RFSOC Radio Frequency Simulation Operations Center

RFSS Radio Frequency Switching Service

RFT Ready For Training • Real Fourier Transform • Request For Technology • Request For Tender • Revisable Form Text • Richest-Field Telescope • Rich-Field Telescope

RFTC Radio Frequency Test Console

RFTCP Request For Test Change Proposal

RFTF Radio Frequency Test Facility

RFTG Reference Frequency Timing Generator

RFTI Radio Frequency Tags and Interrogators

RFTP Request For Technical Proposals • Research for Future Technological Programs

RFTPCE Reduced-Fill Tank Pressure Control Experiment

RFTS Radio Frequency Test Set • Radio Frequency Transmission System • Remote Fiber Test System

RFU Radio Frequency Unit • Ready For Use • Reserved for Future Use

RFV Radiometro de Frequencia Variavel • Request For Visit

RFW Request For Waiver

RFY République Fédérale de Yougoslavie

Rg Rayon Gravitationnel

RG Radialgeschwindigkeit • Radio Galaxy • Radio Guide • Range • Rate Gyro • Rate Gyroscope • Readiness Group • Reconnaissance Group • Reconstitution Group • Recursive Grid • Red Giant (star) • Régimen General • Regular Grammar • Regulatory Group • Renormalization Group • Renseignements Généraux (France) • Reports Generator • Requirement-Gathering • Response Generator • Reticulated Grating • Rio Group • Rive Gauche

RGA Rate Gyro Assembly • Reduction Gas Analyzer • Reflection Grating Array • Relative Growth Rate • Remote Geological Analysis • Reset Group Acknowledgment • Residual-Gas Analyser • Residual-Gas Analysis • Return Goods Authorization • RGS Grating Assembly • Río Grande (Argentina-TF) (location identifier)

RGAI Research Group on Artificial Intelligence

rgb RGB Image (file extension)

RGB RASS - Green Bank (survey) (*) • Red-Giant Branch • Red - Green - Blue • ROSAT - Green Bank (survey)

rgba RGB Image (file extension)

RGBB Red Giant Branch Bump

RGBBL ROSAT - Green Bank BL Lacertae (object)

Rgbf Rangierbahnhof

RGC Red Globular Cluster • Reference for Galaxy Clusters • Reference Great Circle • Remote Global Contact • Repair Group Category

RGCPA Rapporteur Group on Coordination of Profile Activities

RGCS Review of the General Concept of Separation

RGCT Rapporteur Group on Conformance Testing • Reflective Gross Cloud Test

RGC&U Réseau en Génie Civil & Urbain

RGCV Randomized Generalized Cross-Validation

Rgd Registered • Rigid

RGD Ragged • Raleigh - Gans - Debye (criterion) • Rarefied Gas Dynamics • Regular Geophysical Day • Rijksgebouwendienst • Rijks Geneeskundige Dienst • Rijks Geologische Dienst (Netherlands)

Rge Range • Ridge • Rouge

RGE Porgera (Papua New Guinea) (location identifier) • Range

RGF Référentiel Géologique de France • Regional Growth Forecast • Remote Ground Facility

RGH Balurghat (India) (location identifier)

RGI Rangiroa Island (Tuamotu Islands) (location identifier) • Regional-Global Interaction

RGIC Rare Gas Ionization Cell

RGIS Regional Geographic Information System

RGISM Raster Geographic Information System for Mapping

RG&K Reference Guides & Keys

RGK Reference Guides and Keys

RGL Rate Gyroscope Limit • Reading Grade Level • Report Generator Language • Richardson Grating Laboratory (USA-NY) • Rio Gallegos (Argentina-SC) (location identifier)

Rglr Regular

Rglt Regulate

Rgltd Regulated

Rgltr Regulator

RGM Redundant Gyro Monitor • Reflection Grating Module • Remote Geophysical Monitor

RGMS Rrare Gas Mass Spectrometry

RGMT Regiment

Rgn Rangoon (Myanmar) • Region

RGN Rangoon Mingaladon (Myanmar) (location identifier) • Region • Yangon Mingaladon (Myanmar) (location identifier)

RGNP Real Gross National Product

RGO Royal Greenwich Observatory (UK-England) (now closed)

RGOB Royal Greenwich Observatory Bulletin (periodical code)

RGP Raster Graphics Processor • Rate Gyro Package • Reduced Game Property • Reliability Growth Program • Research and Graduate Program • Réseau GPS Permanent

RGPS Radarsat Geophysical Processing System • ROSAT Galactic Plane Survey

RGR Ranger • Ranger (USA-TX) (location identifier) • Recursive Grid Refinement • Regueira • Relative Growth Rate • Repairable Generator Rate

Rgrd Regard

Rgrds Regards

RGRF Repair Generation Rate Forecast

RGRMA Rate Gyro Redundancy Management Algorithm

RGRP Regional Geochemical Reconnaissance Programme

RGS Radio Guidance System • Rate Gyro System • Receiving Ground Station • Reflection Grating Spectrometer • Reflection Grating System • Remote Ground Switching • Requirements Generation System • Restructured General Support • Rocket Guidance System • Royal Geographical Society (UK)

RGSC Ramp Generator and Signal Converter

RGSS Régimen General de la Seguridad Social

Rgstn Registration

Rgstr Register

Rgstrd Registered

Rgstry Registry

RGT Regiment • Relativistic Gravitational Theory • Reliability Growth Test • Reliability Growth Testing • Remote Ground Terminal • Rengat (Indonesia) (location identifier) • Renormalization Group Theory • Requirements Growth Testing • Réseau de Gestion des Télécommunications • Resonant Gate Transistor • Right

Rgtr Register

RGU Rate Gyroscope Unit • Rechner-Gestützter Unterricht • Robert Gordon University (UK-Scotland) • Rot - Gelb - Ultraviolet

RGVTDB Rapidly Generated Virtual Terrain Database

RGWS Remote Graphical Workstation

RGZ Recommended Ground Zero

Rh Rhesus • Rhodium

R-H Rankine-Hugoniot (condition)

RH Haiti (international vehicle registration code) • Radiation-Hardened • Radiation-Hazard • Radiative-Hydrodynamic • Radio-Héliographe • Radiological Health • Rankine-Hugoniot (condition) • Razdow Heliograph • Reentry Home

• Relative Humidity • Remote Handling • Request Header • Response Header • Ressources Humaines • Rheostat • Riemann-Hilbert • Right Half • Right Hand • Right-Handed

RHA Rama Horizontal Azul • Records Holding Area • Reykholar (Iceland) (location identifier)

RHAPIT Rhine/Main Area Project for Integrated Traffic

RHB Red Horizontal Branch

RHBNC Royal Holloway and Bedford New College (UK-England)

RHC Rao - Hartley - Cochran (method) • Remote Hand Controller • Right-Hand Circular • Rotational Hand Controller • Rotation Hand Controller

RHCM Relative Humidity Control/Monitor

RHCP Right-Hand Circularly-Polarized • Right-Hand Circular Polarization

RHCTO Rotational Hand Controller Takeover

Rhd Railhead

RHD Radiation Hydrodynamics • Removable Hard Disk • Response Hold Delay • Right-Hand Drive • Rio Hondo (Argentina-SE) (location identifier)

RHDD Radiation Hardened Disk Drive • Removable Hard Disk Drive

RHDP Right Hand Decimal Point

RHE Radiation Hazards Effect • Raman Holographic Edge • Reims (France) (location identifier)

RHEB Right-Hand Equipment Bay

RHEC Radiation-Hardened Emulating Computer

RHEED Reflected High-Energy Electron Diffraction • Reflection High-Energy Electron Diffraction

Rheinhess Rheinhessen (Germany)

Rheinl Rheinland (Germany)

Rheo Rheostat

RHEO Rheostat

RHESSI Ramaty High-Energy Solar Spectroscopic Imager

Rhet Rhetoric • Rhetorical

RHET Resonant Hot Electron Transistor

RHF Regular High Frequency • Relativistic Hartree-Fock (approximation)

RHFEB Right-Hand Forward Equipment Bay

RHG Residual Halo Gas • Ruhengeri (Rwanda) (location identifier)

RHI Radar Height Indicator • Range-Height Indicator • Relative Humidity Indicator • Rhinelander (USA-WI) (location identifier)

RHIB Rigid Hull Inflatable Boat

RHIC Relativistic Heavy-Ion Collider (*) • Relativistic Heavy-Ion Collision

RHIP Rank Has Its Privileges

RHISE Remote Hitchhiker Investigation Support Environment

RHIT Rose-Hulman Institute of Technology (USA-IN)

RHKO Royal Hong Kong Observatory (now HKO)

RHL Ranked Half-Life • Residual Hazards List • Roy Hill (Australia-WA) (location identifier)

RHM Relative Humidity Meter • Relative Humidity Monitor • Rodent Habitat Module

Rho Rhodesia

RHO Rhodes (Greece) (location identifier) • Rosemary Hill Observatory (USA-FL)

Rhod Rhodesia

RHP Ramechhap (Nepal) (location identifier) • Reconfigurable Hardware Product • Reduced-Hazard Propellant • Revised Harvard Photometry • Riemann-Hilbert Problem • Right-Hand Panel

RHPM Red Hat Package Manager

RHQ Regional Headquarters

r/hr Roentgens per Hour

RHR Radio Horizon Range • Rejectable Hazard Rate • Residual Heat Removal • Right-Hand Rule

RHRI ROSAT High-Resolution Imager

RHRZ Regionales Hochschulrechenzentrum

RHS Radio Hot Spot • Right-Hand Side • Riverview High School (USA-FL) • Rocketdyne Hybrid Simulator • Round-Headed Screw

RHSC Right-Hand Side Console

RHSW Rankine-Hugoniot Shock Wave

RHT Radiant Heat Temperature • Reynolds Hydrodynamic Theory • Right-Hand Turn

RHTW Remote-Handled Transuranic Waste

RHU Radioactive Heater Unit • Radio Heating Unit • Radioisotope Heater Unit (*) • Replacement Holding Unit

RHV Right-Handed Vehicle • San Jose - Reid-Hillview (USA-CA) (location identifier)

RHW Response Header Word • Router Header Word

RHZ Root Hair Zone

Ri Richardson Number

R&I Receiving & Inspection • Regina & Imperatrix • Removal & Installation • Rex & Imperator

R/I Receiving and Inspection • Receiving Inspection

RI Indonesia (international vehicle registration code) • Radiation Intensity • Radio Interference • Radioisotope • Range Instrumentation • Raw Image • Read-In • Real Image • Receiving Inspection • Recherche d'Information • Recomendação Internacional • Referential Integrity • Reflective Insulation • Refractive Index • Regina et Imperatrix • Register Immediate • Relational Indexing • Relative Inertial • Relative Intensity • Reliability Index • Remedial Investigation • Remote Inhibit • Repeated Introduction • Reply Interrogate • Report of Investigation • Représentation Intermédiaire • Request for Information • Research Infrastructure • Resistance Inductance • Restricted Interpretation • Reverberation Index • Rex et Imperator • Rhode Island (USA) • Right Intermediate • Ring In • Ring Indicator • Ring Input • Rockwell International Corporation • Routine Indicator • Routing Identifier • Routing Indicator • Routing Information • Royal Institution

RIA Range Insensitive Axis • Requirements Integration Approval • Robotic Industries Association • Robot Institute of America • Royal Irish Academy • Runway Incursion Alert • Santa Maria (Brazil-RS) (location identifier)

RIAA Recording Industry Association of America (USA)

RIACA Research Institute for Applications in Computer Algebra

RIACS Research Institute for Advanced Computer Science (USA-CA)

RIAD Reply Interrogate Acknowledge Data

RIAGG Regionaal Instituut voor de Ambulante Gestelijke Gezonheidszorg

RIAM Recherche et Innovation en Audiovisuel et Multimedia (project) (France)

RIAO Recherche d'Information Assistée par Ordinateur

RIAR Requirement Inventory Analysis Report

RIAS Radar à Impulsion et Antenne Synthétiques • Radio In the American Sector

RIASS ROSAT-IUE All Sky Survey

RIAV Reply Interrogate Acknowledge Voice

RIB Radar Interface Buffer • Radioactive Ion Beam • Recoverable Item Breakdown • Relevé d'Identité Bancaire

• Research Internet Backbone • Réseau Informatique des Bibliothèques (France) • Riberalta (Bolivia) (location identifier) • Right Inboard • Routing Information Base

RIBE Reactive Ion Beam Etching

RIBOTS Riken-Bisei Optical Transient Seeker

RIBS Radar Interface Buffer System

Ric Ricerca

RIC Radiation-Induced Change • Radiation-Induced Conductivity • Radio Identify Code • Record Interval Counter • Regional Information Center • Relative In Centrality • Remote Indicating Compass • Remote Information Center • Rencontre Inter-Clubs • Request for Instrument Clarification • Requirement Identity Code • Requirements Integration Council • Research and Information Centre • Resistance - Inductance - Capacitance • Resource and Implementation Cooperative • Resource Identification Code • Resource Information Center • Richmond (USA-VA) (location identifier) • Rockwell International Corporation • Routing Identification Code • Routing Identifier Code • Routing Indicator Code • Rx-band Interchange Computer

RICA Radar Image Classification Aid • Regole Italiane di Catalogazione per Autori

RICC Remote Instrument Control Center • Remote Intercomputer Communications • Reportable Item Control Code

RICE Radio Ice Čerenkov Experiment • Relevé d'Identité de Caisse d'Épargne • Remote Inertial and Control Equipment

RICERCAR Reading Initials Clearly Exhibits RACRECIR's Concealed Auto-Reference • Rigid Internal Codes Exclusively Rule Computers and Robots

RICH Ring-Imaging Čerenkov

RICIS Research Institute for Computing and Information System

RICL Roditi International Corporation Ltd. (UK)

RICP Real-time Interactive Computer Program

RICS Range Instrumentation Control System • Remote IMAX Camera System • Royal Institution of Chartered Surveyors (UK)

RICST Real-time Interferometer Control Software Testbed

RID Rack Insertion Device • Radial Immunodiffusion • Radiation-Induced Defect • Radiation-Induced Diffusion • Range Instruments Development • Refractive Index Detector • Remote Isolation Device • Report Identifikationsnummer • Request for Information Disposition • Requirements Integration Directorate • Requirements Interface Document • Reset Inhibit Drive • Review Item Discrepancy (*) • Review Item Disposition • Richmond (USA-IN) (location identifier) • Routing Identification • Routing Identifier • Row Identifier

RIDA Raster Image Device Accelerator • Reverse Isotope Dilution Assay

RIDAC Range Interference Detection And Control

RIDB Readiness Integrated Database

RIDDLE Rapid Information Display and Dissemination in Library Environnements

RIDE Research Issues in Data Engineering

RIDF Random Input Describing Function

RIDGE Ridge Interdisciplinary Global Experiment

RIDI Receiving Inspection Detailed Instruction • Receiving Inspection Detail Instruction

RIDS Radar Image Display System • Radio Information Distribution System • Range Information Display System • Readiness Improvement Data System • Receiving Inspection Data Status • Rules In Database Systems (workshop)

RIDT Raster Imaging and Digital Typography (conference) • Réseau Interrégional de Diffusion Technologique (France)

RIE Radar Interface Equipment • Range of Incentive Effectiveness • Reactive Ion Etched • Reactive Ion Etching (*) • Rice Lake (USA-WI) (location identifier) • Right Inboard Elevon

rif Resource Interchange Format (file extension)

Rif Rifugio

RIF Radio-Influence Field • Radio Interference Filter • Rate Increase Factor • Rate Input Form • Reading Is Fundamental (project) • Recherche dans l'Intérêt des Familles • Reduction In Force • Relative Impact Factor • Relative Importance Factor • Reliability Improvement Factor • Remote Interface • Renormalized Impact Factor • Réseau d'Information sur la Formation • Resource Interchange Format • Richfield (USA-UT) (location identifier) • Risk Information Form • Routing Information Field

RIFA Retry, Ignore, Fail, Abort?

RIFF Raster Image File Format • Resource Interchange File Format

RIFI Radio Interference Field Intensity • Radio-Interference-Free Instrument

RIFIM Radio Interference Field Intensity Meter

RIFL Random Item File Locator

RIFMA Roentgen-Isotope Fluorescent Method of Analysis

RIFS Radioisotope Field Support

RIFT Reactor-In-Flight Test

RIFTS Reactor-In-Flight Test System

RIFU Richtfunk

RIG Radio Inertial Guidance • Rate-Integrating Gyro • Rate Integration Gyro • Regional Information Group • Remote Imaging Group • Research Institute for Fundamental Physics • Reuse Interoperability Group • Rio Grande (Brazil-RS) (location identifier) • Rivista Italiana di Geofisica (periodical code)

RIGA Rate-Integrating Gyro Assembly

RIGES Renewable Intensive Global Energy System

RIGGS Ross Ice Shelf Geophysical and Glaciological Survey

RIGI Receiving Inspection General Instruction

RIGIS Relational Image-based Geographic Information System

RIGS Radio Inertial Guidance System

RIGSA Rivista Italiana di Geofisica e Scienze Affini (periodical code)

RIGTC Research Institute of Geodesy, Topography and Cartography

RIHED Regional Institute of Higher Education and Development

RII Receiving Inspection Instructions • Relevant Information and Intelligence • Request for Intelligent Information • Required Inspection Item • Routing Information Indicator

RIIA Royal Institute of International Affairs

RIIM Recoverable Item Inventory Manager

RIIV Red de Información Industrial de Venezuela

RIIXS Remote Interrogation Information Exchange System

RIJ Rioja (Peru) (location identifier)

RIK Carillo (Costa Rica) (location identifier) • Replacement In Kind

RIKES Raman-Induced Kerr Effect Scattering

RIKZ Rijksinstituut voor Kust- en Zeeonderzoek (Netherlands)

RIL Radio Influence Level • Radio Interference Level • Recherche et Ingénierie Linguistiques • Recoverable Items List • Red Indicator Light • Repairable Items List • Representation Independent Language • Rifle - Garfield County (USA-CO) (location identifier)

RILA Répertoire International de la Littérature sur l'Art

RILB Radio Interférence à Longue Base

RILE Réseau d'Initiatives Locales pour l'Emploi

RILEM Réunion Internationale des Laboratoires d'Essais sur les Matériaux

RILIS Resonance Ionization Laser Ion Source

RILM Répertoire International de Littérature Musicale

RILO Research and Industrial Liaison Officer

RILS Rapid Integrated Logistics System • Regularized Iterative Least Squares • Resident Integrated Logistics Support

RILSA Resident Integrated Logistics Support Activity

RILSS Rapid Integrated Logistics Support System

R/IM Refrigerator/Incubator Module

RIM Mauritania (international vehicle registration code) • Read-In Mode • Read Interrupt Mask • Receipt - Inspection - Maintenance • Regional Ice Model • Rekaleszenz In Matrix • Reliability and Integrity of Media • Remote Installation and Maintenance • Requirements Implementation Manual • Requirements Integration Manager • Resource Inventory Management • Retail Item Manager • Retainable Instructional Material • Rodriguez de Men (Peru) (location identifier)

RIMC Repairable Item Movement Control

RIME Real-time Information Management Environment • Relaynet International Message Exchange

RIMF Remote Integrated Medical Facility

RIMI Research Improvement in Minority Institutions

RIML Rule Identification Markup Language

RIMP Remote Input Message Processor

RIMPAC Rim of the Pacific

RIMS Radiant Intensity Measuring System • Radio Interference Measuring System • Records Information Management System • Rescue Information Management System • Research Institute for Mathematical Sciences • Resonance Ionization Mass Spectrometry • Resource Information Management System • Retarding Ion Mass Spectrometer

RIMU Redundant Inertial Measurement Unit

RIN Rapporteur Group for Internationalization • Record Identification Number • Reference Indicator Number • Regular Inertial Navigator • Relative Intensity Noise • Remote Intelligent Network • Report Identification Number • Research Internet • Resource Information Network • Return Incident Number • Ringi Cove (Solomon Islands) (location identifier) • Routing Instruction Note • Royal Institute of Navigation (UK)

RINAL Radar Inertial Altimeter

RINET Reinsurance Network

RINEX Receiver Independent Exchange (format)

RING Relational Information Grouping

RINT Réseau International de Néologie et de Terminologie

RINU Replacement Inertial Navigation Unit

RIO Radar Intercept Officer • Raytheon Infrared Operations • Reference Identifier Object • Regional Information office • Relative Ionospheric Opacity • Reliable Interoperability • Reporting In and Out • Réseau Intertropical d'Ordinateurs • Rio de Janeiro (Brazil-RJ) (location identifier) • Russian Interface Officer

RIOA Real Instituto y Observatorio de la Armada (Spain)

RioB Observatorio Nacional - Publicações do Serviço Astronômico, Rio de Janeiro (periodical code)

Rioj Riojano

RIOS Radio, Infrared and Optical Sensors • Rotating Image Optical Scanner

RIOT Real-Time Input-Output Transducer • Real-Time Input-Output Translator • Remote Input-Output Terminal • Remotely Independently Operable Transducer • Retrieval of Information by Online Terminal

RIP Radar Image Processor • Raster Image Processing • Raster Image Processor (*) • Raster Input Processor • Readiness Improvement Program • Receiving Inspection Plan • Receiving Inspection Procedure • Recognition Improvement Program • Register Indicator Panel • Regulations of Investigatory Powers • Regulatory and Information Policy • Relevé d'Identité Postal • Reliability Improvement Program • Remain In Place • Remote Imaging Protocol • Repair In Place • Report on Individual Personnel • Requiescat In Pace • Research Initiation Program • Respiratory Impedance Plethysmograph • Rest In Peace • Router Information Protocol • Routine Information Protocol • Routing Information Protocol (*) • Routing Interchange Protocol • Routing Interface Protocol

RIPAS Regional Integrated Protected Areas System

RIPC Remote Information Processing Center • Remote Interprocess Communication

RIPD Remote Input Pointing Device

RIPE RACE Integrity Primitive Evaluation • Range Instrumentation Performance Evaluation • Réseaux IP Européens

RIPFCOMTF Rapid Item Processor to Facilitate Complex Operations on Magnetic-Tape Files

RIPL Remote Initial Program Load • Representation Independent Programming Language

RIPM Répertoire International de la Presse Musicale

RIPP Regional Industrial Property Programme

RIPPL Recoverable Item Provisioning Part List

RIPR Report-In-Process Review

RIPS Radar Impact Prediction System • Radioisotope Power Supply • Radioisotope Power System • Range Instrumentation Planning Study • Raster Image Processing System • Remote Image Processing System • Remote Information Processing System • Road Ice Prediction System

RIPSO Revised Internet Protocol Security Option

RIPT Requirement Integrated Product Team • Réseau d'Informations Pédagogiques et Techniques

RIQ Radio Intermediate QSO • Radio Intermediate Quasar

RIQS Remote Information Query System

RIR Receiving Inspection Report • Recherche Instrumentale en Radioastronomie • Reduction In Requirement • Refractive Index Range • Relative Index Register • Release Initiation Review • Reliability Investigation Request • Reportable Items Report • Riverside Fla-Bob (USA-CA) (location identifier)

RIRIG Reduced-Excitation Inertial Reference Integrating Gyro

RIRMS Remote Information Retrieval and Management System

RIRQ Routed Item Repair Quantity

RIRS Reliability Information Retrieval System

RIRT Russian Institute of Radionavigation and Time (Russia)

RIS Radar Information Service • Radiocommunication Information System • Rail Interface System • Random Input Sampling • Range Information System • Rapid Imaging Software Inc. (USA-NM) • Rapid Imaging System • Receiving Inspection Segment • Record Input Subroutine • Reference Information Store • Reflector In Space • Regional Information Service • Regional Innovation System • Regional Intensive Session • Registered Image Sequence • Regulatory Information Service • Regulatory Information System • Remote Information System • Report Identification Symbol • Reporting Identification Symbol • Research Information Service • Research Information System • Resonance Ionization Spectroscopy • Retroreflector In Space • Revolution Indicating System • Rishiri (Japan) (location identifier) • Rotating Image Scanner • Rural Information System

RISAR Rotation-Invariant Simultaneous Autoregressive

RISC RadioAstron International Science Committee (Russia) • RadioAstron International Scientific Council • Radioastronomy International Science Council • Radioastronomy International Scientific Council (*) • Reduced Instruction Set Chip • Reduced Instruction Set Computer • Reduced Instruction Set Computing • Remote Information Systems Center • Research Institute for Symbolic Computation • Resources, Information and Support Centre • Rockwell International Science Center (USA-CA)

RISD Richardson Independent School District (USA-TX)

RISDE Russian Institute of Space Device Engineering

Rise Rise hvezd Popularne vedecky astronomicky casopis Praha (periodical code)

RISE Radiative Input from the Sun to Earth • Radiative Input of the Sun to Earth (*) • Readiness Improvement Status Evaluation • Readiness Improvement Summary Evaluation • Readiness Improvement through Systems Engineering • Regional Initiative in Science Education • Register for International Service in Education • Relatively Invariant under Strategic Equivalence • Reliability Improvement Selected Equipment • Research Information Service for Education • Resources for Involving Scientists in Education (project)

RISLU Remote Integrated Services Line Unit

RISO Rocket Impacts on Stratospheric Ozone

RISOP Red Integrated Strategic Operations Plan

RISP Robot d'Intervention en Site Planétaire • Robot d'Intervention sur Site Planétaire • Ross Ice Shelf Project

RISRJ Report Ionosphere Space Research Japan (periodical code)

R&ISS R&I SIC&DH Simulator

RISS Real-time IR/EO Scene Simulation • Refractive Interstellar Scintillation • Relational Inquiry and Storage System

RISSAC Research Institute for Soil Science and Agricultural Chemistry

Rist Ristorante

RIST Radioactive Ion Source Testbed

RIT Radio (frequency) Ionization Thruster • Radiofrequency Ion Thruster • Radio-Immunotherapy • Radio Information Test • Radio Interface Terminal • Rate of Information Throughput • Raw Input Thread • Readiness Initiative Team • Real-time Integrated Testbed • Reclamation Insurance Type • Red Interamericana de Telecomunicaciones • Relative Ignition Temperature • Remote Imagery Transceiver • Request for Interface Tool • Request for Interface Tooling • Rio Tigre (Panama) (location identifier) • Rochester Institute of Technology (USA-NY)

RITA Radio-frequency Ionisation Thruster Assembly • Radio-frequency Ion Thruster Assembly • Recognition of Information Technology Achievement (UK) • Recoverable Interplanetary Transport Approach • Reinvented Triple Axis • Reusable Interplanetary Transport Approach

RITC Regional Information Technology Coordinator

RITE Rapid Information Technique for Evaluation • Responsibility - Integrity - Trust - Ethicality

RITN Real-time Information Transfer and Networking

RITS Radiatively Important Trace Substance • Rapid Information Transmission System • Remote Input Terminal System

RITSC Regional Information Technology Service Center

RITSEC Regional Information Technology and Software Engineering Center

RITSI Radiative Ignition and Transition to Spread Investigation

RITSS Raytheon Information Technology and Scientific Services

RITZ Regulatory and Investigative Treatment Zone • Remote Integrated Telescope Zentrum (ESO)

RIU Radar Interface Unit • Radio Interface Unit • Refractive Index Unit • Regulatory Impact Unit • Remote Interface Unit

Riv River

RIV Radio Influence Voltage • Recirculation Isolation Valve • Retrospective Informationsverbreitung • Ricci Inheritance Vector • River • Riverside March (USA-CA) (location identifier) • Riverview College Observatory (Australia-NSW)

RiVas Ridges and Valleys

RIVM Rijksinstituut voor Volksgezondheid en Milieuhygiëne (Netherlands)

RIVR River

RIW Recht der Internationalen Wirtschaft • Reliability Improvement Warranty • Riverton (USA-WY) (location identifier)

RIWG Requirements Integration Working Group

RIX Riga Spilve (Latvia) (location identifier) • Russia Internet Exchange

RIXOS ROSAT International X-ray Optical Survey

RIXT Remote Information Exchange Terminal

RIY Relative Ion Yield • Riyan Mukalla (Yemen) (location identifier)

RIZ Rio Alzucar (Panama) (location identifier)

RIZIV Rijksdienst voor Ziekte- en Invaliditeitsverzekering • Rijksinstituut voor Ziekte- en Invaliditeitsverzekering (Belgium)

RIzZi Riga Izdatel Zinatne (periodical code)

RJ Ramjet • Rayleigh-Jeans • Reaction Jet • Registered Jack • Reject • Rio de Janeiro (Brazil) • Rivet Joint • Road Junction

RJA Rajahmundry (India) (location identifier)

RJB Rajbiraj (Nepal) (location identifier)

RJC Reaction Jet Control

RJD Reaction Jet Device • Reaction Jet Driver • Rejected

RJDA Reaction Jet Driver Aft

RJDF Reaction Jet Driver Forward

RJE Rayleigh-Jeans Equation • Remote Job Entry

RJEC Reaction Jet Engine Control

RJEHM Regional Journal of Energy, Heat and Mass Transfer (periodical code)

RJEP Remote Job Entry Protocol

RJET Remote Job Entry Terminal • Russian Journal of Engineering Thermophysics (periodical code)

RJETS Remote Job Entry Terminal System

RJFC Revised Jagiellonian Field Catalogue

RJGC Revised Jagiellonian General Catalogue

RJH Rajshahi (Bangladesh) (location identifier) • Rigal - Jung - Heck (algorithm)

RJI Rajouri (India) (location identifier)

RJK Rijeka (Croatia) (location identifier)

RJNP Robert J. Novins Planetarium (USA-NJ)

RJO Remote Job Output

RJOD Reaction Jet OMS Driver

RJP Remote Job Processing • Rocket Jet Plume

RJS Remote Job System

RJTAM Russian Journal of Theoretical and Applied Mechanics (periodical code)

RJXB Radio Jet in X Binaries

RK Runge-Kutta (method)

RKA Rekey Agent • Rossiiskoe Kosmicheskoe Agentsvo (Russia) (now RAKA)

RKAS Richmond and Kew Astronomical Society (UK)

RKB Restricted Kinetic Balance

RKC Yreka (USA-CA) (location identifier)

RKD Rockland (USA-ME) (location identifier)

RKE København Roskilde (Denmark) (location identifier)

RKG Receiving Key Generator

RKH Rock Hill (USA-SC) (location identifier)

RKHS Reproducing Kernel Hilbert Space

RKI Rokot (Indonesia) (location identifier)

RKKY Ruderman - Kittel - Kasuya - Yosida (coupling)

RKL Rufklasse

RKM Runge-Kutta Method

RKN Runge - Kutta - Nyström (method)

RKO Range Keeper Operator • Sipora (Indonesia) (location identifier)

RKP Rockport - Aransas County (USA-TX) (location identifier)

RKR Poteau - Robert S. Kerr (USA-OK) (location identifier) • Rydberg - Klein - Rees (method)

RKS Raised Kolmogorov Spectrum • Relational Knowledge Structure • Robotics Kernel System • Rock Springs (USA-WY) (location identifier)

Rkt Rocket

RKT Ras Al Khaimah (UAE) (location identifier) • Rocket

RKU Yule Island (Papua New Guinea) (location identifier)

RKV Remote Keying Variable • Reykjavík (Iceland) (location identifier)

RKVDU Remote Keyboard Visual Display Unit

RKW Rockwood (USA-TN) (location identifier)

Rky Rocky

RKY Rokeby (Australia-QLD) (location identifier)

R-L Richardson-Lucy (algorithm)

R/L Rate Limited • Remote/Local • Return Link

RL Lebanon (international vehicle registration code) • Radiation Laboratory • Radiation Level • Radio-Loud • Radio Location • Radioluminescence • Radioluminescent • Raman Laser • Random Logic • Readiness Level • Reading List • Real Life • Receive Location • Reconfigurable Logic • Record Length • Reduced Level • Reference Library • Reinforcement Learning • Release Line • Remote Loop • Requirements List • Research Laboratory • Residual Layer • Resistance-Inductance • Resistor Logic • Retention Level • Return Link • Return Loss • Richardson-Lucy (algorithm) • Right Line • Roche Lobe • Rocket Launcher • Roll Lift • Runway Light

RLA Remote LAN Access • Remote Link Adapter • Repair-Level Analysis • Repair-Line Agreement • Revealed Literature Advantage • Rolla (USA-MO) (location identifier)

RLAM Relocatable Limited Area Model

RLAN Radio Local Area Network

RLAS Rocket Lunar Attitude System

RLAT Radiation Life Acceptance Testing

RLB Relocatable Binary Code

RLC Radio Launch Control • Radio Link Control • Regional Library Bureau (UK) • Release Complete • Remote Load Controller • Resistance - Inductance - Capacitance • Resistor - Inductor - Capacitor • Restricted Local Cleanest (method) • Roadway Lighting Committee (IESNA) • Run Length Coding

RLCC Remote Launch Control Center

RLCE Request Level Change En-route

RLCM Remote Line Concentrating Module • Remote Line Control Module

RLCS Radio Launch Control System

Rlct Reluctance

Rld Rolled

RLD Ready-to-Load Date • Received Line Detect • Reciprocal Linear Dispersion • Recodable Locking Device • Reload • Remote Link Director • Required Load Date • Richland (USA-WA) (location identifier) • Rijksluchtvaartdienst (Netherlands)

RLDA Relativistic Local Density Approximation

rle Run Length Encoded (file extension)

Rle Ruelle

RLE Request Loading Entry • Research Laboratory for Electronics • Research Laboratory of Electronics • Réseau Local d'Entreprises • Responsible Laboratory Engineer • Run Length Encoded • Run Length Encoding

RLEB Remote Launch Equipment Building

RLEMR Research Laboratory Engineering Materials Report (periodical code)

RLEO Request for Liaison Engineering Order

Rlf Relief

RLF Radio Luminosity Function • Relief • Reusability Library Framework • Robust Luminosity Function • Russian Long-term Flight

RLFOC Regional Laser and Fiber Optics Center

RLG Regional Liaison Group • Research Libraries Group • Ring Laser Gyro • Ring Laser Gyroscope • Rostock-Laage (Germany) (location identifier)

RLGA Ring Laser Gyro Assembly

RLGM Remote Loop Ground Multiplexer • Remote Loop Group Multiplex • Remote Loop Group Multiplexer

RLI Anniston Reilly (USA-AL) (location identifier) • Radiation Level Indicator • Regularized Linear Inversion (algorithm) • Réseau Local d'Information • Right/Left Indicator

RLIN Research Libraries Information Network

RLL Radio in the Local Loop • Radio Local Loop • Right Lower Lobe • Rocket Launcher Locator • Run Length Limited

RLLO Regional Library Liaison Officer

RLLS Runway Lead-in Lighting System

RLM Richardson-Lucy Method • Robert Lee Moore

RLMS Radar Land Mass Simulator

RLN Remote LAN Node

RLNZ Rate-Limited Non-Zero

RLO Relative Light Output • Research Liaison Officer • Returned Letter Office • Roche Lobe Overflow

Rloc Relocate

RLOF Roche Lobe Overflow

RLOGIN Remote Login

RLOS Radio Line Of Site

RLP Radio Link Protocol • Remote Line Printer • Resource Location Protocol • Rosella Plains (Australia-QLD) (location identifier) • Rotatable Log Periodic • Rotating Linear Polarization

RLPA Rotating Log Periodic Antenna

RLPME Research Laboratory on Precision Machinery Electronics (periodical code)

RLPV Recurrent Line Profile Variation

RLQ Radio-Loud Quasar • Right Lower Quadrant

Rlr Roller

RLR Recueil des Lois et Règlements

RLRC Regional Logistics Readiness Center

RLRIU Routing Logic Radio Interface Unit

Rls Release • Rial (currency)

RLS Radio Location Service • Radius of Landing Site • Raman Lidar System • Record-Level Sharing • Recursive Least Squares • Regional Library System • Regularized Least Squares • Release • Restless Legs Syndrome • Restricted Least Squares • Reweighted Least Squares • Rocket Launching System • Roll Limit Switch

RLSAC RAL (Astrophysics cluster) (Starlink internal code)

RLSBB Report Life Sciences - Biomedical Behavioral Science (periodical code)

RLSD Received Line Signal Detected • Released

Rlse Release

RLSE Release

RLSF Regularized Least-Squares Fitting (method)

RLSM Released Message • Responsible Logistics Support Manager

RLSMC Reactive Local Search for Maximum Clique

RLSMSG Released Message

RLSO Regional Logistics Support Office

RLSS Regenerative Life Support System

Rlt Relate

RLT Arlit (Niger) (location identifier) • Relating To • Remote Line Test • Return Line Tether • Rolling Liquid Transporter

Rltn Relation

Rltv Relative

RLU Bornite Upper (USA-AK) (location identifier)

Rlv Relieve

RLV Reusable Launch Vehicle • Ringleitungsverteiler • Roving Lunar Vehicle

RLVAD RAL (Project cluster) (Starlink internal code)

RLVS Recoverable Launch Vehicle Structure

RLW Regie der Luchtwegen • Regularized Long Wave

Rlx Rouleaux

RLX Railway Crossing

Rlxn Relaxation

Rly Railway • Relay

RLY Relay

rm Raummeter • RealAudio (file extension) • Remove • Rhaeto-Romance (ISO language code)

Rm Magnetic Reynolds Number • Ream • Remarks • Room

R&M Redistribution and Marketing • Reliability & Maintainability • Reliability & Maintenance • Reliability & Marketing • Replacement & Modernization • Reports & Memoranda • Requirements & Materiel

R/M Reliability/Maintainability

RM Madagascar (international vehicle registration code) • Radar Mapper • Radial Marking • Radiation Measurement • Radiation Monitor • Radiation Monitoring • Radioactive Material • Radio Maintenance • Radio Modem • Radio Monitoring • Rate Management • Rate Monotonic • Raw Material • Readout Matrix • Receiving Memorandum • Record Management • Record Marking • Recovered Memory • Redundancy Management • Reed-Muller • Reference Material • Reference Method • Reference Mission • Reference Mode • Reference Model • Reference Monitor • Reference Monument • Referent Model • Regional Manager • Regional Meeting • Registre des Métiers • Registro Mercantil • Règlement Mensuel • Reichsmark (currency) • Reject Message • Relational Model • Relationship Management • Relative Mobility • Relay Mirror • Reliability Maintainability • Remedial Maintenance • Remontée Mécanique • Remote Manipulator • Remote Module • Remote Monitoring • Rendezvous Maneuver • Replaceable Module • Reprogrammable Miniprocessor • Requirements Management • Rescue Module • Research Memorandum • Reservation Management • Reset Mode • Residual Map • Résonnance Magnétique • Resource Maintenance • Resource Management • Resource Manager • Resource Module • Resource Monitor • Resupply Module • Retrieval Mode • Richtmeyer-Meshkov (instability) • Right Mid • Risk Management • River Mile • Roman • Room • Rotational Modulation • Rotation Measure • Rotation Module • Routine Maintenance • Royal Mail

R/M/A Reliability/Maintainability/Availability

RM&A Reliability, Maintainability & Analysis • Reliability, Maintainability & Availability

RMA Radio Manufacturers Association (USA) • Rapid Mass Accretion • Rate Monotonic Analysis • Realizable Multiplex Advantage • Rear Maintenance Area • Record Management Application • Relaxation Map Analysis • Reliability - Maintainability - Availability • Remote Manipulation Activity • Remote Manipulator Arm • Return Material Authorization • Return Merchandise Authorization • Return to Manufacturer Authorization • Right Mental Attitude • Risk Management Agency • Risk Management Application • Roma (Australia-QLD) (location identifier) • Roper Mountain Astronomers (USA-SC) • Royal Military Academy

RMAA Revista Mexicana de Astronomía y Astrofísica

RM&AE Reliability, Maintainability & Availability Evolution

RMAE Reliability, Maintainability and Availability Evolution

RMARTS Rate Monotonic Analysis for Real-Time Systems

RMAS Remote Memory Administration System

RMAT Rhein-Main-Astrotreff (Germany)

RMATS Remote Maintenance Administration and Traffic System

Rmb Renminbi (currency)

RMB Buraimi (Oman) (location identifier) • Renminbi (currency) • Request Model Building • Requirements Management Board • Right Mouse Button • Risk Management Board

RMBPT Relativistic Many-Body Perturbation Theory

RMBT Raccord Modulaire de Basse Tension

RMC Radcliffe Magellanic Clouds • Radiative Muon Capture • Radio Meteor-burst Communication • Randolph-Macon College (USA-VA) • Redundancy Management Control • Référentiel Momentanément Comobile • Regional Maintenance Center • Regional Meteorological Centre • Remote Multiplexer Combiner • Resin Matrix Composite • Resource Management Center • Resource Management Committee • Rocket Motor Case • Rockford Machesney (USA-IL) (location identifier) • Rocky Mountain College (USA-MT) • Rod Memory Computer • Role Membership Certificate • Rosette Molecular Cloud • Rotary Mirror Camera • Rotating Modulation Collimator • Rotation Modulation Collimator • Royal Meteorological Centre • Royal Military College

RMCC Remote Monitor Control Center

RMCD Radiation Monitoring Container Device

RMCF Remote Monitor Control Facility

RMCS Range Monitoring Control System

RMCT Records Management Career Track

RMCU Responsable des Moyens Charges Utiles

Rmd Remedy

RMD Ramagundam (India) (location identifier) • Reconnaissance Mutuelle des Diplômes • Regional Meteorological Data • Relationship Management Data • Relationship Management Design • Repair and Modification Directive • Risk Management Document

RMDCN Regional Meteorological Data Communications Network

RMDI Radio-Magnetic-Deviation Indicator

rmdir Remove Directory

RMDM Relationship Management Data Model

Rmdr Remainder

RMDR Remainder

rme Read-Me Text (file extension)

RME Radiation Monitor Experiment • Radiation Monitoring Equipment • Relay Mirror Experiment • Remote Multiplexer Encoder • Request Monitor Entry • Resident Maintenance Engineer • Revenu Minimum Étudiant • Rocket Mission Evaluator • Rome Griffiss (USA-NY) (location identifier)

RMEA Retrieval Mode Electronics Assembly

RMES Relay Mirror Experimental Satellite

RMET Routine Meteorology

RMetA Rivista di Meteorologia Aeronautica (periodical code)

RMF Redistribution Matrix Field • Redistribution Matrix File • Relativistic Mean Field • Remote Management Facility • Resource Management Facility • Resource Measurement Facility • Response Matrix File • Response Matrix Function

RMFT Relativistic Mean-Field Theory

RMG Regularized Multigaussian • Resource Management Group • Right Main Gear • Rome - Richard B. Russell (USA-GA) (location identifier)

RMGA Retrieval Mode Gyro Assembly

R-MHD Radiation-Magnetohydrodynamics

RMHD Radiation-Magnetohydrodynamics • Reduced Magnetohydrodynamics (*) • Resistive Magnetohydrodynamics

rmi RIFF MIDI Data (file extension)

RMI Radio Magnetic Indicator • Reliability Maturity Index • Remote Method Invocation • Resource Manager Interface • Revenu Minimum d'Insertion • Richtmeyer-Meshkov Instability • Rimini (Italy) (location identifier) • Rocket Motor Igniter • Royal Meteorological Institute

RMIF Relative Monthly Impact Factor

RMIP Reentry Measurements Instrumentation Package

R&MIS Resource & Management Information System

RMIS Readiness Management Information System • Remote Multiplexer Instrumentation System • Reprographics Management Information System • Resource and Management Information System • Resource Management Information System

RMIT Royal Melbourne Institute of Technology (Australia-VIC)

Rmk Remark

RMK Remark • Renmark (Australia-SA) (location identifier) • Rotations-Modulations-Kollimator

RMKS Remarks

RML Colombo Ratmalana (Sri Lanka) (location identifier) • Radar Microwave Link • Radio Microwave Link • Range Measurements Laboratory • Read My Lips • Relational Machine Language • Remote Microwave Link • Requirement Modeling Language • Rotating Mirror Laser

RMLR Recueil Méthodique de Lois et Règlements

RMM Mali (international vehicle registration code) • Read-Mostly Memory • Read-Mostly Mode • Relationship Management Methodology • Relative Molecular Mass • Remote Maintenance Monitor • Remote Maintenance Monitoring • Removable Data Manager • Reverberation Mapping Method • Risk Mitigation Model

RMMC Residual Mean Meridional Circulation

RMMP Reliability and Maintainability Management Plan • Risk Management Master Plan

RMMS Remote Maintenance Monitoring System

RMN Remain • Remote Multiplexer Node • Résonnance Magnétique Nucléaire • Réunion des Musées Nationaux (France) • Rumginae (Papua New Guinea) (location identifier)

RMNG Remaining

RMNO Raad voor Milieu- en Natuuronderzoek (Netherlands)

RMNT Recherche en Micro- et Nano-Technologies • Réseau en Micro- et Nano-Technologies (France)

RMO Radio Materiel Officer • Records Management Official • Requirements Management Office • Resources Management Office • Retrieval Mode Operations • Risk Management Office

RMOC Recommended Maintenance Operation Chart • Rocky Mountain Observatories Consortium (USA) • Rosetta Mission Operations Centre

RMODP Reference Model of Open Distributed Processing

RMON Remote Monitor • Remote Monitoring

RMOP Resource Management Operation Plan

RMOS Refractory Metal Oxide Semiconductor

RMOSLM Reflected Magneto-Optic Spatial Light Modulator

RM&P Reliability, Maintainability & Producibility

RMP Rampa • Rampart (USA-AK) (location identifier) • Rate Measuring Package • Rauch Memorial Planetarium (USA-KY) • Reentry Management Program • Reentry Measurement Program • Reliability and Maintainability Plan • Reliability, Maintainability and Producibility • Reliable Multicast Protocol • Remaining Milestone Pageholder • Remote Maintenance Processor • Reprogrammable Microprocessor • Requirements Management Plan • Resource Management Plan • Reviews of Modern Physics • Risk Management Panel • Risk Management Plan • Risk Management Program • Risk Mitigation Plan • Rocket Motor Plume • Rocket Motor Propellant • Royalty Management Program

RMPA Replacement Maritime Patrol Aircraft • Rocky Mountain Planetarium Association (USA)

RMPE Rapid Message Processing Element

RMPM Requirements Management Process Model

RMPS R-Matrix with Pseudo-States (method)

RMPT Reliable Multicast Transfer Protocol

RMQ&A Reliability, Maintainability, & Quality Assurance

RMQA Reliability, Maintainability, and Quality Assurance

RMR Rain Mapping Radiometer • Rapid Memory Reload • Read in Microsoft Reader • Regional Maintenance Representative • Reliability - Maintainability - Recovery • Remote Map Reader • Remote Message Register • Resource Management Review

RMRK Remark

RMRS Recoverable Maintenance Recall System • Repeatable Maintenance and Recall System • Repeatable Maintenance Recall System

RM/S Resources Monitor/Scheduler

RM&S Reliability, Maintainability & Supportability

RMS Radiation and Meteoroid Satellite • Radiation Monitoring System • Railway Mail Service • Ramstein (Germany) (location identifier) • Random Mass Storage • Random Motion Simulator • Range Measuring System • Rate Monotonic Scheduling • Rating Management Service • Reconnaissance Management System • Records Management Service • Records Management System • Receivables Management System • Recoil Mass Separator • Reconnaissance Management System • Record Management Service • Record Management System • Recovery Management Support • Redundancy Management System • Reentry Measurement System • Regulatory Manpower System • Relationship Management System • Reliability and

Maintainability Simulator • Reliability - Maintainability - Supportability • Reliable Multicast Software • Remote Maneuvering System • Remote Manipulating System • Remote Manipulator System • Remote Master Station • Remote Measurement System • Remote Measurement Systems (USA-WA) • Remote Monitoring System • Republic Management Systems Inc. • Requirements Management System • Requirements Mediations Service • Resonant Macrosonic Synthesis • Resource Management Science • Resource Management Service • Resource Management Software • Resource Management System • Resource Monitoring System • Respiratory Monitoring System • Revised Management Scheme • Root Mean Square • Root Mean Squared • Royal Mail Service • Royal Mail Ship • Royal Mail Steamship • Royal Meteorological Society (UK)

RMSAFE Radiation Monitoring System Applicable to Fusion Experiments

RMSC Remote Monitoring System Concentrator • Rochester Museum and Science Center (USA-NY) • Roper Mountain Science Center (USA-SC)

RMSD Root-Mean-Square Difference

RMSDE Root-Mean-Square Distance Error

RMSE Reduced Mean Squared Error • Root-Mean-Square Error

RMSECV Root Mean Square Error of Cross-Validation

RMSEE Root Mean Square Error of Estimation

RMSELC Root Mean Square Error of Leverage-Correction

RMSEP Root Mean Square Error of Prediction

RMSS Range Meteorological Sounding System • Rocky Mountain Star Stare (USA)

RMST Republic Management System Technologies Inc.

RMSVP Remote Manipulator System Verification Plan

Rmt Remote

R/M/T Reliability/Maintainability/Testability

RMT Radiometric Moon Tracer • Rail-Maritime-Transport • Random Matrix Theory • Rapidly Moving Telescope • Rapid Mass Transfer • Rapid Modeling Technique • Real-time Management Team • Recrutement à Mi-Temps • Regional Management Team • Remote • Remote Magnetic Tape • Remote Material Test • Resource Management Team • Resource Management Training • Return Material Test • Ring Management

RMTC Regional Meteorological Training Centre • Russian Meteorological and Training Centre

RMTE Remote

RMTG ROSAT Mission Timeline Generator

RMTME Report Machine Tools Metalworking Equipment (periodical code)

RMTN Regional Meteorological Training Network

RMTP Reliable Multicast Transport Protocol

RMU Radio Maintenance Unit • Radio Management Unit • Radiomessagerie Unidirectionelle • Receive Matrix Unit • Receiver Matrix Unit • Reference Measurement Unit • Remote Maneuvering Unit • Remote Multiplexer Unit • Resource Management Update

Rmv Remove

RMV Reentry Measurement Vehicle • Registry of Motor Vehicles • Regularly Monitored Variable (star)

Rmvbl Removable

Rmvl Removal

RMW Radioactive Mixed Waste • Read - Modify - Write • Roll Maneuver Wheel

RMWC Randolph-Macon Woman's College (USA-VA)

R&MWG Reliability & Maintainability Working Group

RMWG Reliability and Maintainability Working Group • Requirements Management Working Group • Resource Management Working Group • Risk Management Working Group

RMxAA Revista Mexicana de Astronomía y Astrofísica (periodical code)

RMxAC Revista Mexicana de Astronomía y Astrofísica - Conference Series (periodical code)

RMZ Rijksdienst voor Maatschappelijke Zekerheid

rn Kiruni (ISO language code) • Read News

Rn Radon • Rain

R-N Reverse-Normal

RN Niger (international vehicle registration code) • Radio Navigation • Radio Networking • Random Number • Rational Negation • Raumfahrtnutzung • Read News • Rechnernetz • Recovery Net • Red Noise • Reference Noise • Reference Number • Reflection Nebula • Registered Nurse • Registry Number • Reglamento Notarial • Reissner-Nordstrom • Relay Node • Remote Node • Research Note • Resource Node • Revenu National • Reynold's Number • Ring Number • Rio Grande do Norte (Brazil) • Rio Negro (Argentina) • Route Nationale • Routing Number • Royal Navy • Ruta Nacional

RNA Remote Network Access • Réseau de Neurones Artificiels • Ribonucleic Acid

RNAAC Regional Network Associated Analysis Center

RNAC Romanian National Astronomical Committee

RNAGG Residual Non-Aliased Gravity Grid

RNAO Results of National Argentine Observatory (periodical code)

RNAV Radar Navigation

RNB Ronneby (Sweden) (location identifier)

RNBM Radio Noise Burst Monitor

RNC McMinnville - Warren County (USA-TX) (location identifier) • Radio Network Controller • Reflective Null Corrector • Resource Network Committee

RNCC Range Network Control Center • Reference Number Category Code • Report-back Network Control Circuit

RNCH Ranch

Rnd Round

RND Random • San Antonio Randolph (USA-TX) (location identifier)

Rndm Random

RNDS Rounds

RNDZ Rendezvous

RNe Recurrent Novae

RNE Répertoire National des Établissements (France) • Réseau National d'Essais • Roanne (France) (location identifier)

RNET Radio Net

R-Nets Requirements Networks

RNets Requirements Networks

Rnf Renfrewshire (UK-Scotland)

RNF Registry Nomenclature File

RNFC Reference Number Format Code

RNFL Rainfall

RNFR Rename From

Rng Range

RNG Radio Network Gateway • Random Number Generator • Range • Rangely (USA-CO) (location identifier) • Renormalization Group

RNGC Revised New General Catalogue

Rnge Range

Rngg Ringing

Rngr Ringer

RNH New Richmond (USA-WI) (location identifier)

RNI Corn Island (Nicaragua) (location identifier) • Réseau National d'Interconnexion • Reserved Name Indicator

RNID Return Node Identification

RNIPP Répertoire National de l'Identité des Personnes Physiques

RNIS Réseau Numérique à Intégration de Services

RNISZ Rezultaty Nabliudenii Iskusstvennykh Sputnikov Zemli (periodical code)

RNJ Yoronjima (Japan) (location identifier)

rnl Runlength Raster (file extension)

Rnl Renewal

RNL Rennell (Solomon Islands) (location identifier)

RNLAF Royal Netherlands Air Force

RNM Réseau Neuro-Mimétique

RNMG Rosetta Nucleus Modeling Group

RNMI Royal Netherlands Meteorological Institute

RNN Bornholm (Denmark) (location identifier) • Reciprocal Nearest Neighbors • Recurrent Neural Network • Rønne (Denmark) (location identifier)

RNO Red Nebulous Object • Regional Nuclear Options • Reno (USA-NV) (location identifier) • Tahoe (USA-NV) (location identifier)

RNOA Rede Nacional de Observação Astronómica (Portugal)

RNOC Regional Network Operations Center

RNODC Responsible National Oceanographic Data Center

RNP Rede Nacional de Pesquisas • Regional Network Provider • Remote Network Printer • Remote Network Processor • Required Navigation Performance • Rongelap Island (Marshall Islands) (location identifier) • Rotation - Nutation - Precession

RNPC Required Navigation Performance Capability

RNPIC Residual North Pole Ice Cap

RNPP Répertoire National des Personnes Physiques

RNQ Réseau National de la Qualité (France)

RnR Registration and Resolution

RNR Réacteur à Neutrons Rapides • Receive Not Ready • Receiver Not Ready • Robinson River (Papua New Guinea) (location identifier)

RNRT Réseau National de Recherche en Télécommunications (France)

Rns Runs

RNS Radio Navigation Service • Reduced Navier-Stokes • Remote Note Service • Rennes Saint-Jacques (France) (location identifier) • Residue Number System • Reusable Nuclear Shuttle • Reusable Nuclear Stage

RNSC Reference Number Status Code

RNSP Residual Nuclear Strike Plan

RNSS Radio Navigation Satellite Service

RNT Radio Network Termination • Random Numerical Terminal • Renton (USA-WA) (location identifier)

Rntl Rental

RNTL Réseau National en Technologies Logicielles (France)

RNTS Regenerative Non-venting Thermal System

RNU Ranau (Malaysia) (location identifier)

RNV Random Noise Voltmeter • Reusable Nuclear Vehicle

RNVA Réserve Naturelle Volontaire Agréée

RNVC Reference Number Variation Code

Rnwbl Renewable

Rnwy Runway

RNWY Runway

Rny Rainy

RNZ Rensselaer (USA-IN) (location identifier)

RNZAF Royal New Zealand Air Force

RNZN Royal New Zealand Navy

ro Romania • Romanian (ISO language code)

Ro Recto • Río

R&O Routing & Operations

R/O Read Only • Readout • Receive Only • Roll Out • Room Only

R-O Read Only

RO Radar Operator • Radiation Office • Radio Operator • Range Only • Range Operations • Read Only • Readout • Reality Orientation • Real Orden • Receive Only • Receiving Office • Receiving Officer • Record Office • Recovery Operations • Recruiting Officer • Reference Object • Registered Office • Regulated Output • Relative Openness • Remote Observatory • Remote Observe • Remote Observer • Remote Observing • Remote Operation • Reportable Occurrence • Reporting Officer • Requesting Organization • Requirements Objective • Requirements Officer • Research Objective • Reserved Occupation • Reverse Osmosis • Rial Omani (currency) • Ring Output • Risks Officer • Rollover • Romania (international vehicle registration code) • Romania (ISO country code) • Rondônia (Brazil) • Room Only • Roumanie (code ISO de pays) • Route Order • Router • Routine Order • Royal Observatory • Run On • Runout • Runover

RoA Record of Achievement

ROA Raman Optical Activity • Real Instituto y Observatorio de la Armada (Spain) • Record Of Achievement • Reference Optical Alignment • Regional Office for Africa • Reimbursement Obligation Authority • Research Opportunity Award • Restricted Operations Area • Return On Assets • Roanoke (USA-VA) (location identifier)

ROAD Rapid Object Application Development

ROADS Real-time Optical Alignment and Diagnostic System • Resource and Operations Analysis Display System • Resource Organisation And Discovery Service

ROAI Robinet Obturateur Automatique Intégré

RoAJ Romanian Astronomical Journal (periodical code)

ROAM Return On Assets Managed

ROAn Royal Observatory Annals (periodical code)

roAp Rapidly Oscillating Ap

ROAR Rights Of Admission Reserved • Royal Optimizing Assembly Routine

ROB Monrovia Roberts (Liberia) (location identifier) • Recorder Buffer • Recovery Operations Branch • Refocusing Opportunity Bulletin • Remote Order Buffer • Reorder Buffer • Research on Organizational Behavior (periodical code) • Right Outboard • Rigid Optical bench • Royal Observatory of Belgium

ROBCO Readiness Objective Code • Requirements Objective Code

ROBEPS Radar Operating Below Prescribed Standard

ROBIN Rotterdam Bibliotheek en Informatienetwerk (Netherlands)

Robot Robotica (periodical code)

ROBOT Robotic Observatories Basic Operating Tasks

ROBS Rapid Optical Beam-steering System • Rapid Optical Beam System

ROBY Results - Output - Benefits - Yield

ROC Radius Of Curvature • Rappel Opérationnel de Clients • Rate Of Change • Rate Of Climb • Rate Of Coverage • Rayleigh Optical Corp. (USA-MD) • Receiver Operating Characteristics • Reconocimiento Óptico de Caracteres • Reconnaissance Optique de Caractères • Record Of Comments • Regional Operating Co. • Regional Operational Center • Regional Operations Center • Region Of Convergence • Registry Office Code • Relative Out Centrality • Remote Operations Center • Remote Operations Controller • Republic of China (Taiwan) • Request Of Change • Required Operational Capability • Required

Operational Center • Research Organisation Cooperation • Réseau d'Ouverture et de Collaboration • Reserved Observations Catalog • Return On Capital • Reusable Orbital Carrier • Rochester (USA-NY) (location identifier)

ROCC Range Operational Control Center • Range Operations Control Center • Regional Operations Control Center • Region Operational Control Center

ROCE Return On Capital Employed

ROCES Ranger Orbiter Crew Escape System

ROCi Republic Observatory Johannesburg Circular (periodical code)

ROCI Rahim Organizational Conflict Inventory

ROCor Resultados del Observatorio Nacional Argentino en Córdoba (periodical code)

ROCR Remote Optical Character Recognition

ROCS Real-time Operations Control System

ROCU Remote Orderwire Control Unit

Roczn Rocznik Astronomiczny Observatorjum Krakowskiego Kraków (periodical code)

RoD Record of Decision

ROD Radiometric and Orbit Data • Radio On Demand • Record Of Decision • Rate Of Descent • Record Of Discussion • Regional Operational Data • Release Order Directive • Remotely-Operated Door • Report Of Discrepancy • Required Operational Date • Robertson (South Africa) (location identifier) • ROSAT Observational Data-set

RODC Regional Oceanographic Data Center

RODEO Rotating Delivery of Excitation Off-resonance

RoDok Rossijskaia Akademiia Nauk Doklady (periodical code)

RODP Roll-Off Discharge Platform

RODS Record Of Decision System

ROE Readout Electronics • Référentiel d'Objets d'Entreprise • Reflector Orbital Equipment • Report of Expenditures • Return On Equity • Round Off Error • Royal Observatory Edinburgh (UK-Scotland) • Royal Observatory Edinburgh (Starlink internal code)

ROEN Rádio Observátorio Espacial do Nordeste (Brazil)

ROER Remote Operation Error

ROEU Remotely-Operated Electrical Umbilical

ROF Receipt Of Funds • Reliability of Organism Functioning • Réseau Optique Flexible

ROFF Run-Off

ROFL Rolling On the Floor Laughing

ROFT Rapid Optics Fabrication Technology

ROFUS Rapid Optical Follow-Up Spectroscopy

ROG Receipt Of Goods • Rogers (USA-AR) (location identifier) • Royal Observatory Greenwich (UK-England)

ROGUE Radio, Optical, Gamma-ray Undergraduate Experiment • Rapid Off-axis Guider Experiment

ROH Receiver Off Hook • Regular Overhaul • Restricted Overhaul • Robinhood (Australia-QLD) (location identifier)

ROHK Royal Observatory Hong Kong (now HKO)

ROI Radial Orbit Instability • Rádio Observátorio do Itapetinga (Brazil) • Radio - Optical - Inertial • Range Operations Instruction • Rectangle Of Interest • Region Of Influence • Region Of Interest • Rendezvous Orbit Insertion • Report Of Investigation • Return On Investment • Rotating Optical Interferometer • Rules Of Interaction

ROIC Readout Integrated Circuit • Resident Officier In Charge • Return On Invested Capital

ROICC Resident Officier In Charge of Construction

ROID Report Of Item Discrepancy

RoIE Rossiiskaia Akademiia Nauk Izvestiia Energetika (periodical code)

RoIET Rozprawy Inzynierskie Engineering Transactions (periodical code)

RoIMT Rossiiskaia Akademiia Nauk Izvestiia Mekhanika Tverdogo Tela (periodical code)

RoIMZ Rossiiskaia Akademiia Nauk Izvestiia Mekhanika Zhidkosti i Gaza (periodical code)

RoISF Rossiiskaia Akademiia Nauk Izvestiia Seriia Fizicheskaia (periodical code)

RoIn Rozprawy Inzynierskie (periodical code)

ROIS Radio Operational Intercom System

ROIV Remote Operation Invoke

ROK Republic of Korea (international vehicle registration code) • Rockhampton (Australia-QLD) (location identifier) • South Korea (international vehicle registration code)

ROL Leu (Romania) (ISO currency code) • Radiação de Onda Longa • Régie On-Line • Remote Operating Location • Reorder Level • Roosevelt (USA-UT) (location identifier) • Rotate Left

ROLAP Relational On-line Analytical Processing

ROLC Routing Over Large Clouds

ROLD Reorder Level Deficit

ROLM Read Only Line Module

ROLO Robotic Lunar Observatory

ROLOF Roche Lobe Overflow

ROLR Receiving Objective Loudness Rating

ROLRS Reject Online Reentry System

ROLS Recoverable Orbital Launch System

ROLTP Remote On-Line Transaction Processing

ROLun Reports of the Lund Observatory (periodical code)

Rom Roman • Romania • Romanian

ROM Radio Observatório de Maipú (Chile) • Range Of Motion • Read-Only Memory • Reception and Onward Movement • Remote Opacity Monitor • Roma (Italy) (location identifier) • Romania (ISO country code) • Rome (Italy) (location identifier) • Rough Order of Magnitude • Roumanie (code ISO de pays) • Royal Ontario Museum (Canada-ON)

ROMARIN Read-Only Memory of Madrid Actualized Registry Information

ROMBUS Reusable Orbital Module Booster and Utility Shuttle

ROMC Rate Of Manageable Change • Required Operational Messaging Characteristics

RomCo Osservatorio Astronomico di Roma su Monte Mario - Contributi Scientifici (periodical code)

ROME Répertoire Opérationnel des Métiers et des Emplois • Résolution Océanique à Moyenne Échelle

ROMM Read-Only Memory Module

ROMPS Robotic Materials Processing in Space • Robotic Materials Processing System • Robot-Operated Materials Processing System

RON Readout Noise • Remaining Overnight • Remain Overnight • Request Order Number • Research Octane Number • Rondón (Colombia) (location identifier)

RONA Return On Net Assets

RONALD Rocketborne Optical Neutral-gas Analyzer with Laser Diodes

RONAST Royal Nepal Academy of Science and Technology

RONSARD Recherche sur les Orages et les Nuages par un Système à Radars Doppler

ROO Radio Optical Observatory • Rondonôpolis (Brazil-MT) (location identifier)

ROOM Real-time Object-Oriented Modeling

ROOST Reusable One-stage Orbital Space Truck

ROOTS Remotely-Operated Optical Tracking System

ROP Raster Operation • Raster Operations Processor • Read-Only Printer • Receive-Only Printer • Record Of Purchase • Recovery Operating Plan • Reference Observing Proposal • Remote Operator Position • Reorder Point • Republic Of Panama • Republic Of the Philippines • RISC Operation • Rota (Mariana Islands) (location identifier) • Routing Operating Procedure • Run Of Paper

ROPA Realizing Our Potentials Award • Reserve Officer Personnel Act (USA)

ROPE Research on Orbital Plasma Electrodynamics • Rip it Out and Put in Ethernet

ROPL Reorder Point Level

ROPME Regional Organization for the Protection of the Marine Environment

ROPP Receive-Only Page Printer

ROPS Readiness, Operations, and Planning System • Roll Over Protective Structure

ROQ Return On Qualification • Route Order Request

ROR Koror (Palau Islands) (location identifier) • Range Only Radar • Rate Of Read • Rate Of Return • Released on Own Recognizance • Repair Of Repairables • Robust Orthogonal Regression • ROSAT Observation Request • Rotate Right • Route Order Release

RORD Rapid Onset - Rapid Decay

RORF Radio Optical Reference Frame

RORJ Remote Operation Reject

RO-RO Roll-On - Roll-Off

RORO Roll-On - Roll-Off

RORS Remote Operation Result

RORSAT Radar Ocean Reconnaissance Satellite

ROS Radar à Ouverture Synthétique • Radarsat Optical Sensor • Range Operations Supervisor • Rapport d'Ondes Stationnaires • Read-Only Storage • Real-time Operating System • Reconstitution Operational State • Reduced Operational Status • Regulated Oxygen Supply • Relay Optics Structure • Remote Operations Service • Remote Optical System • Removable Overhead Structure • Representative Observations Site • Research Optical Sensor • Retail Operation System • Return from Overseas • Rosario Fisherton (Argentina-SF) (location identifier) • Rotating Optical Scanner

ROSA Report Of Service Activity • Romanian Space Agency

ROSAC ROSAT-Based Search for AGN-Clusters

ROSAR Rotating Synthetic Aperture Radar

ROSAT Radar Ocean Surveillance Satellite • Röntgen Satellite

ROSC Regional Operations and Security Center

ROSDR Read-Only Storage Data Register

ROSE Radar Ocean Sensing • RATS Open Systems Environment • Remotely Operated Special Equipment • Remote-Operations Service Element • Remote-Operations Service Entity • Remote Optical Sensing of Emission • Reporting On the State of the Environment • Request-Oriented Scheduling Engine • Research Open Systems for Europe • Research Open Systems in Europe • Retrieval by Online Search • Rising Observational Sounding Equipment • Rocket and Scatter Experiment • Rural Oxidant in the Southern Environment

ROSI Rome Observatory Schmidt Imager

ROSIG ROSAT Science Interest Group

ROSINA Rosetta Orbiter Spectrometer for Ion and Neutral Analysis

ROSIS Reflective Optics System Imaging Spectrometer

ROSO Review Of Starlink Operations

RosOB Boletín del Observatorio Astronomico Municipal de Rosario (periodical code)

ROSQ Return On Software Quality

ROSS RAND Object Simulation System • Reconstruction Of Serial Sections • Remote Orbital Servicing System • Research Opportunities in Space Sciences

ROSSA Robotics Servicing and Space Assembly

ROSSWG Radarsat Optical Sensor Science Working Group

ROSTA Regional Office for Science and Technology, Africa (UNESCO)

ROSTAS Regional Office for Science and Technology, Arab States (UNESCO)

ROSTLAC Regional Office for Science and Technology, Latin America and Caribbean Islands (UNESCO)

ROSTSCA Regional Office for Science and Technology for South Central Asia (UNESCO)

ROSTSEA Regional Office for Science and Technology for Southeast Asia (UNESCO)

rot Curl

Rót Rótula

Rot Rota • Rotary • Rotating • Rotation • Rotor • Rotule

ROT Rate Of Turn • Refurbish - Operate - Transfer • Remaining Operating Time • Remotely Operated Telescope • Remote Observation Telescope • Requested Observing Time • Reusable Orbital Transport • Right On Time • Rotate • Rotation • Rotorua (New Zealand) (location identifier) • Rotterdam (Netherlands) • Rotunda • Rule Of Thumb • Running Object Table • Runway Occupancy Time

ROTAM Request-Oriented Telecommunications Access Method

ROTAT Robot Observatory Theoretische Astrophysik Tübingen (Germany)

ROTC Reserve Officers' Training Corps

ROTERM Receive Only Terminal

ROTEX Robotic Technology Experiment • Robot Technology Experiment

ROTFL Rolling On The Floor Laughing

Rotg Rotating

ROTHR Relocatable Over-The-Horizon Radar

ROTI Recording Optical Tracking Instrument

ROTL Remote Office Test Line

Rotr Rotator

ROTR Receive-Only Tape Perforator

ROTS Remote Operating Terminals

ROTSCA Regional Office for Science and Technology in South Central Asia (UNESCO)

ROTSE Robotic Optical Transient Search Experiment

ROTV Returnable Orbital Transfer Vehicle • Reusable Orbital Transport Vehicle (*)

ROU Ruse (Bulgaria) (location identifier) • Uruguay (international vehicle registration code)

Roum Roumanian • Roumanie

ROUS Rodent Of Unusual Size

Rout Routine

ROUT Roll Out

ROV Raumordnungsverfahren • Remotely-Operated Vehicle • Remote Operating Vehicle • Remote Optical Viewing • Report Of Visit • Rostov (Russia) (location identifier)

ROVE Radar Observation of Vegetation

ROVOR Remote Observatory for Variable Object Research (BYU)

ROVS Remote Optical Viewing System

ROW Rest Of World • Right Of Way • Roswell (USA-NM) (location identifier)

ROWG RadioAstron Operations Working Group (Russia)

ROWPU Reverse Osmosis Water Purification Unit

ROWS Radar for Observing the Wave Spectrum • Radar Ocean Wave Spectrometer • Remote Observing Weather Sensor • Remote Ocean Wave Spectrometer • Remote Orbital Wave Spectrometer

ROX Right On Target • ROSAT Optical X-ray (survey) • Roseau (USA-MN) (location identifier) • Roxburghshire (UK-Scotland)

ROXS ROSAT Optical X-ray Survey

Roy Royal • Royaume

ROY Rio Mayo (Argentina-CB) (location identifier)

ROYGBIV Read Out Your Good Book In Verse • Red - Orange - Yellow - Green - Blue - Indigo - Violet (*) • Richard Of York Gave Battle In Vain

ROZ Restricted Operations Zone

R° Río

Rp Rappen • Réponse Payée • Rupiah (currency)

R/P Receiver/Processor

R&P Reserve & Process

R-P Roll-Pitch

RP Philippines (international aircraft registration code) • Philippines (international vehicle registration code) • Radiation Pressure • Radiation Protection • Radio Port • Rally Point • Rapid Patch • Rapid Prototyping • Reachable Path • Reaction Product • Reader Printer • Readiness Posture • Readiness Potential • Read Printer • Real Part • Real Property • Received Pronunciation • Reception Poor • Recette Principale • Recommended Practice • Recommended Price • Reconnaissance Party • Reconstitution Priority • Record Position • Recovery Phase • Recreation Park • Recurring Periodical • Redox Potential • Red Phosphorus • Redundancy Payment • Reference Paper • Reference Pattern • Reference Point • Reference Publication • Reference Pulse • Refit Period • Reformed Presbyterian • Regius Professor • Reinforced Plastic • Relative Performance • Relative Pressure • Relativistic Proton • Relativity Principle • Release Point • Reliability Plan • Remote Printer • Rendezvous Point • Repair Parts • Repair Period • Repeated Penetration • Repetitively Pulsed • Repetitive Penetration • Repetitive Pulse • Reply Paid • Réponse Payée • Reporting Post • Représentation Proportionnelle • República Portuguesa • Republic of the Philippines • Request Processing • Research Paper • Reserve and Process • Reserve Personnel • Resolving Power • Response Planning • Restoration Priority • Retained Personnel • Retournement de Pupille • Return Premium • Révérend Père • Reverendus Pater • Review Panel • Revision Proposal • Rocket Projectile • Rocket Propellant • Rocket-Propelled • Roll-Pitch • Rond-Point • Route Package • Route Port • Ruddlesden-Popper • Rules of Procedure • Running Penumbral (wave) • Runtime Process • Ruta Provincial

RPA Radiation Protection Adviser • Random Phase Approximation • Rapid Prototyping Array • Reactive Plate Assembly • Record and Playback Assembly • Record of Personal Achievement • Rème Plasma Analyser • Remotely Piloted Aircraft • Remote Passphrase Authentication • Reply All • Request for Procurement Action • Request for Procurement Authorization • Reserve Pay Account • Reserve Pay and Allowances • Resting Point of Accommodation • Retarded Potential Analyzer • Retarding Potential Analyzer (*) • Retired Pay Accrual • Retired Professional Astronomer • Rolpa (Nepal) (location identifier) • Rotor à Pales Actives • Route Planning Aid

RPAA Rotating Phase Array Antenna

RPAE Random Phase Approximation with Exchange • Retarding Potential Analyzer Experiment

RPAM Recruiting Policy Analysis Model

RPAO Règlement Particulier d'Appel d'Offres

RPAST Raising Public Awareness of Science and Technology

RPB Range Prioritization Board • Recognized Professional Body • Regional Preparedness Board • Reports and Publications Branch • Result Producing Branch • Reverse Path Broadcast • Roper Bar (Australia-NT) (location identifier)

RpBAM USSR Report Space Biology Aerospace Medicine (periodical code)

RPBF Reference Pattern-Based Feature

RPBG Revised Program and Budget Guidance

RPBL Reduced Parameter Bilinear Model

RPBR Result Producing Branch Response

RPC Radio Port Controller • Real Procedure Call • Records Processing Center • Regional Preparedness Committee • Regional Program Committee • Regional Pyrheliometer Comparison • Regras Portuguesas de Catalogação (Portugal) • Relations Publiques et Communication • Remote-Position Control • Remote Power Control • Remote Power Controller • Remote Procedural Call • Remote Procedure Call (*) • Repairable Processing Center • Request the Pleasure of your Company • Report Generator • Resistive Plate Chamber • Resistive Plate Counter • Restoration Priority Code • RGS Power Converter • Ring Plane Crossing • RISC Personal Computer • Root Path Cost • Rosetta Plasma Consortium • Row Parity Check

RPCA Remotely Programmable Conference Arranger

RpCCA USSR Report Cybernetics Computers Automation Technology (periodical code)

RPCL Remote Procedure Call Language

RPCO Reclamation Program Control Officer

RPCU Radio Port Controller Unit

RPCVD Remote Plasma Chemical Vapor Deposition

RPCVE Remote Plasma-enhanced Chemical Vapor Epitaxy

RPD Radar Planning Device • Radiation Pressure Dominated • Rapid • Rapid Program Development • Rapid Prototyping Development • Reconfiguration Production Department • Relative Power Density • Retarding Potential Difference • Rotary Positive Displacement

RPDA Remote Power Distribution Assembly • Robust Partial Discriminant Analysis • ROSAT Public Data Archive

RPDLY Rapidly

RPDS Rapids

RP&E Receive, Process & Exploit

RPE Radial Probable Error • Radio Paging Equipment • Receive, Process and Exploit • Regional Political Ecology • Registered Professional Engineer • Regular Pulse Excited • Relative Pointing Error • Reliability Project Engineer • Remote Peripheral Equipment • Remote Procedure Error • Resource Planning and Evaluation

RPEC Resource Protection Executive Committee

RPECS Rapid Prototyping Electronic Control System

RpEE USSR Report on Engineering Equipment (periodical code)

RpEEE USSR Report on Electronic and Electrical Engineering (periodical code)

RPEG Repartição de Pessoal e Expediente Geral

RpEn USSR Report on Energy (periodical code)

RPEP Register of Planned Emergency Procedures

RPES Residual Polyelastic Sampling

RpESc USSR Report on Earth Sciences (periodical code)

RPF Rapid Prototyping Facility • Raster Product Format • Real Property Facilities • Remote Processor Function • Report Format • Report Program Facility • Reverse Path Forward • ROSAT Proposal Form • Rotatable Pool Factor

RPFITS Radiophysics Flexible Image Transport System

RPFO Resupply Planning Factors Office

RPG Radar Processing Group • Radial Pressure Gradient • Rapid Population Growth • Receiver Processing Group • Receiver Processor Group • Recommended Practices Guide • Relative Publication Growth • Repetitive Pulse Generator • Report Program Generator • Research Planning Grant • Research Planning Guide • Risk Purchasing Group • Rôle-Playing Game

RPGL Report Program Generator Language

rph Revolutions per Hour

RPI Radar Precipitation Integrator • Radio Plasma Imager • Rated Position Identifier • Real Property Inventory • Reduced Price Incentive • Relay Position Indicator • Remote Process Innovation • Remote Process Invocation • Rensselaer Polytechnic Institute (USA-NY) • Repetitive-Motion Injury • Research Performance Indicator • Residual Plan of Instruction • Restoration Priority Indication • Retail Price Index • Retrieval with Probabilistic Indexing • Roll Position Indicator • Rotating Polarizer Interferometer • Runway Point of Intercept • Runway Point of Interception

RPIAS Real Property Inventory Account System

RPIE Real-Property Installed Equipment • Replacement Photographic Image Equipment

RPIF Regional Planetary Image Facility

RPIO Registered Publication Issuing Office

RPK Revenue Passenger Kilometer • Revenue Per Kilometer

rpl Replica (file extension)

RPL Radar Processing Language • Radio-Photo Luminescent • Rated Power Level • Relational Production Language • Remote Procedure Load • Remote Program Load • Repair Parts List • Replaceable Parts List • Requested Privilege Level • Request Parameter List • Resident Programming Language • Reverse Polish LISP • Rocket Propulsion Laboratory • Running Program Language • Runway Parallel Line

RPLC Replace • Reversed-Phase Liquid Chromatography

Rplr Repeller

Rplsn Repulsion

Rplt Repellent

RPLV Reentry Payload Launch Vehicle

rpm Revolutions per Minute (*) • Rotations per Minute • Runs per Minute

RpM Rate per Minute

R&PM Research & Program Management

RPM Ngukurr (Australia-NT) (location identifier) • Radiation Polarization Measurement • Radio Packet Modem • Random Positioning Machine • Rate Per Minute • Read Program Memory • Real Property Maintenance • Recursive Projection Method • Red Hat Package Manager • Reduced Proper Motion • Register of Preservation Microfilms • Relaxation Potential Model • Reliability Performance Measure • Reliability Prediction Module • Remote Payload Monitoring • Remote Performance Monitoring • Reprogrammable Memory • Resale Price Maintenance • Research and Program Management • Research Program Management • Resource Planning and Management • Resupply Provisions Module • Retro-Propulsion Mode • Retro-Propulsion Module • Revenue Passenger Miles • Reverse Path Multicasting • Revolutions Per Minute • Roll Position Mechanism • Rounds Per Minute

RPMA Real Property Maintenance Account • Real Property Maintenance Activity

RPMB Remotely-Piloted Mini-Blimp

RPMD Reduced Proper Motion Diagram

RPMF Real Property Maintenance Floor

RPMG Regional Postmaster General

RPMI Radiant Power Measuring Instrument • Revolutions-Per-Minute Indicator

RPMIO Registered Publication Mobile Issuing Office

RpMP Reports on Mathematical Physics (periodical code)

RPMS Real-time Performance Management System

RpMSM USSR Report on Material Sciences (periodical code)

RPMT Radiant Power Measuring Instrument

RPN Real Page Number • Recherche en Prévision Numérique • Repositionable Note • Reverse Polish Notation • Risk Priority Number • Rosh Pina (Israel) (location identifier)

RpO Revolutions per Orbit

RPO Radiation Protection Officer • Radiological Protection Office • Radiological Protection Officer • Railway Post Office • Recovery Point Objective • Regional Publishing Office • Research Programs Office • Responsible Program Office • Responsible Project Office • Responsible Property Officer • Reuse Project Officer • Revolution Per Orbit

RPOA Recognized Private Operating Agency

RPOC Remote Payload Operations Center

RPOM Real Property Operation and Maintenance

RPOMA Real Property Operation and Maintenance Activity

RPOP Rendezvous and Proximity Operations Program

RPOS Replanned Preferred Observation Schedule • Replanned Preferred Observation Sequence

RPP Radiation Protection Program • Random Phase Plate • Red Patch Panel • Regional Priority Program • Reinforced Pyrolytic Plastic • Reliability Program Plan • Rendezvous Point Position • Repetitive Pulsating Penetration • Représentation des Phénomènes de Physique • Request Parallel Poll • Requisition Processing Point • Resource Protection Program • Respiratory Protection Program • Retrograde Processing Point • Reverse Pulse Polarography • Rotation-Powered Pulsar • Rural Policy Plan

RPPAS Retrograde Processing Point Automated System

RPPE Radiation Protection Program Evaluation • Research, Program, Planning and Evaluation

RPPh Reports of Progress in Physics (periodical code)

RpPhM USSRR Report on Physics and Mathematics (periodical code)

RPPO Regional Printing and Procurement Office

RPPROM Reprogrammable Programmable Read-Only Memory

RPQ Request for Price Quotation

Rpr Repair

RPR Raipur (India) (location identifier) • Rapid Power Reduction • Rassemblement pour la République (France) • Readiness Planning Report • Read Printer • Rear Projection Readout • Rectangular Parallelepiped Resonance • Remote Process Redesign • Repair Product Menu • Resilient Packet Ring • Roll-Pitch Resolver

RPRINTER Remote Printer

RPRSG Resilient Packet Ring Study Group

Rprt Report

RPRV Remotely-Piloted Research Vehicle

rps Revolutions per Second • Runs per Second

RpS Reductions per Second

RPS Radioisotope Power System • Ram Pressure Stripping • Rapid Processing System • Reactor Protection System • Real-time Photogrammetry System • Real-time Processing System • Real-time Programming System • Record and Playback System • Reductions Per Second • Redundant Power System • Regulated Power Supply • Relative Publication Strategy • Remote Printing System • Remote Procedure Service • Remote Processing Service • Remote Processing Station • Remote Proposal Software • Remote

Proposal Submission • Remote Proposal System • Responsable à la Préparation du Satellite • Revolutions Per Second • Rijkspostspaarbank • Ring Parameter Server • Roadway Package System • ROSAT Proposal Submission • Rotational Position Sensing • Royal Photographic Society

RpSBA USSR Reports on Space and Biological Aerospace Medicine (periodical code)

RpScT USSR Reports on Science and Technology (periodical code)

RPSE Real-Property Similar Equipment

RPSEC Ruth Patrick Science Education Center (USA-SC)

RPSF Rotation Processing and Surge Facility

RPSL Routing Policy Specification Language

RPSM Resources Planning and Scheduling Method

RpSpR USSR Reports on Space (periodical code)

RPSS Relative Partial Sum of Squares • Remote Proposal Submission System

RPSTL Repair Parts and Special Tools List

RPS2 Remote Proposal Submission System

rpt Report (file extension)

Rpt Repeat • Report • Reprint

R/Pt Reporting Point

RPT Raw Packet Telemetry • Receive Plain Text • Reference Point Tracking • Regular Public Transport • Repair Parts Transporter • Repeat • Report • Resident Provisioning Team • Rudder Pedal Transducer

RPTA Rudder Pedal Transducer Assembly

RPTC Receive Plain Text Clock

RPTEC Rapid Prototyping, Tooling and Evaluation Center

RPTF Red Patch and Test Facility

RPTGEN Reports Generator

Rptn Repetition

RPTOR Reporting Organization

Rptr Repeater

RpTr USSR Reports on Transportation (periodical code)

RPU Radio Phone Unit • Radio Port Unit • Radio Propagation Unit • Receiver Processing Unit • Receiver-Processor Unit • Rectifier Power Unit • Registered Publication Unit • Remote Positioning Unit • Remote Processing Unit • Research and Publications Unit • Rotor Phasing Unit

RPV Reactor Pressure Vessel • Reactor Pressurized Vessel • Remotely-Piloted Vehicle • Réseau Privé Virtuel • Roper Valley (Australia-NT) (location identifier)

RPW Radio and Plasma Wave • Revenue - Pieces - Weight • Running Penumbral Wave

RPW-E Radio and Plasma Wave - Electric field

RPW-H Radio and Plasma Wave - Magnetic field

RPWS Radio and Plasma Wave Science

RPX Ring Plane Crossing • Roundup (USA-MT) (location identifier)

RPY Roll - Pitch - Yaw

RPZ Runway Protection Zone

RQ Radio-Quiet • Rate Quote • Relational Query • Repeat-Request • Reportable Quantity • Request for Quotation • Requirement Precision • Research Quantum • Respiratory Quotient

R&QA Reliability & Quality Assurance

RQA Reliability and Quality Assurance • River Quality Assessment

RQBE Relational Query By Example

RQC Reliability and Quality Control

R&QE Radiophysics and Quantum Electronics (periodical code)

RQE Relative Quantum Efficiency • Responsive Quantum Efficiency

RQF Rescue Flight

RQFT Relativistic Quantum Field Theory

RQG Rescue Group

RQI Raleigh Quotient Iteration

RQL Reference Quality Level • Revue de Qualification du Lanceur • Rich - Quick - Lean

RQM Radar Quality Monitoring

Rqmt Requirement

RQMT Requirement

Rqmts Requirements

RQMTS Requirements

Rqn Requisition

RQN Radial Quantum Number

RQNS Radio-Quiet Neutron Star

RQPM Recombination Quark Parton Model

RQQ Radio-Quiet Quasar • Really Quick Question

Rqr Require • Requirement

Rqrd Required

RQRPA Renormalized Quasi-Random Phase Approximation

RQS Rate/Quote System • Rate Quoting System • Ready Qualified for Standby • Rescue Squadron

RQT Reliability Qualification Test • Reliability Qualification Testing • Requalification Testing

RQZ Radio Quiet Zone

R-R Roll-Roll • Rolls-Royce

R&R Removal & Repair • Removal & Replacement • Remove & Replace • Rendezvous & Recovery • Repair & Rehabilitation • Repair & Replace • Repair & Return • Rescue & Resuscitation • Rest & Recreation • Rest & Recuperation • Rest & Rehabilitation

R/R Readiness/Reconstitution • Readout and Relay • Record/Retransmit • Remove/Replace

RR Bororo • Radiation Resistance • Radiative Recombination • Radio Range • Radio Regulation • Radio Relay • Radio Resource • Railroad • Rainfall Rate • Range Rate • Readiness Report • Readiness Review • Readout and Relay • Read Restore • Real Reality • Receive Ready • Receiver Ready • Receiving Report • Recurrence Rate • Red Rectangle (nebula) • Reduced Resolution • Redundancy Reduction • Reentry Range • Reference Receiver • Register to Register • Rejection Report • Relative Rate • Relative Ratio • Relative Relevance • Remington Rand • Removal-Replacement • Remove and Replace • Rendezvous Radar • Rendezvous and Recovery • Respiration Rate • Repetition Rate • Reporting Responsibility • Requirements Review • Research Report • Resource Record • Respiration Rate • Retro Rocket • Return Rate • Ridge Regression • Rights Reserved • Rising Rapidly • Risk Reduction • Roll Racheting • Roll-Roll • Rolls-Royce • Root Registry • Roraima (Brazil) • Route Relay • Route Request • Routine Relay • Rural Route • Rush Release

R&RA Research & Related Activities

RRA Radiological Risk Acceptance • Realistic Risk Assessment • Regular Reports of Audit • Research and Related Activities • Resident Research Associate • Retro-Reflective Antenna • ROSAT Result Archive

RRAC Regional Resources Advisory Committee

RRAS Routing and Remote Access Service

RRB Radio Rebroadcast • Radio Regulation Board • Red Rectangle Band • Requirements Review Board • Resource Review Board

RRC Radiative Recombination Continuum • Radio Relay Center • Radio Remote Control • Rapidly Rotating Core

• Ratio of Regression Coefficients • Reentry Rate Command • Regional Repair Center • Regional Research Centre • Reglamento del Registro Civil • Report Review Committee • Requirements Review Committee • Requirements Review Council • Review of Reported Cases • Rocket Research Co. • Roll Ratio Controller

RRCA Radar Reference Coverage Area

RRCT Reflectance Ratio Cloud Test

RRD Reactor Research and Development • Receive, Record and Display • Reclamation Receiving Document • Reformatted Raw Data • Reliability Requirements Directive

RRDA Range-Resolved Differential Absorption • Rendezvous Radar Deployment Actuator • Rendezvous Retrieval, Docking and Assembly

RRDB Recommended Requirements Database

RRDCT Risk Reduction Data Collection Task

RRDE Rotating Ring-Disk Electrode

RRDS Reduced Resolution Data Set

R&RE Radiation & Repair Engineering

RRE Marree (Australia-SA) (location identifier) • Range Rate Error • Reduced-Rank Extrapolation • Remote Range Extension

RREA Rendezvous Radar Electronics Assembly

RREL Risk Reduction Engineering Laboratory

RREU Rendezvous Radar Electronics Unit

RRF Rank Revealing Factorization • Reconnaissance Reporting Facility • Regional Radio Facility • Regional Relay Facility • Relation-Revealing Formula • Retirement Register File

RRFID Radio Reference Frame Image Database

RRG Radio Receiver Group • Remote Receiver Group • Requirements Review Group • Rodrigues Island (Mauritius) (location identifier) • Roll Reference Gyro • Very Red Giant (star)

RRH Relay Radio Hardware • Remote Reading Hygrometer

RRHT Remote Reading Hygrothermometer

RRI Barora (Solomon Islands) (location identifier) • Radio Refractive Index • Radio Research Instrument Inc. (USA-CT) • Raman Research Institute (India) • Range Rate Indicator • Range Response Improvement • Rectangular Region of Interest • Reimbursement and Refund Indicator • Relación Real de Intercambio • Rendezvous Radar Indicator • Response to Request for Information • Riverside Research Institute (USA-CA) • Rocket Research Institute

RRIF Reading Research Impact Factor

RRIS Remote Radar Integration Station

RRIT Réseau de Recherche et d'Innovation Technologique

RRJ Registration Reject

RRK Rourkela (India) (location identifier)

RRL Merrill (USA-WI) (location identifier) • Radio Recombination Line • Radio Relay Link • Radio Research Laboratory • Rayleigh Radiation Law • Reusable Rocket Launcher • Rocket Research Laboratory • Rudder Reference Line • Runway Remaining Lighting

RRLA Reunión Regional Latinoamericana de Astronomía

RRLD Recoverable Rocket Launcher Demonstrator

RRM Marromeu (Mozambique) (location identifier) • Radio Resource Management • Rapid Response Manufacturing • Rayleigh-Ritz Method • Red Resource Monitoring • Reglamento del Registro Mercantil • Relativistic Rotation Measure • Renegociable-Rate Mortgage

RRMD Real-time Radiation Monitoring Device

RRMPA Revue Roumaine de Mathématique Pures et Appliquées (periodical code)

RRN Rapid Response Network • Regional Research Network • Relative Record Number • Serra Norte (Brazil-PA) (location identifier)

rRNA Ribosomal Ribonucleic Acid

RRNS Redundant Residue Number System

RRO Rate and Route Operator • Record Route Object • Reproduction Rights Organisation • Sorrento (Italy) (location identifier)

R&RP Repair & Return Program

RRP Rapid Response Process • Recommended Retail Price • Reconnaissance Reference Point • Response Retrieval Project • Rotating Resonant Pulsator • Rover Robotic Payload • Rudder Reference Plane

RRPA Relativistic Random Phase Approximation

RRPF Réseau de Référence de Polynésie Française

RRPEV Reusable Rocket Powered Experimental Vehicle

RRPL Recommended Repair Parts List

RRQ Registration Request

R/R/R Reassignment/Retargeting/Reconstitution

R&RR Range & Rate Rate

RRR Rapid Runway Report • Read - Reset - Read (mode) • Reconnaissance Rapide des Radiances • Reflectivity and Roughness Radar • Release Readiness Review • Remove - Repair - Replace • Requirements Rational Report • Reset - Read - Read (sampling) • Return Receipt Requested • Rework Removal Rate • Rhône - Poulenc - Rohrer (company) • Risk Reduction Report

RRRA Radar Reflector Resolution Array

RRRP Resource Recovery Recycling Program

RRRQ Routing Repair Replacement Quantity

RRS Radio Receiver Set • Radio Recording Spectrophotometer • Radio Relay Site • Radio Relay Station • Radio Relay System • Radio Repeater Set • Radio Research Station • RAL Research Services • Reaction Research Society (USA-CA) • Reconciliation Sublayer • Record Separator • Regional Research Site • Relay Radio System • Remote Receiving Station • Remote Recovery Support • Remote Relay Station • Remote Relay System • Required Response Spectrum • Research Referral Service • Resource Recovery Service • Restraint Release System • Retrograde Rocket System • Retrorocket System • River and Rainfall Station • Roll Rate Sensor • Røros (Norway) (location identifier) • Rudolf-Römer-Sternwarte (Germany)

RRSF RACF Remote Sharing Facility • Regional Remote Sensing Facility • Repair and Return Status File

RRSP Regional Remote Sensing Programme

RRSSC Regional Remote Sensing Service Centre

RRT Radio Receiver/Transmitter • Radio Relay Terminal • Rail Rapid Transit • Rapid Rail Transit • Rapid Reprogramming Terminal • Raster Ray Tracer • Reentry Reference Time • Remote Ranging Terminal • Rendezvous Radar Transducer • Rendezvous Radar Transfer • Rendezvous Radar Transponder • Requirements Review Team

RRTM Rapid Radiative Transfer Model

RRTS Range-Rate Tracking System

RRU Radio Research Ubit • Remote Readout Unit

RRV Remote Reconnaissance Vehicle • Residual Risk Value • Robinson River (Australia-NT) (location identifier)

RRW Relative Risk Weighting

RRWDS Radar Remote Weather Display System

RRZ Radar Regulation Zone • Restricted Release Zone

RRZK Regionales Rechenzentrum Köln (Germany)

rs Rain and Snow • Sleet

Rs Reis (currency) • Rupees (currency)

R-S Reed-Solomon

R/S Range Safety • Redundant Set • Remote Sensing

R&S Readiness & Sustainability • Reconnaissance & Surveillance • Reliability & Safety • Research & Statistics

RS Radar Set • Radiated Susceptibility • Radiation Status • Radio Science • Radio Scolaire • Radio Set • Radio Shack • Radio Simulator • Radiosondage • Radiosounding • Radio Star • Radio Station • Radio Surveillance • Radio System • Rail Surveillance • Raman Scattering • Random Search • Range Safety • Rate Synthesizer • Rawinsonde • Reactive Search • Reader Service • Reader Stop • Readiness Station • Reading of Standard • Ready Service • Réalisme Subjectif • Received Standard • Receiver Segment • Receiver Sender • Receiver Station • Receive Segment • Reception Station • Recommended Specification • Recommended Standard • Reconnaissance Satellite • Recording Secretary • Record Separator • Record Start • Record Status • Recovery Site • Recovery System • Recruiting Service • Red Straggler • Red Switch • Reduced Slope • Redundancy Status • Redundant Set • Reed-Solomon • Reentry System • Reference Side • Reference Standard • Reformed Spelling • Refurbishment Spare • Register and Storage • Register to Storage • Regulating Station • Relay System • Reliability and Safety • Reliability Summary • Relocation Site • Remote Sensing • Remote Sentry • Remote Single-layer • Remote Site • Remote Station • Remote System • Representative Subset • Request for Support • Request to Send • Requirement Statement • Requirement Submission • Research Scientist • Research Station • Research Summary • Research System • Reset • Reset-Set • Resident School • Resource Scheduler • Resource Server • Respiratory System • Resume Sheet • Retrieval Strategy • Retrieval System • Retrospective Search • Reverse Shock • Reverse Slope • Revised Status • Revised Statutes • Right Side • Rio Grande do Sul (Brazil) • Rocket System • Roll Stabilization • Route Selector • Royal Society (UK) • Russian Segment • Rymdstyrelsen

R&SA Repair & Salvage Authority

RSA Radar with Synthetized Aperture • Rand McNally Statistical Area • Range Safety Approval • Range Standardization and Automation • Rapid Simulation Aid • Rate Sensor Assembly • Rate System Analyst • Regional Security Affairs • Relative Specific Activity • Relative Standard Accuracy • Remote Sensing Application • Remote Standardization and Automation • Remote Storage Activity • Republic of South Africa • République Sud-Africaine • Research Support Agency (South Africa) • Retrograde Storage Area • Reusable Software Asset • Revised Shapley-Ames (catalog) • Revue de Statistique Appliquée • Rheometrics Solids Analyzer • Rivest - Shamir - Adleman (algorithm) • Royal Society of Arts (UK) • Runway Safety Area • Rural Services Area • Russian Space Agency (now RASA) • Santa Rosa (Argentina-LP) (location identifier)

RSAA Remote Sensing Association of Australia • Research School of Astronomy and Astrophysics

RSAC Recreational Software Advisory Council • Remote Sensing Applications Centre • Remote Sensing Applications Committee

RSADU Remote Sensing Applications Development Unit

RSAI Regional Science Association International • Royal Society of Antiquaries of Ireland • R.S. Automation Industrie (France)

RSAL Remote Sensing Applications Laboratory

RSAM Remote Sensing Applications Mission

RSAO Radio Science Atmospheric Occultation

RSAR Risk Status Accounting Report

RSAS Rand Strategic Assessment System • Reusable Software Architecture for Spacecraft

RSAT Rotating Slit Aperture Telescope

RSAW Resistive Shear Alfvén Wave

RSB Radiation Safety Branch • Range Safety Beacon • Rapport Signal sur Bruit • Reference Standards Book • Replica Symmetry Breaking • Research Services Branch • Research Support Branch • Restseitenband • Roseberth (Australia-QLD) (location identifier) • Rudder/Speed Brake

RSBAM USSR Report Space Biological Aerospace Medicine (periodical code)

RSBE Repository-Based Software Engineering

RSBMS Rotating Shadow Band Multispectral Radiometer

RSBR Réseau Synoptique de Base Régional

RSBS Rotating Shadow Band Spectrometer

rsc Resource (file extension)

Rsc Rescue

RSC Radar Scan Converter • Radio Source Coordinates • Range Safety Center • Range Safety Command • Range Safety Computer • Range Safety Control • Reader Service Card • Real-time Schedule Controller • Record Status Code • Regional Support Center • Regional Support Command • Relevance Score Contribution • Remote Sensing Centre • Remote Sensing Committee • Remote Sensing Course • Remote Switching Center • Request System Control • Research School of Chemistry • Reset Circuit • Resource Strategy Committee • Richard Stockton College (USA-NJ) • Riga Skulte (Latvia) (location identifier) • RISC Single Chip • Royal Society of Canada • Royal Society of Chemistry (UK) • Runway Surface Condition

RSCAT Radar Scatterometer

RSCC Range Schedule Control Center • Remote-Site Command Computer • Remote-Site Computer Complex • Roane State Community College (USA-TN) • Russian Segment Cargo Crane

RSCCE Replacement Satellite Configuration Control Element

RSCE Rocket Space Company Energia (Russia)

RSCG Radio Set Control Group

Rsch Research

RScI Review of Scientific Instruments (periodical code)

RSCIE Remote-Station Communication Interface Equipment

RSCL Radioactive Sodium Chemistry Loop • Reactor Simulation and Control Laboratory

RSCN Registered State Change Notice

RSCNJ Richard Stockton College of New Jersey (USA-NJ)

RSCO Range Schedule Control Officer

rscr Resource (file extension)

RSCR Range Safety Command Receiver

RSCS Range Safety Command System • Rate Stabilization and Control System • Remote Source Control Center • Remote Spooling and Connection System • Remote Spooling and Control System • Remote Spooling Communications System (*)

RSCSS Range Safety Command Shutdown System

RSCTC Rough Sets and Current Trends in Computing (conference)

RSCU Regional Soil Conservation Unit

RSD Radiance Spectral Distribution • Radio Science Division • Rainbow Schlieren Deflectometry • Range Support Document • Rapid System Development • Regroupement de Secteurs Disciplinaires • Relative Standard Deviation • Remote Sensing Data • Remote Sensing Division • Repairable Space Division • Requirements and Specifications Document • Residual Standard Deviation • Responsible System Designer • Right Side Down • Rock Sound (Bahamas) (location identifier) • Route Server Daemon

RSDB Remote Sensing Database

RSDC Range Safety Data Controller • ROSAT Science Data Centre (*) • ROSAT Scientific Data Center

RSDD Remote Special Devices Demultiplexer

RSDIS Remote Sensing Data and Information Services

RSDL Reverse Spiral Dual Layer

RSDO Rapid Spacecraft Development Office (NASA)

RSDP Remote-Site Data Processor • ROSAT Science Data Processing • Routine Science Data Processing • Routine Scientific Data Processing

RSDS Radar Signal Detecting Set • Range Safety Display System

RSDT Remote-Station Data Terminal

Rse Remise

R&SE Research & Systems Engineering

RSE Radio Science Experiment • Received Standard English • Receiving Site Equipment • Rees-Sciama Effect • Reference Standards Equipment • Relative State Estimator • Remote Single-layer Embedded • Remote Stream Environment • Request Select Entry • Research Scientists and Engineers • Revised Standard Edition • Rich Structure Element • Royal Society of Edinburgh (UK-Scotland) • Sydney - Au-Rose Bay (Australia-NSW) (location identifier)

RSEA Responsible System Engineering Authority

RSEC Representative Shuttle Environmental Control

RSECS Representative Shuttle Environmental Control System

RSED Research and Scientific Exchanges Division

RSEF Real Sociedad Española de Física (Spain)

RSEM Radio Signal Environment Monitor

RSEMS Remote Sensing Electro Magnetic Spectrum (periodical code)

RSEnv Remote Sensing of Environment (periodical code)

RSEPS Royal Society of Edinburgh Proceedings Section (periodical code)

RSER Remote Sensing of Earth Resources

RSES Research School of Earth Sciences

RSETI Radio Search for Extraterrestrial Intelligence

RSEXEC Resource Sharing Executive

RSF Rapid Star Formation • Receiving-Safing Facility • Referencing Structure Function • Refurbish and Subassemblies Facility • Refurbishment and Subassembly Facility • Region of Star Formation • Remote Site Facility • Remote Support Facility • Reporters Sans Frontières • Retail Stock Fund

RSFG Route Server Functional Group

RSFP Resonant Spin Flavor Precession • Rivet Switch Follow-on Program

RSFQ Rapid Single Flux Quantum

RSG Range Safety Group • Rate Signal Generator • Rate Support Grant • Rate Switching Gyro • Readiness Support Group • Receding Side Gap • Recirculating Steam Generator • Red Supergiant (star) • Reference Signal Generator • Regional Sponsor Group • Remote Sensing Group • Research Steering Group • Research Study Group • Reset Group • Resource Strategy Group • Rising • Serra Pelada (Brazil-PA) (location identifier)

RSGS Remote Sensing Ground Station • Royal Scottish Geographical Society (UK-Scotland)

rsh Remote Shell

RSH Remote Shell • Restricted Shell • Russian Mission (USA-AK) (location identifier)

RSHU Russian State Hydrometeorological University (Russia)

RSI Radiation Systems Inc. • Radio Science Investigation • Rate Stability Indicator • Rationalization - Standardization - Interoperability • Recept - Storage - Issue • Record Status Indicator • Red Spot Hollow (Jupiter) • Relative Specialization Index • Remotely Sensed Imagery • Remote Sensing Institute • Repetitive Strain Injury • Repetitive Stress Injury • Replacement Stream Input • Research Science Institute • Research Studies Institute • Research Systems Inc. • Responsable du Système d'Information • Retrospective Searching Information • Reusable Surface Insulation • Rio Sidra (Panama) (location identifier) • Roll Stability Indicator • Rotational Shearing Interferometer • Rules, Standards and Instructions • Rural Systemic Initiative (USA)

RSIAL Remote Sensing and Image Analysis Laboratory

RSIC Radiation Shielding Information Center • Research Institute for Social Development

RSID Radar Simulator Instrument-on-a-Disk

RSIM Retrospective Single-Ion Monitoring • ROSAT Simulator

RSIMS Remote Sensing Information Management System

RSIOP Red Single Integrated Operations Plan

RSIP Realm-Specific Internet Protocol • Remote Sensing and Image Processing

RSIS Radio Science Investigation for SMART-1

RSITA Règlement du Service International des Télécommunications de l'Aéronautique

RSIU Radio Set Interface Unit

RSJ Rosario (USA-WA) (location identifier)

RSK Ransiki (Indonesia) (location identifier)

RSL Radio Standards Laboratory • RAISE Specification Language • Received Signal Level • Reconciliation Sublayer • Reference Standards Laboratory • Relative Sea Level • Remote-and-Status Link • Remote Sensing Laboratory • Répétition Système Lanceur • Request-and-Status Link • Requirement Specification Language • Research Support and Logistics • Resource Specification Language • Resource Support List • Reusable Software Library • Runtime System Library • Russell (USA-KS) (location identifier)

RSLA Range Safety Launch Approval

RSLP Reentry System Launch Program • Rocket System Launch Program

RSLPS Royal Society of London Proceedings Series (periodical code)

RSLPT Royal Society of London Philosophical Transactions (periodical code)

RSLS Redundant Set Launch Sequence • Runway Status Light System

RSLV Reentry System Launch Vehicle • Resolve

Rslvr Resolver

RSM Radiation Signature Measurement • Radiation Survey Meter • Range Safety Manual • Rapid Search Machine • Reference Sky Map • Regional Sales Manager • Remote Subscriber Multiplexer • Remote Switching Module • Removable Storage Management • Rendezvous Safety Monitoring • Replication Server Manager • Reppublicca di San Marino • Resource Management System • Response Surface Methodology • Resume • Royal Society of Medicine (UK) • San Marino (international vehicle registration code) • San Marino (postal country code)

RSMA Remote Sensing of the Middle Atmosphere

RSMC Regional Specialized Meteorological Centre

RSMCB Rank and Selection and Multiple Comparisons with the Best

RSMD Remote Signal Multiplexer/Demultiplexer

RSMIP Remote Sensing Management Information Program

RSML Rule Structure Markup Language

RSMM Redundant System Monitor Model

RSMO Regional Storage Management Office

RSMSW RVD Safety Monitoring Software

RSMU Remote Satellite Measurement Unit

Rsn Reason

RSN Radio Supernova • Real Soon Now • Record Serial Number • Relative Sunspot Number • Ruston (USA-LA) (location identifier)

RSNC Réseau Synoptique National Complémentaire

RSNe Radio Supernovae

RSNs Remnantes de Supernova

RSNZ Royal Society of New Zealand

RSO Radar à Synthèse d'Ouverture • Radiation Safety Office • Radiation Safety Officer • Radiological Safety Office • Radiological Safety Officer • Radiosonde Observation • Radio-Source Object • Range Safety Office • Range Safety Officer • Range Support Operation • Rankin Science Observatory (USA-NC) • Reconnaissance System Operator • Recreational Sky Observer • Recruiting Staff Office • Recruiting Staff Officer • Red Stellar Object • Regional Security Office • Regional Security Officer • Regional Staff Office • Regional Staff Officer • Research in the Sociology of Organizations • Resident Space Object • Runway Supervisory Office • Runway Supervisory Officer • Rural Suboffice

RSOC Rockwell Space Operations Co. (USA) • Rosetta Science Operations Centre

RSOF Required Special Optional Feature

RSOG RadioAstron Science Operations Group (Russia)

RSOI Radius of Sphere Of Influence

RSOP Range Safety Operational Plan

RSOR Range Safety Operations Requirements

RSP Radar Signal Processor • Radio Switch Panel • Range Solar Panel • Range Sorting Program • Range Support Plan • Rapid Site Preparation • Raspberry Strait (USA-AK) (location identifier) • Rate Sensing Package • Readiness Spare Package • Record Select Program • Recoverable Sensor Package • Red Switch Program • Reinforced Structural Plastic • Remote Sensing Project • Rendezvous Station Panel • Request Serial Poll • Required Space • Research and Survey Programme • Residual Support Plan • Resources Survey Project • Responder Beacon • Response • Roll Stabilization Platform

RSPA Research and Special Programs Administration

RSPAC Remote Sensing Public Access Center

RSPC Range Safety Processing Center • Relevé Systématique du Plan Galactique

RSPCA Royal Society for the Prevention of the Cruelty to Animals (UK)

Rspd Respond

RSPIC Residual South Pole Ice Cap

RSPL Recommended Spare Parts List

RSPP Remote Sensing Preparatory Programme

RSPPG Radio and Space Plasma Physics Group

Rsps Response

RSPS Range Solar Panel Substrate • Research School of Pacific Studies

Rspsb Responsible

RSPT Rapid Screen Prototyping Tool • Real Storage Page Table

Rspv Respective • Responsive

Rsq Rescue

RSQ Remote Sensing Quarterly (periodical code) • Rescue

RSQC Reliability, Safety and Quality Control

RSR Radar Service Request • Rain Scattering Radiometer • Rain Scatterometer and Radiometer • Range Safety Report • Rapid Solidification Rate • Recall - Screen - Repair • Recall - Screen - Return • Release Schedule Record • Release Specification Review • Repair Survival Rate • Request for Scientific Research • Required Supply Rate • Resources Status Report • Resources Status Review • Rohini Sounding Rocket • Rotating Shadowband Radiometer

RSRB Redesigned Solid Rocket Booster • Remote Source Route Bridging

Rsrc Resource

RSRC Resource System Review Committee

RSRE Royal Signals Research Establishment (UK)

RSRF Relative Spectral Response Flux • Relative Spectral Response Function (*)

RSRM Redesigned Solid Rocket Motor • Reusable Solid Rocket Motor (*)

RSRMP Regional Solar Radiation Monitoring Project

RSRP Remote Sensing Research Program

RSRS Range Safety Receiving Station

RSRT Resort

RS&S Receiving, Shipping & Storage

RSS Radar Subsystem • Radiometric Sextant System • Radio Science System • Radio Subsystem • Random Subset Selection • Range Safety Switch • Range Safety System • Rapid Scanning of Spectra • Rapid Structure Search • RDF Site Summary • Reactants Supply System • Readiness System Sustainment • Receiving, Shipping and Storage • Reference Sound Source • Refrigeration Subsystem • Regional Support System • Relaxed Static Stability • Reliability - Supportability - Survivability • Remote Sensing Science • Remote Sensing Society (UK) • Remote Sensing System • Remote Shutdown System • Remote Site System • Remote Slave Set • Remote Slave Station • Remote Speaking System • Remote Subscriber Switch • Remote Switching System • Requirements Status System • Requirements Support System • Réseau Santé Social (France) • Resident Set Size • Residual Sum of Squares • Resource Scheduling Service • Resource Sharing System • Resource Survey Satellite • Restart Selectover System • Return Signal Simulator • Rib Structure Station • Rigid Space Structure • Root-Sum-Square • Root-Sum-Squared • Roseires (Sudan) (location identifier) • Rotating Service Structure • Rotating Shadowband Spectrometer • Routing and Switching System • Routing Subsystem • Royal Statistical Society (UK) • RTG Support Structure • Runway Supervisory Set

RSSA Regional Supply Support Activity

RSSC Refrigerant Self-Sealing Coupling • Regional Satellite Support Cell • Regional Space Support Cell • Regional Space Support Center • Remote Sensing Science Council • Remote-Site Simulator Console

RSSD Research and Scientific Support Department (ESA)

RSSDB RDF Schema-Specific Database

RSSI Receive Signal Strength Indicator • Responsable de la Sécurité des Services Informatiques • Russian Space Science Internet

RSSIA Regional Security Strategy Implementation Analysis

RSSM Russian Segment Service Module (ISS)

RSSN Research Space Surveillance Network

RSSPO Resident Space Shuttle Project Office

RSSS Radio Spectrum Surveillance System • Reusable Space Shuttle System

RSSU Remote-Site Simulation Unit

Rst Restore

RST Radiometric Sun Tracer • Range Search and Track • Read Symbol Table • Readibility - Strength - Tone • Reentry System Technology • Relation Soleil-Terre • Remote Science Terminal • Remote Switching Terminal • Representative Style Table • Reschedule Team • Research on Science and Technology • Research Search Tool • Research Study Team • Reset • Reset-Set Trigger • Resolution Syntax Tree • Resonance Scattering Theory • Resonant Sphere Technique • Rest • Restart • Revue Scientifique et Technique CECLES CERS (periodical code) • Rhetorical Structure Theory • Rochester (USA-MN) (location identifier) • Rough Set Theory • Rough-Surface Tester • Routine Sequence Table

RSTC Remote-Site Telemetry Computer

RSTER Radar Surveillance Technology Experimental Radar

RSTF Remote Sensing Task Force

Rstg Roasting

RSTM Relay Signal Transmitting Monitor

RSTN Radio Solar Telescope Network • Railroad Station • Regional Seismic Test Network

Rstnt Restaurant

RSTP Rapid Spanning Tree Protocol • Remote-Site Telemetry Processor

Rstr Raster • Restrain • Restrict • Restricted

RSTRT Restart

RSTS Resource System Time-Sharing

RSTS-E Resource System Time-Sharing - Enhanced

RSTSE Resource System Time-Sharing - Enhanced

RSTV Real-Time Synthetic Video

RSU Rate Sensing Unit • Rate Sensor Unit • Receiver-Sender Unit • Recommended Service Upgrade • Recovery Storage Unit • Relay Storage Unit • Relay Switching Unit • Remote Sensing Unit • Remote Service Unit • Remote Switching Unit • Rescue Support Umbilical • Ridiculously Small Unit • Rivers State University (Nigeria) • Rostov State University (Russia) • Runway Supervisory Unit • Yosu (Korea-ROK) (location identifier)

Rsv Reserve

RSV Radio-Sondage Vent • Range Safety Van • Ready Storage Vessel • Red Supergiant Variable (star) • Request Service • Reserve • Reservoir • Resolve • Responsable Sauvegarde Vol • Resupply Vehicle • Retrieval Status Value • Revised Standard Version

RSVA Réseaux et Services à Valeur Ajoutée

RSVD Rapid Sequential Visual Display • Reserved • Riemannian Singular Value Decomposition

Rsvn Reservation

RSVP Rapid Serial Visual Presentation • Rare Symmetry Violating Process • Répondez S'il Vous Plaît • Reservation Protocol • Resource Reservation Protocol

Rsvr Reservoir

RSVR Reservoir

RSW Fort Myers - SW Florida (USA-FL) (location identifier)

RSWC Ride Side up With Care

RSWG Response Strategies Working Group

RSWGSC Response Strategies Working Group Steering Committee

RSWK Regeln für den Schlagwortkatalog

RSWP Remote Sensing Working Party

RSX Realistic Sound Experience • Real-time Resource-sharing Executive • Rouses Point (USA-NY) (location identifier)

RSYS Responsible System

Rt Rate • Right

R/T Radio-Telegraph • Radio-Telegraphy • Radio-Telephone • Radio-Telephony • Real Time • Receiver/Transmitter • Receive/Transmit • Record of Trial • Reference Trajectory • Reflection/Transmission • Room Temperature

R&T Recherche & Technologie • Research & Teaching • Research & Technology

R-T Resistance Test

RT Radar Tracking • Radiative Transfer • Radiographic Test • Radiographic Testing • Radio Technician • Radio-Telegraph • Radio-Telegraphy • Radio-Telephone • Radio-Telephony • Radio-Telescope • Radio-Télescope • Radioteleskoop • Radio Tower • Radio Tracking • Radio Transceiver • Radio Transmitter • Range Timing • Range to Target • Range Tracking • Rated Time • Rate Transmitter • Ratio Transformer • Rayleigh-Taylor • Reaction Time • Reading Test • Real Time • Received Text • Receiver-Transmitter • Receive Timing • Reconditioning Test • Record Technician • Record Terminator • Record Transfer • Reduction Table • Reference Trajectory • Regression Testing • Related Term • Relativitätstheorie • Relay Tester • Relay Transmitter • Released Time • Relevance Threshold • Reliable Transfer • Relocatable Target • Remote Terminal • Remote Transmitter • Report Test • Réseau de Télécommunications • Resistance Test • Resource Team • Response Time • Retention Time • Retransmit Message • Retro Table • Return Ticket • Reverse Transcription • Right • Right Turn • Ring Trap • Ring Trip • RISC Technology • Rise Time • Room Temperature • Root Task • Rough Terrain • Round Table • Round Trip • Routing Table • Runtime • Ryle Telescope

RTA Rapid Thermal Annealing • Real-Time Access • Real-Time Accumulator • Real-Time Activity • Real-Time Analyser • Real-Time Analysis • Real-Time Application • Real-Time Assessment • Receiver/Transmitter Antenna • Receiver/Transmitter Assembly • Reciprocal Trade Agreement • Reliability Test Assembly • Remote Trunk Arrangement • Research and Technology Agency • Responsible Test Activity • Responsible Test Agency • Revealed Technological Advantage • Rewriting Techniques and Applications (conference) • Roads and Traffic Authority (Australia) • Rotuma Island (Fiji) (location identifier)

RTAC Real-Time Adaptive Control • Real-Time Atmospheric Compensation • Regional Technical Assistance Center • Research and Technology Advisory Committee • Research and Technology Advisory Council

RTACF Real-Time Auxiliary Computing Facility

RTACS Real-Time Adaptive Control System • Real-Time Automated Control System

RTAD Real-Time Attitude Determination • Remote Tower Alphanumerics Display

RTADS Real-time Telemetry Analysis Data System

R-TAE Resonant Toroidicity-induced Alfvén Eigenmode

RTAE Resonant Toroidicity-induced Alfvén Eigenmode

RTAM Remote Telecommunications Access Method • Remote-Terminal Access Method (*) • Resident Terminal Access Method

RTAP Real-Time Application Platform

RTAS Real-Time Advisory System • Real-time Technology and Applications Symposium

RTB Read Tape Binary • Real-Time Base • Red de Telefonía Bàsica (Spain) • Remote Test Base • Research and Technology Board • Resistance Temperature Bulb • Response/Throughput Bias • Return To Base • Return To Brightest • Roatan (Honduras) (location identifier) • Robot Tour Problem

RTBA Rate To Be Agreed

RTBF Radio Télévision Belge de la Communauté Française (Belgium) • Radio Télévision Belge Francophone (Belgium)

RTBP Restricted Three-Body Problem

RTC Radio Transmission Control • Ratnagiri (India) (location identifier) • Reader Tape Contact • Real-Time Channel • Real-Time Clock • Real-Time Command • Real-Time Compression • Real-Time Computer • Real-Time Control • Real-Time Counter • Red Telefónica Comutada • Regional Training Center • Relative-Time Clock • Relative-Time Command • Remote Terminal Cluster • Remote Terminal Concentrator • Remote Terminal Controller • Removable Top Closure • Report To Congress • Requirements Type Contract • Research and Teaching Committee • Research and Training Centre • Réseau Téléphonique Commuté • Resolution Trust Corporation • Return To Control • Room Temperature Cure • Round Table Conference

RTCA Radio-Technical Commission for Aeronautics • Requirements and Technical Concepts for Aeronautics

RTCAC Regional Tropical Cyclone Advisory Centre

RTCC Real-Time Command Controller • Real-Time Computer Center • Real-Time Computer Command • Real-Time Computer Complex • Research and Technology Coordinating Committee • Rough Terrain Container Crane

RTCE Rotation/Translation Control Electronics

RTCF Real-Time Computer Facility

RTCH Rough Terrain Container Handler

Rtcl Reticle

RTCL Reticle

RTCM Radio-Technical Commission for Maritime services

RTCON Real-Time Connect

RTCP Real-Time Communications Processor • Real-Time Control Protocol • Real-time Transport Control Protocol • Relative Time Command Processor

RTCPS Real-Time Computer Program System

RTCS Real-Time Command Sequence • Real-Time Communications Service • Real-Time Communications System • Real-Time Computer System • Real-Time Computing System • Real-Time Control System

RTCU Real-Time Control Unit

Rtd Returned

RTD Radar Technology Demonstrator • Range Time Decoder • Rate Damping • Read Tape Decimal • Real-Time Data • Real-Time Decision • Real-Time Disk • Real-Time Display • Real-Time Drive • Real-Time Driver • Recherche et Développement Technologique • Reliability Technical Directive • Remote Temperature Detector • Replacement Task Distribution • Research and Technological Development • Research and Technology Development • Research - Technology - Development • Resistance Temperature Detector • Resistance Temperature Device • Resistance Thermal Device • Resonant Tunneling Device • Resonant Tunneling Diode • Returned To Duty • Return To Duty • Rise Time Discriminator • Rotunda (USA-FL) (location identifier) • Round-Trip Delay

Rtda Rotunda

RTDAS Real-Time Data Automation System

RTDBS Real-Time Database System

RTDC Real-Time Data Channel • Real-Time Data Compressor • Real-Time Data Controller

RTD&D Research, Technological Development & Demonstration

RTDD Real-Time Data Distribution • Release To Design Development • Remote Timing and Data Distribution • Research, Technological Development and Demonstration

RTDDC Real-Time Digital Data Conversion

RTD&E Research, Development, Test & Evaluation

RTDE Range Time Data Editor

RTDF Residence Times Density Function

RTDGDS Real-Time Differential Global Positioning System

RTDHS Real-Time Data Handling System

RTDM Real-Time Data Migration • Réseau de Transmission et de Données de la Marine

RTDMP Real-Time Deformation Matrix Production

RTDR Real-Time Data Recording • Reliability Test Data Report

RTDS Real-Time Data System • Reutilization - Transfer - Donation - Sale

RTDT Real-Time Data Translator

Rte Remite • Remitente • Route

RT&E Research, Test & Evaluation

RTE Marguerite Bay (USA-AK) (location identifier) • Radiative Transfer Equation • Real-Time Execution • Real-Time Executive • Receiver Test Equipment • Recovery Techniques Evaluation • Refractory Trace Element • Reliability Test Evaluation • Remote Tele-Excavation • Remote-Terminal Emulation • Remote-Terminal Emulator • Remote-Terminal Equipment • Research, Test and Evaluation • Réseau Transeuropéen • Resident Training Equipment • Resonant Transfer with Excitation • Responsible Test Engineer • Restricted Trace Ensemble • Return To Earth • Route • Runtime Efficiency • Runtime Environment

RTEAM Real-Time Environmental Agriculture Monitor

RTECS Registry of Toxic Effects of Chemical Substances

RTED Return-To-Earth Digital

RTEE Real-Time Engineering Environment

RTEL Reverse Telnet

RTEP Real-Time Event Processor

RTES Real-Time Executive System

rtf Rich-Text Format (file extension)

RTF Radar Test Facility • Radiodiffusion Télévision Française (France) • Radiotelephone • Random Telegraph Function • Real-Time Flow • Real-Time Fortran • Reconnaissance Technical Flight • Regional Task Force • Remote Task Force • Remote Test Facility • Resistance Transfer Factor • Return To Flight • Revision Task Force • Rich-Text File • Rich-Text Format (*) • Robotic Training Facility • Rocket Test Facility • Rotational Test Facility • Run Time File

RTFF Read The FAQ First

RTFM Real-Time Flow Management

RTFSS Real-Time Flight Safety Software

Rtg Rating

RTG Radioisotope Thermal Generator • Radioisotope Thermoelectric Generator • Radiotelegraph • Range To Ground • Reconnaissance Technical Group • Relativistic Theory of Gravitation • Requirements Tape Generator • Ruteng (Indonesia) (location identifier)

RTGC Real-Time Gaze Control

RTH Real-Time Muon Histogram • Regional Telecommunication Hub • Réseau Technique et Humain

RTHC Rotational-Translational Hand Controller • Rotation-Translation Hand Controller

RTHS Real-Time Hybrid System

RTI Radiation Transfer Index • Range-Time Indicator • Range-Time Intensity • Rayleigh-Taylor Instability • Real-Time Image • Real-Time Interface • Real-Time Interrupt • Referred-To-Input • Reflex Transmission Imaging • Regional Training Institute • Relational Technology Inc. • Repetitive Talk Injury • Research - Technology - Innovation • Research Triangle Institute • Rise-Time Indicator • Road Transport Informatics • Robotic Telescope Interface • Roti (Indonesia) (location identifier) • Routing Indicator • Runtime Infrastructure • Runtime Instrumentation • Runtime Interconnect • Runtime Interface

RTIC Real-Time Information to the Cockpit • Real-Time Into the Cockpit • Regional Technical Information Center

RTIF Real-Time Interface

RTIN Return Topology Information

RTIO Radiothérapie Intraopératoire • Real-Time Input/Output • Remote Terminal Input/Output

RTIP Radar Technology Insertion Program • Real-Time Interactive Processor • Real-Time Internet Protocol • Remote-Terminal Interactive Processor

RTIR Real-Time Imaging Reconstruction

RTIRS Real-Time Information Retrieval System

RTIS Raytheon Texas Instruments Systems • Remote Terminal Input System

RTJ Ring Type Joint

RTK Real-Time Kinematic • Recognition Toolkit • Revenue Ton Kilometer • Right To Know

RTL Radio Télévision Luxembourg • Raster Transfer Language • Ready To Launch • Real-Time Language • Register Transfer Language • Register Transfer Level • Resistor-Transfer Logic • Resistor-Transistor Logic • Right To Left • Runtime Library • Spirit Lake (USA-IA) (location identifier)

RTLP Receiver Test Level Point

RTLS Return To Launch Site

RTLT Round Trip Light Time

RTM Radiation Test Model • Radiative Transfer Model • Read The Manual • Ready To Move • Real-Time Metric • Real-Time Mode • Real-Time Monitor • Receiver - Transmitter - Modulator • Recording Tachometer • Recovery Termination Management • Regional Telecommunications Management • Register Transfer Module • Release To Manufacturing • Remote Test Module • Requirements and Traceability Management • Requirements Traceability Management • Requirements Traceability Matrix • Requirements Transition Model • Requirements Translation Model • Research Technical Memorandum • Réserve Technique Minimale • Resident Transport Module • Resin Transfer Molded • Resin Transfer Molding • Resin-Transfer Moulding • Response Time Management • Response Time Monitor • Robot Task Measurement • Rolling Target Mount • Rotterdam (Netherlands) (location identifier) • Running Time Meter • Runtime Manager

RTMB Routing Table Maintenance Protocol

RTMC Riverside Telescope Makers Conference (USA-CA)

RTMD Real-Time Multiplexer Display

RTML Remote Telescope Markup Language (*) • Robotic Telescope Markup Language • Rule Triggering Markup Language

RTMM Reduced Thermal Mathematical Model

RTMOS Real-Time Multiprogramming Operating System

RTMP Real-Time Media Protocol • Routine Table Maintenance Protocol • Routing Table Maintenance Protocol (*)

RTMS Radio Telephone Mobile System • Real-Time Multiprogramming System • Rocket Thrust Measuring System

Rtn Retain • Return

RTN Radial - Tangential - Normal • Radio Telescope Network • Raton Crews (USA-NM) (location identifier) • Received Track Number • Recursive Transition Network • Registered Trade Name • Remote-Telescope Network • Remote-Terminal Network • Remote-Tracking Network • Report Test Number • Research and Training Network • Research Training Network • Resistor Terminating Network • Return • Routing and Transit Number • Rymdteckniknätverk

Rtng Retaining • Returning

Rtnr Retainer

RTNS Retransmission

R&TO Research & Technology Organization

RTO Colton Morrow (USA-CA) (location identifier) • Radiotelephone Officer • Radiotelephone Operator • Rapid Turn-On • Real-Time Object • Real-Time Observation • Real-Time Operation • Real-Time Option • Recovery Time Objective • Referred-To-Output • Reflecting Telescope Optics • Rejected Takeoff • Research and Technology Organization • Responsible Test Organization • Retransmit Time Out

RTOC Real-Time Out of the Cockpit

RTOF Reflecting Time Of Flight (spectrometer)

RTOK Retest OK

RTOL Reduced Takeoff and Landing

RTOP Reformed Teaching Observation Protocol • Research and Technology Objectives and Plans • Research and Technology Operating Plan • Research and Technology Operations Plan

RTOPS Real-Time Operations

RTOS Real-Time Operating System

RTP Rapid Thermal Processing • Rapid Transport Protocol • Rated Thermal Power • Real-Time Peripheral • Real-Time Photogrammetry • Real-Time Photometry • Real-Time Position • Real-Time Positioning • Real-Time Processor • Real-Time Protocol • Real-time Transport Protocol • Remote Terminal Program • Remote-Transfer Point • Request for Technical Proposal • Request To Purchase • Requirement and Test Procedure • Research and Technology Project • Research and Training Programme • Research Triangle Park (USA-NC) • Reservation Transfer Protocol • Return To Pivot • Rijnhuizen Tokamak Project • Room Temperature and Pressure • Room Temperature Phosphorescence • Rutland Plains (Australia-QLD) (location identifier)

RTPE Real-Time Parameter Estimation

RTpH Round Trips per Hour

RTPH Round Trips Per Hour

RTPL Real-Time Procedural Language

RTPM Real-Time Program Management

RTPMS Real-Time Pressure Monitoring System

RTPS Remote Tape Processing System

RTQ Read The Question • Response To Query

RTQC Real-Time Quality Control

Rtr Retour • Rotor

RTR Ready To Run • Real-Time Reconnaissance • Real-Time Record • Receiver-Transmitter • Reel-To-Reel • Reliability Test Requirement • Remote Transmitter/Receiver • Rotor • Router

RTRC Radiotelemetry and Remote Control • Regional Teacher Resource Center

RTRD Remote Tower Radar Display • Retard

RTRE Radiotekhnika Telecommunications Radio Engineering (periodical code)

RTRN Return

RTRO Real-Time Readout

RTRP Reinforced Thermosetting Resin Pipe

RTRSOC Real-Time Reporting System in Oceanic Conditions

RTrTe Revista Transporturilor si Telecomunicatiilor (periodical code)

Rtrv Retrieve

Rtry Rotary

Rts Rights

R&TS Research & Technical Standards

RTS Radar Tracking Station • Radio Télévision Scolaire • Range Time Signal • Range Timing System • Range Tracking Station • Range Tracking System • Rapid Targeting System • Rauch - Tung - Striebel (smoother) • Reactive Tabu Search • Ready To Send • Real-Time Sequencer • Real-Time Server • Real-Time Simulation • Real-Time Simulator • Real-Time Software • Real-Time Supply • Real-Time System • Recorded Time Signal • Regional Training Site • Relative Time Sequence • Relay Telemetry System • Relay Test System • Reliability Test Specimen • Reliable Transfer Service • Reliable Transfer Syntax • Remote Takeover System • Remote Terminal Simulator • Remote Terminal System • Remote Testing System • Remote Thread Server • Remote Tracking Site • Remote Tracking Station • Remote Transfer Switch • Remote Triaxial Sensor • Remote Trunking System • Request To Send • Requirements Tracking System • Research Test Site • Residual Time Stamp • Return To Search • Return To Sender • Return To Service • Risk Theory Society • Rottnest Island (Australia-WA) (location identifier) • Runtime System

RTSA Real-Time Situational Awareness • Real-Time Structure Analysis

RTSC Range Technical Services Contractor • Reuse Technology Support Center

RTSD Resources and Technical Services Division

RTSDS Real-Time Scheduling Display System

RTSE Reliable Transfer Service Element

RTSEE Remote Triaxial Sensor Electronics Enclosure

RTSF Real-Time Simulation Facility

RTSI Real-Time System Integration

RTSJ Real-Time Specification for Java

RTSMP Real-Time Symmetric Multiprocessor

RTSP Real-Time Signal Processor • Real-Time Special Processing • Real-Time Streaming Protocol

RTSQL Real-Time Structured Query Language

RTSS Real-Time Scientific System • Real-Time Simulation System • Real-Time System Simulator

RTSW Real-Time Solar Wind

RTSX Ranger Telerobotic Shuttle Experiment

RTT Radioteletype • Radioteletypewriter • Réduction du Temps de Travail • Régie des Télégraphes et des Téléphones • Regie van Telegraaf en Telefoon • Regie voor Telefoon en Telegrafie • Resonant-Tunneling Transistor • Road Transport Telematics • Round-Trip Time • Round-Trip Timing • Rural Telegraph Terminal

RTTB Radio To The Building

RTTC Regional Technical Transfer Center • Regional Technology Transfer Center

RTTD Real-Time Telemetry Data

RTTDS Real-Time Telemetry Data System

Rtte Restaurante

RTTF Red Taurus Tunable Filter

RTTI Runtime Type Information

RTTM Retransmission Time Measurement

RTTS Real-Time Telemetry System • Real-Time Transmission System

RTTV Real-Time Television

Rtty Radioteletype

RTTY Radioteletype • Radio Teletypewriter

RTU Radio Test Unit • Real-Time Unit • Real-Time Unix • Receiver/Transmitter Unit • Recovery Task Unit • Remote Telemetry Unit • Remote Terminal Unit • Remote Test Unit • Response Test Unit • Retransmission Unit • Right To Use

RTUTE Remote-Terminal Unit Test Equipment

RTV Rapid Terrain Visualization • Raster To Vector • Real-Time Video • Recoverable Test Vehicle • Recovery Test Vehicle • Returned To Vendor • Return To Vendor • Rocket Test Vehicle • Room-Temperature Vulcanized • Room-Temperature Vulcanizer • Room-Temperature Vulcanizing

rtVBR Real-Time Variable Bit Rate

rtVFR Real-Time Variable Frame Rate

RTVM Requirements and Test Verification Matrix • Requirements Traceability Verification Matrix

RTVP Real-Time Video Processing

RTW Ready To Wear • Remote-handled Transuranic Waste • Round The World • Saratov (Russia) (location identifier)

RTWS Raw Tape Write Submodule

RTWUS Research and Technology Work Unit Summary

rtx Rich-Text Format (file extension)

RTX Real-Time Executive • Real-time Telephone Exchange • Real-Time Transmitted • Real-Time Transmitted • Request To Transmit

RTY Merty (Australia-SA) (location identifier)

RTZ Return To Zero

ru Russia • Russian (ISO language code)

Ru Ruthenium • Ruthénium

R&U Repairs & Utilities

RU Are You? • Burundi (international vehicle registration code) • Range Unit • Range User • Receive Unit • Registered User • Reino Unido • Release Unit • Remote Unit • Replaceable Unit • Reporting Unit • Reprocessed Uranium • Reproducing Unit • Request Unit • Rescue Unit • Resource Unit • Response Unit • Restaurant Universitaire • Rijksuniversiteit • Royaume-Uni • Russia (ISO country code) • Russie (code ISO de pays)

RUA Arua (Uganda) (location identifier) • Resource Unit Administrator

RUB Ruhr-Universität Bochum (Germany)

RUC Reporting Unit Code • Roosevelt University Chicago (USA-IL) • Roskilde University Centre (Denmark)

RUCA Rijksuniversitair Campus Antwerpen (Belgium) • Rijksuniversitair Centrum Antwerpen (Belgium)

RuCCS Rutgers Center for Cognitive Science (USA-NJ)

RUCCS Rutgers Center for Cognitive Science (USA-NJ)

Rud Rudder

RUD Rudder

RUDI Resource for Urban design Information

RUE Rational Use of Energy

RUF Raw Units File • Rough

RUFHQ Regroupement des Universités de la Francophonie Hors-Québec

RUFIS Rôle of Universities in the Future Information Society

RUG Regional User Group • Remote User Gateway • Remote User Group • Restricted Users Group • Rijksuniversiteit Gent (Belgium) • Rijksuniversiteit Groningen (Netherlands) • Rugao (China-PRC) (location identifier)

RUH Riyadh - King Khalid (Saudi Arabia) (location identifier) • ROSAT Users' Handbook

RUI Remote Unit Interface • Remote User Interface • Research at Undergraduate Institutions (*) • Research in Undergraduate Institutions • Resource Use Identifier • Royal University of Ireland • Ruidoso (USA-NM) (location identifier)

RUIC Reporting Unit Identification Code

Ruimt Ruimtevaart (periodical code)

RUIP Remote User Information Protocol

RUK Rukumkot (Nepal) (location identifier)

RUL Right Upper Lobe • Rijksuniversiteit Leiden (Netherlands) • Rijksuniversiteit Limburg (Netherlands)

RULE Red de Universidades Latino-Americanas y Europeas

RULER Remaining Useful Life Evaluation Routine

RULLI Remote Ultra-Low Light Imaging

RULR Rigid Ultra-Light Replica

Rum Rumanian

RUM Recinto Universitario de Mayagüez (Puerto Rico) • Remote Underwater Manipulator • Remote-Unit Monitor • Remote User Multiplex • Resource and Unit Monitoring • Rotating Unbalanced Mass • Rumjatar (Nepal) (location identifier)

RUMBLE Requests Users Make By Listings Events

RUMS Range Utilization Measurement System • Resource Use Management Subgroup

RUN Saint-Denis de la Réunion (Reunion Island) (location identifier)

RUNCIBLE Revised Unified New Compiler with Its Basic Language Extended

RUNJOB Russia-Nippon Joint Balloon (program)

RUP Rational Unified Process • Reconnu d'Utilité Publique • Remote User's Package • Rupsi (India) (location identifier)

RUPT Rupture

RUQ Right Upper Quadrant

Rur Rural

RUR Radio Upstream Receiver • Reasoning with Uncertainty in Robotics (workshop) • Resource Utilization Report • Ruble (Russia) (ISO currency code) • Rurutu (Tubuai Islands) (location identifier)

RURE Regional and Urban Restructuring in Europe

RURPIS Resource Utilization and Regional Planning Information System

Rus Russia • Russian

RUS Marau Sound (Solomon Islands) (location identifier) • Rechenzentrum Universität Stuttgart (Germany) • Remote Update System • Resonant Ultrasound Spectroscopy • Russia (international vehicle registration code) • Russia (ISO country code) • Russie (code ISO de pays)

RUSH Remote Use of Shared Hardware • Responsive Urgent Service Handling

Russ Russia • Russian

RUSSWO Revised Uniform Summary of Surveyed Weather Observations

Rúst Rústico

RUSTI Research Unit for Surfaces, Transforms and Interfaces

Rut Rutland

RUT Radio Upstream Transmitter • Reflectance Uniformity Test • Resource Utilization Time • Ruble (Tajikistan) (ISO currency code) • Rutland (USA-VT) (location identifier)

RUTCOR Rutgers Center for Operational Research (USA-NJ)

Rutd Rutland

RUTG Real-Time Users Group

RUU Rijksuniversiteit Utrecht (Netherlands) • Ruti (Papua New Guinea) (location identifier)

RUV Rubelsanto (Guatemala) (location identifier)

RUWC Range Users Work Center

RUWS Remote Underwater Work System • Remote Unmanned Work System • Remote User Workstation

RUY Copan (Honduras) (location identifier)

R/V Range/Value • Research Vessel • Reality/Virtuality

R&V Review & Validation

RV Radar Vector • Radial Velocity • Radio-Vent • Radius of Vulnerability • Random Variable • Random Vibration • Rated Voltage • Réalité Virtuelle • Reality/Virtuality • Rear View • Receive Variable • Recirculation Valve • Recovery Vehicle • Recovery Vessel • Recreational Vehicle • Reentry Vehicle • Reference Voltage • Regionalverband • Rekeying Variable • Relief Valve • Remote Vehicle • Rendezvous • Rendezvous Vehicle • Rescue Vessel • Research Vehicle • Research Vessel • Residual Volume • Retrieval Vessel • Revised Version • Rigveda • Ring Voice • Rotovibrational • Roving Vehicle

RvA Raad van Arbeid • Revista Astronómica de la Asociación Argentina Amigos de la Astronomía (periodical code)

RVA Farafangana (Madagascar) (location identifier) • RAMAC Virtual Array • Reactive Volt-Ampere • Régie des Voies Aériennes • Relative Virtual Address • Reliability Variation Analysis • Remote Voltage Adjustment • Réseau à Valeur Ajoutée • Reuse Value Added • Rijksdienst voor Arbeidsvoorziening (Netherlands)

RvAc Revue d'Acoustique (periodical code)

RVAM Reactive Volt-Ampere Meter

RVAO Reentry Vehicle Associated Object • Remote Vehicle and Associated Objects

RvAqS Reviews in Aquatic Sciences (periodical code)

RVAS Requirements Verification and Accounting System • Roanoke Valley Astronomical Society (USA-VA)

RVB Rendezvous and Berthing • Resonating Valence Bond • Rouge - Vert - Bleu

RvC Raad voor de Certificatie (Netherlands)

RVC Random Vibration Control • Rapidly Varying Component • Relative Velocity Computer • Rivercess (Liberia) (location identifier)

RVCC Rariran Valley Community College (USA-NJ)

RvCF Revista Colombiana de Física (periodical code)

RVCF Remote Vehicle Checkout Facility

RVCN Requirement Verification Completion Notice

RVD Radial Velocity Dispersion • Remote Vehicle Docking • Remote Virtual Disk • Rendezvous and Docking • Rijksvoorlichtingsdienst

RVDISC Rendezvous Discrimination

Rvdo Reverendo

RVDT Remote Video Display Terminal • Remote Visual Display Terminal • Rotary Variable Differential Transducer • Rotary Variable Differential Transformer

RVE Remote Video Encoding • Representative Volume Element • Saravena (Colombia) (location identifier)

RVEC Regional Value Engineering Coordinator

RVES Remote Video Encoding System

RVET Right Ventricular Ejection Time

RvFE Revue Française de l'Électricité (periodical code)

RVFL Random Vector Functional Link

RVG Reference Voltage Generator

RvGeo Reviews of Geophysics (periodical code)

RvGSP Reviews of Geophysics and Space Physics (periodical code)

RVH Rzhevka (Russia) (location identifier)

RVI Ratio Vegetation Index • Reverse Interrupt • Rotation Variant of an Image

RVK Roervík (Norway) (location identifier)

RVL Rolling Vertical Landing

Rvlg Revolving

Rvlv Revolve

RVM Radial-Velocity Measurement • Reactive Voltmeter • Reentry Vehicle Module • Reference Validation Mechanism • Répertoire de Vedettes-Matière • Requirements Verification Matrix • Research Value Mapping • Rotating Vector Model

RvMA Reviews of Modern Astronomy (periodical code)

RvMad Revista Real Academia - Ciencias Exactas - Madrid (periodical code)

RvMin Reviews of Mineralogy (periodical code)

RvMP Reviews of Modern Physics (periodical code)

RVMS Radar Videometric System

RVN Regional Voice Network • Republic of Viet Nam • Requirements Verification Network • Rovaniemi (Finland) (location identifier)

RVO Reichsversicherungsordnung • Reivilo (South Africa) (location identifier) • Russian Virtual Observatory (Russia)

RvOp Reviews of Optics (periodical code)

RVP Radar Verification Programme • Raster-to-Vector Processor • Regional Vice President • Reid Vapor Pressure • Remote Voice Protocol • Rénovation de la Voie Professionnelle • Rotary Vacuum Pump

RvPA Revue de Physique Appliquée (periodical code)

RvPD Revue du Palais de la Découverte (periodical code)

RVQ Recursive Vector Quantization

RVR Green River (USA-UT) (location identifier) • Requirement Verification Revision • River • Runway Visual Range

Rvrb Reverberation

RVRM Runway Visual Range Midpoint

RvRP Revue Roumaine de Physique (periodical code)

RVRR Runway Visual Range Rollout

RVRS Requirements Verification Review Sheet

RvRST Revue Roumaine des Sciences Techniques (periodical code)

RVRT Runway Visual Range Touchdown

Rvs Reverse • Revise

RVS Radial Velocity Scanner • Radial Velocity Spectrometer • Radius Vector Subroutine • Reentry Vehicle Separation • Reentry Vehicle Simulator • Remote Viewing System • Rendezvous Sensor • Reverse • Tulsa - R. Lloyd Jones (USA-OK) (location identifier)

Rvsbl Reversible

Rvse Reverse

RVSM Radial Velocity Sky Mapper • Reduced Vertical Separation Minimum

Rvsn Revision

RVSN Raketniie Voiska Stratigcheskovo Naznacheniia

RVSVP Répondez Vite, S'il Vous Plaît

RvT Revue Technique Thomson (periodical code)

RVT Reliability Verification Test • Remote Video Terminal • Remote Virtual Terminal • Revalidation Testing • RV Tauri

RVTM Requirement Verification Traceability Matrix

RVTO Reentry Vehicle Technology and Observables

RvTS Rivista Tecnica Selenia (periodical code)

RVU Radio Volksuniversiteit • Research Vessel Unit

RVV Runway Visibility Value • Runway Visual Value

Rvw Review

RVW Review

RVWG Reliability and Vulnerability Working Group

RVX River Crossing

RVY Rivera (Uruguay) (location identifier)

rw Kinyarwanda (ISO language code)

Rw Rwanda

R-W Read/Write

R/W Read/Write • Right of Way • Runway

RW Radioactive Waste • Radio-Weak • Random Wander • Reaction Wheel • Read and Write • Read-Write • Real World • Relative Weight • Relative Wind • Resistance Welding • Response Word • Right Wing • Right Worthy • Rijksweg • Robertson-Walker • Rotary Wing • Runway • Rwanda (ISO country code)

RWA Reaction Wheel Assembly • Real-World Application • Regular Workload Assignment • Rotary-Wing Aircraft • Rwanda (international vehicle registration code) • Rwanda (ISO country code)

RWAB Regionale Wissenschaftliche Allgemeinbibliothek

RWB Responsive Workbench • Rowan Bay (USA-AK) (location identifier)

RWBH Records Will Be Handcarried

RWC Radioactive Waste Container • Read - Write - Compute • Read - Write - Continue • Regional Warning Centre • Relative Water Content • Remote Work Center • Robotic Work Crew

RWCC Recycled Water Coordination Committee • Regional Workshop Coordinating Committee

RWCEP Real World Crisis Evaluation Plan

RWCP Real World Computing Partnership (*) • Real World Computing Project

RWD Radioactive-Waste Disposal • Regular World Day • Right Wing Down

RWDC Real World Data Collection

RWE Radio Wave Experiment • Read/Write Electronics

RWF Franc (Rwanda) (ISO currency code) • Redwood Falls (USA-MN) (location identifier)

RwFr Rwanda Franc (currency)

RWG Rechtzentrale Württembergischer Genossenschaften (Germany) • Reliability Working Group

rwho Remote Who

RWI Radio and Wire Integration • Radio Wire Integration • Radio Wire Interface • Read - Write - Initialize • Regular Word Interval • Respective World Interval • Rocky Mount (USA-NC) (location identifier)

RWIC Regional Weather Information Center

RWIF Relative Weekly Impact Factor

RWIN Reynolds Warm Ionized Medium

RWL Rawlins (USA-WY) (location identifier) • Reaction Wheel

R/WM Read/Write Memory

RWM Radioactive-Waste Management • Read-Write Memory • Rectangular wave Modulation

RWMC Reactive Waste Management Committee

RWN Rovno (Ukraine) (location identifier)

RWNBH Records Will Not Be Handcarried

RWND Rewind

RWOD ROSAT WFC Observation Data-set

RWP Radiation Work Permit • Radio Wave Propagation • Rainwater Pipe • Real-time Weather Processor • Real World Processing • Rechts- und Wirtschaftspraxis • Recurring Work Program • Remote Write Protocol

RWPG Real World Problem Generation

RWPO Railway Post Office

RWPV Rotating-Wall Perfused Vessel

RWPWZ Rostock Wilhelm Pieck Universität Wissenschaftliche Zeitschrift (periodical code)

RWR Radar Warning Receiver

RWS Franc (Rwanda) (ISO currency code) • Rat Westdeutscher Sternwarten (Germany) (now RDS) • Reactive Wind Shear • Reconfigurable Workstation • Regional Weather Service • Remote Workstation • Rigid Wall Shelter • Rijkswaterstaat (Netherlands) • Robotic Workstation • Rossby Wave Source

RWT Read-Write Tape

RWTHA Rheinisch Westfälische Technische Hochschule Aerodynamisches Institut Abhandlungen (periodical code)

RWU Robertson-Walker Universe

RWV Read - Write - Verify • Remote Work Vehicle • Rotary-Wing Vehicle • Rotating Wall Vessel

RWVD Real World Visual Display

RWW Rijksgroepsregeling Werkloze Werknemers

Rwy Railway • Runway

RWY Runway

Rx Receive • Receiver

RX Reactor • Receive • Receiver • Remote Exchange • Repairable Exchange • Requisition Express

RXA Raudha (Yemen) (location identifier) • Repairable Exchange Activity

RXC Receive Clock

RxD Receive Data

RXD Received Data

RXF Refracted X-ray Fluorescence

RXLEV Reception Level

RXMT Retransmit

RXO Redundant Crystal Oscillator

RXQUAL Reception Quality

RxR Railroad

RXRC Rank Xerox Research Center

RXS Roxas City (Philippines) (location identifier)

RXTE (Bruno B.) Rossi X-ray Timing Explorer (formerly XTE)

ry Railway • Rydberg

Ry Railway • Runway

R-Y Roll-Yaw

RY Roll-Yaw

RYB Rybinsk (Russia) (location identifier)

RYC Returned Your Call

RYD Real Year Dollars

RYDMAR Reaction-Yield-Detected Magnetic Resonance

RYE Retirement Year Ending

RYG Red - Yellow - Green

RYI Revised Yale Isochrone

RYIF Relative Yearly Impact Factor

RYK Rahim Yar Khan (Pakistan) (location identifier)

RYM Read Your Manual • Reference Your Message

RYN Royan (France) (location identifier)

RYO Rio Turbio (Argentina-SC) (location identifier)

RYTSI RIT-Yale Tip-tilt Speckle Imager

Rz Randziffer

R&Z Range & Zero

RZ Radiation Zone • Range and Zero • Rechenzentrum • Recovery Zone • Recurring Zero • Return to Zero

RZA Santa Cruz (Argentina-SC) (location identifier)

RZB Regulated Zone Boundary

RZE Resonance Zone Effect • Rzeszow (Poland) (location identifier)

RZF Root Zone File

RZG Rechenzentrum Garching (Germany)

RZh Referationyj Zhurnal (periodical code)

RZH Lancaster - Quartz Hill (USA-CA) (location identifier)

RZL Return to Zero Level

RZM Return to Zero Mark • Stichting voor Radiostraling van Zon en Melkweg (Netherlands)

RZN Ryazan (Russia) (location identifier)

RZR Ramsar (Iran) (location identifier)

RZSM Root Zone Soil Moisture

Rzvs Rendezvous

RZZ Roanoke Rapids - Halifax County (USA-NC) (location identifier)

R2 Rapid Response • Reporting Responsibility

R2CSR Rapid Response to Critical Systems Requirements

R2PI Resonant Two-Photon Ionization

R2P2 Rapid Retargeting and Precision Pointing

R2R Run to Run

R2T2 Rapid Response Topographic Team

R3 Redundancy - Resiliency - Robustness • Responder - Relay - Reporter • Rotary Reciprocating Refrigerator

R3B Resources and Requirements Review Board

R3S Rapid Records Retrieval System

S

s/ Su

s Assembly (file extension) • Distance • Path • Path Length • Laplace Variable • Length of Arc • Long-Range Order Parameter • Response Curve • Saint • Satisfactory • Save • School • Scientific • Sculpsit • Sea • Second • Secondary • Seconde • Secret • Section • Sedimentation Coefficient • See • Segundo • Sekunde • Semi • Sensitivity Curve • Series • Set • Shilling • Shock • Siècle • Siège • Siehe • Siglo • Sign • Signature • Signed • Silver • Sin • Sine • Singlet • Singular • Sinister • Sinistra • Sister • Small • Snow • Society • Soft • Solidus • Solo • Son • South • Southerly • Specific Entropy • Spectral Transmission • Spherical • Spin Quantum Number • Steamer • Steel • Stem • Stere • Stock • Stoichiometric Oxidant-to-Fuel Ratio • Strange • Stratus • Strong Absorption • Su • Südlich • Sun • Sunny • sur • Sustantivo • Switch • Symmetrical • Symmetry Number

S/ Su

S Apparent Power • Area • Characteristic Function of Hamilton • Entropy • Flux Density • Linear Stopping Power • Magnetic Solar Daily Variation • Poynting Vector • Probability Current Density • Quantum Number of Electronic Spin • Saint • Salar • Salina • San • Sankt • Santa • Santo • São • Satisfactory • Saturday • Saturn • Satz • Save • Saxon • Schilling (currency) • School • Schwefel • Scientific • Scotland • Scottish • Scout • Sea • Second • Secondary • Secret • Secretary • Sector • See • Segundo • Seite • Senate • Señor • Señora • Sente • Sentence • September • Series • Seven • Seven (Roman numeral) • Seventy • Seventy (Roman numeral) • Shilling • Siècle • Siemens • Sierra (international alphabet) • Siglo • Signature • Signor • Signora • Single • Sinister • Slow • Small • Snow • Society • Socius • Soft • Sol • Sol (currency) • Solar • Solar Constant • Solar Mass • Soufre • South • Southern • Spectral Response • Spectral Type • Spin Quantum Number • Spiral Galaxy • Staff • State • Statute • Stere • Stimulus • Store • Strangeness Quantum Number • Strehl (ratio) • Stress • Strong • Subject • Submarine • Substitution Reaction • Sucre (currency) • Sud • Süd • Sudden • Südlich • Sulfur • Sulphur • Summer • Sun • Sunday • Supplement • Sur • Sweden (international vehicle registration code) • Sweden (postal country code) • Synodic Period • Total Neutron Source Density • 7 (Roman numeral) • 70 (Roman numeral)

s/a Sin Año

sa Sanskrit (ISO language code) • Saudi Arabia • Semi-Annual • siehe auch • Sin Año • Sine Anno • Symmetrical

Sa Samarium (now Sm) • Samstag • Saturday • Serra • Serrania • Sierra • Summa

S-A Safety Arming • Surface-to-Air

S&A Safe & Arm • Safe & Arming • Safety & Arming • Sampling & Analysis • Science & Application • Services & Agencies • Simulation & Analysis • Status & Alert

S/A Safe and Arm • Safing and Arming • Scheduled/Actual • Service/Agency • Site Activation • Situational Awareness • Solar Array • Spacecraft Adapter • Subaccount • Subassembly • Subject to Acceptance • Such As • Supplemental Agreement • Surface-to-Air • Swing Arm

SA Arabia (Saudi_) (ISO country code) • Arabie Saoudite (code ISO de pays) • Safe and Arm • Safety Analysis • Safety and Arming • Safing Area • Sales Analysis • Salta (Argentina) • Samoa • Sandstorm • Sanskrit • Saskatchewan (Canada) • Saudi Arabia (ISO country code) • Saudi Arabian • Scheduled/Actual • Science Archive • Scientific Area • Scintillation Adjust • Seaman Apprentice • Secrétaire Administratif • Section d'Apprentissage • Section of Astronomy • Security Allocation • Security Analysis • Security Architecture • Security Assistance • Security Association • See Also • Select Address • Selected Area • Selected Assembly • Selected Availability • Selective Acquisition • Selective Availability • Semi-Analytic • Semi-Annual • Semi-Annual • Senior Adviser • Sense Amplifier • Sensitivity Analysis • Separated Atom • Sequential Automated • Service Agent • Service Area • Service Assistant • Service Availability • Service d'Aéronomie (France) • Services and Agencies • Set Attribute • Shaft Angle • Shaping Amplifier • Shapley-Ames • Shock-Associated (event) • Shop Aid • Sierra • Signal Accuracy • Signal Analyzer • Signature Analysis • Simple Alert • Simulated Annealing • Simulation and Analysis • Sine Anno • Single Access • Single Aircraft • Single Aisle • Single Article • Site Activation • Situational Analysis • Situational Assessment • Situational Awareness • Situation Analysis • Situation Assessment • Situation Awareness • Sky Atlas • Small Array • Sobolev Approximation • Sociedad Anónima • Sociedade Anónima • Società Anonima • Société Anonyme • Software Application • Software Architecture • Software Assurance • Solar Activity • Solar Array • Source Acquisition • Source Address • South Africa • South African • South America • South American • South Atlantic • South Australia • South Australian • Space Application • Space Available • Spacecraft Adapter • Special Access • Special Agent • Spectral Albedo • Spectrograph Assembly • Spectrum Analysis • Spectrum Analyzer • Spin Axis • Splitting Amplifier • Staging Area • Stand Alone • Star Align • Station Address • Station Automatique • Status and Alarm • Storage Alteration • Storage Area • Store Address • Stress Anneal • Strongly Agree • Storage Analyzer • Strike Assessment • Structural Analyzer • Structured Analysis • Su Alteza • Subaccount • Subassembly • Subject to Approval • Subsistence Allowance • Successive Approximation • Sueldo Anual • Suomen Akademia (Finland) • SuperAGILE • Superassociation • Supplemental Agreement • Supportability Analysis • Supporting Agency • Support Area • Surface Area • Surface-to-Air • Svenska Akademien (Sweden) • Swath Angle • Swing Arm • Symbolic Assembler • System Abort • System Address • Systems Administration • Systems Administrator • Systems Analysis • Systems Architect • Systems Architecture

SAA Safety Assurance Analysis • Same As Above • Saratoga (USA-WY) (location identifier) • Satellite Access Authorization • Satellite Active Archive • Satellite Attitude Acquisition • Security Assistance Automation • Seminario de Astronomía y Astrofísica • Senior Administrative Assistant • Service Académique d'Appui • Service Aspects and Applications • Servo-Actuated Assembly • Shift And Add • Shift-and-Add Algorithm • Single-Access Antenna • Single Agency Action • Single Article Announcement • Single Assignment Approach • Slot Array Antenna • Society of American Archivists (USA) • Solar Aspect Angle • South-Atlantic Anomaly • Space Association of Australia • Speech Association of America • Staff Administrative Assistant • Standards Association of Australia • Stellar Astronomy and Astrophysics • Step Adjustable Antenna • Summary Activity Account • System Aspects Application • Systems Application Architecture

SAAA San Antonio Astronomical Association (USA-TX) • Shoreline Amateur Astronomical Association (USA-MI)

SAAAA San Angelo Amateur Astronomy Association (USA-TX)

SAAAS South African Association of the Advancement of Science

SAAB Svensk Aeroplan Aktiebolag

SAAC Schedule Allocation And Control • Secrétaire Administratif d'Administration Centrale • Security Assistance Accounting Center • Seismic Array Analysis Center • Software for Analyzing Astrometric CCDs • Space Applications Advisory Committee

SAAD Sociedade de Astronomia e Astrofísica de Diadema (Brazil) • South Atlantic Anomaly Detector

SAADS Stellar Autonomous Attitude Determination System

SAAF Small Austere Airfield • South African Air Force

SAAHS Stability Augmentation and Altitude Hold System

SAAL Signaling ATM Adaptation Layer

SAALC San Antonio Air Logistics Center

SAAM Simulation Analysis And Modeling • Software Architectural Analysis Method

SAAMEX Surface and Atmospheric Airborne Microwave Experiment

SAAO South African Astronomical Observatory

SAAOC South African Astronomical Observatory Circulars (periodical code)

SAAP SOFIA Airborne Ambassadors Program • South-Atlantic Anomaly Probe • Subsonic Aircraft Assessment Program

SAAR Seasonally Adjusted at Annual Rates • Strength Accounting and Reporting

SAARC South Asian Association for Regional Cooperation

SAARMS Security Assistance Automated Resource Management System

SAARU Secondary Air and Attitude Reference Unit • Secondary Attitude and Air Reference Unit

SAAS Science Achievement Award for Students • Shuttle Aerospace Actuator Simulator • Shuttle Aerosurface Actuator Simulation • Solent Amateur Astronomers Society (UK-England) • Stand-Alone Analysis Shelter • Stand-Alone Analysis System

SAASE Standard Automated Architecture Support Environment

SAASIN Sociedad Astronómica Amateur de Sinaloa (Mexico)

SAASS South African Airborne Scanner Survey

SAASTEC South African Association of Science and Technology Centers

SAAT Satellite Attitude Acquisition Technique

SAATN Shuttle/Atlas Advanced TIROS Satellite

SAAV System Approach Assistance Visit

Sab Sábado • Sabato • Sabbath

SAB Saba Island (Netherlands Antilles) (location identifier) • Same as Above • Same As Basic • Satellite Applications Branch • Satellite Assembly Building • Science Advisory Board • Scientific Advisory Board • Server Address Book • Shuttle Assembly Building • Shuttle Avionics Breadboard • Simulation of Adaptive Behavior • Site Activation Board • Sociedad de Astronomía Balear (Spain) • Sociedade Astronômica Brasileira (Brazil) • Société Astronomique de Bordeaux (France) • Société Astronomique de Bourgogne (France) • Solar-Array Battery • Source Approval Bulletin • Spacecraft Assembly Building • Standards Activities Board • Storage and Assembly Building • Student Activities Board • Subject as Above • Synoptic Analysis Branch • System Advisory Board

SABA Scope And Budget Authorization

Sabat Sabatique • Sabbatical

SABC South African Broadcasting Corporation

SABCA Société Anonyme Belge de Constructions Aéronautiques (Belgium)

SABENA Société Anonyme Belge d'Exploitation de la Navigation Aérienne (Belgium) (obsolete)

SABER Simplified Acquisition of Base Engineering Requirement • Situational Awareness Beacon with Reply • Sounding of the Atmosphere using Broadband Emission Radiometry (TIMED)

SABEW Society of American Business and Economic Writers (USA)

SABI Secret and Below Information

SABINET South African Bibliographic and Information Network

SABIR Semi-Automatic Bibliographic Information Retrieval • Space-Based Interceptor

SABIRS Semi-Automatic Bibliographic Information Retrieval System

SABIT Special American Business Internship Training

SABLE Simulator Atmospheric Boundary Layer Environment • Systematic Approach to Better Long-range Estimating

SABM Set Asynchronous Balanced Mode

SABME Set Asynchronous Balanced Mode Extended

SABOR Same As Basic Or

SABP Software Acquisition Best Practices

SABRE Sales and Business Reservations done Electronically • Semi-Automatic Business Research Environment • Sweden And Britain Radar Experiment

SABREX Satellite-Borne Radiometer Experiment

SABRO South American Biological Research Organization

SABS Shared-Aperture Boresight Sensor • South African Bureau of Standards • Stanford Automated Bibliographic System • Statistical Analysis of Bibliographical Structure

SABV Suid-Afrikaanse Biblioteekvereniging (South Africa)

S/AC Stabilization/Attitude Control

SAC Sacramento (USA-CA) (location identifier) • Saguaro Astronomy Club (USA-CA) • Satélite de Aplicaciones Científicas • Satellite Applications Centre • Science Abstracting Committee • Science Activity Centre • Scientific Advisory Committee • Scientific Advisory Council • Scientific Analysis Component • Secondary Address Code • Sectional Aeronautical Chart • Seismic Analysis Code • Selected Areas in Cryptography (workshop) • Self-Absorption Curve (method) • Semi-Automatic Coding • Senate Appropriations Committee (USA) • Senior Advisory Council • Sequencing and Application Controller • Service Access Code • Service Access Control • Service Area Code • Service Area Computer • Shannonside Astronomy Club (Ireland) • Shapley Ames Catalog • Short-run Average Cost • Single-Address Code • Single Attached Concentrator • Single-Attachment Concentrator • Slovak Accredition Council • Software Advisory Channel • Solar Activity Cycle • Solar Array Carrier • South Atlantic Coast • Space Activities Commission • Space Advisory Committee • Space and Aviation Control • Space Applications Centre • Space Applications Corporation • Special Access Clearance • Special Agent in Charge • Special Analysis Centre • Special Area Code • Special Area of Conservation • Spherical Aberration Corrector • Stabilization/Attitude Control • State Affairs Committee • Stateful Access Control • Store And Clear • Strategic Air Command • Structural Adjustment Committee • Students Advice Center • Subarray Center • Subject Analysis Committee • Subscriber Acquisition Cost • Supply Administration Center • Support Action Center • Symposium on Applied Computing • System Analysis and Checking • System-Assisted Coordination • System Auxiliary Computer • Système d'Annonce des Crues • Systems Advisory Committee • Systems Automation Course

SACA Société d'Astronomie de Cannes (France)

SACAC South African Council for Automation and Computation

SACAM Ship Acquisition Contract Administration Manual

SACAP Selection, Acquisition, Cataloguing, And Processing

SACAS Security Assistance Central Accounting System

SACC Satellite Archive Coordination Center • Science Action Coordinating Committee • Science Archive Coordination Center • Spatial Analysis Coordinating Committee

SACCH Slow Associated Control Channel

SACCS SAC Automated Command Control System

SACD Super Audio Compact Disk (format)

SACDA System Analysis, Control and Design Activity

SACEM Société des Auteurs, Compositeurs et Éditeurs de Musique

SACEP South Asia Cooperative Environment Program

SACEPO Standing Advisory Committee before the European Patent Office

SACF Single Association Controlling Function

SACG Selective Antenna Coupler Group

Sachs Sachsen (Germany)

Sächs Sächsische (Germany)

Sachs-Anh Sachsen-Anhalt (Germany)

SACI Secondary Address Code Indicator • Special Assistant for Contracting Integrity

SACK Selective Acknowledgment

SACL Standards And Calibration Laboratory

SACM School of Acquisition Management • Security Assistance Case Management • Smart Acoustic Current Meter • Sociedad de Autores y Compositores de México (Mexico) • Statistical Adiabatic Channel Method

SACMA Suppliers of Advanced Composite Materials Association

SACMAPS Selective Automatic Computational Matching and Positioning System

SACMAT Symposium on Access Control Models And Technologies

SACMM Software Acquisition Capability Maturity Model

SACN Secrétaire Administratif de Classe Normale

SACNET Secure Automatic Communications Network

SACO Staff Actions Control Officer

SACP Standing Advisory Committee on Programming

Sacr Sacramento (USA-CA)

SACRD Safety Analysis Computerized Reactor Data

SACRS South Australian Centre for Remote Sensing

SACS Satellite Attitude Control Simulator • Secure Access Control System • Security Access Control System • Software Avionics Command Support • Solar Attitude Control System • State Administration of China for Standardization (China-PRC) • Structure And Composition System • STU Access Control System • Synchronous Attitude Communications Satellite

SACSIR South African Council for Scientific and Industrial Research

SACSIRO South African Commonwealth Scientific and Industrial Research Organization

SACTO Sacramento Test Operations

SACU South African Customs Union

SACV Spacecraft Autonomy Concept Validation

SACY Search for Associations Containing Young (stars) (project)

SACZ South Atlantic Convergence Zone

S&AD Science & Applications Directorate

SAD Safety Analysis Document • Safety Assessment Document • Safety Assurance Diagram • Safford (USA-AZ) (location identifier) • Seasonal Affective Disorder • Security Association Database • Service d'Aide à Domicile • Shuttle Authorized Document • Simple, Average or Difficult • Single Administrative Document • Situation Attention Display • Situation Awareness Display • Software Architecture Document • Solar-Array Drive • Spacecraft Attitude Display • Spaced Antenna Drift • Special Activities Division • Special Adapter Device • Standard Abundance Distribution • System Allocation Document • System Analysis Drawing

SADA Seismic Array Data Analyzer • Solar-Array Drive Assembly • South African Data Archive • Special Assistant for Data Automation • Standard Advanced Dewar Assembly

SADAP Simplified Automatic Data Plotter • Space Astrophysics Data Analysis Program

SADAPTA Solar-Array Drive And Power Transfer Assembly

SADBU Small And Disadvantaged Business Utilization

SADC Sequential Analog-Digital Computer • Southern African Development Community

SADCC Southern Africa Development Coordination Conference

SAdeCV Sociedad Anónima de Capital Variable

SAdeCVyRL Sociedad Anónima de Capital Variable y Responsabilidad Limitada

SADE Signal Analysis and Dissemination Equipment • Solar-Array Deploy Electronics • Solar-Array Drive Electronics (*) • Spark-Assisted Diesel Engine

SADEC Spin Axis Declination

SADEYA Sociedad Astronómica de España y America (Spain)

SADF Semi-Automatic Document Feed • Statistical Analysis of Documentation Files

Sadha Sadhana (periodical code)

SADI Secretariat Automated Document Index

SADIC Solid-state Analog-to-Digital Computer

SADIE San Diego Imaging Echelle • Scanning Analog-to-Digital Input Equipment • Secure Automatic Data Information Exchange • Silicon Anode Detector with Integrated Electronics • Solid-state Analog-to-Digital Equipment

SADIM Stanford Art Directors Invitational Master

SADIO Sociedad Argentina de Investigación Operativa

SADIS Satellite Distribution Service • Southern African Development Information System • Southern African Documentation and Information System

SADL Situation Air Data Link • Situation Awareness Data Link • Sterilization Assembly Development Laboratory • Synchronous Data Link Control

SADM Self-Annihilating Dark Matter • Solar-Array Drive Mechanism (*) • Solar-Array Drive Motor • System Acquisition Decision Memorandum

SADP Synthetic Array Data Processor

SADPT Solar Array Drive and Power Transfer

SADR Selective Address • Spatial Address • System Audit and Design Review

SADRS Simplified Autonomous Data Receiving Station

SADS Solar-Array Drive System

SADT Shared Abstract Data Type • Structured Analysis and Design Technique • Systems Analysis and Design Technique

SADU Solar Array Diode Unit

SADV Semi-Annual Density Variation

SADW Shielded Accretion Disk Wind

SAE Sangir (Indonesia) (location identifier) • Search of Academic Excellence (project) • Self-Addressed Envelope •

Self-Appointed Expert • Service Access Element • Service Acquisition Executive • Shaft Angle Encoded • Simple Arithmetic Expression • Site Acceptance Evaluation • Società fra gli Autori e gli Editori (Italy) • Society for the Advancement of Education (USA) • Society of Automotive Engineers (USA) • Stage d'Accès à l'Entreprise • Stamped Addressed Envelope • Stand Alone Environment • Standard Average European • Steering Angle Error • Stop At Expiration

SAEAC Space and Atmospheric Environment Advisory Committee

SAED Selected Area Electron Diffraction

SAEF Simple Average Ensemble Forecast • Spacecraft Assembly and Encapsulation Facility (USA-FL)

SAEIB Sovietskaia Antarkticheskaia Ekspeditsiia Informatsionnyi Biulleten (periodical code)

SAEL Sensor Atmospheric Effects Library

SAENET South Australia Educational Network

SAEOC Second Alternate Emergency Operations Center

SAERI Seoul Korea Atomic Energy Research Institute (periodical code)

SAERIS South Australian Education Resources Information System

SAES Service d'Allocations d'Études Supérieures • Stand Alone Engine Simulator

SAESAC Space And Earth Sciences Advisory Committee

SAETA Sociedad Anónima Española de Traductores y Autores (Spain)

Saf Safety

SAF San Andreas Fault (USA-CA) • Santa Fe (USA-NM) (location identifier) • Satellite Applications Facility • Science Archive Facility • Search And Find • Sección de Astrofotografía • Secured Access Facility • Security Authorization Facility • See Also From • Semi-Automated Force • Service Access Facility • Société Astronomique de France • Source Acquisition File • Space Agency Forum • Spacecraft Assembly Facility • Special-Access Facility • Special Action Force • Special Applications File • Standard Associative Form • Stavanger Astronomiske Forening (Norway) • Subject to Available Funds • Symmetry-Adapted Function • System Authorization Facility

SAfA Sydney Association for Astrophysics (Australia-NSW)

SAFA Safety Assessment of Foreign Aircraft • Solar-Array Failure Analysis • Southern African Regional Science Initiative

SAFARI South-African Fire-Atmosphere Research Initiative • South African Regional Science Initiative • Swedish And French Auroral Radar Investigation • Système Automatisé pour les Fichiers Administratifs et les Répertoires d'Individus (France)

SAFE Braathens South America and Far East Air Transport SA • Safing and Failure Detection Expert • San Andreas Fault Experiment • Secure Analyst File Environment • Security And Freedom through Encryption • Sequential Analysis for Force Employment • Solar Array Flight Experiment • South Africa - Far East • Stamped Addressed Foolscap Envelope • Supportability Analysis Forecasting and Evaluation • System Analyst File Environment

SAFEA State Administration of Foreign Expert Affairs (India)

SAFENET Survivable Adaptable Fiber Optic Embedded Network

SAFETE Statistical Analysis for Evaluating Theoretical Expectations

S. Afr. South Africa • South African

SAFER Simplified Aid For EVA Rescue • Société d'Aménagement Foncier et d'Établissement Rural • Switch Architecture for Extreme Resiliency

SAFGA Société Astronomique de France - Groupe d'Alsace

SAFI Safety Indicator • Semi-Automatic Flight Inspection

SAFIR Single-Aperture Far-Infrared (observatory) • Sociedad Astronómica de la Faculdad de Ingeniería (Mexico) • Système d'Alerte Foudre par Interféromètre Radioélectrique • Système d'Alerte Foudre par Interférométrie Radioélectrique (*) • Systèmes Algébriques Formels pour l'Industrie et la Recherche

SAFIRE Spectroscopy of the Atmosphere using Far-Infrared Emission • Submillimeter And Far Infrared Experiment

SAFISY Space Agencies Forum for the International Space Year

SAFOC Semi-Automatic Flight Operations Center

SAFM Stuck-At Fault Model

SAFR Schweizerische Arbeitsgemeinschaft für Raumfahrt (Switzerland)

SAFRAS Self-Adaptive Format Retrieval and Storage System

SAFRS Saudi Arabian Facility for Remote Sensing

SAFSA Study Abroad and Foreign Student Adviser

SAFT Safety • Schwäbisches Amateur- und Fernrohrtreffen (Germany)

SAG Sagittarids (meteor stream) • Sagwon (USA-AK) (location identifier) • Schweizerische Astronomische Gesellschaft (Switzerland) • Science Advisory Group • Scientific Advisory Group • Scottish Astronomers Group (UK-Scotland) • Senior Advisory Group • Sociedad Astronómica Granadina (Spain) • Space Astrophysics Group • Special Activities Group • SQL Access Group • Standard Address Generator • Starlink Advisory Group • Street Address Guide • Study Advisory Group • Subactivity Group • Survivability Analysis Group • Systems Administration Group • Systems Analysis Group

SAGA Short Arc Geodetic Adjustment • Solar-Array Gain Augmentation • South Atlantic Geomagnetic Anomaly • Soviet-American Gases and Aerosols (experiment)

SAGAS Southern Area Group of Astronomical Societies (UK)

SAGB Super Asymptotic Giant Branch

SAGD Security Architecture Guidance and Directions • Steam-Assisted Gravity Drainage

SAGE Schéma d'Aménagement et de Gestion de l'Eau • Semi-Automated Ground Environment • Semi-Automatic Ground Environment • Solar-Assisted Gas Energy • Southern Arizona GEMS Educational (center) (USA-AZ) • Soviet-American Gallium Experiment • Sterilization Aerospace Ground Experiment • Strategic Aerosol and Gas Experiment • Stratosphere Aerosol and Gas Experiment (*) • Stratospheric Aerosol and Gas Experiment • Stratospheric Aerosol Gas Experiment • Structuring and Analysis of Gamma-ray Events • Sustainability And the Global Environment (project) • System for Automatic Program Generation • Systems Administrators Guild

SAGEECE Schéma d'Aménagement et de Gestion Écologique des Cours d'Eau

SAGEM Société d'Applications Générales d'Électricité et de Mécanique (France)

SAGENAP Scientific Assessment Group for Experiments in Non-Accelerator Physics

SAGEP Système Automatisé pour la Gestion Électronique de Périodiques

SAGET Satellite-Geodynamical Traverse

SAGGE Synchronous Altitude Gravity Gradient Experiment

SAGMOS Self-Aligning Gate Metal Semiconductor

SAGR Stand-alone Air GPS Receiver

SAGS Semi-Active Gravity Gradient System

SAGSTRAM Standing Advisory Group on the Safe Transport of Radioactive Material

SAGW Schweizerische Akademie der Geistes- und Sozialwissenschaften (Switzerland)

SAH Sana'a - Al'Rahaba (Yemen) (location identifier) • Service d'Action Humanitaire

SAHC South Asia Hydrological Cycle

SAHL Société d'Astronomie du Haut-Léman (Switzerland)

SAHM Société Astronomique de Haute-Marne (France)

SAHR Société Astronomique du Haut-Rhin (France)

SAHRS Standard Attitude and Heading Reference System

SA&I Systems Analysis & Integration

SAI San Marino (location identifier) • Scan Auroral Imager • Scanning Auroral Imager • Service des Affaires Internationales • Serving Area Interface • Single Agency Item • Società Astronomica Italiana (Italy) • Space Application Institute • Spin-scan Auroral Imager • Standards Australia International Ltd. (Australia-NSW) • Sternberg Astronomical Institute (Russia) • Studiecentrum voor Automatische Informatieverwerking (Belgie) • Su Alteza Imperial • Sudden Auroral Intensity • Symbolic Automatic Integrator • Synoptic Atmospheric Imager • Système d'Analyse des Images • Systems Analysis and Integration

SAIC Science Applications International Corporation (USA-VA) • State Administration for Industry and Commerce

SAICSIT South African Institute of Computer Scientists and Information Technologists

SAICYT Sistema Automatizado de Información Cientifica y Técnica

SAID Safety Action Item Database • Safety Analysis Input Data • Secure Association Identifier • Semi-Automatic Integrated Documentation • Shuttle Authorized Item Document • System Analysis Index for Diagnosis • System for Automated Information Dissemination

SAIDAS Sistema de Adquisición e Integración de Datos de Satélite

SAIDET Single-Axis Inertial Drift Erection Test

SAIE Special Acceptance and Inspection Equipment • Special Acceptance Inspection Equipment

SAIF South African Industrial Federation • Spatial Archive Interchange Format • Special Automated Installation File

SAIG Semantics, Applications, and Implementation of Program Generation (conference)

SAIL Scattering from Arbitrarily Inclined Leaves • Serial ASCII Instrumentation Loop • Shuttle Avionics Integration Laboratory • Space and Atmospheric Instrumentation Laboratory • Spacecraft Artificial Intelligence Laboratory • Stanford Artificial Intelligence Laboratory • Structural Analysis Input Language

SAILIS South African Institute of Librarianship and Information Science

SAILS Simplified Aircraft Instrument Landing System • Space Applications of Industrial Laser Systems • Submillimeter Astronomy Investigations of Line Spectra

SAIM Satellite Autonomous Integrity Monitoring • Systems Analysis and Integration Model

SAIMS Selected Acquisition Information and Management System

SAIN Sistema Andino de Información Industrial

SAINT Satellite Array for International and National Telecommunications • Surveillance - Acquisition - Identification - Notification - Tracking • Symbolic Automatic Integrator

SAIO Service Académique d'Information et d'Orientation (France)

SAIP Semi-Automated Imagery Processor • Semi-Automated IMINT Processing • South African Institute of Physics • Spares Acquisition Integrated with Production

SAIR Semi-Annual Supply Report • Synthetic Aperture Interferometric Radiometer

SAIRE Scalable Agent-based Information Retrieval Engine

SAIRS Standardized Advanced Infrared Sensor • Standardized Advanced Infrared System

SAIS School of Advanced International Studies • Science and Applications Information System • Scientific Apparatus Information Service • Selector Automático de Imágenes Solares • Sistema de Adquisición de Imágenes Solares • Standard Aeronautical Indexing System • Standard Automated Information System • Staring Array Imaging Spectrometer • Svenska Artificial Intelligence Sällskapet (Sweden)

SAISA South African Institute for Space Applications (South Africa)

SAIt Società Astronomica Italiana (Italy)

SAIT Samsung Advanced Institute of Technology • Science Applications of Information Technology (program) • Service d'Analyse de l'Information Technologique • South Australian Institute of Technology

SAIV Schedule as an Independent Variable

SAJ Sirajganj (Bangladesh) (location identifier) • Station astronomique Jansky (France) • Subassembly Jig

SAJAC Satellite Japan Corporation (now merged with JC SAT into JSS)

SAJPh South African Journal of Physics (periodical code)

SAJSc South African Journal of Science (periodical code)

SAK Saudarkrókur (Iceland) (location identifier)

Sal Salaire • Salamanca (Spain) • Salarié • Salary • Salizzada

SAL Safeguards Analytical Laboratory • Salida • San Salvador (El Salvador) (location identifier) • Scientific Airlock • Secure Access Level • Semi-Active Laser • Shift Arithmetic Left • Shuttle Avionics Laboratory • Simple Author Language • Société Astronomique de Liège (Belgium) • Société Astronomique de Lyon (France) • Solar Arc Lamp • Solar-Array Leaf • Space Astronomy Laboratory • Space Astrophysics Laboratory • Spread Across Liquids • SQL Application Language • Stellar Absorption Line • Strategic Arms Limitation • Structured Assembly Language • Subject Authority List • Superadiabatic Layer • Supersonic Aerophysics Laboratory • Symbolic Assembly Language • Synthetic Aperture Lidar • Systems Assembly Language

SALA Solar Arc Lamp Assembly

SALAN Sensitivity Analysis of Linear Active Networks

SALAT System for Automatic Language Analysis and Translation

SALC Sacramento Air Logistics Center

SALCC Satellite Airlift Control Center

SALCW Space Astrophysics Laboratory Compound Widget

SALD Security Assistance Logistics Database

SALE Safeguards Analytical Laboratory Evaluation • Simple Algebraic Language for Engineers

SALEN Students Across Europe Language Network

SALG Student Assessment of Learning Gains (project)

SALI Spectral Absorption Line Imager • Standalone Automatic Location Identification • Subnetwork Adjacent Layer Interface • Surface Analysis by Laser Ionization

SALID Solar And Lunar Illumination Data

SALINET Satellite Library Information Network

SALIR Semi-Active Laser Infrared

SaLIS Surveying and Land Information Systems (journal)

SALIS Security Assistance Logistics Information System • Set-Aside Land Information System

SALM Single-Anchor Leg Mooring • Society of Air Line Meteorologists

SALO Security Assistance Liaison Office • Security Assistance Liaison Officer

SALP Special Adult Learning Programme

SALR Saturated Adiabatic Lapse Rate • Synthetic Aperture Laser Radar

SALRO Saudi Arabian Laser Ranging Observatory

SALS Short Approach Lighting System • Small Angle Light Scattering • Statistical Analysis with Least Squares (fitting)

SALSA Spares Acquisition and Logistics Support Analysis • Synthesis Array for Lunar Submillimeter Astronomy

SALT Society for Applied Learning Technology • Southern African Large Telescope • Standard Analysis Language Translator • Strategic Arms Limitation Talks • Streamlined Alternative Logistic Transmission • Symbolic Algebraic Language Translator • System for Analysis of Laboratory Testing

SALTA Snowmass Area Large-scale Time-coincidence Array (USA-CO)

SALTE Semi-Automatic Line Test Equipment

SALTS Streamlined Automated Logistics Transfer System • Streamlined Automated Logistics Transmission System

SALUTE Size - Activity - Location - Unit - Time - Equipment

SALUTER Size-Activity-Location-Unit-Time-Equipment Report

Salv Salvador • Salvage

SALVE Software Aimed at off Line Verifications of ETI (signals)

SALWA Semi-Arid Lowlands of West Africa

sam Symantec Anti-Virus (file extension)

Sam Samedi • Samoa • Samstag

S. Am. South America • South American

SAM Safe Acquisition Mode • Safety Activation Monitor • Safety Analysis Methodology • Salaire Annuel Moyen • Salamo (Papua New Guinea) (location identifier) • Sample Analysis of Mars • Scalable ADSL Modem • Scan Analysis Module • Scanning Auger Microprobe • Scanning Auger Microscope • Scanning Auger Microscopy • School of Aerospace Medicine (USAF) • Script Applier Mechanism • SCSI Architecture Model • Secure Application Module • Security Accounts Manager • Security Application Module • Selective Automatic Monitoring • Selective Automonitoring • Self-Assembled Monolayer • Semantic Analysis Machine • Semantic Analyzing Machine • Semi-Analytical Merger • Semi-Analytical Model • Sensing with Active Microwave • Sequential-Access Memory • Sequential-Access Method • Serial Access Memory • Shared Appreciation Mortgage • Short-Axis Mode • Shuttle Activation Monitor • Shuttle Attachment Manipulator • Shuttle Attachment Mechanism • Signal Analyzing Monitor • Simulated Assignment Model • Simulation of Analogue Methods • Single Agency Manager • Single Application Mode • Single Aviation Market • Sleep/Activity Monitor • Small Angle Maneuver • Smart Acronym Macro • Sociedad Astronómica de México (Mexico) • Société Astronomique de Montpellier (France) • Société d'Astronomie de Montréal (Canada-QC) • Software Acquisition Management • Software Archive Meeting • Solar Activity Monitoring • Sort And Merge • Sourcebook in Applied Mathematics • Southern (hemisphere) Annular Mode • Space Available Mail • Spatial Access Method • Special Advisory Message • Special Air Mission • Spectrum Analyzer Module • Speech Assessment Methodology • SQL Application Message • Squarewave Amplitude Modulation • Stage Assembly and Maintenance • Standard Additive Model • Standard Assembly Module • Static Address Mapping • Stationing Analysis Model • Status And Monitor • Storage/Archive Management • Stratospheric Aerosol Measurement • Stratospheric Aerosol Monitor • Structural Assembly Model • Subsequent Address Message • Subtraction - Addition - Multiplication • Sun Acquisition Mode • Supportability Assessment Model • Sustainability Assessment Model • Symbolic and Algebraic Manipulation • Synchronous Amplitude Modulation • System Accuracy Model • System Activation and Monitoring • System Activity Monitor • System Administration Manager • System Administration Menu • System Analysis Machine • System Assurance Manager • Systems Acquisition Management • Systems Acquisition Manpower • Systems Analysis Module

SAMA Scientific Apparatus Manufacturers Association (USA) • Sociedade Astronômica Maranhense de Amadores (Brazil) • Step-by-step Automatic Message Accounting

SAMAP Security Assistance Management Acquisition Program

SAMAS Security Assistance Manpower Accounting System • Semi-Automated Mesoscale Analysis System • Structure and Manpower Allocation System

SAMBA Satellite for the Measurement of Background Anisotropies • Satellite to Measure Background Anisotropies • Satellite to measure the Microwave Background Anisotropies

SAMBO Simultaneous Auroral Multiballoons Observations

SAMD Security Assistance Management Directorate • Security Assistance Management Division

SAME S-band Angle Measuring Equipment • SQL Ada Module Extension

SAMEC Science And Mathematics Education Center

SAMEDL SQL Ada Module Extension Description Language

S. Amer. South America • South American

SAMEER Society for Applied Microwave Electronics Engineering and Research (India)

SAMER Système Automatique de Mesure de l'Évaporation Régionale

SAMEWG Software Acquisition Management Education Working Group

SAMEX Shuttle Active Microwave Experiment

SAMH Standard Automated Message Handling

SAMHS Standard Automated Message Handling System

SAMI Socially Acceptable Monitoring Equipment • System Acquisition Management Inspection

SAMICS Solar-Array Manufacturing Industry Costing Standards

SAMIR Satellite Microwave Radiometer

SAMIS Security Assistance Management Information System • Structural Analysis and Matrix Interpretation System • Structural Analysis and Matrix Interpretive System

SAML Security Assertion Markup Language • Send And Mail

SAMM Security Assistance Management Manual • Software Acquisition Maturity Model

SAMME Space Active Modular Materials Experiment

SAMMIE Scheduling Analysis Method for Mission Integrated Experiments • System for Aiding Man-Machine Interaction Evaluation

Samml Sammlung

SAMMS Siemens Advanced Motor Master System • Standard Automated Materiel Management System

SAMO School voor Atmosferisch en Marien Onderzoek (Netherlands)

SAMOS Semi-Automatic Meteorological Observing System • Structure Adaptée à la Modélisation Opérationnelle des Systèmes

Samp Sample

SAMP Security Accreditation Management Plan • Security Association Management Protocol • Shuttle Automated Mass Properties • Single Acquisition and Management Plan • System Acquisition Master Plan

SAMPD Science Analysis and Mission Planning Directorate

SAMPE Society for the Advancement of Material Process Engineering

SAMPEX Solar And Magnetospheric Particle Explorer • Solar Anomalous and Magnetospheric Particle Explorer

SAMPIE Solar-Array Module Plasma Interaction Experiment

SAMPJ SAMPE Journal (periodical code)

SAMPLE Single-Assignment Mathematical Programming Language • Survey Analysis Macro Programming Language

SAMPQ SAMPE Quarterly (periodical code)

SAMR Security Assistance Management Review • Structured Adaptive Mesh Refinement

SAMS Sampling Analog Memory System • Satellite Automatic Monitoring System • Segment Alignment and Maintenance System • Segment Alignment Maintenance System • Semi-Autonomous Message Switch • Shuttle Acceleration Measurement System • Shuttle Altitude Measuring System • Shuttle Attached Manipulator System • Shuttle Attachment Manipulator System • Space Acceleration Measurement System • Space Assemblance and Maintenance Study • Spatial Analysis and Modeling System • Standard Automated Maintenance System • Stockroom Analysis and Measures System • Stratospheric Aerosol Mesospheric Sounder • Stratospheric And Mesospheric Sounder • Surface Air Management System

SAMSHA Substance Abuse and Mental Health Services Administration

SAMSI Statistical and Applied Mathematical Sciences Institute (USA-NC)

SAMSON System Analysis of Manned Space Operations

SAMSS Space Assembly Maintenance and Servicing Study

SAMT State-of-the-Art Medium Terminal

SAMU Service d'Aide Médicale d'Urgence

SAMW Schweizerische Akademie der Medizinischen Wissenschaften (Switzerland)

San Sanitary

SAN San Diego - Lindbergh (USA-CA) (location identifier) • Sanitation • Secondary Access Node • Security Assistance Network • Serveur d'Archivage Numérique • Severe Acoustic Noise • Société d'Astronomie de Nantes (France) • Space Age News • Standard Access Number • Standard Account Number • Standard Address Number • Standard Address Numbering • Standard Article Number • Storage-Area Network • Styrene Acrylonitrile • System Advisory Notice • System Area Network • Système d'Archivage Numérique

SANAGA Système d'Acquisition Numérique pour l'Analyse des Grains Africains

SANAMET Servicio Autónomo Nacional de Metrología (Venezuela)

SAnAp Suppléments aux Annales d'Astrophysique (periodical code)

SANAP South African National Antarctic Programme

Sanat Sanatorium

SANCST Saudi National Center for Science and Technology (Saudi Arabia) (now NCST)

Sand Sandwich

SAND Site Activation Need Data (*) • Site Activation Need Date • Sorting and Assembly of New Data • Space-Assisted Network against Desertification • Spectroangular Normalized Difference

SANDS Structural Analysis Numerical Design System

SANE Severe Acoustic Noise Environment • Signaling Area Network Code • Standard Apple Numeric Environment

SANET South African Network

SANIL Système d'Analyse des Images Landsat

Sanit Sanitary • Sanitation

SANMHD Spherical Anelastic Magnetohydrodynamics

SANOS Stabilization Airborne Night Observation System

SANOVA Simultaneous Analysis Of Variance

SANR Subject to Approval - No Risk

SAN REMO Système Analytique de Répartition de Moyens

Sans Sanserif • Sanskrit

SanA Separata Universidad de Chile Departamento de Astronomía (periodical code)

SANS Small-Angle Neutron Scattering • Stellar Autonomous Navigation System • Systems And Network Security

Sánscr Sánscrito

SANT S-band Antenna

SANTA Science Analysis Near-Term Activity

SANU Srpska Akademiia Nauka i Umetnosti

SANUG Srpska Akademiia Nauka i Umetnosti Glas Odeljenje Tekhnichnikh Nauka (periodical code)

SANV Société d'Astronomie du Nord Vaudois (Switzerland)

SANW Schweizerische Akademie der Naturwissenschaften (Switzerland)

SANZ Standards Association of New Zealand (now SNZ)

SAO São Paulo (Brazil-SP) (location identifier) • Saturdays Only • Science Assessment Observation • Security Assistance Office • Security Assistance Officer • Security Assistance Organization • Select Address and Operate • Semi-Annual Oscillation • Shaanxi Astronomical Observatory (China-PRC) • Shanghai Astronomical Observatory (China-PRC) • Smithsonian Astrophysical Observatory (USA-MA) • Sobaeksan Astronomy Observatory (Korea-ROK) (see now SOAO) • Sociedad Astronómica Octante (Uruguay) • Special Access Office • Special Activity Office • Special Adaptive Optics • Special Astrophysical Observatory (Russia) • Spectrographe Acousto-Optique • Spectromètre Acousto-Optique • Staff Action Officer • Statistiques Assistée par Ordinateur • Subject - Action - Object • Surface Airways Observation • Systems Analysis Office

SAOC Smithsonian Astrophysical Observatory Catalog • Space and Astronautics Orientation Course

SAOCo Smithsonian Astrophysical Observatory Contributions to Astrophysics (periodical code)

SAOCRE Stand-Alone Optical Character Recognition Equipment

SAOI Synthetic Aperture Optical Instrument

SAOS Select Address Output Signal

SAOSR Smithsonian Astrophysical Special Report (periodical code)

SAOZ Système d'Analyse par Observations Zénithales

SAp Service d'Astrophysique

SAP Safety Analysis Program • Sampling and Analysis Plan • San Pedro Sula - Villeda Morales (Honduras) (location identifier) • Sans Aucun Problème • Scanned Absolute Power • Science Archive Process • Science Assessment Panel • Scientific Advisory Panel • Secondary Aerosol Particle • Second Audio Program • Security Assistance Program • Senior Analyst and Programmer • Sensor AVE Package • Service Access Point • Service Access Program • Service Access Protocol • Service Access Provider • Service Advertisement Protocol • Service Advertising Protocol (*) • Session Announcement Protocol • Session d'Avant-Projet • Share Assembly Program • Simplified Acquisition Procedure • Site Activation Plan •

Sociedad Astrofísica de Paraguay • Société d'Astronomie Populaire (France) • Software Assurance Plan • Software Automation Panel • Solar Array Paddle • Soon As Possible • Speaker-Adaptation Process • Special Access Program • State Assessment Panel • Strain Arrestor Plate • Strong Anthropic Principle • Structural Adjustment Program • Structural Analysis Program • Subject Access Project • Superadditivity Property • Superbubble Accelerated Particle • Survivable Adaptive Planning • Symbolic Address Program • Symbolic Assembly Program • System Access Protocol • System Acquisition Plan • System Alignment Procedure • Système Automatisé de Production • Systems Assurance Program • Systolic Arterial Pressure

SAPA South African Publishers' Association • System for Analysis of Pyramid Algorithms

SAPAM Society of African Physicists and Mathematicians

SAPC Société Astronomique Populaire du Centre (France) • Standard Asteroid Photometric Catalogue

SAPCO Security Assistance Policy Coordinating Office

SAPE Survivable Adaptive Planning Experiment

SAPES Standardized Advanced Polyelastic Sampling

SAPFU Surpassing All Previous Foul-Ups

SAPI Service Access Point Identifier • Service Access Protocol Interpreter • Speech Application Programming Interface

SAPIR System of Automatic Processing and Indexing of Reports

SAPIU Single-board Adaptive Programmable Interface Unit

SAPL Société d'Astronomie Populaire de Limoges (France)

SAPLA Standing Advisory Panel on Library Automation

SAPM Security Assistance Program Manager • Société d'Astronomie du Planétarium de Montréal (Canada-QC)

SAPMD Stand-Alone Pressure Measurement Device

SApMM North Holland Series on Applied Mathematics and Mechanics (periodical code)

SAPO Space Application Project Office

SAPP Société d'Astronomie Populaire Poitevine (France)

SAPPSC Space Applications and Commercialization Steering Committee (NRC)

SAPR Semi-Annual Progress Report

SAPRG Small-Amplitude Pulsating Red Giant (star)

SAPRR Société des Autoroutes Paris-Rhin-Rhône

SAPRWG Security Assistance Program Review Working Group

SAPT Security Assistance Program Training

SAQ San Andros (Bahamas) (location identifier) • Sociedad Astronómica de Quebradillas (Puerto Rico)

SAQC Statistical Analysis and Quality Control

Sar Sarawak • Sardinia • Sardinian

SAR Riyal (Saudi Arabia) (ISO currency code) • Safety Action Report • Safety Analysis Report • Safety Assessment Report • Satellite Access Request • Scanned Absolute Response • Schedule Add Request • Schedule Allocation Requirements • Science Applications Research • Science Assessment Report • Sealed Authentication System • Search And Release • Search And Rescue • Secondary Angular Resolution • Second Assessment Report • Secrétaire d'Administration de la Recherche • Security Assistance Review • Segment And Reassemble • Segmentation And Reassembly • Selected Acquisition Report • Selected Acquisition Review • Selling, Administrative and Research • Semi-Active Radar • Semi-Annual Report • Shift Arithmetic Right • Significant Activities Report • Simultaneous Autoregressive • Single-Axis Reference • Single-bit Alternation Recording • Site Acceptance Review • Site Action Request • Situation Awareness Report • Slower-than-Average Rotation • Small Active Region (Sun) • Sociedade Astronômica do Recife (Brazil) • Sodium Adsorption Ratio • Software Analysis Room • Solar Array Regulator • Son Altesse Royale • South African Republic • Spacecraft Acceptance Review • Sparta (USA-IL) (location identifier) • Special Access Report • Special Access Required • Special Action Request • Special Action Requirement • Special Administrative Region • Special Analysis Report • Specific Absorption Rate • Specific Activity Report • Specification Approval Report • Stable Auroral Red • Standard Application Report • Stock Appreciation Right • Storage Address Register • Store Address Register • Su Alteza Real • Subauroral Red • Subcommittee on Atmospheric Research • Subsequent Applications Review • Successive Approximation Register • Summary Analysis Report • Supply Assist Request • Support Action Request • Support Analysis Record • Surveillance Assessment Report • Symbol Acquisition Routine • Synthetic Aperture Radar • Syrian Arab Republic • System Acceptance Review • System Acquisition Report • System Acquisition Review • System Activity Rate • System Analysis Report • System Architecture Review • System Assessment Review

SARA Sampled Aperture Receiving Array • Satellite Amateur de Radioastronomie • Satellite pour Astronomie Radio Amateur • Scholarly Articles Research Alerting • Single Aliquot/Regeneration and Added • Society of Amateur Radio Astronomers (USA-NY) • Southeastern Association for Research in Astronomy (USA) • Space Astronomy Research and Analysis • Stichting Rekencentrum Amsterdam (Netherlands) • Superfund Amendments and Reauthorization Act (USA) • Symposium on Abstraction, Reformulation and Approximation • Synthetic Aperture Radar Antenna • Systems Analysis and Resource Accounting

SARAH Search And Range Homing

SARAS Synthetic Aperture Radar Advanced Simulator

SARB Sélecteur Automatique de Références Bibliographiques • Surface and Atmospheric Radiation Budget

SARBE Search And Rescue Beacon Equipment

SARC Satellite Archive Coordination Center • Science Archive Research Center • Scientific and Applied Research Centre (Qatar) • Service Acquisition Review Council • Space and Astronomy Research Center • Surveillance And Reconnaissance Center • Symantec Antivirus Research Center • System Acquisition Review Council

SARD Solar-Array Release and Deployment • Swedish Agency for Research and Development

SARDAPF Synthetic Aperture Radar Data Acquisition and Processing Facility

SARDO Search And Rescue Deputy Office

SARDPF Synthetic Aperture Radar Data Processing Facility

SARDS Situation And Reality Display System

SARE Science Archive Research Environment • Système Automatique de Radiosondage Embarqué

SAREC Swedish Agency for Research Cooperation

SAREF Safety Research Experiment Facility

SAREX Search And Rescue Exercise • Shuttle Amateur Radio Experiment • South American Radar Experiment

SARF Secrétaire d'Administration de Recherche et de Formation • Security Alarm Reporting Function • Special Access Required Facility

SARG Space Assets Review Group • Spettrografo Alta Risoluzione Galileo • System Acquisition Requirement Generator

SARGE Surveillance And Reconnaissance Ground Equipment

SARI Sistema de Almacenamiento y Recuperación de la Información • Subduction Accelerated Research Institute • Synthetic Aperture Radar Industrial

SARIPS Search And Rescue Information Processing System

SARIR Search And Rescue Incident Report

SARIT Swiss Association for Research on Information Technology (Switzerland)

Sarl Société à Responsabilité Limitée

SARL Société À Responsabilité Limitée

SARM Set Asynchronous Response Mode

SARMIS Search And Rescue Management Information System

SAROAD Storage And Retrieval Of Aerometric Data

SAROS Synthetic Aperture Radar for Open Skies

SARP Safety Analysis Report for Packaging • Satellite Acquisition Request Programming • Schedule, Analysis and Review Procedure • Schedule And Resources Procedure • Schedule And Reporting Procedure • Schedule And Review Procedure • Shuttle Astronaut Recruitment Program • Small Autonomous Research Package • Space Allocation Requirement Procedure • Space Allocation Reservation Program • Standards And Recommended Practices

SARPS Standards And Recommended Practices

SARREQ Search And Rescue Request

SARS Segmentation And Reassembly Sublayer • Severe Acute Respiratory Syndrome • Single-Axis Reference System • Solar-Array Reorientation System • Static Automatic Reporting System • Stellar Attitude Reference Study • System Architecture Requirements Specification

SARSAT Search And Rescue Satellite • Search And Rescue Satellite-Aided Tracking (*)

SARSDM State Agency Remote Sensing Data Management

SARSEX Synthetic Aperture Radar Signature Experiment

SARSSDU Segmentation And Reassembly Sublayer Service Data Unit

SART Search And Rescue Transponder • Stimuli Analog Refresh Table

SARTEL Search And Rescue Telephone

SARTF Search And Rescue Task Force

SARTS Switched Access Remote Test System

SARUMAN Secure Access and Restricted Use of Images on the Net

SARUS Search And Rescue Using Satellites

SARV Small-Amplitude Red Variable (star)

SAS Sales Accounting System • Salford Astronomical Society (UK) • Salisbury Astronomical Society (UK) • Salton City (USA-CA) (location identifier) • Sammelanschlußschaltung • Sample Acquisition System • Sample Array System • Science Analysis Software • Science Analysis System • Science and Arts Society • Science Archive System • Scientific Analysis System • Sealed Authentication System • Secondary Alarm Station • Secure Authentication Service • Security Agency Study • Self-Adaptive System • Server-Attached Storage • Service Accounting Segment • Service d'Analyse Stratégique • Sheboygan Astronomical Society (USA-WI) • Shenandoah Astronomical Society (USA-VA) • Sheffield Astronomical Society (UK) • Shropshire Astronomical Society (UK) • Signal Analysis System • Single-Attached Station • Single-Attachment Station • Single-Attach Station • Single Audio System • Skylight Astronomical Society Inc. (USA-MA) • Slovak Astronomical Society • Small-Angle Scattering • Small Applications Satellite • Small Astronomical Satellite • Small Astronomy Satellite • Small Atmosphere Structure • Società in Accomandita Semplice • Société Astronomique de Suisse (Switzerland) • Society for Amateur Scientists (USA-CA) • Society for Applied Spectroscopy (USA-AZ) • Software Approval Sheet • Software Architecture Skeleton • Software Automation Subpanel • Solar-Array Simulator • Solar-Array Structure • Solar-Array System • Solar-Aspect Sensor • Solid-state Array Spectroradiometer • Son Altesse Sérénissime • Source Acquisition Segment • Source Assessment System • Southampton Astronomical Society (UK) • South Astronomical Society (UK) • Southend Astronomical Society (New Zealand) • Southern Astronomical Society (Australia) • Space Activity Suit • Space Adaptation Syndrome • Space Applications Section • Space Architectural Study • Spacecraft Analysis Software • Spacecraft Antenna System • Special Air Service • Spokane Astronomical Society (USA-WA) • Springfield Astronomical Society (USA-MO) • Stability Augmentation System • Stability Augmented System • Stabilization Augmentation System • Staff Administrative Specialist • Stand-Alone Switch • Standard Analysis System • Static Analysis Symposium • Station Automatique Secondaire • Statistical Analysis System • Statistically Assigned Sockets • Stockton Astronomy Society (USA-CA) • Structurally Adaptive System • Studies - Analysis - Simulation • Su Alteza Serenísima • Subarctic Summer • Sun Acquisition Sensor • Supply Analysis System • Supportability Analysis Summary • Surface-Active Substance • Surveillance Architecture Study • Survivable Adaptive System • Survival Avionics System • Svenska Astronomiska Sällskapet (Sweden) • Swiss Accreditation Service • Syndrome d'Apnées du Sommeil • Syracuse Astronomical Society (USA-NY) • System Analysis and Simulation • System Analysis Study • System Application Software • System Architecture Study • System of Aeronomical Satellites • Systems Acquisition School

SASC Science Analysis Support Centre • Semi-Automatic Switch Center

SASCB Shuttle Avionics Software Control Board

SASCOE Solar-Array Special Checkout Equipment

SASDI South African Selective Dissemination of Information

SASE Self-Addressed Stamped Envelope • Society for the Advancement of Socio-Economics • Specific Application Service Element • Standards Analysis, Synthesis and Expression

SASER Sound Laser

SASET Software Architecture Sizing and Estimating Tool

SASF Serveur d'Application Sans Fil

SASG Smoke Aerosol Steering Group

SASHR Senior Assistant Secretary on Human Resources

SASI S-Asteroid Spectral Interpreter • Shugart Associates System Interface • South African Standards Institution • Statistical Analysis Systems Institute (USA-NC) • Sutherland Astronomical Society Inc. (Australia)

SASIDS Stochastic Adaptive Sequential Information Dissemination System

SASIFY Space Agency Forum for the International Space Year

Sask Saskatchewan (Canada)

SASK Saskatchewan (Canada)

SASKR Semi-Active Seeker

SASL Simple Authentication Security Layer • Société d'Astronomie de Saône et Loire (France)

SASLO South African Scientific Liaison Office

SASM Smithsonian Air and Space Museum (USA-DC) • Society for Automation in the Sciences and Mathematics (USA) • Synthetic Aperture Strip Map

SASMO Syrian Arab Organization for Standardization and Metrology (Syria)

SASn Studia Astronomica Sinica (periodical code)

SASN Sedentary-Actor Subnetwork • System-Assigned Serial Number

SASO Saudi Arabian Standards Organization (Saudi Arabia) • Senior Air Staff Officer • Stability And Support Operation

SASP Science and Application Space Platform • Single Acoustic Signal Processor • Single Advanced Signal Processor • Stand Alone Support Program • State Agency for Surplus Property

SASPRSC South African Society for Photogrammetry, Remote Sensing and Cartography (South Africa)

SASPS Standard Automated Small Purchase System

SASPWG System Command Acquisition System Protection Working Group

SAS&R Satellite-Aided Search & Rescue

SASR Satellite-Aided Search and Rescue • Shuttle Avionics Status Review • Shuttle Avionics System Review • Spare Available Space Required

SASRS Satellite-Aided Search and Rescue System

SASS San Antonio Space Society (USA-TX) • Saturn Automatic Software System • Schedules And Status Summary • Scientific Analysis Software System • Seasat Active Scatterometer System • Semi-Automatic Switching System • Shanghai Academy of Social Sciences (China-PRC) • Small Aerostat Surveillance System • Small Airship Surveillance System • Space Asset Support System • Spectral Agile Sensor System • Standard Analysis and Software System • Standard Analysis Software System (*) • Surveillance Analysis Support System • Systems Acquisition and Services Support

SASSE Synchronous Altitude Spin-Stabilized Experiment

SASSIF Self-Adjusting System of Scientific Information Flow

SASSY Selected Areas Starcounts Survey • Supported Activities Supply System

SAST Science Analysis Software Team • Scientific Assessment and Strategy Team • Service and Agency Support Team • Society for the Advancement of Space Travel • South African Standard Time • Statistical Analysis of Sparse Times • Strategic Analysis in Science and Technology • Strategic Assumption Surfacing Technique • Survivable Adaptive System Technology

SASTD Survivable Adaptive System Technology Demonstration

SASTAR Saint Andrews (Starlink internal code)

SASTP Stand-Alone Self-Test Program

SASTU Signal Amplitude Sampler and Totalizing Unit

SASU Secrétaire d'Administration Scolaire et Universitaire (France)

SASY Software Architects' Synthesis

Sat Satellite • Satisfaction • Satisfactory • Satisfied • Satisfy • Saturate • Saturated • Saturday • Saturn

S. At. South Atlantic

SAT Sampler Address Translator • San Antonio (USA-TX) (location identifier) • Satellite • Satisfactory • Saturate • Saturated • Saturday • Scholastic Aptitude Test • Scholastic Assessment Test • Science - Aeronautics - Technology • Science Assessment Team • Security Alert Team • Security Assistance Team • Security Assistance Test • Security Audit Trail • Select At Test • Self-Alignment Technique • Self-Avoiding Trail • Sensor Assessment Technology • Service Autonome de Télédétection • Servicing Aid Tool • Shock Arrival Time • Simplified Acquisition Threshold • Simulation Assessment Team • Simulator Acceptance Test • Site Acceptance Test • Site Activation Team • Site Assignment Time • Situational Awareness Technology • Situational Awareness Terminal • Slit Aperture Telescope • Small Area Triangulation • Società Astronomica Ticinese (Switzerland) • Société Anonyme de Télécommunications • Software Acceptance Test • Software Action Team • South Australian Time • Space Available Travel • Spatial Analysis Toolkit • Special Access Termination • Special Alias Table • Specified Action Table • Stabilization Assurance Test • Standard Assessment Task • Standard Automated Terminal • Static Air Temperature • Stepped Atomic Time • Stochastic Aptitude Test • Strömgren Automatic Telescope • Sulfuric Acid Tetrahydrate • Support Analysis Test • Surface Aerospace Technology • Surveillance - Acquisition - Tracking • Symmetric Axis Transform • Synchronized Acoustic Transmitter • Synthetic Aperture Telescope • System Acceptance Test • System Acceptance Testing • System Access Technique • System Alignment Test • Systems Approach to Training

SATA Safety Arming Test Aid

SATAF Shuttle Activation Task Force • Site Activation Task Force • Site Alteration Task Force

SATAN Satellite Automatic Tracking Antenna • Security Administrator Tool for Analyzing Networks • Sensor for Airborne Terrain Analysis

SATANS Static and Transient Analysis, Non-linear, Shells

SATAR Satellite for Aerospace Research

SATB Strength Aptitude Test Battery

SATCAT Satellite Catalog

SatCo Satellite Communications (periodical code)

SATCO Signal-automatic Air-Traffic Control • Simulated Air-Traffic Control Organization

SatCom Satellite Communications

SATCOM Satellite Communications

SATCON Satellite Condition • Satellite Connectivity

SATDAS Satellite Data Acquisition System

SATE Security Awareness, Training and Education • Semi-Automatic Test Equipment • Special Automated Test Element

Satell Satellite

SATEM Satellite Temperature

SATEX Saturdays Excepted • Semi-Automatic Telegraph Exchange

SATF Satellite Assembly and Test Facility

SATFA Security Assistance Training Field Activity

SatFu Satellitenfunk

SatFv Satellitenfestverbindung

SATH Society for the Advancement of Travel for the Handicapped (USA-NY)

SATI Spectral Airglow Temperature Imager

SATIF Scientific And Technical Information Facility

SATIRE Semi-Automatic Technical Information Retrieval

SATIS Science And Technology In Society • Scientific And Technical Information Service

SATKA Surveillance, Acquisition, Tracking, and Kill Assessment

SATMO Security Assistance Training Management Organization

SATMOS Système d'Archivage et de Traitement Météorologique des Observations Spatiales

Satn Saturation

SATNAV Satellite Navigation

SATNET Satellite Network

SATO Scheduled Airlines Ticket Office • Self-Aligned Thick Oxide • Shuttle-Attached Teleoperator • Supply And Transportation Operations • Système d'Analyse de Textes par Ordinateur

SATOB Satellite Observation

SATODP Satellite Tracking Orbit Determination Program

SATODS Security Assistance Technical Order Data System • Security Assistance Technical Order Distribution System

SATOP Security Assistance Technical Order Program

SATOR Stratospheric And Tropospheric Ozone Research

SATORI Systematic Air Traffic Operations Research Initiative

SATP Security Assistance Training Program • Space Applications Technology Program • Stabilization, Acquisition, Tracking and Pointing

SATRA South African Telecommunications Regulatory Authority • South African Telephone Regulatory Authority

SATRAC Satellite Automatic Terminal Rendezvous And Coupling

SATRAN Satellite Transmission

SATREP Satellite Report

SATRO Science And Technology Regional Organization

SATS Satellite Antenna Test System • Satellite Automated Test System • Shuttle Avionics Test Set • Shuttle Avionics Test Station • Shuttle Avionics Test System • Single-Array Test System • Small Aircraft Transportation System • Small Applications Technology Satellite • Spacecraft Automated Test System • Station Accommodation Test Site • Statistical Analysis of Sparse Times • Supplier Assessment Tracking System • Supply And Transportation Support

SATSI Segmented Aperture Tilted Shearing Interferometer

SATSIM Satellite Simulator

SATSS SEASAT Antenna Temperature Software System

SATSU Science And Technology Studies Unit (UK)

SATT Semi-Automatic Transistor Tester

SATW Schweizerische Akademie der Technischen Wissenschaften (Switzerland)

SAU Arabia (Saudi_) (ISO country code) • Arabie Saoudite (code ISO de pays) • Saint Ambrose University (USA-IA) • Saudi Arabia (ISO country code) • Sawu (Indonesia) (location identifier) • Serial Acquisition Unit • Signal Acquisition Unit • Società Astronomica Urania (Italy) • Spectrum Analysis Unit • Standard Acquisition Unit • Statistical Analysis Unit • Strap-Around Unit

SAUCERS Space And Unexplained Celestial Events Research Society

SAURON Spectrographic Areal Unit for Research on Optical Nebulae

S. Aus. South Australia • South Australian

S. Austral. South Australia • South Australian

sav Saved Data (file extension)

SAV Sale At Valuation • Savannah (USA-GA) (location identifier) • Save • Saved • Saving • Sector Access Velocity • Security Assistance Visit • Service-Appointed Validator • Service Après-Vente • Slovenská Akadémia Vied (Slovakia) • Società Astronomica Versiliese (Italy) • Staff Assistance Visit • Start of Active Video • Stock At Valuation

SAVAR Société d'Astronomie du Valais Romand (Switzerland)

SAVASI Simplified Abbreviated Visual Approach Slope Indicator

SAVB Scientific Applications and Visualization Branch

SAVDM Single Application Virtual DOS Machine

SAVE Satellitenverteildienst • Satellite Validation Experiment • Simulation Assessment Validation Environment • Solstice Atmospheric Variability Experiment • South Atlantic Ventilation Experiment • Special Action for Vigorous Efficiency • Standardized Acceptance and Verification • System for Automatic Value Exchange

SAVER Shuttle Avionics Verification

SAVI Soil-Adjusted Vegetation Index • Space Active Vibration Isolation

SAVIAC Shock And Vibration Information Analysis Center

SAVS Space Analysis and Visualization System • Status And Verification System

SAW Marquette - Sawyer (USA-MI) (location identifier) • Satellite Attack Warning • Scottish Astronomy Weekend • Self-Avoiding Walk • Shear Alfvén Wave • Software Analysis Workstation • Software Audio Workshop • Solar-Array Wing • Sound Acoustic Wave • Space At Will • Spill Analysis Workstation • Subarctic Winter • Surface Acoustic Wave • Surveillance And Warning

SAWB South African Weather Bureau

SAWBOC System Analysis With Blockout Considerations

SAWD Solid Amine Water Desorption • Solid Amine Water-steam Desorption

SAWEGS System for Automatic Wind Extraction from Geostationary Satellite

SAWF Surface-Acoustic-Wave Filter

SAWG Special Advisory Working Group

SAWRS Supplemental Aviation Weather Reporting System • Supplementary Aviation Weather Reporting System

SAWS Solar-Array Wing Simulator • Synoptic Automatic Weather Station

SAW/V Satellite Attack Warning/Verification

SAWV Satellite Attack Warning/Verification

Sax Saxon • Saxony

SAX Sambu (Panama) (location identifier) • Satellite per Astronomia in raggi X • Saturdays Excepted • Simple API for XML

SAXA Slotted Array X-band Antenna

SAXON SAR X-band and Ocean Non-linearity

SAXS Small-Angle X-ray Scattering

SAY Siena (Italy) (location identifier)

SAYE Save As You Earn

SAZ Sasstown (Liberia) (location identifier) • Standards Association of Zimbabwe

SAZU Slovenian Academy of Sciences and Arts

sb Stilb

Sb Antimoine • Antimon • Antimony

S/B Serial Block • Signal/Bruit • Standby

SB Bachelor of Science • Barred Spiral Galaxy • Îles Salomon (code ISO de pays) • Sales Book • Satapatha Brāhmana • Selbstbedienung • Selection Board • Selective Bibliography • Serial Binary • Serial Block • Service Bulletin • Short Bill • Short Block • Sideband • Signal Book • Simultaneous Broadcast • Simultaneous Broadcasting • Slettebak-Brundage (survey) • Sliding Bearing • Small Business • Sodium Borate • Software Bus • Solomon Islands (ISO country code) • Sound Board • South Britain • Southern Boundary • Space Base • Space-Based • Special Bibliography • Special Branch • Special Bureau • Spectroscopic Binary (star) • Speed Brake • Spektralbibliothek • Spiral Galaxy - Barred • Staatsbibliothek • Standby • StarBits (database) • StarBriefs (dictionary) • Starburst • State Beach (USA) • Statsbibliotek • Statute Book • Storage Battery • Straight Binary • Subbranch • Subnegotiation • Subsidiary Body • Subvention Budgétaire • Successful Branch • Sueldo base • Superbubble • Supply Bulletin • Surface Brightness • Switchboard • Synchronization Base • Synchronization Bit

SBA Santa Barbara (USA-CA) (location identifier) • Scene Balance Algorithm • School of Business Administration • Seat-Back Assembly • Secondary Baffle Assembly • Set Buffer Address • Shaped Beam Antenna • Side Band Addressing • Simulation-Based Acquisition • Singapore Broadcast Authority • Small-Beam Approach • Small Business Administration • Spreader Bar Assembly • Standard-Beam Approach • Standards-Based Architecture • Structure-Borne Acoustic • Switchable Beam Antenna • Synchro Booster Amplifier • Synchronous Bandwidth Allocation • Synchronous Bandwidth Allocator

SBAA Sociedade Brasileira dos Amigos da Astronomia (Brazil)

SBAC Safety Data Analysis Unit • Santa Barbara Astronomy Club (USA-CA) • Society of British Aerospace Companies (UK)

SBARM SBARMO Bulletin (periodical code)

SBAS S-Band Antenna Switch • Shreveport-Bossier Astronomical Society (USA-LA) • South Bay Astronomical Society (USA-CA)

SBASS Space-Based Atmospheric Surveillance System

SBAU Santa Barbara Astronomical Unit (USA-CA)

SBB Santa Barbara (Venezuela) (location identifier) • Schweizerische Bundesbahnen (Switzerland) • Stadt- und Bezirksbibliothek • Standard Big Bang (cosmology) • Supplied By Bidder

SBBN Standard Big Bang Nucleosynthesis (*) • Standard Black Body Nucleosynthesis

SBC Sanriku Balloon Center (Japan) • SCSI Block Command • Selbang (Papua New Guinea) (location identifier) • Senate Budget Committee • Service-Based Costing • Signaling Buffer Controller • Simple Bayesian Classifier • Single-Board Computer • Sociedade Brasileira de Computação (Brazil) • Société Bibliographique du Canada • Solar Blind Camera (*) • Solar Blind Channel • Space-Based Communications • Space Biotechnical Complex • Spaceborne Computer • Spectral Brightness Coefficient • Speed Brake Command • Standard Boundary Condition • String Bottom Controller • Sum of Base Cations • Synchro Breaking Concept

SBCA Satellite Broadcasting and Communications Association (USA-VA)

SBCCS Single Byte Command Code Set

SBCDP Small Business Competitiveness Demonstration Program

SBCJ Store Block Control Journal

SBCON Single-Byte Conversion

SBCR Stock Balance and Consumption Report

SBCS Single-Byte Character Set • Southern Boundary Containment System

SBD Dollar (Solomon Islands) (ISO currency code) • San Bernardino - Norton (USA-CA) (location identifier) • S-Band • Scene Block Diagram • Schematic Block Diagram • Schottky Barrier Detector • Schottky Barrier Diode • Sentence Boundary Disambiguation • Silent But Deadly • Simulated Based Design • Simulation-Based Design • Smart Battery Data • Standard Bibliographic Description • Surface Barrier Diode

SBDB Small-Bodies Database

SBDC Small-Business Development Center

SBDEW Space-Based Directed-Energy Weapon

SBDO Space Business Development Operation

SBDS Solid-Based Density Standards • Structure-Based Deduction System

SbE South by East

SBE Scene Balance Algorithm • Simple Boolean Expression • Single Best Estimate • Social - Behavioral - Economic (sciences) • Solar Beam Experiment • Space-Based Element • Suabi (Papua New Guinea) (location identifier) • Subbit Encoder

SBEA Sociedade Brasileira para o Encino da Astronomia (Brazil)

SBED Serial Bit Error Detector

SBENTO Standard Bento Entity for Natural Transport of Objects

SBEP Superbubble Energetic Particle

SBER Subbit Error Rate

SBES Space-Based Experimental System

SBEV Space-Based Experimental Version • Space-Borne Experimental Vehicle

sbf Salvo Buen Fin • Sauf Bonne Fin

SBF Sans Bureau Fixe • Sardeh Band (Afghanistan) (location identifier) • Scientific Balloon Facility • Single Barrier Failure • Société des Bourses Françaises (France) • Southern Base Facility (Gemini) • Surface Brightness Fluctuation

SBFEL Space-Based Free Electron Laser

SBFM Silver Band Frequency Modulation

SBFS Space Business Financial Services

SBFU Standby Filter Unit

SBG Sabang (Indonesia) (location identifier) • Starless Bok Globule

SBH Saint Barthelemy (Leeward Islands) (location identifier) • Schwarzschild Black Hole • Supermassive Black Hole • Switch Busy Hour

SBHC Speed Brake Hand Controller

SBI Koundara Sambailo (Guinea) (location identifier) • Satellite Broadcasting Interface • Shelf Bus Interface • Short Baseline Interferometer • Short Baseline Interferometry • Single Burning Item • Solar Bolometric Imager • Space-Based Interceptor • Speaker Box Interface • Special Background Investigation • Subsidiary Body for Implementation

SBIC Small Business Investment Co.

SBIG Santa Barbara Instrument Group (USA-CA) • Short Baseline Interferometer • Short Baseline Interferometry • Subject-Based Information Gateway

SBIL Schema-Based Interconnection Language

SBIN Shared Bibliographic Input Network

SBIR Small Business Innovation Research • Small Business Innovative Research • Space-Based Infrared

SBIRE Space-Based Infrared Experiment

SBIRP Small Business Innovation Research Program

SBIRS Space-Based Infrared System

SBIS Sustaining Base Information Services

SBIT Startup Built-In-Test

SBITP Santa Barbara Institute of Theoretical Physics (USA-CA)

SBJ São Mateus (Brazil-ES) (location identifier)

SBK Saint-Brieuc (France) (location identifier)

SBKEW Space-Based Kinetic-Energy Weapon

SBKEWS Space-Based Kinetic-Energy Weapon System

SBKKV Space-Based Kinetic Kill Vehicle

SBKKVWS Space-Based Kinetic Kill Vehicle Weapon System

SBL Santa Ana (Bolivia) (location identifier) • Security By Lighting • Sealed Beam Lamp • Space-Based Laser • Stable Boundary Layer • Super Basic Language • Surface Boundary Layer

SBLC S-Band Linear Collider • Standard Base-Level Computer • Sustaining Base-Level Computer

SBLI Savings Bank Life Insurance

SBLIP Space-Based Laser Intercept Program

SBLM Sustaining Base Leadership and Management

SBLP Space-Based Logistic Platform

SBLS Spaceborne Laser Ranging

SB/M Switchable Bandpass/Mirror

SBM Satellite Broadcast Manager • Sciences Biologiques et Médicales • Shared Bandwidth Manager • Sheboygan (USA-WI) (location identifier) • Single Buoy Mooring • Single Buyer Model • Solid Booster Motor • Subnet Bandwidth Manager • Switchable Bandpass/Mirror • System Balance Measure

SBMAC Sociedade Brasileira de Matématica Aplicada e Computacional (Brazil)

SBMB Supervisory Bus Monitor Buffer

SBMD Subscale Beryllium Mirror Demonstrator

SBMS Spaceborne Management Satellite • Store Business Management System

SBN Satellite Broadcast Network • Servizio Bibliotecario Nazionale • Small Bodies Node • South Bend (USA-IN) (location identifier) • Standard Bok Nummer • Standard Book Number (now ISBN) • Standard Book Numbering • Starburst Nucleus

SBNC Small Bodies Names Committee

Sbnd Southbound

SBNPB Space-Based Neutral Particle Beam

SBNPBS Space-Based Neutral Particle Beam System

SBO Salina (USA-UT) (location identifier) • Science-Based Organization • Security By Obscurity • Small Business Office • Sommers-Bausch Observatory (USA-CO)

S'board Starboard

Sboard Starboard

SBOF Standard Book Order Form

SBOLOP Small Board Online Operational Program

SBOS Sound Board Operating System

SBP San Luis Obispo (USA-CA) (location identifier) • Sedimentary Basins Project • Serial Bus Protocol • Service Benefit Plan • Société Belge de Physique (Belgium) • Software Best Practices • Sonic Boom Panel • Space-Based Radar • Spaceborne Programmer • Strategic Business Plan • Summarized Budget Phase • Surface Brightness Profile • Survivor Benefit Plan • Systolic Blood Pressure

SBPA S-Band Power Amplifier • Señal Binaria Pseudo-Aleatoria • Singapore Book Publishers' Association

SBPI Software Best Practices Initiative

SBPM Société Belge des Professeurs de Mathématiques (Belgium)

SBPTC Société Belge de Photogrammétrie, de Télédétection et de Cartographie (Belgium)

SBQ Sibi (Pakistan) (location identifier) • Sociedade Brasileira de Química (Brazil)

s. Br. südliche Breite

SBR Saibai Island (Australia-QLD) (location identifier) • S-Band Receiver • Size-Based Ranking • Space-Based Radar • Space-Borne Radar • Standby Reserve • Statistical Bit Rate • Stellar Birthrate • Styrene Butadiene Rubber • Surface-Based Radiometer

SBRC Santa Barbara Research Center (USA-CA) • Sensor-Based Robot Control

Sbre Septiembre

SBRF Space-Based Radio Frequency

SBRS Sonic Boom Reporting System

SBRTN Satellite-Based Rural Telegraph Network

SBS Satellite Base System • Satellite Business Systems • Satellitenbodenstation • Scientific and Business System • Scientific Business Satellite • Secondary Basic Subject • Second Byurakan Survey • Serially Balanced Sequence • Silicon Bilateral Switch • Simulated Brillouin Scattering • Small Business Specialist • Solar Background Spectrum • Sortie Banalisée Synchrone • Space Bio Separation • Special Broadcasting Service • Steamboat Springs (USA-CO) (location identifier) • Stimulated Brillouin Scattering • Success Breeds Success (principle) • Système Binaire Serré

SBSA Small Business Set Aside

SBSC Supplement to the Bright Star Catalog

SBSD Subside

SBSE Science-Based System Engineering

SBSG Second Byurakan Survey of Galaxies

SBSL Single Bubble-Sonoluminescence

SBSP Space-Based Support Platform

SBSS Sciences Biologiques, Sanitaires et Sociales • Second Byurakan Survey of Stars • Space-Based Space Surveillance • Space-Based Surveillance System • Standard Base Supply System

Sbstr Substrate

SBSTA Subsidiary Body for Scientific and Technological Advice

SBSTTA Subsidiary Body on Scientific, Technical and Technological Advice

SBSWG Small Bodies Science Working Group

SBT San Bernardino - Tri-City (USA-CA) (location identifier) • Scanbeam Table • Space-Based Tug • Spanning Binomial Tree • Statens Bibliotekstjenste • Surface-Barrier Transistor • Sustained Basic Training • System Baseline Testing

SBTA Small Business Technical Advisor

SBTC Speed Brake/Thrust Controller

SBTO Speed Brake Takeover

SBts StarBits (database)

SBTS Shore-Based Tracking System • State Bureau of Technical Supervision

SBTT Small Business Technology Transfer

SBU Sensitive But Unclassified • Springbok (South Africa) (location identifier) • Strategic Business Unit

SBUE Switch-Backup Entry

SBURCS Six-Bit Universal Random Character Set

SBUV Solar and Backscattered Ultraviolet • Solar Backscatter Ultraviolet (*)

SBV Sabah (Papua New Guinea) (location identifier) • Space-Based Visible (sensor) • Space Biospheres Ventures (USA-AZ)

SBVAA San Bernardino Valley Amateur Astronomers (USA-CA)

SbW South by West

SBW Sibu (Malaysia) (location identifier) • Spectral Bandwidth

SBWAS Space-Based Wide-Area Surveillance

SBWR Simplified Boiling Water Reactor

SBX S-Band Transponder • Shelby (USA-MT) (location identifier) • Slow-Burning Explosive • Student Book Exchange

Sby Standby

SBY Salisbury (USA-MD) (location identifier)

SBZ Sibiu (Romania) (location identifier)

s/c Self-Contained • Son Compte • Sous le Couvert • Su Cargo • Su Casa • Su Cuenta

sc Scientific • Scilicet • Scruple • Sculpsit • Small Capitals • Source Code (file extension) • Supercalendered

Sc Scale • Scandinavia • Scandinavian • Scandium • Scène • Scene • Schmidt Number • Science • Scientific • Scientifique • Scotch • Scotland • Scots • Scottish • Screw • Scruple • Sculpsit • Sculpsit Carved • Stratocumulus

S/C Santa Cruz • Security Code • Sensor/Controller • Signal Conditioner • Soft Copy • Software Contractor • Spacecraft • Splitter/Combiner • Stabilization/Control • Stripchart • Subcontract • Subcontractor

S-C Self-Consistent • Serbo-Croat

S&C Search & Clear • Stabilization & Control

SC Safe Custody • Safety Car • Sales Code • Same Case • Same Circular • Same-sense Circular • Santa Catarina (Brazil) • Santa Cruz (Argentina) • SAR Coordinator • Satellite Cell • Satellite Communication • Scale • Scaling Record • Scan Cursor • Scanner Controller • Scheduled Caste • Schmidt-Cassegrain • School Certificate • Science Case • Science Center • Science Coordinator • Sciences Chimiques • Scientific Committee • Scoring Criteria • Scotland • Screen Coordinator • Script • Sculpsit Carved

• Scythia • Sea Current • Search Control • Secondary Channel • Secondary Containment • Sectional Center • Security Code • Security Council • See Copy • Select Committee • Self-Check • Self-Contained • Semi-Classical • Semiconductor • Senatus Consultum • Sending Complete • Sensor Control • Sensor Controller • Separate Cover • Separately Compiled • Separation and Control • Sequence Control • Sequence Count • Sequence Counter • Service Center • Service Certificate • Service Charge • Service Class • Service Component • Service Corporation • Seychelles (ISO country code) • Shaped Charge • Shapley Concentration • Short Camera • Short Circuit • Short Course • Signal Center • Signal Comparator • Signal Conditioner • Significant Characteristics • Silvered Copper • Simulation Coordinator • Simulation Council • Single Card • Single Cartridge • Single Case • Single Channel • Single Chip • Single Column • Single Contact • Single Core • Single Crystal • Slope Code • Small Capitals • Small Coronagraph • Small Craft • Smart Card • Snow Cover • Société Commune de Satellites • Soft Color • Software Center • Software Change • Software Contractor • Solar Cell • Solar Constant • Sorting Carriage • Source • Source Code • Source Count • South Carolina (USA) • Southern Clusters • Space Command • Space Control • Spacecraft • Spain-California • Spatial Coherence • Special Case • Special Command • Special Commission • Specialized Corpus • Specialty Codes • Spectroscopic CCD • Speed Control • Spin Coupling • Splitter/Combiner • Spread Correlation • Spreading Coefficient • Stabilization and Control • Staff Committee • Standards and Coordination • Standing Committee • Standing Conference • Star Cluster • Starlink Cookbook • State and Country (code) • Statement of Capability • Statement of Compatibility • Station Clock • Statistical Control • Statutory Committee • Steering Committee • Stellar Camera • Storage Capacity • Stored Command • Strategic Command • Stratocumulus • Stress Corrosion • String Controller • Strip Chart • Strongly Connected • Structural Change • Structural Computing • Structure Chart • Subcommittee • Subcontract • Subcontractor • Subject Category • Successive Correction • Sudden Commencement • Supercalendered • Supercluster • Supercomputer • Superconductivity • Superimposed Coding • Superimposed Current • Supervisor's Console • Supply Catalog • Support Center • Support Contractor • Support Controller • Support Coordinator • Supreme Court • Surface brightness - Colour • Surface Command • Surface Contamination • Surrogate Carrier • Switching Cell • Switching Center • Symbolic Code • Synchro-Cyclotron • System Characterization • System Code • System Concept • System Control • System Controller

SC&A Security Certification & Accreditation

SCA Santa Catalina (USA-CO) (location identifier) • Schedule Change Authorization • Scientific Computing and its Applications (conference) • Scientific Computing Associates Corporation • SDI Communications Architecture • Security Certification and Accreditation • Sensor Chip Array • Sensor Chip Assembly • Sequence Control Area • Sequence Control Assembly • Sequencer Control Assembly • Serveur de Centre d'Appel • Service Contract Act (USA) • Service Cost Avoidance • Service Cryptologic Agency • Shared Communications Area • Short Code Access • Shuttle Carrier Aircraft • Signal Conditioning Assembly • Simulation Control Area • Simulation Conversion Assembly • Single-Channel Analyzer • Single-Connector Attachment • Sneak Circuit Analysis • Snow-Covered Area • Socio-Cultural Analysis • Software Capability Analysis • Software Command Analysis • Software Control Authorization • Solar-Cell Assembly • Spacecraft Adapter • Speech Communication Association (USA) • Status Control Activity • Student Community Action • Studii si Cercetari de Astronomie Bucureşti (periodical code) • Subcarrier Adapter • Subchannel Adapter • Subcontract Authorization • Subcritical Assembly • Subsidiary Communications Authority • Summary Cost Account • Supporting Contract Administration • Switch Control Assembly • Symposium on Computer Automation • System Communication Architecture • System Comparison Analysis • System Controller Assembly • Système de Contrôle d'Altitude • Système de Contrôle d'Attitude

SCAA School Curriculum and Assessment Authority (UK)

SCAAMP Scientific Applications on Arrays of Multiple Processors

SCAAN Scottish Computer-Assisted Assessment Network (UK-Scotland)

SCAC Santa Cruz Astronomy Club (USA-CA) • Science Computing Advisory Committee • Service Creation and Accounting Center • Service de Coopération et d'Action Culturelle

SCAD Scan Converter And Display • Small Current Amplifying Device

SCADA Supervisory Control And Data Acquisition

SCADC Standard Central Air Data Computer

SCADE Safety Critical Applications Development Environment

SCADS Scanning Celestial Attitude Determination System • Simulation of Combined Analog Digital Systems • Superfund Chemical Analysis Data System

SCAE Society for Computer-Aided Engineers (USA)

SCAEPA Society for Computer Applications in Engineering, Planning and Architecture

SCAFEDS Space Construction Automated Fabrication Experiment Definition Study

SCAFI Studies of Computer Algebra for Industry

SCAFFIP Scanning Fast Field Program

SCALE Scattering Layer Experiment • Shuttle Coherent Atmospheric Lidar Experiment • Space Checkout And Launch Equipment

SCALIR Symbolic and Connectionnist Approach to Legal Information Retrieval

SCALOP Standing Committee on Antarctic Logistics and Operations

SCALP Scenario-Assisted Computer-Assisted Logistics • Solar Calibration Package

SCAM SCSI Configured Automatically • Specific Conservation of Angular Momentum • Spinning Camera • Stanford Copy Analysis Mechanism • Station Control And Monitoring • Support Capability Assessment Model

SCAMA Skewed Circular Arc Method of Analysis • Station Conferencing And Monitoring Agreement • Station Conferencing And Monitoring Arrangement • Switching, Conferencing And Monitoring Arrangement

SCAMP Single-Channel Advanced Man-Portable • Standard Configuration And Modification Program • Supplemental Camera And Maneuvering Platform

SCAMPI Southern California Area Modern Physics Institute (USA-CA)

SCAMPS Small Computer Analytical and Mathematical Programming System

SCAMS Scanning Microwave Spectrometer

Scan Scandinavia • Scandinavian

SCAN Scientific Content Analysis • Selected Current Aerospace Notice • Selected Current Awareness Notification • Self-Correcting Automatic Navigation • Self-Correcting Automatic Navigator • Shipboard Communications Area Network • Shuttle Connector Analysis Network • Spoken Content-based Audio Navigation • Student Career Automated Network • Switched Circuit Automatic Network

Scand Scandinavia • Scandinavian

SCANIQ Système Conjoint d'Analyse Numérique d'Images du Québec (Canada-QC)

SCANS Scheduling and Control by Automated Network System • Scheduling Control and Automation by Network System • Status Configuration and Alarm Notification

System • System Checkout Automatic Network Simulator

SCAO Système de Contrôle d'Attitude et d'Orbite

SCAP Scanning Photometer And Polarimeter • Shuttle Configuration Analysis Program • Silent Compact Auxiliary Power • Solicitation for cooperative Agreement Proposal • Standing Committee of Astronomy Professors • Standing Conference of Astronomy Professors (UK) • States Cooperative Assistance Program

SCAPE Self-Contained Air Pressure Environment • Self-Contained Atmosphere Protective Ensemble • Self-Contained Atmospheric Protective Ensemble • Simulating Composition of Atmospheric Particles at Equilibrium • Slantwise Convective Available Potential Energy

S. Caps. Small Capitals

SCAR Safety Change Assessment Report • Scientific Committee on Antarctic Research (*) • Satellite Capture And Retrieval • Signal Characterization And Recognition • Spacecraft Activity Request • Spacecraft Analysis Room • Spacecraft Assessment Report • Special Committee on Antarctic Research • Starlink Catalogue Access and Reporting • Starlink Catalogue Access and Retrieval • Supplier Corrective Action Request

SCARA Selective Compliance Assembly Robot Arm

ScaRaB Scanner for Radiation Budget

ScaRaBE Scanner for Radiation Budget Experiment

SCARDE Study Committee on Analysis of Research, Development and Engineering

SCARF Standard Collection Asset Request Format

SCARI Self-Compensating All-Reflection Interferometer (*) • Self-Compensating All-Reflective Interferometric

SCARP Skylab Concentrated Atmospheric Radiation Project

SCARS Scientific Computing And Research Support • Serial Control And Record System • Serialized Control And Reporting System • Sneak-Circuit Analysis Report Summary

SCARSD Science Computing And Research Support Division

SCAS Scan String • Seven Color Asteroid Survey • Soil, Crop, and Atmospheric Sciences (Cornell) • Sonoma County Astronomical Society (USA-CA) • Sous-Comité d'Astronomie Spatiale • South Carolina Academy of Sciences (USA-SC) • Southern Cross Astronomical Society (USA-FL) • Stability Control Augmentation System • Steel County Astronomical Society (USA-MN) • Subsystem Computer Application Software

SCASIS Seeing Cell Aperture Synthesis Imaging Spectrometer

SCAST Social Change And Sustainable Transport

SCASU Secrétaire en Chef d'Administration Scolaire et Universitaire

Scat Scatter

SCAT Secondary Combustion Augmented Thruster (*) • Secondary Combustor Augmented Thruster • Security Control of Air Traffic • Self-Contained Automatic Transmitter • Share Compiler-Assembler Translator • Shared Commercial Automobile and Truck • Shared Compiler-Assembler and Translator • Space Communications And Tracking • Storage, Checkout And Transport • Storage, Checkout And Transportation • Supersonic Commercial Air Transport • System Commonality Analysis Tool

SCATE Self-Checking Automatic Testing Equipment

SCATHA Spacecraft Charging At High Altitude

SCATS Satellite Communications Antenna Tracking System • Self-Contained Automatic Test System • Simulation, Checkout And Tracking System • Simulation, Checkout And Training System (*) • Simulation Control And Training System

Scatt Scatterometer

SCATT Scatterable • Scatterometer • Scientific Communication And Technology Transfer • Shared Catalog Accessed Through Terminals

SCAUL Standing Conference of African University Librarians

SCAWG Systems Control Architecture Working Group

ScB Bachelor of Science

SCB Sample Collection Bag • Schedule Change Board • Schedule Control Board • Scheduled Change Board • Scribner (USA-NE) (location identifier) • Scrubbed • Selection Control Board • Selector Control Box • Session Control Block • Signal Conditioning Board • Software Control Board • Space Control Block • Specification Control Board • Steering Control Box • System Control Block • Système de Contrôle d'Attitude

SCBA Self-Contained Breathing Apparatus

ScBC Bachelor of Science in Chemistry

SCBE Seller's Configuration Baseline Document

ScBE Bachelor of Science in Engineering

SCBE Simcoe County Board of Education

SCBP Spectral Centroid Bias Position

SCBU Self-Contained Breathing Unit

scc Single Column Centimeter

SCC Prudhoe Bay - Deadhorse (USA-AK) (location identifier) • Safety Control Center • Sagamihara Control Center (Japan) • Satellite Communications Control • Satellite Communications Controller • Satellite Control Centre • Satellite-Controlled Clock • Scholarly Communication Center • Scientific Computing Center • SCSI Controller Command • Secondary Control Code • Secure Communications Controller • Security Classification Code • Security Control Center • Security Coordination Committee • Selectional Classification Code • Seminole Community College (USA-FL) • Sense of Circular Polarization • Sensor Control Center • Sequence Control Chart • Sequential Control Counter • Serial Communications Controller • Serial Controller Chip • Signal Conditioning Component • Significant Character Code • Simplified Computer Code • Simulation Configuration Control • Simulation Control Center • Single Column Centimeter • Single-Conductor Cable • Single Cotton Covered • Slidell Computer Complex (NASA) • Small Currency Changer • Software Control Center • Sorter Control Computer • South Celestial Cap • South Cusp Cap • Space Commerce Corporation • Space Communications Corporation • Space Components Coordination • Space Control Centre • Spacecraft Clock • Spacecraft Control Centre • Spacecraft Control Computer • Spacecraft Coordinates • Special Computation Center • Special Control Character • Special Coordination Committee • Specialized Commercial Carrier • Specialized Common Carrier • Specialized Communications Carrier • Specific Clauses and Conditions • Standard Commodity Classification • Standard Coordination Committee • Standard Cubic Centimeter • Standards Categories Coordination • Standards Coordinating Committee • Standards Coordination Committee • Standards Council of Canada • Standing Consultative Commission • Star Cluster Candidate • State Consultative Council • Station Control Center • Stress Corrosion Cracking • String Center Controller • Strong Cool Component • Strongly Connected Component • Structured Conceptual Clustering • Submission Control Code • Super Cloud Cluster • Supervisor Control Console • Supply Control Center • Surveillance Command and Control • Sustainable Computing Consortium • Switching Control Center • Synchronous Channel Check • System Command Center • Systems Control Center • Systems Control Console

SCCA Statistical Consulting Center for Astronomy • Subcontract Change Authorization

SCCAC Society for Conceptual and Content Analysis by Computer

SCCB Safety Change Control Board • Service Component Command Center • Simulation Configuration Control Board • Site Configuration Control Board • Software Change Control Board • Software Configuration Control Board • System Configuration Control Board

SCCC Satellite Communications Control Centre • Scottish Consultative Committee on the Curriculum (UK-Scotland) • Seattle Central Community College (USA-WA) • Seward County Community College (USA-KS) • Sociedad Chilena de Ciencia de la Computación (Chile) • South Cusp Cap Collar • Spacecraft Command and Control Center

SCCCB Science Computing Configuration Control Board

SCCE Satellite Configuration Control Element • Scientific Computing in Chemical Engineering • Signal Conditioning Control Electronics

SCCER System Command and Control Equipment Rack

SCCF Satellite Communication Control Facility • Security Clearance Case File • Streamlined Cost Comparison Form

SCCG Space Components Coordination Group • Station Communications Control Group

SCCH Standard Cubic Centimeters per Hour

SCCI Source Code Control Interface

SCCM Standard Cubic Centimeters per Minute • Strongly Connected Compartmental Matrix

SCCN Secure Command and Control Net

SCCNFP Scientific Committee on Cosmetic Products and Non Foodstuff Products (EU)

SCCO Security Classification Control Officer • Soret Coefficient in Crude Oil • Special Circuit Control Officer

SCCP Signaling Connection Control Part • Skinny Client Control Protocol • Standing Committee of Physics Professors

SCCPG Satellite Communications Contingency Planning Group

SCCR Subcontractor Change Request

SCCS Single Channel Collection System • Source Code Control System • Specialized Common Carrier Service • Standard Cubic Centimeters per Second • Switching Control Center System

SCCSB Software Computer Configuration Subboard

SCCT Steering Committee on Conformance and Testing

SC&CU Signal Conditioning & Control Unit

SCCU Signal Conditioning and Control Unit • Spacecraft Command Control Unit

Scd Schedule • Scheduled

Sc. D. Doctor of Science

ScD Doctor of Science

SC&D Stock Control & Distribution

SCD Satélite de Coleta de Dados • Satellite Control Department • Satellite de Coleta Dados • Savings Certificate Division • Science Communications Division • Scientific Capabilities Document • Scientific Computing Division • Scrolling Display • Service Commun de Documentation • Service Creation Environment • Silicon Capacitor Diode • Sociedad Chilena de Derecho (de Autor) (Chile) • Software Conceptual Design • Source Control Document • Source Control Drawing • Space Control Document • SPARC Compliance Definition • Specification Change Drawing • Specification Control Document • Specification Control Drawing • Speed Control Device • Standard Color Display • Stock Control and Distribution • Subcarrier Discrimination • Subcontract Deviation • Subject Cataloguing Division • Sudden Cardiac Death • Sulaco (Honduras) (location identifier) • Synthetic Clock Drive • System Concept Definition • System Conception Definition • System Confidence Demonstration • System Correction Data

SCDA Safing, Cool-down and Decontamination Area

SCDC Source Coding and Data Collection

SCDDB Satellite Climate Diagnostics Database

SCDI Scottish Council for Development and Industry

SCDL Saturated Current Demand Logic • Secure Command Data Link • Stabilized Carbon Dioxide Laser • Structural Dynamics and Control Laboratory • Surveillance and Control Data Link

SCDM Standard Cold Dark Matter (model)

SCDMA Synchronous Code-Division Multiple Access

SCDMS Soil-Climate Database Management System

SCDO Science and Communications Development Office

SCDP Simulation Control Data Package • Specialized Centre for Data Processing

Scdr Screwdriver

SCDR Seller Critical Design Review • Shuttle Critical Design Review • Software Critical Design Review • Special Committee for Disaster Research • Subcontractor Critical Design Review

SCDS Scan Converter Display System • Secure Central Data System • Sensor Communication and Display System • Source Control Data Set • Stock Control and Distribution System

SCDSF Spacecraft Detailed Schedule File

SCDU Satellite Control Display Unit • Signal Conditioning and Display Unit

Sce Scenario • Source

SCE Schedule Compliance Evaluation • Scottish Certificate of Education (UK-Scotland) • Selection Control Element • Service Central d'Exploitation • Service Control Environment • Service Creation Environment • Service Cryptologic Element • Shuttle Cargo Element • Signal Conditioning Equipment • Signal Conversion Equipment • Single Corporate Entity • Sistema de Cuentas Europeo • Situationally Caused Error • Situation-Caused Error • Small Current Element • Société Coopérative Européenne • Society of Computational Economics • Software Capability Evaluation • Software Configuration Engineer • Solar Corona Experiment • Sony Computer Entertainment • Sortie Cargo Element • Space Cabin Environment • Spacecraft Command Encoder • Spacecraft Evaluator • Special Conditioning Equipment • Stabilization Control Electronics • Standard Calomel Electrode • Standard Computing Element • Standard Course of Events • State College (USA-PA) (location identifier) • Structure Chart Editor • Superior Council of Education • Supply Chain Execution • System Checkout Equipment • System Cost Estimate

SCEA Signal Conditioning Electronics Assembly • Society of Cost Estimating and Analysis (USA) • Sont Computer Entertainment America (USA)

SCEAL Standing Conference of East African Librarians

SCEC Small Computer Engineering Center • Southern California Earthquake Center (USA-CA)

SCED Stations and Communications Engineering Division

SCEDI Standing Committee for Electronic Data Interchange

SCEE Scientific Computing in Electrical Engineering (conference)

SCEES Service Central des Enquêtes et Études Statistiques (France)

SCEF Service Creation Environment Function

SCEI Special Committee on Environmental Information

SCEM Service Central d'Exploitation de la Météorologie (*) • Service Central d'Exploitation Météorologique

SCENET Superconductivity European Network

SCEP Service Creation Environment Point • Simple Certificate Enrollment Protocol • Student Career Experience Program

SCEPS Solar-Cell Electric Power System

SCEPTRE System for Circuit Evaluation and Prediction of Transient Radiation Effects • System for Constant Elevation Precipitation Transmission and Recording

SCERT Systems and Computers Evaluation and Review Technique

SCES Satellite Control Earth Station

SCET Scottish Council for Educational Technology • Spacecraft Event Time

scf Standard Cubic Foot

SCF Phoenix - Scottsdale (USA-AZ) (location identifier) • Sampled Channel Filter • Satellite Communications Facility • Satellite Control Facility • Scheduled Change Factor • Science Computer Facility • Science Computing Facility • Scientific Committee on Food • Scientific Computing Facility • Sectional Center Facility • Seeker Calibration Facility • Selective Call Forwarding • Self-Consistent Field (method) • Sequence Compatibility Firing • Sequenced Compatibility Firing • Service Control Facility • Service Control Function • Simplified Contract Format • Simulator Command File • Single Catastrophic Failure • Single Cost Factor • Small Cooling Flow • Sociedad Colombiana de Física (Colombia) • Spacecraft Checkout Facility • Spacecraft Control Facility • Specialized Computing Facility • Spectral Correlation Function • Standard Change Form • Standard Charge Factor • Standard Cubic Feet • Star Catalogue Facility • Statistical Collection File • Steam Condensation Front • Stock and Checkout Facility • Supercritical Fluid • System Control Facility

SCFCS Standing Committee on Freedom in the Conduct of Science (*) • Standing Committee on the Free Circulation of Scientists

SCFD Scientific and Cultural Facilities District

SCFE Special Cassette Flow Experiment • Subcontractor-Furnished Equipment

SCFEL Standard Cryptologic Facility Equipment List

scfh Standard Cubic Feet per Hour

scfm Standard Cubic Feet per Minute

scfph Standard Cubic Feet per Hour

scfpm Standard Cubic Feet per Minute

SCFN Seed Capital Fund Network • Spherical Convergent Flap Nozzle

SCFS Standard Cubic Feet per Second

Scg Scoring

SCG Scaled Conjugate Gradient • Security Classification Guide • Sequential Control Guidance • Single Cell Gel • South Compact Group (of galaxies) • Space Charge Grid • Space Communications Group • Special Circuit Group • Spring Creek (Australia-QLD) (location identifier) • Switching Controller Group • Symposium on Computational Geometry • System Classification Guide

SCGB Steering Committee on Genetics and Biotechnology

SCGD Specification Control Group Directive

SCGG Shakhbazian Compact Group of Galaxies

SCGI Secrétariat Général du Conseil Interministériel

s/cgo Su Cargo

SCGSS Supercritical Gas Storage System

SCGUI Steering Committee for Graphical User Interface

Sch Schedule • Scheduled • Scheduling • Schema • Schematic • Schilling (currency) • Schmidt • Scholar • Scholarship • Scholastic • School • Schule

SCh Sciences Chimiques

SCH Schenectady County (USA-NY) (location identifier) • School • Software Clearinghouse • Space Center Houston (USA-TX) • Spherical Cap Harmonic • Statistics Collection Host • Synchronization Channel

SCHA Spherical Cap Harmonic Analysis

SCHD Schedule

SchDm Schematic Diagram

SCHe Supercritical Helium

Sched Schedule

SCHED Schedule • Scheduled • Scheduling

Schem Schematic • Schematics

SCHEM Schematic • Schematics

SCHG Spectrophotometer Catalog of HII Galaxies • Supercharge

Schill Schilling

SCHINA Schaltanlagen-Informations-Anlagen

Schl Schloß

SCHLT Searchlight

SCHM State Committee on Hydrometeorology

SCHMM Semi-Continuous Hidden Markov Model

Schol Scholar • Scholarship • Scholastic

Schr Schrift

SCHS Small Component Handling System

Schurw Schurwald (Germany)

SchutzbG Schutzbaugesetz

Schw Schwester

Schwäb Schwäbisch

Schwab Schwaben (Germany)

Schwarzw Schwarzwald (Germany)

SchwbG Schwerbehindertengesetz

Sci Science • Science (periodical code) • Scientific • Scientifique • Scientist

sci Single Column Inch

ScI Science Institute

SCI San Cristobal (Venezuela) (location identifier) • Scalable Coherent Interconnect • Scalable Coherent Interface • Science • Science Citation Index (ISI) • Scientific Computing and Imaging • Selected Configuration Item • Seller Configuration Inspection • Sensitive Compartmented Information • Sensitive Compartmented Intelligence • Sensor Chip Imager • Serial Communications Interface • Servicio de Calibración Industrial • Sistema Curricular Interactivo • Société Civile Immobilière (France) • Société de Chimie Industrielle • Society of Chemical Industry • Software Configuration Item • Software Customer Inspection • Space Commerce International Ltd. • Space Craft International (USA-CA) • Special Contol Item • Strategic Computing Initiative • Stratospheric Circulation Index • Structural Composites Interface • Student-Centered Instruction • Supplemental Card Indicator • Switch Closure In • Systemics - Cybernetics - Information • Systems Concepts and Integration

SCIA Smart Card Industry Association

SciAm Scientific American (periodical code)

SCIAMACHY Scanning Imaging Absorption Spectrometer for Atmospheric Cartography

SCIBP Scientific Committee for the International Biological Programme

SCIC Satellite Cloud Information Chart • Semiconductor Integrated Circuit

SciCADE Scientific Computing And Differential Equations (conference)

SCICEX Submarine Science Ice Experiment

SciDAC Scientific Discovery through Advanced Computing (initiative)

SCIDAR Scintillation Detection And Ranging

SCIDF System-Independent Data Format

SciDi Scientific Dimension (periodical code)

SCIDNDR Special Committee for the International Decade on Natural Disaster Reduction

SCIE School on Information Extraction • Science Citation Index Expanded • Servizio Communicazione, Informazione, Educazione

SCIENCE Simulation des Coopérations Internationales et des Échanges Nécessaires aux Chercheurs Européens

SCIES Science Catalog Information Exchange System

ScIF Science Institute Facility

SCIF Sensitive Compartmented Information Facility • Single Console Image Facility

sci. fa. Scire Facias

SCIFER Sounding the Cleft Ion Fountain Energization Region

Sci. Fi. Science Fiction

SCIGBP Special Committee for the International Geosphere-Biosphere Program

SCIGN Southern California Integrated GPS Network (USA-CA)

SCIGY Special Committee for the International Geophysical Year

scil Scilicet

SCIL Simulation Command and Input Language • Software Critical Items List

SCILS School of Communication, Information and Library Studies

SCIM Sample Collection for Investigation of Mars • Shuttle Cargo Integration Manager • Speech Communications Index Meter • Standard Cubic Inches per Minute

SciN Science News (periodical code)

SCINATT Spectra, Composition, and Interactions of Nuclei Avove Ten TeV

SCINS Small Common Inertial Navigation System

SciOp Science Operations

SciOps Science Operations

SCIP Scanning for Information Parameters • Self-Contained Instrument Package • Shuttle Cargo Integration Plan • Site Configuration and Installation Plan • Society of Competitive Intelligence Professionals (USA) • Solar Composition Icy Planetesimal • Systems Control Improvement Program

SCIPP Santa Cruz Institute for Particle Physics (USA-CA)

Scis Sciences (periodical code)

SCIS Satellite Coverage Information System • Science Curriculum Improvement Study • Spacecraft Interface Specification • Standard Cubic Inches per Second • Survivable Communications Integration System

SCISAT Scientific Satellite

SciSim Science Simulator

SciSn Scientia Sinica (periodical code)

SCISRG Semi-Continuous Inferred Stochastic Regular Grammar

SCIT Standard Change Integration and Tracking • Standardization and Control of Industrial Tools • System Concept and Integrated Technology • System Concept and Integration Technology • System Concept Integrating Technology

SCITS Single Channel Injection Transponder System

SCIU Selector Control Interface Unit • Sortie Control Interface Unit • Spacecraft Interface Unit

SCJ Smith Cove (USA-AK) (location identifier)

SCK Stockton (USA-CA) (location identifier) • Studiecentrum voor Kernenergie

Scl Sculptor-Sculptoris (constellation)

SCL Safety Checklist • Santiago - Comodoro Arturo Merino Benitez (Chile) (location identifier) • Scheduling Class Library • Secondary Coolant Line • Secondary Coolant Loop • Separately Controlled Link • Serial Clock • Service Control Level • Sociedad Cooperativa Libre • Software Control Library • SOGS Command Language • Spacecraft Command Language • Specification Change Log • Standard and Calibration Laboratory • Standard Classification List • Standard Conventional Load • Standards and Calibration Laboratory • String Control Language • Symbolic Correction Loader • Systems Command Language • Systems Component List • Systems Control Language

SCLAN Supercomputer Local Area Network

SCLC Space-Charge-Limited Current

SCLERA Santa Catalina Laboratory for Experimental Relativity by Astrometry (USA-CA)

SCLK Spacecraft Clock Kernel

SCLM Software Configuration and Library Management • Software Configuration and Library Manager • Stability, Control, and Load Maneuvers

SCLO Statistical Clearance Liaison Officer

SCLP Signaling Channel Link Protocol

Sclr Scaler

SCLV Spacecraft Launch Vehicle

scm ScreenCam Movie (file extension) • Standard Cubic Meter

ScM Master of Science • Scripta Metallurgica (periodical code)

SCM Satellite Catalog Maintenance • Satellite Communications Module • Scammon Bay (USA-AK) (location identifier) • Scanning Capacitance Microscope • Scanning Capacitance Microscopy • Schedule/Configuration Management • Selected Communications Mode • Service Command Module • Service Control Manager • Shared Calendar Manager • Signal Conditioning Module • Simulator Operations and Maintenance • Single-Column Model • Single-Crystal Meteorite • Site Configuration Message • Small-Core Memory • Sociedad de Crédito Mercantil • Société de Cautionnement Mutuel • Society of Computer Medicine • Software Configuration Management • Software Configuration Manager • Solar-Cell Module • Solar-Constant Monitor • Space Cargo Modification • Space Cargo Modified • Specification Change Memorandum • Spinal Change in Microgravity • Stamp Cancelling Machine • Standard Cosmological Model • Standard Cubic Meter • Station Class Mark • Stochastic Clump Model • Streamline Curvature Method • Subsidies and Countervailing Measures • Superconducting Magnet • Supply Chain Management • Suppressed-Carrier Modulation • System Configuration Management • System Configuration Monitoring • System Control Module

SCMA Société Canadienne des Mathématiques Appliquées (Canada) • Statistical Challenges in Modern Astronomy (conference) • Sustainment Cost Management Annex

SCMAI Société Canadienne de Mathématiques Appliquées et Industrielles (Canada)

SCMB SOGS Configuration Management Bulletin • System Configuration Management Board

SCMFT Self-Consistent Mean-Field Theory

SCMGGI Semi-Continuous Morphic Generator Grammatical Inference

ScMM Scripta Metallurgica et Materialia (periodical code)

SCMM Single-Connection - Multiple-Measurement

SCMO Software Configuration Management Office

SCMOCC Semi-Classical Molecular-Orbital Close-Coupling (method)

SCMP Software Configuration Management Plan • System Console Monitor Printer • System Contractor Management Plan

SCMR Surface Composition Mapping Radiometer

SCMS Serial Copy Management System • Signal Command Management System • Small Cumulus Microphysical Study

• Spacecraft Control and Monitoring System • STScI Code Management System

SCMSS Standing Committee on Membership, Structure and Statutes

SCMT Security Crisis Management Training • Southern Columbia Millimeter Telescope

SCMU Satellite Control and Monitor Unit

SCN Saarbrücken (Germany) (location identifier) • Satellite Communications Network • Satellite Control Network • Schedule Change Notice • Schematic Change Notice • Seattle Community Network (USA-WA) • Self-Checking Number • Self-Contained Navigation • Sensitive Command Network • Show Cause Notice • Software Change Notice • Space Communications Network • Spaceguard Central Node • Spaceguard Control Node • Specification Change Notice • Special Change Notice • Specification Change Notice • Specific Control Number • Student Control Number • Succinonitrile • Suprachiasmatic Nucleus • System Change Notice • System Control Number

SCNA Sudden Cosmic-Noise Absorption

Scng Scanning

SCNP Specification Change Notice Proposal

Scnr Scanner

SCNR Scanner • Sequence Control Number Register • Solid-Core Nuclear Rocket

SCNRC Scanner Controller

SCNS Self-Contained Navigation System

Sco Scenario • Scorpius-Scorpii (constellation)

S/CO Spacecraft Observer

SCO Sabino Canyon Observatory (USA-AZ) • Santa Cruz Operation Inc. • Scheduling Control Office • Senior Civilian Official • Servicing Contracting Office • Shevchenko (Kazakhstan) (location identifier) • Single Copy Order • Slovak Central Observatory (Slovakia) • Smart Card Ontology • Software Change Order • Spacecraft Observer • Spacecraft Operations • Spacecraft Operator • Start Checkout • State Coordinating Office • State Coordinating Officer • Statistical Control Office • Statistical Control Officer • Subcarrier Oscillator • Subcontrol Office • Supercritical Oxygen • Supermassive Compact Object • Switch Closure Out • System Control Officer

SCOA Self-Change Of Address

SCOASO Small Computer/Office Automation Service Organization

SCOC System Control Operations Center • Systems Control and Operations Concept

SCOCE Special Committee On Compromising Emanations

SCOCP State-Constrained Optimal Control Problem

SCODC Subcontract and Other Direct Costs

SCODS Standing Committee on Ocean Data Stations

SCOE Special Checkout Equipment • Specific Checkout Equipment

SCOL Simulation Control Level

SCOLOP System Control Online Operational Program

SCOM Science Operations Manager

SCOMO Satellite Collection Of Meteorological Observations

SCOMP Secure Communications Processor • Secure Communications Protocol

SCONUL Standing Conference of National and University Libraries (UK)

SCOOP Scientific Compilation Of Optimal Programs • Support to Continuity Of Operations Plan • System Concept of Operation Outline Plan

SCOOPS Scheme Object-Oriented Programming System

SCOP Single-Copy Order Plan • Société Canadienne d'Optique et de Photonique (Canada) • Spacecraft Operation • Standard Communications Operating Plan • System Concept of Operation Plan

SCOPE Schedule - Cost - Performance • Scientific Committee On Problems of the Environment • Secure Cooperative Processing Environment • Sequential Customer Order Processing Electronically • Setting - Content - Objectives - Program - Evaluation • Simple Communications Programming Environment • Smart Contract Preparation Environment • Software Certification On Programme in Europe • Solar Connections Observatory for Planetary Environments • Southwest Consortium of Observatories for Public Education (USA) • Standing Committee on Professional Education • Standing Committee on the Problems of the Environment • System for Control of Operations of Payload for EuroMir

SCOPL Simple Checkout-Oriented Program Language

SCOPS Security Concept of Operations

SCOR Scientific Committee for Oceanographic Research • Scientific Committee on Oceanic Research • Self-Calibration Omnirange

SCORE Satellite Computer-Operated Readiness Equipment • Selection Copy and Reporting • Signal Communications by Orbiting Relay Equipment • SIRTF-Cornell Echelle • Spacecraft Operational Readiness Exhibit • System Cost and Operational Resource Evaluation

SCORES Scenario-Oriented Recurring Evaluation System

SCORPIO Spectral Camera with Optical Reducer for Photometric and Interferometric Observations

SCORT Specialized Center Of Research and Training

SCORUS Standing Committee on Regional and Urban Statistics

SCOS Spacecraft Control and Operations System • Spacecraft Control Operations System • Spacecraft Operation System • Subsystem Computer Operating System

SCOST Special Committee on Space Technology

SCOSTEP Scientific Committee on Solar-Terrestrial Physics

Scot Scotland (UK) • Scottish

SCOT Satellite Communication Terminal • Satellite Communications Overseas Terminal • Schéma de Cohérence Territoriale • Small Communication Terminal • Social Construction Of Technology • Standby Compatible On-Tape

SCOTS Surveillance and Control Of Transmission Systems • System Checkout Test Set

SCOTT Signal Channel Objective Tactical Terminal • Synchronous Continuous Orbital Three-Dimensional Tracking

SCOTUS Supreme Court of the United States

SCOUT Serial Checkout Unit

SCOWAR Scientific Committee on Water Research

SCOWL Submillimeter Camera for the Overwhelmingly Large Telescope

scp Secure Copy • Script (file extension)

SCP Saint-Crépin (France) (location identifier) • Satellite Cloud Photograph • Satellite Commissioning Phase • Saunders College Publishing (USA-PA) • Save Cursor Position • Scanner Control Power • Scanning Phased Array • Schedule Change Proposal • Schematic Change Proposal • Science of Computer Programming • Sección de Cielo Profundo • Secure Conferencing Program • Secure Conferencing Project • Security Classification Procedure • Security Computer Program • Service Central des Prêts • Server Control Procedure • Server Control Protocol • Service Connection Point • Service Control Point • Service Cost Position • Session Control Protocol (*) • Signaling Control Point • Signaling Control Point • Simulation Control Plan • Simulator Control Panel • Simulator Converter Program • Single Card Processor • Single-Color Photometry • Single Control Point • Site Configuration Package • Société Civile Professionnelle • Software Change Package • Software Change Proposal • Solar-Cell Panel • South

Celestial Pole • Spacecraft Platform • Spacecraft Processor • Special Contract Provision • Specific Candle Power • Spectral-Curvature Parameter • Standard Change Package • Standard Corporate Protocol • Standards Conformity Program • Station Communications Processor • Status and Controller Panel • Stored Command Processor • Strategic Computing Program • Study of Cosmological Parameters • Subnet Communications Processor • Subsystem Control Port • Supernova Cosmology Project • Supplier Change Proposal • Survey Control Point • Symbolic Conversion Program • System Change Package • System Change Proposal • System Communication Pamphlet • System Concept Paper • System Control Port • System Control Program • System Control Programming • System Coordinating paper

SCPA Semiconductor Protection Act (USA) • Solar-Cell Panel Assembly • Spacecraft Payload Adapter

SCpB Single Channels per Burst

SCPB Single Channels Per Burst • Software Change Proposal Board

SCpC Single Channels per Carrier

SCPC Single Channels Per Carrier (*) • Single Channels Per Connection

SCPD Stanford Center for Professional Development (USA-CA)

SCPE Secure Communications Processing Environment • Simplified Collective Protection Equipment

SCPI Session Control Programmable Interface • Société Civile de Placement Immobilier (France) • Standard Command for Programmable Instrumentation • Standard Command for Programmable Instruments (*) • Standard Command for Programmable Interface

SCPL Symbolic Computation and Programming Language (conference) • System Configuration Provisioning List

SCPM Semi-automatic Circuit Performance Monitor

SCPO Senior Chief Petty Officer

SCPP Seasonal Climate Prediction Program • Sierra Cooperative Pilot Project • Société Civile de Producteurs de Phonogrammes (France)

ScPr Science Progress (periodical code)

SCPR System Concept and Programmatic Review

SCPRI Service Central de Protection contre les Rayonnements Ionisants (France)

SCPS Servo-Controlled Positioning System • Space Communication Protocol Standard • Support Computer Programming System • Survivable Collective Protection Shelter • Survivable Collective Protection System

SCPT Security Control Point • Self-Consistent Perturbation Theory • Single Channel Per Transponder • Strategic Connectivity Performance Test

SCPWG Solar Corona and Particle Working Group

SCQ Santiago de Compostela (Spain) (location identifier) • Shape-Color-based Query

SCQL Structured Card Query Language

SCQM Supersymmetric Classical and Quantum Mechanics

scr Screen Layout (file extension) • Screen Saver (file extension) • Script (file extension)

Scr Screw • Scrip • Scruple

SCR Scanning Control Register • Schedule Change Request • Schedule Confirmation Request • Science Certification Review • Scranton (USA-PA) (location identifier) • Script • Section Cross-Reference • Security Change Request • Segment Completion Review • Selective Catalytic Reaction • Selective Catalytic Reactor • Selective Chopper Radiometer • Semiconductor-Controlled Rectifier • Semiconductor Rectifier • Senior Command Representative • Sequence Checking Routine • Seychelles Rupee (ISO currency code) • Signal Conditioner • Signal-Controlled Rectifier • Signaling Configuration Register • Signal-to-Cross-talk Ratio • Significant Clarification Request • Silicon-Controlled Rectifier • Single-Channel Radio • Single-Channel Reception • Small Circuit Register • Sneak Circuit Report • Software Change Report • Software Change Request • Software Corrective Report • Software Corrective Request • Software Cost Reduction • Solar Corpuscular Radiation • Solar Cosmic Radiation • Solar Cosmic Ray • Source Code Review • Source Cosmic Ray • Space-Constructible Radiator • Special Certification Review • Special Contract Requirement • Speed Change Rate • Standard Context Routing • Stirling Cryogenics and Refrigeration • Strip Chart Recorder • Summary Control Record • Summary Control Report • Support Control Room • Surface Contour Radar • Surveillance Cost Report • Sustainable Cell Rate • Sustained Cell Rate • Symptom - Cause - Remedy • System Change Request • System Circuit Register • System Concept Review • System Control Routine

SCRA Single-Channel Radio access

SCRAM Schedule Risk Assessment Module • Several Compilers Reworked And Modified • Space Capsule Regulator And Monitor • Spares Control, Release And Monitoring • Station Crew Return Alternative Module • Support Center for Regulatory Air Models

SCRAP Series Computation of Reliability And Probability

SCRB Software Change Review Board • Software Configuration Review Board

SCRC Small Computer Requirements Contract

SCRD Service Central de Ressources Documentaires • Symptom-Cause-Remedy Database

SCRE Scottish Council for Research in Education

SCREMS Scientific Computing Environments for the Mathematical Sciences

SCRES Standing Committee on Responsibility and Ethics in Science

SCREWS Solar Cosmic Ray Early Warning System

SCRG Stationary Cosmic Ray Gas • System Change Review Group

SCRI Serviço de Cooperação e Relações Internacionais • Supercomputer Computations Research Institute

SCRIA Supply Chain Relationships In Aerospace

SCRIBE Stratospheric Cryogenic Interferometer Balloon Experiment

SCRIPS Service Center Recognition Image Processing System

Script Scriptural • Scripture

SCRIPT Scientific and Commercial Routine Interpreter and Program Translator • Support for Creative Independent Production Talent

SCRJ Scramjet

SCRL Station Configuration Requirement List

SCRM Safety Control and Risk Management • State Commission of Radio Management (China-PRC)

scrn Screen (file extension)

Scrn Screen

SCRN Screen

SCROD School Cosmic-Ray Outreach Detector

SCROLL String and Character Recording Oriented Logogrammatic Language

SCRP Secure Communications Requirements Panel

SCRS Scalable Cluster of RISC Systems • Serialized Control and Reporting System • Strip Chart Recorder System • Strip Chart Recording System

SCRSY Serialized Control and Reporting System

Scrt Security

SCRT Sealed Cathode Ray Tube • Silicon Compton Recoil Telescope • Single-Channel Transponder Receiver

SCRY Secretary

SCR3S Single Channel Radio Relay and Retransmission System

SCS Safety Control Switch • Sales/Customer Service • Satellite Communications System • Satellite Control Section • Satellite Control System • Satellite Coordinate System • Scan Converter System • Scan Counter System • School of Computational Sciences • Science Commanding System • Scientific Control System • Secondary Control System • Secondary Coolant System • Secret Control Station • Secret Cover Sheet • Security Container System • Security Control Station • Sequence Control System • Serialization Card System • Service and Checkout System • Shetland Islands - Scatsta (UK) (location identifier) • Signal Conditioning System • Signaling Capability Set • Signature Control System • Silicon-Controlled Switch • Simple Classifier System • Simulation Control System • Singapore Computer Society • Single Channel Simplex • Single Control Support • Sira Certification Service • Slaving Control System • Slow Code Scanner • Small Computer System • SNA Character String • Société de Commercialisation de Services • Society for Computer Simulation (USA) • Soil Conservation Service • South China Sea • Space Command Station • Space Communications and Sensors • Space Communications System • Spacecraft Checkout Station • Spacecraft Computer System • Spacecraft Control System • Special Communications System • Stabilization and Control System • Standard Communications System • Standard Coordinate System • State Conservation Strategy • Statistical Control System • Stimulated Compton Scattering • Stock Control System • Suit Communication System • Supervisory Control System • Survivable Control System • Switch Control Server • Syrian Cosmological Society • System Communications Service • System Control Station

SCSA Signal Computing System Architecture • Soil Conservation Society of America • Standard Consolidated Statistical Area

SCSB Small Computer Support Bulletin • Space Components Steering Board (ESA)

SCSBR Southern California Space Business Roundtable (USA-CA)

SCSC Scientific and Commercial Systems Corporation • Service Culturel Scientifique et de Coopération • Small Computer Support Center • South Carolina State College (USA-SC) • Southern Connecticut State College (USA-CT) • Standing Committee on Soil Conservation • Starin Gauge Signal Conditioner • Subcommittee on Standards and Conformance

SCSD Systems Control and Support Directorate

SCSDI Service Central de Statistique et de Documentation Industrielles

SCSE State Commission for Space Exploration

SCSG Satellite Communications Subgroup • Silicon-Chip Strain Gauge • Single-Channel Signal Generator

SCSI Small Computer Standard Interface • Small Computer Systems Interface (*) • Society for Computer Simulation International • Steering Committee for Statistical Information

SCSL SGI/Cray Scientific Library • Standing Lenticular Stratocumulus

SC/SM Spacecraft Systems Monitor

SCSM Spacecraft Systems Monitor • Standard Computer System Manager

SCSMO Systems Control Software Management Office

SCSMP Science China Series Mathematics Physics Astronomy Technological Sciences (periodical code)

SCSN Space Centre for Satellite Navigation (QUT) • Standard Computer Software Number

SCSNT Space Centre for Satellite Navigation Technology

SCSO Space Communications Station Operation • Superconducting-Cavity Stabilized Oscillator

SCSP Standing Committee on Science Policy

SCSPS Standing Committee on the Safeguard of the Pursuit of Science

SCSR Scientific Council on Space Research • Single Channel Signaling Rate

SCSRS Standard Command Supply Review System

SCSS Satellite Communications System Control • Satellite Control Simulation System • Self-Contained Starting System • Service Center Support System • Service Central de Statistiques et de Sondages • Shoe Cove Satellite Station (Canada-NF) • Specialist Computer Systems and Software • Standing Committee for Social Sciences • Standing Committee on Structure and Statutes

SCSSI Service Central de la Sécurité des Systèmes Informatiques (France)

SCSST Standing Conference on School's Science and Technology

SCST State Commission for Science and Technology

SCSU Saint Cloud State University (USA-MN) • South Carolina State University (USA-SC) • Southern Connecticut State University (USA-CT)

SCSW Systems Communication Software

sct Scattered • Screen Capture Text (file extension) • Sculpsit

Sct Scutum-Scuti (constellation)

ScT Spacecraft Time

SCt Supreme Court

SCT Satellite Communications Terminal • Scanning Telescope • Scattered • Scattering Telescope • Scheduled Completion Time • Schmidt-Cassegrain Telescope • Second Coding Theorem • Sentence Completion Test • Sequence Checking Tape • Shipping Control Team • Short Collision Time • Sine-Cosine Transformer • Single-Channel Transponder • Société Canadienne de Télédétection (Canada) • Socotra (Yemen) (location identifier) • Software Certification Test • Software Certification Testing • Spacecraft Team • Spectral Control Technique • Strategic Communications Terminal • Stratus-to-Cumulus Transition • Symposium on Computation Theory • Synchronized Clock and Time • System Capability Test • System Certification Test • System Circuit Test • System Compatibility Test

s/cta Su Cuenta

SCTB Simulator Complexity Testbed

SC/TC Spacecraft Test Conductor

SCTC Small Computer Technical Center • Spacecraft Test Conductor • System Console Transfer Channel

Sctd Scattered

SCTD Scattered • Systems Consultancy Training Development

Sc&Te Sciences & Techniques (periodical code)

SCTE S-band Communication and Tracking Equipment • Society of Cable Telecommunications Engineers • Solar-Cell Technology Experiment • Spacecraft Central Timing Equipment

ScTEn Science of the Total Environment (periodical code)

SCTI Service Central des Télécommunications et de l'Informatique • Special Committee for Technical Information

SCTIS Single Channel Tactical Injection Device • Single Channel Transponder Injector Device

SCTL Small Components Test Loop • Surface Compiler for Temporal Logic

SCTOC Satellite Communications Test Operations Center

ScTP Screened Twisted Pair

SCTP Screened Twisted Pair • Straight Channel Tape Print • Stream Control Transmission Protocol

SCTPC Simple Computer Telephony Protocol Client

SCTPP Straight Channel Tape Print Program

SCTPS Simple Computer Telephony Protocol Server

SCTR Sector • Standing Conference on Telecommunications Research

Sctrd Scattered

SCTRS Single Channel Transponder Receiving Set • Single Channel Transponder Receiving System

SCTS Secondary Clear To send • SFOF Communications Terminal System • Shuttle Cargo Transportation System • Single Channel Transponder System • System Components Test Station

SCTV Separation/Control Test Vehicle

Scty Security

SCU Santiago (Cuba) (location identifier) • Satellite Communications Unit • Scanner Control Unit • SCSI Command Utility • Secondary Control Unit • Selector Control Unit • Sensor Control Unit • Sequence Control Unit • Sequential Control Unit • Service and Cooling Umbilical • Service Command Unit • Servo Control Unit • Setter Control Unit • Shuttle Configuration Unit • Signal Conditioner Unit • Signal Conditioning Unit • Signal Control Unit • Southern Cross University (Australia-NSW) • Special Care Unit • Station Control Unit • Storage Control Unit • Stores Control Unit • Subreflector Control Unit • Supplemental Control Unit • Switching and Control Unit • Synchronous Controller Unit • System Capabilities Upgrade • System Connecting Unit • System Control Unit

SCUBA Self-Contained Underwater Breathing Apparatus • Submillimeter Common-User Bolometer Array

SCUD Structuring Knowledge from Unstructured Data

SCUDS Simplification, Clarification, Unification, Decimalization, Standardization

SCUIO Services Communs Universitaires d'Information et d'Orientation (France)

SCULLS SCUBA-Leiden Lens Survey

Sculp Sculpsit • Sculptor • Sculptural • Sculpture

Sculpt Sculptor • Sculptural • Sculpture

SCUM Small Cumulus Microphysics

SCUMM Script Creation Utility for Maniac Mansion

SCUP Society for College and University Planning

SCUR School Computer Use Plan

SCURL Scottish Confederation of University and Research Libraries (UK-Scotland)

SCV Single-Channel Voice • Stato della Città del Vaticano (Vatican) • Suceava (Romania) (location identifier) • System Component Verification • Vatican (international vehicle registration code)

SCVDR Surface Characteristics Vector Data Record

SCVE Spacecraft Vicinity Equipment

SCVTR Scan Converting Video Tape Recorder

SCVTS Switched Compressed Video Telecommunications Service

SCW Search-Complete Word • Slowly Changing World • State College of Washington (USA-WA) • Syktyvkar (Russia) (location identifier)

SCWA Single Channel Wire Access

SCWAC Sun City West Astronomy Club (USA-AZ)

SCWC Special Commission on Weather Modification

SCWG Satellite Communications Working Group • Science Communications Working Group (NASA) • Security Certification Working Group (NASA) • Star Catalogs Working Group • System Concepts Working Group

SCWO Supercritical Water Oxidation

SCWP Satellite Communications Working Party • Slowly Changing World Policy

SCX Salina Cruz (Mexico) (location identifier) • Single-Charge Exchange • Solar Coronal X-ray • Specialized Communications Exchange • Staff Coordination Exercise

SCXI Signal Conditioning Extension for Instrumentation

SCY Secretary

SCZ Santa Cruz Island (Solomon Islands) (location identifier) • Solar Convection Zone • Solar Convective Zone • Surface Convection Zone

SC2 Sensor Command and Control • Soft Computing and Soft Communications

SC2ISA Standard Command and Control Information Systems Architecture

SC2S Strategic Command and Control Software

SC3 Surveillance - Command - Control - Communications

SC4 System for Command, Control, Communications and Computers

sd Semi-Detached (binary star) • siehe dies! • Sindhi (ISO language code) • Sine Die • Subdwarf

Sd Said • Side • Signed • Sound • Sund

S&D Search & Destroy • Storage & Distribution • Supply & Distribution

S/D School District • Sight Draft • Synchro to Digital

SD Doctor of Science • Safe Dealing • Safe Deposit • Salutem Dicit • Same Date • Sample Data • Sample Delay • Sans Date • Scaling and Display • Scandinavian • Scan Duration • Schematic Diagram • Science Data • Scientific Design • Screening Device • Sea-Damaged • Second Degré • Security Description • Security Directorate • Self-Destroying • Self-Destruction • Self-Dual • Semi-Detached (binary star) • Semi-Diameter • Senatus Decreto • Send Data • Send Direct • Senior Deacon • Sequence Date • Sequenced Data • Several Dates • Shared Disk • Short Delay • Short Delivery • Short Duration • Shut Down • Sicherheitsdienst • Sight Draft • Signal Digitizer • Sine Die • Single Degenerate • Single Density • Single Determination • Single Domain • Sit Down • Situation Display • Size Distribution • Slowdown • Smoke Detector • Software Development • Software Distributor • Solar Dynamic • Solicitation Document • Soudan (code ISO de pays) • South Dakota (USA) • Space Department • Space Dependability • Space Division • Spare Disposition • Spatial Disorientation • Speaker Dependent • Special Delivery • Special Document • Special Duty • Specification Document • Specification for Design • Spectral Distribution • Spectroscopy Detector • Staff Duties • Standard Deviation • Standardization Data • Standardization Directory • Standardization Document • Standards Development • Stardust • Start Delimiter • Starting Delimiter • State Department • Station Director • Statistical Division • Stellar Dynamics • Strongly Disagree • Structural Description • Structural Determinism • Structured Design • Structured Document • Student Director • Study Director • Sudan (ISO country code) • Südliche Durchmusterung • Superdeformed • Super Density • Superdirectory • Supply Depot • Support Directive • Support Document • Supporting Document • Swaziland (international vehicle registration code) • Symmetry Determination • Synchronization Directorate • Synthetic Data • System Demonstration • System Description • System Development • System Dynamics

Sda Senda

SDA Baghdad Saddam (Iraq) (location identifier) • Schutzverband Deutscher Autoren (Germany) • Schweizerische Depeschen Agentur (Switzerland) • Science Data Analysis • Science Data Analyst • Science Data Archive • Scientific Data Analysis • Scientific Data Analyst • Security Design Architecture • Sélection Directe à l'Arrivée • Sequential Degradation Analysis • Service Delivery Arrangement • SGML Document Access • Shaft Drive Axis

• Shock Drift Acceleration • Shot Data Analysis • Significant Digit Arithmetic • Simple Doublet Antenna • Sleeve Dipole Antenna • Slowdown Area • Software Development Activity • Software Disk Array • Solar Depression Angle • Source Data Acquisition • Source Data Automation • Southern Delta Aquarids (meteor stream) • Spaceborne Distributed Aperture • Space Digest Australia • Special Duty Assignment • Standard Descriptive Abstract • Statistical Distribution Analyzer • Step Down Amplifier • Stepwise Discriminant Analysis • Supplier Data Approval • Supporting Data Analysis • Swappable Data Area • Symbolic Disk Address • Symbols - Digits - Alphabetics • System Design Authority • System Display Architecture • Systems Device Annex

SDAA San Diego Astronomy Association (USA-CA)

SDAB Systems Design and Applications Branch

SDAC Satellite Data Analysis Center • Science Data Analysis Center • Scientific Data Analysis Coordinator • Scientific Discovery through Advance Computing (project) • Seismic Data Analysis Center (USA) • Solar Data Analysis Center • Space Data Analysis Center • Surveillance Data Analysis Center

Sdad Sociedad

SDAD Satellite Digital and Analog Display

SDADS Satellite Digital And Display System

SDAE Source Data Automation Equipment

SDAF Science Data Analysis Facility

SDAG Science Data Analysis Group • Space Debris Advisory Group • Strategic Defense Advisory Group

SDAI Standard Data Access Interface • STEP Data Access Interface

SDAIR Symposium on Document Analysis and Information Retrieval

S. Dak. South Dakota (USA)

SDAP SAR Data Acquisition Plan • Science Data Automation Program • Shuttle Descent Analysis Program • Special Duty Assignment Pay • System Development And Performance • System Development and Acquisition Plan • Systems Development Analysis Program

SDAS Science Data Analysis Software • Science Data Analysis System • Scientific Data Automation System • Source Data Automation System • South Dows Astronomical Society (UK) • Stafford and District Astronomical Society (UK) • System Design and Analysis Support • Systems Data Analysis Section

S-DAT Stationary Digital Audio Tape

SDAT Sector Design Analysis Tool • Silicon Diode Array Target • Spacecraft Data Analysis Team • Stanford Diagnostic Arithmetic Test • Stationary Digital Audio Tape

SDAU Safety Data Analysis Unit

sdB Subdwarf B (star)

SDB Saldanha Bay - Langebaanweg (South Africa) (location identifier) • Secondary Data Block • Selective Directed Broadcast • Serial Data Bus • Shallow Draft Barge • Shallow Draft Board • Skill Development Base • Small and Disadvantaged Business • Small Disadvantaged Business • Soil Database • Solnechnye Dannye Byulleten • Source Debugger • Spacecraft Database • Spatial Database • Spectrum Database • Stellar Database • Storage and Data Brokering • Survey Database • Symbolic Debugger • System-Determined Blocksize • Système de Données Bibliographiques • Système de Documentation Bibliographique • Systems Design Board

SDBAS Switched Digital Broadband Access System

SDBC Software Development Baseline Control

SDBL Supportability Design Baseline

SDBM Selective Directed Broadcast Mode

SDBMS Spatial Database Management System

SDBS Scientific Database System • Sodium - Dodecil - Benzene - Sulphonate • Spatial Database System

SDBSA Small Disadvantaged Business Set Aside

SDBUP Small Disadvantaged Business Utilization Program

SDC Sample Data Collection • Sandcreek (Guyana) (location identifier) • Sans Domicile Connu • SAX Data Centre • Scandinavian Data Centre • Science Data Catalogue • Science Data Centre • Science Data Coordinator • Scientific Data Center • Scientific Documentation Center • Secondary Distribution Center • Secondary Domain Controller • Seismological Data Center • Selenoidal Detector Collaboration • Serial Digital Commands • Service Definition Coefficient • Shaft-Driven Compressor • Shipment Detail Card • Signal Data Converter • Single Data Control • Situation Display Console • Slave Display Control • Software Design Definition • Software Design Document • Software Development Center • Software Development Computer • Solenoidal Detector Collaboration • Source Data Collection • Southern Dark Cloud • Space Data Centre • Space Data Corporation • Space Development Corporation • Spares Disposition Code • Spatial Distortion Correction • Stabilization Data Computer • State Distribution Center • Status Display Console • Subreflector Drive Cabinet • Super Data Coder • Support Design Change • SWIFT Data Center • Swiss Development Cooperation • System Decision Paper

SDCA Software Development Capability Assessment

SDCC Small-Diameter Component Cask • Status Display and Control Center • Status Display and Control Console

SDCCC Standard Desktop Computer Companion Contact

SDCCH Stand-alone Dedicated Control Channel

SDCD Secondary Data Carrier Detect • Space Data and Computing Division

SDCE Scientific Data Collection Exercise • Software Development Capability Evaluation

SDCF Sample Data Channel Filter

SDCI Single-Dish Calibration and Imaging

SDCL Supplier Documentation Checklist • Structural Dynamics and Control Laboratory

SDCMO Structural Dynamics and Control MATLAB Operator

SDCN Space Data Communication Network

SDCP Self-Designing Controller Program • Serial Data Control Protocol • Summary Development Cost Plan • Supply Demand Control Point

SDCR Software Development Capacity Review • Software Development Capability Review • Source Data Communication Retrieval

SDCS SAIL Data Communications System • SAR Data Catalog System • Sample Data Control System • Science Data Conditioning System • Simulation Data Conversion System • Single Differential Cross-Section

SD&D Scientific Development & Design

SDD Lubango (Angola) (location identifier) • Scheduled Delivery Date • Secure Data Device • See Detailed Diagram • Segment Design Document • Selective Dissemination of Documentation • Semi-Discrete Decomposition • Shuttle Design Directive • Silicon Drift Detector • Single Data Disable • Single Dish Data • Site Description Database • Software Data Dictionary • Software Data Distribution • Software Description Database • Software Description Document • Software Design Description • Software Design Document • Space Data Division • Standard Delivery date • Stored Data Description • Subscriber Direct Dialing • Synthetized Directivity Diagram • System Definition Directive • System Design Description • System Design Document • System Development and Demonstration • System for Distributed Databases • Systems Development Device

SDDAS SwRI's Data Display and Archival System

SDDBMS Secure Distributed Database Management System

SDDC Science Data Decoder Chassis

SDDD Software Detailed Design Document

SDDF Self-Defining Data Format

SDDi Satellite Data Distribution over the Internet

SDDI Shielded Distributed Data Interface

SDDL Software Design and Document Language • Stored-Data Definition Language

SDDPT Science Digital Data Preservation Task

SDDS Satellite Data Distribution System • Scientific Document Delivery System • Secondary Data Display System • Solid-Discharge Data System

SDDVD Super Density Digital Video Disc

SDE Santiago del Estero (Argentina-SE) (location identifier) • Secure Data Exchange (*) • Security Data Exchange • Service du Développement Économique • Shadow Dose Equivalent • Shared Data Environment • Signal Detection Efficiency • Signal Distribution Equipment • Simulator Development Environment • Smart Device Extension • Software Demonstration Evaluation • Software Development Environment • Space Division Evaluator • Standard Data Element • Standard Data Entry • Stellar Density Enhancement • Stochastic Differential Equation • Submission and Delivery Entity • System Development Environment

SDEB Science Data Evaluation Board

SDECE Service de Documentation Extérieure et de Contre-Espionnage

SDEF Service Development Evaluation Facility

SDEP Software Demonstration Evaluation Procedure

SDER Software Demonstration Evaluation Report

SDES Spatial Data Exchange Standard

SDESG Strapdown Electrically Suspended Gyro

SDEST Societal Dimensions of Engineering, Science and Technology (NSF)

SDEX Standard Executive

SdF Sûreté de Fonctionnement

SDF Louisville Standiford (USA-KY) (location identifier) • Safing and Deservicing Facility • Sans Domicile Fixe • Science Data File • Science Data Formatter • Screen Definition Facility • Screen Definition File • Sensor Data Fusion • Service Data Function • Shared Data Facility • Shore Development Facility • Short-Duration Flare • Shortest Delay First • Shuttle Development Facility • SIG Discretionary Fund (ACM) • Simplified Directional Facility • Simplified Direction Finding • Simulation Data File • Single Degree of Freedom • Slew Data File • Software Description File • Software Development Facility • Software Development File • Software Development Folder • Space Delimited File • Space Delimited Format • Spatial Distance Filter • Special Dispersion Focuser • Spectral Density Function • Standard Data Format • Standard Delay File • Standard Delimited Format • Standard Distance File • Standard Distribution Format • Stored Data File • Structured Display File • Subaru Deep Field • Subdistribution Frame • Sudden Disappearing Filament • Supplementary Data File • Sûreté De Fonctionnement • Synthetic Discriminant Function • System Data Format (*) • System Development Facility

SDFD Software Data Flow Diagram

SDFEM Streamline Diffusion Finite Element Method

SD/FS Smoke Detector/Fire Suppression

SD&FS Smoke Detector & Fire Suppression

SDFS Smoke Detector and Fire Suppression • Standard Disk File System

SDFU Standard Format Data Unit

Sdg Sending • Siding

S/DG Strap-Down Gyro

SDG Sagittarius Dwarf Galaxy • Sample Delivery Group • Sanandaj (Iran) (location identifier) • Scan Display Generator • Signed Directed Graph • Simulated Data Generator • Situation Display Generator • Snow Depth Gauge • Software Development Group • Soli Deo Gloria • Strap-Down Gyro • Supermassive Disk Galaxy • Supplier Documentation Group • Symmetric Derivation Graph • System Development Group • System Documentation Guidelines

SDGE Situation Display Generator Element

SDGW Structural Design Gross Weight • Supernova-Driven Galactic Wind

SDGWM Supernova-Driven Galactic Wind Model

SDH Santa Rosa - Copan (Honduras) (location identifier) • Science Data Header • Seasonal Derated Hour • Select Drive Head • Software Development Handbook • Spatial Data Handling • Synchronous Digital Hierarchy • Synchrotron Digital Handling • System Development Handbook • System Development Hardware

SDHE Spacecraft Data Handling Equipment

SDHOP Surface Développée Hors-Oeuvre Pondérée

SDHS Satellite Data Handling System

SD&I System Development & Integration

SDI Saidor (Papua New Guinea) (location identifier) • Science Discrepancy Item • Selected Descriptive Item • Selected Dissemination of Information • Selective Dissemination of Information (*) • Serial Device Interface • Serial Digital Interface • Service de Documentation et d'Information • Service de Documentation Industrielle • Service de Documentation Interministérielle • Service de Développement Informatique • Serviço de Documentação e Informação • Single-Document Interface • Société de Développement Informatique • Software Development Interface • Source Data Information • Spatial Data Infrastructure • Standard Data Interface • Standard Disk Interconnect • Storage Device Interface • Strategic Defense Initiative • Subscriber Distribution Interface • Support Directive Instruction • System Development and Integration

SDIF SGML Document Interchange Format • Single-Document Interchange Format • Software Development and Integration Facility • Standard Document Interchange Format

S-DIMM Solar Dual Image Motion Monitor

SDIMM Solar Differential Image Motion Monitor • Solar Dual Image Motion Monitor

SDIO Strategic Defense Initiative Office • Strategic Defense Initiative Organization (*)

SDIP Service de Documentation et d'Information Pédagogiques (France) • Simulator Data Integrity Program • Software Development Integrity Program • Strategic Defense Initiative Program

SDIPC Software Digital Image Processing Calculator

SDIR Statement of Detailed Interoperability Requirement

SDIRK Singly Diagonally Implicit Runge-Kutta

SDIS Satellite Data Input System • Simulation Database Interchange Specification • Surveillance and Display Interface System • Switched Digital Integrated Service

SDIT Service de Documentation et d'Information Technique

SDJ Sendai (Japan) (location identifier)

SDK Sandakan (Malaysia) (location identifier) • Skin Decontaminating Kit • Software Developer's Kit • Software Development Kit (*)

sdl SmartDraw Library (file extension)

Sdl Saddle

SDL Samedi - Dimanche - Lundi • Satellite Data Laboratory • Scenario Description Language • Semiconductor Diode Laser • Serpentine Delay Line • Shielded Data Link

• Software Design Language • Software Development Laboratory • Software Development Language • Software Development Library • Space Dynamics Laboratory • Special Duties List • Specification and Description Language • Standard Distribution List • Structural Dynamics Laboratory • Sundsvall (Sweden) (location identifier) • Supporting Document List • Surface Display Library • System Description Language • System Descriptive Language • System Design Language • System Development Language • System Directory List • Systematic Design Language

SDLA Spectrométrie d'Absorption par Diode Laser • Successive Detection Logarithmic Amplifier

SDLC Serial Data Link Control • Software Development Life Cycle • Synchronous Data-Link Control • System Data-Link Control • System Development Life Cycle

SDLM Scheduled Depot Level Maintenance • Standard Depot Level Maintenance

SDLS Satellite Data Link Standard • Software Data Link Standard • Space Data Link Standard

SDLV Shuttle-Derived Launch Vehicle

sdm Standard Deviation of the Mean

SDM Samsonov Density Meter • San Diego - Brown (USA-CA) (location identifier) • Scientific Data Management • Secondary Deployment Mechanism • Secure Digital Modem • Selective Display Model • Selective Dissemination of Microfiche • Semantic Database Model • Semantic Data Model • Sequence Division Multiplexing • Service Decision Memorandum • Service de Détermination des Manoeuvres • Shape Deposition Manufacturing • Short Data Message • Shutdown Mode • Shuttle Data Management • Shuttle Definition Manual • SOC Development Manager • Software Distribution Methodology • Space Disaster Mitigation • Space Division Multiplexing • Spin-Dependent Model • Standardization Design Memorandum • State-Dependent Model • Statistical Data Modulation • Statistical-Dynamical Model • Subrate Data Multiplexer • Subrate Data Multiplexing • Subtractive Double Monochromator • Synchronous Data Mode • System Data Mover • System Decision memorandum • System Definition Manual • System Description Model • System Design and Management • System Design Manual • System Development Methodology • System Development Multitasking

SDMA Shared Direct Memory Access • Space Division Multiple Access

SDMB Shuttle Discrepancy Management Board

SDME Software Development and Maintenance Environment • Synchronous Data Modern Equipment

SDMI Secure Digital Music Initiative

SDMIS Standardization Data Management Information System

SDMJ September - December - March - June

SDMM Solar Dual Motion Monitor

SDMMC Software Development, Modification and Maintenance Contract

SDMO Specification and Data Management Office • Specification and Data Management Officer

SDMP Simulation Development Master Plan • System Development Master Plan

SDMS SCSI Device Management System • Short-Duration Meteor Stream • Shuttle Data Management System • Spatial Data Management System • Supplier Data Management System • Switch Data Management System

SDMTG Space Data Management Task Group

SDMX Space Division Matrix • Space Division Multiplex

Sdn Sedan • Sendirian

SdN Société des Nations (1920)

SDN Safety Discrepancy Notice • Sandane (Norway) (location identifier) • Schedule Deletion Notification • Secure Data Network • Service Delivery Node • Simulator Data Name • Société Des Nations (1920) • Software-Defined Network • Software Delivery Notice • Software Development Note • Software Development Notebook • Software Digital Network • Soudan (code ISO de pays) • Space Data Network • Sudan (ISO country code) • Sustainable Development Networking • System Development Network • System Development Notification

SDNB Software Development Notebook

SDNE Services in Distributed and Networked Environments

SDNRIU Secure Digital Net Radio Interface Unit

SDNS Secure Data Network Service • Secure Data Network System

sdO Subdwarf O (star)

SDO Ryotsu Sado Island (Japan) (location identifier) • Scattered Disk Objects • Scheduled Day Off • Semi-Definite Optimization • Senior Duty Officer • Sensory Data Object • Serial Data Out • Shared Data Object • Shielded Diatomic Orbital • Solar Dynamics Observatory • Solar Dynamics Observer (project) • Spatial Data Option • Spatial Disorientation • Special Development Order • Special Duty Only • Staff Duty Officer • Stamp Distribution Office • Standards-Developing Organization • Standards Development Organization • Station Duty Officer • Subdistrict Office • Synthetic Display Output • System Development Office • System Director's Office

S. Doc. Senate Document

SDOC Science Data Operations Center • Specific Direct Operating Cost

SDOF Single Degree Of Freedom

SDOH SOC Data Operations Handbook

SDOI Spatial Data Organization Information

Sdok Sveriges Doktorander (Sweden)

SDOPR Space Division Office of Primary Responsibility

SDP Pound (Sudan) (ISO currency code) • Safety Data Package • Sand Point (USA-AK) (location identifier) • Science Data Packet • Science Data Plan • Science Data Processing • Secondary Delay Position • Semi-Definite Programming • Sensor Data Processing • Service Delivery Point • Service Discovery Protocol • Servicio de Documentación Periodista • Session Description Protocol • Short-Duration Penetration • Shower Detector Plate • Shuttle Data Processor • Signal Data Processor • Silicon Diode Pellet • Silicon Drift Photodetector • Site Data Processor • Small Distribution Phenomena • Socio-Domestic Pleasure • Software Development Plan • Software Development Program • Solar Disk Photometer • Solar Drum Positioner • Source Data Processing • Space Data Processor • Spectral Diagnostics Package • Standard Data Package • Standard Data Processing • Standard Data Processor • Standard Development Plan • Standard Distance Package • Standards Development Plan • Station Data Processing • Steyr - Daimler - Puch (company) • Storage Distribution Point • Strategic Defense Program • Submission and Delivery Protocol • Supplier Data Package • Supply Distribution Point • Surface Deformation Pattern • Survey Data Processing • System Decision Package • System Decision Paper • System Development Package • System Development Plan

SDPC Science Data Processing Computer • Shuttle Data Processing Complex

SDPE Special Design Protective Equipment

SDPF Satellite Data Processing Facility • Science Data Processing Facility • Sensor Data Processing Facility

SDPL Sensor Data Processing Laboratory

SDPLV Shuttle-Derived-Propellant Launch Vehicle

SDPR System Design and Performance Requirements

SDPS SAR Data Processing System • Science Data Processing Segment • Signal Data Processing System • Spatial Data Production System • Standard Data Processing System

SDPS/W Science Data Processing Software

SDPSW Science Data Processing Software

SDPT South Downs Planetarium Trust (UK-England)

SDPTK Science Data Processing Toolkit

SDQ Santo Domingo - Las Americas (Dominican Rep.) (location identifier)

SDQA Space Division Quality Assurance

SDQL Structured Document Query Language

Sdr Sender

SDR Same Day Return • Santander (Spain) (location identifier) • S-band Data Relay • Science Data Reader • Scientific Data Recorder • Search Decision Rule • Security Deficiency Report • Seismic Detection and Ranging • Sensor Data Record • Sensor Data Recorder • Sensor Data Requirement • Sequential Delivery Service • Serial Data Receiver • Series Dissipating Regulator • Service Detail Record • Service Difficulty Report • Shipment Document Release • Signal Data Record • Signal Data Recorder • Signal Data Reproducer • Signal Distribution Room • Signal-to-Distortion Ratio • Significant Deficiency Report • Smal Development Requirement • Software Design Requirement • Software Design Review • Software Discrepancy Report • Space Division Regulation • Spacelab Disposition Record • Spatial Dynamic Range • Special Drawing Right • Spectral Dynamic Range • Spectrum Data Record • Spoken Document Retrieval • Statistical Data Recorder • Storage Data Register • Streaming Data Request • Subjective-Distance Retrieval • Supply Discrepancy Report • Surrogate Data Radio • Surrogate Digital Radio • Survey of Doctoral Recipients • System Data Record • System Definition Requirement • System Definition Review • System Design Report • System Design Requirement • System Design Review • System Development Requirement • System Discrepancy Report

SDRA Shaped Dual Reflector Antenna • Software Development Resources Architecture

SDRAM Synchronized Dynamic Random Access Memory • Synchronous Dynamic Random Access Memory (*)

SDRB Software Design Review Board • Specification and Data Review Board • Supplier Documentation Review Board

SDRC Structural Dynamics Research Corporation

SDRD Subcontract Data Requirements Document • Supplier Data Requirements Description • Supplier Documentation Review Data • Supportability Design Requirements Document

SDR&I Space Data Resources & Information

SDRL Subcontract Data Requirements List • Subcontractor Data Requirements List • Supplier Data Requirements List

SDRM Safety Design Review Meeting

SDRP Simulated Data Reduction Program

SDRPC Satellite Data Receiving and Processing Center

SDRPS Satellite Data Receiving and Processing System

SDRS Satellite Data Reference System • Shuttle Display Reconfiguration System • Signal Data Recording Set

SDRSI South Dakota Remote Sensing Institute (USA-SD)

SDRT Science Data Restoration Task • Software Development Risk Taxonomy

SDRW Société de Développement Régional pour la Wallonie (Belgium)

SdS Surveillance du Satellite

SDS Sado Shima (Japan) (location identifier) • Safety Data Sheet • Safety Disconnect Switch • Satellite Data Services • Satellite Data System • Science Data Standards • Science Data Store • Scientific Data Set • Scientific Data Store • Scientific Data System • Secondary Data System • Segment Design Specification • Self-defining Data System • Sequential Dynamical System • Service de Documentation Spatiale • Servo Drive System • Shuttle Dynamic Simulation • Shuttle Dynamic Simulator • Signal Data Server • Signal Dispatch Service • Signal Distribution System • Sign-Digit Subtractor • Simulation Data System • Site Data System • Small Digital Switch • Smart Data System • Smoke Detection System • Sodium Dodecyl Sulphate • Software Design Specification • Software Development Specification • Software Development Standard • Software Development System • Solar Diameter Sensor • Solar Disk Sextant • Solar Disk Simulator • Solution Development System • Spacecraft Data Simulator • Spacecraft Design Specification • Space Division Subscriber • Space Documentation Service • Special Docking Simulator • Spectrometer Digital System • Spelda Dedicated Satellite • Spring Damper System • Staff Development Service • Standard Depot System • Steering and Damping System • Steering and Depth Simulator • Steering Damping System • Stored Data Smooth • Strategic Data System • Streaming Data Strobe • Subaru Deep Survey • Surface Distribution System • Surveillance Detection System • Surveillance Direction System • Switched Data Service • Switched Digital Service • Synchronous Data Set • SYSOPs Distribution System • System Data Synthesizer • System Definition Study • System Design Specification

SDSA Société pour la Diffusion des Sciences et des Arts • Software Development and Support Activity

SDSB Satellite Data Services Branch • Systems Development and Sustainment Branch

SDSC San Diego State College (USA-CA) • San Diego Supercomputer Center (USA-CA) • SAR Data Catalog System • Synchronous Data Set Controller

SDSD Satellite Data Services Division • Single Disk Storage Device

SDSF Spool Display and Search Facility • System Display and Search Facility

SDSG Space Division Switching Group

SDSI Shared Data Set Integrity • Short Duration and Specific Date Impact • Simple Distributed Security Information • Simple Distribution Security Information

SDSL Single-line Digital Subscriber Line • Symmetric Digital Subscriber Line (*)

SDSMA Slotted Digital Sense Multiple Access

SDSM&T South Dakota School of Mines & Technology (USA-SD)

SDSMT South Dakota School of Mines and Technology (USA-SD)

SDSP Software Development Support Plan

SDSS Self-Deploying Space Station • Sloan Digital Sky Survey • Space Division Shuttle Simulator • Spatial Decision Support System

SDSSE Science Data System Support Equipment

SDST Science Data Support Team • Small Deep Space Transponder

SDSU San Diego State University (USA-CA) • South Dakota State University (USA-SD)

SdT Service de Traduction

SDT Saidu Sharif (Pakistan) (location identifier) • Scaling and Display Task • Science Data Team • Science Definition Team • Scientific Distribution Technique • Second Destination Transportation • Segment Design Team • Self-Development Test • Serial Data Transmission • Shuttle Data Tape • Shuttle Data Type • Signal Detection Theory • Simulated Data Tape • Single Discipline Team • Skill Development Testing • Software Development Test • Software Development Tool • Source Distribution Technique • Specification-Driven Tool • Spin-Dependent Tunneling • Standard Duty Title • Start Data Traffic • Structural Dynamic Test • Structured Data Transfer • System Development Tool

SDTA Structural Dynamic Test Article

SDTC Standard Duty Title Code

SDTF Standard Duty Title File

SDTI Selective Dissemination of Technical Information • Serial Digital Transmission Interface • System Design and Technology Integration

SDTL Software Design Team Leader

SDTN Synchronous Digital Transmission Network

SDTP Spatial Data Transfer Processor • Startover Data Transfer and Processing • System Demonstration Test Plan

SDTS Satellite Data Transmission System • Spatial Data Transfer Specification • Spatial Data Transfer Standard (*)

SDTT Similarly Dependence Textural Transformation

SDTV Standard-Definition Television • Structural Dynamics Test Vehicle

SDTX Science Data Transmitter

SdU Sciences de l'Univers

SDU Rio de Janeiro - Santos Dumont (Brazil-RJ) (location identifier) • Satellite Data Unit • Satellite Delay Unit • Schéma Directeur d'Urbanisme (France) • Sciences De l'Univers • Scientific Data Unit • Secure Data Unit • Security Device Unit • Sensor Demonstration Unit • Service Data Unit • Service Delivery Unit • Signal Distribution Unit • Slave Display Unit • Spectrum Display Unit • Stand-alone Display Unit • Storage Device Unit • Superhigh Display Unit

SDUC Secondary Data Utilization Centre

SDUK Society for the Diffusion of Useful Knowledge

SDUS Satellite Data Users Station • Secondary Data User Station • Small-scale Data Utilization Station

SDV Sciences De la Vie • Shuttle-Derived Vehicle • Solar Daily Variation • Stella-Della-Valley (weekends) (USA-PA) • Switched Digital Video • Tel Aviv - Yafo Sde-Dov (Israel) (location identifier)

SDVF Software Development and Validation Facility • Software Development and Verification Facility

SDVP System Demonstration Validation Plan • System Demonstration Validation Program

SDVTP System Demonstration Validation/Test Plan

SDW Sandwip (Bangladesh) (location identifier) • Science Data Writer • Serial Digital Word • Shadow • Shared Data Warehouse • Short-Distance Wireless • Spin Density Wave • Spiral Density Wave

SDWA Safe Drinking Water Act (USA) • System Diagnostic Work Area

SDX Satellite Data Exchange • Sedona (USA-AZ) (location identifier) • Selected Diffusion Extended • Storage Data Exchange

SDY Sidney (USA-MT) (location identifier)

SDZ Shetland Islands (UK) (location identifier) • Surface Danger Zone

s/e Sin Editorial

se Standard Error • Sweden

Se Selen • Selenium • Sélénium

S&E Salaries & Expenses • Scheduling & Execution • Science & Engineering • Scientific & Engineering • Scientist & Engineer • Sensors & Effectors • Services & Equipment • Signature Error • Single-Excited

S-E Survivability-Endurance

S/E Service Equipment • Stock Exchange

SE Safety Evaluation • Sanitary Evacuation • Santiago del Estero (Argentina) • Saponification Equivalent • Schrödinger Equation • Scientific Editor • Secondary Eclipse • Secondary Electron • Secretaría de Estado • Secrétariat d'État • Sensors and Effectors • Sequence of Events • Seltene Erden • Semitic • Seneca • September • Septembre • Sergipe (Brazil) • Service Element • Service Engineer • Service Equipment • Session Environment • Shared Everything • Shielding Effectiveness • Signal Excess • Single End • Single-Ended • Single Engine • Single Entry • Société Européenne • Socio-Economical • Socio-Économique • Software Engineer • Software Engineering • Solar Eclipse • Solar Ecliptic • Solar Explorer • Sondereigentum • Son Éminence • Son Excellence • Southeast • Southeastern • Space Exploration • Special Edition • Special Equipment • Special Event • Specialized Engineering • Spectroscopic Ellipsometer • Spherical Equivalent • Spring Equinox • Stable Element • Staff Engineer • Stamped Envelope • Standard Earth • Standard Engine • Standard English • Standard Erection • Standard Error • Statistical Equilibrium • Stock Exchange • Straight Edge • Strategy Evaluation • Stress Engineer • Student Experiment • Subset Entry • Sud-Est • Suède (code ISO de pays) • Su Excelencia • Suppléant Éventuel • Support Element • Support Equipment • Supporter Equipment • Supporting Establishment • Surface Element • Surveillant d'Externat • Survivability-Endurance • Survivability Enhancement • Survival Equipment • Sweden (international aircraft registration code) • Sweden (ISO country code) • Switching Element • Synchrotron Emission • Synthetic Environment • Système Expert • System Effectiveness • System Element • System Engineer • System Engineering • System Evaluator

sea Self-Extracting Archive (file extension)

SE&A Safety Engineering & Analysis

SEA Safety Engineering Analysis • Scanning Electrostatic Analysis • Scientific Exchange Agreement • Scientist's Expert Assistant • SeaTac (USA-WA) (location identifier) • Seattle (USA-WA) (location identifier) • Security Extension Architecture • Self-Extracting Archive • Sensor Electronics Assembly • Service d'Étude des Accélérateurs • Service Educational Activities • Silicon Elastimeter Ablator • Small Earth Approacher • Small Earth-Approaching • Sociedad Española de Astronomía (Spain) • Society for Electronic Access • Society of Engineering Associates (Australia) • Southeast Asia • Spherical Electrostatic Analyzer • Splitting Equilibration Algorithm • Standard Electronic Assembly • Standard Extended Architecture • Standard Extended Attribute • Statistical Energetics Analysis • Statistical Energy Analysis • Strategic Enterprise Architecture • Strategic Environmental Assessment • Sudden Enhancement of Atmospherics • Systems Effectiveness Analyzer • System Engineering Analysis • System Error Analysis • System for Environmental Analysis • Tacoma (USA-WA) (location identifier)

SEAB Secretary of Energy Advisory Board (USA) • Security Education Advisory Board

SeAc Sensors and Actuators (periodical code)

SEAC School Examination and Assessment Council • Société Européenne pour l'Astronomie dans la Culture • Space Exploration Advisory Committee • Standard Eastern Automatic Computer • Standard Eastern Automating Computing

SEACF Support Equipment Assembly and Checkout Facility

SEACOF Support Equipment Assembly and Checkout Facility

SEACS Search and Acquisition System • Support Equipment Acquisition Control System

SEAD Suppression of Enemy Air Defenses

SEADEX Shoreline Environment Atmospheric Dispersion Experiment

SEADS Shuttle Entry Air Data Sensor

SEAES Spacecraft Environmental Anomalies Expert System

SEAGA Selective Employment of Airborne and Ground Alert

SEAI Secrétaire Exécutif aux Affaires Industrielles

SEAIC Southeast Asia Information Center

SEAID Support Equipment Abbreviated Items Description

SEAIG Southeast Asia Information Group

SEAL Screening External Access Link • Sea - Air - Land • Secure Electronic Authorization Laboratory • Segmentation and Assembly Layer • Simple and Efficient Adaptation Layer • Simulated Evolution And Learning (conference) • Standard Electronic Accounting Language

SEALION Sea Level Instrumentation and Observation Network

SEALITE Systematic Evaluation and Analysis of a Laser in a Test Environment

SEALS Security Equipment And Locking System

SEAM Saber Enterprise Applications Manager • Spectral-fitting Expanding Atmosphere Method • System Engineering and Analysis Management

SEAMEC Southeast Asian Ministers of Education Council

SEAMEO Southeast Asian Ministers of Education Organization

SEAMES Southeast Asian Ministers of Education Secretariat

SEAMS Southeast Asian Mathematical Society

SEAN Scientific Event Alert Network (now GVN) • Southeast European Academic Network

SEAOP Sociedade de Estudos Astronômicos de Ouro Preto (Brazil-MG)

SEAP Science and Engineering Apprenticeship Program • Security Education and Awareness Program

SEAPD Scientific and Engineering Professional Development

SEAPG Support Equipment Acquisition Planning Group

SEAQ Stock Exchange Automated Quotation

SEAR System Engineering Analysis Report

SEARCH Study of Earth Rapid Changes

SEARFE Students Exploring Australia's Radio-Frequency Environment (project)

SEARNG Southeast Asian Region Network for Geosciences

SEAS School of Engineering and Applied Science • Shipboard Environmental Acquisition System • Shipboard Environmental Automated System • State Employee Allocation Standard • Stimulated Electron-Acoustic-wave Scattering • Study of the European Arctic Shelf • Synthetic Economy for Analysis and Simulation (project) • Systems, Engineering and Analysis Support

SEASAR Sea Synthetic Aperture Radar

SEASAT Sea Satellite

SEASTARS Satellite Engineering Assisted Shipboard Test and Repair System

SEAT Sociedad Española de Automóviles de Turismo • Status Evaluation And Test • System Evaluation Analysis Tool

SeaTac Seattle-Tacoma (USA-WA)

SEATO South-East Asia Treaty Organization

SEB Scientific Equipment Bay • Sebha (Libya) (location identifier) • Single Event Burnout • Source Evaluation Board • South Equatorial Belt (Jupiter) • State Electricity Board • Strip Electron Beam • Support Equipment Building • Systems Electronic Box • Systems Engineering Branch

SEBA Support Electronics Bay Assembly

SEBC Système Européen de Banques Centrales

SEBD Software Engineering Bibliographic Database • South Equatorial Band Disturbance

SEbE Southeast by East

SEBEX Sahelian Energy Balance Experiment

SEBn South Equatorial Belt - Northern Component (Jupiter)

SEbS Southeast by South

SEBs South Equatorial Belt - Southern Component (Jupiter)

SEBV Soil Environmental Background Value

SEBz South Equatorial Belt - Internal Zone

sec Secant • Secolo

Sec Sección • Second • Secondaire • Secondary • Seconded • Secrétaire • Secrétariat • Secret • Secrétaire • Secretaría • Secretary • Section • Sector • Secular • Secundum • Security

SEC Science Executive Committee • Scientific and Engineering Computation • Scientific Estimates Committee • Second • Secondary • Secondary Electron Conducting • Secondary Electron Conduction • Secondary Emission Conductivity • Secondary Evaluation Committee • Section • Secure • Securities and Exchange Commission (USA) • Security • Security Clearance • Security Exchange Commission • Séjour Éducatif et Culturel • Sensor Emulation Control • Sequential Events Controller • Serre-Chevalier (France) (location identifier) • Simple Electronic Computer • Single-Edge Contact (*) • Single-Edge Contract • Single Electron Capture • Single Error Correction • Size-Exclusion Chromatography • Société Européenne de Culture • Society for Educative Communication • Software Engineering Center • Solar Electric Company • Solar Energy Collector • Solar Energy Concentrator • Solar Event Coincidence • Solid Electrolyte Capacitor • Source Evaluation Committee • South Equator Crossing • Space Education Council • Space Environmental Chamber • Space Environment Center • Space Event Conference • Split Expander Cycle • Sponsor Executive Committee • Standard Error of Calibration • Standard Ethernet Controller • Standard Evaluation Circuit • Sun-Earth Connection • Superelastic Collision • Superexchange • Support Equipment Change • System Essentiality Code

SECAM Séquence de Couleurs Avec Mémoire • Sequential Color And Memory • Systems Engineering Capability Assessment Model

SECAP Société d'Étude et de Construction d'Appareils de Précision (France) • System Experience Correlation and Analysis Program • Systems Experiment Correlation and Analysis Program

SECAS Ship Equipment Configuration Accounting System

SECB Systems Engineering Control Board

Secc Sección

SECC Single Edge Contact Cartridge

SECCHI Sun-Earth Connection Coronal and Heliospheric Investigation

SECDEC Single Error Correction - Double Error Correction

SECDED Single Error Correction - Double Error Detection

SECDS Sun-Earth Connection Data System

SECEF Sun-Earth Connection Education Forum

SECEM Support Equipment Cost Effectiveness Model

SECFO System Engineering and Consensus Formation Office

SECFT Sociedad Española de Cartografía, Fotogrametría y Teledetección (Spain)

Sec. Gen. Secretary General

sech Hyperbolic Secant

SECHO Supervised Extraction and Classification of Homogeneous Objects

SECI Southeast European Cooperation Initiative

SECIF Security Interface Software

SECIS Solar Eclipse Coronal Imaging System

sec. leg. Secundum Legem

SECM Science and Engineering of Composite Materials (periodical code) • Software Engineering Cost Moel • Systems Engineering Capability Model

SECME Southeastern Consortium for Minorities in Engineering

sec. nat. Secundum Naturam

SECO Satellite Ephemeris and Clock • SEC Orthicon • Sun-Earth Connection Observatory • Sustainer Engine Cutoff

SECOIA Systèmes Experts et Conceptions d'Outils pour l'Intelligence Artificielle

SECOM Software Engineering Cost Moel

SECOMO Software Engineering Cost Model

SECOMP Secure Communications Package • Secure Enroute Communications Package

SECP Scientist and Engineer Career Program • Summary Effort Control Package • System Evaluation Control Package

SECPLN Security Plan

Secr Secrétaire • Secretariat • Secrétariat • Secretary

SECR Secretary • Standard Embedded Computer Resource

sec. reg. Secundum Regulam

Secs Seconds • Sections

SECS Sequential Events Control System • Shuttle Events Control System • Simulation and Evaluation of Chemical Synthesis • Simulation/Experiment Control System • Small Experimental Communications Satellite • Solar Electric Communication Satellite • Space Environmental Control System • Standard Embedded Computer Software

Sect Section • Sector

SECT Section

SECTEL Secure Telephone

SECTMA Secretaria de Ciência e Tecnologia e Meio Ambiente

SECVOX Secure Voice

Secy Secretary

Secy.-Gen. Secretary-General

SEC/YR Seconds per Year

SECYT Secretaría de Ciencia Y Tecnologia

sed Stream Editor

Sed Sediment

SED Saisie Électronique de Documents • Saturn Electrostatic Discharge • Science Education Department • Secondary Energy Distribution • Secure Encryption Device • Sedom Min'hat Hashnayim (Israel) (location identifier) • Sensor Evolutionary Development • Service d'Enseignement à Distance • Service d'Étude des Détecteurs • Service d'Études et de Documentation • Smoke-Emitting Diode • Software Engineering Design • Software Engineering Directorate • Solar Energy Density • Space Electronics Division • Space Engineering Document • Space Environment Division • Space Experiments Division • Special Electrical Device • Special Engineering Detachment • Spectral Energy Distribution • Spheroidal Equidensity • Stream Editor • Surface-conduction Emitter Display • Système d'Équations Différentielles • Système d'Exploitation de Disquettes • System Effectiveness Demonstration

SEDA Sociedad Einstein De Astronomía (Spain)

SEDAC Socio-Economic Data and Applications Center

SEDACS Schedule Data Communications System

SEDAR Submerged Electrode Detection and Ranging

SEDAS Satellite Environmental Data Acquisition System • Spurious Emission Detection Acquisition System

SEdB Shielding Effectiveness in Decibels

SEDB Specifically Extragalactic Database • Systems Engineering Database

SEDEE Service d'Enlèvement et de Destruction des Engins Explosifs

SEDF Système d'Équations Différentielles Finies

SEDFB Surface-Emitting Distributed Feedback

SEDI Study of the Earth's Deep Interior

Sedim Sedimentology (periodical code)

SEDIX Selective Dissemination of Indices

SEDL Software Engineering Design Language

SEDN Système Européen de Documentation Nucléaire

SEDO Sociedad Española De Optica (Spain)

SEDPC Scientific and Engineering Data Processing Center (USA)

SEDR Science Experiment Data Record • Single Event Dielectric Burnout • Supplemental Experiment Data Record • Supplementary Experiment Data Record • Supplementary Experimenter Data Record • System Effective Data Rate • Systems Engineering Department Report

SEDRIS Synthetic Environment Data Representation Interchange Specification

SEDS Small Expendable Deployable System • Small Expendable Deployer System • Small Expendable Deployment System • Small Expendable-tether Deployer System • Software Engineering Detailed Schedule • Space Electronics Detection System • Students for the Exploration and Development of Space • Support Equipment Data System • System Effectiveness Data System • System Engineering Detailed Schedule

Sedt Sediment

Sedtn Sedimentation

SEE San Diego - Gillespie (USA-CA) (location identifier) • Satellite Energy Exchange • Science Enterprise Engineering • Secondary Electron Emission • Senior Electrical Engineer • Sensitive Electronic Equipment • Service de l'Énergie de l'État (Luxembourg) • Signing Essential English • Single Event Effect • Small Emplacement Excavator • Small Excavation Equipment • Société des Électriciens et Électroniciens (France) • Society of Environment Engineers • Software Engineering Economics • Software Engineering Environment • Software Engineering Exercise • Software Equality Evaluation • Solar Extreme-ultraviolet Experiment • Solar EUV Experiment • Spectral Energy Emission • Standard End Effector • Standard Error of Estimation • Stimulated Electromagnetic Emission • Study of Expertise and Experience • Sun-Earth Explorer • Supplementary Electronic Equipment • Support Equipment Evaluation • System Exploration Experiment • Systems Effectiveness Engineering • Systems Effectiveness Evaluation • Systems Engineering Environment

SEEA Southeast European Airlines

SEEC Scottish Environmental Education Centre

SEECM Shuttle Environmental Effects on Coated Mirrors

SEED Science Education for Early Development • Science Education through Experiment and Demonstration • Self Electro-optic Effect Device • Standard for the Exchange of Earthquake Data • Support for Eastern European Democracy

SEEDS Spacelab End-to-End Data System • Space-Exposed Experiment Developed for Students

SEEJ Society of Environmental Engineers Journal (periodical code)

SEEK System Evaluation and Exchange of Knowledge

SEEP Scientist and Engineer Exchange Program • Shelf-Edge Exchange Processes • Students to Explore and Experience Physics

SEer Sondereigentümer

SEER Science Education for Equity Reform • Sensor Experimental Evaluation and Review • Sensor Experimental

Evaluation Review • Social and Economic Environmental Research • Software Evaluation and Estimation of Resources • Surveillance, Epidemiology, and End Results • System Engineering, Evaluation and Research • System Evaluation and Estimate of Resources • System for Electronic Evaluation and Retrieval

SEERU Southeast European Research Unit

SEES Students to Experience Engineering and Science

SEET Science End-to-End Test

sef schwer entflammbar

SEF Satellite Evaluation Facilities • Sebring (USA-FL) (location identifier) • Service Emploi Formation (France) • Software Entrepreneurs' Group • Solar Energy Flux • Source-Explicit Forwarding • Spacecraft Event File • Space Environmental Facility • Space Experiment Facility • Stability Enhancement Function • Sustaining Engineering Facility • System Engineering Facility

SEFC Service de l'Enseignement et de la Formation Continue

SEFD Solar Energy Flux Density • System Equivalent Flux Density

SEFF Standardized Electronic File Format

SEFI Service - Formation - Information • Single Event Functional Interrupt

SEFIP Statistical Estimation Fault Isolation Procedure

SEFOR Serviço de Estatísticas e Fomento de Recursos

seg Segment (file extension) • Segue • Segundo

Seg Segment • Segmented • Segovia (Spain) • Segregate • Segregated • Seguente

SEG Scientific Expert Group • Segment • Segregator • Selinsgrove - Penn Valley (USA-PA) (location identifier) • Sequence of Events Generator • Simulation Engineering Group • Society of Economic Geologists (USA) • Society of Exploration Geophysicists (USA) • Software Engineering Group • Solid-Earth Geophysics • Standardization Evaluation Group • System Evaluation Group • Systems Engineering Group

SEGAT Secure Gateway

SEGF Surface Embedded Green Function

SEGPA Section d'Enseignement Général et Professionnel Adapté

SGR Single Event Gate Rupture

SEGS Sequence of Events Generation System

SEGT Section d'Enseignement Général et Technique

SEGway Science Education Gateway (program)

SEH Senggeh (Indonesia) (location identifier) • Shuttle Electronic Hardware • Solar Equivalent Hour • Solar EUVE Hitchhiker • Southeastern Half • Structured Exception Handling

SE&I Systems Engineering & Integration

SEI Salzburger Elektronik Industrie (Austria) • Science Education Initiative (SAAO) • Science Education Institute • Séjours et Échanges Internationaux • Seller End Item • Senhor do Bonfim (Brazil-BA) (location identifier) • Service d'Électronique et d'Informatique • Service des Échanges Internationaux • Service Européen d'Information • Small Engine Instruction • Sobolev with Exact Integration (method) • Software Engineering Institute • Space Exploration Initiative • Special Emphasis Item • Special Engineering Investigation • Special Equipment Item • Special Experience Identifier • Specific Emitter Identification • Statistical Engineering Institute • Stockholm Environment Institute (Sweden) • Support Equipment Illustration • Support Equipment Installation • Systems Engineer and Integrator • Systems Engineering and Integration • System/Equipment Inventory

SEIA Solar-Energy Industries Association • Solar-Energy Institute of America (*) • Systems Engineering Integration Agent

SEIAC Southeastern Iowa Astronomy Club (USA-IA)

SEIC Solar-Energy Information Center • State Educational Information Center (USA) • System Engineering and Integration Contract • System Engineering and Integration Contractor

SEICO Support Equipment Installation and Checkout

SEICT Sistema Estadual de Informação Científica e Tecnológica (Brazil)

SEID Systems Engineering and Integration Division • Systems Engineering, Integration, and Demonstration

SEIDAM System of Experts for Intelligent Data Management

SEII Société d'Échanges Industriels Internationaux (France)

SEIO Systems Engineering and Integration Office

SEIP Standard Engineering Installation Package • Systems Engineering Implementation Plan

SEIR Software and Electronic Information Resources • Software Engineering Information Repository • Southeast Indian Ridge

SEIS Senior Executive Information System • Service des Études Informatiques et Statistiques • Shuttle Experiments Integration Support • Single Element Imaging Spectrograph • Solar-Energy Information Service • Spacelab Experiments Integration Support • Standard End Item Specification • Submarine Escape Immersion Suit

Seismol Seismological • Seismology

SEIST System Engineering, Integration, and Support Team

SEI&T System Engineering, Integration & Test

SEIT Satellite Educational and Informational Television • Software Engineering and Integration Team • System Engineering and Integration Team • System Engineering, Integration and Test • System Evaluation, Integration and Test

SEIU Service Employees International Union

SEJ Seydisfjördur (Iceland) (location identifier) • Society of Environmental Journalists

Sek Sektion • Sekunde

SEK Crown (Sweden) (ISO currency code) • Krona (Sweden) (ISO currency code) • Ksar es Souk (Morocco) (location identifier) • Svenska Elektriska Kommissionen (Sweden)

SEKAS South East Kent Astronomical Society (UK-England)

SEKE Software Engineering and Knowledge Engineering (conference)

Sel Select • Selected • Selecting • Selection • Selectivity • Selector • Selig

Sél Sélection • Sélectionné • Sélectionner

SEL Satellite Experiment Laboratory • School of Engineering and Logistics • Select • Selected Equipment List • Selection • Selector • Self-Extensible Language • Seoul Kimpo (Korea-ROK) (location identifier) • Simulation Engineering Library • Single-Event Latch-up • Software Engineering and Languages (conference) • Software Engineering Laboratory • Space Environment Laboratory • Spontaneously Emitted Light • Stabilized Electromagnetic Levitator • Standard Elektrik Lorentz • Système d'Échange Local • Système d'Équations Linéaires • Système Électrique et Logiciel • Systems Electronics Laboratory

SeLA Seven-Letter Acronym

SELA Sistema Económico Latino-Americano • Système Économique Latinoaméricain

SELC Senior Executive Leadership Course

SELCAL Selective Call • Selective Calling

SELCOM Select Committee • Selection Committee

SELDADS Space Environment Laboratory Data Acquisition and Display System

SELENE Selenological and Engineering Explorer

SELEX Segmented Large (X baryon spectrometer) Experiment

SELF Self-Eject Launch Facility • Solar Electric Light Fund

Selk Selkirk

SELPER Sociedad de Especialistas Latino-americanos en Percepción Remota

SELRAS Space Environment Laboratory Retrieval and Analysis System

SELREL Selective Release

SELSCAN Selective Scanning

SELT Segment Evaluation at Low Temperature • Self-Eject Launch Technique

SELV Separated Extralow Voltage • Small Expendable Launch Vehicle

SELVS Small Expendable Launch Vehicle Services

SELY Southeasterly

Sem Semaine • Semana • Semanal • Semester • Semicolon • Seminar • Seminary • Standard Error of the Mean

Sém Séminaire

SE/M Systems Engineering/Management

SEM Safe Emergency Mode • Scanning Electron Microscope • Scanning Electron Microscopy • Science, Engineering and Mathematics • Secondary Electron Model • Secondary Electron Mirror • Secondary Emission Monitor • Secure Ethernet Multiplexer • Seismometer • Seller's Engineering Memorandum • Selma Craig (USA-AL) (location identifier) • Semi-Empirical Model • Sensitivity Equation Method • Service d'Échange des Médias • Shared Equity Mortgage • Shuttle Experiment Module • Significant Events Message • Simulation - Engineering - Modeling • Simulation Evaluation Methodology • Singularly Expansion Method • Sociedad Espacial Mexicana (Mexico) • Société d'Économie Mixte • Society for Experimental Mechanics • Software Engineering Manual • Software Evaluation Manager • Solar Environmental Monitor • Solar EUV Monitor • Space Environment Monitor • Space Environment Monitoring • Space Experiment Module • Spare Equipment Maintenance • Spectral Evolutionary Model • Standard Electronic Module • Standard Equipment Module • Standard Error of the Mean • Standard Estimating Model • Stray Energy Monitor • String Environmental Module • Structural Econometric Model • Structural Equation Model • Structural Equation Modeling • Structured Entry Model • Subarray Electronics Module • Subcontractor Engineering Memorandum • Sun - Earth - Moon • Superconducting Electrical Machinery • Survivability Evaluation Methodology • Système d'Entretien Mobile • System Effectiveness Model • System Element Management • System Element Model • System Engineering Management • System Environment Monitor • System Error Message

SEMA Sociedad Española de Matematica Aplicada (Spain) • Société d'Encouragement aux Métiers d'Art (France) • Société d'Études de Mathématiques Appliquées (France) • Software Engineering Measurement and Analysis • Special Electronics Mission Aircraft • Special Equipment Mission Aircraft

SEMAA Science, Engineering, Mathematics and Aerospace Academy (USA)

SEMANE Sortie d'Eaux Méditerranéennes en Atlantique Nord-Est

SEMC Standard Electronic Memory Cartridge • Systems Engineering Management Course

SEMD Stray Energy Monitor Device

SEMDP Senior Executive Management Development Program

SEMECCEL Société d'Économie Mixte d'Exploitation de Centres Culturels, Éducatifs et de Loisirs (France)

SEMECO Sección de Meteoros y Cometas

SEMER Schéma d'Ensemble pour la Météorologie Routière

SEMF Semi-Empirical Mass Formula

SEMG Systems Engineering Management Group

SEMI Semiconductor Equipment and Materials Institute • Special Electromagnetic Interference

Semicnd Semiconductor

semih Semilhora

SEMII Special Electromagnetic Interference Investigation

SEMIR Support, Exploitation, Maintenance, Informatique et Réseaux

SEMIRA System of Electronic Marks' Interrogation, Registration and Administration

SEMIRAD Secondary Emission Mixed Radiation Dosimetry

SEMIS Solar Energy Monitor In Space

SEMLAM Semiconductor Laser Amplifier

SEMLAT Semiconductor Laser Array Technique

SEMM Solar Electric Multiple Mission • Systems Engineering Management Meeting

SEMMP Systems and Equipment Maintenance Program • Systems and Equipment Modification Program

SEMMS Solar Electric Multiple-Mission Spacecraft

SEMOS Special Environment Monitoring System

semp Sempre

SEMP Software Engineering Master Plan • Systems Engineering Management Plan • Systems Engineering Master Plan

SEMPA Scanning Electron Microscopy with Polarization Analysis

SEMS Space Environment Monitor System • Standard Environmental Memory System • Stray Energy Monitor System • Systems Engineering Management Schedule • Systems Engineering Management Standard • Systems Engineering Management Support • Systems Engineering Master Schedule

SEMSQ Société des Éditeurs de Manuels Scolaires du Québec (Canada-QC)

SEMTG Simulation Evaluation Methodology Technical Group

Sen Senate • Senator • Senior • Senza

SEN Satellite Education Network • Scottish Enterprise Network (UK-Scotland) • Senator • Senegal (ISO country code) • Sensor • Small Extension Node • Southend (UK-England) (location identifier) • Special Educational Needs • State-Enrolled Nurse • System Engineering Notice • System Evaluation Network

SENAMHI Servicio Nacional de Meteorología e Hidrología (Bolivia)

SENAMI Servicio Nacional de Meteorología e Hidrología (Peru)

SenAR Sendai Astronomiaj Raportoj (periodical code)

SEND Shared Equipment Need Date

SENDA Special Educational Needs and Disability Act (UK)

SENEAM Servicios a la Navegación en el Espacio Aéreo Mexicano

SENEX Senior Executive

SENL Standard Equipment Nomenclature List

Senr Senior

Sens Sensitive • Sensitivity • Sensor

SENS Sensitive • Small Extension Node Switch

Sensi Sensible

Sent Sentido • Sentier

SENTAR Senior Engineering Technology Assessment Review

Sen. VP Senior Vice-President

SEO Satellite for Earth Observation • Sauf Erreur ou Omission • Seguela (Ivory Coast) (location identifier) • Senior Executive Officer • Senior Experimental Officer • Société Européenne d'Optique • Society of Education Officers • Socio-Economic Objective • Software Executive Official • Special Engineering Order • Special Equipment Operator • Survivability Enhancement Option • Sustaining Engineering Organization • Synchronous Equatorial Orbit • Synchronous Equatorial Orbiter • System Engineering Office

SEOCS Sun Earth Observatory and Climatology Satellite

SEON Solar Electronic Observing Network • Solar Electro-Optical Network • Solar Electro-optical Observing Network (*)

SEOO Sauf Erreur Ou Omission

SEOP System Employment and Organizational Plan

S/EOS Standard Earth Observation Satellite

SEOS Stabilized Electro-Optical System • Standard Earth Observation Satellite • Stationary Earth Orbit Satellite • Synchronous Earth Observation Satellite • Synchronous Earth Observatory Satellite

SEOW Single Engineering Order Wire

Sep Separable • Separate • Separated • Separation • September • Septembre

SEP Safety Enhancement Plan • Scientific and Engineering Personnel • Sclérose En Plaques • SCSI Encapsulated Protocol • Section d'Éducation Professionnelle • Security Evaluation Program • Selective Employment Plan • Separate • Separation • September • Septembre • Serial Entry Printer • Service de l'Électronique Physique • Shuttle Electronics Package • Signal Entry Panel • Signaling End Point • Single-Engine Piston • Single-Event Phenomenon • Société Européenne de Propulsion • Software End Product • Software Engineering Process • Software Enhancement Proposal • Software Environment Process • Solar Electric Power • Solar-Electric-Propelled • Solar Electric Propulsion • Solar Energetic Particle • Solar Energetic Proton • Source Evaluation Panel • South Ecliptic Pole • Southern Equatorial Pacific • Space Electronics Package • Space Experiment Plan • Special Embassy Programme • Specific Excess Power • Spherical Error Precision • Spherical Error Probable • Spherical Error Probability • Spheroidal Equipotential • Standard Eigenvalue Problem • Standard Electronic Package • Standard Engineering Practice • Standard Error of Prediction • Standardization Evaluation Program • Star Epitaxial Package • Stephenville Clark (USA-TX) (location identifier) • Strong Equivalence Principle • Supervisor Executive Program • Support Equipment Package • Support Equipment Plan • Surface d'Énergie Potentielle • Synthetic Environment Program • Systematic Evaluation Program • Systems Effectiveness Plan • Systems Engineering Plan • Systems Engineering Process • Systems Enhancement Package • Systems Enhancement Plan • Systems Enhancement Program • Systems Evaluation Plan

SEPA Science Education Partnership Award • Scottish Environment Protection Agency (UK-Scotland) • Site Engineering Project Authorization • Solar Electric Power Association • Software Engineering Process Group • Southeastern Planetarium Association (USA-FL) • State Environmental Protection Agency • System Evaluation Performance Analysis

SEPAC Space Experiment with Particle Accelerators

SEPAP Shuttle Electrical Power Analysis Program

SEPAR Shuttle Electrical Power Analysis Report

Separat Separativo

SEPC Space Exploration Program Council

SEPD Systems Engineering and Programmatics Department

SEPG Software Engineering Process Group

SEPI Social and Economic Policy Institute • Sociedad Estatal de Participaciones Industriales (Spain) • Society for the Exploration of Psychotherapy Integration

SEPIA Structured Elicitation and Processing of Ideas for Authoring

SEPICA Solar Energetic Particle Ionic Charge Analyzer

SEPIL Selective Excitation of Probe Ion Luminescence

SEPL Set-oriented Picture Programming Language

SEPM Solar Electric Propulsion Module

Sepn Separation

SEPO Space Electric Power Office

SEPOL Settlement Problem-Oriented Language

SEPOR Service des Programmes des Organismes de Recherche

SEPP Secure Electronic Payment Protocol (*) • Secure Encryption Payment Protocol • Single Edge Processor Package • SOFIA Education Partners Program • Solar Electric Primary Propulsion • Support Equipment Program Plan • System Effectiveness Program Plan

SEPR South Ecliptic Pole Region

SEPRA Service d'Expertise des Phénomènes de Rentrées Atmosphériques

SEPS Safety and Environmental Protection Service • Service d'Éducation Physique et Sportive • Solar Electric-Propelled Spacecraft • Solar Electric Propulsion System • Solar Electronic Propulsion System • System Engineering and Programming Support

SEPST Solar Electric Propulsion System Technology

SEPSU Science and Engineering Policy Studies Unit

Sept September • Septembre • Septiembre

SEPT Service d'Étude de la Poste et des Télécommunications • Société Européenne de Programmes de Télévision

SEPTR Solar Energetic Particle Tracer

SEPUP Science Education for Public Understanding Program

seq Sequens • Sequente • Sequitur

Seq Sequel • Sequence • Sequential

SEQ Schedule Estimated Quality • Semantically Equivalent Query • Sequence • Sequencer • Sequential • Sungai Pakning (Indonesia) (location identifier)

SEQAS South East Queensland Astronomical Society (Australia-QLD)

seqq Sequentia • Sequentibus

SEQT Sequential Threshold

SEQUAL Seasonal Equatorial Atlantic Experiment

SEQUEL Structured English Query Language

Ser Serbian • Serial • Serie • Series • Sermon • Serpens-Serpentis (constellation) • Service

Sér Série

SER Safety Evaluation Report • Safety Exercise Report • Scanning Emission Radiometer • Scheduling by Edge Reversal • Secrétariat d'État à la Recherche • Section Efficace Radar • Sequential Events Recorder • Serial • Service Employment Redevelopment • Seymour Freeman (USA-IN) (location identifier) • Significant Event Report • Site Evaluation Report • Solar Extinction Radiometer • Special Environmental Report • State of the Environment Support • Stop End of Runway • Support Equipment Requirement • Surface Electrical Resistivity • Symbol Error Rate • System Engineering Report • System Environment Recording • System Evaluation Report • System Evolution Record

SERA Sequential Electrochemical Reduction Analysis • Software Engineering Review and Audits • Southern (extremity of the) Earth's Rotation Axis

SERAD Space Environment Radiation And Debris (mission)

SERAL Southeastern Region of the Astronomical League (USA)

SERAPHIM Segmented Rail Phased Induction Motor

SERAT Structurally-Embedded Reconfigurable Antenna Technology

Serb Serbia • Serbian

SERB Shuttle Engineering Review Board • Software Engineering Review Board • Southeast Regional Bureau • Space Experiment Review Board • Strategic Education Research Board • Study of the Enhanced Radiation Belt • Support Equipment Review Board • Systems Engineering Review Board

SERBS Surface Energy and Radiation Balance System

SERC Satellite Education Resource Consortium • Scientific and Engineering Research Council • Smithsonian Environmental Research Center • Software Engineering Research Centre • Space Engineering Research Centre • State Emergency Response Commission

SERCH Southeast Regional Clearinghouse (USA)

SERCnet SERC Network

SERD Socio-Économie de la Recherche et du Développement • Stored Energy Rotary Drive • Support Equipment Recommendations Data • Support Equipment Recommendations Document • Support Equipment Requirements Data • Support Equipment Requirements Document • System Engineering Requirements Document

SERDP Strategic Environmental Research and Development Program

SERE Spatial Effective Resolution Element

SERENDIP Search for Extraterrestrial Radio Emission from Nearby Developed Intelligent Populations

SEREP System Environment Recording, Editing and Printing

SERF Science and Engineering Research Facility • Solar Electromagnetic Radiation Flux • Space Environment Research Facility • Study of Energy Release in Flares • System Engineering Research Facility

SERG Saint Empire Romain Germanique

SERI Solar Energy Research Institute • System for Earth Resource Information • Systems Engineering Research Institute

SERIES Satellite Emission Radio Interferometric Earth Surveying • Satellite Emission Range Infrared Earth Surveying

SERIN Synthetic Environments Research Initiative

SERIS State of the Environment Reporting Information System

SERIWA Solar Energy Research Institute of Western Australia

SERL System Engineering Reference Library

SERME Sign Error Root Modulus Error

SERN Single-Expansion Ramp-Nozzle • Southeastern

SERNG Saint Empire Romain de la Nation Germanique

serno Serial Number

SERNO Serial Number

SERO System Engineering Release Order

SERP Software Evaluation and Recognition Program • Sustaining Engineering Requirements Plan

SERPA Société d'Enquête et de Recherche sur les Phénomènes Aérospatiaux (France)

SERPIA Service d'Études et de Réalisation des Produits d'Information Avancés

SERR Software Engineering Risk Repository

SERS Shuttle Enhanced Reconfiguration System • Shuttle Equipment Record System • Software Engineering Release System • Support Equipment Requirements Sheet • Surface-Enhanced Raman Spectroscopy

SERT Service d'Études et de Réalisations Techniques • Single-Electron Rise Time • Society of Electronic and Radio Technicians • Soft Excess Radial Trend • Space Electrical Rocket Test • Space Electronic Rocket Test • System Engineering Requirements Team

SERTIT Service Régional de Traitement d'Image et de Télédétection

SERTS Solar EUV Rocket Telescope and Spectrograph

SERURE Service des Énergies Renouvelables et de l'Utilisation Rationnelle de l'Énergie (France)

Serv Service

SERV Service • Space Emergency Reentry Vehicle

S&Es Scientists & Engineers

SES Satellite Earth Station • Scheduling and Execution System • Sciences Économiques et Sociales • Scientific Editorial Service • Scientific Exploration Society • SCSI Enclosure Services • Secondary Electron Scattering • Second Epoch Survey • Section d'Enseignement Spécialisé • Section d'Éducation Spécialisée • Section d'Études Spécialisées • Secure Equipment System • Selma Selfield (USA-AL) (location identifier) • Senior Executive Service • Service Evaluation System • Severely Errored Seconds • Ship Earth Station • Shuttle Engineering Simulation • Shuttle Engineering Simulator • Signal Enhancement Seismograph • Simulation Environment System • Small Extension Switch • Smart Energy System • Social and Economic Sciences • Société Européenne des Satellites • Society of Engineering Science • Socioeconomic Status • Software Evaluation Subpanel • Software Exploitation Segment • Solar Eclipse Sensor • Solar Energy Society • Solar Environment Simulator • Solutions Evaluation System • Source End Station • Space Environment Simulator • Space Erectable Structure • Special-Emphasis Study • Special Exploitation Service • Spin Echo Spectrometer • Spinning Earth Sensor • Stand-alone Edit Station • STJ-based Echelle Spectrograph • Structure of Earnings Statistics • Sun - Earth - Satellite • Super-Eddington Source • Suprathermal Electron Spectrometer • Surface Effect Ship • System Engineering Specification • System Engineering System • System Evaluation Station • Systems Engineering Simulator

SESA Solar-Energy Society of America • Special Equipment Stowage Assembly

SESAC Space and Earth Sciences Advisory Committee

SESAM Secours dans un Établissement Scolaire face à un Accident Majeur • Surface Effective Sample Monitor • Surface Effects Sample Monitor (*) • System for Electronic Support of Academic Material

SESAME Satellite Experiments Simultaneous with Antarctic Measurements • Search for Excellence in Science and Mathematics Education • Second European Stratospheric Arctic and Mid-latitude Experiment • Service, Sort And Merge • Severe Environmental Storms And Mesoscale Experiment • Solid Earth Science and Application Missions for Europe • Structure and Energetics of the Stratosphere and Mesosphere • Synchrotron-light for Experimental Science and Applications in the Middle East

SESC Shuttle Events Sequential Control • Société d'Éditions Scientifiques et Culturelles (France) • Société d'Étude de la Solidification et de la Cristallogénèse (France) • Software Engineering Standards Committee • Space Environment Services Center

SESD Science and Engineering Systems Division • Software Engineering Systems Division

SESE Software Engineering Support Environment • Space Electronics Support Equipment

SESG Systems Engineering Steering Group

SESI System Engineering and System Integration

SESIG Security Special Interest Group

SESIL System Electronics and Software Integration Laboratory

SESL Self-Erecting Space Laboratory • Space Environmental Simulation Laboratory • Space Environment Simulation Laboratory

SESMI Systems Engineering Support and Management Integration

SESNET Space and Earth Science Network

SESP Space Environmental Satellite Program • Space Experiment Support Program

SESPD Space and Earth Science Programs Directorate

SESQUINET Sesquicentennial Network

Sess Session

SESS Space Environmental Support System • Space Environment Sensor Suite

SESSI Service des Études et des Statistiques Industrielles (France)

SeST Semiconductor Science Technology (periodical code)

SEST School of Engineering, Systems and Technologies • Science and Engineering Support Team • Swedish-ESO Submillimeter Telescope

SeSTS Semiconductor Science Technology Supplement (periodical code)

SESTS Security Evaluation Space Transportation System

set Setup Parameters (file extension)

Set Setembro • Settembre • Settlement

SET San Estebán (Honduras) (location identifier) • Satellite Environmental Terminal • Science Engineering and Technology • Science - Engineering - Technology • Scientists - Engineers - Technicians • Section d'Enseignement Technologique • Secured Encrypted Transaction • Secure Electronic Transaction (*) • Secure Encrypted Transaction • Self-Extending Translator • Sensors and Electronics Technology • Setembro • Short Erection Time • Simplified Engineering Technique • Simultaneous Engineering Technology • Simulated Emergency Test • Single-Electron Transistor • Single-Electron Transport • Single-Electron Tunneling • Single Escape Tower • Software Engineering Technology • Software Engineering Terminology • Solar Energy Thermionic • Spacecraft Elapsed Time • Space Education Trust • Space Electronics and Telemetry • Special Engineering Test • Special Evaluation Test • Specialized English Training • Standard d'Échange et de Transfert • Standard Erection Time • Stepped Electrode Transistor • Stress-Energy Tensor • Study of Exceptional Talent • Système d'Échange et de Transfert • System Engineering Testbed • System Evaluation Threat • Systems Engineering Test

SE/TA Systems Engineering/Technical Assistance

SETA Satellite Electrostatic Triaxial Accelerometer • Science Engineering and Technical Analysis • Scientific Engineering and Technical Assistance • Search for Extraterrestrial Activity • Search for Extraterrestrial Artifacts • Security Education, Training, and Awareness • Simplified Electronic Tracking Apparatus • Systems Engineering and Technical Analysis • Systems Engineering and Technical Assistance • Systems Engineering Technical Assistance

SETAC Society of Environmental Toxicology and Chemistry • System Engineering and Technical Assistance Contract • System Engineering and Technical Assistance Contractor • System Engineering Technical Advancement Center

SETAG Software Engineering Technology Advancement Group

SETAMS Systems Engineering, Technical Assistance, and Management Services

SETAR Serial Event Timer And Recorder

SETAS Space Environments and Technology Archive System

SETC Space and Environment Technology Center

SE&TD System Engineering & Technical Direction

SETD Space Environment Test Division • System Engineering and Technical Direction

SETE Support and Electronic Test Equipment

SETEXT Structure Enhanced Text

SETF Small-Engine Test Facility

Setg Setting

SETHI Système Expérimental de Télédétection Hyperfréquence Imageur

SETI Search for Extraterrestrial Intelligence • Semaine Européenne des Technologies de l'Information

SETIM Service des Équipements et des Techniques Instrumentales de la Météorologie

S-et-L Saône-et-Loire (France)

SETL Science Experiment Test Laboratory

Setlg Settling

S-et-M Seine-et-Marne (France)

SETN Science and Engineering Television Network

S-et-O Seine-et-Oise (France)

SETP Society of Experimental Test Pilots • System Engineering Test Phase

SETR Système d'Exploitation en Temps Réel

SETS Seeker Evaluation Test Simulation • Selected Executable Test Suite • Set Equation Transformation System • Shuttle Electrodynamic Tether System • Solar Electric Test Satellite

Sett Settembre

SETT Systems Extension Training Team

SETTA Search for Extraterrestrial Technological Activities

SETWG Science-Engineering-Technology Work Group (USA)

SEU Scan Electronics Unit • Secrétariat d'État aux Universités • Sensor Electronics Unit • Seronera (Tanzania) (location identifier) • Service des Échanges Universitaires • Sight Electronics Unit • Sighting Electronics Unit • Single-Event Upset • Smallest Executable Unit • Soft Error Upset • Software Engineering Unit • Solar Energy Update (periodical) • Source Entry Utility • Structure and Evolution of the Universe • System Electronics Unit

SEUL Support Equipment Utilization List

seuo Salvo Error u Omisión

SEUS Structure and Evolution of the Universe Subcommittee (NASA)

SEUTWG Structure and Evolution of the Universe Technology Working Group (NASA)

Sev Several • Sevilla (Spain)

SEV Sección de Estrellas Variables • Sensor Equivalent Visibility • Severe • Severodoneck (Ukraine) (location identifier) • Special Equipment Vehicle • Status - Error - Validity • Stockpile Emergency Verification • Surface Effects Vehicle

SEVA Secure Voice Access • Surface Extravehicular Activity

SEVAC Secure Voice Access Console

SEVAS Secure Voice Access System

SEVIRI Spinning Enhanced Visible and Infrared Imager

Sevl Several

SEVMS Security-Enhanced Virtual Memory System

SEVOC Secure Voice

SEVOCOM Secure Voice Communications

SEVP Status - Error - Validity - Parity

SEVPCD Syndicats des Entreprises de Vente par Correspondance et à Distance (France)

SEVPEN Service d'Édition et de Vente des Publications de l'Éducation Nationale (France)

SEVS Simulation and Evaluation System

SEVUPS South Eastern Virginia Urban Plume Study

SEW Scientific Engineering Workstation • Siwa (Egypt) (location identifier) • Software Engineering Workshop • Super-Eddington Wind

SEWB System Engineering Workbench

SEWG Scheduling Efficiency Working Group • Sustaining Engineering Working Group • System Effectiveness Working Group • Systems Engineering Working Group

SEWP Scientific and Engineering Workstation Procurement

SEWS Satellite Early Warning System • Scientific Engineering Workstation • Standard Engineering Workstation • Sun-End Workstation

Sex Sextans-Sextantis (constellation)

SEX Saturn Express • Sembach (Germany) (location identifier) • Software Exchange • Sundays Excepted • Système d'Exploitation

SEXSI Serendipitous Extragalactic X-ray Source Identification

Sext Sextant

SEY Secondary Electron Yield • Selibaby (Mauritania) (location identifier)

Seyf Seyfert

SEYM Secondary Electron Yield Measurement

SEYMS Secondary Electron Yield Measurement System

SEYS Secondary Electron Yield System

SEZ Mahe Island (Seychelles) (location identifier) • Special Economic Zone

s/f Sin Fecha

sf Sin Fecha • Sin Foliar • Snow Flurries • Square Foot • Sub Finem

Sf Suriname Guilder (currency)

S&F Store & Forward

S-F Store & Forward

S/F Safety Factor • Scale Factor • Single Flow • Sound Flash • Static Firing • Store and Forward

SF Finland (international vehicle registration code) • Finland (postal country code) (now FIN) • Safe • Safety Factor • Sampled Filter • San Francisco (USA-CA) • Sans Frais • Santa Fe (Argentina) • Scale Factor • Science Fiction • Science File • Screening Factor • Secondary Failure • Secondary Focus • Security Force • Selection Filter • Senior Fellow • Sensor Fusion • Separation Factor • Service Factor • Service Feature • Shift Forward • Short-term Fluctuation • Shrink Factor • Side Frequency • Signal Frequency • Signaling Frequency • Sign Flag • Simulation Facility • Sin Foliar • Single Flow • Single Fraction • Single Frequency • Skip Flag • Sliding Filter • Small Field • Snow Flurries • Software Factory • Solar Flare • Sound and Flash • Source Function • Space Forum • Special Facility • Special Form • Speculative Fiction • Spontaneous Fission • Spot Face • Spotlight Flash • Square Foot • Standard Form • Standard Frequency • Star Field • Star Formation • Star Forming • Star*s Family (resources) • Start Field • Start to Finish • Static Firing • Status Field • Stock Fund • Store and Forward • Strong Field • Structured Field • Structure Function • Subcontractor-Furnished • Subfield • Subfield Factor • Sub Finem • Successful Flight • Suomen Fysikkoseura (Finland) • Superfast • Superframe • Superfund • Support Facility • Surge Filament • Swiss Franc • Synthetic Ferrimagnet

S&FA Shipping & Forwarding Agent

SFA Sales Force Automation • Scientific Film Association (UK) • Securities and Futures Authority (UK) • Security Fault Analysis • Segmented Flow Analysis • Selling Forces Automation • Service de la Formation Administrative • Sfax (Tunisia) (location identifier) • Single Failure Analysis • Single-Frequency Approach • Small Fine Arm • Spaceport Florida Authority (USA-FL) • Spatial Frequency Analyzer • Special Foreign Analysis • Star Formation Activity • Step Frequency Analyser • Stephen F. Austin • Stereo Field Analyzer • Sun Finder Assembly • Supplementary Failure Analysis • Support Facility Annex • Surface Force and Adhesion • Surface Force Assembly • Surface Forces Adhesion • Survey Feedback Action • Sweet Fanny Adams

SFAA San Francisco Amateur Astronomers (USA-CA)

SFAAA South Florida Amateur Astronomers Association (USA-FL)

SFAC Statement of Financial Accounting Concept

SFACI Software Flight Article Configuration Inspection

SFACT Service de la Formation et du Contrôle Technique (France)

SFAF Standard Frequency Action Format

SFAP Standard Field Accounting Procedure

SFAPS Space Flight Acceleration Profile Simulator

SFAR Special Federal Aviation Regulation (USA) • Star*s Family of Astronomy Resources (now SFARR) • System Failure Analysis Report

SFARR Star*s Family of Astronomy and Related Resources (formerly SFAR)

SFAS Safety Feature Activation System • Standard Field Accounting System • Statement of Financial Accounting Standards

SFASU Stephen F. Austin State University (USA-TX)

SFAWG System Functions and Analysis Working Group

SFB Sanford (USA-FL) (location identifier) • Semiconductor Functional Block • Société Française de Biométrie (France) • Sonderforschungsbereich • Spezialforschungsbereich • Sternfreunde Breisgau (Germany) • Super Fast Build

SFBA Société Française de Bibliométrie Appliquée (France)

SFBI Share Frame Buffer Interconnect

Sfc Surface

SFC Saint-François (Guadeloupe) (location identifier) • Secondary Focus Camera • Selection Filter Control • Shopfloor Control • Smaller - Faster - Cheaper • Smart File Cabinet • Société Française de Chimie (France) • Société Francophone de Classification • Solar Feature Catalog • Space-Filling Curve • Space Flight Center • Space Forecast Center • Specific Fuel Consumption • Standard Field Computer • Standing Finance Committee • Star Field Camera • Subject Field Code • Subject Field Coder • Supercritical Fluid Chromatography • Supervisory Function Control • Surface • Synchronized Framing Camera

SFCC Santa Fe Community College (USA-MN)

SFCF Standby Functional Check Flight

SFCG Space Frequency Coordination Group

SFCS Survival Flight Control System

SFCT Service de Formation des Cadres Techniques

SFD San Fernando de Apure (Venezuela) (location identifier) • Simple Formattable Document • Size-Frequency Distribution • Solar Flux Density • Source Follower per Detector • Staff and Faculty Development • Start of Frame Delimiter • Sudden Frequency Deviation • Switching Field Distribution • System Functional Diagram • System Functional Drawing • System Function Description

SFDF Satellite Field Distribution Facility

SFDI Solar Facility Design Integration • Standard Format for Digital Images

SFDK Stepped-Frequency Delta-K

SFDR Supplementary Format Data Record • Supplementary Formatted Data Record

SFdS Société Française de Statistique (France)

SFDS Secondary Flight Display System • System Functional Design Specification

SFDT Site Format Dump Tape

SFDU Standard Format Data Unit • Standard Formatted Data Unit

SFE San Fernando (Philippines) (location identifier) • Scale Factor Error • Schottky Field Effect • Science Formatting Equipment • Seller-Furnished Equipment • Société Française d'Électrophorèse (France) • Solar-Flare Effect • Solid-Fuel Engine • Star Formation Efficiency • Start Field Extended • Static Feed Electrolysis • Stellar Formation Efficiency • Supercritical Fluid Extraction • System Functional Element

SFEC Single Fuel Element Container

SFEF Standard Format Exchange File

SFEL Standard Facility Equipment List

SFEP Society of Freelance Editors and Proofreaders (UK)

SFERE Société Française d'Exploitation des Ressources Éducatives (France)

SFEX Solar Flare X-ray

SFF Self-Forging Fragment • Small Form Factor • Solar Forecast Facility • Solid Freeform Fabrication • Spokane Felts (USA-WA) (location identifier) • Standard File Format • Static Firing Facility • Sternfreunde Franken eV (Germany)

SFFF Sedimentation Field Flow Fractionation

SFFP Small Form-Factor Pluggable

SFfR Svensk Förening för Radiofysik (Sweden)

SFFS Satellite Freeze Forecast System • Satellite Frost Forecast System

SFFT Short Form Functional Test • Superconducting Flux Flow Transistor

SFG Saint-Martin Espérance (French Antilles) (location identifier) • Special Forces Group • Star Formation Group • Sum Frequency Generation • Synthetic Ground Facility

SFGC Space Frequency Coordination Group

Sfgd Safeguard

SFGEP Space Flight Ground Environment Panel

SFGI Syndicat Français des Grossistes Informatiques

SFH San Felipe (Mexico) (location identifier) • Slow Frequency Hop • Slow Frequency Hopping • Star Formation History

SFHA Special Flood Hazard Area

SFHD System Functional Hierarchical Diagram

SFHT Superfluid Helium Tanker

SFI Safi (Morocco) (location identifier) • Santa Fe Institute (USA-NM) • Schweizerisches Forschungsinstitut (Switzerland) • Seller-Furnished Item • Société Fédérale d'Investissement • Solar Flare Index • Space Flight Instrumentation • Special Flight Instrumentation • Step Function Input • Synthetic Final Index

SFIA San Francisco International Airport (USA-CA)

SFIM Space Filter-based Intensity Modulation

SFINX SRON Fabry-Pérot Interferometer Experiment

SFIR Sequential Filter Imaging Radiometer • Swept Frequency Interferometric Radiometer

SFJ Sondre Stromfjord (Greenland) (location identifier)

SFJTI Société Franco-Japonaise des Techniques Industrielles

SFK SmallTalk Frame Kit • So Far as I Know • Special Function Key

SFL São Felipe (Cape Verde) (location identifier) • Secondary Freon Loop • Sequenced Flashing Lights • Short Focal Length • Société Française du Livre (France) • Structured Function Language • Symbolic Flowchart Language

SFLA Stupid Four-Letter Acronym

SFLC Software First Life Cycle

sfm Surface Feet per Minute

SFM Sanford (USA-ME) (location identifier) • Scanning Force Microscope • Scanning Force Microscopy • Senior Functional Manager • Sky Flux Map • Société Française de Mécaniciens (France) • Solar Flux Monitor • Spectrographe à Fentes Multiples • Spool File Manager • Steering Front Mechanism • Stock Fund Materiel • Structure Framework Model • Sustainable Forest Management • Sysplex Failure Management • Systems Flow Model

SFMA Specific Functional Management Area

SFMC San Fernando Meridian Circle • Satellite Facility Management Center

SFMD Storable Fluid Management Demonstration

SFME Storable Fluid Management Experiment

SFMOMA San Francisco Museum Of Modern Art (USA-CA)

SFMR Stepped Frequency Microwave Radiometer

SFMS Superfast Meteor Stream

SfN Society for Neuroscience

SFN Santa Fe (Argentina-SF) (location identifier) • Single Frequency Network • Society For Neuroscience (USA) • Star Formation Newsletter

SFO San Fernando Observatory (USA-CA) • San Francisco (USA-CA) (location identifier) • Sensor de Frente Onda • Serious Fraud Office • Simulated Flameout • Single-Frequency Outlet • Société Française d'Optique (France) • Solicitation For Offer • Space Flight Operations • Sugitani - Fukui - Ogura (catalog)

SFOC Shuttle Flight Operations Contractor • Space Flight Operations Center • Space Flight Operations Complex • Space Flight Operations Contract

SFOCA Single Fiber-Optic Cable Assembly

SFOD Space Flight Operations Director

SFOF Space Flight Operations Facility

SFOM Shuttle Flight Operations Manual • SOC Facility Operations Manual • Space Flight Operations Manual • Space Flight Operations Memorandum

SFOP Space Flight Operating Procedure • Space Flight Operations Plan • Space Flight Operations Procedure

SFOT Solar Flare Observation Technique

SFOV Sounding Field Of View

SFP Scanning Fabry-Pérot • Sensor Fusion Processor • Service de la Formation Professionnelle • Single Failure Point • Single Function Processor • Smooth Field Particle • Société Française de Physique (France) • Solar Flare Proton • Space Flight Participant • Special Fabrication Program • Special Facilitation Program • Standard Format Part • Standard Form of Payment • Standard Front Page • Steady Force Penetration • Strongly Fluctuating Particle • Summary Flight Plan • Surfers Paradise (Australia-QLD) (location identifier) • Sustainer Fuel Pressure

SFPA Simplified Floating-Point Addition • Single Failure Point Analysis • State Family Program Activity

SFPI Scanning Fabry-Pérot Interferometer

SFpM Surface Feet per Minute

SFPM Surface Feet Per Minute

SFPPL Short Form Provisioning Parts List

SFPS Scanning Fabry-Pérot Spectrograph • Secure Fast Packet Switching • Single Failure Point Summary

SFPT Société Française de Photogrammétrie et Télédétection (France)

SFPTB Bulletin de la Société Française de Photogrammétrie et Télédétection (periodical code)

SFQ Sanliurfa (Turkey) (location identifier) • Single Flux Quantum

SFQL Structured Full-text Query Language

SFr Swiss Franc (currency)

SFR San Fernando (USA-CA) (location identifier) • Selective File Retrieval • Shop Floor Reporting • Société Française de Radioprotection (France) • Société Française de Radiotéléphone (France) • Solar Flare Radiation • Star Formation Rate • Star Forming Region • Swiss Franc (currency) • System Functional Review

s/fra Su Factura

SFRA Star Formation Rate per unit Area • System Fielding Readiness Assessment

SFRC Senate Foreign Relations Committee • Short Form Research Contract

SFRD Secret Formerly Restricted Data • Star Formation Rate Derived • System Functional Requirements Description • System Functional Requirements Document

SFREG Summer Faculty Research and Engineering Group

SFREP Summer Faculty Research and Engineering Program

SFRS Service du Film de Recherche Scientifique (France)

SFS Finnish Standards Association (Finland) • Scholarship For Service • Sequential Forward Selection • Shared File System • Shuttle Flight Status • Société Française de Sociologie (France) • Solar Flux Spectrometer • Special Flights Section • Star Field Sensor • Start Frame Sequence • Strong Feature Subset • Sufficient Feature Set • Suomen Standardisoimisliito (Finland) • Synchronous Fluorescence Spectroscopy • System Failure Summary • System File Server • Systems Fielding Support

SFSA Société Française des Spécialistes d'Astronomie (France) (now SF2A)

SFSM South Florida Science Museum (USA-FL)

SFSN Scripta Faculty Science Nat. Ujep Brunensis Physica (periodical code)

SFSS Satellite Field Service Site • Satellite Field Service Station

SFSTC Services Fédéraux Scientifiques, Culturels et Techniques (Belgium)

SFSU San Francisco State University (USA-CA)

Sft Shaft

SFT Shaft • Short Flight Time • Short Functional Test • Simple File Transfer • Simulated Flight Test • Skellefteå (Sweden) (location identifier) • Smart File Transfer • Société Française de Thermique (France) • Special Flight Test • Standard Formatted Tape • Static Firing Test • Stockton Federation of Teachers (USA-NJ) • Strong Flare of Temper • Structural Fatigue Test • Submerged Floating Tunnel • Supercritical Fluid Technology • Supplément Familial de Traitement • System Fault Tolerance • System Fault Tolerant • System File Table

SFTA Structural Fatigue Test Article

SFTF Static Firing Test Facility

SFTP Shielded Foil Twisted Pair • Simple Fax Transfer Protocol • Simple File Transfer Protocol (*) • Société Française de Photogrammétrie et de Télédétection (France)

SFTS Space Flight Test System • Standard Frequency and Time Signals • Synthetic Flight Training System

SFTSS Standard Frequency and Time Signal Satellite • Standard Frequency and Time Signal Service

SFTSSS Standard Frequency and Time Signal Satellite Service

Sftw Software

SFTWE Software

SFU Safia (Papua New Guinea) (location identifier) • Sample Fixation Unit • Secure Frame Unit • Simon Fraser University (Canada-BC) • Solar Flux Unit • Space Flyer Unit • Specimen Fixation Unit • Standard Firing Unit • Suitable For Upgrade • Summary Fault Unit

SFUG Security Features User's Guide

SFUR Self-Filtering Unstable Resonator

SFV Safety of Fishing Vessels • Santa Fé do Sul (Brazil-SP) (location identifier) • Standardfestverbindung

SFVN Secure Fast Virtual Networking

SFW Santa Fe (Panama) (location identifier) • Software

SFWD San Francisco Water District (USA-CA)

SFWE Static Feed Water Electrolysis

SFWK Spatial Framework

SFWP San Francisco Web Professionals (USA-CA)

Sfx Suffix

SFX San Felix (Venezuela) (location identifier) • Sound Effect • Special Effects • Suffix

SFY Springfield (USA-MA) (location identifier) • State Fiscal Year

SFZ Smithfield North Central (USA-RI) (location identifier) • Sonnenfleckenzyklus

SF2A Société Française d'Astronomie et d'Astrophysique (France) (formerly SFSA) (*) • Société Française des Amateurs d'Astronomie (France)

s/g Su Giro

sg Sangho (ISO language code) • Singapore • Singular

s/G Sin Gastos

Sg Sungei

S/G Salary Grade • Strain Gauge

SG Safety Guide • Sa Grâce • Sa Grandeur • Salary Grade • Scenario Generation • Screen Grid • Secrétaire Général • Secrétariat Général • Secretary General • Senior Grade • Seyfert Galaxy • Shain-Gaze (catalog) • Ship and Goods • Signal Generator • Signal Ground • Signaling Gateway • Singapore (ISO country code) • Singapour (code ISO de pays) • Small Gain • Snow Grains • Soldatengesetz • Solicitor General • Special Grade • Specific Gravity • Spinglass • Standing Group • StarGuides (directory) • Starlink Guide • Stern-Gerlach (force) • Stochastic Gradient • Strain Gauge • Student Guide • Study Group • Subgiant • Subgroup • Superconducting Gravimeter • Super Guppy • Supergroup • Surgeon General • Symbol Generator • Systems Generation

SG&A Selling, General & Administrative

SGA Science Goal Analyzer • Security Guard Activity • Shared Global Area • Sheghnan (Afghanistan) (location identifier) • Shocked Graphite Aggregate • Simple Genetic Algorithm • Slave Gyro Assembly • Société de Géologie Appliquée • Spectrometric Gas Analysis • Standards of Grade Authorization • Suppress Go Ahead • Système de Gestion des Abonnés • System Global Area

SGAA Schweizerische Gesellschaft für Astrophysik und Astronomie (Switzerland)

SGAC Space Generation Advisory Council

SGAE Sociedad General de Autores y Editores (Spain)

SGAF Serviço de Gestão Administrativa e Financeira

SGAICO Swiss Group for Artificial Intelligence and Cognitive Science (Switzerland)

SGAM Service de Groupement des Achats de Matériel

SGAP Secrétariat Général pour l'Administration de la Police (France)

SGAPO Space Geodesy and Altimetry Project Office

SGAR Secrétariat Général aux Affaires Régionales

SGASAA Standard Generic Approach to Spacecraft Autonomy and Automation

SGASU Secrétariat Général d'Administration Scolaire et Universitaire

SGATC Second-Generation Advisory Technical Committee

SGB Service Géologique de Belgique (Belgium) • SIG Governing Board • Singaua (Papua New Guinea) (location identifier) • Sozialgesetzbuch • Standard Geographic Base • Subgiant Branch (*) • Supergaussian Beam • Supergiant Branch • Système de Gestion Bord

SGBD Système de Gestion de Base de Données

SGBDOO Système de Gestion de Base de Données Orienté Objet

SGBDR Système de Gestion de Base de Données Relationnelle

SGBDT Système de Gestion de Base de Données Techniques

SGC SCSI Graphic Command • Solar Great Circle • Southern Galaxy Catalog • South Galactic Cap • Space Geodesy Centre • Stabilized Ground Cloud • Stabilizer Gyro Circuit • Superior Geocentric Conjunction • Surgut (Russia) (location identifier)

SGCI Secrétariat Général du Comité Interministériel

SGCM Simplified General Circulation Model

SGCN Secrétariat Général du Comité National

S-GCOS Space-based Global Change Observation System

SGCOS Space-based Global Change Observation System

SGCP Simple Gateway Control Protocol

SGCR Subcommittee on Global Change Research

SGCS Slave Gyro Control System

SGCT Science Goal Capture Tool

Sgd Signed

SGD Dollar (Singapore) (ISO currency code) • Signed • Solar Geophysical Data (journal) • Sønderborg (Denmark) (location identifier) • Symbolized Graphic Data

SGDE Sustainable Gross Domestic Expenditure

SGDF Supergroup Distribution Frame

SGDG Sans Garantie Du Gouvernement

SGDN Secrétariat Général à la Défense Nationale (France)

SGDP Sustainable Gross Domestic Product

SGDT Store Global Descriptor Table • Système de Gestion des Données Techniques

Sge Sagitta-Sagittae (constellation)

SGE Secondary Grid Emission • Simulation Graphics Environment • Standard General Education • Strategische Geschäftseinheit • Système de Gestion de l'Environnement • Système de Gestion Environnementale

SGED Système de Gestion Électronique de Documents

SGEMP Space-Generated Electromagnetic Pulse • System-Generated Electromagnetic Pulse

SGEN Signal Generator • System Generator

SGEOS Second Generation Earth Observation Satellite

SGEP Signaux Géophysiques et Environnements Planétaires

SGEPSS Society of Geomagnetism and Earth, Planetary and Space Sciences (Japan)

SGER Small Grant for Exploratory Research (NSF)

SGES Specialist Group on Expert Systems (BCS)

SGET Spacecraft Ground Elapsed Time

SGF Screen Generation Facility • Second-Generation FLIR • Solution Growth Facility • Space Generation Forum • Space Generation Foundation (USA-CA) • Spaceguard Foundation (USA) • Springfield (USA-MO) (location identifier) • Synthetic Ground Facility • Système de Gestion de Fichiers

SGFGAS Société de Gestion des Fonds de Garantie de l'Accession Sociale

SGFI Schweizerische Gesellschaft zur Förderung der Informatik (Switzerland)

SGFS Special Group on Functional Standardization

SGG Satellite Gravity Gradiometry • Simanggang (Malaysia) (location identifier) • Sozialgerichtsgesetz

SGGA Species-Generating Generic Algorithm

SGGM Superconducting Gravity Gradiometer

SGGS Satellite Geodesy and Geodynamics Systems (JPL)

SGH Simpson Gumpertz & Heyer Inc. • Springfield (USA-OH) (location identifier)

SGHL Schweizerische Gesellschaft für Hydrologie und Limnologie (Switzerland)

sgi Silicon Graphics Inc. Image (file extension)

SGI Sargodha (Pakistan) (location identifier) • Science Group Identifier • Silicon Graphics Inc. • Single Zone Instruction • Small Group Instruction • Specific Gravity Indicator • Sun-Glint Imager

SGIM Shape Globe Invariance Map

SGITS Spacecraft/GOSS Interface Test System

SGJ Sagarai (Papua New Guinea) (location identifier)

SGJP Satellite Graphic Job Processor

SGK Sangapi (Papua New Guinea) (location identifier) • Schweizerische Geologische Kommission (Switzerland) • Schweizerische Gesellschaft für Kristallographie (Switzerland) • Small Generating Kernel

Sgl Single

SGL Manila - Sangley Point (Philippines) (location identifier) • Schweizerische Gesellschaft für Limnologie (Switzerland) • Service Géologique du Luxembourg • Small Group Leader • Société des Gens de Lettre (France) • Space Geodynamics Laboratory • Standard General Ledger • Surface Gridding Library

SGLDM Spatial Grey Level Dependence Matrix

SGLS Satellite-Ground Link System • Space-Ground Link Station • Space-Ground Link System • Space-to-Ground Link System

SGLSY Space Ground Link System

SGM San Ignacio (Mexico) (location identifier) • Schweizerische Gesellschaft für Meteorologie (Switzerland) • Science Goal Monitor • Science Goal Monitoring • Second-Generation Meteosat • Second-Generation Model • Segmentation Message • Shaded Graphics Modeling • Shape Group Method • Special General Meeting • Spherical Grating Monochromator

SGMC Science Goal Monitoring Console • Supergiant Molecular Cloud

SGML Standard Generalized Markup Language (*) • Standard General Markup Language

SGMP Simple Gateway Management Protocol • Simple Gateway Monitoring Protocol

SGMS Satellite Geodetic Measuring Site • Space Geodetic Measurement Site

SGMT Simulated Greenwich Mean Time

SGN Ho Chi Minh City - Tan Son Nhat (Viet Nam) (location identifier) • Service de Géodésie et du Nivellement • Starburst Galactic Nucleus

SGNL Signaling

SGO Saint George (Australia-QLD) (location identifier) • Satellite Geodetic Observatory • Sodankylä Geophysical Observatory (Finland) • Sugar Grove Observatory (USA-IL)

SGOAA Space Generic Open Avionics Architecture

SGOEM Schweizerische Gesellschaft für Optik und Elektronenmikroskopie (Switzerland)

SGOMSEC Scientific Group on Methodologies for the Safety Evaluation of Chemicals

SGOS Shuttle Ground Operations Simulation • Shuttle Ground Operations Simulator

SGP Serviços Geológicos de Portugal • Shay Gap (Australia-WA) (location identifier) • Singapore (international vehicle registration code) • Singapore (ISO country code) • Singapour (code ISO de pays) • Single Gauss Point • Single Grounding Point • Single Ground Point • Southern Great Plains (USA) • South Galactic Pole • Space Geodesy Program • Space Geodesy Project • Spatial General Perturbation • Standard Genetic Programming • Standard Guidance Package • Starlink General paper • Static Ground Point • Structure Ground Point • Supergalactic Plane

SGPBF Schweizerische Gesellschaft für Photogrammetrie, Bildanalyse und Fernerkundung (Switzerland)

SGPI Service de Gestion des Programmes et d'Ingénierie

SGpK Schweizerische Geophysikalische Kommission (Switzerland)

SGPPV Système de Gestion du Pay-Per-View

Sgpr Singapore

SGPS Southern Galactic Plane Survey

SGPW Schweizerische Gesellschaft für Pflanzenbauwissenschfaten (Switzerland)

SGQ Sanggata (Indonesia) (location identifier)

Sgr Sagittarius-Sagittarii (constellation)

SGR Houston Hull (USA-TX) (location identifier) • Science and Government Reports • Seismic Group Recorder • Set Graphics Rendition • Soft Gamma-ray Repeater • Soft Gamma Repeater • Specific Growth Rate • Sugar Land - Hull (USA-TX) (location identifier)

SGRA School of Galactic Radio Astronomy

SGRAM Synchronous Graphics Random Access Memory

SGRBE Swift Gamma-Ray Burst Explorer

SGRS Spaceborne Geodynamics Ranging System

SGS Georgia and South Sandwich Islands (South_) (ISO country code) • Sandwich Islands (South Georgia and Sandwich_) (ISO country code) • Sanga Sanga (Philippines) (location identifier) • Satellite Ground Station • Science Ground Segment • Secondary Ground Station • Segmented Gamma Scanner • Sequence Generation System • Sequence Generator System • Simple Graphics System • Sixth Generation Systems • Smoke Generation System • Société Générale de Surveillance • Société Géophysique Suisse (Switzerland) • Software Generation System • Solar Generator System • South Georgia and South Sandwich Islands (ISO country code) • South Sandwich Islands (South Georgia and_) (ISO country code) • Space Gateway Support (USA-VA) • Statistics Gathering System • Steep Glide Slope • Subgrid Scale • Supergiant Shell • Symbol Generation and Storage • Symbol Generator and Storage

SGSC Strain Gauge Signal Conditioner

SGSE Standard Ground Support Equipment

SGSI Stabilized Glide Slope Indicator

SGSM Subgrid Scale Mixing (model) • Subgrid Scale Modeling

SGSN Serving GPRS Support Node

SGSO Space Ground Support Operations

SGSP Subgroup on Standard Procedures

SGSSA Small Ground Station Segmented Antenna

SGSTS Space-to-Ground System Test Set

SGT Satellite Ground Terminal • Society of Glass Technology • Software-Guided Telescope • Stochastic Growth Theory • Stuttgart (USA-AR) (location identifier) • Subgroup on Time

SGTC Scene Generation Test Capability

SGTE Scientific Group Thermodata Europe

SGtK Schweizerische Geotechnische Kommission (Switzerland)

SGTR Standardized Government Travel Regulations

SGts StarGates (database)

SGU Saint George (USA-UT) (location identifier) • Signal Generator Unit

SGV Sierra Grande (Argentina-RN) (location identifier)

SGW Saginaw Bay (USA-AK) (location identifier)

SGX Songea (Tanzania) (location identifier) • Spacecraft Glow Experiment • Supergalactic X-coordinate

SGY Skagway (USA-AK) (location identifier) • Supergalactic Y-coordinate

SGZ Songkhla (Thailand) (location identifier) • Supergalactic Z-coordinate

s/h Shipping/Handling • Shorthand

sh Hyperbolic Sine • Serbo-Croatian (ISO language code) • Shall • Share • Sheet • Shilling (currency) • Showers

Sh Shall • Share • Sheet • Shilling • Shower

S-H Staff Hour

S/H Sample/Hold

S&H Safety & Health • Sample & Hold • Shipping & Handling

SH Île Sainte-Hélène (code ISO de pays) • Saint Helena (ISO country code) • Sample and Hold • School House • Second Hand • Sequence History • Session Header • Shack-Hartmann • Shift • Short Haul • Shoshoni • Showers • Southern Hemisphere • Speckle Holography • Speech Handler • Staff Hour • StarHeads (database) • State Highway • Subset Header • Sulfhydril • Superhump • Système Harmonisé

sha Unix Shell Archive (file extension)

SHA Sample-and-Hold Amplifier • Secure Hash Algorithm • Secure Hashing Algorithm • Semantic Hypermedia Architecture • Shanghai (China-PRC) (location identifier) • Sidereal Hour Angle • Society for the History of Astronomy (UK-England) • Spherical Harmonic Analysis • Station Housing Allowance • Surface Habitable • System Hazard Analysis

SHAB Schweizerisches Handelsamtsblatt

ShaBaR Shadow Band Ranging

SHABAR Shadow Band Ranging

SHAD Stellar Horizon Atmospheric Dispersion • Stellar Horizon Attitude Determination

SHADE Saharan Dust Experiment • Shared Data Environment • Share Dependency Engineering

SHAEF Supreme Headquarters Allied Expeditionary Force

SHAG Simplified High-Accuracy Guidance

Shak William Shakespeare (1564-1616)

Shaks William Shakespeare (1564-1616)

ShAO Shamakhy Astrophysical Observatory (Azerbaijan)

SHAO Shanghai Astronomical Observatory (China-PRC)

SHAPE Second-generation Hypermedia Application Programming Environment • Second-generation Hypermedia Application Project Environment • Supreme Headquarters of Allied Powers in Europe • Systematic Horizontal Adjustment of Positional Errors

shar Unix Shell Archive (file extension)

SHAR Shell Archive • Shell Archiver • Sriharikota High Altitude Range (India) • Stellar Horizon Atmospheric Refraction

SHARAD Shallow-subsurface (sounding) Radar

SHARC Serendipitous High-redshift Archival ROSAT Cluster • Shared Hierarchical Academic Research Computing (network) • Submillimeter High-Angular Resolution Camera

SHARE Safety and Hazard Assessment Research in Europe • Satellites in Health And Rural Education • Software Help for Applications, Research, and Education • Space-station Heat-pipe Advanced Radiator Element • Study of the Hubble Archives and Reprocessing Enhancements • System for Handicap Accessibility, Recreation and Education (NSAAC)

SHARED Satellite Health Access for Remote Environment Demonstrator

SHARES Sharing of Experience in Space

SHARK Standard HRTP Archiving and Reprocessing Kernel • Standard HRTP Archiving and Retrieving Kernel

SHARMOR Spectrographe Haute-Résolution Meudon-Orsay

SHARP Slender Hypervelocity Aerothermodynamic Research Probe • Spacecraft Health Automated Reasoning Prototype • Space Human Assurance Reliability Program • Standard Hardware Acquisition and Reliability Program • Standard HRTP Archiving and Request Product • Stationary High-Altitude Relay Platform • Summer High-school Apprenticeship Research Program • Super High Altitude Research Project • System for High-Angular-Resolution Pictures

SHARPP Solar Heliospheric Activity Research and Prediction Program

SHASA Servicios Aéreos de Honduras SA • Small High-Altitude Science Aircraft

SHASSA Southern H-Alpha Sky Survey Atlas

SHAVE Shared Virtual Environment

SHB Nakashibetsu (Japan) (location identifier) • Second-Harmonic Band • Spectral Hole

SHBD Spectral Hole Burning Device

SHC Shire Indaselassie (Ethiopia) (location identifier) • Specific Heat Capacity • Spherical Harmonic Coefficient • Spontaneous Human Combustion • Superheterodyne Controller • Supreme High Command

SHCDS Sharing Control Data Set

ShCG Shakhbazian Compact Group (of galaxies)

Shd Should

SHD Shadow • SOCS High-energy Detector • Special Handling Designator • Staunton (USA-VA) (location identifier) • Strahlungshydrodynamik • Superhigh Definition

SHDF Sliding Hinge Double Fold

SHDG Semantic-Head-Driven Generation (algorithm)

SHDM Seeded Hot Dark Matter • Superheavy Dark Matter

SHDN Shutdown

S&HDS Safety & Health Data Sheet

SHDS Safety and Health Data Sheet

S-HDSL Single-line High-bit-rate Digital Subscriber Line

SHDSL Single-line High-bit-rate Digital Subscriber Line • Symmetric High-bit-rate Digital Subscriber Line

s/he She or He

SHe Supercritical Helium

S/HE Sundays and Holidays Excepted

SHE Self-Help Exercise • Service d'Histoire de l'Éducation • Sheffield (Starlink internal code) • Shenyang (China-PRC) (location identifier) • Signal Handling Equipment • Special Handling Equipment • Spherical Harmonic Expansion • Standard Hydrogen Electrode • Supercritical Helium

SHEAL Shuttle High-Energy Astrophysics Laboratory

SHEBA Surface Heat Budget of the Arctic (ocean)

SHED Segmented Hypergraphics Editor

SHEEP Search for the High-Energy Extragalactic Population

Shef Sheffield (UK)

SHEF Standard Hydrologic Exchange Format

SHEFC Scottish Higher Education Funding Council

Sheff Sheffield (UK)

SHELF Superhard Extremely Low Frequency

SHEM Shared Memory

SHEMAT Simulator for Heat and Mass Transport

SHER Spectroscopic High-Energy Resolution

SHERB Sandia Human Error Rate Bank • Spectroscopic High-Energy Resolution Burst

SHERPA Sources de Haute Énergie Relativistes et Processus d'Accrétion-éjection Structurée

Shet Shetland Islands

SHET Soleil, Héliosphère et Environnement Terrestre

Shetl Shetland Islands

SHEVE Southern Hemisphere VLBI Experiment

SHEX Sundays and Holidays Excepted

SHF Shanhaiguan (China-PRC) (location identifier) • Short History File • Short-term History File • Superhandling Facility • Superhigh Frequency

SHFE Space Human Factors Engineering

SHFP Space Human Factors Program

SHFS Superhyperfine Structure

SHFT Shift

SHG Second-Harmonic Generation • Segmented Hypergraphics • Shungnak (USA-AK) (location identifier) • Sozialhilfegesetz • Spectroheliograph

SHGN Service Hydrologique et Géologique National (Switzerland)

SHGS Service Hydrologique et Géologique Suisse (Switzerland)

SHH Shishmaref (USA-AK) (location identifier)

SHHA Safety and Health Hazard Assessment

Shkh Shakhbazian (catalog)

SHI Shimojishima (Japan) (location identifier)

SHIC Social History and Industrial Classification

SHIE Spacecraft Health Insurance Engine

SHIEP Schur-Horn Inverse Eigenvalue Problem

SHIMMER Spatial Heterodyne Imager for Mesopheric Radicals

SHIMS System Hazard Information Management System

SHINE Solar, Heliospheric, and Interplanetary Environment

SHIP Self-Help Issue Point • Single Hazardous Input Program • Standard Hardware Interface Program

SHIPDA Shipping Date

SHIPDTO Ship on Depot Transfer Order

SHIPGO Shipping Order

SHIPIM Ship Immediately

Shipt Shipment

SHIRL Serial Holdings in Irish Libraries

SHIS Scanning High-resolution Interferometer Sounder

SHIVA SDSS Human Interface and Visualization

SHJ Sharjah (UAE) (location identifier)

SHK Schweizerische Hochschulkonferenz (Switzerland) • Sehonghong (Lesotho) (location identifier) • Sternwarte Heiligkreuz

SHL Science Header Line • Semi-Hard Lander • Shift (logical) Left • Shillong (India) (location identifier) • Shoal

SHLB Simulation Hardware Load Box

Shld Shield

SHLD Shield

Shldr Shoulder

SHLDR Shoulder

SHLS Shoals

Shltr Shelter

SHLW Shallow

SHM Nanki Shirahama (Japan) (location identifier) • Safety and Health Manager • Scalar Helium Magnetometer • Semantical Hierarchy Model • Short Hold Mode • Shutter Mechanism • Simple Harmonic Motion • Space Hydrogen Maser • Spectromètre Hétérodyne Submillimétrique • Stage Handling Manual

SHME Single High Magnification Event

SHML Southern Hemisphere Mid-Latitude

SHN Île Sainte-Hélène (code ISO de pays) • Saint Helena (ISO country code) • Servicio de Hidrografía Naval • Shelton Sanderson (USA-WA) (location identifier) • Shorthand Note

SHNT Shunt

SHO Scheduled Order • Service Hydrographique et Océanographique • Simosato Hydrographic Observatory (Japan) • Sokch'o (Korea-ROK) (location identifier)

SHOA Servicio Hidrográfico y Oceanográfico de la Armada (Chile)

SHOALS Scanning Hydrographic Operational Airborne Lidar Survey

SHOB Surface Hors-Oeuvre Brute

SHOE Shadow-Hiding Opposition Effect

SHOFM Système Hydrologique Opérationnel à Fins Multiples

SHOL Ship Helicopter Operational Limitation • Shoal

SHOLAS Simosato Hydrographic Observatory Laser Ranging System

SHOM Service Hydrographique et Océanographique de la Marine (France)

SHON Surface Hors-Oeuvre Nette

SHOOT Shuttle Helium On-Orbit Transfer • Superfluid Helium On-Orbit Transfer (*)

SHORAD Short-range Radar

SHORAN Short-Range Aid to Navigation • Short-Range Navigation

SHOS Segmentation Hiérarchique basée sur l'Optimisation Séquentielle

SHOT Society for History Of Technology • Space Hardware Optimization Technology Inc. (USA-IN)

shp Shaft Horsepower

SHP Pound (Saint Helena) (ISO currency code) • Qinhuangdao (China-PRC) (location identifier) • Science Header Packet • Selective Hydrogenation Process • Separate Heat and Power • Shaft Horsepower • Standard Header Packet • Subharmonically Pumped • Subhost Processor • Subsequent Handling Piece • Sustainer Hydraulic Pressure

SHPE Society of Hispanic Professional Engineers

SHPF Standard Header Packet File

Shpg Shipping

SHPM Scanning Hall Probe Microscope

Shpng Shipping

SHPO State Historic Preservation Officer

Shpt Shipment

SHQ Southport (Australia-QLD) (location identifier)

Shr Share

SHR Sheridan (USA-WY) (location identifier) • Shift (logical) Right • Shore • Solar-Heat Reflecting • Superheavy Relic • Superheterodyne Receiver

SHRD Shroud • Superhigh-Resolution Display • Supplemental Heat Rejection Device

SHRF Ship Regular Freight

SHRI Spectrometer for Heavy Recoil Ions • Stereo High Resolution Imager

SHRIF System for Heuristic Retrieval of Information and Facts

SHRIMP Sensitive High-Mass Resolution Ion Microprobe • Shallow-water Real-time Imaging Polarimeter

SHRIP Stereo High-Resolution Imager Panchromatic

SHRM Society for Human Resource Management

Shrops Shropshire (UK-England)

SHRS Shores

SHRTLY Shortly

SHS Sciences de l'Homme et de la Société • Sciences Humaines et de la Société • Sciences Humaines et Sociales • Secure Hash Standard • Service d'Hygiène et de Sécurité • Shack-Hartmann Sensor • Shashi (China-PRC) (location identifier) • Simulation Hardware System • Spatial Heterodyne Spectrometer • Spatial Heterodyne Spectroscopy • Special Hardware System

SHSA Self-Help Support Agreement

SHSC Safety Hazard Severity Code

SHSG Spherical Harmonic Spatial Grid

SHSM System High Security Mode

SHSU Sam Houston State University (USA-TX)

Sht Sheet

SHT Shepparton (Australia-VIC) (location identifier) • Solar Horizontal Telescope • Space Hand Tool

ShTC Short Time Constant

Shtdn Shutdown

Shtg Shortage

shtml Secure Hypertext (file extension)

SHTR Shutter

SHTTP Secure Hypertext Transfer Protocol (*) • Secure Hypertext Transport Protocol

SHTU Simplified Handheld Terminal Unit

SHU Seton Hall University (USA-NJ) • Smith Point (Australia-NT) (location identifier)

SHUD Smart Head-Up Display

SHUTDN Shutdown

shv Sub Hac Voce • Sub Hoc Verbo

Shv Sheave

SHV Safe High Voltage • Shreveport (USA-LA) (location identifier) • Standard High Volume

SHW Safety - Health - Welfare • Sharurah (Saudi Arabia) (location identifier) • System High Workstation

ShWav Shock Waves (periodical code)

SHWFS Shack-Hartmann Wavefront Sensor

SHWM System High Water Mark

Shwr Shower

SHWR Shower

SHX Shageluk (USA-AK) (location identifier)

SHY Shinyanga (Zimbabwe) (location identifier)

SHZ Seshutes (Lesotho) (location identifier)

SH2 Supercritical Hydrogen

si Sauf Imprévu • Singhalese (ISO language code) • Sin Impresión • Slovenia • SoftImage Image (file extension)

Si Silicium • Silicon

S&I Standardization & Interoperability • Surveys & Investigations

S/I Signal/Interference • Speaker/Interrupt • System Integration

SI Mimas (satellite of Saturn) • Safety Inspection • Sample Implementation • Sample Interval • Sandwich Islands •

Science Instrument • Scientific Information • Scientific Instrument • Search Information • Secondary In • Security Intelligence • Selected Item • Selective Identification • Semi-Insulating • Sensitive Information • Sensitive Intelligence • Sensitivity Index • Senter for Industriforskning • Sequenced Information • Seriously Ill • Seriously Injured • Service de l'Intendance • Service d'Information • Service Indicator • Service Instructeur • Set Indicator • Shetland Islands • Shift-In • Shipping Instructions • Shuttle Integration • Siberia • Side Information • Signal Interface • Signature Invoke • Site Inspection • Situational Interrupt • Skill Identifier • Slovenia (ISO country code) • Slovénie (code ISO de pays) • Smithsonian Institution (USA-DC) • Social Information • Societatis Iesu • Society Island • Society of Indexers (UK) • Software Item • Solar Inertial • Solar Influence • Solubility Index • Source Impedance • Source Index • South Island (New Zealand) • Space Invariant • Spark-Ignited • Spark Ignition • Spark Injection • Sparse Image • Speaker Independent • Special Instruction • Special Intelligence • Special Interest • Special Item • Specific Impulse • Speckle Interferometer • Speckle Interferometry • Spectrum Index • Square Inch • Staff Inspector • Staten Island (USA-NY) • Station Identification • Status Indicator • Statutory Instrument • Stellar Imager • Stewart Island (New Zealand) • Stokes Inversion • Storage Immediate • Straight In • Stream Interface • Sudden Impulse • Sum Insured • Support Installation • Sustainable Indicator • Symplectic Integrator • Syndicat d'Initiative • Système d'Information • Système International • System Information • System Integration • System Integrator

SIA Satellite Industry Association (USA-VA) • Science Image Access • Science Information Association • Self-Interstitial Atom • Semiconductor Industry Association (USA) • Sensor Interface Assembly • Serial Input Adapter • Serial Interface Adapter • Service d'Information Aéronautique • Shuttle Induced Atmosphere • Social Impact Assessment • Société des Ingénieurs de l'Automobile • Software Impact Assessment • Solar Inertial Attitude • Southern Iota Aquarids (meteor stream) • Speaker Intercom Assembly • Special Interface Adapter • Structural Interface Adapter • Subminiature Integrated Antenna • Sustainability Impact Assessment • System Information Agent • System Integration Area • Systems Interconnect Architecture • Xi'An Xiguan (China-PRC) (location identifier)

SIAB Société Industrielle d'Automatisation Bancaire (France)

SiAc Silicon Accelerometer

SIAC Science Instrument Aperture Coordinates • Southeastern Iowa Astronomy Club Inc. (USA-IA) • Spatial Information and Analysis Consortium • Spectrally-Identifiable Ammonia Cloud

SIAD Système Interactif d'Aide à la Décision (*) • Système Interactif d'Aide et de Décision

SIADIS Sistema de Interpretación Automática De Imágenes de Satélite

SIAE Salon International de l'Aéronautique et de l'Espace (France)

SIAF Science Instrument Aperture File

SIAG SIAM Activity Group • Survey Instrument Azimuth Gyro

SIAGL Survey Instrument Azimuth Gyro Lightweight

SIAM Separate Index Access Method • Shipboard Ice Alert and Monitoring • Signal Information And Monitoring • Society for Industrial and Applied Mathematics (USA-PA) • Système d'Information et d'Aide pour les Mutations (France) • System Integrated Access Method

SIAMR Society for Industrial and Applied Mathematics Review (periodical code)

SIAP Single Integrated Air Picture • Standard Instrument Approach Procedure • Statistical Institute for Asia and the Pacific • System Integrity Assurance Program

SIAPP System Integrity Assurance Program Plan

SIAR Scientific Instrument Analysis Room

SIAS Scientific Instrument Aperture System

SIASP Society for Italian-American Scientists and Physicians (USA)

SIASRNR Scientific and Industrial Association for Space Research of Natural Resources

SIAT Scene Image Annotation Tape • Shuttle Independent Assessment Team (NASA) • Special Image Annotation Tape

Sib Siberia • Siberian

SIB Safety Investigation Board • Satellite Information Base • Satellite Integrated Buoy • Satellite Ionospheric Beacon • Science Instrument Background • Science Instrument Branch • Scientometrics - Informatics - Bibliometrics • Security Information Base • Service Independent Block • Ship Information Book • Ship Information Booklet • Sibiti (Congo) (location identifier) • Simulation Interface Buffer • Small Innovative Business • Sociedade Interplanetaria Brasileira (Brazil) • Software Information Base • Solar Indices Bulletin • Special Investigation Branch • Status Indication Busy • Systems Integration Board • Systems Interface Bus

SIBAN Sistema Informativo Biblioteche Ateneo Napoletano • Sofia Izdatel Bolgarskoi Akademii Nauk (periodical code)

SIBEX Submillimeter Infrared Balloon Experiment

SibIZMIR Siberian Institute of Terrestrial Magnetism, Ionosphere and Radio Wave Propagation (Russia)

SIBO Single Board Organiser

sic Siccus

Sic Sicilian • Sicily

SiC Silicon Carbide • Siliziumkarbit

SIC Science Information Council • Scientific Information Center • Second In Command • Semantic Integrity Constraint • Semiconductor Integrated Circuit • Send Interface Clear • Service de l'Information et de la Communication • Service Initiation Charge • Set Initial Condition • Sextant Installation Correction • Silicon Integrated Circuit • Simulated Interface Calibration • Sinop (Turkey) (location identifier) • Software Integration Contractor • Solar Information Center • Spacecraft Identification Code • Special Interest Committee • Specialized Information Center • Specific Impact Contribution • Standard Industrial Classification • Standard Industrial Code • Standard Industry Code • Standard Instrument Controller • Standards Information Center • Subject Identification Code • Subject Index Classification • Subject Index Code • Subject Indicator Code • Subject Information Center • Subscriber Interface Controller • Survey Information Center • Synchronization and Imaging Converter • Synchronization Image Channel • Système d'Informations Commerciales • Systems Integration Contractor

SICA Secondary Inventory Control Activity • Secondary Item Control Activity • Structural Interface Compatibility Assessment

SICAR Sea Ice Chart Archival

SICAS SSR Improvement and Collision Avoidance System

SICAV Société d'Investissement à Capital Variable (France)

SICC Service Item Control Center • Surveying Instrumentation and Coordinate Computation

SICCIM System Input from Command and Control to Information Management

SICD System Interface Control Document

SI/C&DH Science Instrument Control & Data Handling

SICI Serial Item and Contribution Identifier

SICL Standard Instrument Control Library

SICLOPS Scanner Imagery Correction and Line-by-line Orientation Program

SICM Scientific Instrument Central Module • System for Integrated Contract Management

SICMT Space Industry Centre for Microwave Technology (Australia)

SICP Selected Ion Current Profile • Specifying and Implementing Conversation Policies (workshop) • Standard Integrated Command Post

SICR Selected Item Configuration Record • Specific Intelligence Collection Requirement

SICS Scientific Instrument Coordinate System • Scientific Instrument Corrected System • Swedish Institute of Computer Science (Sweden)

SICSP Space Industry Centre for Signal Processing

SICT Serviço de Informação Científica e Técnica

SICWG SOHO Intercalibration Working Group

Sid Sidérurgie

SID Sal (Cape Verde) (location identifier) • Scanned Image Data • Search In Depth • Secondary Imagery Dissemination • Security Identifier • Selective Imagery Dissemination • Sequence Information Data • Serial Input Data • Service d'Information et de Diffusion • Service d'Information et de Documentation • Session Identification • Session Identificator • Session Identifier • Shuttle Integration Device • Silicon Imaging Device • Simulator Interface Device • Single Integrated Database • Singular Isothermal Disk • Situation Information Display • Society for Information Display (USA) • Society for International Development • Software Implementation Directive • Software Interface Drawing • Sound Interface Device • Source Identification • Source Identifier • Source Image Distortion • Spatially Invariant Deconvolution • Specification Interpretation Document • Specification Interpretation Documentation • SRON - IAPG - DEOS (consortium) • Standard Information Display • Standard Instrument Departure • Standard Interface Document • Standard Items Database • Station Identification • Station Identifier • Stored Ion Device • Structure Identifier • Subscriber Identification • Sudden Ionospheric Disturbance • Surface-Induced Disorder • Surface-Induced Dissociation • Symbolic Interactive Debugger • Syntax Improving Device • Système d'Information Distribué • Système d'Information Douanier • System for Information on Demand • System Identification • System Identifier • System Information Directory • System Integration Development • System Interactive Debugger • System Interface Document • System of Information and Documentation

SIDA Swedish International Development Agency (Sweden) • Système Informatisé de Documentation Automatique

SIDAR Selective Information Dissemination And Retrieval

SIDB Simulation Interoperability Database • Single Integrated Database

SIDC Solar Influences Data (analysis) Center • Space Industry Development Centre • Sunspot Index Data Centre (*)

SIDD Scientific Information and Documentation Division

SIDENT Site Identification

SIDES Source Input Data Edit System

SIDF System-Independent Data Format

SIDH Sociedad Interamericana de Derechos Humanos

SIDIP Simple Digital Photogrammetry

SIDL Synchronous Interface Definition Language • System Identification List

SIDM Self-Interacting Dark Matter (model)

SIDN Satellite Imagery Dissemination Network • Selected Item Description Number

SIDP Smaller Interplanetary Dust Particle

SIDPC Service Interministériel de Défense et de Protection Civile

SIDPQ SID Proceedings Quarter (periodical code)

SIDR Secure Intelligence Data Repository

SIDS Satellite Image Display System • Satellite Imagery Dissemination System • Scientific Instrument Detector System • Screening Information Data Set • Secondary Imagery Dissemination System • Self-Inking Data Stamp • Space Investigations Documentation System • Stellar Inertial Doppler System • Support Integrated Data System

SIDT Science Instrument Definition Team • Store Interrupt Descriptor Table

SIE Science Information Exchange • Scientific Information Exchange • Secondary Ion Emission • Selected Information Exchange • Self-Initiated Elimination • Shuttle Interface Equipment • Simple Imaging Extension • Sines (Portugal) (location identifier) • Single Instruction Execute • Society of Industrial Engineers • Sounding Interactive Evaluation • Special Inspection Engineering • Special Inspection Equipment • Standards Improvement Executive • Status Indication Emergency • Surface Ionization Engine • Systems Integration Engineer • Systems Integration Environment • Systems Integration Equipment • Systems Investigation Equipment

SIEA Sensor Interface Electronics Assembly

SIEB Security Impact Evaluation Board

SIEC Service Interacadémique des Examens et Concours (France)

SIECA Summer Institute for Engineering and Computer Applications (USA)

Siegerl Siegerland (Germany)

SIEP Screening Inspection for Electronic Parts • Service d'Information sur les Études et les Professions • Structured Inverse Eigenvalue Problem

SIES Service d'Information Économiques et Statistiques • Supervision - Inspection - Engineering - Services

SIESIP Seasonal to Interannual Earth Science Information Partner

SIESTA Simulation of Interferometric Detector Sensitive to Gravitational Waves

SIF Salon International de la Franchise • Satellite Integration Facility • Science Information Facility • Science Instrument Formatter • Scientific Instrument Formatter • Selective Identification Facility • Selective Identification Feature • Signaling Information Field • Simra (Nepal) (location identifier) • Simulation Information Format • Società Italiana di Fisica (Italy) • Software Integration Facility • SOGS Integration Facility • Sound Intermediate Frequency • Source Input Format • Spannungsintensitätsfaktor • Standard Input Format • Standard Interchange Format • Standards Information Program • Station Information Frame • Status Information Frame • Step Index Fiber • Stress-Intensity Factor • Symbol Information File • Systems Integration Facility

SIFDT Selected Ion Flow Drift Tube

SIFE Second ISLSCP Field Experiment • Stage d'Insertion et de Formation à l'Emploi

SIFES SAR Image Feature Extraction System

SiFoE Siemens Forschungs und Entwicklungsberichte (periodical code)

SIFP System Implementation and Facility Plan

SIFS Short Interframe Space • Standard Industrial Fund System • Supply Interface System

SIFT Selected Ion Flow Tube • Share Internal Fortran Translator • Software Implemented Fault Tolerance • Stanford Information Filtering Tool

SiFTZ Sibirskii Fiziko Tekhnicheskii Zhurnal (periodical code)

SIFUS SOAR Integral Field Unit Spectrograph

sig Signal • Signaler • Signalman • Signature (file extension) • Signetur • Siguiente • Siguientes

Sig Signal • Signal (periodical code) • Signature • Signification • Signifies • Signor • Signore

SIG San Juan - Isla Grande (Puerto Rico) (location identifier) • Senior Interdepartmental Group • SERIN Infrastructure Group • Service d'Information du Gouvernement • Service d'Instrumentation Générale • Signal • Signaling • Signature • Signor • Sistema de Informações Geograficas • Special Interest Group • Strasbourg-Illkirch-Graffenstaden (France) • Système d'Informations de Gestion • Système d'Informations Géographiques • Système d'Informations Géophysiques • Système Intégré de Gestion

SIGACT Special Interest Group on Automata and Computability Theory (ACM)

SIGADA Special Interest Group on Ada (ACM)

SIGAL Système Intégré de Gestion Automatisée du Livre (Belgium)

SIGAPL Special Interest Group on APL (ACM)

SIGAPP Special Interest Group on Applied Computing (ACM)

SIGARCH Special Interest Group on Computer Architecture (ACM)

SIGART Special Interest Group on Artificial Intelligence (ACM)

SIGB Standard Interface Glovebox

SIGBIO Special Interest Group on Biomedical Computing (ACM)

SIGBIT Special Interest Group on Business Information Technology (ACM)

SIGCAPH Special Interest Group on Computers and the Physically Handicapped-Print Edition (ACM)

SIGCAS Special Interest Group on Computers and Society (ACM)

SIGCAT Special Interest Group on CD-ROM Applications and Technology (ACM)

SIGCG Special Interest Group on Computer Graphics (ACM)

SIGCHI Special Interest Group on Computer and Human Interaction (ACM)

SIGCOMM Special Interest Group on Data Communications (ACM)

SIGCPR Special Interest Group on Computer Personnel Research (ACM)

SIGCSE Special Interest Group on Computer Science Education (ACM)

SIGCUE Special Interest Group on Computer Uses in Education (ACM)

SIGC2 Signal Command and Control

SIGDA Special Interest Group on Design Automation (ACM)

SIGDOC Special Interest Group on Documentation (ACM)

SIGFORTH Special Interest Group on Forth (ACM)

SigG Signaturgesetz

SIGGRAPH Special Interest Group on Computer Graphics (ACM)

SIGH Système d'Information Géographique Hydro-océanique

SIGHT Scintillation Imaging Gas-filled Hard X-ray Telescope

SIGICE Special Interest Group on Individual Computing Environments (ACM) (formerly SIGSMALL/PC)

sigill Sigillum

SIGINT Signal Intelligence

SIGIR Special Interest Group on Information Retrieval (ACM)

SIGLE System for Information on Grey Literature in Europe

SIGLINK Special Interest Group on Hypertext/Hypermedia (ACM)

SIGLIT Special Interest Group on Library Information Technology (USA)

SIGMA Science In General Management • Scientist's Intelligent Graphical Modeling Assistant • Stochastic-Integration Global Minimization Algorithm • Système d'Imagerie Gamma à Masque Aléatoire • Système d'Information Géographique Maritime • Système Informatique de Gestion du Matériel

SIGMET Significant Meteorological Phenomenon

SIGMETRICS Special Interest Group on Measurement and Evaluation (ACM)

SIGMICRO Special Interest Group on Microprogramming (ACM)

SIGMOD Special Interest Group on Management of Data (ACM)

Sign Signature

SIGN Special Interest Group Network (ACM)

Signa Signorina

Signal Signalétique

sig. n. pro. Signa Nomine Proprio

SIGNUM Special Interest Group on Numerical Mathematics (ACM)

SIGOIS Special Interest Group on Office Information Systems (ACM)

SIGOPS Special Interest Group on Operating Systems (ACM)

SIGP Scientific Instrument Gas Purge

SIGPLAN Special Interest Group on Programming Languages (ACM)

SigPr Signal Processing (periodical code)

Sigra Signora

SIGRAV Società Italiana di Relatività Generale e Fisica della Gravitazione (Italy)

SIGS Simplified Inertial Guidance System • Stellar Inertial Guidance System • Synthetic Image Generator System • Système d'Information Géographique du SHOM

SIGSAC Special Interest Group on Security, Audit and Control (ACM)

SIGSAM Special Interest Group on Symbolic and Algebraic Manipulation (ACM)

SIGSEC Special Interest Group for Secure Electronic Commerce

SIGSIM Special Interest Group on Simulation (ACM)

SIGSMALL/PC Special Interest Group on Small and Personal Computing Systems and Applications (ACM) (now SIGICE)

SIGSOFT Special Interest Group on Software Engineering (ACM)

SIGTRANS Signal Transmissions

SIGUCCS Special Interest Group on University and College Computing Services (ACM)

SIGWEX Significant Weather

SIG3C Special Interest Group for Computing at Community Colleges • Special Interest Group on Computing in Community Colleges (*)

SIH Silgadi Doti (Nepal) (location identifier) • Système Informatique Hospitalier

SIHI Sampled Images - High dispersion (IUE)

SIHRS Système d'Information Hydrogéologique à Référence Spatiale

SIHSPAI Société Internationale d'Histoire des Sciences et des Philosophies Arabes et Islamiques

SIHT Space Impact Hand Tool

SII Enceladus (satellite of Saturn) • Scalable Information Infrastructure • Science Information Infrastructure • Seiko

Instruments Inc. (Japan) • Seriously Ill or Injured • Servicio de Impuestos Internos • Sidi Ifni (Morocco) (location identifier) • Space Industries Inc. (USA-TX) • Special Instructions Indicator • Special Interest Item • Standards Institution of Israel • Static Invocation Interface • Suprathermal Ion Imager • System Internal Interface

SIIA Software and Information Industry Association (USA-DC)

SIIB Sistema Integrado de Informação Bibliográfica da Região Centro (Portugal)

SIIC Service Item Control Center • Special Interest Item Code

SIICT Sistema Internacional de Información Científica y Ténica

SIID Système Interactif d'Interrogation à Distance

SIIG Security Issue and Insight Group • Service Interuniversitaire d'Informatique et de Gestion • Site Installation/Implementation Guide

SIII Tethys (satellite of Saturn)

SIIMOD Scalable Intelligent Image Module

SIIN Solicitation Instrument Identification Number

SIIP Site Installation and Integration Plan • Site Integration and Installation Plan • Standard Information Interchange Process

SIIR Smithsonian Institution Information Retrieval (system)

SIIRS Smithsonian Institution Information Retrieval System

SIIS Sea Ice Information System

SIIT Stateless Ip/ICMP Translation • Stateless Ip/ICMP Translator

SIJ Siglufjördur (Iceland) (location identifier)

SIK Sikeston (USA-MO) (location identifier)

Sil Silence • Silver

SIL Safety Information Letter • Sample In Liquid • Satellites International Ltd. • Scanner Input Language • Schedule Input Line • Scientific Interchange Language • Selected Item List • Sila (Papua New Guinea) (location identifier) • Silence • Silent • Silver • Simulation Integration Laboratory • Single In Line • Societas Internationalis Limnologiae • Solid Immersion Lens • Sound Interference Level • Speech Interference Level • Sudden Iron Layer • Survivability Integration Laboratory • Systems Implementation Language • Systems Integration Laboratory

SiLA Six-Letter Acronym

SILEX Semiconductor (laser) Intersatellite Link Experiment

SILMA Structur Interne Lancement Multiple Ariane

SILO Sampled Images - Low dispersion (IUE) • Sealed-Interface Lateral Oxidization • Sealed-Interface Local Oxidization • Séjour International Linguistique et Culturel • Signal Intercept from Low Orbit • Signal Location

SILOON Scripting Interface Language for Object-Oriented Numerics

SILS Satellite Interference Location System • School of Information and Library Science • Signal for Interoperable LAN Security • Standard for Interoperability LAN Security

SILTS Shuttle Infrared Leeside Temperature Sensing

sim Simile

Sim Similar • Similarly • Simulate • Simulated • Simulation • Simulator

SIM Satellite Interpretation Message • Scaling Index Methode • Science Instrument Module (*) • Scientific Instrument Module • Scientific Instrumentation Module • Secondary Ion Microscope • Secondary Ion Microscopy • Seismometer • Selected Ion Monitoring • Sensor Interaction Model • Serial Interface Module • Service Information Manager • Simbai (Papua New Guinea) (location identifier) • Simple Inhomogeneous Model • Simulate • Simulation • Simulator • Sistema Interamericano de Metrología • Society for Information Management • Solar Interplanetary Model • Solar Irradiance Measurement • Solar Irradiance Monitor • Space Interferometry Mission • Special Interaction Model • Stellar Image Monitor • Spin-Independent Model • Subscriber Identification Module • Subscriber Identify Module • Subscriber Identity Module (*) • Support Interface Module • Survey Information on Microfilm • Synchronous Interface Module • Systems Integration and Management • Systems Integration Model

SIMA Secondary-mirror Incremental Motor Actuator • SIAM Journal on Mathematical Analysis (periodical code) • Small Incremental Motion Actuator • Systems Integration Management Activity

SIMAI Società Italiana di Matematica Applicata e Industriale (Italy)

SIMAN Simulations Manager • System for Interactive Multispectral Analysis

SIMAP Système d'Information pour les Marchés Publics

SIMAS Shuttle Information Management Accountability System

Símb Símbolo

SIMB Service d'Installation et de Maintenance des Balises

SIMBA SEST Imaging Bolometer Array • Solar Irradiance Monitoring from Balloons

SIMBAD Set of Identifications, Measurements and Bibliography for Astronomical Data • Sistema Municipal de Bases de Datos

SIMBIOS Sensor Intercomparison for Marine Biological and Interdisciplinary Ocean Studies

SIMC Surveillance Infrastructure Management Concept

SIMCEN Simulation Center

SimCom Simulation Complex

SIMD Salon International du Marketing Direct • Single Instruction - Multiple Data

SIMDOG Space Interferometry Mission Dynamical Observations of Galaxies

SIMECA Simulations pour Étoiles Chaudes Actives

SIMEX Simulation and Extrapolation • Simulation Exercise

sIMF Standard Initial Mass Function

SIMFAC Simulation Facility

SIMKO Société Immobilière de Kourou (French Guyana)

SIMM Single In-line Memory Module • SME Internal Market Monitor • Standard In-line Memory Module • Symbolic Integrated Maintenance Manual • System-Integrated Memory Module

SIMNET Simulation Network

SIMO Stage d'Initiation aux Méthodes d'Observation • Systems Integration Management Office

SiMol Single Molecules (periodical code)

SIMOP Simultaneous Operating

SIMOX Separation for Implantation of Oxygen

simp Specific Impulse

SIMP Satellite Information Message Protocol • Systems Integration for Major Projects

SIMPAC Simplified Procurement Acquisition • Simplified Programming for Acquisition and Control

SIMPL Scientific, Industrial, and Medical Photographic Laboratories • Synoptic Interplanetary Monitor Platform

SIMPLE Semi-Implicit Method for Pressure-Linked Equation • Spatial and Inductive Modeling Package for Land Evaluation

SIMPLEX Shuttle Ionospheric Modification with Pulsed Localized Exhaust • Standard Incubator Module for Portable Late-access Experiment

SIMPO Simulation Promotion Office

SIMPR Structured Information Management, Processing and Retrieval

SIMR Systems Integration Management Review

SIMS School of Information Management and Systems • Sea Ice Monitoring Site • Secondary Ion Mass Spectrometer • Secondary Ion Mass Spectrometry • Secondary Ion Mass Spectroscopy • Secondary Ion Microprobe Spectrometer • Selected Item Management System • Selective Interference Modulation Spectrometer • Services Information Management System • Shuttle Imaging Microwave System • Shuttle Inventory Management System • SIAM Institute for the Mathematics and Society • Single Item - Multisource • Stellar Inertial Measurement System • Support Information Management System • Symbolic Integrated Maintenance System • Synthesis, Integration and Modeling Studies

SIMSat Software Infrastructure for Modeling Satellites

SIMSAT Simulation Infrastructure for Modeling Satellites

SIMSWG Space Interferometry Mission Science Working Group

SIMTECH Simulation Technology

SIMTEL Simulation and Teleprocessing

SIMU Simulated Inertial Measurement Unit

Simul Simulation (periodical code) • Simultaneous

SIMULA Simulation

SIMURIS Solar-system Interferometric Mission for Ultrahigh-Resolution Imaging and Spectroscopy • Stellar Interferometric Mission for Ultrahigh-Resolution Imaging and Spectroscopy

SIMVAL Simulation Validation

SIMWG Simulation Working Group • Site Management Working Group

SIMWIN Simulated Workstation Input

SIMWOUT Simulated Workstation Output

sin Sine • Sinecure • Sinestra • Sinus

SiN Silicon Nitride

SIN Sensitive Information Network • Servicio de Inteligencia Nacional • Sinaloa (Mexico) • Sine • Singapore Changi (Singapore) (location identifier) • Status Indication Normal • Study Item Number • Superconducting - Insulator - Normal • Support Information Network • Swedish Ionosonde Network

SINAP Satellite Input to Numerical Analysis and Prediction

SINAPS Système Intégré d'Acquisition par Satellite • System of Integrated Acquisition Procedures for Satellite

SINAS Sir Isaac Newton Astronomical Society (USA-PA)

SINC Site of Important Nature Conservation

SINCHARS Single-Channel Radio System

SIND Satellite Inertial Navigation Determination

SINDA Systems Improved Numerical Differences Analyzer • Systems Improved Numerical Differencing Analyzer • Systems Integrated Numerical Differencing Analyzer

SINE Spacecraft-Induced and Natural Environment • Summer In Northern Europe

SINET Schlumberger Information Network

SINEX Software-Independent Exchange (format) • Solution-Independent Exchange (format)

SINFONI Single Far-Object Near-infrared Investigation

Sing Singapore • Singular • Singulorum

SINGG Search for Ionization in Neutral-Gas Galaxies (project) • Survey for Ionization in Neutral-Gas Galaxies (NOAO)

SINGS SIRTF Nearby Galaxies Survey

sinh Hyperbolic Sine • Sinus Hyperbolic

SiNI Silicon Nitride

SINP Saha Institute of Nuclear Physics (India)

SINS Shipboard Inertial Navigation System • Ships' Inertial Navigation System • Sphère Isotherme Non-Singulière • Supernova Intensive Study

SINSS Shipboard Ice Navigation Support System

SINTEF Stiftelsen for Industriell og Teknisk Forskning (Norway)

SINUM SIAM journal on Numerical Analysis (periodical code)

SiO Silicon Monoxide

S&IO Security & Intelligence Office

SIO Scalable Input/Output • Scheduling Interoperability • Scripps Institution of Oceanography (USA-CA) • Security and Information Operations • Security and Intelligence Office • Senior Intelligence Officer • Serial Input/Output • Service Information Octet • Service Informatique de l'Observatoire • Session d'Information et d'Orientation • Smithton (Australia-TAS) (location identifier) • Special Information Operations • Staged In Orbit • Standard Installation Organization • Supporting Intelligence Office • Systems Integration Office • Systems Integration Organization

SIOE Service Information-Orientation-Emploi

SIOH Supervision - Inspection - Overhead

SION Stichting Informatica Onderzoek in Nederland (Netherlands)

SIOP Selector Input/Output Processor • Single Integrated Operational Plan • Single Integrated Operations Plan (*)

SIORC Senior Information Operations Review Council

SI&P System Integration & Planning

SIP Satellite Information Processor • Scanner Interface Processor • Science Implementation Plan • Scientific Instruction Processor • Scientific Instrument Package • Sección de Investigaciones Policiales • Secondary Improvement Program • Segment Interface Protocol • Semi-Infinite Programming • Separation Instrumentation Package • Separation Instrument Package • Serial Interface Port • Serial Interface Protocol • Service Identification Packet • Session Initiation Protocol • Short Implementation Plan • Short Irregular Pulsation • Signal/Image Processing • Simferopol (Ukraine) (location identifier) • Simulated Input Processor • Simulated Input Profile • Single-Image Pattern • Single Inhibition Process • Single-In-line Package • Single-Investigator Project • Sky Image Processor • SMDS Interface Protocol • SOC Implementation Plan • Software Implementation Plan • Software Instrumentation Package • Software Integration Plan • SOGS Invocable Procedure • Solar and Interplanetary Program • Solar Influence Program • Solar Instrument Probe • Solar Irradiance Photometer • Sputter Ion Pump • Standard Inspection Procedure • Standard Interest Profile • Standard Interface Panel • Standard Interface Plan • Strain Isolation Pad • Strain Isolator Pad • Strategic Information Plan • Structural Integrity Program • Student Internship Program • Submission Information Package • Supplemental Income Plan • Symbolic Input Program • System Improvement Plan • System Improvement Program • System Integration and Planning • System Interface Processing • Systems Implementation Plan • Systems Information Package • Systems Integration and Protection • Systems Integration Plan

SIPAD Système d'Information, de Présentation et d'Accès aux Données

SIPB Student Information Processing Board • Scientific Information Policy Board (CERN)

SIPC Simply Interactive Personal Computer • Simulated Information Processing Center • Stationing and Installations Planning Committee

SIPE Science Instrument Payload Enclosure • Science Instrument Payload Envelope • Science Instrument Protective Enclosure • Science Instrument Protective Envelope • Scientific Instrument Payload Enclosure • Scientific Instrument Payload Envelope • Scientific Instrument Protective Enclosure • Scientific Instrument Protective Envelope

SIPES Society of Independent Professional Earth Scientists

SIPHES Sistema de Predicción de Heladas por Satélites

SIPI Southwestern Indian Polytechnic Institute (USA-NM)

SiPM Silicon Photomultiplier

SIPM Service Integration Program Manager • System Integration Program Manager

SI&PO System Integration & Planning Office

SIPO Serial In - Parallel Out • Spacecraft Integration Project Office • System Integration and Planning Office • System Integration Project Office

SIPOP Satellite Information Processor Operational Program

SIPP Simple Internet Protocol Plus • Single In-line Pinned Package • Single In-line Pin Package • System Information Processing Program

SIPPEREC Syndicat Intercommunal de la Périphérie de Paris pour l'Électricité et les Réseaux de Communication (France)

SIPR Selected Item Program Review • Special In-Progress Review

SIPRI Stockholm International Peace Research Institute (Sweden)

SIPROS Simultaneous Processing Operating System

SIPS Satellite Image Processing System • Satellite Imagery Processing System • Satellite Input Processing System • Simulated Input Preparation System • Small Instrument Pointing System • SOGS Invokable Procedures • Spacelab Input Processing System • Spatial Information Processing System • Spectral Image Processing System • State Implementation Plan System

SIPSF Space-Invariant Point Spread Function

SIPT Société Internationale de Photogrammétrie et de Télédétection • Standard Installation Practice Training

SIPWR System Integrated Pressurized Water Reactor

SIQ Service d'Innovation et de la Qualité • Singkep (Indonesia) (location identifier)

SIQA Sistema Integrado da Qualidade do Ambiente (Portugal)

SIR Safe Integral Reactor • Savings-Investment Ratio • Scientific Information and Retrieval • Scientific Information Retrieval • Selective Image Reconstruction • Selective Item Reporting • Selective Information Retrieval • Semantic Information Retrieval • Semi-annual Inventory Report • Sequential Importance Resampling • Serial Infrared (link) • Serious Incident Report • Service d'Informations Rapides • Service Informatique Régional • Shuttle Imaging Radar • Simultaneous Impact Rate • Signal-to-Interference Ratio • Signature Identification Record • Sion (Switzerland) (location identifier) • Small Income Relief • Soft Insertion Rehearsal • Software Initiated Review • Source-Intermediate Ratio • Spaceborne Imaging Radar • Space Imaging Radar • Special Information Retrieval • Special Inspection Requirement • Special Investigative Requirement • Specialized Information Retrieval • Specific Information Requirement • Spectral Irradiance Response • Standard Interface Record • Standard Interface Requirement • Standardization Interface Record • Statistical Information Retrieval • Stokes Inversion based on Response function • Subsurface Interface Radar • Superconductive Integrated Receiver • Surveys, Investigations, and Research (USGS) • Sustainable Information Rate • Sustained Information Rate • Symbolic Input Routine • Système Informatisé de Réservation • Système International de Référence • System Initialization Routine • System Interface Requirement • System Investigation Request • Systems-Integrated Receiver • Systems Integration Review

SIRA Scientific Instrument Research Association • Stable Isotope Radio Analysis

SIRAS Spaceborne Infrared Atmospheric Sounder

SIRC Science Information Research Center

SIRCA Simulador Reactivo de Conducción de Automoviles • Stockholm Infrared Camera

SIRD SOC Implementation Requirements Document • Support Instrumentation Requirements Document • System Interface Requirements Document

SIRDS Single-Image Random Dot Stereogram

SIRE Satellite Infrared Experiment • Space Imagery Reception and Exploitation • Space Infrared Experiment • Stichting Ideële Reclame (Netherlands)

SIREFF Swenson - Irwin - Rogers - Eggleton - Faulkner - Flannery (formalism)

SIREN Società Italiana Reti Neuroniche (Italy)

SIREV Sector Imaging Radar for Enhanced Vision • SIAM Review

SIRGAS Sistema de Referencia Geocéntrico para América del Sur

SIRH Système d'Information des Ressources Humaines

SIRI Service Intégré de Ressources d'Information

SIRICE Submillimeter Infrared Ice Cloud Experiment

SIRIEL Service Informatique de Renseignements Internationaux Électroniques

SIRIM Synthetic Infrared Imaging Model • Standards and Industrial Research Institute of Malaysia

SIRIS Stratospheric Infrared Interferometric Spectrometer

SIRIUS Simultaneous (three-color) Infrared Imager for Unbiased Surveys

SIRM Saturation Isothermal Remanence Magnetization • Scanning Infrared Microscopy

SIRMO Senior Information Resource Management Officer

SIRN Società Italiana di Reti Neuroniche (Italy)

SIROG Situation-Related Operational Guidance

SIROS Solar and Infrared Radiation Observation Station

SIRP Service de l'Information et des Relations Publiques

SIRQ Session Initiation Request

SIRR Software Integration Readiness Review

SIRRA Sleep Induction - Rapid Reawakening

SIRRM Standard Infrared Radiation Model

SIRS Satellite Infrared Radiation Spectrometer • Satellite Infrared Sounder • Satellite Infrared Spectrometer • Selective Information Retrieval System • Session Initiation Response • Single Integrated Reporting System • Soils Information Retrieval System • Southern International Reference Stars • Statistical Information Retrieval System • Status Inventory Reporting System • Système d'Information à Référence Spatiale • System for Information Retrieval and Storage • System Integration Receiver System

SIRST Shipboard Infrared Search and Track

SIRT Simultaneous Iterative Reconstruction Technique • Synchronous Intelligent Remote Terminal

SIRTF Shuttle Infrared Telescope Facility • Space Infrared Telescope Facility (*)

SIRU Strapdown Inertial Reference Unit

SIRUS Système Intégré de Réception et d'Utilisation des Satellites

Sis Sister

SiS Silicon Sulfide

SIS Safety Interlock System • Satellite Intercept System • Scanning Image Spectrometer • Scanning Imaging Spectrometer • Schengen Information System • School of Information Science • School of Information Systems • Schwer Ionen Synchrotron • Science Information Service • Science Information Systems (NASA) • Scientific Information System • Scientific Instrument System • Secondary Items Study • Selective Inquiry System • Semantic Index System • Semiconducteur - Isolant - Semiconducteur • Semiconductor - Insulator - Semiconductor • Serial Input Special • Shimano Index System • Shock Isolation System • Shorter Interval Scheduling • Short-Interval Scheduler • Shuttle Imaging Spectrometer • Shuttle Information Service • Shuttle Information System • Shuttle Integration Schedule • Shuttle Interface Simulator • Simulation Interface System • Simulator Interface System • Singular Isothermal Sphere • Sishen (South Africa) (location identifier) • Sky Indexation Scheme • Society for Interdisciplinary Studies (UK) • Software Implementation Specification • Software Integrated Schedule • Software Interface Specification • Solar Irradiance Spectrometer • Solar Isotope Spectrometer • Solar Isotope System • Solid-state Imaging Spectrometer • Sound-In-Syncs • Space and Information System • Spatial Information Services • Spatial Information System • Speaker Intercom System • Special Information System • Special Intelligence System • Special Interest Section • Specialized Information System • Specification Information System • Sphère Isotherme Singulière • SPOT Imaging Services Pty. Ltd. (Australia-NSW) • Standard Indexing System • Standard Information System • Standard Interface Specification • Standardiseringskommissionen i Sverige (Sweden) • Statens Institut for Stralehygiejne (Denmark) • Steam Inerting System • Strategic Information System • Strategy Implementation Study • Structure Information Store • Subarcsecond Imaging Spectrograph • Superconducteur - Isolant - Superconducteur • Superconducting - Isolating - Superconducting • Superconductor - Insulator - Superconductor • Super Index System • Super Instrumentation System • Superleiter - Isolator - Superleiter • Swedish Standards Institution (Sweden) • Système d'Information Schengen • Système Informatique de Santé • System Installation Suite • System Integration and Support • Systems Integration Schedule • Systems Integration Support

SISA Science and Infrastructure Support Agency • Science and Infrastructure Support Analysis

SISAC Serials Industry Advisory Committee

SISAM Spectrometer with Interference Selective Amplitude Modulation

SISB Scientific Instrument Support Branch

SISC Science Information System Center • Standard Information System Center

SISD Scientific Information Systems Department • Simulation Interface and Standards Document • Single Instruction - Single Data • Software and Information Systems Division

SISE Science Instrument Systems Engineer • Système d'Information de Suivi des Étudiants

SISEX Shuttle Imaging Spectrometer Experiment

SISFET Semiconductor-Insulator-Semiconductor Field Effect Transistor

SISI Surveillance and In-Service Inspection

SISIC Science Information Systems Interoperability Conference

SISIR Singapore Institute of Standards and Industrial Research

SISM Standard Information Systems Manager

SISMS Standard Integrated Support Management System

SISN Science Information Systems Newsletter (NASA)

SISO Serial In - Serial Out • Shift In - Shift Out • Simulation Interoperability Standards Organization • Single Input - Single Output

SISP Standard Information Systems Panel • Stellar Imager and Seismic Probe • Sudden Increase of Solar Particles • Support Information Systems Panel • Surface Imaging and Sounding Package • System Interconnection Security Policy • System Interface Support Plan

SISPRE Società Italiana per lo Studio della Propulsione a Reazione (Italy)

SISR Selected Item Status Report • Standard Information Systems Review

SISS Scientific Instrument Support Structure • Single Item - Single Source • Standoff Imagery Sensor System • System Integration Support Service

SISSA Scuola Internazionale Superiore di Studi Avanzati (Italy)

SISSR Akademiia Nauk SSSR Sibirskoe Otdelenie Izvestiia Seriia Tekhnicheskie Nauki (periodical code)

SIST Serviceable In Service Time • Society for Imaging Science and Technology • Space Interferometry Study Team • Stellar Interferometry Study Team

SISTER Special Institution for Scientific and Technological Education and Research

SISTERS Space Interferometer for the Search for Terrestrial Exoplanets by Rotation Shearing

SISTO Software and Intelligent Systems Technology Office

SISWG Space Interferometry Science Working Group • STS Integrated Schedule Working Group

SISZ South Iceland Seismic Zone

Sit Sitting • Situate • Situation

SI&T System Integration & Test

SIT Satellite Interactive Terminal • Scientific Imaging Technology • Sección de Instrumentación y Taller • Self-Ignition Temperature • Self-Induced Transparency • Sequential Interval Timer • Service de l'Inspection du Travail • Service d'Information Technique • Service d'Information Topographique • Shuttle Integrated Test • Shuttle Interface Test • Silicon-Intensified Target • Silicon Intensifier Target • Simulation Input Tape • Single Instrument Test • Singular Isothermal Toroid • Site Investigation Team • Sitka (USA-AK) (location identifier) • Situation • Small-to-Intermediate-size Telescope • Small-to-Intermediate Telescope • Society of Industrial Technology • Software-Integrated Test • Software Integration Test • Space Impact Tool • Special Information Tone • Special Interest Target • Spectroimageur Transportable • Spontaneous Ignition Temperature • SSV Integrated Test • Statement of Inventory Transaction • Stopping In Transit • Storing In Transit • Strategic Implementation Team • Structural Integrity Test • Structured Inventive Thinking • Stuff It • Syndicat des Industries de Télécommunication • Système Interbancaire de Télécompensation (France) • System Implementation Team • System Initialization Table • System Integration and Test • System Integration Team • System Integration Test • System Integration Testing • System Integrator Toolkit • Systems Interface Test • Tolar (Slovenia) (ISO currency code)

SITA Société Internationale de Télécommunications Aéronautiques

SITB System Integration and Testing Branch

SITC Signal-Integrated Test Cycle • Single-Integrated Test Cycle • Standard International Trade Classification

SITCEN Situation Center

SITE Satellite Instructional Television Experiment • School In The Exploratorium • School of Information Technology and Engineering • Search Information Tape Equipment • Sensor and Instrument Test and Evaluation • Single Ion Track Electronics • Society for Information Technology and Teacher Education • Spacecraft Instrumentation Test Equipment • Superfund Innovative Technology Education • System Integration, Test and Evaluation

SITEC Scientifique et Technique

SITEF Salon International des Technologies et des Énergies du Futur (France)

SITEL Stage Interdisciplinaire de Télédétection

SITI Schweizerisches Institut für Technische Information (Switzerland) • Search for Intra-Terrestrial Intelligence

sITM Simplest Inductive Turing Machine

SITMAP Situation Map

SITN Stanford Instructional Television Network (USA-CA)

SI&TO System Integration & Test Organization

SITO System Integration and Test Office • System Integration and Test Organization

SITP System Integration Test Plan • System Integration Test Program

SITPB System Integration Test Program Board

SITRB System Integration Test Review Board

SITREP Situation Report

SITS Single Integrated Telecommunications System • System Integration Test Service

SITSUM Situation Summary

Sitt Sitting

SITTEMP Situation Template

SITV Salon International du Tourisme et des Voyages

Sit. Vac. Situation Vacant

SITVC Secondary Injection Thrust Vector Control

SIU Secrétaire d'Intendance Universitaire • Sequence Initiate Update • Serial Interface Unit • Signal Interface Unit • Simulation Interface Unit • Siuna (Nicaragua) (location identifier) • SOGS Interface Unit • Southern Illinois University (USA-IL) • Special Investigation Unit • Standard Interface Unit • Sterrenkundig Instituut Utrecht (Netherlands) • System Interface Unit

SIUAPS Service Interuniversitaire des Activités Physiques et Sportives

SIUC Southern Illinois University, Carbondale (USA-IL)

SIUE Southern Illinois University, Edwardsville (USA-IL)

SIUP Southern Illinois University Press (USA-IL)

SIUZ Seminario Interdisciplinar de la Universidad de Zaragoza (Spain)

Siv Sieve

SIV Dione (satellite of Saturn) • Sample In Vacuum • Selektive Informationsverbreitung • Solar and Interplanetary Variability • Special-Interest Vehicle • Sullivan County (USA-IN) (location identifier) • Systems Integration Van • Systems Integration Vehicle

SIVA Spatial Information, Visualization and Analysis

SIVD Speaker Independent and Vocabulary Dependent

SIVE Shuttle Interface Verification Equipment • Sistema Integrado de Vigilancia Exterior • Spacelab Interface Verification Equipment

SIVIT Siemens Virtual Touchscreen

SIVOM Syndicat Intercommunal à Vocation Multiple (France)

SIVP Stage d'Initiation à la Vie Professionnelle

SIVQ Software Integration, Verification and Qualification

SIVU Syndicat Intercommunal à Vocation Unique (France)

SIV&V Software-Independent Verification & Validation

SIVV Software-Independent Verification and Validation

SIVVP Software-Independent Verification and Validation Plan

SIVVR Software-Independent Verification and Validation Report

SIW Sektion Informationswissenschaft • Sibisa (Indonesia) (location identifier) • Station-In-Wolke • Submission Interworking

SIWG Simulation Interoperability Working Group • Special Interest Working Group • Survivability Integration Working Group

six Sixel Image (file extension)

SIX Phoebe (satellite of Saturn) • Singleton (Australia-NSW) (location identifier)

SIXI Silicon X-ray Imager

SIY Montague - Siskiyou County (USA-CA) (location identifier)

SIZ Seasonal Ice Zone • Sissano (Papua New Guinea) (location identifier)

SIZEX Seasonal Ice Zone Experiment

sj Sub Judice

S/J Signal to Jamming Ratio

SJ Jan Mayen Islands (Svalbard and_) (ISO country code) • Sack Jacket • San Jose (USA-CA) • San Juan (Argentina) • Slip Joint • Supersonic Jet • Svalbard and Jan Mayen Islands (ISO country code)

SJA San Juan (Peru) (location identifier) • Société Jurassienne d'Astronomie (Switzerland)

SJAA San Jose Astronomical Association (USA-CA)

SJAC Society of Japanese Aerospace Companies • South Jersey Astronomy Club (USA-NJ)

SJAM SIAM Journal of Applied Mathematics (periodical code)

SJB San Joaquin (Bolivia) (location identifier)

SJC San Jose (USA-CA) (location identifier) • Silicon Valley (USA-CA) (location identifier) • Slit Jaw Camera • Spacecraft Jettison Controller • Standing Joint Committee • Supreme Judicial Court

SJCO SIAM Journal of Control Optimization (periodical code)

SJD Doctor of Juridical Science • San José del Cabo (Mexico) (location identifier)

SJDM Society for Judgment and Decision Making

SJE San José del Guaviare (Colombia) (location identifier)

SJF Saint John Island (US Virgin Islands) (location identifier) • Scheduler JCL Facility • Shortest Job First

SJG San Pedro - Jagua (Colombia) (location identifier)

SJH San Juan del Cesar (Colombia) (location identifier)

SJI San Jose (Philippines) (location identifier) • San Juan Institute (USA-AZ)

SJJ Sarajevo (Bosnia-Hercegovina) (location identifier)

SJK São José dos Campos (Brazil-SP) (location identifier)

SJL São Gabriel da Cachoeira (Brazil-AM) (location identifier)

SJM Jan Mayen Islands (Svalbard and_) (ISO country code) • San Juan de Maria (Dominican Rep.) (location identifier) • Special Joint Meeting • Svalbard and Jan Mayen Islands (ISO country code) • System Junction Module

SJMA SIAM Journal of Mathematical Analysis (periodical code)

SJMN San Jose Mercury News (USA-CA)

SJN Saint Johns (USA-AZ) (location identifier)

SJNA SIAM Journal of Numerical Analysis (periodical code)

SJO San José - Juan Santamaria (Costa Rica) (location identifier)

SJP São José do Rio Preto (Brazil-SP) (location identifier)

SJPMEA Si Je Puis M'Exprimer Ainsi

SJPO Systems Joint Project Officer

SJQ Sesheke (Zambia) (location identifier)

SJR Salaire Journalier de Référence • San Juan (Colombia) (location identifier) • Software Justification Report

SJS San José (Bolivia) (location identifier)

SJSA Sun - Jupiter - Saturn - Asteroid

SJSP South Jersey Star Party (USA)

SJSU San Jose State University (USA-CA)

SJT San Angelo - Mathis (USA-TX) (location identifier)

SJTI Service Juridique et Technique de l'Information

SJTU Shanghai Jiao Tong University (China-PRC)

SJU Saint John's University (USA-MN) • San Juan - Luis Muñoz Marin (Puerto Rico) (location identifier)

SJUSD San Jose Unified School District (USA-CA)

SJV San Javier (Bolivia) (location identifier)

SJW Shijiazhuang (China-PRC) (location identifier)

SJX Sartaneja (Belize) (location identifier)

SJY Seinäjoki Ilmajoki (Finland) (location identifier)

SJZ São Jorge Island (Azores) (location identifier)

sk Slovak (ISO language code) • Slovakia

Sk Sack • Sanduleak • Sick • Sink • Sketch

SK Sachkatalog • Saskatchewan (Canada) • Service Kit • Sink • Slovakia (ISO country code) • Slovaquie (code ISO de pays) • Station Keeping • Stransky-Krastonov • Super-Kamiokande • Synkronknoten

SKA Skills - Knowledge - Ability • Skills - Knowledge - Attitudes • Spokane Fairchild (USA-WA) (location identifier) • Square Kilometer Array

SKAI Square Kilometer Array Interferometer

SKAMP Station Keeping And Mobile Platform

SKAT Science, Knowledge, And Technology • Smart Knowledge Acquisition Tool

SKB Nevis (Leeward Islands) (location identifier) • Saint Kitts (Leeward Islands) (location identifier) • Skills - Knowledge - Behaviors • Stadt- und Kreisbibliothek

SKBO Scattered Kuiper Belt Objects

SKC Sky Clear • Suki (Papua New Guinea) (location identifier)

SKD Samarkand (Uzbekistan) (location identifier)

SKDR S/Ka-band Data Relay

SKE Schwarzschild Kinematic Explorer • Single-Knife Edge • Skien (Norway) (location identifier) • Software Key Escrow • Station-Keeping Equipment

SKED Schedule

SKEME Secure Key Exchange Mechanism

S. Ken. South Kensington (UK-England)

SKEPP Studiekring voor Kritische Evaluatie van Pseudowetenschap en het Paranormale (Belgium)

SKF San Antonio - Kelly (USA-TX) (location identifier) • Svenska Kullagerfabriken

SKG Stellar Kinematic Group • Thessaloniki (Greece) (location identifier)

SKH Surkhet (Nepal) (location identifier)

SKI Skikda (Algeria) (location identifier)

SkICAT Sky Image Cataloguing and Analysis Tool

SKICAT Sky Image Cataloguing and Analysis Tool

SKIE Second-Kind Integral Equation

SKIP Simple Key for Internet Protocol

SKIRT Shuttle Kinetic Infrared Test • Spacecraft Kinetic Infrared Test

SKJ Sitkinak Island (USA-AK) (location identifier)

SKK Koruna (Slovakia) • Shaktoolik (USA-AK) (location identifier)

SKL Isle of Skye (UK) (location identifier) • Skill Level

Sklt Skylight

Skm Stockholm (Sweden)

SKM Scanning Kelvin (force) Microscopy • Semantic Knowledge Management • Skeldon (Guyana) (location identifier)

SKN Stokmarknes (Norway) (location identifier)

SKO Sets - Kits - Outfits • Sokoto (Nigeria) (location identifier)

SKP Seeker Package • Skopje (Macedonia) (location identifier)

SKPT Skeleton Cursor Table

SKQ Sekakes (Lesotho) (location identifier)

Skr Sanskrit

SKr Swedish Krona (currency)

SKR Saturn Kilometric Radiation • Seeker • Sektorrichtlinie • Shakiso (Ethiopia) (location identifier) • Station Keeping Radar

SKRAMO Skills - Knowledge - Resources - Authority - Motives - Objectives

SKRS Seekers

Skrt Sanskrit

SKRT Square Kilometer Radio Telescope

SKS Skrydstrup (Denmark) (location identifier) • Vojens (Denmark) (location identifier)

Skt Sankt • Sanskrit • Skirt • Socket

SKT Sialkot (Pakistan) (location identifier) • Skill Knowledge Test • Specialty Knowledge Test

Sktd Skirted

SKU Seamless Knowledge Universe • Skiros (Greece) (location identifier) • Sri Krishnadevaraya University (India) • Stock Keeping Unit

SKUB Schweizerische Kommission für Umweltbeobachtung (Switzerland)

SKV Santa Katerina (Egypt) (location identifier) • Seeker Vehicle

SKW Skwentna (USA-AK) (location identifier) • Structured Keyword

SKWOC Structured Keyword Out of Context

SKWY Skyway

SKX Saransk (Russia) (location identifier)

SKY Sandusky Griffing (USA-OH) (location identifier)

SKYLAB Sky Laboratory

SKZ Sukkur (Pakistan) (location identifier)

s/l Su Letra

sl Secundum Legem • Sine Loco • Sin Lugar • Slovenian (ISO language code) • Sus Labores

Sl Sleet • Sliding • Slight • Slightly • Slip • Slovakia • Slovakian

S/L Sea Level • Shop/Laboratory • Space Laboratory • Switch/Light • Synchronization Link

S&L Savings & Loans

SL Safety Level • Safety Limit • Salvage Loss • Same Linear • Same-sense Linear • Sample Laboratory • Sampling Line • Sandqvist-Lindroos • San Luis (Argentina) • Sans Lieu • Savings and Loans • Scanning Slit • Science Linkage • Scripting Language • Seal • Sea Level • Sèche-Linge • Secundum Legem • Security Level • Self-Linearized • Sensitivity Level • Service Layer • Service Lecteurs • Shelf Life • Shoemaker-Levy • Shop Laboratory • Short Landing • Sick Leave • Side Lobe • Sierra Leone (ISO country code) • Signal Level • Signaling Link • Simulation Language • Sine Loco • Single Line • Single Lobe • Skill Level • Slavic • Slavonic • Slovenia (international aircraft registration code) • Small Loop (Sun) • Sociedad Limitada • Soft Landing • Sonoluminescence • Sound Level • Sound Locator • Source Language • Source Level • South Latitude • Spacelab • Space Laboratory • Space Launch • Specialized Lexicon • Special Linear • Squad Leader •

Standard Label • Standard Labelled • Standard of Living • StarLabels (resources) • Star Line • Start Line • Statistical List • Storage Location • Straight Line • Strategic List • Streamline • Strong Line • Superlattice • Superlocal • Support Level • Support Line • Surface Launch • Switch Location • Switch Locator • System Language • System Layer

Sla Slovak

SLA Salta - General Belgrano (Argentina-SA) (location identifier) • Satellite Linear Accelerometer • Satellite Link Adapter • Scanning Laser Altimeter • Scottish Library Association • Second-Language Acquisition • Semiconductor Laser Amplifier • Service Level Agreement • Shuttle Laser Altimeter • Site Layer Aggregation • Société Lorraine d'Astronomie (France) • Software Library Administrator • Special Landscape Area • Special Libraries Association (USA-DC) • Square Loop Antenna • State Licensing Agency • Statistical Local Area • Stereolithography Apparatus • Straight Line Approximation • Superlight Ablative • Support and Logistics Area • Synchronous Line Adapter

SLAC Solar-Lunar Almanac Core • Stanford Linear Accelerator Center (USA-CA) • Supply List Allowance Card • Support List Allocation Card • Support List Allowance Card

SLAD System Logic and Algorithm Development

SLAG Space Launch Advisory Group

SLAHTS Stowage List And Hardware Tracking System

SLAIS Slovenian Artificial Intelligence Society (Slovenia)

SLAK Spacelab Late Accessory Kit

SLALOM Space Laser Low-Orbit Mission

SLAM Scanning Laser Acoustic Microscope • Scanning Laser Acoustic Microscopy • Scanning Low Acquisition Maneuver • Side Load Arrest Mechanism • Simulation Language for Alternative Modeling • Station Logistics And Maintainability

SLAMAP Space Laser Mission Analysis Program

SLAMMR Side-Looking Airborne Multimission Radar

SLAMS Simplified Language for Abstract Mathematical Structures

slan Sine Loco, Anno, vel Nomine

SLAN Secure Local Area Network

SlaOb Slaboproudy Obzor (periodical code)

SLAP Service Life Assessment Programme • Soils Landscape Analysis Project • Solar Lunar Almanac Program • Space Link Access Protocol • Supplemental Low Altitude Photography • SUPREMUM Linear Algebra Package • Symbolic Language Assembly Program

SLAR Side-Looking Airborne Radar • Side-Looking Array Radar

SLART Simultaneous Left And Right Truncation

SLAS Saint Louis Astronomical Society (USA-MO) • Salt Lake Astronomical Society (USA-UT) • Society for Latin American Studies

SLASO Space Licensing And Safety Office (Australia)

S. Lat. South Latitude

SLAT Spacecraft Location/Attitude Tape • Supersonic Low-Altitude Target

SLATE Second Line Automatic Test Equipment • Structural Linguistic Analysis and Text Evaluation

SLATEC Sanida Los Alamos Technical Exchange Committee

Slav Slavic • Slavonia • Slavonian • Slavonic

SLB Îles Salomon (code ISO de pays) • Sea Level Base • Schweizerische Landesbibliothek (Switzerland) • Second Level Buffer • Server Load Balancing • Side Lobe Blanking • Solomon Islands (ISO country code) • Storm Lake (USA-IA) (location identifier) • Strong Light Bridge

SLBD Sea-Light Beam Director

Slbl Soluble

SLBS Saint Lucia Bureau of Standards

SLC Salt Lake City (USA-UT) (location identifier) • Satellite Laser Communications • Scan Line Corrector • Sea Level Change • Side Lobe Canceler • Side Lobe Cancelling • Side Lobe Cancellation • Simple Line Code • Simulated Linguistic Computer • Single Launch Contractor • Single-Layer Ceramic • Single Line Control • Single Look Complex • SLAC Linear Collider • Software Life Cycle • Space Launch Complex • Spacelab Control • Speed of Light Cylinder • Standard Launch Complex • Stanford Linear Collider (USA-CA) • Statute Law Committee • Stockage List Code • Subscriber Line Charge • Subscriber Loop Carrier

SLCA Spacelab Contract Administration

SLCB Single Line Color Bar

SLCD Small Lightweight Chemical Detector

SLCL Spacelab Configuration Layout

SLCM Spacelab Configuration Management • Spacelab Contract Management

SLCMP Software Life Cycle Management Plan

SLCO Senior Laboratory Contracting Official

SLCP Ship Lighterage Control Point • Ship's Loading Characteristics Pamphlet

SLCS Software Library Control System • Survivable Low Communications System

SLCSAT Submarine Laser Communications Satellite

SLCSE Software Life Cycle Support Environment

sld Slide (file extension)

Sld Sailed • Sealed • Slide • Sold • Solid

SLD Satellite Launch Dispenser • Sea Level Datum • Second-Level Destination • Second-Level Domain • Self-Locking Device • Side-Looking Detector • Simulated Launch Demonstration • Single-Load Design • SLAC Large Detector (*) • SLC Large Detector • Slope Line Detection • SOCS Low-energy Detector • Solid • Source Language Debug • Source Language Debugger • Source Language Debugging • Stanford Large Detector • Stiff-Leg Derrick • Supercooled Large Droplet • Superluminescent Diode • System Link designator

SLDC Satellite Launch Dispenser Communications • Synchronous Data Link Protocol

SLDI Sector List Drop Interval

SLDMS SAR/Landsat Data Mapping System • SAR/Landsat Data Merging System

SLDP Spacelab Double Pallet • System Level Diagnostic Program

SLDPF Spacelab Data Processing Facility

Sldr Solder

SLDS Scanning Laser Doppler System • Swedish Land Information System

SLDT Store Local Descriptor Table

SLDV Scanning Laser Doppler Velocimeter

SLE Salem (USA-OR) (location identifier) • Screening Line Editor • Serveur Logique d'Édition • Shapiro - Lightman - Eardley (disk) • Sierra Leone (ISO country code) • Single Loss Expectancy • Size-Luminosity Evolution • Small Lattice Experiment • Spacelab Engineering • Station Liaison Engineer

SLEC Security and Law Enforcement Contractor

SLED SBV Language for Experiment Design • Single Large Expensive Disk • Software Life Cycle Empirical Database • Space Laser Experiment Definition • Stable Large E-mail Database • State Libraries Electronic Doorway • Surface Light-Emitting Diode

SLEE Service de Liaison Étudiants-Entreprise • Service Logic Execution Environment

SLEEP Scanning Low-Energy Electron Probe • Silent Lightweight Electric Energy Plant

SLEL Service Logic Execution Language

SLEM Solution of Linearized Equations of Motion

SLEMSA Soil Loss Estimation Modern for Southern Africa

SLEMU Spacelab Engineering Model Unit

SLEP Service Life Extension Programme

SLES Single Line Entry System

SLF Scientific Laboratory Facility • Service de la Législation Fiscale • Shuttle Landing Facility • Site Listing Filter • Standard Linear Format • Stellar Luminosity Function • Straight Line Frequency • Stress Loading Facility • Student Load Factor • Sulayel (South Africa) (location identifier) • Superlow Frequency • System Library File

SLFA Secondary Level Field Activity

SLFC Selectable Low-Frequency Clock • Supersonic Laminar Flow Control

SLFCS Survivable Low-Frequency Communications System

SLFP Serial Line Framing Protocol

SLG Sloam Springs - Smith (USA-AR) (location identifier) • Symbolized Line Graph • Synthetic Line Generation

SLGR Small Lightweight Ground Receiver

SLGT Slight

SLGTLY Slightly

SLH Smoothed Lagrangian Hydrodynamics • Sola (Vanuatu) (location identifier)

SLHC Standards for Long-Haul Communications

SLHS Sum of Left-Hand Side

SL&I System Load & Initialization

SLI Science Laboratories Infrastructure • Service Logic Interpreter • Solwezi (Zambia) (location identifier) • Spacelab Integration • Space Launch Initiative (NASA) • Strategic List Item

SLIAKI Számitástechnikai és Automatizálási Kutató Intézete (Hungary)

SLIC Simple Low-cost Innovative Concept • Spectral Line Imaging Camera • State Land Information Council • Subscriber-Line Interface Card • Subscriber-Line Interface Circuit • System and Logistics Integration Capability • System-Level Integrated Circuit • System-Level Integration Circuit • System Level Interface Compatibility • System Licensed Internal Code

SLICE Simple and Light ISOCAM Environment

SLID System-Level Integration Demonstration

SLIES Stratospheric Limb Infrared Emission Spectrometer

SLIM Scammed Longwave Infrared Module • Semi-empirical Low-latitude Ionospheric Model • Simpler Legislation for the Internal Market • Simplified Logistics and Improved Maintenance • Software Life Cycle Management • Software Life Cycle Model • System Life Cycle Management

SLIN Standard Line Item Number • Subcontract Line Item Number • Subline Item Number

SLIP Serial Line Interface Protocol • Serial Line Internet Protocol (*) • Serial Link Internet Protocol • Serviceability Level Indication Processing • Simulation Language for Information Processing • Software for Landsat Image Processing • Symmetric List Interpretive Program • Symmetric List Processor • Synchronous Link Internet Protocol • System Level Interconnect Prediction (workshop)

SLIPAR Short Light Pulse Alerting Receiver

SLIPS Scattered Light Intensity Profile Sensor

SLIR Side-Looking Infrared (device)

SLIS School of Library and Information Science • Shared Laboratory Information System

SLISIR Sri Lanka Institute of Scientific and Industrial Research

SLISR Stanford-Lockheed Institute for Solar Research (USA-CA)

SLJ Chandler Stellar (USA-AZ) (location identifier) • Sobre Loja

slk Symbolic Link (file extension)

Slk Slovak

SLK Saranac Lake (USA-NY) (location identifier) • Sichtlochkarte

SLKT Survivability, Lethality, and Key Technologies

SLL Leone (Sierra Leone) (ISO currency code) • Salalah (Oman) (location identifier) • Shelf Life Limit

SLLN Strong Law of Large Numbers

SLLW Solid Low-Level Waste

SLM Salamanca Matacán (Spain) (location identifier) • Senior Level Management • Service Level Management • Service Level Monitor • Servicio Local Medido • Signaling Link Management • Silicon Lightweight Mirror • Simulated Laboratory Module • Single-Longitudinal Mode • Sound Level Meter • Space Laboratory Module • Spatial Light Modulator • Statistical Learning Model • Structure Line Model • Subscriber Line Module • Sul Livello del Mare • System Level Manager • System Library Manager

SLMC Source Locking Microwave Counter

Slmnt Seulement

SLMP Stray Light Modeling Program

SLMR Silly Little Modem Reader

SLMS Single Line Multiservice

Sln Slovene

SLN Salina (USA-KS) (location identifier) • Science Learning Network • Sequence Line Number

slnd Sine Loco Nec Data

SLND Sans Lieu Ni Date

SLNDF Skewed Log Normal Distribution Function

Slo Slovenian

SLO Salem Leckrone (USA-IL) (location identifier) • Saturn-Like Object • Second-Look Observation • Senior Liaison Officer • Service Level Objective • Society of Low-energy Observers (USA-TN) • Spacelab Only • Spacelab Operations • Space Liaison Officer • Swept Local Oscillator

SLOC Sea Lanes Of Communications • Sea Lines Of Communications • Software Lines of Code • Source Lines of Code

SLOCR Single Line Optical Character Reader

SLODB Spacelab Operational Data book

SLOE Special List Of Equipment

SLOG Sine of Laplacian-Of-Gaussian

SLOPE Study of Lunar Orbiter Photographic Evaluation

SLOS Star Line Of Sight • Sun Line Of Sight

SLOSH Sea, Lake and Overland Surge from Hurricane

SLOT Solar Luminosity Oscillations Telescope

SLOTH Strange Loop Or Tangled Hierarchy

Slovak Slovakia • Slovakian

slp Sine Legitime Prole

Slp Slip • Slope

SLP San Luis Potosí (Mexico) (location identifier) • Sea Level Pressure • Segmented Level Programming • Seismic Line Program • Sensor Link Protocol • Service de Lecture Publique • Service Location Protocol • Service Locator Protocol • Service Logic Programme • Significant Lexical Pattern • Silicon Light Pulser • Single Link Procedure • Slope • Slopes • Source Language Processor • Spacelab Pallet • Spacelab Payload • Spacelab Program • Sturm-Liouville Problem • Superlong Play

SLPB Spacelab Program Board

SLPO Spacelab Program Office

SLPOCC Spacelab Payload Operations Control Center

SLPP Summary Level Planning Package

SLPV Stochastic Line Profile Variation

SLPX System level Practical Exercise

SLQ Sleetmute (USA-AK) (location identifier)

SLR Satellite Laser Ranger • Satellite Laser Ranging (*) • Scanning Laser Radar • Sea Level Rise • Side-Looking Radar • Single-Lens Reflex • Solar Laser Ranging • Spacelift Range • Spin - Lattice - Relaxation • State and Local Relations • Sulphur Springs (USA-TX) (location identifier) • Super-Lithium-Rich (star) • Symbolic Logic Retrieval

SLRP Strategic Long-Range Plan

SL Rs Sri Lanka Rupee (currency)

SLRS Satellite Laser Ranging System

SLRU Sequential Least Recently Used

SLRV Shuttle Launch Research Vehicle • Surveyor Lunar Roving Vehicle

SLS Satellite Landing System • Saturn Longitude System • School of Library Science • School of Logistical Science • Sea Level Standard • Secondary Landing Site • Secondary Link Station • Second-Level Source • Selective Laser Sintering • Sequential Least Squares • Service Level Specification • Side Lobe Suppression • Signaling Link Selection • Silicon Light Source • Silistra (Bulgaria) (location identifier) • Society for Literature and Science • Sodium Lauryl Sulphate • Software Loadable System • Solid-state Linear Sensor • Sortie Laboratory Simulator • Source Library System • Southern Loop System • Southern Luminous Stars • Spacelab Life Sciences • Spacelab Simulator • Space Launch System • Space Life Sciences • Specific Living Space • Standard Launch Support • Standard Logistics Supply • Standard Logistics System • Statement Level Simulator • Static Light Scattering • Stereo Line Scanner • Strained Layer Superlattice • Structure of the Lower Stratosphere • Student Learning Service • Submillimeter Limb Sounder • Superhump Light Source • Surface Laboratory System

SLSA Shuttle Logistics Support Aircraft

SLSC Saint Louis Science Center (USA-MO)

SLSG Scanning Laser Slope Gauge • Suit Loop Stimuli Generator

SLSI Sri Lanka Standards Institution • Super Large-Scale Integration

SLSMS Spacelab Support Module Simulator

SLSS Secondary Life Support System • Super Large Scale Structure • Systems Library Subscription Service

SLSTP Space Life Sciences Training Program

Slt Sleet

SLT Salida (USA-CO) (location identifier) • Scripting Language Team • Secondary Logical Unit • Second Level Trigger • Self-Loading Tape • Semi-Lagrangian Transport • Shuttle Loop Transit • Simulated Launch Test • Sleet • Slot • Solid Logic Technique • Solid Logic Technology • Spacelab Technology • Special Launch Trajectory • Specific Launch Trajectory • Subscriber Link Termination • Superlight Telescope • Système de Localisation et de Trajectographie • System Level Test

SLTB System-Level Testbed

Sltd Slotted

SLTDP Special Laser Technology Development Program

SLTE System Level Test Equipment

SLTGT System Level Transaction Group Type

SLTT System Level Technical Test

SLTU Secure Line Termination Unit

SLTV System Level Thermal Vacuum

SLU Saint Lawrence University (USA-NY) • Saint Louis University (USA-MO) • Saint Lucia - Vigie (Windward Islands) (location identifier) • Secondary Logical Unit • Southeastern Louisiana University (USA-LA) • Spoken Language Understanding

SLUC Standard Level User Charge

SLUG Starlink Local Users Group • Superconducting Low-inductance Undulating Galvanometer

SLUGS SCUBA Local Universe and Galaxy Survey

SLUK Spacelab Utility Kit

SLUP Spacelab Utilization Programme

SLURP Simple Linear Unit Reservoir Program

Slv Sleeve

S/L/V Survivability/Lethality/Vulnerability

SLV El Salvador (ISO country code) • Salvador (El_) (ISO country code) • Satellite Launching Vehicle • Saturn Launch Vehicle • Simla (India) (location identifier) • Simulated Launch Vehicle • Soft Landing Vehicle • Southern Launch Vehicle • Space Launch Vehicle • Sport-Luxury Vehicle • Standard Launch Vehicle • Strong-Low velocity

SLVC Super Linear Variable Capacitator

SLVE System Level Validation experiment

Slvg Salvage

SLVG Special Launch Vehicle Group

Slvt Solvent

SLW Saltillo (Mexico) (location identifier) • Slow • Solid Low-level Waste • Spectral Line Width • Stichting Levenwetenschappen (Netherlands) • Store Logical Word • Straight-Line Wavelength • Supercooled Liquid Water

SLWL Straight-Line Wavelength

Slwstr Solwaster (Belgium)

SLWT Side Loading Warping Tug • Super Lightweight Tank

SLX Salt Cay (Turks & Caicos Islands) (location identifier)

Sly Slowly • Southerly

SLY Salehard (Russia) (location identifier)

SLZ São Luís Tirrical (Brazil-MA) (location identifier)

SL9 Shoemaker-Levy 9

sm Samoan (ISO language code) • Seemeile • Small

Sm Small • Samarium (formerly Sa)

S&M Scheduling & Movement • Stock & Machinery • Structural & Mechanical • Structures & Mechanisms • Supply & Maintenance

S/M Scheduled Maintenance • Service/Maintenance • Service Module • Simulation • Structural/Mechanical • Systems Management

SM Master of Science • Safe Mode • Saint Marin (code ISO de pays) • Sales Manager • Samoyed • San Marco • San Marino (ISO country code) • Scan Mechanism • Scheduled Maintenance • Scientific Management • Scientific Memorandum • Secondary Memory • Secondary Mirror • Security Management • Security Manager • Security Monitor • Seeing Monitor • Seismometer • Semi-Monthly • Senior Magistrate • Sensor Module • Sequence Monitor • Serial Magnitude • Service Maintenance • Service Manual • Service Mark • Service Member • Service Mode • Service Module • Servicing Mission • Session Monitor • Set Mode • Shared Memory • Shipment Memorandum • Shopping Mall • Short Message • Short Module • Shuttle Management • Silver Medal • Simulation Model • Simulation Module • Single Manager • Sinus Medii (Moon) • Situation Monitoring • Sky Mapper • Software Manual • Solar Magnet • Solar Magnetic • Solar Magnetospheric • Solid Modeling • Spacecraft Manager • Sparse Mode • Special Memorandum • Special Mission • Specification Memorandum • Stabilized Member • Stable Member • Staff Manager • Staff Memorandum • Staff Month • Standard Model • State Machine • Station Manager •

Status Mile • Statute Mile (*) • Stepper Motor • Storage Manager • Straight Mean • Strange Mode • Structural Mechanics • Structural Model • Subcontract Management • Successful Mission • Summary Memorandum • Super Mongo • Supply Manual • Support Module • Surface Mount • Survey Manual • Suspended Matter • Switching Module • Synchrotron Maser • System Management • System Manager • System Mechanics • System Memory • System Monitor • System Monitoring • Systems Management

SMA Safety and Mission Assurance • Santa Maria (Azores) (location identifier) • S-band Multiple Access • Schweizerische Meteorologische Anstalt (Switzerland) • Science Monitoring Area • Secondary Mirror Actuator • Secondary Mirror Assembly • Semi-Major Axis • Service Management Architecture • Shape Memory Alloy • Sheared Magnetic Arcade • Signal Message Address • Simulation Model Adapter • Site Maintenance Area • Slave Manipulator Arm • Slow Mass Accretion • Small-Angle Maneuver • Sociedad Malagueña de Astronomía (Spain) • Solar-Maximum Analysis • Standardization Management Activity • State Meteorological Administration • State Meteorological Agency • Straight Medium Adaptor • Subject Matter Area • Submillimeter Array • Super Modified Achromat • Support Management Area • Surface Movement Advisor • Survey and Mapping Alliance • Système Mécanique Automatisé • System Management Architecture

SMAAC Southern Michigan Amateur Astronomy Conference

SMAB Solid Motor Assembly Building

SMAC Scene Matching by Area Correlation • Sequencing Multiple Analysis Computer • Sequential Multiple Analysis Computer • Serial Memory Address Counter • Simplified Method for Atmospheric Correction • Simultaneous Multiframe Analytical Calibration • Spacecraft Maximum Allowable Concentration • Space Maximum Allowable Concentration • State Mapping Advisory Committee • State Mapping Advisory Council • Stellar Mass Catalogue • Strategic Member Advisory Committee • Streaming Motions of Abell Clusters • Systèmes Multi-Agents Coopératifs • System Monitor And Control

SMACK Streaming Motions of Abell Cluster Kernels

SMADB Submillimeter Array Database

SMAE Systems Management Application Entity

SMAG Star Magnitude • Supply Management Activity Group

SMAGLE Stationary Magnetic Levitation Experiment

SMAI Société de Mathématiques Appliquées et Industrielles (France)

SMAL Structural Microassembly Language

SMALGOL Small computer Algorithmic Language

SMAN Stage de Mise À Niveau

SMAP Security Management Application Program • Service Management Access Point • Short and Medium-term Action Programme • Software Management and Assurance Program • Station Management Application Program • Systems Management Analysis Project • Systems Management Application Process

SMAQ Shape Memory Alloys for Quarrying

SMARA Safety and Mission Assurance Risk Assessment • Schenectady Museum Amateur Radio Association (USA-NY) • Servicio de Meteorología de la Armada de la Republica Argentina • Surface Mining and Reclamation Act (USA)

SMArc Solid Mechanics Archives (periodical code)

SMARF Solid Motor Assembly and Readiness Facility

SMARG Spectrum Management And RFI-protection Group (NAIC)

SMART Science and Mathematics Achievement Required for To-morrow • Science, Mathematics, Aerospace, Research and Technology Inc. • Science, Mathematics And Relevant Technology • Self-Monitoring, Analysis, and Reporting Technology • Sensors Mounted As Roving Threads • Sequential Mechanism for Automatic Recording and Testing • Simulation and Modeling for Allocation, Requirements, and Training • Simulator for Multiradar Analysis for Realistic Traffic • Simultaneous Multispectral Absolute Radiometer and Transmissometer • Small Ada Run Time • Small Mission for Advanced Research in Technology • Smooth Multiple Additive Regression Technique • Sobolev Modified Approximate Radiative Transfer • Software Market-driven Architecture Trade • Software Metering And Resource Tracking • Space Maintenance And Repair Technique • Spares Management Analysis Review Technique • Storage Modification and Retrieval Transaction • Stereo Mapping with Radar Techniques • Stretched Mesh Attached to Rope Trusses • Structural Maintenance And Repair Team • Submillimeter Array-Receiver for Two (frequencies) • Submillimeter Array of Taiwan (China-ROC) • Supply and Maintenance Assessment and Review Team • Susceptibility Model Assessment and Range Test • System for Management and Allocation of Resources Technique • System for Manipulation And Retrieval of Text • System for Mechanical Analysis and Retrieval of Text • Systems Management Analysis, Research and Test • Systems Management Analysis, Research and Testing

SMARTA Software Market-driven Architecture Trade Association (USA)

SMARTS Shared Memory Asynchronous Runtime System • Simulation, Monitoring, Analysis, Reduction, and Test System • Small and Medium Aperture Research Telescope System • Status Memory And Real-Time System

SMaS Smart Material Structures (periodical code)

SMAS Service Management Application System • Smoky Mountain Astronomical Society (USA-TN) • Southern Maryland Astronomical Society (USA-MD) • Southwest Montana Astronomical Society (USA-MT) • Standard Materiel Accounting System

SMASE Systems Management Application Service Element

SMASS Small Main-belt Asteroid Spectroscopic Survey

SMASSIR Small Main-belt Asteroid Spectroscopic Survey in the Infrared

SMATH Satellite Materials And Target Hardening

SMATS Smashed Target Satellite

SMATV Satellite Master Antenna Television (*) • Small Master Antenna Television

SMB Cerro Sombrero (Chile) (location identifier) • Secondary Mirror Baffle • Server Message Block • Service Message Block • Session Message Branch • Small and Medium Business • Society for Mathematical Biology • Space Meteorology Branch • Submillimeter Background • Surveys and Mapping Branch • System Management Bus • Systems Management Branch

SMBA Shared Memory Buffer Architecture • Supply Management Business Area

Smbdy Somebody

SMBH Supermassive Black Hole

Smthg Something

SMBus System Management Bus

SMC Safe-Mode Command • Safe-Mode Computer • Santa Maria (Colombia) (location identifier) • Santa Monica College (USA-CA) • Scientific Manpower Commission • Secondary Mission Controller • Segmented Maintenance Task • Sequential Machine Controller • Serial Magnitude Command • Service Management Center • SFOC Monitor and Control • Sheet Molding Compound • Site Movement Coordinator • Small Magellanic Cloud • Small Multiuser Computer • Sociedad Matemática de Chile (Chile) • Special Mapping Center • Spectral Management Class • Speech and Music Compression • Spin Muon Collaboration • Staff Message Control • Standard Management Committee • Standard Mean Chord • Standard Microsystems Corporation • Stepping Motor Controller • Stichting

Matematisch Centrum (Netherlands) • Strategic Manufacturing Center • Supermulticoating • Surface Material Class • Switch Memory Control • System Management Center • System Management Code • System Master Catalog • System Monitor and Control • System Monitoring and Coordination • System Monitoring Center • Systems, Man and Cybernetics (conference)

SMCAT Strong Motion Catalog

SMCC Shuttle Mission Control Center • Simulation Monitor and Control Console • Special Material Content Code • State Movement Control Center • Sun Microsystems Computer Corporation • System Monitor Coordination Center

SMCD Simple Code Multiple Data

SmCES Smithsonian Contributions to Earth Sciences (periodical code)

SMCH Standard Mixed Cargo Harness

SMCI System Management Configuration Item

SMCL Secondary Maximum Contaminant Level

SMCM Small Module Core Memory

SMCR Satellite Mode Change Request

SMCRA Surface Mining Control and Reclamation Act (USA)

SMCS Secondary Mirror Control System • Separation Monitor and Control System

SMCT Smart Multicircuit Terminal • Solar Multichannel Telescope

SMCU Separation Monitor Control Unit • Separation Monitoring Control Unit

SMCV Side-Mount Cargo Vehicle

S&MD Science & Mission Design

SMD Fort Wayne - Smith (USA-IN) (location identifier) • Scanning Microdensitometer • Scheduled Maintenance Demand • Scheduling Management Display • Science Mission Division • Scientific Missions Division • Ship Manpower Document • Shop Manufacturing Development • Silicon Microstrip Detector • Singular Multinomial Distribution • Soil Moisture Deficit • Sonntag - Montag - Dienstag • Spacelab Mission Development • Special Measuring Device • Staff Management Division • Standard Magnet Dewar • Standard Microcircuit Drawing • Storage Module Device • Storage Module Drive • Structures and Mechanics Division • Surface Mass Density • Surface Mount Design • Surface Mount Device • Surface-Mounted Device • Surveys and Mapping Division • Switch Multiplexer/Demultiplexer • System Management Device • System Management Directive • System Management Directorate • System Measuring Device • Systems and Mission Division • Systems Management Division • Systems Measuring Device • Systems Monitor Display

SMDB Summary Database

SMDC Shielded Mild Detonation Cord • Superconductive Materials Data Center

SMDD Storage Module Disk Drive

SMDE Static Mercury Drop Electrode

SMDF Soil Moisture Deficit Function

SMDI Simplified Message Desk Interface

SMDL Standard Musical Description Language • Standard Music Description Language

SMDM Small Multiplexer/Demultiplexer

SMDR Station Message Detail Record • Station Message Detail Recording • Structured Manning Decision Review • Subscriber Message Detail Recording • System Message Detail Recording

SMDS Switched Multimegabit Data Service • Systems Management Display System

SMDSM Système Mondial de Détresse et de Sécurité en Mer

SM&E Semiconductor Materials & Equipment

SME Sakharov's Metric Elasticity • School of Mechanical Engineering • Seismic Margin Earthquake • Short Message Entity • Sistema Monetario Europeo • Small and Medium Enterprises • Small and Medium-sized Enterprises • Society of Manufacturing Engineers (USA) • Society of Mining Engineers (USA) • Solar Mesosphere Explorer • Somerset - Pulaski County (USA-KY) (location identifier) • Special Mission Equipment • Stellar Mass Ejection • Storage Management Engine • Structure Mapping Engine • Subject Matter Expert • Supermaximum Entropy • Surface Movement Element • Suriname (international vehicle registration code) • Synchronous Modem Eliminator • Système de Management de l'Environnement • Système de Management Environnemental • Système Monétaire Européen • System Management Environment

SMEA Safe Mode Electronics Assembly

SMEAC Subject Matter Expert Advisory Committee

SMEB Software Management Executive Board

SMEC Stretched Membrane with Electrostatic Curvature

SMEDP Senior Managers Executive Development Program

SMEGMA Sophisticated Modern Editor with Gloriously Magnificent Abilities

SMEI Solar Mass Ejection Imager

SMEO Small End Office

SMES Shuttle Mission Engineering Simulator • Shuttle Mission Evaluation Simulation • Superconducting Magnetic Energy Storage

SMET Science, Mathematics, Engineering and Technology • Shuttle Mission Engineering Simulator • Shuttle Mission Evaluation Simulator • Simulated Mission Endurance Test • Simulated Mission Endurance Testing • Superconducting Magnetic Energy Storage

SMETE Science, Mathematics and Technology Education

SmEX Small Explorer

SMEX Small Explorer

SMF Sacramento (USA-CA) (location identifier) • Simple Message Format • Single Mode Fiber • Sociedad Mexicana de Física (Mexico) • Société Mathématique de France (France) • Société Météorologique de France (France) • Software Maintenance Facility • Solar Magnetic Field • Space Manufacturing Facility • Special Modifying Factor • Standardized Monitoring Framework • Standard Message Format • Static Magnetic Field • Status Message File • Strong Magnetic Field • Submultiframe • Syndrome du Membre Fantôme • System Management Facility • System Management Framework • System Management Function • System Manager Facility • System Measurement Facility • System Message File

SMFA Specific Management Functional Area • System Management Functional Area

SMFG System Management Functional Group

SMFT Solar Magnetic Field Telescope

SMG Santa Maria (Peru) (location identifier) • Secure Mail Guard • Senior Management Group • Spaceflight Meteorology Group • Special Mobile Group • Standard Mail Guard • Submillimeter Galaxy • Super Master Group

SMGC Selective Management Grouping Code • Supply Management Grouping Code • Surface Movement Guidance and Control

SMGCS Surface Movement Guidance and Control System

SMH Sapmanga (Papua New Guinea) (location identifier) • Service de Métrologie Habilité • Service des Monuments Historiques • Short Message Handler • Signaling Message Handling • Spectromètre à Modulation Hétérodyne • Spectromètre Millimétrique Hétérodyne

SMHD Service Multimédia à Haut Débit

SMHE Selected Materiel Handling Equipment

SMHI Sveriges Meteorologiska och Hydrologiska Institut (Sweden)

SMHN Servicio de Meteorología e Hidrología Nacional (El Salvador)

SMHP Summer Microwave Hurricane Program

smi Self-Mounting Image (file extension)

SMI Samos (Greece) (location identifier) • Security Management Information • Security Management Infrastructure • Servo Moteur Intégré • Sistema Monetario Internacional • Slovak Metrology Inspectorate • Solar Magnetism Initiative • Sortie Mission Integrator • Standard Measuring Instrument • Static Memory Interface • Station Management Interface • Strange-Mode Instability • Structure Medium Interaction • Structure of Management Information • Svensk Materialinformation (Sweden) • Swiss Meteorological Institute • System Management Information • System Management Interrupt

SMIA Single Multiplexer Interface Assembly

SMIAC Soil Mechanics Information Analysis Center

SMIB Security Management Information Base

SMIC Salaire Minimum Interprofessionnel de Croissance • Sortie Mission Integration Contractor • Special Maintenance Item Code • Special Materiel Identification Code • Study of Man's Impact on Climate • Surveying and Mapping Industry Council

SMIF Standard Mechanical Interface • Stream-based Model Interchange Format • Système Modulaire Individualisé de Formation

SMIFTS Spatially-Modulated Imaging Fourier Transform Spectrometer

SMIG Salaire Minimum Interprofessionnel Garanti

SMIIS Solar Microwave Interferometer Imaging System

SMIL Statistics and Market Information Library • Synchronized Multimedia Integration Language

SMILE Spherical Micro-Integrated Lens

SMILI Superconducting Magnetic Instrument for Light Isotopes • Superconducting Magnet Instrument for Light Isotopes • Synthetic Modal Interferometric Laser Imaging

SMILS Submillimeter IR Line Survey

SMIM Submillimeter Intermediate Mission

S-MIME Secure Multipurpose Internet Mail Extension

S/MIME Secure Multipurpose Internet Mail Extension

SMIME Secure Multipurpose Internet Mail Extension

SMIO SAR Mission Information Officer

SMIP Spares Management Improvement Program • Specific Management Information Protocol

SMIPS Small Interactive Image Processing System

SMIR Service Météorologique Interrégional • Shuttle Multispectral Infrared Radiometer

SMIRAG Service Météorologique Interrégional Antilles-Guyane (France)

SMIRO Service Météorologique Interrégional Ouest (France)

SMIRR Shuttle Multispectral Infrared Radiometer

S-MIS Single-Mission Integrating System

SMIS Secure Management Information System • Security Management Information System • Service Médical, Infirmier et Social • Ships Management Information System • Single-Mission Integrating System • Single-Mission Integration System • Society for Management Information Systems (USA) • Specific Management Information Service • Standards and Metrology Institute of Slovenia • STDN Management Information System

SMIT Systems Management Information Tool • Systems Management Interface Tool

Smith Smithsonian

SMJ Sim (Papua New Guinea) (location identifier)

Smk Smoke

SMK Saint Michael (USA-AK) (location identifier) • Smoke • Software Migration Kit

Sml Simulate • Simulation • Simulator • Small

SML Science Museum Library • Search Mode Logic • Serials Master List • Service Management Layer • Servicio Médico Legal • Simple Macro Language • Small • Standard Markup Language • Standard Materiel List • Standard Metalanguage • Stella Maris (Bahamas) (location identifier) • Structure Mold Line • Support Material List • Surface Mixed Layer • Symbolic Machine Language

SMLC Software Maintenance Life Cycle • Sterile Massless Component • Suivi Médical Longitudinal Contrôlé

SMLE Suboptimal Maximum-Likelihood Estimator

SMLI Stateful Multilayer Inspection

SMLM Simple-Minded Learning Machine

SMLR Stepwise Multiple Linear Regression • Stepwise Multivariate Linear Regression • Système Mental de Lecture Rapide

SMLS Shelterized Microwave Landing System

SMLTM Spectral Mesosphere/Lower Thermosphere Model

SMLV Standard Memory Load Verifier

SMLWFA Self-Modulated Laser Wakefield Accelerator

S/MM Shuttle MIR Mission

SMM Safe Mercury Mode • Scattering Matrix Method • Science Mission Management • Science Museum of Minnesota (USA-MN) • Semporna (Malaysia) (location identifier) • Server Management Module • Server Manager Module • Shipment Management Module • Shuttle MIR Mission • Sociedad Matemática Mexicana (Mexico) • Société de Mécanique Magnétique (France) • Solar Maximum Mission • Special Mandatory Modification • Spectral Matrix Method • Spectro-Mirror Mechanism • Standard Method of Management • Start of Manual Message • Subcontract Management Manual • Submillimeter • Submillimeter Mission • Subsystem Management Manual • Surface Micromachining • System and Mission Management • System Maintenance Management • System Maintenance Manual • System Management Mode • System Management Module • System Maturity Matrix • System Measurement Management • System Mission Management • System Model Manager • Systems Maturity Matrix

SMMART Society for Mass Media and Resources Technology (Australia)

SmMaS Smart Materials and Structures (periodical code)

SMMC Simultaneous Multiindenture Multiechelon Computation

SMMD Specimen Mass Measurement Device

SMME Small-, Medium- and Micro-Enterprise • Solid Mechanics and Materials Engineering

SMMF Solar Mean Magnetic Field

SMMI Structured Multimedia Information

SMMM Submillimeter Moderate Mission

SMMP Standard Method of Measuring Performance

SMMR Scanning Multichannel Microwave Radiometer • Scanning Multifrequency Microwave Radiometer • Scanning Multispectral Microwave Radiometer

SMMS Support Maintenance Management System

SMMT Society of Motor Manufacturers and Traders • Submillimeter Telescope

SMN Salmon (USA-ID) (location identifier) • SDH Management Network • Sécurité Multiniveau • Servicio Meteorológico Nacional • Software Modification Notice • SONET Management Network

SMO Sample Management Office • Santa Monica (USA-CA) (location identifier) • Senior Management Official • Senior Medical Officer • Senior Meteorological Officer • Sequential Minimal Optimization • Service Module Oxidizer • Servicing Mission Office • Shelter Management

Office • Single-Market Observatory (EU) • Small Magnetospheric Observatory • Small Missions Opportunities • Software Modification Order • So Much Of • Special Management Office • Stabilized Master Oscillator • Strategic Mobility Office • Strategic Mobility Officer • Subcontracts Management Office • Subcontracts Management Organization • Substellar Mass Object • Supermassive Object • Supplementary Meteorological Office • Support Materials Organization (USA-CA) • Systems Management Office • Systems Management Officer • Systems Management Overview

SMOA Single Manager Operating Agency

SMOBC Smolder Mask Over Bare Copper

SMOC Science and Mission Operations Center • Simulation Mission Operation Computer

SMOG Smoke and Fog • Special Monitor Output Generator • Structural Modeling Oriented Graphics • Structured Model Generator • Surface Meteorology and Ozone Generation

SMOH Since Major Overhaul

SMOKA Subaru-Mitaka-Okayama-Kiso Archive

SMOKE Surface Magneto-Optical Kerr Effect

SMON System Monitor

SMOP Satellite Measurement of Oceanographic Parameters • Simple Matter Of Programming • Small Matter Of Programming • So Much Of Paragraph

SMORS Service Mode Optimized Reduction Scheme

SMOS Soil Moisture and Ocean Salinity (experiment) • Surface Meteorological Observation System

SMOV Servicing Mission and Observatory Verification • Servicing Mission Observatory Verification • Servicing Mission Orbital Verification (*)

SMOW Standard Mean Ocean Water

smp Sine Mascula Prole

SMP Safety Management Plan • Sample • Scanning and Measuring Projector • Scheduled Maintenance Program • See Me Please • Selective Message Routing • Self-Maintenance Period • Sensitized Material Print • Service Management Point • Servo Meter Panel • Session Management Protect • Shared Memory Processor • Signaling Mediation Point • Signal Message Processor • Simple Management Protocol • Simulation Model Portability • Simultaneous Membership Program • Small Manufactured Part • Small Mechanical Part • Software Maintenance Plan • Software Management Plan • Software Management Practice • Software Modernization Plan • Soil Management Program • Specialized Mobile Radio • Special Manufacturing Procedure • Spectral Mapping Theorem • Standard Maintenance Procedure • Standards, Methods and Planning • Statutory Maternity Pay • Stockholm (Papua New Guinea) (location identifier) • Stores Management Processor • Stressed Mirror Polishing • Structure Mechanism Policy • Summary Maneuver Plan • Sustained Maintenance Planning • Symbolic Manipulation Program • Symmetrical Multiprocessing • Symmetrical Multiprocessor • Symmetric Multiprocessing • Symmetric Multiprocessor • System Maintenance Procedure • System Management Plan • System Master Plan • System Monitoring Panel

SMPA Solid-Motor Processing Area

SMPC Shared Memory Parallel Computer

SMPG Standardization Management Policy Group

SMPI System Manager Programmer Interface

SM/PM System Management/Performance Monitor

SMPM Structural Materials Property Manual • System Management Performance Monitor

SMPQ Software Management Practice Questionnaire

SMPR Supply and Maintenance Plan and Report

SMPS Sector Maintenance Processor System • Semiconductor Materials Processing System • Service Module Propulsion Stage • Simplified Message Processing Simulation • Switching Mode Power Supply

SMPTE Society of Motion Picture and Television Engineers (USA)

SMPTRB Shuttle Main Propulsion Test Requirement Board

SMPW Superconducting Multipole Wiggler

SMPZ Système Mondial de Prévisions de Zone

SMQ Sampit (Indonesia) (location identifier) • Structure Module Qualification

SMQT Structure Module Qualification Test

SM&R Source, Maintenance & Recoverability

SMR Saint Marin (code ISO de pays) • San Marco Range • San Marino (ISO country code) • Santa Marta (Colombia) (location identifier) • Scanning Microwave Radiometer • Scientific Management Review • Selective Message Router • Service Médical Rendu • Sites and Monuments Records • Software Maintenance Responsibility • Software Management Report • Software Modification Record • Software Modification Report • Solid Moderated Reactor • Source - Maintenance - Recoverability • Spaceborne Meteorological Radar • Specialized Mobile Radio • Special Management Review • Special Mobile Radio • Special Money Regulation • Special Money Requisition • Standardized Mortality Rate • Status Monitoring Routine • Stock Management Report • Submillimeter Radiometer • Super-Metal-Rich (star) • Supply Management Readiness • Supply Management Report • Surface Movement Radar • System Malfunction Report • System Management Review

SMRA Simultaneous Multicomponent Rank Annihilation

SMRAM System Management Random Access Memory

SMRB Senior Material Review Board

SMRC Software Management Review Council • Source Maintenance Recoverability Code

SMRD Science and Mission Requirements Document • Science Modeling Requirements Document • Spin Motor Rotational Detector • Spin Motor Rotation Detector • Spin Motor Rotation Device • Spin Motor Run Detector • Spin Motor Run Direction • Spin Motor Run Discrete • Spin Motor Running Detector • Spin Motor Running Director

SMRG Small Research Grant

SMRI Service Message Routing Indicator

SMRM Solar Maximum Repair Mission

SMRP Supermassive Relic Particle

SMRS Specialized Mobile Radio System

SMRSS Surveys, Mapping and Remote-Sensing Sector

SMRT Single-Message Rate Timing • Software Maintenance Responsibility Transfer

SMRV Small Multiple Reentry Vehicle

SMRWG Science and Mission Requirements Working Group

SM&S Systems Management & Sequencing

SMS Safety Management System • Sainte-Marie (Madagascar) (location identifier) • Sales Management System • Satellite Motion Simulator • Scanning Mirror System • Scenario Modeling System • Schema Management System • School of Management Sciences • School of Mathematical Sciences • School of Mathematics and Statistics • Science Mission Schedule • Science Mission Specification • Sciences Médico-Sociales • Scientific Mission Support • Security Management System • Selective Multicast Server • Semesterwochenstunden • Semi-automatic Meteorological Station • Semiconductor - Metal - Semiconductor • Sensor Monitoring Set • Separation Mechanism System • Sequence Milestone System • Service Management System • Service Manipulator System • Service Médico-Social • Ship Management System • Short Message Service • Short Messaging Service • Short Messaging System • Shuttle Mission Simulator • Signal Measurement System • Single-Molecule Spectroscopy • Site-Managed Software • Sleep Management System • Small Magnetospheric Satellite • Small Messaging System • Société Mathématique Suisse (Switzerland) • Software Maintenance Specialist • Software Modification Sheet • Solar-Mass Sensor • Solar-Maximum Satellite • Sonne - Mond - Sterne • Space Motion Sickness • Spares Management System • Spin Motor Supply • Standard Mobile Segment • Standard Modular System • Standard Molecular System • Stationary Manipulator System • Stationary Meteorological Satellite • Storage Management Service • Storage Management System • Stores Management System • Strategic Management Society (USA) • Subject Matter Specialist • Subscriber Management System • Suprathermal Mass Spectrometer • Surface Management System • Suspended Maneuvering System • Sveriges Mekanstandisering (Sweden) • Swedish Mathematical Society (Sweden) • Switching and Maintenance Set • Switch Management Software • Switch Memory Storage • Synchronous Meteorological Satellite • Synoptic Meteorological Sounding • Systems Maintenance Sector • Systems Maintenance Service • Systems Managed Storage • Systems Management and Sequencing • Systems Management Server • Systems Management Services

SMSA Secondary Mirror Subassembly • Standard Metropolitan Statistical Area • Standard Metropolitan Statistical Average

SMSB System Management Status Briefing

SMSC Semi-Magnetic Semiconductor • Service Module Sequence Controller • Shared-Memory Supercomputer • Short Message Service Center • Software Moding, Sequencing, and Control • Space and Missiles System Center • Standard Modular System Card • Standard Multiuser Small Computer • System Maintenance and Support Center

SMSCC Shuttle Mission Simulator Computer Complex

SMSE Simplified Modified Semi-Empirical (method)

SMSG School Mathematics Study Group • Shuttle Maintenance Steering Group

SMSGEN Science Mission Specification Generation

SMSI Standard Manned Space Flight Initiator

SMSL Standard Multimedia Scripting Language

SMSK Serial Minimum-Shift Keying

SMSP Sodium Metasilicate Peutahydrate

SMSR Side-Mode Suppression Ratio • Survivability Management Status Report

SMSS School of Management and Social Sciences • Shuttle Modular Scanning Spectroradiometer

SMSST Satellite-Measured Skin Surface Temperature

SMSSV Space Mission Scenario Simulation and Visualization

SMST Science Mission Specification Timeline

SMSU Southwest Missouri State University (USA-MO)

SMSW Store Machine Status Word

SMt Station Management

SMT Satellite Mobile Terminal • Science - Mathematics - Technology • Segmented Mirror Telescope • Selective Message Transaction • Service Module Technician • Ship's Mean Time • Short Message Terminal • Simultaneous Multithreading • Slow Mass Transfer • Software Maintenance Team • Specialized Maintenance Training • Special Maintenance Team • Spectral Measurement Toolkit • Standard Measurement Technique • Standards - Measurements - Trials • Static - Mobile - Transportable • Station Management • Steiner Minimum Tree • Submillimeter Telescope • Summit • Sunday - Monday - Tuesday • Sun Moon Lake (China-ROC) (location identifier) • Surface-Mounted Technology • Surface-Mount Technology • Système Modulaire Thermique • Système Mondial de Télécommunications • System Maintenance Test • System Manager Training • System Master Tape

SMTA Subordinate Message Transfer Agent

SMTAS Shuttle Model Test and Analysis System

SMTBCF Space Mission Time Between Critical Failure

SMTBDE Space Mean Time Between Downing Event

SMTC Science, Medicine, and Technology in Culture (program)

SMTF Shuttle Mission Training Facility • Spacecraft Magnetic-Test Facility • State Management Function

SMTH Smooth

SMTI Selective Moving Target Indicator

SMTO Submillimeter Telescope Observatory

SMTP Send Mail Transfer Protocol • Simple Mail Transfer Protocol (*) • Simple Message Transfer Protocol • Standard Mail Transfer Protocol • Standard Mail Transport Protocol

SMTRC Specializing Military Technology Research Center

SMTS Soprotivlenie Materialov i Teoriia Sooruzhenii (periodical code) • Stardust Mission Test System • Synchronous Meteorological Test Satellite

SMTTRF Space Mean Time To Restore Function

SMTTRS Space Mean Time To Restore System

SMU Saint Mary's University (Canada) • Secure Mobile Unit • Sheep Mountain (USA-AK) (location identifier) • Singapore Management University • Single-Motor Unit • Slovensky Metrologicky Ustav (Slovakia) • Software Management Utility • Software Mockup • Source Management Utility • Southeastern Massachusetts University (USA-MA) • Southern Methodist University (USA-TX) • Speaker/Microphone Unit • Special Mission Unit • Spectrum Monitoring Unit • Switch Multiplexer Unit • System Management Utility

SMUR Service Mobile d'Urgence et de Réanimation • Spectrum Management Update and Review

SMURFS Southern Michigan Unorganized Regional Festival of Stargazers (USA-MI)

SMV Saint-Moritz Samedan (Switzerland) (location identifier) • Satellite Mutual Visibility • Science Museum of Virginia (USA-VA) • Symbolic Model verifier

SMVP Shuttle Master Verification Plan

SMVRD Shuttle Master Verification Requirements Document

SMW Smara (Morocco) (location identifier) • Standard Materials Worksheet • Standard Metal Window • System Manager Workstation

SMWHT Somewhat

SMWP Strategic Mobility Work Project

SMWT Self-Managing Work Team

SMWV Science Museum of Western Virginia (USA-VA)

SMX Santa Maria (USA-CA) (location identifier)

SMXP Simple MIME Exchange Protocol

SMY Scientist-Man Year • Simenti (Senegal) (location identifier) • Solar-Maximum Year

SMYA Solar-Maximum Year Analysis

SMZ Stoelmans Eiland (Suriname) (location identifier)

SM2 Second Servicing Mission

s/n Sin Número

sn Sans Nom • Secundum Naturam • Shona (ISO language code) • Sine Nomine • Sin Numerar • Snow • Sub Nomine

Sn Étain • Snow • Tin • Zinn

S-N Stree Number

S/N Serial Number • Service Note • Shipping Note • Signal/Noise • Signal to Noise • Stress Number

SN Sample Name • Scientific Note • Scope Note • Semiconductor Network • Senegal (international vehicle registration code) • Senegal (ISO country code) • Sequence

Number • Serial Number • Series Number • Service National • Service Node • Service Number • Shared Nothing • Shipping Note • Shipping Number • Shot Noise • Side Note • Signaling Network • Small Nephelometer (PV) • Snow • Solar Neutrino • Solution Narrative • South-North • Space Network • Strategic Navigation • Student Number • Subnetwork • Subnote • Supernova • Switching Network • Synchronization Node

SNA Laguna Beach - John Wayne (USA-CA) (location identifier) • Orange County - John Wayne (USA-CA) (location identifier) • Santa Ana - John Wayne (USA-CA) (location identifier) • Satellite Navigation Application • Seltsames Nichtchaotisches Attraktor • Service National de l'Adresse (France) • Simplified Non-standard Acquisition • Simplified Numerical Automatic • Société Neuchâteloise d'Astronomie (Switzerland) • Steerable Null Antenna • Strange Non-chaotic Attractor • Subnetwork Access • Synchronous Network Architecture • Synchronous Numeric Array • System of National Accounts • Systems Network Architecture

SNAcF Subnetwork Access Function

SNAcP Subnetwork Access Protocol

SNACP Systems Network Architecture Control Protocol

SNACRP Société Nouvelle des Ateliers et Chantiers de La Rochelle - La Pallice (France)

SNACS System Network Architecture Character Stream • System Network Architecture Character String

SNADS Systems Network Architecture Distributed Services • Systems Network Architecture Distribution Service

SNAFU Situation Normal, All Fouled Up • Southern Nevada Fantasy and Fiction Union (USA-NV)

SNAIR Smart Navigators to Access and Integrate Resources

SNAME Society of Naval Architects and Marine Engineers

SNAP Selective Niobium Anodization Process • Semantic Network Array Processor • Short-Notice Annual Practice • SIMD Numerical Array Processor • Simplified Numerical Automatic Programmer • Six Node Averaging Program • Small Nuclear Auxiliary Power • Space Nuclear Auxiliary Power • Special Needs Access Program • Standard Network Access Protocol • Steerable Null Antenna Processor • Sterile Nitrogen Atmosphere Processing • Structural Network Analysis Program • Student Needs Assessment Project • Stylus Not A Pen • Subnet Access Protocol • Subnetwork Access Point • Subnetwork Access Protocol • Subnetwork Attachment Point • Supernova Acceleration Probe • Supersonic Non-equilibrium Analysis Program • Switched Network Automatic Profile • Switching Network Analysis Program • System for Nuclear Assistant Power • System for Nuclear Auxiliary Power

SNAPS Space Navigation And Pointing System • Switched Network Automatic Profile System

SNARS Switched Network Automatic Routing System

SNAS Slovak National Accreditation System • Strategic National Alternatives Study

SNASPE Sistema Nacional de Superficies Protegidas por el Estado (Chile)

SNB Seller No Buyer • Snake Bay (Australia-NT) (location identifier) • Surnombre

SNBNK Snowbank

SNBO Supernova Neutrino Burst Observatory

SNC Salinas (Ecuador) (location identifier) • Shergottite - Nakhlite - Chassignite (meteorites) • Shergotty - Nakhla - Chassigny (meteorites) • Side Nodal Controller • Société en Nom Collectif • Space Network Control • Special Nominating Committee • Subnetwork Connection • Système Nerveux Central • System Network Controller

SNCB Société Nationale des Chemins de Fer Belges (Belgium)

SNCC Space Network Control Center

SNCD Syndicat National de la Communication Directe (France)

SNCF Société Nationale des Chemins de Fer Français (France)

SNCI Service National des Champs Intenses (France)

SNCL Serial Number Configuration List

SNCO Staff Non-Commissioned Officer

SNCP Subnetwork Connection Protection • System Node Control Point

SNCs Shergottites - Nakhlites - Chassignites (meteorites)

SNCS Satellite Network Control Station

snd Sound (file extension)

Snd Sand • Sanded • Send • Sound • Sounding

SND Seno (Laos) (location identifier) • Signal Network Distributor • Sound • Standardized Normal Distribution • Street Name Database

SNDCF Subnetwork Dependent Convergence Function

SNDCP Subnetwork Dependent Convergence Protocol

SNDE Society for Non-linear Dynamics and Econometrics

SNDP Signal-to-Noise Data Package • Sustainable National Domestic Product

SNe Supernovae

SNE São Nicolau (Cape Verde) (location identifier) • Severe Noise Environment • Syndicat National de l'Édition (France)

SNEA Société Nouvelle Europe Aéroservice

SNEAP Symposium of Northeastern Accelerator Personnel (USA)

SNECMA Société Nationale d'Études et de Construction de Moteurs d'Avions (France)

SNEPT Space Nuclear Electric Propulsion Test

SNePS Semantic Network Processing Systems (workshop)

SNEWS Supernova Neutrino Early Warning System

SNF San Felipe (Venezuela) (location identifier) • Schweizerischer Nationalfonds (Switzerland) • Second Native Format • Second Normal Form • Secret - No Foreigners • Server Natural Format • Server Normal Format • Schweizerischen Nationalfonds zur Förderung der wissenschaftlichen Forschung (Switzerland) • Spent Nuclear Fuel • System Noise Figure • System Noise Filter

SNFC Semi-National Forwarding Code (UK)

SNFE Standard Network Front End

SNFL Superficial Nerve Fiber Layer

SNFLK Snowflake

SNFR Small Net Flux Radiometer (PV)

Sng Singapore

SNG San Ignacio de Velasco (Bolivia) (location identifier) • Satellite News Gathering • Schweizerische Naturforschende Gesellschaft (Switzerland) • Simulated Natural Gas • Substitute Natural Gas • Synthetic Natural Gas

Sngl Single

SNGR Sin Nuestra Garantia ni Responsabilidad

SNH Stanthorpe (Australia-QLD) (location identifier) • Sternwarte Neanderhöhe Hochdahl eV (Germany)

SNI San Nicolas Island (USA-CA) • Selected Notification of Information • Sequence Number Indicator • Service Node Interface • Siemens Nixdorf Informationsysteme AB • Signal-to-Noise Improvement • Sinoe - R.E. Murray (Liberia) (location identifier) • Sistema Nacional de Información • SMDS Network Interface • SNA Network Interconnect • SNA Network Interconnection • SNC Network Interface • Source Non-Identifiée • Standard Network Interface • Subscriber Network Identifier • Subscriber Network Interface • Supernova of Type I • System Network Interconnect

SNIA Storage Network Industry Association (USA-CA)

SNIAS Société Nationale Industries Aérospatiales (France)

SNICF Subnetwork Independent Convergence Function

SNICP Subnetwork Independent Convergence Protocol

SNICT Sistema Nacional de Informação Cientifica e Tecnológica (Brazil) • Sistema Nacional de Información Científica y Técnica (Uruguay) • Sistema Nacional de Información Científica y Tecnológica (Mexico)

SNID Standard Network Interface Device • Start Note Identification

SNIFTIRS Subtractively Normalized Inter-Facial Fourier Transform Infrared Spectroscopy

SNIG Sistema Nacional de Informação Geografica (Portugal)

SNII Supernova of Type II

SNIMA Service de Normalisation Industrielle Marocaine (Morocco)

SNIP Space Network Interoperability Panel

SNIRD Supposedly Noiseless Infrared Detector

SNIU Secure Network Interface Unit

SNJ San Julián (Cuba) (location identifier)

SNK Snyder Winston (USA-TX) (location identifier)

SNL Sample Noise Level • Sandia National Laboratories (USA-NM) • Shawnee (USA-OK) (location identifier) • Speckle-Noise Limited • Standard Name Line • Standard Nomenclature List

SNLE Sous-marin Nucléaire Lanceur d'Engins

SNLR Services No longer Required

SNM San Ignacio (Bolivia) (location identifier) • Satellite Navigation Map • Seya-Namioka Monochromator • Signaling Network Management • Single Nation Model • Sobre Nivel del Mar • Society of Nuclear Medicine • Special Nuclear Material • Square Nautical Mile • SunNet Manager • System Notification Message

SNMC Système National de Mesure du Canada

SNMP Simple Network Management Protocol (*) • Simple Network Monitoring Protocol

SNMS Sonet Network Management System • Sputtered Neutral Mass Spectrometry

SNMSP Southern New Mexico Star party (USA-NM)

SNN Shannon (Ireland) (location identifier) • Shared Nearest Neighbor • Structure-Nomenclature Notation

SNNE Sustainable Net National Expenditure

SNNS Stuttgart Neural Network Simulator

Sno Snow

SNO Sakon Nakhon (Thailand) (location identifier) • Sierra Nevada Observatory (Spain) • Snow • Sudbury Neutrino Observatory (Canada-ON)

SNOBOL String-Oriented Symbolic Language

SNODAQ Sudbury Neutrino Observatory Data Acquisition

SNOE Smart Noise Equipment • Student Nitric Oxide Explorer

SNOI Signal Not Of Interest • Sudbury Neutrino Observatory Institute (Canada-ON)

SNOK Secondary Next Of Kin

SNOM Scanning Near-field Optical Microscope • Scanning Near-field Optical Microscopy

SNOMED Standard Nomenclature of Medicine

SNOPT Sparse Non-linear Optimizer

SNOS Silicon Nitride Oxide Semiconductor

SNP Saint Paul Island (USA-AK) (location identifier) • Science Network Planning • Secure Network Programming • Sequence Number Protection • Serial Number/Password • Silicon Nanoparticle • Single Nucleotide Polymorphism • Space Nuclear Propulsion • Student Number Projections • SubNet Protocol • Suspected, Not Proved • Synchronous Network Processor • Système Net Protégé • System Network Processor • System Network Protocol

SNPA Subnetwork Point of Attachment • System of Nuclear Auxiliary Power (*)

SNPD Sistema Nacional de Prevención de Desastres (Colombia)

SNPE Société Nationale de Poudres et Explosifs (France)

SNPLI Selected Non-Priority List Item

SNPO Space Nuclear Propulsion Office

SNQ San Quitín (Mexico) (location identifier)

Snr Senior

SNR Saint-Nazaire (France) (location identifier) • Science Networking Representative • Senior National Representative • Short Notice Requirement • Signal-to-Noise Ratio • Sondernutzungsrecht • Supernova Remnant (*) • Supernova Reste • Supplier Non-conformance Report

SNRI Short-Notice Random Inspection

SNRM Set Normal Response Mode

SNRS Sunrise

SNRT Signal-to-Noise Ratio Threshold

SNRW Signal-to-Noise Ratio Weight

SNS Salinas (USA-CA) (location identifier) • Scuola Normale Superiore (Italy) • Secondary Name Server • Secure Network Server • Shared Network Server • Shuttle Network System • Simple Name Service • Simulated Network Simulation • Space Nuclear System • Spallation Neutron Source (ORNL)

SNSF Swiss National Science Foundation (Switzerland)

SNSO Space Nuclear Systems Office

SNSB Swedish National Space Board (Sweden)

SNSH Snow Showers

SNSL Stock Number Sequence List

SNSM Société Nationale de Sauvetage en Mer (France)

SNSO Space Nuclear Systems Office

Snsr Sensor

SNSR Sensor

SNST Sunset

SNT Sabana de Torres (Colombia) (location identifier) • Schmidt Newtonian Telescope • Synchronous Network of Telescopes

SNTP Simple Network Time Protocol • Simple Network Transport Protocol • Space Nuclear Thermal Propulsion

Sntzd Sensitized

SNU Santa Clara (Cuba) (location identifier) • Seoul National University (Korea-ROK) • Solar Neutrino Unit

SNUS Swedish Network Users' Society

SNV Santa Elena (Venezuela) (location identifier) • Schweizerische Normen-Vereinigung (Switzerland) • Sciences de la Nature et de la Vie • Single Network View • Statens Naturvårdsverk (Sweden) • Swiss Association for Standardisation

SNW Sandoway (Myanmar) (location identifier) • Snow • Thandwe (Myanmar) (location identifier)

SNWFL Snowfall

SNWG Supernova Working Group

SNX Sabana de Mar (Dominican Rep.) (location identifier)

Sny Sunny

SNY Sidney (USA-NE) (location identifier)

SNZ Santa Cruz (Brazil) (location identifier) • Standards New Zealand (formerly SANZ)

s/o Shipping Order • Son of • Su Orden

so Seller's Option • siehe oben! • Solución • Somali (ISO language code) • South • Southern

So Sonntag • South • Southern

S-O Shutoff

S/O Shutoff • Switchover

SO Safety Officer • Satellite Oceanography • Saturn Orbiter • Science Operations • Scientific Officer • Secondary Out • Second Overtone • Security Office • Seller's Option • Senior Officer • Serial Output • Service d'Ordre • Service Organization • Shanghai Observatory (China-PRC) • Shift Out • Shipment Order • Shop Order • Shuttle Operations • Shut Off • Siehe Oben! • Significant Other • Slow Operate • Slow Operating • Small Outline • Solar Observing • Solar Orbiter • Somalia (ISO country code) • Somalie (code ISO de pays) • Sorting Office • Southern Oscillation • Soviet • Special Operation • Special Option • Special Order • Spherical Overdensity • Stable Oscillator • Staff Officer • Standing Order • Stationery Office • Statistical Office • Steward Observatory (USA-AZ) • Stockage Objectives • Strike Out • Sud-Ouest • Sulfur Monoxide • Supply Office • Supply Officer • Support Operations • Suroeste • Surveillance Officer • Switching Oscilloscope • Switchover • Synthèse d'Ouverture • System Override

SOA Safe Operating Area • Self-Optimizing and Adaptive • Semiconductor Optical Amplifier • Separate Operating Agency • Service-Oriented Architecture • Session d'Orientation Active • Session d'Orientation Approfondie • Shuttle Orbital Application • Silicon Optical Amplifier • Sky Observers' Association (Hong Kong) • Soc Trang (Viet Nam) (location identifier) • Source Of Assignment • Special Open Allotment • Special Operating Agency • Special Operating Area • Special Operations Aircraft • Speed Of Advance • Start Of Authentication • Start Of Authority • Statement Of Account • State Of Alert • State Of the Art • Status Of Action • Structure Of Array • Switch Off Assembly • Système d'Optique Adaptative

SOAB Special Operation Acquisition Board

SOAC Satellite Ozone Analysis Center • Special Operation Acquisition Center • System On A Chip

SOAF Status Of Action File

SOAL Search Optical Augmentation Laser

SOAO Sobaeksan Optical Astronomy Observatory (Korea-ROK)

SOAP Satellite Orbit Analyzer Predictor • Seal Of Approval • Selective Optical Atmospheric Probe • Simple Object Access Protocol • Simplify Obscure ALGOL Programs • Single Operator Automation System • Spectrometric Oil Analysis Program • Subjective - Objective - Analysis - Plan • Symbolic Optimizing Assembler Program • Symbolic Optimum Assembly Program • Système Opérationnel d'Analyse et de Prévision • Systems Operational Analysis Plan

SOAR Safe Operating Area • Shuttle Orbital Applications and Requirements • Software for Optical Archival and Retrieval • Southern Astrophysical Research (telescope) • Southern Observatory for Astronomical Research (*) • Southern Observatory for Astrophysical Research • State-Of-the-Art Report • Stratospheric Observatory for Astronomical Research (*)

SOARD Steward Observatory Asteroid Relational Database

SOARS Shuttle Operations Automated Reporting System • Significant Opportunities in Atmospheric Research and Science • Statement Of Account Reporting System

SOAS Science Operations Application Server

SOATS Shuttle Operations Automated Training System

SOB Saarmelleek Balaton (Hungary) (location identifier) • Science Observatory Branch • Science Operations Branch • Senate Office Building • Soul On Board • Stage One Booster • Start Of Block • Start Of Business • State Office Building • State Operation Budget • Strap-On Booster

SOBA Scaled OB Association

SOBC SSM Onboard Computer

SOBEPS Société Belge d'Études des Phénomènes Spatiaux (Belgium)

SOBI Second Order Blind Identification

SOBLI Study of Optical Baseline Instrumentation

SOBO Second-Order Bibliometric Operator

SOBOL Second-Order Boundary Layer

SOBP Spread-Out Bragg Peak

SoByu Soobshcheniia Byurakanskoi Observatorii (periodical code)

Soc Social • Sociedad • Società • Société • Society • Sociología • Sociologie • Sociology

S&OC Science & Operations Center

SOC Satellite Operations Center • Satellite Operations Complex • Satellite Orbit Control • Science and Operations Center • Science Observations Center • Science Operations Center • Science Operations Coordinator • Scientific Operations Center • Scientific Organizing Committee • Scientific Oversight Committee • Section Operations Center • Sector Operations Center • Secure Operator Console • Security Operations Center • Self-Organized Critical • Self-Organized Criticality • Self-Organizing Criticality • Service Oversight Center • Set Overrides Clear • Shuttle Operations Center • Simulated Operational Computer • Simulated Operations Center • Simulated Operations Computer • Simulation Operations Center • Simulation Operations Computer • Single-Orbit Computation • Soil Organic Carbon • Solo City (Indonesia) (location identifier) • Space Operations Center • Space Operations Controller • Specialized Oceanographic Center • Special Operations Capability • Special Operations Capable • Special Operations Command • Specific Optimal Control • Specific Optimal Controller • Standard Obstacle Course • Stargazers Observational Course • Start Of Climb • Statement Of Capability • Statement Of Compatibility • State Of Change • State Of Charge • Stationary Overturning Convection • Supply Operation Cost • Surveillance Operations Center • Synthetic Organic Chemical • System On a Chip • System Operational Concept • System Option Controller

SOCA Second-Order Correlation Approximation • Special Operations Communications Assembly

So-Cal Southern California (USA)

SOCAL Southern California (USA)

SOCAR Shuttle Operational Capability Assessment Report • Systems Operational Compatibility Assessment Review

SOCC Satellite Oceanic Control Center • Satellite Operations Command and Control • Satellite Operations Control Center • Science Operations Control Center • Sector Operations Control Center • Service Operation Control Center • Shuttle Operations Computing Center • Space Operations Control Center

SOCCB Secure Operations Configuration Control Board

SOCCE Special Operations Command and Control Element

SOCCER Sample Of Comet Coma Earth Return

SOCCS Summary Of Component Control Status

SoCe Solar Cells (periodical code)

SOCE Standard Office Computing Environment

SOCET Special Operation Cost Estimate Tool

SOCEX Southern Ocean Cloud Experiment

SOCF Spacecraft Operations and Checkout Facility

SoCG Symposium on Computational Geometry

SOCGMR Symposium on Computational Geometry for Mechanics and Applications

SOCH Spacelab Orbiter Common Hardware

Sociol Sociología • Sociological • Sociology

SOCK SPS Outreach Catalyst Kit

SOCKO Systems Operational Checkout

SOCKS Socket Secure

SOCO Switched Out for Checkout

SOCOM Solar Optical Communications (system) • Special Operations Command

SOCP Service d'Orientation et de Consultation Psychologique • Space Oblique Conformal Projection

SOCR Stand-alone Optical Character Recognition • Sustained Operations Control Room • Synchronous Orbit Communications Relay

SOCRATES Strategy for Organisations Concerned in Rural Advanced Telecommunications Experiments • System Of Cellular Radio for Traffic Efficiency and Safety

SOCS Spacecraft Orientation Control System • Sparse Optimal Control Software • Subcommittee On Computer Security • System Operation and Checkout System

Socy Society

Sod Sodium

SoD Service on Demand

SOD Science Operations Database • Science Operations Director • Second Order Design • Separation Of Duties • Serial Output Data • Service d'Orbitographie DORIS (CNES) • Site Occupation Designator • Small Object Detector • Sorocaba (Brazil-SP) (location identifier) • Spectral Optical Depth • Staff Operations Division • Standing Order for Data • Super Oxyde Dismutase • Sustained Operational Date • System Operational Description • System Operational Design

SODA Science Operations and Data Analysis • Small-Object Database • Smart Objects - Dumb Archives (model) • Solar Dynamo Amplitude • Source-Oriented Data Acquisition • Source-Oriented Data Automation • Symposium On Discrete Algorithms

SODALS Simplified Omnidirectional Approach Lighting System

SODAR Sound Detection And Ranging

SODART Soviet-Danish Röntgen Telescope

SODAS Structure-Oriented Description And Automation • Synoptic Oceanographic Data Assimilation System

SODB Science Observations Database • Science Operations Database • Shuttle Operational Data Book • Spacecraft Operational Data Book • Start Of Data Block

SODIM Standard Operational Data Item Manual

SODO Senior Operations Duty Officer

SODRS Synchronous Orbit Data Relay Satellite

SODS Saturn Operational Display System • Shuttle Operational Data Submittal • Shuttle Operational Data System • Skylab Orbit-Deorbit System • Subordinate Operations Data System

SODSIM Strategic Offense-Defense Simulation

SODSSWG Satellite Ocean Data System Science Working Group

SoE Schedule of Events • State of the Environment

SOE Schedule Of Events • Science Operations Element • Sequence Of Events • Server Operating Environment • Significant Operating Experience • Skylab Operational Environment • Slater Orbital Exponent • Software Operating Environment • Sonnenobservatorium Einsteinturm (Germany) • Souanke (Congo) (location identifier) • Special Operations Executive • Standard Operating Environment • State Of the Environment • State-Owned Enterprise • Status Of Equipment • Super Orbit Entry

SOED Shorter Oxford English Dictionary

SOEIS Self-Organization of the European Information Society (research group)

SoEn Solar Energy (periodical code)

SoEnM Solar Energy Materials (periodical code)

SOEP Solar-Oriented Experimental Package • Sozio-Ökonomische Panel

SOER Significant Operating Event Report

SOEST School of Ocean and Earth Science and Technology

SOF Safety Of Flight • Science Operations Facility • Scintillating Optical Fiber • Sofia (Bulgaria) (location identifier) • Soluble Organic Fraction • Sound On a Film • Special Operation Forces • Special Operations Force • Statement Of Functionality

SOFA Selectable Optical Filter Assembly • Sevon-0-Five Astronomers (USA-CA) • Standard Objects For Astronomy • Standards Of Fundamental Astronomy • Statement Of Financial Activities

SOFAR Sound Fixing And Ranging

SOFAS Survivable Optical Forward Acquisition System

S/Ofc Site Office

SOFC Saturn Operational Flight Control • Solid Oxide Fuel Cell

SOFCAL Scintillating Optical Fiber Calorimeter

SOFCS Self-Organizing Flight Control System

S/OFF Sign Off

SoFi Sonnenfinsternis

SOFI Son of ISAAC • Spray-On Foam Insulation • Système d'Ordinateurs pour le Frêt International

SOFIA Stratospheric Observatory For Infrared Astronomy • Surface de l'Océan - Flux et Interactions avec l'Atmosphère • Surface of the l'Ocean Fluxes and Interaction with the l'Atmosphere

SOFIE Scintillating Optical Fiber Isotope Experiment

SOFIN Soviet-Finnish (spectrograph)

SOFM Self-Organizing Feature Map • Self-Organizing Feature Mapping

SOFNET Solar Observing and Forecasting Network

SofR Society of Rheology

SOFRADIR Société Française de Détecteurs Infrarouges (France)

SOFRES Société Française d'Enquêtes et de Sondages (France)

SofS Secretary of State

SOFSEM Software Seminar

SOFT Simple Output Format Translator • Software Operations Focus Team • Solar Flare Telescope • Space Operations and Flight Techniques • Suborbit Flight Test

SOFTE Software Estimator

SOFTEKS Software Engineering and Knowledge-base System

sog sogenannt

SOG Same Output Gate • Scientific Operations Group • Second-Order Gradient • Secteur des Opérations Gouvernementales • Sogndal (Norway) (location identifier) • Special Operations Group • Speed Over Ground • Structure and Organization of Government • Stuck On the Ground

SOGA Schedule Optimizing Genetic Algorithm • SNA Open Gateway Architecture

SoGru Akademiia Nauk Gruzii Soobshcheniia (periodical code)

SOGS Science Operations Ground System

SOH Section Overhead • Since Overhaul • Skylab Operations Handbook • Solita (Colombia) (location identifier) • Start Of Header • Start Of Heading • State Of Health • Suborbital Hopper

SOHF Sense Of Humor Failure

SOHFI Skin-layer Ocean Heat Flux Instrument

SOHMA Servicio de Oceanografía, Hidrografía y Meteorología de la Armada (Uruguay)

SoHo South of Houston (Street) (New York)

SOHO Small Office - Home Office • Solar and Heliospheric Observatory (*) • Solar Heliospheric Observatory

SOHR Solar Hydrogen Rocket Engine

SOI Saturn Orbital Insertion • Saturn Orbit Insertion • Scan Ozone Imager • Scientific and Optical Instruments • Signal Of Interest • Signal Operation Instruction • Silicon On Insulator • Solar Oscillation Imager • Solar Oscillation Investigation • Southern Oscillation Index • South Molle Island (Australia-QLD) (location identifier) • Space Object Identification • Specific Operating Instruction • Sphere Of Influence • Sphere Of Interest • Standard Operating Instruction • Standard Operator Interface • Start Of Installation • Statement Of Intent • Survey Of India

SOIC Small Outline Integrated Circuit

SOICS Summary Of Installation Control Status

SOID Standard Object Identifier

SOIF Structure Object Interchange Format • Summary Object Interchange Format • System Operational Interface Function • System Operations and Integration Function

SOILIMS Soil Laboratory Information and Management System

SOIN Solar Interferometer

SOIO Superoptimal Interoperability

SoIP Storage over Internet Protocol

SOIR Simultaneous Operations on Intersecting Runways

SOIS Space Object Identification System • Spacelab/Orbiter Interface Simulator

SOIWR Simultaneous Operations on Intersecting Wet Runways

SOJ Sorkjosen (Norway) (location identifier)

SOJAAM Société des Jeunes Astronomes Amateurs de Montréal (Canada-QC)

SOJT Supervised On-the-Job Training

SOK Semongkong (Lesotho) (location identifier)

SoKie Sonderdrücke der Sternwarte Kiel (periodical code)

Sol Solaire • Solar • Solenoid • Solicitor • Solid • Soluble • Solution

SOL Schedule Output Line • Science On Line • Scrape-Off Layer • Sequence Order List • Short On Landing • Simulation-Oriented Language • Small Outline • Solar • Solenoid • Solicitation • Solid • Solomon (USA-AK) (location identifier) • Space-On-Line • Strickly Out of Luck • Structure-Oriented Language • System-Oriented Language

SOLA Subtractive Optimally Localized Averages (method) • Subtractive Optimally Localized Averaging (method)

SOLAIR Solomon Islands Airways Ltd.

SOLAR Serialized Online Automatic Recording • SOHO Long-term Archive • Solar On-Line Activity Resources • Solid Application Resource

SOLAS Safety Of Life At Sea

SOLB Start Of Line Block

SOLCON Solar Constant (radiometer)

SOLD Simulation Of Logic Design • Soldering • Source Online Debugger

SOLE Society Of Logistics Engineers (USA) • Special Operations Liaison Element • Star-Obvious Low-level Elongated (PPN)

SOLEM Solar Experiment Module

SOLERS Solar Electromagnetic Radiation Study

SolFDI Solar Flare Dynamics Imager

SOLFEGE Service d'Observation Localisé Favorisant l'Exploitation des Paramètres Géophysiques

SOLID Self-Organizing Large Information Dissemination

SOLIS Sydney Outdoor Lighting Improvement Society (Australia-NSW) • Symbionics Online Information System • Synoptic Optical Long-term Investigations of the Sun

SOLMC Surface-Oriented Locally Monolithic Coordinates

Soln Solution

SOLN Solution

SOLO Selective Optical Lock-On

Sol. Phys. Solar Physics

Solr Solicitor

SOLR Sidetone Objective Loudness Rating

SOLRAD Solar Radiation

SOLSE Shuttle Ozone Limb Sounding Experiment

SOLSPEC Solar Spectrum (radiometer)

SOLSTICE Satellite Online Searching Training Interactive Conferencing Experiment • Solar-Stellar Irradiance Comparison Experiment

SOLT Short - Open - Load - Through • Short - Open - Load - Thru

SOLTIP Solar Transient Interplanetary Phenomenon

Solv Solvent

SOLV Solenoid Valve

SOLVE SAGE III Ozone Loss and Validation Experiment • Solar Variability Enterprise

Soly Solubility

SolZ Solidaritätszuschlag

SolZG Solidaritätszuschlaggesetz

Som Somali • Somerset (UK-England)

SoM School of Management

SOM San Tomé (Venezuela) (location identifier) • Scheme Of Maneuver • Science Operations Manager • Scripting Object Model • Section of Orbital Mechanics • See Our Message • Self-Organized Map • Self-Organizing Machine • Self-Organizing Map • Shift Operations Manager • Ship Operations Manager • Simulation Object Model • Simulator Operation and Maintenance • SIRTF Observer's Manual • Society Of Mind • Software Operation Manual • Software Operators Manual • Somalia (ISO country code) • Somalie (code ISO de pays) • Spacecraft Operations Manager • Spacecraft Operations Manual • Space Oblique Mercator (projection) • Spares Optimization Model • Special Operations Manual • Stage Operating Manual • Standard Operating Manual • Standard Operations Manual • Start Of Message • Structured Object Method • Suborbital Mission • System Object Model • System Operations Manager • Systems Operators Manager • Systems Operators Manual

SOMA Semantic Object Modeling Approach • Signal Organization Mission Assignment • Support Opto-Mechanical Assembly

SOMADA Self-Organizing Multiple-Access Discrete Address

SOMARDS Standard Operations Maintenance And Research Development System

SOMEI Sociedad Mexicana de Estudios Interplanetarios (Mexico)

SOMF Start Of Minor Frame

SOMISS Study Of Management Information Systems Support

SOMI Start Of Message In

SOML Send Or Mail • Steward Observatory Mirror Laboratory (USA-AZ)

SOMM Software Operations and Maintenance Manual • System Organizational Maintenance Manual

SOMMA Single-Operator Multiparameter Metabolic Analyzer

SOMO Space Operations and Management Office • Space Operations Management Office (NASA) • Start Of Message Out

SOMOWG Science Operations Management Operations Working Group

SOMP State Optical and Mechanical Plant (USSR) (obsolete)

SOMPF Special Operations Mission Planning Folder • Special Operations Mission Planning Format

SOMS Shuttle Orbiter Medical System • Start Of Message Sequence • Synchronous Operational Meteorological Satellite

SOMSA Space Object Measurement and Signal Analysis

Som. Sh. Somali Shilling (currency)

SoMue Sonderdrücke Universität Münster Astronomisches Institut (periodical code)

SOMYCE Sociedad de Observadores de Meteoros y Cometas de España (Spain)

S/ON Sign On

SON Espíritu Santo (Vanuatu) (location identifier) • September - October - November • Son Of NAIF • Sonneberg Variable (star) • Sonora (Mexico) • Standards Organization of Nigeria • Statement Of Need • Statement of Operational Need • Stichting voor Scheikundig Onderzoek in Nederland (Netherlands) • Student Observation Network

SONA Society of Neuroscientists of Africa

SONAAS Stratford-on-Avon Astronomical Society (UK)

SONAC Sonar Nacelle

SONAR Sound Navigation And Ranging (*) • Sound Navigation Ranging

Sonderdr Sonderdruck

SONET Synchronous Optical Network

SONEX Subsonic assessment of Ozone and Nitrogen Experiment

SONG Space Oceanography, Navigation and Geodynamics • Stellar Oscillations Network Group • Survey Of Nearby Galaxies

SONIA Southeast Oxidant and Nitrogen Intensive Analysis

SONIC SAO Near-Infrared Camera • Simultaneously Operating Numerical Integration Computer • SPAN Ocean Network Information Center • System Online Network for Information Control • System-wide Online Network for Informational Control

SONITA Société Nigérienne de Transports Aériens

SONNET System Of Neighboring Networks in Ohio (USA-OH)

SONOS Silicon Oxide Nitride Oxide Semiconductor

SONS Self-Organized Nanostructure

SONYR Spin-Orbit N-body Relativistic (model)

SOO Second-Order Operator • Senior Operation Officer • Soderhamn (Sweden) (location identifier) • Söderhamn (Sweden) (location identifier) • Standard Operational Order • Statement Of Objectives

SOOH Suborbital Operations Handbook

SOON Solar Observing Optical Network

SOOS Stacked OSCAR On Scout

Sop Soprano

SOP Safety and Operations Plan • Safety Operating Plan • Safety Operating Procedure • Safety Oversight Program • Saturn Orbiter Probe • Science Operations Plan • Secondary Oxygen Pack • Secure Operations Preboard • Semester Observation Plan • Semester Observing Plan • Senior Officer Present • Service-Object Pair • Ship of Opportunity Programme • Simulated Output Program • Simulation Operations Plan • Small Outline Package • Southern Pines (USA-NC) (location identifier) • Spacecraft Operations Plan • Spacelab Opportunity Payload • Special Observation Period • Special Observing Period • Standard Operating Procedure (*) • Standing Operating Procedure • Statement Of Purpose • Strategic Operating Point • Structural Optimization Procedure • Subject-Object-Protocol • Sum Of Products • Surveillance Opportunities File • Symbolic Optimum Program • Systems Operating Plan • Systems Operating Procedure • Systems Operating Program • Systems Operations Plan

SOPAC Scripps Orbit and Permanent Array Center • South Pacific • South Pacific Applied Geosciences Commission • South Pacific Countries

SOPC Science Operations and Planning Computer • Science Operations Planning Computer • Shuttle Operations and Planning Center • Shuttle Operations Planning Complex

SOPE Simulated Off-the-Pad Ejection • Small ORU Protective Envelope

SOPG Science Operation Planning Group

SoPh Solar Physics (periodical code)

SOPH SOC Operation Procedures Handbook

SOPHI Segmented Optics Phase Integration

SOPHIA SFOC Operational Pilot Hands-on Interactive Activity

SOPI Status Of Planning Indicator

SOPITA South Pacific Isotopic and Thermal Anomaly

SOPM Standard Orbital Parameter Message

SOPRANE Système Océanique de Prévision en Atlantique Nord-Est

SOPRANO Submillimeter Observations of Processes in the Atmosphere Noteworthy for Ozone

SOPRI Space Object Photometry, Radiometry and Identification

SOPS Satellite Output Processing System • Science and Observation Planning System • Southern Objective Prism Survey • Spacecraft Operations Planning Section • Special Operations Power Sources

SOPSA Shuttle Orbit-injection Propulsion System Analysis

SOPSO Space Operations Support Office

SOPT Science Operations Planning Team

SOQ Signe Officiel de Qualité • Sorong (Indonesia) (location identifier) • Statement Of Qualifications • System Optical Quality

SOQAS Statement Of Quality And Support

SOQM STSF Requirements Qualification Matrix

Sor Soria (Spain)

SoR Sale or Return

SOR Al Thaurah (Syria) (location identifier) • Sale Or Return • Sampling Oscilloscope Recorder • Science Operations Room • Senior Officer Review • Set On Receiver • Simultaneous Over-Relaxation • Source Of Repair • Specification Operational Requirement • Specific Operational Requirement • Specific Orders and Requests • Stable Orbit Rendezvous • Standard Operating Report • Standard Operating Rule • Starfire Optical Range (USA-NM) • Start Of Record • Start Of Routing • Statement Of Reasons • Statement Of Requirements • State Of Readiness • Status Of Resources • Struck Off Records • Successive Overrelaxations • Successive Overrelaxing (method) • Surface Of Revolution • Symposium on Operations Research • System Operational Requirement • System Operations Review

SORA Suborganizational Registration Administrator • Suborganizational Registration Authority

SORAP Source Of Repair Analysis Process • Source Of Repair Assignment Process

SORB Subsistence Operations Review Board

SORC Scheduling Operations Request Card • Sound Ranging Control

SORCE Solar Radiation and Climate Experiment

SORD Simplified Operational Requirement Document • Statement of Operational Requirements Document • Submerged-Object Recovery Device • System and Operations Requirements Document • System Operational Requirements Document

SORDAC Special Operations Research, Development and Acquisition Center

SORDC Source Of Repair Decision Criteria

SOREL Sun-Orbiting Relativity

SORIAP Satellite Ocean-Related Imagery Applications Program

SORL Space Optics Research Labs (USA-MA)

SORM System of Efficient Research Measures

SORNG Sound Ranging

SORO Scan On Receive Only

SORP Statement Of Requirement Preliminary

SORPTR South Repeater

SORR Segment Operational Readiness Review

SORS Same Origin Ray Set • Solar Radiospectroscopy • Spacecraft Oscillograph Recording System

SORSA Spatially-Oriented Referencing Systems Association

SORST Solution-Oriented Research for Science and Technology

SORT Solar Orbit Relativity Test • Status of Resource and Training • Structures for Orbiting Radio Telescope

SORTI Solar-Radiance-Transmission Interferometer • Star-Oriented Real-time Teaching Instrument

SORTIE Superorbital Reentry Test Integrated Environment

SORV Surface Operations Research Vehicle

sos Si Opus Sit

SoS Secretary of State

SOS Satellite Observation System • Satellite Operational Services • Save Our Science • Save Our Ship • Save Our Souls • Science Operations Specialist • Science Operations System • Sección de Observación Solar • Secretary Of State • Section d'Optique et de Spectroscopie • Selective Opacity Sampling • Self-Optimized Sector • Self-Organizing System • Serial Output Special • Services Of Supply • Share Operating System • Shilling (Somalia) (ISO currency code) • Short Orbit Spectrometer • Silicon On Sapphire • Software Obsolescence Study • Sophisticated Operating System • Source Of Supply • Southern Oxidants Study • Spacecraft Operations Simulation • Space Object Surveillance • Space Operation Service • Space Option Society (Switzerland) • Space Ordnance System • Special Ordered Set • Spectroscopy and Optics Section • Speed Of Service • Speed Of Sound • SPOCC Operational Status • Stabilized Optical Sight • Standards and Open Systems • Start Of Significance • Statement Of Service • Station Operating Supervisor • Subaru Observatory System • Suborbital Sequence • Sum Of Squares • Supervising Observation Software • Supplemental Oxygen System • Supportability Online System • Surface Of Section • Swedish Optical Society (Sweden) • Symbolic Operating System • Symmetry, Orbitals and Spectra • Synchronous-Orbit Satellite • System Of Systems • System Operational Safety

SOSA Status Of Staff Actions • Suborganizational Systems Administrator • Sustained Operations Support Area • System-Of-Systems Architecture

SoSAO Soobshcheniia Spetsial'noi Astrofizicheskoi Observatorii (periodical code)

SOSC Second-Order Sufficient Condition • Special Operations Support Command • System Operational and Support Capability • Systems Operational Support Center

SOSD Spatial Operational Sequence Diagram

SOSE Science Operations Support Equipment • Special Operations Staff Element • Special Operations Support Element

SOSG Senior Officer Steering Group

SoSh Somali Shilling (currency)

SoShe Soobshchenie Shemakhinskoi Astrofizicheskoi Observatorii (periodical code)

SoSht Soobshcheniia Gosudarstvennogo Astronomicheskogo Instituta (periodical code)

SOSI Serial Out - Serial In • Service de l'Organisation et du Système d'Information (CNRS) • Shift Out - Shift In • Single Output - Single Input • Space Operations Scientific Investigation • System Of Systems Integration

SOSIG Social Science Information Gateway

SOSL Saturn Orbiter Satellite Lander

SOSM Soft-computing, Optimization, Simulations and Manufacturing (conference)

SOSO Safety and Operating Systems Office • Solar Oriented Space Observatory • State Owned - State Operated • Synchronous Orbiting Solar Observatory

SOSP Solar Spectrum • Symposium on Operating System Principles

SoSR Software Specification Review

SOSR Suppress - Observe - Secure - Reduce

SOSRM Strap-On Solid Rocket Motor

SOSS Satellite Optical Surveillance Station • Structurally Oriented Simulation System • Subaru Observation Software System (*) • Subaru Observing Software System

SOSTR Shutoff Starter

SOSUS Sound Surveillance System

SoSy Solar System

SoSyR Solar System Research (periodical code)

SOT Science Observing Team • Science Operations Team • Ship Operability Test • Ship Operability Testing • Small Optical Telescope • Sodankylä (Finland) (location identifier) • Solar Optical Telescope • Solar Orbiting Telescope • Sótão • Special Operational Testing • Special Operations Team • Start Of Tape • Start Of Text • State Of Technology • Strap-On Tank • Syntax-Oriented Translator • System Operability Test

SOTA State Of The Art

SOTACA State-Of-The-Art Contingency Analysis

SOTAS Standoff Target Acquisition System

SOTD Steward Observatory Technical Department (USA-AZ)

SOTE Standard Optical Test Equipment

SOTEP Software Operational Test and Evaluation Plan

SOTF Special Operations Task Force

SOTI Security Operations Training and Intelligence

SOTM Science On The Move

Soton Southampton (UK-England)

SOTON Southampton (Starlink internal code)

SOTP Saturn Orbiter - Titan Probe

SOTS Some Old Tried Stuff • Suborbital Tank Separation • Synchronous Orbiting Tracking Station

SOTT System Organization Tactics and Techniques

SOTUS Sequentially-Operated Teletypewriter Universal Selector

SOTV Solar Orbit Transfer Vehicle

Sou South • Southampton (UK-England) • Southern

SOU Southampton (Starlink internal code) • Southampton Eastleigh (UK-England) (location identifier) • Southern Oregon University (USA-OR)

SOUDAAM Source-Oriented User-Driven Asset-Aware Model

SOULS Site-Originated Utility Library System

SOUP Solar Optical Universal Polarimeter

SOUSY Sounding System

Sov Sovereign • Soviet

SOV Science Opportunity Visualizer • Seldovia (USA-AK) (location identifier) • Shut-Off Valve • Solenoid-Operated Valve • Sovereign • Subject - Object - Verb

SOVA Solar Variability (experiment) • Solar Variations

SovAe Soviet Aeronomii (periodical code)

SOVAFA Sociedad Venezolana de Aficionados a la Astronomía (Venezuela)

SOVI Seguro Obligatorio de Vejez e Invalidez (Spain)

SOVIM Solar Variability and Irradiate Monitor

SOVT System Operations Validation Test

SOW Scope Of Work • Show Low (USA-AZ) (location identifier) • Skylab Orbital Workshop • Small Orbit Width • Start Of Word • Statement Of Work • Subdivision Of Work

SOWETO Southwestern Township

SOWG Science Operations Working Group

SOWICOMS Solar-Wind Composition Mass Spectrometer

SOWP Science Operations Working Group

SOWT Special Operations Weather Team

SOx Sulphur Oxide

SOX Sogamoso (Colombia) (location identifier) • Sound Exchange • Supercritical Oxygen

SOXS Solar X-ray Spectrometer

SOY Stronsay (UK) (location identifier)

SOZ Solenzara (France) (location identifier) • Standard One Zone (approximation) • Stichting Onderzoek der Zee (Netherlands)

SO2 Sulphur Dioxide

SO4 Sulphate Radical

sp Sine Prole

Sp Space • Spain • Spalte • Spaniard • Spanish • Spare • Spatial • Special • Species • Specific • Specimen • Speed • Spelling • Spiral • Spirit • Spitze • Sport • Support

S/P Serial to Parallel • Signal Processor

SP Poland (international aircraft registration code) • Sacramento Peak (USA-NM) • Safety Plan • São Paulo (Brazil) • Satellite Processor • Scalable Parallel • Scaled Point • Schwerpunkt • Scientific Paper • Scientific Processor • Scientific Product • Scientific Publication • Scintillation Profiling • Secteur Postal • Security Plan • Security Policy • Security Protocol • Segment Priority • Self-Propelled • Sensor Processor • Séquence Principale • Sequence Programmer • Serial to Parallel • Service Pack • Service Processor • Service Protocol • Service Provider • Servicio Público • Shore Patrol • Short Period • Short Play • Shoulder Pitch • Shuttle Project • Side Penetration • Signaling Point • Signal Point • Signal Processing • Signal Processor • Single Phase • Single Pole • Single Precision • Single Pulse • Single Purpose • Small Photometer • Smart Push • Smooth Penetration • Software Process • Solar Panel • Solar Physics • Solar Probe • Solid Propellant • Sortie Payload • Source Port • Source Program • Sous-Préfecture • South Pole • Space • Space and Power • Spain • Spare • Spare Part • Sparking Plug • Spark Plug • Special • Specialist • Special Performance • Special Processing • Special Processor • Special Programme • Special Project • Special Publication • Special Purpose • Spectrometer Processor • Spectropolarimeter • Spin • Split Phase • Spontaneous Potential • Stable Platform • Stack Pointer • Staff Planner • Standard and Performance • Standardization Plan • Standardized Profile • Standard or Peculiar • Standard Practice • Standard Procedure • Standard Program • Standing Procedure • StarPages (resources) • Starting Point • Starting Price • Start Point • Statens Provningsantalt (Sweden) • State Park (USA) • Static Pressure • Station Pressure • Stochastic Programming • Stock Point • Stop Payment • Stop Press • Stream Protocol • Strong Point • Structured Programming • Subprocedure • Sudden Penetration • Summus Pontifex • Sun Point • Super-Plössl • Supplement • Supply Point • Supra Protest • Surge Prominence • Surveillance Payload • Surveillance Procedure • Swimming Pool • Switched Power • Switch Panel • Switch Processor • Symbol Programmer • System Parameter • System Processor • System Product • Systems and Procedures • Systems Procedure

Spa Spanish

SpA Società per Azioni

S&PA Safety & Product Assurance

SPA Safety and Product Assurance • S-band Power Amplifier • Scatter Propagation Antenna • Selectable Power Adapter • Self-Phasing Array • Service Provider Agent • Service Provider Application • Servopower Amplifier • Servopower Assembly • Shared Peripheral Area • Shared Program Archive • Shortest Path Algorithm • Signal Preamplifier • Signal Processor Assembly • Single-Parameter Analysis • Single-Particle Approach • Skill Performance Aid • Small-Payload Accommodation • Smooth Potential Approximation • Société Protectrice des Animaux • Society for Popular Astronomy (UK) • Society for Practical Astronomy (USA-IL) (defunct) • Software Performance Analyzer • Software Process Assessment • Software Product Assessment • Software Product Assurance • Software Producers Association (UK) • Software Publishers Association (USA) • Solar-Panel Assembly • Solar-Power Array • Solicitation Package Automation • Solid-state Power Amplifier • South Pole-Aitken (basin) • Space Physics and Aeronomy • Space Processing Application • Spartanburg (USA-SC) (location identifier) • Specially Protected Area • Special Project Agreement • Specification Preparing Activity • Spectral Processing Analysis • Spectrograph Assembly • Spin Phase Angle • Standard de Pouvoir d'Achat • Statement of Program Acceptance • Statistical Processing and Analysis • Steering Position Amplifier • Storage Planning and Allocation • Strategic Posture Analysis • Strategy and Policy Assessment • Subprogramme Area • Sudden Phase Anomaly • System Performance Assessment

SPAA Sociedad Paraguaya de Astronomía Amateur (Paraguay) • Symposium on Parallel Algorithms and Architectures

SPAC Saint Petersburg Astronomy Club (USA-FL) • Scientific Program Advisory Committee • South Plains Astronomy Club (USA-TX) • Space Program Advisory Council • Spacecraft Performance Analysis and Command • Spectrum Planning and Advisory Committee

SP'ACE Space Advanced Communication Engine

SPACE Saint Petersburg Astronomy Club Examiner • Satellite Precipitation And Cloud Experiment • Self-Programming Automatic Circuit Evaluator • Seminario Permanente de Astronomia y Ciencias Espaciales • Sequential Position and Covariance Estimation • Sidereal Polar Axis Celestial Equipment • Simulation de la Propagation Acoustique dans un Conduit avec Écoulement • Single Point of Access to Citizens of Europe • Space Advanced Communication Engine • Spacecraft Prelaunch Automatic Checkout Equipment • Space, Physics and the Advancement of Cosmic Exploration • Space Physics Archive Search Engine • Space, Planetary, and Astronomical Cyberexperience • Space Program American Citizen's Effort • Standards Performance Achievement and Curriculum Enhancement • Strategic Preparedness in Advancing Careers in Engineering • Student Programs for Achievements in Careers and Education • Symbolic Programming Anyone Can Enjoy

SPACECOM Space Communications Corporation

SPACELAB Space Laboratory

SPACEWARN World Warning Agency for Satellites

SPACON Space Control • Spacecraft Controller

SPAD Satellite Position Predictor and Display • Sensor Processing And Display • Shuttle Payload Accommodation Document • Signal Passed At Danger • Single Photon Avalanche Diode • Société pour l'Aviation et ses Dérivés • Space Principles, Applications and Doctrines • Système Portable pour l'Analyse de Données

SPADATS Space Detection And Tracking System

SPADD Smart Passive Damping Device

SPADE Signal Processing And Display Equipment • Small Portable Analyzer Diagnostic Equipment • Spare Parts Analysis, Documentation and Evaluation • Stratospheric Photochemistry, Aerosols, and Dynamics Expedition (NASA) • Stratospheric Photochemistry, Aerosols, and Dynamics Experiment

SPADOC Space Defence Operation Center

SPADS Satellite Position And Display System • Satellite Processing And Display System • Service Performance Analysis Daily System • Shuttle Problem Analysis Data System • Small Parcel Explosive Detection System • Staff Planning And Decision Support (system)

SPADVOS Spaceborne Direct Viewing Optical System

SPAé Service des Programmes Aéronautiques (DGA)

SPAF Simulation Processor And Formatter

SPAG Space Psychology Advisory Group • Standards and Practices Action Group • Standards Promotion and Application Group • Strategic Planning Advisory Group

SPAH Spacelab Payload Accommodation Handbook • Superhydrogenated Polycyclic Aromatic Hydrocarbon

SPALNA Soil and Plant Analytical Laboratories Network of Africa

Sp. Am. Spanish America • Spanish American

SPAM Satellite Processor Access Method • Ship Position and Attitude Measurement • Spectral Analysis Manager • Spectroscopy and Polarimetry At Mees • Stewart Platform Augmented Manipulator • System for Photointerpretation of Airports using Maps

SPAMS Ship Position and Attitude Measurement System

Span Spaniard • Spanish

SPAN Security Policy Automated Network • Security Policy Automation Network • Solar-Particle Alert Network • Sonar Program Analyzer • Spacecraft Analysis • Space Physics Analysis Network • Spiral Anode • Statistical Processing And Analysis • Stored Program Alphanumerics • Successive, Proportionate, Additive Numeration • System Planning and Analysis

SPAND Solar Proton Albedo Neutron Decay

SPANL Spectral Analysis Library

SPANS Simple Protocol for ATM Network Signaling • Spatial Analysis System

SPAP System for Publications Acquisition and Processing

Spar Separata

Sp. Ar. Spanish Arabic

SPAR Shreveport Parks And Recreation (USA-LA) • Shuttle Pointed Autonomous Research • Software Performance Assurance Representative • Space Processing Applications Rocket • Special Prelaunch Analysis Request • Standard Payload Assurance Requirements • Superprecision Approach Radar • Supportability Problem Assessment Report • Symbolic Program Assembly Routine • System Performance and Accuracy Report

SpARC Space Automation and Robotics Center

SPARC Scalable Performance Architecture • Scalable Processor Architecture (*) • Scaled Processor Architecture • Scholarly Publishing and Academic Resources Coalition • Single Processor Architecture • Space Applications and Research Centre • Space Automation and Robotics Center • Space Physics and Aeronomy Research Collaboratory • Spectral Analyzer and Recognition Computer • Standards, Planning And Requirements Committee • Strategic Planning And Research Coordination • Stratospheric Processes And their Role in Climate

SPARCENT Space and Atmospheric Research Center

SPARCLE Spaceborne Aerosols and Cloud Lidar Earthprobe

SPARCS Solar Pointing Aerobee Rocket Control System • Solar Pointing Attitude Rocket Control System

SPARK Student Program for Aeronautics Resource and Knowledge

SPARMO Solar Particles And Radiations Monitoring Organization

SPARO Submillimeter Polarimeter for Antarctic Remote Observations • Submillimeter Polarimeter for Antarctic Remote Observing

SparPG Spar-Prämiengesetz

SPARQ Serving with Pride And Reaching for Quality

SPARRSO Space Research and Remote Sensing Organization (Bangladesh)

SPARS Space and Atmospheric Research Station • Space Precision Altitude Reference System • Space Precision Attitude Reference System (*) • Spatial Analysis and Remote Sensing • System Performance And Readback System

SPARSA Series Pulse, Azimuth, Rate and Spectrum Analyzer

SPARTAN Shuttle Pointed Autonomous Research Tool for Astronomy • System for Personnel Automated Reports, Transactions and Notices

SPARTECA South Pacific Regional Trade and Economic Cooperation Agreement

SPAS Safety Performance Analysis System • Shuttle Pallet Application System • Shuttle Pallet Satellite • Skill Performance Aids • Solar Proton Alpha Spectrometer • Space Power Architecture Study • Standard Procurement Appropriation System

SPASA Servicios Politécnicos Aéreos SA • South Pole Air Shower Array

SpaSci Rev Space Science Reviews

SPASE Small Payload Access to Space Experiment • South Pole Air Shower Experiment • Space Physics Archive Search and Exchange

SPaSM Scalable Parallel Short-range Molecular

SPASM Smithsonian Package for Algebra and Symbolic Mathematics • Space Propulsion Automated Synthesis Modelling • System Performance and Activity Software Monitor

SPASP Small payload Accommodation Switch Panel

SPASS Saint Petersburg Association of Scientists and Scholars (Russia)

Spat Spatial

SPAT Satellite Performance Analysis Team • Silicon Precision Alloy Transistor • Small-Payload Accommodation Terminal • Software Process Action Team

SPATE Space Pointing And Tracking Experiment • Stress Pattern Analysis by Thermal Emission

SPAW Specially Protected Areas and Wildlife

SPB Saint Thomas Island (US Virgin Islands) (location identifier) • Science Planning Branch • Security Policy Board • Shergottite Parent Body • Slowly Pulsating B (star) • Solar-Particle Beam • Space Planning Bureau • Strategic Planning Board • Synchrotron Proton Blazer

SP/BS Small parcel and Bundle Sorter

SPBS Small parcel and Bundle Sorter • Standard Project Book System

SpbSTU Saint Petersburg State Technical University (Russia)

SPC Santa Cruz de la Palma (Canary Islands) (location identifier) • Satellite Processing Center • Scholastic Programming Contest • Science Policy Committee • Science Programme Committee • Scientific Personnel Committee • Scientific Policy Committee • Scientific Programme Committee • Scientific Programmes Committee • Scintillation Proportional Counter • SCSI Primary Command • Security Processing Center • Semi-Permanent Connection • Serial-to-Parallel Converter • Shipping and Packing Cost • Short-Period Comet (*) • Short Periodic Comet • Shuttle Plume Characteristics • Shuttle Processing Contract • Shuttle Processing Contractor • Signaling Point Code • Signal Processing Center • Single-Probe Carrier • Sistema Panamericano de Clasificación • Sky Publishing Corporation (USA-MA) • Signal Processing Center • Small Peripheral Controller • Software Productivity Center • Software Productivity Consortium • Software Publishing Corporation • Solar Pointing Control • South Pacific Commission • South Polar Cap • Space • Space Programs Center • Space Projects Center • Spare Parts Catalog • Special Political Committee • Special-Purpose Committee • Special-Purpose Computer • Spectroscopic and Polarimetric Coronagraph • Staff Planner Course • Standard Products Classification • Starting-Point Code • State Plane Coordinates • Statistical Process Control • Statistical Program Coordinator • Stored Program Command • Stored Program Control • Stored Program Controller • Storm Prediction Center • Strategic Planning Council • Strategy and Planning Committee • Superphysical Contact • Supplementary Protection Certificate • Synoptic Properties Code • System Parameter Control

SPCA Secondary Power Control Assembly • Short Posterior Ciliary Artery • Spectral Principal Component Analysis

SPCBT Self-Paced Computer-Based Training

SPCC Spare Parts Control Center • Stored Program CAMAC Channel • Study Program coordination Committee

SPCD Space Communications Division • Space Data and Computing Division

SPCE Survey Planning and Coordination Element

SPCG Special Circuit Group

Spchg Supercharge

Spcl Special

SPCM Single Photon Counter Module

SPCN Satellite Personal Communication Network • Sciences Physiques, Chimiques, Naturelles • Service de Photogrammétrie et de Cartographie Numérique (IGN)

SpCoB Space Communication Broadcasting (periodical code)

SpCom Space Communications (periodical code)

Spcr Spacer

SPCR Satellite Pointing Change Request • Serial Port Control Register • Software Problem Change Report • Software Problem Change Request

SPCS Satellite Personal Communication Services • Schedule Planning and Control System • Science Programme Communication Service • Selective Paging Communications System • Service and Performance Checkout System • Short Ported Coaxial System • Single Per Carrier System • State Plane Coordinates System • Static Power Conversion System • Stored Program Controlled Switch

SPCU Simulation Process Control Unit • Skylab Process Control Unit

SPCZ South Pacific Convergence Zone

spd Speedo (file extension)

Spd Speed

SPD Safety Program Directive • Saidpur (Bangladesh) (location identifier) • Science Policies Division (STScI) • Science Programs Division • Scientific Passenger Pod • Security Policy Database • Separation Program Designator • Serial Presence Detect • Serviço de Processamento de Dados • Skylab Program Directive • Solar Physics Division • Solid Protoplanetary Disk • South Polar Distance • South Pole Distance • Space Physics Division • Spectral Power Distribution • Standard Practice Directive • Standard Processed Data • Stored Program Decoder • Stored Program Decommutation • Subjective Probability Distribution • Subject to Permission to Deal • Surge Protective Device • Symmetric Positive Definite • System Performance Document • System Preliminary Design • System Program Description • System Program Directive • System Program Director • System Program Directorate • System Program Documentation

SPDA Single-Premium Deferred Annuity

SPDB Science Processing Database • System Power Distribution Box

SPDC Spare Parts Distribution Center • Special Programme for Developing Countries

SPDE Solar Plasma Diagnostic Experiment

Sp. Del. Special Delivery

SPDF Sensor Data Processing Facility • Space Physics Data Facility

SPDIF Sony-Philips Digital Interface • Sony-Philips Digital Interface Format

SPDIR Spectroscopy and Photometry of the Intergalactic Medium's Diffuse Radiation (mission)

Spdl Spindle

SPDL Standard Page Description Language

SPDM Special-Purpose Dexterous Manipulator

SP/DP Signal Processing/Data Processing

SPDP Signal Processing/Data Processing • Society of Professional Data Processors (USA) • Software Preliminary Design Review • Stored Program Data Processor • Symposium on Parallel and Distributed Processing

SPDR Scheduled Planning Drawing Release • Short-Pulse Doppler Radar • Software Preliminary Design Review • System Program Director's Review

SPDS Safe Practice Data Sheet • Scalable Performance Data Server • Sequential Payload Delivery System • Space Physics Data System • Stabilized Payload Deployment System

SPDT Single-Pole - Double-Throw

SPDU Session Protocol Data Unit

SPDVC Special Device Controller

SPE Satellite Precipitation Estimate • Senior Procurement Executive • Senior Project Engineer • Sensor Processing Electronics • Sepulot (Malaysia) (location identifier) • Service Public de l'Emploi • Shrinkage Preliminary Estimator • Signal Processing Element • Single Photoelectron • Single Programming Document • Society of Petroleum Engineers (USA) • Software Performance Engineering • Software Product Evaluation • Software Production Environment • Software Process Engineering • Solar Particle Event • Solar Proton Event • Solid-Phase Epitaxy • Solid-Phase Extraction • Solid Polymer Electrolyte • Solid Polymer Electrolyzer • Space Processing Equipment • Space-tracking and Pointing Experiment • Special-Purpose Equipment • Specifications • Spherical Probable Error • Standardized Polyelastic Sampling • Standard Problem Exercise • Standard Progressive Erection • Static Phase Error • Station Project Engineer • Stored Program Element • Stork Product Engineering BV (Netherlands) • Sun - Planet - Earth • Symbolic Programming Environment • Synchronous Payload Envelope • Systematic Pulsative Erection • System Performance Evaluation • Systems Performance Effectiveness

SPEAR Signal Processing Evaluation And Reporting • Small Payload Ejection And Recovery • Source Performance Evaluation And Reporting • Spaceborne Earth Applications Ranging (system) • Space Plasma Exploration by Active Radar • Space Power Experiments Aboard Rockets • Spectroscopy of Plasma Evolution from Astrophysical

Radiation • Stanford Positron Electron Accelerating Ring (USA-CA) • Support Programme for the Evaluation Activities in Research

SPEARNET South Pacific Education And Research Network

SPEARS Satellite Photoelectric Analog Rectification System • Screener Proficiency Evaluation And Reporting System • Supplier Performance Evaluation And Rating System

Spec Special • Specially • Specialist • Specialization • Specialize • Specialized • Specific • Specifically • Specification • Specified • Specify • Specimen • Spectrum • Speculation

Spéc Spécial • Spécialisé • Spécialiser • Spécialiste • Spécification • Spécifier • Spécifique

SPEC South Pacific Bureau for Economic Cooperation • Space Environmental Compatibility • Specialist • Specification • Specified • Spectrometry • Spectroscopy • Stored Program Educational Computer • Support Programme for Employment Creation • Systems Performance Evaluation Cooperative • Systems Performance Evaluation Corporation

SPECAT Special Category

SPECC Space Enhanced Command and Control

SPECFS Special File System

Special Specialized

SPECIAL Space Processes and Electrical Changes Influencing Atmospheric Layers (network)

Specif Specific • Specifically • Specification

SpecL Spectroscopy Letters (periodical code)

SPECL Surface Plasmon Enhanced Contact Lithography

SPECMAP Spectral Mapping

Specs Specifications

SpECS Special Education of Computational Science

SPECS Solar Physics Element on Columbus/Space Station • Specification and Programming Environment for Communication Software • Specifications • Submillimeter Probe for the Evolution of Cosmic Structure

SPECT Single Photon Emission Computed Tomography (*) • Single Photon Emission Computerized Tomography • Spectrometer • Spectrometry • Spectroscopy • Spectrum

SPECTA Structure-Preserved Error-Correction Tree Automaton

Spectr Spectral • Spectre • Spectrograph • Spectrographe • Spectrometer • Spectromètre • Spectrum

SPECTRA Surface Processes and Ecosystem Changes Through Response Analysis (explorer)

SPECTRE Spectral Radiance Experiment • Student-run Project for Exo-atmospheric Colliding Technologies and Rocket Experiment

SPECTRUM Special Experiment Concerning Typhoon Recurvature and Unusual Movement

SPECTS Symposium on Performance Evaluation of Computers and Telecommunication Systems

SPECVER Specification Verification

Sp. Ed. Specialist in Education

SpEd Space Education (periodical code)

SPED Supersonic Planetary Entry Decelerator

SPEDAC Solid-state Parallel Expandable Differential Analyzer Computer

SPEDE Spacecraft Potential, Electron flux and Dust Experiment • System for Processing Educational Data Electronically • System Procurement by Electronic Data Exchange

SPEDI Small Procurement Electronic Data Interchange

SPEDO Solar-Powered Extended Duration Orbiter

SPEDTAC Stored Program Educational Transistorized Automatic Computer

SPEE Special-Purpose End Effector

SPEEA Society of Professional Engineering Employees in Aerospace (USA)

SPEED Self-Programmed Electronic Equation Delineator • Short-Pulse Experimental Electron Device • Spectral Energy Distribution • Subsistence Preparation by Electronic Energy Diffusion • Systematic Plotting and Evaluation of Enumerated Data • Systems Planning and Effectiveness Evaluation Device • Systems Planning, Engineering and Evaluation Device

SPEEDUP Simulation Program for the Economic Evaluation and Design of Unsteady-state Processes

SPEF Single Program Element Funding • Special Program Element Funding

SPELDA Structure Porteuse Externe pour Lancement Double Ariane

SPELTRA Structure Porteuse Externe pour Lancement Triple Ariane

SPEMS Self-Propelled Elevating Maintenance Stand

SpEnvIS Space Environment Information System

SPEO Space Physics Educational Outreach

SPEPD Space Power and Electric Propulsion Division

SPER Status, Performance and Exception Reporting • Syndicat Professionnel d'Électronique et de Radiocommunications • Syndicat Professionnel Électronique et Radioélectrique

SPERAD Spectrally-scanning Radiometer

SPERT Schedule Performance Evaluation and Review Technique • Simplified Program Evaluation and Review Technique • Special Program Evaluation and Review Technique

SPES Spacecraft Performance and Evaluation Service • Spacecraft Performance Evaluation System • Stimulation Plan for Economic Science • Stored Program Element System

SPET Solid-Propellant Electric Thruster

SPETC Solid-Propellant Electro-Thermal-Chemical

SPETE Special-Purpose Electronic Test Equipment

SPEVAL Spacecraft Evaluation

SPEX Satellite Potential Experiment • Spectral X-ray • Stellar and Planetary Explorer

SPF Schéma Provisionnel de Formation • Science Policy Foundation • Shipper Paid Forwarding • Ship Point Formation • Shortest Path First • Single-Point Failure • Single Project Funded • Single-Project Funding • Site Population Factor • Sociedade Portuguesa de Física (Portugal) • Software Production Facility • South Pacific Forum • South Polar Feature • Space Power Facility • Spacelab Processing Facility • Spearfish (USA-SD) (location identifier) • Special-Purpose Force • Specific Pathogen Free • Spectrofluorometer • Structured Programming Facility • Sun Protection Factor • Superplastically Formed • Superplastic Forming • Survival Probability Function • System Productivity Facility • System Programming Facility

SPFA Single-Point Failure Analysis

SPFD Spectral Power Flux Density

SPFE Special Projects Flight Experiment

SpFl Spaceflight (periodical code)

SPFL Single-Point Failure List

SPFM Single-Point Failure Mode

SPFP Single-Pass Fit Program • Single-Point Failure Potential

SPFPAD Spacecraft Performance and Flight Path Analysis Directorate

SPFR Southern Polarity Flare Ribbon • Spectral Flux Radiometer

SPFS Small Payload Flight System

SPFT Strong Permittivity Fluctuation Theory

Spg Spring

SPG Saint Petersburg - Whitted (USA-FL) (location identifier) • Schweizerische Physikalische Gesellschaft (Switzerland) • Scientific Planning Group • Short-Pulse Generator • Signal Processor Group • Single-Point Ground • Situation Particulière de Gestion • Sort Program Generator • Source Power Gain • Space Group • Special Projects Group • Spring • Système de Préférences Généralisées • System Product Group

SPGAI Surface Position Gimbal Angle Indicator

SPGC Second-Parameter Globular Cluster

Sp. Gr. Specific Gravity

SPGS Springs

SPH Smoothed Particle Hydrodynamics (*) • Smooth Particle Hydrodynamics • Soft-Particle Hydrodynamics • Sopu (Papua New Guinea) (location identifier) • Statement of Personal History

SPhD Soviet Physics Doklady (periodical code)

Spher Spherical

SPHINX Space Plasma High-voltage Interaction Experiment

SPhN Service de Physique Nucléaire

SPHN Société de Physique et d'Histoire Naturelle (Switzerland)

SPHR Sphere

SPhT Service de Physique Théorique

Sp. Ht. Specific Heat

spi Spissatus

SPI Piscids (meteor stream) • Schedule Performance Index • Sciences Physiques pour l'Ingénieur • Sciences Pour l'Ingénieur • SCSI Parallel Interface • Secure Processing Interface • Security Parameters Index • Security Profile Inspector • Security Protocol Identifier • Segmented Pupil Imager • Selected Period Investment • Selective Profile Indexing • Self-Paced Instruction • Serial Peripheral Interface • Serial to Parallel Interface • Service de la Propagande et de l'Information • Service Programming Interface • Service Provider Interface • Single-Process Initiative (*) • Single-Program Initiation • Site Population Index • Software Patent Institute • Software Process Improvement • Software Product Identifier • Space Power Institute (USA-AL) • Special Packaging Instruction • Specific Productivity Index • Spectrographic Particle Imager • Spectrometer on INTEGRAL • Springfield (USA-IL) (location identifier) • Stateful Packet Inspection • Subcontract Program Integrator • Supporting Program Integrator • Surface Position Indicator • Surveillance Principal Investigator • Synchronization Programming Interface • Systèmes et Périphériques Informatiques • System Performance Indicator • System Provider Interface

SPIA Standard Profile for Imagery Archives

SPICA Special Instrument for Cosmic-ray Astrophysics • Spheromak Injector using Conical Accelerator • Support Program for Instructional Competency in Astronomy

SPICE Shuttle Payload Integration Cargo Evaluation • Simulated Program for Integrated Circuit Emphasis • Simulator Program for Integrated Circuit Emulator • Software Process Improvement and Capability • Solar-Particle Intensity Composition Experiment • Spacecraft - Planet - Instrument - C-matrix - Events • Space Integrated Control Experiment • Space Particle Impact Capture Experiment • Space Power Internal Combustion Engine • Spacelab Payload Integration and Coordination in Europe • Suprathermal Plasma Investigation of Cometary Environments • Survey Probe Infrared Celestial Experiment • Survey Program of Infrared Celestial Experiments

SPID Service Profile Identification • Service Profile Identifier • Speckle Imaging by Deconvolution

SPIDAB Space Industrial Database

SPIDASL Space Plasma Interactive Data Analysis and Simulation Laboratory

SPIDER Solar-Powered Ion Driven Eclipsing Rover

SPIDF Support-Planning Identification File

SPIDPO Shuttle Payload Integration and Development Program Office

SPIDR Space Physics Interactive Data Resource • Spectroscopy and Photometry of the IGM Diffuse Radiation (mission)

SPIDS Stabilized Payload Interface Deployment System

SPIE Self-Programmed Individualized Education • Shuttle Plume Impingement Experiment • Simulated Problem - Input Evaluation • Society of Photo-optical Instrumentation Engineers

SPIEL Signal Processing Interactive Extensible Language

SPIF Satellite Payload Integration Facility • Shuttle Payload Integration Facility • Shuttle-POCC Interface Facility • Shuttle Processing Interface Facility • Small Photometric Imaging Facility • Spacecraft Planetary Imaging Facility • Standard Payload Interface

SPIFEX Shuttle Plume Impingement Flight Experiment

SPIFFI Spectrometer for Infrared Fiber-fed Field Imaging

SPIFI South Pole Imaging Fabry-Pérot Interferometer

SPIFR Single-Pilot Instrument Flight Rules

SPIHT Set Partitioning In Hierarchical Trees

SPII Shuttle Program Implementation Instruction

SPIIN Supplemental Procurement Instrument Identification Number

SPIKE Science Planning Interactive Knowledge Environment • Science Planning using Intelligent Knowledge-based Environment • Scientific Planning Interactive Knowledge Expert

SPIL Signal Processing In-the-Loop • Software Prototype and Integration Laboratory

SPILS Spin Prevention and Incidence Limiting System • Stall Prevention and Incidence Limiting System

SPIM Spectrographic Imager

SPIMS Shuttle Program Information Management System

SpIN Space Industry News (CSIRO)

SPIN Scheduling and Planning Information Network • Science Policy Information News • Searchable Physics Information Notices • Software Process Improvement Network • Solution-Provided Information Network • Space Interdisciplinary Network • Space Physics Information Network • Special Instruction • Special-Purpose Inexpensive Satellite Program • Spectropolarimetric Interferometry

SPINE Shared Program Information Network • Space Informatics Network Experiment

SPINES Science Policy Information Exchange System

SPINR Spectrograph for Photometric Imaging with Numeric Reconstruction

SPINS Special Instructions • Stabilized Platform Inertial Navigation System

SPIP Software Process Improvement Plan • System Performance Improvement Plan

SPIPR Sistema de Procesamiento de Imágenes de Percepción Remota

SPIR Spectropolarimeter for Infrared

SPIRAC South Pole Infrared Array Camera

SPIRAL Surface Photometry Interactive Reduction and Analysis Library

SPIRALE Spectromètre Infrarouge par Absorption à Laser Embarqué

SPIRE Sensor Performance in a Radiation Environment • Spatial Inertial Reference Equipment • Spatial Paradigm for Information Retrieval and Exploration • Spectral and Photometric Imaging Receiver • Spectral Infrared Rocket Experiment • Spectrometric and Photometric Imaging Receiver • Stop - Protect - Insulate - Rewarm

SPIREP Special Intelligence Report

SPIRES Standard Personnel Information Retrieval System • Stanford Public Information Retrieval System (USA-CA)

SPIREX South Polar Infrared Explorer • South Pole Infrared Explorer (*)

SPIRIT Service Provider Integrated Requirements for Information Technology • Space Infrared Imaging Telescope • Space Infrared Interferometric Telescope (*) • Space Infrared Interferometry Trailblazer • Spatial Infrared Imaging Telescope • Spatial Infrared Rocket-borne Interferometric Telescope • Special-Purpose Integrated Remote Intelligence Terminal • Spectral Infrared Interferometric Telescope • System Parametric Information Relational Intelligence Tool

SPIRITS Spectral and In-band Radiometric Imaging of Targets and Scenes

SPIRS SuperJANET Project on Information Resources

SPISS Shuttle Polar Ice Sheet Sounder

SPIT Selective Printing of Items from Tapes • Site Performance and Integration Test • Starburst Pixel Interleaving Technique • Strategic Plan for Information Technology

SPIU Sensor Processor Interface Unit

SPJ Service Pages Jaunes • Sparta (Greece) (location identifier)

SPJC Saint Petersburg Junior College (USA-FL)

SPK Sapporo (Japan) (location identifier) • Single-Point Keying • Software Porting Kit • Spacecraft-Planet Kernel

SPKAC Signed Public Key And Challenge

SPKI Simple Public Key Infrastructure

Spkl Sprinkler

SPKL Sprinkle

Spkr Speaker

SPKR Sprinkler

spl Sine Prole Legitima

SPL Science Processing Library • Scratch Pad Line • Sección Planetaria y Lunar • Secondary Payload • Serialized Parts List • Set Priority Level • Signal Processing Language • Simple Programming Language • Software Programming Language • Software Prototyping Laboratory • Sonic Pressure Level • Sound Pressure Level • Space Photography Laboratory • Space Physics Laboratory • Space Programming Language • Spare Parts List • Special-Purpose Language • Spellchecker • Split-Phase Level • Spooler • Student Pilot's Licence • Sun-Pumped Laser • Supplies - Procurement - Logistics • System Programming Language

SPLAASH Spacecraft Protective Landing Area for the Advancement of Sciences and Humanities

SPLAT Simple Performance Lock Analysis Tool • Spatial Planning and Analysis Tool

SPLC Software Product Line Conference • Standard Payload Computer

SPLD South Polar Layered Deposit

SPLE Singular Power Law Ellipsoid

SPLEEN Spin-Polarized Low-Energy Electron Microscope • Spin-Polarized Low-Energy Electron Microscopy

SPLEF Société pour la Propagation des Langues Étrangères en France (France)

SPLINE Scalable Platform for Large Interactive Network Environments

SPLIT Space Program Language Implementation Tool • Spectropolarimetric Lightweight Telescope

SPLL Self-Propelled Launcher Loader

SPLM Space Programming Language Machine

Splmt Simplement

SPLNS South Plains (USA)

SPLOM Scatter Plot Matrix

Splt Spotlight

Spltr Splitter

SPLTS Special Private Line Telephone System

SPLX Simplex

Sply Supply

SPLY Same Period Last Year

spm Sine Prole Mascula

SPM Miquelon (Saint Pierre et_) (code ISO de pays) • Saint Pierre et Miquelon (code ISO de pays) • San Pedro Mártir (Mexico) • Scanning Probe Microscope • Scanning Probe Microscopy • Scanning Proton Microprobe • Scanning Proton Microscope • Scanning Proton Microscopy • Schematic Protection Model • Sciences Physiques et Mathématiques • Scratch Pad Memory • Security Plug-in Module • Security Policy Model • Security Program Manager • Security Protection Module • Seeker and Processor Model • Self Phase Modulation • Self-Propelled Mount • Senior Project Manager • Sequential Processing Machine • Session Protocol Machine • Simple Plasma Monitor • Single Plasma Monitor • Single Port Manager • Single Port Mooring • Skill Projection Model • Sleep Performance Model • Small-Powers Minimization • Smoothed Particle Method • Software Performance Monitor • Software Procedures Manual • Software Programmer's Manual • Software Project Management • Solar Photometer • Solar Physics Meeting • Solar Polar Mission • Solar Probe Mission • Solar Proton Monitor • SONET Physical Mapping • Sonnenphotometer • Source Program Maintenance • Southern Proper Motion (survey) • South Proper Motion • Spangdahlem (Germany) (location identifier) • Spectromagnetograph • Spectrophosphorimeter • Spectrophotomètre Multibandes • Standard Parts Manual • Standard Payload Module • Standard Practice Manual • Standard Practice Memorandum • Standard Procedure Manual • Standard Process Manual • Subordinate Program Manager • Subproject Manager • Sue Proprie Mani • Summary for Policy Makers • Sun Pointing Mode • Sun Point Mode • Sun Probe Mars • Superparamagnetic • Support Program Management • Surveillance Program Manager • Suspended Particulate Matter • Symbol Processing Machine • Symposium • Système Photométrique Multibande • System Performance Model • System Performance Modeling • System Performance Monitor • System Planning Manual • System Program Manager • System Programmer's Manual

SpMar Space Markets (periodical code)

SPMD Single Program/Multiple Data • Single Program/Multiple Data-stream

SPME Solid-Phase Micro Extraction

SPMF Solar Photosphere and Magnetic Fields

SPMG Spectromagnetograph

Spmkt Supermarket

SPMM Schweizerisches Patent, Muster und Markenblatt

SpMME Space Medicine Medical Engineering (periodical code)

SPMN Software Program Managers Network

SPMO San Pedro Mártir Observatory (Mexico)

SPMP Software Project Management Plan

SPMS Software Process and Management System • Software Process Management System • Software Process Modeling System • Solar-Particle Monitoring System • Special-Purpose Manipulator System

SPMV Sparse Matrix Vector Multiply

SPN Saipan (Mariana Islands) (location identifier) • Secrétariat Permanent des Nomenclatures • Servicio de Parques

Nacionales • Shared Processing Network • Shipment Performance Notice • Shuttle Project Notice • Special Program Number • Spectropyranometer • Stop Press News

SPND Self-Powered Neutron Detector

SPNG Spring

SPNGS Springs

SPNFZ South Pacific Nuclear Free Zone

SPNFZT South Pacific Nuclear Free Zone Treaty

SPNP Service Provider Number Portability

Spnr Spanner

SPNR Spanner • Spinner

SPNS Switched Private Network Service

SPNV Special Project Night Vector

SPO Sacramento Peak Observatory (USA-NM) • San Pablo (Spain) (location identifier) • Schmidt Prisme-Objectif • Section des Publications Officielles • Security and Plans Office • Security and Plans Officer • Security - Plans - Operations • Senior Press Officer • Serving Post Office • Short-Period Oscillation • Shuttle Program Office • Shuttle Project Office • Signal Property Office • South Pole Observatory • Spacecraft Program Office • Spacelab Program Office • Spare Parts Order • Special Position Operator • Special Program Office • Special Program Officer • Special Projects Office • Special Projects Officer • Special Projects Operator • Standard Point of Operation • State Planning Office • Support Operations Officer • System Program Office • System Project Office • System Project Organization

SPOA Single Point Of Administration

SPOC Satellite Payload Operations Centre • Science Planning and Operations Center • Search Point Of Contact • Shuttle Payload of Opportunity Carrier • Shuttle Payload Operations Contract • Shuttle Portable Onboard Computer • Single Point Of Contact • Single Point Of Control • Single-Point Orbit Calculator • Singularly Perturbed Optimal Control • Slit Positioning Optimization Code • Spacecraft Payload Operations Centre • SPAS Payload Operations Center

SPOCC SBV Processing, Operations, and Control Center • South Pacific Organizations Coordinating Committee

SPOCT SBV Processing, Operations, and Control Team

SPODP Single-Precision Orbit Determination Program

SPOE Standard Payload Outfitting Equipment

SPOF Science Planning and Operations Facility (*) • Science Planning Operational Facility

SPOI Security - Plans - Operations - Intelligence

SPONG Secrétariat Permanent des Organisations Non-Gouvernementales

SPOOL Simultaneous Peripheral Operations Online

S'pore Singapore

SPORT Small Payload Orbit Transfer

SPORTS Second-generation Portable Remote Telecommunications System • Simplified Payload/Orbiter Thermal Simulator

SPOS Solid-Phase Organic Synthesis

SPOT SADE Photometric Optical Telescope • Satellite Positioning and Tracking • Satellite Pour l'Observation de la Terre • Satellite Probatoire d'Observation de la Terre • Science Planning and Observation Tool • Shared Product Object Tree • SIRTF Planning and Observation Tools • Smithsonian Precision Optical Tracking • South Pole Optical Telescope • Système Probatoire d'Observation de la Terre • Système Probatoire d'Observation Terrestre

SPOTI Service des Programmes d'Observation des Télécommunications et d'Information

SPOUT System Peripheral Output

Spp Species

S/PP System Project Plans

SPP Menongue (Angola) (location identifier) • Scalable Parallel Protocol • Scaled Parallel Protocol • Scenario Preparation Program • Schwerpunktsprogramm • Science and Public Policy • Science Power Platform • Secular Periodic Perturbation • Security Policy and Procedures • Security Practices and Procedures • Separation and Purification Processes • Sequenced Packet Protocol • Service de Physique des Particules • Serviço de Programas e Projectos • Set Packing Problem • Shared Production Program • Signal Preprocessor • Simulation Planning Panel • Single Pearson Precision • Single Process Purge • Site Positioning Problem • Software Portability Plan • Solar Photometry Probe • Solar Physics Payload • Solar Power Platform • Solid Performance Program • Space Plasma Physics • Space Policy Project • Sponsor Program Proposal • Standard Parallel Port • Standard Practices and Procedures • Standard Practices Procedure • Subset Preprocessor • Surface Plasmon Polaritron • Swiss Priority Programme • Switched Poisson Process • System Package Program • System Project Plan

SPPA Système Personnel de Pilotage de l'Apprentissage

SPPD Signal Processing Packaging Design • Signal Processor Package Design • Space Propulsion and Power Division

SPPE Standard Pulsating Progressive Erection • Système Personnel de Pilotage de l'Enseignement

SPPF Société de Producteurs de Phonogrammes Français (France)

SPPG Special Project Planning Group

SPPIL Shuttle Preferred Pyrotechnic Items List

SPPL Spare Parts Provisioning List

SPPM System Performance Prediction Model

SPPMD Stacked-Panel Passive-Matrix Display

SPPO Spacelab Payload Project Office

SpPol Space Policy (periodical code)

SpPow Space Power (periodical code)

SPPPI Secrétariat Permanent pour la Prévention des Pollutions Industrielles (CSG)

SPPS Scalable Power Parallel System • Service de Programmation de la Politique Scientifique • Subsystem Program Preparation Support • System Plot Package Simulator

SPQ San Pedro (USA-CA) (location identifier) • Sistema Português da Qualidade • Sociedade Portuguesa de Química (Portugal)

SPQL Space Physics Query Language

SPQR Senatus Populusque Romani • Small Profit and Quick Return

Spr Spring • Sprinkle

SpR Seconds per Revolution • Seconds per Rotation

SPR Salvage Point Recording • Sample Point Reflector • San Pedro (Belize) (location identifier) • SAR Processor System • Satellite Parametric Reduction • Seconds Per Revolution • Seconds Per Rotation • Secretarial Performance Review • Secure Packet Radio • Sequential Probability Ratio • Service Planning Segment • Simplified Practice Recommendation • Single-Point Resistance • Single Protocol Router • Small Parcels and Rolls • Smoothed Particle Rendering • Software Performance Report • Software Portability Review • Software Problem Record • Software Problem Report • Software Product Report • Software Program Specification • Solid-Propellant Rocket • South Polar Region (Jupiter) • Special Processing Request • Special Process Review • Special Program Requirement • Special Program Review • Special Project Report • Special-Purpose Register • Spezielle Relativitätstheorie • Statistical Pattern Recognition • Strong Penetration Resistance • Subcontractor Performance Review • Supplementary Progress Report • Supply Performance Report • Surface Plasmon Resonance • System Performance Rating • System Performance Report • System

Performance Requirements • System Problem Report • System Program Review

SPRAG STS Payload Requirements and Analysis Group

SpRBu Space Research Bulgaria (periodical code)

SPRC Self-Propelled Robot Craft • Semiconductor Physics Research Center (Korea-ROK)

SPRD Spread • System Performance Requirements Document

SPRDE System Planning, Research, Development and Engineering

SPRDG Spreading

Sprdr Spreader

SPRE Software Process Risk Evaluation • Solid-Propellant Rocket Engine • Spartan Packet Radio Experiment

SPREAD Surface-Promoted Replication and Exponential Amplification of DNA (analogues) • Systems Programming, Research, Engineering And Development

SPREC South Pacific Regional Environment Convention

SPREE Shuttle Potential and Return Electron Experiment

Spreew Spreewald (Germany)

SPREP South Pacific Regional Environmental Programme • South Pacific Regional Environmental Protection • South Pacific Region Environmental Programme • South Pacific Region Environmental Programme

SPREX Spring Runoff Experiment

SPRF Space Propulsion Research Facility • Special-Purpose Receiving Facility

SPRG Special Program Review Group

SPRI Scott Polar Research Institute (UK) • Space Physics Research Institute (South Africa)

SPRING Shared Protection Ring

SPRINTEL Speedy Retrieval of Information on the Telephone

SPRITE Solid-Propellant Rocket Ignition Test and Evaluation • Storage, Processing and Retrieval of Information in a Technical Environment • Stratospheric Perturbation Resulting from Intense Thunderstorm Electrification

Sprl Spiral

SPRL Société de Personnes à Responsabilité Limitée • Space Physics Research Laboratory • Spiral

SPRMD Space, Power Resources, and Manufacturing Development (periodical code)

SPRNG Spring

SPRNGS Springs

SPROB Solid Propellant Space Booster Plant

SprRe Sproul Observatory Reprints (periodical code)

SPRS Single Passenger Reservation System

Sprt Support

SPRT Scientific Programme Review Team • Sequential Probability Ratio Test • Single-Pole Single-Throw • Standard Platinum Resistance Thermometer • Support

SPRU Science Policy Research Unit • Spacecraft Power Regulator Unit • Standard Power Regulator Unit

Sprvsr Supervisor

sps Sine Prole Supersite

SpS Samples per Second • Symbols per Second

SPs Successive Penetrations

S&PS Survivability & Protective Structure

SP&S Special Processes & Sequencing

SPS Samples Per Second • Sanitaire et Phytosanitaire • Sanitary and Phytosanitary • Satellite Power System • School of Physical Sciences • Science Policy Study • Scientific Power Switching • Scratch Pad Store • Secondary Power System • Secondary Propulsion System • Sécurité - Protection - Sûreté • Security Policy Squadron • Self-Provided Summary • Sense Point Scanner • Service Propulsion System • Short Prototype String • Shuttle Procedures Simulator • Shuttle Processing Site • Siam Photon Source (Thailand) • Signal Processing System • Simplified Processing Station • Single-Pole Switch • Society of Physics Students • Soft-Particle Spectrometer • Software Productivity Solution • Software Product Specification • Solar-Panel Substrate • Solar-Physics Section • Solar-Powered Satellite • Solar-Power Satellite • Solar-Power System • Solar-Probe Spacecraft • Source Preparation Segment • South Polar Spot • South Pole Station • Space Peace Sculpture • Space Position System • Space Power System • Spacecraft Propulsion System • Spatial Power Spectrum • Special Processes and Sequencing • Special-Purpose Segment • Special Services • Spectral Planning Subcommittee • Speed Switch • Standard Positioning Service • Standard Positioning System • Standard Process Specification • Standard Procurement System • Standby Power Supply • Standby Power System • Statement of Prior Submission • Static Power System • Statistical Problem Solving • Statistical Programs Specialist • Status Projection System • Stellar Photometer Software • Stellar Photometry Software • Stellar Planet Survey • Stellar Population Synthesis • Stereo Photographic System • Stored Program Simulator • Structured Postscript • Super Proton Synchrotron (CERN) • Survivability and Protective Structure • Swiss Physical Society (Switzerland) • Symbolic Programming System • Symbols Per Second • Synchronous Packet Streaming • Syndiotactic Polystyrene • System Performance Simulation • System Performance Specification • Wichita Falls - Sheppard (USA-TX) (location identifier)

SPSA Signal-Processor Subassembly • Simple Problem Statement Analyzer • Solid Propellant Storage Area

SPSC Science Processing Support Center • Signal Processing and Spectral Control • Special-Purpose Synchronous Control • Statistical Programs Service Center • System Planning and System Control

SPSD Space Power Systems Division • Systems and Project Support Department

SPSE Spacecraft Power System Electronics • Special-Purpose Special Equipment

SPSF Spectral Point Spread Function • Square root of Phase Structure Function

SPSI Society for the Promotion of Scientific Industry (UK)

SPSL Society for the Protection of Science and Learning • Spare Parts Selection List

SPSM Stellar Population Synthesis Model

SPSME Spacelab Payload Standard Modular Electronics

SPSN Submitted Package Sequence Number

SPSO Scale Payment Suboffice • Science Processing Support Office • Science Programs Selection Office • Senior Principal Scientific Officer • South Pole Solar Observatory

SPSP Site Preparation Support Plan • South Pacific Star Party • Spare Parts Support Package

SPSR Service Planning Segment Replacement

SPSS Science Planning and Scheduling System • Science Processing Software Segment • Statistical Package for the Social Sciences

SPST Science Planning and Scheduling Team • Single Pole - Single Throw • South Pole Submillimeter Telescope • Supporting Program Support Team

SPSTD Spectrophotometric Standard

SPSU Southern Polytechnic State University (USA-GA)

Sp. Surf. Specific Surface

SPSW Support Software

Spt Seaport • Support

SpT Sectors per Track

SPT Scaled-Particle Theory • Sectors Per Track • Signal Processing Technique • Silicon Planar Transistor • Silicon-Powered Transistor • Sipitang (Malaysia) (location identifier) • Small-Perturbation Theory • Software Porting Tool • Space Power Tool • Spanning Tree • Special Performance Test • Special Project Team • Special-Purpose Test • Stationary Plasma Thruster • Steering Position Transducer • Support • Symbolic Program Tape • Symbolic Program Translator • System Performance Test • System Planning Team

SPTC Specified Period of Time Contract

SPTCEN Support Center

SPTD Signal Processor Techniques Department • Supplemental Provisioning Technical Documentation • Supplementary Provisioning Technical Documentation

SPTDP South Pacific Telecommunications Development Programme

SPTF Signal Processing Test Facility

SPTI Structural Program Test Input

SPTLB Short-Period Tidally-Locked Binary

SPTMLE Shrinkage Preliminary Test Maximum-Likelihood Estimator

SPTN SPARTAN

SPTP Space Physics Theory Program • Space Power Test Project • Special-Purpose Test Program

SPTR Software Process Tools Registry

SPTS Single Program Transport Stream

SPU Scientific Programs Unit • Self-Provided Underlining • Service Propulsion Unit • Signal Processing Unit • Signal Processor Unit • Single Partition Use • Sky Protection Unit • Spacecraft Processing Unit • Split (Croatia) (location identifier) • Switching Processor Unit • System Power Unit • System Processor Unit

SPUAI Saint Petersburg University of Aerospace Instrumentation (Russia)

SPUC Serial Peripheral Unit Controller

SPUD Solar Power Unit Demonstrator • Superconducting Pop-Up Detector

SPUDS Single Platform Uniform Diagnostic System

SPUIAGG Secção Portuguesa das Uniões Internacionais Astronômica e Geodesica e Geofisica (Portugal)

SPUP Space Platform Users Project

SPUR Separate Parts Usage Record • Space Power Unit Reactor • Support for University Research

SPURC Standard Plume Ultraviolet Radiation Code

SPURM Special-Purpose Unilateral Repetitive Modulation

SPV Semi-Permanente Verbindung • Sepik Plains (Papua New Guinea) (location identifier) • Single Pressure Vessel • Slow Phase Velocity • Solar Photovoltaic • Son - Photo - Vidéo • Sound - Photo - Video • Special-Purpose Vehicle • Subsistence Prime Vendor • System Processing Unit

SPVC Soft Permanent Virtual Circuit

SPVI Subsistence Prime Vendor Interface

SPVM Saddle-Point Variational Method

SPVMG Spectrovideo Magnetograph

Spvn Supervision

SPVPF Shuttle Payload Vertical Processing Facility

Spvw Spaceview (periodical code)

SPW Sheath-Plasma Wave • Signal Processing Worksystem • Spencer (USA-IA) (location identifier) • Systematic Programming Workshop

SpWd Space World (periodical code)

SPWG Security Policy Working Group • Software Planning Working Group • Software Process Working Group

SpWdU Space World U (periodical code)

SpWdW Space World W (periodical code)

SpWdY Space World Y (periodical code)

SPWG Science Planning Working Group • Scientific Priorities Working Group • Spare Parts Working Group

SPX Houston Spaceland (USA-TX) (location identifier) • Sequenced Packet Exchange • Sequential Packet Exchange • Sensor Process Execution • Simplex • Special Power Extension

SPY San Pedro (Ivory Coast) (location identifier) • Spray • Supernova Ia Progenitor Survey (ESO)

SPyCAT Space Physics Catalog

SPYDER System to Provide You Data from Earthquakes Rapidly

SPZ Sovietskaia Peremenna ya Zvezda • Springdale (USA-AR) (location identifier)

SP3T Single Pole Triple Throw (switch)

SP4T Single Pole Quadruple Throw (switch)

sq Albanian (ISO language code) • Sequens • Sequentia

Sq Sequence • Squadron • Square • Squawk

SQ Safety Quotient • Satisfaction Quotient • Service Quality • Shape-based Query • Shocked Quartz • Significant Quantity • Social Question • Squall • Square • Squeezed • Standard Quality • Stephan's Quintet • Stereophonic-Quadraphonic • Student Questionnaire • Super Quick • Survival Quotient • System Qualifier

SQA Shocked Quartz Aggregate • Software Quality Assurance • Subcontractor Quality Assurance • Supplier Quality Assurance • System Queue Area

SQAB Society for Quantitative Analyses of Behavior

SQAE Senior Quality Assurance Evaluator

SQAL Squall

SQAM Software Quality Assessment and Measurement • Software Quality Assurance and Measurement • Staggered Quadrature Amplitude Modulation

SQAMP Software Quality Assessment and Measurement Plan

SQAP Safety and Quality Assurance Plan • Software Quality Assurance Plan • Supplementary Quality Assurance Provision

SQAR Supervisory Quality Assurance Representative

SQAS Supervisory Quality Assurance Specialist

sqb SyQuest Backup (file extension)

SQB Santa Ana (Colombia) (location identifier) • Space-Qualified Booster

SQBC Space-Qualified Booster Charger

SQC Software Quality Control • Southern Cross (Australia-WA) (location identifier) • Specification Quality Control • Statistical Quality Control

sq. cm. Square Centimeter

SQCP Statistical Quality Control Procedure

sqd Squad

SQD Signal Quality Detector • Squad • Standards and Quality Division

Sqdn Squadron

SQE San Luis de Pale (Colombia) (location identifier) • Shared Queue Element • Signal Quality Error • Signal Quality Event • Software Quality Evaluation • Supplier Quality Evaluation

SQEP Software Quality Evaluation Plan

SQF Software Quality Forum • Solano (Colombia) (location identifier) • Solidification Quenching Furnace

SQFD Software Quality Function Deployment

sq. ft. Square Foot

SQG Sintang (Indonesia) (location identifier) • Small Quantity Generator

SQH Supplier Quality History

SQI Rock Falls - Whiteside County (USA-IL) (location identifier) • Sequencing Initialization • Skill Qualification Identifier • Software Quality Investment • Special Qualification Identifier • Statement of Questioned Item • Sterling (USA-IL) (location identifier)

SQID Simultaneous Quad Infrared Detector • Super Quantum Interface Device

SQIDS Satellite Quick-look Interactive Display System

SQIID Simultaneous Quad-color Infrared Imaging Device • Simultaneous Quad Infrared Imaging Device

sq. in. Square Inch

SQK Sidi Barani (Egypt) (location identifier)

sq. km. Square Kilometer

SQL San Carlos (USA-CA) (location identifier) • Sequential Query Language • Simple Query Language • Space-Qualified Laser • Standard Quantum Limit • Standard Query Language • Structural Query Language • Structured Query Language (*) • System Query Language

SQLC Structured Query Language Constructor

SQLDA Structured Query Language Descriptor Area

SQLJ Structured Query Language using Java

SQLN Squall Line

SQLNS Squall Lines

SQLPT Structured Query Language Parser and Translator

sq. m. Square Meter

SQM São Miguel do Araguaia (Brazil-GO) (location identifier) • Software Quality Maintenance • Software Quality Management • Software Quality Maturity • Square Meter • Stochastic Quantization Method • Strange Quark Matter

SQMB Senior Quality Management Board

sq. mi. Square Mile

SQMI Square Mile

sq. mm. Square Millimeter

Sqn Squadron

SQN Sanana (Indonesia) (location identifier) • Spin Quantum Number • Squadron

SQO Storuman (Sweden) (location identifier) • Straight Quality Oversight

SQP Sequential Quadratic Programming • Software Quality Plan • Software Quality Program • Starcke (Australia-QLD) (location identifier)

SQPAT Single-Quasiparticle Photon-Assisted Tunnelling (device)

SQPI Software Quality Profitability Index

SQPN Staggered Quadriphase Pseudo-random Noise

SQPP Software Quality Program Plan

SQPSK Staggered Quadrature Phase Shift Keying

sqq Sequentia • Sequentes

sq. r. Square Rod

SQR Sequel Report • Sequence Relay • Service Request • Signal to Quantizing (noise) Ratio • Software Quality Revenues • Soroako (Indonesia) (location identifier) • Special Qualifications/Requirements • Square • Supplier Quality Representative

sq. rd. Square Rod

SQRE Square

sq. rt. Square Root

SQRT Square Root

SQS San Ignacia (Belize) (location identifier) • Securing Quality System • Skill Qualification Score • Software Quality Standard • Squares • Stochastic Queuing System

SQT Skill Qualification Test • Software Qualification Test • Software Quality Test • Sphere Quadtree • Standard Question Time • Student Quick Take (ACM) • System Qualification Test

SQU Square

SQUARE Specifying Queries As Relational Expressions

SQUEL Spatial Query Language • Structured Query Language

SQUID Simultaneous Quad Imaging Device • Superconducting Quantum Interface Detector • Superconducting Quantum Interface Device • Superconducting Quantum Interference Detector • Superconducting Quantum Interference Device (*)

SQV Sequim (USA-WA) (location identifier)

SQW Single Quantum Well • Square Wave

sq. yd. Square Yard

SQZ Scampton (UK) (location identifier) • Small Quarantine Zone

s/r Su Remesa

sr Serbian (ISO language code) • Steradian • Sun Raster Image (file extension) • Sunrise

Sr Senhor • Senior • Señor • Sentier • Sir • Sister • Strontium • Strouhal Number

S/R Sale or Return • Search/Retrieval • Send and Receive • Shift Register • Stimulus/Response • Subroutine

S&R Search & Rescue • Storage & Retrieval • Surveillance & Reconnaissance

S-R Stimulus-Response

SR Safety Recommendation • Safety Release • Safety Requirement • Safety Review • Sample Rate • Saturable Reactor • Saudi Riyal (currency) • Scanning Radiometer • Scan Ratio • Schedule Result • Scientific Report • Scientific Research • Search and Retrieval • Search and Retrieve • Search Rate • Secondary/Remote • Second Ring • Section Report • Selective Ringing • Selenium Rectifier • Self-Raising • Semantic Reaction • Semi-Regular • Senate Resolution • Send and Receive • Send/Receive • Senhor • Senior • Separate Rations • Server Ring • Service Record • Service Report • Seychelles Rupee (currency) • Shift Register • Shipment Request • Shipping Receipt • Shop Release • Short Range • Short Rate • Shutdown Request • Signature Result • Simulation Report • Size Resolving • Slip Ring • Slow Release • Slow Roll • Socialement Responsable • Software Requirement • Solar Radiation • Solar Reference • Solid Rocket • Sortie Rate • Sound Ranging • Sound Rating • Source Route • Source Routing • Space Reliability • Space Required • Spares Requirement • Special Reconnaissance • Special Regulation • Special Relativity • Special Report • Specific Range • Specification Requirement • Specula Reflection • Speech Recognition • Speed Recorder • Split Ring • Squareness Ratio • Sri Lanka • Staff Report • Staff Requirement • Standard Range • Standard Rate • Standard Repair • Standard Requirement • Standard Resolution • Standardization Report • Star Route • Statement of Requirements • Station Regulation • Status Register • Status Report • Status Review • Steradian • Stimulus-Response • Stochastic Resonance • Stockholders Report • Stoichiometric Ratio • Storage and Retrieval • Storage Ring • Storage Room • Strategic Reconnaissance • Strehl Ratio • Student Records • Subjective Realism • Subroutine • Summary Report • Sunrise • Supplemental Report • Supplementary Regulation • Supporting Research • Support Reaction • Support Request • Support Requirement • Support Room • Suppressed Radiation • Suriname (ISO country code) • Surrogate • Surveillance and Reconnaissance • Surveillance Radar • Surveillance Requirement • Sustainment Rate • Sveriges Radio • Synchronous Radiation • Synchrotron Radiation • Synthetic Rubber • System Ready • System Requirement

Sra Senhora • Señora

SrA Senior Airman

SR&A Supporting Research & Analysis

SRA Santa Rosa (Brazil-RS) (location identifier) • Scanning Radar Altimeter • Science Research Associate • Selected Restricted Availability • Selective Routing Arrangement • Senhora • Separate Reporting Activity • Ship Radio Authorization • Shop Repairable Assembly • Shop Repair Assembly • Shop Replaceable Assembly • Simulation Runtime Agent • Society of Risks Analysis • Solicitation Review Authority • Specialized Repair Activity • Special Repair Activity • Special Rules Area • Spin Reference Angle • Spin Reference Axis • Stability Robustness Analysis • Staff Research Associate • Stage de Réinsertion en Alternance • Standard Radiation Atmosphere • Standard Range Analyzer • Standards of Readiness and Availability • Stanford Radiation Atmosphere • State Recreation Area (USA) • Static Robotic Actor • Stochastic Relaxation with Annealing • Stock Record Account • Subregistration Authority • Sulforicinoleic Acid • Summer Research Assistantship • Supply Readiness Assessment • Supporting Research and Analysis • Support Requirements Analysis • Surveillance and Reconnaissance Aircraft • Surveillance Radar Approach • System Reaction Analysis • System Replacement Assembly • System Requirements Analysis • Systems Research Aircraft • Systems Research and Applications

SRAC Société Royale d'Astronomie du Canada • Student Recruitment and Admissions Committee

SRAD Shuttle Radar Assembly Demonstration (*) • Shuttle Radiator Assembly Demonstration • Space Radiator Assembly Demonstration • Spectral Radiometer

SRAE Solar Radio Astronomy Experiment

SRAEC Scene Radiation and Atmospheric Effects Characterization

SRAG Space Radiation Analysis Group

SRAM Scientific Review of Alternative Medicine • Sequential Random Access Memory • Shadow Random Access Memory • Static Random Access Memory (*)

SRAMP STEP Results, Applications and Modeling Phase

SRAN Stock Record Account Number

SRAO Seoul Radio Astronomical Observatory (Korea-ROK)

SRAP Système de Reconnaissance Automatique de la Parole

SRAS Student Recruitment and Admissions Service • Syndrome Respiratoire Aigu Sévère

SRAT Short-Range Advanced Technology

SRATS Solar Radiation And Thermospheric Structure

SRB Safety Review Board • Santa Rosa (Bolivia) (location identifier) • Schedule Review Board • Service Request Block • Software Release Bulletin • Software Review Board • Solar Radiation Budget • Solar Reflectory Beacon • Solicitation Review Board • Solid-fuel Rocket Booster • Solid Rocket Booster • Source-Route Bridge • Source-Route Bridging • Source Routing Bridging • Specialist Regional Body • Storage Resource Broker • Sulphat-Reducing Bacteria • Surface Radiation Budget • System Readiness Board • System Release Bulletin • System Requirement Baseline • System Review Board

SRBA Société Royale Belge d'Astronomie, de Météorologie et de Physique du Globe (Belgium)

SRBAB Solid-Rocket Booster Assembly Building

SRBDB Solid-Rocket Booster Disassembly Building

SRBF Solid-Rocket Booster Facility

SRBII Société Royale Belge des Ingénieurs et Industriels (Belgium)

SRBL Solar Radio Burst Locator

SRBP Synthetic Resin-Bonded Paper

SRBS Skeletal Reference Baseline Simulation • Skeletal Reference Baseline System • Solar Radio Broadband Spectrometer

SRBSEP Solid Rocket Booster Separation

SRC Safety Review Committee • Sample Recovery Container • Sample Reentry Capsule • Sample Return Canister • Sample Return Container • Sample Rock Container • Saskatchewan Research Council (Canada-SK) • Science Research Center • Science Research Council • Science Review Committee • Searcy (USA-AR) (location identifier) • Semiconductor Research Corporation • Send Register Control • Se Ruega Contestación • Service Régional de Contrôle • Small Rocket Controller • Sociedad Regular Colectiva • Société de Recherche sous Contrat • Software Re-engineering Center • Solvent-Refined Coal • Sound Ranging Control • Source • Source Range Channel • Source Reference Code • Source Requirement Code • Southern Redshift Catalogue • Southwest Research Corporation • Space Regatta Consortium • Space Research Council • Spares Receiving Checklist • Standard Reference Code • Standard Reference Compound • Standard Requirement Code • Station Reliability Coordinator • Stray Radiation Chamber • Student Retention Committee • Student's Representative Council • Supercomputing Research Center • Supply Readiness Center • Support Review Code • Survival and Recovery Cell • Survival Recovery Center • Synchrotron Radiation Center • System Review Committee • System Requirement Code

SRCAP Shanghai Research Center for Applied Physics (China-PRC)

SRCB Simulation Requirements Control Board • Software Requirements Change Board

SRCBD Software Requirements Change Board Directive

SRCD Set-Reset Clocked Data

srCDM Standard Reionized Cold Dark Matter (model)

SRCF Subregional Control Facility

Srch Search

SRCH Search

SRCIM Slovak Research Consortium for Informatics and Mathematics (Slovakia)

SRCHLT Searchlight

SRCM Slow Rotating Centrifuging Microscope

SRCT Standard Recovery Completion Time

SRCU Securable Remote Control Unit • Secure Remote Control Unit

S/RD Secret/Restrictive Data

SRD San Ramón (Bolivia) (location identifier) • Scheduled Release Date • Schnellreferatdienst • Science Requirements Document • Scientific Research Division • Screen Reader • Screen-Reader Device • Secondary Receive Data • Secret Restricted Data • Secret Restrictive Data • Security Requirements Document • Selective Radiation Detector • Selective Retrieval of Documentation • Self-Reading Dosimeter • Service de Règlement Différé • Short-Range Dependence • Short-Range Device • Shuttle Requirements Definition • Shuttle Requirements Document • Significantly Revised Down • Simultaneous Ranging Difference • Small Rigid Dome • Software Requirements Document • Spacecraft Research Division • Standard Reference Data • Standard Reporting Designator • Station de Réception Directe • Step Recovery Diode • Subrate Data • Surplus Release Date • System-derived Requirements Document • Système à Règlement Différé • Système de Recherche Documentaire • Systems Requirements Document

SRDAS Service Recording and Data Analysis System

SRDC Space Research in Developing Countries • Standard Reference Data Center • Subrate Data Crossconnect • Surface Reference Data Centre

SRDCS Simulation Reconfiguration Data Collection System

SRDD Software Release and Distribution Division

SRDE Signals Research and Development Establishment

SRDH Systems Requirements Definition Handbook

SRD&I Space Data Resources & Information

SRDL Signals Research and Development Laboratory

SRDM Subrate Data Multiplexer • Subrate Data Multiplexing

SRDP Standard Reference Data Program

SRDS Single Requirements Determination System • Solar Radio Decimeter Spectrograph • Standard Reference Data System • Symposium on Reliable Distributed Systems

SRE Secondary Reflection Effect • Self Routing Element • Send Remote Enable • Shrinkage Restricted Estimator • Software Reliability Engineering • Software Risk Evaluation • Sounding Rocket Experiment • Standard Revised Edition • Stellar Research and Education • Study of Radar Echos • Sucre (Bolivia) (location identifier) • Surveillance Radar Element • System Research Environment

SR&ED Scientific Research & Experimental Development

s/ref Su Referencia

SREL Space Radiation Effects Laboratory

SREM Software Requirements Engineering Methodology • Software Requirements Evaluation Methodology • Space Radiation Environment Monitor • Standard Radiation Environment Monitor

SREMP Source Region Electromagnetic Pulse

SRES Signed Response • Special Report on Emission Scenarios

SRET Software Reliability Engineered Testing

SRF San Rafael - Hamilton (USA-CA) (location identifier) • Separate Reporting Facility • Service Response Force • Shared Reference Facility • Ship Repair Facility • Shuttle Refurbish Facility • Shuttle Refurbishment Facility • Signaling Relaying Function • Solar Radiation Facility • Solar Radiation Flux • Solar Radiation Function • Special Resource Function • Specifically Routed Frame • Specific Routed Frame • Spectral Response Function • Stable Radio Frequency • Standard Reference File • State Revolving Fund • Status Report Frame • Summary Reference File • Systematic Reconnaissance Flight • System Requirements Handbook

SRFD Service Régional de la Formation et du Développement

SRFF Set-Reset Flip-Flop

SRG Guilder (Suriname) (ISO currency code) • Scientific Review Group • Security Rapporteur Group • Semarang (Indonesia) (location identifier) • Shift-Register Generator • Short and Ring Ground (test) • Small Research Grant • Special Research Group • Spectrum Röntgen Gamma • Spinning Rotor Gauge • SRON Groningen (Netherlands) • Statistical Research Group • Survivability Review Group

SRGP Simple Raster Graphics Package • Small Research Grant Program

SRH Sarh (Chad) (location identifier) • Search • Searching • Systems Requirements Handbook

SRHE Society for Research into Higher Education

SRHIT Small Radar Homing Interceptor Technology

SRHP Space Radiation Health Program

SRHS Sum of Right-Hand Side

SRI Sacrum Romanum Imperium • Samarinda (Indonesia) (location identifier) • Send Routing Information • Service des Relations Internationales • Software Reuse Initiative • Soil Research Institute • Space Research Institute • Spalling Resistance Index • Spatial Reference Information • Standby Request for Information • Standard Request for Information • Standing Request for Information • Stanford Research Institute (USA-CA) • Statistical Reference Index • Surveillance - Reconnaissance - Intelligence • Système de Recherche d'Informations • Système de Référence Inertielle

SRIC Short-Rotation Intensive Culture

SRIF Square Root Information Filter

SRIM Selected Research In Microfiche • Simulated Radar Image Model

SRIP Security Requirements and Implementation Plan • Security Requirements Implementation Plan • Student Record Improvement Project

SRIS Science and Reference Information Service • Science Reference Information Service • Station de Réception des Images Spatiales

SRIT Système de Rectification d'Images en Télédétection

SRIV Station de Réception d'Images pour la Végétation

SRJ San Borja (Bolivia) (location identifier)

SRK Sierra Leone (location identifier)

Srl Società a Responsabilità Limitata

SRl Saudi Riyal (currency)

SRL Santa Rosalia (Mexico) (location identifier) • Scientific Research Laboratory • SGML Reuse Library • Shuttle Research Laboratory • Single-Rack Laboratory • Small-Rocket Launch • Software Reuse Library • Space Radar Laboratory • Space Radiation Laboratory • Speed of Relative Movement • SRON Leiden (Netherlands) • Standard Reference Library • Storage Resource Loader • Structure Return Loss • Study Reference List • Surface Review and Letters (periodical code) • System Reference Library • Systems Requirements Letter

SRLY Series Relay

SR&M Safety, Reliability & Maintainability

SRM Safety, Reliability and Maintainability • Sandringham (Australia-QLD) (location identifier) • Scalable Reliable Multicast • Scanning Radar Microscope • Science Reference Mission • Security Reference Monitor • Shift Register Memory • Solid-fuel Rocket Motor • Solid Rocket Motor • Source Range Monitor • Spares Requirement Model • Specifications Requirement Manual • Specified Risk Material • Spectral Resolving Model • Square Root Mode • Stability Robustness Measure • Standard Reference Material • Standard Reference Module • Storage Resource Management • Strength Ratio Method • Structural Risk Minimization • Supplier Relationship Management • Switch Revision Message • System for Resources Management • System Resources Manager

SRMC Stimulus-Response Measurements Catalog

SRMD Slant Range Measuring Device

SRMF Solid Rocket Motor Facility

SRML Solar Radiation Monitoring Laboratory

SRMLE Shrinkage-Restricted Test Maximum-Likelihood Estimator

SRMP Sounding Rocket Measurement Program

SRM&QA Safety Reliability, Maintainability & Quality Assurance

SRMQA Safety Reliability, Maintainability, and Quality Assurance

SRMS Shuttle Remote Manipulator System

SRMU Solid Rocket Motor Upgrade • Subrate Multiplexer Unit

SRN Serial Reference Number • Serpens Reflection Nebula • Simulation Reference Number • Single Row Nest • Software Release Note • Software Release Notice • Source Recipient Node • Specification Revision Notice • Star in Reflection Nebula • Strahan (Australia-TAS) (location identifier) • Stretched Receptor Neurone

SRNP Secure Remote Network Processor

SRNU Service du Réacteur Nucléaire Universitaire

SRO Santana Ramos (Colombia) (location identifier) • Self-Regulatory Organization • Senior Resource Official • Sensitive Reconnaissance Operation • Sharable and Read Only • Short-Range Order • Single Resonant Oscillator • Single-Room Occupancy • Site Recovery Option • Solar Radio Observatory • Specification Release Order • Standing Room Only • Standing Route Order • Statutory Rules and Orders • Stony Ridge Observatory (USA-CA) • Strategic Research Objective • Superintendent of Range Operations • Supervisor of Range Operations • System Readiness Objective

SROC Senior Readiness Oversight Council • Space Remote Operations Center

SROD Shared Read-Only Data

SROM Spectroradiometer for Ocean Monitoring

SRON Space Research Organization Netherlands • Stichting Ruimte Onderzoek Nederland (Netherlands)

SROSS Stretched Rohini Satellite Series

SROTA Santa Rosa Optical Test Area (USA-CA)

SRP Sale-Recommended Price • School Requirements Package • Science Review Panel • Scientific Research Proposal • Seat Reference Plane • Seat Reference Point • Seaward Recovery Point • Security Requirements Review • Short-Range Plan • Short-Range Planning • Smooth Rolling Plains • Software Rapid Prototype • Software Reuse Program • Solar Radiation Pressure • Solicitation Review Panel • Soluble Reactive Phosphorus • Spatial Reuse Protocol • Special Research Programme • Stabilization Reference Package • Stabilizer Reference Plane • Stabilizer Reference Plane • Stage de Reclassement Professionnel • Standard Repair Procedure • Standard Review Plan • Stord (Norway) (location identifier) • Strategic Reconnaissance Plan • Stratospheric Research Program • Structured Reactive Plan • Suggested Retail Price • Sun Radio Patrol • Surplused Research Professor • Sustained Readiness Program • Switch Route Processor • Système de Retenue Programmée

SR&PA Safety, Reliability & Product Assurance

SRPA Safety, Reliability, and Product Assurance • Small Reclamation Projects Act (USA) • Spherical Retarding Potential Analyzer

SRPI Server-Requester Programming Interface • Swedish Radiation Protection Institute (Sweden)

SRPS Secure Record and Playback System • Sensor-Referenced Positioning System • Subrecursive Programming System

SRPT Small Repair Parts Transporter

SR&Q Safety, Reliability & Quality

SRQ Bradenton (USA-FL) (location identifier) • Sarasota (USA-FL) (location identifier) • Service Request • Service Request Queue

SR&QA Safety, Reliability & Quality Assurance

SRQA Safety, Reliability, and Quality Assurance

SRR Satellite Reconnaissance Radio • Scanned Relative Response • Search and Rescue Region • Serially Reusable Resource • Shift Register Recognizer • Shipment Request Register • Short-Range Radar • Short-Range Radio • Short-Range Recovery • Shuttle Requirements Review • Site Readiness Review • Slow Rotation Room • Software Requirements Review • Solar Radiation Response • Sound Recorder Reproducer • Specification Readiness Review • Split-Ring Resonator • Stradbroke Island - Dunwich (Australia-QLD) (location identifier) • Subcontractor Readiness Review • Survivability - Recovery - Reconstitution • Synchrotron Radiation Research • System Readiness Review • System Requirements Review

SRRB Software Requirements Review Board • System Requirements Review Board

SRRC Synchrotron Radiation Research Center (China-ROC)

SRRF Solar Radiation Response Function

SRRID Site Readiness Review Item Document

SRRP System Requirements Review Panel

SRRPV Short-Range Remotely-Piloted Vehicle

SRRT Site Readiness Review Team

SRS Fellow of the Royal Society • Safety and Reliability Society • Same Runway Separation • San Marcos (Colombia) (location identifier) • SAN Resource Sharing • Satellite Readout Station • Satellite Receiving Station • Satellite Remote Sensing • Schweizerische Radiosonde • Science Requirements Strategy • Science Resource Statistics • Science Resource Studies • Screen Reader System • Search and Rescue Sector • Seismic Recording System • Selective Record Service • Service Registry Service • Shared Registry System • Silicon Readout Sequence • Simple Random Sample • Simulated Remote Site • Simulated Remote Station • Simulation Reconfiguration System • Simulation Recording System • Single Resonance Structure • Situation Reporting System • Small Research Satellite • Smog Regulation System • Societatis Regiae Sodalis • Software Requirements Specification • Solar Radiation Simulator • Solar Radio Spectrograph • Sortie Réservée Synchrone • Sounding Rocket System • Sound Retrieval System • Southern Reference Stars • Southern Reference System • South Research Station • Space Recovery System • Space Research Service • Spatial Reference System • Specification Requirement Sheet • Specification Review Sheet • Specification Revision Sheet • Speed Reference System • Sport and Recreation Service • Standardized Range Security • Standard Range Support • Stanford Research Systems (USA-CA) • Statistical Reporting Service • Stellar Radiation Sensor • Stimulated Raman Scattering • Strict Relevance Scheme • Student Record System • Subscriber Response System • Superluminal Radio Source • Supplemental Restraint System • Support Requirement System • Synchro Retransmission System • Synchrotron Radiation Source • System Requirements Specification • System Requirements Study

SRSA Scientific Research Society of America

SRSC Satellite Remote Sensing Center • Senior Real-time Schedule Controller • Software Review and Scoring Committee • Space Remote Sensing Center

SRSCC Simulated Remote Station Control Center (*) • Simulated Remote Station Control Console

SRSG Science Review Steering Group

SRSI Simulated Reusable Surface Insulation

SRSK Short-Range Station Keeping

SRSL Société Royale des Sciences de Liège (Belgium)

SRSO Senior Range Safety Officer

SRSR Schedule and Resources Status Report

SRSS Shuttle Range Safety System • Simulated Remote Site System • Solar Radiation Simulator System • Square-Root Sum of Squares

Srster Surister (Belgium)

SRSU Sul Ross State University (USA-TX)

SR&T Science Research & Technology • Support Research & Technology • Supported Research & Technology • Supporting Research & Technology

SRT Safety Review Team • Sardinia Radio Telescope (Italy) • Scheduled Return Time • Science Recommendation Team • Science Research and Technology • Secondary Request to Send • Security Response Team • Shuttle Requirements Traceability • Slow Run Through • Small Radiotelescope • Software Review Team • Solar Radiation Test • Solar Radio Telescope • Soroti (Uganda) (location identifier) • Source Route Transparent • Source Routing Transparent • Spacecraft Received Time • Space Radio Telescope • Special Rated Thrust • Special Reaction Team • Special Requirements Test • Specification Requirements Table • Standard Remote Terminal • Station Readiness Test • Step Recovery Transistor • Student Radio Telescope • Sun Reference Sensor • Supporting Research and Technology • Sweeney - Robertson - Tocher (algorithm) • Synthese Radio Teleskoop • System Reaction Time • System Readiness Test • System Reinvention Team • System Reliability Test • System Requirements Test

Srta Senhorita • Señorita

SRTB Source Route Transparent Bridging • Source Routing Translational Bridging

SRTC Special Real-Time Command

SRT&E System Requirements Test & Evaluation

SRTE Scalar Radiative Transfer Equation • System Requirements Test and Evaluation

SRTF Subregional Task Force

SRTG Source-Route Translation Gateway

SRTM Security Requirements Tractability Matrix • Shuttle Radar Terrain Mapping • Shuttle Radar Topography Mission (*)

SRTMU Swami Ramanand Teerth Marathwada University (India)

SRTN Solar Radio Telescope Network

SRToh Sci. Rep. Tohoku Univ. Eighth Ser. (periodical code)

SRTS Short Range Thermal Sight • Status of Resources and Training System • Synchronous Residual Time Stamp

SRU Santa Cruz (USA-CA) (location identifier) • Search and Rescue Unit • Service Replaceable Unit • Shop Repairable Unit • Shop Replaceable Unit • Signal Responder Unit • Significantly Revised Up • Simulation Resource Utilization • Solidarité et Renouvellement Urbains • Space Replaceable Unit • SRON Utrecht (Netherlands) • Stellar Reference Unit • Subassembly Repairable Unit • Subsystem Repairable Unit • Support Resource Unit • System Register Unit • System Replaceable Unit

SRUAV Short-Range Unmanned Aerial Vehicle

Srv Service

SRV Satellite Reentry Vehicle • Scientific Remote Viewing • Semi-Regular Variable (star) • Single Reentry Vehicle • Southern Republic of Viet Nam • Space Rescue Vehicle • Stony River (USA-AK) (location identifier) • Surface Roving Vehicle • System Readiness Verification

SRVIFS Service Installable File System

Srvln Surveillance

SRVM Specification Requirements Verification Matrix

SRVT Suivi des Ressources Végétales par Télédétection • System Readiness Verification Test • System Requirements Verification Test

SRW Salisbury - Rowan County (USA-NC) (location identifier) • Standard Reaction Wheel

SRWBR Short-Range Wideband Radio

SRWCP Short Rotation Woody Crop Program

SRWT Société Régionale Wallonne du Transport (Belgium)

SRX Sert (Libya) (location identifier)

SRY Sary - Dashte Naz (Iran) (location identifier)

SRZ Santa Cruz - El Trompillo (Bolivia) (location identifier) • Special Rules Zone

SRZM Stichting Radiostraling van Zon en Melkweg (Netherlands)

s/s Samples per Second • Symbols per Second

ss Sans • Scilicet • Sections • Seguro Servidor • Semis • Sensu Stricto • Sentencias • Siglos • Siguientes • Siswati (ISO language code) • Sous • Sunset • Supra Scriptum

S-S Surface-to-Surface

S&S Supply & Service

S/S Same Size • Samples per Second • Sector Suite • Service Station • Ship to Shore • Signal Strength • Single Sideband • Sous-Sol • Space Shuttle • Station-Service • Steamship • Subsystem

SS Far South • Far Southern • Sacra Scriptura • Safing System • Saints • Saint-Siège • Same Sense • Same Size • Sandstorm • Santa Sede • Santi • Santissimi • Sa Sainteté • Satellite-Switched • Satellite System • Science Service • Screw Steamer • Secondary School • Seconds • Second Stage • Second Structure • Secretary of State • Secret Service • Sector Suite • Sécurité Sociale • Security Service • Segment Specification • Seguridad Social • Selector Switch • Self-Sustaining • Semana Santa • Sequential Signature • Service Station • Service Subscriber • Set Screw • Seyfert Sextet • Shell Sort • Shift Supervisor • Ship Shore • Shore Station • Short Sleeve • Shuttle System • Side by Side • Signaling System • Signal Strength • Simulator System • Simultaneous Sampling • Single Sideband • Single-Sided • Single Signal • Single Source • Single Star • Single String • Singapore Standard • Sky Screen • Sky Survey • Slow Steady • Social Security • Sodium Sulphate • Software Support • Solar Sentinel • Solar System • Sole Source • Solid State • Source of Supply • Source Selection • Sous-Sol • Space Segment • Space Shuttle • Space Sphere • Space Station • Space Surveillance • Special Series • Special Session • Special Source • Special Study • Speech Sequence • Speech Sequencer • Speech Sequencing • Spent Stage • Spider Squad • Spin-Stabilized • Spiral Structure • Sporadic Server • Spread Spectrum • Stabilization System • Stack Segment • Staff Specialist • Stainless Steel • Standard Sequential • Standard Stage • Standard Size • Star Selector • StarSets (resources) • Start to Start • Statens Strålevern (Norway) • Station Service • Station Set • Statistical Standard • Steady State • Steamship • Steering System • Stephenson-Sanduleak (catalog) • Storage to Storage • Strada Statale • Strain Sensor • Stricto Sensu • Strict Sense • Structure System • Struve-Straka (catalog) • Study Scientist • Study Sponsor • Subordinate System • Submarine • Subsection • Subsolar • Substructure • Subsystem • Summer Solstice • Summing Selector • Sun Sensor • Sunset • Sun Simulator • Sun Synchronous • Supersonic • Supervisory Station • Supplementary Service • Supply Support • Supportability Strategy • Support Segment • Support Software • Support System • Surface-to-Surface • Surveillance Satellite • Surveying Ship • Survey Scientist • Su Señoría • Svensk Standard (Sweden) • Switching Shelter • Symbiotic Star • Symbiotic System • Système Solaire • System Safety • System Segment • System Sensitivity • System Simulation • System Simulator • System Software • System Specification • System Summary • System Supervisor • System Support

SS&A Space Systems & Applications

SSA Salvador - Dois de Julho (Brazil-BA) (location identifier) • S-band Single Access • School of Sciences and Applications • Security Supporting Assistance • Segment Search Argument • Seismological Society of America (USA) • Selective Service Administration • Serial Storage Architecture • Service Statistique Académique • Servicing Security Activity • Shuttle Simulation Aircraft • Simple Shift-and-Add • Single-Service Access • Singular Spectrum Analysis • Slitless Spectra Analysis • Small Science Aperture • Small Scientific Aperture • Social Security Act (USA) • Social Security Administration • Software Safety Analysis • Software Support Activity • Software Support Agency • Solid-State Altimeter • Solid-State Amplifier • Source Selection Authority • Space Suit Assembly • Special Security Agreement • Special Support Activity • Specific Support Action • Stable Solar Analyzer • Staff Supply Assistant • Staff Support Activity • Staff Support Agency • Staging and Support Area • Standard Spending Assessment • Standard System Application • Star Selector Assembly • Strapdown Sensor Assembly • Subsaharan Africa • Subsaharan African • Sun Sensor Assembly • Superconducting Six-axis Accelerometer • Supply Support Activity • Supply Support Area • Supply Support Arrangement • Supporting Support Activity • Symbol Synchronizer Assembly • Synchro Signal Amplifier • Synchrotron Self-Absorption • System Security Analysis • System Supportability Assessment • System Support Analysis

SSAA Société Suisse d'Astrophysique et d'Astronomie (Switzerland) • Space Science Analysis Area • Sus Altezas • System Security Authorization Agreement

SSAAC Space Science and Applications Advisory Committee

SSAAT Sun Sensor Attitude Angle Transducer

SSAC Selective Self-Adaptive Crossover • Source Selection Advisory Committee • Source Selection Advisory Council • Space Science Advisory Committee • Space Science Analysis and Command

SSAD Software System Analysis Document • Space Systems Analysis and Design • Standards and Structures Administrator • System Summary Acceptance Document

SSADC Sofia Sky Archive Data Center (Bulgaria)

SSADM Structured Systems Analysis and Design Method

SSAE Stamped Self-Addressed Envelope

SSAF Standard Single Account File

SSAI Space Systems and Applications Inc. (USA-MD) • Statistical Society of Australia Inc.

SSAIAF Space Systems Automated Integration and Assembly Facility

SSAIP Session Security, Authentication, Initialization Protocol

SSALR Simplified Short Approach Lighting System

SSALT Solid-State Altimeter

SSALTO Segment Sol d'Atltimétrie et d'Orbitographie

SSAM Simple Sequential Access Method

SSAN Social Security Account Number

SSAP Session Service Access Point • Source Service Access Point • Statement of Standard Accounting Practice

SSAR Site Safety Analysis Report • Solid-State Adaptive Rotor • Specifications and Standards Acquisition Reform • Standard Safety Analysis Report

SSARA Scanning Synthetic Aperture Radar Altimeter

SSARR Streamflow Synthesis And Reservoir Regulation

SSAS Signal Security Assessment System

SSASC Space Science and Applications Steering Committee

SSASP Small-Sized Aerospace Plane

SSAT S-band Single Access Transmitter • Shuttle Service and Access Tower • SIRTF Science Activities Timeline • Space Shuttle Access Tower • System Segment Acceptance Test • System Software Acceptance Test

SSB Saint Croix Island (US Virgin Islands) (location identifier) • Science Software Branch • Scientific Software Branch • Script Storyboard • Sea State Bias • Selective Service Board • Silicon Surface Barrier • Single Sideband • Small Solid Booster • Social Security Board • Société Spécialisée de Biotechnologie • Solar Sector Boundary • Solar-System Barycenter • Source Selection Board • Space Science Board • Space Studies Board • Space Support Building • Special Separation Benefit • Spontaneous Symmetry Breaking • State Seismological Bureau • Support Software Bulletin • System Signaling Buffer

SSBAM Single Sideband Amplitude Modulation

SSBC Summary Sheet Bar Chart

SSBCCT Small Solid Booster Critical Component Technology

SSBD Single Sideboard

SSBF Single Sideband Filter

SSBFM Single-Sideband Frequency Modulation

SSBG Single Sideband Generator

SSBI Single Scope Background Investigation

SSBM Single Sideband Modulation

SSBNF Single Sideband Noise Figure

SSBO Single Swing Blocking Oscillator • System Support Buyout

SSBP Self-Supervised Back Propagation

SSBRP Space Station Biological Research Project

SSBS Space-Based Surveillance System

SSBUV Shuttle Solar Backscatter Ultraviolet

SSBW Spread-Spectrum Bandwidth

SSc Social Sciences

S&SC Sized & Supercalendered

SS&C Scope, Sequence & Coordination

SSC Satellite Situation Center • Scanning Shadow Telescope • Scan-to-Scan Correlation • Schéma de Services Collectifs • Science Steering Committee • Science Support Center • Science Survey Consortium • Scientific Steering Committee • Secondary School Certificate • Senior Staff Council • Sensor Signal Conditioner • Sequential System Controller • Serendipitous Survey Catalog • Serial Shift Counter • Serial Stamp Collecting (pipeline) • Service School Command • Set of Station Coordinates • Shapley Supercluster • Shapley Superconcentration • Shop Service Center • Short Segmented Cask • Short Service Commission • Shuttle System Contractor • Simulated Spacecraft • Simulation System Concept • Single-Stage Command • SIRTF Science Center • Site Selection Criterion • Skill Specialty Code • Smaller-Scale Contingency • Small-Saver Certificate • Small Source Catalog • Software Standards Committee • Software Support Center • Solar Stabilization Computer • Solid-Solution Cermet • Solid-state Slit Camera • Solid-State Circuit • Spacecraft System Console • Space Science Center • Space Surveillance Center • Space Surveillance Control • Space Systems Center • Spin Synchronous Clock • Staff Selection Committee • Standardization Status Code • Standard Spacecraft Cooler • Standard Spacecraft Cryocooler • Standard Systems Center • Standards Steering Committee • Starlink Software Change • Starlink Software Collection • Static Standby Computer • Station Selection Code • Station Support Computer • Statistical Society of Canada • STEIS Steering Committee • Stellar Simulation Complex • Stennis Space Center (USA-MS) • Stepping Switch Counter • Stockholm Spectral Catalogue • Storm Sudden Commencement • Subsystem Computer • Subsystem Controller • Subsystem Sequence Controller • Successive Subspace Correction • Sudden Storm Commencement • Sumter Shaw (USA-SC) (location identifier) • Sunspot Count • Superconducting Supercollider • Super Star Cluster (*) • Super Stellar Cluster • Supply Status Code • Supported Software Change • Survey Science Center • Survey Scientist Consortium • Suspended Sediment Concentration • Swedish Space Corporation • SWIFT Science Center • Synchrotron Self-Compton • System Sales Case • System Security Concept • System Sequence Controller • Systems Science and Cybernetics • Systems Support Center • System Support Contractor

SSCA Statistical Consulting Center for Astronomy (USA-PA) • Surface Sampler Control Assembly

SScAC Space Science Advisory Committee (NASA)

SSCAG Space System Cost Analysis Group

SSCB Space Station Control Board

SSCC SATCOM System Control Center • Second-Stage Conduit Container • Space Section Control Center • Space Station Control Center • Space Surveillance Control Center • Special Security Communications Center • Spin-Scan Cloud Camera • Support Services Control Center

SSCCB Support Software Configuration Control Board

SSCD Suspended Solids Concentration Determination

SSCE Solid-Surface Combustion Experiment • Sun - Spacecraft - Earth

SSCF Service-Specific Coordination Function • Space Station Control Facility • Space System Control Facility

SSCFP Senior Service College Fellowship Program

SSCHF Space Shuttle Cargo Handling System

SSCI Social Sciences Citation Index (ISI) • Société de Services et de Conseil en Informatique

SSCL Shuttle System Commonality List • Sunspot Cycle Length • Supercomputing Supercollider Laboratory

SSCM Segment Sol de Contrôle et de Mission • Small Satellite Cost Model • Space Station Cost Model • Sun - Spacecraft - Mars

SSCMP Sequenced Synchronized Clock Multicast Protocol

SSCN Secure Survivable Communications Network

SSCO System Security Control Officer

SSCom Solid State Communications (periodical code)

SSCOP Service-Specific Connection-Oriented Protocol • Service-Specified Connection-Oriented Protocol • Service-Specified Convergence Protocol

SSCP Service Switching and Control Point • Small Self-Contained Payload • Springer Series on Chemical Physics (periodical code) • Sum of Squares and Cross-Products • System Services Control Point

SSCR Standard Systems Center Regulation • Support Software Change Request

SSCS Service Specific Convergence Sublayer • Shipboard Satellite Communications System • Side-Stick Control System • Single Shelter Circuit Switch • Single-Sideband Communications System • Solar-System Creation Sequence • Space Suit Communications System • Special Security Communications System • Stand Satellite Control System • Synchronous Satellite Communications System

SSCSA Society for Student Counseling in South Africa • Standard Small Computer Software Acquisition

SSCSP Space Shuttle Crew Safety Panel

SScT Speculations in Science and Technology (periodical code)

SSCT Science User Resource Expert • Solid-State Celestial Tracker

SSCU Spacecraft Systems Controller Unit • Special Signal Conditioning Unit • Store Station Control Unit

SSCV Solar Sail Cargo Vessel

SSCVR Solid-State Cockpit Voice Recorder

SSCVT Subsystem Communications Vector Table

SSD Sacrae Scripturae Doctor • Safe Separation Device • Sample Server Disk • San Felipe (Colombia) (location identifier) • Satellite Superhighway Demonstrator • Satellite Systems Division • Scaling-Steepest Descent • Science Support Division • Seasat Decimal • Security Support Division • Semiconductor Silicon Detector • Sequence Switch Driver • Shakura-Sunyaev Disk • Single-Sided Disk • Single-Station Doppler • Software Specification Document • Software System Design • Solid-State Detector • Solid-State Device • Solid-State Disk • Solid-state Storage Device • Source Selection Data • Spacecraft Software Division • Space-Scale Decomposition • Space Science Department • Space Science Division • Space Sciences Division • Space Shuttle Display • Space Systems Directorate • Space Systems Division • Split-Screen Display • Sponsor's Study Director • Start of Shell Delimiter • Statistical Subdivision • Subsurface Damage • Sun Shadow Device • Surface Sampler Device • Surveillance Situation Display • Survival Situation Display • Symposium on Spatial Databases • Synthetic State Determination • Systems Sciences Division • Systems Status Display • Systems Summary Display • Systems Support Database • Systems Support Division

SSDA Sequential Similarity Detection Algorithm • Space Systems Development Agreement

SSDAC Space Station Data Acquisition and Control

SSDB Scientific and Statistical Database • Specifically Stellar Database • Spin System Distribution Box • Standard Simulator Database

SSDBM Statistical and Scientific Database Management

SSDBMS Statistical and Scientific Database Management System

SSDC Space Science Data Center

SSDD Single-Side Double-Density • Software System Design Document • Source Selection Decision Document • Subsystem Design Document • System/Segment Design Description • System/Segment Design Document

SSDF Space Science Development Facility

SSDH Subsystem Data Handbook

SSDI Support System Design Integration

SSDL Secondary Standards Dosimetry Laboratories • Single-Side Double-Layer

SSDM Shielding Standard Design Method

SSDO Steady-State Design and Optimization

SSDOO Space Science Data Operations Office (NASA)

SSDP Simple Service Discovery Protocol • Standard Source Data Package

SSDPS Solar System Data Processing System

SSDR Satellite Situation Display Room • Space Systems Division Regulation • Subsystem Design Review • Subsystem Development Requirement • System Safety Domain Report

SSDS Space Sciences Data Service • Space Sciences Data System • Space Shuttle Display and Simulation • Space Station Data System • Software System Development Specification

SSDSAS Space Station Data System Architecture Study

SSDU Session Service Data Unit

SSE Safety System Engineering • Second Starlink Environment • Self-Sustained Emission • Service Social des Élèves • Sholapur (India) (location identifier) • Signal Security Element • Simulation Support Environment • Single Sideband Exciter • Single Site Event • Single Star Evolution • Site-Specific Extractor • Société Suisse des Écrivains (Switzerland) • Society for Scientific Exploration (USA-VA) • Software Support Environment • Software Support Equipment • Software System Environment • Solar-System Exploration • Solid-State Electronics • Sortie Support Equipment • Source Selection Evaluation • South-Southeast • South-Southeastern • Spacecraft Simulation Equipment • Spacecraft Support Equipment • Space Science Enterprise • Space Shuttle Engine • Space Support Equipment • Space Surveillance Equipment • Space Surveillance Experiment • Special Support Equipment • Squared Sum of Errors • Stage Systems Engineer • Star Selector Encoder • Structural Systems and Engineering • Subsystem Element • Sud-Sud-Est • Sum of Squared Errors • Sum-Squared Error • Sun - Spacecraft - Earth • Support System Evaluation • Surface Solar Energy • Surface Support Equipment • Système Statistique Européen • System Safety Engineering • System Security Engineer • System Security Engineering • System Status Evaluation • System Support Engineering • System Support Equipment

S'sea Swansea (UK-Wales)

SSEA Separate Sampling and Excitation Analysis • System Safety Engineering Analysis

SSEAF Space Science Experimenters Analysis Facility

SSEB Source Selection Evaluation Board

SSEC Selective Sequence Electronic Calculator • Solar-System Exploration Committee • Solid-State Electronic Chronograph • Space Science and Engineering Center • Subsystem Essentiality Code

SSECO Second-Stage Engine Cutoff

SSECS Space Shuttle Environmental Control System

S&SED Satellite & Space Electronics Division

SSED Software Systems Engineering Directorate • Solar-System Exploration Division

SSEDF Software Support Environment Development Facility

SSEED Symmetric Self-Electro-optic Effect Device

SSEG Source Selection Evaluation Group • Source Selection Evaluation Guide

SSEL Solid-State Electronics Laboratory • Space Science and Engineering Laboratory

SSEle Solid State Electronics (periodical code)

SSEM Small-Scale Experimental Machine • Space System Effectiveness Model • System Security Engineering Management • System Security Engineering Manager

SSEMC Solar-System Exploration Management Council

SSEMP System Security Engineering Management Plan

SSEO Source Selection Executive Officer

SSEOO Space Shuttle Earth Observation Office • Space Shuttle Earth Observation Optics

SSEOP Space Shuttle Earth Observations Photography • Space Shuttle Earth Observations Photos • Space Shuttle Earth Observations Project (*)

SSEOS Space Shuttle Engineering and Operations Support • Space Station External Operations Simulator

SSEp Springer Series on Electrophysics (periodical code)

SSEP Solar-System Educators Project (NASA/JPL) • Solar-System Exploration Project • Source Selection Evaluation Plan • Space Systems for Economic Purposes • System Safety Engineering Plan • System Security Engineering Phase • System Security Engineering Plan • System Security Engineering Program • System Security Enhancement Program

SSEQ Standard Sequence

SSER System Safety Engineering Report

SSERN South-Southeastern

SSERC Scottish Schools Equipment Research Centre (UK-Scotland)

SSERD Space Science Education Resource Directory

SSES Software System Environment Station • Software System Environment System • Solar-System Exploration Subcommittee

SSET Source Selection Evaluation Team • Source Selection Evaluation Test

SSEU System Selector Extension Unit

SSEWD South-Southeastward

SSF San Antonio - Stinson (USA-TX) (location identifier) • SEA Simulation Facility • Scripps Satellite Facility • Secondary Star Formation • Service Storage Facility • Service Subscriber Field • Service Switching Function • Single-Sideband Filter • Single Solar Flare • Single Stock Fund • Single Suppression Factor • Slit Spread Function • Smooth Source Function • Société Statistique de France • Software Support Facility • Space Simulation Facility • Space Station Freedom • Space Systems Finland Ltd. • Specialized Service Facility • Special Security Facility • Special Security File • Split-Step Fourier • Standard System Failure • Stiff Shipping Frame • Stiftelsen för Strategisk Forskning (Sweden) • Super Skyflux • Supported Solution Forum • Switch Supervisory Function • System Support Facility

SSFBB Small Solid Fast Burn Booster

SSFC Solid-State Frequency Changer • Solid-State Frequency Converter

SSFCS Space Shuttle Flight Control System

SSFD Solid-State Floppy Disk

SSFDC Solid-State Floppy Disk Card

SSFDR Solid-State Flight Data Recorder

SSFF Space Station Furnace Facility • Space Station Future Fighters (USA-TX)

SSFGSS Space Shuttle Flight and Ground System Specification

SSFL Steady-State Fermi Level

SSFLC Surface-Stabilized Ferroelectric Liquid Crystal

SSFM Single-Sideband Frequency Modulation

SSFN Space Station Flight Network

SSFP Space Station Freedom Program

SSFPO Space Station Freedom Project Office

SSFR Solar Spectral Flux Radiometer

SSFS Space Shuttle Functional Simulation • Space Shuttle Functional Simulator • Sydney Space Frontier Society (Australia)

SSFVT Subsystems Functional Verification Test

SSG Malabo (Equatorial Guinea) (location identifier) • Safety Study Group • Science Software Group • Science Steering Group • Scientific Steering Group • Sensor Systems Group • Single-Sideband Generator • Small Signal Gain • Software Strategy Group • South-Seeking Gyroscope • Specialization Subgraph • Special Study Group • Special Support Group • Stonehenge Study Group • Supply Spectrum Generator • Systems Safety Group

SSGA System Support Gate Array

SSGC Short System Ground Check

SSGD Superheated Superconducting Granule Detector

SSGDAC Space Shuttle Guidance, Dynamics And Control

SSGM Synthetic Scene Generation Model

SSG/P Small-Scale Groups and Pairs

SSGP Small-Scale Groups and Pairs

SSGS Solid-State Gamma Switch • Standard Space Guidance System

SSGSR Springer Series on Geophysics and Space Research (periodical code)

SSGT Subsystem Ground Test

SSH Schwartz - Slawsky - Herzfeld (theory) • Sea Surface Height • Second-Stage Hydraulics • Secure Shell • Sharm El Sheik (Egypt) (location identifier) • Simulation Support Hardware • Spatial-Spectral Holography • Starkenburg-Sternwarte Heppenheim eV (Germany)

SSHA Shuttle System Hazard Analysis • Subsystem Hazard Analysis • System Safety Hazard Analysis

SSHB Station Set Handbook

SSHD Single-Side High-Density

SSHF Station Scientifique des Hautes Fagnes (Belgium)

SSHL Société Suisse d'Hydrologie et de Limnologie (Switzerland)

SSHM Society for the Social History of Medicine (USA)

SSHRC Social Sciences and Humanities Research Council

SSHT Sea Surface Height

SSI Brunswick (USA-GA) (location identifier) • San Salvador Island • Schedule Support Impact • Second-Stage Ignition • Sector Scan Indicator • Sécurité des Systèmes d'Information • Sensor System Interface • Serial Signal Interface • Serial Synchronous Interface • Server-Side Includes • Shore Station Interface • Significant Structural Item • Single-Source Integration • Single System Image • Site of Scientific Interest • Sky Survey Instrument • Small-Scale Integration • Social-Security Information • Solid-State Imager • Solid-State Imaging • Solid-State Inverter • Spacecraft System Integration • Spaceport System International • Space Science Institute (USA-CO) • Space Science Instrumentation (periodical code) • Space Segment Integration • Space Services Inc. (USA-TX) • Space Software Italia (Italy) • Space Studies Institute (USA-NJ) • Special-purpose-class Science Instrument (SOFIA) • Special Skill Identifier • Specialty Skill Identifier • Spectral Statistical Interpolation • Spectrum Sciences Inc. (USA-CA) • Standing Signal Instruction • State Systemic Initiative (USA) • Statewide System Initiative • Storage-to-Storage Instruction • Structural Significant Item • Successively Summed Integration • Supplemental Security Income • Supply System Item • Surface Stereo Imager • Sustaining Support Increment • Synchronous Serial Interface • System Software Interface

SSIBD Shuttle System Interface Block Diagram

SSIC Small-Scale Integrated Circuit • Standard Subject Identification Code • System Safety Integration Contractor

SSID Service Set Identification • Standard Software Interface Definition • Survey Science Interface Document

SSIE Smithsonian Science Information Exchange

SSIES Special Sensor for Ions, Electrons, and Scintillations

SSIG Single Signal

SSII Société de Service et d'Ingénierie Informatique • Société de Services en Ingénierie Informatique • Solid-State Image Intensifier

SSIL Security-Sensitivity Items List

SSILS Solid-State Instrument Landing System

SSIM Solar Spectral Irradiance Monitor • Spacecraft Simulator • Statistical Sampling Inventory Method

SSIMU Solid-State Inertial Measurement Unit

SSIP Shuttle Student Involvement Program • Space-science Student Involvement Program • Space Surveillance Interface Processor • Standard Systems Improvement Program • Student Science Investigation Project • Subsystem Integration Plan • Subsystem Integration Program • Surrogate System Implementation Plan • System Safety Implementation Plan • System Setup Indicator Panel • System Software Interface Processing

SSIPF Standard System Information Production Factory

SSIR Sun Shade Interface Ring • Supply System Inventory Report

S/SIS Sales/Services Information System

SSIS Sales/Services Information System • Schedules and Schemes Information System • Solid-State Imaging System • Spacecraft System Integration Support • Space Station Information System

SSISS Spacecraft System Integration Support Service

SSI&T Science Software Integration & Test

SSIT Science Software Integration and Test • Space Science for Illinois Teachers (program)

SSITP Shuttle System Integrated Test Plan

SSIU Singulator-Scan-Induction Unit • Subsystem Interface Unit • Systems Simulator Interface Unit

SSIV Segment Sol Image Végétation

SSIZ Seasonal Sea Ice Zone

SSJ Sandnessjoen (Norway) (location identifier) • Sole Source Justification

SSK Self-Service Kiosk • Sociology of Scientific Knowledge • Strahlenschutz-Kommission • Sturt Creek (Australia-WA) (location identifier)

SSKP Single-Shot Kill Probability

SS/L Space Systems/Loral

SSL Santa Rosalia (Colombia) (location identifier) • Secure Socket Layer • Selected Source List • Shore Station Laboratory • Single Station Locator • Site Support List • Small Space Laboratory • Software Specification Language • Software Solutions Laboratory • Software Support Laboratory • Solid-State Lamp • Solid-State Laser • Solid-State Lighting • Solid-State Logic • Space Science Laboratory • Space Sciences Laboratory • Space Simulation Laboratory • Space Systems Laboratory • Standard Security Label • Storage Structure Layout • Strategic Switching Laboratory • Sudden Sodium Layer • Superscanning Locator • Support Status List • System Software Loader • System Specification Language

SSLCP Secure Socket Layer Control Protocol

SSLDEF Space Shuttle Long-Duration Exposure Facility

SSLO Solid-State Local Oscillator

SSLOR Solar Stellar Limb Occultation Radiometer

SSLP Secure Socket Layer Protocol

SSLS Singapore Synchrotron Light Source • Solid-State Laser System • Standard Space Launch System

SSLSM Single Service Logistics Support Manager

SSLT Scientific Societies' Lecture Theatre (London) • Solid-State Logic Timer

SSLV Standard Small Launch Vehicle • Standard Space Launch Vehicle

SSM Sack Sorting Machine • Samstag - Sonntag - Montag • Satellite Survivability Model • Saturday - Sunday - Monday • Sault Sainte Marie (USA-MI) (location identifier) • Scanning Sky Monitor • Sciences des Structures et de la Matière • Second-Stage Motor • Second Surface Mirror • Signal Strength Monitor • Simulation Support Module • Single Segment Message • Single-Sideband Modulation • Single System Management • Site Status Message • Small Stamp Machine • Société de Secours Mutuel • Soft Systems Methodology • Software Stability Model • Software Structure Model • Software System Manager • Solar Standard Model • Solar Stereoscopic Mission • Solid-State Material • Spacecraft Support Module • Spacecraft Systems Monitor • Space Station and Microgravity • Space Station Module • Special Sensor for Microwave • Special Sensor Microwave • Special Study Module • Spread Spectrum Modulation • Springfield Science Museum (USA-MA) • Standard Solar Model • Station of Solar Monitoring • Station Spatiale et Microgravité • Steady-State Model • Subsystem Manager • Subsystem Model • Supply Support Management • Supply Support Manager • Support System Manager • Support Systems Module • System Safety Management • System Safety Manager • System Security Management • System Security Manager • System Support Management • System Support Manager • System Support Module

SSMA Solid-State Microaccelerometer • Solid-State Microwave Amplifier • Spread-Spectrum Multiple Access

SSMB Space Shuttle Maintenance Baseline • Space Station Manned Base

SSMC Second-Stage Motor Container • Space Shuttle Mission Control

SSME Satellite System Monitoring Equipment • Small-Scale Magnetic Element • Space Shuttle Main Engine • Spread-Spectrum Multiple Equipment

SSMEC Space Shuttle Main Engine Controller

SSMECA Space Shuttle Main Engine Controller Assembly

SSMF Science Studies Methods Forum • Space Station Millimeter Facility • Sun-as-a-Star Magnetic Field

SSMG Société Scientifique de Médecine Générale (Belgium)

SSM/I Special Sensor Microwave/Imager

SSMI Special Sensor Microwave Imager • Special Sensor Microwave Instrument

SSM/IS Special Sensor Microwave Imager/Sounder

SSMIS Special Sensor Microwave Imager/Sounder

SSMM Solid-State Mass Memory • Space Station Mathematical Model

SSMO Southern Skies Mobile Observatory (Australia) • Summary of Synoptic Meteorological Observations

SSMOC SOFIA Science and Mission Operations Center

SSMP Single Support Management Plan • Société Suisse de Minéralogie et de Petrographie (Switzerland) • Stockpile Stewardship and Management Program • System Security Management Plan • System Security Management Program

SSMS Solid-State Mass Spectrometer • Spark Source Mass Spectroscopy • Standard Survivable Message Set • Support Systems Module Simulator

SSM/T Special Sensor Microwave/Temperature

SSMTF Space Station Mockup and Trainer Facility

SSMTG Solid-State and Molecular Theory Group

SSN Seoul (Korea-ROK) (location identifier) • Severely Subnormal • Social-Security Number • Space Surveillance Network (USSC) • Specification Serial Number • Standard Serial Number • Standard Star Newsletter • Standard Study Number • Starlink System Note • Standard

Serial Number • Standard Study Number • Station Serial Number • Sunspot Number • Switched Service Network • System Study Number

SSND Solid-State Neutral Dosimeter

SSNIRS Solid-State Near-Infrared Sensor

SSNS Standard Study Numbering System

SSO Safety Security Office • Safety Security Officer • Safety Signification Operation • Sahara and Sahel Observatory • Salaried Suboffice • Sales Support Officer • São Lourenço (Brazil-MG) (location identifier) • Satellite Systems Operation • Satellite Switching Office • Sayan Solar Observatory (Russia) • Senior Scientific Officer • Siding Spring Observatory (Australia) • Signal Security Office • Signal Security Officer • Single Sign-On • Solar-System Object • Solar-System Observatory • Solid-State Oscillator • Source Selection Office • Source Selection Official • Space Shuttle Orbiter • Special Security Office • Special Security Officer • Special Staff Office • Special Studies Office • Special Studies Officer • Spurious Shut-Off • Staff Security Officer • Steady-State Oscillation • Structure, Sequence and Organization • Substellar Object • Subsystem Operation • Sud-Sud-Ouest • Sun Synchronous Orbit • Swedish Solar Observatory • Swiss Space Office • System Safety Office • System Safety Officer • System Security Office • System Security Officer • System Software Office • System Software Officer • System Staff Office • System Staff Officer

SSOB Senior Scientist Onboard

SSOC Satellite System Operational Center

SSOCR Superscale Optical Character Reader

SSOD Solid-State Optical Detector

SSODB Space Station Operational Data Book

SSODC Steady-State Off-Design Code

SSODCM Space Systems Operational Design Criteria Manual

SSOEG Satellite Systems Operations Evaluation Group

SSOEM Swiss Society for Optics and Electron Microscopy (Switzerland)

SSOF Standby Supervisor Of Flying

SSOG Satellite Systems Operations Guide

SSOI Summary Statement Of Intent

SSOM Société Suisse d'Optique et de Microscopie (Switzerland) • Solid-State Optical Maser

SSOME Société Suisse d'Optique et de Microscopie Électronique (Switzerland)

SSONET Structured Semantic Object Network

SSOP Safety Standing Operating Procedure • Space Systems Operating Procedure

SSOR Symmetric Successive Over-Relaxation

SSOS Severe Storms Observation Satellite • Springer Series on Optical Sciences (periodical code)

SSOSC Strong Second-Order Sufficient Condition

SSP Scanning Spectral Polarimeter • School Science Programme • Scientific Services Program • Scientific Subroutine Package • Security Support Plan • Security System Plan • Seismic Section Profiler • Selective Spectral Processor • Sensor Select Panel • Sensor Signal Processor • Serial Storage Protocol • Service des Soins Palliatifs • Service Social des Personnels • Service-Specific Part • Service Switching Point • Silver Plains (Australia-QLD) (location identifier) • Simple Solar Population • Simple Stellar Population (*) • Simulation Support Plan • Simulation Support Processor • Single-generation Stellar Population • Single-metallicity Stellar Population • Single-Shot Probability • Single Source Processor • Single Stock Point • Small Sortie Payload • Société Statistique de Paris (France) • Société Suisse de Physique (Switzerland) • Society of Scholarly Publishing (USA) • Software Support Package • Soil Survey of Pakistan • Solid-State Photodiode • Solid-State Preamplifier • Source Selection Plan • Source Selection Process • Space and Solar Physics • Space Science Plan • Space Science Program • Space Shuttle Program • Space Station Program • Spread Spectrum • Stable Sensor Platform • Standard Shop Practice • Standard Stability Prediction • Standard Switch Panel • Standby Status Panel • Star/Sky Photometry • Statutory Sick Pay • Stellar Synthesis Population • Stockpile Stewardship Program • Storage Services Provider • Strong Stability Preserving • Subsatellite Point • Subsolar Point • Summer Science Program • Summer Session Program • Summer Star Party • Super Sans Plomb • Superstable Platform • Surface Science Package • Sustained Superior Performance • Swiss Star Party (Switzerland) • Switch-to-Switch Protocol • System Safety Plan • System Security Plan • System Security Policy • System Support Package • System Support Plan • System Support Processor • System Support Program • System Status Panel

SSPA Solid-State Power Amplifier

SSPAIT Société Suisse de Photogrammétrie, d'Analyse d'Images et de Télédétection (Switzerland)

SSPC Self-Service Postal Center • Solid-State Power Controller • Spacelab Stored Program Command • System Safety Program Coordinator

SSPCL System Support Package Components List

SSPD Shuttle System Payload Data • Shuttle System Payload Definition • Shuttle System Payload Description

SSPDA Shuttle System Payload Data Activity

SSPDB Subsystem Power Distribution Box

SSPDS Shuttle System Payload Data Study • Space Shuttle Payload Data Study

SSPE Space Station Program Element • Support System Project Engineer

SSPF Sortie Software Production Facility • Space Shuttle Programming Facility • Space Station Processing Facility (USA-FL)

SSPFA Space Shuttle Program Flight Assignment

SSPI Security Service Provider Interface • Society of Satellite Professionals International (USA-VA) • Student Success Performance Indicator

SSPGSE Space Shuttle Program Ground Support Equipment

SSPII Space Shuttle Program Implementing Instruction

SSPL Steady State Power Level

SSPM Single-Sideband Frequency Modulation • Software Standards and Practices Manual • Software Standards and Procedures Manual • Solid-State Photomultiplier • Space Shuttle Program Manager • System Safety Program Manager

SSPO Space Shuttle Program Office • Space Shuttle Projects Office • Strategic Simulation Planning Office

SSPP Shuttle Small Payloads Project • Solar Sea Power Plant • Space Station Program Participant • System Safety Program Plan

SSPPSG Space Shuttle Payload Planning Steering Group

SSPR Structural and Syntactical Pattern Recognition (workshop)

SSPRO Space Shuttle Program Resident Office

SSPRS Small satellite Power System Regulator

SSPRv Space and Solar Power Review (periodical code)

SSPS Satellite Solar Power Station • Satellite Solar Power System • Sky Survey Prototype System • Solar Satellite Power Station • Solid-State Protection System • Spacecraft Support Planning Section • Space Shuttle Program Schedule

SSPSF Stochastic Self-Propagating Star Formation (model)

SSPTF Santa Susana Propulsion Test Facility

SSQ La Sarre (Canada-QC) (location identifier) • Simple Sinusoidal Quantity • Special Skill Qualification • Star Sequencer • Steep-Spectrum Quasar

SSQM Scanning Superconducting Quantum Microscope

SSR Sara (Vanuatu) (location identifier) • Satélite de Sensoriamento Remoto • Satellite Situation Report • Scanning Spectral Radiometer • Scientific Software Requirement • Secondary Surveillance Radar • Selenium Stack Rectifier • Service Statistique Rectoral • Shop Support Request • Site Suitability Report • Site Survey Report • Software Specification Review • Solid-State Reaction • Solid-State Recorder • Solid-State Relay • Soviet Socialist Republic (now obsolete) • Special Scientific Report • Specification Status Report • Spin-Stabilized Rocket • Staff Support Room • Station Set Requirement • Statistical Sobolev-Rosseland (mean) • Statistical Summary Report • Stock Status Report • Structured Semantic Relationship • Subsurface Radar • Subsynchronous Resonance • Summer Solstice Sunrise • Sum of the Squared Residuals • Supersynthesis Radiometer • Supply Support Request • Support System Requirement • Surface Search Radar • Synchronous Stable Relaying • System Safety Review • System Software Review • System Specification Review • System Status Review • System Study Requirement • System Summary Report • System Support Record

SSRA Subsurface Sounding Radar/Altimeter • System Safety Risk Assessment

SSRAAO Space Station Remote Access Automatic Observatory

SSRAAT Space Station Remote Access Automatic Telescope

SSRAM Synchronous Static Random Access Memory

SSRB Space Science Research Building • Subsystem Review Board

SSRC Social Sciences Research Council (UK) • Standard Software Requirements Contract • Strict Source and Record Route

SSRD Specific Science Requirements Document • Station Set Requirements Document • System Software Requirements Document

SSRE Society for Social Responsibility in Engineering

SSRF Screen Scraper Read File • Shanghai Synchrotron Radiation Facility • Space Systems Research Facility

SSRG Simple Shift Register Generator

SSRI Système de Stockage et de Recherche de l'Information

SSRL Stanford Synchrotron Radiation Laboratory (USA-CA) • Steep-Spectrum Radio-Loud (quasar)

SSRM Scanning Spreading Resistance Microscopy • Second-Stage Rocket Motor

SSRMS Space Station Remote Manipulating System • Space Station Remote Manipulator System (*)

SSRN System Software Reference Number • System Software Requirement Number

SSROP Space Station Resource Optimization Program

SSRP Severe Storms Research Program • Simple Server Redundancy Protocol • Spectral Surface Radiation Program • Summer Student Research Program • System Safety Review Panel

SSRPI Selectable Spectral Resolution Pointing Imager

SSRQ Steep-Spectrum Radio Quasar

SSRR Station Set Requirement Review

SSRS Society for Social Responsibility in Science • Southern Sky Redshift Survey • Standard Service Request Specification • Start-Stop-Restart System

SSRT Siberian Solar Radio Telescope • Single-Stage Rocket Technology • Subsystem Readiness Test

SSRTF Solid Surface Room Temperature Fluorescence

SSRTP Solid Surface Room Temperature Phosphorescence

SSRU Standard Speech Reproducing Unit

SSRv Space Science Reviews (periodical code)

SSRV Solar-Sail Race Vehicle

SSRvE Soviet Scientific Reviews Section E (periodical code)

SSS Safety Status System • Satellite Servicer System • Scanning Star Simulator • Scatterometer System Simulator • Scientific Subroutine System • Sea Surface Salinity • Second Solar Spectrum • Security Surveillance System • Selective Service System • Self-Sustaining System • Sensor Signal Simulator • Sensor Support System • Siassi (Papua New Guinea) (location identifier) • Signal Switching System • Signature Security Service • Silicon-Symmetrical Switch • Simulation Study Series • Single-Screw Ship • Single Support Strategy • SIPE Support Structure • Sira Safety Services • Small-Scale Structure • Small Scientific Satellite • Small Solar Satellite • Small Starlight Scope • Social Studies of Science (periodical) • Software Segment Specification • Software Support Specialist • Solar and Skylight Spectrophotometer • Solar-Stellar Spectrograph • Solar-Stellar Spectrophotometry • Solid-State Spectrometer • Solid-State Switching • Solid-State System • Sortie Support System • Sound Suppression System • Source Selection Sensitive • Spacecraft Software System • Spacecraft System Support • Space Shuttle Segment • Space Shuttle Simulation • Space Shuttle System • Space Station Simulator • Space Surveillance System • Space System Subfile • Special Safety Safeguard • Spinning Space Station • Spinning Sun Sensor • Spin-Stabilized Spacecraft • Staff Summary Sheet • Stage Separation System • Stainless-Steel Sphere • Standard Software System • Standard Supply System • Star Selection Servo • Static Spherically Symmetric • Station Satellite System • Station Set Specification • Steep-Spectrum Source • Storage Serviceability Standard • Strategic Satellite System • Subject Specialization Scheme • Subsystem Segment • Subsystem Specification • Summer Solstice Sunset • Sun Screen Server • Supercosmos Sky Survey • Supersoft Source • Support Software System • Surface Surveillance System • Su Seguro Servidor • Switching Subsystem • Symbolic Shorthand System • Symmetric Spiral Structure • Symposium of System Synthesis • Synchronized Sample Scrambler • System and Segment Specification • System Security Standard • System Segment Specification • System Supervisor Station • System Support Segment

SSSA Software Systems Support Activity • Soil Science Society of America (USA) • Systems Software Support Activity

SSSAAS Space Station Science and Applications Advisory Subcommittee

SSSAJ Soil Science Society of America Journal (periodical code)

SSSB System Source Selection Board

SSSC SAX Scientific Steering Committee • Self-Service Supply Center • Solid State Sciences Committee • Space Science Steering Committee • Space Science Support Center • Sydney Southern Star Catalogue

SSSCB Storage Subsystem Control Block

SSSD Second-Stage Separator Device • Single-Side Single-Density • Solid-State Solenoid Driver • Space Shuttle Simulation Display • State Soil Survey Database

SSSF Surface and Sounding Systems Facility

SSSG Small-Scale System of Galaxies • Space System Support Group • System Simulator Steering Group

SSSI Science of Soft Surfaces and Interfaces • Service de la Sécurité des Systèmes d'Information • Site of Special Scientific Interest • Space Sciences Series of ISSI

SSSK Sociology of Social Scientific Knowledge

SSSMP Scientia Sinica Series Mathematical Physical Technical Sciences (periodical code)

SSSO Specialized Surplus Sales Office • Standard Systems Support Office

SSSP Source Selection Support Plan • Southern Skies Star Party (Bolivia) • Space Shuttle Synthesis Program

SSSR Summer Solstice Sunrise

SSSRD Surrogate System Software Requirements Document

SSSS Society for Social Studies of Science (USA-MA) • Space Shuttle System Specification • Spherically Symmetric Self-Similar

SSSSR Space Shuttle System Safety Review

SSSSS Simulation of the Satellite Signal in the Solar Spectrum • Springer Series on Solid State Sciences (periodical code)

SSST Single Subscriber Terminal • Solar Stokes Spectrum Telescope • Space Station System Trainer

SSSTB South South South Temperate Belt

SSSWG Storage System Standards Working Group

SSSyn Springer Series on Synergetics (periodical code)

SST Safe Secure Trailer • Sample Sound Technology • Santa Teresita (Argentina-BA) (location identifier) • Santé et Sécurité du Travail • Satellite-to-Satellite Tracking • Sauveteur Secouriste du Travail • Scalable Systems Technology • Science Study Team • Seamless Steel Tubing • Search Strategy and Terms • Sea Surface Temperature • Sea Surface Topography • Shuttle System Test • Simple Software Tool • Simulated Structural Test • Simulation and Systems Test • Single-Sideband Transmission • Single Subscriber Terminal • Single System Trainer • Site Survey Team • Social Shaping of Technology • Society of Surveying Technicians • Software Support Team • Software Support Tool • Solar Space Telescope • Solar Submillimeter Telescope • Solid-State Telescope • Solid-State Transmitter • Source Selection Team • Spacecraft Systems Test • Space Solar Telescope • Space Surveillance and Tracking • Spectroscopic Survey Telescope • Spread-Spectrum Technology • Stably Stratified Turbulence • Structural Static Test • Student Science Training • Study Science Team • Submillimeter Solar Telescope • Subsystem Terminal • Subsystem Test • Super Self-aligned Technology • Supersnelle Trein • Supersone Snelheids Tunnel • Supersonic Transport • Superstring Theory • Supplemental Site Test • System Software Test • Systems Services and Technology • System Stress Test • System Subsystem Test • System Support Tape • System Support Team

SSTA Support System Task Analysis

SSTAC Sea Surface Temperature Analysis and Composite

SSTB South South Temperate Belt (Jupiter)

SSTC Single-Sideband Transmitter Carrier • Solid-State Timer-Controller • Spacecraft System Test Console • Space Shuttle Test Conductor • Space System Test Capability • Specialized System Test Contractor • State Science and Technology Commission

SSTCLS Space Shuttle Test, Checkout, and Launch Support

SSTD Space Science and Technology Department (RAL)

SSTDMA Satellite-Switched Time-Division Multiple Access

SSTec Solid State Technology (periodical code)

SSTEP System Support Test Evaluation Program

SSTF Simulation and System Test Facility • Space Shuttle Task Force • Space Shuttle Training Facility • Space Simulation Test Facility • Space Station Task Force • Space Station Training Facility

SSTG Segment Software Test Group • Space Station Task Group • Space Systems and Technology Group

SSTI Small Satellite Technology Initiative • Small Spacecraft Technology Initiative

SSTIMR Sea Surface Temperature Imaging Microwave Radiometer

SSTL Solar System Target List • Solid-State Track Link • Surrey Satellite Technology Ltd. (UK)

SSTM Single Service Training Manager

SSTNET Sociology of Science and Technology Network

SSTO Single-Stage-to-Orbit

SSTOL Supershort Takeoff and Landing

SSTor Studia Societatis Scientiarum Torunensis Sectio F Astronomia (periodical code)

SSTP Screened Shielded Twisted Pair • Small Satellite Technology Program • Subsystem Test Procedure

SSTPC Service Support Transport Protocol Client

SSTR Safing System Tape Recorder

SSTS Satellite Surveillance and Tracking Satellite • Space-based Surveillance and Tracking Satellite • Space-based Surveillance and Tracking System • Space Surveillance and Tracking Satellite • Space Surveillance and Tracking System • Standard Systems Technical Support • Sustainment System Technical Support • Switching System Test Set

SSTT Small Specimen Test Technique

SSTV Slow-Scan Television

SSTWG SETI Science and Technology Working Group

SSTX Solid-State Transmitter

SSTZ South South Temperate Zone

SSU Sample Storage Unit • Secondary Sampling Unit • Self-Service Unit • Semiconductor Storage Unit • Sensor Simulator Unit • Sensor Surveying Unit • Solar-Sail Union • Sonoma State University (USA-CA) • Southwest State University (USA-MN) • Spacecraft Support Unit • Space Science Update • Star Sensor Unit • Station Supervisory Unit • Strategic Services Unit • Stratospheric Sounding Unit • System Services and Utilities • System Support Unit • White Sulphur Springs - Greenbrier (USA-WV) (location identifier)

SSUAS Space Station Utilization Advisory Subcommittee

SSUCS System Services and Utilities Control System

SSUP Space Station User Panel • Space Station Utilization Preparedness

SSUPS Solid-State Uninterruptible Power Source • Solid-State Uninterruptible Power System

SSUR Stringy Spacetime Uncertainty Relation

SSUS Spin-Stabilized Upper Stage • Spinning Solid Upper Stage (*)

SSUSI Special Sensor for Ultraviolet Spectrographic Imaging

SSUSP Spinning Solid Upper Stage Project

SSUT Sous-Système Utilisateur Téléphonique

SSV Satellite Servicing Vehicle • Snittstellenvervielfacher • Set of Stations Velocities • Siasi (Philippines) (location identifier) • Solar System Visualization • Space Shuttle Vehicle • Space Sortie Vehicle • Standard Service Volume • Supersatellite Vehicle

SSVDT Sight, Stabilized, Visual Data Transmitting

SSVN Structure Searching Via Nomenclature

SSVR Silent Suppression Video Rate

SSVS Subsystem Verification Specification

SSVT Swedish Solar Vacuum Telescope

SSW S-band Switch • Schwäbische Sternwarte eV (Germany) • Solar Software (library) • Solar Surface Weather • South-Southwest • South-Southwestern • Stuart Island (USA-WA) (location identifier) • Sudden Stratospheric Warming • Synchro Switch • System Software

SSWC Set of Significant Wavelet Coefficients

SSWG SALT Science Working Group • Servicing Science Working Group • Site Selection Working Group • Site Survey Working Group • Shuttle Security Selection Working Group • Solar-System Working Group • Space Science Working Group • Space Surveillance Working Group •

System Safety Working Group • System Security Working Group

SSWP Space Station Work Package • Springer Series on Wave Phenomena (periodical code)

SSWRD Systems Software Requirements Document

SSWRN South-Southwestern

SSWS Sound Suppression Water System

SSWTI Staatliches System der Wissenschaftlich-Technischen Information

SSWWD South-Southwestward

SSX Samsun (Turkey) (location identifier) • Space Ship Experimental • Subsystem Executive

SSXB Strange Star X-ray Binary • Supersoft X-ray Binary

SSXTF Solar Soft X-ray Telescope Facility

SSY M'Banza Congo (Angola) (location identifier)

S/Sys Subsystem

SSZ Santos (Brazil-SP) (location identifier)

st Sesotho (ISO language code) • Stere • Stet

St Saint • Sanct • Sankt • Sant • Sint • Sitio • Stanton Number • Stanza • State • Statement • Station • Statute • Stem • Stitch • Stokes • Stone • Strait • Straße • Stratus • Street • Stück • Stunde

S/T Short Ton • Source/Type

S&T Science & Technology • Scientific & Technical • Scientific & Technological • Simulation & Training • Single Theodolite • Sky & Telescope (periodical code) • Supply & Transport • Supply & Transportation • Surveillance & Targeting

ST Principe (Sao Tome &_) (ISO country code) • Saint • Sample Tray • Sao Tome & Principe (ISO country code) • Sawtooth • Scalar Totalizer • Scenario Template • Scheduled Tribe • Schmidt Telescope • Schmitt Trigger • Science and Technology • Science Team • Scientific/Technical • Screening Technician • Search and Track • Segment Type • Select Time • Self-Test • Send Timing • Sequence Timer • Sequential Timer • Serial Traceability • Service des Transmission • Service Technique • Service Test • Shipping Ticket • Shock Tunnel • Short Ton • Short Tour • Sidereal Time • Silicon Target • Silicon Tracker • Simplification Task • Simulator Training • Single Throw • Singularity Theorem • Size-Temperature (relation) • Slow Technology • Small Tug • Software Test • Solar-Terrestrial • Sorting Tender • Source/Type • Spacelab Technology • Space Technology • Space Telescope • Space-Time • Spacelab Technology • Special Test • Special Text • Special Tool • Special Tooling • Spécification Technique • Split Training • Spring Tide • Staff Target • Standard Thickness • Standard Time • Stand Talker • Standby Time • Start • Start of Frame • Star Tracker • State • Static Test • Static Thrust • Station • Station Timing • Steam Turbine • Störmer Predictor • Straight Tip • Stratus • Street • Submarine Tender • Substitution Theorem • Subtask • Sudan (international aircraft registration code) • Summer Time • Superconducting Tunnel • Superior Technology • Surveillance Test • Survival Time • Sustainment Training • Switchover Time • Synthesis Telescope • System Tape • System Test

Sta Santa • Station • Stationary • Status

StA Staatsarchiv

STA Saint Andrews (Starlink internal code) • Santa • Satellite Tracking Annex • S-band Test Antenna • Scheduled Time of Arrival • Science and Technology Agency • Science and Technology Agent • Science and Technology Area • Security Threat Analysis • Service Terminal Arrangement • Shuttle Training Aircraft • SIGIR Technology Affiliate • Significant Technical Accomplishment • Significant Technological Achievement • Southern Taurids (meteor stream) • Space Technology Application • Space Transportation Association (USA-VA) • Spanning-Tree Algorithm • Sparing To Availability • Special Temporary Authorization • Staff Training Assistant • Star Tracker Assembly • State Transit Authority • Static Test Article • Status Test Article • Station • Stauning (Denmark) (location identifier) • Strategic Transportation Analysis • Strategy and Technical Assessment • Structural Test Article • Structural Test Assembly • Student Travel Agency • Surveillance and Target Acquisition • System Threat Assessment

STAARTE Scientific Training and Access to Aircraft for Atmospheric Research Throughout Europe

STAAC Space and Terrestrial Applications Advisory Committee • Systems Technology and Advanced Concepts

STAAR Submillimeter Telescope for Astronomy and Atmospheric Research

staatl staatlich

Stab Stabilization • Stabilizer • Stable

StAB Staatliche Allgemeinbibliothek

STAB Space Technology Advisory Board • Stabilize • Stabilized • Stabilizer

STAC Satellite Tracking Angle Calculation • Science and Technical Advisory Committee • Science and Technology Advisory Committee • Science and Technology Agreement Committee • Space Telescope Advisory Committee (*) • Space Telescope Advisory Council • Space Track Auxiliary Computer

STACC Small Telescope Array with CCD Cameras • Standard Telemetry and Command Component • Standard Telemetry and Communications Component

STACE STScI Advisory Council on Education

STACEE Solar Tower Atmospheric Čerenkov Effect Experiment

STACS Serials Tracking And Control System • Subtropical Atlantic Climate Studies • Symposium on Theoretical Aspects of Computer Science

STAD Société de Traitement Automatique des Données (France)

STADAC Station Data Acquisition and Control

STADAD RAL (Starlink internal code) • Satellite Tracking And Data Acquisition Department

STADAN Satellite Tracking And Data Acquisition Network • Space Tracking And Data Acquisition Network

STADAT Starlink Database

STADU System Termination And Display Unit

STAE Specific Task Asynchronous Exit

STAEDT Simulation Tools for Analysis and Evaluation of Data Transmission

STAF Science Team Analysis Facility • Scientific and Technological Applications Forecast • Service de Transfert et d'Archivage de Fichiers • Systems Technical Applications Facility

STAFF Spatio-Temporal Analysis of Field Fluctuations • Stellar-Acquisition Flight Feasibility

Staffs Staffordshire (UK-England)

STAFT Steerable Antenna Focusing Technique

STAG Secure Text And Graphics • Shuttle Turnaround Analysis Group • Space Telescope Advisory Group • Systems Technical Analysis Group

STAGN Stagnation

STAIR Structural Analysis Interpretive Routine

STAIRS Storage and Information Retrieval System (IBM)

STAIS Student Advisory Information Service

STALAS Stationary Laser Site • Stationary Laser Station

STALO Stabilized Local Oscillator • Stable Local Oscillator

STALOC Self-Tracking Automatic Lock-On Circuit

StAM Studies in Applied Mathematics (periodical code)

STAM Standard Tool for Aeronautical Maintenance

STAMO Stabilizer Master Oscillator

STAMOS Sortie Turnaround Maintenance Operations Simulation

STAMP Satellite Telecommunications Analysis and Modeling Program • Space Technology Analysis and Mission Planning • System Testability And Maintainability Program

STAN Space Telescope Analysis Newsletter

STANAG Standardization Agreement

Stand Standard

STANet Science and Technology Awareness Network

StAnpG Steueranpassungsgesetz

STANTEK Standardiseringstekniska Föreningen (Sweden)

StAnz Staatsanzeiger

STA&P System Threat Assessment & Projections

STAP Schmidt Telescope Advisory Panel • Scientific and Technical Advisory Panel • Scientific and Technical Analysis and Program • SmallTalk Apprentice Program • Space-Time Adaptive Processing • Special Technical Assistance Program • System Threat Assessment and Projections

STAPL Standard Test and Programming Language

STAPP Simulation Tape Print Program

STAR Satellite Telecommunications Automatic Routing • Satellites in Telematics Applications and Research (ESA) • Satellite Tracking using RF • Science Teaching Achievement Recognition • Science Teaching through its Astronomical Roots • Science Technology And Research • Science To Achieve Results (program) • Scientific and Technical Aerospace Reports (NASA) • Scientific Toolkit for Astrophysical Research • Sea And Terrain Assessment Radar • Self-defining Text Archival • Self-Test Automated Readout • Self-Test Automatic Readout • Self-Testing And Repairing • Shuttle Turnaround Analysis Report • Significant Technical Achievement in Research • Significant Technical Advancement in Research • Significant Technological Achievement in Research • Significant Technology Achievement through Research • Simulation in Training for Advanced Readiness • Simultaneous Track And Record • Simultaneous Transmission And Reception • SOFIA Terahertz Array Receiver • Software Tools And Requirements • Solar, Terrestrial and Astrophysical Research • Solenoidal Tracker At RHIC • Source Transmission And Response • South Pacific Tectonics and Resources • Space Technology and Advanced Research • Space Telescope Anomaly Room • Space Telescope Audit Report • Space Telescope Axial Replacement • Space Three-Axis Accelerometer for Research • Space Transportation Automated Reconfiguration • Spectral Technology and Applied Research • Spectrograph/Telescope for Astronomical Research • Special Telecommunication Action of Regional (development) • Standard Arrival Route • Standard Telecommunications Automatic Recognizer • Standard Terminal Arrival Route • Statistical Table Assembly and Retrieval • Stellar Attitude Reference • Stockholms Amatörastronomer (Sweden) • Streamer Tubes Analog Readout • String Array • Subsonic Transport Acoustic Research • Surface Target Attenuation Radar • Système Terrestre Automatique de Radiosondage • System Threat Analysis Report • System Threat Assessment Report • Systems Threat Assessment Report

STARAN Stellar Attitude Reference And Navigation

STARBASE Students Training for Achievement in Research Based on Analytical Space-science Experience

STARCAL Star Calibration

STARCAT Space Telescope Archive and Catalog • Space Telescope Archive Catalog

STARCH Stressed Arch

StARe University Observatory St. Andrews Scotland Reprints (periodical code)

STARE Scandinavian Twin Auroral Radar Experiment • Southern Tropical Atlantic Region Experiment • Steerable Telemetry Antenna Receiving Equipment

STARFIRE System To Accumulate and Retrieve Financial Information with Random Extraction

STARLAB Space Technology Applications and Research Laboratory

Starlink Satellite Telemetry And Return Link

STARR Schedule, Technical And Resources Report

STaRS Science Teacher and Research Scientist

STARS Satellite Telemetry Automatic Reduction System • Satellite Tracking And Ranging Station • Satellite TV Airborne Receiving System • Science Teachers Active in Real Science (program) • Seismic Telescope for Astrophysical Research from Space • Software Technology for Adaptable and Reliable Systems • Space Transportation Automated Reconfiguration System • Stabilized Twin-gyro Attitude Reference System • Standard Accounting and Reporting System • Standard Terminal Automation Replacement System • Status Analysis And Reporting System • Stellar Absorption and Refraction Sensor • Stellar Tracking Attitude Reference System • Stored Terrain and Access Retrieval System • Strategic Targeting Activities Reporting System • Structural Target Analysis and Recognition System • Students Taking Astronomy-Related Subjects (USA-CA) • Subaru Telescope Archive System • Support Tracking Analysis Reporting System • Surveillance and Target Attack Radar System • System Test and Astronaut Requirement Simulation

START Selection To Activate Random Testing • Space Test And Reentry Technology • Space Transport And Reentry Test • Strategic Arms Reduction Talks • Strategic Arms Reduction Treaty • System for Analysis, Research and Training • Systemic Tabular Analysis of Requirements Technique

STARTED Strategies, Tools, and Resources for Team-based Early Design

STAS Short-Term Analysis Service • Short-Term Analysis Study • Space Transportation Architecture Study • Stabilized Thermal Imaging System

stat Statim

Stat Statement • Static • Station • Stationary • Statistical • Statistics • Statuary • Statue • Status • Statute

STAT Space Telescope Advisory Team • Space Telescope Astrometry Team • Standards and Technology Assessment Team • Station • Statistical • Statistics • Status • System Testability and Analysis Tool

STATE Structure and Atmospheric Turbulence Environment

STATIS Student Training Analysis and Tracking Information System

STATN Station

STATPAC Statistical Package

STATREC Status Record

STATREQ Status Request

STATS Star Tracker Alignment Test Set

STATSGO State Soil Geographic Database

STAU Standard Acquisition Unit

StAV Steirischer Astronomen Verein (Austria)

STAVF Space Telescope Assembly and Verification Facility

STAWA Science Teachers of Western Australia

Stb Steinbruch

StB Stadtbibliothek

STB Santa Barbara - Zulia (Venezuela) (location identifier) • Save The Bits • Set-Top Box • Shortwave Terrestrial Background • SIM Testbed • Solid Target Boronization • South Temperate Belt (Jupiter) • South Tidal Bridge • South Tropical Belt • Stage Two Booster • Strobe • Surveillance Testbed • System Testbed

StBA Statistiches Bundesamt (Germany)

STBA Selective Top-to-Bottom Algorithm

Stbd Starboard

STBE Space Transportation Booster Engine

StBerG Steuerberatungsgesetz

StBGebV Steuerberatergebührenverordnung

STBL Stable • Stable Boundary Layer • Stratus-topped Boundary Layer

Stbln Stabilization

STBP Sir Thomas Brisbane Planetarium (Australia-QLD)

Stbt Steamboat

Stby Standby

STBY Standby

S&TC Space & Terrestrial Communications

STC Saint Cloud (USA-MN) (location identifier) • Satellite Test Center • Satellite Tracking Center • Science and Technical Committee • Science and Technology Center • Scientific and Technical Committee • Scientific and Technical Computing • Scientific Technical Committee • Secondary Traffic Channel • Security Transition Component • Semi-variogram Textual Classifier • Sensitivity Time Control • Services Techniques Centraux • Set Carry • Short-Time Constant • Short-Title Catalog • Shuttle Test Conductor • Simulation Tape Conversion • Society for Technical Communication (USA-VA) • Software Technology Center • Software Technology Conference • Sound Transmission Class • Spacecraft Technology Center • Spacecraft Test Conductor • Space Technology Center • Space Test Center • Space-Time Continuum • Space Tracking Center • Spécification Technique du Besoin • Specific Transport Cost • Spectral Transfer Coefficient • Standard Telephone Code • Standard Test Condition • Standard Test Configuration • Standard Transmission Code • Started Task Console • Started Task Control • Station Test and Calibration • Stennis Test Center (USA-MS) • Storage Technology Corporation • Stored Time Command • Subtechnical Committee • Suffix Tree Clustering • Supplemental Type Certificate • Supplemental Type Certification • Switching Technical Center • Système de Trajectographie Captive • System Test Complex • System Test Configuration • System Test Console • System Time Clock • System Timing Controller • System Transfer Constant

STCA Short-Term Conflict Alert

STCB Survey Technician Certification Board

STCC Spacecraft Technical Control Center

ST&CD Space, Terrestrial & Communications Directorate

STCD Space, Terrestrial and Communications Directorate

STCDHS Spacecraft Telemetry Command Data Handling System

STCDS System Test Complex Data System

STCE System Test Complex Equipment

StCeF Studii si Cercetari de Fizica (periodical code)

StCeM Studii si Cercetari Matematice (periodical code)

STCG Synchronized Time Code Generator

STCK Store Clock

STCM Service Technique de Cryogénie et de Magnétisme

StCMA Studii si Cercetary de Mecanica Aplicata (periodical code)

STCN Short Title Catalogue Netherlands

STCP Short-Term Cost Plan

ST&CS Supervision, Translation & Control System

STCS Space Transportation and Cargo System • Space Transportation Cargo System • Special Telemetry Conversion System • Supervision, Translation and Control System

STCT Small Transportable Communications Terminal

STCU Science and Technology Center in Ukraine

STCW Standards of Training, Certification and Watchkeeping • System Time Code Word

Std Standard • Standing • Started • Stunde

STD Dobra (Sao Tome & Principe) (ISO currency code) • Salinity - Temperature - Depth • Santo Domingo (Venezuela) (location identifier) • SATCOM Technologies Division • Save The Data • Science and Technology Development • Scientific and Technical Data • Sensor and Test Driver • Servo Tape Display • Set Direction • Shuttle Test Detector • Software Test Description • Software Test Document • Software Transfer Document • Solar-Terrestrial Dispatch • Spacecraft Technology Division • Special Test Device • Standard • Standard Document • Standard Test Device • State Transition Diagram • Subscriber Toll Dialing • Subscriber Trunk Dialing • Supporting Technology Development • Synchronous Time Division • System Training Device

STDA StreetTalk Directory Assistance

STDADS Space Telescope Data Archive and Distribution System

STDAUX Standard Auxiliary

STDB Space Telescope Database • Standard Database

STDBM Spatio-Temporal Database Management (workshop)

STDC Scientific and Technical Data Collection • Single Temperature and Density Calculation • Space Technique Development Center

STDCE Surface Tension Drive Convection Experiment • Surface Tension Driven Convection Experiment (*)

Std. Dev. Standard Deviation

Stde Stunde

STDERR Standard Error

Stdg Standing

STDIN Standard Input

STDIOH Standard Input/Output Header

stdl stündlich

STDL Standard Distribution List

STDM Stadium • Statistical Time-Division Multiplexer • Statistical Time-Division Multiplexing (*) • Synchronous Time-Division Multiplexing

STDMA Self-organizing Time Division Multiple Access

STDMS Space Telescope Data Management System

STDN Spacecraft Tracking and Data Network (*) • Spaceflight Tracking and Data Network • Space Tracking and Data Network

STDO Space Technology Development Office (USRA)

STDOUT Standard Output

STDPRN Standard Printer

STDS Short-Term Data Storage

STDSD Solar-Terrestrial Data Services Division

STDT Science Technology Definition Team

Stdy Steady

STDY Steady

Stdzn Standardization

Ste Sainte • Suite

Sté Société

ST&E Security Test & Evaluation

STE Satellite Tracking Element • Sciences et Techniques Économiques • Section Terminating Equipment • Secure Telephone Equipment • Secure Terminal Equipment • Secure Transmission Equipment • Security Test and Evaluation • Segment Table Entry • Shrinkage Test Estimator • Signaling Terminal Equipment • Simplified Test Equipment • Simulation and Test Equipment • Simulated Telephone • Software Test Environment • Solar-Terrestrial Environment • Spacecraft Test Engineering • Space-Time

Editor • Space-Time Exchange • Spanning Tree Explorer • Special Test Equipment • Staff-years of Technical Effort • Standard Test Environment • Star Tracker Electronics • Station Test Equipment • Statistiques du Travail et de l'Emploi • Stevens Point (USA-WI) (location identifier) • Sudden Temporary Erection • Suitability Test Evaluation • Suite • Support Test Equipment • Surveillance and Tracking Experiment • System Test Engineer • System Test Equipment

STEA System Test, Evaluation and Assembly

STEAM Standard Towing Equipment for Aviation Maintenance

STEAP Simulated Trajectories Error Analysis Program

ST-ECF Space Telescope - European Coordinating Facility

STECF Space Telescope - European Coordinating Facility

STED Science, Technology and Economic Development

STEDI Student Explorer Demonstration Initiative

STEEP STX Terrain Elevation Extraction Package

STEF Short-Term Events File

STEI Scientific, Technical, and Economic Information

STEIN Short-Term Experimental Information Network

STEIS Space Telescope Electronic Information System • STScI Electronic Information Service

STel Stanford Telecommunications

STEL Short-Term Exposure Level • Short-Term Exposure Limit • Solar-Terrestrial Environment Laboratory

STELAB Solar-Terrestrial Environment Laboratory

STELAR Study of Electronic Literature for Astronomical Research

STELLA Satellite Transmission Experiment Linking Laboratories • System To Estimate Latitude and Longitude Astronomically

Stellv Stellvertreten • Stellvertreter • Stellvertreterin

STEM Scanning Transmission Electron Microscope • Scanning Transmission Electron Microscopy • Science, Technology, Engineering and Mathematics • Solar-Terrestrial Environment Model • Space - Time - Energy - Matter • Stay Time Extension Module • Structural-Thermal-Engineering Model • Systems Telecommunications Engineering Manager • System Test Equipment Mission • System Trainer and Exercise Module

STEMPRA Science, Technology, Engineering, and Medicine Public Relations • Association (UK)

STEMS Scenario Test and Evaluation Message System

STEMTP Science, Technology, Engineering and Mathematics Teacher Preparation

Sten Stencil • Stenographer • Stenographic • Stenography

STEN Service Technique de l'Éducation Nationale (France)

Steno Stenographer • Stenographic • Stenography

Stenog Stenographer • Stenographic • Stenography

STENTOR Satellite de Télécommunications pour Expérimenter de Nouvelles Technologies en Orbite

STeP Stanford Temporal Prover

STEP Safety Test and Engineering Program • Satellite Technology Education Program • Satellite Test of the Equivalence Principle • Science and Technology for Environmental Protection • Science, Technology and Economic Policy • Scientific and Technical Exploitation Program • Scientists and Teachers in Educational Partnership • Secure Tunnel Establishment Protocol • Self-Teaching Exportable Package • Sensitivity Temperature Error Program • Simple Transition to Economical Processing • Simple Transition to Electronic Processing • Simulation, Test and Evaluation Process • Society of Trust and Estate Practitioners • Software Technology and Engineering Practice • Software Test and Evaluation Panel • Software Test and Evaluation Plan • Software Test and Evaluation Program • Software Test Estimating Procedure • Software Test Evaluation Program • Solar-Terrestrial Energy Program • Space Technology Experiments Program • Space Telescope Exhibited Preprint • Space Terminal Evaluation Program • Space Test Experimental Program • Space Test Experiment Planner • Space Test Experiment Platform • Space Test Experiment Program • Space Tracking Experiment Program • Space Transportation Engine Program • Special Temporary Employment Programme • Specification To Executable Program • Standard for Engineering Product • Standard for the Exchange of Product • Standardized Tactical Entry Point • Standard Tactical Entry Point • Standard Tape Executive Package • Standard Tape Executive Program • Standard Terminal Program • Standard Test Equipment Procedure • Standard To Exchange Product • Station de Transfert d'Énergie par Pompage • Statistical Trajectory Estimation Program • Stratosphere-Troposphere Exchange Project • Structured Test and Evaluation Process • Supervisory Tape Executive Program • Surveillance and Tracking Experiment Program

STEPAN Science and Technology Policy Asian Network

STEPD Standard for Exchange of Product Data

STEPR Synthetic Training Environment Periodic Review

STEPS Science, Technology, the Economy, Politics, Society • Simulation, Training, Education and Procedural System • Single Telescope Extrasolar Planet Survey • Solar Thermionic Electrical Power System • Statistical Education through Problem Solving • Stellar Planetary Survey • Stored Thermal Energy Propulsion System

STEPSS Survey for Transiting Extrasolar Planets in Solar Systems

Ster Stereophonic • Stereotype • Sterling

STER Space Techniques for Environmental Risks • Stichting Ether Reclame (Netherlands) • Story-averaged Term Error Rate

STERAO Stratospheric-Tropospheric Experiments: Radiation, Aerosols and Ozone

Sterb Der Sternenbote (periodical code)

SteRe Steward Observatory Reprints (periodical code)

Stereo Stereophonic • Stereotype

STEREO Solar-Terrestrial Relations Observatory

STERJ Solar Terrestrial and Environmental Research in Japan (periodical code)

Stern Die Sterne (periodical code)

Sternw Sternwarte

Sterz Sternzeit (periodical code)

STES Single-locker Thermal Enclosure System

STESA Solar-Terrestrial Environment Sensing Assembly

STESOP Space Telescope Elementary Outreach Program

STE&T Special Test Equipment & Tooling

STET Special Test Equipment and Tooling

STETF Solar Total Energy Test Facility

ST&ETR SEcurity Test & Evaluation Test Report

STETR SEcurity Test and Evaluation Test Report

STETS Solar-Terrestrial Energy Transfer Studies

STEVE Space Tool for Extravehicular Emergencies • Stratosphere to Thermosphere Energy Variability Experiment

St. Ex. Stock Exchange

STEX Sensor Technology Experiment • Simulation Test and Exercise • Space Technology Experiment • Static Test and Evaluation Complex

STeZ South Temperate Zone

S&TF Systems & Technology Forum

STF Safety Test Facility • Satellite Test Facility • Satellite Tracking Facility • Save The Frames • S-band Transmit Filter • Sensor Test Facility • Shaping The Future

• Short Title File • Software Test Facility • Space Telescope Fix • Space Telescope Format • Space Track Facility • Specialist Task Force • Special Task Force • Speckle Transfer Function • Spin Test Facility • Source Test File • Spin Test Facility • Standard Task File • Static Test Facility • Stephen Island (Australia-QLD) (location identifier) • Stratiform • Structural Test Facility • Structured File • Structure Function • Summary Tape File • System Test Facility • System Test Framework • System Transfer Function

STFP Standard Time and Frequency Point

STFRM Stratiform

STFS Space Telescope Flight Software • Standard Time and Frequency Signal

STFSDB Space Telescope Flight Software Database

STFSDS Standard Time and Frequency Signal Dissemination Service

STFT Short-Time Fourier Transform

STFV Sensor Total Field of View

STFVS Space Telescope Functional Verification Simulation

Stg Stage • Sterling

STG Saint George Island (USA-AK) (location identifier) • Satellite Terminal Guidance • Scene Transition Graph • Schiffbautechnische Gesellschaft (Germany) • Scientific and Technical Group • Shuttle Test Group • Space Task Group • Space Telescope Guidance • Space Test Group • Special Technology Group • Stage • Steering Task Group • Sterling • Strong

StGB Strafgesetzbuch

Stge Storage

StGG Studia Geophysica et Geodaetica (periodical code)

STGG Superconducting Tensor Gravity Gradiometer

Stgo Santiago

STGP Strongly Typed Genetic Programming

STGPA Special Task Group on Protocol Architecture

Stgt Stuttgart (Germany)

STGT Secondary Target • Second TDRS Ground Terminal (*) • Second TDRSS Ground Terminal

Sth Something • South

STH South • Strathmore (Australia-QLD) (location identifier)

STHD Stored Heading

StHG Staatshaftungsgesetz

Sthn Southern

STHSPAAL Space Telescope High-Speed Photometer Acronym and Abbreviation List

ST&HV Science, Technology & Human Values (periodical)

S&TI Scientific & Technical Intelligence

STI Santiago (Dominican Rep.) (location identifier) • Save The Images • Science and Technology Information • Science and Technology Innovation (program) • Sciences et Technologies Industrielles • Science - Technology - Innovation • Scientific and Technical Information • Scientific and Technological Information • Scientific Technologies Inc. (USA-CA) • Screw Thread Insert • Service Technique Informatique • Set Interrupt • Signal Technology Inc. • Software Testing International • Special Test Instruction • Spine Tip Imager • Standard Tape Interconnect • Standard Terminal Interface • Stationary Target Indicator • Supplemental Technical Instruction • Supplemental Test Instruction • Système de Traitement d'Images • System of Topographic Information • Systolic Time Interval

STIB Scientific and Technical Information Branch • Stratosphere-Troposphere Interactions and the Biosphere

STIC Science and Technology Information Center • Sciences et Techniques de l'Information et de la Communication • Science, Technology and Industry Collection • Scientific and Technical Intelligence Committee • Scintillation Tile Calorimeter • Space Technical Information Control • Space Telescope Institute Council • Space Test Integration Contract

STICA Space Technology Industrial and Commercial Applications (periodical code)

STICS Scalable Transportable Intelligence Communications System • Suprathermal Ion Composition Spectrometer

STID Science and Technology Information and Documentation (EU) • Scientific and Technical Information Dissemination • Scientific and Technical Information Division • Standard Identification • Statistiques et Traitement Informatique des Données • Systems Training Integration Directorate

STIF Science and Technology Information Facility • Scientific and Technical Information Facility • Search Talk Intermediate Frequency • Semantically Transparent Interfacing • Software Test and Integration Facility • Spectral Transmission Interference Filter

STIG Security Technical Implementation Guidelines • Space Technology Interdependency Group

STII Science and Technology Information Institute • Software Test and Integration Instruction

STIISP Scientific and Technical Intelligence Information Services Program

STIK STIS Interface Kit

STIKSCAT Stick Scatterometer

STIL Short Time Interval Launch • Software Test and Integration Laboratory • Statistical Interpretive Language • Streak Tube Imaging Lidar

STILAS Scientific and Technical Information Library Automation System

STILO Scientific and Technical Intelligence Liaison Office

STIM Scanning Transmission Ion Microscope • Scanning Transmission Ion Microscopy • Science - Technology - Information - Medicine • Solar Total Irradiance Monitor • Stimulate • Stimulation

STIMS Scientific and Technical Information Modular System • Shared Technical Information Management System • Shuttle Thermal Infrared Multispectral Scanner

STINET Scientific and Technical Information Network

STINFO Scientific and Technical Information

STINGS Stellar Inertial Guidance System

STINT Scientific and Technical Information Transfer • Standard Interface

STIO Scientific and Technical Information Office (NASA)

Stip Stipend • Stipendiary • Stipulation

STIP Scientific and Technical Information Processing • Scientific and Technical Information Program • Short-Term Instrument Plan • Smart Thermal Interface Plate • Study of Traveling Interplanetary Phenomena

Stir Stirlingshire (UK-Scotland)

STIR Scientific and Technical Intelligence Register • Special Technical Inspection and Repair • Surplus To Immediate Requirements

STIS Science and Technology Information System • Scientific and Technical Information Services • Scientific and Technical Information System • Scientific and Technical Intelligence System • Space Telescope Imaging Spectrograph (*) • Space Telescope Imaging Spectrometer • Stabilized Thermal Imaging System • Standards and Technology Information System

STISS Scientific and Technical Information Systems and Services

s/tít Sin Título

STIT Scientific and Technical Information Team

STITE Scientific and Technical Information Transfer for Education

STIVE Standard Interface Verification Equipment

STJ Saint Joseph - Rosecrans (USA-MO) (location identifier) • Superconducting Tunnel Junction • Superconducting Tunneling Junction

StJB Statistisches Jahrbuch

Stk Stock

S/TK Sectors per Track

STK Satellite Tool Kit • Standard Test Key • Sterling Crosson (USA-CO) (location identifier) • Sternkoppler

STKF Stock Fund

STKFA Stock Fund Accounting

STKFND Stock Fund

STKFS Stock Fund Statement

Stl Steel

STL Saint Louis - Lambert (USA-MO) (location identifier) • Save The Logs • Schottky Transistor Logic • Science and Technology Laboratory • Sciences et Technologies de Laboratoire • Science Trailer Line • Short-term Timeline • Simulated Tape Load • Small Text List • Spacecraft Test Laboratory • Space Technology Laboratory • Space Tissue Loss • Standardized Training Listing • Standardiseringsorganens Ledningsgrupp (Sweden) • Standard Telecommunications Laboratories • Standard Telegraph Level • Standard Template Library • Studio-to-Transmitter Link • Studio-Transmitter Link • Subadiabatic Transition Layer • Synchronous Transistor Logic • Systems Techniques Laboratory • System Test Laboratory • System Test Loop

STLC Small Time Local Controllability • Small Time Local Controllable

STLCC Small Time Local Configuration Controllable

STLDD Software Top Level Design Document

Stlg Sterling

STLI Stockage List Item

STLM Spatial-Temporal Light Modulator

STLO Scientific and Technical Liaison Office • Scientific and Technical Liaison Officer • Scientific Technical Liaison Office • Scientific Technical Liaison Officer

STLP Stochastic-excited Linear Prediction

STLV Slow Turning Lateral Vessel

Stm Steam • Storm

STM Santarém - Eduardo Gomes (Brazil-PS) (location identifier) • Scanning-Tunneling Microscope • Scanning-Tunneling Microscopy • Science and Technology Month (Malaysia) • Scientific and Technical Manuscript • Scientific - Technical - Medical • Send Test Message • Sensor Telemetry Module • Service Test Model • Short-Term Memory • Signal Termination Module • Significant Technical Milestone • Software Test Management • Sonet Transmission Manager • Spatio-Temporal Modulation • Specification Test Material • Spin Tuned Magnetron • Springfield Telescope Makers (USA-VT) • Standard Thermal Model • Static Test Model • Station Management • Storage Management • Store Multiple • Storm • Stream • Structural Test Model • Structural and Thermal Mode • Structural and Thermal Model • Structural Test Model • Structure and Thermal Model • Structure Thermal Model • Synchronous Transfer Mode • Synchronous Transfer Module • Synchronous Transport Module • System Termination Model • System Test Manager • System Test Mode • System Training Mission

STME Space Transportation Main Engine

STMEM Structural/Thermal Model and Engineering Model

STMIS System Test Manufacturing Information System

StMLU Staatministerium für Landesentwicklung und Umweltfragen

STMP Science and Technology Master Plan • Self Test and Monitor Program

STMPLP Stochastic-Multipulse-excited Linear Predictive

STMPM Sonet Transmission Manager Proxy Manager

STMS Sciences et Techniques Médico-Sociales • Short-Term Monetary Support

Stmt Statement

STMU Special Test and Maintenance Unit

STMUX Statistical Data Multiplexer

STMW Subtropical Mode Water

Stmy Stormy

Stn Station • Stazione

STN London Stansted (UK-England) (location identifier) • Satellite Tracking Network • Scientific and Technical Network • Signal To Noise • Software Transfer Notice • Software Trouble Note • Solar Telescope Network • Station • Subaru Telescope Network • Subthalamic Nucleus • Supertwisted Nematic • Supertwist Nematic • Switched Telephone Network • Systemetech Nord GmbH

STNA Service Technique de la Navigation Aérienne

Stng Sustaining

STNLCD Supertwist Nematic LCD

STNMTE Service Technique de la Navigation Maritime et des Transmissions de l'Équipement

STNR Stationary

STNRY Stationary

StNws Standardization News (periodical code)

Sto Santo

STO Santo • Save The Optics • Science and Technology Objective • Senior Technical Officer • Service du Travail Obligatoire (WWII) • Short Takeoff • Short-Term Objective • Slater-Type Orbital • Solar-Terrestrial Observatory • Special Technical Operation • Split Training Option • Standing Order • Stockholm (Sweden) (location identifier) • Store • Strategic Technology Office • Subcontractor Task Order • Subtask Officer • System Test Objective • System Threat Objective

STOA Science and Technology Office of Assessment • Scientific and Technical Options Assessment • Scientific and Technological Options Assessment • Security Trade-Off Analysis

StoAn Stockholms Observatoriums Annaler (periodical code)

STOAT Space Telescope Observatory Assessment Team

STOC Science and Technology Operations Coordination • Special Technical Operations Coordinator • Symposium On the Theory of Computing

STOCC Space Telescope Operations Control Center

STOCS Synchronous Token-Based Communications State

STOD Special Technical Operations Division

STOF Slosh Test Orbital Facility • Special Technical Operations Facility • Superthermal Time Of Flight

STOG Science and Technology Objectives Guide

STOH Since Top Overhaul

STOIC Stratospheric Ozone Intercomparison

STOL Short Takeoff and Landing • Systems Test and Operations Language

STOLS Simulated Telemetry Open Loop System

STOLVCD Short Takeoff and Landing - Vertical Climb and Descent

STOM System Test and Operations Manual

StoMe Stockholms Observatoriums Meddelande (periodical code)

STOMS Space Telescope Observatory Management System

S'ton Southampton (UK-England)

Ston Southampton (UK-England)

StoOR Stockholms Observatoriums Reports (periodical code)

STOP Satellite Targeting Of Polluters • Selected Test Optimization Program • Short-Term Operations Plan • Single-Title Order Plan • Solar Telescope for Operative Prediction

STOPAT Space Telescope Observatory Performance Assessment Team

Stor Storage

STOR Segment Table Origin Register • Store • System for Table-Oriented Reports • System Test and Operations Report

STORE Shuttle Test Of the Relativity Experiment • Storage Technology for Operational Readiness

STORET Storage and Retrieval

STORM Simulation, Testing, Operations and Rehearsal Model • Statistically-Oriented Matrix Program • Storm-scale Operational and Research Meteorology • Storm Tracking and Observation for Rainfall-runoff Monitoring

STORMS Standardized Operation Research Management System

StoS Ship to Shore • Station to Station

STOS Space Test Operations Section • Store String

STOSC Space Telescope Operations Support Cluster

STOU Store Unique

STOVL Short Takeoff and Vertical Landing

Stp S'il te plaît • Stamp

STP Minneapolis Saint Paul (USA-MN) (location identifier) • Principe (Sao Tome &_) (ISO country code) • Sao Tome & Principe (ISO country code) • Save The Plots • Satellite Technology Program • Satellite Test Program • Satellite Tracking Program • Satellite Transport Protocol • Scanning Tunneling Potentiometry • Scheduled Transfer Protocol • Schmidt Telescope Panel • School of Theoretical Physics • Science and Technology Project • Scientifically Treated Petroleum • Secure Transaction Protocol • Secure Transfer Protocol • Security Technical Procedure • Security Test Plan • Selective Tape Print • Self-Test Program • Seminar Traineeship Program • Sensor Tasking Plan • Sensor Track Processor • Service Transfer Point • SGML Target Registry • Shielded Twisted Pair • Short-Term Plan • Short-Term Prediction • Short-Term Project • Shuttle Technology Panel • Signaling Transfer Point • Signal Transfer Point • S'il Te Plaît • Simultaneous Test Procedure • Simultaneous Track Processor • Software Technology Park • Software Technology Programme • Software Test Plan • Software Test Procedure • Software Through Pictures • Soil Temperature Probe • Solar-Terrestrial Physics • Solar-Terrestrial Probe • Space Technology Payload • Space Telescope Project • Space Test Payload • Space Test Program • Spanning Tree Protocol • Special Technical Publication • Special Test Plan • Spectrum of Time Project • Standardized Test Program • Standard Processor • Standard Temperature and Pressure • Standard Terminal Program • Standard Test Procedure • Static Pressure • Static Test Program • Status Test Panel • Stop • Stream Protocol • Structured Test Procedure • Subsurface Transport Program • Supporting Technology Programme • Surveillance Tasking Plan • Synchronized Transaction Processing • Systems Technology Program • Systems Test Plan • Systems Training Plan • Systems Training Program • Systems Transition Plan

STPA Service Technique des Programmes Aéronautiques (France)

STPB Scientific and Technical Publications Branch

STPC Science and Technology Policy Council • Society of Technical Publications Contractors

STPCOE Special-To-Project Checkout Equipment

Stpd Stripped

STPDF Solar-Terrestrial Physics Data Facility

Stpg Stepping

STPL Standard Test Processing Language

STPM Structural/Thermal/Pyro Model

STPMET Solar-Terrestrial Physics - Meteorology

StPO Strafprozeßordnung

STPOA Space Telescope Postoperational Archive

Stpr Stepper

STPR Software Test Procedure

STPRC Space Telescope Proposal Review Committee

STPS Summary Task Planning Sheet

STPX Systems Training Program Exercise

STQ Saint Marys (USA-AK) (location identifier) • Statistiques et Techniques Quantitatives • Stop The Questions

str Stratiformis • streitig

Str Seater • Steamer • Straat • Straight • Strait • Straße • Streamer • Street • Strength • String • Strong • Structural • Structure

STR Satellite Tracking Radar • Satellite Transceiver • Science Tape Recorder • Scientific and Technological Research • Scientific Technical Report • Segment Test Review • Service Test Review • Service Trouble Report • Short-Term Research • Small Tuning Range • Software Tape Recorder • Software Test Report • Software Test Result • Software Trouble Report • Software Turnover Record • Solar-Terrestrial Relation • Source-Target Ratio • Space Test Range • Special Theory of Relativity • Starter • Star Tracker • Steering • Store • Store Task Register • Strait • Street • Strength • Structure • Stuttgart Echterdingen (Germany) (location identifier) • Sub-THz Receiver • Summary Technical Report • Surplus To Requirements • Synchronous Transmitter Receiver • System Test Review • System Trouble Report

STRA Stravenue

STRAC Solar-Terrestrial Relations Advisory Committee

STRAD Signal Transmission, Reception, and Distribution • Signal Transmitting, Receiving, and Distribution

Strai Strain (periodical code)

STRAM Synchronous Transmit Receive Access Method

StrÄndG Strafänderungsgesetz

STRAP Star Tracking Rocket Attitude Positioning • Stellar Tracking Rocket Attitude Positioning • Structured Requirements Analysis Planning • System for Tip Removal with Avalanche Photodiodes

Strat Strategic

STRAT Stratospheric Tracers of Atmospheric Transport

Stratig Stratigraphy

Strato Stratosphere

STRAV Stravenue

STRAVE Stravenue

STRAVEN Stravenue

STRAVN Stravenue

STRB Strobe

STRC Science and Technology Research Center • Scientific and Technology Requirements Committee • Scientific, Technical and Research Commission

STRCH Stretch

STRCHD Stretched

Strd Stranded

STRE Session de Technique de Recherche d'Emploi

STREAM Stratosphere-Troposphere Exchange by Aircraft Measurements

StRefG Steuerreformgesetz

StReL Staub Reinhaltung Luft (periodical code)

STREP Space Trajectory Radiation Exposure Procedure • Specific Targeted Research Project • Standard Repair • System Technology Reentry Experiments Program

STRESS Sediment Transport Events on Shelves and Slopes • Structural Engineering Systems Solver

STRETCH Space Technology Requirements Engineering Test of Component Hardware

STREX Storm Transfer and Response Experiment

STRF SLR Terrestrial Reference Frame

STRFLD Star Field

Strg Steering • Strong

STRG Senior Technical Review Group • Software Technology Reference Guide • Steering • String

STRÍ Icelandic Council for Standardization

STRI Smithsonian Tropical Research Institute

STRICOM Simulation, Training and Instrumentation Command

STRIDE Science and Technology for Regional Innovation and Development in Europe • Science and Technology Reporting Information Distribution Enhancement

STRIM Space Techniques for Risks Management

STRING Soil and Terrain Resources Information Network Generation

STRIP String Processing Language

STRIVE Science Teachers Research Involvement for Vital Education • Standard Techniques for Reporting Information on Value Engineering

Strk Stroke

Strl Structural

Strln Streamline

Strm Storeroom • Stream

STRM Stream

Strn Strainer

STRN Standard Technical Report Number

STROBE Satellite Tracking Of Balloons and Emergencies

STROGANOFF Structured Representation On Genetic Algorithm for Non-linear Function Fitting

Stroj Strojarstvo (periodical code)

StrOp Structural Optimization (periodical code)

STRQ Sensor Termination Request • Session Termination Request

STRR System Test and Readiness Report

STRS Sensor Termination Response • Space-Time Resolved Spectrograph

STRT Start • Strait • Street

STRUC Structure

STRUCT Structure

STRUDL Structural Design Language

STRV Space Technology Research Vehicle

STRVN Stravenue

STRVNUE Stravenue

StrWG Straßen- und Wegengesetz

STrZ South Tropical Zone (Jupiter)

STRZ South Tropical Zone (Jupiter)

Struct Structural • Structure

sts Status

Sts Saints • Streets

STS Santa Rosa (USA-CA) (location identifier) • Satellite To Satellite • Satellite Tracking Station • Satellite Tracking System • Satellite Transmission Systems Inc. • Save The Spectra • S-band Transmitter System • Scheduled Truck Service • Science and Technology Section • Science and Technology Study • Science Talent Search • Science - Technologie - Société • Science - Technology - Society • Scientific Terminal System • Section des Techniciens Supérieurs • Segment Tilt Sensor • Selective Tape Service • Sensor Test Set • Sensor To Service • Servo Test System • Short-Term Schedule • Short-Term Scheduler • Short-Term Scheduling • Shuttle Test Station • Shuttle Transportation System • Shuttle Transport System • Single Trusted System • Skylab Terminal System • Slicing Tree Structure • Socio-Technical System • Soft-Top Shelter • Solar Tracking System • Spacecraft Telecommunications System • Spacecraft Tracking Station • Space Transportation System • Space Transportation System Ltd. • Specialty Training Standard • Stabilized Telescope System • Standardized Teleoperation System • Standard Technical Specification • Static Test Stand • Station To Station • Status Transfer Switch • Streets • Structural Transition Section • Supercooled Ternary Solution • Supersaturated Ternary Solution • Supervisory Training System • Synchronous Transport Signal • Synchronous Transport System • Système de Télécommunication Sol • System Technical Support • System Test Set • System Test Station • System Trouble Shooting

STSA Secure Telephone System Architecture • Space Telescope Solar Array • State Technical Service Act (USA)

STSAM Space Transportation System Analysis Model

STSB Scientific and Technical Services Branch

STSC Satellite Test Support Console • Software Technology Support Center

STSCC Standing Transportation Support Coordinating Committee

STSCHG Status Change

STScI Space Telescope Science Institute (USA-MD)

STScIF STScI Facility

STSCM Space Transportation System Cost Model

STSD Space Transportation System Division

STSDAS Space Telescope Science Data Analysis Software • Space Telescope Science Data Analysis System (*) • Space Telescope Scientific Data Analysis System

STSE Science - Technology - Society - Environment

STSF Science and Technology through Science Fiction

StSky Star Sky (periodical code)

STSM Space Transportation System Management

STSO Senior Technical Staff Officer

STSOC Space Telescope Science Operations Center • Space Transportation System Operations Contract

STSOF Space Telescope Science Operations Facility

STSOP Space Telescope Science and Operations Project

STSOPO Space Transportation System Operations Program Office

STSOS Space Tracking System Options Study

STSP Same Time - Same Place • Secure Telephone System Program • Small Telescope Science Program • Solar-Terrestrial Science Program • Symmetric Traveling Salesman Problem

STSR System Test Summary Report

STSS Sensitive Thrust Stand System • Software Technology Support System • Space Telescope Safing System

STST Short-Term Scheduling Tool

STSU Science and Technology Center in Ukraine

STSV Short Timescale Variability

STSWG Space Telescope Science Working Group

STT Saint Thomas - Cyril E. King (US Virgin Islands) (location identifier) • Schedule Training Time • Sciences et Techniques Tertiaires • Scientific Tools and Techniques • Secure Transaction Technology • Short-Term Training • Single Target Track • Site Testing Telescope • Small Tactical Terminal • Spacecraft Terminal Thrust • Spacelab Transfer Tunnel • Spontaneous Trait Transference • Star Tracker • Strategy To Task • System Technical Test

STTC Service Technique des Technologies Communes

STTE Special Tools and Test Equipment • Spectroscopic Time-Tagged Event

STTF Space Telescope Test Facility • Systems Technology Test Facility

Sttg Starting

STTG Solar-Terrestrial Theory Group (UNH)

STTM Stabilized Tracking Tripod Module

STTO Start - Taxi - Takeoff

STTP SGML Tree Transformation Process • Space Technology Training Program

Sttr Starter • Stator

STTR Short-Term Technology Review • Small-business Technology Transfer

STTT Software Tools for Technology Transfer (periodical code) • Space Telescope Task Team

StTur Studies in Turbulence (periodical code)

Stu Student

STU Santa Cruz (Belize) (location identifier) • Schmidt Telescope Unit • Sciences de la Terre et de l'Univers • Secure Telephone Unit • Secure Terminal Unit • Servo Test Unit • Signal Transfer Unit • Space-Time Unit • Special Test Unit • Special Training Unit • Star Tracker Unit • Static Test Unit • Streaming Tape Unit • Subscriber Terminal Unit • Systems Test Unit • System Timing Unit

StUB Stadt- und Universitätsbibliothek

STUB Stadt- und Universitätsbibliothek

STUC Space Telescope Users Committee

Stud Student

STUTEL Spécifications Techniques d'Utilisation du Téléchange de Logiciel

STV Satellite Television • Scottish Television • Separation Test Vehicle • Short Timescale Variability • Single Transferable Vote • Small-Target Visibility • Small Test Vessel • Solar Thermal Vacuum • Space Test Vehicle • Space Transfer Vehicle • Special Test Vehicle • Standard Test Vehicle • Steerable Television • Stichting Technology Vlaanderen (Belgium) • Structural Test Vehicle • Subscription Television • Surat (India) (location identifier) • Surrogate Teleoperated Vehicle • Surveillance Television

STVF Samoletostroenie Tekhnika Vozdushnogo Flota (periodical code)

StVG Straßenverkehrsgesetz

STVL Short Takeoff and Vertical Landing

STVNS Spherical Tokamak Volume Neutron Source

StVO Straßenverkehrsordnung

STVS Secure Television System

STVSS Space Telescope Verification Software System

STVSW Space Telescope Verification Software

STVT Space Telescope Verification Team

STVTL Space Telescope Verification Team Leader

StVZO Straßenverkehrszulassungsordnung

Stw Steenweg

STW Save The Waves • Standard Typical Waste • Stichting voor de Technische Wetenschappen (Netherlands) • Stravopol (Russia) (location identifier)

STWA Short Trailing Wire Antenna

STWE Society of Technical Writers and Editors

STWG Science and Technology Working Group • Scientific and Technical Working Group • Space Telescope Working Group

STWP Society of Technical Writers and Publishers (USA)

Stwy Stairway

STX Saint Croix - Alexander Hamilton (US Virgin Islands) (location identifier) • Situational Training Exercise • Staff Training Exercise • Start of Text

sty Style Sheet (file extension)

STY Salto (Uruguay) (location identifier) • Style

STZ Santa Terezinha Confresa (Brazil-MT) (location identifier) • South Temperate Zone (Jupiter) • South Tropical Zone

su siehe unten! • Sudanese (ISO language code) • Superuser

Su Sudan • Sudanese • Sunday

S/U Setup

SU Egypt (international aircraft registration code) • Scanner Unit • Scheduling Unit • Search Unit • Seattle University (USA-WA) • Sensor Unit • Service Unit • Service User • Setup • Siehe Unten! • Signaling Unit • Signal Unit • Single User • Sonics and Ultrasonics • Soviet Union (now obsolete) • Space Unit • Special Unitary • Standard Update • Stand Up • Stanford University (USA-CA) • Start Up • Station Unit • Storage Unit • Strontium Unit • Student Union • Submarine • Subscriber • Subscriber Unit • Sudan • Sumerian • Superuser • Supporting Unit • Support Unit • Suppressor • Sustainability • Switching Unit • Syracuse University (USA-NY) • System Upgrade

SUA Satellite Unfurlable Antenna • Sociedad Uruguaya de Astronomía (Uruguay) • State Universities Association • Stuart Witham (USA-FL) (location identifier) • Suppression of Unlawful Acts

SUAOIP Service Universitaire d'Accueil, d'Orientation et d'Insertion Professionnelle

SUAR Start Unload Address Register

SUAS Service Universitaire d'Action Sociale (France) • System for Upper Atmosphere Sounding

SUAV Service Universitaire Audiovisuel

Sub Subida • Subito • Subject • Subjective • Subjectively • Submarine • Submissive • Subordinated • Subscription • Subsidiary • Subsidy • Subsistence • Substitute • Suburb • Suburban • Subvention • Subway

SUB Staats- und Universitätsbibliothek • Subaddressing • Subject • Surabaya Juanda (Indonesia) (location identifier) • Subroutine • Substitute • Subtract • Supplemental Unemployment Benefit

Subd Subdivision

Subj Subject • Subjective • Subjectively • Subjugate • Subjugated • Subjunctive • Subjuntivo

SUBJ Subject

Subl Sublimation

SUBNO Substitutes Not Desired

SUBOK Substitution Acceptable

SUBOP Submersible Operations

Subq Subsequent

SUBR Southern University - Baton Rouge (USA-LA)

Subs Subsequent • Subsequently • Subsidiary • Subsistence

SUBS Single-bus Unidirectional Broadcast System • Subscribers

Subsc Subscription

Subsec Subsecretaría • Subsection

Subseq Subsequent • Subsequently

SUBSIST Subsistence

Subsp Subspecie

Subspp Subspecies

Subst Substantive • Substantively • Substitute

Substand Substandard

Subte Subterráneo

Subtr Subtract

Suc Succeed • Succeeding • Success • Successor • Suction • Sucursal

SUC Satisfied User Criterion • Southern Union College • Starlink Users Committee • Sundance Schloredt (USA-WY) (location identifier)

Succ Succeed • Succeeded • Succeeding • Success • Successful • Succession • Successor • Succursale

SUCCESS Subsonic Aircraft: Contrail and Cloud Effects Special Study

SUCE Spatial Understanding Critical Examination

SUCHII Superultracompact HII (region)

SUCOSIP Subcommittee on Site Properties

Suct Suction

SUD Stroud (USA-OK) (location identifier) • Subjective Unit of Distress

SUDOC Système Universitaire de Documentation

SUDS Software Update and Distribution System • Subjective Unit of Discomfort Scale

SUDSS Spatial Understanding and Decision Supporting System

SUDT Silicon Unilateral Diffused Transistor

SUE Seismic Underwater Explorer • Strontium Unit Equivalent • Stupid User Error • Sturgeon Bay (USA-WI) (location identifier) • Subunit Evaluation • Systematic Usability Evaluation

SUEA Scan Unit Electronics Assembly

SUERC Scottish Universities Environmental Research Centre (UK-Scotland)

Suf Sufficient • Suffix • Sufijo

SUF Lamezia Terme (Italy) (location identifier) • Scientific Update Facility • Service Update Facility • Software Update Facility

SUFCEP Service Universitaire de Formation Continue et d'Éducation Permanente

Suff Sufficient • Suffix • Suffolk (UK-England)

SUFI SUNRISE Filter Imager

SUFOS Section of Unidentified Flying Objects Studies

Sug Suggest • Suggestion

SUG Science Users Group • Starlink User Guide • SUN Users Group • Surigao (Philippines) (location identifier)

SUH Suburban Housing • Sur (Oman) (location identifier)

SUI Sensitive Unclassified Information • Sensory User Interface • Sound User Interface • Speech User Interface • Standard User Interface • State Unemployment Insurance • Sukhumi (Georgia) (location identifier) • System User Interface

SUIC Service Unit Identification Code

SUIM Station Unit Interface Module

SUIO Service Universitaire d'Information et d'Orientation

SUIP Société Unipersonnelle d'Investissement Providentiel (France) • Support Unit Improvement Program

SUIT Simple User Interface Toolkit

Suiv Suivant

SUJ Satu Mare (Romania) (location identifier)

SUK Samchok (Korea-ROK) (location identifier) • Someraj Universitataj Kursoj

SUL Simplified User Logistics • Software Utilization List • Sui (Pakistan) (location identifier)

Sult Sultan

sum Sumat • Sumendum

Sum Summary • Summer

SUM Single User Monitor • Software User's Manual • Statistical Utility Module • Summary • Summer • Sumter (USA-SC) (location identifier)

SUMA Satellite Universal Modulation Adapter

SUMAS Submillimeter Atmospheric Sounder

SUMC Satellite Ultrareliable Modular Computer • Space Ultrareliable Modular Computer

SUMER Solar Ultraviolet Measurement of Emitted Radiation (SOHO)

SUMI Solar Ultraviolet Magnetograph Investigation

Summ Summary • Summer

SUMMIT Submillimeter Tipper • Supervisor of Multiprogramming, Multiprocessing, Interactive Timesharing • Sydney University Multiple Mirror Telescope

SUMO Supermirror Observatory

SUMP Service Universitaire de Médecine Préventive • Submicron Photometer

SUMR Satellite User Mapping Register

SUMS Shuttle Upper-Atmosphere Mass Spectrometer • Specialized Unit Maintenance Support

SUMSS Sydney University Molonglo Sky Survey

SUMT Sequential Unconstrained Minimization Technique

Sun Sunday

SUN Hailey - Sun Valley (USA-ID) (location identifier) • Secure Universal Norm • Single User Network • Solar Ultraviolet Network • Standard Units and Nomenclature • Stanford University Networks (USA-CA) • Starlink User Note • Students Understanding Nature • Sunday • Sun Valley (USA-ID) (location identifier) • Symbols - Units - Nomenclature

SUNA Sociedad Universitaria de Astronomía (Venezuela)

SUNAMCO Symbols, Units, Nomenclature, Atomic Masses and Fundamental Constants

SUNBEAMS Students United with NASA Becoming Enthusiastic About Math and Science

Sund Sunday • Sundial

SUNDIAL Speech Understanding and Dialogue

SUNET Swedish University Network

SUNIST Serveur Universitaire National pour l'Information Scientifique et Technique (France)

SUNNS Survey of Nearby Nuclei with STIS

SUNRS Sunrise

SUNS Small Unit Navigation System

Sunwo Sunworld (periodical code)

SUNY State University of New York (USA-NY)

SUNYA State University of New York at Albany (USA-NY)

SUNYSB State University of New York at Stony Brook (USA-NY)

SUO Small Unit Operations • Sundays Only • Sun River (USA-OR) (location identifier)

SUOA Superunit Of Assessment

SUOC Science User Operation Centre

sup Start-Up Data (file extension) • Supra • Supremum

Sup Superficial • Superfine • Supérieur • Superior • Superlative • Superlativo • Supplement • Supplementary • Supply • Supreme • Suprême

SUP Short/Unstable Penetration • Single Unit Parameter • Soluble Unreactive Phosphorus • Special Utility Program • Standard Unit of Processing • Statistical Utility Program • Sumenep Trunojoyo (Indonesia) (location identifier) • Supervisory • Supplement • Supplemental • Supply • Support • Suspected Unapproved Part • System Utilization Procedure

SUPARCO Space and Upper Atmosphere Research Commission (Pakistan)

Super Superficial • Superfine • Superintendent • Superior • Supernumerary

Superg Supergiant

Superl Superlative • Superlativo

SUPI Simple Universal PMD Interface

SUPIDEN Support Identification

SUPIM Sheffield University Plasmasphere and Ionosphere Model

SUPIR Supplemental Imagery Report

SUPOS SUNRISE Polarimetric Spectrograph

Supp Supplement • Supplementary

SUPPAD Supplemental Address

Suppl Supplement • Supplément • Supplementary • Supplémentaire • Supplied • Supply • Support • Supported • Supporting • Suppress • Suppressed • Suppressing • Suppression

SUPPORT Study to Understand Prognoses and Preferences for Outcomes and Risks of Treatment

SUPPR Suppression

SUPPS Supplemental Procedures

SUPPT Supply Point

Supr Superior • Supervisor • Suppress • Supreme

SUPRA Service Universitaire Pour la Recherche et les Applications en Supraconductivité

SUPS Static Uninterruptible Power Source

SUPSI Scuola Universitaria Professionale della Svizzera Italiana (Switzerland)

Supt Superintendent

SUPT Support

Supv Supervise • Supervisor

SUPV Supervise • Supervisor • Supervisory

Supvr Supervisor

SUQ Sucua (Ecuador) (location identifier)

Sur Surface • Surplus • Surrey (UK) • Surveyor

SUR Rouble (Russia) (ISO currency code) • Shared University Research • Small Unit Radio • Summer Beaver (Canada-ON) (location identifier) • Suriname (ISO country code)

SURA Southeastern Universities Research Association (USA)

SURANET Southeastern Universities Research Association Network

SURAP Summer Undergraduate Research Assistantship Program

SURBAT Simultaneous Unlimited Rigorous Block Analytical Triangulation

SURC Space Utilization Research Center

SURE Sensor Update and Refurbishment Effort • Shuttle Users Review and Evaluation • Simplicity - Usability - Reliability - Economy • Supply Usage Requirements Estimator

Surf Surface

SURF SCUBA User Reduction Facility • Stichting Samenwerkende Universitaire Rekenfaciliteiten (Netherlands) • Summer Undergraduate Research Fellowship • Surface • Synchrotron Ultraviolet Radiation Facility (NIST)

SURFNET Dutch University Network (Netherlands)

SURFSAT Summer Undergraduate Research Fellowship Satellite

SURFRESIDE Surface Reaction Simulation Device

Surg Surgeon • Surgery • Surgical

SURGE Seasat Users Research Group of Europe • Sorting - Updating - Report - Generating - Etc. • Supply User Reports Generator

SURPC Scottish Universities Research Policy Consortium (UK-Scotland)

SURPIC Surface Picture

Surr Surrey (UK)

SURRC Scottish Universities Research and Reactor Centre (UK-Scotland)

SURRVS Standard Umbilical Retraction Retention Vent System

SURS Standard Umbilical Retraction System

SurSc Surface Science (periodical code)

SURTEX Shuttle-Released Target Experiment • Shuttle-Released Technology Experiment

Surv Surveillance • Survey • Surveying • Surveyor • Survival • Survive • Surviving

SURV Standard Underwater Research Vessel • Surveillance • Survey • Survivability

SURVIAC Survivability/Vulnerability Information Analysis Center

Sus Sussex (UK-England)

SUS Silicon Unilateral Switch • Single Unix Specification • Software Unit Specification • Speech-Understanding System • Spirit of Saint Louis (USA-MO) (location identifier) • Suit Umbilical System • Suspend • Sussex (Starlink internal code)

SuScT Superconductor Science Technology (periodical code)

SUSI Stanford University Speckle Interferometry • Structures Under Shock and Impact (conference) • Superb Seeing Imager • Superseeing Imager • Sydney University Stellar Interferometer

SUSIE Satellites for the Understanding of Space Instrumentation and Experimentation

SUSIM Solar Ultraviolet Spectral Irradiance Monitor

Susp Suspected • Suspend • Suspended • Suspending • Suspension

SUSP Suspected • Suspend • Suspended • Suspending • Suspension

Suss Sussex (UK-England)

SUSS Spatial Understanding Support System

SUSSP Scottish Universities Summer School in Physics (UK-Scotland)

SUST Shanghai University of Sciences and Technology (China-PRC)

SUSTAR Sussex (Starlink internal code)

SuStB Staats- und Stadtbibliothek (Germany)

SUSV Small Unit Support Vehicle

SUSY Supersymmetric • Supersymmetry

SUT Satellite Under Test • Science University of Tokyo (Japan) • Science User Tool • Set-Up Time • Small Unit Transceiver • Software Under Test • Sumbawanga (Tanzania) (location identifier) • Suranaree University of Technology (Thailand) • Swinburne University of Technology (Australia) • System Under Test

SUTAGS Shuttle Uplink Text And Graphics Scanner

Suth Sutherland (UK-Scotland)

SUTP Screened Unshielded Twisted Pair

SUU Fairfield Travis (USA-CA) (location identifier) • Secondary User Unit • Southern Utah University (USA-UT) • Suspended Underwing Unit

SuUB Staats- und Universitätsbibliothek (Germany)

SUV Solar Ultraviolet • Suva (Fiji) (location identifier)

SUVB Soft Ultraviolet Background

SUVDDP Space Ultraviolet/Visible Detector Development Program

SUVE Solar Ultraviolet Experiment

SUVO Space Ultraviolet Observatory • Space Ultraviolet Optical Observatory • Space Ultraviolet-Visible Observatory

SUVSR Scandinavian Union of Amateur Astronomers Variable Star Section (periodical code)

SuW Sterne und Weltraum

SUW Superior - Richard I. Bong (USA-WI) (location identifier)

SUX Sioux City (USA-IA) (location identifier) • Sundays Excepted

SUY Sudureyri (Iceland) (location identifier)

SUZ Suria (Papua New Guinea) (location identifier)

SuZIE Sunyaev-Zel'dovich Infrared Experiment

s/v Sub Voce

sv Selbstverlöschend • Sub Verbo • Sub Voce • Swedish (ISO language code)

Sv Sievert

S/V Space Vehicle • Supply Valve • Surface/Volume • Survivability/Vulnerability

SV El Salvador (ISO country code) • Rhea (satellite of Saturn) • Sac Vide • Safety Valve • Sailing Vessel • Salvador (El_) (ISO country code) • Satellite Vehicle • Satellite Virus • Schedule Variance • Science Verification • Sea Vessel • Secondary Verification • Secular Variation • Secure Voice • Security Vehicle • Security Violation • Selenoid Valve • Self-Ventilated • Self-Verification • Short Vector • Shuttle Vehicle • Side Valve • Side View • Significant Vector • Small Volume • Software Visualization • Solenoid Valve • Space Variant • Space Vehicle • Spatially-Variant • Speaker Verification • Spin Vector • Standortverteiler • State Vector • State Violation • Stiftverband • Stop Valve • Stroke Volume • Sub Verbo • Sub Voce • Supply Valve • Support Vector • Surface Vehicle • Surrender Value • Systems View • System Verification

SvA Soviet Astronomy (now obsolete) (periodical code)

SVA Savoonga (USA-AK) (location identifier) • Sectionalized Vertical Antenna • Security Violation Alert • Security Vulnerability Analysis • Service à Valeur Ajoutée • Shared Virtual Area • Skokie Valley Astronomers (USA-IL) • Société Vaudoise d'Astronomie (Switzerland) • Spatially Varying Apodization • Speech-Vision Association • Speech-Vision Associator • Statistical Vibration Analysis • Stator Vane Actuator

SVAA Susquehanna Valley Amateur Astronomers (USA-PA)

SVAAC Sandusky Valley Amateur Astronomy Club (USA-OH)

SVAB Shuttle Vehicle Assembly Building

SVA&C Shuttle Vehicle Assembly & Checkout

SVAC Shuttle Vehicle Assembly and Checkout

SVACO Shuttle Vehicle Assembly and Checkout

SVAFB South Vandenberg Air Force Base (USA-CA)

SvAL Soviet Astronomy Letters (now obsolete) (periodical code)

S/VAR Survivability/Vulnerability Analysis Report

SVAS Sacramento Valley Astronomical Society (USA-CA)

SVAT Soil-Vegetation-Atmosphere Transfer

SVATS Soil-Vegetation-Atmosphere Transfer Scheme

SvB Stichting voor Biofysica (Netherlands)

SVB Sambava (Madagascar) (location identifier) • Shuttle Vehicle Booster • Space Vehicle Booster • Stichting Voor Biofysica (Netherlands)

SVBT Space Vehicle Booster Test

Svc Service

SC/C Survivability-Vulnerability/Crashworthiness

SVC Colón (El Salvador) (ISO currency code) • Science VAX Cluster • Secure Voice Communication • Service • Serviced • Servicing • Serviço • Silver City (USA-NM) (location identifier) • Single Version Charging • Slowly Varying Component • Spacecraft Vehicle Contractor • Space Vehicle Code • Steele Visitor Center (Lowell Obs.) • Store-Value Card • Supervisor Call • Switched Virtual Call • Switched Virtual Circuit (*) • Switched Virtual Connection • Switch Virtual Circuit

SVCC Switched Virtual Channel Connection • Switched Virtual Circuit Connection

SVCCN Secure Voice Command Control Network

SVCDS Space Vehicle Computer Design System • Space Vehicle Control/Display Simulation

Svce Service

SVCMU Space Vehicle Control and Monitor Unit

Svcs Services

SVCS Services

SVD Saint Vincent (location identifier) • Seismic Velocity Discontinuity • Shock Vibration Digest (periodical code) • Silicon Vertex Detector • Simultaneous Voice and Data • Simultaneous Voice/Data • Singular Value Decomposition • Small-Volume Detector • Software Version Description • Space Vehicles Division • Spatially Variant Deconvolution • Supplementary Volume Descriptor • System Verification Diagram

SVDP Skylab Video Documentation Project

SVDS Space Vehicle Dynamic Simulation • Space Vehicle Dynamic Simulator

SVDT Space Vehicle Development Team

SVE SAP Vector Element • Society of Value Engineers (USA) • Space Vehicle Equipment • Survivability - Vulnerability - Endurability • Susanville (USA-CA) (location identifier)

SVEA Slowly Varying Envelope Approximation

SVEB Schweizerische Vereinigung für Erwachsenenbildung (Switzerland)

SVEC Space Vacuum Epitaxy Center (USA-TX)

SVELTE Six Variably-Extended Legs Telescope

SVF Save (Benin) (location identifier) • Simple Vector Format • Software Validation Facility • State Variable Feedback • State Variable Filter

SVFR Special Visual Flight Rule

SVG Scalable Vector Graphics • Silicon Valley Group (USA-CT) • Sozialversicherungsgesetz • Stavanger (Norway) (location identifier)

SVGA Super Video Graphics Adapter • Super Video Graphics Array (*)

Svgs Savings

SVH Statesville (USA-NC) (location identifier)

S-VHS Super Video Home System

SVHS Super Vertical Helical Scan • Super Video Home System

SVI San Vincente del Caguan (Colombia) (location identifier) • Schweizerischer Verband der Informatikorganisationen (Switzerland) • Serveur Vocal Interactif • Single Vendor Integrity • Standardized Virtual Interface • Titan (satellite of Saturn)

SVIC Shock and Vibration Information Center • State Variable Inequality Constraint

SVICB Shock Vibration Information Center Bulletin (periodical code)

SVICD Shock Vibration Information Center Digest (periodical code)

SVICI Shock Vibration Information Center Information (periodical code)

SVICP Shock Vibration Information Center Computer Programs (periodical code)

SVID Satellite-Vehicle Identification • System V Interface Definition • System V Interface Device

SVII Hyperion (satellite of Saturn)

SVIII Japetus (satellite of Saturn)

SVIP Secure Voice Improvement Program

SVIR Scanning Visible and Infrared Radiometer

SVIRI Spinning Visible and Infrared Imager

SVISSR Stretched Visible and Infrared Spin Scan Radiometer

SVIU Space Vehicle Interface Unit

SVJ Svolvaer (Norway) (location identifier)

SvJPP Soviet Journal of Plasma Physics (periodical code)

SVJPS Springer-Verlag Journal Preview Service

SvJQE Soviet Journal of Quantum Electronics (periodical code)

SVK Silver Creek (Belize) (location identifier) • Slovakia (ISO country code) • Slovaquie (code ISO de pays)

SVL Savonlinna (Finland) (location identifier)

SVLA Steered Vertical Line Array

SVLBI Space Very Long Baseline Interferometry

SVLS Spherical with Varied Line Spacing (grating)

SVLTU Service Line Termination Unit

SVM Saint Paul's Mission (Australia-QLD) (location identifier) • Scaling Vector Method • Science Verification Manager • Service Module • Spatial Variance Map • Stamp Vending Machine • Storage Vault Management • Storage Vault Manager • Strong Vertical Mixing • Support Vector Machine • System Validation Model • System Verification Manager • System Virtual Machine

SVMA Space Vehicle Mission Analysis

SVMG Solar Vector Magnetograph

SVML Standard Vehicle Mounted Launcher

SVN Savannah Hunter (USA-GA) (location identifier) • Secure Voice Network • Slovenia (ISO country code) • Slovénie (code ISO de pays) • Space Vehicle Number • Switched Virtual Network

SVO Moscow Sheremetyevo (Russia) (location identifier) • Space Vehicle Operations • Space Volcanic Observatory • Spectroscopic Virtual Observatory • Subject - Verb - Object

SVOD Space Vehicle Operations Director

SVOSA Solar Variability and the Outer Solar Atmosphere

SVP Kuito (Angola) (location identifier) • Saturated Vapor Pressure • Science Verification Phase • Secure Voice Program • Senior Vice-President • S'il Vous Plaît • Software Validation Plan • Software Verification Plan • Software Verification Procedure • Sound Velocity Profile • Space Vehicle Pressure • Special Visibility Program • Spectral Variability Parameter • Star-Vaporizing Pulsar • Support Vector Predictor • Switched Virtual Path

SVPC Switched Virtual Path Connection

SVPCS Springer Series on Physics Chemistry Space (periodical code)

SvPhA Soviet Physics-Astronomy (periodical code)

SvPhD Soviet Physics-Doklady (periodical code)

SvPhU Soviet Physics-Uspekhi (periodical code)

SVPN Secure Virtual Private Network

SvPro Svarochnoe Proizvodstvo (periodical code)

SVPSF Space-Variant Point Spread Function

SVQ Sevilla (Spain) (location identifier)

Svr Severe

SVR Server • Severe • Shop Visit Rate • Smart Video Recorder • Software Verification Report • Space Vehicle Requirement • Spiritus Vini Rectificatus • Supply Voltage Rejection • Svay Rieng (Cambodia) (location identifier) • Synthetic Variable Ratio • System Validation Review • System Verification Review

SVRB Supervisor Request Block

SVRL Several

SVRR Software Verification Readiness Review

SVS Scientific Visualization Studio • Secure Voice System • Service Vision System • Shuttle Vehicle Simulator • Single Virtual Storage • Soviet Variable Stars • Space Vehicle Simulator • Space Vehicle System • Space Vision System • Spinning Vehicle Simulator • Stevens Village (USA-AK) (location identifier) • Still-camera Video System • Suit Ventilation System • Synthetic Vision System

SVSI Secure Voice System Interface

SVSM Semantic Vector Space Model

SVST Swedish Vacuum Solar Telescope

Svt Souvent

SVT Savuti (Botswana) (location identifier) • Sciences de la Vie et de la Terre • Secure Voice Terminal • Silicon Vertex Tracker • Software Validation Test • Solar Vacuum Telescope • Space Vehicle Test • Subscriber Voice Terminal • System Validation Test

SVTC Secure Video Teleconferencing

SVTN Secure Voice Telecommunications Network • Secure Voice Teleconference Network

SVTP Sound - Velocity - Temperature - Pressure

SVTS Secure Video Teleconferencing System

SVU Savusavu (Fiji) (location identifier) • Sri Venkateswara University (India) • Surface Vehicle Unit

SvUNT Sverdlovsk Ural'skii Nauchnyi Tsentr (periodical code)

svv Sub Verbis

SVV San Salvador (Venezuela) (location identifier) • System Vulnerability Value

SVVP Software Verification and Validation Plan

SVVS System V Verification Suite

svw soviel wie

SVW Sparrevohn (USA-AK) (location identifier)

SVWG Science Verification Working Group

SVX Ekaterinburg (Russia) (location identifier) • Silicon Vertex Detector

Svy Survey

SVY Savo (Solomon Islands) (location identifier)

SVZ San Antonio (Venezuela) (location identifier) • Schlüsselverteilzentrale

s/w schwarz/weiß • Sea Water

Sw Swahili (ISO language code) • Sweden • Swedish • Swiss • Switch • Switzerland

S/W Software

S&W Sterne und Weltraum (periodical code) • Surveillance & Warning

SW Sachs-Wolfe (effect) • Salt Water • Schwassmann-Wachmann • Science Watch (journal) • Science Workshop • Sea Water • Semantic Web • Shipper's Weight • Shock Wave • Shortwave • Short Wavelength • Sidewall • Single Weight • Single Wheel • Software • Solar Wind (*) • Solar Wing • South Wales (UK) • Southwest • Southwestern • Space Wing • Spares Warehouse • Specific Weight • Spot Weld • Standard Weight • StarWorlds (database) • Statistical Weight • Status Word • Stellar Wind • Stemming Weight • Strategic Workshop • Subset Word • Superworld • Surface Water • Surface Wave • Swap • Sweden • Switch • Switchboard Wound • Switcher

SWA Namibia (international vehicle registration code) • Scheduler Work Area • Seewetteramt • Shantou (China-PRC) (location identifier) • Slow Wave Activity • Solar-Wind Analyzer • Southwest Africa • Southwest Asia • Superwide Angle • Support Work Authorization • Swiss World Airways

Swab Swabia • Swabian

SWABS Seismic Wide-Angle and Broadband Survey

SWACOM Standard Water Column Model

SWACS Space Warning and Control System

SWAD Semantic Web Advanced Development • Shop Work Authorization Document • Subaccount Work Authorization Document • Subdivision of Work Authorization Document

SWADE Surface Wave Dynamic Experiment

SWADS Scheduler Work Area Data Set

SWAG Simulated Waste Access to Ground

SWAIS Simple Wide-Area Information Server

SWAK Sealed With A Kiss

SWAL Southwest Airlines Inc. (Japan)

SWALK Sealed With A Loving Kiss

SWAMP Science Working Group for the AM Platform • Severe Wind And Moisture Problem

S/WAN Secure Wide-Area Network

SWAN Satellite Wide-Area Network • Secure Wide-Area Network • Solar Wind Anisotropies (SOHO) • Storage Wide-Area Network

SWANA Solid Waste Association of North America

SWAP Severe Weather Avoidance Plan • Shared Wireless Access Protocol • Size, Weight And Power • Southwestern Association of Planetariums (USA-TX)

S/WARE Software

SWARE Software

SWARM Solar Warning And Real-time Monitor • Space Weather Aeronautical Responses Model

SWAS Submillimeter Wave Astronomy Satellite

SWAT Scandinavian Workshop on Algorithm Theory • Short Wavelength Adaptive Technology • Software Analysis and Test • Software Assessment Team • Software Assurance and Test • Software Assurance Team • Solar-system Watching Team • Space Warfare ATM Timing • Special Weapons And Tactics • Stress Wave Analysis Technique • Subjective Workload Assessment Technique • Swift Action Team

SWATEC Short-Wavelength Thermoelectrically Cooled

SWATH Small Water Area Twin Hull (vessel) • Space Weather And Terrestrial Hazards

Swazil Swaziland

SWB Searle - Wilkinson - Bagnuolo (catalog) • Shaw River (Australia-WA) (location identifier) • Solder Waiver Board • Sternwarte Bremen (Germany) • Subtract With Borrow • Switchboard

SWBB Short Wavelength Broadband • Standard Wind - Blown Bubble

Swbd Switchboard

SWBD Switchboard

SWbS Southwest by South

SWBT South Western Bell Telephone (USA)

SWbW Southwest by West

S-W-C Sun-Weather-Climate

S&WC Surveillance & Warning Center

SWC Short Wavelength Channel • Software and Communications • Solar-Wind Compensation • Solar-Wind Compensator • Solar-Wind Composition (experiment) • Space Warfare Center • Stawell (Australia-VIC) (location identifier) • Sun - Weather - Climate • Surveillance and Warning Center • Switch

SWCA Soil and Water Conservation Act (USA)

SWCC Second World Climate Conference • Southwestern Computing Center

SWCDR Software Critical Design Review

SWCG Shane-Wirtanen Clusters of Galaxies

S/WCI Software Configuration Item

SWCI Software Configuration Inspection • Software Configuration Item

SWCNT Single Walled Carbon Nanotube

SWCS Soil and Water Conservation Society

swd Schoolwide Database (file extension)

SWD Schlagwortnormdatei • Seward (USA-AK) (location identifier) • Sun - Wind - Dust • Surface Wave Device • Synchronous Wave Device

SWDA Solid Waste Disposal Act (USA) • Stepwise Discriminant Analysis

SWD&I Software Development & Integration

SWDI Software Development and Integration

SWDMX Software Demultiplexer

SWDP Software Development Plan

SWds StarWords (database)

SWDS Software Development System • Software Distribution System

SWDVS Software Development and Verification System

Swe Swedish

SWE Scalar Wave Equation • Shallow-Water Equation • Siwea (Papua New Guinea) (location identifier) • Snow Water Equivalent • Software Engineering • Solar-Wind Electron • Solar-Wind Experiment • Suède (code ISO de pays) • Sweden (ISO country code)

SWEAP SOTER Water Erosion Assessment Programme

SWEATS Software Engineering And Technical Services • Software Engineering And Technical Support

SWECC Software Engineering Coordinating Committee

Swed Sweden • Swedish

SWEDARP Swedish Antarctic Research Programme

SWEE Software Engineering Environment

SWEEP Structures With Error Expurgation Program

SWEN Space Weather Euro News

SWERVE Sandia Winged Energetic Reentry Vehicle Experiment

SWET Signal Weighted Extraction Technique • Slit-Weighted Extraction Technique

swf Shockwave Flash (file extension)

SwF Swiss Franc (currency)

SWF Newburgh Stewart (USA-NY) (location identifier) • Poughkeepsie Stewart (USA-NY) (location identifier) • Shortwave Fade • Shortwave Fading • Shortwave Fadout • Social Welfare Function • Staff Working File

SWFAS South West Florida Astronomical Society (USA-FL)

SWFC Sliding Window Flow Control

SWFD Selbstwählferndienst

SWFR Slow Write - Fast Read

SWFT Software Fault Tolerant

Swg Sewage

SWG Satwag (Papua New Guinea) (location identifier) • Scenario Working Group • Schedule Working Group • Science Working Group • Security Working Group • Shock Wave Generator • Shuttle Working Group • Simulation Working Group • Sine Wave Generator • Software Working Group • Special Working Group • Squarewave Generator • Standards Working Group • Standard Wire Gauge • Subworking Group • Synchronization Working Group

Swgr Switchgear

SWH Significant Wave Height • Soft Warehouse (USA-HI) • Solar Water Heating • Sternwarte Höfingen (Germany) • Swan Hill (Australia-VIC) (location identifier)

SWHAS South West Herts Astronomical Society (UK)

SWHE Soft Warehouse Europe

Swi Swiss • Switzerland

SWI Schlagwortindex • Sine Wave Inverter • Software Interrupt • Sternwarte Ilsede (Germany) • Swindon (UK-England) (location identifier)

SWICS Solar-Wind Ion Composition Spectrometer

SWIFET Shortwave Infrared Field Effect Transistor • Switched Infrared Field Effect Transistor

SWIFT Selected Words In Full Title • Significant Words In Full Title • Society for the Worldwide Interbank Financial Telecommunications • Southern Wide-Field Telescope • Spectroscopic Wide-Field Telescope • Stratospheric Wind Interferometer for Transport Studies

SWIL Software In the Loop

SwIM Southwest Initiative for Mars

SWIM Solar Wind Interplanetary Measurements • Spherical-Wave Interferometry Module • System-Wide Integrity Management

SWING Stage, Work placement and Internship Networking Group

SWIP Secret Work In Process • Southwestern Institute of Physics

SWIR Shortwave Infrared • Shortwave Infrared Radiometer • Shortwavelength Infrared

SWIRE SIRTF Wide-area Infrared Extragalactic (survey)

SWIRLS Stratospheric Wind Infrared Limb Sounder

SWIRNS Short Wave Infrared Nadir Spectrometer

SWIRT Supernova Watchdogging Infrared Telescope

SWIS Satellite Weather Information System • State-Wide Information System • Straßenzustands- und Wetterinformationssystem

SWISH Simple Web Indexing System for Humans

SWISS Satellite Weather Information for Stratospheric Science

SWISSAR Swiss Interdisciplinary Study with Synthetic Aperture Radar

SWISTO Swingby-Stopover Optimization

Swit Switzerland

SWIT Software Integration and Test

Switz Switzerland

SWJ South West Bay (Vanuatu) (location identifier)

SWK Milano Segrate (Italy) (location identifier) • Schlagwortkatalog

SWL Safe Working Load • Short Wavelength • Spanish Wells (Bahamas) (location identifier) • Sternwarte Langwedel (Germany)

SWLA Short Wavelength Large Aperture

SWLC Salt water Leak Check

SWlds StarWorlds (database)

SWLG Swelling

SWLIB Software Library

SWLS Superconducting Wavelength Shifter

SWLY Southwesterly

SWM Satisfaction With Monitoring • Space Weather Month • Suía-Missu (Brazil-MT) (location identifier)

SWMCT Switch Memory Control

SWMDP Software Management and Development Plan

SWML Stepwise Maximum Likelihood

SWMM Storm Water Management Model

SWMP Solid Waste Management Plan

SWMS Single Wafer Multiple Sensor

SWMST Switch Memory Storage

SWMU Solid Waste Management Unit

SWMUX Switch Multiplexer

SWN Sahiwal (Pakistan) (location identifier) • Scientist Workstation Network

SWNT Single-Walled Nanotube

SWO Space Weather Operations • Squarewave Oscillator • Staff Weather Officer • Standards-Writing Organization • Stillwater Searcy (USA-OK) (location identifier)

SWOB Salaries - Wages - Overhead - Benefits

SWOC Space Weather Operations Center (USA) • Subject-Word of Context

SWOOPS Solar-Wind Observations Over the Poles of the Sun

SWOP Second Window Of Protection • Specifications Web Offset Publication • Star Without Planets • Stereo Wave Observation Project • Structural Weight Optimization Program

SWOT Strengths - Weaknesses - Opportunities - Threats

swp Swap (file extension)

Swp Sweep

SWP Safe Working Pressure • Science Working Panel • Short-Wavelength Primary (camera) • Short-Wavelength Prime (camera) • Software Package • Southwest Pacific • Space - Weight - Power • Special Working Party • Star With Planets • Swakopmund (Namibia) (location identifier) • Swap • Sweep • System of Weight Penalty

SWPA Shuttle Wing Panel Analysis • Southwest Pacific Area

SWPC Shortwavelength Phenomenology Center

SWPDR Software Preliminary Design Review

SWPPP Storm Water Pollution Prevention Plan

SWPWR Switched Power

SWQ Sumbawa (Indonesia) (location identifier)

SWQE Software Quality Engineering

SWQT Software Qualification Test

SWR SAR Wave Mode • Schweizerischer Wissenschaftsrat (Switzerland) • Sea Water Return • Short-Wavelength Radar • Short-Wavelength Radiation • Short-Wavelength Redundant (camera) • Silur (Papua New Guinea) (location identifier) • Sine Wave Response • Standing Wave Ratio • Stepwise Regression

SWRAL Southwest Region of the Astronomical League (USA)

SWRC Strategic Workshop Review Committee

SWRD Software Requirements Document

SwRI Southwest Research Institute (USA-TX)

SWRIMS Southwest Regional Institute for Mathematical Sciences (USA)

SWRR Software Readiness Review • Software Requirements Review

SWRU Software Replaceable Unit

SWS SAS Web Server • Scientific Workstation Support • Semesterwochenstunden • Short-Wavelength Spectrometer • Short-Wave Spectrum • Software System • Solar Wind Spectrometer • Static Water Supply • Swansea Fairwood (UK-Wales) (location identifier) • Switch Scan

SWSA Short Wavelength Small Aperture • Solid Waste Storage Area • Space West Science Activities Inc. (Canada-SK)

SWSL Supplemental Weather Service Location

SWSR Sum of the Weighted Squares of the Residuals

SWSU Southwest State University (USA-MN)

SWT Science Working Team • Shortwave Transmitter • Small Weather Terminal • Southwest Texas (USA) • Sternwarte Trier (Germany) • Strezhevoy (Russia) (location identifier) • Strzhewoi (Russia) (location identifier) • Supersonic Wind Tunnel

SWTP Software Test Plan

Swtr Sea Water

SWTR Surface Water Treatment Rule

SWTSU Southwest Texas State University (USA-TX)

Swtz Switzerland

SWU Separative Work Unit • Su Won City (Korea-ROK) (location identifier) • Switching Unit

SWUIS Southwest Ultraviolet Imaging System

SWV Software Verification

SWW Severe Weather Warning • Space Weather Week • Sweetwater (USA-TX) (location identifier)

SWWAS Southwest Washington Astronomical Society (USA-WA)

SWWBDS Software Work Breakdown Structure

SWWG Software Working Group

SWX Shakawe (Botswana) (location identifier)

Swy Stopway

SWY Sitiawan (Malaysia) (location identifier)

SWys StarWays (database)

SWZ Special Watch Zone • Swaziland (ISO country code)

Sx Sussex (UK-England)

SX Essex (UK-England) • Greece (international aircraft registration code) • Janus (satellite of Saturn) • Saturday Excepted • Science Archive • Simplex • Sioux • Soft X-ray • Space Experimental • Stability Index • Sundays Excepted • Switching Exchange

SXA Sialum (Papua New Guinea) (location identifier) • Stored Index to Address

SXAO Shaanxi Astronomical Observatory (China-PRC)

SXAPS Soft X-ray Appearance Potential Spectrometer • Soft X-ray Appearance Potential Spectroscopy

SXB Soft X-ray Band • Strasbourg Entzheim (France) (location identifier)

SXC Catalina Island (USA-CA) (location identifier) • Soft X-ray Camera (HETE-2)

SXD Self-Examining Device • Sophia Antipolis (France) (location identifier)

SXE Sale (Australia-VIC) (location identifier) • Soft X-ray Experiment

SXES Soft X-ray Electron Spectroscopy

SXF Berlin Schönefeld (Germany) (location identifier) • Solid Exchange Format

SXG Senanga (Zambia) (location identifier) • Spectrum X-Gamma

SXG-CF Spectrum X-Gamma Coordinating Facility

SXGCF Spectrum X-Gamma Coordinating Facility

SXH Sehulea (Papua New Guinea) (location identifier)

SXI Epimetheus (satellite of Saturn) • Sirri Island (Iran) (location identifier) • Solar X-ray Imager • Switching Exchange Interface

SXII Helene (satellite of Saturn)

SXIII Telesto (satellite of Saturn)

SXIV Calypso (satellite of Saturn)

SXJ Shanshan (China-PRC) (location identifier)

SXK Saumlaki (Indonesia) (location identifier)

SXL Sligo (Ireland) (location identifier)

SXLS Superconductivity X-ray Lithography Synchrotron

SXM Self-Examining Mechanism • Sint Maarten - Juliana (Netherlands Antilles) (location identifier)

SXN Section • Suapan (Botswana) (location identifier)

SXNS Sections

SXO São Félix do Araguaia (Brazil-MT) (location identifier)

SXP Sheldons Point (USA-AK) (location identifier) • Solar X-ray Photometer

SXPS Soft X-ray Photoemission Study

SXQ Soldotna (USA-AK) (location identifier)

SXQL Science Archive Query Language

SXR Soft X-Ray • Soft X-ray Region • Srinagar (India) (location identifier)

SXRB Soft X-Ray Background

SXRP Stellar X-Ray Polarimeter

SXRT Soft X-Ray Transient

SXS Sahabat 16 (Malaysia) (location identifier) • Stellar X-ray Spectrum

SXT Sextant • Soft X-ray Telescope • Soft X-ray Transient • Spectroscopy X-ray Telescope • Stable X-ray Transmitter • Taman Negara (Malaysia) (location identifier)

SXTF Satellite X-ray Test Facility

SXU Saint Xavier University (USA-IL) • Soddu (Ethiopia) (location identifier)

SXV Atlas (satellite of Saturn) • Self-Expanding Virus

SXVI Prometheus (satellite of Saturn)

SXVII Pandora (satellite of Saturn)

SXVIII Pan (satellite of Saturn)

SXW Sauren (Papua New Guinea) (location identifier) • Self-Explanatory Wording • Soft X-ray Weak (object)

SXWQ Soft X-ray Weak Quasar

SXX São Félix do Xingu (Brazil-PA) (location identifier)

SXY Sidney (USA-NY) (location identifier)

SXZ Self-Explanatory Zone • Siirt (Turkey) (location identifier)

Sy Seychelles • Seyfert (galaxy) • Supply • Surrey (UK) • Syria

SY School Year • Seasonal Year • Seychelles (international vehicle registration code) • Shoulder Yaw • Square Yard • Syria (ISO country code) • Syrie (code ISO de pays) • System

SYA Shemya (USA-AK) (location identifier) • State Your Answer

SYB Seal Bay (USA-AK) (location identifier)

SYC Seychelles (ISO country code) • Shiringayoc (Peru) (location identifier) • Static Yard Capacity

SYCOM Synchronous Communications • Systems Command

Syd Sydney (Australia-NSW)

S. Yd. Scotland Yard

SYD Sum of the Year's Digits • Sydney - Kingsford Smith Mascot (Australia-NSW) (location identifier)

SYDAS System Data Acquisition System

SydOP Sydney Observatory Papers (periodical code)

SYE Sa Dah (Yemen) (location identifier)

SYG Secretary General

SYH Syangboche (Nepal) (location identifier)

SYI Sediment Yield Index • Shelbyville Bomar (USA-TN) (location identifier) • Synchronization In

SyK Systematischer Katalog

SYK Stykkishólmur (Iceland) (location identifier)

Sy'kat Syarikat

Syl Syllable • Syllabus

SYL San Miguel - Roberts (USA-CA) (location identifier)

SYLDA Système de Lancement Double Ariane

SYLK Symbolic Link

Syll Syllable • Syllabus

Sym Symbol • Symbolic • Symmetrical • Symmetry • Symphony • Symptom

SYM Simao (China-PRC) (location identifier) • Source - Year - Month • Supersymmetric Yang-Mills • Symbol • Symbolic • Symbology

Symb Symbiotic

Symm Symmetrical

SYMM Symmetrical

Symp Symposium

Symph Symphonic • Symphony

Syn Synchronize • Synonym • Synonymous • Synonymy • Synthetic

SYN Stanton Carleton (USA-MN) (location identifier) • Synchronization • Synchronizing • Synchronous • Synonym • Synthetic

Sync Synchronization • Synchronous • Synchronize

SYNC Synchronization • Synchronize • Synchronized • Synchronizing • Synchronous

Synch Synchronous

SYNCOM Synchronous Communications

Synd Syndicate • Syndicated

SYNMOD Synoptic Monitoring of Orbital Debris

Synon Synonymous

Synop Synopsis

SYNOP Synoptic • Synoptic Operation

SYNS Synopsis

Synth Synthèse • Synthesis • Synthetic • Synthétique

SYO Sakata-shi (Japan) (location identifier) • Shonai (Japan) (location identifier) • Synchronization Out

SYP Pound (Syria) (ISO currency code) • Society of Young Publishers (UK)

Syr Syria • Syrian • Syrup

SYR Syracuse (USA-NY) (location identifier) • Syria (international vehicle registration code) • Syria (ISO country code) • Syrie (code ISO de pays)

SYRIANAIR Syrian Arab Airlines

SYRUP System Resource Utilization Package

sys System Module (file extension)

Sys System

SYS Sunchon Yosu (Korea-ROK) (location identifier) • System

SYSADMIN System Administrator

SYSCOM Systems Command

SYSCON Systems Control

SysCoP System for Copyright Protection

SYSEX System Exclusive • System Exercise

SYSGEN System Generation • System Generator

SYSIM System Simulation

SYSIN System Initialization • System Input

SYSLOG System Log

SYSMOD System Modification

SYSOP System Operator

SYSOUT System Output

SYSPAC System Safety Policy Action Committee

SYSPOP System Programmed Operator

SYSREQ System Request

Syst System • Systematic • Systématique • Système

SYST System

SYSTEP System Test and Evaluation Plan

SYSTO System Officer

SYSTRAN Systems Translator

SYSVER System Verification

SYT Saint-Yan - Charolais Bourgogne (France) (location identifier)

SyTec Systems Technology (periodical code)

SYU Warraber Island (Australia-QLD) (location identifier)

SYV Sylvester (USA-GA) (location identifier)

SYW So You Wish

SYX Sanya (China-PRC) (location identifier)

SYY Stornoway (UK-Scotland) (location identifier)

SYZ Shiraz (Iran) (location identifier)

sZ seinerzeit

SZ Size • Sonnenzirkel • Southern Zodiacal • Sternzeit • Süddeutsche Zeitung (Germany) • Sunyaev-Zel'dovich • Surf Zone • Swaziland (ISO country code) • Switzerland

SZA Solar Zenith Angle • Soyo (Angola) (location identifier) • Sunyaev-Zel'dovich (effect) Array • Sun Zenith Angle

SZB Silver-Zinc Battery • Single Zone Boundary • Stadt- und Zentralbibliothek

SZC Santa Cruz - Guanacaste (Costa Rica) (location identifier)

SZE Shared Zone Entry • Sunyaev-Zel'dovich Effect

SZF Single Zero Failure

SZG Salzburg (Austria) (location identifier)

SZH Senipah (Indonesia) (location identifier)

SZJ Single Zone Journey

SZK Skukuza (South Africa) (location identifier)

SZL Lilangeni (Swaziland) (ISO currency code) • Single Zone Licence • Warrensburg Whiteman (USA-MO) (location identifier)

SZM Shared Zone Management • Shielded Zone of the Moon • Stereo Zoom Microscope • Stereo Zoom Microscopy

SZN Santa Barbara - Santa Cruz Island (USA-CA) (location identifier) • Shared Zone Neutrality

SZO Single Zone

SZP Santa Paula (USA-CA) (location identifier)

SZQ Saenz Peña (Argentina-BA) (location identifier)

SZR Stara Zagora (Bulgaria) (location identifier)

SZS Saverni Zavod za Standardizaciju (Yugoslavia-Serbia) • Stewart Island (New Zealand) (location identifier) • Yugoslav Institute of Standardization

SZT Single Zone Territory • Size Threshold

SZTS Stereo Zoom Transferscope

SZU Segou (Mali) (location identifier)

SZV Shared Zone Violation • Suzhou (China-PRC) (location identifier)

SZW Size Weight

SZX Shenzhen (China-PRC) (location identifier)

SZZ Szczecin (Poland) (location identifier)

S1G Seyfert 1 Galaxy

S2 Bangladesh (international aircraft registration code) • Diatomic Sulfur • Security and Survivability • Space Surveillance • Spread Spectrum • Survivability and Security

S2E Seamless Synthetic Environment

S2N2 Space Science Network Northwest (USA)

S2P Station de Planétologie des Pyrénées (France)

S3 Semiconductor Serial Storage • Serial Signaling System • Sortie Support System • Spaceborne Sensor-to-Shooter • Space Surveillance System

S3C Service Central de la Communication et de la Commercialisation

S3I Sensor - Signatures - Signal - Information

S3OS2 Small Solar System Objects Spectroscopic Survey

S3R Sequential Switching Shunt Regulator

S5 Slovenia (international aircraft registration code)

S7 Seychelles (international aircraft registration code)

S9 Sao Tome and Principe (international aircraft registration code)

T

t Age • Celsius Temperature • Cosmic Time • Duration • Hour Angle • Metric Ton • Student's Distribution • Table • Tabulated • Täglich • Taken from • Tare • Tariflich • Technisch • Temperature • Tempo • Temporal • Tempore • Tense • Terminación • Terminal • Territorial • Territory • Tertiary • Thunder • Thunderstorm • Tief • Tiempo • Time • Time Interval • Time Scale • Tome • Tomo • Tomus • Ton • Tonelada • Tonne • Tonneau • Total • Tour • Town • Township • Transit • Transitive • Transport Number • Troy • Tun • Turn • Tutti

T Absolute Temperature • Absolute Time • Couple Moment • Hundred Sixty (Roman numeral) • Internal Transmittance • Kinetic Energy • Kinetic Temperature of Free Particles • Meridian Angle • Moment of a Couple • Orbital Period • Period • Periodic Time • Reactor Time Constant • Reverberation Time • Static Temperature • Surface Tension • Table • Tableau • Tactical • Tactics • Tactique • Tag • Takt • Tal • Tall • Tangage • Tango (international alphabet) • Tank • Tankstelle • Target • Tarif • Tasse • Tat • Tau • Taufe • Tausch • Tausend • Tautomeric Effect • Taxe • Té • Teacher • Technicien • Technik • Teil • Teilung • Telefon • Telegraf • Telegraph • Telegraphic • Telephone • Temperatur • Température • Temperature • Tempo • Temporary • Temps • Tension • Tensor • Tera • Termin • Terminal • Terrain • Terre • Terrestre • Territorial • Territory • Tertiaire • Tesla • Test • Testament • Tête • Text • Thailand (international vehicle registration code) • Thermodynamic Temperature • Thermometer • Threonine • Thrust • Thunderstorm • Thursday • Thymine • Tief • Tiefdruck • Tiefe • Tiefgang • Time • Time Reversal • Tirant • Tireur • Tirol (Austria) • Titel • Titus • Tochter • Toise • Tolérance • Tome • Tomus • Tonikum • Top • Tor • Torpedo • Torque • Torre • Total Light Transmission • Tour • Tourbillon • Touren • Tourist • Touristenklass • Township • Trace • Traceur • Tracté • Tracteur • Traction • Train • Traînée • Tranchée • Transaction • Transducteur • Transformateur • Transformer • Transistor • Transition • Translation • Transmetteur • Transmissibility • Transmission • Transmittance • Transport • Transportation • Transverse • Traveller • Treasury • Tribord • Triebwagenzug • Trillion • Tritium • Tropical • Troupe • True • Tube • Tubulaire • Tuesday • Tür • Turbine • Turboprop • Turbopropulseur • Turbulent • Turkish • Turm • Turnen • Turning • Turnus • Tuyère • Twist • Twisting Number • Typ • Type • Typus • 160 (Roman numeral)

ta Tamil (ISO language code)

Ta Tank • Tanne • Tantal • Tantale • Tantalum • Tanz • Tara • Tarif • Tasche • Tasse • Taste • Température Ambiante • Température d'Arrêt

T/A Temporary Assistant • Training As • Turnaround

T&A Test & Adjustment • Time & Attendance • Traffic & Analysis • Transcription & Analysis

TA Tableau d'Avancement • Tableau des Alarmes • Table of Allowance • Tables Astronomiques • Tanzania • Tape Adapter • Tape Advance • Target • Target Acquisition • Target Analysis • Target Area • Target Audience • Tarifangestellter • Taschenausgabe • Taschenbuchausgabe • Task Analysis • Task Assessment • Task Assignment • Task Authorization • Taux d'Amortissement • Taxe d'Apprentissage • Teaching Assistant • Teal Amber • Team Activity • Technical Advisor • Technical Analysis • Technical Architecture • Technical Area • Technical Assessment • Technical Assessor • Technical Assistance • Technical Assistant • Technicien d'Atelier • Technische Abteilung • Technische Akademie • Technische Anleitung • Technische Anweisung • Technischer Angestellter • Technischer Assistent • Technischer Ausschuß • Technology Assessment • Teilamt • Teilnehmerausweis • Telefon-Anrufbeantworter • Telegrammadresse • Telegraphenamt • Telegraphic Address • Telemetry Address • Téléphone Automatique • Telescope Array • Telescope Assembly • Température Absolue • Temperature Anisotropy • Température d'Arrêt • Temporary Admission • Tension Artérielle • Tension Auxiliaire • Tension d'Asservissement • Terme Associé • Terminal Adapter • Terrain Avoidance • Test Accessory • Test Announcer • Test Article • Text Analysis • Textausgabe • Thermal Analysis • Timbre Amende • Time Actual • Time and Attendance • Timing Advance • Titulaire Académique • Tonabnehmer • Tonne d'Affrètement • Torpille Aérienne • Total Abstinence • Toutes Armes • Trace Analytical • Trachée-Artère • Tracteur Automobile • Traction Avant • Trade Association • Traduction Automatique • Traffic Advisory • Traffic Analysis • Traffic Authority • Trailing Arm (galaxies) • Train Auxiliaire • Train Avant • Train d'Atterrissage • Traînée Aérodynamique • Training Adviser • Training Agency • Training Agent • Traité d'Alliance • Transaction Agent • Transactional Analysis • Transaktionsanalyse • Transfer Agent • Transfer Aisle • Transfer Allowed • Transformateur d'Alimentation • Transit Authority • Transition Altitude • Translation Ambiguity • Transmetteur Automatique • Transport Aérien • Transportability Approval • Transportation Authority • Transverse Acoustic • Travel Allowance • Travel Authorization • Traveling Allowance • Tribunal Administratif • Trip Authorization • Trou d'Air • Troupes Auxiliaires • True Altitude • True Anomaly • Trunk Asynchronous • Trunnion Angle • Turbine Aérienne • Turbulence Amplifier • Turnaround • Turnover Agreement • Type Approval • Type Assistance

TAA Tactical Assembly Area • Tactical Automation Appraisal • Tarapaina (Solomon Islands) (location identifier) • Technical Advisory Area • Technical Area Advisor • Technical Assistance Administration • Technical Assistance Agreement • Temporary Access Authorization • Three-Axis Accelerometer • Total Actual Authority • Trade Agreement Act (USA) • Transatlantic Agreement • Trieste Aurélie Archive • Tulare Astronomical Association (USA-CA)

TAAA Tucson Amateur Astronomy Association (USA-AZ)

TAATS The Australian Advanced Air Traffic System

TAAC Technology Assessment Advisory Council • Triad Action Astronomy Club (USA-NC)

TAACOS Target Acquisition Analysis for Cosmic Origins Spectrograph

TAAD Task Assignment And Directive

TAAF Terres Australes et Antarctiques Françaises • Test, Analyze And Fix

TAAM Total Airspace and Airport Modeller

TAAP Tethered Aerostat Antenna Program • Three-Axis Antenna Positioner • Transient Analysis Array Program

TAAS Telescope Assembly Alignment Simulator • Terminal Advance Automation System • Thames Amateur Astronomical Society (USA-CT) • The Albuquerque Astronomical Society (USA-NM) • Three-Axis Attitude Sensor • Triple Azimuth Angle Scan

TAAU Telescope Assembly Alignment Unit

tab Tabella • Tabulate

Tab Tabelle • Tabla • Table • Tabular • Tabulate • Tabulated • Tabulation • Tabulator

TAB Tabasco (Mexico) • Tab-Automated Bonded • Tabulation • Tabulator • Tape Automated Bonding • Targets And Background • Technical Abstract Bulletin • Technical Advisory Board • Technical Assistance Board • Technology Advisory Board • Technology Assessment Board • Telecommunications Advisory Board • Temporarily Able

Body • Time After Burst • Tobago (Trinidad & Tobago) (location identifier) • Total Allocated Budget • Training Aid Booklet • Transportes Aéreos Bolivianos • Transports Aériens du Bénin

TABD Transatlantic Business Dialogue

TABE Test of Adult Basic Education

TABI Tailorable Advanced Blanket Insulation

Tabl Tablett • Tablette

TABL Tableau

TABLE Tsukuba Atmospheric Boundary Layer Experiment

TABR Targeted Area Background Report

TABS Terminal Access to Batch Service • Time Analysis and Billing System

TABSIM Tabulator Simulator

TABSOL Tabulator-System-Oriented Language

Tabu Taschenbuch

tac Tactical • Tactics

Tac Tactical • Tactics

TAC Tacloban (Philippines) (location identifier) • Tactical • Tactical Advanced Computer • Tactical Air Command • Tactical Air Control • Tactical Air Controller • Tactical Analysis Center • Taiwan Aerospace Corporation • Target Allocation Committee • Target Analysis Center • Technical Advisory Committee • Technical Advisory Council • Technical Assistance Center • Technical Assistance Committee • Technical Assistance Contract • Technical Assistance Council • Technicien de l'Aviation Civile • Technology Advisory Council • Technology Application Center • Technology Assessment Center • Telecommunications Access Controller • Telecommunications Advisory Committee • Telemetry And Command • Telescope Allocation Committee • Tendeur Automatique de Câbles • Terminal Access Center • Terminal Access Circuit • Terminal Access Control • Terminal Access Controller • Terminal Area Chart • Terrain Analysis Center • Test Advisory Committee • Test And Configuration • Tester Analyzer Controller • The Astronomy Club (USA-CA) • Theoretical Astrophysics Center • Tilted Aligned Coupling • Time Allocation Committee • Time Assignment Committee • Time At Completion • Time-to-Amplitude Converter • Timing and Address Counter • Tomografía Axial Computerizada • Tool for Advance Control • Total Access Communication (system) • Total Accumulated Cycles • Total Allowable Catch • Total Average Cost • Traceability And Configuration • Tracking Accuracy Control • Trading Agent Competition • Training - Advising - Counseling • Transformer Analog Computer • Transistorized Automatic Control • Transition Assurance Center • Translator - Assembler - Compiler • Transportation Account Code • Transportes Aéreos Coyhaique • Trapped Air Cushion • Trouble Analysis Chart • Turbulence en Air Clair • Twin Astrograph Catalog • Type of Activity Code • Type of Address Code

TACA Transportes Aéreos Centroamericanos

TACACS Terminal Access Controller Access Control System

TACAMO Take Charge And Move Out

TACAN Tactical Air Control And Navigation • Tactical Air Navigation

TACANS Tactical Air Control And Navigation System

TACAS Tools and Algorithms for the Construction and Analysis of Systems (workshop)

TACC Tactical Air Control Center • Technical Assistance Coordination Committee • Tracking And Control Center

TACCS Time, Age, and Cycle Control System

TACE Technical Analysis and Cost Estimate • Technical Assessment and Cost Estimate • Transmission Alarm and Control Equipment

TACF Temporary Alteration Control Form

Tach Tachometer

TACH Tachometer • Tachymètre

TACHIR Tool for Automatic Construction of Hypertexts for Information Retrieval

Tacho Tachometer

TACIS Technical Assistance to the Commonwealth of Independent States

TACIT Theory and Applications of Continuous Iteration Techniques (project)

TACL Time And Cycle Log

TACMO Take Charge And Move Out

TACO Test And Checkout • Test And Checkout Operation

TACODA Target Coordinate Data

TACOL Thinned Aperture Computed Lens

TACOS Test And Checkout Satellite • Time Allocation Committee Operations Support

TACP Tactical Air Control Party • Technical Analysis of Cost Proposal • Technology Assessment and Control Plan • Training And Certification Plan

TACR Time And Cycle Record

TACS Target Acquisition and Control System • Technical Assignment Control System • Test Assembly Conditioning Station • Theoretical Aspects of Computer Software (conference) • Thruster-Actuated Control System • Thruster Attitude Control System • Time and Attendance Collection System • Total Access Communications System • Transportable Atmospheric Characterization Station

TACT Total Audit Concept Technique • Transactional Analysis in Customer Treatment • Transistor And Component Tester • Transsonic Aircraft Technology • Truc À Côté de la Télé

TACTEC Totally Advanced Communications Technology

TACTIC TeV Atmospheric Čerenkov Telescope with Imaging Camera

TACUP Technical Advisory Committee and University Programs

TACV Tracked Air-Cushion Vehicle

TAD Tabular Application Development • Target Acquisition and Designation • Target Acquisition Data • Target Assembly Data • Target Audience Description • Task Assignment Directive • Task Assignment Drawing • Teaching for Astronomy Development (*) • Teaching in Astronomy Development • Technical Acceptance Date • Technical Analysis Division • Technical Assistance Data • Technology Architecture Division • Technology Area Description • Telephone Answering Device • Temperature And Dewpoint • Temporary Additional Duty • Terrain Analysis Data • Test And Development • Thermal Analysis Data • Throw Away Detector • Timbre À Date • Time Array Detection • Time Available for Delivery • Top Assembly Drawing • Trailing Arm Drive • Traitement Automatique de Documents • Traitement Automatique de Données • Trajectoire d'Alignement de Descente • Transport Automatique de Documents • Traveling Around Drunk • Traveling Atmospheric Disturbance • Trinidad - Las Animas (USA-CO) (location identifier) • Turbo Alternator Drive • Type Abstrait de Données

TADAT Tauern-Dachstein Testite

TADB Terrain Analysis Database

TADC Tunable Atmospheric Dispersion Compensator

TADI Telemetry Acquisition Data Interface

TADIC Telemetry Analog-to-Digital Information Converter

TADIL Tactical Digital Information Link

TADNET Test And Development Network

TADP Terminaison Abonné à Débit Primaire

TADS Target Acquisition and Designation System • Target Acquisition and Detection System • Target Acquisition Designation System • Teletypewriter Automatic Dispatch System • Text Analysis, Description and Synthesis • Tracking And Display System • Trade and Analysis Data System • Training Analysis Data Sheet • Translation Assistant Dictionary System • Typewriter Automatic Dispatch System

TADSS Training Aids, Devices, Simulators and Simulations

TAE Taegu (Korea-ROK) (location identifier) • Tasa Anual Equivalente (Spain) • Techniques d'Applications Énergétiques • Technische Akademie Esslingen (Germany) • Telekommunikationsanschlußeinheit • Test And Evaluation • Text Analysis Module • Toroidal Alfvén Eigenmodes • Toroidicity-induced Alfvén Eigenmodes • Trait d'Atterrissage Escamotable • Transportable Applications Environment • Transportable Applications Executive

TAECG Transportable Applications Environment Code Generator

TAEG Taux Actuariel Effectif Global • Training Analysis and Evaluation Group

TAEHB Terminal-Age Extreme Horizontal Branch

TAEM Terminal Area Energy Management

TAEMm Tokyo Metropolitan College Aeronautical Engineering Memoirs (periodical code)

TAER Telemetry Analog Equipment Room • Time - Azimuth - Elevation - Range

TAES Técnicas Aéreas de Estudios y Servicios SA • Transportes Aéreos de El Salvador SA de CV

TAEWB Transportable Applications Environment Workbench

Taf Tafel

TAF Oran Tafaraoui (Algeria) (location identifier) • Task Analysis Form • Terminal Aerodrome Forecast • Terminal Area Forecast • Test - Analyze - Fix • Training Analysis and Feedback • Trait d'Atterrissage Fixe • Truss Assembly Fixture

TAFE Technical And Further Education

TAFG Two-Axis Free Gyro

TAFIM Technical Architecture For Information Management

TAFM Tunneling Atomic Force Microscopy

TAFT Technical Assistance Field Team • Test - Analyze - Fix - Test

TAFTA Transatlantic Free-Trade Agreement • Transatlantic Free-Trade Area

TAFTIE The Association For Technology Implementation in Europe

TAFTS Tropospheric Airborne Fourier Transform Spectrometer

TAG Tabular And Graphic • Tagbilaran (Philippines) (location identifier) • Technical Advisory Group • Technical Air-to-Ground • Technical Assistance Group • Techniques d'Avant-Garde • Teilnehmeranschlußgerät • Telemetry-Archive Gateway • Telescope Allocation Group • Term Association Graph • Test Analysis Guide • Time Allocation Group • Time Assignment Group • Topic Area Guide • Training Advisory Group • Transient Analysis Generator • Traveling Astronomer's Guide • Tree-Adjoining Grammar • Troupes Auxiliaires du Génie • Two-Axis Gimbal

TAGA Trace Atmospheric Gas Analyser

TAGB Tilt Axis Gearbox

TAGD Technologie de Groupe Assisté par Ordinateur

TAGIU Tracking And Ground Instrumentation Unit

Tägl Täglich

TAGS Technique for Automated Generation of Systems • Technology for Automatic Generation of Systems • Text And Graphics System • Thermal Analysis Graphics System • Total Absorption Gamma-ray Spectrometer • Training And Guided Study

TAGT Theory and Application of Graph Transformations (workshop)

TAH Tanna (Vanuatu) (location identifier)

TAHD Tunnel Aérohydrodynamique

TAHG Target Ad Hoc Group

Tai Taiwan (China-ROC)

TAI Taiz (Yemen) (location identifier) • Target Area of Interest • Targeted Area of Interest • Temps Atomique International • Tiempo Atómico Internacional • Time to Autoignition • Total Active Inventory • Traitement Automatique d'Information • Transmission Automatique Informatique • Transverse Acceleration Instability

TAIDET Triple-Axis Inertial Drift Erection Test

TAIG Test and Analysis Integration Group

TAIMS Three-Axis Inertial Measurement System

TAIR Test And Inspection Record • Test Assembly Inspection Record

TAIS Target Architecture and Implementation Strategy • Technology Application Information System • Transportation Automated Information System

TAJ Aitapa Tadji (Papua New Guinea) (location identifier)

TAK Takamatsu (Japan) (location identifier) • Tierps Astronomiska Klubb (Sweden) • Trainer Appraisal Kit

TAKO Timeline Assembler, Keyword-Oriented

tal Talis

TAL Tanana (USA-AK) (location identifier) • Tandem Accelerator Laboratory • Task Assignment List • Teilnehmeranschlußleitung • Terminal Application Language • Trait d'Atterrissage Largable • Traitement Automatique des Langues • Transatlantic Abort Landing • Transatlantic Landing • Transoceanic Abort Landing • Turbulence en Air Limpide

TalAE Taligent Application Environment

TALAR Tactical Approach and Landing Radar

TALC Text Alignment and Line Consistency • Très À La Carte

TALDT Total Administrative and Logistic Downtime

TALES The Association of Library Equipment Suppliers (UK)

TALI Transfer Adjacent Layer Interface

TALIF Two-photon Atom Light Induced Fluorescence

TALIP Transactions on Asian Language Information Processing (ACM) (periodical code)

TALK Titles Alphabetically Listed by Keyword

TALN Traitement Automatique du Langage Naturel

TALO Time After Lift-Off

TALON Texas - Arkansas - Louisiana - New Mexico (USA)

TALOS Tropical Atmospheric Lidar Observations from Space

Tal. qual. Talis Qualis

Talsp Talsperre

Tam Tamil

TAM Table Assignment Matrix • Taipei Astronomical Museum (China-ROC) • Tampico (Mexico) (location identifier) • Tangent Approximating Manifold • Task Assignment Memorandum • Task Authorization Memorandum • Technical Account Manager • Technicien Agent de Maîtrise • Technology Acceptance Model • Telecommunications Access Method • Telephone Answering Machine • Television Audience Measurement • Temps d'Arrêt Moteur • Thermal Analytical Model • Three-Axis Magnetometer • Throw-Away Maintenance • Thrust Area Manager •

Trainer-Akademie München (Germany) • Trajectory Application Method • Transmission Automatique de Messages • Transportation Aircraft Maintenance • Triangle Amplitude Modulation • Tripartite Agreed Materiel • Tropical Asian Monsoon • Type-Approval Model • Typed Access Matrix

TAMA Technical Assistance and Manufacturing Agreement

TAMAC Three-Axis Manual Attitude Controller

Tamb Tambour

TAMB Température Ambiente

TamCo Tampa Astronomical Contributions (periodical code)

TAMER Testing, Analysis, and Measurement Environment for Robustness

TAMEX Taiwan Area Mesoscale Experiment

TAMMS Transportation Maintenance Management System

TAMOC Theoretical Atomic and Molecular Community

TAMPA Transportes Aéreos Mercantiles Panamericanos SA

TAMPER Tables for Approximation of Midpoints for Exponential Regression

TAMPS Tamaulipas (Mexico)

TAMR Telecommunication Assessment Monitoring Report • Tripartite Agreed Materiel Requirement

TAMS Target Area Meteorological Sensor • Teachers Academy for Mathematics and Science (USA-IL) • Technical and Acquisition Management Support • Technical Assistance and Management Services • Télémétrie Aéroporté des Mouvements du Sol • Telenet Access Management System • Terminal-Age Main Sequence • Termination-Age Main Sequence • Termination Automated Management System • Total Airport Management System • Trade and Analysis Management System • Transient Amplitude Monitor System

TAMSAT Tropical Agricultural Meteorology using Satellite

TAMU Texas A&M University (USA-TX)

TAMUCC Texas A&M University at Corpus Christi (USA-TX)

tan Tangent

TAN Tangalooma (Australia-QLD) (location identifier) • Test Advisory Notice • Total Area Networking • Traité de l'Atlantique Nord • Transaction Number • Twilight All Night

TANAGRA Technique d'Analyse Graphique

tanalt Tangent Altitude

TANEs Tartu Akademiia Nauk Estonskoi SSR (periodical code)

TANet Taiwan Academic Network

Tang Tanganika • Tangente • Tangentiel • Tanger (Morocco)

tanh Hyperbolic Tangent

TANS Terminal-Area Navigation System • Trimble Advanced Navigation Sensor

TANSTAAFL There Are No Such Things As A Free Lunch

TAO Qingdao (China-PRC) (location identifier) • Technical Applications Office • Technical Assistance Operation • Technical Assistance Order • Technology Applications Office • Teilnehmer-Anschlußorgan • Télémanipulation Assistée par Ordinateur • Telephony Application Object • Test Analysis Outline • THEMIS Adaptive Optics • Tokyo Astronomical Observatory (now NAOJ) • Tokyo Atacama Observatory (Chile) • Tomographie Acoustique Océanique • Toyama Astronomical Observatory (Japan) • Track At Once • Tradeable Absorption Obligation • Traduction Assistée par Ordinateur • Transport in the Atmospheres and the Oceans • Trieste Astronomical Observatory (Italy) • Tropical Atmosphere Ocean • TUVA Astronomy Organization (USA-OK)

TAOF Tunable Acousto-Optical Filter

TAOS Taiwanese-American Occultation Survey • Technology for Autonomous Operational Survivability • Thrust-Assisted Orbiter Shuttle

TàP Tirage-à-Part • Tiré-à-Part

TAP Table of Authorized Personnel • Tapachula (Mexico) (location identifier) • Target Analysis and Planning (system) • Target Approach Point • Target Array Payload • Task Area Plan • Team Acquisition Process • Technical Acceptance Plan • Technical Achievement Plan • Technical Advancement Program • Technical Advisory Panel • Technical Assistance Plan • Technical Assistance Program • Technical Assistance Project • Technological Assistance Program • Technology Application Program • Technology Area Plan • Telelocator Alphanumeric Protocol • Telematics Applications Programme • Telemetry Acceptance Pattern • Telemetry Antenna Pedestal • Telemetry Archive Process • Telephone Application Processing • Ter Aangehaalder Plaatse • Terminal Access Point • Terminal Applications Package • Terminal Area Productivity • Terrestrial Auxiliary Power • Test Access Port • Test Administration Plan • Thermal Analysis Program • Thermal Analyzer Program • Thermal-Array Probe • Thermoviscoelastic Analysis Program • Three-Axis Package • Timesharing Assembly Program • Timeslot Access Point • Total Air Pressure • Traitement Automatique de la Parole • Trajectory Analysis Program • Trans-Alaska Pipeline • Transarabian Pipeline • Transboundary Air Pollution • Transit Access Point • Transition Assistance Program • Transports Aériens Privés • Trend Analysis Program • Tripartite Advisory Publication • Troupes Aéroportées

TApA Top. Astrophys. Astron. Space Sci. (periodical code)

TAPAC Tape Automatic Positioning And Control • Transportation, Allocations, Priorities And Controls

TAPAT Tape Programmed Automatic Tester • Technology Application Process Action Team

TAPC Terrestrial Aeronautical Public Correspondence • Traitement Automatique de la Parole Continue

TAPCIS The Access Program for CompuServe Information Service

TAPE Tape Automatic Preparation Equipment • Teaching Aids in Physics Education • Telematics Automated Proposal Editor • Tentative Annual Planning Estimate • Totally Automated Programming Equipment • Tripartite Agreed Plan of Engineering

TAPER Temporary Appointment Pending Establishment of Register

TAPES Tactical Automated Planning and Engineering System

TAPI Technology Application and Promotion Institute • Telephony Applications Program Interface • Telephony Applications Programming Interface (*)

TAPIR Theoretical Astrophysics Including Relativity

TAPM Technology Application Program Management

TAPO Technology Applications Program Office

TAPS Target Analysis and Planning System • Telemetry Automated Processing System • Terminal Applications Processor System • Terrain Analysis Production System • Time Analysis of Program Status • Tracking And Positioning System • Trans-Alaska Pipeline System • Transactions of the American Philosophical Society (periodical code) • Turbo-Alternator Power System • Two-Arm Photon Spectrometer • Two-Axis Pointing System

TAQ Tarcoola (Australia-SA) (location identifier)

TAQL Task Acquisition Language

tar Tape Archive (file extension)

Tar Tara • Tariff

TAR Tape Address Register • Tape Archive • Tape Archiver • Taranto (Italy) (location identifier) • Technical Acceptance Recommendation • Technical Action Request • Technical-Analysis Request • Technical-Assessment Report •

Technical-Assistance Request • Terminal Area Radar • Terrain-Avoidance Radar • Testability Analysis Report • Test Accuracy Ratio • Test Action Requirement • Test Agency Report • Test Analysis Report • Test Authorization Request • Third Assessment Report • Threat Assessment Report • Threshold Autoregression • Thrust-Augmented Rocket • Tibet Autonomous Region • Track Address Register • Train Arrière • Trajectory Adjustment Request • Trajectory Analysis Room • Transportation Acquisition Regulation • Tribunal Administratif Régional • Turbulent Accretion Region

TARA Technology Area Review and Assessment • Technology Assessment/Requirement Analysis • Tools for ACIS Real-time Analysis • Train d'Aterrissage à Roue Avant

TARAN Test And Repair As Necessary • Test And Replace As Necessary

TARC Through Axis Rotational Control • Transmutation by Adiabatic Resonance Crossing

TARD Towed Active Radar Decoy

TARDIS Targeted Delivery of Information and Services • Titles Automated Register and Document Information System

TARE Telegraph Automatic Relay Equipment • Telegraphic Automatic Relay Equipment • Telegraphic Automatic Routing Equipment • Telemetry Automatic Reduction Equipment

TARFOX Tropospheric Aerosol Radiative Forcing Observational Experiment

TARFU Things Are Really Fouled Up

TARGA Truevision Advanced Raster Graphics Adapter

TARGET Training And Rehearsal Generation Toolkit • Transeuropean Automated Real-time Gross-settlement Express Transfer

TARH Terminal Active Radar Guidance

TARL Technical Analytical Research Library

TARM TIROS Atmospheric Radiance Module

TarOT Tartu Astrofuusika Observatoorium Teated (periodical code)

TAROT Télescope à Action Rapide pour les Objets Transitoires

TARP Technical Assistance Research Program • Test And Repair Processor

TarPr Tartuskoi Astrofizicheskoi Observatorii Preprint (periodical code)

TARPS Tactical Air Reconnaissance Pod System

TARS Task Assignment Reporting System • Telemetry Acquisition and Relay System • Terminal Automated Radar Service • Terrain Avoidance Radar System • Tethered Aerostat Radar System • Three-Axis Reference System • Training Activities Reporting System • Transportation Automated Routing System

TART Twin Accelerator Ring Transfer

Tas Tasmania (Australia) • Tasmanian

TAS Table d'Atmosphère Standard • Tacoma Astronomical Society (USA-WA) • Tallahassee Astronomical Society (USA-FL) • Tape Accounting System • Tape Alteration Subroutine • Taree Astronomical Society (Australia) • Target Acquisition System • Tashkent (Uzbekistan) (location identifier) • Tasmania (Australia) • Technical Area Specialist • Technology Applications and Science • Telecommunication Analysis System • Telecommunications Authority of Singapore • Telemetry Acquisition System • Telemetry Antenna System • Telephone Access Server • Telephone Answering Service • Telephone Answering System • Teleprocessing-Auskunftssystem • Temperature-Actuated Switch • Tennessee Academy of Sciences (USA-TN) • Tenure Administration System • Terminal Access System • Terminal Address Selector • Test And Set • Test Article Specification • Texas Astronomical Society (USA-TX) • Thai Astronomical Society (Thailand) • Three-Axis Stabilization • Timeline Analysis System • Total Analysis System • Tracking Adjunct System • Tracking Antenna System • Transportation Addressing System • Transporteur Aérospatial • Triangulum Astronomical Society (USA-VA) • Tritium Analysis System • True Airspeed • Tullamore Astronomical Society (Ireland)

TASA Task And Skills Analysis • Telecomunicações Aeronáuticas SA • Telecommunication Alarm Sequence Analyzer • Television and Audio Support Activity • Test Area Support Assembly • Total Aircraft Status Accounting (system)

TASC Technical Assistance Service Center • Telecommunications Alarm Surveillance and Control • Terminal Area Sequence and Control • The Analytical Science Corporation (USA) • Total Absorption Shower Counter • Touch-Activated Simulator Control • Trailing ASCII Character • Training Aid Service Center • Training and Audiovisual Support Center

TASCA Tololo All-Sky Camera

TASCC Tandem Accelerator Superconducting Cyclotron

TASCON Television Automatic Sequence Control

TASCS Training Analysis Support Computer System

TASE Test Automation Support Environment • Traité de l'Asie du Sud-Est

TASG Technical Architecture Steering Group

TASHE Thermoacoustic Stirling Hybrid Engine

TASI Technical Advisory Service for Images (UK) • Technical Advisory Source for Images • Time Assignment Speech Interpolation

TASK Technologien für Aktive Strukturen und Komponenten • Turnaround Spare Kit

TASKO Task organization

Tasm Tasmania (Australia) • Tasmanian

TASM Turbo Assembler

TASO Terminal Area Security Officer • Training Aids Service Officer • Training Aids Support Organization • Training and Audiovisual Support Officer

TASOS The Astronomical Society Of Singapore

TASP Telemetry Analysis and Simulation Program • Tropical Area Special Product

TASPAC Total Analysis System for Production, Accounting and Control

TASPR Technical And Schedule Performance Report

TASR Telecommunications Assessment Survey Report • Terminal Area Surveillance Radar

TASS Technische Arbeitsgemeinschaft für Schrift und Sprache (Austria) • Telegrafnoie Aguentstvo Sovietskovo Soyuza (USSR) (obsolete) • Telegraph Automatic Switching System • Teleprinter Automatic Switching System • The Amateur Sky Survey • Théorie Analytique des Satellites de Saturne • Towed Array Sonar System • Trinity Astronomy and Space Society (Ireland) • Two Arm Spiral Shock • Two-Axis Sun Sensor

TASSEL Three-Astronaut Space System Experimental Laboratory

TASSF The Atlantic Space Sciences Foundation (Canada-NS)

TAT Tatry Poprad (Slovakia) (location identifier) • Technical Acceptance Team • Technical Approval Team • Technical Area Task • Technical Assessment Team • Technical Assistance Team • Technical Assistance and Training • Technology Application Team • Température d'Air Totale • Thematic Apperception Test • Theoretical Arrival Time • Thrust-Augmented Thor • Total Air Temperature • Total Alias Table • Toxine-Antitoxine • Transatlantic Trunk • Transparent Assembly Test • Turnaround Time • Twin Astrographic Telescope • Two-Axis Tracking

TATC Terminal Air Traffic Control • Transatlantic Telephone Cable

TATCA Terminal Air Traffic Control Automation

TATDL Tabulated Assembly Technical Data List

TATI Total Air Temperature Indicator

TATOM Tracking And Tracing of Overseas Mail

TATP Two-Axis Tracking Pedestal

TATRP Texas Advanced Technology Research Program (USA-TX)

tats tatsächlich

Tats Tatsache

TATS Test And Training Satellite

TATT Total Average Talk Time

Tau Taurus-Tauri (constellation)

TAU Tauramena (Colombia) (location identifier) • Tel Aviv University (Israel) • Terminal Access Unit • Thesaurus Alphabetical Update • Thousand Astronomical Units

TAUMOK Tautenburger Multiobjekt Kamera

TAUN Technical Assistance of the United Nations

TAUP Topics in Astroparticle and Underground Physics

Taut Tautological • Tautology

TAUVEX Tel Aviv Ultraviolet Experiment

t.-à-v. Tout-à-Vous

TAV Target Acquisition and Verification • Tau (American Samoa) (location identifier) • Ten Aanzien Van • Test And Validation • Test And Verification • Total Asset Visibility • Train Avant • Tren de Alta Velocidad

TAVERNS Test And Verification Environment for Remotely Networked Systems

TAVT Terminal Airspace Visualization Tool

TAW Tacuarembo (Uruguay) (location identifier) • Thrust-Augmented Wing • Torsional Alfvén Wave • Twice A Week

TAWDS Transportable Automated Weather Distribution System

TAWG Target Accreditation Working Group

TAWRS Temperature And Wind Rocket Spectrometer

TAWS Terrain Avoidance Warning System • Theater Airborne Warning System

TAWT Torsional Alfvén Wave Train

Tax Taxation

TAX Taliabu (Indonesia) (location identifier)

TAXATA Things Are Exactly As They Appear

TAXI Transparent Asynchronous Transceiver Interface

Taxn Taxation

TAY Tartu (Estonia) (location identifier)

TAZ Tagesanzeiger • Tagesauszug • Tashauz (Turkmenia) (location identifier) • Traffic Analysis Zone

t&b Top & Bottom

Tb Tabelle • Tabulator • Taschenbuch • Tatbestand • Taube • Terabit • Terbium • Tube • Tuberculosis • Turbulence

T&B Turn & Bank

T/B Talkback • Thermal Balance • Thermal Barrier • Title Block • Top/Bottom

TB Tableau de Bord • Tagungsbüro • Tail Bit • Taittirīya Brāhmana • Talk Back • Talking Book • Target Board • Tarifbestimmung • Tätigkeitsbericht • Technical Board • Technical Bulletin • Technische Bestimmungen • Technische Bibliothek • Technischer Beamter • Technischer Bericht • Technisches Büro • Téléphone de Bord • Température de Bruit • Température du Boîtier • Tension aux Bornes • Terabyte • Terminal Base • Terminal Board • Testbed • Text Begin • Thermal Bake • Thermal Balance • Thermal Barrier • Tibet • Time Base • Time of Beginning • Time of Burn • Title Block • Titius-Bode (law) • Toggle Buffer • Token Bus • Tonnage Brut • Train Baladeur • Training Board • Tranquility Base (Moon) • Translation Buffer • Transfer Building • Transformateur Biphasé • Transient Brightening • Transmitter Blocker • Transparent Bridging • Transverse Boom • Treasury Bill • Trial Balance • Tribord • Tropical Band • True Bearing • Tuberculosis • Turbopropulseur • Turn and Bank

TBA Tabibuga (Papua New Guinea) (location identifier) • Teaching By Analogy • To Be Activated • To Be Added • To Be Advised • To Be Agreed • To Be Announced • To Be Arranged • To Be Assigned • To Be Avoided • Triple Bend Achromat

TBAC Task-Based Access Control

TBAG Technical Base Advisory Group

TBAI Temporary Base Activation Instruction

TBAMS Test-Bed Aircraft Multispectral Scanner

TBAN To Be Announced

TBAX Tube Axial

TBB Thick Boxy Bulge • Tuy Hoa (Viet Nam) (location identifier)

TBBS The Bread Board System

Tbc Tuberculosis

TBC Tape Block Channel • Terminal Buffer Controller • The Boeing Company • Thermal Barrier Coating • Time-Base Corrector • Time-Based Corrector • To Be Confirmed • To Be Continued • Token Bus Controller • Transportation Bill Code • Tuba City (USA-AZ) (location identifier)

TBCC Thallium - Barium - Calcium - Copper

TBCCO Thallium-Barium-Calcium-Copper Oxide

TBCF To Be Called For

TBCM Target Background Characterization Model

TBCO To Be Carried Over

TBCP Three-Body Coulomb Problem

TBD Teaching By Demonstration • Temps Barycentrique Dynamique • Timbiqui (Colombia) (location identifier) • Time/Bearing Display • To Be Decided • To Be Defined • To Be Delivered • To Be Determined • To Be Developed • To Be Done • Trade Development Board • Troubleshooting Block Diagram • Turliye Bilisim Dernegi (Turkey)

TBDF Transborder Data Flow

TBE Teaching By Example • Teledyne Brown Engineering (USA-AL) • Tidal Burst Effect • Timbunke (Papua New Guinea) (location identifier) • Time Base Error • Time Base Expansion • Titan Biological Explorer • To Be Evaluated • To Be Expanded • Transient Buffer Exposure • Transmitter Buffer Empty

TBEB Truncated Binary Exponential Backoff

TBED Table Edit

TBESC Technical Base Executive Steering Committee

TBF Tabiteuea North (Kiribati) (location identifier) • Time Between Failure • Test de Bon Fonctionnement • To Be Furnished • Très Basse Fréquence

TBG Tabubil (Papua New Guinea) (location identifier)

TBGA Tape Ball Grid Array

TBGTA Travel By Government Transportation Authorized

TBH Tablas (Philippines) (location identifier)

TBI Ten Bit Interface • Test Bench Installation • The Bight (Bahamas) (location identifier) • Throttle-Body Injection • Through-Bulkhead Initiator • Thru-Bulkhead Initiator • Time Between Inspection • To Be Inactivated • Total Body Irradiation • Traumatic Brain Injury • Turn and Bank Indicator

TbIG Terbium Iron Garnet

TBIG Technicien en Bureautique et Informatique de Gestion

TBIM Trigger-Based Item Management

TBIS Technology Base Investment Strategy

TBJ Tabarka (Tunisia) (location identifier)

tbk Toolbook (file extension)

TBK Timber Creek (Australia-NT) (location identifier) • Toolbook

Tbl Tabelle • Tablette • Tabulator • Titelblatt

TBL Table • Tableland (Australia-WA) (location identifier) • Technology-Based Learning • Télescope Bernard Lyot • The British Library • Through Bill of Lading • Timothy Berners-Lee (1955-) • Turbulent Boundary Layer

Tblr Tumbler

Tblsht Troubleshoot

TBM Table Buffer Management • Temporary Benchmark • Terabit Memory • Training Baseline Materials • Tumbang Samba (Indonesia) (location identifier) • Tunnel Boring Machine • Typenbegleitmannschaft

TBMP Technology Base Master Plan

TBMS Thirteen Band Multispectral Scanner

TBN Fort Leonard Wood (USA-MO) (location identifier) • To Be Negotiated • To Be Nominated • To Be Notified

TBO Tabora (Tanzania) (location identifier) • Time Before Overhauls • Time Between Overhauls (*) • To Be Organized • Total Blackout • Transactions By Others • Tycho Brahe Observatory (Sweden)

TBOI Tentative Basis Of Issue

TBP Technology Business Process • Tetrabenzoporphyrin • To Be Printed • To Be Provided • To Be Published • Tributyl Phosphate • True Boiling Point • Tumbes (Peru) (location identifier) • Tuyauterie de Basse Pression • Two-Body Problem

TBPDD Technology Base Project Definition Document

TbPS Terabits Per Second

TBPS Terabytes Per Second

TBQ Tarabo (Papua New Guinea) (location identifier) • Taxonomy-Based Questionnaire • To Be Questioned

TBR Statesboro (USA-GA) (location identifier) • Technical Basis for Regulations • Ter Beschikking van de Regering • To Be Received • To Be Recommended • To Be Removed • To Be Replaced • To Be Resolved • To Be Reviewed • To Be Revised • Toolbar • Tungsten Bearing Resin

tb&s Top, Bottom & Sides

TBS Talk Between Ships • Tape and Buffer System • Task Breakdown Structure • Tbilisi (Georgia) (location identifier) • Technology-Business Strategy • Test Bench Set • Test Breakdown Structure • Tethered Balloon System • To Be Scheduled • To Be Selected • To Be Specified • To Be Superseded • To Be Supplied • Turner Broadcasting System

TBSL To Be Supplied Later

Tbsp Tablespoon

TBSV Time Between Scheduled Visits

TBT Tabatinga (Brazil-AM) (location identifier) • Technical Barrier to Trade • Technology-Based Training • Time Ball Tower • Tributyl Tin

TBTA To Be Thrown Away

TBTD That'll Be The Day • Thermal Bus Technology Demonstrator

TBU Nuku Alofa - Fua'Amotu (Tonga) (location identifier) • Tape Backup Unit • Time Base Unit • Tongatapu (Tonga) (location identifier)

TBV Tabal (Marshall Islands) (location identifier) • Ten Behoeve Van • Ten Bate Van • To Be Verified

TBW Tambov (Russia) (location identifier) • To Be Withheld • To Be Written • Total Bandwidth

TBx Telebox

TBX Taabo (Ivory Coast) (location identifier)

TBY Tsabong (Botswana) (location identifier)

TBYB Try Before You Buy

Tbyte Terabyte

TBZ Tabriz (Iran) (location identifier)

t/c Thickness/Chord

Tc Critical Temperature • Technetium • Technétium

T/C Telecommunications • Temperature Compensating • Temperature/Conductivity • Termination Check • Test Conductor • Thermal Capacitor • Thermocouple • Thrust Chamber • Training and Certification

T&C Telemetry & Command • Telemetry & Control • Terms & Conditions • Test & Checkout • Tracking & Communication • Tracking & Control

TC Caicos Islands (Turks &_) (ISO country code) • Îles Turks & Caicos (code ISO de pays) • Tableau de Commande • Table de Composition • Tactical Computer • Tagungscenter • Tantalum Capacitor • Target Cluster • Target Complete • Tariff Commission • Taurid Complex • Taux de Charge • Taux de Compression • Taux de Confiance • Taxe Complémentaire • Taxe de Circulation • Teachers College • Technical Center • Technical Characteristics • Technical Circular • Technical College • Technical Committee • Technical Communication • Technical Computing • Technical Control • Technical Cooperation • Technical Coordinator • Technical Council • Technologie-Clé • Technology Center • Telecommand • Télécommande • Telecommunications • Telecommunications Closet • Telecontrol • Telemetry and Command • Telemetry and Control • Telemetry Calibration • Telemetry Combiner • Télévision en Couleurs • Temperature Coefficient • Temperature Compensated • Temperature Compensating • Température Constante • Temperature Control • Température Critique • Temporary Clerk • Temporary Correction • Temps Clair • Tendeur de Câble • Term Combination • Term Coordination • Terminal Computer • Terminal Controller • Termination Check • Terms and Conditions • Test and Checkout • Test Camera • Test Case • Test Cell • Test Change • Test Completion • Test Comprehensiveness • Test Condition • Test Conductor • Test Console • Test Container • Test Control • Test Controller • Test Coordinator • Testing Center • Text Categorization • Text Companion • Thermal Capacitor • Thermal Coefficient • Thermal Conductivity • Thermal Control • Thermal Cycling • Thermocontact • Thermocouple • Thermocurrent • Thermomètre Centigrade • Thiokol Corporation • Thrust Chamber • Tidal Current • Time Check • Time Compensation • Time Constant • Time-Critical • Timed Closing • Time to Circular • Timing Channel • Timing Counter • Tissu Cellulaire • Toll Center • Top Center • Total Capacity • Total Cost • Tour de Contrôle • Town Council • Town Councillor • Traceability Code • Tracking and Communication • Tracking and Control • Tracking Camera • Tracking Console • Traffic Control • Training and Certification • Training Camp • Training Center • Training Circular • Transaction Capability • Transaction Code • Transfer Channel • Transfer Control • Transformateur de Courant • Transit Case • Transit Circle • Transit Corridor • Transition Component • Transition Control • Translation Controller • Transmission Clock • Transmission Control • Transmission Controller • Transmission Convergence • Transmit Clock • Transportation Center • Transportation Coordinator • Transport Carrier • Transport Connect • Travail à la Chaîne • Travel Club • Traveller's Check • Trellis Coding • Trial Counsel • Tribunal Constitutionnel • Tribunal des Conflits • Trinity College • Trip Completed • Tronc Commun • Tropical Cyclone • True Course • Trusteeship Council • Tube Capillaire • Tube Cathodique • Tube de Choc • Turbocharger • Turbocompresseur • Turbulent Concentration • Turkey (international aircraft registration code) • Turks & Caicos Islands (ISO country code) • Type Certificate • Type Certification • Type Classification

TCA Caicos Islands (Turks &_) (ISO country code) • Îles Turks & Caicos (code ISO de pays) • Tachyon Cross-Assembler • Target Class Assessment • Technical Change

Analysis • Technical Contract Administrator • Technology Cooperative Agreement • Télécommande Automatique • Télécommunication Automatique • Télécommunications Association • Telemetering Control Assembly • Temperature Control Amplifier • Temperature Control Assembly • Tennant Creek (Australia-NT) (location identifier) • Terminal Control Area • Terminal Controlled Airspace • Thrust Chamber Assembly • Thrust Control Assembly • Time of Closest Approach • To Come Again • Tour de Contrôle d'Aérodrome • Tower Control Airport • Trace Contamination Analysis • Traffic Control Airport • Translation Control Assembly • Translation Controller Assembly • Transportation Control Agency • Trichloroacetic Acid • Turks & Caicos Islands (ISO country code)

TCAA Taiwan Civil Aeronautics Administration • Twin City Amateur Astronomers (USA-IL)

TCAB Telecommunications Advisory Board (UK) • Temperature of Cabin

TCAC Technical Control and Analysis Center • Tri-City Astronomy Club (USA-WA)

TCACCIS Transportation Coordinator Automated Command and Control Information System

TCAD Technology Computer-Aided Design

TCAE Technical Control and Analysis Element

TCAI Tutorial Computer-Assisted Instruction

TCAL Thermal Calibration

TCAM Telecommunications Access Method • Threat Condition Alerting Message

TCAMS Technical Control, Automation, Maintenance and Support

TCAO Téléconférence Assistée par Ordinateur • Tradeable Carbon Absorption Obligation

T/CAP Thermal Capacitor

TCAP Thermal Capacitor • Trainer Corrective Action Program • Transaction Capabilities Applications Part (*) • Transaction Capabilities Applications Port • Transmission Capability Applications Part

TCAS Technical Control and Analysis Section • Technical Control and Analysis System • Telemetry Consolidation Analysis Study • Terminal Collision Avoidance System • Terminal Controlling Address Space • Traffic-alert and Collision Avoidance System • Traffic-alert Collision Avoidance System • Traffic and Collision Avoidance System (*)

TCAT Tape-Controlled Automatic Testing • Test Coverage Analysis Tool

TCB Take Care of Business • Taking Care of Business • Task Control Block • Technical Coordination Bulletin • Technical Coordinator Bulletin • Temps-Coordonnée Barycentrique • Test Cell Basement • Test Certification Board • Test Connector Buffer • Test Control Board • Time Correlation Buffer • Transmission Control Block • Treasure Cay (Bahamas) (location identifier) • Trouble Came Back • Trusted Computer Base • Trusted Computing Base

TCBL Tradewind Cumulus Boundary Layer

TCC Tarrant County College (USA-TX) • Technical Computing Center • Technical Control Cell • Technical Control Center • Technical Coordination Committee • Technical Coordination Communications • Technological Change Committee • Telecommunications Center • Telecommunications Control Center • Temperature Coefficient of Capacitance • Test Classification Code • Test Conductor's Console • Test Control Center • Test Control Complex • Test Controller Console • Thermal Control Coating • Thrust Control Command • Time Compression Coding • Timing - Clock - Control • Topographic Coordination Communications • Total Change Control • Trace Contaminant Control • Tracking and Control Center • Traffic Control Centre • Transfer Channel Control • Transmission Control Center • Transmission Control Code • Transmit Control Center • Transmitter Control Center • Transport and Communications Commission • Transportation Component Command • Transportation Computer Center • Transportation Control Card • Transportation Control Center • Transportation Coordination Center • Transport Control Center • Trinity College Cambridge (UK-England) • Tropical Cyclone Centre • Tucumcari (USA-NM) (location identifier) • Turbulence par Ciel Clair • Tuyère à Corps Central

TCCA Trace Contaminant Control Assembly

TCCB Technical Configuration Control Board • Trainer Change Control Board

TCCC Technical Committee on Computer Communications • Terrestrial Critical Control Circuit • Tower Control Computer Complex

TCCD Technical Charge-Coupled Device (system)

TCCIS Texas Center for Computational and Information Sciences (USA-TX)

TCCM Technical Committee on Computational Medicine (IEEE) • Temperature-Controlled Crystal Microbalance • Three-Corner Coding Method

TCCP Telecommunications Command Control Program • Telecommunications Command Control Project

TCCS Ticket Consignment Control Sheet • Timed Calculus of Communicating Systems • Trace Contamination Control Subassembly • Transportation Command and Control System

T&CD Timing & Countdown

T/CD Time/Command Distributor

TCD Chad (ISO country code) • Tarapaca (Colombia) (location identifier) • Task Completion Date • Tchad (code ISO de pays) • Teaching Company Directorate • Technical Cooperation for Development • Telemetry and Command Data • Terminal Control Device • Terminal Countdown • Test Completion Date • Test Control Document • Thermal Conductivity Detector • Time/Code Distribution • Time/Command Distributor • Time Compliance Data • Time Correlation Data • Timing and Countdown • Timing Control Distributor • Tour Completion Date • Transistor-Controlled Delay • Transmitted Current Detector • Transport de Chalands de Débarquement • Trinity College Dublin (Ireland) • Two-Colour Diagram • Type Classification Data

TCDC Technical Cooperation among Developing Countries (*) • Technical Cooperation between Developing Countries

TCDD Tetrachlorodibenzodioxin

TCDIS Transportation Coordinator Deployment Information System

TCDL Technical Committee on Digital Libraries (IEEE)

TCDM Tilted Cold Dark Matter (model)

TCDS Tracking and Command Decoding System • Type Certification Data Sheet

TC&E Training Control & Evaluation

TCE Targeted Client Environment • TC Electronics • TEC Control Electronics • Technology Center of Excellence • Telemetry Checkout Equipment • Temperature Coefficient of Expansion • Terminal Control Element • Terrace • Test Cost Estimate • Thermal Canister Experiment • Thermal Coefficient of Expansion • Thermal Control Electronics • Thomson Electronics Consumer • Thread-based Communication Environment • Three-Color Experiment • Tile Control Electronics • Tone Control Equipment • Total Composite Error • Total Cost Estimate • Tout Corps d'État • Training Control and Evaluation • Transaction Cost Economics • Transcoding Equipment • Trichloroethylene • Tulcea (Romania) (location identifier)

TCEA Texas Computer Education Association (USA-TX)

TCEE Temperature-Controlled Emission Enclosure

TCEF Tableau des Consommations d'Énergie en France

TCEM Test des Cycles Élémentaires et Moyens

tcf Trillion Cubic Feet

TCF Tank Checkout Facility • Technical Control Facility • Télévision en Circuit Fermé • Temperature-Compensating Focuser • Test Control Facility • Test Coordination Function • Time-Correction Factor • Tocoa (Honduras) (location identifier) • Touring-Club de France • Transaction Control Function

TCFA Terrain-Corrected Free-Air (anomaly)

TCFH Télescope Canada-France-Hawaii

TCFM Temperature Control Flux Monitor

TCFP Thrust Chamber Fuel Purge

TCFS Threaded-Cell Focusing System

TCG Technical Coordination Group • Technology Coordination Group • Telecommunications Group • Teleport Communication Group • Temps Coordonné Géocentrique • Test Control Group • Time Code Generator • Time Control Gain • Time-Controlled Gain • Transponder Control Group • Tune-Controlled Gain

TCGF T-Cell Growth Factor

TCGR Time Code Generator/Reader

TCGS Tennessee Conference of Graduate Students (USA-TN)

TCH Tchibanga (Gabon) (location identifier) • Threshold Crossing Height • Traffic Channel

Tchg Teaching

Tchr Teacher

TCI Tableau de Composition Illustré • Tape-to-Computer Interface • Technical Critical Item • Telecommunications Inc. • Telemetry Command Interface • Telemetry Components Information • Telnet Command Interpreter • Temperature Control Instrument • Tenerife (Canary Islands) (location identifier) • Terrain Clearance Indicator • Test Control Instruction • Through Check-In • Time Change Item • Total Cost Input • Touring-Club Italiano

TCIC Technical Committee on Industrial Classification

TCID Terminal Computer Identification • Test Configuration Identifier

TCIF Telecommunications Industry Forum • Traveller Contact Information Form

TCIO Task Control and Information Office

TCIP Technical Control Improvement Program

TCIR Two Color Infrared

TCJ Torembi (Papua New Guinea) (location identifier)

TCK Test Clock • Tinboli (Papua New Guinea) (location identifier)

TCL TAE Command Language • Terminal Command Language • Terminal Control Language • Text Composition Language • Thrust Control Lever • Time and Cycle Log • Tool Command Language • Tool Control Language • Transformation Control Language • Transistor Coupled Logic • Transmit Control Logic • Transportable Calibration Laboratory • Tratado de Comercio Libre • Trichloroethylene • Triple Communication Line • Tuscaloosa (USA-AL) (location identifier)

TCLAS Type Classification

TCLC Tri-state College Library Cooperation

TCLE Thermal Coefficient of Linear Expansion

TCLP Toxicity Characteristic Leaching Procedure • Type Classification Limited Procurement • Type Classification Limited Production

TCLT Tentative Calculated Landing Time

TCLZ Texas Learning and Computation Center (USA-TX)

TCM Tableau Climatologique Mensuel • Tacoma McChord (USA-WA) (location identifier) • Target Cost per Mission • Taxiway Collision Monitoring • Technical Content Manager • Technical Control Management • Technical Coordinating Meeting • Technical Coordination Meeting • Technologie de Couches Minces • Télécommande • Télécommunication Manuelle • Telemetry Code Modulation • Telemetry Control and Monitoring • Telephone Control Manager • Temperature Control Model • Terminal Control Mechanism • Terminal-to-Computer Multiplexer • Tetrachloromerkurat • Thermal Conduction Module • Thermal Control Model • Time Compression Multiplexer • Time Compression Multiplexing • Total Corrective Maintenance • Trajectory Control Maneuver • Trajectory Correction Maneuver • Trellis-Coded Modulation • Trellis Code Modulation • Tropical Cyclone Motion

TCMD Transportation Control and Movement Document • Transportation Control Movement Document

TCMF Touch Calling Multifrequency

TCMO Travail Collectif Médiatisé par Ordinateur

TCMP Test Capability Master Plan • Transportation Control and Movement Plan • Trunk Control Message Protocol

TCMS Telecommunications System • Test Control and Monitoring System

TCN Tehuacan (Mexico) (location identifier) • Telecommunications Network • Terrestrial Communications Network • Test Change Notice • Third Country National • Tour à Commande Numérique • Tracing Change Notice • Training Course Number • Transparent Content Negotiation • Transportation Control Number

TCNE Tetracyanoethylene

TCNJ The College of New Jersey (USA-NJ)

TCNQ Tetracyanoquinodimethane

TCNRDI Transmission Corrected Normalized Reflectance Difference Index

TCO Tableau de Contrôle Optique • Taken Care Of • Technical Contracting Office • Technical Contracting Officer • Technical Control Operator • Telecommunications Certification Office • Telecommunications Certification Officer • Telecommunications Control Office • Telecommunications Control Officer • Telemetry and Control • Telephone Control Office • Telephone Control Officer • Terminal Control Office • Termination Contracting Office • Termination Contracting Officer • Test and Checkout • Test Control Office • Test Control Officer • Thomson CSF Optronique (France) • Three College Observatory (USA-NC) • Total Cost of Ownership • Transportation Control Office • Transportation Control Officer • Trinity College Oxford (UK-England) • Tumaco (Colombia) (location identifier)

TCOM Terminal Communications

TCOP Test and Checkout Plan

TCOS Technical Committee on Operating Systems

TCOSS Technical Control Operations Support System

T&CP Test & Checkout Procedure

TCP Tape Carrier Packaging • Tape Conversion Program • Task Change Proposal • Task Control Package • Task Control Packet • Technical Change Proposal • Technical Cooperation Program • Technical Coordination Program • Technical Copyright Protection • Technical Cost Proposal • Technological Capabilities Panel • Telecommunications Protocol • Temporary Change Procedure • Terminal Control Protocol • Test and Checkout Procedure • Test Calibration Panel • Test Case Procedure • Test Checkout Procedure • Test Compare • Test Conductor • Test Constraints Review • Test Control Package • Test Coordination Procedure • Tester Converter Program • TEX Code Page • Thermal Concept Review • Thrust Chamber Pressure • Time Code Processor • Time - Cost - Performance • Timing and Control Panel • Traffic Congestion Protocol • Traffic Control Point • Traffic Control Post • Training Concept Plan • Transfer Control Protocol (*) • Transmission Control Program • Transmission Control Protocol (*) • Transport Connection Protocol •

Transport Control Protocol • Très Courte Portée • Tricalcium Phosphate • Trichlorophenoxyacetic • Tromsø CCD Photometer • Tropical Cyclone Programme

TCPA Telephone Consumer Protection Act (USA) • Trusted Computing Platform Alliance

TCPC Technical Committee for Physics Computing

TCPE Total Cargo Project Engineer

TCPI Technology Consumer Price Index • To Complete Performance Index

TCPP Technical Committee on Parallel Processing

TCPS Technical Control and Processing Section • Technical Copyright Protection System

TCQ Tacna (Peru) (location identifier) • Trellis-Coded Quantization

TCQAM Trellis-Coded Quadrature Amplitude Modulation

TCQO Trellis-Coded Quasi-Orthogonal

tcr Transceiver

TC&R Telemetry, Command & Ranging

TCR Tagged Cell Rate • Tape Cassette Recorder • Task Change Request • T-Cell Receptor • Technical Change Request • Technical Characteristics Review • Technical Control Rebuild • Telemetry Command and Ranging • Temperature Coefficient of Resistance • Test Change Record • Test Compare Result • Test Constraint Review • Text Condition Requirement • Thermal Concept Review • Tonnes of Coal Replacement • Total Contractual Requirements • Training Capabilities Report • Transfer Control Register • Transient Call Record • Traveling Compression Region

TCRB Test Change Review Board • Test Criteria Review Board

T/CRD Telemetry/Command Requirements Document

TC/RD Transmit Combiner/Receiver Divider

TCRD Test and Checkout Requirements Document • Transmit Combiner/Receiver Divider

TCRF Transit Connection Related Function

TCRS Tennessee Consolidated Retirement System (USA-TN) • Training and Certification Records System

TCRSD Test and Checkout Requirements and Specifications Document • Test and Checkout Requirements Specification Document

TCS Teaching Company Scheme • Technical Change Summary • Technical Concurrence Sheet • Technical Control Scanner • Technical Control Shelter • Technical Control Station • Telecommunications Control System • Telecommunications Specialist • Teleconference System • Telemetry and Command System • Telephone Conference Summary • Telephony Control Specification • Telescope Control Software • Telescope Control System • Telescope Coordination Software • Telescopio Carlos Sánchez (IAC) • Television Camera Set • Television Camera System • Temperature Control System • Temporary Change of Station • Terminal Computer System • Terminal Control System • Terminal Countdown Sequencer • Terminal Count Sequence • Test Control Sequence • Test Control Supervisor • Test Control System • Testing Computer Software (conference) • Theoretical Computer Science • Thermal Conditioning Service • Thermal Conditioning System • Thermal Control System • Time-Critical System • TNG Control System • Toroidal Confinement System • Torus Cleanup System • Total Cross-Section • Touting-Club der Schweiz • Touting-Club de Suisse • Tower Communications System • Traffic Conditioning Specification • Traffic Control Station • Traitement Chimique de Surface • Trajectory Control Sensor • Transfer Carry Subtract • Transmission Convergence Sublayer • Transmit Control Sequence • Transportation and Communications Service • Transport Communications Service • Trusted Computer System • Truth or Consequences (USA-NM) (location identifier) • Two Channel Switch

TCSC Technological Cooperation Subcommittee • Trainer Control and Simulation Computer

TCSE Technical Committee on Software Engineering • Technical Council on Software Engineering

TCSEC Trusted Computer Security • Trusted Computer System Evaluation Criteria

TCSI Technical Control System Improvement

TCSP Technical Control Systems Program • Technical Support

TCSPC Time-Correlated Single Photon Counting Technique

TCSS Technical Control Software Support • Telecommunications Service System • Time-table Creation Support System

TCSSS Thermal Control Subsystem Segment

TCSTD Type Classified - Standard

TCSUH Texas Center for Superconductivity at the University of Houston (USA-TX)

TCT Takotna (USA-AK) (location identifier) • Tarski - Church - Turing (theorem) • Technique de Compression du Temps • Technology Cycle Time • Test Coordination Team • Tilted-Component Telescope • Time Code Translator • To Challenge To-morrow • Total Contractor Training • Total Contract Training • Transcontinental Traverse • Translator and Code Treatment • Transmit Cipher Text • Transonic Cryogenic Tunnel • Tribunal Central de Trabajo (Spain)

TCTA Transcontinental Control Area

TCTC Transit Cast Technical Control • Transmit Cipher Text Clock

TCTD Technical Countdown Demonstration Unit

TCTE Telecommand Test Equipment

TCTF Technology Crimes Task Force

TCTI Time Compliance Technical Instruction

TCTO Time-Compliance Technical Order

TCTR Time-Compliance Technical Requirement

TCTS Trans-Canada Telephone System

TCTT Tarski-Church-Turing Theorem

TCTUA Terminal Control Table User Area

TCu Tribunal de Cuentas (Spain)

TCU Tape Control Unit • Teletype Communications Unit • Teletypewriter Control Unit • Temperature Control Unit • Terminal Computer Unit • Terminal Control Unit • Test Computer Unit • Test Control Unit • Texas Christian University (USA-TX) • Thaba Nchu (South Africa) (location identifier) • Thermal Control Unit • Thermal Cuing Unit • Threshold Control Unit • Thrust Control Unit • Time Change Unit • Time Code Update • Timing Control Unit • Towering Cumulus • Tracking and Communications Unit • Tracking Control Unit • Transit Control Unit • Transmission Control Unit • Transportable Computer Unit • Tribal Colleges and Universities (USA) • Trunk Coupling Unit

TCUP Tribal Colleges and Universities Program

TCV Technischer Cartellverband • Temperature-Compensated Voltage • Temperature-Controlled Voltage • Temperature Control Valve • Temperature Control Voltage • Terminal Configured Vehicle • Tete (Mozambique) (location identifier) • Thrust Chamber Valve • Time Code Validity • Tokamak with Variable Configuration • Traveling Convection Vortex

TCVG Technical Committee on Visualization and Graphics (IEEE)

TCVXO Temperature Compensated-Voltage Controlled Crystal Oscillator

TCW Time Code Word • Tocumwal (Australia-NSW) (location identifier) • Track Confirmation Word

TCWG Telecommunications Working Group

TCXO Temperature-Compensated Crystal Oscillator • Temperature-Controlled Crystal Oscillator

TC2S Transportation Command and Control System

td Ter in Die

Td Tausend • Tedeum • Tender • Trend

T/D Time/Date (conversion) • Time Delay • Time Deposit • Touchdown

T&D Test & Diagnostic • Training & Development • Transmission & Distribution • Transport & Diffusion • Transportation & Docking • Tutors & Demonstrators

TD Chad (ISO country code) • Table Data • Table of Distribution • Tabular Data • Tagesdosis • Tangential Discontinuity • Tankdampfer • Target Day • Target Designation • Target Designator • Target Detector • Task Description • Task Design • Task Directive • Taux de Débit • Taux de Dilution • Tchad (code ISO de pays) • Teaching Diploma • Technical Data • Technical Demonstration • Technical Design • Technical Development • Technical Difficulty • Technical Direction • Technical Directive • Technical Director • Technical Directorate • Technical Discussion • Technical Division • Technical Document • Technical Drawing • Technischer Dienst • Technischer Direktor • Technological Determinist • Technology Data • Technology Demonstration • Technology Development • Technology Document • Télédétection • Telemetry Data • Temperature Datum • Temperature Differential • Température Diurne • Temporal Difference • Temporary Duty • Temps de Descente • Tension Disruptive • Terminal Display • Terminal Distributor • Terminating Descriptor • Terminator Deformation • Terrain Data • Test Data • Test Description • Test Design • Test Directive • Test Director • Test Drawing • Testing Device • Thematic Data • Thor-Delta • Threshold Densitometry • Threshold Detection • Tiefendosis • Time Delay • Time Difference • Time-Distance • Time Domain • Time of Departure • Timing Distributor • Topic-Directed • Topographic Division • Topological Defect • Total Depth • Total Distribution • Total Drift • Touchdown • Track Data • Track Display • Traffic Director • Training Developer • Training Development • Training Device • Training Division • Transmission de Données • Transmit Data • Transmitter Distributor • Transportation and Docking • Transportdienst • Transverse Direction • Travail Dirigé • Travaux Dirigés • Treasury Department • Tribord • Tridirectional • Triple Difference • Tropical Depression • Trust Deed • Tube à Décharge • Tunisian Dinar (currency) • Tunnel Diode • Turbodynamo

Tda Tienda

TDA Tables of Distribution and Allowances • Tactical Decision Aid • Target Docking Adapter • Technical Direction Agency • Technical Direction Agent • Technology Development Approach • Telemetric Data Analyzer • Telemetry Data Acquisition • Terminal Data Adapter • Terminal Data Analyzer • Test Data Archive • Text-based Discriminant Analysis • Thermodifferential Analysis • Time Delay Amplifier • Time Division Arrival • Today • Tracking and Data Acquisition • Tracking Data Analysis • Training Device Annex • Transit Detection Algorithm • Trinidad (Colombia) (location identifier) • Trunnion Drive Axis • Tuning Device Assembly • Tunnel Diode Amplifier • Turkheim Detection Algorithm • Type de Données Abstrait

TSAA Training Development and Analysis Activity

TDAAS Two-dimensional Data Assessment and Applications Study

TDAB Terrestrial Digital Audio Broadcasting

TDAC Tactical Data Acquisition and Correlation • Test Data Analysis Center • Training Data Analysis Center • Tycho Data Analysis Consortium

TDAD Table of Distribution and Allowance Designation • Training Development and Analysis Directorate

TDAE Technology Development and Aerospace Environments

TDAF Tracking and Data Analysis Facility

TDAP Technology Development and Acquisition Program • Total Distribution Action Plan • Training Development Action Plan

TDAPWG Test Data Acquisition and Processing Working Group

TDAS Technical Data Automation System • Telemetry Digital Acquisition System • Tracking and Data Acquisition System • Traffic Data Administration System • Tunnel-Diode Amplifier System

TDAT Truncated Disk-Advective Torus

TDATD Total Distribution Advanced Technology Demonstration

TDATS Target Detection, Acquisition, and Tracking System

TDB Target Deployment Bus • Task Database • Technical Database • Technical Directive Board • Temps Dynamique Barycentrique • Terminology Database • Terrain Database • Terrestrial Dust Belt • Tetabedi (Papua New Guinea) (location identifier) • Tiempo Dinámico Baricéntrico • Time Duration of Burn • Total Disability Benefit • Trade and Development Board • Tranet Doppler Beacon • Transportation Database

TDBM Technical Database Management

TDBMS Trusted Database Management System

TDBU Top-Down Bottom-Up

T/DC Time/Date Conversion

TDC Tape Data Controller • Target Designator Control • Tarif Douanier Commun • Taurus Dark Cloud • Technical Data Center • Technical Development Center • Technical Directive Code • Technical Directive Compliance • Technical Documentation Center • Technical Document Center • Telemetry Data Center • Terrestrial Data Circuit • Test Director Console • Text Database • Textes et Documents pour la Classe • Thermal Diffusion Chamber • Thermal Diffusion Coefficient • Time Data Card • Time-Delay Closing • Time Dispersal Closing • Time Dispersed Coding • Time Distance Calculator • Time-to-Digital Converter • Top Dead Centre • Total Distributed Control • Total Distribution Curve • Trade Development Council • Transportation Data Channel • Tribunal de Defensa de la Competencia (Spain) • Tropical Data Centre • True Digital Control • Two-Dimensional Cepstrum

TDCC Training and Development Communications Channel • Transportation Data Coordinating Committee

TDCD Transmitter Duty Cycle Distortion

TDCM Transistor Driver Core Memory

TDCMS Technical Data Configuration Management System

TDCN Technical Data Change Notice

TDCR Technical Data Change Request • Triple to Double Coincidence Ratio (method)

TDCS Traffic Data Collection System

TDD Target Detection Device • Task Description Document • Technical Data Digest • Technical Discussion Document • Techniques Development Department • Technology Development Division • Telecommunications Device for the Deaf • Telemetry Data Digitizer • Telephone Device for the Deaf • Telephonic Device for the Deaf • Temporary Device Departure • Test Definition Document • Test Design Description • Test Development Director • Time Data Dependency • Time Division Duplex • Top Down Design • Training and Development Department • Trinidad (Bolivia) (location identifier)

TDDB Thermodynamic Design Data Block

TDDL Time-Division Data Link

TDDM Time Division Digital Multiplexing

TDDP Trajectory Design Data Package

TDDR Technical Data Department Report • Total Direct-Diffuse Radiometer

TDDS Telemetry Data Decommutation System • Telemetry Data Display System • Telemetry Decommutation and Display System

TDDSG Teledienstdatenschutzgesetz (Germany)

T&DE Test & Diagnostic Equipment

TD&E Tactics Development & Evaluation

TDE Tactics Development & Evaluation • Taux de Demande d'Emploi • Technical Data Engineer • Technical Data Evaluation • Telemetry Displacement Error • Terminal Display Editor • Test and Diagnostic Equipment • Test Design Engineer • Time Displacement Error • Torque Drive Electronics • Total Differential Equation • Total Dose Experiment • Transition Diagram Editor • Two-Dimensional Equilibrium

TDEC Technical Development Evaluation Center

TDED Trade Data Elements Directory

TDEP Tracking Data Editing Program

TDES Tactical Data Encryption System

T/DEV Test/Data Evaluation and Validation

TDEV Test Data Evaluation and Validation

tdf Typeface Definition (file extension)

TDF Tab-Delimited File • Take Data Flag • Task Deletion Form • Télédiffusion De France • Telemetry Data File • Telemetry Data Format • Telemetry Distribution File • Telemetry Distribution Format • Test Data Folder • Test Description Folder • Trace Definition File • Transborder Data Flow • Transmission Decrease Factor • Transport Data File • Two Degrees of Freedom • Typeface Definition File

TDFC Transfert des Données Fiscales et Comptables

TDFCHB Telemetry Data Format Control Handbook

TDFD Telemetry Downlink File Description

TDFR Training Device Fabrication Request

TDFS Terminal Digit Fitting System

TDG Tandag (Philippines) (location identifier) • Telemetry Data Generation • Test Data Generator • Test Date Generator • Test Display Generator • Test Documentation Group • Tidal Dwarf Galaxy • Time Delay Generator • Twist Drill Gauge

TDH Tracking Data Handling

TDHC Time Domain Harmonic Coding

TDHS Tape Data Handling System

TDI Tamper Detection Indicator • Target Data Inventory • Task Description Item • Technical Data Interchange • Technology Demonstration Item • Telecommunications Data Interface • Temporary Disability Insurance • Test Data Interpolation • Time Delay and Integrate • Time Delay and Integration • Thomson Digital Image • Time Delay and Integration • Time-Delayed Integration • Time Delay Integrating • Time Delay Integration • Time-Dependent Integration • Tolerable Daily Intake • Training Development Institute • Traitement Digital des Informations • Transport Device Interface • Transport Driver Interface • Triple Display Indicator • Trusted Database Interpretation • Trusted Data Interpretation • Trusted DBMS Interpretation

TDIC Time Division Interface Controller

TDID Trade Data Interchange Directory

TDIF TOPEX/Poseidon Data Information Facility

TDIG Time Division Interface Group

TDIGM Time Division Interface Group Modified

TDII Technology Demonstration, Implementation, and Integration

TDIM Time Division Interface Module

TDIO Timing Data Input-Output

TDIP Total Disability Income Provisions

TDIPR Test Design In Process Review

TDIS Technical Data Impact Summary • Technical Documentation Information System • Terminal Data Input System • Time Distance • Training Development Information System

TDJ Tadjoura (Djibouti) (location identifier) • Transfer Delay Jitter

TDK Taldy-Kurgan (Kazakhstan) (location identifier)

TDL Tandil (Argentina-BA) (location identifier) • Task-Directed Learning • Technical Data Laboratory • Technical Data Library • Technical Document List • Techniques Development Laboratory • Temporal-Difference Learning • Threshold Detection Level • Transition Documentation Library • Translation Definition Language • Translator Definition Language • Treatment and Disposition Log • Tunable Diode Laser • Tunnel-Diode Logic

TDLAS Tunable Diode Laser Absorption Spectroscopy

TDLHS Tunable Diode Laser Heterodyne Spectrometer

TDLOA Training Device Letter Of Agreement

TDLR Training Device Letter Requirement

TDLRP Training Development Long-Range Plan

TDLS Tower Data-Link Service • Tunable Diode Laser Spectroscopy

Tdm Tandem

TDM Tableau de Définitions et de Mesures • Task Description Memorandum • Tausend Deustche Mark • Technical Data Management • Technical Division Manager • Technical Document Management • Technology Development Mission • Telemetric Data Monitor • Termination Decision Model • Test Data Management • Test Data Memorandum • Thermal Diffusion Method • Thermodynamic Molding • Time-Division Multiplex • Time-Division Multiplexed • Time-Division Multiplexer • Time-Division Multiplexing • Time Duration Modulation • Timing Distributor Module • Transportable - Deployable - Mobile • Transportation Demand Management • Trouble Detection and Monitoring • Tunnel-Diode Mixer

TDMA Tandem Differential Mobility Analyzer • Time-Division Multiple Access (*) • Time-Domain Multiple Access

TDMARN Time-Division Multiple Access Radio Net

TDMAS Teletype Data Multiplex Addressor System • Teletype Data Multiplex Address System

TDMC Technical Data Management Center

TDMD Time Division Multiplex Device • Time Division Multiplex Division

TDMF Time Division Matrix Function • Transparent Data Migration Facility

TDMG Telegraph and Data Message Generator

TDMM Time Division Memory Module

TDMP Technical Data Management Program • Training Device Maintenance Plan

TDMS Technical Data Management System • Telegraph Distortion Measuring Set • Telegraph Distortion Measuring System • Telemetry Data Monitor Set • Terminal Data Management System • Terminal Display Management System • Time Division Multiplex System • Time-shared Data Management System • Transmission Distortion Measuring Set

TDMUX Time-division Data Multiplexer

TDMX Time Division Matrix

TDN Time-Dependent Normalization • Topografische Dienst Nederland • Totally Digestible Nutrient

TDNN Time-Delay Neural Network

TDNO Thermal Detection of Nuclear Orientation

TDNS Training Device Needs Statement

TDO Technical Development Objective • Technical Divisions Office • Technology Development Office • Telemetry Data Officer • Time Delay Opening • Toledo Winlock (USA-WA) (location identifier) • Tornado • Transaction Data Object • Tripartite Development Objective

TDOA Time Delay Of Arrival • Time Difference Of Arrival

TDOP Time Dilution Of Precision

TDOS Tape/Disk Operating System

TdP Turbine de Puissance

TDP Tag Distribution Protocol • Target Data Panel • Target Detection Processor • Target Director Post • Technical Data Package • Technical Development Plan • Technical Dissemination Project • Technical Drawing Package • Technological Demonstration Programme • Technology and Development Program • Technology Demonstration Plan • Technology Demonstration Programme • Technology Demonstrator Programme • Technology Development Plan • Technology Development Programme • Teledata Processing • Telelocator Data Protocol • Telemetry Data Processing • Temperature Density Plotter • Temperature and Dewpoint • Test Data Package • Test Design Plan • Three Dipole Problem • Tidal Deformation Polarization • Time-Dependent Processor • Total Development Plan • Total Distribution Program • Tracking Data Processor • Training Development Plan • Transmitter Documentation Project • Transmitting Data Programming

TDPAC Time Differential Perturbed Angular Correlation

TDPAD Time Differential Perturbed Angular Distribution

TDPD Technical Data Program Document

TDPFO Temporary Duty Pending Further Orders

TDPI Treasury Distributed Processing Infrastructure

TDPL Technical Data Package List

TDPM Training Device Program Manager

TDPMIS Technical Data Package Materiel Information System

TDPMP Technical Data Package Management Plan

TDPS Tokyo Declaration of Physical Societies • Tracking Data Processing System

TDPSK Time Differential Phase Shift Keying

TDQAS Training Development Quality Assurance System

TDQM Total Data Quality Management

Tdr Tender

TDR Tape Data Register • Target Data Receiver • Teardown Deficiency Report • Technical Data Relay • Technical Data Report • Technical Data Request • Technical Deficiency Report • Technical Design Report • Technical Design Review • Technical Development Requirement • Technical Dispatch Reliability • Technical Documentary Report • Technical Documentation Report • Test Data Report • Test Discrepancy Report • Theodore (Australia-QLD) (location identifier) • Time Delay Relay • Time Division Reflectometer • Time-Domain Reflectometer • Time-Domain Reflectometry • Torque Differential Receiver • Total Defect Rate • Tous Droits Réservés • Tracking and Data Relay • Traffic Distribution Record • Training Device Requirements • Transistorized Digital Readout • Transmission de Données par Radio • Transmit Data Register • Transportation Discrepancy Report • Treasury Deposit Request

TDRB Test Data Review Board • Tool Design Review Board

T&DRE Tracking & Data Relay Equipment

TDRE Tracking and Data Relay Equipment

TDRL Temporary Disability Retired List • Terminal Distinguishable Regular Language

TDRR Test Data Recording and Retrieval

TDRRB Technical Data Requirement Review Board

TDRRC Training Device Requirements Review Committee

TDRS Telemetry Downlink Receiving Site • Text Data Retrieval System • Tracking and Data Relay Satellite

TDRSGS Tracking and Data Relay Satellite Ground Station

TDRSS Tracking and Data Relay Satellite System

TDRSSGT Tracking and Data Relay Satellite System Ground Terminal

TDRST Tracking and Data Relay Satellite Terminal

tds Ter Die Sumendum

TDS Tabular Data Stream • Tailored Delivery System • Tape Data Selector • Tape Duplication System • Teacher Diagnostics System • Technical Data System • Technical Directive System • Technical Documentation Standard • Technology Demonstration Satellite • Telemetry Distribution Switch • Telescope Data System • Terrestrial Data Services • Test Data Sheet • Test Data System • TEX Directory Structure • Thermal Degradation Sample • Thermal Design System • Thermal Desorption Spectroscopy • Thermal Diffuse Scattering • Time Delay Switch • Time-Dependent Sensitivity • Time - Distance - Speed • Time Distribution System • Time Division Subscriber • Time-Domain Spectroscopy • Timing Distribution System • Total Dissolved Solids • Total Distribution System • Track Data Simulator • Track Data Storage • Tracking and Data System • Tracking Display System • Training Development Study • Training Device Study • Training Device System • Transaction-Driven System • Transfert des Données Sociales • Translation and Docking Simulator • Transputer Development System

TD&SA Telephone, Data & Special Audio

TDSA Technical Data Status Accounting • Technical and Data Signals Analyzer • Telephone, Data and Special Audio • Tribal-Designated Statistical Area

TDSB Test Data Scoring Board

TDSCC Tidbindilla Deep Space Communications Complex

TDSE Task Data Space Element

TDSG Time Division Switching Group

TDSGM Time Division Switching Group Matrix

TDSI Traitement du Signal et de l'Image

TDSM Time Division Switching Matrix

TDSP Technical Data Support Package • Time-Dependent Signal Processor • Top-Down Structured Programming

TDSR Transmitter Data Service Request

TDSS Technical Data Storage System • Technical Document Support Software • Texas Deep Sky Survey • Transportation Decision Support System

TdT Traitement de Texte

TDT Tanda Tula (South Africa) (location identifier) • Target Designator Transmitter • Target Docking Trainer • Telemetry Data Tape • Temps Dynamique Terrestre • Terrestrial Dynamical Time • Terrestrial Dynamic Time • Test Data Team • Test Direction Team • Test Dwell Time • Tiempo Dinámico Terrestre • Topic Detection and Tracking • Tower Disconnect Technician • Transonic Dynamics Tunnel • Tunnel-Diode Transducer

TDTC Test, Development, and Training Center

TDTG Time Division Transmission Group • True Date-Time Group

TDTL Test Data Team Leader • Tunnel-Diode Transistor Logic

TDU Telemetry Decommutation Unit • Terminal Display Unit • Third Dredge-Up (episode) • Time Delay Unit • Time Display Unit • Time Distribution Unit • Tracking Display Unit

TDUM Tape Dump and Utility Monitor

TDUP Technical Data Usage Program

TDV Tanandava (Madagascar) (location identifier) • Technology Development Vehicle • Test Data Van • Tool Design Variance • Touchdown Velocity

TDVS Telemetry Data Verification System • Telemetry Decommutation Validation System

TDW Amarillo Tradewind (USA-TX) (location identifier) • Training Development Workload • Twisted Double Shielded

TDWG Taxonomic Database Working Group • Technology Development Working Group • Test Development Working Group

TDWM Training Development Workload Management

TDWO Test and Development Work Order

TDWP Training Development Workload Planner

TDWR Terminal Doppler Weather Radar

TDWT Transonic Dynamic Wind Tunnel

TDX Thermal Demand Transmitter • Time Division Multiplexing

TDy Temporary Duty

TDY Temporary Duty

TDZ Toledo (USA-OH) (location identifier) • Total Danger Zone • Touchdown Zone

TDZE Touchdown Zone Elevation

TDZL Touchdown Zone Lights

te Tegulu (ISO language code)

Te Tellur • Tellure • Tellurium • Température d'Équilibre • Tetanus

T/E Table of Equipment • Technical Evaluation • Transporter/Erector

T&E Test & Evaluation • Testing & Evaluation • Training & Education • Travel & Entertainment • Trial & Error

TE Talk Enable • Tangent Elevation • Task Element • Taux d'Erreurs • Teacher Enhancement • Technical Engineer • Technical Evaluation • Technical Exchange • Technicien d'Études • Technicien d'Exploitation • Technische Einheit • Tecklenburg (Germany) • Teileigentum • Teileinheit • Teilentladung • Telecommunications Engineering • Template Element • Temps d'Érection • Temps d'Escale • Temps d'Établissement • Temps d'Exploitation • Temps Estimé • Tension à l'Entrée • Tension d'Essai • Tension Électrique • Terminal Emulator • Terminal Equipment • Terminal Event • Terrain d'Envol • Terrestrial Ecosystem • Test and Engineering • Test and Evaluation • Test Edition • Test Engineer • Test Engineering • Test Equipment • Tête d'Enregistrement • Text End • Thermal Efficiency • Thermal Element • Thermal Expansion • Thermodynamic Equilibrium • Thermoelectric • Thermomètre Enregistreur • Tiempo de las Efemérides • Time Earliest • Time Expected • Time of End • Timing Electronics • Tirant d'Eau • Topographic Engineer • Total Effect • Total Energy • Totally Enclosed • Trace Element • Traction Électrique • Trading Expert • Traffic Engineering • Trägerfrequenzerzeuger • Trailing Edge • Training and Education • Training Effectiveness • Training Emphasis • Training Equipment • Trajectory Engineer • Trans-Earth • Transequatorial • Transformateur Électrique • Transfusioneinheit • Transmission Error • Transporter/Erector • Transverse Electric • Transverse Électrique • Transverse Electrostatic • Trennungsentschädigung • Trial and Error • Turbine à Engrenages • Turbine Engine • Tuyère d'Échappement • Twin Engine

TEA Task Equipment Analysis • Technical Exchange Agreement • Tela (Honduras) (location identifier) • Temperate East Asia • Temps d'Entretien Actif • Test Environment Assessment • Test Equipment Accessory • Test Equipment Analysis • Test and Evaluation Agency • Thermal Energy Analysis • Thermal Energy Analyzer • Torque Equilibrium Attitude • Track Error Angle • Training Effectiveness Analysis • Transferred Electron Amplifier • Translation Exception Address • Transportability Engineering Analysis • Transportation Engineering Agency • Transportation Engineering Agreement • Transverse Electric Atmospheric • Transversely Excited Atmospheric • Triethylamonium • Tunnel Emission Amplifier

TEAC Technical Educational Advisory Council • Tribunal Económico Administrativo Central (Spain) • Turbine Engine Analysis Check

Teach Teacher • Teaching

TEAF Test and Evaluation Analysis Facility

TEAL Transversely Excited Atmospheric Laser

TEAM Technique for Evaluation and Analysis of Maintainability • Technology-based Enhanced Autonomous Machine • Technology Exchange Assessment Methodology • Test - Evaluation - Analysis - Modeling • Top European Advertising Media • Total Environment Analysis and Management

TEAMS Technology Experiments for Advancing Missions in Space • Terrestrial Ecosystem Analysis and Modeling System • Test Evaluation And Monitoring System • Time-of-flight Energy Angle Mass Spectrometer

TEAMUP Test, Evaluation, Analysis, and Management Uniformity Plan

TEANS Test and Evaluation Analysis Network System

TEAP Terminating Extended-Area Descriptor • Trajectory Error Analysis Program • Transversely Excited Atmospheric Pressure

TEAR Transportability Engineering Analysis Report

TEARRR Times, Elevations, Azimuths, Ranges and Range Rates

TEARS Telecommunications Equipment Automated Retrieval System

TEAS Technical and Engineering Acquisition Support • Test and Evaluation Analysis System • Test and Experiment Activity Summary

TEAT Technical Engineering Analysis Team

TEB Technical Evaluation Board • Tender Evaluation Board • Test and Evaluation Board • Teterboro (USA-NJ) (location identifier) • Thread Environment Block • Tone-Encoded Burst • Tropical Experiment Board

TEBALDI Télécommande Ballon Longue Distance

TEBC Time - Ephemeris - Barycenter Correction

TEC Teaching Excellence Center • Technical • Technical Evaluation Committee • Technology Exchange Co. • Technology - Employment - Competitiveness • Temperature Electric Cooler • Temporary Engineering Change • Terminal Equipment Configuration • Test and Evaluation Center • Test and Evaluation Committee • Test Environment Certification • Test Equipment Center • Test Equipment Committee • Theory of Electroconformational Coupling • Thermal Electric Cooler • Thermal Expansion Coefficient • Thermionic Energy Conversion • Thermo-Electric Cooled • Thermo-Electric Cooler • Tokyo Electronics Corporation • Tonne d'Équivalent Charbon • Topographic Engineering Center • Total Electric Content • Total Electron Content (*) • Total Environment Centre • Total Estimated Cost • Tower En-route Control • Training and Enterprise Council • Training Evaluation and Control • Training Extension Course • Trans-Earth Coast • Transistor à Effet de Champ • Transports En Commun • Transports Européens Combinés • Type Equipment Code

TECCB Training Equipment Change Control Board

TECCS Tactical Engineer Command and Control System

TECD Training Equipment Change Directive

TECDOC Technical Documentation

TECE Teleprinter Error Correction Equipment

TECEVAL Technical Evaluation

TECG Test and Evaluation Coordination Group

Tech Technical • Technician • Technics • Technique • Technische • Technological • Technologie • Technologique • Technology • Technology (periodical code)

TECH Team Encouraging Career Horizons • Technical • Technician • Technology

TECHDOC Technical Document • Technical Documentation

TECHEVAL Technical Evaluation

Techn Technical • Technician • Technics • Technique • Technische • Technological • Technologie • Technologique • Technology

Technol Technological • Technologie • Technologique • Technology

TECHREP Technical Report • Technical Representative

TECL Test Equipment Configuration Log

TECM Test Equipment Commodity Manager

Tecn Tecnicismo

TecN Technical News (periodical code)

TECNET Technical Network

Tecnol Tecnología

TECO Tape Editor and Corrector • Test and Evaluation Coordination Office • Test and Evaluation Coordination Officer • Text Editor and Corrector • Turbine Engine Checkout

TECOMAC Technical Conference on Observation and Measurements of Atmospheric Contaminants

TECP Trainer Engineering Change Proposal

TECR Technical Requirement • Test Equipment Change Requirement • Training Equipment Change Request

TECRM Technology-Enabled Customer Relation Management

TECRO Taiwan's Economic and Cultural Representative Office

TecRv Technology Review (periodical code)

TECS Tether Elevator/Crawler System • Total Energy Control System • Total Environment Control System

Tecto Tectonophysics (periodical code)

TECU Total Electron Content Unit

TED Tableau Effectif-Dotation • Tailored Edge-ray Design (method) • Technological and Economic Dynamics • Tenders Electronic Daily • Test, Evaluation and Development • Test and Evaluation Directorate • Test and Evaluation Division • Testability and Embedded Diagnostic • Test Evaluation Directorate • Thermoelectric Device • Thermoionic Emission Detector • Thisted (Denmark) (location identifier) • Threat Environment Description • Tiny Editor • Total Energy Detector • Towards Electronic Democracy (project) • Tracking Error Detector • Trailing Edge Down • Training Equipment Development • Traitement Électronique des Données • Transferred-Electron Device • Translation Error Detection • Translation Error Detector • Trunk Encryption Device • Turbine Engine Diagnostics

TEDB Test Engineering Database

TEDH Tribunal Européen des Droits de l'Homme • Tribunal Europeo de Derechos Humanos

TEDI Terminology Editor

TEDIS Trade Electronic Data Interchange System

TEE Tape Editing Equipment • Tbessa (Algeria) (location identifier) • Telecommunications Engineering Establishment • Test Equipment Engineering • Trans Europ Express • Triaxial Earth Ellipsoid

TEECP Training Equipment Engineering Change Proposal

TEEM Techno-Economico-Environmental Model • Test and Evaluation Estimating Model • Trans Europ Express Marchandises • Two-phase Extended Evaluation in Microgravity

TEEP Training Effectiveness Evaluation Plan

TEer Teileigentümer

TEES Texas Engineering Experiment Station (USA-TX)

TEF Technology Experiment Facility • Technology Exposure Facility • Telfer (Australia-WA) (location identifier) • Thermoelectric Fan • Total Environment Facility • Toxicity Equivalence Factor

TEFB Test and Evaluation Facility Base

TEFF Turbine Engine Fatigue Facility

TEFL Teaching English as a Foreign Language

TEFLON Polytetrafluoroethylene

TEG Taux Effectif Global • Technical Evaluation Group • Tenkodogo (Burkina Faso) (location identifier) • Test and Evaluation Group • Test Evaluation Group • Thermoelectric Generator • Top Edges Gilt • Tuyère d'Éjection des Gaz

TEGa Triethyl Gallium

TEGA Thermal and Evolved Gas Analyzer

TEGD Technical Enforcement Guidance Document

TEGFET Two-dimensional Electron Gas Field Effect Transistor

TEGP Theory and Experimentation in Gravitational Physics

Teh Tehran (Iran)

TEH Tetlin (USA-AK) (location identifier)

TEHM Terrestrial Ecosystem Hydrology Model

TEHR Tension Elliptical Hollow Rod

TEI Technical and Economic Information • Technical Economic Item • Technical Engineering Item • Terminal Endpoint Identifier • Terminal Equipment Identifier • Text Electronic Interchange • Text Encoding Initiative • Tezu (India) (location identifier) • The Education Initiative • Thermal Electron Imager • Thermoelectric Incubator • Thermostat Électronique Intelligent • Traitement Électronique de l'Information • Trans-Earth Injection • Tube à Entretien d'Images • Tusas Engine Industries

TEIFU Thousand Element Integral Field Unit

teilw teilweise

Teilz Teilzahlung

TEIn Triethyl Indium

TEIN Test and Evaluation Identification Number

TEIPT Test and Evaluation Integrated Product Team

TEISC Test and Evaluation Information Support Capability

TEK Tatitlek (USA-AK) (location identifier) • Traffic Encryption Key

TekEl Tekhnicheskaia Elektrodinamika (periodical code)

Tél Télécommunications • Télémètre • Télémétrie • Téléphone • Télescope • Télescopique

Tel Telecommunications • Telegram • Telegraph • Telegraphic • Telemeter • Telemetry • Telephone • Telescope • Telescopic • Telescopium-Telescopii (constellation)

TEL Technische Einsatzleitung • Telecommunications • Telephone • Telupid (Malaysia) (location identifier) • Transporter - Erector - Launcher

TELAPO Terkán Lajos Public Observatory (Hungary)

TELB Test Engineering Logbook

TelCo Telephone Company

TELCON Telephone Conversation

TELE Telescope • Total Extinction Level Event

Telec Telecommunicación

Télécom Télécommunication • Télécommunications

Telecom Telecommunicate • Telecommunicating • Telecommunications

TELECON Telephone Conference • Telephone Conversation

TELEDAC Telemetric Data Converter

TELEFAX Telecommunications Facsimile

Télég Télégramme • Télégraphe • Télégraphie • Télégraphique

Teleg Telegram • Telegraph • Telegraphic • Telegraphy

Télém Télémètre • Télémétrie

Telem Telemeter • Telemetry

TELEMED Telemedicine

TELENET Telecommunications Network • Teletypewriter Network

Téléph Téléphone • Téléphonie

Teleph Telephone • Telephony

TELERJ Telecomunicações do Rio de Janeiro (Brazil)

Télesc Télescope • Télescopique

Telesc Telescope • Telescopic

TELESAT Telecommunications Satellite

Telet Telettra (periodical code)

Télév Télévision

Telev Television

TELEX Teleprinter Exchange • Teletypewriter Exchange

TELH Telheiro

TELIAS Telidon Image Analysis System

TelIn Telematics and Informatics (periodical code)

Tell Tellus (periodical code)

TellB Tellus Series B - Chemical and Physical Meteorology (periodical code)

TELMS Telecommunications Management System

TELNET Telecommunications Network • Teletype Network

Tel. No. Telephone Number

TeLoA Technika Lotnicza i Astronautyczna (periodical code)

TelOp Teleconference Operator

TELOPS Telemetry On-line Processing System

TELOS The Electronic Library Of Science

TELSAT Teledetection by Satellite

TELSEC Telecommunications Security

TELUS Telemetric Universal Sensor

TEM Tableau Effectif-Matériel • Target Engine Module • Target Experiment Module • Technical Exchange Meeting • Temora (Australia-NSW) (location identifier) • Terrain Evaluation Module • Terrestrial Ecosystem Model • Test Evaluation Matrix • Tête Électromécanique • Thermoelectric Module • Tot En Met • Transformed Eulerian Mean • Transform Engine Motor • Transition Electron Microscope • Transition Electron Microscopy • Transmission Electron Microscope (*) • Transmission Electron Microscopy (*) • Transportability Engineering Management • Transverse Electric Mode • Transverse Electromagnetic • Travail Électrique des Métaux

TEMA Test and Evaluation Management Agency • Test Evaluation Matrix • Training, Educational and Mutual Assistance • Training, Education and Mutual Assistance

TEMAC Test and Evaluation Manager Committee • Trailing Edge Mean Aerodynamic Chord

TEMC Test and Evaluation Management Course

TEMEX Telemetry Exchange

TEMIC Telecommunications Executive Management Institute of Canada

TEML Turbo Editor Macro Language

TEMO Test and Evaluation Management Office

Temp Temperate • Temperature • Température • Tempered • Template • Tempo • Temporal • Temporary • Tempore

TEMP Temperature • Test and Evaluation Master Plan • Test Evaluation Master Plan • Thermal Energy Management Process • Transportability Engineering Management Program

TeMPEST Texas McDonald Photometric Extrasolar Search for Transits

TEMPEST Transient Electromagnetic Emanations Standard

TEMPO Technical Electronic Management Planning Organization • Time and Earth Motion Precision Observation • Timing Experiment Modules for PAMELA Observations • Tyrrhenian Eddy Multiplatform Observation

TEMPUS Tiegelfreies Elektro-Magnetisches Prozessieren Unter Schwerelosigkeit • Trans European Mobility Programme for University Studies

TEMS Technical Engineering and Management Support • Technical Engineering Management Support • Test and Evaluation Mission Simulator • Test Equipment Maintenance Set • Tethered Magnetometer System • Transport Environment Monitoring System • Turbine Engine Monitoring System

TEMSI Temps Significatif

TEMSS Test Equipment Mass Storage System

Ten Tenant • Tenement

TEN Technicien de l'Éducation Nationale • Tenente • Tongren (China-PRC) (location identifier) • Transeuropean Network • Trans-Europ-Nacht (railcar) • Trans-Europ-Nuit (railcar)

TENCAP Tactical Exploitation of National Capabilities

Tency Tenancy

Tend Tendenz • Tender • Tendeur

TENIS Terminal Environment Network Information System

Tenn Tennessee (USA)

Tens Tension

TENS Transcutaneous Electrical Nerve Stimulation • Transcutaneous Electrical Nerve Stimulator • Trans European Networks

T&EO Training & Evaluation Outline

TEO Technology Enhancement Office • Technology Executive Officer • Terapo (Papua New Guinea) (location identifier) • Terrestrial Ephemeris Origin • Test and Evaluation Objective • Test Equipment Operator • Theory of Equipment Operations • Training and Evaluation Outline • Transferred Electron Oscillator

TEOA Test and Evaluation Objective Annex

TEOC Trømsø Earth Observation Center

TeoEl Teoreticheskaia Elektrotekhnika (periodical code)

Teol Teología

TEOM Tapered Element Oscillating Microbalance • Transformer Environment Overcurrent Monitor

TEOP Two-Electron One-Photon (transition)

TeoPM Teoreticheskaia i Prikladnaia Mekhanika (periodical code)

TEOS Tetraethoxysilane

TeoVP Teoriia Veroiatnostei i ee Primeneniia (periodical code)

tep Top Edges Gilt

TEP Tabletop Enhancement Program • Tactical Electronic Plot • Tactical Entry Point • Task Execution Plan • Teaching Electronic Publishing • Technical Evaluation Panel • Technical Evaluation Program • Technology and Environmental Policy • Technology Enhancement Program • Temps d'Entretien Préventif • Temps Écoulé Prévu • Teptep (Papua New Guinea) (location identifier) • Terminal Edge Pixel • Terminal Emulation Processor • Test and Evaluation Plan • Test Evaluation Plan • Thermal Exchange Pump • Thermoelectric Power • Thermostat

Électronique Programmable • Tonne Équilavent Pétrole • Traded Endowment Policy • Traffic Engineering Program • Transatlantic Economic Partnership • Transformable Edge Pair • Transit of Extrasolar Planet • Transit of Extrasolar Planets (network) • Trust and Estate Practitioner • Turbine Exhaust Pressure

TEPA Taiwan Environmental Protection Agency

TEPC Tissue Equivalent Proportional Counter

TEPCO Tokyo Electronic Power Co.

TEPG Thermoionic Electrical Power Generator

Teplo Teploenergetika (periodical code)

TEPP Test and Evaluation Program Plan • Tetraethyl Pyrophosphate • Transportation Emergency Preparedness Program

TEPS Test and Evaluation Plan Summary

Tepsg Teplofizicheskie svoistva gazov (periodical code)

T/EPSCG Time/Event-Phased Security Classification Guide

TepT Teplofizika i Teplotekhnika (periodical code)

TepVT Teplofizika Vysokikh Temperatur (periodical code)

TEQ Toxicity Equivalence Quotient

TEQB Thermal Equilibration

Ter Terrace • Terrain • Territorial • Territory • Tertiary • Teruel (Spain)

TER Technical Expert Representative • Terceira Island - Lajes (Azores) (location identifier) • Term Error Rate • Terrace • Terrain • Test and Evaluation Report • Test Effectiveness Ratio • Test Equipment Readiness • Test Evaluation Report • Thermal Eclipse Reading • Time Estimating Relationship • Total Expense Ratio • Train Express Régional (France) • Transmission Equivalent Resistance • Travel Expense Report • Triple Ejection Rack

TERA Temporary Early Retirement Authority • Terminal Effects Research and Analysis

TERAPIX Traitement Élémentaire de Réduction et Analyse de Pixels

Terat Teratology

TERB Time En-route Bulletin

TERC Technical Education Research Center • Test and Evaluation Resources Committee • Total Environmental Response Contract • Total Environmental Restoration Contract • Turbine Engine Research Center

TerCo Osservatorio Astronomico di Colluriana Teramo Contributi (periodical code)

TERCOM Terrain Contour Matching

TERENA Transeuropean Research and Education Networking Association

TEREZA Température Rotationnelle en Zone Aurorale

TERI Tata Energy Research Institute (India) • Terrestrial Ecosystem Research Initiative

TERIB Test and Evaluation Resource and Investment Board

TERINT Terrain Intelligence

TERL Test Equipment Readiness List

TERLS Thumba Equatorial Rocket Launching Station (India)

Term Terminación • Terminal • Terminate • Terminated • Terminateur • Terminating • Termination • Terminator • Terminological • Terminologie • Terminologique • Terminology

TERM Tackling Environmental Resource Management • Temporary Equipment Recovery Mission • Terminal • Terminate • Transport and Environment Reporting Mechanism • Transport and Environment Research Mechanism

TerMm Osservatorio Astronomico di Colluriana Teramo Memorie ed Osservazioni (periodical code)

TERMPWR Terminator Power

TERMS Test Equipment Reporting Management System

TerNC Osservatorio Astronomico di Colluriana Teramo Note e Comunicazioni (periodical code)

TERP Technical Evaluation Review Panel • Terrain Elevation Retrieval Program

TERPS Terminal Enquiry/Response Programming System

Terr Terrace • Terrain • Terrestre • Terrestrial • Territoire • Territorial • Territory

TERR Terrace • Terreiro

TERRA Technique of Reproducing the Response for ASCA • Tecniche di Elaborazione e Rilevamento delle Risorse Ambientali • Terrestrial Ecosystems Regional Research and Analysis

TERRIERS Tomographic Experiment using Radiative Recombinative Ionospheric EUV and Radio Sources

Territ Territorial

TERS Tropical Earth Resources Satellite • Tropical Environments Remote Sensing

TERSE Tunable Etalon Remote Sounder for the Earth

TERSS Tasmanian Earth Resources Satellite Station (Australia-TAS)

TERSSE Total Earth Resources System for the Shuttle Era

TERSWG Tropical Environments Remote Sensing Working Group

Tert Tertiaire • Tertiary

TERU Training and Employment Research Unit

Terz Terrazzo

TES Telemetry Evaluation Station • Telephony Earth Station • Terminal Encounter System • Terrain Edit System • Tessenei (Ethiopia) (location identifier) • Text Editing System • Thermal Emission Spectrometer • Thermal Enclosure System • Thermal Energy Storage • Thermophysikalische Eigenschaften Schmelzen • Throughput Enhancement Supervisor • Tidal Erection Surge • Tilting Etalon Spectrometer • Time-Enriched Statechart • Training of European Statisticians • Transition Edge Sensor • Transition-Edge Superconductor • Travel Expert System • Tropospheric Emission Spectrometer

TESA Table Équatoriale Solaire Automatisée • Technical Expert Status Accreditation • Technology, Economic and Social Applications

TESAC Temperature - Salinity - Currents

TESCAS Tropospheric Environmental Studies by Laser Soundings

TESH Technical Shop

TESL Teaching English as a Second Language • Teaching of English as a Second Language

Tesla TESLA Electronics (periodical code)

TESLA TeV Energy Superconducting Linear Accelerator

TESLAS Tropospheric Environmental Studies by Laser Sounding

TESOL Teachers of English to Speakers of Other Languages • Teaching English to Speakers of Other Languages

TESOS Telecentric Solar Spectrograph • Telecentric Solar Spectrometer (*)

TESP Test and Evaluation Security Plan

TESPEL Tracer Encapsulated Solid Pellet

TESR Test Equipment Status Report • Torsion Elliptical Solid Rod

TESRE Technologie E Studio delle Radiazioni Extraterrestri (CNR)

TESRP Test and Evaluation Support Resource Plan

TESS Tactical Environmental Support System

TESSA Testbed for Evaluating Spatial Similarity Algorithms • Texas Severe Storm Association

TESSE Test Environment Support and System Enhancement

Test Testament • Testator • Testatrix • Testigo • Testimonial • Testimonio • Testimony

TEST Thesaurus of Engineering and Scientific Terms • Time-Efficient Sequencer of Tests • Total Elapsed Sidereal Time • Transferable European Space Technology

Testa Testament • Testamento

TESTA Transeuropean Services for Telematics between Administrations

Testis Testigos

TESTS Technical European Support for Testing Standards

TESZ Transeuropean Suture Zone

Tet Tetrahedron

TET Technical Evaluation Test • Technicien d'Études et de Travaux • Test Environment Toolkit • Test Equipment Team • Test Equipment Tool • Test Evaluation Team • Tete Matunda (Mozambique) (location identifier) • Theoretical Evolutionary Track • Thermionic Emission Technique • Thermometric Enthalpimetric Titration • Total Elapsed Time • Trajectory Events Timeline • Transistor Evaluation Test • Turbine Entry Temperature • Turbine Exhaust Temperature

TETB Thyssen Edelstahl Technische Berichte (periodical code)

TETM Thermal Effects Test Model

Tetr Tetragonal

TETRA Terrestrial Trunked Radio • Transeuropean Trunked Radio Access • Transeuropean Trunk Radio

TEU Te Anau (New Zealand) (location identifier) • Télé-Enseignement Universitaire • Telemetry Electronics Unit • Telemetry Equipment Unit • Treaty on European Union • Twenty-foot Equivalent Unit

TEUE Training Effectiveness User Evaluation

Teut Teuton • Teutonic

Teutón Teutónico

TeV Teraelectronvolt

TEV Target Engine Vehicle • Taux d'Entrée en Virage • Thermoelectric Voltage

TEW Total Equivalent Weight • Threat Early Warning

TEWG Technical Executive Working Group • Test and Evaluation Working Group

TEWS Tactical Electronic Warfare System

tex TEX or LaTEX Document (file extension)

Tex Texan • Texas (USA)

TEx Transition Experiment

TEX Telluride (USA-CO) (location identifier) • Text Executive Processor • Transit Exchange

TEXES Texas Echelon-Cross-Echelle Spectrograph

TEXnet Texas Network

TEXONO Taiwan experiment On Neutrinos

TexOx Texas-Oxford (survey)

TEXOX Texas-Oxford (survey)

text ASCII Text (file extension)

TEXT Texas Experimental Tokamak

TEXUS Technologische Experiment Unter Schwerelosigkeit

TEY Thingeyri (Iceland) (location identifier)

TEZ Tezpur (India) (location identifier) • Total Exclusion Zone

t/f True/False

tf Term Frequency • Time Frame • Tonne-Force

T/F Terrain Following • True/False • Turbofan

T&F Time & Frequency

TF French Southern Territories (ISO country code) • Iceland (international aircraft registration code) • Southern Territories (French_) (ISO country code) • Tabulating Form • Target Field • Task Force • Taux de Fiabilité • Tax Free • Technical fault • Telefax • Telemetry Format • Temperature Factor • Temperature Function • Temps de Fonctionnement • Tenue en Fatigue • Term Frequency • Terrain Following • Territoire Français Austral (code ISO de pays) • Test Facility • Test Fixture • Test Flight • Text Flow • Thermomètre Fahrenheit • Thin Film • Throughfall • Tidal Field • Tidal Friction • Tierra del Fuego (Argentina) • Time and Frequency • Time Frame • To Fill • To Follow • Trägerfrequenz • Traînée de Frottement • Train Fixe • Transaction Format • Transfer Function • Transformateur • Transformée de Fourier • Transportfunktionsteil • Très Fort • Très Forte • Triple Fraction • Triple Frequency • Trust Framework • Trust Fund • Tube à Flamme • Tully-Fisher (relation) • Tunable Filter • Turbofan • Turbo-Fusée • Tuyau Flexible

TFA Target Factor Analysis • Taux de Fausse Alarme • Technology Forecasting and Assessment • Tilfalmin (Papua New Guinea) (location identifier) • Transfer Function Analyzer • Transistor Feedback Analyzer • Transparent File Access • Trifluoroacetic Acid • Trust Fund Account

TFAC Taux de Fausse Alarme Constant • To Facility

TFAP Tropical Forestry Action Plan

Tfc Traffic

TFC Telemetry File Catalog • Telemetry Format and Control • Telescopic Field Center • Time From Cutoff • Total Fixed Cost • Traffic • Transfer Function Computer • Transistorized Frequency Converter • Trigonometric Function Computer • Tunable Frequency Converter

TFCS Triplex Flight Control System

TFCV Theoretical Foundations of Computer Vision (workshop)

TF/D Time Frequency Dissemination

TFD Tactical Feasibility Decision • Television Feasibility Demonstration • Termination For Default • Terrain-Following Device • Test Flow Diagram • Thermo Field Dynamics • Thin-Film Detector • Thin-Film Diode • Thomas - Fermi - Dirac (model) • Time Frequency Digitizer • Time-Frequency Display • Time Frequency Dissemination • Total Frequency Deviation • Transcription Factor Database • Transformée de Fourier Discrète

TFDD Text File Device Driver

TFDM Technical Feasibility Demonstration Model

TFE Teflon • Tetrafluoroethylene • Thermionic Fuel Element • Time From Event • Transportation Feasibility Estimator • Türfreisprecheinrichtung

TFEL Thin-Film Electroluminescent

TFF Tefé (Brazil-AM) (location identifier) • Telescope Filter Function • Time of Free Fall

TFFAP Teoriia Funktsii Funktsional nyi Analiz i ikh Prilozheniia (periodical code)

TFFF Thermal Field Flow Fractionation

TFG Telecommunications Functional Group • Tracking Frequency Gauge • Transportation Facility Guide

TFGS Time and Frequency Generation System

TFH Technische Fachhochschule (Germany) • Timing Failure Handler

TFHR Task Force Human Resources

TFI Technology Flight Instrumentation • Temporal Fractional Integral • Thin-Film Inductive • Time From Ignition • Tufi (Papua New Guinea) (location identifier)

TFIDF Term Frequency times Inverse Document Frequency

TFizG Teploobmen i Fizicheskaia Gazodinamika (periodical code)

TFK Turbulent Flare Kernel

TFL Telemetry Format Load • Teófilo Otoni (Brazil-MG) (location identifier) • Time Filtering • Time From Launch

TFLES Time-Filtered Large Eddy Simulation

TFLOPS Tera Floating Point Operations Per Second • Trillion of Floating Point Operations Per Second

tfm Tagged Font Metrics (file extension)

TFM Tagged Font Metrics • Telefomin (Papua New Guinea) (location identifier) • TeX Font Metric • Traffic Flow Management • Transmit Frame Memory • Transmitter Frequency Multiplier • Treasury Financial Manual • Trusted Facility Manual

TFMRP Transfer Function Mode Reduction Package

TFMS Text and File Management System • Time-of-Flight Mass Spectrometer

TFN Tenerife Norte (Canary Islands) (location identifier) • Till Further Notice • Tribe Flood Network

Tfno Teléfono

TFO Tactical Fiber Optics • Technology Flight Opportunity • Trainer Fabrication Organization • Transaction For Others • Transferring From Original

TFOCA Thin Fiber Optic Cable Assembly

TFODM Task Force on Observations and Data Management

TFOL Tape File Octal Load • Thermo Field Dynamics

TFOM Testability Figure Of Measure

TFOV Target Field Of View • Telescope Field Of View • Total Field Of View

TFP Test Facility Program • Thermal Front Parameter • Time and Frequency Processing • TOPS Filing Protocol • Total Factor Productivity • Tunable Fabry-Pérot

TFPC Thin-Film Polarizing Cube • Toll Fraud Prevention Committee

TFPF Table Formats and Parameters File

TFQMR Transpose-Free Quasi-Minimal Residual

Tfr Transfer

TFR Tarif Forfaitaire de Remboursement • Tarif Forfaitaire de Responsabilité • Temporary Flight Restriction • Terrain-Following Radar • Test Facilities and Resources • Tightly Folded Resonator • Time From Receipt • Tokamak de Fontenay-aux-Roses (France) • Total Final Report • Total Final Reporting • Transfer Function Response • Transformación de Fourier Rápida • Transformée de Fourier Rapide • Transverse Field Region • Traveller Failure Report • Trouble and Failure Report • Tully-Fisher Relation • Tunable Frequency Range • Twisted Flux Rope

TFRD Test Facilities Requirement Document

TFS Table File System • Taux de Formation Stellaire • Teaching From Space (program) • Telemetry Format Selection • Temporary File System • Tenerife Sur - Reina Sofia (Canary Islands) (location identifier) • Terminal Fix Sensor • Thermal and Fluid Systems • Thursday - Friday - Saturday • Time and Frequency Standard • Traffic Flow Security • Translucent File Service • Translucent File System • Trusted File Server • Tunable Frequency Source

TFSS Technical Facilities Subsystem

TFT Taftan (Pakistan) (location identifier) • Technical Feasibility Test • Technical Feasibility Testing • Technology For Technology • Tensor Field Theory • Thermal Fatigue Test • Thin-Film Technology • Thin-Film Transistor • Thought Field Therapy • Time From Transmission • Tit For Tat

TFTEO Thin Film Transistor Electro-optical

TFTOA Technical Feasibility and Trade Off Analysis

TFTP Trivial File Transfer on PCs • Trivial File Transfer Protocol (*)

TFTR Tokamak Fusion Test Reactor

TFTS Terrestrial Flight Telecommunications System • Terrestrial Flight Telephone System (*)

TFTT Thanks For The Thoughts

TFU Test Facility Utilization • Theoretical First Unit • Typical Follow-Up

tFWM Temporal Four Wave Mixing

TFxIDF Term Frequency and Inverse Document Frequency

TFY Tarfaya (Morocco) (location identifier)

TFYT Thanks For Your Time

TFZ Tropospheric Frontal Zone

tg Tajik (ISO language code) • Tangent • Tangente

Tg Tanjong • Tanjung

TG Guatemala (international aircraft registration code) • Tail Gear • Task Group • Technical Group • Technical Guidance • Technical Guide • Technology Gap • Telegiornale • Telegramme • Téléguidage • Telemetry Group • Terme Générique • Terminal Guidance • Test Group • Test Guide • Test Guideline • Test Guaranteed • Thermogravimetry • Tiefgarage • Timing Gate • Togo (international vehicle registration code) • Togo (ISO country code) • Trace Gas • Tracking and Guidance • Trainer's Guide • Training Group • Transaction Group • Transformational Generative • Transformational Grammar • Transform Group • Transmission Group • Trésorerie Générale • Triglyceride • Tube de Guidage • Tuned Grid • Turbine à Gaz • Turbine Générateur • Turbogénérateur

tga Targa Graphics (file extension)

TGA Tandy Graphics Adapter • Task Group on AXAF • Télégestion des Abonnés • Thermal Gravimetric Analysis • Thermogravimeter Analyzer • Thermogravimetric Analysis • Total gas Age • Trace Gas Analyzer • Transient Gas Assessment • Transportable Ground Antenna • Trusted Global Advisor • Turbogénérateur d'Air

TGAO Technologie de Groupe Assistée par Ordinateur

TGB Tagbita (Philippines) (location identifier) • Task Generator Bulletin • Track Group Block • Très Grande Bibliothèque

TGBE Track Group Block Entry

TGBL Through Government Bill of Lading

TGBNTOO Task Group on the BMDO New Technology Orbital Observatory

TGC Task Group Chief • Test Generator Controller • Tradable Green Certificate • Transmission Group Control • Transmit Gain Control • Trunk Group Cluster • Trunk Group Communications

TGCM Thermospheric General Circulation Model

TGCP Test Ground Control Point

TGCR Thermal Gross Cloud Restoral

TGCT Thermal Gross Cloud Test

TGD Timing Group Delay • Titograd (Yugoslavia-Montenegro) (location identifier) • Total Gas Distribution • Trajectory and Guidance Data

TGDDIS Trace Gas Dynamics Data Information System

TGDF Turbulent Gasjet Diffusion Flame

TGE Tidal Giant Erection • Transverse Gravimeter Experiment • Très Grand Équipement • Turbogénérateur Externe • Tuskegee - Sharpe Field (USA-AL) (location identifier)

TGEA Task Group on Education in Astronomy (AAS)

TGEN Table Generate

TGF Terrain Feature Generation • Tignes (France) (location identifier) • Transforming Growth Factor

TGG Kuala Terengganu (Malaysia) (location identifier) • Third-Generation Graphics • Third-Generation Gyro • Transformational-Generative Grammar • Turbogénérateur de Gaz

TGGPB Task Group on Gravity Probe B

TGGRS Telemetry /GPS Ground Receiving System

TGH Tongoa (Vanuatu) (location identifier)

TGI Target Group Identifier • Technique Graphique Industrielle • Temperature Growth Index • Tingo Maria (Peru) (location identifier) • Transaction-Generated Information • Tribunal de Grande Instance

TGID Trunk Group Identification Number

TGIF Transportable Ground Interface Facility

TGIR Test Generation with Infrared Reasoning • Turning Goals Into Reality (conference)

TGIRS Two Grating Infrared Spectrometer

TGJ Tiga (Loyalty Islands) (location identifier)

Tgl Toggle

TGL Tagula (Papua New Guinea) (location identifier) • Task Group Leader • Touch and Go Landing

TGM Timetable Generation Module • Tirgu Mures (Romania) (location identifier) • Trace Gas Model • Transmission Group Module • Trunk Group Multiplexer

TGMA Tortugas Mountain Geothermal Area (USA-NM)

TGME Total Generalized M-Estimation

TGMOW Transmission Group Module Orderwire

TGMT Temperature Gradient Mass Transport

tgn Tangent

TGN Traralgon (Australia-VIC) (location identifier)

TGO Time To Go • Togo (ISO country code) • Tongliao (China-PRC) (location identifier)

TGOESS Task Group on Exploring Organic Environments in the Solar System (NRC)

Tgov Tegenover

TGOWG Teleoperator Ground Operations Working Group

TGP Tidal Generating Potential • Très Gros Plan

TGPl Tiefgaragenplatz

TGR Timing Gap Record • Touggourt (Algeria) (location identifier)

TGRA Très Grand Rayon d'Action

TGRAP Task Group on Research and Analysis Programs

TGRD Turbine à Gaz à Réaction Directe

TGRISS Task Group on Research on the International Space Station

TGRS Transient Gamma-Ray Spectrometer

TGS Chokwe (Mozambique) (location identifier) • Telemetry Ground Station • Telemetry Ground System • Terminal Guidance Sensor • Terminal Guidance System • Text and Graphics Server • The Great Spitter • Thermogravimetric System • Thermogravimetry System • Ticket-Granting Server • Ticket-Granting Service • Time Generation System • TOPEX Ground System • Translator Generator System • Transmission Grating Spectrometer • Transportable Ground Station • Triglycine Sulfate

TGSAA Task Group on Space Astronomy and Astrophysics

TGSE Telemetry Ground Support Equipment • Test Ground Support Equipment

TgSF Télégraphie Sans Fil

TGSS Terminal Guidance Sensor System

Tgt Target

TGT Tanga (Tanzania) (location identifier) • Target • Temperature Gradient Technique • Ticket-Granting Ticket • TDRSS Ground Terminal • Transaction Group Type • Turbine Gas Temperature

TGTM Transportation Guidance Technical Manual

TGU Target Generation Unit • Technical Guidance Unit • Tegucigalpa Toncontín (Honduras) (location identifier) • Théorie de Grande Unification • Trunk Compatibility Unit

TGV Targovishte (Bulgaria) (location identifier) • Ten Gevolge Van • Test Generation with Verification • Train à Grande Vitesse (*) • Tren de Gran Velocidad • Très Grande Vigilance • Très Grande Vitesse

TGW Tropospheric Gravity Wave

TGX Tingrela (Ivory Coast) (location identifier)

TGZ Tuxtla Gutierrez - Llano San Juan (Mexico) (location identifier)

th Hyperbolic Tangent • Tangente Hyperbolique • Thai (ISO language code) • Thailand

Th Thai • Theoretical • Theoretische • Théorie • Théorique • Theory • Thermal • Thorium • Thursday

T&H Test & Handling • Transportation & Handling

T/H Track/Hold

TH Table Header • Technische Hochschule • Technische Hogeschool • Temporary Hold • Territory of Hawaii • Thailand (ISO country code) • Thailande (code ISO de pays) • Thesaurus • Third Harmonic • Threshold • Time Hopping • Titre Homologué • Totalisateur Horaire • Town Hall • Track and Hold • Train Haut • Transmission Header • Transportation and Handling • Transport Header • Traufhöhe • Triple Hydrant • True Heading • Turbine à Hélice • Turbine Hydraulique • Turbo-Hélice

THA Test Hazard Analysis • Thailand (ISO country code) • Thailande (code ISO de pays) • Thermal Hydraulic Analyzer • Thread Hazard Analysis • Tullahoma (USA-TN) (location identifier)

THAAD Theater High-Altitude Area Defense

ThAFM Theoretical and Applied Fracture Mechanics (periodical code)

Thai Thailand

THAMA Toxic and Hazardous Materials Agency

THAP Thermal Hydraulic Analyzer Program

THAR Test Hazard Analysis Review • Test Hazard Assessment Review

THB Bath (Thailand) (ISO currency code) • Technische-Hochschule-Bibliothek • Thaba-Tseka (Lesotho) (location identifier) • Third-Harmonic Band

THC Tchien (Liberia) (location identifier) • Telecommunications Host Processor • Temperature and Humidity Control • Thermohaline Circulation • Through Hole Connection • Thrust Hand Controller • Total Hydrocarbon • Translational Hand Control • Translation Hand Controller

ThCFD Theoretical and Computational Fluid Dynamics (periodical code)

THCS Temperature and Humidity Control System • Total Heading Control System

Thd Thread

ThD Doctor of Theology

THD Tag Hierarchy Diagram • Technische Hochschule Darmstadt (Germany) (now TUD) • Technische Hogeschool Delft (Netherlands) • Thread • Thunderhead • Total Harmonic Distortion • Très Haut Débit • Tunnel Hydrodynamique

THDA Temperature of Camera Head Amplifier

THDB Technology History Database

THDG True Heading

THD+N Total Harmonic Distortion plus Noise

THDN Total Harmonic Distortion plus Noise

THDR Thunder • Transport Layer Header

THDS Threads • Time Homogeneous Data Set

THE Tape-Handling Equipment • Technische Hogeschool Eindhoven (Netherlands) • Teresina (Brazil-PI) (location identifier)

THEA Thousand Element Array

Theat Theater • Theatrical

THEL Tactical High-Energy Laser

THEMA Thermal Mapper

THEMIS Télescope Héliographique pour l'Étude du Magnétisme et des Instabilités Solaires • Thermal Emission Imaging System • Time History of Events amd Macroscale Interactions during Substorms (mission)

THEMISTOCLE Tracking High-Energy Muons In Showers Triggered On Čerenkov Light Emission

THENet Texas Higher Education Network (USA-TX)

THEO Theodolite • Theoretical

Theol Theologian • Theological • Theology

Theor Theorem • Theoretical • Theory

Théor Théorique

Theoret Theoretical

Theos Theosophical • Theosophy

Therm Thermal • Thermique • Thermometer • Thermometry • Thermostat • Thermostatic

THERM Thermal

Thermochem Thermochemistry

Thermodyn Thermodynamics

Thermodynam Thermodynamics

Thermom Thermometer • Thermometry

THERP Technique for Human Error Rate Prediction

Thes Thesis

THESEO Third European Stratospheric Experiment on Ozone

THEX Thursdays Excepted

THF Berlin Tempelhof (Germany) (location identifier) • Token Holding Frame • Tremendously High Frequency • Très Haute Fréquence

THG Thangool Biloela (Australia-QLD) (location identifier) • Third Harmonic Generation

THH Taharoa (New Zealand) (location identifier)

ThI Thallium Iodide

THI Temperature-Humidity Index • Tichitt (Mauritania) (location identifier) • Time Handed In

THIA Toledo Heavy Ion Accelerator

THIR Temperature/Humidity Infrared Radiometer • Thermal Infrared

THIS Telescope Human Interface System • Total Health Information System • Transportable Hemispherical Illumination System • Tunable Heterodyne Infrared Spectrometer

THISBE Telescope of Heidelberg for Infrared Studies by Balloon-borne Experiments

THISTLE Tracking Heavy Isotope Spectrometer Telescope for Low-Energies • Tracking Heavy Isotope Spectrometer Telescope for Low-Energy (cosmic rays) (*)

THJM Thermal Joining of Metals

Thk Thick

THK Thakhek (Laos) (location identifier) • Thick

ThkF Thick Film

Thkns Thickness

Thl Thermal

THL Tachilek (Myanmar) (location identifier) • Transhybrid Loss

ThLA Three-Letter Acronym

Thld Threshold

THLF Thermal Hydraulic Load Flow

ThM Master of Theology

THM Thomsonfield (USA-MO) (location identifier) • Traveling Heater Method • Tricode Hexaphase Modulation • Trihalomethane

Thml Thermal

Thmom Thermometer

Thms Thermistor

thn Thumbnail Data (file extension)

THN Thin • Trollhättan (Sweden) (location identifier)

ThnF Thin Film

Thnr Thinner

THO Thórshöfn (Iceland) (location identifier) • Thursdays Only

THOLD Threshold

THOPS Tape Handling Operational System

THOR Tandy High-intensity Optical Recording • Tandy High-performance Optical Recording • Tape-Handling Option Routine • Technology by Highly Oriented Research • Thickness from Off-beam Returns (project)

THORP Thermal Oxide Reprocessing Plant

THOTS Technical Hands-On Training System

Thou Thousand

THOUS Thousand

Thp Thrust Horsepower

THP Terminal-to-Host Protocol • Thermopolis - Hot Springs (USA-WY) (location identifier) • Thrust Horsepower • Turbine à Haute Pression

Thr Their • Threshold • Through • Thrust

THR Tehran Mehrabad (Iran) (location identifier) • Terahertz Heterodyne Receiver • Threshold • Thrust • Thruster • THz Receiver • Transmit Holding Register

THRFTR Thereafter

Throt Throttle

THROT Throttle

THRSTR Thruster

Thrt Throttle

THRU Through

THRUT Throughout

THS Test Host Substitute • Thesaurus • Traitement Hormonal Substitutif • Treuhandstelle • Trimmable Horizontal Stabilizer

THSD Thousand

Thstm Thunderstorm

THT Tamchakett (Mauritania) (location identifier) • Token Holding Time • Token Holding Timer • Token Hold Time • Très Haute Tension

THTN Threaten

THTR Thorium High-Temperature Reactor

Thu Thursday

THU Thule (Greenland) (location identifier) • Thursday

THUDD Thermal Uplink Data Display

Thur Thursday

Thür Thüringen (Germany)

Thüringer W Thüringer Wald (Germany)

Thurs Thursday

THV York (USA-PA) (location identifier)

THW Technisches Hilfswerk

Thwy Throughway

THX Thanks • Thursdays Excepted

THY Thohoyandou (South Africa) (location identifier)

Thyr Thyristor

THz Terahertz

THZ Tahoua (Niger) (location identifier)

ti Tigrinya (ISO language code)

Ti Titan • Titan • Titanium

T&I Test & Integration

TI Costa Rica (international aircraft registration code) • Tape Inverter • Target Identification • Target Indicator • Taux d'Incidents • Technical Information • Technical Insertion • Technical Inspection • Technical Institute • Technical Instruction • Technical Instructor • Technical Integration • Technical Intelligence • Technical Interchange • Technical Interface • Technical Investigation • Technicien d'Instruments • Technique Industrielle • Technologies de l'Information • Technology Insertion • Temperature Indication • Temperature Indicator • Temporary Installation • Temps d'Indisponibilité • Terminal Initiation • Terminator Irregularity • Test Incident • Test Increment • Test Initialization • Test Instruction • Test Instrumentation • Texas Instruments Inc. • Thermal Imager • Thermal Imagery • Thermal Imaging • Thermal Infrared • Tiempo Internacional • Time Interval • Timing Interface • Title • Title Information • Topographic Instruction • Total Immersion • Trademark Information • Traditional Instruction • Traînée d'Interaction • Traînée Induite • Training Instructor • Training Integrator • Traitement de l'Information • Transfer Impedance • Transfer Initiate • Translation Invariant • Transmetteur d'Impulsion • Transmission Identification • Transmission Identifier • Transparency Index • Transport-Independent • Turn In • Type d'Institution

TIA Task Item Authorization • Telecommunications Industries Association • Telecommunications Industry Association (USA) • Territorial and International Affairs • Test Interface Assembly • Thanks In Advance • The Internet Adapter • Time Interval Analyzer • Tirana (Albania) (location identifier) • Total Information Awareness • Training Impact Analysis • Transimpedance Amplifier

TIAC Technical Information Analysis Center • Texas Instruments Automatic Computer

TIACA The International Air Cargo Association

TIADD Texas Instruments Automatic Data Display System

TIALD Thermal Imaging Airborne Laser Designator

TIATA Technical Information and Training Authoring

Tib Tibet • Tibetan

TIB Technical Information Branch • Technical Information Bulletin • Technical Information Bureau • Technische Informations-Bibliothek • Telescope and Instruments Branch • Telescope Instrument Branch • Thread Information Block • Tibu (Colombia) (location identifier) • Traitement Indiciaire Brut

TIBAQ Text-and-Reference-Based Answering of Questions

TIBL Thermal Internal Boundary Layer

TIBM Time-Integration Block Marching

TIBOE Transmitting Information By Optical Electronics

TIBR Truly Integrated Baseline Review

TIBS Tactical Information Broadcast Service • Tactical Information Broadcast System • Through-Ice Bathymetry System • Transcribed Information Briefing Service • Trends in Biochemical Sciences

TIC Taken Into Consideration • Tape Identification Card • Tape Index Count • Task Identification Code • Technical Implementation Committee • Technical Information Capability • Technical Information Center • Technical Integration Center • Technical Interface Certification • Technical Interface Committee • Technical Interface Concept • Technologie de l'Information et de la Communication • Tecnología de la Información y de la Comunicación • Technology Insertion Candidate • Technology Integration Center • Telecommunications Information Center • Temperature Indicating Controller • Temperature Indicator Controller • TermInfo Compiler • Testing and Interoperability Center • Test Investment Committee • Thermal Imaging Camera • Thermionic Integrated Circuit • Thin Ionization Calorimeter • Timed Calculus • Time Interval Counter • Timing Interface Controller • Tinak Island (Marshall Islands) (location identifier) • Types In Compilation (workshop) • Token-ring Interface Coupler • Total Inorganic Carbon • Total Item Change • Tourist Information Centre • Trade Information Committee • Transaction Identification Code • Transducer Information Center • Translation Information Center • Transmission Interval Code • Trigger Interface Controller • Tuned Integrated Circuit • Turbulence Instrument Cluster

TICA Technical Information Center Administration

TICCE Time-band Capture Cell Experiment (EURECA)

TICCIT Time-Shared Interactive Computer-Controlled Information Television

TICE Technologies de l'Information et de la Communication pour l'Enseignement • Time Integral Cost Effectiveness

TICER Tilburg Innovation Center for Electronic Resources BV (Netherlands)

TICES Type-In Coding and Editing System

TICF Transient Installation Confinement Facility

TICHS Thermal Ion Capped Hemisphere Spectrometer

TICL Transport Interface Compatibility Layer

TICLER Technical Input Checklist/Evaluation Report

TICM Test Interface and Control Module

TICR Tycho Input Catalogue Revision

TICS Teacher Interactive Computer System • Test Interferometer Control Software

TICT Twisted Intramolecular Charge Transfer

tid Ter in Die

TID Target Identification • Target Information Display • Technical Implementation Directive • Technical Information Division • Technical Information Document • Technical Interface Description • Ter In Die • Test Instrument Development • Test Introduction Document • Thermal Ion Detector • Thread Identifier • Tiaret (Algeria) (location identifier) • Time Integration Delay • Time-Intensity Diagram • Total Ionizing Dose • Traitement Informatique des Données • Traitement Intégré des Données • Transaction Identification • Traveling Ionospheric Disturbance

TIDAR Time Delay Array Radar

TIDAS Totally Integrated Data System

TIDB Technical Information Dissemination Bureau

TIDDB Thermal Interface Design Data Block

TIDE Tachyon Integrated Development Environment • Telematics Initiative for the Disabled and Elderly • Thermal Infrared Differential Emission • Thermal Ion Dynamics Experiment

TIDER Technical-Information Deficiency Report

TIDI TIMED Doppler Interferometer

TIDMA Tape Interface Direct Memory Access

TIDP Technical Interface Design Plan • Technical Interface Development Plan • Telemetry and Image Data Processing

TIDS Target Information Display System • Technical Information Distribution Service

TIDU Technical Information and Documents Unit

TIDY Teletypewriter Integrated Display

TIE Targeted Image Extraction • Taux d'Intérêt Équivalent • Technical Idea Exchange • Technical Independent Evaluation • Technical Independent Evaluator • Technical Information Exchange • Technical Integration and Evaluation • Technology Information Exchange • Technology Insertion Environment • Technology Integration Experiment • Teleform Interpretation Engine • Telescopes In Education • Thermosphere-Ionosphere Electrodynamics • Time Interval Error • Tippi (Ethiopia) (location identifier) • Tool for Image Encryption • Training Instrumentation Evaluation • Translated-Image Environment

TIED Test Independent Evaluation Directorate

TIE-GCM Thermosphere/Ionosphere/Electrodynamic General Circulation Model

TIEGCM Thermosphere/Ionosphere/Electrodynamic General Circulation Model

TIER Technical Independent Evaluation Report

TIERCE Transport des Ions Et des Rayonnements dans les Cibles Épaisses

TIES Technical Integration Engineering Support • Technological Information Exchange System • Telecom Information Exchange Service • Terrain Information Extraction System • The Interactive Encyclopaedia System (now HyperTIES) • Time-Independent Escape Sequence • Transmission and Information Exchange System

TIEU Technical Information and Educational Unit

tif Tagged Image (file extension)

TIF Tagged Image File • Taif (Saudi Arabia) (location identifier) • Tape Inventory File • Task Initiation Form • Technical Information File • Technology Integration Facility • Telephone Influence Factor • Telephone Interference Factor • Test Incident Form • Text Interchange Format

TIFDL TOPEX Inter-Facility Data Link

tiff Tagged Image File Format (file extension)

TIFF Tagged Image File Format • Tag Image File Format

TIFR Tata Institute of Fundamental Research (India)

TIFS Total In-Flight Simulator

TIG The Inspector General • Time In Grade • Time of Ignition • Time until Ignition • Tingwon (Papua New Guinea) (location identifier) • Transmission Identification Generator • Travail d'Intérêt Général • Tungsten Inert Gas

TIGA Texas Instruments Graphics Architecture • Tide Gauge

TIGCM Thermosphere/Ionosphere General Circulation Model

TIGER Tasman International Geospace Environment Radar • Terrestrial Initiative in Global Environment Research (*) • Terrestrial Initiative on Global Environment Research • Thermal Infrared Ground Emission Radiometer • Topologically Integrated and Geophysically Encoded Reference • Topologically Integrated Geographic Encoding and Referencing • Traitement Intégral des Galaxies par l'Étude de leurs Raies • Transiron Galactic Element Recorder

TIGO Transportable Integrated Geodetic Observatory (*) • Transportable Integrated Geophysical Observatory

TIGR The Institute for Genomic Research (USA-MD) • Time Inertial Guidance and Radar • TOVS Initial Guess Retrieval

TIGRE Tracking and Imaging Gamma-Ray Experiment

TIGRIS Televised Images of Gaseous Region in Interplanetary Space • Topologically Integrated Geographically Referenced Information System • Topologically Integrated Geographic Resource and Information System

TIH Tikehau Atoll (Tuamotu Islands) (location identifier)

TIHS Torino Impact Hazard Scale

TII Technology Independent Interface • Texas Instruments Inc. • Tirinkot (Afghanistan) (location identifier) • Tooling Inspection Instrumentation • Total Inactive Inventory

TIIAP Telecommunications and Information Infrastructure Assistance Program (now TOP)

TIIP Topographic Imagery Integration Prototype

TIIPS Technically-Improved Interference Prediction System

TIIS Thermal Infrared Imaging Spectrometer

TIJ Tijuana (Mexico) (location identifier) • Tribunal Internacional de Justicia

TIK Oklahoma City - Tinker (USA-OK) (location identifier)

TIL Inverlake (USA-CA) (location identifier) • Targeted Institutional Link • Technical Information Library • Technology Insertion Laboratory • Transportation Information Library • Travail d'Initiative Locale • Until

TILL Total Initial Lamp Lumens

TILO Technical Industrial Liaison Office • Technical Information Liaison Office

TILS Technical Information and Library Service

TILT Teaching and Learning Technology Programme • Technique d'Images Laser Tridimensionnelle

TIM Technical Information Management • Technical Information Manual • Technical Information Memorandum • Technical Information on Microfilm • Technical Integration Management • Technical Integration Manager • Technical Interchange Meeting • Technical Interface Meeting • Technology and Innovation Management • Telescope and Image Modeling • Telescope Image Model • Telescope Image Modeling • Telescope Imaging Model • Telescope Instrument Model • Telescope Instrument Modeling • Telescopio Infrarojo Mexicano • Tembagapura (Indonesia) (location identifier) • Temperature Indicator Monitor • Ternary Ionic Mixture • Test Item Malfunction • Thermal Imaging Module • Ticket-Issuing Machine • Time Interface Module • Time Interval Measurement • Time Interval Meter • Time Interval Monitor • Time Is Money • Topic Indexing Matrix • Total Information Management • Tracking Information Memorandum • Tracking Instruction Manual • Tracking Instrument Mount • Transcardiac Impedance Measurement • Transmission Interface Module

TIMA Techniques de l'Informatique et de la Microélectronique pour l'Architecture (d'ordinateurs) • Tool Inventory Management Application

TIMC Techniques de l'Imagerie, de la Modélisation et de la Cognition

TIME Technique for Information Management and Employment • Thermosphere-Ionosphere-Mesosphere Electrodynamics • Total Inventory Management in Engineering

TIMED Thermosphere-Ionosphere-Mesosphere Electrodynamics • Thermosphere-Ionosphere-Mesosphere Energetics and Dynamics (*)

TIME-GCM Thermosphere/Ionosphere/Mesosphere/Electrodynamic General Circulation Model

TIMEGCM Thermosphere/Ionosphere/Mesosphere/Electrodynamic General Circulation Model

TIMEREQ Time Request

TIMERESP Time Response

TIMES Thermoelectric Integrated Membrane Evaporation System • Transportation Information Management and Evaluation System

TIMEX Time Multiplexer

TIMI Technical Information Maintenance Instruction • Technology-Independent Machine Interface

TIMIG Time In Grade

TIMINT Time Interval

TIMIS Totally Integrated Management Information System

TIMM Task Identification Management Matrix • The INTEGRAL Mass Model • Thermally-Integrated Micromodule • Toshiba Integrated Multimedia Monitor

TIMMI Thermal Infrared Multimode Imager • Thermal Infrared Multimode Instrument

TIMO Temporary Installation Manufacturing Order

TIMP Test Investment Master Plan

TIMS Teaching Integrated Math and Science (program) • Technical Information Management System • Telecommunications Information Management System • Test Interactive Management System • Thaad Information Management System • The Institute of Management Sciences (now merged with ORSA into INFORMS) • Thermal Imaging Spectrometer • Thermal Infrared Mapping Spectrometer • Thermal Infrared Multispectral Scanner • Thermal

Ionization Mass Spectrometry • Total Information Management System • Traditional Interactive Multiuser System • Training Integration Management System • Transmission Impairment Measurement Set • Transmission Impairment Measuring Set

TIMSS Third International Mathematics and Science Study

TIM3 Techniques de l'Informatique, des Mathématiques, de la Microélectronique et de la Microscopie

TIN Task Implementation Notice • Taxpayer Identification Number • Temporary Identification Number • Temporary Instruction Notice • Tindouf (Algeria) (location identifier) • Transaction Identification Number • Triangular Irregular Network • Triangulated Irregular Network • Turn In

TINA Telecommunication Information Networking Architecture • There Is No Alternative • Truth In Negotiation Act (USA)

TINS Thermal Imaging Navigation Set • Transinertial Navigation System

TINTIN Terahertz Integrated Technology Initiative

TIO Target Insertion Orbit • Task Input/Output • Technical Information Office • Technical Information Officer • Technology Insertion Office • Telescope and Instrument Operator • Tilin (Myanmar) (location identifier) • Time Interval Optimization • Total Interoperability • Training Integration Office • Transistorized Image Orthicon

TIOA Tomsk Institut Optiki Atmosfery (periodical code)

TIOE Test In Other Environment

TIOM Terminal Input/Output Module

TIOP Taux Interbancaire des Opérations á Paris • Taux Interbancaire Offert á Paris

TIOT Task Input/Output Table

TIP Tailored Instruction Program • Target Instrument Payload • Task Implementation Plan • Task Initiation and Prediction • Technical Implementation Plan • Technical Improvement Plan • Technical Improvement Panel • Technical Improvement Program • Technical Improvement Project • Technical Information Package • Technical Information Panel • Technical Information Pilot • Technical Information Processing • Technical Information Program • Technical Information Project • Technical Information Publishing • Technical Insertion Program • Technical Integration Panel • Technical Interest Profile • Technical Interest Profiling • Technology Insertion Plan • Technology Insertion Project • Technology Integration Plan • Telecommunications Implementation Plan • Teletype Input Processing • Template-controlled Image Processor • Tenerife Infrared Polarimeter • Terminal Interface Package • Terminal Interface Processor • Terminal Interface Program • Terrestrial Interface Port • Test Integration Plan • Textual Information Processing • The Industrial Physicist (periodical) • Thermospheric and Ionospheric Physics • Thesis In Progress • Thrust Inlet Pressure • Tilted Payload • TIROS Information Processor • Titre Interbancaire de Paiement • To Insure Promptitude • Total Information Processing • Tracking and Impact Prediction • Tracking Impact Prediction • Track Initiation Processor • Trainer Improvement Program • Transaction Internet Protocol • Transit Improvement Program • Transition and Integration Plan • Transition Implementation Plan • Treaty Implementation Plan • Tripoli (Libya) (location identifier) • True Image Projector • Truly Important Person • Truncated Inverted Pyramid • Turbine Inlet Pressure • Turn-In Point

TIPA TeX International Phonetic Alphabet • Treaty Implementation Plan Agency

TIPACS Texas Instruments Planning And Control System

TIPE Travail d'Initiative Personnelle Encadré

TIPEX Test-In-Progress Executive

TIPG Telephone Interface Planning Guide

TIPIL Turn-In Point Item List

TIPL Teaching Information Processing Language

TIPM Transport in Porus Media (periodical code)

TIPMEX Tibetan Plateau Meteorology Experiment

TIPP Taxe Intérieure sur les Produits Pétroliers (France) • Test Investment Plan and Program • Test Investment Planning and Programming • Test Investment Planning Process • Test Investment Program Plan • Time-Phasing Program

TIPPS Total In-house Publication Production System

TiPS Tether Physics and Survivability

TIPS Teaching Information Processing System • Technical Information Processing Service • Technical Information Processing System • Technological Information Pilot System • Telemetry Integrated Processing System • Terminal Information Processing System • Test Information Planning Sheet • Tether Physics and Survivability (experiment) • Text Information Processing Service • Text Information Processing System • Thematic Mapper Image Processing System • Theory of Inventive Problem Solving • Thruster Instability Protection System • Tile Information Processing System • Tools for Innovative Publishing in Science (project) • Total Integrated Pneumatic System • Traitement d'Images Perceptifs Spécifiques • Transportation Information Processing System • Truevision Image Processing Software • Tutorial Image Processing System

TIPSH Translation Invariant Poisson Smoothing using Haar (wavelets)

TIPSY Theoretical Image Processing System

TIPT Technology Integrated Product Team • Transition Integrated Product Team

TIPTOP Tape Input - Tape Output • The Internet Pilot To Physics

TIQ Tinian Island (Mariana Islands) (location identifier)

Tir Tirage

TIR Target-Intermediate Ratio • Technical Information Release • Technical Information Retrieval • Technical Interface Recertification • Technical Interface Report • Technical Interface Requirement • Technical Interoperability Requirement • Technology Insertion Recapitalization • Telescopio Infrarosso • Temporary Installation Record • Terminal Imaging Radar • Test Incident Report • Thermal Imaging Radar • Thermal Infrared • Thermal Infrared Radiometer • Time In Rate • Tirupati (India) (location identifier) • Total Indicator Reading • Total Internal Reflection • Total Item Record • Transportation Item Reporter • Transport International Routier

TIRA Tracking and Imaging Radar

TIRACS Telecommanded Inertially Referenced Attitude Control System

TIRAD Tool for Infrared Algorithm Development

TIRAS Task Identification and Resource Allocation System • Technical Information Retrieval and Analysis System

TIRC Technical Information Retrieval Center • Technical Information Retrieval Committee

TIRDOC Test Integration Requirement Document

TIREC TIROS Ice Reconnaissance

TIREX TIROS Ice Experiment

TIRFM Total-Internal-Reflection Fluorescence Microscopy

TIRGO Telescopio Infrarosso del Gornergrat

TIRL Telematique International Research Laboratories • Trapped Instantaneous Residual Liquid

TIRLS Thermal Infrared Linescan Imagery

TIRM Training for Information Resources Management • Transparent Infrared Material

TIROS Television and Infrared Observation Satellite (*) • Television and Infrared Observing Satellite • Television Infrared Observation Satellite

TIRP Tenerife Infrared Polarimeter • Total Internal Reflection Prism

TIRPC Transport-Independent Remote Procedure Call

TIRV Tuned Inter-Range Vector

TIRPF Total Integrated Radial Peaking Factor

TIRS Text Information Retrieval System • The Integrated Reasoning Shell • Thermal Infrared Scanner

TIRSS Technical Information Retrieval and Storage System

TIS Teacher In Space • Technical Information Service • Technical Information Sheet • Technical Information System • Technical Inspection and Safety • Technical Interface Specification • Telemetry Input System • Temperature Indicating Switch • Test Information Sheet • Test Instruction Set • Test Instruction Sheet • Test Instrumentation System • Test Interface Summary • Test Interface System • Test Investment Strategy • Test Item Simulator • The Information Society • Thermal Imaging Scanner • Thermal Imaging System • Thermal Insulation System • Thematic Information System • Thursday Island (Australia-QLD) (location identifier) • Time In Service • Total Information System • Total Integrated Scatter • Tracking and Injection Station • Tracking Instrumentation System • Training Integration System • Truncated Isothermal Sphere • Trusted Information System • Turkish Informatics Society (Turkey)

TISA Technical Information Support Activity • Technique for Interactive Systems Analysis

TISAB Total Ionic Strength Adjustment Buffer

TISAP Technical Information Support Activity Project

TiSAT Time Series Analysis Tool

TISB Technical Information Services Branch

TISC Technical Information Steering Committee • Technology Integration Steering Committee

TISCA Technical Information Storage and Control Application

TISCM Thermal Imaging System Common Module

TISE Technical Information Service Extension

TISG Technical Interoperability Standards Group

TISI Thai Industrial Standards Institute (Thailand) • Thermal Infrared Spectral Index

TISN Tokyo International Science Network

TISO Technical Information Services Office • Terrestrial Impact of Solar Output • Tropical Intraseasonal Oscillation

TISP Teacher In Space Program • Test Investment Strategic Plan

TISR Technical Information Statistical Report

TISS Terrestrial Impacts and Spherules Symposium • Thermal Imaging Sensor System • Transportation Information Subsystem

TISSEC Transactions on Information and Systems Security (ACM) (periodical code)

TISSS Tester Independent Support Software System

TISTR Thailand Institute for Scientific and Technical Research

Tít Título

Tit Titel • Title • Titre • Titular • Titulaire

TIT Tecnologias da Informação e Telecomunicações • Thermal Inactivation Time • Thermoinsulation Technique • Total Insertion Time • Turbine Inlet Temperature

TITas Tsirkulyar Astronomicheskogo Instituta Akademiia Nauk Uzbekskoi SSR (periodical code)

TITF Test Item Transmittal Form

Titr Titration

TITUS Traitement Interactif du Texte à l'Usage des Scientifiques

TIU Tape Identification Unit • Technical Improvements Update • Technical Informatioh Unit • Telemetry Interface Unit • Terminal Interface Unit • Timaru (New Zealand) (location identifier) • Time Interval Unit • Time Isolation Unit • Time In View • Timing Insertion Unit • Timing Interface Unit • Trunk Interface Unit • Trusted Interface Unit

TIV Tableau Indicateur de Vitesse • Threat Impact Value • Torus Interface Valve • Total Immersion Videogramme

TIVA Transmission d'Images par Voie Acoustique

TIW Tacoma (USA-WA) (location identifier) • Test Issue Workout • Topic Index Word

TIWE Tropical Instability Wave Experiment

TIWG TDRSS Interface Working Group • Technical Interface Working Group • Technology Integration Working Group • Test Integration Working Group

TIX Titusville Space Center (USA-FL) (location identifier)

TIY Tidjikja (Mauritania) (location identifier)

TIZ Tari (Papua New Guinea) (location identifier)

TIzSS Tbilisi Izdatel Sabchota Sakartvelo (periodical code)

tj Trajectory

T/j Tonne par Jour

T/J Turbojet

TJ Cameroon (international aircraft registration code) • Tajik • Tajikistan (ISO country code) • Technical Journal • Telephone Jack • Temperature Junction • Terajoule • Test Jack • Thermal Junction • Tower Jettison • Trajectoire • Trajectory • Transmetteur de Jaugeur • Turbojet

TJA Tarija (Bolivia) (location identifier) • True Jupiter Analog

TJAu Telecommunication Journal of Australia (periodical code)

TJB Tanjung Balai (Indonesia) (location identifier)

TJC Trajectory Chart • Tyler Junior College (USA-TX)

TJD Trajectory Diagram • Truncated Julian Day

TJCM Télescope James Clerk Maxwell

TJG Tanjung Warukin (Indonesia) (location identifier)

TJH Toyooka (Japan) (location identifier)

TJI Tonneau de Jauge International • Trujillo Capiro (Honduras) (location identifier)

TJJPT Tuijin Jishu Journal of Propulsion Technology (periodical code)

TJK Tajikistan (ISO country code)

TJM Tower Jettison Motor • Tyumen (Russia) (location identifier)

TJNAF Thomas Jefferson National Accelerator Facility (USA-VA) (formerly CEBAF)

TJPh Turkish Journal of Physics (periodical code)

TJQ Tanjung Pandan (Indonesia) (location identifier)

Tjrs Toujours

TJS Tanjung Selor (Indonesia) (location identifier)

TJSA Tribal Jurisdiction Statistical Area

TJV Thanjavur (India) (location identifier)

tk Turkmen (ISO language code)

Tk Taka (currency) • Teluk • Telukan • Track

TK Tank • Teknisk Kommitté • Telekommunikation • Tokelau (ISO country code) • Tool Kit • Tonne-Kilomètre • Track • Turkmenistan

TKA Talkeetna (USA-AK) (location identifier)

TKB Tekadu (Papua New Guinea) (location identifier)

TKBS Token Bus Controller

TKC Tiko (Cameroon) (location identifier)

TKD Takoradi (Ghana) (location identifier)

TKE Tenakee Springs (USA-AK) (location identifier) • Terminology and Knowledge Engineering • Thermodynamic Kinetic Equilibrium • Track Error • Turbulence Kinetic Energy • Turbulent Kinetic Energy (*)

TKF Takeoff • Truckee (USA-CA) (location identifier)

TKG Bandar Lampung (Indonesia) (location identifier) • Telekommunikationsgesetz • Terminal Key Generator • Transmitting Key Generator

TKGA Task Generation Advisor

TKH Takhli (Thailand) (location identifier)

TKI Tokeen (USA-AK) (location identifier) • Trail Kit Installation

TKJ Tok (USA-AK) (location identifier)

TKK Truk (Caroline Islands) (location identifier)

TKL Taku Lodge (USA-AK) (location identifier) • Tokelau (ISO country code)

TKM Tikal - El Peten (Guatemala) (location identifier) • Turkmenistan (ISO country code)

TKN Tokunoshima (Japan) (location identifier)

TKO Technical Knockout • Tlokoeng (Lesotho) (location identifier)

Tkof Takeoff

TKP Takapoto (Tuamotu Islands) (location identifier)

TKQ Kigoma (Tanzania) (location identifier)

TKR Terrestrial Kilometric Radiation • Thakurgaon (Bangladesh) (location identifier)

Tks Thanks

TKS Task Knowledge Structure • Tokushima (Japan) (location identifier)

TKSC Tsukuba Space Center (Japan)

Tkt Ticket

TKT Tak (Thailand) (location identifier) • Threshold Knowledge Test

TKTS Thermodynamic Kelvin Temperature Scale

TKU Turku (Finland) (location identifier)

TKV Tatatoko (Tuamotu Islands) (location identifier) • Telekommunikationsverordnung

TKW Tekin (Papua New Guinea) (location identifier)

TKX Takaroa (Tuamotu Islands) (location identifier)

TKY Turkey Creek (Australia-WA) (location identifier)

TKZ Tokoroa (New Zealand) (location identifier)

tl Tagalog (ISO language code) • Thunderstorm

Tl Teil • Thallium

T-L Temperature-Luminosity

T/L Talk/Listen • Timeline • Time Loan • Total Loss • Trace by Lot • Transformer Load • Transport/Loader

TL Central African Rep. (international aircraft registration code) • Tableau Lumineux • Table de Lancement • Table de Logarithmes • Talk and Listen • Talk-Listen • Tape Library • Target Language • Target List • Task Leader • Task List • Taxe Locale • Taxi Lane • Team Leader • Technical Letter • Technical Library • Telecommunication Laboratories • Temps Local • Ten Laatste • Termination Liability • Terminological Logic • Terminological Logics • Test Laboratory • Test Link • Test Load • Test Log • Tête de Lecture • Thermoluminescence • Thermoluminescent • Thrust Level • Thrust Line • Thrust Link • Tie Line • Time Latest • Time Length • Time Limit • Timeline • Tired Light • Title List • Tool • Total Length • Total Load • Total Loss • Total Luminescence • Tour de Lancement • Trace by Lot • Trade List • Transequatorial Loop • Transformée de Laplace • Transformer Load • Transient Load • Transition Level • Transit Level • Transmission Level • Transmission Line • Transmittal Letter • Travel and Living • Treuil de Levage • Truck Load • Trusted Label • Tube de Lancement • Turbine Libre • Turkish Lira (currency) • Until

TLA Tableau Lumineux des Alarmes • Teller (USA-AK) (location identifier) • Temporal Logic of Actions • Temporary Living Allowance • Temporary Lodging Allowance • Thin-Layer Activation • Three-Letter Abbreviation • Three-Letter Acronym • Timeline Analysis • Top Layer Aggregation • Top Level Aggregation • Top Level Aggregator • Training Level Assignment • Transmission Line Adapter • Transmission Line Assembly • Travel and Living Allowance

TLAP Token-ring Link Access Protocol

TLAPS Tropical Limited Area Prediction Scheme

TLAX Tlaxcala (Mexico)

tlb Table Lookaside Buffer

TLB Table Lookaside Buffer • Tarbela (Pakistan) (location identifier) • Tidally-Locked Binary • Top Level Budget • Translation Lookaside Buffer (*) • Transmission Lookaside Buffer

TLC Talker - Listener - Controller • Tangent Latitude Computer • Teacher-Librarian Certificate • Teaching - Learning - Computing • Telecommand • Télécommande • Telecommunications • Telecommunications Line Center • Telecommunications Live Controller • Telemetry Limit Check • Thin-Layer Chromatography • Timeline Controller • Toluca (Mexico) (location identifier) • Total Life Cycle • Traffic Load Control • Translunar Coast • Tratado de Libre Comercio

TLCA Tangent Latitude Computer Amplifier • Typed Lambda Calculi and Applications (conference)

TLCC Texas Learning and Computation Center (USA-TX)

TLCE Transmission Line Conditioning Equipment

TLCF Team Leader Computing Facility • Teleconference • Teleconferencing

TLCN Test Logic and Constraint Network

TLCPC Top Level Computer Program Component

TLCSC Top Level Computer Software Component

TLC2 Texas Learning and Computation Center (USA-TX)

TLD Target Level Debugger • Technical Logistics Data • Thermal Luminescence Detection • Thermal Luminescence Detector • Thermoluminescence Dosimeter • Thermoluminescent Detector • Thermoluminescent Device • Thermoluminescent Dosimeter • Thermoluminescent Dosimetry • Top Level Demonstration • Top Level Domain • Traffic Load Device • Transport and Logistics Data • Tuli Lodge (Botswana) (location identifier)

TLDI Technical Logistics Data and Information • Types in Language Design and Implementation (workshop)

Tle Teile

TLE Target Location Error • Taxe Locale d'Équipement • Technical Liaison Engineer • Temperature-Limited Emission • Temporal Lobe Epilepsy • Thin-Layer Electrode • Thin-Layer Electrophoresis • Thin Leading Edge • Top Loading Electrical • Total Lunar Eclipse • Tower Lighting Equipment • Tracking Light Electronics • Transfert Linéaire d'Énergie • Treaty-Limited Equipment • Tulear (Madagascar) (location identifier) • Two-Line Element

TLF Telida (USA-AK) (location identifier) • Temporary Lodging Facility • Terminal Launch Facility • Theoretical Luminosity Function • Timeline Form • Top Level Function • Trésors de la Langue Française

TLFIRD Top Level Functional Interface Requirements Document

Tlg Telegraph

TLG Tail Landing Gear • Tulagi Island (Solomon Islands) (location identifier)

TLH Tallahassee (USA-FL) (location identifier)

TLHSQ Transactions of the Literary and Historical Society of Quebec (periodical code)

TLI Tape Library Interface • Telephone Line Interface • Terrain Line Interpolation • Thermionics Laboratory Inc. (USA-CA) • Three-Letter Identifier • Time-Life International • Tolitoli (Indonesia) (location identifier) • Translunar Injection • Transport Layer Interface (*) • Transport Level Interface • Type Line Item

TLIC Type Line Item Code

TLIEF Thin-Layer Isoelectric Focusing

TLJ Tatalina (USA-AK) (location identifier)

TLK Talk • Talknafjördur (Iceland) (location identifier) • Terrace - Ledge - Kink (model) • Test Link

TLL Tallinn Ulemiste (Estonia) (location identifier) • Telelink • Television Laser Link

TlLd Total Load

TLM Tagish Lake Meteorite • Target Locator Module • Technical Liaison Memorandum • Telemeter • Télémesure • Télémètre • Telemetry • Télémétrie • Tlemcen (Algeria) (location identifier) • Top Loading into Mixture • Topographic Line Map • Transition Layer Model • Transmission Line Method • Transmission Line Modeling • Trouble Locating Manual

TLMA Telematic Agent

TLMS Tape Library Management System

Tlmy Telemetry

TLN Toulon Hyères (France) (location identifier) • Toulon - Le Palyvestre (France) (location identifier) • Trunk Line Network

TLO Technical Liaison Officer • Terminal Learning Objective • The Last One • Tol (Papua New Guinea) (location identifier) • Total Loss Only • Transmittal Letter Only

TLOD Technical Letter Of Delegation

TLOS Target Location Observation System

TLOTOS Timed Language Of Temporal Ordering Specification

TLP Technology Leadership Programme • Telescope Local Processor • Test Level Point • Thread Level Parallelism • Threshold Learning Process • Top Load Pad • Total Language Processor • Total Loss of Pay • Transient Lunar Phenomenon • Transmission Level Point • Transparent Language Processing • Transport Layer Connection • Triangular Libration Point • Trouble Location Problem • Tumolbil (Papua New Guinea) (location identifier)

TLR Tape Loop Recorder • Technical Letter Report • Telecommunications Leasing Request • Termination Liability Reserve • Top Level Requirement • Trailer • Tulare (USA-CA) (location identifier) • Twin Lens Reflex

TLRBSE Teacher Leaders in Research-Based Science Education (program)

TLRN Technical Logistics Reference Network

TLRO Top Level Requirement Overview

TLRP Track Last Reference Position

TLRS Total Logistics Readiness and Sustainability • Transportable Laser Ranging System

TLS Taiwan Light Source (China-ROC) • Target Level of Safety • Teaching and Learning Service • Technical Library Service • Telecommunications Liaison Staff • Temporal Linking Service • Terminal Landing System • Thailand Landsat Station • Three-Line-Scanner • Through Life Support • Thüringer Landessternwarte (Germany) • Time Limited Signal • Timeline Sheet • Top Left Side • Top Level Specification • Total Least Squares • Total Luminescence Spectroscopy • Toulouse Blagnac (France) (location identifier) • Training Launch Station • Transponder Landing System • Transport Layer Security • Tunable Laser Spectrometer • Two-Level System

TLSA Torso Limb Suit Assembly

TLSAB Teaching and Learning Service Advisory Board

TLSC Total Logistics Support Cost

Tlscp Telescope

TLSCP Telescope

TLSP Transport Layer Security Protocol • Twin Lakes Star Party (USA-KY)

TLSS Technical Library Services Section • Texas Large-scale Structure Survey

TLT Target Launch Time • Test Loop Translator • Tilt • Trademark Law Treaty • Traffic Loop Test • Tuluksak (USA-AK) (location identifier)

TLTA Two-Loop Test Apparatus

TLTP Teaching and Learning Technology Program • Télétype • Temporal Linking Transfer Protocol

Tltr Translator

TLTSN Teaching and Learning Technology Support Network (UK)

TLU Table Lookup • Texas Lutheran University (USA-TX) • Threshold Logic Unit • Time of Last Update • Tolu (Colombia) (location identifier) • Transportable Laser Unit

TLv Transition Level

TLV Target Launch Vehicle • Tel Aviv Yafo - Ben Gurion (Israel) (location identifier) • Threat Likelihood Value • Threshold Limit Value • Transporter-Loader Vehicle • Type - Length - Value

TLVO Type Length Value Object

tlw teilweise

TLW Talasea (Papua New Guinea) (location identifier) • Termination Liability Worksheet

TLX Talca (Chile) (location identifier) • Teletype • Telex (*)

TLZ Catalão (Brazil-GO) (location identifier) • Titanium - Lead - Zinc • Transfer on Less than Zero

t/m Tot en Met

tm Team • Telemetry • Time • Trademark

Tm Thulium • Tonelada Métrica

T&M Testing & Maintenance • Time & Maintenance • Time & Materials

T/M Telemetry • Trace by Member

TM Table de Mortalité • Table de Multiplication • Table Maintenance • Talking Machine • Tambour Magnétique • Tangent Mechanism • Tape Mark • Tape Module • Tasking Memorandum • Task Management • Task Master • Task Memory • Taux de Modulation • Team Mate • Team Member • Technical Maintenance • Technical Management • Technical Manager • Technical Manual • Technical Memorandum • Technical Minutes • Technical Monograph • Technicien de la Météorologie • Technology Mission • Telemeter • Telemetered • Télémètre • Télémétrie • Telemetry • Télémétrie • Telescope Maker • Telescope Model • Teleskopischer Meteor • Temperature Meter • Temperature Monitor • Température Moyenne • Template Matching • Temps de Maintien • Temps de Mission • Temps de Montée • Temps Moyen • Terminal Man • Terminal Mode • Terminal Multiplexer • Terminological Model • Terminological Module • Territory Manager • Test Maintenance • Test Manager • Test Manual • Test Message • Test Model • Tête Motrice • Thematic Mapper • Thermal Mapper • Thermal Model • Threat Manager • Time and Materials • Time Management • Time Mission • Time Monitor • Time of Maximum • Tone Modulation • Tonne Métrique • Topic Map • Top Management • Torque Motor • Tot en Met • Tour de Montage • Trace by Member • Tracer Model • Trade Mark • Traffic Management • Traffic Manager • Traffic Model • Training Manual • Transaction Manager • Transcendental Meditation • Transfer Memorandum • Transition Metal • Transmission and Multiplexing • Transmission Matrix • Transverse Magnetic (field) • True Mean • Turing Machine • Turkmenistan (ISO country code) • Type Model

TMA Target Motion Analysis • Technology Mirror Assembly • Telecommunications Management Association (USA) • Telescope Makers Association • Tense Mood Aspect • Terminal Managing Application • Terminal Maneuvering Area • Terminal Maneuver Area • Thermomechanical Analysis • Thermomechanical Analyzer • Three-Mirror Anastigmat • Tierce Maintenance Applicative • Tifton -

Henry Tift Myers (USA-GA) (location identifier) • Time-Modulated Antenna • Traffic Management Advisor • Traffic Management Agency • Traffic Management Analysis • Traffic Management Analyst • Training Mission Area • Tycho Magnetic Anomaly

TMAC Technical Mapping Advisory Council (USA)

TMACC Transition Management And Control Center

TMAG Travel More Advantageous to the Government

TMAH Tetramethyl Ammonium Hydroxide

TMAP Table Mountain Astrometry Project • Telecommunications Monitoring and Assessment Program • Telemetry Analysis Plan

TMAS Transaction Monitoring and Analysis System

TMAW Technical Managers Advanced Workshop

TMB Miami - Tamiami (USA-FL) (location identifier) • Task Maintenance Burden • Technical Management Board • Terminating Message Buffer • Tetramethylbenzene • Traitement Mensuel Brut • Transient Microwave Brightening • Transportation Management Bulletin

TMBF Temps Moyen de Bon Fonctionnement

TMBPS Thematic Mapper Bulk Processing System

Tmbr Timber

TMBU Table Maintenance Block Update

TMC Tambolaka (Indonesia) (location identifier) • Tape Management Catalog • Task Management Component • Taurus Molecular Cloud • Taux de Masse Corporelle • Taxe de Mise en Circulation • Technical Management Committee • Technical - Management - Cost (review) • Technical Manual Code • Technology Management Consultants • Telemetry Controller • Test Management Council • Test Mass Chamber • Test Monitoring Console • The Map Collector (journal) • Thermal Management Control • Thinking Machines Corporation • Third Mediterranean Country • Thrust Magnitude Control • Titanium Matrix Composite • Total Manufacturing Cost • Traffic Management Center • Traffic Management Coordinator • Traffic Message Channel • Triple Magnetic Configuration • Triple Molecular Collision

TMCA Thrust Management Control Analysis • Transportation Management Control Agency • Transportation Movement Control Agency

TMCC Test Monitoring and Control Center • Time-Multiplexed Communications Channel • Traffic Management Computer Complex

TMCF Team Member Computing Facility

TMCP Two-Megabit Channelized Port

TMCR Technical Manual Contract Requirement

TMCRL Tailored Master Cross-Reference List

TMCS Tactical Maintenance Control System • Telecommunications Management and Control System

Tmd Timed

TMD Technical Management Database • Technical Management Division • Telemetry Doppler • Temps Moyen de Disponibilité • Thematic Mapper Derivative • Theoretical Maximum Density • Timbedra (Morocco) (location identifier) • Transient Mass Distribution • Turk Matematik Dernegi (Turkey)

TMDB Technical Management Database

TMDC Technical Manual Data Card

TMDE Test - Measurement - Diagnostic Equipment

TMDF Trunk Main Distribution Frame

TMDL Technical Manual Data List

TMDR Technical Manual Deficiency Report • Telemetry Data Reduction

TMDS Test - Measurement - Diagnostic System

TMDT Total Mean Downtime

Tme Time

TME Tame (Colombia) (location identifier) • Test and Measurement Equipment • Test Maintenance Equipment • Tidal Mass Exchange • Tivoli Management Environment • Tokyo-Mitaka catalog of Equatorial stars • Tropical Mountain Ecosystem

TMEC Tempo Medio dell'Europa Centrale • Temps Moyen de l'Europe Centrale

TMECO Time of Main Engine Cuttoff

TMED Temps Moyen Entre Défaillances

TMEP Temps Moyen Entre Pannes

TMER Temps Moyen Entre Révisions

TMET Technical Management Evaluation Team • Transportation Medium Earth Terminal

TMF Table Mountain Facility (USA-CA) • Telemetry Module Facility • Time Management Facility • Time Marker Frequency • Traffic Management Facility • Training Mission Facility • Transporter Maintenance Facility

TMFD Technical Manual Functional Description

TMFE Time multiplexed Frequency Expansion

TMFG Telemetry Format Generator

TmFl Time of Flight

Tmg Timing

TMG Temps Moyen de Greenwich • Test Message Generator • Thermal Meteoroid Garment • Thermal Model Generator • Time Mark Generator • Timing • Tomanggong (Malaysia) (location identifier) • Traffic Management

TMGE Thermomagnetic-Galvanic Effect

T/MGS Transportable Mobile Ground Station

TMGS Terrestrial Magnetic Guidance System • Transportable Mobile Ground Station • Two Micron Galactic Survey

TMH Tanahmerah (Indonesia) (location identifier)

TMHA Trimethylhexanoic Acid

TMI Technical Management Instruction • Technical Management Item • Technical Measurements Inc. • Technical Meetings Information • Telesat Mobile Inc. • Temps Moyen d'Indisponibilité • Test Method Instruction • Three Mile Island (USA) • Too Much Information • Tranche Marginale d'Imposition • Trans-Mars Injection • TRMM Microwave Imager • Tumling Tar (Nepal) (location identifier)

TMIC Technical Meetings Information Center

TMICP Topographic Map Inventory Control Point

TMICS Transportation Management Information Control System

TMIG Telemetered Inertial Guidance

TMIS Technical Management Information System • Technical Meetings Information Service • Technician's Maintenance Information System • Telecommunications Management Information System • Total Management Information System • Training and Maintenance Information System

TMJ Temporomandibular Joint • Termez (Uzbekistan) (location identifier)

TMK Tamky (Viet Nam) (location identifier)

TMKF Technische Mitteilungen Krupp Forschungsberichte (periodical code)

Tmkpr Timekeeper

Tml Terminal

TML Tamale (Ghana) (location identifier) • Television Microwave Link • Temps Moyen Local • Terminal • Tetramethyllead • TEX Macro Language • Three-Mile Limit • Total Mass Loss • Total Material Loss • Truck-Mounted Launcher • Trusted Multilevel • Turbulent Mixing Layer

TMLB Terminating Message Line Block

TMLI Television Microwave Link Indicator

TMLR Television Microwave Link Repeater

TMLS Tactical Microwave Landing System

TMLT Television Microwave Link Terminal

TMM Manat (Turkmenistan) (ISO currency code) • Tamatave (Madagascar) (location identifier) • Tape Mount Management • Télescope du Mont Mégantic (Canada-QC) • Testing Maturity Model • Test Message Monitor • Theory of Moon's Motion • Thermal Mathematical Model • Thermal Math Model

TMME Tidal Massive Mass Ejection

TMMP Technical Manual Management Program

TMMS Technical Manual Management System • Telephone Message Management System

TMMSH Texas Memorial Museum of Science and History (USA-TX)

TMMT Technical Manual Management Team

TMN Tamana Island (Kiribati) (location identifier) • Tatebayashi - Matsuzaki - Newman (protocol) • Technical and Management Note • Technologie de Micro- et Nanostructures • Telecommunications Management Network • Time Management Networking • Tokyo Meteor Network (Japan) • Transmission • True Match Number

TM&O Telecommunications Management & Operations

TMO Table Mountain Observatory (USA-CA) • Tabor Mountain Observatory (Canada-BC) • Target Mount • Technique Magnéto-Optique • Telecommunications Management and Operations • Telegraphic Money Order • Télémesure Opérationnelle • Telemetry Monitor Officer • Telemetry Observer • Telemetry Officer • Telephone Money Order • Température Maximale Opérationnelle • Tenant Management Organisation • Test Management Order • Test Manufacturing Order • Tokyo Meteorological Observatory (Japan) (now JMA) • Tool Manufacturing Order • Traffic Management Office • Training Material Outline • Transition Mass Object • Transportation Management Office • Transportation Management Officer • Transportation Movement Office • Transportation Movement Officer • Tumeremo (Venezuela) (location identifier)

TMOB Tidal Mass Outburst

TMOD Telecommunications and Mission Operations Directorate

TMOO Test Mission Operations Order

TMOP Transitional Meteosat Operational Programme

tmp Temporary (file extension)

TMP East Timor (ISO country code) • Tampere (Finland) (location identifier) • Target Master Plan • Taux Moyen Pondéré • Technical Management Plan • Technical Manual Part • Technical Manual Plan • Technology Master Process • Technology - Maths - Physics • Telecommunications Management Program • Telecommunications Modem Processor • Telemetry Processing • Telemetry Processor • Temperature • Temporary • Terminal Message Processor • Terminal Monitor Program • Test Maintenance Panel • Test Management Plan • Test Management Protocol • Test Methods and Procedures • Tetramethylpentane • Theodolite Measuring Point • Thermal Modeling Program • Thermomechanical Processing • Thermomechanical Pump • Time Management Processor • Timor (East_) (ISO country code) • Timor Orientale (code ISO de pays) • Traffic Management Processor • Training Management Plan • Transistor Mounting Pad • Transportable Measurement Package • Transportation Motor Pool • Transversely Magnetized Plasma • Turbomolecular Pump

TMPA Traffic Management Program Alert

TMPFS Temporary File System

TMPGO Tagung über Mathematische Probleme Geodäsie Oberwolfach (periodical code)

Tmpl Template

TMPL Temps Moyen de Préparation au Lancement

TMPO Terrain Modeling Project Office

Tmpry Temporarily • Temporary

TMPRY Temporary

TMPS Temperature Monitoring Power Supply • Test Maintenance Panel Subassembly

TMPSU Telescope Module Power Supply Unit

TMPV Torque Motor Pilot Valve

TMQ Tambao (Burkina Faso) (location identifier)

Tmr Timer

TMR Tamanrasset (Algeria) (location identifier) • Technical Memorandum Report • Télémesure Rapide • Temperature Minimum Region • Time Meter Reading • Timer • TOPEX Microwave Radiometer • Total Materiel Requirement • Training and Mobility of Researchers • Transient Memory Record • Transportation Movements Release • Triple Modular Redundant • Triple Modular Redundancy • Tunneling Magnetoresistance

TMRA Telemetry Reduction and Analysis

TMRAO Table Mountain Radio Astronomy Observatory (USA-CA)

TMRC Transportation Modeling Research Center

TMRF Technical Meeting Request Form

TMRM Transformation Remanent Magnetization

TMRS Temps Moyen avant Remise en Service • Test Management Reporting System • Traffic Measurement and Recording System • Traffic Metering Remote System

TMS Sao Tome (location identifier) • Tactical Message System • Tanzanian Meteorological Society (Tanzania) • Tape Management System • Target Monitoring System • Task Management System • Tax Modernization System • Technical and Management Support • Technical Management System • Telecommunications and Microgravity Section • Telecommunications Management System • Telemarketing Management System • Telemetry Modulation System • Telemetry Multiplexer System • Telemobiloskop • Teleoperator Maneuvering System • Telephone Management System • Telescope Management System • Temperature Management Station • Temporary Mobile Station • Test and Measurement Station • Test and Monitoring Station • Test and Monitor Station • Test Monitor Software • Test Monitor System • Tetramethylsilane • Thematic Mapper Simulator • Thermal Maneuvering System • Thermomechanical System • Thrust Measuring System • Time-Multiplexed Switching • Time of Maximum Strangeness • Time-shared Monitor System • Topographic Mapping Service • Tracing Management System • Traffic Management Specialist • Traffic Management System • Traffic Monitoring System • Trainee Management System • Transcranial Magnetic Simulation • Transmission Measuring Set • Transportation Management System • Tray Management System • Trend Monitoring System • Trouble Management System • Truth Maintenance System • Turbulence Measuring System • Twist Monitoring System • Two-Micron Survey

TMSC Talcott Mountain Science Center (USA-CT) • Transportation Management System Center

TMSI Temporary Mobile Station Identity • Temporary Mobile Subscriber Identity

TMSL Test and Measurement Systems Language

TMSO Telecommunications Management and Services Office • Telecommunications Management Service Office

TMSP Table Mountain Star Party (USA-WA)

TMSPA Table Mountain Star Party Association (USA-WA)

TMSS Technical Manuals Specifications and Standards • Two-Micron Sky Survey

TMT Technique Moderne de Transformation • Technology - Media - Telecommunications • Telecommunications - Media - Technologies • Télémesure Technologique • Ten-Meter Telescope • Terminal Monitor Program • Thermal Measurement Treatment • Thirty-Meter Telescope

(*) • Training Management Team • Transition Management Team • Transportation Movement Team • Trombetas (Brazil-PA) (location identifier) • Turbo Motor Train • Twenty-Meter Telescope

TMTC Technical Manual Technical Council

TMTE Telemetry Test Equipment

TMTR Thermistor • Transmetteur

TMTS Thematic Mapper Transcription System

TMTSF Tetramethyltetraselenofulvalene

TMU Table Management Utility • Tambor (Costa Rica) (location identifier) • Temperature Measurement Unit • Test Maintenance Unit • Thermal-Mechanical Unit • Thermomechanical Unit • Time Measurement Unit • Tokyo Metropolitan University (Japan) • Traffic Management Unit • Transmission Message Unit

TMUNER Tokyo Metropolitan University Network for Education and Research (Japan)

TMUX Transmultiplex • Transmultiplexer

TMV Temps Moyen de Vol • True Mean Value • Tube à Modulation de Vitesse

TMVS Telemetry Validation System

TMW Tamworth (Australia-NSW) (location identifier) • To-morrow • Transverse Magnetic Wave

TMWG Techniques and Methodology Working Group

TMX Timimoun (Algeria) (location identifier)

TMXRT Three-Mirror X-Ray Telescope

TMY Tiom (Indonesia) (location identifier)

TMZ Thames (New Zealand) (location identifier)

tn Setswana (ISO language code) • Tunisia

Tn Thoron • Ton • Tonne • Town • Train • Transportation • Ton • Tonne • Town • Train • Transportation

TN Congo (international aircraft registration code) • Tactical Note • TDRSS Network • Technical Needs • Technical Note • Technology Need • Telephone Number • Telnet • Tempérament Nerveux • Température Nocturne • Température Normale • Tennessee (USA) • Tension Normale • Terminal Node • Test Number • Tête Nucléaire • Thermonuclear • Tonnage Net • Trade Name • Traineeship Number • Transaction Number • Transfer on Negative • Translator's Note • Transmittal Notice • Transport Network • Trou Noir • True Negative • True North • Tunisia (international vehicle registration code) • Tunisia (ISO country code) • Tunisie (code ISO de pays) • Twisted Nematic

TNA Jinan (China-PRC) (location identifier) • Task Notification Area • Taux Nominal Annuel • Telematics Network between Administrations • Télémètre de Nuages Analogique • Thermal Neutron Activation • Thermal Neutron Analysis • Transient Network Analyzer

TNAD Trusted Network Access Device

TNAU Tamil Nadu Agricultural University (India)

TNB Tanah Grogot (Indonesia) (location identifier) • Terra Nova Bay • Transneptunian Binary • Trinitrobenzene • Trusted Network Base

TNBS Taiwan National Bureau of Standards (China-ROC)

TNC Tekniska Nomenklaturcentralen • TERENA Networking Conference • Terminal Node Controller • Threaded-Nut Connector • Tin City (USA-AK) (location identifier) • Too Numerous to Count • Total Number of Citations • Total Numerical Control • Trade Negotiating Committee • Transnational Corporation • Trunk Node Control

TND Dinar (Tunisia) (ISO currency code) • Telecommunications Network for the Deaf • The Next Definition • The Next Dimension • Training Need Date • Trinidad (Cuba) (location identifier)

TNDCY Tendency

TNDS Total Network Data System

TNDT Thermal Non-Destructive Testing

TNE Tanega Shima (Japan) (location identifier)

TNEF Transport Neutral Encapsulation Format

TNEK Teplovye Napriazheniia Elementakh Konstruktsii (periodical code)

TNF Third Normal Form • Timing Negative Film • Toussus-le-Noble (France) (location identifier)

TNFE Twisted Nematic Field Effect

Tng Tongue • Training • Turning

TNG Tanger Boukhalef (Morocco) (location identifier) • Telescopio Nazionale Galileo (Italy) • The Next Generation

TNGT Tonight

TNH Temperate Northern Hemisphere • Tonghua (China-PRC) (location identifier)

TNI Satna (India) (location identifier) • Tennelec/Nucleus Inc. • Total Normalized Impact • Trusted Network Interpretation

TNIEG Trusted Network Interpretation Environment Guideline

TNIU Trusted Network Interface Unit

TNJ Tanjung Pinang (Indonesia) (location identifier)

Tnk Trunk

TNK Tank • Tununak (USA-AK) (location identifier)

TNKS Tochnost i Nadezhnost Kiberneticheskikh Sistem (periodical code)

Tnl Tunnel

TNL Technological Research Programme • Télémètre de Nuages à Laser • Terminaison Numérique de Ligne • Terminal Net Loss • Ternopol (Ukraine) (location identifier) • Traitement en Langage Naturel • Tunnel

TNM Telecommunications Network Management • Teniente R. Marsh (Antarctica) (location identifier) • Tetranitromethane • Transmission Network Management

TNMR Tritium Nuclear Magnetic Resonance

TNN Tainan (China-ROC) (location identifier) • Topological Neural Network • Transport Network Node • Trunk Network Number

TNO Toegepast-Natuurwetenschappelijk Onderzoek (Netherlands) • Tamarindo (Costa Rica) (location identifier) • Transnational Operation • Transneptunian Object • TUBITAK National Observatory (Turkey)

TNOP Total Network Operation Plan

TNP Total Number of Publications • Traité de Non-Prolifération • Transportation Network Profile • Twentynine Palms (USA-CA) (location identifier)

TNPC Taiwan New PC Consortium (China-ROC) • Traffic Network Planning Center

TNPDS Torque Noise Power Density Spectrum

Tnpk Turnpike

TNPN Traité de Non-Prolifération Nucléaire • Tratadi de No-Proliferación Nuclear

TNR Antananarivo (Madagascar) (location identifier) • Terminaison Numérique de Réseau • Thermonuclear Runaway • Transit Nuclear Radiation

TNRE Transit Nuclear Radiation Effect

TNRIS Texas Natural Resources Information System

TNRSP Thailand National Remote Sensing Programme

TNS Technology Need Statement • The Next Step • Transit Network Selection • Tungsten (Canada-NT) (location identifier)

TNSA Technical Nuclear Safety Analysis

TNSC Tanegashima National Space Center (Japan) • Trailside Nature and Science Center (USA-NJ)

TNSE Total Normalized Stress Error

Tnsn Tension

Tnt Torrent

TNT Miami - Dade Collier (USA-FL) (location identifier) • Télévision Numérique de Terre • Télévision Numérique Terrestre • Theoretical Nose Tip • Tightened - Normal - Tightened • Timer No Token • Timmi Navigator Terminal • Trinitrotoluene • Turbulent/Non-Turbulent • Typographical Number Theory

TNTC Too Numerous To Count

TNTCAM Ten and Twenty Micron Camera

TNTDL Tabulated Numerical Technical Data List

Tntv Tentative

TNU Newton (USA-IA) (location identifier)

TNV Ten name Van • Total Net Value • Total Network Visibility

TNX Stung Treng (Cambodia) (location identifier) • Thanks

t&o Taken & Offered

to Tonga (ISO language code)

To Togo

T/O Table of Organization • Takeoff • Takeover • Target of Opportunity • Turnover

T&O Test & Operations

TO Table of Organization • Takeoff • Takeover • Target of Opportunity • Task Order • Technical Observer • Technical Officer • Technical Operation • Technical Order • Technical Office • Technical Officer • Technical Order • Teide Observatory (Canary Islands) • Telecommunications Operator • Telecommunications Outlet • Telegraphic Order • Télégraphie Optique • Telegraph Office • Téléobjectif • Telephone Office • Telephone Order • Telescope Operation • Telescope Operator • Température Optimale • Teraoctet • Terminal Ordinaire • Term-Ordered • Test and Operations • Thermo-Optic • Tiempo Oficial • Timeout • Tocantins (Brazil) • Togo • Tonga (ISO country code) • Tonnage Officiel • Tool Order • Torsional Oscillation • Train d'Ondes • Trained Operator • Training Objective • Training Observation • Transfer Optics • Transfer Order • Transistor Outline • Transition Object • Transmitter Oscillator • Transportation Office • Transportation Officer • Transporteur Orbital • Transport Officer • Transverse-Optical • Transverse Optics • Traveling Office • Travel Order • Turnoff • Turnout • Turnover • Tuyère Orientable • Two-Oscillator (model)

TOA Table Of Allowance • Table of Organization and Allowance • Tarifordnung für Angestellte • Terms Of Agreement • Time Of Arrival • Time Off Award • Top Of the Atmosphere • Torrance (USA-CA) (location identifier) • Total Obligational Authority • Tradeoff Analysis • Transfer Of Authority • Transportation Operating Agency • Tropical Ocean-Atmosphere

TOAD THz Optical Asymmetrically-positioned Demultiplexer • Tremendous Outburst Amplitude Dwarf (star)

TOADS Transportable Optical Atmospheric Data System

TOAE Terre - Océan - Air - Espace • Terre - Océan - Atmosphère - Espace (*)

TOAM Titan Orbiter Aerover Mission

TOAR Top Of Atmosphere Radiation

TOAS Technology Opportunities Analysis System

ToASC Torino Accademia delle Scienze Classe di Scienze Fisiche Matematiche e Naturali Atti (periodical code)

TOAST Task Organization And Status Tool • The Orpington Astronomical Society Times

TOASTE Transport of Ozone and Stratosphere/Troposphere Exchange

TOB Telemetry Output Buffer • Temporary Office Building • Tobruk (Libya) (location identifier)

toc Table of Contents (file extension)

ToC Table of Contents • Time of Coincidence

TOC Table Of Contents • Tactical Operations Center • Takeover Code • Technical Options Committee • Technical Order Compliance • Telemetry Operations Control • Telescope Operations Control • Telescope Operations Crew • Test Operations Center • Test Operations Change • Texas-Oxford Cluster (survey) • TexOx Cluster (survey) • Theory Of Computing • Theory Of Constraint • Time Of Coincidence • Time Of Completion • Toccoa (USA-GA) (location identifier) • Total Organic Carbon • Total Organic-carbon Content • Total Ownership Cost • Total Ozone Content • Training Observation Center • Training Operations Center • Transfer Of Control • Transportation Operating Command • Travel Order Claim • Tropospheric Ozone Column • Trouble Obsessionnel Convulsif • Trousse d'Outils de Communication

TOCC TDRS Operations Control Center • Telescope Operations Control Center • Test Operations Control Center • Type Of Change Code

TOCE Tonnes Of Carbon Equivalent

TOCHI Transactions On Computer Human Interaction (ACM) (periodical code)

TOCL Transactions On Computational Logic (ACM) (periodical code)

TOCN Technical Order Change Notice

TOCOTOX Too Complicated To Explain

TOCR Total Ownership Contract Requirement • Total Ownership Cost Reduction

TOCS Terminal Operations Control System • Terminal-Oriented Computer System • Term-Oriented Classification System • Transactions On Computer Systems (ACM) (periodical code)

TOCU Technical Order Control Unit

ToD Tour of Duty

TOD Takeoff Distance • Task Order Directive • Technical Objective Directive • Technical Objective Document • Technical Operations Department • Technical Order Data • Technical Order Directive • Technical Order Distribution • Temporally-Ordered Data • Test Operations Directorate • Thermal Observation Device • Ticket On Departure • Time Of Day • Time Of Delivery • Time Of Departure • Time Of Dispatch • Time-Ordered Data • Time-Ordered Domain • Tioman (Malaysia) (location identifier) • TIROS Oceanographic Drifter • Top Of Descent • Total Oxygen Demand • Tour Of Duty • Tradeoff Determination • Triangle Orientation Dissemination • Turnover Device

TODA Takeoff Distance Available

TODAES Transactions on Design Automation of Electronic Systems (periodical code)

TODARS Terminal-Oriented Data Analysis and Retrieval System

TODC Tine-Of-Day Clock

TODCOR Two-Dimensional Correlation (technique)

TODD Telescope Observation Designer and Driver • Telescope Observation Designer Driver

TODO Technical Order Distribution Office

TODR Takeoff Distance Required

TODS Transaction On Database Systems (ACM)

TO&E Table of Organization & Equipment

TOE Table of Organization and Equipment • Theory Of Everything • Time Of Event • Ton Oil Equivalent • Tozeur (Tunisia) (location identifier)

TOEFL Testing Of English as a Foreign Language • Test Of English as a Foreign Language

TOEIC Test Of English for International Communication

TOEL Time-Ordered Event List

TOES Telephone Order Entry System

TOET Test Of Elementary Training

ToF Time of Flight

TOF Test Operations Facility • Theoretical Orthogonal Function • Time Of Filing • Time Of Flight • Timeout Factor • Tomsk (Russia) (location identifier) • To Order From • Top Of Form

TOFC Trailer On Flat Car

TOFCN Technical Order Field Change Notice

TOFMS Time-Of-Flight Mass Spectrometer • Time-Of-Flight Mass Spectrometry • Time-Of-Flight Mass Spectroscopy

TOFS Time-Of-Flight Spectrometer • Time-Of-Flight Spectrometry

ToFU To FITS Utilities

Tog Together

TOG Technical and Operations Group • Toggle • Togiak Village (USA-AK) (location identifier) • Transactions On Graphics (ACM) (periodical code)

TOGA Takeoff/Go-Around • Tropical Oceans and the Global Atmosphere

Togr Together

TOGW Takeoff Gross Weight

TOH Time Of Handover • Top Of Hill • Torres (Vanuatu) (location identifier)

tohm Terohmmeter

TOHO Time Of Handover

TOI Technical Operation Instruction • Term Of Induction • Top Of the Instrument • Track Of Interest • Transfert Orbit Insertion • Troy (USA-AL) (location identifier)

TOIMDR Total organizational Intermediate Maintenance Demand Rate

TOIPT Technical Oversight Integrated Product Team

TOIPTM Technical Oversight Integrated Product Team Meeting

TOIS Technical Order Improvement System • Transactions On Information Systems (ACM) (periodical code)

TOIT Transactions On Internet Technology (ACM) (periodical code)

ToIzL Technisch oekonomische Informationen zivilen Luftfahrt (periodical code)

TOJ Madrid Torrejón (Spain) (location identifier)

Tok Tokyo (Japan)

TOK Torokina (Papua New Guinea) (location identifier)

TokAB Tokyo Astronomical Bulletin (periodical code)

TokRe Tokyo Astronomical Observatory Reprints (periodical code)

Tol Toledo (Spain) • Tolerance

TOL Temporary Occupation Licence • Test-Oriented Language • Time Of Liftoff • Toledo Express (USA-OH) (location identifier) • Tolerance • Tololo (catalog) • Topographic Orbiting Lidar

TOLD Takeoff and Landing Data

TOLIP Trajectory Optimization and Linearized Pitch

TOLIS Total Online Information System

TOLO Time Of Liftoff

TOLR Transmitting Objective Loudness Rating

TOLTS Total Online Test System

Tom Tomus

TOM Target Object Map • Task Ordering Management • Technical Operations Manager • Territoire d'Outre-Mer (France) • Thermal Optical Modeling • Thermo - Optical - Mechanical • Tilt-Over Mode • Time Of Month • Time of Orbital Midnight • Tombouctou (Mali) (location identifier) • Totally Ordered Mesh • Total Organic Matter • Transitioning Operations and Maintenance • Trunk Operations Monitor

TOMA Technical Order Management Activity • Technical Order Management Agency • Telescope Optics Manufacturers' Association • Training Operations and Management Activity

TOMACS Transactions On Modeling And Computer Graphics (ACM) (periodical code)

TOMar Travaux de l'Observatoire de Marseille (periodical code)

TOMCAT Telemetry Online Monitoring Compression And Transmission • Toulouse Off-line Model of Chemistry And Transport

TOMIS Task-Oriented Multipurpose Information System

TOMP Technical Order Management Plan • Transition Operations and Maintenance Plan

TOMS Total Ozone Mapping Spectrometer (*) • Total Ozone Mapping System • Transactions On Mathematical Software (ACM) (periodical code)

Ton Tonantzintla (survey) • Tonelaje • Tonne

TON Taiwan Oscillations Network • Telescope Operations News • Test Operation Notice • Tonga (ISO country code) • Tonu (Papua New Guinea) (location identifier) • Total Organic Nitrogen • Transactions on Networking (ACM) (periodical code)

TONIC Technique Optronique Nouvelle pour l'Identification des Cibles

Tonn Tonnage

TONS TDRSS Onboard Navigation System

ToO Target of Opportunity

TOO Target Of Opportunity • Target of Opportunity Observation • Test Operations Order • Thousand Oaks Optical (USA-MI) • Threshold Of Odor • Time Of Origin • To Order Only

TOOIS Transactions On Office Information Systems (ACM) (periodical code)

TOOL Test-Oriented Operated Language

TOOLS Technology for Object-Oriented Linking and Sharing

TOOT TexOx One Thousand (survey) • TexOx-1000 (survey)

Top Topographic • Topological • Topology

TO&P Technical Objectives & Plans

TOP Pa'anga (Tonga) (ISO currency code) • Tactical Operations Plan • Take Or Pay • Technical and Office Protocol • Technical Objectives and Plans • Technical Office Protocol • Technical Operating Procedure • Technical Operations Procedure • Technology Opportunities Program (formerly TIIAP) • Telecoms Operating Protocol • Temporarily Out of Print • Terrestrial Observation Panel • Tertiary Operation • Test and Operations Plan • Tester Operational Plan • Test Operational Plan • Thunderstorm Overflight Program • Time-averaged Orbiting Potential • Time Of Preparation • Time Of Processing • Topeka (USA-KS) (location identifier) • Topic • Topocentric • Tracker Optical Package • Transfer Orbit Phase • Tube à Ondes Progressives • Turbine Outlet Pressure

TOPAS Terra Observation Pattern Analysis System • Tropospheric Optical Absorption Spectroscopy

TOPAZ Traffic Organization and Perturbation Analyzer

TOPEX Topographic Experiment • (ocean) Topography Experiment • Topography Explorer • Typhoon Operational Experiment

TOPIAS Topographic Information and Archiving Software

TOPIC Tactical Operations and Increment Centre • Teletext Output Price Information Computer • Time-Ordered Programmer Integrated Circuit

TOPL Top Left

TOPLAS Transactions On Programming Languages And Systems (ACM) (periodical code)

TOPM Takeoff Performance Monitor

TOPMS Takeoff Performance Monitoring System

Topo Topographic • Topography

Topog Topographer • Topographical • Topography

Topogr Topografía

Topol Topological • Topology

TOPP Terminal-Operated Production Program • Torino Observatory Parallax Program

TOPR Top Right

TOPS Tape Order Processing System • Teaching Opportunities in the Plasma Sciences • Technical Order Page Supplements • Technology Opportunities Showcase • Telecommunications Orbiting and Pricing System • Telemetry Online Processing System • Teletype Optical Projection System • Tera Operations Per Second • Terascale Optimal PDE Simulation (center) • Terminal Operating System • Text-Oriented Processing System • Thermodynamic Ocean Prediction System • The Operational PERT System • Total Object Programming System • Total Operating paperless System • Total Operations Processing System • Total Organization Productivity Support • Towards Observation of Planetary Systems • Towards Other Planetary Systems (*) • Traffic Operator Position System • Transcendental Operating System • Transfer of Optical Processing to Systems • Transfer Orbit Perigee Stage • Transistorized Operational Phone System • Transition of Optical Processing to Systems • Transition of Optical Processors into Systems • Trends in Optics and Photonics (OSA)

TOPSAR Topographic Synthetic Aperture Radar

TOPSEC Top Secret

TOPSIM Torino Politecnico Simulator

TOPSIS Technique for Order Preference by Similarity to Ideal Solutions

TOPSS Tenerife Observatory for Planetary and Stellar Science (Canary Islands)

TOPTS Test-Oriented Paper-Tape System

TOQ Table d'Observations Quotidiennes • Tocopilla Barriles (Chile) (location identifier)

Tor Toronto (Canada-ON)

TOR Technical Operating Report • Technical Operations Research • Technical Override • Technical Oversight Representative • Technological Opportunities Report • Tentative Operational Requirement • Terminal Owning Region • Terms Of Reference • Test Of Reasonableness • Test Operation Report • Test Operating Report • Time Of Receipt • Time Of Reception • Time Of Release • Torrington (USA-WY) (location identifier) • Torsion • Torsional • Tout Ou Rien • Tropospheric Ozone Research • Tube à Onde Régressive

TORA Takeoff Run Available • Test Operations Risk Assessment

TORB Technical Order Review Board

TORES Text-Oriented Editing System

TORF Time of Retrofire

TORI Tables Of Representative Images

Torn Tornado

TORPEDO The Optical Retrieval Project: Electronic Documents Online

TORR Takeoff Run Required

TORS Tables Of Representative Spectra • Technical Order Status Report • Time-Ordered Reporting System • Torsion

TOS Tachyon Operating System • Tape Operating System • Taux d'Ondes Stationnaires • Technical and Operational Support (*) • Technical Operational Support • Teleoperator System • Temporarily Out of Service • Temporarily Out of Stock • Terminal-Oriented Software • Terminal-Oriented System • Terms Of Service • Test Operating System • Text Organizing System • The Oceanography Society • Time On Station • Time-Ordered System • TIROS Operational Satellite • Top Of Stack • Towarzystwo Observatorów Slonca (Poland) • Traffic Orientation System • Training Objective Statement • Transfer Orbital Stage • Transfer Orbit Stage (*) • Transition Operational State • Tromsø (Norway) (location identifier) • Trusted Operating System • Type Of Service • Type Of Shipment

TOSCA Testing Operational Scenarios for Concepts in ATM • Text and Office Systems Color Architecture • Text-Oriented System Controller for ALIS

TOSCANE Travaux d'Océanographie Spatiale dans l'Atlantique Nord-Est

TOSEM Transactions On Software Engineering and Methodology (ACM) (periodical code)

TOSP Taux d'Ondes Stationnaires en Puissance

TosRv Toshiba Review (periodical code)

TOSS Television Optical Scoring System • Test Operation Support Segment • TIROS Operational Satellite System • Transient Or Steady State

TOST Taux d'Ondes Stationnaires en Tension

Tot Total

TOT Task-Oriented Training • Telescope Optical Throughput • Time Of Transmission • Time On Target • Time Over Target • Total • Total Operating Cost • Totness Coronie (Suriname) (location identifier) • Transfer Of Technology • Triumph Over Tragedy • Turbine Outlet Temperature • Turn-On Time

TOTAL Toroidal Transport Analysis Linkage (code)

TOTE Tropical Ozone Transport Experiment

TOTM Total Message

TOTO Totable Tornado Observatory

TOTOVS Total Ozone from TOVS

TOTS Tower Operator Training System • Transportable Orbital Tracking Station

Tou Toulon (France)

TOU Touho (New Caledonia) (location identifier) • Transmission Operating Unit

TouCE Toulouse Cepaduès Éditions (periodical code)

TOUFPAC Tohoku University Focal Plane Array Controller

Tour Tourism • Tourist

TOUS Test On Understanding Science • Transmission Oscillator Ultrasonic Spectrometer

TOV Telemetering Oscillator Voltage • Teleoperated Vehicle • Ten Opzichte Van • Ten Overstaan Van • Tolman - Oppenheimer - Volkoff (equation) • Tortola West End (British Virgin Islands) (location identifier)

TOVP Technical Order Verification Plan

TOVS TIROS Operational Vertical Sounder

TOW Takeoff Weight • Tank and Orbiter Weight • Time Of Week • Toledo (Brazil-PR) (location identifier)

TOWA Terrain and Obstacle Warning and Avoidance

TOWARD Tower Ocean Wave and Radar Dependence

TOWG Technical Order Working Group • Turnover Working Group

Tox Toxicology

TOX Tobolsk (Russia) (location identifier) • Toxic • Toxicology

TOY Time Of Year • Toyama (Japan) (location identifier)

TOYal Transactions of the Astronomical Observatory of Yale University (periodical code)

TOZ Total Ozone (product) • Touba (Ivory Coast) (location identifier)

t&p Theft & Pilferage

Tp Telephone • Township • Troop • Type

T/P TOPEX/Poseidon

T&P Tracking & Processing • Transportation & Packaging

TP East Timor (ISO country code) • Tape Punch • Target Practice • Task Proposal • Taux de Pannes • Taux de Pression • Taxe Professionnelle • Taxpayer • Taylor Polynomial • Teaching Practice • Team pack • Technical Pamphlet • Technical Pan (Kodak) • Technical Paper • Technical Performance • Technical Preprint • Technical Problem • Technical Proposal • Technical Publication • Technographic Publication • Technology Parameter • Telemetry Processor • Telephone • Teleprinter • Teleprocessing • Telescope Project • Temperature Pod • Temperature Probe • Temperature Profile • Tempo Primo • Temps de Partage • Temps de Préparation • Temps de Propagation • Temps Partiel • Temps Passé • Tension de Polarisation • Terminal Portability • Terminal Processing • Term Probability • Test Panel • Test Period • Test Plan • Test Point • Test Port • Test Position • Test Pressure • Test Procedure • Test Program • Text Processing • Theorem Proving • Thermally Pulsing • Thermal Pulse • Thermo-Plastic • Third Party • Time Pulse • Time to Perigee • Timor (East_) (ISO country code) • Timor Orientale (code ISO de pays) • Title Page • Toll Point • Tool Permit • To Pay • Top Priority • Totally Positive • Total Package • Total Power • Total Pressure • Tout Payé • Tracking and Processing • Tracking Program • Tracking Processor • Trading Partner • Traffic Post • Traffic Program • Traînée de Pression • Traînée de Profil • Training Package • Training Parameter • Training Partner • Training Period • Training Plan • Training Practice • Traîté de Paix • Traitement Préventif • Trajectographie Passive • Trajectory Prediction • Transaction Processing • Transaction Processor • Transformation Procedure • Transition Period • Transition Plan • Transition Point • Translation Process • Translation Program • Transmission Priority • Transmission Protocol • Transpondeur • Transportation Priority • Transport Protocol • Transputer • Travail Pratique • Travaux Pratiques • Travaux Publics • Trésor Public (France) • Trial Period • Trigonometric Point • Triple Play • Triple Pole • Tropical Pacific • True Position • True Positive • Trusted Process • Turbopompe • Turbopropulseur • Turbo Pump • Turning Point • Turn Point • Twisted Pair • Type

TPA Clearwater (USA-FL) (location identifier) • Saint Petersburg (USA-FL) (location identifier) • Tampa (USA-FL) (location identifier) • Tape Pulse Amplifier • Technical Publications Announcement • Technology Planning Annex • Technology Program Annex • Telemetry Power Amplifier • Telemetry Processing Assembly • Temporary Program Area • Terephthalic Acid • Terminal Processor Assembly • Test Planning Analysis • Test Preparation Area • Test Process Archive • Thermal Plasma Analyzer • Third-Party Access • Third-Party Application • Tissue Plasminogen Activator • Total Package Approach • Trade Promotion Authority • Traffic Pattern Altitude • Transaction Processing Activity • Transistor Power Amplifier • Transpolar Arc • Tunable Parametric Amplifier • Turbopump Assembly • Two-Photon Absorption

TPAC Technology Program Area Committee • Telescope Precision Angle Counter • Treasury Program Advisory Committee (STScI)

TPAD Trunnion Pin Attachment Device

TPADM Thermal Plasma Analyzer Development Model

TPAG Test Plan Assessment Guide

TPAGB Thermally Pulsating Asymptotic Giant Branch (star)

TPAM Teleprocessing Access Method

TPAP Transaction Processing Application Program

TPAR Technology Producibility Assessment Review

TPAS Technical Publication Automated System • The Phoenix Astronomical Society (New Zealand)

TPAT Test Point Algorithm Technique

TPAW Tiger Parallel Architecture Workbench

TPB Tape Playback Bit • Test Prioritization Board • Tetraphenyl-Butadiene • Theory of Planned Behavior

TPBVP Two-Point Boundary Value Problem

TP&C Thermal Protection & Control

TPC Tactical Pilotage Chart • Tarapoa (Ecuador) (location identifier) • Technical Performance Council • Technical Performance Criterion • Technical Prime Contractor • Technical Processing Council • Technical Program Committee • Telecommunications Policy Committee • Telemetry Preprocessor Computer • The Peace Corps • Thermal Protection and Control • Time Polarity Control • Time Projection Chamber • Total Program Concept • Total Program Cost • Trade Practices Commission • Training Path Chart • Transmission Power Control • Tropical Prediction Center • Two-Person Control

TPCA Task Performance Capability Analysis • Test Procedure Change Authorization

TPCC Third-Party Call Control • Trade Promotion Coordinating Committee

TPCD Theorem Provers in Circuit Design (conference)

TPCE Tank Pressure Control Experiment • Test Program Cost Estimate

TPCF Two-Point Correlation Function

TPCN Task Plan Change Notice • Technical Program Change Negotiation • Test Plan Change Notice • Test Procedure Change Notice

TPCP Third-Party Collection Program • Two-Particle Coulomb Problem

TPCR Task Plan Change Request

TPCS Technical Procedure Control System

TPC3 Technology Panel for Command, Control and Communications

tpd Tons per Day

Tpd Tapped

TpD Tons per Day

TPD Technical Physics Department • Technical Product Documentation • Technical Products Division • Technisch Physische Dienst • Temperature Programmed Desorption • Terminal Protection Device • Test Planning Directive • Test Planning Document • Test Point Data • Test Procedure Drawing • Test Program Directive • Third-Party Developer • Time Pulse Distributor • Tons Per Day • Total Power Detector • Trailing Packet Discard • Train Principal Droit

TPDA Tapada

TPDAD Time Differential Perturbed Angular Distribution

TPDC Trade Point Development Centre • Training Performance Data Center

TPDDI Twisted Pair Distributed Data Interface

TPDK Teleform Programmer's Developer Kit

TPDMP Training Program Development Master Plan

TPDR Technical Publications Deficiency Report

TPDS Test Procedures Development System

TPDU Transport Protocol Data Unit

TPE Escudo (East Timor) (ISO currency code) • Tableau de Protection Électrique • Tableau Périodique des Éléments • Taipei - Chiang Kai Shek (China-ROC) (location identifier) • Teacher Preparation and Enhancement • Terminal de Paiement Électronique • Test Planning and Evaluation • Test Project Engineer • Threshold Photoelectron • Total Potential Energy • Tracking and Pointing Experiment • Transaction Processing Executive • Travail Personnel Encadré • Très Petite Entreprise

TPEB Technical Proposal Evaluation Board

TPED Tasking - Processing - Evaluation - Dissemination

TPEG Technical Planning and Evaluation Group

TPEM Technology Program Element Manager

TPEP Trusted Product Evaluation Program

TPES Technology for Performance-Engineered Systems • Threshold Photoelectronic Spectrum • Topological Polyelastic Sampling

TPESP Technical Panel on the Earth Satellite Program

TPF Tampa - Peter O'Knight (USA-FL) (location identifier) • Technology Performance Financing • Telemetry Processing Facility • Terminal Phase Final • Terminal Phase Finalization • Terminal Phase Finish • Terminal Presentation Facility • Terrestrial Planet Finder • Thermal Phase Finish • Tibetan Plateau Field (experiment) • Transaction Processing Facility • Transfer Phase Final • Triple Point Failure • Tug Processing Facility • Two-Phase Flow

TPFE Two-Phase Flow Experiment

TPFLEX Two-Phase Fluid Loop Experiment

TPFO TOPEX-Poseidon Follow-On

Tpg Tapping

TPG Taiping (Malaysia) (location identifier) • Technology Planning Guide • Test Pulse Generator • Thermal Plasma Generator • Thermionic Power Generator • Time Pattern Generator • Timing Pulse Generator • Topping • Train Principal Gauche • Trésorier-Payeur Général (France) • Trigger Primitive Generator

TPGID Terminal Process Group Identification • Terminal Process Group Identificator • Terminal Process Group Identifier

tph Tons per Hour

TPH Tonopah (USA-NV) (location identifier) • Total Piece Handling • Transmetteur de Pression Hydraulique

TPHOL Theorem Proving in Higher-Order Logic (conference)

tpi Teeth per Inch • Tracks per Inch • Turns per Inch

TPI Tape Phase Inverter • Tape Position Indicator • Tapini (Papua New Guinea) (location identifier) • Target Position Indicator • Tax and Price Index • Technical Proficiency Inspection • Technical Proposal Instructions • Tennessee Polytechnic Institute (USA-TN) • Terminal Phase Ignition • Terminal Phase Initiate • Terminal Phase Initiation • Terminal Phase Insertion • Test Planning Integration • Test Process Improvement • Test Program Instruction • Thermal Protection Investigation • Threads Per Inch • Title Page and Index • Tons Per Inch • Tracks Per Inch • Transport Provider Interface • Tribunal de Première Instance • Tribunal Pénal International (ONU) • Tritium Pellet Injector • Turns Per Inch • Two-Person Integrity

TPID Temporal Proposal Identification

TPIN Trading Partner Identification Number

TPIO Technical Program Integration Office

TPIP Time-Phased Implementation Plan

TPIPT Technical Planning Integrated Product Team

TPITS Two-Phase Integrated Thermal System (*) • Two-Phase Integration Thermal System

TPJ Taplejung (Nepal) (location identifier)

Tpk Turnpike

TPK Tapaktuan (Indonesia) (location identifier) • Turnpike

TPKE Turnpike

Tpl Triple

TPL Table Producing Language • Tapered Power Law • Teacher Programming Language • Technical Publications Library • Technical Publications List • Telephone Private Line • Temple (USA-TX) (location identifier) • Terminal Processing Language • Test Parts List • Test Plan • Test Planning Letter • Test Point Logic • Thrust Power Lever • Transaction Processing Language • Transpacific Landing • Tunable Pulsed Laser • Two-Phase Line

TPLS Terminal Position Location System • Tunable Pulsed Laser System

tpm Tons per Minute • Tours par Minute • Transactions per Minute

TPM Tape Preventive Maintenance • Technical Performance Measure • Technical Performance Measurement • Technical Performance Module • Technical Presentation Metafile • Technology Planning and Measurement • Technology Planning Measure • Telemetry Pipeline Manager • Telemetry Processor Module • Telescope Performance Monitor • Ten-Point Maneuver • Terminal Phase Maneuver • Terminal Phase Midcourse • Terrain Profile Matching • Test Planning Manager • Third-Party Maintenance • Tide-Predicting Machine • Token Passing Model • Tons Per Minute • Tons Per Month • Torsion Photoelectric Magnetometer • Total Power Mode • Total Productive Maintenance • Tours Par Minute • Transactions Per Minute • Transfer Phase Midcourse • Transition Probability Matrix • Transmission and Processing Model • Tree-Particle Mesh (code)

TPMA Thermodynamics Properties of Metals and Alloys

TPMC Tokyo Photoelectric Meridian Circle

TPMP Total Performance Measurement Plan

TPMR Training Program Management Review

TPMS Technical Performance Measurement System

TPN Température et Pression Normales • Tiputini (Ecuador) (location identifier) • Titulaire de Physique Nucléaire

TPNS Teleprocessing Network Simulator

TPO Tanalian Point (USA-AK) (location identifier) • Tank Pressurizing Orifice • Technical Planning Office • Technical Planning Officer • Technical Project Office • Technical Project Officer • Technical Publications Office • Telecommunications Program Objective • Tentative Program Objective • Test Plan Outline • Test Program Outline • Transfer Post Office • Traveling Post Office • Tube à Propagation d'Onde

TPOCC Transportable Payload Operations Control Center

TPOCP Turbopump Oxidizer Cavity Purge

TPOD Test Plan Of the Day

TPOM Tube à Propagation d'Onde Magnétique

TPON Telephony on a Passive Optical Network

TPoP Traditional Point of Presence

TPOP Traditional Point Of Presence

TPP Tarapoto (Peru) (location identifier) • Technical Performance Parameter • Technical Performance Plan • Technical Program Plan • Technical Publication Plan • Technology and Policy Program • Telecommunications Preparatory Programme • Telemetry Preprocessor • Terminal Procedures Publication • Test Participation Plan • Test Point Pace • Test Project Plan • Tetraacrylporphyrin • Thermal Power Plant • Thermal Protection Panel • Total Package Procurement • Total Program Planning • Total Program Procurement • Trans-Pluto Probe • Two-Phase Principle • Two-Photon Propagation

TPPC Total Package Procurement Concept • Transaction Processing Performance Council

TPPD Twisted Pair - Physical-Media Dependent

TPPG Test Plan Preparation Guide

TPPMD Twisted pair Physical Media Dependent

TPPP Theory and Practice of Parallel Programming (workshop)

TPPS Tape Postprocessing System

TPQ Tepic (Mexico) (location identifier) • Terms Per Query

Tpr Taper • Teleprinter

TPR Tape Programmed Row • Technical Progress Report • Technical Progress Review • Technical Proposal Requirement • Teleprinter • Telescopic Photographic Recorder • Temperature Profile Recorder • Température - Pouls -

Respiration • Temperature - Pulse - Respiration • Test Performance Recorder • Test Phase Report • Test Problem Report • Test Procedure Report • Third-Party Relay • Token Passing Ring • Tom Price (Australia-WA) (location identifier) • Total Projected Requirement • Trained Personnel Requirement • Training Personnel Requirement • Transaction Processing Routine • Trappe de Positionnement de Réception

TPRA Task Performance Requirements Analysis • Test Planning Requirement Analysis

TPRL Transport Properties Research Laboratory

tps Topspeed Data (file extension) • Tours par Seconde • Transactions per Second

TPS Tape Plotting System • Task Parameter Synthesizer • Tautenburg Plate Scanner • Technical Performance Specification • Technical Publishing Society • Technical Publishing Software • Technology Protection System • Telemetry Processing Station • Telemetry Processing System • Telephone Preference Service • Teleprocessing Services • Television Par Satellite • Terminal Performance Specification • Test Package Set • Test Plotting System • Test Preparation Sheet • Test Procedure Specification • Test Program Set • Test Program Software • The Planetary Society • Thermal Protection System • Thin-Plate Splines • Total Parameter Space • Total Personal Score • Total Pressure Sensor • Training Path System • Training Planning System • Transaction Processing System • Transactions Per Second • Transfer Propulsion System • Translator Processing Station • Translator Processing System • Translator Processor Station • Translunar Propulsion Stage • Transportation Protection Service • Trapani (Italy) (location identifier) • Tritium Purification System • Turbine-Powered Simulation • Turbo-Powered Simulator • Twisted Pair Shielded

TPSA Telekommunikacji Polskiej SA (Poland)

TPSB Telemetry Processing System Buffer

TPSC Task Performance Support Code

TPSDB Test Program Set Database

TPSE Thermal Protection System Experiment

TPSG Terrestrial Planets Study Group

TPSI Time Position Space Information

TPSK Tri-Phase Shift Keying

TPSO TOPEX-Poseidon Follow-On

TPSS Test Preparation Sheet Summary

Tpt Transport

TPT Tapeta (Liberia) (location identifier) • Technicien Principal Nucléaire • The Physics Teacher (periodical code) • Time Priority Table • Training Planning Team • Transmit Plain Text • Transport

TPTA Transient Pressure Test Article

TPTB The Powers That Be

TPTEP Training Program and Training Equipment Plan

TPTG Tuned-Plate Tuned-Grid

TPU Tank Pump Unit • Tape Preparation Unit • Taux de Panne en Utilisation • Taxe Professionnelle Unique (France) • Terminal Processor Unit • Text Processing Utility • Tikapur (Nepal) (location identifier) • Time Processing Unit • Tomsk Polytechnic University (Russia) • Transient Processing Unit

TPUN Test Procedure Update Notice

TPV Terminal au Point de Vente • Terminal Punto de Venta • Thermophotovoltaic

TPVS Transaction Program Verb Set

TpW Tons per Week

TPW Tactical Planning Workstation • Tons Per Week • Total Precipitable Water • Training Program Worksheet • True Polar Wander • True Polar Wandering • Turbo Pascal for Windows

TPWG Test Planning Working Group

TPX Tokamak Physics Experiment • Two-Phase Experiment

TpY Tons per Year

TPY Tons Per Year

TP2 Transmission Protocol Class 2

TP3 Transmission Protocol Class 3

TP4 Transmission Protocol Class 4

TQ Tel Quel • Topographic Quadrangle • Total Quality • Transparent Query

T&QA Test & Quality Assurance

TQA Teaching Quality Assessment • Test and Quality Assurance • Transformational Question Answering

TQAE Test Quality Assurance Evaluator

TQC Théorie Quantique des Champs • Time - Quality - Cost • Total Quality Commitment • Total Quality Control • Three-Quark Cluster

TQCM Temperature-controlled Quartz Crystal Microbalance

TQDM Total Quality Data Management

TQE Timer Queue Element • Torque • Total Quality Education

TQF Threshold Quality Factor

TQFP Thin Quad Flat Pack

TQFT Topological Quantum Field Theory

TQI Total Quality Involvement

TQL Total Quality Leadership

TQM Technical Quality Measurement • Total Quality Management • Total Quality Movement

TQMC Total Quality Management Course

TQMP Total Quality Management Plan

TQMS Total Quality Management System • Triple-Quadrupole Mass Spectrometer

TQMSC Total Quality Management Steering Committee

TQMW Total Quality Management Workshop

TQN Taluqan (Afghanistan) (location identifier)

TQO Transformation to Quality Organization

TQP Total Quality Program • Transistor Qualification Program

TQR Torque Rotor • Training Quality Report

TQS Term Question Specificity • Total Quality of Service • Total Quality System • Tres Esquinas (Colombia) (location identifier)

TQT Transistor Qualification Test

TQTP Transistor Qualification Test Program

tr Translucidus • Turkey • Turkish (ISO language code)

Tr Tare • Trace • Track • Train • Tranche • Transaction • Transitif • Transitive • Transitivo • Translate • Translated • Translation • Translator • Transpose • Transposition • Treasurer • Trench • Triassic • Troop • Through • Trust • Trustee

T-R Transmit-Receive

T&R Tip & Ring

T/R Tape Recorder • Technical Report • Transformer/Rectifier • Transmit/Receive • Transmitter/Receiver • Transportation Request • Travel Request • Turnaround Requirements

TR Gabon (international aircraft registration code) • Table Row • Tape Reader • Tape Recorder • Tape Register • Tariff Reform • Taux des Retards • Technical Reference • Technical Regulation • Technical Report • Technical Representative • Technical Requirement • Technical Review • Technische Richtlinie • Teknologirådet • Telecommunications Request • Teleoperator Retrieval • Temperature Range • Temperature Recorder • Tempore Regis •

Temps de Ravitaillement • Temps de Réaction • Temps de Roulage • Temps Réel • Terminal Ready • Terms of Reference • Terres Rares • Test Register • Test Report • Test Request • Test Response • Test Retest • Test Rocket • Test Routine • Test Run • Tête de Rotor • Text Retrieval • Theodore Roosevelt (1858-1919) • Thermomètre Réaumur • Thermorégulation • Threat Reduction • Thrust Reverser • Time for Retrofire • Time Record • Time Remaining • Time Resolution • Time-Resolved • Time to Retrofire • Time to Retrograde • Titulaire Remplaçant • Token Ring • Tons Registered • Tool Request • Top Return • Total Regulation • Total Revenue • Tracking Radar • Trail • Trailer • Training Report • Training Requirement • Training Review • Train Radio • Transaction • Transaction Register • Transfer • Transfer Register • Transformation Ratio • Transformer-Rectifier • Transient Response • Transition Radiation • Transition Region • Translation • Transmetteur-Récepteur • Transmission-Réception • Transmit-Receive • Transmitter-Receiver • Transportability Report • Transportation Report • Transportation Request • Transpose • Trip Report • Trouble Report • Trust Receipt • Tuesday-Thursday • Turboréacteur • Turkey (international vehicle registration code) • Turkey (ISO country code) • Turquie (code ISO de pays) • Turkey (postal country code) • Turnaround Requirements

TrA Triangulum Australe - Trianguli Australis (constellation)

TRA Tape Recorder Amplifier • Taramajima (Japan) (location identifier) • Targeted Research Action • Teacher Resource Agent • Technical Review Authority • Telecommunications Regulatory Agency • Teledyne Ryan Aeronautical • Temporary Reserved Airspace • Temporary Restricted Area • Test Requirements Analysis • Theory of Reasoned Action • Token Ring Adapter • Traffic Routing Administration • Training Requirements Analysis • Triaxial Recording Accelerometer • Turnaround Requirements Analysis

TRAA Tule River Amateur Astronomers (USA-CA)

Trac Tractatus • Tracto

TRAC Targeting and Reflective Alignment Concept • Tar River Astronomy Club (USA-NC) • Teaching, Reflection And Collaboration • Technical Recommendations Application Committee • Technical Report Awareness Circular • Technological Research of Advanced Concepts • Teléfono Rural de Acceso Celular • Text Reckoning And Compiling • Total Record Access Control • Transaction Reporting and Control System • Transient Radiation Analysis by Computer

TRACALS Traffic Control And Landing System

TRACE Task Reporting And Current Evaluation • Technical Report Analysis, Condensation, Evaluation • Teleprocessing Recording for Analysis by the Customer • Test Equipment for Rapid Automatic Checkout Evaluation • Time-shared Routine for Analysis, Classification, and Evaluation • Tolls Recording and Computing Equipment • Total Resource and Cost Evaluation • Total Risk Assessment Cost Estimate • Trace Remote Atmospheric Chemical Evaluation • Trajectory and Constraint Evaluation • Transition Region And Coronal Explorer • Transportable Automated Control Environment • Transport and Atmospheric Chemistry near the Equator • Transport And Chemistry near the Equator • Transregional Academic mobility and Credential Evaluation

TRACECA Transport Corridor Europe-Caucasus-Asia

TRACER Tropospheric Radiometer for Atmospheric Chemistry and Environmental Research

TRACON Terminal Radar Approach Control

TRACS Tool-Record Accountability System • Transportation Cost System • Trouble Adviser for Customer Services

TRACT Transportable Remote-Area Communication Terminal

Trad Tradition • Traditional • Traducción • Traducteur • Traduction • Traductor • Traduit • Traduttore • Traduzione

TRAD Training Requirements Analysis Document

TRADE Transatlantic Research and Development Environment

TRADEX Target Resolution and Discrimination Experiment

TRADIC Transistorized Airborne Digital Computer

TRADOC Traitement de Données Océanographiques

TRAE Time - Range - Azimuth - Elevation

TraFIX Trade Facilitation Information Exchange

Trafo Transformator

TRAG Technical Regulations Application Group

TRAGB Tritium Receiving and Analytical Glove Box

TraGe Travaux Géophysiques (periodical code)

TRAGEX Trace Gas Exchange

TRAI Telecom Regulatory Authority of India

TRAIL Technical Report Awareness Internet Link

TRAIN Training Resources Access Information Network

TRAIPR Trade-Related Aspects of Intellectual-Property Rights

Traj Trajectory

TRAJ Trajectory

TRAK Track

TrAlm Trudy Astrofizicheskogo Instituta Alma-Ata (periodical code)

TRAM Technology and Risk Assessment Meeting • Tilt-Rotor Aeroacoustics Model • Trajectory Reconstruction Analysis Method • Transputer Module

TRAMAR Tropical Rain Mapping Radar

TRAMAS Transportable Microwave Active Spectrometer

TRAMP Temperature Regulation And Monitor Panel • Testing, Reporting, and Maintenance Program • Test Retrieval And Memory Print • Three-dimensional Radiation-hydrodynamical Modeling Project

Tran Transient

TRAN Transfer

TRANET Tracking Network

TRANID Transaction Identification

Trans Transaction • Transfer • Transferred • Transform • Transformation • Transformed • Transformer • Transit • Transition • Transitive • Transitory • Translate • Translated • Translation • Translator • Transmission • Transmit • Transmittance • Transmitted • Transparence • Transparent • Transport • Transportation • Transpose • Transposition • Transvaal (South Africa) • Transverse

TRANS Transfer • Transition • Transitory • Translation • Transmitter • Transportation

Transa Transaction

TRANSAC Transistorized Automatic Computer

Transcr Transcribed • Transcription

Transf Transfer

Transfd Transferred

Transfg Transferring

Transl Translate • Translated • Translation • Translator

TRANSL Translation

Transp Transparent • Transport • Transportation

TRANSV Transversal

Trany Transparency

TRAO Taeduk Radio Astronomy Observatory (Korea-ROK) (formerly DRAO) • Torun Radio Astronomy Observatory (Poland)

TRAP Tactical Recovery of Aircraft and Personnel • Thermal Radiation Program • Training Resources Adjudication Panel • Transient Response Analysis Program • Transmettre - Ranger - Agir - Poubelle

TRAPS Template Review and Analysis for Performance Specifications

TRAS Training Requirements Analysis System

TRASYS Thermal Radiation Analysis System • Thermal Radiation Analyzer System

TRAU Transcoder and Rate Adaptor Unit • Transcoding Rate Adaption Unit

Trav Travail • Travaux • Travel • Traveller

TRAWL Tape Read And Write Library

TRAWMAR Targeted Research Action on Waste Minimization And Recycling

Trb Trebe

TRB Technical Review Board • Test Requirements Bulletin • Test Review Board • Three-Ring Binder • Translation Rule Builder • Transparent Bridging • Transportation Research Board • Turbo (Colombia) (location identifier)

TRBL Troubleshooting

TRC Tape Reader Calibrator • Tape Reader Control • Tape Reader Coordinator • Tape Record Coordinator • Teacher Resource Center (NASA/JSC) • Technical Reference Center • Technical Reference Code • Technical Repair Center • Technical Reports Center • Technical Research Center • Technical Resources Center • Technical Review Committee • Technological Repair Center • Technology Repair Center • Technology Resources Center • Telemetry and Remote Control • Test Readiness Certificate • Thermodynamics Research Center (USA-TX) • Token Ring Connection • Torreón (Mexico) (location identifier) • Transition Region Camera • Transmission Release Code • Transverse Redundancy Check • Tsukuba Research Consortium • Tube à Rayons Cathodiques • Tycho Reference Catalogue

TRCE Thermionic Reactor Critical Experiment • Trace

Tr. Co. Trust Co.

TRCO Technical Representative of the Contracting Officer

TRCP Tape Recorder Control Panel

TRCR Traceur

TRCS Toronto Red-sequence Cluster Survey

Trd Tread • Trinidad

TR&D Technology Research & Development

TRD Technical Requirements Data • Technical Requirements Document • Technology Research and Development • Technology Resources Department • Telemetry Requirements Document • Test Requirements Data • Test Requirements Document • Transition Radiation Detector • Trondheim (Norway) (location identifier)

TRDA Technological Research and Development Authority

TRDI Technical Research and Development Institute

TRDL Technical Reference Digital Library

TRDP Technology Research and Development Program • Thousand Research and Development Personnel

TRDPS Topographic Retrieval and Data Presentation System

TRDS Text Retrieval Data System

TrDus Trudy Instituta Astrofiziki Dushanbe (periodical code)

TRE Target-Rich Environment • Task Responsible Engineer • Taux de Reprise à l'Emploi • Technique de Recherche d'Emploi • Timing Read Error • Tiree (UK) (location identifier) • Transient Radiation Effect • True Radiation Emittance

Treas Treasurer • Treasury

Treasr Treasurer

TREAT Trouble Report Evaluation and Analysis Tool

TREC Text Retrieval Conference • Tracking Radar Echoes by Correlation

TRECAMS Tritium Remote Control And Monitoring System

T-RECS Thermal-Region Camera System

TRECS Thermal-Region Camera System

TreDS Trend sin Distributed Systems (workshop)

Tree Trustee

TREE Transeuropean Employment • Transient Radiation Effects on Electronics • Transition Region Explosive Event

TREES Tropical Ecosystem Environment Satellite

TREF Transient Radiation Effects Facility • Tripartite Employment Conditions Forum (CERN)

TREIS Tropical Radar Environmental Information System

TREK Transmit Record Electronically and Quickly

TREM Tropical Rainfall Explorer Mission

TREMI Tracking Radar Echoes by Moment Invariants

TREMP Trapped Radiation Environment Modeling Program

TREND Technical Requirements Engineering Network Database • Terrestrial Radiation Environment Development

TreP Osservatorio Treviso Pubblicazione (periodical code)

TREP Trusted Replica

TREPS Trusted Replicas

Trés Trésor • Trésorier

TRES Time-Resolved Emission Spectrum

TRESR Time-Resolved Electron Spin Resonance

TRESTLE Text Retrieval, Extraction and Summarization Technologies for Large Enterprises

TRET Telecommunications and Radio Engineering Telecommunications (periodical code)

Trf Tariff • Transfer

TRF Sandefjord (Norway) (location identifier) • Technical Reference File • Term Relevance Feedback • Terrestrial Reference Frame • Transfer • Transformer - Rectifier - Filter • Tribunal Regional Federal • Tube à Refroidissement Forcé • Tuned Radio Frequency • Turbine Research Facility • Two-Ribbon Flare

Trfc Traffic

TRFS Transparent Remote File System

TRFY Trafficway

Trg Training • Triangle

TRG Tailed Radio Galaxy • Target • Tauranga (New Zealand) (location identifier) • Teaching, Research and Development • Technical Review Group • Threshold Reporting Group • Tonrufgerät • Training Group • Training Requirements Generator • Transport Routier Guidé • Trigger

TRGB Tip of the Red Giant Branch

TRGD Technical Requirements and Goals Document

Trgf Traffähigkeit

TrGRC Transactions of the Geothermal Resources Council (periodical code)

TrGru Tiflis Izdatel Metsniereba Akademiia Nauk Gruzinskoi SSR (periodical code)

TRGS Technische Regelung Gefahrtstoff

TRGUI Technical Reports Graphical user Interface

TRH Trona (USA-CA) (location identifier)

Tri Triangulum-Trianguli (constellation) • Tributary • Triode

TRI Bristol (USA-VA) (location identifier) • Johnson City (USA-TN) (location identifier) • Kingsport (USA-TN) (location identifier) • Tape Recorder Interface • Taux de Retour sur Investissement • Technical Report Instruction

• Time Reversal Invariance • Toxic Release Inventory • Training Requirements Integration • Transaction Routing Index • Tributary • Tri-Cities (USA-TN) (location identifier)

TrIA Trainable Information Assistant

TRIAD Timesharing Research Image Analysis and Display • Tucson Revised Index of Asteroid Data

TRIAL Technique to Retrieve Information from Abstracts of Literature

Trib Tributary

TRIB Transfer Rate of Information Bits • Tributary

TRIBECA Triangle Below Canal (Street) (New York)

TRIC Transaction Identification Code • Transition Radiation and Ion Calorimeter

TRICON Triple Container

TRICS Trajectory Incremental Correction System

trid Triduum

TRID Training Requirement Identification Display

TRIDAC Three-Dimensional Analog Computer

TRIF Tiled Raster Interchange Format

TRIFAR Three Frequency Airborne Radar

Trig Trigger • Trigonometría • Trigonometric • Trigonometrical • Trigonometry

TRIG Trigger

Trigon Trigonometría • Trigonometric • Trigonometrical • Trigonometry

TRIL Training and Research in Italian Laboratories (ICTP)

TRILD Temps Réduit Indemnisé de Longue Durée

Trim Trimester

TRIM Terrain Resource Information Management • Testing for Reliability by Incremented Maximization • Time Resolution Imaging Mode • Time Resolved Imaging • Time Resolved Imaging Mode • Trade-Related Investment Measure • Transport of Ions in Matter

TRIMS Technical Risk Identification and Mitigation System • Time-Related Instruction Management System • Time-Resolved Ion Momentum Spectrometry • Trade-Related Investment Measures • Transportation-Integrated Management System

Trin Trinidad

Trip Triple • Triplicate

TriP Osservatorio Astronomico di Trieste Pubblicazioni (periodical code)

TRIP Tape-Recorded Interview Project (AMS) • Technology Readiness and Implementation Plan • Telephony Routing Information Protocol • Thunderstorm Research International Program • Thunderstorm Research International Project • Trade-Related aspects of Intellectual Property • Trade-Related Intellectual Property

Tripl Triplet • Triplicate

TRIPS Trade-Related Intellectual Property Rights (agreement) • Trillion Instructions Per Second

TRIS Test and Reliability Information System • Transportation Reporting and Inquiry System • Trapped Ions in Space • Tropical Region Imaging Spectrometer

TrITA Institut Teoreticheskoi Astronomii Trudy (periodical code)

TRIUMF Tri-University Meson Facility (Canada-BC)

TRIZ Theory of Inventive Problem Solving (Russian)

TRJ Tarakbits (Papua New Guinea) (location identifier)

Trk Track • Truck

TRK Tarakan (Indonesia) (location identifier) • Track • Truck • Trunk

TrKaz Trudy Kazanskaia Gorodkoi Astronomicheskoi Observatorii (periodical code)

Trkg Tracking

TRKG Tracking

Trkhd Truckhead

Trkmtd Truck-mounted

Trkr Tracker

TRKR Tracker

TrKra Seminar Kraevym Zadacham Trudy (periodical code)

TKS Tracks

Trl Trail

TRL Lira (Turkey) (ISO currency code) • Tarasko - Richardson - Lucy (method) • Target Recommendation List • Teacher Resource Laboratory • Technical Report Literature • Technology Readiness Level • Terrell (USA-TX) (location identifier) • Test Readiness List • Thermodynamics Research Laboratory • Through-Reflect Line • Thru-Reflect Line • Trail • Trailer • Transistor-Resistor Logic

TrLen Trudy Astronomicheskoi Observatorii Leningrad (periodical code)

TRLF Time-Resolved Lidar Fluorosensor

Trlg Trailing

TrLit Trudy Akademiia Nauk Litovskoi (periodical code)

Trlr Trailer

TRLR Trailer

TRLS Trails

Trly Trolley

trm Terminal Parameters (file extension)

TRM Team Risk Management • Technical Reference Model • Technology Reference Mission • Telecommunications Resource Manager • Terminal • Test Request Message • Test Requirements Manual • Thermal (USA-CA) (location identifier) • Thermal Remanent Magnetisation • Thermal Resistance Measurement • Thermoremanence • Thermoremanent Magnetism • Thermoremanent Magnetization • Time Ratio Modulation • Time-Reversal Mirror • Totally Reflective Mirror • Track Recovery Module • Trademark • Training and Research Mobility • Training Manual • Training Resource Model • Transmission Resource Manager • Trust-Region Method

TRMD Tableau de Répartition des Moyens par Discipline

TRMH Taux de Répartition des Moyens Horaires

Trml Terminal

TRML Terminal

TRMM Tropical Rainfall Measuring Mission

TrMMO Trudy Moskovskoe Matematicheskoe Obshchestvo (periodical code)

TrMO Tabor Mountain Observatory (Canada-BC)

TRMP Test Resource Management Plan • Test Resource Master Plan

Trmr Trimmer

TRMS Technical Requirements Management System • Test Resources Management System

Trmt Treatment

TRMTR Transmitter

Trn Train

TRN Task Requirements Notice • Technical Requirements Notice • Technical Research Note • Temporary Record Number • Terrain-Referenced Navigation • Threaded Read News • Time Rule Name • Token-Ring Network • Torino (Italy) (location identifier) • Train • Trunnion • Turin (Italy) (location identifier)

tRNA Transfer Ribonucleic Acid

Trnbkl Turnbuckle

Trnd Turned

Trng Training

TRNG Training

Trngl Triangle

TRNL Triangle Research Libraries Network

TRNPK Turnpike

Trnr Trainer

Trnsp Transportation

Trnspn Transportation

TRO Taree (Australia-NSW) (location identifier) • Technical Reviewing Office • Temporary Restraining Order • Test Requirements Outline • Thermal Relaxation Oscillation • Training and Readiness Oversight • Transparent Remote Operation • Truck Route order

TROFF Text Run Off

TROI Titania - Thea - Oberon - Iapetus

Trol Troland

TROLL Transmission and Receiving Of Laser Light

TROMEX Tropical Meteorology Experiment

TRON The Real-time Operating-system Nucleus • Time Rationale Organizational Notation

TROOP Transportation Operational and Organizational Plan

Trop Tropic • Tropical

Trop. Can. Tropic of Cancer

Trop. Cap. Tropic of Capricorn

TROP Tactical Resource and Operations Plan • Tropopause

TROPIC Token-Ring Protocol Interface Controller

TROPICS Transparent Object-oriented Parallel Information Computing System

Tropo Troposphere

TROPO Troposphere

TROPOZ Tropospheric Ozone

TROS Tape-Resident Operating System • Televisie Radio Omroep Stichting

TROV Telepresence Remotely Operated Vehicle

Trp Troop

TRP Target Reference Point • Technical Requirement Package • Technical Research Program • Technical Review Panel • Technological Research Program • Technology Reinvestment Program • Technology Reinvestment Project • Technology Research Plan • Technology Research Program • Terminal Rendezvous Phase • Test Resource Plan • Three Point (USA-AK) (location identifier) • Thunderstorm Research Project • Tiepoint Registration Processor • Time Reduction Processor • Traffic Regulation Point • Training Requirements Report • Transformation - Ring - Projection • Trip

TRPC Transaction Remote Procedure Call

TRPCL Tropical

TRPD Test Requirements and Plan Documentation

TrPet Trudy Seminar imeni G. Petrovskogo (periodical code)

TRPI Telecommunications-Related Personal Information

TRPK Turnpike

TRPS Temperature Regulating Power Supply

TrPul Trudy Glavnoi Astronomicheskoi Observatorii v Pulkovo (periodical code)

Trq Torque

TRQ Tarauacá (Brazil-AC) (location identifier) • Tariff Rate Quota • Torque

TRR Tape Recorder Reproducer • Teacher Retention and Renewal (program) • Technical Readiness Review • Technical Requirements Review • Test Readiness Review • Test Requirements Review • Theoretical Research Report • Time-Resolved Reflectivity • Topical Report Request • Topical Report Review • Torre • Training Requirements Report • Transformer - Rectifier - Regulator • Trincomalee - China Bay (Sri Lanka) (location identifier) • Tropical Report Request • Tropical Report Review

TRRB Test Readiness Review Board

TrRig Trudy Astrofiz. Lab. Riga (periodical code)

TRRMS Training Requirements and Resource Management System

TRRN Terrain

TRRU Tape Recorder Reproducer Unit

Trs Transfer • Transpose • Trustees

TRS Tandy Radio Shack • Tape Recorder System • Technical Repair Standard • Technical Report Service • Technical Requirements Specification • Technical Research Support • Telemetry Receiver Site • Telemetry Receiving Site • Telemetry Receiving Station • Teleoperator Retrieval System • Telephone Relay Service • Telerobotics Servicer • Terapin Reading Society (USA-MD) • Terminal Reservation System • Term-Rewriting System • Terrestrial Reference System • Test Record Sheet • Test Requirements Sheet • Test Requirements Specification • Test Requirements Summary • Test Research Service • Test Research Station • Test Resource Status • Test Response Spectrum • Test Resume Sheet • Thermal Radiation Source • Time Reference Signal • Time Reference System • Time-Resolved Spectrometry • Time-Resolved Spectroscopy • Tone-Ranging System • Top Right Side • Total Routhian Surface • Tower Research Simulator • Township Range System • Traceability and Reporting System • Tracking and Reporting System • Training Record System • Transpose • Trieste (Italy) (location identifier) • Tritium Receiving System • Tropical Revolving Storm • Troubleshooting Record Sheet • Trunked Radio System • Tug Rotational System

TRSA Terminal Radar Service Area

TRSB Time Reference Scanning Beam

Trsbr Transcriber

TRSC Thailand Remote Sensing Centre

Trscb Transcribe

Trsd Transferred • Transposed

TRSD Test Requirements and Specification Document • Test Requirements/Specification Document • Total Radiance Spectral Distribution

TRSE Token Ring Switching Engine

Trsfr Transfer

Trsfrt Transfert

TrSht Trudy Gosudarstvennogo Astronomicheskogo Instituta Shternberga (periodical code)

TRSL Test Requirements Specification Language

Trsn Transition

TRSN Torsion

Trsp Transport

TRSP Total Radiance Spectral Polarization

TRSR Taxi and Runway Surveillance Radar • Trieste Solar Radio System

TrSSR Trudy Akademiia Nauk SSSR Fizicheskii Institut (periodical code)

TrSta Trudy Instituta Astrofiziki Stalinabad (periodical code)

Trstn Transition

TRSU Tropical Remote Sensing Unit

TRSV Time-Resolved Surface Viscometer

Trt Turret

TRT Technical Review Team • Test Reference Tool • Token Rotation Time • Total Running Time • Tremonton (USA-UT) (location identifier) • Triply Reflecting Telescope • Triply Reflection Telescope

TrTas Trudy Tashkentskoi Astronomicheskoi Observatorii (periodical code)

Trtd Treated

TrTIM Trudy Tiflis Izdatel Metsniereba (periodical code)

TRTL Transistor-Resistor-Transistor Logic

TRTM Test Requirements Traceability Matrix

Trtmt Treatment

TrTsA TsAGI Trudy (periodical code)

TrTsi TsIAM Trudy (periodical code)

TRU Test Repairable Unit • Test Replaceable Unit • Time Release Unit • Transceiver Unit • Transformer-Rectifier Unit • Transmit-Receive Unit • Transmitter-Receiver Unit • Transportable Radio Unit • Transuranic • Transuranium • Trujillo (Peru) (location identifier)

TRUCE Tropical Urban Climate Experiment

TRUE Task-Related User Exit

TRUFO True Unidentified Flying Object

TRUMP Teaching Resources Unit for Modern Physics • Technical Review Updated Manuals and Publications • Total Revision and Upgrading of Maintenance Procedures

TRUST TDRSS Resource User Support Terminal

Trv Travail • Travesia

TRV Terminal Rendezvous • Transient Recovery Voltage • Trivandrum (India) (location identifier)

TRVEH Tracked Vehicle

TRVS Text - Radio - Video - Speech

Trvx Travaux

TRW Tarawa (Kiribati) (location identifier) • Thompson Raine Woolridge Inc.

TRWG Trajectory Reconstruction Working Group

TRWY Throughway

Trx Triplex

TRX Trenton (USA-MO) (location identifier)

Try Truly

TRY Tororo (Uruguay) (location identifier)

TRZ Tiruchirapally (India) (location identifier)

t&s Toilet & Shower

ts Tsonga (ISO language code)

Ts Tosyl Group • Tous

T&S Timing & Synchronization • Transport & Supply

T/S Temperature/Salinity • Test Station • Thermal/Structural • Time Sharing • Traceability by Serial • Troubleshoot • Troubleshooting • Turboshaft

TS Taittirīya Samhitā • Targeted Search • Target Search • Target Start • Target Strength • Target Suppression • Target System • Tasmania (Australia) • Team Supervisor • Technical Secretariat • Technical Service • Technical Specialist • Technical Specification • Technical Standard • Technical Support • Technicien Supérieur • Telecommunications System • Telescope Scientist • Telescope Section • Temperature Switch • Temporary Storage • Tensile Strength • Tentative Specification • Terme Spécifique • Terminal Server • Terminal Service • Termination Shock • Terminatorschatten • Terminator Shadow • Term Specificity • Test Schedule • Test Scheduling • Test Set • Test Site • Test Specification • Test Stand • Test Station • Test Summary • Test Surveillance • Text Split • Thermal Spray • Third Stage • Thunderstorm • Tiempo Sideral • Time Scale • Time Scheduled • Time Series • Time Service • Time Sharing • Time Stamp • Time Switch • Timing Selector • Timing System • Tool Steel • Top Secret • Total Strength • Traceability by Serial • Tracking Scope • Tracking Station • Tracking System • Traffic Shaping • Trainer System • Training Sequence • Training Simulator • Training Support • Trajectory Sensor • Transaction Server • Transfer Set • Transfer Syntax • Transfer System • Transient Source • Transient State • Transient Steady • Transition State • Transit Storage • Transmission Set • Transmitter Segment • Transparent Substrate • Transport Service • Transport Stream • Transport Support • Transport System • Très Secret • Tropical Storm • Troubleshoot • Troubleshooting • Trunk Synchronous • Tunisia (international aircraft registration code) • Tuplespace • Turboshaft • Twin Screw • Type of Signaling • Type Specification • Typescript

TSA Taipei - Sung Shan (China-ROC) (location identifier) • Target Service Agent • Target Signature Analysis • Target Signature Array • Taxiway Safety Area • Teacher-Scientists Alliance • Technical and Safety Assessment • Technical Support Activity • Technical Support Alliance • Technical Support Area • Technique des Systèmes Automatisés • Technologie de Systèmes Automatisés • Technology and Safety Assessment • Technology Security Analysis • Temperature Swing Absorption • Test and Simulation Assembly • Test Site Activation • Test Start Approval • Thermal Suprathermal Analizer • Time Scale Analysis • Time Series Analysis • Time-Shared Amplifier • Time Study Analysis • Toronto Sidewalk Astronomers (Canada-ON) • Total Sunspot Area • Total Surface Area • Trainer Support Activity • Training Site Agreement • Training Situation Analysis • Training Support Agent • Transportation Security Administration • Travel Security Advisory • Tripartite Standardization Agreement • Two-Step Antenna

TSAC Time of Spectrum Acquisition • Title, Subtitle And Caption • Total System Analysis and Cost • Tracking System Analytical Calibration

TSACC Telecommunications Standards Advisory Council of Canada

TSACS Terminal Server Access Control System

TSAD Trade Study and Analysis Database

TSAG Telecommunications Standardization Advisory Group

TSAP Technical Standards Activities Program (NIST) • Time Series Analysis Package • Transport Service Access Point

TSAPI Telephony Services Application Programming Interface

TSAPTS The Southern Automatic Patrol Telescope Service

TSAR Télécommande de Sauvegarde Ariane • Test Schedule And Review • Time-series Storage And Retrieval

TSARC Test Schedule And Review Committee

TSAVI Transformed Soil Adjusted Vegetation Index

TSB Task Selection Board • Technical Sector Board • Technical Service Bulletin • Telecommunications Standardization Bureau • Telecommunications Standards Bureau • Telematics Supervisory Board • Temporary Stowage Bag • Terminal Status Block • Termination Status Block • Total Surface Brightness • Transportation Safety Board • Trunk Signaling Buffer • Tsumeb (Namibia) (location identifier) • Twin Sideband

TSBF Tropical Soil Biology and Fertility

TSC Taisha (Ecuador) (location identifier) • Tandem Switch Center • Tangential Spherical Cube • Taxes and Social Contributions • Technical Service Center • Technical Standards Committee • Technical Subcommittee • Technical Support Center • Technical Support Contractor • Telecommunications Support Contract • Teledyne Systems Corporation • Telescope Supervising Controller • Television Scan Converter • Tension Sous Charge • Terminal Sterilization Chamber • Terrestrial Science Chamber • Test Score Category • Test Set Computer • Test Set Connection • Test Setup Complete • Test Support Complex • Test Support Controller • Test Support Coordinator • Thermally-Stimulated Current • Time-Sharing Control • Tooling Supplemental Control • Training Support Center • Transmission System Component • Transmitter Start Code • Transportation Systems Center • Triangular-Shape Cloud • Trimble System Controller • Triple Store Carrier • Trunked System Controller (*) • Trusted Software Capability • Twin-assigned Synchronous Composite • Two-Stage Command

TSCA Toxic Substances Control Act (USA)

TSCC Test Support Control Center • Tilt-Scan CCD Camera

TSCCS Terminal Server Access Controller System

TSCD Test Specification and Criteria Document • Test Specification Control Document

TSCDP Technical Service Center Development Plan

TSCF Task Schedule Change Form

TSCLT Transportable Satellite Communications Link Terminal

TSCM Taylor Series Correction Method • Technical Survey Countermeasure • Test Station Configuration Model

TSCMM Trusted Software Capability Maturity Model

TSCO Test Support Coordination Office • Test Support Coordinator • Top Secret Control Office • Top Secret Control Officer

TSCP Training Simulator Control Panel

TSCPF Time Switch and Call Processor Frame • Time Switch and Central Processor Frame

TSCS Test Support Computer System

TSCT Transportable Satellite Communications Terminal

TSCTR Transportable Single Channel Transponder Receiver

Tsd Tausend

TSD Technical Services Division • Technical Support Document • Test Sequence Document • Test Software Development • Test Start Date • Text - Speech - Dialogue • Thermionic Specific Detector • Total Spectral Density • Total System Design • Traffic Situation Display • Transient Signal Detector • Transmission Systems Directorate • Treatment - Storage - Disposal • Tshipise (South Africa) (location identifier)

TSDA Theory of Signal Detection Analysis

TSDCS Tritium Storage and Delivery Cleanup System

TSDF Telemetry Subset Definition File • Time Slot Duty Factor

TSDGB Tritium Storage and Delivery Glovebox

TSDIA Two-Scale Direct Interaction Approximation

TSDIS TRMM Science Data and Information System

TSDL Test Software Development Laboratory

TSDM Temporal and Spatial Data Mining • Time-Shared Data Management (system) • Trusted Software Development Methodology

TSDMO Tellus Series on Dynamic Meteorology and Oceanography (periodical code)

TSDMS Time-Shared Data Management System

TSDN Transfer Syntax Description Notation

TSDOS Time-Shared Disk Operating System

TSDS Tritium Storage and Delivery System

TSDU Transport Service Data Unit

TSE Team Supportability Element • Technical Support Effort • Technicien Supérieur d'Exploitation • Telemetry Support Equipment • Tender Support Equipment • Terminal Server Edition • Test of Spoken English • Test Scoring Equipment • Test Set Electrical • Test Support Equipment • Thermoelastic Stress Analysis • Toronto Stock Exchange (Canada-ON) • Total Solar Eclipse • Transmissible Spongiform Encephalopathy • Transmission Secondary Emission • Transportation Support Equipment • Tselinograd (Kazakhstan) (location identifier) • Türk Standardlari Enstitüsü (Turkey)

TSEC Telecommunications Security • Top Secret • Translinear Scanning Electro-optical Camera • Transmission Security

TSE&E Test, Systems, Engineering & Evaluation

TSEE Test Support Equipment Evaluation • Test, Systems, Engineering & Evaluation

TSEEC Transmission Systems Engineering Evaluation Center

TSEEF Transmission Systems Engineering Evaluation Facility

TSEF Technicien Supérieur d'Étude et de Fabrication

TSEG Telemetry Support Equipment Group

TSEM Transmission Secondary Emission Multiplier

TSEP Telecommunications System Engineering Plan

TSEQ Time Sequence

TSER Targeted Socio-Economic Research

TSES Technical Simulation and Evaluation System • Transportable Satellite Earth Station

TS&ET Test Support & Evaluation Team

TSET Test Support and Evaluation Team

TSEU Technology Seeker Evaluation Unit

TSF Tactical Support Facility • Télégraphie Sans Fil • Telemetry Scalar File • The Space Foundation • Thin Solid Films (periodical code) • Topographic Shell Fabrication • Traffic Service Function • Travail Sans Fin • Typical Shape Function • Venice Treviso (Italy) (location identifier)

TSFAR The Star*s Family of Astronomy Resources

TSFC Thrust-Specific Fuel Consumption

Tsfr Transfer

TSG Table de Situation Générale • Tanacross (USA-AK) (location identifier) • Target Scene Generator • Technical Specialty Group • Technical Study Group • Technical Subgroup • Technical Support Group • Technology Study Group • Technology Support Group • Température de Sortie des Gaz • Term Support Graph • Test Signal Generator • Test Support Group • Threat Scenario Generator • Time Signal Generator • Timing Systems Group • Transversely-adjusted Group

TSGC Texas Space Grant Consortium (USA-TX)

TSGP Test Sequence Generator Program

TSGS Time Series Generation System

TSh Tanzanian Shilling (currency)

TSH Telephony Signal Handler • Télescope Spatial Hubble • Temperate Southern Hemisphere • Tshikapa (Zaire) (location identifier)

TsHUJ Tsin Hua University Journal (periodical code)

TSHWR Thundershower

tsi Tons per Square Inch

T/SI Turn and Slip Indicator

TSI Tarjeta de Identificación Sanitaria (Spain) • Technical Standardization Inspection • Technicien Supérieur en Instruments • Technique et Science Informatiques • Teinte - Saturation - Intensité • Temporal Status Interval • Terminal System Initialization • Test Support Instruction • Threshold Signal-to-Interference (ratio) • Time Slice Interval • Time Slot Interchange • Time Slot Interchanger • Time Status Interval • Total Solar Irradiance • Tracking Source Indicator • Traitement du Signal et des Images • Transportation and Safety Institute • Transport Service Interpreter • Tsili Tsili (Papua New Guinea) (location identifier) • Turn and Slip Indicator

TSIC Transaction Set Implementation Convention

TSID Target System Interface Drawing

TSIG Trusted Systems Interoperability Group

TSIM Total Solar Irradiance Mission

TSIO Time-Shared Input/Output

TSIP Telescope System Instrumentation Program

TSIPS Teachers and Students Investigating Plants in Space

TSIR Time-Shared Information Retrieval • Total System Integration Responsibility

TSIT Technical Service Intelligence Team

TSJ Tribunal Superior de Justicia • Tsushima (Japan) (location identifier)

TSK Tagaki - Sugeno - Kang (model) • Task • Taskul (Papua New Guinea) (location identifier) • Time Shift Keying • Transmission Security Key

TSL Tamuin (Mexico) (location identifier) • Technical Scripting Language • Technical Services Laboratory • Temporal Surface Language • Temps Sidéral Local • Test Set Logic • Test Source Library • Test Support Language • The Software Link (USA-GA) • Three Searching Language

TSLC Taijuan Satellite Launch Center (China-PRC)

TSLD Troubleshooting Logic Diagram

TSLS Two-Stage Least Squares

TSM Tableau de Suivi des Moyens • Tail Service Mast • Taos (USA-NM) (location identifier) • Tax System Modernization • Technical Staff Member • Technology - Science - Mathematics • Température de Surface de la Mer • Temporary Storage Memory • Temps Sidéral à Minuit • Terminal Security Matrix • Terminal Support Module • Test Standards Module • Test Support Manager • Test Support Methodology • Textron Specialty Materials • Time Scheduled Maintenance • Time-Shared Monitor • Time Sharing Monitor • TIROS Stratospheric Mapper • Tivoli Storage Manager • Total Suspended Matter • Toyama Science Museum (Japan) • Trade Study Management • Training Support Material • Transportation Simulation Model • Tray Sorting Machine • Trunk Signaling Model • Trusted Software Methodology • Trusted System Methodology

TSMC Taiwan Semiconductor Manufacturing Co. Ltd. (China-ROC)

TSMD Time-Stress Measurement Device

TSMG Tiempo Sideral Medio en Greenwich

TSMIS Test Support Management Information System

TSMT Technical Support Management Team

TSMU Tbilisi State Medical University (Georgia)

TSN Tape Serial Number • Technical Support to Negotiations • Term Semantic Network • Test Support Network • Tianjin (China-PRC) (location identifier) • Time Since New

Tsnt Transient

TSNU Tajik State National University (Tajikistan)

TSO Isles of Scilly - Tresco (UK) (location identifier) • Table Structure Overview • Technical Service Order • Technical Staff Officer • Technical Standard Order • Technical Standing Order • Technical Support Office • Technical Support Online • Technical Support Order • Technisch Secundair Onderwijs • Telecommunications Service Order • Terminator Sensor Output • Test Site Office • Test Support Office • Test Support Operation • Time-Sharing Operations • Time-Sharing Option • Time Since Overhaul • Town Suboffice • Transportation Supply Officer • Turtle Star Observatory (Germany)

TSOC Triana Science and Operations Center

TSO/E Time-Sharing Option Extended

TSOE Time-Sharing Option Extended

TSOP Technical Standard Operating Procedure • Thin Small Outline Package • Tripartite Standard Operating Procedure

TSOR Tentative Specific Operational Requirement

TSORT Transmission System Optimum Relief Tool

TSOS Time-Sharing Operating System

TSOSC Test Set Operational Signal Converter

TSOT Total System Operational Test

TSOW Technical Scope Of Work

Tsp Teaspoon

TSP Team Software Process • Technical Support Package • Technical Support Plan • Tehachapi - Kern County (USA-CA) (location identifier) • Telecommunications Service Priority • Telecommunications Service Provider • Telemetry Simulation Package • Telemetry Support Package • Teleprocessing Services Program • Temperature-Sensitive Parameter • Temperature-Sensitive Period • Temporary Standard Practice • Temporary Standard Procedure • Terminal Support Processor • Test Site Position • Test Software Program • Test Status Panel • Test Supervisor • Test Support Package • Test Support Plan • Test Support Point • Test Support Position • Test Support Program • Texas Star Party (USA-TX) • Thermospray • Timely Spares Provision • Time Series & Polarimetry • Time Series Polarimetry • Time Sorting Program • Time Synchronization Protocol • Titanium Sublimation Pump • Total Soluble Phosphorus • Total Suspended Particles • Trade Security Policy • Traffic Service Position • Traffic Supervision Position • Training Support Package • Training Support Plan • Trajectory Shaping Program • Trajectory Synthesis Program • Transaction and Service Provider • Traveling Salesman Problem • Triple Super Phosphate • Trisodium Phosphate • Trusted Software Project • Twisted Shielded Pair

TSPA Total System Performance Assessment

TSPC Time Stored Program Command

TSPD Technology and Space Program Development

tspi Tons per Square Inch

TSPI Time, Space, Position Information

TsPK Tsentr Podgotovska Kosmonavti (Russia)

TSPM Total Suspended Particulate Matter

TSPR Total System Performance Requirement • Total System Performance Responsibility

TSPS Time-Sharing Programming System • Traffic Service Position System

TSPU Test, Synchronization and Pulsargate-Unit • Test, Synchronization and Pulsar-gating Unit

TsPul Tsirkulyary Glavnoi Astronomicheskoi Observatorii i Pulkove (periodical code)

TSQLS Thundersqualls

TSR Technical Safety Requirement • Technical Software Review • Technical Status Review • Technical Study Report • Technical Summary Report • Technical Support Request • Technical Support Requirement • Telecommunications Service Request • Télévision de la Suisse Romande (Switzerland) • Temps Significatif Routier • Terminal and Stay Resident • Terminal Specified Routing • Terminate and Stay Resident (*) • Terminate Stay Resident • Test Schedule Report • Test Schedule Request • Test Status Report • Test Storage Ring • Test Strategy Report • Test Summary Report • Test Support Request • Test Support Requirement • Thermal and Solar Radiometer • Time-Sensitive Reporting • Time Status Register • Timişoara (Romania) (location identifier) • Torsion Solid Rod • Total Stress Range • Trade Study Report • Training Status Report • Training Support Requirement • Transport Service Request • Tripartite Staff Requirement • T-shirts over Suits Ratio

TSRA Total System Requirements Analysis • Training System Requirements Analysis

TSRB Termination Settlement Review Board

TSRD Technology Systems Research Development • Telecommunications Security Requirements Document

TSRM Third-Stage Rocket Motor

TSRO Temps Significatif Routier Observé

TSRP Temps Significatif Routier Prévu • Training System Resource Plan

TSRPC Transport-Specific Remote Procedure Call

TSRS Trieste Solar Radio System

TSRSG Tohoku University Science Reports Series Geophysics (periodical code)

TSRT Transportable Standard Remote Terminal

TSRV Transport Systems Research Vehicle

TSS New York - East 34th Street (USA-NY) (location identifier) • Tactical Satellite Support • Tank Subsystem • Tape Search System • Targeted Search System • Target Selection Standard • Target Sensing System • Task State Segment • TDRSS Service Session • Technical Specification Sheet • Technical Staff Surveillance • Technical Support Service • Technical Support Software • Telecommunications Security System • Telecommunications Segment Specification • Telecommunications Standardization Sector • Telecommunications Standards Sector • Telecommunication Switching System • Télémanipulateur de la Station Spatiale • Telemetry Simulation System • Telescope Software System • Terminal Security System • Terminal Support System • Test Set Simulator • Test Software Support • Test Support System • Tethered Satellite System • Time-Sharing System • Time Signal Set • Time-Slot Shifter • Topographic Support System • Top Side Sounder • Toroidal Space Station • Total Surveying System • Total Suspended Solids • TPS Summary Sheet • Tracked Suspension System • Trading Support System • Traffic Separation Scheme • Traffic Service Section • Traffic Service Station • Training Support System • Training System Software • Trajectory Subsystem • Transport Support System • Trapezoid Self-Scheduling • Travel Support System • Tromsø Satellite Station (Norway) • Tropospheric Scatter System • Tug Structural Support • Turbine Steamship • Twin-Screw Steamer

TSSA Test Scorer and Statistical Analyzer • Test Site Support Activity • Training Systems Support Activity

TSSC Technical Services Support Contract • Training System Support Center • Transportable Space Support Center

TSSD Typesetting System for Scientific Documents

TSSF Tape Staging and Storage Facility

TSSG Top-Seeded Solution Growth

TsShe Tsirkulyar Shemakhinskoi Astrofizicheskoi Observatorii (periodical code)

TSSI Top secret Special Intelligence

TSSLW Travaux de la Société des Sciences et des Lettres de Wroclaw (periodical code)

TSSM Thruster Subsystem Module

TSSMS Time Sharing Service Management System

TSSN Trailer Station Serial Number

TSSO Time Sharing System Option

TSSP Tactical Satellite Signal Processor • Task-Specific Safety Plan

TSSR Troposcatter Satellite Support Radio • Troposcatter Satellite Support Relay

TSSS Time-Sensitive Support System

TSSSO Time Sharing Subsystem Option

TSST Time - Space - Space - Time

TsSta Tsirkulyar Stalinabadskoi Astronomicheskoi Observatorii (periodical code)

TSSU Test Signal Switching Unit

Tst Test

TST Tab-Separated Table • Target Selector Telescope • Technical Support Team • Telefunken Systems Technik (Germany) • Test • Test Status Tracker • Time-Sharing Terminal • Time - Space - Time • Trang (Thailand) (location identifier) • TRMM Science Team

TSTA Technology Security Technical Assessment • Tritium System Test Assembly

TsTas Tsirkulyar Tashkentskoi Astronomicheskoi Observatorii (periodical code)

TSTC Technical Specialist Training Course

TSTE Training System Test and Evaluation

Tstg Testing

TSTM Thunderstorm • Tropical Science Team Member

TSTN Triple Supertwisted Nematic

TSTO Two-Stage-To-Orbit

TSTOL TRMM System Test and Operations Language

TSTP Test of Selected Topics in Physics • Total System Training Plan • Transaction Switching and Transport Protocol

Tstr Tester

TSTR Transistor

TSTRF Terrestrial Space-Time Reference Frame

Tstrz Transistorize

TSTS Third Stage Test Set • Time - Space - Time - Space • Tracking System Test Set • Transaction Switching and Transport Service

TSTT Terascale Simulation Tools and Technologies (center)

TSTV Two-Stage Transport Vehicle

TSU Tabiteuea South (Kiribati) (location identifier) • Tajik State University (Tajikistan) • Tape Search Unit • Technical Service Unit • Technical Support Unit • Telebriefing Switching Unit • Telescopic Sight Unit • Tennessee State University (USA-TN) • Texas Southern University (USA-TX) • Thermal Structural Unit • This Side Up • Time-Sharing User • Time Standard Unit • Towson State University (USA-MD) • Transfer Switch Unit • TRMM Science User • Truman State University (USA-MO)

TSUM Troy State University Montgomery (USA-AL)

TsUP Tsentr Upravileniia Polyoti (Russia)

tsv Tabbed-Separated Values (file extension)

TSV Tabbed-Separated Values • Tab-Separated Values • Telefonsteckverbinder • Term Selection Value • Townsville (Australia-QLD) (location identifier)

TSVD Truncated Singular Value Decomposition

TSVP Tournez S'il Vous Plaît

TSVQ Tree-Structured Vector Quantizer

TSVS Time-Sharing Virtual System

TSW Technical Software • Test Software • Test Switch • Tsewi (Papua New Guinea) (location identifier)

TSWA Temporary Secure Working Area

TSWG Technical Support Working Group • Test Support Working Group • Training Support Working Group • TRMM Science Working Group

TSX Tanjung Santan (Indonesia) (location identifier)

TSY Tasikmalaya (Indonesia) (location identifier) • Toxic Support Yard

tSZ Truncated Sunyaev-Zel'dovich (effect)

TSZ Thermal Sunyaev-Zel'dovich (effect) • Tsetserleg (Mongolia) (location identifier)

TS3 Time Sensitive Support System • Top Secret Support System

T&T Transportation & Transportability • Travel & Tourism • Trinidad & Tobago

tt Tatar (ISO language code) • Teletypewriter • Tomos • Tout

T/T Terminal Timing • Timing/Telemetry

TT Chad (international aircraft registration code) • Tartar • Task Team • Task Type • Team Terminal • Technical Task • Technical Team • Technical Test • Technical Testing • Technical Training • Technical Translation • Technicien Titulaire • Technology Transfer • Technology Transition • Telegraphic Transfer • Telescope Team • Teletype • Teletypewriter • Temps Terrestre • Terminal Timing • Terminal Transfer • Terrestrial Time • Tertiary • Test Temperature • Therapeutic Touch • Think Tank • Thrust

Termination • Tidal-Torque (theory) • Tiefe Temperatur • Time Trial • Timing/Telemetry • Tip-Tilt • Tobago (Trinidad &_) (ISO country code) • Top Term • Total Temperature • Total Time • Tracking Telescope • Transit Temporaire • Transmit Timing • Travail Temporaire • Travel and Tourism • Traveling Telescope • Tremblement de Terre • Trinidad & Tobago (international vehicle registration code) • Trinidad & Tobago (ISO country code) • Trouble Ticket • True Track • True Type • Turn Time • Typewriter Text

TTA Tailored Training Approach • Tan Tan (Morocco) (location identifier) • Technological/Technical Assistance • Technology Teachers Association • Technology Trends Assessment • Telecommunications Technology Association • Telemetry and Telecommand Assembly • Thermomechanical Test Area • Time To Apogee • Transmission Temporelle Asynchrone • Transport-Triggered Architecture • Travel Time Authorized • Travel Trust Association (UK) • Triplet-Triplet Annihilation

TTAB Technical Training Acceptance Board

TTAE Total Time Aircraft Engine

TTAF Total Time Aircraft Frame

TTAP Training Technology Applications Program

TTAPS Turco - Toon - Ackerman - Pollack - Sagan (model)

TTAS Technology Transfer Annual Survey • Transportable Telemetry Acquisition System

TTAU Teletex Access Unit

TTB Technology Testbed • Telescience Testbed • Tetragonal Tungsten Bronze • Tip-Tilt Box • Tortoli Arbatax (Italy) (location identifier) • Train - Tram - Bus

TTBS Trinidad and Tobago Bureau of Standards

ttc TrueType Compressed Font (file extension)

TT&C Telemetry, Tracking & Command • Telemetry, Tracking & Commanding • Telemetry, Tracking & Communications • Telemetry, Tracking & Control • Tracking, Telemetry & Command • Tracking, Telemetry & Control

TTC Taltal (Chile) (location identifier) • Technical Training Center • Telecommunications Technology Committee • Technical Training Centre • Technology Transfer Center • Telecommunications Techniques Corporation • Telemetry and Telecommand • Telemetry, Telecommand and Control • Telemetry Test Console • Telemetry, Tracking and Command • Telemetry, Tracking and Control • Thompson Techniques de Communication • Time To Circularize • Time To Control • Time To Customer • Timing - Trigger - Control • Toutes Taxes Comprises • Tracking - Telemetry - Command • Track-to-Track Correlator • Transportation Technology Center • Tropical Test Center • TrueType Compressed (font) • Tunnel Thermal Control

T/TCA Thrust Translation Controller Assembly

TTCA Thrust Translation Controller Assembly

TTC&M Telemetry, Tracking, Control & Monitoring

TTCM Telemetry, Tracking, Control and Monitoring

TTCN Tree and Tabular Combined Notation

TTCP Technology Transfer Control Plan • Test Training Certification Plan • Toppo Three-Channel Photometer

TTCS Target Tracking and Control Station • Target Tracking and Control System • Telemetry, Tracking, and Command System

TTCU Teletypewriter Control Unit

TTCV Tracking Telemetry, Command, and Voice

TTD Dollar (Trinidad & Tobago) (ISO currency code) • Tactical Terrain Data • Technical Task Directive • Technical Training Device • Technology Transfer Division • Technology-Transparent Device • Temporary Text Delay • Test Technique Demonstrator • Touch Tone Director • Training and Technical Data • Transit Time Damping • Troutdale (USA-OR) (location identifier)

TTDL Terminal-Transparent Display Language

TTDR Tracking Telemetry Data Receiver

Tte Tante

TT&E Technical Test & Evaluation • Test, Teardown & Evaluation

TTE Technical Test and Evaluation • Technical Training Equipment • Technology Transfer Institute • Telephone Terminal Equipment • Temporary Test Equipment • Ternate (Indonesia) (location identifier) • Test, Teardown and Evaluation • Time-Tagged Event • Time To Event • Trace Table Entry

TTEG Technical Task Execution Group

TTEL Tools and Test Equipment List

Ttes Toutes

TTESP Test Time Event Sequencer Plan

ttf TrueType Font (file extension)

TTF Taurus Tunable Filter • Temporary Test Facility • TESLA Test Facility • Tetrathiofulvalene • Transition Task Force • TrueType Font

TTFF Time To First Fix

TTFN Tah Tah, For Now • Ta-Ta For Now

Ttfs Toutefois

TTF&T Technology Transfer, Fabrication & Test

TTFT Technology Transfer, Fabrication and Test

TTFX Time To First Fix

TTG Tartagal (Argentina-SA) (location identifier) • Technical Translation Group • Tiling Task Group • Time To Go • Trainer Test Group • Transition Task Group

TTH Tactical Transport Helicopter • Thumrait (Oman) (location identifier)

TTHS Transients - Trainees - Holdees - Students

TTI TechTrans International Inc. (USA-TX) • Télécommunications et Traitement de l'Information • Telecommunication Training Institute • Teletypewriter Test Instruction • Term-based Topic Identification • Test Thoroughness Index • Tetiaroa Island (French Polynesia) (location identifier) • Theme-Topic Interface • Tide-Thermal Instability • Time-Temperature Indicator • Toyota Technological Institute (Japan) • Transitional Technology Inc.

TTIC Technology Transfer In Cryogenics

TTIP Test Team Initial Planning

TTIS Translation and Technical Information Service

TTJ Tottori (Japan) (location identifier)

TTK Tottenham - Hale Station (UK-England) (location identifier)

Ttl Total

TTL Telescope Technologies Ltd. (UK-England) • Test Tone Level • Thermal Test Laboratory • Through The Lens • Time To Live • Time-Transformed Leapfrog • Total Time to Launch • To Take Leave • Training Task List • Transfer Tube - Low-loss • Transistor-to-Transistor Logic • Transistor-Transistor Logic • Turtle Island (Fiji) (location identifier)

TTLM Through-The-Lens Metering

Ttlmnt Totalement

TTM Tablón de Tamara (Colombia) (location identifier) • Temperature Test Model • Temps de Traversée Moyenne • Thermal Test Model • Thick-Target Model • Time To Market • Tip-Tilt Mirror • Trigger Timing Module • Two-Tone Modulation

TTMM True Tape Motion Monitor

TTMOP Technical Test Measure Of Performance

TTMP Transit Time Magnetic Pumping

TTMT Thermal-Timescale Mass Transfer

TTN Technology Transfer Network • Technology Transfer Node • TEX and TUG News • Transfer Technology Node

• Transfer Tune - Normal-loss • Trenton (USA-NJ) (location identifier)

Tto Toronto (Canada-ON)

TTO Britton (USA-SD) (location identifier) • Tailored Technical Order • Technology Transfer Office • Tobago (Trinidad &_) (ISO country code) • Total Toxic Organic (compound) • Transient Tracers in the Ocean • Trinidad & Tobago (ISO country code)

TTOC Tape Table Of Contents

TTOE Tentative Table of Organization and Equipment • Touch Tone Order Entry

TTOES Technology Task Order Engineering Service

TTOF Thermally Tunable Optical Filter

TT&P Tactics, Techniques & Procedures

TTP Tactics, Techniques and Procedures • Tape-To-Print • Technical Task Plan • Technical Test Plan • Tecnología de Telescopios Pequeños • Technology Transfer Program • Telescience Testbed Project • TERENA Technical Programme • Test Transfer Port • Thermal-Transfer Printing • Thermistor Test Program • Time To Perigee • Time To Protection • Time-Triggered Protocol • Trainer Test Procedure • Transport and Thermal Processing • Trusted Third Party

TTPP Telescience Testbed Pilot Program

TTPRR Trainer Test Procedure and Results Report

TTP&S Trainees, Transients, Patients & Students

TTPS Trainees - Transients - Patients - Students

TTQ Tortuquero (Costa Rica) (location identifier)

TTR Tana Toraja (Indonesia) (location identifier) • Target Tracking Radar • Technical Test Report • Time-Temperature Recorder • Time Token Rotation • Transient Thermal Radiation • Triangle-To-Rectangle (transformation)

TTRB Telescope Time Review Board

TTRP Timed Token Rotation Protocol

TTRR Technical Test Readiness Review

TTRT Target Token Rotation Time

TTRTO Time-Triggered Real-Time Object

TTS Teachers Touch the Sky (project) • Technical Test Specification • Technology Tracking System • Telemetry Transmission System • Telemetry Tracking System • Teletypesetter • Teletypesetting • Temperature Test Set • Temporary Threshold Shift • Test and Training Satellite • Text-To-Speech • Thrust Termination System • Time To Service • Time To Spill • Time To Station • Total Training System • Training and Training Support • Transaction-Tracking System • Transmission Test Set • Transponder Test Set • Transportable Telemetry System • Trouble Ticketing System • True To Scale • Tsaratanana (Madagascar) (location identifier) • T Tauri Star

TTSARB Technology Transfer Security Assistance Review Board

TTSF Time To Subsequent Fix

TTSI Tandem Telecommunications Systems Inc.

TTSN Taiwan Telemetered Seismographic Network (China-ROC) • Total Time Since New

TTSP Training Test Support Package

TTT Taitung (China-ROC) (location identifier) • Tasks - Tools - Time • Team Time Trial • Teleteaching and Training • Through The Thickness • Tidal-Torque Theory • Trunk-to-Trunk Transfer • Trust - Transparency - Timing

TTTS Tanker/Transport Training System

TTU Tape Transport Unit • Technology Transfer Unit • Teletex-Telex-Umsetzer • Tennessee Technological University (USA-TN) • Texas Tech University (USA-TX) • Terminal Time Unit • Terminal Timing Unit • Tetuan (Morocco) (location identifier) • Time Tag Unit • Time Terminal Unit • Timing Terminal Unit • Très Très Urgent

TTUC Technical Teachers Upgrading Centre

TTV Tartu Variable (star) • Termination, Test, and Verification • Time To View • Total Television

TTVC Test Termination and Verification Contractor

T&TW Test & Training Wing

TTW Test and Training Wing

TTWG Technology Transfer Working Group • Turnover and Transfer Working Group

Ttx Teletex

TTX Teletext • Tracking Transducer

TTY Teleprinter • Teletype • Teletypewriter

TTYB Teletype Buffer

TTYC Teletype Controller

TTYL Talk To You Later

TTYU Teletype Unit

ttz 't is te zeggen

Tu Thulium (now Tm) • Tuesday • Turkish

TU Côte d'Ivoire (international aircraft registration code) • Ivory Coast (international aircraft registration code) • Table Update • Tape Unit • Task Unit • Technical University • Technical Utilization • Technische Universität • Technology Utilization • Temps d'Utilisation • Temps Universel • Terminal Unit • Teutonic • Thermal Unit • Tiempo Universal • Timing Unit • Toamotu • Towson University (USA-MD) • Trade Union • Traffic Unit • Training Unit • Transfer Unit • Transition Unit • Transmission Unit • Transport Unit • Traversing Unit • Très Urgent • Tributary Unit • Tucumán (Argentina) • Tuner Unit • Tungu • Turkey • Type Unit

TUA Tulcan (Ecuador) (location identifier)

TUAAM Tuber Unit Automatic Antenna Matching

TUAID Tübingen Universität Astronomisches Institut Diplomarbeit (periodical code)

TUAM Tuber Unit Antenna Matching

TUASB Tokyo University Institute of Space and Aeronautical Science Bulletin (periodical code)

TUASR Tokyo University Institute of Space and Aeronautical Science Report (periodical code)

tub Tuba

Tub Tubular

TUB Technische Universität Berlin (Germany) • Tubuai Island (location identifier)

TUBA TCP/UDP with Bigger Addresses • Transition Unit Box Assembly • Turkiye Bilimler Akademisi (Turkey)

TUBSAT Technical University of Berlin Satellite

Tuc Tucana-Tucanae (constellation)

TUC Temps Universel Coordonné • Tiempo Universal Coordinado • Time of Useful Consciousness • Total User Cell • Trade Union Congress (UK) • Travail d'Utilité Collective • Tucumán - Benjamin Matienzo (Argentina-TU) (location identifier) • Type Unit Characteristic • Type Unit Code

TUCA Taiwan University Consortium in Astronomy

TUCHA Type Unit Characteristic

Tud Tubmányos

TUD Tambacounda (Senegal) (location identifier) • Technische Universität Darmstadt (Germany) (formerly THD) • Technische Universiteit Delft (Netherlands) • Technology University of Denmark • Trunk Up Down

TUDE Teletype Users Data Entry

Tue Tuesday

TUE Technische Universiteit Eindhoven (Netherlands) • Traité de l'Union Européenne • Tratado de la Unión Europea • Tuesday • Tupile (Panama) (location identifier)

Tues Tuesday

TUES Tübingen Ultraviolet Echelle Spectrograph

TUEX Tuesdays Excepted

TUF Technology Utilization Foundation • Terminal User Function • Tours (France) (location identifier)

TUFA Telexumsetzer für Fernsprechnebenstellenanlagen

TUFEJ Tokyo University Faculty of Engineering Journal Series (periodical code)

TUFER Tokyo Denki University Faculty of Engineering Research Reports (periodical code)

TUFI Toughened Unipiece Fibrous Insulation

TUG Technical Users Group • Technische Universität Graz (Austria) • TeX Users Group • TRANSAC Users Group • Tributary Unit Group • TUBITAK Ulusal Gozlemevi (Turkey) • Tuguegarao (Philippines) (location identifier) • Turbo User Group

TUH Tullahoma Arnold (USA-TN) (location identifier)

TUI Tangible User Interface • Text-based User Interface • Transport Urbain Intermodal • Turaif (Saudi Arabia) (location identifier)

TUISR Tokyo University Institute Industrial Science Report (periodical code)

TUJ Tum (Ethiopia) (location identifier)

TUK Turbat (Pakistan) (location identifier)

TUL Tulsa (USA-OK) (location identifier)

TULE Transistorized Universal Logic Element

TULIP The University Licensing Programme

TULIPS Terrain Understanding and Land Information Processing System

TULS Toroidal with Uniform Line Spacing (grating)

TUM Table Universelle des Marées • Technische Universität München (Germany) • Tumut (Australia-NSW) (location identifier) • Tuning Unit Member

TUMS Table Update and Management System

Tun Tuning • Tunisia • Tunnel

TUN Tunis Carthage (Tunisia) (location identifier) • Tunisia (ISO country code) • Tunisie (code ISO de pays)

TUNA Transition Unit Nest Assembly

Tung Tungsten

TUnGG Technische Univ. Geodesy Global Geodyn. (periodical code)

TUNIS Toronto University System (Canada-ON)

TUNL Triangle Universities Nuclear Laboratory (USA-NC) • Tunnel

TUNS Technical University of Nova Scotia (Canada-NS)

TUO Taupo (New Zealand) (location identifier) • Technology Utilization Office • Technology Utilization Officer • Tuesdays Only

TUP Teknologiske Udviklingsprogram (Denmark) • Telephony User Part • Temple University Press (USA) • Titre Universel de Paiement • Tsinghua University Press (China-PRC) • Tupelo (USA-MS) (location identifier)

TUPR Total Uplink Power Requirement

TUQ Tougan (Burkina Faso) (location identifier)

Tur Turret

TUR Teaching Unit of Resource • Thermal Uniformity Restoral • Tucuruí (Brazil-PA) (location identifier) • Turkey (ISO country code) • Turquie (code ISO de pays)

TURA Thematic Uniform Resource Agent

Turb Turbine • Turbulence

TURB Turbine • Turbulence

TURBC Turbulence

TURBT Turbulent

TURF Tuning Unit Radio Frequency

Turk Turkey • Turkish

Turkmen Turkmenistan

TURN Turnaround

TurTe Turbulentnye Techeniia (periodical code)

TUS The Untold Story • Transmission Unattended Site • Tucson (USA-AZ) (location identifier)

TUSA Tactical Unmanned Surveillance Aircraft

TUT Tauta (Papua New Guinea) (location identifier) • Telemetry User Table • Thermal Uniformity Test • Transistor Under Test • Tutorial

TUTT Tropical Upper Tropospheric Through

TUU Tabuk (Saudi Arabia) (location identifier)

TUUD Total Unit Uplight Density

TÜV Technischer Überwachungsverein

TUV Tucupita (Venezuela) (location identifier) • Tuvalu (ISO country code)

TUVR Total Ultraviolet Radiation

TUW Technische Universität Wien (Austria) • Tubala (Panama) (location identifier)

TUX Tuesdays Excepted • Tumbler Ridge (Canada-BC) (location identifier)

TUY Tulum (Mexico) (location identifier)

TUZ Tucumã (Brazil-PA) (location identifier)

T/V Testing Verification • Thermal Vacuum • Threat/Vulnerability

TV Target Vehicle • Technical View • Téléviseur • Television (*) • Télévision • Temps de Vol • Temps Variable • Temps Vrai • Tension de Vapeur • Terminal Velocity • Terre Végétale • Test Vehicle • Textverarbeitung • Thermal Vacuum • Thrust Vector • Thrust Vectoring • Time Variable • Time-Varying • Tokyo Variable (star) • Tololo-Vienna • Total Variation • Total Vision • Total Volume • Trajectoire de Vol • Transport Vehicle • Travel Voucher • Traverse • Travessa • Tuva • Tuvalu (ISO country code)

TVA Morafenobe (Madagascar) (location identifier) • Target Value Analysis • Target Value Assessment • Taxe à la Valeur Ajoutée • Taxe sur la Valeur Ajoutée • Temporary Variance Authority • Tennessee Valley Authority (USA) • Thrust Vector Actuator • Thrust Vector Alignment

TVAC Taxe sur la Valeur Ajoutée Comprise

TVAR Test Variance

TVASI Tee Visual Approach Slope Indicator

TVB Television Broadcast • Top Versus Bottom • Total Variation Bounded (method) • Tout Va Bien

TVBS Television Broadcast Satellite

TVC Television Camera • Télévision Communautaire par Câble • Télévision en Couleurs • Terminal Velocity Curve • Thermal Vacuum Chamber • Thermal Voltage Converter • Thrust Vector Control • Time Variable Constant • Time-Varying Coefficient • Trapped Vortex Combustor • Traverse City (USA-MI) (location identifier) • Troisième Vitesse Cosmique

TVCA Thrust Vector Control Actuator • Thrust Vector Control Assembly

TVCD Thrust Vector Control Driver

TVCO Television Converter

TVCS Thrust Vector Control System • Thrust Vectoring Control System • Tritium Vault Cleanup System

TVCU Thrust Vector Nozzle Control

TVD Torque Velocity Dynamometer • Total Variation Diminishing (scheme)

TVDPM Total Variation Diminishing Particle Mesh

TVDS Transportable VLBI Data System

TVDU Television Display Unit

TVE Technical Validation Experiment • Technology Validation Experiment • Technology Verification Experiment • Temps de Vol des Équipements • Test and Verification Equipment • Test Vehicle Engine • Tête de Visée

Électronique • Théorie des Valeurs Extrêmes • Thermal Vacuum Environment • Thermo-Visco-Elasticity

TVF Television Video File • Thief River Falls (USA-MN) (location identifier) • TOC Verbosely from File

TVFI Television Frame International

TVFS Thermal Video Frame Scanner

TVG Tarifvertragsgesetz (Germany) • Television Guiding • Test Vector Generator • Threshold Voltage Generator • Time Variation of Gain

TVGDHS Television Ground Data Handling System

TVH Taxe de Vente Harmonisée (Canada)

TVHD Télévision à Haute Définition

TVI Technical Validation Inspection • Télévision Industrielle • Television Interference • Terminal de Vente Image • Thomasville (USA-GA) (location identifier) • Transformed Vegetation Index • Translation Variant of an Image • Tutored Videotape Instruction

TVIG Television and Inertial Guidance

TVK Tagesverkehrskurve

Tvl Transvaal (South Africa) • Transversal • Travel

TVL South Lake Tahoe (USA-CA) (location identifier) • Transversal - Vertical - Longitudinal

Tvlg Traveling

Tvlr Traveller

TVLS Toroidal Varied-Line Space

TVM Tachometer Voltmeter • Television Memory • Temporal Variance Map • Test Verification Matrix • Thermal Vorticity Mesoscale (model) • Tower Vector Magnetograph • Transistor Voltmeter

TVMA True Visual Magnitude Atlas

TVMP Terrain Visualization Master Plan

TVN Televisión Nacional • Test Verification Network

TVOC Television Operations Center

TVOL Television Online

TVOR Terminal VHF Omnidirectional Range

TVP Taxe de Vente Provinciale (Canada) • Test Verification Program • Time-Variable Parameter • Topological Vector Profile • Trombosis Venosa Profunda

TVPS Torus Vacuum Pumping System

TVQ Taxe de Vente du Québec (Canada-QC)

TVr Temps Vrai

TVR Télétel à Vitesse Rapide • Television Rating • Television Recorder • Temperature Variation of Resistance

TVRO Television Receive Only

TVS Telecommunicatie- en Verkeersbeleidingssystemen • Telemetry Validation System • Television Sensor • Television System • Temporal Variance Spectrum • Term Vector Space • Thermal Video System • The Virtual School • Thrust Vectoring System • Thrust Vector System • Topological Vector Space • Tornado Vortex Signature • Toxic Vapor Suit • Transient Voltage Suppressor • Tri-Valley Stargazers (USA-CA)

TVSA Thrust Vector Servoamplifier

TVSD Time-Varying Spectral Display

TVSL Thin Viscous Shock Layer

TVSO Television Space Observatory

TVSt Teilnehmervermittlungsstelle

TVT Television Tape • Television Trainer • Thermal Vacuum Test

TVTA Thermal Vacuum Test Article

TVTB Tout Va Très Bien

TVU Taming Virtual University • Taveuni (Fiji) (location identifier)

TVY Dawe (Myanmar) (location identifier) • Tavoy (Myanmar) (location identifier)

TV3D Télévision en Relief

tw Taiwan (China-ROC) • Te Weten • Twi (ISO language code)

T/W Tailwind • Thrust/Weight

TW ROC (ISO country code) • Tactical Warning • Tail Wind • Taiwan (ISO country code) • Target Week • Taxiway • Teamwork • Terawatt • Termination Word • Thermal Windowing • Thermal Wire • Threat Warning • Thrust-Weight • Thumbwheel • Time Words • Total Work • Traveling Wave • Turbulent Wake (region) (Jupiter) • Typewriter

TWA Time-Weighted Average • Trailing Wire Antenna • Transaction Work Area • Traveling-Wave Amplifier • TW Hya Association • Twin Hills (USA-AK) (location identifier) • Two-Way Alternate

TWAIN Technology Without An Important Name • Technology Without An Interesting Name • Toolkit Without An Important Name • Toolkit Without An Interesting Name

TWAMP Time-Weighted Average Month Program

TWAN Third World Astronomy Network

TWARS Third World Association for Remote Sensing

TWAS Third World Academy of Sciences (Italy)

TWASI Two-Wire Analog Subscriber Interface

TWAT Traveling-Wave Amplifier Tube

TWB Terrestrial Wideband • Toowoomba (Australia-QLD) (location identifier) • Traveling-Wave Beam

TWC Tank Waste Characterization • The Weather Channel • Third-World Country • Three-Way Catalytic • Total Water Content

TWCRT Traveling-Wave Cathode-Ray Tube

TWD New Dollar (China-ROC) (ISO currency code) • Port Townsend (USA-WA) (location identifier) • Tail Wags Dog • Total World Domination • Toward • Transparente Wärmedämmung

TWDS Twisted Double Shielded

TWE Taylor (USA-AK) (location identifier) • Test of Written English • Threat Warning Equipment

TWEB Transcribed Weather Broadcast • Transcribed Weather En-route Briefing

TWERLE Tropical Wind, Energy and Reference Level Experiment • Tropospheric Wind Earth Radio Location Experiment

TWF Twin Falls (USA-ID) (location identifier)

TWG Target Working Group • Technical Working Group • Technology Working Group • Telecommunications Working Group • Telescope Working Group • Test Working Group • TEXnical Working Group • Transfermium Working Group • Transfer Working Group • Transition Working Group • Tripartite Working Group • Turnover Working Group

TWGMLC Technical Working Group on Multiple Language Coordination

TWh Terawatt Hour

TWH Catalina Island - Two Harbors (USA-CA) (location identifier)

TWI Trade Weighted Index • Training With Industry • Training Within Industry

TWICE Traveling Wave-tube Interactive Computational Environment

TWICOLS Twelfth International Conference on Laser Spectroscopy

TWIG Technology Watch Information Group • Test Working Integration Group

TWIMC To Whom It May Concern

TWIN Third-World Information Network

TWINS Two Wide-angle Imaging Neural-atom Spectrometer

TWK Too Well Known • Typewriter Keyboard

TWL Threat Wavelength • Total Weight Loss • Traveling-Wave Laser

TwLA Two-Letter Acronym

TWLT Twente Workshop on Language Technology

TWM Tab Windows Manager • Traveling-Wave Maser • Two-Way Mirror

TWMBK Traveling-Wave Multiple-Beam Klystron

TWN ROC (ISO country code) • Taiwan (China-ROC) (ISO country code) • Teleprinter Weather Network • Tewantin (Australia-QLD) (location identifier) • Third-World Network

TWNC Two-Way Non-Coherent

TWNSO Third World Network of Scientific Organizations

TWOM Traveling-Wave Optical Maser

TWOV Transit Without Visa

Twp Township

TWP Task Work Package • Team Work Plan • Telescope Working Platform • Torwood (Australia-QLD) (location identifier) • Township • Tropical Western Pacific • Twisted Wire Pair

TWPA Traveling-Wave Parametric Amplifier

TWPB Total Work Package Budget

TWPL Teletypewriter Private Line

TWPO Tropical Western Pacific Ocean

Twr Tower

TWR Temporary Work Request • Total Weighted Ranks • Tower

TWRG Towering

TWS Task Work Specification • Telefonisch Waarschuwings-Systeem • Telemanipulation Workstation • Test Workstation • Thermal Wavelength Shift • Track While Scan • Trader Workstation System • Trusted Workstation • Two-Way Simultaneous

TWSB Twin Sideband

TWSTFT Two-Way Satellite Time and Frequency Transfer

TWSTT Two-Way Satellite Time Transfer

TWT Tawitawi (Philippines) (location identifier) • Transonic Wind Tunnel • Traveling-Wave Tube • Trisonic Wind Tunnel • Tuesday - Wednesday - Thursday • Two-Way Time • Two-Way Traveltime

TWTA Traveling-Wave Tube Amplifier! (* • Traveling-Wave Tube Assembly

TWU Tawau (Malaysia) (location identifier) • Technical Writing Unit • Transportation Workers Union • Transport Workers Union • Trinity Western University (Canada-BC)

TWV Ten Waarde Van

TWX Teletypewriter Exchange • Telex

Twy Taxiway • Twenty

TWY Tawa (Papua New Guinea) (location identifier) • Taxiway

TWZ Mount Cook - Pukaki-Twizel (New Zealand) (location identifier) • Mt. Cook - Pukaki-Twizel (New Zealand) (location identifier)

Tx Tax • Taxation • Teleprinter Exchange • Tonneaux • Transmit

TX Telex • Texas (USA) • Transmission • Transmit • Transmitter

TXC Transmission Control • Transmit Clock

TxD Transmit Data

TXD Transmit Data • Transmitted Data

TXE Telephone Exchange Electronics • Transmission Experiment

TXF Tax Exchange Format

TXG Taichung (China-ROC) (location identifier)

TXI Tunable X-ray Imager

TXK Texarkana (USA-TX) (location identifier)

TXL Berlin Tegel (Germany) (location identifier)

TXM Teminabuan (Indonesia) (location identifier)

TXN Tunxi (China-PRC) (location identifier)

Txp Transponder

TXP Transmission Problem • Transmit Plain (text)

Txpdr Transponder

TXQ Transparent Question

TXR Tanbar (Australia-QLD) (location identifier)

TXRF Total X-Ray Fluorescence

TXS Texas Survey

txt ASCII Text (file extension)

TXT Text

TXTHDR Text Header

Txtl Textile

TXT2STF Text To Structured File

TXU Tabou (Ivory Coast) (location identifier)

Ty Territory • Truly

TY Benin (international aircraft registration code) • Thank You • Then Year • Total Yield • Typhoon

TYA Tula (Russia) (location identifier)

TYB Tibooburra (Australia-NSW) (location identifier)

TYC Two-Year College

TYD Tynda (Russia) (location identifier)

TYDAC Typical Digital Automatic Computer

TYE Tyonek (USA-AK) (location identifier)

TYF Torsby (Sweden) (location identifier)

TYG Thylungra (Australia-QLD) (location identifier)

TYL Talara (Peru) (location identifier)

TYM Staniel Cay (Bahamas) (location identifier)

TYN Taiyuan (China-PRC) (location identifier)

TYNIAC Tiny Almost Automatic Computer

TYO Tokyo (Japan) (location identifier)

Typ Typical • Typing • Typist • Typographer • Typographic • Typographical • Typography

TYP Ten-Year Plan • Tobermorey (Australia-NT) (location identifier) • Typical • Typist

Typh Typhoon

Typo Typographer • Typographic • Typographical • Typography

Typog Typographer • Typographic • Typographical • Typography

Typstg Typesetting

Typw Typewriter • Typewriting • Typewritten

TYR Tyler (USA-TX) (location identifier)

TYS Knoxville (USA-TN) (location identifier)

TYT Treinta-y-Tres (Uruguay) (location identifier)

TYVM Thank You Very Much

TYZ Taylor (USA-AZ) (location identifier)

Tx Teilziffer • Textziffer

TZ Mali (international aircraft registration code) • Tanzania (ISO country code) • Tanzanie (code ISO de pays) • Time Zero • Time Zone • Truncated Zel'dovich (approximation)

TZA Belize City (Belize) (location identifier) • Tanzania (ISO country code) • Tanzanie (code ISO de pays) • Truncated Zel'dovich Approximation

TZB Technische Zentralbibliothek

TZD True Zenith Distance

TZE Total Zone Exclusion

TZK Temperatur-Zeit-Kristallisation • Transportable Zenitkamera

TZM Titanium - Zirconium - Molybdenum • Tizimin (Mexico) (location identifier) • Traveling Zone Method

TZN South Andros (Bahamas) (location identifier)

TZO Thorne-Zytkow Object

TZP Time Zero Pulse

TZS Shilling (Tanzania) (ISO currency code)

TZT Te Zijner Tijd

TZW Technologiezentrum Wasser (Germany)

TZX Trabzon (Turkey) (location identifier)

T1M Télescope de 1m

T1T Treburer-1m-Teleskop (Germany)

T2 Technology Transfer • Tuvalu (international aircraft registration code)

T2IWG Team Test Integration Working Group

T2L2 Time Transfer by Laser Link

T2M Télescope de 2m

T2S2 Translation/Transcription Support System

T3 Kiribati (international aircraft registration code) • Trust - Transparency - Timing

T4C Termination for Convenience

T4D Termination for Default

T7 San Marino (international aircraft registration code)

T9 Bosnia-Hercegovina (international aircraft registration code)

U

ú Úsase

u Atomic Mass Unit • Control Vector • Displacement of Ion • Energy Density • Instantaneous Potential Difference • Lethargy • Micro • Molecular Velocity • Object Distance • Relative Velocity • Sound Particle Velocity • Specific Internal Energy • Speed • Ugly Threatening Sky • Und • Underfloor • Ungerade • Unified Atomic Mass Unit • Uniform • Unit • Unsatisfactory • Unter • Up • Upper • Utility • Velocity Component

Ú Úsase

U Electric Tension • Gas Ionization Parameter • Internal Energy • Potential Difference • Rate of Heat Loss • Tension • Total Energy • Umbrella • Umlagen • Unclassified • Uncle • Und • Unidirectional • Unidirectionnel • Uniform (international alphabet) • Union • Unit • Unite • Unité • United • Universe • Université • University • Unix • Unknown • Unsatisfactory • Unterer • Untergrundbahn • Up • Upper • Uranium • Uranus • Uruguay (international vehicle registration code) • Utilisation • Velocity Component • Wind Speed • You

uä und ähnlich

ua Ukraine • und andere • unter anderen

UA Ukraine (ISO country code) • Unauthorized Absence • Unbalanced Asynchronous • Uncertainty Avoidance • Und Andere • Und Anderes • Under Age • Unexplained Anomaly • Unified Abstract • Uniform Allowance • Union Africaine (formerly OUA) • Unique Application • Unit Administration • Unité Administrative • Unité Associée • Unité Astronomique • Unité d'Acquisition • Universitätsarchiv • University of Alabama (USA-AL) • University of Alaska (USA-AK) • University of Arizona (USA-AZ) • Unnumbered Acknowledgment • Unter Anderem • Unter Anderen • Upper Air • Urbanisation et Aménagement • Urbanized Area • User Agency • User Agent • User Area • User Availability • Usine d'Aviation • Usque Ad

UA&A User Authentication & Authorization

UAA Ukrainian Astronomical Association • Umpqua Amateur Astronomers (USA-OR) • Undeliverable As Addressed • Universal Administered Address • Universidad Autónoma de Aguascalientes (Mexico) • University Aviation Association (USA) • University of Alaska in Anchorage (USA-AK) • Upper Advisory Area • Uppsala Amatörastronomer (Sweden) • User Authentication and Authorization

UAAARA Unión de Asociaciones y Aficionados a la Astronomía de la República Argentina

UAAG User Agent Accessibility Guidelines

UAB Unione Astrofili Bresciani (Italy) • Unit Accreditation Board • Unitary Awarding Body (UK) • Universitat Autònoma de Barcelona (Spain) • University of Alabama in Birmingham (USA-AL) • Until Advised By • User Address Book

UAC San Luis - Rio Colorado (Mexico) (location identifier) • Unattended Assembly Call • Unbalanced Asynchronous Class • Uniform Annual Cost • Unit Advisory Council • Universele Artikelcodering • Universele Auteursrechtconventie • Unseen Astrometric Companion • Upper Area Control • Upper Atmosphere Composition • User Access and Control • User Access Component

UACC Upper Area Control Center

UACNJ United Astronomy Clubs of New Jersey (USA-NJ)

UACPyP Unidad Académica de los Ciclos Profesional y de Posgrado

UAD Unit Assembly Drawing • Unit Authorization Document • Universal Antennas Division • Updated Authorization Document • User Accommodation Document

UADP Uniform Automated Data Processing • Uniform Automatic Data Processing

UADPS Uniform Automated Data Processing System • Uniform Automatic Data Processing System

UADS User Attribute Data Set

UADSL Universal Asymmetric Digital Subscriber Line

UAE Mount Aue (Papua New Guinea) (location identifier) • United Arab Emirates • Uninterruptible Application Error • Unrecoverable Application Error • User Agent Entity

UAEU United Arab Emirates University

UAF Unidentified Absorber Fraction • Unit Authorization File • University-Affiliated Facility • University of Alaska in Fairbanks (USA-AK) • Upper Atmosphere Facility • User Account File • User Authorization Facility • User Authorization File

UAG Hryvna (Ukraine) (ISO currency code) • Unité Administrative de Gestion • Upper-Atmosphere Geophysics • User Advisory Group

UAH Ua Huka (Marquesas Islands) (location identifier) • Universidad de Alcalá de Henares (Spain) • University of Alabama in Huntsville (USA-AL)

UAHCI Universal Access in Human-Computer Interaction (conference)

UAI Suai (Indonesia) (location identifier) • Uncertainty Avoidance Index • Union Académique Internationale • Union Astronomique Internationale • Union des Associations Internationales • Unione Astrofili Italiani (Italy) • Universal Access to Information • Universal Astronomy Index • Universal Azimuth Indicator • Universitair Instelling Antwerpen (Belgium) • Université Al Akhawayn à Ifrane (Morocco) • Use As Is

UAIS Universal Access on the Information Society (journal)

UAJ Unité Administrative Juridique

UAK Narssarssuaq (Greenland) (location identifier)

UAL Luau (Angola) (location identifier) • Unfounded Actuarial Liability • Unit Area Loading • Unit Authorization List • Universal Assembly Language • Upper Acceptance Limit • User Access Line • User Adaptive Language • User Agent Layer • User Authorization List • User Authorization Listing • Usinage Assisté par Laser

UALI Unit Authorization List Item

UALR University of Arkansas at Little Rock (USA-AR)

uam und andere mehr • und anderes mehr

UAM Guam - Anderson (Guam) (location identifier) • Unified Access Management • Unité d'Apprentissage Mobile • Universidad Autónoma de Madrid (Spain) • Universidad Autónoma Metropolitana • University of Arkansas at Monticello (USA-AK) • User Access Manager

UAMHS User-oriented Automated Message Handling System

UAM-I Universidad Autónoma Metropolitana - Iztapalapa (Mexico)

UAMI Universidad Autónoma Metropolitana - Iztapalapa (Mexico)

UAMS Upper-Atmosphere Mass Spectrometer

UAMU User Assistance and Marketing Unit

UAN User Access Node

UÄndG Unterhaltsänderungsgesetz

UAO Unexplained Aerial Object • University of Arizona Observatories (USA-AZ) • Upper Air Observatories • Upper Airway Obstruction • Uppsala Astronomical Observatory (Sweden) • Urumqi Astronomical Observatory (China-PRC) • Usinage Assisté par Ordinateur • Ussurijsk Astrophysical Observatory (Russia)

UAOB Upper Air Observation

UAOM Upper Air Observatory - Mobile

UAON Upper Air Observatory - North

UAOPP Upper Air Observatory - Pillar Point

UAOS Upper Air Observatory Site • Upper Air Observatory - South • User Alliance for Open Systems

UAP Ua Pou (Marquesas Islands) (location identifier) • Unidentified Atmospheric Phenomenon • Union des Assurances de Paris • Universal Access Point • Universal Availability of Publications • Universidad Autónoma de Puebla (Mexico) • University-Affiliated Program • Unmanned Aerial Platform • User Acceptance Program • Usine d'Aviation Pilote

UAPD Universe of All POssible Discourse

UAPDU User Agent Protocol Data Unit

UAPI Uniform Application Programming Interface

UAPT Union Africaine des Postes et Télécommunications • Union of African Posts et Telecommunications

UAPV Université d'Avignon et des Pays de Vaucluse (France)

UAQ San Juan (Argentina-SJ) (location identifier)

UAQPSK Unequal Asynchronous Quadrature Phase Shift Key • Unequal Asynchronous Quadrature Phase Shift Keying

UAR Unconventional Assisted Recovery • United Arab Republic • Upper Air Route • Upper-Atmosphere Research • Use As Required • User Action Routine

UARC Upper-Atmosphere Research Collaboration • Upper-Atmosphere Research Collaboratory

UARM Unconventional Assisted Recovery Mechanism

UARP Upper-Atmosphere Research Program

UARPP Upper-Atmosphere Rocket Research Panel

UARS Unmanned Aerial Reconnaissance System • Upper-Atmosphere Research Satellite • Upper-Atmosphere Research Spacecraft • Upper Atmospheric Research Satellite (*)

UART Universal Asynchronous Receiver/Transmitter

UARTE Unidade de Apoio à Rede Telemática Educativa (Portugal)

UARV Unmanned Air Reconnaissance Vehicle

UAS Samburu (Kenya) (location identifier) • Unavailable Seconds • Uniform Accounting System • Union of African States • Unit Activation Schedule • Unit Approval System • United Arab States • University Admissions System • University of Agricultural Sciences • Upper Airspace • Upper-Atmospheric Sounder • Urban-Area Study • Urgent Action Survey • User Application Software

UASB Upflow Anaerobic Sludge Blanket

UASIF Union des Associations Scientifiques et Industrielles Françaises (France)

UASLP Universidad Autónoma de San Luis Potosí (Mexico)

UAST Unité Administrative du Service de Transmission

UAT Ultraviolet Acquisition Technique • Union Africaine des Télécommunications • Universal Access Transceiver • Universal Alias Table • User Acceptance Test

UATI Union des Associations Techniques Internationales

UATL Unité d'Action, de Temps et de Lieu • Universal Ada Test Language

UATE Universal Automatic Test Equipment

UATI Union des Associations Techniques Internationales

UATM Unité d'Acquisition de Télémétrie

UATR Universal Attenuated Total Reflectance

UAV Uncrewed Aerial Vehicle • Uninhabited Aerial Vehicle • Uninhabited Aeronautical Vehicle • Universally Accepted Value • Unmanned Aerial Vehicle (*) • Unmanned Aerospace Vehicle • Unmanned Airborne Vehicle • Unmanned Air Vehicle • Upper Atmosphere Vehicle • User Antenna View

UAVGCS Unmanned Aerial Vehicle Ground Control Station

UAVGCM Unmanned Aerial Vehicle Ground Control Module

UAW United Astronomers of the World • United Auto Workers • Universal Access Workstation • Upper-Atmosphere Wind

UAwg Um Antwort wird gebeten

UAWG Universal ADSL Working Group

UAX Uaxactun (Guatemala) (location identifier) • Unit Automatic Exchange

UAYF Universe At Your Fingertips (project)

UAZ Universidad Autónoma de Zacatecas (Mexico)

U/B Usage Block

UB Ulaan Baatar (Mongolia) • Ultraviolet-Blue • Undistributed Budget • Unified Build • Unité de Bataille • Universitat de Barcelona (Spain) • Universitätsbibliothek • Universiteitsbibliotheek • Unterbegriff • Upper Brace • Usage Block • Utility Bridge • User Background

UBA Uberaba (Brazil-MG) (location identifier) • Umweltbundesamt (Germany) • Underwater Breathing Apparatus • Unibus Adapter • Union Bank of Africa

UBAI Ulug Beg Astronomical Institute (Uzbekistan)

UBB Mabuiag Island (Australia-QLD) (location identifier)

UBBM Unendorsed Bulk Business Mail

UBC Unified Byte Count • Uniform Building Code • Universal Bibliographic Control • Universal Buffer Controller • University of British Columbia (Canada-BC)

UBCIM Universal Bibliographic Control and International MARC

UBD User Brain Damage

UBE Universal Bus Exercisor • Unsolicited Bulk E-mail • Upper Band Edge

UBER Undetected Bit Error Rate

UBF Universal Birefringent Filter

UBFFP Universal Birefringent Filter Fabry-Pérot

UBI Buin (Papua New Guinea) (location identifier)

UBIC Universal Bus Interface Controller

UBIDEP Universal Building Information and Data Exchange Protocol

UBIS Universitätsbibliothek Informationsstelle • Users of Book Industry Standards

UBITRON Undulating Beam Interaction Electron

UBJ Ube (Japan) (location identifier) • Univariant Box-Jenkins

UBK Universelle Bibliographische Kontrolle

UBL Unblocked • Unblocking • Unit Basic Load

UBM Unité de Bande Magnétique • Universität des Bundeswehr München (Germany) • Unpressurized Berthing Mechanism

UBMTA Uniform Biological Materials Transfer Agreement

UBO Unidentified Burning Object • Université de Bretagne Occidentale (France)

UBP Ubon Ratchathani (Thailand) (location identifier) • Umweltbelastungspunkt • Upper Boundary of the Plate

UBQS US Bright QSO Survey • US Bright Quasar Survey

UBR Uniform Business Rate • Unspecified Bit Rate • Upper Brazing Region

UBRD Usage-Based Requirements Determination

UBS Columbus (USA-MS) (location identifier) • Unclassified But Sensitive • Université de Bretagne-Sud (France) • US Biological Survey

UBT Ubatuba (Brazil-SP) (location identifier)

UBU Kalumburu (Australia-WA) (location identifier)

UBV Ultraviolet - Blue - Visual

UBVRI Ultraviolet - Blue - Visual - Red - Infrared

u/c Under Charge

uc Upper Case

U/C Under Current • Upconverter • User Code

UC Ultracompact • Ultracompression • Ultracompressor • Umbral Core • Under Construction • Undercover • Under Current • Unemployment Compensation • Union Catalogue • Unit Call • Unit Chairman • Unit Cost • Unité Capitalisable • Unité Centrale • Unité de Cadre • Unité de Compte • University of California (USA-CA) • University of Chicago (USA-IL) • University College • Universidad Complutense (Spain) • Unofficial Collaborator • Unoperated Control • Unsatisfactory Condition • Unseen Companion • Update Code • Uplink Command • Upper Canada • Upper Case • Upper Cassette • Usable Control • Use Case • User Class • User Code • User Committee • Users' Committee • Using Command • Utilization Control

UCA Undefinitized Contract Action • Undefinitized Contractual Action • Uniform Chart of Accounts • Unitary Convolution Approximation • University of Central Arkansas (USA-AR) • Upper-Case Alphabet • Universal Calibration Adapter • Upper Control Area • USNO CCD Astrograph • Utica (USA-NY) (location identifier)

UCAC USNO CCD Astrograph Catalog

UCAC-S USNO CCD Astrograph Catalog - South

UCAFR Universitas Comeniana Acta Facultatis Rerum Naturalium Physica (periodical code)

UCAID University Corporation for Advanced Internet Development

UCAMIS Uniform Chart of Accounts Management Information System

UCAMS Undefinitized Contractual Action Management System

UCANSS Union des Caisses Nationales de Sécurité Sociale (France)

UCAR University Corporation for Atmospheric Research (USA) • Unsatisfactory Condition Action Report

UCARE Unité Capitalisable d'Adaptation Régionale

UCAS Union of Central African States • Universities and Colleges Admissions Service (UK)

UCASS University Computer-Assisted Search Service

UCB Ufficio Centrale Brevetti (Italy) • Ultraviolet Cosmic Background • Union Chimique Belge (Belgium) • Unique Consumer Benefit • Unit Control Block • Universal Channel Bank • University Campus Box • University of California, Berkeley (USA-CA) • Unless Caused By • Utilization Control Board

UCBL Université Claude Bernard de Lyon (France)

UCBT Universal Circuit Board Tester

UCC Ultracompact Core • Umbilical Checkout Cable • Unadjusted Contractual Cable • Unemployment Compensation Coordinator • Unified Classification Code • Uniform Commercial Code • Union Carbide Corporation • Unit Control Center • Unité de Chaleur Centésimale • Unité de Conditionnement Centrale • Unité de Contrôle Capitalisable • Universal Checkout Console • Universal Copyright Convention (1952) • University College Cork (Ireland) • Upper Control Center • Utility Control Console • Yucca Flat (USA-NV) (location identifier)

UCCD Undergraduate Course and Curriculum Development

UCCEGA Union des Chambres de Commerce et des Établissements Gestionnaires d'Aéroports (France)

UCCFE Unemployment Compensation Coordinator for Federal Employees (USA)

UCCIFE Union des Chambres de Commerce et d'Industrie Françaises à l'Étranger

UCCP Unité Capitalisable pour Contrôle Ponctuel

UCCS Universal Camera Control System • University of Colorado at Colorado Springs (USA-CO)

UCCSN University and Community College System of Nevada (USA-NV)

UCD Ultra-Compact-Dwarf (galaxy) • Ultra-Cool-Dwarf (star) • Unified Column Descriptor • Unified Concept Descriptor • Unified Content Descriptor • Uniform Call Distribution • Uniform Call Distributor • Uniform Content Descriptor • University College Dublin (Ireland) • University of California, Davis (USA-CA) • Unstructured Cell Data • Upper Critical Depth • Urine Collection Device • User-Centered Design

UCDL University of California Digital Library

UCDP Uncorrected Data Processor

UCE Eunice (USA-LA) (location identifier) • Ultimate Computer Environment • Unió de Consumidores Españoles (Spain) • Unit Checkout Equipment • Unit Correction Entry • Unité de Compte Européenne • Unité de Contrôle Espace • Unsolicited Commercial E-mail

UCEC University of Chicago Engineering Center (USA-IL)

UCEI Union Culturelle pour les Échanges Internationaux

UCERC Union des Centres d'Études Rurales par Correspondance

UCET Universities Council for the Education of Teachers

UCF Uniform Contract Format • Union Catalogue File • Union Circulation File • University of Central Florida (USA-FL)

UCFM Unitree Central File Manager

UCG Unit Cost Goal • Unité de Calcul et de Gestion • Uppsala Catalog of Galaxies

UCGIS University Consortium for Geographic Information Science

UCh Universidad de Chile

UCHII Ultra-Compact HII (region)

UCHIIR Ultra-Compact HII Region

UCHR Ultra-Compact HII Region

UCI Unclassified Controlled Information • Unidad Central de Inteligencia • Unidad de Cuidados Intensivos • Union Cycliste Internationale • Unit Construction Index • University of California, Irvine (USA-CA) • Updated Coordinating Instructions • Utility Card Input

UCIS Uniterm Coordinate Index System • University Center for International Studies (USA) • Uprange Computer Input System

UCITA Uniform Computer Information Transactions Act (USA)

UCITS Undertaking for Collective Investment in Transferable Securities

UCJC Universidad Camilo José Cela (Spain)

UCK Lutsk (Ukraine) (location identifier)

Uckerm Uckermark

UCL Universal Communications Language • Université Catholique de Louvain (Belgium) • University College London (Starlink internal code) • University College London (UK-England) • Update Control List • Upper Centaurus-Lupus (association) • Upper Confidence Level • Upper Control Limit • User Control Language

UCLA University of California, Los Angeles (USA-CA)

UCLAN University of Central Lancashire (UK-England)

UCLAP Astronomical Papers University of California, Los Angeles (periodical code)

UCLES University College London Echelle Spectrograph

UCLGES Unified Cartographic Line Graph Encoding System

UCLJ University of California, La Jolla (USA-CA)

UCLSI Unité de Cours en Libre Service Informatique

UCLXE University College of London Experiment

UCM Unified Cloudy Model • Unité de Calcul Miniaturisée • Universal Cable Module • Universal Communications Module • Universidad Complutense de Madrid (Spain) • User Context Module

UCMD Unified Command

UCMS Unit Capability Measurement System

UCN Buchanan (Liberia) (location identifier) • Ultra-Centrifuge Nederland BV • Uniform Control Number • Unit Command Net • Universidad Católica del Norte (Chile) • Unmanned Communications Network

UCNI Unclassified Controlled Nuclear Information

UCNW University College of North Wales (UK-Wales)

UCO Ultra Compact Object • Uncommitted Obligation • Undefinitized Change Order • Unidad Control Operativa • University of California Observatories (USA-CA) • Unstructured Complex Object • Upper Critical Orbit • Use Case Object • User Consultancy Office

UCOAC University of California Observatories Advisory Council

UCOIF University of California Observatories Instrumentation and Facilities

UCOS Universal Checkout Station • Uprange Computer Output System

UCOSP University of California Observatories Science Policy Committee

UCOT Uncompensated Overtime

UCOWR Universities Council on Water Resources

UCP Unified Command Plan • Unified Control Plane • Uniform Custom and Practice • Unité de Commande du Processeur • Universal Communications Processor • Universal Computer Protocol • University of Chicago Press (USA-IL) • User Change Proposal • Usinage Chimique Progressif • Utility Control Program

UCPA Union des Centres de Plein Air

UCPTE Union for the Coordination of the Production and Transport of Electricity

UCPU Urine Collection and Pretreatment Unit

UCR Unconditioned Reflex • Under-Color Removal • Uniform Crime Report • Unit Card Reader • Unit Cost Report • Unit Cost Reporting • Unit Cost Resourcing • Unité de Circulation Routière • Unité de Commande du Réseau • Unité de Contrôle de Transmission • Universidad de Costa Rica • University of California, Riverside (USA-CA) • Unsatisfactory Condition Report • User Control Routine

UCRC University Cooperative Research Center

UCRL University of California Radiation Laboratory

UCRP University Collaborative Research Programme

UCS Unconditioned Stimulus • Unconfined Compressive Strength • Unicode Character Set • Unicode Conversion Support • Unified Communications Services • Uniform Chromaticity Scale • Uniform Color Scale • Union of Concerned Scientists (USA) • Unit Count System • Universal Cabling System • Universal Card Scanner • Universal Character Set • Universal Classification System • Universal Command System • Universal Component System • Universal Control System • University Computer Services • Urine Collection System • User Coordinate System • Utilities Control System

UCSB University of California, Santa Barbara (USA-CA)

UCSC Universal Communications Switching Device • University of California, Santa Cruz (USA-CA)

UCSD University of California, San Diego (USA-CA)

UCSF University of California, San Francisco (USA-CA)

UCSR Unit Control Supply Rate

UCSTRI Unified Computer Science Technical Report Index

UCSW University College of South Wales (UK-Wales)

UCT Ukhta (Russia) (location identifier) • Uncorrelated Target • Underwater Construction Team • Unit Cost Target • Units Compatibility Test • Universal Coordinated Time • University of Cape Town (South Africa) • User Check Test • User Communication Tracking

UCTA Urine Collection Transfer Assembly

UCTE Union pour la Coordination des Transport de l'Électricité

UCTM Unité de Contrôle de la Télémétrie • Unité de Contrôle et de Télémesure

UCU Ultra Capacity Unit

UCV Unidade Ciência Viva (Portugal) • Universidad Católica de Valparaiso (Chile) • Universidad Central de Venezuela

UCW Ulysses Comet Watch • Union of Communications Workers • Unit Control Word • University College of Wales (UK-Wales)

UCWE Upper Coupling of Wave Energy • Upward Coupling of Wave Energy

UCX Universal Component Extension

UCY Union City - Everett-Stewart (USA-TN) (location identifier)

ud Ut Dictum

Ud Usted

U/D Under Deed • Uni-Directional • Update

UD Udmurtian • Ultra-Dense • Umbral Dot • Unavoidable Delay • Undesirable Discharge • Unidirectional • Uniform Disk • Uniform Distribution • Uniquely Decipherable • Unit Designation • Unit Director • Unité à Disque • Unité de Documentation • Unité Dynamique • Universal Dipole • Universitair Docent • Update • Usage Data • User Data • User Dependability • Ut Dictum

UDA Unité Décentralisée d'Apprentissage • Urgent Deployment Acquisition • User Data Area

UDAPAC Uccle Direct Astronomical Plate Archive Centre (Belgium)

UDAS UAO-DLR Asteroid Survey • Unified Direct-Access System

UDB Unified Database • Universal Database • Up-Data Buffer • Update Buffer • Urban Development Board • User Database

UDC Undefinitized Document Control • Uniform Document Count • Unit Descriptor Code • Universal Decimal Classification • Universal Digital Channel • University of The District of Columbia (USA-DC) • Up-Down Counter • Urban Data Center • Urban District Council • User-Defined Command • Utility Distribution Center

UDCF Unbinned Discrete Correlation Function

UDD Palm Springs - Bermuda Dunes (USA-CA) (location identifier)

UDDF Updrafts/Downdrafts

UDDI Universal Description, Discovery and Integration (protocol) • Universal Directory Discovery Interface

UDE Uden Volkel (Netherlands) (location identifier) • Universal Data Exchange

UDEAC Union Douanière des États de l'Afrique Centrale • Union Douanière et Économique de l'Afrique Centrale

UDEAO Union Douanière des États d'Afrique de l'Ouest

UdeC Universidad de Chile (Chile) • Universidad de Concepción (Chile)

UDEC Unitized Digital Electronic Calculator

UDES User Operational Evaluation System

UDESC Unit Description

UDF UAV Demonstration Flight • UHF Direction Finder • UHF Direction Finding • Ultra Deep Field • Unducted Fan • Unit Data File • Unit Development Folder • Universal Data Format • Universal Disk Format • User-Defined Function • Utility and Data Flow

UdeG Universidad de Guadalajara (Mexico)

UDFAM User-Defined File Access Memory

UDFOP Union Départementale pour la Formation et l'Orientation Professionnelle (France)

UdG Universitat de Girona (Spain) • User-Defined Gateway

u. dgl. und dergleichen

UDHD User Defined Data Field

UDHII Ultra-Dense HII (region)

UDHR UNO Declaration on Human Rights

UDI Uberlândia - Eduardo Gomes (Brazil-MG) (location identifier) • Uniform Driver Interface • Unique Data Item • Universal Digital Interface • Universal Document Identifier

UDID Universal Data Identification • Unique Data Item Description • User Defined Image Data Field

UDJ Uzhgorod (Ukraine) (location identifier)

UDK Universal Decimalklassifikationssystem • User Defined Key • User Domain Knowledge

UDL Unclassified Data Link • Uniform Data Language • Uniform Data Link • Unique Data Log • Up-Data Link • Update Link

UDLC Universal Data Link Control

UDLR Uni-Directional Link Routing

udM unter dem Meeresspiegel

üdM Über dem Meeresspiegel

UDM Under Data Module • Universal Docking Module • University of Detroit Mercy (USA-MI) • User Data Module

UDMH Unsymmetrisches Dimethylhydrazine • Unsymmetrical Dimethylhydrazine

UDMU Universal Decoder Memory Unit

UDN Udine (Italy) (location identifier) • User Densification Network

UDO Udomxay (Laos) (location identifier) • Ultra Density Optical • Undefinitized Delivery Order • Unrelated Deep-Sky Object • Upper Domain Ontology

UDOP Ultrahigh Doppler

UDP Unicast Discovery Protocol • Uniform Data Processing • Unitary Development Plan • Unit Data Processing • Unit Deployment Program • Universal Datagram Protocol • Unix Datagram Protocol • Uplink Data Processing • Upper Deck Pressure • User Datagram Protocol (*) • User Data Packet • User Data Protocol • User Development Program

UDR Udaipur (India) (location identifier) • Ultra-Dipole Radiation • Universal Data Repository • Universal Digital Readout • Universal Document Reader • Unprocessed Data Record • Urgent Data Request • Usage Data Report • Utility Data Reduction

UDRC Utility Data Retrieval Control

UDRO Utility Data Retrieval Output

UDRP Uniform Dispute Resolution Policy

Uds Ustedes

UDS Ultra Deep Survey • Ulysses Data System • Unified Data System • Unité de Détection au Sol • Universal Digital Switch • Universal Documentation System • Universal Document System • Unresolved Double Star • US Differential System • User Database Searching • User Data System • User Display System • Utility Distribution System

UDSC Usuda Deep Space Center (Japan)

UDSL Universal Digital Subscriber Line

UDSPL Ultrasonics and Digital Signal Processing Laboratory

UdSSR Union der Sozialistischen Sowjetrepubliken

udT unter dem Titel

UDT Uniform Data Transfer • Unit Detail Testing • Universal Data-flow and Telecommunications • Universal Detection Technology • Unstructured Data Transfer • User-Defined Type • User Display Terminal

Udtd Updated

UDTS Universal Data Transfer Service • Universal Data Transmission System

UDU Underwater Dive Unit

UDW University of Durban-Westville (South Africa)

UDZ Ustav Dozimetrie Zâreni (Czech Republic)

uE unseres Erachtens

UE Unfaulted Edge • Unified Endeavor • Unión Europea • Union Européenne (formerly CEE) • Unité d'Enseignement • Unité Électronique • Unit Entry • Unit Equipment • Unit Establishment • Univers Émotionnel • University Entrance • University Extension • University of the East (Philippines) • Unrestricted Estimator • Until Exhausted • Usability Engineer • Useful Elevation • User Element • User Enterprise • User Environment • User Equipment

UEA Unattended Equipment Area • University of East Anglia (UK-England) • University of East Asia

UEAC Union des États de l'Afrique Centrale

UEB Upper Equipment Bay

UEBL Union Économique Belgo-Luxembourgeoise (1921)

UEBS Union des Entreprises de la Base Spatiale

UEC Unité d'Échange et de Couplage • Universitas Europae Centralis (Austria) • User Environment Component • Users Executive Committee • Users Executive Council

UECBu University Electro Communications Bulletin (periodical code)

UED University Education Diploma

UEE Queenstown (Australia-TAS) (location identifier) • Unit Essential Equipment

UEET Ultra Efficient Engine Technology

UEF Uniform Electric Field • United Engineering Foundation (USA-NY) • Universal Extrafine

UEG Union Européenne des Géosciences

UEGN Unified European Gravity Network

UEI Union of Educational Institutions (UK) • United Electronic Industries

UEIP Ural Electrochemical Integrated Plant

UEL Quelimane (Mozambique) (location identifier)

UELF Union des Éditeurs de Langue Française (France)

UELV Ultralight Expandable Launch Vehicle • Ultra-lite Expandable Launch Vehicle

UEM Unión Económica y Monetaria • Union Économique et Monétaire • Union Européenne Monétaire • Unité Électromagnétique • Universal Electron Microscope • Universal Electron Microscopy

UEMMA Unsolicited Electronic Mail with Multiaddressing

UEMOA Union Économique et Monétaire Ouest-Africaine

UEN Universität Erlangen-Nürnberg (Germany)

UEO Kume Jima (Japan) (location identifier) • Union de l'Europe Occidentale (*) • Unione dell'Europa Occidentale • Union Européenne Occidentale

UEP Unequal Error Protection • Uniform External Pressure • Union Européenne des Paiements • Unplanned Event Pickup

UEPR Unsatisfactory Equipment Performance Report

UER Union Européenne de Radiodiffusion • Unique Equipment Register • Unité d'Enseignement et de Recherche • University Entrance Requirements • Unplanned Event Record • Unsatisfactory Equipment Report

UERE User Equivalent Range Error

UERL Unplanned Event Record Log

UES Uniform Emission Standard • Union Européenne et Suisse • Unité Électrostatique • Utrecht Echelle Spectrograph • Waukesha (USA-WI) (location identifier)

UESAC Uppsala-ESO Survey of Asteroids and Comets

UET Quetta (Pakistan) (location identifier) • Unité Élementaire de Travail • Université Européenne de Tourisme • User Earth Terminal

UETA Uniform Electronic Transactions Act (USA)

UEU Universal Exciter Upgrade

UEX Underexposed

U/F Unique File

UF Ultrafilter • Ultrafiltration • Ultrafine • Ultrasonic Frequency • Umbral Flash • Unavailability Factor • Unidad de Fomento • Unique File • Unité de Force • Unité de Formation • Universal Format • University of Florida (USA-FL) • Unreleasable Frontier • Used For • User Foreground • Usine Frigorifique • Utilization Factor • Utilization Flight

UFA Ufa (Russia) (location identifier) • Unité de Formation par l'Apprentissage • Université Franco-Allemande • University Faculty Award (program) • University Film Association (USA) • Until Further Advised • Upper (tropospheric) Formation of Aerosols • Use Frequency Analysis

UFACNC Ultrafine Aerosol Condensation Nucleus Counter

UFAG Union Française des Associations Graphiques (France)

UFAM Universal File Access Method

UFAS Unified File Access System • Unified Flight Analysis System • Uniform Federal Accessibility Standard

UFBE Union Francophone des Belges à l'Étranger

UFC Unified Flight Classification • Uniform Freight Classification • Universal Flight Computer • Universities Funding Council

UFCC Uniform Freight Classification Code

UFCV Union Française des Centres de Vacances (France)

UFD Unit Functional Description • User File Directory • User Functional Description

UFDWI University of Florida Digital Words Institute (USA-FL)

UFE Undergraduate Faculty Enhancement

UFEP Union de France des Étudiants en Préparations (France)

UFES Universidade Federal de Espíritu Santo (Brazil-ES)

Uff Ufficiale

UF-F Universal Flip-Flop

UFF Universal Flats Feeder • Universal Flip-Flop • Universality of Free Fall • Upper Forward Fuselage

UFFS Ultrafast Fine Structure

UFG Universidade Federal de Goiás (Brazil-GO)

UFI Uniform File Identifier • Union des Foires Internationales • Universal Fermi Interaction • User-Friendly Index • User-Friendly Interface

UFIR Ultraluminous Far-Infrared

UFIRS Universal Far-Infrared Sensor

UFJF Universidade Federal de Juiz de Fora (Brazil-MG)

UFL Upper Flammability Limit

UFM Ultrasonic Force Microscopy • Unintentional Frequency Modulation

UFMG Universidade Federal de Minas Gerais (Brazil-MG)

UFN Until Further Notice

UFO UHF Follow-On • Unbekanntes Flugobjekt • Unidentifiable Format Object • Unidentified Flying Object • Universal Fiber Optic • Universitair Functie Ordenen (Netherlands)

UFOD Union Française des Organismes de Documentation (France)

UFOE Upgraded Fiber Optic Echelle (spectrograph)

UFOIRC Unidentified Flying Object Information Retrieval Center

UFOP Universidade Federal de Ouro Preto (Brazil-MG)

UFP Ultra-Fine Particle • Ultra-Fine Powder • Unit Focal Point • Université Française du Pacifique • Unix File Transfer

UFPR Universidade Federal de Parana (Brazil)

UFR Ultra-Fast Rotator • Underfrequency Relay • Unfinanced Requirement • Unfunded Requirement • Unité de Formation et de Recherche (France) • Unsatisfactory Feature Report

UFRF University of Florida Research Foundation (USA-FL)

UFRGN Universidade Federal do Rio Grande do Norte (Brazil-RN)

UFRGS Universidade Federal do Rio Grande do Sul (Brazil-RS)

UFRJ Universidade Federal de Rio de Janeiro (Brazil-RJ)

UFRO University of Florida Radio Observatory (USA-FL)

UFS Umweltforschungssatellit • Unconnected Fine Structure • United Feeder Service • Unix File System • Unresolved Fine Structure

UFSA Union Française des Sociétés d'Auteurs (France)

UFSC Universidade Federal de Santa Catarina (Brazil-SC)

UFSM Universidade Federal de Santa Maria (Brazil-RS) • Upgraded Flat Sorting Machine

UFSS Unified Flexible Spacecraft Simulation

UFT Ultrasonic Frequency Transformer • Ultrasonic Frequency Translator • Unified Field Theory • Unified File Transfer • United Federation of Teachers

UFTI UKIRT Fast Track Imager

UFTP Upper Feedthru Plate

UFTS Union des Formations de Travailleurs Sociaux (France)

UFV Universidad Federico Villareal (Peru)

UFVA University Film and Video Association (USA)

UFVC Ultrafast Varying Component

UFZ Umweltforschungszentrum

u/g Underground

Ug Uganda

UG Uganda (ISO country code) • Undergraduate • Underground • Unité de Gestion • Universal Gateway • Universal Generalization • Universal Grammar • Universidad de Guanajuato (Mexico) • Untergeschoß • User Group

UGA Bulgan (Mongolia) (location identifier) • Uganda (ISO country code) • Unity Gain Amplifier • University of Georgia at Athens (USA-GA) • User Global Area

UGAMP Universities Global Atmospheric Modeling Programme (*) • Universities Global Atmospheric Modeling Project

UGAP Union des Groupements d'Achats Publics (France)

UGB Pilot Point - Ugashik Bay (USA-AK) (location identifier) • Urban Growth Boundary

UGBW Unity Gain Bandwidth

UGC Nova Acta Regiae Soc. Sci. Upsaliensis Ser. V (periodical code) • University Grants Commission • University Grants Committee • Uppsala General Catalogue • Urgench (Uzbekistan) (location identifier)

UGCG Uppsala General Catalogue of Galaxies

UGE Union des Grandes Écoles (France)

UGGI Union Géodésique et Géophysique Internationale

UGHW Universal Ground Handling Wheel

UGI Uganik (USA-AK) (location identifier) • Union Géographique Internationale

UGID Universal Geographic Identity

UGLI Universal Gate for Logic Implementation

UGM Universitas Gadjah Mada (Indonesia) • Unusual Ground Marking • User Group Maintenance

UGN Waukegan Memorial (USA-IL) (location identifier)

Ugnd Underground

UGO Uige (Angola) (location identifier) • User - Group - Other

UGR Unified Glare Rating

UGS Unattended Ground Sensor • Undergraduate Student • Uniaxial Gyrostabilizer

UGSJR US Geological Survey Journal Research (periodical code)

UGT Underground Test • Underground Testing • Upgrade Training

UGUI Universal Graspable User Interface

UGV Unmanned Ground Vehicle • Usinage à Grande Vitesse

UGX Shilling (Uganda) (ISO currency code)

UH Ultra Heavy • Unit Head • University of Hawaii (USA-HI) • University of Houston (USA-TX) • Upper Half • Usine Hydraulique • Utility Helicopter

UHA Ultrahigh Altitude • Université de Haute-Alsace (France) • Upper Half Assembly • Uppsala H-Alpha

UHAO University of Hawaii Adaptive Optics

UHB Ultrahigh Bandwidth • Ultrahigh Bypass • Urban Haute Bourgeoisie • User Home Base

UHC Ultrahigh Compact • Ultrahigh Contrast • Ultrahigh-density Core • Unburnt Hydrocarbon • Under Honorable Conditions • Unit Hardware Cost • University Healthsystem Consortium • Upper Hatch Closed

UHCA Ultrahigh Capacity Aircraft

UHCI Universal Host Controller Interface

UHCL University of Houston - Clear Lake (USA-TX)

UHCR Ultrahigh-energy Cosmic-Ray

UHCRE Ultrahigh Cosmic-Ray Experiment

UHCS Ultrahigh Capacity Storage

UHD Universitair Hoofddocent

UHDM Ultra-Heavy Dark Matter

UHE Uherské Hradiště (Czech Rep.) (location identifier) • Ultrahigh Energy

UHEA Utah Home Educator's Association (USA-UT)

UHEC Ultrahigh-Energy Cosmic (ray)

UHECR Ultrahigh-Energy Cosmic Ray

UHF Ultra-Haute Fréquence • Ultrahigh Frequency • Unrestricted Hartree-Fock • Upper Harkless Flat (USA-CA) • Upper Heyford (UK) (location identifier)

UHFF Ultrahigh Frequency Filter

UHFG Ultrahigh Frequency Generator

UHFO Ultrahigh Frequency Oscillator

UHFR Ultrahigh Frequency Receiver

UHI Urban Heat Island

UHM Ultrahigh Modular • Ultrahigh Modulus • Universal Horizontal Metroscope

UHMS Undersea and Hyperbaric Medical Society • Urban Hydrology Monitoring System

UHMW Ultrahigh Molecular Weight

UHO University of Hertfordshire Observatory (UK-England) • Upper Hatch Open

UHP Ultrahigh Power • Ultrahigh Pressure • Ultrahigh Purity

UHPTRS Ultrahigh Precision Time Reference Station

UHR Ultra-Harmonic Resonance • Ultrahigh Reduction • Ultrahigh Resolution • Upper Hybrid Resonance

UHRF Ultrahigh Resolution Facility

UHRR Ultrahigh Resolution Radar

UHRS Ultrahigh-Resolution Spectroscopy

UHRXS Ultrahigh Resolution XUV Spectroheliograph

UHS Ultrahigh Speed • Ultramedia Host Support • University Health Service

UHT Ultra-Haute Température • Ultra-Heat Treated • Ultrahigh Temperature • Umbilical Handling Technician • Universal Hand Tool

UHTC Ultrahigh Temperature Ceramic • Ultrahigh Transmission Coating

UHV Ultrahigh Vacuum • Ultrahigh Voltage

UHVC Ultrahigh Vacuum Chamber

UHVCVD Ultrahigh Vacuum Chemical Vapor Deposition

UHVS Ultrahigh Vacuum System

u/i Under Instruction

ui Ut Infra

U/I Unidentified • Unit of Issue • Urban/Industrial • User Interface

U-I Union-Intersection

UI Ariel (satellite of Uranus) • Understemming Index • Unemployment Insurance • Union-Intersection • Unique Identifier • Unité Intermédiaire • Unité Internationale • Unit Interval • Unit of Issue • Universal Identifier • Universal Instantiation • University of Ibadan (Nigeria) • University of Illinois (USA-IL) • Unix International • Unnumbered Information • Upgren Intermediate • Upper Intercept • User Institution • User Interaction • User Interface

UIA Ukraine International Airlines • Unión Industrial Argentina • Union Internationale des Architectes • Union of International Associations • Universidad Iberoamericana • Universitaire Instelling Antwerpen (Belgium) • Usable Inside Area • User Interface Agent

UIAA Union Internationale des Astronomes Amateurs

UIAS University of Illinois Astronomical Society (USA-IL)

UIB Quibdo (Colombia) (location identifier) • Unidentified Infrared Band • Unidentifiziert Infrarotband • Unité d'Interface Bus • Universidad de las Islas Baleares (Spain)

UIC Underground Injection Control • Union des Industries Chimiques (France) • Union Internationale des Chemins de fer • Unit Identification Code • University of Illinois at Chicago (USA-IL) • Urban Industrial Complex • Urban Information Center • Usenet Information Center • User Identification Code

UICA Union of Independent Colleges of Art (USA)

UICC Union Internationale Contre le Cancer

UICN Union Internationale pour la Conservation de la Nature

UICP Uniform Inventory Control Program

UICPA Union Internationale de Chimie Pure et Appliquée

UID Unique Identifier • Universal Identifier • User Identification • User Identificator • User Identifier • User Identity • User Interface Description

UIDE User Interface Design Environment

UIDL User Interface Definition Language

UIDS User Interface Design System

UIE UNESCO Institute for Education • Union Internationale d'Électrothermie • Union Internationale des Éditeurs • Union Internationale des Étudiants

UIEO Union of International Engineering Organizations (now UITA)

UIEV Université Internationale d'Été de Versailles (France)

UIF Unfavorable Information File • Union Interlinguistique de France (France) • User Interface

UIFG Unser Interface Functional Group

UIFN Universal International Freephone Number

Uig Uighur

UIG User Instruction Group

UIH Qui Nhon (Viet Nam) (location identifier)

UIHPS Union Internationale d'Histoire et de Philosophie des Sciences

UII Umbriel (satellite of Uranus) • Utila (Honduras) (location identifier)

UIIG Union Internationale de l'Industrie du Gaz

UIII Titania (satellite of Uranus)

UIIP User Integration Implementation Plan

UIJ Union Internationale des Journalistes

UIJPLF Union Internationale des Journalistes de la Presse de Langue Française

UIK Ust-Ilimsk (Russia) (location identifier)

UIL Quillayute State (USA-WA) (location identifier) • Undergraduate Instructional Laboratory • Undergraduate Instruction Language • User Interface Language

UILE Union Internationale pour la Liberté de l'Enseignement

UIM Unité d'Implantation des Moyens • Unit Information Module • Universal Interface Module • User Interaction Module • User Interface Module

UIMM Union des Industries Minières et Métallurgiques (France)

UIMP Universidad Internacional Menéndez Pelayo (Canary Islands)

UIMS User Interface Management System

UIN Quincy (USA-IL) (location identifier) • Unit Identification Number • User Identification Number • User Information

UIO Quito - Mariscal Sucre (Ecuador) (location identifier) • Unified Interoperability • Unique Input/Output • User Interface Object • Utility Iterative Operation

UIOAR User Interface Off-line Analysis Request

UIOD User Input/Output Device

UIOM Usine d'Incinération des Ordures Ménagères

U/IP User/International Partner

UIP Quimper Cornouaille (France) (location identifier) • Quimper Pluguffan (France) (location identifier) • Union Internationale de Physique • Union Interparlementaire • Universal International Pictures • Université Interdisciplinaire de Paris (France) • User Interaction Protocol • User Interface Processor • User/International Partner • Utility Incentive Program • Utility Interface Panel

UIPE United Institute for Physics of the Earth (Russia)

UIPPA Union Internationale de Physique Pure et Appliquée

UIPPI Union Internationale pour la Protection de la Propriété Industrielle

UIPRE Union Internationale de la Presse Radiotechnique et Électronique

UIPRN Universal International Premium Rate Number

UIPT Ukrainian Institute of Physics and Technology • Utilization Integrated Product Team

UIQ Quine Hill (Vanuatu) (location identifier)

UIR Quirindi (Australia-NSW) (location identifier) • Unidentified Infrared (feature) (*) • Uniform Infrared (feature) • University Industrial Research • University/Industry Research • Upper Information Region • Upwelling Infrared • User Interface Requirement

UIRB Unidentified Infrared Band

UIRK Unique Interswitch Rekeying Key

UIRV Unique Interswitch Rekeying Variable

UIS Ubiquitous Information System • Union Internationale de Spéléologie • Unit Identification System • University of Illinois at Springfield (USA-IL) • Upper Information System • Urban Information System • User Information Service • User Interface Service • User Interface Specification • User Interface System • User Interface Specification

UISB Union Internationale des Sciences Biologiques

UISL User Interface Specification Language • User Interface Standard Library

UISM User Interface Standards Manual

UISRM User Interface System Reference Model

UIST UKIRT Image Spectrometer • User Interface Software and Technology

UISTR Univelt Inc Science Technology Series (periodical code)

UIT Jaluit Island (Marshall Island) (location identifier) • Ultraviolet Imaging Telescope • Union Internationale des Télécommunications • Unit Investment Trust • User Interface Technology • User Interface Terminal

UITA Union of International Technical Associations

UITC User Interface Transit Case

UITF Union des Ingénieurs et Techniciens utilisant la langue Française

UITP Union Internationale des Transports Publics

UITS University Information Technology Services

UIU User Interface Unit

UIUC University of Illinois at Urbana-Champaign (USA-IL)

UIV Oberon (satellite of Uranus)

UIVS Universal Infrared Viewing System

UIW University of the Incarnate Word (USA-TX)

UIWG User Interface Working Group

UIX Cressida (satellite of Uranus)

UIW Usable Inside Width

UIZ Utica Berz-Macomb (USA-MI) (location identifier)

UJ Union Jack • Universal Joint

UJC Urgency Justification Code

UJCL Universal Job Control Language

UJE Ujae Island (Marshall Island) (location identifier)

UJF Université Joseph Fourier (France)

UJJEF Union des Journaux et Journalistes d'Entreprises de France

UJO Unidentified Jumping Object

UJT Unijunction Transistor

UJTL Universal Joint Task List • Universidad Jorge Tadeo Lozano (Colombia)

uk Ukrainian (ISO language code) • United Kingdom

UK Ukraine • Underkommitté • United Kingdom • University of Kansas (USA-KS) • Uzbekistan (international aircraft registration code)

UKAC United Kingdom Automation Council

UKAEA United Kingdom Atomic Energy Authority

UKAF United Kingdom Astrobiology Forum

UKAFF United Kingdom Astrophysical Fluids Facility

UKAIS United Kingdom Academy for Information Systems

UKAPE United Kingdom Association of Professional Engineers

UKAS United Kingdom Accreditation Service

UKATC United Kingdom Astronomy Technology Centre

UKB Kobe (Japan) (location identifier) • Universal Keyboard

UKCIS United Kingdom Chemical Information Service

UKCOSA United Kingdom Council for Overseas Student's Affairs

UKCRS United Kingdom Controlled Release Society

UKCSU University of Kent at Canterbury Students' Union (UK-England)

UKDC United Kingdom Data Centre

UKDMC United Kingdom Dark Matter Collaboration

UKERNA United Kingdom Education and Research Networking Association

UKESRU United Kingdom Earth Satellite Research Unit

UKFCC United Kingdom Federation for Culture Collections

UkFiZ Ukrainskii Fizicheskii Zhurnal (periodical code)

UKgal United Kingdom Gallon

UKGOC United Kingdom Guest Observer Centre

UKGSG United Kingdom Gemini Support Group

UKHG Ultrakompakte H-II Gebiete

UKI Ukiah (USA-CA) (location identifier)

UKIDSS United Kingdom Infrared Deep Sky Survey

UKIRT United Kingdom Infrared Telescope

UKIS Universal Knowledge-based Imaging System

UKISC United Kingdom Industrial Space Committee

UKK Ust-Kamenogorsk (Kazakhstan) (location identifier)

UKL Pound Sterling (currency)

UKLT United Kingdom Large Telescope

UKM Universiti Kebangsaan Malaysia • User Keying Material

UkMaZ Ukrainskii Matematicheskii Zhurnal (periodical code)

UKMO United Kingdom Meteorological Office

Ukn Unknown

UKN Unknown • Waukon (USA-IA) (location identifier)

UKNCPRS United Kingdom National Committee for Photogrammetry and Remote Sensing

UKOL United Kingdom On Line

UKOLN United Kingdom Office for Library Networking

UKOLUG United Kingdom Online User Group

UKOP United Kingdom Official Publications

Ukr Ukraine • Ukrainian

UKR Mukeiras (Yemen) (location identifier) • Ukraine (ISO country code)

UKRDAC United Kingdom ROSAT Data Archive Centre

UKS United Kingdom Satellite • United Kingdom Schmidt

UKSC United Kingdom Survey Centre

UKSEDS United Kingdom Students for the Exploration and Development of Space

UKSS United Kingdom Simulation Society

UKST United Kingdom Schmidt Telescope

UKSTU United Kingdom Schmidt Telescope Unit

UKT Quakertown - Upper Bucks (USA-PA) (location identifier)

UKTS United Kingdom Treaty Series

UKU Nuku (Papua New Guinea) (location identifier)

UKW Ultrakurzwellen

UKX Ust-Kut (Russia) (location identifier)

UKY Kyoto (Japan) (location identifier)

U/L Uplink

UL Ulci • Ultraléger • Ultralinear • Ultraluminous • Underload • Underwriters Laboratories • Unit Level • Unit of Length • Universitat de Lleida (Spain) • Universiteit Leiden (Netherlands) • Université Laval (Canada-QC) • University Libraries • University Library • Unordered List • Upgren Late • Uplink • Upload • Upper Left • Upper Level • Upper Limb • Upper Limit • Upper Lobby • User Language

ULA San Julian (Argentina-SC) (location identifier) • Uncommitted Logic Array • Unit Level Automation • Universidad de Los Andes (Venezuela) • Upper Layer Architecture

ULANA Unified Local-Area Network Architecture

ULAS Union List of Astronomy Serials

ULB Ulei (Vanuatu) (location identifier) • Unit Log Book • Universal Logic Block • Universitäts- und Landesbibliothek (Germany) • Université Libre de Bruxelles (Belgium)

u&lc Upper & Lower Case

ULC Santiago - Los Cerrillos (Chile) (location identifier) • Underwriters' Laboratories of Canada • Uniform Language of Composition • Unitary Launch Concept • Unitary Launch Control • Unité Logique de Contrôle • Unit Level Code • Unit Level Computer • Unit Load Container • Unit Location Code • Universal Logic Circuit • Unpressurized Logistics Carrier • Upper Left Caption

ULCAS Unpressurized Logistics Carrier Attach System

ULCC Ultra Large Crude Carrier • University of London Computer Centre • Utility Location and Coordination Council

ULCE Unified Life-Cycle Engineering

ULCI Unpressurized Logistics Carrier Interface

ULCS Unit Level Circuit Switch

ULD Ulundi (South Africa) (location identifier) • Unit Load Device • Unit Logic Device • Upload • Uploaded • Upper Level Discriminator

ULDA Uniform Low-Dispersion Archive

ULDB Ultra-Long-Duration Balloon

ULDEST Ultimate Destination

UL/DL Uplink/Downlink

ULDL Uplink-Downlink

ULDMI Ultraprecise Laser Distance Measuring Instrument

ULDS Unit Level Digital Switch

ULE Sule (Papua New Guinea) (location identifier) • Ultra-Low Expansion • Unit Location Equipment

ULECA Ultra-Low-Energy Charge Analyzer

ULEIS Ultra-Low-Energy Isotope Spectrometer

ULET Ultra-Low Energy Telescope

ULEV Ultra-Low Emission Vehicle

ULF Ultralow Field • Ultra-Low Frequency • Upper Limiting Frequency • Upward Lighting Flux • Utilization and Logistics Flight

ULg Université de Liège (Belgium)

ULG Ulgit (Mongolia) (location identifier)

ULI Ulithi (Caroline Islands) (location identifier) • Universal Laboratory Interface • Universal Language Interface • Universal Logic Implementer

ULIBG Ultra-Luminous Infrared Bright Galaxy

ULICS University of London Institute of Computer Science

ULIG Ultraluminous Infrared Galaxy

ULIPC Upper Layer Interprocess Communication

ULIRG Ultra-Luminous Infrared Galaxy (*) • Ultra-Luminous IRAS Galaxy

ULK Uplink

ULL Mull (UK) (location identifier) • Unit Level Logistics • Unit Local Loading • Universidad de La Laguna (Spain)

ULLC Unit Level Logistics Computer

ULLS Unit Level Logistics System

ULLV Unmanned Lunar Logistics Vehicle

ULM New Ulm (USA-MN) (location identifier) • Ultra-Léger Motorisé • Ultra-Light Motorized • Ultrasonic Light Modulator • Universal Logic Module

ULMS Unit Level Message Switch

ULN Ulaanbataar (Mongolia) (location identifier) • Unité de Liaison Numérique • Unit Line Number • Universal Link Negotiation

ULO Ulaangom (Mongolia) (location identifier) • University of London Observatory (UK-England) • Unliquidated Obligation • Unmanned Launch Operations • Unmanned Lunar Orbiter

ULOR Upward Light Output Ratio

ULP Quilpie (Australia-QLD) (location identifier) • Unfair Labor Practice • Unit in Last Place • Université Louis Pasteur (France) • University of London Press (UK-England) • Upper-Layer Protocol (*) • Upper-Level Protocol

ULPGC Universidad de Las Palmas de Gran Canaria (Spain)

ULQ Tulua Farfán (Colombia) (location identifier)

ULS Mulatos (Colombia) (location identifier) • Union List of Serials • Union List System • Unit Level Switch • Upload Station • Upward-Looking Sonar • User Location Server • User Location Service

ULSA Ultra-Low Sidelobe Antenna

ULSB Ultra-Low Surface Brightness

ULSD Ultra-Low-Sulphur Diesel

ULSDM Ultra-Light Scalar Dark Matter

ULSDS Ultra-Large Screen Display System

ULSI Ultra-Large-Scale Integration • United Launch Systems International (Australia)

ULSS User Logistic Support Summary

ULST Uniform Low Stratus Test

ULSV Unmanned Launch Space Vehicle

Últ Último

Ult Ultimate • Ultimately • Ultimo

ULT Ultimate • Ultra-Low Temperature

ULTD Unit Level Training Device

UltIm Ultrasonic Imaging (periodical code)

Ultmi Ultramicroscopy (periodical code)

Ulto Ultimo

Ultra Ultrasonics (periodical code)

ULTU Universal Line Termination Unit

ULU Gulu (Uganda) (location identifier) • Uplink Unit

ULUT UCB Lookup Table

ULV Ultralow Volume • Unmanned Launch Vehicle

ULX Ultraluminous X-ray (source) • Ulusaba (South Africa) (location identifier)

ULY Ulyanovsk (Russia) (location identifier)

ULZ Uliastai (Mongolia) (location identifier)

u/m Undermentioned

uM unter dem Meeresspiegel

üM Über dem Meeresspiegel

U/M Unit of Measure • Unmanned • Unscheduled Maintenance

UM Mauritanian Ouguiya (currency) • Minor Outlying Islands (United States_) (ISO country code) • Umbilical Mast • Under Mentioned • Unified Messaging • Unified Model • Unité de Mesure • United States Minor Outlying Islands (ISO country code) • Unité Marine • Unité Motrice • Unit Manager • Unit of Measure • Unit of Measurement • Universal Modem • Universal Monitor • Universal Multifractal • Universiteit van Maastricht (Netherlands) • University of Maryland (USA-MD) • University of Massachusetts (USA-MA) • University of Miami (USA-FL) • University of Michigan (USA-MI) • University of Minnesota (USA-MN) • Unmanned • Unscheduled Maintenance • User Manual • User Model • User's Machine • Usine Mécanique • Utilization Management

UMa Ursa Major - Ursae Majoris (constellation)

uma Unité de Masse Atomique

UMA Punta de Maisi (Cuba) (location identifier) • Unified Memory Access • Unified Memory Architecture • Union du Maghreb Arabe • Unión Matemática Argentina (Argentina) • Unité de Masse Atomique • Universal Mail Agent • Universal Measurement Assembly • Universal Measurement Amplifier • Universal Measuring Amplifier • Universal Multimedia Access • Universal Multiple Access • Unscheduled Maintenance Action • Upper Memory Area • Uppsala Model Atmosphere

UMAB University of Maryland at Baltimore (USA-MD)

UMaCh Universidad Marítima de Chile

UMaG Ursa Major Group

UMAI Union des Maîtres assistants et Assistants Indépendants (France)

UMALCA Unión Matemática de América Latina y el Caribe

UMAM Uspekhi Mekhaniki Advances Mechanics (periodical code)

UMARF Unified Meteorological Archive and Retrieval Facility

UMASS Unlimited Machine Access from Scattered Sites

Umb Umbilical

UMB Umbilical • Umnak Island (USA-AK) (location identifier) • Université Marc Bloch (France) • Upper Memory Block

Umbc Umbilical

UMBC Umbilical Cord • University of Maryland - Baltimore County (USA-MD)

Umbl Umbilical

UMC Umba (Papua New Guinea) (location identifier) • Unassigned Multiplex Channel • Unconventional Model of Computation • United Microelectronics Corporation • Unit Manning Report • Unit Movement Code • Unit Movement Coordinator • Universal Multizone Crystallizator • Universidade de Mogi das Cruzes (Brazil) • University of Missouri at Columbia (USA-MO) • Upper Mantle Committee

UMCP Unit Maintenance Collection Point • University of Maryland at College Park (USA-MD)

UMCS University Marie Curie-Sklodowska (Poland)

UCMT Unit Maintenance Collection Team

UMD Uniform Mode Distribution • Unit Manning Document • Unit Manpower Document • University of Massachusetts at Dartmouth (USA-MA) • University of Michigan-Dearborn (USA-MI) • University of Minnesota at Duluth (USA-MN) • Unmatched Disbursement • Uummannaq (Greenland) (location identifier)

UMDL University of Michigan Digital Library (USA-MI)

UME Ulusal Metroloji Enstitüsü (Turkey) • Umeå (Sweden) (location identifier) • Undergraduate Mathematics Education • UNI Management Entity • Union des Maisons d'Étudiants (France) • Union Monétaire Européenne •

Unité Monétaire Expérimentale • Unit Management Entity • Unit Mission Equipment • University of the Middle East

UMEB Undergraduate Monitoring in Environmental Biology

UMES University of Maryland Eastern Shore (USA-MD)

UMF Ultra-Microfiche • Uniform Magnetic Field • User Master File • User Message Format

UMG University's Management Group

UMH Unité Mobile Hospitalière • Université de Mons-Hainaut (Belgium)

UMi Ursa Minor - Ursae Minoris (constellation)

UMI Minor Outlying Islands (United States_) (ISO country code) • Quincemil (Peru) (location identifier) • UMIST (Starlink internal code) • Unione Matematica Italiana • United Microelectronics Inc. • United States Minor Outlying Islands (ISO country code) • University Microfilms International • University of Manchester Institute of Science and Technology (Starlink internal code) • User MAN Interface

UMIACS University of Maryland's Institute for Advanced Computer Studies (USA-MD)

UMIB Unione Matematica Italiana Bollettino (periodical code) • Urgent Marine Information Broadcast

UMIF Upper Mechanical Integration Frame

UMIP Université Musicale Internationale de Paris (France)

UMIS Urban Management Information System

UMIST University of Manchester Institute of Science and Technology (UK-England)

UMJ Urgence Médico-Judiciaire

UMK Universal Meßkamer

UMKC University of Missouri - Kansas City (USA-MO)

Uml Umlage

UML Unified Message Language • Unified Modeling Language • Uniform Modeling Language • Universal Mission Load • University of Massachusetts Lowell (USA-MA)

UMLA Universal Multiline Adapter

UMLE Unrestricted Maximum-Likelihood Estimator

UMLS Unified Medical Language System

UMM Summit (USA-AK) (location identifier) • Unidirectional Multipoint-to-Multipoint • Universal Measuring Machine

UMMS Unified Multimedia System

UMO Unmanned Orbital

UMOL Unmanned Orbital Laboratory

UMOS Uniform Mission-Oriented Structure

UMP Ultra-Metal-Poor (star) • Unité Modulaire de Pilotage • Unité Modulaire Programmable • Upper Mantle Programme

UMPA Unité de Mathématiques Pures et Appliquées

UMPDU User Message Protocol Data Unit

UMPS Unit Mission Planning System • Universal Multiprogramming System • Up Mission Processing Start

UMR Unité Mixte de Recherche • Unit Manning Report • Unsatisfactory Materiel Report • Woomera (Australia-SA) (location identifier)

UMRAO University of Michigan Radio Astronomy Observatory (USA-MI)

UMS Ultrasound Monitoring Experiment • UNESCO Member State • Unified Messaging Service • Unité Mixte de Service • Universal Modem System • University of Massachusetts System • Unmanned Multifunction Satellite • Unresolved Multiple Star • Urine Monitoring System • User Management System

UMSA Universidad Mayor de San Andrés (Bolivia)

UMSI University of Minnesota Supercomputer Institute (USA-MN)

UMSL University of Missouri in Saint Louis (USA-MO)

UMSM Universidad Mayor de San Marcos (Peru)

UMT Umiat (USA-AK) (location identifier) • Unachieved Merge Total • Underground Muon Telescope • Unit Maintenance Technician • Unit Mobility Tasking • Universal Mercator Time

UMTA Urban Mass Transportation Administration

UMTBCF User Mean Time Between Critical Failures

UMTBDE User Mean Time Between Downing Events

UMTE Unmanned Threat Emitter

UMTF User Migration Task Force

UMTRAP Uranium Mill Tailings Remedial Action Program

UMTRCA Uranium Mill Tailings Radiation Control Act (USA)

UMTS Universal Maintenance Training System • Universal Mobile Telecommunications System (*) • Universal Mobile Telephone System

UMTTRF User Mean Time To Restore Function

UMTTRS User Mean Time To Restore System

UMU Umuarama - Ernesto Geisel (Brazil-PR) (location identifier)

UMV Unbiased Minimal Variance • Unrestricted Motion Vector

UMVE Uniformly Minimum Variance Estimator

UMVF Université Médicale Virtuelle Francophone • Unmanned Vertical Flight

UMW Ultramicrowave

UMWD Ultra-Massive White Dwarf (star)

UmwG Umwandlungsgesetz

UmwStG Umwandlungs-Steuergesetz

UMY Sumy (Ukraine) (location identifier)

un Undulatus • Unit • Unité

Un Unified • Union • United • Unsatisfactory

ÜN Übermittlungsnetz

UN Unbalanced Normal • Uncertainty • Union • United Nations • Update Number

UNA United Nations Association • Universitäts-Netz • University of North Alabama (USA-AL) • Upstream Neighbor Address

UNAAPE Union Nationale des Associations Autonomes de Parents d'Élèves (France)

Unab Unabridged

UNAC United North American Conference

Unacc Unaccompanied

UNACC United Nations Administrative Committee and Coordination

Unaccomp Unaccompanied

UNADS UNIVAC Automated Documentation System

UNAEDE Union Nationale des Éducateurs de l'Enfance (France)

UNAer Università di Napoli Aeritalia S (periodical code)

UNAF Union Nationale des Associations Familiales (France)

UnAG Université des Antilles et de Guyane (Martinique)

UNAH Universidad Nacional Autónoma de Honduras

UNALM Universidad Nacional Agraria La Molina (Peru)

Unalot Unalotted

UNAM Universidad Nacional Autónoma de México

UNAMACE Universal Automatic Map Compilation Equipment

Unan Unanimous

UNAP Universidad Nacional de la Amazonia Peruviana (Peru)

UNAPEI Union Nationale d'Associations de Parents d'Enfants Inadaptés (France)

UNAS Universal Network Architecture Services

Unasgd Unassigned

Unasged Unassigned

Unatnd Unattended

Unatt Unattached

Unattrib Unattributed

Unauth Unauthorized

Unauthd Unauthorized

Unavbl Unavailable

UNAVCO University NAVSTAR Consortium (USA)

UNAVF Union Nationale des Accueils des Villes Françaises (France)

Unb Unbound

UNB Universal Navigation Beacon • University of New Brunswick (Canada-NB)

UNBAL Unbalance

Unbd Unbound

UNBIS United Nations Bibliographic Information System

UNBS Uganda National Bureau of Standards

unc Uncinus

Unc Uncertain • Uncirculated

UNC Unbalanced Normal Class • Unclassified • Uncoated • Unencoded Netnews Collator • Unguia (Colombia) (location identifier) • Unified Coarse • Universal Naming Convention • Universal Navigation Computer • Universidad Nacional de Cuyo (Argentina) • University of North Carolina (USA-NC) • University of Northern Colorado (USA-CO) • Uranyl Nitrate Concentrate • Uuencoded Netnews Collator

UNCA University of North Carolina at Asheville (USA-NC)

UNCAST United Nations Conference on the Applications of Science and Technology

UNCC United Nations Cartographic Commission

UNCCC United Nations Convention on Climate Change

UNCDF United Nations Capital Development Fund

UNCE Union Nationale des Coopératives d'Étudiants (France)

UNCED United Nations Conference on Environmental Development • United Nations Conference on Environment and Development (*)

UNCG University of North Carolina at Greensboro (USA-NC)

UNCHE United Nations Conference on Human Environment

UNCHS United Nations Centre for Human Settlements

UNCIO United Nations Conference on International Organizations

UNCITRAL United Nations Commission on International Trade Law

UNCJIN United Nations Crime and Justice Information Network (*) • United Nations Criminal Justice Information Network

UNCLA Unclassified

Unclas Unclassified

UNCLAS Unclassified

Unclass Unclassified

UNCLE United Nations Command for Law Enforcement • Unsteady Computation of Linear Equations

UNCLOS United Nations Conference on the Law of the Sea

UNCOD United Nations Conference On Desertification

UNCOL Universal Computer-Oriented Language

Uncond Unconditional

Unconst Unconstitutional

UNCOPUOS United Nations Committee On the Peaceful Uses of the Outer Space

Uncor Uncorrected

UNCPIE Union Nationale des Centres Permanents d'Initiation à l'Environnement (France)

UNCRD United Nations Centre for Regional Development

UNCSTD United Nations Centre for Science and Technology for Development • United Nations Conference on Science and Technology for Development

UNCTAD United Nations Conference on Trade And Development

UNCTC United Nations Centre for Transnational Corporations

UNCU Union Nationale des Clubs Universitaires (France)

UNCW University of North Carolina at Wilmington (USA-NC)

Und Under

UND Kunduz (Afghanistan) (location identifier) • University of Natal at Durban (South Africa) • University of North Dakota (USA-ND) • University of Notre Dame (USA-IN) • User Need Date

Undc Undercurrent

UNDC United Nations Disarmament Commission

UNDCP United Nations Drug Control Programme

Undef Undefined

UNDEO Union Nationale des Diplômés d'État en Organisation (France)

Undetm Undetermined

UNDEX Underwater Explosive

Undf Underfrequency

UNDHA United Nations Department of Humanitarian Affairs

UNDK Undock

Undld Underload

UNDP United Nations Development Program

UNDRO United Nations Disaster Relief Organization

Undsgd Undersigned

UNDTCD United Nations Department of Technical Cooperation for Development

Undtkr Undertaker

Undv Undervoltage

UNDV Undervoltage

UNE Qachas Nek (Lesotho) (location identifier) • Underground Nuclear Explosion • Universidad Nacional de Educación (Peru) • University of New England (Australia-NSW)

UNEC United Nations Education Conference

UNECA United Nations Economic Commission for Asia

UNECE United Nations Economic Commission for Europe

UNECLAC United Nations Economic Commission for Latin America and the Caribbean

UnECR University Electro Communications Reports (periodical code)

UNED Universidad Nacional de Educación a Distancia (Spain) • Université Nationale d'Enseignement à Distance

UNEDA United Nations Economic Development Administration

UNEDIC Union Nationale pour l'Emploi Dans l'Industrie et le Commerce (France)

UNEF Unified Extrafine • Union Nationale des Étudiants de France

UNEP United Nations Environmental Program • United Nations Environment Program (*)

UNES Union Nationale des Étudiants en Sciences (France) • Union Nationale des Étudiants Salariés (France)

UNESCAP United Nations Economic and Social Commission for Asia and the Pacific

UNESCO United Nations Educational, Scientific and Cultural Organization

UNESIS United Nations Economic and Social Information System

UNESP Universidade Estadual Paulista (Brazil-RJ)

UnESS University Earth System Science (project)

UNETP Union Nationale de l'Enseignement Technique Privé (France)

UNETPSA United Nations Educational and Training Program for Southern Africa

UNEX University-class Explorer (*) • University Explorer

Unexpl Unexplained • Unexplored

UNF Unified Fine • University of North Florida (USA-FL)

UNFAO United Nations Food and Agriculture Organization

Unfav Unfavorable

UNFB United Nations Film Board

UNFCC United Nations Framework for Climate Change

UNFCCC United Nations Framework Convention on Climate Change

Unfin Unfinished

UNFPA United Nations Fund for Population Activities

UNFSTD United Nations Fund for Science and Technology Development

UNFURNOTE Until Further Notice

ung Unguentum

UNG Kiunga (Papua New Guinea) (location identifier)

UNGA United Nations General Assembly

UNGASS United Nations General Assembly Special Session

UNGCRP United Nations Global Change Research Program

UNGEGN United Nations Group of Experts on Geographical Names

ungt Unguentum

Unh Unnilhexium

UNH University of New Hampshire (USA-NH) • University of New Haven (USA-CT) • Uranyl Nitrate Hexahydrate

UNHCR United Nations High Commission for Refugees

UNHQ United Nations Headquarters

Uni Universität

UNI Ente Nazionale Italiano di Unificazione (Italy) • Union Island (Grenadine Islands) (location identifier) • Union Nationale Interuniversitaire (France) • Unité de Navigation Inertielle • University of Northern Iowa (USA-IA) • User-Network Interface • User-to-Network Interface (*)

UNIC United Nations Information Centre

UNICC United Nations International Computing Center

UNICE Union des Industries de la Communauté Européenne • Union of Industrial and Employers' Confederation of Europe

UNICEF United Nations International Children's Emergency Fund

UNICOM Universal Integrated Communication

UNICOMP Universal Compiler

UNIDO United Nations Industrial Development Organization

UNIDRV Universal Driver

UNIECO Union Internationale d'Écoles par Correspondance

Unif Uniform

UNIFET Unipolar Field-Effect Transistor

UNII Unlicensed National Information Infrastructure

UNINETT Nordic University Network

UNIPEDE Union Internationale des Producteurs et Distributeurs d'Électricité

UNIPOL Universal Problem-Oriented Language • Universal Procedure-Oriented Language

UNIPS Unité d'Indicateurs de Politique Scientifique

UNIS Unified Information System

UNISA University of South Africa

UNISE Uganda National Institute for Special Education

UNISIST Universal System for Information in Science and Technology

UNISTAR United Nations Institute for Scientific Training And Research • UNIVAC Storage And Retrieval • User Network for Information Storage, Transfer Acquisition and Retrieval

UNIT Instituto Uruguayo de Normas Técnicas (Uruguay) • Universal Numerical Interchange Terminal

UNITAR United Nations Institute for Training and Research

Univ Universal • Universally • Universe • Universel • Universidad • Universidade • Università • Universität • Université • Universiteit • Universiti • University

UNIV Universal

UNIVAC Universal Automatic Computer

UNJSPF United Nations Joint Staff Pension Fund

Unk Unknown

UNK Unalakleet (USA-AK) (location identifier) • University of Nebraska at Kearney (USA-NE) • Unknown

Unkn Unknown

Unl Unloading

UNL Universal Networking Language • Universidad Nacional del Litoral (Argentina) • University of Nebraska at Lincoln (USA-NE) • Unlimited

UNLCH Unlatch

Unlgtd Unlighted

Unlim Unlimited

Unliq Unliquidated

UNLK Unlock

Unlkg Unlocking

UNLP Universidad Nacional de La Plata (Argentina-BA)

UNLV University of Nevada at Las Vegas (USA-NV)

Unm Unmarried

UNM University of New Mexico (USA-NM)

UNMA Unified Network Management Architecture

UNMC United Nations Mediterranean Commission

Unmkd Unmarked

Unmon Unmonitored

UNMRI Union Nationale des Mutuelles de Retraites des Institutions (France)

Unrelbl Unreliable

UNMS Slovak Office of Standards, Metrology and Testing

UNMSM Universidad Nacional Mayor de San Marcos (Peru)

Unmtd Unmounted

UNN Ranong (Thailand) (location identifier) • University of Nigeria at Nsukka

UNNE Universidad Nacional del Nordeste (Argentina)

UNO Underground Nucleon Observatory • United Nations Organization • University of Nebraska at Omaha (USA-NE) • University of New Orleans (USA-LA)

UNOASD United Nations Outer Space Affairs Division

UNODIR Unless Otherwise Directed

UNOG United Nations Office in Geneva (Switzerland)

UNOLS University National Oceanographic Laboratory System

Unop Unopposed

UNOPS United Nations Office for Project Services

UNOV United Nations Office in Vienna (Austria)

Unp Unnilpentium • Unpaged • Unpaid

Unpd Unpaid

UNP Uluru National Park (Australia-NT) • University of Natal at Pietermaritzburg (South Africa)

UNPA United Nations Postal Administration

UNPS Universal Power Supply

Unpub Unpublished

Unpubl Unpublished

Unq Unnilquadium

Unqual Unqualified

UNR Underkhaan (Mongolia) (location identifier) • Université Nationale du Rwanda • University of Nevada in Reno (USA-NV)

UNREP Union Nationale Rurale d'Éducation et de Promotion (France)

UNRFNRE United Nations Revolving Fund for Natural Resources Exploration

Unrgltd Unregulated

UNRISD United Nations Research Institute for Social Development

UNRRA United Nations Relief and Rehabilitation Administration

UNRSTD Unrestricted

UNRWA United Nations Relief and Works Agency

Uns Unserviceable

UNS Umnak Island (USA-AK) (location identifier) • Unions • Universidad Nacional del Sur (Argentina) • Université Numérique de Strasbourg (France)

UNSA Unconsolidated Sedimentary Aquifer • Universidad Nacional San Agustín (Peru)

UNSAC United Nations Scientific Advisory Committee

Unsat Unsatisfactory • Unsaturated

UNSBL Unseasonable

UNSC United Nations Security Council

UNSCC United Nations Standards Coordinating Committee

UNSCCUR United Nations Scientific Conference on the Conservation and Utilization of Resources

UNSCEAR United Nations Scientific Committee on Effects of Atomic Radiations

UNSD United Nations Statistical Division

UNSE Union Nationale des Séjours à l'Étranger (France)

UNSG United Nations Secretary-General

UNSIS United Nations Statistical Information System

UNSJ Universidad Nacional de San Juan (Argentina)

UNSL Universidad Nacional de San Luis (Argentina)

UNSM United States National Museum (USA)

UNSO United Nations Statistical Office

UNSR United Nations Space Registry

UNSTBL Unstable

UNSTDY Unsteady

UNSTL Unstettle

UNSTLD Unstettled

Unstpd Unstamped

Unsvc Unserviceable

UNSW University of New South Wales (Australia-NSW)

UNSWIRF University of New South Wales Infrared Fabry-Pérot

UNSWPWR Unswitched Power

Unsym Unsymmetrical

Unt Unter

UNT Undergraduate Navigation Training • Unified Networking Technology • Universidad Nacional de Trujillo (Peru) • Universidad Nacional de Tucumán (Argentina) • University of Newcastle-on-Tyne (UK) • University of North Texas (USA-TX) • Unst - Shetland Island - Baltasound (UK) (location identifier)

UNTAA United Nations Technical Assistance Administration

UNTAB United Nations Technical Assistance Board

UNTAG United Nations Transition Assistance Group

UNTAM United Nations Technical Assistance Mission

UNTAP United Nations Technical Assistance Programme

UNTCD United Nations Technical Cooperation for Development

Unterfr Unterfranken (Germany)

UNTPDC United Nations Trade Point Development Centre

Untrtd Untreated

UNTSFA United Nations Trust Fund for Southern Africa

UNTSO United Nations Truce Supervision Organization

UNU Juneau - Dodge County (USA-WI) (location identifier) • United Nations University

UNUIST United Nations University Institute for Software Technology

UNUIIST United Nations University International Institute for Software Technology

Unusbl Unusable

UNV United Nations Volunteers

UNX Underground Nuclear Explosion

uö und öfter

u/o Used on

U&O Use & Occupancy

U/O Used On

UO Undefinitized Option • Undelivered Order • University of Oregon (USA-OR) • Used On

UoA University of Arizona (USA-AZ)

UOA Unit Of Assessment

UOC Ultimate Operating Capability • Ultimate Operational Configuration • Unequilibrated Ordinary Chondrite • Universal Output Computer • Universitat Oberta de Catalunya (Spain) • Unusual Occurrence Control • Uranium Ore Concentrate • Usable On Code

UoD Universe of Discourse

UOD Ultimate Oxygen Demand

UODDL User-Oriented Data Display Language

UOE Unit Of Error • User Operational Evaluation

UOES User Operational Evaluation System

UOF Ultimate Operating Frequency • Unplanned Outage Factor • User Operations Facility

UofA University of Alaska (USA-AK)

UofNC University of North Carolina (USA-NC)

UofS University of Orange Free State (South Africa) • University of Saskatchewan (Canada-SK)

UofT University of Toronto (Canada-ON)

UofW University of Washington

UOG Universitätsorganisationsgesetz (Austria) • University Of Guam

UOH Unplanned Outage Hour

UOHC Under Other than Honorable Conditions

UOI Uranus Orbital Insertion • Uranus Orbit Insertion • User Online Interaction

UoL Universe of Language

UOL Buol (Indonesia) (location identifier) • Utility Octal Load • Utility-Oriented Language

UOLI User-Oriented Language Interface

UoM University of Mauritius

UOMS Unmanned Orbital Multifunction Satellite

UON Muong Sai (Laos) (location identifier) • Uniform Object Notation

UOO Undelivered Order Outstanding

UOP UCAR Office of Programs

UOR Unplanned Outage Rate • Unusual Occurrence Report • Urgent Operational Requirement

UoS Universe of Structure

UOS Sewanee - Franklin County (USA-TN) (location identifier) • Ultraviolet Ozone Spectrometer • Undelivered Order Schedule • Unless Otherwise Specified • Unmanned Orbital Satellite

UoSAT University of Surrey Satellite

UoT Universe of Thought

UOT Universal Oxidation Test • Universal Oxidation Testing • User-Oriented Test • User-Oriented Testing

UOTS Untouched Off-The-Shelf

UOV Unit Of Variance

UoW University of Wales (UK-Wales) • University of Wollongong (Australia)

UOX Oxford University (USA-MS) (location identifier)

Up Upper

U/P Unit Price

UP Ultra-Puissant • Uncertainty Principle • Under Proof • Under Provision of • Uniprocessor • Unité de Production • United Press • Unité Pédagogique • Unit Pack • Unit Price • Universidade do Porto (Portugal) • Universidad Panamericana • Université de Paris (France) • University of Pennsylvania (USA-PA) • University of Pittsburgh (USA-PA) • University of the Philippines • University Press • Unknown Precipitation • Unpublished • Unsaturated Polyester • Unsolicited Proposal • Unusual Programme • Update Processor • Upper • Upper Peninsula • Urban Planning • User Part • User Plane • User Profile • Utility Path • Uttar Pradesh (India)

UPA Punta Alegre (Cuba) (location identifier) • Ultrafine Particle Analyzer • Ultra Port Architecture • Uncooled Parametric Amplifier • Underwriting Profitability Analysis • Union Postale Arabe • Union Professionnelle Artisanale (France) • Unitary Pole Approximation • Unpaired Contract Action

UPAAE UNESCO Pilot African Academic Exchange

UPADI Unión Panamericana de Asociaciones De Ingenieros

UPAJ Union Panafricaine des Journalistes

UPAMS Unpriced Action Management System

UPAR Uninstantiated Parametrized Action Representation • Unit Public Affairs Representative

UPAS Underpass • Unit Performance Assessment System • Universal Pilot Application Service

UPB Uniform Pixon Basis • Unit Price Book

UPC Puerto la Cruz (Spain) (location identifier) • UKIRT Programme Committee • Unidata Program Center • Uniform Price Code • Unit Processing Code • Unit Product Code • Unit Production Cost • Universal Postal Convention • Universal Product Code • Universidad Pontificia Comillas (Spain) • Universitat Politecnica de Catalunya (Spain) • Unutilized Plant Capacity • Uplink Power Control • Usage Parameter Control • User Parameter Control • User Permit Code • Utilized Plant Capacity

UPCA Uniform Phase Evolution and Chaotic Amplitude

UPCC Uniform Product Code Council (USA)

UPCE University of the Philippines College of Engineering

UPCHUK University Program for the Comprehensive Handling and Utilization of Knowledge

Upd Unpaid

UPD University Police Department • Update

UPDF Universal Probability Distribution Function

UPDFTS Updrafts

Updt Update

UPE Unitary Pole Expansion • Université Populaire Européenne • University of Port Elizabeth (South Africa) • User Portability Extension

UPEP Unité de Première Expérience Professionnelle

UPEPI Union des Practiciens Européens en Propriété Industrielle

UPERC Unsolicited Proposal Evaluation Review Committee

UPES Universal Polyelastic Sampling

UPF Pforzheim (Germany) (location identifier) • Universal Patch Format • Universitat Pompeu Fabra (Spain) • User Productivity Facility

UPG Ujung Pandang (Indonesia) (location identifier) • Upgrade • Usine de Propergol de Guyane

UPGD Upgrade • Upgraded

UPGMA Unweighted Pair Group Method using arithmetic Averages

UPGMC Unweighted Pair Group Method using Centroids

UPGWA United Plant Guard Workers of America

UPH Unaccompanied Personnel Housing

Uphd Uphold

UPI Uncertain Prior Information • United Press International • Universal Presentation Interface

UPIG User Product Information Group

UPIO Undefinitized Provisioned Item Order

UPJV Université de Picardie Jules Verne (France)

UPK Uganda Polytechnic Kyambogo

UPL Unapproved Parts List • Universal Programming Language • Upala (Costa Rica) (location identifier) • User Program Language • User Programming Language

UPLIFT Utrecht Project: Linguistic Information Free Text

UPLK Uplink

UPM Unidirectional Point-to-Multipoint • Universal Permissive Module • Universal Processor Module • Universidad Politécnica de Madrid (Spain) • University Pertanian Malaysia • Unix Programmer's Manual • Urban Pollution Management • User Profile Management

UPMA Universal-Primärmultiplexanschluß

UPMAIL Uppsala Programming Methodology and Artificial Intelligence Laboratory (Sweden)

UPMC Université Pierre et Marie Curie (France)

UPMR Unit Personnel Management Roster

UPN Unified Personal Number • Unique Program Number • Unique Project Number • Universidad Pedagógica Nacional (Mexico) • Uruapan (Mexico) (location identifier)

UPNE University Press of New England (USA-NH)

UPnP Universal Plug and Play

UPO Unit Personnel Officer • Unpaired Purchase Order • Unstable Periodic Orbit • Usage Printout

UPOCC Ulysses Payload Operations Control Centre

UPOS University Partnership for Operational Support • Utility Program Operating System

Upp Upper

UPP Unidirectional Point-to-Point • Universal Payment Preamble • Universal Plug and Play • Universal Procedure Pointer • University of Pennsylvania Press (USA-PA) • Unliquidated Progress Payment • Upolu Point (USA-HI) (location identifier) • User Parameter Processing • User Participation Program • User Partnership Program • User Preference Profiling • Utility Print Punch

UppAn Uppsala Astronomical Observatory Annals (periodical code)

UPPB Union de la Presse Périodique Belge (Belgium)

UppOR Uppsala Astronomical Observatory Reports (periodical code)

UPPR Upper

UPPS Universal Projection Plotting System

Upr Upper

UPR Ultrasonic Parametric Resonance • Unité Propre de Recherche • Universidad de Puerto Rico • University of Puerto Rico • Upiara (Papua New Guinea) (location identifier) • Upper • User Preferred Routing

UPRA University Postgraduate Research Award

UPRM University of Puerto Rico at Mayaguez (Puerto Rico)

UPRS Unique Property Reference Number

Uprt Upright

UPS Ultraviolet Photoelectron Spectroscopy • Ultraviolet Photoemission Spectroscopy • Unattended Power Source • Unauthorized Product Substitution • Under Provision of Section • Unified Propulsion System • Uniform Procurement System • Uninterrupted Power Source • Uninterrupted Power Supply • Uninterrupted Power System • Uninterruptible Power Supply (*) • Uninterruptible Power System • Unité Propre de Service • Universal Polar Stereographic (projection) • Universal Profiling Schema • Università Popolare Sestrese (Italy) • Université de Paris-Sud (France) • Université Paul Sabatier (France) • University of Puget Sound (USA-WA) • Upright Perigee Stage • User Planning System • User Profile System

UPSI User Program Switch Indicator

UPSLP Upslope

UPSO Uttar Pradesh State Observatory (India)

UPSR Unidirectional Path Switching Ring

UPSTAIRS Universal Processing System for Treasuring Up and Analyzing Images in Remote Sensing

UPSTM Upstream

UPT Undergraduate Pilot Training • Universal Personal Telecommunications • Universal Protocol Translator • User Profile Table

UPTCAGP University of the Philippines Training Center for Applied Geodesy and Photogrammetry

UPTEC Unité Pluridisciplinaire des Techniques d'Expression et de Communication

UPTLM Uplink Telemetry

UPTP Universal Package Test Panel

UPU Union Postale Universelle • Unión Postal Universal • Universal Postal Union

UPV Universidad del País Vasco (Spain) • Upavon (UK-England) (location identifier)

uPVC Unplasticized Polyvinyl Chloride

UPWARDS Unified Procedures for World-wide Access to Research Databases

Upwd Upward

UPWD Upward

UPWT Unitary Plan Wind Tunnel

UQ Ultimo Quarto • University of Queensland (Australia-QLD) • Upper Quartile

UQ&A Users' Questions & Answers

UQA Users' Questions and Answers

UQAC Université du Québec à Chicoutimi (Canada-QC)

UQAH Université du Québec à Hull (Canada-QC)

UQAM Université du Québec à Montréal (Canada-QC)

UQAR Université du Québec à Rimouski (Canada-QC)

UQAT Université du Québec en Abitibi-Témiscamingue (Canada-QC)

UQCP Uniform Quality Control Program

UQE Queen (USA-AK) (location identifier) • Uniform Query Engine

UQEI Uniform Query Engine Interface

UQF Uniform Query Format • Users' Question Forum

UQG Uniform Query Gateway

UQGT Unified Quantum Gravitational Theory

UQP Users' Questions Procedure

UQPSK Unbalanced Quadrature Phase-Shift Keying

UQT Users' Questions Time

UQTR Université du Québec à Trois-Rivières (Canada-QC)

UQTSM Unit Quality and Technical Support Manager

ur Urdu (ISO language code)

uR Upwind Roughness

Ur Urdu • Uruguay

U/R Underrange • Up Range • Urban/Rural

UR Ukraine (international aircraft registration code) • Ultra-Rapide • Unconditioned Reflex • Unfinanced Requirement • Unfunded Requirement • Uniform Regulations • Unité de Recherche • Unité de Reconnaissance • Unité de Réserve • Unit Record • Universal Rectifier • Universal Reference • Universal Relation • University of Rochester (USA-NY) • Unsatisfactory Report • Unspecified Reception • Upper Right • Urkundenrolle • Uruguay • Usage Restreint • User Reliability • User Requirement • Utility Room • Utilization Rate

Ura Uranus

URA Uniform Resource Agent • Uniformly Redundant Array • Unité de Raccordement d'Abonnés • Unité de Recherche Associée • Universities Research Associates • Universities Research Association Inc. (USA) (*) • Ural'sk (Kazakhstan) (location identifier) • Uranova (catalog) • Urine Receptacle Assembly • User Radar Altimeter • User Requirement Analysis

URACA Union Rhône-Alpes des Clubs d'Astronomie (France)

Uran Uranus

Urani Urania (periodical code)

URANIA Universal Research Archive of Networked Information in Astronomy

URAP Ulysses Radio And Plasma • Unified Radio And Plasma-wave (project) (*)

URAV Uninhabited Reconnaissance Aerial Vehicle

Urb Urban • Urbanism • Urbanisme • Urbanismo • Urbanización

URB Universal Radio Background • Urbanização • Urubupunga - Ernesto Pochler (Brazil-SP) (location identifier)

UR&C Usual, Reasonable & Customary

URC Uniform Resource Characteristic • Uniform Resource Citation • Unit Record Card • Unit Repair Cost • University Research Center • Urumqi (China-PRC) (location identifier) • Usual, Reasonable and Customary

URCN Unité de Raccordement de Circuits Numériques

URD University Resources Database (USRA) • User Requirements Document

URDB User Requirements Database

URDU Urban Resource Development Unit

URE Undergraduate Record Examination • Unintentional Radiation Exploitation • User Range Error • Utilisation Rationnelle de l'Énergie

UREC Unité Réseaux du CNRS

UREF Université des Réseaux d'Expression Française

UREP Unix RSCS Emulation Program • Unix RSCS Emulation Protocol

URETI University Research, Engineering and Technology Institute

URF Unité de Responsabilité Fonctionnelle • Unit Risk Factor

URFIST Unité Régionale de Formation pour l'Information Scientifique et Technique (France)

URFPIST Unité Régionale de Formation et de Promotion pour l'Information Scientifique et Technique

Urg Urgence • Urgent

URG Urgent • Uruguaiana - Ruben Berta (Brazil-RS) (location identifier)

URGENT Upper Rhine Graben Environmental Tectonics

UrhG Urheberrechtsgesetz

URI Unfavorable Rotation Index • Uniform Resource Identifier • Unité de Recherche et d'Innovation • Universal Resource Identifier • Universal Resource Indicator • University of Rhode Island (USA-RI) • University Research Initiative • Uribe (Colombia) (location identifier)

URIF Universal Research Incentive Fund

URIFT University of Regina Image File Translator

URISP University Research Infrastructure Support Program

URJ Uraj (Russia) (location identifier)

url Bookmarked URLs (file extension)

URL Unfinanced Requirements List • Uniform Record Locator • Uniform Resource Locator (*) • Unité de Recherche Linguistique • Unité Remplaçable en Ligne • Universal Resource Language • Universal Resource Locator • Universidad Ramón Llull (Spain) • Upper Range Limit • User Requirements Language

URLA Unité de Recherche Linguistique Associée

U/RM Utilities/Recurring Maintenance

URM Ulysses Reference Mission • Unit Record Manager • Unlimited Register Machine • Urban-to-Regional Multiscale • Uriman (Venezuela) (location identifier)

URMA Union Régionale des Métiers d'Art

UrMad Urania Madrid (periodical code)

URMS Universal Reproducing Matrix System

URMSG Your Message

URM3 Università di Roma 3 (Italy)

URN Uniform Random Numerator • Uniform Resource Name (*) • Uniform Resource Naming • Uniform Resource Number • Unit Requirement Number • Universal Resource Name • Universal Resource Number • Urgoon (Afghanistan) (location identifier) • Urine

URO Rouen (France) (location identifier) • Unidentified Running Object • User Readout

UROP Undergraduate Research Opportunities Program

URP Unité Remplaçable sur Piste • Unit Record Processor • University Research Program • Unsolicited Research Proposal

URPA Undergraduate Research Program in Astronomy

URQ Unregistered Request • Unsatisfactory Report Questionnaire

URR Ultra-Rapid Reader • Ultra-Reliable Radar • Unit Readiness Record • Unresourced Requirement • Urrao (Colombia) (location identifier) • User Requirement Review

URS Kursk (Russia) (location identifier) • Unit Reference Sheet • Universal Reference System • Universal Regulating System • Université Robert Schuman (France) • University Research Support • Unrestricted Random Sample • Ursids (meteor stream) • User Requirements Specification • Utilization Reporting System

URSA Union Romande des Sociétés d'Astronomie (Switzerland)

URSI Union Radio Scientifique Internationale

Urspr Ursprünglich

URSS Union des Républiques Socialistes Soviétiques

URSSAF Union de Recouvrement pour la Sécurité Sociale et les Allocations Familiales (France)

Urt Urteil

URT Surat Thani (Thailand) (location identifier) • Universal Remote Terminal • University Research and Training • Upright

URTI Université Radiophonique et Télévisuelle Internationale

URTNA Union des Radiodiffusions et Télévisions Nationales d'Afrique

Uru Uruguay

URU Universal Research Unit

Urug Uruguay

URUPABOL Uruguay - Paraguay - Bolivia

URV Underwater Recovery Vehicle • Unité à Réponse Vocale • Unit Risk Value • Universitat Rovira i Virgili (Spain)

URY Gurayat (Saudi Arabia) (location identifier) • Uruguay (ISO country code)

URZ Universitätsrechenzentrum • Uruzgan (Afghanistan) (location identifier)

u/s Unserviceable • Useless

us Ubi Supra • United States of America • Ut Supra

Us Usuel

U/S Unserviceable • Useless

U&S Unified & Specified

US États-Unis d'Amérique (code ISO de pays) • Ufficio Stampa • Ultrasoft • Ultrasonic • Ultrason • Ultrasound • Ultrasonic Scanning • Ultrasound Scanning • Ultraspectral Sensor • Uncle Sam • Unconditioned Stimulus • Unified Scheme • Uniform System • United Service • United States • United States of America (ISO country code) • Unités et Symboles • Unit Separator • Unit Specification • University of Singapore • Untreated Sample • Update Server • Upper Stage • USA (ISO country code) • User Segment • User Station • Usher (survey)

USA États-Unis d'Amérique (code ISO de pays) • Unconventional Stellar Aspect • Unified Solutions Algorithm • Union of South Africa • United Space Alliance • United States of America (international vehicle registration code) • United States of America (ISO country code) • Universal System of Addressing • University of South Alabama (USA-AL) • USA (international vehicle registration code) • User Support Astronomer • Utilization - Support - Accountability

USAA United Services Automobile Association

USAAF United States Army Air Force

USAATS United States of America Air Traffic Service

Usáb Usábase

USABC United States of America Battery Consortium

USACA United States Advanced Ceramics Association (USA)

USACE United States Army Corps of Engineers (USA)

USACOM United States Atlantic Command (USA)

USAE Unconventional Stellar Aspect Experiment

USAEC United States Atomic Energy Commission (USA)

USAETL United States Army Engineering Topographic Laboratories (USA)

USAF United States Air Force (USA)

USAFA United States Air Force Academy (USA-CO)

USAFE United States Air Force in Europe

USAFETAL United States Air Force Environmental Technical Applications Center

USAID United States Agency for International Development (USA)

USAFE United States Air Forces in Europe

USAFM United States Air Force Museum

USAFR United States Air Force Reserve (USA)

USAM Unique Sequential Access Method

USAMO United States of America Mathematical Olympiad

USAN University Satellite Network

USAO United States Attorneys Office

USAP United States Antarctic Program

USARP United States Antarctic Research Program

USART Universal Synchronous-Asynchronous Receiver-Transmitter

USARV Ultra-Small-Amplitude Red Variable (star)

USAS United States of America Standard

USASCSOCR United States of America Standard Character Set for Optical Character Recognition

USASDC United States Army Strategic Defense Command (USA)

USASDG United States Army Space Defense Group

USASI United States of America Standards Institute

USAT Ultra-Small Aperture Terminal

USATS United States Air Traffic Service (USA)

USAXS Ultra-Small-Angle X-ray Scattering

usb Universal Serial Bus (file extension)

U-SB Unified S-Band

USB Unified S-Band • Universal Serial Bus • Universidad Simon Bolívar (Mexico + Venezuela) • Universitäts-Sternwarte Bonn (Germany) • University Science Books (USA-CA) • Unterer-Sonderkanal-Bereich • Unused Serial Bus • Upper Sideband • User Support Branch

USBC United States Bureau of Census (USA) • Universal Standard Book Code

USBCF United States Bureau of Commercial Fisheries (USA)

USBE Unified S-Band Equipment • United States Book Exchange (USA) • Universal Serials and Book Exchange

USBI United Space Boosters Inc.

USBLS United States Bureau of Labor Statistics (USA)

USBM United States Bureau of Mines (USA)

USBP United States Border Patrol (USA)

USBR United States Bureau of Reclamation (USA)

USBS Unified S-Band System • United States Bureau of Standards (USA) • User Signaling Bearer Service

USBUCEN United States Bureau of Census (USA)

USC Ultrasonic Storage Cell • Under Separate Cover • Unified Soil Classification • United States Code (USA) • United States Congress (USA) • Unit Security Code • Universal Service Code • University of San Carlos (Philippines) • University of Santa Clara (USA-CA) • University of South Carolina (USA-SC) • University of Southern California (USA-CA) • Upper Stage Contractor • User Security Component • User Spacecraft

USCA University of South Carolina Aiken (USA-SC)

USCAR United States Council for Automotive Research (USA)

USCC United States Chamber of Commerce (USA) • User Support Coordinating Committee

USCE United States Corps of Engineers

USCF United States Chess Federation (USA)

USCG United States Coast Guard (USA)

USCGNET United States Communications Grid Network (USA)

USC&GS United States Coast & Geodetic Survey (USA)

USCGS United States Coast and Geodetic Survey (USA)

USCM United States Conference of Mayors (USA)

USCMI United States Commission on Mathematical Instruction (USA)

USCO United States and Canada Only

USCS United States Civil Service (USA) • United States Customary System (USA) • Urine Sampling and Collection System

USCW Understanding Space, Climate and Weather

USD Dollar (USA) (ISO currency code) • Ultimate Strength Design • Unified School District • United States Dollar (currency) • Universal Synchronous Data • University of South Dakota (USA-SD)

USDA United States Department of Agriculture (USA)

USDC United States Department of Commerce (USA) • United States Digital Cellular (protocol)

USDD United States Department of Defense (USA)

USDE Undesired Signal Data Emanation • United States Department of Education (USA) • United States Department of Energy (USA)

USDHEW United States Department of Health Education and Welfare (USA)

USDHUD United States Department of Housing and Urban Development (USA)

USDI United States Department of the Interior (USA)

USDJ United States Department of Justice (USA)

USDL Ultra-Stable Diode Laser • United States Department of Labor (USA)

USDMA United States Defense Mapping Agency (USA)

USDOC United States Department Of Commerce (USA)

USDOD United States Department Of Defense (USA)

USDOE United States Department Of Energy (USA)

USDOI United States Department Of the Interior (USA)

USDP United States Disclosure Policy

USDT United States Department of Transportation (USA)

USDW Underground Source of Drinking Water • Usable Supply of Drinking Water

Use Usine

USE Undersea Scientific Expedition • Unified Stochastic Engine • Uniform Stochastic Engine • Unique Support Equipment • Unit Support Equipment • UNIVAC Scientific Exchange • University Space Experiment • Unmanned Surveillance Equipment • Unobtrusive Sensors and Effectors • User Support Environment

USEC United System of Electronic Computers

USEIT User Station Evaluation Integration and Test • User System Evaluation and Integration Tool

USEMAP Urban Survey Experimental Method for Analyzing Photo-data

USENET Users' Network

USEPA United States Environment Protection Agency (USA)

USER User Systems Ergonomics Research (IBM)

USERID User Identification

USERP United Scientists for Environmental Responsibility and Protection

USES United States Employment Service (USA)

USESSA United States Environmental Science Service Administration

usf und so fort

USF Universal Service Fund • University of San Francisco (USA-CA) • University of South Florida (USA-FL)

USFFT Unequally Spaced Fast Fourier Transform

UsFiN Uspekhi Fizicheskikh Nauk (periodical code)

USFS United States Forest Service (USA) • United States Frequency Standard (USA)

USFWS United States Fish and Wildlife Service (USA)

USG Ultrasonic Space Grating • United States Government • Unité Secondaire de Gestion • Universitäts-Sternwarte Göttingen (Germany) • Unterhaltssicherungsgesetz • User Support Group

USgal United States Gallon

USGCA United States Government Contract Awards (USA)

USGCRP United States Global Change Research Program (USA)

USGCRPO United States Global Change Research Program Office (USA)

USGEID United States Government Equipment Identification

USGLI United States Government Life Insurance

USGP United States Gemini Program

USGPO United States Gemini Project Office (USA) • United States Government Printing Office (USA)

USGR United States Government Report

USGRDR United States Government Research and Development Report

USGS United States Geological Service (USA) • United States Geological Survey (USA) (*)

Ush Ushuaia (Argentina-TF)

USh Uganda Shilling (currency)

USH User Shell • Ushuaia (Argentina-TF) (location identifier)

USHA United States Housing Authority (USA)

USHDI United States Historical Documents Institute (USA)

US-HEPNET United States High-Energy Physics Network

USHS United States Home Systems (USA) • Université des Sciences Humaines de Strasbourg (France)

USI Mabaruma (Guyana) (location identifier) • Ultrasound Imaging • Ultraviolet Spectroheliographic Instrument • Unité de Soins Intensifs • Università delle Svizzera Italiana (Switzerland) • Update Software Identity • Urban Systemic Initiative (USA) • User-Specific Interface • User-System Interface • Usine de Séparation Isotopique

USIA United States Information Agency (USA)

USIBA United States International Book Association (USA)

USIC Unité de Soins Intensifs Coronariens • United States Information Center (USA) • United States International Carrier (USA)

USICA United States International Communication Agency (USA)

USIM Universal Subscriber Identity Module

USIS Ultraviolet Stratospheric Imaging Spectrometer • United Society In Space (USA-CO) • United States Imagery System (USA) • United States Information Service (USA)

USIT University System for Information Technology

USITC United States International Trade Commission (USA)

USJ Universitäts-Sternwarte Jena (Germany)

USJGOFS United States Joint Global Ocean Flux Study

USJPRS United States Joint Publications Research Service (USA)

USK Usinsk (Russia) (location identifier)

USL Unit Source Listing • Universal Sign Language • University of Southern Louisiana (USA-LA) • Unix Systems Laboratories Inc. • Useless Loop (Australia-WA) (location identifier) • User Specification Limit

USLC United States Liaison Committee

USLE Universal Soil Loss Equation

USLO United States Liaison Office • United States Liaison Officer

USM Koh Samui (Thailand) (location identifier) • Ultimate Safe Mode • United States Mail (USA) • United States Mint • Unit Security Manager • Universal Service Market • Universitäts-Sternwarte München (Germany) • Universiti Sains Malaysia • University of Southern Maine (USA-ME) • University of Southern Mississippi (USA-MS) • Unscheduled Maintenance • Upravliaiushchie Sistemy i Mashiny (periodical code) • User Security Manual • User Support Management

USMA United States Military Academy (USA) • Universidad Santa María La Antigua (Honduras)

USMAPS United States Map Analysis Package System

USMARC United States Machine Readable Cataloguing Standard

USMC United States Marine Corps (USA) • United States Maritime Commission (USA)

USMES Unified Science and Mathematics for Elementary Schools

USMG Unité Scientifique et Médicale de Grenoble (France)

USMI Universal Software Market Identifier

USML United States Microgravity Laboratory (USA)

USMM Upper Semi-Midmean

USMP United States Microgravity Payload

USMTF Uniform Services Message Text Format • Uniform Services Message Transfer Format • United States Message Text Format • United States Message Transfer Format

USN Ulsan (Korea-ROK) (location identifier) • United States Navy (USA) • USA Dollar (next day) (ISO currency code) • User Subnetwork

USNA United States National Army (USA) • United States Naval Academy (USA-MD)

USNARS United States National Archives and Record Service (USA)

USNBS United States National Biological Survey (USA)

USNC United States National Committee (USA)

USNCHC United States National Committee on Scientific Hydrology (USA)

USNDC United States Nuclear Data Committee (USA)

USNMAS United States National Map Accuracy Standard

USNO United States Naval Observatory (USA-DC)

USNOC United States Naval Observatory Circulars (periodical code)

USNOFS United States Naval Observatory Flagstaff Station (USA-AZ)

USNOR United States Naval Observatory Reports (periodical code)

USNOS United States National Ocean Service (USA) • United States National Ocean Survey (USA)

USNPS United States National Parks Service (USA)

USNR United States Navy Reserve (USA)

USNRC United States National Research Council (USA)

USNRL United States Naval Research Laboratory (USA-DC)

USNS United States Navy Ship

USO Udaipur Solar Observatory (India) • Ultrastable Oscillator • Unidentified Superconducting Object • Unidentified Supraconducting Object • United Service Organization • United States Only • Universal Service Obligation • Universal Service Order • Unix Software Operation • Unmanned Seismological Observatory • User Support Office • User Support Organization • Usino (Papua New Guinea) (location identifier)

USOC Unified System for Orbit Computation • Unified System for Orbit Computation (periodical code) • Uniform Service Order Code • Universal Service Order Code • Universal Service Ordering Code • User Support and Operations Centre

USOE United States Office of Education (USA)

USofA United States of America

USOP United States Office Products (USA)

USP Uniform Specification Program • Unique Selling Proposition • United States Patent (USA) • United States Pharmacopoeia (USA) • United States Property (USA) • Unit Sale Price • Universal Sampling Plan • Universal Signal Processor • Universidade de São Paulo (Brazil-SP) • University of the South Pacific • Unsolicited Proposal • Unusual Stellar Pattern • Urgent Special Report • User-Sensitive Pricing • Utility Summary Program

USPAS United States Particle Accelerator School

USPat United States Patent

Uspd Underspeed

USPDI United States Professional Development Institute (USA)

USP&FO United States Property & Fiscal Office (USA) • United States Property & Fiscal Officer

USPFO United States Property and Fiscal Office (USA) • United States Property and Fiscal Officer

USPHS United States Public Health Service (USA)

USPLSS United States Public Land Survey System (USA)

USPO United States Patent Office (USA) • United States Post Office (USA)

USPOD United States Post Office Department (USA) (now USPS)

USPS United States Postal Service (USA)

USPSUQC United States Postal Service Uniform Quality Control (USA)

USPTO United States Patent and Trademark Office (USA)

USQ Universal Structured Query Language • University of Southern Queensland (Australia-QLD) • Unsqueezed

USQBR United States Quarterly Book Review

Usr User

USR Unité de Service et de Recherche • Unité de Surveillance et de Reconfiguration • Unit Status Report • Universal Space Rectangular (projection) • Unsatisfactory Service Report • User • User Service Request • User Service Routine

USRA Universities Space Research Association (USA)

USRDA United States Recommended Daily Allowance

USRLVC United States Reusable Launch Vehicle Corporation (USA)

USRN Unité de Synchronisation de Réseau Numérique

USRP Undergraduate Student Research Programme (NASA)

USRSDC United States ROSAT Science Data Center (USA)

USRT Universal Synchronous Receiver/Transmitter

USS Sancti Spiritus (Cuba) (location identifier) • Ultrasteep Spectrum • Umbilical Separation System • Unattended Scoring System • Understanding Support System • Uniform Symbology Specification • Unique Support Structure • United States Senate (USA) • United States Service • United States Ship • United States Standard • United States Steamer • United States Steamship • Unit for Space Sciences • Unix System Service • Unmanned Space Satellite • Unstructured Software System • Upper Stage System • USA Dollar (same day) (ISO currency code) • User Support System • Utility Support Structure

USSA Uniaxial Split Sphere Apparatus • Unit for Space Sciences and Astrophysics • User Supported Software Association (UK)

USSB United States Satellite Broadcasting • United States Shipping Board (USA)

USSC United States Space Command (USA) (formerly NORAD) • Unix Starlink Software Collection

USSCS United States Soil Conservation Service (USA)

USSCt United States Supreme Court (USA)

USSD Unstructured Supplementary Services Data

USSDTS United States Spatial Data Transfer Specification

USSF United States Space Foundation (USA)

USSI United States Standards Institution (USA)

USSIE Ultraviolet Solar Spectral Irradiance Experiment

USSP ULDA Software Support Package • Universal Sensor Signal Processor • User Support Software Package

USSPA United States Student Press Association (USA)

USSPACECOM United States Space Command (USA)

USSR Union of Soviet Socialist Republics (now obsolete)

USSS United States Secret Service (USA) • United States SIGINT System • United States Steamship • Unmanned Sensing Satellite System • User Services and Systems Support

USSSP United States Science Support Program (USA)

USt Umsatzsteuer

UST Saint Augustine (USA-FL) (location identifier) • Ultra Surface Treatment • Undergraduate Space Training • Underground Storage Tank • Undersea Technology • Uniform Specification Tree • Universal Servicing Tool • Universal Standard Time • University of Science and Technology (Ghana)

USTA United States Telephone Association (USA) • United States Trademark Association (USA) (now INTA)

UStB Universitäts- und Stadtbibliothek

UStÄR Umsatzsteuer-Änderungsrichtlinien

USTC United States Tariff Commission (USA) • University of Science and Technology of China (PRC)

UStDV Umsatzsteuer-Durchführungsverordnung

UStG Umsatzsteuergesetz

USTIS Ubiquitous Scientific and Technical Information System

USTL Université des Sciences et Techniques de Lille (France) • Université des Sciences et Techniques du Languedoc (France)

U/STP Unscreened Shielded Twisted Pair

UStR Umsatzsteuer-Richtlinien

USTR United States Trade Representative

USTS United States Travel Service (USA)

USTTA United States Travel and Tourism Administration (USA) • United States Travel and Tourism Association (USA)

Usu Usual • Usually

USU Busuanga (Philippines) (location identifier) • Ural State University (Russia) • Usually • Utah State University (USA-UT)

USUHS Uniformed Services University of the Health Sciences

USV Universal Stamp Vendor • Unterbrechungsfreie Stromversorgung

USVI United States Virgin Islands

USVMS Urine Sample Volume Measuring System

usw und so weiter

USW Ultrashort Wave • Unterschlagwort

USWB United States Weather Bureau (USA)

USWG User Services Working Group

USWI United States West Indies

USWRP United States Weather Research Program (USA)

USX Ultra-Soft X-ray

Usz Undersize

Ut Utah (USA) • Utility

U/T Under Trust

UT Umbilical Tower • Uncontrolled Term • Underground Temperature • Unité de Traitement • Unité de Transport • United Technologies • Unité Terminale • Unité Territoriale • Unit Telescope • Unit Terminator • Unit Test • Unit Tester • Unit Testing • Unit Trainer • Universal Time • Universiteit Twente (Netherlands) • Université Technologique • University of Tasmania (Australia-TAS) • University of Tennessee (USA-TN) • University of Texas (USA-TX) • University of Tulsa (USA-OK) • Unorganized Territory • Unspecified Temperature • Unterer Tubus • Upper Tester • Upper Troposphere • User Terminal • User Test • Usine Thermique • Utah (USA)

UTA Mutare (Zimbabwe) (location identifier) • Undergraduate Teaching Assistant • Unidad Tributaria Anual • Unified Tracking Antenna • Unité de Transport Automobile • Unit Tax Advisor • Unit Training Area • Unit Training Assembly • Université du Troisième Âge • University of Texas at Arlington (USA-TX) • Unresolved Transition Array • Upper Terminal Area

UTB Muttaburra (Australia-QLD) (location identifier) • Universal Testbed

UTBM Université de Technologie de Belfort-Montbéliard (France)

UTC Coordinated Universal Time • Undergraduate Teaching Committee • Unified Transportation Command • Unité de Traitement Centrale • United Technologies Corporation • United Technology Centre • Unit Test Case • Unit Time Coding • Unit Type Code • Universal Test Console • Universal Time Code • Universal Time Compensated • Universal Time Constant • Universal Time Coordinated (*) • Université de Technologie de Compiègne (France) • University of Tennessee at Chattanooga (USA-TN) • User Time Clock • Utrecht Soesterberg (Netherlands) (location identifier)

UTCF Universal Time Correlation Factor

UTCQ Universal Trellis-Coded Quantization

UTCSR University of Texas Center for Space Research (USA-TX)

Utd United

UTD Nutwood Downs (Australia-NT) (location identifier) • United • Unit Training Device • University of Texas at Dallas (USA-TX) • User Terminal and Display

UTDF Universal Tracking Data Format

ut dict Ut Dictum

UT&E User Test & Evaluation • User Test & Experimentation

UTE Butterworth (South Africa) (location identifier) • Union Technique de l'Électricité • Unión Temporal de Empresas (Spain) • Unité Technologique Élémentaire • Unit Test Equipment • Universal Test Equipment • User Terminal Element • User Test and Evaluation • User Test and Experimentation

UTEC Universal Test Equipment Compiler

UTEL University of Toronto English Library (Canada-ON)

utend Utendum

UTEP University of Texas at El Paso (USA-TX)

UTES Unit Training Equipment Site

UTF UCS Transformation Format • Underwater Test Facility

UTG Quthing (Lesotho) (location identifier)

UT&GS Uplink Test & Graphics System

UTGS Uplink Test and Graphics System

UTH Udon Thani (Thailand) (location identifier) • Unité de Travail Homme • Upper Tropospheric Humidity • Up The Hill

UTI Kouvola Utti (Finland) (location identifier) • Unidad de Tratamiento Intensivo • Unité de Traitement d'Images • Universal Text Interchange • Universal Text Interface • Unordered Tree Inclusion

UTIAS University of Toronto Institute for Aerospace Studies (Canada-ON)

UTICEF Utilisation des Technologies de l'Information et de la Communication dans l'Éducation et la Formation

UTIG University of Texas Institute of Geophysics (USA-TX)

Util Utilisateur • Utilisation • Utility • Utilization

UTIL Utility

ut inf Ut Infra

UTIP User Test Instrumentation Plan

UTK University of Tennessee at Knoxville (USA-TN) • Utirik Island (Marshall Islands) (location identifier)

UTL Torremolinos (Spain) (location identifier) • Under The Limit

UTLAS University of Toronto Library Automation Systems

UTLB Unified Translation Look-aside Buffer

UTLM Up Telemetry

UTLS Upper Troposphere/Lower Stratosphere

UTM Unidad Tributaria Mensual • Unified Transport Model • Unified Type Model • Unit Training Mission • Universal Testing Machine • Universal Test Message • Universal Turing Machine • Universal Transverse Mercator (projection) • Université de Toulouse-le-Mirail (France) • Utility Technical Manual

UtMat Utilitas Mathematica (periodical code)

UTMC United Technologies Microelectronics Center

UTMOSST University of Texas McDonald Observatory Supernovae Search Team

UTMS User Total Management System

UTN Upington (South Africa) (location identifier)

UTO Unité de Traitement Obligatoire • Unité de Transformation Opérationnelle • United Towns Organization • Universal Time Observed • Utopia Creek - Indian Mountain (USA-AK) (location identifier)

UTOG University Thesis Online Group (UK)

UTOL Universal Translator-Oriented Language

UTOPIA Universal Test and Operations Interface for ATM • University of Tsukuba Online Processing of Information Archives

UTOPIO University of Texas Orbit Processor

UTP Unified Test Plan • Uniform Training Plan • Unit Test Plan • Universal Tape Processor • Unscreened Twisted Pair • Unshielded Twisted Pair (*) • Untwisted Pair • Utapao (Thailand) (location identifier) • Utility Tape Processor

UTPA University of Texas Pan American (USA-TX)

UTR Ukrainian T-shape Radiotelescope • Unité de Turboréacteur • Unprogrammed Transfer Register • User Terminal Room • Uttaradit (Thailand) (location identifier)

UTRA UMTS Terrestrial Radio Access

UTRAN UMTS Terrestrial Radio Access Network

UTRAO University of Texas Radioastronomical Observatory (USA-TX)

UTRC United Technologies Research Center (USA-CT)

UtrOv Utrechtse Sterrekundige Overdrukken (periodical code)

UTRS Universal Trouble Reporting System

UTS Ultimate Tensile Strength • Umbilical Test Set • Unified Transfer System • Unité de Travail Séparé • Universal Terminal System • Universal Test Set • Universal Test Station • Universal Thrust Stand • Universal Time-sharing System • Universal Time Standard • Universal Transport System • University of Technology in Sydney (Australia-NSW) • Unmanned Teleoperator Spacecraft • Urine Transfer System

UTSA University of Texas at San Antonio (USA-TX)

UTSI University of Tennessee Space Institute (USA-TN)

UTSO University of Toronto Southern Observatory (Chile)

ut sup Ut Supra

UTT Umtata (South Africa) (location identifier) • Université de Technologie de Troyes (France) • University of Tennessee at Tullahoma (USA-TN)

UTTC Universal Tape-to-Tape Converter

UTTR Utah Test and Training Range (USA-UT)

UTTS Universal Target Tracking Station

UTU Undergraduate Teaching Unit • Ustupo (Panama) (location identifier)

UTW Queenstown (South Africa) (location identifier)

UTX Unit Training Exercise

uu UU-encoded Compression (file extension)

uU unter Umständen

UU Universal Usability • Universiteit Utrecht (Netherlands) • University of Ulster (Northern Ireland) • Uppsala University (Sweden) • Usenet University • User Unit • Utrecht University (Netherlands)

UUA Bugul'ma (Russia) (location identifier) • UNIVAC Users Association

UUB Uppsala Universitetsbibliothek (Sweden)

UUC UKIRT Users' Committee • University of Ulster at Coleraine (Northern Ireland) • Unix-to-Unix Copy

UUCICO Unix-to-Unix Copy Incoming Copy Outgoing

UUCP Unix-to-Unix Copy Program

uud UU-encoded Compression (file extension)

UUD Ulan-Ude (Russia) (location identifier) • Unix-to-Unix Decoding

uue UU-encoded Compression (file extension)

UUE Unix-to-Unix Encoding • Use Until Exhausted

UUG Unix User Group

UUI Uniform User Interface • User-to-User Information • User-to-User Interface

UUID Unique User Identifier • Universally Unique Identification • Universally Unique Identifier • Universal Unique Identifier

UUJ University of Ulster at Jordanstown (Northern Ireland)

UUK Kuparuk (USA-AK) (location identifier)

UUN Baruun-Urt (Mongolia) (location identifier)

UUR Unit Under Review

UUS User-to-User Signaling • Yuzhno-Sakhalinsk (Russia) (location identifier)

UuStB Universitäts- und Stadtbibliothek

UUT Unit Under Test

U/UTP Unscreened Unshielded Twisted Pair

UUU Manumu (Papua New Guinea) (location identifier)

uüV unter üblichem Vorbehalt

UUV Unauthorized Use of Vehicle • Unmanned Underwater Vehicle

UUX Unix-to-Unix Execute

U/V Under Voltage

UV Miranda (satellite of Uranus) • Ultraviolet • Ultravoltage • Under Voltage • Unité de Valeur • Unit Value • Universitäts-Verlag • Unmanned Vehicle

uva und viele andere • und vieles andere

UvA Universiteit van Amsterdam (Netherlands)

UVa University of Virginia (USA-VA)

UVA Ultraviolet A • Unidad Vecinal de Absorción • Uvalde Garner (USA-TX) (location identifier)

UVAA Utah Valley Astronomy Association (USA-UT)

UVAI Ultraviolet Auroral Imager

UVAISIS USNO/ATNF VLBI Astrometry and Imaging of Southern ICRF Sources

UVAL Ultraviolet Argon Laser

UVAS Ultraviolet Astronomical Satellite

UVASER Ultraviolet Amplification by Stimulated Emission of Radiation

UVAT Ultraviolet Automatic Tracker

UVB Ultraviolet B • Ultraviolet Background (radiation)

UVBS Ultraviolet Bright Star

UVC Ultrahigh Vacuum Chamber • Ultraviolet C • Ultraviolet Camera • Ultraviolet Converter • Ultraviolet Coronagraph • Ultraviolet Court • Ultraviolet to Visible Converter

UVCI Ultraviolet and Visible-light Coronagraph Imager

UVCS Ultraviolet Coronal Spectrometer • Ultraviolet Coronagraph and Spectrometer (*) • Ultraviolet Coronagraph Spectrometer

UVD Ultraviolet Detector • Under Voltage Device

UVDD Ultraviolet Detector Development

UVDDP Ultraviolet Detector Development Program

UVE Ouvea (Loyalty Islands) (location identifier) • Ultraviolet Excess • Usine de Valorisation Énergétique

UVES Ultraviolet Echelle Spectrograph • Ultraviolet/Visible Echelle Spectrograph • Ultraviolet/Visual Echelle Spectrograph (*)

UVF San Lucia - Hewanorra (Windward Islands) (location identifier) • Ultraviolet Filter • Unmanned Vertical Flight

UVGA Ultraviolet, Visible, and Gravitational Astrophysics

UVI Cordelia (satellite of Uranus) • Ultraviolet Irradiation • Unidad de Vigilancia Intensiva (Spain) • University of the Virgin Islands

UVic University of Victoria (Canada-BC)

UVII Ophelia (satellite of Uranus)

UVIII Bianca (satellite of Uranus)

UVIL Ultraviolet Ion Laser

UVIR Ultraviolet-Infrared

UVIS Ultraviolet Imaging Spectrograph • Ultraviolet Imaging Spectrometer

UVISI Ultraviolet Visible Imager and Spectrographic Imager

UVISS Ultraviolet Italian Sky Surveyor

UVIT Ultraviolet Imaging Telescope

UVK Universal Vehicle Kit

UVL New Valley (Egypt) (location identifier) • Ultraviolet Lamp • Ultraviolet Laser • Ultraviolet Light • Ultraviolet Long

UVLIM Ultraviolet Limb Imaging

UVLMB Ultraviolet Limb

UVLS Ultraviolet Light Stabilizer • Ultraviolet Limb Spectrometer

UVM Ultraviolet Moyen • Universal Vendor Marking • Urban Vector Map • User's Virtual Machine

UVMCS Ultraviolet Multiple Channel Spectrometer

UVMS Urine Volume Measuring System

UVNO Ultraviolet Nitric Oxide Spectrometer

UVO Ultraviolet/Optical • Uvol (Papua New Guinea) (location identifier)

UVOIR Ultraviolet - Optical - Infrared

UVOT Ultraviolet/Optical Telescope

UVOWG Ultraviolet-Optical Working Group

UVP Ultrahigh Vacuum Pump • Ultraviolet Photometry • Umweltverträglichkeitsprüfung

UVPE UV/Video Processing Electronics

UVPG Umweltverträglichkeitsprüfunggesetz

UVPI Ultraviolet Plume Imager • Ultraviolet Plume Instrument

UVPROM Ultraviolet Programmable Read-Only Memory

UVPS Ultrahigh Vacuum Pumping Station

UVR Ultraviolet Radiation • Ultraviolet Radiometer • Ultraviolet Receiver • Ultraviolet Rocket • Under Voltage Relay

UVRSS Ultraviolet Rotating Shadowband Spectroradiometer

UVS Ultraviolet Sensor • Ultraviolet Spectrometer • Under Voltage Sensing • User Visiting Session

UVSC Ultraviolet Solar Constant • Utah Valley State College (USA-UT)

UVSE Ultraviolet Small Explorer • Ultraviolet Survey Explorer

UVSOR Ultraviolet Synchrotron Orbital Radiation

UVSP Ultraviolet Spectral Photometer • Ultraviolet Spectrometer and Polarimeter • Ultraviolet Spectropolarimeter • Ultraviolet Surveillance Program

UVSQ Université de Versailles/Saint-Quentin-en-Yvelines (France)

UVSST Ultraviolet Sky Survey Telescope

UVSTAR Ultraviolet Spectrograph Telescope for Astronomical Research

ÜVt Übergabeverteiler

UVT Ultraviolet Transmission • Ultraviolet Tube • Universal Voltage Tester

UVV Upward Vertical Velocity

UVVC Unit Volume Variable Cost

UVVIS Ultraviolet-Visible • Ultraviolet-Visual

UVVS Ultraviolet Visible Spectrometer

UVX Ultraviolet Excess

UVXS Ultraviolet Excess

u/w Used With

U/W Under Will • Underwriter • Used With

UW Unique Word • University of Wales (UK-Wales) • University of Washington (USA-WA) • University of Wisconsin (USA-WI) • Upstream Wave • Usable Width • Used With

UWa University of Washington (USA-WA)

UWA Ultra-Wide Angle • University of Wales at Aberystwyth (UK-Wales) • University of Western Australia (Australia-WA) • Ware (USA-MA) (location identifier)

UWB Ultrawide Band

UWC Underwater Communications • Universal Wireless Communications • University of Wales at Cardiff (UK-Wales) • University of Western Cape (South Africa)

UWCC Universal Wireless Communications Consortium • University of Wales - College of Cardiff (UK-Wales) • University of Wisconsin Computer Center (USA-WI)

UWCG University of Wisconsin Controls Group

UWE University of the West of England (UK-England) • Wiesbaden (Germany) (location identifier)

UWEC University of Wisconsin, Eau Claire (USA-WI)

UWF University of West Florida (USA-FL)

UWG User Working Group

UWGB University of Wisconsin, Green Bay (USA-WI)

UWi University of Wisconsin (USA-WI)

ÜWI Überwachungsinstrument

UWI Uniform Window Interface • University of West Indies (Jamaica)

UWIST University of Wales Institute of Science and Technology (UK-Wales)

UWL Underwater Launch

UWLR Upward Waste Light Ratio

UWM Universal Windows Manager • University of Wisconsin, Madison (USA-WI) • University of Wisconsin, Milwaukee (USA-WI)

UWNDS Upper Winds

UWO University of Western Ontario (Canada-ON)

UWP University of Wisconsin, Platteville (USA-WI)

UWPC United World Press Cooperative (USA)

UWRF University of Wisconsin, River Falls (USA-WI)

UWS University of Wales at Swansea (UK-Wales) • University of Western Sydney (Australia-NSW) • Unmanned Weather Station • User Workstation

UWSM University of Western Sydney Macarthur (Australia-NSW)

UWSP University of Wisconsin, Stevens Point (USA-WI)

UWT Unabbreviated Wavelet Transform • Undecimated Wavelet Transform • User's Working Team

Uwtr Underwater

UWY Upper Airway

ux Uxor

UX Desdemona (satellite of Uranus) • Unité X

UXI Juliet (satellite of Uranus)

UXII Portia (satellite of Uranus)

UXIII Rosalind (satellite of Uranus)

UXIV Belinda (satellite of Uranus)

UXOR UX Orionis (type)

UXV Puck (satellite of Uranus)

uy Uruguay

UY Unit Year • Uruguay (ISO country code)

UYAA Ukrainian Youth Aerospace Association (Ukraine)

UYL Nyala (Sudan) (location identifier)

UYN Yulin (China-PRC) (location identifier)

UYP Peso (Uruguay) (ISO currency code)

UYSO Unidentified Young Stellar Object

uz Uzbek (ISO language code)

Uz Uzbekistan

UZ University of Zimbabwe (Zimbabwe) • Uzbekistan (ISO country code)

UZB Uzbekistan (ISO country code)

Uzbek Uzbekistan

UZC Updated Zwicky Catalog

UZGOST Uzbek State Centre for Standardization, Metrology and Certification (Uzbekistan)

UZH Unayzah (Saudi Arabia) (location identifier)

UZM Unofficial Zone Monitoring

UZS Sum (Uzbekistan) (ISO currency code)

UZU Curuzu Cuatia (Argentina-CR) (location identifier)

u. zw. und zwar

U2 Unclassified Unlimited (distribution)

U3A Université du Troisième Age • University of the Third Age

U3M Université du Troisième Millénaire (France)

U3P Union pour la Promotion de la Propulsion Photonique (France)

V

v/ Vista • Visto

v Flight Velocity • Image Distance • Instantaneous Potential Difference • Instantaneous Voltage • Quantum Number of Vibrational Mode • Specific Volume • Speed • Vacuum • Vale • Valley • Valve • Van • Variable • Variable Absorption • Véase • Vedi • Vel • Velocity Component • Velocity Vector • Verb • Verbal • Vers • Version • Verso • Versus • Vertical • Very • Via • Vibrational Quantum Number • Vice • Vicinal • Vide • Village • Visibility • Vision • Voice • Voir • Volcano • Volt • Voltage • Volume • Volumen • vom • von • vor • Votre • Voyez • Vuelta • Vuestra • Vuestro

V Dwarf • Electric Potential • Five (Roman numeral) • Luminous Efficiency • Macroscopic Speed • Main-Sequence • Potential Difference • Potential Energy • Potential of Conservative Forces • Speed • Stokes Parameter • Tension • Usted • Vacuum • Val • Validated • Valle • Valley • Valve • Vanadin • Vanadium • Variable • Varient • Vatican (international vehicle registration code) • Vatican (postal country code) • Véase • Vector • Vehicle • Velocity • Velocity Component • Venerable • Ventana • Venus • Verb • Vers • Verse • Version • Versus • Vertical • Very • Veterinary • Via • Vice • Victor (international alphabet) • Victoria (Australia) • Vide • Ville • Virtual • Visa • Vision • Visual Acuity • Vitesse • Vocative • Voice • Voir • Volcán • Volt • Voltage • Voltmeter • Volume • Vortex • Vortrag • Vowel • Voyez • 5 (Roman numeral)

Va Vanadium • Villa • Virginia (USA)

V-A Vibroacoustic • Volt-Ampere

V/A Video/Analog • Visualization/Analysis • Volt/Ampere • Voucher Attached

VA Valeur Absolue • Valeur Ajoutée • Value Added • Value Analysis • Variable Aléatoire • Variance Analysis • Vatican City State (ISO country code) • Véhicule Aérien • Velocity at Apogee • Verwaltungsanordnung • Veteran Astronomer • Veterans Affairs • Vibroacoustic • Video Amplifier • Vie Artificielle • Virginia (USA) • Virtual Address • Visiting Astronomer • Vistas in Astronomy (periodical code) • Visual Accounting • Visual Accumulation • Visual Acuity • Visual Aid • Visualization-Analysis • Vital Area • Vitesse Absolue • Vitesse Accélérée • Vitesse Acquise • Vitesse Angulaire • Vitesse d'Atterrissage • Voice of America • Voie Aérienne • Volcanic Ash • Volt-Ampere • Vor Allem • Vostra Altezza • Votre Altesse • Vuestra Alteza • Vulnerability Assessment

VAA Vaasa (Finland) (location identifier) • Vehicle Assembly Area • Vehicular Amplifier Adapter • Veteran Amateur Astronomer • Viersener Astronomischer Arbeitskreis (Germany) • Viewpont Adapter Assembly • Virtual and Augmented Architecture (conference) • Visible Acquisition Aid • Vorarlberger Amateur Astronomen (Austria)

VAAC Vectored-thrust Aircraft Advanced Control • Volcanic Ash Advisory Centre

VAACP Visible Acquisition Aid Control Panel

VAAE Vereinigte Amateur-Astronomen Eschwege 1975 eV (Germany)

VAAP Vulnerability Analysis and Assistance Program

VAAS Virginia Association of Astronomical Societies (USA-VA)

VAB Van Allen Belts • Vehicle Assembly Building • Vertical Assembly Building • Voice Answer Back • Yavarate (Colombia) (location identifier)

VABD Van Allen Belt Densitometer

VABI Variable Area Bypass Injection

VABM Vertical Angle Benchmark

VABR Vehicle Assembly Building Repeater

Vac Vacancy • Vacant • Vacation • Vacuum • Vacuum (periodical code)

VAC Vacant • Vacuum • Value-Added Carrier • Value-Added Center • Variance At Completion • Vector Analog Computer • Vehicle Assembly and Checkout • Versatile Advection Code • Video Access Center • Vidicon Alignment Coil • Visiting Advisory Committee • Visual-Aid Console • Voies de l'Aviation Civile (periodical code) • Voltage/Alternate Current • Volts Alternating Current

VACA Volatile Aerosol Component Analysis • Volatile Aerosol Composition Analysis

Vac. Dist. Vacuum Distilled

VACF Velocity Auto-Correlation Function

VACM Vector Averaging Current Meter

VACO Vehicle Assembly and Checkout

VACP Virtual Active Control Point

Vac.Pmp. Vacuum Pump

VACR Visual Aircraft Recognition

VACTL Vertical Assembly Component Test Laboratory

VAD Valdosta Moody (USA-GA) (location identifier) • Validation Difficulty • Value-Added Dealer • Value-Added Detailer • Value-Added Distributor • Vapor-phase Axial Deposition • Velocity-Azimuth Display • Ventricular-Assist Device • Voice Activity Detection

VADAS Voice-Activated Domestic Appliance System

VADC Venting Analysis Digital Computer • Voice Analyzer and Data Converter

VADD Value-Added Disk Driver

VADE Vandenberg Automatic Decommutation Equipment • Versatile Automatic Data Exchange

VADER Video Analysis, Digitization, Enhancement and Retransmission

VADIS Value-Added Distributed-Information System

VADM Vice-Admiral • Virtual Axial Dipole Moment

VADMT VKI New Approaches in the Description and Modeling of Turbulence (periodical code)

VADS Verdix Ada Development System • Visual Assist Drive System

VADSL Very-high-rate Asymmetric Digital Subscriber Line

VADVS VKI Advanced Design of Ventilation Systems (periodical code)

VAE Validation des Acquis de l'Expérience • Vereinigte Arabische Emirate • Visible Airglow Experiment

VAEP Variable - Attributes - Error Propagation • Verify And Enter Program

VAES Voice-Activated Electronic System

VAF Valence Chabeuil (France) (location identifier)

VAFB Vandenberg Air Force Base (USA-CA)

VAFC VESA Advanced Feature Connector • VKI Axial Flow Compressors (periodical code)

VAFTAD Volcanic Ash Forecast Transport And Dispersion

VAG Varginha - Maj. Brig. Trompowsky (Brazil-MG) (location identifier) • Versicherungsaufsichtsgesetz • Vierteljahresschrift der Astronomischen Gesellschaft (periodical code) • VRML Architecture Group

VAGI VESA Advanced Graphics Interface

VAGO Vssessoiusnoie Astronomo-Geodesicheskoie Obtshestvo (Russia)

VAH Vallegrande (Bolivia) (location identifier) • Vanne d'Alimentation en Hydrogène

VAHT Valeur Ajoutée Hors Taxes

VAI Vanimo (Papua New Guinea) (location identifier) • Video-Assisted Instruction • Vol Aux Instruments • Vorticity Area Index

VAIB Vehicle Assembly and Integration Building

VAID Variable Air Inlet Duct

VAIO Video Audio Input/Output • Video Audio Integration Operation

VAIP Vietnamese Association for Information Processing (Viet Nam)

VAIS Value Added Information Service

Vaiss Vaisseau

VAK Chevak (USA-AK) (location identifier) • Vertical Assembly Kit

Val Valencia (Spain) • Valeur • Valid • Validate • Validated • Validating • Validation • Validity • Valley • Valuation • Value • Valued • Valuta • Wert

VAL Valença (Brazil-BA) (location identifier) • Valid • Validated • Validation • Validity • Valley • Value • Variable Absorption Line • Véhicule Automatique Léger • Visit Authorization Letter • Visual Approach and Landing • Voice Application Language • Vulnerability Assessment Laboratory (USA-NM)

VALD Vienna Atomic Line Database

Valdtn Validation

Valid Validate • Validation

VALID Validation

Vall Valladolid (Spain)

Vallad Valladolid (Spain)

VALM Vice-Almirante

Valn Valuation

Valpo Valparaiso (Chile)

Valpso Valparaiso (Chile)

VALSAS Variable Length Word Symbolic Assembly System

VALT Variable-stress Accelerated Life-testing Trial

VALUE Valorisation et Utilisation pour l'Europe

Vam Voltammeter

VAM Value-Added Machinery • Variational Analysis Method • Vector Airborne Magnetometer • Vertical Angle Measurement • Virtual-Access Method • Vogel's Approximation Method • Voltammeter

VAMAS Versailles Project on Advanced Materials and Standards

VAMOS Velocity And Magnetic Observations of the Sun

VAMP Validation of Model Prediction • Valise Automatique Modulaire Programmable • Vector Arithmetic Multiprocessor • Vibroacoustic Modeling and Prediction • Visual-Acoustic-Magnetic Pressure

Van Vancouver (Canada-BC)

VAN Value-Added Network • Van (Turkey) (location identifier) • Vehicle Area Network • Virtual Active Network

Vanc Vancouver (Canada-BC)

VANS Value-Added Network Service

VANT Virginia Network for Technology (USA-VA)

VANTS Voprosy Atomnoi Nauki i Tekhniki Seriia Fizika Plazmy i Problemy Upravliaemykh Termoiadernykh Reaktsii (periodical code)

VAO Suavanao (Solomon Islands) (location identifier) • Vanishing Astronomical Observers • Vie Assistée par Ordinateur • Virtual Aberration Oblivion • Virtual Activities Organization • Virtual Astronomical Observatory

Vap Vapeur

VAP Validation des Acquis Professionnels • Valparaiso (Chile) (location identifier) • Value-Added Process • Value-Added Product • Value-Added Provider • Value-Added Publishing • Vapor • Vector Analysis Power • Vehicle Antenna Position • Véhicule Automatique Planétaire • Velocity Analysis Program • Vendor Assistance Phase • Verification Acceptance Procedures • Verification and Acceptance Program • Verlagsabgabepreis • Videotext Access Point • Visitor Academic Programme • Volume Arrival Profile • Voluntary Assistance Programme (now VCP)

VAPC Variable Absolute Pressure Controller • Vector Adaptive Predictive Coding

VAPCS Verification and Acceptance Program Computer System

VAPEPS Vibroacoustic Payload Environment Prediction System

VAPGS Verification and Acceptance Program Ground System

VAPI Virtual Application Programming Interface • Visual Approach Path Indicator

Vapor Vaporization

Vapp Vitesse d'Approche

VAPP Vector And Parallel Processing (conference)

Vap. Prf. Vapor Proof

VAPS Virtual Applications Prototyping System • Virtual Avionics Prototyping System • Volume - Article - Paragraph - Sentence

var Volt-Ampere Reactive

Var Variable • Variant • Variante • Variation • Variety • Variometer • Various

VAR Value-Added Remarketer • Value-Added Reseller • Value-Added Retailer • Variable • Variable Angle Reflectance • Variance • Variance Analysis Report • Variant • Variation • Varna (Bulgaria) (location identifier) • Verification Analysis Report • Verification Analysis Review • Vertical Air Rocket • Visual Aural Range • Volt-Ampere Reactive • Votre Altesse Royale • Vuestra Alteza Real

VARCS Vehicle Application and Replacement Categorical System

VARIS Virginia Resource Information System (USA-VA)

VARMM Variables - Maths - Macros

VARR Variable Range Reflector • Visual Aural Radio Range

VARS Vending Activity Reporting System • Vertical Azimuth Reference System

VARZ Várzea

VAS Sivas (Turkey) (location identifier) • Validation Status • Value-Added Service • Vectis Astronomical Society (Isle of Wight) • Véhicule Aérospatial • Vehicle Allocation and Summary • Verein der Amateurastronomen des Saarlandes (Germany) • Vermont Astronomical Society (USA-VT) • Vibration Analysis System • Video Analysis Software • Villanova Astronomical Society (USA-PA) • Visible Acquisition Sensor • Visible Atmospheric Sounder • Visiting Astronomers Section • VISSR Atmospheric Sounder • Visual Analogue Scale • Visual Augmentation System • Vlašim Astronomical Society (Czech Rep.) • Voice-Activated System • Vulnerability Assessment System

Vasc Vascuense

VASC Virginia Air and Space Center (USA-VA)

VASCAR Visual Average Speed Computer And Recorder

VASCO Vibro-Acoustique des Structures Composées

VASE Variable Angle Spectroscopic Ellipsometer • Variable Angle Spectroscopic Ellipsometry

VASG VHDL Analysis and Standardization Group

VASI Visual Approach Slope Indicator

VASIMR Variable Specific Impulse Magnetoplasma Rocket

VASIS Visual Approach Slope Indicator System

VASOST Vacuum Solar Spectral Telescope

VASP Value-Added Service Provider • Variable Automatic Synthesis Program

VASS Visual Analysis Subsystem

VAST Variable Array Storage Technology • Versatile Automatic Specification Tester • Versatile Avionics Shop Test • Versatile Avionics System Tester • Virtual Advanced Switch Test • VLAN Advanced Switch Test

VASTC Virtual Advanced Software Technology Consortium

Vat Vatican

VAT Validation Assessment Team • Value-Added Tax • Vatican (ISO country code) • Vatomandry (Madagascar) (location identifier) • Vibration Acceptance Test • Vibroacoustic Test • Video Alias Table • Video Audio Tool • Virtual-Address Translation • Virtual-Address Translator • Visual Audio Tool • Voice-Activated Typewriter • Volontaire d'Aide Technique • Vulnerability Analysis Team

VATA Vertical Assembly and Test Area • Vibroacoustic Test Article

VatAR Specola Astronomica Vaticana Annual Reports (periodical code)

VATB Visibility - Amount - Top - Base

VatCo Specola Astronomica Vaticana Comunicazione (periodical code)

VATE Value-Added Tax Excluded • Versatile Automatic Test Equipment

VATF Vibration and Acoustic Test Facility

VATI Value-Added Tax Included

VATLS Visual Airborne Target Locator System

VatMA Specola Astronomica Vaticana Miscellanea Astronomica (periodical code)

VatOP Vatican Observatory Publications (periodical code)

VATP Vector Adaptive Transform Processing

VATPOL Vatican Polarimeter

VatRA Specola Astronomica Vaticana - Ricerche Astronomiche (periodical code)

VatRS Specola Astronomica Vaticana - Ricerche Spettroscopiche (periodical code)

VATS Vertical Assembly Test Stand • Vibration Analysis Test Set • Video-Augmented Tracking System

VATT Vatican Advanced Technology Telescope

VAU Vatukoula (Fiji) (location identifier)

VAUB Vehicle Authorization Utilization Board

VAULT Very-high-Angular-resolution Ultraviolet Telescope

VAUO Vehicle Authorization Utilization Officer

VAUS Value-Added Utilization System

VAUT Very Advanced Ultraviolet Telescope

VAV Variable Air Volume • Vava'u Lupepau'u (Tonga) (location identifier)

VAW Vardø (Norway) (location identifier)

VAWP Voice-Activated Word Processor

VAWR Vector Averaging Wind Recorder

VAWT Vertical-Axis Wind Turbine

Vax Vitesse Axiale

VAX Variable-Access Exchange • Virtual-Address Executive • Virtual-Address Extended • Virtual-Address Extension • Virtual-Array Executive

VAXBI VAX Bus Interconnect

VAZ Val d'Isère (France) (location identifier)

vb Volledige Betrekking

Vb Verb • Verbal

VB Vacío de Boyero • Variable Block • Variable (star) - Bologna • Véhicule Blindé • Verb • Verhandlungsbasis • Violet Belly • Virtual Basic • Visual Basic • Visual Binary • Vitesse Balistique • Vitesse de Balayage

VBA Variable Balance Architecture • Variable Block Architecture • Veterans Benefits Administration • Vibrating Beam Accelerometer • Visual Basic Application • Visual Basic for Applications (*)

VBAS Von Braun Astronomical Society (USA-AL)

VBB Verkehrsgemeinschaft Berlin-Brandenburg (Germany) • Very Broad Band

VBC Valid Bid Controller • Very Broad Component • Video Bandwidth Compressor • Visualization in Biomedical Computing (conference)

VBD Voice Band Data

VBE VESA BIOS Extension • Visual Basic Editor

Vbf Verschriebebahnhof

VBG Lompoc Vandenberg (USA-CA) (location identifier) • Very Blue Galaxy

VBH Volkshochschule Bad Homburg (Germany)

VBI Vertical Blanking Interrupt • Vertical Blanking Interval (*) • Video-Based Instruction • Virginia Bioinformatics Institute (USA-VA) • Virtual Binary Interface • Vital Bus Inverter

VBITS Video-Based Information Transmission System

Vbl Verbal

VBL Vertical Block Line • Voyager Biological Laboratory

VBLR Very Broad Line Region

VBLT VKI Boundary Layers in Turbomachines (periodical code)

VBM Value-Based Management • Variable Block Management • Volatile Bulk Memory

VBN Vermittelndes Breitbandnetz • Visitor-Based Network

vBNS Very-high-performance Backbone Network Service

VBNS Very-high-speed Broadband Network Services • VHS Broadband Network Services

VBO Vainu Bappu Observatory (India) • Velocity at Burnout • Verbond van Belgische Ondernemingen (Belgium) • Vertical Burnout • Views Brothers Observatory (USA-CO)

vbp Visual Basic Project (file extension)

VBP Value-Based Pricing

VBPS Variable Bandpass Filter

VBR Variable Bit Rate

vbs Visual Basic Script (file extension)

VBS Verteiltes Betriebssystem • Very Blue Star • Very Broadband Structure • Very Broad Structure • Visual Basic Script • Visual Basic Scripting • Voice Broadcast System

VBSMCVC Variable Block Size Motion Compensated Video Color

VBSP Video Baseband Signal Processor

VBSS Visit - Board - Search - Seizure

VBT Vainu Bappu Telescope (India) • Virtual Browsing Tool

Vbtm Verbatim

VBV Vanuabalavu (Fiji) (location identifier)

VBW Very Bad Wording • Video Bandwidth

VBWL Variable Block-Word-Length

vbx Visual Basic Extension (file extension)

VBX Virtual Basic Extension • Visual Basic Extension

VBY Visby (Sweden) (location identifier)

v/c Votre Compte

V/C Vector Control • Velocity Counter

VC Grenadine Islands (Saint Vincent and_) (ISO country code) • Saint Vincent and Grenadine Islands (ISO country code) • Vaisseau Capillaire • Vapor Compression • Variable Cataclísmica (estrella) • Variable Cataclysmique (étoile) • Variable Cost • Vatican City • Vector Character • Vector Coding • Vector Control • Vehicular Communication • Velocity Counter • Velocity Curve • Venture Capital • Venture Capitalism • Venture Capitalist • Version Control • Vice-Chairman • Vice-Chamberlain • Vice-Chancellor • Vice-Consul • Vicinity • Video Camera • Video Centroid • Video Communication • Video Conferencing • Video Converter • Video Correlator • Vinyl Chloride • Virgo Cluster • Virtual Call • Virtual Channel • Virtual Circuit • Virtual Connection • Virtual Container • Visiting Committee • Visual Communication • Vital Capacity • Vitesse Calibrée • Vitesse Corrigée • Voice Coil • Voice Compression • Volt-Coulomb • Voltage Comparator • Voltage Compensator • Volume Control • Votre Compte

VCA Can Tho (Viet Nam) (location identifier) • Vehicle Checkout Area • Véhicule à Coussin d'Air • Velocity Channel Analysis • Very Compact Array • Vinyl Carbonate • Virtual Channel Access • Voice Communication Assembly • Voltage-Controlled Amplifier • Voltage Control of Amplification

VCAN Vótice Ciclônico de Altos Níveis

VCAP Vehicle Charging and Potential Experiment

VCAS VCOS Application Server • Ventura County Astronomical Society (USA-CA)

VCATS Vendors Costing And Tracking System • Visually Coupled Acquisition and Targeting System

VCB Vertical of Center of Buoyancy • Video Converter Board • View Cove (USA-AK) (location identifier) • Virtuelle Campusbibliothek • VISTA Consortium Board • Visual Circuit Board • Visual Control Board

VCC Variable Ceramic Capacitor • Variable Command Count • Vehicle Command Count • Vehicle Crew Chief • Velocity Code Counter • Verification Code Counter • Very Close Call • Very Cold Cloud • Vervoer Coördinatie Centrum • VHSIC Central Computer • Victoria (Cameroon) (location identifier) • Video Coaxial Connector • Video Compact Cassette • Video Compression Code • Video Conferencing Communication • Video Conferencing Connection • Virgo Cluster Catalog • Virtual Channel Connection • Virtual Circuit Connection • Virtual Concurrent Copy • Vitrociset-CRI Consortium • Voice Communications Circuit • Voice Communications Control • Voice Control Center • Voice Control Circuit • Voltage Change Controller • Voltage Coefficient of Capacitance • Voltage-Controlled Capacitor

VCCA Video Connect Card Adapter

VCCPP VKI Combined Cycles for Power Plants (periodical code)

VCCR Vienna Convention on Consular Relations

VCCS Vega Command Control System • Voltage-Controlled Current Source

VCD Vapor Compression and Distillation • Vapor Compression Distillation • Variable-Capacitance Diode • Verification Control Document • Vernier Engine Cutoff • Vibrational Circular Dichroism • Victoria River Downs (Australia-NT) (location identifier) • Video Compact Disk • Virtual Communications Driver • Virtual COM-port Driver • Voltage Crossing Detector

VCDS Voice Communications Distribution System

VCDU Virtual Channel Data Unit

VCDX Very Compact Digital Exchange

Vce Venice (Italy)

VCE Vanne à Commande Électrique • Variable-Cycle Engine • Venezia - Marco Polo (Italy) (location identifier) • Venice - Marco Polo (Italy) (location identifier) • Virginia Center of Excellence (USA-VA) • Virtual Cluster Exploratory • Voice

VCF Valcheta (Argentina-RN) (location identifier) • Variable Crystal Filter • Velocity Correlation Function • Visuomotor Coordination Facility • Voltage-Controlled Frequency

VCFD VKI Computational Fluid Dynamics (periodical code)

VCFDI VKI Computational Fluid Dynamics for Industrial Flows (periodical code)

VCG Vapor Crystal Growth • Vector Cardiogram • Vertical at Centre of Gravity • Vertical (location) of Center of Gravity • Vice-Consul-General • Vice-Consul-Général

VCGFV VKI Computer Graphics Flow Visualization (periodical code)

VCGL Versatec Color Graphics Language

VCGS Vapor Crystal Growth System

VCH Vichadero (Uruguay) (location identifier) • Vinyl Cyclohexene

VCHAD VGA Change Display

VCHCI Vienna Conference on Human-Computer Interaction

VCHP Variable Conductance Heat Pipe

v. Chr. Vor Christus

VCI Valence Configuration Interaction • Vegetation Condition Index • Velocity Change Indicator • Verband der Chemischen Industrie (Germany) • Virtual Channel Identification • Virtual Channel Identifier • Virtual Channel Indicator • Virtual Circuit Identifier • Volatile Corrosion Inhibitor

VCID Virtual Channel Identifier • Voice-Controlled Interactive Device

VCIF Vehicle and Cargo Integration Facility

VCITE Vertical Cargo Integration Test Equipment

Vcl Vehicle

VC/L Vehicle Container/Launcher

VCL Vegetation Canopy Lidar • Vehicle Centerline • Vehicle Container/Launcher • Vertical Center Line • Vertical Component of Lift • Virtual Channel Link • Visual Component Library • Visual Computing Library

VCLC Virtual Channel Link Control

V/cm Volt/Centimeter

VCM Value Chain Management • Variable-Curvature Mirror • Vector Correlation Map • Vehicle Control Module • Ventilation Control Module • Version Compatibility Matrix • Vibrating Coil Magnetometer • Video Control Module • Vinyl Chloride Monomer • Virtual Call Manager • Virtual Channel Multiplexer • Voice Coil Motor • Volatile Condensable Material • Volatile Condensed Materials • Volatile Condensible Materials • Voronezh Crystal Massif

VCMC Vocational Curriculum Management Center (USA)

VCN Vehicle Configuration Number • Verification Change Notice • Verification Completion Notice • Virtual Circuit Number • Virtual Community Network • Visual Communications Network • Voice Communications Network

VCNO Verkenningscommissie Natuurkundig Onderzoek (Netherlands)

VCNR Voltage-Controlled Negative Resistance

Vcnty Vicinity

VCO Value Concepts Office • Vehicle Control Office • Vehicle Control Officer • Venus Climate Orbiter • Virtual Circuit Operation • Voltage-controlled Crystal Oscillator • Voltage-Controlled Oscillator • Voltage Control Operator • Voltage Control Oscillator • Voluntary Conservation Organisation

VCON Voltage-Controlled Oscillator-model Neutron

VCOS Vehicle Control and Operating System • Vehicle Control Operating System • Virgo CO Survey • Visible Catching Operating System • Visual Caching Operating System

VCP São Paulo - Viracopos (Brazil-SP) (location identifier) • Vandenberg Contract Report • Vehicle Collecting Point • Vehicle Control Point • Vehicle Control Processor • Vehicle Control Program • Velocity Control Program • Velocity Control Programmer • Venus Composition Probe • Vertical Complementarity Problem • VideoRISC Compression Processor • Virtual Channel Processor • Virtual Circuit Packet • Visual Context Processing • Visual Context Processor • Visual Control Position • Voluntary Cooperation Programme (formerly VAP)

VCPC Vienna Center for Parallel Processing (Austria)

VCPI Virtual Control Program Interface

VCPS Velocity Control Propulsion System • Verification Completion Planning Sheet

VCQ Vibrant Color Quality

Vcr Vancouver (Canada-BC)

VCR Carora (Venezuela) (location identifier) • Video Cassette Recorder (*) • Video Cassette Recording • Virtual Control Room • Visual Control Room • Volume Conflict Rename

VCRI Verification Cross-Reference Index

VCRM Verification Cross-Reference Matrix

VCRT Variable Contrast Resolution Test

Vcs Voices

VCS Vacuum Control Switch • Vane Control System • Vapor Coating System • Vapor Control System • Vapor-Cooled Shield • Vapor Cooling System • Variable Correlation Synchronization • Vehicle Cargo Ship • Vehicle Cargo System • Vehicle Computer Simulation • Vehicle Control System • Vehicular Communications System • Velocity Control System • Velocity Cutoff System • Ventilation Cutoff System • Verification Control Sheet • Vernier Control System • Versatile Console System • Video Camera System • Video Coding System • Video Communications System • Video Computer System • Video Conference System • Vidicon Camera System • VILSPA Computer Services (ESA) • Virtual Circuit Switch • Virtual-Circuit System • Virtual Command Sphere • Virtual Content Store • Visual Control Station • Visually-Coupled System • VLT Common Software • VLT Control Software • VLT Control System • Voice Command System • Voice Communication System • Voice Compression Server • Voice Controlled System • Voice Control Switch • Voltage Calibration Set

VCSEL Vertical-Cavity Surface-Emitting Laser (*) • Vertical-Cavity Surface-Emitting Laser-diode

VCSU Valley City State University (USA-ND)

VCT Grenadine Islands (Saint Vincent and_) (ISO country code) • Saint Vincent and the Grenadines (ISO country code) • Victoria (USA-TX) (location identifier) • Voice Code Translation • Voice Congestion Threshold • Voltage Control Transfer

VCTC Virtual Channel To Channel

Vctr Vector

VCTR Vector

VCU Vehicular Communication Unit • Versatile Computer Unit • Video Control Unit • Virginia Commonwealth University (USA-VA) • Voltage Control Unit

VCV Variance-Covariance

VCVS Voltage-Controlled Voltage Source

VCW Vertical Component of Weight

VCX Virtual Component Exchange

VCXO Voltage-Controlled Crystal Oscillator

V. Cz. Vera Cruz (Mexico)

VC3 Victoria Computational Cosmology Conference (Canada-BC)

vd van de • vor dem • vor der

Vd Usted • Vend • Void

V/D Voice/Data

VD Democratic Viet-Nam (ISO country code) • Vapor Density • Various Dates • Vector Displacement • Vent Debout • Video Disk • Viet-Nam (Democratic_) (ISO country code) • Virtual Data • Virtual Destination • Voltage Detector • Voltage Drop

Vda Viuda

V&DA Video & Data Acquisition • Video & Data Assembly

VDA Ovda (Israel) (location identifier) • Valve Driver Assembly • Vapor-Deposited Aluminium • Vapor Diffusion Apparatus • Variable Data Area • Verband Deutscher Autoren (Germany) • Verein für das Deutschtum im Ausland • Video and Data Acquisition • Video and Data Assembly • Video Data Assembly • Video Distribution Amplifier • Video Drawer Assembly • Visual Discriminatory Acuity

VDADC Virtual Domain Application Data Center

VDAL Variable Data-length Assembly Language

VDAM Virtual Data Access Method

VDAS Vibration Data Acquisition System

VDASD Virtual Direct Access Storage Device

VDb Virtual Database

VDB Fagernes (Norway) (location identifier) • Verification Database • Virtual Database • VISSR Database

VDBG Van den Bergh Galaxies • Van Dors Best Glory

VDBMS Video Database Management System

VDC Video Display Control • Virgo Data Center • Virtual Device Coordinates • Virtual Display Coordinates • Vitória da Conquista (Brazil-BA) (location identifier) • Voice Data Control • Voltage Direct Current • Voltage-to-Digital Converter • Volts Direct Current

VDCP Video Data Collection Program

VDCS Virtual Data Center Services

VDCT Viaduct

VDD Vehicle Detection Device • Verification Description Document • Version Description Document • Vienna Danubepier (Austria) (location identifier) • Virtual Device Driver • Virtual Display Driver • Visual-Display Data • Voice Digital Display • Wien Danubepier (Austria) (location identifier)

VDDI Voyager Data Detailed Index

VDDL Voyager Data Detailed List

VDDM Virtual Device Driver Manager

VDDP Video Digital Data Processing

VDDS Voyager Data Description Standard

VDE Valve Drive Electronics • Valverde (Canary Islands) (location identifier) • Vehicle Drive Electronics • Verband Deutscher Elektroniker (Germany) • Verband Deutscher Elektrotechniker (Germany) • Verein Deutscher Elektroniker (Germany) • Vertical Doping Engineering • Video Display Editor • Virtual Directory Engine • Visual Development Environment • Visual Display Equipment

VDEh Verein Deutscher Eisenhüttenleute (Germany)

VDEM Velocity Differential Emission Measure

VDES Voice Data Entry System

VDET Voltage Detector

VDETS Voice Data Entry Terminal System

VDEW Vereinigung Deutscher Elektrizitätswerke (Germany)

VDF Validation Data File • Variable Digital Filter • Velocity Distribution Function • VHF Direction Finder • VHF Direction Finding

VDFM Virtual Disk File Manager

VdG Van de Graaf

VDG Vertical and Direction Gyro • Video Display Generator • Virtual Data Grid

VDI Vector Display Indicator • Vendeur à Domicile Indépendant • Vendor Documentation Inventory • Verein Deutscher Ingenieure (Germany) • Vertical Display Indicator • Vidalia (USA-GA) (location identifier) • Video Device Interface • Video Display Interface • Virtual Data Interchange • Virtual Device Interface • Voice - Data - Image • Voix - Données - Image

VDIF VDI Forschungsheft (periodical code)

VDIG Vertical Display Indicator Group • Visualization and Digital Information Group

VDIN Voice - Data - Image - Navigation

VDIR Video Data Insertion and Retrieval

VDISK Virtual Disk

VDIZF VDI Zeitschriften Fortschritt Berichte Reihe Strömungstechnik (periodical code)

VDJ Verein der Deutschen Journalisten (Germany)

Vdkhr Vodokhranilishche

VDL Van Diemen's Land (Australia-TAS) • Velocity Dividing Line • VHF Data Link • Video Direct Line • Vienna Definition Language • Voice Direct Line

VDLF Variable Depth Launch Facility

VDM Vague Discrete Modeling • VAS Digital Multiplexer • Vector Dominance Model • Vector Drawn Map • Video Display Monitor • Viedma (Argentina-RN) (location identifier) • Vienna Development Method • Virtual Device Machine • Virtual Device Metafile • Virtual Display Manager • Virtual Document Management • Virtual DOS Machine • Visual Development Method

VDMA Verband Deutscher Maschinen und Anlagenbau eV (Germany)

VDMAD Virtual Direct Memory Access Device • Virtual Direct Memory Access Driver

VDME VOR with Distance Measuring Equipment

VDMS Vocal Data Management System

VDN Vector Directory Network • Virtual Domain Network

VDO Video • Virtual Dumb Observatory

VDOP Vertical Dilution Of Precision

VDP Validation Demonstration Phase • Valle de Pascua (Venezuela) (location identifier) • Variable-Delivery Pump • Vehicle Down for Parts • Velocity Dispersion Profile • Ventura Database Publisher • Verband Deutscher Papierfabriken (Germany) • Verband der Deutschen Presse (Germany) • Vertical Data Processing • Vibration Diagnostic Program • Vibration Dissociation Process • Video Datagram Protocol • Video Data Processor • Video Decompression Processor • Video Display Processor

VDPI Voyager Data Processing Instruction

VDPM Variable-Probability Detection Model

VDPS Voice Data Processor System

VDR Validation Data Repository • Variable-Diameter Rotor • VHF Data Radio • Video-Disc Recorder • Video-Disc Recording • Video-Disk Recorder • Video-Disk Recording • Villa Dolores (Argentina-CD) (location identifier) • Virtual Data Recovery • Voice Digitization Rate • Voltage-Dependent Resistor

VDRA Voice and Data Recording Auxiliary

VDRT Visual Display Research Tool

VDRZ Verband Deutscher Rechenzentren (Germany)

Vds Ustedes

VdS Vereinigung der Sternfreunde (Germany)

VDS Vadsø (Norway) (location identifier) • Variable-Depth Sonar • Vehicle Detection System • Vehicle Dynamics Simulator • Vehicular Data System • Verband Deutscher Schriftsteller (Germany) • Versatec Data Standard • Vertex Detector System • VHDL Development System • Video Display System • Video Distribution System • Viewfinder Detector System • Virtual DMA Service • Virtual Dome System • Visual Display System

VDSÄ Verein Deutscher Strahlenschutzärzte (Germany)

VDSD Virtual Distress Signaling Device

VDSI Video Distribution Subsystem Interface

VDSL Very-high-data-rate Digital Subscriber Line • Very-high-speed Digital Subscriber Line

VDSS Vehicle Dynamic Simulator System

VDT Vehicle Data Table • Video Data Terminal • Video Dial Tone • Video Display Terminal • Video Display Tube • Virtual Data Tool • Virtual Data Toolkit • Virtual Dumb Terminal • Visual Database Tool • Visual-Display Terminal

VDTOP Video Display Terminal Operator • Visual Display Terminal Operator

VDTS Video Display Supervisor • Visual Display Supervisor

VDU Refugio - Tom O'Connor (USA-TX) (location identifier) • Video Display Unit • Visual Display Unit

VDUC VAS Data Utilization Console • Video Display Unit Controller • Visual Display Unit Controller

VDV Vorbestellte Dauerwählverbindung

vdW van der Waals (interaction)

VDW van der Waals (interaction) • Venus Departure Window • Verband Deutscher Wissenschaftler (Germany) • Vibrational Distorted Wave

VDWQT Vertical Deferred Write Queue Threshold

VDZ Valdez (USA-AK) (location identifier) • Verband Deutscher Zeitschriftenverleger (Germany)

ve van een • Venezuela • Vertebratus

Ve Veuve • Voie

VE Valence Electron • Valeur Énergétique • Value Engineering • Vehicle Engineering • Velocity Error • Venezuela (ISO country code) • Vent Effectif • Verification • Vertex Element • Victory in Europe • Virtual Enterprise • Virtual Environment • Visual Efficiency • Visualisation Électronique • Vitesse d'Écoulement • Vitesse d'Éjection • Vitesse d'Entrée • Vitesse d'Évasion • Vitesse Équivalente • Vocational Education • Vol d'Essai • Voltage Efficiency • Vostra Eccellenza • Votre Éminence • Votre Excellence • Vuestra Excelencia

VEA Validation and Edit Acceptance • Validation Edit Addition

VeARI Veröffentlichungen des Astronomischen Rechen-Instituts Heidelberg (periodical code)

VEB Bolivar (Venezuela) (ISO currency code) • Variable Elevation Beam • Vehicle Enclosure Building • Vehicle Engineering Building • Vehicle Equipment Bay • Venus Entry Body • VISTA Executive Board • Volkseigener Betrieb (Germany) • Volkssternwarte Erich Bär (Germany)

VeBAB Veröffentlichungen der Sternwarte Babelsberg (periodical code)

VeBam Veröffentlichungen der Remeis-Sternwarte zu Bamberg (periodical code)

VeBB Veröffentlichungen der Universitätssternwarte zu Berlin-Babelsberg (periodical code)

VeBD Veröffentlichungen des Astronomischen Rechen-Instituts zu Berlin-Dahlem (periodical code)

VeBer Veröffentlichungen des Königlichen Astronomischen Rechen-Instituts zu Berlin (periodical code)

VeBoc Veröffentlichungen des Astronomischen Instituts der Ruhr-Universität Bochum (periodical code)

VeBon Veröffentlichungen des Astronomischen Instituts der Universität Bonn (periodical code)

Vec Vector

VEC Value Engineering Coordinator • Vector • Vocational Educational Committee

VECAM Veille Européenne et Citoyenne sur les Autoroutes de l'Information et le Multimédia

VECC Variable Energy Cyclotron Centre (India)

VECI Vehicular Equipment Complement Index • Virtual Environment Control Interface

VECIB Vehicle Engineering Change Implementation Board

VECM Vocational Education Curriculum Material

VECO Vernier Engine Cutoff

VECOS Vehicle Checkout Set • Vehicular Checkout Set

VECP Value Engineering Change Proposal

VECTA Virtual Environment Configurable Training Aid

Ved Vedova

VED Vacuum Electron Device • Verification Difficulty • Virtual Environment Display • Volontaire Européen de Développement

VEDA Validation of European Design Advances

VEDS Vandenberg Elevated Dispersion Study • Vehicle Emergency Detection System

VEE Venetie (USA-AK) (location identifier)

VEEGA Venus-Earth-Earth Gravity-Assist • Venus-Earth-Earth Gravity-Assisted

VEEI Vehicle Electrical Engine Interface

VEELR Very Extended Emission-Line Region

VEESS Vehicle Engine Exhaust Smoke System

VEF Variable Eddington Factor • Variable Electronic Filter • Variable Explanation Format • Virtual Event File

VEFCO Vertical Functional Checkout • Vertient Functional Checkout

VEFI Vector Electric Field Instrument

VEFR Vehicle Frame

VeFra Veröffentlichungen des Astronomisches Institute der Universität Frankfurt (periodical code)

Veg Vegetable • Vegetarian • Vegetation

VEG Maikwak (Guyana) (location identifier) • Virtual Environment Group • Vitesse d'Éjection des Gaz • VLBI Expert Group

VEGA Validation Expérimentale en Grille Annulaire • Vector-Evaluated Genetic Algorithm • Venus-Earth Gravity-Assist • Video Enhanced Graphics Adapter • VLBI Experiment for Geodetic Applications

VEGF Vascular Endothelial Growth Factor

VeGoe Veröffentlichungen der Universitäts-Sternwarte zu Goettingen (periodical code)

Véh Véhicule

Veh Vehicle • Vehicular

VEH Vehicle • Vehículo Eléctrico Hibrido

VeHei Veröffentlichungen des Sternwarte Heidelberg-Königstuhl (periodical code)

VEI Value Engineering Incentive • Vehicle End Item • Vocabulário Electrotécnico Internacional • Volcanic Explosivity Index • Volcano Explosivity Index

VEIL Variable Encryption and Intelligence Labelling

VEIN Vocational Education Information Network

VEIS Vector Electron and Ion Spectrometer • Vocational Education Information System

VEJ Vejle (Denmark) (location identifier)

VeJen Veröffentlichungen der Universitäts-Sternwarte zu Jena (periodical code)

VeKar Veröffentlichungen der Grossherzoglichen Sternwarte zu Karlsruhe (periodical code)

VeKaz Akademiia Nauk Kazakhskoi SSR Vestnik (periodical code)

VeKie Veröffentlichungen der Universitäts-Sternwarte Kiel (periodical code)

VeKoe Veröffentlichungen der Universitäts-Sternwarte Königsberg (periodical code)

vel Velum

Vel Vela-Velorum (constellation) • Vellum • Velocidad • Velocity • Velvet

VEL Velocity • Vernal (USA-UT) (location identifier) • Virtual Electronic Library

VELCRO Velours et Crochets

VeLei Veröffentlichungen der Universitätssternwarte zu Leipzig (periodical code)

VeLen Leningradskii Universitet Vestnik Matematika Mekhanika Astronomiia (periodical code)

VELF Velocity Filter

VEM Vasoexcitor Material • Vertical Equivalent Muon • Virtual Electron Microscope • Viscoelastic Material • Volume Emission Measure • VSCS Electronic Module

VEMASID Vehicle Magnetic Signature Duplicator

VEMM Virtual Expanded Memory Manager

VEMMI Versatile Multimedia Interface • Videotex Enhanced Multimedia Interface

VEMPS Vertical Electromagnetic Pulse Simulator

VEMR Virtual Electronic Medical Record

VeMun Veröffentlichungen der Sternwarte München (periodical code)

Ven Vendredi • Venerable • Venerdi • Venezuela • Venezia (Italy) • Ventral • Venus

VEN Venerable • Venezuela (ISO country code) • Virtual Enterprise Network

VeNav Akademiia Nauuk BSSR Vestsi Seriia Fizika Tekhnichnykh Nauuk (periodical code)

Venez Venezuela

Vent Ventilate • Ventilateur • Ventilation • Ventilator • Venture

VENUS Vehicle for Education of the Network of Universities in Space

VEO Virtual Earth Observatory • Voluntary Environmental Organisation

VEOS Versatile Electro-Optical System • Virtual-Environment Operating System

VEP Value Engineering Plan • Value Engineering Proposal • Vehicle Evaluation Payload • Video Entrance Panel • Vieillissement Exceptionnellement Prolongé • Violation of Equivalence Principle • Visually Evoked Potential

VEPIS Vocational Education Program Information System

VEPM Value Engineering Program Manager

VEPR Value Engineering Program Requirement

VEQ Vertically-loaded External Quench

Ver Verein • Vereinigung • Verification • Verified • Verify • Verse • Version • Vertical

VER Vera Cruz (Mexico) • Vera Cruz - La Bajadas (Mexico) (location identifier) • Verified • Verify • Version • Vertical

VERA Versatile Reactor Assembly • Vision Electronic Recording Apparatus • VLBI Exploration of Radio Astrometry

verb verbessert

Verb. et Lit. Verbatim et Literatim

VerbrKrG Verbraucherkreditgesetz

VERDI Victoria Explorer for Redshift Databases on the Internet

Verf Verfasser • Verfassung

VerglO Vergleichsordnung

verh verheiratet

Verif Verification • Verified • Verify

Vérif Vérification • Vérifié • Vérifier

VERI Virtual Environments Research Institute (USA-TX)

VERIF Verification

VERITAS Very Energetic Radiation Imaging Telescope Array System

Verl Verlag

VermBG Vermögensbildungsgesetz

VermG Vermögensgesetz

Vern Vernacular

VERN Vernier

Veröff Veröffentlichung

VERONICA Very Easy Rodent-Oriented Net-wide Index to Computerized Archives

VERR Verify Read • Verrière

Verrou Verrouillage

Vers Version

Versch Verschieden

Verst Versatile

Vert Vertebra • Vertebrate • Vertica (periodical code) • Vertical

VERT Venture Evaluation Review Technique • Vertical

Vertal Vertical

VERTEX Vertical Transport Experiment

Vertf Vertiflite (periodical code)

VERTIS Vehicle, Road and Traffic Intelligence Society

VERVIS Vertical Visibility

Verw Verwaltung • Verwitwet

VERW Verify Write

VerwZG Verwaltungszustellungsgesetz

Ves Vespere • Vessel

VES Vacuum Exhaust System • Vatican Emission Stars • Vertical Equivalent Size • Villa Elisa Survey

VESA Video Electronics Standards Association • Visualization and Earth Science Applications

VESAT Venus Environmental Satellite

VeSon Veröffentlichungen des Sternwarte Sonneberg (periodical code)

VESP Value Engineering Study Proposal

VESPER Venus Sounder for Planetary Exploration

VESS Vehicular Smoke System • Velocity Extraction Subsystem • Vending Equipment for Sales and Services • Visual Environment Simulation System

VeSSR Akademiia Nauk SSSR Vestnik (periodical code)

VESSS Villa Elisa Southern Sky Survey

VEST Vehicle Electrical System Test

VESTA Vecteur à Statoréacteur

Vet Veteran

VET Veteran • Video Editing Terminal • Virtual Environment for Training • Vocational Education and Training

VETA Verification Engineering Test Article • Virginia Educational Technology Alliance

VETC Vapor-Exposure Test Cell

VETT Voprosy Elektroniki Tverdogo Tela (periodical code)

VEU Virtual European University

VeUSB Veröffentlichungen der Universität Sternwarte zu Bonn (periodical code)

VEV Barakoma (Solomon Islands) (location identifier) • Vacuum Expectation Valve • Vernier Engine Vibration • Voice-Excited Voicecoder

VEVI Virtual Environment Vehicle Interface

VEW Vereinigte Edelstahlwerke

VeWFS Veröffentlichung der Wilhelm Foerster Sternwarte (periodical code)

VEx Venus Express

VEX Tioga (USA-ND) (location identifier) • Venus Express • Voice Encryption Accessory

VEXUVIO Visible Explorer and Ultraviolet Infrared Observer

VEY Vestmannaeyjar (Iceland) (location identifier)

Vf Verfasser • Veuf

V/F Voltage to Frequency

VF Vacuum Fluorescent • Vanne de Flux • Variable Frequency • Variance Factor • Variation de Force • Vector Feature • Vector Field • Velocity Failure • Velocity Field • Ventricular Fibrillation • Version Française • Vertical Flight • Very Fair • Very Fast • Very Fine • Video Frequency • Virtual Factory • Virtual Filestore • Virtual Floppy • Visiting Fellowship • Visual Field • Voice Frequency • Voie Ferrée

VFA Variation Flow Analysis • Vertical Feed Arm • Victoria Falls (Zimbabwe) (location identifier) • Video Frequency Amplifier

VFAT Virtual File Allocation Table (*) • Virtual File Attribute Table

VFBF Vite Fait - Bien Fait

VFC Variable Frequency Control • Vector Function Chainer • Version First Class • Vertical Flats Casing • V Fast Class • Video Feature Connector • Video Frequency Carrier • Video Frequency Channel • Voice Frequency Carrier • Voice Frequency Channel • Voltage-to-Frequency Converter

VFCS Vehicle Flight Control System • Vienna Fortran Compilation System

VFCT Variable-Frequency Carrier Tone • Voice Frequency Carrier Telegraph • Voice Frequency Carrier Terminal

VFD Vacuum Fluorescent Display • Variable Field Data • Verified Free Distribution

VFDR Variable Flow Ducted Rocket

VFE Vendor-Furnished Equipment

VFF Valence Force Field • Voice Frequency Filter

VFFC Variable Field Fluid Coupling

VFFH Very Fast Frequency Hopping

VFG Video Frame Grabber

Vfg Verfügung

Vfg. v. Verfügung von

VFI Verification Flight Instrumentation

VFIR Very Fast Infrared

VFK Variable Function Key

VFL Variable Field Length • Variable Focal Length

VFLFS Very Fast and Large File System

VFLO Variable Frequency Local Oscillator

VFM Value For Money • Variable Frequency Modem • Viewfinder Monitor

VFMED Variable Format Message Entry Device

VFMP Vehicle Fleet Management Program

VFN Vendor Feature Node • Verification Change Notice • Virtual File Name

VFO Variable-Frequency Oscillator

VFOAR Vandenberg Field Office of Aerospace Research

VFP Variable-Factor Programming • Variance Frequency Processor • Visiting Fellows Program

VFPA Visible Focal-Plane Assembly

VFPC Vertical Feet Per Minute

VfR Verein für Raumschifffahrt (Germany) (obsolete)

VFR Vector Relational Format • Very Far Radiation • Visual Flight Rules

VFS Variable Filter System • Vertical Full Scale • Virtual File Storage • Virtual File Store • Virtual File System • Visual Flight Rule • Volume and File Structure

VFSC Very Faint Source Catalogue

VFSS VAP Flight Simulation Software

VFST Vacuum Fine Structure Telescope (China-PRC)

VFT Verification Flight Test • Vertical Flight Test • Very Fast Train • Very Forward Tracker • Visual Function Test • Visual Function Tester • Voice Frequency Telegraph • Voice Frequency Terminal

VFTG Voice Frequency Telegraph Group

VFU Vertical Format Unit • Vertical Forms Unit

VFV Venus Flyby Vehicle

VFW Variable/Fixed Wavelength • Veterans of Foreign Wars (USA) • Video For Windows

Vfy Verify

VFY Verify

vg Verbi Gratia • vorgenannte

V/G Variable Geometry

VG British Virgin Islands (ISO country code) • Îles Vierges Britanniques (code ISO de pays) • Variable Geometry • Variable Gradient • Velocity Generator • Velocity/Gravity • Verlagsrechtsgesetz • Vertical Gyro • Verwaltungsgericht • Verwertungsgesellschaft • Very Good • Voice Grade • Voie de Garage • Voltage Gain • Volume Group • Votre Grâce • Votre Grandeur • Virgin Islands (British_) (ISO country code) • Vraie Grandeur

VGA Variable Gain Amplifier • Variable Graphics Adapter • Vertical Gyro Alignment • Very General Algorithm • Video Graphics Adapter • Video Graphics Array • Vijayawada (India) (location identifier) • Visual Graphics Array

VGB British Virgin Islands (ISO country code) • Îles Vierges Britanniques (code ISO de pays) • Virgin Islands (British_) (ISO country code)

VGC Very Good Condition • Video Graphics Controller

VGCF Voronoi Galaxy Cluster Finder

VGCS Voice Group Call System

VGD Vologda (Russia) (location identifier)

VGE Visual Gross Error • Voluntary Group of Experts

VGF Vapor Growth Facility • Voice Grade Facility

VGG Vangrieng (Laos) (location identifier) • Visual Graphics Generator

VGH Velocity - Ground range - Height • Velocity - Gust - Height • Verwaltungsgerichtshof • Very Good Health

VGI Vertical Gyro Indicator

VGIS Virtual Graphic Information System

vgl vergleiche

Vgl Vergelijk • Vergleiche • Vergleichung

VGLIS Video Guidance, Landing and Imaging System

VGML Vegetarian Meal

VGNA Video Graphics Network Adapter

VGO Vigo (Spain) (location identifier)

VGOR Vandenberg Ground Operations Requirement • Vehicle Ground Operations Requirement

VGP Vehicle Ground Point • Virtual Geomagnetic Pole

VGPI Visual Glide Path Indicator

VGPO Velocity Gate Pull Off

VGPS VLA Galactic Plane Survey

v. gr. Verbi Gratia

VGR Voyager

VGRF Variable-Gravity Research Facility

VGS General Villegas (Argentina-BA) (location identifier) • Vector and Symbol Generator • Vereinigung Gandersheimer Sternfreunde (Germany) • Versatec Graphics Software

VGSA Visitor Group Security Agreement

VGSR Vacuum Gas-Solid Reaction

VGSW VAP Ground Software

VGT La Vegas North (USA-NV) (location identifier) • Vacuum Gregory Telescope • Vegetation • Vehicle Ground Test

VGTET VKI Gas Turbine Engine Transient Behavior (periodical code)

VGTM Variable Geometry-Truss Manipulator

vgu vorgelesen-genehmigt-unterschrieben

VGV Variable Guide Vane • Véhicule à Grande Vitesse

VGVI Vereniging voor Geografische Informatie en Vastgoedinformatie

VGVT Vertical Ground Vibration Test

VGVTA Vertical Ground Vibration Test Article

VGW Variable-Geometry Wing • Veiligheid - Gezondheid - Welzijn

VGWS Video Graphic Workstation

VGZ Mocoa (Colombia) (location identifier) • Villagarzón (Colombia) (location identifier)

vh van het

vH vom Hundert

Vh Véhicule

V/H Velocity to Height

V&H Vertical & Horizontal • Vulnerability & Hardness

VH Australia (international aircraft registration code) • Vertical Height • Very Heavy • Very High • Virginia Hydrodynamics (code) • Visibilité Horizontale • Vitesse Horaire • Vitesse Horizontale • Vitesse Hyperbolique • Vol Horizontal • Volume Holographic • Vorderhaus

VHA Variable Housing Allowance • Very High Altitude

VHAA Very-High-Altitude Abort

VHARC Very High-Angular-Resolution Camera • Visible High-Angular-Resolution Camera

VHB Vergabehandbuch

VHC Saurimo (Angola) (location identifier) • Very Highly Commended • Virgo - Hydra - Centaurus (sky area) • Volatile Halocarbon Compound

VHD Video High Density

VHDA Very High Density Amorphous (ice)

VHDCI Very-High-Density Cable Interconnect

VHDL VHSIC Hardware Description Language • Video High Density Logic

VHDU Visual Hand Display Unit

VHE Very High Energy • Virtual Home Environment

VHEGR Very High Energy Gamma Ray

VHEGRA Very High Energy Gamma-Ray Astronomy

VHENTW Very-High-Energy Neutrino Telescope Workshop

VHET Very-High-Energy Telescope

VHF Very High Fidelity • Very High Frequency

VHFDF Very-High-Frequency Direction Finder

VHFF Very-High-Frequency Filter

VHFG Very-High-Frequency Generator

VHFO Very-High-Frequency Oscillator

VHFR Very-High-Frequency Receiver

VHH Volatile Halogenated Hydrocarbon

VHLL Very-High-Level Language

VHLW Vitrified High-Level Waste

VHM Vector Helium Magnetometer • Vehicle Health Maintenance • Vehicle Health Management • Vehicle Health Monitoring • Vilhelmina (Sweden) (location identifier) • Virtual Hardware Monitor

VHMS Vehicle Health Monitoring System

VHN Van Horn - Culberson County (USA-TX) (location identifier) • Vickers Hardness Number

VHO Very High Output

VHOL Very High Order Language

VHP Vacuum Hot Pressing • Very High Performance • Very High Pressure • Video-to-Hardcopy Recorder

VHPBNS Very-High-Performance Backbone Network Service

VHR Very High Reduction • Very High Resolution

VHRA Very-High-Resolution Apodizer

VHRFC Very-High-Resolution Framing Camera

VHRG Very High Ratio Gas

VHRR Very-High-Resolution Radiometer

Vhs Véhicules

VHS Variable Hard Sphere • Vertical Head Synchronization • Very High Speed • Very High State • Video Home System • Virtual Host Storage • Vitesse Hypersonique • Volkshochschule

VHSA Variable Hydrologic Source Area

VHSD Very-High-Speed Data

VHSI Very-High-Speed Integrated

VHSIC Very-High-Speed Integrated Circuit

VHST Very-High-Speed Transit

V&HT Vulnerability & Hardness Technology

VHT Very High Temperature

vHTS Virtual High Screening Throughput

VHU Véhicule Hors d'Usage

VHV Very High Vacuum

VHVC Very-High-Velocity Cloud

VHY Vichy (France) (location identifier)

VHZ Vahitahi (Tuamotu Islands) (location identifier)

vi Vietnamese (ISO language code) • Visual Editor

Vi Vitesse

VI Îles Vierges Américaines (code ISO de pays) • Subdwarf • US Virgin Islands (ISO country code) • Vancouver Island (Canada-BC) • Variable Interval • Vegetation Index • Vertical Incidence • Vertical Interval • Very Important • Victoria (Australia) • Vide Infra • Video Input • Video Integrator • Virgin Islands (US_) (ISO country code) • Virtual Impactor • Virtual Instrument • Virtual Interface • Viscosity Index • Visual Identification • Visual Impairment • Visual Information • Visual Inspection • Visual Interactive • Vitesse d'Impact • Vitesse d'Injection • Vitesse Indiquée • Vitesse Initiale • Vitesse Intermédiaire • Vocabulary Independent • Volontaire International • Volontariat International • Volume Indicator

Via Viaduct

VIA Vendir-Independent Ada • Vendors ISDN Association • Versatile Integrated Avionics • Viaduct • Video Indicate Active • Virtual Interface Architecture • Vistas In Astronomy

VIABLE Vertical Installation Automation Baseline

VIAEA Vienna International Atomic Energy Agency (periodical code)

VIAS Video Image Analyzer System

Vib Vibrant • Vibrate • Vibrating • Vibration • Vibrer

VIB Vast Interface Bulletin • Vehicle Integration Building • Vertical Integration Building • Vibraphone • Vibrate • Vibration • Villa Constitución (Mexico) (location identifier) • VTOC Index Block

VIBE Visual Information Browsing Environment

VIBES Vibration Isolation Box Experiment System • Vibration Isolation in a Box Experiment System

Vic Vicinity • Victoria (Australia)

VIC Vacuum-Insulated Contained • Variable Instruction Computer • Vast Interface Change • Vector In Command • Very Important Customer • Vehicle Identification Code • Vicenza (Italy) (location identifier) • Victoria (Australia) • Video Conferencing • Vienna International Centre (Austria) • Visitor Information Center • Vitesse Indiquée Corrigée • VLT Instrument Consortium • Vortex In Cell • Vulnerability/Incident Control • VXI Interactive Control

VICAR Video Image Communication And Retrieval (*) • Video Information Communication And Retrieval

VICD Vendor Item Control Drawing

Vice-Prés Vice-Président

Vice-Pres Vice-President

VICFD VKI Introduction to Computational Fluid Dynamics (periodical code)

VICI Velocity Indicating Coherent Integrator • Video Input Code Identifier

VICOS Verification, Integration and Checkout Software

VICS Vehicle Information and Communications System • Voluntary Interindustry Communications Standards

Vict Victoire • Victoria • Victorian • Victory

Vid Vidange • Vide • Video • Videtur • Vidua

VID Vertical Indication Display • Video • Vidin (Bulgaria) (location identifier) • Virtual Image Display • Visual Identification • Visual Information Design • Visual Interactive Debugger • Visual Interactive Debugging

VIDA VLT Imager Densified Array

VIDAC Visual-Information Display And Control

VIDAMP Video Amplifier

VIDAP Vibration Data Accuracy Program • Video Data Processor

VIDAR Velocity Integration, Detection And Ranging

VIDC Video Controller

VIDD Vertical Interval Data Detector

Vide Videtur

VIDEO Visual-Data Entry Online

VidF Video Frequency

VIDF Video Frequency

VIDI Visual-Input Detection Instrumentation

VIDINET Video on Digital Network

VIDOC Visual Information Documentation

VIDR Visual Information Display and Retrieval

VIDS VICAR Interactive Display System • Visual Information Display System • Visual Integrated Display System

VIDSMAF Visual Information Distribution System Multiaction Forum

VIE Vienna Schwechat (Austria) (location identifier) • Virtual Information Environment • Virtual Integration Exercise • Visual Information Engineering • Volontaire International en Entreprise • Volontariat International en Entreprise • Wien Schwechat (Austria) (location identifier)

VIEF Volume Integration Equation Function

VIEL Viela

Vien Vienna (Austria)

VIEP Visual Information Equipment Plan

VIES VAT Information Exchange System

VIEW Video for Exploring the World • Virtual and Interactive Environments for Workplaces (project) • Virtual Interface Environment Workstation

VIEWS Video Intelligence Enhanced Workstation System • Virtual Interactive Environment Workspace

vif Visualization Image (file extension)

VIF Valence (point) Interaction Formula • Variable Import Fee • Variance Inflation Factor • Vehicle Integration Facility • Virtual Image Facility • Virtual Interface • Virtual Interrupt Flag • Visualization Image File

VIFC Vertical Interface Frame Code

VIFIS Variable Interference Filter Imaging Spectrometer

Vig Vignette

VIG El Vigia (Venezuela) (location identifier) • Video Image Generator

ViGeo Vissha Geodeziia (periodical code)

Vign Vignette

VIGRE Vertical Integration of Research and Education • Vertically Integrated Grants for Research and Education (*)

VIGS Visual Image Generating System

VIH Vichy Rolla (USA-MO) (location identifier)

VIHE VKI Industrial Heat Exchangers (periodical code)

VII White Dwarf

VIICS Vehicular Inter/Intra Communications System

VIIRS Visible Infrared Imaging Radiometer Suite

VIJ Virgin Gorda (British Virgin Islands) (location identifier)

VIK Kavik (USA-AK) (location identifier) • Value-In-Kind • Voltage/temperature Improvement Kit

Vil Village

VIL Dakhla (Morocco) (location identifier) • Vehicle Integration and Launch • Vertical-Injection Logic • Very Important Lady • Very Important Launch • Village • Volume Imaging Lidar

VilCo Villanova University Observatory Contributions (periodical code)

Vill Villa • Village

VilOB Vilnius Astronomijos Observatorijos Biuletenis (periodical code)

VILSPA Villafranca del Castillo Satellite Tracking Station (ESA)

VILT Visible Light Tomography

VIM Vacuum Induction Melting • Variability-Induced Mover • Vendor-Independent Mail • Vendor-Independent Messaging • Vertical Improved Mail • Vertically-Integrated Mail • Video Interface Module • Virtual Interconnected Management • Vision Immersion Module • Voyager Interstellar Mission

VIMA VLT Interferometer Main Array (*) • VLT Interferometric Main Array • VLT Interferometric mode using the Medium Array

VIMAD Voice Interactive Maintenance Aiding Device

vIMF Variable Initial Mass Function

ViMF Vierer in Metallfolie

VIMIS Vertically Integrated Metal Insulated Semiconductor

VIMOS Visible Multiobject Spectrograph

VIMS Vehicle Integrated Management System • Visible Imaging Spectrometer • Visual and Infrared Mapping Spectrometer

VIMSARS Video Image Map Storage And Retrieval System

VIMT VKI Introduction to Modeling Turbulence (periodical code)

Vin Vinum

VIN Vehicle Identification Number • Vinnica (Ukraine) (location identifier)

VINCE Vendor-Independent Network Control Entity

VINCI VLT Interferometer Commissioning Instrument

VINES Virtual Networking Software • Virtual Network Software • Virtual Network System

VINT Video Integration

VIO Véhicule Interorbital • Vertical Interoperability • Video Input/Output • Virtual Input/Output

VIOC Variable Input-Output Code

VIOS Voltage Imaging Optical System

ViP Visual Programming

VIP Vacuum Image Pipeline • Vacuum-Insulated Panel • Variable Information Processing • Variable Input Phototypesetter • Vehicle Improvement Program • Vehicle Interaction Program • Vehicle Interface Processor • Versatile Information Processor • Very Important Paper • Very Important Particle • Very Important Passenger • Very Important Person • Video Inertial Pointing • Video Information Processor • Video Information Provider • Video Integrator and Processor • Video Integrator Processor • Video Interrupt Pending • Virtual Interface Processor • Virtual Internet Protocol • Virtual Interrupt Pending • Visual Information Processing • Visual Information Processor • Visual Information Projection • Visual Interaction Platform • Visual Interactive Preview • Vocabulary Improvement Project • Volcanisme Intraplaque

VIPC Vegetation Image Processing Centre

VIPER Verifiable Integrated Processor for Enhanced Reliability • Video Processing and Electrical Reduction

VIPOR Visual Investigation Program on Orbiter Operations

VIPP Variable Information Processing Process

VIPR Validation-In-Process Review • Versatile Intelligence Portable Receiver • Virtual Internet Protocol Routing

VIPS Vandenberg Impact Prediction System • VDUC Interactive Processing System • Vehicle Impact Prediction System • Video Ice Particle Sampler • Video Image Processing System • VIZIR Image Processing System • Voice Interruption Priority System

VIQ Variation In Quantity • Viqueque (Indonesia) (location identifier)

vir Virus-type Module (file extension)

Vir Virage • Virgo-Virginis (constellation) • Viridis

VIR Durban Virginia (South Africa) (location identifier) • Îles Vierges Américaines (code ISO de pays) • US Virgin Islands (ISO country code) • Victoria Imperatrix Regina • Video Indication Ready-to-active • Virginids (meteor stream) • Virgin Islands (US_) (ISO country code) • Visible and Infrared Radiometer • Visual Information Retrieval • Vulnerability/Incident Report

VIRA Venus International Reference Atmosphere

VIRAC Ventspils International Radioastronomy Centre (Latvia) (formerly VIRC)

VIRAF Visible Infrared Reflectance Absorbance Fluorescence

VIRC Ventspils International Radioastronomy Centre (Latvia) (now VIRAC)

VIRCUI Visible and Infrared Close-Up Imager

VIRD Verification and Integration Requirements Document

Virg Virginia (USA) • Vírgula

VIRGO Variability of solar Irradiance and Gravity Oscillations (SOHO) • Velocity Imager for Resolving Galaxy Origins

VIRGS VISSR Interactive Registration and Gridding System

VIRI Visible/Infrared Imager • Visual Information Retrieval Interface

VIRIS Visible-Infrared Intelligent Spectrometer

VIRL Visible and Infrared Lidar

VIRMOS Visible Infrared Multiobject Spectrograph

VIRNS Velocity Inertia Radar Navigation System

VIRR Visible and Infrared Radiometer

VIRS Vegetation Image Receiving Station • Visible and Infrared Scanner • Visible Infrared Scanner

VIRSR Visible and Infrared Scanning Radiometer

VIRT Visible to Infrared Transducer

VIRTIS Visible and Infrared Thermal Imaging Spectrometer

VIRUS Visible Replicable Ultradeep Spectrograph

Vis Verbis • Viscosity • Viseur • Visibilité • Visibility • Visible • Visit • Visite • Visité • Visited • Visiteur • Visiting • Visitor • Vista • Visual • Visuel

VIS VAT Information System • Vector Instruction Set • Vehicle Information System • Vehicular Intercommunications System • Véhicule d'Intervention et de Sauvetage • Vendor Information System • Verification Information System • Video Information System • Virtual Information Storage • Virtual Instruction Set • Visalia (USA-CA) (location identifier) • Visibility • Visible • Visible Imaging Spectrometer • Visible Imaging System • Visite d'Intégrité de Structure • Visitor Information Station • Vista • Visual • Visual Imagery System • Visual Imaging System • Visual Information Seeking • Visual Information Specialist • Visual Information Storage • Visual Information System • Visual Instruction Set • Visual Instrumentation System • Vital Information System • Voice Information System • Voice Interactive System

VISA VLT Interferometric mode using the Small Array (*) • VLT Interferometer Subarray (*) • VLT Interferometric Subarray • Voluntary Intermodal Shipping Agreement

VisAD Visualization for Algorithm Development

VISAM Virtual Index Sequential Access Method

VISAR Velocity Interferometer System for Any Reflector

VISAS Visiting Astronomers Section (ESO)

VISBD Vibration Inst Shock Vibration Digest (periodical code)

Visc Viscosity • Viscount

ViSC Visualization in Scientific Computing

VISC Visconde • Visual Information Support Center • VSOP International Science Council

ViSCA Video System Control Architecture

VISCA Video System Control Architecture

VISCSS Viscondessa

VISG Visual Information Services Group

Visi Visibility

Visigót Visigótico

VISIR VLT Imager/Spectrometer in the Infrared • VLT Imaging and Spectrometry in the mid-Infrared

VISIT Virtual Institute for Satellite Integration Training (NWS)

VISITT Vendor Information System for Innovative Treatment Technologies

VISN Virtual Integrated Sky Network

ViSP Visible Spectropolarimeter

VISP Video Imagery Standards Profile • Visual Information Systems Program

VISPA Väisälä Institute for Space Physics and Astronomy (Finland)

VISR Virginia Institute for Scientific Research (USA-VA)

VISS VHS Index Search System

VISSR Visible and Infrared Spin Scan Radiometer • Visible/Infrared Spin Scan Radiometer • Visual Infrared Spin Scan Radiometer

VIST Vista

VISTA Variable In-flight Stability Test Aircraft • Verbal Information Storage and Text Analysis • Visible and Infrared Survey Telescope for Astronomy • Visible Interferometer with Separate Telescope Assemblies • Visual Imagery Simulation Training Aid • Visual Information for Satellite Telemetry Analysis • Visual Inspection Software for Technical Applications • Visual Interface for Space and Terrestrial Analysis • Volunteers In service to America

VisUV Visible-Ultraviolet

VIS-UV Visible-Ultraviolet

VISYON Variables Intelligentes Synchronisches Optisches Netz

Vit Vitesse • Vitreous

VIT Vehicle Integration Test • Very Important Traveller • Vitoria (Spain) (location identifier)

VITA Virginia Institute for Theoretical Astronomy (USA-VA) • Volunteer Income Tax Assistance • Volunteer In Technical Assistance

VITAC Voice Input To Automatic Computer

VITAL Variably Initialized Translator for Algorithmic Languages • Virtually Integrated Technical Architecture Lifecycle

VITB Vocational Industrial Training Board

VITC Vertical Interval Time Code

VITD Vector Interim Terrain Data

Vitr Vitreous • Vitreum

VITS Vertical Interval Test Signal • Videoconferencing Interaction for Teaching Science • Video Image Tracking System

Vit.Stat. Vital Statistics

VITT Vehicle Integration Test Team

VIU Vehicle Interface Unit • Venice International University (Italy) • Video Interface Unit • Viru (Solomon Islands) (location identifier) • Voiceband Interface Unit • Voice Interface Unit

ViUkr Akademiia Nauk Ukrains'koi RSR Visnik (periodical code)

VIV Vivigani (Papua New Guinea) (location identifier)

VIVA Viajes Internacionales de Vacaciones SA • Virtual Library of Virginia

VIX Vitória - Eurico Sales (Brazil-ES) (location identifier)

VIXL Video Cross-Link

VIXS Video Information Exchange System • Video Information Transfer System

viz Namely • Videlicet

Viz Vizcaya (Spain)

VIZ Vizag (India) (location identifier)

Vizc Vizcaya (Spain)

vJ vorigen Jahres

VJ Vacuum-Jacketed • Vedāṅga Jyotisha • Victory over Japan • Visite Journalière • Vitesse de Jet

VJB Xai Xai (Mozambique) (location identifier)

VJQ Gurue (Mozambique) (location identifier)

VJS Vierteljahresschrift • Virginia Journal of Science (periodical code)

VJTI Veermata Jijabai Technological Institute (India)

VJUS Virtual Journal of Ultrafast Science

VK Verbundkatalog • Voyageur-Kilomètre

VKA Volatile Keying Assembly

VKB Visual Knowledge Builder

VKD Virtual Keyboard Device • Virtual Keyboard Driver

VKE Von Karman Equation

VKF Variable Key Function • Von Karman Facility

VKG Rach Gia (Viet Nam) (location identifier)

VKH Vitesse en Kilomètres par Heure

VKha Vestnik Khar'kovskogo Universiteta (periodical code)

VKI Verteilte Künstliche Intelligenz • Von Kármán Institute

VKie Vestnik Kievskogo Universiteta Seriia Astronomii (periodical code)

VKIFD Von Karman Institute for Fluid Dynamics

VKILV VKI Laser Velocimetry (periodical code)

VKIMT VKI Measurement and Techniques (periodical code)

VKIRT VKI Radial Turbines (periodical code)

VKISP VKI Spacecraft Propulsion (periodical code)

VKL Vegetation Killing Liquid

VKO Moscow Vnukovo (Russia) (location identifier)

VKS Vicksburg (USA-MS) (location identifier) • Visuelles Kommunikationssystem

VKT Vorkuta (Russia) (location identifier)

VKW West Kavik (USA-AK) (location identifier)

VkV Verkaufsstättenverordnung

vl Varia Lectio

VL Véhicule de Liaison • Véhicule Léger • Vent Latéral • Vertical Landing • Very Low • VESA Local • Video Logic • Viking Lander • Vila • Ville • Virtual Laboratory • Virtual Language • Virtual Life • Virtual Liquidity • Visible Light • Visual Language • Vitesse de la Lumière • Vitesse de Lancement • Vitesse de Libération • Vitesse Linéaire • Voie Lactée • Vorlesung • Vulgar Latin

VLA Vandalia (USA-IL) (location identifier) • Verification Loads Analysis • Vertical Line Array • Very Large Aircraft • Very Large Array • Very Low Altitude • Video Logarithmic Amplifier • Visual Landing Aid • Volume Licensing Agreement

VLACC Vlaamse Geautomatiseerde Centrale Catalogus

Vlad Vladivostok (Russia)

VLAGES Very-Low-Altitude Gravity Extraction System

VLAIS Very Large Array Information System

VLAM Variable Level Access Method

vLAN Virtual Local Area Network

VLAN Virtual Local Area Network

VLAT Very Large Array Telescope

vlb Ventura Library (file extension)

VLB Very Long Baseline • Verzeichnis Liefbarer Bücher • VESA Local Bus • VESA Logical Bus • Visual Laser Beam

VLBA Very Long Baseline Array

VLB Vertical-Lift Bridge

VLBC Very Large Bulk Carrier

VLBI Very Long Baseline Interferometry • Viking Lander Biological Instrument

VLBP Very Long Baseline Polarimetry

VLBW Very Low Birth Weight

VLC Valencia Manises (Spain) (location identifier) • Variable-Length Code • Variable-Length Coder • Variable-Length Coding • Verification Load Cycle • Very Large Corpus • Very Long Camera • Very Low Clearance • Very Low Cost • Viking Lander Capsule • Volcano • Vulpecula - Lyra - Cygnus

VLCC Very Large Cargo Carrier • Very Large Crude Carrier

VLCS Voltage - Logic - Current - Switching

VLCT Very Large Capacity Transport • Very Large Commercial Transport

VLCTY Velocity

VLD Vacuum Leak Detector • Valdosta (USA-GA) (location identifier) • Variable-Length Decoder

VLDB Very Large Database

VLDC Variable Length Don't Care

VLDL Very-Low-Density Lipoprotein

Vle Viale

VLE Valeur Limite d'Exposition • Valle (USA-AZ) (location identifier) • Vapor-Liquid Equilibrium • Very Large Event • Very-Low Excitation • Virtual Learning Environment • Visible Light Emission

VLED Visible-Light-Emitting Diode

VLF Variation Linéaire de Fréquence • Vertical Laminar Flow • Vertical Launch Facility • Very Large Format • Very Low Frequency • Virtual Lookaside Facility

VLFA Very Low-Frequency Array • Very Low-Frequency Astronomy

VLFN Very-Low-Frequency Noise

VLFR Very-Low-Frequency Receiver

Vlg Volgend

VLG Village • Villa Gesell (Argentina-BA) (location identifier)

VLGS Villages

VLGT Very Large Gravitational Telescope

VLHC Very Large Hadron Collider

VLI Port Vila (Vanuatu) (location identifier) • Variable Length Integer • Violation of Lorentz Invariance

VLIRG Very Luminous Infrared Galaxy

VLIS Viking Lander Imaging System

VLISM Very Local Interstellar Medium

VLIW Very Long Instruction Word

VLK Volgodonsk (Russia) (location identifier)

VLL Valladolid (Spain) (location identifier) • Very Low Luminosity • Video Layout Terminal

VLLRD Virtual Law Library Reference Desk

VLLW Very Low-Level Waste

VLLY Valley

VLM Variable Length Multiply • Very Large Memory • Very Low Mass • Villamontes (Bolivia) (location identifier) • Virginia Living Museum (USA-VA) • Virtual Loadable Module (*) • Virtual Load Module • Vortex Lattice Method

VLMC Very Low-Mass Companion

VLMF Very Low Magnetic Field

VLML Lacto-ovo Vegetarian Meal

VLMS Very Low Mass Star

Vlmtrc Volumetric

VLN Valencia (Venezuela) (location identifier) • Violin

vlnr van links naar rechts • von links nach rechts

VLNS Vehicle Land Navigation System

VLNT Violent

VLO Vallejo Stolport (USA-CA) (location identifier) • Vernieuwd Lager Onderwijs • Very Low Observable

VLOS Vertical Line Of Sight

VLOT Very Large Optical Telescope

VLP Vertical Long Period • Very Long Period • Very Long Play • Very Long Player • Vila Rica (Brazil-MT) (location identifier) • Visiting Lecturer Programme

VLPC Visible Light Photon Counter

VLPE Very-Long-Period Experiment

VLPRT Very-Low-Power Radio Transmitter

VLPS Vandenberg Launch Processing System • Visible Light Photon Source

VLR Vallenar (Chile) (location identifier) • Very Long Range • Very Low Range • Visiting Location Register • Visitor Location Register

VLRA Véhicule de Liaison, de Reconnaissance et d'Appui

VLS Valesdir (Vanuatu) (location identifier) • Vandenberg Launch Site • Vapor - Liquid - Solid • Variable Line Spacing • Véhicule Logistique Spatial • Veículo Lançador de Satélites • Velocity Lowest Selectable • Vertical Launching System • Vertical Launch System • Very Low Speed • Viking Lander System • Visible-Light Sensor • Visible-Light System

VLSAL Very-Large-Scale Application of Logic

VLSB Very Low Surface Brightness

VLSBG Very Low Surface Brightness Galaxy

VLSI Very-Large-Scale Integrated • Very-Large-Scale Integration (*)

VLSIC Very-Large-Scale Integrated Circuit (*) • Very-Large-Scale Integration Circuit

VLSIPS Very Large Scale Immobilized Polymer Synthesis

VLSISC Very Large Scale Integration Socketing Capability

VLSM Variable-Length Subnet Mask

VLSN Very Large Scale Neighborhood

VLSR Velocity of Local Standard of Rest

VLSS Visible-Light Sensor System • Visible-Light Surveillance System

VLSST Vandenberg Launch Site Support Team

VLST Vertical Launch/Landing Space Transport

vlt Vliet

Vlt Violet

VLT Variable List Table • Very Large Telescope • Very Low Threshold • Video Layout Terminal • Video Lookup Table

VLTC Very Long Time Coming

Vltg Voltage

VLTI Very Large Telescope Interferometer • Very Large Telescope Interferometry

VLTICS Very Large Telescope Interferometer Control Software

VLTP Very Late Thermal Pulse • Very Long Tensor Perturbation

VLU Velikiye Luki (Russia) (location identifier) • Video Logic Unit

VLV Valera (Venezuela) (location identifier) • Value • Valve

VLVS Voltage - Logic - Voltage - Switching

VLWI Very Long Word Instruction

VLWIR Very Long-Wave Infrared • Very Long-Wavelength Infrared

VLY Valley

VLYS Valleys

vm Vormittags

vM vorigen Monats

V/m Volts per Meter

V/M Velocity Meter

VM Vacuum Module • Value Management • Velocity Meter • Velocity Modulation • Vertical Magnet • Video Machine • View Manager • Virtual Machine • Virtual Manufacturing • Virtual Memory • Visitor Mode • Vitesse de Montée • Vitesse Maximum • Vitesse Moyenne • Vlasov-Maxwell • Voice Mail • Voice Messaging • Voice Modulation • Volatile Matter • Voltmeter • Von Neumann-Morgenstern • Voormiddag • Vorstandsmitglied • Voyager Mars

VMA Valid Memory Address • Vehicle Maintenance Area • Virtual Memory Address

VMAG Visual Magnitude

VMAP Vector Map

VMARS Video Metric Analysis and Reduction System

VMAS Vehicle Management Accounting System

VMATP VKI Modeling and Applications of Transport Phenomena in Porous Media (periodical code)

VMB Video Memory Buffer • Virtual Machine Boot • Voice Mail Box

VMBF Ventilation Mécanique Bi-Flux

VMC Velocity Modulating Camera • Velocity of Minimum Control • Ventilation Mécanique Continue • Ventilation Mécaniquement Contrôlée (*) • Venus Monitoring Camera • Vertical Microcode • VESA Media Channel • Video-Mediated Communication • Visual Meteorological Conditions • Visual Monitoring Camera • Vitesse Maximum de Contrôle • VLT Master Catalogue

VMCF Virtual Machine Communications Facility

VMCM Vector Measuring Current Meter

VMD Video Motion Detector • Virtual Manufacturing Device • Virtual Mouse Driver

VMDB Visual Meteor Database

VMDK VCOS Multimedia Development Kit

VME Valeur Moyenne d'Exposition • Versabus-Module-European • Versa Modular Eurocard • Versa Module Eurocard • Versa Module Europe • Versa Multibus Eurostandard • Vertical Memory Expansion • Villa Mercedes (Argentina-SL) (location identifier) • Virial Mass Estimator • Virtual Machine Environment • Virtual Memory Environment • Virtual Memory Extended • Virtual Memory Extension • Visibility Management File

VMF Vacuum Melting Furnace • Variable Message Format • Vehicle Maintenance Facility • Vehicle Manufacturing Facility • Vertical Maintenance Facility • Voenno-Morskoy Flot (Russia) • Voice Message Format

VMG Vector Magnetograph • Video Magnetograph (*)

VMH Vitesse Moyenne à l'Heure

VMHT VKI Methodology Hypersonic Testing (periodical code)

VMI Variable Moment of Inertia • Vegetation Monitoring Instrument • Vendor-Managed Inventory • Vibration Measurement Integrator • Virginia Military Institute (USA-VA) • Visual and Mechanical Inspection

VMIR Vandenberg Management Integration Review

VMIS VAS Model Impact Study

VMIU Virtual Memory Interface Unit

VML Vector Markup Language • Virtual Machine Language

VMM Veille Météorologique Mondiale • Virtual Machine Management • Virtual Machine Manager • Virtual Machine Monitor • Virtual Memory Management • Virtual Memory Manager • Virtual Memory Monitor • Vlaamse Milieumaatschappij (Belgium)

VMMM Virtual Machine Memory Manager

VMMPS Vehicle Management and Mission Planning System

VMO Value Management Office • Velocity-Modulated Oscillator • Very Massive Object • Vitesse de Mise en Orbite • Vitesse Maximale Opérationnelle

VMOA Valley of the Moon Observatory Association (USA-CA)

VMOIP Voprosy Metrologicheskogo Obespecheniia Izmereniia Parametrov Tekhnologicheskikh Lazerov (periodical code)

VMOS Vertical Metal Oxide Semiconductor • Virtual Memory Operating System

VMP Value Management Program • Valve de Mise en Pression • Virtual Modem Protocol • Voice Message Processor • Vol de Mise au Point

VMPM Venus Multiprobe Mission

VMQ Vote à la Majorité Qualifiée

VMR Variance-to-Mean Rate • Vela Molecular Ridge • Vertical Market Reseller • Video Mail Retrieval • Virtual

Meeting Room • Voltmeter Reverse • Volume Mixing Ratio • Volume Movement Report

VMRR Vendor Material Review Report

VMRS Voice Message Relay System

V/M/S Vision/Mission/Strategy

VMS Variable Mass System • Variable Message Sign • VAX Management System • Vehicle Management System • Vehicle Motion Sensor • Velocity Measurement System • Velocity Measuring System • Vertical Motion Simulator • Very Massive Star • Vessel Monitoring System • Vibration Measuring System • Video Modulation System • Virtual Machine System • Virtual Memory System • Visual Management System • Visual Memory System • Visual Motion Simulator • Vital Memory Storage • Voice Mail Server • Voice Mail System • Voice Management System • Voice Message System • Voice Messaging System • Voyage Management System

VMSC Voice Mail Server Center

VMSG Vening Meinesz School for Geodynamics

VMT Validate Master Tape • van den Bergh - Marschner - Terzian (catalog) • Variable Mass Theory • Very Many Thanks • Virtual Memory Technique • Volcano Mitigation Team

VMTA VKI Measurement Techniques in Aerodynamics (periodical code)

VMTF Vertical Magnet Test Facility

VMTP Versatile Message Transaction Protocol

VMTSS Virtual Machine Time Sharing System

VMU Baimuru (Papua New Guinea) (location identifier) • Voice Message Unit • Volume Maintenance Utility

VMV Vision - Mission - Value • Vitesse Maximum de Vol

VMVM Vector Measuring Current Meter

VMX Voice Message Exchange

VMZ Verkehrsmanagementzentrum

vn Viet Nam

Vn Volcán

VN Vacancy Notice • Valeur Nominale • Verbal Noun • Vereinte Nationen • Verenigde Naties • Verification Note • Version Numérique • Viet Nam (international vehicle registration code) • Viet Nam (ISO country code) • Virtual Network • Vitesse Normale • Voie Navigable • Voltmètre Numérique • Von Neumann

Vna Vienna (Austria)

VNA Saravane (Laos) (location identifier) • Vector Network Analyzer • Video Network Adapter • Video Noise Analyzer • Virtual Network Architecture

VNAP Vertical Noise Abatement Procedure

VNAS Vehicle Navigation Aiding System

VNAV Vertical Navigation

VNB Very Narrow Band

VNC Valeur Nette Comptable • Venice (USA-FL) (location identifier) • Virtual Network Computing

VNCA VTAM Node Control Application

VNCC Vandenberg Network Control Center

VND Dong (Viet Nam) (ISO currency code)

VNE Vannes (France) (location identifier)

VNESSA Voyager Neptune Encounter Science Support Activity

VNET Virtual Network

VNF Virtual Network Feature

VNG Viengxay (Laos) (location identifier)

VNGG VKI Numerical Grid Generation (periodical code)

VNIIFTRI Institute for Physical Technical and Radio-Technical Measurements (Russia)

VNIM Voyager Neptune/Interstellar Mission

VNIP Virtual Network Interactive Protocol

VNIR Very Near Infrared • Visible and Near Infrared • Visible and Near Infrared Radiometer

VNIRP Visible and Near Infrared Polarimeter

vnl Voornamelijk

VNL Variable Name Length • Via Net Loss • Virtual National Laboratory

VNLF Via Net Loss Factor

VNM Viet Nam (ISO country code) • Vitesse au Niveau de la Mer

VNMFT VKI Numerical Methods for Flows in Turbomachinery (periodical code)

VnMP Vychislitel nye Metody i Programmirovanie (periodical code)

VNMRF Virtual Nuclear Magnetic Resonance Facility (USA-WA)

VNO Value Not Obtained • Verbond van Nederlandse Ondernemingen • Vilnius (Lithuania) (location identifier) • Vitesse Normale Opérationnelle

VnPM Vychislitel naia i Prikladnaia Matematika (periodical code)

VNR Van Nostrand Reinhold • Vanrook (Australia-QLD) (location identifier) • Video News Release • Vol Non Régulier

VNRT Very Near Real Time

VNS Varanasi (India) (location identifier) • Vehicle Navigation System • Virtual Network Switching • Virtual Notebook System

VNSU Vereniging van Samenwerkende Nederlandse Universiteiten (Netherlands)

VNT Vacuum Newton Telescope • Vakuum Newton Teleskop

VNU Volontaire des Nations-Unies

vnw Voornaamwoord

VNX Vilanculos (Mozambique) (location identifier)

VNY Los Angeles - Van Nuys (USA-CA) (location identifier)

v/o Votre Compte

vo Volapuk (ISO language code) • von oben

Vo Verbo • Verso

VO Validation Office • Vanishing Observer • Vatican Observatory • Vehicle Operations • Véhicule d'Occasion • Verbal Order • Verb Object • Verordnung • Versión Original • Version Originale • Very Old • Video Output • Viking Orbiter • Virtual Object • Virtual Observatory • Virtual Organization • Visualization Option • Visual Object • Vitesse Orbitale • Voice • Votre Ordre

VOA Variable Optical Attenuator • Velocity Of Arrival • Voice Of America

VÖAA Vereinigung der Österreichischen Amateurastronomen (Austria)

VoATM Voice over Asynchronous Transfer Mode

VOB Vacuum Optical Bench • Verdingungsordnung für Bauleistungen • Versioned Object Base • Version Object Base • Vertical Optical Bench • Virtual Optical Bench • Voice Of the Business

Vobf Vorbahnhof

V°B° Visto Bueno

Voc Vocal • Vocation

VOC Verenigde Oostindische Compagnie • Vereniging voor Ordinatie en Classificatie (Netherlands) • Video Output Card • Video Output Control • Vocative • Voice Of the Customer • Voice-Operated Coder • Volatile Organic Carbon • Volatile Organic Chemical • Volatile Organic Compound (*) • Volts Open Circuit

VOCA Voice Communications Assembly

Vocab Vocabulary

VOCAR Variation Of Coastal Atmospheric Refractivity

Vocat Vocativo

VOCC Venous Occlusion Cuff and Controller • Voltage-Operated Calcium Channel

VOCOM Voice Communication

VOCP Vehicle Out-of-Commission Part

Voctl Vocational

VOCU Voice Orderwire Control Unit

VoD Video on Demand • Voice over Data

VOD Vertical Onboard Delivery • Video On Demand • Virtual Optical Disk • Voice Over Data

VODAS Voice Over Data Access Station

VODAT Voice-Operated Device for Automatic Transmission

VoDSL Voice over Digital Subscriber Line

VOE Venus Orbit Ejection • Voice Of the Employee

VOF Vatican Observatory Foundation • Virtual Observatory of the Future • Volume Of Fluid

VoFR Voice over Frame Relay

VOFR Voice Over Frame Relay

VOG Video-Oculographic • Video-Oculography • Video-Okulographie • Video-Okulographisch • Voice Over Frame • Volgograd (Russia) (location identifier)

VOGAA Voice-Operated Gain-Adjusting Amplifier

VOGAD Voice-Operated Gain-Adjusting Device

Vogtl Vogtland (Germany)

VOH Vohemar (Madagascar) (location identifier)

VOI Venus Orbital Insertion • Venus Orbit Insertion • Virtual Observatory - India • Voinjama (Liberia) (location identifier) • Vol Opérationnel aux Instruments

VOICE Vertical Observations Involving Convective Exchange Experiment

Voil Voilure

VOILA Variable Object Identification, Location and Acquisition

VoIP Voice over Internet Protocol

VOIR Veille Omnidirectionnelle Infrarouge • Venus Orbiting Imaging Radar (now Magellan) • Visualization Of Information Retrieval

VOIS Visual-Observation Instrumentation System • Visual-Observation Integration System • Voice-Operated Information System

VOK Camp Douglas - Volk (USA-WI) (location identifier)

Vol Volans-Volantis (constellation) • Volant • Volatería • Volatile • Volcán • Volcanic • Volcánico • Volcano • Volet • Voltige • Volume • Volumen • Volunteer

VOL Variable Orientation Launcher • Variations Of Latitude • Volos - Nea Anchialos (Greece) (location identifier) • Volume

Volc Volcanic • Volcánico • Volcano

VOLCAS Voice-Operated Loss Control And Suppression

Vols Volúmenes • Volumes

VOLS Vertical Optical Landing System

VOLT Virginia Outdoor Lighting Task force (USA-VA) • Visual Observations Layout Tool

VOLTAN Voltage Amperage Normalizer

Volum Volumetric

VOM Vacuum Oscillation Mechanism • Versioned Object Manager • Vitesse Optimale de Montée • Volt-Ohm-Meter

VOMD Vandenberg Operations and Maintenance Documentation

VOMP Vandenberg Operational Management Plan

VON Video Over the Net • Virtual On Net (service) • Voice Over the Net

VONA Vehicle Of the New Age

VOO Vehicle Operation Office • Virtual Observatory Object

VOOQL Virtual Observatory Object Query Language

voorpl Voorplein

VoP Voice over Packet

VOP Vertical Ozone Profile

VopDP Voprosy Dinamiki i Prochnosti (periodical code)

VopGA Voprosy Gidrodinamiki Atmosfery (periodical code)

VopK Voprosy Kibernetiki (periodical code)

VopTP Voprosy Teorii Plazmy (periodical code)

VOQ Visiting Officer Quarters

VOR Vestibular-Ocular Reflex • VHF Omnidirectional Radiorange • VHF Omnidirectional Range • VHF Omnirange • Vision Of Reality • Visual Omnirange Radius • Visual Operating Instructions • Voice-Operated Relay

VORAD Vehicle Onboard Radar

VORG Vatican Observatory Research Group (USA-AZ)

vorm vormals • vormittags

Vors Vorsitzender

VORTAC VHF Omnirange TACAN

VORTEX Venus Orbiter Radiometric Temperature sounding Experiment • Verification of the Origins of Rotation in Tornadoes Experiment

VOS Verbal Operating System • Very Old System • Viking Orbiter System • Virtual Operating System • Voice-Operated Switch • Voice Operating System • Voluntary Observing Ship (WMO) • Volunteer Observer Ship

VOSA Vistas on Organizations and Strategies in Astronomy

VOSS Vatican Observatory Summer School

VoSt Vorsteuer

VOST Version Originale Sous-Titrée

Vot Voto

VOT Verification of Operational Test • VHF Omnidirectional Test • Voice - Onset - Time • VOR Test • Votuporanga (Brazil-SP) (location identifier)

VOTA Vibration Open Test Assembly

VoTAS Voprosy Teorii Atomnykh Stolknovenii (periodical code)

VOTE Vortex Ozone Transport Experiment

Vou Voucher

VOU Virtual Online University

VOV Video Output Voltage

VOWB Vlaams Overlegorgaan voor Wetenschappelijk Bibliotheekwerk

VOX Voice • Voice Actuation • Voice Extension • Voice-Operated Transmission • Voice-Operated Transmit • Voice-Operated Transmitter

VOXEL Volume Picture Element

VOY Voyage

VOZ Voronezh (Russia) (location identifier)

vp Ventura Publisher (file extension)

Vp Vitesse de Périgée

VP Falkland Islands (international aircraft registration code) • Islas Malvinas (international aircraft registration code) • Vacuum Pump • Vaglia Postale • Valve à Pointeau • Vanishing Point • Vanne Pneumatique • Vapor Phase • Vapor Pressure • Variable Physique • Variable Pitch • Variable Procedure • Variation Périodique • Vāyu Purāna • Vector Process • Vector Processing • Vector Processor • Vector Program • Véhicule Porteur • Velocity Profile • Vent Pipe • Verb Phrase • Verification Plan • Verification Polarization • Vertex Point • Vertical Pipe • Vertical Parallelism • Vertical Polarization • Very Poor • Vía Preferente • Vice President • Vice-Président • Vice-Principal

• Video Processor • Viewing Period • View Point • Ville de Paris (France) • Virgin Islands (international aircraft registration code) • Virtual Path • Virtual Pool • Virtual Processor • Virtual Prototyping • Vision Processing • Visite Périodique • Visual Programming • Vital Point • Vitesse de Pointe • Vitesse Parabolique • Vitesse Périphérique • Vitesse Propre • Vlasov-Poisson • Volume Partiel • Vuestra Paternidad • Vulnerable Point

VPA Vanne de Prélèvement d'Air • Variable Profile Antenna • Vehicle Power Adapter • Vendor Payment Authorization • Veteran Professional Astronomer • Virtual Personal Assistant • Virtual Physics and Astronomy (project) • Virtual Population Analysis

VPAA Vice President for Academic Affairs

VPAM Virtual Partitioned Access Method

VPB Variable Procrustean Bed • Version Planning Board

VPBA Varipolarization Beacon Antenna

VPC Vacuum Pump Chamber • Vapor-Phase Chromatography • Vente Par Catalogue • Vente Par Correspondance • Video Processor Control • Viewport Coordinate • Virgo Photometric Catalogue • Virtual Path Connection • Virtual Physics Centre • Virtual Physics Council

VPCA Video Prelaunch Command Amplifier

VPCAR Vapor Phase Catalytic Ammonia Removal

VpD Vehicles per Day

VPD Vapor-Phase Decomposition • Vapor Pression Deficit • Variable Power Divider • Vector-Point Diagram • Vehicles Per Day • Version Planning Document • Vertically Polarized Dipole • Virtual Printer Device • Vital Product Data • Volume Partition Descriptor

VPDN Virtual Private Data Network

VPDS Virtual Private Data Service

VPE Ongiva (Angola) (location identifier) • Vapor-Phase Epitaxy • Vernetztes Polyäthylen • Video Port Extension • Video Processing Equipment • Virtual Prototyping Environment • Visualization for Planetary Exploration • Visual Programming Environment

VPES Virtual Polyelastic Sample

VP-F Falkland Islands (international aircraft registration code)

VPF Falkland Islands (international aircraft registration code) • Vacuum Pump Filter • Variable Phase Filter • Vector Product Format • Vertical Processing Facility • Void Probability Function • Voigt Profile Fitting

VPFG Variable Phase Function Generator

VPG Virtual Proving Ground

VpH Vehicles per Hour

VPH Vehicles Per Hour • Vertical Photography • Volume-Phase Holographic

VPHD Vertical Payload Handling Device

VPHG Volume Phase Holographic Grating

VPI Valve Position Indicator • Vapor-Phase Inhibitor • Versatec Parallel Interface • Vibration Pattern Imager • Virginia Polytechnic Institute (USA-VA) • Virtual Path Identifier • Virtual Path Indicator • Virtual Private Intranet

VPICD Virtual Programmable Interrupt Controller Driver

VPIDV VKI Particle Image Displacement Velocimetry (periodical code)

VPIM Virtual Personal Information Manager • Voice Profile for Internet Mail

VPI&SU Virginia Polytechnic Institute & State University (USA-VA)

VpK Vehicle per Kilometer

VPK Vehicle Per Kilometer • Volts Peak

VPL View Program Listings • Virtual Path Link • Virtual Physics Laboratory • Virtual Planetary Laboratory • Virtual Private Line • Virtual Processing Laboratory • Virtual Programming Language • Visual Process Language • Volkssternwarte und Planetarium Laupheim eV (Germany)

VP-LA Anguilla (international aircraft registration code)

VPLCC Vehicle Propellant Loading Control Center

VP-LMA Montserrat (international aircraft registration code)

VP-LV Virgin Islands (international aircraft registration code)

VpM Vehicles per Mile • Vibrations per Minute • Volts per Meter

VPM Vacuum Pumping Module • Variability/Proper Motion (survey) • Vehicle Project Manager • Vehicles Per Mile • Vertical Polarization Mode • Vibrations Per Minute • Video Port Manager • Virtual Parallel Machine • Virtual Project Management • Volts Per Meter (*) • Volts Per Mile

VPMS Variability/Proper Motion Survey

VPN Virtual Page Number • Virtual Private Network • Virtual Private Networking • Vopnafjördur (Iceland) (location identifier)

Vpo Valparaiso (Chile)

VPO Vapor Pressure Osmometry • Vernonia Peak Observatory (USA) • Virtual Projection Origin • Vivienda de Protección Oficial

vPoP Virtual Point of Presence

VPP Value Payable Post • Vehicle Processing Phase • Volume Purchase Plan • Voluntary Protection Program

VPPAW Variable-Polarity Plasma Arc Welding

VPPSEQT Virtual Pool Parallel Sequential Threshold

VPR Variable Processing Register • Video Production Recorder • Voluntary Price Reduction

V. Pres. Vice-President

VPRF Variable Pulse Repetition Frequency

VPRN Virtual Private Remote Networking

VpS Vibrations per Second

VPS Destin (USA-FL) (location identifier) • Fort Walton Beach (USA-FL) (location identifier) • Vacuum Plasma Spraying • Vacuum Pumping System • Vakuum - Plasma - Spritzen • Valparaiso (USA-FL) (location identifier) • Variable Pump System • Variable Parameter System • Variable Phase Shifter • Vehicle Peculiar Supplement • Vernier Propulsion System • Versatec Plot Server • Vibrations Per Second • Vibrator Power Supply • Video Processing System • Video Program System • Voice Processing System

VPSA Vehicle Processing Storage Area

VPSB Virtual Parts Supply Base

VPSC Vault - Process - Structure - Configuration

VPSEQT Virtual Pool Sequential Threshold

VPT Vente Par Téléphone • Violence Prevention Team • Virtual Path Terminator • Virtual Printer Technology • Virtual Print Technology • Virtual Private Trunking

VPTS Voice Processing Training System

VPU Vacuum Penetration Unit • Video Processing Unit • Video Production Unit • Vision Processing Unit • VME Pattern Unit

VPUS Vacation Programme for University Students

VPW Vente Par WWW • Vertically Polarized Wave

VPX Video Pixel

VPZ Valparaiso - Porter County (USA-IN) (location identifier)

VQ Saint Helena and Ascension (international aircraft registration code) • Turks and Caicos (international aircraft registration code) • Vector Quantization • Vector Quantizer • Ventilateur de Queue

VQA Vibrating Quartz Accelerometer • Voice Quality Analyzer

VQC Vector Quantization Coding

VQF Vector Quantization Format

VQL Variable Quantizing Level • Virtual Query Language • Visual Query Language

VQM Valeur Quadratique Moyenne • Virginia Quilt Museum (USA-VA)

VQP Variable Quadratic Propagator

VQS Vapor Quality Sensor • Vieques (Puerto Rico) (location identifier) • Visual Query System

VQ-T Turks and Caicos Islands (international aircraft registration code)

VQT Turks and Caicos Islands (international aircraft registration code)

v/r Valor Recibido

vr Vrouwelijk

Vr Rotation Speed

V-R Vibrational-Rotational

V/R Very Respectfully • Virtual Reality

VR Bermuda (international aircraft registration code) • Cayman Islands (international aircraft registration code) • Gibraltar (international aircraft registration code) • Hong Kong (international aircraft registration code) (obsolete) • Valeur Résiduelle • Valve de Remplissage • Vanne de Régulation • Variability Reduction • Variable Reluctance • Variable Response • Vedi Retro • Veer • Véhicule de Reconnaissance • Velocity Ratio • Vent Relatif • Verlagsrecht • Vetenskapsrådet • Via Rápida • Video Recorder • Virtual Reality • Virtual Route • Visiting Researcher • Visual Reference • Vitesse de Rafale • Vitesse de Rapprochement • Vitesse de Rentrée • Vitesse de Rotation • Vitesse de Roulage • Vitesse Radiale • Vitesse Relative • Voice Recognition • Voie Routière • Voiture Radio • Vol de Réception • Vol Rectiligne • Vol Régulier • Voltage Reference • Voltage-Regulated • Voltage Regulator • Voltage Relay • Vorstandsrat

VRA Varadero (Cuba) (location identifier) • Vertical Reference Altitude • Vertical Resolution Advisory • Viking Radar Altimeter • Virtual Resources Association • Voltage Reference Amplifier

VRAM Variable Random Access Memory • Video Random Access Memory • Virtual Random Access Machine • Virtual Random Access Memory

VRAP Visiting Research Assistant Professorship

VRAS Verification Requirements Analysis Sheet

VRASP Virtual Reality Alliance of Students and Professionals

VR-B Bermuda (international aircraft registration code)

VRB Variable • Vero Beach (USA-FL) (location identifier) • VHF Recovery Beacon

VRBL Variable

VR-C Cayman Islands (international aircraft registration code)

VRC Cayman Islands (international aircraft registration code) • Variable Rate Coil • Vehicular Radio Communications • Vertical Red Clump • Vertical Redundancy Character • Vertical Redundancy Check • Viewing Reference Coordinate • Virac (Philippines) (location identifier) • Visual Record Computer • Voice Response Controller

VRCS Vernier Reaction Control System

VRD Vanne Régulatrice de Débit • Virtual Retinal Display • Vitesse de Rotation au Décollage • Voies et Réseaux Divers

VRDA Vereda

VRDI Virtual Raster Display Interface

VRE Virtual Reality Environment • Visual Recognition Engine • Voltage-Regulated Extended • Vredendal (South Africa) (location identifier)

VREO Visiting Researcher Exchange and Outreach (program) (USRA)

VRF Versatec Random Format • Vertical Removal Fixture • Vestibular Research Facility • Visual-Recording Facility • Visual-Reproduction Facility

VRFD Virtual Reality Flight Deck

Vrfy Verify

VRFY Verify

VR-G Gibraltar (international aircraft registration code)

VRG Gibraltar (international aircraft registration code) • Variable Report Generation • Vorruhestandsgesetz

VRGC Voucher Register and General Control

VR-H Hong Kong (international aircraft registration code) (obsolete)

VRH Hong Kong (international aircraft registration code) (obsolete) • Variable Range Hopping

VRI Vanguard Research Inc. (USA-CA) (formerly JSE) • Virtual Radio Interferometer • Visual - Red - Infrared • Visual Rule Instrument

VRIFS Vector Recurrent Iterated Function System

VRISL Vancouver Island (Canada-BC)

VRK Varkaus (Finland) (location identifier) • Video Recorder Kit

VRL Vertical Recovery Line • Vertical Reference Line • Vila Real (Portugal) (location identifier)

VRLA Valve-Regulated Lead-Acid (battery)

VRLI Virtual Reality Laboratories Inc. (USA-CA)

VRLSA Virtual-Reality Large-Scale Application

VRM Variable-Rate Mortgage • Variance Review Meeting • VCOS Resource Manager • Venus Radar Mapper (formerly VOIR - now Magellan) • Verification Requirements Matrix • Video Record Machine • Virtual Register Machine • Virtual Resource Management • Virtual Resource Manager • Voltage Regulator Module

VRML Virtual Reality Markup Language (*) • Virtual Reality Modeling Language

VRMP Voltage Reference Maintenance Program

VRMS Voltage Root Mean Square

VRN Vernier • Verona (Italy) (location identifier)

VRNCC Vandenberg Range Network Control Center

vrnl van rechts naar links • von rechts nach links

VRNR Vernier

Vro Ventisquero

VRO Vehicle Registration Office • Vermilion River Observatory • Very Red Object

VROM Vocabulary Read-Only Memory • Volkshuisvesting Ruimtelijke Ordening en Milieubeheer (Netherlands)

VROOMM Virtual Real-time Object-Oriented Memory Manager

VRP Validation RDF Parser • Variability Reduction Program • Vector-to-Raster Processor • Vehicle Routing Problem • Viewing Reference Point • Visual Reporting Point

VRPM Visible Radiation Polarization Monitor

VRPS Voltage-Regulated Power Supply

VRPV Voie Rapide du Piémont des Vosges (France)

VRR Verification Readiness Review • Visual Radio Range • Voice Recognition Response

VRRP Virtual Router Redundancy Protocol

vrs Versus

VRS Vacuum Resources System • Vehicle Registration System • Velocity Response Shape • Vendor Rating System

• Verkehr Raumfahrt und Systemtechnik GmbH (Germany) • Versailles (USA-MO) (location identifier) • Verse • Video Reception System • Video Recognition System • Video Recording System • Video Reference Service • Video Relay System • Video Response System • Virtual Reality and Simulation • Virtual Reality Society • Virtual Route Switching • Visual Reference System • Voice Recording System • Voice Response System

VRSA Voice-Reporting Signal Assembly

VRSD Verification Requirements and Specifications Document

VRSS Virginia Remote Sensing Study

VRST Virtual Reality Software and Technology

Vrsx Versoix (Switzerland)

Vrt Vert

VRT Video Recording Tape • Voltage Reduction Technology • Voltage Reference Tube • Voltage Regulation Technology (*) • Voltage Regulator Tube

VRTD Verification Requirements Traceability Document

VRTM Vacuum Resin Transfer Molding

VRTMOTN Vertical Motion

VRU Velocity Reference Unit • Version - Release - Update • Vertical Reference Unit • Voice Response Unit • Voltage Readout Unit • Vryburg (South Africa) (location identifier)

VRUI Virtual-Reality User Interface

VRUMI Virtual Reality User Machine Interface

VRY Vaerøy (Norway) (location identifier)

vs Verse • Versus • Vide Supra

Vs Vous

V/S Vertical Speed

VS Vacuum Switch • Vaisseau Sanguin • Vanne de Sécurité • Vanne de Surpression • Vapeur Saturée • Variable Speed • Variable Sweep • Variant Selector • Vehicle Station • Véhicule de Sauvetage • Véhicule de Service • Véhicule Spatial • Velocity Search • Ventilateur de Sustentation • Ventilation System • Vent Solaire • Verbal Scale • Verenigde Staten • Verification Specification • Versus • Verteilte Systeme • Vertical Speed • Vertical Stereoscopic • Vertical Structure • Vertical Sync • Very Slow • Vestigial Sideband • Video Switch • Vide Spatial • Vide Supra • Vie de Stockage • Vie Scolaire • Vieux Style • Virialized System • Virtual Scheduling • Virtual Simulation • Virtual Single • Virtual Source • Virtual Storage • Viseur Stabilisé • Visite de Sécurité • Visite Systématique • Visual Storage • Vitesse à la Sortie • Vitesse de Sécurité • Vitesse de Sortie • Vitesse du Son • Vitesse-Sol • Vocational Studiescessing • Voice Secure • Voile Solaire • Volatile Solid • Vol Spatial • Vol Stationnaire • Voltmeter Switch • Volumetric Solution • Vuestra Señoría

VSA Vacuum Society of Australia • Validity and Statistical Analysis • Variable-Stability Aircraft • Velocity Sensor Antenna • Vendor Security Analyst • Verenigde Staten van Amerika • Verification Site Approval • Vernier Solo Accumulator • Vertical Sensor Assembly • Very Small Array • Villahermosa (Mexico) (location identifier) • Virtual Server Architecture • Virtual Storage Access • Volkssterrenwacht Amsterdam (Netherlands) • Voltage-Sensitive Amplifier • Volume Stimulus Actuator

VSAM Video Surveillance and Activity Monitoring • Virtual Sequential Access Method • Virtual Storage Access Method • Virtual System Access Method

VSAR Variable Synthetic Aperture Radar • Versatile Synthetic Aperture Radar

VSASSY Very Small Array Sky Survey

VSAT Very Small Angle Tagger • Very Small Aperture Terminal

VSB Vast System Bulletin • Vereinigten Staatsbibliotheken (Germany) • Vertical Sideband • Vestigial Sideband (*) • Visible • VME Serial Bus • Volkssternwarte Bonn (Germany) • Volkssterrenwacht Bussloo (Netherlands)

Vsbl Visible

VSBR Visual Surface Brightness Ratio

Vsby Visibility

VSBY Visibility

VSC Variable Speed Coupling • Vast System Change • Vehicle Support Contractor • Very Small Cell • Vibration Safety Cutoff • Video Standards Council • Virtual Science Centre • Visiting Scientists Committee (ESO) • VISTA Science Committee • Volatile Sulphur Compound • Voltage-Saturated Capacitor • Volumes - Seats - Circulation

VSCE Variable Stream Control Engine

VSCF Variable Speed Constant Frequency

VSCS Vertical Stabilizer Control System • Voice Switching and Control System

VSD Variable Spark Duration • Variable-Speed Drive • Vendor's Shipping Document • Vendredi - Samedi - Dimanche • Vertical Situation Display • Video Symbology Display • Visualisation - Saisie - Déport • Volkssternwarte Darmstadt eV (Germany)

VSE Vector Symbol Editor • Velocity - Silence - Efficiency • Virtual Storage Extended • Virtual Storage Extension • Virtual System Environment • Virtual System Extended • Viseu (Portugal) (location identifier) • Visite Selon État • Vol Sans Escale

VSEL Vickers Shipbuilding & Engineering Ltd.

VSELP Vector-Sum Excited Linear Prediction

VSEP Visiting Students Enrichment Program • Visiting Summer Enrichment Program

VSEPR Valency Shell Electron Pair Repulsion

VSF Springfield (USA-VT) (location identifier) • Variable Star Field • Vertical Scanning Frequency

VSG Lugansk (Ukraine) (location identifier) • Vernier Step Gauge • Very Small Gain • Very Small Grain • Very Steep Gradient

VSGC Virginia Space Grant Consortium (USA-VA)

VSH Vaisseau Spatial Habité • Variable Spring Hanger

V/SHM Vector/Scalar Helium Magnetometer

VSHM Vector-Scalar Helium Magnetometer

VSHPS Vernier Solo Hydraulic Power System

VSI Vents Stellaires en Interaction • Vertical Speed Indicator • Very Seriously Ill • Video Simulation Interface • Virtual Socket Interface • VLBI Standard Interface • Volcanological Survey of Indonesia • Voluntary Separation Incentive • Voluntary Severance Incentive • Vuestra Señoría Ilustrísima

VSII Very Seriously Ill or Injured

VSIMM Video Single Inline Memory Module

VSIO Virtual Serial Input/Output

VSIP Valence State Ionization Potential • Voluntary Separation Incentive Pay • Voluntary Separation Incentive Program

VSIPL Vector/Signal/Image Processing Laboratory (DARPA) • Vector/Signal/Image Processing Library

VSK Volkssternwarte Köln (Germany)

VSL Van Swinden Laboratorium (Netherlands) • Variable Safety Level • Variable Speed of Light • Véhicule Sanitaire Léger • Very Strong Line • Very Strong-Lined • Virtual Shareware Library • Viscous Shock Layer

VSLAN Verdix Secure Local Area Network

VSLED Vibration Structural Life and Engine Diagnostics

VSM Variable Screening Model • Vector Space Model • Vector Spectromagnetograph • Véhicule Sous-Marin • Verein Schweizerischer Machinen-Industrieller (Switzerland)

• Vestigial Sideband Modulation • Vibrating Sample Magnetometer • Video Switching Matrix • Virtual Shared Memory • Virtual Storage Management • Virtual Storage Manager • Virtual Storage Memory • Visual System Management • Volkssternwarte Mainz (Germany) • Vol Sans Moteur

VSMF Vendor Specification Microfile • Visual Search Microfilm File

VSMSC Virtual Short Message Service Center

VSN Video Switching Network • Virtual Station Number • Vision • Volkssternwarte Norderstedt eV (Germany) • Volume Serial Number

VSNA Volontaire du Service National Actif

VSNE Volontaire du Service National à l'Étranger

VSNET Variable Star Network

VSNL Videsh Sanchar Nigam Ltd. (India)

VSNS Virtual School of Natural Sciences

VSNU Vereniging van Samenwerkende Nederlandse Universiteiten (Netherlands)

VSO Phuoclong (Viet Nam) (location identifier) • Verb - Subject - Object • Vernieuwd Secundair Onderwijs • Very Stable Oscillator • Very Superior Old • Virtual Solar Observatory • Virtual Storage Option • Visual Satellite Observer

VSOLJ Variable Star Observing League of Japan

VSOP Very Superior Old Pale • VLBI from Space Orbiting Platform • VLBI Science Observatory Program • VLBI Space Observatory Program (now HALCA) (*)

VSOS Virtual Storage Operating System

VSP Vacation Students Programme • Vehicle Scheduling Program • Vehicle Synthesis Program • Vertical Seismic Profile • Vertical Speed • Very Short Period • Video Signal Processor • Vienna Software Publishing • Vie Socio-Professionnelle • Visible Sensor Package • Visiting Scientist Program • Visualization Service Provider • Voice Selection Panel • Vol Spatial Piloté

VSPC Virtual-Storage Personal Computing

VSPN Virtual Secure Private Network

VSPS Vernier Solo Power Supply

VS&PT Vehicle Summary & Priority Table

VSPT Vehicle Summary and Priority Table

VSPX Vehicle Scheduling Program Extended

VSR Vaisseau Spatial Réutilisable • Very Short Range • Very Special Reserve • Vibration Sensitive Relay • Voice Storage and Retrieval • Voltage-Sensing Relay

VSRA V/STOL System Research Aircraft

VSRF Very-Short-Range Forecasting

VSRP Vacation Student's Research Programme

VSRR Verein Sternwarte Rotgrueb Rümlang (Switzerland)

vss Versions

VSS Variable Stability System • Variable Stars Section • Vehicle Support System • Vehicle System Simulator • Velocity Sensor System • Velocity Spectrum Stack • Very Steep Spectrum • Video Signal Simulator • Video Storage System • Virtual Simulation Software • Virtual Storage System • Visual Scene System • Visual Source Save • Vitesse Supersonique • Vocabulary Switching System • Voice Storage System • Vol Supersonique • Voltage-Sensing Switch • Voyager Spacecraft System

VSSC Vikram Sarabhai Space Centre (ISRO)

VSSCi RASNZ Variable Star Section Circulars (periodical code)

VSSHG Video Spectra-Spectroheliogram • Video Spectra-Spectroheliograph

VSSM Video Scanner Switch System

VSSRe RASNZ Variable Star Section Reprints (periodical code)

VSSSC RASNZ Variable Star Section Special Circulars (periodical code)

VSST Vacuum Solar Spectrum Telescope (China-PRC)

Vst Vermittlungstelle

VST Variable Speed Technology • Västerås (Sweden) (location identifier) • Velocity-to-Space Transformation • Very Small Telescope • Vibration and Shock Test • Vista • VLT Survey Telescope

VSTAG Vandenberg Shuttle Turnaround Analysis Group

VSTAR Vandenberg Shuttle Turnaround Assessment Report

VStG Vermögensteuergesetz

V/STOL Vertical/Short Takeoff and Landing

VSTOL Vertical/Short Takeoff and Landing

VSTP Visual Satellite Tracking Program

VStR Vermögensteuer-Richtlinien

VSU Valdosta State University (USA-GA)

VSV Variable Stator Vane • Véhicule à Stabilité Variable • Vol Sans Visibilité

VSW VAP Software • Verification Software • Very Shallow Water

VSWR Variable Standing Wave Ratio • Visual Standing Wave Ratio • Voltage Standing Wave Ratio (*)

VSYNC Vertical Synchronization

VSZ Verzeichnis Schweizerischer Zeitschriften

vT vom Tausend

Vt Vermont (USA) • Vietnamese

V&T Verification & Test

V/T Voltage/Temperature • Voltage-limited/Temperature-compensated

V-T Vibrational to Translational • Voltage-Temperature

VT India (international aircraft registration code) • Vacuum Technology • Vacuum Tube • Vaisseau de Transport • Vanne de Transfert • Variable Temperature • Variable Thrust • Variable Time • Variable Transmission • Variation Thermique • Vatu (currency) • Véhicule de Transport • Vent • Vermont (USA) • Verteiler • Vertical Tab • Vertical Tabulation • Vibration Testing • Videotape • Videotext • Viet Nam • Vie Totale • Virginia Tech (USA-VA) • Virial Theorem • Virtual Terminal • Virtual Tributary • Vitesse de Tangage • Vocational Training • Voice Tube • Voie de Transmission • Vol de Transition • Vol en Translation • Vol Télécommandé • Volume Total • Voronoi Tessellation

Vta Vuelta

VTA Vacuum Test Article • Vacuum Tube Amplifier • Valley Transportation Authority (USA-CA) • Variable Transfer Address • Vertex Time of Arrival • Victoria (Honduras) (location identifier) • Voie du Train d'Atterrissage

VTAM Virtual Telecommunications Access Method (*) • Virtual Teleprocessing Access Method • Virtual Terminal Access Method

VTAP Vehicle Tracking Analysis and Performance

VTAPS Vehicle Tracking Analysis and Performance System

VTAR Verification Test/Analysis Report

VTAS Visual Target Acquisition System

VTB Vitebsk (Belarus) (location identifier) • Voltage Time to Breakdown

VTBC Vertical Time Bar Chart

VTC Vehicular Technology Conference • Video Tape Camera • Video Teleconference • Video Teleconference Center • Video Teleconferencing • Video Teleconferencing Capability • Video Teleconferencing Center • Video Time Code • Vidicon Television Camera • Virtual Tape Control • Virtual Tape Controller • Virtual Terminal Control • Virtual Training Centre • Vol Télécommandé • Voting Trust Certificate

VTCEIG Video Teleconferencing Executive Interest Group

VTCN Video Telecommunications Conferencing Network

VTCS Vehicular Traffic Control System • Virtual Tape Control System

VTCW Vehicle Time Code Word

VTD Vakuumtechnik Dresden GmbH (Germany) • Vandenberg Test Director • Variable Time Delay • Variation Totale Décroissante • Vertical Tape Display • Virtual Tape Drive • Virtual Timer Device • Virtual Timer Driver • Voting District

VTDA Verification Team Data Analyst

VTDB Virtual Terrain Database

VTDP Virtual Takeoff Departure Point

Vte Variante

VTE Vapor Transport Equilibrated • Variable Thrust Engine • Véhicule Transporteur-Érecteur • Vendor Trading Exchange • Vibration Test Equipment • Vicarious Trial and Error • Vientiane (Laos) (location identifier) • Virtual Terminal Environment • Virtual Transmit Engine • Viscous Transonic Equation

VTEC Vertical Total Electron Content

VTELN Vapor Transport Equilibrated Lithium Niobate

VTES Variable Thrust Engine System • Vinyltriethoxysilane

VTEST Virtual Test

VTF Vacuum Test Furnace • Variable Temperature Facility • Vatulele (Fiji) (location identifier) • Vertical Test Facility • Vertical Test Flight • Vibration Test Facility • Voltage Transfer Function

Vtg Voting

VTG Video Timing Generator • Video Tracking Group

VTGCM Venus Thermospheric General Circulation Model

VTGV Véhicule à Très Grande Vitesse

VTH Virtual Town Hall

VTHD Vraiment Très Haut Débit

VTHL Vertical Takeoff and Horizontal Landing

VTI Variable Temperature Insert • Vidéotransmission Interactive • Virtual Terminal Interface

VTIR Visible and Thermal Infrared • Visible and Thermal Infrared Radiometer (*)

VTJ Vitro Technical Journal (periodical code)

VTK Visualization Toolkit

VTL Vacuum-Tube Launcher • Variable Threshold Logic • Verification Test Laboratory • Video Tape Lecture • Virtual Tape Library • Virtual Training Library • Vittel (France) (location identifier)

VTLS Virginia Tech Library System

VTM Vacuum-Tube Module • Valve Test Manifold • Verification Team Manager • Verification Test Matrix • Versatile Tracking Mount • Vibration Test Module • Voltage Tunable Magnetron

VTMIS Vessel Traffic Management and Information Service

VTML Versioned Text Markup Language

VTMO Voltage Tunable Microwave Oscillator

VTMS Vinyltrimethysilane

VTN Valentine Miller (USA-NE) (location identifier) • Verification Test Network • Video Tape Network

Vto Vencimiento • Vuelto

VTO Véhicule de Transfert Orbital • Vertical Takeoff • Voltage Tunable Oscillator • Voltage Tuned Oscillator

VToA Voice and Telephony over ATM

VTOA Voice and Telephony Over ATM

VTOC Visual Table Of Contents • Volume Table Of Contents (*)

VTOGW Vertical Takeoff Gross Weight

VTOHL Vertical Takeoff - Horizontal Landing

VTOI Verification Team Operating Instruction

VTOL Vertical Takeoff and Landing

VTOVL Vertical Takeoff - Vertical Landing

VTP Vehicle Test Plan • Verification Test Plan • Verification Test Program • Virtual Tape Protocol • Virtual Telescope Program • Virtual Terminal Protocol • Virtual Training Program • Virtual Tunneling Protocol • Voronoi Tessellation and Percolation

VTPR Vertical Temperature Profile Radiometer

VTPS Vibration Test Plotting System • Vibroacoustic Test Plan Summary

VTQ Voice Transmission Quality

VTR Verification Test Report • Vertical Test Range • Video Tape Recorder • Video Tape Recording • Voice Tape Recorded • Voltage Transformation Ratio

VTRE Vented Tank Resupply Experiment

VTRF VLBI Terrestrial Reference Frame

VTRS Vandenberg Telemetry Receiver Site • Vandenberg Telemetry Receiving Site • Vandenberg Telemetry Receiving Station • Vandenberg Tracking Station (USA-CA) • Videotape Recording System • Videotape Response System

VTS Vandenberg Tracking Station • Variable Time Step • Variable Tracking Strategy • Vault Tracking System • Vereinigung Technischer Schriftsteller (Austria) • Verification Test Station • Versatile Training System • Vertical Test Site • Vertical Test Stand • Vertical Test System • Vessel Traffic Service • Vessel Traffic System • Vibration Test Specification • Vibration Test System • Vibroacoustic Test Summary • Video Teleconferencing System • Video Teleseminar • Video Test Set • Video Timesharing Service • Video Tracker System • Viewfinder Tracking System • Virtual Tape Server • Virtual Tape Solution • Virtual Tape System • Virtual Terminal Service • Virtual Terminal System • Visual Telemetry System • Vitesse Transsonique • Voice Telecommunications System

VTSAU Voprosy Teorii Sistem Avtomaticheskogo Upravileniia (periodical code)

VTSF VKI Turbulent Shear Flows (periodical code)

VTSS Values, Technology, Science and Society • Virginia Tech Spectral-line Survey • Virtual Tape Subsystem

VTT Vacuum Thermal Testing • Vacuum Tower Telescope (now RBDST) • Vacuum-Tube Transmitter • Vakuum Turm-Teleskop • Validation Test Suite • Valtion Teknillinen Tutkimuskeskus (Finland) • Variable Threshold Transmitter • Véhicule Tout Terrain • Video Teletraining • Virtual Teletraining • Visual Target Tuner • Vol Tous Temps • Voyage Through Time

VTU Las Tunas (Cuba) (location identifier) • Video Teleconferencing Unit • Virtual Time Unit • Vital Telemetry Unit

VTV Verband für Textverarbeitung (Germany) • Verification Test Vehicle • Virtual Tape Volume

VTVL Vertical Takeoff and Vertical Landing

VTVM Vacuum-Tube Voltmeter

VTW Variable Transmission Window • Vehicle Time Word

Vtx Videotex

VTX Vertex • Videotext

VTZ Vishakhapatnam (India) (location identifier)

vu Volume Unit • von unten

V/U Vehicle Utility

VU Vanuatu (ISO country code) • Vehicle Unit • Vehicle Utility • Véhicule Utilitaire • Vertical Unit • Very Urgent • Virtual Universe • Virtual University • Vitesse Uniforme • Voice Unit • Voie Unique • Volume Unit • Vrije Universiteit

VUA Virtual Unit Address • Vrije Universiteit Amsterdam (Netherlands)

VUB Vrije Universiteit Brussel (Belgium)

VUBIS Vrije Universiteit Brussel Information System

VUCA Volatility - Uncertainty - Complexity - Ambiguity

VUCC VHF/UHF Century Club (award)

VUE Visible Ultraviolet Experiment • Visual User Environment

VUI Video User Interface • Visual User Interaction • Visual User Interface • Voice User Interface

VUIM Voyager Uranus/Interstellar Mission

Vul Vulpecula-Vulpeculae (constellation)

Vulg Vulgaire • Vulgar

VUN Virtual University Network

VUP Valledupar (Colombia) (location identifier) • VAX Unit of Performance

VUPS VAX Units of Performance

VUS Velikij Ustyug (Russia) (location identifier) • Verbindungsunterstützungssystem • Versatile Upper Stage

VUSW Vehicle Utility Software

VUT Vanuatu (ISO country code) • Vervroegde Uitreding

VUTS Verification Unit Test Set

V/UV Visible/Ultraviolet

VUV Vacuum Ultraviolet • Vatu (Vanuatu) (ISO currency code) • Visible-Ultraviolet

VUVES Vacuum Ultraviolet Echelle Spectrograph

VUW Victoria University of Wellington (New Zealand)

v/v Volume in Volume

vv Varios • Vice Versa • Volúmenes

V/V Validation & Verification • Vent Valve • Vertical Velocity

V&V Validation & Verification • Verification & Validation

V-V Vibrational-Vibrational

VV Ustedes • Validation and Verification • Vanne de Ventilation • Varios • Vatican Variable (star) • Velocity Vector • Vent Valve • Verification and Validation • Vertical Velocity • Vertical-Vertical • Vertical Visibility • Vertreterversammlung • Verwaltungsvorschrift • Vibrational-Vibrational • Vice Versa • Visibilité Verticale • Visibility Value • Vitesse de Virage • Vitesse de Vol • Vitesse du Vent • Vitesse Verticale • Vitesse Vraie • Vivenda • Vol à Vue • Vorontsov-Vel'yaminov (catalog)

VV&A Verification, Validation & Accreditation • Verification, Validation & Authentication • Verify, Validate & Accredit

VVA Verein für Volkstümliche Astronomie • Verification, Validation and Accreditation • Verification, Validation and Authentication • Verify, Validate and Accredit • Verstoring Van Aardsystemen

VV&A/C Verification, Validation & Authentication/Certification

VVB Mahanoro (Madagascar) (location identifier)

VV&C Verification, Validation & Certification • Verify, Validate & Certify

VVC Variable Vacuum Capacitor • Verification, Validation and Certification • Verify, Validate and Certify • Vertical-Velocity Console • Villavicencio (Colombia) (location identifier) • Voltage Variable Capacitor • Voltage Versatile Capacitor

VVCC Victor Valley Community College (USA-CA)

VVCCC Version Virtuelle du Catalogue Collectif Canadien (Canada)

VVCD Voltage Variable Capacitance Diode

VVCS Vernier Velocity Correction System

VVD Voltage Variable Diode

VVDS VIRMOS VLT Deep Survey

Vve Veuve

VVE Very Very Easy

VVH Very Very Hard • Very Very Heavy

VVI Santa Cruz - Viru Viru (Bolivia) (location identifier) • Vertical Velocity Indicator • Vitesse Verticale Instantanée • Voltage Variation Indicator

VVK Västervik (Sweden) (location identifier) • Verwaltung Vermessungs- und Kartenwesen

VVM Vector Voltmeter • Velocity Vector Measurement

VVMS Velocity Vector Measurement System

VVO Van Vleck Observatory (USA-CT) • Very Very Old • Vladivostok (Russia) (location identifier)

VVORe Van Vleck Observatory Reprints (periodical code)

VVP Vitesse de Variation de Pression

VVR Variable Voltage Rectifier

VVRD VKI Vibration Rotor Dynamics (periodical code)

VVS Vereniging voor Sterrenkunde, Meteorologie, Geophysica en Aanverwante Wetenschappen (Belgium) • Very Very Slightly • Very Very Superior • Voenno-Vosdushniie Sili (Russia) • Vol à la Vitesse du Son

VVSA Velocity Vector Sensor Assembly

VVSR Vegetation Visible Spectral Reflectance

VVSS Vertical Volute Spring Suspension

VVTP Verification and Validation Test Plan

VVV Vereneging voor Vreemdelingen Verkeer

VVZ Illizi (Algeria) (location identifier)

V&W Verkeer & Waterstaat

VW Vector Warning • Verkeer en Waterstaat • Very Weak • View • Virtually Worthless • Virtual World • Visual Worth

VWA Vacuum Window Assembly

VWB Visual Workbench

VWC Vector Wavelet Coding

VWFC Very Wide Field Camera

VWG Validation Working Group • Verification Working Group

VwGO Verwaltungsgerichtsordnung

VWL Variable Word Length

VWO Voorbereidend Wetenschappelijk Onderwijs

VWP Variable Width Pulse • Visa Waiver Program (USA) • Visible Web Presence

VWRS Variable Wire Rate Sensor

VWS Vertical Wind Shear • Vibrating Wire Stressmeter • Views

VWSI Vegetation Water Stress Index

VwVfG Verwaltungsverfahrensgesetz

VwZG Verwaltungszustellungsgesetz

Vx Vertex • Vieux

VXC Lichinga (Mozambique) (location identifier)

vxd Virtual Device Driver (file extension)

VxD Virtual Extended Driver

VXD Virtual X Device

VXDS VETA X-ray Detector System

VXE São Vicente (Cape Verde) (location identifier)

VXI Virtual Extension for Instrumentation • Virtual Extension Interface • VME Extension for Instrumentation

VXML Voice Extensible Markup Language

VXO Variable Crystal Oscillator • Växjö (Sweden) (location identifier) • Vaxjo (Sweden) (location identifier)

Vy Various Years • Very

VYD Vryheid (South Africa) (location identifier)

VyGeo Vyssha Geodeziia (periodical code)

VYS Peru (USA-IL) (location identifier)

Vz Voorzitter

VZ Virtual Zero • Vollzug

Vzk Verzweigungskabel

vzm Voor Zover Mogelijk

vZM von Zeipel Method

VZM Vereniging Zonder Winstoogmerk

VZOG Vermögenzuordnungsgesetz

V2 Antigua and Barbuda (international aircraft registration code)

V3 Belize (international aircraft registration code)

V4 Saint Kitts and Nevis (international aircraft registration code)

V5 Namibia (international aircraft registration code)

V6 Micronesia (international aircraft registration code)

V7 Marshall Islands (international aircraft registration code)

V8 Brunei (international aircraft registration code)

W

w/ With

w Axial Velocity • Coefficient of Concordance • Concordance • Dew • Electromagnetic Energy Density • Mass Fraction • Radiant Energy Density • Sound Energy Density • Velocity Component • Warden • Warehouse • Water • Weak Absorption • Walk • Warm • Water • Weather • Week • Weight • West • Western • Westlich • Wet • White • Wide • Width • Wife • Wind • Wire • With • Withdrawn • Withdrew • Withheld • Won • Word • Work • Wrong

W/ With

W Energy • Load • Mechanical Work • Radiant Energy • Radiant Flux • Tungsten • Tungstène • Velocity Component • Wagon • Wald • Wales (UK) • Warden • Warehouse • Warm • Warning • Washington • Water • Watt • Weak • Wednesday • Weekly • Weight • Well • Welsh • Wesleyan • West • Westen • Western • Westerhout • Westlich • Whisky (international alphabet) • White • Width • Wide • Widow • Widowed • Width • Wien Constant • Window • Windows • Wire • With • Withdrawn • Withdrew • Withheld • Wohnung • Wolfram • Won (currency) • Work

wa With Answers

W/A Work Around

WA Wales • Warm Absorber • Washington (USA) • Watt Angle • Weather Advisory • West Africa • Western Arctic • Western Australia • Wetteramt • Wide Angle • With Average • Withholding Agent • Work Around • Work Authorization

WAA Wales (USA-AK) (location identifier) • Washington Area Astronomers (USA-DC) • Westchester Amateur Astronomers (USA-NY) • Wide-Angle Astrograph • Wide-Aperture Array • Wiener Arbeitsgemeinschaft für Astronomie (Austria) • Wireless Advertizing Association (USA) • Wisconsin Alumni Association (USA-WI) • With Abject Apologies

WAAC Wollongong Amateur Astronomy Club (Australia-NSW)

WAAE World Association for Adult Education

WAAL Wisconsin Association of Academic Librarians (USA-WI)

WAAS Wide-Area Augmentation System • World Academy of Art and Science

WAAT Wide-Area Application Testbed

wab Windows Address Book (file extension)

WAB Wabag (Papua New Guinea) (location identifier) • When Authorized By • Wissenschaftliche Allgemeinbibliothek

WABA Western Australia Booksellers' Association (Australia-WA)

WABCR Wide-Area Bar Code Reader

WABI Windows Application Binary Interface

WABS Wide-Angle Broadband Spectrometer

WAC Waca (Ethiopia) (location identifier) • Wide-Angle Camera • Wide-Area Center • Wide-Area Communications • Work Accomplishment Code • World Aeronautical Chart • World Aeronautical Corporation • Wyoming Academic Consortium (USA-WY)

WACA Weeks After Contract Award • Write Access Connection Acceptor

WACC Work Activities and Coordination and Collaboration

WACCC Worldwide Air Cargo Commodity Classification

WACG Workshop on Applied Computational Geometry

WACI Write Access Connection Initiator

WACK Wait Acknowledgment • Wait before Acknowledge

WACL Worcester Area Cooperating Libraries (USA-MA)

WACO Whatcom Association of Celestial Observers (USA-WA) • Written Advice of Contracting Officer

WACS Wire Automated Check System • Wireless Access Communications System

WACT Wide-Angle Čerenkov Telescope

Wad Wadham

WAD Andriamena (Madagascar) (location identifier) • Wide-Area Development • Wohnungsanschlußdose • Work Authorization Document • Work Authorizing Document • Workload Assignment Document • World Academic Database • World Astronomy Day • World Aviation Directory

WADA World Anti-Doping Agency

WADB West African Development Bank

WADC Wright Air Development Center

WADD Wright Air Development Division

WADE World Association of Document Examiners

WADEX Word and Author Index

WADGNSS Wide-Area Differential Global Navigation Satellite System

WADGPS Wide-Area Differential Global Positioning System

WADM Wavelength Add-Drop Multiplexer

WAdO Wing Administrative Officer

WADS Wide-Angle Display System • Wide-Area Data Service • Wide-Area Differential System • Workshop on Algorithms and Data Structures • Write-Ahead Data Set

WADT Workshop on Algebraic Development Techniques

WAE Wadi ad Dawasir (Saudi Arabia) (location identifier) • Wheel Abrasion Experiment • When Actually Employed • Wireless Applications Environment • Workshop on Algorithm Engineering

WAEA World Airline Entertainment Association

WAES Web Attribute Evaluation System

WAF Wana (Pakistan) (location identifier) • Width Across Flats • Wiring Around Frame • Word Address Format • Work Approval Form

WAFC World Area Forecast Centre

WaFIRS Waveguide Far-Infrared Spectrometer

WAFL Write Anywhere File Layout

W. Afr. West Africa • West African

WAFS World Area Forecast System

WAG Gambia (international vehicle registration code) • Wanganui (New Zealand) (location identifier) • Waste Area Grouping

WAGS Worldwide Acoustic Gravity-wave Study • Worldwide Atmospheric Gravity-wave Study

WAH Wahpeton (USA-ND) (location identifier) • Wide-Area Hypertext

WAHA Wide-Angle High Aperture

WAI Antsohihy (Madagascar) (location identifier) • Web Accessibility Initiative • Wörterbuchentwicklung für Automatisches Indexing

WAIF Weighted Average Impact Factor

WAIL Wide-Angle Imaging Lidar

WAIM Web-Age Information Management (conference)

WAIN Wide-Area Integrated Network

WAIOP Will Accept, If Offered, the Position

WAIPA World Association of Investment Promotion Agencies

WAIS Wechsler Adult Intelligence Scale • West Antarctic Ice Sheet • Western Antarctic Ice Sheet • Wide-Area Imaging System • Wide-Area Information Server • Wide-Area Information Service • Wide-Area Information System

WAIT Western Australia Institute of Technology

WAITIO World Association of Industry and Trade Information Offices

WAITRO World Association of Industrial and Technological Research Organizations

WAITS Wide-Area Information Transfer System

WAJ Wawoi Falls (Papua New Guinea) (location identifier)

WAK Ankazoabo (Madagascar) (location identifier)

WAKS Wiener Allgemeines Kategorienschema

Wal Wallon • Wallonie (Belgium)

WAL Chincoteaque Wallops (USA-VA) (location identifier) • Sierra Leone (international vehicle registration code) • Wide-Angle Lens • Wohnungsanschlußleitung

WALA West African Library Association

WALC Washington Accelerated Learning Center (USA-DC)

WALES Water (vapor) Lidar Experiment in Space (project)

WahlO Wahl Ordnung

WALIS Western Australia Land Information System

WALL Western Association of Law Libraries (USA-CA)

WALN Western Australia Library Network (Australia-CA)

WALP Wide-Array Linear Planar

WALS Wide-Angle Light Scattering

WALT Warning Assessment Logic Terminal

WALTA Washington Large-area Time-coincidence Array (USA-WA)

WAM Ambatondrazaka (Madagascar) (location identifier) • Warren Abstract Machine • Work Analysis and Measurement • Wraparound Mortgage

WAMA Weight After Mars Arrival

WAMAP Waste Management Advisory Programme

WAMDII Wide-Angle Michelson Doppler Imaging Interferometer

WAMET Warped Modified Envelope Tracing (method)

WAMEX West African Monsoon Experiment

WAMI Washington - Alaska - Montana - Idaho • World Association for Medical Informatics

WAML Western Association of Map Libraries

WaMot Wave Motion (periodical code)

WAMP Weight And Mass Property

WAMS Wide-Area Master Station

WAN Nigeria (international vehicle registration code) • Waverney (Australia-QLD) (location identifier) • Wide-Area Network • Work Authorization Number • World Association of Newspapers

WAND Wide-Area Neutralization Device

WANO World Association of Nuclear Operators

WAO Wabo (Papua New Guinea) (location identifier) • Wallace Astrophysical Observatory (USA-MA) • Wet op de Arbeidsongeschiktheidsverzekering • Wissenschaftliche Arbeitsorganisation • Work Assignment Order

WAOS Wide-Angle Optical System

WAOSS Wide-Angle Optical Stereo Scanner • Wide-Angle Optoelectronic Stereo Scanner (*)

WAP Alto Palena (Chile) (location identifier) • Waste Analysis Plan • Weak Anthropic Principle • Wide-Area Paging • Wireless Application Protocol • Wissenschaftlerarbeitsplätze • Work Analysis Program • Work Assignment Plan • Work Assignment Procedure

WAPI World Aerial Photographic Index • World Air Photography Index

WAPOR World Association for Public Opinion Research

WAPP Wideband Arecibo Pulsar Processor

WAQ Antsalova (Madagascar) (location identifier)

War Warrant • Warsaw Okecie (Poland) • Warszawa Okecie (Poland) • Warwickshire (UK-England)

WAR Waris (Indonesia) (location identifier) • Weekly Activity Report • Wide-Area Routing • Work Authorization Report • Work Authorization Request • Write After Read

WARAL World Amateur Radio Astronomers League

WARC World Administrative Radio Conference (*) • World Allocations Radio Conference

WARCO Warranty Control Office

WARDA West Africa Rice Development Association

WARF Wide-Aperture Research Facility

Warks Warwickshire (UK-England)

WARLA Wide-Aperture Radio Location Array

Warn Warning

WARN Warning • Worker Adjustment and Retaining Notification

WaRP Wavelength Routing Protocol

WARP Weather And Radar Processor • Web Academic Resource Publisher • Weight Associate Rule Processor • Wisconsin Alumni Research Foundation (USA-WI) • Work And Recreation Party

WARPAC Warsaw Pact

WARPS Wide-Angle ROSAT Pointed Survey

Warr Warranty

WARR Wide-Angle Reflection and Refraction

WarRe Warsaw University Reprints (periodical code)

Warrty Warranty

WARS Wedge Absorption Remote Sensor • Wide-Area Reference Station • Wide-Area Remote Sensor

WARSIDEC Western Australia Remote Sensing Industry Development and Education

Warwicks Warwickshire (UK-England)

WAS Wadhurst Astronomical Society (UK-England) • Walsall Astronomical Society (UK-England) • Warren Astronomical Society (USA-MI) • Warsaw Astronomical Society (USA-IN) • Washington (USA-DC) (location identifier) • Washington Academy of Sciences • Wasilewski (catalog) • Wehr Astronomical Society (USA-WI) • Wellington Astronomical Society (New Zealand) • Weltastronomie Studienreisen (Germany) • Wessex Astronomical Society (UK-England) • Westminster Astronomical Society (USA-MO) • Westport Astronomical Society (USA-CT) • Wide-Angle Sensor • Wide-Angle Survey (INT) • Wide-Area Search • Wide-Area System • Wideband Antenna System • Willingboro Astronomical Society (USA-NJ) • Worthing Astronomical Society (UK) • Wycombe Astronomical Society (UK)

WASA West African Science Association • Wide Angle Star Acquisition

WASAR Wide-Angle Synthetic Aperture Radar

WASC Washington Analytical Service Center (EG&G) • Western Association of Schools and Colleges (USA)

Wash Washington (USA)

WASL Windows Aspect Script Language

WasOA Washburn Observatory Astrophysics (periodical code)

WAsP Wind Atlas Analysis and Application Programme

WASP Water Analysis Simulation Program • Water And Steam Program • Weather-Atmospheric Sounding Projectile • Weightless Analysis Sounding Probe • Wideband Autocorrelation Spectrometer • Wien Automatic System Planning • Window Atmosphere Sounding Projectile • Workshop Analysis and Scheduling Programming • Workshop on Agent/Swarm Programming • Wrap Around Simulation Program • Wyoming-Arizona Search for Planets

WASPAM Wasser- und Spurengasmessung in der Atmosphäre mit Mikrowolken

WASS Wide-Area Surveillance System

WAST Western Australia Standard Time

WASTAC Western Australian Satellite Technology Application Consortium

Wat Waterford (Ireland)

WAT Waterford (Ireland) (location identifier) • Wavelet Alias Table • Web Action Time • Weight - Altitude - Temperature • Weight And Temperature • Weighted-Average Toxicity • Wide-Angle Tail • Wide-Angle Tailed (source) • Word Association Test

WATA World Association of Travel Agencies

WATCH Wide-Angle Telescope for Cosmic Hard (X-rays)

WATERS Wide-Area Technical Report Server • Wide-Area Technical Report Service

WATIS Work and Training Information System

WATM Wide-area Asynchronous Transfer Mode • Wireless Asynchronous Transfer Mode

WATOX Western Atlantic Ocean Experiment

WATS Website Activity Tracking Statistics • Wide-Area Telecommunications Service • Wide-Area Telephone Service • Wide-Area Telephone System • Wide-Area Tracking System • Wind And Temperature Spectrometer • World Air Transport Statistics

WAU Wauchope (Australia-NSW) (location identifier) • Wavelet Audio Unit

WAUBA West African University Booksellers Association

W. Aus. Western Australia

wav Windows Sound (file extension)

WAV Waveform • Wave Hill - Kalkgurung (Australia-NT) (location identifier)

WAVE Water Vapor Experiment • Web Analysis and Visualization Environment • Wide-Area Variability Experiment • Wide-Area Voice Exchange • World Association of Video-Makers and Editions

WAVECAL Wavelength Calibration

WavEl Wave Electronics (periodical code)

WAVES Waveform And Vector Exchange Specification

WAW Warszawa (Poland) (location identifier) • Weaving the Astronomy Web • Write After Write

WAWD Write-After-Write Dependency

WAWG Where Are We Going?

WAX Warroad (USA-MN) (location identifier)

WAXD Wide-Angle X-ray Diffraction

WAXS Wide-Angle X-ray Scattering • Wide-Area X-ray Survey

WAY Waynesburg - Green County (USA-PA) (location identifier)

WAZ Warwick (Australia-QLD) (location identifier) • Westdeutsche Allgemeine Zeitung (Germany)

W/B Waybill • Weight & Balance • Wideband

W&B Weight & Balance

WB Wage Board • Warehouse Book • Water Ballast • Water Boiler • Waveband • Waybill • Weather Bureau • Wechsler-Bellevue • Weekly Bulletin • Weight and Balance • West Berlin (ISO country code) (now obsolete) • Westbound • Wet Bulb • Wet Bulk • Wetenschappelijke Bibliotheek • Wheel Base • White Belly • Wideband • Will Be • Wissenschaftliche Bibliothek • Word-Based • Word Before • Workbench • World Bank • Write Back • Write Buffer

WBA Welsh Border Astronomers (UK-Wales) • Wideband Amplifier • Worn By Astronaut

WBAFC Weather Bureau Area Forecast Center

WBAWS Weather Briefing And Warning Service

WBB Stebbins (USA-AK) (location identifier) • Wiener Brückenbau (Austria) • Wind-Blown Bubble

WBC Weight and Balance Computer • West Balkan Country • Western Boundary Current • White Blood Cell • Wideband Channel • Wideband Coupler • Wide-Body Cargo • Wien Bridge Circuit • Wire Bridge Circuit

WBCO Waveguide Below Cutoff

WBCS Western Boundary Containment System

WBCSD World Business Council for Sustainable Development

WBCT Wideband Current Transformer

WBCV Wideband Coherent Video

WBD Befandriana (Madagascar) (location identifier) • Whiteboard • Whole Body Dose • Wideband • Wideband Data • Wire-Boom Drive • Wire-Bound

WBDA Wideband Data Assembly

WBDCS Wideband Data Collection System

WBDI Wideband Data Interleaver

WBDL Wideband Data Link

WBDTS Wideband Data Transmission System

WBDX Wideband Data Switch

WBE Bealanana (Madagascar) (location identifier) • Wideband Electronics • Wire-Boom Electronics

WBEM Web-Based Enterprise Management

WBF Wegener - Bergeron - Findeisen (mechanism)

WBFDM Wideband Frequency Division Multiplexer

WBFM Wideband Frequency Modulation

Wbg Webbing

WBGT Wet-Bulb Globe Temperature

WBGU Wissenschaftlicher Beirat der Globale Umweltveränderungen

WBHF Wideband High Frequency

WBHFR Wideband High Frequency Radio

WBI Boulder - Broker Inn (USA-CO) (location identifier) • Web-Based Instruction • Web Browser Intelligence • Whole-Body Irradiation • Will Be Issued

WBIF Wideband Intermediate Frequency

WBIZ Wissenschaftliches Bibliotheks- und Informationszentrum

wbk Word Backup (file extension)

WBL Wideband Laser • Wideband Limiting

WBM Wapenamanda (Papua New Guinea) (location identifier)

WBMC Weight Before Mars Capture

WBMCR Wideband Multichannel Receiver

WBMOD Wideband Model

WBMS Work Breakdown Management Structure

WbN West by North

WBN Woburn - Cummings Park (USA-MA) (location identifier)

WBND Westbound

WBNL Westbound Noise Limiting

WBNV Westbound Noise Voltage

WBO Beroroha (Madagascar) (location identifier) • Wideband Oscilloscope • Wideband Overlap • Wien Bridge Oscillator

WBPS Web-based Business Procurement System • Web-based Business Purchasing System

WBQ Beaver (USA-AK) (location identifier)

WBR Big Rapids (USA-MI) (location identifier) • Whole Body Radiation • Wideband Radio • Wideband Radiometer • Wideband Receiver • Word Buffer Register • Workbench Rack

WBRR Weather Bureau Radar Remote

WBRS Wideband Remote Switch

WbS West by South

WBS Walking Beam Suspension • Weight and Balance System • Welsh Bibliographical Society (UK-Wales) • Whole-Body Scan • Wideband Spectrometer • Wideband Spectrum • Wideband System • Without Benefit of Salvage • Work Breakdown Schedule • Work Breakdown Structure • Work Breakdown System

WBSA Wideband Spectrum Analyzer

WBSAP Workload-Based Staffing Analysis Program

WBSC Wideband Signal Channel • Wideband Signal Conditioner

WBSCB Work Breakdown Structure Control Board

WBSO Wet Bevordering Speur- en Ontwikkelingswerk (Netherlands)

WBSS Wideband Digital Switching System • Wide-Binary Supersoft Source • Workshop for Basic Space Science

WBST Wideband Secure Terminal • Wideband Subscriber Terminal

WBSV Wideband Secure Voice

WBSVN Wideband Secure Voice Network

WBSVVS Wideband Secure Voice Via Satellite

WBT Wait Before Transmitting • Web-Based Training • Wet-Bulb Temperature • Wideband Terminal • Wideband Transformer • Wideband Transmitter • Windows-Based Terminal

WBTCH Water Boiler Thermal Control Hydraulics

WBTS Wideband Transmission System

WBTV Weather Briefing Television

WBU Boulder (USA-CO) (location identifier)

WBV Wideband Voltage

WBVCO Wideband Voltage-Controlled Oscillator

WBVCXO Wideband Voltage-Controlled Crystal Oscillator

WBVT Wideband Video Tape

WBVTR Wideband Video Tape Recorder

WBW Wilkes-Barre - Wyoming Valley (USA-PA) (location identifier) • Word By Word

wc Without Charge • Word Count

W&C Wire & Cable

W/C Workmen's Compensation

WC Carbon Wolf-Rayet (star) • Tungsten Carbide • Wagon-Citerne • Watch Committee • Water Clock • Water Closet • Weather Center • Web Club • Wideband Communications • Wind Current • Wire and Cable • Wiring Closet • Withdrawal Credit • Without Charge • Word Collocation • Word Contribution • Word Count • Work Center • Work Component • Work Control • Work Credit • Working Capital • Workmen's Compensation • World Coordinates • Write and Compute

WCA Castro Gamboa (Chile) (location identifier) • Water Clearance Authority • Wideband Cassegrain Antenna • Wind Correction Angle • Worst-Case Analysis

WCAC Western Colorado Astronomy Club (USA-CO) • Wilderness Center Astronomy Club (USA-OH) • Williamson County Astronomy Club (USA-TX)

WCAFS Wideband Cassegrain Antenna Feed System

WCAG Web Content Accessibility Guidelines

WCAP World Climate Applications Programme

WCASP World Climate Applications and Services Program

WCAT Web Capacity Analysis Tool

WCB World Council for the Biosphere • WWW Catalogue Browser

WCC Washington Comprehensive Catalog • Western Computing Center • Wilson Cloud Chamber • Wideband Communications Capsule • Wire Chamber Conference • Work Control Center • World Computer Congress

WCCE World Computer Conference on Education

WCCP Web Cache Control Protocol

WCCS Window Contamination Control System • Wireless Communications and Control System

WCCSAS Western Connecticut Chapter of the Society for Amateur Scientists (USA-CT)

WCCU Wireless Crew Communication Unit

WCD Wind-Compressed Disk • Work Center Description • Work Control Document • World Commission on Dams

WCDA Western Canada Deformation Array

WCDB Work Control Database

WCDDT Wet Countdown Demonstration Test

WCDMA Wideband Code Division Multiple Access

WCDMP World Climate Data and Monitoring Programme

WCDP World Climate Data Programme

WCE Early-type Carbon Wolf-Rayet (star) • Wavefront Control Experiment • Write Cache Enable

WCED World Commission on Environment and Development

WCEFA World Conference on Education For All

WCET Worst-Case Execution Time

WCF Working Capital Fund

WCG Washington Conservation Guild (USA-DC) • Water-Cooled Garment • Water-Cooled Generator • Workstation Computer Graphics • Workstation Coordinated Graphics • World Cyber Games

WCGM Writable Character Generation Module

WCGR Waterloo Center for Groundwater Research (Canada-ON) (now CRESTech)

WCH Chaiten (Chile) (location identifier)

WCHR Wheelchair

WCI White Light Coronagraph • Wildlife Conservation International

WCIP World Climate Impacts Programme

WCIRP World Climate Impact and Response Programme

WCISP World Climate Impact Studies Program

WCL Water Coolant Line • Water Coolant Loop • Work Control Log • World Confederation of Labor

WCLS Water Coolant Loop System

WCM Web Content Management • Wired-Core Matrix • Wired-Core Memory • Word Combine and Multiplexer • Workshop on Conceptual Modeling • Writable Control Memory

WCMC World Conservation Monitoring Centre

WCN Worksheet Control Number • Worldwide Communications Network

WCNDR World Conference on Natural Disaster Reduction (UN)

WCNN World Conference on Neural Networks

WCO World Customs Organization

WCOT Wall-Coated Open Tubular

WCOTP World Confederation of Organisations of the Teaching Professions

WCP Waste Characterization Plan • Weapon Control Platform • Weather Communications Processor • Weight Control Program • Wing Chord Plane • Work Control Plan • Working Capital Productivity • World Climate Programme

WCPT West Coast Precision Traverse

WCR Chandalar (USA-AK) (location identifier) • Water-Cooled Reactor • Wire Contact Relay • Word Control Register

WCRA Weather Control Research Association

WCRP World Climate Research Programme

WCS Waste Collection System • Waste Collector System • Waveguide Communications System • Wavelength Coincidence Statistics • Wing Center Section • Wireless Communication Service • Work Control System • Workload Control System • Workstation Control and Synchronization • World Communications System • World Conference of Science • World Conservation Strategy • World Coordinate System • Writable Control Storage • Writable Control Store

WCSAR Wisconsin Center for Space Automation and Robotics (USA-WI)

WCSC Walker County Science Center (USA-GA) • West Coast Switching Center • Western Commercial Space Center (USA-CA)

WCSI World Centre for Scientific Information

WCSMP World Climate System Monitoring Programme

WCSU Western Connecticut State University (USA-CT)

WCT Wavefront Control Testbed • World Class Test

WCU Water Cooling Unit • West Chester University (USA-PA) • Western Carolina University (USA-NC) • World Customs Union

WCY World Communications Year

WCZ Wind Collision Zone • Wind-Compressed Zone

WC3 Warning, Command, Control and Communications

w/d Warranted • Well Developed • Withdrawn

wd Ward • When Distributed • White Dwarf (star) • Width • Wind • Withdrawn • Wood • Word

Wd Warranted • Wind • Wood • Word • Would

W/D Weight/Drag • Withdrawal

WD Dominica (international vehicle registration code) • Wage Determination • Walking Distance • Warm Disk • Warnier Diagram • Warning Display • Wave Drag • Western Digital • Westward Drift • White Dwarf (star) • Wide Distribution • Wilson-Devinney • Wind Direction • Wired Discrete • Wiring Data • Wiring Diagram • Withdrawn • Work Day • Work Distribution • Working Day • Working Distance • Working Document

WDA Wadi Ain (Yemen) (location identifier) • Wavelet Decomposition Algorithm • Wavelet Detection Algorithm • Weather Decision Aid • Wheel Drive Assembly

Wdb Woordenboek

wdb Works Database (file extension)

WDB Deep Bay (USA-AK) (location identifier) • World Data Bank • World Database

WDC Work Distribution Chart • World Data Center

WDCGG World Data Center for Greenhouse Gases

WDCM World Data Center on Microorganisms

WDCO World Data Center for Oceanography

WDCRS World Data Center for Rockets and Satellites

WDCSA World Data Center for Solar Activity

WDD White Dwarf (star) Database • Windows Disk Defragmenter • World Data Depository

WDDES World Digital Database for Environmental Sciences

WDDHRS Wideband Digital Decimeter High-Resolution Spectroscope

WDF Wave Distribution Function • Weather Data Facility

Wdg Winding

WDG Enid Woodring (USA-OK) (location identifier)

WDH Windhoek - J.G. Strijdom (Namibia) (location identifier)

WDI Wind Direction Indicator • Wondai (Australia-QLD) (location identifier)

WDIMF White-Dwarf-heavy Initial Mass Function

WDL Weak Double Layer • Western Development Laboratory • Wien Displacement Law • Windows Description Language • Windows Driver Library • Wireless Data Link • Won - Drawn - Lost

WDLC Wireless Data Link Control

WDLF White Dwarf (star) Luminosity Function

Wdlg Widerlager

WDLY Widely

WDM Warm Dark Matter (*) • Warm Disk Matter • Wave Division Multiplexer • Wave Division Multiplexing • Wavelength Division Multiplex • Wavelength Division Multiplexing (*) • Weight after Departure from Mars • Win32 Driver Model

WDMF Weak Disordered Magnetic Field

WDMO Weight before Departure from Mars Orbit

WDMS Watershed Data Management System

WDN Waldron Island (USA-WA) (location identifier)

Wdo Window

WDP Wavelength-Dependent Polarization • Wireless Datagram Protocol • Work Distribution Policy

WDPC Western Data Processing Center (USA-CA)

WDR Winder (USA-GA) (location identifier)

WDRAM Windows Dynamic Random Access Memory

WDRC World Data Referral Centre

WDRR Winchester Disk Recorder-Reproducer

WDS Wake Deluge Supply • Washington Double Star (catalog) • Wave Dynamic System • Wavelength-Dispersive Spectrometer • Wetterdienstschule • Wire Data Service

WDSD Water Data Sources Directory

WDSH Work- and Data-Sharing

WDSL Wireless Digital Subscriber Line

WDSPR Widespread

WDSPRD Widespread

WDSH-PTc Work- and Data-Sharing Parallel Tree-code

WDSHPTc Work- and Data-Sharing Parallel Tree-code

WDT Warning Display Terminal • Watch Dog Timer

Wdth Width

WDTT World Directory of Think Tanks

WDV Wissenschaftliche Datenverarbeitung • Written-Down Value

WDVL Web Developers Virtual Library

WDVI Weighted Difference Vegetation Index

WDY Kodiak (USA-AK) (location identifier)

w/e Week End • Week Ending

We Weber Number

W/E Wing Elevon • Write/Erase

WE Weather Effects • West-East • Westinghouse Electric • Wing Elevon • With Equipment • Withholding Exemptions • Wohneinheit • Wohnungseigentum • Write Enable • Write-Erase • Writing Environment • Wyld Equation

Wea Weather

WEA Warm Electronics Assembly • Weather • Weatherford - Parker County (USA-TX) (location identifier) • Workers' Educational Association (UK)

WEAAP Western European Association of Aviation Psychologists

WEAG Western European Armaments Group

WEAO Western European Armament Organization

WEAP World Environment Action Plan

WEAR WBS Element Audit Review

WEAVE Water and Energy: Atmospheric, Vegetative and Earth (interactions)

WEB Warm Electronic Box • Whole Earth Blazar

WEBB Water, Energy, and Biogeochemical Budget

WEBS Woven Electronic Book System

WEBT Whole Earth Blazar Telescope

WEC Wave and Electric-field Consortium • Wave Energy Converter • Wave Experiment Consortium (Cluster) • Weak Energy Condition • Wind Energy Converter • World Energy Conference • World Energy Council • World Environment Centre

WECA Wireless Ethernet Compatibility Alliance

WECC Western European Calibration Cooperation

WECo Western Electric Co.

WECOM Western Electric Satellite Communications

Wed Wednesday

WED Wedau (Papua New Guinea) (location identifier) • Wednesday • World Environment Day

WEDC West Engineering for Developing Countries

WEDOS World Environment Disaster Observation System

Weds Wednesday

WEDSS Whole Earth Decision Support System

WEE What Everyone Eats • Wind Erosion Equation

WEED Windows Extracted from Earthquake Data

WEEE Waste Electrical and Electronic Equipment

WEEG Wednesday Evening Evaluation Group

WEENIAC Weeny Almost Automatic Computer

wef With Effect From

WEF Water Environment Federation • Weifang (China-PRC) (location identifier) • With Effect From • World Economic Forum • World Electronic Forum

WEFAX Weather Facsimile

WEFT Wings-Engine-Fuselage-Tail

WEG Washington Energy Group • Wohnungseigentumsgesetz

WEGA Wind Energy Groß-Anlagen

WEGENER Working group of European Geoscientists for the Establishment of Networks for Earthquake Research

WEGS Western European Geological Survey

WEH Weihai (China-PRC) (location identifier)

WEHI Walter and Elisa Hall Institute

WEI Weipa (Australia-QLD) (location identifier)

WeiEn Weight Engineering (periodical code)

WEIZAC Weizmann Automatic Computer

WEJ Westward Electrojet

Wel Welsh

WEL Welkom (South Africa) (location identifier)

WELAC Western European Laboratory Accreditation Cooperation

Weld Welding

WeldJ Welding Journal (periodical code)

Weled Weledele

Weledel Weledelgeboren

Well Wellington (New Zealand)

WELL Whole Earth 'Lectronic Link

WELMEC Western European Legal Metrology Cooperation

WELS Weak Emission-Line Star

WEM West Malling (UK-England) (location identifier) • World Event Management

WEMC Western European Metrology Club

WeMFAS West Midlands Federation of Astronomical Societies (UK)

WEN Write Enable

WEND Worldwide Electronic Navigational Database

WENDS World Energy Data System

WENET World Energy Network

WENO Weighted Essentially Non-Oscillatory

WENSS Westerbork Northern Sky Survey

WEO Western European Organization

WEP Weak Equivalence Principle • Weam (Papua New Guinea) (location identifier) • Western Equatorial Pacific • Wired Equivalent Privacy • Wireless Equivalent Privacy • Wisconsin Experiment Package • Work Execution Plan

WEPA Wireless Equivalent Privacy Algorithm

WEPT Week on Electronic Publishing and Typography

WER Weight Estimating Relationship • Word Error Rate

WERT Wide Energy Range Telescope

WERTID West European Round Table on Information and Documentation

Wes Wesleyan

WES Waterways Experiment Station • Weasua (Liberia) (location identifier) • What Everyone Sees • World Economic Survey

West Western

WEST West European Science and Technology • Wideband European Satellite Telecommunication (*) • Wideband European Space Telecommunication

WESTAR Westinghouse Space Transportation And Recovery

Westerw Westerwald (Germany)

Westf Westfalen (Germany)

WESTHEM Western Hemisphere

Westm Westminster

WestPac Western Pacific

WESTPAC Western Pacific

WET Wagethe (Indonesia) (location identifier) • Waste Extraction Test • Western European Time • West European Time • Wetland Evaluation Technique • Whole Earth Telescope

Wetb Wetboek

WETEP Weekend Training Equipment Pool

WETF Weightless Environment Training Facility

WETN Whole Earth Telescope Network

WEU Western European Union (*) • West European Union

WEV Water - Energy - Vegetation

WEVS Wetland Ecosystem Vulnerability Study

WEW Wee Waa (Australia-NSW) (location identifier) • What Everyone Wants

WEX Wednesdays Excepted • Wexford Castlebridge (Ireland) (location identifier)

WEZ Westeuropäische Zeit

wf Wrong Font

wF wahrscheinlicher Fehler

WF Futuna Islands (Wallis &_) (ISO country code) • Wallis & Futuna Islands (ISO country code) • Wave Function •

Weighting Function • Wiener Filter • Wide Field • Wind Force • WOW Factor • Write Forward

WFA Wavefront Adding • Wavefront Analyzer • Weak-Field Approximation • Weighted Factor Analysis • Wide-Field Astronomy • Wide-Frequency Antenna

WFAG Wideband Frequency Agile Radar

WFAU Weiterbildender Fernstudiengang Angewandte Umweltwissenschaften • Wide-Field Astronomy Unit

WFB Ketchikan Waterfront (USA-AK) (location identifier)

WFC Wallops Flight Center (NASA) • Washington Fundamental Catalog • Wave Front Curvature • Wide-Field Camera • Wide-Field Channel • Windows Foundation Class • World Food Council

WFCAM Wide-Field Camera

WFCC World Federation for Culture Collections

WFCCI Wide Field Camera Consortium Institutes (ROSAT)

WFCMV Wheeled Fuel Consumption Motor Vehicle

WFCS Workshop on Factory Communications Systems

WFE Wavefront Error

WFEA World Federation of Educational Associations

WFEO World Federation of Engineering Organizations

WFF Wallops Flight Facility (NASA) • Well-Formed Formula

WFG Wait For Graph • Waveform Generator

WFHRI Wide-Field High-Resolution Imaging

WFI Fianarantsoa (Madagascar) (location identifier) • Wide-Field Imager • Wide-Field Imaging • Wide-Field Instrument • Wide-FOV Instrument • Work Flow Information

WFIC Wide-Field Imaging CCD

WFIR Wide-Field Imaging in the Red • Wide-Field Imaging Radiometer

WFIS Wide-Field Imaging Spectrograph • Wide-Field Imaging Spectrometer

WFK Frenchville (USA-ME) (location identifier)

Wfl Wohnfläche

WFL Work Flow Language

WFM Wired For Management • Work Flow Management • Work Flow Manager

WFMOS Wide-Field Multiobject Spectrograph • Wide-Field Multiobject Spectroscopy

WFMP Work Force Motivation Plan

WFMS Work Flow Management System

WFMU Weather Fixed Map Unit

WFNA White Fuming Nitric Acid

WFNDEC World Federation of Non-Destructive Evaluation Centers

WFO Weather Forecast Office • Wide-Field Optics • Work For Others

WFOF Wide-Field Optical Filter

WFOV Wide Field Of View

WFP Waiver Focal Point • Warm Frontal Passage • Warm Front Passage • Waste Form Producer • World Food Programme (UNO)

WFPA Wide-Field Plate Archive

WF/PC Wide-Field/Planetary Camera

WFPC Wide-Field and Planetary Camera

WFPCB Wavefront Power Control Board

WFPC2 Second Wide-Field and Planetary Camera

WFPDB Wide-Field Plate Database

WFQ Waste Form Qualification • Weighted Fair Queuing

wfr Wafer

WFR Waveform Receiver

WfS Workgroup for Spectrography

WFS Wavefront Sensor • Wide-Field Survey • Wilhelm-Foerster-Sternwarte (Germany) • WORM File System

WFS&C Wavefront Sensing & Control

WFSC Wavefront Sensing and Control

WFSW World Federation of Scientific Workers

WFT Wavefront Testing • Windowed Fourier Transform

WFTA Winograd Fourier Transform Algorithm

WFTU World Federation of Trade Unions

WFU Wake Forest University (USA-NC) • Withdrawn From Use

WfW Word for Windows

WFW Windows For Workgroups • World Forest Watch

WFWG Windows For Workgroups

WFXT Wide-Field X-ray Telescope

WFZ Weapons Free Zone

Wg Wing

W-G Water-Glycol

W/G Water/Glycol • Within Guidance

WG Grenada (international vehicle registration code) • Wage Grade • Water Gauge • Water-Glycol • Waveform Generator • Waveguide • Wechselgesetz • Weight Guaranteed • West Germany • Wigner Distribution • Windgeschwindigkeit • Wing • Wired Glass • Wire Gauge • Working Group

WGA Wagga Wagga - Forrest Hill (Australia-NSW) (location identifier) • Waveguide Assembly • Weekly Government Abstracts • Writer's Guild of America

WGADBRA Working Group on Archiving and Databases for Radioastronomy

WGAE Working Group on Astronomy Education (AAS)

WGAI Working Group Agenda Item

WGAIT Working Group on Astronomical Imaging Technology

WGAR Working Group on Astronomical Refraction

WGAS Working Group on Astronomical Software

WGB Bahawalnagar (Pakistan) (location identifier)

WGBC Waveguide (operating) Below Cutoff

WGC Warrangal (India) (location identifier)

WGCA Working Group on Computational Astrophysics

WGCV Working Group on Calibration and Validation

WGD Working Group on Data

WGDA Working Group for the Development of Astronomy

WGDAA Working Group for Development of Antarctic Astronomy

WGDC Waveguide Directional Coupler

WGDL Waveguide Delay Line

WGDM Working Group on Data Management

WGE Walgett (Australia-NSW) (location identifier) • Working Group of Experts

WGEN Weather Generator

WGEP Working Group on Energy Policy

W. Ger. West Germany

WGES World's Greatest Environment Statement

WGESP Working Group on Extrasolar Planets (IAU)

WGET Working Group for Encryption and Telecommunications

WGF Waveguide Filter

WGFLSF Working Group on Future Large-Scale Facilities (IAU)

WGGC Working Group on Global Change

WGGCR Working Group on Global Change Research

WGGGI Working Group on Geodesy and Geographic Information

WGH Working Group on Hydrology

WGI World Geophysical Interval • World Glacier Inventory

WGII Working Group on Internal Instrumentation

WGISS Working Group on Information Systems and Services

W. Gl. Wired Glass

WGL Wavefront Generation Language • Waveguide Load • Windows Glyph List • Wissenschaftsgemeinschaft Gottfried (Wilhelm) Leibniz (Germany)

WGLAM Working Group on Limited Area Modeling

WGM Waveguide Meter • Weighted Guideline Method • Working Group Meeting

WGMAM Working Group on Modern Astronomical Methodology

WGmc West Germanic

WGMS World Glacier Monitoring Service

WGN Waitangi (New Zealand) (location identifier) • Working Group News

WGNE Working Group on Numerical Experimentation

WGNEO Working Group on (natural) Near-Earth Objects

WGO What's Going On? • Winchester (USA-VA) (location identifier)

WGP Waingapu (Indonesia) (location identifier) • Working Group on Policy • Working Group Policy

WGPAC Working Group on Professional-Amateur Collaboration (AAS)

WGPSN Working Group on Planetary System Nomenclature

WGR Waste Generation Rate • Waveguide Grating Router • Working Group on Reform • Working Group Report

WGRA Working Group on Resource Assessment

WGRF Working Group on Radiation Fluxes • Working Group on Reference Frames (IAU)

WGRO Whipple Gramma-Ray Observatory

WGRS Working Group on Reference Systems

WGS Weak Gravitational Shear • Weather Graphics System • Work Group Server • Work Group System • Working Group on Software • World Geodetic Spheroid • World Geodetic Survey • World Geodetic System (*)

WGSIM Working Group on Satellite Ionospheric Measurements

WGSPR Working Group for Space Physics Research

WGSS Working Group on Sky Surveys (IAU) • Working Group on Standard Stars (IAU)

WGSSA Working Group on Space Sciences in Africa

WGSS2 Working Group on Space Science in Zambia

WGSU Water-Glycol Service Unit

Wgt Weight

WGT Wangaratta (Australia-VIC) (location identifier) • Weight

WGTC Working Group on Tracking and Computation

WGU Wagau (Papua New Guinea) (location identifier) • Western Governors University (USA)

WGWFI Working Group on Wide-Field Imaging

WGWDA Working Group for the Worldwide Development of Astronomy

WGWWDA Working Group for the Worldwide Development of Astronomy

WGY Wagny (Gabon) (location identifier)

w/h Withholding

W/h Watt/Hour

Wh Watt-Hour • White

W/H Wire Harness

WH Warehouse • Warhead • Warm Halo • Water Heater • White House (USA-DC) • Widrow-Hoff (classifier) • Wire Harness • Wirtshaus • Withholding • Word Hypothesis • World Heritage

WHA Tungsten Heavy Alloy • Waste Handling Area • World Heritage Area

WHA/FH Waste Handling Area Fume Hood

WHAFH Waste Handling Area Fume Hood

WHALIGN Wheel Alignment

WHAM Waveform Hold And Modify • Wisconsin H-Alpha Mapper • Wisconsin Hydrogen-Alpha Mapper

W'hampton Wolverhampton (UK)

WHAT Wind Height And Temperature

WHAT A CAT West Hellenic Arc Tectonics And Calabrian Arc Tectonics

WHB Wash-Hand Basin

WHC Western Hemisphere Cooperation

WHCA War Hazard Compensation Act (USA) • White House Communications Agency (USA-DC)

WHCCAMP White House Commission on Complementary and Alternative Medicine Policy (USA)

WGCF White House Central Files (USA)

WHD Hyder (USA-AK) (location identifier) • Warm Halo and Disk

WHDF William Herschel Deep Field

WHDM Watt Hour Demand Meter

WHDTV Wide High-Definition Television

WHEPP Workshop on High-Energy Physics Phenomenology

Whf Wharf

WHF Wadi Halfa (Sudan) (location identifier)

Whg Wohnung

WHG Walter-Horn-Gesellschaft (Germany)

WHH Boulder - Hiltons Har (USA-CO) (location identifier)

WHI Hwange (Zimbabwe) (location identifier) • Werner-Heisenberg-Institut (Germany)

WHIG Warm Highly Ionized Gas

WHIM Warm Highly Ionized Medium • Warm-Hot Intergalactic Medium

WHIP Wormhole Induction Propulsion

WHIPS Widebeam High-Intensity Pulsed Source

WHIRCAM William Herschel Infrared Camera

WHIRL Wide-angle High-Resolution Line

WHISP Westerbork HI Spiral-galaxy Project • Westerbork HI (survey of) Irregular and Spiral (galaxies)

WHISPER Waves of High frequency and Sounder for Probing of Electron density by Relaxation

WHK Welty - Hobbs - Kulkarni (survey) • Whakatane (New Zealand) (location identifier)

Whl Wheel

WHL Welshpool (Australia-VIC) (location identifier)

WHM Watt Hour Meter • Work Hour Management

WHO Franz Josef (New Zealand) (location identifier) • White House Office • Wilfred Hall Observatory (UK-England) • World Health Organization • Worth Hill Observatory (UK-England)

WHOI Woods Hole Oceanographic Institute (USA-MA)

Whol Wholesale

WHOSIS World Health Organization Statistical Information System

WHP Los Angeles - Whiteman (USA-CA) (location identifier) • WOCE Hydrographic Programme

WHPT While-light and H-alpha Patrol Telescope

WHQ Weak Headed Quasar • Wing Headquarters • World Headquarters

Whr Watt-Hour • Whether

WHR Vail Avon Stolport (USA-CO) (location identifier) • Watt Hour

Whs Warehouse

WHS Whalsay (UK) (location identifier) • White Sands (USA-NM) • World Heritage Site

WHSC White House Science Council (USA)

Whse Warehouse

Whsle Wholesale

Whsmn Warehouseman

Wht White

WHT Wharton (USA-TX) (location identifier) • William Herschel Telescope

WHU Wuhu (China-PRC) (location identifier)

WHYCOS World Hydrological Cycle Observing System

w/i Within

W&I Weighting & Inspection

WI Wallops Island (USA-VA) • Warning Installation • Web Intelligence • Weizmann Institute (Israel) • West Indian • West Indies • When Issued • Windward Islands • Wisconsin (USA) • Workstation Interface • Wrought Iron

WIA Warned In Absentia • Workshop on Implementing Automata • World Institute of Achievement

WIAS Weierstraß-Institut für angewandte Analysis und Stochastik (Germany)

WIB When Interrupt Block • Wissenschaftliche Internationale Bibliographie

WIBNI Wouldn't It Be Nice If

WIC Warning Information Correlation • Westindische Compagnie • Whole Internet Catalog • Wick (UK-Scotland) (location identifier) • Wideband Imaging Camera • WSO Implementation Committee

WICA With-In Commuting Area

WICAT World Institute of Computer-Aided Teaching

WICEM World Industry Conference on Environment Management

WICHE Western Interstate Commission for Higher Education (USA)

WICN Work Item Completion Notice

WICS Warning Information and Correlation System • Worldwide Intelligent Communications System

WICSA Working IFIP/IEEE Conference on Software Architecture

WICU Weather Intercept Control Unit

Wid Widow • Widower

WID Wildenrath (Germany) (location identifier) • Wireless Information Device

WIDA Weather Impact Division Aids

WIDAR Wideband Interferometric Digital Architecture

WIDE Wiring Integration Design

WIDER World Institute for Development Economics Research

WIDL Web Interface Definition Language

WIDM Web Information and Data Management (conference)

WIE Wiesbaden (Germany) (location identifier) • With Immediate Effect

WIEBS Winter Ice Experiment in Beaufort Sea

WIESS Workshop on Industrial Experiences with Systems Software

WIF Water Immersion Facility • Web Impact Factor • Web Interface Facility • Weekly Impact Factor • What If

WIFI Wireless Fidelity

WiFS Wide Field Sensor

WIFS Wide Field Sensor

WIG Wing In Ground (effect)

WIGE Wing In Ground Effect

WIGM Warm Intergalactic Medium

WIHG Wadia Institute of Himalayan Geology (India)

WII Worldwide Information Infrastructure

WIIT Wide-field Imaging Interferometry Testbed

WIK Surfdale (New Zealand) (location identifier)

WIL Nairobi Wilson (Kenya) (location identifier)

WILCO Will Comply

WILMA Wire List Management and Accounting

Wilts Wiltshire (UK-England)

WIM WAP Identity Module • Warm Interstellar Medium • Warm Ionized Medium (*) • Wideband Interconnection Module • Windows - Icons - Mouse • Wireless Identification Module

WIMC Whom It May Concern

WIMP Weakly Interactive Massive Particle • Window - Icon - Mouse - Pointer • Window - Icon - Menu - Pop up • Windows - Icons - Menus - Pointer • Windows - Icons - Mouse - Pull-down

WIMPS Windows - Icons - Menus - Pointer - Scroll

WIMS Weather Information Management System • Work Information Management System

win Windows Backup (file extension)

WIN Walden Information Network • Wallonie Internet • Weather Information Network • Welfare Information Network • Western Information Network • Westlaw Is Natural • Winter • Winton (Australia-QLD) (location identifier) • Wisconsin - Indiana - NOAO (consortium) • Wissenschaftsnetz (Germany) • Work Incentive

WINC Wavelength Insensitive Coupler

WINCE Winter Cloud Experiment

Winch Winchester (UK-England)

W. Ind. West Indian • West Indies

WIND Weather Information Display • Weather Information Network and Display (*)

WINDEX Window Experiment

Wind.I. Windward Islands

WINDII Wind Imaging Interferometer

WINDO Wide Information Network for Data Online

WINDS Weather Information Network and Display System (*) • Weather Information Network Display System

WINDII Wind Imaging Interferometer

WINDSAT Wind Satellite

WINE Windows Emulator • Winter In Northern Europe

WinEn Wind Engineering (periodical code)

WING Western Pacific Integrated Network of GPS • Work and Internship Networking Group

WINGS Web Interactive Network of Government Services • Wireless Internet Gateway System

WinHEC Windows Hardware Engineering Conference

Winn Winnipeg (Canada-MB)

WINNERSS Wisconsin Idea National Network for Education and Research in Space Sciences (USA-WI)

WINS Weather Information Network Service • Windows Internet Name Service • Windows Internet Naming Service (*) • Wireless Integrated Network Sensor • Wiskunde - Informatica - Natuurkunde - Sterrenkunde

WINSOCK Windows Socket

Wint Winter

WINT Winter

WINTER Web Internationalization

WIO Wilcannia (Australia-NSW) (location identifier) • Work It Out

WIP Waste Incineration Plant • Wissenschaftler-Integrations-Programm • Wissenschaftler-Integrations-Projekt • Work In Place • Work In Process • Work In Progress

WIPERS Weakly Ionized Powerful Extragalactic Radio Source

WIPG West Information Publishing Group

WIPICS Work-In-Proces Inventory Control System

WIPIS Whi Is Publishing In Science

WIPLA World Intellectual Private Law Agency (France)

WIPO World Intellectual Property Organization

WIPP Waste Isolation Pilot Plant (USA-NM)

WIPS Weather Image Processing System

WIPT Working Integrated Process Team • Working Integrated Product Team

WIR Wet op de Investeringsrekening • Wide Irradiance Response • Woomera Instrument Range

WIRAC Wideband Radio Access

WIRC Wide-Field Infrared Camera

WIRCAM Wide-field Infrared Camera

WIRDS Weather Information Reporting and Display System

WIRE Web/Internet Resource Executive • Wide-Field Infrared Explorer

WIRESCRIPT Web Interactive Review of Scientific Culture, Research, Innovation Policy and Technology

WIRL West India Reference Library

WIRO Wyoming Infrared Observatory (USA-WY)

WIRP Watershed Integrated Research Program

Wis Wisconsin (USA)

WIS Wallops Island Station (USA-VA) • WAN Interface Sublayer • Wavelength-Independent Switch • Weather Information System • Web-based Information System • Wedge Imaging Spectrometer • Wideband Instrumentation System • Wireless Intercommunications System • Work Information System • WWMCCS Information System

WISARD Web Interface for Searching Archival Research Data • Wideband System for Acquiring and Recording Data

WisBT Wissenschaftliche Berichte (periodical code)

Wisc Wisconsin (USA)

WISC Wechsler Intelligence Scale for Children • Writable Instruction Set Computer

WISE Washington Information Systems for Education • Web Information Systems Engineering (conference) • Whole Information System Expert • Wisconsin Initiative for Space Education (USA-WI) • WordPerfect Information System Environment • Workshop on Information Systems and Economics • World Information Service for Energy • World Inventory of Soil Emissions • Worldwide Information System for Engineering

WISH Westerbork (survey) In the Southern Hemisphere

WISI World Index of Space Imagery

WISIWYS What I See Is What You See

WISLAD World Information System for Library Automation Development

WISP Waves In Space Plasma • Wide-field Imaging Survey Polarimeter • Wide-range Imaging Spectrometer • Winter Icing and Storms Project • Workstation Integrated Security Program

WISPI Wide-field Spectral Imager

Wiss Wissenschaft • Wissenschaftlich

WISS World Institute on Space and Society

Wissensch Wissenschaft • Wissenschaftlich

WiStG Wirtschaftsstrafgesetz

WisZe Wissenschaftliche Zeitschrift (periodical code)

Wit Witness

WIT Washington Interactive Television • Waterford Institute of Technology (Ireland) • Web Interactive Talk • Wireless Internetworking Testbed • Wittenoom (Australia-WA) (location identifier) • Work Improvement Team

WITFOR World Information Technology Forum

Withdrl Withdrawal

Witht Without

WITID Wissenschafliche und Technische Information und Dokumentation

WITL Wireless In The Loop

WITS Washington Interagency Telecommunications System (USA-DC) • Web Infrared Toolshed • Windows Integrated Test Suite • Workshop on Internet Technology and Systems • Workshop on Issues in the Theory of Security • World Ionosphere Thermosphere Study • Worldwide Information and Trade System

WIU Web Information Unit • Western Illinois University (USA-IL) • Witu (Papua New Guinea) (location identifier)

WIVA Wireless Internet Venture Association

WIX Wait for Index

WIXC Wavelength Interchanging Crossconnect

WIYN Wisconsin - Indiana - Yale - NOAO (consortium)

wiz Word Wizard (file extension)

WJA Woja (Marshall Islands) (location identifier) • Working Joint Arrangement

WJEC Welsh Joint Education Committee (UK-Wales)

WJF Lancaster - William J. Fox (USA-CA) (location identifier)

WJR Wajir (Kenya) (location identifier)

WJU Wheeling Jesuit University (USA-WV)

Wk Walk • Weak • Week • Work

WK Weak • Well Known • Wissenschaftskolleg

WKA Wanaka (New Zealand) (location identifier)

WKB Warracknabeal (Australia-VIC) (location identifier) • Wentzel - Kramers - Brillouin (theory) • Williams - Kenderdine - Baldwin (catalog) • Workbook

WKAA Western Kentucky Amateur Astronomers (USA-KY)

WKBCH Workbench

WKBJ Wentzel - Kramers - Brillouin - Jeffreys (method)

Wkdays Weekdays

Wkds Weekdays

WKEND Weekend

Wkends Weekends

WKF Well-Known Factor • Worksheet

Wkg Working

WKG Working

WKJ Wakkanai (Japan) (location identifier)

WKK Aleknagik (USA-AK) (location identifier)

WKL Waikoloa (USA-HI) (location identifier)

WKLD Workload

Wkly Weekly

WKM Wechselkursmechanismus • Hwange National Park (Zimbabwe) (location identifier)

WKN Wakunai (Papua New Guinea) (location identifier) • Weaken • Weakening

WKQ Spreadsheet

Wkr Worker

WKR Walkers Cay (Bahamas) (location identifier)

wks Works Module (file extension)

Wks Weeks • Works • Workshop

WKS Weeks • Well-Known Service • Werner-Keller-Schinke • Worksheet

WkSchG Wohnraumkündigungsschutzgesetz

Wkshp Workshop

WKU Western Kentucky University (USA-KY)

WKZ Compressed Spreadsheet

wl Workload

wL westliche Länge

WL Saint Lucia (international vehicle registration code) • Wagon-Lit • Waiting List • Water Line • Waterloopkundig Laboratorium • Wavelength • Well • Werkstofflabor • Western Laboratories • White Light • Whiteshell Laboratories • Will • Wing Left • Wing Load • Wire List • Word Line • Working Level • Workload

WLA Wallal (Australia-WA) (location identifier) • Wire Line Adapter

wLAN Wireless Local Area Network

WLAN Wireless Local Area Network

WLANA Wireless Local Area Network Association

Wlb Wallboard

WLB Labouchere Bay (USA-AK) (location identifier)

WLC Walcha (Australia-NSW) (location identifier) • White-Light Channel • White-Light Cooling • White-Light Coronagraph • Winnerless Competition • Workload License Charge

WLCS Workload and Cosr Schedule

Wld Welded • Would

WLD Wärmeleitfähigkeitsdetektor • West-Longitude Date • Winfield - Arkansas City (USA-KS) (location identifier) • Won - Lost - Drawn

WLDC Wall-Less Drift Chamber

WLDD World-Level Decision Diagram

Wldr Welder

WLE Weak Localization Effect • Wing Leading Edge

WLF Futuna Islands (Wallis &_) (ISO country code) • Wallis & Futuna Islands (ISO country code) • White-Light Flare • White-Light Fringe

WLG Wellington (New Zealand) (location identifier)

WLH Walaha (Vanuatu) (location identifier)

WLI Workload Index

WLIF Wireless LAN Interoperability Forum

Wlk Walk

WLK Selawik (USA-AK) (location identifier)

WLL Wireless Local Loop • Wollogorang (Australia-QLD) (location identifier)

WLLN Weak Law of Large Numbers

WLM Waltham (USA-MA) (location identifier) • Water Line Monitor • Wavelength Modulation • Wolf - Lundmark - Melotte (catalog) • Wolf - Lundmark - Melotte (galaxy) (*) • Working-Level Month • Workload Manager

Wln Wellington (New Zealand)

WLN Little Naukati (USA-AK) (location identifier) • Washington Library Network • Western Library Network • Wiswesser Line Notation

WLO Waterloo (Australia-NT) (location identifier) • Windows Library for OS/2

W. Long. West Longitude

WLOS Window Line Of Sight

WLR Loring (USA-AK) (location identifier) • Wind-Luminosity Relationship • Wind (momentum) - Luminosity Relationship • Wire Line Receiver • Wrong Length Record

WLRG Weak Line Radio Galaxy

WLRS Wetzels Laser Ranging System

WLS Wallis Island (location identifier) • Wavelength Scanning • Wavelength Shifter • Weighted Least Squares • Wells • Werkgroep Leidse Sterrewacht (Netherlands) • Western Launch Site • Wide Latitude Storm • World Library Service • World Listening Service

WLSE Weighted Least Squares Estimation

WLSP World List of Scientific Periodicals

WLT White-Light Telescope • Wins - Losses - Ties

WLU Wilfrid Laurier University (Canada-ON)

WLV Wide Load Vehicle

WLW Willows - Glenn County (USA-CA) (location identifier)

Wly Westerly

WLY Westerly

Wm William

W/m Words/Minute

W/M Weight/Measurement

WM Wagon de Marchandises • Waste Management • Wattmeter • Webmaster • Western Modular • Whisker Mapping • Window Manager • Workload Manager • Work Manager • Work Measurement

WMA Mandritsara (Madagascar) (location identifier) • Windows Media Audio • Winsurf Mainframe Access

WMAA Western Montana Astronomical Association (USA-MT)

WMAC Windows Media Access Control • Wireless Media Access Control

WMAN Wireless Metropolitan Area Network

WMAP Wilkinson Microwave Anisotropy Probe

WMASH Wireless Mobile Applications and Services on Hitspots (conference)

WMB Warrnambool (Australia-VIC) (location identifier)

WMC Washington Multiplicity Catalog • Waste Management Compartment • Ways and Means Committee • Western Maryland College (USA-MD) • Winnemucca (USA-NV) (location identifier) • World Manufacturing Congress • World Meteorological Centre

WMCI Weighted Median Citation Index

WMCSA Workshop on Mobile Computing Systems and Applications

WMD Mandabe (Madagascar) (location identifier) • Water Management District • Weapon of Mass Destruction • World Meteorological Day

WMDB Waste Management Database

WME Mount Keith (Australia-WA) (location identifier) • Waste Management Engineer • Window Meteoroid Experiment

WMES Workload Management Expert System

wmf Windows Media Format (file extension) • Windows Metafile (file extension)

WMF Windows Media Format • Windows Metafile • Window Metafile Format • World Monuments Fund

WMH Mountain Home (USA-AR) (location identifier) • Welty - Morton - Hobbs (survey)

WMI Windows Management Instrumentation • World Metal Index • World Meteorological Interval

Wmk Watermark

WMK Meyers Chuck (USA-AK) (location identifier)

WMKO W.M. Keck Observatory (USA-HI)

WML Malaimbandy (Madagascar) (location identifier) • Wind Momentum-Luminosity (relation) • Wireless Markup Language

WMLR Wind-Momentum-Luminosity Relation

WMM World Magnetic Model

WMN Maroantsetra (Madagascar) (location identifier)

WMO White Mountain (USA-AK) (location identifier) • World Meteorological Organization

WMOP Weather Observation Master Plan

WMP Waste Management Plan • Weather Modification Program • WindowsMaker Professional • With Much Pleasure

WMPI Weakly Mean Percentage Increase

WMQ Why More Questions?

WMR Mananara (Madagascar) (location identifier) • Weekly Management Review

WMRA Write Many Read Always

WMRF Wide-sense Markov Random Field

WMS Warehouse Management System • Waste Management System • Watchdog Monitoring System • Wind Measuring System • Workload Management System • Work Management System • World Magnetic Survey • World Mapping System

WMSC Weather Message Switching Center

WMSCR Weather Message Switching Center Replacement

WMT Web Marketing Tool • Weighting More Than • West-Meridian Time • Windows Media Technologies • Wrongly Merge Total

WMU Western Michigan University (USA-MI) • World Maritime University

WMV Madirovalo (Madagascar) (location identifier) • Windows Media Video

WMX Wamena (Indonesia) (location identifier) • Windows Manager for X

WMY World Mathematical Year (2000)

WN Nitrogen Wolf-Rayet (star) • Warning Notice • Wideband Non-secure • Will not

WNA Napakiak (USA-AK) (location identifier) • Wireless Network Access

WNAA Workshop on Numerical Analysis and its Applications

WNAS Wireless Network Access System

WNB Will Not Be

WNC Nichin Cove (USA-AK) (location identifier) • Tuxekan Island - Nichen Cove (USA-AK) (location identifier)

Wnd Waarnemend • Wound

WND Wind • Windarra (Australia-WA) (location identifier)

WNDP With No Down Payment • World-wide Numbering and Dialing Plan

Wndr Winder

WNE Wora na Ye (Gabon) (location identifier)

WNG Warning

WNGO Weakly Non-linear Geometric Optics

WNIC Wide-area Network Interface Coprocessor

WNILS WHOIS and Network Information Lookup service

WNIM Wide-area Network Interface Module

WNK Weitverkehrs-Netzknoten

WNL Late-type Nitrogen Wolf-Rayet (star) • Within Normal Limits

WNLRT Weakly Non-Linear Ray Theory

WNM Warm Neutral Material • Warm Neutral Medium (*) • Warm Nuclear Medium

WNN Wunnummin Lake (Canada-ON) (location identifier)

WNP Naga (Philippines) (location identifier) • Will Not Proceed • Will Not Process • World Numeracy Program

Wnr Weeknummer

WNr Week Number

WNR Windorah (Australia-QLD) (location identifier)

WNRC Washington National Records Center (USA-DC)

WNS Nawabshah (Pakistan) (location identifier) • World News Service • World-wide Navigation System

WNSL Wright Nuclear Structure Laboratory (USA-CT)

WNSSG Workshop on Network and System Support for Games

WNU Wanuma (Papua New Guinea) (location identifier)

WNV World New Vistas

WNW West-Northwest • West-Northwestern • What's New on the Web

WNWRN West-Northwestern

WNWWD West-Northwestward

WNY Wynyard (Australia-TAS) (location identifier)

WNZ Wenzhou (China-PRC) (location identifier)

w/o Without • Written Off

wo Waaronder • wie oben • Without • Wolof (ISO language code)

W/O Without • Written Of

WO Oxygen Wolf-Rayet (star) • Wait Order • Warning Order • Warrant Officer • Wednesdays Only • Westerlund-Olander (catalog) • Western Observatorium (USA-CA) • Wireless Operator • Wise Observatory (Israel) • Work Order • Write Once • Write Only • Write Out • Written Order

WOA Web Object Application • Without Answers • Wonerara (Papua New Guinea) (location identifier) • World Ocean Atlas

WOB Suttonheath Woodbridge (UK-England) (location identifier)

WoBauFG Wohnungsbauförderungsgesetz

WoBauG Wohnungsbaugesetz

WoBindG Wohnungsbindungsgesetz

WOC Weekly Operating Cost • Without Compensation • Work Order Control

WOCAT World Overview of Conservation Activities and Technologies

WOCE World Ocean Circulation Experiment

WOCS WIYN Open Cluster Study • WIYN Open Cluster Survey (*) • Workshop on Optics and Computer Science

WOD Wood River (USA-AK) (location identifier) • Word Of Day

WODC World Ozone Data Center

WODD World Oceanographic Data Display

WODDIN Worldwide Online Data and Documents Intelligence System

WODES Workshop On Discrete Event Systems

WOE Weight Of Evidence • Without Enclosure • Without Equipment

WOFOST World Food Study

WOFS Write-Once File System

WOG Water - Oil - Gas • Weekly Orbit Geometry (file) • With Other Goods • Woodgreen (Australia-NT) (location identifier)

WoGG Wohngeldgesetz

WOGS World Organization for Cybernetics and Systems

WoGV Wohngeldverordnung

WohneigFG Wohneigentumsförderungsgesetz

WOI Wind Outflow Inflow • Wologissi (Liberia) (location identifier)

WOIEP World Office of Information on Environmental Problems

WoK Web of Knowledge (ISI)

WOK Wonken (Venezuela) (location identifier)

WOL Wollongong (Australia-NSW) (location identifier)

WOLF Work Order Logistics File

Woll Wollongong (Australia-NSW)

WOM Web Object Management • Wheel Off-loading Mode • Word Of Month • Write-Only Memory • Write Optical Memory

WOMBAT Waste Of Money, Brains, And Time • Wavelength-Oriented Microwave Background Analysis Team

WON Wondoola (Australia-QLD) (location identifier) • Work Order Number

WONG Weight On Nose Gear

WoO Werk ohne Opuszahl

WOO Western Operations Office • Woodchopper (USA-AK) (location identifier) • World Oceanographic Organization

WOOL Words Out of the Ordinary Language

WOON White-Object-Oriented Night (Russia)

WOP Web Offensive Presence • Weekly Observing Plan • Weekly Operating Plan • Wing Outer Panel • With Other Property • Without Papers • Without Passport • Without Payment • Without Personnel • Without Preference • Without Priority

WOPE Without Personnel and Equipment • Without Personnel or Equipment

WoPG Wohnungsbauprämiengesetz

WOPPLOT Workshop On Parallel Processing: Logic, Organization, and Technology

WOR Weekly Operating Report • Without Our Responsibility • Work Order Record • Work Order Release

WORAM Word-Oriented Random Access Memory

WORCC World Optical depth Research and Calibration Center

Worcs Worcestershire (UK-England)

WORF Window Observational Research Facility • WWW Observing Remote Facility

WO/RM Write Once/Read Many

WORM Write Once - Read Many

WORMTOMS WORM To Mass Storage System

WOROM Write-Only Read-Only Memory

WoS Web of Science (ISI)

WOS White Oval Spot (Jupiter) • Workplace Operating System • Workstation Operating System

WOSA Windows Open Services Architecture • Windows Open Systems Architecture

WOSAC Worldwide Synchronization of Atomic Clocks

WOSB Weather Observation Site Building

WOSC World Organization on Systemics and Cybernetics

WOSE Weather Officer in Space Experiment

WOSP Workshop On Software and Performance

WoStd Wochenstunde

WOT Wide Open Throttle • Wonan (China-ROC) (location identifier)

WOTRO Wetenschappelijk Onderzoek van de Tropen in Ontwikkelingslanden

WOTS Wallops Optical Tracking Station • Worldwide Origin Thread System

WOU Western Oregon University (USA-OR)

WoVermittG Wohnvermittlungsgesetz

WOVO World Organization of Volcano Observatories

WOW Waiting On Weather • Web - Oracle - Web • Weight On Wheels • Willow (USA-AK) (location identifier) • Wisconsin Observer's Weekend • World Of Wonders • Worst On Worst

WoWMoM Workshop on Wireless Mobile Multimedia (ACM)

WOY Week Of Year

WOYDOY Week Of Year/Day Of Year (calendar)

WOZ Wahre Ortszeit

wp WordPerfect Document (file extension)

W/P Waypoint

WP Wagon Particulier • Wagon Planeur • Wagon-Poste • Warsaw Pact • Waste Paper • Water Plane • Water Point • Water Processor • Water Pump • Waterproof • Way Point • Weapons Platform • Weather Permitting • Weatherproof • Wegener-Panda • Western Province (South Africa) • White Paper • White Phosphorus • Will Proceed • Wind Profile • Wind Profiles • Wire Payment • Without Prejudice • WordPerfect • Word Processing • Word Processor • Working Package • Working Paper • Working Party • Working Pressure • Working Protocol • Work Package • Work Plan • Work Program • Work Progress • Wrist Pitch • Write Protect • Write Protected

WPA Puerto Aisen (Chile) (location identifier) • Waiting Point Approximation • Water Pump Assembly • With Particular Average • Work Package Address • Work Package Agreement • Works Progress Administration • Works Projects Administration (*) • World Patents Abstracts • World Press Archives

WPABX Wireless Private Automatic Branch Exchange

WPAFB Wright-Patterson Air Force Base (USA-OH)

WPAN Wireless Personal Area Network

WPANL Wright-Patterson Aeronautical Laboratory

WPB Port Berge (Madagascar) (location identifier) • Waste Paper Basket • Write Printer Binary

Wpc Watts per Candle

WPC Pincher Creek (Canada-AB) (location identifier) • Washington Photoelectric Catalogue • Wave-Parallel Computing • Widening Participation Committee • Wired Program Computer • Wood-Plastic Composite • Word Processing Center • Work Package Concept • World Power Conference

WPCF Water Pollution Control Federation

wpd Windows Printer Description (file extension) • WordPerfect Document (file extension)

WPD WordPerfect under DOS • Work Package Description • Work Product Description • Work Program Directive • Write Printer Decimal

WPDA Wireless Personal Digital Assistant

WPDL Workflow Process Definition Language

WPDN Wind Profile Demonstration Network

WPE Word Processing Equipment

WPES Wideband Polyelastic Sampling

WPESS Within Pulse Electronic Sector Scanning

WPF Wissenschaftliche Projektführung • Work Process Flow

Wpfl Wahlpflicht • Wahlpflichtfach

WpflG Wehrpflichtgesetz

wpg WordPerfect Graphics (file extension)

Wpg Waterproofing

WPG WordPerfect Graphics • Work Package Grouping

WPGMA Weighted Pair Group Method using arithmetic Averages

WPGMC Weighted Pair Group Method using Centroids

WPHD Write-Protect Hard Disk

WPI Wholesale Price Index • Worcester Polytechnic Institute (USA-MA) • Work Process Indicator • Work Progress Indicator • World Patent Index • World Patent Information • World Press Institute

WPIE Warning Plan Interface Equipment

WPK WordPerfect Keyboard • Wrotham Park (Australia-QLD) (location identifier)

WPKI Wireless Public Key Infrastructure

WPL Powell Lake (Canada-BC) (location identifier) • Warning-Point Level • Wave Propagation Laboratory • Wesenheit - Period - Luminosity • Windows Personal Librarian

WPLS Western Pacific Laser Satellite

WPLTN Western Pacific Laser Tracking Network

WPLTO Western Plateau

wpm WordPerfect Macro (file extension) • Words per Minute

WPM Weighted Product Model • Wipim (Papua New Guinea) (location identifier) • WordPerfect Macro • Words Per Minute • Work Package Management • Work Package Manager • Write Program Memory

WPMCP Work Package Manpower and Cost Plan

WPMPR Work Package Milestone Progress Report

Wpn Weapon

WPO Paonia - North Fork Valley (USA-CO) (location identifier) • Wirtschaftsprüfungsordnung • Woomera Project Office • World Packaging Organization

WPOM Word Processing Output Microform

WPOP Winds Product Oversight Panel

WPOS Workplace Operating System

WPP Water Pump Package • Weibull Probability Paper • White Pages Pilot • White Patch Panel • Work Package Plan

WPPG Workshop on Programming Paradigm in Graphics

WPPS Work Package Planning Sheet

WPPSI Wechsler Preschool and Primary Scale of Intelligence

WPR Porvenir (Chile) (location identifier) • Web-based Peer Review

WPRAC Working Program Resource Advisory Committee

wps Word Processor Module (file extension) • Words per Second

WpS Watts per Steradian • Words per Second

WPS Wallops Island Tracking Station (NASA) • Watts Per Steradian • Weather Protection System • Windows Printing System • With Prior Service • Word Processing Specialist • Word Processing System • Words Per Second • Working Prototype System • Work Package Structure • Workplace Shell • World Patent Service

WPSF Waste Processing and Storage Facility

wpt WordPerfect Template (file extension)

Wpt Waypoint

WPT Windows Processing Tools • Wireless Power Transmission • World Press Trends

WPU Puerto Williams (Chile) (location identifier) • Well Processing Unit • William Patterson University (USA-NJ) • William Penn University (USA-IA)

WPVM Windows Parallel Virtual Machine

WPW WordPerfect under Windows

WPWOD Will Proceed Without Delay

WQ Water Quality • Winter Quality

WQA Water Quality Act (USA) • Water Quality Association

WQAS Water Quality Analysis System

WQC Water Quality Criteria • Waveform Quality Center

WQCD Water Quality Criteria Document

WQD Water Quality Database

WQF Workstation Query Facility

WQL While Quantities Last

WQM Water Quality Management • Water Quality Monitor

WQP Water Quality Parameter

WQR Waste Qualification Report

WQS Water Quality Standard

wr Water Rail

Wr Washroom • Writer

W/R Water/Rock • Write/Read

WR Wagon-Réservoir • Wagon-Restaurant • Waiver Request • Warehouse Receipt • Water Resources • Weather Reconnaissance • Weber (variable star) • Western Range • Western Region • Wet Runway • Whole Rock • Windrichtung • Wing Right • Wire Receive • Wire Recorder • Wire Reply • Wissenschaftsrat • Wolf-Rayet • Word Recognition • Word Restoration • Work Request • World Record • Wrinkle Ridge • Wrist Roll • Write

WRA Warder (Ethiopia) (location identifier) • Web Research Alliance • Whiteshell Research Area

WRAM Window Random Access Memory

WRAP Water Resources Assessment Program • Western Region Applications Program • Winter Rainfall Acceptance Proposal • Work Risk Analysis Procedure

Wrb Wardrobe

WRB Macon Robins (USA-CA) (location identifier) • Warranty Review Board • Web Request Broker • World Reference Base

WRBP W. Russell Blake Planetarium (USA-MA)

WRC Water Research Commission • Water Resources Council • Weather Relay Center • Weekly Relative Citedness • World Radiation Centre • World Radiocommunication Conference

WRCB Water Resources Control Board

WRCI Weighted Relative Citation Index

WRD Water Resources Department • Water Resources Division

WRDC World Radiation Data Centre • Wright Research and Development Center

WRDD Water Resources Development Department

WRE Whangarei (New Zealand) (location identifier)

WREBUS Water Reactor Extended Burnup Study

WRED Weighted Random Early Detection • Weighted Random Early Discord

W. Ref. With Reference to

WREN World Renewable Energy Network

WRF Waste Retrieval Facility • Weather Research Forecast

WRG Warm Residual Gas • Wrangell (USA-AK) (location identifier)

WRGP Weak Reduced Game Property • Western Regional Graduate Program (USA)

WRH World Radio Handbook

wri Microsoft Write Module (file extension)

WRI Fort Dix - McGuire (USA-NJ) (location identifier) • Water Resources Information • Web Report Interface • World Resources Institute (USA-DC) • Write

WRIST Weather Radar Identification of Severe Thunderstorms

Wrk Work

WRK Windows Resource Kit • Wreck

Wrkg Working

WRL Willow Run Laboratory (NASA) • Wing Reference Line • Worland (USA-WY) (location identifier)

WRLC Washington Research Library Consortium

WRM Warm • Water Recovery and Management • Water Resources Management • Waves Random Media (periodical code) • Word Recognition Module • Work Rate Master

WRMAG Water Resources Management Action Group

WRMFNT Warm Front

WRMLTK What the Respondent Might Like To Know

wrmn Wireman

WRMS Weighted Root Mean Square

WRMSR Write Machine Specific Register

Wrn Warning • Western

WRN Wideband Radio Networking • World Radio Network

WRNG Warning

Wrnt Warrant

WRO Warren Rupp Observatory (USA-OH) • Work Release Order • Wroclaw (Poland) (location identifier)

WroRe Wroclaw Astronomical Observatory Reprints (periodical code)

WRP Water Resource Planning • Weight Reduction Proposal • Wiener Random Process • Wing Reference Plane

WRQ Walker - Richer - Quinn (company)

WRR Water Resources Research (periodical code) • Weighted Round Robin • World Radiometric Reference • World Resources Report

WRRC Water Resources Research Center

WRS WAAS Reference Station • Warning and Report System • Waveguide Raman Spectroscopy • Wavelet Redshift • Wave Radiometer System • Wide-Range Sensor • Word Recognition System • World Reference System • Worldwide Reference System

WRSIC Water Resources Scientific Information Center

WRSP World Register of Scientific Periodicals

Wrt Wrought

WRT With Regards To • With Request To • With Respect To • Write

Wrtr Writer

WRU Weighting and Rating Unit • Who Are You?

WRW Warrawagine (Australia-WA) (location identifier) • Waste Reduction Workshop

WRY Westray (UK-Scotland) (location identifier)

W/S Workstation

W&S Wärme & Stoffübertragung (periodical code) • Whisky & Soda

WS Samoa Occidentale (code ISO de pays) • Samoa (Western_) (international vehicle registration code) • Samoa (Western_) (ISO country code) • Wait State • Walk Sequence • Water Servicer • Water Soluble • Water Supply • Water Surface • Water System • Weather Service • Weather Ship • Weather Station • Weather Stripping • Western Samoa (international vehicle registration code) • Western Samoa (ISO country code) • Windshear • Wind Shield • Wind Speed • Wind Sonde • Wind Speed • Winter Solstice • Withholding Statement • Word Send • Word Spotting • WordStar • Working Storage • Work Statement • Workstation • World Service

WSA Wasua (Papua New Guinea) (location identifier) • Wire Subscriber Access • Workshop on Static Analysis

WSAA Waveguide Slot Array Antenna

WSAAG Western Sydney Amateur Astronomy Group (Australia-NSW)

WSAL Winston-Salem Astronomical League (USA-NC)

W. Sam. Western Samoa

WSAPI Web Site Application Programming Interface

WSAR Weekly Significant Activity Report

WSB Steamboat Bay (USA-AK) (location identifier) • Water Spray Boiler • Weak Stability Boundary

WSBS Wide-Sweep Backscatter Sounder

WSc Wind Scatterometer

WSC Water Survey of Canada • Wavelength-Selective Coupler • Wayne State College (USA-NE) • Web Servlet Component • Westfield State College (USA-MA) • White Sands Complex (USA-NM) • Whole Sun Catalog • Wideband Signal Conditioner • Wind Scatterometer • Winter Simulation Conference • World Solar Challenge (Australia) • World Space Congress

WSCC Workstation Control Center

WSCG West Superclusters of Galaxies

WSCMO Weather Service Contract Meteorological Observatory

WSCOM Workstation Communications

WSD Water Saturation Deficit • Wet Server Director • White Sands - Condron (USA-NM) (location identifier) • Wholly Significant Difference • Wieland System Design • Wind Speed Detector • Working Stress Design • Work Standard Day

WSDA Web Services Discovery Architecture • World Storage, Documentation and Abstracting

WSDB Working Survey Database

WSDDM Worldwide Solutions Design and Delivery Methods

WSDL Web Services Definition Language (*) • Web Services Description Language • Web Structure Definition Language

WSDRN Wide Satellite Data Relay Network

WSDS Working Space Data Set

WSE Web Search Engine • Workstation Environment • Workstation Support Environment

WSEN Western Space Education Network (Canada-SK)

WSF Sarichef (USA-AK) (location identifier) • Wake Shield Facility • Water Soluble Fraction • Workstation Function • World Space Foundation

WSFA Workstation Functional Analysis

WSFL Web Services Flow Language

WSFO Weather Service Forecast Office

WSFS World Science Fiction Society

WSG Washington County (USA-PA) (location identifier) • Wehrsoldgesetz • Weltstandardgruppe • World Standard Group

WSGC Wisconsin Space Grant Consortium (USA-WI)

WSGR White Sands Ground Relay

WSGS White Sands Ground Station (USA-NM)

WSGT White Sands Ground Terminal

w/sh Workshop

Wsh Workshop

WSH Shirley Brookhaven (USA-NY) (location identifier) • Windows Script Host • Windows Scripting Host • World Spectra Heritage

WSHFT Wind Shift

Wshg Washing

Wshld Windshield

Wshr Washer

WSI Wafer-Scale Integration • Whole-Sky Imager • Wind Speed Indicator • World Synoptic Interval

WSID Workstation Identification

WSIL Web Services Inspection Language

WSIO Web Services Interoperability Organization

WSIR Wide Swath Imaging Radar

WSIS World Space Industry Survey

WSJ San Juan (USA-AK) (location identifier) • Wall Street Journal (periodical code)

WSJE Wall Street Journal Europe

WSK Webmaster Starter Kit

WSL Wallonia Space Logistics (Belgium) • Westerbork Source List

WSM Samoa (Western) (ISO country code) • Samoa Occidentale (code ISO de pays) • Watershed Model • Weighted Sum Model • Western Samoa (ISO country code) • Wet Satellite Model • Whole Sun Month • Wigner-Seitz Method • Wiseman (USA-AK) (location identifier)

WSMGR Workstation Manager

WSMO Weather Service Meteorological Observatory

WSN South Naknek (USA-AK) (location identifier) • Washington State Network (USA-WA) • Water-Soluble Nitrogen

WSNA Wireless Sensor Networks and Applications (conference)

WSO Washabo (Suriname) (location identifier) • Water Servicer Operator • Weapon System Officer • Weather Service Office • Western Support Office (NASA) • Wilcox Solar Observatory (USA-CA) • Wing Staff Officer • World Space Observatory

WSOF Wavelength-Shifting Optical Fibre

WSOLOP Workstation Online Operational Program

WSOM Weather Service Operations Manual

WSP Wagon-Salon Pullman • Waspam (Nicaragua) (location identifier) • Water Supply Paper • Water Supply Point • Whirpool Star Party (Ireland) • Wideband Signal Processor • Winter Star Party (USA-FL) • Wireless Session Protocol • Work Sequence Plan • Work Simplification Program • Workstation Processor • Workstation Security Package • World Scientific Publishing

WSPC World Scientific Publishing Co.

WSPEC Workshop on Security and Privacy in Electronic Commerce

WSPR Weapon System Problem Report

WSPS Worldwide Standard Port System

WSQ Wavelet-packet Scalar Quantization • Web-Supported Query

WSQJ Webb Society Quarterly Journal

WSR Wasior (Indonesia) (location identifier) • Weak Signal Reception • Weather Search Radar • Weather Service Radar • Weather Surveillance Radar • Weekly Summary Report • Weltstrahlungsreferenz • Winter Solstice Sunrise • Workstation Replacement

WSRD Worldwide Software Resource Directory

WSRH Weather Service Regional Headquarters

WSRT Westerbork Synthesis Radio Telescope (Netherlands)

WSS Wall Shear Stress • Wavelength-Selective Switch • Wide-field Spectroscopic Survey • Winter Solstice Sunset • Workflow Suspense System • Workstation Server • Workstation Software System • Work Summarization System

WSSD World Summit on Sustainable Development

WSSI World Seismic Safety Initiative

WSSL Washington Select Star List

WSSN World Standards Services Network

WSSP White Sands Star Party (USA-NM)

WSSR Winter Solstice Sunrise

WSSU Weather Service Support Unit • Winston-Salem State University (USA-NC)

Wst Wetterstation

WST Tala (Samoa) (ISO currency code) • Weak Shock Theory • Weapon System Trainer • Westerly (USA-RI) (location identifier) • Wholesale Sales Tax • Write Symbol Table

WSTA Weapon System Testability Analyzer

WSTAG West of Scotland Training and Accreditation Group (UK-Scotland)

WSTAWG Weapon Systems Technical Architecture Working Group

WSTB Water Science and Technology Board • Westerbork (Netherlands)

WStDV Wechselsteuer-Durchführungsverordnung

WSTF White Sands Test Facility (USA-NM)

WStG Wechselsteuergesetz

WSTN World Service Television News

WSTV World Service Television • W-Sensor Test Vehicle

WSU Washington State University (USA-WA) • Wasu (Papua New Guinea) (location identifier) • Water Servicing Unit • Wayne State University (USA-MI) • Weather Support Unit • Weber State University (USA-UT) • Wichita State University (USA-KS) • Wright State University (USA-OH)

WSUIPA Washington State University International Phonetic Alphabet

WSV Weapon System Vehicle • Wideband Secure Voice

WSVN Wideband Secure Voice Network

WSW West-Southwest • West-Southwestern

WSWRN West-Southwestern

WSWWD West-Southwestward

WSX Westsound (USA-WA) (location identifier)

WSXC Wavelength-Selective Crossconnect

WSY Airlie Beach - Whitsunday (Australia-QLD) (location identifier)

WSZ Westport (New Zealand) (location identifier)

Wt Warrant • Weight • Without

W&T Wear & Tear

W/T Walkie-Talkie • Wind Tunnel

WT Waist Tether • Walkie-Talkie • Watchdog Timer • Watch Time • Water Tank • Watertight • Waveguide Transmission • Wavelet Transform • Weather Trend • Weight • Wind Tunnel • Winter Time • Wireless Telegraphy • Wireless Telephony • Wireless Transmitter • Wire Ticket • Wire Transmit • With Tape • Write Through

WTA Tambohorano (Madagascar) (location identifier) • West Texas Astronomers (USA-TX) • Westwälder Tag der Astronomie (Germany) • Willingness To Avoid • Winner Takes All • Wireless Telephony Application • Wire Traceability and Accountability

WTAI Wireless Telephony Application Interface

WTB Wire To Board • Wissenschaftliche Taschenbücher

WTC New York - World Trade Center (USA-NY) (location identifier) • Workload Technician Code • World Trade Center

Wtd Wanted • Warranted

WTD West End (Bahamas) (location identifier) • Wind Tunnel Data

WTDC World Telecommunication Development Conference

WTDM Wavelength Time Division Multiplexing

WTE Eing Trailing Edge • World Tapes for Education • Wotje Island (Marshall Islands) (location identifier)

Wtf Waterford

WTF Wednesday - Thursday - Friday • Wideband Tunable Filter

Wtg Waiting • Weighting

WTG Way To Go • Wind Turbine Generator

WTH Whole-Tree Harvest

Wthr Weather • Weather (periodical code)

WTHR Weather

WTI Wetenschappelijke en Technische Informatie • Wissenschaftliche und Technische Information • World Translation Index

WTIC World Trade Information Center

WTID Wissenschaftlich-Technische Information und Dokumentation

WTIE Wastewater Treatment and Information Exchange

WTIS World Trade Information Service • World Trade Information System

WTK Noatak (USA-AK) (location identifier)

WTL Tuntutuliak (USA-AK) (location identifier) • Won - Tied - Lost

WTLO Water Terminal Logistic Office

WTLS Wireless Transport Layer Security

WTM Waste Tank Maintenance • Wind Tunnel Model

WTMC Wetenschap, Technologie en Moderne Cultuur

WTN Waddington (UK) (location identifier) • Worldwide Television Network • Worldwide Television News

WTO Warsaw Treaty Organization • World Tourism Organization • World Trade Organization (formerly GATT) • Wotho Island (Marshall Islands) (location identifier) • Write To Operator

WTOR Write To Operator with Reply

WTP Willingness To Pay • Wireless Transaction Protocol • Woitape (Papua New Guinea) (location identifier)

WTPF World Telecommunication Policy Forum

Wtr Water • Winter • Writer

WTR Water • Weighted Total Rank • Western Test Range (USA) • White River (USA-AZ) (location identifier) • Wireless Technology Research • Work Transfer Record • Work Transfer Request

WTROD Western Test Range Operations Division (NASA)

Wtrprf Waterproof

WTRS Waters

Wtrtt Watertight

WTS Tsiroanomandidy (Madagascar) (location identifier) • Wallops Tracking Station (USA-VA) • Wavelet Transform Space • Westward Traveling Surge • Windows Terminal Server • Wind Test Site • Wind Tunnel Study • Word Terminal Synchronous • Workload Tracking System • World Terminal Synchronous

WTSA World Telecommunications Standardization Assembly

WTSC Wet Tantalum Slug Capacitor • World Telecommunication Standardization Conference

WTSPT Waterspout

wTT Weak-line T Tauri (star)

WTT Wantoat (Papua New Guinea) (location identifier) • Weak-emission T Tauri (star) • Weak-lined T Tauri (star) • Weapons and Tactics Trainer • Wide Terrestrial Telescope (now OWL) • Wind Tunnel Test • Wind Tunnel Testing

WTTC World Travel and Tourism Council

WTTM WIYN Top-Tilt Module

WTTS Weak-lined T Tauri Star (*) • Weak-line T Tauri Star • Weak T Tauri Star • Weak-wind T Tauri Star

WTU Wastewater Treatment Unit • Wuhan Technical University (China-PRC)

WTUC World Trade Union Conference

WTUSM Wuhan Technical University of Surveying and Mapping (China-PRC)

WTW Wire To Wire • Worldwide Transaction Web

WTWIWYH What They Want Is What You Have

WTWS Windshear and Turbulence Warning System

WTX Workstation Technology Extended (format)

WTZ Whitianga (New Zealand) (location identifier) • Wissenschaftlich-Technisches Zentrum

Wu Puissance Utile

WU Washington University (USA-MI) • Wave Universe • Wesleyan University (USA-CT) • Western Union • Work Unit

WUA Weighted Usable Area

WUBR Weighted Unspecified Bit Rate

WUC Wake-Up Call • Wave Universe Concept • Work Unit Code • Writers Union of Canada

WUCF Work Unit Code File

WUD Wudinna (Australia-SA) (location identifier)

WUE Water Use Efficiency • Western Undergraduate Exchange (USA) • Western University Exchange (USA)

WUF Woke Up Fast

WUFF WIYN Universal Fiber Feed

WUG Wau (Papua New Guinea) (location identifier) • Webmasters User Group

WuH Wohnung und Haus

WUH Wuhan (China-PRC) (location identifier)

WUI Web-based User Interface

WUIS Work Unit Information Summary • Work Unit Information System

WUL Woke Up Late

WUM Work Unit Manager

WUN Wiluna (Australia-WA) (location identifier)

WÜP Wohnungs-Übergabepunkt

WUPL Where-Used Parts List

WUPPE Wisconsin Ultraviolet Photo-Polarimeter Experiment

WUR Wageningen Universiteit en Researchcentrum (Netherlands)

Württ Württemberg (Germany)

WUS World University Service

WUSCI Western Union Space Communications Inc.

WUStL Washington University in Saint Louis (USA-MO)

WUSTL Washington University in Saint Louis (USA-MO)

WUT Warm Up Time • Warsaw University of Technology (Poland) • Washburn University of Topeka (USA-KS)

WUTC Washington Utilities and Transportation Commission (USA-DC)

WUTS Weekend Under The Stars (USA-WY)

WUU Wau (Sudan) (location identifier)

WUUA World Union for a Universal Alphabet

WUV Wuvulu Island (Papua New Guinea) (location identifier)

WUX Wuxi (China-PRC) (location identifier)

WUZ Wuzhou Changzhoudao (China-PRC) (location identifier)

w/v Weight in Volume

W/V Wind Vector

WV Saint Vincent and the Grenadines (international vehicle registration code) • Wagon de Voyageurs • Water Valve • Water Vapor • Wave • Wellenlängenvielfach • Westdeutscher Verlag • West Virginia (USA) • Wind Velocity • Working Voltage

W. Va. West Virginia (USA)

WVA Warenverkehrsanmeldung

WVAS Wabash Valley Astronomical Society (USA-IN) • Wake Vortex Avoidance System

WVB Walvis Bay (South Africa) (location identifier)

WVC Water Vapor Column • Water Vapor Content • West Valley College (USA-CA)

Wvd Waived

WVD Waived • Wavelet-Vaguelette Decomposition

WVDC Working Voltage/Direct Current

wve Psion Sound Module (file extension)

WVE Water Vapor Electrolysis • Wave

WVEH Wheel Vehicle

Wvfm Waveform

WVI Watsonville (USA-CA) (location identifier) • Work Volume Indicator

WVIOP Water Vapor Intensive Observing Period

WVIS Work Volume Indicator System

WVK Manakara (Madagascar) (location identifier)

wvl Wavelet-compressed Bitmap (file extension)

WVL Waterville (USA-ME) (location identifier)

WVM Water Vapor Monitor

WVMCCD West Valley Mission Community College District (USA-CA)

WVN Wilhelmshaven (Germany) (location identifier)

WVNET West Virginia Network for Educational Telecomputing (USA-WV)

WVO Wet Verontreiniging Oppervlaktewateren (Netherlands)

WVP Waste Volume Projection

WVR Water Vapor Radiometer • Water Vapor Radiometry • Wide Vehicle Roadway

WvS Wetboek van Strafrecht

WVS Weekly Visits Statistics • Window Value Selection • World Vector Shoreline • Worst Variable Selection

WVSC West Virginia State College (USA-WV)

WVSGC West Virginia Space Grant Consortium (USA-WV)

WVT Water Vapor Transfer • Water Vapor Transmission

WVTM Wharton Visual Test Market

WVU West Virginia University (USA-WV)

WVWV Water-Vapor Wind Vector

w/w Weight for Weight • Weight in Weight

W/W Wall to Wall • Weather • Wheel Well • Wheel Wing

WW Wall to Wall • Warehouse Warrant • Water Waste • Werkloosheidswet • Wheel Wing • Who's Who • Wire Way • Wire Wound • Wire Wrap • With Warrants • With Wind • World War • Worldwide

WWA Wasilla (USA-AK) (location identifier) • Weighted Wavelet Amplitude • Western Writers of America • World Warning Agency

WWAN Wireless Wide-Area Network

WWAR World-Wide Arts Resources

WWAS World Warning Agency for Satellites

WWB Warehouse Workshop Building • Wavelet Workbench • Westerly Wind Burst

WWBI Wireless Web-Based Instruction

WWC Wind-Wind Collision • Wing Work Center • World Water Council • World-Wide Computer

WWCA World-Wide Computing and its Applications (conference)

WWD Weather Working Day • Westward • Wildwood (USA-NJ) (location identifier)

WWDGPS Worldwide Differential Global Positioning System

WWDM Wide Wavelength Division Multiplex

WWDMS World-Wide Data Management System

WWDSA World-Wide Digital Systems Architecture

Wwe Witwe

WWF Woodrow Wilson Fellowship • World Wildlife Fund

WWFPP Weit-Winkel-Flächen-Photometer-Polarimeter

WWI Woodie Woodie (Australia) (location identifier) • World War I • World Watch Institute

WWII World War II

WWIMS Worldwide Warning Indicator Monitoring System

WWIO Worldwide Inventory Objective

WWIR Wind-Wind Interacting Region

WWIS World-Wide Information Services • World-Wide Information System

WWIZ Wind-Wind Interaction Zone

WWK Wewak (Papua New Guinea) (location identifier)

WWLC Worldwide Launching Campaign

WWM Weizsäcker-Williams Method • Welded Wire Matrix • Wireless Wellness Monitor

W/WMS Water and Waste Management System

WWMS Waste Water Management System • Water and Waste Management System

ww/o With or Without

WWO Waste Water Outflow • World-Wide Organization

WWOLS World-Wide Online System

WWP Whale Pass (USA-AK) (location identifier) • World Weather Program • Write Without Program

WWPS World-Wide Port System

WWR West Woodward (USA-OK) (location identifier)

WWRE World-Wide Retail Exchange

WWRP World Weather Research Programme

WWRS Wash-Water Recovery System

WWS Water and Waste System • Wildwood (USA-AK) (location identifier) • World Weather System

WWSE Who's Who in Science and Engineering

WWSN Worldwide Seismology Net

WWSSN Worldwide Standardized Seismograph Network • Worldwide Standard Seismograph Network

WWST Who's Who in Science and Technology

WWSV World-Wide Secure Voice

WWSVA World-Wide Secure Voice Architecture

WWSVCS World-Wide Secure Voice Conferencing System

WWT Newtok (USA-AK) (location identifier) • Waste Water Treatment

WWTCIP World-Wide Technological Control Improvement Plan • World-Wide Technical Control Improvement Program

WWTP Waste Water Treatment Plan

WWU Western Washington University (USA-WA) • Westfälische Wilhelms-Universität (Germany) • Wirtschafts- und Währungsunion

WWW Warm Wet Weather • Web-Weit Welt • Who's Who in the World • Who Was Who • World Weather Watch • World-Wide Waste (of time) • World-Wide Wealth (of data) • World-Wide Web (*)

WWWC World-Wide Web Consortium

WWWI First World-Wide Web Conference

WWWO World-Wide Web Organization

WWWS World Who's Who in Science

WWWT Worldwide Windows on TEX

WWWVL World-Wide Web Virtual Library

WWWW Who - Where - When - Why • World-Wide Web Worm

WWW1 First World-Wide Web Conference

WWW2 Second World-Wide Web Conference

WWW3 Third World-Wide Web Conference
WWY West Wyalong (Australia-NSW) (location identifier)
WWZ Weighted Wavelet Z-transform
Wx Weather • Weight
WX Weather • Wednesdays Excepted
WXC Wide-field X-ray Camera
WXF Braintree Wether (UK-England) (location identifier)
WXFCST Weather Forecast
WXM Wide-field X-ray Monitor (HETE-2)
WXN Wanxian (China-PRC) (location identifier)
WXR Weather Radar
WXTM Weather Team
Wy Wyoming (USA)
WY Way • Working Year • Work Year • Wyoming (USA)
WYA Whyalla (Australia-SA) (location identifier)
WYAIO Will You Accept, If Offered
WYAS West Yorkshire Astronomical Society (UK)
WYB Yes Bay (USA-AK) (location identifier)
WYE Yengema (Sierra Leone) (location identifier)
WYFFOS Wide-Field Fibre Optics Spectrograph
WYG What You Got
WYN Wyndham (Australia-WA) (location identifier)
Wyo Wyoming (USA)
WYP What's Your Point
WYS West Yellowstone (USA-MT) (location identifier)
WYSBYGI What You See Before You Get It
WYSIAWYG What You See Is Almost What You Get
WYSIAWYH What You See Is (Almost) What You Hear
WYSIAYG What You See Is All You'll Get
WYSIWES What You See Is What Everyone Sees
WYSIWIS What You See Is What I See
WYSIWYG What You See Is What You Get
WYSIWYP What You See Is What You Print
WYWIWYG What You Want Is What You Get
Wz Warenzeichen
WZ Weißer Zwerg • Weltzeit
WZG Warenzeichengesetz
WZN Weltzentrum für Niederschlagsklimatologie
WZW Wess - Zumino - Witten (model)
WZY Nassau (Bahamas) (location identifier)
W3 World-Wide Web
W3A World-Wide Web Applet
W3C World-Wide Web Consortium
W3DS World-Wide Web Data System
W3J World-Wide Web Journal
W3O World-Wide Web Organization
W3S World-Wide Web Server
W4 What Works With What • World-Wide Web Wanderer
W4G World-Wide Web Working Group
W4WG Windows For Workgroups

X

x Algebraic Variable • Cartesian Abcissa • Cartesian Coordinate • Choice • Coordinate Axis • Cross • Crossed with • Degree of Ionization • Distance • Ex • Excess • Experimental • Extra • Hoar Frost • Kiss • Magnification • Mole Fraction • Multiplication • Optimization Parameter • Out of • Path Length • Space Coordinate • State Vector • Times • Travel Distance • Without

X Air Mass • Anonyme • Anonymous • Cloud Depth • Cloud Diameter • Correlation • Cross • Crossing • École Polytechnique (France) • Error • Exchange • Experimental • Explosive • Exposure • Exposure Dose • Extension • Extra • Extraordinary • Fractional Mass Abundance of Hydrogen • Halogen • Hydrogen Abundance • Inconnu • Index • Initial Position Error • Location • Mole Fraction • Multiplication • North Component of the Magnetic Vector • Reactance • Ten (Roman numeral) • Times • Unknown • Unspecified • Xenon (now Xe) • X-ray (international alphabet) • Xylonite • 10 (Roman numeral)

xa Ex All

XA Extended Architecture • Extended Attribute • Mexico (international aircraft registration code)

Xaa Unknown Amino Acid

XAA Extended Attribute Area

XADC Extended Atlantic Data Coverage

XADR Transmit Address

XAF Extended Attribute Field • Franc (CFA/Banque des États d'Afrique Centrale) (ISO currency code)

XAFS X-ray Absorption Fine Structure

XAIU Multiplex Access Interface Unit

XAL Alamos (Mexico) (location identifier)

Xals Crystals

XALS Extended Application Layer Structure

XAML Transaction Authority Markup Language

Xan Xanthene • Xanthic

XANADU X-ray Analysis and Data Utilization

XANES X-ray Absorption Near-Edge Structure

XAO Ingénierie Assistée par Ordinateur

XAP Chapecó (Brazil-SC) (location identifier)

XAPIA X.400 Application Program Interface Association

XAPWG X-ray Astronomy Program Working Group (NASA)

XAR Aribinda (Burkina Faso) (location identifier) • Extended Attribute Record

Xarm Cross-Arm

XAS X-ray Absorption Spectroscopy • X-ray Analysis System

XAT Extended Alias Table

XAU Saul (French Guyana) (location identifier)

XAX Excellence Above Expectation

XAY Xayabury (Laos) (location identifier)

xb Ex Bonus

XB Mexico (international aircraft registration code) • X-ray Binary • X-ray Bright • X-ray Burst • X-ray Burster

XBAC X-ray Brightest Abell-type Cluster (catalog)

Xbar Cross-Bar

XBB Blubber Bay (Canada-BC) (location identifier)

XBE Bearskin Lake (Canada-ON) (location identifier)

Xber December

XBG Bogande (Burkina Faso) (location identifier) • X-ray Background

XBGN X-ray Bright Galactic Nucleus

XBJ Birjand (Iran) (location identifier)

XBL Buno Bedelle (Ethiopia) (location identifier) • X-ray BL Lacertae (object) • X-ray-loud BL Lacertae (object) (*)

xbm X-Windows Bitmap (file extension)

XBM Extended Basic Mode • X Bitmap

XBN Biniguni (Papua New Guinea) (location identifier)

XBO Boulsa (Burkina Faso) (location identifier) • X-ray Blow Out

XBP X Bright Point • X-ray Bright Point

XBR Brockville (Canada-ON) (location identifier) • Cross-Band Radar

Xbra Cross-Bracing

Xbre Décembre

XBSA X/Open Backup Service API

XBSS X-ray Background Survey Spectrometer

xbt Exhibit

XBT Cross-Bar Tandem • Expandable Bathythermograph

XBW Port Burwell (Canada-QC) (location identifier)

xc Ex Capitalization • Ex Coupon

XC Crossconnect • Cross-Correlation • Cross-Country • Mexico (international aircraft registration code) • XOR Character

XCA Extreme Circumstances Allowance

XCAL Cross-Calibration

XCAT X-ray Calibration And Test

XCD Dollar (East Caribbean) (ISO currency code)

XCES XML Corpus Encoding Standard

XCF Cross-Communication Facility • Cross-Coupling Facility • Cross-system Coupling Facility • Experimental Computing Facility • X-ray Crystallography Facility

XCH Christmas Island (location identifier)

XCHANGR Exchanger

XCHG Exchange

XCL Cluff Lake (Canada-SK) (location identifier) • Excess Current Liabilities

XCLF X-ray Cluster Luminosity Function

XCM Chatham (Canada-ON) (location identifier) • Extended Computer Modern

XCMD External Command

XCN Coron (Philippines) (location identifier)

XCO Colac (Australia-VIC) (location identifier) • Crystal-Controlled Oscillator

XCOM Executive Committee

Xconn Crossconnection

XCONN Crossconnection

XCOPY Extended Copy

xcp Ex Coupon

Xcp Except

Xcpt Except

Xcptd Excepted

Xcptn Exception

XCS Xerox Computer Services • XMM-Newton Cluster Survey • X-ray Cluster Survey

Xct Exact

XCT X-ray Computed Tomography

XCTD Extended Conductivity-Temperature-Depth

Xctly Exactly

XCVR Transceiver

XCY Cross-Country

X'd Executed

xd Ex Dividend

XD Cross-Disperser • Cross-Dispersion • Exothermic Dispersion • Without Dividend • X-ray Diffraction • X-ray Dominant

XDA X-band Drive Amplifier • X-ray Detector Aperture

XDACS X-ray Data Acquisition and Control System

XDB Extended Database

XDC Cross-point Data Conflict

XDCR Transducer

XDCS X-ray Dark Cluster Survey

XDDCMP External Digital Data Communications Message Protocol

XDE Diebougou (Burkina Faso) (location identifier)

XDF Exchange Data Format • Extensible Data Format • Extended Density Format • Extended Distance Feature • XML Data Format • XRM Decrease Factor • XTE Data Finder

xdiv Ex Dividend

XDJ Djibo (Burkina Faso) (location identifier)

XDL Cross-Delay Line • Crossed-Delay Line • X-ray Dividing Line

XDLB Extended Digital Line Backend

XDM Extended Data Management • X Display Manager • Xerox Dry Microfilm

XDMCP X Display Manager Control Protocol

XDP Extended Data Processing • X-ray Density Probe

XDR Exchange Data Representation • Extended Data Representation • Extended Dynamic Range • External Data Representation • X-ray Dissociation Region • X-ray Dominated Region

XDS Dollar (Saint Christopher) (ISO currency code) • X Display Station • Xerox Data Systems • X/Open Directory Server • X/Open Directory Service • X-ray Detection System • X-ray Detector System

XDT Xenon Discharge Tube • XUV Doppler Telescope

XDUP Extended Disk Utilities Program

Xe Xenon • Xénon

XED Extended Electronic Document

XEG X-ray Emission Gauge

XEN Xingcheng (China-PRC) (location identifier)

XEQ Extra Easy Question

Xer Xerox

Xerogr Xerographie

Xerok Xerokopie

XES Lake Geneva Municipal (USA-WI) (location identifier) • Xerox Escape Sequence • X-ray Emission Spectrum • X-ray Emission Spectrometry • X-ray Emission Spectroscopy

XEU European Currency Unit (ISO currency code) • Xerox Encryption Unit

XEUS X-ray (mission) for Evolving Universe Spectroscopy

XF Extra Fine • Transfer Frame

XFA Extremely Fast Access • X-ray Fluorescence Absorption • X-ray Fluorescence Analysis

XFC X-band Frequency Converter

XFCN External Function

XFD X-ray Flow Detection

XFDB XTE FITS Database

XFDS XMM Flight Dynamics System

XFEL X-ray Free Electron Laser

Xfer Transfer

XFER Transfer

XFF Extended Feature Function • XTE FITS Format (*) • XTE FITS Formatter

XFG Transfer Frame Generator

XFH X-ray Fluorescence Holography

XFM X-ray Flux Monitor

Xfmr Transformer

XFMR Transformer

XFN Xiangfan (China-PRC) (location identifier) • X/Open Federated Naming

Xformer Transformer

Xfr Transfer

XFR Transfer

XFREQ Transmit Frequency

XFS Extra Fast System

XFT Extremely Fast Tracker

XFTS XMM File Transfer System

XFW Finkenwerder (Germany) (location identifier)

XFX Crosstalk Network Simulator

Xg Crossing

XGA Extended Graphics Array • Gaoua (Burkina Faso) (location identifier)

XGAUI Ten Gigabit Attachment Unit Interface

XGG Gorom-Gorom (Burkina Faso) (location identifier)

XGKS X-Windows Graphic Kernel System

XGL Granville Lake (Canada-MB) (location identifier)

XGMII Ten Gigabit Medium-Independent Interface

XGN Xangongo (Angola) (location identifier)

XGP Xerox Graphic Printer

XGR Kangiqsualujjuaq (Canada-QC) (location identifier)

XGRS X and Gamma-Ray Spectrometer

XGT External Gamma-ray Target

xh Xhosa (ISO language code)

XHC Extra High Ceiling

XHD Cross-Head • Extended Header Data

XHF Extrahigh Frequency

XHI Extra High Intensity

XHL Extensible Hyperlinking Language

XHR Extrahigh Reliability

Xhst Exhaust

XHTML Extensible Hypertext Markup Language

XHV Extreme High Vacuum

Xhvy Extra Heavy

xi Ex Interest

XI X-flare Index • X-ray Imaging

XIA X-ray Instrumentation Associates (USA-CA)

XIC Xichang (China-PRC) (location identifier)

XICS Xerox Integrated Composition System

XID Cross-Identify • Cross-Identity • Exchange Identification • Exchange Identifier • X-ray Identification • X-window Identification • X-window Identifier

XIDB Extended Integrated Database • Extended Intelligence Database

XIDT X-ray Identification Team

XIE Xienglom (Laos) (location identifier) • X Image Extension • X-ray Imaging Experiment

xif Xerox Image Format (file extension)

XIF Xerox Image Format

XIG Xinguara (Brazil) (location identifier)

XIL Xilinhot (China-PRC) (location identifier) • X Imaging Library

xin Ex Interest

XIN Xingning (China-PRC) (location identifier)

XING Crossing

XIO Extended Interoperability

XIOS Extended Input/Output System

XIP Execute In Place • Execution In Place

XIPC X Interprocess Communication

XIPS Xenon-Ion Propulsion System

XIR X Irradiance Response at Wavelength X

XIRQL Extended Information Retrieval Query Language

XIS X-ray Imaging Spectrometer

XIT Exit • Exit Information Table

XITE X-ray Imaging Telescope

XIWG Cross-Industry Working Group

XIWT Cross-Industry Working Team

XIY Xi'An - Xianyang (China-PRC) (location identifier)

XJT Expansion Joint

XKA Kantchari (Burkina Faso) (location identifier)

XKH Xieng Khouang (Laos) (location identifier)

XKO Kemano (Canada-BC) (location identifier)

XKS Kasabonika (Canada-ON) (location identifier)

XKY Kaya (Burkina Faso) (location identifier)

xl Excel Spreadsheet (file extension)

XL Extended Language • Extra Large • Extra Long

XLAB Experimental Laboratory

XLAT Translate

XLB Lac Brochet (Canada-MB) (location identifier)

xlc Excel Chart (file extension)

XLD Experimental Laser Device • X-ray Luminosity Distribution

XLDF X-ray Luminosity Distribution Function

XLF Leaf Bay (Canada-QC) (location identifier) • X-ray Luminosity Function

XLFD X Logical Font Description

xlk Excel Backup (file extension)

XLL Extensible Linking Language

xlm Excel Macro (file extension)

XLM Excel Macro Language • Extended Local Memory

XLMU Extended Local Memory Unit

Xlnt Excellent

XLO Long Xuyen (Viet nam) (location identifier)

XLPE Cross-Linked Polyethylene

xls Excel Spreadsheet (file extension)

XLS Excel • Saint-Louis (Senegal) (location identifier)

xlt Excel Template (file extension)

XLT Extremely Large Telescope

XLU Leo (Burkina Faso) (location identifier)

xlw Excel Workspace (file extension)

XLW Lemwerder (Germany) (location identifier)

XLWB Extra Long Wheel base

XLWIR Extra Longwave Infrared

XM Extended Memory

XMA Extended Middleware Architecture • Maramag (Philippines) (location identifier) • X-ray Microanalysis • X-ray Mirror Assembly • XRT Mirror Assembly

X-MAC Cross-Multiplier/Accumulator

XMAP Extended Master Analysis Package • X-ray (instrument for) Mineralogical Analysis of Planetary (material)

XMB M'Bahiakro (Ivory Coast) (location identifier)

XMC Mallacoota (Australia-VIC) (location identifier)

XMCL Extensible Media Commerce Language

XMCS XMM Mission Control System

XMD Extremely Metal-Deficient (star) • Madison (USA-SD) (location identifier)

XMG Mahendranagar (Nepal) (location identifier)

XMH Manihi (Tuamotu Islands) (location identifier)

XMI Exchange Marketing Information • Masasi (Tanzania) (location identifier) • XML Metadata Interchange

XMISSION Transmission

Xmit Transmit

XMIT Transmit

XMIT/RCV Transmit/Receive

Xmits Transmits

X-ml Ex Mill

Xml Ex Mill

XML Extended Markup Language • Extensible Markup Language • Minlaton (Australia-SA) (location identifier)

X-mll Ex Mill

Xmll Ex Mill

XMM Extended Memory Manager • X-ray Multimirror • X-ray Multimirror Mission • X-ray Multiple Mirror

XMN Xiamen (China-PRC) (location identifier)

XMP Extended Multiprocessing • Extended Multiprocessor • Extensible Metadata Platform • McMillan Pass (Canada-YT) (location identifier) • Xanthosine Monophosphate • X/Open Management Protocol • X-ray Microprobe

XMS Extended Memory Services • Extended Memory Specification (*) • Extended Memory System • Macas (Ecuador) (location identifier) • X-ray Microcalorimeter Spectrometer

Xmsn Transmission

XMSN Transmission

Xmt Transmit

XMT Transmit

Xmtd Transmitted

XMTD Transmitted

Xmtg Transmitting

XMTG Transmitting

XMTL Transmittal

Xmtr Transmitter

XMTR Transmitter

XMY Yam Island (Australia) (location identifier)

xn Ex New

XNC Extranuclear Cloud

XNG Quang Ngai (Viet Nam) (location identifier)

XNN Xining (China-PRC) (location identifier)

XNS Crosstalk Network Simulator • Xerox Network Services (*) • Xerox Network System

X/nt Excellent

XNT Excellent • Xingtai (China-PRC) (location identifier)

XNU Nouna (Burkina Faso) (location identifier)

X/O X-ray/Optical (flux ratio)

XO Crystal Oscillator • Exactly Once • Exactly One • Executive Office • Executive Officer • Expenditure Order • Xylenol Orange

XOC Xanadu Operating Co.

XOF Franc (CFA/Banque Centrale des États de l'Afrique de l'Ouest) (ISO currency code)

XOFF Transmitter Off

XOL X Open Look

XOM External Output Management • X/Open Object Management • X/Open OSI-abstract-data Manipulation

XON Transmitter On

XOR Chi Orionids (meteor stream) • Exclusive Or

XOS Experimental Operating System • Extended Operating System • Xerox Operating System

XP Cross-Polar • Exchange Point • Expendable Processor • Exprès Payé • Express • Express Paid • Extra Person • Extreme Programming • Transponder

XPA Pama (Burkina Faso) (location identifier) • X-band Power Amplifier • X Public Access • X Public Address

XPARS External Research Publication and Retrieval System

XPC Expect

Xpct Expect

XPD Cross-Polar Discrimination

Xpdr Transponder

XPDR Transponder

XPE Cross-Platform Extension • X-ray Polarimeter Experiment

XPES Extended Polyelastic Sampling • X-ray Photoelectron Spectroscopy

XPF Franc (CFP) (ISO currency code)

XPG Xerox Publishing Group • X/Open Portability Guide

XPI Xanadu Parameter Interface

XPK Pukatawagan (Canada-MB) (location identifier)

Xpl Explosive

XPL Comayagua Palmerola (Honduras) (location identifier) • Explosive

XPLRS X-Windows Position Location and Reporting System

Xplt Exploit

xpm X-Windows Pixel Map (file extension)

XPM Cross-Phase Modulation • X11 Print Manager

Xpn Expansion

XPN Extragalactic Planetary Nebula • X-ray Photoionized Nebula

Xpnd Transponder

Xpndr Transponder

XPORT Transport

XPP Express Pallet with Payloads • Poplar River (Canada-MB) (location identifier)

XPr Without Privileges

XPR Pine Ridge (USA-SD) (location identifier) • Power Response at Wavelength X

XPRM Xerox Print Resources Manager

XPRMNTL Experimental

XPRS Experimental Precision Resupply System

Xprt Transport

XPS X-ray Photoelectron Spectroscopy • X-ray Photoemission Spectroscopy • X-ray Point-like Source

Xprtn Transportation

XPt Cross-Point

XPT Cross-Point

XPU West Kuparuk (USA-AK) (location identifier)

XQ Cross-Question • Excuse • Relative Concentration

XQC X-ray Quantum Calorimeter

XQL Extensible Query Language • XML Query Language

XQP Quepos (Costa Rica) (location identifier)

XQSO X-ray Quasar • X-ray Quasi-Stellar Object

XQU Qualicum (Canada-BC) (location identifier)

XQV Extra Quiet Vehicle

xr Ex Rights

Xr Examiner

XR X-Ray

XRAP X-Windows Radar Analysis Processing

XRASE X-Ray Spectroscopic Explorer

XRB X-Ray Background • X-Ray Beacon • X-Ray Binary • X-Ray Buster

XRCF X-Ray Calibration Facility

XRD Cross-Redistribution • Cross-Road • X-Ray Diffraction • X-Ray Diffractometry

Xrds Cross-Roads

XRE X-Ray Emission

XREA X-Ray Event Analyzer

XRef Cross-Reference

XREF Cross-Reference

XRF Experimental Reproduction Film • X-Ray Flash • X-Ray Fluorescence

XRFA X-Ray Fluorescence Analysis

XRFS X-Ray Fluorescence Spectrometer • X-Ray Fluorescence Spectrometry

XRL X-Ray Laser

XRM X-Ray Microanalyzer • X-Ray Monitor

XRMA X-Ray Microprobe Analysis

XRML Extensible Rule Markup Language

XRN X-Ray Nova • X-ray Reflection Nebula

XRP X-Ray Polychromator • X-Ray Pulsar

XRPM X-Ray Projection Microscope

XRPS XMM Remote Proposal Submission

XRR Ross River (Canada-YT) (location identifier) • X-Ray Reflectivity

XRS X-Ray Sensor • X-Ray Spectrometer

XRT Exit Routine Table • Extension for Real-Time • X-Ray Telescope • X-Ray Topography

xrts Ex Rights

XRY Jerez de la Frontera (Spain) (location identifier) • X-Ray

Xs Expenses

XS Excess • Exit Shadow • Extra Small • Multispectral

XSA Sinoe (Liberia) (location identifier) • X-band Satellite Antenna • XMM-Newton Science Archive

X-SAR X-band Synthetic Aperture Radar

XSAR X-band Synthetic Aperture Radar

XSAT Exceptional Satellite

XSBL X-ray Selected BL Lac (object)

XSC Extended Source Catalog • South Caicos (Turks & Caicos Islands) (location identifier)

XSCS XMM Science Control System

XSD Halogen-Selective Detector • Tonopah (USA-NV) (location identifier) • XML Schema Definition • XML Schema Description

XSDC XTE Science Data Center

XSE Sebba (Burkina Faso) (location identifier)

XSECT Cross-Section

XSF X-ray Scattering Facility

X-ship Ex Ship

Xship Ex Ship

X-shp Ex Ship

Xshp Ex Ship

XSI South Indian Lake (Canada-MB) (location identifier)

XSIL Extensible Scientific Interchange Language

XSL Experimental Space Laboratory • Extensible Style Language • Extensible Stylesheet Language • XML Schema Language

XSLC Xichang Satellite Launch Center (China-PRC)

XSLT Extensible Stylesheet Language Transform • Extensible Stylesheet Language Transformation • XML Stylesheet Language Transformation

XSM Saint Marys (USA-MD) (location identifier) • X/Open System Management

XSMD Extended Storage Module Drive

XSO Siocon (Philippines) (location identifier)

XSP Extrasolar Planet • Extrasolar Planetary • Singapore Selatar (Singapore) (location identifier)

XSPEC X-ray Spectral Analysis Package

XSS X-ray Source System

XSSI Extended Server Side Included

XST Exist

XSTA X-band Satellite Tracking Antenna

Xstg Extra Strong

Xstr Transistor

XSTR Transistor

X-stre Ex Store

Xstre Ex Store

xSU Former Soviet Union

XT Burkina Faso (international aircraft registration code) • Crosstalk • Extended • Extended Technology • X-ray Telescope

XTA Crosstalk Attenuation • X-band Tracking Antenna

XTACACS Extended Terminal Access Controller Access Control System

XTAL Crystal

Xtalk Crosstalk

XTC External Transmit Clock

XTCLK External Transmit Clock

XTD Cross-Track Distance

XTE Cross-Track Error • X-ray Timing Explorer (now RXTE) (*) • X-ray Transient Explorer

XTEN Xerox Telecommunications Network

X/Term X-Windows Terminal

XTF X-ray Temperature Function

XTG Thargomindah (Australia-QLD) (location identifier)

X-tgd Cross-Tongued

Xtgd Cross-Tongued

XTI Extended Transport Interface • X/open Transport Interface

XTION Transition

XTK Cross-Track

XTL Extraterrestrial Life • Tadoule Lake (Canada-MB) (location identifier)

XTLO Crystal Oscillator

XTM Experimental Test Model • Extensible Topic Map • X-ray Tomography Microscopy

XTND Extend

XTO Taroom (Australia-QLD) (location identifier)

XTP Express Transfer Protocol

XTR Tara (Australia-QLD) (location identifier)

Xtra Extra

XTRAN Cross-Transfer • Extended Fortran

XTRM Extreme

Xtry Extraordinary

XTV Experimental Test Vehicle • X Teleconferencing and Viewing

XU Cambodia (international aircraft registration code)

XUI X-windows User Interface

XUL Xavier University of Louisiana (USA-LA)

XUV Expendable Unmanned Vehicle • Extreme Ultraviolet • X-ray and Ultraviolet

XUVI Extreme Ultraviolet Imager

XUVT Extreme Ultraviolet Telescope

XUZ Xuzhou (China-PRC) (location identifier)

XV Viet Nam (international aircraft registration code)

XVA X-ray Vidicon Analysis

XVL Vinh Long (Viet Nam) (location identifier)

XVMX Xerox Voice Message Exchange

XVQ X-Windows Vector Quantization

Xvs Transverse

XVSAT Switched Very Small Aperture Terminal

XVT Executive Vector Table • X Virtual Terminal

xw Ex Warrants

X/W Cross-Wind

XWA Extended Wide Area • Wakeham Bay (Canada-QC) (location identifier)

xwd X-Windows Dump (file extension)

XWD X-Window Dump

X-whse Ex Warehouse

Xwhse Ex Warehouse

XWIB X-Windows Interface Builder

X-wks Ex Works

Xwks Ex Works

XWM X-Windows Manager

XWS Extended Web Site

XX Heavy

XXI Twenty First

XXL Extra Extra Large

XXS Extra Extra Small

XY Myanmar (international aircraft registration code)

XYA Yandina (Solomon Islands) (location identifier)

XYE Ye (Myanmar) (location identifier)

Xyl Xylographie • Xylometer • Xylophon

XYR Yellow River (Papua New Guinea) (location identifier)

XYT Toulouse-Montaudran (France) (location identifier)

XZ Myanmar (international aircraft registration code)

XZA Zabre (Burkina Faso) (location identifier)

X2B Hexadecimal to Binary

X2C Hexadecimal to Character

X2D Hexadecimal to Decimal

Y

y Algebraic Variable • Altitude • Cartesian Coordinate • Cartesian Ordinate • Coordinate Axis • Dry Air • Lateral Axis • Space Coordinate • Yard • Year • Yellow • Yen (currency) • Yes • Young • Youngest

Y Admittance • East Component of the Magnetic Vector • Fractional Mass Abundance of Helium • Helium Abundance • Hundred Fifty (Roman numeral) • Hypercharge • Planck Function • Time Period of Orbit • Yankee (international alphabet) • Yaw • Yen (currency) • Yes • Yttrium • Yuan (currency) • 150 (Roman numeral)

YA Afghanistan (international aircraft registration code) • Yao • Yaw Axis • Yet Another • Young Adult • Young Astronomer

YAA Anahim Lake (Canada-BC) (location identifier) • Yet Another Addition • Young Amateur Astronomer

YAB Arctic Bay (Canada-NT) (location identifier)

Yac Yacimiento

YAC Cat Lake (Canada-ON) (location identifier) • Youth Activities Committee

YACC Yet Another Compiler-Compiler

YAD Moose Lake (Canada-MB) (location identifier)

YAE Alta Lake (Canada-BC) (location identifier)

YAF Asbestos Hill (Canada-QC) (location identifier)

YAFIYGI You Asked For It, You Got It

YAG Fort Frances (Canada-ON) (location identifier) • Yttrium-Aluminium Garnet

YAH Lagrande 4 (Canada-QC) (location identifier)

YAHOO Yet Another Hierarchically Officious Oracle • Yet Another Hierarchical Officious Oracle (*)

YAI Chillan (Chile) (location identifier)

YAJ Lyall Harbour (Canada-BC) (location identifier)

YAK Yakutat (USA-AK) (location identifier) • Youth And Knowledge

YAL Alert Bay (Canada-BC) (location identifier)

YALO Yale - AURA - Lisbon - Ohio (consortium) • Yttrium Aluminate

YAM Sault Sainte Marie (Canada-ON) (location identifier) • Yaw Acquisition Mode • Yet Another Modem • Young Astronomers Meeting

YAMM Yet Another Menu Manager

YAN Yangambi (Zaire) (location identifier)

YANA Yet Another Bloated Application

YAO Yaounde (Cameroon) (location identifier) • Yunnan Astronomical Observatory (China-PRC)

YAP Yap (Caroline Islands) (location identifier) • Yield Analysis Pattern • Young Aspiring Professional • Yttrium - Aluminium - Perowskite

YAPA Youngstown Area Physics Alliance (USA-OH)

YAQ Maple Bay (Canada-BC) (location identifier)

YAR Lagrande 3 (Canada-QC) (location identifier) • Yearly Activity Report • York-Antwerp Rules

YAS Yasawa Island (Fiji) (location identifier) • Yaw Attitude Sensor • York Astronomical Society (UK)

YaST Yet Another Setup Tool

YAT Attawapiskat (Canada-ON) (location identifier) • Your Alias Table

YAUN Yet Another Unix Nerd

YAV Miner's Bay (Canada-BC) (location identifier)

YAVIS Young - Attractive - Verbal - Intelligent - Successful

YAW Halifax-Shearwater (Canada-NS) (location identifier)

YAWYSYA You Are Who You Say You Are

YAX Angling Lake (Canada-ON) (location identifier)

YAY Saint Anthony (Canada-NF) (location identifier)

YAZ Tofino - Metropolitan Area (Canada-BC) (location identifier)

Yb Ytterbium

YB Yearbook • Yellow Belly • Yottabyte

YBA Banff (Canada-AB) (location identifier)

YBB Pelly Bay - Townsite (Canada-NT) (location identifier)

YBC Baie Comeau (Canada-QC) (location identifier) • Yttrium - Barium - Copper (oxide)

YBCO Yttrium-Barium-Copper Oxide

YBD Young Brown Dwarf

YBE Uranium City (Canada-SK) (location identifier) • Yang-Baxter Equation

YBF Bamfield (Canada-BC) (location identifier)

YBG Bagotville (Canada-QC) (location identifier)

YBH Bull Harbour (Canada-BC) (location identifier)

YBI Black Tickle (Canada-NF) (location identifier)

YBJ Baie Johan Beetz (Canada-QC) (location identifier)

YBK Baker Lake (Canada-NT) (location identifier)

YBL Campbell River (Canada-BC) (location identifier)

YBM Bronson Creek (Canada-BC) (location identifier)

YBN Borden (Canada-ON) (location identifier)

YBO Bobquinn Lake (Canada) (location identifier)

YBP Yibin (China-PRC) (location identifier)

YBQ Telegraph Harbour (Canada-BC) (location identifier)

YBR Brandon (Canada-MB) (location identifier)

YBS Yale Bibliographic System

YBT Brochet (Canada-MB) (location identifier)

YBV Berens River (Canada-MB) (location identifier)

YBW Bedwell Harbour (Canada-BC) (location identifier)

YBX Blanc Sablon (Canada-QC) (location identifier)

YBY Bonnyville (Canada-AB) (location identifier)

YBZ Toronto Downtown (Canada-ON) (location identifier)

Y/C Luminance/Chrominance

Y-C Yale-Columbia

YC Yachtclub • Yale Catalogue • Yale College • Yale-Columbia • Young Cluster • Youth Council

YCA Courtenay (Canada-BC) (location identifier)

YCAS York County Astronomical Society (USA-PA)

YCB Cambridge Bay (Canada-NT) (location identifier)

YCbCr Brightness-Chrominance/Blue-Chrominance/Red

YCC Cornwall (Canada-ON) (location identifier) • Lightness - Colour - Chromatics • Youth Computing Club

YCD Nanaimo Cassidy (Canada-BC) (location identifier)

YCE Centralia (Canada-ON) (location identifier) • You Cannot Enter

YCF Cortes Bay (Canada-BC) (location identifier)

YCG Castlegar (Canada-BC) (location identifier)

YCH Chatham (Canada-NB) (location identifier)

YCI Caribou Island (Canada-ON) (location identifier)

YCJ Cape Saint James (Canada-BC) (location identifier)

YCK Colville Lake (Canada-NT) (location identifier)

YCL Charlo (Canada-NB) (location identifier)

YCM Saint Catharines (Canada-ON) (location identifier) • Yellow - Cyan - Magenta

YCMOU Yashwantrao Chavan Maharashtra Open University (India)

YCN Cochrane (Canada-ON) (location identifier)

YCO Coppermine (Canada-NT) (location identifier) • York College Observatory (USA-NY)

YCP Co-op Point (Canada-MB) (location identifier) • York College of Pennsylvania (USA-PA)

YCPS Yale Center for Parallel Supercomputing (USA-CT)

YCQ Chetwynd (Canada-BC) (location identifier)

YCR Cross Lake (Canada-MB) (location identifier) • Youth Council Representative

YCS Chesterfield Inlet (Canada-NT) (location identifier)

YCSP Yale Catalogue of Stellar Parallaxes

YCT Coronation (Canada-AB) (location identifier)

YCTSP Yale Catalogue of Trigonometric Stellar Parallaxes

YCU Cullaton Lake (Canada-NT) (location identifier)

YCV Cartierville (Canada-QC) (location identifier)

YCW Chilliwack (Canada-BC) (location identifier) • You Can't Win

YCX Gagetown (Canada-NB) (location identifier)

YCY Clyde River (Canada-NT) (location identifier)

YCZ Creston (Canada-BC) (location identifier)

yd Yard

YD Democratic Yemen (ISO country code) • Yaw Deviation • Yemen (Democratic_) (ISO country code) • Yemeni Dinar (currency)

YDA Dawson City (Canada-YT) (location identifier)

YDAC Yohkoh Data Archive Center

Y'day Yesterday

YDB Burwash Landings (Canada-YT) (location identifier)

YDC Drayton Valley (Canada-AB) (location identifier)

YDD Dinar (Yemen) (ISO currency code)

YDE Paradise River (Canada-NF) (location identifier)

YDF Deer Lake (Canada-NF) (location identifier)

ydg Yarding

YDG Digby (Canada-NS) (location identifier)

YDH Daniels Harbour (Canada-NF) (location identifier)

YDI Davis Inlet (Canada-NF) (location identifier)

YDK Main Duck Island (Canada-ON) (location identifier)

YDL Dease Lake (Canada-BC) (location identifier)

YDN Dauphin (Canada-MB) (location identifier)

YDO Dolbeau - Sainte Méthode (Canada-QC) (location identifier)

YDP Nain (Canada-NF) (location identifier)

YDQ Dawson Creek (Canada-BC) (location identifier)

YDR Broadview (Canada-SK) (location identifier)

yds Yards

YDS Desolation Sound (Canada-BC) (location identifier)

YDT Vancouver Boundary Bay (Canada-BC) (location identifier)

YDX Doc Creek (Canada-BC) (location identifier)

Y-E Younger-Elder

YE Yemen (ISO country code) • Young Engineer • Younger-Elder • Your Excellency

YEA Edmonton Metropolitan (Canada-AB) (location identifier) • Yaw Error Amplifier • Year-End Appraisal Report

Yearb Yearbook

YEAST Year of Science and Technology (1998)

YEC Yechon (Korea-ROK) (location identifier) • Young Earth Creationist • Young Engineer Club

YED Edmonton Namao (Canada-AB) (location identifier)

YEG Edmonton (Canada-AB) (location identifier)

YEI Ennadai Lake (Canada-NT) (location identifier)

YEK Arviat (Canada-NT) (location identifier) • Eskimo Point (Canada-NT) (location identifier)

Yel Yellow

YEL Elliot Lake (Canada-ON) (location identifier) • Yellow

YEM Yemen (ISO country code)

YEN Estevan (Canada-SK) (location identifier)

YEO Yeovilton (UK-England) (location identifier) • Your Eyes Only

YEP Estevan Point (Canada-BC) (location identifier)

YEQ Yenkis (Papua New Guinea) (location identifier)

YER Fort Severn (Canada-ON) (location identifier) • Rial (Yemen) (ISO currency code) • Yearly Effective Rate

YERAC Young European Radio Astronomers Conference

YerOB Yerkes Observatory Bulletin (periodical code)

YerPhI Yerevan Physics Institute (Armenia)

YES Young Engineers and Scientists • Young Engineers' Satellite • Youth Education Series • Youth Engineering and Science • Youth Extension Service

YESS Yaw Earth Sensor System

Yesty Yesterday

YET Edson (Canada-AB) (location identifier)

YETA Yet Another

YEU Eureka (Canada-NT) (location identifier)

YEV Inuvik (Canada-NT) (location identifier)

YEY Amos (Canada-QC) (location identifier)

YFA Fort Albany (Canada-ON) (location identifier)

YFB Iqaluit (Canada-NT) (location identifier)

YFC Fredericton (Canada-NB) (location identifier)

YFD Yard Floating Dock

YFE Forestville (Canada-QC) (location identifier)

YFF Your Friend Forever

YFG Fontanges (Canada-QC) (location identifier) • Yttrium-Ferrite Garnet

YFH Fort Hope (Canada-ON) (location identifier)

YFL Fort Reliance (Canada-NT) (location identifier)

YFO Flin Flon (Canada-MB) (location identifier)

YFPES Yawfree Polyelastic Sampling

YFR Fort Resolution (Canada-NT) (location identifier)

YFS Fort Simpson (Canada-NT) (location identifier) • Photopic Flux Sensitivity

YFX Fox Harbour - Saint Lewis (Canada-NF) (location identifier)

YG Year Group

YGA Gagnon (Canada-QC) (location identifier)

YGB Gillies Bay (Canada-BC) (location identifier)

YGC Grande Cache (Canada-AB) (location identifier) • Young Globular Cluster

YGE Gorge Harbour (Canada-BC) (location identifier)

YGG Ganges Harbour (Canada-BC) (location identifier)

YGH Fort Good Hope (Canada-NT) (location identifier)

YGJ Yonago (Japan) (location identifier)

YGK Kingston (Canada-ON) (location identifier)

YGL La Grande (Canada-QC) (location identifier)

YGM Gimli (Canada-MB) (location identifier)

YGN Greenway Sound (Canada-BC) (location identifier)

YGO Gods Narrows (Canada-MB) (location identifier)

YGP Gaspé (Canada-QC) (location identifier)

YGQ Geraldton (Canada-ON) (location identifier)

YGR Îles de la Madeleine (Canada-QC) (location identifier)

YGS Germansen (Canada-BC) (location identifier)

YGT Igloolik (Canada-NT) (location identifier) • Young Graduate Trainee

YGV Havre Saint-Pierre (Canada-QC) (location identifier)

YGW Kuujjuarapik (Canada-QC) (location identifier)

YGX Gillam (Canada-MB) (location identifier)

YGY Déception (Canada-QC) (location identifier)

YGZ Grise Fiord (Canada-NT) (location identifier)

YH Younger Halo

YHA Port Hope Simpson (Canada-NF) (location identifier) • Youth Hostels Association

YHB Hudson Bay (Canada-SK) (location identifier)

YHC Hakai Pass (Canada-BC) (location identifier) • Young Harris College (USA-GA)

YHD Dryden (Canada-ON) (location identifier)

YHE Hope (Canada-BC) (location identifier)

YHF Hearst (Canada-ON) (location identifier)

YHG Charlottetown (Canada-NF) (location identifier)

YHH Campbell River Harbour (Canada-BC) (location identifier)

YHI Holman Island (Canada-NT) (location identifier)

YHK Gjoa Haven (Canada-NT) (location identifier)

YHL You Have Lost

YHM Hamilton (Canada-ON) (location identifier)

YHN Hornepayne (Canada-ON) (location identifier)

YHO Hopedale (Canada-NF) (location identifier) • Youth Hostels Organization

YHP Poplar Hill (Canada-ON) (location identifier)

YHR Chevery (Canada-QC) (location identifier) • Harrington Harbour (Canada-QC) (location identifier)

YHS Sechelt (Canada-BC) (location identifier)

YHT Haines Junction (Canada-YT) (location identifier)

YHU Montréal Saint-Hubert (Canada-QC) (location identifier)

YHY Hay River (Canada-NT) (location identifier)

YHZ Halifax (Canada-NS) (location identifier)

YI Iraq (international aircraft registration code) • Yale Isochrone • Young Investigator

YIB Atikokan (Canada-ON) (location identifier)

YIC Isachsen (Canada-NT) (location identifier)

YIF Pakuashipi (Canada-QC) (location identifier) • Yearly Impact Factor

YIG Big Bay - Marina (Canada-BC) (location identifier) • Yttrium-Iron Garnet

YIH Yichang (China-PRC) (location identifier)

YIK Ivujivik (Canada-QC) (location identifier)

YIL Yellow Indicating Light

YIN Standard Lamp Intensity • Yining (China-PRC) (location identifier)

YINS Photopic Intensity Sensitivity

YIO Pond Inlet (Canada-NT) (location identifier) • Yielding Interoperability

YIP Detroit - Willow Run (USA-MI) (location identifier) • Young Investigator Program

YIQ Intensity - Inphase - Quadrature

YIR Year In Review

YIS Photometric Illuminance Sensitivity

YITP Yukawa Institute for Theoretical Physics (Japan)

YIV Island Lake - Garden Hill (Canada-MB) (location identifier)

YIW Yiwu (China-PRC) (location identifier)

YIX Port Radium (Canada-NT) (location identifier)

YJ Vanuatu (international aircraft registration code)

YJA Jasper Hinton (Canada-AB) (location identifier)

YJD Young Jong Do (Korea-ROK) (location identifier)

YJF Fort Liard (Canada-NT) (location identifier)

YJN Saint Jean (Canada-QC) (location identifier)

YJO Johnny Mountain (Canada-BC) (location identifier)

YJT Stephenville (Canada-NF) (location identifier)

YK Syria (international aircraft registration code) • Yakutian

YKA Kamloops (Canada-BC) (location identifier)

YKC Collins Bay (Canada-SK) (location identifier)

YKD Kincardine (Canada-ON) (location identifier)

YKE Knee Lake (Canada-MB) (location identifier)

YKF Kitchener - Waterloo-Wellington (Canada-ON) (location identifier)

YKG Kangirsuk (Canada-QC) (location identifier)

YKI Kennosao Lake (Canada-MB) (location identifier)

YKJ Key Lake (Canada-SK) (location identifier)

YKK Kitkatla (Canada-BC) (location identifier)

YKL Schefferville (Canada-QC) (location identifier)

YKM Yakima (USA-WA) (location identifier)

YKN Yankton (USA-SD) (location identifier) • Yukon (Canada)

YKQ Waskaganish (Canada-QC) (location identifier)

YKS Yakutsk (Russia) (location identifier)

YKT Klemtu (Canada-BC) (location identifier)

YKU Chisasibi (Canada-QC) (location identifier)

YKX Kirkland Lake (Canada-ON) (location identifier)

YKY Kindersley (Canada-SK) (location identifier)

YKZ Toronto Buttonville (Canada-ON) (location identifier)

YL Latvia (international aircraft registration code) • Yield Limit

YLA Langara (Canada-BC) (location identifier)

YLB Lac Biche (Canada-AB) (location identifier)

YLC Lake Harbour (Canada-NT) (location identifier)

YLD Chapleau (Canada-ON) (location identifier)

YLE Lac La Martre (Canada-NT) (location identifier)

YLF LaForges (Canada-QC) (location identifier) • Yttrium Lithium Fluoride

YLG Yalgoo (Australia-WA) (location identifier) • Youth Library Group

YLH Lansdowne House (Canada-ON) (location identifier)

YLI Ylivieska (Finland) (location identifier)

YLJ Meadow Lake (Canada-SK) (location identifier)

YLL Lloydminster (Canada-AB) (location identifier)

YLM Clinton Creek (Canada-YT) (location identifier)

YLN Yilan (China-PRC) (location identifier)

YLP Mingan (Canada-QC) (location identifier)

YLQ La Tuque (Canada-QC) (location identifier)

YLR Leaf Rapids (Canada-MB) (location identifier)

YLS Lebel-sur-Quevillon (Canada-QC) (location identifier) • Photometric Luminance Sensitivity

YLSTN Yellowstone (USA)

YLT Alert (Canada-NT) (location identifier)

YLW Kelowna (Canada-BC) (location identifier) • Young - Lada - Wilking (catalog)

YLX Long Point (Canada-ON) (location identifier)

YLY Lytton (Canada-BC) (location identifier)

YM Yang-Mills • Yield Management • Your Message

YMA Mayo (Canada-YT) (location identifier)

YMB Merritt (Canada-BC) (location identifier)

YMC Maricourt (Canada-QC) (location identifier) • Young Massive Cluster

YMCA Young Magellanic Cloud Association

YMD Mould Bay (Canada-NT) (location identifier) • Year - Month - Day

YME Matane (Canada-QC) (location identifier) • Younger - Middle-ager - Elder

YMF Montagne Harbour (Canada-BC) (location identifier)

YMG Manitouwadge (Canada-ON) (location identifier)

YMH Mary's Harbour (Canada-NF) (location identifier) • Yang - Mills - Higgs (system)

YMI Minaki (Canada-ON) (location identifier) • Why Am I ...

YMJ Moose Jaw (Canada-SK) (location identifier)

YML Murray Bay - Charlevoix (Canada-QC) (location identifier)

YMM Fort McMurray (Canada-AB) (location identifier)

YMMV Your Mileage May Vary

YMN Makkovik (Canada-NF) (location identifier)

YMO Moosonee (Canada-ON) (location identifier)

YMP Port McNeil (Canada-BC) (location identifier)

YMPI Yearly Mean Percentage Increase

YMQ Montréal Metropolitan (Canada-QC) (location identifier)

YMR Merry Island (Canada-BC) (location identifier) • Yearly Management Review • Young Merger Remnant

YMS Young Micro System • Yurimaguas (Peru) (location identifier)

YMT Chibougamau (Canada-QC) (location identifier)

YMU Yamanashi Medical University (Japan)

YMW Maniwaki (Canada-QC) (location identifier) • Yaw Maneuver Wheel

YMX Montréal Mirabel (Canada-QC) (location identifier)

YMY Montréal - Victoria Stol (Canada-QC) (location identifier)

Y/N Yes/No

YN Nicaragua (international aircraft registration code)

YNA Natashquan (Canada-QC) (location identifier)

YNB Yanbu (Saudi Arabia) (location identifier)

YNC Wemindji (Canada-QC) (location identifier)

YND Gatineau (Canada-QC) (location identifier)

YNE Norway House (Canada-MB) (location identifier)

YNG Youngstown (USA-OH) (location identifier)

YNH Hudson Hope (Canada-BC) (location identifier)

YNI Nichequon (Canada) (location identifier)

YNJ Yanji (China-PRC) (location identifier)

YNK Nootka Sound (Canada-BC) (location identifier)

YNL Points North Landing (Canada-SK) (location identifier)

YNM Matagami (Canada-QC) (location identifier) • You and Me

YNO North Spirit Lake (Canada-ON) (location identifier)

YNR You have No Return

YNS Nemiscau (Canada-QC) (location identifier)

YNT Yantai (China-PRC) (location identifier)

YNW You are Not Welcome

YNZ Yancheng (China-PRC) (location identifier)

yo Yoruba (ISO language code)

YO Years Old • Yunnan Observatory (China-PRC)

YOA Yale Observatory Archives

YOB Year Of Birth • Your Office in the Box

YOC Old Crow (Canada-YT) (location identifier) • Young Open Cluster

YOD Cold Lake (Canada-AB) (location identifier) • Year Of Death

YOE Falher (Canada-AB) (location identifier)

YOG Ogoki (Canada-ON) (location identifier) • Youth Observing Group

YOH Oxford House (Canada-MB) (location identifier)

YOJ High Level (Canada-AB) (location identifier)

YOK Yokohama (Japan) (location identifier)

YOL Yola (Nigeria) (location identifier)

YOLT You Only Live Twice

YOO Oshawa (Canada-ON) (location identifier)

YOP Rainbow Lake (Canada-AB) (location identifier) • Youth Opportunity Program

YOQ Your Own Question

YOR Yearly Operating Report

Yorks Yorkshire (UK-England)

YORP Yarkovsky - O'Keefe - Radzievskii - Paddack (effect)

YOS Years Of Service

YOTO Year Of The Ocean

YOW Ottawa McDonald Cartier (Canada-ON) (location identifier)

YOY Valcartier (Canada-QC) (location identifier)

Y-P Yaw-Pitch

YP Year's Purchase • Yellow Pages • Yield Point • Young Person

YPA Prince Albert (Canada-SK) (location identifier) • Yaw Precession Amplifier • Yearbook Printers Association • Young Professional Astronomer

YPB Port Alberni (Canada-BC) (location identifier)

YPC Paulatuk (Canada-NT) (location identifier) • Yale Parallax Catalog

YPD Parry Sound (Canada-ON) (location identifier) • Yaw Phase Detector

YPE Peace River (Canada-AB) (location identifier)

YPF Esquimalt (Canada-BC) (location identifier)

YPG Portage - La Prairie (Canada-MB) (location identifier) • Yuma Proving Ground

YPH Inukjuak (Canada-QC) (location identifier)

YPI Port Simpson (Canada-BC) (location identifier)

YPJ Aupaluk (Canada-QC) (location identifier)

YPL Pickle Lake (Canada-ON) (location identifier)

YPM Pikangikum (Canada-ON) (location identifier)

YPMO Young Planetary Massive Object

YPN Port Menier (Canada-QC) (location identifier)

YPO Peawanuck (Canada-ON) (location identifier)

YPOP Yohkoh Public Outreach Project

YPP Pine Point (Canada-NT) (location identifier) • Yellow Patch Panel

YPQ Peterborough (Canada-ON) (location identifier)

YPR Prince Rupert - Digby Island (Canada-BC) (location identifier)

yps Yards per Second

YPS Port Hawkesbury (Canada-NS) (location identifier) • Yellow-Page Service

YPT Pender Harbour (Canada-BC) (location identifier)

YPW Powell River (Canada-BC) (location identifier)

YPX Povungnituk (Canada-QC) (location identifier)

YPY Fort Chipewyan (Canada-AB) (location identifier)

YPZ Burns Lake (Canada-BC) (location identifier)

YQA Muskoka (Canada-ON) (location identifier)

YQB Québec (Canada-QC) (location identifier)

YQC Quaqtaq (Canada-QC) (location identifier)

YQD The Pas (Canada-MB) (location identifier)

YQE Kimberley (Canada-BC) (location identifier)

YQF Red Deer (Canada-AB) (location identifier)

YQG Windsor (Canada-ON) (location identifier)

YQH Watson Lake (Canada-YT) (location identifier)

YQI Yarmouth (Canada-NS) (location identifier)

YQK Kenora (Canada-ON) (location identifier)

YQL Lethbridge (Canada-AB) (location identifier)

YQM Moncton (Canada-NB) (location identifier)

YQN Nakina (Canada-ON) (location identifier)

YQQ Comox (Canada-BC) (location identifier)

YQR Regina (Canada-SK) (location identifier)

YQS Saint Thomas - Pembroke (Canada-ON) (location identifier)

YQT Thunder Bay (Canada-ON) (location identifier)

YQU Grande Prairie (Canada-AB) (location identifier)

YQV Yorkton (Canada-SK) (location identifier)

YQW North Battleford (Canada-SK) (location identifier)

YQX Gander (Canada-NF) (location identifier)

YQY Sydney (Canada-NS) (location identifier)

YQZ Quesnel (Canada-BC) (location identifier)

Yr Year • Younger • Your

Y-R Yaw-Roll

YR Romania (international aircraft registration code) • Yaw-Roll • Year

YRA Rae Lakes (Canada-NT) (location identifier)

YRB Resolute (Canada-NT) (location identifier)

Yrbk Yearbook

YRC Yearly Relative Citedness

YRD Dean River (Canada-BC) (location identifier)

YRDST Year-Round Daylight Saving Time

YRE Resolution Island (Canada-NT) (location identifier)

YREC Yale Rotating Evolution Code

YRF Cartwright (Canada-NF) (location identifier)

YRG Rigolet (Canada-NF) (location identifier)

YRI Rivière du Loup (Canada-QC) (location identifier)

YRIF Yuri Rubinsky Insight Foundation

YRJ Roberval (Canada-QC) (location identifier)

YRl Yemen Ryal (currency)

YRL Red Lake (Canada-ON) (location identifier)

Yrly Yearly

YRM Rocky Mountain House (Canada-AB) (location identifier) • You Are Matched

YRMODA Year - Month - Day

YRN River's Inlet (Canada-BC) (location identifier)

YRO Ottawa - Rockcliffe Street (Canada-ON) (location identifier) • Year Round Operations

YRQ Trois-Rivières (Canada-QC) (location identifier)

YRR Stuart Island (Canada-BC) (location identifier)

Yrs Years • Yours

YRS Red Sucker Lake (Canada-MB) (location identifier) • Years • Young Researchers Symposium • Ysgarth Rules System

YRT Rankin Inlet (Canada-NT) (location identifier) • Young Researchers Track

YRV Revelstoke (Canada-BC) (location identifier)

YS El Salvador (international aircraft registration code) • Yardstick • Yellow Sky • Yield Strength • Young Scholar

YSA Sable Island (Canada-NS) (location identifier)

YSAG Yttrium-Scandium-Aluminium Garnet

YSAR Yearly Significant Activity Report

YSB Sudbury (Canada-ON) (location identifier)

YSC Sherbrooke (Canada-QC) (location identifier) • Yearly Spares Cost • Yokohama Science Center (Japan) • Young Star Cluster • Young Stellar Cluster

YSD Suffield (Canada-AB) (location identifier)

YSE Swan River (Canada-MB) (location identifier) • Yaw Steering Error

YSF Stony Rapids (Canada-SK) (location identifier) • Yield Safety Factor • Youth Science Foundation

YSG Snowdrift (Canada-NT) (location identifier) • Yellow Supergiant (star) • Young Star Group • Young Star Grouping • Young Supergiant (star)

YSH Smith Falls (Canada-ON) (location identifier)

YSI Sans Souci (Canada-ON) (location identifier) • Young Springs Instruments (USA)

YSJ Saint John (Canada-NB) (location identifier)

YSK Sanikiluaq (Canada-NT) (location identifier)

YSL Saint Leonard (Canada-NB) (location identifier)

YSLF Yield Strength Load Factor

YSM Fort Smith (Canada-NT) (location identifier) • Yale Survey Machine

YSN Salmon Arm (Canada-BC) (location identifier) • Young Scientists' Network

YSO Postville (Canada-NF) (location identifier) • Yale Southern Observatory (Argentina-SJ) • Young Stellar Object

YSP Marathon (Canada-ON) (location identifier)

YSPU Yaroslavl State Pedagogical University (Russia)

YSQ Spring Island (Canada-BC) (location identifier)

YSR Nanisivik (Canada-NT) (location identifier)

YSS State Island (Canada) (location identifier) • Youth Support Scheme

YSSA Young Scorpius-Sagittarius Association

YSSP Young Scientists' Summer Program

Yst Youngest

YST Sainte Therese Point (Canada-MB) (location identifier) • Young Scientist Training • Yukon Standard Time (Canada)

YSTAR Yonsei Survey Telescopes for Astronomical Research (ROK)

YSU Summerside (Canada-PE) (location identifier) • Yerevan State University (Armenia) • Youngstown State University (USA-OH)

YSV Saglek (Canada-NF) (location identifier)

YSW Salluit (Canada-QC) (location identifier)

YSX Shearwater (Canada-BC) (location identifier)

YSY Sachs Harbour (Canada-NT) (location identifier)

YSZ Squirrel Cove (Canada-BC) (location identifier) • Yttrium-Stabilized Zirconium

Y-T Yam-Trim

YT Mayotte (code ISO de pays) • Yaw Trim • Yukon Territory (Canada) • Yukon Time

YTA Pembroke (Canada-ON) (location identifier) • Yaw Trim Angle

YTB Hartley Bay (Canada-BC) (location identifier) • Yearly Training Brief

YTC Sturdee (Canada-BC) (location identifier)

YTD Thicket Portage (Canada-MB) (location identifier) • Year To Date

YTE Cape Dorset (Canada-NT) (location identifier)

YTF Alma (Canada-QC) (location identifier)

YTG Sullivan Bay (Canada-BC) (location identifier) • Yearly Training Guidance • Yukon Territorial Government (Canada-YT)

YTH Thompson (Canada-MB) (location identifier)

YTI Triple Island (Canada-BC) (location identifier)

YTJ Terrace Bay (Canada-ON) (location identifier)

YTK Tulugak (Canada-QC) (location identifier) • Year Two Thousand

YTL Big Trout Lake (Canada-ON) (location identifier)

YTN Rivière au Tonnerre (Canada-QC) (location identifier)

YTO Toronto Metropolitan (Canada-ON) (location identifier)

YTP Tofino (Canada-BC) (location identifier)

YTQ Tasiujuaq (Canada-QC) (location identifier)

YTR Trenton (Canada-ON) (location identifier)

YTS Timmins (Canada-ON) (location identifier) • Youth Training Scheme

YTT Tisdale (Canada-SK) (location identifier)

YTU Tasu (Canada-BC) (location identifier)

YTV Yaw Thrust Vector

YTX Telegraph Creek (Canada-BC) (location identifier)

YTZ Toronto Island (Canada-ON) (location identifier)

YU Yale University • York University (UK-England) • Yugoslavia (international vehicle registration code) • Yugoslavia (ISO country code) • Yugoslavia (postal country code) • Yugoslavie (code ISO de pays)

YUA Yuanmou (China-PRC) (location identifier)

YUAC York University Astronomy Club (Canada-ON)

YUB Tuktoyaktuk (Canada-NT) (location identifier)

YUC Yucatan (Mexico)

YUCL Young and Upcoming Leaders

YUD Umiujaq (Canada-QC) (location identifier)

YUE Yuendumu (Australia-NT) (location identifier)

YUF Pelly Bay - Dewline Site (Canada-NT) (location identifier)

YUFEM Yamaguchi University Faculty of Engineering Memoirs (periodical code)

YUG Yugoslavia (ISO country code) • Yugoslavie (code ISO de pays)

Yugo Yugoslavia

YUJ Lady Franklin (Canada-NT) (location identifier)

Yuk Yukon

YUL Montréal Dorval (Canada-QC) (location identifier) • Yale University Library

YUM Yuma (USA-AZ) (location identifier)

YUN New Dinar (Yugoslavia) (ISO currency code) • Johnson Point (Canada-NT) (location identifier)

YUO Yonsei University Observatory (ROK)

YUP Yale University Press

Yuppie Young Upwardly-mobile Professional People • Young Urban Professional

YUT Repulse Bay (Canada-NT) (location identifier)

YUTR Yamaguchi University Technology Reports (periodical code)

YUX Hall Beach (Canada-NT) (location identifier)

YUY Rouyn (Canada-QC) (location identifier)

YV Venezuela (international aircraft registration code) • Venezuela (international vehicle registration code)

YVA Moroni (Comoros) (location identifier)

YVAA Yellowstone Valley Amateur Astronomers (USA-MT)

YVAC Yakima Valley Astronomy Club (USA-WA)

YVB Bonaventure (Canada-QC) (location identifier)

YVC La Ronge (Canada-SK) (location identifier)

YVD Yeva (Papua New Guinea) (location identifier)

YVE Vernon (Canada-BC) (location identifier) • Youth Versus Experience

YVG Vermilion (Canada-AB) (location identifier)

YVM Broughton Island (Canada-NT) (location identifier)

YVN Cape Dyer (Canada-NT) (location identifier)

YVO Val d'Or (Canada-QC) (location identifier)

YVP Kuujjuaq (Canada-QC) (location identifier)

YVQ Norman Wells (Canada-NT) (location identifier)

YVR Vancouver (Canada-BC) (location identifier)

YVS Yearly Visits Statistics

YVT Buffalo Narrows (Canada-SK) (location identifier)

YVV Wiarton (Canada-ON) (location identifier)

YVW You are Very Welcome

YVZ Deer Lake (Canada-ON) (location identifier)

YWA Petawawa (Canada-ON) (location identifier)

YWB Kanjiqsujuaq (Canada-QC) (location identifier)

YWF Halifax - Downtown Waterfront (Canada-NS) (location identifier)

YWG Winnipeg (Canada-MB) (location identifier)

YWH Victoria - Inner Harbour (Canada-BC) (location identifier)

YWIA You're Welcome In Advance

YWJ Deline (Canada-NT) (location identifier) • Fort Franklin (Canada-NT) (location identifier)

YWK Wabush (Canada-NF) (location identifier)

YWL Williams Lake (Canada-BC) (location identifier)

YWM Williams Harbour (Canada-NF) (location identifier)

YWN Winisk (Canada-ON) (location identifier)

YWP Webequie (Canada-ON) (location identifier)

YWR White River (Canada-ON) (location identifier)

YWS Whistler (Canada-BC) (location identifier)

YWY Wrigley (Canada-NT) (location identifier)

YXC Cranbrook (Canada-BC) (location identifier)

YXD Edmonton (Canada-AB) (location identifier)

YXE Saskatoon (Canada-SK) (location identifier)

YXF Snake River (Canada-YT) (location identifier)

YXG Youth Examination Group

YXH Medicine Hat (Canada-AB) (location identifier)

YXI Killaloe (Canada-ON) (location identifier)

YXJ Fort Saint John (Canada-BC) (location identifier)

YXK Rimouski (Canada-QC) (location identifier)

YXL Sioux Lookout (Canada-ON) (location identifier)

YXN Whale Cove (Canada-NT) (location identifier)

YXP Pangnirtung (Canada-NT) (location identifier)

YXQ Beaver Creek (Canada-YT) (location identifier)

YXR Earlton (Canada-ON) (location identifier)

YXS Prince George (Canada-BC) (location identifier)

YXT Terrace (Canada-BC) (location identifier)

YXU London (Canada-ON) (location identifier)

YXX Abbotsford (Canada-BC) (location identifier)

YXY Whitehorse (Canada-YT) (location identifier)

YXZ Wawa (Canada-ON) (location identifier)

Y-Y Yaw-Yaw

YY Yaw-Yaw

YYA Big Bay - Yacht Club (Canada-BC) (location identifier)

YYB North Bay (Canada-ON) (location identifier)

YYC Banff (Canada-AB) (location identifier) • Calgary (Canada-AB) (location identifier)

YYD Smithers (Canada-BC) (location identifier)

YYDDD Year/Julian day

YYE Fort Nelson (Canada-BC) (location identifier)

YYF Penticton (Canada-BC) (location identifier)

YYG Charlottetown (Canada-PE) (location identifier)

YYH Spence Bay (Canada-NT) (location identifier) • Taloyak (Canada-NT) (location identifier)

YYI Rivers (Canada-MB) (location identifier)

YYJ Victoria (Canada-BC) (location identifier)

YYL Lynn Lake (Canada-MB) (location identifier)

YYM Cowley (Canada-AB) (location identifier)

YYMMDD Year - Month - Day

YYN Swift Current (Canada-SK) (location identifier)

YYP You and Your Partner

YYQ Churchill (Canada-MB) (location identifier)

YYR Goose Bay (Canada-NF) (location identifier)

YYT Saint John's (Canada-NF) (location identifier)

YYU Kapuskasing (Canada-ON) (location identifier)

YYW Armstrong (Canada-ON) (location identifier)

YYY Mont Joli (Canada-QC) (location identifier)

YYZ Toronto - Lester B. Pearson (Canada-ON) (location identifier)

YZ Yale Zones • Youth Redshift

YZA Ashcroft (Canada-BC) (location identifier)

YZC Beatton River (Canada-BC) (location identifier) • Yale Zone Catalogue

YZE Gore Bay (Canada-ON) (location identifier)

YZF Yellowknife (Canada-NT) (location identifier)

YZG Salluit (Canada-QC) (location identifier)

YZH Slave Lake (Canada-AB) (location identifier)

YZL Liard River (Canada-BC) (location identifier)

YZM Buchans (Canada-NF) (location identifier)

YZP Sandspit (Canada-BC) (location identifier)

YZR Sarnia (Canada-ON) (location identifier)

YZS Coral Harbour (Canada-NT) (location identifier)

YZT Port Hardy (Canada-BC) (location identifier)

YZU Whitecourt (Canada-AB) (location identifier)

YZV Sept-Îles (Canada-QC) (location identifier)

YZW Teslin (Canada-YT) (location identifier)

YZX Greenwood (Canada-NS) (location identifier)

Y10K Year 10,000

Y2K Year 2000

Y2KWG Year 2000 Working Group

Z

z Algebraic Variable • Altitude • Cartesian Coordinate • Cartesian Third Axis • Charge Number of Ion • Coordinate Axis • Depth • Distance from the Galactic Plane • Doppler Shift • Effort • Haze • Height • Ion Charge Number • Redshift • Space Coordinate • Vertical Component of the Magnetic Vector • Vertical Distance • Zenith Angle • Zenith Distance • Zero • Zip-compressed Archive (file extension) • Zone • zu • zum • zur

Z Altitude • Atomic Number • Azimuth Angle • Charge Number of Atoms • Coriolis Correction • Ensemble des Entiers Relatifs • Flash Precedence • Fractional Mass Abundance of Heavy Elements • Fractional Mass Abundance of Metals • Greenwich Mean Time • Impedance • Mechanical Impedance • Metallicity • Normal Acceleration • Partition Function • Proton Number • Radar Reflectivity Factor • Two Thousand (Roman numeral) • Zahl • Zahlung • Zaïre (currency) • Zambia (international vehicle registration code) • Zeichen • Zeile • Zeit • Zeitschrift • Zeitung • Zenith • Zenith Distance • Zenith Pointing • Zentralblatt • Zentrale • Zentrifuge • Zentrum • Zero • Zéro • Zettel • Zeuge • Zeugnis • Ziffer • Zimbabwe (international aircraft registration code) • Zimmer • Zip-compressed Archive (file extension) • Zitat • Zoll • Zone • Zoulou Time • Zuchthaus • Zucker • Zug • Zulu (international alphabet) • Zulu Time • Zurücknahme • 2000 (Roman numeral)

za South Africa • zirka

zA zur Ansicht • zur Anstellung

ZA Afrique du Sud (code ISO de pays) • Albania (international aircraft registration code) • South Africa (international vehicle registration code) • South Africa (ISO country code) • Zahlungsabkommen • Zahlungsanweisung • Zahnarzt • Zaire • Zeitschrift für Astrophysik (periodical code) • Zel'dovich Approximation • Zenith Angle • Zentralabteilung • Zentralamt • Zentralanstalt • Zentralarkiv • Zero Adjusted • Zero Age • Zero and Add • Zollamt • Zoneamt

ZAA Alice Arm (Canada-BC) (location identifier) • Zeeman Atomic Absorption • Zeeman-effect Atomic Absorption • Zentralanmeldeamt

ZAB Zollamt am Bahnhof • Zone Absolute Barrier

ZAbfO Zollabfertigungsordnung

ZAbfSt Zollabfertigungsstelle

ZAC York Landing (Canada-MB) (location identifier) • Zacatecas (Mexico) • Zone d'Aménagement Concerté (France)

ZACV Zero-Age Cataclysmic Variable (star)

ZAD Zadar (Croatia) (location identifier) • Zenith Angle Distribution • Zone d'Aménagement Différé (France)

ZaDN Zagadnienia Drgan Nieliniowych (periodical code)

ZAE Zone d'Activité Économique (France) • Zone of Absolute Exclusion

ZAED Zentralstelle für Atomkernenergie-Dokumentation

ZAEHB Zero-Age Extreme Horizontal Branch

ZAF Afrique du Sud (code ISO de pays) • South Africa (ISO country code)

ZAG Zagreb Pleso (Croatia) (location identifier) • Zeitschriftenverlag Aktien-Gesellschaft • Zentralarbeitsgemeinschaft

ZAH Zahedan (Iran) (location identifier)

ZAHB Zero-Age Horizontal Branch

Zahl Zahlung

Zahlm Zahlmeister

zahlr Zahlreich

Zahlst Zahlstelle

ZAI Zero Address Instruction • Zero Administration Initiative • Zero Administrative Initiative

ZAIK Zentrum für Angewandte Informatik Köln (Germany)

ZAJ Zaranj (Afghanistan) (location identifier)

ZAK Zeitschriftenaufsatzkatalog • Zentralarbeitskreis • Zero Administration Kit

ZAL Financial South African Rand (ISO currency code) • Valdivia Pichoy (Chile) (location identifier) • Zona de Actividades Logísticas

ZaLab Zavodskaia Laboratoriia (periodical code)

Zam Zamora (Spain)

ZAM Zamboanga (Philippines) (location identifier) • Z-Axis Modulation • Zentralinstitut für angewandte Mathematik (Germany) • Zentrum für angewandte Mikroelektronik

ZAMG Zentralanstalt für Meteorologie und Geodynamik (Austria)

ZaMM Zeitschrift für angewandte Mathematik und Mechanik (periodical code)

ZAMO Zero Angular Momentum Observer

ZaMP Zeitschrift für angewandte Mathematik und Phyik (periodical code)

ZAMS Zero-Age Main Sequence

z. Anst. zur Anstellung

ZAnst Zahlungsanweisung

ZAnwO Zollanweisungsordnung

ZAO Cahors Laberandie (France) (location identifier)

ZAP Zone Axis Pattern • Zone d'Aménagement Protégé (France)

ZAPA Zambia Press Agency

ZAPE Zonal Available Potential Energy

ZAPI Zone d'Action Prioritaire Intégrée (France)

ZAPO Zulassungs-, Ausbildungs- und Prüfungsordnung

Zar Zaragoza (Spain)

ZAR Rand (South Africa) (ISO currency code) • Zaire (ISO country code) • Zaria (Nigeria) (location identifier) • Zentral-Archiv Retrieval • Zone d'Action Rurale (France)

ZARM Zentrum für Angewandte Raumfahrttechnologie und Mikrogravitation (Germany)

ZAS Zaire Aero Service • Zambia Astronomical Society (Zambia) • Zentrum für Auslandsstudien • Zero Access Storage

ZASt Zentrale Auskunftstelle • Zinsabschlagsteuer • Zollaufsichtstelle

ZAT Zero Alias Table • Zhaotong (China-PRC) (location identifier) • Zone d'Activités Tertiaires (France)

ZaTsA Zapiski TsAGI Uchenye (periodical code)

ZAV Zentralstelle für Arbeitsvermittlung (Germany)

ZAW Zero Administration for Windows • Zero Administration Workstation

ZAZ Zaragoza (Spain) (location identifier)

zB zum Beispiel • zur Beachtung

ZB Zahlungsbefehl • Zahlungsbewilligung • Zentralbatterie • Zentralbibliothek • Zentralbild • Zentralblatt • Zentralbüro • Zero Beat • Zettabyte • Zuiderbreedte • Zwischenbericht

ZBA Zero Bias Anomaly • Zoning Board of Appelas

ZBB Zero Base Budget • Zero Base Budgeting • Zimbabwe Booksellers Association

ZBCP Zero-Bias Conductance Peak

ZBE Zabreh - Dolni Benesov (Czech Rep.) (location identifier) • Zone Backend

ZBF Bathurst (Canada-NB) (location identifier) • Zentralbereich Förderung • Zugbahnfunk

ZBI Zentralverband Berufsständischer Ingenieurvereine • Zero Baseline Interferometer • Zero Baseline Interferometry

ZBK Zabljak (Yugoslavia-Montenegro) (location identifier)

Zbl Zentralblatt

ZBL Biloela (Australia-QLD) (location identifier) • Ziegler - Biersack - Littmark (potential)

ZBM Bromont (Canada-QC) (location identifier) • Zentralbibliothek der Medizin (Germany) • Zone Barrier Monitoring

ZBO Bowen (Australia-QLD) (location identifier)

ZBP Baltimore Penn (USA-MD) (location identifier)

ZBR Chah-Bahar (Iran) (location identifier) • Zero-Based Review • Zero-Base Review • Zone Bit Recording

ZBS Zebra Pattern Structure

ZBT Zero Balance Transfer

ZBTSI Zero Byte Time Slot Interchange

zbV zur besonderen Verwendung

ZBV Beaver Creek (USA-CO) (location identifier)

ZBX Branson Bus (USA-MO) (location identifier)

ZBY Sayaboury (Laos) (location identifier)

ZC Zero Crossing • Zodiacal Catalogue • Zone Convective • Zwicky Cluster

ZCA Zone Central ACID • Zone Control Authority

ZCAS Zona de Convergência do Atlântico Sul

Z-CAV Zone Constant Angular Velocity

ZCAV Zone Constant Angular Velocity • Zoned Constant Angular Velocity

ZCD Zero Cross-Detector • Zone Core Distribution

ZCG Impedance Cardiogram • Zeolite Crystal Growth

Zch Zürich (Switzerland)

ZCH Zone Continuellement Habitable

Zchg Zeichnung

Zchn Zeichnen

Zchng Zeichnung

ZCIT Zona de Convergencia Intertropical • Zone de Convergence Intertropicale

ZCK Zero Check

ZCL Zacatecas (Mexico) (location identifier)

ZCN Zinc-Coated Nut

ZCO Temuco Manquehue (Chile) (location identifier)

ZCPS Zona de Convergência do Pacífico Sul

ZCR Zero-Crossing Rate • Zero-temperature Coefficient Resistor

ZCS Zero Code Suppression

ZCV Zone Center Value

ZCW Zinc-Coated Washer

ZCY Zirconium - Cerium - Yttrium

zD zur Disposition

ZD Zeeman Doppler (technique) • Zeitschriftendienst • Zeitungsdienst • Zener Diode • Zenithal Distance • Zenith Distance • Zentraldienst • Zentraldirektion • Zentrale Dokumentation • Zero Defects • Ziff-Davis • Zimmer mit Dusche • Zoned Decimal • Zusatz-Deskriptor

zdA zu den Akten

ZDA Zentrale Direktionsabteilung

ZDB Zeitschriftendatenbank

ZDBOp Ziff-Davis Benchmark Operation

ZDC Zero-Degrees Calorimeter

ZDCF Z-transformed Discrete Correlation Function

ZDDSE Zentrale Dokumentation für Entwicklungshilfe und Entwicklungsländer

ZDE Zentrale Dokumentation Elektrotechnik • Zentralstelle Dokumentation Elektrotechnik

ZDEC Two-Dimensional Electron Gas

ZDF Zweites Deutsches Fernsehen (Germany)

Zdg Zündung

ZDG Zentrale Dokumentation für Geographie • Zentralen Dominierenden Galaxien • Zivildienstgesetz

Zdh Zündhölzer

ZDH Zentralverband des Deutschen Handwerks (Germany)

ZDI Zeeman-Doppler Imager • Zeeman-Doppler Imaging • Zentralstelle für Dokumentation und Information

ZDIWD Zentrale Dokumentations- und Informationsstelle für Werkstoffdaten

ZDJ Berne Main Station (Switzerland) (location identifier) • Bern Hauptbahnhof (Switzerland) (location identifier) • Bern Hbf. (Switzerland) (location identifier)

ZDK Zentrale Driftkammer

ZDL Zero Delay Lockout

ZDN Zentrum zur Dokumentation für Naturheilverfahren

ZDO Zentrum für Datenverarbeitung und Organisation • Zero Differential Overlap

ZDR Zentrale Dokumentation für Raumordnung

ZDS Zenith Data Systems

ZDT Zero-Ductility Transition

Zdw Zündwaren

ZDW Zeitung-Druck-Wirtschaft • Zentralamt des Deutschen Wetterdienstes

zE zu Ehren • zum Exempel

ZE Zahlungseinheit • Zahlungseinstellung • Zeiteinheit • Zentraleinheit • Zugkrafteinheit

ZEAT Zone d'Études et d'Aménagement du Territoire (France)

ZEB Zenith-Enhanced Backscatter

ZEBRA Zero-Energy Breeder-Reactor Assembly

ZEC Secunda (South Africa) (location identifier)

ZEDE Zone d'Études Démographiques et d'Emploi

ZEE Zona Económica Exclusiva • Zone Économique Exclusive

ZEEP Zero-Energy Experimental Pile

ZEFU Zentrum Für Umweltforschung (Germany)

ZEG Senggo (Indonesia) (location identifier) • Zero Economic Growth • Zero Entropy Gradient

ZeIE Zeitschrift für elektrische Information und Energietechnik (periodical code)

ZeiPh Zeitschrift für Physik (periodical code)

zeitgen zeitgenössisch

zeitl zeitlich • zeitlos

Zeitschr Zeitschrift

zeitw zeitweilig • zeitweise

Zeitw Zeitwert • Zeitwort

ZEKE Zero Kinetic Energy

ZEL Bella Bella (Canada-BC) (location identifier) • Zell am Mosel (Germany)

Zem Zemlia

ZEM Eastmain (Canada-QC) (location identifier) • Zero-Effort Miss • Zero-Energy Mode

ZeMet Zeitschrift für Meteorologie (periodical code)

ZemVS Zemlia i Vselennaia (periodical code)

Zen Zenith

ZEN Zenag (Papua New Guinea) (location identifier) • Zero-Effort Network • Zero-Effort Networking

ZENITH Zero-Energy Nitrogen-Heated Thermal Reactor

Zent Zentral

zentr zentral • zentralisiert • Zentralistisch • zentriert • zentrisch

Zentr Zentrale • Zentrum

zentrif zentrifugal

ZEP Zone d'Échanges Préférentiels • Zone d'Éducation Prioritaire (France) • Zone d'Environnement Protégé (France)

ZePAN Zeitschrift Physik Atomic Nuclei (periodical code)

ZER Zero (India) (location identifier)

Zerst Zerstörung

zerstr zerstreut • zerstritten

ZES Zone Économique Spéciale

ZESA Zodiac Española SA (Spain)

ZETA Zero-Energy Thermonuclear Assembly

ZETR Zero-Energy Thermal Reactor

ZEU Zentrum für Europäische Forschung

ZEV Zero-Emission Vehicle

ZEVGA Zeitschrift Eisenbahnwesen und Verkehrstechnik Glasers Annalen (periodical code)

ZEVIS Zentrales Verkehrsinformationssystem

zF zu Fuß • zur Folge

Zf Zufahrt • Zugführer

ZF Zentrifuge • Zermelo-Fraenkel • Zero Flag • Zero Forcing • Zero Frequency • Zona Franca • Zone Franche • Zone of Fire • Zwischenfrequenz

ZFA Faro (Canada-YT) (location identifier)

ZFB Old Fort Bay (Canada-QC) (location identifier) • Zentralinstitut für Film und Bild

ZfBB Zeitschrift für Bibliothekswesen und Bibliographie

ZFC Zermelo - Fraenkel - Choice (axiom) • Zero Failure Criteria • Zero-Field-Cooled • Zero-Field Cooling

ZfD Zeitschrift für Datenverarbeitung

ZFD Fond du Lac (Canada-SK) (location identifier) • Zwei-Farben-Diagramm

zFdW zur Förderung der Wissenschaften

Zfg Zufügung • Zusammenfassung

ZfG Zeitschrift für Geschichtswissenschaft

ZFH Zweifamilienhaus

ZFL South Trout Lake (Canada-ON) (location identifier)

ZFlu Zeitschrift Flugwissenschaften (periodical code)

ZFlWe Zeitschrift Flugwissenschaften und Weltraumforschung (periodical code)

ZfM Zentralamt für Mobilfunk

ZFM Fort McPherson (Canada-NT) (location identifier)

ZFN Fort Norman (Canada-NT) (location identifier) • Zero (order), Fixed (aperture), Non(-redundant)

zfr zinsfrei • zollfrei

ZfS Zentralstelle für Standardisierung

ZFS Zero Field Splitting

ZFT Fort Lauderdale - Railway Station (USA-FL) (location identifier)

ZfU Zentrum für Unternehmenskommunikation (Germany)

ZFU Zentralstelle für Fernunterricht

ZfV Zeitschrift für Vermessungswesen

ZFV Zentrale für Fremdenverkehr

ZFW Fairview (Canada-AB) (location identifier) • Zeitschrift für Flugwissenschaften und Weltraumforschung • Zero Fuel Weight

Zg Zeitung • Zeuge • Zeugung • Zug • Zugabe • Zugang

ZG Zivilgericht • Zollgesetz • Zoologischer Garten • Zug

ZGB Zivilgesetzbuch

ZGC Lanzhou Zhongchuan (China-PRC) (location identifier)

ZGCS Compressed Guide Star Catalog

ZGD Zollgrenzdienst

ZGDV Zentrum für Graphische Datenverarbeitung eV (Germany)

zge ziemlich gut erhalten

ZGE Zero Gravity Effect

ZGeo Zeitschrift für Geophysik (periodical code)

ZGET Zero-Gravity Expulsion Technique

Zgf Zugfolge • Zugführer

ZGF Grand Forks (Canada-BC) (location identifier) • Zero-Gravity Facility

ZGG Zero-Gravity Generator

Zgh Zugehörigkeit

ZGH Zero-Gravity Harmonic • Zona Galáctica Habitable • Zone Galactique Habitable

ZGI Gods River (Canada-MB) (location identifier) • Zentrales Geologisches Institut

Zgkw Zugkraftwagen

zgl zugleich

ZGL South Galway (Australia-QLD) (location identifier)

ZGM Ngoma (Zambia) (location identifier)

Zgn Zeugen • Zeugnis

ZGP Zeiss-Grossplanetarium Berlin (Germany)

ZGPU Zabaikalsky State Pedagogical University (Russia)

ZGR Little Grand Rapids (Canada-MB) (location identifier) • Zugangsrechner

zgs zusammengesetzt

ZGS Gethsemani (Canada-QC) (location identifier) • Zeichengabesystem • Zero-Gradient Synchrotron • Zero-Gravity Shower • Zero-Gravity Simulator

zgst zusammengestellt

ZGT Zero-Gravity Trainer

ZGU Gaua (Vanuatu) (location identifier)

ZGWBS Zero-Gravity Whole-Body Shower

zh Chinese (ISO language code)

zH zu Händen • zur Hälfte

Zh Zhurnal

ZH Zentralheizung • Zero Halogen • Zero Hour • Zijne Heiligheid • Zijne Hoogheid

ZHA Zhangjiang (China-PRC) (location identifier)

z. Hd. zu Händen

z. Hdn. zu Händen

z. Hdn. v. zu Händen von

z. Hd. v. zu Händen von

ZhETF Zhurnal Eksperimental'noi i Teoreticheskoi Fiziki (periodical code)

ZHF Zodiacal History File

ZHK Zusatzheilkosten

zhlr zahlreich

ZHM Shamshernagar (Bangladesh) (location identifier)

ZHO Houston (Canada-BC) (location identifier)

ZHP High Prairie (Canada-AB) (location identifier)

ZhPhy Zhurnal Physik (periodical code)

ZhPmR Zhurnal Pis ma Redaktsiiu (periodical code)

ZhPS Zhurnal Prikladnoi Spektroskopii (periodical code)

ZHR Zenithal Hourly Rate • Zenith Hourly Rate

ZHSA Zeiss Historica Society of America (USA-NY)

ZhTFi Zhurnal Tekhnicheskoi Fiziki (periodical code)

zHv zu Händen von

ZHV Zoll- und Handelsvertrag

Zi Ziffer • Zimmer

ZI Zentralinkasso • Zentralinstitut • Zero Intensity • Zielinformation • Zimbabwe • Zinner (variable star) • Zonal Index • Zone Industrielle • Zone of the Interior

ZIA Zentralinstitut für Automation • Zentralinstitut für Automatisierung • Zinc Ion Availability

ZIAP Zentralinstitut für Astrophysik Potsdam, Germany

ZIB Zentralinformationsbank • Zentralinstitut für Bibliothekswesen • Zentrum für Informationstechnik Berlin (Germany) • Zero-Impulse Bolt

ZIC Victoria (Chile) (location identifier) • Zentrale Informationsverarbeitung Chemie (Germany) • Zone Intertropicale de Convergence

ZID Zeitschrifteninhaltsdienst • Zentraler Informationsdienst • Zentralstelle für Information und Dokumentation • Zone of Initial Dilution

ZIDIS Zentrale Information und Dokumentation der Isotopen- und Strahlentechnik

ZIE Zone Locale Élargie

ZiF Zentrum für interdisziplinäre Forschung (Germany)

ZIF Zero Insertion Force

Ziff Ziffer

Zig Zigarre • Zigarette • Zigarillo

ZIG Zählimpulsgeber • Ziegenhain (Germany) • Ziguinchor (Senegal) (location identifier) • Z39.50 Implementation Group

ZIH Ixtapa Zihuatanejo (Mexico) (location identifier)

ZIHD Zentrale Informationsstelle für Hochschuldidaktik

ZIID Zentralinstitut für Information und Datenverarbeitung • Zentralinstitut für Information und Dokumentation

ZIIS Zentrum für Internationale und Interdisziplinäre Studien (Austria)

ZIL Zentrale Informationsleitstelle • Zone d'Intervention Localisée

Zim Zimmer

ZIMPOL Zürich Infrared Polarimeter (*) • Zürich Imaging Polarimeter

ZIO Zero Interoperability

zip Zip-compressed Archive (file extension)

ZIP Zero-Inflated Poisson • Zigzag In-line Package • Zigzag In-line Pin • Zodiacal Infrared Probe • Zodiacal Infrared Project • Zone Improvement Plan • Zone Improvement Program • Zone Information Protocol • Zoning Improvement Plan • Zukunftinvestigationsprogramm • Zukunftinvestitionsprogramm

ZIPE Zentralinstitut für Physik der Erde (Germany)

Zirk Zirkular

ZIRST Zone pour l'Innovation et la Recherche Scientifique et Technique

ZIS Scotopic Illuminance Sensitivity • Seis Station (Switzerland) (location identifier) • Zentrales Informationssystem • Ziekehuis Informatie System • Zone Information Socket • Zweiginformationssystem

zit zitieren • zitiert

ZISC Zero-Instruction-Set Computer • zitiert

Zit Zitadelle • Zitat

ZIT Zone Information Table • Zone Information Table

zil zivil • zivilisiert

Zil Zivil • Zivilisation

ZIV Inverness Scotrail (UK-Scotland) (location identifier) • Zentrale Informationsstelle für Verkehr

ZIZ Zweigorientiertes Informationszentrum

ZiZa Ziran Zazhi (periodical code)

ZJG Jenpeg (Canada-MB) (location identifier)

ZJID Zentralstelle für Journalistische Information und Dokumentation

ZJN Swan River (Canada-MN) (location identifier)

zk zurück

zK zur Kenntnis

Zk Zahlkarte • Zuschlagkarte

ZK New Zealand (international aircraft registration code) • Zeitkontrolle • Zentrale Kommission • Zentralkatalog • Zentralkommission • Zentralkomitee • Zollkontrolle • Zugkraft • Zuschlagkarte • Zwischenklärung

ZKA Zentraler Kreditausschuß • Zollkriminalamt

ZKB Kasaba Bay (Zambia) (location identifier) • Zimmer - Küche - Bad

ZKBB Zimmer - Küche - Bad - Balkon

ZKE Kaschechewan (Canada-ON) (location identifier) • Zonal Kinetic Energy

Zkft Zukunft • Zusammenkunft

zKg zur Kenntnis genommen

ZKG Kegaska (Canada-QC) (location identifier)

ZKH Zijne Keizerlijke Hoogheid • Zijne Koninklijke Hoogheid

ZKK Zentrale Kontrolkommission

ZKL Steenkool (Indonesia) (location identifier)

ZKM Sette Cama (Gabon) (location identifier) • Zentrum für Kunst und Medientechnologie (Germany) • Zweikoordinatenmeßgerät

ZKP Kasompe (Zambia) (location identifier)

Z. Kts. Zur Kenntnis

Zkw Zugkraftwagen

ZKW Zero Kilowatt

ZKZ Zentralkatalog der Zeitschriften

Zł Złoty (currency)

Zl Zahl • Zeile • Ziel • Zugleitung

ZL New Zealand (international aircraft registration code) • Zero Lift • Zodiacal Light • Zwischenlandung

ZLA Zambia Library Association • Zimbabwe Library Association • Zone with Limited Access

ZLC Zero Length Column

ZLD Zero Level Drift • Zentralstelle für Luftfahrtdokumentation • Zero Lift Drag • Zodiacal Light Device

ZDLI Zentralstelle für Luftfahrtdokumentation und Information

Zle Zeile • Ziele

ZLE Zero Latency Enterprise

ZLEA Zone de Libre-Échange des Amériques

Zlg Zahlung • Zulassung • Zuleitung

ZLG El Gouera (Mauritania) (location identifier) • Zero Line Gap • Zero-lock Laser Gyro

ZLISP Zilog List Processor

ZLKB Zentrale Leit- und Koordinierungstelle der Bibliotheken

ZLM Zone de Logistique et de Maintenance • Zweifrequenz-Laserwegmeßsystem

ZLO Manzanillo (Mexico) (location identifier) • Zone de Logistique et d'Opérations

ZLR Zentrum für Luft- und Raumfahrttechnik (Germany)

ZLS Scotopic Luminance Sensitivity • Zeeman Laser Scattering • Zentralstelle der Länder für Sicherheitstechnik

ZLSM Zeiss Light Section Microscope

ZLT La Tabatière (Canada-QC) (location identifier)

ZLV Zero-Length Vector

Zm Zahlmeister • Zeitmaß • Zeitmesser • Zeitmessung • Zugmaschine

ZM New Zealand (international aircraft registration code) • Zambia (ISO country code) • Zambie (code ISO de pays) • Zeitmessung • Zeitmultiplex • Zentralmeridian • Zipf-Mandelbrot (distribution) • Zone Marker • Zwischenmahlzeit • Zwischenmodulation

ZMA Zone Multicast Access • Zone Multicast Address

ZMB Zambia (ISO country code) • Zambie (code ISO de pays) • Zero Momentum Bias

ZMD Sena Madureira (Brazil-AC) (location identifier) • Zambia Meteorological Department

ZME Metro Park - Railway Station (USA-NY) (location identifier) • Zeolite-Modified Electrode

ZMete Zeitschrift für Meteorologie (periodical code)

ZMetl Zeitschrift für Metallkunde (periodical code)

ZMH 108 Mile Ranch (Canada-BC) (location identifier) • Zone Mail Hour

ZMI Napoli - Mergellina Stazione (Italy) (location identifier)

ZMK Kwacha (Zambia) (ISO currency code)

ZML Milwaukee Railway Station (USA-WI) (location identifier)

ZMM Zamora (Mexico) (location identifier)

ZMMAS Zodiacal Microparticle Multiparameter Analysis System

ZMMD Zürich - Mainz - München - Darmstadt (collaboration)

ZMMF Zentrale Medizin-Meteorologische Forschungsstelle

ZMP Zentralmeridianpassage • ZIP Mail Translator

ZMR Zone Melt Recrystallization

ZMS Firenze - S.M. Novella Stazione (Italy) (location identifier) • Zero Mechanical State

ZMT Masset (Canada-BC) (location identifier)

zn zu nehmen • zu nennen

Zn Zeichen • Zeitnahme • Zinc • Zink • Zone • Zoon • Zunahme • Zuname

ZN Zahlungsnachweis • Zimmernachweis • Zusatz-Nichtdeskriptor

ZNA Nanaimo Harbour (Canada-BC) (location identifier) • Zone of No Access

ZNat Zeitschrift für Naturforschung (periodical code)

ZNatA Zeitschrift für Naturforschung - Teil A (periodical code)

ZNC Nyac (USA-AK) (location identifier)

ZND Zentrale Namendatei • Zinder (Niger) (location identifier)

ZNE Newman (Australia-WA) (location identifier)

ZNG Negginan (Canada-MB) (location identifier)

ZNIEFF Zone Naturelle d'Intérêt Écologique, Faunistique et Floristique (France)

ZNK Zentrale Namenkartei

ZNP Zero Knowledge Proof

ZNPFK Zhurnal Nauchnoi i Prikladnoi Fotografii i Kinematografii (periodical code)

ZNS Zentrales Nervensystem • Zentralnervensystem

ZnSe Zinc Selenide

ZNTH Zenith

ZNU Namu (Canada-BC) (location identifier)

ZNZ Zanzibar (Tanzania) (location identifier)

ZO Zahlungsordnung • Zero Output • Zollort • Zuidoost • Zulassungsordnung • Zuständigkeitsordnung

ZOA Zone Of Avoidance

ZOAG Zone Of Avoidance Galaxy (catalog)

ZOB Zentral-Omnibus-Bahnhof

Zod Zodiac

ZOD Zero-Order Design • Zero-Order Detector

ZOE Zone Of Entry • Zone Of Exchange • Zone Of Exclusion

ZOF Ocean Falls (Canada-BC) (location identifier) • Zero-Order Field • Zone Of Fire

ZOHF Zodiacal Observation History File

ZOI Zone Of Incorporation

ZOL Zero Overlap

ZOO Zone Of Occupation

Zoochem Zoochemistry

Zool Zoología • Zoological • Zoologist • Zoology

Zoot Zootecnia

ZOP Zero Overpricing • Zeroth-Order Prediction

ZOPI Zero Order Polynomial Interpolator

ZOPP Zero Order Polynomial Predictor

ZOS Osorno - Cañal Bajo (Chile) (location identifier) • Zone Of Separation • Zoom Optical System

zoz Zie Ommezijde

ZOZ Zie Ommezijde

zP zur Person

ZP Paraguay (international aircraft registration code) • Zero Pressure • Zielpunkt • Zimmerpreis • Zivilprozeß • Zone of Protection • Zusatzpatent • Zusatzprovision

ZPA Zero Period Acceleration

ZPB Sachigo Lake (Canada-ON) (location identifier)

ZPC Pucón (Chile) (location identifier) • Zero Point of Change • Zero Point of Charge

ZPD Zenithal Point Determination • Zenith Path Delay

ZPE Zero-Point Energy

ZPES Zagalic Polyelastic Sampling

zPf zur Pferd

ZPF Zero Power Finder

ZPG Zero Population Growth

ZPH Zephyrhills (USA-FL) (location identifier)

ZPhy Zeitschrift für Physik (periodical code)

ZPID Zentralstelle für Pädagogische Information und Dokumentation

ZPL Zero Phonon Line

ZPM Zone de Pénétration Massive • Zone-Plate Mirror

ZPN Impedance Pneumogram • Zenithal Polynomial Projection • Zentralstern eines Planetarischen Nebels

ZPO Pine House (Canada-SK) (location identifier) • Zivilprozeßordnung

ZPP Zinc Protoporphyrin

ZPR Zentrum für Paralleles Rechnen • Zivilprozeßrecht

Zprav Zpravodaj VZLU (periodical code)

ZPRSN Zürich Provisional Relative Sunspot Number

ZPS Zone de Protection Spéciale

ZPT Zero Power Test

ZPV Zoomed Port Video

ZQC Zero Quantum Coherence

ZQN Queenstown (New Zealand) (location identifier)

ZQS Queen Charlotte Island (Canada-BC) (location identifier)

zr zahlreich

zR zu Recht • zur Revision • zur Ruhe

Zr Zimmer • Zirconium

ZR Zaire (ISO country code) • Zeeman Regime • Zentralrat • Zivilrecht • Zone de Remplacement • Zone Rurale

ZRA Zone of Restricted Access

ZRD Buffalo Depew (USA-NY) (location identifier)

ZRE Zaire (international vehicle registration code) • Zero Rate Error

ZRF Rockford - Clock Tower (USA-IL) (location identifier)

ZRFG Zonenrandførderungsgesetz

ZRG Bratislava Bus Station (Slovakia) (location identifier)

ZRH Zürich Kloten (Switzerland) (location identifier)

ZRI Serui (Indonesia) (location identifier) • Zero Resistance Penetration • Zimbabwe Research Index

ZRJ Round Lake (Canada-ON) (location identifier)

ZRK Zentrale Regionenkonferenz

ZRL Albany-Rensselaer Railway Station (USA-NY) (location identifier) • Zero Risk Level

ZRM Sarmi (Indonesia) (location identifier)

ZRO Zero Rate Offset

ZRP Newark Penn (USA-NJ) (location identifier)

ZRR Arctic Red River (Canada-NT) (location identifier)

ZRS Zivilrechtssache

ZRT Zero Reaction Tool

ZRU Zone de Redynamisation Urbaine (France)

ZRV Zero Relative Velocity

ZRZ Zaire (Zaire) (ISO currency code) • Zentrales Rechenzentrum

zs zusammen

zS zur Sache • zur See

Zs Zeitschrift • Zusage • Zusatz • Zusammensetzung

ZS South Africa (international aircraft registration code) • Zahlungsstelle • Zeichensynchronisation • Zenith Star • Zentralsekretariat • Zentralstation • Zentralstelle • Zero and Subtract • Zivilschutz • Zivilsenat

ZSA San Salvador (Bahamas) (location identifier) • Zambia Science Abstracts

ZSB Zero Surface Brightness • Zweiseitenband

ZSC Zero Semantic Comment

Zschr Zeitschrift • Zuschrift

ZSD Ziviler Sicherheitsdienst

ZSE Saint-Pierre de la Réunion (Reunion Island) (location identifier) • Zero Stress Everywhere

ZSEE Zoneamento Socio-Economico-Ecologico

ZSF Springfield Railway Station (USA-MO) (location identifier)

Zsfg Zusammenfassung • Zusammenfügung

Zshg Zusammenhang

ZSI Zambia Standards Institute • Zentralstelle für Sicherheit in der Informationstechnik (Germany)

ZSJ Sandy Lake (Canada-ON) (location identifier)

ZSL Zentrale Sendeleitung • Zero Slot LAN

zsm Zo Spoedig Mogelijk

ZSM Zavod za Standardizacijai Metrologija (Macedonia) • Zone Specific Maintenance

ZSMA Zodi-Subtracted Mission Average

Z-SN Zwicky Supernova

ZSN Zwicky Supernova

ZSP Saint Paul (Canada-QC) (location identifier) • Zone de Solidarité Prioritaire (France)

ZSPN Zentralstern eines Planetarischen Nebels

ZSR Zugsteuerrechner

Zss Zeitschriften

ZSS Sassandra (Ivory Coast) (location identifier)

Zst Zustand • Zusammenstellung • Zustellung

ZSt Zentralstelle

ZST Stewart (Canada-BC) (location identifier) • Zone Standard Time

ZStA Zentrales Staatarchiv

Zstg Zusammenstellung • Zustellung

ZSV Saint Louis - Railway Station (USA-MO) (location identifier) • Zone Settling Velocity

ZSW Prince Rupert - Seal Cove (Canada-BC) (location identifier) • Zentrum für Sonnenenergie- und Wasserstoffforschung (Germany)

zT zum Teil • zum Termin

Zt Total Available Effort • Zeit • Zitat • Zuteilung

ZT South Africa (international aircraft registration code) • Zahlungstermin • Zenith Telescope • Zone Time

ZTA Tureia (Tuamotu Islands) (location identifier) • Zirconia Toughened Alumina • Zone de Télécommunication Avancée

ZTB Tête-à-la-Baleine (Canada-QC) (location identifier)

ZTC Zero Tolerance to Circumstances

Ztg Zeitung • Zuteilung

Ztgn Zeitungen

ZTH Zakynthos Island (Greece) (location identifier)

ZTIBC Zero Torque Inner Boundary Condition

ztl zeitlich • zeitlos

ZTLP Zero Transmission Level Point

Ztm Zeitmaß • Zeitmesser • Zeitmessung

ZTM Shamattawa (Canada-MB) (location identifier)

Ztn Zeitnahme • Zeitnehmer

ZTO Zero Time Outage • Zone Transportation Office • Zone Transportation Officer

Ztr Zeitrechnung • Zentner • Zentrale • Zentrum • Zutritt

ZTR Zhitomir (Ukraine) (location identifier)

ZTS Tahsis (Canada-BC) (location identifier) • Zoom Transfer Scope

Ztschr Zeitschrift

ZTU Zero Tolerance Unit

ztw zeitweilig • zeitweise

Ztw Zeitwert

ZtW Zeitwert

ZTY Philadelphia - 30th Station (USA-PA) (location identifier)

zu Zulu (ISO language code)

zU zur Unterschrift • zur Untersuchung

ZU South Africa (international aircraft registration code) • Zentralunternehmen • Zuni • Zustellungsurkunde

Zub Zubehör

Zubr Zubringer

ZUC Ignace (Canada-ON) (location identifier)

ZUD Ancud (Chile) (location identifier)

ZUE Zuenoula (Ivory Coast) (location identifier)

zuf zufällig • zufolge

Zuf Zufall • Zufuhr

zug zugängig • zugänglich

Zug Zugabe • Zugang

zugel zugelassen

zuget zugeteilt

zugew zugewandert • zugewiesen

zugl zugleich

ZUH Zhuhai (China-PRC) (location identifier)

ZUI Zooming User Interface

zul zulässig • zuletzt

ZUL Zilfi (Saudi Arabia) (location identifier)

ZUM Churchill Falls (Canada-NF) (location identifier)

ZUMA Zeitung für Umfragen, Methoden und Analysen

ZÜN Zentraler Übersetzungsnachweis

ZUN Chicago Union (USA-IL) (location identifier)

ZUP Zone à Urbaniser en Priorité (France)

zur zurück

Zus Zusammen • Zusatz • Zusätzlich

zust zustimmend

ZUVAD University College London (Starlink internal code)

zuw zuweilen

Zuz Zuzüglich

zv zur verkaufen • zu vermieten • zu vernichten

zV zum Verkauf

ZV Zeitungsverlag • Zentralvorstand • Zoomed Video • Zoom Video

ZVA Miandrivazo (Madagascar) (location identifier) • Zeitungsvertriebsamt • Zero (order), Variable (aperture), Adjusted

ZVC Zero-Velocity Contour • Zero-Velocity Curve

ZVE Zentrale Verarbeitungseinheit

ZVEI Zentralverband der Elektrotechnischen Industrie eV (Germany)

z.Vg. zum Vorgang

ZVG Springvale (Australia-WA) (location identifier)

ZVIS Zentralverwaltung Informationssysteme

ZVK Savannakhet (Laos) (location identifier)

ZVMMF Zhurnal Vychislitel'noi Matematiki i Matematicheskoi Fiziki (periodical code)

ZVSt Zentralvermittlungsstelle

ZVZV Zeitungsverlag und Zeitschriftenverlag

zw zwar • zwecks • zwei • zwischen

Zw Zahlungsweise • Zahlwort • Zeitwort • Zuwachs • Zwang • Zweck • Zwicky • Zwillinge

ZW Zimbabwe (international vehicle registration code) • Zimbabwe (ISO country code) • Zinner-West (scale) • Zuidwest • Zwicky

ZWA Andapa (Madagascar) (location identifier) • Zentralstelle für Wissenschaftliche Altbestände

ZwCl Zwicky Clusters

ZwD Zweigbibliothek

ZWD Dollar (Zimbabwe) (ISO currency code)

ZWE Zimbabwe (ISO country code)

ZWer Zeitschrift für Werkstofftechnik (periodical code)

ZWG Zwicky (catalog of) Galaxies

ZWI Wilmington - Railway Station (USA-DE) (location identifier)

ZWL Wollaston Lake (Canada-SK) (location identifier) • Zentralstelle für Wissenschaftliche Literatur

ZWO Organisatie voor Zuiverwetenschappelijk Onderzoek (Netherlands)

ZWP West Palm Beach - Railway Station (USA-FL) (location identifier)

ZWS Stuttgart Hauptbahnhof (Germany) (location identifier) • Stuttgart Hbf. (Germany) (location identifier) • Stuttgart Main Station (Germany) (location identifier) • Zero Wait State

ZWTI Zentrum für Wissenschaftlich-Technische Information

ZWTR Zentraler Wissenschaftlich-Technischer Rat

ZWU Washington Union (USA-DC) (location identifier)

ZXS Buffalo - Exchange Street (USA-NY) (location identifier)

ZYI Zunyi (China-PRC) (location identifier)

Zyl Zylinder

ZYL Sylhet (Bangladesh) (location identifier)

ZYP New York - Penn (USA-NY) (location identifier)

zz Ginger

zZ zur Zeit

Zz Ginger

ZZ Zero-Zero • Zig-Zag • Zonenzeit

ZZB Zirkumzenitalbogen

ZZBB Zentralkatalog der Zeitungsbestände in Berliner Bibliotheken

ZZC Zone Zero Conditions

ZZD Zeitungs- und Zeitschriftendienst • Zig-Zag Diagram

ZZF Zentralamt für Zulassungen im Fernmeldewesen (Germany)

zzgl zuzüglich

ZZK Zeitschriftenzentralkatalog • Zentral-Zeichengabe-Kanal

ZZO Zuidzuidoost

ZZP Zweizentrenproblem

z. Zt. zur Zeit

ZZU Mzuzu (Malawi) (location identifier)

ZZV Zanesville (USA-OH) (location identifier)

ZZW Zone Zero Weather • Zuidzuidwest

Special Sections

Numerical Beginnings

0 Supersupergiant Star
0D Zero-Dimensional
0g Null Gravity
0OI Zeroth-Order Ignorance
0th Zeroth
1 Monday
1A First Ariel V Catalogue
1ASE First American Science and Engineering Catalogue
1C First Cambridge Catalogue of Radio Sources
1CG First COS B Catalogue
1D One-Dimensional
1dF One-degree Field
1E EINSTEIN Observatory Catalog • HEAO-3 Catalog
1G First Generation
1hT One Hectare Telescope
1HT One Hectare Telescope
1M Massachusetts OSO-7 Catalogue
1NF First Normal Form
1OI First-Order Ignorance
1OT First Outer Track • First Overtone
1Q First Quarter
1st First • Readme First (file extension)
1U First Uhuru Catalogue
1XRS First X-ray Sources Catalogue • SAS-3 Catalogue
10GE 10 Gigabits/Ethernet
10GFC 10 Gigabits/Fiber Channel
12MISS Twelve Micron ISOCAM Serendipitous Survey
2 Tuesday
2a Segunda-feira (Monday/Portuguese)
2A Second Ariel V Catalogue
2ASE Second American Science and Engineering Catalogue
2AT Second Alias Table
2BMS Two-Bed Molecular Sieve
2BSD Second Berkeley Software Distribution
2B1Q Two Binary One Quaternary
2C Second Cambridge Catalogue of Radio Sources
2CG Second COS B Catalogue
2D Two-Dimensional • Two-Poisson
2DAG Two-Dimensional Atom Gas
2DD Two-Dimensional and Dynamic (method)
2DEG Two-Dimensional Electron Gas
2DES Two-Dimensional Electron System
2dF Two-Degree Field
2dFGRS Two-Degree Field Galaxy Redshift Survey
2DKS Two-Dimensional Kolmogorov-Smirnov (test)
2EG Second EGRET Catalog
2G Second Generation
2GL Second-Generation Language
2H Double Hard
2MASS Two-Micron All-Sky Survey
2MRS 2MASS Redshift Survey
2MTRT Two-Mirror Three-Reflection Telescope
2nd Second
2NF Second Normal Form
2OI Second-Order Ignorance
2OT Second Outer Track • Second Overtone
2P Two-Poisson
2PC Two-Phase Commit
2PCF Two-Point Correlation Function
2PS Two-Phase Search
2Q Second Quarter
2QZ 2dF QSO Redshift (survey) • 2dF Quasar Redshift (survey)
2SCS Second Satellite Control Squadron
2SLS Two-Stage Least Squares
2SOPS Second Satellite Operations Squadron
2T Two-Temperature (mode)
2U Second UHURU Catalogue
2WD Two-Wheel Drive
2XRS Second X-ray Sources Catalogue
3 Wednesday
3a Terça-feira (Tuesday/Portuguese)
3A Monaco (international aircraft registration code) • Third Ariel V Catalogue
3B Mauritius (international aircraft registration code)
3C Care - Companionship - Clothing • Equatorial Guinea (international aircraft registration code) • Third Cambridge Catalogue of Radio Sources
3CR Revised Third Cambridge Catalogue of Radio Sources
3D Désétatisation - Défiscalisation - Dérégulation • Dimension - Durée - Dispersion • Swaziland (international aircraft registration code) • Three-Dimensional
3dF Three-Degree Field
3DTV Three-Dimensional Television
3DUP Third Dredge-Up
3F Form - Fit - Function
3FMS Free Flight Flight Management System
3G Third Generation
3GL Third-Generation Language
3GPP Third Generation Partnership Project
3H Hot - High - Humid • Treble Hard
3I Improved Initialization Inversion
3M Maintenance and Material Management • Minnesota Mining and Manufacturing Co. (USA-MN)
3NF Third Normal Form
3OI Third-Order Ignorance
3OT Third Outer Track • Third Overtone
3PCF Three-Point Correlation Function
3PN Third Postnewtonian
3PS Three-Phase Search
3PT Third-order Perturbation Theory
3PTY Three Party (service)
3Q Third Quarter
3R Rapid Radiance Recognition • Reading - Writing - Arithmetic • Reconnaissance Rapide des Radiances • Reference and Research Resources
3rd Third
3RD Réseau Radioélectrique à Ressources pour Données
3RP Réseau Radioélectrique à Ressources Partagées

3S Satellite Surface Stress • Sols - Solides - Structures

3T Twente Technology Transfer BV (Netherlands)

3U Third UHURU Catalogue

3WC Three-Way Calling

3X Guinea (international aircraft registration code)

4 For • Thursday

4a Quarta-feira (Wednesday/Portuguese)

4A Association Andromède Astronomie Aveyronnaise (France) • Atlas Automatisé des Absorptions Atmosphériques • Automated Atmospheric Absorption Atlas

4BMS Four-Bed Molecular Sieve

4C Computer - Communication - Control - CRT • Fourth Cambridge Catalogue of Radio Sources

4D Four-Dimensional

4dF Four-Degree Field

4G Fourth Generation

4GL Fourth-Generation Language

4H Head - Heart - Hands - Health

4K Azerbaijan (international aircraft registration code)

4MAPS 4-Meter Active Primary Support

4MASS Four-Meter All-Sky Survey

4NF Fourth Normal Form

4OI Fourth-Order Ignorance

4OT Fourth Overtone

4P Prévision Par Petits Pavés

4PS Four-Phase Search

4Q Fourth Quarter

4QC Four Quadrants Coronagraph

4R Sri Lanka (international aircraft registration code)

4S Society for the Social Studies of Science

4th Forth (file extension) • Fourth

4to Quarto

4U Fourth UHURU Catalogue

4WD Four-Wheel Drive

4X Israel (international aircraft registration code)

5 Friday

5a Quinta-feira (Thursday/Portuguese)

5A Libya (international aircraft registration code)

5As Amherst Area Amateur Astronomers Association (USA-MA)

5B Cyprus (international aircraft registration code)

5C Fifth Cambridge Catalogue of Radio Sources

5D Five-Dimensional

5dF Five-Degree Field

5G Fifth Generation

5GL Fifth-Generation Language

5H Tanzania (international aircraft registration code)

5J Private Jet Expeditions Inc. (location identifier)

5N Nigeria (international aircraft registration code)

5OT Fifth Overtone

5R Madagascar (international aircraft registration code)

5S Satellite Signal Simulation in the Solar Spectrum

5T Mauritania (international aircraft registration code) • Time - Turn - Twist - Throttle - Talk

5th Fifth

5U Niger (international aircraft registration code)

5V Togo (international aircraft registration code)

5W Samoa (international aircraft registration code)

5WD Five-Wheel Dive

5X Uganda (international aircraft registration code)

5Y Kenya (international aircraft registration code)

6 Saturday

6a Sexta-feira (Friday/Portuguese)

6C Sixth Cambridge Catalogue of Radio Sources

6dF Six-degree Field

6dFAG Six-degree Field Advisory Group

6dFGRS Six-Degree Field Galaxy Redshift Survey

6dFGS Six-Degree Field Galaxy Survey

6dFSAG Six-degree Field Science Advisory Group

6G Sixth Generation

6O Somalia (international aircraft registration code)

6OT Sixth Overtone

6th Sixth

6V Senegal (international aircraft registration code)

6W Senegal (international aircraft registration code)

7 Sunday

7G Seventh Generation

7O Yemen (international aircraft registration code)

7OT Seventh Overtone

7P Lesotho (international aircraft registration code)

7QY Malawi (international aircraft registration code)

7T Algeria (international aircraft registration code)

7th Seventh

8G Eighth Generation

8OT Eighth Overtone

8P Barbados (international aircraft registration code)

8Q Maldives (international aircraft registration code)

8R Guyana (international aircraft registration code)

8th Eighth

8to Octavo

9A Croatia (international aircraft registration code)

9G Ghana (international aircraft registration code) • Ninth Generation

9H Malta (international aircraft registration code)

9J Zambia (international aircraft registration code)

9K Kuwait (international aircraft registration code)

9L Sierra Leone (international aircraft registration code)

9M Malaysia (international aircraft registration code)

9N Nepal (international aircraft registration code)

9OT Ninth Overtone

9Q Zaire (international aircraft registration code)

9th Ninth

9U Burundi (international aircraft registration code)

9V Singapore (international aircraft registration code)

9XR Rwanda (international aircraft registration code)

9Y Trinidad and Tobago (international aircraft registration code)

Greek Alphabet

α .... Acoustic Absorption Coefficient
alpha
Angle of Attack
Angle of Incidence
Angular Acceleration
Attenuation Coefficient
Cubic Expansion Coefficient
Dissociation Degree
Effective Thermal Diffusivity
Fine-Structure Constant
Internal-Conversion Coefficient
Internal-Conversion Factor
Lagrangian Multiplicator
Mach Angle
Madelung Constant
Molecular Electrical Polarizability
Payload Ratio
Plane Angle
Polarizability
Radio Spectral Index
Recombination Coefficient
Right Ascension
Thermal Accomodation Coefficient
Thermal Diffusivity
Thermal Expansion Coefficient
Townsend Electron Ionization Coefficient
α^h .... Absorptance
α_i .... Electron-Ion Recombination Coefficient
α_i .... Ion-Ion Recombination Coefficient
α_l .... Linear Expansion Coefficient
α_p .... Relative Pressure Coefficient
$\alpha_x\ \alpha_y\ \alpha_z$ ... Dirac (4×4) Matrices
α_T .... Thermal Diffusion Factor
$\alpha(A)$ .... Nullity Index of A
A .... Alpha
β .... Ballistic Coefficient
beta
Coefficient of Thermal Expansion
Dirac (4×4) Matrix
Dry Mass Ratio
Ecliptic Latitude
Magnetic Pressure Ratio
Phase Coefficient
Plane Angle
Pressure Coefficient
Ratio of Gas Pressure to Total Pressure
Townsend Ion Ionization Coefficient
$\beta(A)$ .... Deficiency Index of A
B .... Beta
γ .... Acceleration
Aspect Ratio of the Duct
Conductivity
Cubic Expansion Coefficient
Damping Constant
Electrolytic Conductivity
Flight Path Angle
gamma
Grüneisen Parameter
Gyromagnetic Ratio
Photon
Plane Angle
Polarizability
Propagation Coefficient
Propellant Fraction
Ratio of Specific Heat Capacities
Secondary Electron Emission Coefficient
Shear Strain
Slope of Characteristic Curve of Photographic Emulsion
Specific Heat Capacities Ratio
Surface Tension
γ_B .... Activity Coefficient
Γ .... Boudary of A
Gamma
Grüneisen Parameter
Level Width
Specific Gamma-Ray Constant
Van den Kerkhove Factor
δ .... Boundary Layer Thickness
Damping Coefficient
Declination
Deflection Angle
delta
Dissipation Coefficient
Flow Angle
Loss Angle
Thickness
Wedge Angle
δ_e .... Pitch Control Deflection
δ^i_j .... Kronecker Symbol
δ_{ij} .... Kronecker Symbol
δ_T .... Throttle Setting
$\delta(x)$ .... Dirac Delta Function
Δ .... Area of Triangle
Delta
Difference of
Distance to the Earth in AU
Geocentric Distance of Perihelion
Mass Excess
Superconductor Energy Gap
$\Delta\epsilon$ .... Nutation in Ecliptic Latitude
$\Delta\lambda$ .... Doppler Wavelength Shift
Wavelength Bandwidth
$\Delta\nu$ .... Frequency Bandwidth
$\Delta\varphi$ .... Nutation in Ecliptic Longitude
ΔS .... Preston's Metallicity Parameter (RR Lyrae stars)
ΔT .... Ephemeris Time Difference
Universal Time Difference
ϵ .... Emissivity
Energy of a Particle
epsilon
Linear Strain
Molecular Attraction Energy Parameter
Permittivity
Relative Elongation
ϵ_F .... Fermi Energy
ϵ_{ab} .... Seebeck Coefficient for Substances a and b
ϵ_o .... Permittivity of Vacuum
ϵ_r .... Relative Permittivity
ε .... Asymmetry Parameter
Collision Parameter (azimuth angle)
Ecliptical Obliquity
Emission Coefficient
Emissivity
Energy per Unit of Mass
epsilon

Error
Expansion Ratio
Fast Fission Factor
Inclination of the Galactic Plane on the Equatorial One
Nuclear Energy Generation
Permittivity
Sign Factor
Structural Fraction
Turbulence Dissipation

ε_P .... Potential Energy

ε_T .... Thrust Vector Angle

ε_i .... Energy of Level i
Energy of State i

ε_o .... Dielectric Constant of Vacuum

$\varepsilon(t)$ .... Unit Step Function

E .... Epsilon

ζ .... Apparent Zenithal Distance
Modal Damping Ratio
Normalized Pressure Drop
zeta

$\zeta(\lambda)$ .... Interstellar Extinction Law

Z .... Zeta

η .... Charge Density
Efficiency
Efficiency Rate
eta
Modal Coordinates
Neutron Yield per Absorption
Propulsion Efficiency
Sector-Triangle Ratio
Viscosity

η_m .... Magnetic Diffusivity

H .... Eta

θ .... Bulk Strain
Dimensionless Temperature
Scattering Angle
theta
Plane Angle
Regression Coefficient
Volume Strain

ϑ .... Angular Coordinate (galactocentric system)
Bragg Angle
Celsius Temperature
Elongation
Euler Angle
Flow Angle
Pitch Angle
Plane Angle
Polar Angle
Position Angle
Scattering Angle
theta
Volume Strain

Θ .... Characteristic Temperature
Current Linkage
Diffraction Angle
Local Sidereal Time
Sidereal Time
Thermodynamic Temperature
Theta

Θ_D .... Debye Temperature

Θ_E .... Einstein Temperature

Θ_W .... Characteristic Weiss Temperature

Θ_r .... Rotational Temperature

Θ_v .... Vibrational Temperature

ι .... iota

I .... Iota

κ .... Absorption Coefficient
Bulk Compressibility
Compressibility
Epicyclic Frequency
Isentropic Exponent
kappa
Landau-Ginzburg Parameter
Magnetic Susceptibility
Molar Absorption Coefficient
Opacity
Ratio of Specific Heat Capacities
Rosseland Mean Absorption Coefficient
Thermal Diffusivity

K .... Kappa

λ .... Decay Constant
Desintegration Constant
Ecliptic Longitude
Geographical Longitude
Lagrangian Multiplier
lambda
Mean Free Path
Mean Molecular Free Path
Payload Fraction
Planetographic Longitude
Thermal Conductivity
Wavelength

$\lambda\lambda$ .... Wavelength Range

λ_B .... Absolute Activity of Substance B

λ_C .... Compton Wavelength

λ_D .... Debye Length

λ_L .... London Penetration Depth

λ_i .... Quantum Number of Component of Electronic Orbital Angular Momentum Vector Along Symmetry Axis

λ_{ik} .... Thermal Conductivity Tensor

Λ .... Boundary Layer Shape Factor
Characteristic Diffusion Length
Collision Parameter (polar angle)
Cosmological Constant
Lambda
Logarithmic Decrement
Mean Free Path of Phonons
Non-Leakage Probability
Phonon Mean Free Path
Quantum Number of Component of Electronic Orbital Angular Momentum Vector Along Symmetry Axis
Shock Wave Thickness

μ .... Atomic Attenuation Coefficient
Chemical Potential
Dynamic Viscosity
Earth/Moon Mass Ratio
Electromagnetic Moment
Friction Coefficient
Friction Factor
Joule-Thomson Coefficient
Linear Attenuation Coefficient
Magnetic Moment of Particle
Mass Flow Ratio
Mass Fraction
Micro
Micron
Mobility
mu
Muon
Nucleus Magnetic Moment
Particle Magnetic Moment
Permeability
Poisson Ratio
Proper Motion
Reduced Mass
Surface Mass Density

Thompson Coefficient
Viscosity
μ_B .... Bohr Magneton
Chemical Potential of Substance B
μ_N .... Nuclear Magneton
μ_a .... Atomic Attenuation Coefficient
Bank Angle
μ_e .... Magnetic Moment of Electron
μ_l .... Linear Attenuation Coefficient
μ_m .... Mass Attenuation Coefficient
μ_n .... Magnetic Moment of Neutron
μ_o .... Magnetic Permeability of Vacuum
Permeability of Vacuum
μ_r .... Relative Permeability
M .... Mu
ν .... Frequency
Kinematic Viscosity
Neutrino
Neutron Yield per Fission
nu
Poisson Number
Sound Particle Velocity
Specific Volume
True Anomaly
ν_B .... Stoichiometric Number of Substance B
ν_c .... Collision Frequency
ν_{coll} .... Collision Frequency
ν_i .... Ionization Frequency
N .... Nu
ξ .... Average Logarithmic Energy Decrement
Coherence Length
Extent of Reaction
Inertial Parameter
Microturbulent Velocity
Molecular Velocity
Mole-Mass Ratio
Sound Particle Displacement
xi
Ξ Xi
o .... omicron
O .... Omicron
π .... Equatorial Horizontal Parallax
Parallax
pi (see also ϖ)
Pion
Ratio of a Circle Circumference to its Diameter
$\pi_\odot$ .... Solar Parallax
ϖ .... Longitude of Perihelion
pi (see also π)
Π .... Linear Polarization Fraction
Osmotic Pressure
Pi
Product
Π_{ab} .... Peltier Coefficient for Substances a and b
ρ .... Atmospheric Density
Charge Density
Charge Volume Density
Density
General Secular Precession in Longitude
Geocentric Distance
Linear Strain
Mass Density
Reactivity
Reflection Coefficient
Resistivity
rho
ρ^h .... Reflectance
ρ_R .... Residual Resistivity
ρ_{ik} .... Resistivity Tensor
$\rho(A)$ ...... Resolvent Set of A
ϱ .... rho
P .... Rho
σ .... Charge Surface Density
Conductivity
Cross Section
Dispersion
Leakage Coefficient
Momentum Accomodation Coefficient
Normal Stress
Population Standard Deviation
Short-Range Order Parameter
sigma
Specific Fuel Consumption
Standard Deviation
Stefan-Boltzmann Constant
Stress Components
Surface Charge Density
Surface Tension
Symmetry Factor
Wave Number
Wave Vector
σ_i .... Quantum Number of Component of Electronic Spin Along Symmetry Axis
σ_{ik} .... Electric Conductivity Tensor
σ_x, σ_y σ_z .. Pauli Matrices
σ_1, σ_2 σ_3 .. Pauli Matrices
$\sigma(A)$ ...... Spectrum of A
$\sigma_{as}(A)$ ..... Absolutely Continuous Spectrum of A
$\sigma_c(A)$ ...... Continuous Spectrum of A
$\sigma_d(A)$ ...... Discrete Spectrum of A
$\sigma_e(A)$ ...... Essential Spectrum of A
$\sigma_p(A)$ ...... Point Spectrum of A
$\sigma_r(A)$ ...... Residual Spectrum of A
$\sigma_{sc}(A)$ ..... Singular Continuous Spectrum of A
σ_T .... Thomson Cross Section
ς .... sigma
Σ .... Covariance Matrix
Macroscopic Cross Section
Quantum Number of Component of Electronic Spin Along Symmetry Axis
Sigma
Sum
Summation
τ .... Age
Carrier Lifetime
Containment Time Scale
Dead Time
Eigenvalue
Heat Release Parameter
Hour Angle
Interval
Mean Life
Molecular Cross Section
Opacity
Optical Depth
Reacting Time
Relaxation Time
Shear Stress
tau
Thompson Coefficient
Time Constant
Transmission Coefficient
Viscous-Stress Tensor
τ^h .... Transmittance
τ_A .... Light Time per Astronomical Unit
τ_c .... Mean Time Interval between Collisions
τ_{coll} .... Mean Time Interval between Collisions
T .... Tau
υ .... Neutron Speed
upsilon

Υ .... Upsilon
Vernal Equinox
ϕ .... Constraint Mode
Interaction Potential
Latitude
Merit Function
phi
Porosity
φ .... Absorption Profile
Argument of z
Azimuthal Angle
Bank Angle
Composite Mass Fraction
Eigenmode
Electric Potential
Euler Angle
Geographic Latitude
Heat Current Density
Neutron Flux Density
Osmotic Pressure
Parallactic Angle
Particle Flux Density
Phase (of z)
Phase Difference
Phase Displacement
phi
Plane Angle
Planetographic Latitude
Potential
Radiant Flux Density
Scattering Angle
Solution Osmotic Coefficient
Work Function
φ_B .... Volume Fraction of Substance B
φ_{ij} .... Intercation Energy between Molecules i and j
Φ .... Angular Diameter of a Source
Heat Flow Rate
Luminous Flux
Magnetic Flux
Maxwell-Boltzmann Velocity Distribution Function
Phi
Polarization Angle
Potential Energy
Radial Phase (density wave)
Radiant Flux
Work Function
Φ_o .... Magnetic Flux Quantum
χ .... chi
Potential
Susceptibility
χ_c .... Electric Susceptibility
χ_m .... Magnetic Susceptibility
X .... Chi
ψ .... Degeneracy Parameter
Energy Flux Density
Euler Angle
Heading
psi
Radiant Flux Density
Stream Function
Vertex Deviation
Ψ .... Electric Flux
Electron Wave Function
Galactic Gravitational Potential
Psi
ω .... Angular Frequency
Angular Velocity
Argument of Perigee
Argument of Perihelion
Circular Frequency
Longitude of Periastron
omega
Oscillation Frequency
Prandtl Angle
Pulsatance
Rotational Rate
Solid Angle
ω_D .... Debye Circular Frequency
ω_L .... Larmor Circular Frequency
ω_{cc} .... Electron Cyclotron Circular Frequency
ω_{ci} .... Ion Cyclotron Circular Frequency
ω_i .... Quantum Number of Total Electronic Angular Momentum Along Symmetry Axis
ω_{pc} .... Electron Plasma Circular Frequency
Ω .... Density Parameter
Electric Resistance
Longitude of the Ascending Node
Ohm
Omega
Open Set in a Topological Space
Pattern Speed (density wave)
Quantum Number of Total Electronic Angular Momentum Vector Along Symmetry Axis
Solid Angle
Volume in γ Phase Space

Mathematical Signs

alphabetical on the explanations)

$\sigma_{as}(A)$	Absolutely Continuous Spectrum of A
$\lvert a \rvert$	Absolute Value of a
A^*	Adjoint of A
$\aleph$	Aleph
$\dim X$	Algebraic Dimension of X
$\xrightarrow{**}$	Almost Sure Convergence
$\angle$	Angle
$\measuredangle$	Angle
$\sphericalangle$	Angle
U°	Annihilator of Subspace U
$[A,B]_+$	Anticommutator of A and B
$\rightarrow$	Approaches
$\approx$	Approximately Equal to
$\cong$	Approximately Equal to
$\arg z$	Argument of z
$\arg(z)$	Argument of z
$*$	Asterisk Operator
$\simeq$	Asymptotically Equal to
$\asymp$	Asymptotical to
e	Base of Natural Logarithms
$\in$	Belongs to
X''	Bidual Space of X
$\mathrm{lb}\, x$	Binary Logarithm of x
$\log_2 x$	Binary Logarithm of x
$\binom{n}{p}$	Binomial Coefficient
$a_\circ$	Bohr Radius
Ω	Boundary of A
∂A	Boundary of A
$:=$	By Definition
$\stackrel{\mathrm{def}}{=}$	By Definition Equal to
$\stackrel{\Delta}{=}$	By Definition Equal to
$\widehat{=}$	By Definition Equal to
$\cap$	Cap
$\aleph$	Cardinal Number
$A \times B$	Cartesian Product of A and B
pv	Cauchy Principal Value (of an integral)
U_A	Cayley Tranform of A
∇	Characteristic Boundary Layer Length
∇	Characteristic Boundary Thickness
$r_\circ$	Classical Electron Radius
$\bar{A}$	Closure of A
$\lg x$	Common Logarithm of x
$\log_{10} x$	Common Logarithm of x
$[A,B]$	Commutator of A and B
$[A,B]_-$	Commutator of A and B
$F \subset\subset E$	Compact Injection of the Set F in E
A^c	Complement of A
$A \backslash B$	Complement of B in A
$\complement_A^B$	Complement of B in A
$\bar{z}$	Complex Conjugate of z
z^*	Complex Conjugate of z
A^*	Complex Conjugate of A
$g \circ f$	Composite Function of f and g
$f \circ g$	Composite Mapping of $f: F \rightarrow G$ and $g: E \rightarrow F^{(2)}$
f^n	Composite Mapping of Order n of $f: E \rightarrow E$
&	Conjunction
$\wedge$	Conjunction
$\supseteq$	Contains
$\ni$	Contains as Element or Member
$\supset$	Contains Properly
$\sigma_c(A)$	Continuous Spectrum of A
$\xrightarrow{d}$	Convergence in Distribution
$\xrightarrow{L^i}$	Convergence in L^i
$\xrightarrow{P}$	Convergence in Probability
$f * g$	Convolution Product of f and g
D_i	Cook's Influence Measure
$\amalg$	Coproduct
r_{xy}	Correlation Coefficient (linear) between x and y
$\Rightarrow$	Corresponds to
$\stackrel{\Delta}{=}$	Corresponds to
$\operatorname{cosec} x$	Cosecant of x
$\cos x$	Cosine of x
$\cot x$	Cotangent of x
$\operatorname{ctg} x$	Cotangent of x
$\cup$	Cup
$\nabla \times \mathbf{A}$	Curl of $\mathbf{A}$
$\operatorname{curl} \mathbf{A}$	Curl of $\mathbf{A}$
$\operatorname{rot} \mathbf{A}$	Curl of $\mathbf{A}$
$\Box f$	Dalembertian of f
$a_n \downarrow$	Decreasing Suite
$\beta(A)$	Deficiency Index of A
$\dot{f}$	Derivative of f
f'	Derivative of f
$\mathrm{d}f/\mathrm{d}x$	Derivative of f
$\frac{\mathrm{d}f}{\mathrm{d}x}$	Derivative of f
$f^{(n)}$	Derivative of Order n of f
$\det A$	Determinant of A
$\operatorname{diam} A$	Diameter of A
$\backslash$	Difference
dm	Differential Mass
dt	Differential Time
$\partial/\partial\mathbf{r}$	Differential Vector Operator
dv	Differential Velocity
$\dim X$	Dimension of X
$\delta(x)$	Dirac Delta Function
δ_o	Dirac Distribution
$\delta(x)$	Dirac Distribution
$\delta_o(x)$	Dirac Distribution
$\alpha_x\ \alpha_y\ \alpha_z$	Dirac (4×4) Matrices

β .... Dirac (4×4) Matrix
$\sigma_d(A)$ .... Discrete Spectrum of A
$A \oplus B$ .... Direct Sum of Vector Spaces A and B
$\vee$ .... Disjunction
$d(A,B)$ .... Distance between A and B
$\nabla \cdot f$ .... Divergence of f
div f .... Divergence of f
$\neq$ .... Diversity Relation
$\not\approx$ .... Diversity Relation
$/$ .... Divides
$:$ .... Divides
$\div$ .... Division
$\notin$ .... Does not Belong to
$\not\ni$ .... Does not Contain
$\neq$ .... Does not Equal
$\nprec$ .... Does not Precede
$\npreceq$ .... Does not Precede or not Equal to
$\nsucc$ .... Does not Succede
$\nsucceq$ .... Does not Succede or not Equal to

$D(A)$ .... Domain of A
$\dot{x}$ .... Dot (here on x)
$\doteq$ .... Dot Equal
$\ddot{x}$ .... Double Dot (here on x)
X' .... Dual Space of X
ab .... Dyadic Product of **a** and **b**
$a \circ b$ .... Dyadic Product of a and b
A_{ij} .... Element ij of A
$\emptyset$ .... Empty Set
□ .... End of a Logical Sequence
▯ .... End of a Logical Sequence
■ .... End of a Logical Sequence
▮ .... End of a Logical Sequence
$\stackrel{\text{def}}{=}$ .... Equal by Definition to
$\stackrel{\Delta}{=}$ .... Equal by Definition to
$\widehat{=}$ .... Equal by Definition to
$:=$ .... Equal by Definition to
$=$ .... Equal to
$\doteq$ .... Equal up to a Certain Order
$\leftrightarrow$ .... Equivalence
$\Leftrightarrow$ .... Equivalence
$=$ .... Equivalence Relation
$\approx$ .... Equivalence Relation
$\sigma_e(A)$ .... Essential Spectrum of A
$\hat{a}$ .... Estimated Value of a
$|x|$ .... Euclidean Norm of x
ψ .... Euler Angle
ϑ .... Euler Angle
φ .... Euler Angle
e .... Euler Number
$\underline{\vee}$ .... Exclusive Disjunction
$\underline{\vee}$ .... Exclusive or
$\exists$ .... Existential Quantifier
$< a >$ .... Expectation Value of a
$\bar{a}$ .... Expectation Value of a
E(a) .... Expectation Value of a
exp x .... Exponential of x
e^x .... Exponential of x
$\bigoplus_{n=1}^{\infty}$ .... Exterior Hilbert Sum of a Sequence H_n of Hilbert Spaces
$n!$ .... Factorial n
Δx^b .... Finite Increase of x
Fp .... Finite Part (of an integral)
$\forall$ .... For All
$\forall$ .... For Every
$\exists$ .... For Some
$:$ .... For which
$|$ .... For which
$\hat{f}$ .... Fourier Transform of f
Df .... Fréchet Derivative of f
$D^n f$ .... Fréchet Derivative of order n of f
∞ .... Freestream
$f(\cdot,t)$ .... Function Defined by $x \mapsto f(x,t)$ for fixed t
$f(x)$ .... Function of x
f(x) .... Function of x
$\Rightarrow$ .... Gives
∇f .... Gradient of f
grad f .... Gradient of f
$G(A)$ .... Graph of A
$>$ .... Greater than
$\geq$ .... Greater than or Equal to
$\geqq$ .... Greater than or Equal to
$\geqslant$ .... Greater than or Equal to
ent a .... Greatest Integer $\leq a$
$\Rightarrow$ .... Has for an Image
$A^\dagger$ .... Hermitian Conjugate of A
$\equiv$ .... Identically Equal to
I_E .... Identity Mapping in E
I .... Identity Operator
$=$ .... Identity Relation
$\approx$ .... Identity Relation
iff .... If and only if
$\Longleftrightarrow$ .... If and only if
Im A .... Image of A
$R(A)$ .... Image of A
Im $f = f(E)$ ...... Image of $f : E \rightarrow F$ in $F^{(3)}$
$\Im z$ .... Imaginary Part of z
Im z .... Imaginary Part of z
Im(z) .... Imaginary Part of z
z'' .... Imaginary Part of z
$\Rightarrow$ .... Implication
$\rightarrow$ .... Implication
$\Rightarrow$ .... Implies
$\rightarrow$ .... Implies
$\subseteq$ .... Inclusion
$\vee$ .... Inclusive Disjunction
$\vee$ .... Inclusive or
$a_n \uparrow$ .... Increasing Suite
inf A .... Infimum Bound of A
∞ .... Infinity
$\int$ .... Integral
$\mathring{A}$ .... Interior of A
$\cap$ .... Intersection
$G^s(A)$ .... Inverse Graph of A
f^{-1} .... Inverse of f
A^{-1} .... Inverse of A
$\in$ .... Is a Member of
$\in$ .... Is an Element of
$\approx$ .... Is Approximately Equal to
$\leftrightarrow$ .... Is Approximately Equal to
$\subseteq$ .... Is Contained as Subset in
$: =$ .... Is Defined to be
$=$ .... Is Equal to
$=$ .... Is Equivalent to
$\leftrightarrow$ .... Is Equivalent to
$\Leftrightarrow$ .... Is Equivalent to
$\approx$ .... Is Equivalent to
$=$ .... Is Identical with
$\approx$ .... Is Identical with
$\subseteq$ .... Is Included in
$\notin$ .... Is not a Member of
$\notin$ .... Is not an Element of
$\not\approx$ .... Is not Approximately Equal to

$\nleftrightarrow$ ... Is not Approximately Equal to
$\not\subset$ ... Is not a Proper Subset of
$\neq$ ... Is not Equal to
$\neq$ ... Is not Equivalent to
$\not\approx$ ... Is not Equivalent to
$\nleftrightarrow$ ... Is not Equivalent to
$\neq$ ... Is not Identical with
$\not\approx$ ... Is not Identical with
$\not\subset$ ... Is not Properly Included in
$\subset$ ... Is Properly Contained as Subset in
$\subset$ ... Is Properly Included in
$\Rightarrow$ ... Is Transformed into
$\vdash$ ... It is Asserted that
J_f ... Jacobian of f
$\cup$ ... Join
$\ker A$ ... Kernel of A
$N(A)$ ... Kernel of A
δ_{ij} ... Kronecker Symbol
δ^i_j ... Kronecker Symbol
$\nabla^2\varphi$ ... Laplacian φ
Δf ... Laplacian of f
$>$ ... Larger than
$\geq$ ... Larger than or Equal to
$\geqslant$ ... Larger than or Equal to
$\geqq$ ... Larger than or Equal to
$<$ ... Less than
$\leq$ ... Less than or Equal to
$\leqslant$ ... Less than or Equal to
$\leqq$ ... Less than or Equal to
$\lim_{x\to a}\ f(x)$ ... Limit of $f(x)$
$\lim\limits_{x\to a} f(x)$ ... Limit of $f(x)$
r_{xy} ... Linear Correlation Coefficient between x and y
L ... Linear Integral Operator
$\log_a x$ ... Logarithm to the base a of x
$\cap$ ... Logical Product
$\cup$ ... Logical Sum
$\inf A$ ... Lower Bound of A
$<$ ... Lower than
$\leq$ ... Lower than or Equal to
$\leqslant$ ... Lower than or Equal to
$|a|$ ... Magnitude of a
$f:^{A\to F}_{x\mapsto f(x)}$ ... Mapping of A into F such that to the elements $x \in A$, there corresponds $f(x) \in F^{(1)}$
$x \in A \to f(x) \in F$ Mapping of A into F such that to the elements $x \in A$, there corresponds $f(x) \in F^{(1)}$
$\imath$ ... Mathematical i
$\jmath$ ... Mathematical j
ℓ ... Mathematical l
a_{ij} ... Matrix Element
(a_{ij}) ... Matrix whose ij-th Element is a_{ij}
$\hat{x}$ ... Maximal Value of x
$\bar{x}$ ... Mean Value of x
$< x >$ ... Mean Value of x
$-$ ... Minus
$\mp$ ... Minus-Plus
$|z|$ ... Modulus of z
$\gg$ ... Much Greater than
$\gg$ ... Much Larger than
$\ll$ ... Much Less than
$\ll$ ... Much Smaller than
$\times$ ... Multiplies
$\bullet$ ... Multiplies

∇ ... Nabla
$\ln x$ ... Natural Logarithm of x
$\log_e x$ ... Natural Logarithm of x
$\neg$ ... Negation
$\sim$ ... Negation
' ... Negation
$|a|$ ... Norm of a
$\| A \|$ ... Norm of A
$\neg$ ... Not
$\sim$ ... Not
' ... Not
$\not\asymp$ ... Not Asymptotical to
$\neq$ ... Not Equal to
$\not\equiv$ ... Not Equivalent to
$\ngtr$ ... Not Greater than
$\ngeq$ ... Not Greater than or not Equal to
$\ngtr$ ... Not Larger than
$\ngeq$ ... Not Larger than or not Equal to
$\nless$ ... Not Less than
$\nleq$ ... Not Less than or not Equal to
$\not\sim$ ... Not of the Order of
$\ncong$ ... Not Similar or not Equal to
$\not\simeq$ ... Not Similar or not Equal to
$\nless$ ... Not Smaller than
$\nleq$ ... Not Smaller than or not Equal to
$\nsubseteq$ ... Not Subset of or not Equal to
$\not\supset$ ... Not Superset of
$\nsupseteq$ ... Not Superset of or not Equal to
$\alpha(A)$ ... Nullity Index of A
$\emptyset$ ... Null Set
$\sim$ ... Of the Order of
$\leftrightarrow$ ... One-to-One Reciprocal Correspondence
$B(x, \alpha)$ ... Open Ball with Centre x and Radius $\alpha > 0$
Ω ... Open Set in a Topological Space
$\vee$ ... Or
$H^{\perp}$ ... Orthogonal Complement of H in a Pre-Hilbert Space
$\|$ ... Parallel
f_x ... Partial Derivative of f
$\partial_x f$ ... Partial Derivative of f
$\frac{\partial f}{\partial x}$ ... Partial Derivative of f
$\partial f/\partial x$ ... Partial Derivative of f
$D_k f$ ... Partial Derivative of f
$\partial_k f$ ... Partial Derivative of f
s ... Path Length
σ_x, σ_y σ_z ... Pauli Matrices
σ_1, σ_2 σ_3 ... Pauli Matrices
$\perp$ ... Perpendicular
$\arg z$ ... Phase of z
$+$ ... Plus
$\pm$ ... Plus-Minus
$\pm$ ... Plus or Minus
$\sigma_p(A)$ ... Point Spectrum of A
a^n ... Power n (or n-th power) of a
$\prec$ ... Precedes
$\preceq$ ... Precedes or Equal to
$\hat{a}$ ... Predicted Value of a
$\prime$ ... Prime
Π ... Product
$*$ ... Product
$\mathbf{S} \cdot \mathbf{T}$ ... Product of Tensor $\mathbf{S}$ and Vector $\mathbf{A}$
$\subset$ ... Proper Inclusion
$\subset$ ... Proper Subset of
$\propto$ ... Proportional to
$\sim$ ... Proportional to
A/B ... Quotient Space of Vector Spaces A and B

rank A .... Rank of A
rk(A) .... Rank of A
$\hbar$ .... Rationalized Planck constant
π .... Ratio of a Circle Circumference to its Diameter
$\Re z$ .... Real Part of z
Re z .... Real Part of z
Re(z) .... Real Part of z
z' .... Real Part of z
$\sigma_r(A)$ .... Residual Spectrum of A
$R(\lambda, A)$ .... Resolvent Operator of A
$R(\lambda)$ .... Resolvent Operator of A
$\rho(A)$ .... Resolvent Set of A
$f\,|_A$ .... Restriction of $f : E \to F$ to the subset A of E
$\sqrt[n]{a}$ .... Root n of a
$\bar{x}$ .... Root-Mean-Square Value of x
R_∞ .... Rydberg Constant
$\bullet$ .... Scalar Product
$\mathbf{S} : \mathbf{T}$ .... Scalar Product of Tensors $\mathbf{S}$ and $\mathbf{T}$
sec x .... Secant of x
$\ddot{x}$ .... Second Derivative (here of x)
$\mathbf{A}$ .... Second-Order Tensor
$\{a_k\}$ .... Sequence a_1, a_2, ..., a_k, ...
(a_k) .... Sequence a_1, a_2, ..., a_k, ...
C .... Set of Complex Numbers
$\mathbb{C}$ .... Set of Complex Numbers
$\{a_1, a_2, ...\}$ .... Set of Elements
$\mathbb{Z}$ .... Set of Integers
$\mathbb{N}$ .... Set of Natural Numbers
$[a, b]$ .... Set of Objets x such that $a \leq x \leq b$
$[a, b)$ .... Set of Objets x such that $a \leq x < b$
$(a, b]$ .... Set of Objets x such that $a < x \leq b$
(a,b) .... Set of Objets x such that $a < x < b$
Q .... Set of Rational Numbers
$\mathbb{Q}$ .... Set of Rational Numbers
Q^+ .... Set of Rational Numbers ≥ 0
Q^{+*} .... Set of Rational Numbers > 0
R .... Set of Real Numbers
$\mathbb{R}$ .... Set of Real Numbers
$A \backslash B$ .... Set-Theoric Difference $A \cap (B^c)$
sgn a .... Signum a
$\cong$ .... Similar or Equal to
$\simeq$ .... Similar or Equal to
sin x .... Sine of x
$\sigma_{sc}(A)$ .... Singular Continuous Spectrum of A
sing supp$_a f$ .... Singular Analytic Support of f
sing supp f .... Singular Support of f
$<$ .... Smaller than
$\leq$ .... Smaller than or Equal to
$r(A)$ .... Spectral Radius of A
$S(A)$ .... Spectrum of A
$\sigma(A)$ .... Spectrum of A
$\surd a$ .... Square Root of a
$\sqrt{a}$ .... Square Root of a
$a^{1/2}$ .... Square Root of a
s_y .... Standard Deviation of y
x_n .... Subscript n on x
$\subseteq$ .... Subset of or Equal to
$\{x \in A; B\}$ .... Subset of A Consisting of the Elements with Property B
$\succ$ .... Succedes
$\succeq$ .... Succedes or Equal to
$|$ .... Such that
$:$ .... Such that
x^n .... Superscript n on x
$\supset$ .... Superset of
$\supseteq$ .... Superset of or Equal to
Σ .... Sum
Σ .... Summation
$A + B$ .... Sum of Subsets A and B in a Vector Space
supp f .... Support of f
sup A .... Supremum of A
tan x .... Tangent of x
tg x .... Tangent of x
$\to$ .... Tends to
$f \otimes g$ .... Tensor Product of f and g
$\mathbf{S} \cdot \mathbf{T}$ .... Tensor Product of Tensors $\mathbf{S}$ and $\mathbf{T}$
$A \otimes B$ .... Tensor Product of Vector Spaces A and B
$\exists$ .... There Exists at Least One
$\exists! v$ .... There Exists Exactly One v such that
$\exists\,|$ .... There Exists Uniquely
df .... Total Differential of f
tr A .... Trace of A
Tr A .... Trace of A
$\tilde{A}$ .... Transpose of A
${}^tA$ .... Transpose of A
M^T .... Transpose of Matrix M
a^T .... Transpose of Vector a
$\cup$ .... Union
$\exists! v$ .... Uniqueness Quantifier
E .... Unit Matrix
I .... Unit Matrix
$\varepsilon(t)$ .... Unit Step Function
$\forall$ .... Universal Quantifier
sup A .... Upper Bound of A
δ .... Variational Operator
δf .... Variation of f
$\mathbf{a}$ .... Vector $\mathbf{a}$
$\mathbf{a} \wedge \mathbf{b}$ .... Vector Product of $\mathbf{a}$ and $\mathbf{b}$
$\mathbf{a} \times \mathbf{b}$ .... Vector Product of $\mathbf{a}$ and $\mathbf{b}$
V^* .... Vector Space Dual to Vector Space V
$\wp$ .... Weierstraß Function
$\wr$ .... Wreath Product

Other Signs

(alphabetical on the explanations)

@ .... About

$\mathfrak{N}$ .... Absorption (total-to-selective ratio)

á .... Acute Accent (here on a)

& .... Ampersand

& .... And

+ .... And

∠ .... Angle

∡ .... Angle

⊲ .... Angle

Å .... Ångström

' .... Apostrophe

* .... Asterisk

@ .... At

\ .... Backslash

$\bar{a}$ .... Bar Accent (here on a)

$\hbar$ .... Barred h

a̲ .... Bar-Under Accent (here under a)

♣ .... Black Club Suit

♠ .... Black Spade Suit

¶ .... Blind P

a_o .... Bohr Radius

m_{bol} .... Bolometric Magnitude

⊥ .... Bottom Sign

⋈ .... Bowtie Sign

$\breve{a}$ .... Breve Accent (here on a)

• .... Bullet

ç .... Cedilla Accent (here under c)

¢ .... Cent Sign

č .... Check Accent (here on c)

√ .... Check Mark

â .... Circumflex Accent (here on a)

r_o .... Classical Electron Radius

♣ .... Clubsuit Sign

: .... Colon

$\mathfrak{N}$ .... Column Density

, ... Comma

ˆ .... Control Character

© .... Copyright

† .... Dagger

° .... Degree Sign

Ø .... Diameter

⋄ .... Diamond

♢ .... Diamondsuit Sign

ä .... Dieresis (here on a)

† .... Diesis

<| .... Dirac Bra Vector

|> .... Dirac Ket Vector

" .... Ditto Mark

$ .... Dollar

/ .... Done (after making an e-mail remark)

$\dot{a}$ .... Dot Accent (here on a)

ạ .... Dot-Under Accent (here under a)

‡ .... Double Dagger

‡ .... Double Obelisk

// .... Double Prime

" .... Double Quote

↓ .... Down Arrow

√ .... Earmark

⊕ .... Earth

□ .... End of a Logical Sequence

▯ .... End of a Logical Sequence

▪ .... End of a Logical Sequence

▮ .... End of a Logical Sequence

ß .... Es-zet

+ .... Et

! .... Exclamation Point

D^a .... Feminine Ordinal Indicator (here after D)

♭ .... Flat Sign

' .... Foot

⌢ .... Frown

$\Re$ .... Gas Constant

à .... Grave Accent (here on a)

č .... Háček Accent (here on c)

.... Hash Mark

ˆ .... Hat

♡ .... Heartsuit Sign

? .... Help

↩ .... Hooked Left Arrow

↪ .... Hooked Right Arrow

... . Horizontal Ellipsis

ő .... Hungarian Umlaut (here on o)

" .... Inch

¡ .... Inverted Exclamation Mark

¿ .... Inverted Question Mark

⟨ .... Left Angle Bracket

← .... Left Arrow

{ .... Left Brace

[.......... Left Bracket

⌈ .... Left Ceiling Bracket

⌊ .... Left Floor Bracket

(.......... Left Parenthesis

“ .... Left Double Quote

≪ .... Left Pointing Guillemet

‘ .... Left Quote

↔ .... Left Right Arrow

□ .... Logical-Sequence End

▯ .... Logical-Sequence End

▪ .... Logical-Sequence End

▮ .... Logical-Sequence End

$\mathfrak{L}$ .... Luminosity Density

$L_\odot$ .... Luminosity of the Sun

$\mathfrak{L}_\odot$ .... Luminosity of the Sun

$\bar{a}$ .... Macron Accent (here on a)

✠ .... Maltese Cross

D^o	Masculine Ordinal Indicator (here after D)
$\mathfrak{M}$	Mass
$M_\oplus$	Mass of the Earth
$\mathfrak{M}_\oplus$	Mass of the Earth
$M_\odot$	Mass of the Sun
$\mathfrak{M}_\odot$	Mass of the Sun
'	Minute of Arc
...	More to Come
♮	Natural Sign
#	Number
†	Obelisk
¡	Open Exclamation Mark
¿	Open Question Mark
§	Paragraph
¶	Paragraph
%	Percentage
.	Period
m_{pg}	Photographic Magnitude
m_{pv}	Photovisual Magnitude
.	Point
£	Pound Sign
′	Prime
?	Question Mark
"	Quotation Mark
$\mathfrak{L}$	Radiant Flux
$R_\oplus$	Radius of the Earth
$\mathfrak{R}_\oplus$	Radius of the Earth
$R_\odot$	Radius of the Sun
$\mathfrak{R}_\odot$	Radius of the Sun
$\hbar$	Rationalized Planck constant
®	Registered Mark
‘	Reverse Apostrophe
$\rangle$	Right Angle Bracket
$\rightarrow$	Right Arrow
}	Right Brace
]	Right Bracket
$\rceil$	Right Ceiling Bracket
”	Right Double Quote
$\rfloor$	Right Floor Bracket
)	Right Parenthesis
$\gg$	Right Pointing Guillemet
’	Right Quote
R_∞	Rydberg Constant
”	Second of Arc
§	Section
;	Semi-Colon
ß	Sharp S
#	Sharp Sign
♯	Sharp Sign
/	Slash
č	Slavic Háček Accent (here on c)
$\smile$	Smile
$L_\odot$	Solar Luminosity
$\mathfrak{L}_\odot$	Solar Luminosity
$M_\odot$	Solar Mass
$\mathfrak{M}_\odot$	Solar Mass
$R_\odot$	Solar Radius
$\mathfrak{R}_\odot$	Solar Radius
♠	Spadesuit Sign
□	Square
□°	Square Degree
□ '	Square Feet
□ "	Square Inches
□ '	Square Minute of Arc
□ "	Square Second of Arc
˜	Squiggle Accent
★	Star
$\odot$	Sun
∇	Temperature Gradient
&&&	Temporary (file extension)
$M_\oplus$	Terrestrial Mass
$\mathfrak{M}_\oplus$	Terrestrial Mass
$R_\oplus$	Terrestrial Radius
$\mathfrak{R}_\oplus$	Terrestrial Radius
✓	Tick-off Sign
o͡o	Tie-After Accent (here linking oo)
ñ	Tilde (here on n)
×	Times
$\top$	Top Sign
$\mathfrak{R}$	Total-to-Selective Absorption Ratio
ö	Umlaut (here on o)
+	Und
_	Underscore
$\uparrow$	Up Arrow
$\perp$	Up Tack
m_v	Visual Magnitude
¥	Yen Sign
+	&

E-mail Smileys

(alphabetical on the explanations)

=):-)= .... Abraham Lincoln
:-o .... Amazed
(:-& .... Angry
:-|| .... Angry
:-| .... Apathetic
(-: .... Australian
:~-(........... Bawling
:-X .... Big Kiss
(:+) .... Big Nose
!-(............. Black Eye
):-(............ Bull-Headed
}:-(............ Bull-Headed
:-X .... Closed Mouth
%-) .... Confused
%-(............ Confused
%-) .... Cross-Eyed
;(.............. Crying
;-(............. Crying
:'-(............ Crying
:-@ .... Cursing
>:-) .... Devil
]:-) .... Devil
]:-> .... Devil
%*} .... Drunk
:-(............. Frowning
(:-(............ Frowning
:-| .... Grim
<g> .... Grin
:-} .... Grin
:-) .... Happy
[x] .... Hugs to x
(:-$ Ill
:-* .... Kiss
:-x .... Kiss
:-D .... Laughing
(-: .... Left-Handed
=):-)= .... Lincoln (Abraham_)
:-# .... Lips Sealed
:-X .... Mouth Closed
:-O .... Mouth Open
:-{ .... Mustache
:-(............. Not Happy
:-Q .... Not Inhaling!
:-o .... Oh, no!
:-O .... Open Mouth
+-(:-) .... Pope
+<|:-) .... Pope
:-{ .... Pouting
:-? .... Pouting
+-:-) .... Priest
+:-) .... Priest
:-(............. Sad
:-] .... Sarcastic
:-[............. Sarcastic
:-x .... Say Nothing
:-# .... Sealed Lips
(-_-) .... Secret Smile
8-o .... Shocked
:-@ .... Shouting
(-::-) .... Siamese Twins
(-:|:-) .... Siamese Twins
:) .... Smiling
:-) .... Smiling
:-,, ... Smirking
:-] .... Smirking
:-i .... Smoking
:-Q .... Smoking
:-v .... Speaking
[:-] .... Square Head
:-p .... Striking Out Tongue
&-| .... Tearful
:-& .... Tied Tongue
:-p .... Tongue Striking Out
:-& .... Tongue Tied
=):-) .... Uncle Sam
:-(............. Unhappy
<g ... g> ...... Very Wide Grin
? .... Want to Ask a Question
! .... Want to Make a Comment
:-|) .... Wearing a Moustache
<ggg> .... Wide Grin
8-) .... Wide-Eyed
'-) .... Wink
;-) .... Wink
-=:-)\ .... Wizard
:-o .... Wow!
:O .... Yelling
:-O .... Yelling
$-) .... Yuppie

GPSR Compliance
The European Union's (EU) General Product Safety Regulation (GPSR) is a set of rules that requires consumer products to be safe and our obligations to ensure this.

If you have any concerns about our products, you can contact us on

ProductSafety@springernature.com

In case Publisher is established outside the EU, the EU authorized representative is:

Springer Nature Customer Service Center GmbH
Europaplatz 3
69115 Heidelberg, Germany

www.ingramcontent.com/pod-product-compliance
Ingram Content Group UK Ltd.
Pitfield, Milton Keynes, MK11 3LW, UK
UKHW051358190926
475689UK00005B/278
9781402019258